Multiplication Notations for SI Units

MULTIPLICATION FACTOR	PREFIX	SYMBOL
10^9	giga	G
10^6	mega	M
10^3	kilo	k
10^{-2}	centi	c
10^{-3}	milli	m
10^{-6}	micro	μ

Note that conversion factors are approximations. The given multiplication factors are those commonly used in engineering mechanics. For further details, see ASTM Metric Practice, E380-76.

TABLE 2

Approximate Coefficients of Static Friction

MATERIALS IN CONTACT		μ_s
Aluminum	Metal (dry)	0.2
	Metal (greasy)	0.1
Asbestos (brake lining)	Cast iron (dry)	0.4
	Cast iron (wet)	0.2
Asphalt (dry)	Rubber tires	0.8
Cast iron	Cast iron	0.2
	Leather	0.2
	Rubber	0.2
	Wood	0.2
Leather	Metal	0.4
	Wood	0.5
Nylon	Metal	0.3
Rubber	Metal	0.4
	Slippery roads	0.1
	Wood	0.4
Steel	Clay	0.6
	Graphite	0.1
	Ice	0.1
	Metal (dry)	0.2
	Metal (greasy)	0.1
	Teflon	0.1
Teflon	Metal	0.1
Wood	Metal (dry)	0.4
	Metal (wet)	0.2

ENGINEERING MECHANICS: STATICS AND DYNAMICS

ENGINEERING MECHANICS: STATICS AND DYNAMICS

PRENTICE-HALL, INC., ENGLEWOOD CLIFFS, N.J.07632

COMBINED VOLUME

BELA I. SANDOR

Library of Congress Cataloging in Publication Data

SANDOR, BELA IMRE.
Engineering mechanics.

Includes index.
Contents: v. 1. Statics — v. 2. Dynamics — v. 3.
Statics and dynamics.
1. Mechanics, Applied. I. Title.
TA350.S29 1983 620.1 82-13136
ISBN 0-13-278945-0 (v. 3)

ENGINEERING MECHANICS: STATICS AND DYNAMICS
Bela I. Sandor

Printed in the United States of America

10 9 8 7 6 5 4 3 2 1

Editorial/Production Supervision by Virginia Huebner
Interior & Cover Design by Janet Schmid
Art Supervision by Janet Schmid
Cover: The space shuttle Columbia's first flight; launching 1981.
Photograph courtesy NASA/Johnson Space Flight Center, Houston, Texas.
Page Layout by Diane Korohmas & Jenny Markus
Interior Illustrations by J & R Technical Services, Inc.
Manufacturing Buyers: Joyce Levatino & Anthony Caruso
Acquisitions Editor: Chuck Iossi

ISBN 0-13-278945-0

Prentice-Hall International, Inc., *London*

Prentice-Hall of Australia Pty. Limited, *Sydney*

Editora Prentice-Hall do Brazil, LTDA, Rio de Janeiro

Prentice-Hall Canada Inc., *Toronto*

Prentice-Hall of India Private Limited, *New Delhi*

Prentice-Hall of Japan, Inc., *Tokyo*

Prentice-Hall of Southeast Asia Pte. Ltd., *Singapore*

Whitehall Books Limited, Wellington, *New Zealand*

CONTENTS

DYNAMICS—VOLUME 2

PREFACE

PREFACE TO THE STUDENT

Statics and dynamics are fundamental engineering subjects; their concepts and methods are used in subsequent courses and in engineering practice. It is essential for the student to develop a working knowledge in these areas. The material in this text facilitates the learning of statics and dynamics in the following ways:

Perspectives. Concise statements at the beginning of each chapter describe the need for learning the material in that chapter. A few sections that may be omitted in elementary statics and dynamics courses are included to serve ambitious students and to provide for special assignments by the instructor if time permits.

Examples. The text is interspersed with groups of worked-out examples to show the applications of the equations and methods developed in the preceding sections. Realistic examples show the relevance of the subjects to engineering practice. Additional examples are presented in *Learning and Review Aid for Statics* by Sandor and Schlack, and in *Learning and Review Aid for Dynamics* by Schlack and Sandor, both by Prentice-Hall.

Judgment of the results. Many example problems are concluded by a brief section showing judgment of the results. The student should exercise similar evaluation in all problems because the numerical answer to a problem may or may not be reasonable. Quick and correct assessment of an answer for acceptability requires thought and experience. Whenever possible, the judgment should be based on calculations, preferably using a different method of solution. Even apparently simplistic judgments (for example, knowing that the answer makes sense for the given problem) are important for the student and in engineering practice.

Homework problems. Problem sets of a wide range and of increasing difficulty follow each group of examples. A few of the problems require the plotting of the results as a function of a given parameter. Although this is time consuming, it is important for two reasons. First, a graphic display of the results enhances the learning process. The patterns of the data may quickly reinforce one's knowledge or provoke thought. Second, engineers often work and communicate in graphic ways. The practice of plotting results and sketching reasonable diagrams is useful preparation for the student. Answers are given to approximately two-thirds of the problems.

Mathematics. A working knowledge of elementary calculus is necessary. Vector methods are introduced in the text and are used in many problems.

Units. The SI and U.S. customary units are used approximately to the same extent throughout. There are enough examples and homework problems to allow a total concentration on either system of units if desired. Engineers in the United States will need to know both systems of units in the foreseeable future.

Summary sections. Each chapter is concluded with a graphically illustrated summary of the essential methods and equations. These summaries are very useful at all stages in the process of learning and in preparation for examinations. It is recommended that the student look at the summaries even at the start of new topic study. For a final review of the whole course, it may be convenient to remove the summary sheets from the book and use them separately.

Appendices. The four Appendices at the end of the text should be read as they become appropriate. The following comments about these are worth noting:

Appendix A. Vector derivatives. This Appendix is devoted to the derivation of an important equation (Eq. A-7). The details of the derivation are not necessary to remember, but Eq. A-7 and its special cases are useful in the discussion of several topics.

Appendix B. Comments on the use of units. The two major systems of units to be used are discussed in the main body of the text.

In addition, the reader should be aware that other units, even incorrect ones, may be encountered in practice. It is necessary to mention these units, but they are removed from the mainstream of topics to reduce the chance of confusion about them.

Appendix C. Inertia force and dynamic equilibrium. All students of dynamics should be aware of this topic. However, for practical reasons (which are given in App. C), this topic is also removed from the mainstream of elementary engineering mechanics.

Appendix D. Coriolis effects. The understanding of Coriolis acceleration is rather difficult for many students, although its calculation is quite simple. A detailed discussion of Coriolis effects is presented in this Appendix in order to streamline the main body of the text where several accelerations must be considered simultaneously.

PREFACE TO THE INSTRUCTOR

The main purpose in writing this text was to offer a useful learning aid to students of a wide range of abilities and ambitions. Instructors who read the book will notice the educational value of its large-scale structure and the attention given to details. Nevertheless, it is also useful to summarize the book's main features in this space.

Large-Scale Structure

1. *Organization of chapters.* Many topics in elementary mechanics can be studied in various orders, each with some advantages and disadvantages for the student. The compromises in this text were made with an effort to ease the student's overall progress through the course. For example, bending moments in beams are first discussed in Ch. 5, but bending moments in beams under distributed loadings are presented in Ch. 7, because the latter topic is rather difficult for many students. Therefore, they benefit from the early presentation of the basic concept (which is simple) and from the considerable additional practice with free-body diagrams in Ch. 6. (trusses and frames) before studying distributed loadings. However, such an order of topics is by no means sacred. Those instructors who strongly wish to rearrange some of the presentations can do so without great difficulty for themselves or harm for the students.

2. *Organization within chapters.* Each chapter has three distinct parts:

a. The introductory part (always one section) describes the need for learning the material in that chapter. Concise statements give perspectives to each section in the central, major part of the chapter. The comments about emphasizing, or, in the other extreme, possibly omitting some sections are based on my experiences and observations. Some instructors may have different views for serving their own students, and I would most likely accept their reasons and opinions.

b. The central part of each chapter presents new material to the student. Generally, in this part every few sections are followed by several worked-out example problems and a set of homework problems.

c. The summary section at the end of each chapter presents the essential methods and equations, graphically illustrated wherever appropriate. The instructor should encourage the students to use these summaries at all stages in the process of learning and in preparation for examinations.

Details

1. *Minimal role of dynamic equilibrium.* The method of dynamic equilibrium is mentioned but not used anywhere in the main body of the text; it is described in some detail in App. C. Students of dynamics should be aware of this topic, but I am convinced that it should not be in the mainstream of elementary dynamics for several reasons:

a. The application of dynamic equilibrium is confusing to most students because it is fictitious.

b. Imagining dynamic equilibrium is not necessary to solve any problems.

c. The method of dynamic equilibrium becomes increasingly cumbersome (even mindboggling) as the complexity of the problems increases.

d. Most frequently it is best to use the basic method of drawing free-body diagrams with real forces and applying the equations of motion according to these drawings.

2. *Use of other appendices.* The instructor should encourage the students to read App. A, B, and D as they become appropriate. App. A presents useful vector expressions, App. B describes some common but incorrect usages of units, and App. D provides physical concepts for understanding the Coriolis acceleration.

3. *Photographs.* The text is interspersed with photographs to enhance the realistic aspects of the material. In many instances the photographs also arouse the curiosity of the student toward the subject, an important part of educational motivation.

4. *Examples.* Each set of homework problems is preceded by one or more worked-out example problems. The total number of these examples in Statics and Dynamics is 214. It is not reasonable to put many more of them in the text, although some students could benefit from seeing more examples. Such students should be encouraged to obtain *Learning and Review Aid for Statics* by Sandor and Schlack, and *Learning and Review Aid for Dynamics* by Schlack and Sandor, both published by Prentice-Hall. These books are devoted to worked-out examples of the most important topics in elementary courses.

5. *Judgment of the results.* Many example problems are concluded by a brief section showing judgment of the results. The instructor should encourage students to exercise similar evaluation in solving all problems because judgment is an important part of problem-solving in engineering practice.

6. *Homework problems.* Problem sets of a wide range and of increasing difficulty follow each group of examples. The instructor occasionally may encourage the students to use a vector method of solution for practice even when a scalar method is simpler. Answers are given to approximately two-thirds of the 1743 problems.

7. *Mathematics.* The student should have a working knowledge of elementary calculus. Vector methods are used in the text in the discussion of many topics.

8. *Units.* The SI and U.S. customary units are used approximately to the same extent throughout. There are enough examples and homework problems to allow a total concentration on either system of units if desired.

It is hoped that the instructor will encourage some independent effort (other than homework) on the part of the students for enhancing their learning experience and to better prepare them for the realities of engineering work. This book will aid both the teacher and the student in such endeavors. For example, students should study independently some topics of special interest; the instructor may play a role as a counselor in that work. Another useful way for the students to acquire independence is to develop the judgments of the results of problem solutions. This may involve different methods of solution to a given problem, different qualitative judgments, or relevant thought experiments. I believe that instructors would give recognition and credit to students both for effort and ability in this kind of work. Good luck to everyone!

ACKNOWLEDGEMENTS

I am indebted to many people for giving invaluable assistance in the writing of this book. Concerning the Statics part, I should first mention Prof. Robert D. Cook and former graduate students Keith R. Loss and Bruce A. Dale, all of the Department of Engineering Mechanics at the University of Wiconsin—Madison. I also wish to express my gratitude to the external reviewers, especially Prof. Leonard Berkowitz, Cal Poly Pomona, Prof. William Clausen, Ohio State University, Prof. G. W. May, University of New Mexico, Prof. David McGill, Georgia Institute of Technology, and Capt. Robert Schaller, United States Air Force Academy. Most of their suggestions are incorporated in the text.

In the Dynamics part I have received outstanding assistance from graduate students Elizabeth A. Fuchs, and Keith R. Loss of my Department. I am also grateful to the external reviewers, especially Prof. Reese Goodwin, Brigham Young University, Prof. J. Huston, Iowa State University, Prof. Richard K. Kuntz, Georgia Institute of Technology, Prof. G. W. May, University of New Mexico, Prof. Robert McKee, University of Nevada, and Prof. Drew Nelson, Stanford University. I have followed the majority of their suggestions in revising the text.

Margaret Lynch and Jan McCreary were patient and efficient in typing the manuscript.

Special thanks are due Karen J. Richter, graduate student in the Department of Engineering Mechanics at the University of Wisconsin—Madison, and Professor William M. Lee of the Mechanical Engineering Department at the U.S. Naval Academy—Annapolis, for intensive reviews prior to publication.

BELA I. SANDOR

VOLUME 1

STATICS

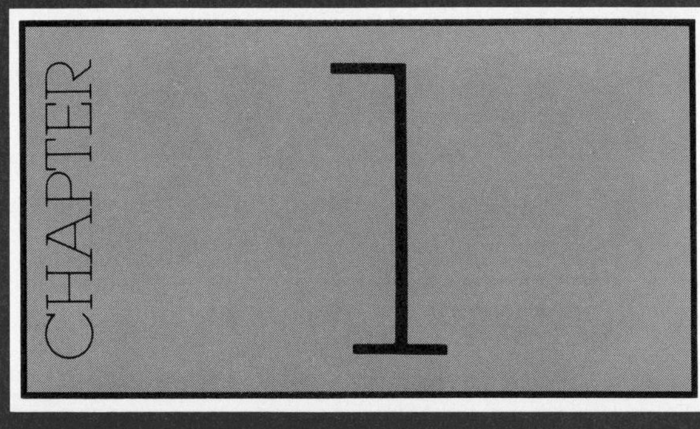

INTRODUCTION TO ENGINEERING MECHANICS

OVERVIEW

Students of several engineering specialties begin their engineering studies in the sophomore year by taking a course in statics, one of the fundamental subjects in engineering mechanics. These courses are based on the mechanics concepts of physics and provide the foundation for many other courses and engineering practice. Statics is the prerequisite for two common courses, dynamics and mechanics of materials (also called strength of materials). These in turn open the door to many other courses, as illustrated in the following flow diagram.

Statics
→ Dynamics
  Courses in
  Advanced dynamics
  Mechanics of fluids
  Machine design
  Vibrations
  etc.
→ Mechanics of materials
  Courses in
  Advanced mechanics of materials
  Machine design
  Structural mechanics
  Experimental stress analysis
  etc.

Both dynamics and mechanics of materials are more difficult than statics for the average student. Therefore, a firm foundation in statics is important to assure satisfactory progress in subsequent courses.

STUDY GOALS

The remaining sections of this introductory chapter present definitions, basic concepts, systems of measurement, and some practical suggestions that are important for all fundamental or advanced courses in engineering mechanics. An overview of these sections is given here.

SECTION 1-1 defines the major branches of engineering mechanics. Most engineers know the fundamentals of all of these subjects, but at the advanced level people tend to specialize in the dynamics, the materials, or the fluids area.

SECTIONS 1-2 and 1-3 present the terminology and basic definitions of Newtonian mechanics, which is the foundation of engineering mechanics. The student should try to remember these sections because a concise review of the basic concepts is desirable from time to time throughout statics and dynamics.

SECTION 1-4 defines the two kinds of quantities used in mechanics, scalars, and vectors. Various aspects and details of manipulations of vectors are presented throughout statics and dynamics. At all levels, these should be studied with intensity to facilitate the solution of complex problems.

SECTION 1-5 describes the SI system of units for the specific measure of the quantities used in engineering mechanics. The student should read this section to understand the scheme of basic and derived units and to note the statements concerning proper practices. For solving problems, the student will find the table inside the front cover the most useful to check the formulas of units and conversion factors between SI and the U.S. customary units.

SECTION 1-6 introduces the method of analytical modeling, which is used throughout statics and dynamics to facilitate the solving of problems. In many cases the modeling will appear very simple. However, sufficiently accurate and yet practical analytical models may require careful thought and should always have high priority in setting up solutions and checking the results.

SECTION 1-7 introduces recurring practical examples which show that the fundamental concepts of engineering mechanics and the necessary simplifications in modeling are indispensable in working on small or large projects. These examples will give the student a perspective view of engineering analysis in general.

1-1 ENGINEERING MECHANICS

Mechanics is a branch of physics, and it has the longest history among the physical sciences. Archimedes (287–212 B.C.) is considered the first systematic investigator of this subject, mainly for his work on the principles of levers and of buoyancy. The most prominent of numerous other scientists in this area are Galileo (1564–1642), who made the first fundamental analyses and experiments in dynamics, and Newton (1642–1727), who formulated the laws of motion and of gravitation.

Engineering mechanics mainly concerns the practical applications of the fundamental principles of mechanics. It is the foundation of many engineering sciences and is used daily by many people. The vocabulary of engineering mechanics is part of the language that most engineers must learn. The large area of engineering mechanics is divided into three parts:

1. ***Mechanics of rigid bodies.*** Here it is assumed that all solids are perfectly rigid and do not deform even under large forces.
   a. *Statics* deals with rigid bodies at rest or under conditions that are static for certain practical purposes.
   b. *Dynamics* deals with rigid bodies in motion.
2. ***Mechanics of deformable bodies*, or *mechanics of materials*, or *strength of materials.*** This area deals with real solids which are never perfectly rigid. Even very small forces cause deformations in the strongest materials.
3. ***Mechanics of fluids.*** This field concerns the behavior of liquids and gases.

The student should appreciate that often, the distinctions between the various areas are not sharp. For example, mechanics of materials may be considered in the realms of statics or dynamics.

1-2 BASIC CONCEPTS OF MECHANICS

Engineering mechanics is based on Newtonian mechanics, in which relativistic effects are negligible. In Newtonian mechanics the basic concepts are space, time, and mass. These are absolute concepts because they are independent of each other. They are also quite abstract. It is best to accept them in engineering work on the basis of personal experiences of the physical world, and intuition if necessary. Other concepts, such as force, are dependent on the basic concepts. *Space* is a geometric region in which events involving bodies occur. The region of interest may be one-, two-, or three-dimensional. The bodies may be *particles* of negligible dimensions, or rigid bodies, or deformable bodies of finite dimensions. *Reference* frames are used to describe the relative positions or coordinates of objects with respect to each other and in the space under consideration. Frequently, the frame of reference chosen is stationary with respect to the earth. Sometimes it is necessary to use an astronomical frame of reference in which the earth itself is moving. *Time* is an absolute quantity in Newtonian mechanics, and in this system it requires no special definition. *Mass* is the quantitative measure of *inertia*, which is the property of all matter. Inertia is the resistance to change in motion of matter. *Force* is the action of a body on another. Each body tends to move in the direction of the external force acting on it.

1-3 NEWTON'S LAWS

Sir Isaac Newton formulated the laws that are basic in statics and dynamics analysis. These may be stated as follows.

FIRST LAW

A particle remains at rest (if originally at rest) or moves straight with a constant velocity if the net force on it is zero.

SECOND LAW

A particle with a force acting on it has an acceleration proportional to the magnitude of the force and in the direction of that force.

This law is formulated as

$$\boxed{\mathbf{F} = m\mathbf{a}} \qquad \text{1-1*}$$

where **F**, m, and **a** are force, mass, and acceleration, respectively. The boldface letters indicate vectors, which are described in Sec. 1-4. Equation 1-1 is first used extensively in Ch. 13 of the associated Dynamics volume.

THIRD LAW

The forces of action and reaction between interacting bodies are equal in magnitude, opposite in direction, and have the same line of action.

This principle is valid at all times and is essential in understanding the nature of forces. For example, consider a rope-pulling contest between two persons, as shown in Fig. 1-1. Assume that person A is pulling the rope to the left with a force F. Intuitively, person B must be pulling the rope to the right with a force of equal magnitude F if there is a stalemate. Next, assume that person B slips and pulls the rope with a different force F_1 at a particular

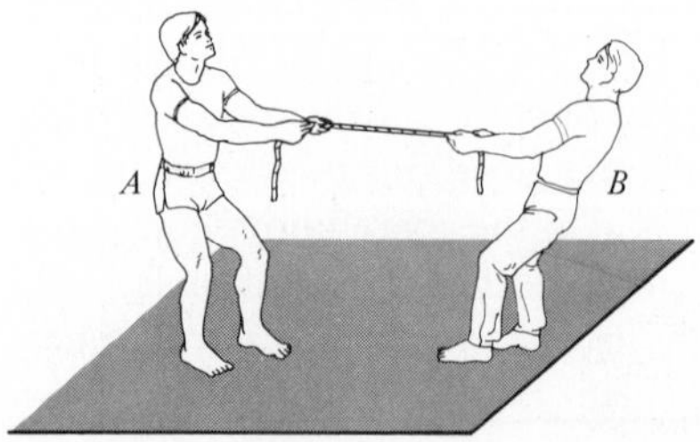

FIGURE 1-1

Rope-pulling contest illustrates Newton's Third Law of action and reaction instant. According to Newton's Third Law, person A pulls the rope in the opposite direction with the same force F_1 at that particular instant. This is true regardless of who or what caused person B to slip.

Law of Gravitation

Newton expressed the mutual gravitational attraction between two particles by the equation

$$F = G\frac{Mm}{r^2} \tag{1-2}$$

where F = magnitude of force of attraction
M and m = masses of the particles
G = universal constant of gravitation
r = distance between the centers of the particles

The mutual forces F determined from this equation must satisfy all of the conditions of the Third Law (equal, opposite, collinear forces).

The force of attraction between the earth and any body near its surface is called the *weight* W of that body. This force causes the acceleration of falling bodies. The *gravitational acceleration* is denoted by g, and it can be determined by experiments with freely falling bodies in vacuum. Alternatively, it can be obtained from Eqs. 1-1 and 1-2, using the experimental value of the constant G. The average value of g is 9.81 m/s^2 or 32.2 ft/s^2. It varies slightly depending on location on the surface of the earth because the earth is not a perfect, homogeneous sphere.

Equation 1-1 can be written using the quantities W and g,

$$W = mg \tag{1-3}$$

This formula is frequently used in practice and is worth remembering.

1-4 SCALARS AND VECTORS

Two kinds of quantities are distinguished in mechanics. *Scalar* quantities have only magnitudes. Examples are density, energy, mass, speed, temperature, time, and volume. *Vector* quantities have magnitude and direction. Examples are acceleration, displacement, force, moment, momentum, and velocity. Vectors are commonly represented by arrows in drawings, with letters or numbers by the arrows indicating the quantity of them. In this text boldface letters indicate vectors. The following classification of vectors is useful.

A *free vector* is not uniquely associated with any given point or line in space. For example, consider the body of a car that moves in a straight path. At any given instant the velocity vector for point A is the same as for point B, or any other point in the car (Fig. 1-2). Thus, a single vector drawn anywhere in the body of the car is adequate to describe its motion, which is convenient in analysis. Note that a free vector is not entirely free; only parallel lines of action are acceptable for choice.

A *sliding vector* may be moved along its line of action but not to another line even if it is parallel to the original line of action. For example, consider the laboratory test of an automobile whose wheels are on rigid supporting platens. It makes no difference whether a sideways force F is pushing or F' of equal magnitude is pulling the platen as in Fig. 1-3. This is called the *principle of transmissibility*. The same force applied to the spindle ($P = F$ in magnitude) would make a significant difference in the vehicle's response because of the relatively flexible tire between the platen and the spindle. This does not rule out using P instead of F with proper corrections, but it has to be recognized that these vectors are not the same in effect.

A *fixed vector* has the most restrictions, with both its line of action and its point of application specified. Such a vector must be used when the deformation of a nonrigid body has to be involved. For example, consider the identical, collinear forces V (acting on the bottom of the tire) and V' (acting on the spindle) in Fig. 1-3.

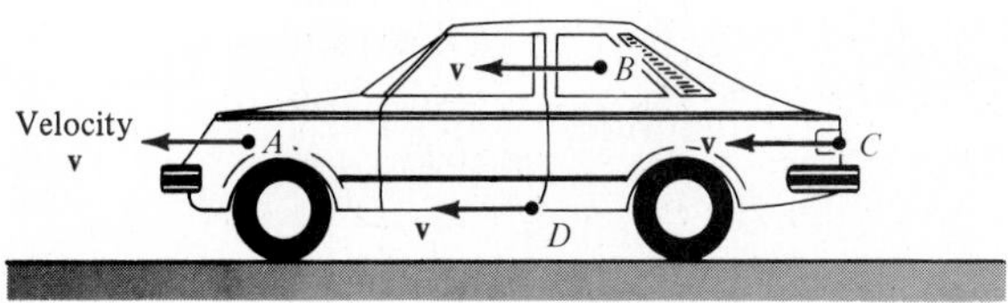

FIGURE 1-2

A free vector may be placed anywhere in a parallel position in a rigid body

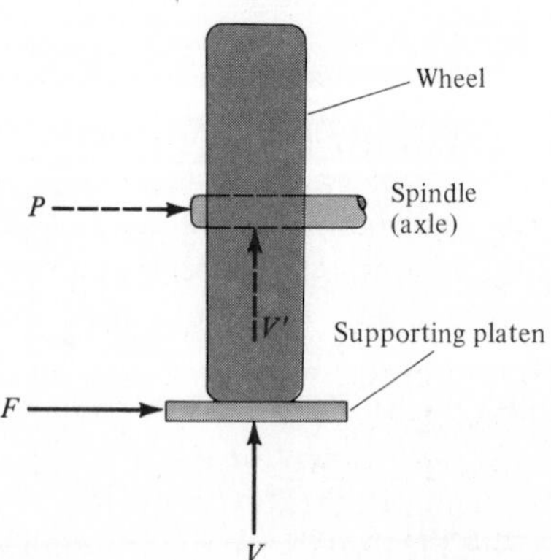

FIGURE 1-3

One wheel of an automobile in a laboratory setup for simulating forces from the road

These vectors could be considered as sliding vectors if the tire were rigid. As it is, the point of application of the force makes a large difference in the vehicle's response to a force.

More detailed considerations and manipulations of vectors are presented in Ch. 2.

UNITS 1-5

Several systems have evolved for the specific measure of the basic and derived quantities. Engineers have traditionally preferred to use a gravitational system with length, time, and force as fundamental quantities, and mass as a derived quantity. In physics mass is taken as fundamental and force is derived; there is no practical difference between the systems if they are used properly.

The system of measurement that is favored now is the International System of Units (SI). This is a modern version of the MKS (meter, kilogram, second) system. Its details are established and controlled by an international treaty organization. Many agencies and technical societies publish these in full or in part. One example is the Standard for Metric Practice E380, which is published by the American Society for Testing and Materials. This standard presents the SI system and a small number of non-SI units that are strongly needed and have been widely accepted.

The SI system of units is intended for worldwide standardization, and it is the basic system in this text. A reasonably detailed description of the SI system is presented here. However, inevitably, engineers will encounter other systems of units for a long time, especially the customary English system, which may persist in the United States. This system is described only briefly because of its familiarity to Americans. Since nearly half of the problems herein are on the U.S. customary units, a conversion table for these is given inside the front cover.

SI Units

SI units are divided into three classes: base, supplementary, and derived units. The three base units are arbitrarily defined, using the *meter* (m) for length, the *kilogram* (kg) for mass, and the *second* (s) for time. For example, the kilogram is the mass of a platinum standard kept in France (one kilogram is approximately equal to the mass of 0.001 m^3 of water). The most important derived unit is that of force, called the *newton* (N). It is expressed from Eq. 1-1 using the unit magnitudes of mass and acceleration,

$$1\ \text{N} = (1\ \text{kg})(1\ \text{m/s}^2) = 1\ \text{kg}\cdot\text{m/s}^2 \qquad \boxed{1\text{-}4}$$

The most relevant SI units for statics and dynamics are presented here.

	QUANTITY	UNIT	SYMBOL	FORMULA
Base units	Length	meter or metre	m	
	Mass	kilogram	kg	
	Time	second	s	
Supplementary unit	Plane angle	radian	rad	
Derived units	Energy, work	joule	J	N·m
	Force	newton	N	$kg \cdot m/s^2$
	Frequency (of a periodic phenomenon)	hertz	Hz	1/s
	Power	watt	W	J/s
	Pressure, stress	pascal	Pa	N/m^2

Other commonly used and acceptable derived units of SI are as follows:

QUANTITY	UNIT	SYMBOL
Acceleration	meter per second squared	m/s^2
Angular acceleration	radian per second squared	rad/s^2
Angular velocity	radian per second	rad/s
Area	square meter	m^2
Density, mass	kilogram per cubic meter	kg/m^3
Moment of force	newton·meter	N·m
Velocity	meter per second	m/s
Volume	cubic meter	m^3

The following non-SI units may be used with the SI system according to the ASTM standard E380:

QUANTITY	UNIT	SYMBOL	DEFINITION
Time	minute	min	min = 60 s
	hour	h	1 h = 60 min
	day	d	1 d = 24 h
Plane angle	degree[a]	°	$1° = \frac{\pi}{180}$ rad
Volume	liter or litre	ℓ	$1\,\ell = 10^{-3}\,m^3$
Mass	metric ton	t	$1\,t = 10^3\,kg$

[a] Decimal submultiples are permissible, but minutes and seconds are discouraged.

Correct usage of the SI system is important to avoid a gradual degradation that has occurred in the older measurement systems. The following aspects of the proper practices should be noted.

Prefixes are used to denote orders of magnitude, and spaces are used in long strings of digits as shown for some common numbers in engineering mechanics.

MULTIPLICATION FACTOR	PREFIX	SYMBOL
10^9 = 1 000 000 000	giga	G
10^6 = 1 000 000	mega	M
10^3 = 1000	kilo	k
10^{-2} = 0.01	centi	c
10^{-3} = 0.001	milli	m
10^{-6} = 0.000 001	micro	μ

It is best to choose prefixes so that the numerical value is between 0.1 and 1000. Prefixes representing powers of 1000 are preferred. In calculations the powers-of-10 notation should be used instead of the prefixes.

The meanings of the words *mass*, *force*, and *weight* should be clarified, especially since SI has a principal difference from the gravitational system of metric units with respect to these quantities. Gravity is involved in measuring mass with a spring scale. The result of such a measurement is called *weight*, but it would be better to call it *force of gravity*, which is mass times the local acceleration of gravity. For this reason, the term *weight* should be avoided in technical practice (which is rather difficult because of widespread habits) and kilograms for *mass* or newtons for *force* used correctly. Most students will need plenty of practice with numerical problems before fully understanding and accepting these units and the various derived ones.

U.S. Customary Units

The base units of the English system still widely used in American industry are the *foot* (ft) for length, the *pound* (lb) for force, and the *second* (s) for time. It is convenient to define these in terms of SI units. The foot is defined as equal to 0.3048 m. The pound is the *weight* of a platinum standard (at the National Bureau of Standards, Washington, D.C.) whose mass is 0.453 592 43 kg. The second for time is the same in both systems of units.

The derived unit of mass in the U.S. customary system is the *slug*. This is the mass that is accelerated at 1 ft/s^2 when a force of 1 lb is applied to it. Mass can be expressed from Eq. 1-1 using the unit magnitudes of force and acceleration,

$$1 \text{ slug} = \frac{1 \text{ lb}}{1 \text{ ft/s}^2} = 1 \text{ lb}\cdot\text{s}^2/\text{ft} \qquad \text{(1-5)}$$

Comments on the SI and U.S. Customary Systems

Conversions between these major systems of units are readily made, but their differences in points of view should be emphasized.

1. The SI system is an *absolute* system of units because the three base units are chosen to be independent of the locations where the quantities are measured. This is true for any location in the known universe.

2. The U.S. customary system is *not* an absolute system but a *gravitational* system of units because the weight of a body depends on the gravitational attraction of the earth, which varies somewhat with location even on the surface of the earth. The variation of weight of a given mass with location is negligible in many engineering problems, but the possibility of even large variations should be kept in mind. If there is any doubt concerning the accuracy of weight or mass using U.S. customary units, the value of the acceleration of gravity, g, should be verified for the given location.

Conversions of Units

Engineers occasionally have to convert numerical quantities from one system of units to another. The conversions are straightforward using the table given inside the front cover. Some of the most common conversions are presented in detail here to illustrate the procedures.

Length. By definition,

$$1 \text{ ft} = 0.3048 \text{ m} \qquad \text{(1-6)}$$

The U.S. customary unit of *mile* (mi) is defined in terms of feet and converted into SI units as follows:

$$1 \text{ mi} = 5280 \text{ ft} = 5280(0.3048 \text{ m}) = 1609 \text{ m}$$

or

$$1 \text{ mi} = 1.609 \text{ km} \qquad \text{(1-7)}$$

Similarly for the unit of *inch* (in.),

$$1 \text{ in.} = \frac{1}{12} \text{ ft} = \frac{1}{12}(0.3048 \text{ m}) = 0.0254 \text{ m}$$

or

$$1 \text{ in.} = 2.54 \text{ cm} = 25.4 \text{ mm} \qquad \text{(1-8)}$$

Force. The U.S. customary unit of pound is defined as the weight of a mass of approximately 0.4536 kg at a location where $g = 9.807$ m/s^2. Thus, from Eq. 1-3,

$$W = mg$$

$$1 \text{ lb} = (0.4536 \text{ kg})(9.807 \text{ m/s}^2) = 4.448 \text{ kg}\cdot\text{m/s}^2$$

From Eq. 1-4,

$$1 \text{ lb} = 4.448 \text{ N} \qquad \boxed{1\text{-}9}$$

A force of 1000 lb is called the *kilopound* (kip), and 2000 lb equals the *ton* in the U.S. customary system. It should be noted that the *metric ton* in the SI system represents a mass of 1000 kg. Thus, ton can be an ambiguous unit if not specified precisely.

Mass. The U.S. customary unit of mass, the slug, can be converted into kilograms as follows. From Eqs. 1-5, 1-6, and 1-9,

$$1 \text{ slug} = 1 \text{ lb}\cdot\text{s}^2/\text{ft} = \frac{4.448 \text{ N}}{0.3048 \text{ m/s}^2} = 14.59 \text{ N}\cdot\text{s}^2/\text{m}$$

From Eq. 1-4,

$$1 \text{ slug} = 14.59 \text{ kg} \qquad \boxed{1\text{-}10}$$

Moment of a force. The moment of a force has units of force times length in both systems of units. For example, *foot·pound* (ft·lb) or *pound·foot* (lb·ft) is used in the U.S. customary system, and *newton·meter* (N·m) in the SI system. The conversion is written using Eqs. 1-6 and 1-9,

$$1 \text{ ft}\cdot\text{lb} = (0.3048 \text{ m})(4.448 \text{ N}) = 1.356 \text{ N}\cdot\text{m} \qquad \boxed{1\text{-}11}$$

To illustrate the reverse procedure in a more complex conversion, assume that 500 N·m should be converted into inch·pounds. Again using Eqs. 1-6 and 1-9, and the fact that 1 ft = 12 in.,

$$500 \text{ N}\cdot\text{m} = 500 \text{ N}\cdot\text{m}\left(\frac{1 \text{ lb}}{4.448 \text{ N}}\right)\left(\frac{1 \text{ ft}}{0.3048 \text{ m}}\right)\left(\frac{12 \text{ in.}}{1 \text{ ft}}\right)$$

Note that each quantity in parentheses has a magnitude of unity; their purposes are to change the units of the original quantity to the desired final set of units. After the computations and the cancelation of newtons, meters, and feet, the result is

$$500 \text{ N}\cdot\text{m} = 4426 \text{ in}\cdot\text{lb}$$

Need for Displaying Units

The student should always remember that in most cases numbers have meaning only if the correct units are displayed with the numbers (exceptions are ratios, which have no units). In performing calculations there is often a temptation to work only with the numbers and write the "expected" units with the final result. This method of economizing in the computation may be reasonable in some cases, but in general it increases the chances of error. The ideal method is to show the correct units with each number throughout the solution. This helps in detecting errors both in the concepts and details of the solution.

1-6 MODELING, APPROXIMATIONS, REQUIRED ACCURACIES

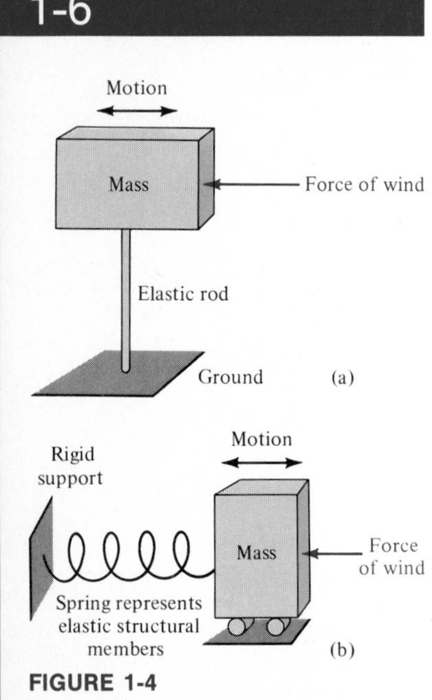

FIGURE 1-4

Simple models of skyscrapers

Support represents car's body
Spring
Shock absorber
Mass of wheel assembly
Force from the road

FIGURE 1-5

Simple model of a suspension system of an automobile

Modeling for analysis means that a real system of bodies, no matter how complex, is assumed to have only a few parts, that these are of simple shape, and that only a few major forces are acting on them. This is always necessary to do in analyzing any system because it is not possible to include everything that has an effect on the system. For example, the gravitational effects of the other planets at the surface of the earth are real but negligible in comparison with the effect of the earth.

It is always best to start with the simplest possible model that reasonably represents the system under consideration. Sometimes this may involve difficult choices regarding the inclusion of certain items. Even then it is best to start with a simple model and gradually improve it as experience is gained concerning the given problem.

For examples of analytical models of complex systems, consider skyscrapers and suspension systems of cars. Figure 1-4 shows the simplest reasonable models for skyscrapers. Of course, other models of similar or greater complexity may be drawn. The model for the suspension system in Fig. 1-5 includes a shock absorber because its effect is quite significant on the spring and mass system.

In making analytical models the important assumptions must be explicitly stated, preferably in writing. This allows rapid reevaluation and changes for improvements when the results from the original model appear (or are proved by experiment) to be wrong. Work in modeling complex systems is quite challenging and interesting mainly because of the judgments and evaluations that are necessary.

Since assumptions are always involved in solving engineering problems, the numerical accuracy of the answers must be in some doubt. In most cases, especially in course work, it is reasonable to round the answers within a few percent. Accuracies within 1% are easily obtained from electronic calculators but may be meaningless in practice. Rounding the answers reasonably requires understanding the practical aspects of engineering work. The student should constantly strive to develop a physical sense of the problem beyond obtaining a facility in manipulating symbols.

RECURRING EXAMPLES FOR PROGRESSIVE ANALYSIS

1-7

Many students in introductory courses such as Statics find it difficult to visualize how the basic concepts, models, and methods form an integral part of solving more complex and interesting problems in engineering work. To assist the student in developing a coherent view of engineering analysis, several examples in this text reappear from time to time for the analysis of different problems. It is hoped that these examples will gradually become familiar to the student so that the various analyses might make sense as necessary parts in the task of completing a project. Two recurring examples are introduced in this section to establish the context for progressive analysis of all such projects.

Skyscrapers with Tuned Mass Dampers

Very tall buildings can vibrate in a wind as a reed does. The amplitude (maximum displacement of the building during oscillations) depends on the geometry and materials of construction of the building, and on the wind. Modern steel-and-glass buildings are relatively susceptible to large displacements in wind, with amplitudes exceeding 1m possible. These displacements would not be dangerous from a structural point of view, but are undesirable with respect to the comfort of the occupants.

There are several ways to keep the oscillations of a building within tolerable limits, although the motions cannot be completely eliminated. One of the viable solutions to the problem uses a *tuned mass damper* system. Such dampers have been used extensively to reduce the vibrations of bridges, tall antennas, power line cables, automotive and aircraft piston engines, and others. They are new in the area of skyscraper design.

The essence of a tuned mass damper in a building is illustrated in Fig. 1-6. A large mass M_2 is located near the top of the building of mass M_1. The damper mass M_2 can move only in a horizontal plane and has a spring and other constraining devices attached to it. The damper system's mass, frictional resistance to motion, and spring characteristics are designed to make the damper mass M_2 move opposite to the direction of displacement of the main mass M_1 at every instant. This reactive oscillating motion is called out-of-phase vibration, which can be adjusted or "tuned" by choosing the system's components to minimize the displacements of the building. An application of this device is in the John Hancock Tower in Boston. This building has two damper masses A and B as illustrated in Fig. 1-7. Each mass has a weight of 6×10^5 lb and a maximum displacement of ± 81 in. relative to the building.

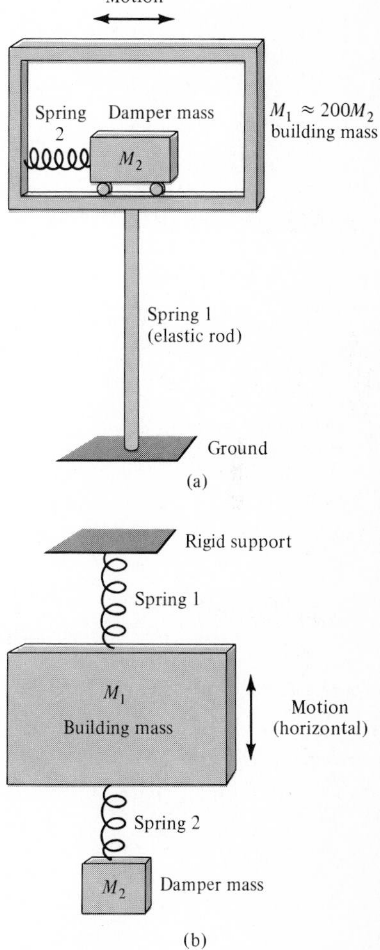

FIGURE 1-6

Models of a skyscraper with a tuned mass damper

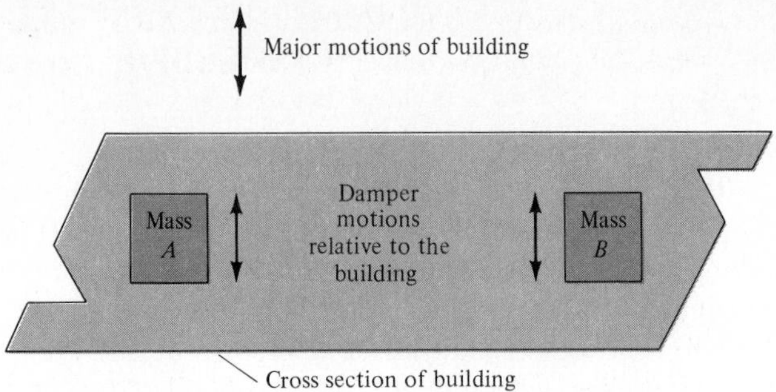

FIGURE 1-7

Tuned-mass-damper locations on 58th floor in John Hancock Tower in Boston

The design of a tuned mass damper is mainly in the realm of vibrations, an advanced area of dynamics, but the design also involves statics and elementary dynamics. There are numerous aspects of such a system that can be analyzed progressively as the student's knowledge of engineering mechanics gradually increases. A few relatively simple problem areas of a tuned mass damper are outlined as follows.

1. The damper mass M_2 in Fig. 1-6 is supported by linear bearings that allow it to move horizontally. Concentrated vertical forces are applied through the bearings to the mass even when it is stationary. Elementary statics is useful in analyzing these forces.

2. The constraining devices of the damper mass must apply moments, and these can also be analyzed early in a statics course.

3. There is friction involved in the motion of the damper mass. The conditions of interest are among the most common in statics and dynamics.

4. The energy methods of elementary engineering mechanics can be readily exemplified with the tuned mass damper.

5. The complete system of a tuned mass damper includes overtravel bumpers for safety. The elastic impacts of the damper mass with these bumpers are among the common problems in elementary dynamics.

6. The oscillating motion of the tuned mass damper (ignoring motion of the building) can be easily characterized using the basic concepts of vibrations which are included in most dynamics texts (although seldom covered in elementary dynamics courses).

Road Simulation in the Laboratory

The complete road testing of cars and trucks is very expensive and time consuming, and the results are often difficult to evaluate. The modern method is to use road simulation under the controlled conditions of the laboratory. Simulators are used to study the endurance life of components, subassemblies, and the vehicle as a whole. Other testing is aimed at studying shock and vibration characteristics, safety, and ride comfort.

Two kinds of simulators are common. Figure 1-8 shows a firmly held vehicle frame, with three electronically controlled *hydraulic actuators* (hydraulic rams) directly coupled to a *wheel spindle* (axle) *A*. These actuators apply forces to the spindle on command. Three independently controlled actuators can simulate a wide range of vertical, fore-and-aft, and sideways forces on the wheel. Figure 1-9 shows an arrangement for independently controlled vertical force input through each tire. In this case the vehicle can bounce, twist, and vibrate on top of the hydraulic actuators.

FIGURE 1-8

Automobile in a laboratory test. The vehicle's frame is firmly fixed while three independent forces may be applied to the wheel spindle *A*. (Courtesy MTS Systems Corporation, Minneapolis, Minnesota.)

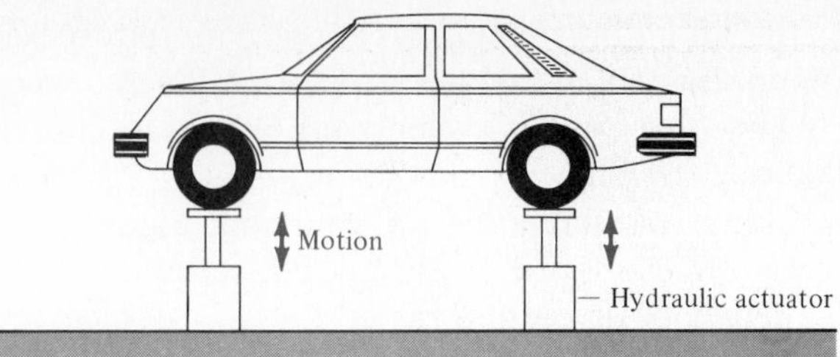

FIGURE 1-9

Automobile in a laboratory test. Hydraulic actuators apply realistic vertical forces to each wheel while the vehicle's frame is not constrained. (Courtesy MTS Systems Corporation, Minneapolis, Minnesota.)

A road simulator for automotive vehicles offers numerous examples for progressive analysis involving elementary engineering mechanics, as outlined in the following.

1. The minimal requirement of each hydraulic actuator that applies the vertical force to a wheel is to hold the vehicle stationary. These requirements can be established using elementary statics.

2. The force applied by each hydraulic actuator can be measured quite accurately and easily. Thus, each vertical actuator in Fig. 1-9 can be considered as an electronic scale for measuring weight, which allows for the experimental determination of the center of gravity of the vehicle. This topic is covered in statics.

3. The linkages and brackets between the hydraulic actuators and the wheel spindle in Fig. 1-8 provide simple examples for vector manipulations, moments of forces, equilibrium, and two-force members, all topics of statics.

4. When a vehicle is assumed to be a rigid body, the road simulation provides examples for dynamics analysis.

5. The basic equipment and techniques of road simulation for automotive vehicles are readily extended to exemplify other kinds of important engineering analysis, such as the impact testing of packages, locomotives, or electronic equipment. Thus, the student will find that the principles and methods of elementary engineering mechanics have wide-ranging practical application, regardless of how complex the overall project may be.

CHAPTER 2

FORCES ON PARTICLES

OVERVIEW

Basic Concepts in Modeling Forces on Bodies

Forces are generally described by their *magnitudes*, *directions*, and *points of application*. The body being studied is called a *particle* if it has negligible dimensions compared to other dimensions in the same problem, or if its dimensions are not important in the solution. Particles are useful models in many practical problems. Since a particle has negligible dimensions, all forces on it act at the same point, which is another useful simplification.

STUDY GOALS

SECTIONS 2-1 AND 2-2 present the methods for adding and subtracting vectors or resolving vectors into components. Force vectors are used to exemplify the methods, but these operations are also essential for other vector quantities of statics and dynamics.

SECTION 2-3 introduces unit vectors, which provide a concise scheme of notation for all vector quantities. Unit vectors are especially useful in complex vector operations.

SECTIONS 2-4 AND 2-5 extend the methods of Secs. 2-1 and 2-2 to three-dimensional force systems. Some advantages of using unit vectors are demonstrated here. Other advantages will become obvious in several areas of statics and especially in dynamics.

SECTION 2-6 defines the scalar product of two vectors and illustrates several useful applications of scalar products in mechanics.

SECTION 2-7 presents the definition of equilibrium of a particle and the method of modeling using free-body diagrams. The concept of equilibrium is extremely important in engineering mechanics because it is used in determining unknown forces. Free-body diagrams are indispensable in solving most problems, so they should be mastered in this section. Students are strongly advised to practice the use of free-body diagrams as much as possible.

2-1 TWO-DIMENSIONAL FORCE SYSTEMS. RESULTANTS OF FORCES

Consider a body that is modeled as a particle at point O in Fig. 2-1a. Assume that two separate forces **F** and **P** are acting on the particle as shown.* It is useful in practice that the two forces **F** and **P** may be replaced by a single resultant force **R** that has the same effect on the particle. The resultant force is the diagonal through point O of the parallelogram constructed using the original two vectors as in Fig. 2-1a. The procedure is expressed in a *vector equation* as $\mathbf{F} + \mathbf{P} = \mathbf{R}$. This method is called the *parallelogram law* for the addition of two vectors, which can be proved only experimentally. In the case of forces, the proof of this law requires force measuring devices; in other cases such as for displacement or velocity vectors, the proof may be more intuitively obvious to some people. The parallelogram method is valid for *any* two vectors that are *concurrent* (acting at the same point). The resultant vector is always the diagonal of the parallelogram, as illustrated in Fig. 2-2a.

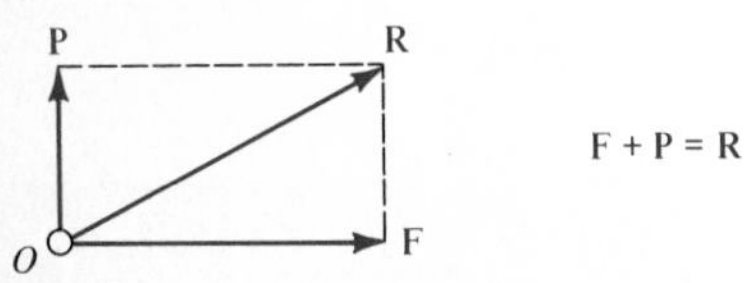

F + P = R

(a) Parallelogram law

The parallelogram law has led to an alternative method of vector addition, the *triangle rule*. In this method one of the vectors is redrawn parallel to its original position, with its tail at the arrowhead of the other vector. The vector that completes the triangle is the resultant of the original two vectors. The various possibilities are shown in Fig. 2-lb and c and in Fig. 2-2b and c. The results are identical from both routes in each figure.

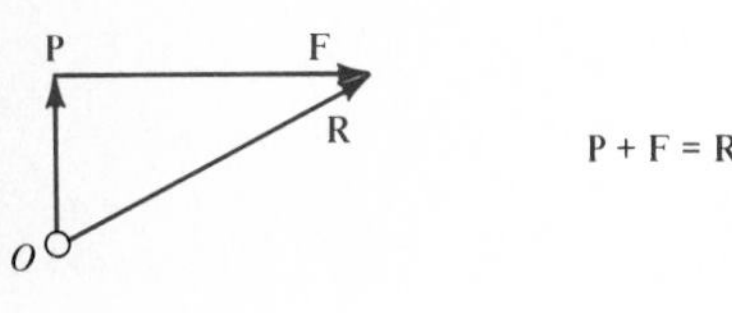

P + F = R

(b) Triangle rule

The vector addition of two vectors is readily extended to that of any number of vectors, as shown for three, **A**, **B**, and **C**, in Fig. 2-3a. **P** is the resultant of vectors **A** and **B**, and **R** is the resultant of **P** and **C**. The addition can be started with any pair of these. The final resultant **R** should be the same for a given set of vectors. Thus, the

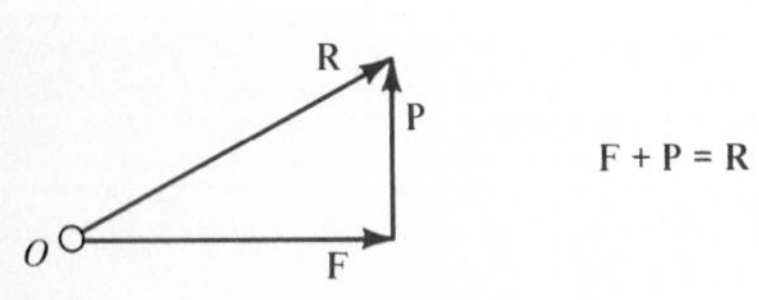

F + P = R

(c) Triangle rule

FIGURE 2-1

Addition of perpendicular, concurrent vectors **P** and **F**

* Vectors are distinguished here by boldface type (**P**). Common notations for vectors in longhand writing are $\vec{P}$, $\underline{P}$, and $\mathbb{P}$.

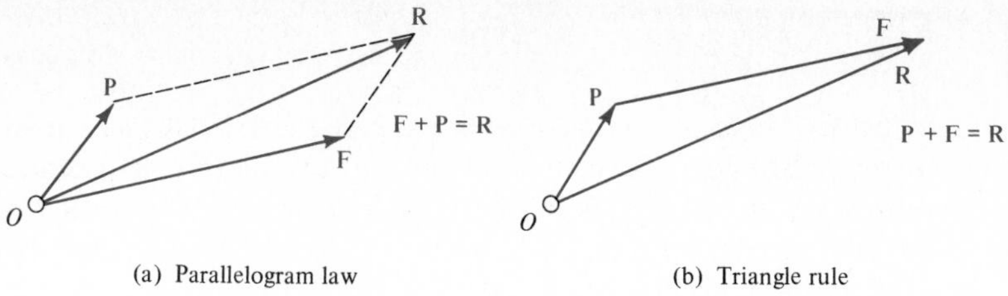

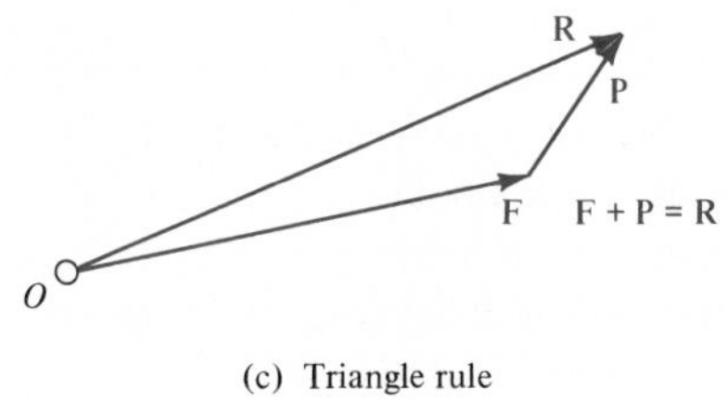

FIGURE 2-2

Addition of arbitrary, concurrent vectors **P** and **F**

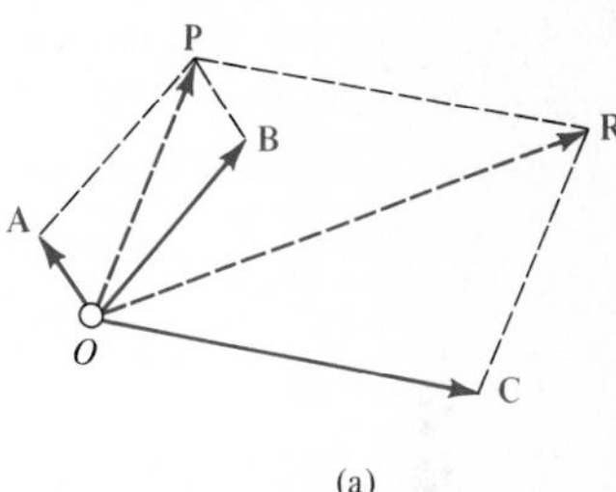

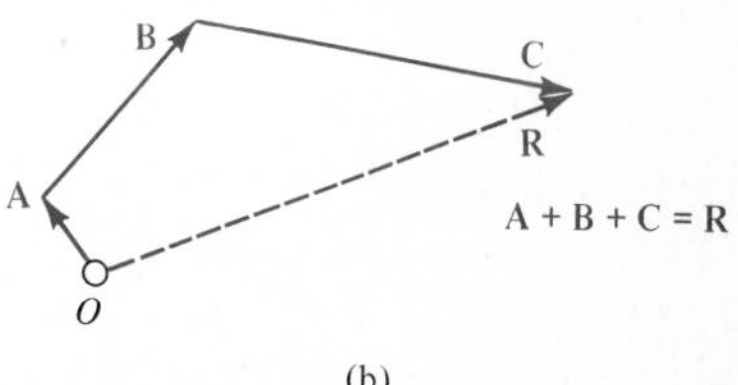

FIGURE 2-3

Addition of concurrent, coplanar vectors **A**, **B**, and **C**

addition of vectors is *commutative*. Formally, with respect to Fig. 2-3a,

$$\mathbf{A} + \mathbf{B} = \mathbf{B} + \mathbf{A} = \mathbf{P} \qquad \text{2-1a}$$

$$\mathbf{A} + \mathbf{C} = \mathbf{C} + \mathbf{A} \qquad \text{2-1b}$$

$$\mathbf{B} + \mathbf{C} = \mathbf{C} + \mathbf{B} \qquad \text{2-1c}$$

$$\begin{aligned} \mathbf{A} + \mathbf{B} + \mathbf{C} &= (\mathbf{A} + \mathbf{B}) + \mathbf{C} = \mathbf{A} + (\mathbf{B} + \mathbf{C}) \\ &= (\mathbf{B} + \mathbf{C}) + \mathbf{A} = \mathbf{B} + (\mathbf{C} + \mathbf{A}) \\ &= \mathbf{R} \end{aligned} \qquad \text{2-1d}$$

The last of these, which could be further rearranged, shows that vector addition is also *associative.*

The triangle rule leads to the convenient *polygon rule* for graphically adding more than two concurrent, *coplanar* vectors (they must be in the same plane). This involves rearranging the vectors in a head-to-tail sequence as in Fig. 2-3b, and drawing the resultant arrow from the tail of the sequence to close the polygon.

The subtraction of a vector from another vector can be visualized as the summation of a positive and a negative vector. For appreciating this method, note that vectors **A** and −**A** have the same magnitude but opposite directions. These facts are used in the illustration of vector addition and subtraction in Fig. 2-4. Part a of this figure shows the addition of vectors **P** and **F**, **P** + **F** = **R**. Part b shows the subtraction of vector **F** from vector **R**, as **R** + (−**F**) = **P**.

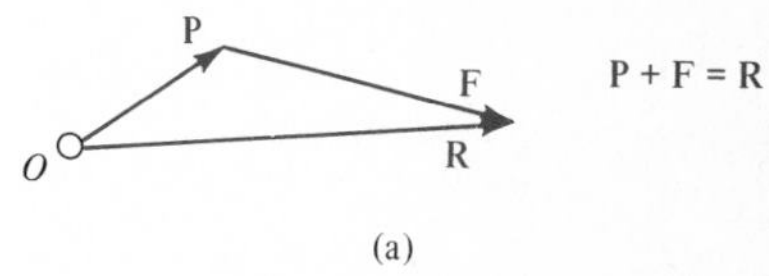

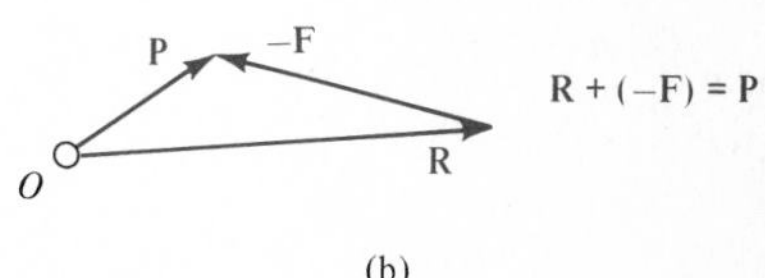

FIGURE 2-4

Addition and subtraction of two vectors

2-2 COMPONENTS OF A FORCE

Frequently, there is a need to replace a given force by other forces that have the same effect on the body but are more conveniently oriented for analysis. This *resolving of a force into components* is done as a reversal of the vector addition procedure. For any force there is an infinite number of possible sets of components, but in practice only a few specific sets may be of interest. Two frequently encountered cases of specifying components are worth noting:

1. *The lines of action of the components are given.* For an illustration of this, assume that the components of a given force **F** are required along lines m and n in Fig. 2-5. There is only one set of two components that satisfies the parallelogram law for **F** and the two lines. These components are $\mathbf{F}_m$ and $\mathbf{F}_n$. Their resultant in vector addition must be **F**. Note that the magnitude of the components can be found graphically or by using the law of sines if **F** and the angles α and β are given.

2. *One component is completely specified.* For example, assume that force **F** in Fig. 2-6 is given and that one of its two components must be force $\mathbf{F}_1$. The other component $\mathbf{F}_2$ is determined by applying the parallelogram law or the triangle rule. Using the triangle rule, $\mathbf{F}_2'$ is a vector from the tip of $\mathbf{F}_1$ to the tip of **F**, but both components $\mathbf{F}_1$ and $\mathbf{F}_2$ are applied at point O.

There are many other possibilities in specifying conditions for the components of a vector, but *any set of components must satisfy the parallelogram law*. In most engineering problems the components of interest are along orthogonal coordinates.

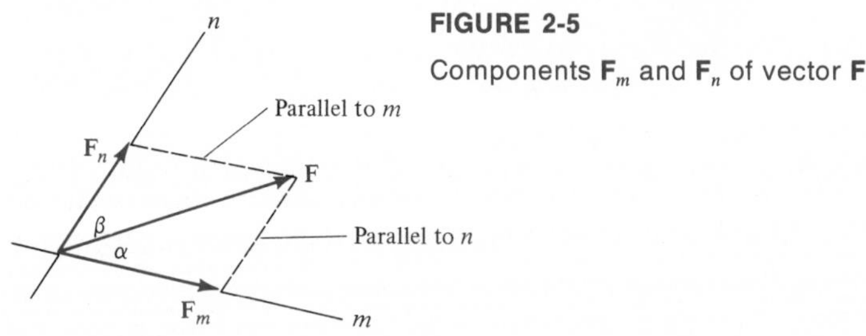

FIGURE 2-5

Components $\mathbf{F}_m$ and $\mathbf{F}_n$ of vector **F**

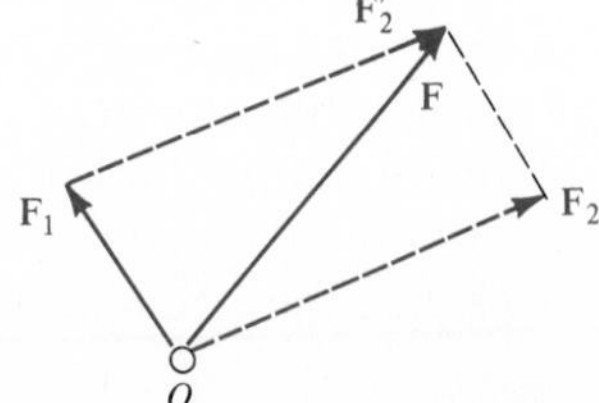

FIGURE 2-6

Determination of component $\mathbf{F}_2$ if force **F** and one of its components $\mathbf{F}_1$ are given

EXAMPLE 2-1

Determine the resultant **R** of forces **F** and **P** applied concurrently to the bracket in Fig. 2-7a. Use a graphical and a trigonometric method of solution.

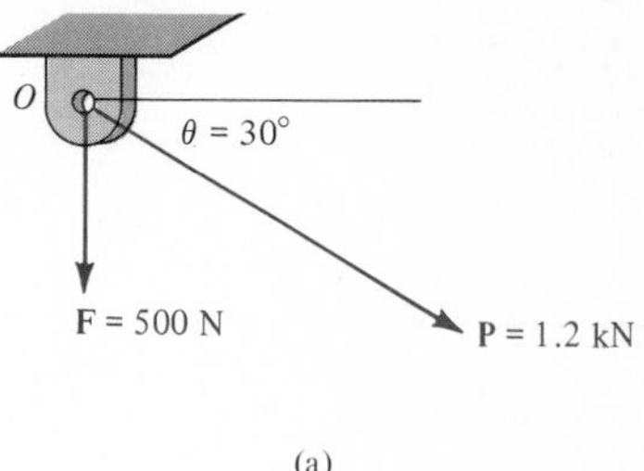

(a)

SOLUTION

GRAPHICAL METHOD: A parallelogram is constructed with sides **F** and **P** drawn to scale in Fig. 2-7b. The magnitude and direction of the resultant **R** are measured from the drawing. The results can be stated in scalar or vector forms as follows.

Answer in scalar form: $R = 1500\text{ N}, \alpha = 43°.$

Answer in vector form: $\mathbf{R} = 1500\text{ N}$ 43°

The same results can be obtained using the triangle rule, as shown in Fig. 2-7c.

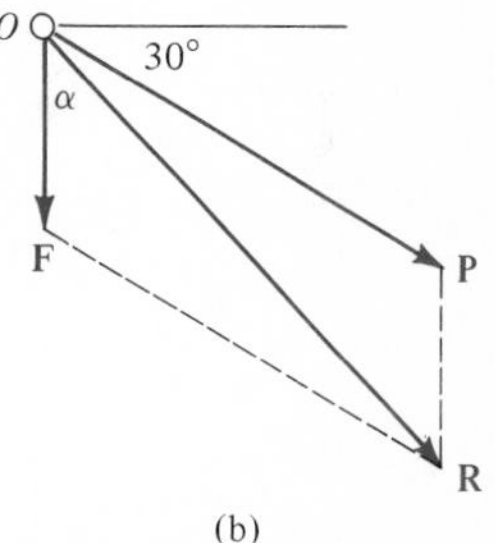

(b)

TRIGONOMETRIC METHOD: The law of cosines can be applied to the triangle of the three forces in Fig. 2-7d (from the triangle rule).

$$R^2 = F^2 + P^2 - 2FP\cos\beta$$

$$= (500\text{ N})^2 + (1200\text{ N})^2 - 2(500\text{ N})(1200\text{ N})(\cos 120°)$$

$$R = 1513\text{ N}$$

The direction of **R** is obtained by applying the law of sines.

$$\frac{\sin\beta}{R} = \frac{\sin\gamma}{P} = \frac{\sin\alpha}{P}$$

$$\sin\alpha = \frac{P}{R}\sin\beta = \frac{1200\text{ N}}{1513\text{ N}}\sin 120°$$

$$\alpha = \arcsin(0.6869) = 43.4°$$

Answer in vector form: $\mathbf{R} = 1513\text{ N}$ 43°

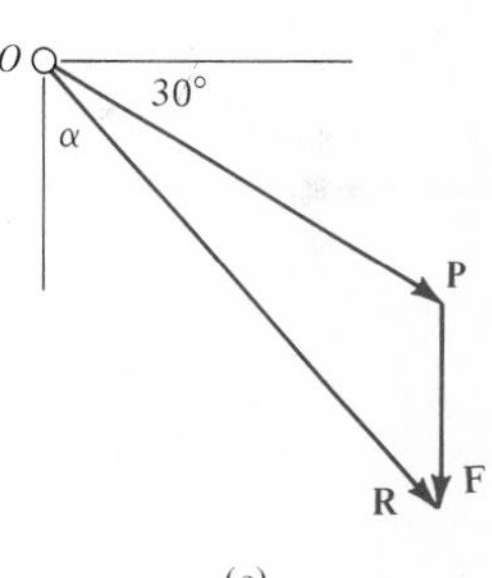

(c)

JUDGMENT OF THE RESULTS

The magnitude and direction of the resultant **R** obtained trigonometrically agree with those from the graphical construction. In practice, the trigonometric solution is often preferable since it is more accurate. The values of the answer are judged reasonable at a glance (or from a freehand sketch) using Fig. 2-7a.

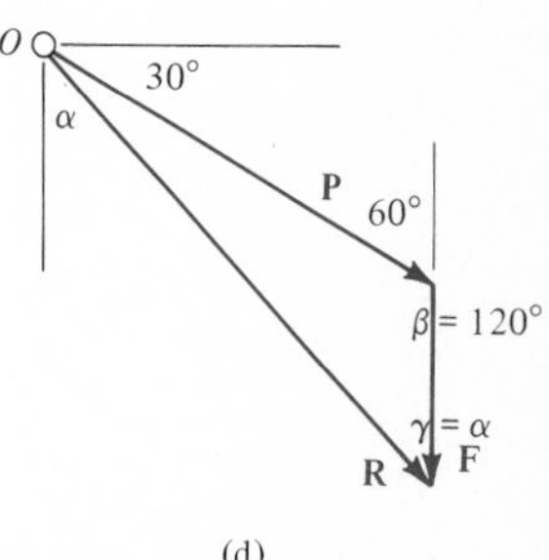

(d)

FIGURE 2-7

The resultant **R** of vectors **F** and **P**

EXAMPLE 2-2

In a mechanism of a road simulator (for a description of this, see Sec. 1-7) the maximum intended force to the wheel spindle is **P**, as modeled in Fig. 2-8a. What is the required force capacity **F** of the hydraulic actuator that applies the force **P** through pin *B* of the rigid rocker arm *AB*? What is the component **A** of the force **F** toward pin *A*? Note that **F** is applied to cause a specified force **P** at the same point, so **P** is a known component of the unknown force **F**.

SOLUTION

GRAPHICAL METHOD: From Fig. 2-8b,

$$\mathbf{F} = 5500 \text{ lb}$$

$$\mathbf{A} = 1900 \text{ lb}$$

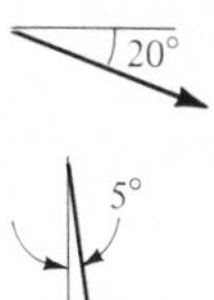

TRIGONOMETRIC METHOD: Using the law of sines,

$$\frac{\sin 20^\circ}{A} = \frac{\sin 65^\circ}{P}$$

$$A = \frac{\sin 20^\circ}{\sin 65^\circ}(5000 \text{ lb}) = 1887 \text{ lb}$$

Using the law of sines again yields

$$\frac{F}{\sin 95^\circ} = \frac{5000}{\sin 65^\circ}$$

$$F = 5500 \text{ lb}$$

Thus, the forces **F** and **A** agree reasonably with those obtained from the graphical solution.

FIGURE 2-8

Determination of a required force **F** to produce a given component **P**

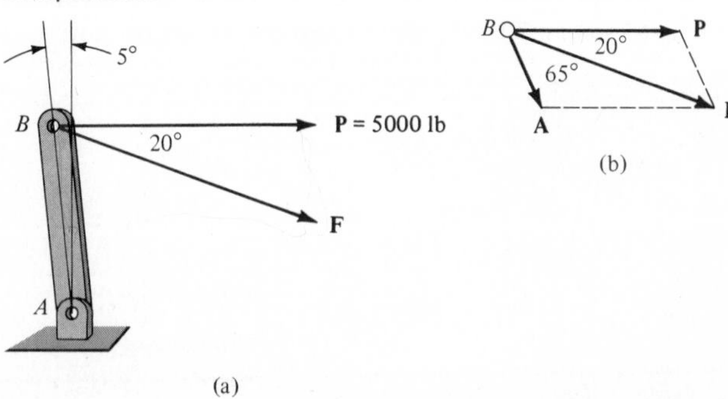

EXAMPLE 2-3

Consider a very simple model of eight radial spokes with equal tensions (pulling forces) at the hub of a stationary bicycle wheel. The magnitudes of the tensions are $T_1 = T_2 = T_3 = \ldots = T_8$ in Fig. 2-9a. Determine graphically the net force on the hub by adding one force at a time to $\mathbf{T}_1$. Check the result by using the polygon rule for the eight forces at once.

SOLUTION

The resultants $\mathbf{R}_1, \mathbf{R}_2, \ldots$ for the successive pairs of vectors are drawn in Fig. 2-9b. Note that the last addition is of $\mathbf{T}_8$ and $\mathbf{R}_6$, which are equal in magnitude but opposite in direction, so the net resultant is zero. The same is shown by the polygon construction in Fig. 2-9c. The lesson from the latter is that *whenever the polygon of the given forces is closed* (*the last arrowhead touches the tail of the first vector*), *the net force is zero.*

JUDGMENT OF THE RESULTS

It is reasonable that the resultant force on the hub is zero in the absence of external forces on the wheel, since the spokes pull on the stationary hub evenly in all directions.

FIGURE 2-9

Summation of eight concurrent forces

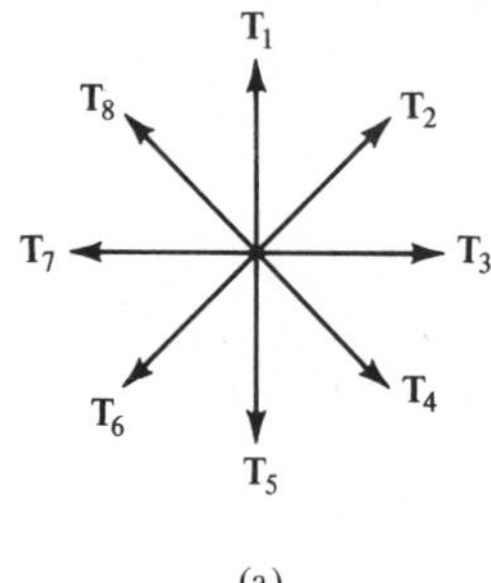

(a)

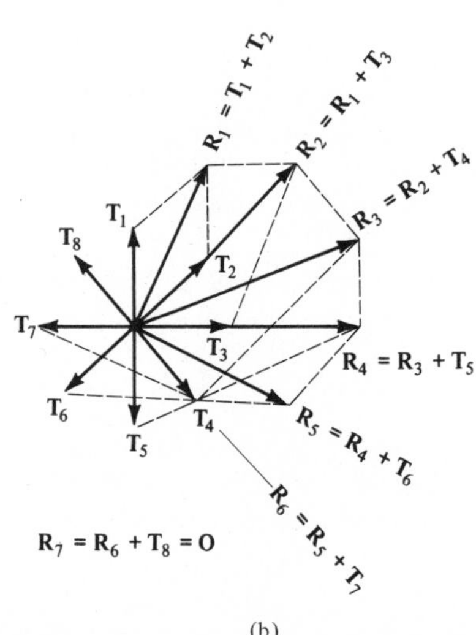

(b)

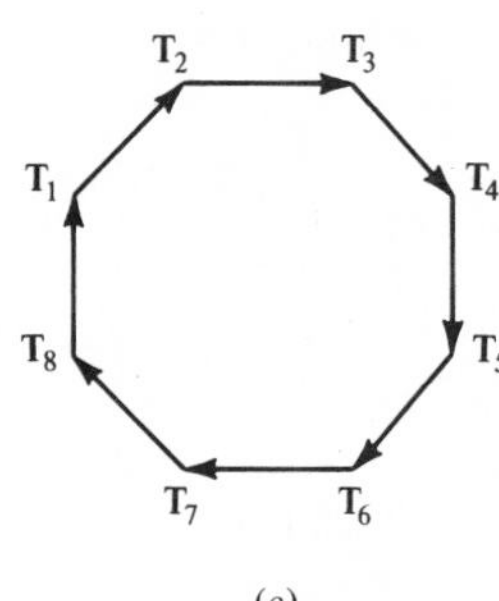

(c)

PROBLEMS

FIGURE P2-1

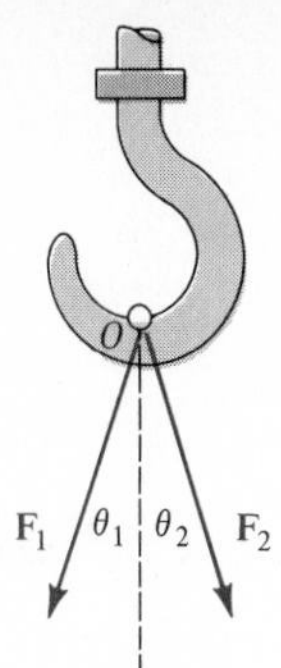

2–1 Determine graphically the resultant of the two forces if $F_1 = F_2 = 3$ kN and $\theta_1 = \theta_2 = 20^\circ$. Use the parallelogram law and the triangle rule.

2–2 In Fig. P2-1, $F_1 = 500$ lb, $F_2 = 800$ lb, and $\theta_1 = \theta_2 = 30^\circ$. Determine graphically the resultant of $\mathbf{F}_1$ and $\mathbf{F}_2$.

FIGURE P2-3

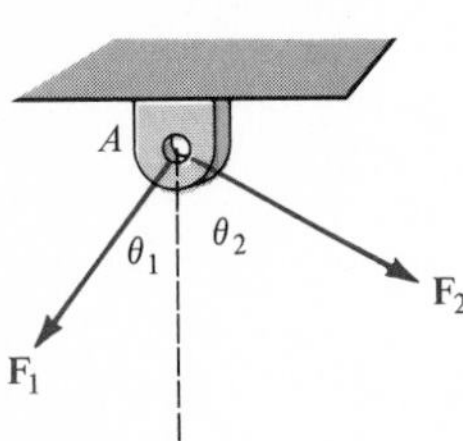

2–3 Determine by trigonometry the resultant of the two forces if $F_1 = 800$ N, $F_2 = 1.4$ kN, $\theta_1 = 30^\circ$, and $\theta_2 = 45^\circ$.

2–4 In Fig. P2-3, $F_1 = 300$ lb at $\theta_1 = 25^\circ$. Determine graphically the magnitude of force $\mathbf{F}_2$ at $\theta_2 = 15^\circ$ to make the resultant of $\mathbf{F}_1$ and $\mathbf{F}_2$ vertical. What is the magnitude of the resultant in this case?

FIGURE P2-5

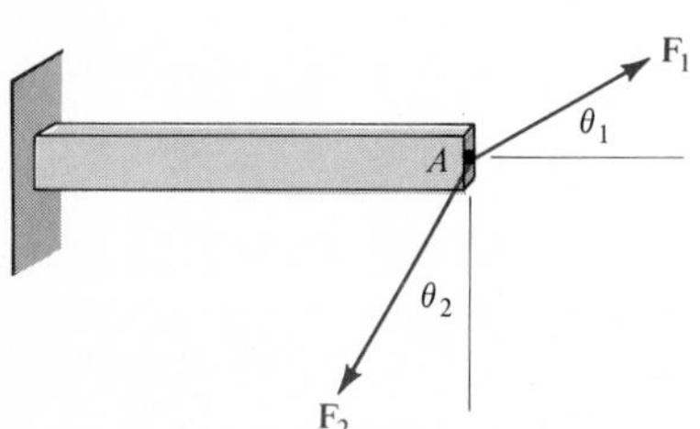

2–5 Determine by trigonometry the angle θ_1 to make the resultant of $\mathbf{F}_1$ and $\mathbf{F}_2$ horizontal if $F_1 = 5$ kN, $F_2 = 1$ kN, and $\theta_2 = 30^\circ$. What is the magnitude of the resultant force?

FIGURE P2-6

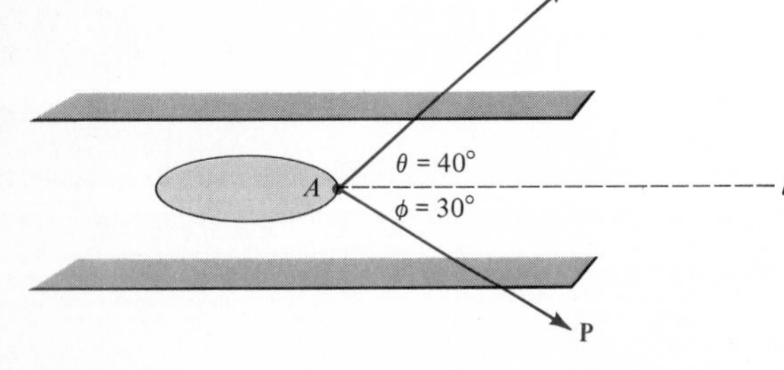

2–6 A boat is towed in a narrow channel by applying forces **F** and **P** through ropes attached to the boat. Determine by trigonometry the ratio P/F for which the resultant force is along line l. Write an expression for the magnitude of the resultant.

FIGURE P2-7

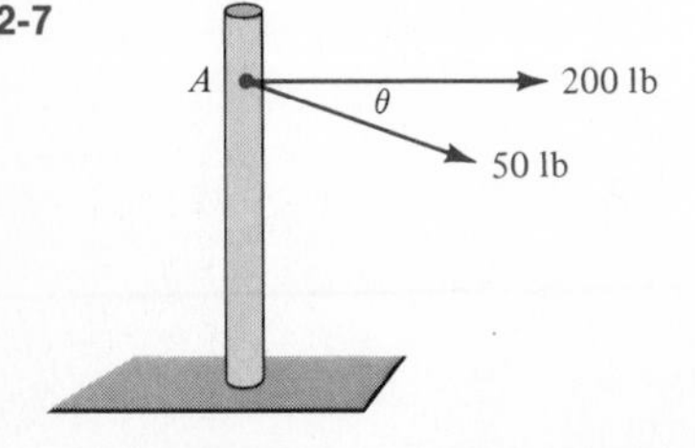

2–7 Two ropes are attached to a post at point A. For the given tensions in the ropes, determine the range of the angle θ between the ropes to limit the maximum resultant force at A to 200 lb.

2–8 Determine the resultant of the three coplanar forces using a graphical method. $F_1 = 500$ N, $F_2 = 800$ N, $F_3 = 900$ N, $\theta = 30°$, and $\phi = 50°$.

FIGURE P2-8

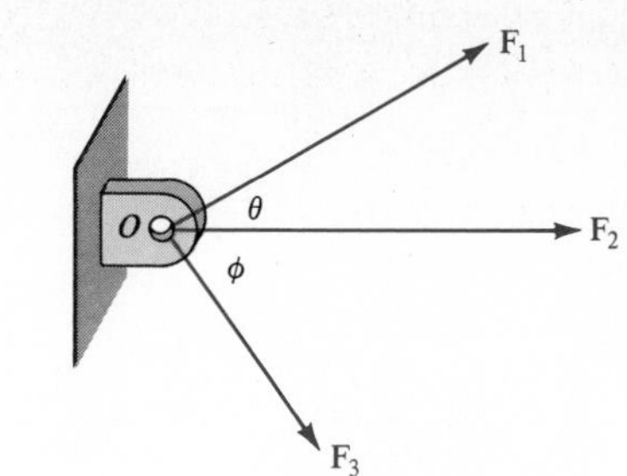

2–9 Determine graphically the additional force $\mathbf{F}_4$ to make the resultant of the three coplanar forces in Fig. P2-8 zero ($\mathbf{R} = 0$). $F_1 = 200$ lb, $F_2 = 250$ lb, $F_3 = 400$ lb, $\theta = 20°$, and $\phi = 70°$.

FIGURE P2-10

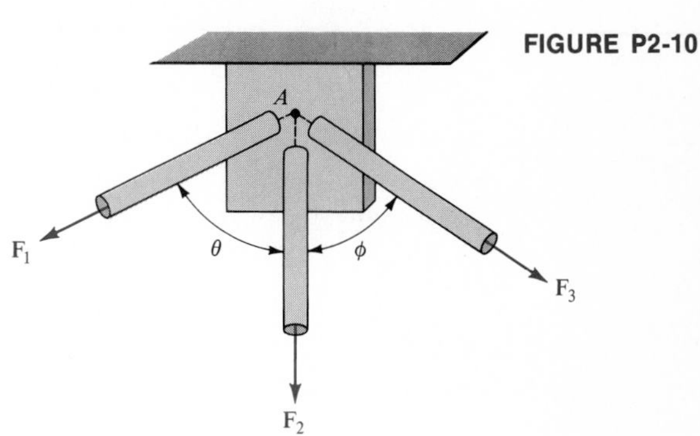

2–10 Determine by trigonometry the resultant of the three coplanar forces acting on the welded bracket. $F_1 = 4$ kN, $F_2 = 5$ kN, $F_3 = 7.5$ kN, $\theta = 45°$, and $\phi = 60°$.

2–11 Assume that force $\mathbf{F}_2$ is reversed in direction in Fig. P2-10 while $\mathbf{F}_1$ and $\mathbf{F}_3$ act as shown. Determine by trigonometry the resultant of the three coplanar forces. $F_1 = 1000$ lb, $F_2 = 1500$ lb, $F_3 = 800$ lb, $\theta = 90°$, and $\phi = 20°$.

FIGURE P2-12

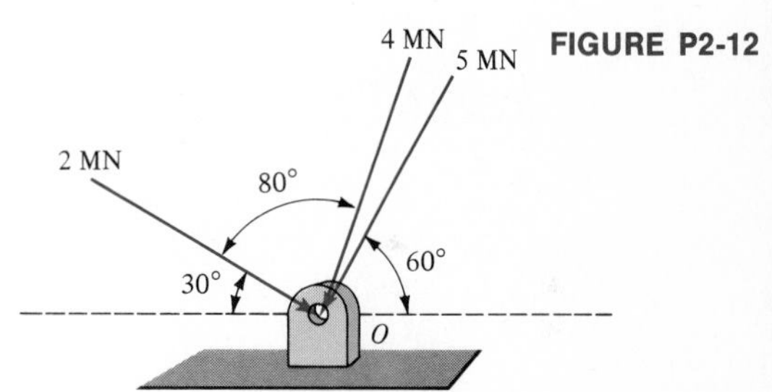

2–12 Determine by trigonometry the resultant of the three coplanar forces.

2–13 Determine graphically the additional force **P** that would make the resultant force on the bracket in Fig. P2-12 zero ($\mathbf{R} = 0$).

FIGURE P2-14

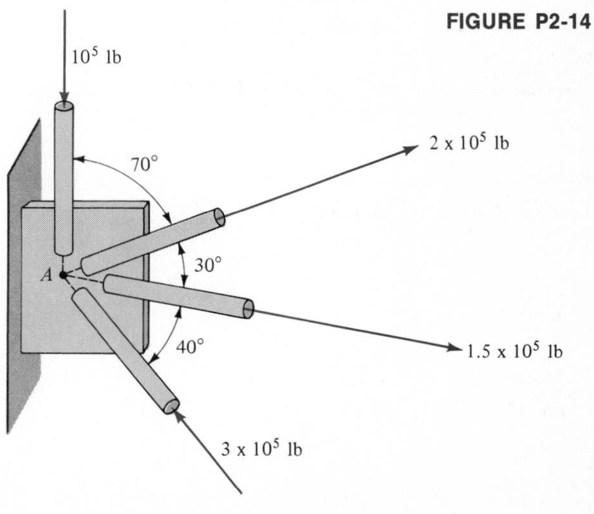

2–14 Determine by trigonometry the resultant of the four coplanar forces acting on the welded bracket.

2–15 Draw the polygon for the four forces and their resultant in Fig. P2-14. Assume that the lowest bar suddenly fails so that the force of 3×10^5 lb becomes zero. Show the new resultant of the remaining three forces on the same diagram.

FIGURE P2-16

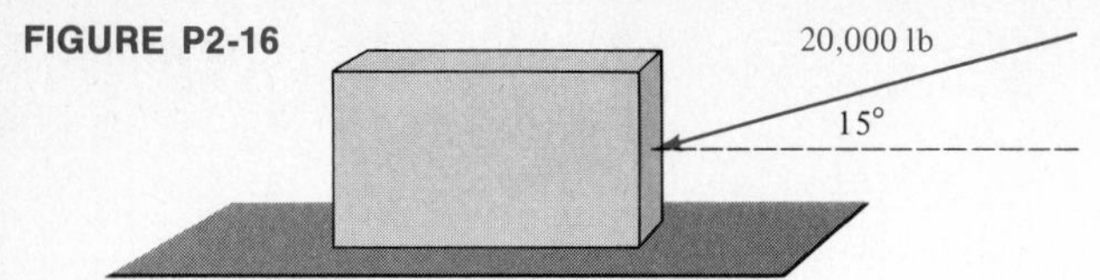

2–16 One of the constraining forces acting on the block of a tuned mass damper (see Sec. 1-7) is inclined with the horizontal as shown. Determine the horizontal and vertical components of this force.

FIGURE P2-17

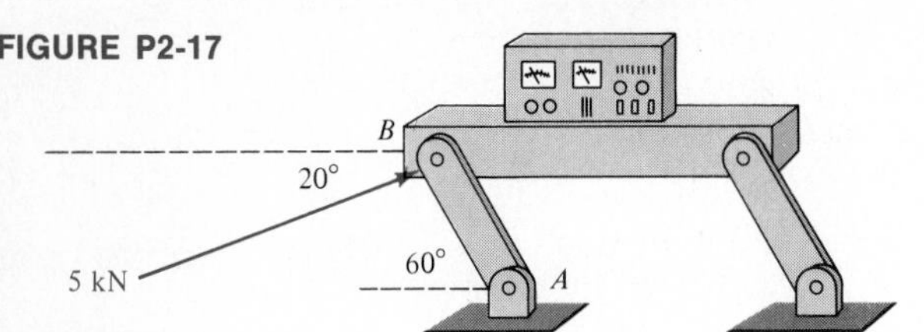

2–17 Determine the horizontal and vertical components of the force acting on the test platform. Also determine the component perpendicular to the arm *AB*.

FIGURE P2-18

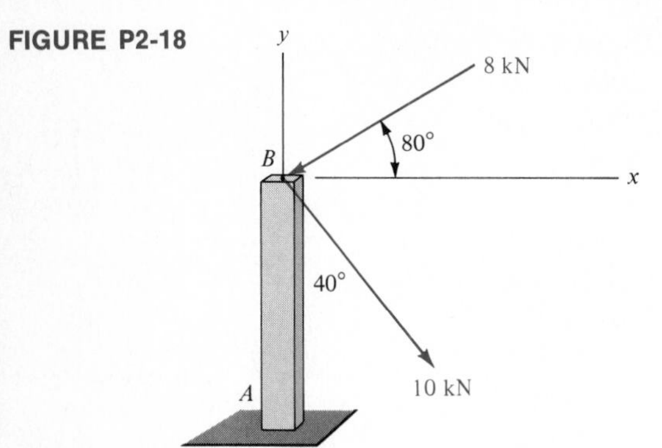

2–18 Determine the *x* and *y* components of each force.

FIGURE P2-19

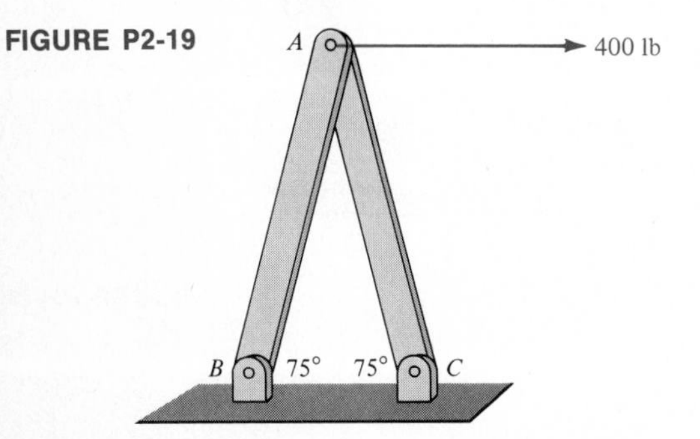

2–19 Resolve the horizontal force into components along lines *AB* and *AC*.

FIGURE P2-20

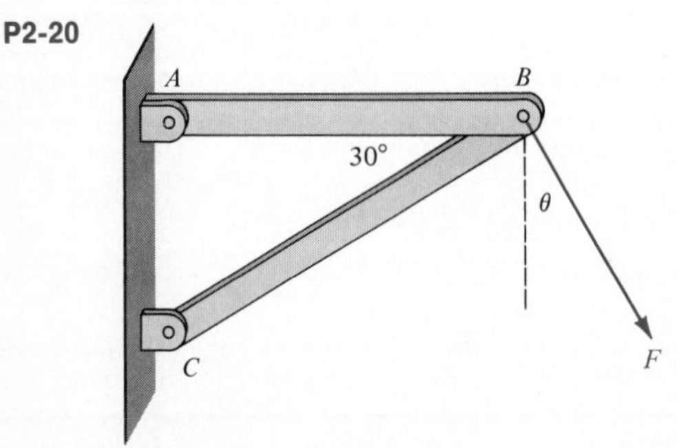

2–20 Plot the components of the force along bars *AB* and *BC* for $\theta = 0$, 30°, and 60°. Enter all data on the same diagram and sketch smooth curves through the appropriate points for each bar.

There are many situations where the directions of forces remain constant but their magnitudes change. For example, a force of magnitude $0.9P$ increases to a magnitude of $2.2P$ while acting in the same direction. This change can be illustrated as a *uniaxial* vector addition (meaning that there is only one line of action) as in Fig. 2-10a. The change of magnitude in the uniaxial vector addition is equivalent to a scalar addition. Thus, the operation $0.9P + 1.3P = 2.2P$ completely describes the change of the vector in the example. The same change can be expressed as a multiplication of the magnitude of the vector by a scalar, $2.444(0.9P) = 2.2P$.

A decrease in the magnitude of a vector can be considered as a uniaxial vector subtraction (Fig. 2-10b). This can be described by a scalar subtraction or a division of the magnitude of the vector by a scalar. For example, the above vector of magnitude $2.2P$ decreases to $1.7P$. In scalar subtraction this is stated as $2.2P - 0.5P = 1.7P$. Using the scalar division procedure, $2.2P \div 1.294 = 1.7P$.

The multiplication or division of a vector by a scalar is important in the application of *unit vectors*. By definition, *a unit vector* **n** *is a quantity that has a direction; its magnitude is unity.* It is very useful that any vector **F** may be expressed as the unit vector **n**, taken in the same direction as **F**, multiplied by the scalar quantity F. This is illustrated in Fig. 2-11, where **n** is simply superimposed on **F** so that $\mathbf{F} = F\mathbf{n}$. For writing this expression, it is not important where **n** is shown on the line of action of **F**, but the two vectors must have the same direction.

The concept of the unit vector can be extended to the components of vectors. The unit vectors along rectangular coordinates are so frequently used that they have widely accepted names, **i**, **j**, and

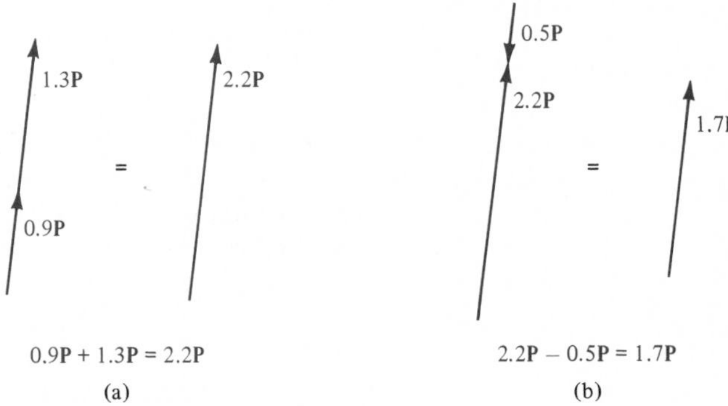

FIGURE 2-10

Uniaxial vector addition and subtraction

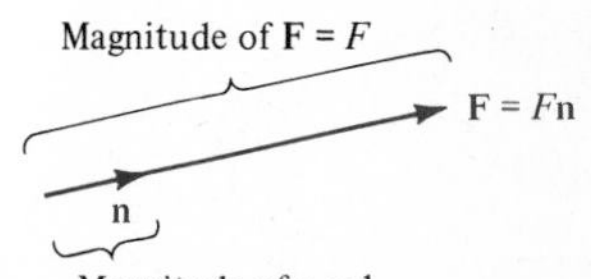

FIGURE 2-11

Representation of a vector **F** by its own magnitude F multiplied by the unit vector **n** of the same direction as **F**

k, as shown in Fig. 2-12. Each one of these has a magnitude of unity. These unit vectors provide a concise, written scheme of notation for vectors. The scheme is illustrated for a force with two-dimensional components in Fig. 2-13. The *vector component* $\mathbf{F}_x$ can be written as its magnitude F_x times the unit vector **i**, and similarly for the y component,

$$\mathbf{F}_x = F_x\mathbf{i} \qquad \mathbf{F}_y = F_y\mathbf{j} \tag{2-2}$$

The original force **F** is completely stated by

$$\mathbf{F} = F_x\mathbf{i} + F_y\mathbf{j} \tag{2-3}$$

The magnitudes F_x and F_y are the *scalar components* of the vector **F**. They are calculated from the magnitude of **F** and the angle θ in Fig. 2-13,

$$F_x = F\cos\theta \qquad F_y = F\sin\theta \tag{2-4}$$

The angle θ can be readily calculated if the components are known,

$$\tan\theta = \frac{F_y}{F_x} \tag{2-5}$$

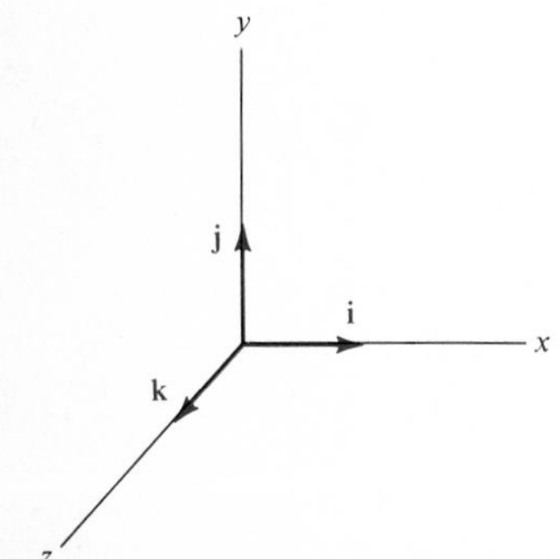

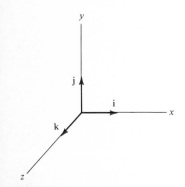

FIGURE 2-12

Unit vectors in rectangular coordinates

The magnitude F of **F** can be determined either from Eq. 2-4 or from

$$F = \sqrt{F_x^2 + F_y^2} \tag{2-6}$$

which is derived from Fig. 2-13 using the Pythagorean theorem.

The unit vector notation is convenient in vector summation. Since the addition or subtraction of uniaxial vectors can be performed as scalar addition or subtraction (Fig. 2-10), all components denoted by the same unit vector can be summed as scalars. Thus, if **A**, **B**, and **C** are coplanar vectors and **R** is their resultant, $\mathbf{A} + \mathbf{B} + \mathbf{C} = \mathbf{R}$ can be written as $A_x\mathbf{i} + A_y\mathbf{j} + B_x\mathbf{i} + B_y\mathbf{j} + C_x\mathbf{i} + C_y\mathbf{j} = (A_x + B_x + C_x)\mathbf{i} + (A_y + B_y + C_y)\mathbf{j} = R_x\mathbf{i} + R_y\mathbf{j}$. Using summation notation for any number of vectors,

$$\textit{Scalar components:}\quad R_x = \sum (x \text{ components}) \qquad R_y = \sum (y \text{ components}) \tag{2-7a}$$

$$\textit{Vector components:}\quad \mathbf{R}_x = \sum \mathbf{F}_x = \sum F_x\mathbf{i} \qquad \mathbf{R}_y = \sum \mathbf{F}_y = \sum F_y\mathbf{j} \tag{2-7b}$$

Note that in practice the intermediate step such as $A_x\mathbf{i} + B_x\mathbf{i} + \dots$ is frequently skipped and $(A_x + B_x + \dots)\mathbf{i}$ is written directly.

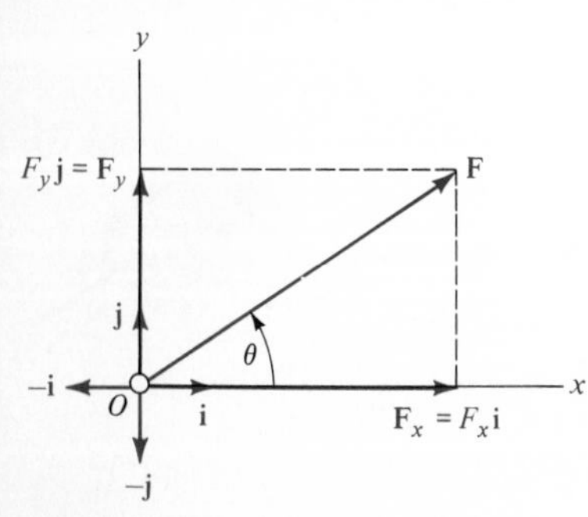

FIGURE 2-13

Unit vector notation for the components of vector **F**

Sign conventions. It is necessary to discuss the sign conventions for unit vectors, vector components, and scalar components of vectors.

1. *The unit vectors* **i**, **j**, *and* **k** are positive when directed along the corresponding positive coordinate axes, and vice versa (**i** along x, $-\mathbf{i}$ along $-x$, etc.).

2. *Rectangular vector components* are positive when directed along the positive coordinate axes, and vice versa. Since each vector component is given in terms of a scalar component times a unit vector, the sign of the product (scalar component) · (unit vector) gives the direction of the vector component.

3. *Scalar components* are given signs according to the signs that the corresponding vector components would have (F_x corresponds to $F_x\mathbf{i}$, $-F_x$ corresponds to $-F_x\mathbf{i}$).

The application of signs is illustrated with respect to Fig. 2-13 as follows. Note in particular the use of negative signs such as in $-\mathbf{i}$ vs. $-F_x\mathbf{i}$ vs. $-F_x$ (the vector components can be directly written from the appropriate scalar components if the unit vectors are always taken as positive!).

QUANTITY	INDICATES
$F_x\mathbf{i}$	Vector component $\mathbf{F}_x$ of magnitude F_x is directed in positive x direction (defined by **i**)
$-F_x\mathbf{i}$ (not shown)	Vector component $\mathbf{F}_x$ of magnitude F_x is directed in negative x direction (defined by $-\mathbf{i}$)
F_x	Scalar component of vector $F_x\mathbf{i}$ which is directed in positive x direction
$-F_x$	Scalar component of vector $-F_x\mathbf{i}$ which is directed in negative x direction
$-\mathbf{i}$	Vector **i** of magnitude 1 is directed in negative x direction
$\mathbf{j}$	Vector **j** of magnitude 1 is directed in positive y direction
$F_y\mathbf{j}$	Vector component $\mathbf{F}_y$ of magnitude F_y is directed in positive y direction
$-F_y\mathbf{j}$ (not shown)	Vector component $\mathbf{F}_y$ of magnitude F_y is directed in negative y direction

Unit Vectors in Parallel Coordinate Systems

The **ijk** unit vectors of an xyz coordinate system (such as in Fig. 2-12) are also valid for any other $x'y'z'$ coordinate system if the respective axes are parallel to one another (x is parallel to x', etc.), because the unit vectors indicate directions in space. This is especially useful in situations where the forces are not concurrent.

EXAMPLE 2-4

Write an expression for the resultant **R** of the two forces in Fig. 2-14a using force components with unit vectors.

SOLUTION

First the x and y components of the forces are written as functions of the given angles.

From Fig. 2-14b and c,

$$\mathbf{F}_{1x} = F_1 \cos\theta\,\mathbf{i} \qquad \mathbf{F}_{2x} = -F_2 \sin\phi\,\mathbf{i}$$

$$\mathbf{F}_{1y} = F_1 \sin\theta\,\mathbf{j} \qquad \mathbf{F}_{2y} = F_2 \cos\phi\,\mathbf{i}$$

The resultant force has two components, which are determined using Eq. 2-7,

$$\mathbf{R}_x = \sum \mathbf{F}_x = F_1 \cos\theta\,\mathbf{i} - F_2 \sin\phi\,\mathbf{i}$$

$$\mathbf{R}_y = \sum \mathbf{F}_y = F_1 \sin\theta\,\mathbf{j} + F_2 \cos\phi\,\mathbf{j}$$

The resultant is the vector sum of the components,

$$\mathbf{R} = R_x\mathbf{i} + R_y\mathbf{j} = (F_1 \cos\theta - F_2 \sin\phi)\mathbf{i} + (F_1 \sin\phi + F_2 \cos\phi)\mathbf{j}$$

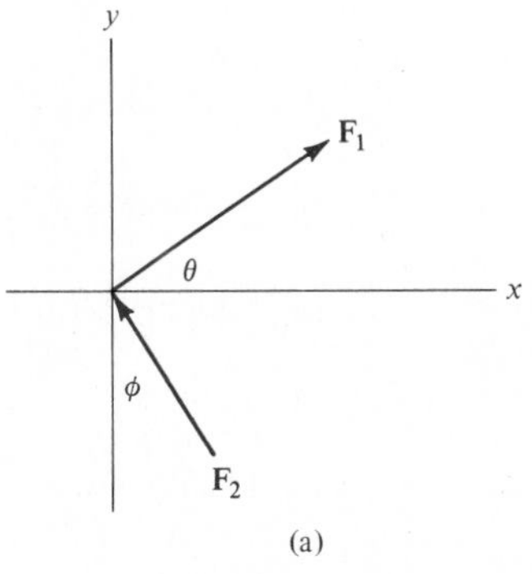

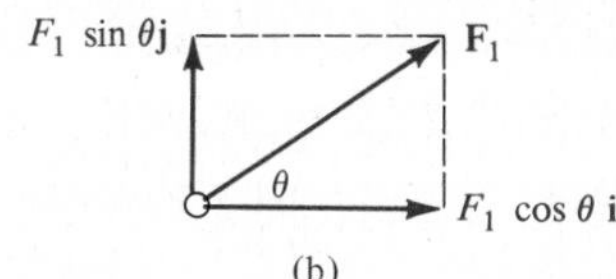

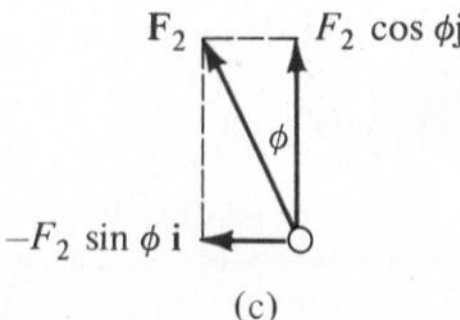

FIGURE 2-14
Summation of two forces using components and unit vectors

EXAMPLE 2-5

Consider an anchoring block of a tall antenna tower that holds six guy wires for steadying the tower (Fig. 2-15). The initial tension in each wire is 23 kN when there is no wind. Assume a simple model of this junction with the forces concurrent. Determine the resultant force of the six tensions in terms of its magnitude R and angle θ with respect to the horizontal.

SOLUTION

First measure the angle of each wire with the horizontal in Fig. 2-15. The resultant force is $\mathbf{R} = R_x\mathbf{i} + R_y\mathbf{j}$, where the components R_x and R_y are determined using Eq. 2-7.

$$\mathbf{R}_x = \sum \mathbf{F}_x = 23 \text{ kN}(\cos 15^\circ + \cos 30^\circ + \cos 40^\circ + \cos 50^\circ + \cos 57.5^\circ + \cos 65^\circ)\mathbf{i}$$

$$\mathbf{R}_x = 23 \text{ kN}(4.20)\mathbf{i} = 96.6 \text{ kN } \mathbf{i}$$

$$\mathbf{R}_y = \sum \mathbf{F}_y = 23 \text{ kN}(\sin 15^\circ + \sin 30^\circ + \sin 40^\circ + \sin 50^\circ + \sin 57.5^\circ + \sin 65^\circ)\mathbf{j}$$

$$\mathbf{R}_y = 23 \text{ kN}(3.92)\mathbf{j} = 90.1 \text{ kN } \mathbf{j}$$

Using Fig. 2-13, we obtain

$$R = \sqrt{R_x^2 + R_y^2} = 132.1 \text{ kN} \qquad \tan\theta = \frac{R_y}{R_x} = \frac{90.1 \text{ kN}}{96.6 \text{ kN}} = 0.933$$

$$\theta = 43.0^\circ$$

JUDGMENT OF THE RESULTS

The magnitude and direction of the resultant force appear reasonable by inspection of Fig. 2-15.

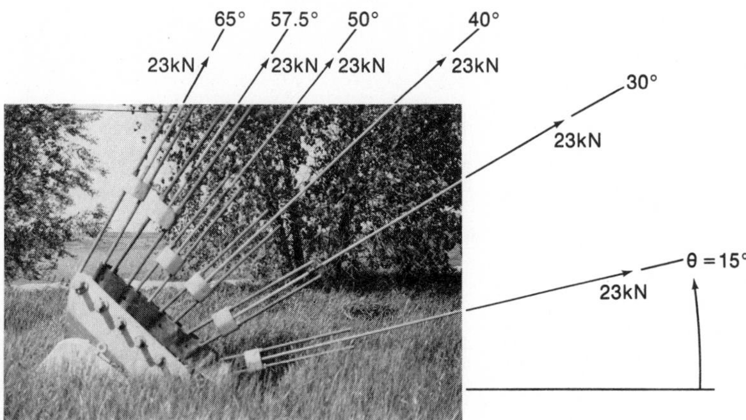

FIGURE 2-15

Concurrent tensions in the guy wires of an antenna tower

EXAMPLE 2-6

Reconsider the wheel hub in Ex. 2-3 and determine the net force **R** on it using the addition of force components with unit vectors.

SOLUTION

Use the notation in Fig. 2-13. Denote the magnitude of each tension by T, and go clockwise starting from the top.

$$\mathbf{R} = (O\mathbf{i} + T\mathbf{j}) + \left(\frac{\sqrt{2}}{2}T\mathbf{i} + \frac{\sqrt{2}}{2}T\mathbf{j}\right) + (T\mathbf{i} + O\mathbf{j})$$

$$+ \left(\frac{\sqrt{2}}{2}T\mathbf{i} - \frac{\sqrt{2}}{2}T\mathbf{j}\right) + (O\mathbf{i} - T\mathbf{j}) + \left(-\frac{\sqrt{2}}{2}T\mathbf{i} - \frac{\sqrt{2}}{2}T\mathbf{j}\right)$$

$$+ (-T\mathbf{i} + O\mathbf{j}) + \left(-\frac{\sqrt{2}}{2}T\mathbf{i} + \frac{\sqrt{2}}{2}T\mathbf{j}\right)$$

$$\mathbf{R} = \left(\frac{\sqrt{2}}{2} + 1 + \frac{\sqrt{2}}{2} - \frac{\sqrt{2}}{2} - 1 - \frac{\sqrt{2}}{2}\right)T\mathbf{i}$$

$$+ \left(1 + \frac{\sqrt{2}}{2} - \frac{\sqrt{2}}{2} - 1 - \frac{\sqrt{2}}{2} + \frac{\sqrt{2}}{2}\right)T\mathbf{j} = O\mathbf{i} + O\mathbf{j} = \mathbf{0} = 0$$

JUDGMENT OF THE RESULTS

It agrees with the previous result in Ex. 2-1. Since **R** is a vector, it is correct to give the answer as the *zero vector* **0**. However, since the resultant of zero magnitude cannot have a direction, it is also reasonable to use the scalar zero 0 instead of the zero vector.

PROBLEMS

2–21 Determine the resultant **R** of the two forces using unit vectors. $F_1 = F_2 = 3$ kN and $\theta_1 = \theta_2 = 20°$.

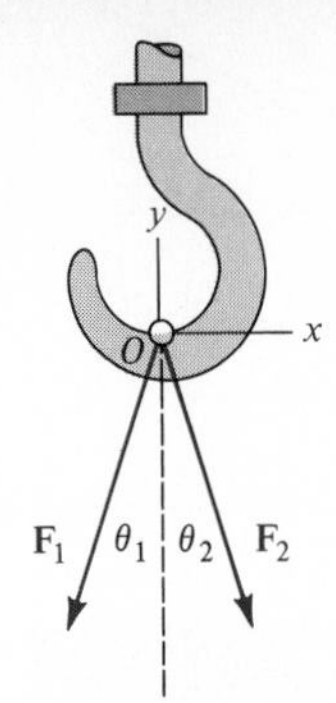

FIGURE P2-21

2–22 In Fig. P2-21, $F_1 = 500$ lb, $F_2 = 800$ lb, and $\theta_1 = \theta_2 = 30°$. Express the resultant **R** in terms of unit vectors.

2–23 Express the resultant **R** of the two forces in terms of the magnitude **R** and the angle ϕ of **R** with the x axis. $F_1 = 800$ N, $F_2 = 1.4$ kN, $\theta_1 = 30°$, and $\theta_2 = 45°$.

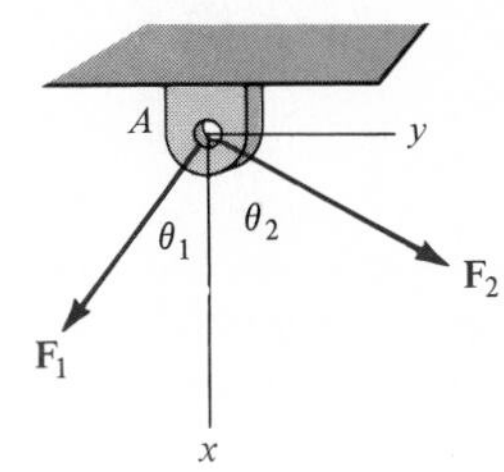

FIGURE P2-23

2–24 In Fig. P2-23, $F_1 = 300$ lb at $\theta_1 = 25°$. Determine the magnitude of force $\mathbf{F}_2$ at $\theta_2 = 15°$ to make the resultant of $\mathbf{F}_1$ and $\mathbf{F}_2$ vertical. Use unit vectors.

2–25 Express the resultant **R** of the two forces using unit vectors. $F_1 = 5$ kN, $F_2 = 1$ kN, $\theta_1 = 25°$, and $\theta_2 = 30°$.

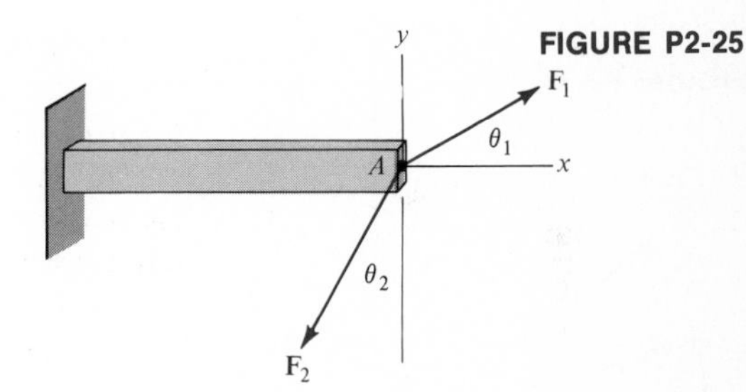

FIGURE P2-25

2–26 A boat is towed in a narrow channel by applying forces $F = 4$ kN and $P = 6$ kN through ropes attached to the boat. Express the resultant force using unit vectors.

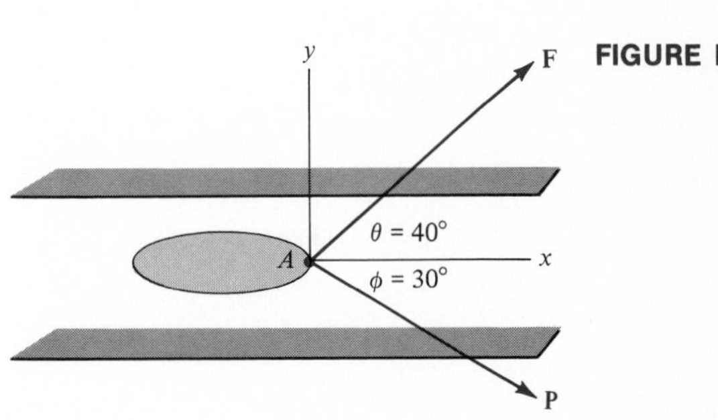

FIGURE P2-26

2–27 Two ropes are attached to a post at point A. For the given tensions in the ropes, plot the x component of the resultant force **R** as a function of $\theta = 0, 10°$, and $20°$ (plot R_x vs. θ).

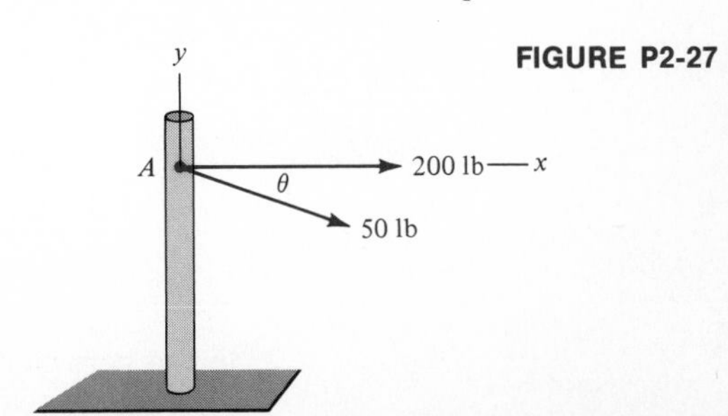

FIGURE P2-27

FIGURE P2-28

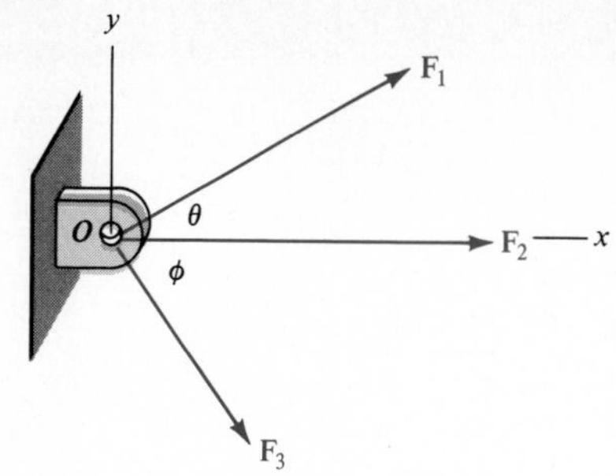

2–28 Determine the resultant of the three coplanar forces using unit vectors. $F_1 = 500$ N, $F_2 = 800$ N, $F_3 = 900$ N, $\theta = 30°$, and $\phi = 50°$.

2–29 Determine by rectangular components the smallest additional force $\mathbf{F}_4$ to make the resultant component $R_y = 0$ in Fig. P2-28. $F_1 = 200$ lb, $F_2 = 250$ lb, $F_3 = 400$ lb, $\theta = 20°$, and $\phi = 70°$.

FIGURE P2-30

2–30 Determine the resultant of the three forces in terms of its magnitude R and the angle α of $\mathbf{R}$ with the x axis. $F_1 = 4$ kN, $F_2 = 5$ kN, $F_3 = 7.5$ kN, $\theta = 45°$, and $\phi = 60°$. Use rectangular components.

2–31 Determine by rectangular components the smallest additional force $\mathbf{F}_4$ to make the resultant component $R_y = 0$ in Fig. P2-30. $F_1 = 1000$ lb, $F_2 = 1500$ lb, $F_3 = 800$ lb, $\theta = 40°$, and $\phi = 50°$.

FIGURE P2-32

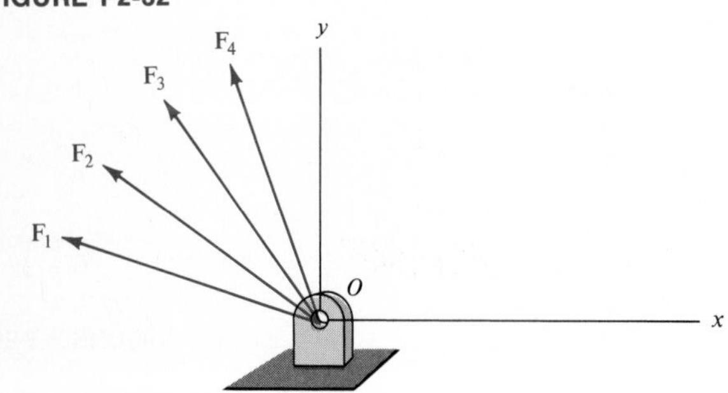

2–32 The anchoring block for four guy wires of a radio antenna tower is modeled as a simple bracket with concurrent forces. Determine the resultant force $\mathbf{R}$ if $\mathbf{F}_1 = (-15\mathbf{i} + 5\mathbf{j})$ kN, $\mathbf{F}_2 = (-12\mathbf{i} + 7\mathbf{j})$ kN, $\mathbf{F}_3 = (-10\mathbf{i} + 9\mathbf{j})$ kN, and $\mathbf{F}_4 = (-8\mathbf{i} + 10\mathbf{j})$ kN.

FIGURE P2-33

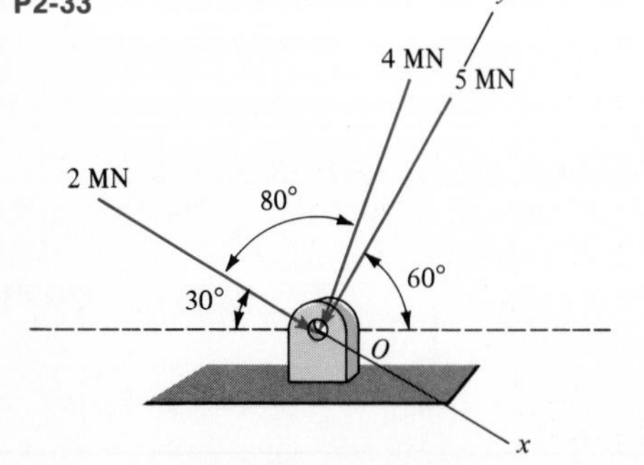

2–33 Determine the resultant force $\mathbf{R}$ using unit vectors in the xy coordinate system shown.

2–34 Determine the smallest additional force $\mathbf{P}$ that would make the resultant component $R_x = 0$ in Fig. P2-33.

2–35 Determine the resultant **R** of the four forces using unit vectors in the xy coordinate system shown.

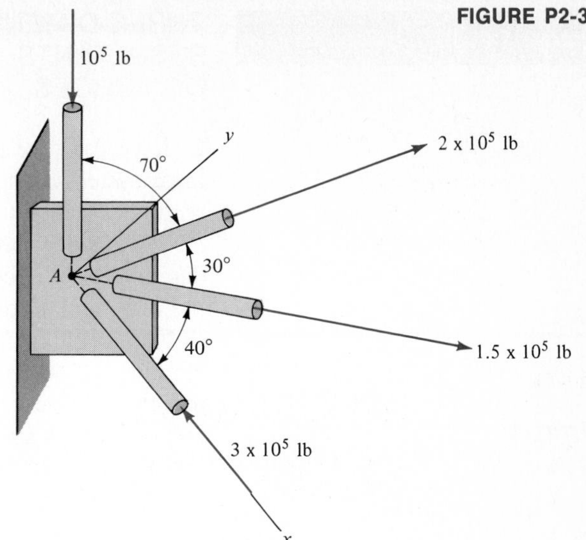

FIGURE P2-35

2–36 Solve Prob. 2-35 as stated. Double the magnitude of each force and solve for the new resultant **R**′. Write the ratio R/R'.

2–37 Express the resultant **R** of the three forces in terms of unit vectors of the xy coordinate system.

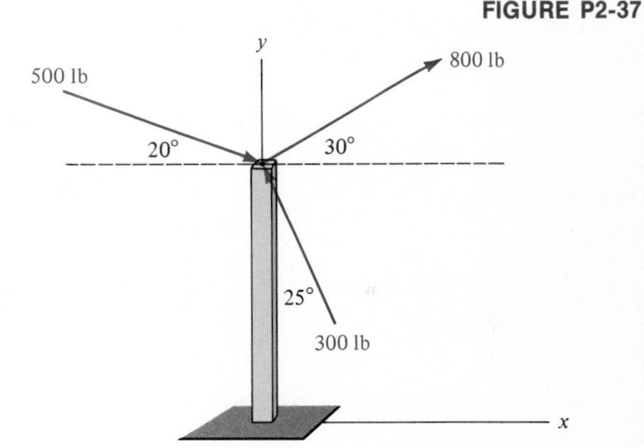

FIGURE P2-37

2–38 Two forces are acting on the bracket of a road simulator (see Sec. 1-7). Express each force using unit vectors of the xy coordinate system. Determine the resultant **R** of the two forces assuming that they are concurrent.

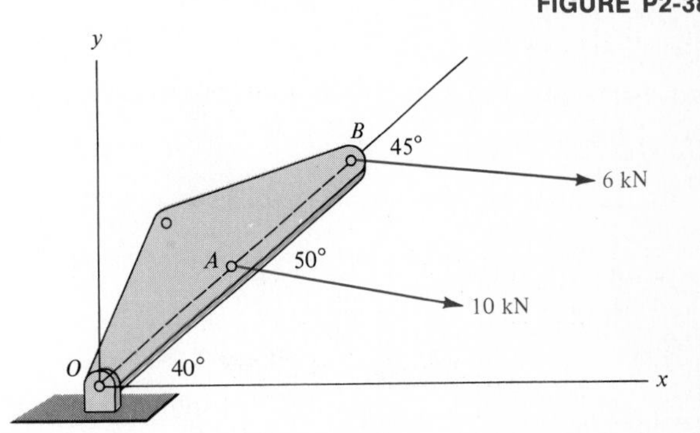

FIGURE P2-38

2-4 THREE-DIMENSIONAL FORCE SYSTEMS. RESULTANTS AND COMPONENTS OF FORCES

Unit vectors are most useful when applied to three-dimensional force systems. The analysis of a force in two dimensions can be readily extended to three dimensions.

Consider force **F** in Fig. 2-16 and a box for which **F** is a diagonal line. Each edge of the box is along or parallel to one of the rectangular coordinate axes, x, y, and z. The three components of **F** are equal in magnitude to the respective dimensions of the box. From Fig. 2-16, $F_x = OH$, $F_y = OC$, and $F_z = OE$. A detailed explanation of these components can be made using two-dimensional components in succession. First, the vector components $\mathbf{F}_y$ and $\mathbf{F}_G$ are drawn in the plane $OGAC$. $\mathbf{F}_y$ is one of the major components sought, so it is completely determined. Next, $\mathbf{F}_G$ is seen as the diagonal of the rectangle $OEGH$ in the xz plane. $\mathbf{F}_x$ and $\mathbf{F}_z$ are the components of $\mathbf{F}_G$ in this plane. The force **F** can be written in terms of the scalar components and the unit vectors,

$$\mathbf{F} = F_x\mathbf{i} + F_y\mathbf{j} + F_z\mathbf{k} \tag{2-8}$$

The scalar components are also given by

$$F_x = F\cos\theta_x \qquad F_y = F\cos\theta_y \qquad F_z = F\cos\theta_z \tag{2-9}$$

The angles θ_x, θ_y, and θ_z are defined in Fig. 2-17. θ_x is between **F** and the x axis, θ_y is between **F** and the y axis, and θ_z is between **F** and the z axis. The cosines of these angles are the *direction cosines* of the force **F**. The angles are always between 0 and 180°. They are measured from the positive coordinate axes as in Fig. 2-17. The

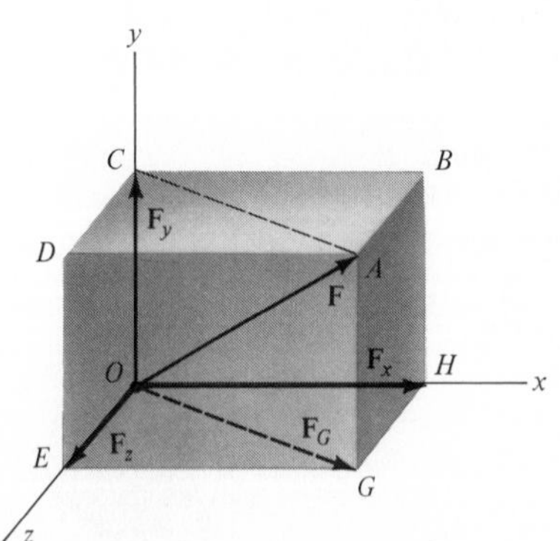

FIGURE 2-16

Three-dimensional components of vector **F**

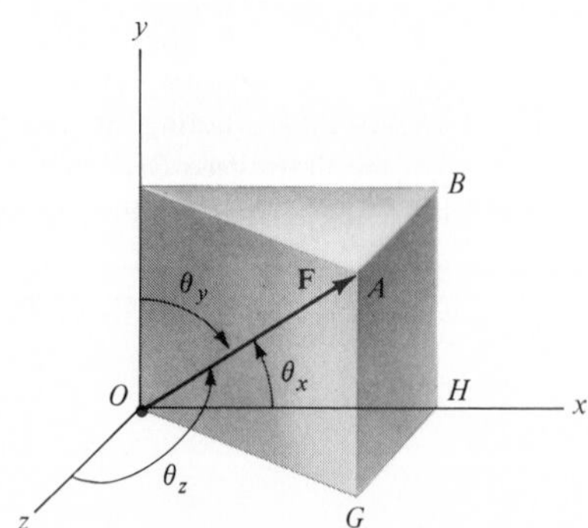

FIGURE 2-17

Angles used to define the direction of vector **F**

direction cosines are positive or negative depending on the size of the angles.

Another relation between the magnitude of **F** and its rectangular scalar components is obtained from the triangles OAC and OGH in Fig. 2-16. According to the Pythagorean theorem,

$$F^2 = F_y^2 + F_G^2$$

$$F_G^2 = F_x^2 + F_z^2$$

From these, the magnitude of the vector **F** is

$$\boxed{F = \sqrt{F_x^2 + F_y^2 + F_z^2}} \qquad \text{2-10}$$

The substitution of Eq. 2-9 into Eq. 2-8 gives

$$\mathbf{F} = F(\cos\theta_x\,\mathbf{i} + \cos\theta_y\,\mathbf{j} + \cos\theta_z\,\mathbf{k}) = F\mathbf{n} \qquad \text{2-11}$$

where $\cos\theta_x$, $\cos\theta_y$, and $\cos\theta_z$ are the scalar components of the vector **n** whose magnitude is equal to 1. The unit vector **n** is collinear with **F** as in Fig. 2-18. The components of the unit vector are equal to the corresponding direction cosines of **F**,

$$n_x = \cos\theta_x \qquad n_y = \cos\theta_y \qquad n_z = \cos\theta_z \qquad \text{2-12}$$

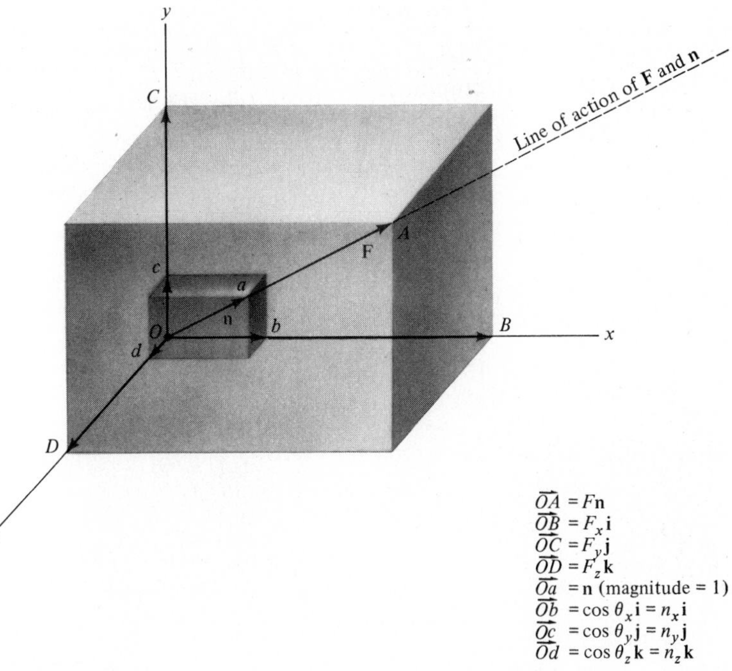

$\overrightarrow{OA} = F\mathbf{n}$
$\overrightarrow{OB} = F_x\mathbf{i}$
$\overrightarrow{OC} = F_y\mathbf{j}$
$\overrightarrow{OD} = F_z\mathbf{k}$
$\overrightarrow{Oa} = \mathbf{n}$ (magnitude = 1)
$\overrightarrow{Ob} = \cos\theta_x\mathbf{i} = n_x\mathbf{i}$
$\overrightarrow{Oc} = \cos\theta_y\mathbf{j} = n_y\mathbf{j}$
$\overrightarrow{Od} = \cos\theta_z\mathbf{k} = n_z\mathbf{k}$

FIGURE 2-18

Rectangular components of force **F** and unit vector **n**

These can be combined as in Eq. 2-10. Noting that the magnitude of **n** is 1, the result is

$$n_x^2 + n_y^2 + n_z^2 = 1 \qquad \text{2-13}$$

It is seen in Fig. 2-18 that the respective sides and diagonals of the two boxes are proportional to one another. These can be expressed formally as

$$\frac{n_x}{F_x} = \frac{n_y}{F_y} = \frac{n_z}{F_z} = \frac{1}{F} \qquad \text{2-14}$$

In practice it is not necessary to draw either of the boxes to exactly define a vector **F** as in Fig. 2-18. The analysis can be performed if the coordinates of two points on the line of action are known. This is demonstrated for a two-dimensional vector for simplicity, and can be extended for a three-dimensional vector.

Consider the problem of defining a force **F** from point A to C on the line AB in Fig. 2-19. Assume that the coordinates of points A and B are known and that $\mathbf{F}'$ is an imaginary vector from A to B. The differences d_x and d_y between the coordinates are the scalar components of $\mathbf{F}'$, so

$$\mathbf{F}' = d_x\mathbf{i} + d_y\mathbf{j} = d\mathbf{n} \qquad \text{2-15}$$

where d is the distance AB, $d = \sqrt{d_x^2 + d_y^2}$. The real force **F** is given in terms of its scalar components,

$$\mathbf{F} = F_x\mathbf{i} + F_y\mathbf{j} = F\mathbf{n} \qquad \text{2-16}$$

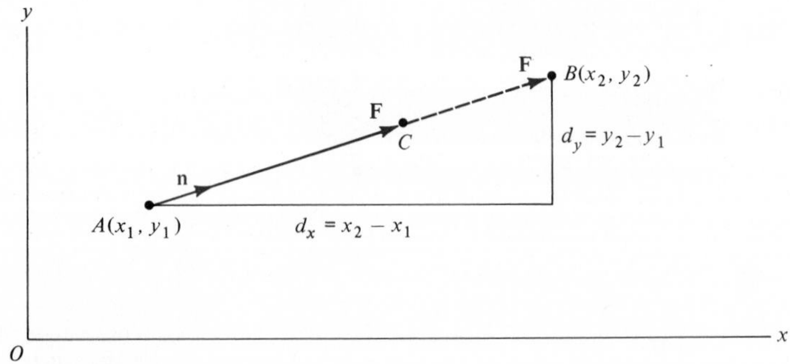

FIGURE 2-19

Direction of force **F** defined by points *A* and *B* on its line of action

The respective terms in Eq. 2-15 and 2-16 are proportional, and the results are also valid for the third dimension z (on the basis of Fig. 2-18). Therefore,

$$\frac{F_x}{d_x} = \frac{F_y}{d_y} = \frac{F_z}{d_z} = \frac{F}{d} \qquad \boxed{2\text{-}17}$$

Finally, from Eqs. 2-14 and 2-17,

$$\frac{n_x}{d_x} = \frac{n_y}{d_y} = \frac{n_z}{d_z} = \frac{1}{d} \qquad \boxed{2\text{-}18}$$

Addition of Forces

The unit vector notation is the most convenient for the addition of concurrent forces. The method of addition that was already used for two-dimensional force systems is equally valid and even more advantageous for three-dimensional forces. In general, each component of the resultant force equals the sum of the corresponding individual components. As in Eq. 2-7,

Scalar components:

$$R_x = \sum F_x$$
$$R_y = \sum F_y \qquad \boxed{2\text{-}19a}$$
$$R_z = \sum F_z$$

Vector components:

$$\mathbf{R}_x = \sum \mathbf{F}_x = \sum F_x\mathbf{i}$$
$$\mathbf{R}_y = \sum \mathbf{F}_y = \sum F_y\mathbf{j} \qquad \boxed{2\text{-}19b}$$
$$\mathbf{R}_z = \sum \mathbf{F}_z = \sum F_z\mathbf{k}$$

2-5 METHOD FOR DETERMINING FORCE VECTOR FROM SCALAR INFORMATION

In some cases a force may be given in terms of its magnitude F, sense of direction, and the coordinates of two points on the line of action of the force. It is often advantageous in the solution of problems to express such a force as a vector quantity. The transformation of the scalar quantities into the appropriate vector can be made using the following procedure based on Fig. 2-19 as a simple model (the third dimension z is added for completeness).

Given scalar information:

1. F, magnitude of force
2. (x_1, y_1, z_1), coordinates of point A on line of action of force $\mathbf{F}$

3. (x_2, y_2, z_2), coordinates of point B on line of action of force **F**
4. Direction of force **F** from point A to B

Unknown to be determined: Vector expression

$$\mathbf{F} = F_x\mathbf{i} + F_y\mathbf{j} + F_z\mathbf{k}$$

Method of solution:

1. From Eq. 2-11,

$$\mathbf{F} = F\mathbf{n} \qquad \boxed{2\text{-}20}$$

where F is known.

2. Define a vector $\mathbf{F}'$ between the given points A and B as in Eq. 2-15,

$$\mathbf{F}' = d_x\mathbf{i} + d_y\mathbf{j} + d_z\mathbf{k} = d\mathbf{n} \qquad \boxed{2\text{-}21}$$

where $d = \sqrt{d_x^2 + d_y^2 + d_z^2}$.

3. Express the unit vector

$$\mathbf{n} = n_x\mathbf{i} + n_y\mathbf{j} + n_z\mathbf{k} = \frac{d_x\mathbf{i} + d_y\mathbf{j} + d_z\mathbf{k}}{d} \qquad \boxed{2\text{-}22}$$

4. Substitute Eq. 2-22 into Eq. 2-20,

$$\mathbf{F} = F\frac{(d_x\mathbf{i} + d_y\mathbf{j} + d_z\mathbf{k})}{d} \qquad \boxed{2\text{-}23}$$

5. With $F_x = Fd_x/d$, etc.,

$$\mathbf{F} = \frac{Fd_x}{d}\mathbf{i} + \frac{Fd_y}{d}\mathbf{j} + \frac{Fd_z}{d}\mathbf{k} \qquad \boxed{2\text{-}24}$$

The desired vector **F** can be expressed either by Eq. 2-23 or 2-24.

Notations for Units of a Vector

There are several acceptable ways to write the units of a vector when the vector is given in terms of its rectangular components and the corresponding unit vectors. It is only necessary to use the same units for all components. For example,

$$\begin{aligned}\mathbf{F} &= 50\ \text{N}\ \mathbf{i} - 80\ \text{N}\ \mathbf{j} + 130\ \text{N}\ \mathbf{k}\\ &= (50\mathbf{i} - 80\mathbf{j} + 130\mathbf{k})\ \text{N}\\ &= 50\mathbf{i} - 80\mathbf{j} + 130\ \mathbf{k}\ \text{N}\end{aligned}$$

The last expression is the simplest in writing but the least precise of the three. It is used because its meaning is generally understood. The student will see and practice these different notations throughout Statics and Dynamics.

EXAMPLE 2-7

Consider the mast of a sailboat with three guy wires as shown in Fig. 2-20a. Each initial tension in the wires at B and C is planned to be 200 lb. Determine whether the same tension in the third wire would make the horizontal resultant R_h at A equal to zero.

SOLUTION

From Fig. 2-20b,

$$\mathbf{R} = \overrightarrow{AB} + \overrightarrow{AC} + \overrightarrow{AD}$$

The three tensions of known magnitude (200 lb) must be written as vectors. Using Eqs. 2-22 and 2-23,

$$\overrightarrow{AB} = (\text{tension } AB)(\text{unit vector } A \text{ to } B)$$

$$= 200 \text{ lb } \mathbf{n}_{AB} = 200 \text{ lb} \frac{(d_x\mathbf{i} + d_y\mathbf{j} + d_z\mathbf{k})}{d}$$

With the data in Fig. 2-20a,

$$\overrightarrow{AB} = \frac{200 \text{ lb}}{\sqrt{5^2 + 4^2 + 10^2}}(-5\mathbf{i} - 4\mathbf{j} - 10\mathbf{k})\frac{\text{ft}}{\text{ft}}$$

$$= -84.22 \text{ lb } \mathbf{i} - 67.37 \text{ lb } \mathbf{j} - 168.4 \text{ lb } \mathbf{k}$$

Similarly for the other two tensions,

$$\overrightarrow{AC} = \frac{200 \text{ lb}}{11.87 \text{ ft}}(5\mathbf{i} - 4\mathbf{j} - 10\mathbf{k}) \text{ ft} = 84.22 \text{ lb } \mathbf{i} - 67.37 \text{ lb } \mathbf{j} - 168.4 \text{ lb } \mathbf{k}$$

$$\overrightarrow{AD} = \frac{200 \text{ lb}}{11.66 \text{ ft}}(0\mathbf{i} + 6\mathbf{j} - 10\mathbf{k}) \text{ ft} = 102.9 \text{ lb } \mathbf{j} - 171.5 \text{ lb } \mathbf{k}$$

The resultant of the three tensions is written according to Eq. 2-19b,

$$\begin{aligned}\mathbf{R} &= \sum F_x\mathbf{i} + \sum F_y\mathbf{j} + \sum F_z\mathbf{k} \\ &= (-84.22 + 84.22 + 0) \text{ lb } \mathbf{i} + (-67.37 - 67.37 + 102.9) \text{ lb } \mathbf{j} \\ &\quad + (-168.4 - 168.4 - 171.5) \text{ lb } \mathbf{k} \\ &= 0 \text{ lb } \mathbf{i} - 31.8 \text{ lb } \mathbf{j} - 508.3 \text{ lb } \mathbf{k}\end{aligned}$$

The horizontal resultant of interest is obtained from

$$R_h = (R_x^2 + R_y^2)^{1/2} = (0^2 + 31.8^2)^{1/2} \text{ lb} = 31.8 \text{ lb} \neq 0$$

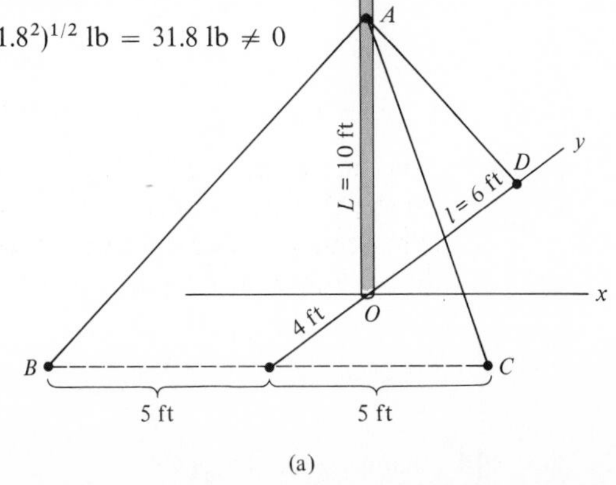

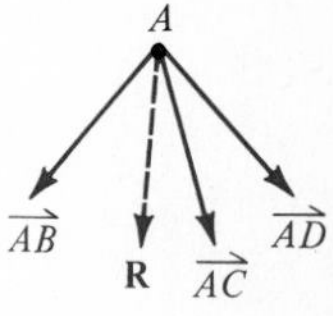

FIGURE 2-20
Model for mast of a sailboat with steadying guy wires

EXAMPLE 2-8

Figure 2-21a and b are the side and top views of a tall antenna tower, showing only the three guy wires AA', BB', and CC'. The origin O of the coordinate system is at the bottom of the tower. Anchoring points A, B, and C are in the xy plane (the ground), and the z axis is vertical along the tower. The distances on the ground from the tower to the anchoring points are $OA = OB = OC = 100$ m. Determine the angles θ_x, θ_y, θ_z of wires AA' and BB' with respect to the coordinate axes. Use unit vector notation, and assume that A', B', C' are practically at the same point.

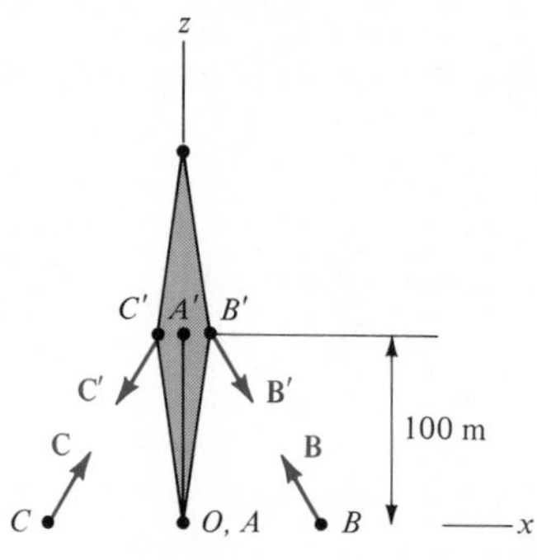

(a) Side view of tower

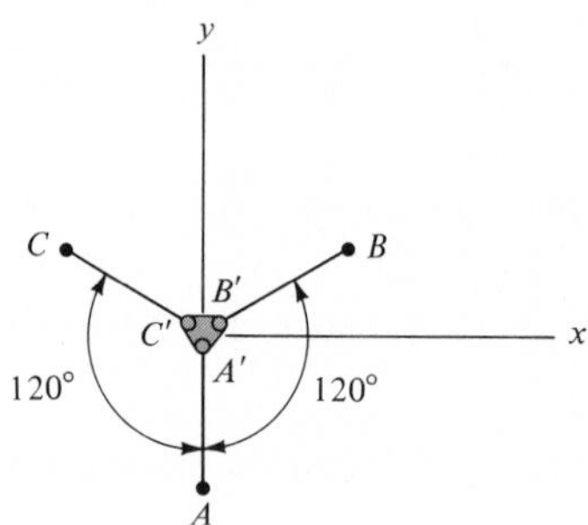

(b) Top view of tower

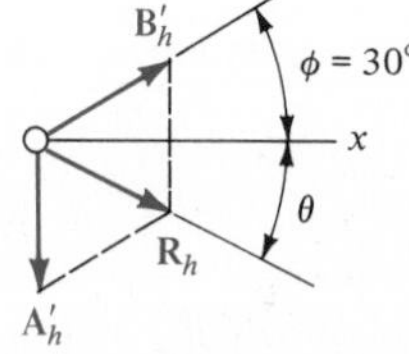

(c) Resultant of A_h' and B_h'

FIGURE 2-21

Three guy wires of an antenna tower

SOLUTION

The angles of the wires with respect to the axes are determined using Eqs. 2-12 and 2-22. Placing a unit vector from A toward A',

for wire AA'

$$n_x = \cos\theta_x = \frac{d_x}{d} = \frac{x_{A'} - x_A}{\sqrt{d_y^2 + d_z^2}}$$

$$= \frac{(0 - 0)\text{ m}}{\sqrt{100^2 + 100^2}\text{ m}} = 0, \qquad \theta_x = 90^\circ$$

$$n_y = \cos\theta_y = \frac{d_y}{d} = \frac{y_{A'} - y_A}{d}$$

$$= \frac{0 - (-100\text{ m})}{141.4\text{ m}} = 0.7071, \qquad \theta_y = 45^\circ$$

$$n_z = \cos\theta_z = \frac{d_z}{d} = \frac{z_{A'} - z_A}{d}$$

$$= \frac{100\text{ m} - 0}{141.4\text{ m}} = 0.7071, \qquad \theta_z = 45^\circ$$

Placing a unit vector from B toward B',

for wire BB'

$$n_x = \cos\theta_x = \frac{d_x}{d} = \frac{x_{B'} - x_B}{d}$$

$$= \frac{0 - 100\text{ m}\cos 30^\circ}{141.4\text{ m}} = -0.6125, \qquad \theta_x = 127.8^\circ$$

$$n_y = \cos\theta_y = \frac{d_y}{d} = \frac{y_{B'} - y_B}{d}$$

$$= \frac{0 - 100\text{ m}\sin 30^\circ}{141.4\text{ m}} = -0.3536, \qquad \theta_y = 110.7^\circ$$

$$n_z = \cos\theta_z = \frac{d_z}{d} = \frac{z_{B'} - z_B}{d}$$

$$= \frac{100\text{ m} - 0}{141.4\text{ m}} = 0.7071, \qquad \theta_z = 45^\circ$$

JUDGMENT AND INTERPRETATION OF THE RESULTS

The values for θ_y of wire AA' and for θ_z in both cases are correct by inspection. For wire BB', the complementary angles could be stated as the final results for θ_x and θ_y ($\theta_x = 52.2^\circ$, $\theta_y = 69.3^\circ$). For interpreting the angle between two nonintersecting lines such as BB' and the x axis, imagine a line x' that is parallel to x but intersects line BB'.

EXAMPLE 2-9

Assume a tension of 18 kN in each wire of the tower described in Ex. 2-8. (a) Express the forces applied by the wires at points A, B, A', and B' using unit vector notation. (b) Determine the horizontal resultant of the two forces at A' and B' (this would be the horizontal force acting on the tower immediately after wire CC' broke).

SOLUTION

(a) The force vectors at A and B are directed toward the tower, while those at A' and B' are directed toward A and B on the ground, respectively. Denote each force by the letter assigned to the point where the force acts. The force vectors all have the same magnitude, so each is expressed correctly when 18 kN is multiplied by the appropriate unit vector along the line of action. The required forces **A**, **B**, **A**′, and **B**′ are determined according to the following outline of major steps, using the results of Ex. 2-8.

FORCE	MAGNITUDE	DIRECTION	COMPONENTS OF UNIT VECTOR	FORCE = (MAGNITUDE)(UNIT VECTOR)
A	18 kN	A to A'	$0\mathbf{i} + 0.707\mathbf{j} + 0.707\mathbf{k}$	$\mathbf{A} = 12.73\mathbf{j} + 12.73\mathbf{k}$ kN
B	18 kN	B to B'	$-0.613\mathbf{i} - 0.354\mathbf{j} + 0.707\mathbf{k}$	$\mathbf{B} = -11.03\mathbf{i} - 6.37\mathbf{j} + 12.73\mathbf{k}$ kN
A′	18 kN	A' to A	$0\mathbf{i} - 0.707\mathbf{j} - 0.707\mathbf{k}$	$\mathbf{A}' = -12.73\mathbf{j} - 12.73\mathbf{k}$ kN
B′	18 kN	B' to B	$0.613\mathbf{i} + 0.354\mathbf{j} - 0.707\mathbf{k}$	$\mathbf{B}' = 11.03\mathbf{i} + 6.37\mathbf{j} - 12.73\mathbf{k}$ kN

(b) The horizontal resultant $\mathbf{R}_h$ of force **A**′ and **B**′ is the vector sum of their **i** and **j** components.

$$\mathbf{R}_h = 11.03\mathbf{i} + (-12.73 + 6.37)\mathbf{j}\text{ kN} = 11.03\mathbf{i} - 6.36\mathbf{j}\text{ kN}$$

JUDGMENT OF THE RESULTS

By Newton's law of action and reaction, forces **A** and **A**′ (and also **B** and **B**′) should be equal in magnitude and opposite in direction. The vector components of the pairs of forces satisfy these requirements. By inspection, $\mathbf{R}_h$ should make an angle $\theta = 30°$ with the x axis as in Fig. 2-21c. This can be proved in several ways. For example, the **i** components of **B**′ and $\mathbf{R}_h$ are equal, and the **j** components of **B**′ and $\mathbf{R}_h$ are equal in magnitude but opposite in direction, so $\theta = \phi = 30°$.

SECTIONS 2-4, 2-5 PROBLEMS

2–39 The wire under tension has xyz coordinates (a, b, c) at point A. Determine the unit vector at point O in the direction of A.

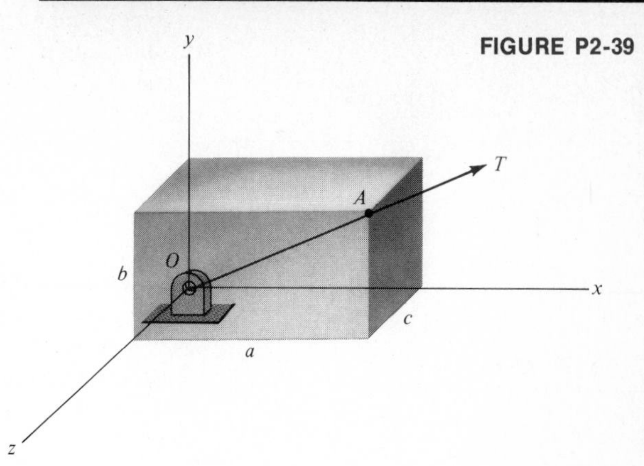

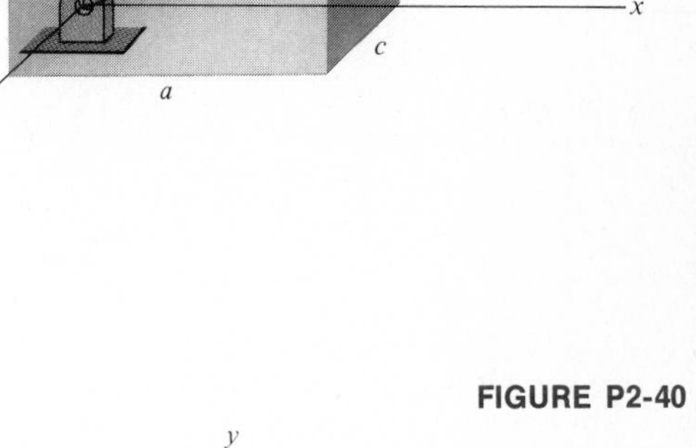

FIGURE P2-39

2–40 The chain under tension has xyz coordinates $(-a, -b, O)$ at point A. Determine the unit vector at point O in the direction of A.

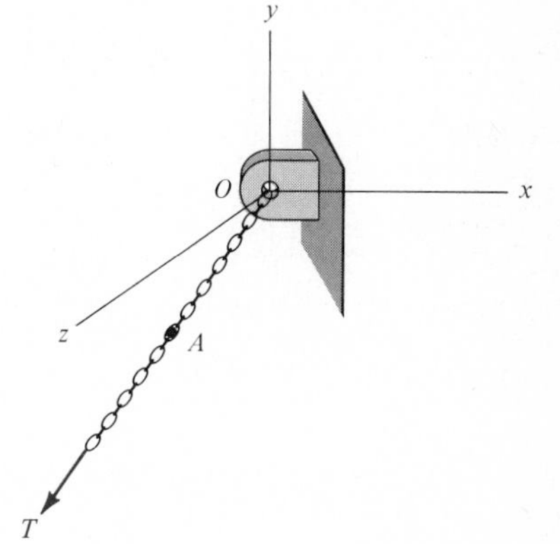

FIGURE P2-40

2–41 The rope under tension is in the xy plane. Determine its direction cosines with respect to the x, y, and z axes.

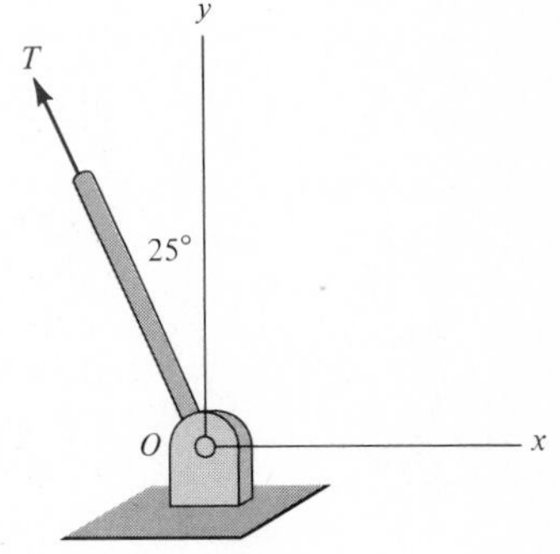

FIGURE P2-41

FIGURE P2-42

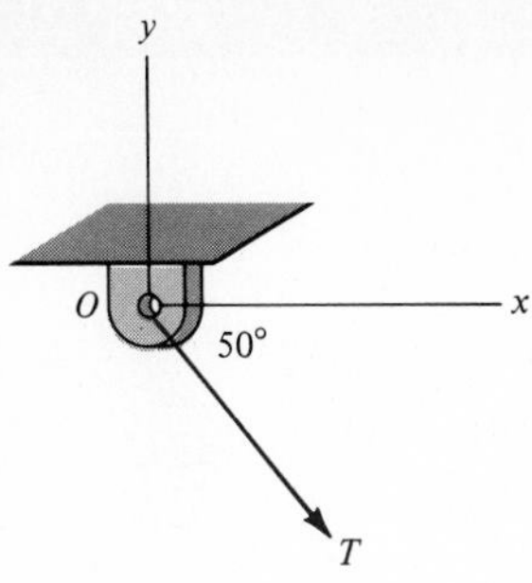

2–42 The wire under tension is in the xy plane. Determine its direction cosines with respect to the three coordinate axes.

FIGURE P2-43

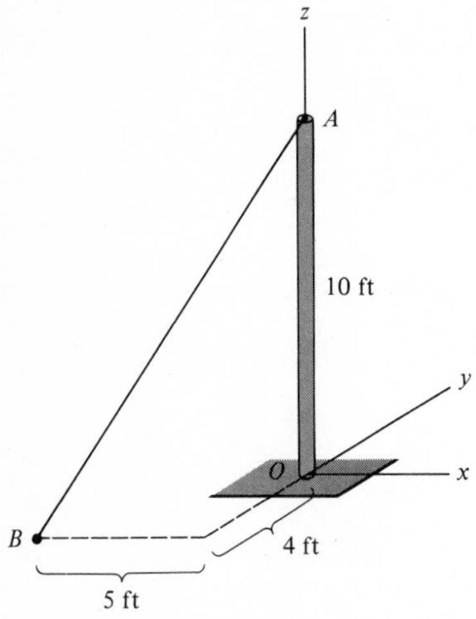

2–43 Determine the angles θ_x and θ_z of wire AB. (This is part of Ex. 2-7 and Fig. 2-20.)

FIGURE P2-44

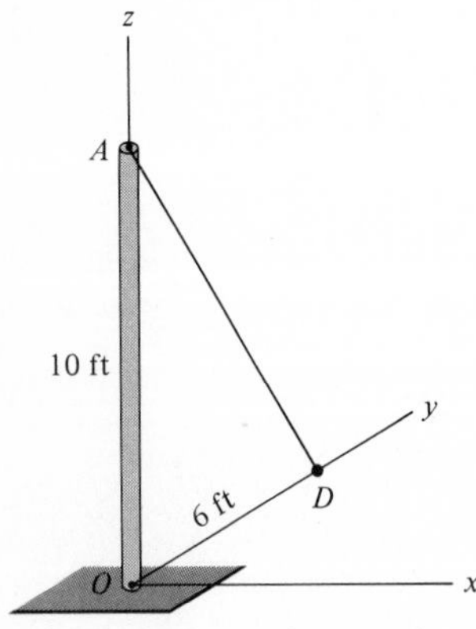

2–44 Determine the three direction cosines of wire AD. (This is part of Ex. 2-7 and Fig. 2-20.)

2–45 Determine both unit vectors that may be drawn for line AC. (This is part of Ex. 2-7 and Fig. 2-20.)

FIGURE P2-45

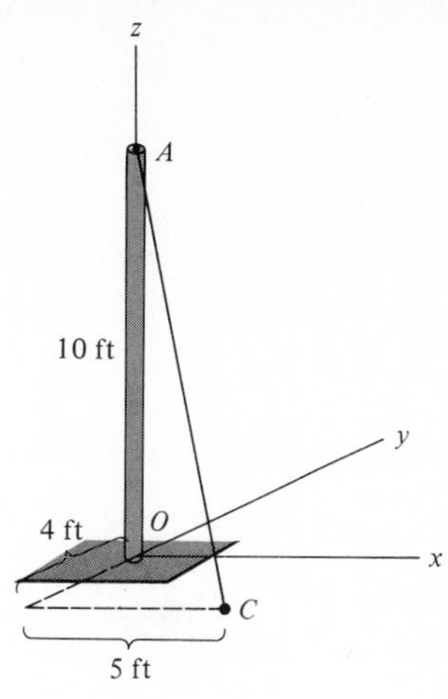

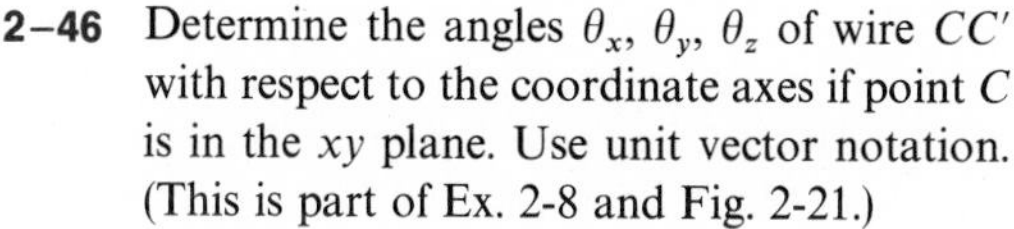

2–46 Determine the angles θ_x, θ_y, θ_z of wire CC' with respect to the coordinate axes if point C is in the xy plane. Use unit vector notation. (This is part of Ex. 2-8 and Fig. 2-21.)

FIGURE P2-46

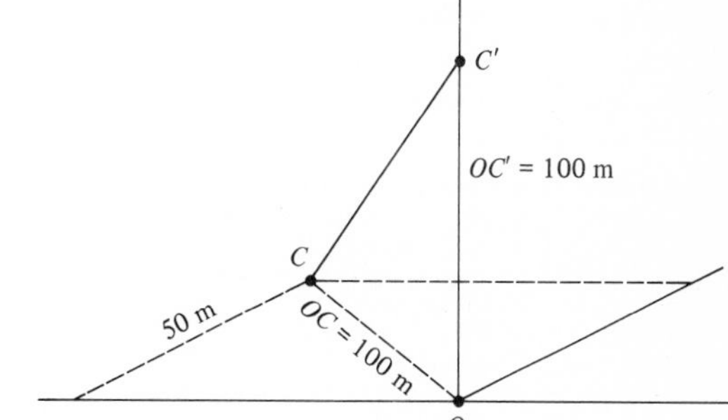

2–47 Determine the angles θ_x, θ_y, θ_z of wires AB and AC using unit vector notation and any reasonable shortcuts. The coordinates of the three points are A: (2 m, 0, 0), B: (0, 1.5 m, 1 m), C: (0, 1.5 m, −1 m).

2–48 Express unit vectors in terms of xyz components for all of the directions indicated with respect to Prob. 2-47: A to B, B to A, A to C, C to A, O to A, A to O, B to C, and C to O.

FIGURE P2-47

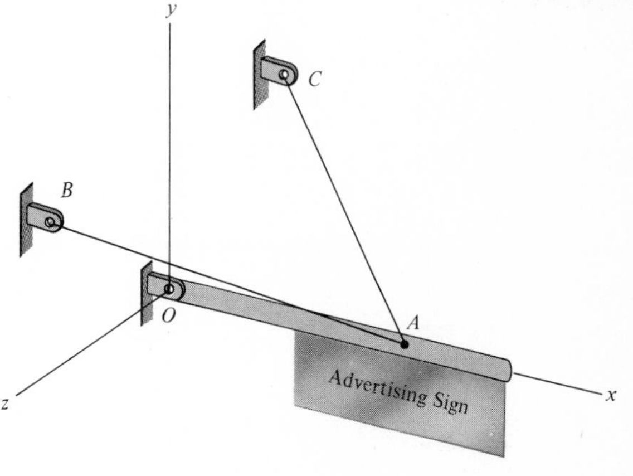

2–49 The magnitude of the tension in Prob. 2–39 is 6 kN. Express the force **T** in terms of vector components if $a = 50$ cm, $b = 30$ cm, and $c = 15$ cm.

2–50 In Fig. P2-47, the coordinates of A, B, and C are A: (7 ft, 0, 0) B: (0, 4 ft, 3 ft), C: (0, 4 ft, −3 ft). The tension in each wire (AB and AC) is 250 lb. Determine the vector components of the resultant force **R** that the two wires apply at point A.

FIGURE P2-51

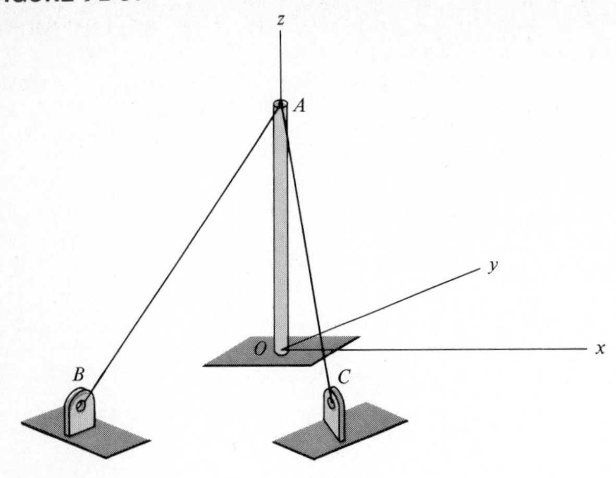

2–51 The coordinates of points A, B, and C are A: (0, 0, 5 m), B: (−2 m, −1.5 m, 0), C: (2 m, −1.5 m, 0). The tension in each wire (AB and AC) is 2 kN. Determine the horizontal and vertical components of the resultant force that the two wires apply at point A. The xy axes are horizontal.

2–52 In Fig. P2-51, the coordinates are A: (0, 0, 20 ft), B: (−8 ft, −6 ft, 0), C: (7 ft, −5 ft, 2 ft). The tension is 400 lb in wire AB, and 500 lb in wire AC. Determine the vector components of the resultant force $\mathbf{R}$ that the two wires apply at point A.

FIGURE P2-53

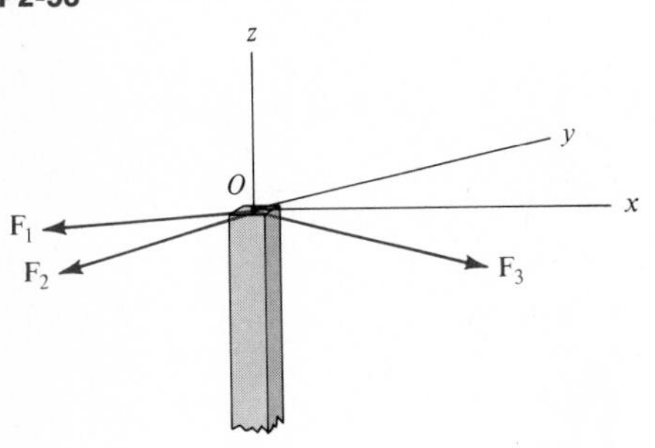

2–53 Three forces are concurrent at the top of an electric utility pole. Determine the horizontal (xy plane) components of the resultant force $\mathbf{R}$ if $\mathbf{F}_1 = -0.8F\mathbf{i} - 0.1F\mathbf{k}$, $\mathbf{F}_2 = -0.7F\mathbf{i} + 0.1F\mathbf{j} - 0.25F\mathbf{k}$, $\mathbf{F}_3 = 1.2F\mathbf{i} - 0.05F\mathbf{j} - 0.2F\mathbf{k}$.

2–54 Reconsider Prob. 2–53. Determine the smallest force $\mathbf{F}_4$ that should be added at point O in Fig. P2-53 to make the component R_x of the resultant zero.

FIGURE P2-55

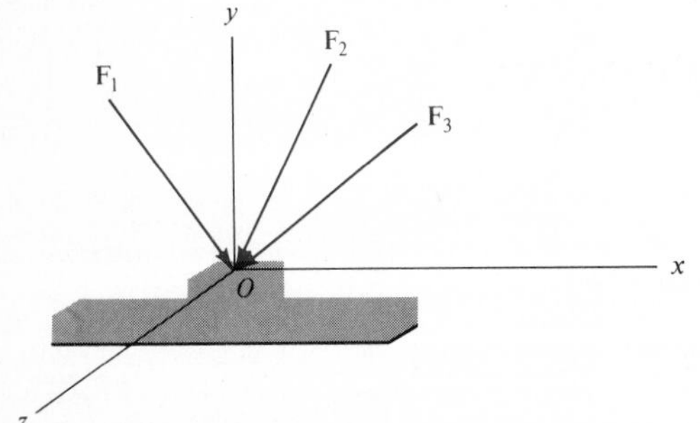

2–55 Three forces are concurrent at the top of a foundation block. Determine the horizontal (xz plane) and vertical components of the resultant force $\mathbf{R}$ if $\mathbf{F}_1 = a\mathbf{i} - b\mathbf{j}$, $\mathbf{F}_2 = -c\mathbf{i} - d\mathbf{j} + e\mathbf{k}$, $\mathbf{F}_3 = -f\mathbf{i} - g\mathbf{j} - h\mathbf{k}$.

2–56 Determine the scalar components of the force $\mathbf{F}_4$ that should be added to $\mathbf{F}_1$, $\mathbf{F}_2$, and $\mathbf{F}_3$ in Prob. 2–55 to make all components of the resultant force $\mathbf{R}$ zero.

SCALAR PRODUCT OF TWO VECTORS. ANGLES BETWEEN VECTORS. PROJECTION OF A VECTOR

2-6

The *scalar product* of two concurrent vectors **A** and **B** is, *by definition,* the product of the magnitudes of **A** and **B** and the cosine of the angle between them. This product of **A** and **B** is denoted by $\mathbf{A} \cdot \mathbf{B}$ and is also called the *dot product.* The formal expression for the vectors in Fig. 2-22 is

$$\boxed{\mathbf{A} \cdot \mathbf{B} = AB \cos \phi} \tag{2-25}$$

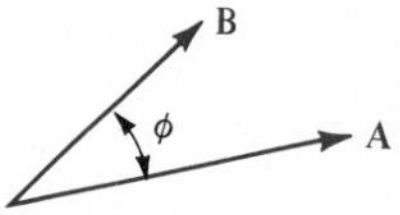

FIGURE 2-22

The scalar product of two vectors: $\mathbf{A} \cdot \mathbf{B} = AB \cos \phi$

The scalar product of two vectors is *commutative,*

$$\mathbf{A} \cdot \mathbf{B} = \mathbf{B} \cdot \mathbf{A} \tag{2-26}$$

since $AB \cos \phi = BA \cos \phi$.

The scalar product is also *distributive,*

$$\mathbf{A} \cdot (\mathbf{B} + \mathbf{C}) = \mathbf{A} \cdot \mathbf{B} + \mathbf{A} \cdot \mathbf{C} \tag{2-27}$$

The proof of this is left as an exercise for the student (Prob. 2-59).

There is no basis to discuss the *associative* property of scalar products since $\mathbf{A} \cdot \mathbf{B}$ is a scalar, so $(\mathbf{A} \cdot \mathbf{B}) \cdot \mathbf{C}$ is a scalar dotted with a vector, which is undefined.

The various possible scalar products of unit vectors are simple and worth remembering. Since $\cos 0 = 1$ and $\cos 90° = 0$, the basic set of products is

$$\begin{aligned} \mathbf{i} \cdot \mathbf{i} = \mathbf{j} \cdot \mathbf{j} = \mathbf{k} \cdot \mathbf{k} = 1 \\ \mathbf{i} \cdot \mathbf{j} = \mathbf{j} \cdot \mathbf{k} = \mathbf{k} \cdot \mathbf{i} = 0 \end{aligned} \tag{2-28}$$

The other possibilities, such as $\mathbf{j} \cdot \mathbf{i}$, give the same results because of the commutative property.

Following are several useful applications of scalar products in mechanics.

Magnitude of a Vector

Equation 2-10 can also be proved using the scalar product. In general,

$$\mathbf{A} \cdot \mathbf{B} = (A_x\mathbf{i} + A_y\mathbf{j} + A_z\mathbf{k}) \cdot (B_x\mathbf{i} + B_y\mathbf{j} + B_z\mathbf{k})$$

This can be reduced using Eqs. 2-27 and 2-28,

$$\mathbf{A} \cdot \mathbf{B} = A_xB_x + A_yB_y + A_zB_z$$

For the special case when $A = B$,

$$\mathbf{A} \cdot \mathbf{A} = A_x^2 + A_y^2 + A_z^2 = A^2$$

which is equivalent to Eq. 2-10.

Angle between Vectors

The angle between two vectors can be easily calculated when the rectangular components of both are known. From

$$\mathbf{A} \cdot \mathbf{B} = AB \cos \phi = A_x B_x + A_y B_y + A_z B_z$$

$$\cos \phi = \frac{A_x B_x + A_y B_y + A_z B_z}{AB} \qquad \boxed{2\text{-}29}$$

Projection of a Vector

It is frequently necessary to investigate the effect of a vector in a different direction. This "side effect" can be determined by making the "nearest-distance projection" of the vector onto the line of interest. This is shown for a two-dimensional vector **F** in Fig. 2-23. The *projection* **F**′ has a magnitude

$$F' = F \cos \phi \qquad \boxed{2\text{-}30}$$

In all such cases the projection is easily obtained using the graphical or trigonometric method. Note that in practice the term "projection" may mean a vector or a scalar quantity.

In three-dimensional situations it may be very difficult to determine the angle ϕ between two lines. The scalar product is convenient to use in such problems. The idea is to place a unit vector on the line of interest as in Fig. 2-24. $\mathbf{F} \cdot \mathbf{n} = F(1) \cos \phi = F'$, as in Eq. 2-25. The advantage is that the angle ϕ is not involved in the calculation if the rectangular coordinates of **F** and **n** are known. With $\mathbf{n} = \cos \theta_x\, \mathbf{i} + \cos \theta_y\, \mathbf{j} + \cos \theta_z\, \mathbf{k}$,

$$F' = F_x \cos \theta_x + F_y \cos \theta_y + F_z \cos \theta_z \qquad \boxed{2\text{-}31}$$

where $\cos \theta_x, \ldots$ are the direction cosines of the unit vector along the line of interest (they are not related to the angle ϕ between **F** and **n**).

FIGURE 2-23

Projection of vector **F** on line *l*

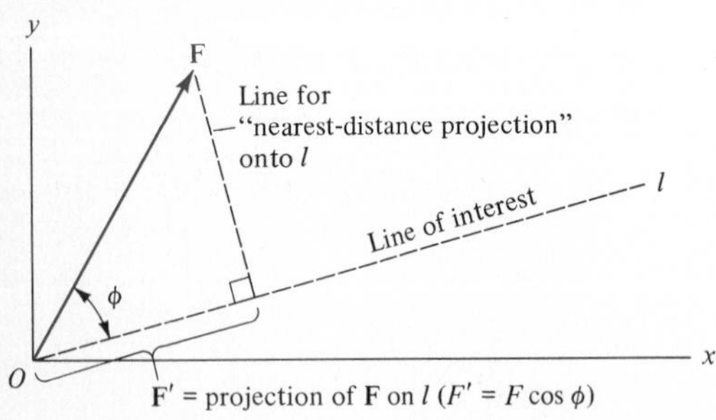

FIGURE 2-24

Scalar product of **F** and **n** for determining angle ϕ

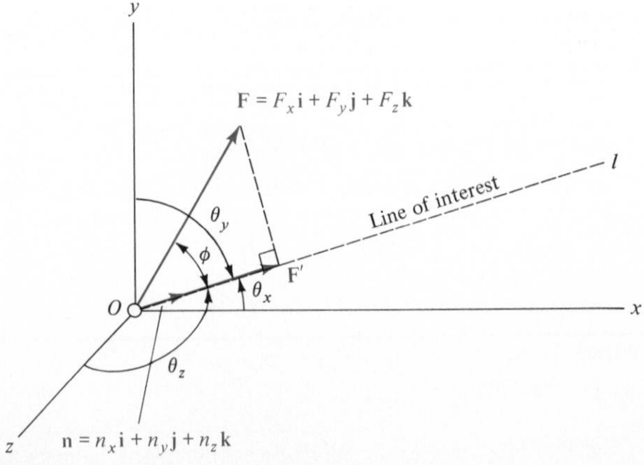

EXAMPLE 2-10

The laboratory test of a structure is modeled in Fig. 2-25. Forces $\mathbf{F}_1$ and $\mathbf{F}_2$ are applied by hydraulic actuators (see Sec. 1-7) at point O. Determine the force vectors and the angle between them if $F_1 = 1.4$ kN and $F_2 = 5.6$ kN. Point A is in the yz plane, and B is in the xy plane.

SOLUTION

First, the forces must be expressed in vector notation. This is done using unit vectors $\mathbf{n}_{AO}$ and $\mathbf{n}_{BO}$ as follows.

$$\mathbf{F}_1 = F_1\mathbf{n}_{AO} = F_1 \frac{\text{vector } \overrightarrow{AO}}{\text{magnitude of } \overrightarrow{AO}}$$

$$= (1.4\text{ kN})\frac{(0\mathbf{i} - 0.6\mathbf{j} - 0.8\mathbf{k})\text{ m}}{\sqrt{0.6^2 + 0.8^2}\text{ m}}$$

$$= (-0.84\mathbf{j} - 1.12\mathbf{k})\text{ kN}$$

$$\mathbf{F}_2 = F_2\mathbf{n}_{BO} = F_2 \frac{\text{vector } \overrightarrow{BO}}{\text{magnitude of } \overrightarrow{BO}}$$

$$= (5.6\text{ kN})\frac{(0.9\mathbf{i} - 0.7\mathbf{j} + 0\mathbf{k})\text{ m}}{\sqrt{0.9^2 + 0.7^2}\text{ m}}$$

$$= (4.42\mathbf{i} - 3.44\mathbf{j})\text{ kN}$$

The angle θ between $\mathbf{F}_1$ and $\mathbf{F}_2$ is calculated using Eq. 2-29,

$$\theta = \arccos\left(\frac{F_{1x}F_{2x} + F_{1y}F_{2y} + F_{1z}F_{2z}}{F_1F_2}\right)$$

$$= \arccos\left[\frac{(0)(4.42) + (-0.84)(-3.44) + (-1.12)(0)}{(1.4)(5.6)}\right]$$

$$= 68.4^\circ$$

JUDGMENT OF THE RESULT

The angle appears reasonable by inspection since it should be a little less than 90° (which would be the angle if the distances AA' and BB' were both zero).

FIGURE 2-25

Determination of two force vectors and the angle between them from scalar information

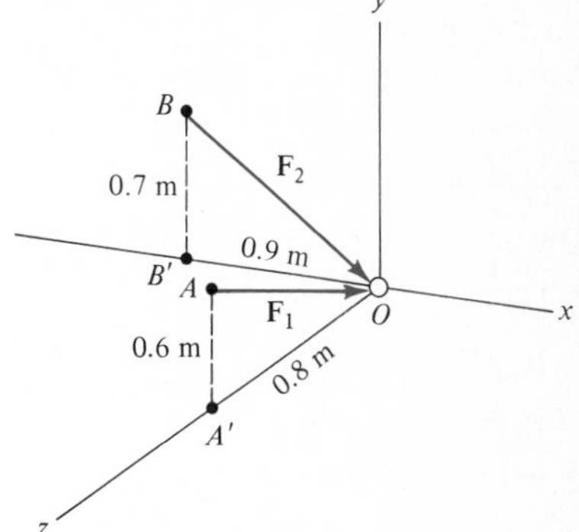

EXAMPLE 2-11

Part of a mechanism consists of slider S on rod OA, as shown in Fig. 2-26. The coordinates of point A are (10, 7, -4) in. Force $\mathbf{F} = 3\mathbf{i} - 0.5\mathbf{j} - 2\mathbf{k}$ lb acts on the slider. Determine the magnitude of the projection $\mathbf{F}'$ of force $\mathbf{F}$ on line OA (this is the effective force for moving the slider along line OA).

SOLUTION

First, a unit vector $\mathbf{n}_{OA}$ along line OA is determined using Eq. 2-22,

$$\mathbf{n}_{OA} = \frac{10\mathbf{i} + 7\mathbf{j} - 4\mathbf{k}}{\sqrt{10^2 + 7^2 + 4^2}} = 0.78\mathbf{i} + 0.55\mathbf{j} - 0.31\mathbf{k}$$

The projection F' is calculated using Eq. 2-30 or 2-31.

$$\begin{aligned} F' &= (3)(0.78) + (-0.5)(0.55) + (-2)(-0.31) \text{ lb} \\ &= 2.69 \text{ lb} \end{aligned}$$

JUDGMENT OF THE RESULT

Because the projection F' is the nearest-distance projection, F must be greater than F'. Indeed, $F = \sqrt{3^2 + 0.5^2 + 2^2}$ lb $= 3.64$ lb > 2.69 lb.

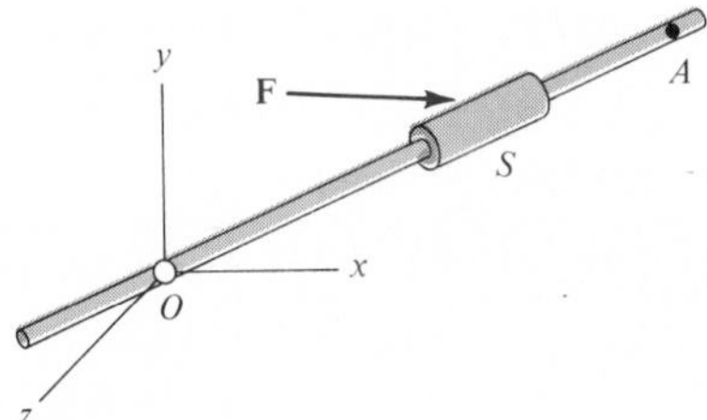

FIGURE 2-26

Calculation of projection of vector **F** on line *OA*

PROBLEMS

2–57 Determine the angle between lines AC and BC using vector methods.

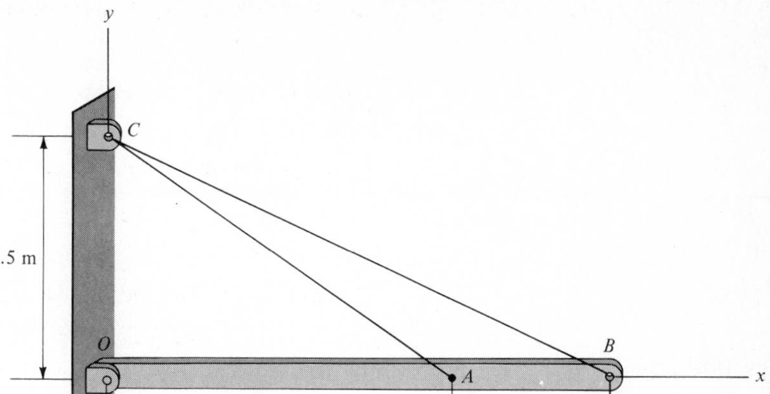

FIGURE P2-57

2–58 The tension in wire AC in Fig. P2-57 is 2 kN. Calculate the projection of force $\overrightarrow{CA}$ on line CB using the scalar product of two vectors.

2–59 Prove Eq. 2-27.

2–60 Use vector methods to determine the angle ϕ between force $\mathbf{F} = (200\mathbf{i} + 170\mathbf{j})$ lb and arm OA of the testing platform.

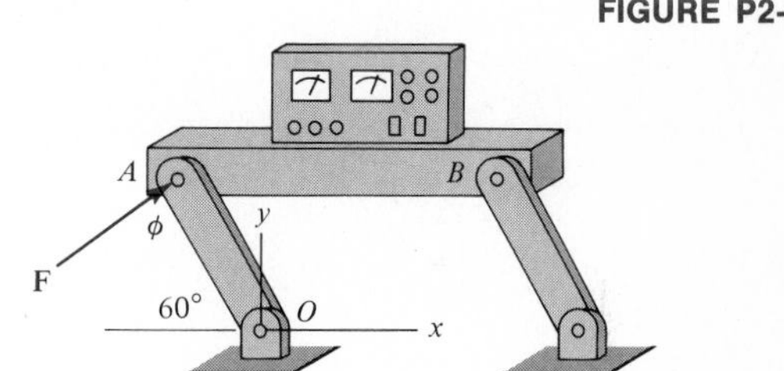

FIGURE P2-60

2–61 Calculate the projection of force **F** in Prob. 2–60 on line AB, which is parallel to the x axis. Use the dot product of two vectors.

2–62 Determine the angle between lines AB and AC.

2–63 The tension in wire AB in Fig. P2-62 is 1.5 kN. Calculate the projection of force $\overrightarrow{AB}$ on line AC.

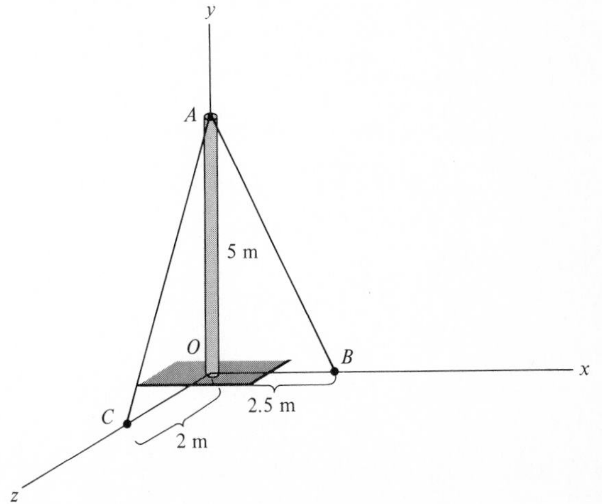

FIGURE P2-62

FIGURE P2-64

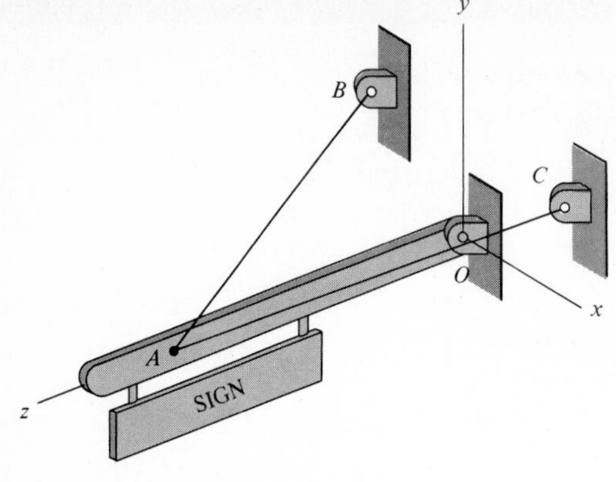

2–64 The coordinates of A, B, and C are (0, 0, 2) m, (−1, 1.5, 0) m, and (1.2, 1.5, 0.2) m, respectively. Calculate the angle between lines AB and AC.

2–65 In Prob. 2–64, the tension in each wire is 900 N. Determine the projections of the tensions in AB and AC on line BC.

FIGURE P2-66

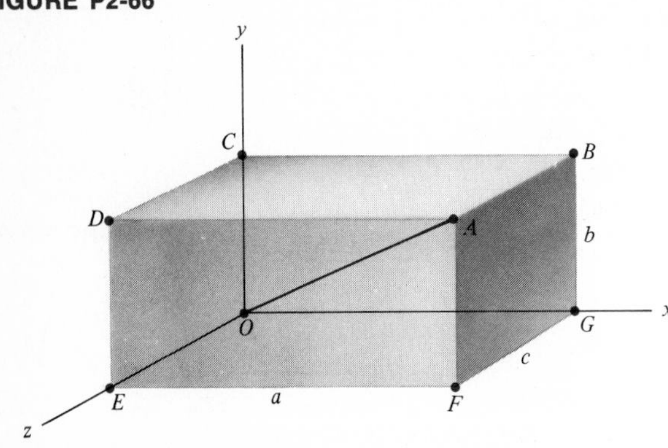

Problems 2–66 to 2–70 are based on Fig. P2-66. Use the scalar product of two vectors to determine the required quantity in each problem.

2–66 Determine the angle between lines OA and OE.

2–67 Determine the angle between lines OA and BG.

2–68 A force **F** is acting from O to A. Determine the projection of this force on line AB.

2–69 A force **P** is acting from D to G. Determine the projection of this force on line EF.

2–70 A 100-lb force is acting from O to A. Determine the projection of this force on the x axis in two ways:

(a) directly, going from OA to OG

(b) first going from OA to OF, then from OF to OG (that is, obtain the projection of another projection)

In both cases, the dimensions in Fig. P2-66 are $a = 2b = 1.5c$.

EQUILIBRIUM OF A PARTICLE. FREE-BODY DIAGRAMS

2-7

In previous discussions the resultant of the force system was occasionally zero. Such a situation is defined as *equilibrium*. According to strict definition:

> A particle is in equilibrium when the resultant of all forces acting on it is zero

The concept of equilibrium is enormously useful in engineering because conditions of equilibrium are very common. These conditions are discussed in the following for uniaxial, biaxial, and triaxial force systems acting on a particle.

Uniaxial Equilibrium

At most two forces are acting on a particle that is in uniaxial equilibrium. These forces must be equal and opposite, either compressing the particle or pulling it apart, as in Fig. 2-27. An alternative statement of this condition is often useful. When it is known that *only two forces* are acting on a particle that is in equilibrium, those forces must have the same line of action and magnitude, and be opposite in direction. This is the only way that the resultant of two forces can be zero.

Biaxial Equilibrium

A particle that is free to move in two dimensions may have an infinite number of forces on it and still be in equilibrium if the resultant of the forces is zero. The condition can be stated or checked graphically using the polygon rule of vector addition. The algebraic method of summing all components of the forces provides the necessary and sufficient conditions for biaxial equilibrium of a particle:

$$\sum F_x = 0 \qquad \sum F_y = 0 \tag{2-32}$$

The orthogonal xy axes of reference may be chosen arbitrarily without changing the mathematical statement of equilibrium.

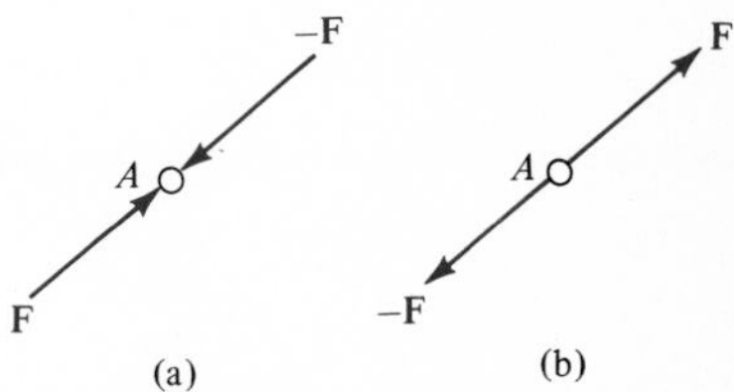

FIGURE 2-27

Two forces in equilibrium must be collinear, equal in magnitude, and opposite in direction

Triaxial Equilibrium

The conditions of equilibrium of a particle in three dimensions are extensions of the simpler cases. The major practical difference is that the polygon rule is not convenient to use for a three-dimensional problem. The algebraic method of summing components is valid and convenient to use with the following formal statement of the necessary and sufficient conditions of equilibrium:

$$\sum F_x = 0 \qquad \sum F_y = 0 \qquad \sum F_z = 0 \tag{2-33}$$

Determination of Unknown Forces

Knowledge of the existence of equilibrium is very useful in determining unknown forces. The polygon rule is easy to use for two-dimensional systems of forces. For example, all of the known forces $\mathbf{F}_1$ to $\mathbf{F}_4$ that act on particle A in Fig. 2-28a are drawn first in Fig. 2-28b. The resultant of the four original forces is **R**. Equilibrium is possible only if there is a fifth force **P** on the particle. **P** must be equal and opposite to **R** to close the polygon and obtain a zero resultant force on the particle. A force, such as **P**, which opposes the resultant of any number of other forces is called a *reaction* to those forces.

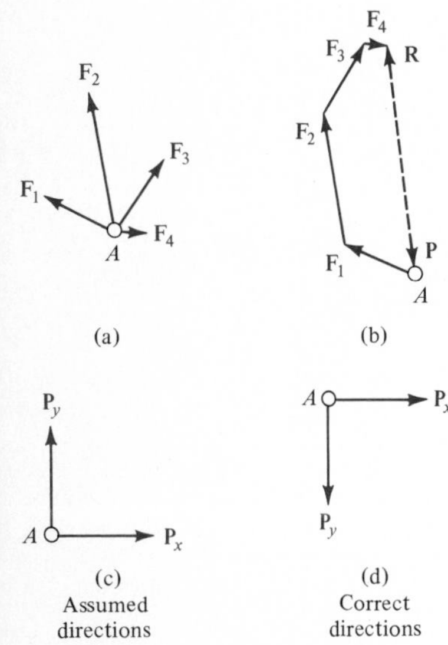

FIGURE 2-28

Graphical determination of an unknown force **P** for equilibrium of a particle

The unknown force **P** in Fig. 2-28 can also be determined algebraically if the components of the other forces are known. The force summations are simply made, including the unknown components of **P**. The solutions of Eq. 2-32 give the correct magnitudes of P_x and P_y. Positive answers for the unknowns indicate that the assumptions regarding their directions (Fig. 2-28c) were correct. Negative answers indicate that the assumed directions were wrong. It appears from comparing Fig. 2-28b and c that P_x was assumed correctly but P_y should be downward. At this stage the correct directions may be drawn, but usually it is better to leave the sketch in its original form and let the negative number show the real direction. (If the directions of any forces are changed, their algebraic signs in the equations must also be changed.)

Free-Body Diagrams

Simple sketches of the physical situations relevant to a problem are often drawn in the process of solution. More commonly, however, a bare minimum of a sketch is made of the body, part of the body, or system of bodies involved. On this diagram *all* significant, relevant forces must be shown to allow a proper determination of unknown forces. The effects of other bodies on the body of interest are simply shown by drawing the forces that they cause. Thus, one draws a *free-body diagram*, which is the simplest possible model of the body under consideration. This model is imagined to be isolated from all other bodies. The effects of the latter are represented by showing appropriate arrows on the free-body diagram. For example, the free-body diagram for the anchoring block of six guy wires shown in Fig. 2-15 is given in Fig. 2-29. Such diagrams are essential in the solution for unknown quantities.

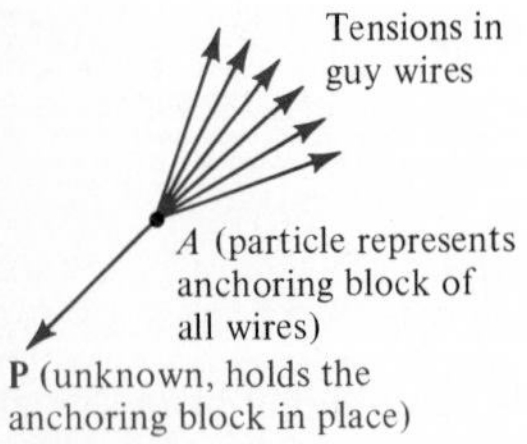

FIGURE 2-29

Free-body diagram of an anchoring block of six guy wires

EXAMPLE 2-12

Determine the magnitude and direction of the force **P** that must oppose the tensions in the guy wires shown in Fig. 2-15 since the block is in equilibrium.

SOLUTION

Recall from Ex. 2-5 that the resultant of the forces due to the wires was $\mathbf{R} = 96.6\ \text{kN}\ \mathbf{i} + 90.1\ \text{kN}\ \mathbf{j}$. Since there is equilibrium, $\sum \mathbf{F} = 0$. Thus, $\mathbf{P} + \mathbf{R} = 0$, from which $\mathbf{P} = -\mathbf{R}$, and

$$\mathbf{P} = -96.6\ \text{kN}\ \mathbf{i} - 90.1\ \text{kN}\ \mathbf{j}$$

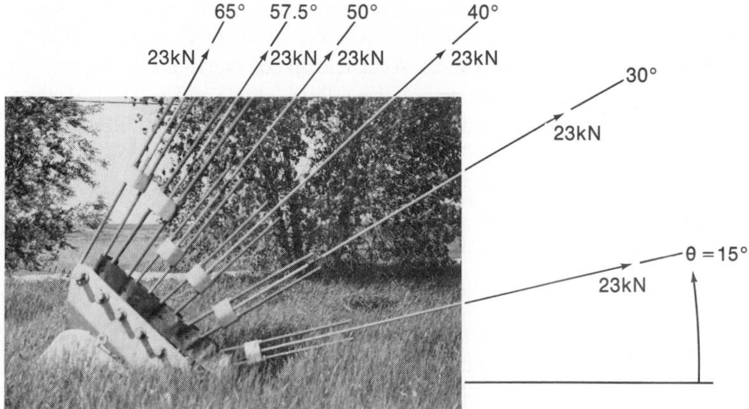

EXAMPLE 2-13

Suppose that you wish to hold a cart of weight $W = 1960$ N on a 10° incline by applying a force **F** to the cart parallel to the incline. The cart is modeled in Fig. 2-30a. Determine the required magnitude of **F** along line l to hold the cart in equilibrium. An important and reasonable assumption in such a problem is that all forces between the wheels and the inclined plane are *perpendicular* to the incline. These are commonly called *normal forces*.

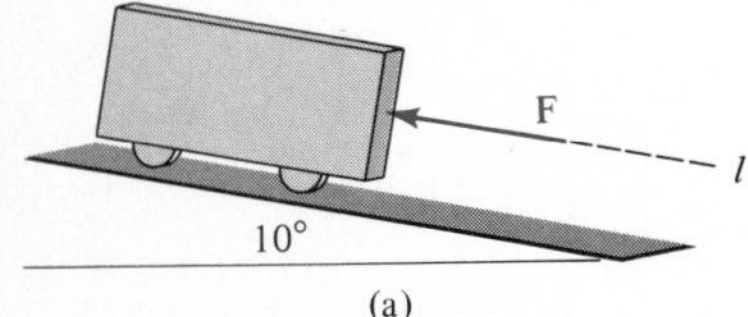

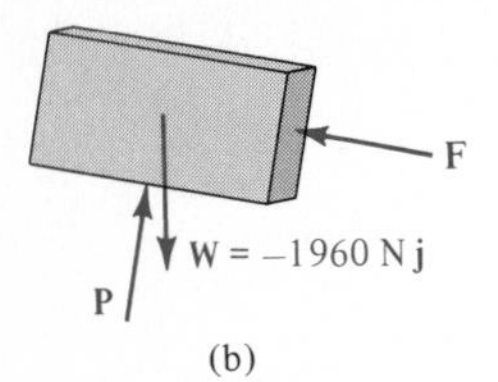

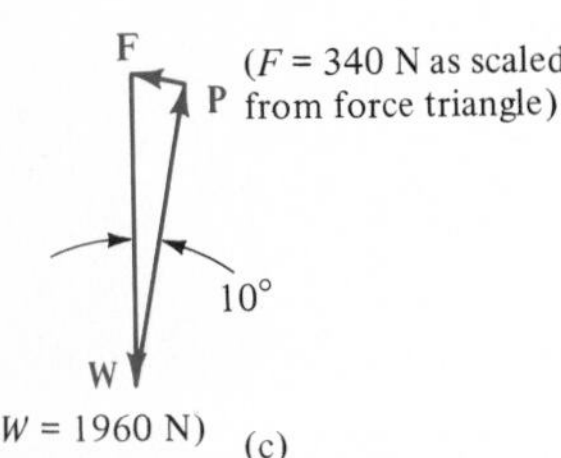

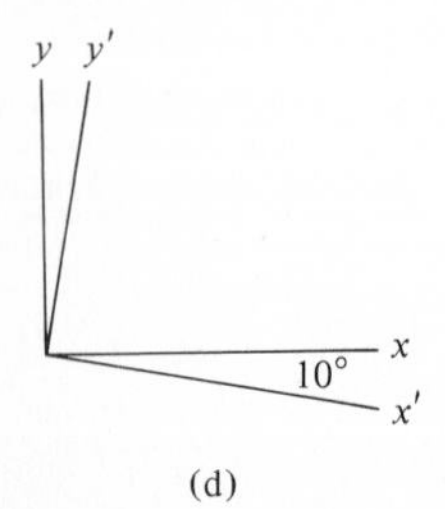

FIGURE 2-30

Determination of unknown force F that holds the cart in equilibrium

SOLUTION

One graphical and two analytic methods of solution are presented for comparison:

(a) First draw the free-body diagram in Fig. 2-30b. Assuming that the cart is a particle, a graphical solution is rapid for the three forces. The force **F** has a magnitude of 340 N in the force triangle of Fig. 2-30c.

(b) The analytic solution for the x (horizontal) and y (vertical) components of the forces is obtained from the two scalar equilibrium equations that can be written,

$$\sum F_x = 0$$

$$P \sin 10° - F \cos 10° = 0$$

$$\sum F_y = 0$$

$$P \cos 10° - W + F \sin 10° = 0$$

From these,

$$F = P \tan 10°$$

$$P(\cos 10° + \sin 10° \tan 10°) = W$$

$$P = \frac{W}{1.02} = 1930 \text{ N}$$

$$F = 340.3 \text{ N}$$

(c) There is a simpler analytic solution if the coordinate axes are chosen more advantageously. It is generally best to orient the coordinates parallel and perpendicular to the unknown force as x' and y' in Fig. 2-30d. With these,

$$\sum F_{x'} = 0$$

$$W \sin 10° - F = 0$$

$$F = 340.4 \text{ N}$$

JUDGMENT OF THE RESULT

The force F must approach zero as the angle of the incline becomes zero, so a small F is reasonable. Furthermore, the three different solutions lead to the same answer.

EXAMPLE 2-14

A body of weight W is lifted by a rope going over two pulleys as shown in Fig. 2-31a. Determine the tension T in the rope and the force F_B on the axle of pulley B for (a) $\theta = 0$, and (b) $\theta = 60°$. Assume that the tension in a rope is the same on the two sides of each pulley. This can be proved using the concept of moment, which is presented in the next chapter.

SOLUTION

Assume that each pulley is a particle with three concurrent forces acting on it. In each case, the solution should start at pulley A because a known force, **W**, is acting there.

(a) $\theta = 0$. The free-body diagram of pulley A is drawn in Fig. 2-31b. Since the rope tensions **T** are concurrent, parallel, and equal in magnitude, they are shown as a single force 2**T**, which must be equal in magnitude and opposite in direction to **W**. Thus, $T = W/2$.

A similar procedure is applied at pulley B as shown in Fig. 2-31c. Here the combined tensions 2**T** in the ropes are acting downward, and $T = W/2$ as at pulley A, by Newton's law of action and reaction. The other required force is $F_B = W$ by inspection (this is the magnitude of the force vector $\mathbf{F}_B = -\mathbf{W}$).

(b) $\theta = 60°$. The free-body diagram of pulley A in Fig. 2-31b is still applicable since the center of pulley B is fixed and the ropes at A remain parallel. Thus, $T = W/2$ at pulley A.

To determine F_B, consider Fig. 2-31d. The tensions $\mathbf{T}_1$ and $\mathbf{T}_2$ have equal magnitudes, T, so their resultant **R** bisects the 60° angle between them. $\mathbf{F}_B$ must be equal and opposite to **R**, making a 30° angle with the vertical as shown. For equilibrium,

$$\sum F_x = 0$$

$$-F_B \sin 30° + T_2 \cos 30° = 0$$

Substituting $T_2 = T = W/2$,

$$F_B = \frac{W \cos 30°}{2 \sin 30°} = W \cos 30° = 0.866W$$

JUDGMENT OF THE LAST RESULT

The biaxial forces in Fig. 2-31 can be reduced to uniaxial ones by summing forces along the line of action of $\mathbf{F}_B$. Then $F_B = T_1 \cos 30° + T_2 \cos 30° = 2T \cos 30° = W \cos 30°$, in agreement with the previous result.

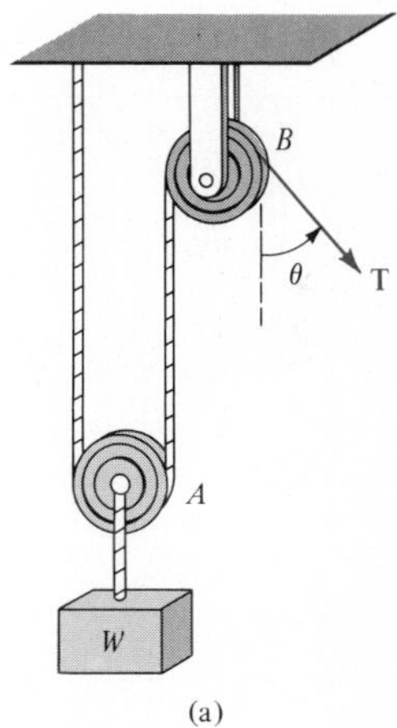

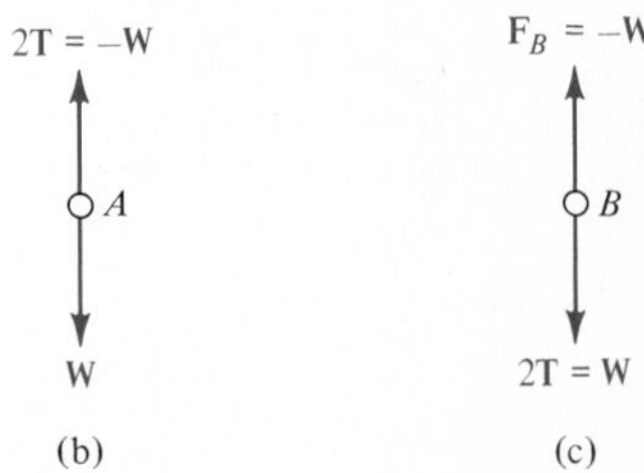

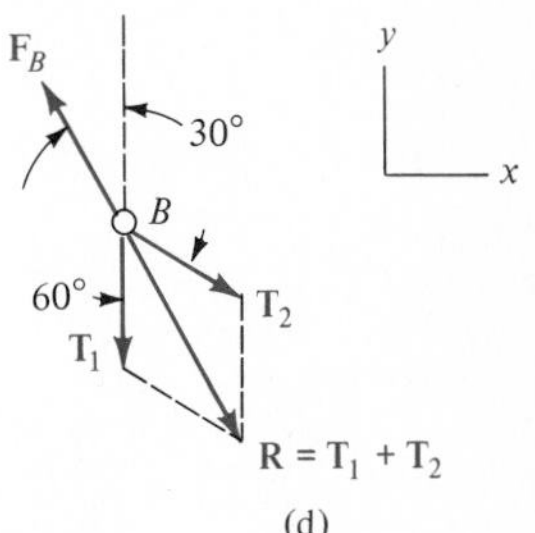

FIGURE 2-31

Determination of unknown forces T and F_B in a pulley system

EXAMPLE 2-15

A helicopter is to lift a cargo of 9 kN weight. The cables are attached as in Fig. 2-32, with point A on the y axis and points B, C, and D in the xz plane. Determine the tension in each cable.

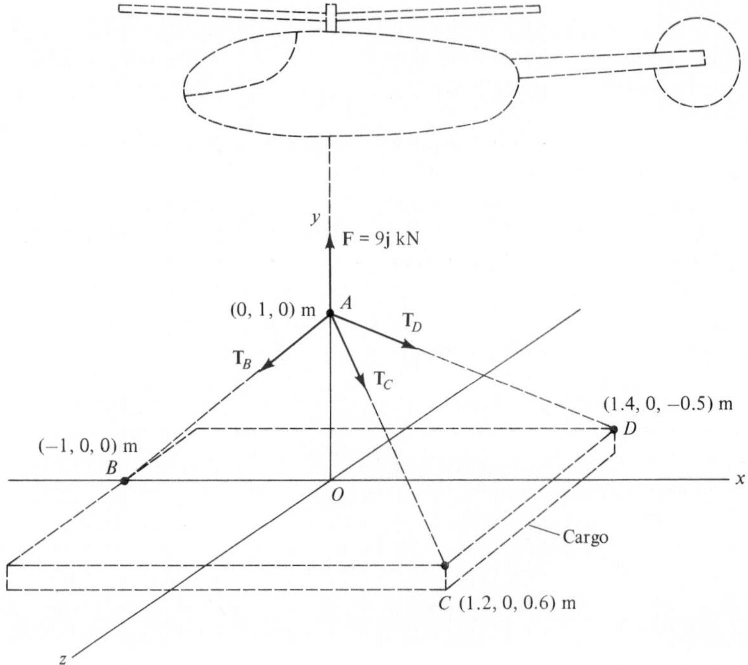

FIGURE 2-32

Determination of unknown tensions T_B, T_C, and T_D

SOLUTION

First represent all forces in vector notation. From Eqs. 2-11 and 2-22,

$$\mathbf{T}_B = T_B\mathbf{n}_B \qquad \mathbf{T}_C = T_C\mathbf{n}_C \qquad \mathbf{T}_D = T_D\mathbf{n}_D$$

$$\mathbf{n}_B = \frac{\text{vector } \overrightarrow{AB}}{\text{magnitude of } \overrightarrow{AB}} = \frac{-\mathbf{i} - \mathbf{j} + 0\mathbf{k}}{\sqrt{1^2 + 1^2 + 0^2}}$$

$$= -0.707\mathbf{i} - 0.707\mathbf{j}$$

$$\mathbf{n}_C = \frac{\text{vector } \overrightarrow{AC}}{\text{magnitude of } \overrightarrow{AC}} = \frac{1.2\mathbf{i} - \mathbf{j} + 0.6\mathbf{k}}{\sqrt{1.2^2 + 1^2 + 0.6^2}}$$

$$= 0.717\mathbf{i} - 0.598\mathbf{j} + 0.359\mathbf{k}$$

$$\mathbf{n}_D = \frac{\text{vector } \overrightarrow{AD}}{\text{magnitude of } \overrightarrow{AD}} = \frac{1.4\mathbf{i} - \mathbf{j} - 0.5\mathbf{k}}{\sqrt{1.4^2 + 1^2 + 0.5^2}}$$

$$= 0.781\mathbf{i} - 0.558\mathbf{j} - 0.279\mathbf{k}$$

For equilibrium, $\mathbf{T}_B + \mathbf{T}_C + \mathbf{T}_D + \mathbf{F} = 0$. Summing scalar components according to Eq. 2-33 is convenient,

$$\sum F_x = 0$$

$$-0.707T_B + 0.717T_C + 0.781T_D = 0$$

$$\sum F_y = 0$$

$$-0.707T_B - 0.598T_C - 0.558T_D + 9 = 0$$

$$\sum F_z = 0$$

$$0T_B + 0.359T_C - 0.279T_D = 0$$

From the simultaneous solution of these three equations

$$T_B = 7.23 \text{ kN} \qquad T_C = 2.97 \text{ kN} \qquad T_D = 3.82 \text{ kN}$$

JUDGMENT OF THE RESULTS

The tension values are acceptable since the sum of the vertical components of T_B, T_C, and T_D is 9 kN.

PROBLEMS

SECTION 2-7

UNIAXIAL EQUILIBRIUM

In Probs. 2–71 to 2–80, assume that all the ropes are essentially vertical and that the system is in equilibrium.

FIGURE P2-71

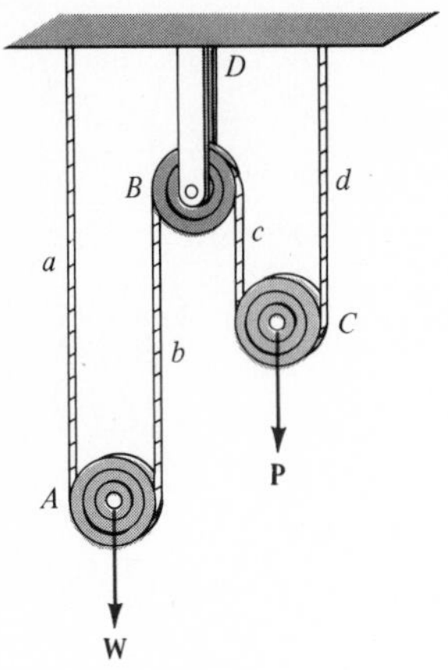

2–71 Determine the tension in the continuous parts a, b, c, and d of the rope if $W = 2$ kN.

2–72 Determine the force P acting on pulley C and the force in bar BD if $W = 600$ lb in Fig. P2-71.

FIGURE P2-73

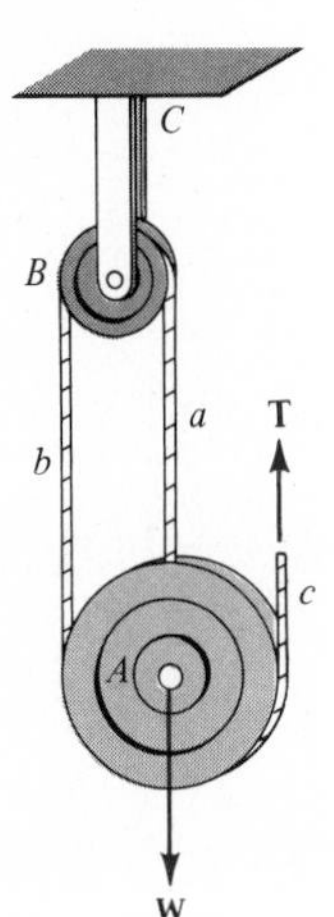

2–73 The rope segment a is attached to the center shaft of pulley A, where the force **W** is also acting. The rope is continuous from part a through b to c. Calculate the tension T if $W = 1$ kN.

2–74 Determine the force in bar BC in Prob. 2–73.

FIGURE P2-75

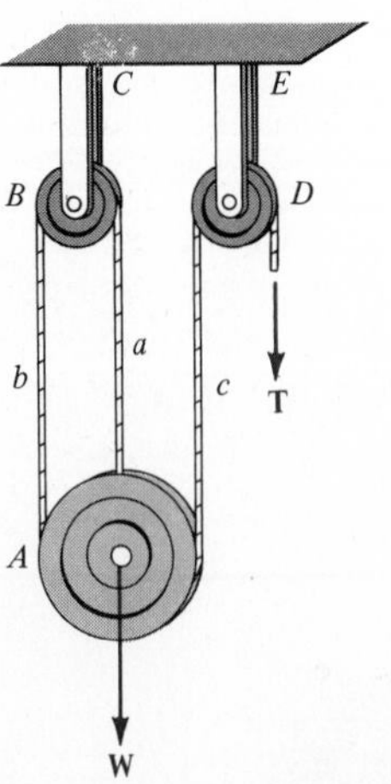

2–75 The rope segment a is attached to the center shaft of pulley A, where the force W is also acting. The rope is continuous from part a through d. Calculate the tension T if $W = 5$ kN.

2–76 Determine the forces in bars BC and DE in Prob. 2–75.

2–77 The rope segment a is attached to the center shaft of pulley A, where the force W is also acting. The rope is continuous from part a through d. Calculate the tension in bar BC if $W = 4$ kN.

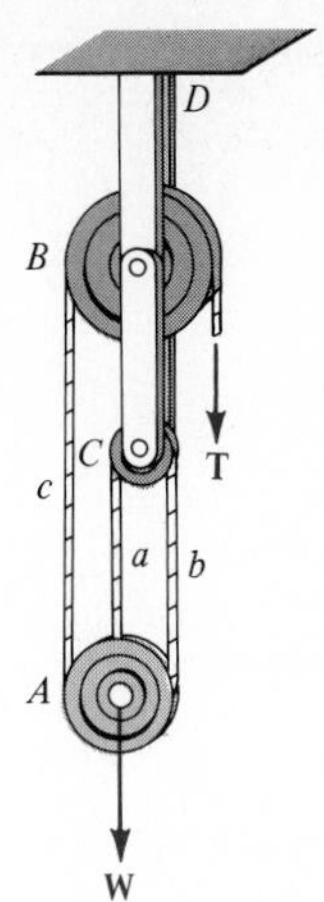

FIGURE P2-77

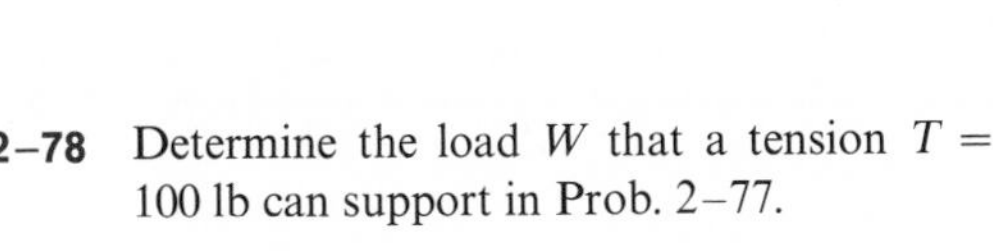

2–78 Determine the load W that a tension $T = 100$ lb can support in Prob. 2–77.

2–79 The rope segment e is attached to the center shaft of pulley D, where bar AD is also connected. Determine the forces in bars AD, BC, and BE if a tension $T = 400$ N is required to support a load W.

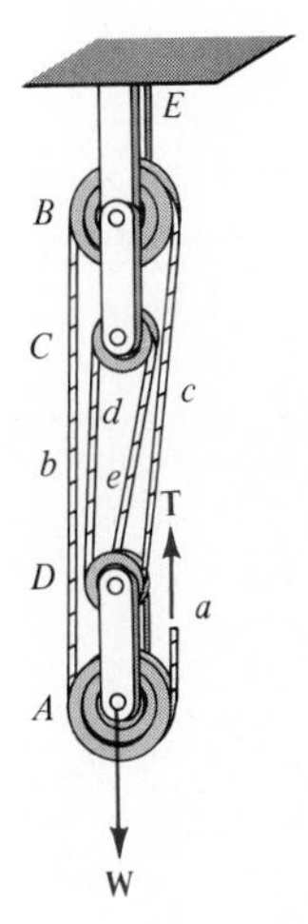

FIGURE P2-79

2–80 Calculate the tension in the rope segment e in Prob. 2–79 if $W = 500$ lb.

BIAXIAL EQUILIBRIUM

2–81 A wire is supported at points A and B. Determine the tensions in segments AC and BC if $W = 500$ N, $\theta = \phi = 20^\circ$, and $L = 4$ m.

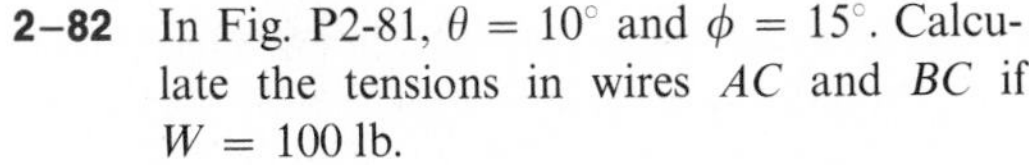

2–82 In Fig. P2-81, $\theta = 10^\circ$ and $\phi = 15^\circ$. Calculate the tensions in wires AC and BC if $W = 100$ lb.

2–83 In Fig. P2-81, $\theta = \phi$. Plot the tension T in the wire AB as a function of $\theta = 5^\circ$, 10°, and 15° if the load is W. Draw a smooth curve through the three points and extrapolate the curve to $\theta = 1^\circ$.

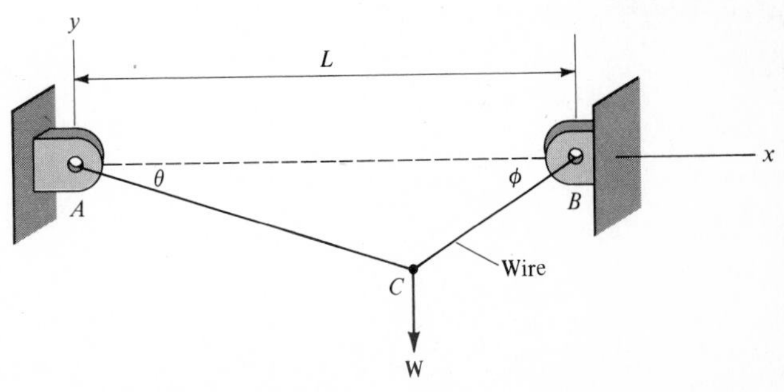

FIGURE P2-81

FIGURE P2-84

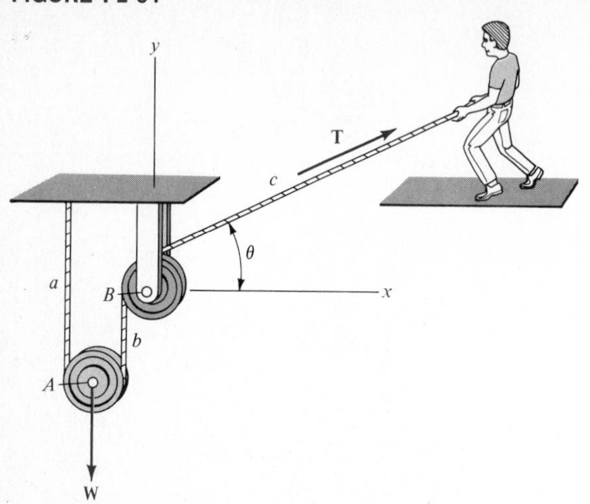

2–84 Determine the tension T and the reactions R_x and R_y at axle B if $W = 500$ N and $\theta = 30°$.

FIGURE P2-85

(a)

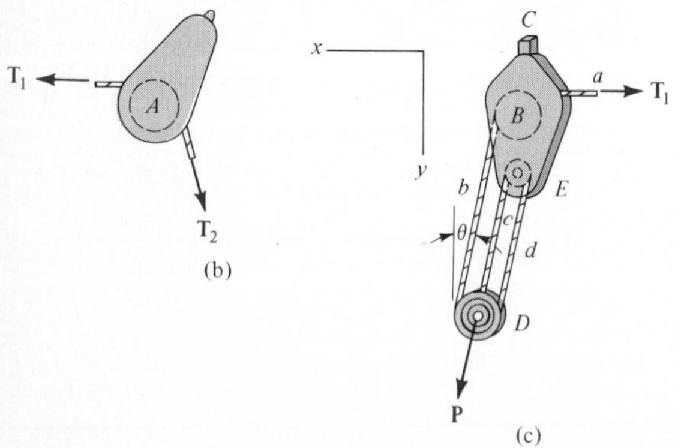

(b)

(c)

2–85 Determine the supporting reactions R_x and R_y on pulley A if $\mathbf{T}_1$ is horizontal and $\mathbf{T}_2 = (-5\mathbf{i} + 40\mathbf{j})$ lb.

2–86 In Fig. P2-85c, the rope is continuous from a through d as it passes over three pulleys. Segments b, c, and d are parallel to one another and make an angle $\theta = 5°$. Determine the supporting reactions at point C if $\mathbf{T}_1 = 300\mathbf{i}$ N.

2-87 A 100-lb block is supported by force **F** on an inclined plane. The block is essentially frictionless, which means that only a normal force can act between the block and the incline. Plot the force F necessary to maintain equilibrium of the block as a function of the angle, using $\theta = 60°, 90°$, and $120°$.

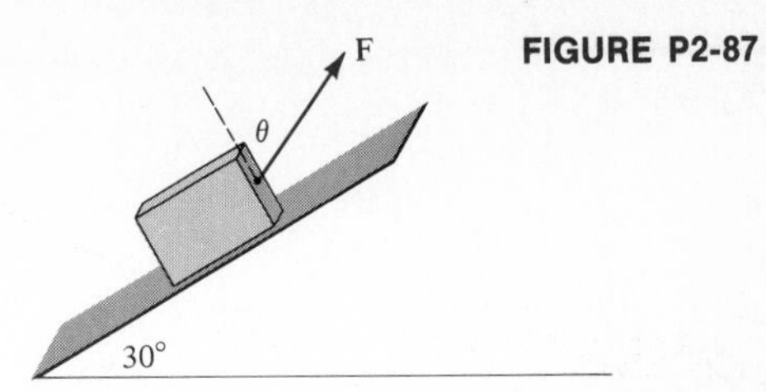

FIGURE P2-87

FIGURE P2-88

Network of horizontal cables of electric trolley bus system in Vancouver, B.C. Photo is from the ground, in the vertical direction

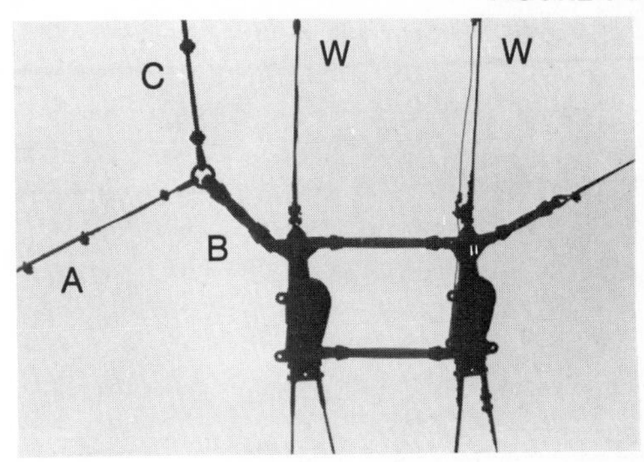

FIGURE P2-88

2-88 The cable junction structure of an electric trolley bus system is essentially in the horizontal plane. The buses pass below this network and obtain electric power through arms that rub against the wires (w). The tensions in two supporting cables are $T_A = 1.5$ kN and $T_C = 1.2$ kN. Determine the tension in cable B. All angles seen in the picture are true angles.

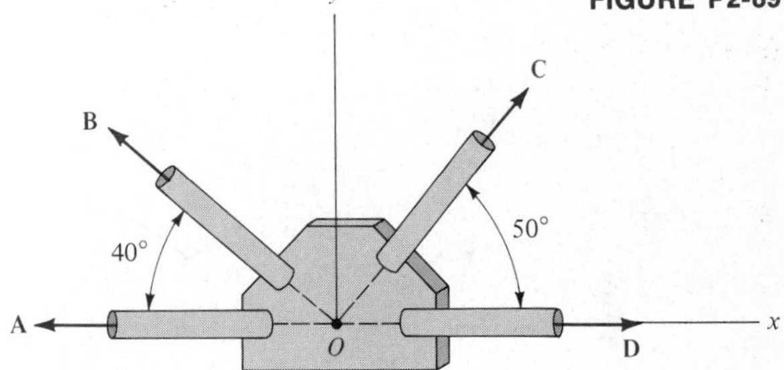

FIGURE P2-89

2-89 Four bars are welded to a bracket. Calculate the unknown reaction **R** if $A = 1000$ lb, $B = 2000$ lb, $C = 500$ lb, and $D = 6000$ lb.

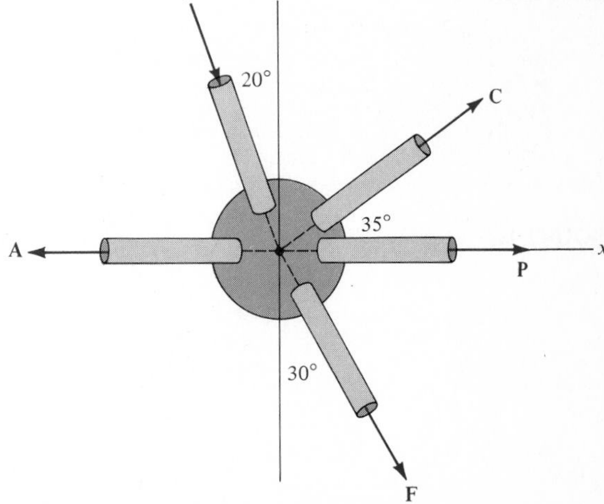

FIGURE P2-90

2-90 The following forces on a welded bracket have fixed directions as shown: $A = 2$ kN, $B = 3$ kN, and $C = 5$ kN. Forces **F** and **P** have unknown magnitudes and directions, but their lines of action are fixed as shown. Determine **F** and **P** in terms of vector components.

FIGURE P2-91

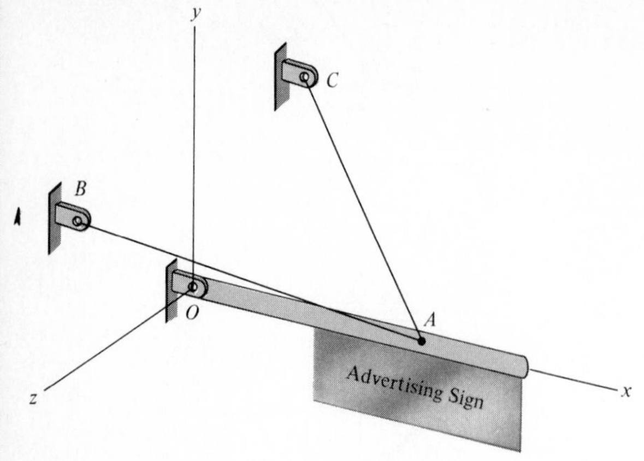

TRIAXIAL EQUILIBRIUM

2–91 The two wires are attached to a single bolt at A. The coordinates of the three points are A: (2 m, 0, 0), B: (0, 1.5 m, 1 m), C: (0, 1.5 m, -1 m). The tension in AB is 800 N; in AC it is 850 N. Determine the reaction **R** of the beam acting on the bolt.

FIGURE P2-92

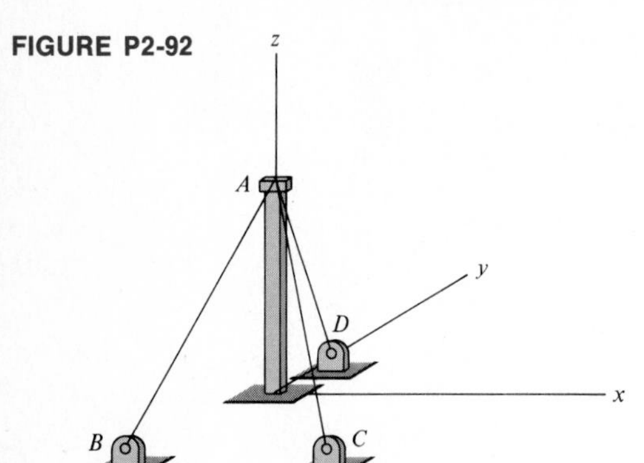

2–92 The three wires are attached to a small collar A on the pole. The tension in each wire is 300 lb, and the coordinates are A: (0, 0, 15) ft, B: (-5, -4, 0) ft, C: (5, -4, 0) ft, and D: (0, 7, 0) ft. Determine the reaction **R** of the pole acting on the collar.

FIGURE P2-93

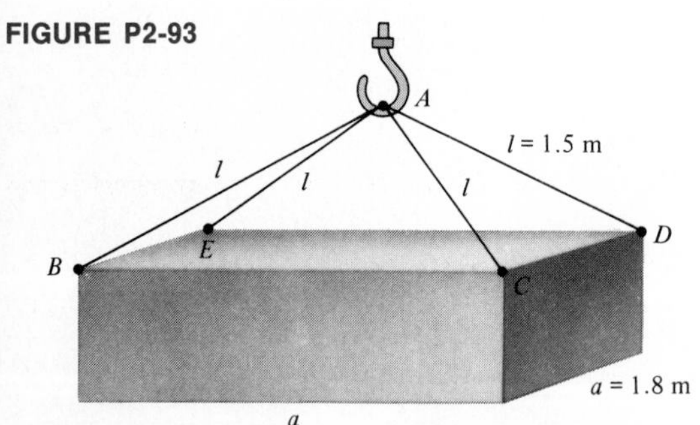

2–93 The rectangular crate is lifted by four ropes of equal length l. Determine the net force applied by the crane hook to the junction A of the four ropes if the tension in each rope is 2.5 kN. What is the weight of the crate?

FIGURE P2-94

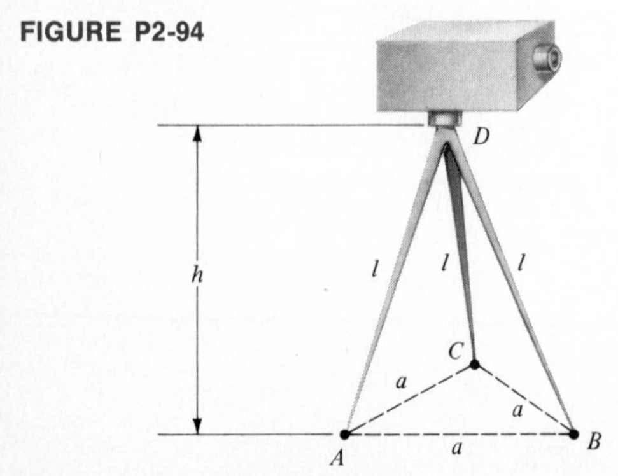

2–94 A tripod is designed to support a camera of weight W. The camera is at height h above the ground when the legs form an equilateral triangle ABC of side a. Determine the forces in the legs assuming that they share the load equally, and that the forces are axial (that is, longitudinal) in the legs.

2–95 Four identical wires are suspended from a horizontal ceiling at points A, B, C, and D. The wires are concurrent at E and support a load $W = 2$ kN. Calculate the tension in the wires if $a = 2$ m, $b = 4$ m, and $l = 3$ m. Assume that the wires share the load symmetrically.

FIGURE P2-95

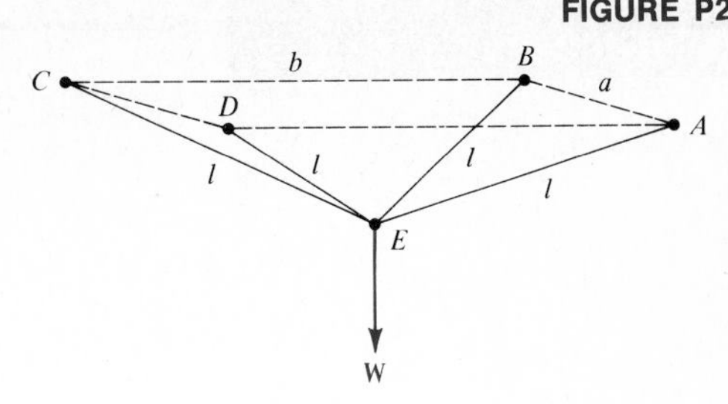

2–96 Solve Prob. 2–95 using $a = 4$ m, 2 m, and 0. There are four wires in each case. Plot the tension T in wire BE as a function of a, and explain the observed general tendency.

2–97 The tensions in chains AB and AC and the force **F** support the load **W**. The coordinates of the three points are A: (10, 0, 0) ft, B: (0, 7, −4) ft, and C: (0, 7, 4) ft, and **F** is in the xy plane. Determine F if $W = 500$ lb and θ must be zero.

FIGURE P2-97

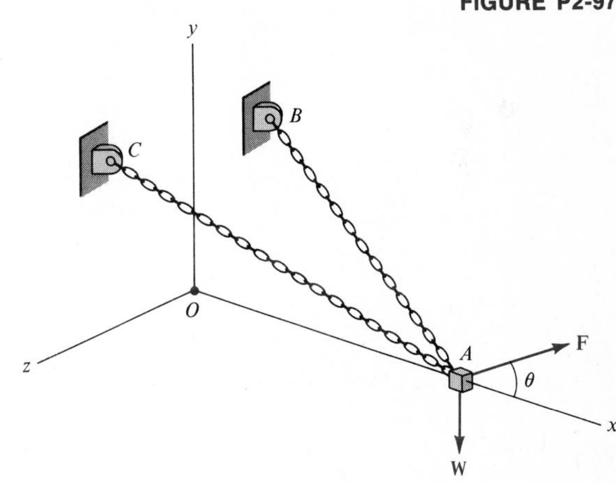

2–98 Solve Prob. 2–97 using $\theta = 0$, 45°, and 90°. Plot F as a function of θ and estimate the angle at which F is a minimum.

2–99 A proposed light fixture for a theater is to weigh 800 N. It is to be suspended by 16 wires of length $l = 10$ m. The wires are evenly spaced at the supports, which are on a horizontal circle of 2.5-m radius on the ceiling. Determine the tension in the wires if they share the load uniformly.

FIGURE P2-99

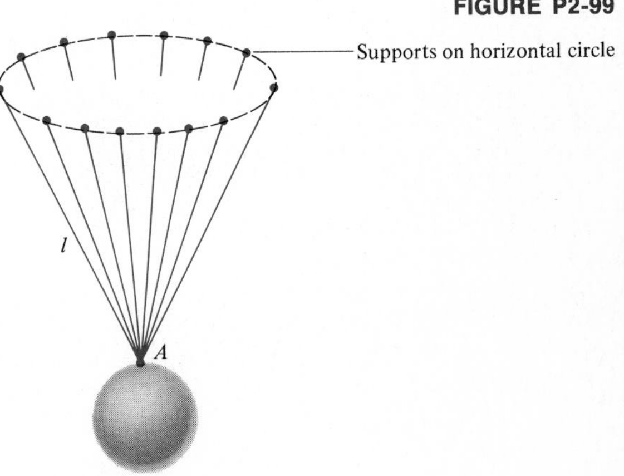

SUMMARY

Section 2-1

PARALLELOGRAM LAW:

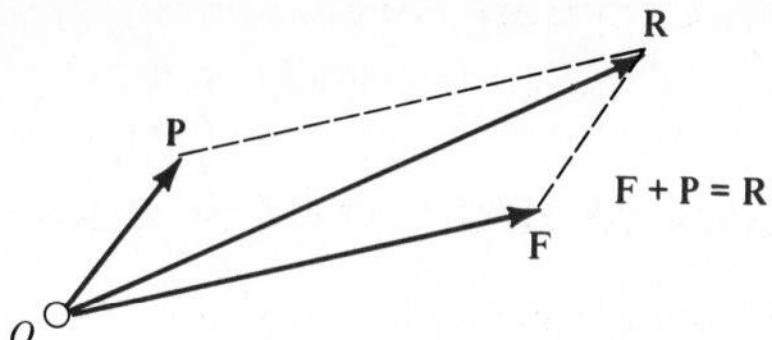

TRIANGLE RULE:

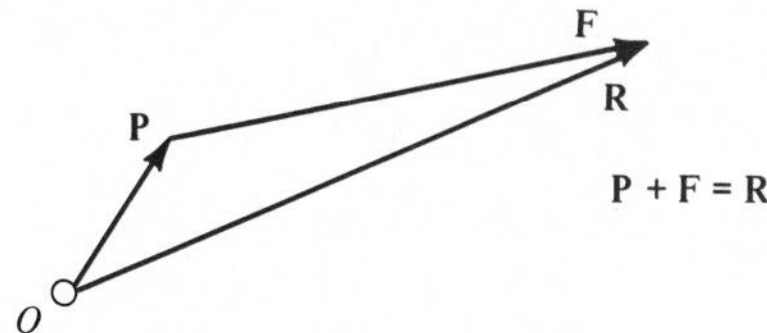

ADDITION OF VECTORS F AND P: $\mathbf{F} + \mathbf{P} = \mathbf{R}$

SUBTRACTION OF VECTOR F FROM R: $\mathbf{R} + (-\mathbf{F}) = \mathbf{P}$

POLYGON RULE:

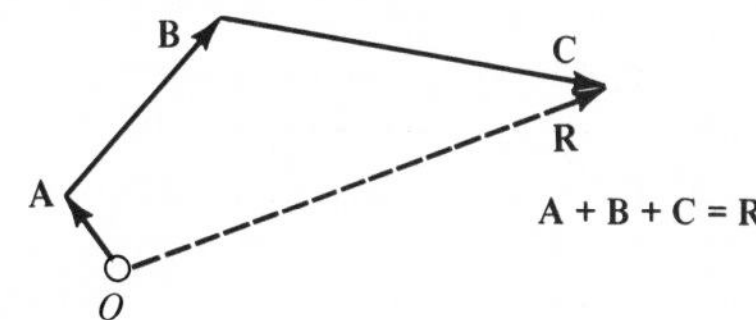

Section 2-2

COMPONENTS $\mathbf{F}_m$ AND $\mathbf{F}_n$ OF FORCE F:

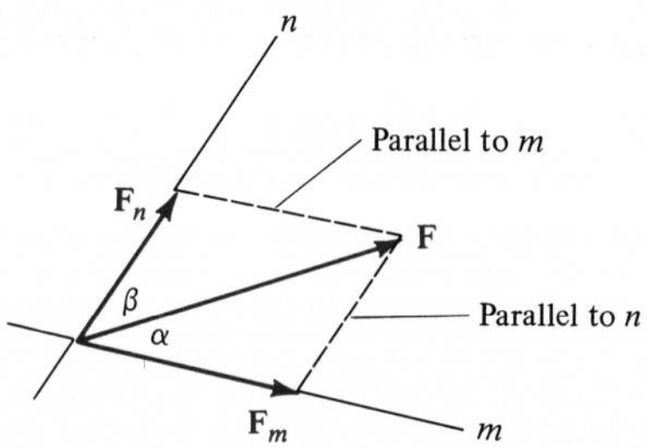

Sections 2-3, 2-4

UNIT VECTORS n AND i, j, k:

1. Each has a magnitude of 1.

2. Vector **F** of magnitude F can be written as

$\mathbf{F} = F\mathbf{n}$ where **F** and **n** have the same direction

In terms of xyz components,

$$\mathbf{F} = F_x\mathbf{i} + F_y\mathbf{j} + F_z\mathbf{k}$$

VECTOR SUMMATION TO OBTAIN RESULTANT $\mathbf{R} = \mathbf{R}_x\mathbf{i} + \mathbf{R}_y\mathbf{j} + \mathbf{R}_z\mathbf{k}$:

Scalar Components: $R_x = \sum F_x \quad R_y = \sum F_y$

$R_z = \sum F_z$

Vector Components: $\mathbf{R}_x = F_x\mathbf{i} \quad \mathbf{R}_y = F_y\mathbf{j}$

$\mathbf{R}_z = F_z\mathbf{k}$

SIGN CONVENTIONS:

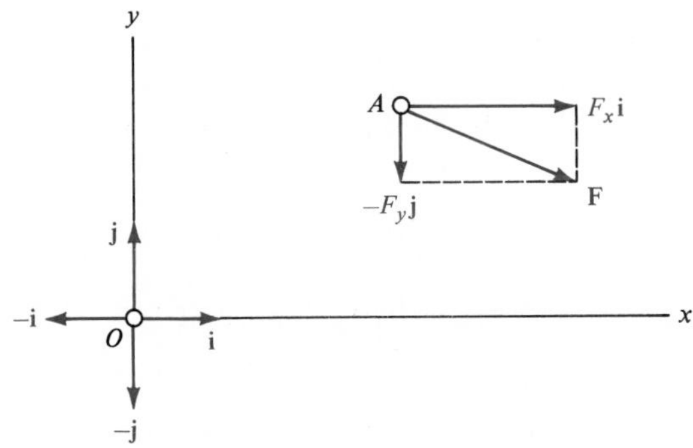

MAGNITUDE F OF VECTOR $\mathbf{F} = \mathbf{F}_x\mathbf{i} + \mathbf{F}_y\mathbf{j} + \mathbf{F}_z\mathbf{k}$:

$$F = \sqrt{F_x^2 + F_y^2 + F_z^2}$$

SCALAR COMPONENTS OF VECTOR F IN TERMS OF ANGLES:

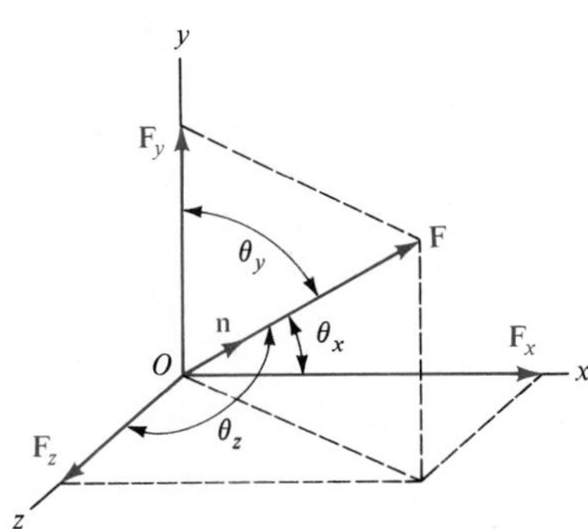

$$F_x = F\cos\theta_x \qquad F_y = F\cos\theta_y \qquad F_z = F\cos\theta_z$$

DIRECTION COSINES OF COLLINEAR VECTORS F AND n:

$$n_x = \cos\theta_x \qquad n_y = \cos\theta_y \qquad n_z = \cos\theta_z$$

USEFUL EXPRESSIONS:

$$\mathbf{n} = n_x\mathbf{i} + n_y\mathbf{j} + n_z\mathbf{k}$$

$$n_x^2 + n_y^2 + n_z^2 = 1$$

Section 2-5

FORCE VECTOR F FROM SCALAR INFORMATION:

Necessary information:

1. F, magnitude of force
2. Direction of force from point A to B
3. Coordinates of points A and B on the line of action of force

The required force vector **F** is expressed as

$$\mathbf{F} = F(n_x\mathbf{i} + n_y\mathbf{j} + n_z\mathbf{k}) = F\,\frac{d_x\mathbf{i} + d_y\mathbf{j} + d_z\mathbf{k}}{d}$$

where $n_x, \ldots$ are direction cosines of line A to B
$d_x, \ldots$ are coordinate increments from A to B
d = distance A to B = $\sqrt{d_x^2 + d_y^2 + d_z^2}$

Section 2-6

SCALAR PRODUCT (OR DOT PRODUCT) OF VECTORS A AND B

$$\mathbf{A} \cdot \mathbf{B} = AB\cos\phi$$

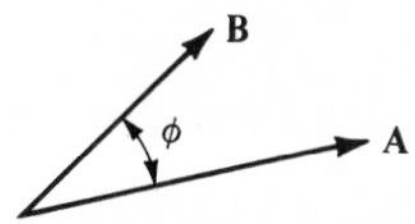

USEFUL PROPERTIES:

$$\mathbf{A} \cdot \mathbf{B} = \mathbf{B} \cdot \mathbf{A}$$

$$\mathbf{A} \cdot (\mathbf{B} + \mathbf{C}) = \mathbf{A} \cdot \mathbf{B} + \mathbf{A} \cdot \mathbf{C}$$

$$\mathbf{i} \cdot \mathbf{i} = \mathbf{j} \cdot \mathbf{j} = \mathbf{k} \cdot \mathbf{k} = 1$$

$$\mathbf{i} \cdot \mathbf{j} = \mathbf{j} \cdot \mathbf{k} = \mathbf{k} \cdot \mathbf{i} = 0$$

ANGLE ϕ BETWEEN VECTORS A AND B:

$$\phi = \text{arc cos} \frac{A_xB_x + A_yB_y + A_zB_z}{AB}$$

PROJECTION F′ OF VECTOR F ON LINE l:

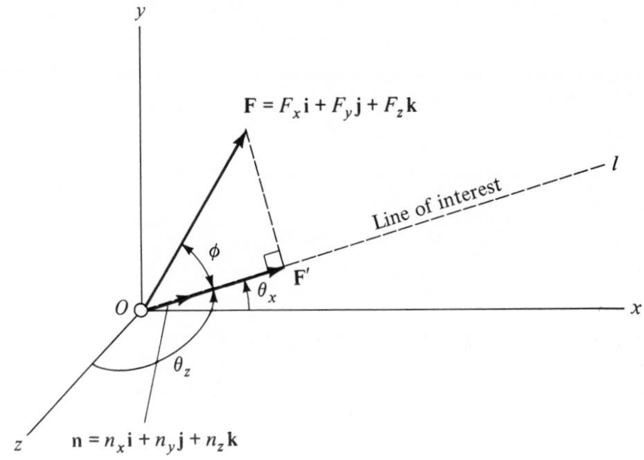

$$\begin{aligned} F' &= F_xn_x + F_yn_y + F_zn_z \\ &= F_x \cos \theta_x + F_y \cos \theta_y + F_z \cos \theta_z \end{aligned}$$

Section 2-7

SCALAR EXPRESSIONS FOR EQUILIBRIUM OF A PARTICLE:

$$\sum F_x = 0 \qquad \sum F_y = 0 \qquad \sum F_z = 0$$

where the *xyz* axes may be chosen for convenience.

FREE-BODY DIAGRAMS: Make a simple sketch of a body (or part of a body, or system of bodies) to be analyzed. Imagine this body to be isolated from all other bodies. Show *all* significant, relevant forces on the free body, including unknown forces. Use the equilibrium equations to solve for unknown forces.

REVIEW PROBLEMS

FIGURE P2-100

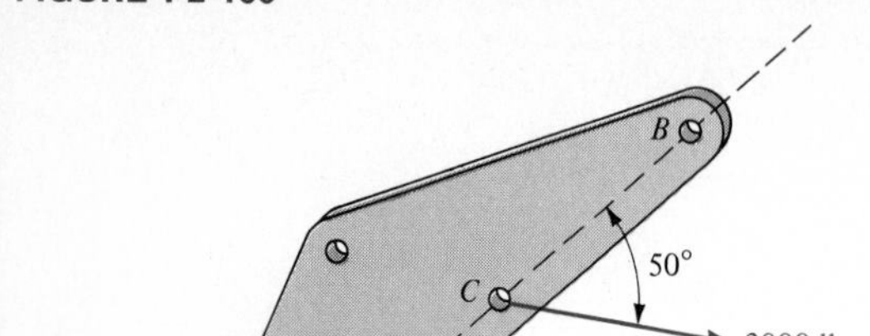

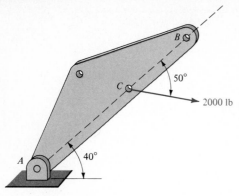

2–100 Calculate the horizontal and vertical components of the force acting on the bracket of a road simulator (see Sec. 1-7). Also calculate the components normal and tangent to line *AB*.

FIGURE P2-101

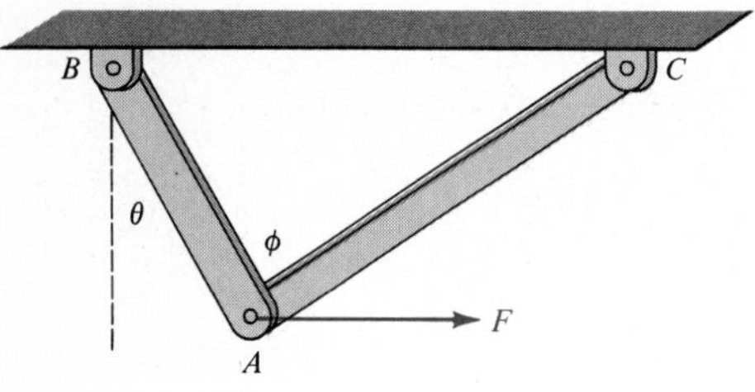

2–101 Plot the components of the horizontal force along bars *AB* and *AC* for the following pairs of angles:

θ	ϕ
0	40°
10°	50°
20°	60°

Enter all data on the same diagram and sketch smooth curves through the appropriate points for each bar.

FIGURE P2-102

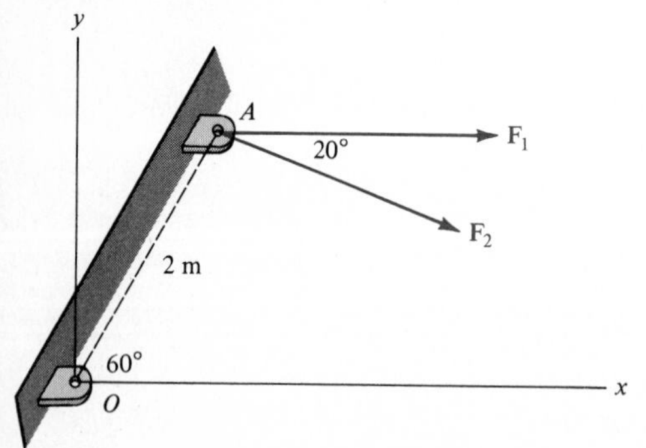

2–102 Express the resultant **R** of the two forces in terms of unit vectors of the *xy* coordinate system. $F_1 = 30$ kN and $F_2 = 45$ kN, both in the *xy* plane.

2–103 Three forces are acting on the frame of a simple crane. Express the resultant **R** using unit vectors of the xy coordinate system and assuming that the forces are concurrent.

FIGURE P2-103

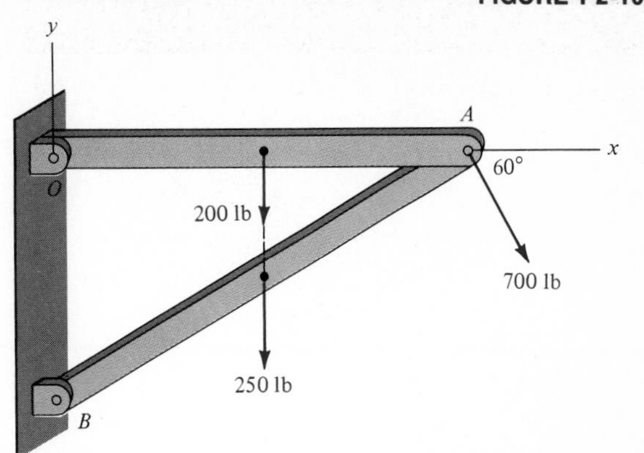

2–104 A rectangular crate is lifted by four ropes of equal length l. It is estimated that the tension in each rope is 2.5 kN, simultaneously. Determine the horizontal and vertical components of the resultant force applied by the four ropes at point A.

FIGURE P2-104

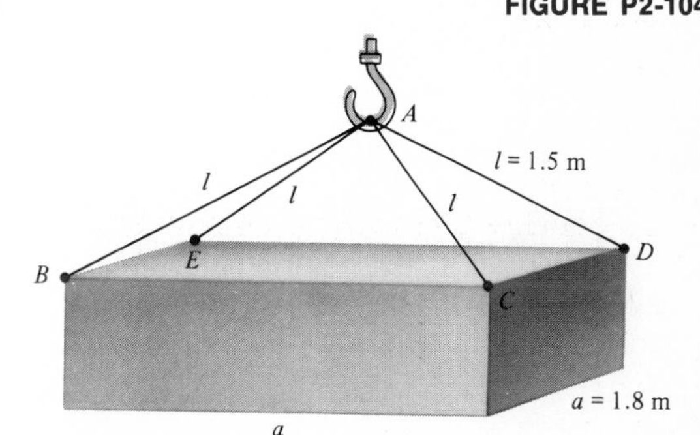

2–105 A proposed light fixture for a theater is to be suspended by 16 wires of equal length l. The wires are evenly spaced at the supports, which are on a horizontal circle of radius r on the ceiling. The wires are concurrent at point A, and each has tension T. Determine the horizontal and vertical components of the resultant of the 16 tensions at A.

FIGURE P2-105

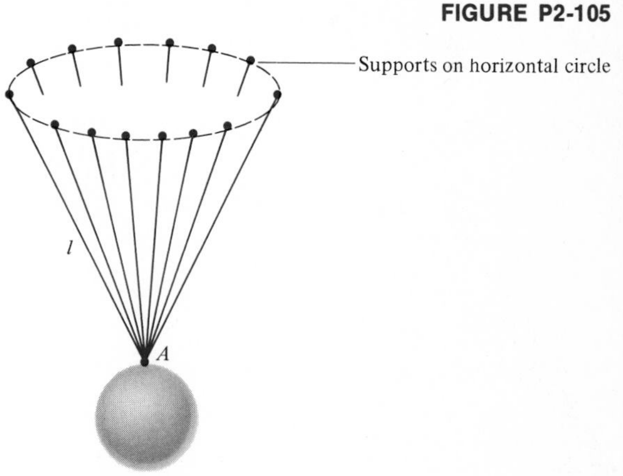

FIGURE P2-106

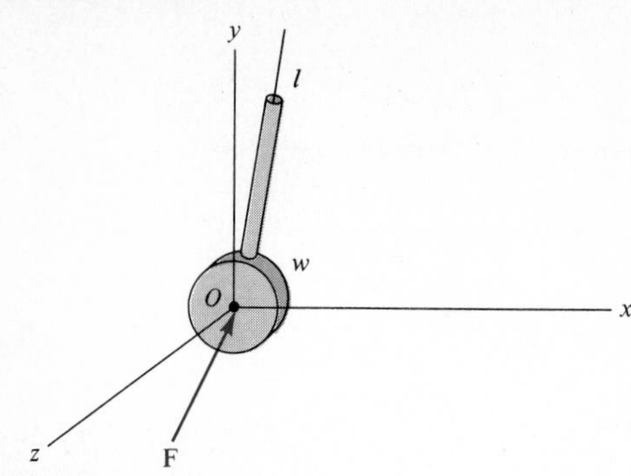

FIGURE P2-108

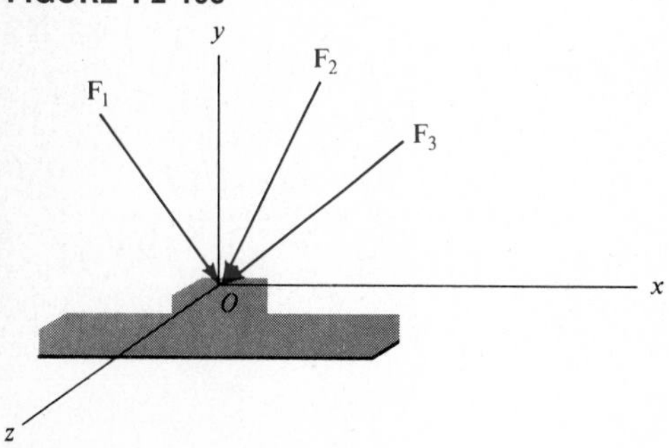

2–106 In a laboratory test of the strut l and wheel w of an aircraft a force $\mathbf{F} = (500\mathbf{i} + 700\mathbf{j} - 200\mathbf{k})$ lb is applied at point O. The orientation of the strut is given by the vector $\mathbf{l} = \mathbf{i} + 8\mathbf{j} - 2\mathbf{k}$. Calculate the angle between $\mathbf{F}$ and $\mathbf{l}$.

2–107 Determine the projection of force $\mathbf{F}$ along $\mathbf{l}$ in Prob. 2–106.

2–108 Three concurrent forces are applied at the top of a foundation block: $\mathbf{F}_1 = 10^4(3\mathbf{i} - 4\mathbf{j} - \mathbf{k})$ lb, $\mathbf{F}_2 = 10^4(-2\mathbf{i} - 3\mathbf{j} + 2\mathbf{k})$ lb, $\mathbf{F}_3 = 10^4(-3\mathbf{i} - 2.5\mathbf{j} - \mathbf{k})$ lb. Determine the reaction $\mathbf{R}$ that must act on the block to hold it in place.

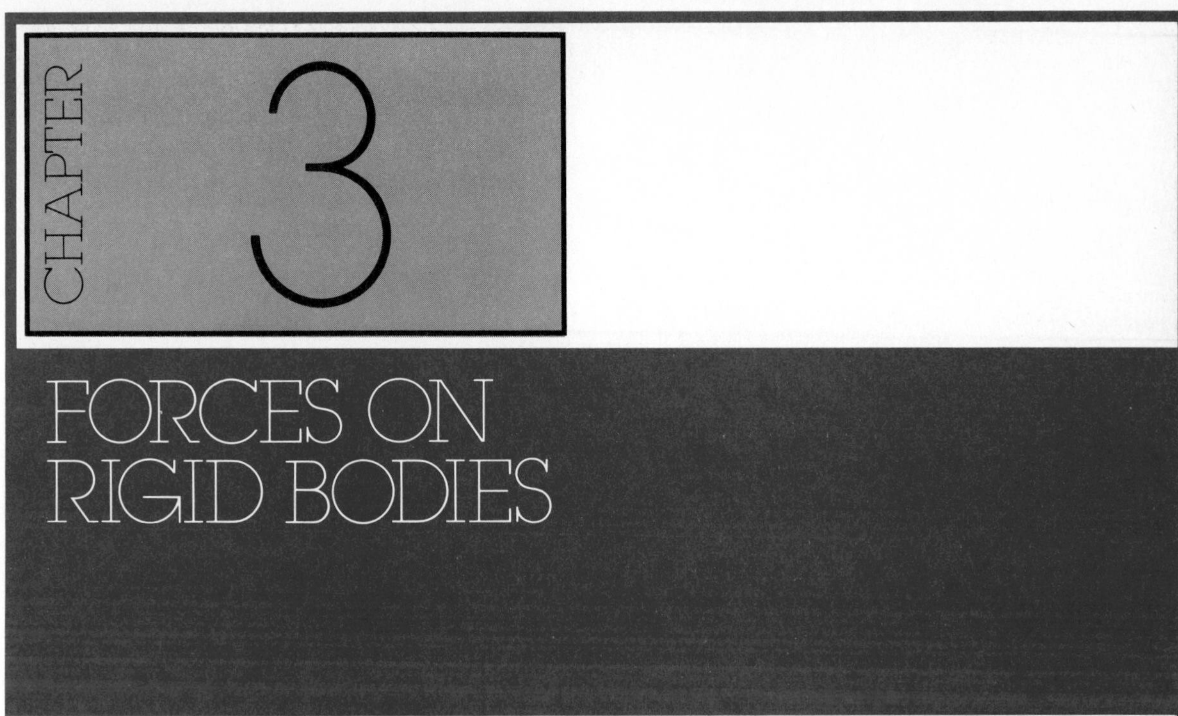

OVERVIEW

Basic Concepts in Modeling Rigid Bodies

There are several important concepts and principles of engineering mechanics that must be kept in mind when dealing with sizable members. The most commonly used concept is the assumption of rigidity. All solid materials deform when forces are applied to them, but frequently the deformations are negligible in practice. A *rigid body* is a useful model in the solution of many different problems in mechanics. This is true at least in the early part of the analysis, which eventually may have to involve the deformability of the material. The various other principles discussed in this chapter will be applied to rigid-body models.

For an example of modeling rigid bodies, consider the movable *wave board* of a complex ocean-wave-simulating facility, shown in Fig. 3-1a and b. Such a board is the end wall of a long channel of water, and its controlled motion about hinges *A* can generate a wide variety of waves. The two-piece wave board shown has two hinges at the bottom (*A*) and two hinges (*B*) between the two boards. Motion of the boards is caused by the two *hydraulic actuators* (hydraulic rams) shown in Fig. 3-1b. Hydraulic actuator *C* applies a nearly horizontal force to the lower board. Hydraulic actuator *D*

(a)

(b)

FIGURE 3-1

Two-piece wave board for the generation of water waves for research purposes. (Courtesy MTS Systems Corporation, Minneapolis, Minnesota.)

applies a force to move the upper board relative to the lower board. Depending on the displacements of these actuators, the wave board may be considered either as a single rigid body pivoted at the fixed axis A, or as two rigid bodies that are connected at axis B and supported at axis A. Such wave boards will be useful examples in numerous fundamental mechanics analyses.

STUDY GOALS

SECTION 3-1 presents the important distinction between external and internal forces that may act on a rigid body. Understanding these forces is essential in drawing free-body diagrams of rigid bodies.

SECTION 3-2 defines equivalent forces for the purpose of altering the statements of some problems to advantage.

SECTIONS 3-3 AND 3-4 introduce the concept of moment of a force. A moment is a "turning effect" of a force. It is one of the most frequently used concepts in engineering mechanics and should be thoroughly studied. The vector definition and evaluation of moment are particularly useful in solving complex problems.

SECTIONS 3-5 AND 3-6 present the method of evaluating the moment of a force about a line. In the most general case this is the most complex problem involving moments. Vector methods are especially useful for solving such problems.

SECTIONS 3-7 TO 3-9 present couples (a special case of moments) and the basic operations concerning them. These are advantageous in certain computations.

SECTION 3-10 shows the procedures for transforming a force into a force–couple system, or vice versa. These possibilities facilitate solving problems.

SECTIONS 3-11 AND 3-12 extend the force–couple transformations for the simplified analysis of any complex system of forces and moments. These are advantageous both in the theoretical and experimental investigations of mechanical systems.

3-1 EXTERNAL AND INTERNAL FORCES. FREE-BODY DIAGRAMS

The two-piece wave board shown in Fig. 3-1 can be used in two distinct configurations. When actuator *D* is "locked" in any particular position, the upper board does not move relative to the lower one, so the two boards may be analyzed as a single rigid body. There are a number of *external forces* acting on this member at any time, as shown in the free-body diagram of Fig. 3-2. Here **W** is the weight of the complete board and attached mechanisms that move as a unit about the hinges on line *A–A*. This force **W** is always vertical. Force **C** is from the horizontal actuator and force **F** is the net resultant of the forces caused by the water. Reaction **R** is from the two hinges on the bottom.

Figure 3-3 shows a more general configuration of the single wave board. The magnitudes and directions of the forces (except **W**) depend on the shape, position, and movement of the board. The points of application of forces **C** and **R** are fixed by the geometries of the actuator and lower hinge connections. **F** can move up or down depending on the level and movement of the water. **W** can move left or right depending on the relative position θ of the two boards; it always acts at the *center of gravity* of the whole body shown in the free-body diagram. This is discussed in detail in Ch. 7.

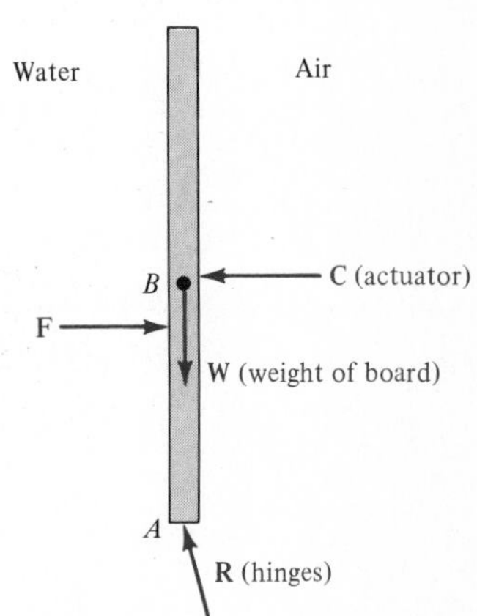

FIGURE 3-2

Free-body diagram of a rigid wave board in a simple configuration

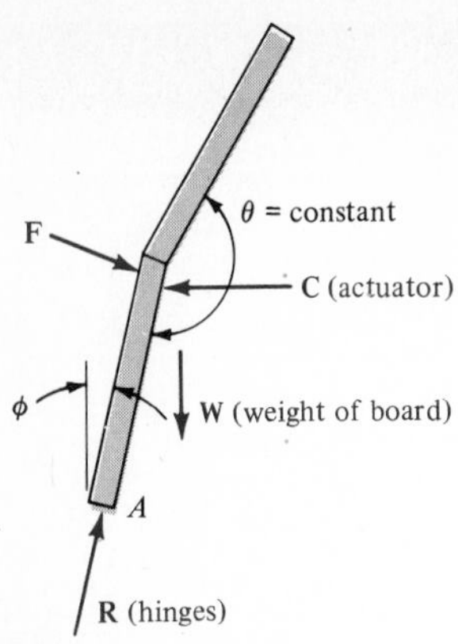

FIGURE 3-3

Free-body diagram of a rigid wave board in a general configuration

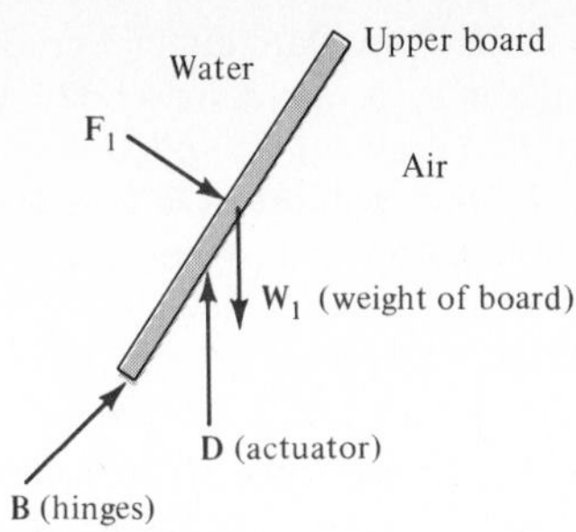

FIGURE 3-4

Upper part of a two-piece wave board

The forces acting between the upper and lower parts of the wave board are *internal forces* when the whole board is considered as a rigid body. These internal forces completely cancel each other and must be ignored when drawing the whole free-body diagram as in Fig. 3-2. However, the free-body diagram may be drawn for any body of interest (for example, the upper piece alone of the wave board). In that case the forces from the hinges at line *B–B* and from actuator *D* are external forces and must be shown as in Fig. 3-4. Here $\mathbf{F}_1$ and $\mathbf{W}_1$ are the parts of **F** and **W** acting only on the upper board, while **C** and **R** are not shown because they are not directly acting on this member. The details of analyzing interrelated free-body diagrams are presented in Ch. 4.

3-2 EQUIVALENT FORCES

Sliding vectors and the *principle of transmissibility* have been introduced in Sec. 1-5. These concepts are especially important in dealing with rigid bodies, but the assumption of rigidity must be carefully checked for these. For example, it does not make any difference in theory whether the actuator force is pushing (**D**) or pulling (**D**′) on the wave board in Fig. 3-5. However, it may make a significant difference in practice, depending on the deformations of the tubular structure through which the force is applied to the board (Fig. 3-1b).

Two forces **D** and **D**′ acting at different points are *equivalent* if they have the same magnitude, direction, line of action, and the same effect for all practical purposes on a given rigid body. A practical limitation of the concept of equivalent forces is from the magnitude of the deformations. Further limitations may be applicable since certain failures depend on tensile or compressive loading. These may involve the nature of the material or the geometry of the member (see texts on strength of materials or mechanics of materials).

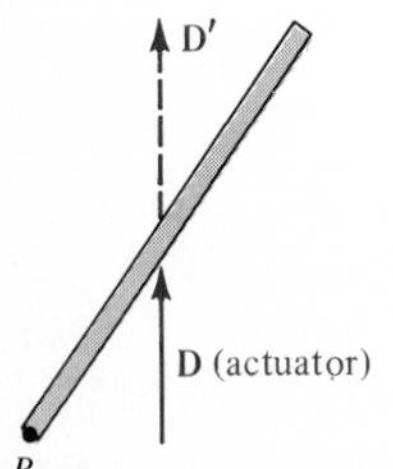

FIGURE 3-5

Equivalent force **D** and **D**′ acting on a rigid body ($D = D'$)

MOMENT OF A FORCE

The *moment* M_A of a force **F** (also called *torque*) about a point A is defined as a scalar quantity

$$M_A = Fd \tag{3-1}$$

where d is the nearest distance from A to the line of action of **F** as in Fig. 3-6. The terms *lever arm* or *moment arm* are often used for the distance d.

The concept of the moment describes the turning effect of the force with respect to the point of interest. The sense of the rotational tendency gives the sense of the moment, clockwise or counterclockwise. The *right-hand rule* is commonly used to describe a moment: with the fingers of the right hand curled in the sense of the rotational tendency, the pointed thumb shows the axis about which rotation tends to occur.

The *scalar definition of moment* given in Eq. 3-1 is valid for rigid bodies of any shape and in three dimensions, as well, but care must be taken that d is the nearest distance to the line of action. The unit of moment is force times length.

The statement in Sec. 3-2 concerning the equivalence of two forces can be expanded to include a formal statement of equivalence with respect to turning effects. Forces **D** and **D**′ are equivalent if

$$\mathbf{D} = \mathbf{D}' \quad \text{and} \quad M_A = M'_A \tag{3-2}$$

These are the necessary and sufficient conditions for equivalence.

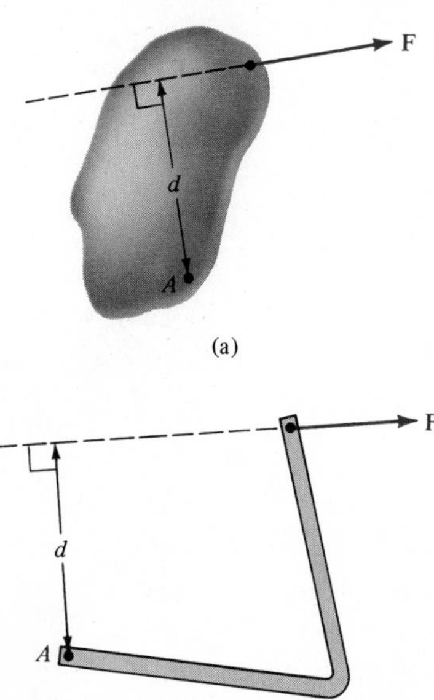

FIGURE 3-6

Force **F** has a moment of magnitude $M_A = Fd$ with respect to point A

EXAMPLE 3-1

A relatively small, one-piece wave board is modeled in Fig. 3-7a. The maximum force C from the hydraulic actuator is always the same 150 kN, regardless of the angular positions of the board. The distance AB is 0.8 m. Determine the moment of **C** about point A for (a) $\theta = \phi = 0$, and (b) $\theta = 4°$, $\phi = 2°$.

SOLUTION

(a) $M_A = Fd$

For $\theta = \phi = 0$, $d = 0.8$ m, so

$$M_A = (150 \text{ kN})(0.8 \text{ m}) = 120 \text{ kN}\cdot\text{m}$$

(b) From Fig. 3-7b,

$$d = 0.8 \cos(\phi + \theta) = 0.8 \cos(6°) = 0.796 \text{ m}$$

$$M_A = Fd = (150 \text{ kN})(0.796 \text{ m}) = 119.3 \text{ kN}\cdot\text{m}$$

JUDGMENT OF THE RESULTS

As expected, the moment of the force applied at a given point depends on its angle of application. The difference here is small because the change in angle is small.

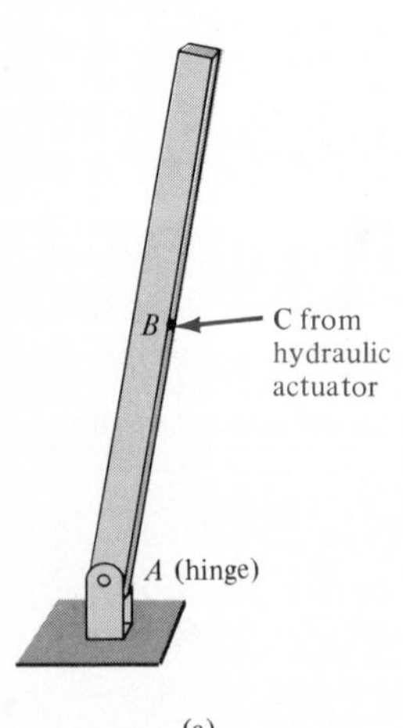

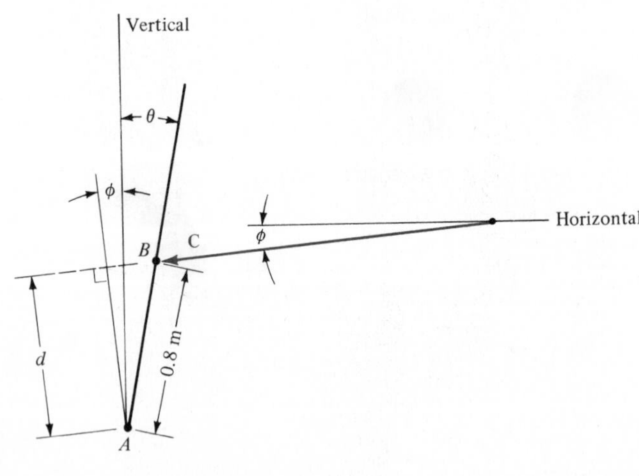

FIGURE 3-7

Model of a simple wave board

EXAMPLE 3-2

Figure 3-8a shows a test setup for applying vertical forces to the ball joint of a car. Determine the moments of force F about points A, B, C, and D. The locations of these points are given in Fig. 3-8b using the xy coordinate system for reference.

SOLUTION

$M = Fd$ where d is a horizontal distance in each case since $\mathbf{F}$ is vertical

$M_A = F\,(14 \text{ in.})$

$M_B = F\,(19 \text{ in.})$

$M_C = F\,(21 \text{ in.})$

$M_D = F\,(40 \text{ in.})$

JUDGMENT OF THE RESULTS

Point D is farthest from the applied force, hence the moment about D is largest as expected; $M_D > M_C > M_B > M_A$. It is emphasized that the vertical distances to a vertical force were not needed (law of transmissibility).

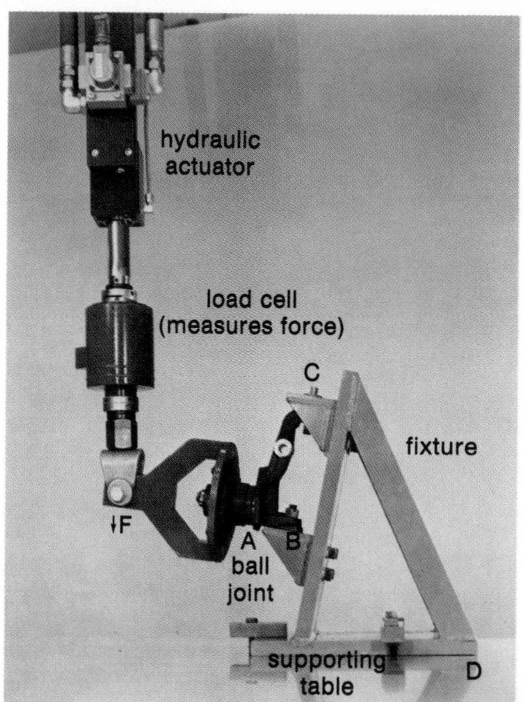

(a)

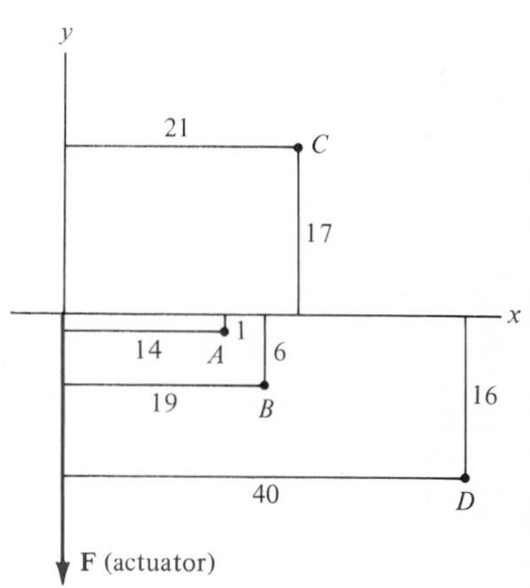

The numbers represent distances in inches.

(b)

FIGURE 3-8

Test setup of a ball joint of a car. (Photograph courtesy MTS Systems Corporation, Minneapolis, Minnesota.)

PROBLEMS

SECTIONS 3-1, 3-2, 3-3

FIGURE P3-1

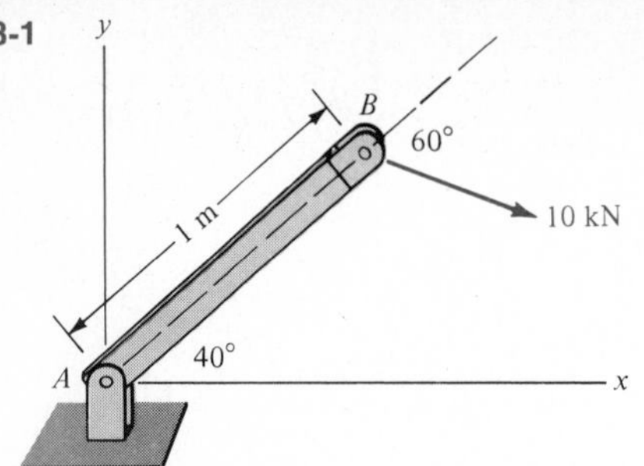

3–1 Draw a free-body diagram for the bar that is hinged at A. Include unknown forces. Determine the moment of the force about point A.

FIGURE P3-2

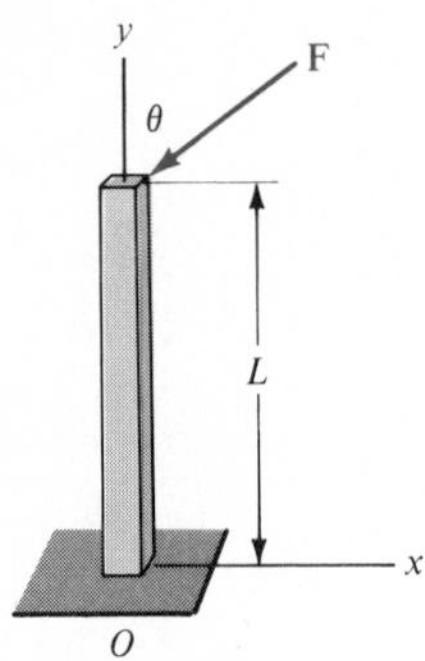

3–2 Determine the moment of $\mathbf{F} = (-300\mathbf{i} - 100\mathbf{j})$ lb about point O if $L = 5$ ft.

3–3 Plot M_O for the pole in Fig. P3-2 using F and L as constants and $\theta = 0$, 30°, 60°, and 90°.

FIGURE P3-4

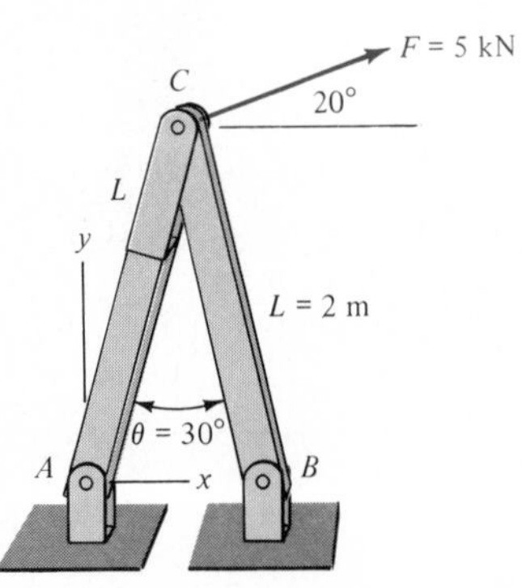

3–4 The simple frame consists of two bars hinged at A, B, and C. Draw a free-body diagram of the frame without its supports at A and B. Include unknown forces.

3–5 Determine the moments M_A and M_B of the force F in Fig. P3-4.

3–6 Plot the moments M_A and M_B of the force F in Fig. P3-4. Let $F = 500$ lb, $L = 4$ ft, and $\theta = 20°$, 30°, and 40°. Use a single diagram for showing both moments.

FIGURE P3-7

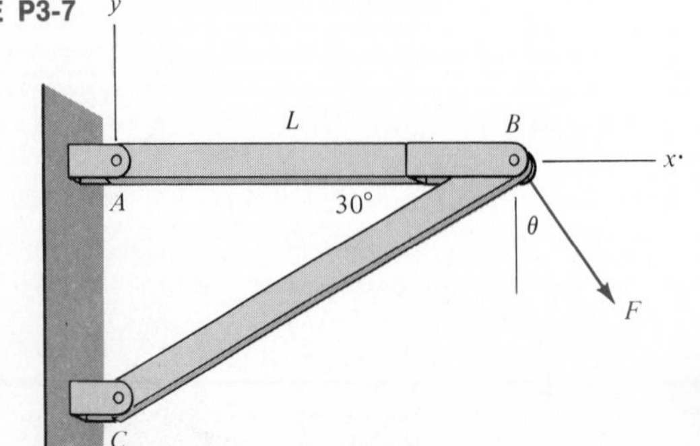

3–7 Determine the moments M_A and M_C of the force $F = 3$ kN if $L = 4$ m and $\theta = 20°$.

3–8 Determine the angle θ in Fig. P3-7 for which the moments of the force F are $M_A = M_C$.

3–9 Calculate the moments M_A and M_B of the force $F = 2000$ lb if $L = 5$ ft, $\theta = 70°$, $\phi = 50°$, and $\alpha = 60°$.

3–10 Determine the angle α in Fig. P3-9 for which the moments of the force $F = 10$ kN are $M_A = M_B$. $L = 1.5$ m, $\theta = 70°$, and $\phi = 45°$.

FIGURE P3-9

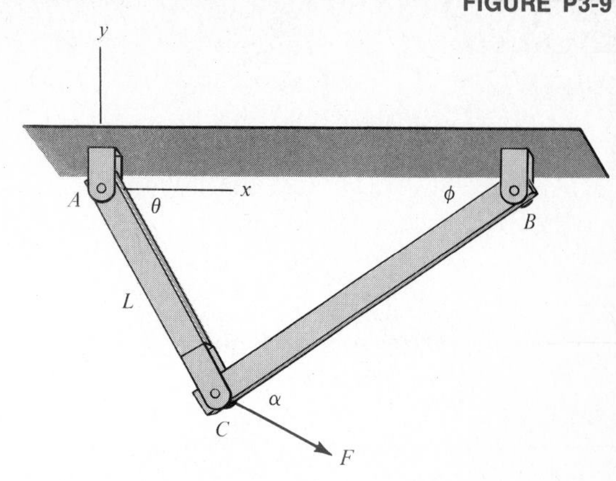

3–11 Calculate the moments M_1 and M_2 of forces F_1 and F_2 about A. Also determine the resultant F_R of F_1 and F_2, and the moment of F_R about A. $F_1 = 2$ kN, $F_2 = 3$ kN, $L = 1$ m, $\theta = 60°$, and $\phi = 40°$.

3–12 Assume that $F_1 = 500$ lb is always horizontal and to the right in Fig. P3-11. Determine the angle ϕ and the smallest value of F_2 to apply a moment M_{2A} which is equal in magnitude but opposite in direction to M_{1A}. $L = 4$ ft and $\theta = 50°$.

FIGURE P3-11

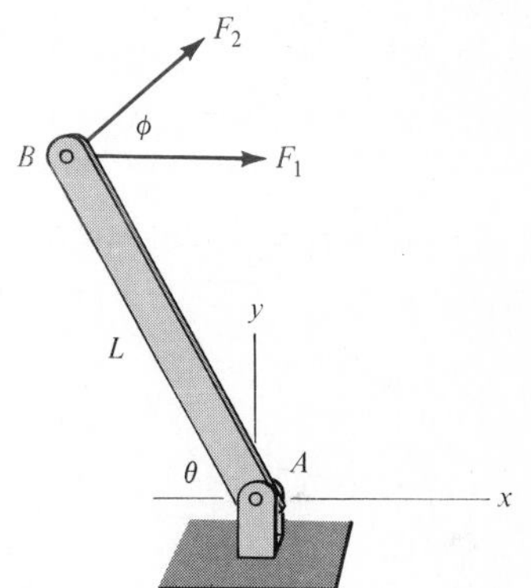

3–13 The street lamp A weighs $W = 400$ N. The force of the wind is $F = 300$ N. Determine the moments of F and W about O. Also calculate the resultant F_R of F and W, and the moment of F_R about O. $d = 2$ m and $h = 5$ m.

3–14 Assume that the force of the wind $F = 40$ lb on lamp A in Fig. P3-13 is directly into the paper. $W = 50$ lb as shown. Determine the moments of F and W about O. $d = 7$ ft and $h = 15$ ft.

FIGURE P3-13

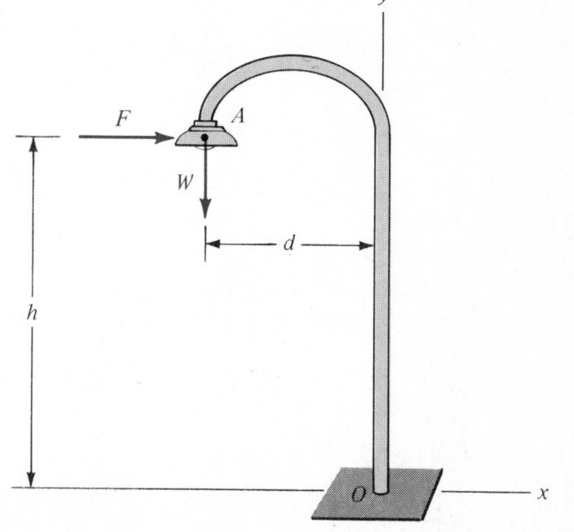

FIGURE P3-15

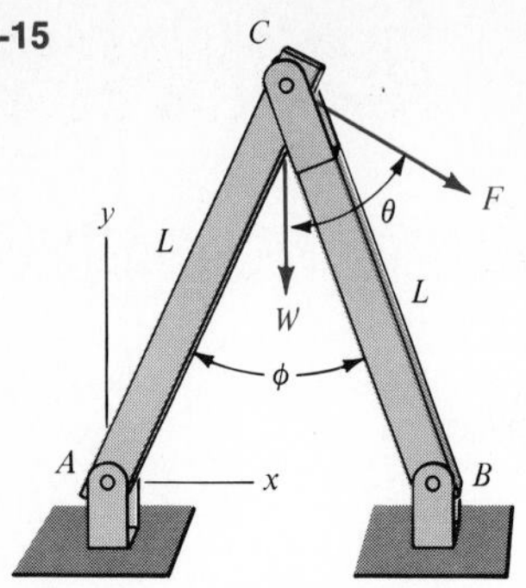

3–15 Determine the moments of F and W about A and B individually (four moments). Calculate the resultant F_R of F and W, and the moments of F_R about A and B. $F = 800$ N, $W = 500$ N, $L = 3$ m, $\theta = 70°$, and $\phi = 30°$.

FIGURE P3-16

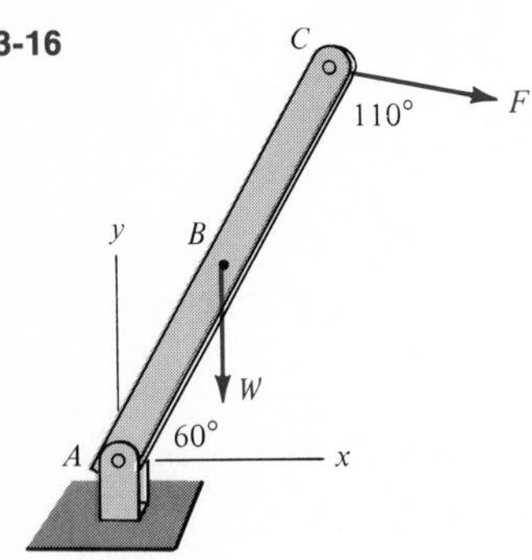

3–16 Determine the moments of F and W about A. Also calculate the resultant F_R of F and W, and the moment of F_R about A. $F = 200$ lb, $W = 40$ lb, and $AB = BC = 4$ ft.

FIGURE P3-17

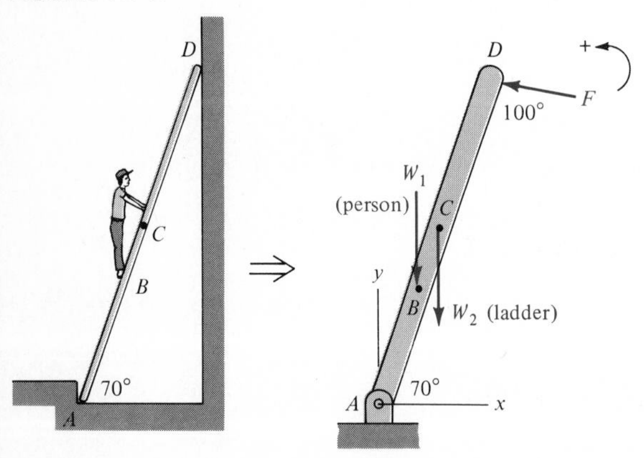

3–17 Write an expression for the sum of the moments M_A of the three forces. Use the distances AB, AC, and AD.

FIGURE P3-18

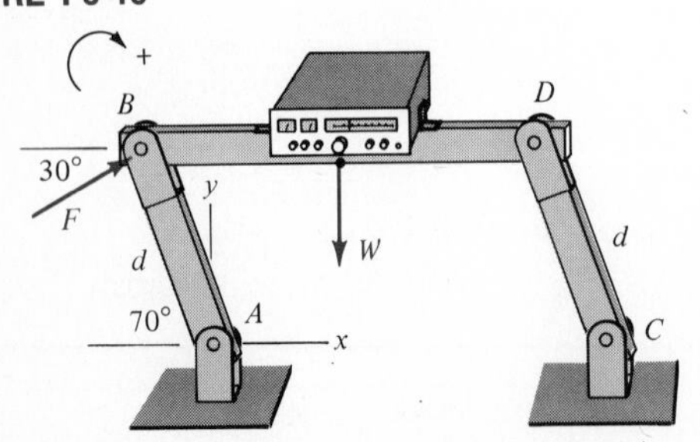

3–18 The test platform and the test object have a total weight $W = 3$ kN, with W acting at a horizontal distance of 1 m from B. Determine the sum of the moments of F and W about A if $F = 6$ kN and $d = 80$ cm.

VECTOR PRODUCT OF TWO VECTORS. MOMENT OF A FORCE ABOUT A POINT

3-4

The scalar method of determining the moment of a force is the best in solving simple problems where the nearest distance d to the line of action is easily found. The solutions of more complex problems are simplified by using additional mathematical tools.

The *vector product* of two concurrent vectors **A** and **B** is, by definition, the vector **V** that has the following properties:

1. **V** is perpendicular to the plane of vectors **A** and **B** as in Fig. 3-9a.
2. The sense of **V** obeys the right-hand rule: **A** rotated into **B** gives **V**, and **B** rotated into **A** gives **V**′ in Fig. 3-9a.
3. The magnitude of **V** is the product of the magnitudes of **A**, **B**, and sin θ, where θ is the angle between **A** and **B**:

$$V = AB \sin \theta \tag{3-3}$$

The commonly used mathematical notation for the vector product is

$$\mathbf{V} = \mathbf{A} \times \mathbf{B} \tag{3-4}$$

which is also called the *cross product* of **A** and **B**.

According to the second property given above, vector products are *not commutative*, so

$$\mathbf{A} \times \mathbf{B} \neq \mathbf{B} \times \mathbf{A} \tag{3-5}$$

but

$$\mathbf{A} \times \mathbf{B} = -(\mathbf{B} \times \mathbf{A}) \tag{3-6}$$

Vector products are *distributive*, so for three vectors **A**, **B**, and **C**,

$$\mathbf{A} \times (\mathbf{B} + \mathbf{C}) = \mathbf{A} \times \mathbf{B} + \mathbf{A} \times \mathbf{C} \tag{3-7}$$

This is intuitively acceptable to most people. A rigorous proof is left as an exercise for the reader.

The distributive property is the basis for quantitative determination of the vector product in terms of vector components. For the general three-dimensional case,

$$\begin{aligned}\mathbf{V} = \mathbf{A} \times \mathbf{B} &= (A_x\mathbf{i} + A_y\mathbf{j} + A_z\mathbf{k}) \times (B_x\mathbf{i} + B_y\mathbf{j} + B_z\mathbf{k}) \\ &= (A_x\mathbf{i}) \times (B_x\mathbf{i}) + (A_x\mathbf{i}) \times (B_y\mathbf{j}) + \ldots + (A_z\mathbf{k}) \times (B_z\mathbf{k})\end{aligned} \tag{3-8}$$

This can be simplified because all the possible vector products of the unit vectors are, according to definition,

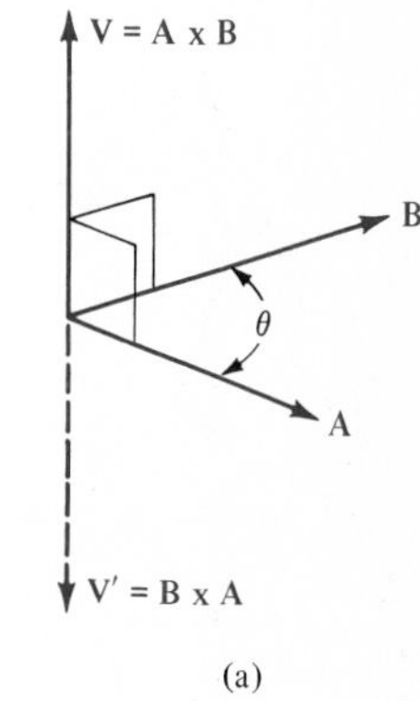

(a)

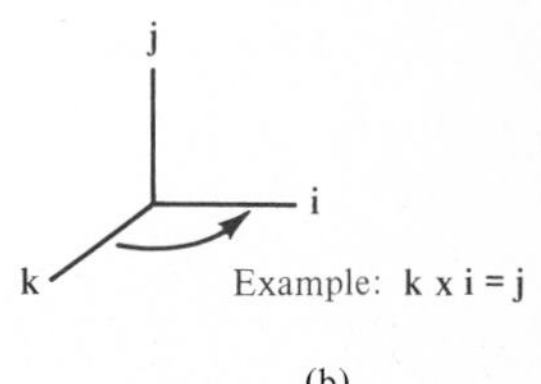

(b)

FIGURE 3-9
Vector product **V** of vectors **A** and **B**

$$\mathbf{i} \times \mathbf{i} = \mathbf{j} \times \mathbf{j} = \mathbf{k} \times \mathbf{k} = 0 \qquad \text{since } \sin 0 = 0$$

$$\mathbf{i} \times \mathbf{j} = \mathbf{k} \qquad \text{since } \sin 90^\circ = 1$$

$$\mathbf{i} \times \mathbf{k} = -\mathbf{j}$$

$$\mathbf{j} \times \mathbf{i} = -\mathbf{k} \tag{3-9}$$

$$\mathbf{j} \times \mathbf{k} = \mathbf{i}$$

$$\mathbf{k} \times \mathbf{i} = \mathbf{j}$$

$$\mathbf{k} \times \mathbf{j} = -\mathbf{i}$$

The positive or negative signs of these are obtained from the right-hand rule as in Fig. 3-9b. With these, Eq. 3-8 becomes

$$\mathbf{V} = (A_yB_z - A_zB_y)\mathbf{i} + (A_zB_x - A_xB_z)\mathbf{j} + (A_xB_y - A_yB_x)\mathbf{k} \tag{3-10}$$

The terms in parentheses are the rectangular components of the vector **V**,

$$\begin{aligned} V_x &= A_yB_z - A_zB_y \\ V_y &= A_zB_x - A_xB_z \\ V_z &= A_xB_y - A_yB_x \end{aligned} \tag{3-11}$$

The vector product can also be calculated from a determinant. This method should be favored in routine solutions because it is the easiest to remember.

$$\mathbf{V} = \begin{vmatrix} \mathbf{i} & \mathbf{j} & \mathbf{k} \\ A_x & A_y & A_z \\ B_x & B_y & B_z \end{vmatrix} \tag{3-12}$$

This can be evaluated as follows.* Write the sum of the products along the solid diagonals, and subtract from this the sum of the products along the dashed diagonals. Each term must be the product

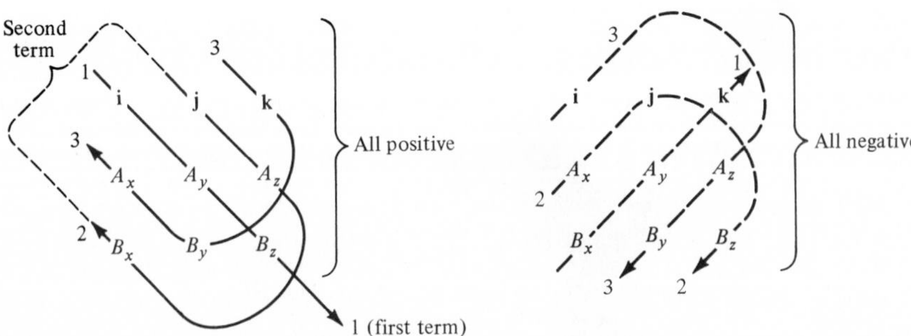

* Note that this particular method is easy to remember after a little practice, but it is not valid for higher-order determinants.

of two different scalar components and a unit vector. In practice it is not necessary to write these as done here to illustrate the method. It is possible to write directly from Eq. 3-12,

$$\mathbf{V} = A_yB_z\mathbf{i} + A_zB_x\mathbf{j} + A_xB_y\mathbf{k} - A_yB_x\mathbf{k} - A_xB_z\mathbf{j} - A_zB_y\mathbf{i} \tag{3-13}$$

This can be rearranged to give Eq. 3-10 or 3-11.

The vector product has great practical utility in determining moments of forces. It has been found that the moment of a force **F** about a point O is the vector product of the position vector **r** (from O to *any* point A on the line of action of **F**) and **F**,

$$\boxed{\mathbf{M}_O = \mathbf{r} \times \mathbf{F}} \tag{3-14}$$

This is the *vector definition of moment*. It is denoted by a double-headed arrow in drawings in this text.

For a simple example of validating the vector definition of moment with the intuitively more acceptable scalar definition, consider Fig. 3-10, which shows a two-dimensional arrangement. The force **F** is applied at point A that has coordinates a and b. From the scalar definition,

$$M_O = Fa \qquad \text{counterclockwise}$$

From the vector product,

$$\mathbf{M}_O = \mathbf{r} \times \mathbf{F} = (a\mathbf{i} + b\mathbf{j}) \times F\mathbf{j} = Fa\mathbf{k}$$

The unit vector **k** is correct for $\mathbf{M}_O$ since it means "counterclockwise" in this diagram according to the right-hand rule.

Note that the definition leading to Eq. 3-14 states that **r** is a position vector from point O to *any* point A on the line of action of **F**. To check this, the moment $\mathbf{M}_O$ is recalculated using $\mathbf{r}'$ in Fig. 3-10.

$$\mathbf{M}_O = \mathbf{r}' \times \mathbf{F} = \left(a\mathbf{i} + \frac{b}{3}\mathbf{j}\right) \times F\mathbf{j} = Fa\mathbf{k}$$

It is clear that for any r_y component $r_y\mathbf{j} \times F\mathbf{j} = 0$, and M_O is always

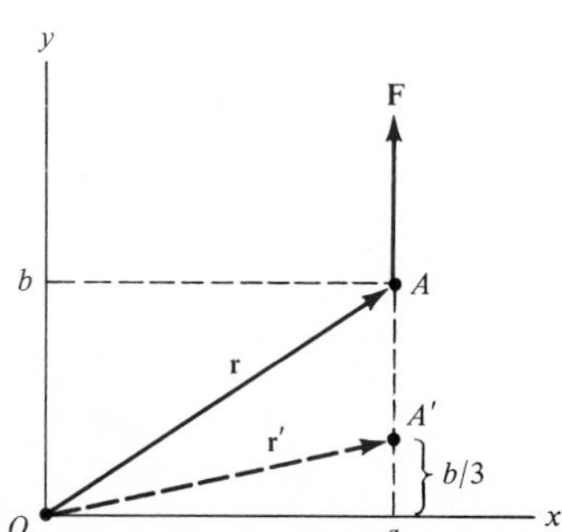

FIGURE 3-10

To determine the moment $\mathbf{M}_O$ of force **F** use a position vector from point O to *any* point on the line of action **F**

$F a\mathbf{k}$. The vector method is valid for more complex problems than this. Generally, *the more complex the arrangement, the more advantageous the vector method becomes.*

Rectangular Components of a Moment

In the general case where both **r** and **F** have three-dimensional components, the moment of **F** about a point also has three components. From Fig. 3-11,

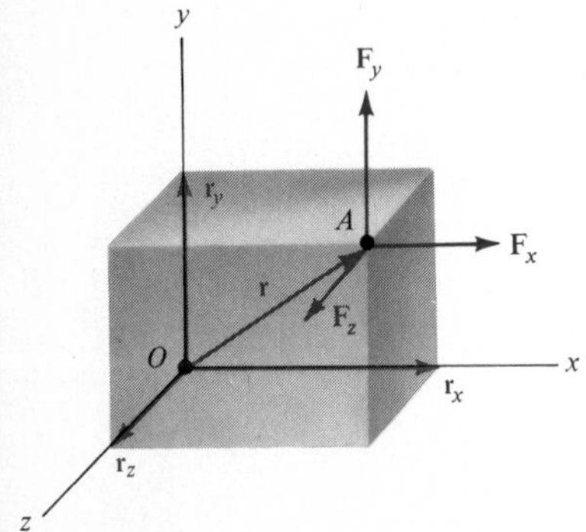

FIGURE 3-11

Rectangular components of a moment $\mathbf{M}_O = \mathbf{r} \times \mathbf{F}$

$$\begin{aligned} \mathbf{M}_O &= \mathbf{r} \times \mathbf{F} = (r_x\mathbf{i} + r_y\mathbf{j} + r_z\mathbf{k}) \times (F_x\mathbf{i} + F_y\mathbf{j} + F_z\mathbf{k}) \\ &= (r_yF_z - r_zF_y)\mathbf{i} + (r_zF_x - r_xF_z)\mathbf{j} \\ &\quad + (r_xF_y - r_yF_x)\mathbf{k} \\ &= M_x\mathbf{i} + M_y\mathbf{j} + M_z\mathbf{k} \end{aligned} \tag{3-15}$$

The moment components M_x, M_y, and M_z show the turning effect of the force **F** about the corresponding coordinate axes. Note that a single force **F** causes only one moment about a given point. However, since the moment is a vector, it can have three components with respect to the chosen coordinate system.

Varignon's Theorem

The distributive property of vector products is sometimes advantageous in calculating the resultant moment of several concurrent forces as in Fig. 3-12. On the basis of Eq. 3-7,

$$\boxed{\mathbf{M}_O = \mathbf{r} \times (\mathbf{F}_1 + \mathbf{F}_2 + \ldots) = \mathbf{r} \times \mathbf{F}_1 + \mathbf{r} \times \mathbf{F}_2 + \ldots} \tag{3-16}$$

This procedure is related to the preceding one, where the three rectangular components of a force were used to determine the moment (Eq. 3-15). The statement of Eq. 3-16 is useful for dealing with any number of concurrent forces or for resolving a given force into components that are not parallel to the chosen coordinate axes.

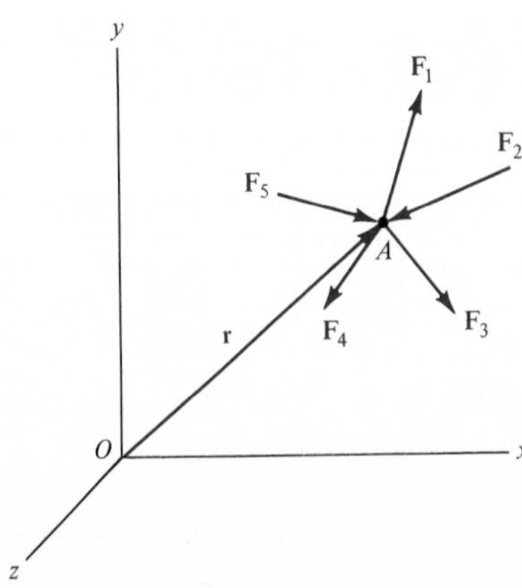

FIGURE 3-12

Resultant moment of several concurrent forces: $\mathbf{M}_O = \mathbf{r} \times \mathbf{F}_1 + \mathbf{r} \times \mathbf{F}_2 + \ldots$

EXAMPLE 3-3

Particle A and point O in Fig. 3-13 are in a vertical plane. Use the scalar and vector methods to determine the moment of the particle's weight W about point O for the position given.

SOLUTION

SCALAR METHOD:

$$M = Fd$$

$$d = r \sin 50^\circ$$

$$M = Wr \sin 50^\circ \quad \text{(clockwise)}$$

VECTOR METHOD:

$$\mathbf{M}_O = \mathbf{r} \times \mathbf{F}$$

$$\mathbf{F} = -W\mathbf{j} \qquad \mathbf{r} = r \sin 50^\circ\,\mathbf{i} - r \cos 50^\circ\,\mathbf{j}$$

$$\begin{aligned} \mathbf{M}_O &= (r \sin 50^\circ\,\mathbf{i} - r \cos 50^\circ\,\mathbf{j}) \times (-W\mathbf{j}) \\ &= -Wr \sin 50^\circ\,(\mathbf{i} \times \mathbf{j}) \\ &= -Wr \sin 50^\circ\,\mathbf{k} \quad \text{(clockwise by right-hand rule)} \end{aligned}$$

JUDGMENT OF THE RESULTS

Both methods yield the same result. For this problem the scalar method is the easiest since the calculation of d is simple.

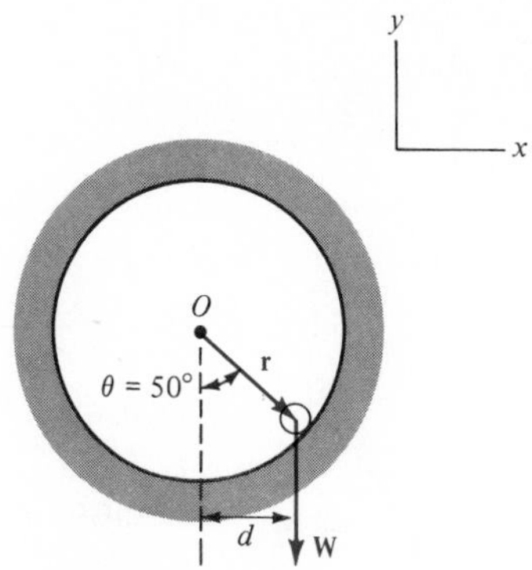

FIGURE 3-13
Moment of a particle's weight

EXAMPLE 3-4

The torque wrench in Fig. 3-14a has an arm of constant length L but a variable socket length d because of interchangeable tool sizes. Determine how the moment applied at point O depends on the length d for a constant force $\mathbf{F}$ from the hand. Use the coordinate system shown in Fig. 3-14b.

SOLUTION

$$\mathbf{M}_O = \mathbf{r} \times \mathbf{F} \qquad \mathbf{r} = L\mathbf{i} + d\mathbf{j} \qquad \mathbf{F} = F\mathbf{k}$$

$$\mathbf{M}_O = (L\mathbf{i} + d\mathbf{j}) \times F\mathbf{k} = Fd\mathbf{i} - FL\mathbf{j}$$

JUDGMENT OF THE RESULT

According to a visual analysis the wrench should turn clockwise, so the $-\mathbf{j}$ component of the moment is justified. Looking at the wrench from the positive x direction, point A has a tendency to rotate counterclockwise. Thus, the $\mathbf{i}$ component is correct using the right-hand rule.

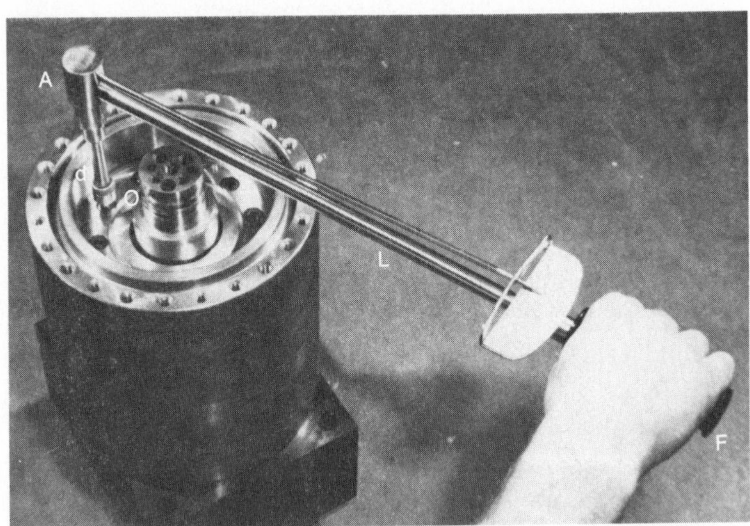

(a)

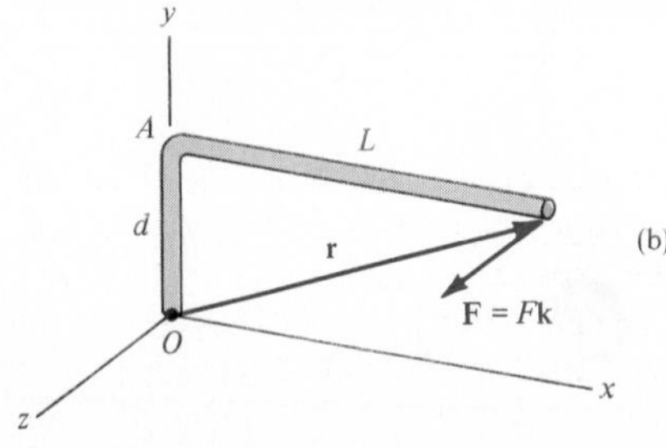

(b)

FIGURE 3-14
Moment applied using a torque wrench. (Photograph courtesy MTS Systems Corporation, Minneapolis, Minnesota.)

EXAMPLE 3-5

Consider two sets of three guy wires on a tall antenna tower as shown in Fig. 3-15. The initial tension is 50,000 lb in each wire. The ground distance OA to each anchor A is 1000 ft. The wires are attached to the tower at elevations $B = 150$ ft and $C = 1200$ ft. Assume that a wire breaks at B *or* C (but not both at once). In each case the remaining two wires cause a net moment about point O. Determine which failure causes the larger moment about O.

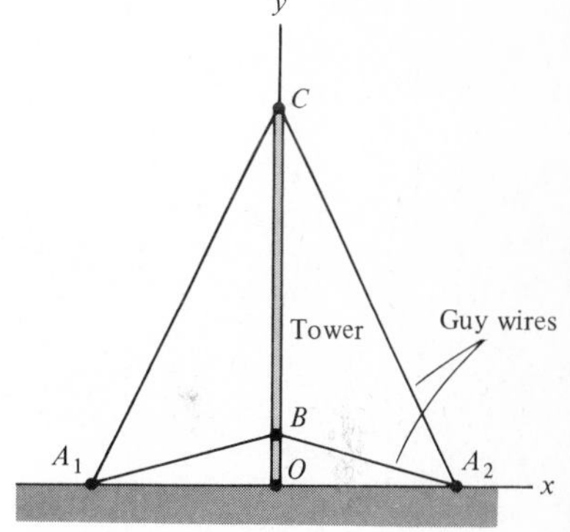

(a) Side view

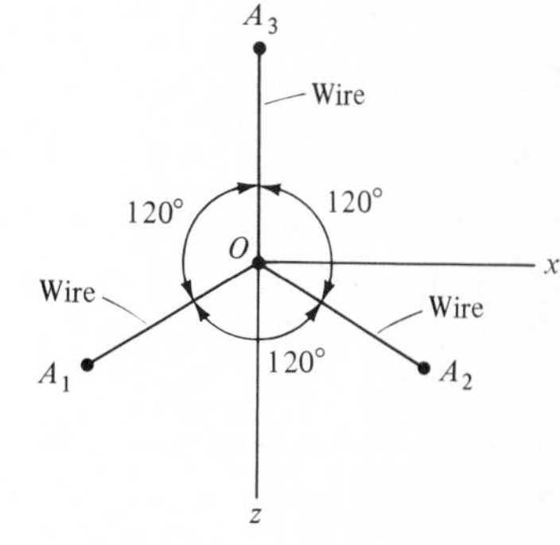

(b) Top view

FIGURE 3-15
Model of antenna tower

SOLUTION

There is no resultant moment about O when all six cables are intact. Since all tensions are equal and the cable distribution is symmetric, there is no resultant force in the x or z directions and the resultant force in the y direction goes through the point O; hence, $M_O = F_y d = 0$ $(d = 0)$.

CASE 1: Assume that cable BA_3 breaks.

$$\begin{aligned}\sum \mathbf{M}_O &= \sum (\mathbf{r} \times \mathbf{F}) = \mathbf{r}_{OB} \times (\mathbf{T}_{A_1} + \mathbf{T}_{A_2}) \\ &= \mathbf{r}_{OB} \times T(\mathbf{n}_{BA_1} + \mathbf{n}_{BA_2}) \\ &= (150 \text{ ft } \mathbf{j}) \times (50{,}000 \text{ lb})(-0.856\mathbf{i} - 0.148\mathbf{j} + 0.494\mathbf{k} \\ &\quad + 0.856\mathbf{i} - 0.148\mathbf{j} + 0.494\mathbf{k}) \\ &= (150 \text{ ft } \mathbf{j}) \times (50{,}000 \text{ lb})(0\mathbf{i} - 0.296\mathbf{j} + 0.988\mathbf{k}) \\ &= 7.41 \times 10^6 \text{ ft}\cdot\text{lb } \mathbf{i}\end{aligned}$$

CASE 2: Assume that cable CA_3 breaks.

$$\begin{aligned}\sum \mathbf{M}_O &= \sum (\mathbf{r} \times \mathbf{F}) = \mathbf{r}_{OC} \times (\mathbf{T}_{A_1} + \mathbf{T}_{A_2}) \\ &= \mathbf{r}_{OC} \times T(\mathbf{n}_{CA_1} + \mathbf{n}_{CA_2}) \\ &= (1200 \text{ ft } \mathbf{j}) \times (50{,}000 \text{ lb})(-0.554\mathbf{i} - 0.768\mathbf{j} \\ &\quad + 0.32\mathbf{k} + 0.554\mathbf{i} - 0.768\mathbf{j} + 0.32\mathbf{k}) \\ &= (1200 \text{ ft } \mathbf{j}) \times (50{,}000 \text{ lb})(0\mathbf{i} - 1.536\mathbf{j} + 0.64\mathbf{k}) \\ &= 38.4 \times 10^6 \text{ ft}\cdot\text{lb } \mathbf{i}\end{aligned}$$

JUDGMENT OF THE RESULTS

If the cable breaks at C, there is a larger moment caused about O. The cause for this larger moment is the much larger moment arm to point C. The moment is not as large as might be expected because there is a smaller resultant force component in the z direction which is the only component of the force causing a moment about O. There is no x component and the y component goes through O.

EXAMPLE 3-6

The designers of a new goal post (American football) wish to prevent damage to the post by overly excited spectators. It is estimated that at most about 10 people can hang onto a rope thrown over the post. The rope can be pulled in many different directions, but the maximum force is assumed to be 2500 lb regardless of the direction. What is the moment of this force about the base O of the post for the situation sketched in Fig. 3-16? Use the vector method.

SOLUTION

$$\mathbf{M}_O = \mathbf{r}_{OD} \times \mathbf{F}_D = (8\mathbf{i} + 12\mathbf{j} - 2\mathbf{k}) \times (1768\mathbf{i} - 1768\mathbf{j})$$

$$= \begin{vmatrix} \mathbf{i} & \mathbf{j} & \mathbf{k} \\ 8 & 12 & -2 \\ 1768 & -1768 & 0 \end{vmatrix}$$

$$= (-3536\mathbf{i} - 3536\mathbf{j} - 35{,}360\mathbf{k})\ \text{ft}\cdot\text{lb}$$

The magnitude of the moment is

$$M = 35{,}700\ \text{ft}\cdot\text{lb}$$

JUDGMENT OF THE RESULT

The relative magnitudes of the moment's components are intuitively correct. The goal post would tend to bend around the z axis much more than it would tend to bend about the x or y axes. A check with the right-hand rule shows that the sign of each component is also correct. It is left as an exercise for the reader to check the result using scalars (perpendicular distances and force components).

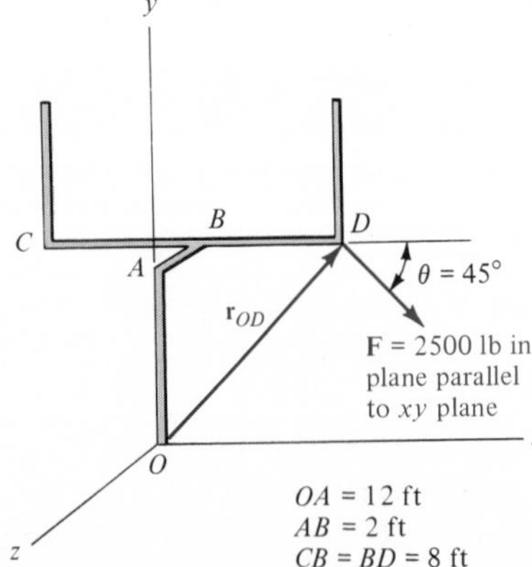

FIGURE 3-16
Model of a goal post

Use the vector product in each problem.

3–19 Prove Eq. 3-7.

3–20 Show that the magnitude of the vector product of vectors **A** and **B** equals the area of the parallelogram with sides A and B. Let $A = 3$ and $B = 2$ arbitrary units and
(a) $\theta = 90^\circ$
(b) $\theta = 30^\circ$
(c) $\theta = 150^\circ$

FIGURE P3-20

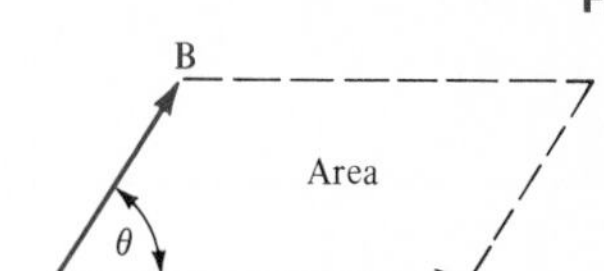

FIGURE P3-21

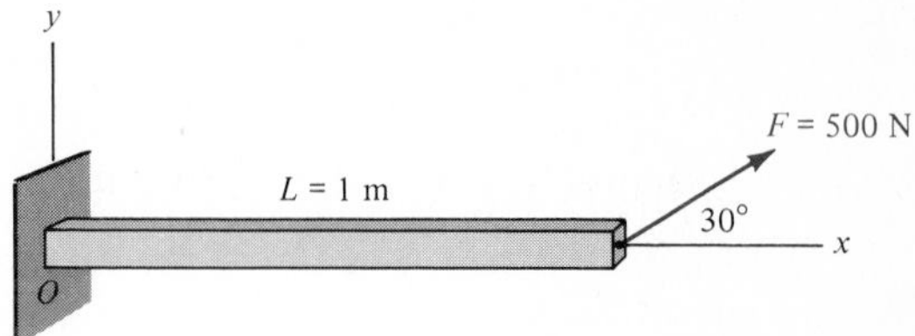

3–21 Determine the moment of force F about O.

FIGURE P3-22

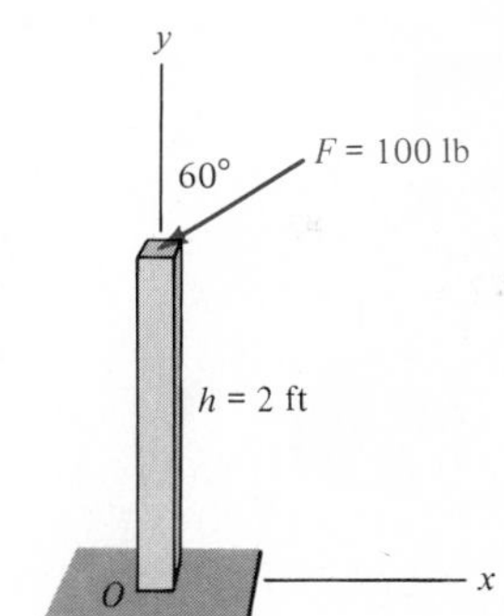

3–22 Calculate the moment of force F about O.

FIGURE P3-23

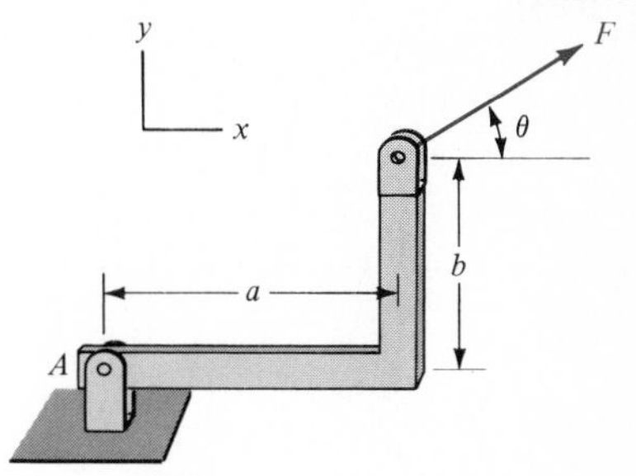

3–23 Determine the moment $\mathbf{M}_A$ of force $F = 10$ lb at $\theta = 20^\circ$ if $a = 4$ in. and $b = 3$ in.

FIGURE P3-25

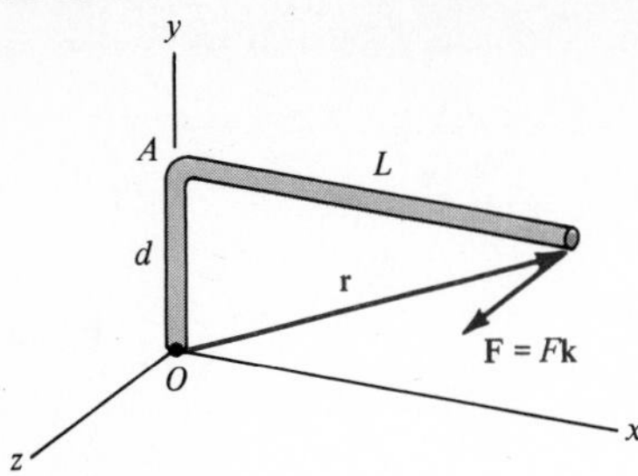

FIGURE P3-27

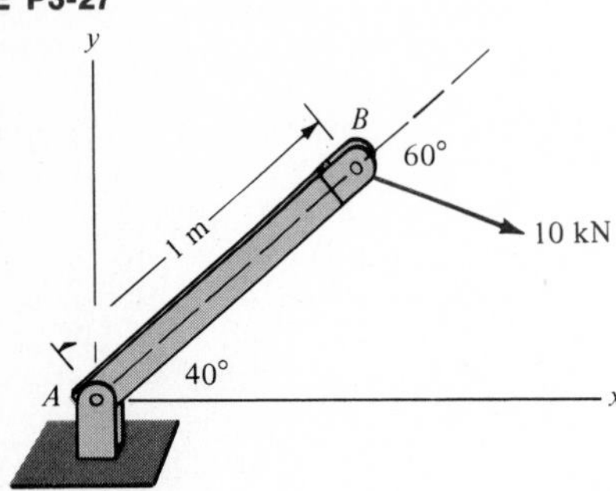

FIGURE P3-28

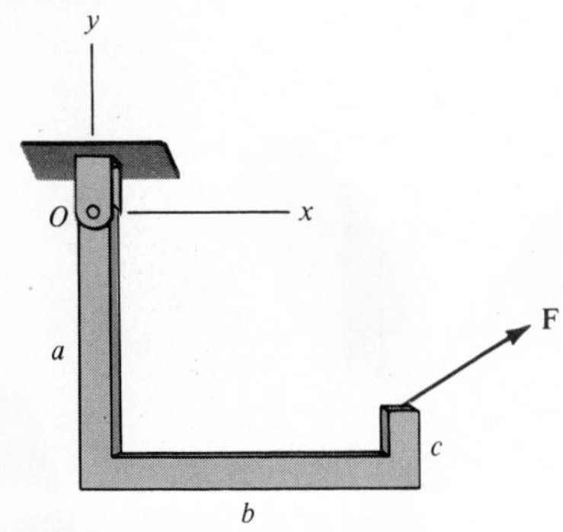

FIGURE P3-29

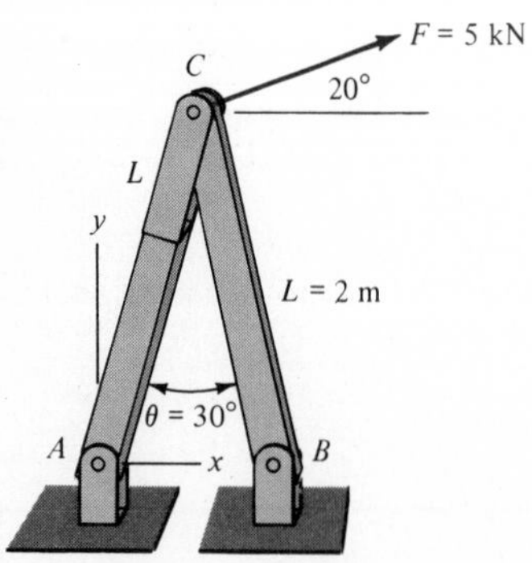

3–24 Calculate the moment $\mathbf{M}_A$ of force $F = 200$ N in Fig. P3-23 if $\theta = 120°$, $a = 50$ cm, and $b = 30$ cm.

3–25 Determine the moment $\mathbf{M}_O$ of force $\mathbf{F} = F\mathbf{k}$ using the dimensions d and L (this is from Fig. 3-14).

3–26 In Fig. P3-25 $d = 3$ in., $L = 12$ in., and $\mathbf{F} = (2\mathbf{i} - 3\mathbf{j} + 30\mathbf{k})$ lb. Calculate the moment $\mathbf{M}_O$ of $\mathbf{F}$.

3–27 Calculate the moment $\mathbf{M}_A$ of the 10-kN force.

3–28 Calculate the moment $\mathbf{M}_O$ using $a = b = 3c = 1$ ft, and $\mathbf{F} = (120\mathbf{i} + 70\mathbf{j})$ lb.

3–29 Determine the moment $\mathbf{M}_B$ of the force F.

3–30 Calculate the moment $\mathbf{M}_C$ of the force $F = 400$ lb using $\theta = 45^\circ$ and $L = 6$ ft.

FIGURE P3-30

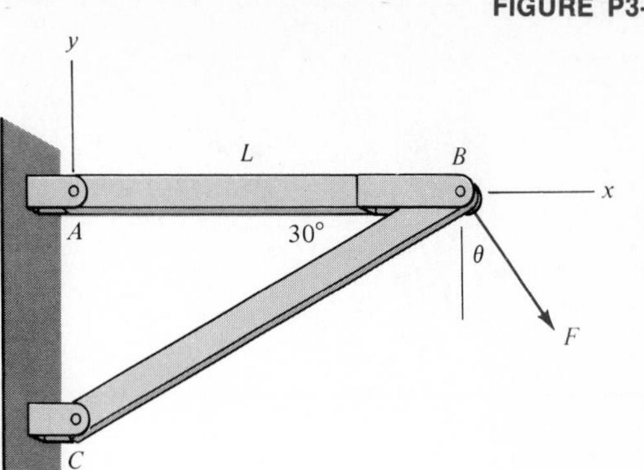

3–31 The street lamp A weighs $W = 400$ N. The force of the wind is $\mathbf{F} = 300$ N $\mathbf{i}$. Calculate the moment $\mathbf{M}_O$ of the resultant of the two forces. $d = 2$ m and $h = 5$ m.

FIGURE P3-31

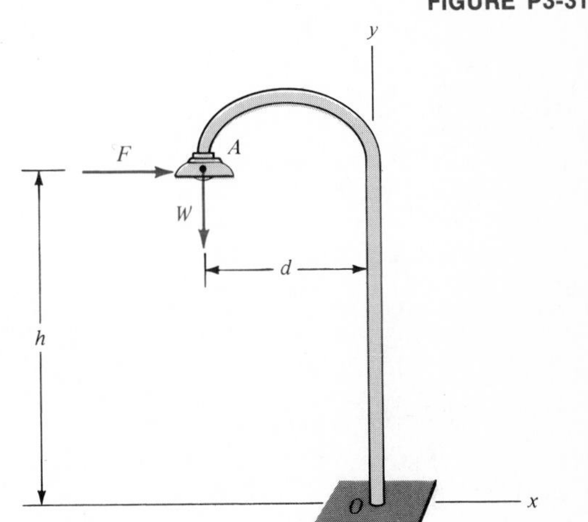

3–32 Assume that the force of wind on lamp A in Fig. P3-31 is $\mathbf{F} = -40$ lb $\mathbf{k}$. $\mathbf{W} = -50$ lb $\mathbf{j}$. Determine the moment $\mathbf{M}_O$ of the resultant of the two forces. $d = 7$ ft and $h = 15$ ft.

3–33 The boom of the simple crane is supported by a single wire AB in which the tension is 2 kN. Calculate the moment $\mathbf{M}_O$ caused by the resultant force of the tension and $W = 1$ kN.

FIGURE P3-33

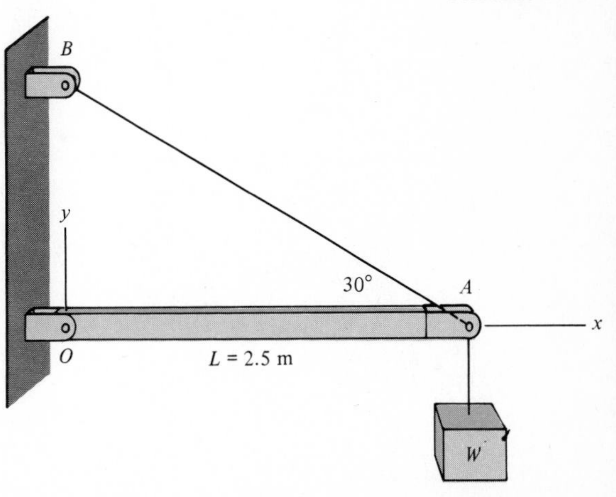

FIGURE P3-34

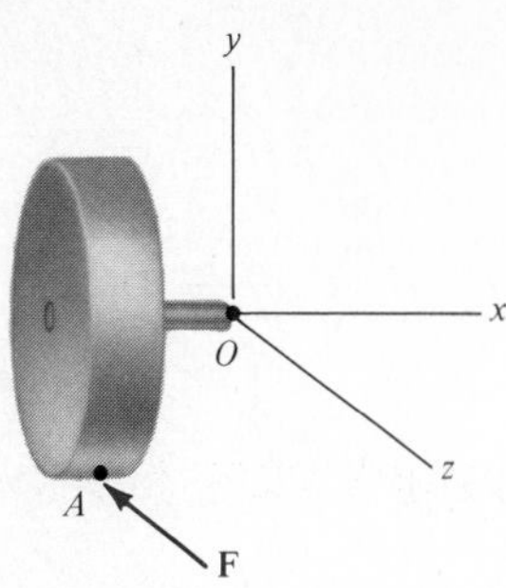

3–34 In a road simulator for cars a force $\mathbf{F} = (-300\mathbf{i} + 1000\mathbf{j} - 250\mathbf{k})$ lb is applied to a wheel at point A, which has coordinates $(-1, -1.3, 0.4)$ ft. Determine the moment $\mathbf{M}_O$ of the force.

FIGURE P3-35

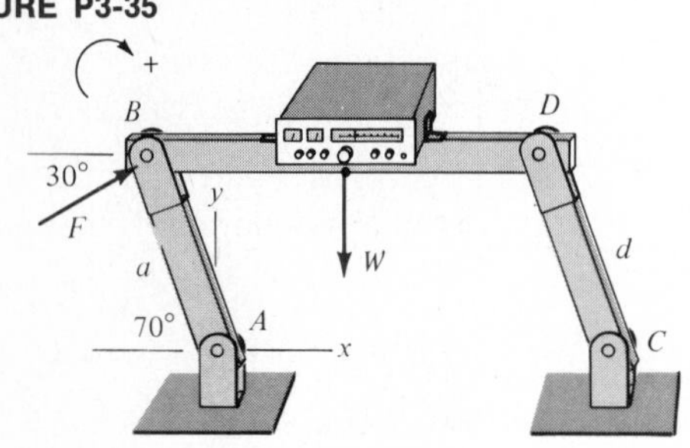

3–35 Assume that the test platform, arms AB and CD, and all forces are in the xy plane. Determine the moments $\mathbf{M}_{F_A}$ and $\mathbf{M}_{W_A}$ if $F = 5$ kN, $W = 2$ kN at 1 m horizontally from B, and $d = 0.7$ m.

FIGURE P3-36

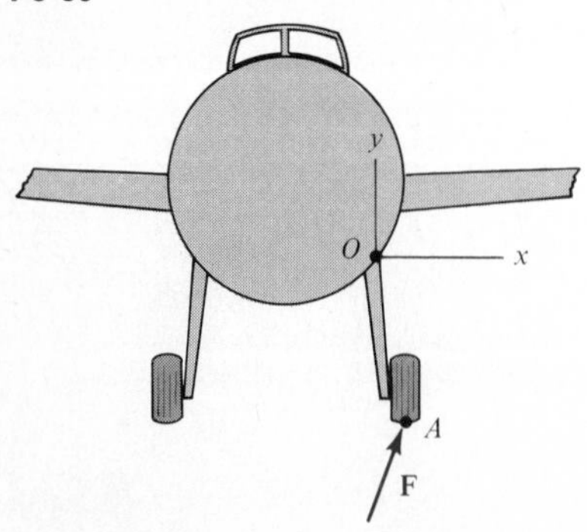

3–36 In a laboratory test for the landing gear of a small aircraft a force $\mathbf{F} = (300\mathbf{i} + 200\mathbf{j} - 300\mathbf{k})$ lb is applied at point A, which has coordinates $(1, -4, 0)$ ft. Calculate the moment $\mathbf{M}_O$ of the force.

3–37 A mast is supported by several guy wires. Calculate the moment $\mathbf{M}_O$ of a 5-kN tension in wire AB. The coordinates of A and B are (10, 0, 5) m, and (0, 12, 0) m, respectively.

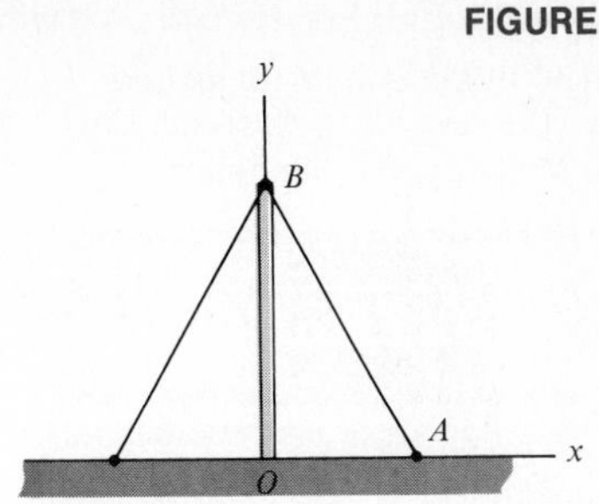

FIGURE P3-37

3–38 A jet aircraft is flying horizontally when the left engine fails. Determine the moment $\mathbf{M}_C$ of the thrust $\mathbf{T}$ of the right engine. $\mathbf{T}$ is applied at point A, which has coordinates (18, 2, −4) ft.

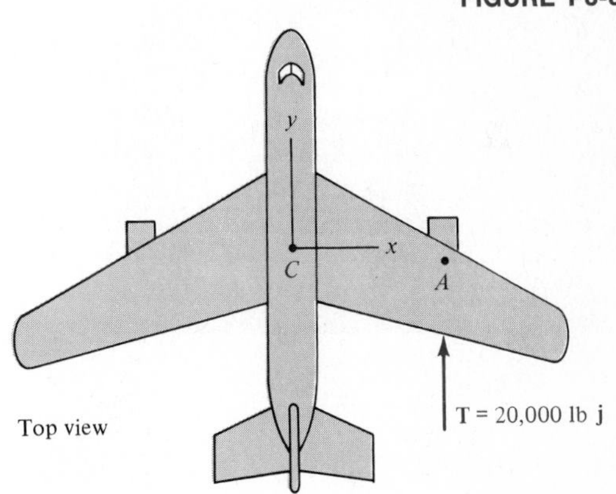

FIGURE P3-38

3–39 A force $\mathbf{F} = (100\mathbf{i} - 40\mathbf{j} + 20\mathbf{k})$ N is acting at the middle of handle BC (10 cm from B) of the crank. Determine the moments $\mathbf{M}_O$, $\mathbf{M}_A$, and $\mathbf{M}_B$ of the force when the crank is in the yz plane.

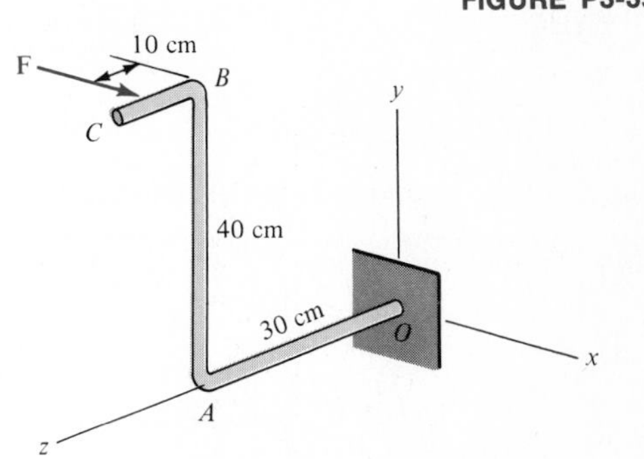

FIGURE P3-39

FIGURE P3-40

Fixture *ABC* in a mechanical testing machine. (Courtesy MTS Systems Corporation, Minneapolis, Minnesota.)

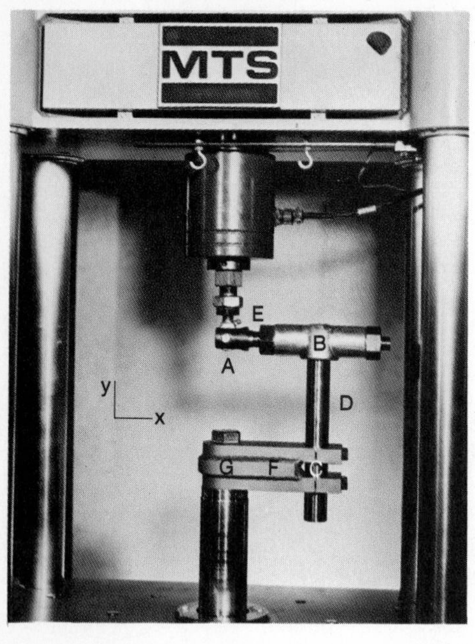

3–40 A T-shaped fixture *ABC* is tested in tension in a testing machine. An upward force $F = 200$ lb is applied at *A*, while a downward force *F* is applied at *G*. Determine the moments $\mathbf{M}_A$, $\mathbf{M}_B$, $\mathbf{M}_C$, $\mathbf{M}_D$, $\mathbf{M}_E$, $\mathbf{M}_F$, and $\mathbf{M}_G$. $AB = GC = 2FC = 3AE = 8$ in. and $BC = 10$ in.

FIGURE P3-41

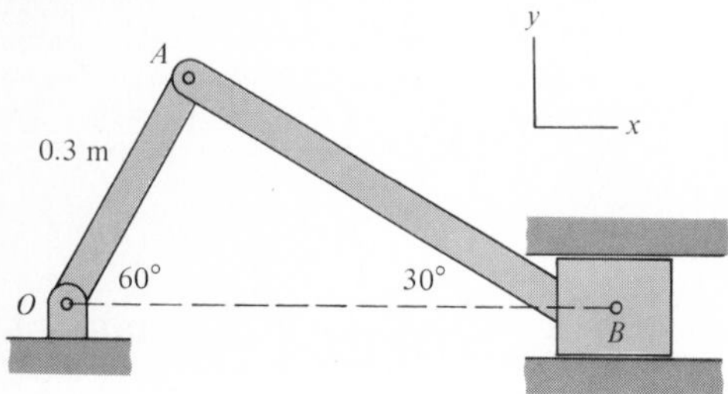

3–41 The piston rod applies a force $F = 5$ kN directed from *B* to *A*. Calculate the moment $\mathbf{M}_O$ acting on the crank arm.

FIGURE P3-42

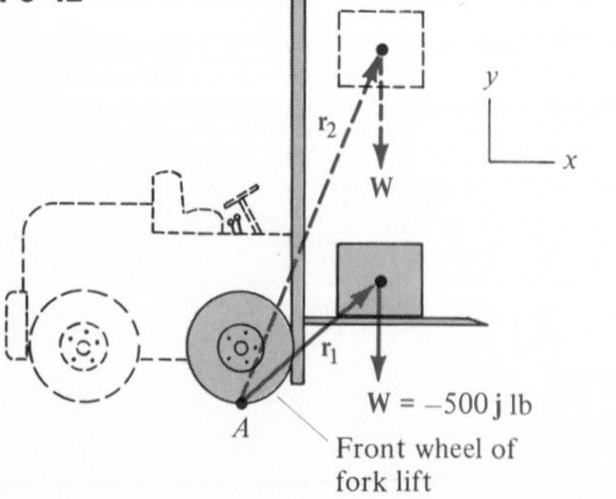

y
x
$\mathbf{r}_2$
W
$\mathbf{r}_1$
W = −500 j lb
A
Front wheel of fork lift

3–42 A crate is lifted by a fork lift. Calculate the moments $\mathbf{M}_A$ for the two extreme positions $\mathbf{r}_1 = (3\mathbf{i} + 2\mathbf{j})$ ft and $\mathbf{r}_2 = (2\mathbf{i} + 7\mathbf{j})$ ft.

TRIPLE PRODUCTS OF THREE VECTORS 3-5

Various products of three vectors may be considered. The *vector triple products* are mathematically valid because $\mathbf{A} \times \mathbf{B}$ is a vector, so$(\mathbf{A} \times \mathbf{B}) \times \mathbf{C}$ can be calculated. However, the associative property applies to this product only for unit vectors, and the result is trivial,

$$(\mathbf{i} \times \mathbf{j}) \times \mathbf{k} = \mathbf{i} \times (\mathbf{j} \times \mathbf{k}) = 0 \tag{3-17}$$

since

$$\mathbf{i} \times \mathbf{j} = \mathbf{k} \qquad \text{and} \qquad \mathbf{j} \times \mathbf{k} = \mathbf{i}$$

In other cases,

$$(\mathbf{A} \times \mathbf{B}) \times \mathbf{C} \neq \mathbf{A} \times (\mathbf{B} \times \mathbf{C}) \tag{3-18}$$

The vector triple product is used in dynamics analysis.

A useful product of three vectors in the area of statics is the *mixed triple product*, which is the scalar product $\mathbf{A}$ with the vector product of $\mathbf{B}$ and $\mathbf{C}$,

$$\mathbf{A} \cdot (\mathbf{B} \times \mathbf{C}) \tag{3-19}$$

The result of this triple product is a scalar.

The mixed triple product is calculated by first performing the vector product part of it.

$$\mathbf{A} \cdot (\mathbf{B} \times \mathbf{C}) = \mathbf{A} \cdot \mathbf{D} \tag{3-20}$$

where $\mathbf{D}$ is $\mathbf{B} \times \mathbf{C}$. In terms of rectangular components,

$$\mathbf{A} \cdot (\mathbf{B} \times \mathbf{C}) = A_x D_x + A_y D_y + A_z D_z$$

On the basis of Eq. 3-10, the complete details are

$$\mathbf{A} \cdot (\mathbf{B} \times \mathbf{C}) = A_x(B_y C_z - B_z C_y) + A_y(B_z C_x - B_x C_z) + A_z(B_x C_y - B_y C_x) \tag{3-21}$$

Equation 3-21 is also the expansion of the determinant

$$\boxed{\mathbf{A} \cdot (\mathbf{B} \times \mathbf{C}) = \begin{vmatrix} A_x & A_y & A_z \\ B_x & B_y & B_z \\ C_x & C_y & C_z \end{vmatrix}} \tag{3-22}$$

which is the easier form to remember.

3-6 MOMENT OF A FORCE ABOUT A LINE

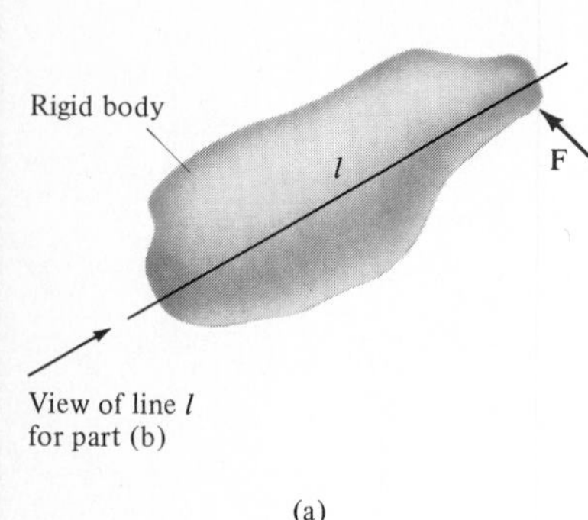

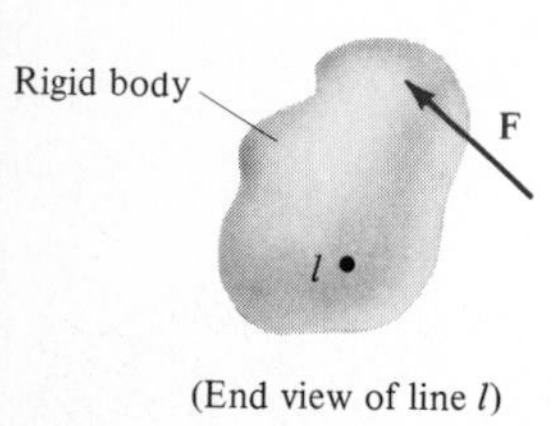

FIGURE 3-17

Moment of a force about a line

Commonly in practice, a body is constrained to rotate about a given axis regardless of the directions of forces acting on it. If the resultant of the forces acts in a plane that is perpendicular to the axis, the concept of moment about a point can be used directly. This was implicit in dealing earlier with the wave board, for example. The restriction on using this simple approach is explained with the aid of Fig. 3-17. A force **F** tends to cause rotation of a rigid body about the axis l. The axis may be fixed or moving. Imagine that the situation in Fig. 3-17a is viewed from one end of the line l so that the line appears as a point (Fig. 3-17b). In this end view the force may or may not appear in its full length. The concept of moment about a point is valid and easy to use for any line l if **F** is seen in its full length in Fig. 3-17b. The situation is more complicated if **F** is not in the plane of the paper in Fig. 3-17b. This is discussed in detail as follows.

By definition, the moment of an arbitrarily oriented force about a line is the projection, on the line, of the force's moment about any point on the same line. Consider Fig. 3-18 for a more formal description of this statement. **F** is an arbitrary force whose line of action does not intersect the line of interest l. $\mathbf{M}_O$ is the moment of **F** about an arbitrary point O on the line l. $\mathbf{M}_O$ is perpendicular to the parallelogram $OABC$ according to the vector product of **r** and **F**. $\mathbf{M}_l$ is the projection of $\mathbf{M}_O$ on the line l. This projection is the moment of **F** about l. There may be other components of $\mathbf{M}_O$, but these components do not have a moment about the line l, only about point O. For example, if $\mathbf{M}_O$ is resolved into rectangular components $\mathbf{M}_l$ and $\mathbf{M}'$, the latter would tend to rotate l itself, not have a moment *about* l. M_l alone is what tends to cause rotation about l.

The mixed triple product is a convenient mathematical tool for calculating the moment of a force about a line. Using **n**, the unit vector along the line l in Fig. 3-18, the magnitude of the projection of $\mathbf{M}_O$ on the line l is

$$\boxed{M_l = \mathbf{n} \cdot \mathbf{M}_O = \mathbf{n} \cdot (\mathbf{r} \times \mathbf{F})} \qquad \text{3-23}$$

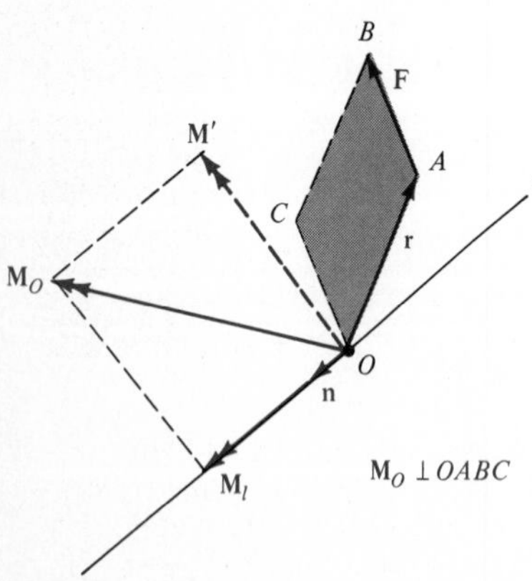

FIGURE 3-18

Moment $\mathbf{M}_O$ of force **F** about line l

This can be written as a determinant as in Eq. 3-22,

$$M_l = \begin{vmatrix} n_x & n_y & n_z \\ r_x & r_y & r_z \\ F_x & F_y & F_z \end{vmatrix} \qquad \text{3-24}$$

where n_x, n_y, n_z = rectangular components of **n** (also called *direction cosines* of **n**)

r_x, r_y, r_z = components of position vector from O to point A where **F** is applied

F_x, F_y, F_z = rectangular components of force **F**

Special Cases

1. The moment about a line is zero when the line of action of the force intersects the line of interest l as force $\mathbf{F}_1$ does in Fig. 3-19. This can be appreciated by inspection since the moment arm is zero in this case.

2. The moment about a line is zero when the line of action of the force is parallel to the line of interest l as is the case for $\mathbf{F}_2$ in Fig. 3-19. In this case the moment $\mathbf{M}_O$ is perpendicular to the plane of l, r, and $\mathbf{F}_2$, and the projection of $\mathbf{M}_O$ on l is a point of zero length.

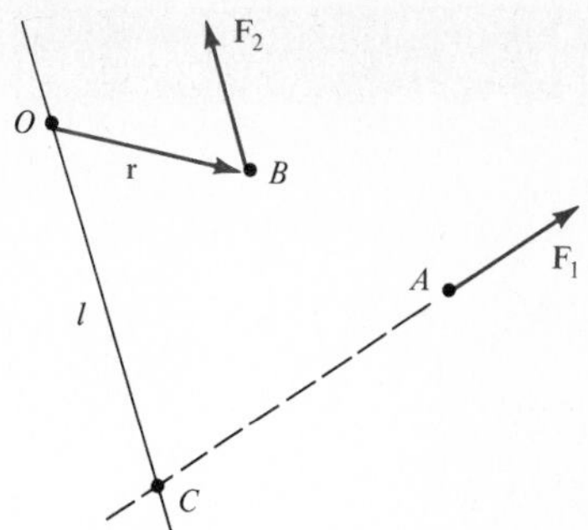

FIGURE 3-19

A force has no moment about an intersecting line or one parallel to its line of action

The various possibilities of moments about lines are illustrated for a door in Fig. 3-20.

(a) Large moment about line through hinges because of large moment arm; the door turns easily

(b) Zero moment about line through hinges because the force has no moment arm; the door does not turn

(c) Zero moment about line through hinges because the force is parallel to that line; the door does not turn

FIGURE 3-20

Illustration of moments about lines

EXAMPLE 3-7

The wheel assembly tester shown in Fig. 3-21a allows changing the position of the wheel in many ways with respect to the large driving drum as in part b of the figure. The position of the wheel affects the force **F** applied to the supporting structure (or the vehicle in real life) because of dimensional changes and tire sliding mixed with rolling. Moments about two lines are of interest: x and y are the transverse and longitudinal center lines of the vehicle (Fig. 3-21c). Determine M_x and M_y for $\mathbf{F} = (-300\mathbf{j} + 1000\mathbf{k})$ lb applied at $A: (-3, 6, -2)$ ft.

FIGURE 3-21

Test equipment for automotive wheel assemblies. (Photographs courtesy MTS Systems Corporation, Minneapolis, Minnesota.)

(a)

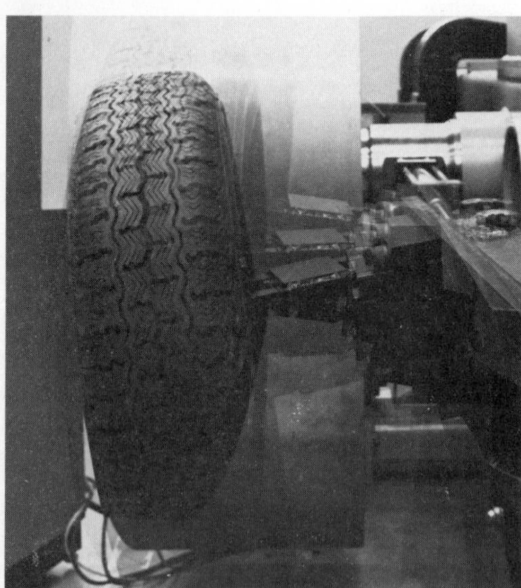

(b)

SOLUTION

Using Eq. 3-14,

$$\mathbf{M}_O = \mathbf{r}_{OA} \times \mathbf{F}$$

$$\mathbf{r}_{OA} = (-3\mathbf{i} + 6\mathbf{j} - 2\mathbf{k}) \text{ ft}$$

$$\mathbf{F} = (-300\mathbf{j} + 1000\mathbf{k}) \text{ lb}$$

$$\mathbf{M}_O = \mathbf{r}_{OA} \times \mathbf{F} = \begin{vmatrix} \mathbf{i} & \mathbf{j} & \mathbf{k} \\ -3 \text{ ft} & 6 \text{ ft} & -2 \text{ ft} \\ 0 & -300 \text{ lb} & 1000 \text{ lb} \end{vmatrix}$$

$$= (5400\mathbf{i} + 3000\mathbf{j} + 900\mathbf{k}) \text{ ft}\cdot\text{lb}$$

From Eq. 3-23,

$$M_x = \mathbf{M}_O \cdot \mathbf{i} = 5400 \text{ ft}\cdot\text{lb}$$

$$M_y = \mathbf{M}_O \cdot \mathbf{j} = 3000 \text{ ft}\cdot\text{lb}$$

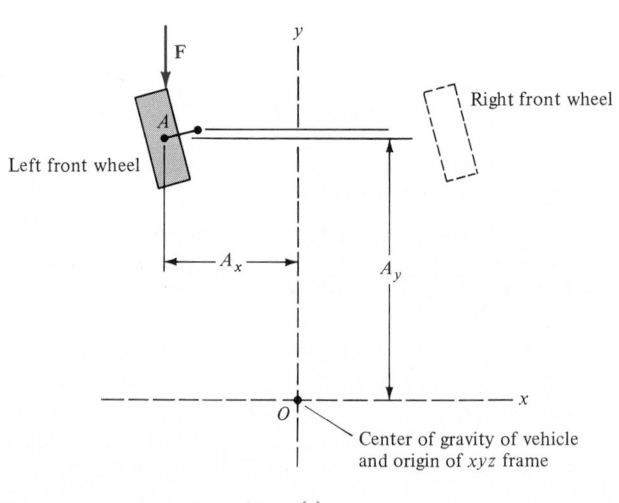

(c)

EXAMPLE 3-8

In bending and torsion tests of pipe systems a hydraulic actuator applies the desired force **F** through an end plate as shown in Fig. 3-22a. $\mathbf{F} = 500\mathbf{i}$ kN, and it is applied at A: (4, 1, 2) m with respect to the chosen coordinate system. Determine the moment of **F** about a line l defined by two points, B: (0, 2, 0) m, C: (1, 4, 0.5) m. The situation is sketched in Fig. 3-22b.

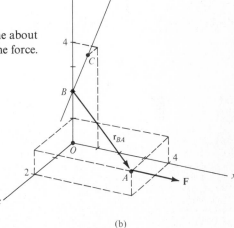

(a)

(b)

FIGURE 3-22

Test setup of a large pipe. (Photograph courtesy MTS Systems Corporation, Minneapolis, Minnesota.)

SOLUTION

From Eq. 3-23,

$$M_{BC} = \mathbf{n}_{BC} \cdot (\mathbf{r} \times \mathbf{F})$$

From Eq. 2-22,

$$\mathbf{n}_{BC} = \frac{1}{d_{BC}}(d_x\mathbf{i} + d_y\mathbf{j} + d_z\mathbf{k})$$

$$d_{BC} = [(x_C - x_B)\mathbf{i} + (y_C - y_B)\mathbf{j} + (z_C - z_B)\mathbf{k}] \text{ m}$$
$$= [(1 - 0)\mathbf{i} + (4 - 2)\mathbf{j} + (0.5 - 0)\mathbf{k}] \text{ m}$$

$$\mathbf{n}_{BC} = \frac{(-\mathbf{i} - 2\mathbf{j} - 0.5\mathbf{k})}{[(1)^2 + (2)^2 + (0.5)^2]^{1/2}}$$
$$= 0.436\mathbf{i} + 0.873\mathbf{j} + 0.218\mathbf{k}$$

The criterion for choosing vector **r** is that it join any point on the line about which the moment is taken and any point on the line of action of the force. Choosing points B and A on these lines, respectively,

$$\mathbf{r}_{BA} = (x_A - x_B)\mathbf{i} + (y_A - y_B)\mathbf{j} + (z_A - z_B)\mathbf{k}$$
$$= (4 \text{ m} - 0)\mathbf{i} + (1 \text{ m} - 2 \text{ m})\mathbf{j} + (2 \text{ m} - 0)\mathbf{k}$$
$$= (4\mathbf{i} - \mathbf{j} + 2\mathbf{k}) \text{ m}$$

$$\mathbf{F} = 500\mathbf{i} \text{ kN}$$

$$M_{BC} = \mathbf{n}_{BC} \cdot (\mathbf{r}_{BA} \times \mathbf{F}) = \begin{vmatrix} 0.436 & 0.873 & 0.218 \\ 4 \text{ m} & -1 \text{ m} & 2 \text{ m} \\ 500 \text{ kN} & 0 & 0 \end{vmatrix}$$

$$= [873 - 109] \text{ kN} \cdot \text{m} = 764 \text{ kN} \cdot \text{m}$$

Note that if the unit vector going from C to B ($\mathbf{n}_{CB} = -\mathbf{n}_{BC}$) had been chosen, an answer of -764 kN·m would have been obtained. However, both answers mean the same, remembering that by the right-hand rule the thumb is pointed along the *positive* sense of the axis about which the moment is taken. In the first case, pointing the thumb along the line going from B to C, the curl of the fingers gives the sense of rotation about this line (since the result was positive). In the second case, with the thumb pointing from C to B, the fingers curl in the opposite direction, but the negative sign of the answer means that this is opposite to the direction of rotation, hence both results yield the same sense of rotation about the line BC.

PROBLEMS

SECTIONS 3-5, 3-6

FIGURE P3-47

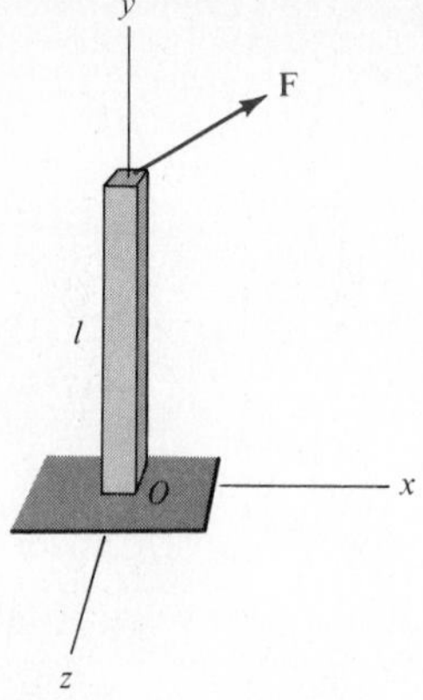

FIGURE P3-49

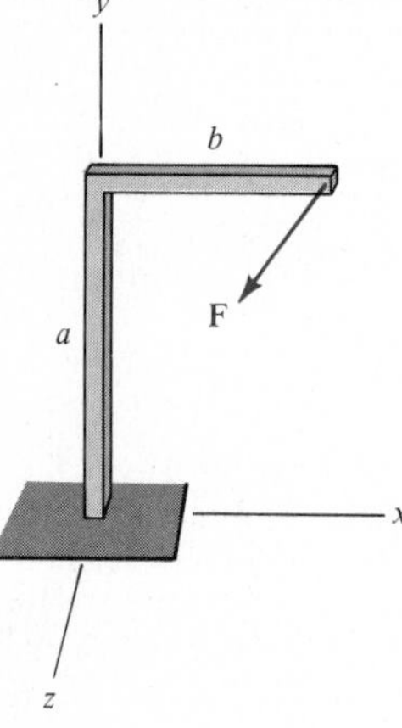

FIGURE P3-51

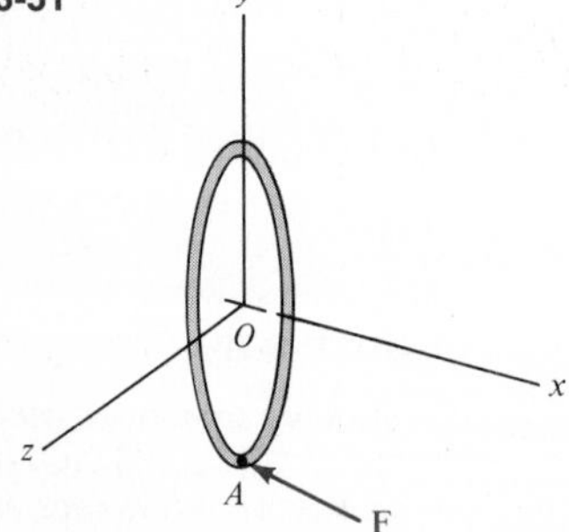

FIGURE P3-53

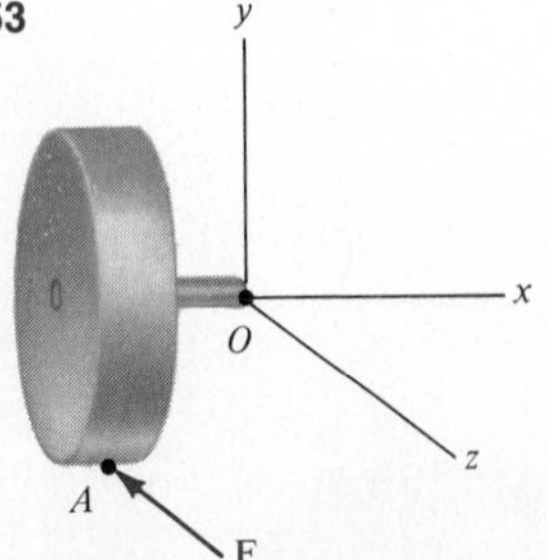

3–43 Calculate the vector triple product $\mathbf{V} = (\mathbf{A} \times \mathbf{B}) \times \mathbf{C}$ of $\mathbf{A} = 2a\mathbf{i} + 4b\mathbf{j} - 3c\mathbf{k}$, $\mathbf{B} = -b\mathbf{j} + 5c\mathbf{k}$, and $\mathbf{C} = 6a\mathbf{i} - 8b\mathbf{j}$.

3–44 Calculate the vector triple product $\mathbf{V} = (\mathbf{A} \times \mathbf{B}) \times \mathbf{C}$ of $\mathbf{A} = 10\mathbf{i} - 5\mathbf{j} - 12\mathbf{k}$, $\mathbf{B} = -\mathbf{i} + 10\mathbf{j}$, and $\mathbf{C} = 7\mathbf{k}$. Show that $(\mathbf{A} \times \mathbf{B}) \times \mathbf{C} \neq \mathbf{A} \times (\mathbf{B} \times \mathbf{C})$.

3–45 Calculate $\mathbf{A} \cdot (\mathbf{B} \times \mathbf{C})$ if $\mathbf{A} = \mathbf{i} + \mathbf{j} + \mathbf{k}$, $\mathbf{B} = 3\mathbf{i} + 5\mathbf{k}$, and $\mathbf{C} = -4\mathbf{i} - 3\mathbf{j}$.

3–46 Calculate $\mathbf{A} \cdot (\mathbf{B} \times \mathbf{C})$ if $\mathbf{A} = -5\mathbf{i} - 4\mathbf{j} - 10\mathbf{k}$, $\mathbf{B} = -12\mathbf{i} + 10\mathbf{j} + 6\mathbf{k}$, and $\mathbf{C} = 20\mathbf{i} - 15\mathbf{j} + 30\mathbf{k}$.

3–47 Determine the moments M_x, M_y, and M_z of the force $\mathbf{F} = (600\mathbf{i} + 150\mathbf{j})$ N if $l = 2$ m.

3–48 Calculate the moments M_x, M_y, and M_z of the force $\mathbf{F} = (100\mathbf{i} + 60\mathbf{j} - 80\mathbf{k})$ lb if $l = 4$ ft in Fig. P3-47.

3–49 Calculate the moments M_x, M_y, and M_z of the force $\mathbf{F} = (-70\mathbf{j} + 100\mathbf{k})$ N if $a = b = 0.2$ m.

3–50 Determine the moments M_x, M_y, and M_z of the force $\mathbf{F} = (40\mathbf{i} - 35\mathbf{j} + 10\mathbf{k})$ lb in Fig. P3-49 where $a = 8$ in. and $b = 6$ in.

3–51 In a laboratory test of a bicycle wheel a force $\mathbf{F} = (-300\mathbf{i} + 700\mathbf{j})$ N is applied at point A of coordinates (0, −0.38 m, 0). Determine the moments M_x, M_y, and M_z of the force.

3–52 Solve Prob. 3–51 with $\mathbf{F} = (50\mathbf{i} + 300\mathbf{j} + 40\mathbf{k})$ lb and point A at (0, −13 in., 0).

3–53 The wheel of a car in a test is supported at point O. Calculate M_x, M_y, and M_z using $\mathbf{F} = (4\mathbf{j} + 6\mathbf{k})$ kN acting at point A, which has coordinates (−0.3, −0.4, 0) m.

3–54 Solve Prob. 3–53 using $\mathbf{F} = (500\mathbf{i} + 1500\mathbf{j} - 300\mathbf{k})$ lb acting at point A of coordinates $(-10, -13, 3)$ in.

3–55 In a laboratory test of the landing gear of a small aircraft a force $\mathbf{F} = (1.5\mathbf{i} + 20\mathbf{j} - 0.8\mathbf{k})$ kN is applied at point A of coordinates $(0.3, -1, 0.25)$ m. Determine the moments M_x, M_y, and M_z of the force.

3–56 The rectangular plate is hinged at the x axis and lies in the xz plane. Determine the moments M_x, M_y, and M_z of the force $\mathbf{F} = (-50\mathbf{i} + 180\mathbf{j} + 30\mathbf{k})$ lb.

3–57 What is the smallest force F that can be applied at point B in Fig. P3-56 to have the same moment M_x that the force $\mathbf{F}$ has in Prob. 3–56?

3–58 Force $F = (-20\mathbf{i} - 30\mathbf{j} + 200\mathbf{k})$ N is applied to the rigid linkage $OACB$, which is hinged at O and B. Determine the moment M_x of $\mathbf{F}$ if θ in the yz plane is 30°.

3–59 Determine the moment M_l of a force $\mathbf{F} = (100\mathbf{i} + 300\mathbf{j} - 600\mathbf{k})$ N in Fig. P3-58. Line OA makes an angle $\theta = 20°$ with the y axis. The line of interest l is through points O and D; the latter has coordinates $(1, 0, 0.5)$ m.

3–60 The force acting on the crank arm is $\mathbf{F} = (120\mathbf{i} - 30\mathbf{j} - 20\mathbf{k})$ N. Calculate the moments M_x, M_y, and M_z of the force, assuming that the crank is in the yz plane.

3–61 Define a line l through points O and D in Fig. P3-60; D has coordinates $(1, 1, 0)$ m. Determine the moment M_l of $\mathbf{F}$ if the crank is in the yz plane.

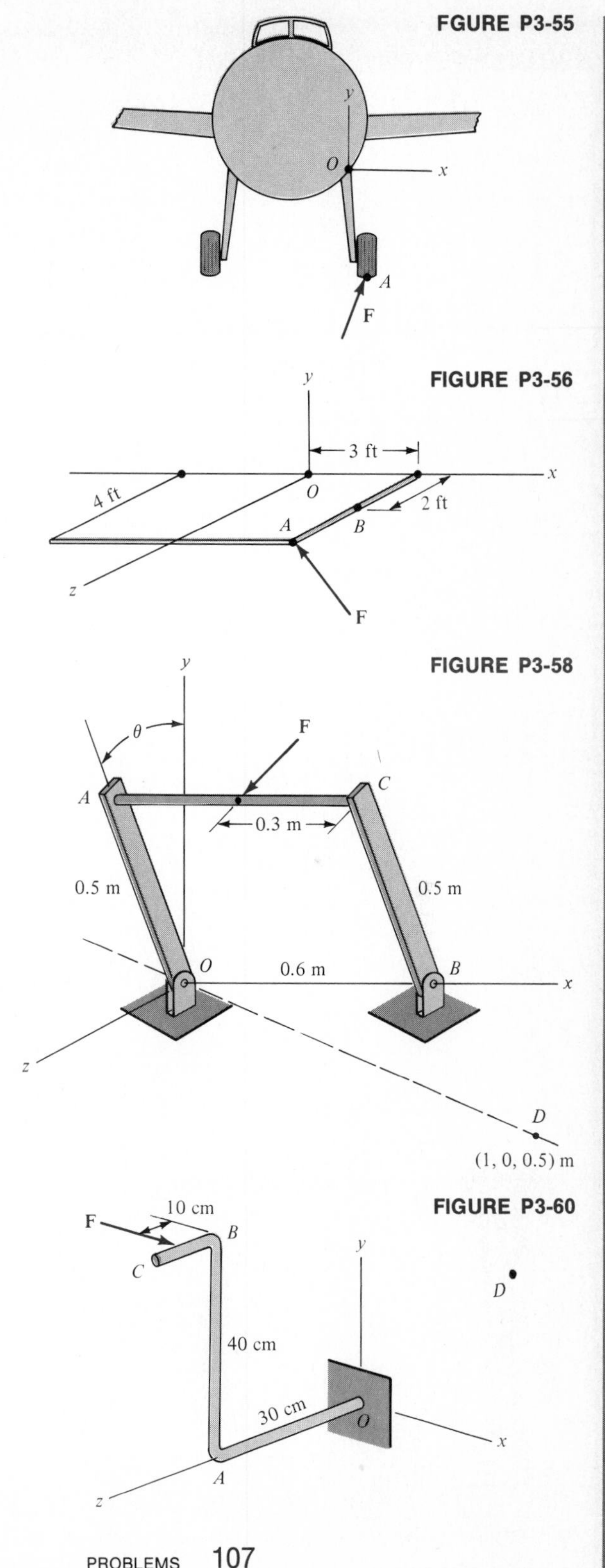

FGURE P3-55

FIGURE P3-56

FIGURE P3-58

FIGURE P3-60

3-7 MOMENT OF A COUPLE

A special case of moment is when it is caused not by a single force but by two forces that are equal in magnitude, parallel in lines of action, and opposite in direction. Such a pair of forces is called a *couple*. An example is the moment applied to the test specimen by the elevated-temperature extensometer (sensitive device for the measurement of deformations) shown in Fig. 3-23a. On the basis of intuition, the weight of the device would cause it to rotate clockwise, and this causes tension at section A on the specimen and compression at B (Fig. 3-23b). These forces are equal in magnitude (this is proved in Ch. 4), parallel, and opposite in direction. Thus, they have no net force on the specimen as a whole, but slightly tend to rotate it in the clockwise direction (at high temperatures even this small moment may be important).

The magnitude of the moment of a couple is given by

$$M = Fd \tag{3-25}$$

where d is the distance between the lines of action of the forces F. To prove this, sum the moments about any of the three points in Fig. 3-23b:

For A, $\quad M_A = Fd$ (only the lower force has a moment)

For B, $\quad M_B = Fd$ (only the upper force has a moment)

For C, $\quad M_C = 2\left(F\dfrac{d}{2}\right)$

$= Fd$ (both forces have moments but the moment arm is $d/2$)

The distinction between moments caused by a single force or a couple is straightforward. For example, the single force **F** causes a moment of $Fd/2$ on the steering wheel of diameter d in Fig. 3-24. It also acts as a net force **F** to the left on the steering column assembly. Exactly the same turning effect about the steering column is obtained

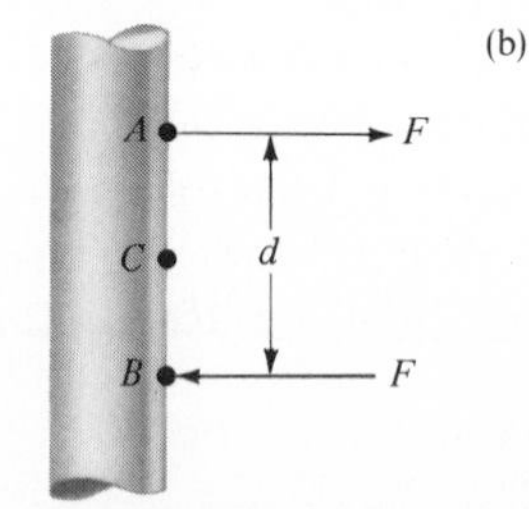

FIGURE 3-23

Extensometer attached to a metal specimen. (Photograph courtesy MTS Systems Corporation, Minneapolis, Minnesota.)

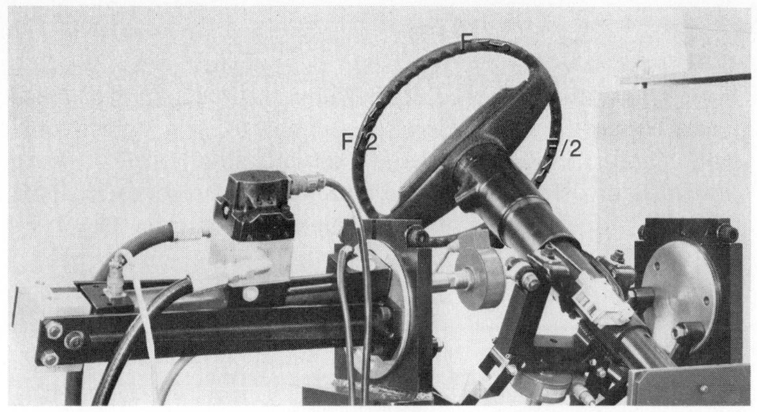

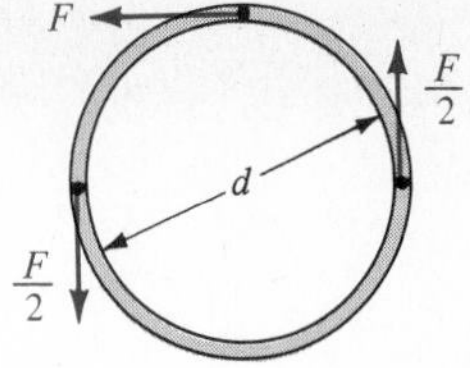

(a) (b)

FIGURE 3-24

Moment of a single force *F* and moment of a couple of *F*/2. (Photograph courtesy MTS Systems Corporation, Minneapolis, Minnesota.)

by applying a couple of $F/2$ at a distance d as shown. In this case the net force on the steering column is zero, and the turning effect is not dependent on where the forces are applied on the wheel.

For a simple vector analysis of a couple, consider Fig. 3-25. Taking moments about the arbitrary point O, with all points and vectors in the xy plane,

$$\begin{aligned}\mathbf{M}_O &= \mathbf{r}_A \times \mathbf{F} + \mathbf{r}_B \times (-\mathbf{F}) = (\mathbf{r}_A - \mathbf{r}_B) \times \mathbf{F} \\ &= \mathbf{d} \times \mathbf{F} \qquad (\text{since } \mathbf{r}_A = \mathbf{r}_B + \mathbf{d})\end{aligned} \tag{3-26}$$

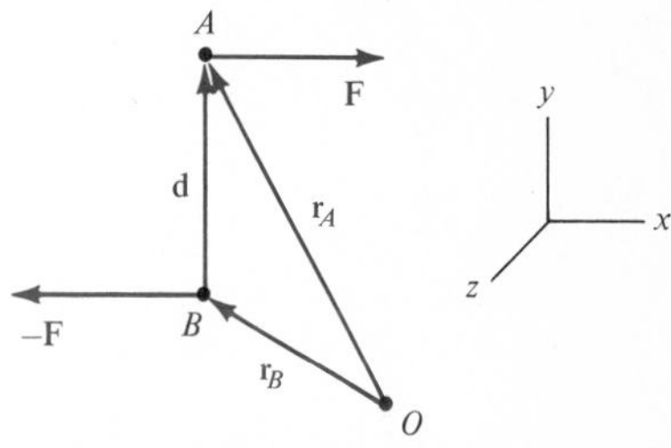

FIGURE 3-25

Vector analysis of a couple

For the given arrangement,

$$\mathbf{M}_O = \begin{vmatrix} \mathbf{i} & \mathbf{j} & \mathbf{k} \\ 0 & d & 0 \\ F & 0 & 0 \end{vmatrix} = -Fd\mathbf{k}$$

The sign is correct according to the right-hand rule. Note that the cross product should involve a force lying at the head of the position vector.

The same result is obtained if three-dimensional components are used for the forces and position vectors. An important aspect of this analysis is that the position of the chosen reference point O does not affect the moment of the couple, which is thus a *free vector* and can be applied anywhere in the given plane.

It should be noted that a single vector **M** representing a couple (sometimes called a *couple vector*) is perpendicular to the plane of the two forces as in Fig. 3-26a. In general, the vector **M** may have three components, $\mathbf{M}_x$, $\mathbf{M}_y$, and $\mathbf{M}_z$. The turning effect of **M** or any of its components is established using the right-hand rule. For example, $\mathbf{M}_z$ in Fig. 3-26 means a counterclockwise rotation when viewed from the positive z direction. Care must be taken in each problem so that the moment vector is not confused with any other vector quantity. Sometimes it is helpful to use a curled arrow in the plane of the two forces instead of the couple vector **M**, as in Fig. 3-26b.

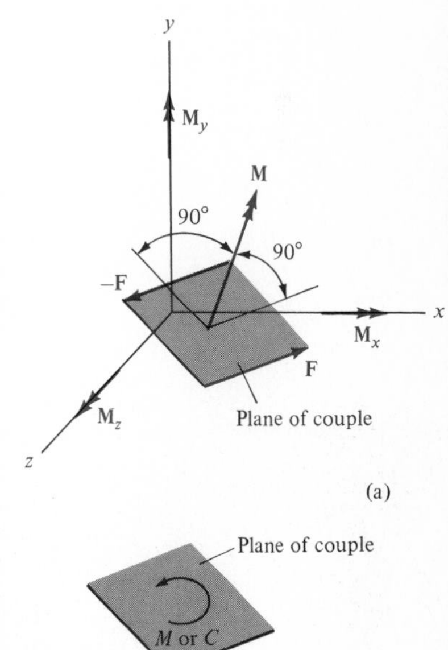

FIGURE 3-26

Rectangular components of a couple vector

3-8 EQUIVALENT COUPLES

For considerations of space requirements and strengths of members, it may be necessary to decide where and how to apply a given couple to a member. Evaluation of a few possible alternatives shows the principles involved and the most reasonable directions in design.

Figure 3-27 shows three of the infinite number of ways of

FIGURE 3-27
Equivalent couples

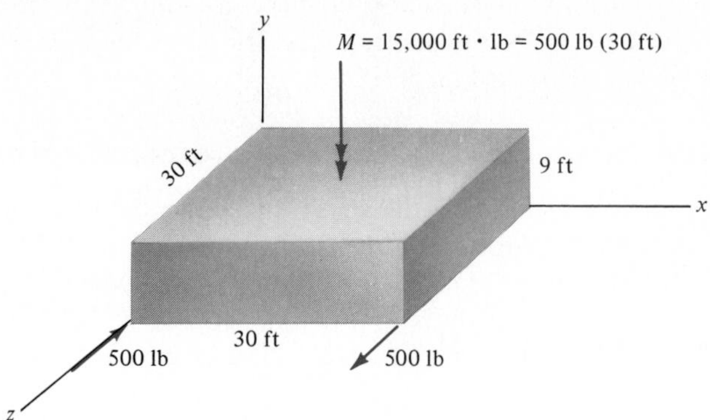

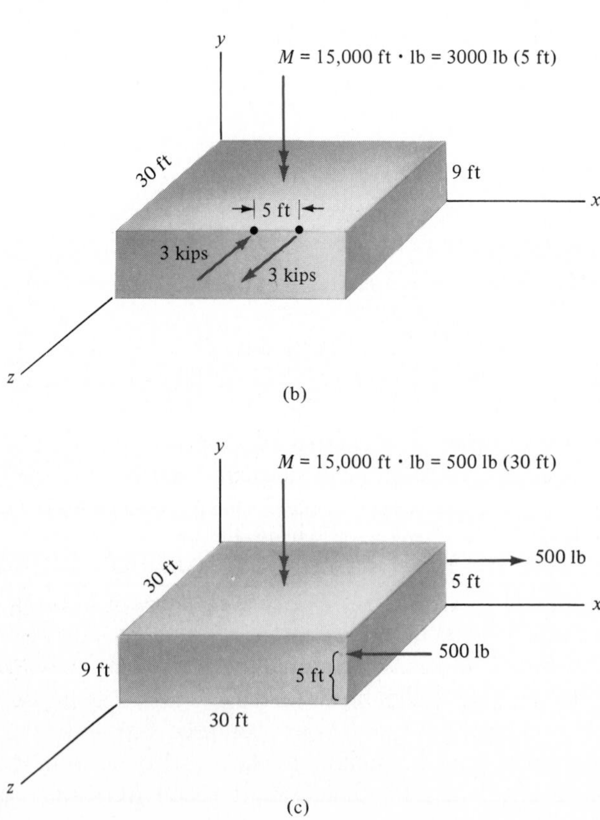

applying the same moment of a couple to a block. These are *equivalent* in theory since they have the same turning effect, and the net force is also the same, zero in each case. Of course, they are different to the extent that those in Fig. 3-27a and c may require considerable space for a mechanism, while that in part b of the figure requires much stronger linkages to apply the forces. A compromise is often made as shown in Fig. 3-28, where bars *A* and *B* apply the force couple to the large block *C*.

The generalization about the equivalence of couples is based on the previous vector analysis associated with Fig. 3-25, and is summarized with respect to Fig. 3-29:

1. A couple can be moved anywhere in its own plane without changing the moment acting on a given rigid body (Fig. 3-29a).

2. A couple can be moved to any other plane that is parallel to its original plane in the same rigid body (Fig. 3-29b). Conversely: a couple cannot be moved to change the position of its plane without changing its effect on the rigid body.

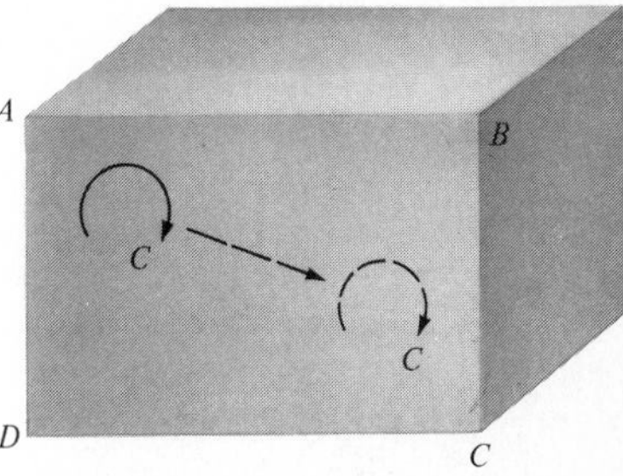

(a) Couple *C* can be moved anywhere in plane *ABCD*

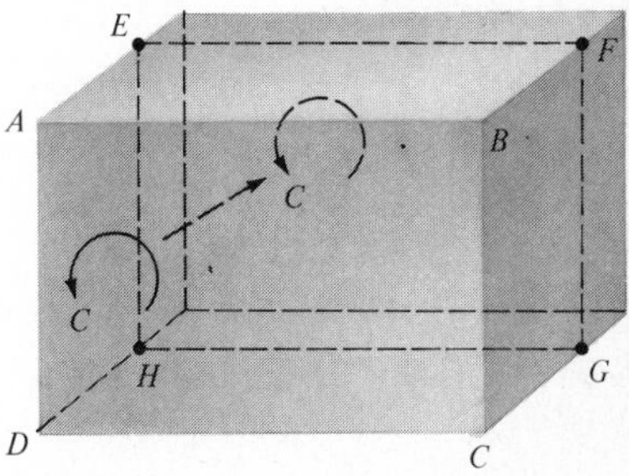

(b) Couple *C* can be moved from plane *ABCD* to plane *EFGH;* it cannot be moved to planes *AFGD, BDGC,* etc.

FIGURE 3-29

Equivalence of couples in the same plane or in parallel planes

FIGURE 3-28

Antiyaw (antirotation) mechanism of a tuned mass damper (see Sec. 1-7). (Courtesy MTS Systems Corporation, Minneapolis, Minnesota.)

3-9 COMPONENTS AND ADDITION OF COUPLES

The vector representing a couple can be resolved into components as can be done for any other vector. For example, the antiyaw mechanism shown in Fig. 3-28 applies a moment of variable direction to the block as illustrated in Fig. 3-30. The moment **M** is always perpendicular to the plane of the two linkage bars represented by *BC* (this is based on the concept of two-force members which is proved in Ch. 4). The angle θ varies as the block moves left and right. The block can move only horizontally, so the effective antiyaw moment is $\mathbf{M}_y$ in Fig. 3-30. Of course, $\mathbf{M}_x$ is present, tending to cause rotation about the x axis. This moment is felt by the horizontal plane on which the block glides (M_x is small and does not tip the block).

The addition of couples obeys the rule of vector addition. The moment vectors of couples in parallel planes can be moved to have a single line of action, so such couples can be algebraically added. Couples that are in different, nonparallel planes can be resolved into rectangular components to facilitate their addition. The simple procedure is illustrated in Fig. 3-31. In general, the resultant **M** of couples $\mathbf{M}_1, \mathbf{M}_2, \ldots$ is obtained by vector addition,

$$\mathbf{M} = \mathbf{M}_1 + \mathbf{M}_2 + \ldots \tag{3-27}$$

Any number of couples can be reduced to a single resultant **M** that may have up to three rectangular components, $\mathbf{M}_x$, $\mathbf{M}_y$, and $\mathbf{M}_z$.

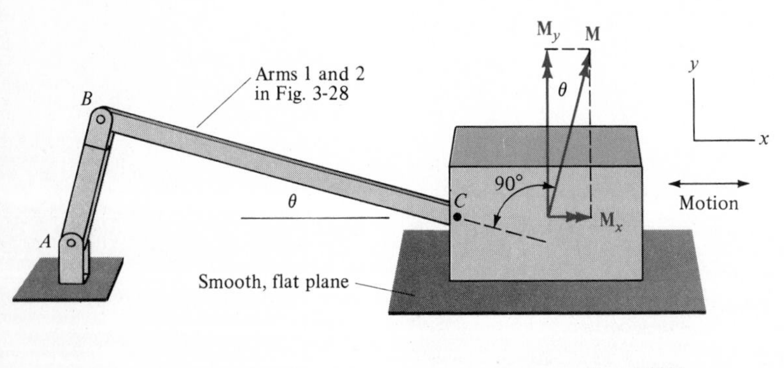

FIGURE 3-30

Rectangular components of a moment acting on a tuned mass damper (see Sec. 1-7)

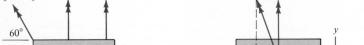

FIGURE 3-31

Addition of couples by vector addition. Moment vectors $\mathbf{M}_1$, $\mathbf{M}_2, \ldots$ are all in the *xy* plane.

TRANSFORMATION OF A FORCE INTO A FORCE-COUPLE SYSTEM

3-10

Consider the use of a torque wrench for bolts arranged on a circle as in Fig. 3-14. The general situation is depicted in Fig. 3-32a. The force **F** may have distinct effects at two points of interest, at a bolt A, and at center C of the whole rigid body. These effects are most clear if the force is transformed into an equivalent system consisting of a couple and a force acting at the point of interest.

To deal with point A first, add two collinear but opposite forces of magnitude F at A. This does not change the situation in any way, regardless of the orientation of the line of action of these forces, but it is advantageous to put the forces parallel to the original force at B as in Fig. 3-32b. Here the upward force at A and the downward force at B make a couple of magnitude FL. Thus, the single force at B can be replaced by an identical force acting at A and a couple **M** which can be moved anywhere in the plane of **F** and L (Fig. 3-32c).

A similar procedure can be applied in going from point A to C resulting in an identical force **F** but a different moment **M′** (Fig. 3-32d). Several of these interesting aspects of moments can be demonstrated using common T-shaped and L-shaped tire tools on the wheel-lug bolts of a jacked-up car. In such experiments the small motions of the wheel should be noted as the tool is used to loosen or tighten the bolts.

The manipulations of moments shown in Fig. 3-32 can be stated in vector notation as follows.

$$\mathbf{M}_A = \mathbf{r}_{AB} \times \mathbf{F} \quad \text{and} \quad \mathbf{M}_C = \mathbf{r}_{CB} \times \mathbf{F} = \mathbf{M}_A + \mathbf{r}_{CA} \times \mathbf{F} \qquad \boxed{3\text{-}28}$$

The force–couple systems described here are such that the force and moment vectors are mutually perpendicular in each case. For example, the moment vector **M** is into the paper in Fig. 3-32c, while **M′** is out of the paper in part d of the figure.

The procedures described above can be applied in the reverse order to replace a force–couple system by a single force that has the same total effect on a rigid body. The idea is to move the force **F** in its plane perpendicular to the moment vector **M** until it causes the same moment as **M** would cause. For example, starting with Fig. 3-32c, **F** is moved to the right a distance L, parallel to itself. At this stage **F** alone must be shown as in Fig. 3-32a.

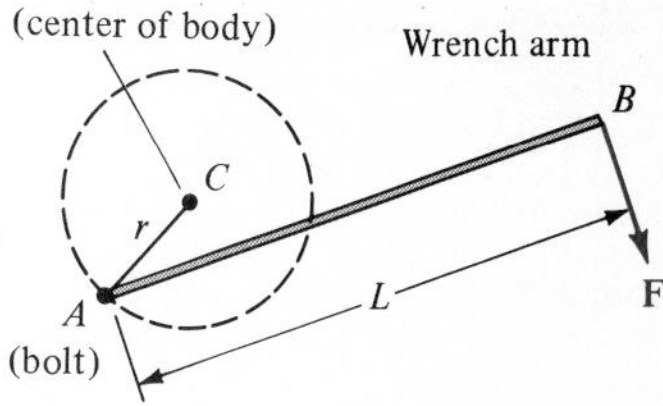

(a)

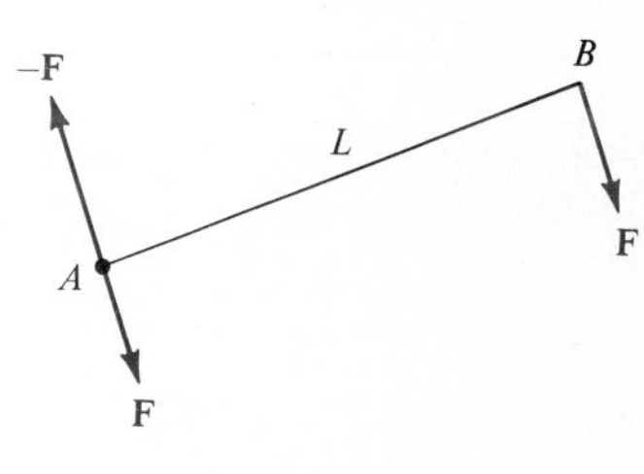

(b)

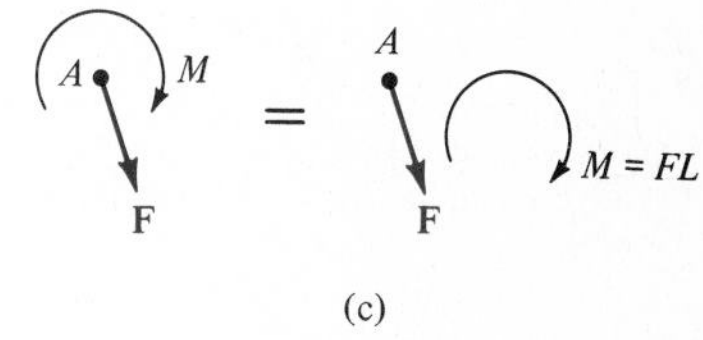

(c)

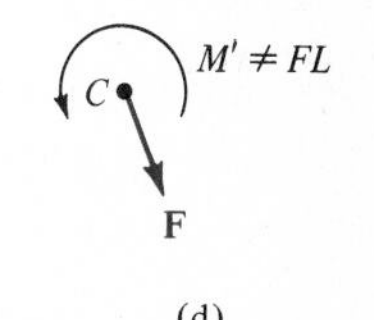

(d)

FIGURE 3-32

Transformation of a force into a force–couple system

EXAMPLE 3-9

Prove that a couple Fd acting on a rigid body can be moved anywhere in the same plane along line L, even to make the angle θ with L as in Fig. 3-33, and its moment about point A remains the same if F and d are constant.

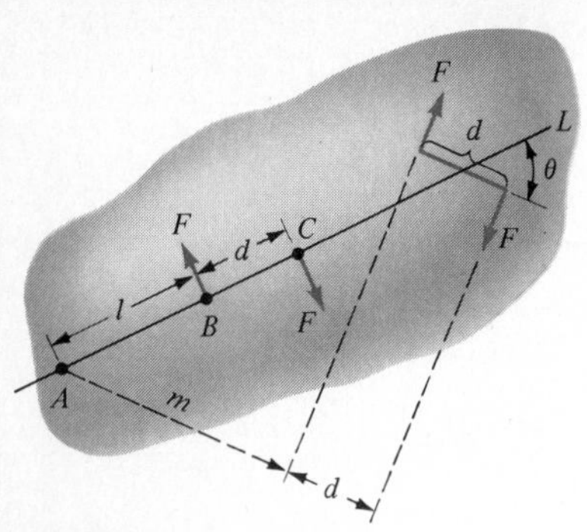

FIGURE 3-33

Equivalence of couples *Fd*

SOLUTION

With the forces perpendicular to L and taking clockwise as positive, the net moment about A is $M_A = F(l + d) - Fl = Fd$, independent of l. For the position with the angle θ, $M_A = F(m + d) - Fm = Fd$, independent of m or θ.

Note that in this relatively simple problem the scalar method of calculating moments is the easiest and allows visualizing that the results are correct. Using vector products is slightly more complicated in this case but provides the same results. This is left as an exercise for the reader.

EXAMPLE 3-10

In a machine the access hole for tightening a bolt is misaligned because of a mistake in fabrication (Fig. 3-34). The moment on the handle of the screwdriver is applied as shown. Determine the effective moment that turns the bolt as a function of θ. At what angle is the effective moment 90% of the applied moment?

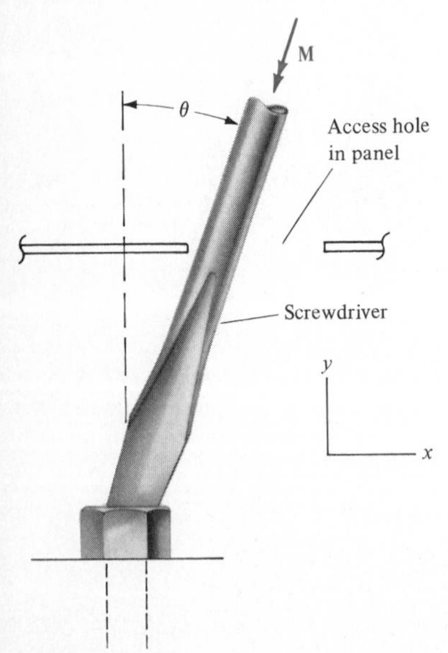

FIGURE 3-34

Effective moment M_y for tightening a bolt

SOLUTION

The effective moment is the M_y component of **M**, and

$$M_y = M \cos\theta$$

$$M_y = 0.9M = M\cos\theta \qquad \text{when } \theta = 25.8^\circ$$

3–62 Use vector products to prove that the results obtained in Ex. 3-9 are correct.

3–63 Determine the moment of the two forces F with respect to
(a) any point along the x axis
(b) any point along the y axis
(c) any point along the z axis

FIGURE P3-63

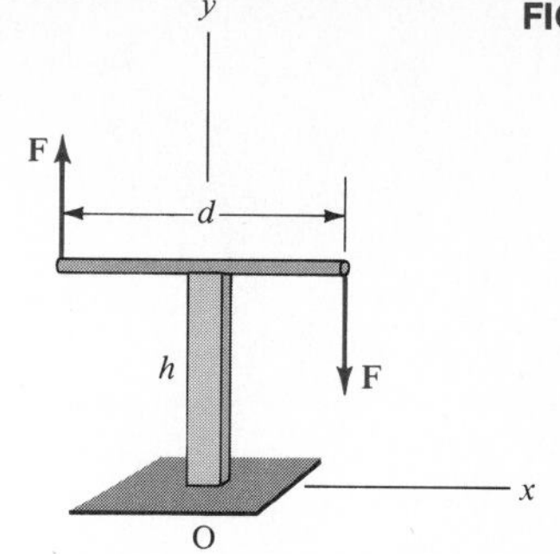

3–64 Determine the net moment **M** caused by the two couples on the rectangular block. All forces are parallel to the x axis.

FIGURE P3-64

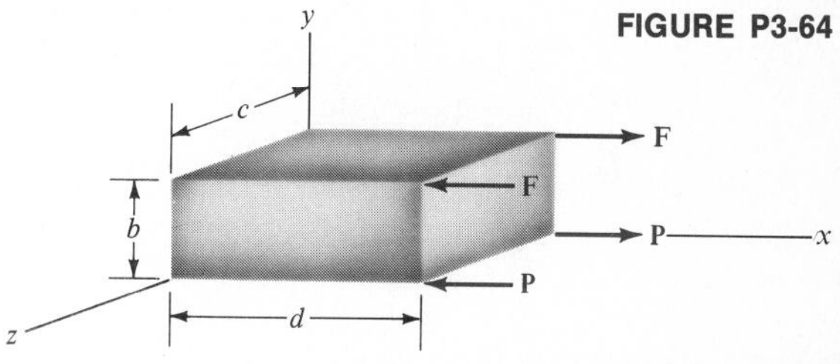

3–65 Determine the net moment **M** caused by the two couples on the rectangular block. All forces are parallel to the y axis.

FIGURE P3-65

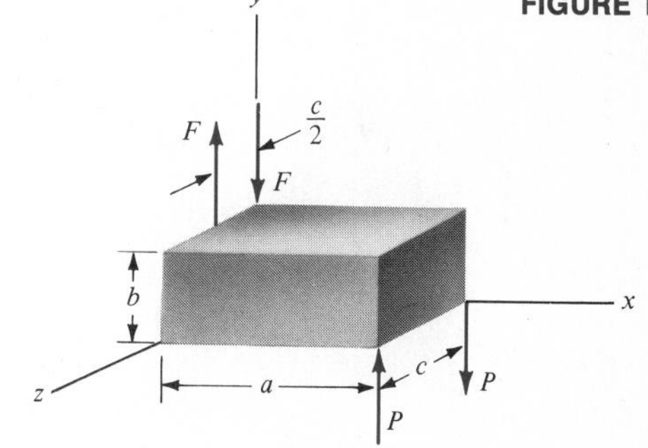

3–66 Determine the net moment **M** caused by the two couples on the rectangular block. All forces are parallel to the z axis.

FIGURE P3-66

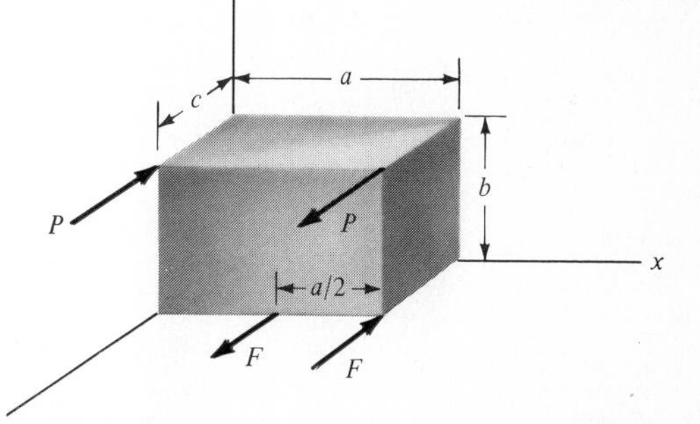

FIGURE P3-67

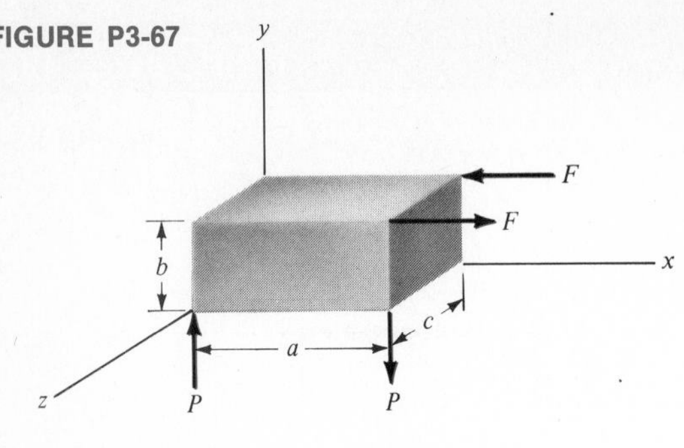

3–67 Two couples are acting on the rectangular block. Determine the resultant moment **M** of these couples.

FIGURE P3-68

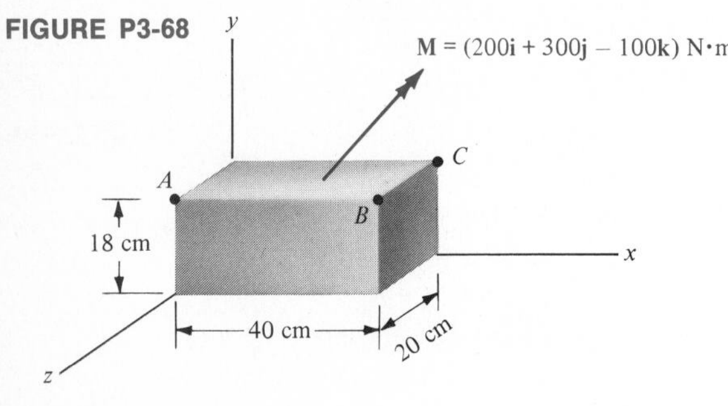

3–68 A moment $\mathbf{M} = M_x\mathbf{i} + M_y\mathbf{j} + M_z\mathbf{k}$ is acting on the rectangular block. Determine the directions and magnitudes of horizontal and vertical forces acting at points A, B, C to have the same effect on the block as **M** does. Is your set of forces the only set possible?

FIGURE P3-69

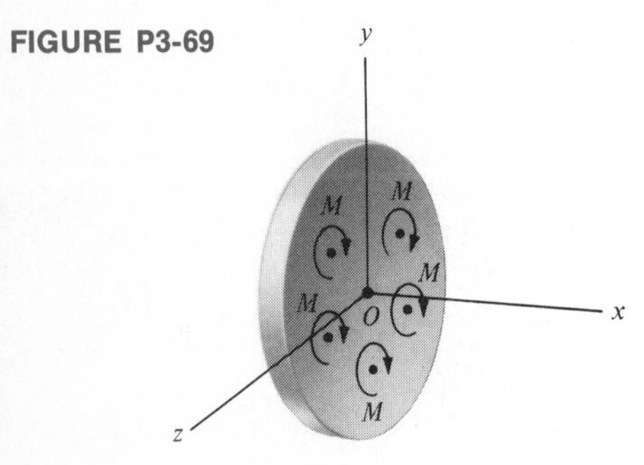

3–69 In an assembly operation five lug nuts on a car's wheel are tightened simultaneously. The same torque M is applied to each nut. Calculate the net moment $\mathbf{M}_O$ acting on the wheel if x is perpendicular to the plane of the nuts.

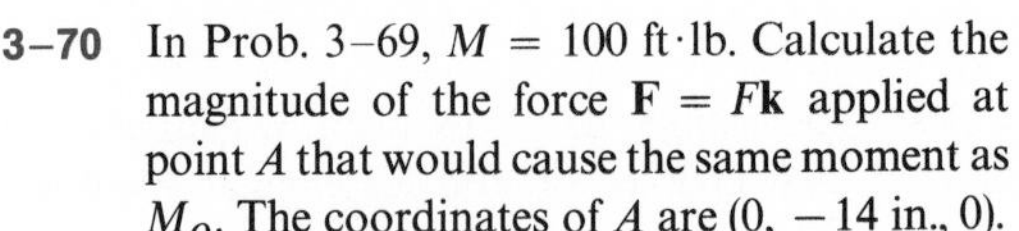

3–70 In Prob. 3–69, $M = 100$ ft·lb. Calculate the magnitude of the force $\mathbf{F} = F\mathbf{k}$ applied at point A that would cause the same moment as M_O. The coordinates of A are (0, −14 in., 0).

FIGURE P3-71

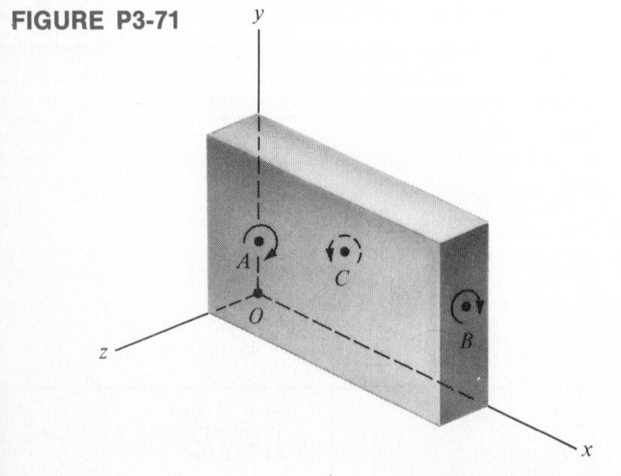

3–71 In an assembly operation the following moments are applied simultaneously: $\mathbf{M}_A = -200$ N·m **k**, $\mathbf{M}_B = -250$ N·m **i**, $\mathbf{M}_C = 150$ N·m **k**. Determine the net moment $\mathbf{M}_O$ acting on the block. Calculate the smallest force F that could cause the same moment when applied at a distance of 0.5 m from O.

3–72 The turning effect of the 10-kN force on the bar is to be duplicated by a couple of magnitude Fd. What is F if the distance d between the forces must be 20 cm?

3–73 Determine the equivalent force and moment of **F** at point O if $F = 200$ lb, $\theta = 80°$, and $L = 3$ ft.

3–74 Determine the equivalent force and moment of each of the two forces at point A if $F_1 = 500$ N, $F_2 = 300$ N, $L = 0.6$ m, $\theta = 70°$, and $\phi = 40°$.

3–75 The turning effects of forces F and W on the light pole with respect to point O are to be represented by a single couple of magnitude Pl. If the distance l between the forces P is given, determine P in terms of the other quantities.

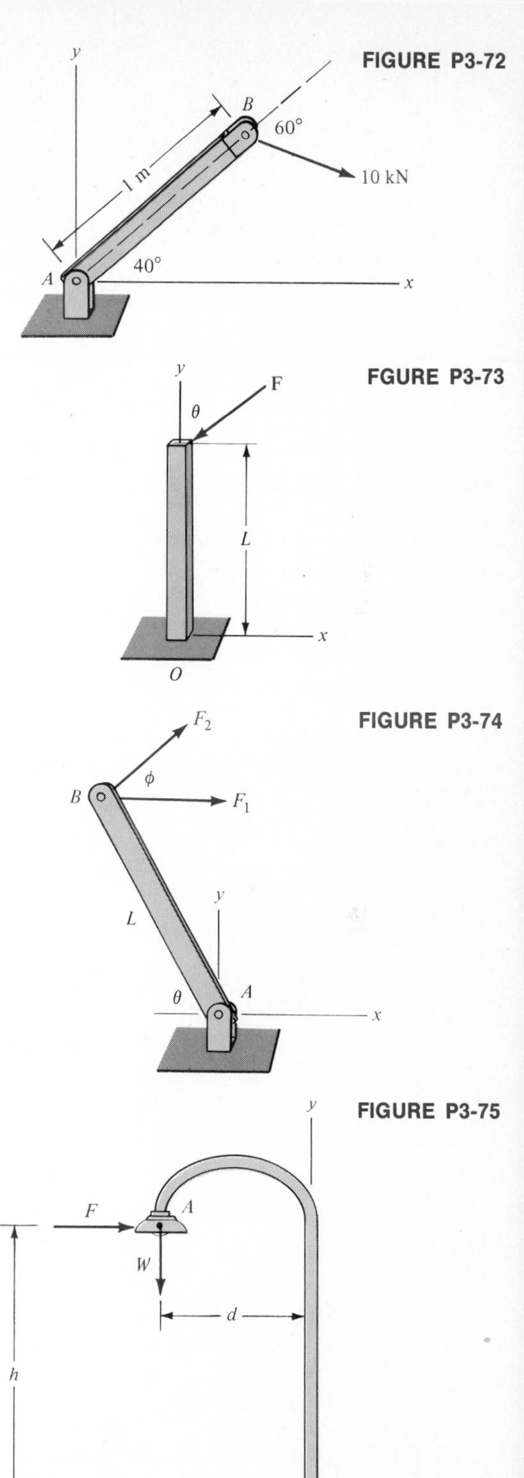

FIGURE P3-72

FGURE P3-73

FIGURE P3-74

FIGURE P3-75

FIGURE P3-76

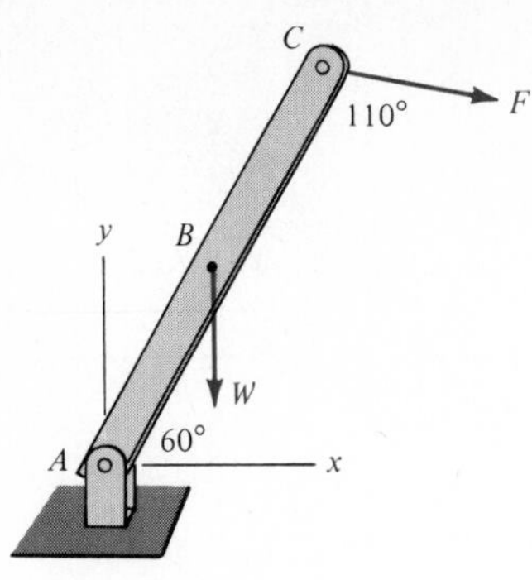

3–76 Assume that $F = 100$ lb and $AC = 4$ ft. The turning effect of this force about point A is to be represented by a couple of magnitude Pd. If P must be 200 lb, determine the distance d of the couple.

FIGURE P3-77

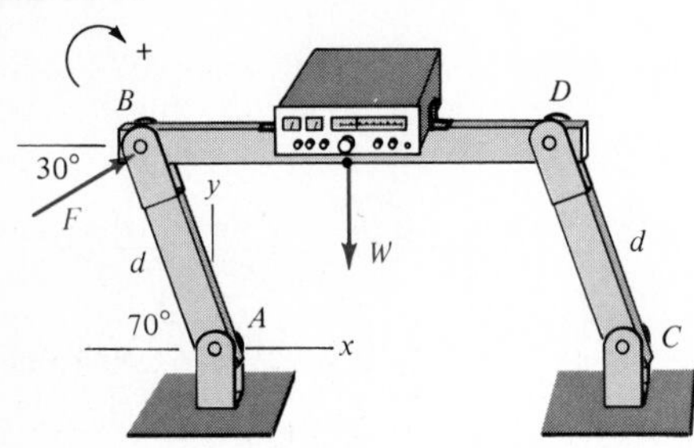

3–77 Assume that $F = 5$ kN and $d = 0.7$ m. The turning effect of F about point A is to be duplicated by a couple of magnitude Pl. Calculate P if l must be 0.1 m.

FIGURE P3-78

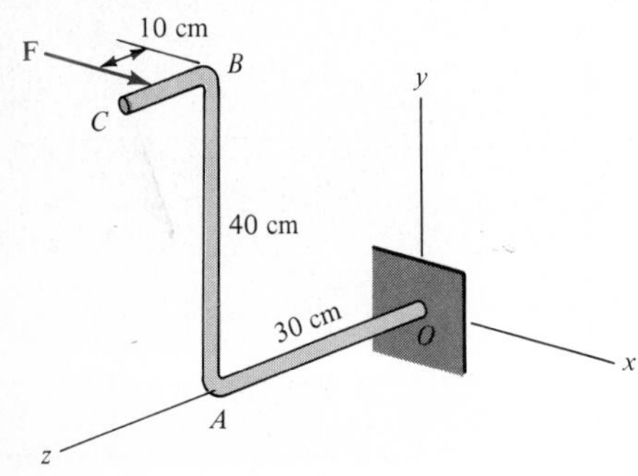

3–78 Force $\mathbf{F} = (200\mathbf{i} - 30\mathbf{j} - 20\mathbf{k})$ N. Determine the equivalent force and moment of $\mathbf{F}$ at point O.

FIGURE P3-79

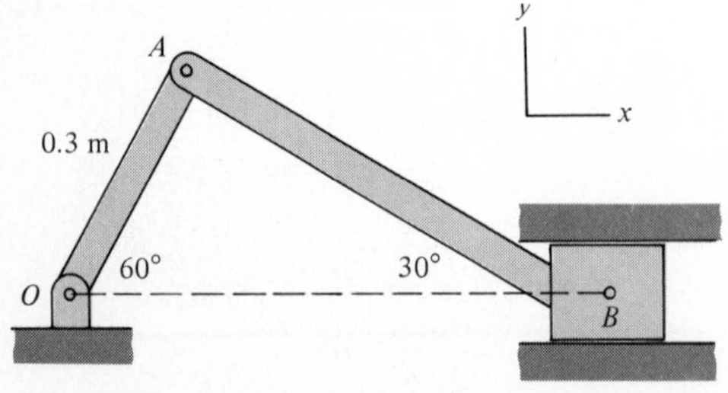

3–79 A 2-kN·m moment is applied to the crank arm OA. What is the equivalent force along piston rod AB to cause the same moment?

(a)

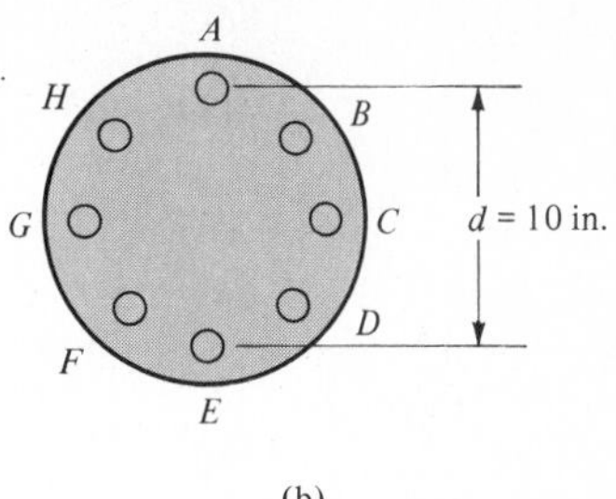

(b)

FIGURE P3-80

Torsion test setup of a shift. (Photograph courtesy MTS Systems Corporation, Minneapolis, Minnesota.)

3–80 The shaft S is subjected to a torsion test. The rotary hydraulic actuator A applies a moment $M = 2000$ in·lb through the bolts A to H. Determine the force P acting on each bolt assuming that the bolts share the total load equally.

FIGURE P3-81

Test equipment for tires. (Courtesy MTS Systems Corporation, Minneapolis, Minnesota.)

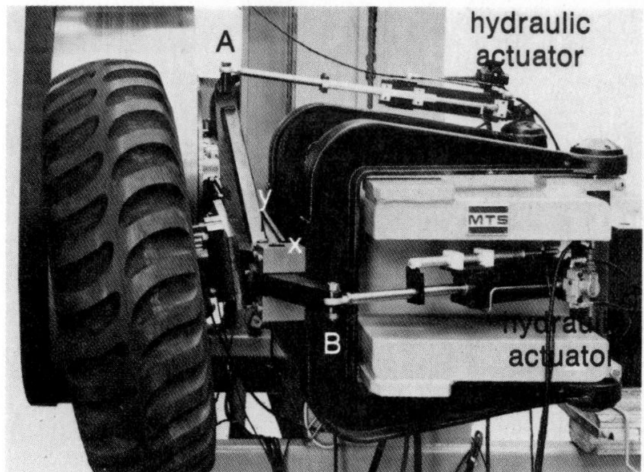

3–81 Two linear hydraulic actuators of a tire testing machine are visible in the photograph. The maximum force from the upper actuator is $\mathbf{F}_1 = (-3\mathbf{i} + 0.5\mathbf{j})$ kN acting at A of coordinates $(-0.05, 0.8, 0)$ m. Determine the equivalent force and moment of this force acting at point O.

3–82 Part of a tire testing machine is shown in Fig. P3-82. A rotary hydraulic actuator is considered to apply a moment at A to replace the linear actuator BC, which can apply a force of 500 lb. BC and the wheel's spindle are in the same plane. What is the required capacity (moment) of the rotary actuator? For scale, the diameter of the wheel is 4 ft.

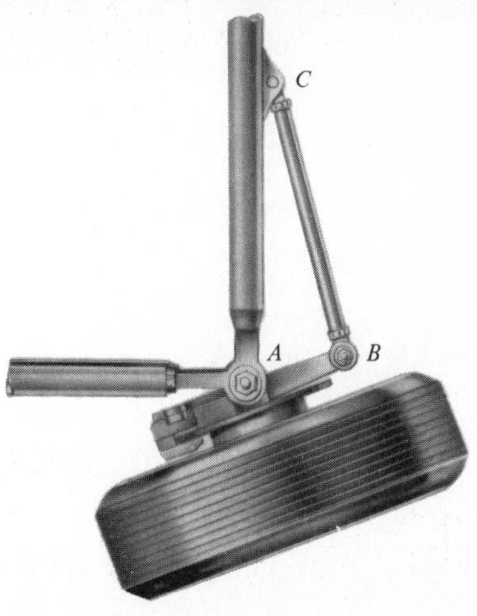

FIGURE P3-82

Drawing of a tire testing machine. (Courtesy MTS Systems Corporation, Minneapolis, Minnesota.)

3-11 SIMPLIFICATION OF MANY FORCES TO A FORCE AND A COUPLE

There have been some instances in the preceding discussions where more or less complex systems of forces were simplified without altering the net effects on rigid bodies. These procedures can be formally stated at this stage to cover all possible systems of any complexity. The following statements are for a three-dimensional system of forces such as the one in Fig. 3-35a.

In an attempt to simplify this complex system, it is necessary to establish a point of reference. The center of mass, point O, may be convenient. Some forces may have a moment about O, and each of these may be replaced by an identical force at O and the appropriate couple (for example, $\mathbf{M}_5 = \mathbf{r}_5 \times \mathbf{F}_5$). With all the forces made concurrent at O, they have a resultant $\mathbf{F}_R$ in Fig. 3-35b. Also, all the individual moments of the forces in their original positions ($\mathbf{M}_i = \mathbf{r}_i \times \mathbf{F}_i$) can be reduced to a single resultant moment, $\mathbf{M}_R$. Thus, the equivalent force–couple system for Fig. 3-35a is shown schematically in part b of the figure, and the corresponding equations are

$$\mathbf{F}_R = \sum_{i=1}^{7} \mathbf{F}_i \qquad \text{and} \qquad \mathbf{M}_R = \sum_{i=1}^{7} (\mathbf{r}_i \times \mathbf{F}_i) \tag{3-29}$$

Note that $\mathbf{M}_R$ is calculated with respect to the arbitrarily chosen point O. The result would be different for another reference point O', but it could be calculated equally easily.

In the most general case of a system of forces a resultant force $\mathbf{F}_R$ and resultant moment $\mathbf{M}_R$ can be determined, but these will not be mutually perpendicular. This limits the chances for further simplification of the equivalent system.

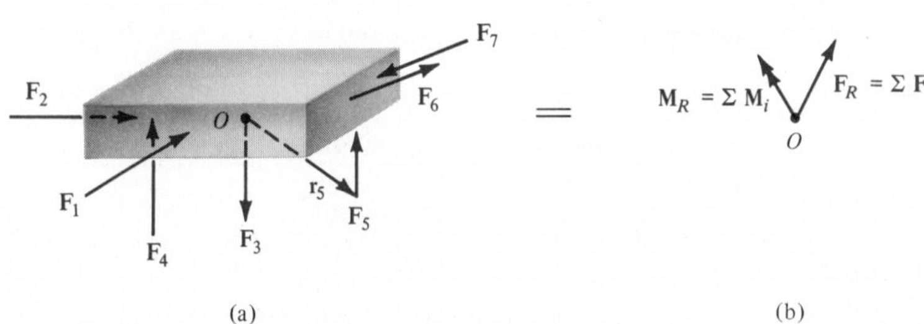

FIGURE 3-35

Simplification of many forces of a tuned mass damper to a force and a couple. F_1 and F_2 from hydraulic actuators; F_3 = weight; $F_4 \neq F_5$, track reactions; F_6 and F_7, antiyaw couple

Equivalent Force Systems

Equation 3-29 is the basis for checking the equivalence of two force systems. For all cases involving rigid bodies, *two systems of forces are equivalent if they can be reduced to the same resultant force and resultant moment with respect to any arbitrary point.* In concise notation, a system of forces $\mathbf{F}_1, \mathbf{F}_2, \ldots, \mathbf{F}_i \ldots$ is equivalent to another system consisting of $\mathbf{P}_1, \mathbf{P}_2, \ldots, \mathbf{P}_j \ldots$ if

$$\sum_{i=1}^{m} \mathbf{F}_i = \sum_{j=1}^{n} \mathbf{P}_j \quad \text{and} \quad \sum_{i=1}^{m} (\mathbf{r}_i \times \mathbf{F}_i) = \sum_{j=1}^{n} (\mathbf{r}_j \times \mathbf{P}_j) \qquad \boxed{3\text{-}30}$$

for any point of reference. Here i and j are not unit vectors but subscripts to distinguish individual vectors.

3-12 COMMON SIMPLE FORCE SYSTEMS

It is of considerable advantage in analyzing complex problems of rigid bodies that any force system can be reduced to an equivalent system of a resultant force $\mathbf{F}_R$ and a resultant moment $\mathbf{M}_R$. There are special cases of these resultant systems that are quite common. They are classified and listed here to allow a convenient overview and comparison of them.

1. ***Resultant force is zero.*** A nontrivial case is when $\mathbf{F}_R$ is zero but there is a resultant moment $\mathbf{M}_R$. Sometimes this $\mathbf{M}_R$ is called the *resultant couple* of the system.

2. ***Concurrent forces.*** When all the forces act at the same point O, they have a resultant $\mathbf{F}_R$ but the moment $\mathbf{M}_R$ is zero with respect to O.

3. ***Coplanar forces.*** When all the forces act in the same plane, they can be reduced to a resultant force $\mathbf{F}_R$ in the same plane and a resultant moment $\mathbf{M}_R$ about a chosen reference point O. $\mathbf{M}_R$ is perpendicular to the plane of the forces as in Fig. 3-36a. The force–couple system could be replaced by the force $\mathbf{F}_R$ alone by moving $\mathbf{F}_R$ from point O parallel to itself in the xy plane through a distance d until its moment about O equals $\mathbf{M}_R$. In this situation $\mathbf{M}_R$ should not be shown since $\mathbf{F}_R$ (dashed line in Fig. 3-36b) represents the total effect of $\mathbf{F}_1$, $\mathbf{F}_2$, and $\mathbf{F}_3$ with respect to point O.

4. ***Parallel forces.*** All such forces have parallel lines of action but their directions may or may not be the same as in Fig. 3-37a. In this example their resultant $\mathbf{F}_R$ is also parallel to the y axis (Fig. 3-37b). The resultant moment $\mathbf{M}_R$ about point O is perpendicular to $\mathbf{F}_R$, as shown in Fig. 3-37c.

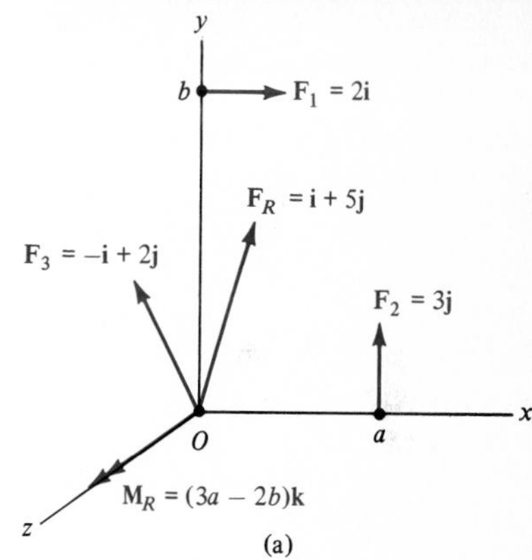

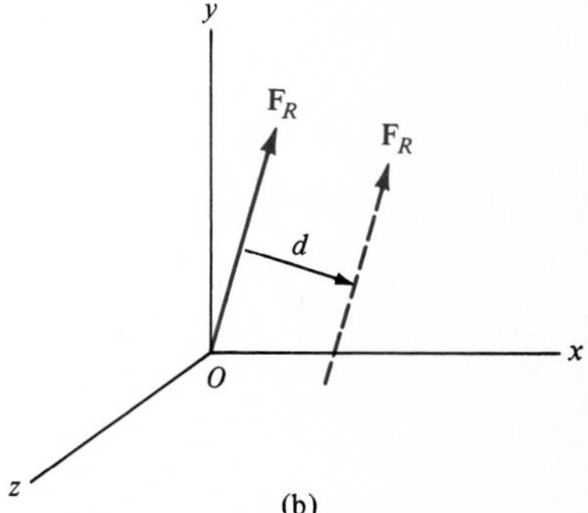

FIGURE 3-36

Reduction of a coplanar force system

The magnitude of $\mathbf{F}_R$ is obtained by algebraic summation. Its location can be determined in several ways without introducing any new concepts. Essentially, it amounts to writing the same moment in two different ways, with one of them involving an unknown coordinate, and then equating the two expressions. This is illustrated with finding coordinate x of $\mathbf{F}_R$ in Fig. 3-37b. This coordinate is the moment arm for $\mathbf{F}_R$, inducing rotation about the z axis. Equating the moments about the z axis in Fig. 3-37a and b,

$$F_1 a_x + F_2 b_x - F_3 c_x = F_R x$$

$$3(2) + 2(4) - 1.5(1) = 3.5x$$

$$x = \frac{12.5}{3.5} = 3.57$$

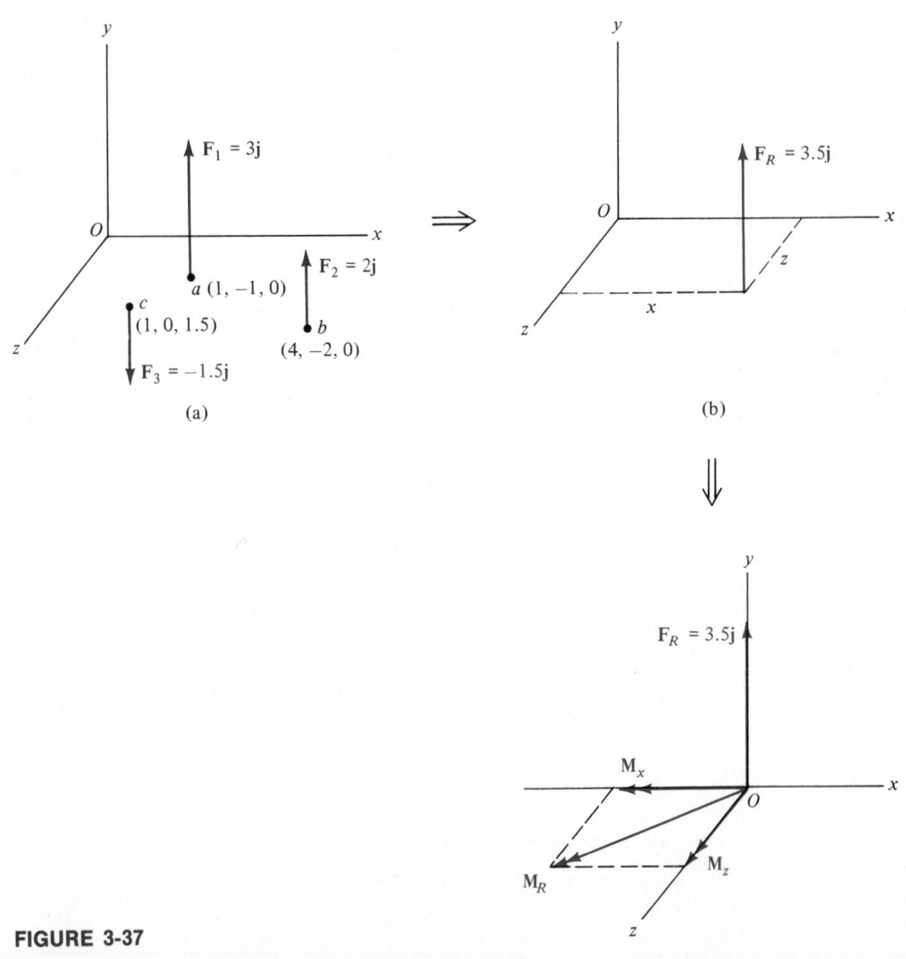

FIGURE 3-37

Reduction of a general parallel force system

EXAMPLE 3-11

Assume that the major forces acting on the tall crane modeled in Fig. 3-38 are coplanar. Force **F** on the right is normally vertical, but it may make an angle up to 5° with the vertical. The other major loads are the counterweight **P** at the top left and the weight **W** of the vertical column whose centerline is the y axis. Determine the force–couple system acting at the base O of the column for a given height L of the crane, with **F** making a 5° angle with the vertical and acting in the xy plane as shown.

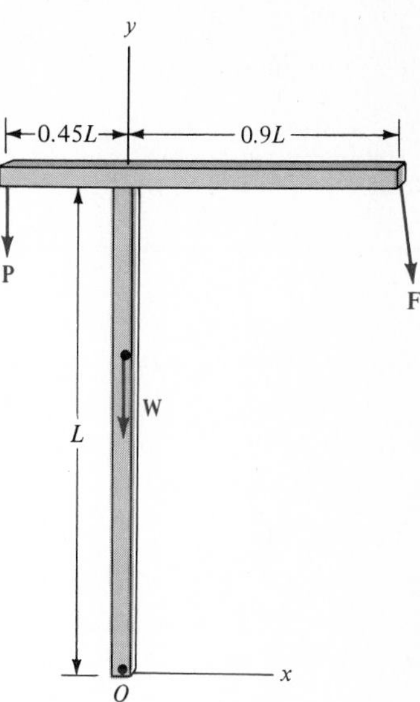

FIGURE 3-38

Model of a crane

SOLUTION

When **F** acts in the xy plane,

$$\mathbf{F} = F(\sin 5^\circ\,\mathbf{i} - \cos 5^\circ\,\mathbf{j}) = 0.087F\mathbf{i} - 0.996F\mathbf{j}$$

$$\sum \mathbf{F} = 0.087F\mathbf{i} - 0.996F\mathbf{j} - P\mathbf{j} - W\mathbf{j}$$

$$\begin{aligned}\sum \mathbf{M}_O &= \mathbf{M}_R = \mathbf{M}_{FO} + \mathbf{M}_{PO} \\ &= (0.9L\mathbf{i} + L\mathbf{j}) \times (0.087F\mathbf{i} - 0.996F\mathbf{j}) \\ &\quad + (-0.45L\mathbf{i} + L\mathbf{j}) \times (-P\mathbf{j}) \\ &= -0.983FL\mathbf{k} + 0.45PL\mathbf{k}\end{aligned}$$

Thus, the force–couple system with respect to point O is

$$\mathbf{F}_R = 0.087F\mathbf{i} - (0.996F + P + W)\mathbf{j}$$

$$\mathbf{M}_R = (0.45P - 0.983F)L\mathbf{k}$$

EXAMPLE 3-12

A rail test setup such as in Fig. 3-39a can be imagined as a bar in Fig. 3-39b with coplanar forces. Determine the equivalent force–couple system at O, where there is a weldment. Also determine the location of the single resultant force that would have the same effect as the force–couple system.

SOLUTION

$$\mathbf{F}_R = \sum_{i=1}^{4} \mathbf{F}_i = (-50\text{ kN} + 10\text{ kN} + 20\text{ kN} - 5\text{ kN})\mathbf{j} = -25\text{ kN}\,\mathbf{j}$$

The scalar method is convenient for calculating the resultant moment since all forces are coplanar and parallel. Taking counterclockwise moments as positive, the moments about O are

$$\begin{aligned}M_R &= \sum_{i=1}^{4} F_i d_i \\ &= (50\text{ kN})(1.5\text{ m}) - (10\text{ kN})(0.8\text{ m}) \\ &\quad - (20\text{ kN})(2.3\text{ m}) + (5\text{ kN})(2.5\text{ m}) \\ &= 33.5\text{ kN}\cdot\text{m} \quad \text{counterclockwise}\end{aligned}$$

$$\mathbf{M}_R = 33.5\text{ kN}\cdot\text{m}\,\mathbf{k} \quad \text{by inspection}$$

To obtain the location of the resultant force to cause the same resultant moment, assume that $\mathbf{F}_R$ acts at a distance $l\mathbf{i}$ with respect to point O. Equating the moment $\mathbf{M}_R$ (from the individual forces) with the moment of the resultant force $\mathbf{F}_R$, the unknown distance l is determined.

$$\mathbf{M}_R = l\mathbf{i} \times \mathbf{F}_R$$

$$33.5 \text{ kN}\cdot\text{m } \mathbf{k} = l\mathbf{i} \times (-25 \text{ kN } \mathbf{j}) = -25l \text{ kN } \mathbf{k}$$

$$l = -1.34 \text{ m}$$

The negative sign of l indicates that $\mathbf{F}_R$ acts to the left of point O, as expected.

(a)

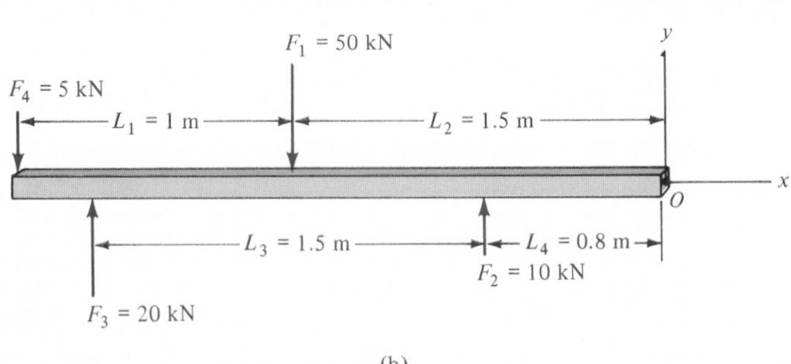

(b)

FIGURE 3-39

Test setup of railroad tracks. (Photograph courtesy MTS Systems Corporation, Minneapolis, Minnesota.)

EXAMPLE 3-13

Determine the equivalent force–couple system at O and the components M_x, M_y, and M_z caused by the following forces acting on the crane shown in Fig. 3-40.

$$\mathbf{F}_A = -2\mathbf{j} \text{ MN at } A, \quad \mathbf{r}_A = 50\mathbf{i} \text{ m}$$
$$\mathbf{F}_B = -2.9\mathbf{j} \text{ MN at } B, \quad \mathbf{r}_B = (65\mathbf{i} + 10\mathbf{k}) \text{ m}$$
$$\mathbf{F}_C = -2.9\mathbf{j} \text{ MN at } C, \quad \mathbf{r}_C = (65\mathbf{i} - 10\mathbf{k}) \text{ m}$$
$$\mathbf{F}_D = -2.9\mathbf{j} \text{ MN at } D, \quad \mathbf{r}_D = 85\mathbf{i} \text{ m}$$
$$\mathbf{F}_E = -0.7\mathbf{j} \text{ MN at } E, \quad \mathbf{r}_E = (90\mathbf{i} + 15\mathbf{j} - 19\mathbf{k}) \text{ m}$$

SOLUTION

$$\begin{aligned}\mathbf{F}_R &= \sum \mathbf{F} = \mathbf{F}_A + \mathbf{F}_B + \mathbf{F}_C + \mathbf{F}_D + \mathbf{F}_E \\ &= -2 \text{ MN}\,\mathbf{j} - 2.9 \text{ MN}\,\mathbf{j} - 2.9 \text{ MN}\,\mathbf{j} - 2.9 \text{ MN}\,\mathbf{j} - 0.7 \text{ MN}\,\mathbf{j} \\ &= -11.4 \text{ MN}\,\mathbf{j}\end{aligned}$$

$$\begin{aligned}\mathbf{M}_R &= \sum \mathbf{M}_R = (\mathbf{r}_A \times \mathbf{F}_A) + (\mathbf{r}_B \times \mathbf{F}_B) \\ &\quad + (\mathbf{r}_C \times \mathbf{F}_C) + (\mathbf{r}_D \times \mathbf{F}_D) + (\mathbf{r}_E \times \mathbf{F}_E) \\ &= (50\mathbf{i} \text{ m}) \times (-2\mathbf{j} \text{ MN}) + [(65\mathbf{i} + 10\mathbf{k}) \text{ m} \times (-2.9\mathbf{j} \text{ MN})] \\ &\quad + [(65\mathbf{i} - 10\mathbf{k}) \text{ m} \times (-2.9\mathbf{j} \text{ MN})] + (85\mathbf{i} \text{ m} \times (-2.9\mathbf{j} \text{ MN}) \\ &\quad + [(90\mathbf{i} + 15\mathbf{j} - 19\mathbf{k}) \text{ m} \times (-0.7\mathbf{j} \text{ MN})] \\ &= -100\mathbf{k} \text{ MN}\cdot\text{m} - 188.5\mathbf{k} \text{ MN}\cdot\text{m} + 29\mathbf{i} \text{ MN}\cdot\text{m} \\ &\quad - 188.5\mathbf{k} \text{ MN}\cdot\text{m} - 29\mathbf{i} \text{ MN}\cdot\text{m} - 246.5\mathbf{k} \text{ MN}\cdot\text{m} \\ &\quad - 63\mathbf{k} \text{ MN}\cdot\text{m} - 13.3\mathbf{i} \text{ MN}\cdot\text{m} \\ &= -13.3\mathbf{i} \text{ MN}\cdot\text{m} - 786.5\mathbf{k} \text{ MN}\cdot\text{m}\end{aligned}$$

$$M_x = -13.3 \text{ MN}\cdot\text{m}$$

$$M_y = 0$$

$$M_z = -786.5 \text{ MN}\cdot\text{m}$$

Note that there is no moment component about the y axis as should be expected since all forces act parallel to the y axis.

FIGURE 3-40

Force system acting on a crane. (Courtesy General Dynamics Quincy Shipbuilding Division.)

PROBLEMS

SECTIONS 3-11, 3-12

3–83 Solve Ex. 3-11 assuming that the force **F** makes a 5° angle with the vertical, acting in the z direction in the yz plane.

FIGURE P3-84

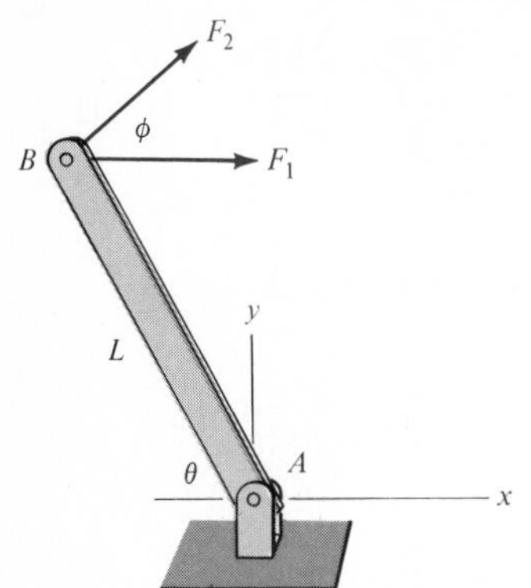

3–84 Determine the equivalent force and moment of the two forces at point A if $F_1 = 500$ N, $F_2 = 300$ N, $L = 0.6$ m, $\theta = 70°$, and $\phi = 40°$. First work with the individual forces, then check the results by working directly with the resultant $\mathbf{F}_R$ of $\mathbf{F}_1$ and $\mathbf{F}_2$.

FIGURE P3-85

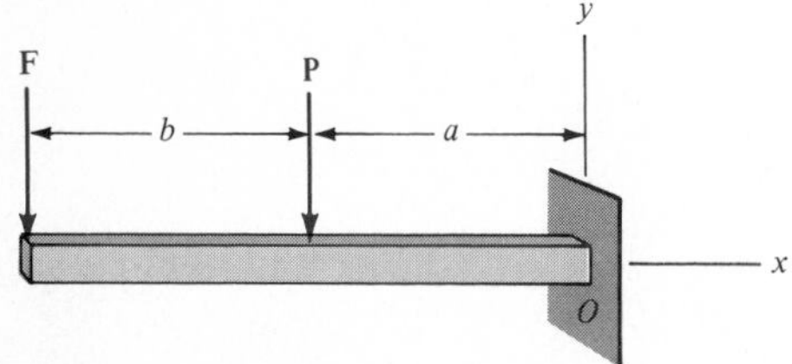

3–85 Determine the equivalent force and moment of the two forces at point O of the cantilever beam. **F** and **P** are parallel to the y axis.

FIGURE P3-86

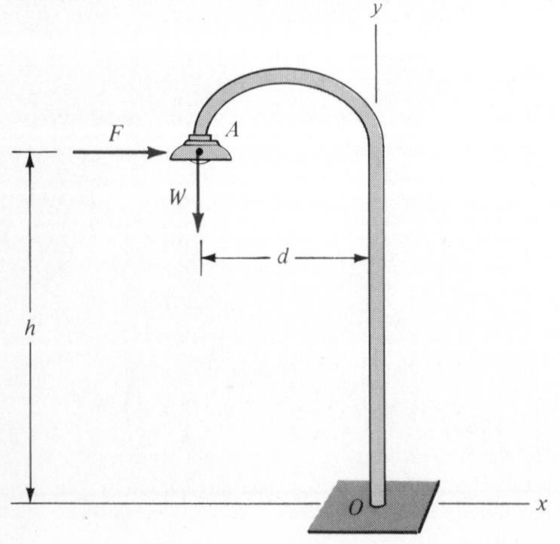

3–86 Determine the equivalent force and moment of the two forces at point O on the light pole. First use F and W individually, then check the results using the resultant $\mathbf{F}_R = \mathbf{F} + \mathbf{W}$.

3–87 Calculate the equivalent force and moment of forces F and P at point O on the column.

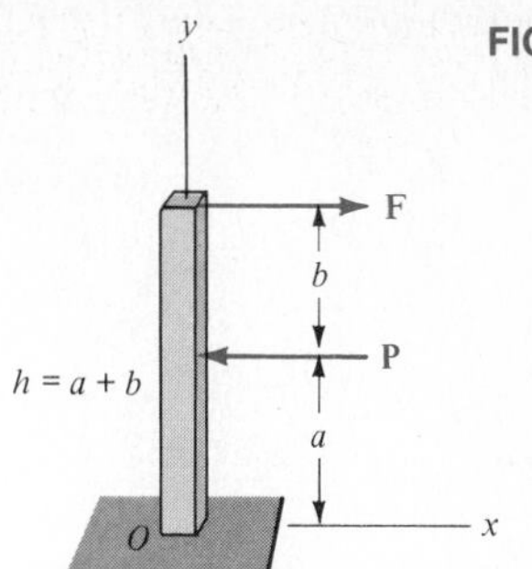

FIGURE P3-87

3–88 Consider Fig. P3-87 again. Let the equivalent force and moment at point O be $\mathbf{F}_O$ and $\mathbf{M}_O$, respectively. Prove that if $\mathbf{F}_O = 0$, $\mathbf{M}_O \neq 0$, and if $\mathbf{M}_O = 0$, $\mathbf{F}_O \neq 0$.

FIGURE P3-89

3–89 Assume that $W = 30$ lb, $F = 100$ lb, and $AB = BC = 2$ ft. The turning effects of these forces about point A are to be represented by a single couple of magnitude Pd. If d must be 2 in., determine the force P.

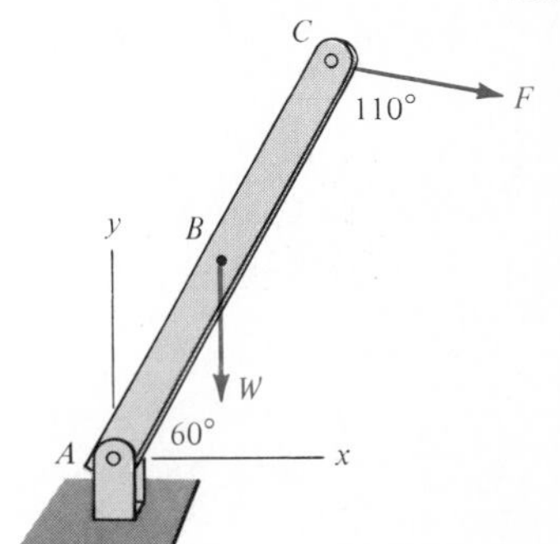

3–90 Determine the equivalent force and moment at point O if all forces are in the xy plane on the beam.

FIGURE P3-90

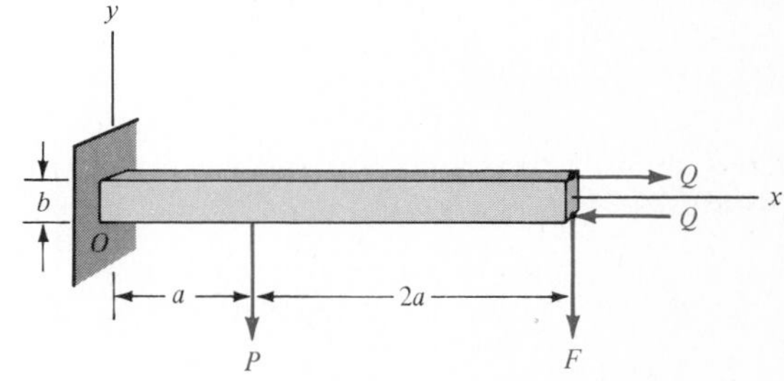

3–91 The vertical forces acting on an airplane wing are approximated by four forces given in pounds. The forces intersect the x axis at the indicated distances which are given in feet. Calculate the equivalent force and moment of these forces at O.

FIGURE P3-91

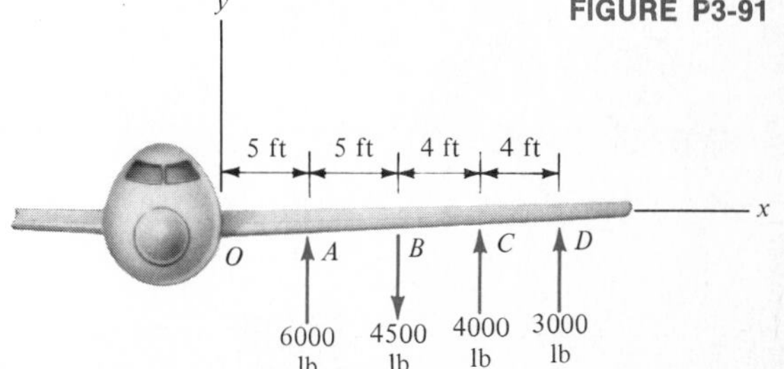

FIGURE P3-92

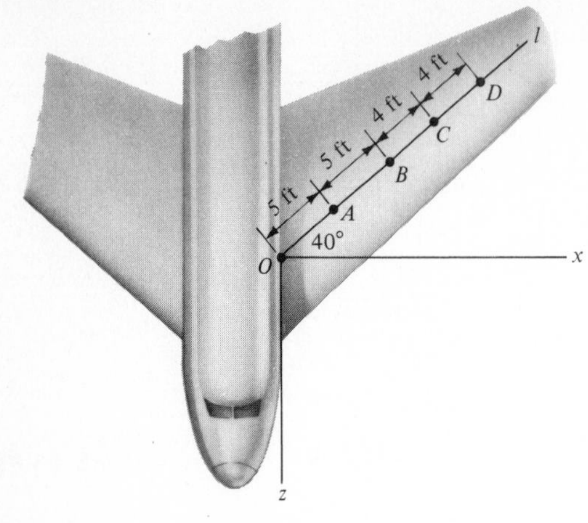

3–92 Consider a variation of Prob. 3–91 using a swept wing. In Fig. P3-92, the forces are perpendicular to the plane of the paper while the distances are along line l in the xz plane. Determine the equivalent force and moment at point O.

3–93 In Prob. 3–91, determine the location d measured from point O along the x axis where the resultant force F_R of the four forces is acting.

FIGURE P3-94

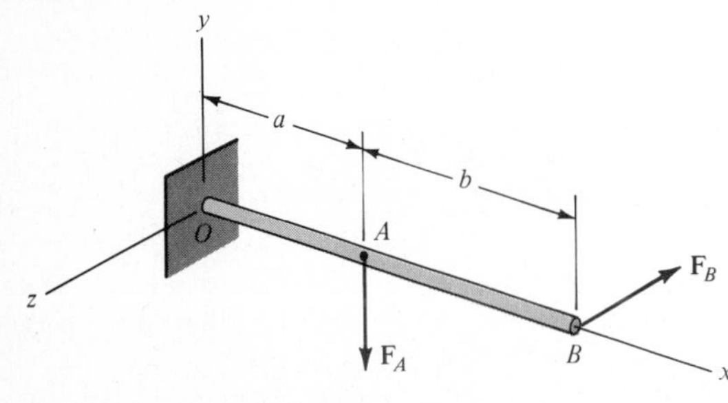

3–94 Calculate the equivalent force and moment of $\mathbf{F}_A = -400\text{ N }\mathbf{j}$ and $\mathbf{F}_B = -300\text{ N }\mathbf{k}$ at point O; $a = 0.3$ m and $b = 0.4$ m. Determine the distance d from O, where the resultant force $\mathbf{F}_R$ of $\mathbf{F}_A + \mathbf{F}_B$ appears to be acting.

FIGURE P3-95

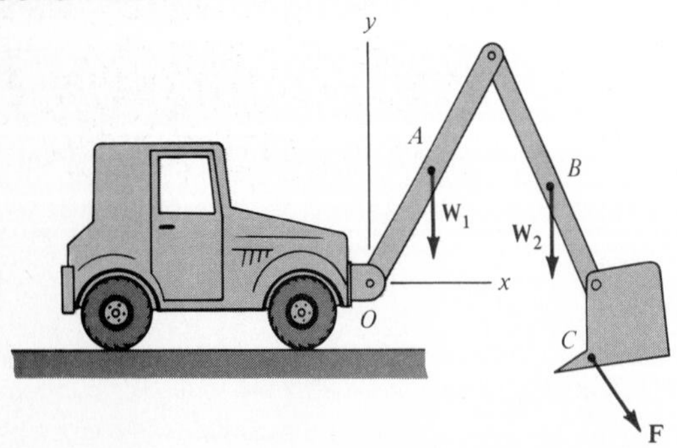

3–95 A mechanical hoe is modeled with forces $\mathbf{W}_1 = -1.2\text{ kN }\mathbf{j}$ at A of coordinates (0.5, 1.5, 0) m, $\mathbf{W}_2 = -1.6\text{ kN }\mathbf{j}$ at B of coordinates (1.4, 1.2, 0) m, and $\mathbf{F} = (800\mathbf{i} - 900\mathbf{j})$ N at C of coordinates (1.7, −0.6, 0) m. Determine the location where the resultant of the three forces intersects the x axis.

3–96 In Prob. 3–95, use $\mathbf{F} = (800\mathbf{i} - 900\mathbf{j} + 400\mathbf{k})$ N acting at C of coordinates (1.7, −0.6, 0.3) m. Calculate the distance d from point O, where the resultant of the three forces appears to be acting.

3–97 In a laboratory test of a motorcycle three forces are applied simultaneously: $\mathbf{F}_1 = 500 \text{ lb } \mathbf{j}$, $\mathbf{F}_2 = 450 \text{ lb } \mathbf{j}$, $\mathbf{F}_3 = -200 \text{ lb } \mathbf{j}$. The coordinates are A: $(-2.5, -1.7, 0)$ ft, B: $(2.3, -1.7, 0)$ ft, and C: $(0.7, 1, 0)$ ft. Calculate the x intercept of the resultant force $\mathbf{F}_R$.

FIGURE P3-97

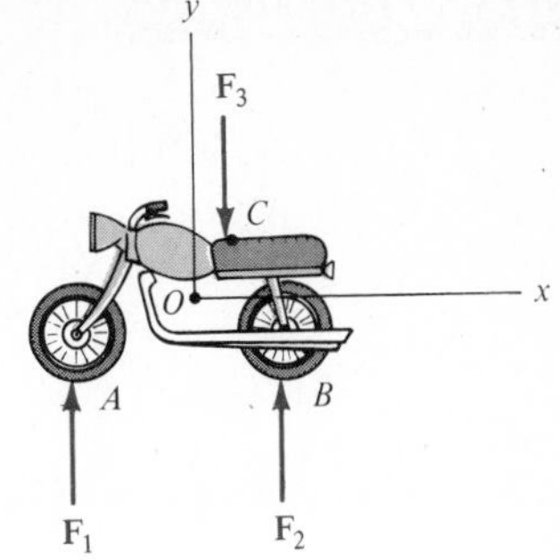

3–98 In Prob. 3–97, an additional force $\mathbf{F}_4 = 50 \text{ lb } \mathbf{i} + 40 \text{ lb } \mathbf{k}$ is applied at point A. Determine the equivalent force and moment of the four forces at point O.

3–99 Consider the tractor-trailer truck in a laboratory test. The origin O of the xyz frame is at the pivot connection of the tractor and the trailer. Determine the equivalent force–couple system at point O for the given forces (this is important for making the pivot connection strong enough).

$\mathbf{F}_A = 40\mathbf{j}$ kN at A, $(2, -0.7, 1.4)$ m

$\mathbf{F}_B = 0$ (temporarily)

$\mathbf{F}_C = 120\mathbf{j}$ kN at C, $(-0.5, -0.7, 1.3)$ m

$\mathbf{F}_D = 80\mathbf{j}$ kN at D, $(-0.5, -0.7, -1.3)$ m

$\mathbf{F}_E = -40\mathbf{j}$ kN (weight of tractor) at E, $(2, 0, 0)$ m

$\mathbf{F}_F = -160\mathbf{j}$ kN (load in trailer) at F, $(0, 2, 1)$ m

FIGURE P3-99

Laboratory test of a tractor-trailer truck. (Courtesy MTS Systems Corporation, Minneapolis, Minnesota.)

3–100 In Prob. 3–99, an additional force $\mathbf{P} = -10\mathbf{k}$ kN is acting at point A. Determine the components M_x, M_y, and M_z of the equivalent moment of all of the forces (these are important in dealing with the "ride characteristics" of the cab).

SUMMARY

Section 3-1

FORCES IN FREE-BODY DIAGRAMS: Show all significant external forces (such as weight) on a free-body diagram of a rigid body or system of bodies. The forces shown must act directly on the parts drawn. Do not show any internal forces between members that are included in the diagram.

Section 3-2

EQUIVALENT FORCES: Two forces acting at different points on a given rigid body are equivalent if they have the same magnitude, line of action, and direction. Note certain practical limitations concerning the equivalence of forces.

Section 3-3

MOMENT OF A FORCE, SCALAR DEFINITION:

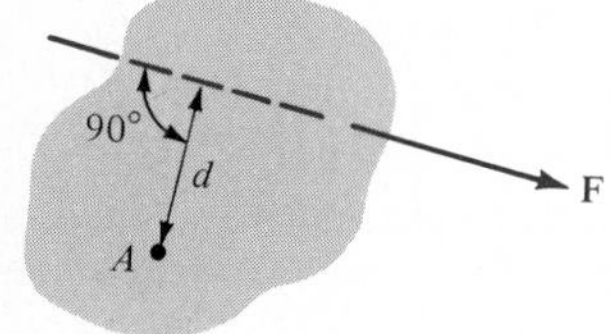

$$M_A = Fd$$

Section 3-4

VECTOR PRODUCT V OF TWO VECTORS A AND B:

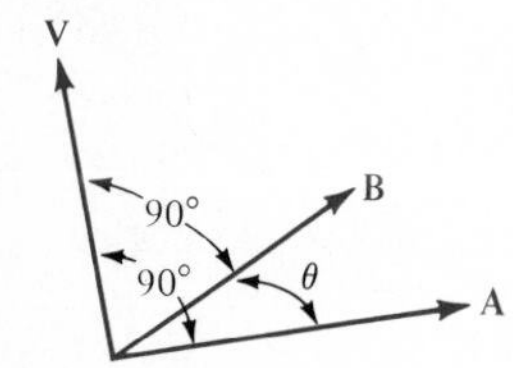

$$\mathbf{V} = \mathbf{A} \times \mathbf{B}$$

$$V = AB \sin \theta$$

USEFUL PROPERTIES:

$$\mathbf{A} \times \mathbf{B} \neq \mathbf{B} \times \mathbf{A}$$

$$\mathbf{A} \times \mathbf{B} = -(\mathbf{B} \times \mathbf{A})$$

$$\mathbf{A} \times (\mathbf{B} + \mathbf{C}) = \mathbf{A} \times \mathbf{B} + \mathbf{A} \times \mathbf{C}$$

$$\mathbf{i} \times \mathbf{i} = \mathbf{j} \times \mathbf{j} = \mathbf{k} \times \mathbf{k} = 0$$

$$\mathbf{i} \times \mathbf{j} = \mathbf{k} \qquad \mathbf{j} \times \mathbf{i} = -\mathbf{k} \qquad \mathbf{k} \times \mathbf{i} = \mathbf{j}$$

$$\mathbf{i} \times \mathbf{k} = -\mathbf{j} \qquad \mathbf{j} \times \mathbf{k} = \mathbf{i} \qquad \mathbf{k} \times \mathbf{j} = -\mathbf{i}$$

$$\mathbf{V} = \mathbf{A} \times \mathbf{B} = \begin{vmatrix} \mathbf{i} & \mathbf{j} & \mathbf{k} \\ A_x & A_y & A_z \\ B_x & B_y & B_z \end{vmatrix}$$

MOMENT OF A FORCE, VECTOR DEFINITION:

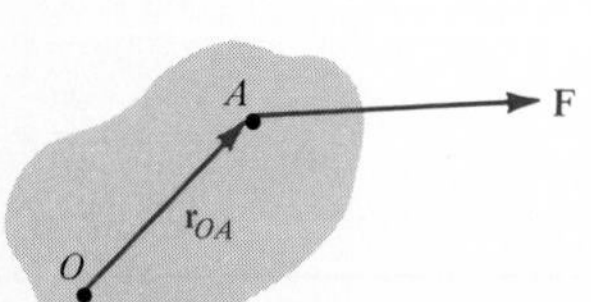

$$\mathbf{M}_O = \mathbf{r}_{OA} \times \mathbf{F} \qquad (A \text{ is any point on the line of action of } \mathbf{F})$$

RECTANGULAR COMPONENTS OF A MOMENT:

$$\mathbf{M}_O = M_x\mathbf{i} + M_y\mathbf{j} + M_z\mathbf{k}$$

Section 3-5

VECTOR TRIPLE PRODUCT: $\mathbf{V} = (\mathbf{A} \times \mathbf{B}) \times \mathbf{C}$ (used in dynamics)

MIXED TRIPLE PRODUCT:

$$\mathbf{A} \cdot (\mathbf{B} \times \mathbf{C}) = \begin{vmatrix} A_x & A_y & A_z \\ B_x & B_y & B_z \\ C_x & C_y & C_z \end{vmatrix}$$

Section 3-6

MOMENTS OF A FORCE ABOUT A LINE:

$$M_l = \mathbf{n} \cdot \mathbf{M}_O = \mathbf{n} \cdot (\mathbf{r} \times \mathbf{F}) = \begin{vmatrix} n_x & n_y & n_z \\ r_x & r_y & r_z \\ F_x & F_y & F_z \end{vmatrix}$$

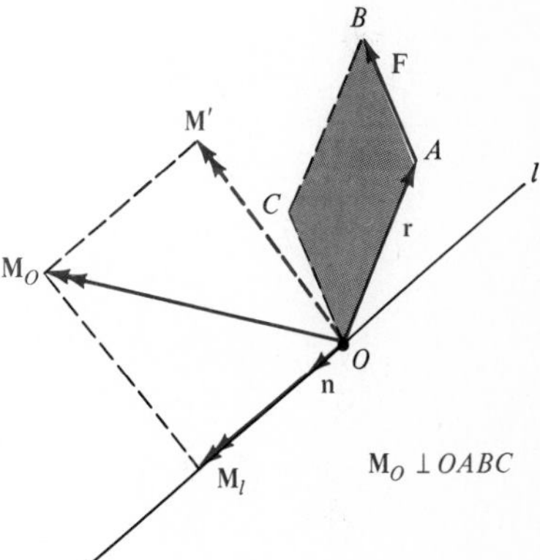

Sections 3-7, 3-8

MOMENT OF A COUPLE:

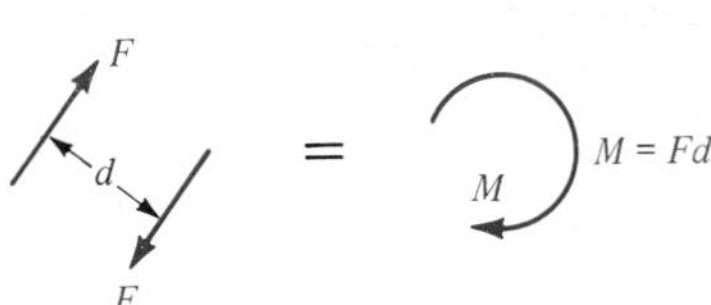

USEFUL PROPERTIES:

1. A couple can be moved anywhere in its own plane in the same rigid body.
2. A couple can be moved to any other parallel plane in the same rigid body.

Section 3-9

COMPONENTS AND ADDITION OF COUPLES:

1. Couples in the same plane or in parallel planes in a rigid body can be summed algebraically.
2. In general, couples are resolved into rectangular components for vector addition.
3. Any number of couples $\mathbf{M}_1, \mathbf{M}_2, \ldots$ can be reduced to a single resultant $\mathbf{M}$ with xyz components,

$$\mathbf{M}_1 + \mathbf{M}_2 + \cdots = \mathbf{M} = M_x\mathbf{i} + M_y\mathbf{j} + M_z\mathbf{k}$$

Section 3-10

FORCE–COUPLE TRANSFORMATIONS:

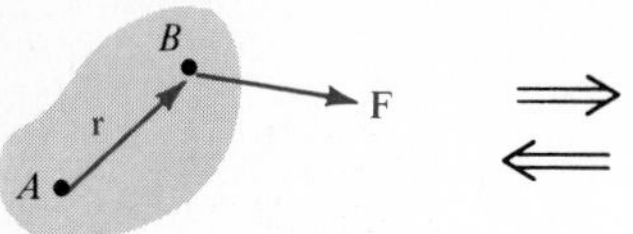

$\Rightarrow$
$\Leftarrow$

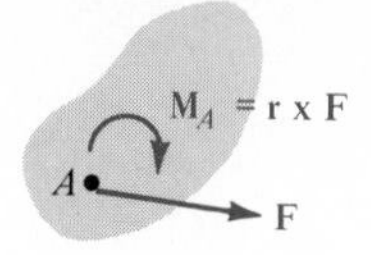

1. A force **F** acting at B can be replaced by the same force **F** acting at A and a moment $\mathbf{M}_A = \mathbf{r} \times \mathbf{F}$ about A.
2. A force **F** and moment $\mathbf{M}_A$ acting at A can be replaced by a force **F** acting at B for the same total effect with respect to A.

Section 3-11

REDUCTION OF COMPLEX FORCE SYSTEMS: Any system of forces acting on a rigid body can be reduced to a resultant force $\mathbf{F}_R$ and a resultant moment $\mathbf{M}_R$,

$$\mathbf{F}_R = \sum_{i=1}^{n} \mathbf{F}_i \qquad \mathbf{M}_R = \sum_{i=1}^{n} \mathbf{M}_i$$

EQUIVALENT FORCE SYSTEMS: A system of forces $\mathbf{F}_1, \mathbf{F}_2, \ldots, \mathbf{F}_i$ is equivalent to another system of forces $\mathbf{P}_1, \mathbf{P}_2, \ldots, \mathbf{P}_i$ if

$$\sum_{i=1}^{m} \mathbf{F}_i = \sum_{j=1}^{n} \mathbf{P}_i \qquad \text{(equivalence of forces)}$$

$$\sum_{i=1}^{m} (\mathbf{r}_i \times \mathbf{F}_i) = \sum_{j=1}^{n} (\mathbf{r}_j \times \mathbf{P}_j) \qquad \text{(equivalence of moments)}$$

Section 3-12

COMMON FORCE SYSTEMS:

1. Resultant force is zero, but there is a resultant moment,

$$\mathbf{F}_R = 0 \qquad \mathbf{M}_R \neq 0$$

2. Concurrent forces have a resultant force, but there is no resultant moment,

$$\mathbf{F}_R \neq 0 \qquad \mathbf{M}_R = 0$$

3. Coplanar forces may have a resultant force and a resultant moment about a chosen point,

$$\mathbf{F}_R \neq 0 \qquad \mathbf{M}_R \neq 0$$

$\mathbf{M}_R$ is perpendicular to the plane of the forces.

4. Parallel forces may have a resultant force and a resultant moment about a chosen point

$$\mathbf{F}_R \neq 0 \qquad \mathbf{M}_R \neq 0$$

$\mathbf{M}_R$ is perpendicular to $\mathbf{F}_R$.

REVIEW PROBLEMS

3–101 A garage door is modeled with three forces acting on it: $F = 70$ lb, $P = 200$ lb, and $W = 180$ lb. Determine the sum of the moments of F, P, and W about O if $OA = 2$ ft, $OB = 4$ ft, and $OC = 8$ ft.

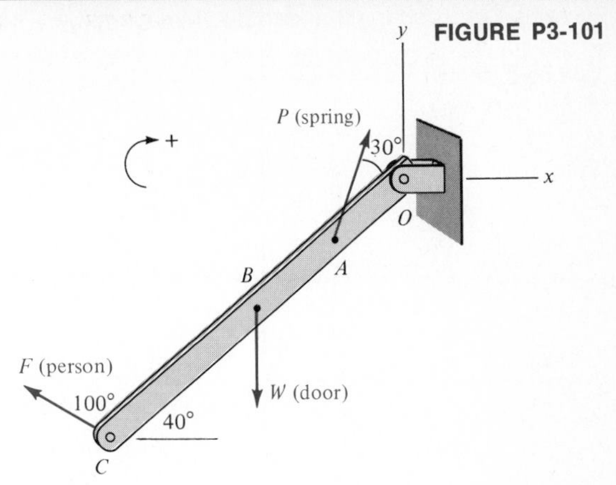

FIGURE P3-101

3–102 Three coplanar forces acting at the top of the pole are parallel to the xz plane. Calculate the moment of each force about O. Also determine the resultant F_R of the three forces, and the moment of F_R about O. Compare the sum of the three individual moments with the moment of the resultant force F_R.

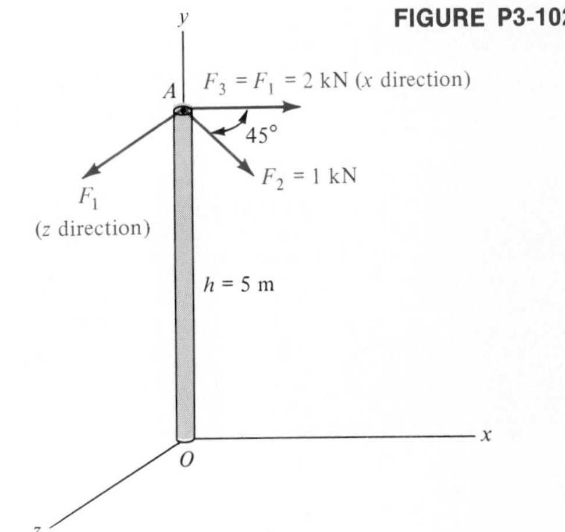

FIGURE P3-102

3–103 The shake table for testing electronic equipment has a hydraulic actuator A attached to it. Pivot B of the actuator and the lower pivots of the four supporting arms are attached to the same base plate. For a simple modeling of an arm, consider the arrangement in Fig. P3-103, where l and r are constant. Determine $\mathbf{M}_O$ of the force $\mathbf{F}$ for $\theta = 40°, 60°$, and $80°$. F is constant.

(a)

FIGURE P3-103

Shake table for shock and vibration testing of electronic equipment. (Photograph courtesy MTS Systems Corporation, Minneapolis, Minnesota.)

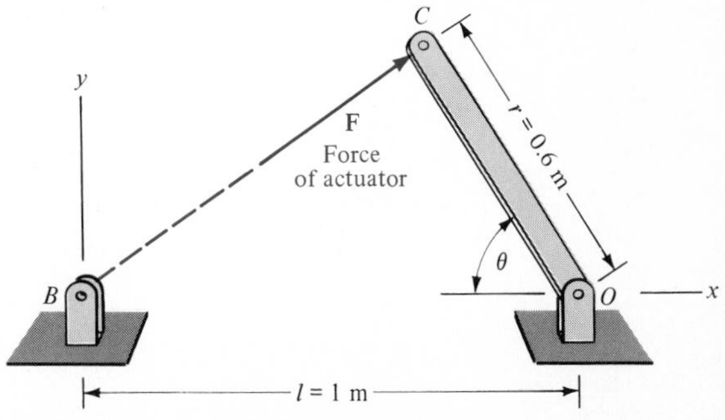

(b)

FIGURE P3-105

Test fixture attached to a car axle. (Courtesy MTS Systems Corporation, Minneapolis, Minnesota.)

3–104 For any angle θ in Prob. 3–103, prove that only the component of $\mathbf{F}$ perpendicular to r causes a moment $\mathbf{M}_O$.

3–105 The test fixture shown is attached to the right front wheel spindle A of a car. A has coordinates (5, −0.3, 3) ft with respect to point O at the center of the car. Three forces are applied to the spindle A: $\mathbf{F}_1 = 1500$ lb $\mathbf{j}$, $\mathbf{F}_2 = -700$ lb $\mathbf{i}$, and $\mathbf{F}_3 = -1000$ lb $\mathbf{k}$. Determine the moments M_x, M_y, and M_z of these simultaneous forces. x is the longitudinal centerline of the car.

3–106 Assume that in Prob. 3–105 each force can vary randomly from zero to the given value. Determine the largest possible moment M_x about the x axis of the car as the three forces vary randomly.

3–107 Represent the turning effect of force P at supports A and B by a couple of magnitude Fd. Calculate F.

FIGURE P3-107

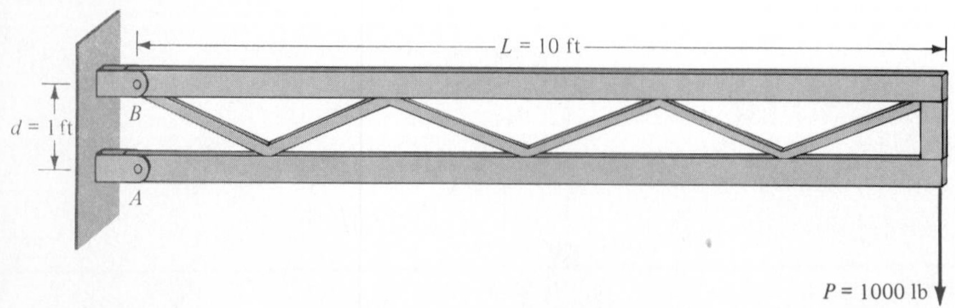

3–108 A force $\mathbf{F} = (300\mathbf{i} + 2000\mathbf{j} - 300\mathbf{k})$ lb is applied to the landing gear of the airplane. The coordinates of point A are $(1, -4, 0)$ ft. Determine the equivalent force and moment of $\mathbf{F}$ at point O.

FIGURE P3-108

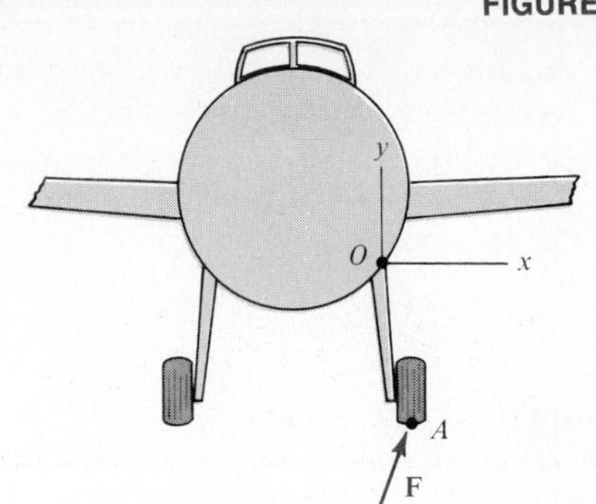

3–109 Three tensions in the wires of an antenna tower are $\mathbf{T}_1 = (-50\mathbf{i} - 60\mathbf{j} + 50\mathbf{k})$ kN, $\mathbf{T}_2 = (50\mathbf{i} - 55\mathbf{j} + 53\mathbf{k})$ kN, and $\mathbf{T}_3 = (-70\mathbf{j} - 65\mathbf{k})$ kN. Determine the equivalent force and moment at point O if $OA = 30$ m.

FIGURE P3-109

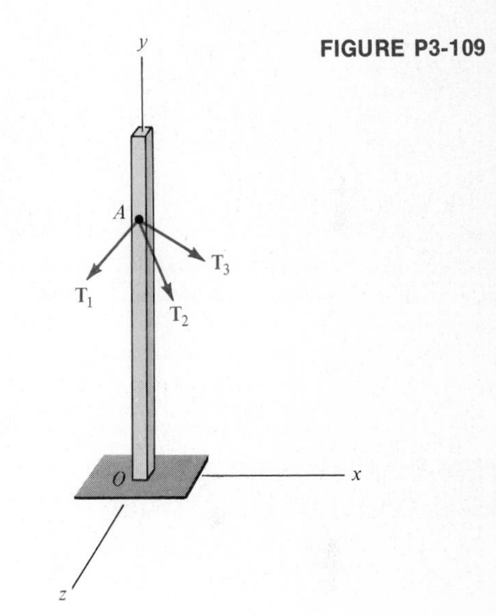

3–110 Consider the proposal for a perpetuum mobile sketched in Fig. P3-110. The wheel rotates on a horizontal axle. It is divided into eight identical chambers and has an identical ball in each chamber. As shown in the figure, there are more balls on the left side of the vertical centerline than on the right side, so the wheel tends to rotate counterclockwise. Note that the chamber dividers are not radial lines, a key element in the concept. Determine the horizontal moment arm for the resultant of the eight small weights with respect to the center of the wheel. Scale distances from the drawing.

FIGURE P3-110

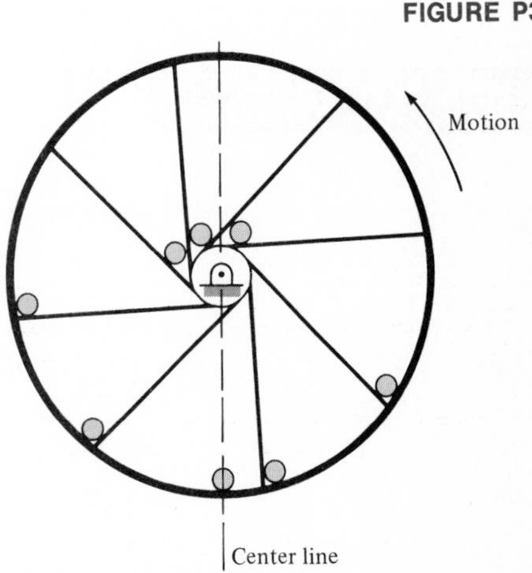

CHAPTER 4

EQUILIBRIUM OF RIGID BODIES

OVERVIEW

Advantages of Using Equilibrium Conditions

The concept of equilibrium is enormously useful in analyzing engineering problems. The reason is that equilibrium conditions are the basis for determining unknown forces and moments. The unknowns may be external or internal to a rigid body.

STUDY GOALS

SECTION 4-1 presents the most useful set of equations in statics. The equilibrium equations express the necessary and sufficient conditions for equilibrium of any rigid body. They are essential in mechanical design.

SECTION 4-2 describes the method of drawing proper free-body diagrams that are essential in the practical application of the equilibrium equations.

SECTION 4-3 introduces the application of equilibrium equations for determining unknown forces and moments. This method is extremely useful in engineering mechanics, and it should be practiced in as many different problems as possible. Such preparation is important for solving new problems.

SECTION 4-4 describes two-force members, which are the simplest of common structural members. The recognition of some components as two-force members in any mechanical system simplifies the analysis of that system.

SECTION 4-5 presents the method of analyzing connected rigid bodies. This method is based on Newton's Third Law (action and reaction). It is particularly useful for determining unknown forces that are internal to a rigid body or a system of bodies.

SECTION 4-6 gives several additional examples with a minimum of detail for the practice of drawing free-body diagrams and writing equilibrium equations. These are such important topics that the readers are urged to study every example and eventually make examples of their own.

COMPLETE EQUILIBRIUM CONDITIONS 4-1

The equilibrium equations that describe the state of balance for forces are valid for particles and for rigid bodies. In general, equilibrium means that no motion of the body occurs. For analyzing such a condition, it is convenient to state that a rigid body not in equilibrium *could* have six distinct motions relative to an arbitrary xyz reference frame. There could be linear displacements along the x, y, or z axes (this is called *translation*), and there could be rotations (angular displacements) about the x, y, or z axes. Any two or more (up to six) of these motions may occur simultaneously. Equilibrium means that none of these six motions are occurring.

For example, the helicopter in Fig. 4-1 may translate with respect to each of the xyz coordinates, and it may rotate with respect to each of these axes, as well. Complete equilibrium of the helicopter's fuselage means that it hovers in place without any rotation (of course, rotors and internal machinery rotate). Such a situation is possible if the external forces have neither a resultant force nor a resultant moment on the body.

The formal statement of equilibrium is that *a rigid body is in equilibrium when the equivalent force–couple system of the external forces acting on it is zero.* The equations that express the necessary

FIGURE 4-1

A helicopter may have six distinct motions in the *xyz* frame. (Courtesy MTS Systems Corporation, Minneapolis, Minnesota.)

and sufficient conditions for equilibrium of any rigid body are

$$\sum \mathbf{F} = 0 \quad \text{and} \quad \sum \mathbf{M}_O = \sum (\mathbf{r} \times \mathbf{F}) = 0 \tag{4-1}$$

where O is an arbitrary point of reference.

In practice it is most convenient to write Eq. 4-1 in terms of rectangular components,

$$\begin{aligned} \sum F_x &= 0 & \sum M_x &= 0 \\ \sum F_y &= 0 & \sum M_y &= 0 \\ \sum F_z &= 0 & \sum M_z &= 0 \end{aligned} \tag{4-2}$$

These scalar equilibrium equations are easy to use, but occasionally it is necessary to be cautious and judiciously check the results as shown in Ex. 4-1.

Useful Notes Concerning the Equilibrium Equations

Problems concerning equilibrium of rigid bodies may involve one, two, or three dimensions. In each case it is necessary to start the solution of the problem by drawing the free-body diagram. The equilibrium equations are written for each rigid body as follows.

1. ***One-dimensional problem.*** Only one force summation is necessary using Eq. 4-1 or 4-2,

$$\sum \mathbf{F} = 0 \quad \text{or} \quad \sum F = 0$$

It makes no difference what subscript (x, y, or z) is used for the force F.

2. ***Two-dimensional problem.*** At most two force summations and one moment summation can be used from Eq. 4-2. If axes x and y are chosen for the forces, the z axis is the one for moments,

$$\sum F_x = 0 \qquad \sum F_y = 0 \qquad \sum M_z = 0$$

The other two possible combinations can be made using the same pattern. Note that taking moments about any point A in the xy plane is equivalent to taking moments about the z axis through point A.

3. ***Three-dimensional problem.*** At most all six of Eq. 4-2 can be used. The moments must be taken about the appropriate *axes* rather than about a single, arbitrary point A.

The details of these methods are exemplified in subsequent sections.

FREE-BODY DIAGRAMS

The equilibrium equations are necessary to determine unknown forces and moments. The first step in this process is the drawing of a proper *free-body diagram* as was done in Ch. 2. Such a diagram is for a mechanical system (a body or group of bodies, rigid or nonrigid) that is imagined to be isolated from all other bodies. All significant effects caused by the other bodies *on* the free body must be shown in the free-body diagram. Internal forces and moments are ignored because they occur in pairs and cancel one another. *Unknown forces and moments* must be shown at the appropriate places on the diagram. The directions of the unknowns may be assumed arbitrarily, but it is best to show them in the correct directions if these are known.

Common Forces and Constraints

Drawing free-body diagrams always involves making simplifying assumptions. These are facilitated by knowledge of the common characteristics of mechanical systems. The following list and the schematic drawings in Fig. 4-2 are representative of frequently encountered two-dimensional problems. The basic ideas can be extended to three-dimensional situations.

1. The *weight* of the body is often included in the diagram. Its direction is always toward the center of the earth. The weight can be neglected only if it is much smaller than the other forces acting on the body.

2. *Forces with known lines of action* are common in several special cases for bodies in contact.

- **a.** *Cables, chains, belts, or ropes* are flexible, so they can have only axial tension in them (Fig. 4-2a).
- **b.** *Two-force members* are rigid members with all external forces applied at two separate, freely hinged pin connections. For such a member the net force at each connection has a line of action that goes through both connections (Fig. 4-2b). This is proved in Sec. 4-4.
- **c.** *Rolling contacts* allow motion parallel to the surfaces in contact, so there is only one force between the bodies and it is perpendicular to these surfaces (Fig. 4-2c). The force is called a *normal force*.
- **d.** Some sliding members have very low friction, so these tend to have only normal forces as rolling members do (Fig. 4-2d).

3. *Pin connections* allow no translation of a body, but rotation about the axis of the pin is possible. In a freely hinged connection there is no moment acting between the members. The force at the

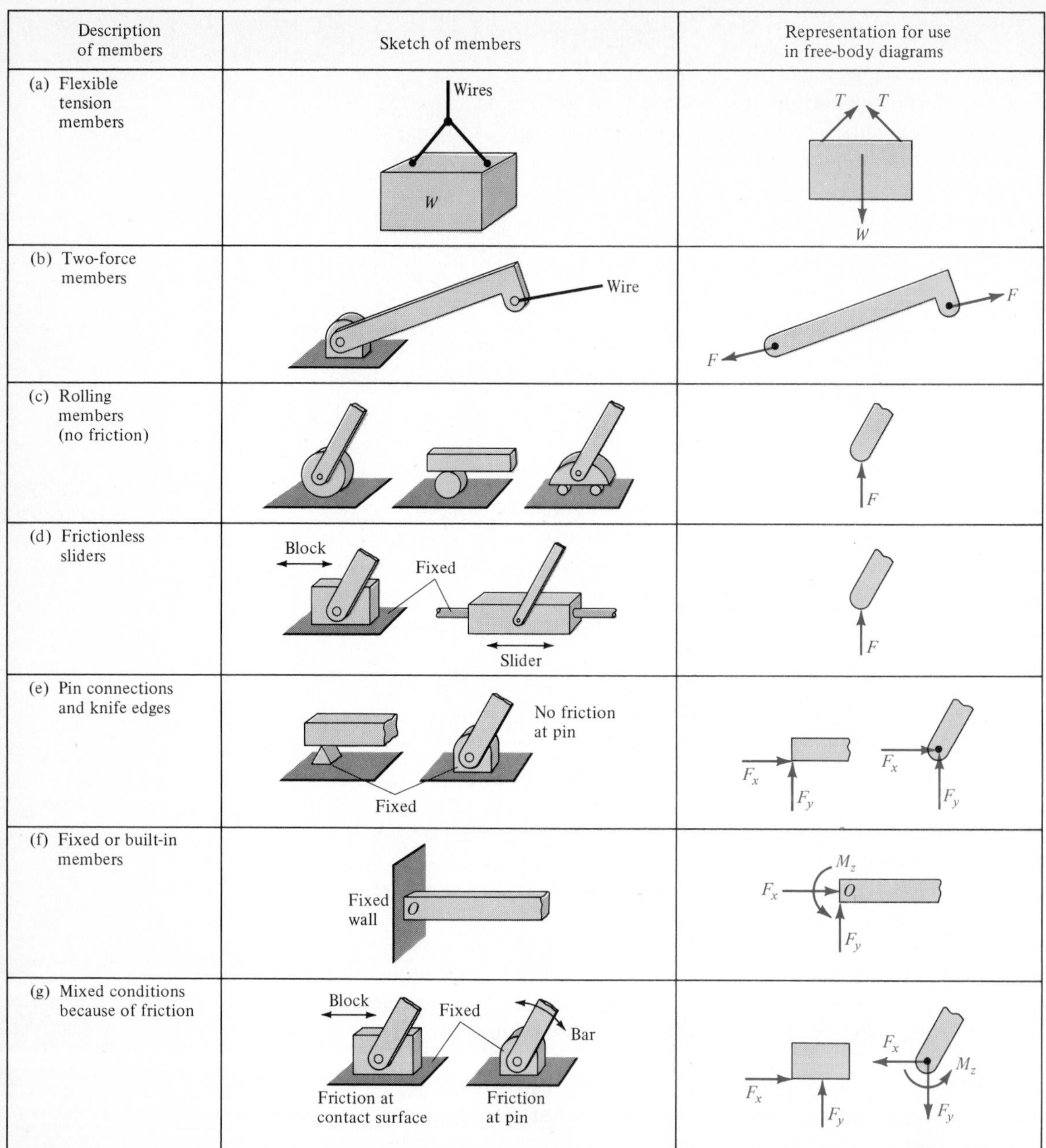

Description of members	Sketch of members	Representation for use in free-body diagrams
(a) Flexible tension members	Wires; W	T; T; W
(b) Two-force members	Wire	F; F
(c) Rolling members (no friction)		F
(d) Frictionless sliders	Block; Fixed; Slider	F
(e) Pin connections and knife edges	No friction at pin; Fixed	F_x; F_y; F_x; F_y
(f) Fixed or built-in members	Fixed wall; O	M_z; F_x; O; F_y
(g) Mixed conditions because of friction	Block; Fixed; Bar; Friction at contact surface; Friction at pin	F_x; F_y; F_x; M_z; F_y

FIGURE 4-2

Common forces, supports, and their representations in free-body diagrams

pin has two components in general (Fig. 4-2e) when the problem involves two-dimensional loading.

4. *Fixed or built-in members* are restrained from translation and rotation. There may be up to three components of the force and three components of the moment at the connection. The two-dimensional case is illustrated in Fig. 4-2f.

5. *Mixed conditions* are quite common. For example, a member may slide with friction opposing the motion; here there are normal and tangential forces at the point of contact. A pin connection with friction has a force and a moment acting at the connection (Fig. 4-2g).

The constraining forces or moments acting on a body are often called *reactions* or *supports*. In the majority of the problems these are unknown and must be determined analytically.

4-3 CALCULATION OF UNKNOWN FORCES AND MOMENTS

The number of unknown quantities that can be determined in a problem is the same as the number of independent equilibrium equations that can be written using the available information. The student should realize that any number of equilibrium equations can be written for a given problem, but only a few of them are independent, and it is only fruitful to work with these. However, there are many different sets of independent equations, and each can be used to obtain a solution. This is illustrated for the beam in Fig. 4-3a.

The beam is called *simply supported* at B because the roller can apply only a normal-force reaction R_B as shown in the free-body diagram in Fig. 4-3b. The pin at A could apply both a horizontal

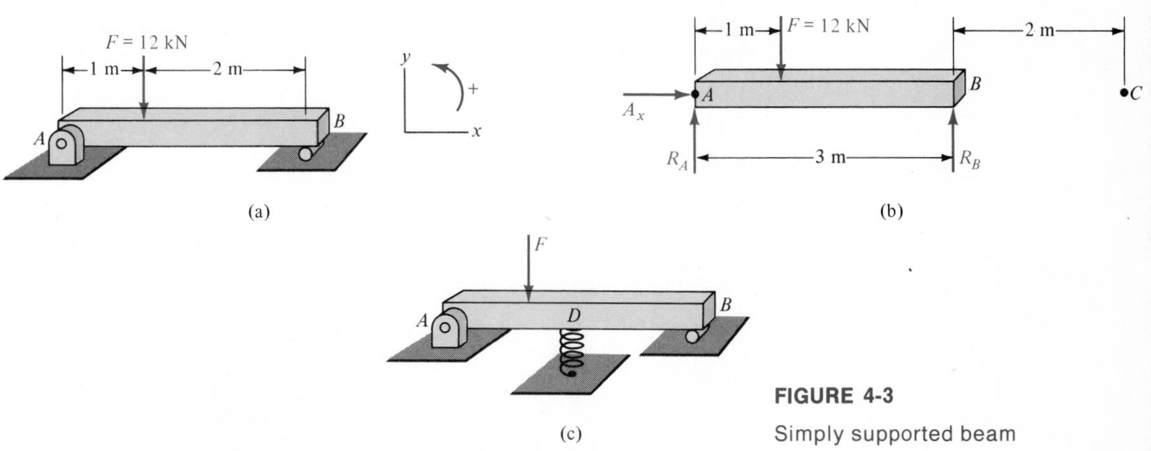

FIGURE 4-3
Simply supported beam

force A_x and a vertical force R_A. However, from Fig. 4-3b, equilibrium in the x direction is possible only if

$$\sum F_x = A_x = 0$$

Thus, there are two unknown forces, R_A and R_B, and two equations are needed to solve for them. Two independent equations are obtained from Eq. 4-2,

$$\sum F_y = 0 \qquad \text{and} \qquad \sum M_A = 0 \qquad \boxed{4\text{-}3}$$

$$R_A + R_B - 12\text{ kN} = 0 \qquad (3\text{ m})R_B - (1\text{ m})(12\text{ kN}) = 0$$

From these, $R_B = 4$ kN, $R_A = 8$ kN.

An equally useful set of equations is

$$\sum F_y = 0 \qquad \text{and} \qquad \sum M_B = 0 \qquad \boxed{4\text{-}4}$$

from which the same results are obtained.

An equally valid but slightly less convenient set of equations is

$$\sum F_y = 0 \qquad \text{and} \qquad \sum M_C = 0 \qquad \boxed{4\text{-}5}$$

because the moment equation also has two unknowns. Generally, equations with one unknown in each are the best to use, so taking moments about a point where an unknown force acts is usually advantageous.

The two supports shown in Fig. 4-3a are the minimum number of supports with which this beam can be in equilibrium. The beam would fall down if either R_A or R_B were removed. Members in such situations, with the minimum number of constraints for equilibrium, are called *statically determinate.*

Any extra supports beyond the minimum are *redundant* in theory; they could be removed without changing the state of equilibrium. It may not be reasonable to remove redundant supports in practice since they may be necessary for extra stiffness or safety of the structure. The solution of problems is more difficult when there are redundant supports. For example, consider the addition of a spring support to the beam at midspan as in Fig. 4-3c. Three independent equations are required to solve for the three reactions. One equation is the force summation

$$\sum F_y = R_A + R_B + R_D - 12\text{ kN} = 0$$

No other force summation can be written. Then, it may be tempting to write two moment summations because they look different if two reference points are used. For example, write

$$\sum M_A = (3\text{ m})R_B + (1.5\text{ m})R_D - (1\text{ m})F = 0$$

and

$$\sum M_B = -(3\text{ m})R_A + (2\text{ m})F - (1.5\text{ m})R_D = 0$$

The equations

$$\sum F_y = 0 \qquad \sum M_A = 0 \qquad \sum M_B = 0 \tag{4-6}$$

are valid but cannot be solved for the three unknown forces. The reason is that the two moment equations are not independent (the equivalent couple of the given forces can be moved anywhere in the same plane).

Unfortunately, the same is true for any other set of correct equations. Such members with any redundant supports are called *statically indeterminate*. They can be solved only with the methods of statics and mechanics of materials together.

Another configuration that is common is the built-in member. The simplest built-in member is the cantilever beam (Fig. 4-4a), which is statically determinate. The free-body diagram for this beam is drawn on the basis that the external force F tends to cause a translation and a rotation at the support A. These effects of the force F are shown by the equivalent force–couple system at point A in Fig. 4-4b. The wall at A resists the applied force and couple with an equal and opposite force and couple if there is equilibrium. The complete free-body diagram is shown in Fig. 4-4c. Here the reactions V and M_O are considered as an unknown, external force–couple system to the beam. Their directions may be drawn correctly in the free-body diagram, knowing that they oppose the applied force–couple system shown in Fig. 4-4b. The equations of equilibrium are written for Fig. 4-4c,

$$\begin{aligned} &\sum F_y = 0 && \sum M_A = 0 \\ &V - F = 0 && M_O - LF = 0 \\ &V = 3000 \text{ lb} && M_O = 15{,}000 \text{ ft}\cdot\text{lb} \end{aligned} \tag{4-7}$$

Note that changing the directions of M_A and V while they are considered as unknowns is permissible. Applying the given sign convention to Fig. 4-4d, the equilibrium equations become

$$\begin{aligned} &-V - F = 0 && -M_O - LF = 0 \\ &V = -3000 \text{ lb} && M_O = -15{,}000 \text{ ft}\cdot\text{lb} \end{aligned} \tag{4-8}$$

The negative signs indicate that the directions of the unknown force and couple were assumed incorrectly. This is such a simple matter that it is not reasonable to spend time in figuring out the correct directions before the calculations are made.

The methods presented here are valid for three-dimensional problems. In that case up to six equations are available (Eq. 4-2), whereas in a two-dimensional problem at most three equations can be used (two for force summations and one for moment summation).

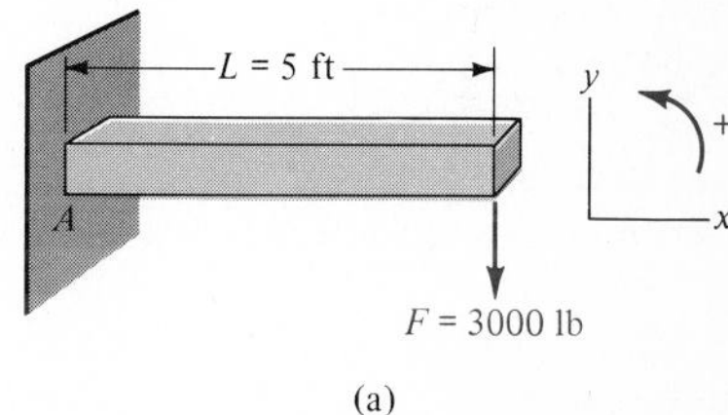

(a)

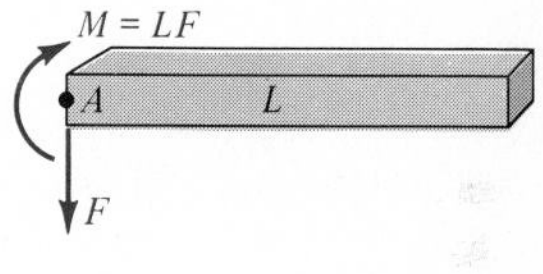

(b)

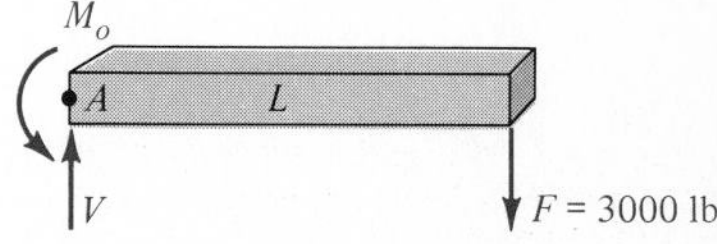

(c) Free-body diagram

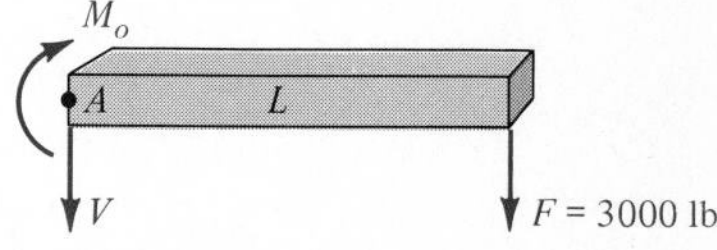

(d) Free-body diagram with unknowns reversed with respect to (c)

FIGURE 4-4
Cantilever beam

EXAMPLE 4-1

The selection of a reference point for taking moments is arbitrary when there is equilibrium, but occasionally a mistake in another part of the analysis may cause problems with the moments. For a simple example, imagine an attempt to analyze how the moment reaction is created at the supporting wall of a cantilever beam. Somebody proposes that an internal force F caused by the wall resists the moment of the applied load W as in Fig. 4-5a. Thus, taking moments about O,

$$F = \frac{WL}{d}$$

Is this the correct view of the situation?

SOLUTION

It is not entirely correct. For one thing, the force F is the only force in the horizontal direction, so the beam could not be in equilibrium in that direction. For another, there is no equilibrium with respect to moments, either. This is more subtle than the matter of force summation since it appears correct to take moments about O. However, it is also correct to take moments about any other point, for example, point A in Fig. 4-5a. Now, it is clear that both F and W cause clockwise moments, so the beam could not be in equilibrium.

The proper construction of the free-body diagram could be initiated as in Fig. 4-5a because the moment of W must indeed have a moment reaction. But it is necessary to go further and make sure that equilibrium is attained in all respects. Figure 4-5b is a reasonable free-body diagram for the cantilever beam.

JUDGMENT OF THE RESULTS

Figure 4-5b is acceptable because the force summations are zero (by inspection), and taking moments about *any* point gives the same result. Note that taking moments about two different points is often enough for checking the result, but it does not necessarily reveal a mistake. For example, taking moments about points B and O in Fig. 4-5a are equally inadequate.

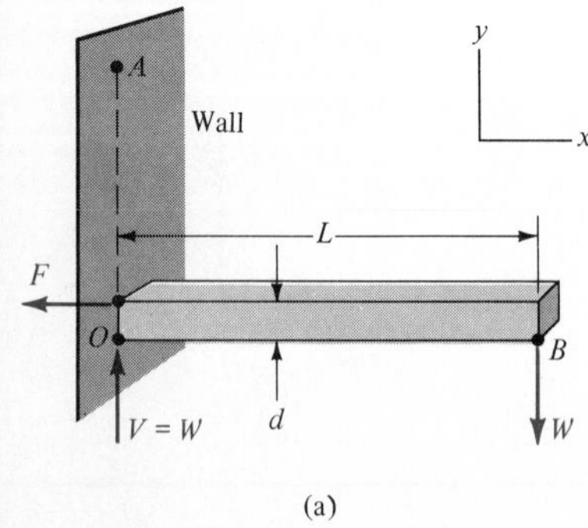

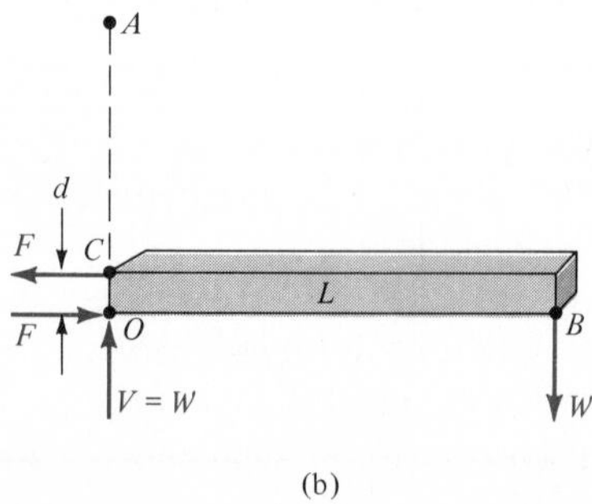

FIGURE 4-5

Analysis of the supporting reactions of a cantilever beam

EXAMPLE 4-2

Consider a pulley (Fig. 4-6) which has a frictionless axle at its center O. Prove that the magnitude of the tension in a rope or wire is the same on the two sides of the pulley. Assume that tension T is known.

SOLUTION

Figure 4-6 is the free-body diagram of the pulley. The required proof is obtained from the equilibrium equations. From Eq. 4-2, $\sum F_x = 0, \sum F_y = 0$, and $\sum M_O = 0$ are applicable. Note that the unknown reaction **R** at point O must be included in the analysis. Thus, there are three unknowns, T_1, R_x, and R_y, and three independent equations,

$$\sum F_x = 0 \qquad \sum F_y = 0 \qquad \sum M_O = 0$$

$$-T + T_1 \sin\theta + R_x = 0 \qquad R_y - T_1 \cos\theta = 0 \qquad Tr - T_1 r = 0$$

$$R_x = T(1 - \sin\theta) \qquad R_y = T\cos\theta \qquad T_1 = T$$

It is remarkable that the tension T_1 is independent of the angle θ but the reaction **R** depends on θ.

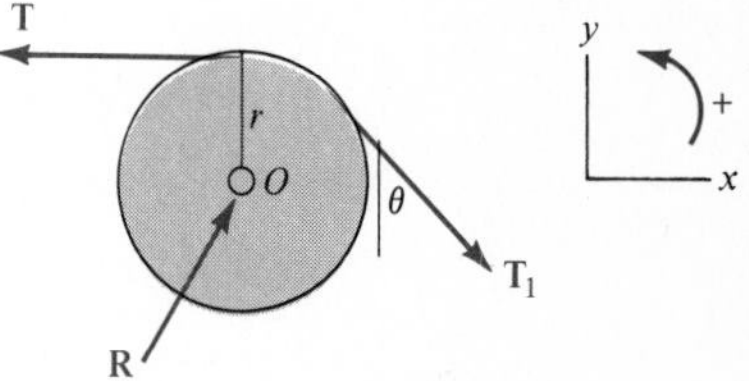

FIGURE 4-6

Equilibrium of a pulley

EXAMPLE 4-3

Susie weighs 120 lb, and when she walks on her stilts (Fig. 4-7a) the whole weight is periodically applied to a triangular peg. Each peg is attached to the pole by a nail set 3 in. above corner B of the peg. Determine the forces F_A and V in the nail and F_B on the peg, using the free-body diagram in Fig. 4-7b for the peg. Assume that the peg is touching the stilt leg only at points A and B and that force F_B must be horizontal (this implies that the peg is free to slide at B).

SOLUTION

Since the free-body diagram is given, the applicable equations can be written at once from Eq. 4-2,

$$\sum F_x = 0$$

$$-F_A + F_B = 0 \qquad \textbf{(a)}$$

$$\sum F_y = 0$$

$$V - 120 \text{ lb} = 0 \qquad \textbf{(b)}$$

$$\sum M_A = 0$$

$$-(120 \text{ lb})(2 \text{ in.}) + F_B(3 \text{ in.}) = 0 \qquad \textbf{(c)}$$

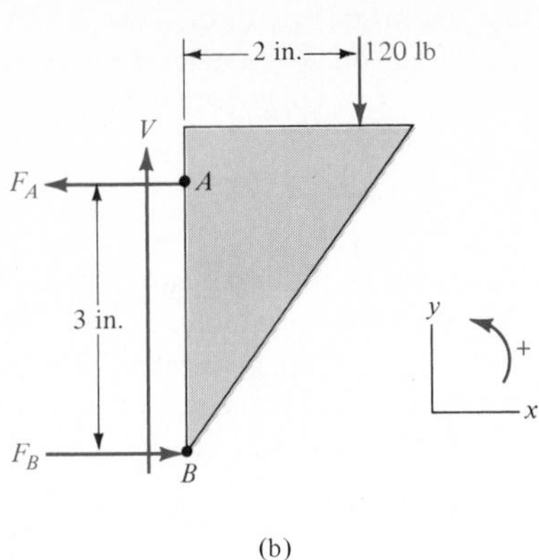

FIGURE 4-7
Analysis of a supporting peg on a stilt (a) (b)

Both Eqs. (b) and (c) can be solved directly, giving

$$V = 120 \text{ lb}$$

$$F_B = 80 \text{ lb}$$

From Eq. (a),

$$F_A = F_B = 80 \text{ lb}$$

The positive answers for V, F_A, and F_B indicate that their directions were correct as assumed in Fig. 4-7b.

JUDGMENT OF THE RESULTS

Since forces F_A and F_B have a larger distance between them than that between the vertical forces, the 80-lb force on the nail is reasonable. This also shows that increasing the vertical dimension of the peg may improve the reliability of the stilts by decreasing the force F_A.

EXAMPLE 4-4

Consider the model of a crane in Fig. 4-8a. The weights of the column and boom are neglected. Assume that the reactions from the column to the boom are entirely horizontal at points A and B. Determine the forces F_A and F_B at A and B, respectively. Also calculate the tension T in the cable which passes over a frictionless pulley at the top of the crane.

SOLUTION

First draw the free-body diagram in Fig. 4-8b. The applicable equilibrium equations are written from Eq. 4-2,

$$\sum F_x = 0$$

$$T\cos 30^\circ - T\cos 30^\circ + F_A - F_B = 0 \qquad \textbf{(a)}$$

$$\sum F_y = 0$$

$$2T\sin 30^\circ - 1\text{ MN} = 0 \qquad \textbf{(b)}$$

$$\sum M_A = 0$$

$$(1\text{ MN})(16\text{ m}) - T\sin 30^\circ(8.5\text{ m}) + T\sin 30^\circ(6.5\text{ m}) - F_B(1.5\text{ m}) = 0 \qquad \textbf{(c)}$$

Equation (b) can be solved first, giving

$$T = 1\text{ MN}$$

From Eq. (c),

$$F_B = 10\text{ MN}$$

From Eq. (a),

$$F_A = F_B = 10\text{ MN}$$

JUDGMENT OF THE RESULTS

The moment of the load about the pulley is 15 MN·m. This is opposed by the couple $F_A(1.5\text{ m}) = 15\text{ MN·m}$, so the results appear to be correct.

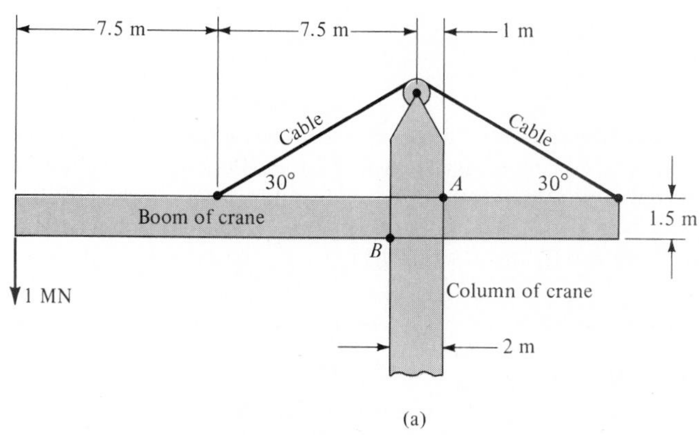

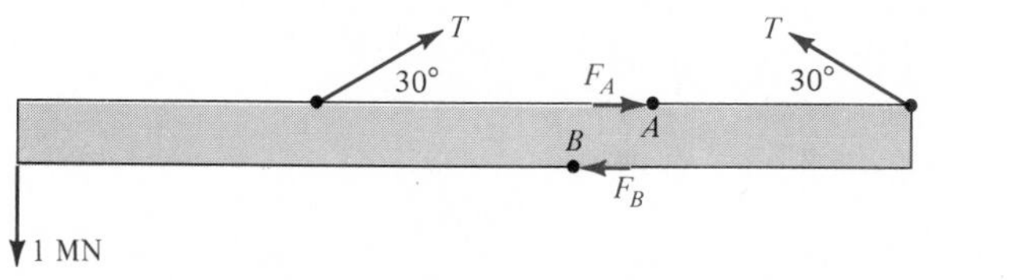

FIGURE 4-8
Model of a crane

EXAMPLE 4-5

Some street-light fixtures are poorly designed. Such a fixture is shown and modeled in Fig. 4-9a and b, respectively. Find the tension T in the cable BC and the reactions at point A in order to evaluate the fixture. Assume that B is 0.75 ft above A, and C is 1 ft above B. BC makes an angle 12.5° with the horizontal.

SOLUTION

From the free-body diagram of the bar and cable together in Fig. 4-9b, and using Eq. 4-2,

$$\sum F_x = 0$$

$$A_x - T\cos 12.5^\circ = 0 \quad \textbf{(a)}$$

$$\sum F_y = 0$$

$$A_y - T\sin 12.5^\circ - 75\text{ lb} = 0 \quad \textbf{(b)}$$

$$\sum M_A = 0$$

$$(T\cos 12.5^\circ)(0.75\text{ ft}) - (75\text{ lb})(5\text{ ft}) = 0 \quad \textbf{(c)}$$

Equation (c) can be solved first, giving

$$T = 512\text{ lb}$$

From Eq. (a),

$$A_x = 500\text{ lb}$$

From Eq. (b),

$$A_y = 185\text{ lb}$$

JUDGMENT OF THE RESULTS

The magnitudes of the calculated forces appear to be correct since the forces and the moment arms are inversely related by inspection. Generally, the longer the distance BC, the larger the tension in the cable and the reaction at A. The design is poor because the cable is pulling *down* on the light in this arrangement, contrary to the purpose of having a cable. The result is an unnecessarily large compressive force (which could cause buckling) on the tube AC.

FIGURE 4-9

Street-light fixture

(a)

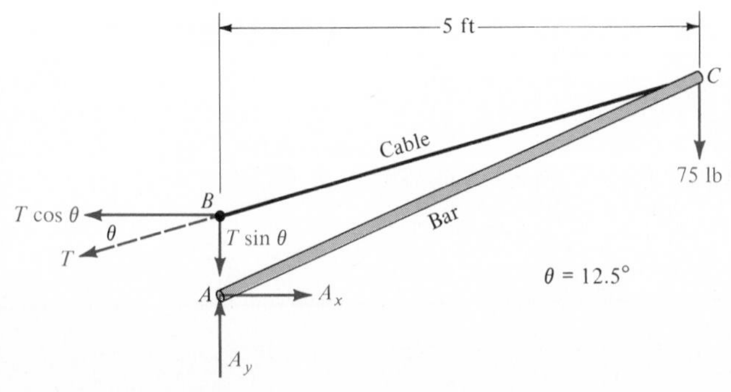

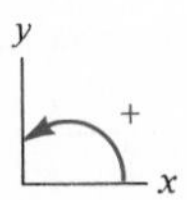

(b)

EXAMPLE 4-6

The bracket *ABC* in Fig. 4-10 is part of a road simulator for automobiles (see Sec. 1-7 for a description of these). It is desired to apply a force F_C from a hydraulic actuator to hold the bracket when a force $P = 6$ kN is applied to it as shown in Fig. 4-10a. (a) Determine the force F_C and the reactions A_x and A_y at the frictionless pin *A*. (b) Where should the actuator be attached if it could be anywhere between *C* and *E* in Fig. 4-10b?

SOLUTION

(a) Figure 4-10a is the free-body diagram of the bracket. The necessary equilibrium equations are written from Eq. 4-2,

$$\sum F_x = 0$$

$$A_x - 6\text{ kN} + \frac{4}{5} F_C = 0 \qquad \textbf{(a)}$$

$$\sum F_y = 0$$

$$A_y - \frac{3}{5} F_C = 0 \qquad \textbf{(b)}$$

$$\sum M_A = 0$$

$$(6\text{ kN})(0.7\text{ m}) - F_C(0.5\text{ m}) = 0 \qquad \textbf{(c)}$$

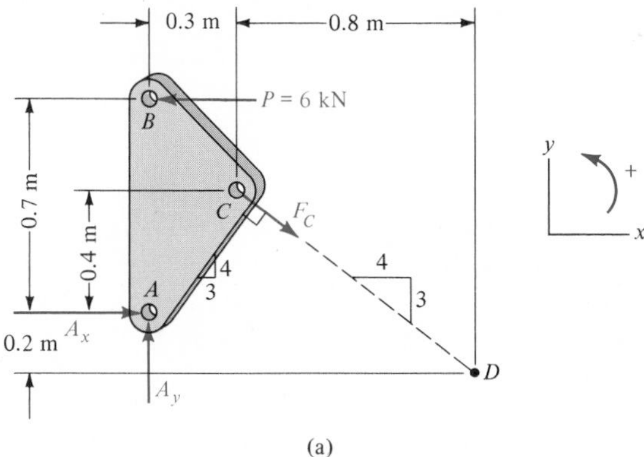

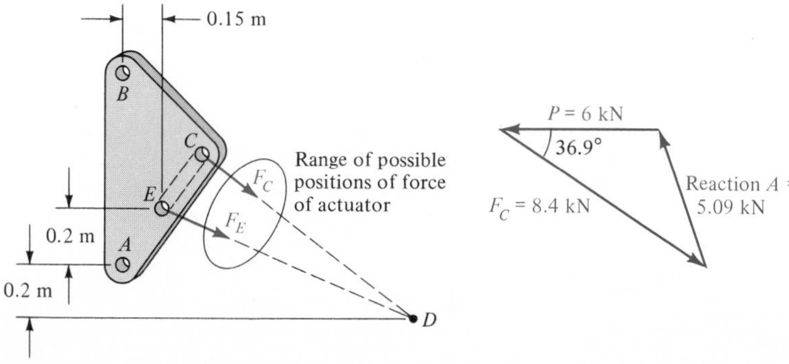

FIGURE 4-10

Bracket of a laboratory testing mechanism

Equation (c) can be solved first, giving

$$F_C = 8.4 \text{ kN}$$

From Eq. (a),

$$A_x = -0.72 \text{ kN}$$

The negative sign indicates that A_x is actually to the left in Fig. 4-10a. From Eq. (b),

$$A_y = 5.04 \text{ kN}$$

(b) For the possibility of changing the attachment of the actuator, it is enough to calculate the required force F_E in the extreme position at E. Here the actuator is not perpendicular to the line AC, so it is easiest to use the vector method of summing moments. From Eq. 4-1,

$$\sum \mathbf{F} = 0 \qquad \sum \mathbf{M}_A = \mathbf{r} \times \mathbf{F} = 0$$

Here it is sufficient to solve the moment equation, which has only one unknown, $\mathbf{F}_E$. The force summation should be checked by the readers for an exercise.

First the unknown force $\mathbf{F}_E$ is expressed in vector form using Eq. 2-23,

$$\mathbf{F}_E = F\mathbf{n}_{ED} = F\left(\frac{\overrightarrow{ED}}{|ED|}\right) = F\left(\frac{0.95\mathbf{i} - 0.4\mathbf{j}}{1.031}\right)$$

$$= 0.92F\mathbf{i} - 0.388F\mathbf{j}$$

The moment summation is written as

$$\sum \mathbf{r} \times \mathbf{F} = \mathbf{r}_p \times \mathbf{P} + \mathbf{r}_E \times \mathbf{F}_E = 0$$

$$(0.7 \text{ m } \mathbf{j}) \times (-6 \text{ kN } \mathbf{i}) + (0.15 \text{ m } \mathbf{i} + 0.2 \text{ m } \mathbf{j}) \times (0.92\mathbf{i} - 0.388\mathbf{j})F_E = 0$$

from which

$$F_E = \frac{4.2 \text{ kN}\cdot\text{m}}{0.242 \text{ m}} = 17.34 \text{ kN}$$

Comparing forces F_C and F_E, it is clear that attaching the actuator at C is the most efficient.

JUDGMENT OF THE RESULTS

Forces F_C and F_E appear correct in comparison with force P by inspection. The reaction $\mathbf{A} = A_x\mathbf{i} + A_y\mathbf{j}$ can be determined from the force triangle in Fig. 4-10c and from $A = \sqrt{A_x^2 + A_y^2}$. $A = 5.09$ kN using both methods, so A_x and A_y appear correct.

PROBLEMS

4–1 Assume that tension T_1 and angles θ and ϕ are given for the pulley. Determine the tension T_2 and the reactions R_x and R_y at the center C.

FIGURE P4-1

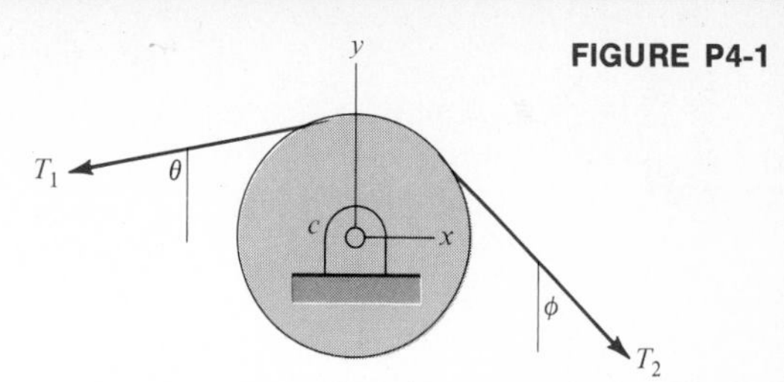

4–2 In Fig. P4-1, $T_1 = 5$ kN, $\theta = 70°$, and $\phi = 0$. Calculate the tension T_2 and the reactions on the pulley at point C.

4–3 The largest allowable vertical reaction R_y at center C of the pulley in Fig. P4-1 is 1000 lb. Determine the maximum tension in the belt if $\theta = 60°$ and $\phi = 40°$.

FIGURE P4-4

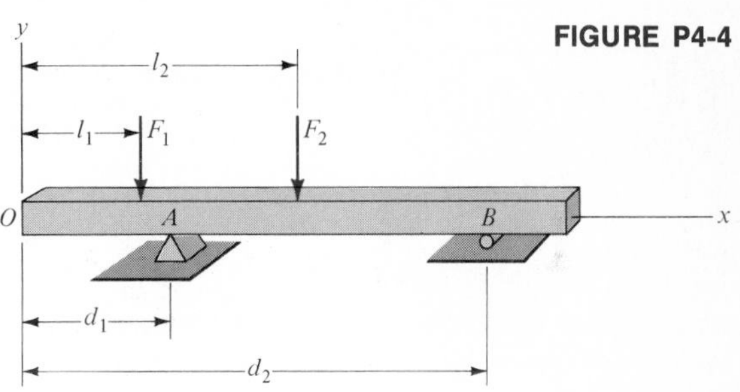

4–4 The weightless beam is loaded by forces $F_1 = 2$ kN at $l_1 = 0.5$ m and $F_2 = 5$ kN at $l_2 = 2$ m. Determine the reactions at A and B if $d_1 = 0$ and $d_2 = 3$ m.

4–5 Force F_2 on the beam is 2000 lb at $l_2 = 5$ ft in Fig. P4-4. The supports are at $d_1 = 2$ ft and $d_2 = 7$ ft. Determine the force F_1 at $l_1 = 1$ ft that would make the vertical reactions at A and B equal.

FIGURE P4-6

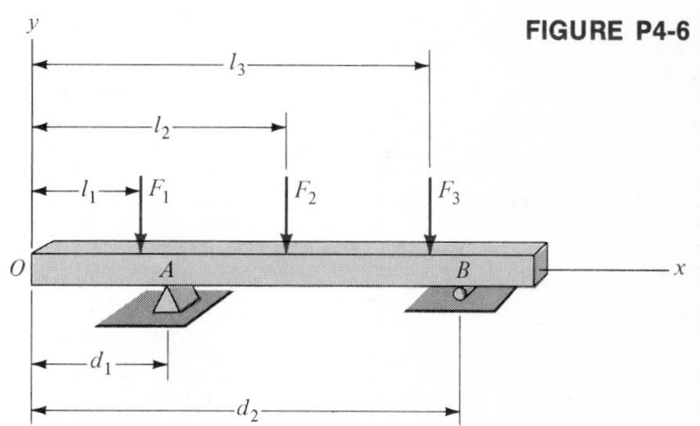

4–6 The downward forces on the beam are $F_1 = 1$ kN at $l_1 = d_1 = 0.4$ m, $F_2 = 3$ kN at $l_2 = 1$ m, and $F_3 = 2$ kN at $l_3 = 2$ m. Calculate the reactions at A and B if $d_2 = 2.5$ m.

4–7 Assume that the largest allowable reactions $R_A = R_B = 5000$ lb on the beam in Fig. P4-6 are specified, and the dimensions are given as $d_1 = 0$, $d_2 = 10$ ft, $l_1 = 2$ ft, $l_2 = 5$ ft, and $l_3 = 9$ ft. Determine F_1, F_2, and F_3. If this is not possible, state the reason for the difficulty and determine the minimum additional information for solving the problem.

FIGURE P4-8

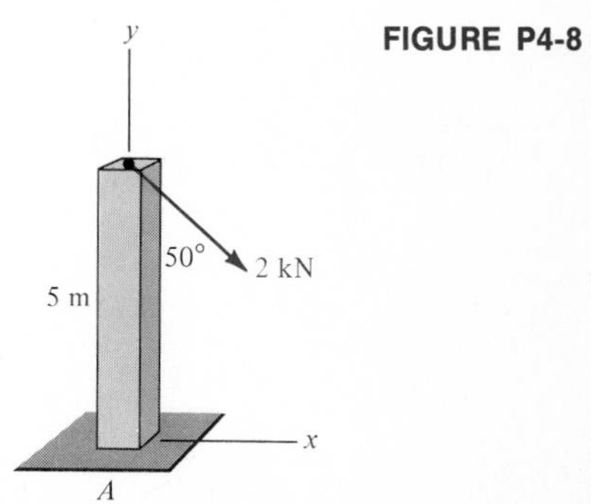

4–8 Calculate the reactions at point A of the pole.

FIGURE P4-9

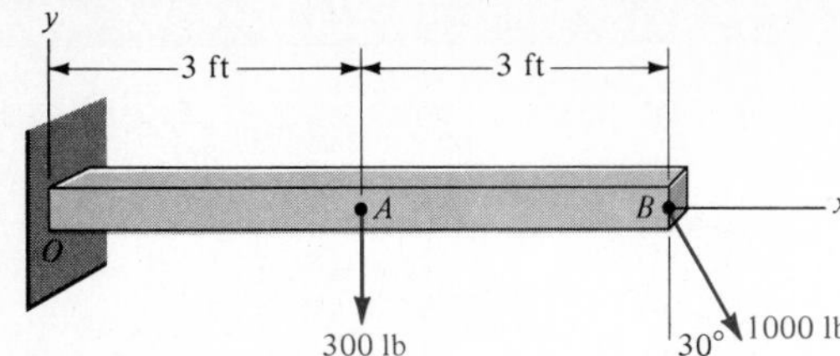

4–9 Determine the supporting reactions of the cantilever beam.

FIGURE P4-10

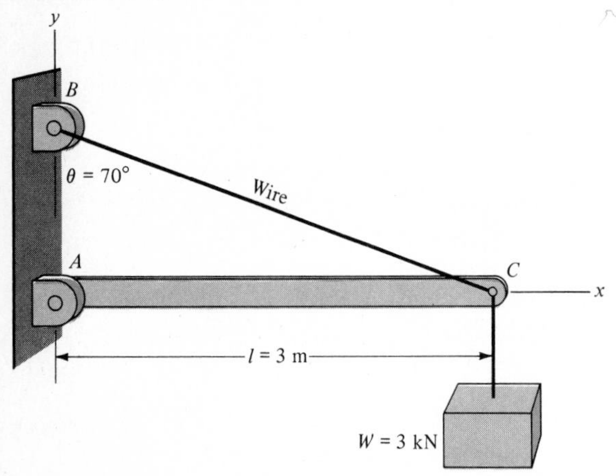

4–10 Calculate the reactions at hinge A of the simple crane, assuming that the tension in the wire must be along line BC.

4–11 Solve Prob. 4–10 using $W = 1000$ lb, $l = 8$ ft, and $\theta = 45°$, $60°$, and $75°$. Plot A_x and A_y on the same diagram as a function of θ. The wire is always attached to the end of the beam.

FIGURE P4-12

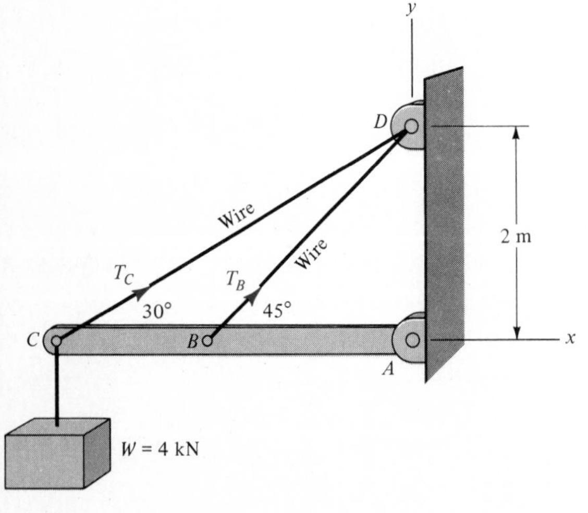

4–12 Prove that the beam which is hinged at A is statically indeterminate. The tension in each wire is along the wire. Consider the unknown forces T_B, T_C, A_x, and A_y.

4–13 Assume that the tension in wire CD in Fig. P4-12 is measured with an instrument and found to be $T_C = 5$ kN. Calculate the tension T_B and the reactions A_x and A_y.

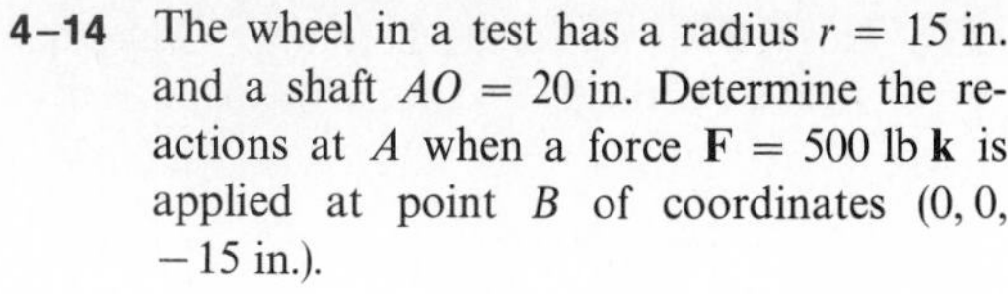

4–14 The wheel in a test has a radius $r = 15$ in. and a shaft $AO = 20$ in. Determine the reactions at A when a force $\mathbf{F} = 500 \text{ lb } \mathbf{k}$ is applied at point B of coordinates (0, 0, −15 in.).

FIGURE P4-14

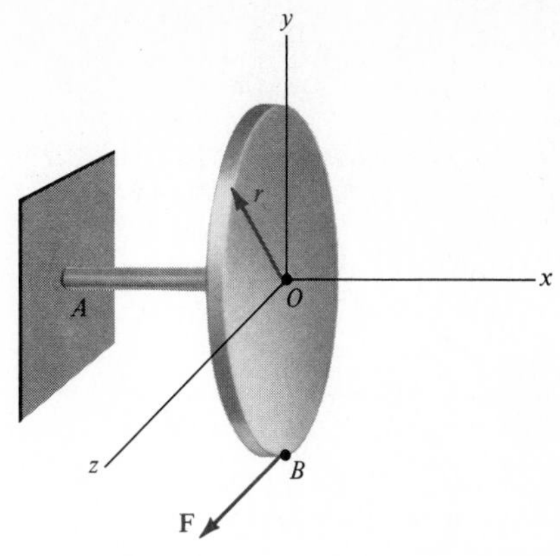

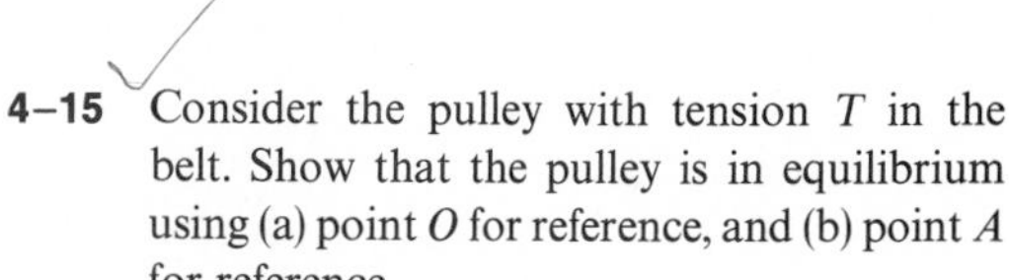

4–15 Consider the pulley with tension T in the belt. Show that the pulley is in equilibrium using (a) point O for reference, and (b) point A for reference.

FIGURE P4-15

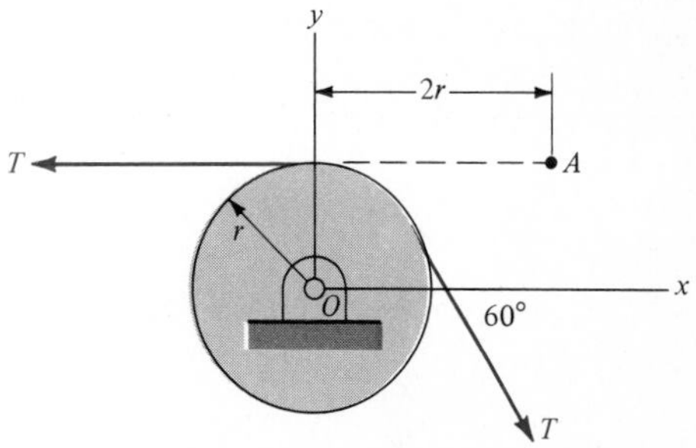

4–16 Prove that Eq. 4-5 for forces and moments are valid and usable to determine the two unknowns in the situation of Fig. 4-3b.

4–17 Prove that Eq. 4-6 are valid but not independent, and cannot be solved for the three unknowns in Fig. 4-3c.

4–18 Use vector methods to determine the force $\mathbf{F}$ if $\mathbf{P} = (-5\mathbf{i} - 4\mathbf{j})$ kN. The coordinates are (−0.2, 0.4, 0) m for point A, and (0, 0.5, 0) m for point B.

FIGURE P4-18

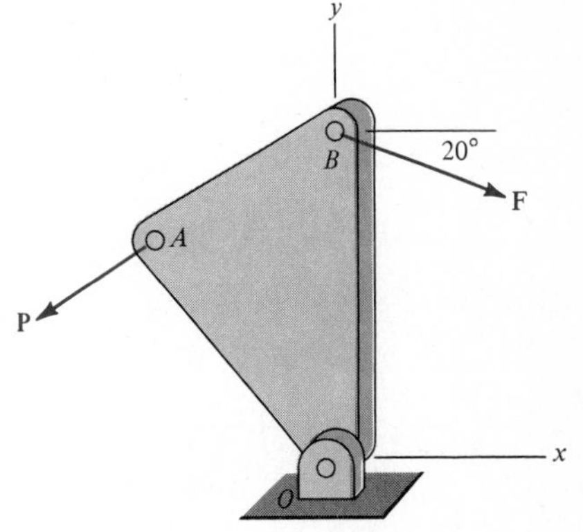

4–19 Calculate the reactions R_x and R_y acting on the bracket in Fig. P4-18 if $\mathbf{P} = (-500\mathbf{i} - 350\mathbf{j})$lb. The coordinates are (−4, 9, 0) in. for A, and (0, 12, 0) in. for B.

FIGURE P4-20

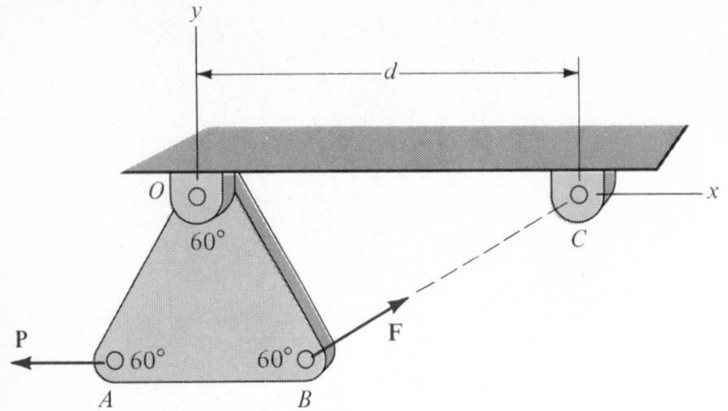

4–20 Force $\mathbf{P} = -5$ kN $\mathbf{i}$ on the bracket. Calculate the reactions R_x and R_y at point O if $d = 0.5$ m and $OA = OB = 0.35$ m.

4–21 Consider different positions for point C along the x axis in Prob. 4–20. Force $\mathbf{F}$ must always be directed to point C (an actuator is located between B and C). Plot the reactions R_x and R_y as functions of $d = 0.3$ m, 0.5 m, and 1 m. Plot both curves on the same diagram.

FIGURE P4-22

4–22 Assume that the grain chute of length l is pinned at A and simply supported at B. Its weight W can be assumed to act at the center of the chute. Determine the reactions at A and B for a chute that makes an angle θ with the horizontal.

FIGURE P4-23

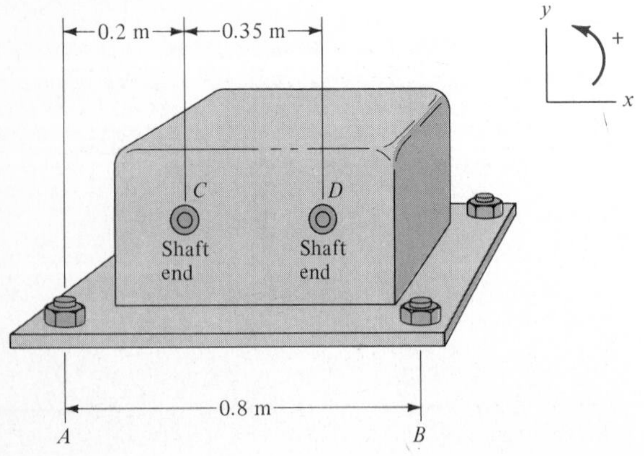

4–23 Consider the sketch of a gearbox. The external torques applied at the shafts are $\mathbf{C} = 200$ N·m $\mathbf{k}$ and $\mathbf{D} = -100$ N·m $\mathbf{k}$. Calculate the vertical forces at bolts A and B to resist the applied torques.

4–24 Consider the model of a tall crane where ABC is assumed to be a single rigid member pivoted at A. Determine the tension T in the supporting wire and the reactions at A.

FIGURE P4-24

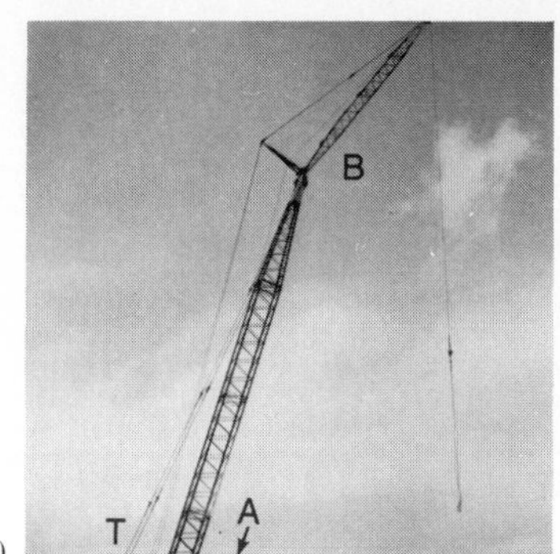

(a)

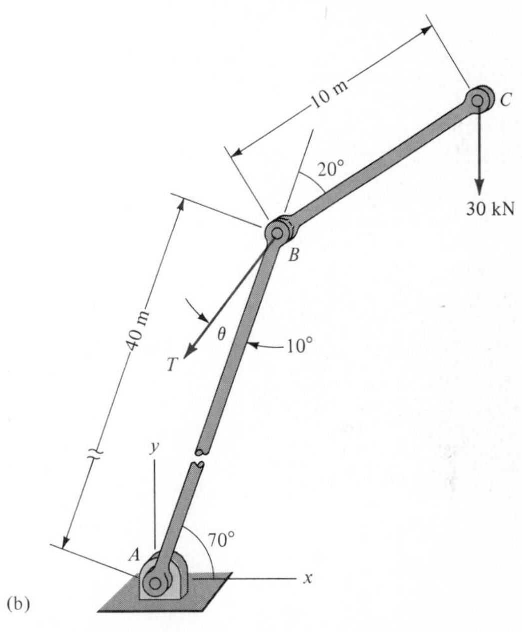

(b)

4–25 Plot T, A_x, and A_y as functions of $\theta = 6°, 8°$, and $10°$ from Prob. 4–24. Use a single diagram.

FIGURE P4-26

4–26 The 80-in.-diameter saw for road repair work rotates clockwise about its axis A. The whole assembly pivots about B according to the positioning of the actuator CD. The equivalent weight at A is 300 lb. Estimate the dimensions and angles involving points A, B, C, and D using the circular saw as the basis for measurements. Determine the force F from D to C which is large enough to lift the saw blade off the road.

FIGURE P4-27

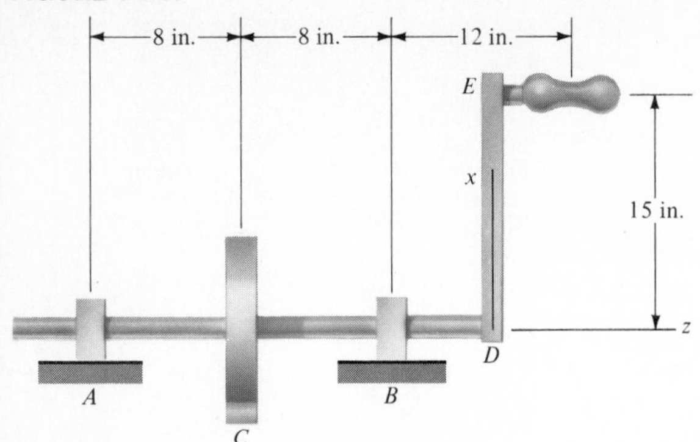

(a) Top view (*ADE* is horizontal)

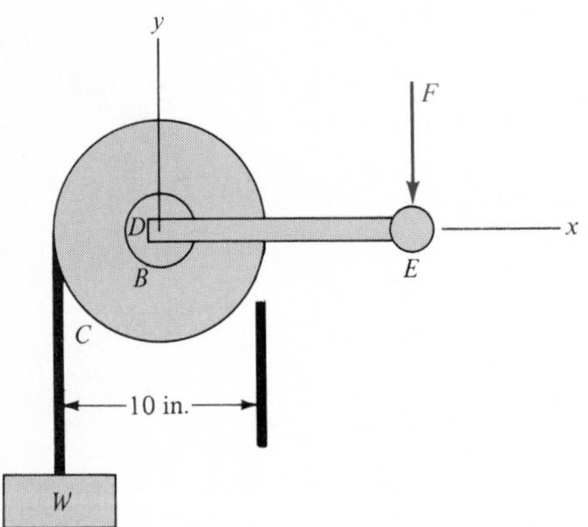

(b) End view from right side

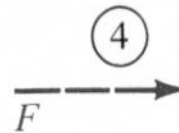

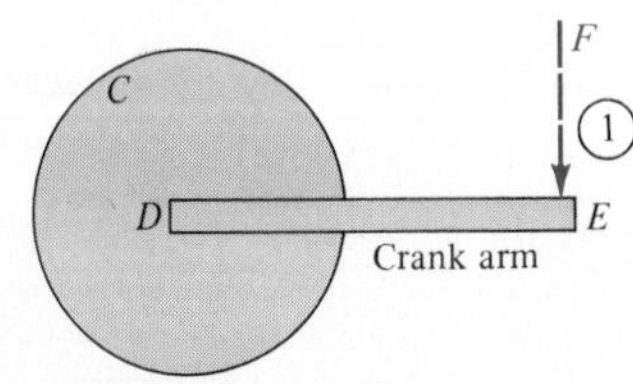

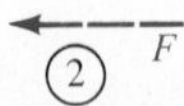

(c) End view from right side

4–27 Consider the hand-crank hoist shown, where *A* and *B* are fixed sleeve bearings, and the wire holding the 100-lb weight is wound on the drum *C*. *F* is the minimum force required to hold the weight *W*. Calculate the reactions A_x and A_y when the crank arm *DE* is horizontal in position 1 (Fig. P4-27c).

4–28 Determine the reactions B_x and B_y in Prob. 4–27 when the crank arm *DE* is vertical in position 2 (Fig. P4-27c).

4–29 Determine the reactions at *A* and *B* in Prob. 4–27 when the crank arm *DE* is (a) horizontal in position 3, and (b) vertical in position 4 (Fig. P4-27c).

TWO-FORCE MEMBERS 4-4

In many engineering problems particular conditions of constraint must be created. An interesting example is the tire-coupled road simulation of any vehicle (see Fig. 1-9 in Sec. 1-7). In typical tests only vertical forces are intentionally applied to each wheel, but the vehicle could suddenly or gradually move horizontally during a test if not constrained properly. There are stories of vehicles that moved off their actuator platens and crashed in the laboratory. Such horizontal motions can be prevented but it must be done with a minimum of interference with the vertical motions. The best solution is to attach horizontal wires or pinned rods to the vehicle at several points. Wire A and frame B (and others in the background) in Fig. 4-11 show such a system.

(a)

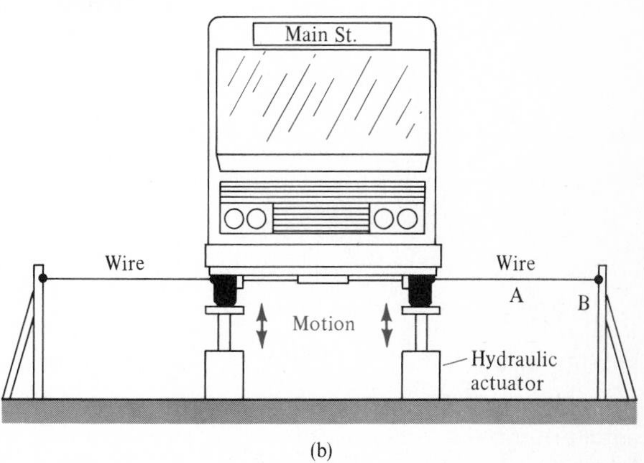

(b)

FIGURE 4-11

Laboratory test setup of a bus. Horizontal wires such as *A* apply horizontal constraint while allowing vertical motion of the bus. (Photograph courtesy MTS Systems Corporation, Minneapolis, Minnesota.)

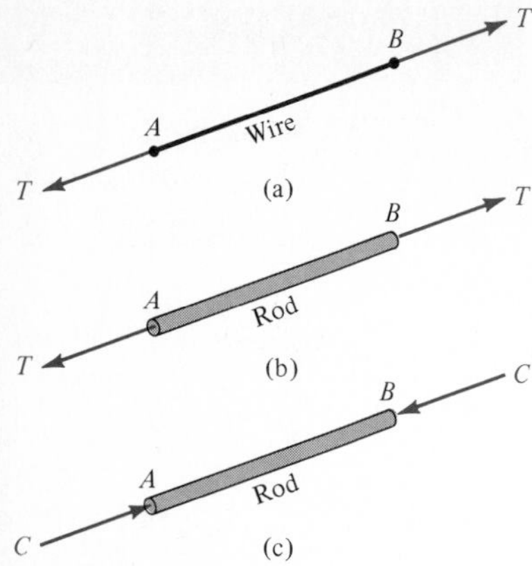

FIGURE 4-12

Two-force members

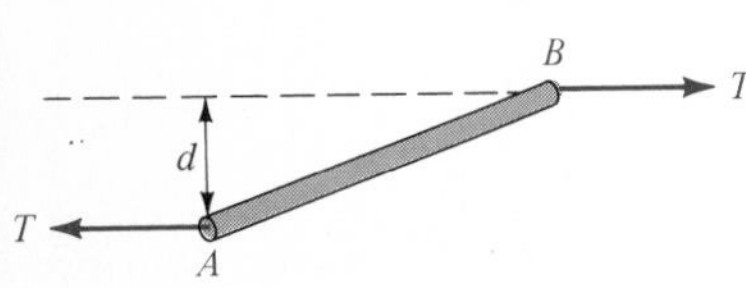

FIGURE 4-13

Two-force member not in equilibrium because the forces are not collinear

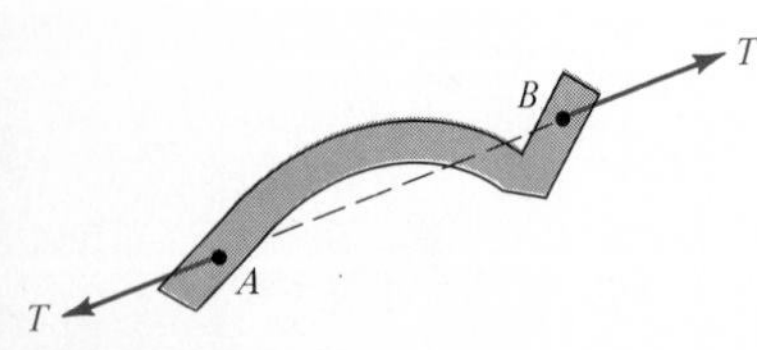

FIGURE 4-14

Member of arbitrary shape is in equilibrium if the forces are collinear

The constraining wires and pinned rods are the simplest structural components because they have only axial forces acting on them. The wires can have only tension, and the rods either tension or compression on them. Both members are called *two-force members* because their free-body diagrams can have only two forces shown as in Fig. 4-12. These forces must be collinear if the member is in equilibrium. The situation in Fig. 4-13 is not possible because only the summation of forces is zero. There is a net moment of magnitude Td with respect to point A or B, so the member cannot be in equilibrium.

The two-force member does not have to be straight or simple in shape. For example, the component in Fig. 4-14 must have collinear forces acting on it to be in equilibrium. These findings lead to the general statement that *a body subjected to forces applied at two points is in equilibrium only when the* (*resultant*) *forces have the same line of action and are opposite in direction.*

The numerous possibilities of constraining a body with two-force members are illustrated with several two-dimensional models in Fig. 4-15. Assume that the two-force member is a pinned rod in each case. The situations in Fig. 4-15a, b, and c represent *partial constraint* because the body could move in the x direction (a and b) or rotate a little in the xy plane (c). *Complete constraint* is achieved by a rearrangement of only three members, as in Fig. 4-15d. Naturally, it is easy to overdo the fixation of a member in theory, and have *redundant constraints* as in Fig. 4-15e. Certain aspects of such statically indeterminate bodies have been mentioned in Sec. 4-2.

Generally it is found that partial constraint (sometimes called *improper constraint*) exists when all the restraining links are parallel (Fig. 4-15a and b) or concurrent (Fig. 4-15c). For the latter, it is not necessary to have mutually perpendicular lines.

The concepts of constraint described above can be extended to three-dimensional cases and to bodies with simple roller or pivot joints (these can be considered as two-force links of zero length). In general, the adequacy of constraint is closely related to the number of ways a body could move, which is the number of independent coordinates that completely specify the position of the body. For short, this is called the number of *degrees of freedom* of the body. A single degree of freedom exists for a body sliding along a fixed axis or rotating about a fixed axle. Motion in a plane may be characterized by a maximum of three degrees of freedom (two linear coordinates and one angular coordinate). Three-dimensional problems may involve up to six degrees of freedom (linear motion in three directions and rotation about three axes). The constraints required for equilibrium must prevent all possible motion of the body.

Finally, the practical difference between rods (rigid) and chains or wires (flexible) is worth remembering. Rigid rods are often the

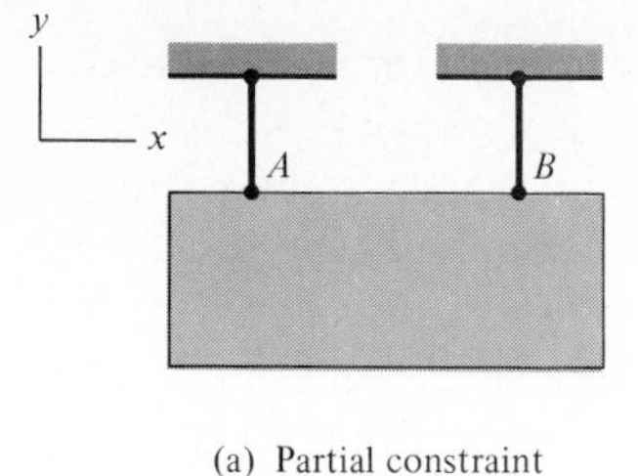

(a) Partial constraint

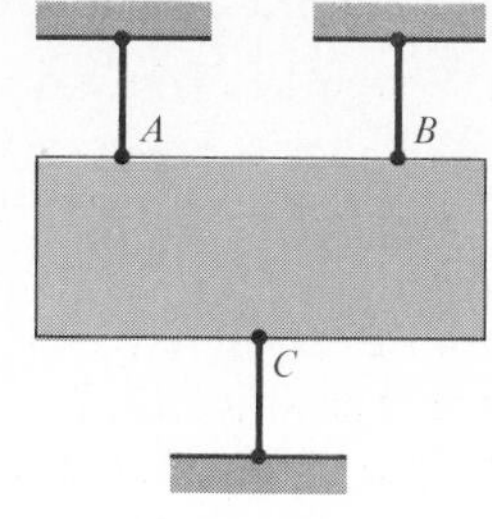

(b) Partial constraint

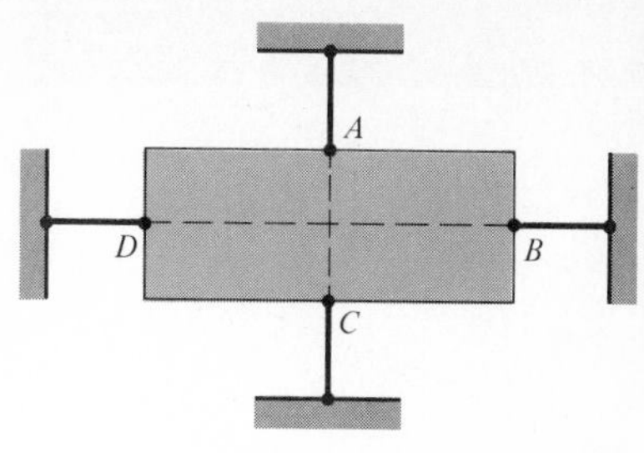

(c) Partial constraint

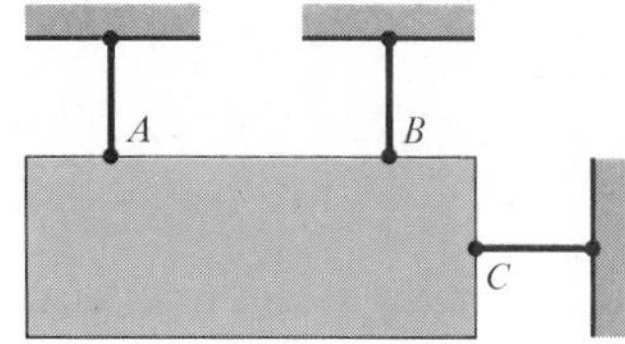

(d) Complete constraint

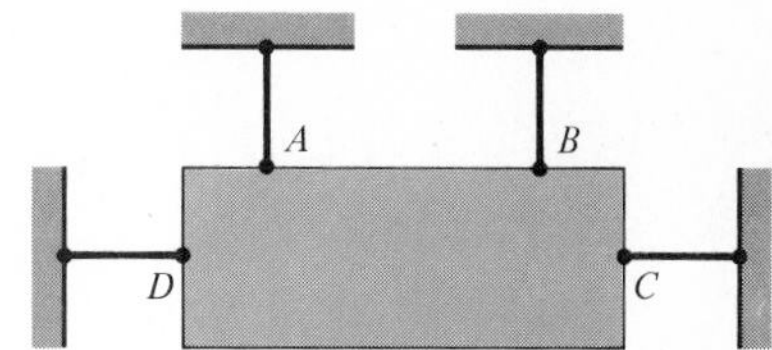

(e) Redundant constraint

FIGURE 4-15

Various constraints of a body by two-force rods

most desirable since they transmit tension and compression (sometimes these forces alternate as in bars *A*, *B*, and *C* in Fig. 4-16). Thus, two flexible members would be required to replace one rigid rod. Nevertheless, wires and chains are often used (such as wire *A* in Fig. 4-11) because they are self-aligning and have simple end connections.

FIGURE 4-16

Rods *A*, *B*, and *C* apply forces to the right front axle of a car in a test. (Courtesy MTS Systems Corporation, Minneapolis, Minnesota.)

4-5 INTERRELATED FREE-BODY DIAGRAMS. NEWTON'S THIRD LAW

An important concept in analyzing the equilibrium of rigid bodies is that parts of those rigid bodies are also in equilibrium. Consequently, complete free-body diagrams can be drawn for all parts that are of interest. Neighboring parts may cause forces and moments on each other, and these must be shown on the individual diagrams. The internal forces and moments are unknown when the whole body is first analyzed; of course, they are not shown on the free-body diagram of the whole body. The forces and moments caused by neighboring parts on each other are related according to Newton's Third Law: the forces are equal in magnitude and opposite in direction; thus, they cancel in pairs when the whole body is considered. These concepts are most easily appreciated with the aid of drawings.

Reconsider the two-piece wave board that was first discussed in Ch. 3. Assume that the two boards act as a single, weightless rigid body, with actuator D "locked in position." Forces $\mathbf{W}_1$ and $\mathbf{W}_2$ in Fig. 4-17a are the net forces from the water on each board. They could be reduced to a single net force ($\mathbf{W} = \mathbf{W}_1 + \mathbf{W}_2$) when the board is considered as a single rigid body. The actuator D is a two-force member between the two boards, so it applies two equal

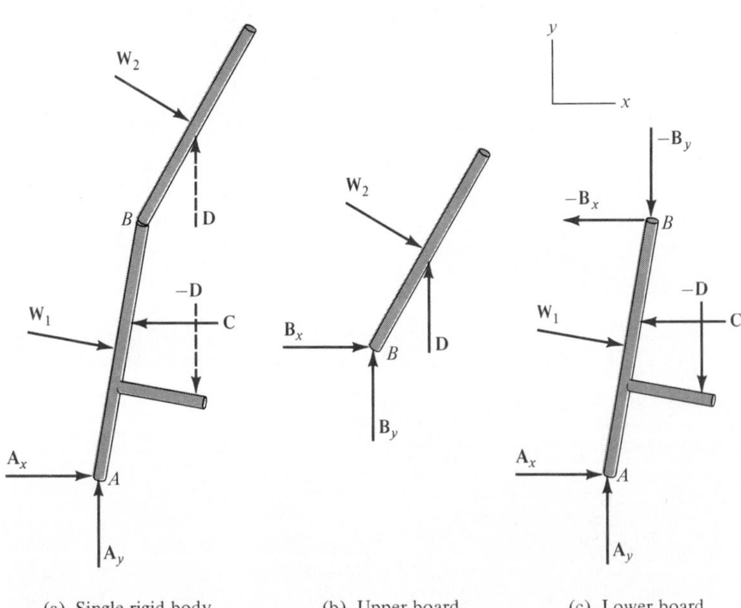

FIGURE 4-17
Interrelated free-body diagrams using vector notation

and opposite forces to the whole rigid body. Thus, the forces **D** and $-\mathbf{D}$ are not necessary in Fig. 4-17a, but at this stage it is helpful to see them in proceeding to the individual free-body diagrams.

The free-body diagram of the upper board is drawn in Fig. 4-17b. Here the forces **D** and $\mathbf{W}_2$ must be shown since they are external forces acting directly on this member. There is an unknown force at *B*, and this is conveniently shown by its rectangular components. The directions of $\mathbf{B}_x$ and $\mathbf{B}_y$ are arbitrarily assumed since they are not known.

The free-body diagram of the lower board is readily drawn, but here the existence of the diagram for the upper board must be taken into account. The forces $-\mathbf{D}$ and $\mathbf{W}_1$ are external forces for the lower board, so they must be shown as in Fig. 4-17c. $\mathbf{A}_x$ and $\mathbf{A}_y$ are simply taken from Fig. 4-18a because at point *A* it does not matter which free body is analyzed. The force at *B* on the lower board is caused by the upper board; therefore, $-\mathbf{B}_x$ and $-\mathbf{B}_y$ in Fig. 4-17c must be equal in magnitude and opposite in direction to those in Fig. 4-17b. The concept of action and reaction (Newton's Third Law) must be used in each case when free-body diagrams are related to one another.

Frequently, a scalar notation is used with the arrows representing forces. In that case negative signs are not used to distinguish forces of action and reaction. The arrows in opposite directions

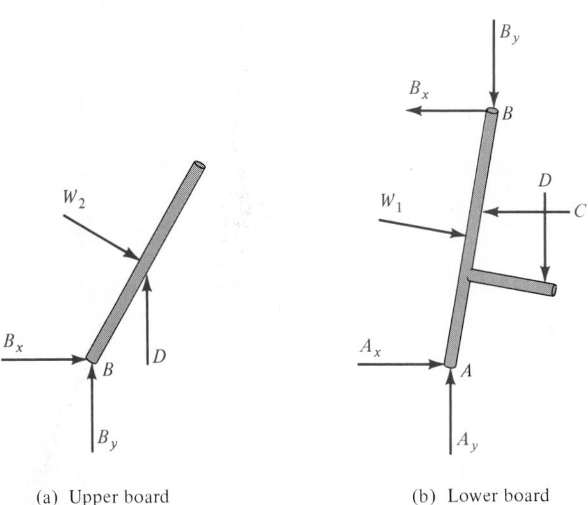

FIGURE 4-18

Interrelated free-body diagrams using scalar notation

represent such differences, while the symbols by the arrows show only the magnitudes of the forces. For example, Fig. 4-17b and c are repeated with scalar notation in Fig. 4-18a and b.

For another example of interrelated free-body diagrams, consider the test setup for a seat belt in Fig. 4-19a. The system to analyze is the *load frame* (the part of the equipment that applies forces to

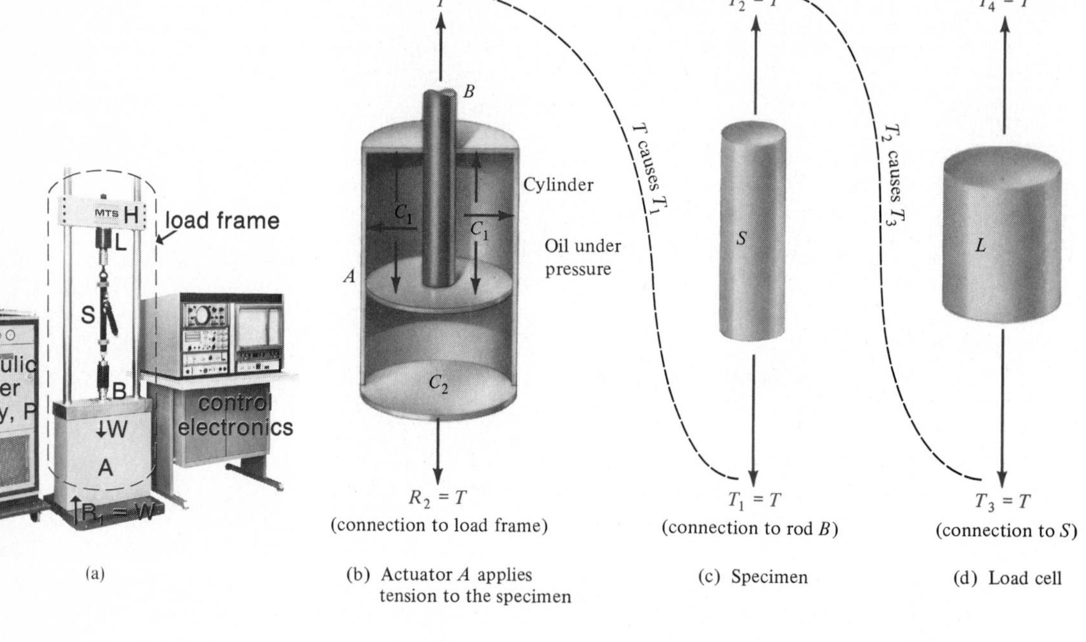

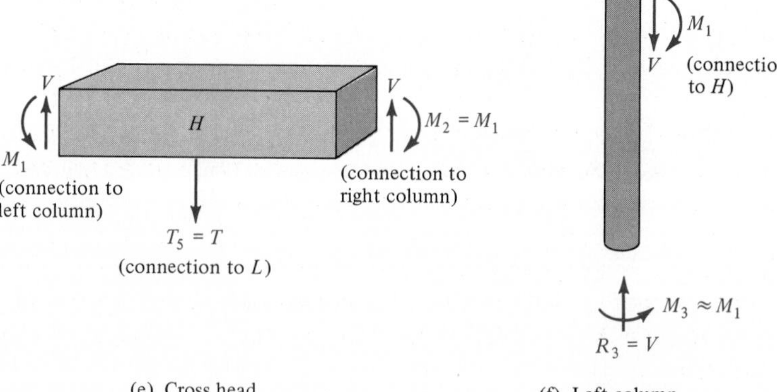

FIGURE 4-19

Test setup of a seat belt. (Photograph courtesy MTS Systems Corporation, Minneapolis, Minnesota.)

test specimens) shown in Fig. 4-19a. The only external forces on the load frame are its weight W and the reaction R_1 from the floor. Note that this diagram is completely independent of any force that may be applied to a test specimen S in the load frame.

The test force in the load frame is applied by a hydraulic actuator A that is not visible in Fig. 4-19a. The pump P delivers oil under pressure to the upper chamber C_1 of the actuator shown in Fig. 4-19b. This forces the piston rod B downward against the tension T developed in the specimen. The upward force T on the actuator is resisted by the reaction R_2 at the place where the actuator is attached to the load frame. Naturally, if pressure is applied in chamber C_2, the rod B is forced upward, applying compression to the specimen. In this case the upper force T and R_2 would reverse directions to point toward each other.

The free-body diagram of the test specimen is drawn in Fig. 4-19c on the basis of part b of the figure. The lower force T_1 is drawn first because it must be equal and opposite to the upward force T in Fig. 4-19b. The upper force T_2 is readily drawn for equilibrium of the specimen.

Above the specimen there is a so-called *load cell L* for the measurement of forces using electronic equipment. The free-body diagram of the load cell is drawn in Fig. 4-19d. Force T_3 on the bottom of the load cell is drawn first on the basis of the force T_2 on top of the specimen. The upper force T_4 on the load cell is again readily drawn for equilibrium. The dashed lines between Fig. 4-19b, c, and d show the progression in drawing interrelated free-body diagrams.

The load cell is attached to the horizontal *cross-head H* of the load frame. The free-body diagram of the cross head is relatively complex, but advantage is taken of its obvious symmetry in Fig. 4-19e. The force T_5 is from the load cell, and it must be opposed by vertical forces V where the cross head is attached to the columns; for equilibrium, $T_5 = 2V$. The cross-head cannot rotate with respect to the columns, so it is necessary to show unknown moments M on the free-body diagram. Essentially, the cross-head acts as a beam built in on both ends, which makes it a statically indeterminate member (one of the columns is redundant in theory).

For a last item in the sequence of major, related free-body diagrams (and many more are possible), consider one of the columns. V and M_1 in Fig. 4-19f are caused by the cross-head on the column, while R_3 and M_3 are the reactions from the lower platen of the load frame.

Note that the whole load frame is quite heavy, so its weight is included in Fig. 4-19a. However, the weights of individual parts may be ignored in their free-body diagrams since often they have much larger external forces acting on them.

EXAMPLE 4-7

Determine the possible conditions for equilibrium of a two-dimensional *three-force member*, which means that forces act on the body at three points. Consider the concurrent, parallel, and general force systems in Fig. 4-20.

SOLUTION

CONCURRENT FORCES: It is clear that equilibrium is possible, and the only equations that should be checked are for the summation of forces. With respect to Fig. 4-20a,

$$\sum F_x = 0 \qquad \sum F_y = 0$$

There are no moments about point O. This condition of equilibrium is valid for more than three concurrent forces, as well.

PARALLEL FORCES: Again, equilibrium is possible. The only requirements with respect to Fig. 4-20b are

$$\sum F_y = 0$$

$$\sum M_O = 0 \qquad \text{where 0 is an arbitrary point}$$

This condition is valid for more than three parallel forces, as well.

GENERAL FORCES (Not Concurrent, Not Parallel): Complete equilibrium is not possible, as seen after a simple change from Fig. 4-20a to c. Equilibrium of the forces is still conceivable, but not of moments. Taking moments about a point O where the lines of action of any two forces intersect,

$$\sum M_O \neq 0$$

because the third force must have a moment about point O (Fig. 4-20c). For more than three forces, equilibrium may be possible. In conclusion, *a body subjected to three forces may be in equilibrium only when the forces are concurrent or parallel.*

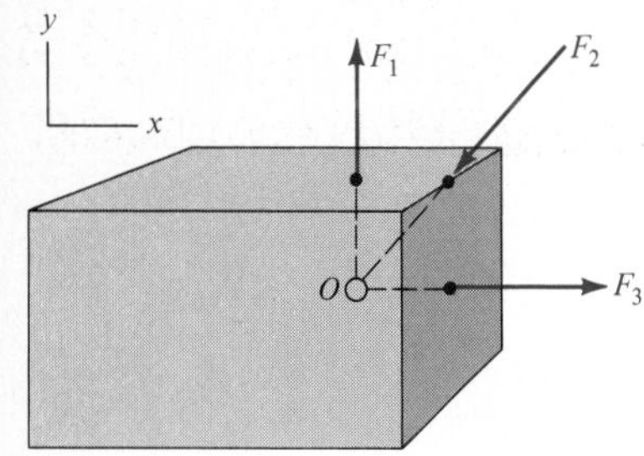

(a) Concurrent forces

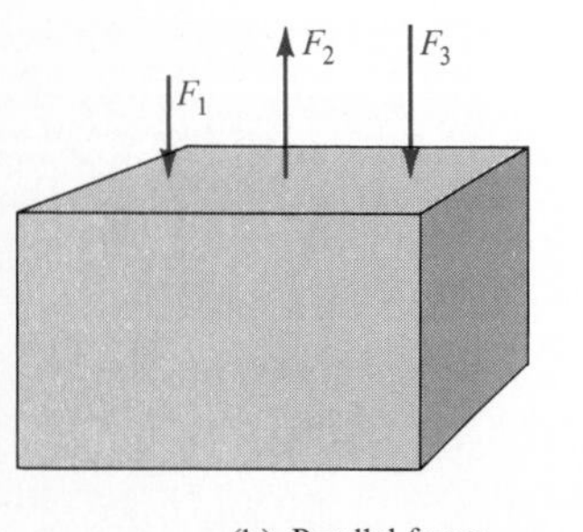

(b) Parallel forces

O O

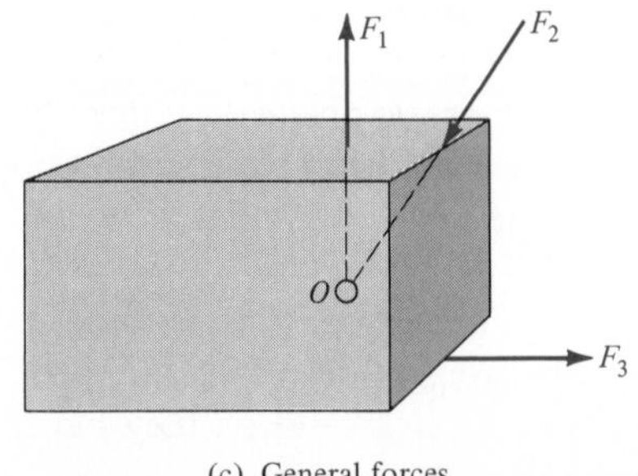

(c) General forces

FIGURE 4-20

Analysis of a two-dimensional three-force member

EXAMPLE 4-8

The boat winch of Fig. 4-21a is drawn schematically in Fig. 4-21b. The tension in the cable is up to 2000 lb. Determine the force F needed at the handle to start winding the cable up. Also determine the force F_s that the stop must exert to keep the cable from unwinding. Handle H is fixed on gear B.

SOLUTION

The free-body diagrams of the gears are in Fig. 4-21c and d. From these and Eq. 4-2 (noting that only moment equations are useful since force directions are not given),

$$\sum M_A = 0$$

$$F_B(3 \text{ in.}) - (2000 \text{ lb})(1 \text{ in.}) = 0$$

$$F_B = 667 \text{ lb}$$

$$\sum M_B = 0$$

$$(667 \text{ lb})(1 \text{ in.}) - (F)(12 \text{ in.}) = 0$$

$$F = 56 \text{ lb}$$

The force on the stop is found from Fig. 4-21d with $F = 0$,

$$\sum M_B = 0$$

$$(F_s)(1 \text{ in.}) - (667 \text{ lb})(1 \text{ in.}) = 0$$

$$F_s = 667 \text{ lb}$$

JUDGMENT OF THE RESULTS

The magnitudes of F_B and F are reasonable, and F_s must equal F_B when F is zero (hands off). Note that F can be reduced by decreasing the radius of gear B, and by increasing the length of the handle arm.

FIGURE 4-21

Analysis of a boat winch

(a)

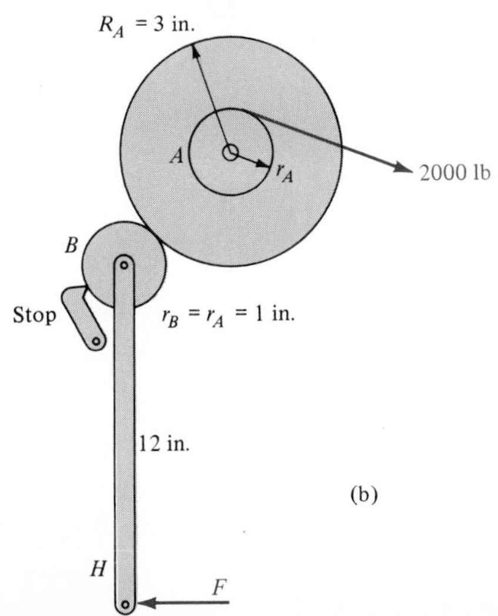

(b)

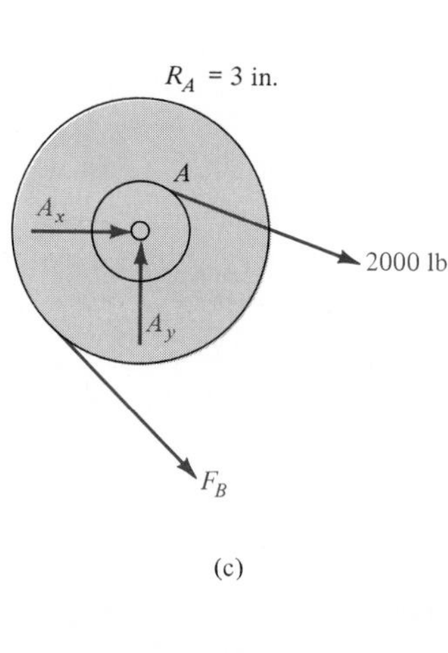

(c)

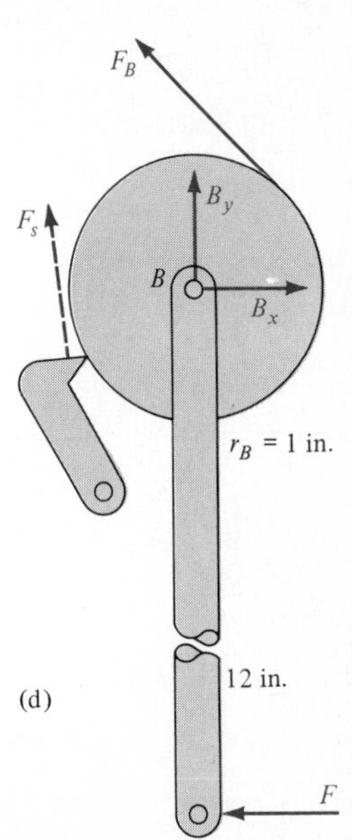

(d)

EXAMPLE 4-9

The rotary actuator A in Fig. 4-22a applies a clockwise torque of $T = 5 \text{ kN}\cdot\text{m}$ to the specimen S as viewed from point 0. Draw free-body diagrams of the specimen S, torsion load cell L, end fixture F, and base platform B, ignoring the weights of the members. Make the views from the z direction.

SOLUTION

The individual free-body diagrams are drawn in Fig. 4-22b through e. The arrows represent moments. Forces are not shown because weights are negligible and no other forces are given or can be assumed as significant.

JUDGMENT OF THE RESULTS

Action and reaction are properly followed from diagram to diagram. As a final check, the free-body diagram of the actuator A in Fig. 4-22f fits that for the specimen S. This is drawn with the idea that the specimen S and the base B apply equal and opposite couples to the actuator A.

FIGURE 4-22

Test setup for torsion of a shaft. (Photograph courtesy MTS Systems Corporation, Minneapolis, Minnesota.)

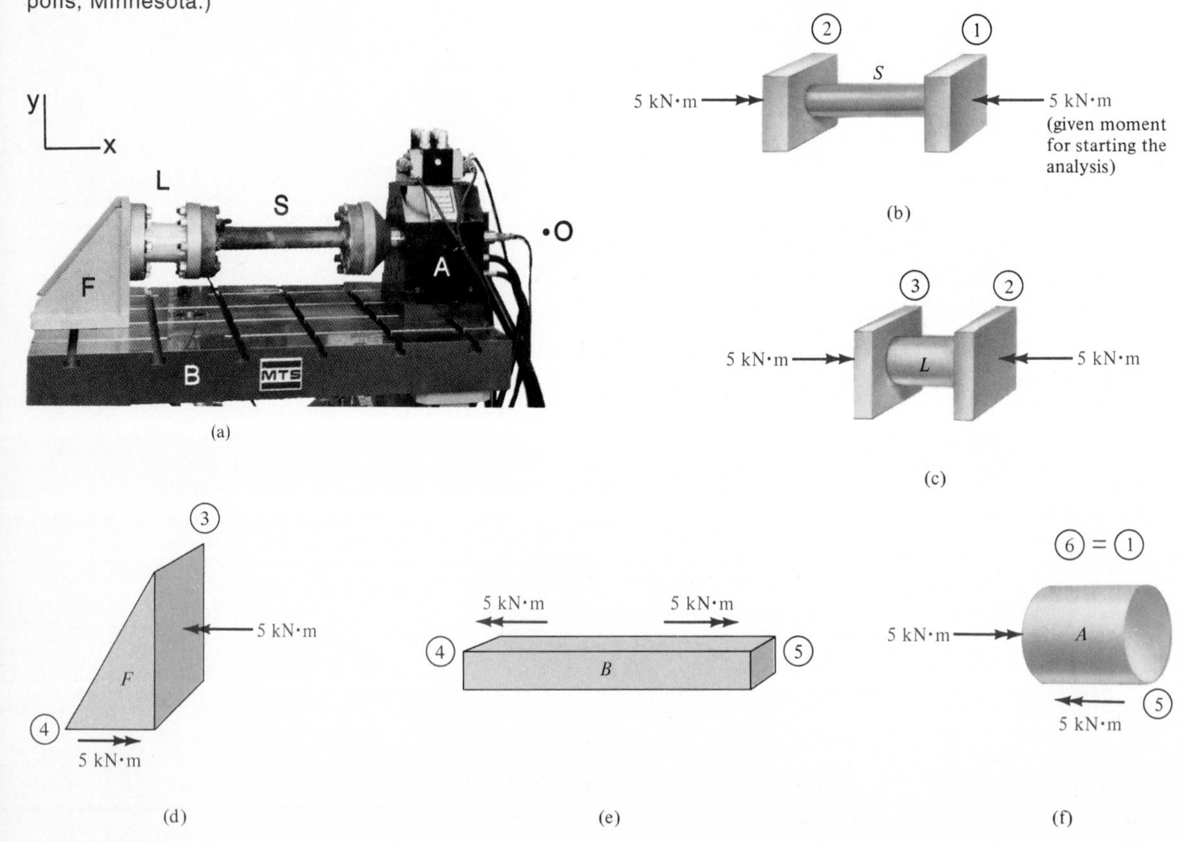

EXAMPLE 4-10

A biaxial test setup of a shock absorber S is shown in Fig. 4-23a. The tensile forces from the testing machine are $F_y = 15$ kN and $F_x = 10$ kN when line AB makes a 5° angle with the horizontal. Draw and analyze the free-body diagrams for the specimen S, load cell L, and member ACD. Calculate the reactions at E, the base of the horizontal actuator. Assume that the specimen is pinned at its ends.

SOLUTION

Arbitrarily, start with the load cell in Fig. 4-23b. The vertical force of 15 kN is given (this is known quite accurately because it is measured by the load cell itself). The horizontal force H_1 is not known precisely; it is caused by the specimen S being pulled to the left by the horizontal actuator AE. The moment M_1 is expressed using Eq. 4-2, $\sum M_G = M_1 - H_1 l = 0$. Thus, $M_1 = H_1 l$.

The maximum total tension T in cables AB can be determined accurately because the force of the horizontal actuator is given as 10 kN. T alone resists this, but at a 5° angle. So, from Fig. 4-23c, and $\sum F_x = T\cos 5° - 10\text{ kN} = 0$,

$$T = \frac{10\text{ kN}}{\cos 5°} = 10.04\text{ kN}$$

From $\sum F_y = T\sin 5° - V_1 - V_2 = 0$, the vertical force at C and D is $V = V_1 + V_2 = T\sin 5°$. V is shared by the two clamps unpredictably since the fixture is statically indeterminate here (either clamp alone could be enough for equilibrium). There are also horizontal forces at C and D that create a couple M_2. From $\sum M_I = M_2 - T\sin 5°(0.4\text{ m}) = 0$,

$$M_2 = T\sin 5°(0.4\text{ m}) = 0.35\text{ kN·m}$$

The free-body diagram for the specimen S is in Fig. 4-23d; F and T are known here. From $\sum F_y = 0$,

$$P = F - T\sin 5° = 15 - 0.88 = 14.12\text{ kN}$$

For equilibrium in the horizontal direction, $\sum F_x = 0$, and

$$T\cos 5° = H_1 + H_2$$

Since $\sum M_B = 0$,

$$H_2 b - H_1 a = 0$$

The reactions at base E of the horizontal actuator are obtained using Fig. 4-23e and Eq. 4-2,

$$\sum F_x = 0 \qquad H_3 = T\cos 5° = 10\text{ kN}$$

$$\sum F_y = 0 \qquad V_3 = T\sin 5° = 0.88\text{ kN}$$

$$\sum M_E = 0 \qquad M_3 = T\sin 5°\,(0.25\text{ m}) = 0.22\text{ kN·m}$$

Note that the reactions at E could be obtained by drawing the free-body diagram of part AE directly and realizing that $H_3 = 10$ kN is given.

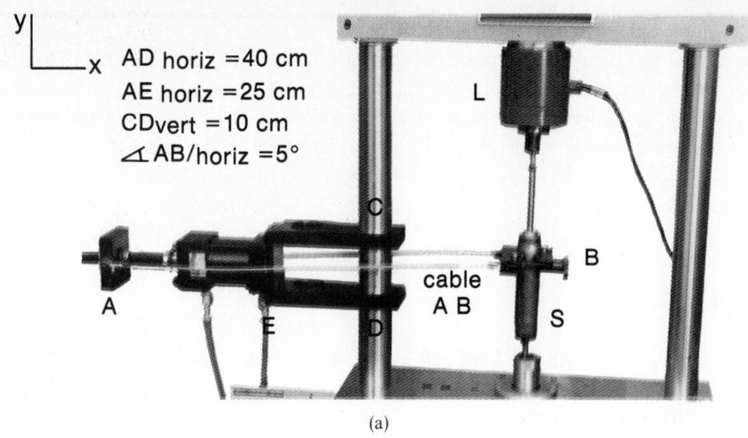

(a)

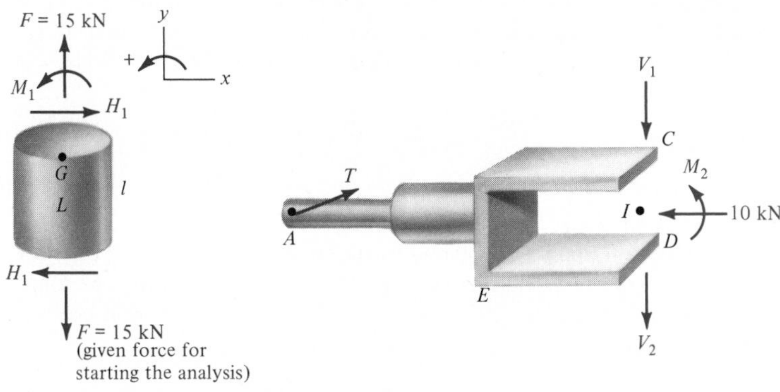

(b) (c)

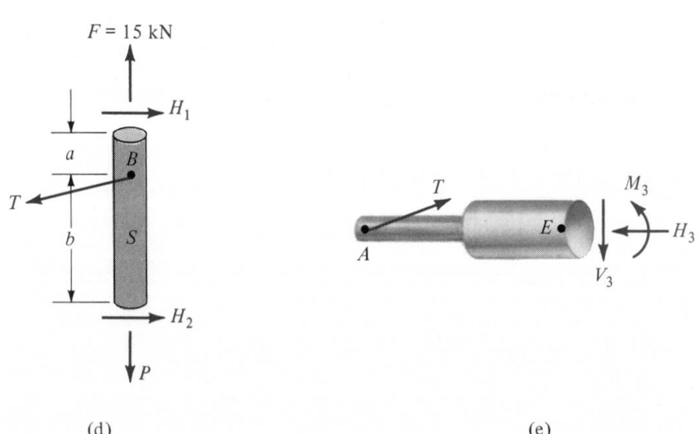

(d) (e)

FIGURE 4-23

Biaxial test setup of a shock absorber. (Photograph courtesy MTS Systems Corporation, Minneapolis, Minnesota.)

PROBLEMS

In Probs. 4–30 to 4–38, determine whether the rectangular plate is completely or partially constrained. Compute the reactions where possible.

FIGURE P4-30

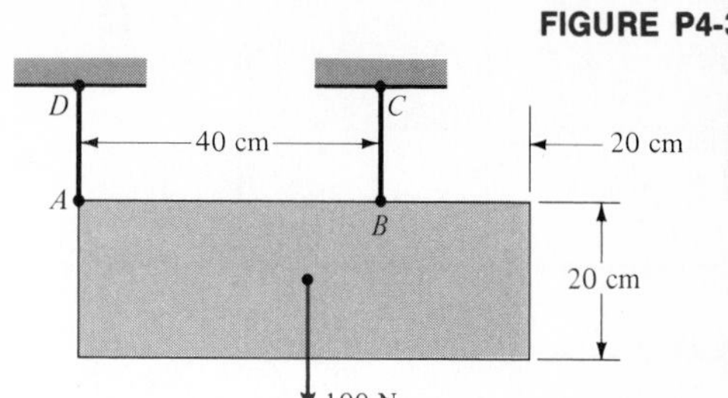

4–30 There are pin joints at *A*, *B*, *C*, and *D*. The force acts at the center of the plate.

FIGURE P4-31

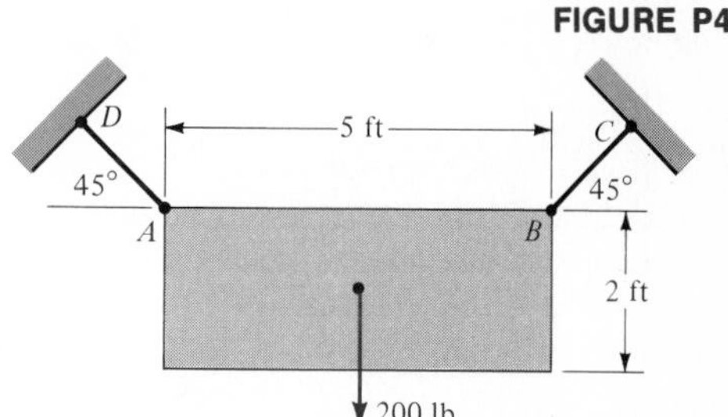

4–31 There are pin joints at *A*, *B*, *C*, and *D*. The force acts at the center of the plate.

FIGURE P4-32

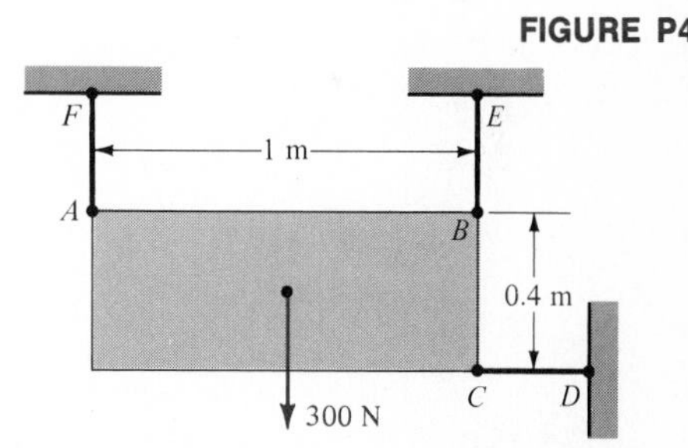

4–32 There are pin joints at *A*, *B*, *C*, *D*, *E*, and *F*. The force acts at the center of the plate.

FIGURE P4-33

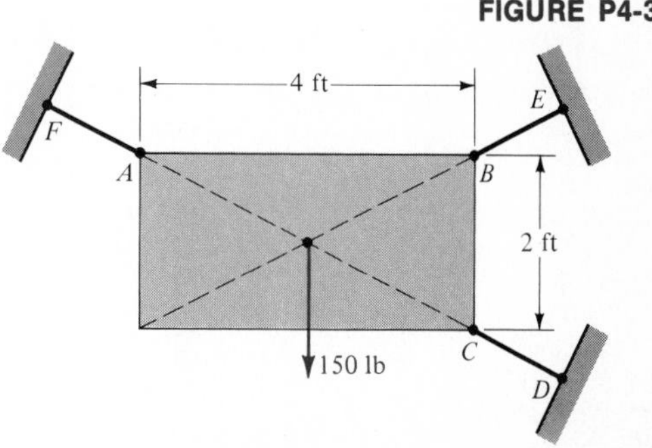

4–33 There are pin joints at *A*, *B*, *C*, *D*, *E*, and *F*.

FIGURE P4-34

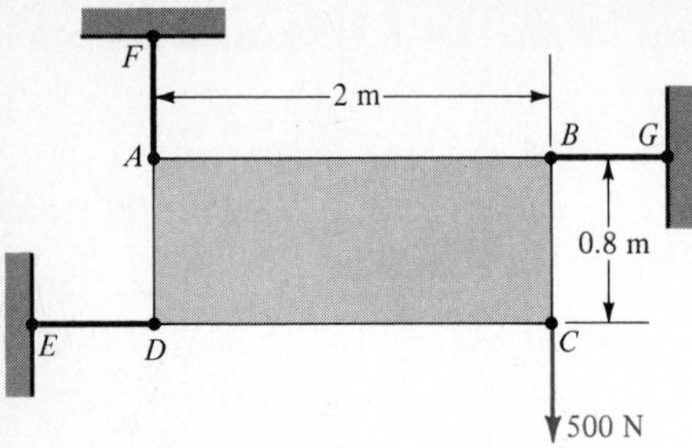

4–34 There are pin joints at *A*, *B*, *D*, *E*, *F*, and *G*.

FIGURE P4-35

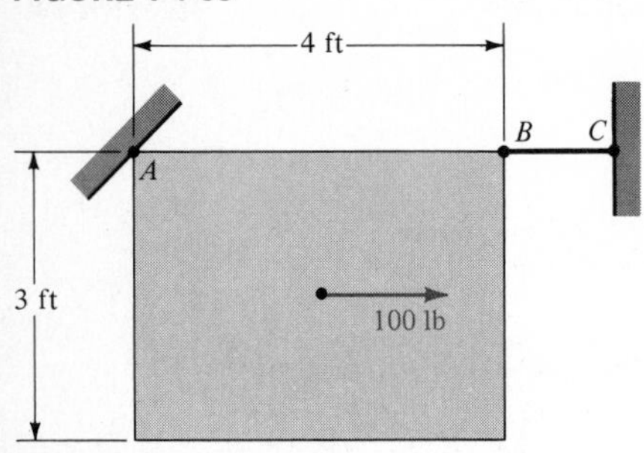

4–35 There are pin joints at *A*, *B*, and *C*. The force acts at the center of the plate.

FIGURE P4-36

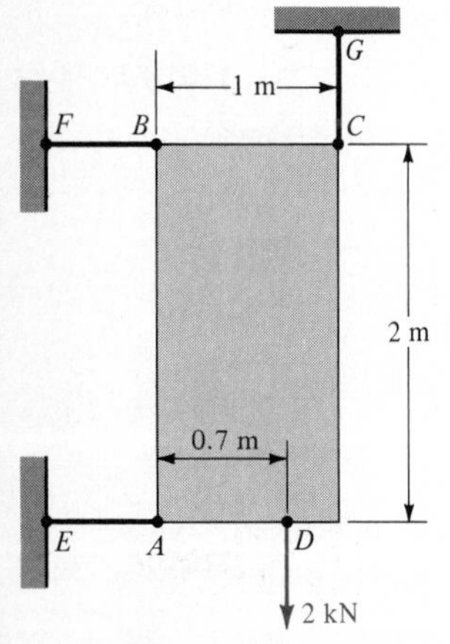

4–36 There are pin joints at *A*, *B*, *C*, *E*, *F*, and *G*.

4–37 There are pin joints at *A*, *B*, *C*, *D*, *E*, *F*, *G*, and *H*.

4–38 There are pin joints at *A*, *B*, *C*, *D*, *E*, and *F*. The force acts at the center of the plate.

FIGURE P4-37

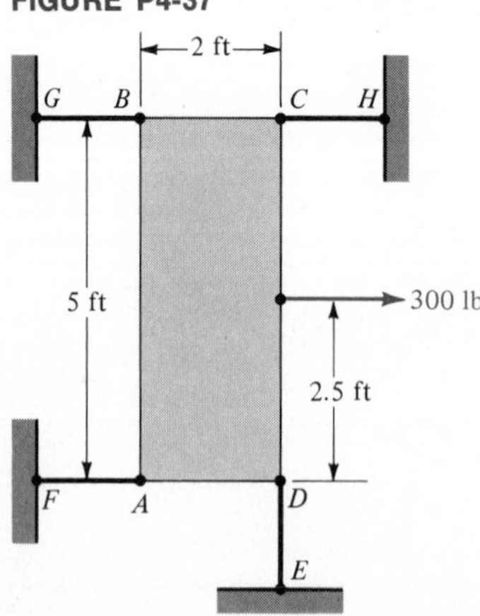

FIGURE P4-38

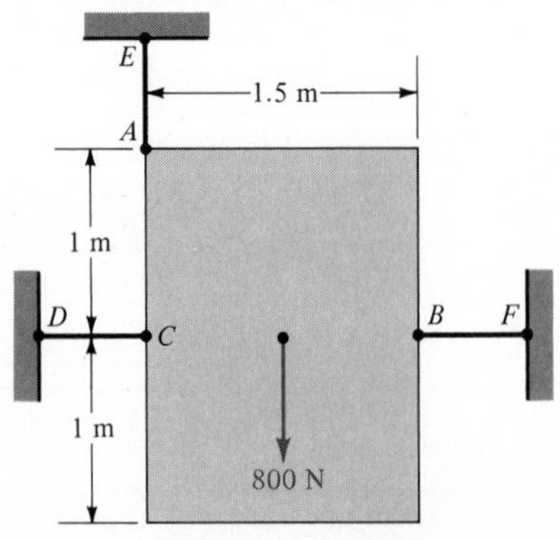

4–39 The weightless bars are pinned at A, B, and C. Determine the reactions at A and B if $W = 5\text{ kN}$, $l = 3\text{ m}$, and $d = 1\text{ m}$. W acts directly on bar AC at C.

FIGURE P4-39

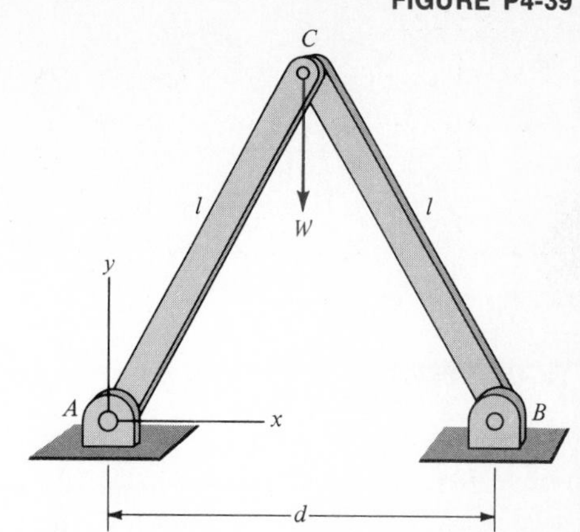

4–40 In Fig. P4-39, $W = 1000\text{ lb}$ and $l = 10\text{ ft}$. Plot the reactions A_x, A_y, B_x, and B_y as functions of $d = 3\text{ ft}$, 5 ft, and 7 ft. Use a single diagram. W acts directly on bar BC at C.

4–41 The simple crane has two weightless members which are pinned at A, B, and C. The small pulley at B is directly attached to bar AB. Calculate the reactions at A and C if $W = 2\text{ kN}$, $\theta = 0$, and $l = 2.5\text{ m}$.

FIGURE P4-41

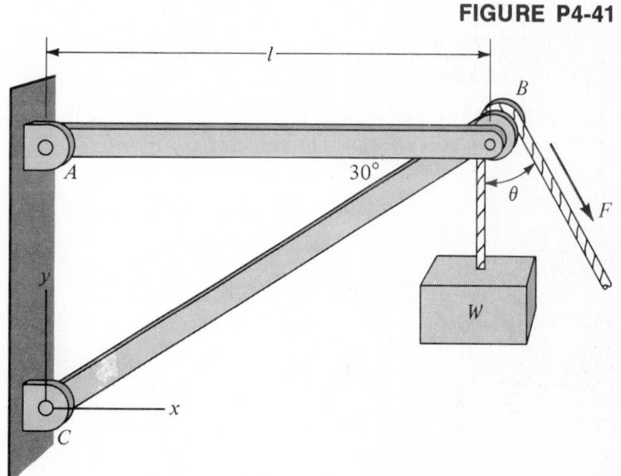

4–42 Solve Prob. 4–41 using $W = 500\text{ lb}$, $\theta = 40^\circ$, and $l = 8\text{ ft}$.

4–43 Assume that $W = 800\text{ N}$, acting on bar AC at pin C. Determine the reactions at A, B, and C if the distance AB is $d = 2\text{ m}$, and l_1 and $l_2 = 1.5\text{ m}$.

4–44 In Fig. P4-43, $W = 500\text{ lb}$ acting on bar BC at pin C. The distance AB is $d = 7\text{ ft}$, $l_1 = 4\text{ ft}$, and $l_2 = 8\text{ ft}$. Calculate the reactions at A, B, and C.

4–45 Pins A and B of the bar can slide without friction in the narrow grooves. Determine the complete systems of forces required at A and B if a force F at A is given and if the bar is to be in equilibrium.

FIGURE P4-43

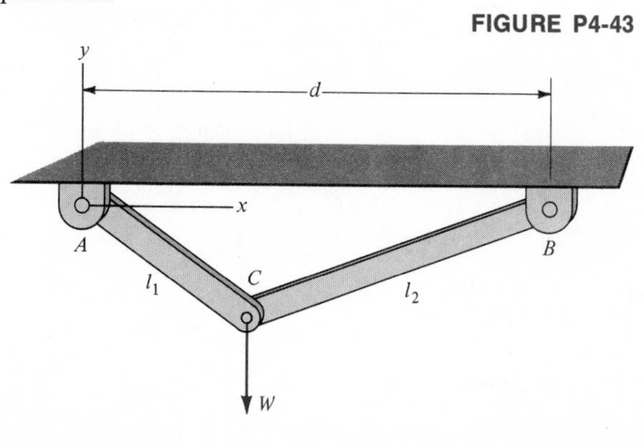

FIGURE P4-45

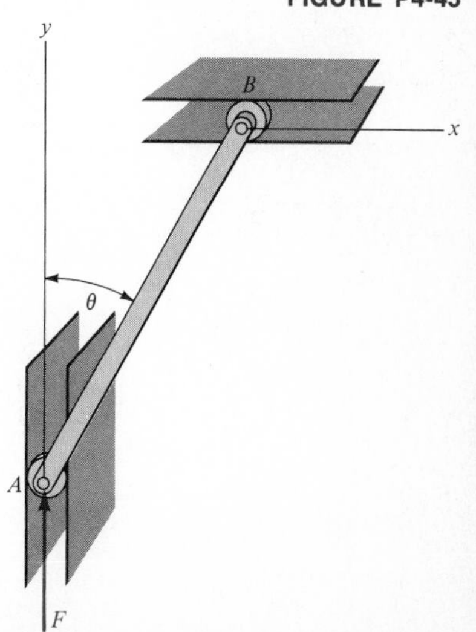

FIGURE P4-47

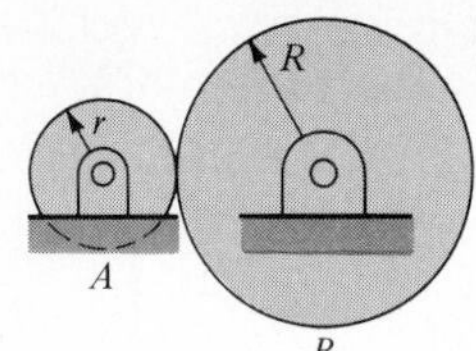

4–46 Plot the required forces B_x and B_y in Prob. 4–45 as functions of $\theta = 10°$, 30°, and 50°.

4–47 A clockwise moment M is applied to gear A. Determine the vertical reactions on both axles if the gears are in equilibrium.

FIGURE P4-48

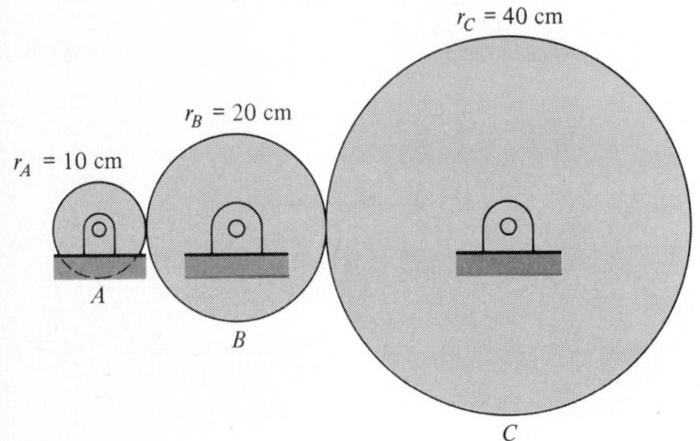

4–48 A clockwise moment $M = 600$ N·m is applied to gear A. Determine the moment acting on gear C and the vertical reaction on its axle in equilibrium.

4–49 In Fig. P4-48, $r_A = 2$ in., $r_B = 5$ in., and $r_C = 12$ in. Determine the largest moment that can be applied to gear A if the vertical reaction should not exceed 1000 lb on any of the axles.

FIGURE P4-50

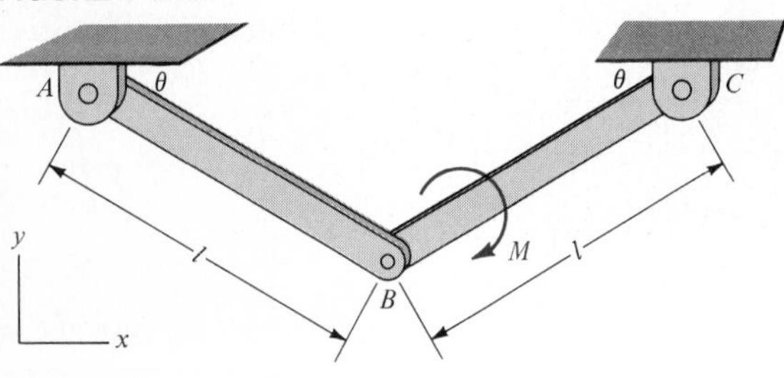

4–50 A mechanism includes two identical bars of negligible weight. Determine the reactions at pins A, B, and C.

4–51 Determine the reactions at A and B for the situation shown.

FIGURE P4-51

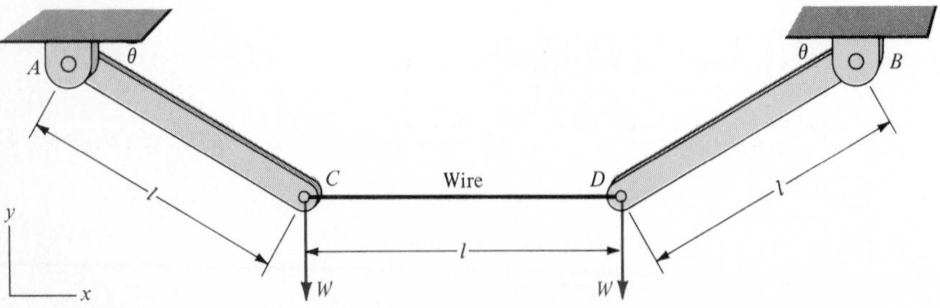

4–52 A ladder is modeled as a uniform bar of weight $W = 300$ N and length $l = 4$ m. W acts at the center of the bar. Calculate the reactions at A and B assuming that all contacting surfaces are frictionless.

FIGURE P4-52

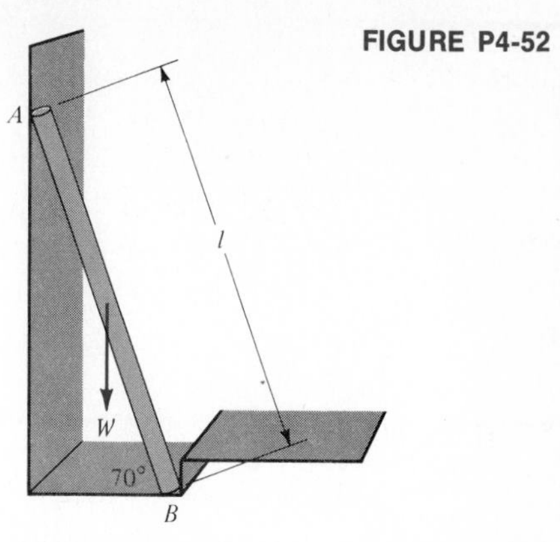

4–53 The piston B can slide without friction in the horizontal cylinder. Determine the reaction at A if the pinned rod AB is to be in equilibrium when a force $F = 300$ lb is applied.

4–54 Determine the force and moment reactions at point O in Prob. 4–53 if $F = 400$ lb is acting to the right on the piston.

FIGURE P4-53

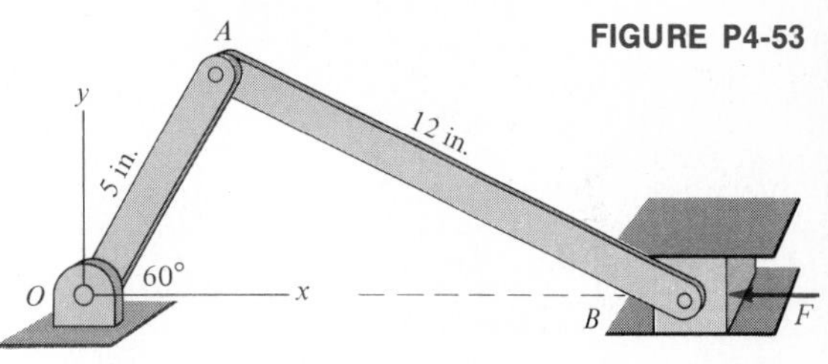

4–55 The dimensions of the boat-trailer suspension are $AB = BC = 15$ in. and $CD = 6$ in. Estimate the forces and moments at A, B, C, and D for a load of 1000 lb acting at B. What is the purpose of the member CD, which makes a 5° angle with the vertical?

FIGURE P4-55

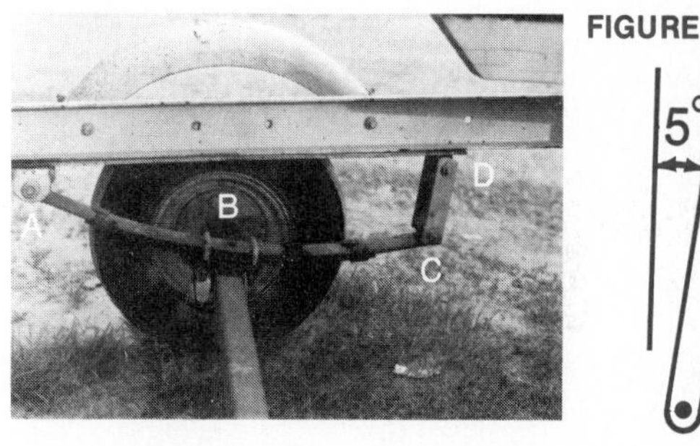

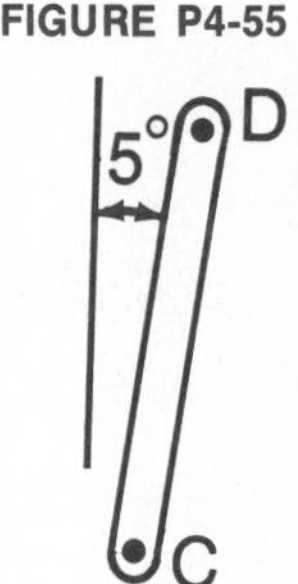

4–56 A load of 7 kN may be placed anywhere within A and B in the trailer. Determine the reactions at the wheels C, D, E, and the force on the hitch F that is mounted on the car, for the extreme position A of the load.

FIGURE P4-56

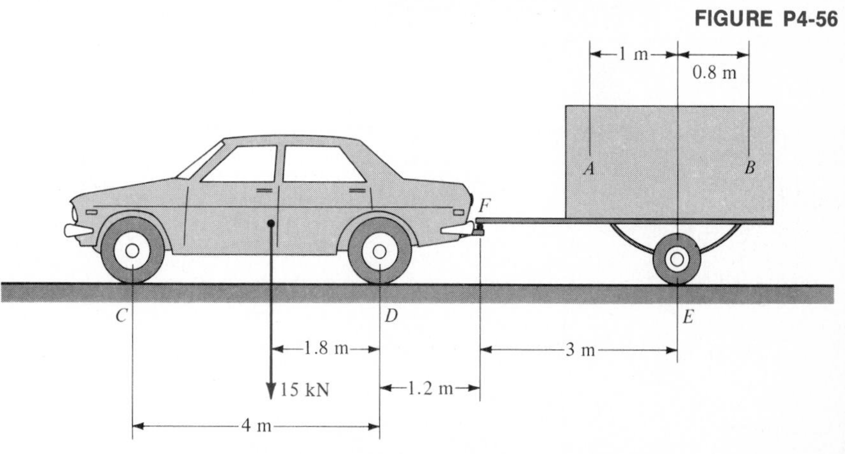

FIGURE P4-58

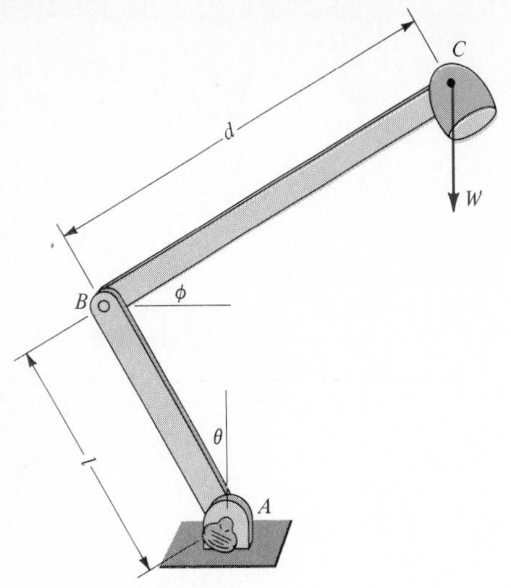

FIGURE P4-60

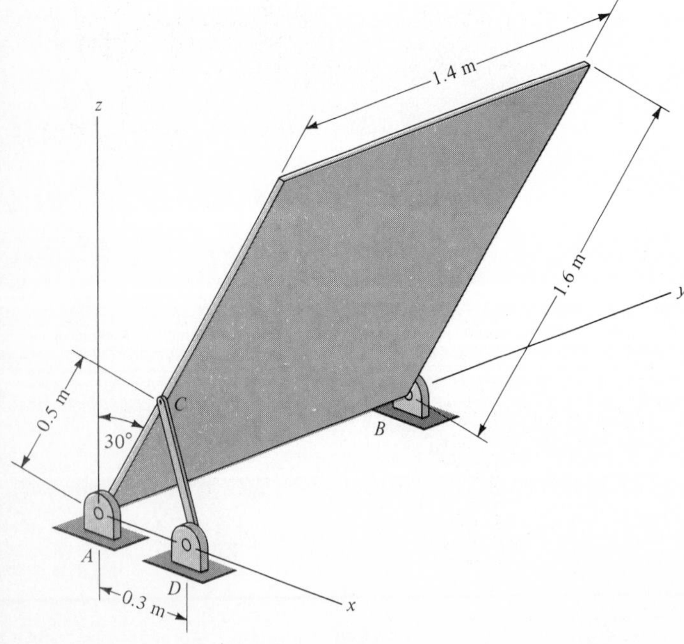

4–57 Solve Prob. 4–56 assuming that the load in the trailer is at *B*.

4–58 Determine the reactions at *A* and *B* of the adjustable lamp to hold it in equilibrium.

4–59 Determine the complete force vectors **C** and **D** acting at points *C* and *D*, respectively, in Fig. 4-23c (Ex. 4-10).

4–60 The hood of a car is modeled as a uniform rectangular plate of 150-N weight (acts at the center of the plate). Determine the reactions at hinges *A* and *B* and the force in bar *CD*, which is pinned at both ends.

4–61 Solve Prob. 4–60 as stated and also with the hinges *A* and *B* moved to $y_A = 0.3$ m and $y_B = 1.1$ m, respectively. Which are the best positions for the hinges to minimize the supporting forces of the plate?

ADDITIONAL EXAMPLES 4-6

This section is devoted entirely to the practice of basic skills in drawing free-body diagrams and writing equilibrium equations. All of these are done using symbols rather than numbers, because the correct setting up of problems is of utmost importance. Generally, the numerical solutions of the equations are quite simple.

EXAMPLE 4-11

Block A is compressed between pads B of the clamp C in Fig. 4-24a. (a) Draw the free-body diagrams of A, B_1, B_2, and C with D. (b) Draw the free-body diagram of the screw D of clamp C in Fig. 4-24a.

SOLUTION

(a) The diagrams are shown in Fig. 4-24b where F is unknown.

(b) The diagram is shown in Fig. 4-24c. Forces f represent the reactions from a few threads in C. All the forces f together oppose force F.

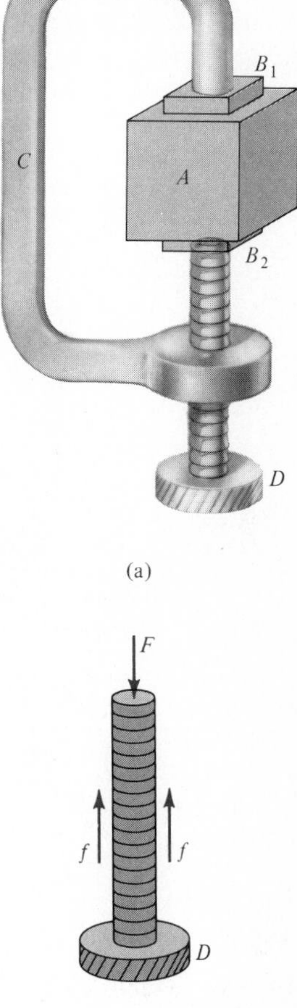

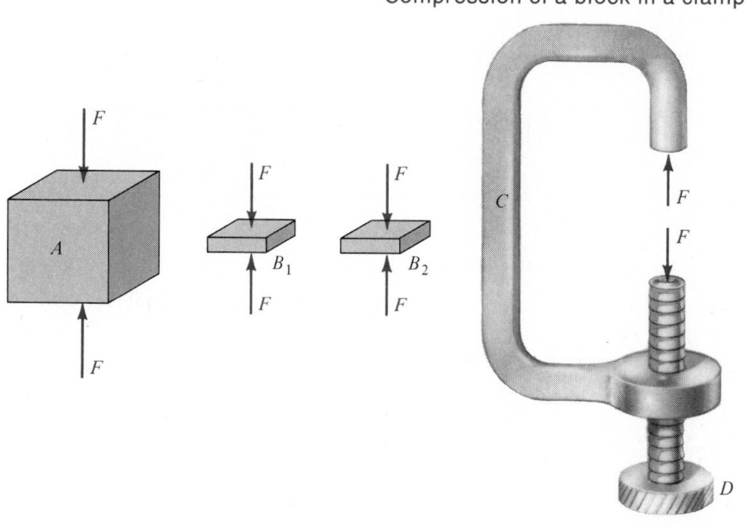

FIGURE 4-24
Compression of a block in a clamp

EXAMPLE 4-12

Force P is applied to block A, which is gripped in clamp C as shown in Fig. 4-25a. Draw the free-body diagrams and write the equilibrium equations for A and C. The face plates in Fig. 4-25a are integral to clamp C, and may be ignored (for clarity) when drawing the free-body diagram of C.

SOLUTION

From Fig. 4-25b,

$$\sum F_x = P - 2Q = 0$$
$$\sum F_y = F - F = 0$$

From Fig. 4-25c,

$$\sum F_x = 2Q - P = 0$$
$$\sum F_y = F - F = 0$$

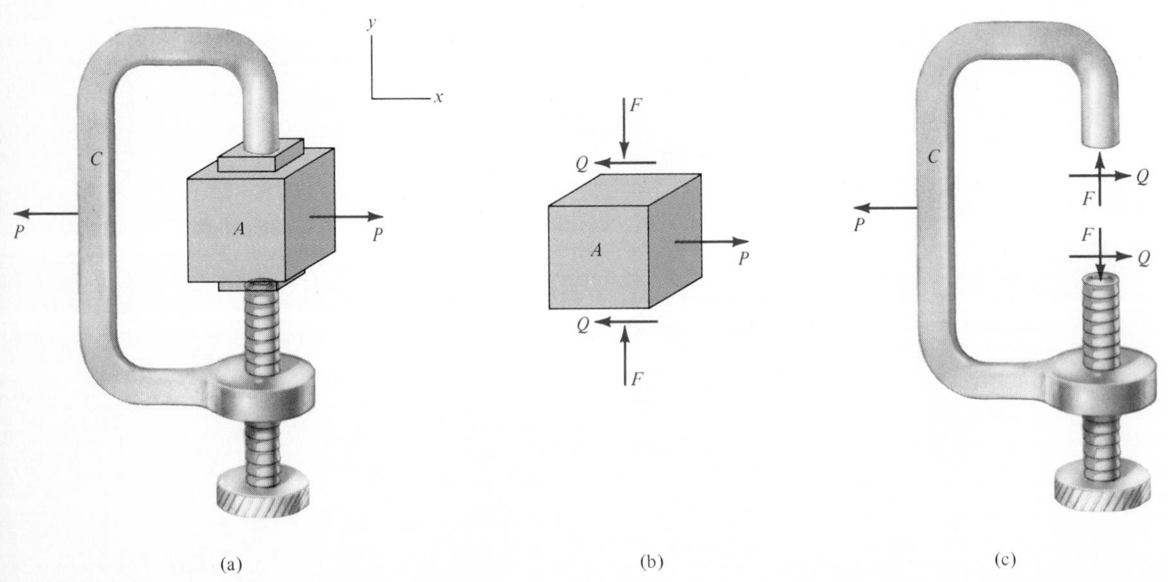

FIGURE 4-25
Biaxial loading of a block in a clamp

EXAMPLE 4-13

A person of weight W walks on the weightless diving board from A to C (Fig. 4-26a). Determine the forces A_y and B_y for $x = a/3$, $2a/3$, and $3a/2$.

SOLUTION

From Fig. 4-26b,

$$x = \frac{a}{3}: \quad \sum F_y = A_y - W + B_y = 0$$

$$\sum M_A = aB_y - \frac{a}{3} W = 0$$

$$B_y = \frac{W}{3}$$

$$A_y = W - \frac{W}{3} = \frac{2W}{3}$$

$$x = \frac{2a}{3}: \quad \sum F_y = A_y - W + B_y = 0$$

$$\sum M_A = aB_y - \frac{2a}{3} W = 0$$

$$B_y = \frac{2W}{3}$$

$$A_y = W - \frac{2W}{3} = \frac{W}{3}$$

$$x = \frac{3a}{2}: \quad \sum F_y = A_y - W + B_y = 0$$

$$\sum M_A = aB_y - \frac{3a}{2} W = 0$$

$$B_y = \frac{3W}{2}$$

$$A_y = W - \frac{3}{2} W = -\frac{W}{2}$$

(A_y is downward)

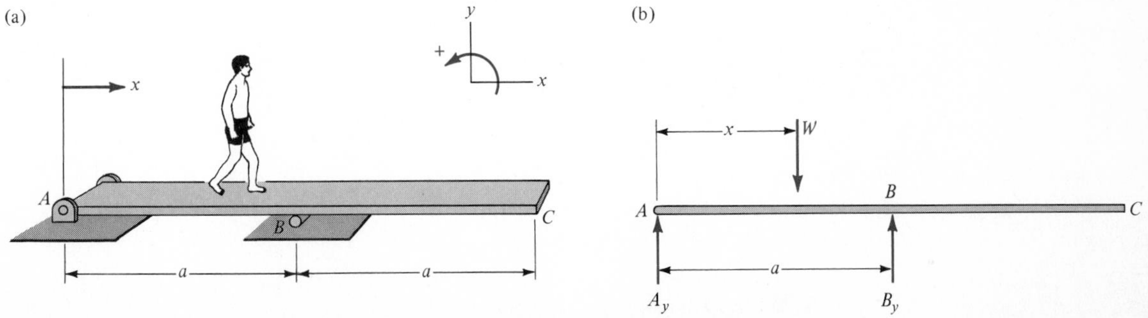

FIGURE 4-26
Analysis of a diving board

EXAMPLE 4-14

Hinge C of the boom is directly above wheel B of the crane in Fig. 4-27a. A and B represent two wheels each. W_1, W_2, P, and the dimensions are known. (a) Write equations from which the forces A_y and B_y can be determined. (b) Assume that the wire supporting the boom might break. Determine the reactions at C that would be required to hold the boom.

SOLUTION

(a) From Fig. 4-27b,

$$\sum F_y = A_y + B_y - W_1 - W_2 - P = 0$$

$$\sum M_A = -W_1 a + B_y(a + b) - W_2(a + b + c) - P(a + b + c + d) = 0$$

(b) From Fig. 4-27c,

$$\sum F_x = C_x = 0$$

$$\sum F_y = C_y - W_2 - P = 0$$

$$C_y = W_2 + P$$

$$\sum M_C = M_C - W_2 c - P(c + d) = 0$$

$$M_C = W_2 c + P(c + d)$$

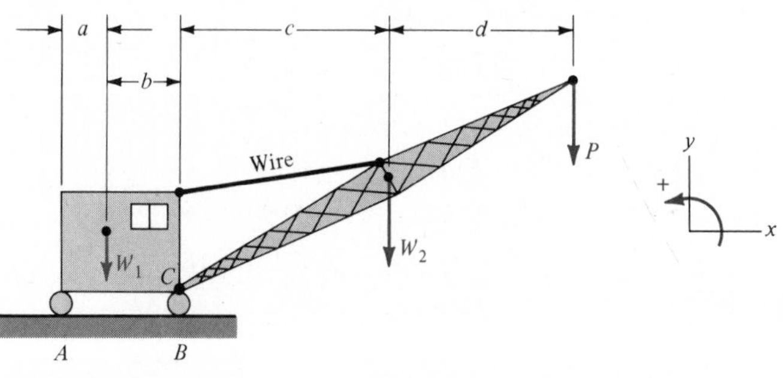

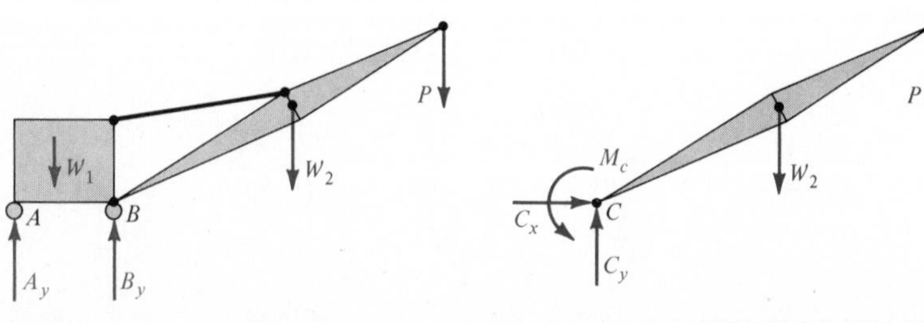

FIGURE 4-27
Model of a mobile crane

EXAMPLE 4-15

The drill press applies a downward force P and a clockwise moment M_o (as viewed from the top) to the work piece in Fig. 4-28a. Express the reactions $\mathbf{F}_A$ and $\mathbf{M}_A$ acting at point A.

SOLUTION

From Fig. 4-28c,

$$\sum \mathbf{F} = P\mathbf{j} + \mathbf{F}_A = 0 \qquad \mathbf{F}_A = -P\mathbf{j}$$

$$\begin{aligned}\sum \mathbf{M} &= \mathbf{M}_A + (\mathbf{r} \times P\mathbf{j}) + M_o\mathbf{j} \\ &= \mathbf{M}_A + (-a\mathbf{i} + b\mathbf{j}) \times P\mathbf{j} + M_o\mathbf{j} = 0 \\ \mathbf{M}_A &= -M_o\mathbf{j} + Pa\mathbf{k}\end{aligned}$$

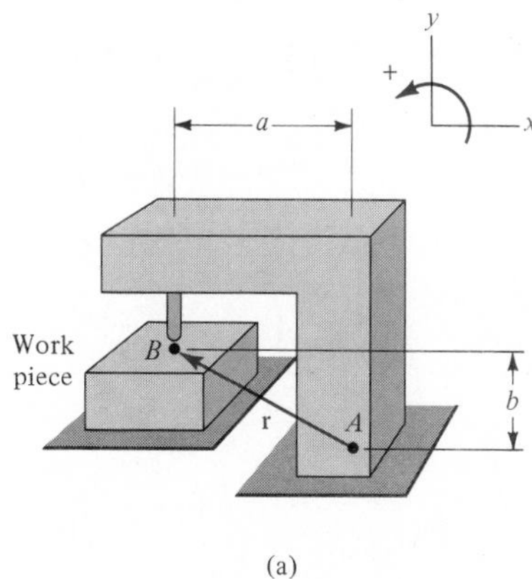

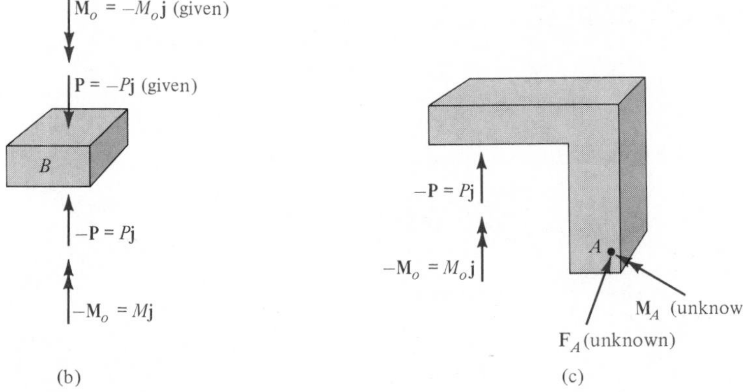

FIGURE 4-28
Model of a drill press

EXAMPLE 4-16

A motorcycle of weight W_1 pulling a one-wheeled trailer of weight W_2 is modeled in Fig. 4-29a. Bar BD is hinged at axle B but is rigidly attached at D. Determine the forces A_y, B_y, C_y, and F, the force applied by bar BD to axle B.

SOLUTION

From Fig. 4-29b,

$$\sum F_y = A_y + B_y - W_1 - F = 0 \qquad \textbf{(a)}$$

$$\sum M_A = -W_1 a + B_y(a + b) - F(a + b) = 0 \qquad \textbf{(b)}$$

From Fig. 4-29c,

$$\sum F_y = F + C_y - W_2 = 0 \qquad \textbf{(c)}$$

$$\sum M_B = C_y(c + d) - W_2 c = 0 \qquad C_y = \frac{W_2 c}{c + d} \qquad \textbf{(d)}$$

From Eq. (c),

$$F = \left(1 - \frac{c}{c + d}\right) W_2$$

From Eq. (b),

$$B_y = \frac{W_1 a}{a + b} + \left(1 - \frac{c}{c + d}\right) W_2$$

From Eq. (a),

$$A_y = W_1 + F - B_y = \left(1 - \frac{a}{a + b}\right) W_1$$

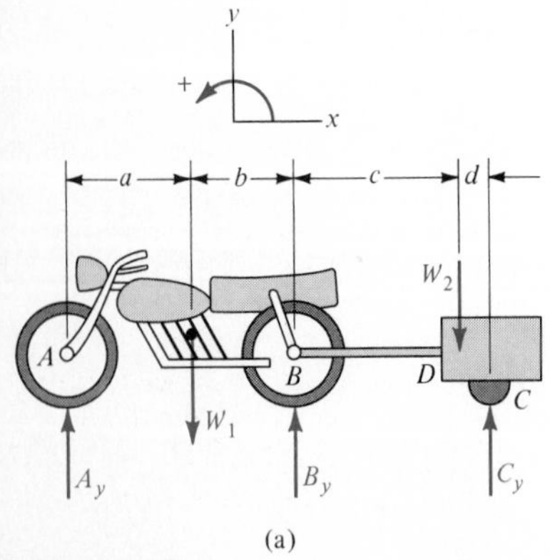

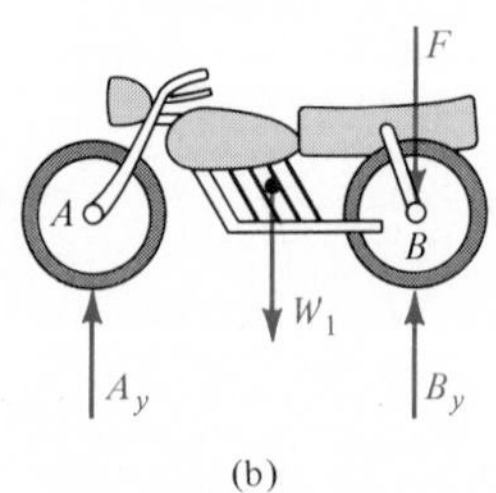

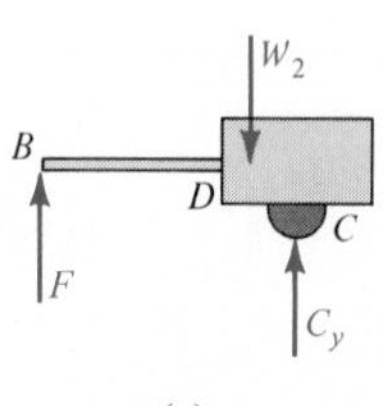

FIGURE 4-29
Analysis of a motorcycle with trailer

EXAMPLE 4-17

An airplane of weight W is resting on three wheels (Fig. 4-30), which are located at point O, point A of coordinates $(a, 0, b)$, and point B of coordinates $(a, 0, -b)$. The weight acts at point C of coordinates $(0.7a, h, 0)$. Determine the forces acting on the wheels.

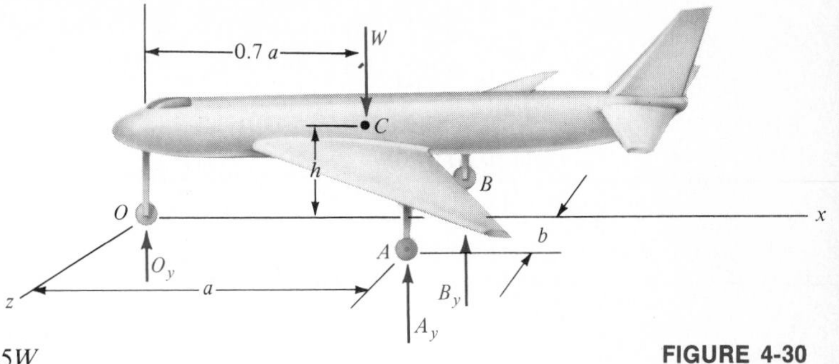

FIGURE 4-30

Model of an airplane resting on three wheels

SOLUTION

From Eq. 4-2,

$$\sum F_y = O_y + A_y + B_y - W = 0$$

$$\sum M_x = B_y b - A_y b = 0, \; A_y = B_y$$

$$\sum M_z = A_y a + B_y a - W(0.7a) = 0$$

$$2A_y a - W(0.7a) = 0, \; A_y = 0.35W$$

$$O_y = W - 2A_y = 0.3W$$

EXAMPLE 4-18

A helicopter of weight W in a test is attached to a fixture at A, which allows rotations about the y axis with negligible friction but prevents other significant displacements (Fig. 4-31a). The main rotor applies a vertical lift force L while overcoming a moment M_D of air drag forces acting on the blades, which rotate clockwise when viewed from above. The tail rotor at B is in the xy plane and applies a thrust P to oppose the moment M_D. Determine the reactions at A if the helicopter's body should not move.

FIGURE 4-31

Test setup of a helicopter

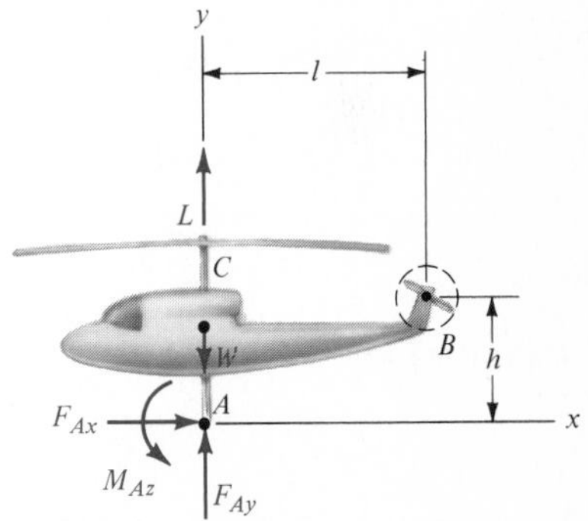

(a) Side view

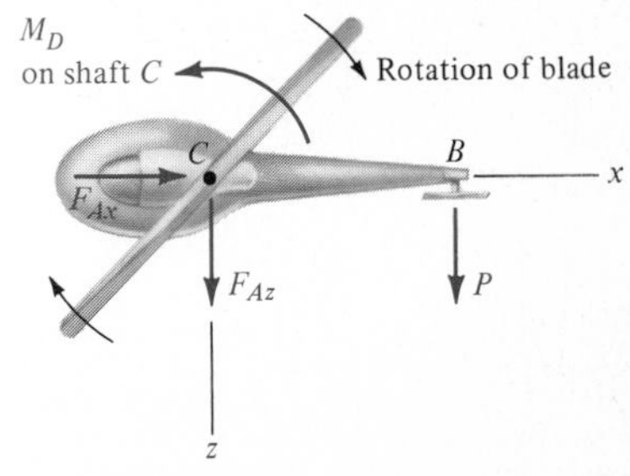

(b) Top view

SOLUTION

Force P is perpendicular to xy, so it cannot be shown in Fig. 4-31a. From the top view in Fig. 4-31b, P must be in the z direction to oppose M_D (here it is not possible to show the forces L, F_{Ay}, and W). From Fig. 4-31b,

$$\sum M_A = M_D - Pl = 0 \qquad P = \frac{M_D}{l}$$

The reactions F_{Ax}, F_{Ay}, F_{Az}, M_{Ax}, M_{Ay}, and M_{Az} are obtained using Eq. 4-2,

$$\sum F_x = F_{Ax} = 0$$

$$\sum F_y = F_{Ay} + L - W = 0 \qquad F_{Ay} = W - L$$

$$\sum F_z = F_{Az} + P = 0 \qquad F_{Az} = -P$$

$$\sum M_x = Ph + M_{Ax} = 0 \qquad M_{Ax} = -Ph$$

$$\sum M_y = 0 \quad \text{by statement of the problem}$$

$$\sum M_z = M_{Az} = 0$$

PROBLEMS

SECTION 4-6

FIGURE P4-62

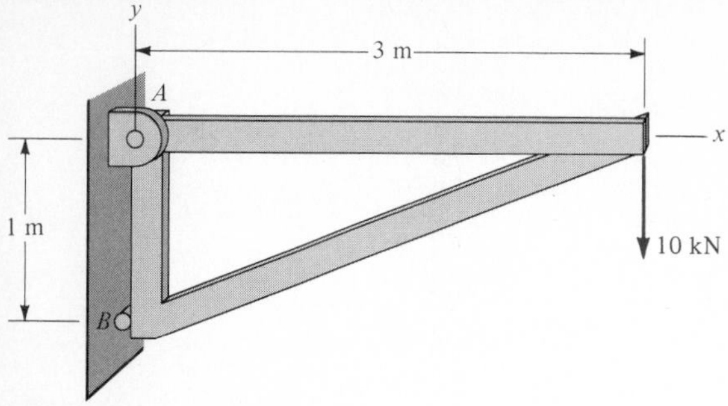

4–62 The boom of the light crane is supported by a smooth pin at A and a roller at B. Calculate the reactions at A and B.

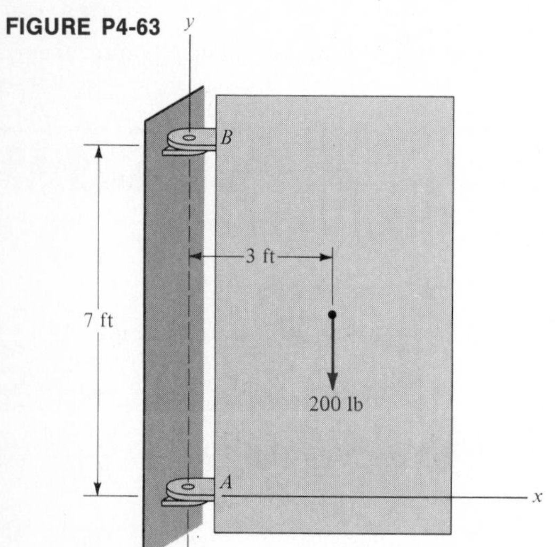

FIGURE P4-63

4–63 The door is hinged at A and B, with the latter not supporting any vertical load. Determine the reactions at the two hinges.

FIGURE P4-64

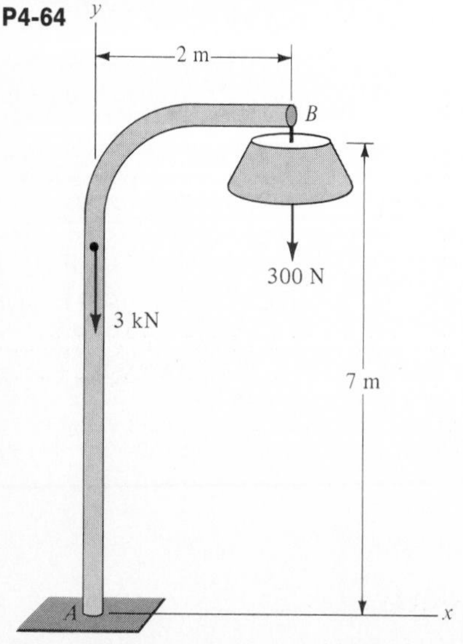

4–64 Determine the reactions at A for the street light shown.

4–65 The effect of wind on the street light B in Fig. P4-64 is represented by a concentrated force $\mathbf{F} = -500\text{ N }\mathbf{k}$. Calculate the reactions at A due to all three forces.

4–66 Determine the reactions at A and O due to the three forces acting on the highway sign. $\mathbf{F} = -200 \text{ lb } \mathbf{i}$, $\mathbf{P} = -130 \text{ lb } \mathbf{k}$, and $\mathbf{W} = -300 \text{ lb } \mathbf{k}$.

FIGURE P4-66

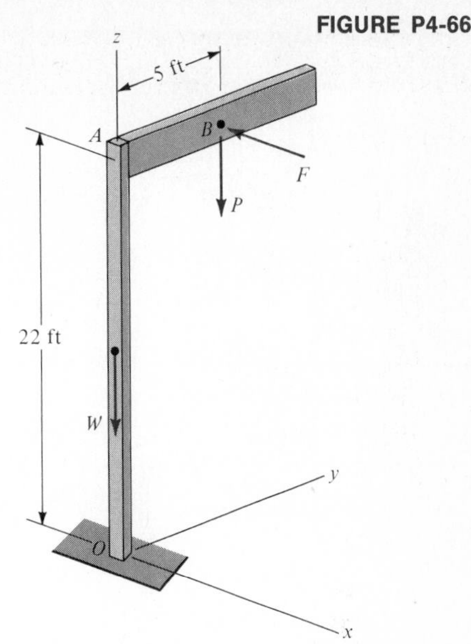

4–67 Consider the addition of a guy wire at point A to the structure described in Prob. 4–66. The wire is also attached to point C of coordinates $(0, -6 \text{ ft}, 0)$. The tension in the wire is adjusted to eliminate the moment of force P about point O. Calculate the resulting reactions at O.

4–68 The gross weight of a cart is $W = 8$ kN acting at point C, halfway between the wheels. Calculate the tension in the cable that holds the cart and the reactions on wheels A and B.

FIGURE P4-68

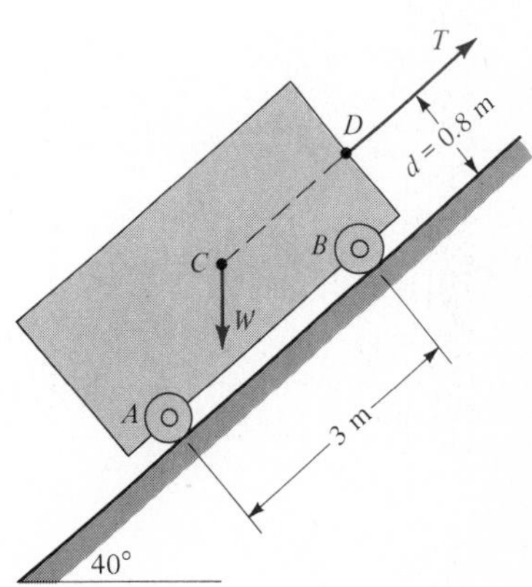

4–69 Plot the reactions on wheels A and B for $d = 0.5$ m, 0.8 m, and 1.1 m of the cable in Prob. 4–68. Assume that C remains at a distance of 0.8 m from the track.

FIGURE P4-70

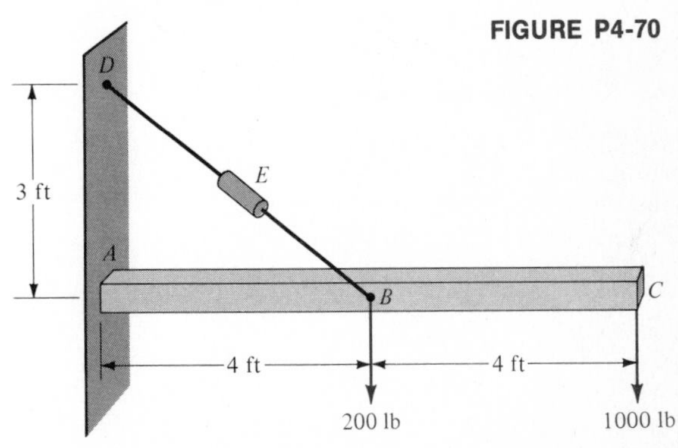

4–70 The beam AC is held rigidly at A, and it is also supported by the cable BD. The tension in the cable is 300 lb after the turnbuckle E is adjusted. Determine the reactions at A.

FIGURE P4-71

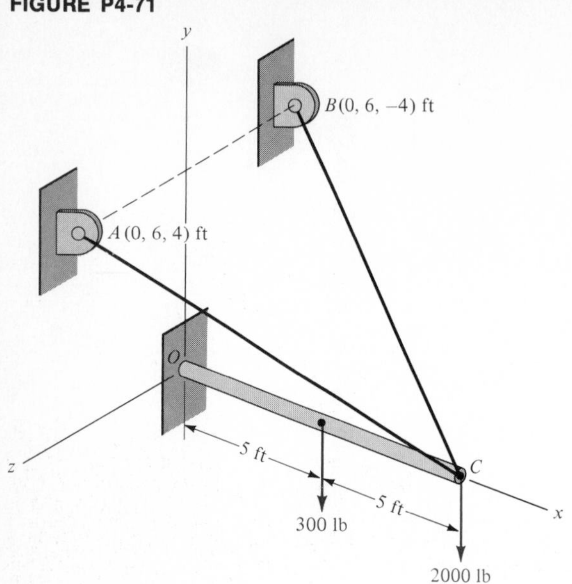

4–71 The horizontal boom of the crane is hinged at O and is held by cables at A and B. Determine the reactions at O.

4–72 A member of a mechanism is supported by smooth pins A and B in vertical and horizontal tracks, respectively. Calculate the reactions on the pins.

4–73 Plot the reactions A_x, A_y, B_x, and B_y in Prob. 4–72 for $\theta = 40^\circ, 50^\circ$, and 60°. Assume that force P is always in the given direction.

4–74 The aircraft rests on landing gears at A, B, and D. The total weight is represented by a concentrated force $W = 50{,}000$ lb acting at C (in the negative z direction in the given coordinate system). Determine the reactions at wheels A, B, and D.

FIGURE P4-72

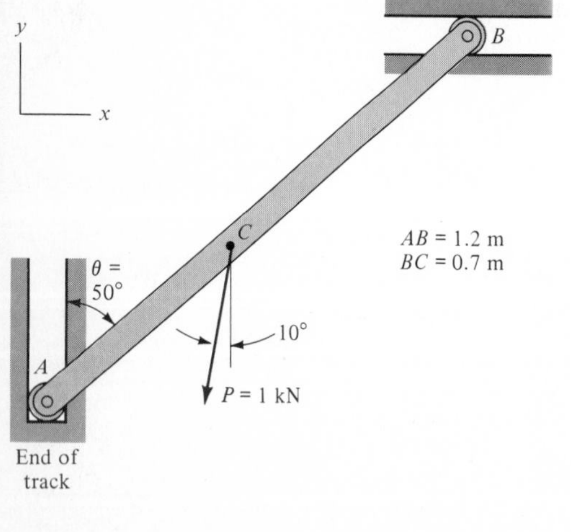

FIGURE P4-74

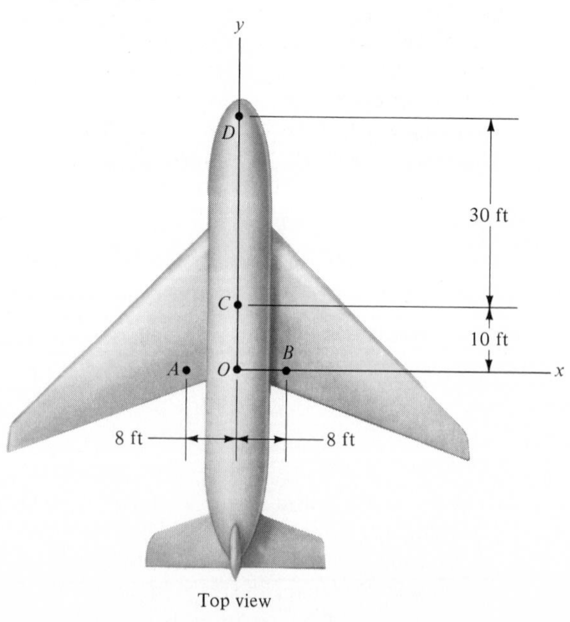

4–75 The piping system ABC delivers concrete from the mixer truck. The pipe weighs 1.5 kN per meter along the whole pipe. AB is 15 m long and makes a 50° angle with the horizontal. In the same vertical plane, the part BC is 12 m long and makes an 80° angle with the horizontal. Determine the reactions at A.

FIGURE P4-75

4–76 An antenna tower is modeled as a rigid bar pinned at the bottom and supported by three identical guy wires (top view). The effect of wind is represented by a concentrated force that causes wire C to become slack. Determine the reactions at O, A, and B.

FIGURE P4-76

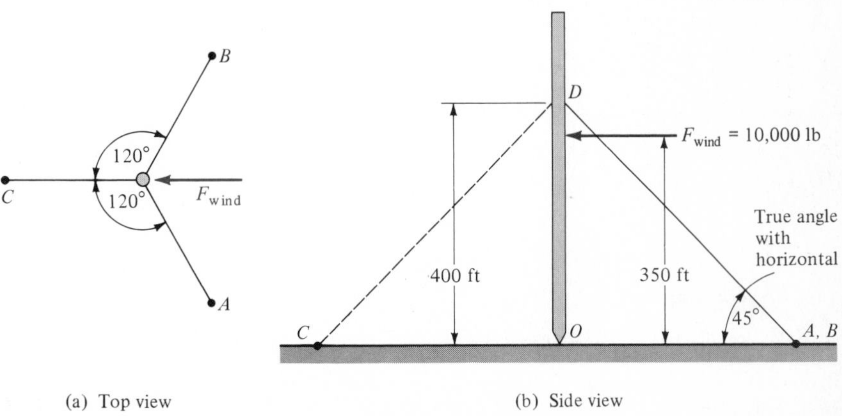

FIGURE P4-77

Loading frame for testing an airplane. (Photograph courtesy The Boeing Company, Seattle, Washington.)

(a)

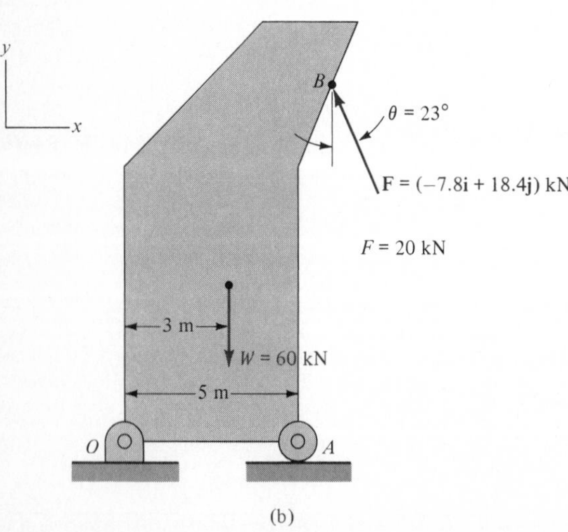

(b)

4–77 Assume that the load reaction frame in Fig. P4-77a is a two-dimensional structure. Point O is the origin of the chosen coordinate system and the location of braked supporting wheels. Wheels A are at $x_A = 5$ m and are free to roll. The structure has weight $W = 60$ kN acting at $x_W = 3$ m. The actuator force at point B ($x_B = 7.4$ m, $y_B = 13.5$ m) is $\mathbf{F} = (-7.8\mathbf{i} + 18.4\mathbf{j})$ kN. Determine the reactions at wheels O and A.

4–78 Draw a free-body diagram of the rear view of the helicopter in Fig. 4-31 (Ex. 4-18). Write the equilibrium equations relevant to this diagram.

SUMMARY

Section 4-1

CONDITIONS OF EQUILIBRIUM

VECTOR NOTATION:

$$\sum \mathbf{F} = 0$$
$$\sum \mathbf{M}_O = \sum (\mathbf{r} \times \mathbf{F}) = 0$$

where O is an arbitrary point of reference.

SCALAR NOTATION:

$$\sum F_x = 0 \qquad \sum M_x = 0$$
$$\sum F_y = 0 \qquad \sum M_y = 0$$
$$\sum F_z = 0 \qquad \sum M_z = 0$$

MAXIMUM NUMBER OF EQUATIONS FOR ONE BODY:

1. One-dimensional problem: $\sum F = 0$
2. Two-dimensional problem: $\sum F_x = 0$, $\sum F_y = 0$, $\sum M_z = 0$ (or permutations of subscripts)
3. Three-dimensional problem: $\sum F_x = 0$, $\sum F_y = 0$, $\sum F_z = 0$
   $\sum M_x = 0$, $\sum M_y = 0$, $\sum M_z = 0$

Section 4-2

FREE-BODY DIAGRAMS: In solving for unknown forces or moments, always draw the free-body diagram first. The directions of unknowns may be assumed arbitrarily.

Section 4-3

CALCULATION OF UNKNOWNS: The number of unknowns that can be determined is the same as the number of independent equilibrium equations available. Only statically determinate members can be analyzed using the methods of statics.

Section 4-4

TWO-FORCE MEMBERS:

1. Wires, chains, ropes: only tension may be applied to these
2. Rigid members: tension or compression may be applied to these

GENERAL RULE: A body subjected to forces applied at two points is in equilibrium only when the resultant forces at those points have the same line of action and are opposite in direction.

Section 4-5

INTERRELATED FREE-BODY DIAGRAMS: When two or more bodies are in contact, separate free-body diagrams may be drawn for each body. The mutual forces and moments between the bodies are related according to Newton's Third Law (action and reaction). The directions of unknown forces and moments may be arbitrarily assumed in one diagram, but these initial choices affect the directions of unknowns in all other related diagrams. The number of unknowns and of usable equilibrium equations both increase with the number of related free-body diagrams.

SCHEMATIC EXAMPLE:

Given: F_1, F_2, F_3, M (all in the xy plane)

Unknowns: P_1, P_2, P_3, and forces and moments at joint A (rigid connection)

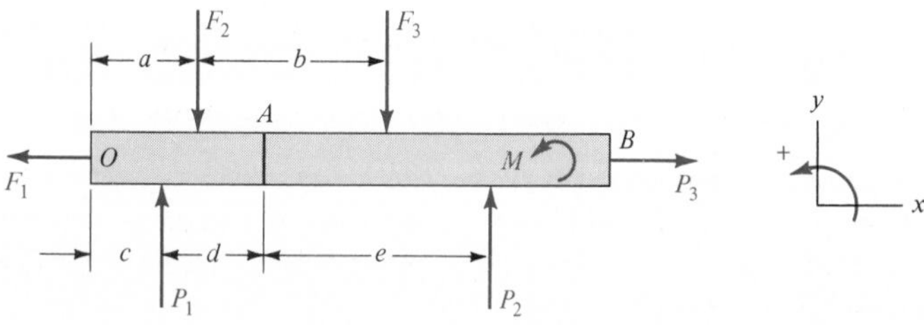

EQUILIBRIUM EQUATIONS:

$$\sum F_x = -F_1 + P_3 = 0$$
$$\sum F_y = P_1 + P_2 - F_2 - F_3 = 0$$
$$\sum M_O = P_1 c + P_2(c+d+e) + M - F_2 a - F_3(a+b) = 0$$

Three unknowns (P_1, P_2, P_3) in three equations.

RELATED FREE-BODY DIAGRAMS:

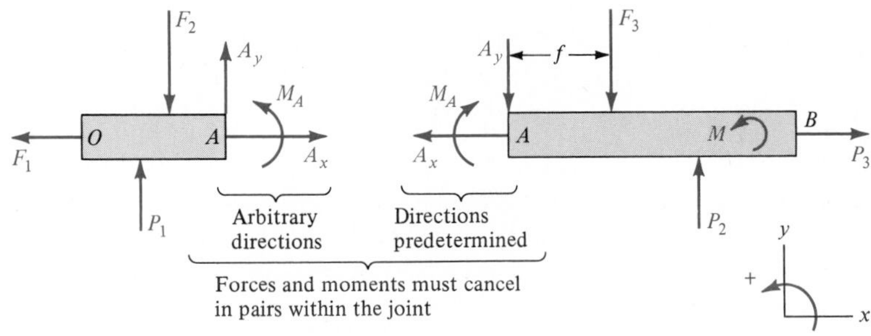

Dimensions a, b, c, d and e are also valid here.

NEW SET OF EQUILIBRIUM EQUATIONS:

Left part: (OA)

$$\sum F_x = -F_1 + A_x = 0$$
$$\sum F_y = P_1 + A_y - F_2 = 0$$
$$\sum M_O = P_1 c + A_y(c+d) + M_A - F_2 a = 0$$

Right side: (AB)

$$\sum F_x = -A_x + P_3 = 0$$
$$\sum F_y = P_2 - A_y - F_3 = 0$$
$$\sum M_A = -M_A + P_2 e + M - F_3 f = 0$$

Six unknowns (P_1, P_2, P_3, A_x, A_y, M_A) in six equations.

NOTE: In the first diagram, the couple M may be moved anywhere from O to B. M is not shown in the second diagram because it is shown in the third diagram (in which it may be moved anywhere from A to B).

REVIEW PROBLEMS

FIGURE P4-79

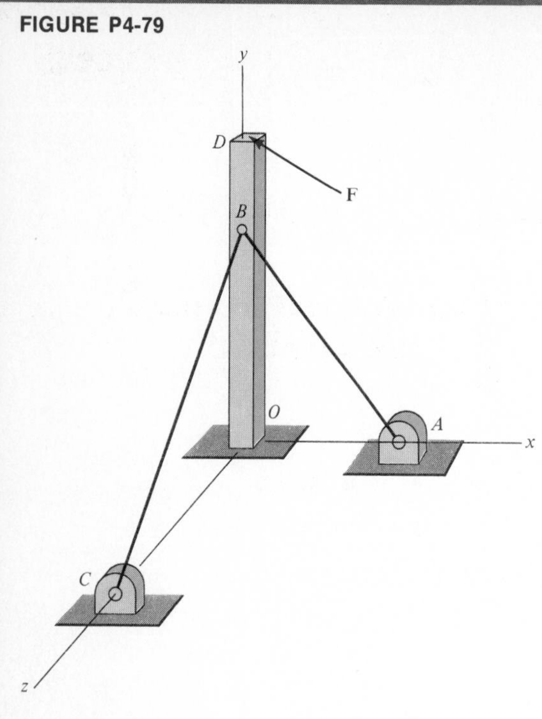

4–79 The mast is hinged at O to allow rotation in any direction. The tensions in the wires are along BA and BC. Calculate the tensions T_A and T_C if $\mathbf{F} = (-200\mathbf{i} - 100\mathbf{k})$ lb, $OA = OC = 8$ ft, $OB = 12$ ft, and $OD = 15$ ft.

4–80 A tool is modeled as a rigid member OA. Determine the reactions at O if $\mathbf{F} = -F\mathbf{k}$ and $\mathbf{M} = M\mathbf{j}$ are applied at point A. Arm d is parallel to the x axis.

4–81 A 20-ft ladder is set against a wall. It is on grass at A and acts as if it were pinned. The wall is smooth and essentially frictionless at B. Calculate the changes in reactions at A and B when a 200-lb person moves from C to D, if $AC = 7$ ft and $AD = 14$ ft.

FIGURE P4-80

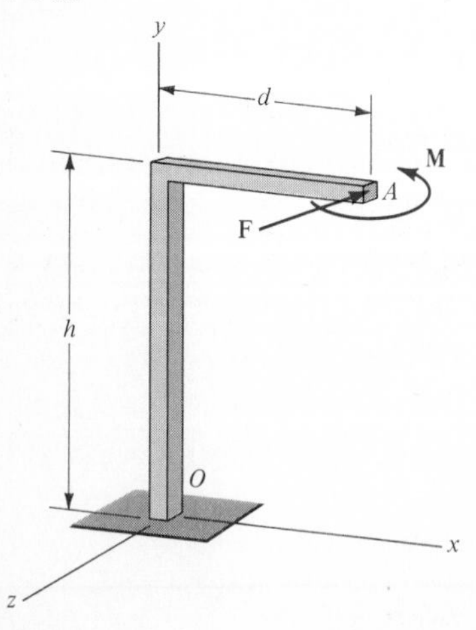

FIGURE P4-81

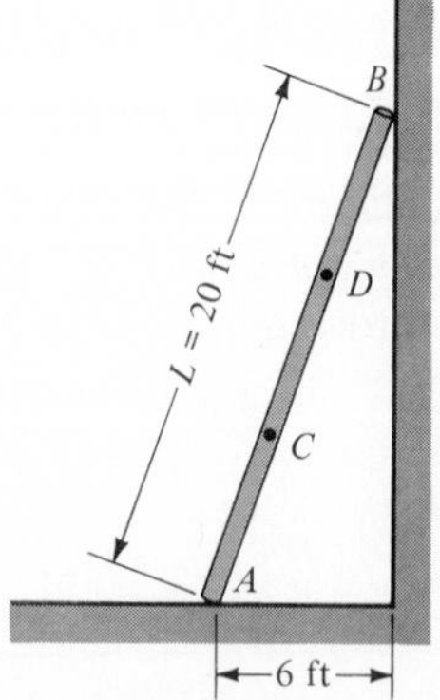

4–82 A hand-crank hoist is designed to have a ratcheting crank arm DE with a 60° range of motion. A and B are fixed sleeve bearings, and the wire holding the weight W is wound on the drum C. Determine the reactions A_z and A_y when the crank arm DE is parallel to the y axis (position 2). Assume that F is always perpendicular to DE, and the ratchet applies no force.

FIGURE P4-82

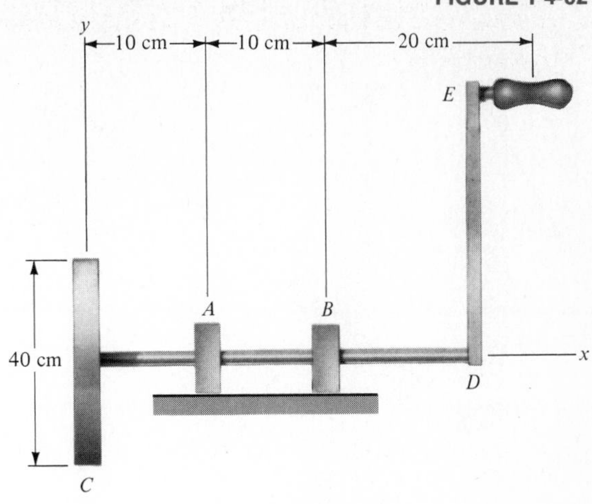

(a) Side view

4–83 Calculate the reactions at A and B in Prob. 4–82 when arm DE is in position 1 (Fig. P4-82c).

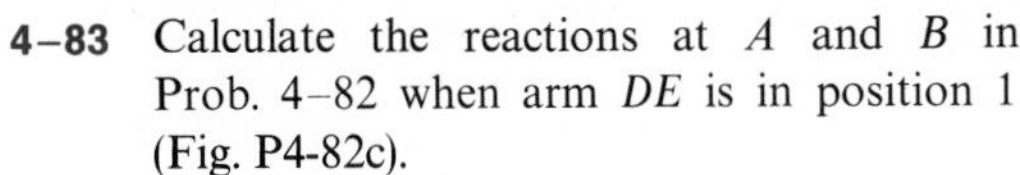

4–84 Assume that in Prob. 4–33 the positions of B and C are fixed but bearing A could be placed ± 5 cm from the position indicated. If the bearing reactions should be minimized, where should bearing A be placed on the x axis?

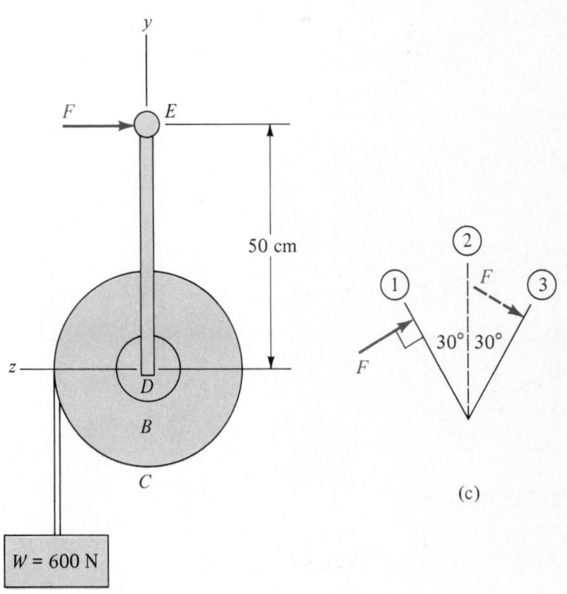

(b) End view

(c)

FIGURE P4-85

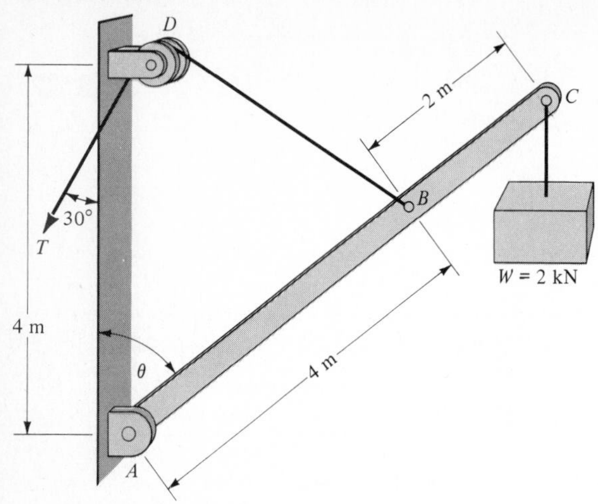

4–85 The boom of a light crane is hinged at A and is raised or lowered by a rope going over the pulley at D. Determine the reactions at A and D for $\theta = 60°$.

4–86 The medical X-ray apparatus has adjustable joints at the following points, with 0 at the origin of the coordinates:

A: $x_A = 0$, $y_A = 0.1$ m, $z_A = 0.5$ m

B: $x_B = -0.1$ m, $y_B = 0.2$ m, $z_B = 1.0$ m

C: $x_C = -0.1$ m, $y_C = 0.6$ m, $z_C = 1.0$ m

D: $x_D = 0.55$ m, $y_D = 1.1$ m, $z_D = 0.9$ m

E: $x_E = 1.1$ m, $y_E = 0.85$ m, $z_E = 0.9$ m

F: $x_F = 1.1$ m, $y_F = 0.4$ m, $z_F = 0.9$ m

Assume weightless members and a 100-N weight for the device at F. Determine the force and moment reactions at joints O, A, B, C, D, and E.

FIGURE P4-86

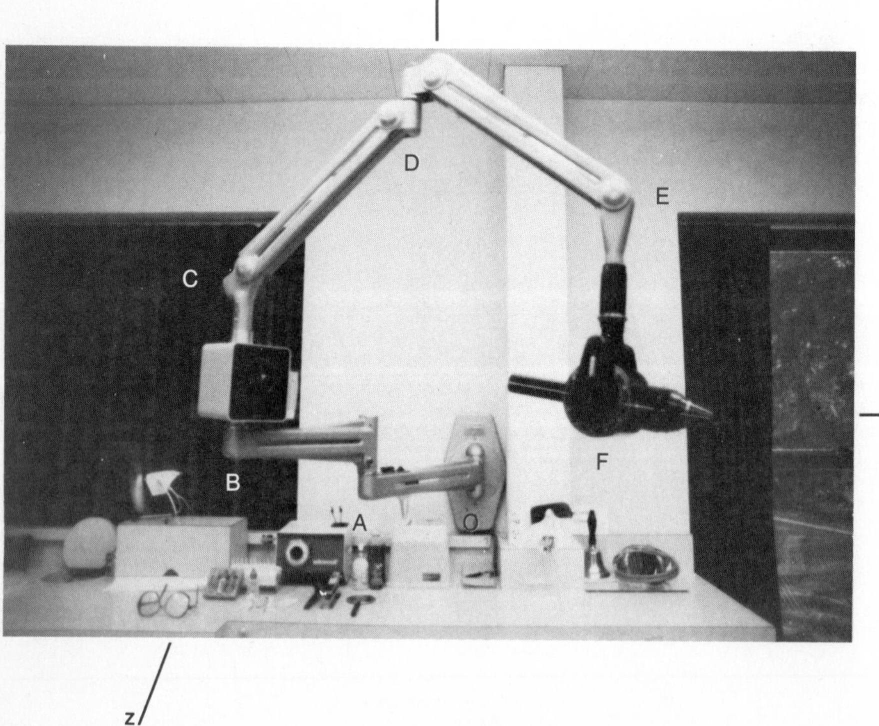

CHAPTER 5

FORCES AND MOMENTS IN BEAMS

OVERVIEW

Basic Concepts in Modeling Beams

Beams are among the most common structural members. Frequently, beams are rather long in comparison with their lateral dimensions, and at least some of the loads acting on them cause bending of the beams. The bending of beams is often very important even if the deformations are not readily noticed. The following analyses of beams can be made using the methods of statics: (1) The external (supporting) reactions acting on a statically determinate beam can be calculated using the methods of Ch. 4; and (2) subsequently, the internal reactions (forces and moments) at any cross section in the beam can also be determined using the methods of Ch. 4. These analyses are essential background for investigating the geometric changes or safety aspects of real beams using the methods of mechanics of materials. Interestingly, the more advanced analysis of a beam (such as of its deformability) would not be possible without first assuming that it is perfectly rigid.

STUDY GOALS

SECTION 5-1 describes several common categories of supports. The reasonable modeling of the supports is indispensible in solving engineering problems. Quite frequently this can be done by inspection.

SECTIONS 5-2 AND 5-3 extend the methods developed in Ch. 4 for determining internal forces and moments at any section in a beam. These formalize the methods to make them efficient and uniform for the analysis of many kinds of beams.

SECTION 5-4 presents shear force and bending moment diagrams, which show at a glance how the internal reactions vary along a beam. These diagrams are important in locating the critical sections in a beam, and are used extensively in mechanics of materials.

NOTE: This chapter is devoted mainly to the analysis of beams on which concentrated forces are acting. Occasionally, couples acting on beams are also considered. The important but more complex problems of distributed loads on beams are analyzed in Ch. 7. This allows the readers to obtain a firm background in solving problems of beams and those relevant to beams.

5-1 CLASSIFICATION OF SUPPORTS

Several common supports for beams are shown in Fig. 5-1. The nature of the support determines the unknown reactions that must be shown at the appropriate place in the free-body diagram. The complete set of possible unknowns for two-dimensional loading of the beams in Fig. 5-1 is tabulated in Fig. 5-2. Various combinations of these supports are also common. For example, a roller may be placed under the right end of the cantilever beam in Fig. 5-lb.

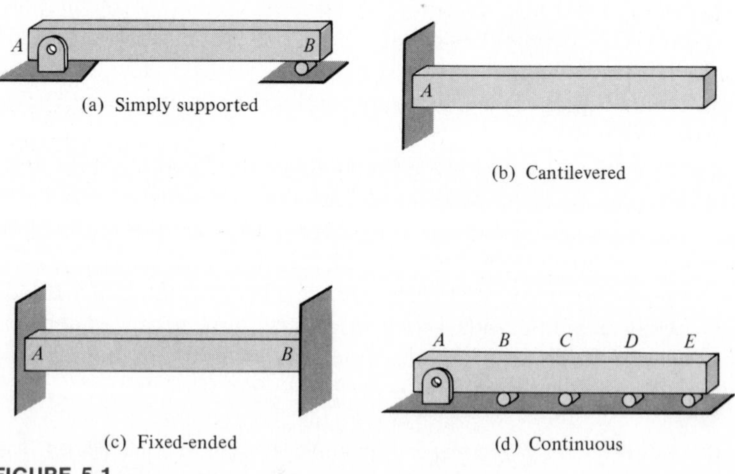

FIGURE 5-1

Common supports for beams

Beam	Possible reactions					
	R_{Ax}	R_{Ay}	M_A	R_{Bx}	R_{By}	M_B
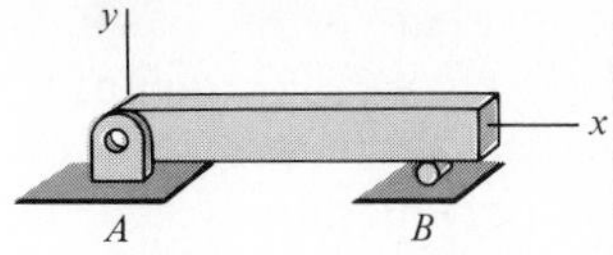 A, B (pin at A, roller at B)	√	√			√	
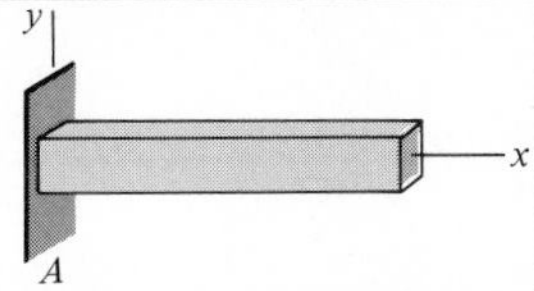 A (fixed at A)	√	√	√			
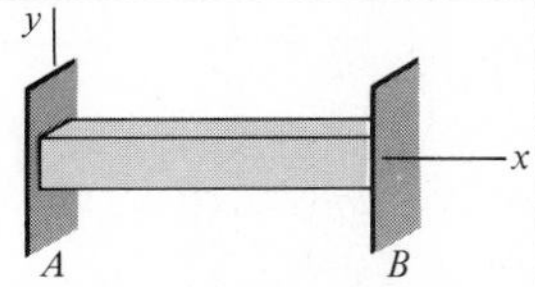 A, B (fixed at A and B)	√	√	√	√	√	√
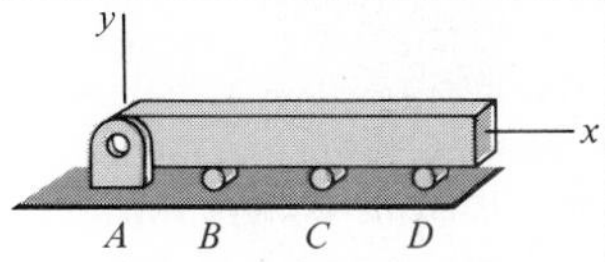 A, B, C, D (pin at A, rollers at B, C, D)	√	√			√ similarly for C, D	

FIGURE 5-2

Possible unknowns for common beams

INTERNAL FORCES AND MOMENTS 5-2

The internal forces and moments (also called *internal reactions*) are essential for analyzing the strength of a beam. These reactions are determined using the concept that *if a beam is in equilibrium, any part of it is also in equilibrium.* Consider a cantilever beam with an external force F for a simple example, as in Fig. 5-3a. The free-body diagram is shown in part b of the figure. The reactions V_O (called *vertical shear force,* or *transverse shear force,* in general) and M_O are required for equilibrium, but their directions are arbitrarily assumed. From Eq. 4-2,

$$\sum F_y = 0$$

$$-F + V_O = 0, \qquad V_O = F$$

$$\sum M_O = 0$$

$$M_O - FL = 0, \qquad M_O = FL$$

Next, consider a portion of the beam, that to the right of section A. This part is also in equilibrium, and its free-body diagram is drawn in Fig. 5-3c. Note that V_O and M_O are not shown here because they are not acting directly on this part of the beam. Of course, there must be reactions at A to keep the part in equilibrium. At this stage the internal reactions V_A (vertical shear) and M_A (often called *bending moment**) are unknown, but they are readily calculated from the equilibrium equations,

$$\sum F_y = 0$$

$$-F + V_A = 0, \qquad V_A = F$$

$$\sum M_A = 0$$

$$M_A - Fl = 0, \qquad M_A = Fl$$

Clearly, any other arbitrary length x may be substituted for l, so the internal reactions as functions of x are $V(x)$ and $M(x)$,

$$V(x) = F \qquad \text{and} \qquad M(x) = Fx \tag{5-1}$$

which are valid for $0 < x < L$. Note that it is best to take moments about a point in the section of interest (point A here) because the shear force on the section has no moment about that point.

The free-body diagram of the part to the left of A is drawn in Fig. 5-3d. The internal reactions here are equal in magnitude and opposite to those in Fig. 5-3c by the principle of action and reaction. One could say that V_A and M_A in part c are caused by the left part of the beam (in a sense, they are the remote effects of V_O and M_O). Similarly, V_A and M_A in part d are caused by the right part of the beam (the remote effects of F).

It is reasonable to check if the piece $L - l$ is in equilibrium. Working with point O,

$$\sum F_y = V_O - V_A = F - F = 0$$

$$\sum M_O = M_O - M_A - V_A(L - l)$$

$$= FL - Fl - F(L - l) = 0$$

so the part of Fig. 5-3d is in equilibrium. It is equally reasonable to work with point A on the piece $L - l$, which leads to the same results,

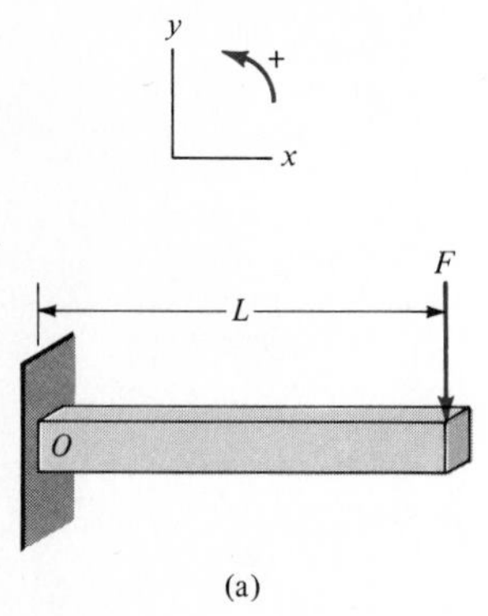

(a)

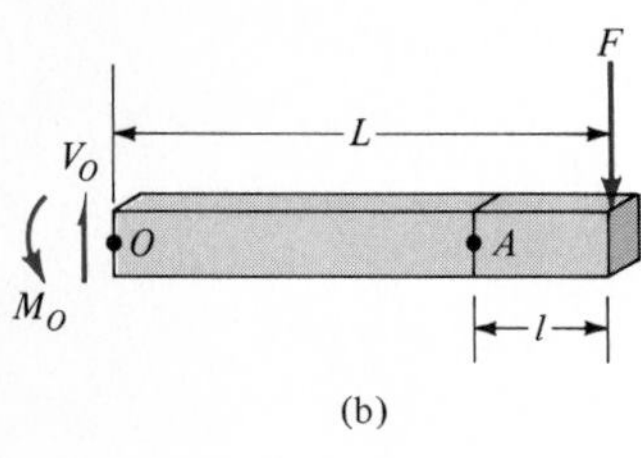

(b)

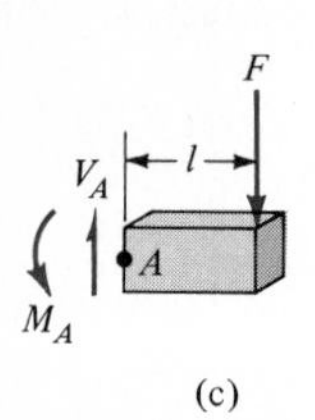

(c)

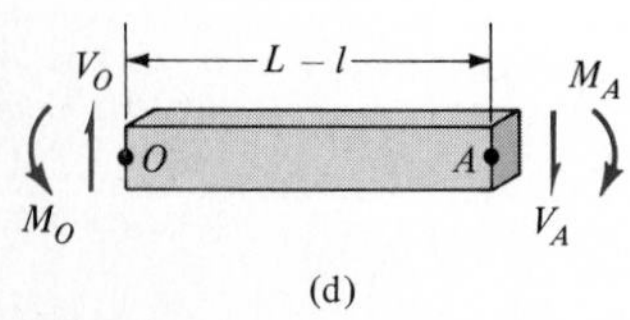

(d)

FIGURE 5-3

Internal forces and moments in a cantilever beam

* Sometimes it is necessary to distinguish a *bending moment* from a *twisting moment*. All bending moment vectors are perpendicular to the longitudinal axis of the beam, while twisting moment vectors are along that axis. These can be demonstrated by bending or twisting a pencil by hand.

$$\sum F_y = V_O - V_A = F - F = 0$$

$$\sum M_A = M_O - M_A - V_O(L - l)$$
$$= FL - Fl - F(L - l) = 0$$

The cantilever beam is interesting since the internal reactions on any transverse section along the beam may be calculated directly, without first calculating the reactions at the fixed end. The simply supported beam is slightly more complicated in this respect. Its reactions must be calculated first; otherwise, there are too many unknowns in the equilibrium equations. For an example, consider the beam in Fig. 5-4a. From Fig. 5-4b and Eq. 4-2,

$$\sum F_y = 0$$

$$R_A + R_B - F = 0$$

$$\sum M_A = 0$$

$$2LR_B - LF = 0, \quad R_B = \frac{F}{2}, \quad \text{so } R_A = R_B = R = \frac{F}{2}$$

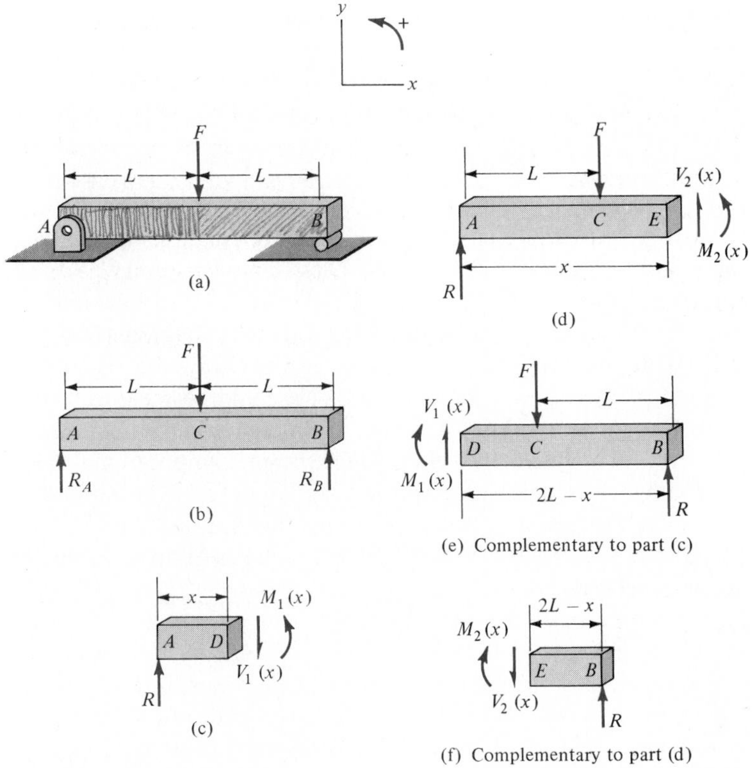

(a)

(b)

(c)

(d)

(e) Complementary to part (c)

(f) Complementary to part (d)

FIGURE 5-4

Internal forces and moments in a simply supported beam

Next consider an element of length x at the left end of the beam. Its free-body diagram is shown in Fig. 5-4c. Here everything is unknown if the reaction R is not computed first, and nothing can be calculated. If R is known, $V_1(x)$ and $M_1(x)$ are readily calculated for any given x up to the length L.

$$\sum F_y = 0$$

$$R - V_1(x) = 0, \qquad V_1(x) = R$$

$$\sum M_D = 0$$

$$-Rx + M_1(x) = 0, \qquad M_1(x) = Rx$$

If x were to get larger than L, a new free-body diagram and new equilibrium equations would be needed to include the force F, as shown in Fig. 5-4d. Here $V_2(x)$ is clearly upward because F is larger than R. The magnitudes of the internal reactions are obtained from the new equilibrium equations. Again using Eq. 4-2,

$$\sum F_y=0$$

$$R+V_2(x)-F=0, \qquad V_2(x)=F-R=\frac{F}{2}$$

$$\sum M_E=0$$

$$-Rx+F(x-L)+M_2(x)=0, \quad M_2(x)=Rx-F(x-L)$$

$$M_2(x)=-\frac{Fx}{2}+FL$$

It is worthwhile to draw the free-body diagrams of the parts adjoining those in Fig. 5-4c and d. These complementary parts are shown in Fig. 5-4e and f, respectively. Newton's Third Law must be obeyed when drawing the interrelated free-body diagrams.

Note that in Fig. 5-4 the dimension x is always measured from point A to the right. This is arbitrary; x could be measured from point B to the left without making any difference in the final results. However, there may be a notable difference in computational speed depending on which way a coordinate is chosen. This should be clear after thoughtful comparison of parts c, d, e, and f in Fig. 5-4. Experience with numerous free-body diagrams helps in choosing the simplest procedures.

Beams can also have axial forces acting on them, and these are easily added to the internal reactions already discussed. For example, an axial tension T is added to the beam in Fig. 5-4a, as shown in Fig. 5-5a. Now the free-body diagram of an element at the left end of the beam includes the tension, as shown in Fig. 5-5b. The presence of this tension does not affect any of the other external or internal reactions.

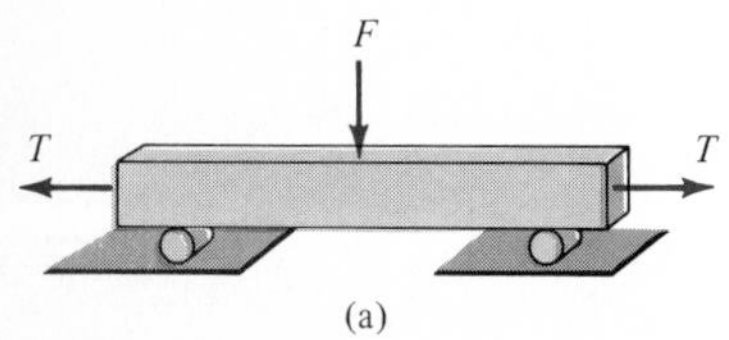

(a)

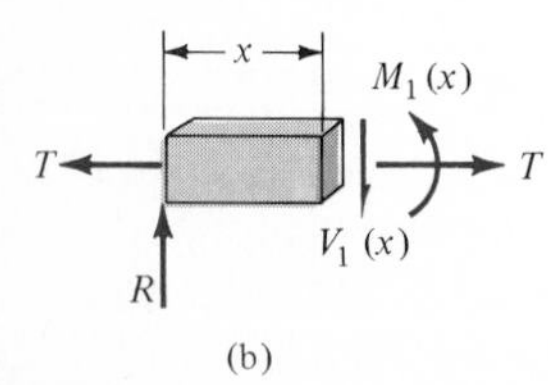

(b)

FIGURE 5-5

Internal reactions in a biaxially loaded beam

5-3 SIGN CONVENTIONS

At the beginning of the analysis of any problem the signs for the internal force and moment may be arbitrarily chosen. It is only essential to adhere to the chosen system of signs throughout the solution of that particular problem. Nevertheless, most people prefer to follow a custom in analyzing beams.

According to custom, a moment that makes a beam concave upward (not necessarily visible to the naked eye) is taken as positive. This is illustrated in Fig. 5-6. The element in part a of the figure is in equilibrium with the moments M acting on it. These moments are shown as force couples in Fig. 5-6b. The top fibers of the element are shortened because of the compressive force on them, and the bottom fibers are lengthened because they are in tension. A bending distortion of the element results as shown in Fig. 5-6c. This deformation is often negligible, so the assumption of ideal rigidity is reasonable. The magnitudes of such deformations are analyzed with the methods of strength of materials.

The sign convention includes that a shear force acting downward on the right side of a section, or upward on the left side, is positive. Thus, both shear forces V are positive in Fig. 5-6d.

The signs of the moment and shear acting on a given section may or may not be the same. Figure 5-7 shows the possible combinations according to the common notation.

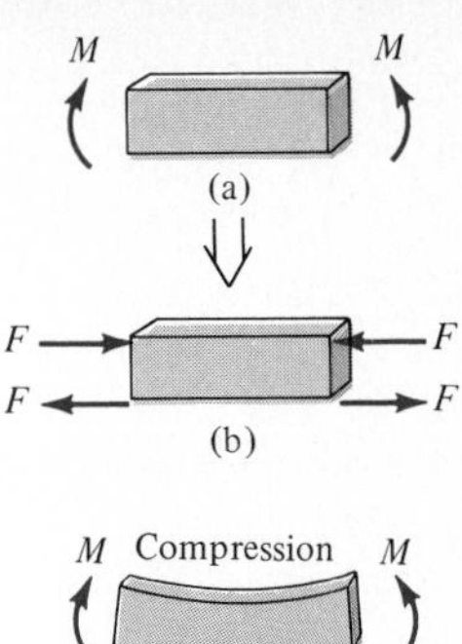

(c) Beam concave upward; moments are positive here

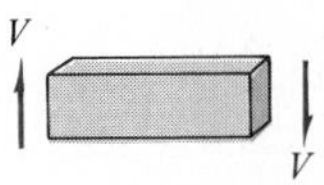

(d) Shear forces are positive here

FIGURE 5-6
Sign convention for shear force and bending moment

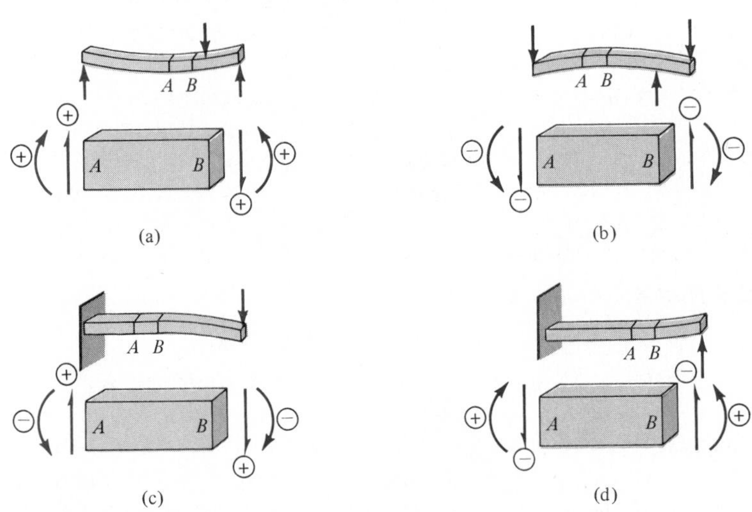

FIGURE 5-7
Possible combinations of signs of shear force and bending moment

EXAMPLE 5-1

Assume that a skyscraper with wind acting on it is modeled for convenience as in Fig. 5-8a. Determine the vertical shear and the bending moment at sections *A*, *B*, *C*, *D*, and *E* in the cantilever beam.

SOLUTION

Free-body diagrams are drawn for each section of interest. For each calculation, either the left or right portion of the beam can be considered. In this problem the right-hand section makes for easiest calculations. Note that all internal shear forces and bending moments are assumed positive. Equation 4-2 is repeatedly used as follows.

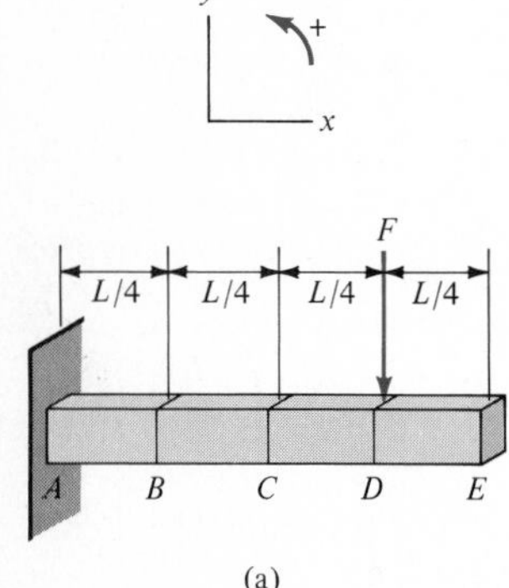

SECTION *A*: From Fig. 5-8b,

$$\sum F_y = 0 \qquad V_A - F = 0 \qquad V_A = F$$

$$\sum M_A = 0 \qquad -F\left(\frac{3L}{4}\right) - M_A = 0 \qquad M_A = -\frac{3}{4}FL$$

SECTION *B*: From Fig. 5-8c,

$$\sum F_y = 0 \qquad V_B - F = 0 \qquad V_B = F$$

$$\sum M_B = 0 \qquad -F\left(\frac{L}{2}\right) - M_B = 0 \qquad M_B = -\frac{1}{2}FL$$

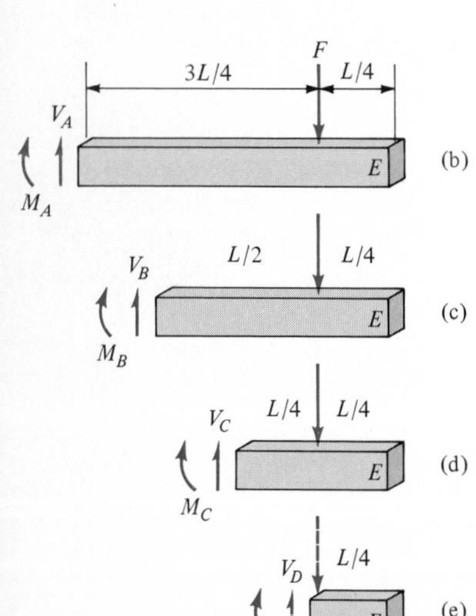

SECTION *C*: From Fig. 5-8d,

$$\sum F_y = 0 \qquad V_C - F = 0 \qquad V_C = F$$

$$\sum M_C = 0 \qquad -F\left(\frac{L}{4}\right) - M_C = 0 \qquad M_C = -\frac{1}{4}FL$$

SECTION *D*: From Fig. 5-8e,

$\sum F_y = 0$ $\quad$ *V* is undefined in the immediate neighborhood of a concentrated load (it is not known precisely how *V* and *F* act at the same section *D*)

$$\sum M_D = 0 \qquad -F(0) - M = 0 \qquad M_D = 0$$

SECTION *E*: From Fig. 5-8f,

$$\sum F_y = 0 \qquad V_E - 0 = 0 \qquad V_E = 0$$

$$\sum M_E = 0 \qquad M_E + 0 = 0 \qquad M_E = 0$$

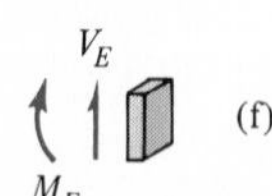

FIGURE 5-8
Analysis of a cantilever beam

JUDGMENT OF THE RESULTS

As expected from Fig. 5-8b, c, and d, $V_A = V_B = V_C = F$. The moment did vary by reasonable increments with the section considered, however. The obtained signs for *V* and *M* can be accepted as final since all internal reactions were assumed positive.

EXAMPLE 5-2

Determine the internal forces and moments at sections A, B, and C in the simply supported beam shown in Fig. 5-9a.

SOLUTION

First the reactions at supports 1 and 2 must be determined. From Fig. 5-9b for the whole beam, and using Eq. 4-2,

$$\sum F_x = 0 \qquad R_{1x} = 0$$

$$\sum F_y = 0 \qquad R_{1y} + R_2 - 2F - F = 0$$

$$\sum M_1 = 0 \qquad -2F\left(\frac{L}{3}\right) - F\left(\frac{2L}{3}\right) + R_2(L) = 0$$

$$R_2 = \frac{4}{3}F$$

From $\sum F_y = 0$, $R_{1y} = \frac{5}{3}F$.

SECTION *A*: Choosing the section to the left of A (Fig. 5-9c), and taking V_A and M_A as positive,

$$\sum F_y = 0 \qquad \frac{5}{3}F - V_A = 0 \qquad V_A = \frac{5}{3}F$$

$$\sum M_A = 0 \qquad M_A - \left(\frac{5}{3}F\right)\left(\frac{L}{4}\right) = 0 \qquad M_A = \frac{5}{12}FL$$

SECTION *B*: From Fig. 5-9d,

$$\sum F_y = 0 \qquad \frac{5}{3}F - 2F - V_B = 0 \qquad V_B = -\frac{1}{3}F$$

$$\sum M_B = 0 \qquad M_B - \left(\frac{5}{3}F\right)\left(\frac{L}{2}\right) + 2F\left(\frac{L}{6}\right) = 0 \qquad M_B = \frac{1}{2}FL$$

SECTION *C*: Here it is best to use the section to the right of C (this reduces the number of forces that must be considered). From Fig. 5-9e,

$$\sum F_y = 0 \qquad V_C + \frac{4}{3}F = 0 \qquad V_C = -\frac{4}{3}F$$

$$\sum M_C = 0 \qquad \frac{4}{3}F\left(\frac{L}{4}\right) - M_C = 0 \qquad M_C = \frac{1}{3}FL$$

JUDGMENT OF THE RESULTS

The same results can be obtained by using complementary sections to those used here. To realize the facility of using either the left or right section for calculating internal reactions, the student should compute the bending moments and shear forces at sections A and C using sections other than those used in the example. Remember to assume the appropriate positive shear force and bending moment at each section.

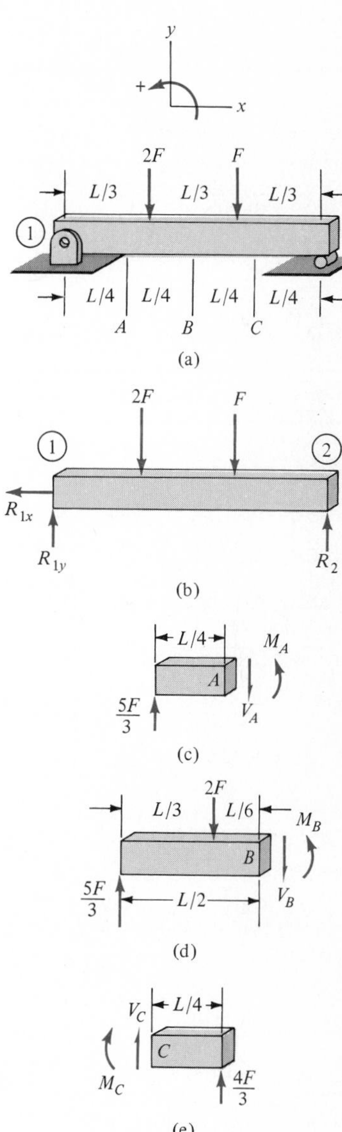

FIGURE 5-9

Analysis of a simply supported beam

EXAMPLE 5-3

Determine the internal forces and moments at sections A, B, and C in the cantilever beam shown in Fig. 5-10a. Determine the new reactions at A, B, and C after adding a downward force $F = 3$ kN acting at the middle of the beam.

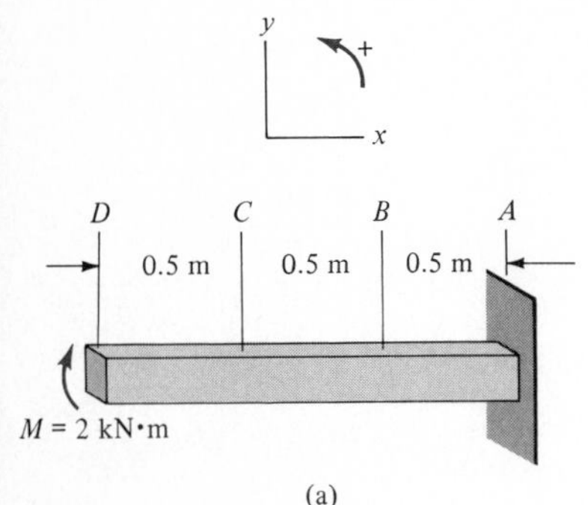

(a)

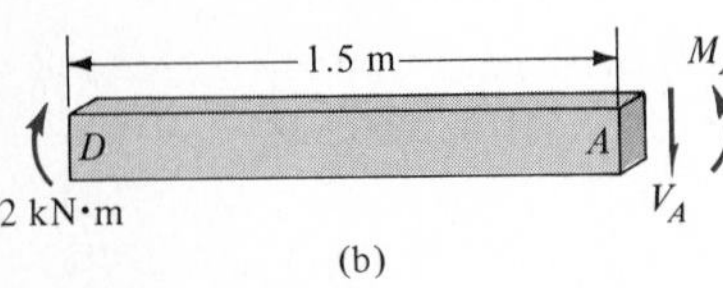

(b)

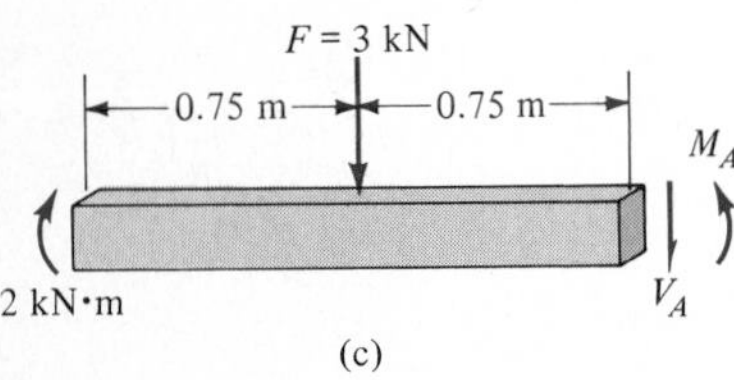

(c)

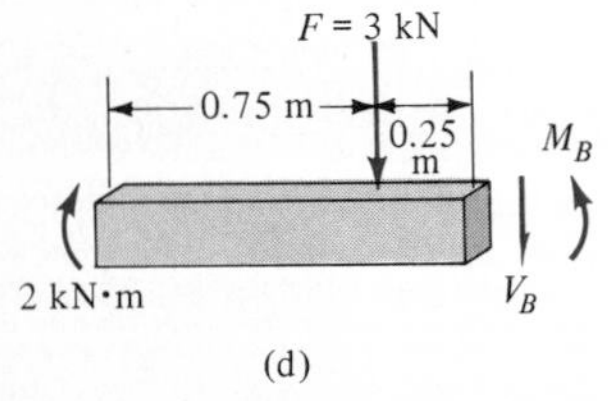

(d)

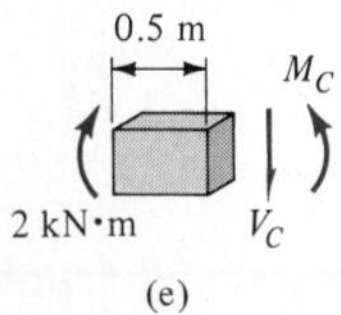

(e)

SOLUTION

CASE 1. SECTION *A*: From Fig. 5-10b and using Eq. 4-2,

$$\sum F_y = 0 \qquad V_A = 0$$

$$\sum M_A = 0 \qquad M_A - 2\text{ kN}\cdot\text{m} = 0 \qquad M_A = 2\text{ kN}\cdot\text{m}$$

It can be realized that the shear force is zero throughout the beam since there are no vertical forces applied. Also, the internal bending moment is equal to 2 kN·m everywhere since the moment calculation does not depend on position for a couple.

CASE 2. SECTION *A*: From Fig. 5-10c, for the whole beam,

$$\sum F_y=0 \qquad -3\text{ kN}-V_A=0 \qquad V_A=-3\text{ kN}$$

$$\sum M_A=0 \qquad M_A-2\text{ kN}\cdot\text{m}+3\text{ kN}\,(0.75\text{ m})=0 \qquad M_A=-0.25\text{ kN}\cdot\text{m}$$

SECTION *B*: From Fig. 5-10d,

$$\sum F_y=0 \qquad -3\text{ kN}-V_B=0 \qquad V_B=-3\text{ kN}$$

$$\sum M_B=0 \qquad M_B-2\text{ kN}\cdot\text{m}+3\text{ kN}\,(0.25\text{ m})=0 \qquad M_B=1.25\text{ kN}\cdot\text{m}$$

SECTION *C*: From Fig. 5-10e,

$$\sum F_y = 0 \qquad V_C = 0$$

$$\sum M_C = 0 \qquad M_C - 2\text{ kN}\cdot\text{m} = 0 \qquad M_C = 2\text{ kN}\cdot\text{m}$$

JUDGMENT OF THE RESULTS

The same results are obtained when the complementary sections are used.

FIGURE 5-10
Analysis of a cantilever beam

EXAMPLE 5-4

(a) Determine the forces and moments at A, B, C, and D for the simply supported beam shown in Fig. 5-11a. **(b)** Determine the forces and moments at A, B, C, and D after moving the force F to B.

SOLUTION

(a) First the reactions must be determined from Fig. 5-11b for the whole beam. Using Eq. 4-2,

$$\sum F_x = 0 \qquad (600\text{ lb})\cos 60^\circ - A_x = 0 \qquad A_x = 300\text{ lb}$$

$$\sum F_y = 0 \qquad A_y + E - (600\text{ lb})\sin 60^\circ = 0$$

$$\sum M_A = 0 \qquad -(600\text{ lb})(\sin 60^\circ)(30\text{ in.}) + E(60\text{ in.}) = 0, \qquad E = 260\text{ lb}$$

From $\sum F_y = 0$, $A_y = 260$ lb.

SECTION *A*: From Fig. 5-11c,

$$\sum F_x = 0 \qquad P_A - 300\text{ lb} = 0 \qquad P_A = 300\text{ lb}$$

$$\sum F_y = 0 \qquad 260\text{ lb} - V_A = 0 \qquad V_A = 260\text{ lb}$$

$$\sum M_A = 0 \qquad 260\text{ lb}(0) + M_A = 0 \qquad M_A = 0$$

SECTION *B*: From Fig. 5-11d,

$$\sum F_x = 0 \qquad P_B - 300\text{ lb} = 0 \qquad P_B = 300\text{ lb}$$

$$\sum F_y = 0 \qquad 260\text{ lb} - V_B = 0 \qquad V_B = 260\text{ lb}$$

$$\sum M_B = 0 \qquad M_B - (260\text{ lb})(15\text{ in.}) = 0 \qquad M_B = 3900\text{ in}\cdot\text{lb}$$

SECTION *C*: From Fig. 5-11e,

$$\sum F_x = 0 \qquad P_C \text{ is undefined (see Ex. 5-1)}$$

$$\sum F_y = 0 \qquad V_C \text{ is undefined (see Ex. 5-1)}$$

$$\sum M_C = 0 \qquad M_C - (260\text{ lb})(30\text{ in.}) = 0 \qquad M_C = 7800\text{ in}\cdot\text{lb}$$

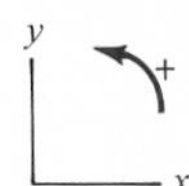

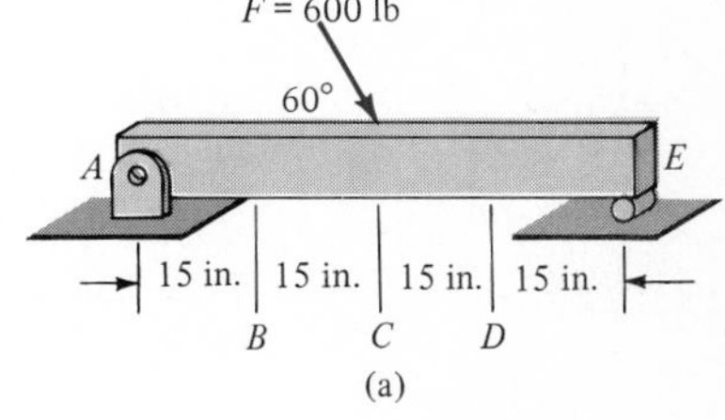

(a)

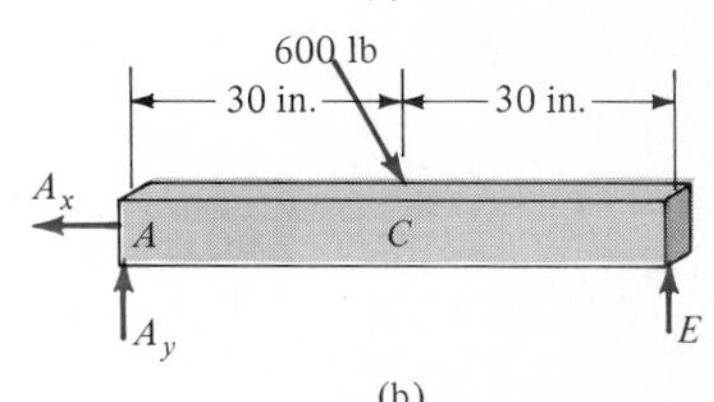

(b)

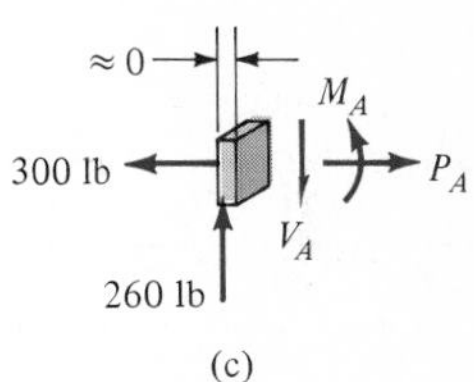

(c)

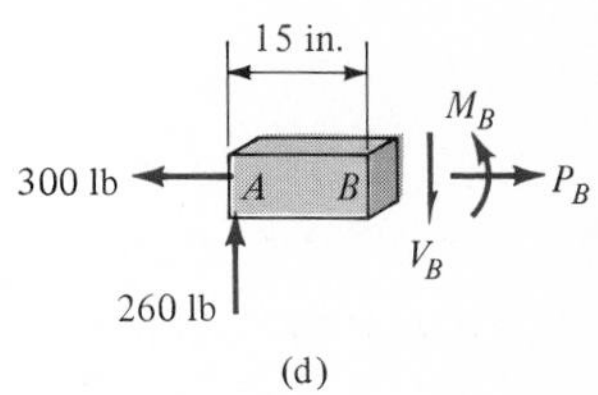

(d)

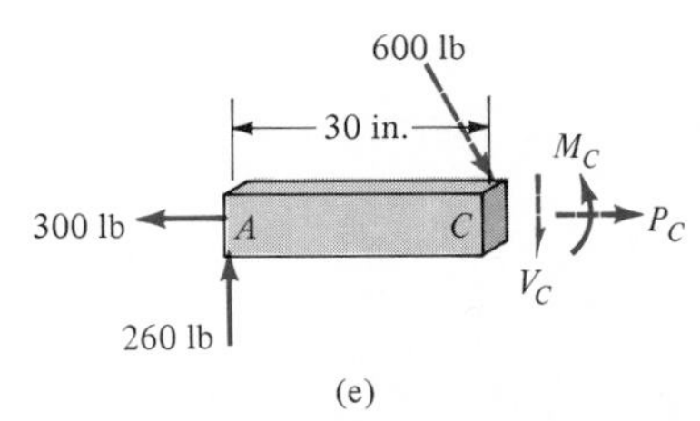

(e)

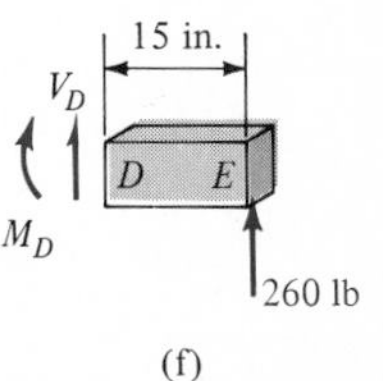

(f)

FIGURE 5-11
Analysis of a biaxially loaded, simply supported beam

SECTION *D*: From Fig. 5-11f,

$$\sum F_x=0 \qquad P_D=0$$

$$\sum F_y=0 \qquad V_D+260\text{ lb}=0 \qquad V_D=-260\text{ lb}$$

$$\sum M_D=0 \qquad (260\text{ lb})(15\text{ in.})-M_D=0 \qquad M_D=3900\text{ in}\cdot\text{lb}$$

(b) For the case when F is applied at B, the reactions must be recalculated:

$$\sum F_x=0 \qquad (600\text{ lb})\cos 60^\circ-A_x=0 \qquad A_x=300\text{ lb}$$

$$\sum F_y=0 \qquad A_y+E-(600\text{ lb})\sin 60^\circ=0$$

$$\sum M_A=0 \qquad (-600\text{ lb})(\sin 60^\circ)(15\text{ in.}) + E(60\text{ in.})=0, \qquad E=130\text{ lb}$$

From $\sum F_y = 0$, $A_y = 390$ lb

SECTION *A*:

$$\sum F_x = 0 \qquad P_A - 300\text{ lb} = 0 \qquad P_A = 300\text{ lb}$$

$$\sum F_y = 0 \qquad 390\text{ lb} - V_A = 0 \qquad V_A = 390\text{ lb}$$

$$\sum M_A = 0 \qquad 390\text{ lb}\,(0) + M_A = 0 \qquad M_A = 0$$

SECTION *B*:

$\sum F_x=0$ $\quad P_B$ is undefined (see Ex. 5-1)

$\sum F_y=0$ $\quad V_B$ is undefined (see Ex. 5-1)

$$\sum M_B=0 \qquad M_B-(390\text{ lb})(15\text{ in.})=0 \qquad M_B=5850\text{ in}\cdot\text{lb}$$

SECTION *C* (Using the Right Hand Section):

$$\sum F_x=0 \qquad P_C=0$$

$$\sum F_y=0 \qquad 130\text{ lb}+V_C=0 \qquad V_C=-130\text{ lb}$$

$$\sum M_C=0 \qquad (130\text{ lb})(30\text{ in.})-M_C=0 \qquad M_C=3900\text{ in}\cdot\text{lb}$$

SECTION *D*:

$$\sum F_x=0 \qquad P_D=0$$

$$\sum F_y=0 \qquad 130\text{ lb}+V_D=0 \qquad V_D=-130\text{ lb}$$

$$\sum M_D=0 \qquad (130\text{ lb})(15\text{ in.})-M_D=0 \qquad M_D=1950\text{ in}\cdot\text{lb}$$

JUDGMENT OF THE RESULTS

It is seen that by moving the force F from the midpoint to the left the internal forces become intensified on the left and diminished on the right as expected. The changes in reactions at A and E are acceptable by inspection.

EXAMPLE 5-5

Show that the weightless cantilever beam in Fig. 5-12a is statically indeterminate for F acting in the range $0 \leq x \leq 1.2$ m. Next, assume that the beam is cut into two equal parts and that these are connected with a hinge at point E. Determine the reactions at A, B, C, D, and E for the hinged beam.

SOLUTION

The free-body diagram of the beam in Fig. 5-12b shows there are four unknowns. From Eq. 4-2, only three independent equations of equilibrium are available, $\sum F_x = 0$, $\sum F_y = 0$, and $\sum M_A = 0$. Hence, the beam is indeterminate. If the beam is hinged at E, it is seen from Fig. 5-12c that all reactions at D and E must both be zero (otherwise, part DE would experience a resultant moment or force); the problem is now determinate.

SECTION *A*: From Fig. 5-12d,

$$\sum F_y = 0 \qquad V_A - 3\text{ kN} = 0 \qquad V_A = 3\text{ kN}$$

$$\sum M_A = 0 \qquad -3\text{ kN}(0.4\text{ m}) - M_A = 0 \qquad M_A = -1.2\text{ kN}\cdot\text{m}$$

SECTION *B*: From Fig. 5-12e,

$$\sum F_y = 0 \qquad V_B \text{ is undefined (see Ex. 5-1)}$$

$$\sum M_B = 0 \qquad M_B = 0 \qquad M_B = 0$$

Since there is no force at E or D, everything to the right of B experiences no force, and all internal forces and moments are zero in this part of the beam.

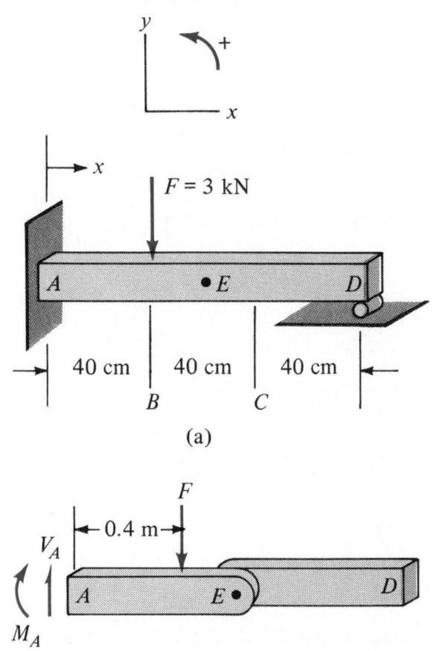

(a)

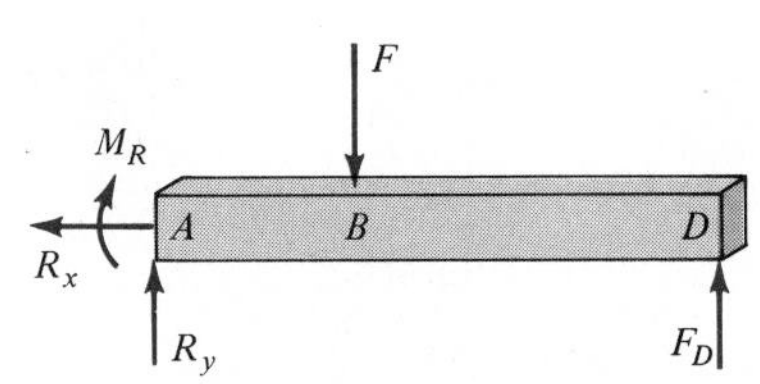

(b) Continuous beam from A to D

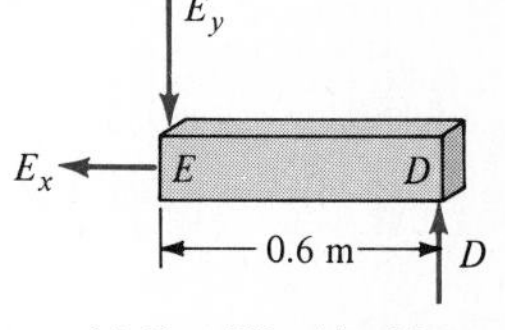

(c) Part ED with a hinge at E and a roller at D

(d) Assume that the beam is cut at E and a hinge is installed there

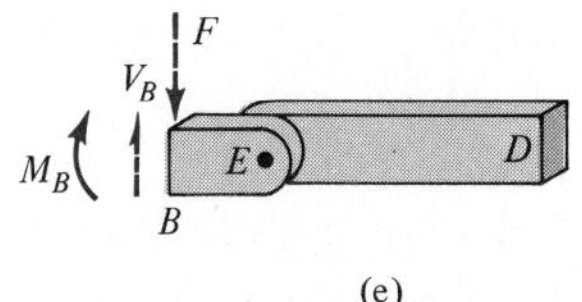

(e)

FIGURE 5-12

Statically indeterminate beam made into a determinate beam

PROBLEMS

SECTIONS 5-1, 5-2, 5-3

FIGURE P5-1

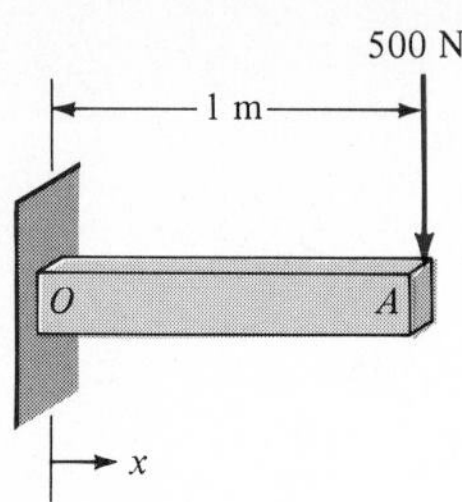

5–1 Determine the shear force and bending moment at $x = 0$, 0.2 m, and 0.4 m.

FIGURE P5-2

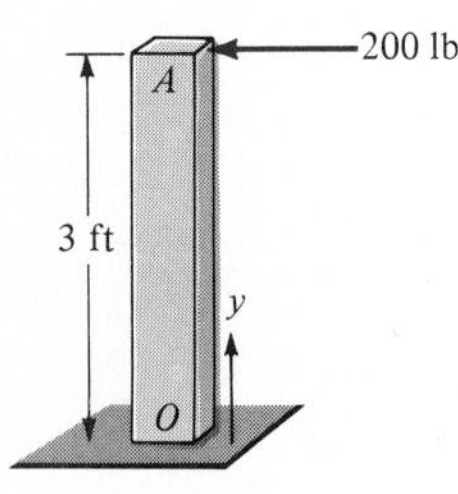

5–2 Determine the shear force and bending moment at $y = 0$, 1 ft, and 2 ft.

FIGURE P5-3

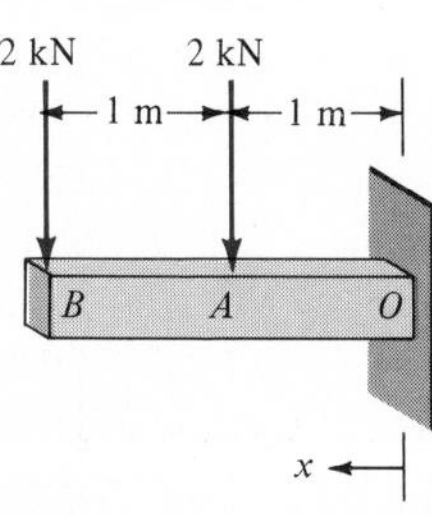

5–3 Calculate the shear force and bending moment at $x = 0$, 0.5 m, and 1.5 m.

FIGURE P5-4

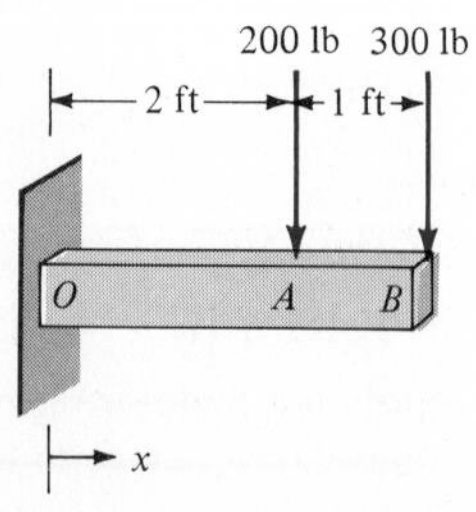

5–4 Determine the shear force and bending moment at $x = 0$, 1 ft, and 2.5 ft.

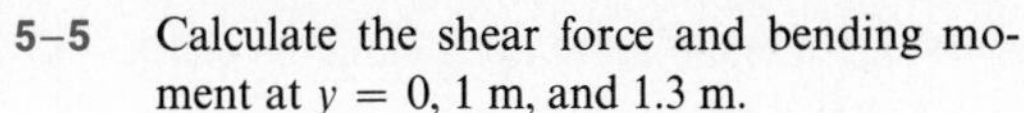

5–5 Calculate the shear force and bending moment at $y = 0$, 1 m, and 1.3 m.

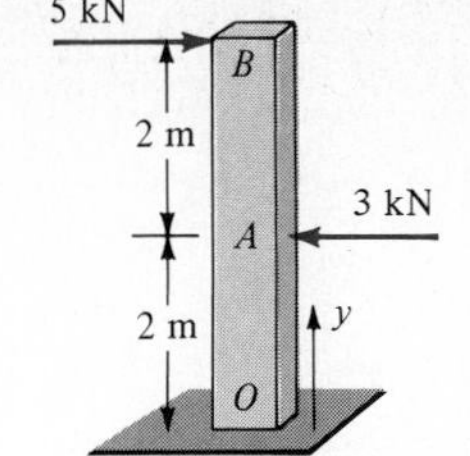

FIGURE P5-5

5–6 Calculate the shear force and bending moment at $x = 0$, 0.5 ft, 1.2 ft, and 2.5 ft.

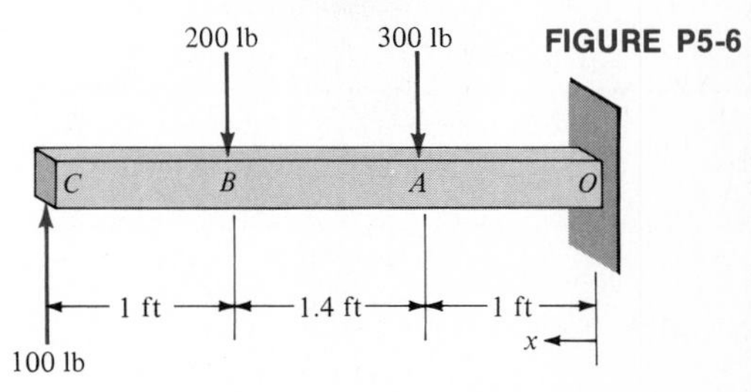

FIGURE P5-6

5–7 Force W always acts at point A on the simple crane, but force F has variable positions. Determine the shear force and bending moment at support 0 for $x_C = 1$ m, 2 m, and 4 m.

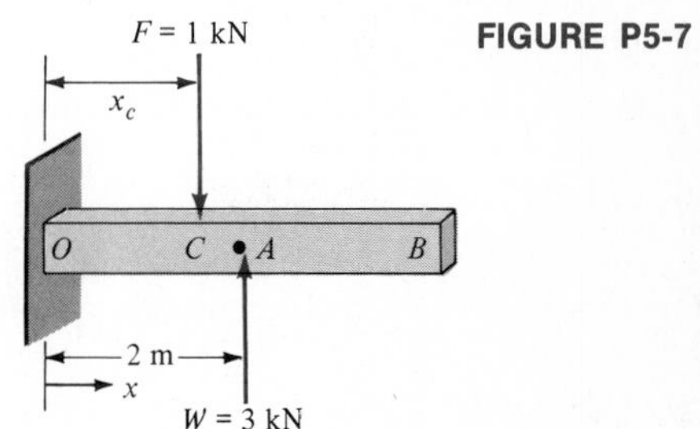

FIGURE P5-7

5–8 A simple crane is modeled as a weightless cantilever beam. The load F is applied through a symmetric carriage which has two wheels, A and B. Calculate the shear force and bending moment at support 0 for $x_C =$ 3 ft, 6 ft, and 9 ft.

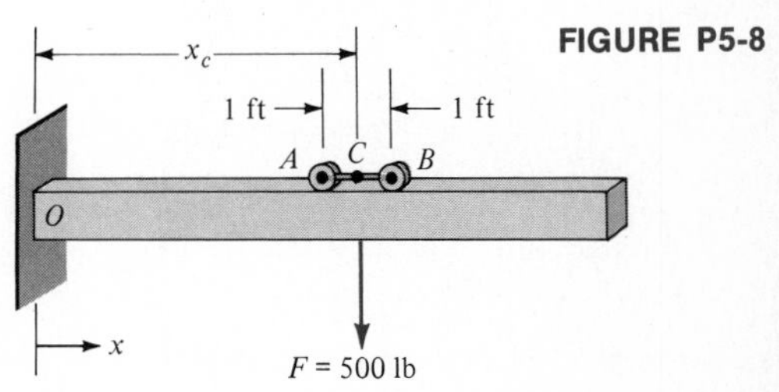

FIGURE P5-8

5–9 Calculate the shear force and bending moment at $x = 1$ m and 2.5 m. Give the values of the bending moment at A and C.

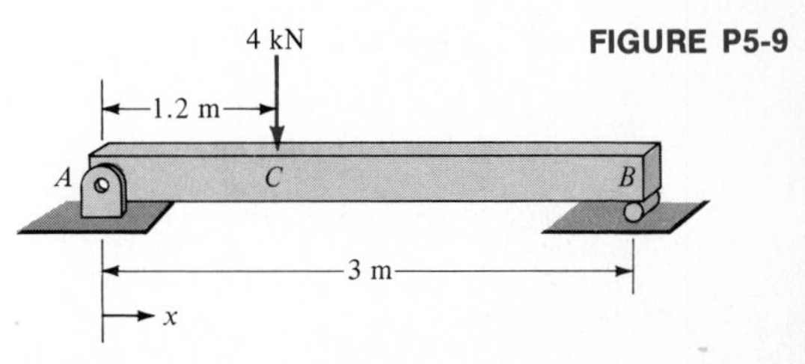

FIGURE P5-9

FIGURE P5-10

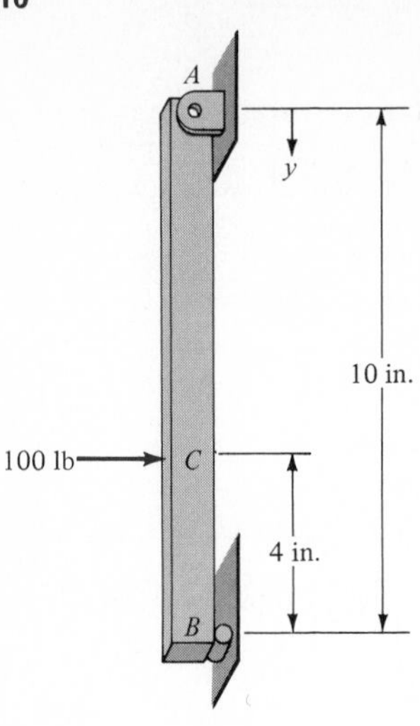

5–10 Determine the shear force and bending moment at $y = 2$ in. and 8 in. Give the values of the bending moment at *A*, *B*, and *C*.

FIGURE P5-11

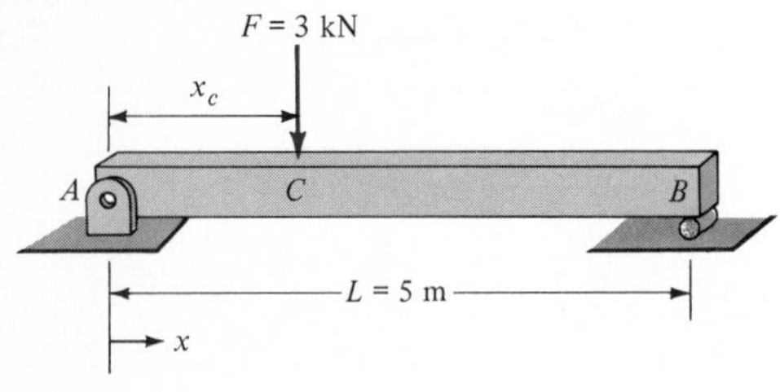

5–11 The force *F* acts at various places along the beam *A–B*. Calculate the bending moment at section *C* for $x_C = 1$ m, 2.5 m, and 4 m.

5–12 Let $F = 500$ lb and $L = 10$ ft in Fig. P5-11. Consider a section *D* at $x_D = 5$ ft. Calculate the shear force and bending moment at *D* for *F* acting at $x_C = 1$ ft, 2.5 ft, and 4 ft (*F* is at one place at a time).

FIGURE P5-13

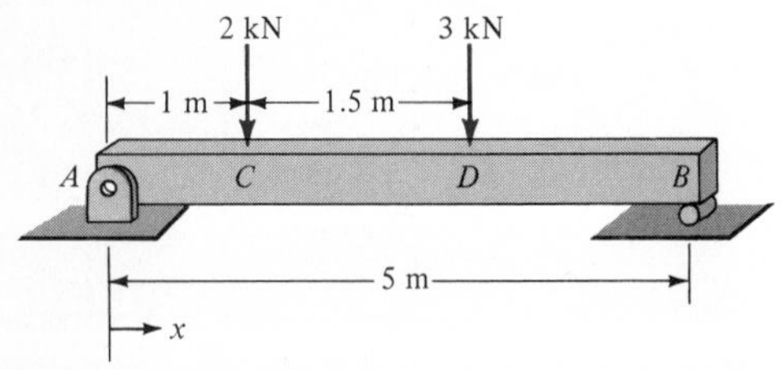

5–13 Determine the bending moments at sections *C* and *D*. Also calculate the shear force and the bending moment at section *E*, where $x_E = 4$ m.

FIGURE P5-14

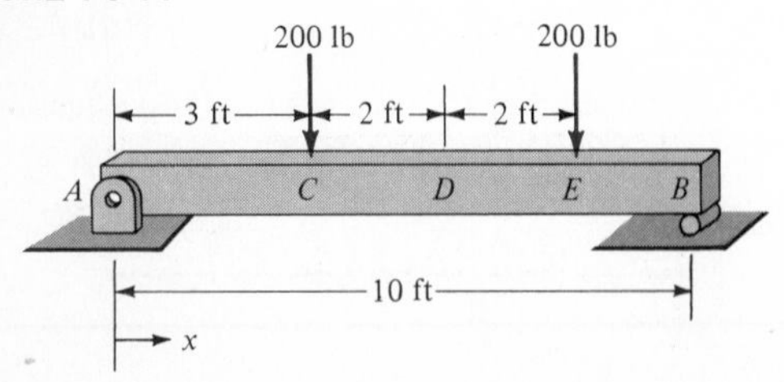

5–14 Calculate the bending moments at sections *C*, *D*, and *E*.

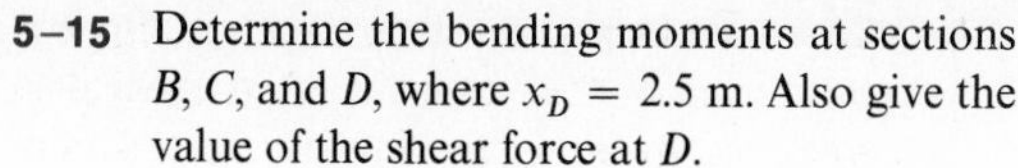

5–15 Determine the bending moments at sections B, C, and D, where $x_D = 2.5$ m. Also give the value of the shear force at D.

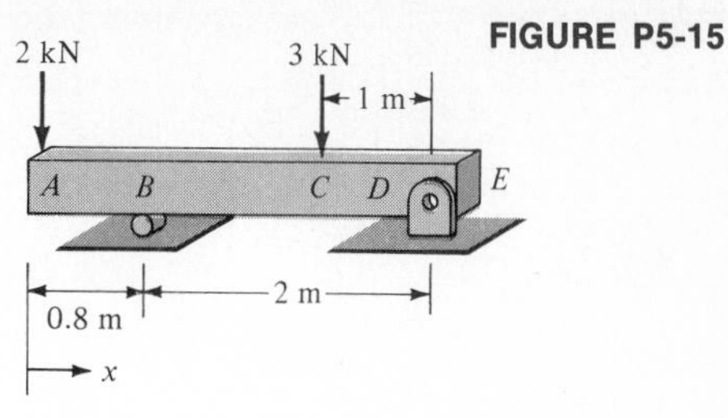

FIGURE P5-15

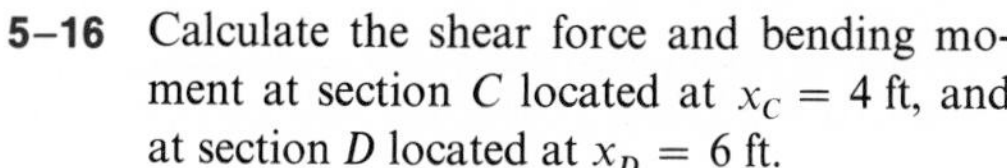

5–16 Calculate the shear force and bending moment at section C located at $x_C = 4$ ft, and at section D located at $x_D = 6$ ft.

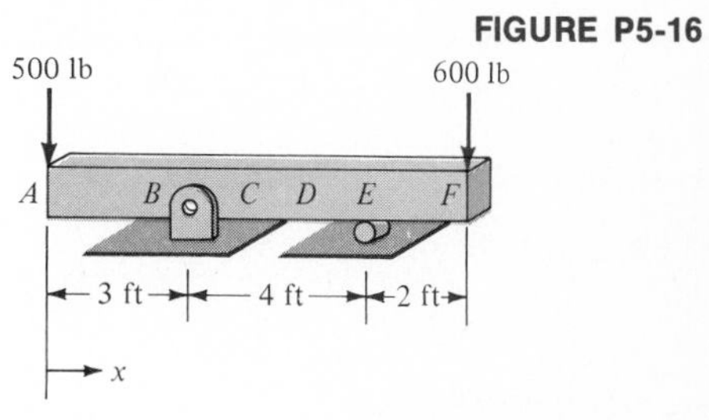

FIGURE P5-16

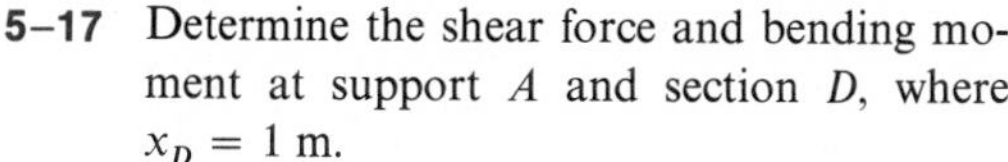

5–17 Determine the shear force and bending moment at support A and section D, where $x_D = 1$ m.

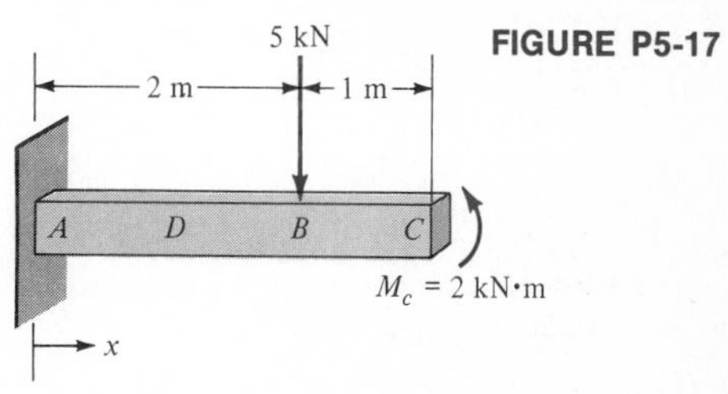

FIGURE P5-17

5–18 Calculate the shear force and bending moment at A, B, and C.

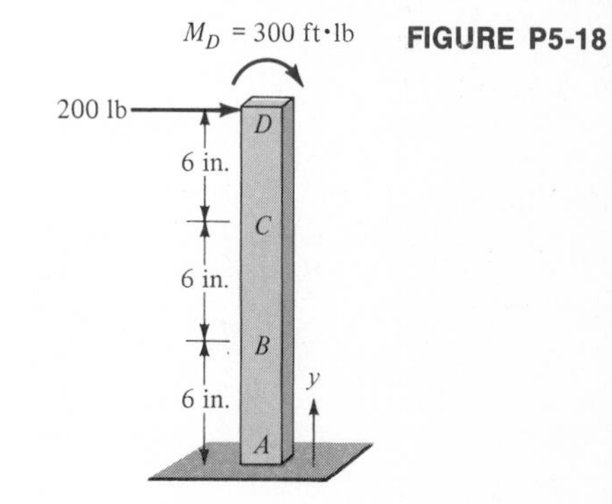

FIGURE P5-18

5–19 Determine the shear force and bending moment at section C.

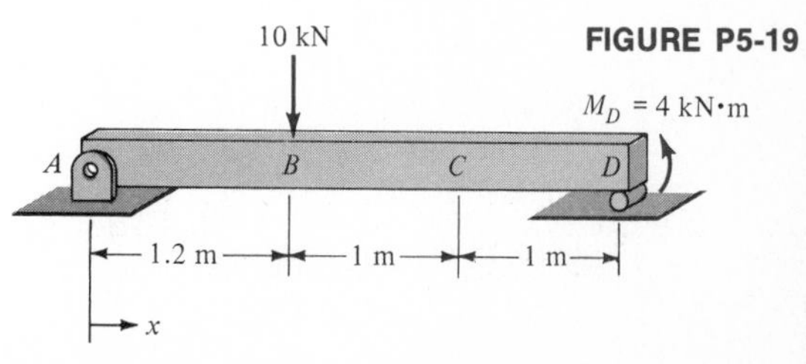

FIGURE P5-19

FIGURE P5-20

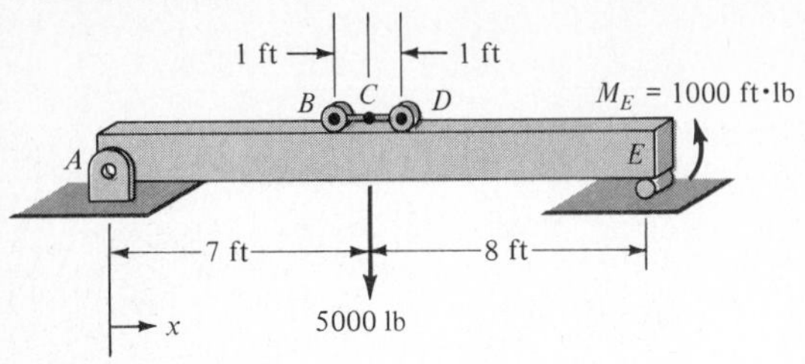

FIGURE P5-24

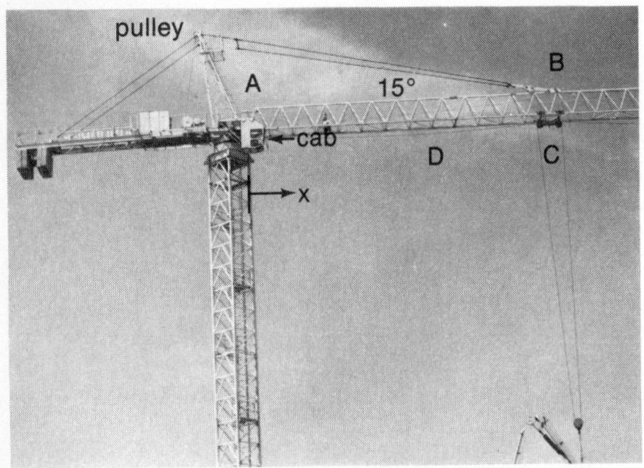

FIGURE P5-26

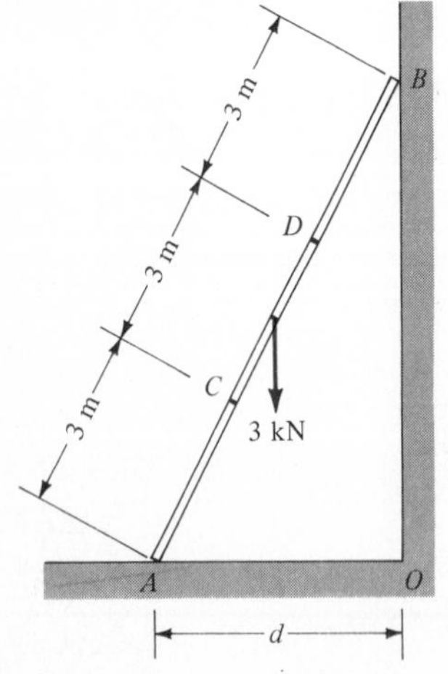

5–20 When the two-wheeled carriage of a crane is positioned at C, a moment M_E is applied to track AE by a neighboring structure. Calculate the bending moments at B, C, and D in the beam. Also give the shear force at C.

5–21 Solve Ex. 5-1 using the complementary parts of the beam to those used in the example. Thus, use the left-hand parts at sections B, C, D, and E.

5–22 Solve Ex. 5-2 using the complementary parts of the beam to those used in Fig. 5-9.

5–23 Solve Ex. 5-3 using the complementary parts of the beam to those used in Fig. 5-10. Thus, use the right-hand parts at sections B and C.

5–24 Consider the tall crane shown in the photograph. The horizontal boom is pivoted at A, which is behind the operator's cab in the picture. The permanently attached cables at B make a 15° angle with the boom. The horizontal distance between A and B is 80 ft. The load carriage C can move along the 120-ft-long boom, which is assumed weightless. Calculate the reactions at A for a 10,000-lb vertical load on the carriage C at the position x_C = 120 ft. Also calculate the shear force and bending moment in the boom at x_D = 50 ft assuming that the boom is a homogeneous beam.

5–25 It may be expected that the internal reactions in the boom described in Prob. 5–24 depend on the position x_C of the carriage C. Determine the horizontal reaction, the vertical shear, and the bending moment at x_E = 5 ft when C is at x_C = 50 ft, 80 ft, and 110 ft (at one place at a time).

5–26 Consider a proposed 9-m-long ladder. The possible internal reactions in the ladder must be known to minimize its weight. For an initial estimation, determine the transverse shear force and bending moment at C and D for a 3-kN weight applied at the center of the ladder, which is positioned at d = 4 m. Assume that the wall is frictionless at B, and that the ladder acts as if it were pinned at A.

5–27 It may be expected that the placement distance d makes a difference for the ladder described in Prob. 5–26. Plot the transverse shear force and the bending moment at C and D as functions of $d = 2$ m, 4 m, and 6 m. Use one diagram for the shear forces and one for the bending moments.

FIGURE P5-28

Suspension system of a railroad car. (Courtesy MTS Systems Corporation, Minneapolis, Minnesota.)

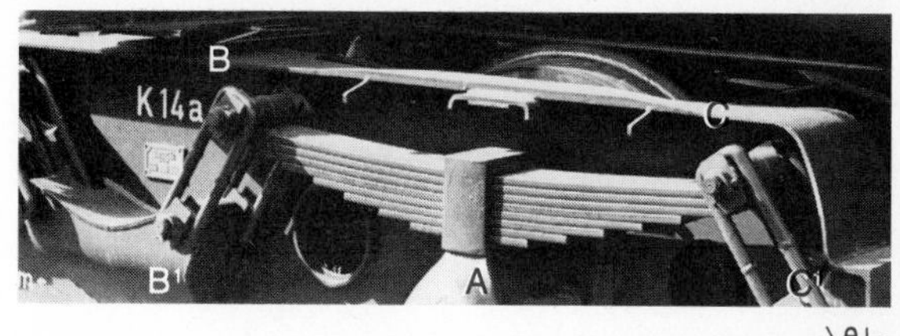

5–28 Consider the spring BC of a railroad tank car. The wheel axle A applies a maximum upward force of 20,000 lb to the spring. The distance BC is 30 in. Determine the maximum bending moment and vertical shear in the spring for the angle $\theta = 30°$ that the links BB' and CC' make with the vertical.

5–29 Assume that the angle θ of the links BB' and CC' in Prob. 5–28 could be chosen between 0 and 30°. Determine the shear force and bending moment at the center A as functions of $\theta = 0$, 15°, and 30°.

FIGURE P5-30

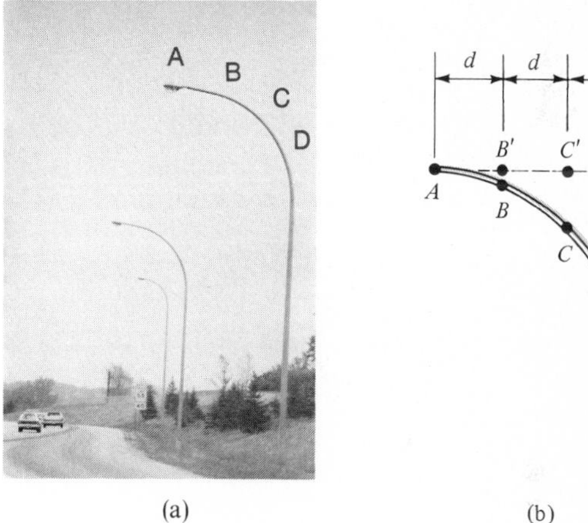

(a) (b)

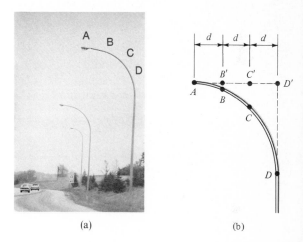

5–30 The light poles in the photograph are quite appealing esthetically. To investigate part of their technical merits, consider the bending moments in the curved poles and in poles made of straight members. To be specific, compare the bending moments at points B and B', C and C', and D and D' in Fig. P5-30b. Assume that a weight W acts at A, and that the pole is weightless for any geometry.

FIGURE P5-31

5–31 Consider the model of the large earth mover used in strip mining. The long arm AD of the conveyor is pivoted at A and has a roller at B. Determine the transverse shear force and bending moment at points E and F.

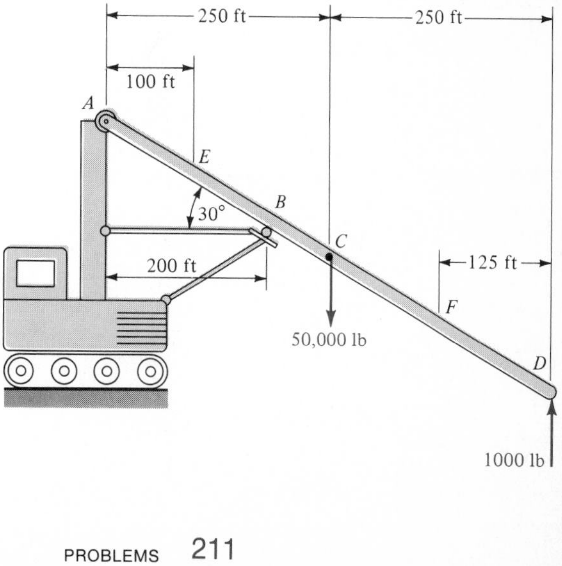

FIGURE P5-32

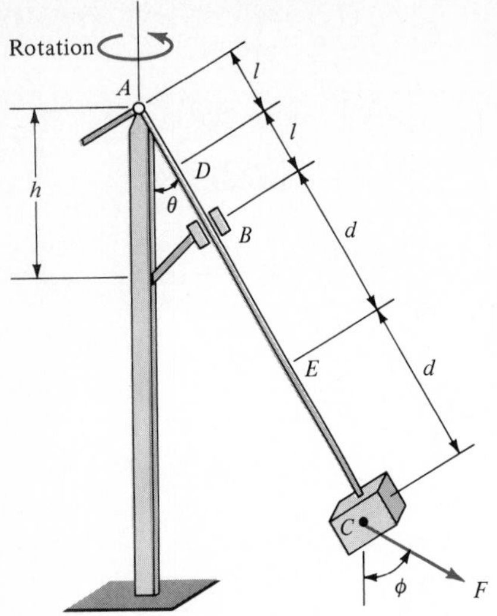

5–32 Part of an amusement ride is modeled as a rigid rod AC, which is pivoted at A. It has a frictionless sleeve at B. The magnitude and direction of force F is obtained from dynamics analysis. Determine the transverse shear force and bending moment at D and E in terms of the general quantities indicated.

FIGURE P5-33

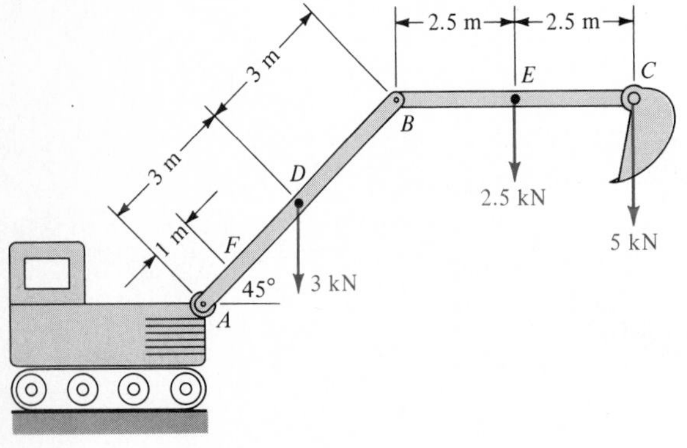

5–33 An excavator is modeled with rigid arms AB and BC, which can be locked in a variety of positions. Calculate the transverse shear force and bending moment at point F.

FIGURE P5-34

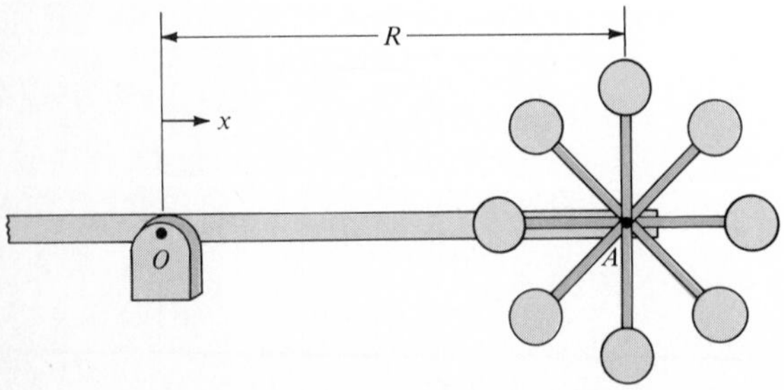

5–34 An amusement ride is to be modeled as a cantilever beam supported at O. The weights are assumed to be concentrated, W_1 acting at $x = R/2$, and W_2 acting at A. Furthermore, a clockwise couple of magnitude M_A is acting at A (because of motion of the small arms). Determine the vertical shear force and bending moment at $x = 0$ and $x = R/4$ in the main arm.

FIGURE P5-35

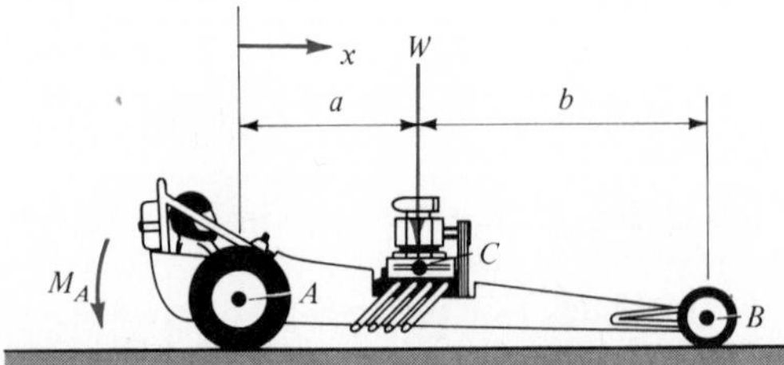

5–35 A drag racing car at the start is to be modeled as a weightless, horizontal beam pinned at A and supported on a roller at B. The weight W acts at C, and a counterclockwise moment M_A is acting on AB at point A (an opposite moment is acting on the wheels at A; a large M_A causes the occasional lifting of the front end). Determine the vertical shear force and bending moment at $x = a/2$ and $x = a + b/2$, assuming a normal force of $0.1W$ acting on wheel B.

SHEAR FORCE AND BENDING MOMENT DIAGRAMS

A common problem in the design of beams is to find the critical sections where the shear force and bending moment are the largest. The signs of the bending moment are also of great interest because they show where there is tension or compression in a beam. The signs of the shear forces are of smaller practical significance.*

The critical locations in some beams may be determined by inspection after a little experience. In any case, it is most satisfactory to plot shear and moment diagrams for the whole length of each beam. These are excellent pictorial views of the variations in shear force and bending moment, and are essential when the loading is complex.

It is best to draw the shear and moment diagrams directly below the complete free-body diagram of the beam to obtain an overall view of the details of the problem. The construction of the diagrams is illustrated in Fig. 5-13. The free-body diagram of a simply supported beam is drawn in part a of the figure. From this, using Eq. 4-2, $P = R_1 + R_2$, $R_1 a = R_2 b$, and

$$R_1 = \frac{P}{1 + \frac{a}{b}}, \qquad R_2 = \frac{P}{1 + \frac{b}{a}}$$

Then vertical lines are drawn at the major points of interest such as where $\mathbf{R}_1$, $\mathbf{R}_2$, and $\mathbf{P}$ act. Next, the coordinate axes of the shear (V) and moment (M) diagrams are drawn.

The plotting of the shear and moment diagrams is done after the calculation of shear and moment at a few judiciously selected sections in the beam. These are shown in Fig. 5-13d through j. If the beam is in equilibrium, each part of it is also in equilibrium. Thus, from the diagrams of the individual parts,

$$\text{Fig. 5-13d:} \quad V_1 = R_1 = \frac{P}{1+\frac{a}{b}}, \quad M_1 \approx 0.$$

$$\text{Fig. 5-13e:} \quad V_2 = R_1 \qquad M_2 = R_1 \frac{a}{3}$$

$$\text{Fig. 5-13f:} \quad V_3 = R_1 \qquad M_3 = R_1 \frac{2a}{3}$$

$$\text{Fig. 5-13g:} \quad V_4 = R_1 \qquad M_4 \approx R_1 a$$

* See B. I. Sandor, *Strength of Materials* (Englewood Cliffs, N.J.: Prentice-Hall, Inc., 1978).

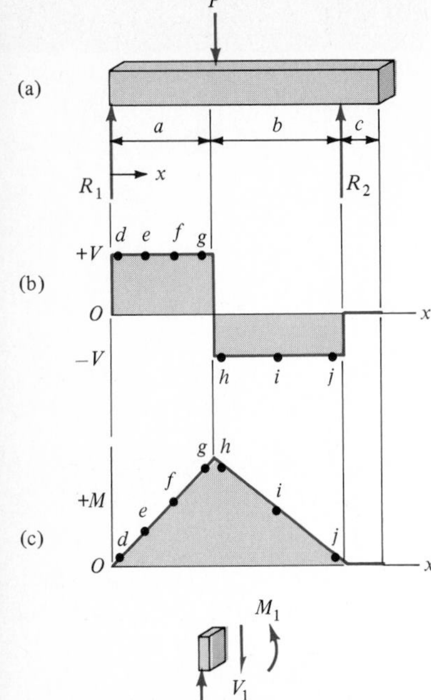

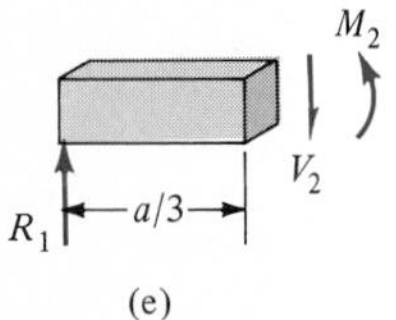

(d) Horizontal length of element is near zero

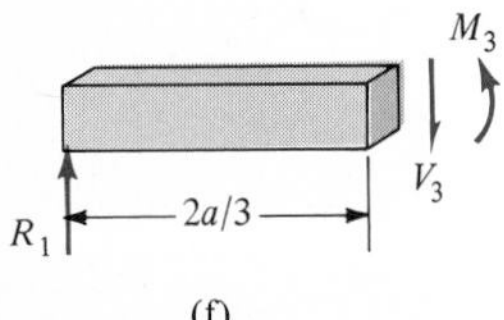

(e)

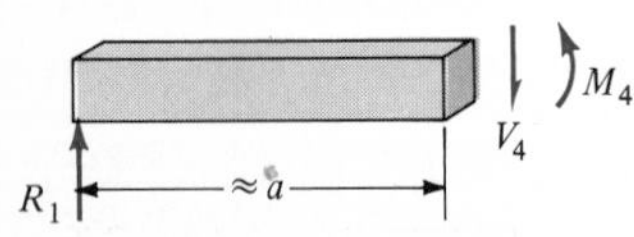

(f)

Fig. 5-13h: $V_5 = P - R_1 = R_2$ $\qquad M_5 \approx R_1 a$

Fig. 5-13i: $V_6 = R_2$ $\qquad M_6 = R_1\left(a + \frac{b}{2}\right) - P\frac{b}{2}$

Fig. 5-13j: $V_7 = R_2$ $\qquad M_7 \approx R_1(a+b) - Pb \approx 0$

These values are plotted in Fig. 5-13b and c as points *d* through *j*, respectively. The points are connected with reasonable lines; the areas between these lines and the *x* coordinate are crosshatched to help in the visual evaluation of the plots. The shear and moment are zero in region *C* (to the right of R_2).

The sharp breaks in the plots are characteristic for sections where concentrated forces are acting. Such regions cannot be analyzed precisely. In theory, a force can act at a point, but in reality, even such forces are distributed over small but finite areas or volumes. Thus, the shear and moment diagrams are not precise at $x \approx 0$, $x \approx a$, and $x \approx a + b$. However, the approximations with the simple lines as shown in the diagrams are reasonable in practice.

The procedure described above can be formalized by writing functions of the shear and moment in terms of the external loads and the distance *x* along the beam. Each function is based on a single free-body diagram that is valid for a part of the beam. The beam in Fig. 5-13a has three unique parts, and there is a shear and moment function for each part as follows:

$$\left.\begin{aligned} V_a(x) &= R_1 \\ M_a(x) &= R_1 x \end{aligned}\right\} \qquad 0 < x < a$$

$$\left.\begin{aligned} V_b(x) &= P - R_1 = R_2 \\ M_b(x) &= R_1 x - P(x - a) \end{aligned}\right\} \qquad a < x < a + b$$

$$\left.\begin{aligned} V_c(x) &= 0 \\ M_c(x) &= 0 \end{aligned}\right\} \qquad a + b < x < a + b + c$$

These functions can be evaluated at selected values of *x*. The points are plotted on the appropriate diagram and reasonable lines are drawn through the points as before. Much practice with these equations and diagrams is recommended.

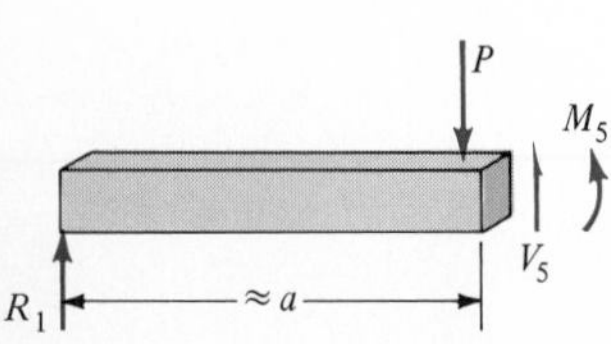

(h) Length is slightly larger than *a*

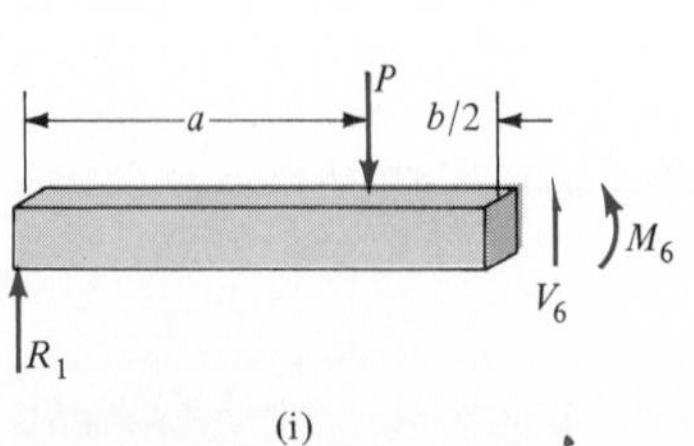

(i)

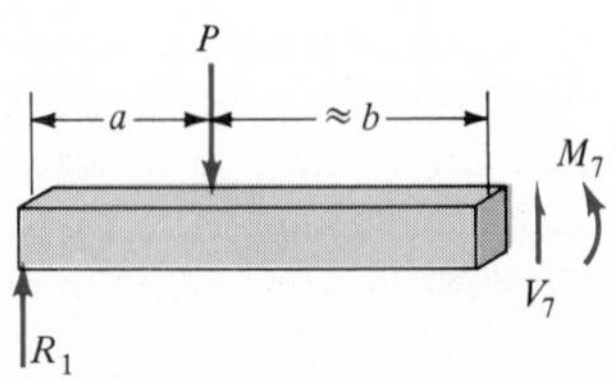

(j) Length is slightly less than *a* + *b*; R_2 is not acting on this part

(g) Length is slightly less than *a*; *P* is not acting on this part

FIGURE 5-13

Shear force and bending moment diagrams of a simply supported beam

EXAMPLE 5-6

Write the shear and moment equations and draw the shear and moment diagrams for the cantilever beam shown in Fig. 5-14a.

SOLUTION

First the reactions at end A are obtained (Fig. 5-14b) using Eq. 4-2,

$$\sum F_x = 0 \qquad R_x = 0$$

$$\sum F_y = 0 \qquad R_y - 5\text{ kN} = 0 \qquad R_y = 5\text{ kN}$$

$$\sum M_A = 0 \qquad M_R - 5\text{ kN}(2\text{ m}) = 0 \qquad M_R = 10\text{ kN}\cdot\text{m}$$

To obtain the general shear and moment equations a section K at a general distance x from the left end is analyzed. Assuming positive shear and moment reactions at section K in Fig. 5-14e,

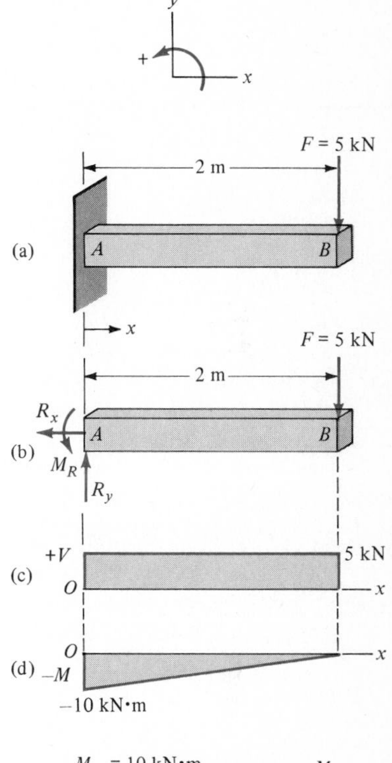

PART *A–K*:

$$\sum F_y = 0 \qquad V_K = 5\text{ kN}$$

$$\sum M_K = 0 \qquad 10\text{ kN}\cdot\text{m} - 5\text{ kN}(x) + M_K = 0$$

$$M_K = 5\text{ kN}(x) - 10\text{ kN}\cdot\text{m}$$

Plotting the equations above, Fig. 14c and d are obtained.

A check at the end of the beam shows (Fig. 5-14f):

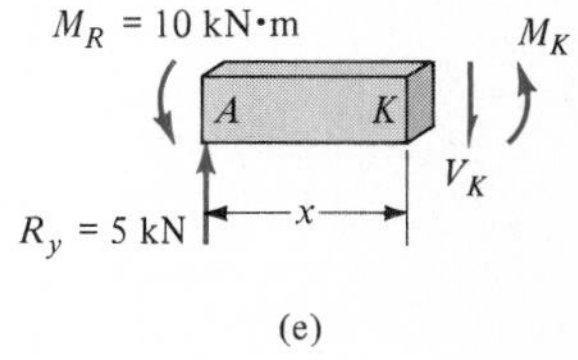

PART *A–B* (*K* IS AT *B*):

$$\sum F_y = 0 \qquad 5\text{ kN} - 5\text{ kN} - V_K = 0,$$

so $V_K = 0$ as it should be

$$\sum M_K = 0 \qquad 10\text{ kN}\cdot\text{m} - 5\text{ kN}(2\text{ m}) + M_K = 0,$$

so $M_K = 0$ as it should be

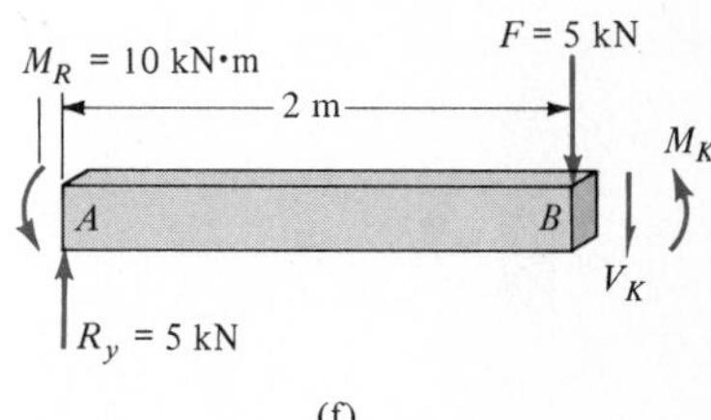

FIGURE 5-14

Shear force and bending moment diagrams of a cantilever beam

JUDGMENT OF THE RESULTS

The results seem logical. Since there is no change in the applied load along the length of the beam, the shear force, which resists the vertical load, should not change either. The linear moment distribution seems logical after studying the free-body diagram in Fig. 5-13e. Using only one general section is reasonable since no other sections would show a different loading situation.

EXAMPLE 5-7

Write the shear and moment equations and draw the shear and moment diagrams for the cantilever beam shown in Fig. 5-15a.

SOLUTION

First the reactions at A are obtained (Fig. 5-15b) using Eq. 4-2,

$$\sum F_x = 0 \qquad R_x = 0$$

$$\sum F_y = 0 \qquad R_y = 0$$

$$\sum M_A = 0 \qquad 12{,}000 \text{ in}\cdot\text{lb} - M_R = 0$$

$$M_R = 12{,}000 \text{ in}\cdot\text{lb} = 1000 \text{ ft}\cdot\text{lb}$$

Analyzing a general section at distance x, again assuming positive internal reactions, the shear and moment equations are obtained.

PART *A–K*:

$$\sum F_y = 0 \qquad -V_K = 0 \qquad V_K = 0$$

$$\sum M_K = 0 \qquad M_K - 1000 \text{ ft}\cdot\text{lb} = 0 \qquad M_K = 1000 \text{ ft}\cdot\text{lb}$$

Plotting these equations, Fig. 5-15c and d are obtained.

JUDGMENT OF THE RESULTS

The results appear logical. Since the resultant of the applied vertical forces is zero, no internal shear forces need exist to resist them. Also, since the only applied load is a couple, a free vector, the internal moment should be the same regardless of position along the beam.

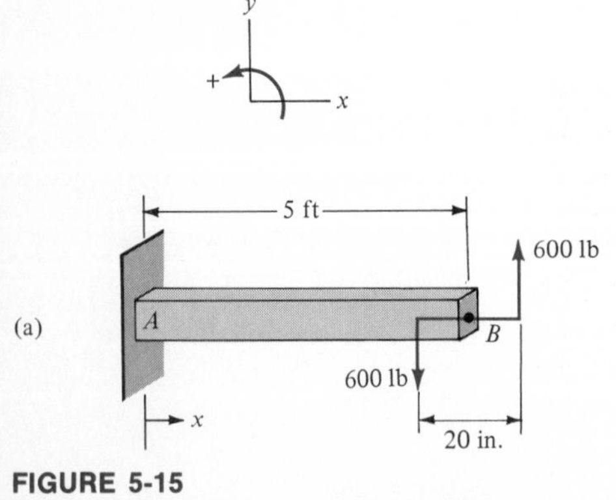

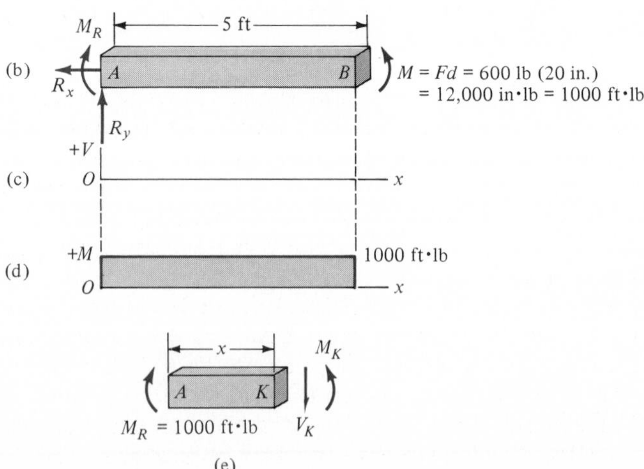

FIGURE 5-15

Shear force and bending moment diagrams of a cantilever beam

EXAMPLE 5-8

Write the shear and moment equations and draw the shear and moment diagrams for the beam shown in Fig. 5-16a.

SOLUTION

The reactions are given in Fig. 5-16b. These should be checked by the reader. Free-body diagrams are drawn to place a general section K in each section between neighboring vertical forces (between A and B, B and C, and C and D). Equations 4-2 are used for each part.

PART *A–K* (Fig. 5-16e): For the region $0 \le x \le 1$ m,

$$\sum F_y = 0 \qquad 2.43\text{ kN} - V_{K1} = 0 \qquad V_{K1} = 2.43\text{ kN}$$

$$\sum M_K = 0 \quad M_{K1} - 2.43\text{ kN}(x) = 0 \quad M_{K1} = 2.43(x)\text{ kN}\cdot\text{m}$$

PART *A–B–K* (Fig. 5-16f): The equations are written for the region $1\text{ m} \le x \le 2.5$ m,

$$\sum F_y = 0 \qquad 2.43\text{ kN} - 2\text{ kN} - V_{K2} = 0 \qquad V_{K2} = 0.43\text{ kN}$$

$$\sum M_K = 0 \qquad M_{K2} - 2.43\text{ kN}(x) + 2\text{ kN}(x - 1\text{ m}) = 0$$

$$M_{K2} = 0.43(x)\text{ kN}\cdot\text{m} + 2\text{ kN}\cdot\text{m}$$

PART *A–B–C–K* (Fig. 5-16g): The equations are written for the region $2.5\text{ m} \le x \le 3.7$ m,

$$\sum F_y = 0$$

$$2.43\text{ kN} - 2\text{ kN} - 3\text{ kN} - V_{K3} = 0 \qquad V_{K3} = -2.57\text{ kN}$$

$$\sum M_K = 0$$

$$M_{K3} - 2.43\text{ kN}(x) + 2\text{ kN}(x - 1\text{ m}) + 3\text{ kN}(x - 2.5\text{ m}) = 0$$

$$M_{K3} = (2.43 - 2 - 3)\text{ kN}(x) + 2\text{ kN}(1\text{ m}) + 3\text{ kN}(2.5\text{ m})$$
$$= -2.57(x)\text{ kN}\cdot\text{m} + 9.5\text{ kN}\cdot\text{m}$$

These equations are plotted on the appropriate axes for the intervals for which they are valid (see Fig. 5-16c and d). Plotting is facilitated by evaluating each equation at the end points of the interval for which it is valid and plotting the values at those points. A line is drawn connecting the neighboring points. It is seen that two equations are valid where each force is acting, one for the section to the left and one for the section to the right of the point where the force is applied. If the reaction that is being sketched (shear or bending moment) is discontinuous at a section (for example, the shear force at points B and C), then there are two different values plotted at that section and a vertical line is assumed to join the two points (see Fig. 5-16c). If the reaction being plotted is *not* discontinuous (for example, the bending moment throughout the beam), then the two equations should yield the same point at the interval end points. This helps to serve as a check of the obtained equations.

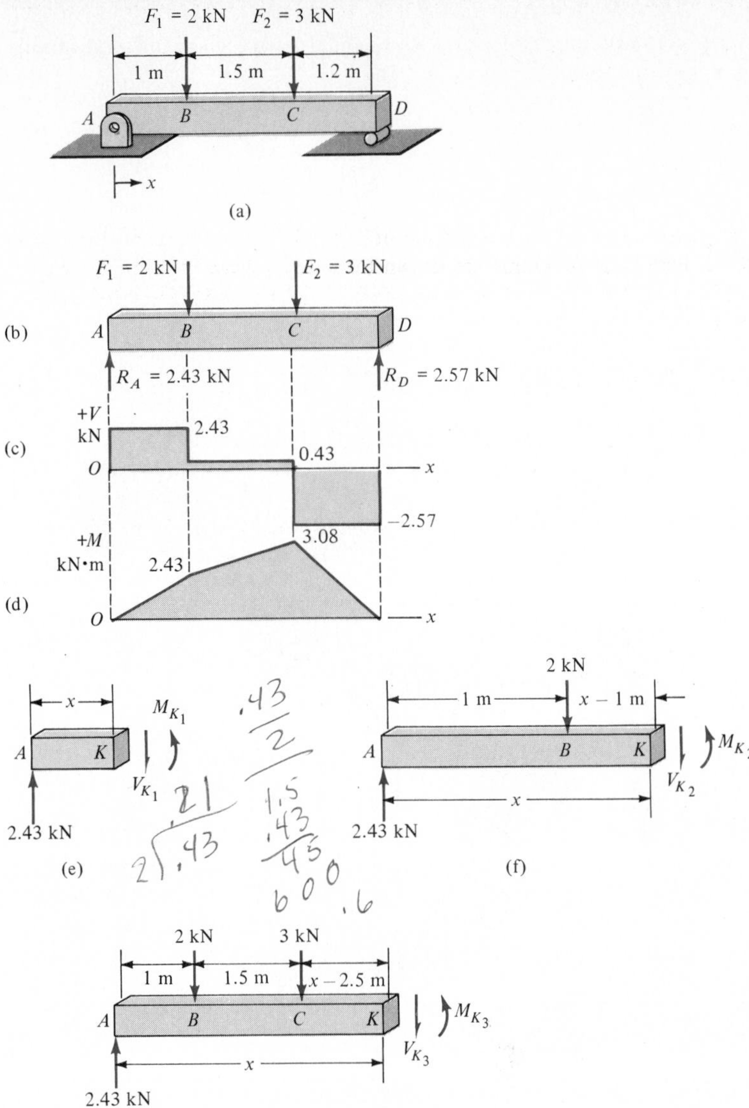

FIGURE 5-16

Shear force and bending moment diagrams of a simply supported beam

JUDGMENT OF THE RESULTS

The shapes of the sketched plots appear reasonable when remembering the discussion above. Since the vertical loads are discontinuous at A, B, C, and D, the shear force diagram is expected to be discontinuous at these points, as well. When calculating the internal bending moments at B and C, the same values are obtained using either of the two equations which are valid for each point.

EXAMPLE 5-9

Write the shear and moment equations and draw the shear and moment diagrams for the beam shown in Fig. 5-17a.

SOLUTION

The reactions at B and D are given in Fig. 5-17b. The reader should check these. Free-body diagrams are drawn to place a general section K in each section between neighboring vertical forces. Equations 4-2 are used for each part.

PART *A–K* (Fig. 5-17e): For the region $0 \le x \le 2$ ft,

$$\sum F_y = 0 \qquad -400\text{ lb} - V_{K1} = 0 \qquad V_{K1} = -400\text{ lb}$$

$$\sum M_K = 0 \qquad M_{K1} + 400\text{ lb}(x) = 0 \qquad M_{K1} = -400(x)\text{ ft}\cdot\text{lb}$$

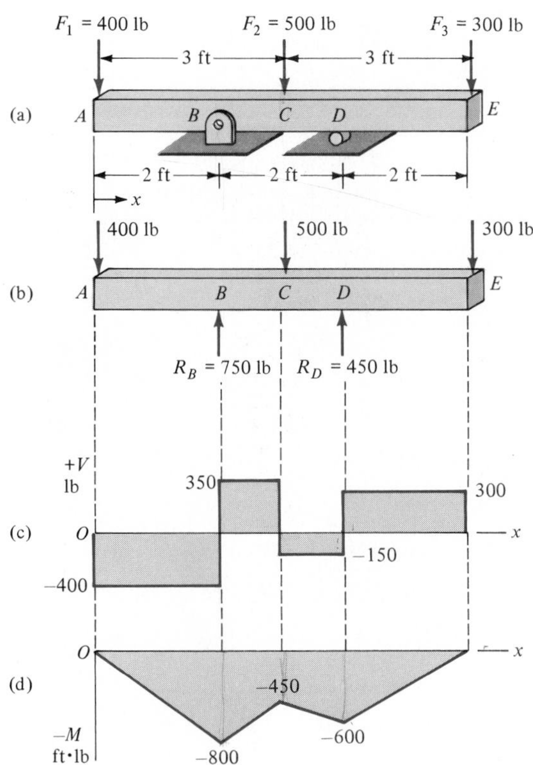

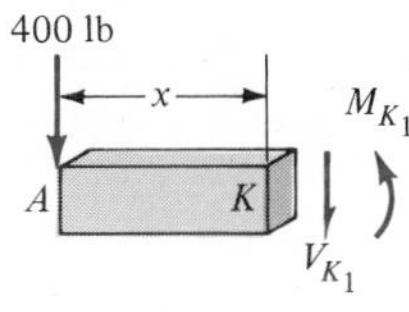

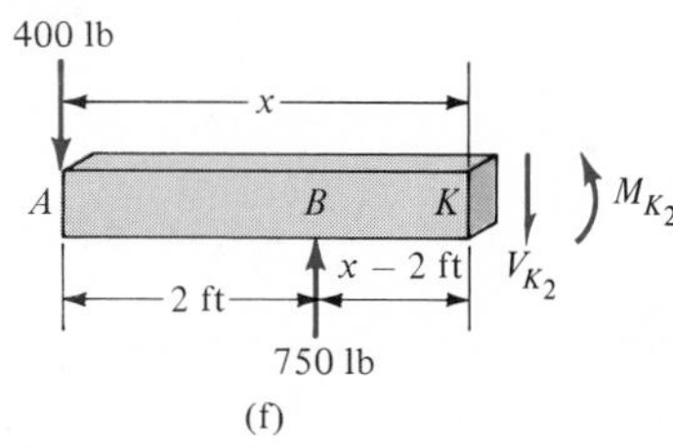

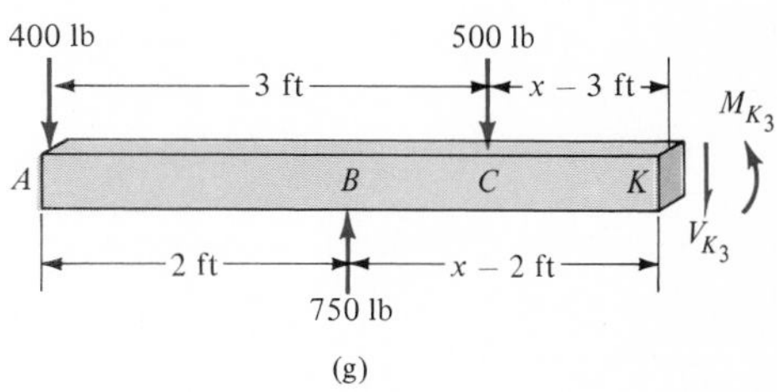

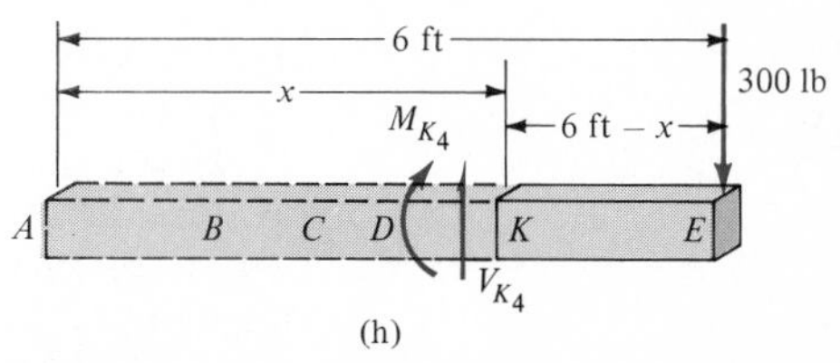

FIGURE 5-17

Shear force and bending moment diagrams of a simply supported beam

PART A–B–K (Fig. 5-17f): The equations are written for the region $2 \text{ ft} \leq x \leq 3 \text{ ft}$,

$$\sum F_y = 0 \qquad -400 \text{ lb} + 750 \text{ lb} - V_{K2} = 0 \qquad V_{K2} = 350 \text{ lb}$$

$$\sum M_K = 0 \qquad M_{K2} + 400 \text{ lb}(x) - 750 \text{ lb}(x - 2 \text{ ft}) = 0$$

$$M_{K2} = 350(x) \text{ ft} \cdot \text{lb} - 1500 \text{ ft} \cdot \text{lb}$$

PART A–B–C–K (Fig. 5-17g): The equations are written for the region $3 \text{ ft} \leq x \leq 4 \text{ ft}$,

$$\sum F_y = 0$$

$$-400 \text{ lb} + 750 \text{ lb} - 500 \text{ lb} - V_{K3} = 0 \qquad V_{K3} = -150 \text{ lb}$$

$$\sum M_K = 0$$

$$M_{K3} + 400 \text{ lb}(x) - 750 \text{ lb}(x - 2 \text{ ft}) + 500 \text{ lb}(x - 3 \text{ ft}) = 0$$

$$M_{K3} = (750 - 400 - 500) \text{ lb}(x) - 1500 \text{ ft} \cdot \text{lb} + 1500 \text{ ft} \cdot \text{lb}$$
$$= -150(x) \text{ ft} \cdot \text{lb}$$

PART A–B–C–D–K: For the region $4 \text{ ft} \leq x \leq 6 \text{ ft}$, it is seen that the calculation would become long and tedious because of the many forces involved. Remembering that the same internal reactions can be obtained using either the right or left section, it is advantageous to use the right-hand section for section DE. However, the equations must remain consistent, with x being measured from the left end (point A); hence the general distance shown in Fig. 5-17h is $(6 \text{ ft} - x)$. It must also be remembered to assume the proper positive shear and bending moments for a right section as shown. Thus, the last part used, instead of A–B–C–D–K, is

PART K–E (Fig. 5-17h): For the region $4 \text{ ft} \leq x \leq 6 \text{ ft}$,

$$\sum F_y = 0 \qquad V_{K4} - 300 \text{ lb} = 0$$

$$V_{K4} = 300 \text{ lb}$$

$$\sum M_K = 0 \qquad -300 \text{ lb}(6 \text{ ft} - x) - M_{K4} = 0$$

$$M_{K4} = 300(x) \text{ ft} \cdot \text{lb} - 1800 \text{ ft} \cdot \text{lb}$$

These equations are plotted for the appropriate intervals (as outlined in Ex. 5-8) in Fig. 5-17c and d.

JUDGMENT OF THE RESULTS

The bending moments are checked at sections B, C, and D by using both moment equations (for the part to the left and the part to the right from the point considered). It is seen that the same moments are obtained. Hence the results can be stated with confidence. The shear force diagram is correct by inspection.

EXAMPLE 5-10

Write the shear and moment equations and draw the shear and moment diagrams for the beam shown in Fig. 5-18a.

SOLUTION

The reactions at A are given in Fig. 5-18, the free-body diagram of the whole beam. The reader should check these. There are two distinct parts of the beam, A–K and A–B–K. Equations 4-2 are used for each of these parts.

PART *A*–*K* (Fig. 5-18e): For the region $0 \leq x \leq 1$ m,

$$\sum F_y = 0 \qquad 2\text{ kN} - V_{K1} = 0$$

$$V_{K1} = 2\text{ kN}$$

$$\sum M_K = 0 \qquad M_{K1} + 0.6\text{ kN}\cdot\text{m} - 2\text{ kN}(x) = 0$$

$$M_{K1} = 2(x)\text{ kN}\cdot\text{m} - 0.6\text{ kN}\cdot\text{m}$$

PART *A*–*B*–*K* (Fig. 5-18f): The equations are written for the region $1\text{ m} \leq x \leq 2.3$ m,

$$\sum F_y = 0 \qquad 2\text{ kN} - V_{K2} = 0 \qquad V_{K2} = 2\text{ kN}$$

$$\sum M_K = 0 \qquad M_{K2} + 0.6\text{ kN}\cdot\text{m} + 4\text{ kN}\cdot\text{m} - 2\text{ kN}(x) = 0$$

$$M_{K2} = 2(x)\text{ kN}\cdot\text{m} - 4.6\text{ kN}\cdot\text{m}$$

The equations are plotted (by the method outlined in Ex. 5-8) in Fig. 5-18c and d. The discontinuities at A and B in the bending moment curve are caused by the concentrated moments applied at these points. Plotting the moment equations near the end points of each part yields two different values at B, which are connected by a vertical line. It should be noted that the effect of a concentrated moment does not show up in the shear diagram. This appears to be logical after reconsidering the free-body diagrams of the general sections.

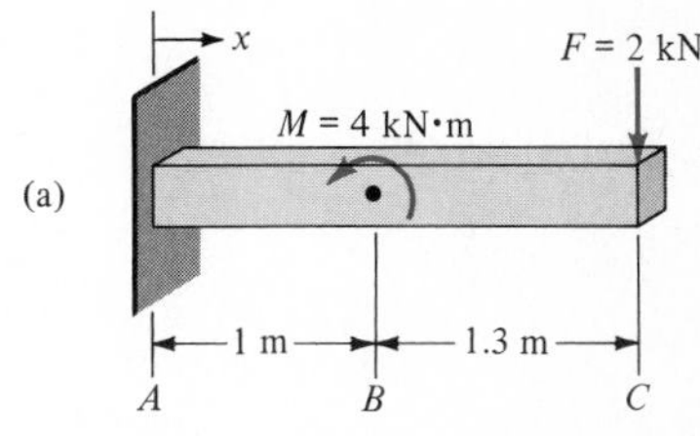

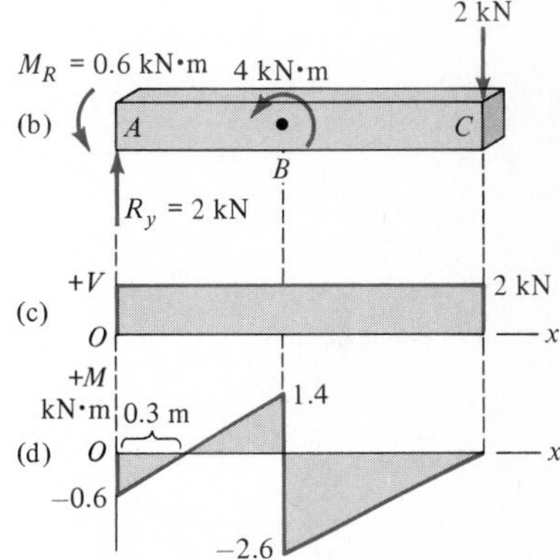

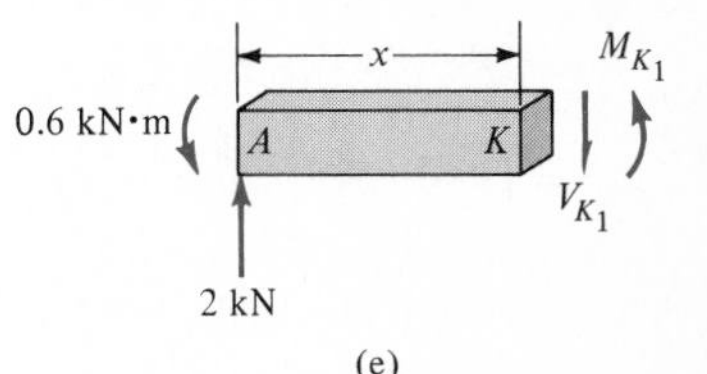

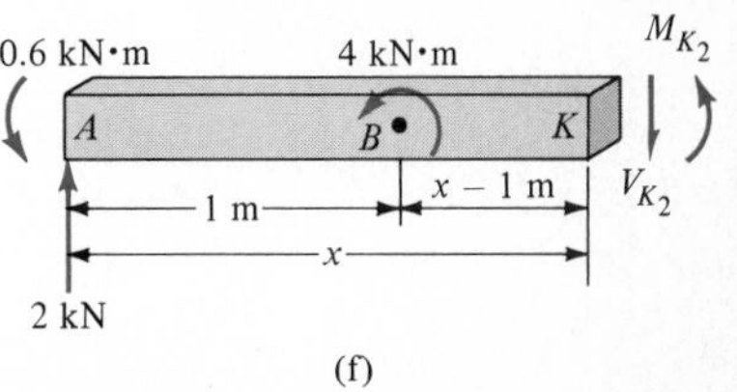

FIGURE 5-18

Shear force and bending moment diagrams of a cantilever beam

EXAMPLE 5-11

Write the shear and moment equations and draw the shear and moment diagrams for the beam shown in Fig. 5-19a.

SOLUTION

The reactions at A and C are given in Fig. 5-19b, the free-body diagram of the whole beam. The reader should check these. There are two distinct parts of the beam, A–K and A–B–K. Equations 4-2 are used for each of these parts.

PART *A–K* (Fig. 5-19e): For the region $0 \leq x \leq 3$ ft,

$$\sum F_y = 0 \qquad -V_{K1} - 714\ \text{lb} = 0 \qquad V_{K1} = -714\ \text{lb}$$

$$\sum M_K = 0 \qquad M_{K1} + 714\ \text{lb}(x) = 0 \qquad M_{K1} = -714(x)\ \text{ft}\cdot\text{lb}$$

PART *A–B–K* (Fig. 5-19f): The equations are written for the region $3\ \text{ft} \leq x \leq 7$ ft,

$$\sum F_y = 0 \qquad -V_{K2} - 714\ \text{lb} = 0 \qquad V_{K2} = -714\ \text{lb}$$

$$\sum M_K = 0 \qquad M_{K2} + 714\ \text{lb}(x) - 5000\ \text{ft}\cdot\text{lb} = 0$$

$$M_{K2} = 5000\ \text{ft}\cdot\text{lb} - 714(x)\ \text{ft}\cdot\text{lb}$$

The equations are plotted (by the method outlined in Ex. 5-8) in Fig. 5-19c and d. Since there is a concentrated moment applied to the beam, the bending moment diagram is discontinuous. The shear diagram is unaffected at the point of the concentrated moment and would be the same regardless of the point where the moment is applied.

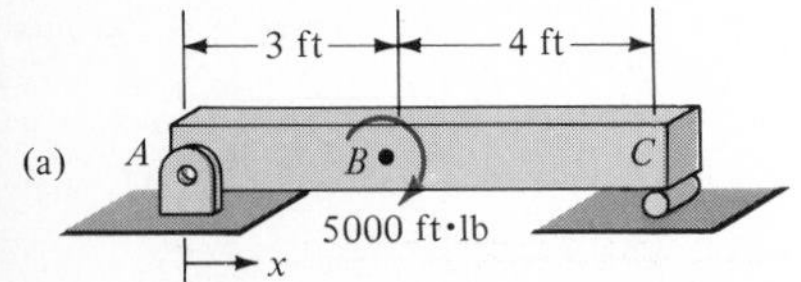

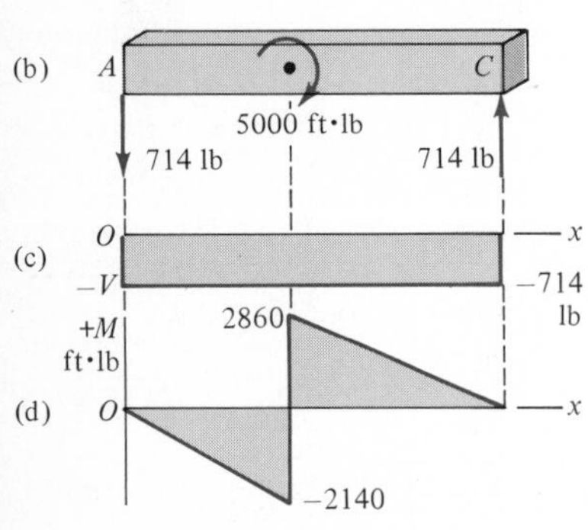

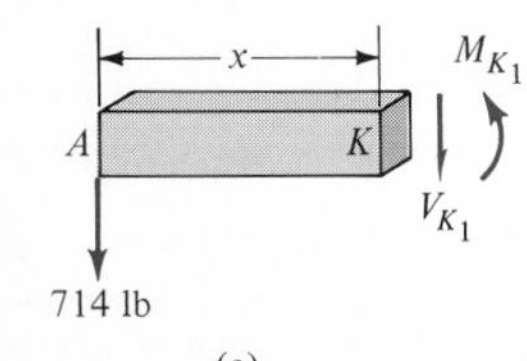

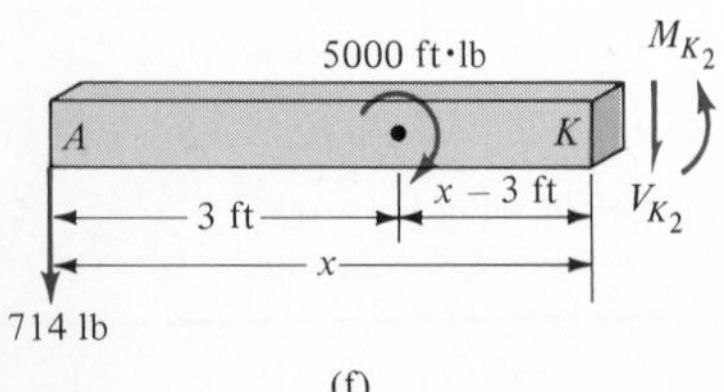

FIGURE 5-19

Shear force and bending moment diagrams of a simply supported beam

In Probs. 5–36 to 5–56, write the equations of transverse shear force and bending moment for the beams shown. Sketch the complete shear force and bending moment diagrams.

FIGURE P5-36

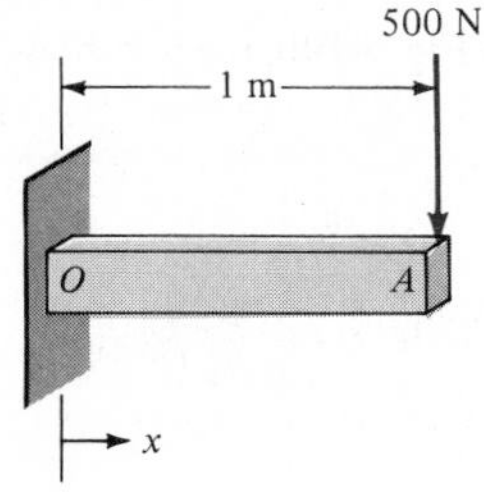

FIGURE P5-37

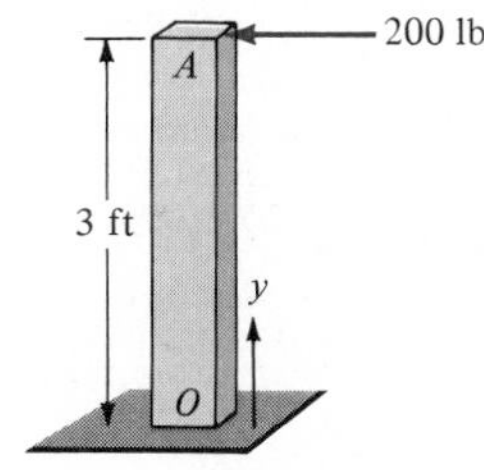

FIGURE P5-38

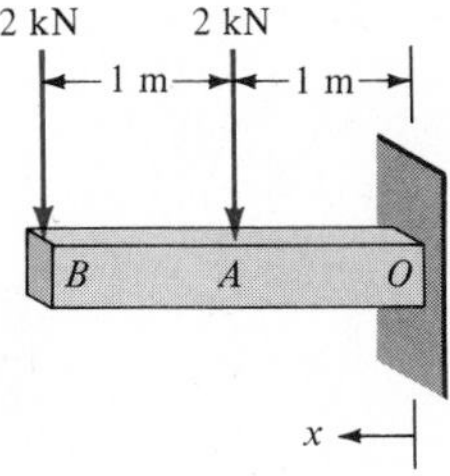

FIGURE P5-39

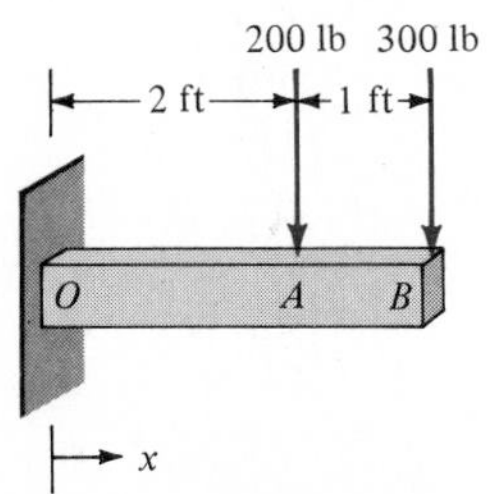

FIGURE P5-40

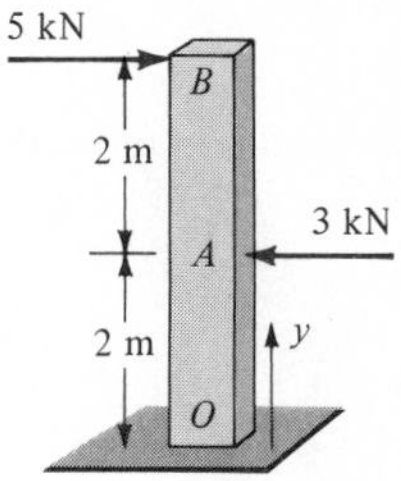

FIGURE P5-41

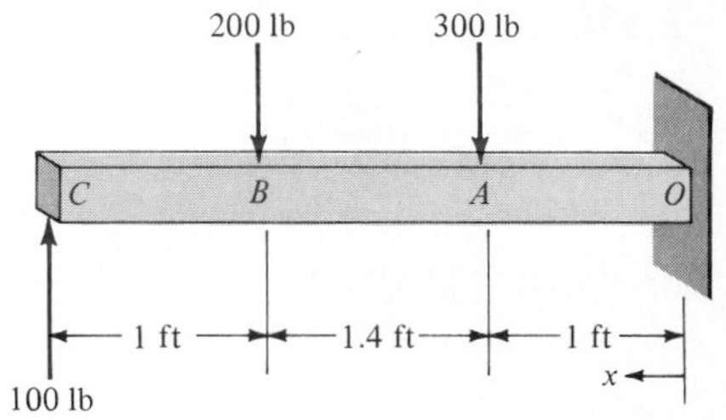

FIGURE P5-42

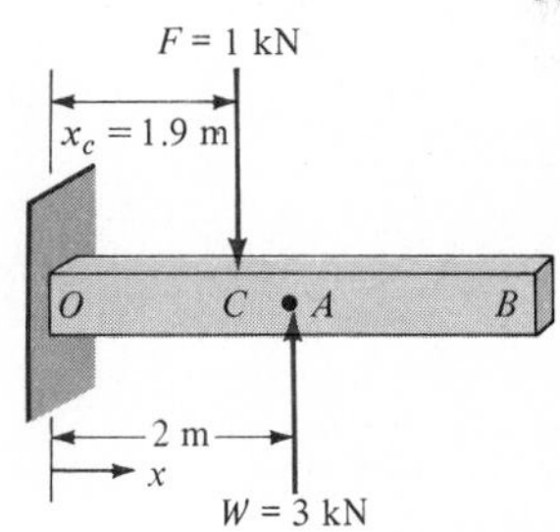

FIGURE P5-43

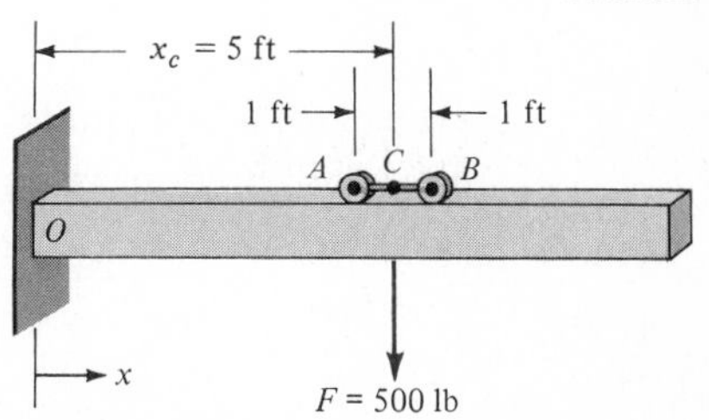

FIGURE P5-44

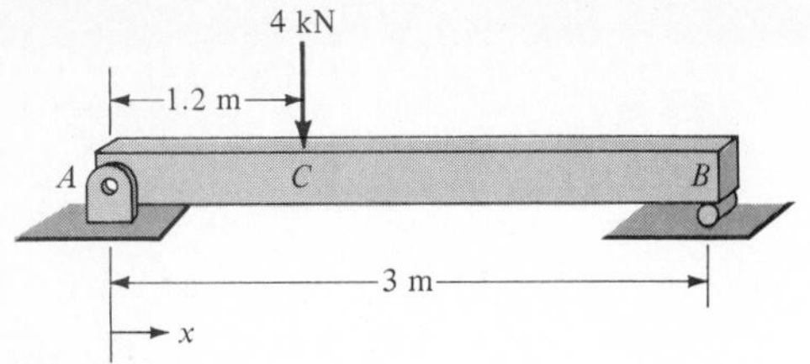

FIGURE P5-48

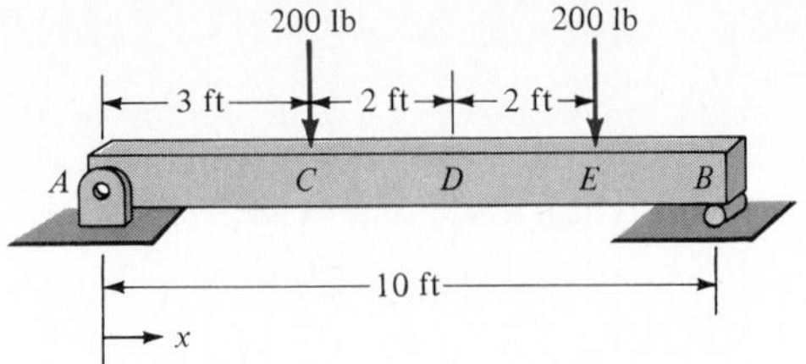

FIGURE P5-45

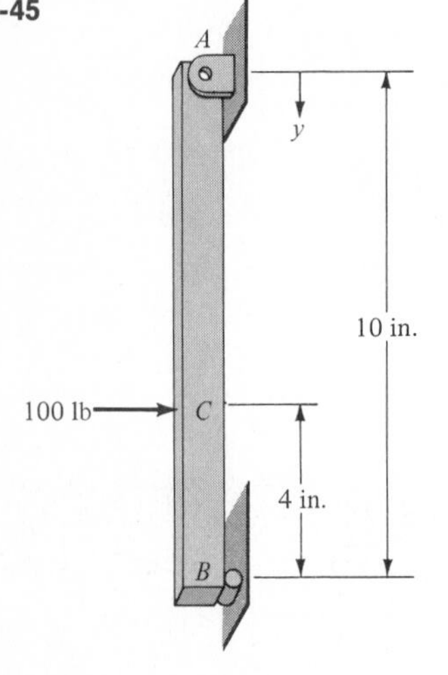

FIGURE P5-49

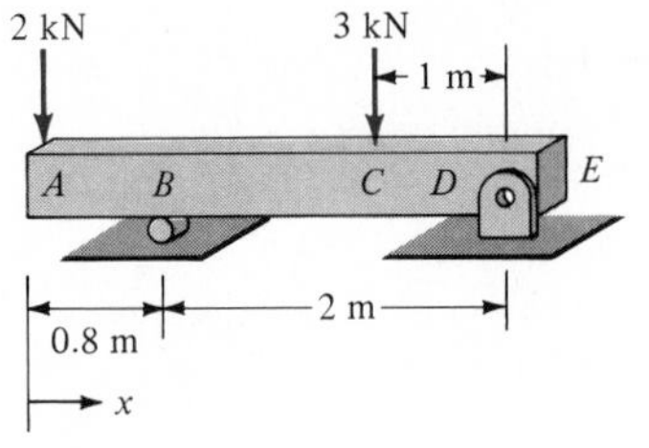

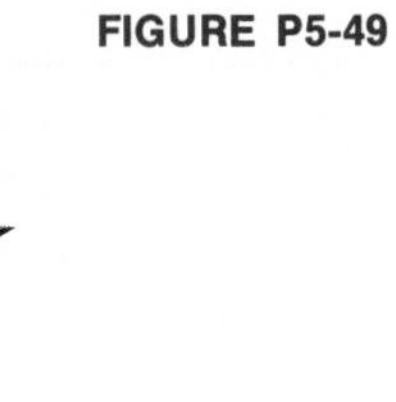

FIGURE P5-50

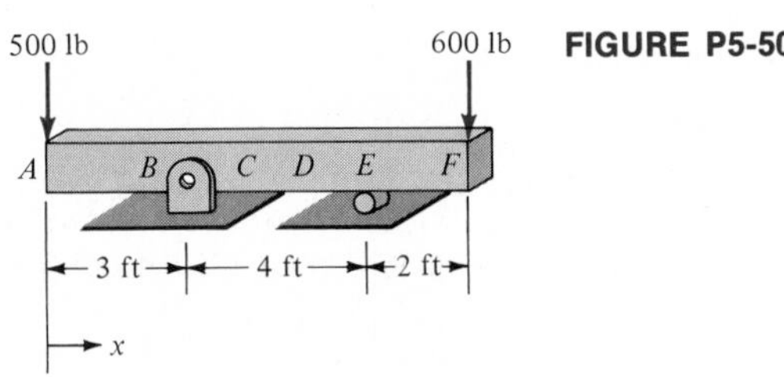

FIGURE P5-46

F = 3 kN

x_c = 4.8 m

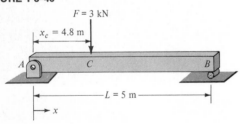

A C B

L = 5 m

x

FIGURE P5-51

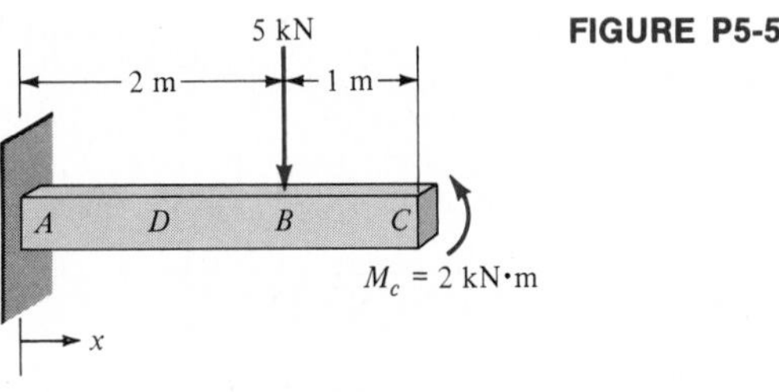

FIGURE P5-52

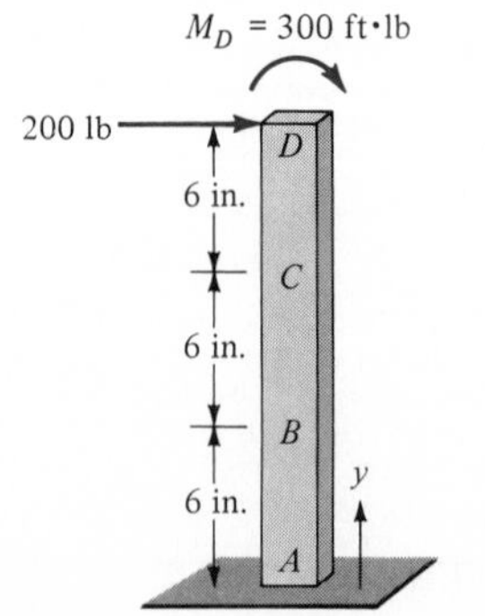

FIGURE P5-47

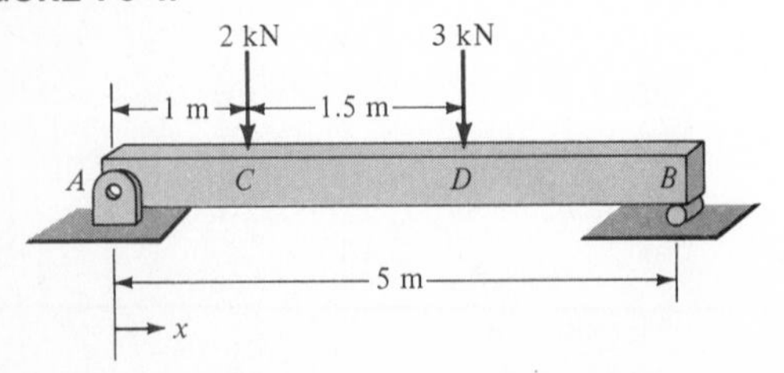

FIGURE P5-53

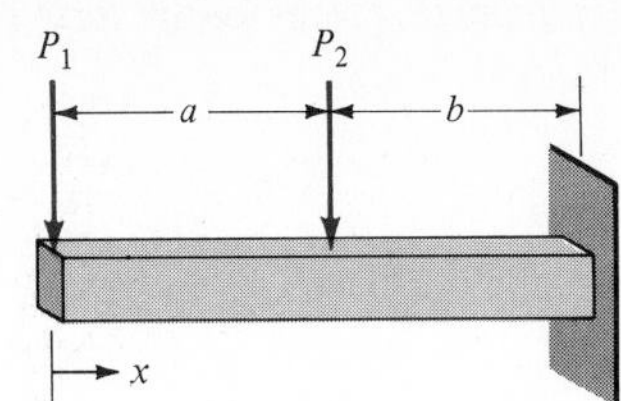

FIGURE P5-55

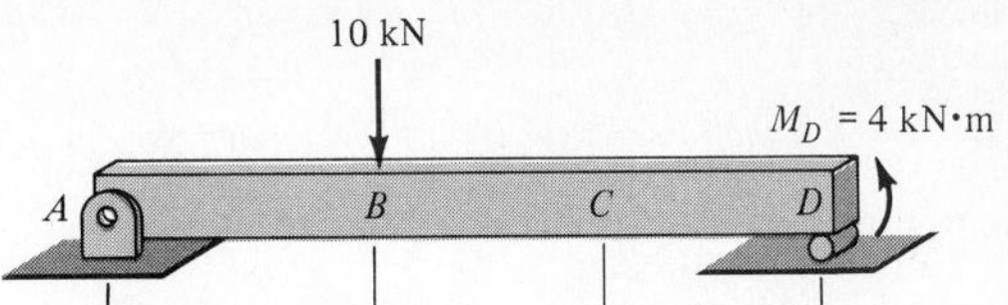

FIGURE P5-54

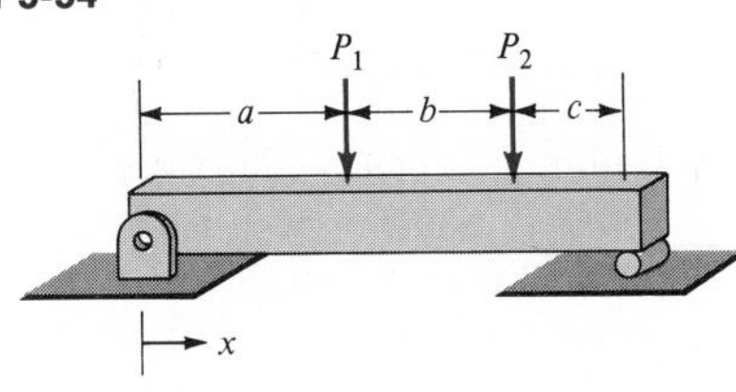

FIGURE P5-56

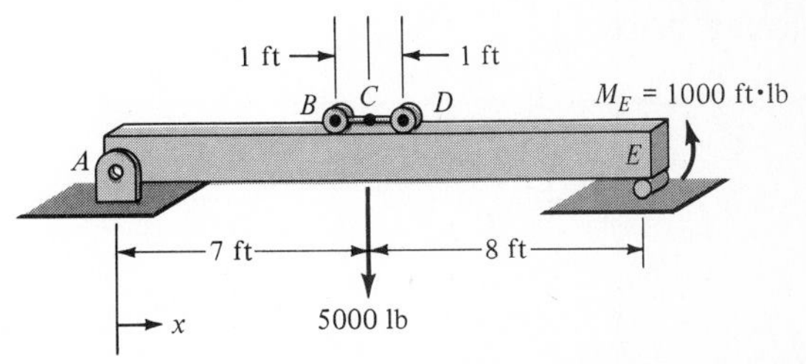

5–57 Sketch the shear force and bending moment diagrams for the boom of the crane described in Prob. 5–24. Make the plots for $0 < x < 120$ ft (the load is at $x_C = 120$ ft).

FIGURE P5-57

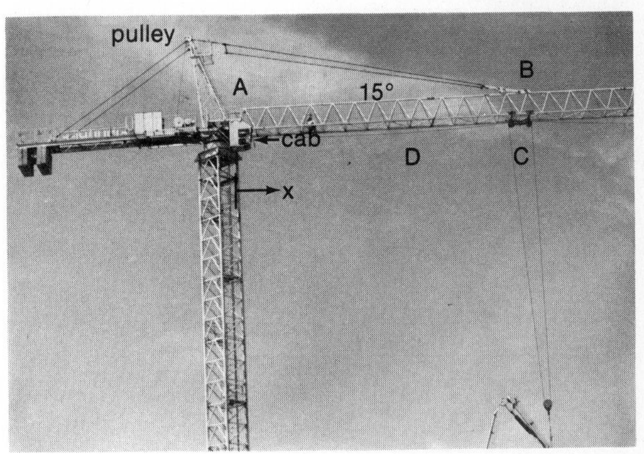

5–58 Sketch the transverse shear force and bending moment diagrams for the ladder described in Prob. 5–26.

FIGURE P5-58

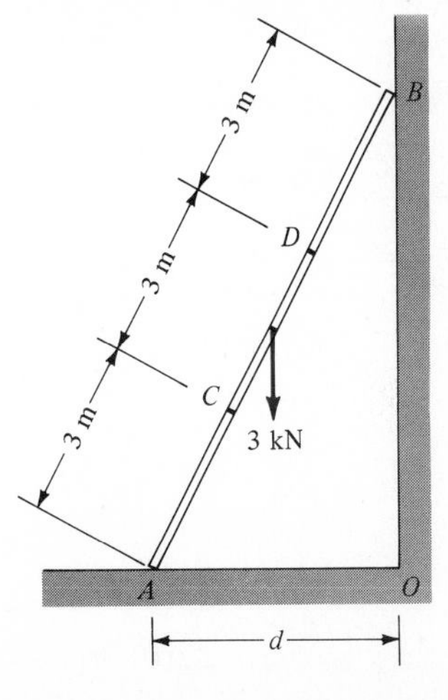

FIGURE P5-59

5–59 Sketch the transverse shear force and bending moment diagrams for the spring BC described in Prob. 5–28.

FIGURE P5-60

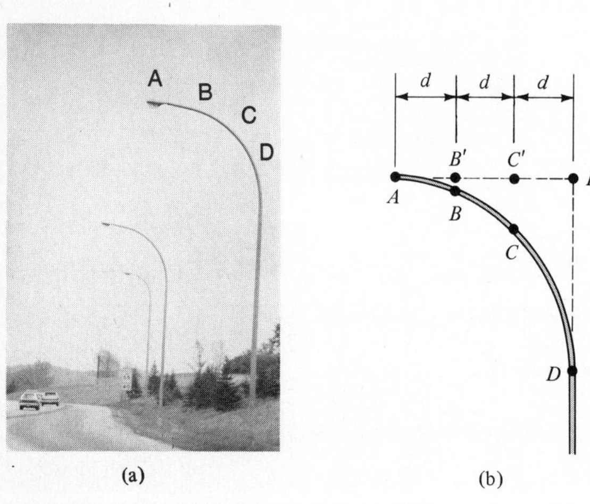

(a) (b)

5–60 Consider a light pole described in Prob. 5–30. Assume that section AD is a quarter circle of radius $3d$. Calculate the bending moments at A, B, C, and D, and plot these as functions of the linear distance l along the pole. Thus, make a standard bending moment diagram with M on the vertical axis, and l on the horizontal axis (l increases from A to D).

FIGURE P5-61

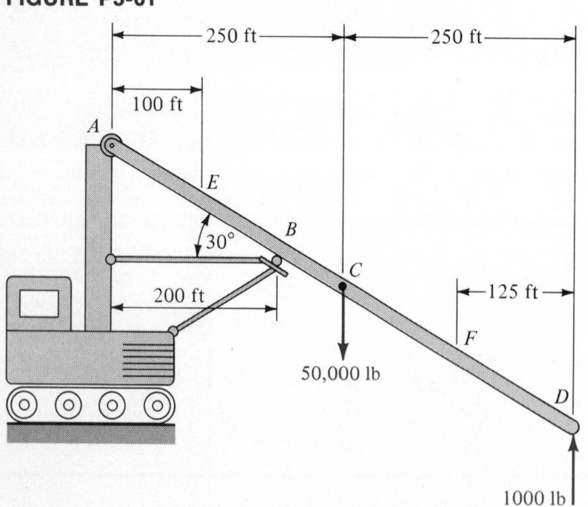

5–61 Sketch the transverse shear force and bending moment diagrams for the arm AD of the excavator in Prob. 5–31.

5–62 Draw the transverse shear force and bending moment diagrams for the arm AC of the amusement ride described in Prob. 5–32. Assume that $\theta = 60^{\circ}$, $\phi = 70^{\circ}$, $F = 10$ kN, and $d = 3l = 3$ m.

FIGURE P5-62

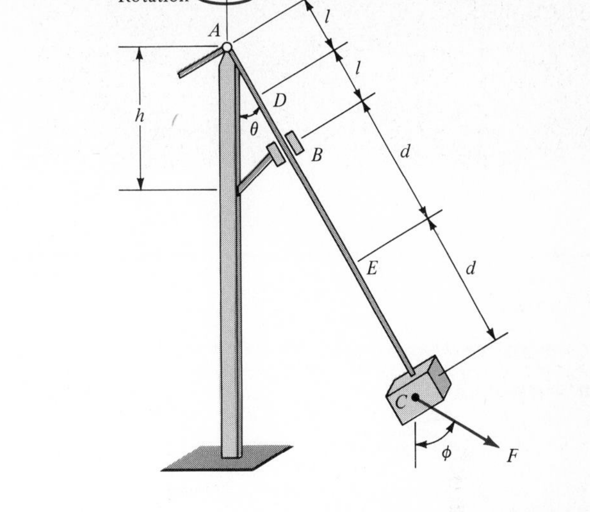

5–63 In an assembly operation four bolts are tightened simultaneously as shown. Draw the shear force and bending moment diagrams for the member AF.

FIGURE P5-63

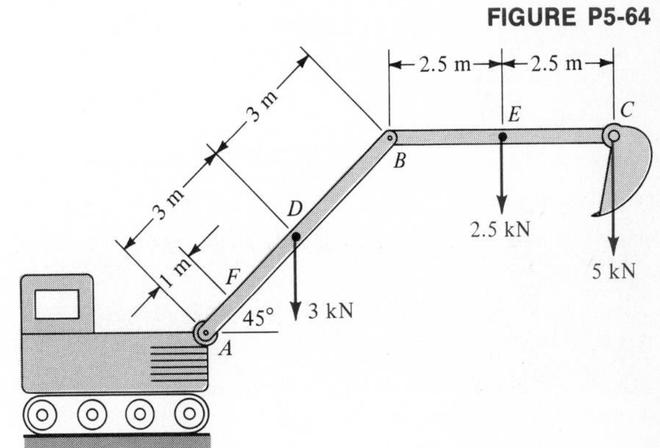

5–64 Consider the excavator described in Prob. 5–33 and neglect the forces acting at D and E. Sketch the bending moment diagrams for arms AB and BC.

5–65 A 5-ft-long cantilever beam is modeled with five equal forces equally spaced along the beam. The forces represent the beam's weight and other loads. Draw the shear force and bending moment diagrams.

FIGURE P5-65

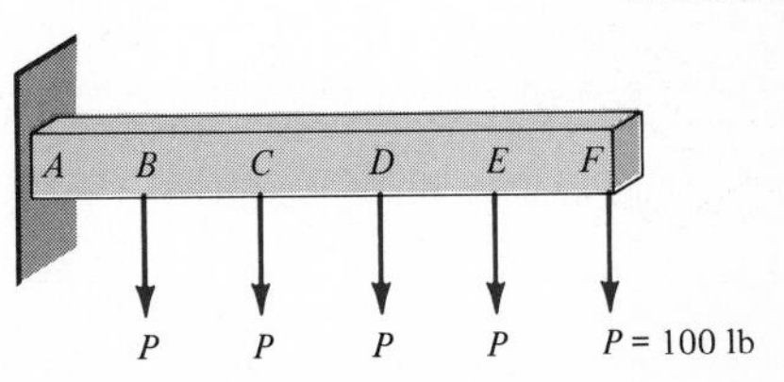

FIGURE P5-66

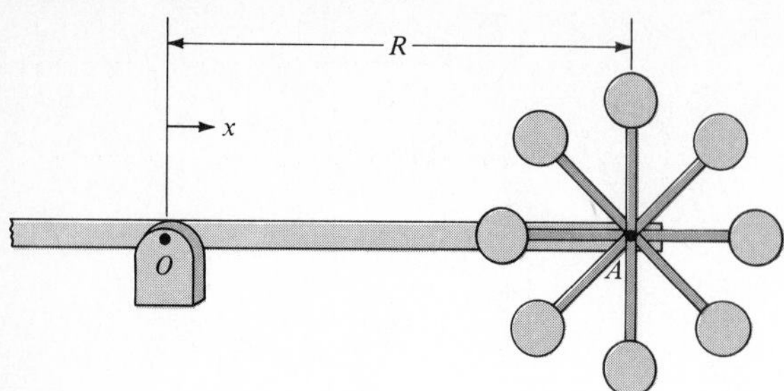

5–66 An amusement ride is modeled as a cantilever beam supported at O. The weight W_1 of arm OA is represented by three forces, $P = W_1/3$. A force P is acting at $x = R/4$, $x = R/2$, and $x = 3R/4$. The weight $W_2 = 2W_1$ of the small arms is acting at A. A clockwise couple of magnitude $M_A = PR$ is also acting at A (because of motion of the small arms). Sketch the bending moment diagram of arm OA.

5–67 Draw the shear force and bending moment diagrams for beam AB of the drag racing car described in Prob. 5–35.

5–68 Assume that the end pipe of an agricultural sprinkler system is horizontal and is hinged at A. Its distributed weight is represented by nine concentrated forces as shown. Draw the shear force and bending moment diagrams assuming that only the wire at B is holding the pipe up ($T_C = T_D = 0$). For optional work, assume that either $T_B = T_D = 0$, or $T_B = T_C = 0$.

FIGURE P5-67

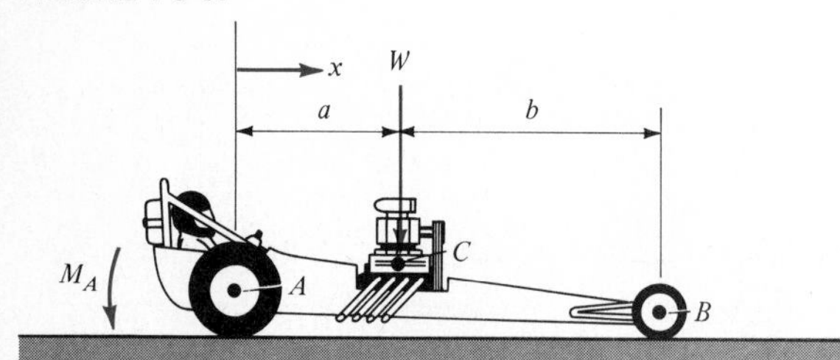

FIGURE P5-68

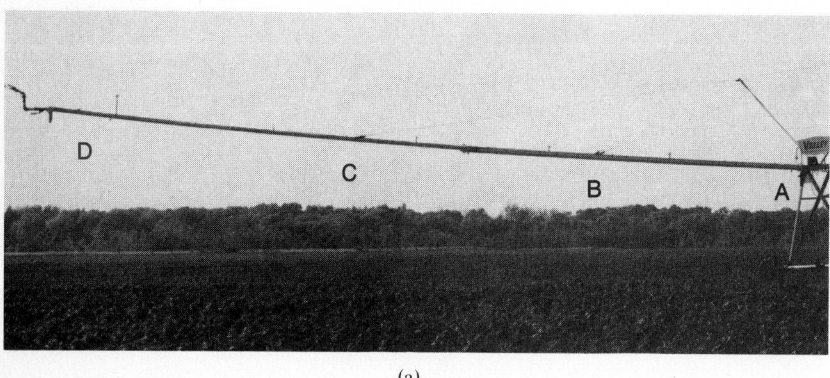

(a)

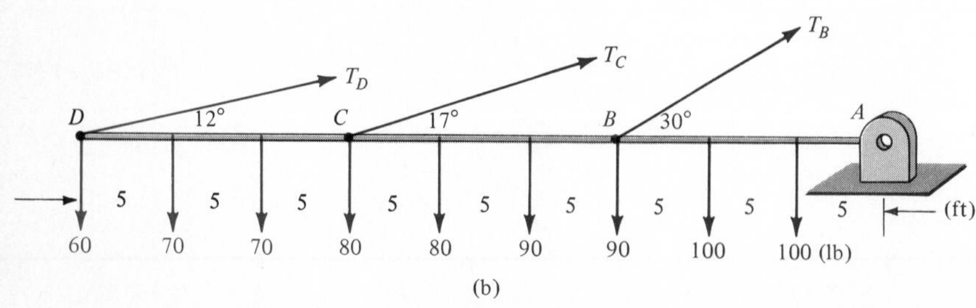

(b)

5–69 Each jet engine on the wing weighs 50 kN. The resultant of the aerodynamic lift forces is 2 MN. Plot the moment diagram for the whole wing using the concentrated forces. Sketch on the same diagram your estimate of a more realistic moment variation. Assume that the wing is weightless and that L is the resultant of many parallel forces that are largest at O and smallest at D (the correct way of doing this is covered in Ch. 7; here, rely on simple techniques and intuition).

FIGURE P5-69

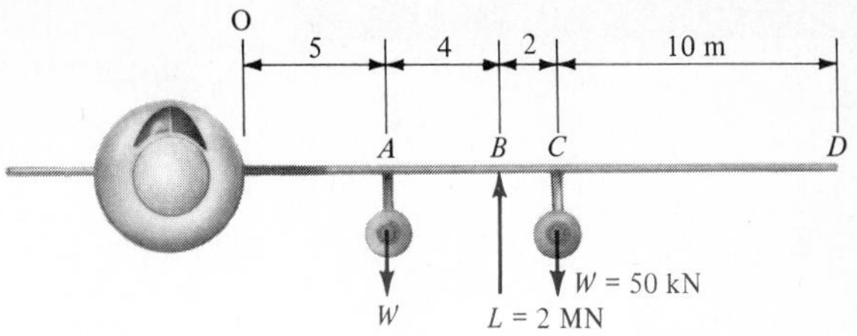

5–70 Consider the boom of the 100-metric-ton crane whose model is shown. The boom is hinged at A and supported by wires at B that make a 30° angle with it. Sketch on a single moment diagram two curves that show the bending moments in the boom with the 1-MN load at C and later at B. $AB = 15$ m, $BC = 30$ m.

FIGURE P5-70

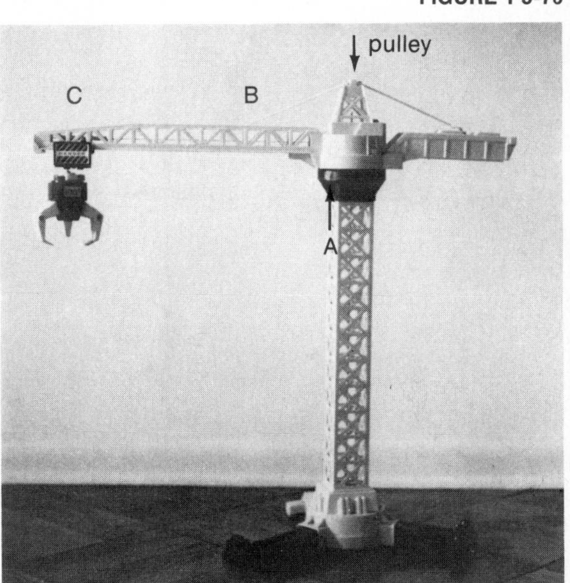

SUMMARY

Section 5-1

COMMON SUPPORTS OF BEAMS AND POSSIBLE REACTIONS:

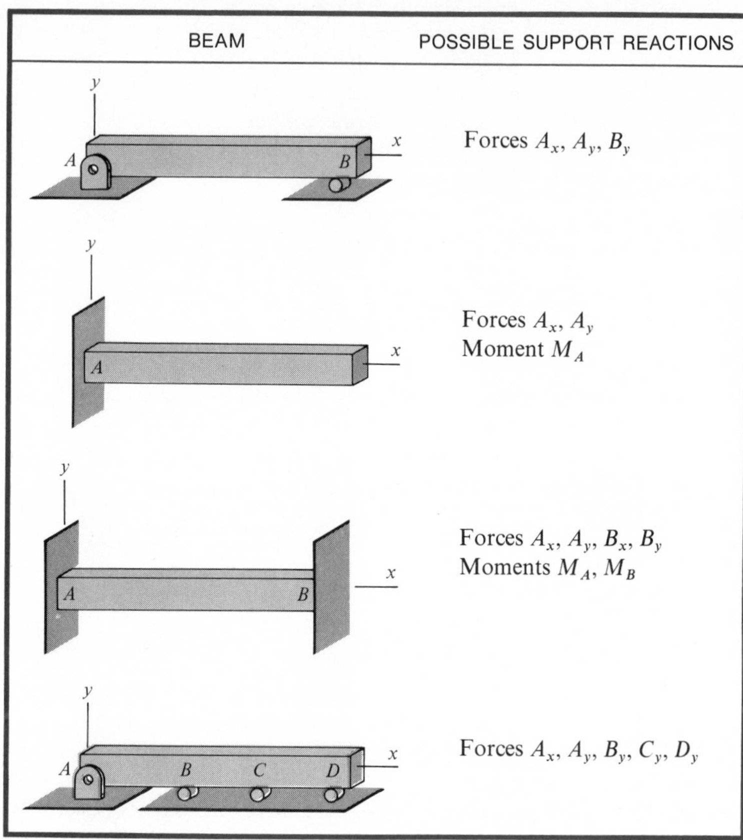

BEAM	POSSIBLE SUPPORT REACTIONS
(pin at A, roller at B)	Forces A_x, A_y, B_y
(fixed at A)	Forces A_x, A_y Moment M_A
(fixed at A and B)	Forces A_x, A_y, B_x, B_y Moments M_A, M_B
(pin at A, rollers at B, C, D)	Forces A_x, A_y, B_y, C_y, D_y

Section 5-2

INTERNAL FORCES AND MOMENTS—EXAMPLE:

Given loading of a beam:

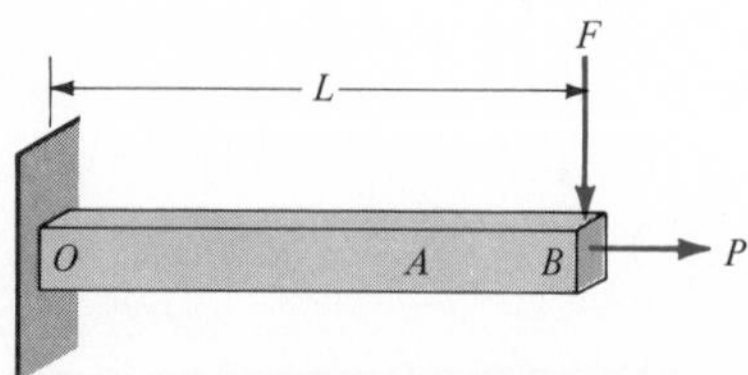

Free-body diagram of whole beam:

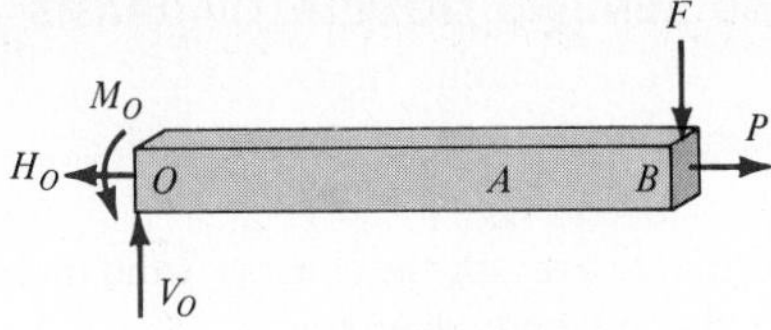

$$H_O = \text{horizontal reaction} = P$$

$$M_O = \text{bending moment} = FL$$

$$V_O = \text{vertical shear} = F$$

Internal reactions at arbitrary section A:

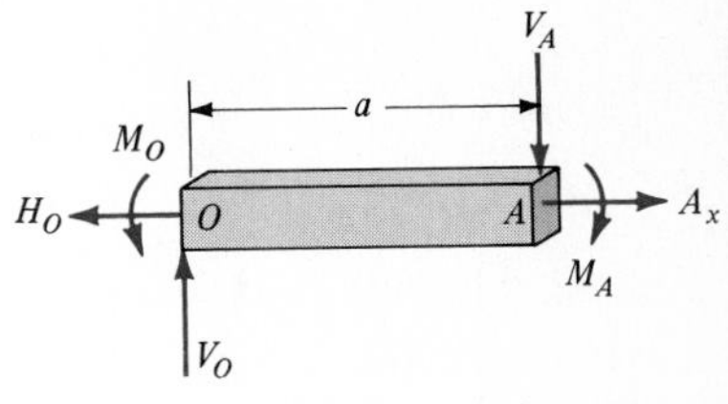

$V_A = F$

$A_x = P$

$$M_A = \underbrace{Fb}_{\text{from right piece}} = \underbrace{M_O - Fa}_{\text{from left piece}}$$

$$\left(\begin{array}{ll}\text{check:} & \text{as } a \to L,\ M_A \to 0 \\ & \text{as } a \to 0,\ M_A \to M_O\end{array}\right)$$

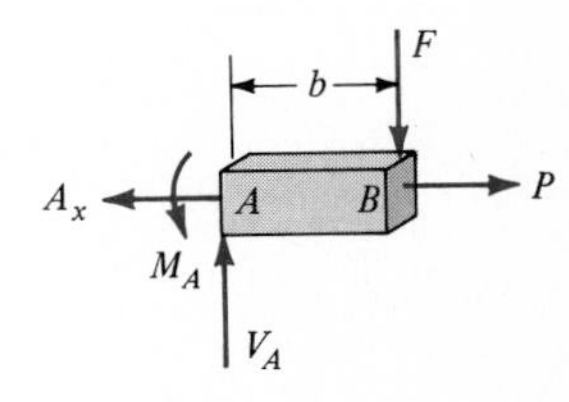

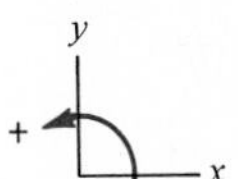

Section-5-3

SIGN CONVENTIONS FOR SHEAR FORCE AND BENDING MOMENT:

Distinguish the directions of forces and moments on the right side of an element from those on the left side of the same element. The following combinations of signs are possible if no other forces or moments act on element AB:

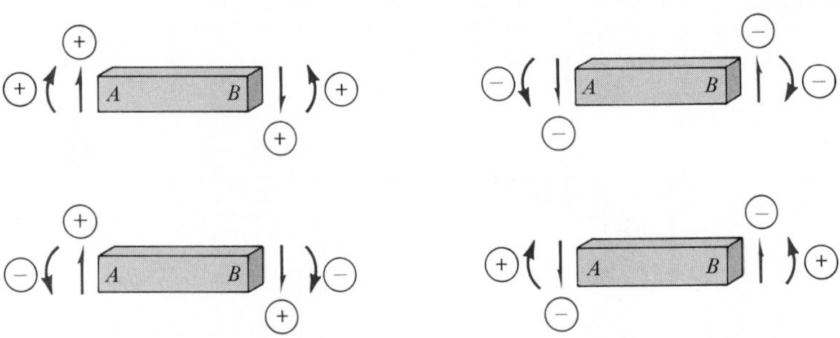

Section 5-4

SHEAR FORCE AND BENDING MOMENT DIAGRAMS—PROCEDURE:

1. Draw the free-body diagram of the whole beam and determine all reactions.
2. Draw the coordinate axes for the shear (V) and moment (M) diagrams directly below the free-body diagram.
3. Immediately plot those values of V and M that can be determined by inspection (especially where they are zero).
4. Calculate and plot as many additional values of V and M as are necessary for drawing reasonably accurate curves through the plotted points.
5. Draw the complete curves of V and M and check a few points on them by using free-body elements that are complementary to those used in the original calculations.

EXAMPLE:

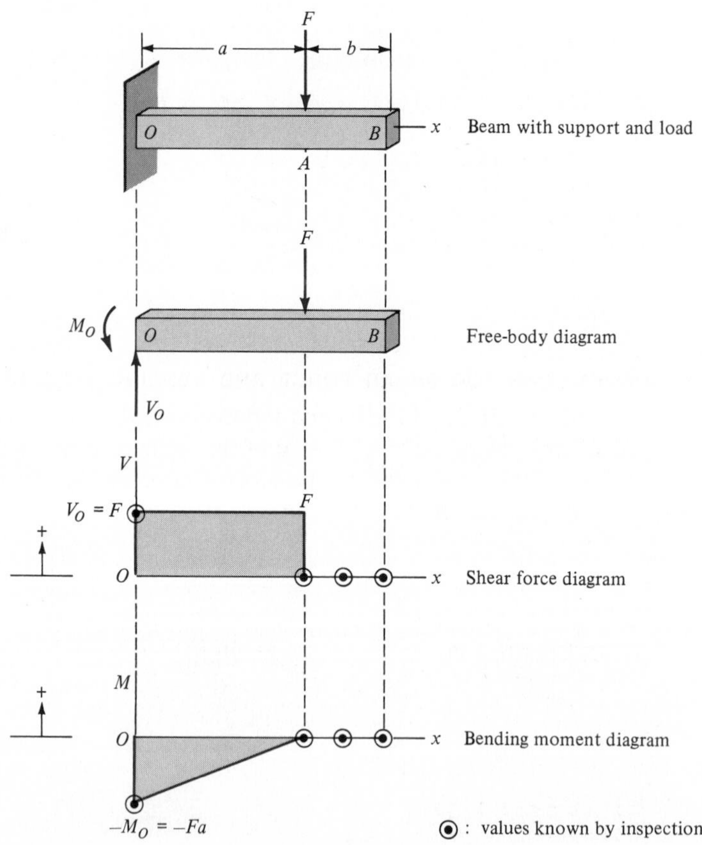

In Probs. 5–71 and 5–72 write the equations of transverse shear force and bending moment for the beams shown. Sketch the complete shear force and bending moment diagrams.

FIGURE P5-71

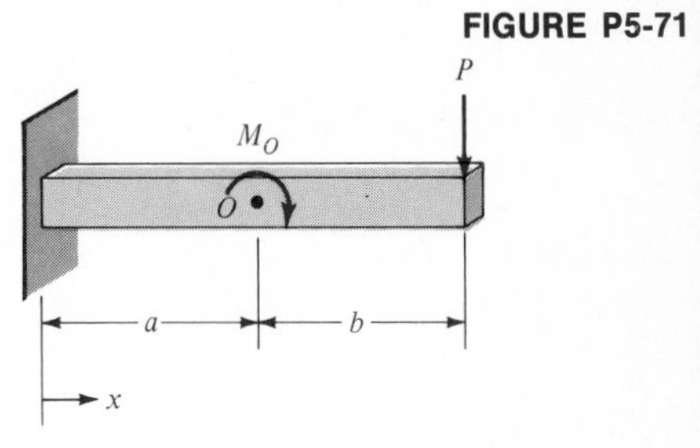

FIGURE P5-72

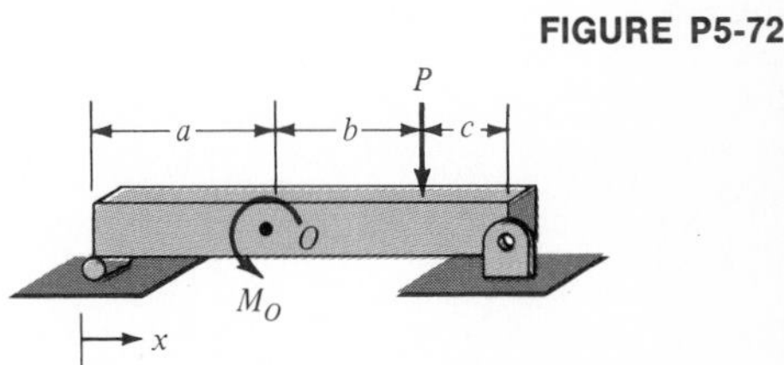

5–73 Consider the model of a racing car mounted in a wind tunnel. The car is attached by two supports (each consisting of three parts, A, B, C) to the bottom of the tunnel (the horizontal line visible at the bottom of the wheels is on the tunnel wall behind the car; the heavy bar below the car is also irrelevant to the car and its supports). Each segment A, B, and C of the supports is in the yz plane and thus perpendicular to the direction of air flow. The supports ABC are instrumented (note the wires) to measure air drag and upward or downward lift on the car. For simplicity, assume that there is only one support ABC, with partial lengths $A = 5$ cm, $B = 8$ cm, and $C = 40$ cm. Ignore the weight of the car. Assume that a drag force $\mathbf{D} = -10\text{ N }\mathbf{i}$ is acting at the top of rod A while the model is adjusted to have no lift force on it. Calculate the transverse shear force and bending moment at the middle (E) and bottom (F) of rod A. Give the answers in vector form.

5–74 Determine the transverse shear force and bending moment at the top (G) and bottom (H) of rod C described in Prob. 5–73. Distinguish bending from twisting by using an equivalent force–couple system.

FIGURE P5-73

(a)

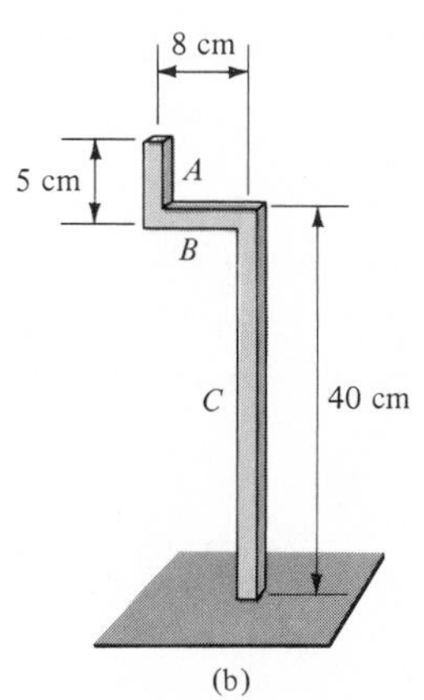

(b)

5–75 Calculate the transverse shear force and bending moment at the middle (I) of rod B described in Prob. 5–73. Assume that besides force **D** a lift force **L** = 2N **j** is also acting at the top of rod A. Distinguish bending from twisting by using an equivalent force–couple system.

5–76 Determine the transverse shear force and bending moment at the middle (J) of rod C described in Prob. 5–73. Assume that besides force **D** a downward aerodynamic force (negative lift) is also acting at the top of rod A, **L** = −4N **j**. Distinguish bending from twisting by using an equivalent force-couple system.

5–77 Consider Prob. 5–73 assuming that the drag force is **D** = −12N **i** while a negative lift force **L** = −3N **j** is also acting at the top of of rod A. For purposes of instrumentation, the bending moments in the three rods should be compared. Calculate the bending moment at the middle of each rod: at point E in rod A, point I in rod B, and point J in rod C. (These points agree with those in the preceding related problems.) Distinguish bending from twisting by using an equivalent force–couple system.

5–78 Review Prob. 5–73 and plot the bending moment diagram for rod A under the action of the drag force **D**.

5–79 Sketch the bending moment diagram for rod B described in Prob. 5–73. Assume that a lift force **L** = 2 N **j** is also acting at the top of rod A. Distinguish bending from twisting by using an equivalent force–couple system.

5–80 Sketch the bending moment diagram for rod C described in Prob. 5–73. Assume that a negative lift force **L** = −4 N **j** is also acting at the top of rod A. Distinguish bending from twisting by using an equivalent force–couple system.

CHAPTER 6

ANALYSIS OF SIMPLE STRUCTURES

OVERVIEW

Basic Concepts of Structures

Engineering structures are made of individual rigid members for the purpose of carrying loads safely and efficiently. For example, airplane wings must be strong to prevent failure but also minimum in weight for economical flying. Even stationary structures such as bridges and buildings must be designed to save materials.

Fortunately, it is possible to design beams and other members that are as strong but much lower in weight than the simplest solid members conceivable for the same purpose. The idea is to shape each member and the whole structure in such a way that most of the material carries reasonably high loads. This is the ultimate challenge in the engineering design of any structure: the amount of material with small loads in it should be minimized, but excessive loads are not permissible even in small regions of the whole structure. Knowledge of statics and strength of materials are important in such design work.

STUDY GOALS

SECTION 6-1 describes trusses, the most common structures, which have high strength and relatively low weight. Simple trusses are statically determinate and can be analyzed using the methods developed in Ch. 4. This section describes the simplifying assumptions necessary for solving basic problems.

SECTION 6-2 presents the method of joints, which is based on equilibrium of a particle. This method provides substantial practice in the application of Newton's Third Law and interrelated free-body diagrams.

SECTION 6-3 presents special cases of joints which are worth remembering to simplify the analysis of trusses made of two-force members.

SECTIONS 6-4 AND 6-5 will be omitted in most statics courses, but they should be of interest to those students who expect to work regularly with complex trusses. Section 6-4 describes a formal procedure for setting up problems, and Sec. 6-5 shows a graphic method of analyzing trusses on the basis of force polygons. This graphic method can be used with or without the formal notation described in Sec. 6-4.

SECTION 6-6 presents the method of sections, which is another method for determining forces in a truss. This method is especially useful when it is required to calculate the forces only in a few members of a truss. It involves a simple procedure which is ideal for practicing basic skills with free-body diagrams and equilibrium equations.

SECTION 6-7 extends the methods of analysis of plane trusses to three-dimensional trusses. This topic is often omitted in regular statics courses because, while it generalizes the basic methods, it requires rather laborious analyses.

SECTION 6-8 presents the important method of analyzing frames and machines which involve multiforce members. This is also based on concepts of equilibrium and interrelated free-body diagrams (Ch. 4). Proficiency in solving problems in this area is indicative of satisfactory knowledge of equilibrium principles.

SECTION 6-9 describes a simple method of analysis for cables which are assumed to be weightless and have only concentrated forces acting on them. Such a cable can be analyzed as a set of two-force members without introducing any new concepts. This topic provides additional practice in solving for unknowns.

6-1 TRUSSES

The most common structures with high strength-to-weight ratios are called *trusses.* They are made in many different sizes and shapes, as can be seen in certain roof and bridge structures. They have straight, relatively slender members whose ends are connected at joints. Two-dimensional trusses (also called *plane trusses*) are designed to carry loads acting in their planes, and they are often connected to form three-dimensional trusses (also called *space trusses*).

The individual members of a truss are bolted, pinned, riveted, or welded at the connections. A major simplification in the analysis

FIGURE 6-1

Part of a crane's boom

of any truss is to assume that all members are joined by frictionless pins at their ends. This way each member can be treated as a two-force member, which of course can only have axial tension or compression in it. A consequence of this assumption is that in the analysis all forces must be applied at the joints. Thus, all members are assumed to be weightless, or their weights are divided and assumed to act at the joints.

Naturally, real structural frameworks are more complex than is implied in most analyses. For example, the boom of the crane in Fig. 6-1 is pinned only at sections *A* and *B*, and loads may be applied between the joints, laterally to some members. Still, the simplifying assumptions may be quite reasonable. The load on the boom is shared and applied by eight small wheels, so the local lateral load on a member is reduced. The numerous welded joints between sections *A* and *B* make the structure somewhat more rigid than if it were pinned at every connection, so the assumption of pinned joints is a conservative one for practical purposes.

Rigid Trusses

The basic element of a rigid plane truss is the triangle formed by three straight bars as modeled in Fig. 6-2a. This forms a rigid frame to the extent that the three bars and the connections themselves are rigid. In contrast, four or more bars connected to make polygons with as many sides are nonrigid frames. For example, the frame of four bars in Fig. 6-2b collapses to the right when a force **F** is applied at *D*. The addition of a single bar *BD or AC* would create triangular elements and make the frame rigid. This is called a *simple truss*. The addition of *both* bars *BD* and *AC* makes the frame rigid, but one of these members is redundant, so the truss is statically indeterminate in the analysis.

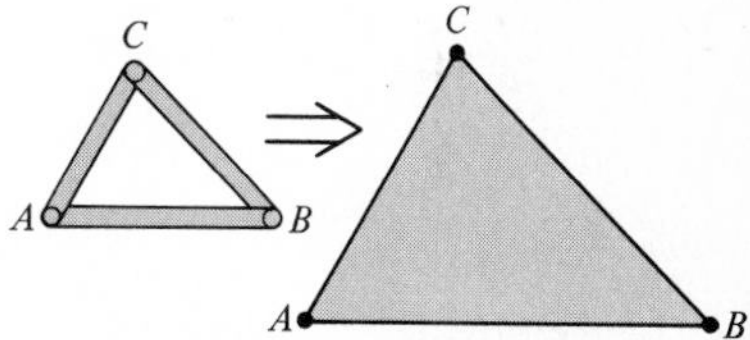

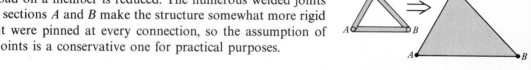

(a) Simplest drawing of truss element

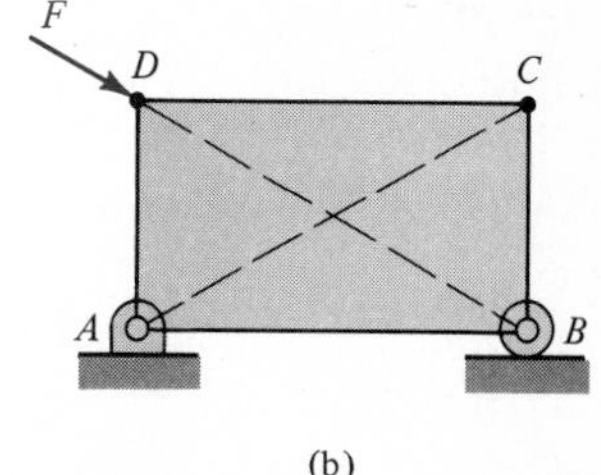

(b)

FIGURE 6-2

Three pinned bars form the basic element of a rigid plane truss

6-2 METHOD OF JOINTS

A truss consists only of pins and two-force members in the simplest cases, and is simply supported similar to beams. It is remarkable that all the forces on the pins and bars can be determined even when there are many such components. These calculations are based on the equilibrium of the whole truss and of its parts. The method is illustrated using the simple truss shown in Fig. 6-3a. The truss is pinned to a fixed bracket at A, while the roller at joint C allows motion in the horizontal direction. This is a typical arrangement for the prevention of undesirable forces that would be caused by thermally or mechanically induced deformations in a redundantly supported truss.

The free-body diagram of the truss is in Fig. 6-3b. The external reactions A_y and C_y are obtained from the equilibrium equations of the truss. By inspection, an equation for moments can be written to solve for one unknown at once. Thus, from Eq. 4-2,

$$\sum M_A = 0 \qquad \sum F_y = 0$$

$$2.5lC_y - lP = 0 \qquad A_y + C_y - P = 0$$

$$C_y = 0.4P \qquad A_y = 0.6P$$

The solution of a truss problem should be started at a joint where only two unknown forces act. This way the two equilibrium

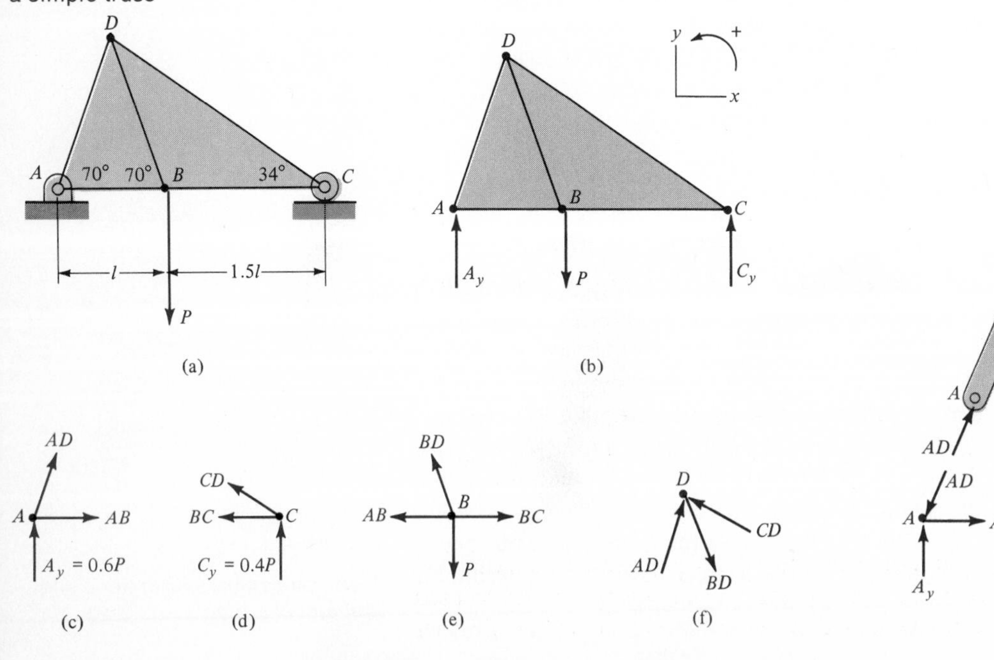

FIGURE 6-3

Method of joints for the analysis of a simple truss

equations for forces at a joint ($\sum F_x = 0, \sum F_y = 0$) can be solved immediately for two unknowns. Joints A and C satisfy this criterion in the present problem. To start with A, draw the free-body diagram of this joint in Fig. 6-3c. The unknown forces $\overrightarrow{AB}$ and $\overrightarrow{AD}$ have known lines of actions. They may be arbitrarily directed away from A or toward A in this diagram. To use a simple convention, assume that all unknown forces are tensions which are indicated by arrows directed away from the joints. The unknown forces are determined using the equilibrium equations for forces at the joint A. From Eq. 4-2,

$$\sum F_x = 0 \qquad \sum F_y = 0$$

$$AB + AD \cos 70^\circ = 0 \qquad 0.6P + AD \sin 70^\circ = 0$$

$$AB = -AD \cos 70^\circ \qquad AD = -\frac{0.6P}{\sin 70^\circ}$$

$= -0.64P$ (the negative sign indicates that $\overrightarrow{AD}$ is actually toward joint A) (compression)

Using the results for AD,

$AB = 0.22P$ (the positive answer indicates that the assumed direction of $\overrightarrow{AB}$ is correct) (tension)

The same procedure is applied for joint C with the aid of Fig. 6-3d. Again using Eq. 4-2,

$$\sum F_x = 0 \qquad \sum F_y = 0$$

$$-CD \cos 34^\circ - BC = 0 \qquad 0.4P + CD \sin 34^\circ = 0$$

$$BC = -CD \cos 34^\circ \qquad CD = -\frac{0.4P}{\sin 34^\circ} = -0.72P$$

(compression)

Using the result for CD,

$BC = 0.6P$ (tension)

The analysis of joint B is made using the free-body diagram in Fig. 6-3e,

$$\sum F_x = 0 \qquad \sum F_y = 0$$

$$BC - AB - BD \cos 70^\circ = 0 \qquad BD \cos 20^\circ - P = 0$$

$0.6P - 0.22P - BD \cos 70^\circ = 0 \qquad BD = 1.06P$ (tension)

$BD = 1.11P$ (tension)

The two answers for the same force BD are caused by round-off errors. Thus, one of the last two equilibrium equations can be used to check the result obtained from the other equation. Further checking can be done with the free-body diagram of joint D, where all of the forces are known at this stage. From Fig. 6-3f (noting the actual directions of the forces),

$$\sum F_x: \quad AD \sin 20^\circ + BD \sin 20^\circ - CD \cos 34^\circ$$
$$= 0.64P(0.34) + 1.06P(0.34) - 0.72P(0.83) \cong 0$$

$$\sum F_y: \quad AD \sin 70^\circ - BD \sin 70^\circ + CD \sin 34^\circ$$
$$= 0.61P - 1.00P + 0.40P \cong 0$$

These results show that the forces were calculated correctly.

Note that Newton's Third Law is essential in using the method of joints. For example, the directions of forces in Fig. 6-3f are determined by the directions established in Fig. 6-3c, d, and e. Consider the joints A and D and the bar between them in detail. It is this bar that applies the force AD on joint A as shown in Fig. 6-3g. In reaction, the joint A applies the force AD at the bottom of this bar as shown. The bar is in equilibrium, so an equal and opposite force must be applied by the joint D at the top of the bar as in Fig. 6-3g, putting the bar in compression. Again by Newton's law, the force AD acts on joint D as shown in Fig. 6-3f and g.

The interrelated free-body diagrams must be drawn with complete consistency, but this is easily done after a little experience. Generally, it is sufficient to work only with the joints in the sketches, with a mental picture of the members between the joints such as AD in Fig. 6-3g. It is helpful to draw the arrows at the joints on that side of the pin where the appropriate member is. For these, an arrow away from the pin indicates tension in the bar, and an arrow toward the pin indicates compression in the bar.

For a plane truss that contains n pins it is possible to write $2n$ independent equilibrium equations (for summations in the x and y directions at each pin), so the maximum number of unknown forces that could be determined is $2n$. Another way to look at this is with respect to the total number of connecting members, m. In a simple truss $m = 2n - 3$ (to prove this, start with three members and three joints; there is one additional joint for every two new members that are added). Accordingly, the maximum number of unknown forces that can be determined with the method of joints is $m + 3$ for a truss having m members. These considerations are valid for rigid plane trusses that are statically determinate externally and internally.

The procedure in conjunction with Fig. 6-3 typifies a simple and commonsense approach to understanding and solving truss problems. The emphasis in this approach is on using interrelated free-body diagrams of the joints, equilibrium equations for the diagrams, and on judgment of selecting the joints for efficient analysis. This procedure is ideal and sufficient for most students.

SPECIAL CASES OF JOINTS 6-3

A few special arrangements of members at joints are worth remembering because they can save time and effort in the analysis. In all cases the special properties of the arrangement are the results of having only two-force members connected at the joints.

1. *Two members* connected at a joint must be collinear to transmit any force as in Fig. 6-4a. If there are forces in these members, the forces must be equal for equilibrium. Both tension and compression are possible in theory, but only tension is possible in reality. Since perfect alignment is difficult to achieve in practice, even small compressive loads would cause a collapse at a two-member joint. On the other hand, tension causes self alignment of the members. When the two members are not collinear, as in Fig. 6-4b, the forces in them must be zero for equilibrium.

2. *Three members* with two of them collinear are possible only if the third member has no force in it, as in Fig. 6-4c. The reason is that members A and B could not apply a lateral force to oppose the component of the force in C that is perpendicular to A and B. This is true for any angle θ.

In that case, why have a member such as C in Fig. 6-4c? The answer is again found after considering the difference between theory and practice. In theory, there is absolutely no reason to have the member C when A and B are collinear. In practice, A and B may make a small angle, and member C is important to maintain the geometry at the joint. This can be accomplished with a relatively small force in C. In other words, the so-called *zero-force member* C may have a theoretically negligible force in it, which may even allow compressive loading of members A and B. Of course, C can be ignored in much of the analysis. An exception to this is when an external load is applied at the joint of A, B, and C, as in Fig. 6-4d. Another practical reason for having apparently unnecessary members is that most structures are subjected to changing loading conditions.

3. *Four members* that are collinear in pairs at a single joint must have equal and opposite forces in pairs, as shown in Fig. 6-4e. The force polygon for such a joint is a parallelogram, as in Fig. 6-4f.

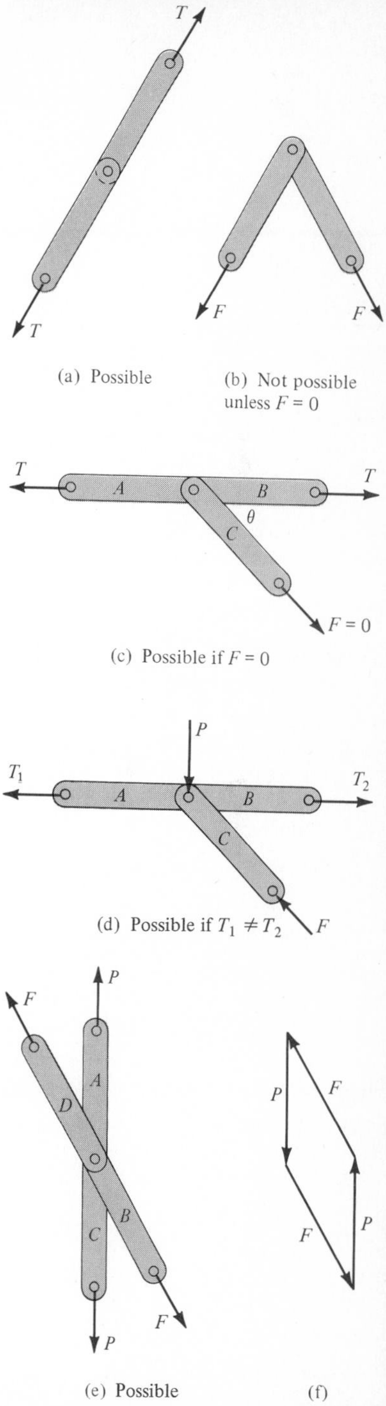

(a) Possible

(b) Not possible unless $F = 0$

(c) Possible if $F = 0$

(d) Possible if $T_1 \neq T_2$

(e) Possible

(f)

FIGURE 6-4
Special cases of joints

EXAMPLE 6-1

Determine the force in each member of the truss in Fig. 6-5a using the method of joints. Take advantage of symmetry.

SOLUTION

Because of symmetry, it is sufficient to determine the forces at joints A, B, C, G, and H. The support reactions are determined first. By symmetry, $R_1 = R_2 =$ total load/2 = 75 kN. The joint-by-joint analysis is begun at a joint where there are only two unknown forces.

JOINT *A*, Fig. 6-5b:

$$\sum F_x = 0 \qquad \sum F_y = 0$$

$$AH + AB \cos 60^\circ = 0 \qquad 75 \text{ kN} + AB \sin 60^\circ = 0$$

$$AB = -86.6 \text{ kN (bar } AB \text{ is in compression)}$$

Using the result for AB,

$$AH = 43.3 \text{ kN} \quad \text{(tension, as assumed)}$$

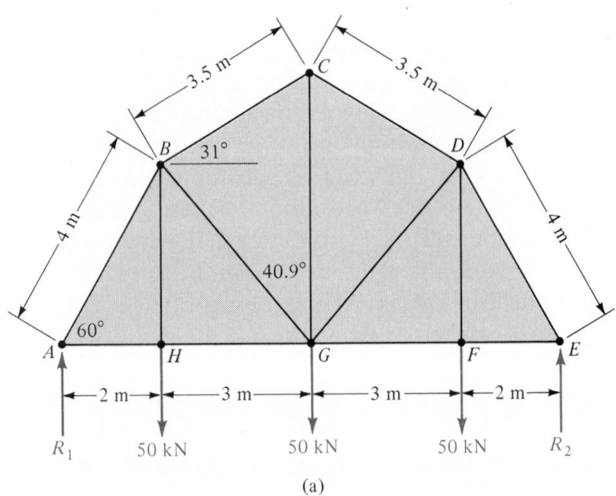

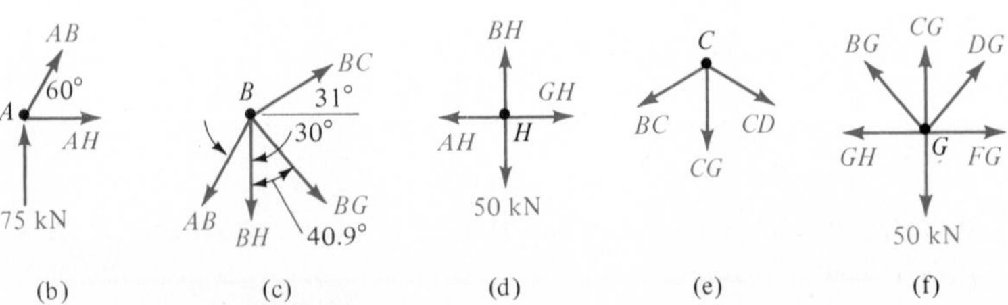

FIGURE 6-5
Analysis of a symmetric plane truss

JOINT *B*, Fig. 6-5c: A quick check shows there are three unknown forces and only two equations of equilibrium to use for solving. Hence this joint is not solvable yet.

JOINT *H*, Fig. 6-5d:

$$\sum F_x = 0 \qquad\qquad \sum F_y = 0$$

$$GH - AH = 0 \qquad\qquad BH - 50\text{ kN} = 0$$

$$GH = 43.3\text{ kN} \quad \text{(tension)} \qquad\qquad BH = 50\text{ kN} \quad \text{(tension)}$$

JOINT *B*: Joint B may now be analyzed. From Fig. 6-5c,

$$\sum F_x = 0$$

$$BC \cos 31^\circ + BG \sin 40.9^\circ - AB \sin 30^\circ = 0$$

$$0.857BC + 0.655BG + 43.3\text{ kN} = 0$$

$$\sum F_y = 0$$

$$BC \sin 31^\circ - AB \cos 30^\circ - BH - BG \cos 40.9^\circ = 0$$

$$0.515BC + 75\text{ kN} - 50\text{ kN} - 0.756BG = 0$$

Solving the two equations simultaneously yields

$$BG = -0.88\text{ kN} \quad \text{(compression)}$$

$$BC = -49.8\text{ kN} \quad \text{(compression)}$$

JOINT *C*, Fig. 6-5e: From symmetry, $CD = BC = -49.8$ kN (compression),

$$\sum F_y = 0$$

$$-CG - BC \sin 31^\circ - CD \sin 31^\circ = 0$$

$$CG = -2BC \sin 31^\circ = -2(-49.8\text{ kN})(0.515)$$

$$= 51.3\text{ kN} \quad \text{(tension)}$$

This completes the analysis of the symmetric truss.

JUDGMENT OF THE RESULTS

A summation of the known forces in the y direction at joint G serves as a check on the other calculations. From Fig. 6-5f,

$$\sum F_y = BG \cos 40.9^\circ + CG + DG \cos 40.9^\circ - 50\text{ kN}$$

With $BG = DG$ by symmetry,

$$-0.88\text{ kN}(0.756) + 51.3\text{ kN} - 0.88\text{ kN}(0.756) - 50\text{ kN}$$

$$= -0.03\text{ kN}$$

The deviation from zero in the final summation is caused by round-off errors. The results are accepted as correct.

EXAMPLE 6-2

Determine the forces acting at joints A, B, C, H, and I of the truss in Fig. 6-6a. The angles are $\alpha = 56.3°$, $\beta = 38.7°$, $\phi = 39.8°$, and $\theta = 36.9°$.

SOLUTION

First the reactions at the supports are determined and are shown in Fig. 6-6b. A joint at which only two unknown forces act is the best as the starting point for the solution. Joints A and F satisfy this criterion. Choosing joint A, the solution is progressively and thoughtfully developed, always seeking the next joint with only two unknowns. In each diagram circles indicate the quantities that are known from the preceding analysis.

JOINT *A*, Fig. 6-6c:

$$\sum F_x = 0 \qquad \sum F_y = 0$$

$$AI = 0 \qquad AB - R_A = 0$$

$$AB - 50 \text{ kips} = 0$$

$$AB = 50 \text{ kips} \quad \text{(tension)}$$

FIGURE 6-6

Analysis of a plane truss

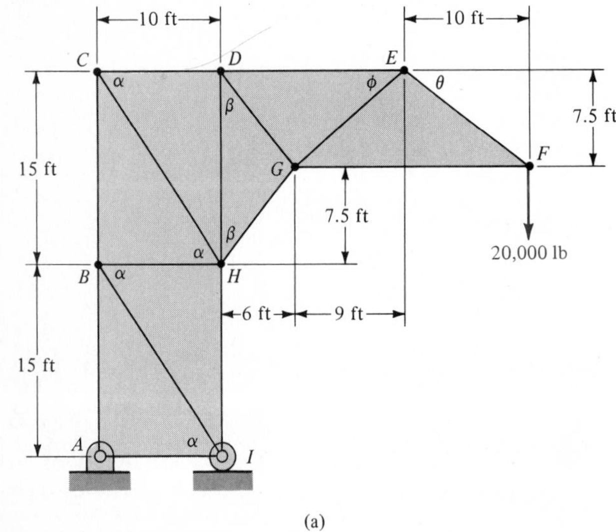

(a)

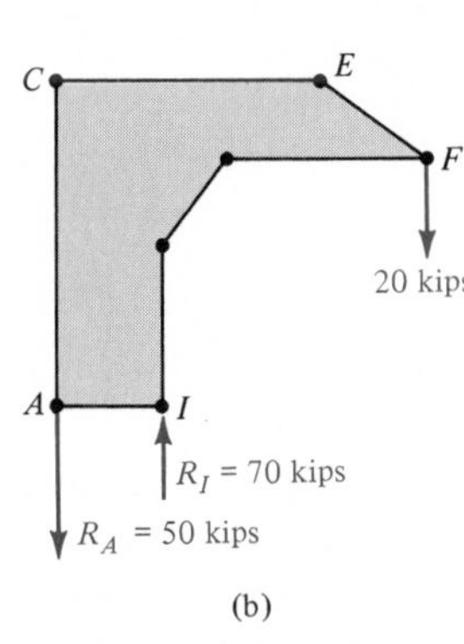

(b)

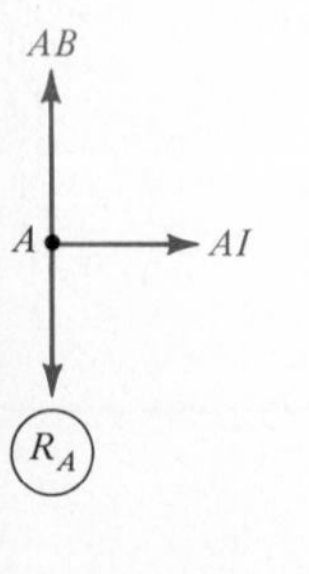

(c)

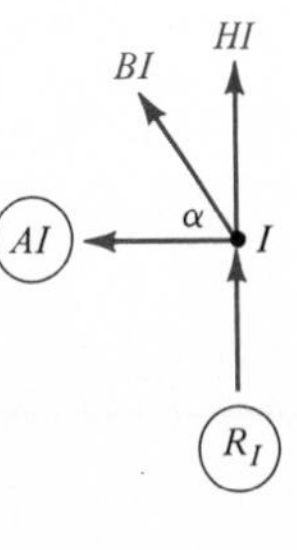

(d)

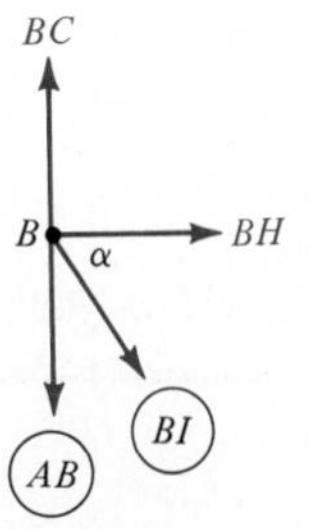

(e)

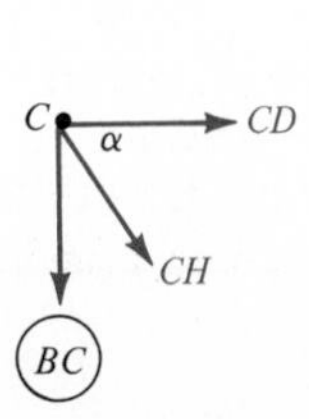

(f)

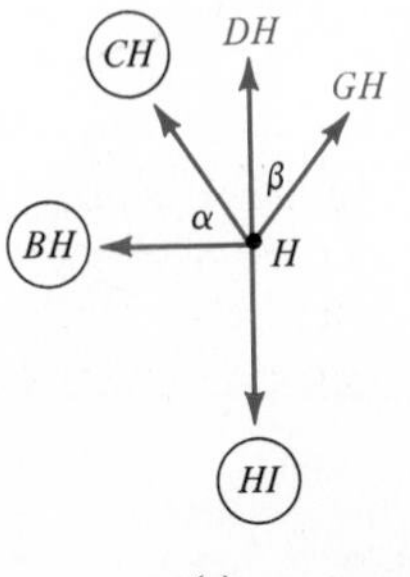

(g)

JOINT *I*, Fig. 6-6d:

$\sum F_x = 0$

$-BI \cos \alpha - AI = 0$

$BI \cos \alpha + 0 = 0$

$BI = 0$

$\sum F_y = 0$

$R_I + HI + BI \sin \alpha = 0$

$70 \text{ kips} + HI + 0 \sin \alpha = 0$

$HI = -70 \text{ kips}$ (compression)

JOINT *B*, Fig. 6-6e:

$\sum F_x = 0$

$BH + BI \cos \alpha = 0$

$BH + 0 \cos \alpha = 0$

$BH = 0$

$\sum F_y = 0$

$BC - BI \sin \alpha - AB = 0$

$BC - 0 \sin \alpha - 50 \text{ kips} = 0$

$BC = 50 \text{ kips}$ (tension)

JOINT *C*, Fig. 6-6f:

$\sum F_x = 0$

$CD + CH \cos \alpha = 0$

$CD - 60.1(0.555) = 0$

$CD = 33.3 \text{ kips}$ (tension)

$\sum F_y = 0$

$-BC - CH \sin \alpha = 0$

$-50 \text{ kips} - CH(0.832) = 0$

$CH = -60.1 \text{ kips}$ (compression)

JOINT *H*, Fig. 6-6g:

$\sum F_x = 0$

$GH \sin \beta - CH \cos \alpha - BH = 0$

$GH(0.625) + 60.1 \text{ kips}(0.555) - 0 = 0$

$GH = -53.4 \text{ kips}$ (compression)

$\sum F_y = 0$

$CH \sin \alpha + DH + GH \cos \beta - HI = 0$

$-60.1 \text{ kips}(0.832) + DH - 53.4 \text{ kips}(0.780) + 70 \text{ kips} = 0$

$DH = 21.7 \text{ kips}$ (tension)

JUDGMENT OF THE RESULTS

The magnitudes of the forces in the two-force members appear reasonable considering the applied force and the reactions at the supports. The senses of most of these forces, namely tension or compression in the members, are intuitively correct for the given situation. A specific check on the results can be made as homework by starting the analysis at joint *F* of the truss.

PROBLEMS

SECTIONS 6-1, 6-2, 6-3

FIGURE P6-1

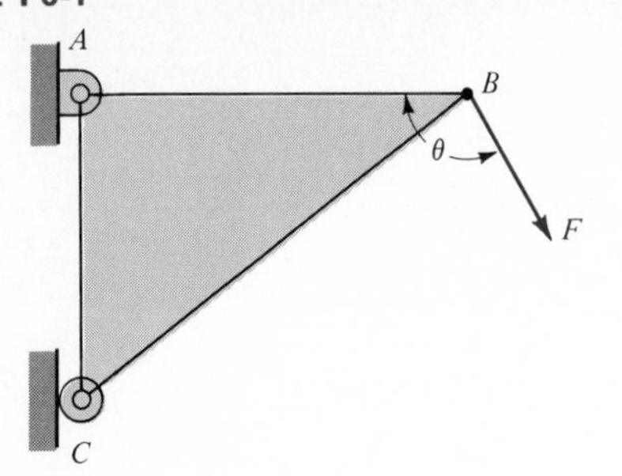

6–1 Determine the force in member AC if $F = 3$ kN, $\theta = 70°$, $AB = 3$ m, and $AC = 1.4$ m.

6–2 Determine the forces in the three members in Fig. P6-1 if $F = 1000$ lb, $\theta = 100°$, $AB = 8$ ft, and $BC = 12$ ft.

FIGURE P6-3

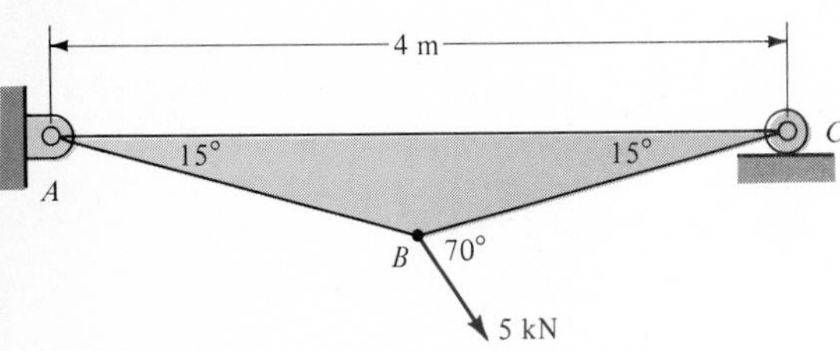

6–3 Calculate the forces in each member of the truss.

FIGURE P6-4

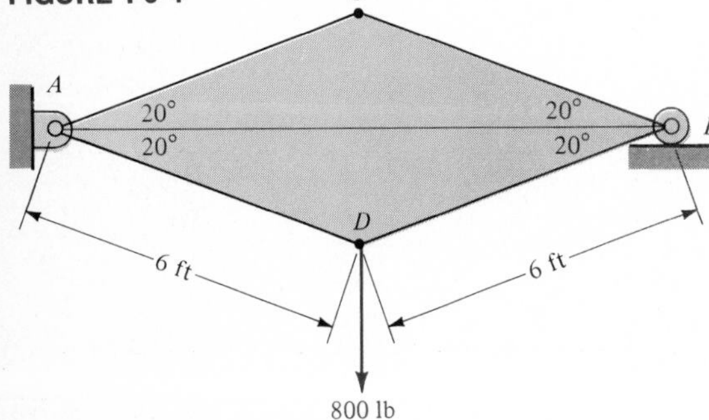

6–4 Determine the forces in the five members of the truss.

FIGURE P6-5

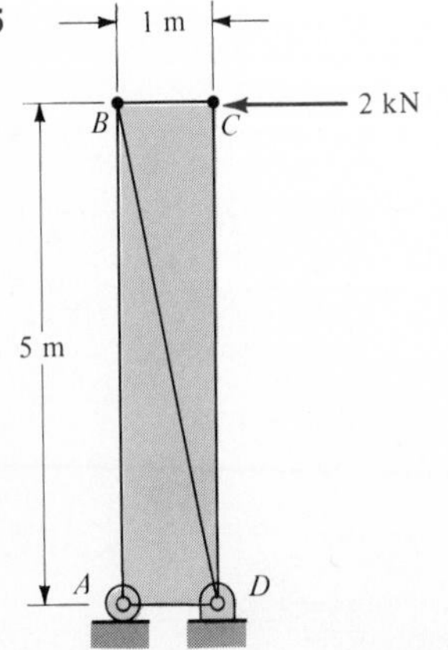

6–5 The effect of wind on a structure is modeled as a concentrated force acting at B. Calculate the forces in the five members of the truss.

6–6 Consider Fig. P6-5. (a) Calculate the force in member BD. (b) Replace member BD with member AC and determine the force in member AC.

6–7 Calculate the forces in the five members of the truss.

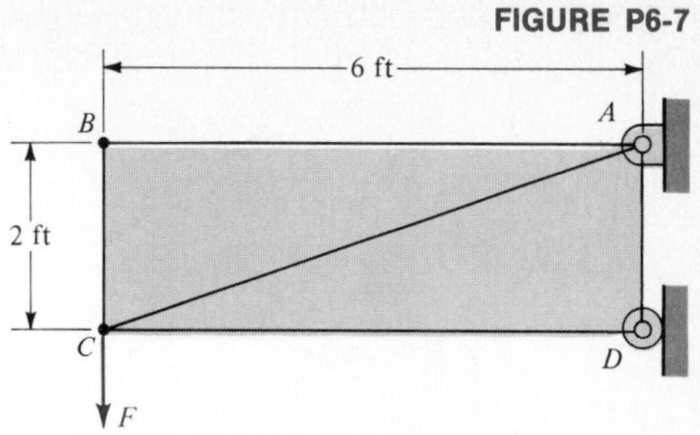

FIGURE P6-7

6–8 Solve Prob. 6–7 as stated and also after replacing member *AC* with member *BD*. Which configuration appears to be the best to minimize the forces, especially to minimize compression in long members?

6–9 Determine the forces in the five members if $\mathbf{F}_1 = (2\mathbf{i} - 1.5\mathbf{j})$ kN and $\mathbf{F}_2 = 4\mathbf{i}$ kN.

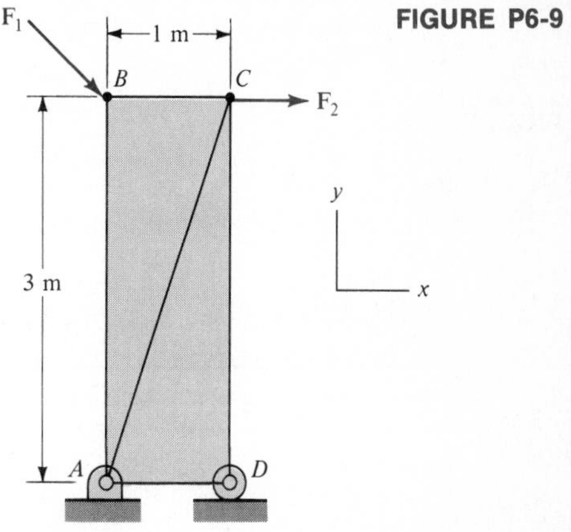

FIGURE P6-9

6–10 Calculate the forces in the five members of the truss.

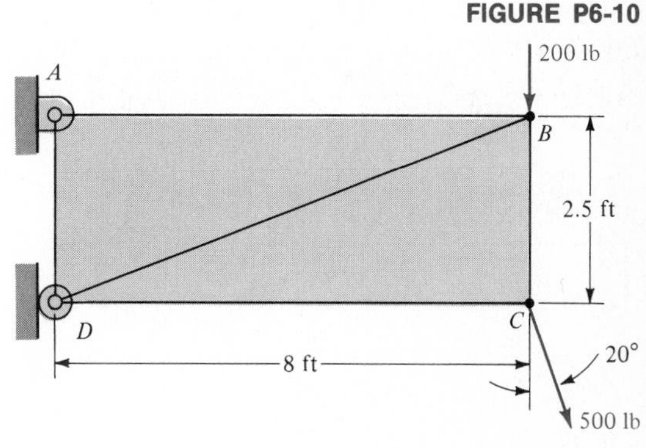

FIGURE P6-10

6–11 Determine the forces in members *AB*, *BE*, and *EF* if $P = 5$ kN and $l = h = 1.5$ m.

6–12 Consider Fig. P6-11 with the following changes: replace member *BF* with *AE*, and member *CE* with *BD*. Calculate the forces in every member if $P = 1000$ lb and $l = 5$ ft, $h = 4$ ft.

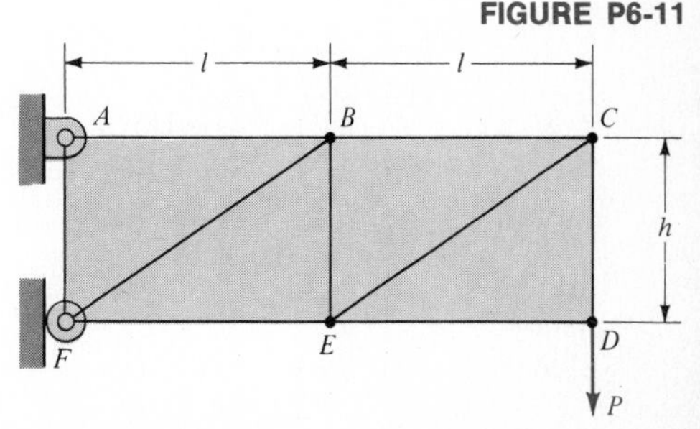

FIGURE P6-11

FIGURE P6-13

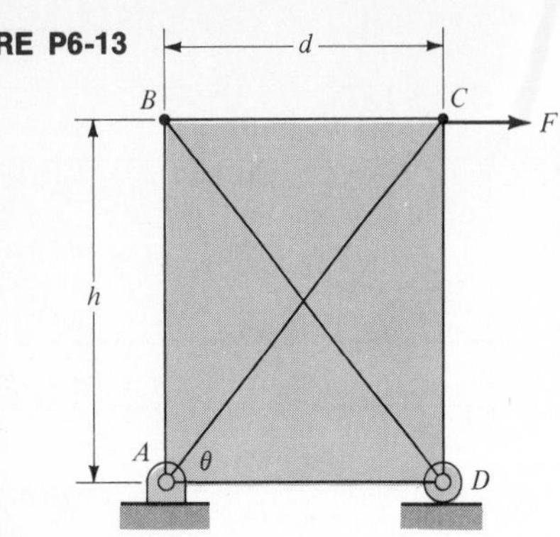

6–13 Assume that members *AC* and *BD* of the truss have no common pin connection. List all the unknown forces and independent equilibrium equations for the truss.

FIGURE P6-14

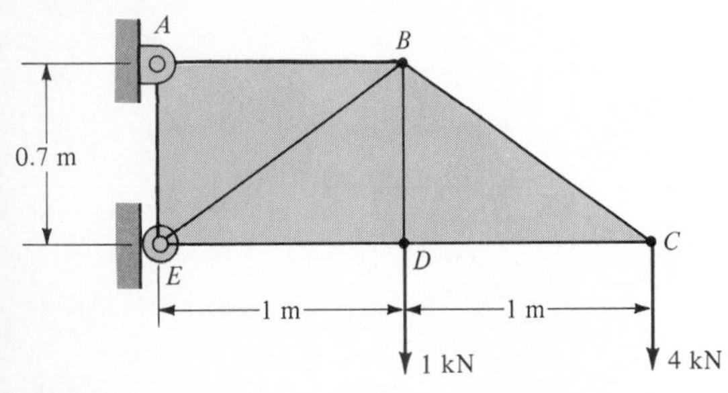

6–14 Determine the forces in members *AB*, *BD*, and *BE* in the truss.

FIGURE P6-15

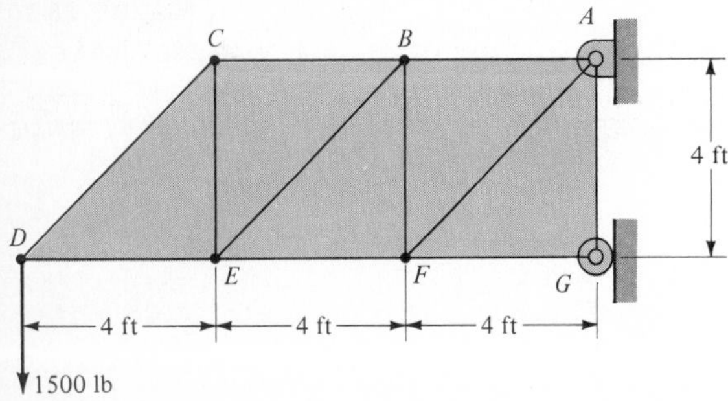

6–15 Calculate the forces acting at joint *F* in the truss.

6–16 The weight of the truss in Fig. P6-15 is modeled by a 200-lb force acting at *E*, and a 250-lb force acting at *F*. Calculate the other forces acting at *F*.

6–17 Solve Ex. 6-1 by starting the analysis at joint *E* (Fig. 6-5).

FIGURE P6-19

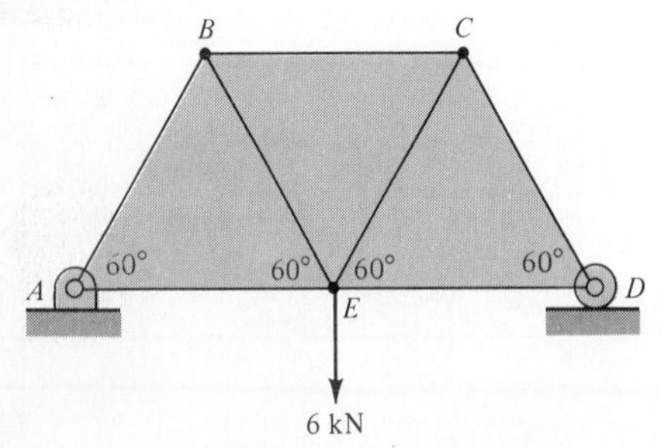

6–18 Solve Ex. 6-2 by starting the analysis at joint *F* (Fig. 6-6).

6–19 Determine the force in every member of the truss. Each member is 2 m long.

6–20 Add 1-kN downward forces at joints *B* and *C* in Prob. 6–19, and determine the forces in the members.

FIGURE P6-21

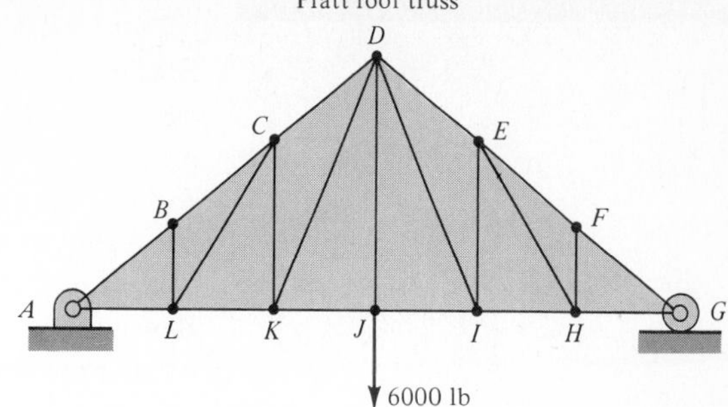

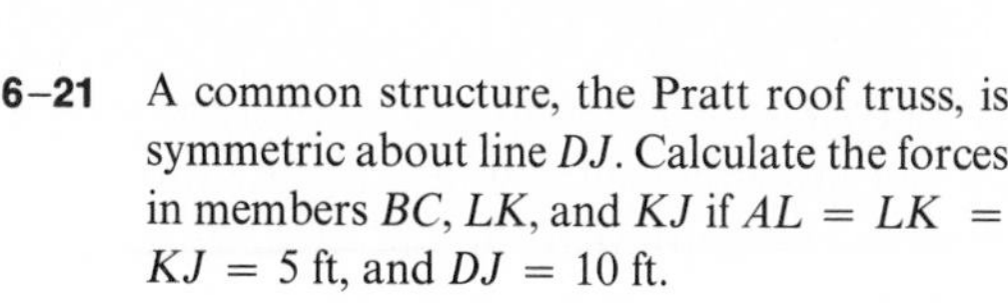

6–21 A common structure, the Pratt roof truss, is symmetric about line *DJ*. Calculate the forces in members *BC*, *LK*, and *KJ* if $AL = LK = KJ = 5$ ft, and $DJ = 10$ ft.

FIGURE P6-22

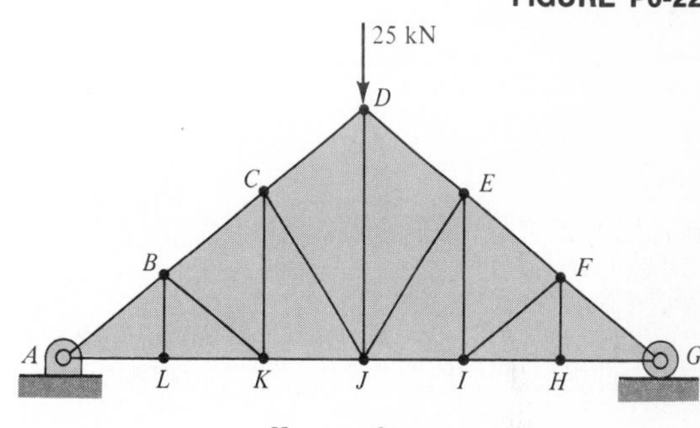

Howe roof truss

6–22 The Howe roof truss is symmetric about line *DJ*. Each horizontal member is 2 m long, and $DJ = 3$ m. Calculate the forces acting at joint *J*.

FIGURE P6-23

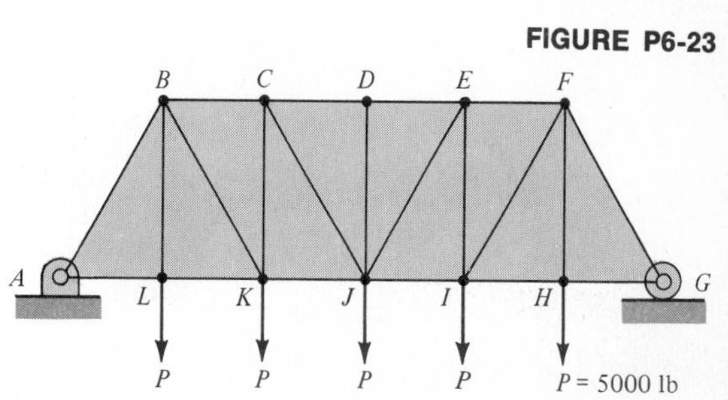

6–23 The bridge truss is symmetric about line *DJ*. Each horizontal member is 8 ft long, and the structure is 10 ft high. Determine the forces acting on joints *L*, *K*, and *J*.

6-4 FORMAL PROCEDURE IN JOINT ANALYSIS

Students who expect to work regularly with complex trusses may be interested in a formal procedure for setting up problems for analysis. The formal method of dealing with members, joints, and forces makes use of *Bow's notation*. This is illustrated in Fig. 6-7 for a simple plane truss.

First, every region between the external loads and reactions is denoted by a lowercase letter, going clockwise around the truss, as in Fig. 6-7b. The lettering is continued for the areas inside the truss, still going clockwise. These letters are used to denote the members, joints, and external and internal forces as follows.

1. *Members* are named by the letters of the two adjacent areas. For example, the diagonal member is *gh* or *hg*, the top member is *bh* or *hb*.

2. *Joints* are denoted by numbers or by listing the letters of areas at the joint in clockwise order. For example, the joint at $\mathbf{R}_2$ is *bci*, or *cib*, or *ibc*.

3. *External forces* (loads and reactions) are named by the letters of the two areas adjacent to them, written in clockwise order with capital letters and vector notation with an arrow. For example, the force **P** is written as $\overrightarrow{AB}$, the reaction $\mathbf{R}_2$ as $\overrightarrow{BC}$.

4. *Internal forces* (exerted by members on pins) are named by the letters of the two adjacent areas to the member. Capital letters with an arrow are used with respect to a particular joint. The arrows are omitted when the letters denote only magnitudes of forces. For example, the force exerted by member *af* on joint 1 is $\overrightarrow{AF}$.

The whole scheme is illustrated in Fig. 6-8 for three joints of the truss in Fig. 6-7. Approximate force polygons are also given for the three joints to show an alternative method to the algebraic calculations.

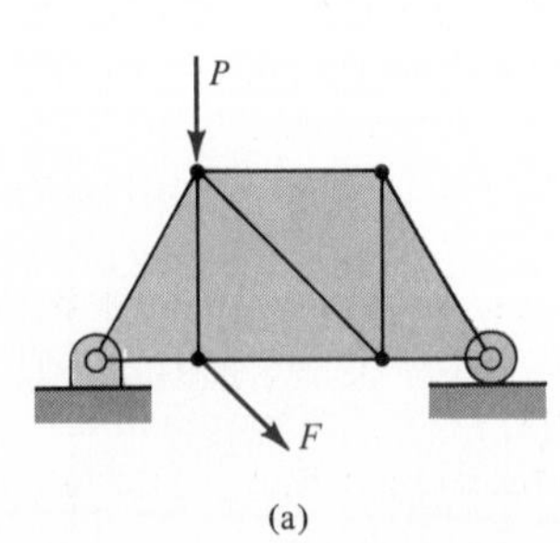

(a)

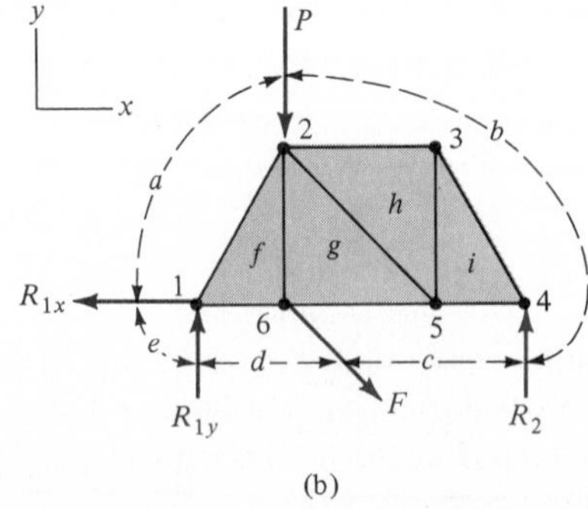

(b)

FIGURE 6-7
Formal procedure in joint analysis

Joint	Free-body diagram (all forces are known here)	Force polygons
1 or *afde*	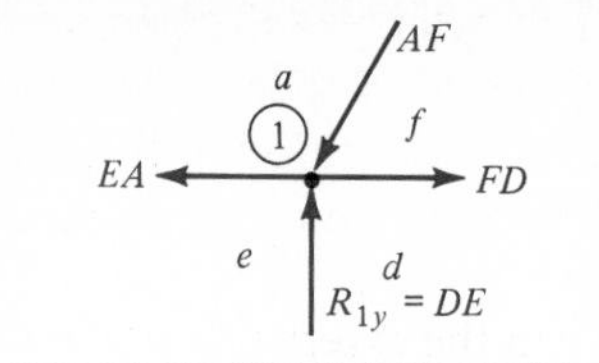 *AF*, *a*, *f*, (1), *EA*, *FD*, *e*, *d*, $R_{1y} = DE$	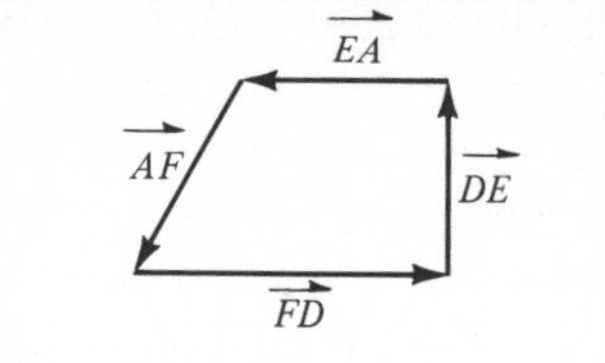 $\overrightarrow{EA}$, $\overrightarrow{AF}$, $\overrightarrow{DE}$, $\overrightarrow{FD}$
3 or *bih*	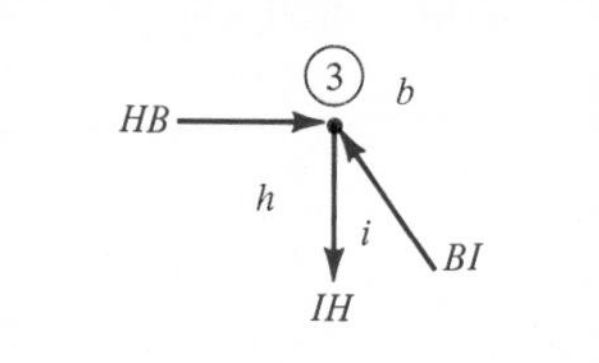 (3), *b*, *HB*, *h*, *i*, *BI*, *IH*	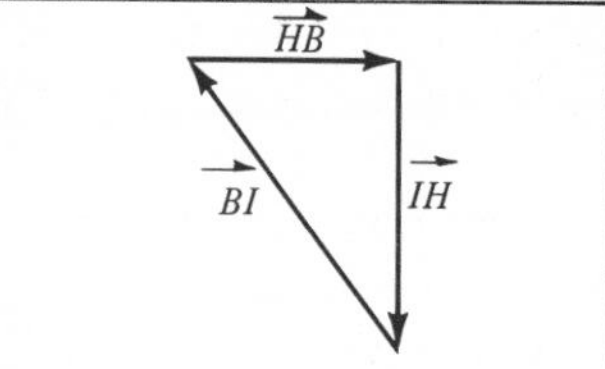 $\overrightarrow{HB}$, $\overrightarrow{BI}$, $\overrightarrow{IH}$
4 or *bci*	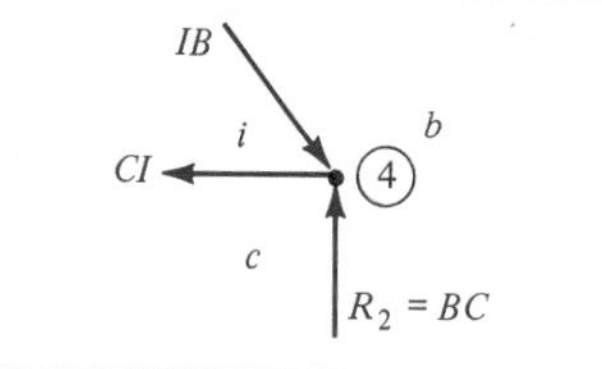 *IB*, *i*, *b*, *CI*, (4), *c*, $R_2 = BC$	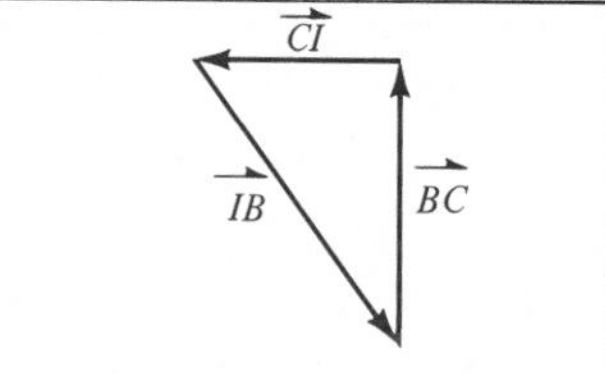 $\overrightarrow{CI}$, $\overrightarrow{IB}$, $\overrightarrow{BC}$

FIGURE 6-8

Joint analysis schemes for Fig. 6-7

6-5 GRAPHICAL ANALYSIS OF TRUSSES. MAXWELL DIAGRAM

The force polygons for a truss may be superimposed to make a composite figure called the *Maxwell diagram*. Construction of this diagram is based on the equilibrium of each joint and on the correct relationship of the individual free-body diagrams of the joints. The detailed, step-by-step procedure is shown in Fig. 6-9.

At first the external reactions must be determined from the equilibrium equations for the whole truss. Assume that the external load P causes the reactions R_1 and R_2 shown in Fig. 6-9a. The graphic construction can be started and continued at any joint where the unkown forces can be determined directly from a force polygon (at most two unknown forces at a joint). For example, start with joint 1 in Fig. 6-9b. Here $\overrightarrow{CA}$ is known and must be drawn to scale. The force triangle is drawn knowing the lines of action of the other forces at the same joint. Note that the lettering sequence for the forces must be correct.

Consider joint 2 next, where force $\overrightarrow{DA}$ is known (it must be equal and opposite to force $\overrightarrow{AD}$) and there are only two unknown forces. Force $\overrightarrow{DA}$ is added as a dashed line in Fig. 6-9c. Successively, forces $\overrightarrow{AE}$ and $\overrightarrow{ED}$ are drawn in Fig. 6-9c since their lines of action are known.

The procedure is continued similarly in Fig. 6-9d and e. All previously drawn lines are retained at each step. New forces are added as solid arrows when first considered, and as dashed arrows when opposing already drawn forces. Figure 6-9e shows at a glance that each joint and the whole truss is in equilibrium since there is no net force acting anywhere.

The solid and dashed arrows are useful to illustrate the composite graphic construction of the Maxwell diagram, but they are not necessary in routine analysis. Simple lines are used in practice as in Fig. 6-9f to obtain a view of the force polygons. The magnitude of each force is measured from the length of the corresponding line. Tension or compression in a member is determined by considering whether the member is pushing or pulling a joint.

FIGURE 6-9

Construction of a Maxwell diagram

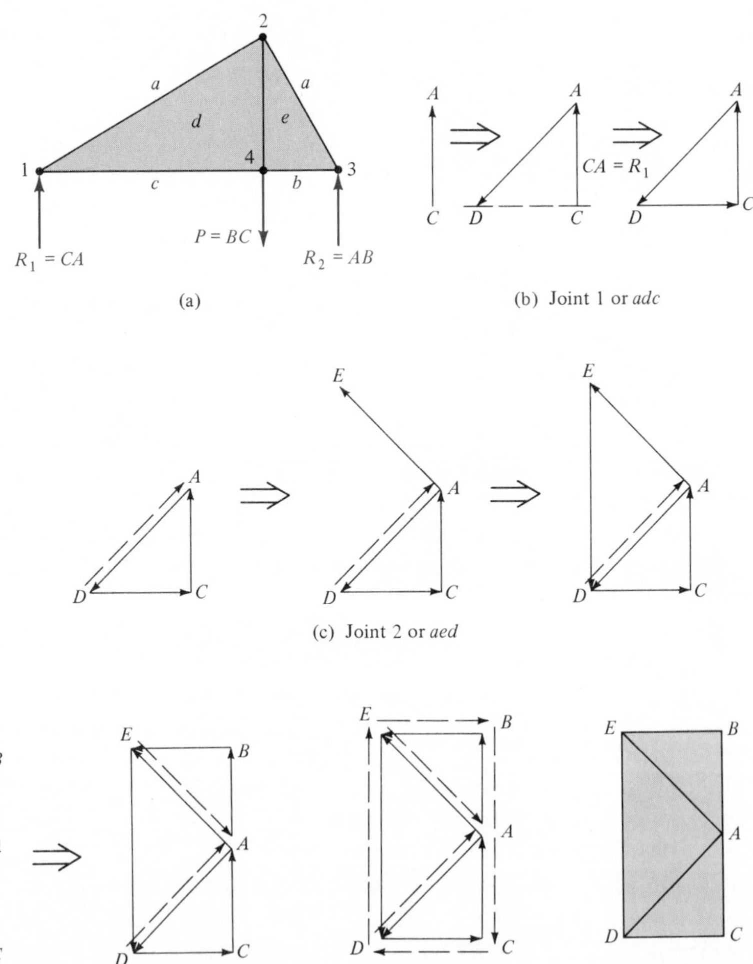

EXAMPLE 6-3

Construct the Maxwell diagram for the truss in Fig. 6-10a. Use informal notation, solid and dashed arrows, and several diagrams to show the major steps in the construction. Each member is 3 m long.

SOLUTION

First the reactions are determined. Since there is symmetry, $R_A = R_D = 5\text{ kN}/2 = 2.5\text{ kN}$. Beginning with joint A, a force triangle is drawn.

JOINT *A*, Fig. 6-10b:

$$AB = \frac{R_A}{\sin 60^\circ} = \frac{2.5\text{ kN}}{0.866} = 2.89\text{ kN}$$

$$AE = \frac{R_A}{\tan 60^\circ} = \frac{2.5\text{ kN}}{1.732} = 1.44\text{ kN}$$

JOINT *B*, Fig. 6-10c: Retaining the force triangle from joint A, the forces acting at joint B are added. Forces AB, BC, and BE form an equilateral force triangle, hence their magnitudes are all equal: $AB = BC = BE = 2.89\text{ kN}$.

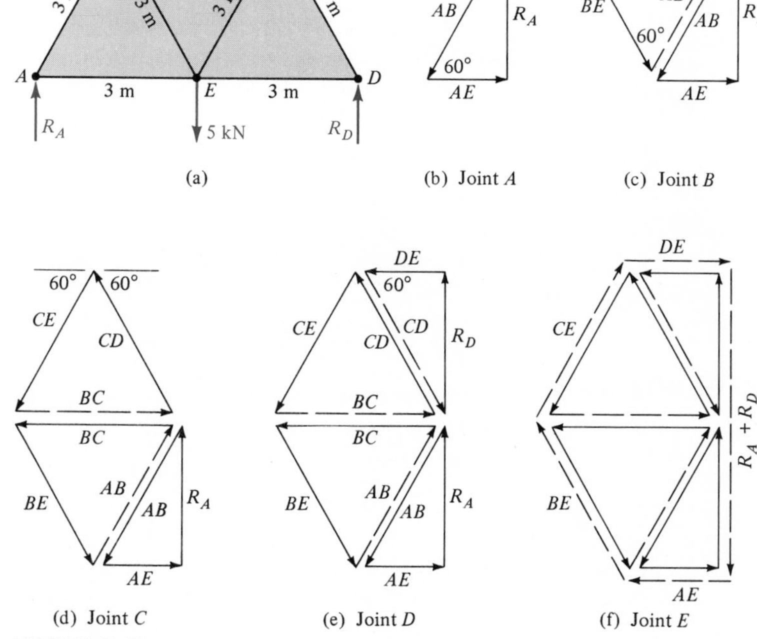

FIGURE 6-10

Construction of a Maxwell diagram

JOINT *C*, Fig. 6-10d: Continuing to add on to the previous diagram, solid arrows are shown for members being considered for the first time and dashed arrows are shown for members being considered the second time. Forces *CD*, *CE*, and *BC* form an equilateral force triangle, hence their magnitudes are all equal: $CD = CE = BC = 2.89$ kN

JOINT *D*, Fig. 6-10e: Reaction R_D is known already but may be used as a check on the other calculations.

$$R_D = CD \cos 30^\circ = 2.89 \text{ kN}(0.866) = 2.5 \text{ kN} \quad \text{(check)}$$

$$DE = CD \sin 30^\circ = 2.89 \text{ kN}(0.5) = 1.45 \text{ kN}$$

JOINT *E*, Fig. 6-10f: Realizing that the applied force (5 kN) is equal and opposite to the sum of the reactions R_A and R_D, it is seen that the Maxwell diagram is indeed closed.

JUDGMENT OF THE RESULTS

There should be considerable confidence in the results obtained from force triangles. However, it should be noted that the sense (tension or compression) of the forces is not apparent from these diagrams. In order to obtain that extra information from the analysis, a combination of the methods discussed so far has to be used.

EXAMPLE 6-4

Construct the Maxwell diagram for the truss in Fig. 6-11a. Use Bow's notation, solid and dashed arrows, and several diagrams to show the major steps in the construction. The angles are $\theta = 53.1^\circ$ and $\phi = 63.4^\circ$.

SOLUTION

It is noted that the vertical members at joints 3 and 6 are zero-force members, so these joints can be ignored in the analysis. First the reactions are calculated by considering the truss as a whole (Fig. 6-11b). Choosing joint 4 as the starting point, a force triangle is drawn.

JOINT 4, Fig. 6-11c:

$$BG = \frac{BA}{\sin \theta} = \frac{10 \text{ kips}}{\sin 53.1^\circ} = 12.5 \text{ kips}$$

$$GA = BG \cos 53.1^\circ = 7.5 \text{ kips}$$

JOINT 5, Fig. 6-11d: Since member *BG* is being considered for the second time, it is shown as a dashed arrow in the opposite direction originally shown.

$$FG = \frac{BA}{\sin \phi} = \frac{10 \text{ kips}}{\sin 63.4^\circ} = 11.2 \text{ kips}$$

$$BF = GA + FG \cos 63.4^\circ = 7.5 \text{ kips}$$
$$+ 11.2 \text{ kips} \cos 63.4^\circ = 12.51 \text{ kips}$$

JOINT 2, Fig. 6-11e: From symmetry, $EF = FG = 11.2$ kips,

$$AE = BF + EF \cos \phi = 12.51 \text{ kips}$$
$$+ 11.2 \text{ kips} \cos 63.4^\circ = 17.5 \text{ kips}$$

JOINT 7, Fig. 6-11f:

$$EC = BA = 10 \text{ kips}$$

$$CB = EA = 17.5 \text{ kips}$$

JOINT 1, Fig. 6-11g: It is seen that the only way for the force diagram to be consistent with the lines of action of the reactions and members is for point D to be the same as point B. It is also seen that the Maxwell diagram does indeed form a closed polygon, so the whole structure is in equilibrium.

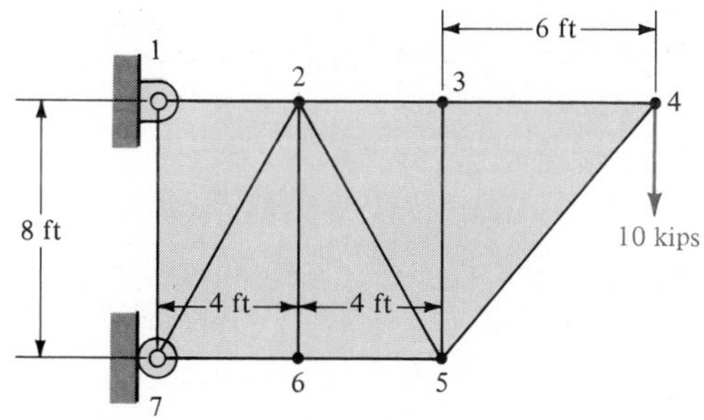

(a)

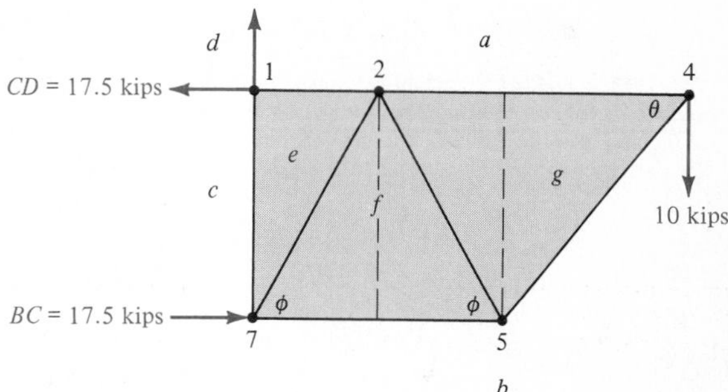

(b) Note: a, b, c, and d denote regions between forces

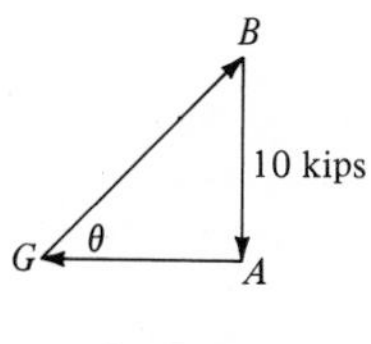

(c) Joint 4

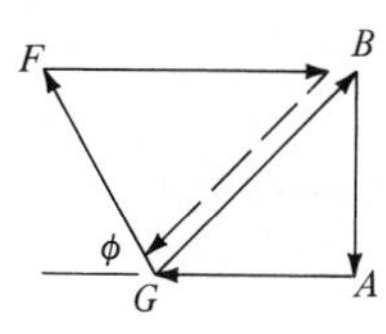

(d) Joint 5

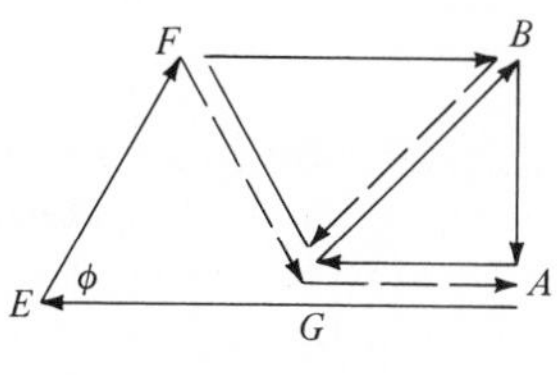

(e) Joint 2

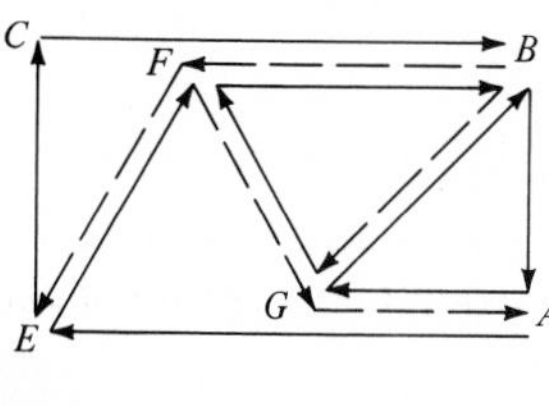

(f) Joint 7

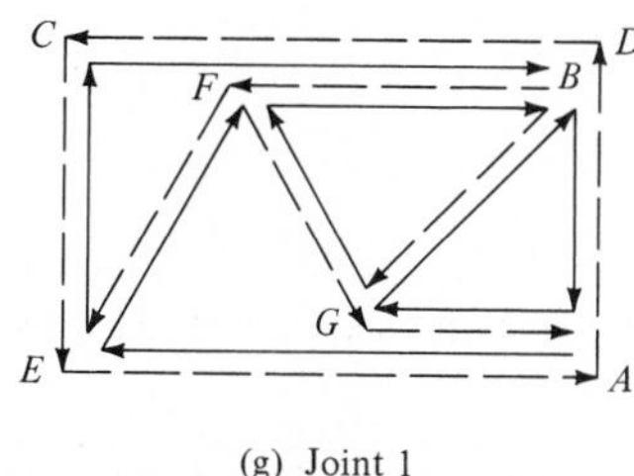

(g) Joint 1

FIGURE 6-11

Construction of a Maxwell diagram

PROBLEMS

FIGURE P6-24

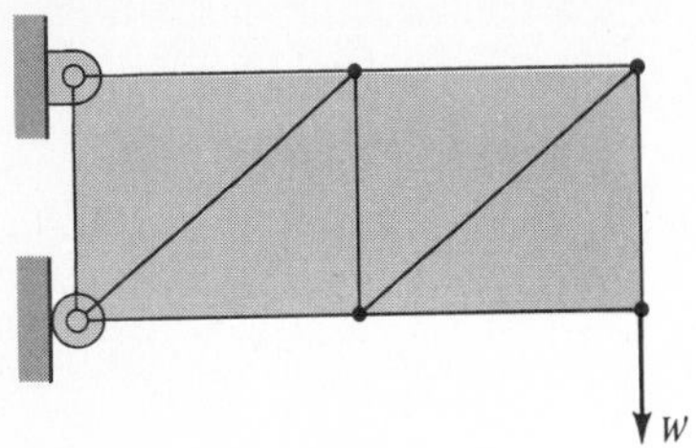

FIGURE P6-25

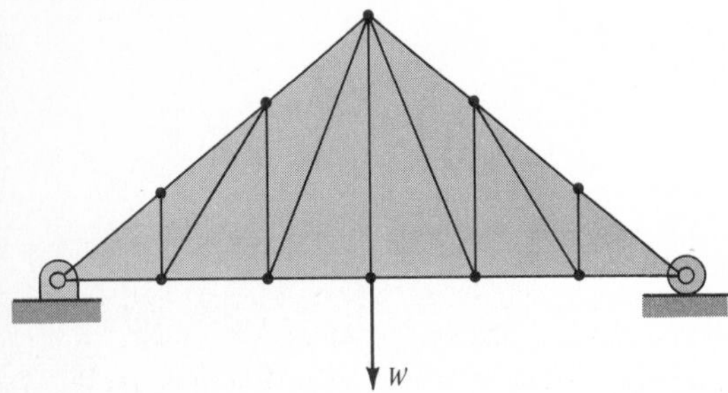

SECTIONS 6-4, 6-5

6–24 Establish a formal notation for the cantilever truss starting with the upper support.

6–25 Establish a formal notation for the roof truss, starting with the left support.

In Probs. 6–26 to 6–37, construct the Maxwell diagram for the truss or part of the truss indicated. Use solid and dashed arrows and several diagrams to show the major steps in the construction. Determine the required forces from the diagrams.

6–26 Prob. 6–1. $F_{AB} = ?$

6–27 Prob. 6–2. $F_{AB} = ?$ $F_{BC} = ?$

6–28 Prob. 6–3. $F_{AC} = ?$

6–29 Prob. 6–4. $F_{AB} = ?$

6–30 Prob. 6–5. $F_{AB} = ?$

6–31 Prob. 6–7. $F_{AC} = ?$

6–32 Prob. 6–9. $F_{CD} = ?$

6–33 Prob. 6–10. $F_{AB} = ?$

6–34 Prob. 6–11. Construct the Maxwell diagram for section *BCDE* of the truss. $F_{BC} = ?$

6–35 Prob. 6–11. Give the force in every member.

6–36 Prob. 6–14. Give the force in every member.

6–37 Prob. 6–15. $F_{AB} = ?$

METHOD OF SECTIONS 6-6

It is generally necessary to determine the forces in all members of a truss during its design. The method of joints and Maxwell's diagram are the most useful for such detailed analysis. There are other situations where only a few members must be analyzed regardless of the size and complexity of the whole structure (for example, checking the critical regions in a truss). In such cases the *method of sections* is the most efficient because it involves the members of interest at once without requiring many intermediate steps in the calculation.

The method of sections is based on the idea that any portion of a truss may be considered as a free body in equilibrium. The chosen portion may have any number of joints and members in it, but the number of unknown forces acting on it should not exceed three in most cases because only three equilibrium equations can be written for each portion of a plane truss.

To illustrate the possible choices of sections in a truss, assume that only the force in member *CD* in Fig. 6-12a must be known. A section at line *l* would involve the member *CD*, but both the upper and the lower portions of the truss have the unknown forces *BA*, *BI*, *CI*, and *C*H also acting on them as shown in Fig. 6-12b. Thus, this particular imaginary sectioning is useless in the analysis because there are too many unknowns.

The useful sections for the truss in Fig. 6-12a are those that cut member *CD* and only two other members. Lines *m* or *n* are equally good choices, and either the left or the right portions of the truss can be used as shown in Fig. 6-12c and d for the section *m*. The unknown forces *CD*, *CH*, and *IH* have arbitrarily assumed directions at this stage, but these have to be consistent in the interrelated diagrams in Fig. 6-12c and d.

Judgment is again necessary to expedite the solution. Since only the force *CD* must be determined, it is best to take moments about point *H* in Fig. 6-12d. This gives an equation with only one unknown.

$$\sum M_H = 0$$

$$CD(b) + P(2a) = 0$$

$$CD = -\frac{2a}{b}P$$

The negative sign indicates that *CD* is not a tensile force as assumed, but a compressive force (pushing joint *C* to the left and joint *D* to the right).

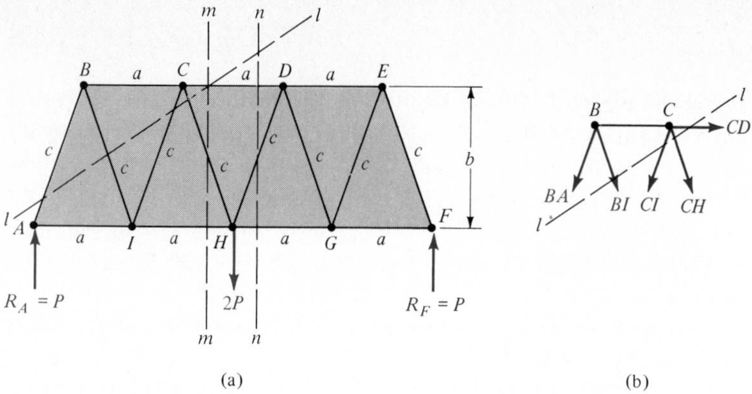

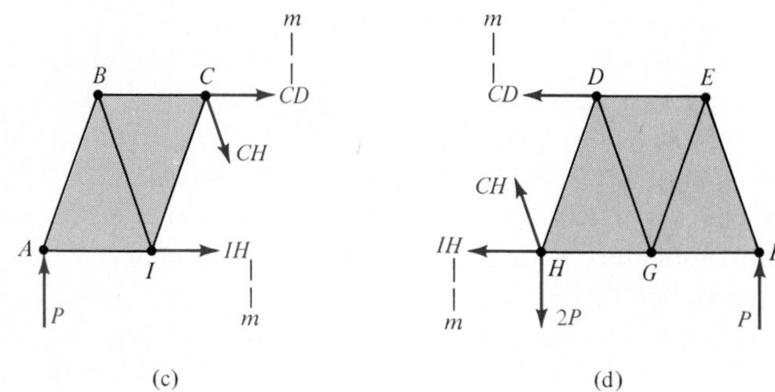

FIGURE 6-12

Method of sections for the analysis of selected members

In general, equilibrium equations for horizontal and vertical forces can also be used, so three independent equations are available for a section of a plane truss. Note that the approach taken in simplifying the analysis of member CD (taking moments about point H) somewhat relaxes the requirement for cutting only three members in drawing a section of the truss. In certain cases the sectioning is allowed to result in more than three unknowns if it is not necessary to solve for every unknown. For example, there may be any number of unknowns at point H in Fig. 6-12d if only the force CD must be determined.

SPACE TRUSSES 6-7

The basic elements of the plane truss may be combined in three-dimensional structures called *space trusses*. The idealized, statically determinate space truss is made of rigid bars connected at their ends by ball-and-socket joints. These joints cannot apply moments to the bars, so each bar can be treated as a two-force member. The supports of the ideal space truss also apply only forces to the structure.

The basic rigid unit of space trusses has six bars connected to form the edges of a tetrahedron as shown in Fig. 6-13a. Each additional rigid unit requires three more members attached at different existing joints and connected to each other at a new joint as in Fig. 6-13b. Here AE, BE, and DE are the new members connected at the old joints A, B, and D, and at the new joint E. It can be established from Fig. 6-13 that $m = 3n - 6$, where m is the total number of members and n is the total number of joints in a space truss.

A space truss can be analyzed with the method of joints or with the method of sections. For each joint, there are three scalar equilibrium equations, $\sum F_x = 0$, $\sum F_y = 0$, and $\sum F_z = 0$. The analysis must begin at a joint where there are at least one known force and no more than three unknown forces. The solution must progress to other joints in a similar fashion.

There are six scalar equations available when the method of sections is used,

$$\sum F_x = 0 \qquad \sum F_y = 0 \qquad \sum F_z = 0$$
$$\sum M_x = 0 \qquad \sum M_y = 0 \qquad \sum M_z = 0$$

Therefore, the number of unknowns at a section generally should not exceed six. This method for space trusses is less useful in practice than it appears at first. The reason is in the difficulty of finding moment axes involving only one unknown in each moment equation.

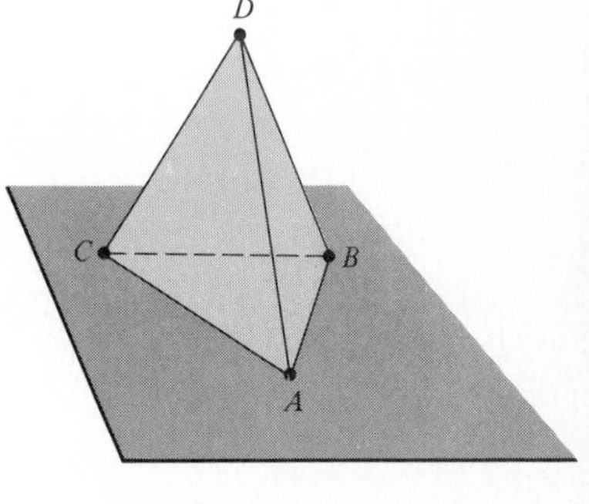

(a)

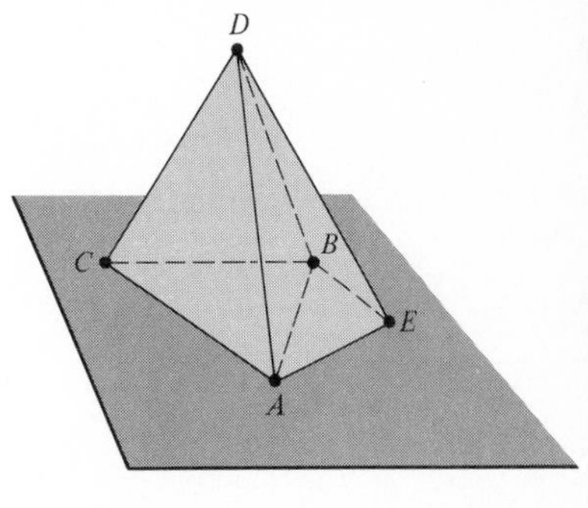

(b)

FIGURE 6-13

Six bars form the basic element, a tetrahedron, of a rigid space truss

EXAMPLE 6-5

Use the method of sections to determine the forces in members *CJ* and *DJ* of Fig. 6-14a.

SOLUTION

First, the support reactions are found; because of the symmetry, $R_A = R_G = 1200$ lb. Next an appropriate section must be chosen such that forces *CJ* and

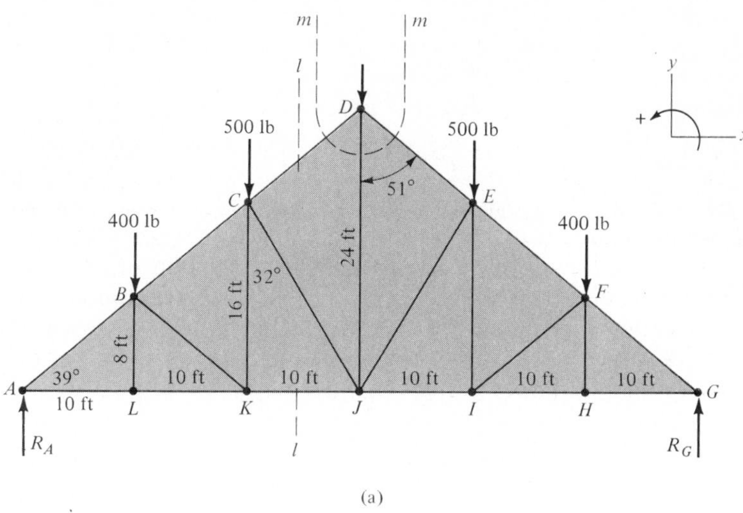

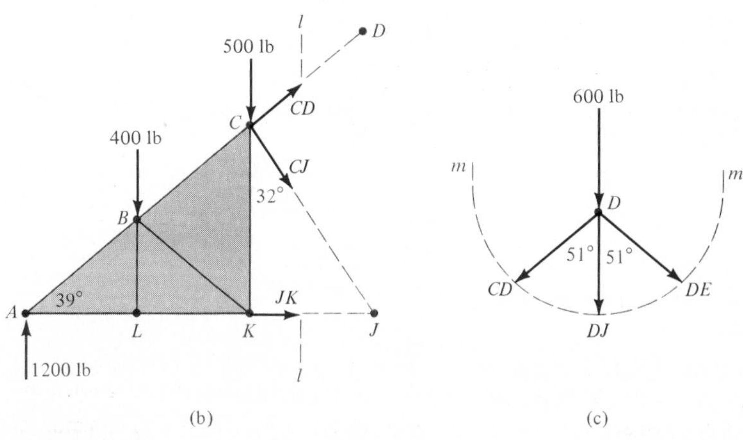

FIGURE 6-14

Analysis of a plane truss using the method of sections

DJ are unknowns, and there are no more than three unknowns involved. It is seen that no one section fulfills all these criteria, hence two separate sections need to be analyzed.

Choosing the section to the left of line l–l, the force CJ can be solved for (Fig. 6-14b). Since the other unknown forces are both directed through joint A, this is a convenient point to sum moments about.

$$\sum M_A = 0$$

$$-400\text{ lb}(10\text{ ft}) - 500\text{ lb}(20\text{ ft}) - CJ\cos 32^\circ(20\text{ ft}) - CJ\sin 32^\circ(16\text{ ft}) = 0$$

$$CJ = \frac{-14{,}000\text{ lb}\cdot\text{ft}}{(20\cos 32^\circ + 16\sin 32^\circ)\text{ ft}} = -550\text{ lb}\quad\text{(compression)}$$

The calculation of force DJ is not as straightforward. The only section that cuts through only three members including DJ is m–m. This, however, is essentially the same as considering joint D alone (Fig. 6-14c). Since all forces at a joint are concurrent, the moment equation becomes trivial and there are only two equations of equilibrium available ($\sum F_x = 0, \sum F_y = 0$). Joint D has three unknown forces acting on it, hence more information is needed before force DJ may be solved. It is seen that force CD can be obtained from the previous section (Fig. 6-14b). By summing moments about joint J an equation involving CD as the only unknown is obtained since forces CJ and JK are both directed through joint J.

$$\sum M_J = 0$$

$$-1200\text{ lb}(30\text{ ft}) + 400\text{ lb}(20\text{ ft}) + 500\text{ lb}(10\text{ ft}) - CD\cos 39^\circ(16\text{ ft}) - CD\sin 39^\circ(10\text{ ft}) = 0$$

$$CD = \frac{-23{,}000\text{ lb}\cdot\text{ft}}{(16\cos 39^\circ + 10\sin 39^\circ)\text{ ft}} = -1230\text{ lb}\quad\text{(compression)}$$

Reconsidering Fig. 6-14c, by symmetry $CD = DE = -1230$ lb. Now force DJ is the only unknown at joint D. From Fig. 6-14c,

$$\sum F_y = 0$$

$$-CD\cos 51^\circ - DE\cos 51^\circ - 600\text{ lb} - DJ = 0$$

$$DJ = 1230\text{ lb}\cos 51^\circ + 1230\text{ lb}\cos 51^\circ - 600\text{ lb} = 950\text{ lb}\quad\text{(tension)}$$

It is seen that the method of sections allows for much faster calculation of forces if only a few forces are desired. While calculation of the force DJ might seem lengthy, it should be remembered that a solution by the method of joints would be even more complicated. In that case joints A, L, B, K, C, and D would have to be analyzed.

EXAMPLE 6-6

Assume that the triangular structure on top of the tall crane in Fig. 6-15a consists of simple plane trusses as modeled in Fig. 6-15b. Determine the forces in members BC, CJ, and IJ using the method of sections.

SOLUTION

First it is decided where to "cut" the truss for easiest analysis. Section l–l shown in Fig. 6-15b involves the three members of concern and no others. A decision must now be made whether to use the top section or the bottom section. By choosing the top section calculation of the support reactions is avoided, so a free-body diagram of this section is drawn (Fig. 6-15c).

From trigonometry the angle that member CJ makes with the horizontal is found to be 43.6°. By summing moments about joint C an equation involving only force IJ is obtained.

$$\sum M_C = 0$$

$$-50\text{ kN}\cos 15^\circ(2.8\text{ m}) + 50\text{ kN}\sin 15^\circ(6\text{ m}) + IJ\sin 5^\circ(0\text{ m}) - IJ\cos 5^\circ(2.27\text{ m}) = 0$$

$$IJ = \frac{-57.6\text{ kN}\cdot\text{m}}{(\cos 5^\circ)(2.27\text{ m})} = -25.5\text{ kN}\quad\text{(compression)}$$

By summing moments about joint J an equation involving only force BC is obtained since CJ and IJ are both directed through this joint. Note that it is valid to sum moments about a point that is not part of the section being considered because moment equilibrium is valid about *any* point in space.

$$\sum M_J = 0$$

$$-50\text{ kN}\cos 15^\circ(0.7\text{ m}) + 50\text{ kN}\sin 15^\circ(8\text{ m}) + BC\cos 25^\circ(2.1\text{ m}) + BC\sin 25^\circ(2\text{ m}) = 0$$

$$BC = \frac{-69.7\text{ kN}\cdot\text{m}}{\cos 25^\circ(2.1\text{ m}) + \sin 25^\circ(2\text{ m})}$$

$$= -25.4\text{ kN}\quad\text{(compression)}$$

By summing moments about joint F an equation involving only force CJ is obtained. It is seen that the external force acts through this joint as well as IJ and BC; hence the moment equation is trivial and $CJ = 0$.

A check may be made by summing forces in the x and y directions for the whole section.

It should be noted that by choosing certain points to sum moments about, equations involving only one unknown are often available. Summing forces in certain directions can also reduce the number of unknowns. This has obvious advantages and should be used whenever possible.

(a)

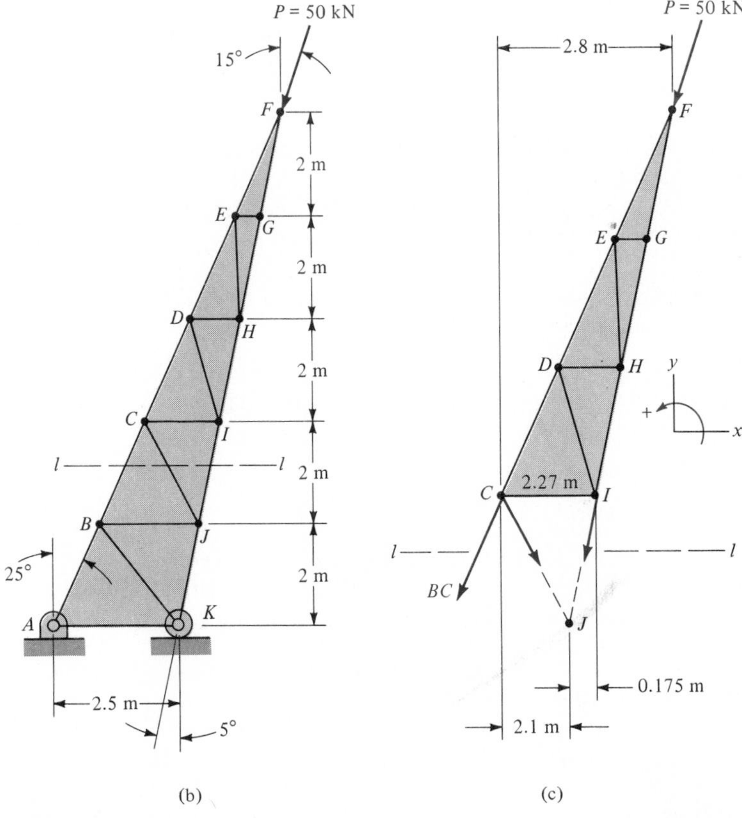

FIGURE 6-15

Analysis of a plane truss using the method of sections

EXAMPLE 6-7

The boatyard crane shown in Fig. 6-16a has a basic space truss element as its stationary structure, $ABCD$ in Fig. 6-16b. The position of joint A is fixed, while joints B and C (in the same horizontal plane as A) are only constrained to maintain their positions with respect to A in the horizontal plane. Determine the forces in members AD and BD when $L = 6\text{m}$ and $W = 10$ kN. The other dimensions, coordinates, and angles are given as constants or quantities that are valid for the particular situation.

SOLUTION

Since the supports at B and C resist motion in the y direction only, y component support reactions only are shown. From symmetry it is seen that $R_B = R_C = R$. These are determined by taking moments about the x axis,

$$\sum M_x = 0$$

$$W(5.69\text{ m}) - 2R(3\text{ m}\cos 30°) = 0$$

$$R = \frac{(10\text{ kN})(5.69\text{ m})}{2(3\text{ m})(\cos 30°)} = 10.95\text{ kN} = R_B = R_C$$

The reactions at joint A are determined by making summations of forces along x, y, and z.

$$\sum F_x = 0$$

$$A_x = 0$$

$$\sum F_y = 0$$

$$2R - A_y - W = 0$$

$$A_y = 2(10.95\text{ kN}) - 10\text{ kN} = 11.9\text{ kN}$$

$$\sum F_z = 0$$

$$A_z = 0$$

The best joint to analyze next is joint E, which has only two unknowns. Here vector notation is initiated to facilitate the solution. The unknown forces are given as vectors of known directions using Eq. 2-23.

JOINT *E*, Fig. 6-16c:

$$\overrightarrow{ED} = ED\frac{(0\mathbf{i} - 1.9\mathbf{j} - 5.69\mathbf{k})\text{ m}}{6\text{ m}} = -0.32ED\mathbf{j} - 0.95ED\mathbf{k}$$

$$\overrightarrow{EA} = EA\frac{(0\mathbf{i} - 8.22\mathbf{j} - 5.69\mathbf{k})\text{ m}}{10\text{ m}} = -0.82EA\mathbf{j} - 0.57EA\mathbf{k}$$

(a)

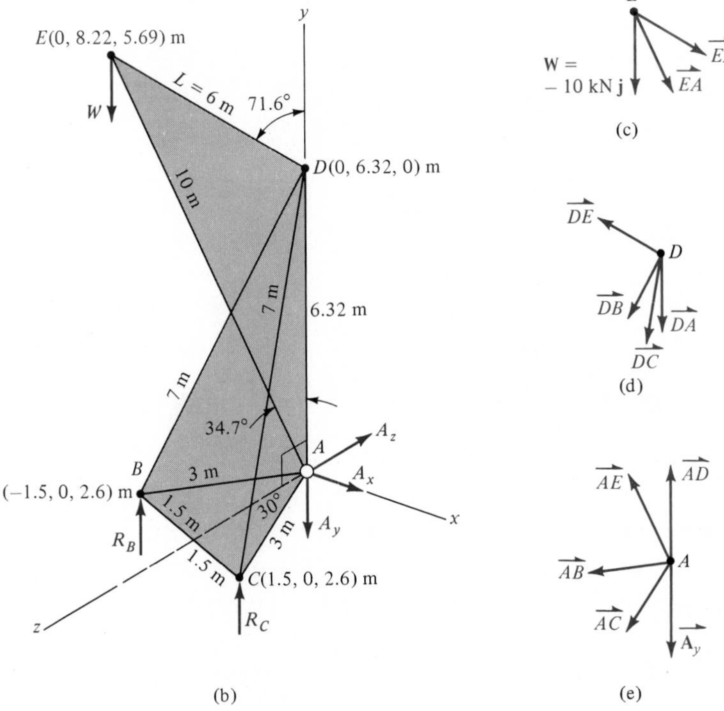

(b)

(c)

(d)

(e)

FIGURE 6-16

Analysis of a space truss

Summing the scalar components of the forces acting at joint E,

$$\sum F_y = 0 \qquad \sum F_z = 0$$

$$-10\text{ kN} - 0.32ED - 0.82EA = 0 \qquad -0.95ED - 0.57EA = 0$$

$$-10\text{ kN} - 0.32(-0.6EA) - 0.82EA = 0 \qquad ED = -0.6EA$$

$$EA = -15.9\text{ kN} \quad \text{(compression)}$$

$$ED = 9.5\text{ kN} \quad \text{(tension)}$$

JOINT *D*, Fig. 6-16d:

$$\overrightarrow{DB} = DB\frac{(-1.5\mathbf{i} - 6.32\mathbf{j} + 2.6\mathbf{k})\text{ m}}{7\text{ m}}$$

$$= -0.21DB\mathbf{i} - 0.9DB\mathbf{j} + 0.37DB\mathbf{k}$$

$$\overrightarrow{DC} = DC\frac{(1.5\mathbf{i} - 6.32\mathbf{j} + 2.6\mathbf{k})\text{ m}}{7\text{ m}}$$

$$= 0.21DC\mathbf{i} - 0.9DC\mathbf{j} + 0.37DC\mathbf{k}$$

From symmetry, $DB = DC$. By Newton's law, $\overrightarrow{DE} = -\overrightarrow{ED} = 9.5(0.32\mathbf{j} + 0.95\mathbf{k})$ kN. Summing forces in the z direction at joint D, only one unknown (force $DB = DC$) is involved,

$$\sum F_z = 0$$

$$9.5(0.95)\text{ kN} + 2(0.37)\ DB = 0$$

$$DB = DC = -12.2\text{ kN} \quad \text{(compression)}$$

Summing forces in the y direction at joint D gives the solution for the unknown force DA,

$$\sum F_y = 0$$

$$0.32DE - 2(0.9)DB - DA = 0$$

$$0.32(9.5)\text{ kN} - 2(0.9)(-12.2)\text{ kN} - DA = 0$$

$$DA = 25\text{ kN} \quad \text{(tension)}$$

JUDGMENT OF THE RESULTS

All of the forces in the y direction at joint A are known, and a summation of these serves as a check. From Fig. 6-16e,

$$\sum F_y = AD - A_y + AE(0.82)$$

$$= 25\text{ kN} - 11.9\text{ kN} - 15.9(0.82)\text{ kN} = 0.06\text{ kN}$$

The small net force calculated is attributed to round-off errors. The results are accepted as correct.

PROBLEMS

6–38 Determine the forces in members *AB*, *AC*, and *CD*.

FIGURE P6-38

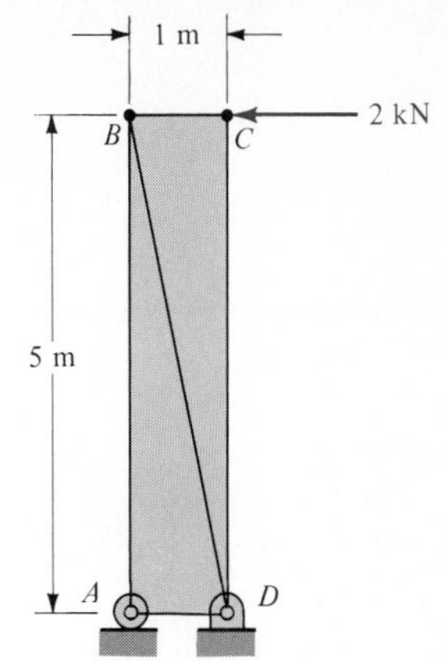

6–39 Determine the forces in members *AB*, *AC*, and *CD*.

FIGURE P6-39

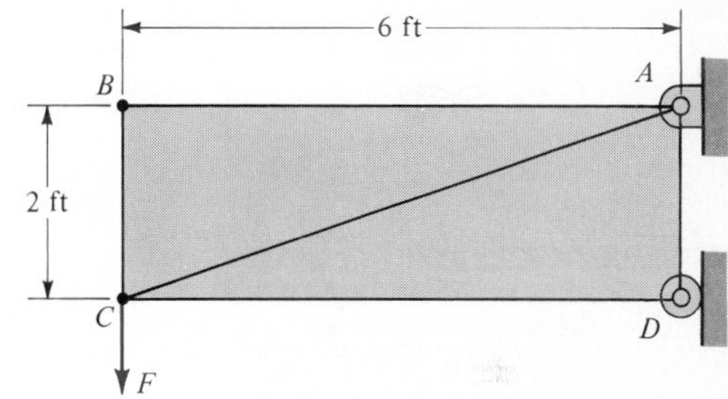

6–40 Calculate the forces in members *AB*, *AC*, and *CD*.

FIGURE P6-40

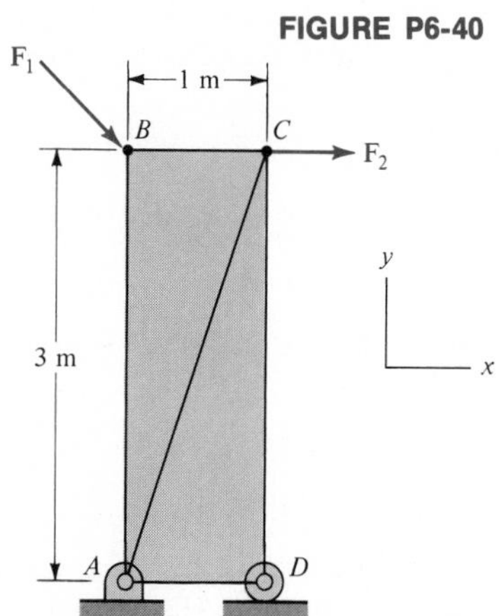

6–41 Determine the forces in members *AB*, *BD*, and *CD*.

FIGURE P6-41

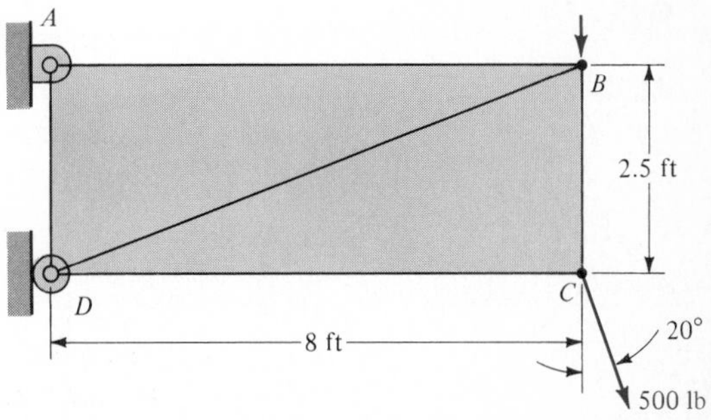

FIGURE P6-42

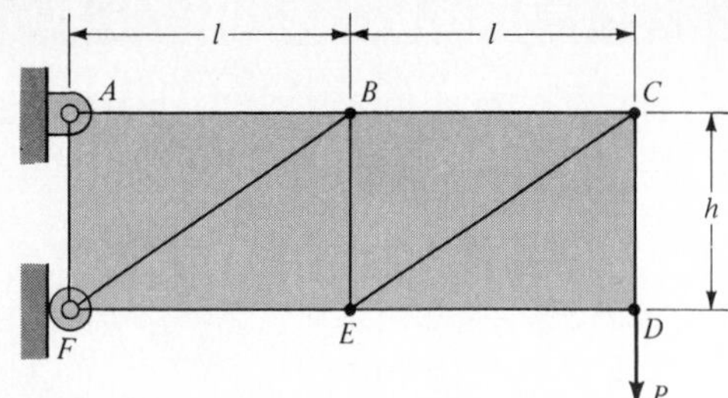

6–42 Calculate the forces in members AB, BF, and EF. $P = 4$ kN, $l = 1.5$ m, and $h = 1$ m.

6–43 Consider Fig. P6-42, assuming that $P = 600$ lb, $l = 7$ ft, and $h = 5$ ft. Calculate the forces in members BC, BE, and EF.

FIGURE P6-44

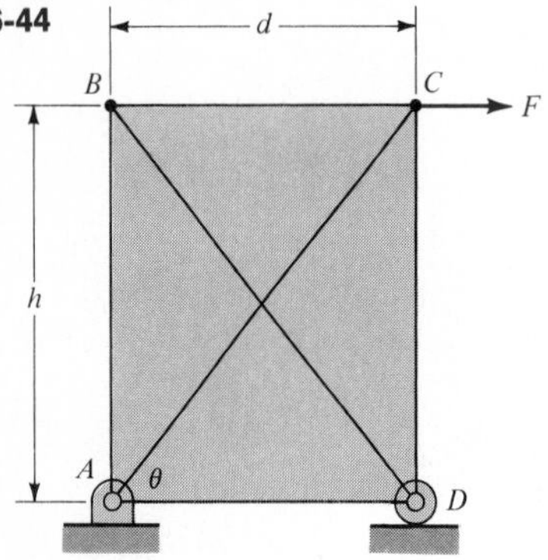

6–44 Take a section through members AB, AC, BD, and CD. Show that there are not enough independent equations of equilibrium available to solve for the unknown forces in these members. Consider both the upper and lower sections.

FIGURE P6-45

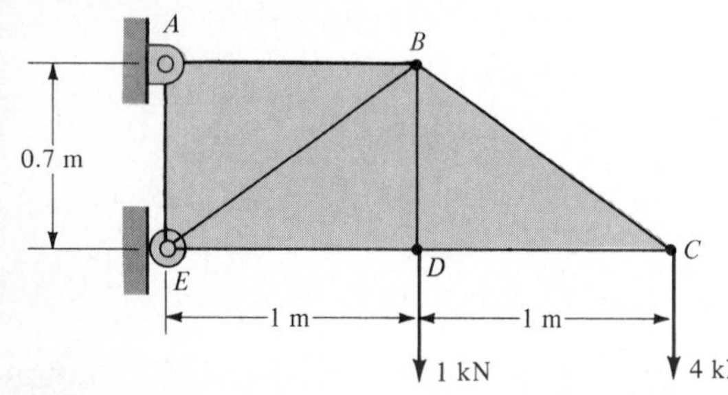

6–45 Determine the forces in members AB, BE, and DE.

FIGURE P6-46

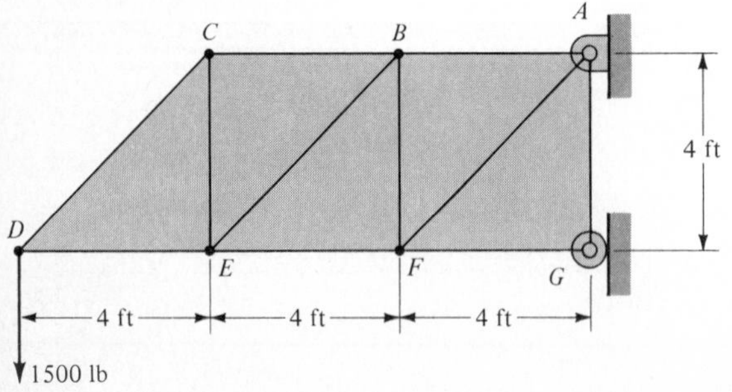

6–46 Calculate the forces in members BC, BE, and EF.

6–47 Consider Fig. P6-46 and determine the force in member AB using two different sections: (a) through AB, AF, and FG, and (b) through AB, BF, and EF.

6–48 Calculate the forces in members *BC*, *CE*, and *DE*.

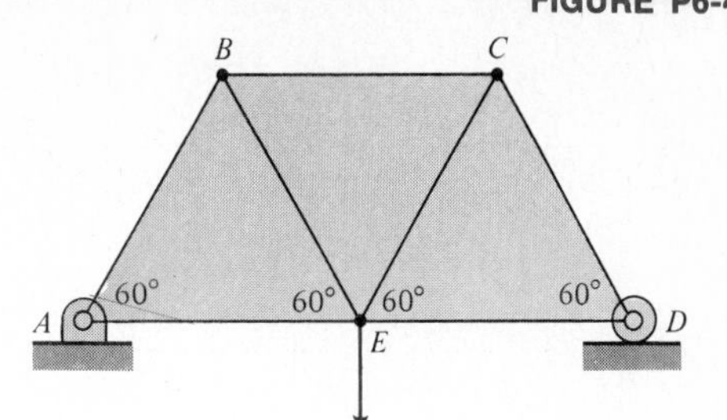

FIGURE P6-48

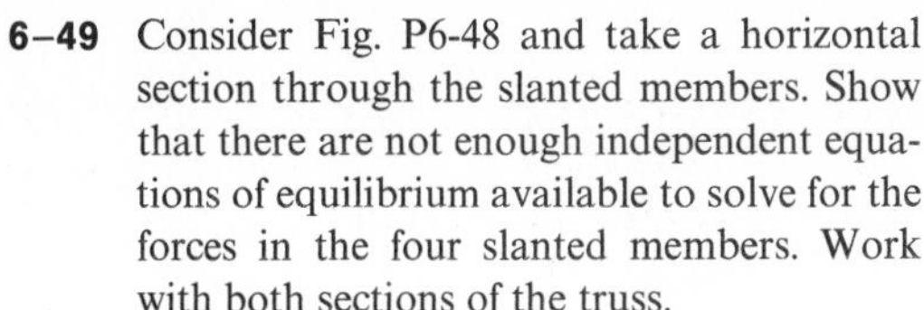

6–49 Consider Fig. P6-48 and take a horizontal section through the slanted members. Show that there are not enough independent equations of equilibrium available to solve for the forces in the four slanted members. Work with both sections of the truss.

6–50 Determine the forces in members *CD*, *DK*, and *JK*. $AL = LK = KJ = 5$ ft and $DJ = 10$ ft. The truss is symmetric.

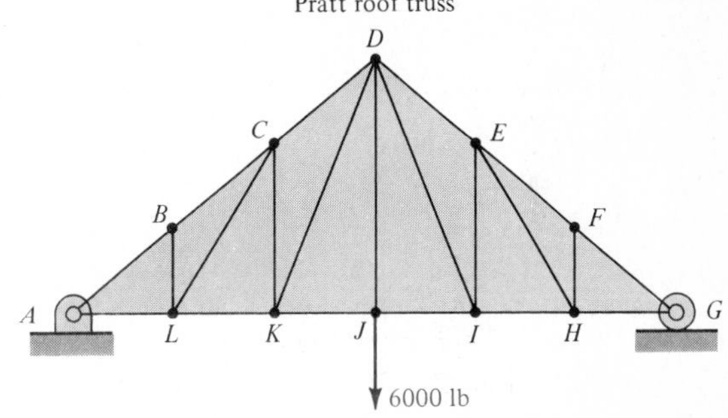

FIGURE P6-50

6–51 Solve Prob. 6–50 as stated, then replace member *DK* with *CJ*, and member *DI* with *EJ*. Calculate the forces in *CD*, *CJ*, and *JK* after the change.

6–52 Calculate the forces in members *DE*, *EJ*, and *IJ*. $AL = LK = KJ = 2$ m, $DJ = 3$ m. The truss is symmetric.

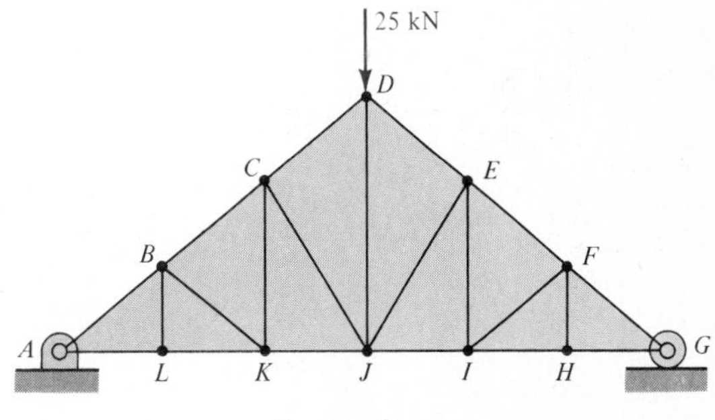

FIGURE P6-52

6–53 Solve Prob. 6–52 as stated, then replace all internal slanted members with oppositely slanting bars (for example, *BK* is replaced by *CL*). Determine the forces in *DE*, *DI*, and *IJ* after the change.

6–54 In the symmetric bridge truss $AL = LK = KJ = 8$ ft, and $DJ = 10$ ft. Determine the forces in members *CD*, *CJ*, and *KJ*.

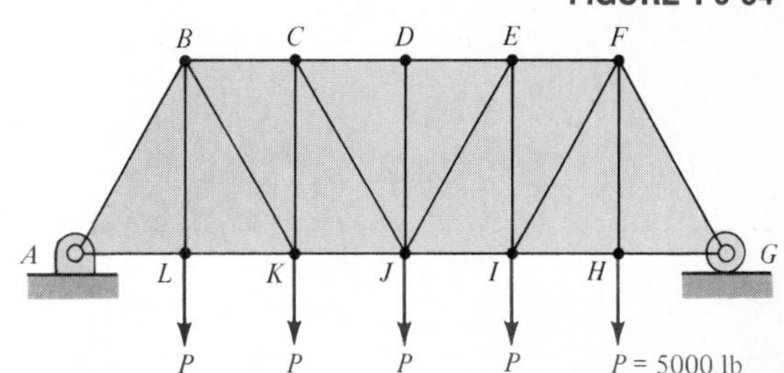

FIGURE P6-54

6–55 Check the force *KJ* obtained in Prob. 6–54 by using a section through *BC*, *CK*, and *KJ*.

FIGURE P6-56

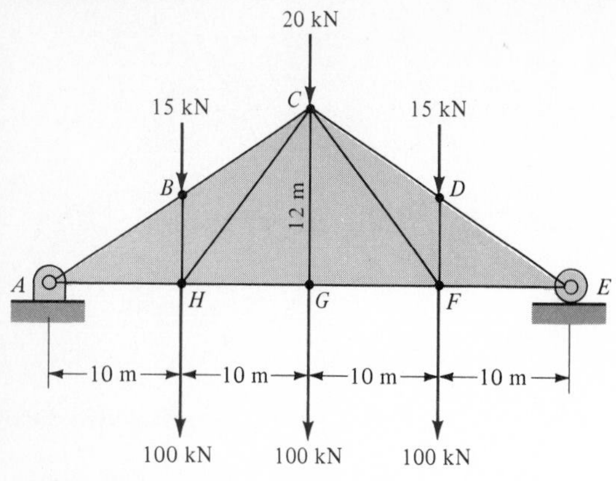

6–56 Calculate the forces in members *BC*, *CH*, and *GH*.

FIGURE P6-57

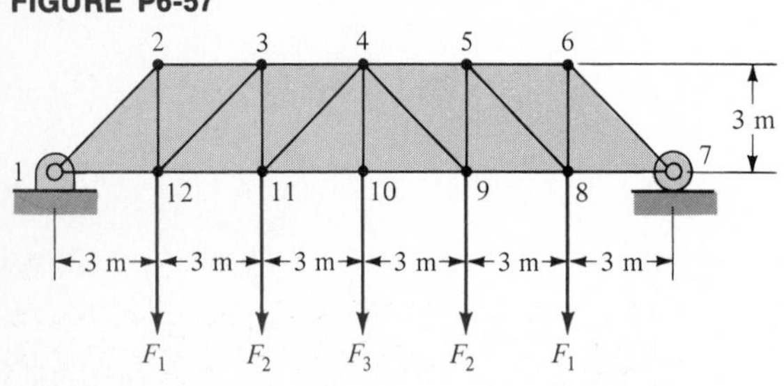

6–57 Calculate the forces acting at joint 9 if $F_1 =$ 20 kN, $F_2 =$ 25 kN, and $F_3 =$ 30 kN.

FIGURE P6-58

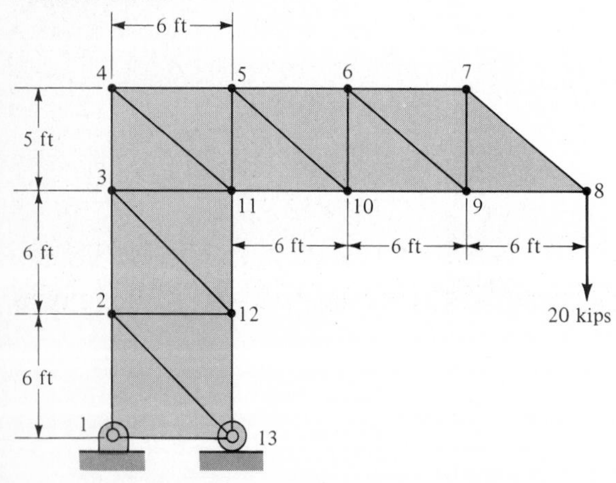

6–58 Calculate the forces acting at joint 11.

6–59 Determine the forces in the four vertical members of the truss.

FIGURE P6-59

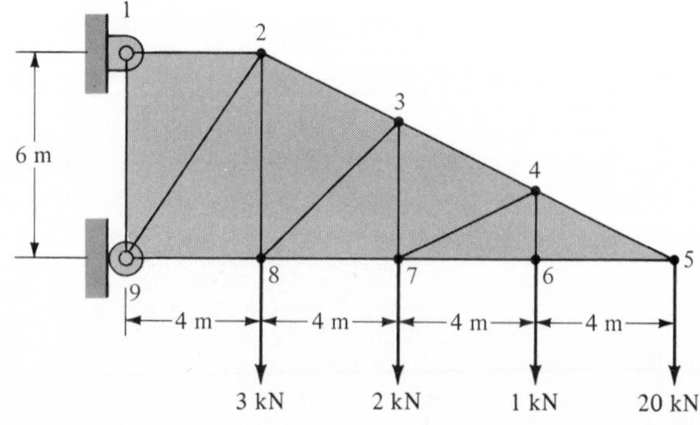

6–60 Determine the forces acting at joint 18.

FIGURE P6-60

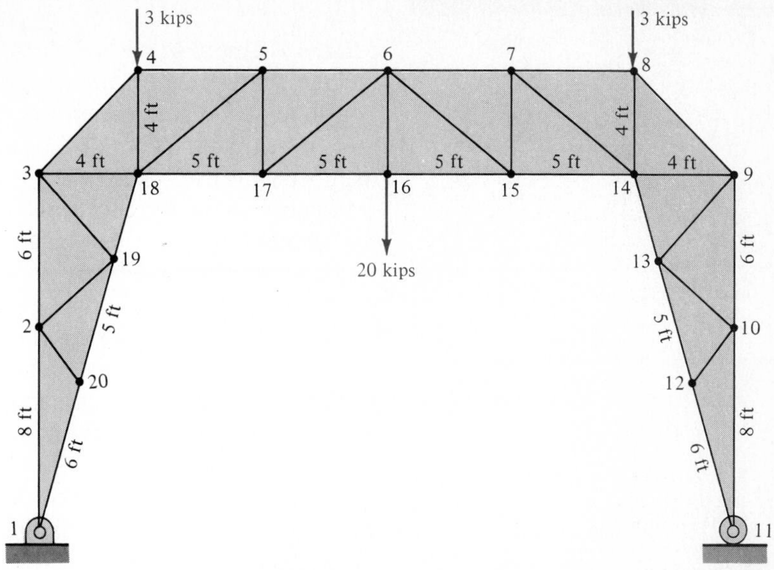

6–61 Consider the container-loading derrick and assume that the supporting cables at joint 11 become slack. Determine the forces acting at joints 2 and 15.

FIGURE P6-61

(a)

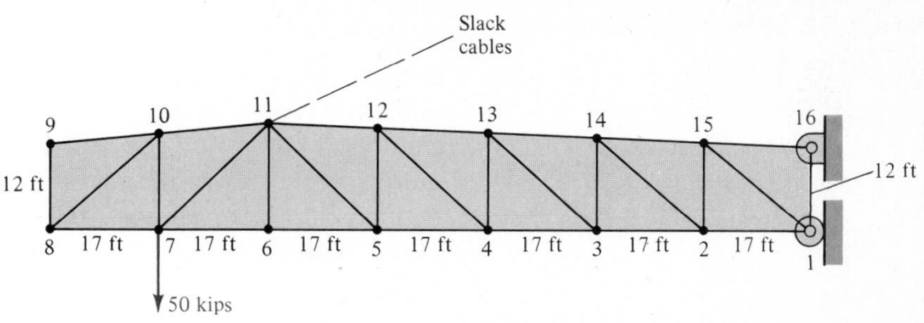

(b) Plane truss model of one side of boom

6-8 FRAMES AND MACHINES

Multiforce members (with three or more forces acting on each member) are common in structures. In these cases the forces are not directed along the members, so they are a little more complex to analyze than the two-force members in simple trusses. Multiforce members are used in two kinds of structure. *Frames* are usually stationary and fully constrained. *Machines* always have moving parts, so the forces acting on a member depend on the location and orientation of that member.

The analysis of multiforce members is based on the consistent use of interrelated free-body diagrams. The solution is often facilitated by representing forces by their rectangular components. Scalar equilibrium equations are the most convenient for two-dimensional problems, and vector notation is advantageous in three-dimensional situations.

To illustrate the analysis of frames, consider the folding lawn chair shown in Fig. 6-17a. Assume that the two-dimensional frame in Fig. 6-17b is a reasonable model of the chair. Here *ABC*, *BH*, *CDF*, *DIJ*, *EFG*, and *GHI* are rigid members connected by smooth pins, and the supporting legs *A* and *J* are on a rough surface. The members are connected only where dots are shown to represent pins in Fig. 6-17b.

The free-body diagrams of the members are drawn in Fig. 6-17c through h; the *principle of action and reaction* must be carefully observed in this process. For example, in Fig. 6-17c, forces A_x, A_y, B_x, B_y, C_x, and C_y are all unknown and may be drawn arbitrarily in the positive or negative x and y directions. After this drawing is made, B_x and B_y have fixed directions in Fig. 6-17d, as shown. Here force P is also fixed, but H_x and H_y are assumed arbitrarily. The process is continued with the same consistency, considering the previously drawn diagrams in making each new diagram. Note that the actual direction of each unknown force remains to be determined from the sign of the result.

The next task is to write equilibrium equations and determine the unknown forces. It is worth noting that the total number of equations required to solve the whole problem is equal to the number of unknowns. Two force summations and one moment summation may be written for each two-dimensional free-body diagram as follows. The moment equation should be chosen to have only one unknown in it, or as few as possible. Careful choice of these equations is necessary to reduce the need to solve simultaneous equations.

(a)

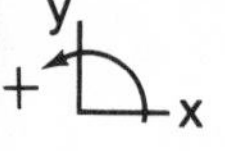

FIGURE 6-17

A folding chair modeled as a frame

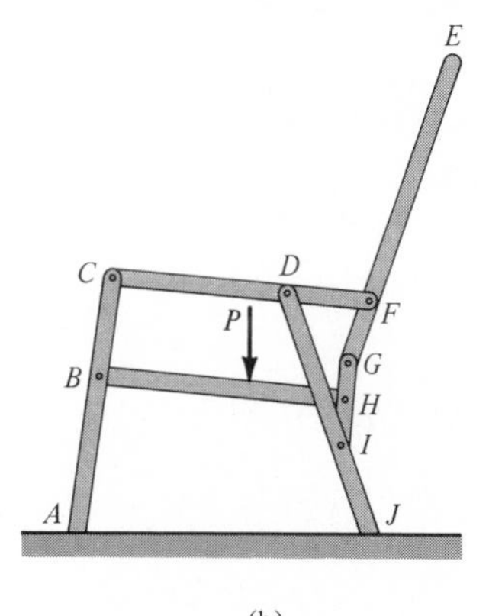

(b)

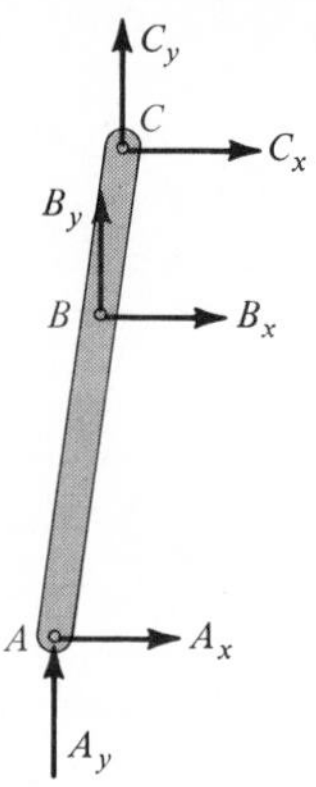

(c)

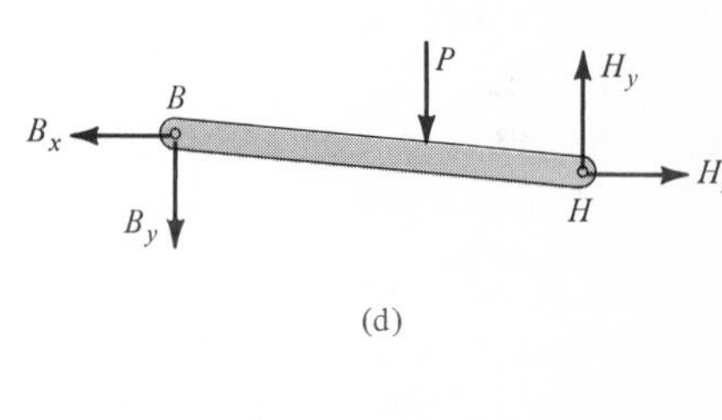

(d)

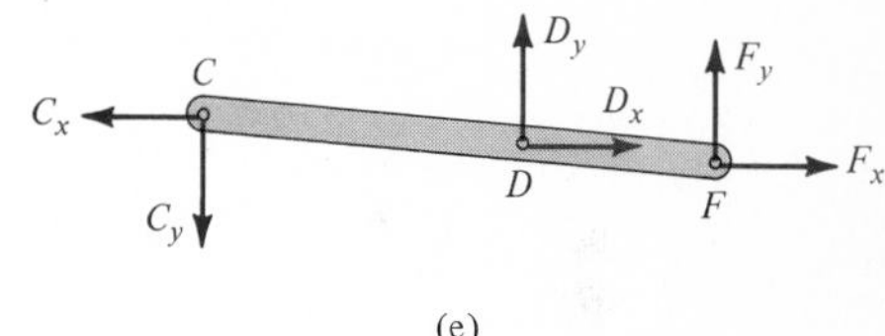

(e)

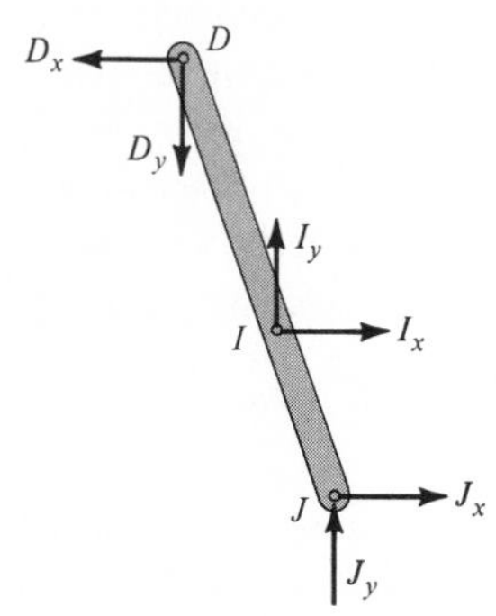

(f)

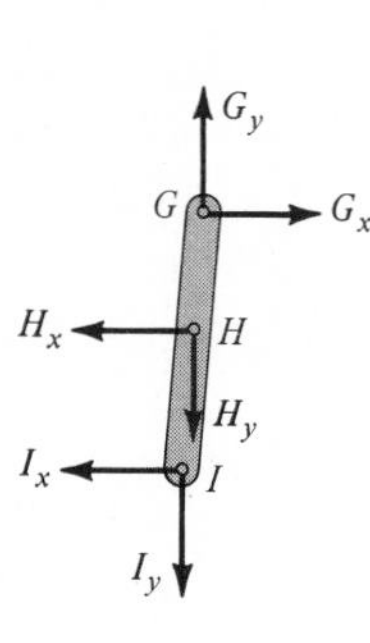

(g)

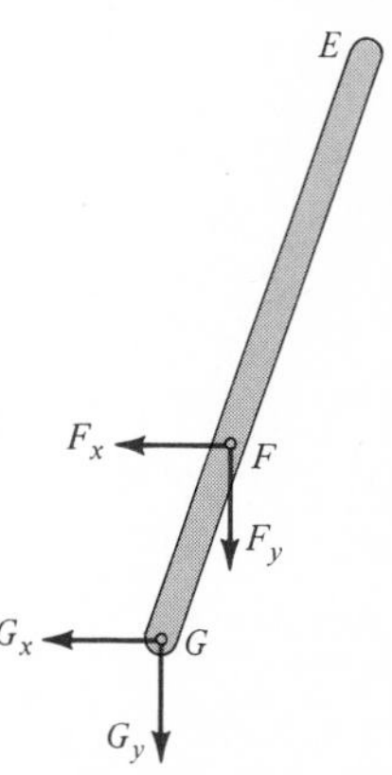

(h) Note that EG is a two-force member, so the directions of **F** and **G** are along EG, and force $F = G$

$$\text{Fig. 6-17c:}\quad \sum F_x=0\rightarrow A_x+B_x+C_x=0$$
$$\sum F_y=0\rightarrow A_y+B_y+C_y=0$$
$$\sum M_B=0 \quad or \quad \sum M_C=0$$
$$\text{Fig. 6-17d:}\quad \sum F_x=0\rightarrow H_x-B_x=0$$
$$\sum F_y=0\rightarrow H_y-B_y-P=0$$
$$\sum M_B=0 \quad or \quad \sum M_H=0$$
$$\text{Fig. 6-17e:}\quad \sum F_x=0\rightarrow D_x+F_x-C_x=0$$
$$\sum F_y=0\rightarrow D_y+F_y-C_y=0$$
$$\sum M_C=0 \quad or \quad \sum M_D=0$$
$$or \quad \sum M_F=0$$
$$\text{Fig. 6-17f:}\quad \sum F_x=0\rightarrow I_x+J_x-D_x=0$$
$$\sum F_y=0\rightarrow I_y+J_y-D_y=0$$
$$\sum M_D=0 \quad or \quad \sum M_I=0$$
$$\text{Fig. 6-17g:}\quad \sum F_x=0\rightarrow G_x-H_x-I_x=0$$
$$\sum F_y=0\rightarrow G_y-H_y-I_y=0$$
$$\sum M_G=0 \quad or \quad \sum M_H=0$$
$$or \quad \sum M_I=0$$
$$\text{Fig. 6-17h:}\quad \left.\begin{aligned}\sum F_x&=0\rightarrow -F_x-G_x=0\\ \sum F_y&=0\rightarrow -F_y-G_y=0\\ \sum M_F=0 \quad &or \quad \sum M_G=0\end{aligned}\right\}\ \text{Note that } FG \text{ is a two-force member}$$

These equations are available to calculate the unknown forces if the geometry of the chair and the loading are specified. A positive number for a result indicates that the direction of that force was assumed correctly in the beginning. A negative answer indicates that the force is acting in the direction opposite to the assumed direction. Generally it is not necessary to correctly redraw any of the forces in the free-body diagrams since the signs are easy to interpret.

Special Cases of Frames

The number and arrangement of members and supports in a frame can be quite varied, so a few distinct possibilities are worth mentioning.

1. *Rigid frames* may be treated as rigid bodies whether or not they are attached to supports. Figure 6-18a shows such a frame that remains rigid in any orientation, under any loading (as long as the individual bars remain rigid), and with or without being attached to fixed supports.

2. *Semirigid frames* may be treated as rigid bodies only when they are attached to fixed supports. For example, the member *BE* in Fig. 6-18a may be removed only if ends *A* and *F* are attached to fixed supports as in Fig. 6-18b. An interesting aspect of this frame is that all the external reactions at *A* and *F* cannot be determined from the free-body diagram of the whole frame consisting of members *AD* and *CF*, since this frame is not rigid by itself. The individual free-body diagrams are needed to complete the analysis.

3. *Statically determinate and rigid frames* have an equal number of unknown force components and independent equilibrium equations. All the unknown forces can be determined, and all the equations can be satisfied (Fig. 6-18a). If there are more equations than unknowns, the frame is called *nonrigid*. This seems to be the case for the chair in Fig. 6-17 if equilibrium equations for the whole chair are added to the equations already written. However, these new equations, although valid, would not be independent of the original equations because they could be obtained by superimposing all of the individual free-body diagrams (this cancels all internal forces).

4. *Statically indeterminate and rigid frames* have more unknown forces than independent equilibrium equations. For example, member *DE* in Fig. 6-18c is redundant in theory, although it may be useful in practice. A statically indeterminate frame may also be nonrigid if there are more equations than unknown forces.

5. *Machines* are used to transmit and modify forces and motion. They consist of two or more main rigid parts that are relatively constrained. The simplest devices (hand tools such as pliers) and large and complex mechanisms can all be analyzed using the basic equilibrium conditions. Machines always have moving parts, so they are nonrigid on the whole even if they include some rigid structures.

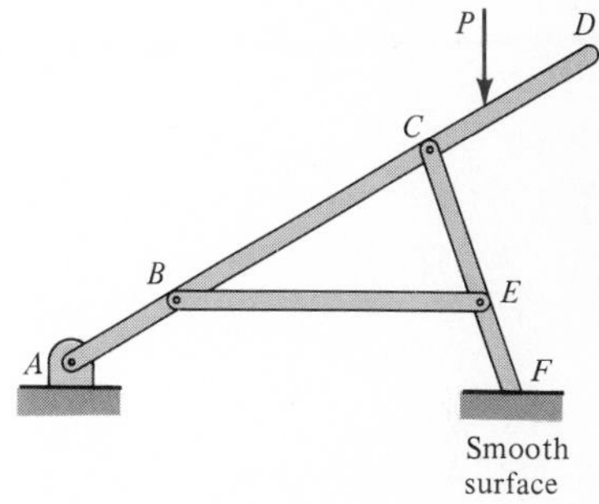

(a) Rigid frame of three rigid bars (*AD* and *CF* are continuous)

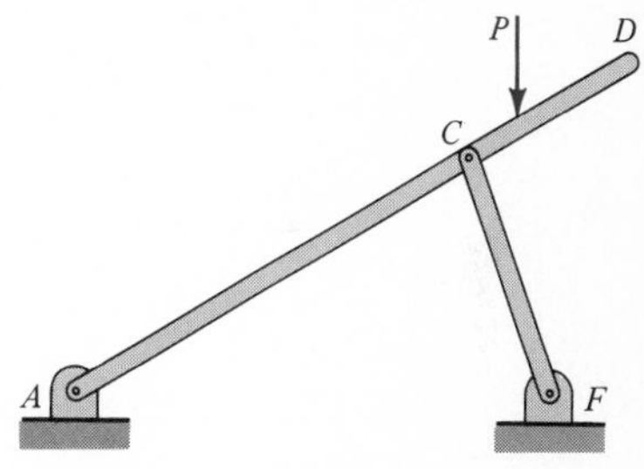

(b) Semirigid frame

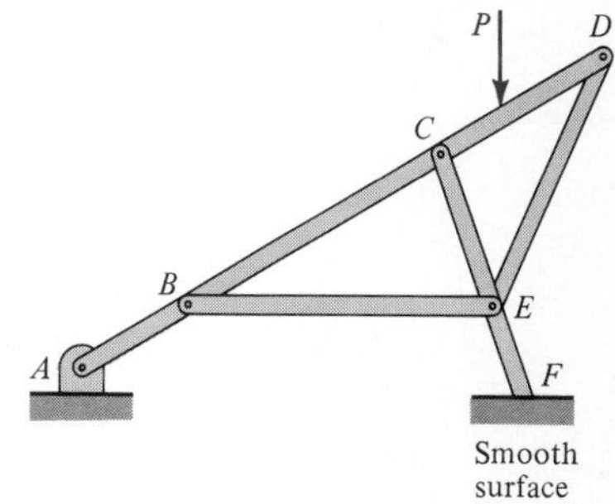

(c) Statically indeterminate rigid frame

FIGURE 6-18

Special cases of frames

EXAMPLE 6-8

The folding chair shown in Fig. 6-19a is represented by the two-dimensional model of the front or rear legs in Fig. 6-19b. The floor supporting the chair is assumed to be frictionless. Determine the reactions at the bottom of the legs and the forces at the joints B, C, and E.

SOLUTION

Consider the whole chair's free-body diagram (Fig. 6-19b) to determine the reactions A_y and H_y. Sum the moments about either point A or H. This yields an equation with only one unknown.

$$\sum M_A = 0$$

$$-P\cos 30^\circ(0.5\text{ m}) + P\sin 30^\circ(0\text{ m}) + P\cos 30^\circ(0.5\text{ m})$$

$$-P\sin 30^\circ(0.714\text{ m}) + H_y(0.714\text{ m}) = 0$$

$$H_y = P\sin 30^\circ = 600\text{ N}(0.5) = 300\text{ N}$$

By symmetry, $A_y = H_y = 300$ N.

It is now necessary to work with free-body diagrams of the individual members.

MEMBER *DCBH*, Fig. 6-19c:

$$\sum F_x = 0$$

$$P\cos 30^\circ + C_x + B_x = 0$$

$$520\text{ N} + C_x + B_x = 0 \qquad \textbf{(a)}$$

$$\sum F_y = 0$$

$$-P\sin 30^\circ + C_y + B_y + H_y = 0$$

$$-300\text{ N} + C_y + B_y + 300\text{ N} = 0$$

$$C_y = -B_y \qquad \textbf{(b)}$$

It is best to take moments about point B or C, where two unknown forces act. Choosing C,

$$\sum M_C = 0$$

$$-P\cos 30^\circ(0.15\text{ m}) + P\sin 30^\circ(0.214\text{ m}) + B_y(0.143\text{ m})$$

$$+ B_x(0.1\text{ m}) + H_y(0.5\text{ m}) = 0$$

$$-77.9\text{ N}\cdot\text{m} + 64.2\text{ N}\cdot\text{m} + B_y(0.143\text{ m})$$

$$+ B_x(0.1\text{ m}) + 150\text{ N}\cdot\text{m} = 0 \qquad \textbf{(c)}$$

There are four unknowns (B_x, B_y, C_x, and C_y) and only three equations, so no unknowns can be solved. Equilibrium equations must be written for other members.

MEMBER *CE*, Fig. 6-19d:

$$\sum M_C = \sum M_E = 0 \qquad \text{(member } CE \text{ is a two-force member)}$$

(a)

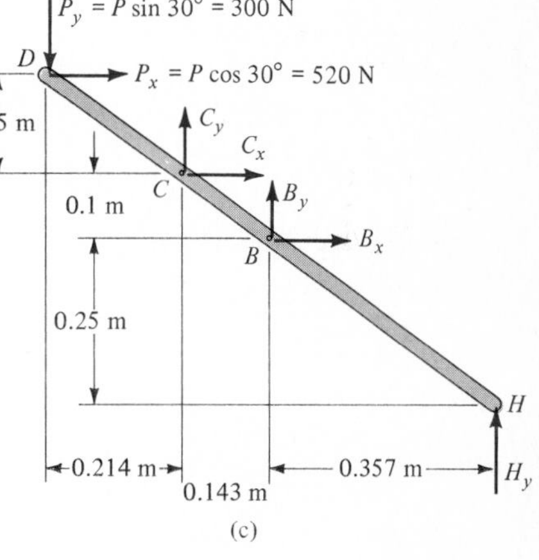

(c)

Hence,

$$C_y = E_y = 0 \qquad \text{and} \qquad C_x = E_x \qquad \textbf{(d)}$$

By Eq. (b),

$$C_y = -B_y = 0$$

Equation (c) now has only one unknown, and this is readily calculated,

$$B_x = -\frac{136.3\ \text{N}\cdot\text{m}}{0.1\ \text{m}} = -1360\ \text{N}$$

From Eq. (a), C_x can now be solved,

$$C_x = -520\ \text{N} - B_x = -520\ \text{N} - (-1360\ \text{N}) = 840\ \text{N}$$

From Eq. (d), $C_x = E_x = 840$ N. From Fig. 6-19e,

$$G_x = E_x = 840\ \text{N}$$

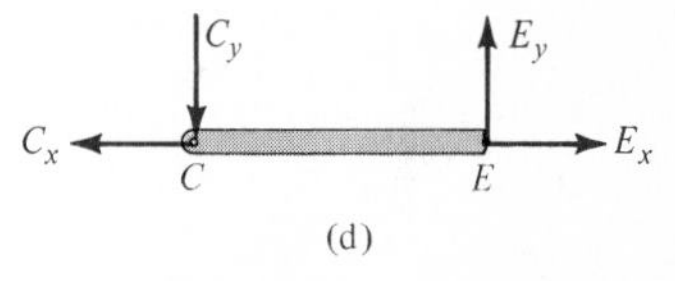

(d)

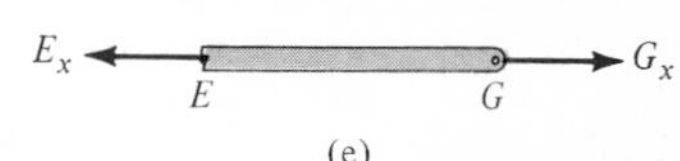

(e)

SUMMARY:

$$A_x = H_x = 0 \qquad A_y = H_y = 300\ \text{N}$$
$$B_x = -1360\ \text{N} \qquad B_y = 0$$
$$C_x = E_x = G_x = 840\ \text{N} \qquad C_y = E_y = G_y = 0$$

JUDGMENT OF THE RESULTS

The forces calculated can be checked using Fig. 6-19f, where every force is known. Summing moments about point F is convenient,

$$\begin{aligned} \sum M_F &= -300\ \text{N}(0.714\ \text{m}) + 1360\ \text{N}(0.25\ \text{m}) - 840\ \text{N}(0.15\ \text{m}) \\ &= -0.2\ \text{N}\cdot\text{m} \end{aligned}$$

The net moment calculated is negligible, so the results are accepted as correct.

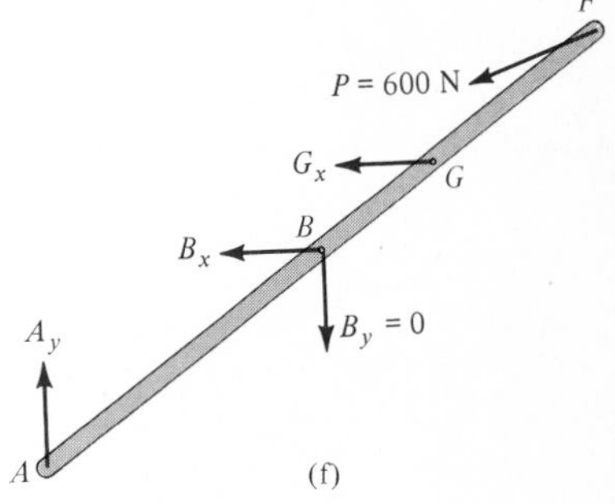

(f)

FIGURE 6-19

Analysis of a folding chair

EXAMPLE 6-9

A winch and mast-crutch frame of a boat trailer is shown in Fig. 6-20a. Its simplified model is shown in Fig. 6-20b, where the given forces are estimated for a braking situation. Calculate the forces at points *A*, *B*, and *D*.

SOLUTION

First the reaction components at the supports *A* and *D* are sought by analysis of the frame as a whole (Fig. 6-20c).

$$\sum F_x = 0$$

$$-50 \text{ lb} \cos 10^\circ - 500 \text{ lb} \cos 20^\circ + A_x + D_x = 0$$

$$A_x + D_x = 470 \text{ lb} + 49 \text{ lb} = 519 \text{ lb} \quad \textbf{(a)}$$

$$\sum F_y = 0$$

$$-50 \text{ lb} \sin 10^\circ - 500 \text{ lb} \sin 20^\circ + A_y + D_y = 0$$

$$A_y + D_y = 180 \text{ lb} \quad \textbf{(b)}$$

$$\sum M_A = 0$$

$$D_y(0.91 \text{ ft}) - 500 \text{ lb} \sin 20^\circ(1 \text{ ft} \sin 20^\circ)$$
$$+ 500 \text{ lb} \cos 20^\circ(1.66 \text{ ft} \cos 20^\circ) + 50 \text{ lb} \cos 10^\circ(5 \text{ ft}) = 0$$

$$D_y = \frac{59 \text{ lb}\cdot\text{ft} - 733 \text{ lb}\cdot\text{ft} - 246 \text{ lb}\cdot\text{ft}}{0.91 \text{ ft}} = -1011 \text{ lb}$$

From Eq. (b),

$$A_y = 180 \text{ lb} + 1011 \text{ lb} = 1191 \text{ lb}$$

After these results it is necessary to analyze the individual members. From Fig. 6-20d,

$$\sum F_x = 0$$

$$A_x + B_x - 50 \text{ lb} \cos 10^\circ = 0 \quad \textbf{(c)}$$

$$\sum F_y = 0$$

$$A_y + B_y - 50 \text{ lb} \sin 10^\circ = 0$$

$$B_y = -1191 \text{ lb} + 9 \text{ lb} = -1182 \text{ lb}$$

$$\sum M_B = 0$$

$$A_x(2.5 \text{ ft}) + 50 \text{ lb} \cos 10^\circ(2.5 \text{ ft}) = 0$$

$$A_x = -49 \text{ lb}$$

From Eq. (c),

$$B_x = 50 \text{ lb} \cos 10^\circ - A_x = 98 \text{ lb}$$

From Eq. (a),

$$D_x = 519 \text{ lb} - A_x = 568 \text{ lb}$$

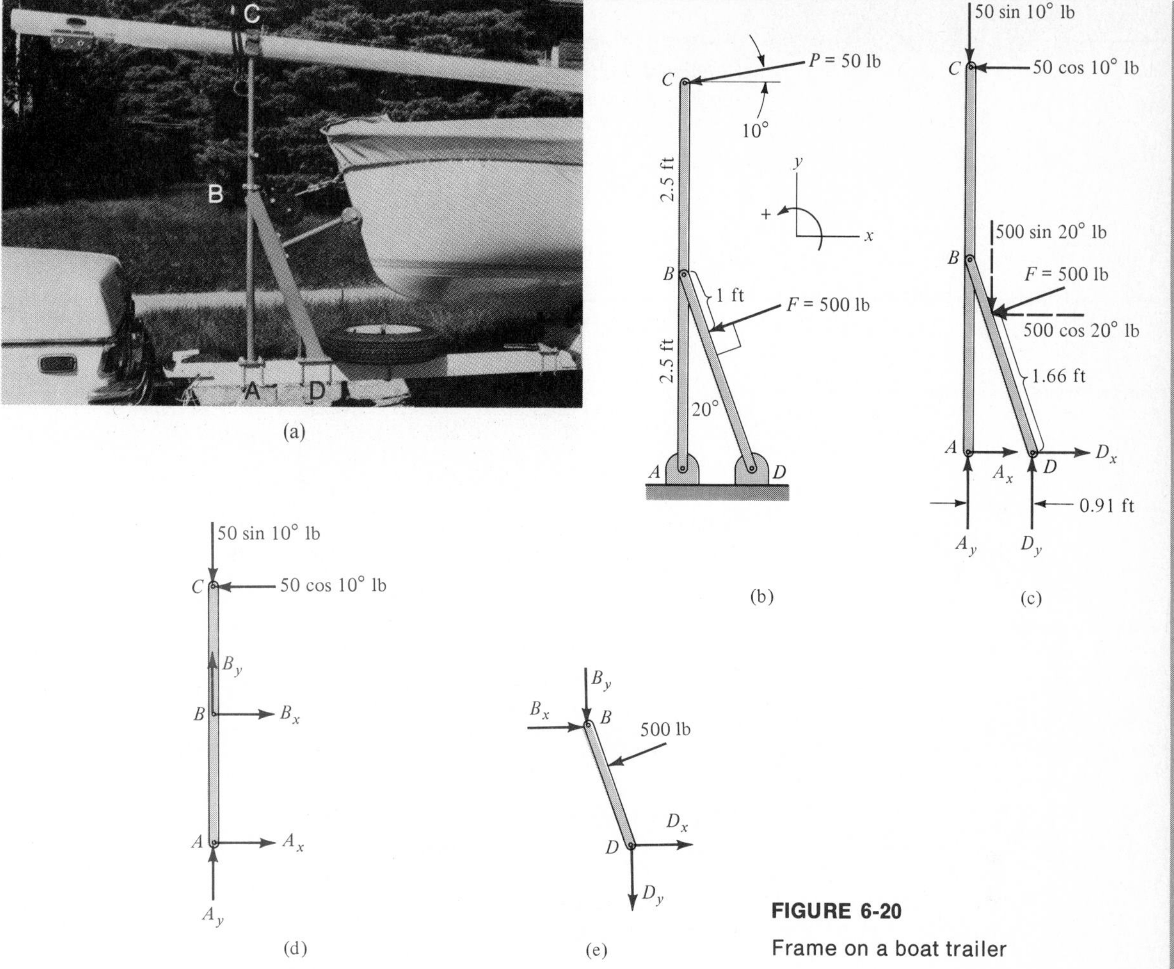

FIGURE 6-20
Frame on a boat trailer

SUMMARY:

$A_x = -49$ lb	$A_y = 1191$ lb
$B_x = 98$ lb	$B_y = -1182$ lb
$D_x = 568$ lb	$D_y = -1011$ lb

JUDGMENT OF THE RESULTS

The solution may be checked using Fig. 6-20e, where each force is known. Taking moments about point B,

$$\sum M_B = D_x(2.5 \text{ ft}) + D_y(0.91 \text{ ft}) - 500 \text{ lb}(1 \text{ ft}) = -0.01 \text{ ft}\cdot\text{lb}$$

The net moment calculated is negligible, so the results are accepted as correct.

EXAMPLE 6-10

The hood support mechanism of a car is shown in Fig. 6-21a, with the details of its extended position given in part b of the figure. (a) First determine the force S in the spring required to hold the position of the linkage. (b) Also calculate the forces at pin A.

SOLUTION

(a) The solution is best started by analyzing the member that has the known force acting on it.

MEMBER *CD*, Fig. 6-21c: This is a two-force member, hence the reaction CD at C is 200 N (by inspection) acting at an angle of 6° with the vertical as shown.

MEMBER *ABC*, Fig. 6-21d (Using a Simplified, Equivalent Shape for the Bar): First the geometry of this member must be determined.

$$|AB|^2 = (0.04\text{ m})^2 + (0.18\text{ m})^2 - 2(0.04\text{ m})(0.18\text{ m})\cos 10^\circ$$

$$|AB| = 0.141\text{ m}$$

$$\frac{\sin\alpha}{0.18\text{ m}} = \frac{\sin 10^\circ}{0.141\text{ m}} = \frac{\sin\beta}{0.04\text{ m}}, \qquad \alpha = 167.2^\circ, \quad \beta = 2.8^\circ$$

$$\gamma = 90^\circ - 70^\circ - 2.8^\circ = 17.2^\circ$$

By summing moments about point A, an equation is obtained involving the spring force, S, as the only unknown.

$$\sum M_A = 0$$

$$S\sin 33^\circ(AB\sin\gamma) - S\cos 33^\circ(AB\cos\gamma)$$
$$+ 200\text{ N}\cos 6^\circ(0.18\text{ m}\cos 20^\circ)$$
$$-200\text{ N}\sin 6^\circ(0.18\text{ m}\sin 20^\circ) = 0$$

$$S\sin 33^\circ(0.141\text{ m}\sin 17.2^\circ)$$
$$-S\cos 33^\circ(0.141\text{ m}\cos 17.2^\circ) + 32.4\text{ N}\cdot\text{m} = 0$$

$$S = \frac{32.4\text{ N}\cdot\text{m}}{0.09\text{ m}} = 360\text{ N}$$

(b)

$$\sum F_x = 0$$

$$200\text{ N}\sin 6^\circ + S\cos 33^\circ - A_x = 0$$

$$A_x = 20.9\text{ N} + 360\text{ N}\cos 33^\circ = 323\text{ N}$$

$$\sum F_y = 0$$

$$-200\text{ N}\cos 6^\circ - S\sin 33^\circ + A_y = 0$$

$$A_y = 199\text{ N} + 360\text{ N}\sin 33^\circ = 395\text{ N}$$

JUDGMENT OF THE RESULTS

The magnitudes and directions of the reaction forces appear reasonable. A check on the results may appear to be possible by making sure the whole system is in equilibrium, but this would not be a bona fide check. Since member *CD* is a two-force member, it merely transmits the 200 N force along line *CD*. Hence, by the principle of transmissibility, analyzing the whole free body is the same as analyzing member *ABC* alone.

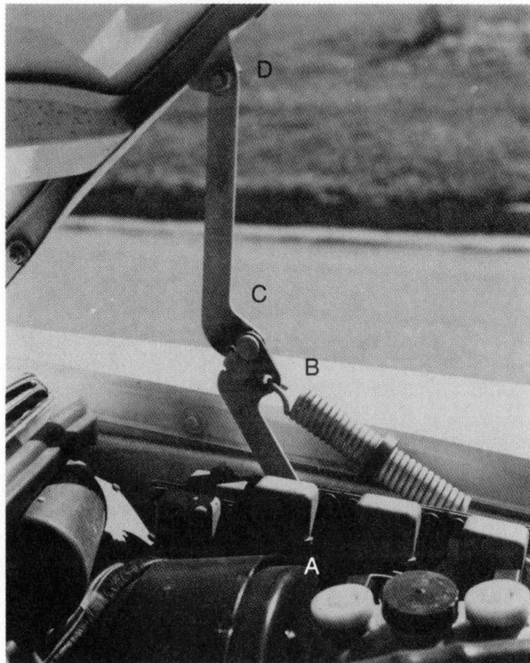

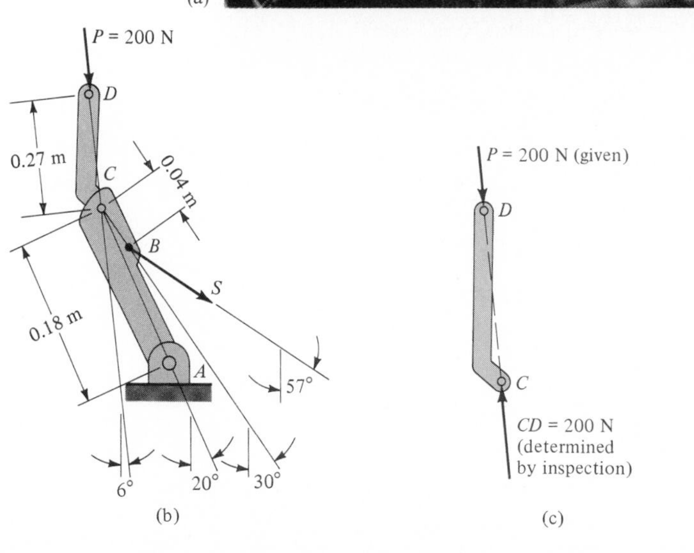

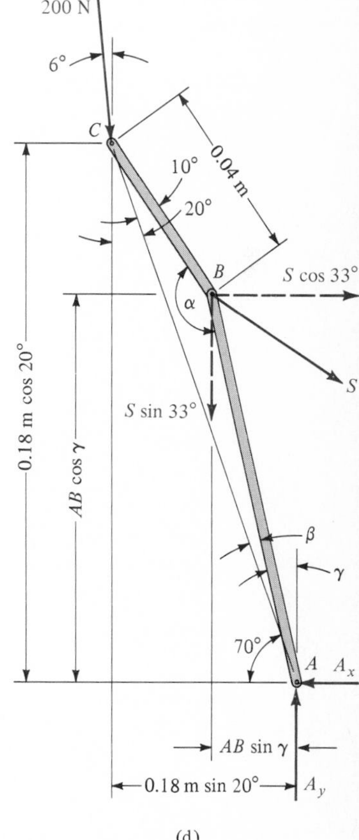

FIGURE 6-21
Hood support mechanism

EXAMPLE 6-11

Determine the forces at points B, C, D, and E in the fireplace tongs shown in Fig. 6-22a and b.

SOLUTION

First a free-body diagram of member ABC is drawn and analyzed.

MEMBER *ABC*, Fig. 6-22c:

$$\sum F_x = 0 \qquad \sum F_y = 0$$

$$B_x + C_x = 0 \qquad 30\text{ lb} + B_y + C_y = 0$$

$$\sum M_B = 0$$

$$-30\text{ lb}(5\text{ in.}) + C_y(7\text{ in.}) - C_x(7\text{ in.}\tan 15^\circ) = 0$$

There are three equations and four unknowns (B_x, B_y, C_x, and C_y); hence no forces may be solved for yet. Member CDE is drawn next (Fig. 6-22d). It is noticed that the total number of unknowns increases by four, so analyzing this member does not help at this time. From symmetry, it is found that free-body diagrams of members $A'BC'$ and $C'DE'$ yield the same equations as for ABC and CDE, thus still no forces can be determined. A free-body diagram of the whole system is then drawn (Fig. 6-22e). It is seen from symmetry considerations that $E_x = E'_x$, but from equilibrium of forces in the x direction $E_x = -E'_x$. Hence, $E_x = 0 = E'_x$. Free-body diagrams of combinations of members should then be considered. Looking at member $ABCDE$ (Fig. 6-22f) it is seen that there are three additional unknowns (D_x, D_y, and E_y), to make a total of seven unknowns and six equations of equilibrium when used in conjunction with Fig. 6-22c. The combination of ABC and $A'BC'$ is then analyzed (Fig. 6-22g), and this turns out to be fruitful.

$$\sum F_x = 0$$

$$C_x + C'_x = 0$$

$$\sum M_{C'} = 0$$

$$30\text{ lb}(12\text{ in.}) - 30\text{ lb}(12\text{ in.}) - C_x(2)(7\text{ in.}\tan 15^\circ) = 0$$

$$C_x = 0, \qquad C'_x = 0$$

Now member ABC may be solved (Fig. 6-22c),

$$\sum F_x = 0 \qquad \sum F_y = 0$$

$$B_x = 0 \qquad 30\text{ lb} + B_y + C_y = 0$$

$$\sum M_B = 0$$

$$C_y(7\text{ in.}) - 30\text{ lb}(5\text{ in.}) = 0$$

$$C_y = \frac{150\text{ lb}\cdot\text{in.}}{7\text{ in.}} = 21.4\text{ lb}$$

$$B_y = -C_y - 30\text{ lb} = -21.4\text{ lb} - 30\text{ lb} = -51.4\text{ lb}$$

MEMBER *CDE*, Fig. 6-22d:

$$\sum F_x = 0 \qquad \sum F_y = 0$$

$$D_x = 0 \qquad -C_y + D_y - E_y = 0$$

$$\sum M_D = 0$$

$$C_y(7 \text{ in.}) - E_y(7 \text{ in.}) = 0$$

$$E_y = C_y = 21.4 \text{ lb}$$

$$D_y = E_y + C_y = 21.4 \text{ lb} + 21.4 \text{ lb} = 42.8 \text{ lb}$$

SUMMARY:

$$B_x = 0 \qquad B_y = 51.4 \text{ lb}$$

$$C_x = 0 \qquad C_y = 21.4 \text{ lb}$$

$$D_x = 0 \qquad D_y = 42.8 \text{ lb}$$

$$E_x = 0 \qquad E_y = 21.4 \text{ lb}$$

FINAL COMMENTS: Persistence pays off! It was known that this problem was solvable since there were 12 unknowns (two unknowns at each of *B*, *C*, *C*′, *D*, *E*, and *E*′) and there were 12 equations available (three equations for each member). The magnitudes of the forces appear reasonable from an intuitive appeal.

FIGURE 6-22

Fireplace tongs

(a)

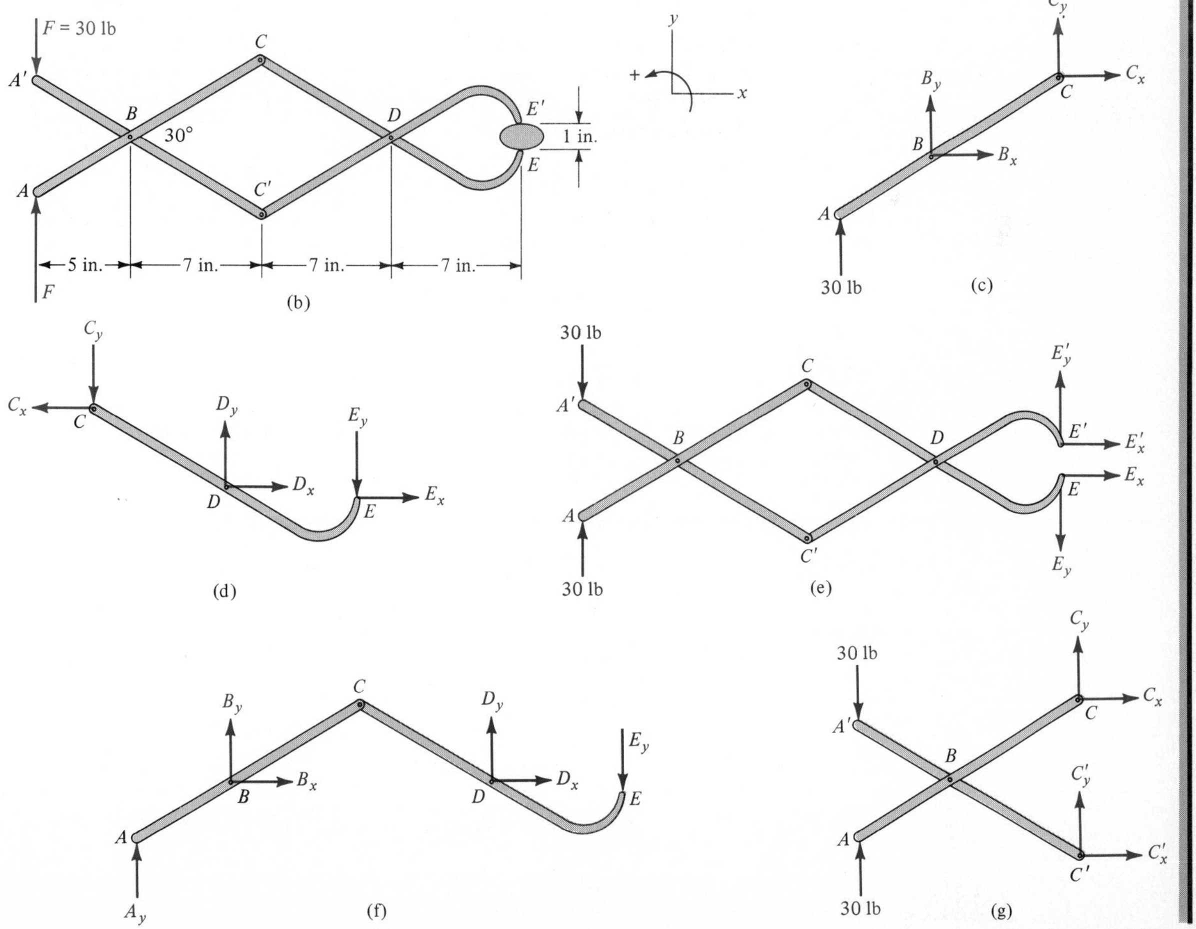

(b)

(c)

(d)

(e)

(f)

(g)

PROBLEMS

SECTION 6-8

FIGURE P6-62

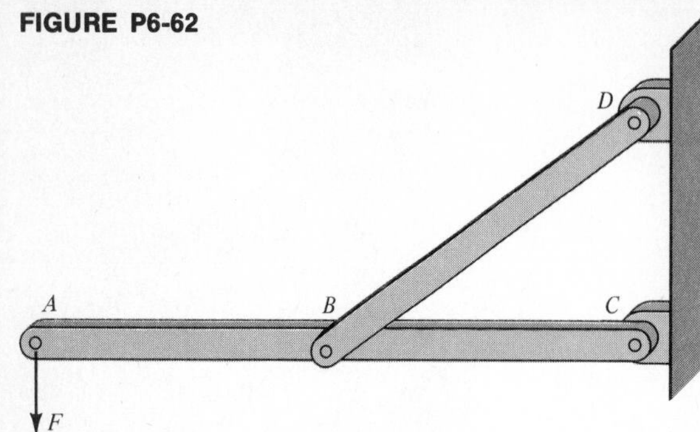

6–62 The rigid boom AC of a simple crane is pinned at B and C. Determine the forces at B, C, and D if $F = 3$ kN, $AB = BC = 1.5$ m, and $CD = 1$ m.

6–63 In Fig. P6-62, $F = 1200$ lb, $AB = 6$ ft, $BC = 5$ ft, and $CD = 4$ ft. The weight of the frame is represented by a force $P = 200$ lb acting at B. Calculate the forces acting at B, C, and D.

FIGURE P6-64

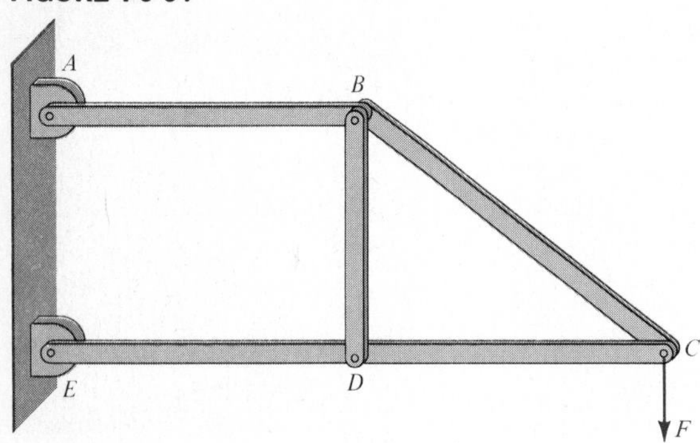

6–64 The rigid boom EC of the crane is supported by three other bars as shown. Determine the forces acting at A, B, and E if $F = 5$ kN, $ED = DC = 1.3$ m, and $AE = 0.8$ m.

6–65 In Fig. P6-64, $F = 1500$ lb, $ED = 5$ ft, $DC = 7$ ft, and $BD = 4.5$ ft. The weight of the structure is represented by a force $W = 250$ lb acting at D. Calculate the forces in members BC and BD.

FIGURE P6-66

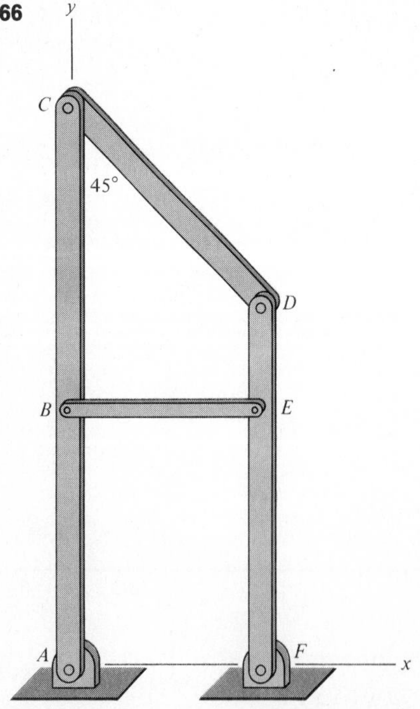

6–66 A force $\mathbf{F} = 6$ kN $\mathbf{i}$ is acting at joint C of the structure. Determine the forces at B and E if $AB = BE = 2$ m, and $AC = 4$ m.

6–67 Consider Fig. P6-66 with a force $\mathbf{F}_C = 800$ lb $\mathbf{i}$ at C, and a force $\mathbf{F}_B = (-300\mathbf{i} - 350\mathbf{j})$ lb at B. Calculate the forces at joints D, E, and F if $AF = 4$ ft, $AB = 6$ ft, and $AC = 15$ ft.

6–68 The pontoon struts of the hydroplane in Fig. P6-68a are modeled in part b of the figure. The small member DE serves as a step to the cockpit and gives no extra stiffness to the structure. Determine the forces in the rigid bars BF and CF if $P = 1$ kN, $AB = 1.2$ m, $BC = 1$ m, and $EF = 0.6$ m. Bar AB carries no load horizontally.

6–69 Solve Prob. 6–68 after replacing bar BF by a member BE. Let $P = 200$ lb, $AB = 5$ ft, $BC = 4.5$ ft, and $EF = 2.5$ ft.

FIGURE P6-68

(a)

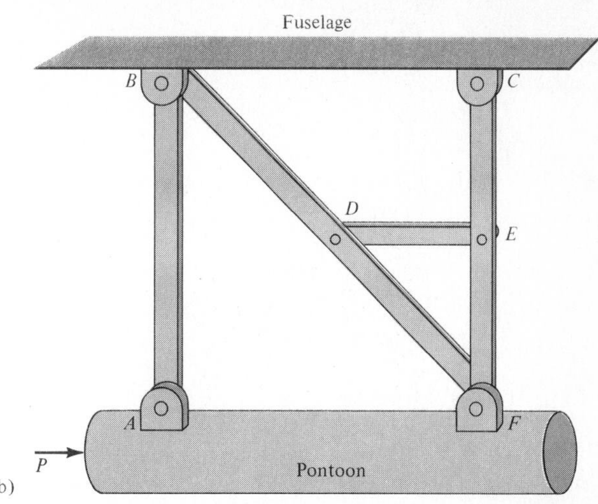

(b)

6–70 The hinged frame on the utility pole in Fig. P6-70 appears simple to analyze, but it requires a lot of work. Write down all the equations that are necessary to determine the force at pin A. Discuss the cause of any major difficulty if the solution were attempted.

FIGURE P6-70

(a)

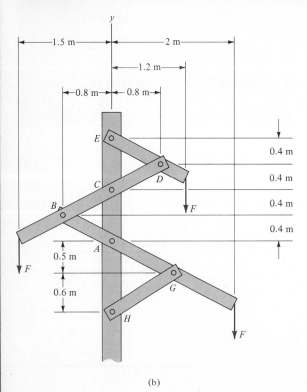

(b)

6–71 Consider the model of an ironing board. The two legs are pinned to the board at A and B and stand on a smooth, frictionless floor. $EC = DC = 32$ in., $CA = CB = 16$ in., and $ED = 37$ in. The weight of the board is $W = 15$ lb, assumed concentrated directly above joint C. Determine the forces acting at A, C, and D.

6–72 Solve Prob. 6–71 assuming that an additional downward force $F = 10$ lb is acting on the board directly above point E.

6–73 A ladder is modeled as a two-dimensional structure standing on a smooth, frictionless floor. Calculate the forces acting on the rigid leg CF if a force $W = 200$ lb is applied to rung 5.

6–74 Solve Prob. 6–73 with W acting on rung 3, and an additional downward force $P = 50$ lb acting at joint B.

6–75 A tripod supports a load of 30 N that acts at the center of the vertical pole. The legs are 120° apart. A typical leg AC is 40 cm long, and BC is 10 cm. The distance CD is also 10 cm. Pins C and D are both 2 cm from the vertical centerline of the tripod. Determine the forces in members AC and BD assuming a smooth, frictionless floor.

6–76 Consider the tripod described in Prob. 6–75, and assume that the force W may act on an arm at a distance $d = 20$ cm away from the vertical centerline, directly over the bar AC. Calculate the forces acting on leg AC.

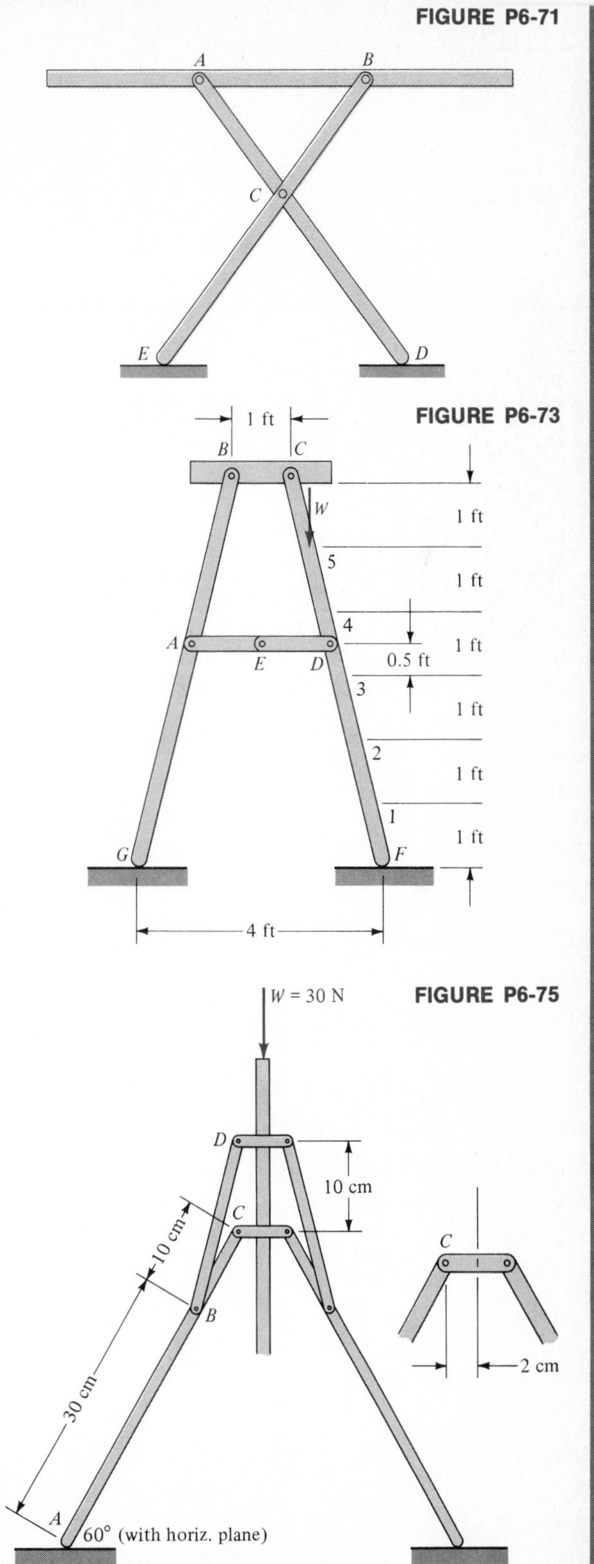

6–77 An adjustable lamp arm has rigid bars *AJ*, *BC*, *CI*, *JD*, *DH*, *IE*, *EG*, and *HF*. A typical element $DH = 6$ in. is pinned at its center. Determine the forces acting at joints *B*, *D*, and *F* if $\theta = 90°$.

FIGURE P6-77

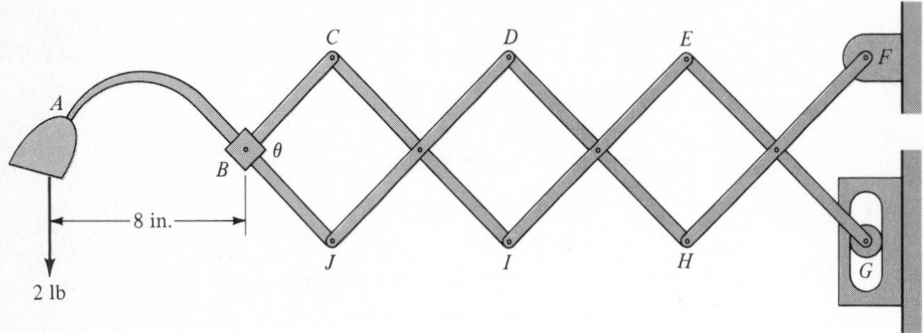

6–78 Solve Prob. 6–77 after adding three more forces, 0.5 lb acting at *H*, *I*, and *J* each.

6–79 Determine the force applied by the bolt cutter to the bolt at *A*.

FIGURE P6-79

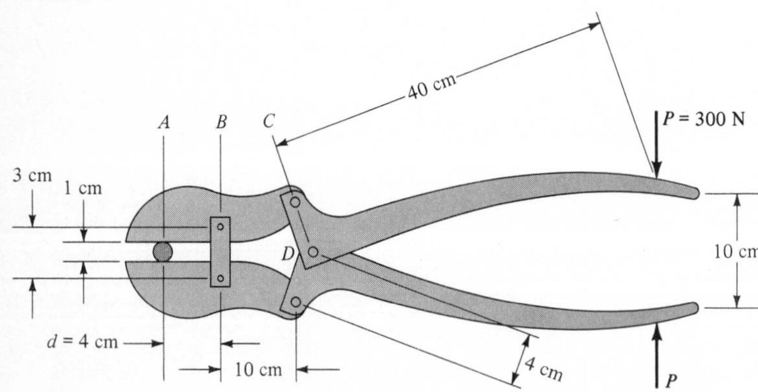

6–80 It is assumed that the forces in the bolt cutter of Fig. P6-79 depend on where the bolt is placed in the jaws. Plot the force acting at pin *D* as a function of $d = 3$ cm, 4 cm, and 5 cm.

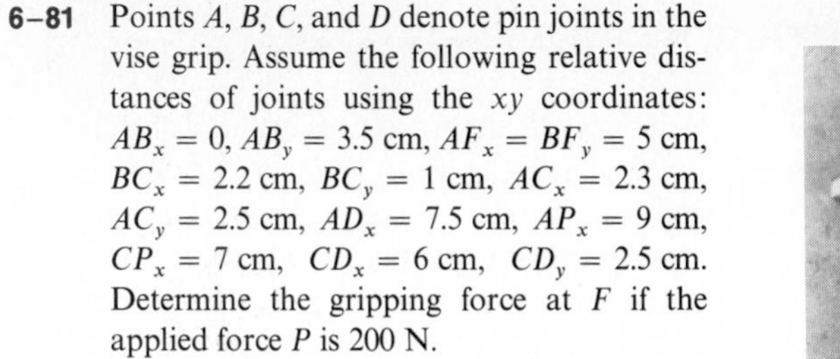

6–81 Points A, B, C, and D denote pin joints in the vise grip. Assume the following relative distances of joints using the xy coordinates: $AB_x = 0$, $AB_y = 3.5$ cm, $AF_x = BF_y = 5$ cm, $BC_x = 2.2$ cm, $BC_y = 1$ cm, $AC_x = 2.3$ cm, $AC_y = 2.5$ cm, $AD_x = 7.5$ cm, $AP_x = 9$ cm, $CP_x = 7$ cm, $CD_x = 6$ cm, $CD_y = 2.5$ cm. Determine the gripping force at F if the applied force P is 200 N.

FIGURE P6-81

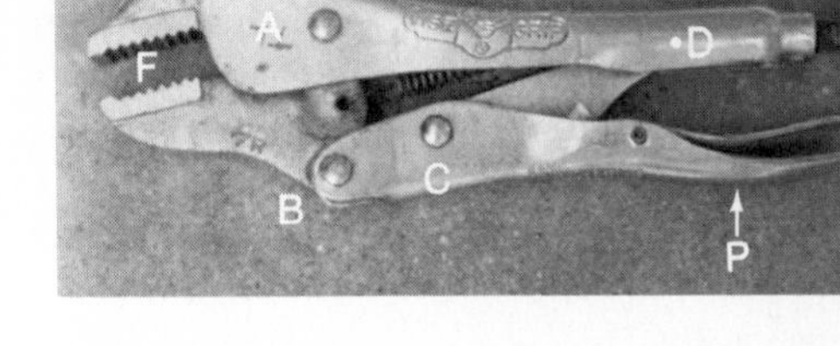

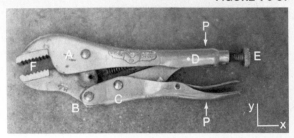

6–82 Discuss the change in the force at F in Prob. 6–81 that would be caused by shortening the distance AD (this is done using the screw E).

6–83 Determine the forces acting on member DF of the front loader. Members BC and EG are hydraulic actuators which can be treated as two-force members of adjustable length. Their lengths are constant in Fig. P6-83b for maintaining the given configuration.

6–84 Calculate the forces acting on member AI in Prob. 6–83.

FIGURE P6-83

(a)

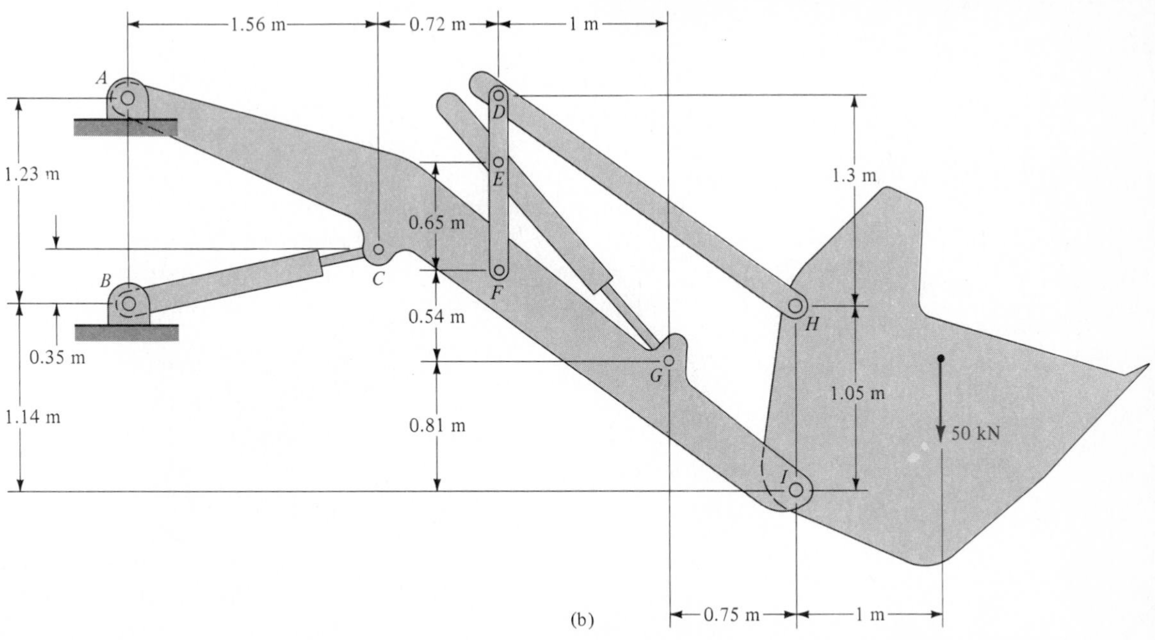

(b)

6-9 CABLES WITH CONCENTRATED LOADS

Cables are commonly used in certain structures and machines such as suspension bridges, cranes, transmission lines, and antenna towers. In the simplest situations cables can be analyzed as two-force members. It is only slightly more complicated to analyze a cable with concentrated loads acting on it. For example, the suspension cable in Fig. 6-23a can be modeled as in part b of the figure. Here the weight of the upper cable can be neglected in comparison with the large loads applied to it by the vertical cables. The more complex problem of including distributed loads on a cable (its own weight, ice, or wind) is discussed in Ch. 7.

Another assumption made in the analysis is that the cable is flexible and consequently in pure axial tension between neighboring points where the loads are applied. This way the cable is treated as a set of two-force members. The only unknown forces are at the external supports and the tensions in the individual segments of the cable. The solution for these unknowns is straightforward if the loads and the geometry are known, but it must be remembered that the cable is not a rigid body, so changes in the loading would change its shape.

The method of solution is illustrated using Fig. 6-23b, where the loading and the geometry are symmetric with respect to D ($P_B = P_F$, $y_B = y_F$, etc.). First, the free-body diagram of the whole cable is drawn in Fig. 6-23c. The directions of the unknown reactions at A and G can be correctly drawn at once since there is only tension in the cable. From symmetry, $A_y = G_y = W/2$ where $W = P_B + P_C + P_D + P_E + P_F$. A_x and G_x are equal in magnitude by inspection, but they cannot be calculated from Fig. 6-23c. It is useful to draw another diagram such as that in Fig. 6-23d. If the coordinates of A, B, and Q are known, A_x can be calculated by taking moments about point Q, writing $\sum M_Q = 0$.

With the external reactions A_x and A_y known, the tension in each segment of the cable can be calculated using the force summations $\sum F_x = 0$ or $\sum F_y = 0$ for the appropriate portion of the cable. For example, T_{CD} is determined from Fig. 6-23e using $\sum F_x = 0$,

$$T_{CD} = \frac{A_x}{\cos \theta_{CD}}$$

Since A_x is a constant, the horizontal component of the tension is also constant for the entire cable, so

$$T_{AB} \cos \theta_{AB} = T_{BC} \cos \theta_{BC} = \dots$$

Another useful observation from these results is that the tension in the cable is largest where $\cos \theta$ is smallest. This has to be at a support such as A and G in Fig. 6-23b.

(a)

FIGURE 6-23

Suspension cable with concentrated loads

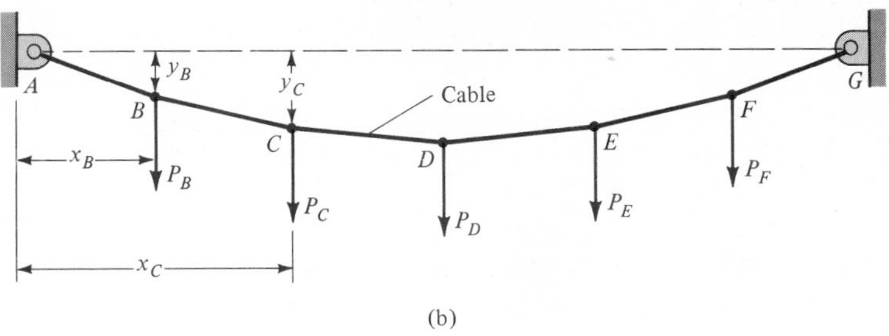

(b)

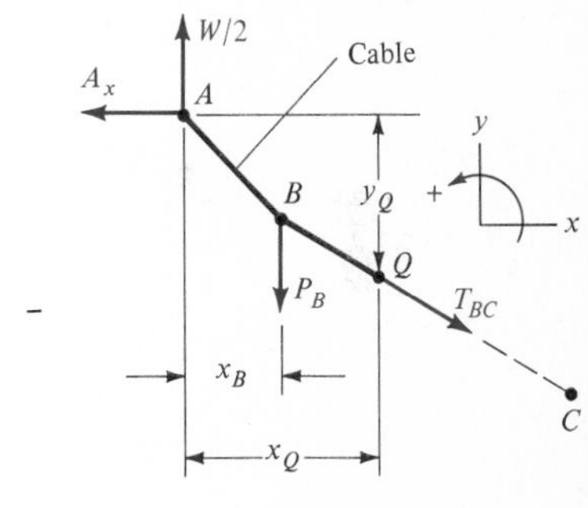

(d)

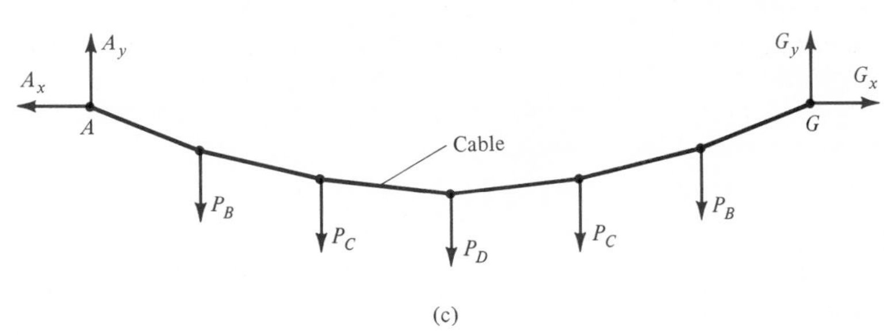

(c)

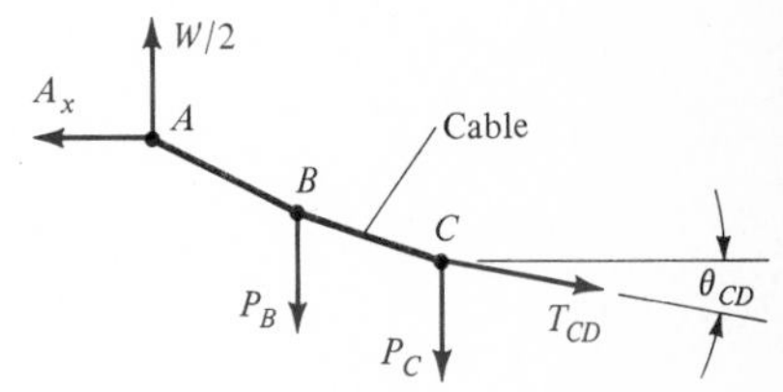

(e)

EXAMPLE 6-12

A cable supports three loads as shown in Fig. 6-24a. What is the required maximum tension in the cable if point C should be 2 m below point A? What are the coordinates y_B and y_D in this configuration?

SOLUTION

First a free-body diagram of the whole cable is drawn (Fig. 6-24b) and the y components of the support reactions are determined.

$$\sum F_x = 0 \qquad A_x = E_x \tag{a}$$

$$\sum F_y = 0$$

$$A_y - 2\text{ kN} - 3\text{ kN} - 1\text{ kN} + E_y = 0$$

$$\sum M_E = 0$$

$$-A_y(43\text{ m}) + 2\text{ kN}(33\text{ m}) + 3\text{ kN}(21\text{ m}) + 1\text{ kN}(10\text{ m}) = 0$$

$$A_y = \frac{(66 + 63 + 10)\text{ kN}\cdot\text{m}}{43\text{ m}} = 3.23\text{ kN}$$

$$E_y = 6\text{ kN} - A_y = 6\text{ kN} - 3.23\text{ kN} = 2.77\text{ kN}$$

The x components of the support reactions are obtained by considering a section of the cable and summing moments about a point that has known coordinates (point C in this case). From Fig. 6-24c,

$$\sum M_C = 0$$

$$A_x(2\text{ m}) - 3.23\text{ kN}(22\text{ m}) + 2\text{ kN}(12\text{ m}) = 0$$

$$A_x = \frac{(71.1 - 24)\text{ kN}\cdot\text{m}}{2\text{ m}} = 23.5\text{ kN}$$

From Eq. (a), $E_x = A_x = 23.5$ kN. The largest tension in a cable occurs where the angle between the horizontal and the cable is largest (where $\cos\theta$ is smallest since $T_{\text{segment}} = A_x/\cos\theta$). For this case segments AB and DE must be considered.

SEGMENT *AB*: Since AB is considered a two-force member, the force $\mathbf{A} = -A_x\mathbf{i} + A_y\mathbf{j}$ must be collinear with cable segment AB. Therefore, θ_{AB} equals the angle that $\mathbf{A}$ makes with the negative horizontal axis (Fig. 6-24d).

$$\tan\theta_{AB} = \frac{A_y}{A_x} = \frac{3.23\text{ kN}}{23.5\text{ kN}}, \qquad \theta_{AB} = 7.8^\circ$$

$$T_{AB} = A = \frac{A_x}{\cos\theta_{AB}} = 23.7\text{ kN}$$

Also,

$$\frac{y_B}{10\text{ m}} = \tan\theta_{AB} = 0.137, \qquad y_B = (10\text{ m})(0.137) = 1.37\text{ m}$$

SEGMENT *DE*: By analogy to segment *AB*,

$$\tan \theta_{DE} = \frac{E_y}{E_x} = \frac{2.77 \text{ kN}}{23.5 \text{ kN}}, \qquad \theta_{DE} = 6.7^\circ$$

Since θ_{DE} is smaller than θ_{AB}, the tension in segment *AB* is the largest. Also,

$$\frac{y_D}{10 \text{ m}} = \tan \theta_{DE} = 0.118, \quad y_D = 10 \text{ m}(0.118 \text{ m}) = 1.18 \text{ m}$$

JUDGMENT OF THE RESULTS

The largest tension occurred in the expected segment. Since the largest loads were applied closer to the left end it was expected that this portion of the cable should make the largest angle with the horizontal and have the largest tension. The large tensions caused by relatively small loads also seem reasonable from an intuitive standpoint (the smaller the sag, the larger the tensions).

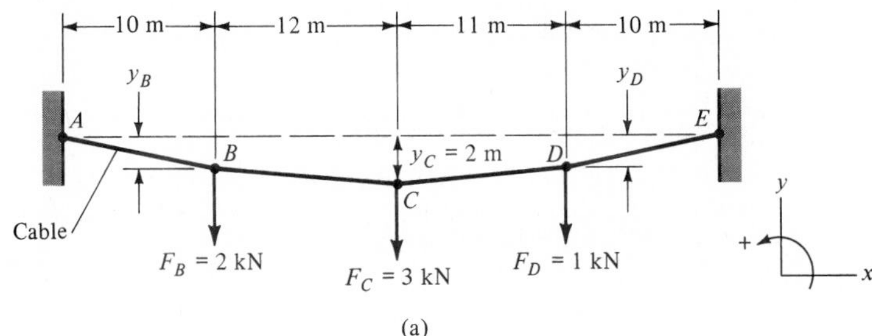

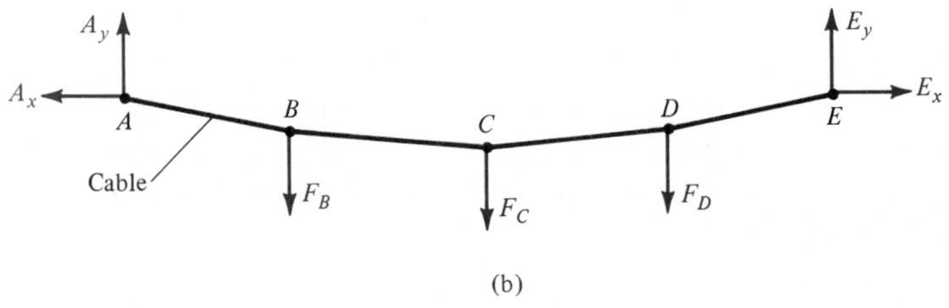

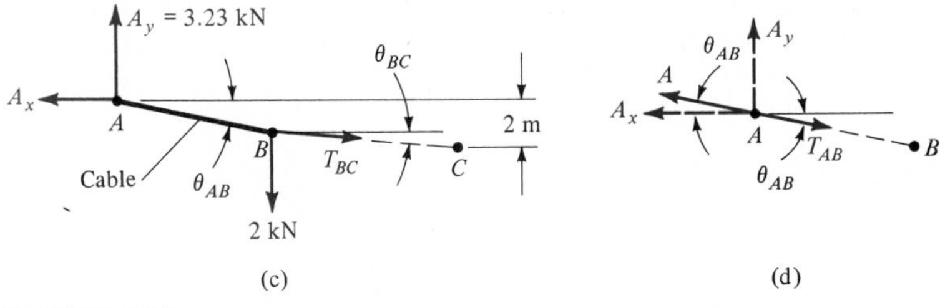

FIGURE 6-24

Suspension cable with three loads supported at the same elevation

EXAMPLE 6-13

A cable supports three loads as shown in Fig. 6-25a. Determine the tension in each segment of the cable and the coordinates y_B and y_D of points B and D.

SOLUTION

Considering the whole cable as a free body, equations involving the support reaction components are obtained. From Fig. 6-25b,

$$\sum F_x = 0$$

$$A_x = E_x \tag{a}$$

$$\sum F_y = 0$$

$$A_y + E_y - 800\text{ lb} - 500\text{ lb} - 400\text{ lb} = 0$$

$$A_y + E_y = 1.7\text{ kips} \tag{b}$$

$$\sum M_E = 0$$

$$-A_x(20\text{ ft}) - A_y(78\text{ ft}) + 800\text{ lb}(53\text{ ft}) + 500\text{ lb}(33\text{ ft}) + 400\text{ lb}(15\text{ ft}) = 0$$

$$20A_x + 78A_y = 64.9\text{ kips} \tag{c}$$

Three equations involving four unknowns are obtained since none of the x or y components of the support reactions are collinear. Therefore, a section of the cable involving at most two additional unknowns must be analyzed. Cutting the cable through section BC only two more unknowns are involved, the tension T_{BC} of unknown direction and magnitude (Fig. 6-25c).

$$M_C = 0$$

$$-A_y(45\text{ ft}) + A_x(2\text{ ft}) + 800\text{ lb}(20\text{ ft}) = 0$$

$$-2A_x + 45A_y = 16\text{ kips} \tag{d}$$

Solving Eqs. (c) and (d) simultaneously,

$$A_x = 1590\text{ lb}$$

$$A_y = 426\text{ lb}$$

From Eqs. (a) and (b),

$$E_x = 1590\text{ lb}$$

$$E_y = 1270\text{ lb}$$

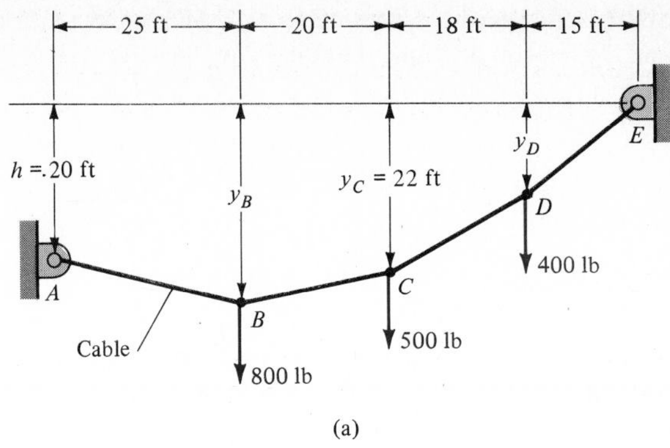

(a)

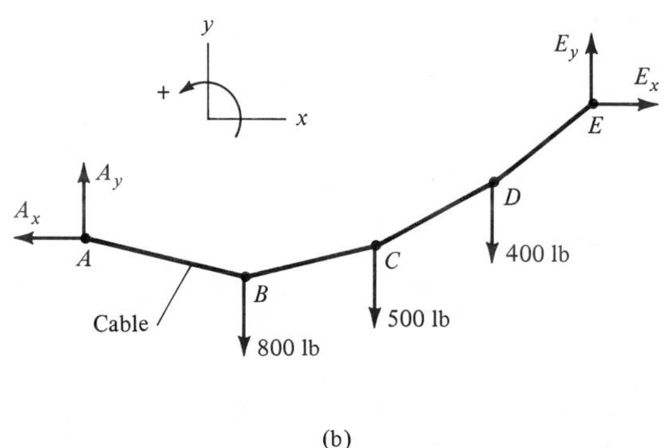

(b)

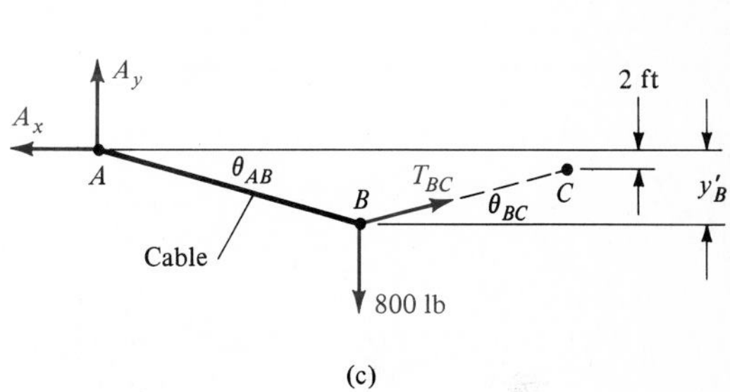

(c)

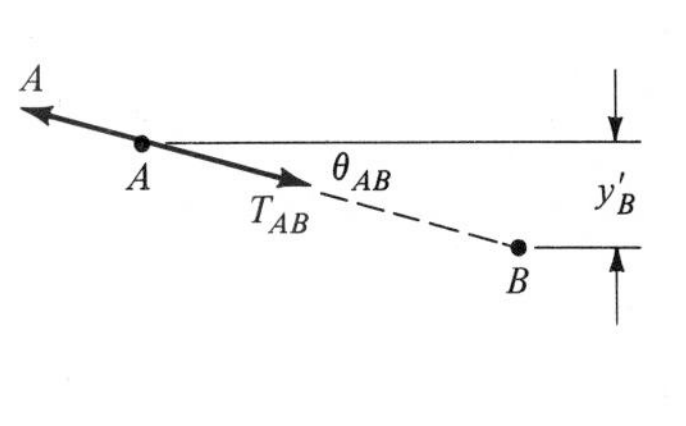

(d)

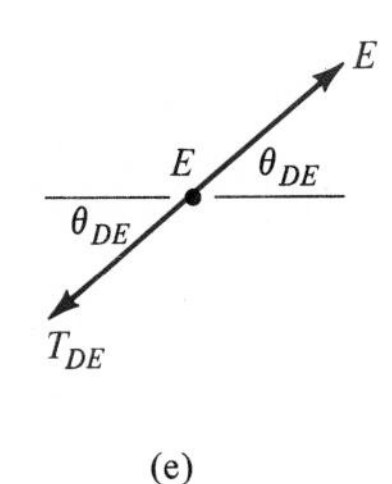

(e)

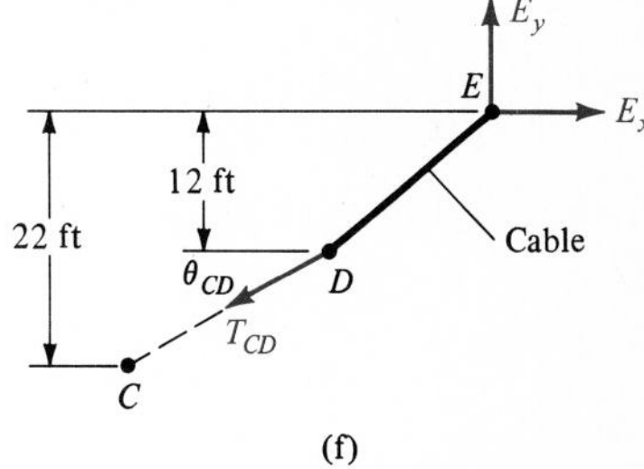

(f)

FIGURE 6-25

Suspension cable with three loads supported at different elevations

Since each segment is considered a two-force member, the reaction force A and the tension T_{AB} must be equal and opposite (Fig. 6-25d).

$$T_{AB} = (A_x^2 + A_y^2)^{1/2} = 1650 \text{ lb}$$

$$\theta_{AB} = \tan^{-1}\frac{A_y}{A_x} = \tan^{-1}\left(\frac{426}{1590}\right) = 15^\circ$$

$$\tan\theta_{AB} = \frac{y'_B}{25 \text{ ft}} = 0.27, \quad y'_B = (25 \text{ ft})(0.27) = 6.7 \text{ ft (Fig. 6-25c)}$$

$$y_B = 20 \text{ ft} + 6.7 \text{ ft} = 26.7 \text{ ft}$$

For the tension in segment BC, Fig. 6-25c:

$$\tan\theta_{BC} = \frac{y'_B - 2 \text{ ft}}{20 \text{ ft}} = \frac{6.7 \text{ ft} - 2 \text{ ft}}{20 \text{ ft}} = 0.24, \qquad \theta_{BC} = 13.2^\circ$$

$$\sum F_x = 0$$

$$T_{BC}\cos\theta_{BC} - A_x = 0$$

$$T_{BC} = \frac{A_x}{\cos\theta_{BC}} = \frac{1590 \text{ lb}}{\cos 13.2^\circ} = 1630 \text{ lb}$$

For the tension in segment DE, Fig. 6-25e:

$$T_{DE} = (E_x^2 + E_y^2)^{1/2} = 2030 \text{ lb}$$

$$\theta_{DE} = \tan^{-1}\frac{E_y}{E_x} = \tan^{-1}\left(\frac{1270}{1590}\right) = 38.6^\circ$$

$$\frac{y_D}{15 \text{ ft}} = \tan\theta_{DE} = 0.8, \qquad y_D = 15 \text{ ft}(0.8) = 12 \text{ ft}$$

For the tension in segment CD, Fig. 6-25f:

$$\tan\theta_{CD} = \frac{22 \text{ ft} - 12 \text{ ft}}{18 \text{ ft}} = 0.56, \qquad \theta_{CD} = 29.1$$

$$\sum F_x = 0$$

$$-T_{CD}\cos\theta_{CD} + E_x = 0$$

$$T_{CD} = \frac{E_x}{\cos\theta_{CD}} = \frac{1590 \text{ lb}}{\cos 29.1^\circ} = 1820 \text{ lb}$$

JUDGMENT OF THE RESULTS

The largest tension (T_{DE}) occurred in a segment connected to a support as it should be. It should be expected that T_{DE} be greater than T_{AB} since the angle that DE makes with the horizontal is much larger than the angle of AB. The magnitudes of the other tensions appear reasonable for the given loading situation.

PROBLEMS

6–85 Determine the tension in each segment of the cable if $F_B = F_C = 1$ kN, $d = 12$ m, and $y_B = 1.5$ m.

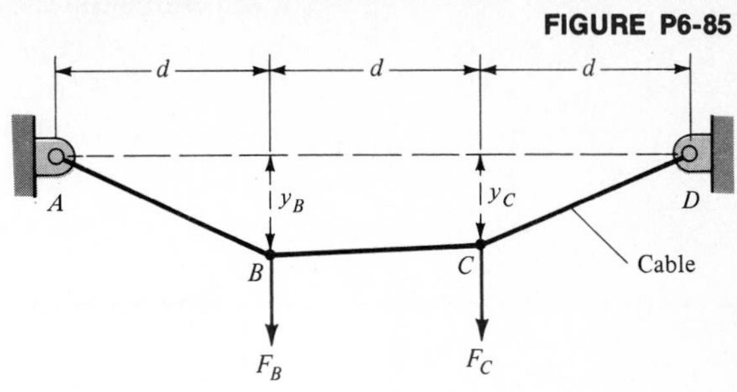

FIGURE P6-85

6–86 In Fig. P6-85, $F_C = 2F_B = 400$ lb, $d = 15$ ft, and $y_C = 3$ ft. Calculate y_B and the tension in each segment of the cable.

6–87 Calculate the tension in each segment of the cable if $F_C = 800$ N, $F_B = F_D = 600$ N, $d = 15$ m, and $y_C = 2$ m.

6–88 In Fig. P6-87, $F_B = 300$ lb, $F_C = 100$ lb, $F_D = 50$ lb, $d = 10$ ft, and $y_B = 4$ ft. Calculate y_C and y_D.

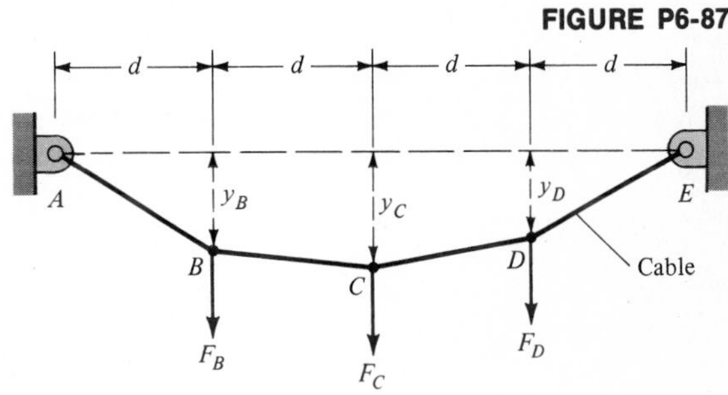

FIGURE P6-87

6–89 In Fig. P6-87, $F_B = F_C = F_D = F$. Plot the maximum tension in the cable as a function of $y_C = 0.1d$, $0.2d$, and $0.3d$.

6–90 The forces acting on the cable are $F_B = 500$ N, $F_C = 600$ N, $F_D = 900$ N. Calculate the maximum tension in the cable if $a = b = c = d = 5$ m, $y_D = 10$ m, and $y_E = 8$ m.

6–91 In Fig. P6-90, $F_B = F_C = F_D = 200$ lb. Determine the tension in each segment of the cable if $a = 10$ ft, $b = 14$ ft, $c = 16$ ft, $d = 15$ ft, $y_C = 17$ ft, and $y_E = 13$ ft.

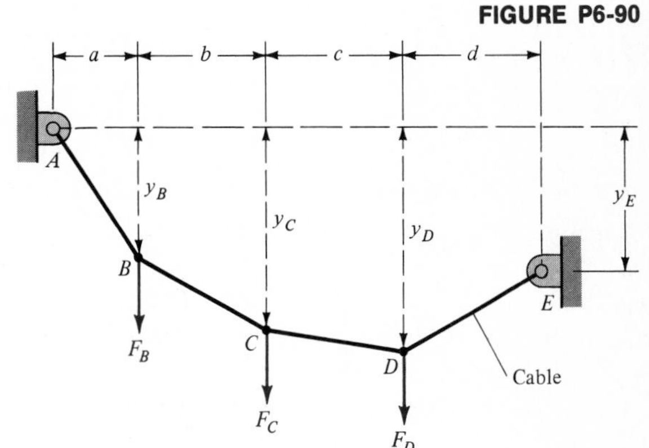

FIGURE P6-90

6–92 In Fig. P6-90, $F_B = 800$ N, $F_C = 700$ N, and $F_D = 500$ N; $a = 3$ m, $b = 5$ m, $c = 6$ m, $d = 5.5$ m, $y_B = 4$ m, and $y_E = 7$ m. Calculate y_C and y_D.

6–93 In Fig. P6-90, $F_C = 3F_B = 4F_D = 360$ lb; $a = 12$ ft, $b = 15$ ft, $c = 20$ ft, $d = 17$ ft, $y_C = 23$ ft, and $y_E = 18$ ft. Determine the tensions in BC and CD.

SUMMARY

Section 6-1

TRUSSES: Assume that each member of a truss is a two-force member connected at frictionless pins at its ends. The members are assumed to be weightless, with forces applied at some of the joints. The basic element of a simple, rigid truss is the triangle formed by three bars.

Section 6-2

METHOD OF JOINTS: Two equilibrium equations may be written for each joint of a two-dimensional truss:

$$\sum F_x = 0 \qquad \sum F_y = 0$$

PROCEDURE FOR ANALYSIS:

1. Determine the support reactions of the truss.
2. Start the detailed analysis at a joint where only one or two unknown forces act. Draw the free-body diagram of this joint, assuming that the unknown forces are tensions (arrows directed away from the joint).
3. Draw free-body diagrams for the other joints to be analyzed, using Newton's Third Law (action and reaction) consistently with respect to the first diagram.
4. Write the equilibrium equations for the forces acting at the joints and solve them. Attempt to progress from joint to joint in such a way that each equation contains only one unknown (to simplify the calculations). Positive answers indicate that the assumed directions of the unknown forces were correct, and vice versa.

Section 6-3

TWO MEMBERS AT A JOINT:

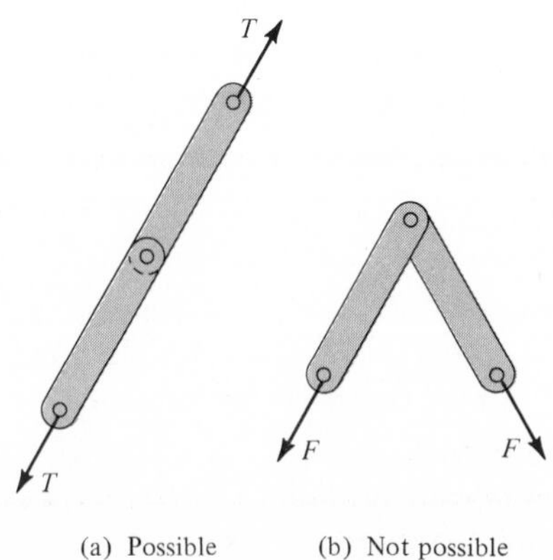

(a) Possible (b) Not possible unless $F = 0$

THREE MEMBERS AT A JOINT:

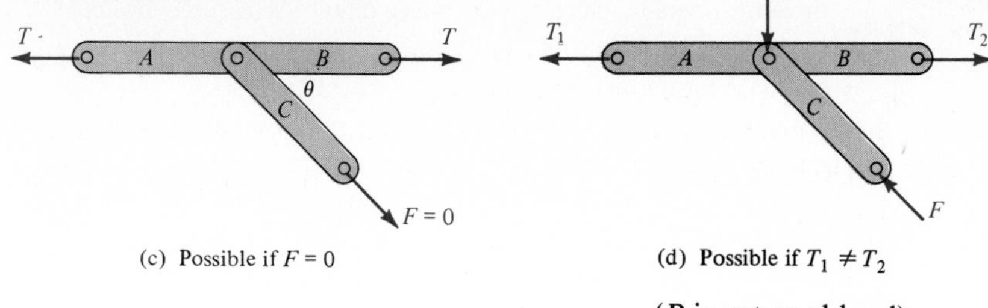

(c) Possible if $F = 0$ (d) Possible if $T_1 \neq T_2$

(P is external load)

FOUR MEMBERS: Collinear in pairs with equal and opposite forces in pairs:

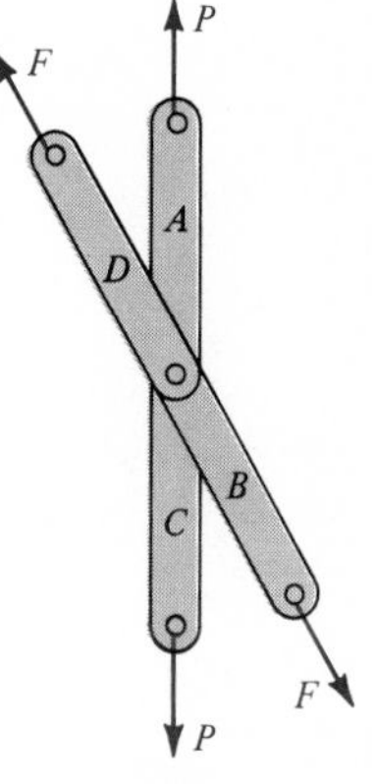

(e) Possible

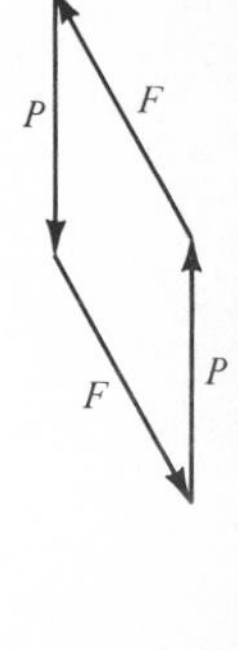

(f)

Section 6-4

FORMAL PROCEDURE USING BOW'S NOTATION: For examples of lettering regions of a truss, its members, and external or internal forces, see Figs. 6-7 and 6-8.

Section 6-5

PROCEDURE FOR GRAPHICAL CONSTRUCTION OF TRUSSES:

1. Determine the support reactions for the truss.
2. Start the graphical construction at any joint where the unknown forces can be determined from a force polygon.
3. Add force triangles using the known lines of action at neighboring joints.
4. Each part of the diagram should have closed polygons showing equilibrium.
5. The magnitude of each force is measured from the length of the corresponding line, based on the scale established in the beginning for the known forces.

Section 6-6

METHOD OF SECTIONS: This method is most valuable when only a few forces must be determined in a complex truss. An imaginary cut is made through the members of interest. Three independent equations of equilibrium may be written for each portion of a plane truss (if the part has two or more joints in it):

$$\sum F_x = 0 \qquad \sum F_y = 0 \qquad \sum M_A = 0$$

where A is an arbitrary point of reference.

PROCEDURE FOR ANALYSIS

1. Determine the support reactions if the section used in the analysis includes the joints supported.

2. Make the cut through the members of interest, preferably to cut only three members in which the forces are unknown (assume tensions).
3. Write the equations of equilibrium. Choose a convenient point of reference for moments to simplify the calculations.
4. Solve the equations. If necessary, use more than one cut in the vicinity of interest to allow writing more equilibrium equations. Positive answers indicate that the assumed directions of the unknown forces were correct, and vice versa.

Section 6-7

SPACE TRUSSES: For analyzing a three-dimensional truss, six scalar equations of equilibrium are available when using the method of sections:

$$\sum F_x = 0 \qquad \sum F_y = 0 \qquad \sum F_z = 0$$
$$\sum M_x = 0 \qquad \sum M_y = 0 \qquad \sum M_z = 0$$

Only the three force summations are available when using the method of joints.

Section 6-8

FRAMES AND MACHINES: In general, a member may have two or more forces acting on it. Three equations of equilibrium are available for each member or combination of members in two-dimensional loading:

$$\sum F_x = 0 \qquad \sum F_y = 0 \qquad \sum M_A = 0$$

where A is an arbitrary point of reference.

PROCEDURE FOR ANALYSIS

1. Determine the support reactions if necessary.
2. Draw the free-body diagram of the first member assuming that the unknown forces are tensions.
3. Draw the free-body diagrams of the other members using Newton's Third Law (action and reaction) consistently with respect to the first diagram.
4. Write the equilibrium equations for the members or combinations of members and solve them. Positive answers indicate that the assumed directions of the forces were correct, and vice versa.

Section 6-9

CABLES WITH CONCENTRATED LOADS: These are to be analyzed as two-force members for which the method of joints is applicable. Note that cables are not rigid, so any change in the loading changes the configuration of a cable. The horizontal component of the tension in each segment of a given cable is a constant.

6–94 Calculate the forces acting at joints G and H in the truss.

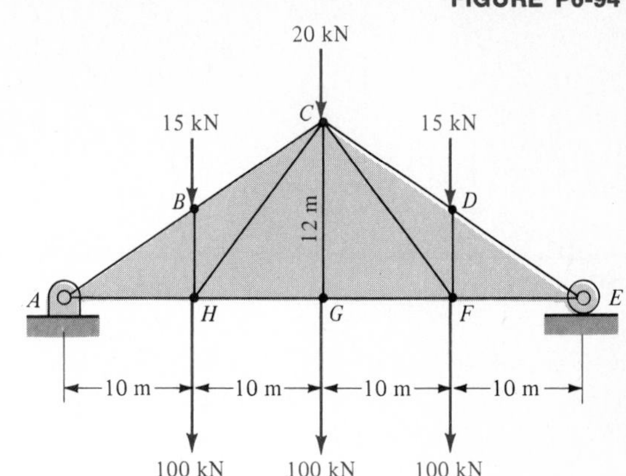

FIGURE P6-94

6–95 Determine the forces acting at joints 17, 18, and 19 in the truss.

6–96 Construct the Maxwell diagram for the truss in Prob. 6–19. Determine the forces F_{AE} and F_{BC}.

6–97 Calculate the forces acting at joints 2 and 9.

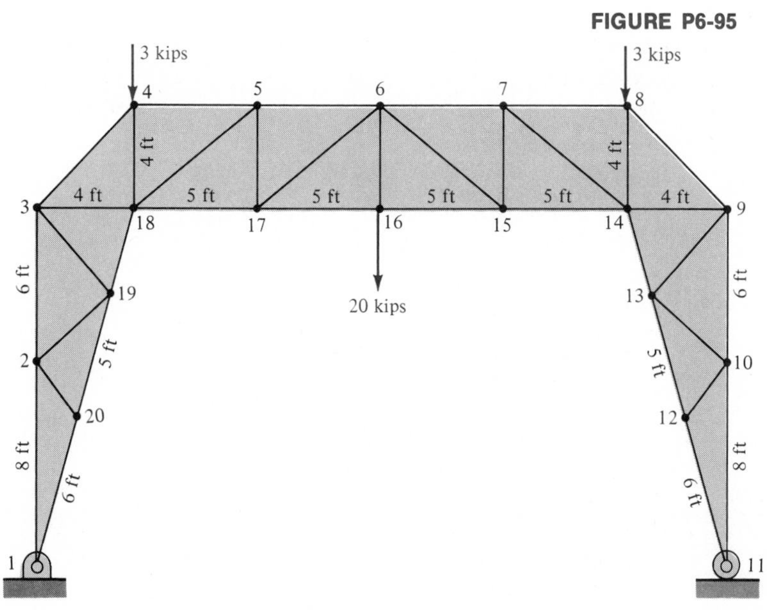

FIGURE P6-95

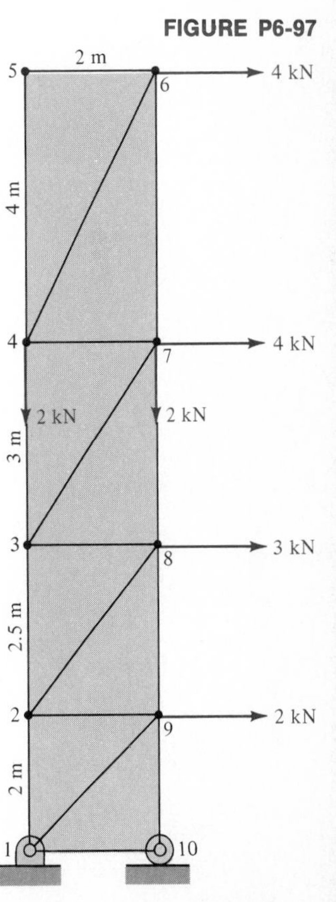

FIGURE P6-97

FIGURE P6-98

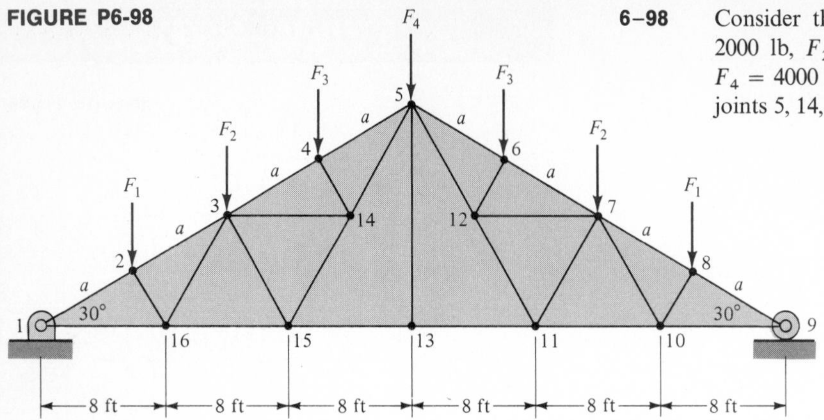

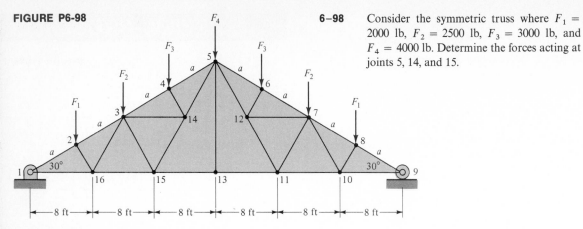

6–98 Consider the symmetric truss where $F_1 = 2000$ lb, $F_2 = 2500$ lb, $F_3 = 3000$ lb, and $F_4 = 4000$ lb. Determine the forces acting at joints 5, 14, and 15.

FIGURE P6-99

T
20 in.
A
15 in.
B
B
2 ft
C
4 ft
D
2 ft
D
$W = 500$ lb

6–99 The symmetric lifting tongs hold the box by friction at the pads D. Determine the forces acting at joints B and C.

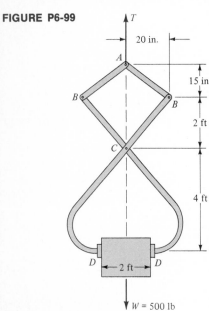

6–100 Determine the forces acting on arm BF of the excavator. Joint D connects three rigid members as shown in Fig. P6-100c.

FIGURE P6-100 a and b

(a)

(b)

FIGURE P6-100 (*continued*)

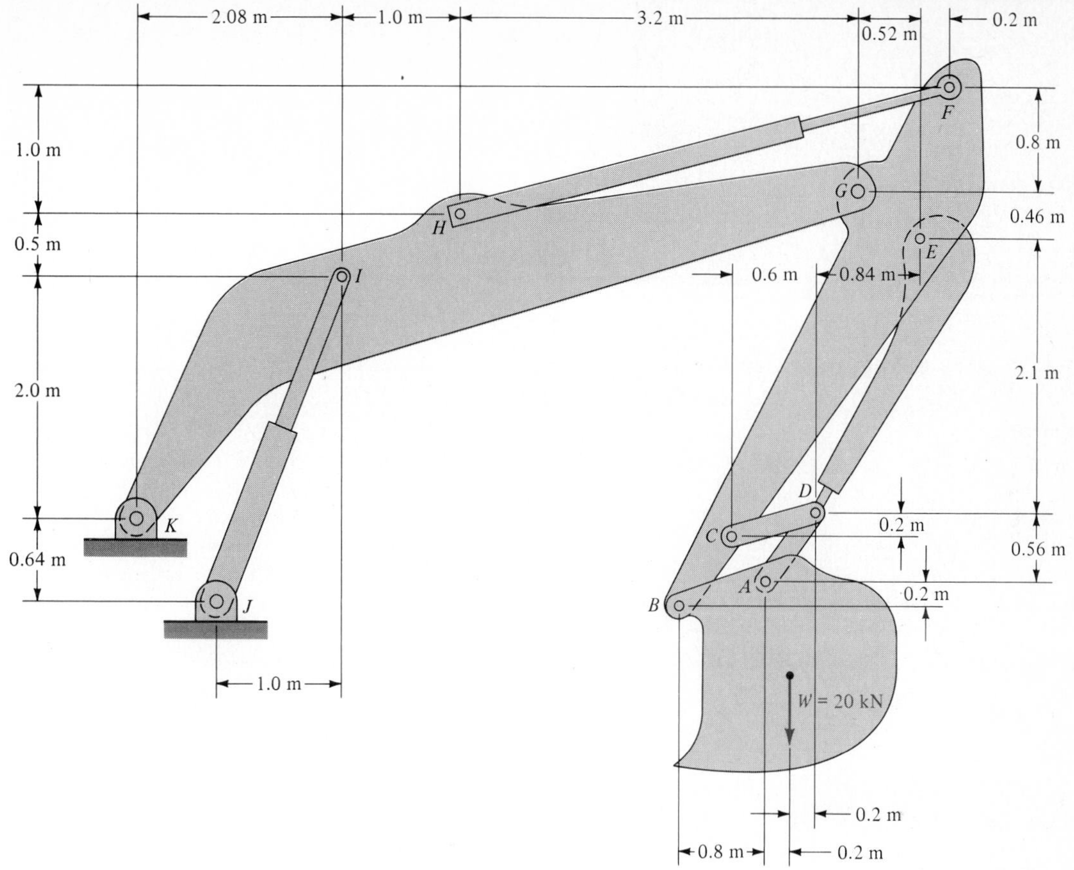

(c)

6–101 Determine the forces acting on member KG in Prob. 6–100.

6–102 The same downward force F is applied at points B, C, and D; $a = b = c = d$, and $y_E = 2a$. Plot the maximum and minimum tensions in the cable as functions of $y_C = 2.5a$, $2.75a$, and $3a$. (The maximum tension is of interest to assure safety; the minimum tension is relevant to considerations of efficient use of the cable.)

FIGURE P6-102

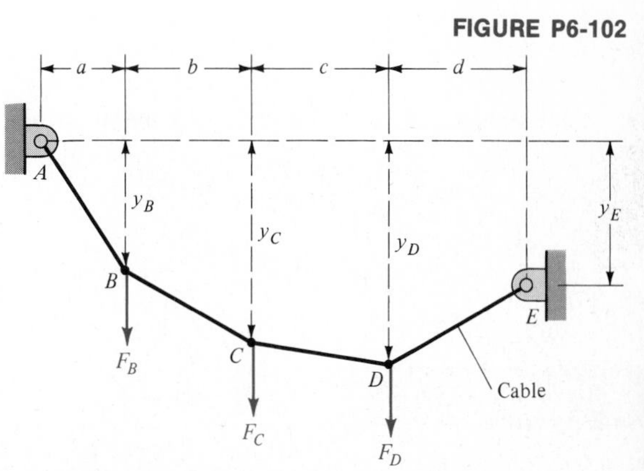

CHAPTER 7

DISTRIBUTED FORCES

OVERVIEW

Basic Concepts of Distributed Forces

So far in this text forces have always been assumed to be concentrated and acting at distinct points on bodies. This idealization is very useful even in advanced analysis, but it is not immediately possible or adequate in certain practical situations. For example, the force of gravity acts as a distributed force on a rigid beam. The distribution of the force over the beam depends on the distribution of matter in the body. The distributed force (a parallel force system here) can be represented by an equivalent concentrated force which is useful in the analysis, but both the magnitude and location of the equivalent force must be determined before the support reactions on the beam can be calculated. In more advanced analyses, such as determining the load-carrying ability of the beam, the local magnitude of the distributed internal force is of greater importance than the equivalent force for the given distribution. Several important special cases of distributed forces are presented in this chapter.

STUDY GOALS

SECTION 7-1 presents the concept and method of determination of the center of gravity. This topic is essential for engineers who will analyze dynamics of rigid bodies, stability of structures, or mechanics of materials.

SECTIONS 7-2, 7-3, AND 7-4 describe purely geometrical approaches to determine centers of gravity. These methods are useful because most rigid bodies are sufficiently homogeneous so that the center of gravity coincides with the centroid, the geometric center of the body. The topics of Secs. 7-2 and 7-3 are used in several courses and in engineering practice, but the theorems of Sec. 7-4 are mainly for academic exercise.

SECTION 7-5 presents the analysis of beams under distributed loading. It is an extension of the methods used for beams under concentrated loading, but considerably more difficult for many students. This topic is frequently encountered in mechanics of materials and structural analysis, so it should be studied with emphasis.

SECTION 7-6 introduces the relatively complex analysis of cables under distributed loads. This topic is important in the design of certain bridges and power lines, but it is omitted from most statics courses.

SECTIONS 7-7 AND 7-8 present the important basic concepts of fluid statics. These can be readily learned after Sec. 7-5. However, sometimes these topics are omitted from statics because they are also covered in fluid mechanics courses.

SECTION 7-9 is a brief introduction to stress analysis to show another area involving distributed forces. This very important topic is often omitted from statics because it is covered in much more detail in mechanics of materials courses.

CENTER OF GRAVITY 7-1

The most common distributed force is the gravitational force of the earth acting on a body. It is essentially a parallel force system distributed all over the body. The resultant of this distributed force is called the *weight* of the body and it acts at an imaginary point called the *center of gravity*. In other words, the center of gravity is a point where the mass of the whole body may be imagined to be concentrated as far as gravitational attraction is concerned. The concept of *center of mass* is based on a similar imaginary concentration of mass, but independent of a particular gravitational field. The centers of mass and gravity are at the same point for most practical purposes. They are not precisely the same for extremely large bodies because the magnitude and direction of the gravitational attraction (toward the center of the earth) varies from one part of the body to other parts.

The concept of the center of gravity is illustrated with the aid of Fig. 7-1. Assume that two very small bodies A and B of different masses are situated along the x axis. These *point masses* may be separate or connected by a rigid rod of negligible mass. The center of gravity of these masses is the point where the equivalent resultant force caused by gravity is acting. In other words, the two weights balance each other with respect to their center of gravity. Clearly, the magnitude of the resultant force is $R = \sum W_i = 4W$. The location of its line of action is determined by setting up the moments for the given system of bodies in two different ways. Since the net moment about any chosen point must be the same regardless of how it is calculated in detail, the moments written in different ways can be equated. For example, the summation of moments for bodies A and B with respect to point O in Fig. 7-1a is

$$\sum M_O = \sum W_i x_i = Wa + 3W(a + b) = W(4a + 3b)$$

This is equivalent to working with the resultant force **R** in Fig. 7-1b where $\bar{x}$ is the unknown location of the line of action of **R**. The moment of R is $\sum M_O = R\bar{x}$. Since $\sum W_i x_i$ must equal $R\bar{x}$,

$$W(4a + 3b) = 4W\bar{x}$$

and

$$\bar{x} = \frac{4a + 3b}{4} = a + \frac{3}{4}b$$

It makes sense that the center of gravity C is closer to the larger of the masses. To check the details, take moments about point A as in Fig. 7-1c. Here $3Wb$ is equivalent to $4W\bar{x}'$, where $\bar{x}' = \bar{x} - a$,

$$\bar{x}' = \frac{3}{4}b \qquad \text{and} \qquad \bar{x} = a + \bar{x}' = a + \frac{3}{4}b$$

The same procedure can be applied to more complex systems of point masses, and using other coordinate axes if necessary.

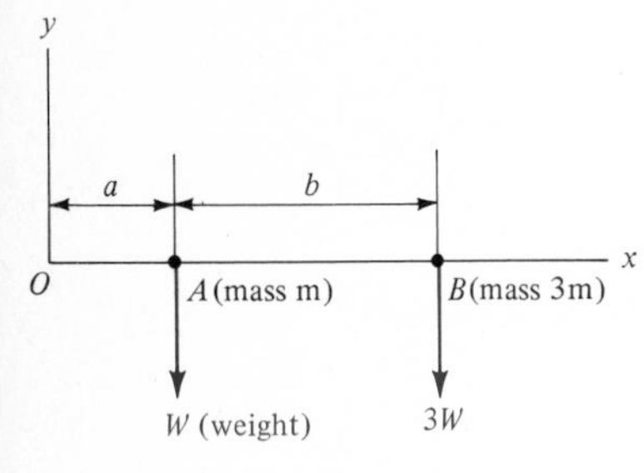

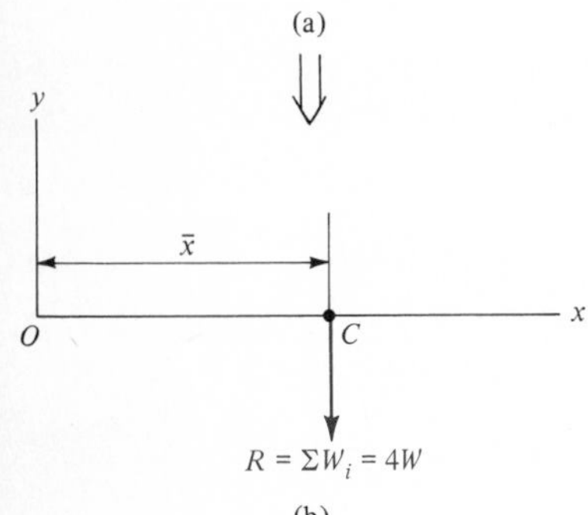

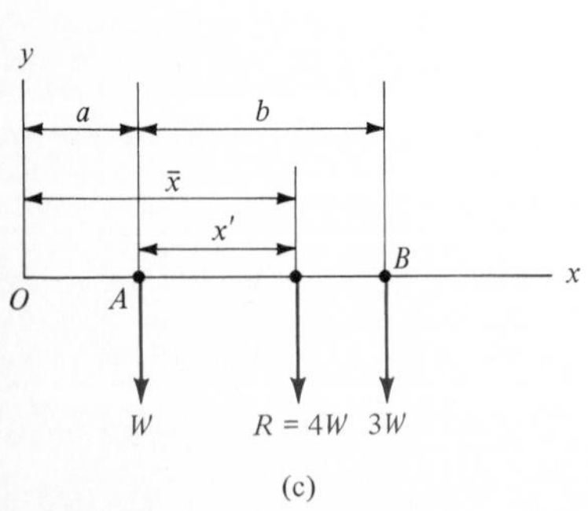

FIGURE 7-1.

Center of gravity C of masses A and B

Center of Gravity of a Three-Dimensional Body

The concepts used for point masses can be extended to deal with bodies of substantial dimensions. For example, consider the homogeneous plate of constant thickness t in Fig. 7-2a. Here every infinitesimal part dAt of the total volume At has the same force dW acting on it, but the moments caused by them with respect to the y axis are not necessarily the same. The x coordinate of the center of gravity is established as follows:

$$R = \int dW = W, \quad \text{the total weight of the plate} \tag{7-1}$$

The moments with respect to point O are written in two ways using Fig. 7-2a and b, respectively,

$$\sum M_O = \int x\, dW \qquad \text{and} \qquad \sum M_O = \bar{x}W \tag{7-2}$$

The last two summations are equivalent, so

$$\bar{x} = \frac{\int x\,dW}{W} \tag{7-3}$$

To determine the y coordinate of the center of gravity, imagine that the plate and the coordinate axes are rotated 90° to obtain Fig. 7-2c. The two ways of writing the moments with respect to point O are now

$$\sum M_O = \int y\,dW \qquad \text{and} \qquad \sum M_O = \bar{y}W \tag{7-4}$$

from which

$$\bar{y} = \frac{\int y\,dW}{W} \tag{7-5}$$

In practice this can be written directly from Fig. 7-2a without drawing Fig. 7-2c.

Similarly, the method is applicable to the third dimension, and

$$\bar{z} = \frac{\int z\,dW}{W} \tag{7-6}$$

The elemental force dW can be expressed as

$$dW = \gamma\,dV \tag{7-7}$$

where γ = specific weight of the material (weight per unit volume)

dV = elemental volume

With this expression, Eqs. 7-3, 7-5, and 7-6 can be written as

$$\bar{x} = \frac{\int x\gamma\,dV}{\int \gamma\,dV} \qquad \bar{y} = \frac{\int y\gamma\,dV}{\int \gamma\,dV} \qquad \bar{z} = \frac{\int z\gamma\,dV}{\int \gamma\,dV} \tag{7-8}$$

Having presented these derivations, the general meaning of the center of gravity can be stated: any body suspended or supported at the center of gravity would be perfectly balanced regardless of the orientation of the body. This is the basis for rapid and accurate experimental determinations of the center of gravity. In general, the balance of a body is checked in three orthogonal orientations.

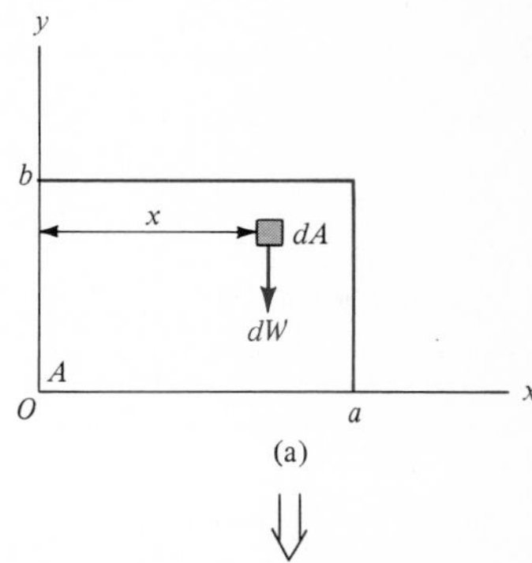

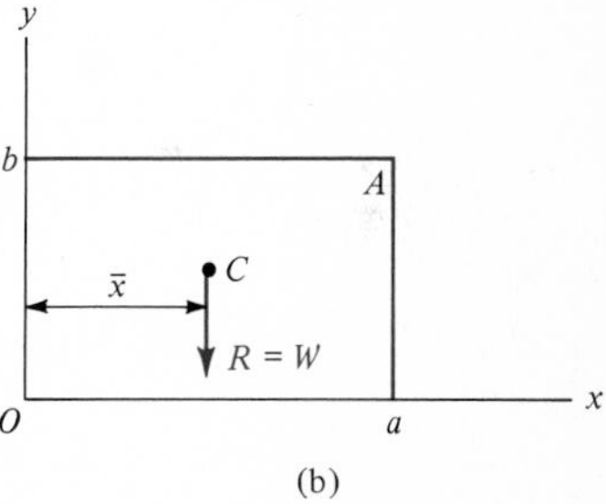

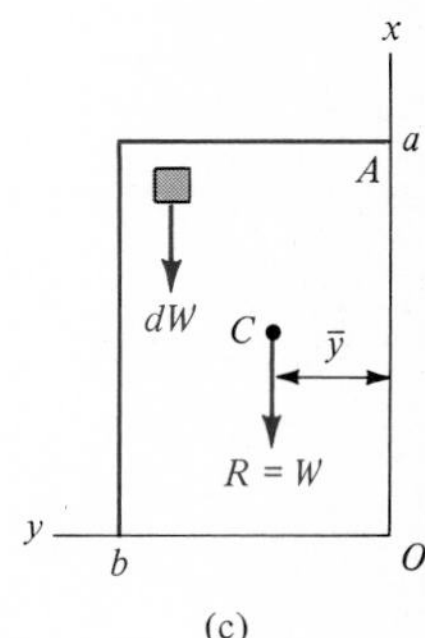

FIGURE 7-2

Center of gravity C of a uniform plate

7-2 CENTROIDS OF LINES, AREAS, AND VOLUMES

Equations 7-8 are generally valid even when the specific weight γ varies throughout the body. Of course, one can hope to solve these equations only when γ can be expressed as simple functions of the x, y, and z coordinates. Fortunately, in most bodies γ is a constant and can be canceled from Eq. 7-8. In such cases the center of gravity coincides with the *centroid*, which is a purely geometrical property of the body.

The centroid is determined in a manner similar to that used for the center of gravity in homogeneous bodies. It is assumed that line segments, areas, or volumes have moments with respect to the origin of arbitrary coordinate axes, depending on their positions in the reference system.

Centroids of Lines

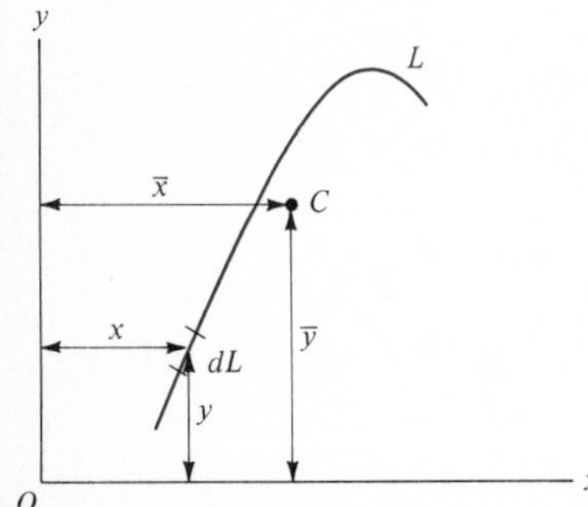

FIGURE 7-3

Centroid C of line L in the xy plane

Slender bodies of uniform cross-sectional areas A can be treated simply as line segments. For illustration, consider the two-dimensional line L in Fig. 7-3. It is good practice to guess and show the approximate location of the center of gravity C, which is also the centroid when γ and A are constant over the whole length of the line. The coordinates of the centroid are obtained after writing $dW = \gamma\, dV = \gamma A\, dL$,

$$\bar{x} = \frac{\int x\, dL}{L} \qquad \bar{y} = \frac{\int y\, dL}{L} \tag{7-9}$$

The method is extended to deal with a three-dimensional line in Fig. 7-4. Equations 7-9 are also valid for this line, and a similar equation can be added for the z coordinate,

$$\boxed{\bar{x} = \frac{\int x\, dL}{L} \qquad \bar{y} = \frac{\int y\, dL}{L} \qquad \bar{z} = \frac{\int z\, dL}{L}} \tag{7-10}$$

Note that the centroid C seldom lies on the line L unless the line is entirely straight.

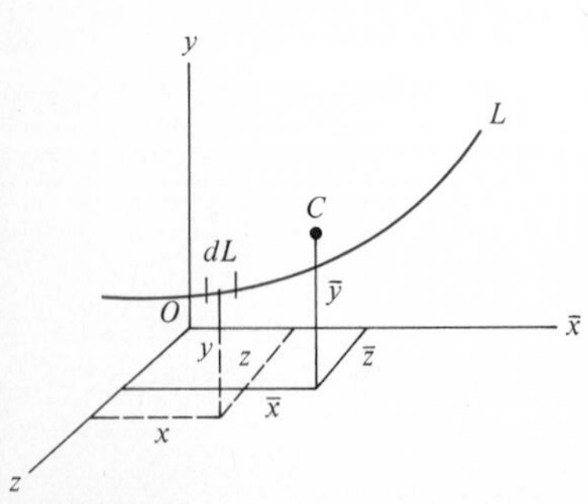

FIGURE 7-4

Centroid C of line L in three dimensions

Centroids of Areas

Bodies of small thickness t can be treated as surface areas. The method of determining the centroid is illustrated with the three-dimensional sheet in Fig. 7-5. Assuming that γ and t are constant over the entire area A, and using $dW = \gamma\, dV = \gamma t\, dA$,

$$\boxed{\bar{x} = \frac{\int x\, dA}{A} \qquad \bar{y} = \frac{\int y\, dA}{A} \qquad \bar{z} = \frac{\int z\, dA}{A}} \tag{7-11}$$

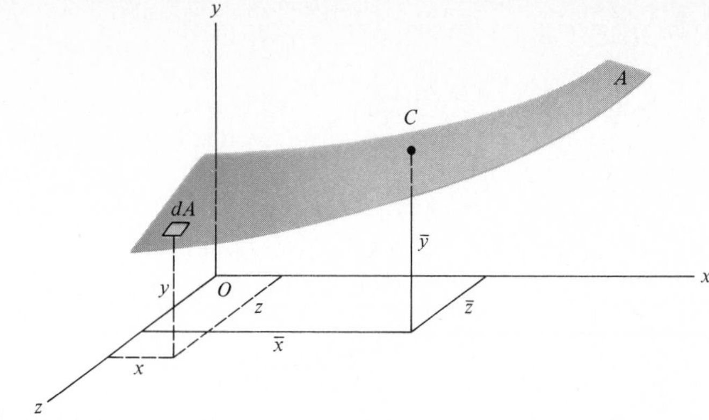

FIGURE 7-5

Centroid C of a three-dimensional area

Note again that the centroid seldom lies in the body except when the entire surface is a single flat area. The integral $\int x\,dA$ is called the *first moment of the area A with respect to the z axis*. Similarly, first moments can be defined with respect to any line, in each case multiplying areas by the appropriate distances which are analogous to moment arms.

Centroids of Volumes

The concepts described above can be readily extended to determine the coordinates of the centroid of a general body of volume V. Assuming that γ is constant over the whole volume, and using $dW = \gamma\,dV$,

$$\bar{x} = \frac{\int x\,dV}{V} \qquad \bar{y} = \frac{\int y\,dV}{V} \qquad \bar{z} = \frac{\int z\,dV}{V} \tag{7-12}$$

The integral $\int x\,dV$ is called the *first moment of the volume V with respect to the yz plane*. Similarly, first moments can be defined with respect to the other planes.

Volumes with a plane of symmetry have their centroids in that plane. Volumes with two planes of symmetry have their centroids on the line of intersection of those planes. Volumes with three planes of symmetry that intersect at a single point have the centroid at that point. Since these properties can be determined by inspection, the centroids are easily found for cubes, spheres, ellipsoids, and other symmetric bodies.

The centroidal coordinates of several common lines, areas, and volumes are given in Table 3, inside the back cover. In practice these are used directly rather than calculating integrals.

7-3 CENTROIDS OF COMPOSITE OBJECTS

Frequently, the centroid of a complex body is much easier and more accurate to determine experimentally (see the last paragraph in Sec. 7-1) than by calculations. Nevertheless, such calculations provide useful practice with the analytic methods and also may facilitate the experimentation by locating the centroid approximately before the actual tests.

The centroids of complex objects are determined in a manner similar to dealing with point masses. The object is broken down into simple parts, aiming to determine the centroid of each part by inspection. The method is illustrated for areas in which case the ideal simple parts are circles and rectangles. The composite centroid is determined by equating the first moment of the whole object with the sum of the first moments of the individual parts, using the same reference system. For the centroidal coordinate $\bar{x}$ of an area A consisting of discrete parts A_i,

$$\underbrace{\bar{x}A}_{\text{first moment of whole area}} = \underbrace{\sum \bar{x}_i A_i}_{\text{sum of first moments of individual parts}} \qquad \text{7-13a}$$

where $\bar{x}_i$ are the centroidal coordinates of parts A_i and $A = \sum A_i$. Note the analogy of this expression with $\bar{x}A = \int x\, dA$ in Eq. 7-11. Similarly for the other coordinates,

$$\bar{y}A = \sum \bar{y}_i A_i \qquad \bar{z}A = \sum \bar{z}_i A_i \qquad \text{7-13b}$$

The method, which is valid for objects of any shape, is illustrated for the area A in Fig. 7-6a.

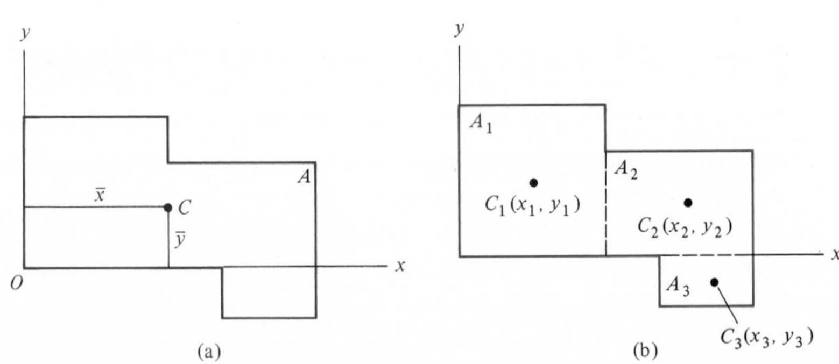

FIGURE 7-6

Centroids: C_1 of coordinates x_1 and y_1 is centroid of area A_1, and so on; C of coordinates $\bar{x}$ and $\bar{y}$ is centroid of whole area $A = A_1 + A_2 + A_3$

First choose a convenient coordinate system along two sides of the area. Guess and show the approximate location of the centroid C as in Fig. 7-6a. Next the area is divided into rectangles; one possibility is shown in Fig. 7-6b. The centroid of each rectangle (C_1, etc.) is at the intersection of its diagonals (this makes sense but can also be proved using integration). The first moment of area A about point O is written in two different ways for each axis of reference. Working with the x coordinates, the resulting equation is

$$\underbrace{A\bar{x}}_{\text{first moment of whole area}} = \underbrace{A_1x_1 + A_2x_2 + A_3x_3}_{\text{sum of first moments of individual parts}} \quad \boxed{7\text{-}14a}$$

$$\bar{x} = \frac{A_1x_1 + A_2x_2 + A_3x_3}{A}$$

Similarly, but noting that A_3 is on a different side of the x axis than A_1 and A_2,

$$\bar{y} = \frac{A_1y_1 + A_2y_2 - A_3y_3}{A} \quad \boxed{7\text{-}14b}$$

where $A = \sum A_i = A_1 + A_2 + A_3$.

Whenever the complex body cannot be broken down into simple geometrical shapes, it is often reasonable to make approximations using rectangles and other simple shapes to facilitate the calculations. The error resulting from the approximation can be reduced by increasing the number of simple parts used. When the number of parts is large, it is advantageous to make a systematic table to list all areas (or lines, or volumes) and their centroidal distances with the proper magnitudes and signs. This is illustrated for Fig. 7-6 as follows.

AREA PART A_i	CENTROIDAL COORDINATE $\bar{x}_i$	PRODUCT $\bar{x}_iA_i$	CENTROIDAL COORDINATE $\bar{y}_i$	PRODUCT $\bar{y}_iA_i$
$+A_1$	$+x_1$	$+x_1A_1$	$+y_1$	$+y_1A_1$
$+A_2$	$+x_2$	$+x_2A_2$	$+y_2$	$+y_2A_2$
$+A_3$	$+x_3$	$+x_3A_3$	$-y_3$	$-y_3A_3$
$\sum A_i = A$		$\sum \bar{x}_iA_i = \bar{x}A$		$\sum \bar{y}_iA_i = \bar{y}A$

The areas considered are positive if they are real parts of the body analyzed; they are negative for imaginary parts, such as a circular hole in a plate.

EXAMPLE 7-1

Determine the centroid of the circular arc shown in Fig. 7-7 using Table 3, inside the back cover.

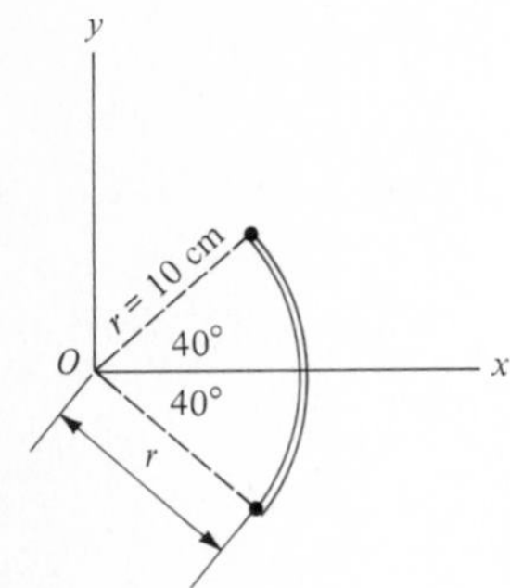

FIGURE 7-7

Circular arc in the *xy* plane

SOLUTION

From symmetry it is seen that $\bar{y} = 0$. From Table 3,

$$\bar{x} = \frac{r \sin \theta}{\theta}$$

where θ must be stated in radians.
$\theta = 40^\circ = 0.698$ rad, so

$$\bar{x} = \frac{(10 \text{ cm})(\sin 0.698)}{0.698}$$

$$= 9.21 \text{ cm}$$

JUDGMENT OF THE RESULTS

The answer is reasonable since $r \cos 40^\circ < \bar{x} < r$, as should be expected.

EXAMPLE 7-2

The centroid C of the rectangle ab in Fig. 7-8 has coordinates $(a/2, b/2)$ by inspection. Prove this using integration.

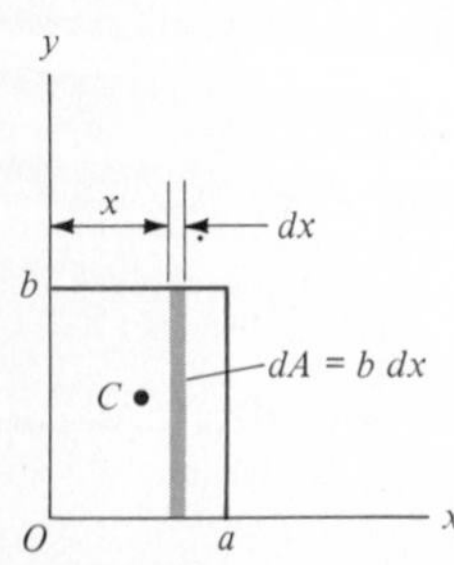

FIGURE 7-8

Centroid C of a rectangle in the *xy* plane

SOLUTION

From Eq. 7-11 and Fig. 7-8,

$$\bar{x} = \frac{\int x\, dA}{A} = \frac{\int xb\, dx}{ab} = \frac{1}{a}\left[\frac{x^2}{2}\right]_0^a = \frac{a}{2}$$

Similarly for the y coordinate, but using a horizontal elemental area $dA = a\,dy$,

$$\bar{y} = \frac{\int y\, dA}{A} = \frac{\int ya\, dy}{ab} = \frac{1}{b}\left[\frac{y^2}{2}\right]_0^b = \frac{b}{2}$$

EXAMPLE 7-3

(a) State the centroidal coordinates $\bar{x}, \bar{y}$ of the right triangle in Fig. 7-9a. Use Table 3, inside back cover.

(b) Apply the results from part (a) to the composite triangle in Fig. 7-9b using Eq. 7-13.

SOLUTION

(a) Since the relation $\bar{y} = h/3$ in Table 3 is for a general triangle, it is valid for a right triangle. This result is applied to the triangle in Fig. 7-9a, using the y and x axes for reference to determine $\bar{x}$ and $\bar{y}$, respectively. The centroidal coordinates (which are worth remembering) are

$$\bar{x} = \frac{b}{3} \qquad \bar{y} = \frac{h}{3}$$

(b) The results from part (a) are readily applied to triangles 1 and 2 in Fig. 7-9b after making adjustments (by inspection) for the particular configuration.

$$\bar{x}_1 = \frac{2}{3}\left(\frac{b}{5}\right) = \frac{2b}{15}$$

$$\bar{y}_1 = \frac{h}{3}$$

$$\bar{x}_2 = \frac{b}{5} + \frac{1}{3}\left(\frac{4}{5}b\right) = \frac{7b}{15}$$

$$\bar{y}_2 = \frac{h}{3}$$

The centroidal coordinates $\bar{x}_C$ and $\bar{y}_C$ of the composite triangle are obtained using Eq. 7-13,

$$\bar{x}_C = \frac{\bar{x}_1 A_1 + \bar{x}_2 A_2}{A_1 + A_2} \qquad \bar{y}_C = \frac{\bar{y}_1 A_1 + \bar{y}_2 A_2}{A_1 + A_2}$$

where

$$A_1 = \frac{1}{2}\left(\frac{bh}{5}\right) = \frac{bh}{10} \quad \text{and} \quad A_2 = \frac{1}{2}\left(\frac{4b}{5}h\right) = \frac{2bh}{5}$$

Thus,

$$\bar{x}_C = \frac{\dfrac{2b}{15}\left(\dfrac{bh}{10}\right) + \dfrac{7b}{15}\left(\dfrac{2bh}{5}\right)}{\dfrac{bh}{10} + \dfrac{2bh}{5}} = \frac{2b}{5}$$

$$\bar{y}_C = \frac{\dfrac{h}{3}\left(\dfrac{bh}{10}\right) + \dfrac{h}{3}\left(\dfrac{2bh}{5}\right)}{\dfrac{bh}{10} + \dfrac{2bh}{5}} = \frac{h}{3}$$

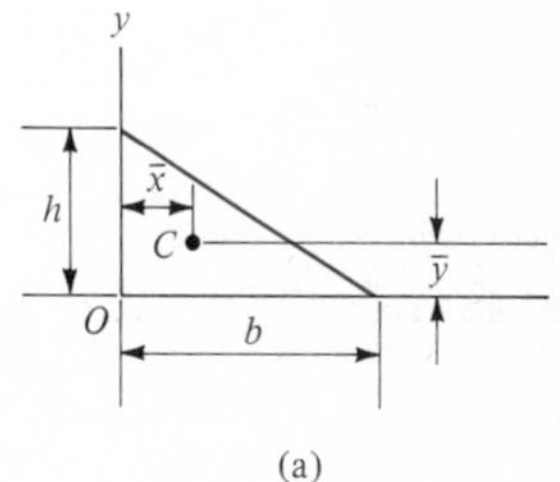

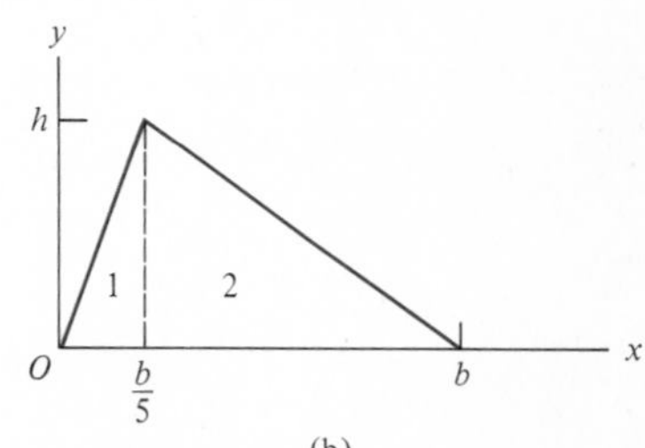

FIGURE 7-9
Centroids of triangles

EXAMPLE 7-4

Determine the first moment M_x of the rectangle ab in Fig. 7-10 with respect to the x axis. Use two methods: (a) Break the rectangle into four areas as divided by the x and y axes. (b) Work directly with the centroid of the whole area. (c) Use the simpler of the methods in parts (a) and (b) to calculate the first moment M_y of the rectangle with respect to the y axis.

SOLUTION

(a) The first moment M_x of the whole area is the sum of the first moments of the individual parts, $M_x = \sum \bar{y}_i A_i$, where $\bar{y}_i$ is the y coordinate of the centroid of area A_i.

$$\begin{aligned} M_x &= \sum \bar{y}_i A_i = \bar{y}_1 A_1 + \bar{y}_2 A_2 + \bar{y}_3 A_3 + \bar{y}_4 A_4 \\ &= \left(\frac{1}{3}b\right)\left[\left(\frac{5}{6}a\right)\left(\frac{2}{3}b\right)\right] + \left(-\frac{1}{6}b\right)\left[\left(\frac{5}{6}a\right)\left(\frac{1}{3}b\right)\right] \\ &\quad + \left(\frac{1}{3}b\right)\left[\left(\frac{1}{6}a\right)\left(\frac{2}{3}b\right)\right] + \left(-\frac{1}{6}b\right)\left[\left(\frac{1}{6}a\right)\left(\frac{1}{3}b\right)\right] \\ &= \frac{18}{108}ab^2 = \frac{1}{6}ab^2 \end{aligned}$$

(b) The moment may also be calculated by considering the whole area A and the vertical distance of its centroid from the x axis:

$$M_x = \bar{y}A = \left(\frac{2}{3}b - \frac{1}{2}b\right)(ab) = \frac{1}{6}ab^2$$

(c) Realizing the facility of the method in part (b) the moment of the area about the y axis is found to be

$$M_y = \bar{x}A = \left(\frac{5}{6}a - \frac{1}{2}a\right)(ab) = \frac{1}{3}a^2 b$$

JUDGMENT OF THE RESULTS

The identical results from parts (a) and (b) indicate that the calculations were performed correctly.

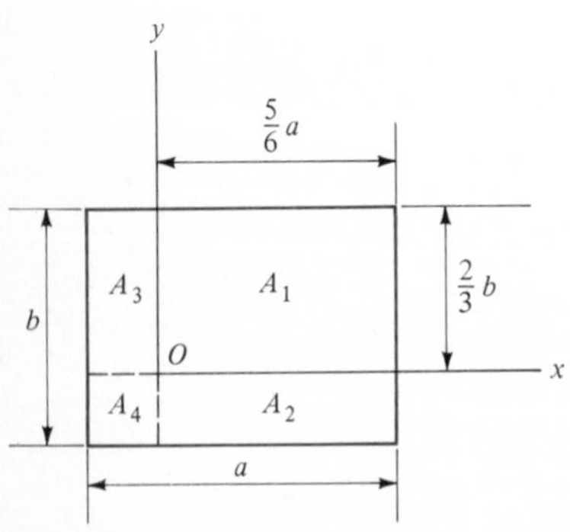

FIGURE 7-10

First moments of a rectangular area

EXAMPLE 7-5

Determine the centroid of area A in Fig. 7-6 if it consists of three squares arranged similarly to the rectangles shown. $A_1 = 5 \times 5$, $A_2 = 3.5 \times 3.5$, $A_3 = 1.4 \times 1.4$ units of length. Determine the new centroid after a circular hole of radius 0.8 is cut at point C_1.

SOLUTION

It is helpful to set up a table for the three regions as discussed in Sec. 7-3.

PART	A_i	$\bar{x}_i$	$\bar{x}_i A_i$	$\bar{y}_i$	$\bar{y}_i A_i$
A_1	25	2.5	62.5	2.5	62.5
A_2	12.25	6.75	82.69	1.75	21.44
A_3	1.96	7.8	15.29	−0.7	−1.37
Summations:	39.21		160.48		82.57

$$\bar{x}A = \sum \bar{x}_i A_i, \qquad \bar{x} = \frac{\sum \bar{x}_i A_i}{A} = \frac{160.48}{39.21} = 4.09$$

$$\bar{y}A = \sum \bar{y}_i A_i, \qquad \bar{y} = \frac{\sum \bar{y}_i A_i}{A} = \frac{82.57}{39.21} = 2.11$$

Next the area with a hole at C_1 is analyzed. The hole may be considered a negative area, $A_4 = -2.01$ with centroid $(\bar{x}_4, \bar{y}_4) = (2.5, 2.5)$. The problem is now considered that of a composite area consisting of two regions, the original area A and A_4.

$$\bar{x}' = \frac{\bar{x}A + \bar{x}_4 A_4}{A + A_4} = \frac{(4.09)(39.21) + (2.5)(-2.01)}{39.21 + (-2.01)} = 4.18$$

$$\bar{y}' = \frac{\bar{y}A + \bar{y}_4 A_4}{A + A_4} = \frac{(2.11)(39.21) + (2.5)(-2.01)}{39.21 + (-2.01)} = 2.09$$

JUDGMENT OF THE RESULTS

The results appear reasonable since the centroid of the final area is to the right and below the centroid of the original area. This was expected as a larger percent of the solid area is now to the right of the original centroid. Similarly, the centroid also moved down because a larger percent of the solid area is now below the original centroid.

EXAMPLE 7-6

Locate the centroid of the composite solid hemisphere and attached hollow cylinder shown in Fig. 7-11.

SOLUTION

It is seen from symmetry that $\bar{x} = \bar{z} = 0$. Labeling the hemisphere V_1 and the cyclinder V_2, the centroid of the composite piece is found using Eq. 7-13 and relations from Table 3, inside the back cover,

$$\bar{y} = \frac{\bar{y}_1 V_1 + \bar{y}_2 V_2}{V_1 + V_2}$$

$$= \frac{\left(\frac{3}{8}r\right)\left(\frac{2}{3}\pi r^3\right) + (-2r)(4r\pi)\left[r^2 - \left(\frac{3}{4}r\right)^2\right]}{\frac{2}{3}\pi r^3 + 4r\pi\left[r^2 - \left(\frac{3}{4}r\right)^2\right]}$$

$$= \frac{\frac{\pi r^4}{4} - 8\pi r^4 + \frac{9}{2}\pi r^4}{\frac{2}{3}\pi r^3 + 4\pi r^3 - \frac{9}{4}\pi r^3} = \frac{-\frac{13\pi r^4}{4}}{\frac{29\pi r^3}{12}} = \frac{-39r}{29}$$

JUDGMENT OF THE RESULTS

The sign and magnitude of the result appear reasonable. It should be expected that the centroid be on the left side of the xz plane since the larger portion of the material is on this side.

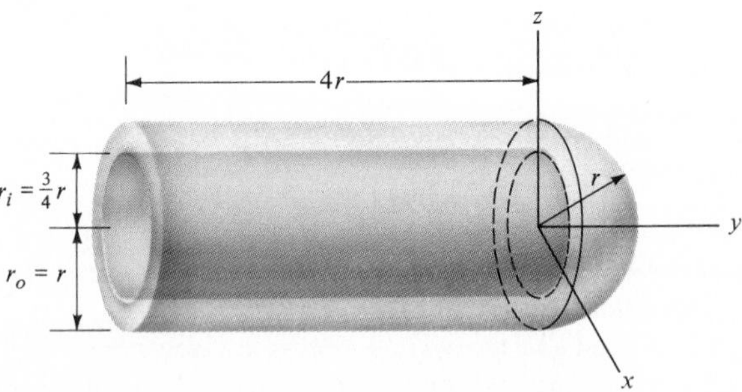

FIGURE 7-11

Centroid of a composite body

SECTIONS 7-1, 7-2, 7-3

PROBLEMS

7–1 Two thin, uniform wires are soldered at A to form the T shape. Determine the centroid of the line $OABC$.

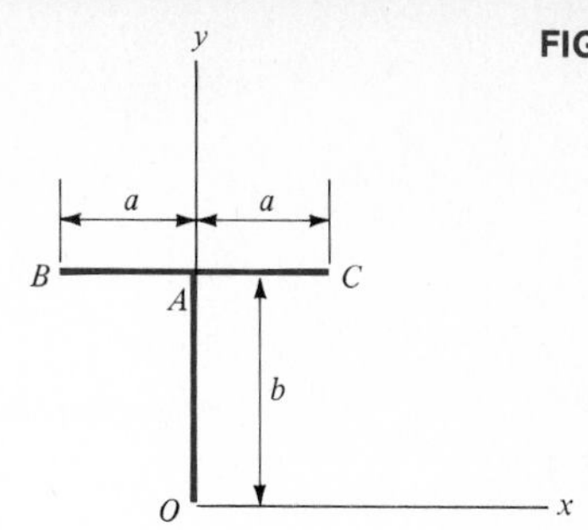

FIGURE P7-1

7–2 Determine the centroid of the thin, uniform wire OAC in Fig. P7-1.

7–3 Determine the centroid of the thin, uniform wire $AOBC$.

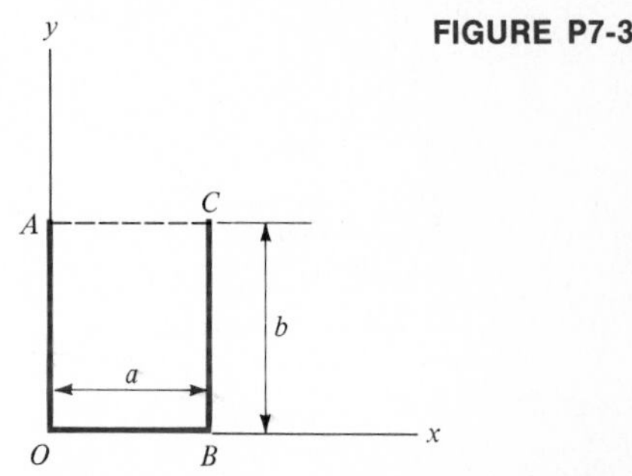

FIGURE P7-3

7–4 Solve Prob. 7–3 assuming that the segment OA is doubled in length to $2b$.

7–5 Determine the centroid of the thin, uniform wire $AOBC$.

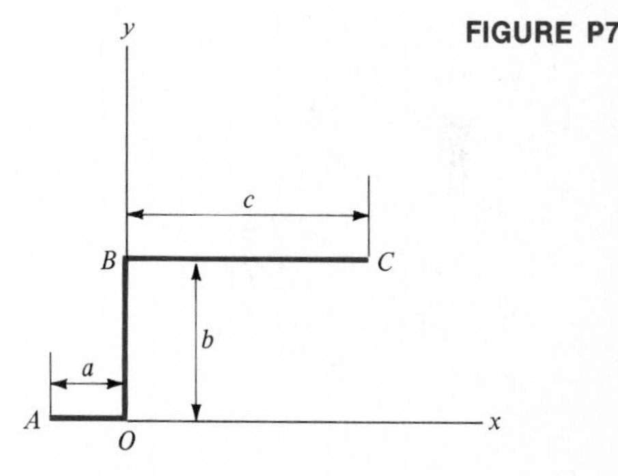

FIGURE P7-5

7–6 Determine the centroid of the thin, uniform wire AOB.

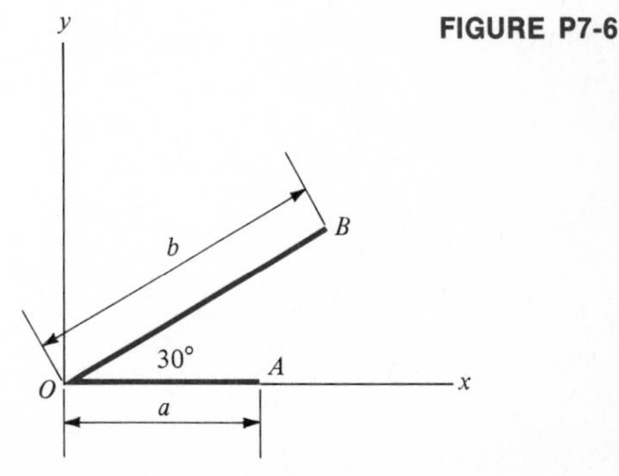

FIGURE P7-6

FIGURE P7-7

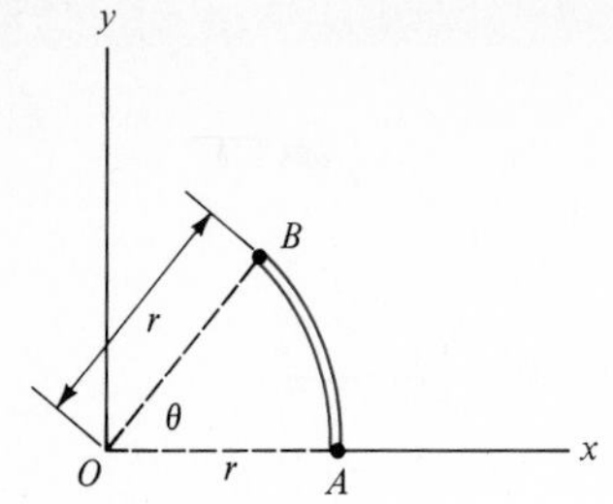

7–7 Determine the centroid of the circular arc AB for $\theta = 40^\circ$ using integration.

7–8 Determine the centroid of the circular arc AB in Fig. P7-7 for $\theta = 90^\circ$. Use Table 3, inside the back cover.

FIGURE P7-9

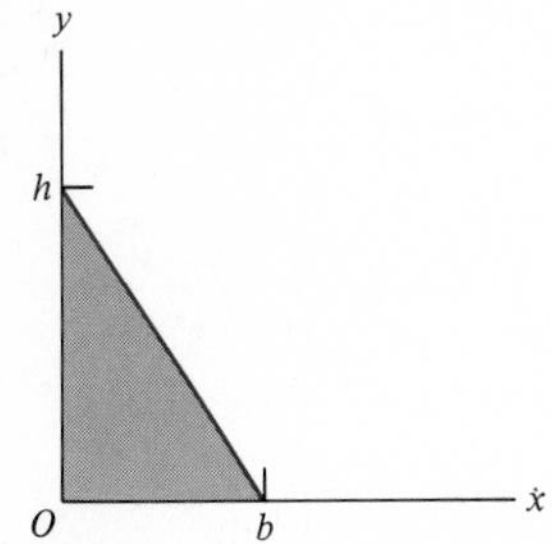

7–9 Determine the centroid of the triangular area by integration.

7–10 Assume that the triangle in Fig. P7-9 is truncated horizontally at $y = h/2$. Determine the centroid of the resulting trapezoid. Use the centroids and areas of the original triangle and of the smaller triangle that is removed.

FIGURE P7-11
Plane area

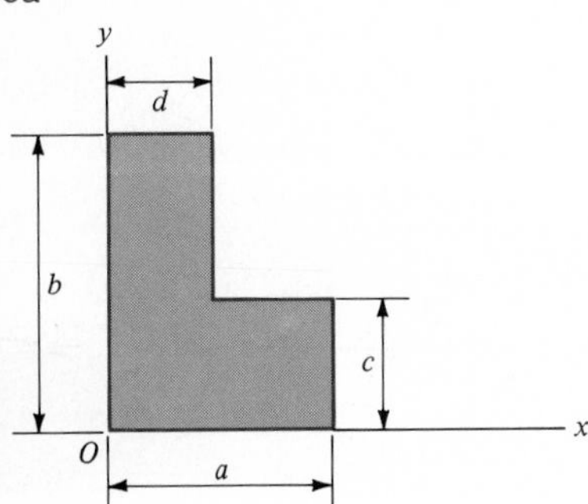

7–11 Determine the centroid of the L-shaped area working with two rectangles: one of area ac, the other of area $(b - c)d$.

7–12 Solve Prob. 7–11 working with two imaginary rectangles: one of area ab, the other of area $(a - d)(b - c)$.

FIGURE P7-13
Plane area

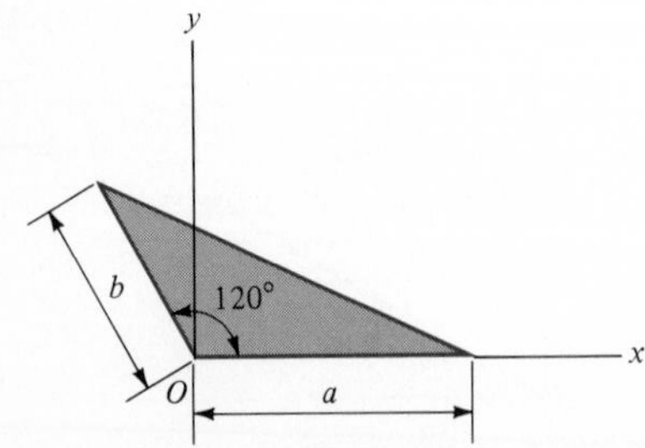

7–13 Determine the centroid of the triangular area.

7–14 Determine the centroid of the trapezoidal area working with the rectangle of area ab and the triangle above it.

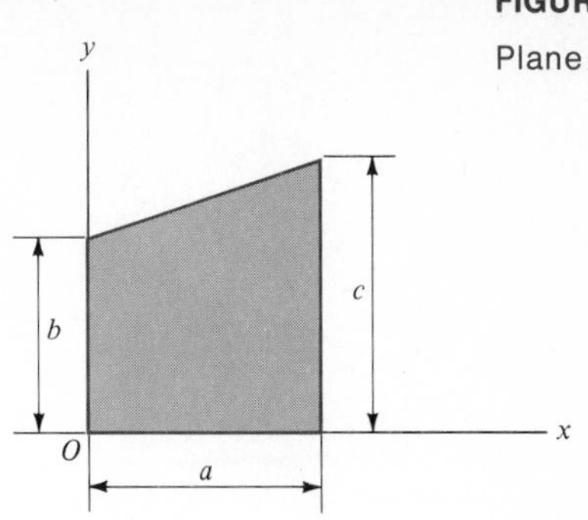

FIGURE P7-14

Plane area

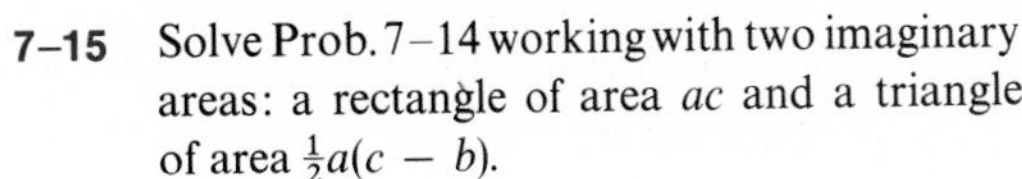

7–15 Solve Prob. 7–14 working with two imaginary areas: a rectangle of area ac and a triangle of area $\frac{1}{2}a(c - b)$.

7–16 Determine the centroid of the shaded area.

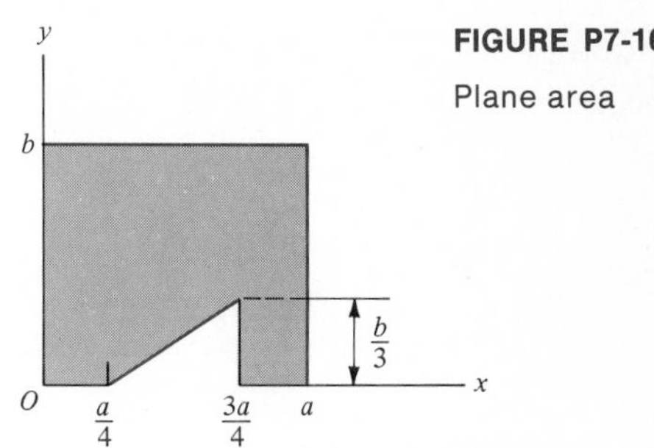

FIGURE P7-16

Plane area

7–17 Determine the centroid of the shaded area if the inner radius is $r_i = r_o/2$.

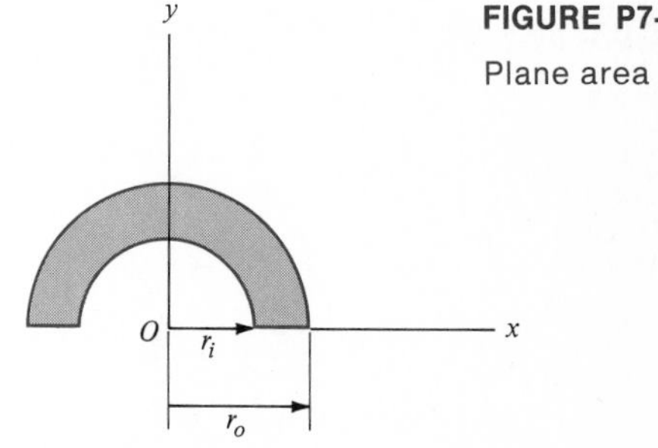

FIGURE P7-17

Plane area

7–18 Assume that in Fig. P7-17 the inner radius is $r_i = 0.95r_o$. Determine the centroid of this area. For comparison, determine the centroid of a semicircular line of radius r_o.

FIGURE P7-19

Plane area

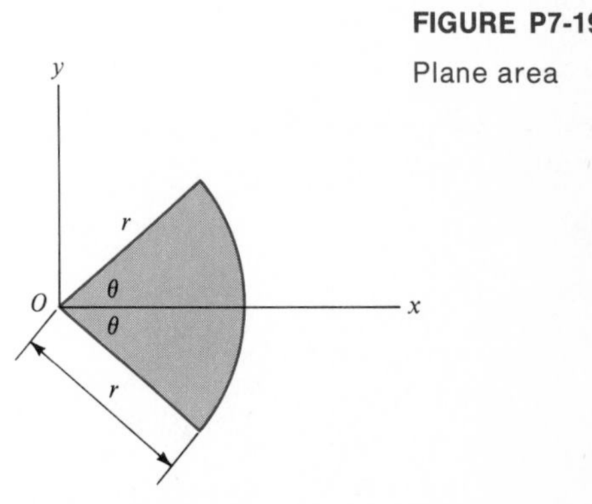

7–19 Determine the centroid of the circular segment by integration.

FIGURE P7-20
Plane area

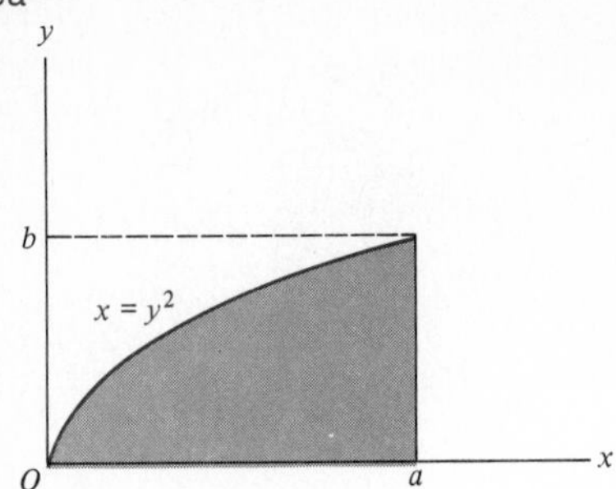

7–20 Determine the centroid of the shaded area.

7–21 Determine the centroid of the area between the dashed line and the curve $x = y^2$ in Fig. P7-20.

FIGURE P7-22
Plane area

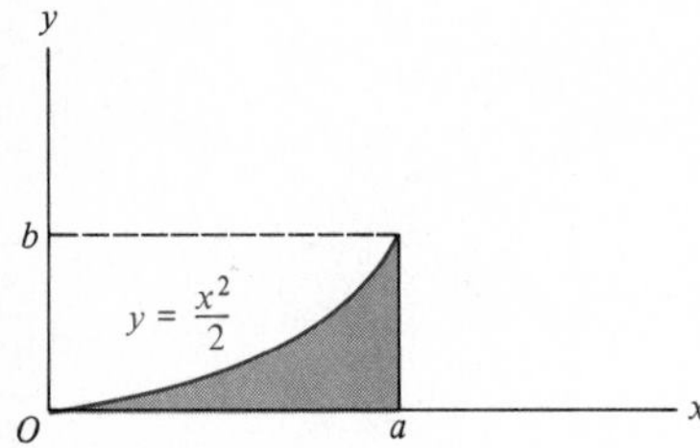

7–22 Determine the centroid of the shaded area.

7–23 Determine the centroid of the area between the dashed line and the curve $y = x^2/2$ in Fig. P7-22.

FIGURE P7-24
Plane area

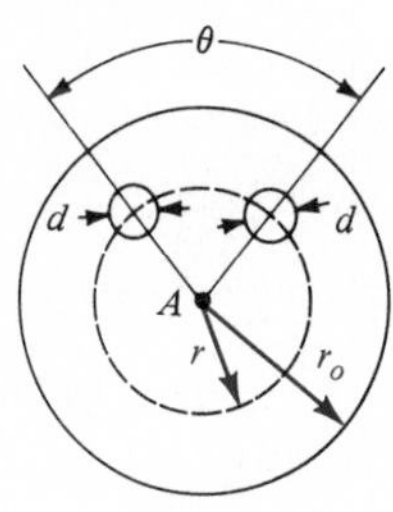

7–24 Determine the centroid of a circular area of radius r_o after one hole of diameter $d = 0.2r_o$ is drilled at a distance $r = r_o/2$ from the center A. (The second hole is for Prob. 7–25.)

7–25 Solve Prob. 7–24 assuming that two identical holes of diameter d are drilled (Fig. P7-24) at $\theta = 60°$.

FIGURE P7-26
Plane area

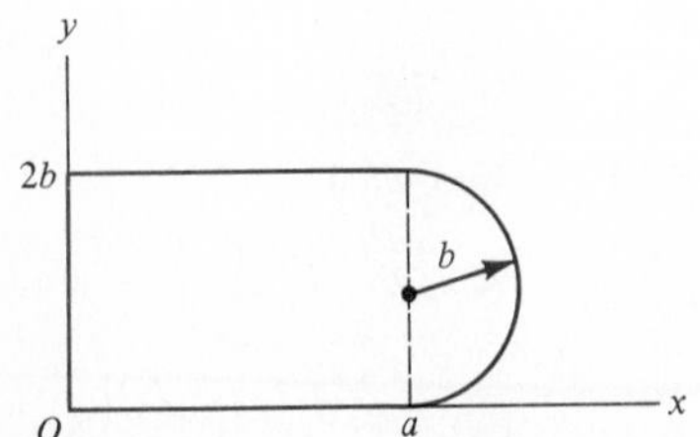

7–26 Determine the centroid of the area composed of a rectangle $a(2b)$ and a semicircle of radius b.

7–27 Determine the centroid of the segment of a solid spherical volume. The height of the segment is $h = r/3$.

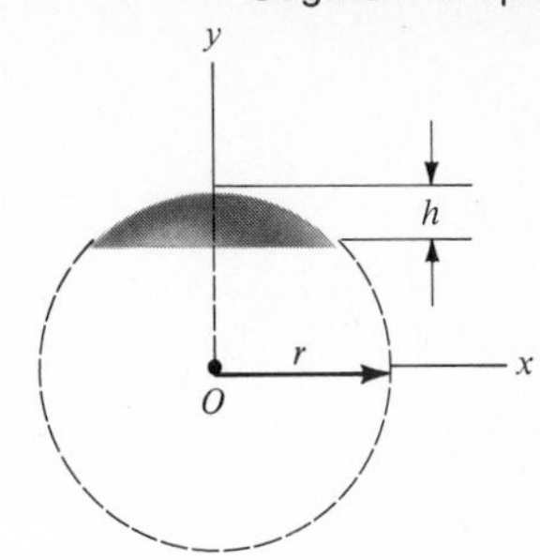

FIGURE P7-27

Segment of spherical volume

7–28 Determine the centroid of the truncated pyramid's volume. The height of the small pyramid that is cut off at the top is $h = a/4$.

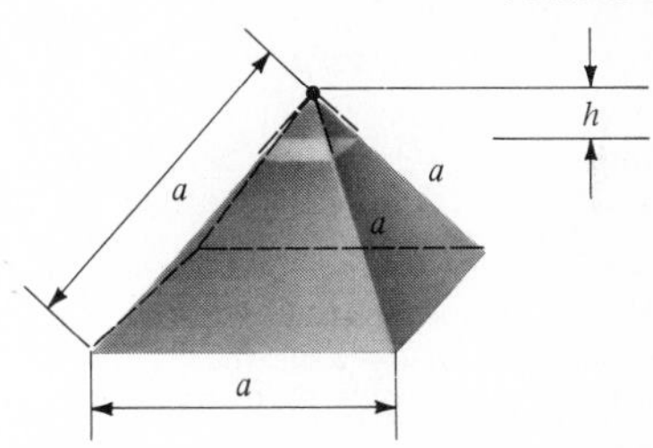

FIGURE P7-28

Volume of pyramid

7–29 Determine the centroid of the conical volume using integration. (The dimension d is for Prob. 7–30.)

FIGURE P7-29

Volume of circular cone

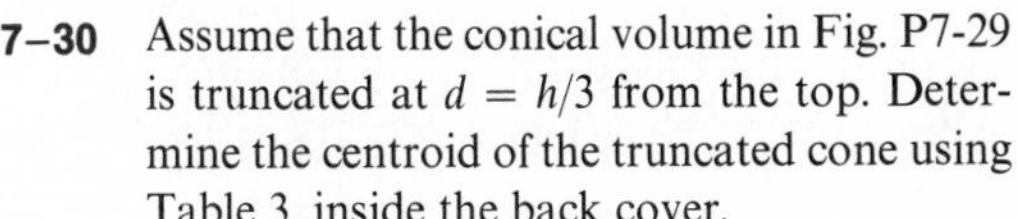

7–30 Assume that the conical volume in Fig. P7-29 is truncated at $d = h/3$ from the top. Determine the centroid of the truncated cone using Table 3, inside the back cover.

7–31 Determine the centroid of the composite volume.

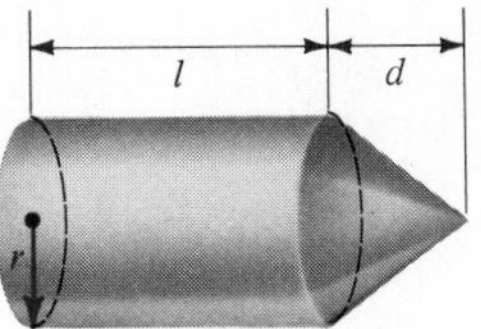

FIGURE P7-31

Volume of cylinder and cone

7–32 A cylindrical piece is machined to produce an internal cone shape as shown in the cross-sectional view. Determine the centroid of the final volume if $r = 1$ cm, $d = 0.4$ cm, and $h = 1.5$ cm.

FIGURE P7-32

Cross section of volume of cylinder minus cone

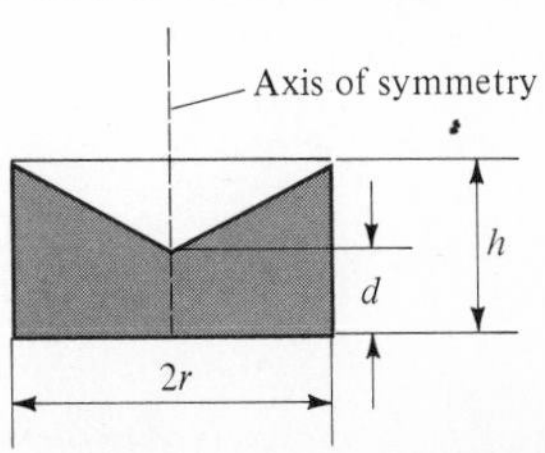

7-4 THEOREMS OF PAPPUS

Bodies with *surfaces of revolution* are commonly used in engineering. A surface of revolution is defined as a surface generated by rotating a plane curve about a fixed axis. Conversely, the body may be rotated with respect to a fixed cutting edge as in a lathe. For example, consider the hourglass-shaped specimen S and the grip G in Fig. 7-12a. The specimen in Fig. 7-12b is rotated about the x axis during machining while the cutting tool swings in a plane about a fixed point to generate the surface line L. The grip is made similarly.

An element of the surface of revolution generated is always a ring of width dL, as in Fig. 7-12. The outer surface area of this ring is

$$dA = 2\pi y \, dL$$

The total area of the surface of revolution is

$$A = 2\pi \int y \, dL \tag{7-15}$$

The volume dV of an elemental ring of revolution with cross section dA is obtained using Fig. 7-12c,

$$dV = 2\pi y \, dA$$

The total volume within the surface is

$$V = 2\pi \int y \, dA \tag{7-16}$$

Simpler and more useful methods for calculating the surface areas and volumes of bodies of revolution were formulated by Pappus, a Greek geometer, in the third century A.D. The name of Guldinus, a Swiss mathematician of the seventeenth century, is also mentioned occasionally since he has restated these concepts. The two theorems of Pappus are based on the definitions of centroidal coordinates of lines and areas. These are given in concise form as follows.

THEOREM I. Since $\int y \, dL = \bar{y}L$,

$$\boxed{A = 2\pi\bar{y}L} \tag{7-17}$$

where L = length of the generating curve
$\bar{y}$ = centroid of line L with respect to the axis of revolution

THEOREM II. Since $\int y \, dA = \bar{y}A$,

$$\boxed{V = 2\pi\bar{y}A} \tag{7-18}$$

where A = area of the generating surface
$\bar{y}$ = centroid of area A with respect to the axis of revolution

Naturally, Eqs. 7-17 and 7-18 can also be used to determine the centroids when the surface area A and the volume V of a body of revolution are known.

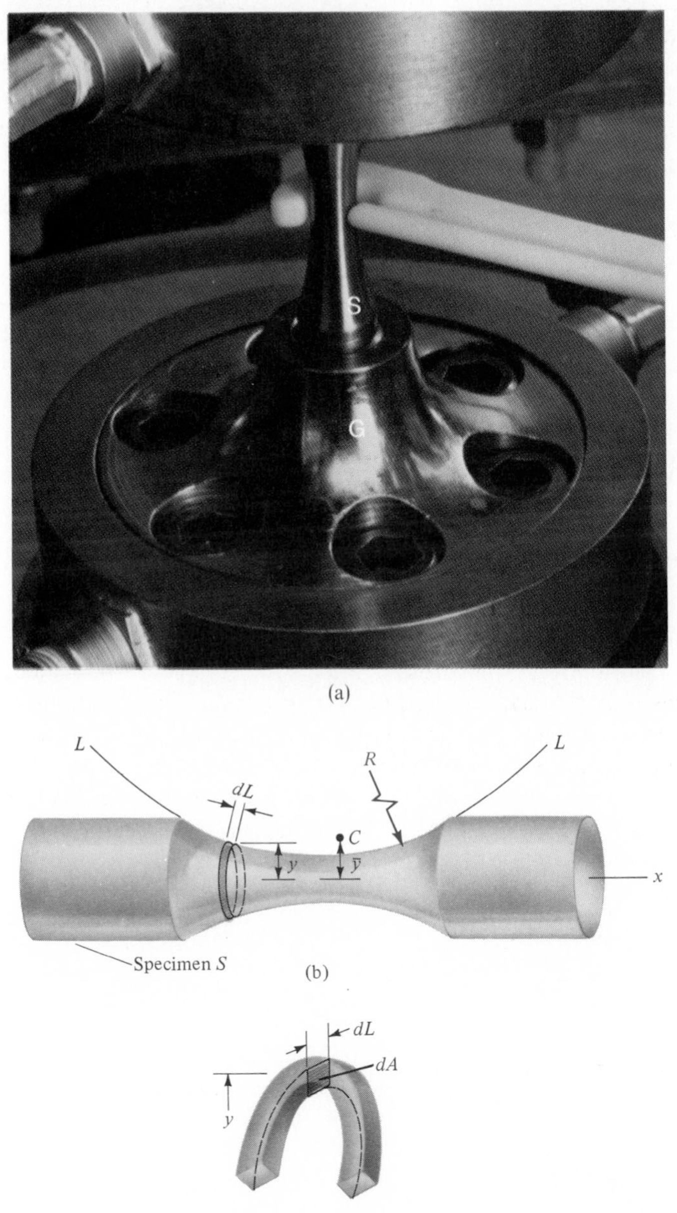

FIGURE 7-12

Hourglass-shaped specimen. (Photograph courtesy MTS Systems Corporation, Minneapolis, Minnesota.)

EXAMPLE 7-7

Use the theorems of Pappus to determine the centroids of (a) a semicircular arc, and (b) a semicircular area of radius r. The well-known formulas for the surface area and volume of a sphere can be used.

SOLUTION

(a) The semicircular arc in Fig. 7-13a would generate a spherical shell when rotated about the x axis. The total surface area is $4\pi r^2$. From Eq. 7-17, $A = 2\pi \bar{y} L$,

$$4\pi r^2 = 2\pi \bar{y}(\pi r)$$

$$\bar{y} = \frac{2r}{\pi}$$

(b) The semicircular area in Fig. 7-13b would generate a solid sphere when rotated about the x axis. The total volume of the sphere is $\frac{4}{3}\pi r^3$. From Eq. 7-18, $V = 2\pi \bar{y} A$,

$$\frac{4}{3}\pi r^3 = 2\pi \bar{y}\left(\frac{\pi r^2}{2}\right)$$

$$\bar{y} = \frac{4r}{3\pi}$$

JUDGMENT OF THE RESULTS

Both results are reasonable by inspection. Note that the centroidal distances are different in the two cases even though the same symbol is used for them (this is common practice) and r is the same in both cases.

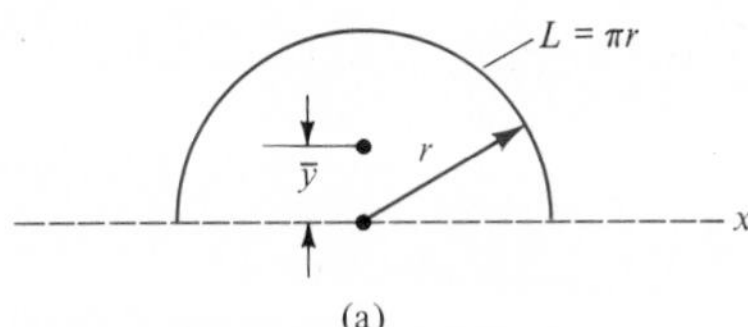

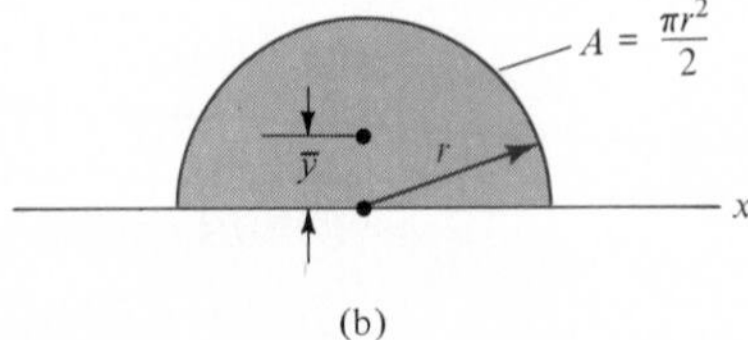

FIGURE 7-13

Line segment L and area A for the use of the theorems of Pappus

EXAMPLE 7-8

Determine the surface area and volume of the pulley whose cross section is shown in Fig. 7-14.

SOLUTION

To find the surface area of the pulley the centroid of the boundary line *abcdef* must be found. Dividing the region into segments *ab*, *bc*, *cd*, *de*, *ef*, and *fa*, $\bar{y}$ is found for the composite area:

$$\bar{y}=\frac{\sum \bar{y}_i l_i}{\sum l_i}$$

$$=\frac{2(25\,\text{cm})(0.5\,\text{cm})+\left[25-\frac{2(1.5)}{\pi}\right]\text{cm}(1.5\pi\,\text{cm})+2(14\,\text{cm})(22\,\text{cm})+(3\,\text{cm})(4\,\text{cm})}{2(0.5\,\text{cm})+1.5\pi\,\text{cm}+2(22\,\text{cm})+4\,\text{cm}}$$

$$=\frac{766.3\,\text{cm}^2}{53.7\,\text{cm}}=14.3\,\text{cm} \quad \text{(relative to the axis of rotation)}$$

From Eq. 7-17,

$$A = 2\pi\bar{y}L = 2\pi(14.3\text{ cm})(53.7\text{ cm}) = 4820\text{ cm}^2$$

Similarly, the centroid of the area *abcdef* is needed to calculate the volume of the pulley. Dividing the area into the rectangle *adef* and semicircular hole *bc* the centroid of the composite area is found.

$$\bar{y} = \frac{\sum \bar{y}_i A_i}{\sum A_i}$$

$$= \frac{(14\text{ cm})(88\text{ cm}^2) + \left[25 - \frac{4(1.5)}{3\pi}\right]\text{cm}\left[-(1.5\text{ cm})^2\frac{\pi}{2}\right]}{88\text{ cm}^2 - (1.5\text{ cm})^2\left(\frac{\pi}{2}\right)}$$

$$= \frac{1146\text{ cm}^3}{84.5\text{ cm}^2} = 13.6\text{ cm} \quad \text{(relative to the axis of rotation)}$$

From Eq. 7-18,

$$V = 2\pi\bar{y}A = 2\pi(13.6\text{ cm})(84.5\text{ cm}^2) = 7220\text{ cm}^3$$

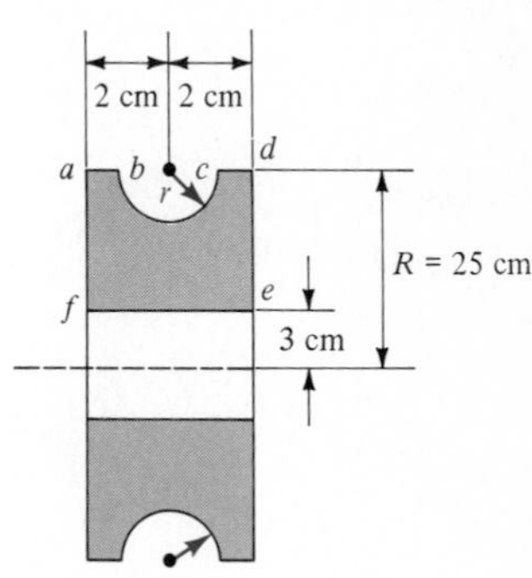

FIGURE 7-14
Cross section of a pulley

EXAMPLE 7-9

The removable seat of an O-ring consists of two halves as illustrated in Figure 7-15. Determine the centroid and the volume of the two halves put together to form the complete circular seat.

SOLUTION

Using the relations of Table 3, inside the back cover, the centroid of the boundary of a cross-sectional area is found.

$$\bar{y}' = \frac{\sum \bar{y}_i l_i}{\sum l_i}$$

$$= \frac{(1\text{ cm})(1\text{ cm}) + (0.9\text{ cm})(0.2\text{ cm}) + (0.5\text{ cm})(1\text{ cm})}{1\text{ cm} + 0.2\text{ cm} + 1\text{ cm} + 0.2\text{ cm} + \dfrac{\pi(0.8\text{ cm})}{2}}$$

$$+ \frac{(0\text{ cm})(0.2\text{ cm}) + \dfrac{2(0.8\text{ cm})}{\pi}\left(\dfrac{0.8\pi}{2}\text{ cm}\right)}{1\text{ cm} + 0.2\text{ cm} + 1\text{ cm} + 0.2\text{ cm} + \dfrac{\pi(0.8\text{ cm})}{2}}$$

$$= \frac{2.32\text{ cm}^2}{3.66\text{ cm}} = 0.63\text{ cm}$$

$$\bar{y} = R + \bar{y}' = 10\text{ cm} + 0.63\text{ cm} = 10.63\text{ cm}$$

From Eq. 7-17,

$$A = 2\pi\bar{y}L = 2\pi(10.63\text{ cm})(3.66\text{ cm})$$
$$= 244\text{ cm}^2$$

Similarly, by first finding the centroid of the cross-sectional area the volume of revolution can be found.

$$\bar{y}' = \frac{\sum \bar{y}_i A_i}{\sum A_i} = \frac{(0.5\text{ cm})(1\text{ cm}^2) + \dfrac{4(0.8\text{ cm})}{3\pi}\left[-\dfrac{\pi(0.8\text{ cm})^2}{4}\right]}{1\text{ cm}^2 - \dfrac{\pi(0.8\text{ cm})^2}{4}}$$

$$= \frac{0.33\text{ cm}^3}{0.50\text{ cm}^2} = 0.66\text{ cm}$$

$$\bar{y} = R + \bar{y}' = 10\text{ cm} + 0.66\text{ cm} = 10.66\text{ cm}$$

From Eq. 7-18,

$$V = 2\pi\bar{y}A = 2\pi(10.66\text{ cm})(0.50\text{ cm}^2) = 33.3\text{ cm}^3$$

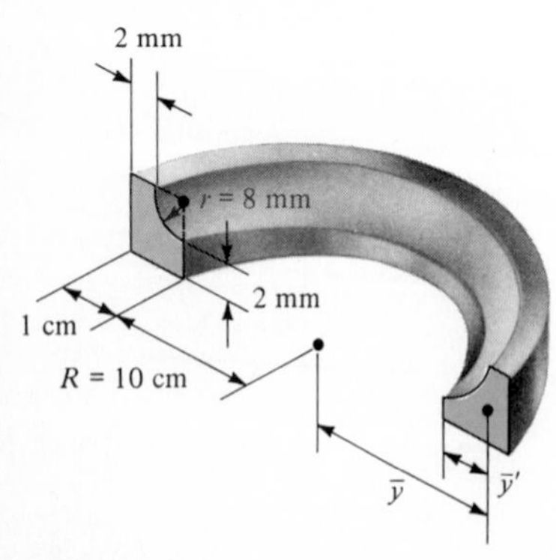

FIGURE 7-15
Circular seat of an O-ring

PROBLEMS

7–33 The solid ring has a circular cross section of radius r. Determine the surface area of the ring.

FIGURE P7-33

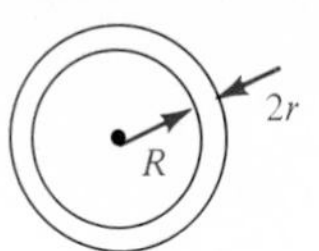

7–34 Determine the volume of the ring in Prob. 7–33.

7–35 The segment of a solid circular ring is cut at an angle $\theta = 120°$. Determine the total surface area of the segment.

FIGURE P7-35

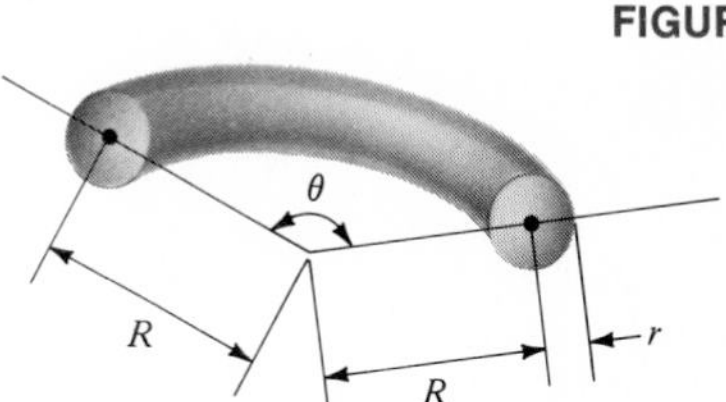

7–36 Determine the volume of the ring segment described in Prob. 7–35.

7–37 Determine the total surface area of the solid half ring.

FIGURE P7-37

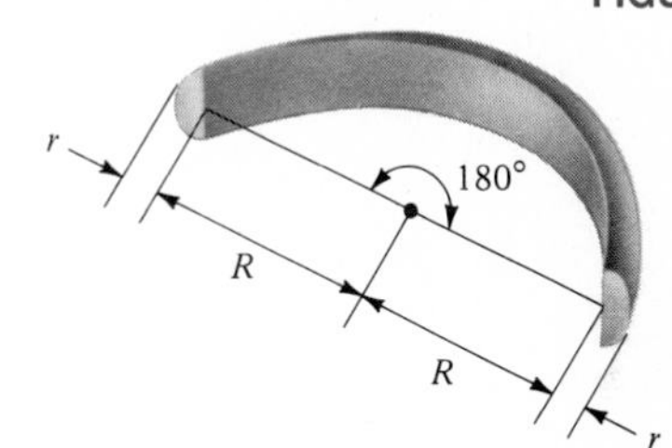

7–38 Determine the volume of the half ring described in Prob. 7–37.

7–39 Determine the total surface area and volume of a solid hemisphere of radius r.

7–40 Determine the total surface area and volume of a solid circular cone of base radius r and height h.

7–41 Determine the total surface area of the composite solid body shown in Fig. P7-31.

7–42 Determine the volume of the solid body described in Prob. 7–32.

FIGURE P7-43

Parabolic dish antenna

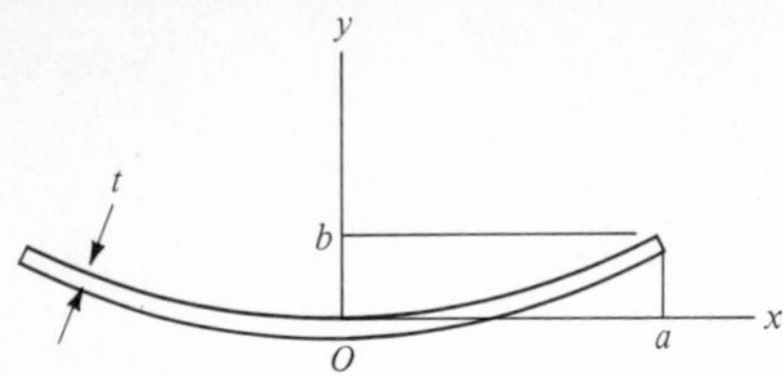

7–43 Calculate the inside surface area (toward the y axis) of the parabolic antenna if $a = 2$ m, $b = 0.5$ m. The antenna appears as a circle when viewed from the y direction. Use integral tables.

7–44 Calculate the volume of the parabolic antenna dish itself in Fig. P7-43 if $a = 8$ ft, $b = 2$ ft, and $t = 1$ in. The antenna appears as a circle when viewed from the y direction. Use integral tables.

FIGURE P7-45

Hollow shaft

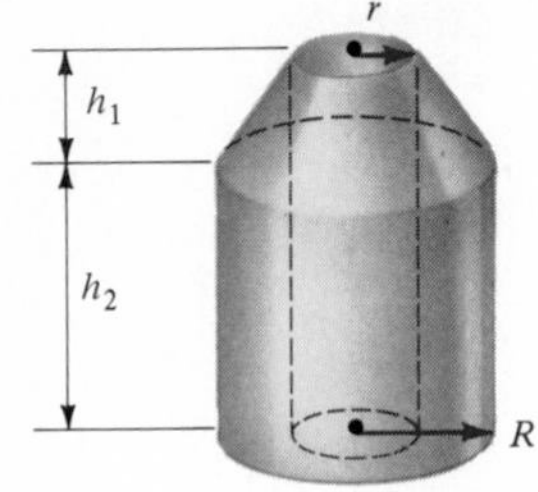

7–45 Determine the total surface area of the tapered, hollow shaft if $2r = R$ and $h_1 = h_2 = h$.

7–46 Calculate the volume of the tapered, hollow shaft in Prob. 7–45 if $r = 1$ cm, $R = 1.6$ cm, $h_1 = 20$ cm, and $h_2 = 5$ cm.

FIGURE P7-47

Concrete cooling tower, open on ends

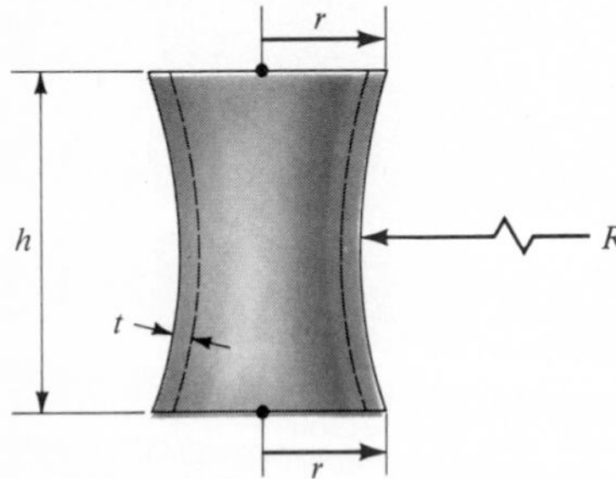

7–47 The concrete cooling tower of a power plant is modeled as an open-ended body, with the center of curvature at half height. Calculate the area of the external curved surface if $R = 300$ m, $r = 15$ m, and $h = 50$ m.

7–48 Calculate the volume of the concrete wall of the cooling tower in Prob. 7–47. Let $R = 1000$ ft, $r = 50$ ft, $h = 160$ ft, and $t = 1.7$ ft.

Beams with distributed loads on them frequently have to be analyzed. The most common such loading is from the weight of the beam itself. This and all other distributed loads acting simultaneously on a beam can be superimposed vectorially to determine the resultant force distribution along the beam. The purpose of the analysis is to determine the internal shear and bending moment at any cross section of the beam. At various stages in the solution the concept of centroid can be applied to the distributed loading to simplify the calculations, as shown in the following.

If the external reactions are required in the solution, they should be calculated first. This is illustrated for an airplane wing in Fig. 7-16a. Assume that the airplane is stationary on the ground so that the distributed load on it is caused by weight alone. The loading $w(x)$ (in terms of force per unit length) is continuous but irregular because of the numerous components in the wing. The support reactions at the fuselage are the shear force V_O and the moment M_O. The shear V_O is evaluated by integrating the load $w(x)$ on the element dx over the whole length l of the wing. For equilibrium of the wing in the vertical direction,

$$V_O = \int_0^l w(x)\,dx \qquad \text{7-19}$$

The moment M_o is evaluated by integrating the elemental moment $xw(x)\,dx$ over the length l of the wing. For the equilibrium of moments with respect to point O,

$$M_O = \int_0^l xw(x)\,dx \qquad \text{7-20}$$

Another way to state these is that V_O is equal to the total area A under the load distribution curve [since $A = \int dA = \int w(x)\,dx$], and M_O is equal to the first moment of the area A with respect to point O. In short, $V_O = A = W$, and

$$\boxed{M_O = \bar{x}A} \qquad \text{7-21}$$

where $A = W$ = resultant of the distributed forces
$\bar{x}$ = x coordinate of centroid of area A under the load distribution curve

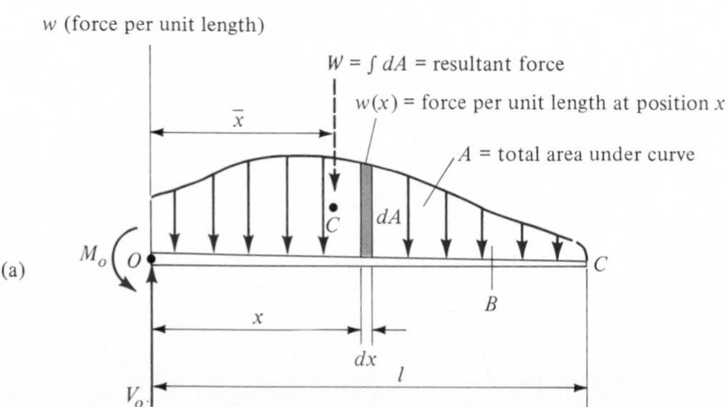

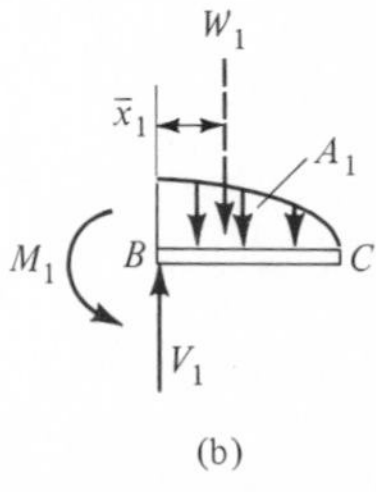

FIGURE 7-16
Distributed load on an airplane wing

Accordingly, the distributed load may be replaced by a concentrated load whose magnitude is equal to the area under the load curve and whose line of action is at the centroid of that area. This result is very useful in calculating the support reactions of a beam.

The concepts described above are also useful in determining internal forces and moments (such as V_1 and M_1 at section B in Fig. 7-16) when applied with caution. This is illustrated in Fig. 7-16b, where V_1 and M_1 can be evaluated by considering only the part of the wing that is to the right of section B. Here the area A_1 under the load distribution curve, the centroidal dimension $\bar{x}_1$ of the area A_1, and the resultant force W_1 are only for part BC of the wing and the load on it (Fig. 7-16b). Alternatively, the segment from O to B could be considered to calculate V_1 and M_1. In that case the distributed loading from O to B *and* the reactions V_O and M_O (which were obtained for the whole wing) must be used in the equilibrium equations. In any case, the correct free-body diagram for the part considered must be used.

The distributed external load on a beam causes variations in the internal shear V and moment M as functions of position along the beam. For the illustration of this, consider a small part of a beam shown in Fig. 7-17. It is assumed that the load $w(x)$ is constant for a differential element dx even if $w(x)$ varies in a macroscopic way. Assume that the shear V and moment M at section A are known from calculations involving a large segment of the beam. There must be a different shear $V + dV$ and a different moment $M + dM$ at section B since these are caused by a different load distribution than that used for section A. In other words, looking at the differential element alone, dV and dM represent the necessary difference in internal reactions between sections A and B because of the presence of $w(x)$ on the element.

The element in Fig. 7-17 is in equilibrium so the summation of vertical forces on it is zero,

$$\boxed{V - (V + dV) - w(x)\,dx = 0}$$

and

$$\boxed{w(x) = -\frac{dV}{dx}} \qquad \text{7-22}$$

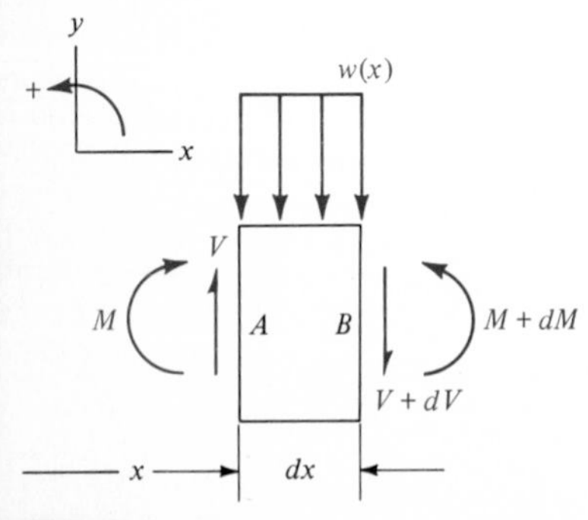

FIGURE 7-17
Free-body element of a beam

Thus, the magnitude of the slope dV/dx is equal to the negative value of the load function at that location. It should be noted that the negative sign here results from the fact that a downward load is acting on the element. This sign convention should be remembered in using Eq. 7-22. It should also be noted that in the analysis above, x is measured from the left end of the beam, and V and M are assumed to increase to the right. For x increasing to the left, $w(x) = dV/dx$.

The difference in shear forces at any two sections A and B is equal in magnitude to the area under the load distribution curve between A and B as shown in the following. Integrating Eq. 7-22,

$$\int_{V_A}^{V_B} dV = \int_{x_A}^{x_B} -w(x)\,dx$$

$$V_B - V_A = -\int_{x_A}^{x_B} w(x)\,dx,$$

or

$$V_A - V_B = \int_{x_A}^{x_B} w(x)\,dx \qquad \text{7-23}$$

$$= \text{area under } w(x)$$

The element in Fig. 7-17 is also in equilibrium with respect to moments. For analyzing this, the resultant of $w(x)$ is put at the centroid of the $w(x)$ distribution, which is halfway between A and B. Taking moments about section B,

$$-M + (M + dM) + w(x)\,dx\,\frac{dx}{2} - V dx = 0$$

Ignoring the differential of higher order (dx^2), the following simple relation is obtained for x increasing to the right,

$$\boxed{V = \frac{dM}{dx}} \qquad \text{7-24}$$

Thus, the shear at any section is equal to the slope of the moment curve. Conversely, the difference in moments at two sections A and B is the integral of the shear in that region,

$$\int_{M_A}^{M_B} dM = M_B - M_A = \int_{x_A}^{x_B} V\,dx \qquad \text{7-25}$$

where M_A is the moment at the location x_A, and so on. Another way to state Eq. 7-25 is that

$$M_B - M_A = \text{area under shear diagram from } x_A \text{ to } x_B \qquad \text{7-26}$$

Equation 7-26 provides a simple and practical way of constructing the moment diagram in some cases.

Another relationship involving the shear and the moment in a beam under continuous loading is obtained by differentiating V with respect to x. From Eq. 7-24,

$$\boxed{\frac{dV}{dx} = \frac{d^2M}{dx^2}} \qquad \text{7-27}$$

The positive signs in Eqs. 7-24 through 7-27 result from the particular loading used in Fig. 7-17. Negative signs in these equations are possible for different loadings.

EXAMPLE 7-10

Determine the shear force V and bending moment M as functions of x for the cantilever beam shown in Fig. 7-18a. Draw the shear force and bending moment diagrams.

SOLUTION

The load function $w(x)$ varies linearly with x and has a slope $\Delta w/\Delta x = w_B/l$, hence $w(x) = (w_B/l)x$. Consideration of a section K at a general position x (Fig. 7-18d) yields the shear force and internal moment as functions of x. The effect of the distributed load is found by replacing that portion (and only that portion) of the load applied to the section with its equivalent concentrated load W applied at the centroid of its area. The area of the triangular load distribution is $\frac{1}{2}bh$, where $b = x$ and $h = (w_B/l)x$, and its centroid is at $x/3$ measured from section K. Since part AK is in equilibrium,

$$\sum F_y = 0$$

$$-W - V = 0$$

$$V = -\frac{1}{2}\left(\frac{w_B}{l}x\right)x = -\frac{w_B x^2}{2l}$$

$$\sum M_K = 0$$

$$W\left(\frac{x}{3}\right) + M = 0$$

$$M = -\frac{w_B x^2}{2l}\left(\frac{x}{3}\right) = -\frac{w_B x^3}{6l}$$

It is seen that the shear curve is parabolic and the moment curve is cubic. The appropriate curves are sketched in Fig. 7-18b and c.

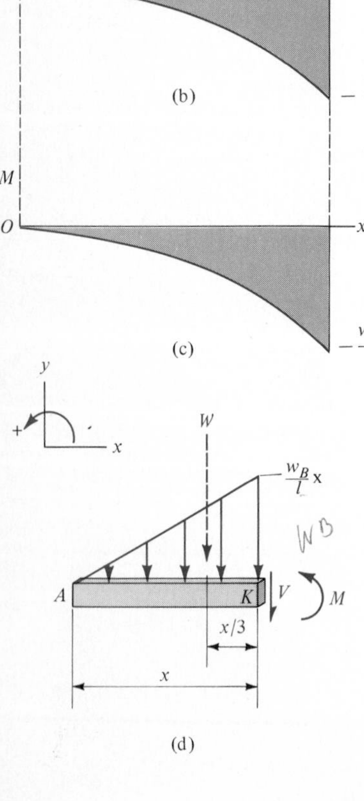

JUDGMENT OF THE RESULTS

As a check it is seen that indeed $w(x) = -dV/dx$ and $V = dM/dx$, which increase one's confidence in the results. A further check could be made by calculating the reactions at the right end using the whole beam's equilibrium and comparing them to values that are obtained using the developed shear and moment equations.

FIGURE 7-18

Shear force and bending moment diagrams of a cantilever beam

EXAMPLE 7-11

Determine the shear force V and bending moment M as functions of x for the cantilever beam shown in Fig. 7-19a. Draw the shear force and bending moment diagrams.

SOLUTION

Since the right end of the beam is unsupported, consideration of sections measured from this free end avoids the necessity of calculating the reaction at the wall. Evaluating a general section, $0 \le x \le l/2$ (Fig. 7-19d), it is seen that only that portion of distributed load up to the general location K_1 is considered. The effect of this portion of the load is handled by replacing the distributed load by its equivalent concentrated load equal to the area under the load curve located at the centroid of the area. Part BK_1 of the beam is in equilibrium, so

$$\sum F_y = 0$$

$$V_1 - w_B x = 0, \qquad V_1 = w_B x$$

$$\sum M_{K_1} = 0$$

$$M_1 + w_B x\left(\frac{x}{2}\right) = 0, \qquad M_1 = -w_B\left(\frac{x^2}{2}\right)$$

A quick check shows that $V_1 = -dM_1/dx$ and $w(x) = w_B = dV_1/dx$, which have the opposite signs than they should have by Eqs. 7-22 and 7-24. These are due to the fact that x is being measured from the right end in this example rather than from the left, which was used for the derivation of these equations. For $l/2 \le x \le l$ (Fig. 7-19e), the load curve is considered as a composite area as shown. The load W_2 is a triangle of area $\frac{1}{2}bh$. The height of the triangle h at a general position x is determined by realizing that the load curve has a linear slope $\Delta w/\Delta x = w_B/(l/2)$ and that this portion of the load magnitude varies directly with $b = x - l/2$. Hence,

$$h = \frac{w_B}{l/2}(x - l/2) \qquad \text{and} \qquad W_2 = \frac{1}{2}bh = \frac{w_B}{l}(x - l/2)^2$$

Part BK_2 is in equilibrium, so

$$\sum F_y = 0$$

$$V_2 - w_B x - \frac{w_B}{l}(x - l/2)^2 = 0, \qquad V_2 = \frac{w_B x^2}{l} + \frac{w_B l}{4}$$

$$\sum M_{K_2} = 0$$

$$-M_2 - \frac{w_B}{3l}(x - l/2)^2(x - l/2) - w_B x\left(\frac{x}{2}\right) = 0$$

$$M_2 = -\frac{w_B x^3}{3l} - \frac{w_B x l}{4} + \frac{w_B l^2}{24}$$

As a check it is found that $V_2 = -dM_2/dx$.

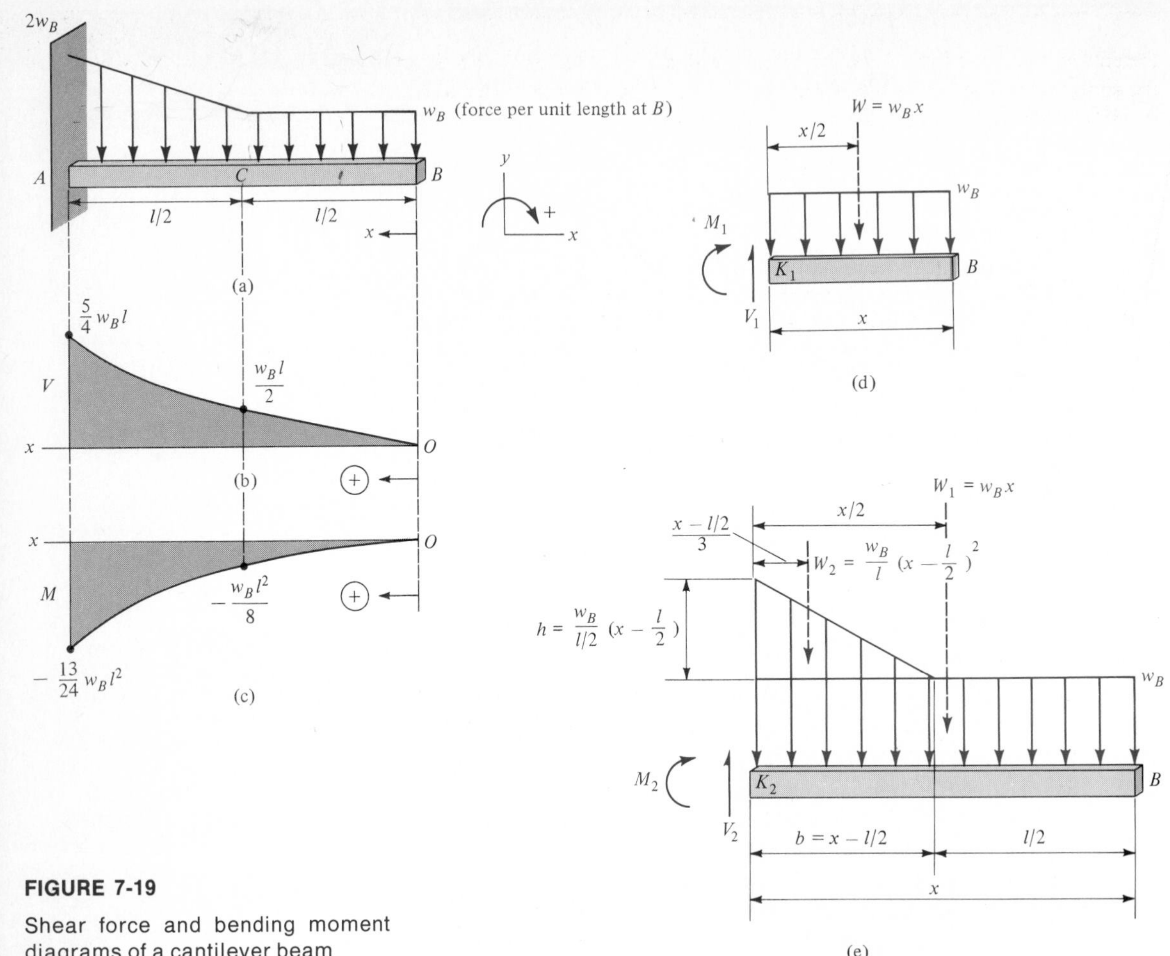

FIGURE 7-19

Shear force and bending moment diagrams of a cantilever beam

With these, relations for both the internal shear V and moment M as they vary with x have been determined for the total length of the beam. The V and M diagrams are drawn for the proper intervals as shown in Fig. 7-19b and c. Graphing is facilitated by plotting values at the end points of the intervals where each equation is valid and sketching in a curve of reasonable shape between points. For example, the V curve for the interval $l/2 \leq x \leq l$ is quadratic and is sketched appropriately.

JUDGMENT OF THE RESULTS

The checks made along the way give one confidence that the solution was proceeding correctly. One further check would be to determine the reactions at the wall to see if the shear and moment as calculated from the equations are consistent. It is seen from the equations that both the shear and moment at the free end are zero, as they should be.

EXAMPLE 7-12

Determine the shear force V and bending moment M as functions of x for the simply supported beam shown in Fig. 7-20a. Draw the shear force and bending moment diagrams.

SOLUTION

Before a general section can be considered the support reactions must be determined. However, the total weight of the distributed load must be calculated before the support reactions can be solved. It is easiest to consider the load as a composite area as shown in Fig. 7-20d.

$$W = W_1 + W_2 = (400\text{ lb/ft})(8\text{ ft}) + \frac{1}{2}(800 - 400)\text{ lb/ft}(8\text{ ft})$$

$$= 3200\text{ lb} + 1600\text{ lb} = 4800\text{ lb}$$

From symmetry,

$$R_A = R_B = \frac{1}{2}W = 2400\text{ lb}$$

A general section AK at a position x, for $0 \leq x \leq 4$ ft, is then drawn (Fig. 7-20e). The loads W'_1 and W'_2 on this section must be found first as functions of x. For the rectangular distribution, $W'_1 = 400x$ lb. The load W'_2 is the area of a triangle ($\frac{1}{2}bh$) with $b = x$ and

$$h = \frac{\Delta w}{\Delta x}x = \frac{(800 - 400)\text{ lb/ft}}{4\text{ ft}}x$$

For equilibrium of part AK,

$$\sum F_y = 0$$

$$R_A - W'_1 - W'_2 - V = 0$$

$$2400\text{ lb} - (400\text{ lb/ft})x - \frac{1}{2}\left(\frac{400\text{ lb/ft}}{4\text{ ft}}\right)(x)(x) - V = 0$$

$$V = 2400 - 400x - 50x^2\text{ (lb)}$$

$$\sum M_K = 0$$

$$-(R_A)(x) + W'_1\left(\frac{x}{2}\right) + W'_2\left(\frac{x}{3}\right) + M = 0$$

$$-2400(x) + 400\left(\frac{x^2}{2}\right) + 50\frac{x^3}{3} + M = 0$$

$$M = 2400x - 200x^2 - \frac{50}{3}x^3\text{ (ft}\cdot\text{lb)}$$

These equations are valid only for $0 \leq x \leq 4$ ft but due to the symmetry of loading only this half of the beam need be considered. The appropriate shear and moment equations are sketched in Fig. 7-20b and c.

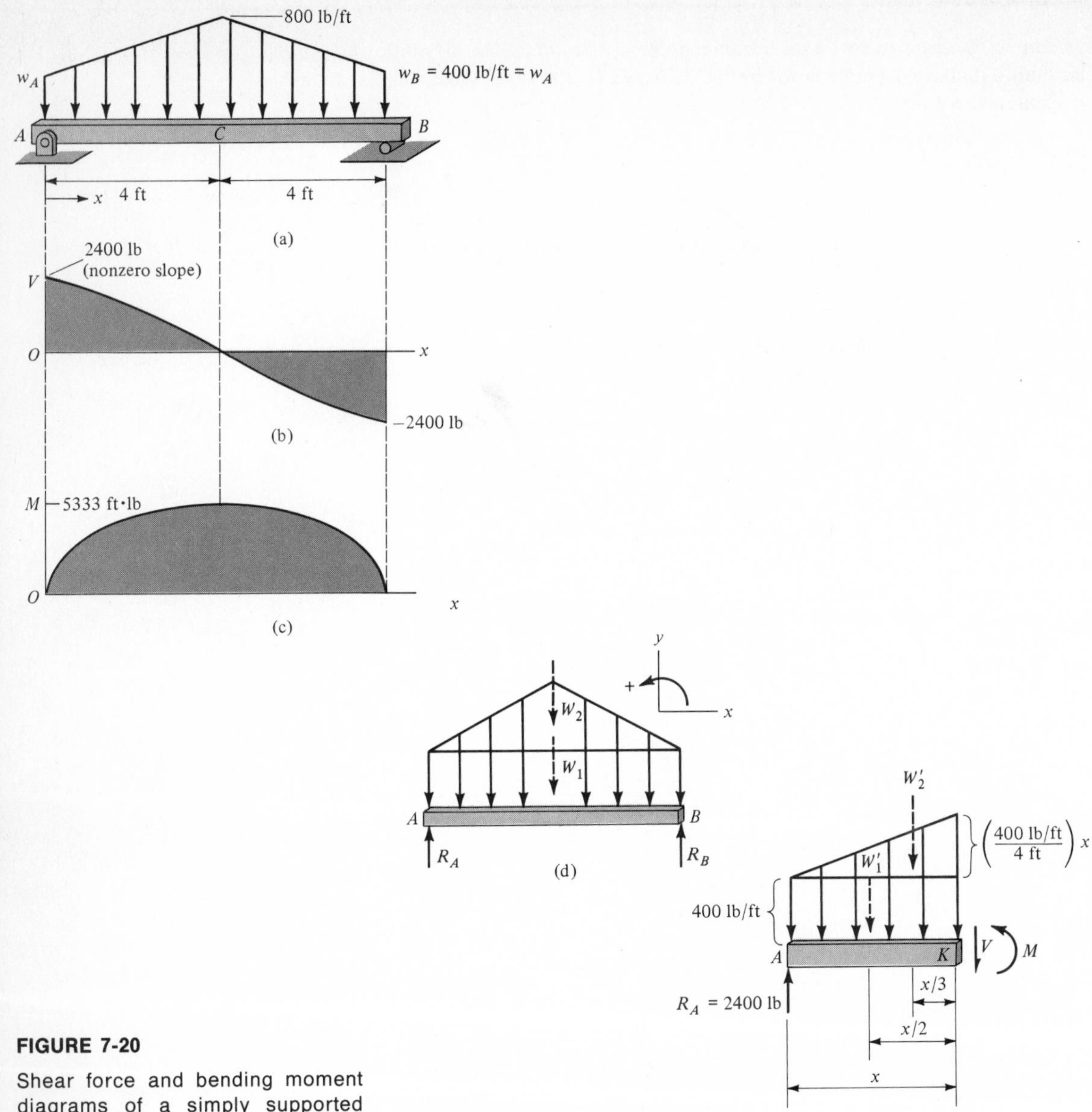

FIGURE 7-20

Shear force and bending moment diagrams of a simply supported beam

JUDGMENT OF THE RESULTS

The shapes and magnitudes of the shear force and moment diagrams appear reasonable near points A, B, and C. A check to see that $V = dM/dx$ and $w(x) = -dV/dx$ lends further confidence to the results. It should be noted that the moment equation shows that there is no moment at either end, as it should be, since simple supports such as rollers cannot apply a moment.

EXAMPLE 7-13

Determine the shear force V and bending moment M as functions of x for the beam shown in Fig. 7-21a. Draw the shear force and bending moment diagrams. $P = 2wa$, $M_O = 0.3Pa = 0.6wa^2$.

SOLUTION

First the distributed load is replaced by its equivalent concentrated load W at the centroid of the load area and the support reactions are calculated. From Fig. 7-21d,

$$\sum M_A = 0$$

$$-0.6wa^2 + 2wa(0.6a) + R_B a = 0$$

$$R_B = -0.6wa$$

$$\sum M_B = 0$$

$$-0.6wa^2 + 2wa(1.6a) + w(2a)(a) - R_A a = 0$$

$$R_A = 4.6wa \qquad \text{(for a check, } R_A + R_B = P + W\text{)}$$

General sections between the loading discontinuities are sketched and the equations of equilibrium used to find the internal shear and moment as functions of x.

REGION 1, $0 \le x \le 0.4a$ (Fig. 7-21e):

$$\sum Fy = 0$$

$$-wx - V_1 = 0$$

$$V_1 = -wx$$

$$\sum M_{K_1} = 0$$

$$-0.6wa^2 + wx\left(\frac{x}{2}\right) + M_1 = 0$$

$$M_1 = 0.6wa^2 - w\left(\frac{x^2}{2}\right)$$

REGION 2, $0.4a \le x \le a$ (Fig. 7-21f):

$$\sum F_y = 0$$

$$-wx - 2wa - V_2 = 0$$

$$V_2 = -w(2a + x)$$

$$\sum M_{K_2} = 0$$

$$-0.6wa^2 + wx\left(\frac{x}{2}\right) + 2wa(x - 0.4a) + M_2 = 0$$

$$M_2 = -w\left(\frac{x^2}{2} + 2ax - 1.4a^2\right)$$

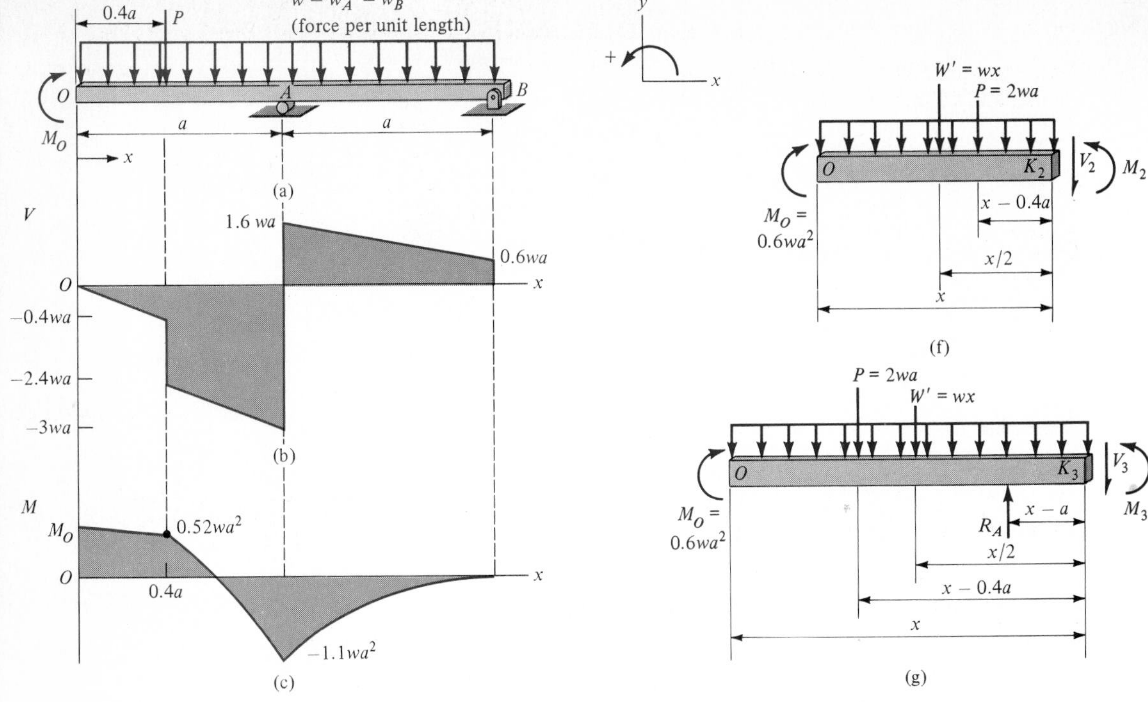

REGION 3, $a \leq x \leq 2a$ (Fig. 7-21g):

$$\sum F_y = 0$$

$$-2wa - wx + 4.6wa - V_3 = 0$$

$$V_3 = 2.6 - wx$$

$$\sum M_{K_3} = 0$$

$$-0.6wa^2 + 2wa(x - 0.4a) + wx\left(\frac{x}{2}\right) - 4.6wa(x - a) + M_3 = 0$$

$$M_3 = -w\left(\frac{x^2}{2} - 2.6ax + 3.2a^2\right)$$

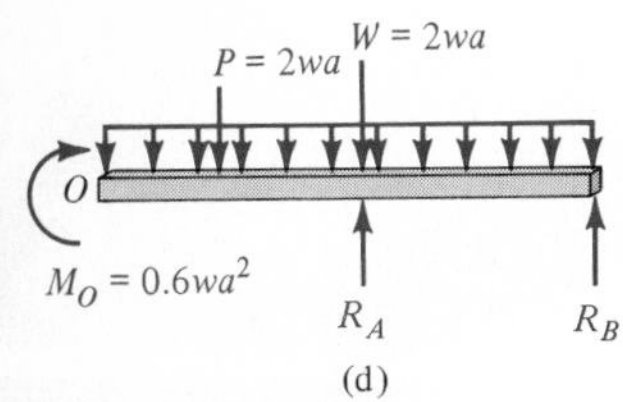

The curves are drawn by plotting the shear and moment values at the end points of each interval for which each equation is valid and sketching a curve of appropriate order between the plotted points (Fig. 7-21b and c).

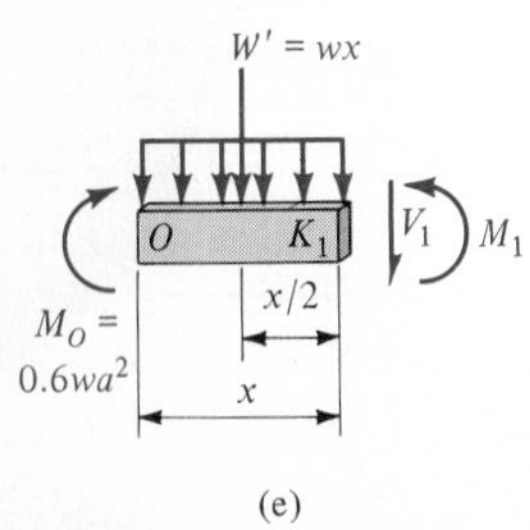

FIGURE 7-21

Analysis of a beam under complex loading

JUDGMENT OF THE RESULTS

Several specific checks give one confidence in the results. The shear curve ends at $V = 0.6wa$, which is equal but opposite in sign to the reaction at B, as it should be. The moment curve goes to zero at the right end, as it should, since no moment can be applied by a pin. One further check would be to see if Eq. 7-26 holds.

SECTION 7-5 PROBLEMS

7–49 Determine the shear force V and bending moment M as functions of x.

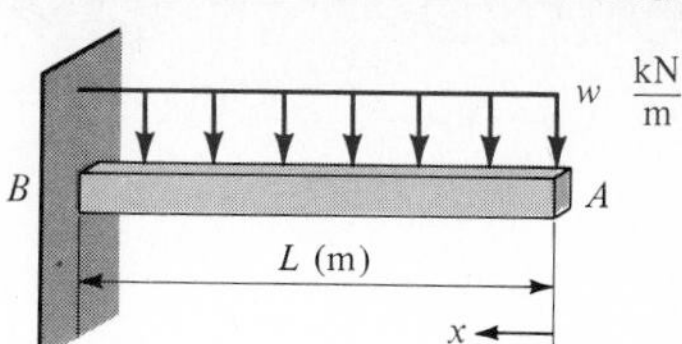

FIGURE P7-49

7–50 Draw the shear and moment diagrams for the cantilever beam in Fig. P7-49.

7–51 Determine the shear force V and bending moment M as functions of x.

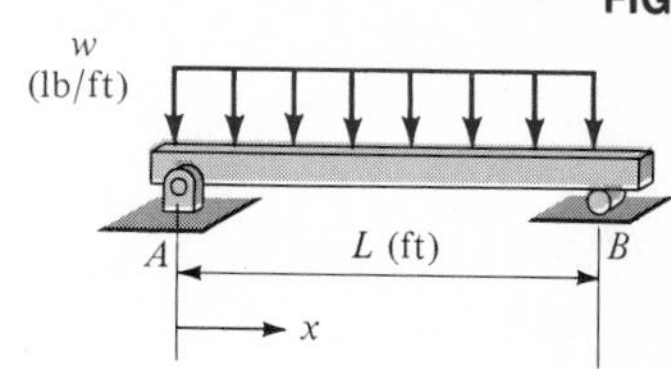

FIGURE P7-51

7–52 Draw the shear and moment diagrams for the beam in Fig. P7-51.

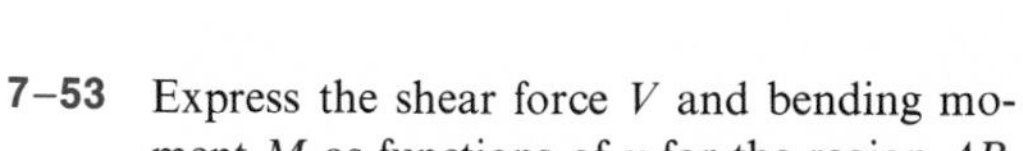

7–53 Express the shear force V and bending moment M as functions of x for the region AB.

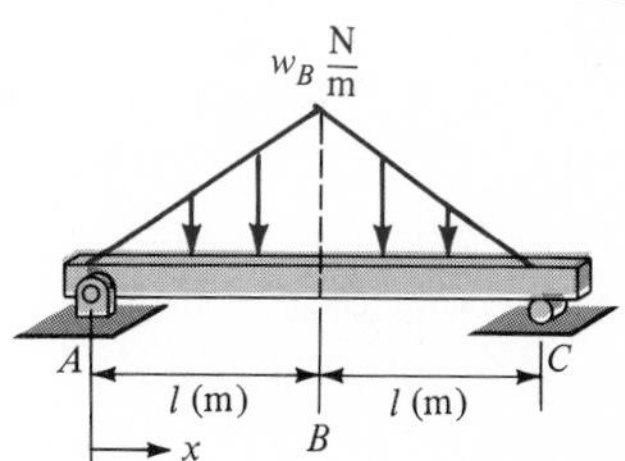

FIGURE P7-53

7–54 Draw the complete shear and moment diagrams for the beam in Fig. P7-53.

7–55 Determine the shear force V and bending moment M as functions of x.

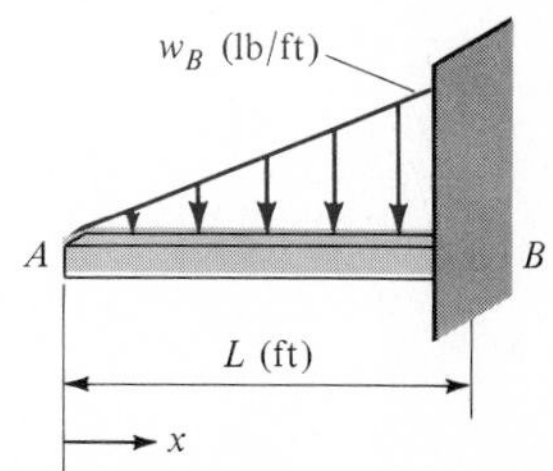

FIGURE P7-55

7–56 Draw the shear and moment diagrams for the beam in Fig. P7-55.

FIGURE P7-57

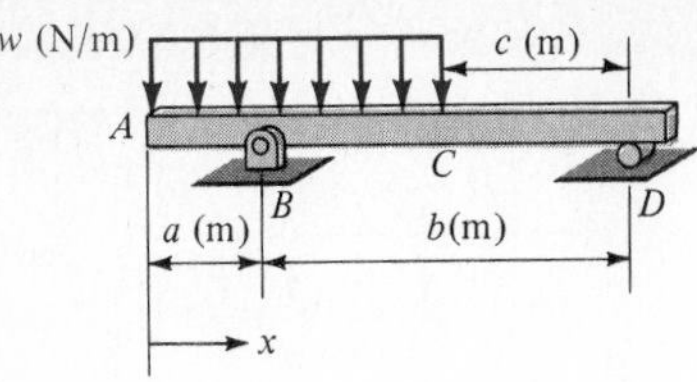

7–57 Determine the shear force V and bending moment M as functions of x for the region BC.

7–58 Draw the complete shear and moment diagrams for the beam in Fig. P7-57.

FIGURE P7-59

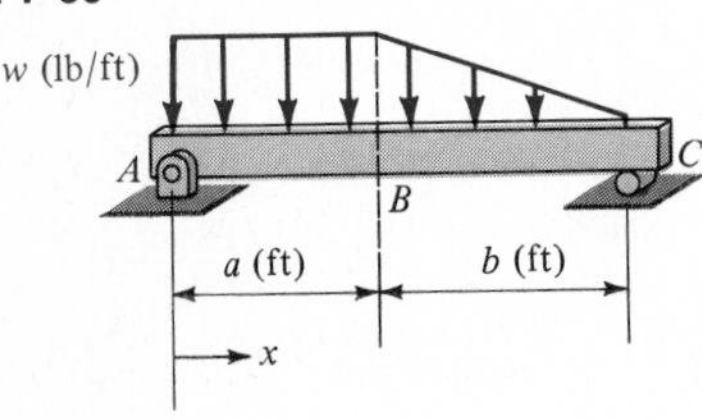

7–59 Express the shear force V and bending moment M as functions of x for the region BC.

7–60 Draw the complete shear and moment diagrams for the beam in Fig. P7-59.

FIGURE P7-61

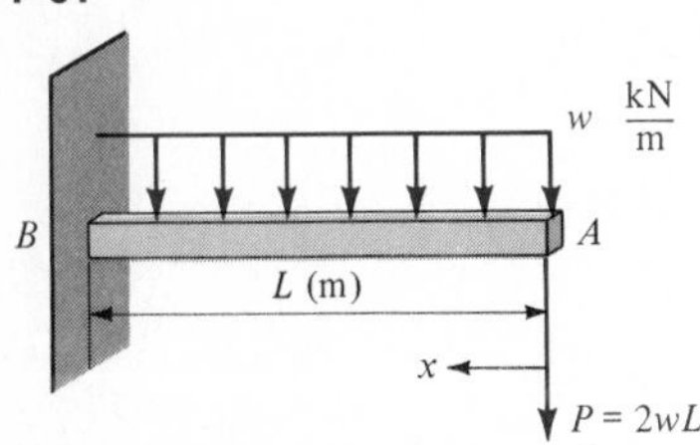

7–61 Determine the shear force V and bending moment M as functions of x.

7–62 Draw the shear and moment diagrams for the beam in Fig. P7-61.

FIGURE P7-63

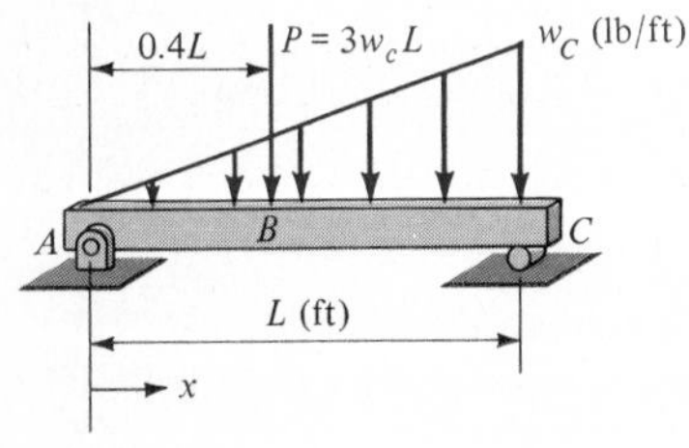

7–63 Express the shear force V and bending moment M as functions of x for the region AB.

7–64 Draw the complete shear and moment diagrams for the beam in Fig. P7-63.

FIGURE P7-65

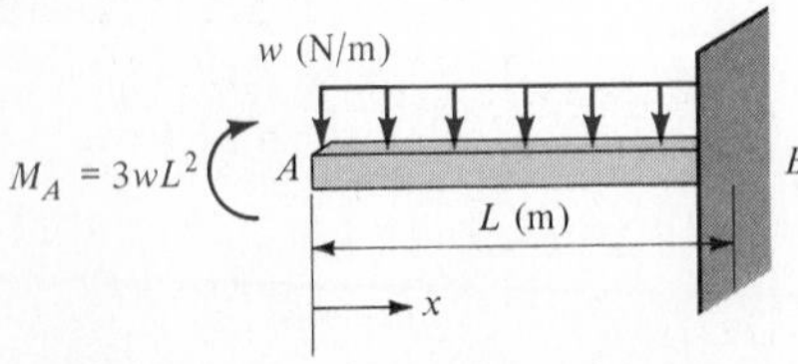

7–65 Determine the shear force V and bending moment M as functions of x. M_A is an external moment.

7–66 Draw the shear and moment diagrams for the beam in Fig. P7-65.

7–67 Express the shear force V and bending moment M as functions of x for the region AB. M_B is an external moment.

FIGURE P7-67

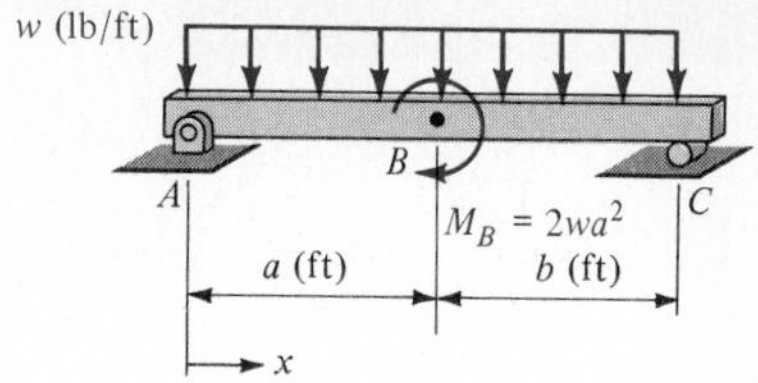

7–68 Draw the complete shear and moment diagrams for the beam in Fig. P7-67.

7–69 An airplane wing is modeled as a uniform beam with a linearly varying net upward force on it. Determine the shear force V and bending moment M as functions of x.

FIGURE P7-69

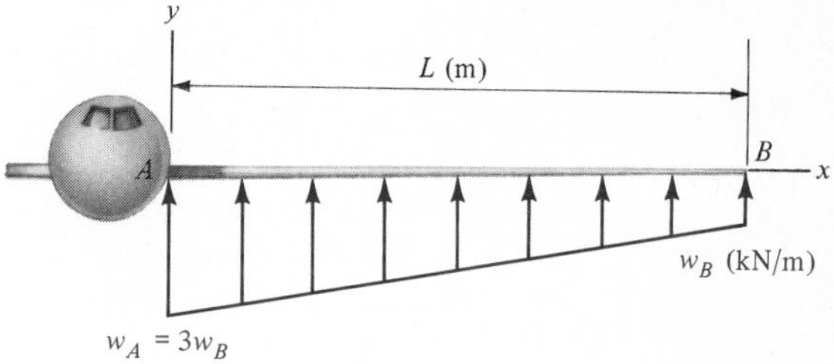

7–70 Draw the shear and moment diagrams for the wing in Fig. P7-69.

7–71 The boom of a crane is modeled as a uniform beam of 1.2-kN weight. It is hinged at A and supported by a wire at B. Draw the bending moment diagram for the beam.

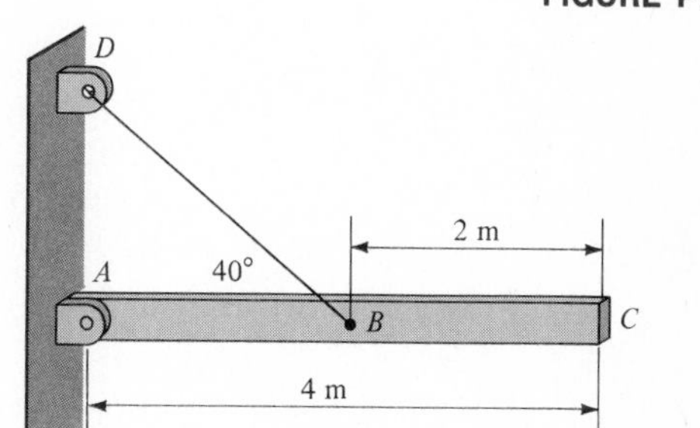

FIGURE P7-71

7–72 Solve Prob. 7–71 assuming that a concentrated vertical load of 3 kN is applied at point C.

7–73 A uniformly distributed force of 200 lb caused by wind is acting normal to the highway sign BC. Draw the bending moment diagram for the arm AC assuming that the wind causes no force acting directly on part AB. Neglect weight.

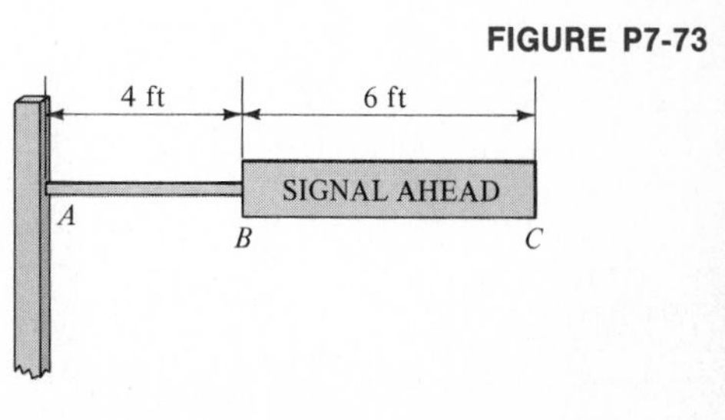

FIGURE P7-73

7-6 DISTRIBUTED LOADS ON CABLES

The mechanics analysis of flexible cables with distributed loads is somewhat complex, but it is important to understand the effects of self weight, ice, or wind on such members. The basic assumptions in the analysis are that a flexible cable has practically no resistance to bending and that the internal force at any point is tangent to the cable at that point. In the following discussions $w(x)$ denotes a continuous, but possibly variable, distributed load applied to a cable. The units of $w(x)$ are always force per unit length.

General Considerations

Assuming that the whole cable is in equilibrium, a differential element of it shown in Fig. 7-22 is also in equilibrium. Here $w(x)\,dx$ is the resultant force of $w(x)$ distributed over the distance dx of the cable. On the basis of equilibrium in all directions, two independent equations can be written for the small element of the cable.

$$\sum F_x = 0$$

$$(T + dT)\cos(\theta + d\theta) - T\cos\theta = 0 \qquad \text{7-28}$$

$$\sum F_y = 0$$

$$(T + dT)\sin(\theta + d\theta) - T\sin\theta - w(x)\,dx = 0 \qquad \text{7-29}$$

These can be simplified using the trigonometric identities for the sum of two angles and the approximations $\sin d\theta = d\theta$ and $\cos d\theta = 1$ for the very small angle $d\theta$. The resulting two equations are, respectively,

$$d(T\cos\theta) = 0 \qquad \text{7-30}$$

$$d(T\sin\theta) = w(x)\,dx \qquad \text{7-31}$$

Equation 7-30 shows that the horizontal component of the tension is constant along the cable, as can be expected by inspection of the element. This component is often denoted by T_o. Using $T\cos\theta = T_o$, $\tan\theta = dy/dx$, and Eq. 7-31 allows one to write

$$\frac{d^2y}{dx^2} = \frac{d}{dx}\left(\frac{\sin\theta}{\cos\theta}\right) = \frac{d}{dx}\left(\frac{T\sin\theta}{T_o}\right)$$

$$\boxed{\frac{d^2y}{dx^2} = \frac{w(x)}{T_o}} \qquad \text{7-32}$$

The final result is the most useful differential equation of the cable. Integrating Eq. 7-32 gives the equation for the slope dy/dx of the cable, and integrating the latter gives the equation for the position y

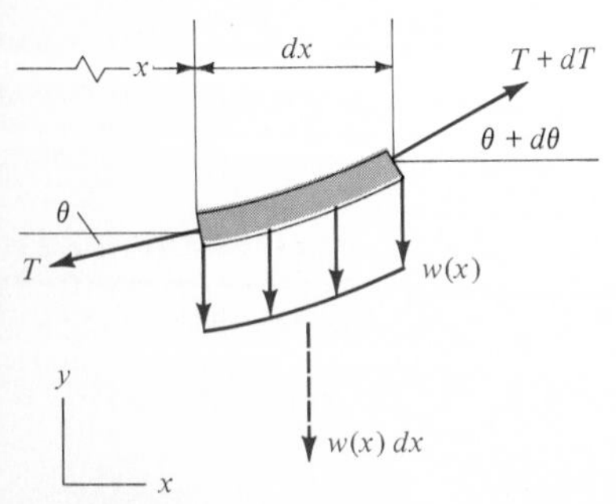

FIGURE 7-22
Free-body element of a cable

as a function of x. These are illustrated for two important classes of cables in the following.

Parabolic Cables

Sometimes interesting reversals are made in the modeling process. Clearly, a distributed load on a cable may be represented by numerous concentrated forces for all practical purposes. Conversely, it is advantageous to substitute (in the analysis) a simple function of a distributed load for a large number of concentrated forces that would be cumbersome to use in detail. Common examples of such problems are suspension bridges (Fig. 6-23a).

In a suspension bridge the weight of the cable is negligible compared to the weight of the roadway. The latter is often a constant load w per unit length measured horizontally in the x direction. It is convenient to place the origin of the coordinate axes at the lowest point on the cable (Fig. 7-23a), which must be halfway between the supports if they are of the same height. Since w and T_o are constant, Eq. 7-32 is readily integrated to give

$$\frac{dy}{dx} = \frac{wx}{T_o} + C$$

With the slope of the cable equal to zero at the lowest point, C is also zero, so the slope at any position x is

$$\frac{dy}{dx} = \frac{wx}{T_o} \qquad \text{7-33}$$

Integration of Eq. 7-33 gives

$$y = \frac{wx^2}{2T_o} + C_1$$

where $C_1 = 0$ because $y = 0$ at $x = 0$. Thus,

$$\boxed{y = \frac{wx^2}{2T_o}} \qquad \text{7-34}$$

which is the equation of a parabola.

Designers are most interested in the magnitudes of the tension in the cable. A simple formula that gives the tension T as a function of the horizontal position x along the cable can be obtained with the aid of Fig. 7-23b. The segment of the cable to the right of the origin may be of any finite length x. The resultant of the distributed load on this segment is wx. Since $T_y = wx$,

$$\boxed{T = \sqrt{T_o^2 + w^2x^2}} \qquad \text{7-35}$$

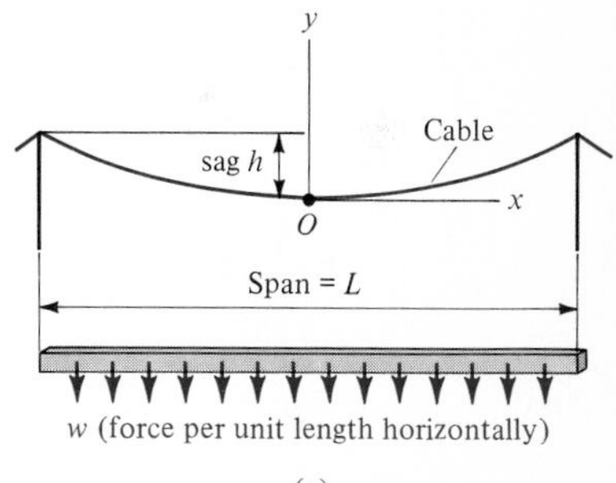

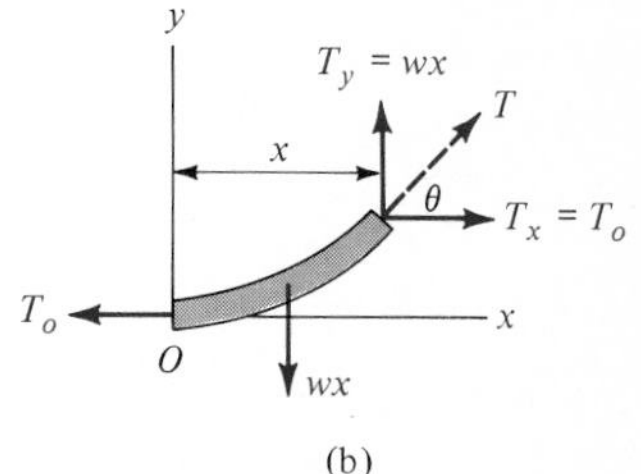

FIGURE 7-23

Parabolic cable of a suspension bridge

Clearly, the minimum tension is at the lowest point of the cable ($x = 0$), and the maximum tension is at the supporting towers ($x = L/2$). Putting the values of $x = L/2$ and $y = h$ in Eq. 7-34,

$$T_{\min} = T_o = \frac{wL^2}{8h} \tag{7-36}$$

Note that T_o can be obtained by using Eq. 7-34 for any point on the curve except $x = y = 0$. The maximum tension is obtained from Eqs. 7-35 and 7-36,

$$T_{\max} = \sqrt{\left(\frac{wL^2}{8h}\right)^2 + w^2\left(\frac{L}{2}\right)^2} = \frac{wL}{2}\sqrt{1 + \frac{L^2}{16h^2}} \tag{7-37}$$

The concepts and results discussed above are also relevant to unsymmetric cables where the supports are at different elevations, as in Fig. 7-24. Using the appropriate values of x and y for each support in Eqs. 7-34 and 7-35, the tensions at points A and B are

$$T_A = wa\sqrt{1 + \frac{a^2}{4h_A^2}} \quad \text{and} \quad T_B = wb\sqrt{1 + \frac{b^2}{4h_B^2}} \tag{7-38}$$

The lengths S_A or S_B of the cable from the lowest point to the supports can be expressed as in elementary calculus for the length of a curve. For example, referring to Fig. 7-24,

$$S_A = \int_0^a \sqrt{1 + \left(\frac{dy}{dx}\right)^2}\, dx = \int_0^a \sqrt{1 + \frac{w^2x^2}{T_o^2}}\, dx$$

Substituting $T_o = wa^2/2h_A$ from Eq. 7-34,

$$S_A = \int_0^a \sqrt{1 + \frac{4h_A^2x^2}{a^4}}\, dx \tag{7-39}$$

This is most easily evaluated by using the binomial theorem to get an infinite series whose first few terms can be integrated, resulting in

$$S_A = a\left(1 + \frac{2h_A^2}{3a^2} - \frac{2h_A^4}{5a^4} + \ldots\right) \tag{7-40}$$

This series converges for all values of $h_A/a < 0.5$, and usually the first two or three terms give a reasonable approximation to the cable length. The segment S_B can be evaluated similarly.

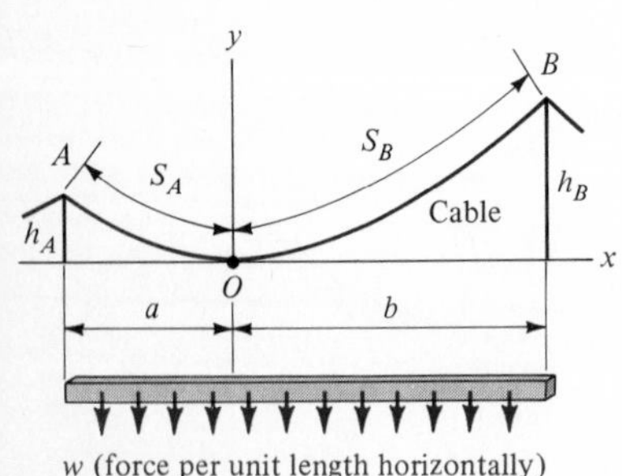

FIGURE 7-24
Unsymmetric suspension cable

Catenary Cables

Frequently, a cable has to support a weight that is uniformly distributed along the cable itself (self weight, ice). This is in contrast to the parabolic cable, whose own weight is ignored and which supports a load uniformly distributed in the horizontal direction, not along the cable. The shape of a cable under its own weight is called a *catenary*. For a tight cable (with a small ratio of sag to span) this shape is nearly the same as that of a parabolic cable and can be analyzed accordingly. More deeply sagging cables must be analyzed differently, as illustrated with the aid of Fig. 7-25.

In Fig. 7-25a the origin of the coordinate axes is placed at the lowest point on the cable, as before. A small part of the cable to the right of the origin is drawn in Fig. 7-25b. The length of this piece is s, so the resultant of the distributed load on it is ws. For equilibrium of any infinitesimal element ds in the y direction, Eq. 7-32 can be rewritten (after multiplying it by dx/dx) as

$$\frac{d^2y}{dx^2} = \frac{w}{T_o}\frac{ds}{dx} \qquad \boxed{7\text{-}41}$$

Since s is a function of x and y, ds can be expressed in terms of dx and dy, using $(ds)^2 = (dx)^2 + (dy)^2$. On this basis, Eq. 7-41 becomes

$$\frac{d^2y}{dx^2} = \frac{w}{T_o}\sqrt{1 + \left(\frac{dy}{dx}\right)^2} \qquad \boxed{7\text{-}42}$$

which is the differential equation of the catenary curve. To solve this equation, let $dy/dx = f$. With this substitution,

$$\frac{df}{\sqrt{1 + f^2}} = \frac{w}{T_o}\,dx$$

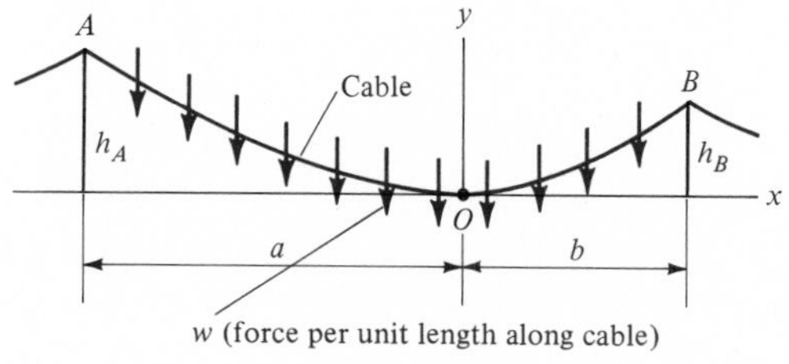

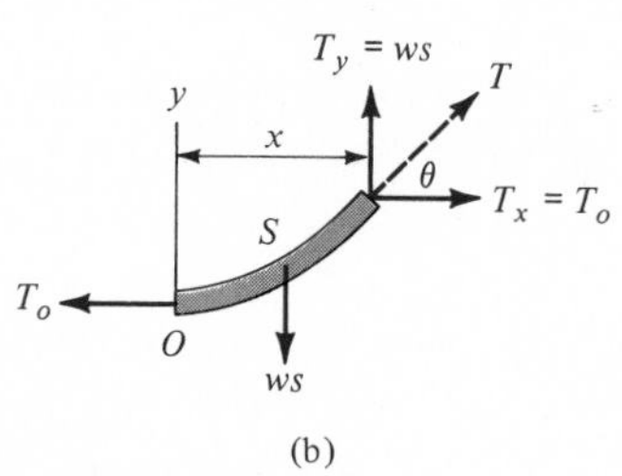

FIGURE 7-25

Model for a catenary cable

This can be integrated at once, and

$$\ln(f + \sqrt{1 + f^2}) = \frac{w}{T_o}x + C_1 \tag{7-43}$$

Since the slope $dy/dx = 0$ at $x = 0$, the constant $C_1 = 0$. Equation 7-43 can be further manipulated to advantage as follows:

$$f + \sqrt{1 + f^2} = e^{(w/T_o)x}$$

$$1 + f^2 = (e^{(w/T_o)x} - f)^2$$

$$= e^{2(w/T_o)x} - 2e^{(w/T_o)x}f + f^2$$

$$f = \frac{dy}{dx} = \frac{e^{(w/T_o)x} - e^{-(w/T_o)x}}{2} = \sinh\frac{wx}{T_o} \tag{7-44}$$

The position y of the cable is obtained by integrating Eq. 7-44,

$$y = \frac{T_o}{w}\cosh\frac{wx}{T_o} + C_2$$

Since $y = 0$ at $x = 0$, the constant $C_2 = -T_o/w$, so

$$y = \frac{T_o}{w}\left(\cosh\frac{wx}{T_o} - 1\right) \tag{7-45}$$

This is the equation of the shape of the catenary curve.

The length s of the cable in Fig. 7-25b (measured from point O) is obtained from Eq. 7-44. With $\tan\theta = dy/dx = ws/T_o$ from Fig. 7-25b,

$$s = \frac{T_o}{w}\sinh\frac{wx}{T_o} \tag{7-46}$$

The tension T in the cable is also obtained from Fig. 7-25b,

$$T^2 = T_x^2 + T_y^2 = T_o^2 + w^2s^2$$

Using Eq. 7-46,

$$T^2 = T_o^2\left(1 + \sinh^2\frac{wx}{T_o}\right) = T_o^2\cosh^2\frac{wx}{T_o}$$

The result can be finally simplified using Eq. 7-45,

$$\boxed{T = T_o\cosh\frac{wx}{T_o} = T_o + wy} \tag{7-47}$$

The equations for catenary cables given above can be solved graphically or by trial and error (where computers are helpful).

EXAMPLE 7-14

Determine the position of the cable shown in Fig. 7-26a at $x = 5$ m for two different loadings: (a) $w_1 = 100$ N/m = constant load per unit length measured horizontally, and (b) $w_2 = 100$ N/m = constant load per unit length measured along the cable. The sag $h = 5$ m and the span $L = 20$ m are the same for both loadings. Sketch the two cable shapes on the same diagram using the given and calculated few positions. Calculate the minimum and maximum tensions for both loadings.

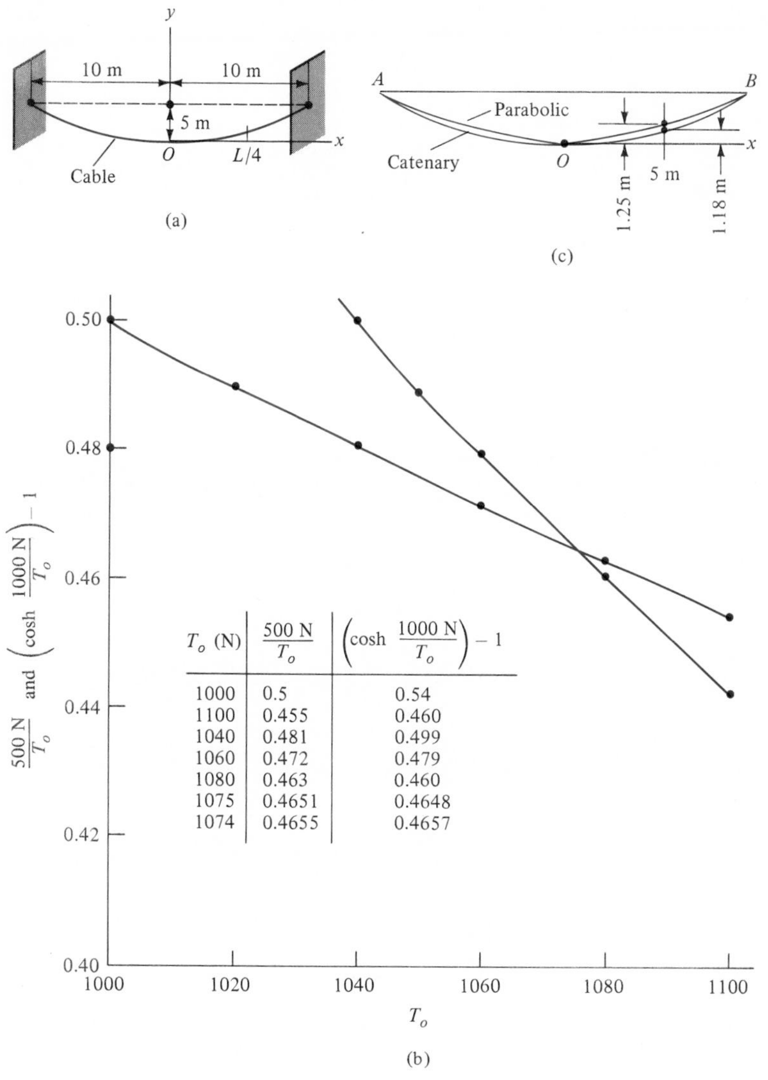

T_o (N)	$\frac{500\text{ N}}{T_o}$	$\left(\cosh \frac{1000\text{ N}}{T_o}\right) - 1$
1000	0.5	0.54
1100	0.455	0.460
1040	0.481	0.499
1060	0.472	0.479
1080	0.463	0.460
1075	0.4651	0.4648
1074	0.4655	0.4657

FIGURE 7-26

Cable shapes under two different loadings

SOLUTION

(a) The cable is parabolic and Eqs. 7-33 to 7-40 apply. The minimum tension occurs at the midpoint of the cable. From Eq. 7-36,

$$T_o = \frac{wL^2}{8h} = \frac{(100\text{ N/m})(20\text{ m})^2}{8(5\text{ m})} = 1000\text{ N} \quad \text{(at midpoint)}$$

The maximum tension occurs at the supports (at $x = 10$ m). From Eq. 7-37,

$$T_{\max} = \frac{wL}{2}\sqrt{1 + \frac{L^2}{16h^2}}$$

$$= \frac{(100\text{ N/m})(20\text{ m})}{2}\sqrt{1 + \frac{(20\text{ m})^2}{16(5\text{ m})^2}} = 1410\text{ N}$$

The position y at $x = 5$ m is obtained using Eq. 7-34,

$$y = \frac{wx^2}{2T_o} = \frac{(100\text{ N/m})(5\text{ m})^2}{2(1000\text{ N})} = 1.25\text{ m}$$

(b) The cable makes a catenary curve and Eqs. 7-41 to 7-47 apply. By knowing the coordinates of any point on the cable, Eq. 7-45 can be used to evaluate the magnitude of T_o. The point (0, 0) is known but is of no help since Eq. 7-45 yields the trivial result that $0 = 0$ for this point. The point (5, 10), however, does allow for determination of T_o, but it can only be solved by trial and error. Solution is facilitated by rearranging Eq. 7-45 as shown below and solving graphically in Fig. 7-26b.

$$\frac{yw}{T_o} = \left(\cosh\frac{wx}{T_o}\right) - 1 = \frac{(5\text{ m})(100\text{ N/m})}{T_o}$$

$$= \left(\cosh\frac{(100\text{ N/m})(10\text{ m})}{T_o}\right) - 1$$

$$\frac{5000\text{ N}}{T_o} = \left(\cosh\frac{1000\text{ N}}{T_o}\right) - 1, \qquad T_o = 1075\text{ N}$$

From Eq. 7-47, $T = T_o + wy$.

Therefore,

$$T_{\min} = T_o = 1075\text{ N at } y = 0 \qquad \text{(midpoint of cable)}$$

and

$$T_{\max} = 1075 + \left(100\frac{\text{N}}{\text{m}}\right)(5\text{ m}) = 1575\text{ N} \qquad \text{(at support)}$$

The height at $x = 5$ m is obtained from Eq. 7-45,

$$y = \frac{1075\text{ N}}{100\text{ N/m}}\left(\cosh\frac{(100\text{ N/m})(5\text{ m})}{1075\text{ N}} - 1\right) = 1.18\text{ m}$$

Sketches of the cables are shown in Fig. 7-26c.

It is seen in this case that the maximum tensions in parabolic and catenary cables do not differ much. However, cables with larger sag (h/L) ratios or larger distributed loads will have larger differences between the maximum tensions in parabolic and catenary cables under similar loading.

PROBLEMS

7–74 The cable supports a load of $w = 1$ kN/m (uniformly distributed horizontally). Calculate the tensions in the cable at the supports if $L = 50$ m and $h = 5$ m.

FIGURE P7-74

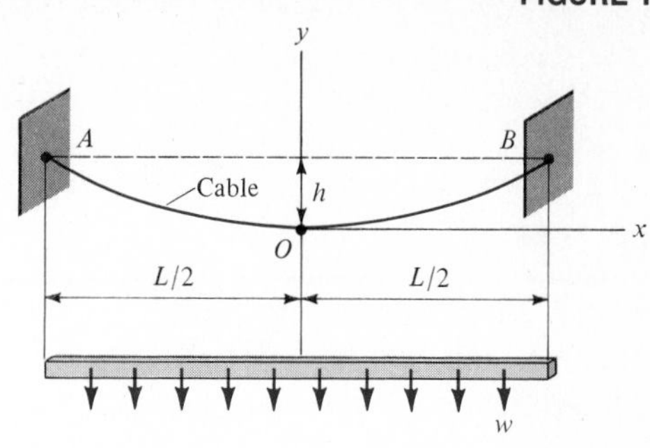

7–75 Determine the maximum tension in the cable of Fig. P7-74 and the position of the cable at $x = L/4$ if $w = 300$ lb/ft (uniformly distributed horizontally), $L = 200$ ft, and $h = 15$ ft.

7–76 Consider Prob. 7–74 and sketch the cable shapes on a single diagram using $h = 3$ m, 5 m, and 7 m.

7–77 The cable supports a load of $w = 2$ kN/m (uniformly distributed horizontally). Determine the tensions in the cable at the supports if $a = 40$ m, $b = 60$ m, and $h_A = 5$ m.

FIGURE P7-77

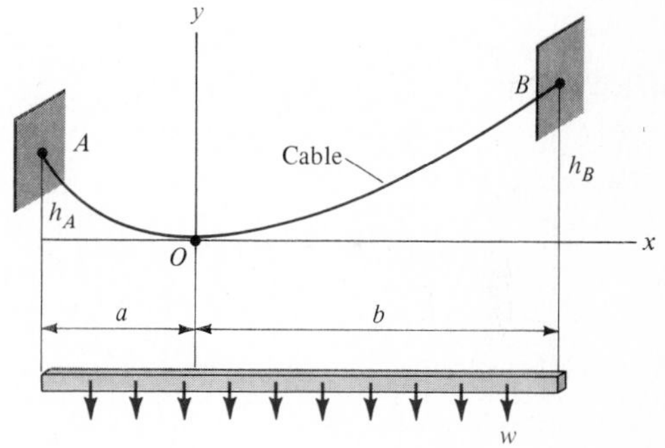

7–78 Determine the tensions at points A, O, and B in the cable in Fig. P7-77. Let $w = 200$ lb/ft (uniformly distributed horizontally), $a = 400$ ft, $b = 150$ ft, and $h_B = 10$ ft.

7–79 Plot the maximum tension in Prob. 7–78 as a function of $h_B = 10$ ft, 20 ft, and 30 ft.

FIGURE P7-80

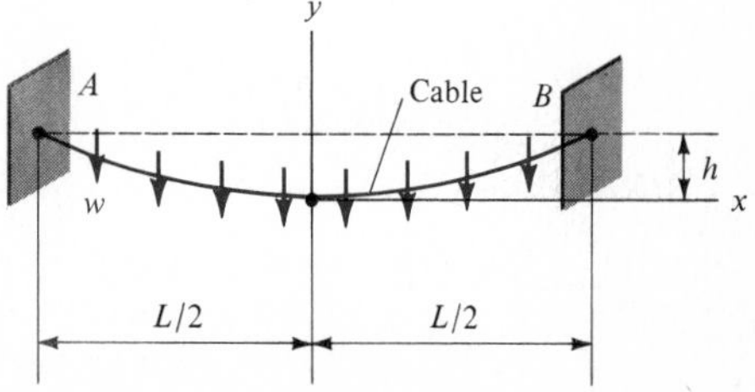

7–80 The cable supports a load of $w = 10$ N/m uniformly distributed along the cable. Calculate the tensions in the cable at the supports if $L = 70$ m and $h = 8$ m.

7–81 Determine the maximum and minimum tensions in the cable in Fig. P7-80. Also calculate the position of the cable at $x = L/4$. Let $w = 2$ lb/ft distributed uniformly along the cable, $L = 200$ ft, and $h = 18$ ft.

FIGURE P7-83

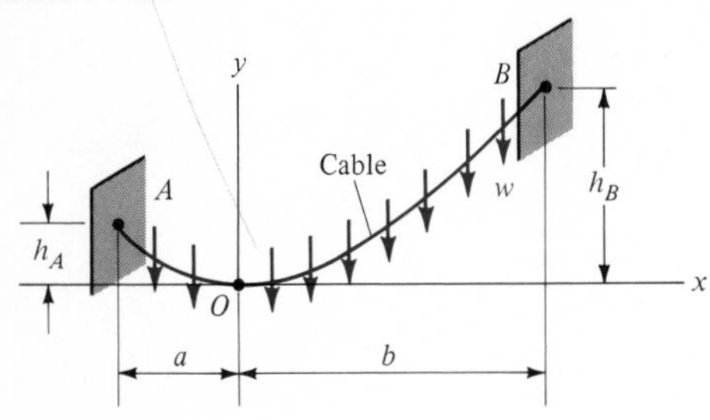

7–82 Plot the maximum tension in Prob. 7–80 as a function of $h = 8$ m, 10 m, and 12 m.

7–83 Determine the maximum tension in the cable if $w = 5$ N/m uniformly distributed along the cable, $a = 30$ m, $b = 60$ m, and $h_A = 4$ m.

FIGURE P7-84

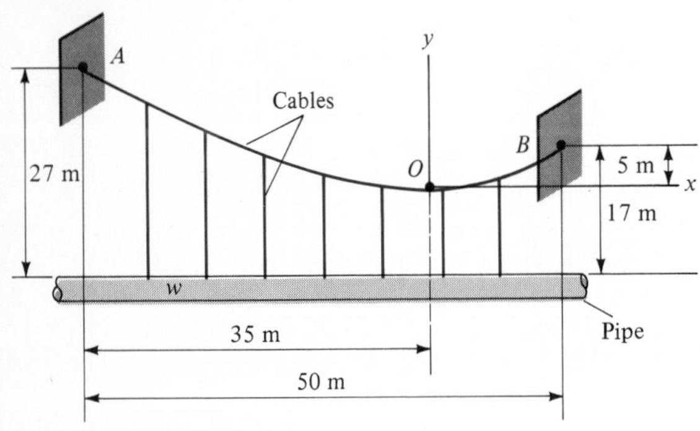

7–84 A proposed pipeline must cross a ravine and is to be supported by a cable. The pipe weighs $w = 5$ kN/m (uniformly distributed horizontally). Calculate the maximum tension in the cable AB.

7-7 FLUID STATICS

A common problem area involving distributed forces is fluid mechanics. Fluids are classified as *liquids* and *gases*, or, on the basis of relative constancy of volume under the action of forces, as *incompressible* and *compressible fluids.* An important characteristic of all fluids is that at rest they can support only normal forces and not shear forces. This is the basis for Pascal's law, according to which *the pressure is the same in all directions at any given point in a fluid at rest.* Pressure is defined as the force per unit area, p = force/area, and it always implies compression in fluid statics. Pascal's law is intuitively obvious to many people (the pressure on the ear in deep water is independent of body orientation).

To illustrate Pascal's law, consider an infinitesimal element of fluid. The rectangle in Fig. 7-27a represents a three-dimensional element of dimensions dx, dy, dz, and weight dW. Pressures $p_1, \ldots$ act on the element, while it is assumed that no shear forces are acting on it. For equilibrium in the x direction,

$$\sum F_x = 0$$

$$\underbrace{\underbrace{p_1}_{\text{pressure}}\ \underbrace{dy\,dz}_{\text{area}}}_{\text{force}} - p_2\,dy\,dz = 0, \qquad \text{so } p_1 = p_2$$

A similar statement can be made for the pressures acting in the z direction. In the y direction,

$$\sum F_y = 0$$

$$p_3\,dx\,dz - p_4\,dx\,dz - dW = 0$$

where $dW = \gamma\,dx\,dy\,dz$

γ = specific weight (weight per unit volume)

Here the weight term dW can be neglected since it is of higher order and approaches zero more rapidly than the terms involving pressures. Thus, $p_3 = p_4$.

To show that the pressure is the same on nonparallel planes, consider a wedge from the rectangular element in Fig. 7-27a. The pressures p_1, p_3, and p_5 in Fig. 7-27b are potentially different from

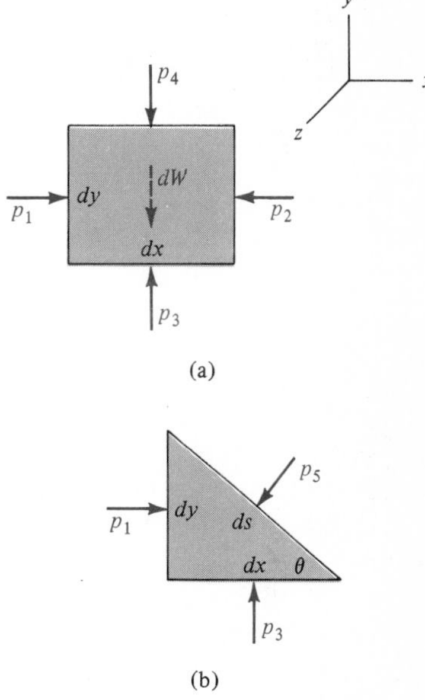

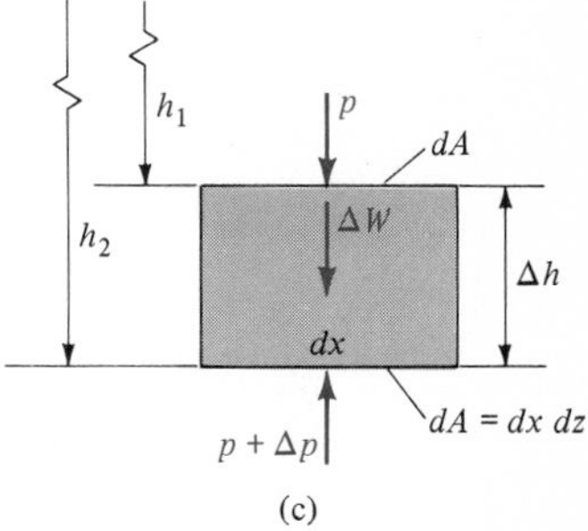

FIGURE 7-27
Infinitesimal elements of fluid

one another at this stage in the analysis. For equilibrium in the x and y directions,

$$\sum F_x = 0$$

$$p_1\, dy\, dz - p_5\, ds \sin\theta\, dz = 0$$

$$\sum F_y = 0$$

$$p_3\, dx\, dz - p_5\, ds \cos\theta\, dz = 0$$

Since $ds \sin\theta = dy$, and $ds \cos\theta = dx$, $p_1 = p_5 = p_3 = p$, where p is the pressure in any direction at the point considered.

For an important development based on the concepts discussed above, consider an element with an infinitesimal cross-sectional area ($dA = dx\, dz$) and a small height Δh that makes the element's weight ΔW significant (Fig. 7-27c). In this case, for equilibrium of the element in the y direction,

$$-p\, dx\, dz + (p + \Delta p)\, dx\, dz - \Delta W = 0$$

Finally, using $\Delta W = \gamma\, dA\, \Delta h$,

$$\boxed{\Delta p = \gamma\, \Delta h} \qquad \text{7-48}$$

This shows that the pressure varies linearly with the depth h in the fluid (downward h is positive). The relation is valid for liquids and gases.

When γ is a constant (as in liquids, which are essentially incompressible for most purposes), the reference for pressure and depth measurement is often taken at the open surface of the liquid, where the pressure is caused by the atmosphere above that surface. The pressure at such a surface is 1 atmosphere and the depth h is zero. At any other depth the measured pressure shows only the increment above atmospheric pressure. This increment in pressure is called *gage pressure*. With these in mind, Eq. 7-48 can be written simply as

$$p = \gamma h \qquad \text{7-49}$$

where p is the gage pressure and h is the depth from the open surface of the liquid to the point where p is measured. This is depicted in Fig. 7-28, where the horizontal lines show how the pressure (magnitude only) varies with the depth h. The small triangle denotes the free surface of the liquid.

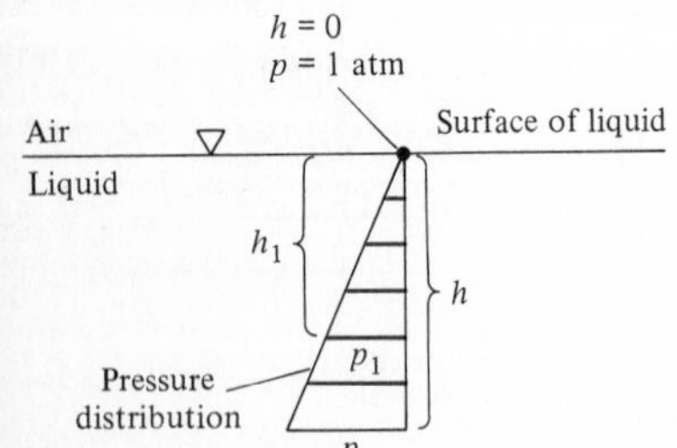

FIGURE 7-28

Pressure distribution in a liquid with a free surface

Hydrostatic Forces on Plane Surfaces

It is frequently necessary to know the magnitude, direction, and point of action of the resultant force of *hydrostatic pressure* on a solid surface submerged in a liquid (the term *aerostatics* is used when a gas is involved). For example, consider a two-piece wave board (Fig. 3-1) in a stationary position shown in Fig. 7-29a. The pressure p linearly increases to the bottom of the channel. The force of the water on each board is analyzed individually, keeping in mind the linear variation in pressure and that the pressure acts normal to any surface.

The upper board is drawn in Fig. 7-29b with the triangular pressure distribution; the reactions R_B and M_B at joint B complete the free-body diagram. P_B is the resultant force caused by the distributed pressure; its magnitude equals the average pressure $(p_B/2)$ times the total submerged area of board BC. Assume a board of constant width b. Using Eq. 7-49 and an elemental area $dA = b\,dh$,

$$P_B = \int p\,dA = \int_0^{l_B} \gamma hb\,dh$$

$$= \gamma b\frac{l_B^2}{2} = \underbrace{\gamma\left(\frac{l_B}{2}\right)}_{\substack{\text{average pressure,}\\ \text{occurs at } l_B/2}} l_B b$$

Note that a constant pressure p is acting on the elemental strip of area dA.

The resultant force P_B acts at a distance $\bar{h}$ below the surface. This distance is determined as for a beam with distributed loading, by writing the moments about a point in two different ways and equating them. Working with point C in Fig. 7-29b,

$$P_B\bar{h} = \int hp\,dA, \qquad \text{so} \quad \bar{h} = \frac{\int hp\,dA}{\int p\,dA} \tag{7-50}$$

After reviewing Eq. 7-11, it is found that P_B acts at the centroid of the pressure distribution, called the *center of pressure* (not at the centroid of the plate), which simplifies the calculations conveniently. For the triangular distribution, $\bar{h} = \frac{2}{3}l_B$.

A similar procedure is applied to the lower board with the aid of Fig. 7-29c, but this board has a more complex trapezoidal pressure distribution on it. For convenience, divide the area of the arrows into a rectangular distribution with the resultant force P_{A_1} acting halfway between points A and B, and a triangular distribution with the resultant force P_{A_2} acting at $l_A/3$ from point A. Thus, the resultant forces are

$$P_{A_1} = p_B(\text{area of board } AB)$$

$$P_{A_2} = \frac{1}{2}(p_A - p_B)(\text{area of board } AB)$$

$$P_A = P_{A_1} + P_{A_2}$$

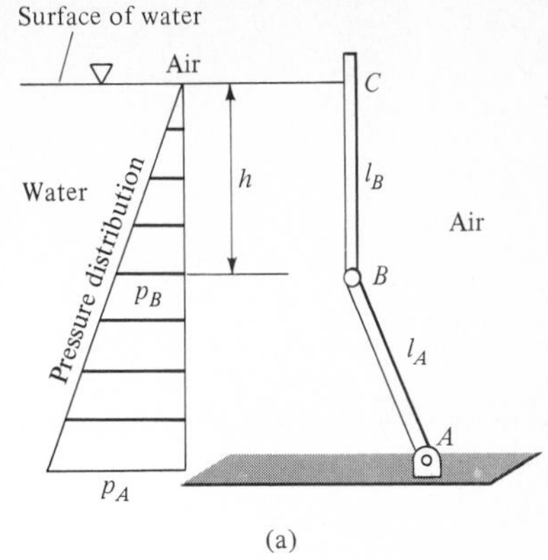

(a)

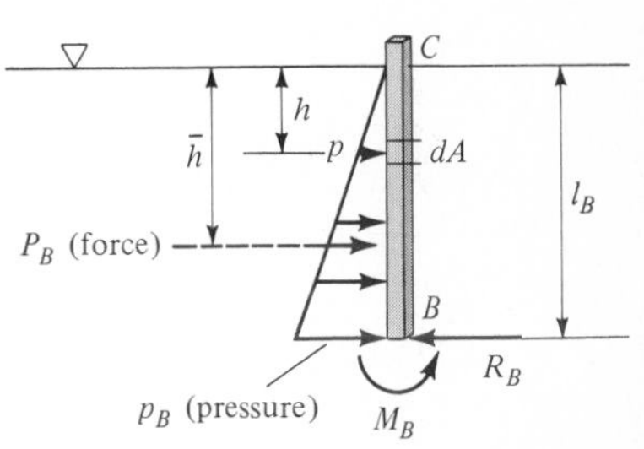

(b) Upper board

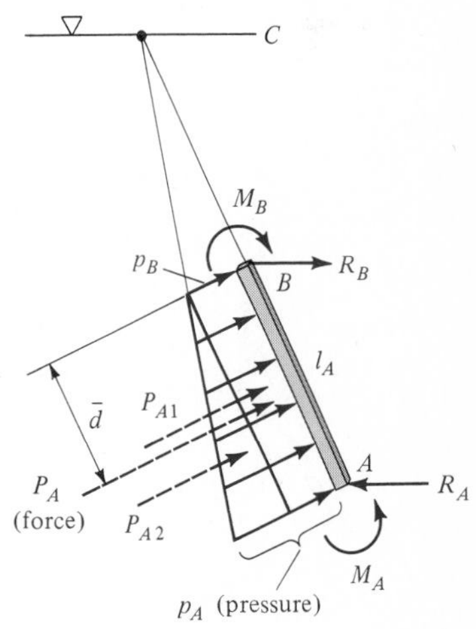

(c) Lower board

FIGURE 7-29
Pressure distribution on a two-piece wave board

The location $\bar{d}$ where P_A is acting is found by taking moments about a convenient point such as B,

$$P_A\bar{d} = P_{A_1}\frac{l_A}{2} + P_{A_2}\frac{2l_A}{3}$$

where $\bar{d}$ is the only unknown.

At this stage it is useful to restate and summarize several aspects of forces caused by hydrostatic pressure on vertical or inclined plane surfaces. The following are true for any plane area and are worth remembering:

1. *The magnitude of the total force* of hydrostatic pressure on a plane surface is equal to the area of the surface times the pressure at the center of gravity of the surface area. This force is independent of the inclination of the plane if the centroid of the plane area remains at a constant depth.

2. *The center of pressure* is where the resultant force of hydrostatic pressure is acting on the plane area. This point is the perpendicular projection of the centroid of the pressure distribution on the plane area. The center of pressure is always below the centroid of the area, and its location is independent of the specific weight of the liquid and of the inclination of the plane at a given depth.

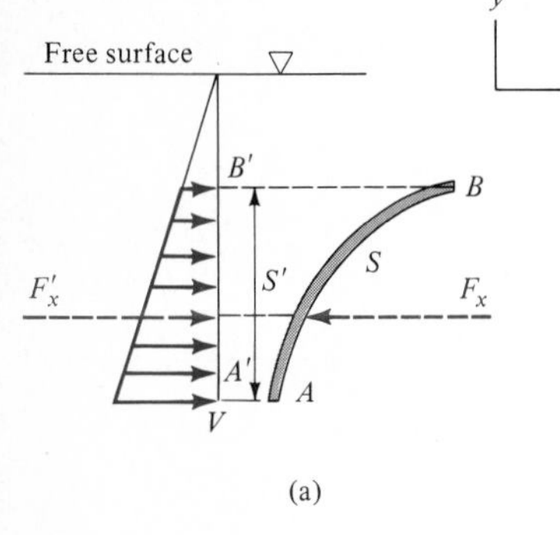

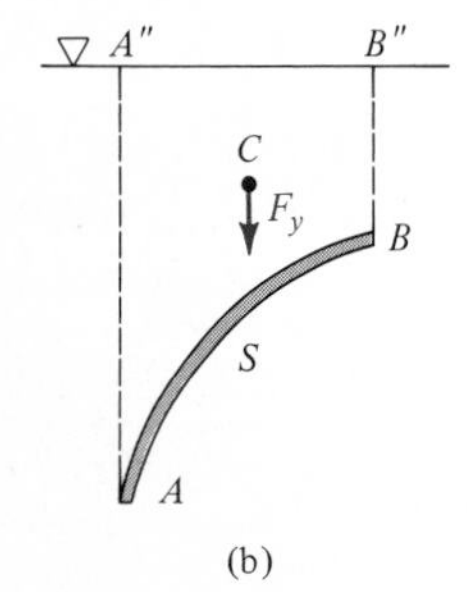

FIGURE 7-30

Hydrostatic forces on a curved surface S

Hydrostatic Forces on Curved Surfaces

An algebraic summation of elemental hydrostatic forces on a curved surface is impractical because the direction of pressure is not constant on the whole area. Fortunately, the resultant forces in the horizontal and vertical directions can be relatively easily determined even for curved surfaces. For example, consider the surface S in Fig. 7-30a. This area is projected as area S' on an arbitrary vertical plane V. The pressure distribution and the resultant force F'_x on the imaginary area S' can be determined exactly as for any plane area. For horizontal equilibrium of the imaginary volume represented by $AA'BB'$, $F_x = F'_x$ with identical lines of action. Thus, *the horizontal force on a submerged surface of any shape is equal to the force on a projected vertical area of the same depth.* The vertical plane must be normal to the desired direction of the force.

The vertical force on a submerged surface of any shape is equal to the weight of the volume of fluid above that surface, extending to the free surface. The line of action of this force (F_y in Fig. 7-30b) must pass through the center of gravity C of the volume represented by $AA''BB''$.

The theory of flotation of bodies was first formulated by Archimedes. A simple statement of Archimedes' principle is that a submerged body is subject to an upward buoyant force equal to the weight of the displaced fluid. The buoyant force B acts at the *center of buoyancy* (C.B. in Fig. 7-31), which is the center of gravity of the displaced fluid. Note that the center of gravity of the body (C.G. in Fig. 7-31) does not necessarily coincide with the center of buoyancy. The relative magnitudes and points of action of the weight W and the buoyant force B affect the position and stability of the submerged body in the following distinct ways.

1. $B > W$: the body rises in the fluid.

2. $B = W$: the body floats in equilibrium.

3. $B < W$: the body sinks in the fluid.

4. C.B. is above C.G. (Fig. 7-31a): stable position with respect to rotation because B and W form a couple that tends to bring C.B. directly above C.G.; an increase in the angle θ increases the moment of the *righting couple* of B and W, and this tends to return the body to its original position.

5. C.B. is below C.G. (Fig. 7-31b): unstable position with respect to rotation because B and W form a couple that tends to increase the angle θ; as a result, the body overturns until C.B. is above C.G., where it becomes stable.

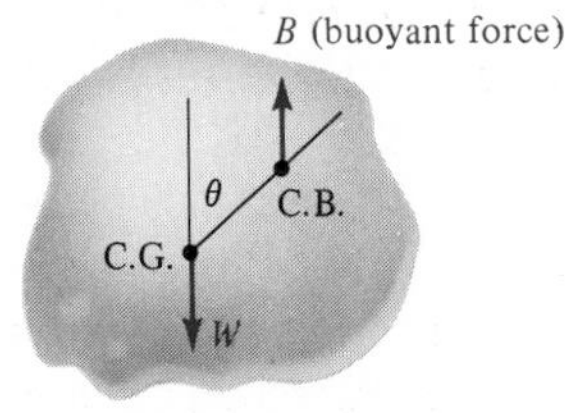

(a) Stable position

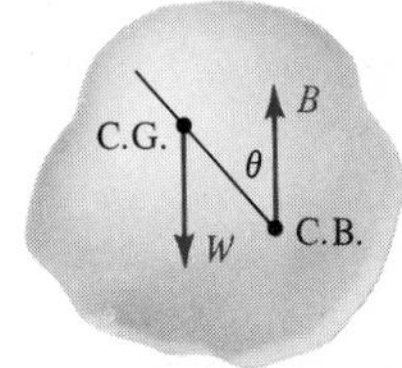

(b) Unstable position

C.G. = center of gravity of solid body
C.B. = center of buoyancy
= center of gravity of fluid displaced by the solid body

FIGURE 7-31
Center of buoyancy

EXAMPLE 7-15

Determine the horizontal and vertical hydrostatic forces (magnitudes and locations) on each part of the two-piece wave board shown in Fig. 7-32a. The wave board is 6 m wide, and the specific weight of water is 9.8 kN/m³.

SOLUTION

Top board: The maximum pressure, at B, is calculated first (Fig. 7-32b),

$$p_B = \gamma h_B = (9.8 \text{ kN/m}^3)(2.5\text{m} \sin 50^\circ) = 18.8 \text{ kN/m}^2$$

The resultant force of the distributed load is equal to the area of the pressure distribution times the width of the plate,

$$P_B = \frac{1}{2}\left(18.8 \frac{\text{kN}}{\text{m}^2}\right)(2.5 \text{ m})(6 \text{ m}) = 141 \text{ kN}$$

Therefore,

$$P_{B_H} = 141 \text{ kN} \cos 40^\circ = 108 \text{ kN}$$

$$P_{B_V} = -141 \text{ kN} \sin 40^\circ = -91 \text{ kN}$$

The location of this resultant force is at $\frac{1}{3}$ the distance along the board:

$$\bar{h} = \frac{BC}{3} = \frac{2.5 \text{ m}}{3} = 0.83 \text{ m}$$

Bottom board: The pressure distribution is trapezoidal as shown in Fig. 7-32c. The force caused by the pressure is found by considering the pressure distribution in the two parts shown.

$$p_A = \gamma h_A = (9.8 \text{ kN/m}^3)(4.4 \text{ m}) = 43.1 \text{ kN/m}^2$$

$$P_{A_1} = p_B(A) = (18.8 \text{ kN/m}^2)(2.5 \text{ m})(6 \text{ m}) = 282 \text{ kN}$$

$$P_{A_2} = \frac{1}{2}(p_A - p_B)(A) = \frac{1}{2}(43.1 - 18.8)\frac{\text{kN}}{\text{m}^2}(2.5 \text{ m})(6 \text{ m})$$
$$= 182 \text{ kN}$$

$$P_A = P_{A_1} + P_{A_2} = 282 \text{ kN} + 182 \text{ kN} = 464 \text{ kN}$$

Therefore,

$$P_{A_H} = 464 \text{ kN} \cos 10^\circ = 457 \text{ kN}$$

$$P_{A_V} = -464 \text{ kN} \sin 10^\circ = -81 \text{ kN}$$

The location of this resultant force is at the centroid of the trapezoid,

$$\bar{d} = \frac{P_{A_1}\bar{h}_1 + P_{A_2}\bar{h}_2}{P_{A_1} + P_{A_2}}$$
$$= \frac{(282 \text{ kN})(1.25 \text{ m}) + (182 \text{ kN})(0.83 \text{ m})}{282 \text{ kN} + 182 \text{ kN}} = 1.09 \text{ m}$$

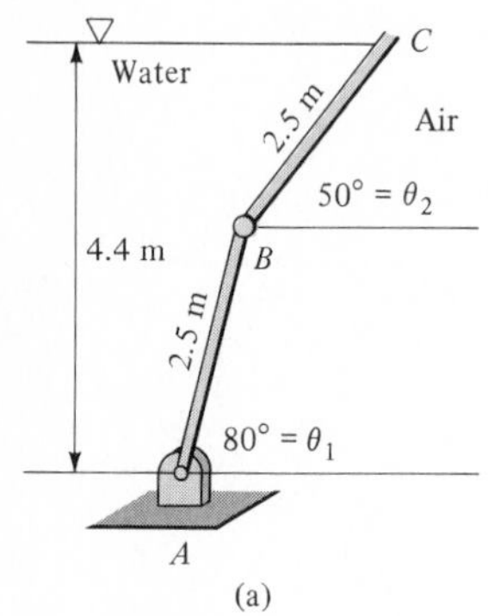

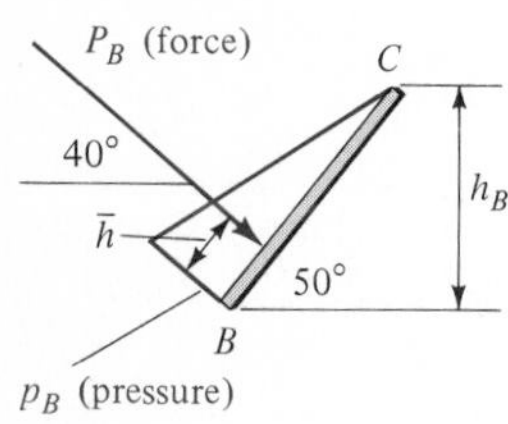

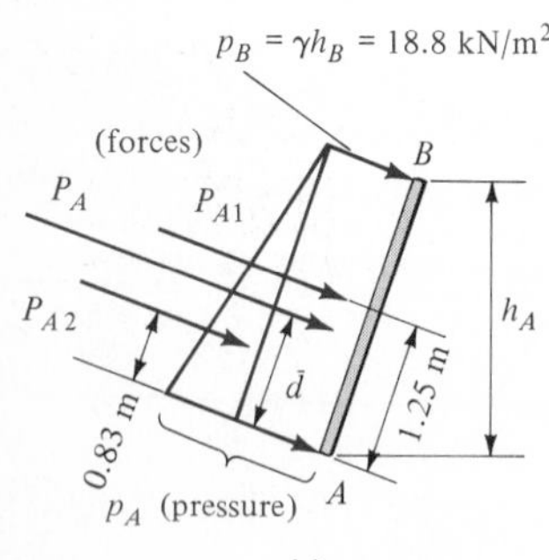

FIGURE 7-32
Hydrostatic forces on a two-piece wave board

EXAMPLE 7-16

The buoy shown in Fig. 7-33a is a circular cylinder 5 ft in diameter and 6 ft in length. It is made of steel with a uniform wall thickness of 0.4 in. What is the tension in the cable at the bottom of the buoy if a 4-ft-long section of the buoy is submerged? Assume that no ship is tied at the buoy. Steel weighs 0.283 lb/in^3.

SOLUTION

Since the buoy is in equilibrium,

$$\sum F_y = 0$$

$$-W + B - T = 0$$

where
$$\begin{aligned} W &= \gamma_{st} V_B = \gamma_{st}(2\pi r t h + 2\pi r^2 t) \\ &= (0.283\ \text{lb/in}^3)[2\pi(30\ \text{in.})(0.4\ \text{in.})(72\ \text{in.}) \\ &\quad + \pi(30\ \text{in.})^2(0.4\ \text{in.})2] \\ &= 2180\ \text{lb} \end{aligned}$$

$$\begin{aligned} B &= \text{buoyant force} = \text{weight of water displaced} \\ &= \gamma_w V_w = \gamma_w(\pi r^2 h) = (62.4\ \text{lb/ft}^3)[\pi(2.5\ \text{ft})^2](4\ \text{ft}) \\ &= 4900\ \text{lb} \end{aligned}$$

Therefore,

$$T = B - W = (4900 - 2180)\ \text{lb} = 2720\ \text{lb}$$

JUDGMENT OF THE RESULTS

The large magnitude of the results might cause one to reconsider the calculations. These magnitudes are reasonable, however, due to the large volume of water displaced by a relatively small amount of steel. A large tension is needed to maintain the buoy in this position.

(a)

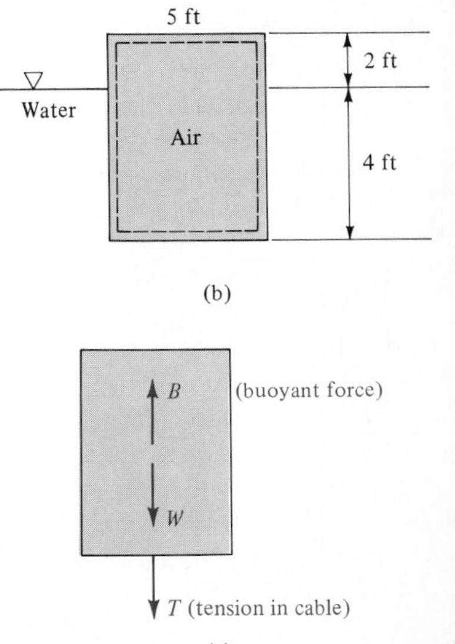

FIGURE 7-33
Analysis of an anchored buoy

PROBLEMS

FIGURE P7-85

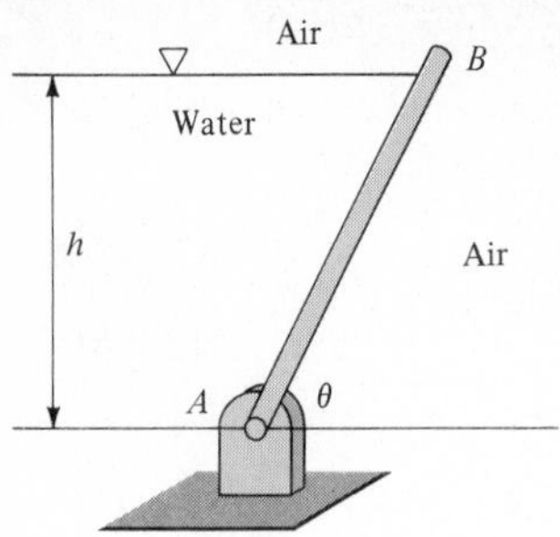

7–85 Consider a single-piece wave board in a stationary position (see Sec. 3-1 for an introduction to these devices). Determine the magnitudes and locations of the horizontal and vertical hydrostatic forces acting on the 3-m-wide board if $\theta = 70^\circ$, $h = 2$ m, and $\gamma = 9.8$ kN/m^3.

7–86 Plot the required supporting moment M_A in Prob. 7–85 as functions of $\theta = 50^\circ$, 70°, and 90°.

FIGURE P7-87

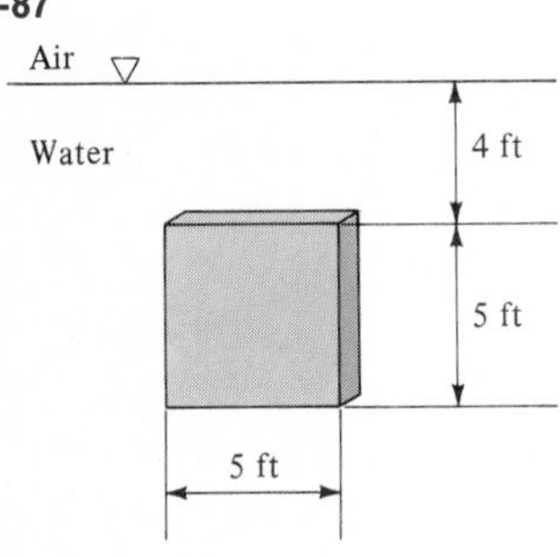

7–87 The square plate is a model for a gate under water. Determine the magnitude and location of the hydrostatic force acting on one side of the plate if there is air on the other side. $\gamma = 62.4$ lb/ft^3.

FIGURE P7-88

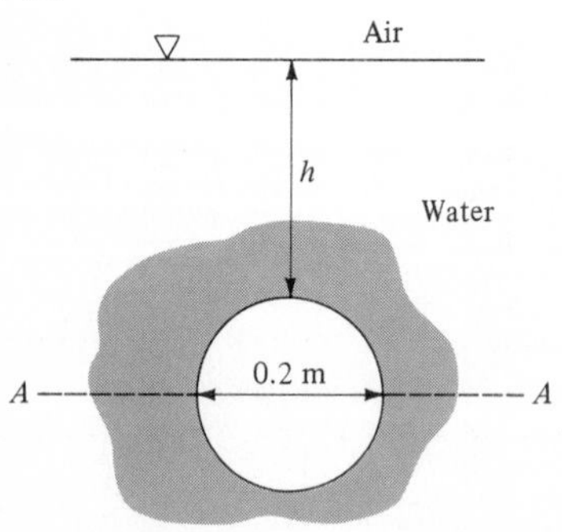

7–88 Determine the magnitude of the hydrostatic force on the circular window of a small submarine if $h = 50$ m and $\gamma = 9.8$ kN/m^3.

7–89 Plot the distance d of the center of pressure from the centroidal line AA of the window in Prob. 7–88. Let $h = 0$, 50 m, and 100 m.

7–90 Assume that the stationary rudder $ABCD$ is under water. Determine the horizontal hydrostatic force and the center of pressure on the rudder. Discuss the possible effect of this force on (a) a solid metal rudder, and (b) a hollow, lightweight rudder. Salt water weighs 64 lb/ft^3.

FIGURE P7-90

(a)

Air
2 ft
6 ft
A
B
6 ft
D
C
5 ft

(b) Symmetric rudder

7–91 Consider a proposed curved wave board with a 10-m length in the z direction. Determine the magnitudes and locations of the horizontal and vertical hydrostatic forces on the board if $\gamma = 9.8$ kN/m^3.

FIGURE P7-91

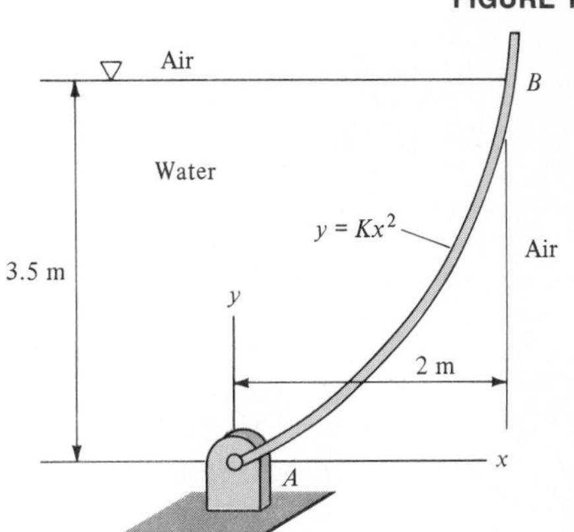

FIGURE P7-92

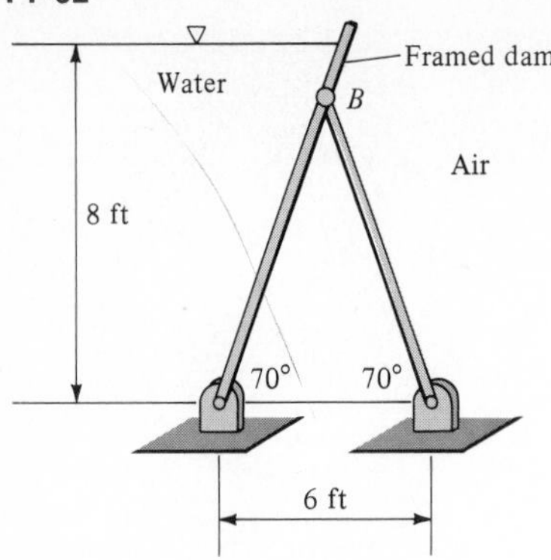

7–92 Estimate the forces at the hinged joints A, B, and C in the framed dam. There is a strut BC every 5 ft along the dam. Water weighs 62.4 lb/ft^3.

FIGURE P7-93

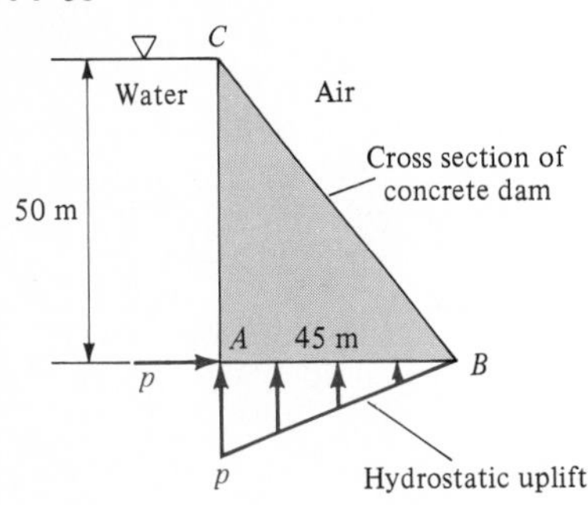

7–93 The cross section of a proposed *gravity dam* is approximated by a triangle ABC. The *hydrostatic uplift* between A and B is caused by water seeping under the dam. The stability of the gravity dam depends on its weight (for concrete, $\gamma = 23.5\ \text{kN/m}^3$) and on the horizontal and vertical hydrostatic forces; the stability is independent of the total length of the dam. Is there any danger of this dam overturning? Water weighs 9.8 kN/m^3.

FIGURE P7-94

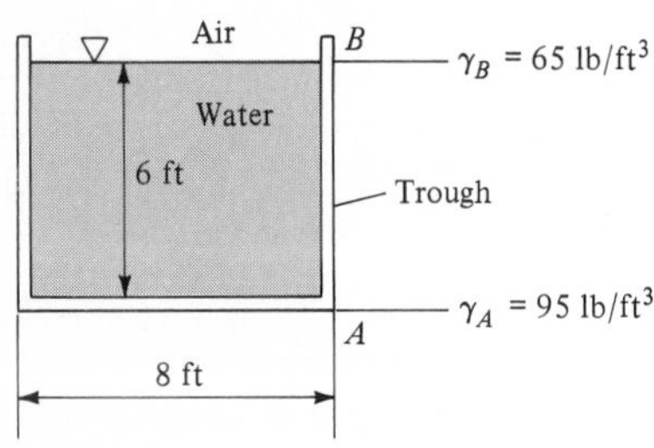

7–94 The trough shown may carry muddy water whose specific weight varies linearly between the values given. Determine the internal reactions at corner A per unit length of the trough.

FIGURE P7-95

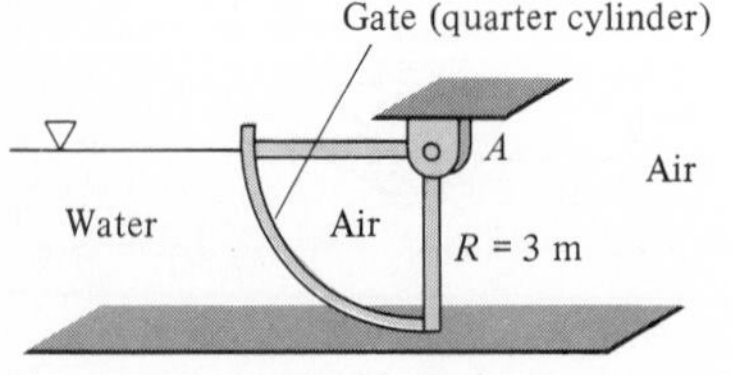

7–95 The gate of a spillway is pivoted at point A. Determine the total force and moment at A for a 5-m-wide gate. $\gamma = 9.8\ \text{kN/m}^3$.

7–96 A balloon is designed to carry a load of 200 lb using no more than 20,000 ft^3 of helium. What is the maximum allowable magnitude of the balloon's own weight? Helium weighs 0.0112 lb/ft^3 and air weighs 0.0807 lb/ft^3 at the same pressure and temperature.

FIGURE P7-96

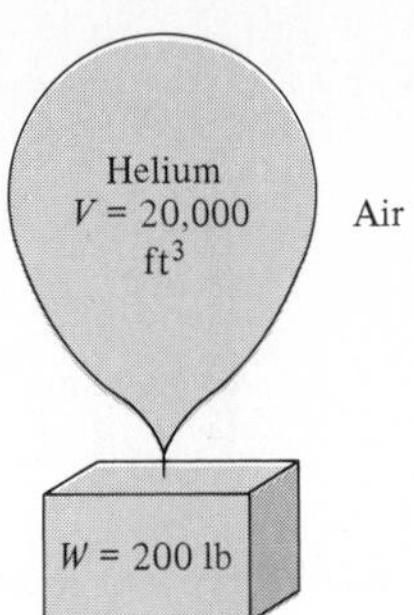

7–97 Consider solid cylinders whose specific gravity is half of that of water. Discuss why a short cyclinder (a disk with $d \gg h$) is stable in water in the position shown, while a long rod (with $h \gg d$) is unstable in the same position. Assume small displacements from the equilibrium position where the cylinder's axis is vertical.

FIGURE P7-97

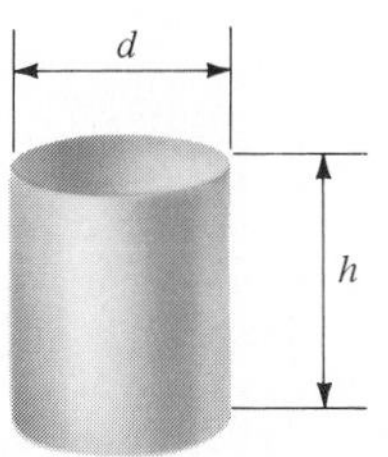

7–98 A deep-sea drill rod is assembled from uniform pieces of steel pipe that has an 8-cm inside diameter and a 10-cm outside diameter. There is seawater (specific weight 10 kN/m^3) inside and outside the pipe. Determine the force that supports the hanging pipe at the ship. Assume that the drill is not touching bottom at a depth of 5000 m, the length of the pipe. Steel weighs 76.8 kN/m^3.

7–99 A small raft is designed with a square platform 6 ft × 6 ft in size. It has three buoyant elements, wooden boards 2 in. × 12 in. in cross section, 6 ft long, and weighing 0.02 lb/in^3. Determine the distance d of the platform above the water.

FIGURE P7-99

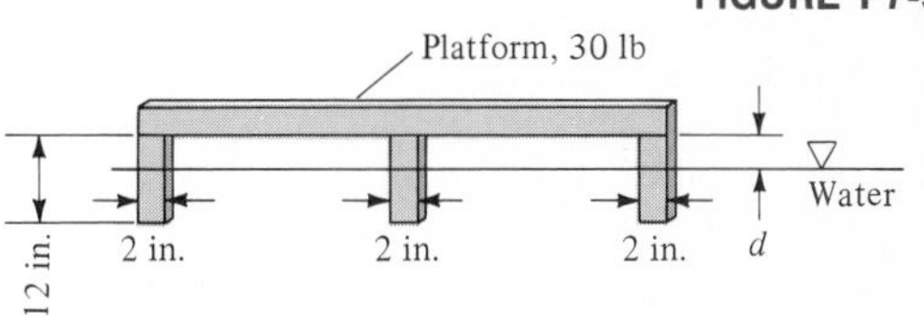

7–100 Determine h for the solid hemisphere shown in Fig. P7-100a. The specific weight of the solid is half of that of water.

FIGURE P7-100

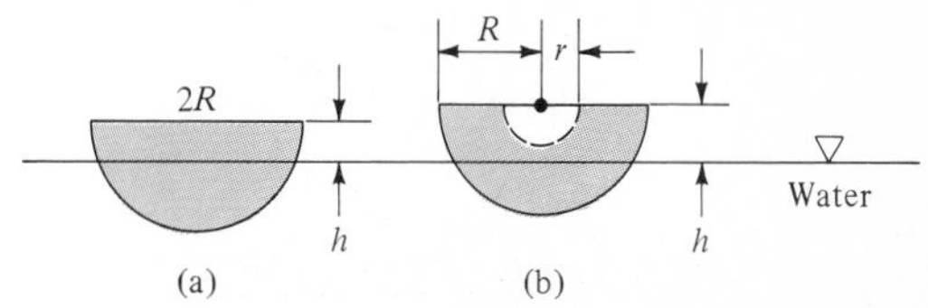

FIGURE P7-103

(a)

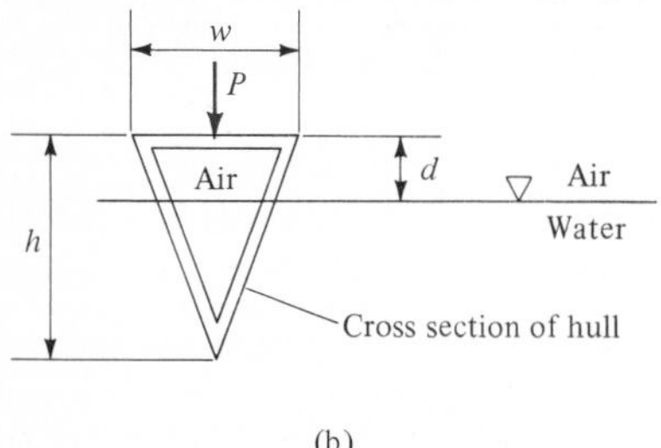

(b)

7–101 Determine h for the hemispherical shell shown in Fig. P7-100b if $r = 0.6R$. The specific weight of the solid is half of that of water.

7–102 Plot h as a function of $r = 0$, $0.3R$, $0.6R$, and $0.9R$ in Fig. P7-100. R is the same radius in each case, and the specific weight of each solid is half of that of water. Estimate r to maximize h.

7–103 The thin-walled hulls of a catamaran are approximately triangular in cross section and 3 m in length. Assume that $d = h/2 = 15$ cm and $w = 20$ cm. Ignore any changes in the cross-sectional shape and area along the hull. Calculate the force P.

ELEMENTS OF STRESS ANALYSIS 7-9

In solid materials there are internal distributed forces somewhat analogous to the pressure in a fluid. The main difference is that a solid element may have normal and shear forces acting on it as shown in Fig. 7-34, while only normal forces act on a fluid at rest. Internal normal and shear forces can be generated by axial tension or compression, bending, and torsion of solid materials. These are studied in detail in the area of *mechanics of materials* (also called *strength of materials*). A brief introduction to this subject is given here to show the basic concepts and their similarities or differences to other kinds of distributed forces.

The magnitudes of the internal forces are of great practical interest because these forces can be directly related to the strength of any solid. This relationship is important in design to prevent failures of load-carrying members. The best way to determine the extreme values of internal forces is to calculate their local intensities.

The intensity of an internal force is called *stress* and is defined as the force per unit area. The common symbol for normal stress is the lowercase Greek letter σ, and for shear stress the lowercase Greek letter τ. They are calculated by dividing the total force acting on an area by the size of that area,

$$\sigma = \frac{F_{\text{normal}}}{A} \quad \text{and} \quad \tau = \frac{F_{\text{shear}}}{A} \tag{7-51}$$

The units for both σ and τ are pascals (Pa $=$ N/m^2) or pounds per square inch (psi).

The stresses calculated from Eq. 7-51 are the average stresses on the given area. The magnitudes of the stresses can vary from position to position in a member even if the external loading is constant. Thus, a more satisfactory definition of stress requires working with a differential load dF acting on a differential area dA,

$$\sigma = \frac{dF_{\text{normal}}}{dA} \quad \text{and} \quad \tau = \frac{dF_{\text{shear}}}{dA} \tag{7-52}$$

If the functional relation for the variation of stress is known, the total force on an area is calculated by integration of Eq. 7-52,

$$F_{\text{normal}} = \int \sigma \, dA \quad \text{and} \quad F_{\text{shear}} = \int \tau \, dA \tag{7-53}$$

In words, the total force acting on an area is equal to the stress integrated over the area it acts upon. It should be emphasized in this regard that *force and stress are not the same* even though they are directly related.

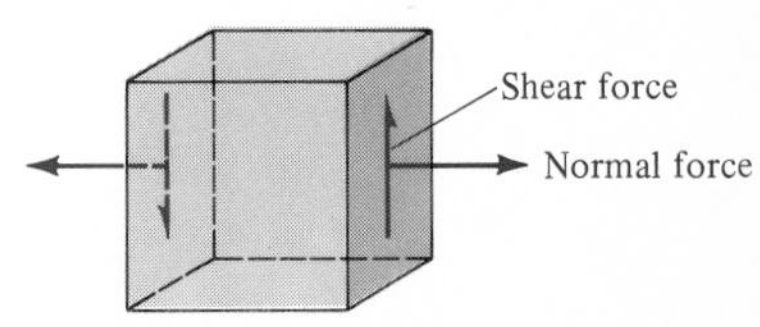

FIGURE 7-34
Element of a solid body

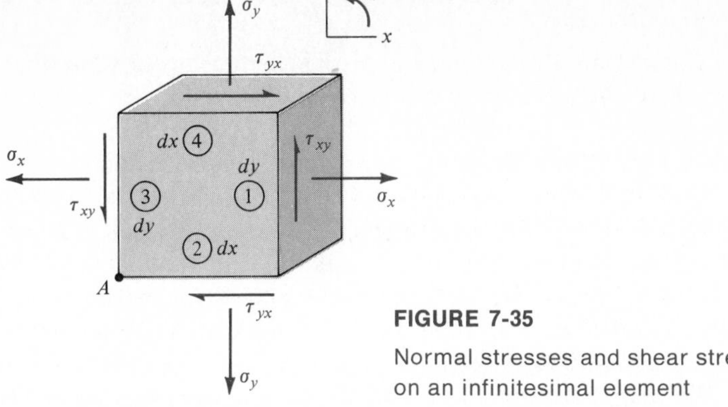

FIGURE 7-35

Normal stresses and shear stresses on an infinitesimal element

Normal and shear stresses on an infinitesimal element are commonly shown as in Fig. 7-35, where each arrow represents a distributed load. Three-dimensional states of stress (which have stresses acting in the x, y, and z directions) frequently exist, but the two-dimensional case (called plane stress) is the most illustrative at this stage. The notations of the normal stresses (which may be tensile or compressive) are self-explanatory. For shear stress notation, the first subscript denotes the face on which the stress is acting: the subscript shows the coordinate axis normal to the face. The second subscript denotes the direction of the shear stress. For example, τ_{xy} is acting on a face normal to the x axis, and in the y direction (positive or negative).

There are several aspects of the magnitudes and directions of the stresses on a given element that are noteworthy, assuming a thickness of unity for the element. Also assume that σ and τ may act simultaneously, or either one may act without the other. From the equilibrium of forces in Fig. 7-35,

σ_x on face 1 is equal and oppsite to σ_x on face 3

σ_y on face 2 is equal and opposite to σ_y on face 4

τ_{xy} on face 1 is equal and opposite to τ_{xy} on face 3

τ_{yx} on face 2 is equal and opposite to τ_{yx} on face 4

From the equilibrium of moments taken about any corner of the element in Fig. 7-35, $\tau_{xy} = \tau_{yx}$ in magnitude, and they cause opposite moments (this is true even when $dx \neq dy$). When considering equilibrium of internal stresses, it is important to remember that *equilibrium equations can be written only for forces and moments of forces*, and not for stresses directly. For example, take moments about point A in Fig. 7-35: the normal forces cancel each other, two shear forces have zero moment arms, so only the shear forces on faces 1 and 4 cause moments,

$$\sum M_A = \underbrace{\tau_{xy}\underbrace{(dy)(1)}_{\text{area}}}_{\text{force}}\;\underbrace{(dx)}_{\substack{\text{moment}\\\text{arm}}} - \underbrace{\tau_{yx}\underbrace{(dx)(1)}_{\text{area}}}_{\text{force}}\;\underbrace{(dy)}_{\substack{\text{moment}\\\text{arm}}} = 0 \qquad \boxed{7\text{-}54}$$

from which $\tau_{xy} = \tau_{yx}$.

The equilibrium equations for forces based on known stresses are useful to determine the stresses on other planes in a given element. For example, consider the element in Fig. 7-36a, where the stresses on plane AB must be found. Next draw only the triangular element ABC on which the unknown stresses σ_θ and τ_θ are acting. Note that σ_x and τ_{xy} are redrawn with the original magnitudes since the concept of stress is independent of the area in a given small neighborhood. For convenience, assume that the plane AB has an area of unity, and sum forces along axes m and n in Fig. 7-36b. For equilibrium in the m direction,

$$\sum F_m = \underbrace{\tau_\theta\underbrace{(1)}_{\text{area}}}_{\substack{\text{force in}\\ m \text{ direction}}} + \underbrace{\underbrace{\sigma_x \underbrace{\sin\theta}_{\text{area}}}_{\text{force}} \cos\theta}_{\substack{\text{component of}\\ \text{force of } \sigma_x\\ \text{in } m \text{ direction}}} + \underbrace{\underbrace{\tau_{yx} \underbrace{\cos\theta}_{\text{area}}}_{\text{force}} \cos\theta}_{\substack{\text{component of}\\ \text{force of } \tau_{yx}\\ \text{in } m \text{ direction}}} \qquad \boxed{7\text{-}55}$$

$$- \underbrace{\underbrace{\tau_{xy} \underbrace{\sin\theta}_{\text{area}}}_{\text{force}} \sin\theta}_{\substack{\text{component of}\\ \text{force of } \tau_{xy}\\ \text{in } m \text{ direction}}} = 0$$

For additional clarification of the components of stresses, consider Fig. 7-36c, d, and e.

For equilibrium in the n direction,

$$\sum F_n = \sigma_\theta(1) - \sigma_x \sin\theta \sin\theta - \tau_{xy} \sin\theta \cos\theta \qquad \boxed{7\text{-}56}$$
$$- \tau_{yx} \cos\theta \sin\theta$$
$$= 0$$

Since the only unknowns are σ_θ and τ_θ in these equations, they are readily determined. This method of stress analysis is called the *wedge method* because in each case a triangular element is used.

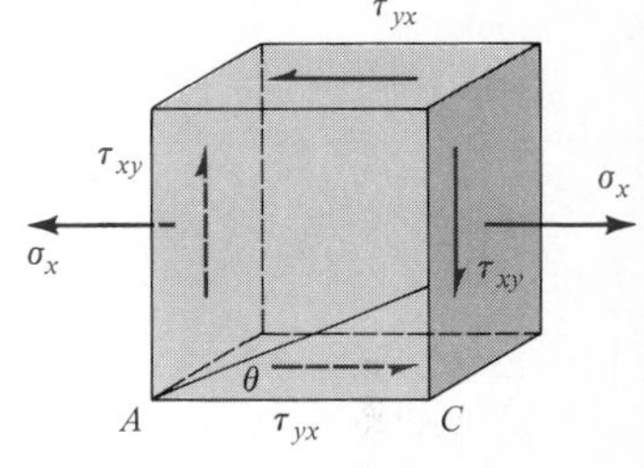

(a)

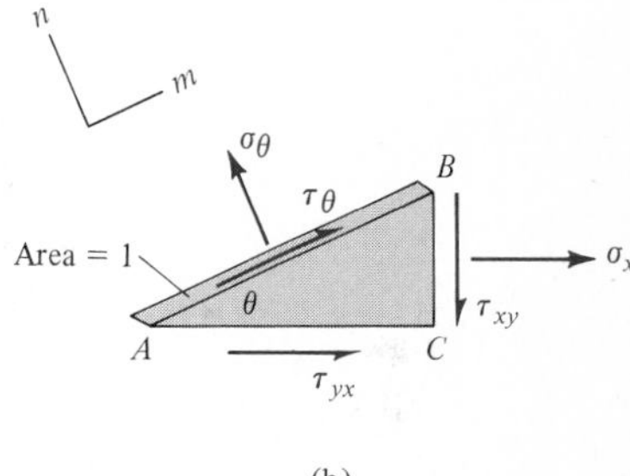

(b)

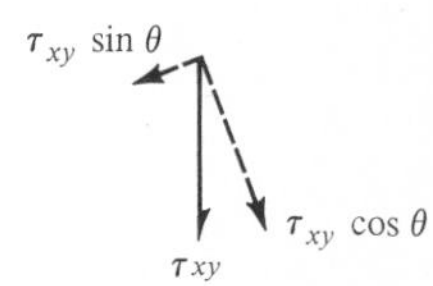

(c) Components of τ_{xy} on face BC in m and n directions

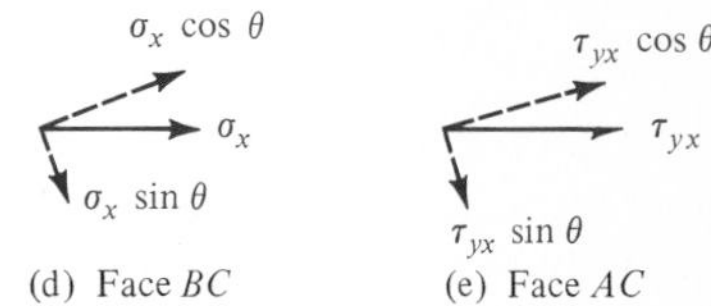

(d) Face BC (e) Face AC

FIGURE 7-36

Wedge method of stress analysis

EXAMPLE 7-17

In a graphite–epoxy composite plate the fibers are oriented as shown in Fig. 7-37a. Determine the stresses parallel and normal to the fibers.

SOLUTION

The stresses σ_θ and τ_θ are found by realizing that the wedge in Fig. 7-37b is in equilibrium. It is most convenient to sum forces in the directions of the unknown stresses to avoid the necessity of solving simultaneous equations. Therefore, since each force is obtained by multiplying the stress by the area it acts upon,

$$\sum F_m = 0$$

$$\tau_\theta A - (30 \text{ MPa} \sin 20^\circ)(A \cos 20^\circ) = 0$$

$$\tau_\theta = 30 \text{ MPa} \sin 20^\circ \cos 20^\circ = 9.6 \text{ MPa}$$

$$\sum F_n = 0$$

$$\sigma_\theta A - (30 \text{ MPa} \cos 20^\circ)(A \cos 20^\circ) = 0$$

$$\sigma_\theta = 30 \text{ MPa} \cos^2 20^\circ = 26.5 \text{ MPa}$$

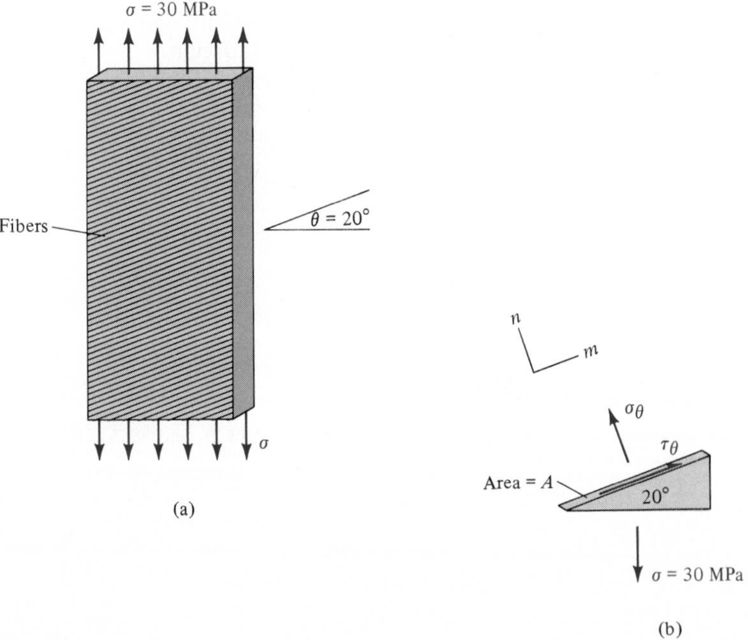

FIGURE 7-37

Stress analysis of a composite plate

EXAMPLE 7-18

The element in Fig. 7-38a is from a large glued plate. Determine the shear and normal stresses on plane AB of the joint.

SOLUTION

Resolving the known stress components into components in the m and n directions (Fig. 7-38b),

$$\sum F_m = 0$$

$$\tau_\theta A + (\tau \sin 30^\circ)A \sin 30^\circ - (\sigma_x \cos 30^\circ)A \sin 30^\circ - (\tau \cos 30^\circ)A \cos 30^\circ - (\sigma_y \sin 30^\circ)A \cos 30^\circ = 0$$

$$\begin{aligned}\tau_\theta &= \tau \cos^2 30^\circ - \tau \sin^2 30^\circ + \sigma_x \cos 30^\circ \sin 30^\circ + \sigma_y \sin 30^\circ \cos 30^\circ \\ &= 500 \text{ psi}(\cos^2 30^\circ - \sin^2 30^\circ) + (1000 \text{ psi} + 2000 \text{ psi}) \sin 30^\circ \cos 30^\circ \\ &= 1550 \text{ psi}\end{aligned}$$

$$\sum F_n = 0$$

$$\sigma_\theta A + (\tau \cos 30^\circ)A \sin 30^\circ + (\sigma_x \sin 30^\circ)A \sin 30^\circ + (\tau \sin 30^\circ)A \cos 30^\circ - (\sigma_y \cos 30^\circ)A \cos 30^\circ = 0$$

$$\begin{aligned}\sigma_\theta &= -\tau \cos 30^\circ \sin 30^\circ - \tau \sin 30^\circ \cos 30^\circ - \sigma_x \sin^2 30^\circ + \sigma_y \cos^2 30^\circ \\ &= -500 \text{ psi}(2 \cos 30^\circ \sin 30^\circ) - 1000 \text{ psi} \sin^2 30^\circ + 2000 \text{ psi} \cos^2 30^\circ \\ &= 820 \text{ psi}\end{aligned}$$

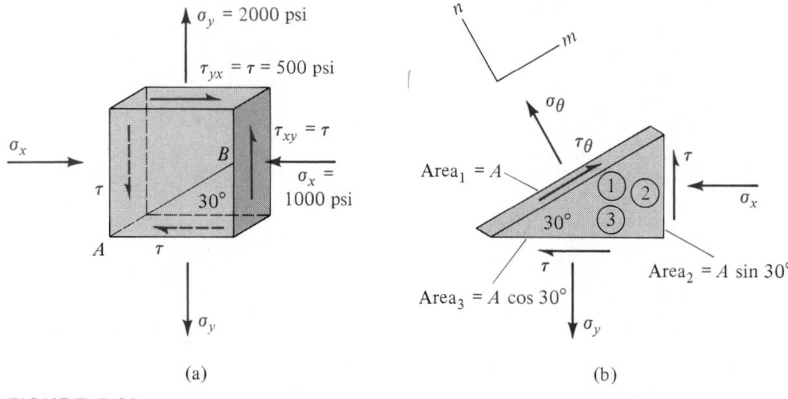

FIGURE 7-38

Stress analysis of a glued plate

PROBLEMS

SECTION 7-9

In Probs. 7–104 to 7–113, use the wedge method of analysis to determine the shear and normal stresses on plane AB in the appropriate figure.

FIGURE P7-104

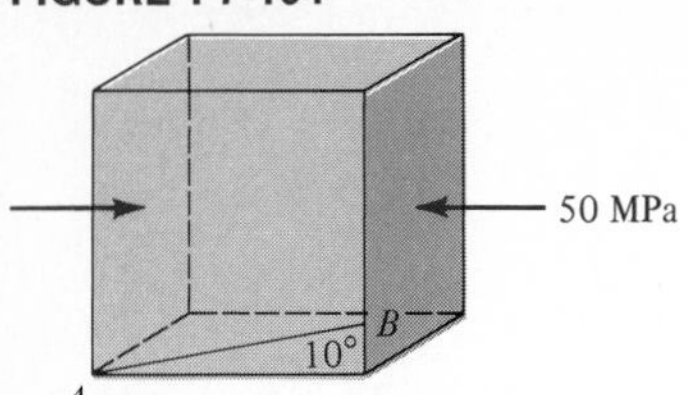

FIGURE P7-105

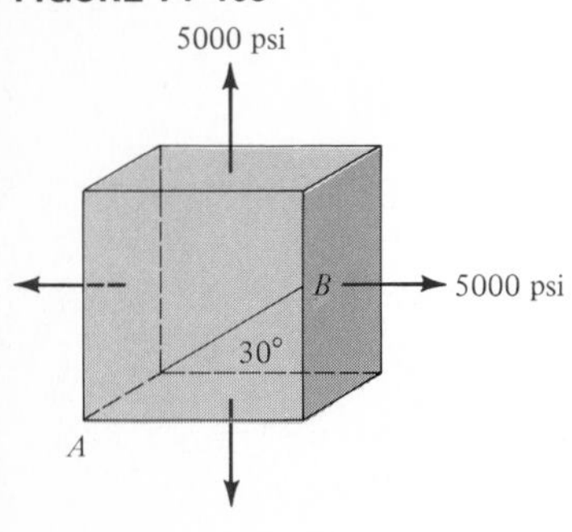

FIGURE P7-108

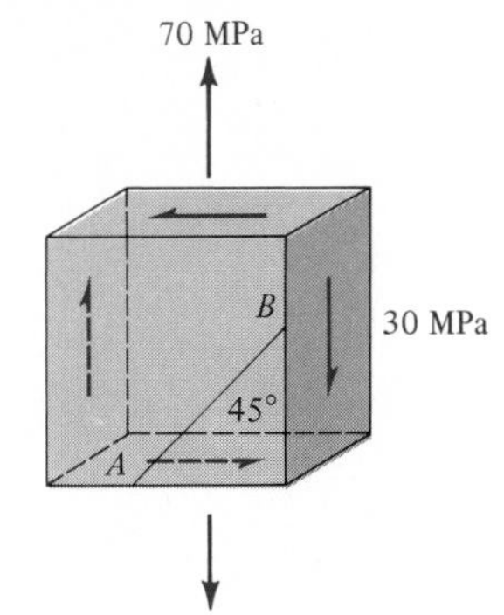

FIGURE P7-111

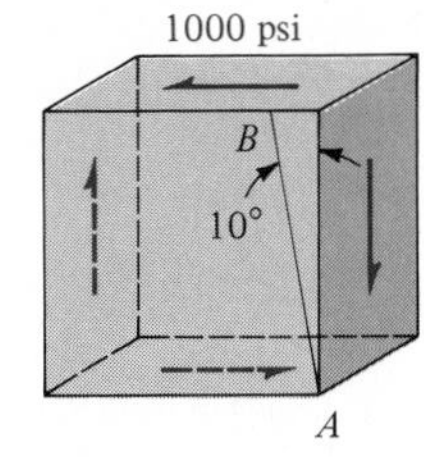

FIGURE P7-106

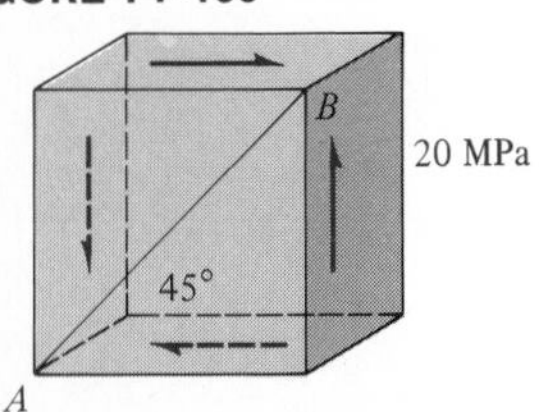

FIGURE P7-109

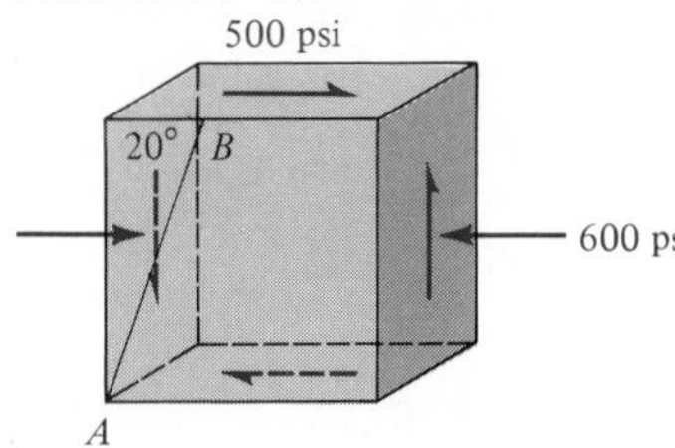

FIGURE P7-112

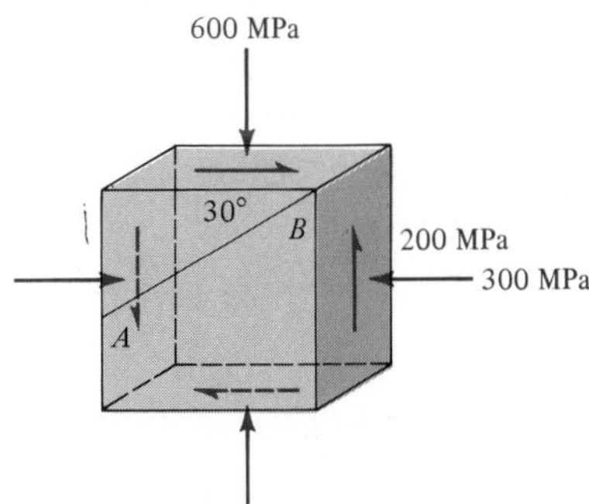

FIGURE P7-107

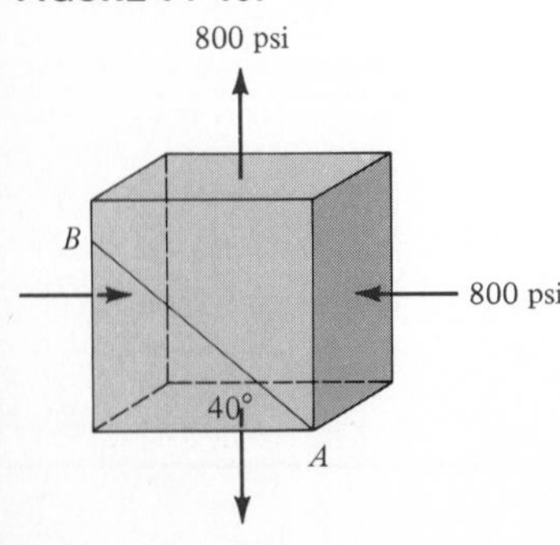

FIGURE P7-110

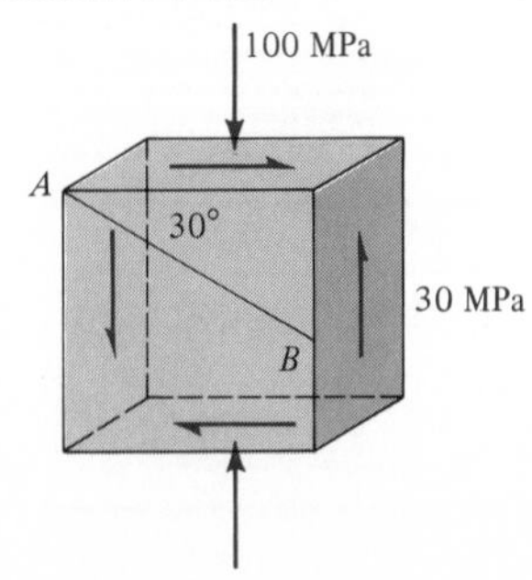

FIGURE P7-113

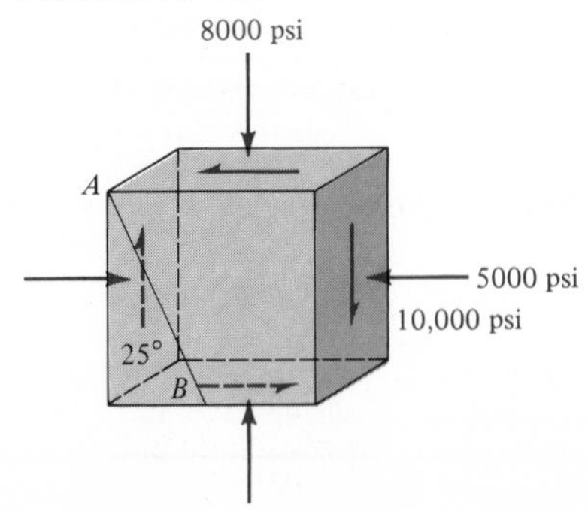

SUMMARY

Section 7-1

COORDINATES OF CENTER OF GRAVITY:

$$\bar{x} = \frac{\int x\, dW}{W} \qquad \bar{y} = \frac{\int y\, dW}{W} \qquad \bar{z} = \frac{\int z\, dW}{W}$$

where $dW = \gamma\, dV$
γ = specific weight of the material (weight per unit volume)
dV = volume element
$W = \int \gamma\, dV$ = total weight of the body

Section 7-2

CENTROIDS:

LINES: $$\bar{x} = \frac{\int x\, dL}{L} \qquad \bar{y} = \frac{\int y\, dL}{L} \qquad \bar{z} = \frac{\int z\, dL}{L}$$

where dL = line element
L = total length of line

AREAS: $$\bar{x} = \frac{\int x\, dA}{A} \qquad \bar{y} = \frac{\int y\, dA}{A} \qquad \bar{z} = \frac{\int z\, dA}{A}$$

where dA = area element
A = total area

VOLUMES: $$\bar{x} = \frac{\int x\, dV}{V} \qquad \bar{y} = \frac{\int y\, dV}{V} \qquad \bar{z} = \frac{\int x\, dV}{V}$$

where dV = volume element
V = total volume

Section 7-3

CENTROIDS OF COMPOSITE OBJECTS: Equate the first moment of the whole object with the sum of the first moments of the individual parts, using the same reference system.

LINES: $$\bar{x}L = \sum \bar{x}_i L_i, \qquad \bar{x} = \frac{\sum \bar{x}_i L_i}{L}$$

$$\bar{y}L = \sum \bar{y}_i L_i, \qquad \bar{y} = \frac{\sum \bar{y}_i L_i}{L}$$

$$\bar{z}L = \sum \bar{z}_i L_i, \qquad \bar{z} = \frac{\sum \bar{z}_i L_i}{L}$$

where $\overline{x}$, $\overline{y}$, $\overline{z}$ are the centroidal coordinates of the whole line L, and $\overline{x}_i$, $\overline{y}_i$, $\overline{z}_i$ are the centroidal coordinates of the line element L_i; $L = \sum L_i$.

Similarly for areas and volumes,

AREAS: $\overline{x} = \dfrac{\sum \overline{x}_i A_i}{A} \qquad \overline{y} = \dfrac{\sum \overline{y}_i A_i}{A} \qquad \overline{z} = \dfrac{\sum \overline{z}_i A_i}{A}$

VOLUMES: $\overline{x} = \dfrac{\sum \overline{x}_i V_i}{V} \qquad \overline{y} = \dfrac{\sum \overline{y}_i V_i}{V} \qquad \overline{z} = \dfrac{\sum \overline{z}_i V_i}{V}$

Section 7-4

THEOREMS OF PAPPUS:

THEOREM I: $A = 2\pi \overline{y} L$

where A = total area of surface of revolution
L = length of the generating curve
$\overline{y}$ = centroid of line L with respect to the axis of revolution

THEOREM II: $V = 2\pi \overline{y} A$

where V = total volume within the generating surface of revolution
A = area of the generating surface
$\overline{y}$ = centroid of area A with respect to the axis of revolution

Section 7-5

DISTRIBUTED LOADS ON BEAMS: Assume the following distributed load $w(x)$ on the element AB of a beam:

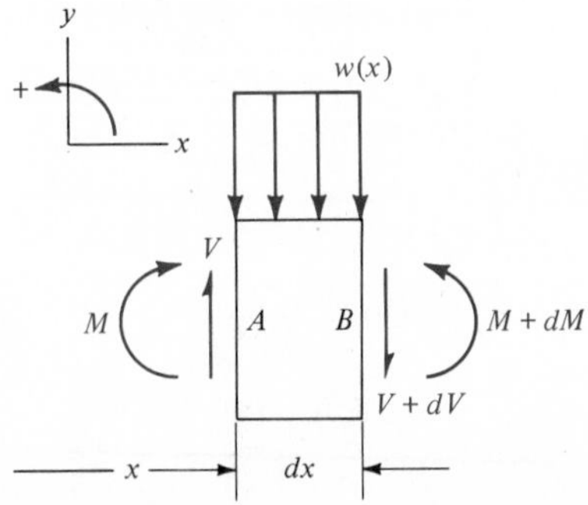

The load $w(x)$, shear force V, and bending moment M are related as

$$w(x) = -\frac{dV}{dx} \qquad V = \frac{dM}{dx}$$

where the signs are for the situation shown here. Other useful expressions for any two sections A and B of a beam:

$$V_A - V_B = \int_{x_A}^{x_B} w(x)\,dx = \text{area under } w(x)$$

$$M_B - M_A = \int_{x_A}^{x_B} V\,dx = \text{area under shear diagram}$$

EXAMPLE:

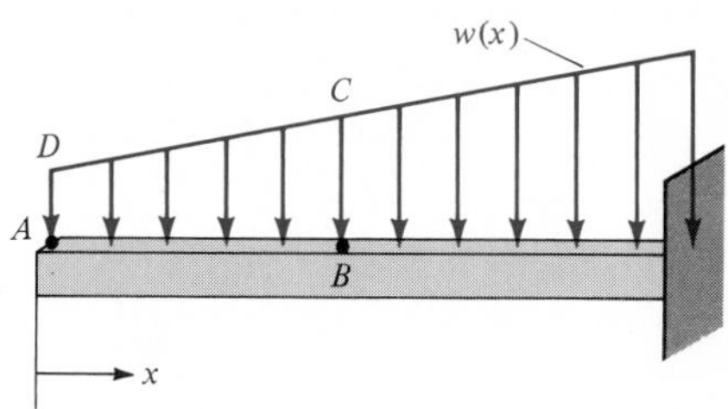

Free-body diagram for AB:

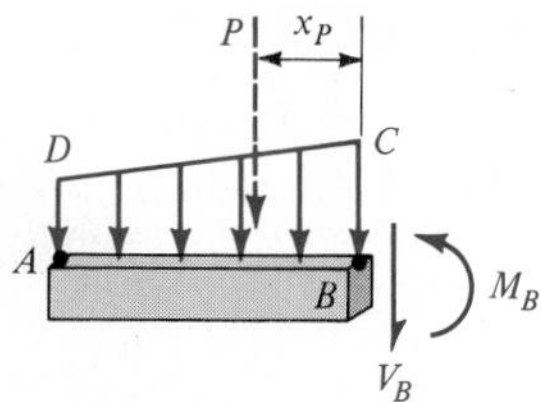

where P = resultant external force on part AB (known from $w(x)$ and length of AB) = area $ABCD$

x_P = location where force P acts to make P equivalent to load $ABCD$ = centroid of area $ABCD$ as measured horizontally from section B

V_B = shear force at B (unknown, assumed positive)

$V_B = -P$ from equilibrium of part AB

M_B = bending moment at B (unknown, assumed positive)

$M_B = -Px_P$ from equilibrium of part AB

Section 7-6

DIFFERENTIAL EQUATION OF A CABLE:

$$\frac{d^2y}{dx^2} = \frac{w(x)}{T_o}$$

where $w(x)$ = distributed load as a function of x (x is horizontal axis)
T_o = constant = horizontal component of the tension in the cable

PARABOLIC CABLES: The cable supports a load w which is uniformly distributed horizontally. The shape of the cable is given by

$$y = \frac{wx^2}{2T_o} \qquad (x = 0 \text{ at lowest point})$$

The tension in a symmetric cable is $T = \sqrt{T_o^2 + w^2x^2}$

CATENARY CABLES: The cable supports a load w which is uniformly distributed along the cable. The shape of the cable is given by

$$y = \frac{T_o}{w}\left(\cosh \frac{wx}{T_o} - 1\right)$$

The tension in the cable is $T = T_o + wy$.

Section 7-7

FLUID STATICS:

$$\text{Pressure} \quad p = \frac{\text{force}}{\text{area}}$$

PASCAL'S LAW

The pressure is the same in all directions at any given point in a fluid at rest.

The pressure varies linearly with the depth h in a fluid,

$$\Delta p = \gamma\, \Delta h$$

where γ is the specific weight of the fluid (weight per unit volume).

Gage pressure is measured from the open surface of a liquid,

$$p = \gamma h$$

Hydrostatic force = resultant of hydrostatic pressure on a solid surface
= area under pressure distribution curve

Center of pressure = location where the hydrostatic force acts
= centroid of pressure distribution (not at the centroid of the solid surface on which the pressure acts)

Section 7-8

ARCHIMEDES' PRINCIPLE

A submerged body is subject to an upward buoyant force equal to the weight of the displaced fluid. The buoyant force acts at the center of buoyancy, which is the center of gravity of the displaced fluid (not necessarily the center of gravity of the body that displaces the fluid).

Section 7-9

GENERAL DEFINITIONS OF STRESS:

$$\text{Stress} = \frac{\text{force}}{\text{area}}$$

$$\text{Normal stress} \quad \sigma = \frac{\text{normal force}}{\text{area}}$$

$$\text{Shear stress} \quad \tau = \frac{\text{shear force}}{\text{area}}$$

For precise definitions, use a differential force acting on a differential area,

$$\sigma = \frac{dF_{\text{normal}}}{dA} \qquad \tau = \frac{dF_{\text{shear}}}{dA}$$

To calculate force from stress,

$$F_{\text{normal}} = \int \sigma \, dA \qquad F_{\text{shear}} = \int \tau \, dA$$

WEDGE METHOD OF STRESS ANALYSIS: Draw a triangular element with all known ($\sigma_x, \ldots$) and unknown (σ_θ, τ_θ) stresses on it:

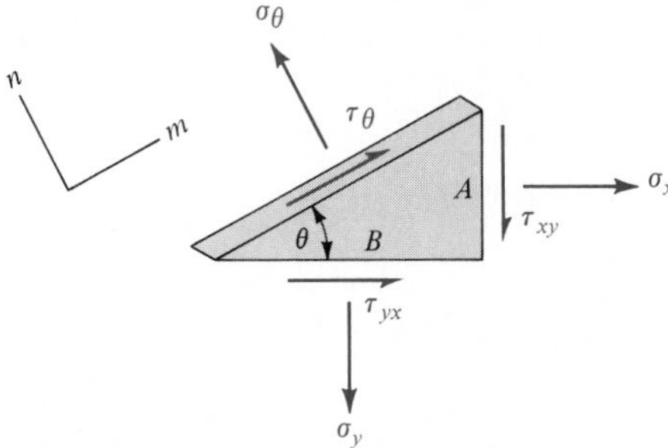

Write two equilibrium equations, one for the m direction, the other for the n direction. These equations are for *forces*, so each term must involve a stress times the area on which that stress is acting. Solve the equations for the unknown stresses.

REVIEW PROBLEMS

FIGURE P7-114

Plane area

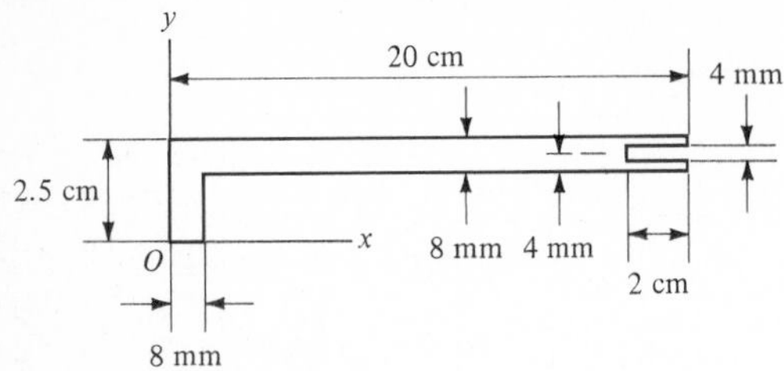

7–114 The probe of a mechanical testing instrument is modeled as a flat plate of negligible thickness. Determine its centroid assuming that the slot on the right end is not made.

7–115 Solve Prob. 7–114 as stated, then make the correction for the 2-cm long slot as shown in Fig. P7-114.

7–116 The shield of a small electronic measuring device is made of sheet metal of negligible thickness. Determine the centroid of the five-sided box shape.

FIGURE P7-116

Box shape made of plane areas

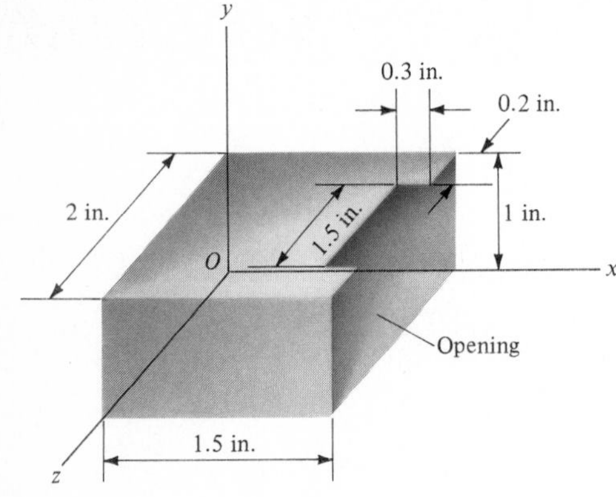

7–117 Determine the external surface area of the bellows which appears as a circle when viewed from the right or left.

7–118 Determine the volume enclosed by the bellows in Fig. P7-117.

7–119 An airplane wing is modeled as a uniform beam with a distributed and a concentrated load on it. Express the shear force V and bending moment M as functions of x for the region AB.

7–120 Draw the complete shear and moment diagrams for the wing in Fig. P7-119.

FIGURE P7-117

Thin-walled bellows with ends closed

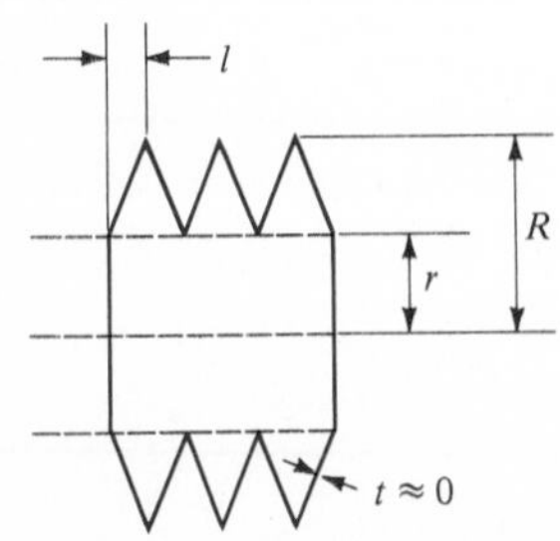

FIGURE P7-119

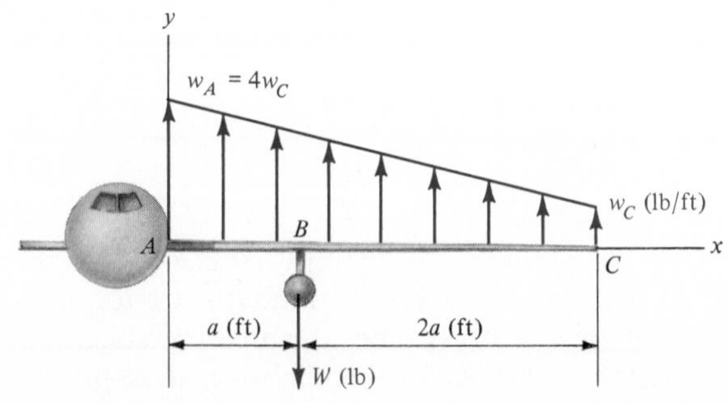

7–121 Calculate the maximum and minimum tensions in the cable in Fig. P7-83. Let $w = 0.7$ lb/ft uniformly distributed along the cable, $a = 100$ ft, $b = 350$ ft, and $h_A = 15$ ft.

FIGURE P7-123

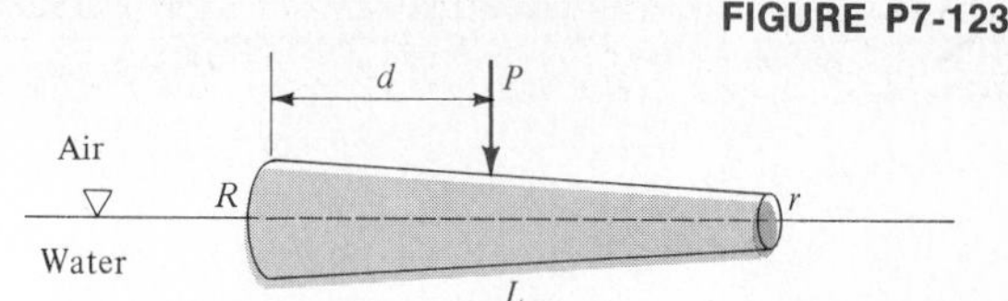

7–122 Plot the maximum tension in Prob. 7–121 for $h_A = 10$ ft, 15 ft, and 20 ft.

7–123 The pontoons of a seaplane are modeled with a uniformly varying circular cross section. What is the load P and its location x to place the centerline of the pontoon at the surface of the water? Assume that the pontoon itself is weightless.

FIGURE P7-124

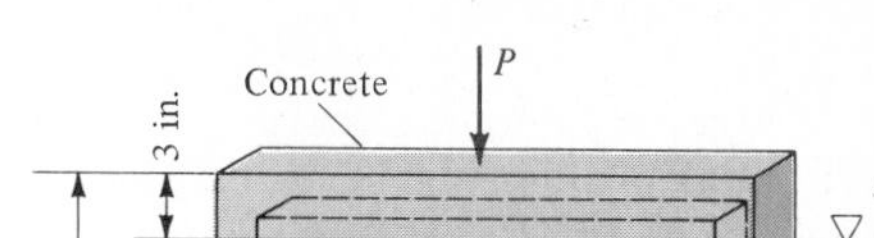

(a) Floating Dock

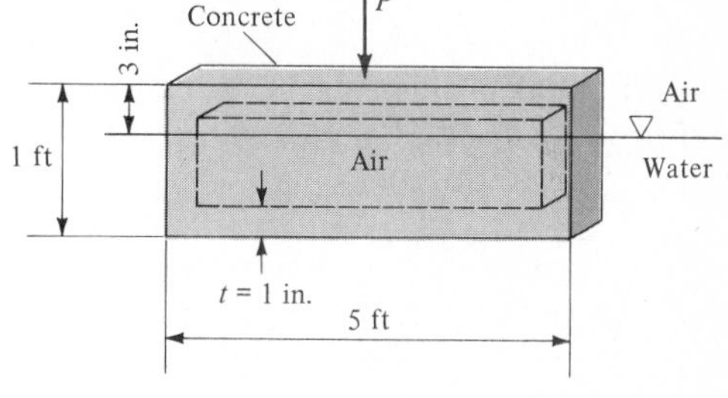

(b) Width of chamber = 2.5 ft

7–124 The concrete flotation chambers of the floating dock are 5-ft long, 1-ft high, and 2.5-ft wide. The uniform wall thickness is $t = 1$ in. of each box. What is the external load P that allows the box to extend 3 in. above the water? Concrete weighs 150 lb/ft^3, and sea water weighs 64 lb/ft^3.

In Probs. 7–125 and 7–126 determine the shear and normal stresses on plane AB.

FIGURE P7-125

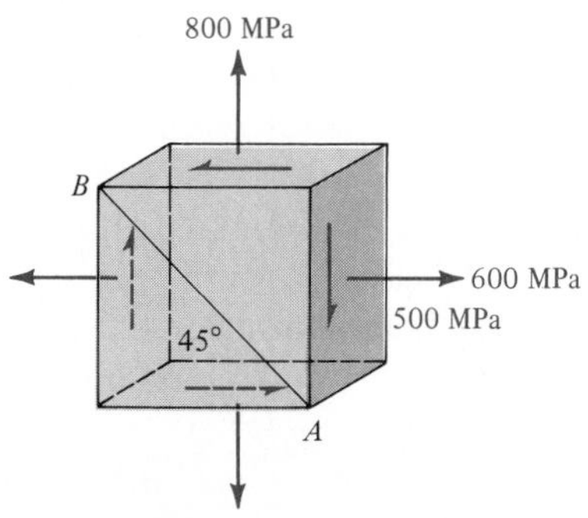

FIGURE P7-126

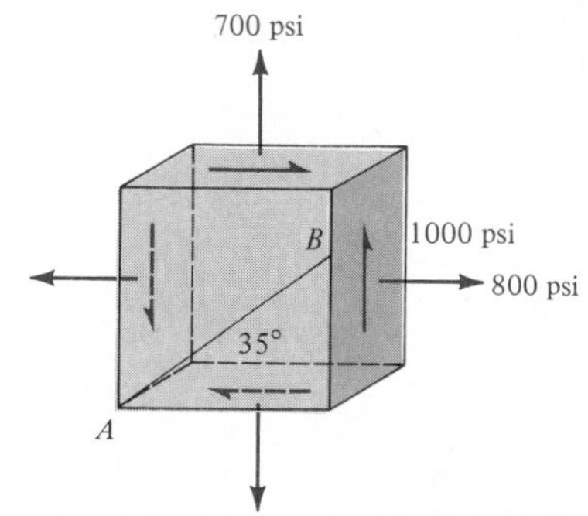

CHAPTER 8

FRICTION

OVERVIEW

Basic Concepts in Analyzing the Effects of Friction

Some phenomena involving friction are frequently observed and are naturally understood by most people. For example, consider the heat generated by rubbing dry hands together, the skid marks made by automobile tires sliding on dry pavements, and the action of bicycle brakes. It appears from these and other phenomena that a *friction force* acts between contacting bodies when one slides relative to the other or when sliding tends to occur. The friction force is tangential to each body at the point of contact, as evidenced by the displacement of skin when the hands are slowly rubbed. It is also clear from the common examples that the friction force depends on the *normal force* pressing the bodies together and on the condition of the contacting surfaces.

These concepts can be extended to qualitatively explain more complex phenomena of friction. For example, there is *internal friction* in solid materials when deformed, as evidenced by the heat generated in the sidewalls of rolling tires or in repeatedly bent wires. Such friction involves atoms or molecules moving tangentially relative to one another in the material. *Fluid friction* or *drag* involves a fluid moving past a solid body. These phenomena can be briefly introduced in statics after an extensive analysis of friction forces between solid bodies.

STUDY GOALS

SECTION 8-1 presents the fundamental aspects of dry friction between solid bodies. These include the concepts of static and kinetic friction (the latter for moving bodies). These concepts are essential in solving many statics and dynamics problems and should be mastered.

SECTION 8-2 describes the principles for experimentally determining the coefficient of static friction, which is assumed a constant in most problems. Engineers rarely do such experiments since approximate values are available from handbooks, but the method is also illustrative of friction problems in general.

SECTION 8-3 presents a useful classification of problems involving dry friction and procedures for solving them.

SECTION 8-4 extends the basic concepts of static friction to the analysis of wedges and screws, the simplest machines. These topics are very useful for the exercise of skills in modeling real situations with free-body diagrams and solving equilibrium equations.

SECTIONS 8-5 AND 8-6 present the application of the concepts of dry friction to disk bearings, disk clutches, disk brakes, and journal bearings for axles. These components of machines and vehicles provide for useful practice in statics analyses. They also build a background for more advanced courses in machine design.

SECTION 8-7 introduces the topic of rolling resistance of wheels and balls. This is sufficiently specialized and different from those involving simple dry friction that it is often omitted from statics courses. The topic is presented after Sec. 8-6 because of similarities in the analytical approaches used.

SECTION 8-8 presents the method of analysis and simple formulas for the friction problems of belts, ropes, and wires on pulleys or drums. This topic is relevant to machine design.

SECTION 8-9 presents the elementary concepts of fluid friction. This topic is often omitted from statics since it is covered in fluid mechanics.

8-1 DRY FRICTION

Assume that the driver of a moving vehicle on a steep, dry road applies enough braking force to lock every wheel. The vehicle slides on the road and may come to a complete stop without hitting anything. For analyzing these situations, both the sliding and the stationary vehicle (Fig. 8-1) can be modeled simply as a block on an inclined plane, as in Fig. 8-2a. The next logical step is to consider the free-body diagram of the block, which is shown in Fig. 8-2b. Here W is the weight of the block and for convenience it is resolved

FIGURE 8-1

Friction may be involved in the analysis of both moving or stationary bodies

into components parallel and perpendicular to the inclined plane, W_m and W_n, respectively.

If the body in Fig. 8-2b is stationary, there must be a reaction from the inclined plane acting on the body that is equal and opposite to W. This reaction can be resolved into convenient components $\mathscr{F}$ and N that act in opposite directions to W_m and W_n, respectively. The normal force N always acts perpendicular to the surface and prevents motion in that direction. Thus, N is equal to $W \cos \theta$ whether or not there is motion parallel to the surface. The tangential force $\mathscr{F}$ is the friction force, which may or may not prevent motion. It is often denoted by a script letter to distinguish it from other forces. The friction force always acts at the surfaces of contact between bodies, parallel (or tangent) to the surfaces, and it acts in the direction opposite to motion or impending relative motion. The friction force does not depend on the area of the contact patch if there are no large deformations there. However, the location of the resultant normal force is not known unless there is a distinct point of contact.

The magnitude of the friction force may have the following values with respect to $W \sin \theta$, the force that tends to cause motion along the incline.

1. $\mathscr{F} = W \sin \theta$, there is no motion, or there is motion at constant velocity.
2. $0 < \mathscr{F} < W \sin \theta$, the body accelerates downhill.
3. $\mathscr{F} > W \sin \theta < 0$, the body initially moving downhill has a decreasing velocity and eventually stops; after stopping, $\mathscr{F} = W \sin \theta$.

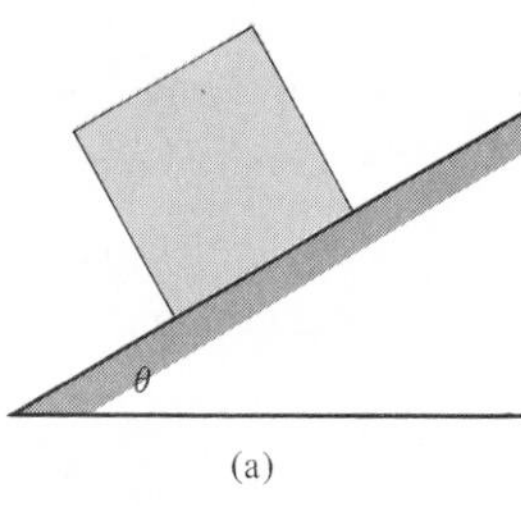

(a)

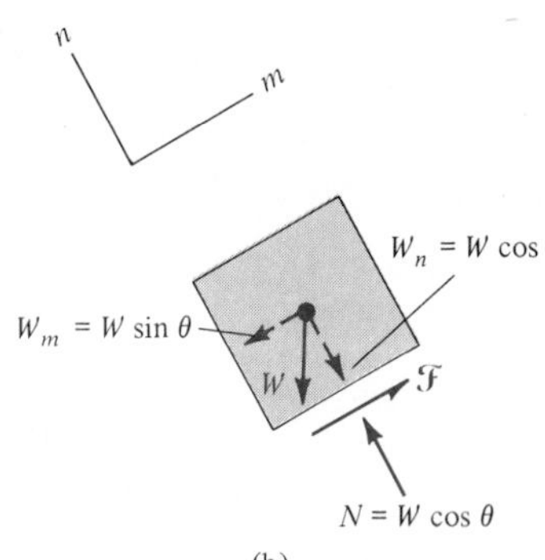

(b)

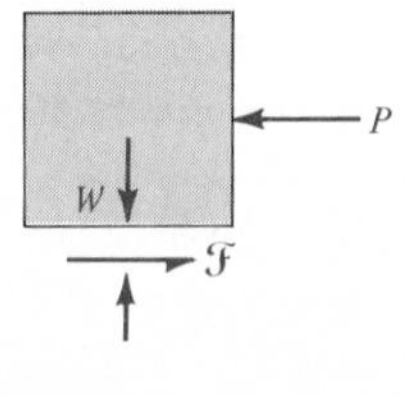

(c)

FIGURE 8-2

A vehicle on a sloping road is modeled as a block on an incline

These generalizations should make sense even without much knowledge of dynamics. They can also be extended to situations such as that in Fig. 8-2c where $W \sin\theta = 0$, but another external force P tends to cause motion parallel to the supporting surface.

The magnitude of the friction force may vary from zero to a maximum value for any given pair of bodies. $\mathscr{F}$ is always zero when there is no force for the friction force to oppose, as in Fig. 8-2c if $P = 0$. The friction force can be equal to the applied force up to a limiting value $\mathscr{F}_{max}$ shown in Fig. 8-3. The maximum friction force depends on the materials involved, on the geometry and condition of the surfaces in contact, and on the normal force N that presses the bodies together. The material and surface properties are lumped together and represented in analysis by the *coefficient of friction* μ. It is found that under most conditions the limiting friction force is proportional to the normal force between the bodies,

$$\boxed{\mathscr{F}_{max} = \mu N \quad (0 \leq \mathscr{F} \leq \mathscr{F}_{max})} \qquad \boxed{8\text{-}1}$$

The coefficient of friction is largest for a given pair of materials when there is static equilibrium; the *coefficient of static friction* is denoted by μ or μ_s. The coefficient decreases when there is relative sliding of the bodies; this is called the *coefficient of kinetic friction* (or sliding friction, or dynamic friction) and is denoted by μ_k. The coefficient of kinetic friction ($\mu_k < \mu_s$) is approximately constant at moderate rubbing speeds. At higher speeds the coefficient of friction decreases further because of the heat generated by friction (Fig. 8-3). This is the reason for the phenomenon of brake fading at high speeds or during long descents. Clearly, locking the wheels in a

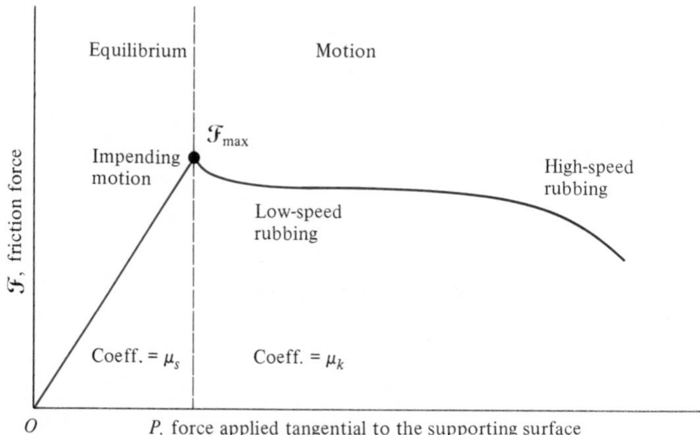

FIGURE 8-3

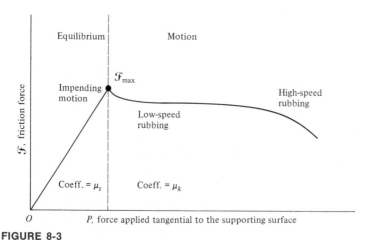

Variation of the friction force between two bodies

moving vehicle immediately reduces the braking force at the road surface ($\mu_k < \mu_s$), and may cause lateral sliding of the vehicle since sliding in any direction facilitates sliding in other directions. The heat generated at the sliding contact patch of each tire may cause further changes in the friction force acting on a vehicle.

8-2 ANGLES OF FRICTION

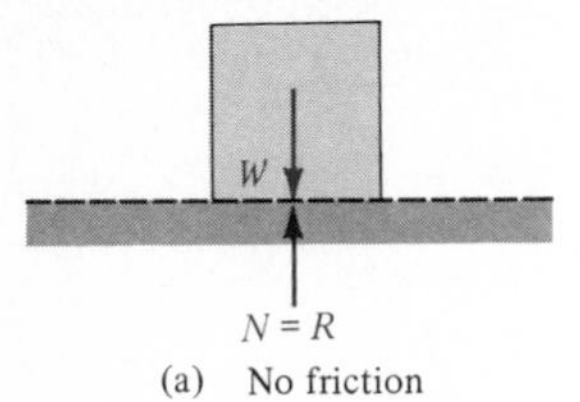

(a) No friction

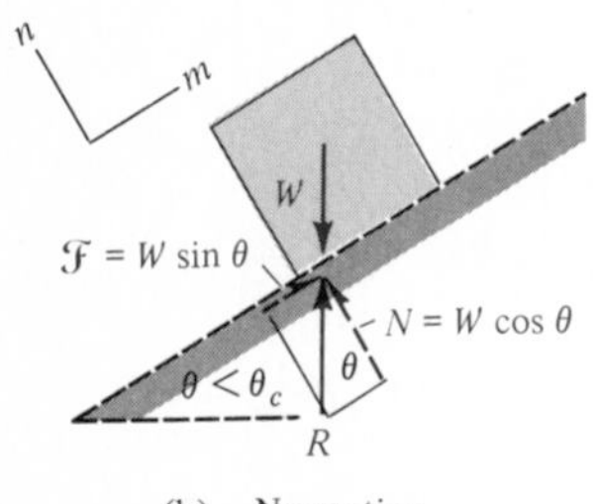

(b) No motion

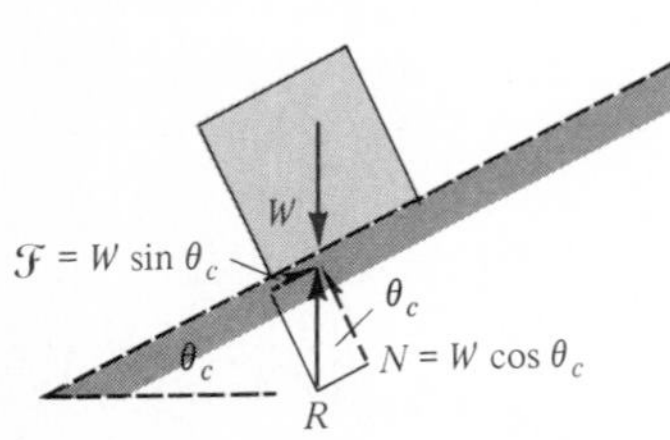

(c) Motion impending or beginning

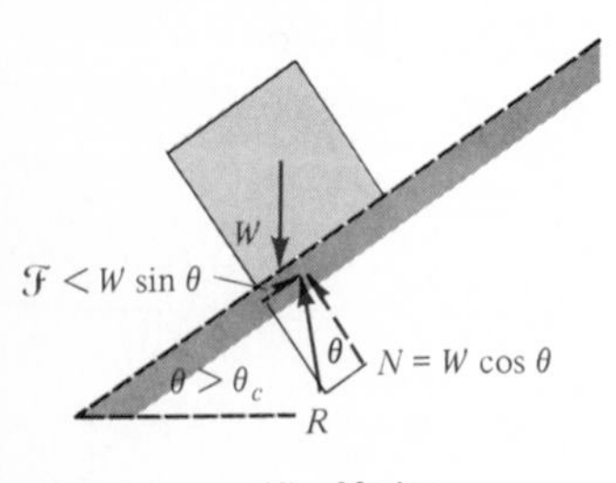

(d) Motion

FIGURE 8-4
Angles of friction

The experimental determination of the coefficient of static friction is normally an easy matter. To illustrate the principles used in this work, consider Fig. 8-4. Here a block of weight W is on a plane that is represented by a dashed line. In Fig. 8-4a the plane is horizontal, so the entire reaction R acting on the block is normal to the plane. Next assume that the plane is inclined at a small angle θ with the horizontal, as in Fig. 8-4b. The reaction R now is the resultant of the friction force $\mathscr{F}$ and the normal force N. As the angle θ is further increased, it reaches a critical value θ_c, at which angle the block begins to slide (Fig. 8-4c). The reaction R is again the resultant of forces $\mathscr{F}$ and N. The largest possible value of the friction force $\mathscr{F}$ for the given bodies occurs when motion is impending, which is at the critical angle where motion begins, for all practical purposes. In fact, there is no way to prove that motion is impending without letting the motion begin. At angles larger than the critical angle, the block accelerates down the incline. In these cases the reaction **R** makes an angle with the weight **W** since $\mathscr{F} < W \sin\theta$.

The critical angle θ_c at which motion is impending is called the *angle of repose*. The friction force is at its maximum at this angle, so from Fig. 8-4c and Eq. 8-1,

$$\tan\theta_c = \frac{\mathscr{F}}{N} = \frac{\mu_s N}{N} = \mu_s \qquad \boxed{8\text{-}2}$$

Thus, measurement of the angle of repose is the only major step required in determining the coefficient of static friction.

The concepts discussed above can be extended to situations where the weight of a body does not tend to cause motion. For example, the block in Fig. 8-5 is on a horizontal plane, so it can move only if a large enough force P is applied. Here motion is impending or beginning when the lateral force reaches a critical value P_c. At that instant the friction force is at its maximum, so from Fig. 8-5c and Eq. 8-1,

$$\tan\theta_c = \frac{\mathscr{F}}{N} = \frac{\mu_s N}{N} = \mu_s$$

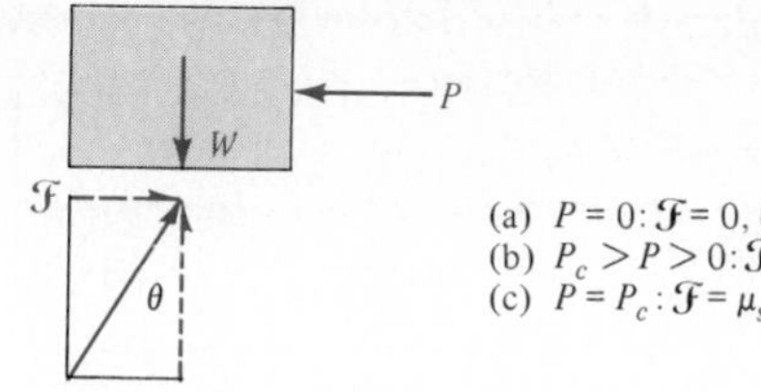

(a) $P = 0: \mathcal{F} = 0, \theta = 0$, no motion
(b) $P_c > P > 0: \mathcal{F} > 0, \theta > 0$, no motion
(c) $P = P_c: \mathcal{F} = \mu_s N, \theta = \theta_c$, impending or beginning motion

FIGURE 8-5

Critical value of a lateral force for causing relative motion between two bodies

which is Eq. 8-2. In this case the angle θ_c is called the *angle of static friction*, which is equivalent to the angle of repose.

The coefficients of friction in dry friction (also called Coulomb friction) are normally sensitive to humidity, temperature, dust, contamination, oxide films, surface finish, and chemical reactions. The contact area and the normal force affect the coefficients only when they cause large deformations of one or both bodies in contact. Published values of the coefficients of friction are all empirically determined, and most often they are for static friction. As a rule of thumb, high values of the coefficient of friction are between 0.6 and 1.0, and low values are about 0.1 or less. Approximate values of μ_s for many materials are given inside the back cover of this book for easy reference. For rough or sticky surfaces, μ_s may be larger than unity.

8-3 CLASSIFICATION OF PROBLEMS OF DRY FRICTION

There are numerous ways to classify problems involving dry friction. Generally, the first step is to decide whether Eq. 8-1 is applicable or not. It is essential to remember that $\mathcal{F} = \mu N$ *when there is motion or impending motion.* There are many situations where there is friction that is less than the maximum possible friction for the given conditions; in such cases $\mathcal{F} < \mu N$ and cannot be used to determine the friction force.

The major steps in solving problems of dry friction are listed in three categories in Table 8-1. Note that at least the directions of unknown friction forces are easily determined by inspection. The correct directions should be used in the solutions. The details of the procedures are illustrated in several examples.

TABLE 8-1

Procedures for Solving Friction Problems

CLASS OF PROBLEM	GIVEN INFORMATION	PROCEDURE
A	Bodies, forces, coefficients of friction are known. Impending motion is not assured; $\mathscr{F} \neq \mu_s N$.	To determine if equilibrium is possible: 1. Construct the free-body diagram. 2. Assume that the system is in equilibrium. 3. Determine the friction and normal forces necessary for equilibrium. 4. Results: (a) $\mathscr{F} < \mu_s N$, the body is at rest. (b) $\mathscr{F} > \mu_s N$, motion is occurring, static equilibrium is not possible. Since there is motion, $\mathscr{F} = \mu_k N$. Complete solution requires principles of dynamics.
B	Bodies, forces are given. Impending motion is specified, $\mathscr{F} = \mu_S N$ is valid.	To determine the coefficients of friction: 1. Construct the free-body diagram. 2. Write $\mathscr{F} = \mu_s N$ for all surfaces where motion is impending. 3. Determine μ_s from the equation of equilibrium.
C	Bodies, forces, coefficients of friction are known. Impending motion is specified, but the exact motion is not given. The possible motions are sliding, tipping, or rolling. Alternatively, the forces or coefficients of friction may have to be determined to produce a particular motion from several possible motions.	To determine the exact motion that may occur, or unknown quantities required: 1. Construct the free-body diagram. 2. Assume that motion is impending in one of the two or more possible ways. Repeat this for each possible motion, and write the equation of equilibrium. 3. Compare the results for the possible motions and select the likely event. Determine the required unknowns for any preferred motion.

EXAMPLE 8-1

A car that weighs 10 kN is to be parked on the same 30° incline year round. The static coefficient of friction between the tires and the road varies between the extremes of 0.05 and 0.9. Is it always possible to park the car at this place? Assume that the car can be represented by a single block on an incline.

SOLUTION

The problem is in class A of Table 8-1. It is most reasonable to consider the lowest coefficient of friction first since the larger the coefficient of friction, the larger the resistance to sliding. The situation considered is modeled in Fig. 8-4b. For equilibrium,

$$\sum F_n = 0$$

$$N - W\cos 30^\circ = 0$$

$$N = W\cos 30^\circ = 8.66 \text{ kN}$$

$$\sum F_m = 0$$

$$\mathscr{F} - W\sin 30^\circ = 0$$

$$\mathscr{F} = W\sin 30^\circ = 5 \text{ kN}$$

$$= \text{friction force necessary for equilibrium}$$

The smallest friction force expected is

$$\mathscr{F}_{\text{max }1} = \mu_{s1}N = 0.05(8.66 \text{ kN}) = 0.433 \text{ kN}$$

Since $\mathscr{F}_{\text{max }1} < \mathscr{F}$, the car would slide and cannot be parked year round on the same incline.

JUDGMENT OF THE RESULTS

The result can be checked by determining the angle of repose θ_c for the smallest coefficient of friction. From Eq. 8-2,

$$\theta_c = \tan^{-1}\mu_{s1} = \tan^{-1} 0.05 = 3^\circ$$

Since $\theta_c < 30^\circ$, the car would slide.

EXAMPLE 8-2

Consider a wheel of a car on an incline as in Fig. 8-6a. Determine the moment M that must be applied at the wheel spindle to hold the wheel at rest. Note that two different motions, sliding and spinning of the wheel, must be prevented.

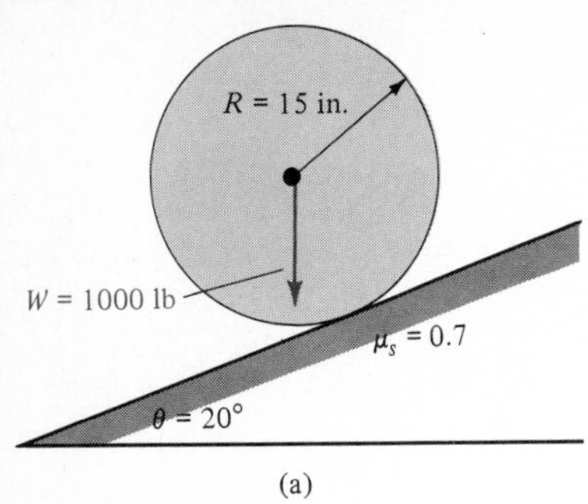

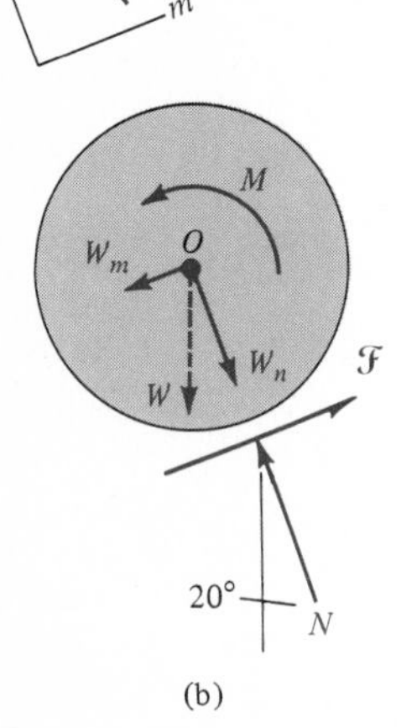

FIGURE 8-6
Wheel of a car parked on an incline

SOLUTION

The problem is in class A of Table 8-1. To check the possibility of sliding, the free-body diagram is drawn (Fig. 8-6b). The normal and friction forces are unknown. For equilibrium in the n and m directions,

$$\sum F_n = 0$$

$$-W_n + N = 0$$

$$N = W \cos 20^\circ = 1000 \text{ lb} \cos 20^\circ = 940 \text{ lb}$$

$$\sum F_m = 0$$

$$-W_m + \mathscr{F} = 0$$

$\mathscr{F} = W \sin 20^\circ = 1000 \text{ lb} \sin 20^\circ = 340 \text{ lb} =$ friction force necessary for equilibrium. The available friction force is $\mathscr{F}_{\max} = \mu N = (0.7)(940 \text{ lb}) = 660$ lb, so $\mathscr{F} < \mathscr{F}_{\max}$. Therefore, there is no sliding.

For no spinning,

$$\sum M_o = 0$$

$$\mathscr{F}R + M = 0$$

$$M = -(340 \text{ lb})(15 \text{ in.}) = -5100 \text{ in}\cdot\text{lb}$$

Thus, the moment M must be clockwise.

EXAMPLE 8-3

A car is to be parked with all wheels locked on an incline as in Fig. 8-7. Can the car remain at rest? If not, what is the additional force P required parallel to the incline for equilibrium? Assume that the forces are symmetric with respect to the longitudinal axis of the car.

SOLUTION

The problem is in class A of Table 8-1. Figure 8-7 is the free-body diagram. Assuming symmetry of the left and right sides of the car, the normal and friction forces at the two back tires and two front tires are represented by their sums N_A, $\mathscr{F}_A$, and N_B, $\mathscr{F}_B$, respectively (Fig. 8-7). These four forces are unknown. First, it is necessary to calculate the normal forces. This is started by taking moments about either A or B.

$$\sum M_A = 0$$

$$W \sin 10°(0.3 \text{ m}) - W \cos 10°(2 \text{ m}) + N_B(3.5 \text{ m}) = 0$$

$$N_B = \frac{10 \text{ kN}(2 \text{ m} \cos 10° - 0.3 \text{ m} \sin 10°)}{3.5 \text{ m}} = 5.5 \text{ kN}$$

$$\sum F_n = 0$$

$$N_A + N_B - W \cos 10° = 0$$

$$N_A = 10 \text{ kN} \cos 10° - 5.5 \text{ kN} = 4.4 \text{ kN}$$

In investigating equilibrium parallel to the incline, the total friction force necessary for equilibrium, $\mathscr{F}_T = \mathscr{F}_A + \mathscr{F}_B$, must be determined. Summing forces along m,

$$\sum F_m = 0$$

$$\mathscr{F}_A + \mathscr{F}_B - W \sin 10° = 0$$

$$\mathscr{F}_T = \mathscr{F}_A + \mathscr{F}_B = 10 \text{ kN} \sin 10° = 1.7 \text{ kN}$$

The available friction force is

$$\mathscr{F}_{\text{max}} = (\mathscr{F}_A + \mathscr{F}_B)_{\text{max}} = \mu N_A + \mu N_B = (0.1)(N_A + N_B)$$

$$= (0.1)(W \cos 10°) = 1 \text{ kN}$$

$\mathscr{F}_T > \mathscr{F}_{\text{max}}$; therefore, the car will slide. The additional force P required parallel to the incline for equilibrium is obtained from

$$\sum F_m = 0$$

$$P + \mathscr{F}_{\text{max}} - W \sin 10° = 0$$

$$P = 1.7 \text{ kN} - 1 \text{ kN} = 0.7 \text{ kN}$$

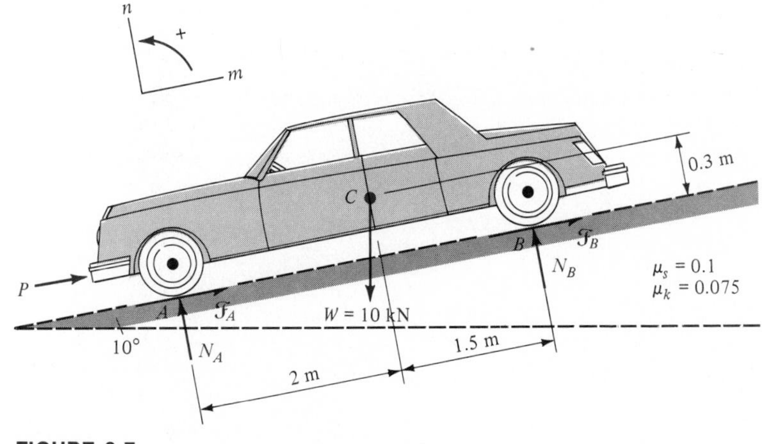

FIGURE 8-7

Car parked on an incline

EXAMPLE 8-4

Consider the crate shown in Fig. 8-8a. (a) Determine the largest force F that can be applied without the crate sliding or tipping. (b) What are the limiting values of F and μ_s such that sliding occurs first? This is preferable to tipping.

3 ft
3 ft
2 ft
Mass center
F
4 ft
W = 800 lb
$\mu_s = 0.3$

(a)

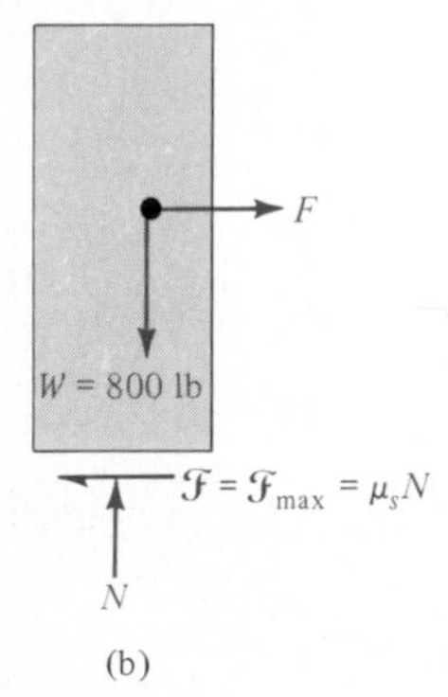

(b)

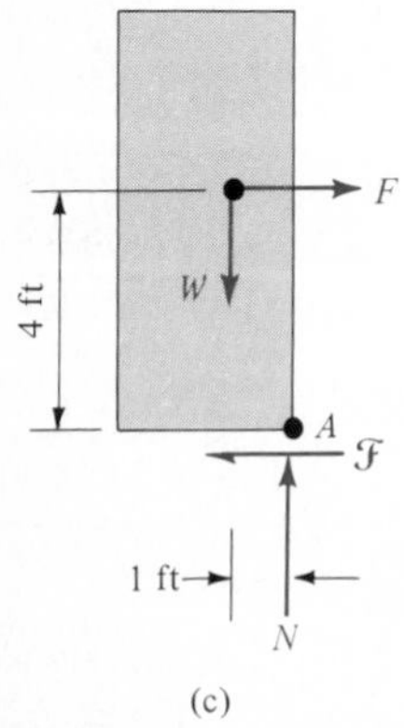

(c)

FIGURE 8-8

Sliding or tipping of a crate

SOLUTION

(a) This is a situation as discussed in Table 8-1, class C. The possible motions must be considered separately.

MOTION 1, SLIDING (Motion Impending): From Fig. 8-8b,

$$\sum F_x = 0$$
$$F - \mathscr{F}_{max} = 0$$
$$F = \mathscr{F}_{max} = \mu_s N = 0.3W = 0.3(800 \text{ lb}) = 240 \text{ lb}$$

MOTION 2, TIPPING (Fig. 8-8c): If tipping is impending, the only point of contact is corner A, hence the location of the normal force N is known. Therefore,

$$\sum M_A = 0$$
$$(800 \text{ lb})(1 \text{ ft}) - F(4 \text{ ft}) = 0$$
$$F = 200 \text{ lb}$$

The smallest value of F is the governing value, so

$$F_{max} = 200 \text{ lb}$$

(b) If slipping is the desired first motion, then a smaller coefficient of friction is required. Now the problem is in class B, Table 8-1: determine μ_s for slipping and tipping to be equally likely; then for all $\mu < \mu_s$ sliding occurs first. From equilibrium in the x direction when slipping is impending,

$$F = \mathscr{F} = \mathscr{F}_{max} = \mu_s N = \mu_s W$$

When tipping is impending, Fig. 8-8c is applicable:

$$\sum M_A = 0$$
$$W(1 \text{ ft}) - F(4 \text{ ft}) = 0$$
$$F = \frac{1}{4} W$$

Setting the two values of F equal makes slipping or tipping equally likely,

$$\mu_s W = \frac{1}{4} W, \qquad \mu_s = 0.25$$

Hence, for slipping to occur first, the coefficient of static friction μ must be less than 0.25.

EXAMPLE 8-5

Consider a wheel at rest on an incline as in Fig. 8-9a. The forces F press the internal brake pads B against the wheel. The brake pads are fixed with respect to the vehicle frame and always remain parallel to the inclined plane. (a) Determine F that prevents the wheel from rolling downhill. (b) For this F, at what value of μ_A does the wheel slide down without rolling?

SOLUTION

(a) The problem is in class C of Table 8-1. The free-body diagram is drawn in Fig. 8-9b. By summing moments about point A, calculation of the unknown $\mathscr{F}_A$ and N_A is avoided. For impending motion, the friction forces at the brakes are $\mathscr{F}_B = \mathscr{F}_{\max} = \mu_B F$. Realizing that the braking forces form a couple (a free moment vector),

$$\sum M_A = 0$$

$$W_m R - \mathscr{F}_B(2R) = 0$$

$$\mathscr{F}_B = \frac{1}{2} W_m = \frac{1}{2}(10 \text{ kN} \sin 20^\circ) = 1.7 \text{ kN} = \mu_B F = 0.5F$$

Hence, $F = 1.7 \text{ kN}/0.5 = 3.4 \text{ kN}$.

(b) Now the problem is in class B, Table 8-1. For impending sliding,

$$\mathscr{F}_A = \mathscr{F}_{\max} = \mu_A N_A$$

For equilibrium,

$$\sum F_n = 0$$

$$N_A - W_n = 0$$

$$N_A = W \cos 20^\circ$$

$$\sum F_m = 0$$

$$-W_m + \mathscr{F}_A = 0$$

$$-W \sin 20^\circ + \mu_A W \cos 20^\circ = 0$$

$$\mu_A = \tan 20^\circ = 0.36$$

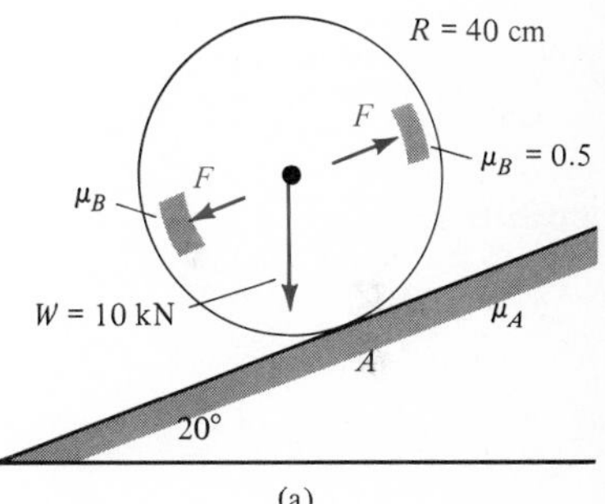

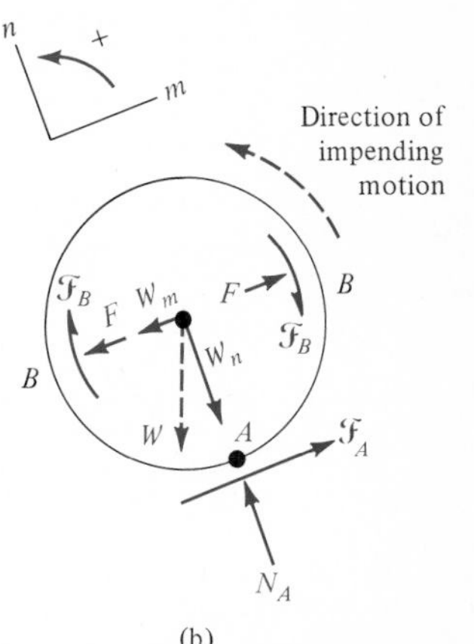

FIGURE 8-9
Wheel at rest on an incline

PROBLEMS

SECTIONS 8-1, 8-2, 8-3

FIGURE P8-1

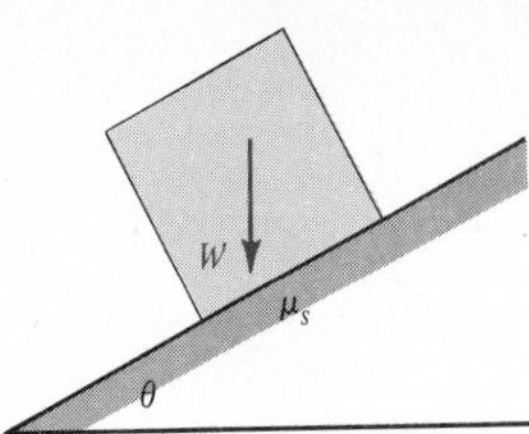

8–1 Determine whether the block can remain at rest on the inclined plane if $W = 300$ N, $\theta = 10^\circ$, and $\mu_s = 0.1$.

8–2 Consider a block on an incline (Fig. P8-1), where $\mu_s = 0.2$. Plot the angle θ at which the block begins to slide as a function of $W = 10$ lb, 50 lb, and 100 lb.

FIGURE P8-3

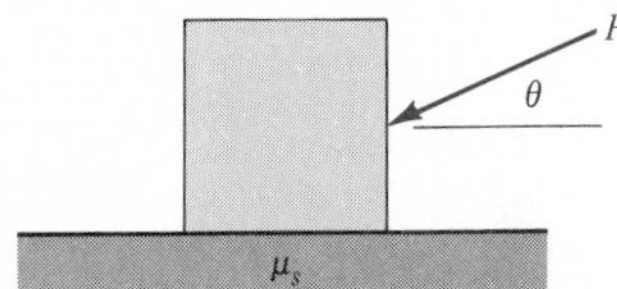

8–3 Assume that the block is weightless, $P =$ 500 N, $\theta = 15^\circ$, and $\mu_s = 0.3$. Does the block begin to slide when P is applied?

8–4 Solve Prob. 8–3 using $P = 10$ N, 500 N, and 10 kN. What is the general conclusion from the results?

8–5 The block in Fig. P8-3 weighs 100 lb, $P =$ 80 lb, $\theta = 10^\circ$, and $\mu_s = 0.25$. Does the block begin to slide when P is applied?

FIGURE P8-6

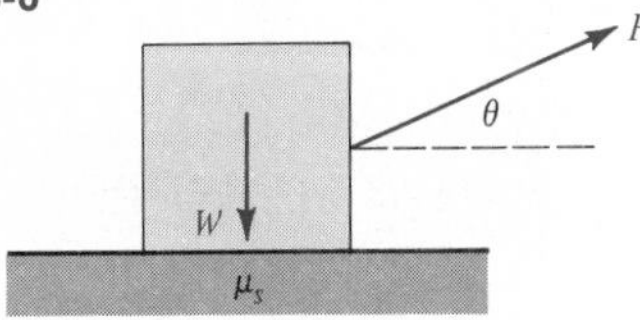

8–6 The block weighs $W = 1$ kN and $\mu_s = 0.4$. Determine the force P at $\theta = 20^\circ$ to start moving the block.

8–7 The block in Fig. P8-6 weighs 200 lb and $\mu_s = 0.5$. Determine the angle θ at which a force $P = 90$ lb causes the block to slide.

8–8 In Fig. P8-6 $\mu_s = \mu_k = 0.7$ and $P = W/2$. Determine if the block would move at $\theta = 0$, 40°, and 80°.

FIGURE P8-9

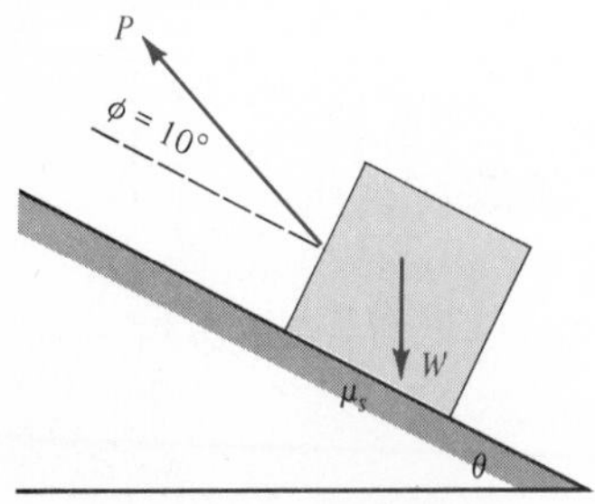

8–9 Determine the smallest force P (in terms of W) that causes the block to start moving up the incline if $\theta = 25^\circ$ and $\mu_s = 0.3$.

8–10 Assume that in Prob. 8–9 the force P has a fixed direction and magnitude. Determine the weight W (if any) to make the block slide down the incline in spite of the force P.

8–11 The ladder AB is supported by a smooth wall at B. Determine whether the 5-m-long ladder can remain at rest with a person of weight $W = 800$ N standing at $h = 4$m if $\theta = 70^\circ$ and $\mu_A = 0.4$.

8–12 Determine the angle θ in Prob. 8–11 at which the ladder starts to slide.

8–13 A uniform board AB of weight $W = 50$ lb is leaning against a smooth wall at B. Determine whether the board can remain at rest if $\theta = 65^\circ$ and $\mu_A = 0.3$.

8–14 Assume that $\theta = 70^\circ$ and $\mu_A = 0.8$ in Prob. 8–13. Calculate the vertical force P acting on the board at B that would start end B to move upward.

8–15 A uniform ladder AB is supported by a smooth wall at B. It is to be adjusted by applying a horizontal force F to the right at point C. Determine the force F required to cause motion if $W = 500$ N, $L = 7$ m, $h = 1.3$ m, $\theta = 75^\circ$, and $\mu_A = 0.3$.

8–16 Determine if the 60-lb, 8-ft-long beam can remain in equilibrium on the two steps. If yes, calculate the friction forces at A and B.

FIGURE P8-11

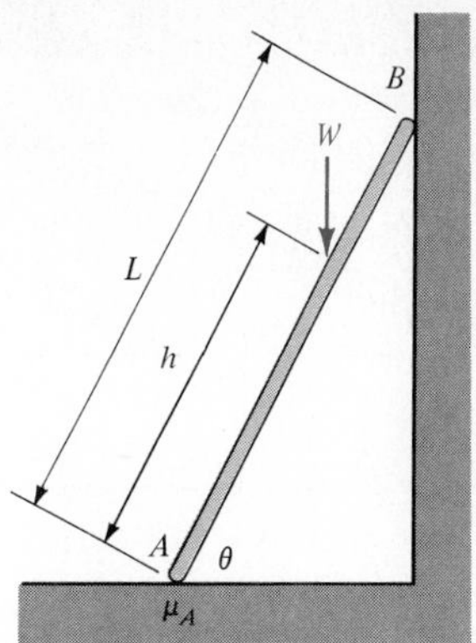

FIGURE P8-13

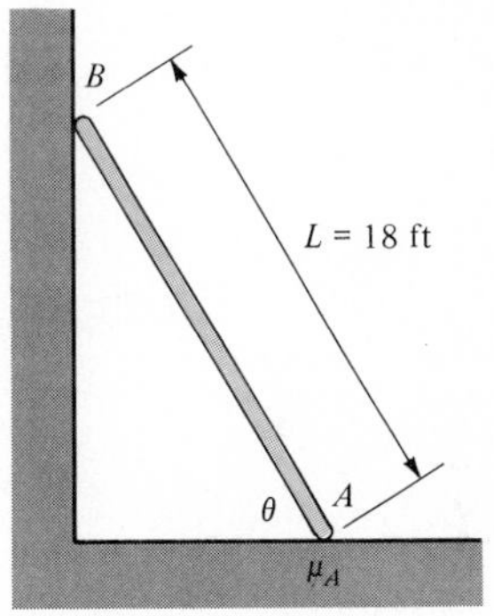

FIGURE P8-15

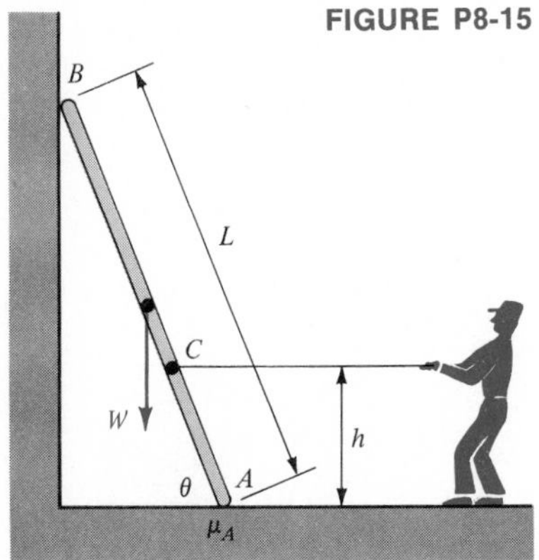

FIGURE P8-16

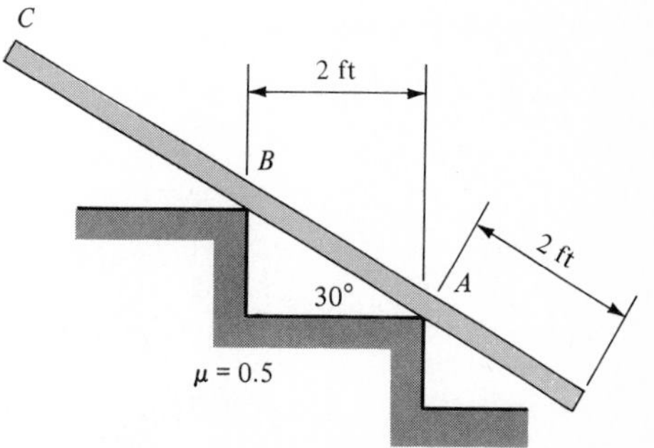

FIGURE P8-18

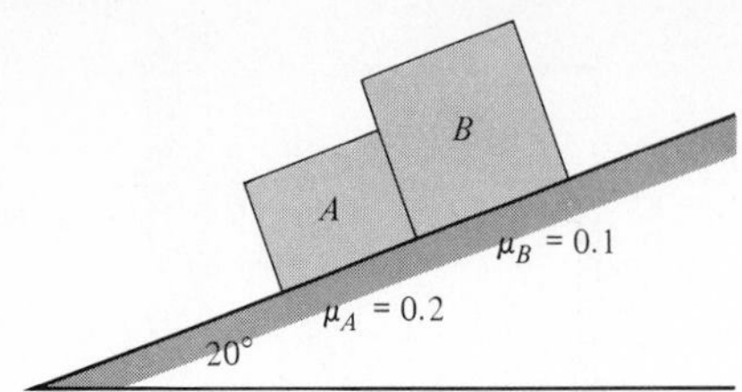

8–17 In Prob. 8–16 determine the horizontal force F at point C that would cause the beam to start sliding upward.

8–18 Determine if boxes A and B can remain in equilibrium as shown. The boxes have weights $W_A = 200$ N and $W_B = 300$ N.

8–19 Calculate the force F acting on box A parallel to the incline in Prob. 8–18 to cause impending upward motion of the boxes.

FIGURE P8-20

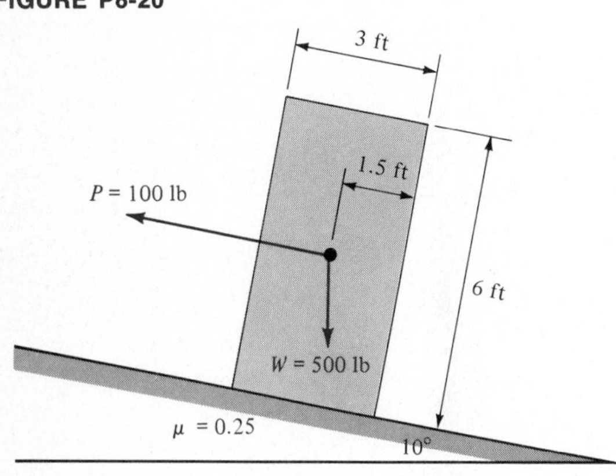

8–20 Determine if the uniform rectangular console can remain in equilibrium as shown. If yes, calculate the friction force and the location of the normal force on it.

8–21 Consider a static test of brake pads B of a bicycle wheel. Determine the normal force F that must act on each pad to hold the wheel at rest on a horizontal road if a horizontal force $P = 0.5W = 150$ N is applied to the axle of the wheel.

8–22 Determine the clockwise moment M_C that would start wheel C to rotate while a braking force $P = 500$ N is applied to arm AB of the simple brake. $r = 20$ cm, $a = 30$ cm, $b = 70$ cm, and $c = 30$ cm.

FIGURE P8-21

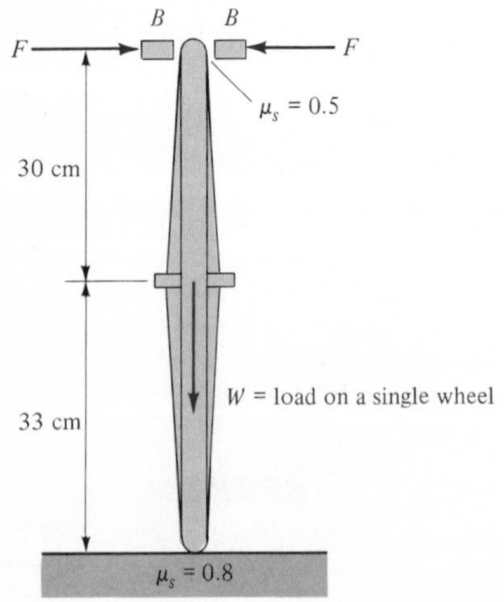

FIGURE P8-22

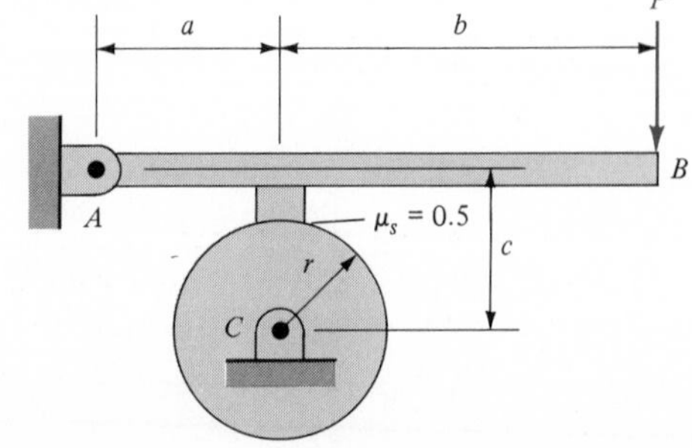

8–23 Determine the counterclockwise moment M_C that would start wheel C in Fig. P8-22 to rotate while a braking force $P = 100$ lb is applied to arm AB of the brake. $r = 8$ in., $a = 10$ in., $b = 40$ in., and $c = 13$ in.

8–24 Solve Prob. 8–22 with $c = 30$ cm and 40 cm. Plot the clockwise and counterclockwise starting moments M_C (magnitudes only) for the two values of c using the same diagram.

8–25 Determine the clockwise moment M_C that would start wheel C to rotate while a braking force $P = 500$ N is applied to arm AB of the brake. $r = 20$ cm, $a = 25$ cm, $b = 75$ cm, and $c = d = 30$ cm.

8–26 Determine the counterclockwise moment M_C that would start wheel C in Fig. P8-25 to rotate while a braking force $P = 100$ lb is applied to arm AB of the brake. $r = 8$ in., $a = 12$ in., $b = 35$ in., and $c = d = 13$ in.

8–27 Solve Prob. 8–26 with $c = 13$ in., $d = 10$ in., 13 in., and 16 in. Plot the clockwise and counterclockwise starting moments M_C(magnitudes only) for the three values of d. Make a single diagram.

8–28 There is a so-called *skating force* acting on the tone arms of most record players. This force F_s appears to push the arm toward the center C of the turntable. Explain this phenomenon by considering friction on the pickup stylus.

8–29 The launching ramp of a ship is modeled as a V-block, and the ship is assumed to have a circular cylindrical hull. Determine the force P necessary to start a ship of weight W to move if $\theta = 0$ and $\phi > 0$.

FIGURE P8-25

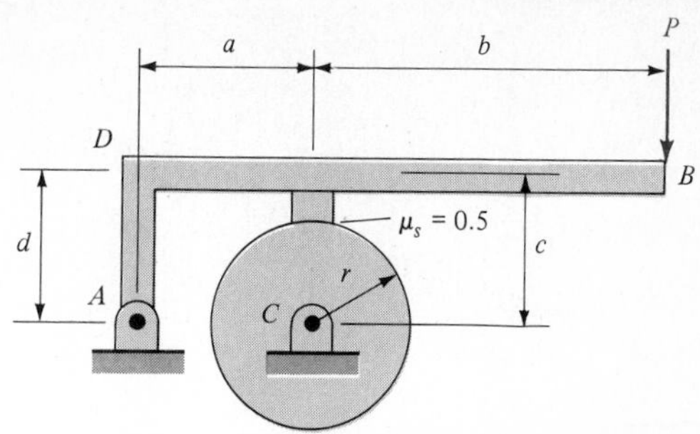

FIGURE P8-28

FIGURE P8-29

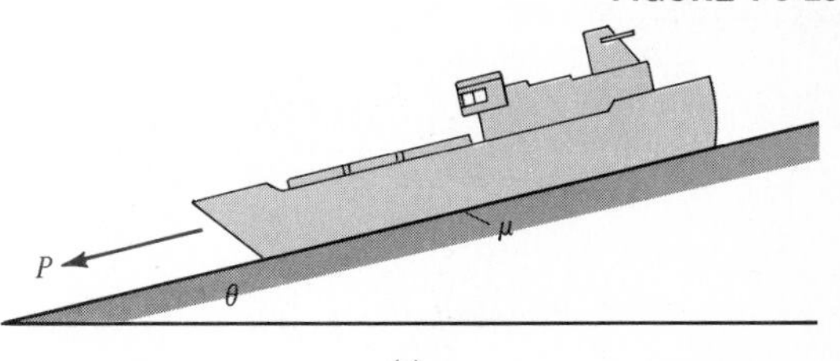

(a)

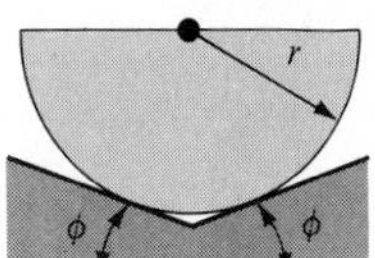

(b) Cross section of hull and ramp

FIGURE P8-31

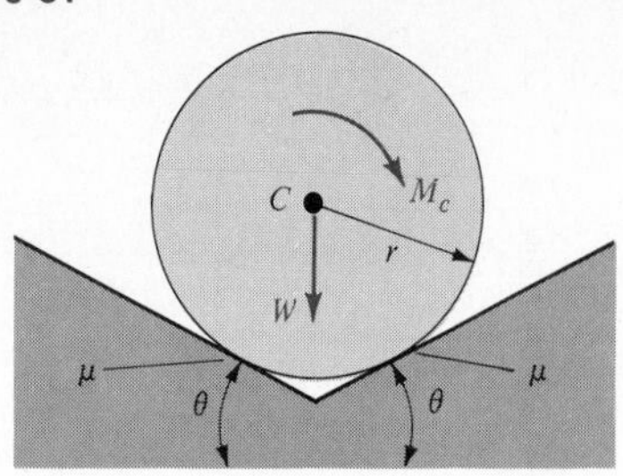

8–30 Solve Prob. 8–29 if $W = 50{,}000$ lb, $r = 10$ ft, $\theta = 5^\circ$, $\phi = 15^\circ$, and $\mu = 0.1$.

8–31 An exercise wheel is modeled as resting on a V-block. Write an expression for the couple M_C that would cause the wheel to rotate.

FIGURE P8-32

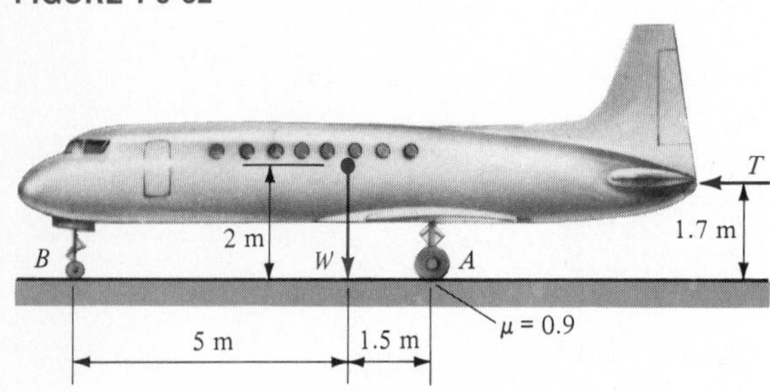

8–32 A jet aircraft weighs 70 kN. At the beginning of the takeoff run the aircraft is at rest with the brakes applied to the two main wheels A but not to the nose wheel B. Determine the total horizontal thrust T during the final check of the engines that would cause the locked wheels to slide.

FIGURE P8-33

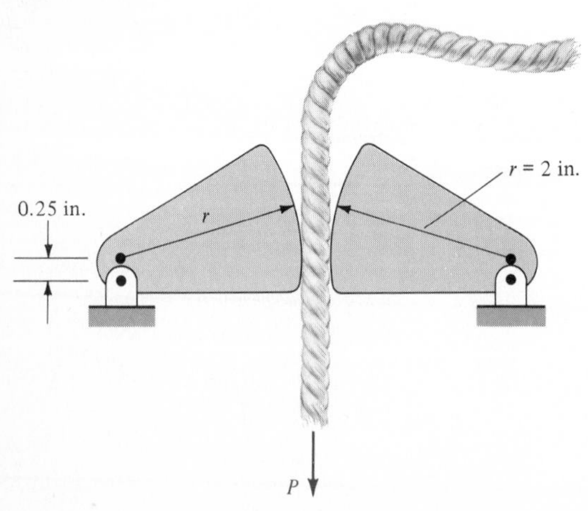

8–33 The cam cleat (commonly used on small sailboats) tightly grips the rope when it is pulled downward (and releases it when pulled upward). Determine the maximum force P that allows holding the rope without slipping if $\mu = 0.2$.

WEDGES AND SCREWS 8-4

Wedges are among the oldest and simplest machines. They are still very useful in amplifying the effects of small forces or making precise, small adjustments in the positions of bodies. Friction plays an important role in the behavior of every wedge as illustrated in the following.

Figure 8-10a shows a body A of weight W that is to be raised by driving the wedge B of negligible weight to the left by the force F. Assume that the coefficient of static friction is μ_s at all surfaces where motion is possible. Two directions of motion must be considered for the given arrangement, with the friction forces always opposing any impending or actual motion.

First assume that wedge B is used to raise body A, and the force F that initiates motion must be determined. The free-body diagram of the wedge under this condition is shown in Fig. 8-10b. The corresponding diagram of body A is shown in Fig. 8-10c. Note that force $\mathscr{F}_3$ is downward because A would move upward if B were driven to the left. Four equations of equilibrium can be written for the forces in the x and y directions from the two free-body diagrams. The unknown forces F, N_1, N_2, and N_3 can be determined in terms of W, θ, and μ_s from these equations. The problem can also be solved by drawing force polygons (these become force triangles if each pair of normal and friction forces is replaced by their resultant).

FIGURE 8-10

Analysis of a wedge

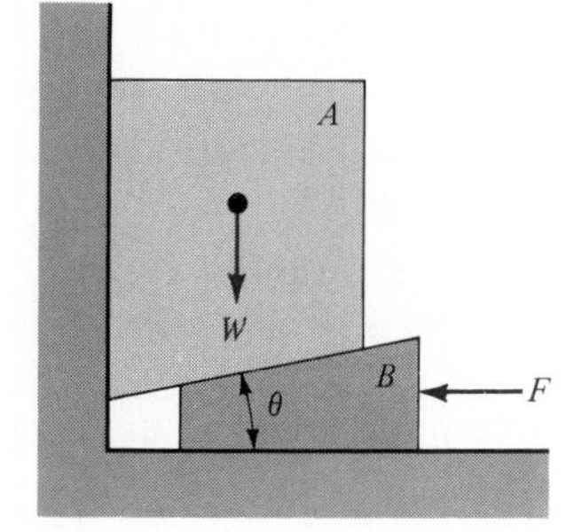

(a)

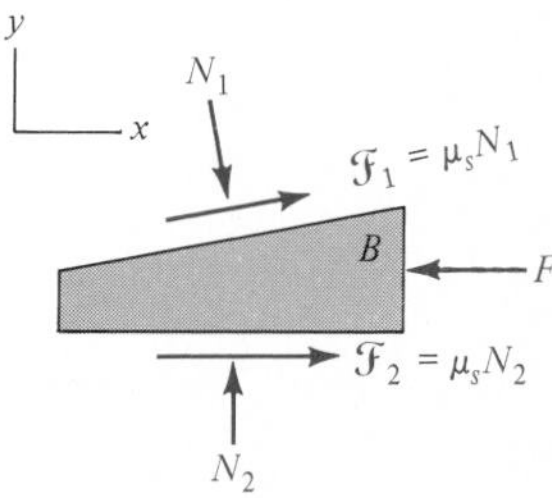

Diagram for raising A

(b)

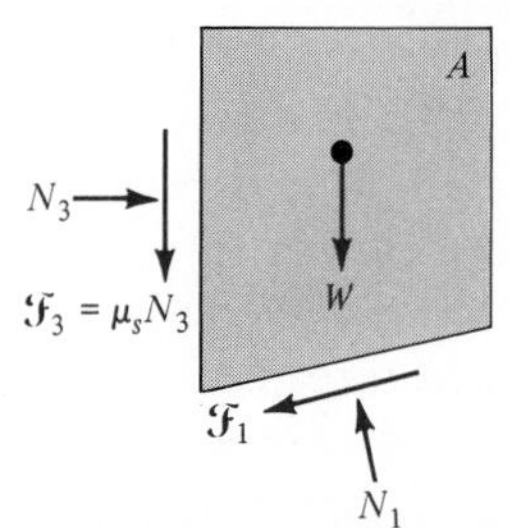

(c) Diagram for raising A

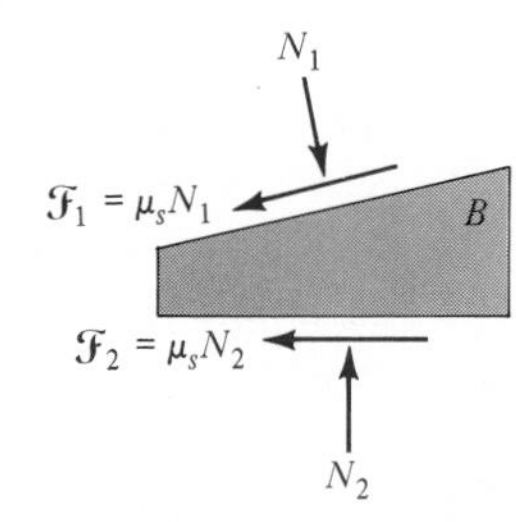

(d) Diagram to check self-locking

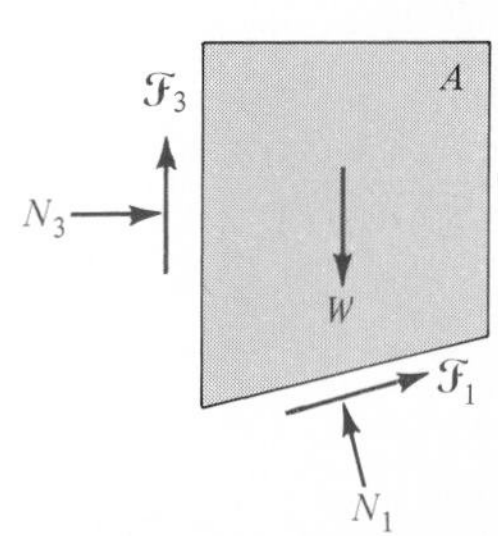

(e) Diagram to check self-locking

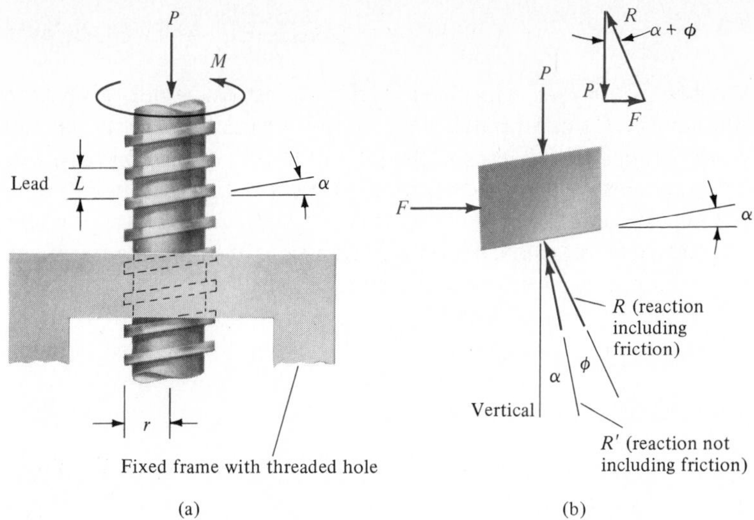

FIGURE 8-11

Square-threaded screw

It is also worthwhile to check whether the arrangement in Fig. 8-10a is *self-locking*, which means that the wedge *B* remains in place when the force *F* is removed. This situation is characterized by the free-body diagrams in Fig. 8-10d and e. Note the differences in the directions of the friction forces with respect to those in Fig. 8-10b and c. The solution of a specific problem is again straightforward using either the analytical or the graphical method.

Screws represent a most clever application of the concept of wedges: each screw thread may be considered a wedge wrapped on a circular shaft. Screw threads are made in a variety of shapes (square, trapezoidal, V-shape, etc.) for numerous purposes. Of these, square threads are the easiest to analyze since they are the most similar to wedges or to blocks moving on inclined planes. Square-threaded screws are commonly used in jacks, presses, and for transmitting power in machines.

For the general analysis of a square-threaded screw, consider Fig. 8-11a. Assume that a moment M turns the screw in its fixed frame so that the screw moves out of the frame against the axial load P. The situation is modeled in Fig. 8-11b, where the element represents the screw in the fixed frame. The angle α depends on the mean radius r and of the *lead* L (advancement per revolution) of the screw. For single-threaded screws L is equal to the *pitch* p, the distance between similar points on adjacent threads (multiple-threaded screws are also available; for them, $L = np$, where n is the multiplicity of threads). Thus, from a right triangle formed by unwrapping a complete turn of the single thread,

$$\alpha = \tan^{-1} \frac{L}{2\pi r} = \tan^{-1} \frac{np}{2\pi r} \tag{8-3}$$

In Fig. 8-11b the force F represents the turning effect of the moment M since $M = Fr$. The reaction of the fixed frame on the screw is the force **R**. If the screw could move without friction, the reaction would be **R**′, normal to the thread. The angle ϕ of **R** with the normal to the thread is the angle of friction, so $\tan \phi = \mu$, the appropriate coefficient of friction. For equilibrium of the element in the horizontal and vertical directions,

$$F = R \sin (\alpha + \phi) = \frac{M}{r}$$

$$P = R \cos (\alpha + \phi)$$

From these two equations, the moment required to move the screw against the axial load is

$$\boxed{M = Pr \tan (\alpha + \phi)} \tag{8-4}$$

The relative magnitudes of α and ϕ determine whether the screw is self-locking. Three possibilities are shown in Fig. 8-12. If $\phi > \alpha$, the screw can support an axial load without unwinding (Fig. 8-12a). In that case a moment must be applied causing a force F opposite to that in Fig. 8-11b if the screw should move in the direction of the force P. The moment required to move a self-locking screw in the direction of the axial load is $M_1 = Pr \tan (\phi - \alpha)$. At the opposite extreme, a moment $M_2 = Pr \tan (\alpha - \phi)$ must be applied to the screw modeled in Fig. 8-12c to hold the screw stationary, making it equivalent to the model in Fig. 8-12b.

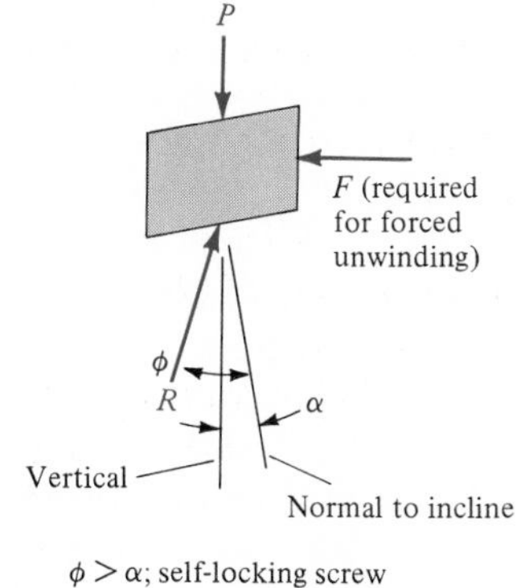

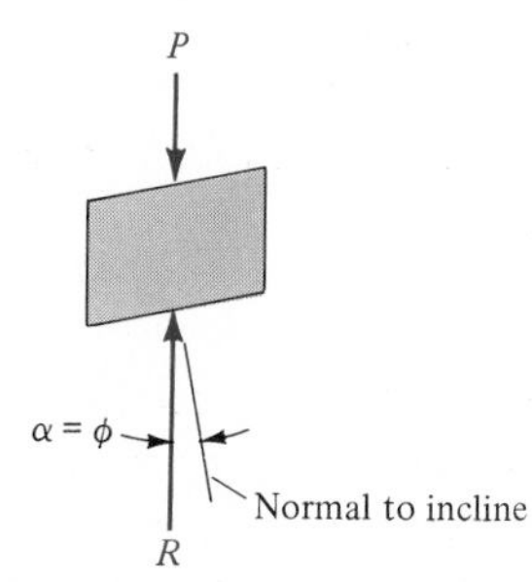

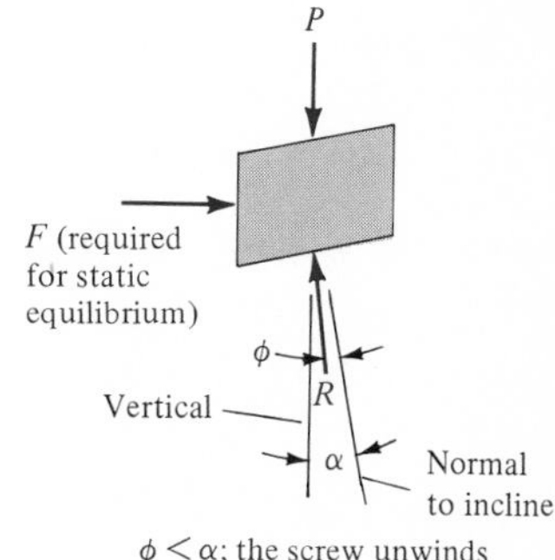

FIGURE 8-12

Special cases of modeling screws

EXAMPLE 8-6

In Fig. 8-10, $\theta = 5^\circ$, $\mu_s = 0.3$, and $W = 10$ kN. Determine the forces F and N_3 during the process of slowly raising block A. What is the horizontal force F required for lowering block A?

SOLUTION

Since the block is raised slowly, the equations for static equilibrium are used. From Fig. 8-10b,

$$\sum F_x = 0$$

$$-F + \mu_s N_2 + \mu_s N_1 \cos\theta + N_1 \sin\theta = 0 \qquad \textbf{(a)}$$

$$\sum F_y = 0$$

$$N_2 - N_1 \cos\theta + \mu_s N_1 \sin\theta = 0 \qquad \textbf{(b)}$$

From Fig. 8-10c,

$$\sum F_x = 0$$

$$N_3 - \mu_s N_1 \cos\theta - N_1 \sin\theta = 0 \qquad \textbf{(c)}$$

$$\sum F_y = 0$$

$$-\mu_s N_3 - W - \mu_s N_1 \sin\theta + N_1 \cos\theta = 0 \qquad \textbf{(d)}$$

From Eq. (c),

$$N_3 = N_1(\mu_s \cos\theta + \sin\theta) = 0.39 N_1$$

Substituting into Eq. (d),

$$N_1 = \frac{W}{\cos\theta - \mu_s \sin 5^\circ - \mu_s(0.39)} = \frac{10 \text{ kN}}{0.85} = 12 \text{ kN}$$

Substituting into Eq. (b),

$$N_2 = 12 \text{ kN} \cos 5^\circ - 0.3(12 \text{ kN}) \sin 5^\circ = 11.6 \text{ kN}$$

Substituting into Eq. (a),

$$F = 0.3(11.6 \text{ kN}) + 12 \text{ kN}(0.3 \cos 5^\circ + \sin 5^\circ) = 8.1 \text{ kN}$$

For lowering the block, the directions of friction forces change. From Fig. 8-10e,

$$\sum F_x = 0$$

$$N_3 + \mu_s N_1 \cos\theta - N_1 \sin\theta = 0$$

$$N_3 = N_1(\sin\theta - \mu_s \cos\theta) = -0.21 N_1$$

A negative value for N_3 implies that the vertical wall must apply a pulling force, which it cannot do. Hence, it is seen that the block A is self-locking. This result could have also been found by checking to see if block A has a tendency to slip down the inclined plane of the wedge. A quick check shows that $\tan^{-1} \mu_s = \tan^{-1} 0.3 = 16.7^\circ > 5^\circ$. Thus, block A does not tend to slip, even if the wall were not there, and cannot be lowered simply by changing F on block B.

EXAMPLE 8-7

In Fig. 8-10 $W = 3000$ lb and $\mu_s = 0.2$. Determine the range of the angle θ for which the wedge is self-locking.

SOLUTION

The system is self-locking if it remains still after removal of the force F ($F = 0$). The upper limit on θ is obtained by considering the case of motion impending for both blocks. From Fig. 8-10d,

$$\sum F_x = 0$$

$$-\mu_s N_2 - \mu_s N_1 \cos\theta + N_1 \sin\theta = 0 \qquad \textbf{(a)}$$

$$\sum F_y = 0$$

$$N_2 - N_1 \cos\theta - \mu_s N_1 \sin\theta = 0 \qquad \textbf{(b)}$$

From Fig. 8-10e,

$$\sum F_x = 0$$

$$N_3 - N_1 \sin\theta + \mu_s N_1 \cos\theta = 0 \qquad \textbf{(c)}$$

$$\sum F_y = 0$$

$$\mu_s N_3 - W + N_1 \cos\theta + \mu_s N_1 \sin\theta = 0 \qquad \textbf{(d)}$$

Adding Eq. (b) and (d),

$$N_2 + \mu_s N_3 - W = 0 \qquad \textbf{(e)}$$

Adding Eq. (a) and (c),

$$N_3 - \mu_s N_2 = 0 \qquad \textbf{(f)}$$

Solving (e) and (f) simultaneously yields

$$N_2 = 2885 \text{ lb} \qquad N_3 = 577 \text{ lb}$$

From Eq. (b), assuming $N_2 \simeq N_1 \cos\theta$ and substituting into Eq. (c),

$$N_3 - N_1 \sin\theta \left(\frac{\cos\theta}{\cos\theta}\right) + \mu_s N_2 = 0$$

$$N_3 - N_2 \tan\theta + \mu_s N_2 = 0$$

$$\tan\theta = \frac{N_3 + \mu_s N_2}{N_2} = \frac{577 + 0.2(2885)}{2885} = 0.4 \qquad \theta = 21.8^\circ$$

The lower limit on θ is of course $\theta = 0$.

EXAMPLE 8-8

A square-threaded screw has a mean diameter of 6 cm, a pitch of 0.7 cm, and a coefficient of static friction of 0.2. Determine the moment required to raise the load when the axial force on the screw is 5 kN (Fig. 8-11a).

SOLUTION

Referring to Fig. 8-11b, the angle of inclination is

$$\alpha = \tan^{-1}\frac{np}{2\pi r} = \tan^{-1}\frac{(1)(0.7\text{ cm})}{2\pi(3\text{ cm})} = 2.13^\circ$$

When motion is just beginning the reaction R is its maximum and

$$\phi = \phi_s = \tan^{-1}\mu_s = 11.3^\circ$$

Using a force triangle such as in Fig. 8-11b, the necessary pushing force F is found,

$$F = P\tan(\phi + \alpha) = 5\text{ kN}\tan(11.3^\circ + 2.13^\circ) = 1.19\text{ kN}$$

The moment caused by F which must be overcome by the applied moment M is

$$Fr = M = (1.19\text{ kN})(0.03\text{ m}) = 35.7\text{ N}\cdot\text{m}$$

The solution shows that the necessary moment to raise the load decreases as the coefficient of friction decreases and also as the pitch decreases.

EXAMPLE 8-9

The jackscrew for a sports car is designed with a mean diameter of 0.6 in. The maximum load on the jack is 1200 lb. What is the required number of threads per inch on the single-threaded screw if unwinding is to be prevented even with the minimum coefficient of friction of 0.1?

SOLUTION

From Fig. 8-12b, a screw is just self-locking if the angle of inclination equals the angle of repose or when

$$\alpha = \tan^{-1}\frac{np}{2\pi r} = \tan^{-1}\mu_s = \tan^{-1}(0.1) = 5.71^\circ$$

$$\frac{np}{2\pi r} = \frac{(1)p}{2\pi(0.3\text{ in.})} = 0.1$$

$$p = 2\pi(0.3\text{ in.})(0.1) = 0.188\text{ in.}$$

Since p is the distance between adjacent threads, the number of adjacent threads per inch equals $1/p = 5.31$ threads/in. It should be noted that the result was independent of the magnitude of the load.

PROBLEMS

8–34 Block A is to be raised using wedges B and C, which have negligible weight. The coefficient of friction is 0.4 at all surfaces. Determine the horizontal force F that can move the block A upward if $\theta = 8°$.

8–35 Determine the horizontal force F acting on wedge C in Fig. P8-34 to lower block A. $\theta = 4°$, and $\mu = 0.4$ on all surfaces.

8–36 Plot the force F as a function of $\theta = 4°$, $8°$, and $12°$ in Prob. 8–35. Estimate from the diagram the angle θ at which the wedge ceases to be self-locking.

FIGURE P8-34

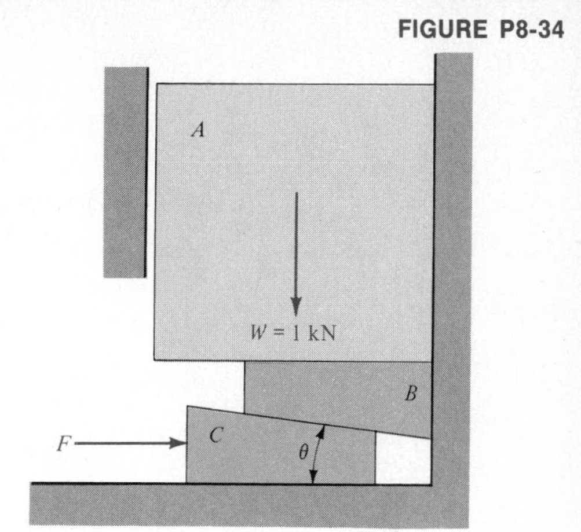

8–37 Determine whether the force F acting on wedge B is sufficient to raise block A if $\mu = 0.3$ at all surfaces $\theta = 6°$.

8–38 Plot the force F acting on wedge B in Fig. P8-37 to lower block A. Use $W = 500$ lb, 1000 lb, and 2000 lb. $\mu = 0.3$ at all surfaces.

FIGURE P8-37

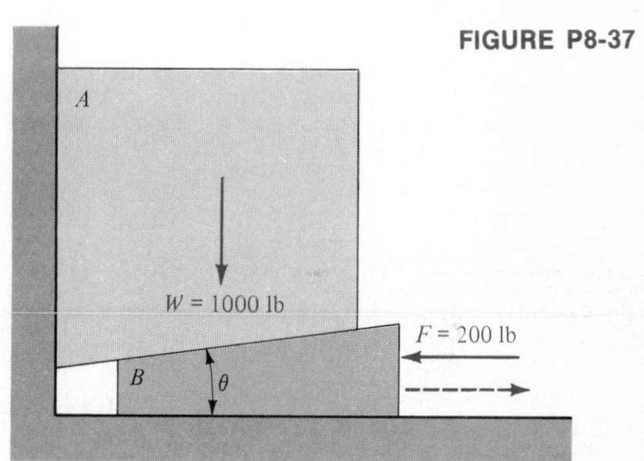

8–39 Block A of weight 3 kN is to be moved away from the wall using wedge B. Calculate the required force P if the coefficient of friction is 0.3 at all surfaces.

FIGURE P8-39

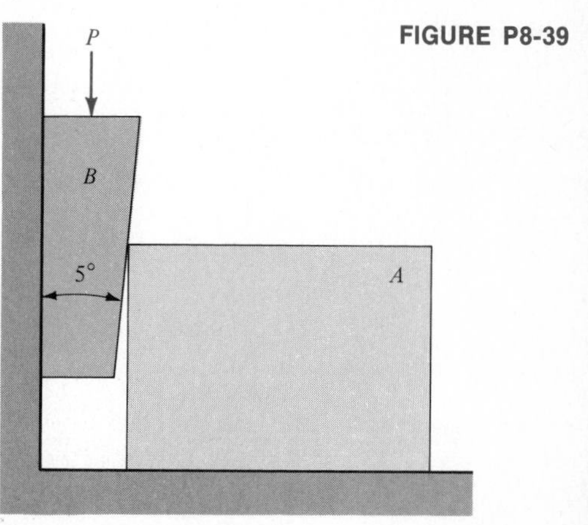

FIGURE P8-40

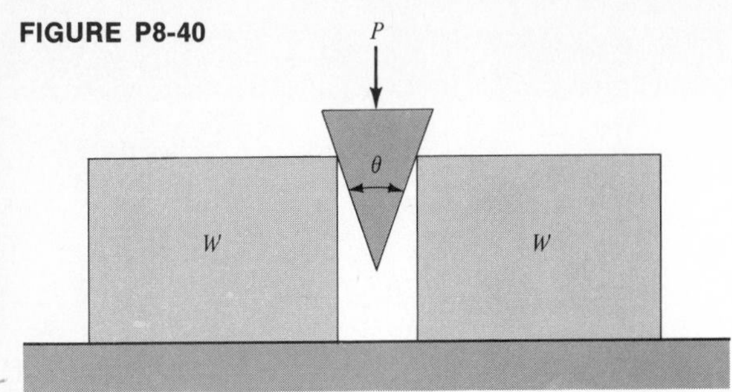

8–40 Two equal blocks are to be separated by driving a wedge between them. Write an expression for the required force P, assuming that the coefficient of friction is μ at all surfaces, and that the two blocks start moving simultaneously.

FIGURE P8-41

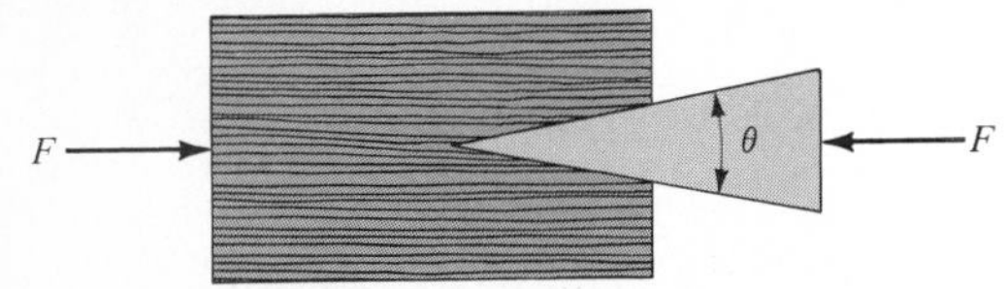

8–41 A log-splitting machine is modeled in Fig. P8-41. Plot the required driving force F as a function of $\theta = 3°$, $6°$, and $9°$. Assume that in all cases the friction force from the wood on the wedge is 5 kN and the coefficient of friction is 0.5.

FIGURE P8-42

8–42 The hull of the icebreaker shown is tapered so that horizontal crushing forces from thick ice would tend to lift the whole ship without damaging it. What is the minimum angle that the hull must have with the vertical for this purpose? The ship weighs 10^6 lb, and its coefficient of friction on ice is 0.05.

FIGURE P8-43

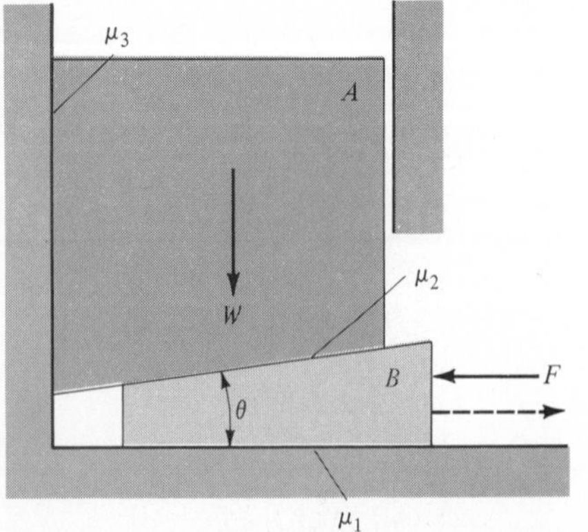

8–43 Determine the force F acting on the weightless wedge B to raise block A if $W = 10$ kN, $\theta = 5°$, $\mu_1 = 0.2$, $\mu_2 = 0.3$, and $\mu_3 = 0.5$.

8–44 Determine the force F acting on wedge B in Fig. P8-43 to lower block A if $W = 2000$ lb, $\theta = 4°$, $\mu_1 = 0.2$, $\mu_2 = 0.4$, and $\mu_3 = 0.3$.

8–45 Consider the model of a mechanism in which wedges A and B have negligible weight. Determine the required coefficient of friction μ_{AB} to make the system self-locking (if possible) when a force F_A is applied.

FIGURE P8-45

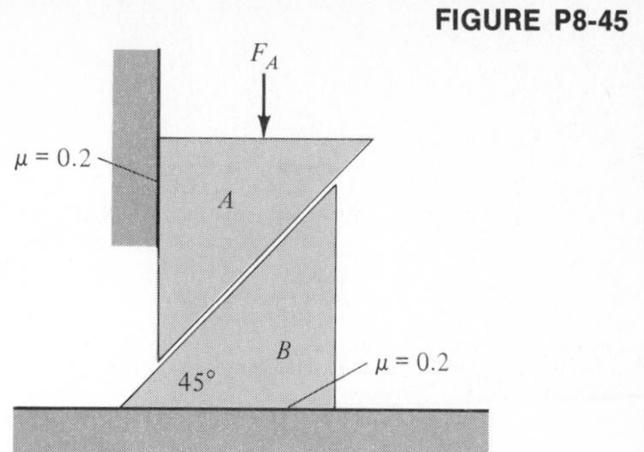

8–46 Consider the two square-threaded screw columns of a mechanical testing machine. Each column has a mean diameter of 8 cm, a coefficient of friction of 0.25, and a maximum axial load of 100 kN. Determine the necessary moment M for turning each column to raise the cross head if the pitch of the single thread is $p = 0.8$ cm.

FIGURE P8-46

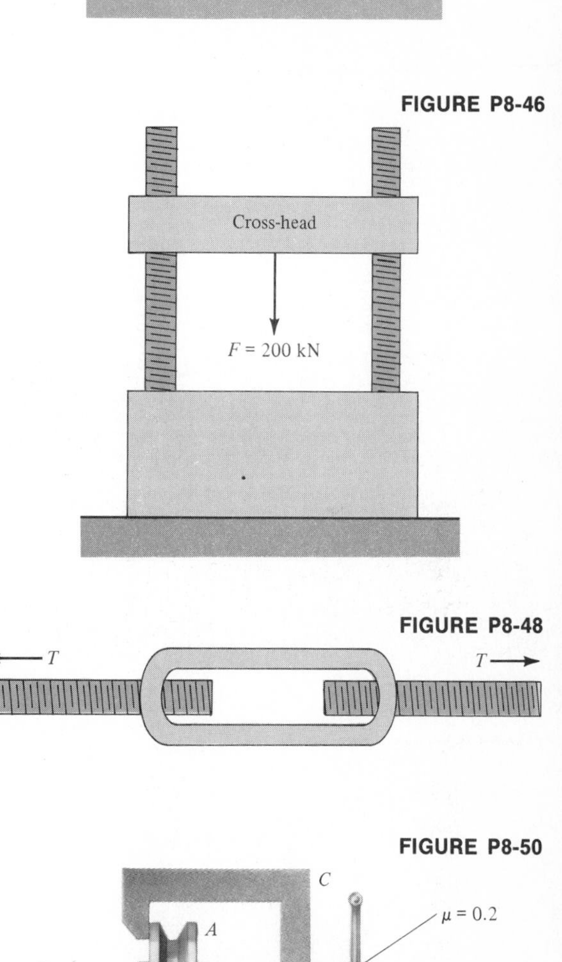

FIGURE P8-48

8–47 Plot the moment M vs. the pitch $p = 0.8$ cm, 1 cm, and 1.2 cm in Prob. 8–46. Discuss, on the basis of the diagram, whether the pitch of the screws is an important factor in choosing a motor for turning the columns.

8–48 One threaded rod has right-handed threads, the other left-handed threads in the turnbuckle. The mean diameter of the screws is $d = 1$ in., the pitch of the single threads is $p = 0.08$ in., and the coefficient of friction is 0.2. Determine the moments with which the turnbuckle could be slightly tightened or loosened if $T = 10{,}000$ lb. The two rods cannot rotate.

8–49 The tension in a guy wire is to be set approximately using a turnbuckle (Fig. P8-48). The mean diameter of the screws is $d = 2.4$ cm, the pitch of the single threads is $p = 2$ mm, and the coefficient of friction is 0.25. Determine the tension in the nonrotating rods that is set by applying a torque of 300 N·m to the turnbuckle.

8–50 Pulley A is to be removed from the shaft B by turning the double-threaded screw in the clamp C. The mean diameter of the screw is $d = 0.8$ in., the pitch is $p = 0.05$ in., and the coefficient of friction is 0.2. The screw has negligible friction at point D. Calculate the total force that the clamp applies to the pulley when the screw is tightened with a moment of 150 ft·lb.

FIGURE P8-50

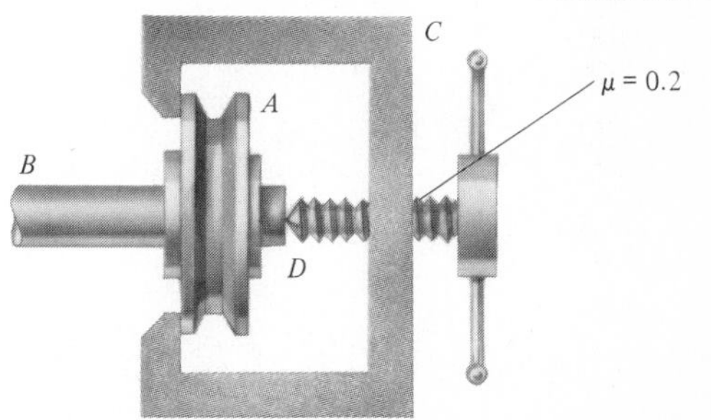

FIGURE P8-51

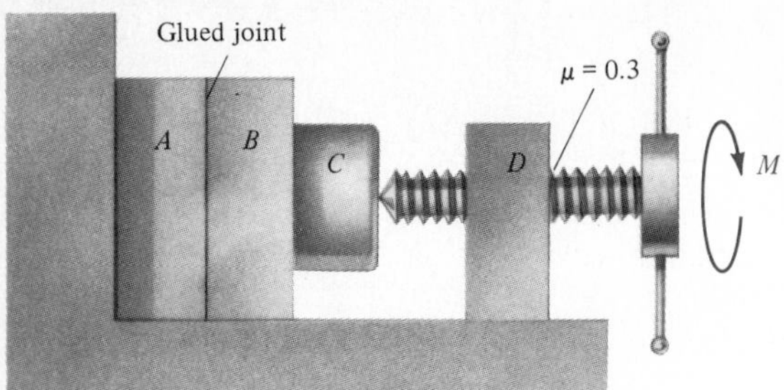

8–51 Blocks A and B are pressed together in a single-threaded vise. The mean diameter of the screw is $d = 2$ cm, the pitch is $p = 1.9$ mm, and the coefficient of friction is 0.3. Determine the moment necessary to apply a compressive force of 3 kN to the blocks. There is negligible friction at D.

8–52 At what pitch would the screw in Prob. 8–51 not be able to compress the blocks if M were zero?

FIGURE P8-53

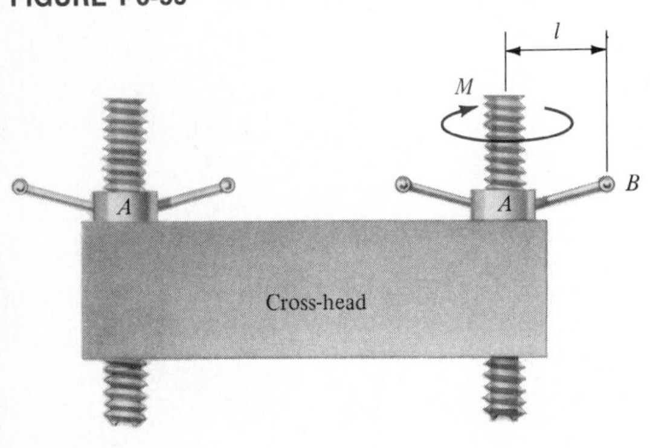

8–53 The large locking nuts A of a mechanical testing machine are designed to be tightened by hand. The mean diameter of a single-threaded column is $d = 3$ in., the pitch is $p = 0.2$ in., the coefficient of friction is 0.25 at the threads and assumed negligible at the cross-head's surface. Calculate the axial force on a nut when a 100 ft·lb moment is applied to tighten it.

8–54 Determine the smallest force at point B in Fig. P8-53 that would loosen the locking nut A if $l = 8$ in. Assume that the nut was tightened as described in Prob. 8–53.

FIGURE P8-55

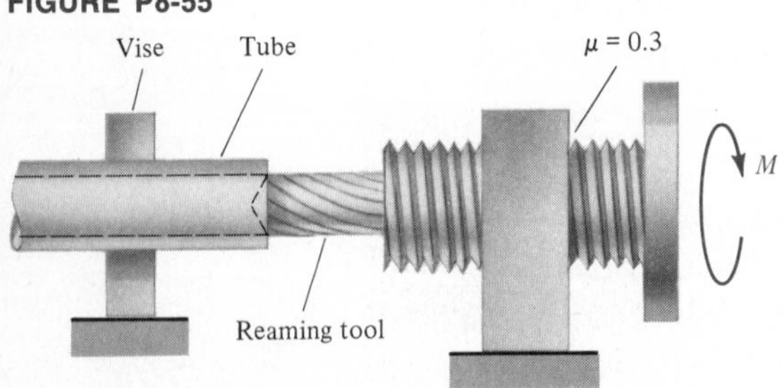

8–55 A tube-reaming device is designed to have a single-threaded screw of $d = 4$ cm mean diameter, pitch $p = 5$ mm, and coefficient of friction of 0.3. Estimate the axial force on the tube when the net moment M driving the tool into the tube is 400 N·m.

8–56 Estimate the axial forces in Prob. 8–55 for single-, double-, and triple- threaded screws, with everything else being the same.

DISK FRICTION 8-5

The friction between flat surfaces in relative rotary motion must be analyzed in a number of devices. These include end bearings and collar bearings of shafts for axial support, disk clutches, and disk brakes.

For a general analysis of disk friction, consider a flat-ended hollow shaft that is rotated by a moment M while being pressed against another flat surface as in Fig. 8-13. Assume that the same coefficient of friction μ is valid for the whole ring-shaped area (outer radius R_o, inner radius R_i), and that the pressure is constant over the whole area. The normal force dN on an elemental area dA is

$$dN = \frac{P}{A}\,dA = \frac{P}{\pi(R_o^2 - R_i^2)}(r\,d\theta\,dr)$$

The frictional force on the area dA is $\mu\,dN$, and the moment of this frictional force about the axis of the shaft is approximately $dM = \mu\,dN\,r$. The total moment caused by friction on the whole area is

$$M = \int \mu\,dN\,r = \frac{\mu P}{\pi(R_o^2 - R_i^2)}\int_0^{2\pi}\int_{R_i}^{R_o} r^2\,dr\,d\theta$$

$$= \frac{2}{3}\mu P\frac{R_o^3 - R_i^3}{R_o^2 - R_i^2} \qquad \text{8-5}$$

For a solid shaft with radius R, Eq. 8-5 becomes

$$M = \frac{2}{3}\mu PR \qquad \text{8-6}$$

Equations 8-5 and 8-6 may be used in static or dynamic situations with the appropriate coefficient of friction, μ_s or μ_k. In the static case they give the largest torque that can be transmitted without slippage. In the case of relative motion, they give the moment required to maintain motion against the frictional resistance. It has been found that the frictional moment decreases for well-worn surfaces to about 75% of its original value for new surfaces (solid shafts).

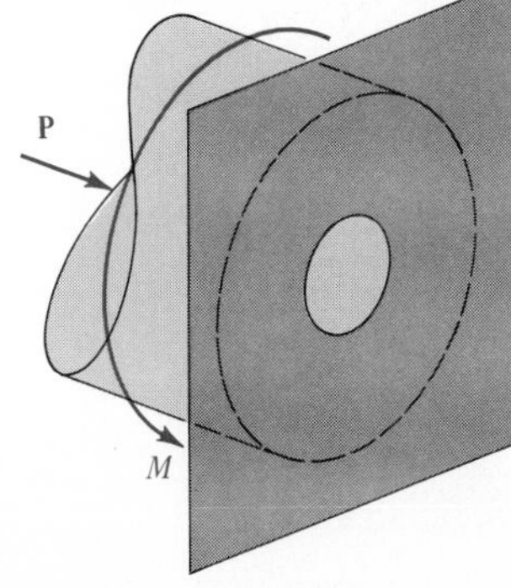

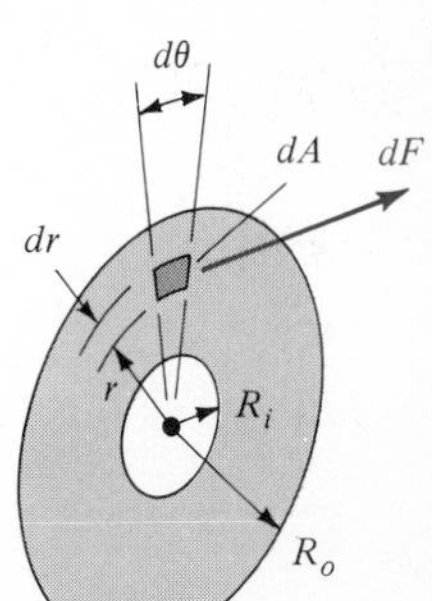

FIGURE 8-13

Model for analyzing disk friction

8-6 AXLE FRICTION

The lateral support of rotating axles is provided by bearings. The simplest of these are called *journal bearings* or *sliding bearings.* Such a bearing consists of a stationary member with a hole that is slightly larger than the shaft it has to support, as shown in Fig. 8-14. In dry or partially lubricated bearings the shaft makes contact with the bearing on a long and narrow area parallel to the centerline of the shaft. Such bearings can be analyzed using the principles of dry friction.

The rotating shaft in a journal bearing climbs a little as shown by the point of contact A in the cross-sectional view in Fig. 8-14a. The height of the climb is limited by the inevitable (and desirable) slippage that occurs. The frictional resistance of the bearing can be determined from the climbing displacement of the shaft in the bearing.

Assume that the vertical load applied to a shaft is P in Fig. 8-14. The free-body diagram of the shaft is drawn in Fig. 8-14b. Here M is the torque required to maintain steady motion with slippage occurring at point A. For equilibrium of forces in the vertical direction, the reaction R acting on the shaft must be equal and opposite to P, with parallel lines of action. The applied torque M must be large enough to overcome the resisting couple formed by P and R with the distance r_f between them. If M is reversed to act counterclockwise, the reaction R' acts at point A' in Fig. 8-14b, at the same distance r_f from the line of action of P. The circle that can be drawn with the radius r_f is called the *circle of friction* which is a constant for a given journal bearing.

The reaction R of the bearing can be conveniently resolved into the normal force N and the friction force $\mathscr{F}$ as in Fig. 8-14b. The angle ϕ between **R** and **N** is related to the coefficient of kinetic friction μ_k according to Eq. 8-2, $\tan\phi = \mu_k$. For the equilibrium of moments about the center of the shaft C,

$$M = Rr_f = \mathscr{F}r = Rr\sin\phi \qquad \text{8-7}$$

Properly constructed bearings have low coefficients of friction, so $\sin\phi \simeq \tan\phi = \mu_k$. This leads to the following approximate for-

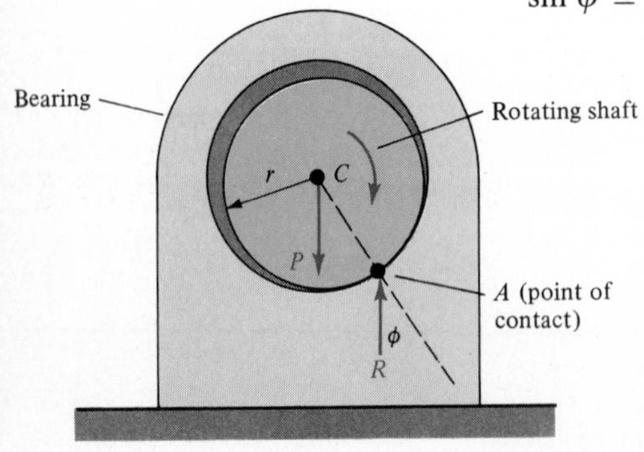

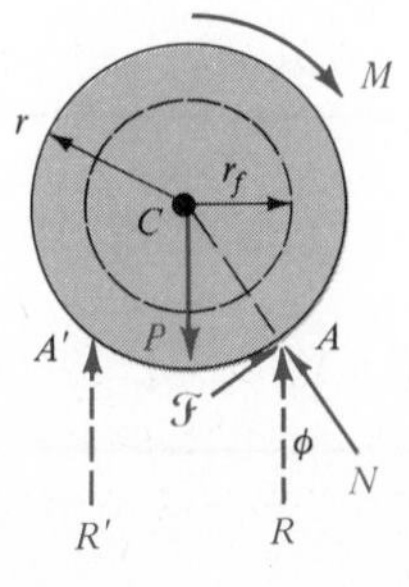

FIGURE 8-14

Cross section of an axle in a sliding bearing

mulas for the friction moment M and the radius r_f of the circle of friction,

$$M = Rr\mu_k = Pr\mu_k \qquad \text{8-8}$$

$$r_f = r\mu_k \qquad \text{8-9}$$

Thus, the friction moment depends on the applied load P and the properties of the bearing, whereas the circle of friction is independent of the load.

ROLLING RESISTANCE 8-7

Wheels and balls are very useful in minimizing the resistance to motion of bodies. This does not mean, however, that the friction of a rolling wheel or ball can be zero. To give a qualitative example, the rolling resistance of tires is the major part of the total losses in cars traveling at speeds up to about 70 km/h (at higher speeds air drag becomes the largest resistance; the size and shape of the vehicle may greatly affect the speed at which the transition occurs). Steel rolling on steel is much more efficient than rubber tires, but it still cannot eliminate friction.

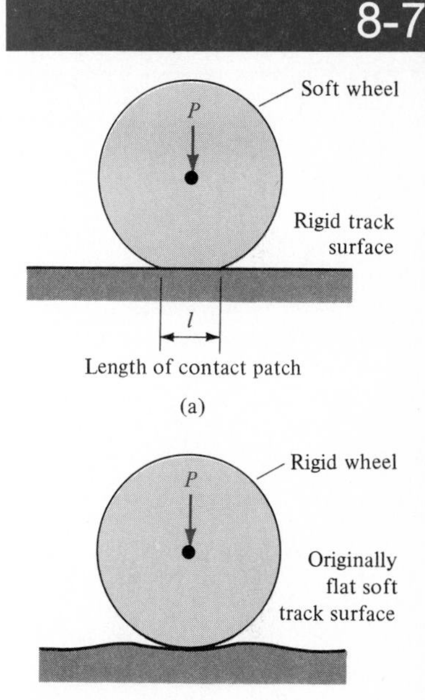

FIGURE 8-15

Deformations of wheel or track cause resistance to rolling

The rolling resistance of any wheel, not counting axle friction, is caused by deformations of the wheel and its track at the region of mutual contact. In the extreme, a soft wheel may deform considerably on a relatively rigid track (Fig. 8-15a), or a relatively rigid wheel may cause much deformation of a soft track (Fig. 8-15b). The deformations of rigid wheels and tracks may be extremely small and therefore difficult to measure. In any case, the internal friction of the deforming materials is the cause of the rolling resistance. Internal friction is somewhat similar to dry friction between bodies in contact. Both generate heat when displacement occurs as can be easily demonstrated by rapidly rubbing the hands together (surface friction) or bending soft wires back and forth (internal friction).

A wheel with rolling resistance can be modeled as in Fig. 8-16. Here P is the vertical load on the wheel, and F is the minimum horizontal force that can keep the wheel rolling at constant speed. The reaction R from the track is the resultant of all distributed forces between the track and the wheel. Assume that the length l of the contact patch (Fig. 8-15) is much shorter than the radius r of the wheel, and that, in Fig. 8-16, the moment arm of F is approximately r. Thus,

$$\sum M_A = 0, \qquad Pa \simeq Fr \qquad \text{8-10}$$

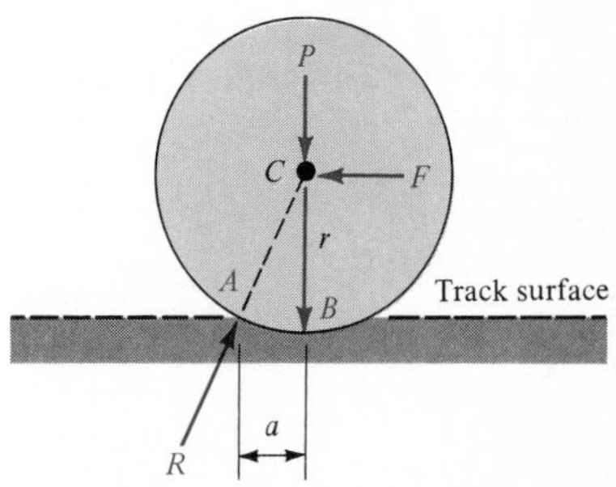

FIGURE 8-16

Model for analyzing rolling resistance

The distance a is taken as the *coefficient of rolling resistance*; it has units of length. The values of a, which must be experimentally determined, range upward from the low of about 0.005 mm for hardened steel on hardened steel.

EXAMPLE 8-10

Determine the maximum torque that can be transmitted without slippage by the disk clutch shown in Fig. 8-17. The disk plates are new and they are solid circles. How much would the maximum torque change if the disk plates were hollow circles with $R_o = 14$ cm and $R_i = 7$ cm?

SOLUTION

The disks transmit the maximum torque when slipping is impending. From Eq. 8-6, for solid shafts, the torque that may be applied is

$$M_S = \frac{2}{3}\mu_s P R_o = \frac{2}{3}(0.7)(4\text{ kN})(0.14\text{ m}) = 0.26\text{ kN}\cdot\text{m}$$

For the hollow shaft, from Eq. 8-5,

$$M_H = \frac{2}{3}\mu_s P\frac{R_o^3 - R_i^3}{R_o^2 - R_i^2} = \frac{2}{3}(0.7)(4\text{ kN})\frac{(0.14^3 - 0.07^3)\text{ m}^3}{(0.14^2 - 0.07^2)\text{ m}^2}$$

$$= 0.3\text{ kN}\cdot\text{m}$$

$$\text{Percent change} = \frac{M_H - M_S}{M_S} \times 100\%$$

$$= \frac{0.3 - 0.26}{0.26} \times 100\% = 15.4\%$$

JUDGMENT OF THE RESULTS

It is surprising that the hollow rings transmit a larger torque than the solid circles. After checking the calculations, the results are still the same. Upon further thought, these results make sense. The torque caused by friction increases with the moment arm from the center and also with normal force exerted at each point on the surfaces in contact. By reducing the area in contact, the normal force at each point is increased. Hence, even though some resistance is lost due to loss of material, the maximum torque is increased due to the larger normal force at those points that also have the largest moment arms.

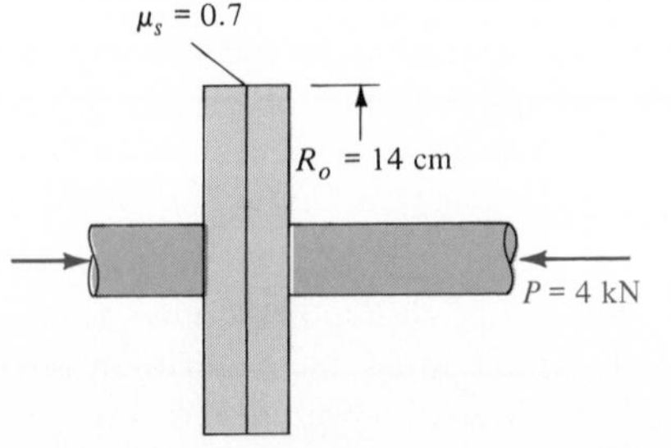

FIGURE 8-17
Disk clutch

EXAMPLE 8-11

A small hoist is supported by two 10-in.-diameter wheels, with each wheel rotating on a 1.2-in.-diameter shaft. Assuming that the major friction is at the wheel axles with $\mu_s = 0.2$ and $\mu_k = 0.17$, determine the force F required to start the hoist moving horizontally (Fig. 8-18a).

SOLUTION

From Fig. 8-18b it is seen that the friction moment $M_{\mathscr{F}}$ is overcome by the moment caused by F and $\mathscr{F}_s$, the friction force at the surface. Summing moments about point A and using Eq. 8-8,

$$\sum M_A = 0 \qquad \text{(at impending motion)}$$

$$M_{\mathscr{F}} - FR = 0$$

$$Pr\mu_s - FR = 0$$

$$F = \frac{(3000\text{ lb})(0.6\text{ in.})(0.2)}{5\text{ in.}} = 72\text{ lb}$$

The actual force applied must be slightly greater than 72 lb.

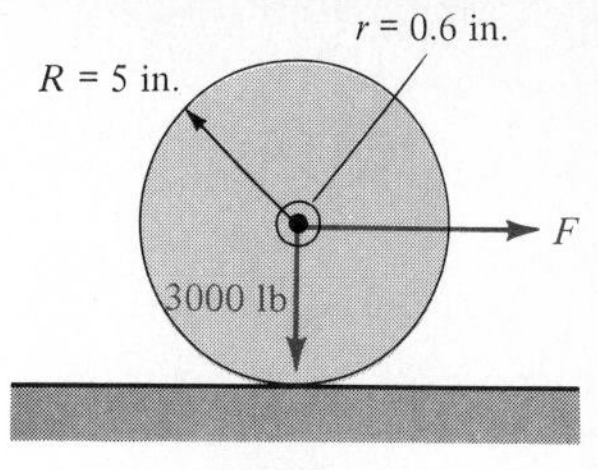

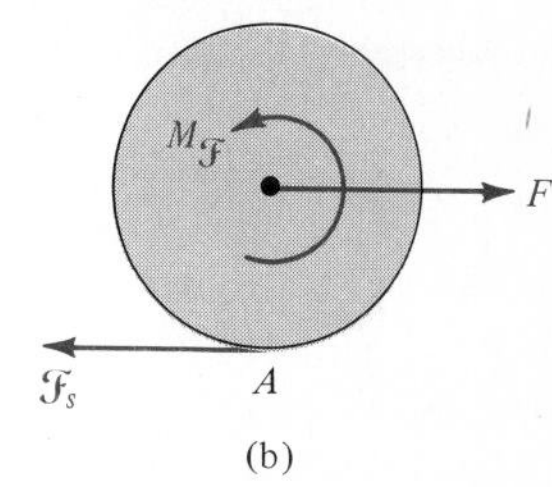

FIGURE 8-18
Analysis of axle friction

EXAMPLE 8-12

The uniform wheel shown in Fig. 8-19 has a rolling coefficient of friction $a = 2$ mm on a steel surface. Determine the largest angle θ with which the wheel can remain at rest.

SOLUTION

Using Fig. 8-19b, the weight of the wheel can be broken into components that are analogous to forces P and F in Fig. 8-16.

$$\sum M_O = 0$$

$$W \sin\theta(R) - W\cos\theta(a) = 0$$

$$R\sin\theta = a\cos\theta$$

$$\frac{\sin\theta}{\cos\theta} = \tan\theta = \frac{a}{R} = \frac{2\text{ mm}}{300\text{ mm}}$$

$$\theta = 0.38^\circ$$

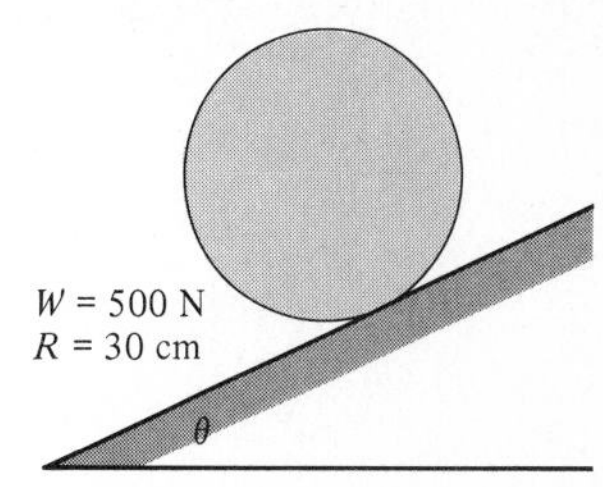

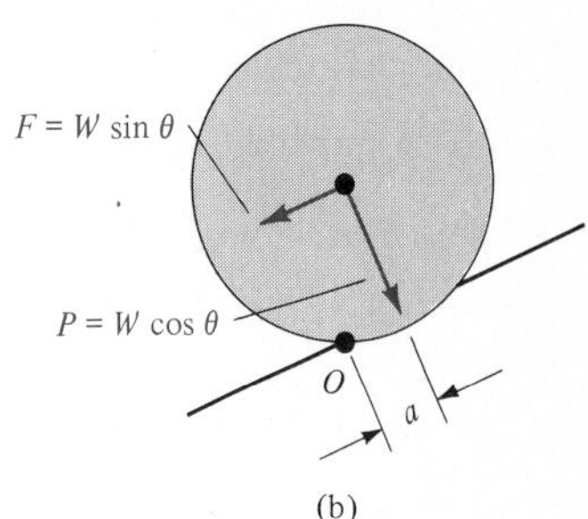

FIGURE 8-19
Analysis of rolling resistance

PROBLEMS

SECTIONS 8-5, 8-6, 8-7

FIGURE P8-57

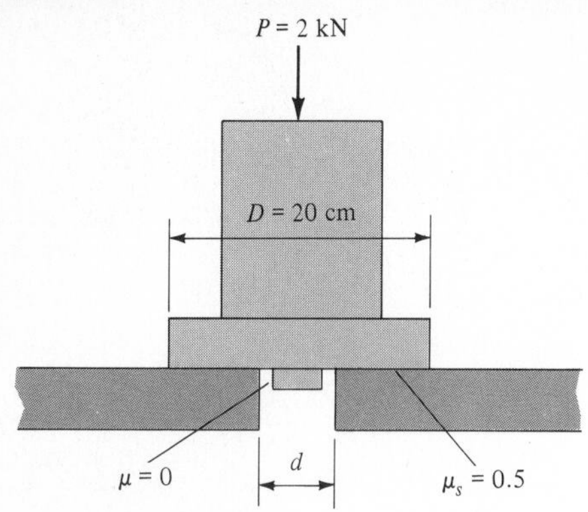

8–57 Determine the torque required to start the vertical shaft rotating if $d = 2$ cm.

8–58 Plot the calculated torque vs. the diameter d in Prob. 8–57 for $d = 2$ cm, 5 cm, and 10 cm. Extrapolate the curve and estimate the torque at $d = 0$. Make a calculation to check the estimation.

8–59 In Fig. P8-57, $d = 2$ in. and $P = 3000$ lb. Determine the largest allowable value of D if the starting torque should not exceed 350 ft·lb.

FIGURE P8-60

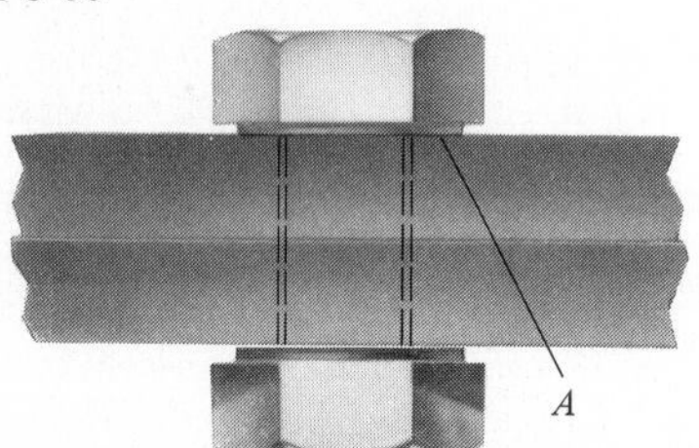

8–60 The load-bearing face A of a bolt has an outer diameter $D = 4$ cm, an inner diameter $d = 2.5$ cm, and a coefficient of friction of 0.3. Calculate the original tension in the bolt if it is loosened with a torque of 230 N·m.

8–61 In Prob. 8–50, the tip of the screw at D is flat with a 0.3-in. diameter and a coefficient of friction of 0.25. Determine the additional moment necessary to overcome the friction at D. Assume that the same force is applied to the pulley in both problems.

8–62 In Prob. 8–51, the tip of the screw at D is flat with a 1.6-cm diameter and a coefficient of friction of 0.3. Determine the required moment in this case, with everything else being the same.

8–63 The load-bearing flat face of the locking nut A in Prob. 8–53 has an outer diameter $D_o = 4.4$ in., an inner diameter $D_i = 3.2$ in., and a coefficient of friction of 0.25. Determine the total moment necessary for applying an axial force of 1000 lb by tightening the nut.

8–64 A flat buffing disk of diameter $D = 12$ cm is gripped in a small electric drill and rotated clockwise. The disk is pressed uniformly against a flat surface with a force $P = 100$ N. Determine the moment necessary at the handle of the drill to hold it steady if the coefficient of friction at the surface being polished is 0.7.

FIGURE P8-64

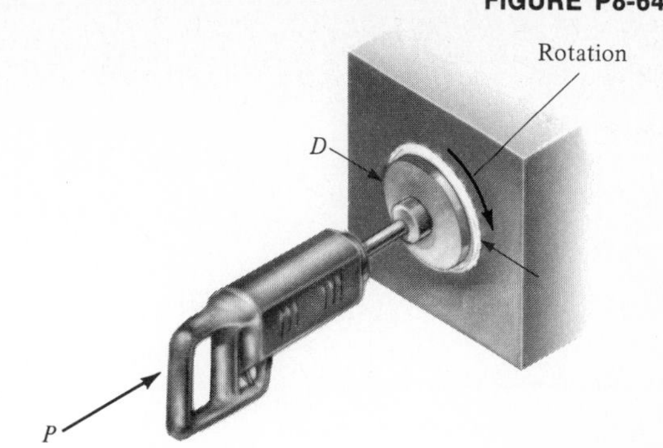

8–65 Each brush of the street sweeper has an outer diameter $D = 4$ ft and an inner diameter $d = 1$ ft at the ground. The vertical load on a brush spindle is 50 lb, and the coefficient of friction at the ground is 0.5. Calculate the moment reaction on the spindle arm A caused by rotation of the brush, assuming that it has uniform contact with the ground.

FIGURE P8-65

8–66 An inventor wishes to improve pointe shoes used in ballet. To analyze the frictional resistance of a shoe during a turn as shown, it is assumed to have a circular, flat tip at A of diameter $D = 4$ cm and coefficient of friction of 0.3. Calculate the moment M that a dancer weighing 500 N must apply to start turning on this shoe.

8–67 Plot the required starting moment M as a function of $D = 3$ cm, 4 cm, and 5 cm in Prob. 8–66. If possible, state the ideal diameter D on the basis of the diagram, assuming that a small moment is advantageous for turning easily and gracefully, while a large diameter D is desirable for stability in any elevated position on the toes.

8–68 A disk brake is modeled as a flat disk A sandwiched between circular pads B and C. There are four clamping forces $P = 100$ lb, but the normal force per unit area is uniform on each side of disk A. Determine the moment M that can turn disk A relative to plates B and C if $D = 8$ in. and $d = 2$ in.

FIGURE P8-66

FIGURE P8-68

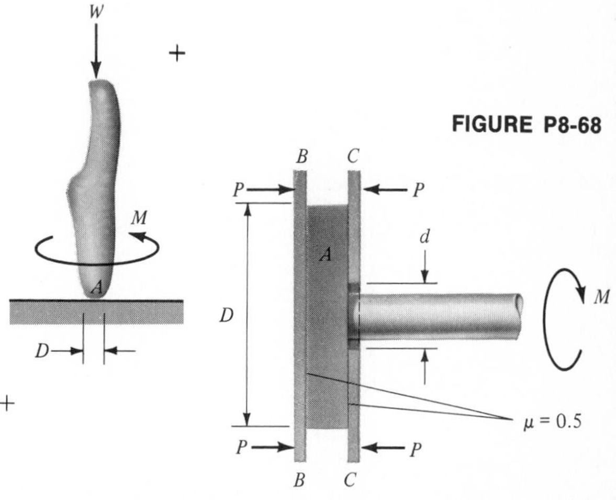

FIGURE P8-69

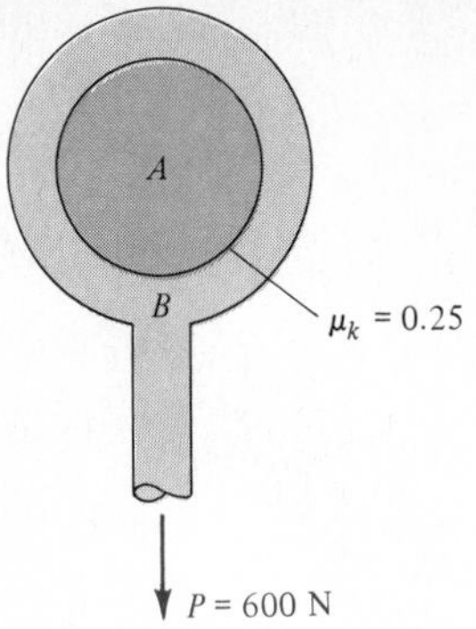

8–69 Determine the moment that must be applied to the bar B to keep it rotating about the axle A, which has a diameter $d = 3$ cm.

8–70 Plot the moment as a function of $d = 2$ cm, 3 cm, and 4 cm in Prob. 8–69. Use two values of the lateral load, $P = 300$ N and 600 N, and show all results on the same diagram.

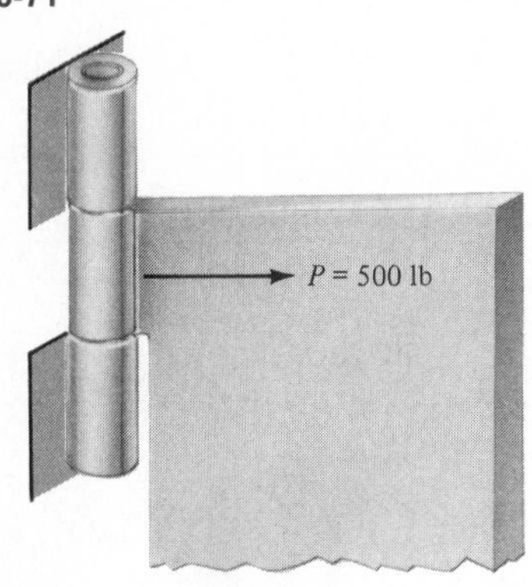

FIGURE P8-71

8–71 The upper hinge of a heavy door has a 0.5-in.-diameter vertical shaft and a coefficient of friction of 0.3. This hinge is designed to support the door only horizontally. Determine the moment caused by friction in the hinge as the door is turned.

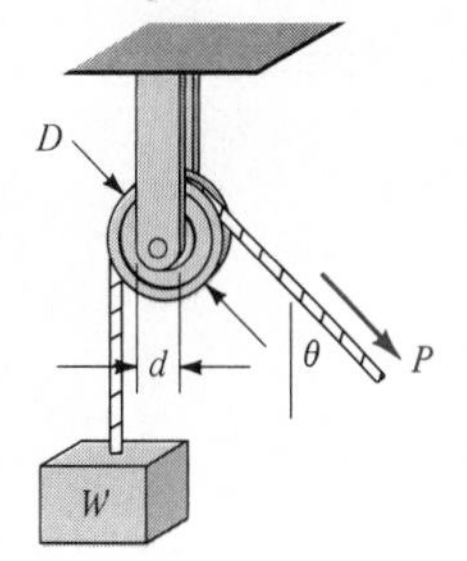

FIGURE P8-72

8–72 A pulley of diameter $D = 30$ cm is supported by a fixed axle of diameter $d = 3.2$ cm. The coefficient of kinetic friction at the axle is 0.3. Determine the smallest force P at $\theta = 0$ for raising the load of $W = 2$ kN.

8–73 Determine the smallest force P in Prob. 8–72 if the load is to be slowly lowered. Assume that $\theta = 30°$.

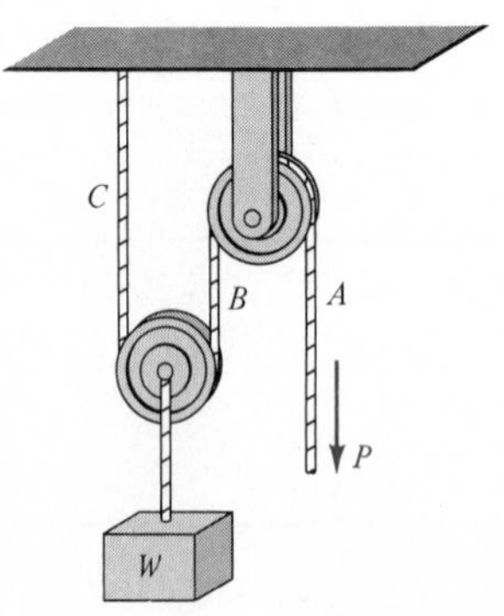

FIGURE P8-74

8–74 Each pulley has a 12-in. diameter and rotates on a 0.75-in.-diameter shaft, where the coefficient of friction is 0.2. Determine the tension in segments B and C of the rope when the load $W = 400$ lb is slowly raised.

8–75 Determine the tension in segments A, B, and C of the rope in Prob. 8–74 when the load is slowly lowered.

8–76 In a remote region a hand-operated well includes a drum of radius $R = 15$ cm on a steel shaft of radius $r = 1.3$ cm. The wooden bearings A and B are 1 m apart with the load $W = 250$ N midway between them. The coefficient of friction in each bearing is 0.3. Determine the moment required to slowly raise the load.

FIGURE P8-76

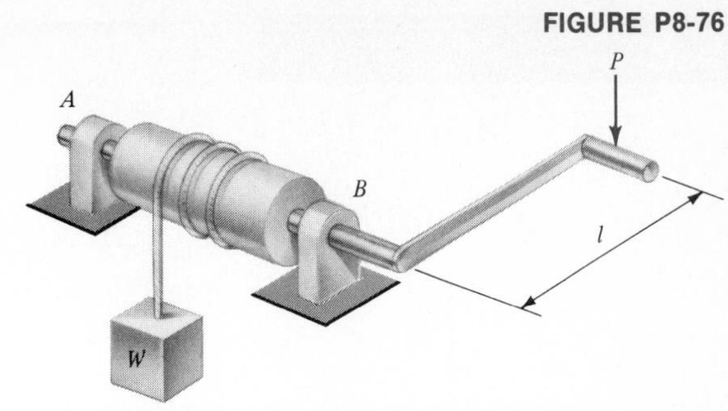

8–77 Solve Prob. 8–76 assuming that the last turn of the rope on the drum (that is, the vertical plane in which the load moves) is at a distance of 30 cm from bearing A.

8–78 A stepped shaft loosely fits in a hole drilled through a flat plate. Axial and lateral forces act on the shaft as shown (assume that the shaft has no tendency to tip away from the vertical). Calculate the minimum torque required to keep the shaft turning at constant speed. $\mu_k = 0.3$ at all surfaces.

FIGURE P8-78

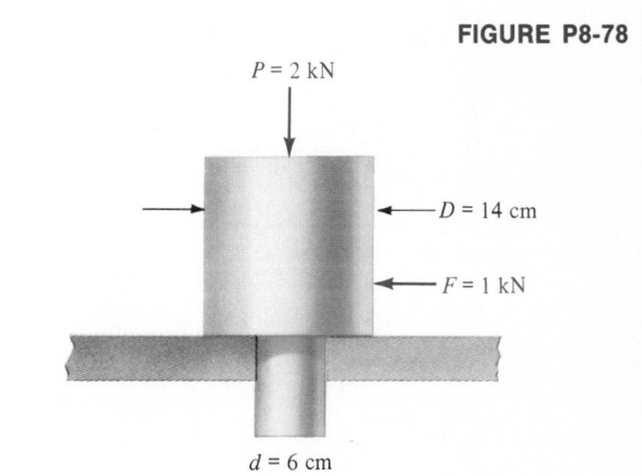

8–79 A uniform wheel of radius $r = 10$ cm and weight $W = 100$ N is tested for its rolling resistance on a slightly inclined track. Determine the smallest angle θ of the incline with the horizontal on which the wheel rolls under the force W alone. The coefficient of rolling friction is constant, $a = 0.5$ mm.

8–80 Plot the angle θ vs. the radius r in Prob 8–79. Let $r = 10$ cm, 15 cm, and 20 cm.

8–81 Plot the angle θ vs. the weight of the wheel in Prob 8–79. Let $W = 50$ N, 100 N, and 150 N.

8–82 In Prob. 8–111, the coefficient of rolling friction is $a = 0.02$ in. for each wheel. Determine the component of the horizontal force F required to overcome the rolling friction.

8–83 A horizontal force $F = 1$ kN is required to start a car of weight $W = 15$ kN rolling on a horizontal road. The tires are 60 cm in diameter. Determine the coefficient of rolling resistance assuming that other frictions are negligible.

8-8 BELT FRICTION

The friction properties of belts, ropes, and wires on pulleys and drums are important in the design of belt drives, band brakes, and gripping devices for flexible members. The main features of belt friction are illustrated in Fig. 8-20. A belt under tensions T_1 and T_2 is laid on a drum making contact from A to B. Generally, $T_1 \neq T_2$. A force R and a moment M act on the drum to oppose the resultant force and moment of T_1 and T_2 as shown in Fig. 8-20a. Part b of the figure shows the free-body diagram of the belt with distributed normal forces N and friction $\mathscr{F}$ caused by the drum. The friction forces $\mathscr{F}$ are counterclockwise here if $T_2 > T_1$.

For a detailed analysis of belt friction and the tensions T_1 and T_2, assume that the belt is on the verge of sliding to the right on the drum shown in Fig. 8-20a. The free-body diagram of an infinitesimal element dl of the belt is drawn in Fig. 8-20c. Here the differential friction force $d\mathscr{F}$ applies for impending motion under the obvious net tension to the right. For equilibrium in the x direction, $\sum F_x = 0$,

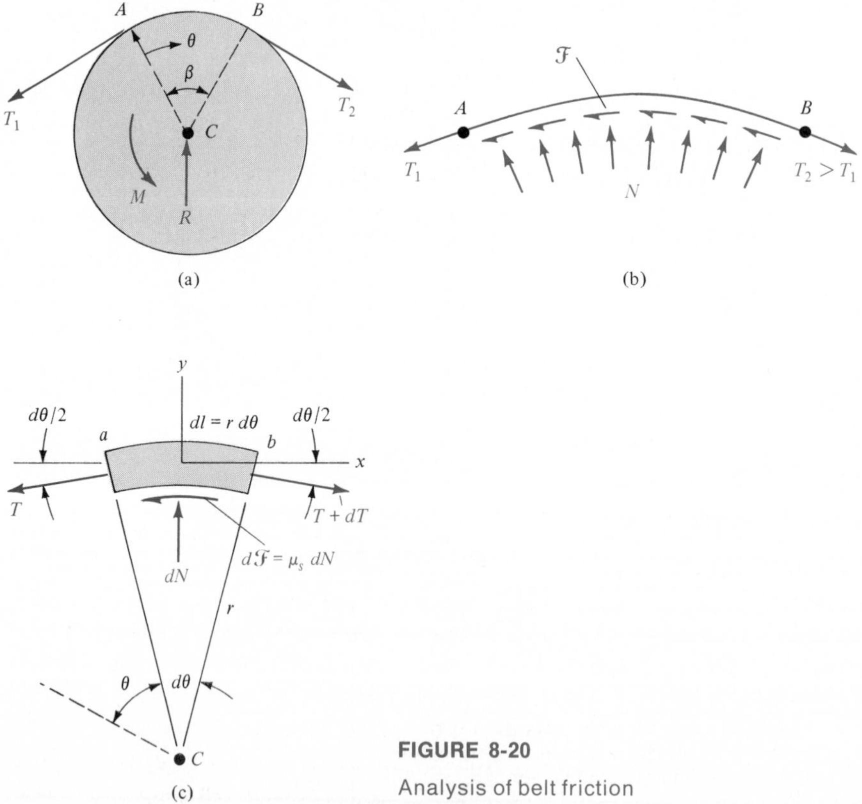

FIGURE 8-20
Analysis of belt friction

$$(T + dT)\cos\frac{d\theta}{2} - T\cos\frac{d\theta}{2} - \mu_s\,dN = 0$$

This can be greatly simplified since $d\theta$ is infinitesimally small, so $\cos(d\theta/2) \simeq 1$. Thus,

$$dT = \mu_s\,dN \tag{8-11}$$

For equilibrium in the y direction, $\sum F_y = 0$,

$$dN - T\sin\frac{d\theta}{2} - (T + dT)\sin\frac{d\theta}{2} = 0$$

This can be reduced using the following approximations that are valid for differential quantities:

$$\sin\frac{d\theta}{2} \simeq \frac{d\theta}{2} \qquad \text{and} \qquad dT\sin\frac{d\theta}{2} \simeq 0.$$

With these,

$$dN = T\,d\theta \tag{8-12}$$

Equations 8-11 and 8-12 can be combined to eliminate the quantity dN, resulting in

$$\frac{dT}{T} = \mu_s\,d\theta \tag{8-13}$$

The ultimate aim of relating the belt friction to the tensions T_1 and T_2 can be achieved by integrating Eq. 8-13 between the appropriate limits of tension and angle of contact,

$$\int_{T_1}^{T_2}\frac{dT}{T} = \int_0^{\beta}\mu_s\,d\theta$$

$$\ln\frac{T_2}{T_1} = \mu_s\beta \tag{8-14}$$

or

$$\boxed{T_2 = T_1 e^{\mu_s\beta}} \tag{8-15}$$

The total angle of belt contact β must be expressed in radians. For a member wrapped around a drum n times, the angle β is $2\pi n$. It should be noted that the derivations above are valid for $T_2 > T_1$ and for impending slipping. If slipping occurs, μ_s may be replaced by the coefficient of kinetic friction μ_k. The equations are not valid when there is no slipping or impending slipping.

The most common belts in belt drives are V belts, which make contact on two slanted sides with the pulleys as shown in cross section in Fig. 8-21. The relation of the high and low tensions in

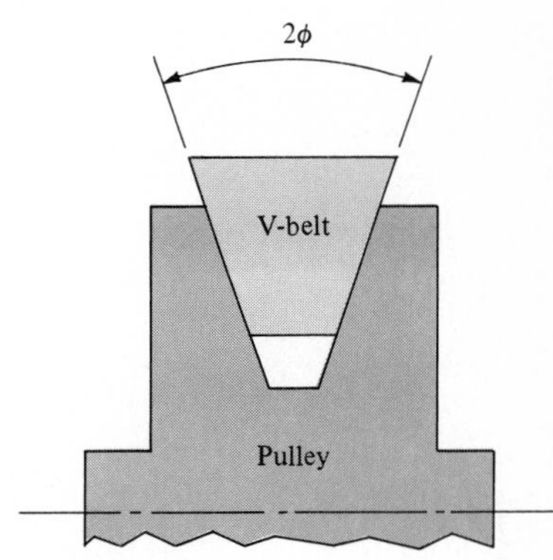

FIGURE 8-21
Cross section of a V-belt on a pulley

such a belt is derived in a similar procedure to that used above. The details of the derivation are left as an exercise for the students. For a V belt of angle 2ϕ, the final result is

$$T_2 = T_1 e^{\mu_s \beta / \sin \phi} \tag{8-16}$$

This equation has the same limitations as those for flat belts.

8-9 FLUID FRICTION

There is always a *drag force* (or resistance) on a body when there is relative motion between the body and a surrounding fluid. The details of analyzing drag are somewhat complex and should be learned in fluid mechanics. It is useful, however, to learn the basic qualitative aspects of fluid friction at this stage since practical problems often involve this kind of friction along with others.

In simplistic form, the drag force D is expressed as

$$D = CV^2 \tag{8-17}$$

where C is a constant depending mainly on the size and shape of the body and the properties of the fluid. V is the relative velocity between the body and the fluid (either one may be stationary while the other is moving).

It is useful to distinguish between *form* (or *pressure*) *drag* and *friction* (or *surface*) *drag*. Form drag depends strongly on the projected area of the body normal to the direction of the flow. Friction drag depends on the friction of the surface in a given fluid and on the surface area parallel to the flow. In both cases the coefficients in the equation for the drag must be determined experimentally.

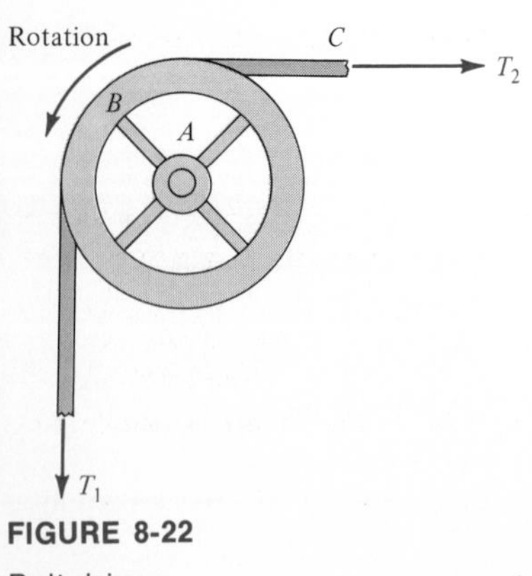

FIGURE 8-22
Belt drive

To illustrate the difference between the form and friction drags, consider the belt drive in Fig. 8-22. The spokes A of the pulley "cut" through the air, so their projected areas normal to the flow should be minimized to reduce the drag. The same spokes also have surface friction, so their areas parallel to the flow should also be minimized. The outer rim B of the pulley and the belt C have only friction drag, which means that their shapes are much less important than their surface areas and the quality of their surfaces.

It is important that any kind of drag depends on the square of the velocity of relative fluid motion. In most automobiles this becomes a very important factor at speeds of about 70 km/h or higher. Clearly, even small reductions in the operating speeds of vehicles and other machines are highly effective in reducing the energy losses caused by drag.

EXAMPLE 8-13

Determine the resisting tension T_1 in the rope shown in Fig. 8-23 if the pulling tension is $T_2 = 500$ N. Assume that $\mu_s = 0.2$ and that slippage of the rope is impending. Also assume that both T_1 and T_2 are horizontal.

FIGURE 8-23

Rope wrapped on a small post of a boat

SOLUTION

From Eq. 8-15, the angle β is all that is needed to solve for T_1. The rope is wrapped $1\frac{1}{4}$ times around the post; hence $\beta = 2n\pi = 2\pi(1.25) = 7.85$ rad. Therefore,

$$T_1 = \frac{T_2}{e^{\mu_s \beta}} = \frac{500 \text{ N}}{e^{(0.2)(7.85)}} = 104 \text{ N}$$

EXAMPLE 8-14

(a) Determine which arrangement shown in Fig. 8-24 is best to prevent clockwise slippage of the rope around the posts. Thus, the minimum resisting tension T_1 is sought. **(b)** What would T_1 have to be in each case to cause counterclockwise slippage of the rope (with T_1 pulling and T_2 opposing it)? $\mu_s = 0.3$.

SOLUTION

(a) ONE POST: $\beta = 4\pi$ rad (since the rope is wrapped around the post twice),

$$T_1 = \frac{T_2}{e^{\mu_s \beta}} = \frac{2000 \text{ lb}}{e^{(0.3)(4\pi)}} = 46.1 \text{ lb}$$

TWO POSTS: This case should be split into two calculations, one for calculation of T_1' on the left side of the right post and then using this as the pulling tension (T_2) for the left post.

$$\beta_1 = \beta_2 = 2\pi \text{ rad}$$

$$T_1' = \frac{T_2}{e^{\mu_s \beta_1}} = \frac{2000 \text{ lb}}{e^{(0.3)(2\pi)}} = 304 \text{ lb}$$

$$T_1 = \frac{T_1'}{e^{\mu_s \beta_2}} = \frac{304 \text{ lb}}{e^{(0.3)(2\pi)}} = 46.1 \text{ lb}$$

Both cases yield the same result.

(b) For the case of slippage to the left, let T_1'' be the tension to the left and remember that the larger tension must be in the numerator. Now Eq. 8-15 yields

$$T_1'' = T_2 e^{\mu_s \beta} = (2000 \text{ lb}) e^{(0.3)(4\pi)} = 86{,}750 \text{ lb}$$

The same result is valid for both cases on the basis of the results in part (a).

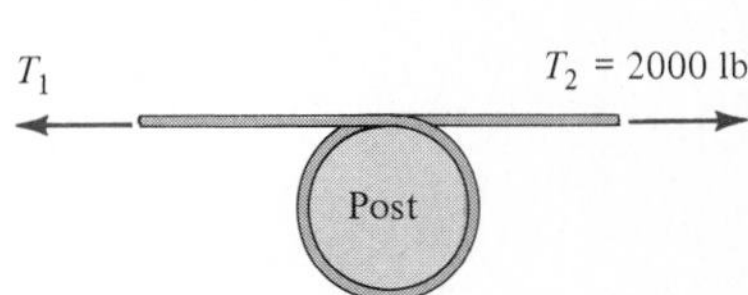

(a) Rope is wrapped around post two times

JUDGMENT OF THE RESULTS

The fact that using one or two posts yield the same result should not be too surprising. From a physical point of view the two cases are the same in that both involve a rope wrapped two full revolutions around the same material. From a mathematical point of view it is seen that the calculations for two posts reduce to

$$T_1 = \frac{T_1'}{e^{\mu_s \beta_2}} = \frac{T_2 / e^{\mu_s \beta_1}}{e^{\mu_s \beta_2}} = \frac{T_2}{e^{(\mu_s \beta_1 + \mu_s \beta_2)}} = \frac{T_2}{e^{\mu_s (\beta_1 + \beta_2)}}$$

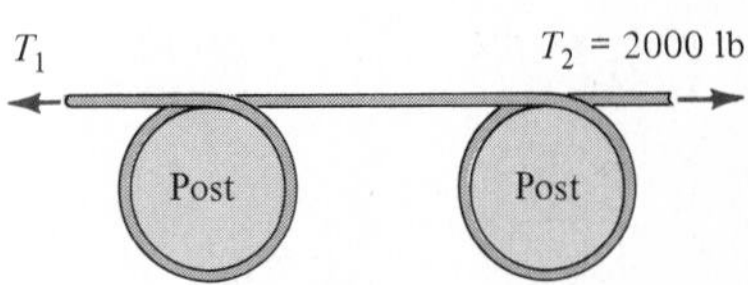

(b) Rope is wrapped around each post once

FIGURE 8-24

Cross sections of posts with ropes wrapped around them

PROBLEMS

SECTIONS 8-8, 8-9

FIGURE P8-84

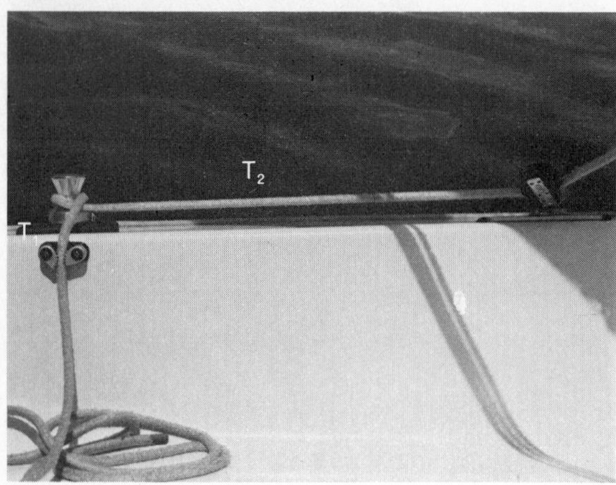

8–84 Determine the resisting tension T_1 if the pulling tension T_2 is 400 N, and $\mu_s = 0.2$. Assume that both T_1 and T_2 act horizontally and that slipping of the rope is impending. The angle of contact on the post is 270°.

FIGURE P8-85

8–85 Calculate the required pulling tension T_2 (before inserting the rope in the cam cleats) if the resisting tension T_1 is 50 lb. Assume that both T_1 and T_2 act horizontally, $\mu_k = 0.2$, and slippage is occurring. Estimate the angle of contact to the nearest quarter turn around the post.

8–86 A rope is to be wound around a mooring post. The pulling tension T_2 from the ship is 50 kN, and $\mu_s = 0.3$. Determine the required resisting tension T_1 (applied by hand of dock workers) for two complete turns ($n = 2$) of rope on the post if slippage is impending.

FIGURE P8-86

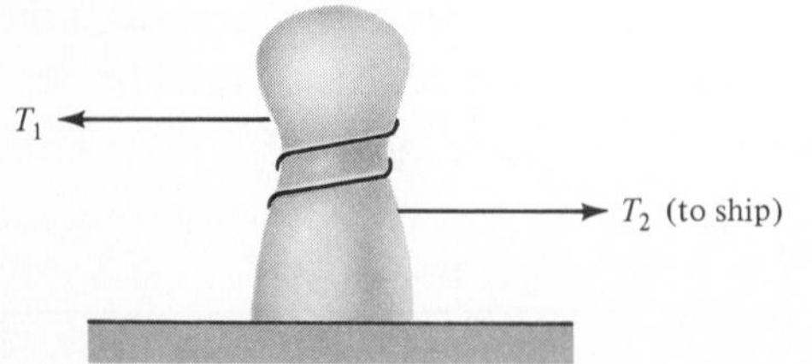

8–87 Plot the tension T_1 vs. the number of turns $n = 0.75$, 2, and 3 in Prob. 8–86. Extrapolate the curve to $n = 0$ and $n = 10$. Discuss the practical significance of the results, assuming that a worker can easily apply a force of $T_1 \simeq 500$ N.

8–88 A broken locking mechanism prevents the pulley from rotation. Determine the required tension T to raise the load by sliding the rope over the pulley if $\theta = 90^\circ$ and $\mu_k = 0.3$.

FIGURE P8-88

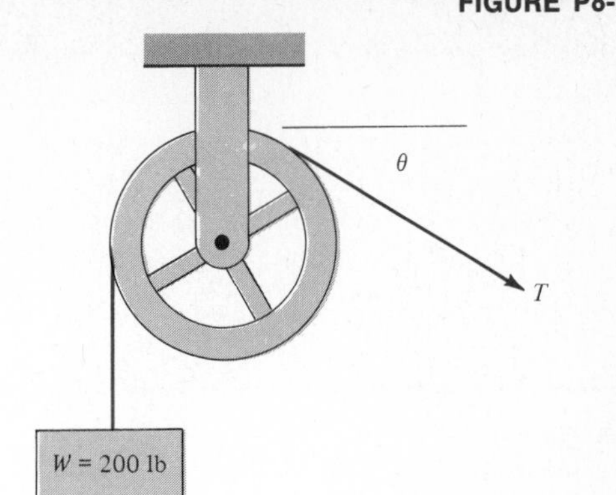

8–89 Plot the tension T as a function of $\theta = 0, 45^\circ$, and 90° in Prob. 8–88.

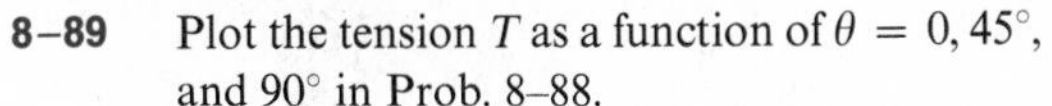

8–90 Consider a magnetic tape passing over a curved tape head. The minimum tension T_1 must be 0.2 lb. Determine the pulling tension T_2 if $\mu_k = 0.07$.

8–91 Plot the tension T_2 vs. $\mu_k = 0.04, 0.07$, and 0.1 in Prob. 8–90. What is your conclusion from this diagram concerning the significance of friction in attempting to minimize the pulling tension T_2 in the machine?

FIGURE P8-90

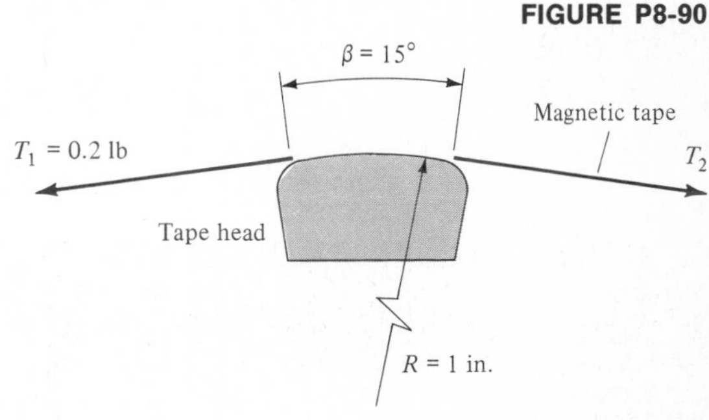

8–92 Prove that the tape tension in Prob. 8–90 can be analyzed the same way even if the tape head has a noncircular contour provided that β is the angle as shown in Fig. P8-92 and μ_k is the same.

FIGURE P8-92

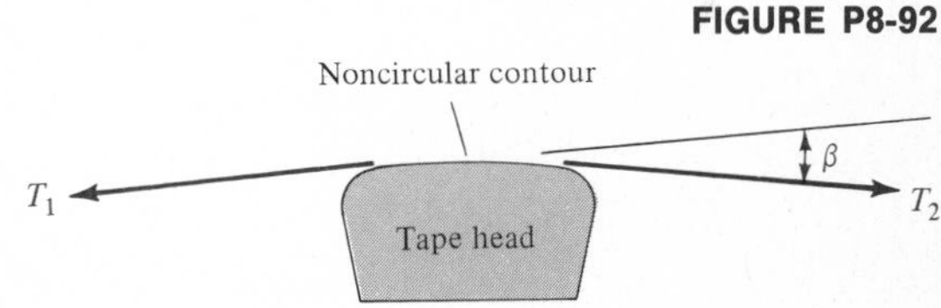

8–93 Consider a planned V-belt drive. The desired minimum tension is T_1, the angle of contact is $\beta = 130^\circ$, and $\mu_s = 0.5$. Express the pulling tension T_2 if the belt angle is $2\phi = 50^\circ$ (Fig. 8-21).

8–94 Plot the tension T_2 vs. the belt angle $2\phi = 20^\circ, 50^\circ$, and 90° in Prob. 8–93. Extrapolate the curve to $2\phi = 180^\circ$ and discuss whether this extrapolation is reasonable or not.

8–95 Derive Eq. 8-16.

8–96 Determine the tensions at A and B in the band brake. Consider both clockwise and counterclockwise rotations of the drum.

FIGURE P8-96

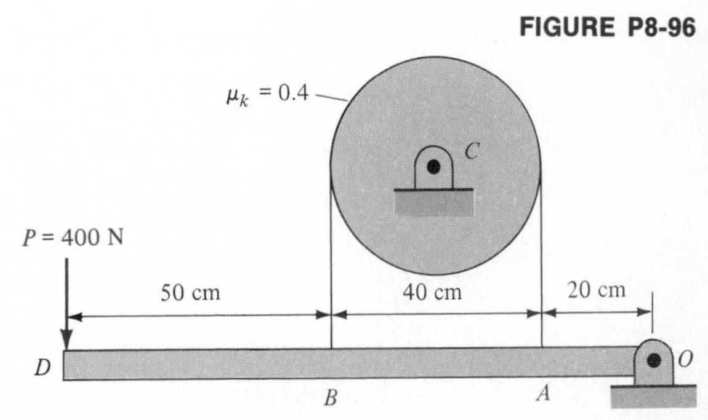

FIGURE P8-97

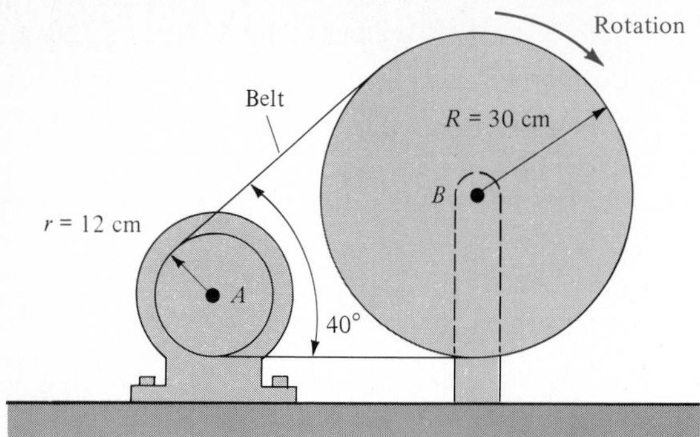

8–97 Motor A must deliver a torque of 2 kN·m to pulley B. Determine the largest tension in the flat belt if the coefficient of static friction is 0.3 at both pulleys.

8–98 Solve Prob. 8–97 assuming that the flat belt drive is replaced by a V-belt drive with a belt angle of $2\phi = 60^\circ$.

8–99 In Prob. 8–97 an idler pulley is proposed to make the belt's angle of contact 180° on pulley A and 270° on pulley B. Determine the horizontal and vertical bearing reactions of the idler pulley if all friction at this pulley can be ignored.

FIGURE P8-100

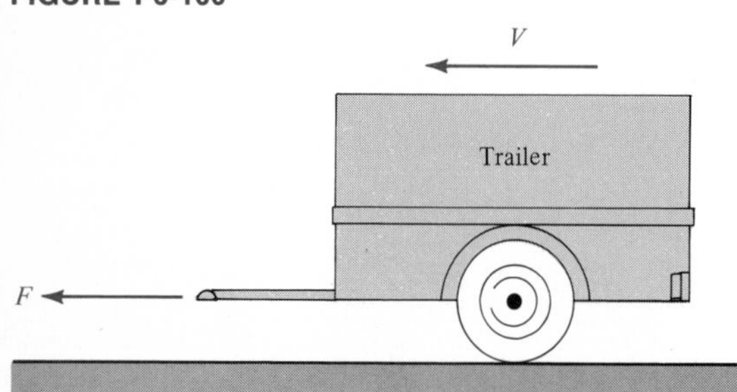

8–100 The horizontal force F on the drawbar of a trailer traveling at a constant speed of 40 mph is 500 lb. Sketch a reasonable curve of the force F vs. the speed V for speeds ranging from zero to 70 mph.

8–101 A tugboat is towing two large container barges. Discuss whether the relative cable distances a and b are reasonable or not with respect to total fluid friction on the two barges. This is a complex but important practical problem. Base your discussion on concepts presented in this chapter and on your intuition.

FIGURE P8-101

Section 8-1

SUMMARY

DRY FRICTION: A friction force $\mathscr{F}$ may act at the surfaces of contact between bodies. This force is always tangent to the surfaces in contact, and acts in the direction opposite to motion or impending relative motion. The magnitude of the friction force may vary from zero to a maximum value, depending on the external force system and the condition of the bodies in contact. $\mathscr{F}$ is always zero when there is no force for the friction force to oppose. The maximum value is

$$\mathscr{F}_{\max} = \mu N$$

where μ = coefficient of friction (μ_s or μ_k)
N = normal force between the bodies

In general, $0 \leq \mathscr{F} \leq \mathscr{F}_{\max}$.

Section 8-2

ANGLES OF FRICTION:

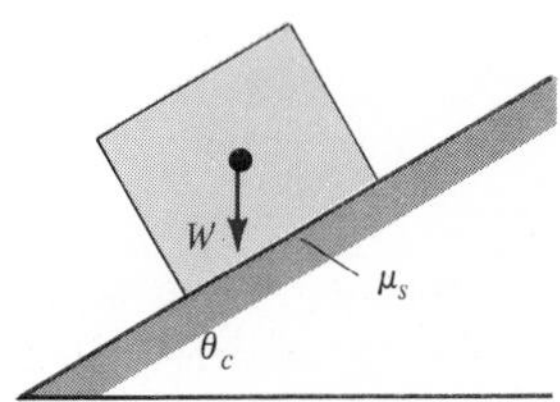

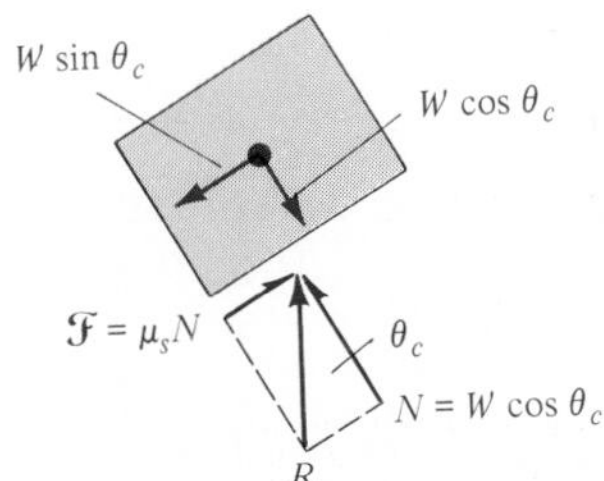

The angle of repose is the critical angle θ_c of the incline at which motion of the block is impending because of the weight of W.

$$\tan \theta_c = \frac{\mathscr{F}}{N} = \frac{\mu_s N}{N}$$

$$= \mu_s = \text{coefficient of static friction}$$

On a horizontal plane where force P_c causes impending motion,

$$\tan \theta_c = \frac{\mathscr{F}}{N} = \frac{\mu_s N}{N} = \mu_s$$

Here θ_c is the angle of static friction, which is equivalent to the angle of repose for inclines.

Section 8-3

PROCEDURES FOR SOLVING FRICTION PROBLEMS

CLASS OF PROBLEM	GIVEN INFORMATION	PROCEDURE
A	Bodies, forces, coefficients of friction are known. Impending motion is not assured; $\mathscr{F} \neq \mu_s N$.	To determine if equilibrium is possible: 1. Construct the free-body diagram. 2. Assume that the system is in equilibrium. 3. Determine the friction and normal forces necessary for equilibrium. 4. Results: (a) $\mathscr{F} < \mu_s N$, the body is at rest. (b) $\mathscr{F} > \mu_s N$, motion is occurring, static equilibrium is not possible. Since there is motion, $\mathscr{F} = \mu_k N$. Complete solution requires principles of dynamics.
B	Bodies, forces are given. Impending motion is specified, $\mathscr{F} = \mu_S N$ is valid.	To determine the coefficients of friction: 1. Construct the free-body diagram. 2. Write $\mathscr{F} = \mu_s N$ for all surfaces where motion is impending. 3. Determine μ_s from the equation of equilibrium.
C	Bodies, forces, coefficients of friction are known. Impending motion is specified, but the exact motion is not given. The possible motions are sliding, tipping, or rolling. Alternatively, the forces or coefficients of friction may have to be determined to produce a particular motion from several possible motions.	To determine the exact motion that may occur, or unknown quantities required: 1. Construct the free-body diagram. 2. Assume that motion is impending in one of the two or more possible ways. Repeat this for each possible motion, and write the equation of equilibrium. 3. Compare the results for the possible motions and select the likely event. Determine the required unknowns for any preferred motion.

Section 8-4

WEDGES: Draw a free-body diagram for each wedge and body to be supported or moved by the wedge. Write equilibrium equations for each body and solve these for the unknown quantities. Note that all friction forces must oppose the impending relative motions.

SCREWS (Square Threads):

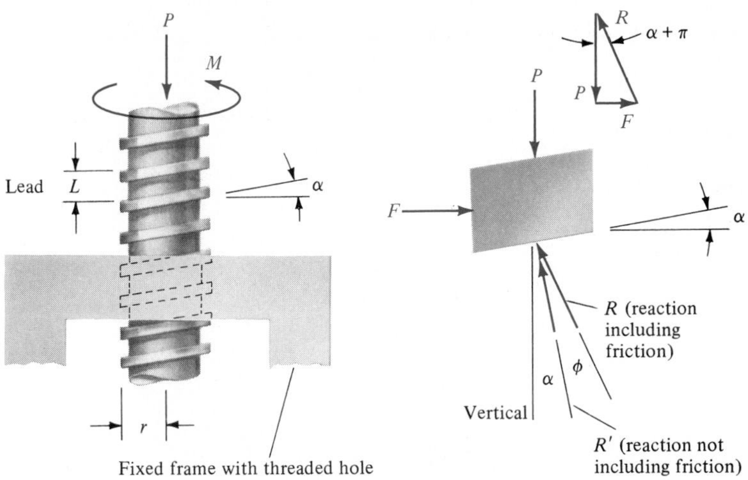

The moment M required to move the screw against the axial load P is

$$M = Pr \tan(\alpha + \phi)$$

where $\alpha = \tan^{-1} L/2\pi r = \tan^{-1} np/2\pi r$

L = lead (advancement per revolution)

p = pitch (distance between similar points on adjacent threads)

n = multiplicity of threads

$\phi = \tan^{-1} \mu$

μ = coefficient of friction (μ_s or μ_k)

Section 8-5

DISK FRICTION:

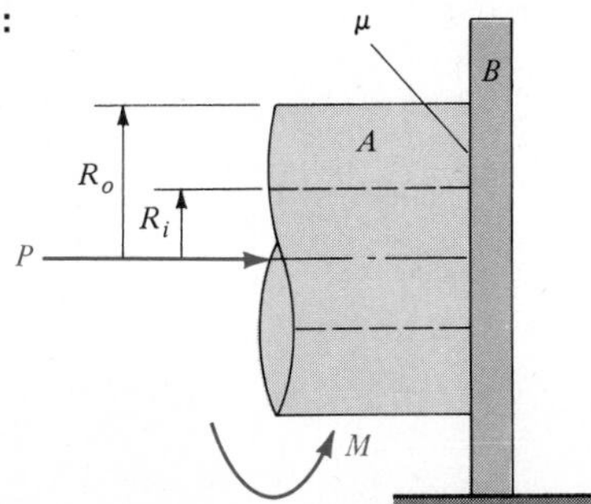

For the hollow member A pressed against the flat member B, the moment M to cause relative rotation is

$$M = \frac{2}{3}\mu P \frac{R_o^3 - R_i^3}{R_o^2 - R_i^2}$$

For a solid shaft of radius R,

$$M = \frac{2}{3}\mu P R$$

where P = normal force between the bodies
μ = coefficient of friction (μ_s or μ_k)

Section 8-6

AXLE FRICTION:

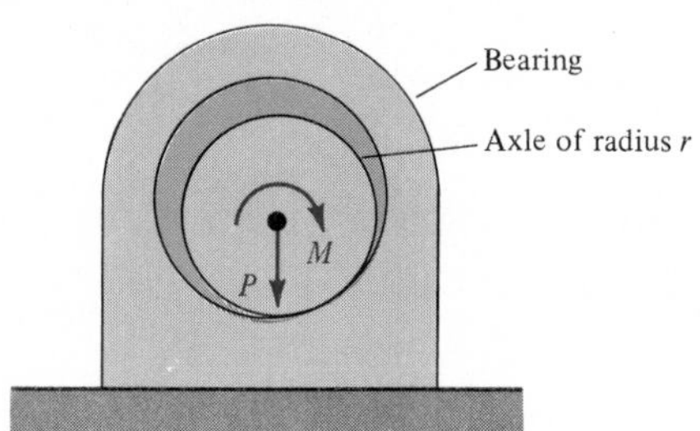

The moment M required to overcome friction in steady rotation of the axle is approximately

$$M = Pr\mu_k$$

where P = lateral force on the axle
μ_k = coefficient of kinetic friction

Section 8-7

ROLLING RESISTANCE:

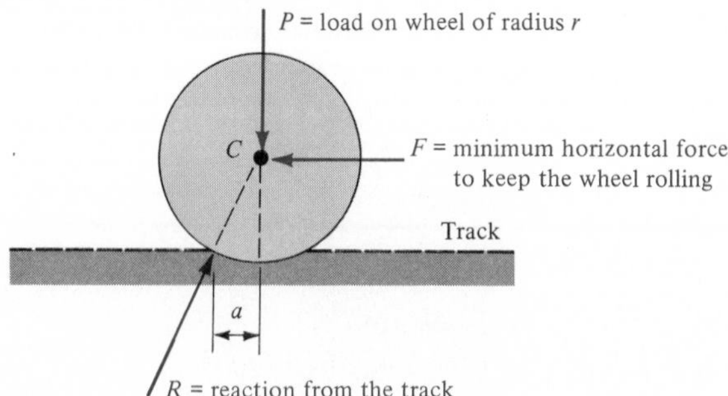

The coefficient of rolling resistance is

$$a \simeq \frac{Fr}{P} \text{ in units of length}$$

Section 8-8

BELT FRICTION:

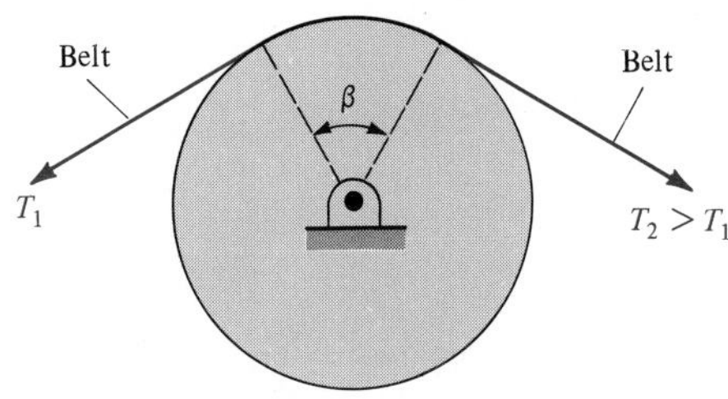

For impending slipping to the right,

$$T_2 = T_1 e^{\mu_s \beta}$$

where β = angle of belt contact in radians
μ_s = coefficient of static friction (may be replaced by μ_k if there is slipping)

For a V belt of belt angle 2ϕ,

$$T_2 = T_1 e^{\mu_s \beta / \sin \phi}$$

Section 8-9

FLUID FRICTION: The drag force D caused by fluid friction is approximately

$$D = CV^2$$

where C = constant depending on size and shape of the body and the properties of the fluid
V = relative velocity of the body and the fluid

REVIEW PROBLEMS

FIGURE P8-102

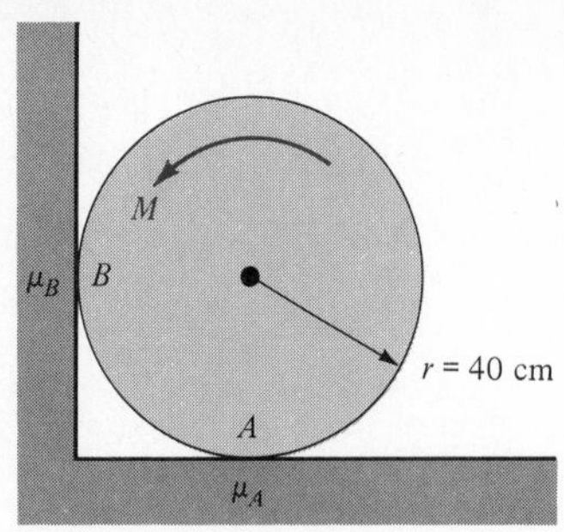

FIGURE P8-103

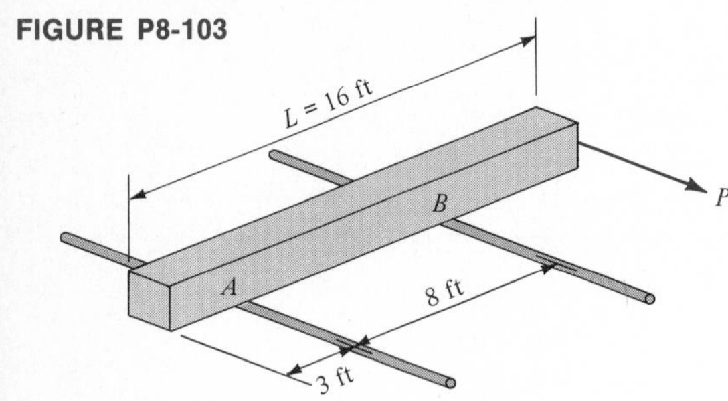

FIGURE P8-104

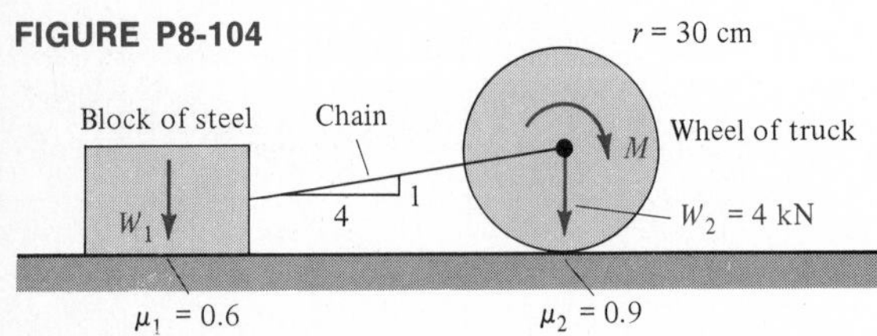

FIGURE P8-106

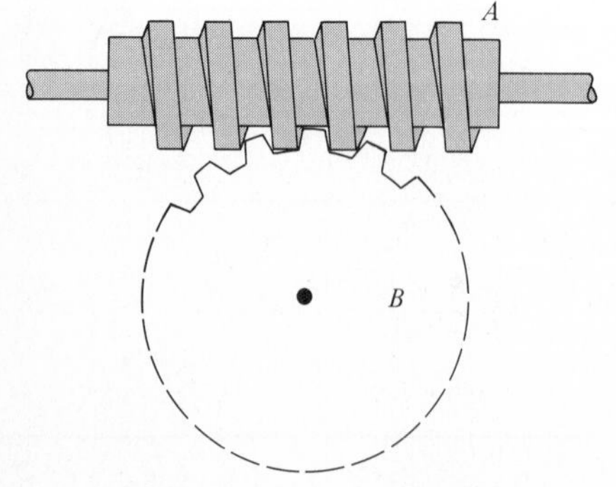

8–102 An exercise machine is modeled as a uniform wheel of weight $W = 600$ N. Determine whether a moment $M = 300$ N·m would cause slippage of the wheel if $\mu_B = 0$.

8–103 A uniform, 2500-lb concrete beam is resting on two rails. Calculate the horizontal force P parallel to the rails that would start the beam to slide if $\mu_A = \mu_B = 0.3$.

8–104 Consider the single-wheel model of a small electric truck used in a foundry to drag blocks of steel on the floor. Of course, a block may be so heavy that it will not slide while the wheel spins. Determine the smallest torque M that would cause slipping of the block or of the wheel if $W_1 = 3$ kN.

8–105 In Prob. 8–104 determine the smallest load W_1 that causes the wheel to slip first (this is where the method becomes useless with the given truck).

8–106 The mean diameter of the single-threaded worm gear A is 1.6 in., and of gear B is 10 in. The coefficient of friction between the gear teeth is 0.25. Ignoring friction elsewhere in the mechanism, determine the required moment on gear A to turn gear B clockwise against a 500 ft·lb moment on gear B. The pitch is $p = 0.2$ in.

8–107 Plot the required moment as a function of the pitch in Prob. 8–106 for $p = 0.2$ in., 0.25 in., and 0.3 in. Is the pitch an important factor in designing the machine that includes this worm gear?

8–108 Determine the possibility of unwinding of the mechanism in Prob. 8–106 if the maximum resisting moment acting on gear A is 10 ft·lb while a 500 ft·lb moment is acting on gear B.

8–109 For cooling purposes, a disk brake has radial slots in one disk which is pressed against a full circle of the same radius R_o. Determine the torque that can start one disk to slip on the other when they are pressed together by a force $P = 1$ kN. $R_o = 15$ cm, $r_i = 2$ cm, $r_o = 13$ cm, and $\mu_s = 0.6$. The slots remove half of the area between r_i and r_o.

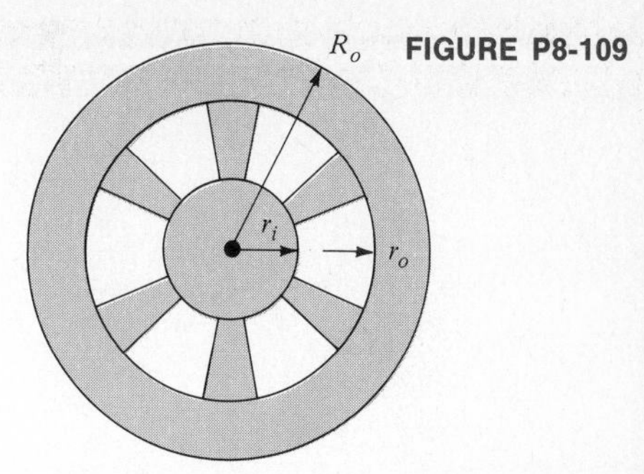

FIGURE P8-109

8–110 Derive a formula to determine the frictional resisting torque of the conical pivot.

FIGURE P8-110

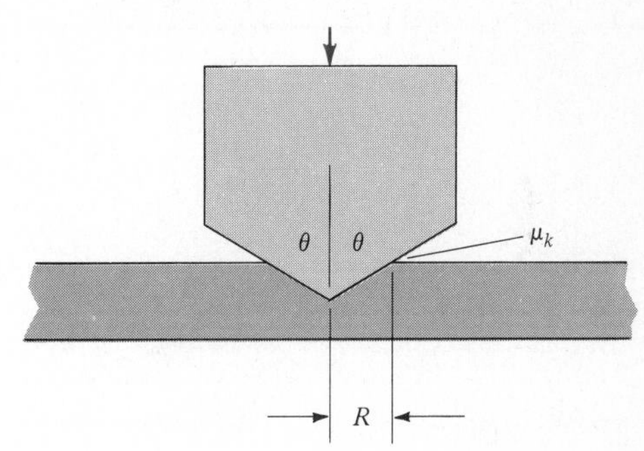

8–111 The four wheels of a crane's carriage share equally in supporting the 40,000 lb load. The wheel diameters are 10 in., and the axles have 2 in. diameters with coefficient of kinetic friction of 0.2. Calculate the horizontal force F required to maintain steady motion of the carriage if rolling friction is negligible.

FIGURE P8-111

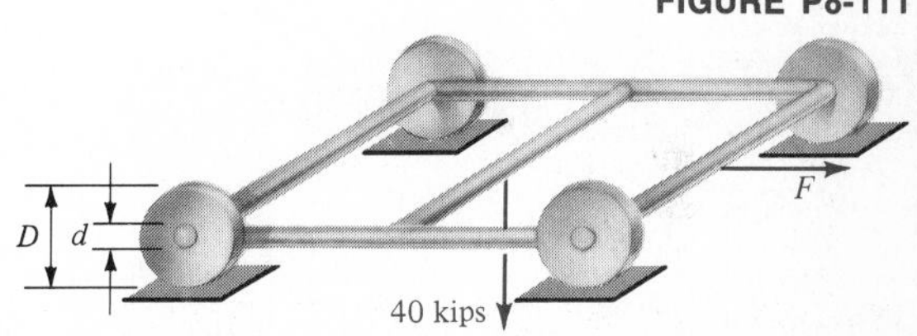

8–112 Calculate the required coefficient of friction in the proposed band saw. Are the required tensions practical to move the band saw by friction alone?

FIGURE P8-112

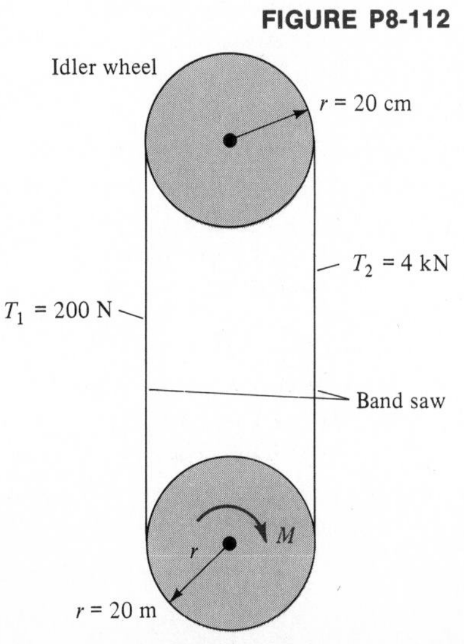

8–113 Determine the tension in both segments of the band brake attached to the brake arm at B. The drum is rotating clockwise.

FIGURE P8-113

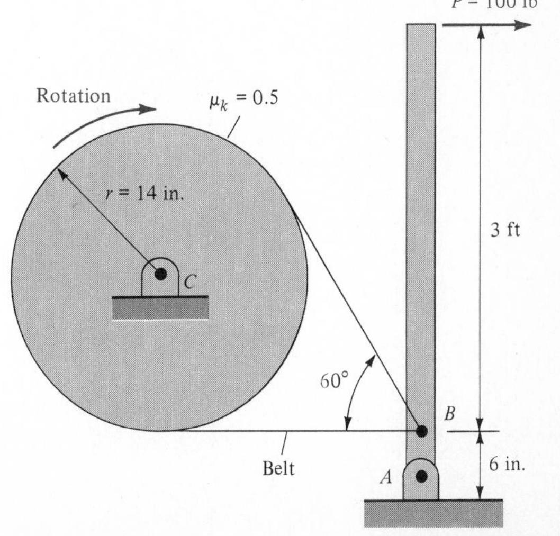

CHAPTER 9

MOMENTS OF INERTIA

OVERVIEW

Reasons for Studying the Topics of Inertia in Statics

The concepts and methods of inertia are used in the areas of dynamics and mechanics of materials. As far as subject matter is concerned, these topics have relatively little to do with statics since they do not involve the analysis of equilibrium. The reasons for presenting the topics of inertia in statics are twofold: First, the methods of first moments (Secs. 7-2 and 7-3) can be readily extended to analyze inertias. Second, in most academic situations it is easiest to schedule the detailed introduction of topics of inertia in statics, requiring only that they be rapidly reviewed in dynamics or mechanics of materials. This enhances the understanding and facility of using the methods of inertia even if they do not seem entirely natural in statics.

Basic Concepts of Inertia

Superficially, there are two ways of viewing the concepts of inertia. From the mathematical point of view, they involve the methods of elementary calculus, and they can be readily used without trying to understand the relevant physical phenomena. A satisfactory physical view of the concepts of inertia is more difficult to attain because the concepts are rather abstract. Basically, the term *inertia* is used in dynamics to describe the tendency of matter to remain at rest or continue moving with a constant velocity. Inertia is also used in other areas to show resistance to any change. These are illustrated in a qualitative discussion of lifting a barbell from the floor as follows.

Consider an end view of a barbell in Fig. 9-1. If the disks were fixed on the bar, each disk would rotate counterclockwise when the hands rotate the bar during a lift (Fig. 9-1b and c). This rotation would require an extra effort from the lifter, especially if the disks were large. The inertia, or resistance of a disk to rotation, depends on its mass and how it is distributed with respect to the axis of rotation A. In practice, the disks are relatively free to rotate on the bar, so the line AB tends to remain horizontal because of inertia even though the bar rotates (this minimizes the effort of the lifter).

Next consider a side view of the barbell during a lift (Fig. 9-2). If the disks are heavy, the bar bends noticeably. The bending of the bar allows another rotation of each disk, that shown in Fig. 9-2b. The inertia of the disk in this rotation depends on its mass and how it is distributed with respect to the axis of rotation, which is perpendicular to the bar in this case. The two inertias discussed are not the same since the mass distributions with respect to the two axes of rotation are different.

A more abstract concept of inertia is relevant to the bending of the bar itself in Fig. 9-2. Here the resistance to change is taken as a property of the cross-sectional area of the bar (note that an originally vertical cross section becomes slanted as the bar bends). This property is used extensively in mechanics of materials.

Since the inertia of a mass is easier to understand intuitively than the inertia of an area, they are first presented in that order. Mathematically they are similar, so in problems the difference between masses and areas is a difference of symbols and units.

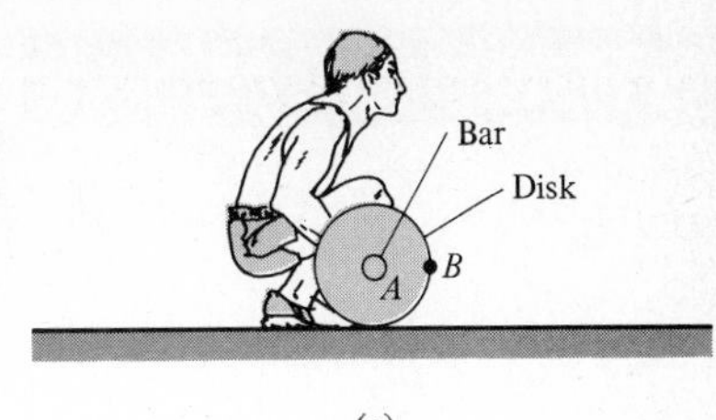

(a)

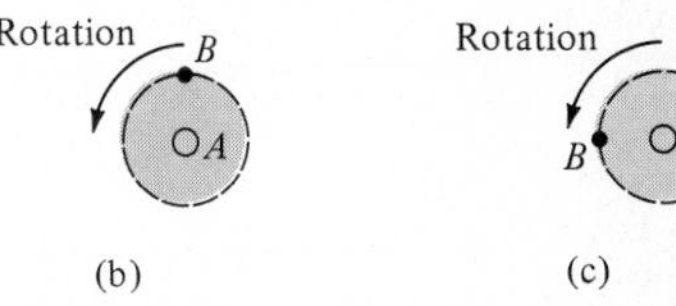

(b) (c)

FIGURE 9-1

Weightlifting is used to illustrate the concepts of inertia. End view of a barbell

(a)

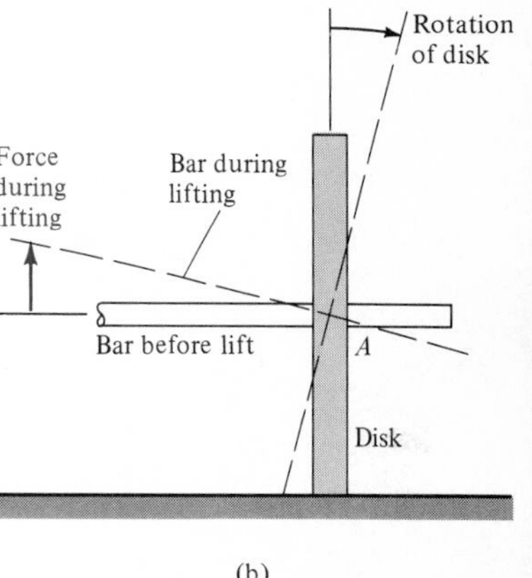

(b)

FIGURE 9-2

Side view of a barbell

STUDY GOALS

SECTION 9-1 presents the definition of moment of inertia of a mass with respect to an axis of reference. This topic is essential for studying the dynamics of rigid bodies.

SECTION 9-2 presents the definition of moment of inertia (or second moment) of an area with respect to an axis of reference. This topic is essential for studying the mechanics of materials (bending of beams and columns).

SECTION 9-3 extends the concept of moment of inertia of an area to the polar moment of inertia of the area with respect to a point of reference. This topic is essential for studying the mechanics of materials (torsion of cylindrical shafts).

SECTION 9-4 presents the parallel-axis theorem for a mass or an area. This useful theorem facilitates the solving of problems in dynamics and mechanics of materials. The simple formulas for the transformation of moments of inertia from a centroidal axis to a parallel axis are worth remembering.

SECTION 9-5 shows that the moment of inertia of a complex area or body about a specified axis may be determined in several ways using the parallel-axis theorem. The results of the different calculations should be the same in a given situation, but they are not necessarily equal in computational effort. It is demonstrated that an efficient approach can be found with a little experience and judgment.

SECTIONS 9-6, 9-7, AND 9-8 present products of inertia, principal axes, and transformations of moments of inertia. These are advanced topics and are seldom discussed in statics courses because they are normally used only in advanced courses in dynamics and mechanics of materials. They are included in this chapter to facilitate studying them when required in a later course.

9-1 MOMENT OF INERTIA OF A MASS

The moment of inertia dI_x of a differential mass dM about the x axis (Fig. 9-3) is defined as

$$dI_x = \rho^2\, dM = (y^2 + z^2)\, dM \qquad \text{9-1}$$

where ρ is the nearest distance from the element of mass to the x axis. The moment of inertia of an extended body is obtained by integrating Eq. 9-1 over the whole mass M while always referring to the same axis,

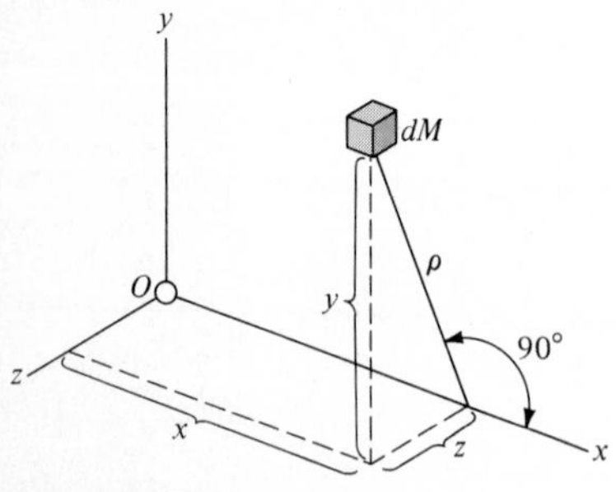

FIGURE 9-3
Mass element dM at distance ρ from the x axis

$$I_x = \int \rho^2\, dM = \int (y^2 + z^2)\, dM \qquad \text{9-2a}$$

Similarly, for using the y or z axis for reference in Fig. 9-3,

$$I_y = \int (x^2 + z^2)\, dM \qquad \text{9-2b}$$

$$I_z = \int (x^2 + y^2)\, dM \qquad \text{9-2c}$$

Radius of Gyration

Imagine that a body of mass M is replaced by a point mass M (same mass) at a distance r_g from a given x axis. If the moments of inertia I_x are the same for the body and the point mass, $I_x = r_g^2 M$ from Eqs. 9-1 and 9-2, and

$$r_g = \sqrt{I_x/M} \qquad \text{9-3}$$

The distance r_g is called the *radius of gyration* and it can be similarly defined for any other axis of reference. A point mass may be replaced by a thin strip or shell to obtain Eq. 9-3 if all masses are essentially at a constant distance r_g from the axis of reference.

Because the moment of inertia of a mass depends on ρ^2 of differential masses at distance ρ from a common axis of reference, the geometry of a body strongly influences the moment of inertia. This is true for any axis of reference such as those in Figs. 9-1 and 9-2 for the same body. The moments of inertia of masses have units of mass times length squared.

9-2 MOMENT OF INERTIA OF AN AREA

The moment of inertia dI_x of a differential area dA about the x axis (Fig. 9-4) is defined as

$$dI_x = \rho^2\, dA \qquad \text{9-4}$$

where ρ is the nearest distance from the element of area to the x axis. The moment of inertia I_x of an area A is obtained by integrating Eq. 9-4 over the whole area A while always referring to the same axis,

$$I_x = \int \rho^2\, dA \qquad \text{9-5}$$

The term *second moment* of the area A is sometimes used in connection with Eq. 9-5. The radius of gyration of an area is defined the same way as it is for a mass in Eq. 9-3,

$$r_g = \sqrt{I_x/A} \qquad \text{9-6}$$

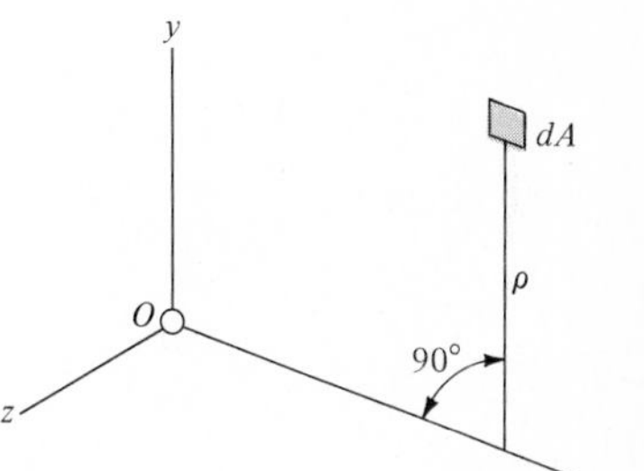

FIGURE 9-4
Area element dA at distance ρ from the x axis

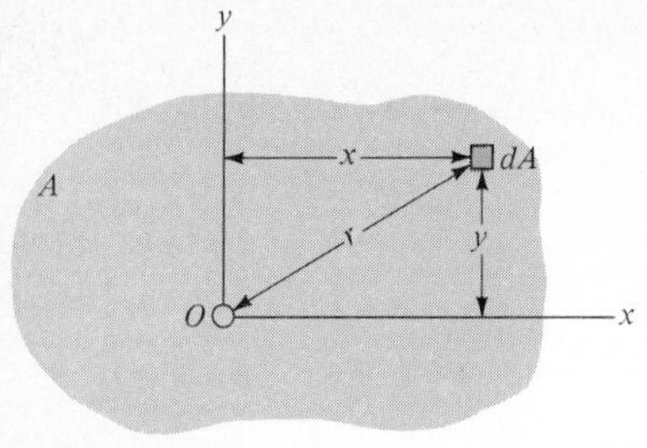

FIGURE 9-5
Area A in the xy plane

The moment of inertia of areas is an abstract concept. To obtain a feeling for what it means physically, assume that it represents the resistance of an area to rotational displacement from its original plane. For example, the bending of the bar in Fig. 9-2 would cause any vertical cross section in this part of the bar to rotate a little clockwise. The imaginary resistance of areas to such rotation is a useful concept in analyzing the strength of members even in static situations.

As in the case of masses, geometry strongly affects the moments of inertia of areas. Thus, the choice of the reference axis is important in any calculation of moments of inertia. For example, the area A in Fig. 9-5 is likely to have different moments of inertia with respect to the x and y axes,

$$I_x = \int y^2 \, dA, \qquad I_y = \int x^2 \, dA \tag{9-7}$$

Note that in this case (Fig. 9-5) a moment of inertia I_z cannot be defined in the same way as I_x or I_y (I_z would involve in-plane rotation while I_x and I_y imply out-of-plane rotations).

9-3 POLAR MOMENT OF INERTIA OF AN AREA

The *polar moment of inertia of an area*, J_O, is defined as the (imaginary) resistance of an area A to rotational displacements in its own plane. Using the concepts discussed above, and the xy plane and the origin O for reference in Fig. 9-5,

$$J_O = \int r^2 \, dA \tag{9-8a}$$

In other words, the polar moment of inertia is calculated with respect to an axis perpendicular to the area considered. It is easy to show that

$$J_O = I_x + I_y \tag{9-8b}$$

since $r^2 = x^2 + y^2$ in Fig. 9-5.

It should be noted that the moment of inertia is always positive regardless of the position of the body or area with respect to the reference axis. Also, the moments of inertia are additive; this is advantageous in working with bodies of complex shape. The moment of inertia of an area has units of length raised to the fourth power, m^4 or in^4.

EXAMPLE 9-1

Derive the equations for the moments of inertia of a rectangular area with sides a and b with respect to axes x, x_1, and y, y_1 shown in Fig. 9-6.

SOLUTION

The moments of inertia about the centroidal xy axes are determined first. Using the elemental area of width dy and length a, at a distance y from the x axis ($\rho = y$),

$$I_x = \int dI_x = \int y^2\,dA = \int_{-b/2}^{b/2} y^2 a\,dy = \frac{1}{12}ab^3$$

Using the elemental area of width dx and length b, at a distance x from the y axis ($\rho = x$),

$$I_y = \int dI_y = \int x^2\,dA = \int_{-a/2}^{a/2} x^2 b\,dx = \frac{1}{12}ba^3$$

The same procedure is used for I_{x1} and I_{y1}, but with different limits of integration,

$$I_{x1} = \int dI_{x1} = \int y^2\,dA = \int_0^b y^2 a\,dy = \frac{1}{3}ab^3$$

$$I_{y1} = \int dI_{y1} = \int x^2\,dA = \int_0^a x^2 b\,dx = \frac{1}{3}ba^3$$

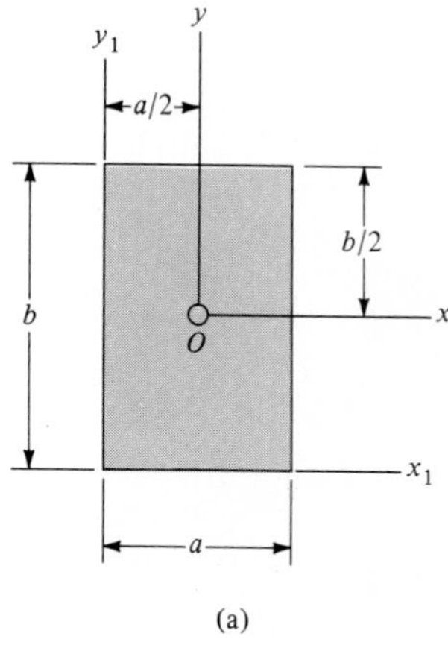

(a)

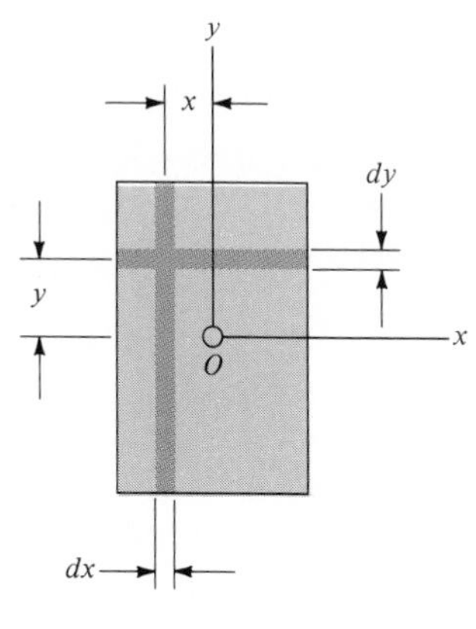

(b)

FIGURE 9-6
Area *ab* in the *xy* plane

EXAMPLE 9-2

(a) Determine the moments of inertia I_x and I_y of the semicircular area shown in Fig. 9-7. **(b)** Calculate the mass moment of inertia I_x of a thin plate with a semicircular area shown in Fig. 9-7. The plate has thickness $t = 0.1$ in. and density $\delta = 0.001$ slug/in^3. Assume that I_x is not a function of z.

SOLUTION

(a) It is easiest to use polar coordinates for this problem with an elemental area $dA = \rho\, d\theta\, d\rho$ as shown in Fig. 9-7. From Eq. 9-7,

$$\begin{aligned} I_x &= \int y^2\, dA = \int_{\theta=0}^{\theta=\pi} \int_{\rho=0}^{\rho=r} (\rho \sin \theta)^2\, \rho\, d\theta\, d\rho \\ &= \int_{\theta=0}^{\theta=\pi} \left(\frac{r^4}{4} \sin^2 \theta\right) d\theta = \frac{r^4}{4} \int_0^\pi \left(\frac{1}{2} - \frac{\cos 2\theta}{2}\right) d\theta \\ &= \frac{r^4}{4}\left(\frac{\pi}{2}\right) = \frac{\pi r^4}{8} \\ &= 245 \text{ in}^4 \end{aligned}$$

$$\begin{aligned} I_y &= \int x^2\, dA = \int_{\theta=0}^{\theta=\pi} \int_{\rho=0}^{\rho=r} (\rho \cos \theta)^2 \rho\, d\theta\, d\rho \\ &= \frac{r^4}{4} \int_0^\pi (\cos^2 \theta)\, d\theta = \frac{r^4}{4} \int_0^\pi \left(\frac{1}{2} + \frac{\cos 2\theta}{2}\right) d\theta \\ &= \frac{r^4}{4}\left(\frac{\pi}{2}\right) = \frac{\pi r^4}{8} \\ &= 245 \text{ in}^4 \end{aligned}$$

(b) For the mass moment of inertia, Eq. 9-2 is used,

$$I_x = \int y^2\, dM$$

$dM = \delta\, dV$, where dV is an elemental volume $dV = t\, dA$. Hence,

$$\begin{aligned} I_x &= \int y^2 \delta\, dV = \int y^2\, \delta t\, dA = \delta t \int y^2\, dA = \delta t\, I_x \\ &= \left(0.001 \frac{\text{slug}}{\text{in}^3}\right)(0.1 \text{ in.})(245 \text{ in}^4) = 0.0245 \text{ slug}\cdot\text{in}^2 \end{aligned}$$

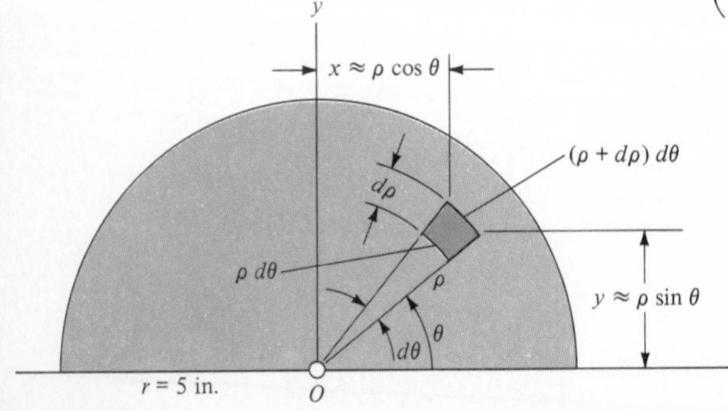

FIGURE 9-7
Semicircular area in the xy plane

SECTIONS 9-1, 9-2, 9-3

PROBLEMS

9–1 Determine the mass moment of inertia I_z of the uniform slender rod of mass M.

FIGURE P9-1

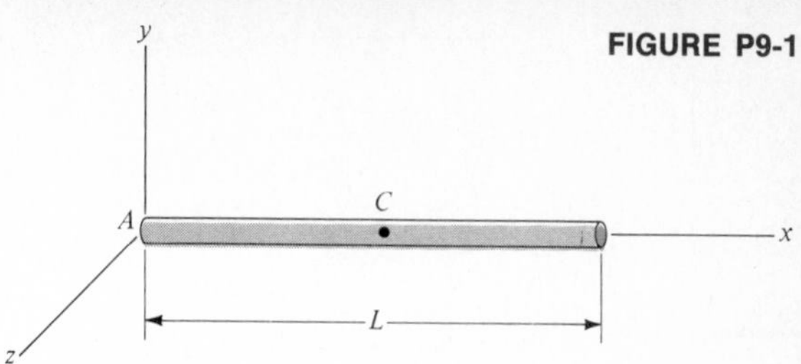

9–2 Determine the mass moment of inertia $I_{y'}$ of the uniform slender rod in Fig. P9-1. The axis y' is parallel to y at point C, $x = L/2$. The rod has mass M.

FIGURE P9-3

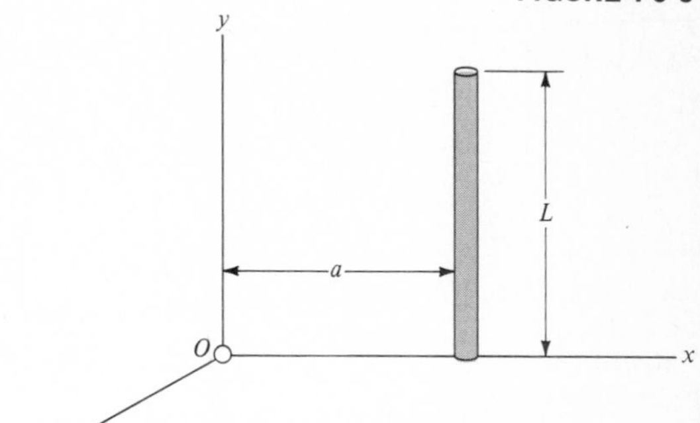

9–3 A uniform slender bar of mass M is parallel to the y axis at $x = a$. Determine the mass moments of inertia I_x and I_y, and the radii of gyration.

FIGURE P9-4

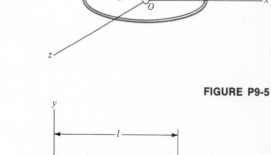

9–4 A thin hoop of radius R and mass M is in the xz plane. Determine the mass moments of inertia I_x and I_y of the hoop.

FIGURE P9-5

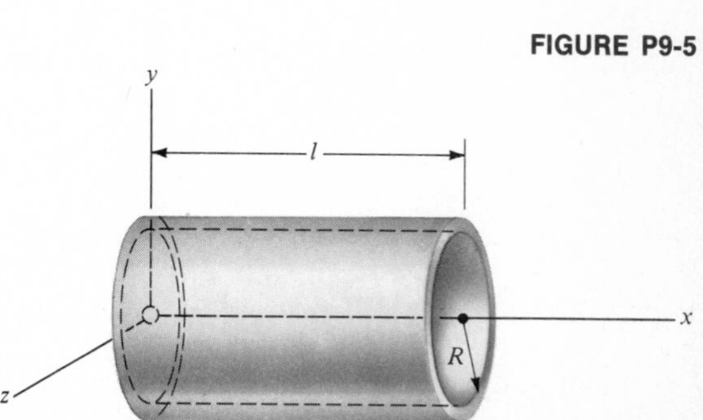

9–5 A thin cylindrical shell of radius R and mass M is centered on the x axis. Determine the mass moment of inertia I_x.

FIGURE P9-6

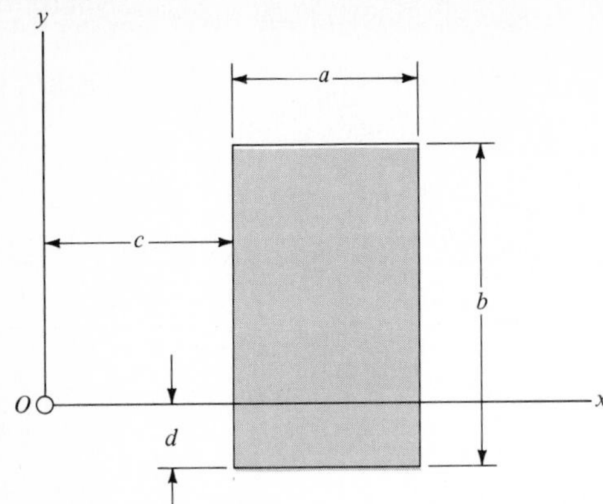

FIGURE P9-9

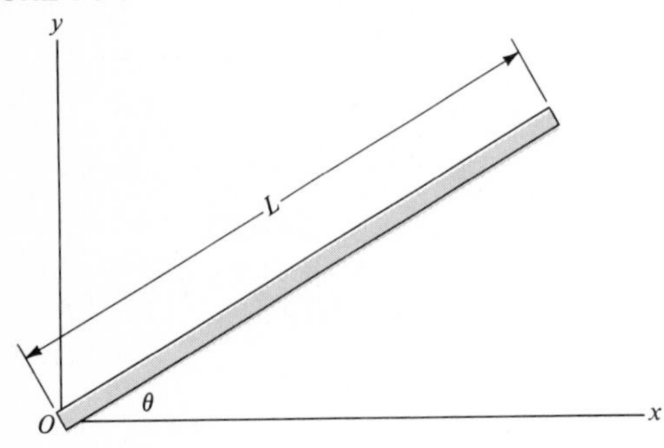

FIGURE P9-10

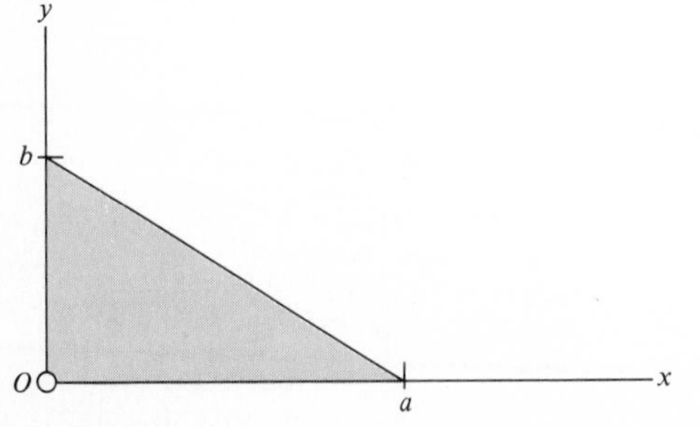

9–6 The thin plate *ab* has thickness t and density δ. Determine the mass moments of inertia I_x and I_y assuming that they are not functions of z (since t is small).

9–7 Determine the moments of inertia I_x and I_y of the area *ab* in Fig. P9-6. Express the radius of gyration with respect to the x axis.

9–8 Denote by C the centroid of the area *ab* in Fig. P9-6. Place centroidal axes x' and y' at C, parallel to x and y, respectively. Determine the polar moment of inertia J_C of the area *ab*.

9–9 Calculate the mass moments of inertia I_x and I_y of the uniform slender rod of mass M.

9–10 Determine the mass moment of inertia I_x of the thin triangular plate of thickness t and density δ. Assume that I_x is not a function of z (since t is small).

9–11 Determine the moment of inertia I_y of the triangular area in Fig. P9-10.

9–12 Determine the polar moment of inertia J_O of the triangular area in Fig. P9-10.

9–13 Calculate the mass moment of inertia I_y of the thin triangular plate of thickness t and density δ. Assume that I_y is not a function of z (since t is small).

FIGURE P9-13

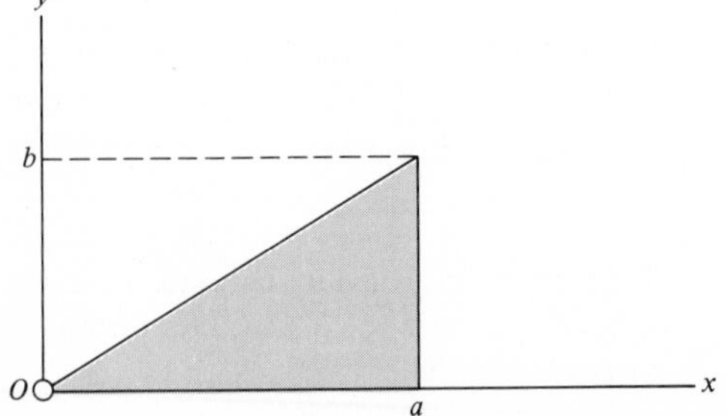

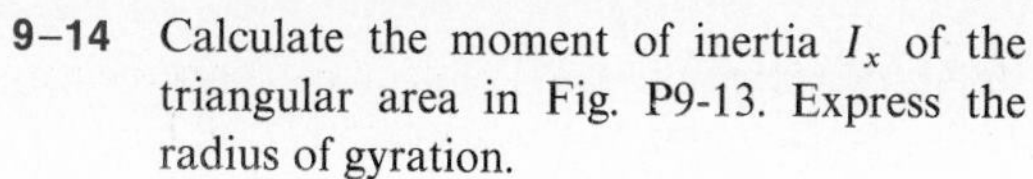
9–14 Calculate the moment of inertia I_x of the triangular area in Fig. P9-13. Express the radius of gyration.

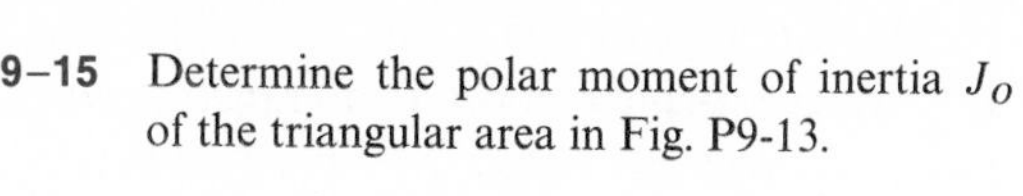
9–15 Determine the polar moment of inertia J_O of the triangular area in Fig. P9-13.

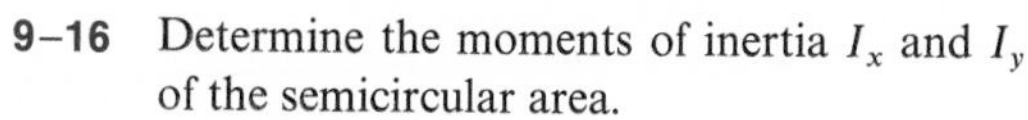
9–16 Determine the moments of inertia I_x and I_y of the semicircular area.

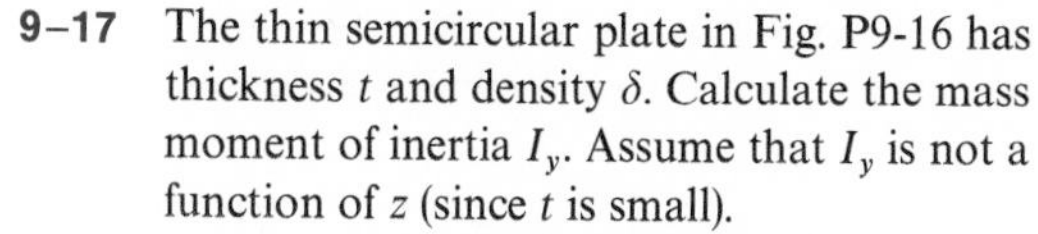
9–17 The thin semicircular plate in Fig. P9-16 has thickness t and density δ. Calculate the mass moment of inertia I_y. Assume that I_y is not a function of z (since t is small).

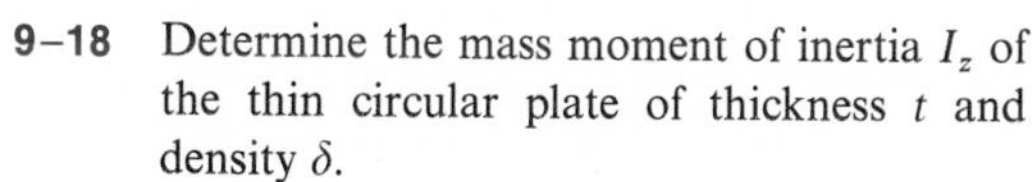
9–18 Determine the mass moment of inertia I_z of the thin circular plate of thickness t and density δ.

9–19 Calculate the moments of inertia I_x and I_y of the circular area in Fig. P9-18.

9–20 Determine the polar moment of inertia J_O of the circular area in Fig. P9-18 using two methods: (a) direct integration (Eq. 9-8a), and (b) using Eq. 9-8b.

9–21 Calculate the mass moment of inertia I_x of the thin, symmetric plate of thickness t and density δ. Assume that I_x is not a function of z (since t is small).

9–22 Determine the moment of inertia I_y of the shaded area in Fig. P9-21. Express the radius of gyration.

FIGURE P9-16

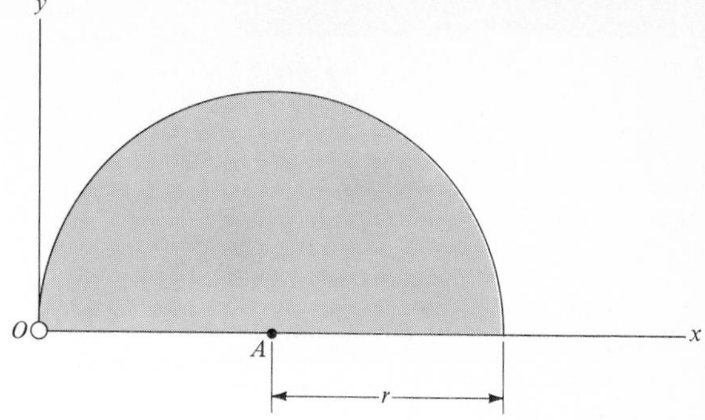

FIGURE P9-18

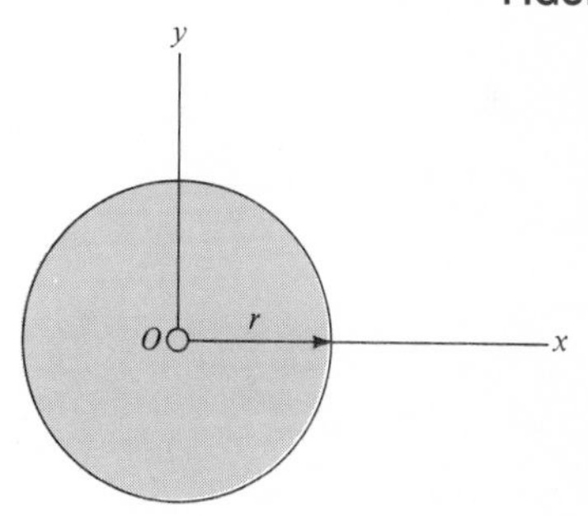

FIGURE P9-21

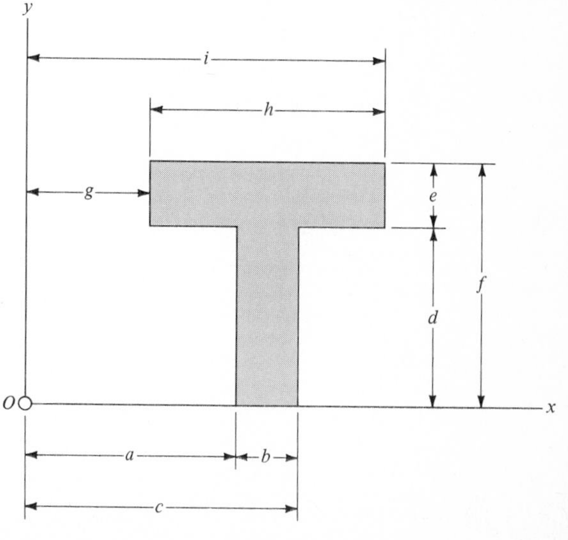

9-4 PARALLEL-AXIS THEOREM

In problems involving moments of inertia it is often convenient to first calculate the moment of inertia about a centroidal axis and then transform this to a moment of inertia about a required axis which is parallel to the given centroidal axis. The transformation can be done using a simple formula based on the *parallel-axis theorem*,

$$I = I_C + Md^2 \qquad \text{for a mass } M \tag{9-9}$$

$$I = I_C + Ad^2 \qquad \text{for an area } A \tag{9-10}$$

where I = moment of inertia of mass M or area A about any line l
I_C = moment of inertia of mass M or area A about a line through the centroid of M or A and parallel to line l
d = perpendicular distance between the two parallel lines

The parallel-axis theorem is proved in general terms for the rectangular area A shown in Fig. 9-8. Here the axes x' and y' are centroidal axes of the area A. Assume that the moments of inertia with respect to the axes x and y must be determined. Considering the x axis first, $dI_x = (y' + b)^2\, dA$ by definition of the moment of inertia (Eq. 9-4). For the whole area,

$$\begin{aligned} I_x &= \int_A (y' + b)^2\, dA \\ &= \int_A y'^2\, dA + 2b \int_A y'\, dA + b^2 \int_A dA \end{aligned}$$

Note that $\int_A y'\, dA = 0$ since it is the first moment of the area A about the x' axis which is a centroidal axis (the first moment is always zero about any centroidal axis). Thus,

$$I_x = I_{x'} + Ab^2 \tag{9-11}$$

Using the same procedure in reference to the y axis,

$$I_y = I_{y'} + Aa^2 \tag{9-12}$$

The method is also applicable to transformations of the polar moment of inertia. With respect to Fig. 9-8,

$$J_O = J_{O'} + Ac^2 = I_x + I_y \tag{9-13}$$

The proof of this equation is left as an exercise for the reader. It is important to note that Eqs. 9-11, 9-12, and 9-13 are valid only if one of the two axes in each equation is a centroidal axis since this was used to advantage in the derivation.

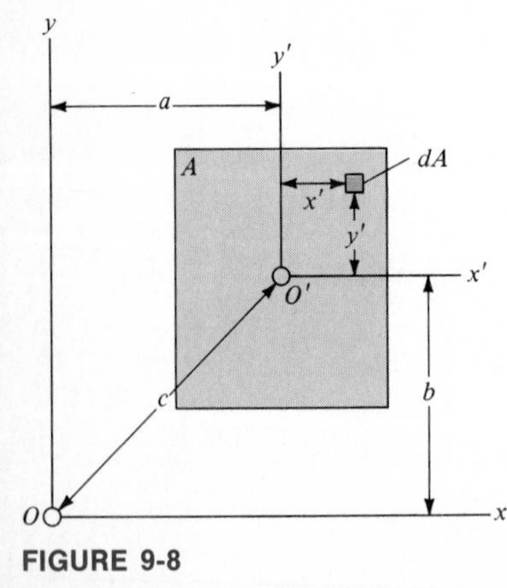

FIGURE 9-8
Area A in the xy plane

9-5 MOMENTS OF INERTIA OF COMPLEX AREAS AND BODIES

In some problems it is necessary to determine the moments of inertia for masses or areas of complex shape. The procedure for these is reasonably easy if the object can be broken down into simple parts in the analysis. For each part, the location of the centroid should be obtained by inspection, and the centroidal moment of inertia should be calculated using a readily available formula (such as that for rectangles). Finally, the centroidal moments of inertia may have to be transformed to a common axis of interest using the parallel-axis theorem. The method is illustrated schematically for the moment of inertia of a complex area. Generally, the details of the calculations can be done in several different ways to obtain the same result.

Assume that it is required to determine the moments of inertia of the channel section about the x and y axes (I_x and I_y) in Fig. 9-9. I_x is calculated in two ways to illustrate that judgment is necessary in finding the most efficient procedure. First break the area down into three rectangles (A and two B) as shown by the dashed lines in Fig. 9-9. For area A, I_{x_A} is its centroidal moment of inertia. For each area B, using Eq. 9-10,

$$I_{x_B} = I_{x'_B} + (ad)\left(b + \frac{d}{2}\right)^2$$

For the whole channel section (since the moments of inertia are additive),

$$I_x = I_{x_A} + 2I_{x_B}$$

In an alternative procedure the channel section is imagined to be the difference of two rectangles, area $C = a(2b + 2d)$ and area $D = c(2b)$. Since the x axis is centroidal to both of these rectangles, the moment of inertia can be calculated without using the parallel-axis theorem. Thus,

$$I_x = I_{x_C} - I_{x_D}$$

Clearly, the latter method is the simpler of the two since it avoids the calculation of $(ad)(b + d/2)^2$.

In the calculation of I_y it is necessary to use the parallel-axis theorem for every part of the total area, so again it is best to minimize the number of parts used. Working with area $C = a(2b + 2d)$ and area $D = c(2b)$,

$$I_y = I_{y_C} + a(2b + 2d)\left(e + \frac{a}{2}\right)^2 + I_{y_D} + c(2b)\left(e + d + \frac{c}{2}\right)^2$$

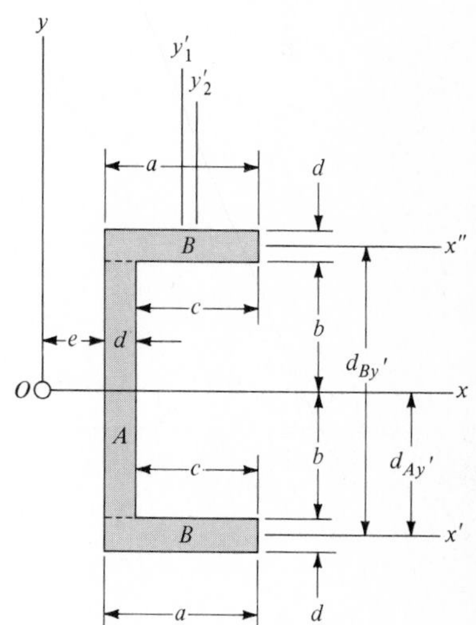

FIGURE 9-9

Channel cross section in the xy plane

EXAMPLE 9-3

Determine the moments of inertia of the rectangle ab in Fig. 9-10 with respect to axes x_1, y_1, and x_2, y_2. Example 9-1 is relevant to this problem.

SOLUTION

Using Eq. 9-10 and the results of Ex. 9-1 for the centroidal moments of inertia I_x and I_y,

$$I_{x_1} = I_x + Ad^2 = \frac{1}{12}ab^3 + ab\left(\frac{b}{2}\right)^2$$

$$= \frac{1}{12}ab^3 + \frac{1}{4}ab^3 = \frac{1}{3}ab^3$$

$$I_{y_1} = I_y + Ad^2 = \frac{1}{12}ba^3 + ab\left(\frac{a}{2}\right)^2$$

$$= \frac{1}{12}ba^3 + \frac{1}{4}ba^3 = \frac{1}{3}ba^3$$

$$I_{x_2} = I_x + Ad^2 = \frac{1}{3}ab^3 + ab\left(\frac{b}{2}\right)^2 = \frac{1}{3}ab^3$$

$$I_{y_2} = I_y + Ad^2 = \frac{1}{3}ba^3 + ab\left(\frac{a}{2}\right)^2 = \frac{1}{3}ba^3$$

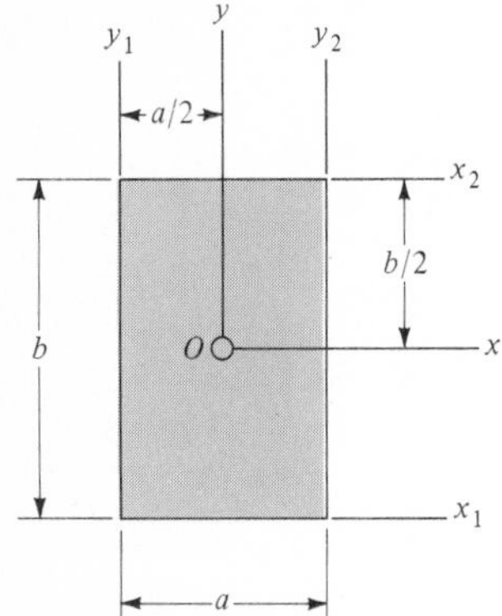

FIGURE 9-10
Area ab in the xy plane

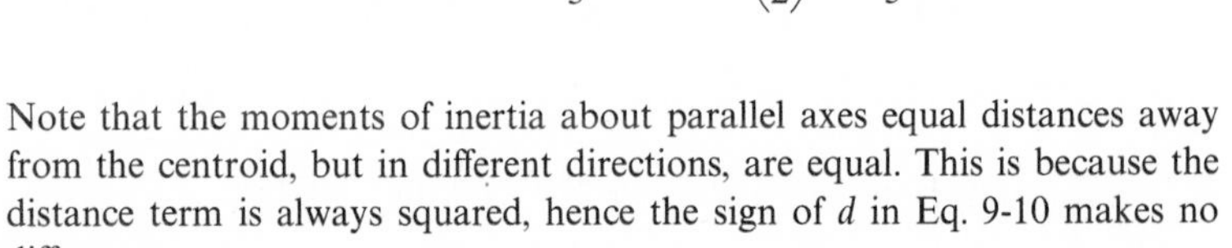

Note that the moments of inertia about parallel axes equal distances away from the centroid, but in different directions, are equal. This is because the distance term is always squared, hence the sign of d in Eq. 9-10 makes no difference.

EXAMPLE 9-4

Determine the moments of inertia of the whole channel section area shown in Fig. 9-9 with respect to axes x, x', and y. $a = 10$ cm, $b = 8$ cm, $c = 9$ cm, $d = 2$ cm, and $e = 4$ cm. Also compute the polar moment of inertia J_O.

SOLUTION

Let I_A and I_B be centroidal moments of inertia of each of the areas A and B.

$$I_x = I_{A_x} + A_A d_{A_y}^2 + 2(I_{B_x} + A_B d_{B_y}^2)$$

$$= \frac{1}{12}(2b)^3 d + 2bd(0) + 2\left[\frac{1}{12}d^3 a + ad\left(b + \frac{d}{2}\right)^2\right]$$

$$= \frac{1}{12}(16 \text{ cm})^3(2 \text{ cm}) + 0 + 2\left[\frac{1}{12}(2 \text{ cm})^3(10 \text{ cm})\right.$$

$$\left. + (10 \text{ cm})(2 \text{ cm})(8 \text{ cm} + 1 \text{ cm})^2\right]$$

$$= 683 \text{ cm}^4 + 2(6.7 \text{ cm}^4 + 1620 \text{ cm}^4)$$

$$= 3936 \text{ cm}^4$$

$$I_y = I_{A_y} + A_A d_{A_x}^2 + 2(I_{B_y} + A_B d_{B_x}^2)$$

$$= \frac{1}{12}(2b)d^3 + 2bd\left(e + \frac{d}{2}\right)^2$$

$$+ 2\left[\frac{1}{12}da^3 + ad\left(e + \frac{a}{2}\right)^2\right]$$

$$= \frac{1}{12}(16 \text{ cm})(2 \text{ cm})^3 + 2(8 \text{ cm})(2 \text{ cm})(4 \text{ cm} + 1 \text{ cm})^2$$

$$+ 2\left[\frac{1}{12}(2 \text{ cm})(10 \text{ cm})^3 + (10 \text{ cm})(2 \text{ cm})(4 \text{ cm} + 5 \text{ cm})^2\right]$$

$$= 10.7 \text{ cm}^4 + 800 \text{ cm}^4 + 2(166.7 \text{ cm}^4 + 1620 \text{ cm}^4)$$

$$= 4384 \text{ cm}^4$$

$$I_{x'} = I_{A_x} + A_A d_{A_{y'}}^2 + I_{B_{x''}} + A_B d_{B_{y'}}^2 + I_{B_{x'}}$$

$$= \frac{1}{12}(16 \text{ cm})^3(2 \text{ cm}) + (16 \text{ cm})(2 \text{ cm})(8 \text{ cm} + 1 \text{ cm})^2$$

$$+ \frac{1}{12}(10 \text{ cm})(2 \text{ cm})^3 + (10 \text{ cm})(2 \text{ cm})$$

$$\times (16 \text{ cm} + 2 \text{ cm})^2 + \frac{1}{12}(10 \text{ cm})(2 \text{ cm})^3$$

$$= 683 \text{ cm}^4 + 2592 \text{ cm}^4 + 6.7 \text{ cm}^4$$

$$+ 6480 \text{ cm}^4 + 6.7 \text{ cm}^4$$

$$= 9768 \text{ cm}^4$$

$$J_O = I_x + I_y = 3936 \text{ cm}^4 + 4384 \text{ cm}^4 = 8320 \text{ cm}^4$$

JUDGMENT OF THE RESULTS

The large magnitude of $I_{x'}$ in relation to I_x might be surprising. It means that rotating the whole channel cross-sectional area about the x' axis would be much more difficult than rotating it about the x axis. This is reasonable considering the r^2 dependence of moments of inertia.

PROBLEMS

SECTIONS 9-4, 9-5

Use the parallel-axis theorem in Probs. 9–23 to 9–44. Take advantage of tabulated values of moments of inertia when appropriate.

9–23 Determine the mass moment of inertia I_z of the uniform slender rod of mass M.

FIGURE P9-23

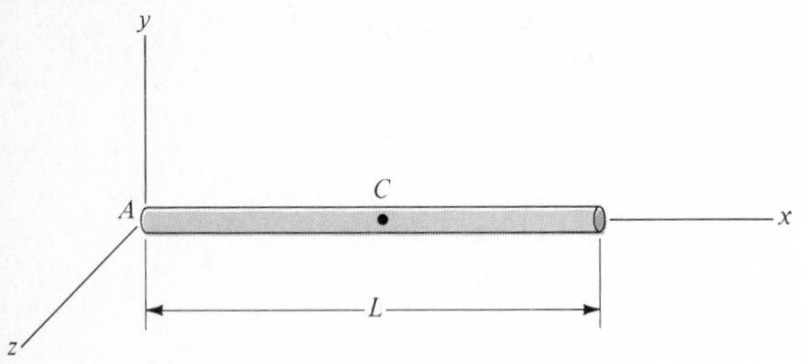

9–24 Determine the mass moment of inertia I_x of the uniform slender rod of mass M.

FIGURE P9-24

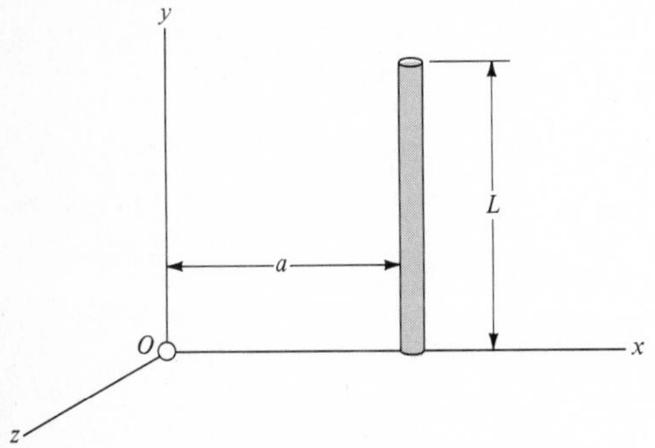

9–25 Calculate the moments of inertia I_x and I_y of the rectangular area *ab*.

FIGURE P9-25

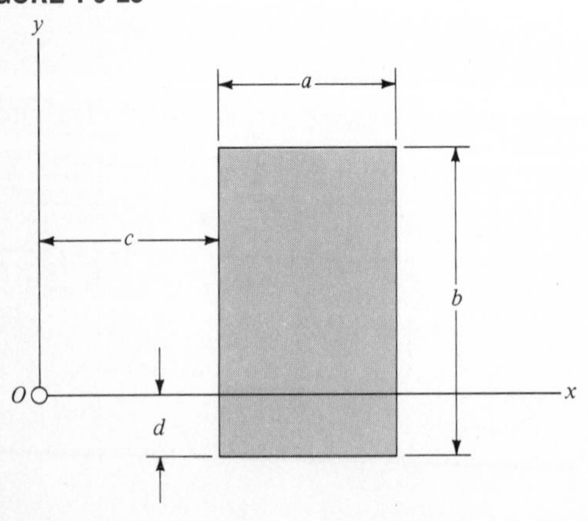

9–26 Determine the polar moment of inertia J_O of the rectangular area *ab* in Fig. P9-25. Use the centroidal polar moment of inertia J_C.

9–27 Calculate the mass moments of inertia I_x, I_y, and I_z of the uniform slender rod of mass M.

FIGURE P9-27

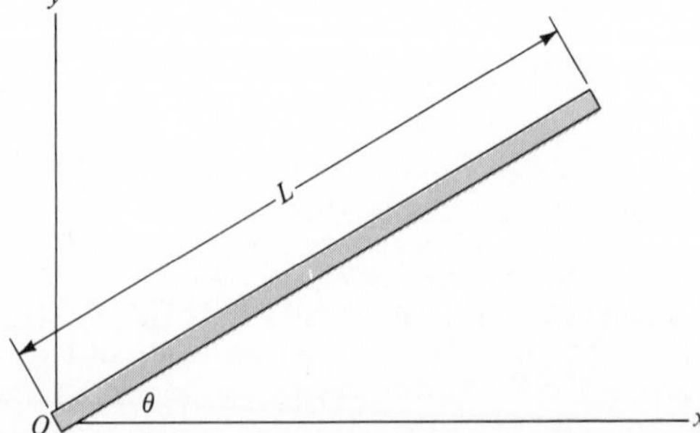

9–28 A thin hoop of radius R and mass M is in the xz plane. Determine the mass moment of inertia I_{y_1} where y_1 is parallel to y at $x = R$.

9–29 The thin cylindrical shell of radius R and mass M is centered on the x axis. Calculate the mass moment of inertia I_{x_1} where x_1 is parallel to x at $y = R$.

9–30 Determine the moments of inertia I_x and I_y of the triangular area.

9–31 Calculate the polar moment of inertia J_O of the triangular area in Fig. P9-30. Determine and use the centroidal polar moment of inertia J_C.

9–32 Calculate the moment of inertia $I_{x'}$ of the semicircular area. Axis x' is parallel to x through the centroid C of the area.

9–33 The polar moment of inertia J_A of the area in Fig. P9-32 is $\frac{1}{4}\pi r^4$. Determine the polar moment of inertia J_O. Note a limitation of the parallel-axis theorem which is relevant to this problem.

9–34 Consider a disk of mass M, radius r, and negligible thickness. Calculate the mass moment of inertia I_{y_1}, where y_1 is parallel to y at $x = 2r$.

9–35 Determine the moment of inertia I_{x_1} of the area in Fig. P9-34. Axis x_1 is parallel to x at $y = r/2$.

FIGURE P9-32

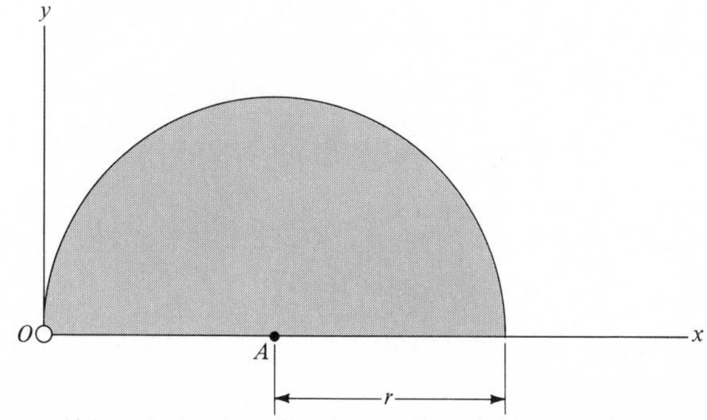

FIGURE P9-28

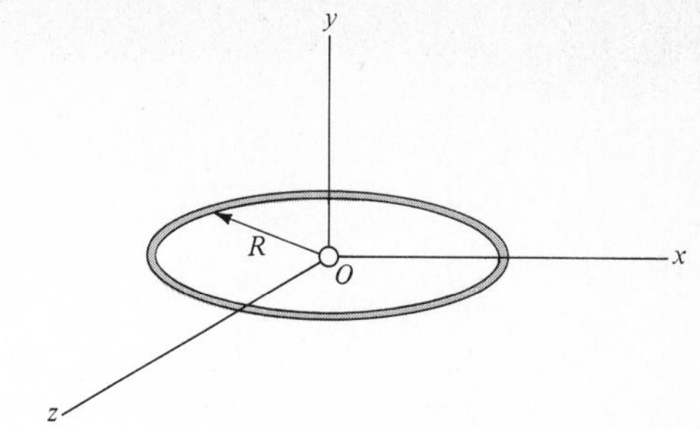

FIGURE P9-29

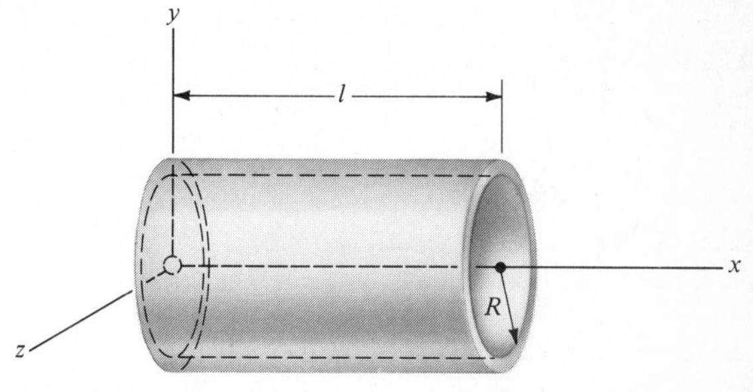

FIGURE P9-30

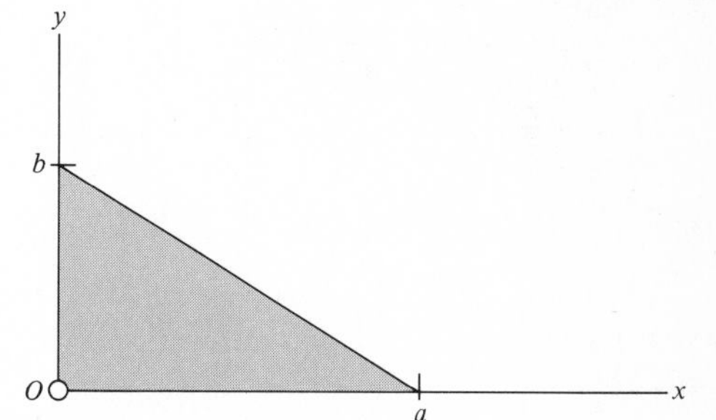

FIGURE P9-34

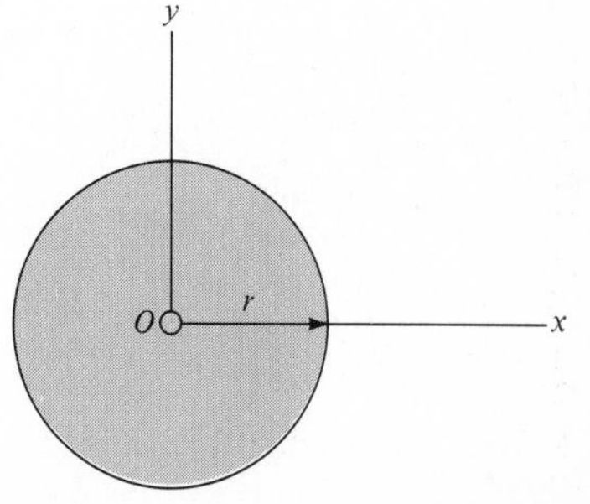

9–36 Calculate the moment of inertia $I_{x'}$ of the shaded area. Axis x' is parallel to x through the centroid of the area.

9–37 A rectangular prism has edges of lengths a, b, and c, with $a = 3b = 4c$. Calculate the mass moment of inertia I_l of the prism if l coincides with one of the edges of length c.

9–38 Determine the moment of inertia I_x of the shaded area. Work with a rectangle ac, from which a triangular piece is removed.

9–39 Determine the moment of inertia I_y of the shaded area in Fig. P9-38. Work with a rectangle ac from which a triangular piece is removed.

9–40 Calculate the moment of inertia I_x of the shaded area.

9–41 Calculate the moment of inertia I_y of the shaded area in Fig. P9-40.

9–42 Assume that the shaded area in Fig. P9-40 represents a thin plate of density δ. The thickness is t in the lower part bd, and $2t$ in the upper part he. Determine the mass moment of inertia I_x.

9–43 Determine the moment of inertia I_x of the shaded area in terms of the dimensions given without units.

9–44 Calculate the moment of inertia I_y in Prob. 9–43.

FIGURE P9-36

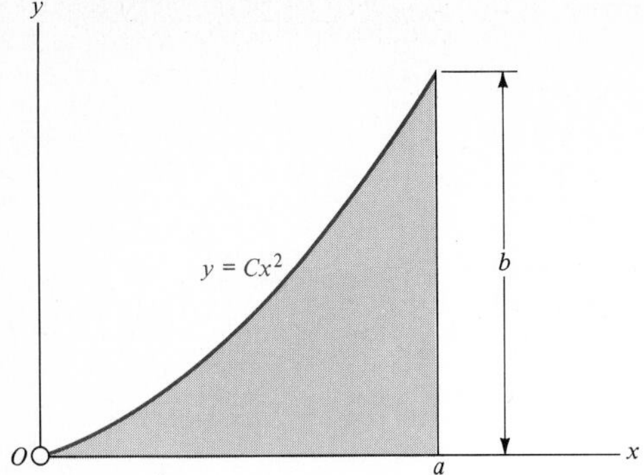

FIGURE P9-38

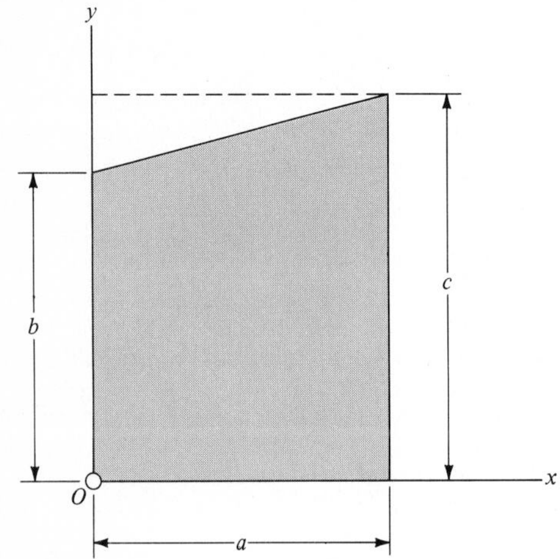

FIGURE P9-40

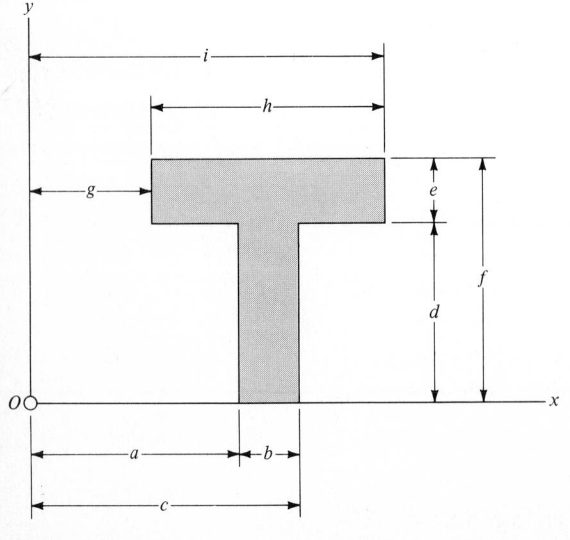

FIGURE P9-43

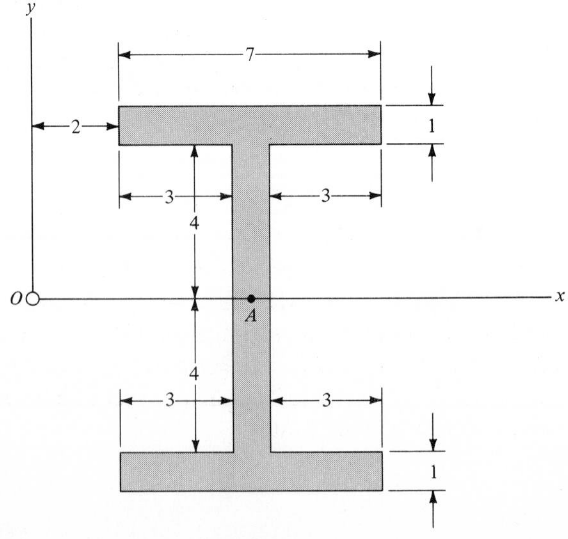

In certain problems of mechanics of materials and dynamics it is necessary to determine the maximum and minimum moments of inertia for areas or masses and the orientations of the corresponding reference axes. For this purpose it is useful to define the *product of inertia* I_{xy}, which is calculated with respect to the x and y axes simultaneously. This is illustrated using areas. For the element of area dA in Fig. 9-11, the product of inertia is defined as $dI_{xy} = xy\, dA$. For the entire area A,

$$I_{xy} = \int xy\, dA \tag{9-14a}$$

Similarly for the other pairs of axes,

$$I_{yz} = \int yz\, dA \tag{9-14b}$$

$$I_{xz} = \int xz\, dA \tag{9-14c}$$

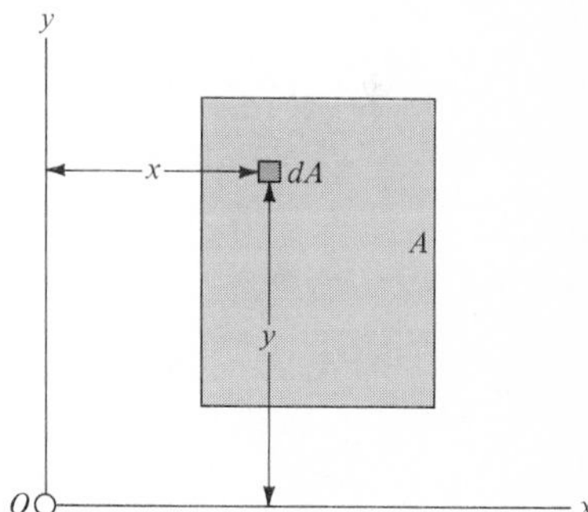

FIGURE 9-11
Area A in the xy plane

The product of inertia has the same units as the moment of inertia of an area, length raised to the fourth power, m^4 or in^4. In contrast to the moment of inertia, which is always positive, the product of inertia may be positive, negative, or zero, since x and y may be zero or of various signs.

The product of inertia I_{xy} is zero if either the x or y axis (or both) is an axis of symmetry of the area. For example, consider the rectangle in Fig. 9-12 that has the y axis as an axis of symmetry. Here any element dA with coordinates $+x$ and $\pm y$ has a corresponding image element dA' with coordinates $-x$ and $\pm y$. The products of inertia for the elements shown in Fig. 9-12 are $xy\, dA$ and $-xy\, dA'$, where the areas dA and dA' are equal. Thus, the sum of the products of inertia for the pair of elements, dA and dA' is zero. On this basis the product of inertia for the whole area A in Fig. 9-12 is zero.

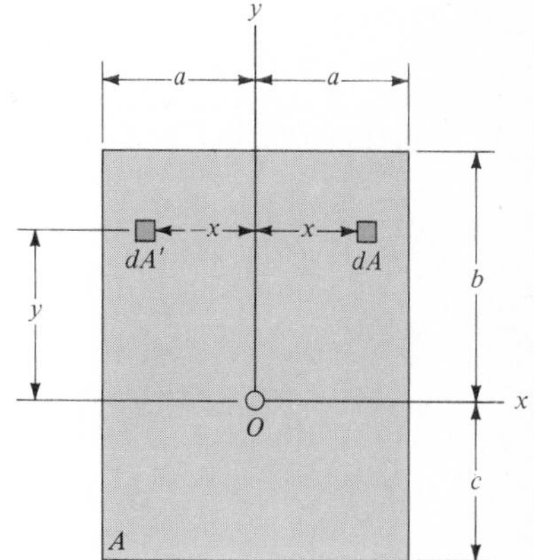

FIGURE 9-12
Area A in the xy plane

Parallel-Axis Theorem

A parallel-axis theorem that is valid for products of inertia clearly must take into account the simultaneous reference to *two* new axes. To derive the appropriate formula, consider the area A in Fig. 9-13. Here the x' and y' axes are centroidal axes of the nonsymmetric area A, and it is desired to find the product of inertia with respect to the set of parallel axes x and y.

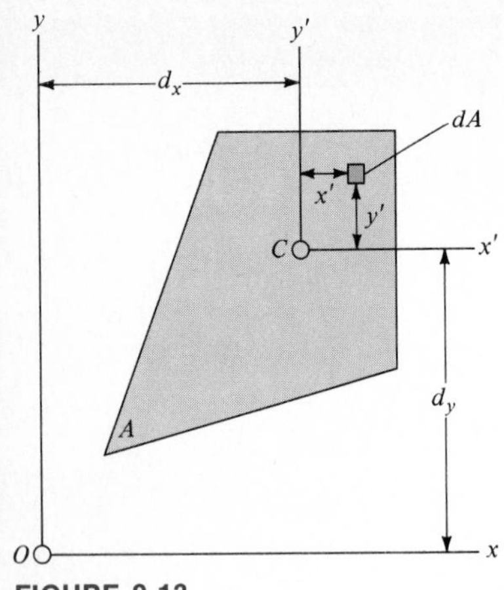

FIGURE 9-13

Area A in the xy plane

The product of inertia of the element dA in Fig. 9-13 can be defined with respect to both the $x'y'$ and the xy system of axes,

$$dI_{x'y'} = x'y'\,dA$$

$$dI_{xy} = (x' + d_x)(y' + d_y)\,dA$$

These can be integrated to obtain the products of inertia for the whole area A,

$$I_{x'y'} = \int x'y'\,dA$$

$$\begin{aligned} I_{xy} &= \int (x' + d_x)(y' + d_y)\,dA \\ &= \int x'y'\,dA + d_x \int y'\,dA \\ &\quad + d_y \int x'\,dA + d_x d_y \int dA \end{aligned}$$

The second and third terms in the last equation are zero because the first moment of any area about a centroidal axis is zero. The fourth integral is equal to the area A, so

$$\boxed{I_{xy} = I_{x'y'} + Ad_x d_y} \qquad \text{9-15a}$$

Similarly for the other pairs of axes,

$$I_{yz} = I_{y'z'} + Ad_y d_z \qquad \text{9-15b}$$

$$I_{xz} = I_{x'z'} + Ad_x d_z \qquad \text{9-15c}$$

where d_x, d_y, and d_z have signs depending on their locations with respect to the centroidal reference system. This equation is similar to the parallel-axis theorem for moments of inertia and it is convenient to use in dealing with composite areas.

Mass Products of Inertia

Mass products of inertia are defined similarly to those for areas. For a body of mass M three products of inertia may be calculated with respect to axes x, y, and z,

$$I_{xy} = \int xy\,dM \qquad I_{yz} = \int yz\,dM \qquad \text{9-16}$$

$$I_{xz} = \int xz\,dM$$

Mass products of inertia may be zero when the body has symmetry, as in the case of areas. The parallel-axis theorem for mass products of inertia is also similar to that for areas, resulting in

$$I_{xy} = I_{x'y'} + Md_xd_y \quad \text{(9-17a)}$$

$$I_{yz} = I_{y'z'} + Md_yd_z \quad \text{(9-17b)}$$

$$I_{xz} = I_{x'z'} + Md_xd_z \quad \text{(9-17c)}$$

where x', y', and z' are axes through the center of gravity, and x, y, and z are parallel to them, respectively. d_x, d_y, and d_z are the apparent distances between the pairs of parallel axes as viewed from another coordinate direction (see Ex. 9-7). $I_{x'y'}$, $I_{y'z'}$, and $I_{x'z'}$ are products of inertia with respect to the indicated centroidal axes. As before, Eqs. 9-17 are convenient in dealing with composite bodies.

9-7 PRINCIPAL AXES AND MOMENTS OF INERTIA OF AN AREA

It is sometimes necessary to transform the known moments and product of inertia I_x, I_y, and I_{xy} to refer them to other axes u and v that are inclined to the x and y axes. The transformation can be performed with the aid of Fig. 9-14. The u and v coordinates can be expressed in terms of x, y, and θ,

$$u = x \cos \theta + y \sin \theta$$

$$v = y \cos \theta - x \sin \theta$$

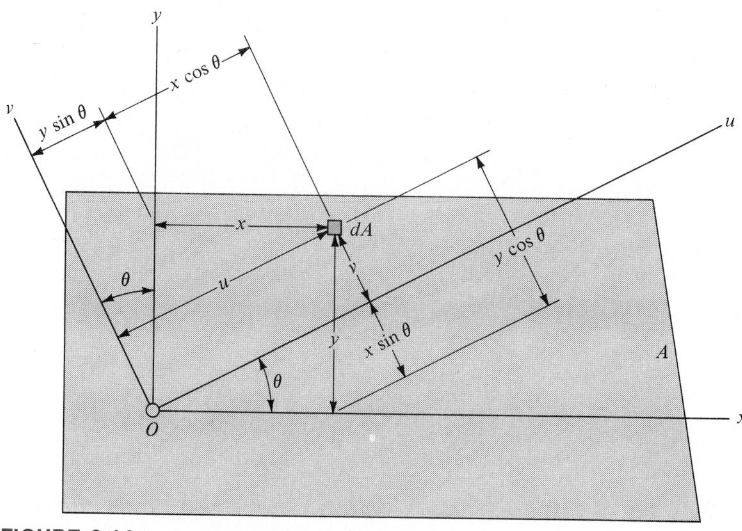

FIGURE 9-14

Area A in the xy plane

Using these transformation equations, the moments and product of inertia of the area A can be written as

$$I_u = \int v^2\,dA = \int (y\cos\theta - x\sin\theta)^2\,dA$$
$$= \cos^2\theta \int y^2\,dA - 2\sin\theta\cos\theta \int xy\,dA$$
$$+ \sin^2\theta \int x^2\,dA$$

Finally, using the fundamental definitions for inertias,

$$I_u = I_x\cos^2\theta - 2I_{xy}\sin\theta\cos\theta + I_y\sin^2\theta \tag{9-18}$$

Following a similar procedure for I_v and I_{uv} yields

$$I_v = I_x\sin^2\theta + 2I_{xy}\sin\theta\cos\theta + I_y\cos^2\theta \tag{9-19}$$

$$I_{uv} = I_x\sin\theta\cos\theta + I_{xy}(\cos^2\theta - \sin^2\theta) - I_y\sin\theta\cos\theta \tag{9-20}$$

After adding Eqs. 9-18 and 9-19 it is found that

$$I_u + I_v = I_x + I_y = J_O \tag{9-21}$$

This is reasonable since the polar moment of inertia J_O for a given area should depend only on its position with respect to the reference point O (Eq. 9-8) and not on the orientation of any axes through that point.

It is worthwhile to change Eqs. 9-18, 9-19, and 9-20 using the trigonometric identities $\sin 2\theta = 2\sin\theta\cos\theta$ and $\cos 2\theta = \cos^2\theta - \sin^2\theta$. The resulting equations are

$$I_u = \frac{I_x + I_y}{2} + \frac{I_x - I_y}{2}\cos 2\theta - I_{xy}\sin 2\theta \tag{9-22}$$

$$I_v = \frac{I_x + I_y}{2} - \frac{I_x - I_y}{2}\cos 2\theta + I_{xy}\sin 2\theta \tag{9-23}$$

$$I_{uv} = \frac{I_x - I_y}{2}\sin 2\theta + I_{xy}\cos 2\theta \tag{9-24}$$

These can be further manipulated to advantage if the functions of θ are eliminated. A simple but less than obvious procedure involving Eqs. 9-22 and 9-24 involves the following major steps. From Eq. 9-22,

$$\left(I_u - \frac{I_x + I_y}{2}\right)^2 = \left(\frac{I_x - I_y}{2}\cos 2\theta - I_{xy}\sin 2\theta\right)^2$$

Adding this to the square of Eq. 9-24 and noting that $\sin^2 2\theta + \cos^2 2\theta = 1$,

$$\left(I_u - \frac{I_x + I_y}{2}\right)^2 + I_{uv}^2 = \left(\frac{I_x - I_y}{2}\right)^2 + I_{xy}^2 \tag{9-25}$$

Here I_x, I_y, and I_{xy} may be assumed known constants or relatively easily calculated quantities. It is required to determine the unknown quantities I_u and I_{uv}. This task is facilitated by substituting in Eq. 9-25

$$\frac{I_x + I_y}{2} = I_{ave} \quad \text{and} \quad \left(\frac{I_x - I_y}{2}\right)^2 + I_{xy}^2 = R^2 \tag{9-26}$$

Here I_{ave} is the average of I_x and I_y. With these, Eq. 9-25 becomes

$$\boxed{(I_u - I_{ave})^2 + I_{uv}^2 = R^2} \tag{9-27}$$

which is the equation of a circle of radius R in the I_u and I_{uv} coordinate system as shown in Fig. 9-15. This is called *Mohr's circle* after the German engineer Otto Mohr (1835–1918). Since I_x, I_u, and so on, are related and their units are all the same, these quantities may be plotted in the same diagram. I_{ave} is the location of the center of the circle on the I_u axis, and it is always on that axis according to Eq. 9-27 (since I_{uv} is squared directly). The moments of inertia labeled I_{max} and I_{min} in Fig. 9-15 are called the *principal moments of inertia* of the area A with respect to point O in Fig. 9-14. Their magnitudes can be obtained from Eq. 9-27 after setting $I_{uv} = 0$. In detail, the extreme values of I_u are

$$\boxed{I_{\substack{max \\ min}} = \frac{I_x + I_y}{2} \pm \sqrt{\left(\frac{I_x - I_y}{2}\right)^2 + I_{xy}^2}} \tag{9-28}$$

Another item of interest is the location of the principal axes x' and y', which are the reference axes for I_{max} and I_{min}. The simplest case is when the area has at least one axis of symmetry. Since $I_{uv} = 0$ was used in deriving Eq. 9-28, and since the product of inertia is zero with respect to any axis of symmetry, *any axis of symmetry is a principal axis of the area.* The reverse is not necessarily true, so a principal axis may or may not be an axis of symmetry.

The problem of locating the principal axes is more complex when the area has no symmetry. For the analysis of such a case, assume that I_x, I_y, and I_{xy} are relatively easy to calculate for the area A in Fig. 9-16a, so these are considered as known quantities. The circle representing Eq. 9-27 can be drawn with the following considerations involving I_x, I_y, and I_{xy}.

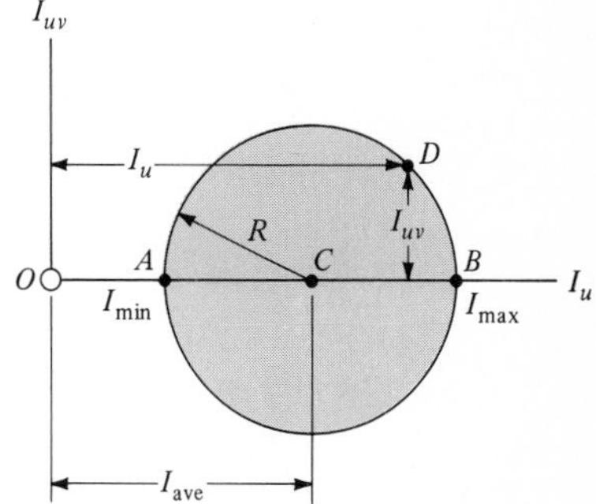

FIGURE 9-15

Mohr's circle in the plane of the I_u and I_{uv} axes

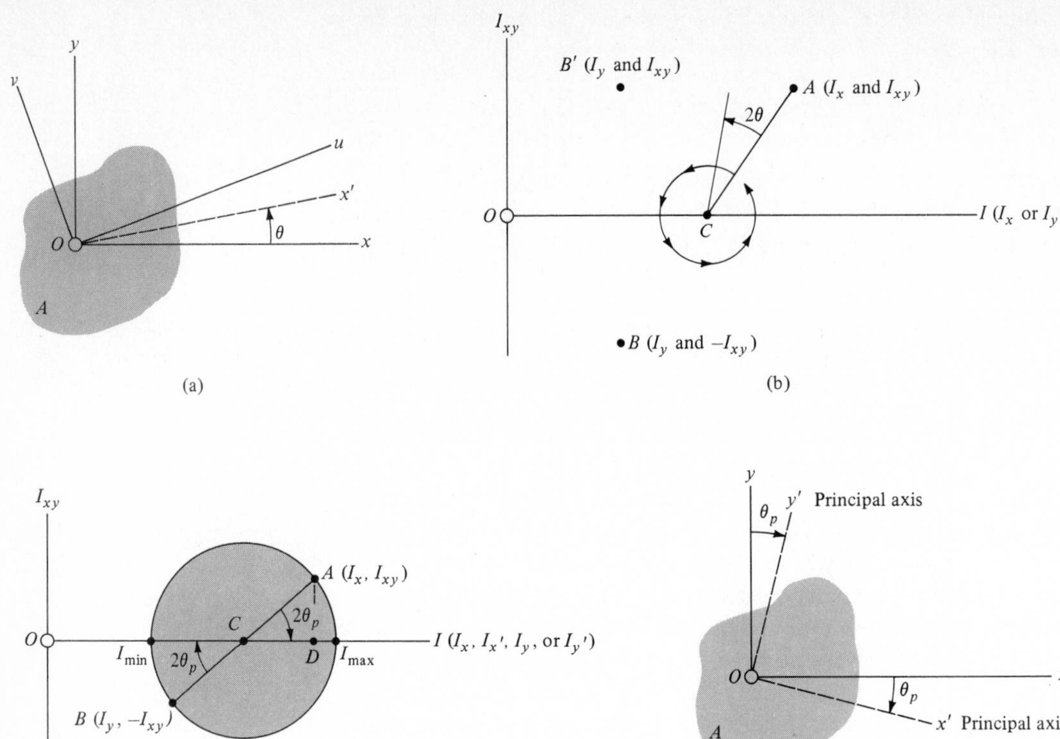

FIGURE 9-16

Principal axes and moments of inertia of area A in the xy plane

1. The center of the circle must be on the horizontal I axis, and its location is at $C = (I_x + I_y)/2$ in Fig. 9-16b.

2. I_x and I_{xy} must be plotted at a single point (assume point A in Fig. 9-16b) since they must simultaneously satisfy Eq. 9-27 as a point on the circle.

3. I_y and I_{xy} must also be at a single point, and on the same circle as I_x and I_{xy}, since I_x, I_y, and I_{xy} refer to the same area. It is reasonable to assume at this stage that point B' in Fig. 9-16b could represent I_y and I_{xy}. However, careful thought leads to the finding that point B with coordinates I_y and $-I_{xy}$ is the correct point on the circle. The reasons are outlined as follows.

a. I_y and $-I_{xy}$ satisfy Eq. 9-27 just as I_y and I_{xy} do since the product of inertia is squared.

b. The x and y axes are 90° apart in Fig. 9-16a, but after careful consideration they "appear" to be 180° apart in the Mohr's circle representation of Fig. 9-16b. Clearly, a rotation through 360° would be required in Fig. 9-16b to go along the circle from A and return to A. However, a 180° rotation of the x axis in Fig. 9-16a produces the same reference axis for I_x and I_{xy}. In other words, I_x and I_{xy} are not changed if the reference axis x is rotated 180° about the origin of the coordinate system. Thus, any rotation θ in Fig. 9-16a must be represented by a rotation 2θ in the construction of Mohr's circle (Fig. 9-16b). With these considerations, point B is the logical choice for I_y and I_{xy} in Fig. 9-16b.

The location of the principal axes with respect to the known axes x and y is now straightforward using Fig. 9-16c. The principal axes are reached by a clockwise rotation through an angle $2\theta_p$ in Mohr's circle. From Fig. 9-16c,

$$\tan 2\theta_p = \frac{AD}{CD} = \frac{2I_{xy}}{I_x - I_y} \qquad \boxed{9\text{-}29}$$

The locations of the principal axes with respect to area A are obtained in Fig. 9-16d by a rotation from the x and y axes through an angle θ_p. The direction of this rotation is the same as in Mohr's circle.

9-8 TRANSFORMATION OF MOMENTS OF INERTIA OF A MASS

The dynamics analysis of bodies in some situations involves the concept of *principal moments of inertia of a body*. This is similar to the principal moments of inertia for an area, and it is reasonable to use here in preparation for a course in dynamics. The basic problem again is to transform known (or relatively easily calculated) moments of inertia to refer them to other axes of interest.

The transformation procedure is explained with the aid of Fig. 9-17. A body of total mass M is outlined here with respect to the xyz coordinate system. Assume that the moment of inertia of the body should be determined with respect to the line l. By definition, $I_l = \int \rho^2 \, dM$, where ρ is the shortest distance from the elemental mass dM to the line l. It is advantageous to replace ρ by the vector product $\mathbf{n} \times \mathbf{q}$, where $\mathbf{n}$ is a unit vector along the line l (this is allowed since $\rho = q \sin \theta$, the magnitude of $\mathbf{n} \times \mathbf{q}$). In terms of the rectangular components n_x, n_y, and n_z of $\mathbf{n}$, and x, y, and z of the mass dM,

$$I_l = \int [(n_x y - n_y x)^2 + (n_y z - n_z y)^2 + (n_z x - n_x z)^2] \, dM \qquad \boxed{9\text{-}30a}$$

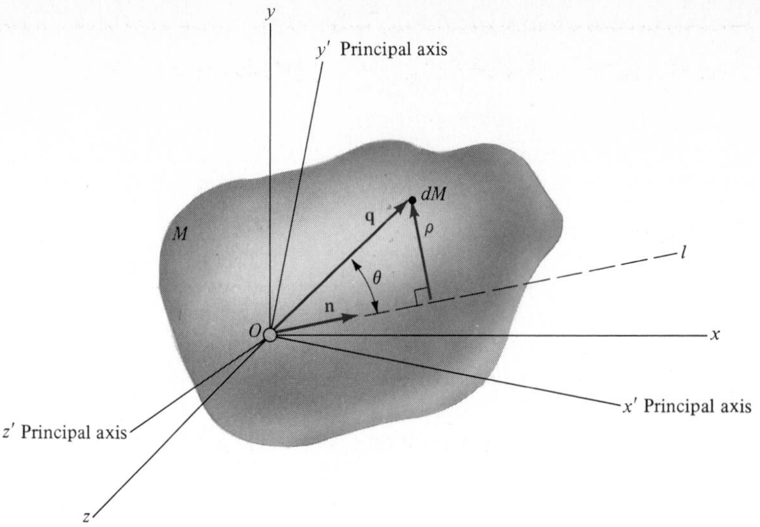

FIGURE 9-17

Transformation of moments of inertia of mass M in the xyz frame

This can be written as

$$\begin{aligned} I_l = n_x^2 \int (y^2 + z^2)\, dM + n_y^2 \int (z^2 + x^2)\, dM \\ + n_z^2 \int (x^2 + y^2)\, dM - 2n_x n_y \int xy\, dM \\ - 2n_y n_z \int yz\, dM - 2n_z n_x \int zx\, dM \end{aligned}$$

With $\int (y^2 + z^2)\, dM = I_x$, $\int xy\, dM = I_{xy}$, and so on,

$$\begin{aligned} I_l = I_x n_x^2 + I_y n_y^2 + I_z n_z^2 \\ - 2I_{xy} n_x n_y - 2I_{yz} n_y n_z - 2I_{zx} n_z n_x \end{aligned} \qquad \boxed{9\text{-}30b}$$

For the case where principal moments of inertia about principal axes x', y', and z' are used, $I_{x'y'} = I_{y'z'} = I_{z'x'} = 0$. Thus, Eq. 9-30b reduces to

$$\boxed{I_l = I_{x'} n_{x'}^2 + I_{y'} n_{y'}^2 + I_{z'} n_{z'}^2} \qquad \boxed{9\text{-}31}$$

This equation is valid for bodies of any shape since even irregular bodies have maximum and minimum values of their moments of inertia. Naturally, it is easiest to work with symmetric bodies for which the principal axes can be established by inspection. These equations are also valid for the moment of inertia of an area about a line l where the appropriate quantities for the area are used. Note that the principal moments of inertia for objects of simple shape are available in Table 4, inside the back cover.

EXAMPLE 9-5

Determine the product of inertia I_{xy} of the triangle shown in Fig. 9-18.

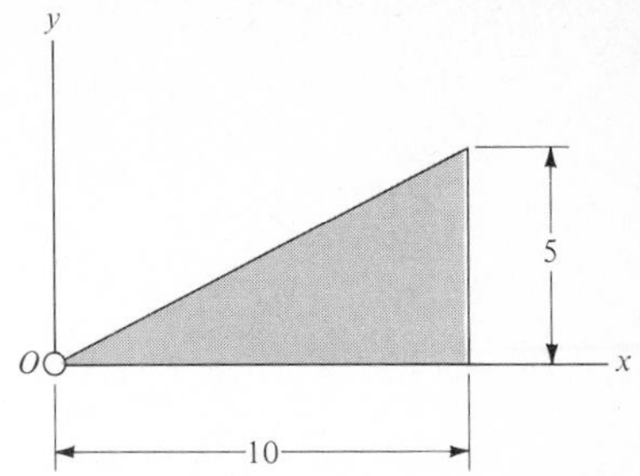

FIGURE 9-18

Triangular area in the *xy* plane

SOLUTION

Choosing an elemental area of width dx,

$$I_{xy} = \int xy\, dA = \int_0^{10} x\left(\frac{y}{2}\right) y\, dx$$

From Fig. 9-18, $y = \frac{1}{2}x$; hence

$$I_{xy} = \int_0^{10} x\left(\frac{x}{4}\right)\left(\frac{x}{2}\right) dx = \frac{x^4}{32}\bigg|_0^{10} = \frac{10{,}000}{32} = 312.5$$

EXAMPLE 9-6

Calculate the principal moments of inertia of the area shown in Fig. 9-19. Determine the orientation of the principal axes.

SOLUTION

The moments and product of inertia of the composite area are calculated with respect to the given xy axes. Partitioning the area as shown and using the parallel-axis theorem for each area,

$$\begin{aligned} I_x &= I_{x_1} + I_{x_2} \\ &= \frac{1}{12}(10\text{ cm})(1\text{ cm})^3 + (10\text{ cm}^2)(0.5\text{ cm})^2 \\ &\quad + \frac{1}{12}(1\text{ cm})(11\text{ cm})^3 + (11\text{ cm}^2)(6.5\text{ cm})^2 \\ &= 579\text{ cm}^4 \end{aligned}$$

$$\begin{aligned} I_y &= I_{y_1} + I_{y_2} \\ &= \frac{1}{12}(10\text{ cm})^3(1\text{ cm}) + (10\text{ cm}^2)(5\text{ cm})^2 \\ &\quad + \frac{1}{12}(1\text{ cm})^3(11\text{ cm}) + (11\text{ cm}^2)(0.5\text{ cm})^2 \\ &= 337\text{ cm}^4 \end{aligned}$$

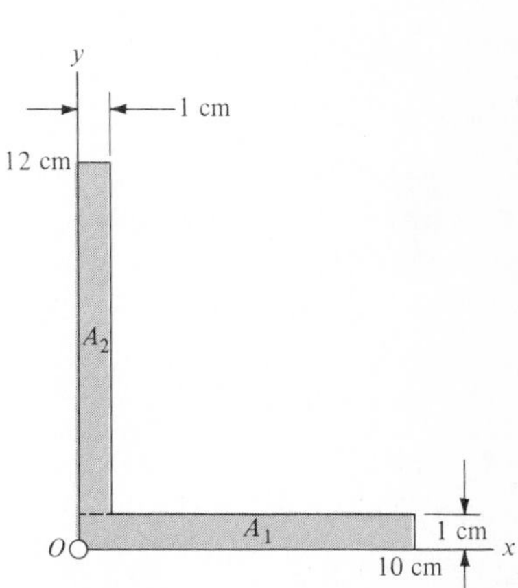

FIGURE 9-19

The product of inertia I_{xy} is found by realizing both composite areas are symmetric about their centroidal axes; hence, the products of inertia about the centroids of these areas are zero for these reference axes. By applying Eq. 9-15 to each area,

$$I_{xy} = I_{xy_1} + I_{xy_2}$$
$$= 0 + (10\text{ cm}^2)(5\text{ cm})(0.5\text{ cm}) + 0$$
$$+ (11\text{ cm}^2)(0.5\text{ cm})(6.5\text{ cm})$$
$$= 60.75\text{ cm}^4$$

Substituting into Eq. 9-28 yields

$$I_{\substack{\max\\\min}} = \frac{I_x + I_y}{2} \pm \sqrt{\left(\frac{I_x - I_y}{2}\right)^2 + I_{xy}^2}$$

$$= \frac{579 + 337}{2} \pm \sqrt{\left(\frac{579 - 337}{2}\right)^2 + (60.75)^2}$$

$$I_{\max} = 593.4\text{ cm}^4 \qquad I_{\min} = 322.6\text{ cm}^4$$

Referring to Fig. 9-16b and Eq. 9-29,

$$\tan 2\theta_p = \frac{I_{xy}}{\dfrac{I_x - I_y}{2}} = \frac{60.75}{\dfrac{579 - 337}{2}} = 0.502$$

Hence, $2\theta_p = 26.7°$ or $-153.3°$ and $\theta_p = 13.3°$ or $-76.7°$, where positive θ is measured clockwise to the I_x axis. By analogy to Fig. 9-16 c and d, rotation of the x axis 13.3° clockwise gives the axis of maximum moment of inertia. The equally rotated y axis is the axis of minimum moment of inertia.

EXAMPLE 9-7

Determine the principal axes and moments of inertia about the centroid of the area shown in Fig. 9-20. Also determine the moments and product of inertia with respect to the uv axes. Use the equations of Mohr's circle.

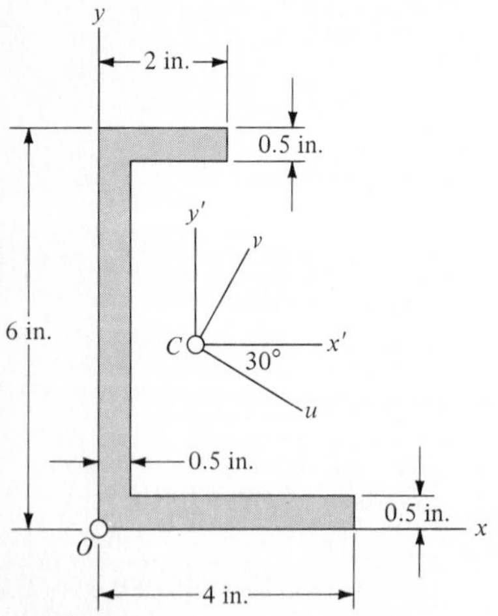

FIGURE 9-20
Area in the xy plane

SOLUTION

It is necessary to determine the moments and product of inertia about the x' and y' axes (parallel to the x and y axes) with the origin at C. First C must be located, however.

$$\bar{x}_C = \frac{\sum \bar{x}_i A_i}{\sum A_i} = \frac{(1\text{ in.})(1\text{ in}^2) + (0.25\text{ in.})(2.5\text{ in}^2) + (2\text{ in.})(2\text{ in}^2)}{1\text{ in}^2 + 2.5\text{ in}^2 + 2\text{ in}^2}$$
$$= 1.02\text{ in.}$$

$$\bar{y}_C = \frac{\sum \bar{y}_i A_i}{\sum A_i} = \frac{(5.75\text{ in.})(1\text{ in}^2) + (3\text{ in.})(2.5\text{ in}^2) + (0.25\text{ in.})(2\text{ in}^2)}{1\text{ in}^2 + 2.5\text{ in}^2 + 2\text{ in}^2}$$
$$= 2.5\text{ in.}$$

Using the parallel-axis theorem (with I_{x_i} as centroidal moments of inertia of subareas A_i),

$$
\begin{aligned}
I_{x'} &= \sum (I_{x_i} + A_i d_{y_i}^2) \\
&= \frac{1}{12}(0.5 \text{ in.})^3(2 \text{ in.}) + (1 \text{ in}^2)(5.75 - 2.5)^2 \text{ in}^2 \\
&\quad + \frac{1}{12}(5 \text{ in.})^3(0.5 \text{ in.}) + (2.5 \text{ in}^2)(3 - 2.5)^2 \text{ in}^2 \\
&\quad + \frac{1}{12}(0.5 \text{ in.})^3(4 \text{ in.}) + (2 \text{ in}^2)(0.25 - 2.5)^2 \text{ in}^2 \\
&= 26.6 \text{ in}^4
\end{aligned}
$$

Similarly,

$$
\begin{aligned}
I_{y'} &= \sum (I_{y_i} + A_i d_{x_i}^2) \\
&= \frac{1}{12}(0.5 \text{ in.})(2 \text{ in.})^3 + (1 \text{ in}^2)(1 \text{ in.} - 1.02 \text{ in.})^2 \\
&\quad + \frac{1}{12}(5 \text{ in.})(0.5 \text{ in.})^3 + (2.5 \text{ in}^2)(0.25 \text{ in.} - 1.02 \text{ in.})^2 \\
&\quad + \frac{1}{12}(0.5 \text{ in.})(4 \text{ in.})^3 + (2 \text{ in}^2)(2 \text{ in.} - 1.02 \text{ in.})^2 \\
&= 6.6 \text{ in}^4
\end{aligned}
$$

$$
\begin{aligned}
I_{x'y'} &= \sum (I_{xy_i} + A_i d_{y_i} d_{x_i}) \\
&= 0 + (1 \text{ in}^2)(5.75 \text{ in.} - 2.5 \text{ in.})(1 \text{ in.} - 1.02 \text{ in.}) \\
&\quad + 0 + (5 \text{ in}^2)(3 \text{ in.} - 2.5 \text{ in.})(0.25 \text{ in.} - 1.02 \text{ in.}) \\
&\quad + 0 + (2 \text{ in}^2)(0.25 \text{ in.} - 2.5 \text{ in.})(2 \text{ in.} - 1.02 \text{ in.}) \\
&= -0.07 \text{ in}^4 + (-1.93 \text{ in}^4) + (-4.41 \text{ in}^4) \\
&= -6.4 \text{ in}^4
\end{aligned}
$$

The principal moments of inertia are obtained by Eq. 9-28,

$$
I_{\substack{\max \\ \min}} = \frac{26.6 + 6.6}{2} \pm \sqrt{\left(\frac{26.6 - 6.6}{2}\right)^2 + (-6.4)^2}
$$

$$
I_{\max} = 28.5 \text{ in}^4, \qquad I_{\min} = 4.73 \text{ in}^4
$$

From Eq. 9-29,

$$
\tan 2\theta = \frac{2I_{x'y'}}{I_{x'} - I_{y'}} = \frac{2(-6.4)}{26.6 - 6.6} = -0.64
$$

$$
2\theta = 147.4^\circ \quad \text{or} \quad -32.6^\circ
$$

$$
\theta = 73.7^\circ \quad \text{or} \quad -16.3^\circ
$$

From Mohr's circle it is seen that the axis of maximum moment of inertia is obtained by rotating the x' axis counterclockwise 16.3°. Using Eqs. 9-22, 9-23, and 9-24, for $\theta = -30^\circ$ (a clockwise rotation from axis x' for the transformation equations),

$$I_u = \frac{26.6 + 6.6}{2} + \frac{26.6 - 6.6}{2} \cos(-60°) - (-6.4) \sin(-60°)$$

$$= 16.1 \text{ in}^4$$

$$I_v = \frac{26.6 + 6.6}{2} - \frac{26.6 - 6.6}{2} \cos(-60°) + (-6.4) \sin(-60°)$$

$$= 17.1 \text{ in}^4$$

$$I_{uv} = \frac{26.6 - 6.6}{2} \sin(-60°) + (-6.4) \cos(-60°)$$

$$= -11.9 \text{ in}^4$$

JUDGMENT OF THE RESULTS

It is important to be careful with the signs of d_x and d_y when computing products of inertia. The correct signs are obtained by going *from* the centroidal axes to the other axes. The correct sign of θ is obtained by remembering that positive θ is measured clockwise from the x' axis going *from* the original axis *to* the new axis (this θ is for the transformation equations).

Check:

$$I_u + I_v = (16.1 + 17.1) \text{ in}^4 = 33.2 \text{ in}^4$$

$$= (26.6 + 6.6) \text{ in}^4 = I_{x'} + I_{y'}$$

EXAMPLE 9-8

A railroad freight car on a test track (Fig. 9-21a) is modeled as a rectangular body of uniform density δ in Fig. 9-21b. Determine the mass moments of inertia of the body with respect to the x axis, line AB, and the diagonal OF. Use Table 4, inside the back cover.

SOLUTION

From Table 4 (inside back cover), the moment of inertia about the x' axis through the centroid of the body is $I_{x'} = (M/12)(b^2 + c^2)$. Using the parallel axis theorem, and d_1 as the nearest distance between the x and x' axes, the mass moment of inertia with respect to the x axis is

$$I_x = I_{x'} + Md_1^2 = \frac{M}{12}(b^2 + c^2) + M\left[\left(\frac{b}{2}\right)^2 + \left(\frac{c}{2}\right)^2\right]$$

$$= \frac{M}{3}(b^2 + c^2)$$

where $M = \delta abc$. Similarly,

$$I_{AB} = I_{y'} + Md_2^2 = \frac{M}{12}(a^2 + c^2) + M\left[\left(\frac{a}{2}\right)^2 + \left(\frac{c}{2}\right)^2\right]$$

$$= \frac{M}{3}(a^2 + c^2)$$

Equation 9-30b is used to calculate the moment of inertia with respect to line OF,

$$I_{OF} = I_x n_x^2 + I_y n_y^2 + I_z n_z^2 - 2I_{xy} n_x n_y - 2I_{yz} n_y n_z - 2I_{xz} n_x n_z$$

(a)

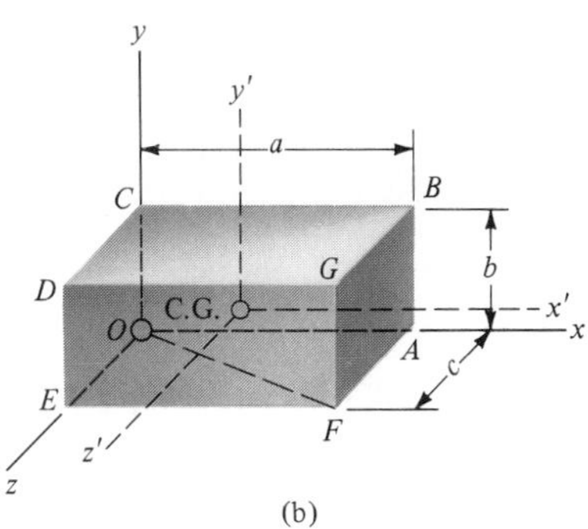

(b)

FIGURE 9-21

Modeling of a railroad freight car for determining its moments of inertia. (Photograph courtesy MTS Systems Corporation, Minneapolis, Minnesota.)

Several of these terms must be calculated yet.

$$I_y = I_{AB} = \frac{M}{3}(a^2 + c^2)$$

$$I_z = I_{z'} + Md_3^2 = \frac{M}{12}(a^2 + b^2) + M\left[\left(\frac{a}{2}\right)^2 + \left(\frac{b}{2}\right)^2\right]$$

$$= \frac{M}{3}(a^2 + b^2)$$

$$I_{xy} = I_{x'y'} + Md_xd_y = 0 + M\left(-\frac{b}{2}\right)\left(-\frac{a}{2}\right) = \frac{M}{4}ab$$

where d_x is the vertical distance in going from the x' to the x axis, as viewed from the z axis, and so on (the sign is taken going *from* the CG).

$$I_{yz} = I_{y'z'} + Md_yd_z = 0 + M\left(-\frac{c}{2}\right)\left(-\frac{b}{2}\right) = \frac{M}{4}bc$$

$$I_{xz} = I_{x'z'} + Md_xd_z = 0 + M\left(-\frac{c}{2}\right)\left(-\frac{a}{2}\right) = \frac{M}{4}ac$$

$$I_{x'y'} = I_{y'z'} = I_{x'z'} = 0$$

because the body is symmetric with respect to the centroidal $x'y'z'$ axes. The unit vector in the direction of OF is

$$\frac{\overrightarrow{OF}}{OF} = \frac{a\mathbf{i} + c\mathbf{k}}{(a^2 + c^2)^{1/2}}, \qquad \text{so} \quad n_x = \frac{a}{(a^2 + c^2)^{1/2}},$$

$$n_y = 0, \qquad n_z = \frac{c}{(a^2 + c^2)^{1/2}}$$

Finally,

$$I_{OF} = \frac{M}{3}(b^2 + c^2)\frac{a^2}{a^2 + c^2} + \frac{M}{3}(c^2 + a^2)(0)$$

$$+ \frac{M}{3}(a^2 + b^2)\frac{c^2}{a^2 + c^2} - 2\left(\frac{M}{4}ab\right)(0)$$

$$- 2\left(\frac{M}{4}bc\right)(0) - 2\left(\frac{M}{4}ac\right)\left(\frac{ac}{a^2 + c^2}\right)$$

$$= \frac{M}{3}\left(\frac{a^2b^2 + b^2c^2}{a^2 + c^2}\right) + \frac{M}{6}\left(\frac{a^2c^2}{a^2 + c^2}\right)$$

$$= \frac{M}{6}\left(2b^2 + \frac{a^2c^2}{a^2 + c^2}\right)$$

JUDGMENT OF THE RESULTS

The units obtained are correct.

PROBLEMS

In Probs. 9–45 to 9–66, calculate the indicated moments of inertia and products of inertia of the area A or mass M. The masses are equal to the volume times the density δ. The thickness of each body is t.

9–45 Area $A = ab$ in Fig. P9-45. $I_{xy} = ?$

FIGURE P9-45
Area A or volume $V = At$

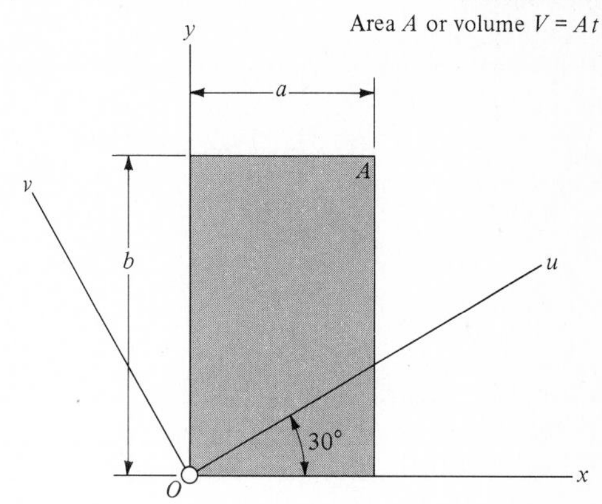

9–46 Body $V = abt$ in Fig. P9-45. $I_{xy} = ?$

9–47 Area $A = ab$ in Fig. P9-45. $I_{uv} = ?$

9–48 Area $A = ab$ in Fig. P9-45. Plot Mohr's circle showing I_x, I_y, I_u, I_v, I_{xy}, I_{max}, and I_{min}. Use simple formulas to calculate any of these quantities when possible.

9–49 Area A in Fig. P9-49. $I_{xy} = ?$

FIGURE P9-49
Area A or volume $V = At$

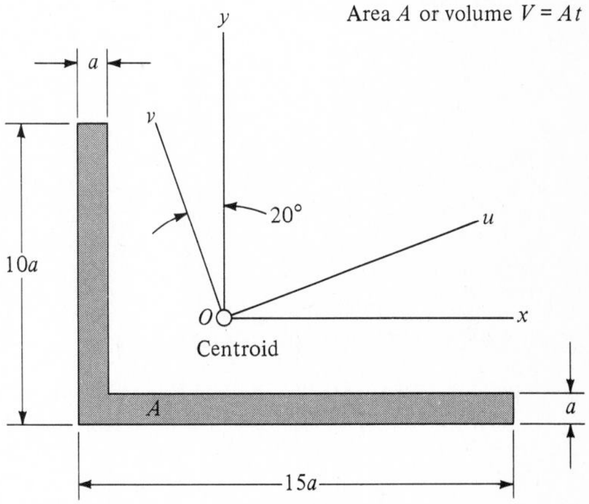

9–50 Body V in Fig. P9-49. $I_{xy} = ?$

9–51 Area A in Fig. P9-49. $I_{uv} = ?$

9–52 Body V in Fig. P9-49. $I_{uv} = ?$

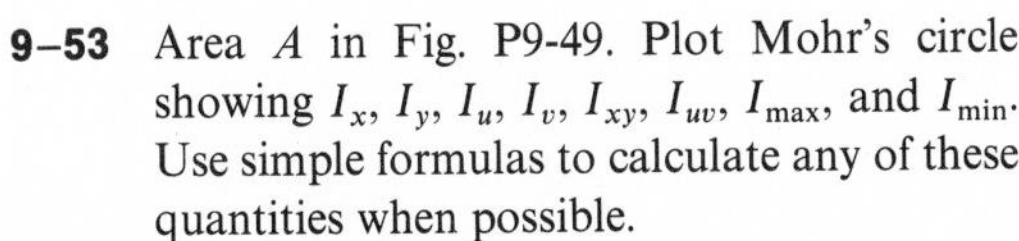
9–53 Area A in Fig. P9-49. Plot Mohr's circle showing I_x, I_y, I_u, I_v, I_{xy}, I_{uv}, I_{max}, and I_{min}. Use simple formulas to calculate any of these quantities when possible.

FIGURE P9-54

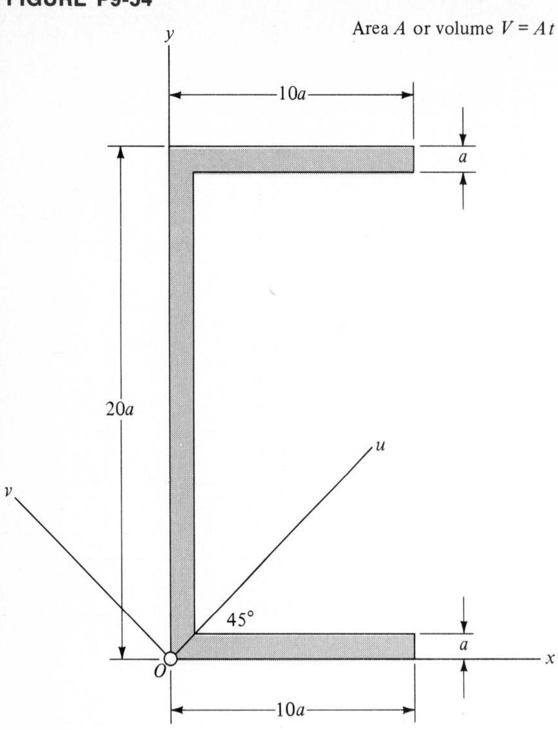

9–54 Area A in Fig. P9-54. $I_{xy} = ?$

9–55 Body V in Fig. P9-54. $I_{uv} = ?$

9–56 Area A in Fig. P9-54. Plot Mohr's circle showing I_x, I_y, I_u, I_v, I_{xy}, I_{uv}, $I_{\max}$, and $I_{\min}$. Use simple formulas to calculate any of these quantities when possible.

9–57 Area A in Fig. P9-57; $\theta = 40°$. $I_{uv} = ?$

9–58 Body V in Fig. P9-57; $\theta = 30°$. $I_{uv} = ?$

9–59 Area A in Fig. P9-59. $I_{xy} = ?$

9–60 Body V in Fig. P9-59. $I_{uv} = ?$

FIGURE P9-57

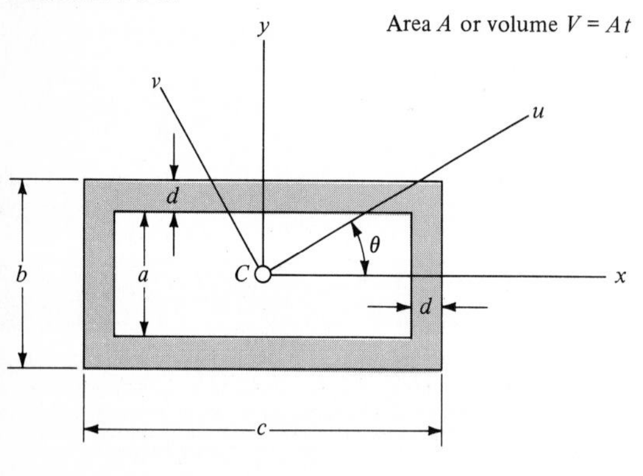

FIGURE P9-59

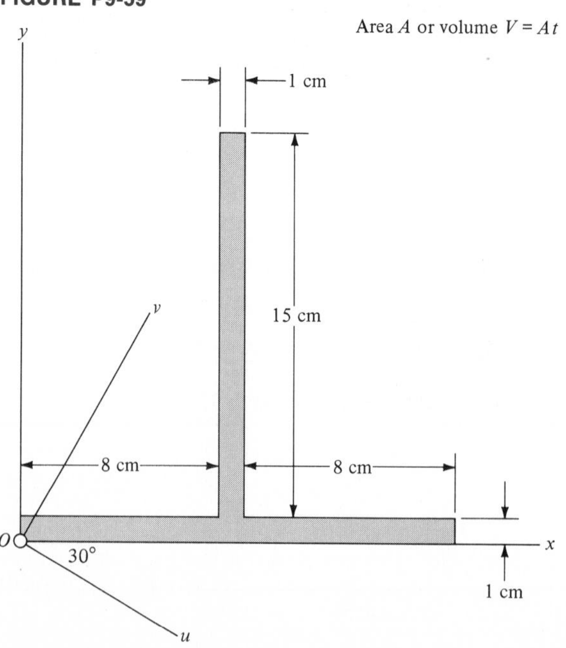

9–61 Area A in Fig. P9-61. $I_{xy} = ?$

9–62 Body V in Fig. P9-61. $I_{xy} = ?$

9–63 Area A in Fig. P9-61. $I_{uv} = ?$

9–64 The thin hoop in the xz plane and the two pegs on it are used as a simple gear. The material weighs 10 N/m, and r is 3 cm. Determine the principal moments of inertia of this gear.

9–65 The thin-walled cylindrical shell is made of steel and weighs 0.283 lb/in^3. Determine the principal moments of inertia of the shell. The origin of the coordinates is at the centroid of the shell.

9–66 A hole of radius $r = 0.2$ in. is drilled at the positive x axis in the shell shown in Fig. P9-65. Calculate the mass products of inertia of the new member. Estimate the principal moments of inertia. Discuss the assumptions and limitations of your analysis.

FIGURE P9-61

Area A or volume $V = At$

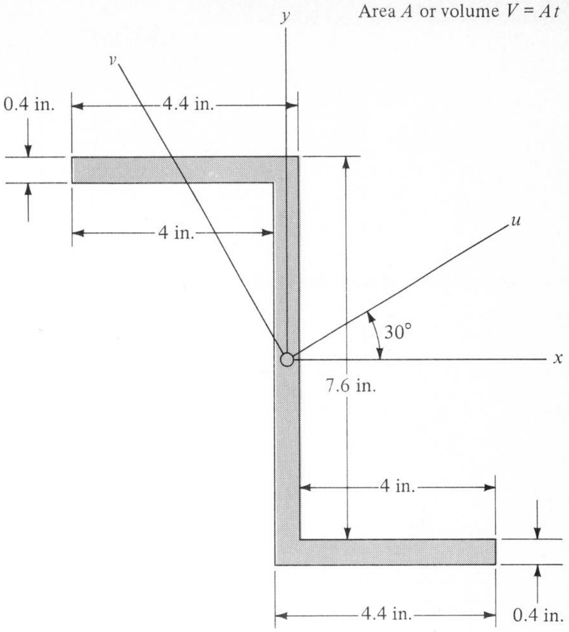

FIGURE P9-64

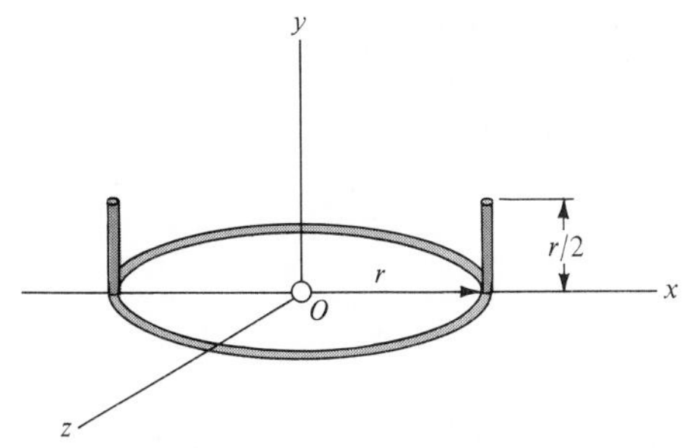

FIGURE P9-65

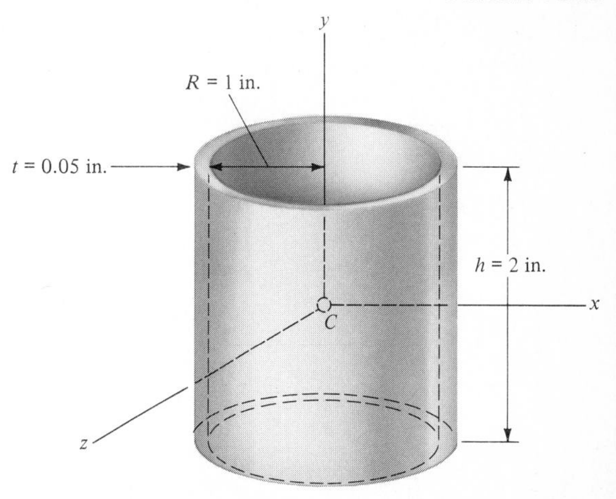

SUMMARY

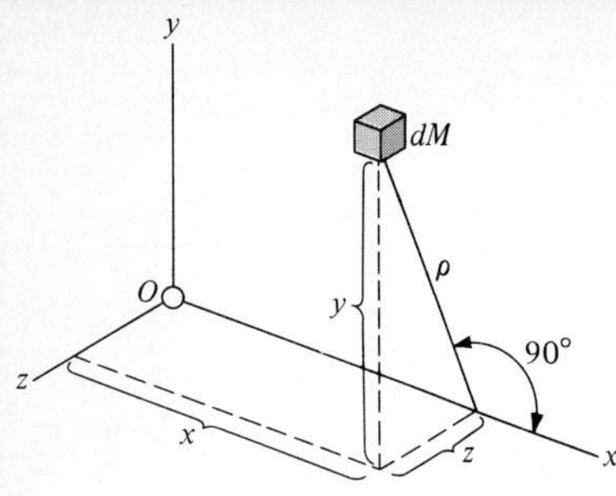

Section 9-1

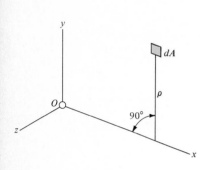

MOMENT OF INERTIA OF A MASS *M*:

$$dI_x = \rho^2\, dM = (y^2 + z^2)\, dM$$

$$I_x = \int \rho^2\, dM = \int (y^2 + z^2)\, dM$$

$$I_y = \int (x^2 + z^2)\, dM$$

$$I_z = \int (x^2 + y^2)\, dM$$

RADIUS OF GYRATION: $r_g = \sqrt{I/M}$, where r_g and I are with respect to the same axis.

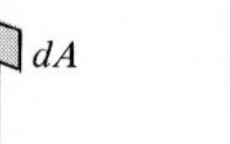

Section 9-2

MOMENT OF INERTIA OF AN AREA *A*:

$$dI_x = \rho^2\, dA$$

$$I_x = \int \rho^2\, dA = \int y^2\, dA \qquad I_y = \int x^2\, dA$$

RADIUS OF GYRATION: $r_g = \sqrt{I/A}$, where r_g and I are with respect to the same axis.

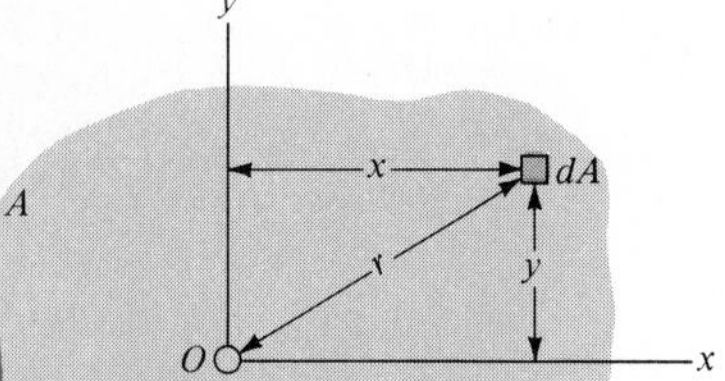

Section 9-3

POLAR MOMENT OF INERTIA OF AN AREA *A*:

$$J_O = \int r^2\, dA = I_x + I_y$$

Section 9-4

PARALLEL-AXIS THEOREM:

For a mass M: $I = I_C + Md^2$

For an area A: $I = I_C + Ad^2$

$$J_O = J_{O'} + Ac^2$$

where I or J_O = moment of inertia of mass M or area A about any appropriate line l

I_C or $J_{O'}$ = moment of inertia of mass M or area A about a line through the centroid of M or A and parallel to line l

c or d = perpendicular distance between the two parallel lines

Note: Each equation of the parallel-axis theorem is valid only if one of the two axes is a centroidal axis.

Section 9-5

PROCEDURE TO DETERMINE MOMENTS OF INERTIA OF COMPLEX AREAS AND BODIES:

1. Break the object down into simple parts for which the centroids can be determined by inspection.
2. Calculate the centroidal moments of inertia of each part. Use tabulated formulas when possible.
3. Transform the individual centroidal moments of inertia to the required common axis using the parallel-axis theorem.

Note: Use judgment in the initial choosing of parts to simplify the calculations throughout the solution.

Section 9-6

PRODUCTS OF INERTIA:

FOR AN AREA *A*:

$$I_{xy} = \int xy\,dA$$

$$I_{yz} = \int yz\,dA$$

$$I_{xz} = \int xz\,dA$$

Parallel-axis theorem:

$$I_{xy} = I_{x'y'} + Ad_x d_y$$

$$I_{yz} = I_{y'z'} + Ad_y d_z$$

$$I_{xz} = I_{x'z'} + Ad_x d_z$$

FOR A MASS *M*:

$$I_{xy} = \int xy\,dM$$

$$I_{yz} = \int yz\,dM$$

$$I_{xz} = \int xz\,dM$$

Parallel-axis theorem:

$$I_{xy} = I_{x'y'} + Md_x d_y$$

$$I_{yz} = I_{y'z'} + Md_y d_z$$

$$I_{xz} = I_{x'z'} + Md_x d_z$$

where $x'y'z'$ are centroidal axes, and xyz are arbitrary axes parallel to $x'y'z'$, respectively. d_x, d_y, and d_z are the apparent distances between the pairs of parallel axes as viewed from another coordinate direction (see Ex. 9-7).

Section 9-7

PRINCIPAL AXES AND MOMENTS OF INERTIA OF AN AREA: The principal axes x' and y' are the reference axes for $I_{\max}$ and $I_{\min}$. The angle θ_p between the principal axes $x'y'$ and given axes xy is obtained from

$$\tan 2\theta_p = \frac{2I_{xy}}{I_x - I_y}$$

The principal moments of inertia $I_{\max}$ and $I_{\min}$ are obtained from

$$I_{\substack{\max \\ \min}} = \frac{I_x + I_y}{2} \pm \sqrt{\left(\frac{I_x - I_y}{2}\right)^2 + I_{xy}^2}$$

Section 9-8

TRANSFORMATION OF MOMENTS OF INERTIA OF A MASS:

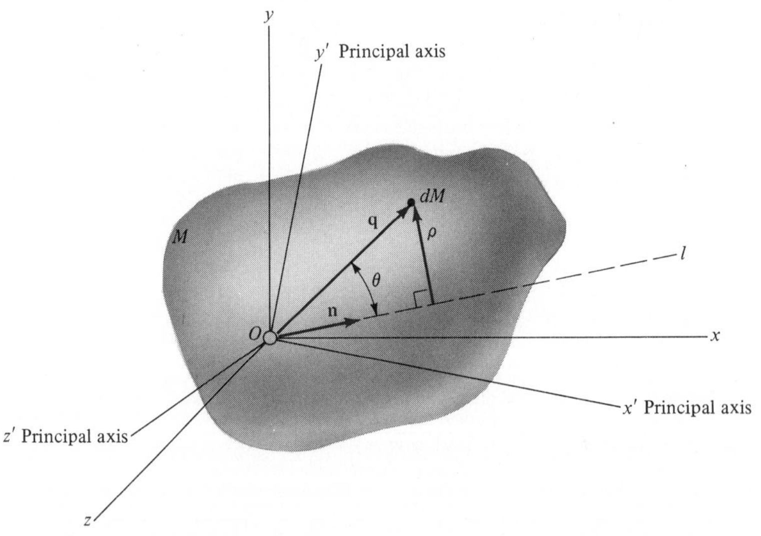

For an arbitrary line l,

$$I_l = I_x n_x^2 + I_y n_y^2 + I_z n_z^2 - 2I_{xy} n_x n_y - 2I_{yz} n_y n_z - 2I_{zx} n_z n_x$$

When principal moments of inertia about principal axes $x'y'z'$ are used,

$$I_l = I_{x'} n_{x'}^2 + I_{y'} n_{y'}^2 + I_{z'} n_{z'}^2$$

REVIEW PROBLEMS

9–67 The thin plate has thickness t and density δ. Determine the mass moment of inertia I_y, assuming that it is not a function of z (since t is small).

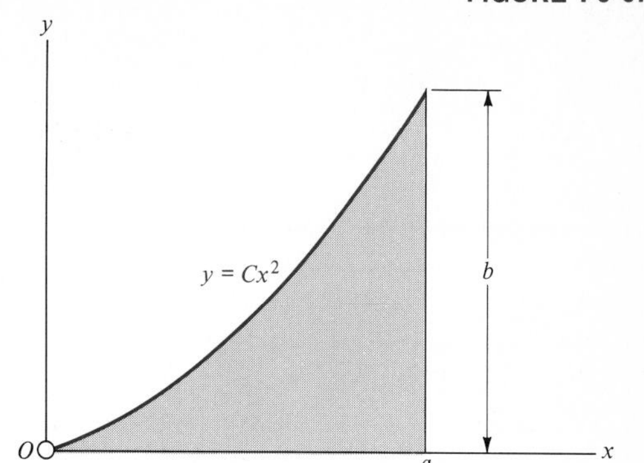

FIGURE P9-67

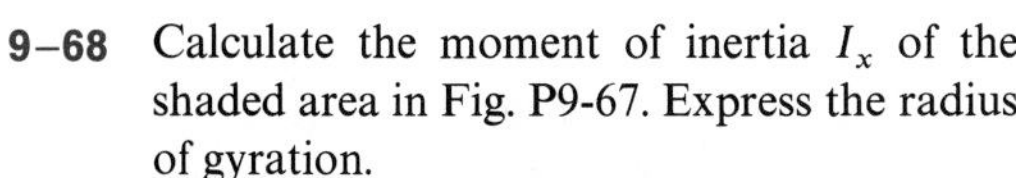

9–68 Calculate the moment of inertia I_x of the shaded area in Fig. P9-67. Express the radius of gyration.

9–69 Determine the polar moment of inertia J_O of the shaded area in Fig. P9-67.

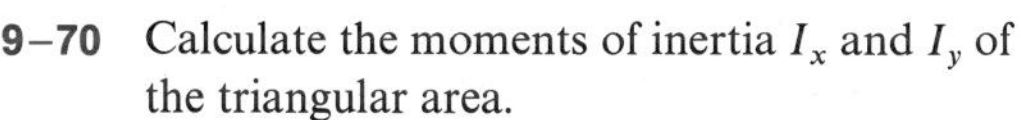

9–70 Calculate the moments of inertia I_x and I_y of the triangular area.

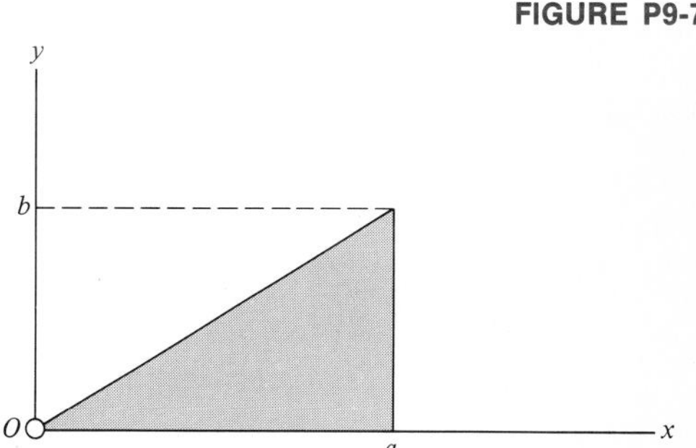

FIGURE P9-70

9–71 Determine the polar moment of inertia J_O of the triangular area in Fig. P9-70. Calculate and use the centroidal polar moment of inertia J_C.

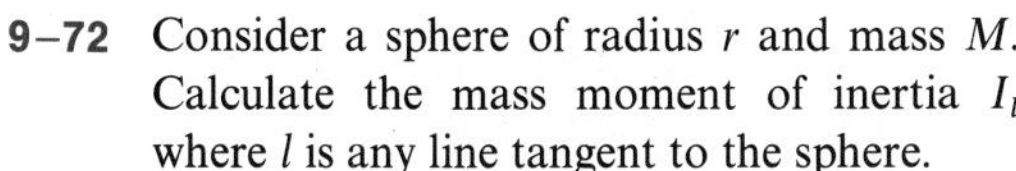

9–72 Consider a sphere of radius r and mass M. Calculate the mass moment of inertia I_l where l is any line tangent to the sphere.

In Probs. 9–73 to 9–75 calculate the indicated moments of inertia and products of inertia of the area A or mass M. The masses are equal to the volume times the density δ. The thickness of each body is t.

9–73 Area A in Fig. P9-73; $\theta = 20°$. $I_{uv} = ?$

9–74 Body V in Fig. P9-73; $\theta = 30°$. $I_{uv} = ?$

9–75 Area A in Fig. P9-73; $\theta = 20°$. Plot Mohr's circle showing I_x, I_y, I_u, I_v, I_{xy}, I_{uv}, I_{max}, and I_{min}. Use simple formulas to calculate any of these quantities when possible.

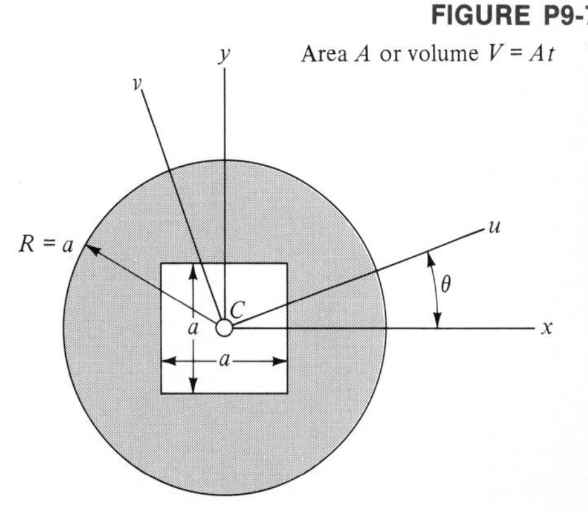

FIGURE P9-73
Area A or volume $V = At$

CHAPTER 10

WORK AND POTENTIAL ENERGY

OVERVIEW

Basic Concepts of Work and Energy

The term *work* has an exact, quantitative meaning when used in the context of mechanics. Quantitatively, *work is the product of the force and the corresponding displacement*. This simple definition has to be carefully interpreted and used. The term *energy* is used to describe the capacity of a system to do work on another system.

These concepts are advantageous for analyzing the equilibrium of complex systems in statics, a variety of motions in dynamics, and the behaviors of engineering members in mechanics of materials. Several practical problem areas of work and energy are introduced here using the flight simulation of aircraft structures. In the old days flight simulation was done by putting people on the wings or applying dead weights at the regions of interest. The modern method is to use numerous hydraulic actuators controlled by electronic equipment to produce a wide range of realistic forces that can vary in magnitude and repetitions. This requires an elaborate reaction-frame structure around the aircraft, as shown in Fig. 10-1.

The strength and performance characteristics of the test equipment must be determined in the design almost as carefully as those of the aircraft to be tested. For example, some wings are highly

FIGURE 10-1

Boeing 747 aircraft in a test setup. (The Boeing Company, Seattle, Washington.)

FIGURE 10-2

Static deflections of a long wing in a test. The maximum upward deflection is 22 ft and the downward deflection is 10 ft (double exposure) from the undeflected position. (The Boeing Company, Seattle, Washington.)

flexible, as shown in Fig. 10-2. The minimum requirement of the test equipment is to apply static forces to the wing over the full range of expected deformations. A more severe requirement is to apply the forces dynamically, even repeatedly (since the wing may vibrate in service). Thus, the concepts of work (force times displacement) and the rate of doing work are needed in designing such tests. In all cases the strength properties of the cantilever wing are also involved in the design of the test equipment.

STUDY GOALS

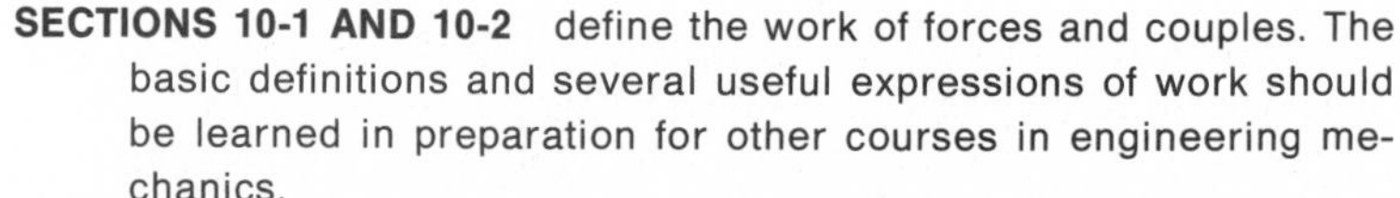

SECTIONS 10-1 AND 10-2 define the work of forces and couples. The basic definitions and several useful expressions of work should be learned in preparation for other courses in engineering mechanics.

SECTION 10-3 presents the concept and method of virtual work. In this analysis, displacements do not actually take place. The use of imaginary small displacements is very appealing philosophically and powerful practically for investigating the equilibrium of complex systems. However, this topic is sometimes omitted from regular statics courses because it is not essential for most subsequent courses.

SECTION 10-4 defines the work of friction forces when relative motion between bodies occurs. The distinction of useful work and frictional losses is used in the concept of mechanical efficiency. This is important in designing machines.

SECTION 10-5 presents the concept of potential energy (the capacity to do work) of a mechanical system because of the effect of gravity on the system or elastic deformability of its members. This topic is especially useful in the study of dynamics.

SECTIONS 10-6 AND 10-7 apply the relationship of work and potential energy to investigate the equilibrium and the stability of a system. These topics are often omitted from standard statics courses because they are not used in subsequent elementary courses.

10-1 WORK OF A FORCE

By definition, work is done by a force when there is a corresponding displacement of a body. There are several aspects of this concept that must be distinguished in engineering analyses.

A simple case is illustrated in Fig. 10-3a where the constant force **F** slowly moves the block from point 1 to 2 along a frictionless inclined plane. If **F** is parallel to the incline, the work U of the force is defined as

$$U = Fs \qquad \text{10-1}$$

where s is the displacement of the block. If **F** is at an angle θ with the plane as in Fig. 10-3b,

$$U = F \cos \theta \, s \qquad \text{10-2}$$

The reason for Eq. 10-2 is that only the component $F \cos \theta$ is associated with the displacement of the block. The other component, $F \sin \theta$, presses the block to the inclined plane but is not involved in the displacement (assuming that the plane is rigidly fixed) and therefore does no work on the block according to the definition stated above.

A more generally useful form of Eq. 10-2 is necessary because both the magnitude of **F** and its direction with respect to the path of the displacement may vary during the displacement. This can be obtained by considering the infinitesimal work dU of a force **F** during an infinitesimal displacement. Work on a particle (such as P in Fig. 10-4) can be defined vectorially as the scalar product of two vectors, which agrees in principle with Eq. 10-2,

$$dU = \mathbf{F} \cdot d\mathbf{r} = F\, dr \cos \theta \qquad \text{10-3}$$

The total work U_{12} in displacing particle P from point 1 to 2 is obtained by integrating Eq. 10-3. Using the rectangular components of **F** and $d\mathbf{r}$,

$$\begin{aligned} U_{12} &= \int_1^2 \mathbf{F} \cdot d\mathbf{r} = \int_1^2 F \cos \theta \, dr \\ &= \int_1^2 (F_x\, dx + F_y\, dy + F_z\, dz) \end{aligned} \qquad \text{10-4}$$

Work is a scalar quantity with units of force times distance, N·m or lb·ft (dimensionally the same as for moments). Work may be

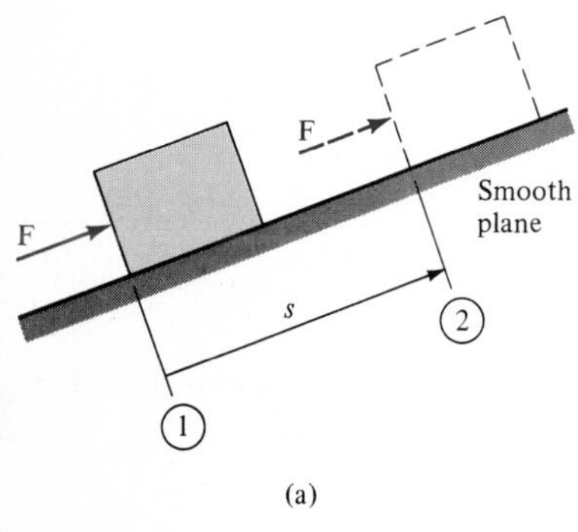

(a)

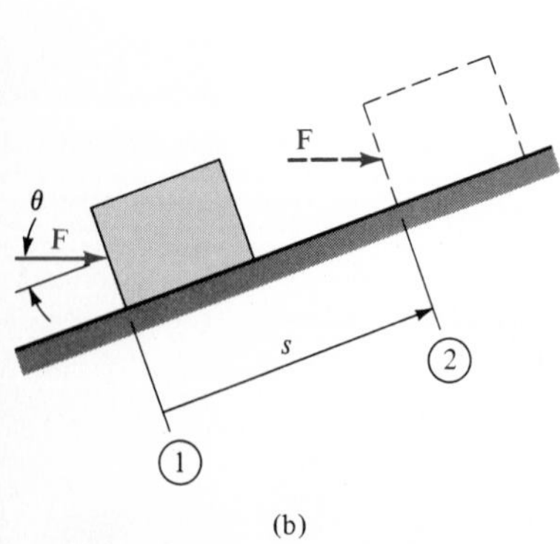

(b)

FIGURE 10-3
Displacement of a block along an incline

positive or negative according to the cosine function in Eq. 10-3. Forces that do no work are common. For example, a force applied to a fixed particle causes no displacement, so its work is zero. Similarly, a force acting perpendicular to the displacement has no component in the direction of the displacement, and its work is also zero.

There are several special aspects of total work that are worth remembering. It is easy to prove that the total work of concurrent forces is equal to the work of the resultant of those forces. This proof is left as an exercise for the reader. Another aspect that should be noted is that the work of internal forces in an inextensible member cancel out. For example, assume that the segment dl of a rigid rod in Fig. 10-5 is displaced to the right through a distance s. If $T_A = T_B$, the positive work of T_B is canceled by the negative work of T_A since they are equal in magnitude.

An interesting and important aspect of total work must be considered when the displacements go through one or more complete cycles. In such a case the total work is zero when each cycle is completed, with no frictional losses during the displacements. For example, the total work done on an object is zero in the complete cycle of lifting it up from the floor and putting it back on the floor. Getting tired in the process is caused by internal losses in the human body and not by doing net work on the object that was moved.

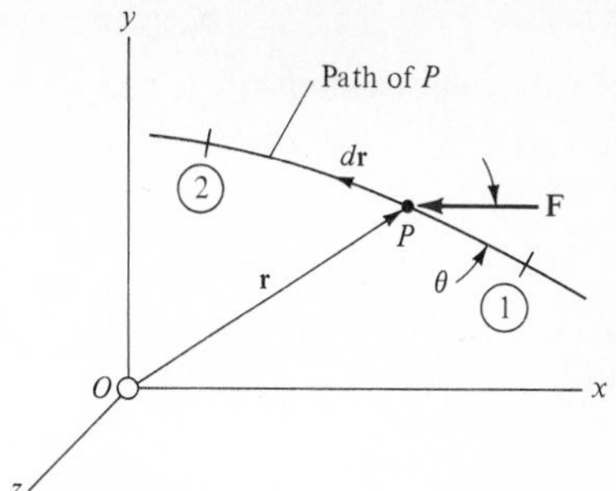

FIGURE 10-4

Infinitesimal displacements of a particle

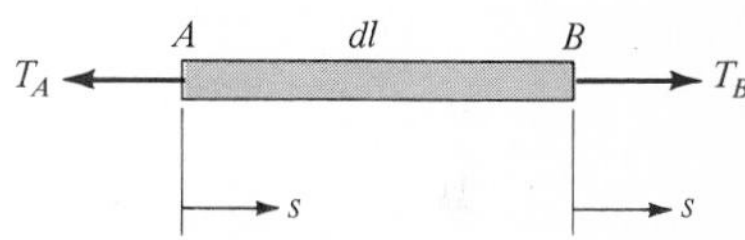

FIGURE 10-5

The work of internal forces T_A and T_B cancel each other if the member displaced is rigid

Qualitative Example of Major Concepts of Work

For a synthesis of the concepts discussed above, consider several test loads on the wing of an airplane. Two views of the wing are shown in Fig. 10-6, where the coordinate system is fixed at the root of the wing, a region of relatively high rigidity. The forces are applied by the hydraulic actuators A and B, whose piston rods move axially, rod A in the y direction, and rod B in the z direction.

Assume at first that actuator B is not connected to the wing. An aerodynamic lift force is simulated by rod A moving upward. The force F_A required for this increases with increasing deflection of the wing. Actuator A does positive work $\int F_A \, dy_A$ as the wing is raised. Note that this work is simple to calculate since F_A is only a function of y_A, no matter how the wing deflects. As the wing is slowly lowered back to its neutral position, actuator A resists the motion and does negative work $\int F_A \, dy_A$ on the wing. It is equally valid to say that now the wing does positive work $\int F_A \, dy_A$ on the actuator. The net work of the actuator on the wing (also of the wing on the actuator) in the complete cycle is zero, neglecting frictional effects.

Assume next that actuator B is connected to the wing to simulate drag forces, but actuator A is not connected. Since the force

FIGURE 10-6
Model of a statically tested wing

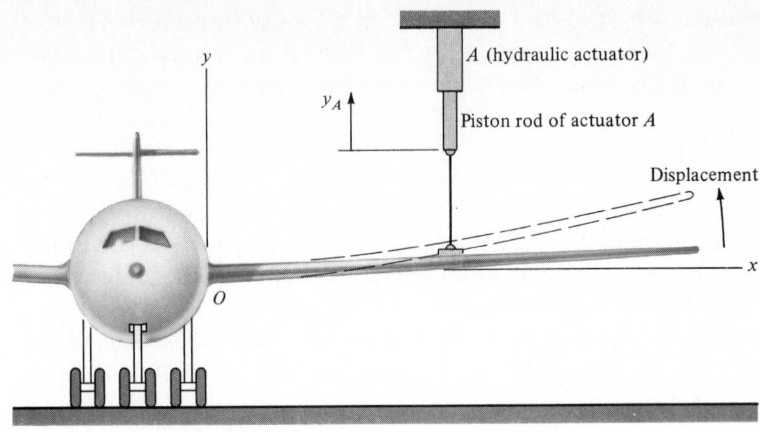

(a)

y
A
B (Hydraulic actuator)
Piston rod
z
O
z_B

(b)

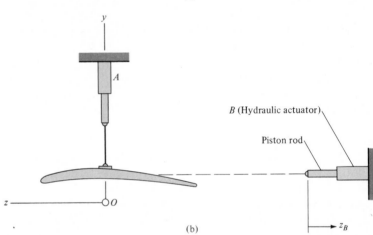

F_B is a function of the displacement z_B, the work of F_B is calculated using $\int F_B\, dz_B$ and the same consideration for signs as discussed above for F_A.

The situation is more complex when actuators A and B cause simultaneous displacements y_A and z_B. If B applies only a horizontal force to the wing (which can be done approximately if necessary), the work of force F_B does not affect the work of force F_A, and vice versa. Otherwise, each actuator can do work not only on the wing but also on the other actuator. In all cases, the work of internal forces cancel out (as was discussed for Fig. 10-5) in the relatively rigid piston rods, connecting linkages, and pin joints.

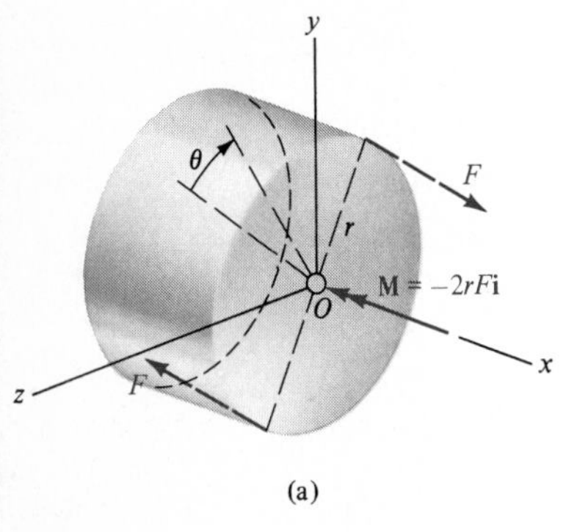

(a)

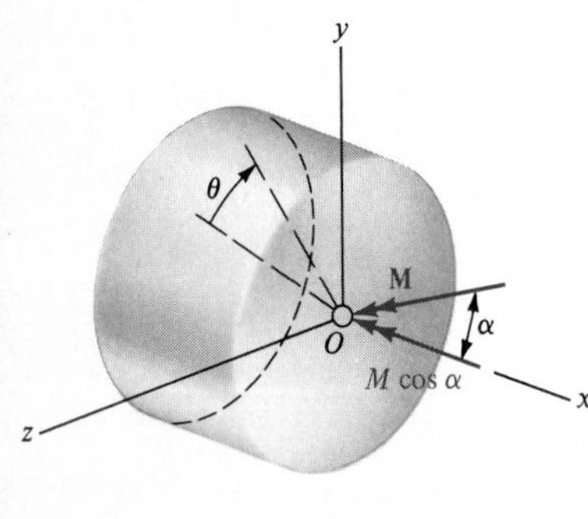

(b)

FIGURE 10-7
Work of a couple

10-2 WORK OF A COUPLE

The work of a couple of moment M is defined in a similar way as that of a force. The only essential difference is that a couple does work on a body when there is a corresponding angular displacement. In the simplest case, the work of a constant couple of moment **M** acting on a rigid body is defined with respect to Fig. 10-7a as

$$U = M\theta \tag{10-5}$$

where the moment $M = 2rF$ (in units of force times distance) and the angular displacement θ (in radians) are measured in the same plane (the moment vector **M** is perpendicular to the plane of the displacement). The derivation of Eq. 10-5 using the work of the two forces F is left as an exercise for the reader.

In the case where **M** is not perpendicular to the plane of the angular displacement, only the perpendicular component of **M** does work on the body. Thus, referring to Fig. 10-7b, the work of the constant couple of moment **M** is defined as

$$U = M\theta \cos \alpha \tag{10-6}$$

In the general case the magnitude of the couple and its orientation with respect to the angular displacement may vary, so it is useful to consider differential quantities. Expressing both the moment **M** and the infinitesimal rotation $d\boldsymbol{\theta}$ with respect to the same coordinate system as in Fig. 10-8 (where the body is not shown), the incremental work is

$$dU = \mathbf{M} \cdot d\boldsymbol{\theta} \tag{10-7}$$

which has the units of moment since the rotations are measured in radians.

The total work U in a rotation from angular position β to γ can be expressed using Eqs. 10-6 and 10-7,

$$U = \int_{\beta}^{\gamma} M \cos \alpha \, d\theta = \int_{\beta}^{\gamma} \mathbf{M} \cdot d\boldsymbol{\theta} \tag{10-8}$$

These are analogous to the expressions defining the work of a force. The work of a couple is a scalar, and it may be positive or negative.

The work of a couple should be computed using Eq. 10-8 rather than Eq. 10-5 when successive rotations about different axes occur. The reason is that *finite rotations are not commutative*, but *differential rotations are commutative*. This is illustrated in Fig. 10-9. First consider $\Delta\theta_x = -90°$ and $\Delta\theta_z = +90°$ rotations about the x and z axes, respectively. These are shown according to the right-hand rule in Fig. 10-9a. Point P rotates to different positions P_1 and P_2 depending on the order of the $\Delta\theta_x$ and $\Delta\theta_z$ rotations. P_1 is obtained when $\Delta\theta_x$ is followed by $\Delta\theta_z$ (the latter does not move the point to a new position). P_2 is the final location when $\Delta\theta_z$ is followed by $\Delta\theta_x$ (the latter does not move the point to a new position).

The differential rotations are shown in Fig. 10-9b. Clearly, P moves to the same point P_2 whether the sequence of rotations is $d\theta_x + d\theta_z$ or $d\theta_z + d\theta_x$. Thus, these rotations are commutative.

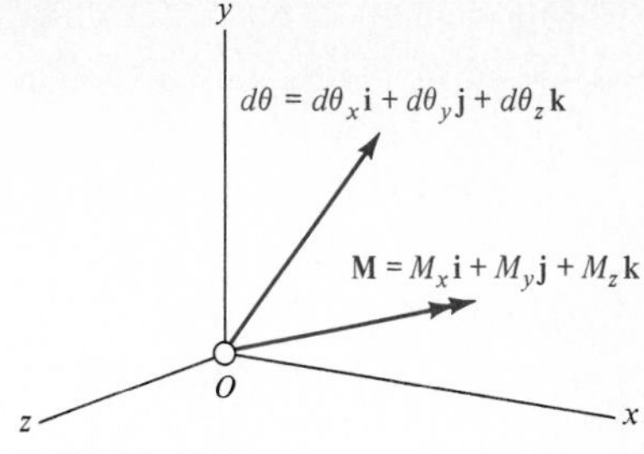

FIGURE 10-8

Detailed expressions for a moment and an infinitesimal rotation

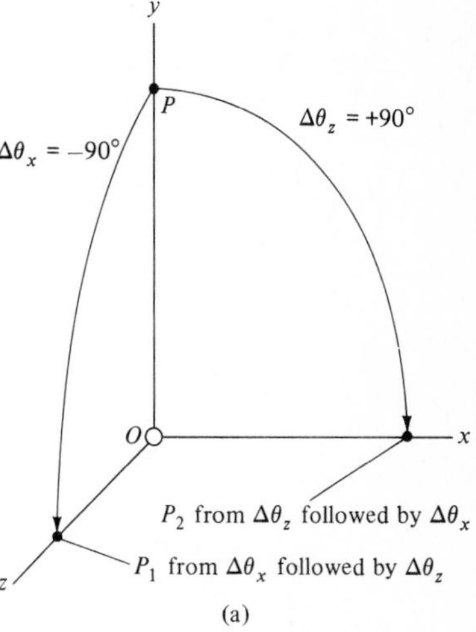

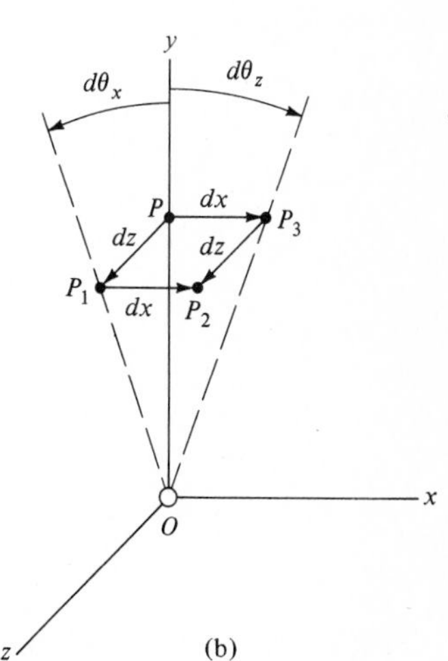

FIGURE 10-9

Distinction of differential and finite rotations

EXAMPLE 10-1

Assume that a block of weight W has to be moved from position A to B, and two different paths are reasonable to consider, as shown in Fig. 10-10. Evaluate the total input work in moving the block from A to B for the two paths. (a) First, assume that all paths are frictionless. (b) Next, assume the same coefficient of friction μ for all surfaces of contact. Ignore the differences in angular orientation of the block.

FIGURE 10-10

Work in moving a block to a given elevation using different paths

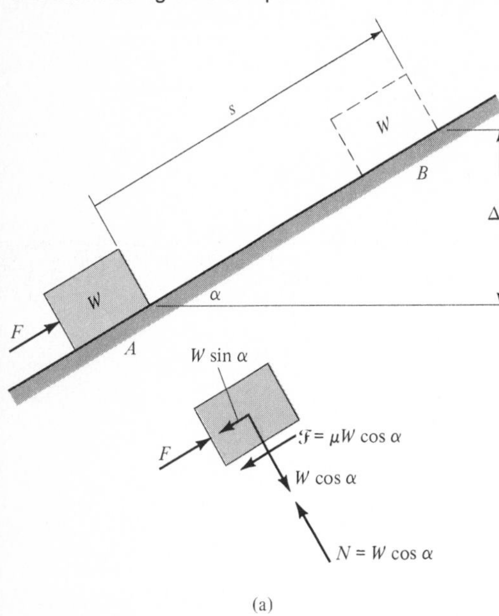

(a)

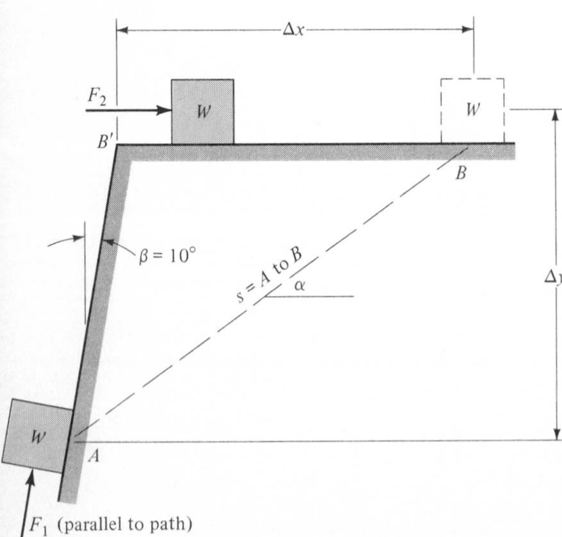

(b)

SOLUTION

(a) Frictionless ($\mu = 0$):

STRAIGHT PATH: The force F necessary to begin motion is $F = W \sin\theta$, hence the work done on the block is

$$U_{AB} = Fs = Ws \sin\alpha$$

LONG PATH: From A to B' the force F_1 necessary to begin motion is $F_1 = W \cos 10^\circ$, and the distance F_1 acts through is

$$\frac{\Delta y}{\cos 10^\circ} = \frac{s \sin\alpha}{\cos 10^\circ}$$

$$U_{AB'} = F_1 \frac{s \sin\alpha}{\cos 10^\circ} = W \cos 10^\circ \left(\frac{s \sin\alpha}{\cos 10^\circ}\right) = Ws \sin\alpha$$

From B' to B the force $F_2 \simeq 0$ since the surface is frictionless, hence $U_{B'B} = 0$ and the total work done is

$$U_{AB} = Ws \sin\alpha$$

In the frictionless situation the work done is independent of the path.

(b) With coefficient of friction μ:

STRAIGHT PATH: The force F necessary to cause motion is

$$F = W \sin\alpha + \mu N = W \sin\alpha + \mu W \cos\alpha$$

Hence, the work done is

$$U_{AB} = Fs = (W \sin\alpha + \mu W \cos\alpha)s$$

LONG PATH: From A to B', the force F_1 necessary to cause motion is

$$F_1 = W \cos 10^\circ + \mu N_{AB'} = W \cos 10^\circ + \mu W \sin 10^\circ$$

Thus, the work done is

$$U_{AB'} = (W \cos 10^\circ + \mu W \sin 10^\circ) \frac{s \sin\alpha}{\cos 10^\circ}$$

$$U_{AB'} = Ws \sin\alpha + \mu Ws \tan 10^\circ \sin\alpha$$

The force F_2 necessary to cause motion from B' to B is $F_2 = \mathscr{F} = \mu W$ and the work done is

$$U_{B'B} = \mu W \, \Delta x \approx \mu Ws \cos \alpha$$

The total work on the body is

$$\begin{aligned} U_{AB} &= U_{AB'} + U_{B'B} \\ &= Ws \sin \alpha + \mu Ws \tan 10^\circ \sin \alpha + \mu Ws \cos \alpha \end{aligned}$$

JUDGMENT OF THE RESULTS

The form and magnitude of the results seem reasonable. It is seen that for small angles β ($\tan \beta \simeq 0$) and small coefficients of friction, the work done is almost the same for either case.

EXAMPLE 10-2

(a) Calculate the work of the couple C in Fig. 10-11 for raising a weight $W = 1000$ N through a vertical distance $h = 50$ m. Compare the results for two drum radii, $r_1 = 20$ cm and $r_2 = 40$ cm. Ignore frictional losses. (b) What is the total work of C in a complete cycle of raising and lowering the cable if the weight raised is always W and the one lowered is always $0.1W$?

SOLUTION

(a) From Eq. 10-5, the work done by the couple C is $C\theta$, where θ is the angle through which the drum rotates. In a frictionless system the work U_C of the couple is equal in magnitude and opposite in sign to the work U_W of the weight, so

$$C\theta = Wh = (1000 \text{ N})(50 \text{ m}) = 50 \text{ kN}\cdot\text{m}$$

It is seen that the work done by C does not depend on the radius of the drum.

(b) The total work U_T of the couple C in a complete cycle is

$$\begin{aligned} U_T &= U_{\text{up}} + U_{\text{down}} = Wh - 0.1\,Wh = 0.9\,Wh \\ &= (0.9)(1000 \text{ N})(50 \text{ m}) = 45 \text{ kN}\cdot\text{m} \end{aligned}$$

JUDGMENT OF THE RESULTS

The result of part (a) can be checked using a different approach. Summing moments about point 0 for a static situation yields

$$\sum M_O = 0$$

$$C - Wr = 0, \qquad C = Wr$$

The angle θ through which C acts can be determined. Since $h = r\theta$,

$$\theta = \frac{h}{r} \quad \text{and} \quad U_C = C\theta = (Wr)\left(\frac{h}{r}\right) = Wh = 50 \text{ kN}\cdot\text{m}$$

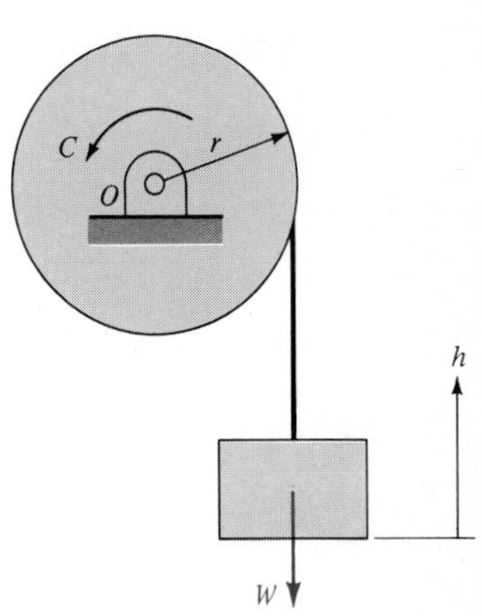

FIGURE 10-11

Work of a couple C in raising a block

PROBLEMS

SECTIONS 10-1, 10-2

FIGURE P10-1

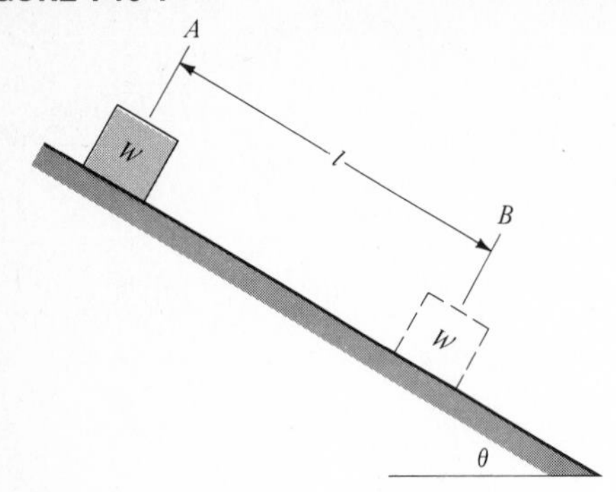

FIGURE P10-4

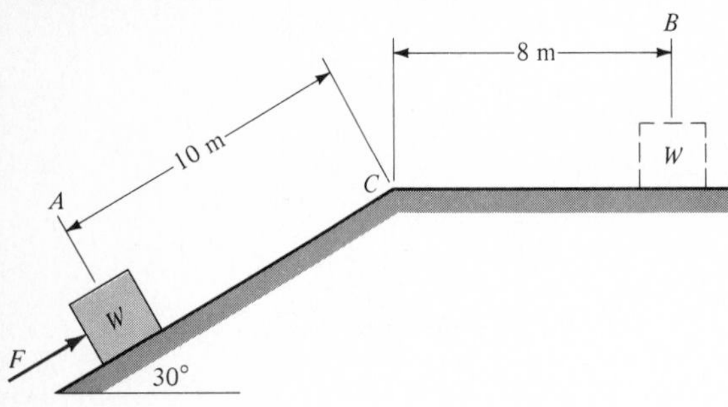

FIGURE P10-7

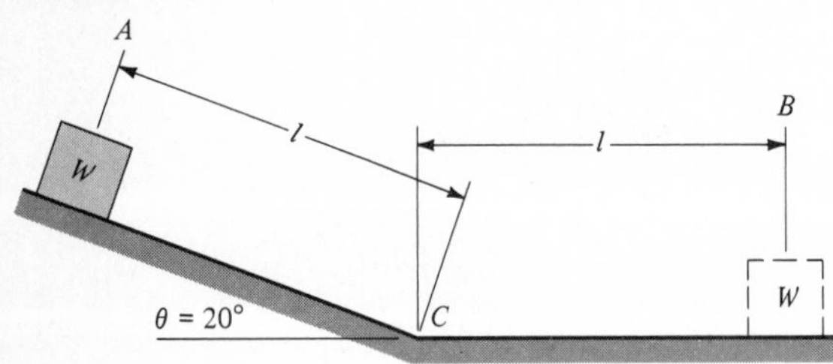

FIGURE P10-9

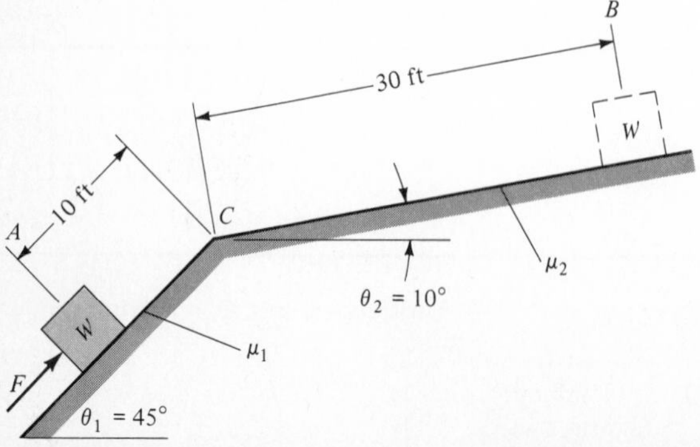

10–1 A block of weight W is slowly moved by a machine from A to B. Determine the total work done by the machine if friction is negligible.

10–2 Solve Prob. 10–1 using $W = 500$ N, $l = 10$ m, $\theta = 30^\circ$, and a coefficient of friction $\mu_k = 0.1$.

10–3 The block in Fig. P10-1 weighs $W = 100$ lb. A machine slowly moves the block through a complete cycle A to B to A. Calculate the total work done by the machine in the cycle of displacement $A - B - A$ if $l = 8$ ft, $\theta = 25^\circ$, and the coefficient of friction is $\mu_k = 0.3$.

10–4 A block of weight $W = 100$ N is slowly moved by a machine from A to B. Calculate the total work done by the machine if friction is negligible.

10–5 Solve Prob. 10–4 assuming that the coefficient of friction is $\mu_k = 0.3$ everywhere.

10–6 The block in Fig. P10-4 has weight $W = 1$ kN. It is slowly moved by a machine from A to B and back to A. Determine the total work done by the machine in the complete cycle of displacement A–B–A if the coefficient of friction is $\mu_k = 0.4$ everywhere.

10–7 A machine slowly moves the block of weight W from A to B. Write an expression for the total work done by the machine if the coefficient of kinetic friction is μ_1 on the incline and μ_2 on the horizontal path. The block could rest on the incline.

10–8 Assume that in Prob. 10–7, $\mu_1 = \mu_2 = \mu$. Express the ratio of the work U_{AB} to U_{BA} done by the machine during a complete cycle $A - B - A$ of the displacements.

10–9 A block of weight $W = 200$ lb is slowly moved by a machine from A to B. Calculate the total work done by the machine if friction is negligible.

10–10 Solve Prob. 10–9 assuming that the coefficients of kinetic friction are $\mu_1 = 0.2$ and $\mu_2 = 0.3$.

10–11 The block in Fig. P10-9 represents a container used in a mine. Its total weight is 3000 lb as it moves from A to B, and 400 lb as it moves from B to A. Determine the total work done by the slowly moving hoisting machine during a complete cycle $A - B - A$ of the motion if $\mu_1 = \mu_2 = 0.1$.

10–12 A machine slowly moves the block of weight $W = 2$ kN from A to B. Calculate the total work done by the machine if the coefficient of friction is $\mu_k = 0.3$ and $l = 10$ m.

10–13 Consider the model of a dump truck. Two hydraulic actuators A are mounted on the sides of the truck bed as shown. Assume that the location of the center of gravity C of the truck bed is fixed. Calculate the total work of the actuators when the truck bed is raised from $\theta = 0$ to 30°. The weight of the truck bed is $W = 100$ kN.

10–14 A uniform slender bar of weight W is hinged at point O. Determine the work of the moment M_O when the bar is slowly raised from $\theta = 0$ to 90°.

10–15 An amusement ride is modeled as a cage of weight $W_1 = 800$ lb fixed on a uniform arm of weight $W_2 = 300$ lb, which is hinged at A. Determine the work of the moment M_A when the arm is slowly raised from $\theta = 0$ to 60°. $AB = 20$ ft and $BC = 3$ ft.

FIGURE P10-14

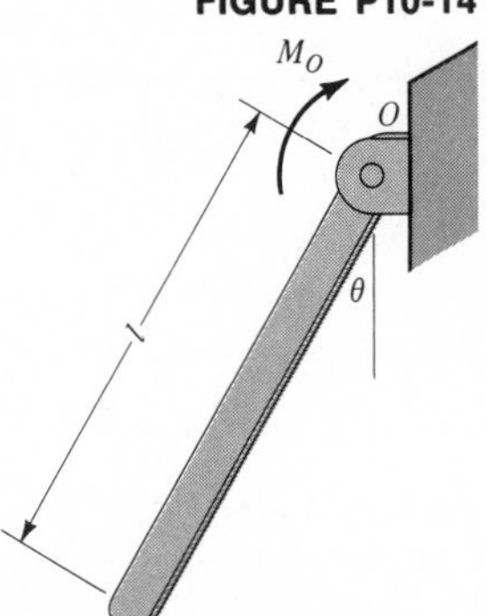

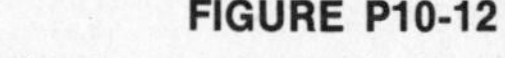
FIGURE P10-12

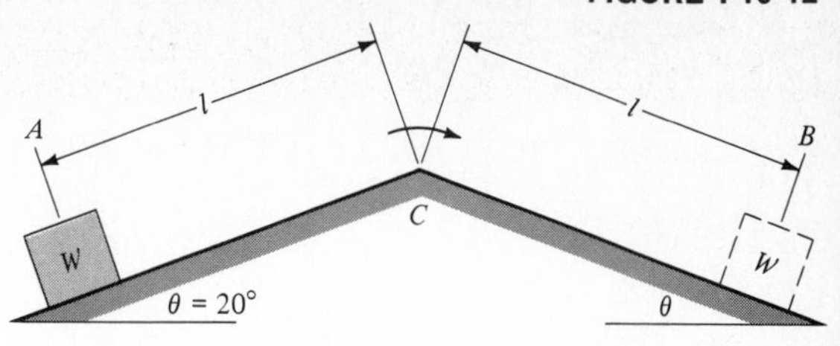

FIGURE P10-13

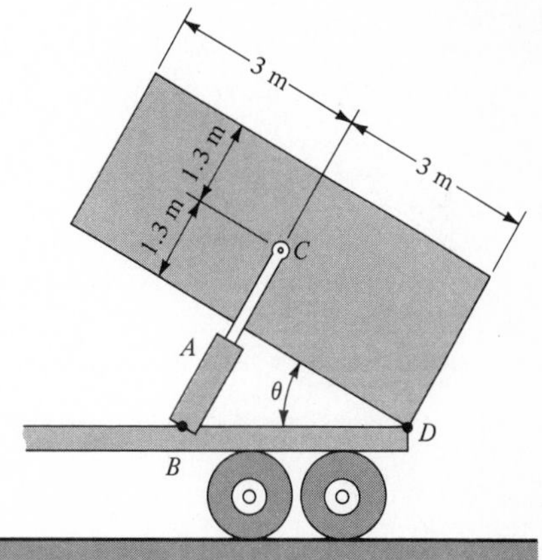

FIGURE P10-15

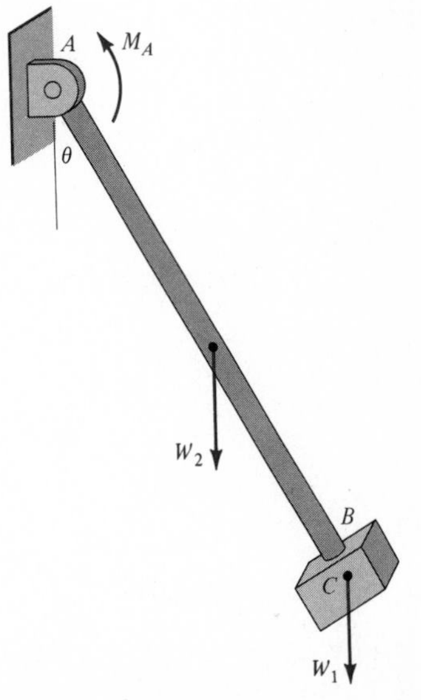

FIGURE P10-16

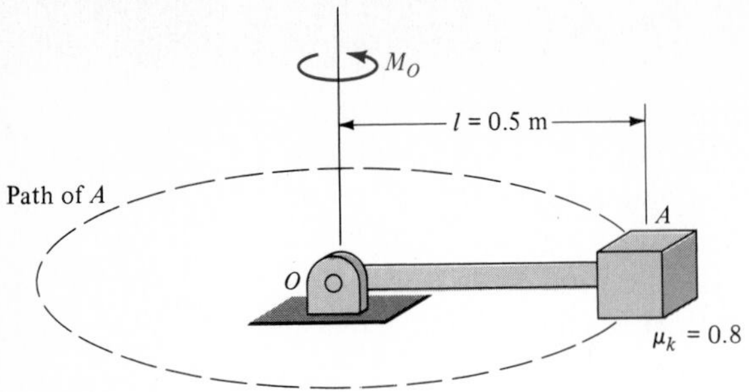

FIGURE P10-17

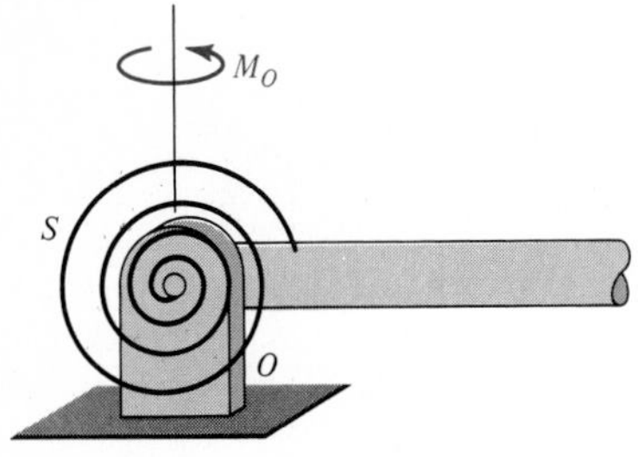

FIGURE P10-18

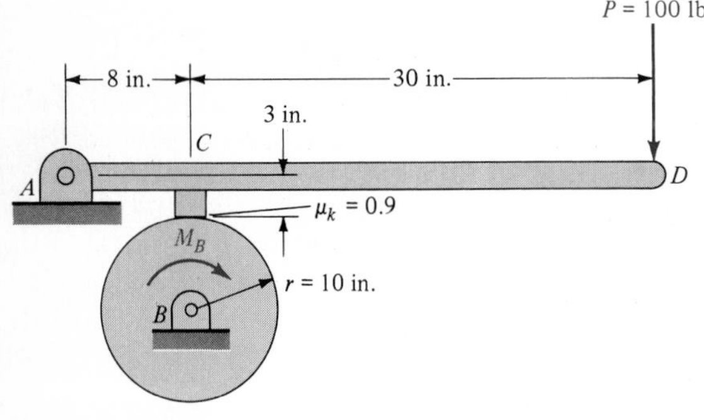

FIGURE P10-20

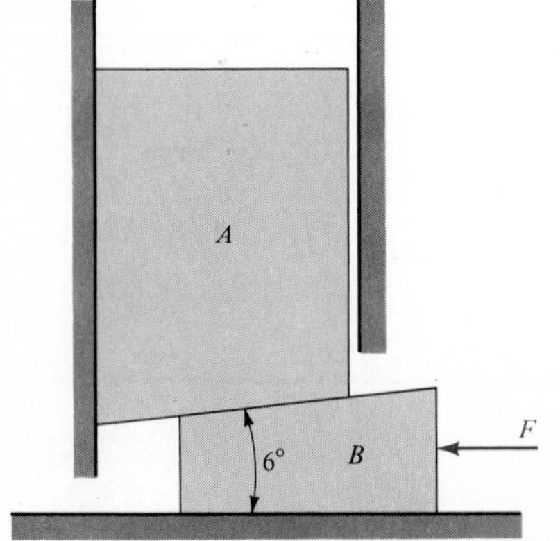

10–16 A polishing machine is modeled as a small block A of weight $W = 100$ N mounted on a weightless arm OA. The arm is hinged at O to freely swing vertically. A moment M_O slowly moves the arm in the horizontal plane. The motion is opposed by friction acting on the block. Calculate the work of M_O in a complete revolution.

10–17 In Prob. 10–16, a spring S is added to the arm at hinge O. The spring is in the vertical plane and doubles the normal force between block A and the supporting surface. Determine the work of M_O and the work of the spring in a complete revolution.

10–18 Determine the work of the moment M_B which slowly rotates the wheel through a complete revolution.

10–19 Assume that the wheel in Fig. P10-18 is alternately moving in the clockwise and counterclockwise directions. Determine the total work of the driving moment M_B in two complete but opposite turns of the wheel.

10–20 Part of a machine is modeled as block A of weight $W = 3$ kN, which is raised using the weightless wedge B. Calculate the total work of force F in raising block A through a distance of 5 cm. The coefficient of friction on all surfaces is $\mu_k = 0.3$.

10–21 Determine the work of force F in Prob. 10–20 if block A is to be lowered through a distance of 5 cm.

10–22 A car is stuck with one of its driving wheels slowly spinning. Calculate the work required to make one complete turn of the wheel if its radius is 15 in. and the axle load is 600 lb.

FIGURE P10-22

VIRTUAL WORK 10-3

The concept of work is useful in the analysis of certain problems even when displacements do not actually take place. The idea is to imagine that infinitesimal displacements occur within the constraints of a given system (that is, the displacements should be possible), and examine the resulting work done by the external forces acting on the system. The method can be used to investigate the equilibrium of the system. This is somewhat analogous to testing the strength of ice on a pond by gently pressing it with an extended foot. It is not necessary to break the ice to prove that it is unsafe; one can imagine from its response to small loads that the ice would break under the full body weight.

The imaginary small displacement is called a *virtual displacement* and is denoted by $\delta\mathbf{r}$, which is to be distinguished from a similarly small but real displacement $d\mathbf{r}$. The imaginary work done by a force $\mathbf{F}$ during a virtual displacement of a point where the force is acting is called *virtual work* and defined as

$$\boxed{\delta U = \mathbf{F} \cdot \delta\mathbf{r}} \qquad \text{10-9}$$

Similarly for the virtual work of a couple of moment $\mathbf{M}$,

$$\boxed{\delta U = \mathbf{M} \cdot \delta\boldsymbol{\theta}} \qquad \text{10-10}$$

where $\delta\boldsymbol{\theta}$ is a virtual angular displacement of infinitesimal magnitude.

It is reasonable to assume that both forces and couples remain constant during the virtual displacements. This is because any change in a force or a couple caused by an infinitesimal displacement must itself be infinitesimal. The resulting difference in work is a second-order differential and can be neglected.

The concept of virtual work can be used to analyze the equilibrium of particles and rigid bodies because the sum of all virtual work done by a system of forces or couples is zero when the system is in equilibrium. This is true for any system, and it is easiest to prove for a particle or single rigid body under the action of two forces. Consider the particle A in Fig. 10-12 for a simple example. If the particle is in equilibrium, the forces F_1 and F_2 must be equal in magnitude, collinear, and opposite in direction. Thus, F_1 does positive work during a virtual displacement to the right, while F_2 does negative work of equal magnitude. Formally, with $F_1 = F_2$,

$$\delta U = \sum_{i=1}^{n=2} \mathbf{F}_i \cdot \delta\mathbf{r} = F_1\, \delta r + (-F_2\, \delta r) = 0$$

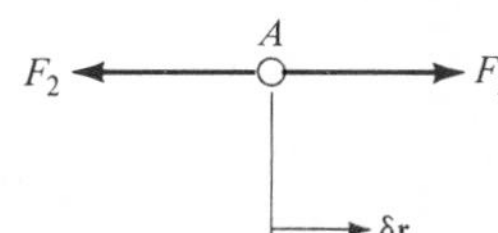

FIGURE 10-12

Virtual work is done in the imaginary displacement $\delta\mathbf{r}$ of a particle

This procedure can be extended to deal with any number of forces or couples acting in any direction. The general statement of equilibrium on the basis of virtual work defined by Eqs. 10-9 and 10-10 is

$$\delta U = \sum_{i=1}^{m} \mathbf{F}_i \cdot \delta \mathbf{r}_i + \sum_{j=1}^{n} \mathbf{M}_j \cdot \delta \boldsymbol{\theta}_j = 0 \qquad \text{10-11}$$

where the subscript i refers to an individual force and the corresponding virtual displacement, while j refers to an individual couple and the corresponding virtual rotation. Equation 10-11 is valid for particles (with the moment terms naturally absent) or for rigid bodies. This equation includes the work of all forces and moments that act through a distance or rotation under a virtual displacement. The work may be of applied forces or moments, including body forces (such as weight or magnetic force), and friction forces. Equation 10-11 is not advantageous in comparison with the fundamental equations of equilibrium unless the system consists of at least several interconnected rigid bodies (see Ex. 10-3). A particular advantage of the method of virtual work is in the determination of the equilibrium configuration for a system of rigid bodies.

There are several items of practical interest in solving a problem using the method of virtual work. First, it is not necessary to draw the interrelated free-body diagrams for the members of the given system of rigid bodies. The required diagram has to show only those *active forces and couples* that have virtual displacements associated with them. Thus, the reactions at any fixed joint do not have to be shown. The number of equations that has to be solved in a given problem equals the number of degrees of freedom n for that system. The reason is that the virtual work according to Eq. 10-11 can be written in n different ways, one for each particular possible virtual displacement.

10-4 MECHANICAL EFFICIENCY OF REAL SYSTEMS

The concepts of work and virtual work are extremely useful because of their simplicity when frictional effects are ignored. However, real mechanical systems always operate with frictional losses, so a more precise analysis may have to be done. This is simple if the magnitude of the friction force is known. The work of the friction force is always taken as negative for a sliding body because the direction of the friction force is always opposite to the direction of the displacement of the body.

To illustrate the work of friction in a simple case, consider the block of weight W on the inclined plane in Fig. 10-13a. The force F

required to push the block up the incline can be determined using Eq. 10-11,

$$\delta U = \underbrace{F\,\delta s}_{\substack{\text{work by}\\ \text{applied}\\ \text{force}}} - \underbrace{W \sin\theta\,\delta s}_{\substack{\text{work by}\\ \text{weight}}} - \underbrace{\mu W \cos\theta\,\delta s}_{\substack{\text{work by}\\ \text{friction force}}} = 0$$

$$F > W(\sin\theta + \mu\cos\theta) \qquad \text{to cause upward motion}$$

Of course, this could also be found without using the concept of work.

It is useful to rewrite the expression for the virtual work as follows,

$$F\,\delta s = W \sin\theta\,\delta s + \mu W \cos\theta\,\delta s$$

which is stated in words as

$$\text{input work} = \text{useful work} + \text{work of friction} \qquad \boxed{10\text{-}12}$$

The useful work is also called *output work*; in the case of a machine this is the work available to be done on another body or system of bodies.

The ratio of output work to input work is defined as the *mechanical efficiency* η of a machine,

$$\boxed{\eta = \frac{\text{output work}}{\text{input work}} = \frac{\text{useful work}}{\text{total work required}}} \qquad \boxed{10\text{-}13}$$

The mechanical efficiency of an ideal machine is $\eta = 1$, and of any real machine $\eta < 1$.

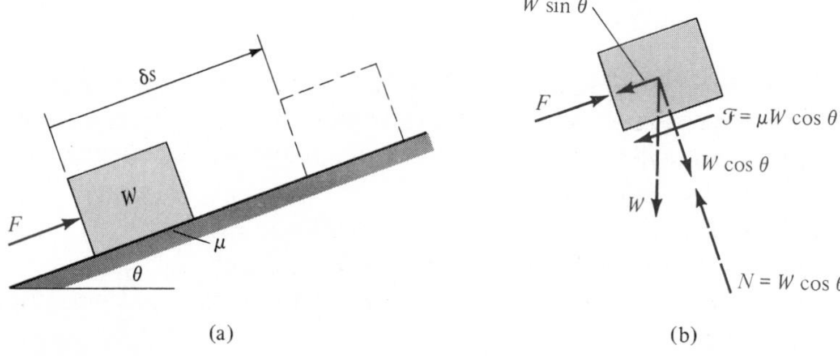

FIGURE 10-13

Model of a real system with friction

EXAMPLE 10-3

The two uniform booms A–B of the electric trolley bus shown in Fig. 10-14a are 18 ft long, weigh 30 lb each, and have essentially frictionless hinges at A. The normal force at B needed for proper contact is 10 lb and the coefficient of friction is 0.6 for each contact slider. Use the method of virtual work to determine the couple M required at joint A of each boom to maintain proper contact at B. The nearly vertical strings on the booms are slack during normal operation. $\theta = 30^\circ$.

SOLUTION

A reference system that moves with the trolley car with its origin at A is considered as shown in Fig. 10-14b. The position of the boom is totally defined by the angle θ.

$$y_B = (18 \text{ ft}) \sin \theta$$

$$\frac{dy_B}{d\theta} = (18 \text{ ft}) \cos \theta$$

$$\delta y_B = (18 \text{ ft}) \cos \theta \, \delta\theta$$

$$x_B = (18 \text{ ft}) \cos \theta$$

$$\frac{dx_B}{d\theta} = -(18 \text{ ft}) \sin \theta$$

$$\delta x_B = -(18 \text{ ft}) \sin \theta \, \delta\theta$$

For the center of gravity G,

$$y_G = (9 \text{ ft}) \sin \theta$$

$$\frac{dy_G}{d\theta} = (9 \text{ ft}) \cos \theta$$

$$\delta y_G = (9 \text{ ft}) \cos \theta \, \delta\theta$$

A virtual displacement $\delta\theta$ is imagined as shown in Fig. 10-14b, and Eq. 10-11 is used. Note that the signs are obtained by deciding if the force or moment being considered acts in the direction of its displacement (positive work) or against it (negative work).

$$\begin{aligned} \delta U &= M\,\delta\theta - N\,\delta y_B - \mathscr{F}\,\delta x_B - W\,\delta y_G = 0 \\ &= M\,\delta\theta - 10 \text{ lb}(18 \text{ ft} \cos \theta)\,\delta\theta - 6 \text{ lb}(18 \text{ ft} \sin \theta)\,\delta\theta \\ &\quad - 30 \text{ lb}(9 \text{ ft} \cos \theta)\,\delta\theta = 0 \end{aligned}$$

$$\begin{aligned} M &= 180 \text{ ft}\cdot\text{lb} \cos 30^\circ + 108 \text{ ft}\cdot\text{lb} \sin 30^\circ + 270 \text{ ft}\cdot\text{lb} \cos 30^\circ \\ &= 444 \text{ ft}\cdot\text{lb} \end{aligned}$$

JUDGMENT OF THE RESULTS

A check using the equations of equilibrium confirms this result. Note that the check is much faster than using the method of virtual work. The example is worked out here for practice with Eq. 10-11.

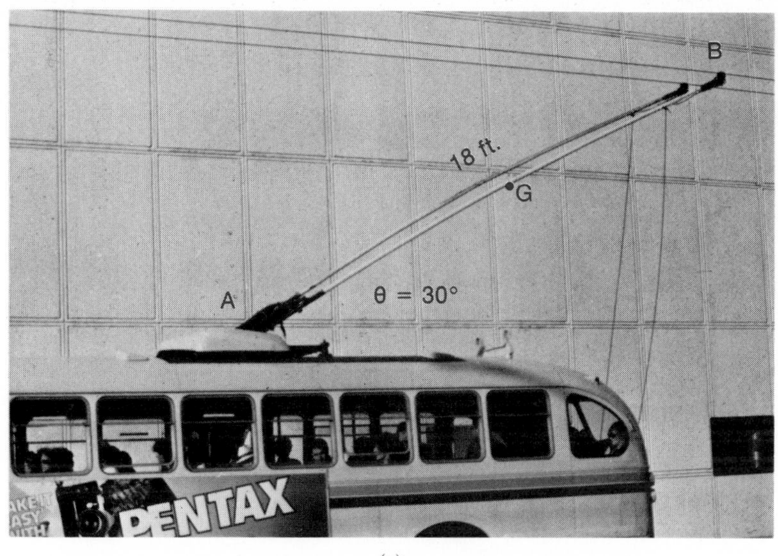

(a)

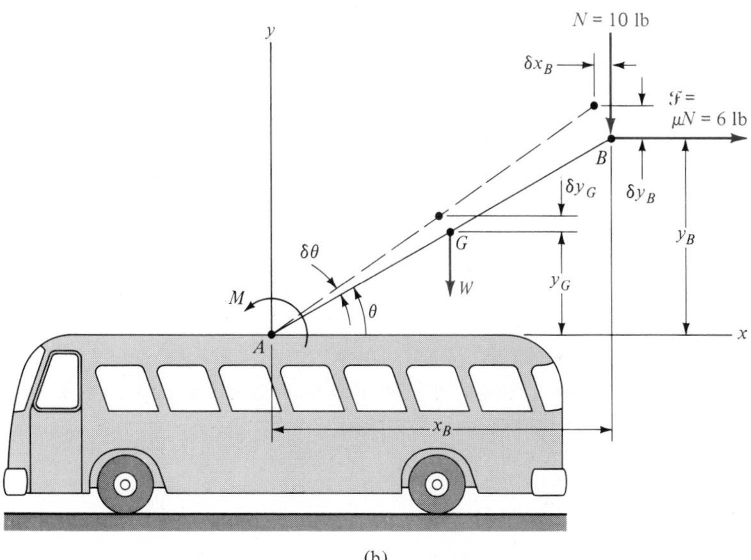

(b)

FIGURE 10-14

Analysis of the booms of an electric trolley bus using virtual displacements

EXAMPLE 10-4

Consider the wing-loading mechanism with the hydraulic actuator A shown in Fig. 10-15a. Use the method of virtual work to determine the actuator force F required to apply a vertical force P to the wing first upward, then downward. Assume that all joints are frictionless and that the mechanism is weightless.

(a)

(b)

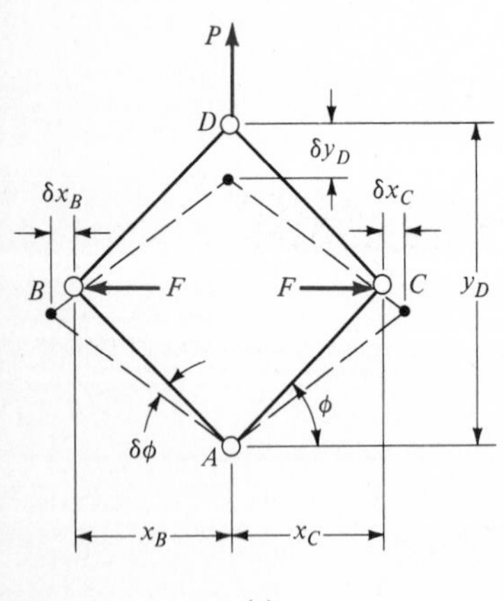

(c)

FIGURE 10-15

Virtual displacements of a mechanism

SOLUTION

First the mechanism is sketched showing virtual displacements. All of the virtual displacements of interest are determined in terms of a single parameter, the angle ϕ. For P acting upward on the wing (Fig. 10-15b),

$$x_B = -l\cos\phi \qquad \delta x_B = l\sin\phi\,\delta\phi$$

$$x_C = l\cos\phi \qquad \delta x_C = -l\sin\phi\,\delta\phi$$

$$y_D = 2l\sin\phi \qquad \delta y_D = 2l\cos\phi\,\delta\phi$$

Equation 10-11 is written after carefully checking the resultant signs in each term,

$$\begin{aligned}\delta U &= F\,\delta x_B + (-F)\,\delta x_C - P\,\delta y_D = 0\\ &= 2F(l\sin\phi)\,\delta\phi - P(2l\cos\phi)\,\delta\phi = 0\end{aligned}$$

$$2Fl\sin\phi = 2Pl\cos\phi$$

$$F = P\cot\phi = P\cot 45^\circ = P$$

For P acting down on the wing, Fig. 10-15c,

$$\begin{aligned}\delta U &= 2F\,\delta x_B - P(2l\cos\phi)\,\delta\phi = 0\\ &= 2Fl\sin\phi - 2Pl\cos\phi = 0\end{aligned}$$

$$F = P\cot\phi = P\cot 45^\circ = P$$

It should be noted that the signs of the work terms are checked by deciding if the force being considered acts in the direction of its displacement (positive work) or against it (negative work). Also note that the vertical deflections of the forces F are never considered since the forces always act horizontally, hence F does not work through the vertical displacement.

JUDGMENT OF THE RESULTS

This problem could be solved using the equations of equilibrium but it would be a much longer solution. In this case, using the method of virtual work is fastest. The results appear reasonable because of the geometry ($|\delta AD| = |\delta BC|$), but they are valid only for $\phi = 45^\circ$.

PROBLEMS

10–23 Determine the force P required to maintain equilibrium of the weightless bar supported by a frictionless hinge.

10–24 Assume that in Fig. P10-23 force $P = 4F$ and the uniform bar's weight is $W = 2F$. Determine the moment M_O required to maintain equilibrium.

10–25 Assume that the rigid bar AB is weightless and has a frictionless pin at A. Derive an expression for the force P required to maintain equilibrium.

10–26 Assume that in Fig. P10-23 force $F = 3P$ and the uniform bar's weight is $W = 0.5F$. Determine the moment M_O required to maintain equilibrium.

10–27 A uniform bar of weight W is supported by frictionless surfaces and a force F. Determine the force F required to maintain equilibrium.

10–28 Solve Prob. 10–27 after moving force F from point A to point B, where it acts in the vertical direction.

10–29 Determine the force P required to maintain equilibrium of the weightless bars, which have frictionless connections.

FIGURE P10-23

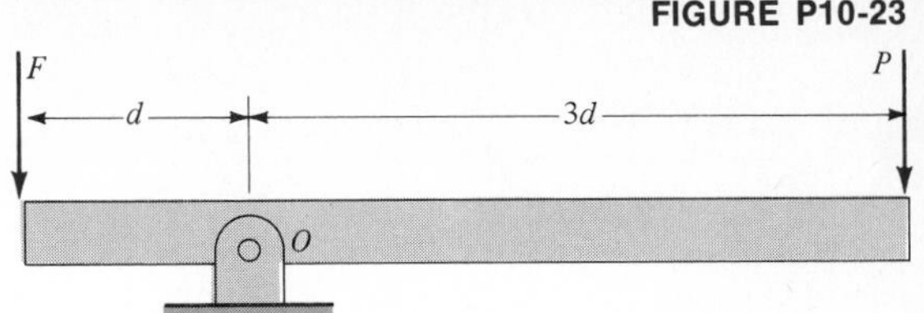

FIGURE P10-25

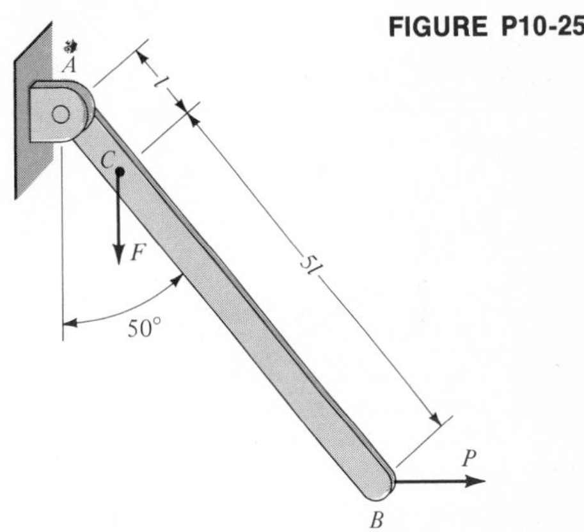

FIGURE P10-27

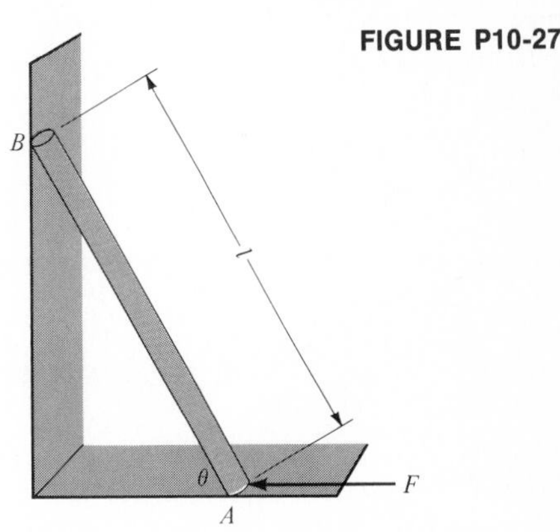

FIGURE P10-29

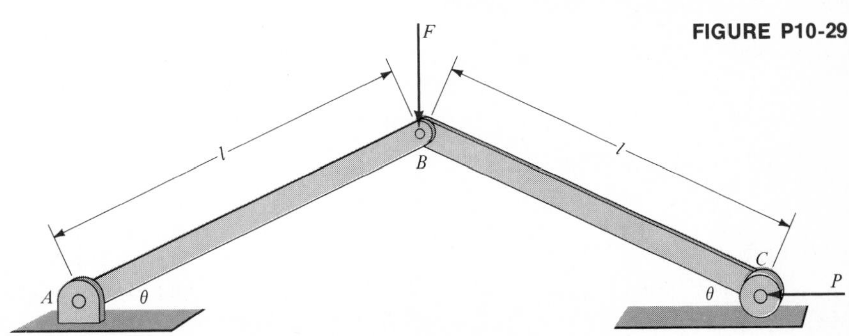

FIGURE P10-31

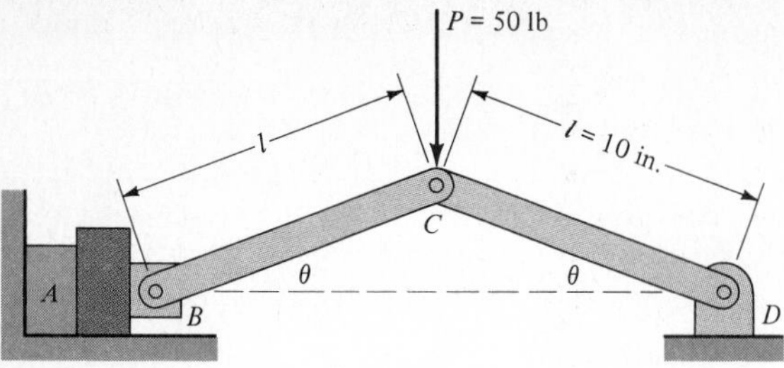

FIGURE P10-33

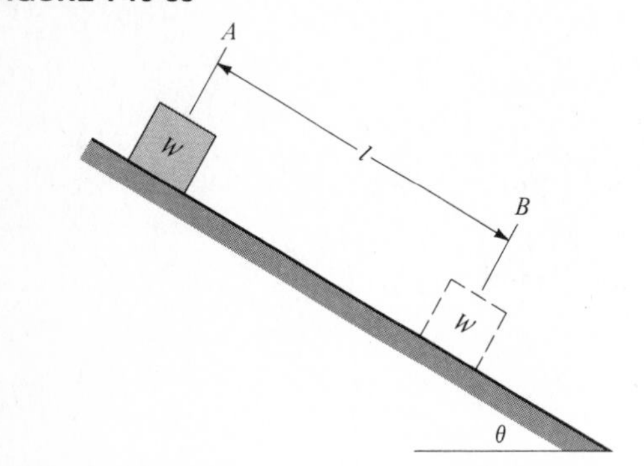

FIGURE P10-35

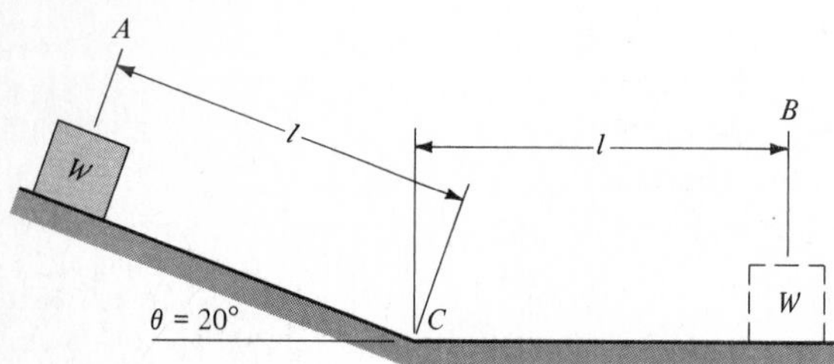

FIGURE P10-36

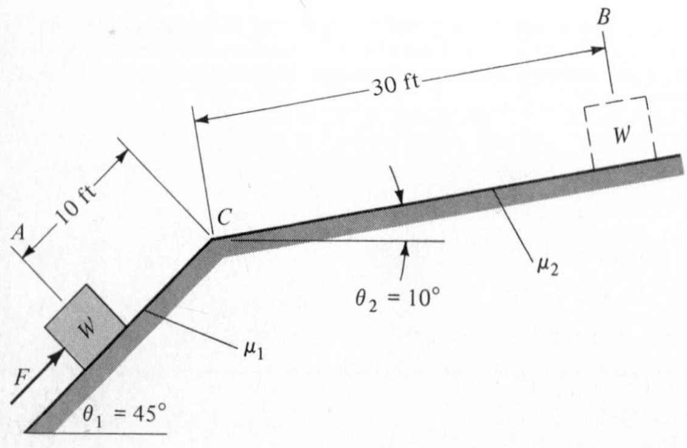

10–30 Solve Prob. 10–29 assuming that each bar is uniform and has weight $W = 0.5F$.

10–31 Calculate the horizontal force F that the toggle vise applies to block A if friction is negligible and $\theta = 10°$.

10–32 Plot the force F vs. the angle $\theta = 5°$, $10°$, and $15°$ in Prob. 10–31.

10–33 Determine the mechanical efficiency of the system which slowly moves the block from A to B. $W = 600$ N, $l = 10$ m, $\theta = 20°$, and $\mu_k = 0.3$. Other frictional losses are ignored.

10–34 The block in Prob. 10–33 is moved through a complete cycle of displacements A–B–A. Calculate the mechanical efficiencies in the distinct motions A–B and B–A.

10–35 Write an expression for the mechanical efficiency of the system which moves the blqck from A to B. The coefficient of friction is μ_k everywhere. The block could rest on the incline.

10–36 The block represents a container used in a mine. Its total weight is 3000 lb as it moves from A to B and 400 lb as it moves from B to A. Calculate the mechanical efficiency of the slowly moving container system in a complete cycle of displacements A–B–A if $\mu_1 = \mu_2 = 0.1$.

10–37 The mechanical efficiency is 0.85 in the tilting action of the dump truck described in Prob. 10–13. Calculate the work of friction in a tilt from $\theta = 0$ to $30°$.

The method of potential energy in engineering has an interesting historical background. The majority of inventors of perpetuum mobiles have tried for centuries to find clever ways to get continuously available and cheap work from the force of gravity. The basic concept is illustrated in Fig. 10-16. When the weight W is allowed to go from level 1 to a lower level 2 (with respect to the center of the earth), the weight does useful work on the system that lowers it (drives a clock, pumps water, etc.). In an ideal system this work, according to Eq. 10-4, is

$$U_{12} = \int_1^2 W\,dy = W(y_2 - y_1) = Wh$$

The potential of a weight W to do work because of its relative height h is defined as its *potential energy*. The term "potential energy" comes from the perception that a weight situated at any level could do useful work if allowed to descend to a lower level. Potential energy thus depends on the levels chosen. The weight W has potential energy U_{23} at level 2 with respect to level 3 in Fig. 10-16,

$$U_{23} = \int_2^3 W\,dy = Wl$$

Naturally,

$$U_{13} = \int_1^3 W\,dy = W(h + l)$$

Many inventors have tried to make a machine that does work in lowering a weight from level 1 to level 2 (Fig. 10-16), then uses part of this work to return the weight to level 1 so that the cycle can be repeated. The problem, of course, is that the work required to move the weight from level 2 to level 1 in an ideal system is

$$U_{21} = \int_2^1 -W\,dy = -Wh$$

The minus sign is present since gravity acts in opposing the motion upward. Thus, in a complete cycle from 1 to 2 to 1, the net work is

$$U_{12} + U_{21} = Wh - Wh = 0$$

In a real machine some work is dissipated by friction (in the form of heat), so the work available in the descent from 1 to 2 is never enough to completely raise the weight from 2 to 1.

The concept of potential energy is very useful in judging most of the proposed perpetuum mobiles (but not all of them). The idea is to ignore the paths of descent and ascent of all members of the machine and concentrate on the changes in their elevations. If the net change in potential energy in a complete cycle is zero when friction is ignored, the machine is proved to be unable to run perpetually on its own. Often there are other methods also for

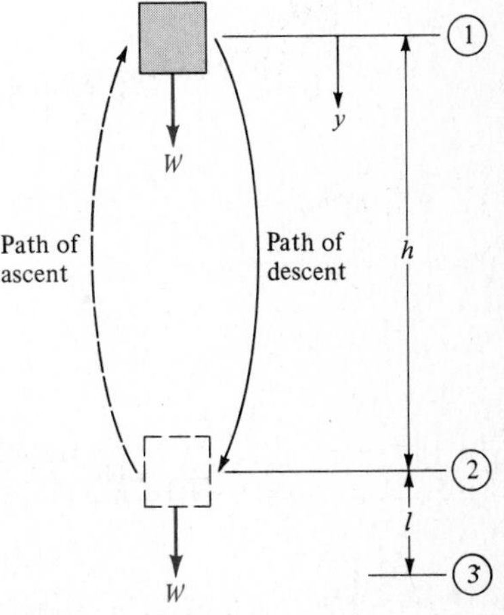

FIGURE 10-16
Potential energy of a weight

proving the same result, but the method of potential energy is generally the simplest.

Elastic Potential Energy

There are many other forms of potential energy besides that associated with the force of gravity. The potential energy of elastic members is the most important of these in elementary mechanics.

The background for analyzing any elastic member is the *theory of springs* discovered by Robert Hooke in 1642. This important concept is easy to understand with the aid of Fig. 10-17. Consider a helical spring which is loaded with progressively larger weights W as in Fig. 10-17a through d. If the deformations x are not excessive, they are proportional to the applied force. Consequently, a plot of force F vs. deformation x is a straight line as in Fig. 10-17e. The slope k of this line is called the *spring constant*. It has units of N/m or lb/in., and it is the property of each particular spring (depends on material, length of spring, wire diameter, and coil diameter). The formal relation of the force and deflection in a spring, most commonly given in terms of the spring constant, k, is

$$F = kx \tag{10-14}$$

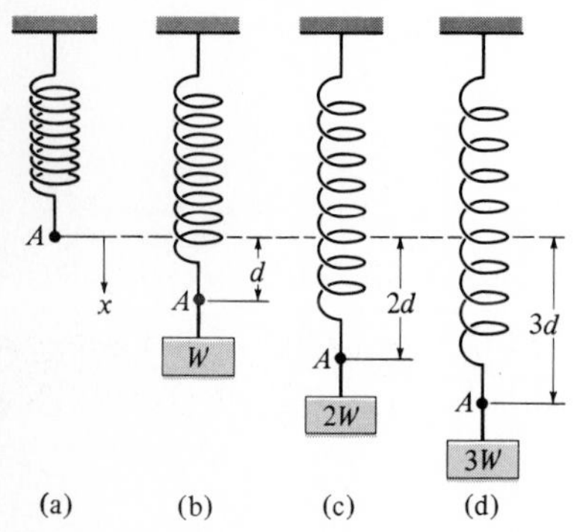

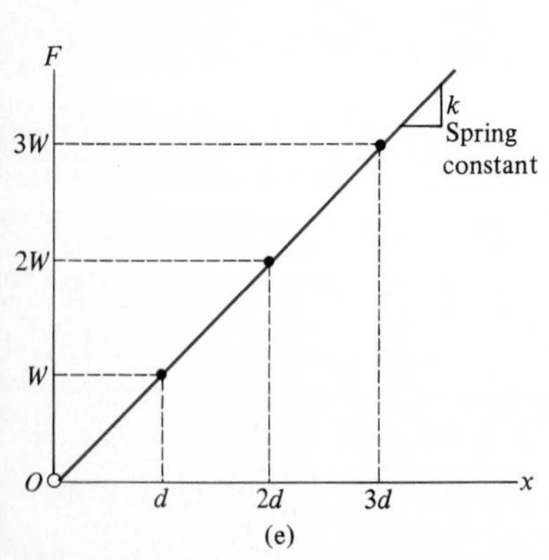

FIGURE 10-17

Theory of springs for an ideally elastic (linearly deforming) spring

The work of a force F on a spring is obtained from Eq. 10-4. With respect to Fig. 10-18, where the force F_1 causes a deformation x_1 from the undeflected position,

$$U_1 = \int_0^{x_1} kx\,dx = \frac{1}{2}kx_1^2$$

which is also equal to the shaded area under the force vs. deflection line in Fig. 10-18. The result is generally true for a linearly deforming spring, so the work U of a force F on the spring is

$$U = \frac{1}{2}kx^2 \tag{10-15a}$$

where x is the change in length (from the unstretched length of the undeformed spring) caused by the force. Care must be taken to use consistent units for k and x. In a more general case, as in going from x_1 to x_2 in Fig. 10-18,

$$\boxed{U = \int_{x_1}^{x_2} kx\,dx = \frac{1}{2}k(x_2^2 - x_1^2)} \tag{10-15b}$$

Torsion Springs

The concepts described above are readily extended to analyze the deformation and energy in a member which is twisted through an angle θ (measured in radians) when a couple of moment M is applied to it. In this case Eq. 10-14 becomes

$$M = k\theta$$

where k is the torsional spring constant in units of N·m/rad or ft·lb/rad.

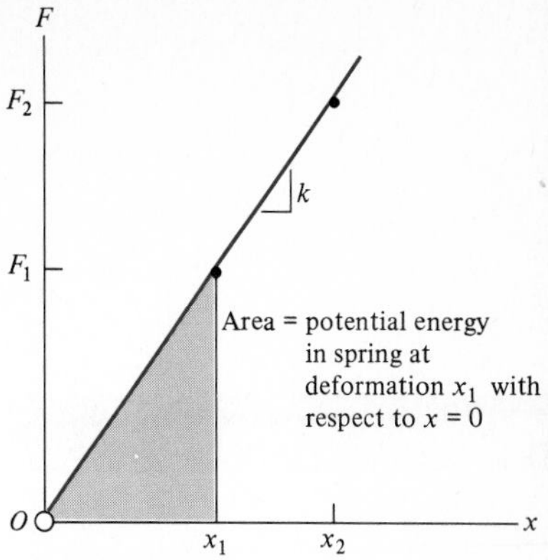

Notation for Potential Energy

The units of potential energy are the units of work, N·m or ft·lb, but it is useful to have a special notation for potential energy as shown in Secs. 10-6 and 10-7. One of the most common symbols for potential energy is V. The relation of work U done on a mechanical system by an external force and the *resulting* potential energy V of the system is generally expressed as

$$U = -V \tag{10-16}$$

This equation shows that negative work is done by a system while its own potential energy is increased through the action of an external force or moment. Of course, the external agent does positive work in this process since it acts in the same direction as that of the resulting motion.

Equation 10-16 is valid for any ideal system based on gravity, springs, and so on. These ideal systems are called *conservative*, and the forces involved are called *conservative forces.* In reality there is always some friction, so Eq. 10-16 is not strictly correct. Even in the best case, there is internal friction in the spring material. Nevertheless, Eq. 10-16 is quite useful in practice when idealized systems are analyzed.

FIGURE 10-18

Potential energy of a deformed spring

10-6 POTENTIAL ENERGY AT EQUILIBRIUM

It was shown in Sec. 10-3 that the concept of virtual work can be used to determine the equilibrium of forces and moments in a system. This method can be altered advantageously using the relation of work and potential energy. Assume that Eq. 10-16 is valid for virtual displacements, so $\delta U = -\delta V$. For equilibrium, $\delta U = 0$, and consequently, $\delta V = 0$. The physical interpretation of this is that small variations δ in the configuration of a system do not change the potential energy V of that system. Assuming that the virtual variations are equivalent to taking the differential with respect to an independent parameter q, the condition of equilibrium on the basis of total potential energy V of the system is

$$\frac{dV}{dq} = 0 \tag{10-17}$$

The independent parameter q should be an independent coordinate along which displacement of the system is possible. There is only one such coordinate for a system with one degree of freedom. For a system with n degrees of freedom, the partial derivatives of the potential energy with respect to each independent parameter q_i must be zero for the system to be in equilibrium. The mathematical statement of this is

$$\frac{\partial V}{\partial q_i} = 0 \tag{10-18}$$

where $i = 1, 2, \ldots, n$. The resulting n equations are independent, and they can be used to check the equilibrium of the system.

10-7 STABILITY OF EQUILIBRIUM

An important aspect of the condition of equilibrium is the stability of the configuration. Various possibilities of equilibrium are illustrated in Fig. 10-19. A ball in a bowl (Fig. 10-19a) is in *stable equilibrium* because when it is displaced from its resting position, it tends to return there. In the position of *neutral equilibrium* (Fig. 10-19b) the ball has no tendency to move away from or toward any particular position. In unstable equilibrium (Fig. 10-19c and d) a small displacement can cause the ball to move even farther from its resting position.

The condition of stability can be analyzed on the basis of potential energy of the system as outlined in the following.

1. *Stable:* Displacement causes an increase in potential energy
2. *Neutral:* Displacement causes no change in potential energy
3. *Unstable:* Displacement causes a decrease in potential energy

A graphical illustration of each of these is given in Fig. 10-20, where the potential energy V is sketched as a function of q, the independent coordinate along which displacement of the system is possible. Since the maximum or minimum values of a function can be determined by considering its second derivative, the conditions of equilibrium and stability can be summarized for a system with one degree of freedom as follows.

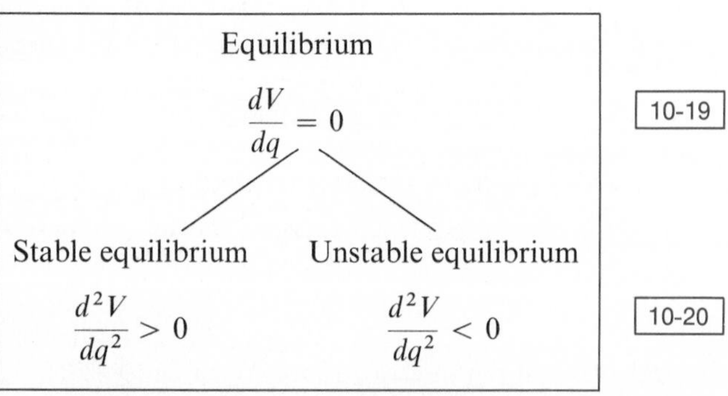

The possible result that $dV/dq = 0$ and $d^2V/dq^2 = 0$ does not indicate neutral equilibrium for certain because derivatives of higher order may be positive, negative, or zero. The higher-order deriva-

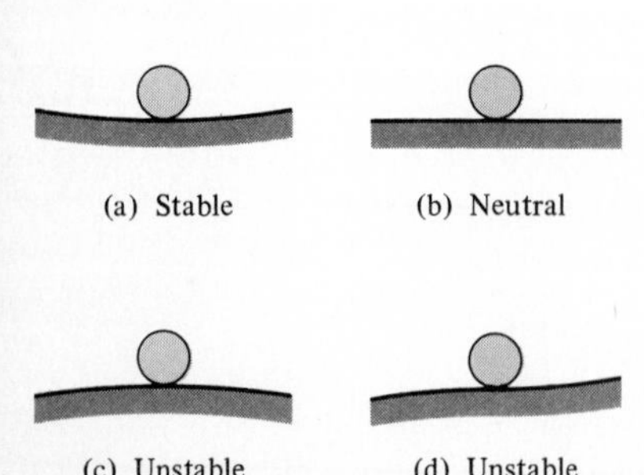

FIGURE 10-19
Stability of equilibrium

tives, while small in magnitude, may affect the variation of the potential energy near the equilibrium position. This variation is the basis of the criterion for stability. In general (without giving proof, because of the subtle details):

1. *Neutral equilibrium:* All derivatives are zero
2. *Stable equilibrium:* The first nonzero derivative is positive and of even order
3. *Unstable equilibrium:* The first nonzero derivative is of odd order (positive or negative), or
   The first nonzero derivative is negative and of even order

The complexities of these statements indicate that in some problems both the qualitative aspects of the stability (stable, neutral, or unstable) and the degree of extent of that condition must be considered. To illustrate this, imagine a small dip at the peak of the curve in Fig. 10-20c. The dip indicates stability, but it is small and is superimposed on a large-scale instability. Naturally, many important problems can be solved without having to consider many derivatives of the potential energy function.

In systems with n degrees of freedom the potential energy depends on n variables, and it may be difficult to visualize or analytically determine the extreme values of the potential energy function. For example, in the case of a system with two degrees of freedom the variation of potential energy is represented as three-dimensional surfaces of V vs. x and y (analogous to Fig. 10-20). The analysis of such a system would involve checking the following:

For *equilibrium* (stable or unstable),

$$\frac{\partial V}{\partial x} = \frac{\partial V}{\partial y} = 0 \tag{10-21}$$

For *stability* (minimum potential energy),

$$\left(\frac{\partial^2 V}{\partial x\, \partial y}\right)^2 - \left(\frac{\partial^2 V}{\partial x^2}\right)\left(\frac{\partial^2 V}{\partial y^2}\right) < 0 \tag{10-22}$$

and

$$\frac{\partial^2 V}{\partial x^2} + \frac{\partial^2 V}{\partial y^2} > 0 \tag{10-23}$$

For *instability* (maximum potential energy),

$$\left(\frac{\partial^2 V}{\partial x\, \partial y}\right)^2 - \left(\frac{\partial^2 V}{\partial x^2}\right)\left(\frac{\partial^2 V}{\partial y^2}\right) < 0 \tag{10-24}$$

and

$$\frac{\partial^2 V}{\partial x^2} + \frac{\partial^2 V}{\partial y^2} < 0 \tag{10-25}$$

The proof of these criteria are beyond the scope of this text but can be found in texts on advanced calculus.

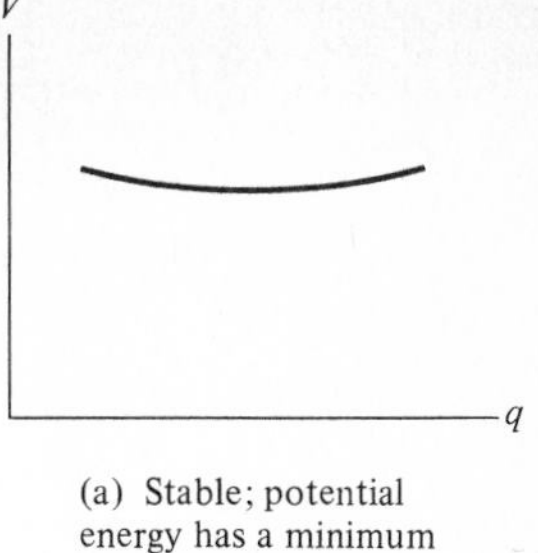

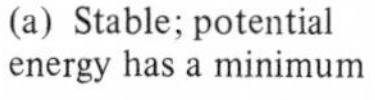

(a) Stable; potential energy has a minimum

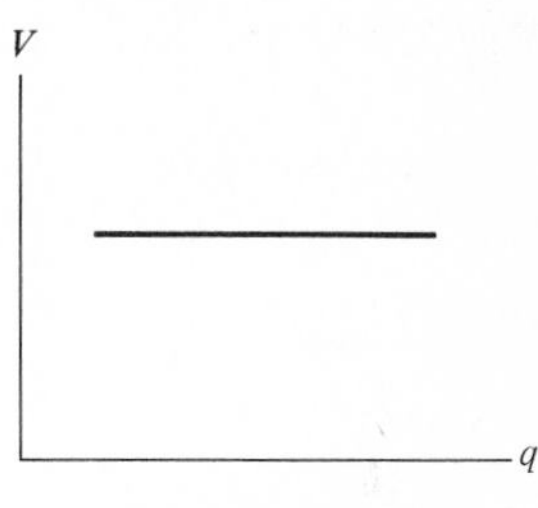

(b) Neutral; potential energy is constant

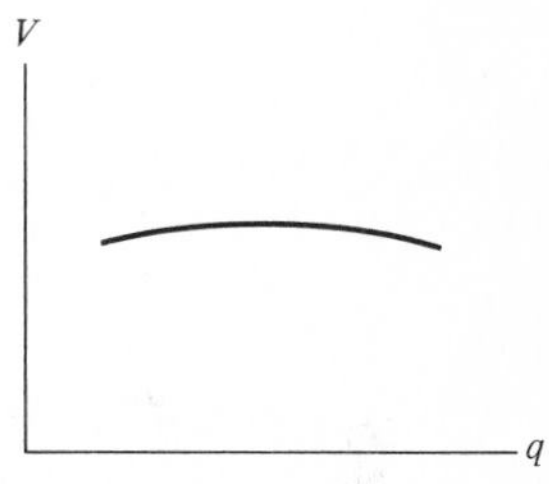

(c) Unstable; potential energy has a maximum

FIGURE 10-20

Potential energy V as a function of coordinate q for one degree of freedom

EXAMPLE 10-5

Show that the scale mechanism shown in Fig. 10-21a is always in equilibrium regardless of the angle θ and the positions of the weights W on the horizontal arms. The bars AB, CD, EF, CE, and DF are rigid members. B, C, D, E, F, and G are smooth pin joints. $EG = GF = l$. What reaction moment is necessary at point A?

SOLUTION

The potential energy of the system arises from the elevation of the weights only. An expression for the potential energy of the system with respect to point A (arbitrarily chosen datum) is written in terms of the parameter θ,

$$\begin{aligned} V &= W_{y_1} + W_{y_2} \\ &= W(h + l\sin\theta + a) + W(h - l\sin\theta + a) \end{aligned}$$

Using Eq. 10-17,

$$\frac{dV}{d\theta} = Wl\cos\theta - Wl\cos\theta = 0$$

$$\frac{d^2V}{d\theta^2} = 0$$

All derivatives of V with respect to θ are zero, hence the mechanism is always in neutral equilibrium and no position is preferred to another. Summing moments about point A (Fig. 10-20b) yields

$$\sum M_A = 0$$

$$M_A + W(l\cos\theta + d_1) - W(l\cos\theta + d_2) = 0$$

$$M_A = W(d_1 - d_2)$$

The resultant moment caused by the two weights is counteracted by the reaction moment M_A at the base. This moment depends only on d_1 and d_2, not on θ. This moment reaction is transferred to the mechanism through the horizontal reactions B_x and G_x (not shown), which form a couple.

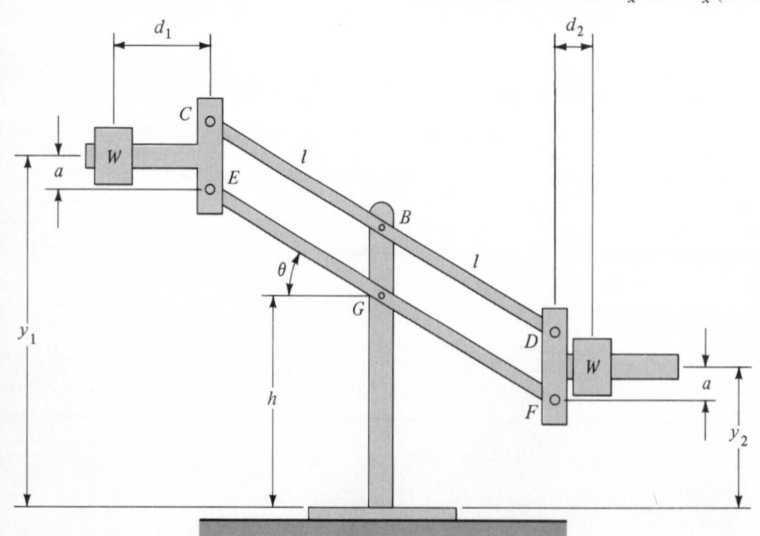

FIGURE 10-21

Scale mechanism with two horizontal arms for the movable weights *W*

EXAMPLE 10-6

The double-decker bus shown in Fig. 10-22a weighs 100 kN when empty. The wheels have a radius of 40 cm and the track width of the wheels is 2.2 m. The center of gravity is halfway between the wheels, 1.2 m above the ground. (a) Determine the angle θ (Fig. 10-22b) where the bus is in equilibrium in a tipped position. Check for stability in this position. (b) Determine whether or not the bus would be stable on a 20-cm-high curb.

SOLUTION

(a) First the bus is sketched in the tipped position, as shown in Fig. 10-22b. The potential energy of the system is written in terms of the displacement parameter, in this case θ.

$$V = Wy = W(1.2\text{ m}\cos\theta + 1.1\text{ m}\sin\theta)$$

For equilibrium, using Eq. 10-17,

$$\frac{dV}{d\theta} = 0 = W(-1.2\text{ m}\sin\theta + 1.1\text{ m}\cos\theta)$$

$$\frac{\sin\theta}{\cos\theta} = \tan\theta = \frac{1.1\text{ m}}{1.2\text{ m}}$$

$$\theta = 42.5^\circ$$

Check for stability for $\theta = 42.5^\circ$,

$$\frac{d^2V}{d\theta^2} = W(-1.2\text{ m}\cos\theta - 1.1\text{ m}\sin\theta)$$

$$= W(-1.2\text{ m}\cos 42.5^\circ - 1.1\text{ m}\sin 42.5^\circ)$$

$$= -1.63\text{ m } W$$

Since $d^2V/d\theta^2 < 0$, the position $\theta = 42.5^\circ$ is an unstable equilibrium, as would be expected.

(b) With the bus on a 20-cm-high curb,

$$\theta = \sin^{-1}\frac{0.2\text{ m}}{2.2\text{ m}} = 5.2^\circ < 42.5^\circ$$

Hence the bus is stable when tipped up on the curb.

(a)

1.1 m sin θ
1.1 m
1.2 m
y
1.2 m cos θ
W = 100 kN
θ
20 cm

(b)

FIGURE 10-22

Equilibrium and stability of a tilted bus

PROBLEMS

SECTIONS 10-5, 10-6, 10-7

FIGURE P10-38

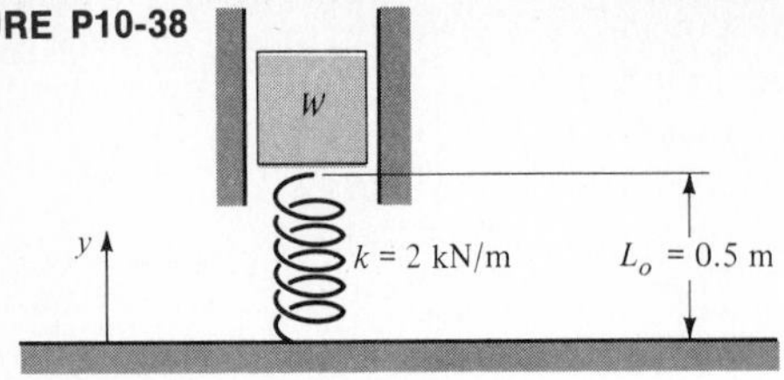

10–38 A block of weight $W = 100$ N is placed (between frictionless guides) on a helical spring whose original length is L_o. Calculate the potential energy of the spring and the position of the block.

10–39 Assume that a second block of weight W is placed on top of the first block in Prob. 10–38. Calculate the change in potential energy of the spring caused by the second block.

FIGURE P10-40

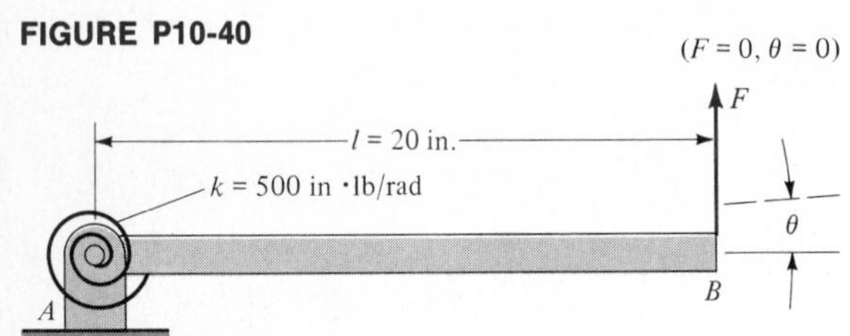

10–40 A torsion spring of spring constant k is attached to the hinged bar AB. The spring is undeformed at $\theta = 0$. Determine θ and the potential energy in the spring when $F = 60$ lb.

10–41 Assume that the force F in Prob. 10–40 increases. Calculate the change in potential energy of the spring as F changes from 60 lb to 150 lb.

FIGURE P10-42

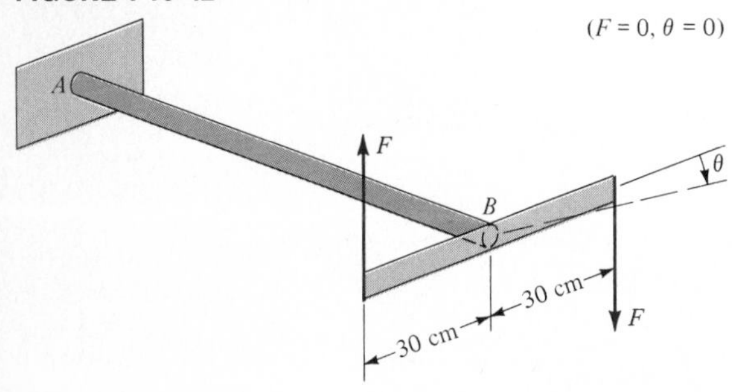

10–42 A shaft is fixed at A, and a couple is applied to it at end B. The torsional spring constant of the shaft is $k = 5$ kN·m/rad (the angular displacement θ is that of end B relative to end A). Calculate the force F required to cause a rotation of $\theta = 1°$.

10–43 Determine the change in potential energy of the shaft in Prob. 10–42 as the angle θ increases from 1° to 2°.

FIGURE P10-44

FIGURE P10-45

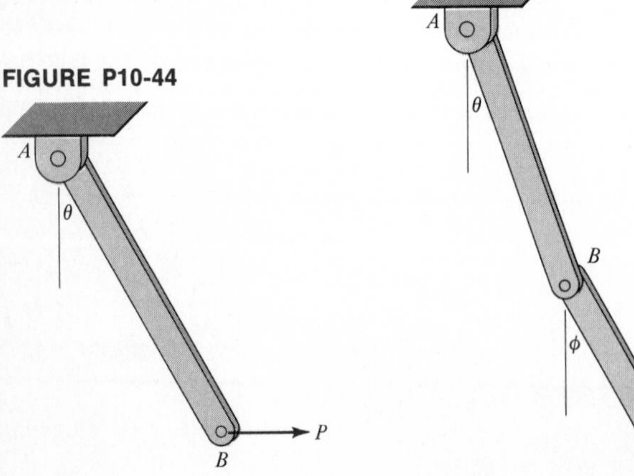

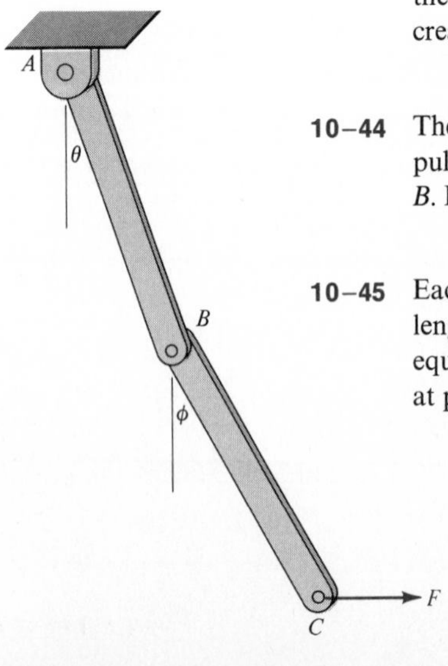

10–44 The uniform bar of weight W and length l is pulled to the right by force $P = 3W$ at end B. Determine the angle θ at equilibrium.

10–45 Each of the uniform bars has weight W and length l. Determine the angles θ and ϕ at equilibrium after a force $F = W$ is applied at point C.

10–46 Each of the uniform bars has weight W and length l. Determine the angle θ at equilibrium after a force $P = 2W$ is applied at point C. Is this a stable position?

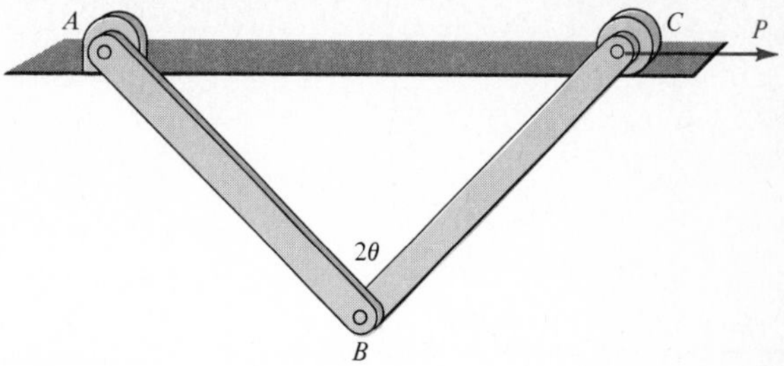

10–47 The rope supporting the weight $W = 50$ N is attached to the unstretched spring when A is at $x = 20$ cm. Calculate the position of A after the spring is deformed and check the system for stability.

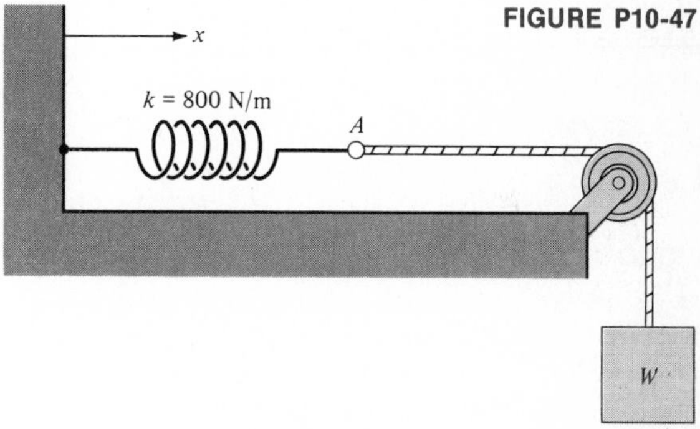

10–48 The weightless bar of length $l = 6$ in. is supported by a spring at $d = 2$ in. The bar is horizontal when $P = 0$. Determine the angle θ of the bar when $P = 3$ lb is applied. Is this a stable position?

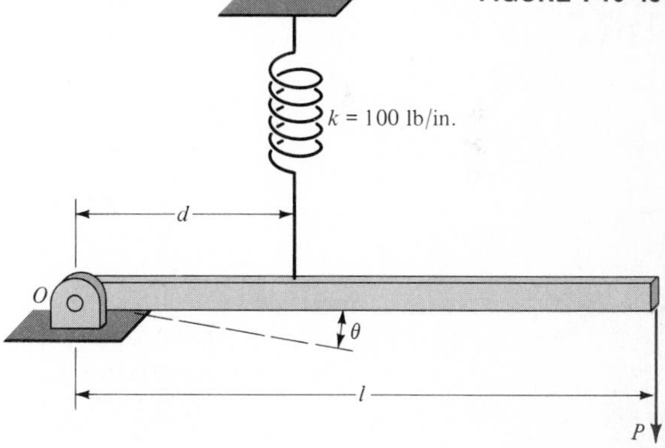

10–49 Consider the model of a compact scale mechanism. The dimensions a and b are measured when P and the force in the spring AB are both zero. Determine the angle θ for $P = 100$ N and check the stability of the system. Solve by trial and error if necessary.

10–50 An engine throttle mechanism is modeled as a hinged bar. The torsional spring has a constant $k = 200$ in·lb/rad; it is unstrained when $\theta = 30°$ counterclockwise from the horizontal. Determine the angle θ for $P = 5$ lb (which is always vertical) and check the stability of the system. Solve by trial and error if necessary.

FIGURE P10-49

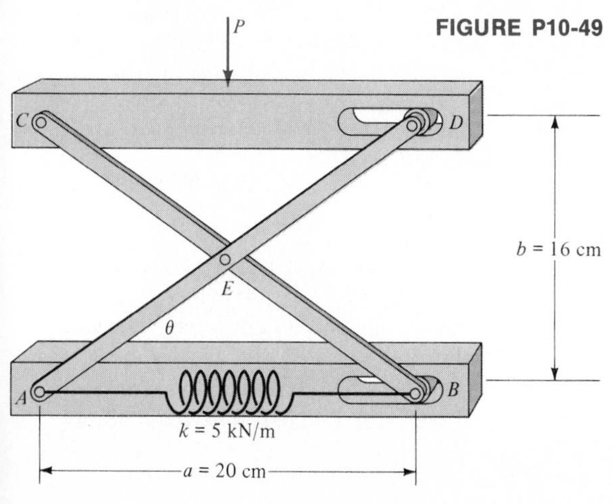

FIGURE P10-46

FIGURE P10-47

FIGURE P10-48

FIGURE P10-50

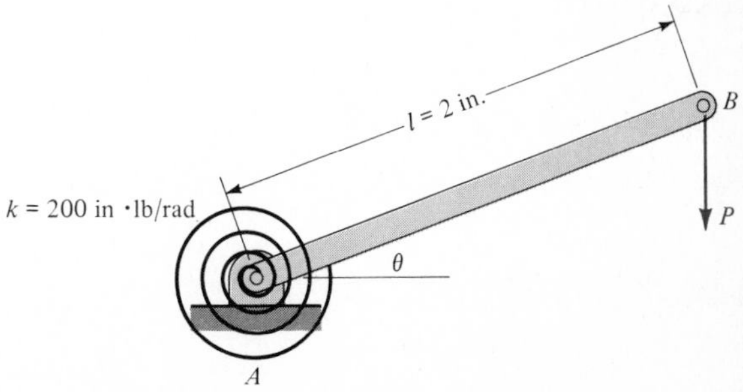

FIGURE P10-51

FIGURE P10-55

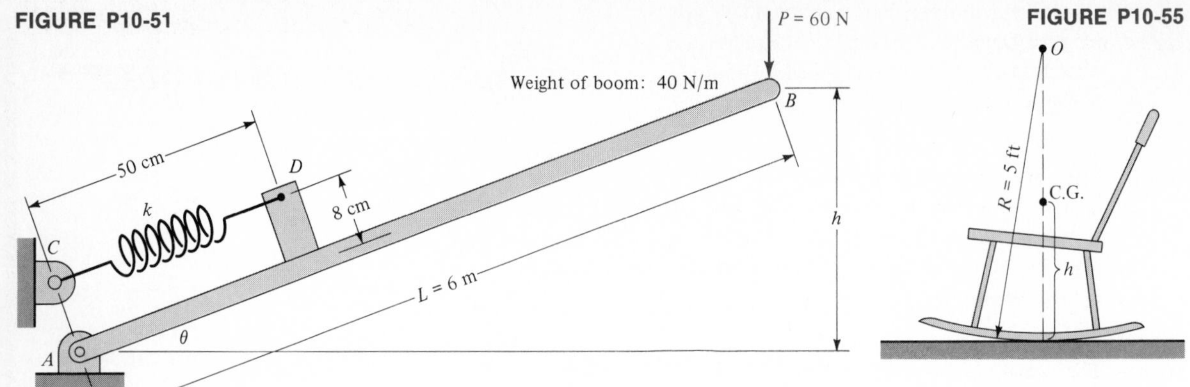

FIGURE P10-52

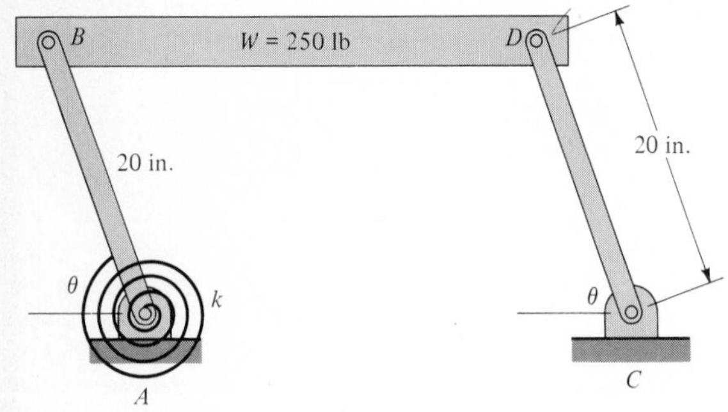

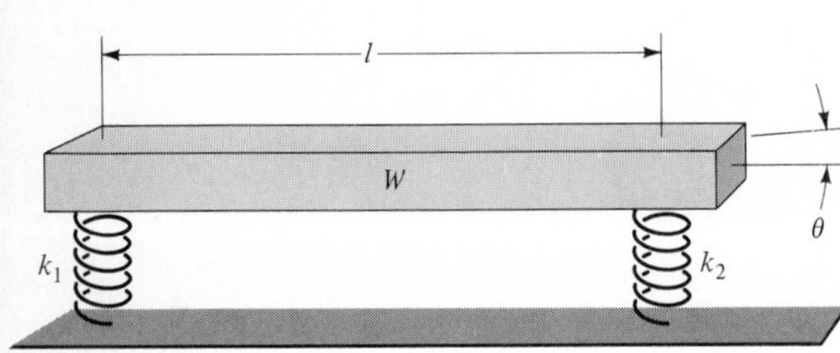

FIGURE P10-54

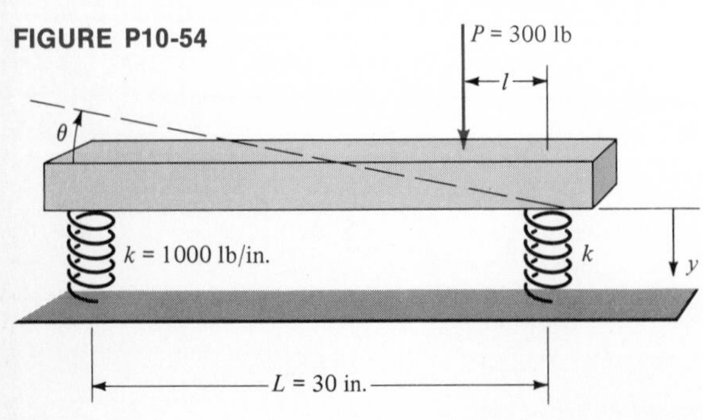

10–51 Consider the boom of an electric trolley (Fig. 10-14). Assume that the weight of each boom is uniformly distributed, 40 N/m. Determine the spring constant k (there are two springs on each boom) for $\theta = 30^\circ$ and a spring elongation of 2 cm (from 48 cm to 50 cm). Check the equilibrium using the potential energy in the given configuration. Use statics, small deflections, and assume that the spring is parallel to the boom in the deformed state.

10–52 A proposed shake table for vibration testing is modeled as a heavy bar supported by two hinged arms of negligible weight. A torsion spring of stiffness $k = 500$ ft·lb/rad is attached to arm AB. Determine the angle θ at equilibrium if the spring applies no moment to the arm at $\theta = 90^\circ$. Solve by trial and error if necessary.

10–53 A uniform bar of weight $W = 200$ N is placed on two springs of identical initial height $h = 10$ cm. Calculate the angle θ of the bar as it rests on the springs if $l = 1$ m, $k_1 = 4$ kN/m, and $k_2 = 6$ kN/m.

10–54 A weightless bar is placed on two identical springs. Determine the angle θ of the bar after the force P is applied at $l = 8$ in.

10–55 Consider a rocking chair of simple construction. Determine if there is stable equilibrium for a person of weight $W = 200$ lb with center of gravity at $h = 2$ ft.

SUMMARY

Section 10-1

WORK OF A FORCE:

$$U = Fs$$

where U = work
F = constant force
s = displacement of a body in the same direction as F is acting

For an arbitrary path from point 1 to 2,

$$U = \int_1^2 \mathbf{F} \cdot d\mathbf{r} = \int_1^2 (F_x\, dx + F_y\, dy + F_z\, dz)$$

There is no work when:

1. A force is applied to a fixed, rigid body

$$(dr = 0, \text{ so } dU = 0)$$

2. A force acts perpendicular to the displacement ($\mathbf{F} \cdot d\mathbf{r} = 0$)

Section 10-2

WORK OF A COUPLE:

$$U = M\theta$$

where M = moment of the couple
θ = angular displacement (in radians) in the same plane as the couple is acting

In a rotation from angular position α to β,

$$U = \int_\alpha^\beta \mathbf{M} \cdot d\boldsymbol{\theta} = \int_\alpha^\beta (M_x\, d\theta_x + M_y\, d\theta_y + M_z\, d\theta_z)$$

Section 10-3

VIRTUAL WORK δU OF FORCE F OR MOMENT M:

$$\delta U = \mathbf{F} \cdot \delta\mathbf{r}$$

$$\delta U = \mathbf{M} \cdot \delta\boldsymbol{\theta}$$

where $\delta\mathbf{r}$ and $\delta\boldsymbol{\theta}$ are imaginary infinitesimal displacements.

A general statement of equilibrium on the basis of virtual work is

$$\delta U = \sum_{i=1}^{m} \mathbf{F}_i \cdot \delta\mathbf{r}_i + \sum_{j=1}^{n} \mathbf{M}_j \cdot \delta\boldsymbol{\theta}_j = 0$$

where i refers to an individual force and the corresponding virtual displacement, and j refers to an individual couple and the corresponding virtual rotation.

Section 10-4

MECHANICAL EFFICIENCY η OF REAL SYSTEMS

$$\text{input work} = \text{useful work} + \text{work of friction}$$

$$\eta = \frac{\text{output work}}{\text{input work}} = \frac{\text{useful work}}{\text{total work required}}$$

For an ideal machine: $\eta = 1$

For any real machine: $\eta < 1$

Section 10-5

GRAVITATIONAL WORK AND POTENTIAL ENERGY:

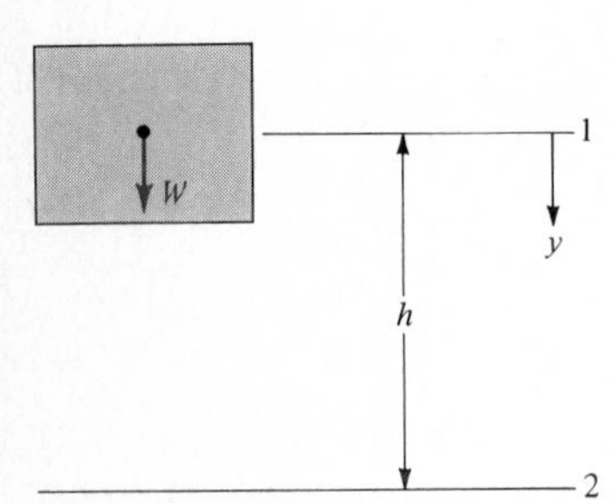

The work of weight W as the block moves from level 1 to 2 is

$$U_{12} = \int_1^2 W\,dy = Wh$$

$$= \text{potential energy of the block at level 1 with respect to 2}$$

The work of weight W as the block moves from level 2 to 1 is

$$U_{21} = \int_2^1 -W\,dy = -Wh$$

$$= \text{potential energy of the block at level 2 with respect to 1}$$

RELATION OF WORK U AND RESULTING POTENTIAL ENERGY V: The work U done on a mechanical system by an external force and the resulting potential energy V of the system are expressed as

$$U = -V$$

The change in potential energy as the block moves from level 2 to 1 is

$$V_{21} = -U_{21} = Wh$$

This is an alternative way to express the potential energy of the block at level 1 with respect to 2.

ELASTIC POTENTIAL ENERGY: The work of a force ($F = kx$) on a spring is

$$U = \frac{1}{2}kx^2$$

$= $ potential energy of the deformed spring with respect to the undeformed spring

where $k = $ spring constant of the spring (N/m or lb/in.)
$x = $ change in length (from the length of the undeformed spring)

In deforming a spring from length x_1 to x_2,

$$U = \int_{x_1}^{x_2} kx\,dx = \frac{1}{2}k(x_2^2 - x_1^2)$$

$= $ potential energy at length x_2 with respect to length x_1

Section 10-6

POTENTIAL ENERGY AT EQUILIBRIUM: For equilibrium,

$$\frac{dV}{dq} = 0$$

where $q = $ an independent coordinate along which displacement of the system is possible

For a system with n degrees of freedom,

$$\frac{\partial V}{\partial q_i} = 0 \qquad \text{where} \quad i = 1, 2, \ldots, n$$

Section 10-7

STABILITY OF EQUILIBRIUM:

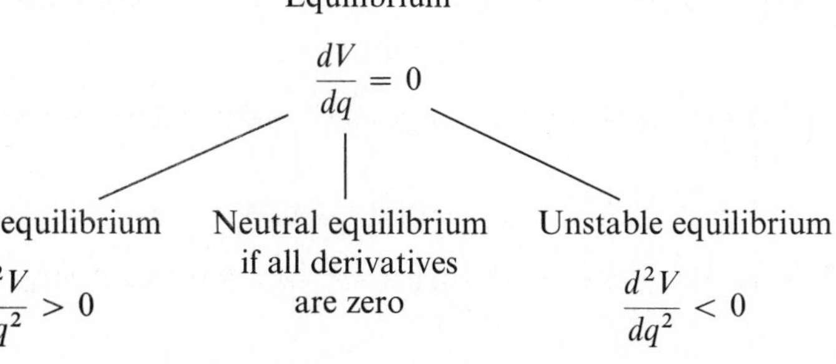

Note that the conditions $dV/dq = 0$ and $d^2V/dq^2 = 0$ do not necessarily indicate neutral equilibrium because derivatives of higher order may be other than zero. In complex configurations the derivatives of higher order and their signs must be considered as described in Sec. 10-7 to determine the stability of the system.

REVIEW PROBLEMS

FIGURE P10-56

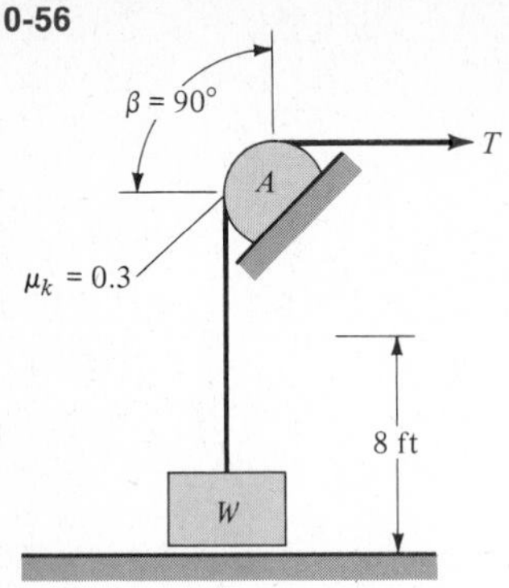

FIGURE P10-58

FIGURE P10-60

10–56 A block of weight $W = 200$ lb is raised to a height of 8 ft by a rope which passes over the fixed member A. Calculate the total work done by the tension T in raising the block.

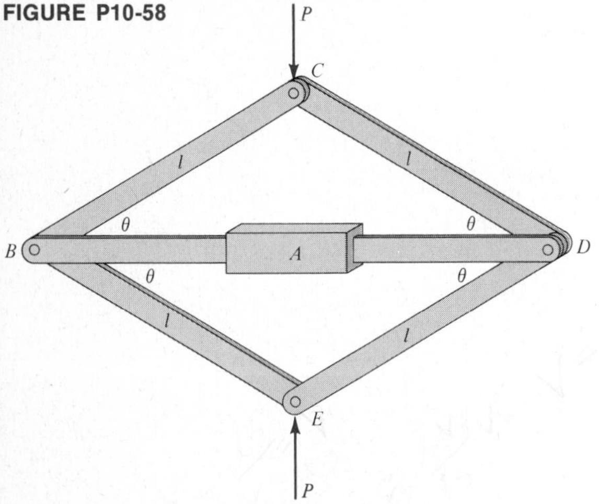

10–57 Four identical bars of length $l = 2$ ft and negligible weight are pinned at B, C, D, and E. A hydraulic actuator A (two-force member) applies a force F at B and D. Determine the force F of the actuator required to maintain equilibrium if $P = 10{,}000$ lb, $l = 3$ ft, and $\theta = 30^\circ$.

10–58 Plot the force F vs. the angle $\theta = 30^\circ$, 45°, and 60° in Prob. 10–57. State the range of angles θ for which $F \leq P$.

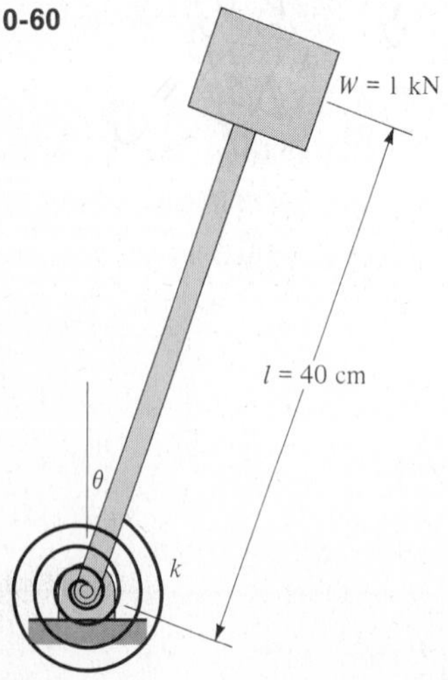

10–59 Determine the mechanical efficiency of the polishing machine described in Prob. 10–16.

10–60 The mechanism shown is a model for a typical office chair. Determine the required spring constant k for equilibrium at $\theta = 30^\circ$ and check the stability of equilibrium. The spring applies no force when $\theta = 0$.

VOLUME 2

DYNAMICS

CHAPTER 11

KINEMATICS OF PARTICLES: SCALAR METHODS

OVERVIEW

Dynamics is the analysis of moving bodies. The fundamental concepts of this area were first investigated and stated by Galileo Galilei (1564–1642) and Isaac Newton (1642–1727). Since that time, dynamics has been applied in many areas of engineering work. It is indispensable in the analysis of all moving vehicles and high-speed machinery. Dynamics is frequently used in the design of massive and apparently static structures such as bridges and high-rise buildings. There are also applications of dynamics in biomechanics work for aerospace travel, and in medical and sports investigations.

Dynamics consists of two major areas. *Kinematics* deals with the geometry and the time-dependent aspects of motion without considering the forces causing motion. In these studies forces may or may not be associated with the motion. The parameters of interest in kinematics are position, displacement, velocity, acceleration, and time. *Kinetics* is based on kinematics and it includes the effects of forces on masses.

Further divisions of dynamics are made according to the sizes and rigidities of the masses whose motions are analyzed. *Particles* include those masses whose dimensions can be ignored when considering their motion. Bodies that are not rotating may be considered as particles in many cases. The criterion for this assumption is that only the motion of the center of mass should be important. On this basis, particles may range in size from submicroscopic to astronomical. Often a mass cannot be treated as a particle. In that case, it should be analyzed either as a *rigid body*, a *deformable solid*, or a *fluid*. In this text the emphasis is on the dynamics of particles and rigid bodies.

STUDY GOALS

In simple problems of dynamics the scalar methods are easily understood by the average student and are adequate to obtain solutions. Many scalar methods are special cases of vector methods which are essential to solve complex problems. There are pedagogical advantages and disadvantages of starting a study of dynamics with the scalar methods. Those who prefer a heavy emphasis on vector methods should rapidly progress through this chapter and devote the time gained to Chap. 12. The topics of this chapter are introduced in the following overview.

SECTION 11-1 presents definitions and mathematical relationships for the position, velocity, and acceleration of a particle that moves linearly in a rectangular coordinate system. These statements are part of the basic vocabulary in dynamics.

SECTION 11-2 has simple examples in which either the position, or the velocity, or the acceleration are given as a function of time, and the other two quantities are determined by differentiation or integration.

SECTION 11-3 presents several useful expressions for common problems where the acceleration is given as a function of another quantity. Some of the equations derived in this section are frequently used in solving problems where initial conditions of position and velocity are given.

SECTION 11-4 introduces the concept of relative motion of two particles, which may be separate or belong to the same body. Here the scalar method of analysis is limited to the case where both particles move along the same path or when modeling with a single path is reasonable.

11-1 RECTILINEAR POSITION, VELOCITY, AND ACCELERATION

The motion of a particle can be analyzed by considering its position and the rates of change of its position. These are most conveniently described in terms of rectangular coordinates. The simplest motion of a particle is along a straight line which is also conveniently specified by rectangular coordinates. Such motion is called *rectilinear motion.* To understand all aspects of this simple motion it is sufficient to consider motion along one axis, as in Fig. 11-1.

Assume that a particle is at point A at time $t = 0$. The particle is at position B at time t, and at position C at time $t + \Delta t$. In general, the position coordinate x will have a positive or negative sign to describe the direction of the displacement from the original point of reference.

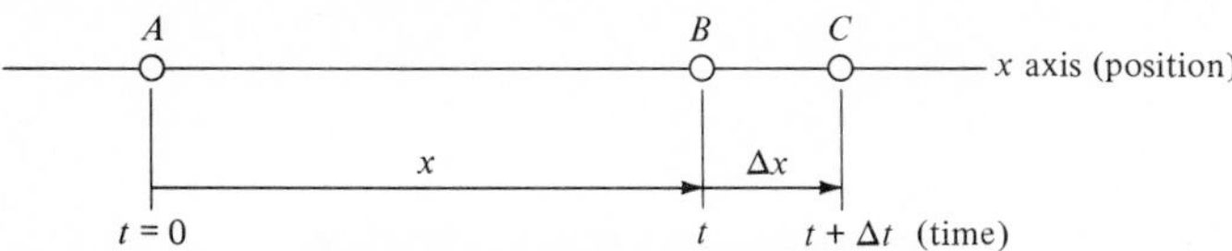

FIGURE 11-1

Rectilinear motion of a particle along the x axis

The next item of interest concerning the motion of the particle is the time rate of change of its position. The quotient of displacement and time interval at the limit is called the *instantaneous velocity v.* For the particle at point B in Fig. 11-1,

$$\text{instantaneous velocity} = v = \lim_{\Delta t \to 0} \frac{\Delta x}{\Delta t} = \frac{dx}{dt} = \dot{x} \tag{11-1}$$

When the displacement is finite, the average velocity may also be of interest. The average velocity is the quotient of the net displacement to the total time interval. For the particle moving from point B to C in Fig. 11-1,

$$\text{average velocity} = v_{\text{ave}_{BC}} = \frac{\Delta x}{\Delta t}$$

The units of velocity are always in terms of distance divided by time. Typically, they are meters per second (m/s) in the SI terminology, and feet per second (ft/s) in the U.S. customary system. Velocity may have a positive or negative sign, denoting the direction of the displacement in the chosen coordinate system. The magnitude of the velocity is known as the *speed* of the particle. Velocity is actually a vector quantity, as shown in Chap. 12, but the scalar analysis of magnitude and sense of direction (sign) is adequate for rectilinear motion.

The instantaneous velocity is usually not constant during a displacement. For analyzing the variation of velocity, the change in instantaneous velocity divided by the corresponding infinitesimal time interval is defined as the

$$\text{instantaneous acceleration} = a = \lim_{\Delta t \to 0} \frac{\Delta v}{\Delta t} = \frac{dv}{dt} = \dot{v} \qquad \text{11-2}$$

Also, from Eq. 11-1,

$$a = \frac{dv}{dt} = \frac{d(dx/dt)}{dt} = \frac{d^2x}{dt^2} = \ddot{x}$$

When the displacement is finite, the average acceleration is determined as the quotient of the net change in velocity Δv to the corresponding total time interval Δt,

$$\text{average acceleration} = a_{\text{ave}} = \frac{\Delta v}{\Delta t}$$

The units of acceleration are always in terms of distance divided by time squared. Typically, they are m/s^2 in SI units, and ft/s^2 in the U.S. customary system. Acceleration may be positive or negative, denoting an increase or a decrease in velocity, respectively. Sometimes a negative acceleration is called *deceleration*. For numerical analysis, let $\Delta v = |v_{\text{final}} - v_{\text{initial}}|$. Acceleration is a vector quantity, as shown in Chap. 12. The scalar analysis of acceleration must be limited to rectilinear motion.

In many problems it is advantageous to eliminate the time dt from the equations. Using the chain rule,

$$a = \frac{dv}{dt} = \frac{dv}{dx}\frac{dx}{dt} = \frac{dv}{dx}v$$

gives

$$v\,dv = a\,dx \qquad \text{11-3}$$

Equations 11-1, 11-2, and 11-3 are the differential equations of motion of a particle. Similar equations can be written using the y or z coordinates when the motion is rectilinear.

11-2 MEASURED AND DERIVED QUANTITIES AS FUNCTIONS OF TIME

In practice it is possible to measure either position, or velocity, or acceleration as a function of time. From any of these three, the other two quantities may be calculated. The three basic possibilities in rectilinear motion are discussed in the following.

Position measured, $x = f(t)$. When the position of a particle is measured continuously as a function of time, the position can be expressed as a mathematical function. For example, consider the position x given by

$$x = t^3 - 4t^2 + 5 \qquad \textbf{(a)}$$

It is seen from Eq. 11-1 that the velocity of the particle can be obtained by differentiating Eq. (a) with respect to time t,

$$v = \frac{dx}{dt} = 3t^2 - 8t \qquad \textbf{(b)}$$

The velocity can be calculated directly for chosen times t, or for positions x using Eq. (a).

It is seen from Eq. 11-2 that acceleration of the particle is determined by differentiating Eq. (b) with respect to time t,

$$a = \frac{dv}{dt} = \frac{d^2x}{dt^2} = 6t - 8 \qquad \textbf{(c)}$$

The position, velocity, and acceleration of the particle can be plotted on separate graphs using the same time base for a coordinate, as shown in Fig. 11-2. Such graphs are valuable in showing the various aspects of the motion at a glance.

The three graphs in Fig. 11-2 have several notable features that are typical for all such plots. The slope of the position vs. time curve at any time t is the velocity v of the particle at that particular time. The slope of the velocity vs. time curve at any time t is the acceleration a at that time. The differential area under the velocity curve (Fig. 11-2b) is $dA = v\,dt$, which is equal to the differential displacement dx. Integrating the velocity between two arbitrary times t_1 and t_2 is thus equivalent to integrating dx between the corresponding positions x_1 and x_2. The area A under the velocity curve from time t_1 to t_2 is expressed as

$$A = \int_{t_1}^{t_2} v\,dt = \int_{x_1}^{x_2} dx = x_2 - x_1 \qquad \boxed{11\text{-}4}$$

This means that the area under the velocity–time curve in a specified time interval is equal to the net displacement during that time interval. The velocities and accelerations can be related in a similar procedure. From Fig. 11-2c, $dA' = a\,dt = dv$, and the area A' under the acceleration curve is expressed as

$$A' = \int_{t_1}^{t_2} a\,dt = \int_{v_1}^{v_2} dv = v_2 - v_1 \qquad \boxed{11\text{-}5}$$

This means that the area under the acceleration–time curve during a specified time interval is equal to the net change in velocity during that time.

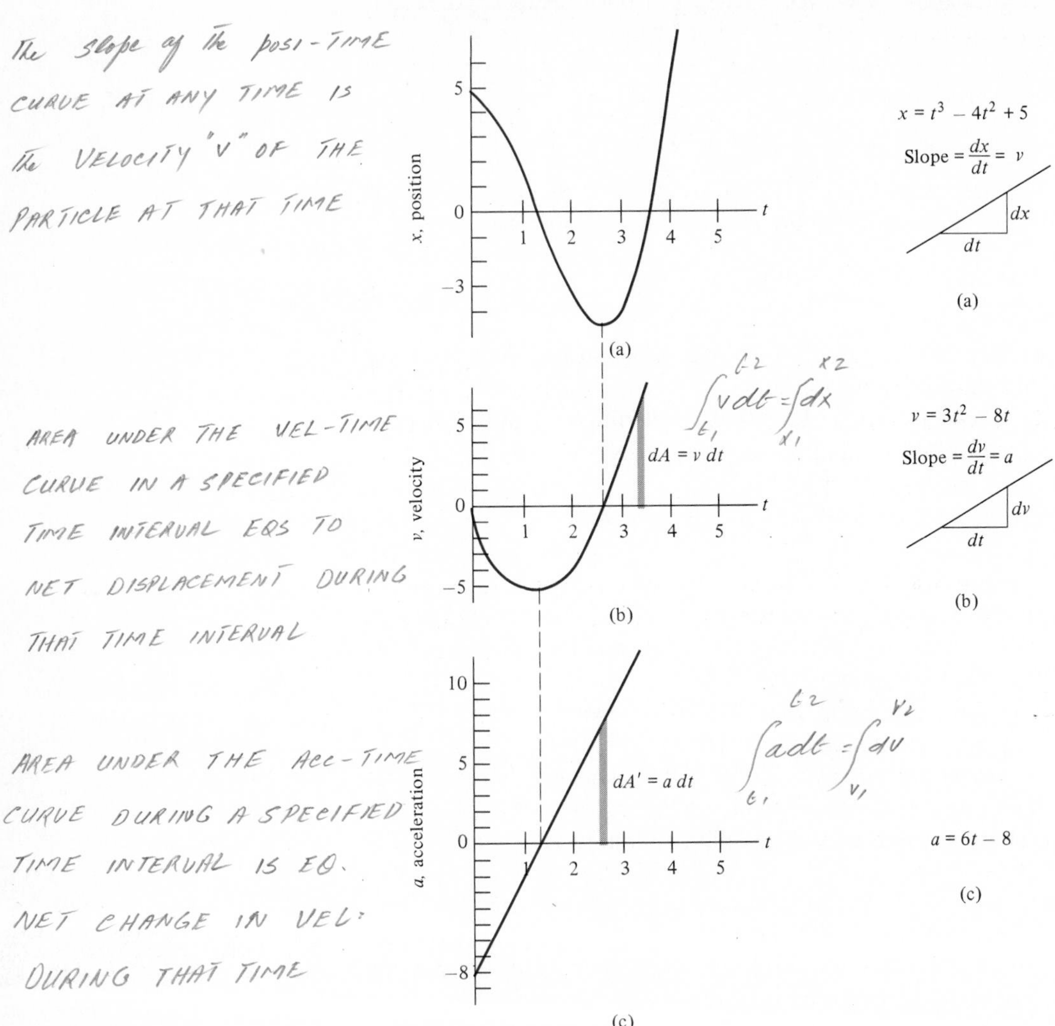

FIGURE 11-2

Plots of a particle's position, velocity, and acceleration

Velocity measured, $v = f(t)$. Sometimes the velocity is the parameter determined directly as a function of time. For example, assume that the velocity of a particle in rectilinear motion is estimated from a plot of data in the time interval from $t_0 = 0$ to t as

$$v = \sin 2t + 3 \tag{d}$$

From this function the position x and the acceleration a at time t can be determined by integration and differentiation, respectively. From Eqs. 11-1 and 11-2,

$$x = \int_{t_0}^{t} v\,dt = \left[\frac{-1}{2}\cos 2t + 3t\right]_0^t$$

$$= -\frac{\cos 2t}{2} + 3t - \frac{1}{2} \tag{e}$$

$$a = \frac{dv}{dt} = 2\cos 2t \tag{f}$$

Acceleration measured, $a = f(t)$. Acceleration as a function of time is frequently the most convenient to measure in modern experimental analysis of motion. For example, assume that the acceleration of a particle in rectilinear motion is measured from time $t_0 = 0$ to t, and it is expressed as

$$a = 2t - 3 \tag{g}$$

with $x = v = 0$ at $t = 0$. The velocity and position at time t can be determined from this by successive integration. From Eqs. 11-2 and 11-1,

$$v = \int_{t_0}^{t} a\,dt = [t^2 - 3t]_0^t = t^2 - 3t \tag{h}$$

$$x = \int_{t_0}^{t} v\,dt = \left[\frac{1}{3}t^3 - \frac{3}{2}t^2\right]_0^t = \frac{t^3}{3} - \frac{3t^2}{2} \tag{i}$$

which satisfy the given initial conditions.

11-3 USEFUL EXPRESSIONS BASED ON ACCELERATION

Acceleration is often the basic quantity in the analysis of motion because it is easily measured or is specified in the problem. Therefore, it is useful to learn to work with expressions based on functions of acceleration. These involve time, velocity, displacement, or a combination of some of these quantities, as shown in the following. It should be noted that there are a few basic expressions (these are the most important to remember) which can be applied in several special cases.

Basic Expressions

When time is used as a parameter, $a = dv/dt$ (Eq. 11-2) is rearranged and integrated:

$$\int_{v_0}^{v} dv = \int_0^t a\,dt \qquad \boxed{11\text{-}6}$$

When displacement is used as a parameter, the expression $v\,dv = a\,dx$ (Eq. 11-3) is integrated:

$$\int_{v_0}^{v} v\,dv = \int_{x_0}^{x} a\,dx \qquad \boxed{11\text{-}7}$$

In these equations the initial conditions $t = 0$, $x = x_0$, and $v = v_0$ are used as the lower limits of integration. The upper limits are the arbitrary time, position, and velocity of interest. Note the relationship between Eqs. 11-6 and 11-7 since Eq. 11-3 was obtained from Eq. 11-2 using the chain rule.

Special Cases

Constant acceleration. The simplest and also one of the most common cases is when the acceleration is constant. From Eq. 11-6,

$$\int_{v_0}^{v} dv = a\int_0^t dt$$

Thus, at an arbitrary time t,

$$\boxed{v = at + v_0} \qquad \boxed{11\text{-}8}$$

The velocity can also be obtained as a function of the distance x. From Eq. 11-7,

$$\int_{v_0}^{v} v\,dv = a\int_{x_0}^{x} dx$$

From this, at an arbitrary displacement x,

$$\boxed{v^2 = 2a(x - x_0) + v_0^2} \qquad \boxed{11\text{-}9}$$

A useful expression for the position x is derived from $v = dx/dt$ (Eq. 11-1) and Eq. 11-8,

$$\int_{x_0}^{x} dx = \int_0^t (at + v_0)\,dt$$

The result is in terms of constant acceleration a, initial velocity v_0, and arbitrary time t,

$$x = \frac{1}{2}at^2 + v_0t + x_0 \qquad \text{11-10}$$

Note that Eqs. 11-8, 11-9, and 11-10 are valid *only* for constant acceleration; this includes the possibility of $a = 0$.

Variable acceleration given as a function of time. Assume that $a = f(t)$. From Eqs. 11-6 and 11-1,

$$\int_{v_0}^{v} dv = \int_0^t f(t)\,dt \qquad v = \int_0^t f(t)\,dt + v_0$$

$$\int_{x_0}^{x} dx = \int_0^t v\,dt \qquad x = \int_0^t v\,dt + x_0$$

These expressions can be further evaluated only when $f(t)$ is specified.

Variable acceleration given as a function of displacement. Assume that $a = f(x)$. From Eq. 11-7,

$$\int_{v_0}^{v} v\,dv = \int_{x_0}^{x} f(x)\,dx \qquad v^2 = 2\int_{x_0}^{x} f(x)\,dx + v_0^2$$

This can be solved for x as a function of t when $f(x)$ is specified and v is replaced by dx/dt.

Variable acceleration given as a function of velocity. Assume that $a = f(v)$. From $a = dv/dt$,

$$dt = \frac{dv}{f(v)}$$

$$t = \int_0^t dt = \int_{v_0}^{v} \frac{dv}{f(v)}$$

After integrating this expression, v is obtained as a function of t. Another integration results in the relation for the displacement x. Equation 11-3 may also be used, giving

$$a = f(v) = v\frac{dv}{dx} \qquad dx = \frac{v\,dv}{f(v)}$$

$$\int_0^x dx = x = \int_{v_0}^{v} \frac{v\,dv}{f(v)}$$

Note that the solution of problems involving variable acceleration as a function of velocity is complex and intractable except in trivial cases.

EXAMPLE 11-1

A particle P moves along a straight line and its position is given by $x = 2t^3 - 4t^2 + 3$, where x is in meters and t is in seconds. Calculate the times when the velocity and the acceleration are zero. Sketch x vs. t, v vs. t, and a vs. t, and check the numerical answers using the graphs.

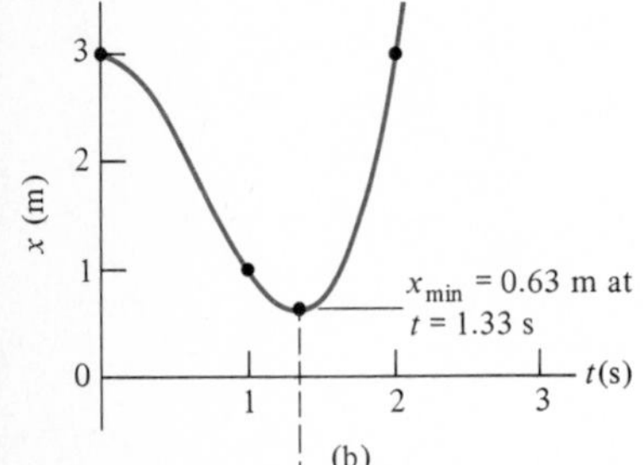

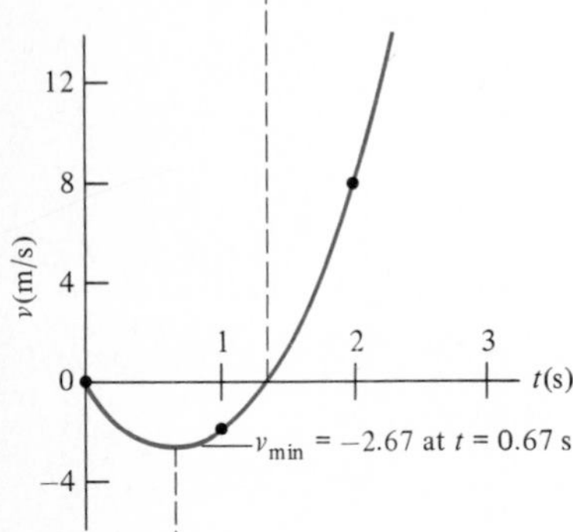

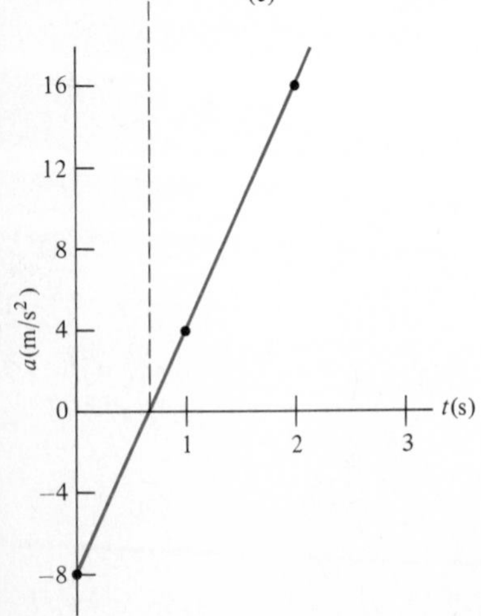

SOLUTION

A relation for the velocity as a function of time is determined using Eq. 11-1,

$$v = \frac{dx}{dt} = 6t^2 - 8t = t(6t - 8)$$

Setting this equation to zero yields that the velocity is zero at times $t = 0$ and $t = \frac{4}{3}$ s.

The acceleration is found using Eq. 11-2,

$$a = \frac{dv}{dt} = 12t - 8$$

Setting this equation to zero yields $a = 0$ at $t = \frac{2}{3}$ s.

The velocity curve should be a sketch of the slope of the $x(t)$ curve at each point and hence the velocity should be zero when the position is at a local maximum or minimum (when the slope of the tangent to the curve is zero). Similarly, the acceleration curve should be a graph of the slope of the velocity curve and should be zero when the velocity is a local maximum or minimum. Using these principles and calculations with several values of t, the three curves are sketched in Fig. 11-3. The results are consistent.

FIGURE 11-3

Plots of a particle's position, velocity, and acceleration

EXAMPLE 11-2

The velocity of a large-displacement piston (Fig. 11-4a) is measured from $t = 0$ as $v = \sin(t/2) + 1/\pi$, where v is in feet per second and t is in seconds. The piston starts at the position $x_0 = 0$ at $t = 0$. Calculate the maximum acceleration and sketch x vs. t, v vs. t, and a vs. t for a complete cycle after which the motion is repeated.

SOLUTION

The position x as a function of time is determined first. From Eq. 11-1,

$$\int_{x_0=0}^{x} dx = \int_{t=0}^{t} v\,dt = \int_0^t \left(\sin\frac{t}{2} + \frac{1}{\pi}\right) dt$$

$$x = -2\cos\frac{t}{2} + \frac{t}{\pi} + 2$$

The acceleration a is determined from Eq. 11-2,

$$a = \frac{dv}{dt} = \frac{1}{2}\cos\frac{t}{2}$$

The maximum acceleration occurs when $\cos(t/2) = 1$ or when $t = 0, 4\pi, 8\pi$, and so on. The acceleration at these times is $a = \frac{1}{2}\ \text{ft/s}^2$. The sketches of x vs. t, v vs. t, and a vs. t are based on the following table of values and are shown in Fig. 11-4.

t (s)	x (ft)	v (ft/s)	a (ft/s^2)
0	0	$\frac{1}{\pi}$	$\frac{1}{2}$
π	3	$1 + \frac{1}{\pi}$	0
2π	6	$\frac{1}{\pi}$	$-\frac{1}{2}$
3π	5	$-1 + \frac{1}{\pi}$	0
4π	4	$\frac{1}{\pi}$	$\frac{1}{2}$
6.93	6.10	0	-0.47
11.92	3.90	0	0.47

The local maximum and minimum values of x and the times at which they occur are found by solving for the times when the velocity is zero and substituting these times into the expression for x.

$$v = \sin\frac{t}{2} + \frac{1}{\pi} = 0 \qquad \sin\frac{t}{2} = -\frac{1}{\pi}$$

Therefore, $v = 0$ at $t = 2\sin^{-1}(-1/\pi)$,

$$t = 6.93\ \text{s} \quad \text{or} \quad 11.92\ \text{s}$$

Only positive values of t less than 4π are used. These values are then substituted in the expression for $x(t)$ and plotted together with several other points.

JUDGMENT OF THE RESULTS

The velocity and acceleration curves cross the zero axis when the position and velocity are local maximums or minimums, respectively. Thus, the results appear to be correct. Note that the oscillatory component of the position x is found to be small relative to the total displacement when time is large in the equation for x.

FIGURE 11-4

Plots of a piston's position, velocity, and acceleration

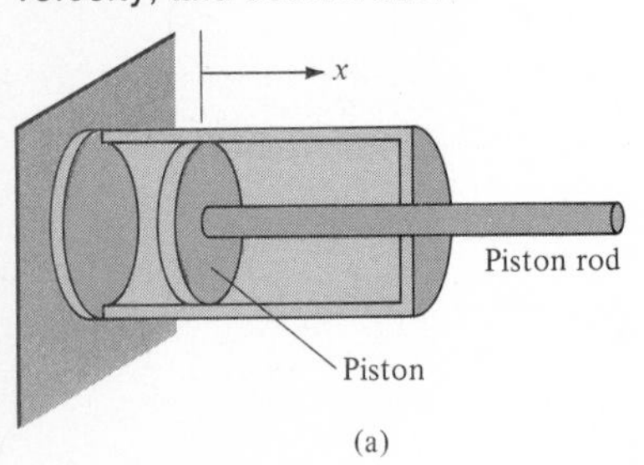

(a)

x (ft)

$x_{max} = 6.1$ at $v = 0$

$x_{min} = 3.9$ at $v = 0$

$t = 6.93$ s

$t = 11.92$ s

t (s)

(b)

v (ft/s)

$v_{max} = 1.32$ at $a = 0$

$v = \frac{1}{\pi}$

$v_{min} = -0.68$ at $a = 0$

t (s)

(c)

a (ft/s^2)

t (s)

(d)

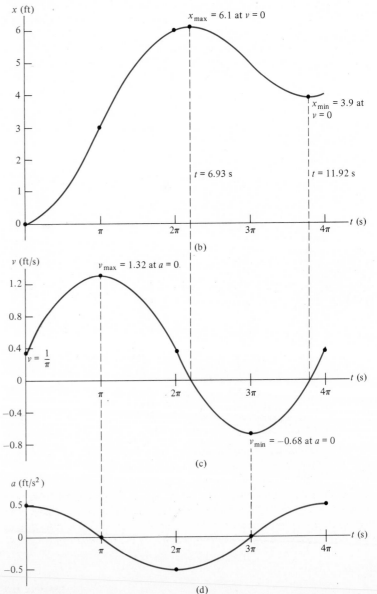

EXAMPLE 11-3

In the laboratory simulation of impacts on hard hats (Fig. 11-5) a hammer is dropped from a height of 3 m. Determine the hammer's velocity just before impact if during the free fall the hammer is subject to a constant acceleration of 9.81 m/s^2 downward. What is the required initial velocity of the hammer at the height of 3 m to simulate a free fall from a height of 20 m?

SOLUTION

The relations for velocity and position when the acceleration is constant are derived in case 1 in Sec. 11-3. First a reference system must be established. Arbitrarily defining the origin to be at the point of impact and positive x being up, $a = -9.81$ m/s^2, and the initial conditions are $x_0 = 3$ m and $v_0 = 0$. The velocity v is required at position $x = 0$. From Eq. 11-9,

$$\begin{aligned} v^2 &= 2a(x - x_0) + v_0^2 \\ &= 2(-9.81 \text{ m/s}^2)(0 - 3 \text{ m}) + 0^2 \\ &= 58.9 \text{ m}^2/\text{s}^2 \\ v &= 7.7 \text{ m/s} \end{aligned}$$

The velocity of the hammer after falling 20 m under an acceleration of 9.81 m/s^2 is

$$\begin{aligned} v^2 &= 2(-9.81 \text{ m/s}^2)(0 - 20 \text{ m}) + 0^2 \\ &= 392 \text{ m}^2/\text{s}^2 \\ v &= 19.8 \text{ m/s} \end{aligned}$$

To obtain the same final velocity v after falling only 3 m, the initial velocity v_0 is expressed from Eq. 11-9,

$$\begin{aligned} v_0^2 &= v^2 - 2a(x - x_0) \\ &= 392 \text{ m}^2/\text{s}^2 - 2(-9.81 \text{ m/s}^2)(0 - 3 \text{ m}) \\ &= 333 \text{ m}^2/\text{s}^2 \\ v_0 &= 18.3 \text{ m/s} \end{aligned}$$

JUDGMENT OF THE RESULTS

The absolute and relative magnitudes of the results are reasonable by inspection considering the height of fall and the magnitude of the acceleration. Note that a reference system must always be defined and the given information must be made consistent with that system before the problem can be solved.

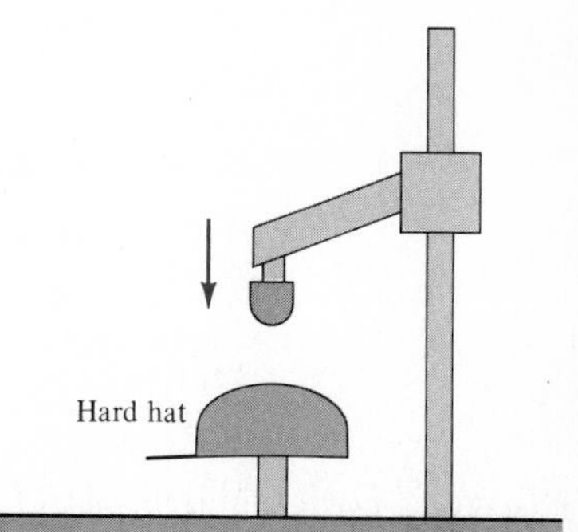

FIGURE 11-5
Impact testing of hard hats

EXAMPLE 11-4

Worker A throws a small tool vertically toward worker B (Fig. 11-6a). The tool's initial velocity is $v_0 = 50$ ft/s upward at the initial height of $y_0 = 6$ ft. The acceleration of the tool is constant, $a = 32.2$ ft/s^2 downward. Determine (a) the highest elevation y_{max} that the tool could reach, and (b) the velocity of the tool at the ground if it falls back without interference. Use integration of basic equations and draw the y vs. t and v vs. t curves.

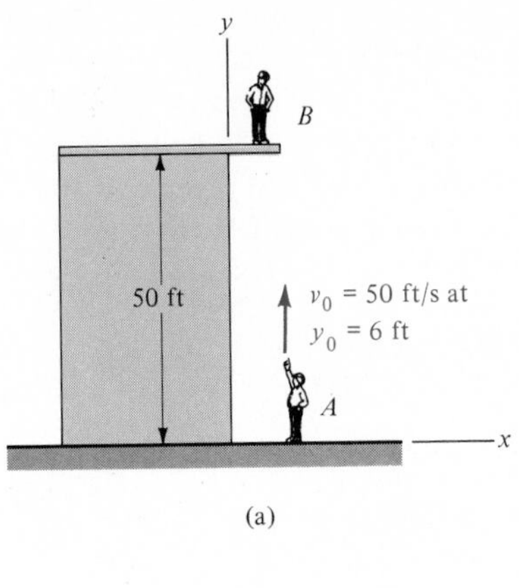

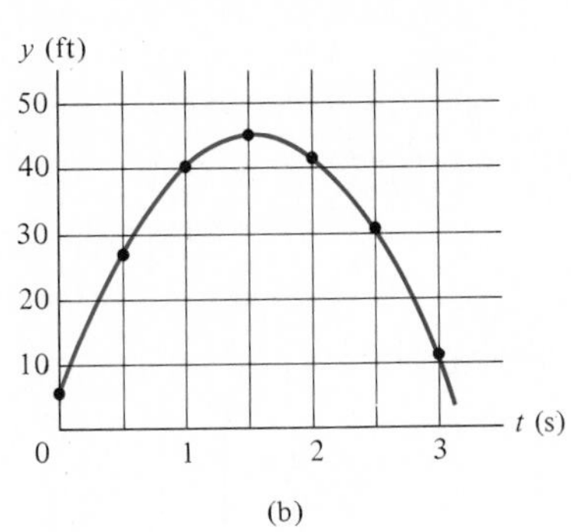

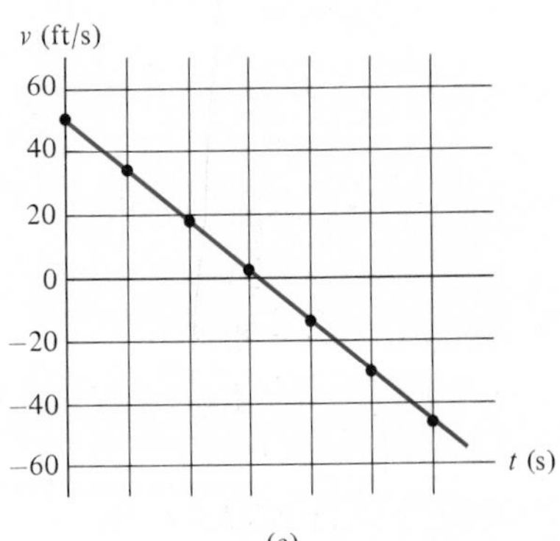

FIGURE 11-6

Plots of a vertically thrown object's position and velocity

SOLUTION

Using the given reference system, $a = -32.2$ ft/s^2, and at $t = 0$, $y_0 = 6$ ft and $v_0 = 50$ ft/s.

(a) First it is necessary to obtain expressions for the velocity and elevation at any time t. From Eq. 11-6,

$$\int_{v_0=50}^{v} v = -\int_0^t 32.2\, dt$$

$$[v]_{50}^{v} = -[32.2t]_0^t$$

$$v = 50 - 32.2t \qquad \textbf{(a)}$$

From Eq. 11-1 and using $y_0 = 6$ ft at $t = 0$,

$$v = \frac{dy}{dt} = 50 - 32.2t$$

$$\int_{y_0=6}^{y} dy = \int_0^t (50 - 32.2t)\, dt$$

$$[y]_6^y = [50t - 16.1t^2]_0^t$$

$$y = 6 + 50t - 16.1t^2 \qquad \textbf{(b)}$$

At the highest elevation $v = 0$. Substituting into Eq. (a), the time t_1 of upward flight is obtained,

$$50 - 32.2t_1 = 0$$

$$t_1 = 1.553 \text{ s} \qquad \textbf{(c)}$$

Substituting Eq. (c) into Eq. (b) yields

$$y_{max} = 6 + (50)(1.553) - (16.1)(1.553)^2$$
$$= 44.8 \text{ ft}$$

(b) When the tool hits the ground, $y = 0$. Substituting into Eq. (b), the total time t_2 of upward and downward flight is obtained,

$$6 + 50t_2 - 16.1t_2^2 = 0$$

$$t_2 = 1.553 \pm 1.668$$

Using the positive root for the total flight measured from $t = 0$ yields

$$t_2 = 3.221 \text{ s} \qquad \textbf{(d)}$$

The velocity of the tool at the ground is obtained from Eqs. (a) and (d),

$$v = 50 - (32.2)(3.221) = -53.7 \text{ ft/s}$$

The y vs. t and v vs. t plots are obtained from Eqs. (b) and (a), respectively, and are drawn in Fig. 11-6b and c.

JUDGMENT OF THE RESULTS

The final velocity is reasonable considering the initial velocity, the initial and final elevations, and the magnitude of the acceleration.

PROBLEMS

SECTIONS 11-1, 11-2, 11-3

11–1 The position x of a particle is defined by the expression $x = 3t + 4$, where x is in meters and t is in seconds. Determine the velocity and acceleration when $t = 4$ s.

11–2 The position y of a particle is defined by the relationship $y = 2t^2 - 5$, where y is in feet and t is in seconds. Calculate the velocity and acceleration at $t = 3$ s.

11–3 The position z of a particle is defined by the expression $z = t^3 - 2t$, where z is in meters and t is in seconds. Determine the position and acceleration when the velocity is zero.

11–4 The position x of a particle is defined by the relationship $x = 5 \sin 2t + 4$, where x is in inches and t is in seconds. Calculate the velocity and acceleration at $t_1 = 1$ s and $t_2 = 2$ s.

11–5 The position y of a particle is defined by the expression $y = t^4 - 4t^2 + t + 2$, where y is in meters and t is in seconds. Determine the velocity when the acceleration is zero.

11–6 The velocity of a particle is defined by the expression $v = 2t$, where v is in feet per second and t is in seconds. The particle is at $x_0 = 2$ ft when $t = 0$. Calculate the position and acceleration at $t = 3$ s.

11–7 The velocity of a particle is defined by the expression $v = 3 \cos t + 1$, where v is in meters per second and t is in seconds. The particle is at $y_0 = 2$ m when $t = 0$. Determine the position and acceleration at $t = 10$ s.

11–8 The velocity of a particle is estimated from data and expressed as $v = t^2 + 3t - 2$, where v is in inches per second and t is in seconds.

The particle is at $x_0 = -2$ in. at $t = 0$. Determine the position and acceleration at $t = 5$ s.

11–9 The velocity of a particle is defined by the relationship $v = 3x$, where v is in meters per second and x is in meters. Determine the position, velocity, and acceleration at $t = 0.5$ s if initially $v_0 = 0$ at $x_0 = 0.1$ m.

11–10 The velocity of a particle is defined by the expression $v = ky^2$, where v is in feet per second, y is in feet, and k is a constant. Determine the acceleration at $y = 100$ ft if initially $v_0 = 2$ ft/s at $y_0 = 5$ ft.

11–11 The acceleration of a particle is given as $a = 6\text{ m/s}^2$. At time $t = 0$ the position is $x_0 = 0$ and the velocity is $v_0 = 0$. Calculate the position and velocity at time $t = 5$ s.

11–12 The acceleration of a particle is given as $a = 4t\text{ ft/s}^2$, where t is in seconds. At time $t = 0$ the position is $x_0 = 0$ and the velocity is $v_0 = 5$ ft/s. Determine the position and velocity at $t = 2$ s.

11–13 The acceleration of a particle is defined by the expression $a = 4 \sin t$, where a is in m/s^2 and t is in seconds. At time $t = 0$ the position is $x_0 = 0$ and the velocity is $v_0 = 8$ m/s. Determine the first position x_1 where the velocity is maximum.

11–14 The acceleration of a particle is defined by the relationship $a = kx\text{ in./s}^2$, where k is a constant and x is in inches. At $x_0 = 0$, $v_0 = 0$, and at $x_1 = 2$ in., $v_1 = 10$ in./s. Determine the velocity v_2 at $x_2 = 5$ in.

11–15 The acceleration of a particle is defined by the expression $a = 100 - x^2$, where a is in m/s^2 and x is in meters. At $x_0 = 0$, $v_0 = 0$. Determine the position x where the velocity is maximum.

11–16 The acceleration of a particle is defined by the expression $a = -kx^{-2}$ ft/s^2, where x is in feet and k is a constant. At $x_0 = 0.1$ ft, $v_0 = 0$, and at $x_1 = 1$ ft, $v_1 = 10$ ft/s. Determine the velocity v_2 at $x_2 = 5$ ft.

11–17 The acceleration of a particle is defined by the relationship $a = -v$, where a is in m/s^2 and v is in m/s. At $t = 0, v_0 = 500$ m/s. Determine the distance traveled by the particle as it comes to rest from the initial velocity.

11–18 The acceleration of a particle is defined by the expression $a = 400 - 5v$, where a is in ft/s^2 and v is in ft/s. At $t = 0$, $v_0 = 200$ ft/s. Determine the time required to reduce the velocity (a) to half its initial value, and (b) to zero.

11–19 The acceleration of a particle is given as $a = -v^2$, where a is in m/s^2 and v is in m/s. Determine the distance d traveled by the particle as it nearly comes to rest from its initial velocity v_0. Plot v/v_0 vs. d as a function of v_0, $2v_0$, and $3v_0$.

11–20 The acceleration of a particle falling through the atmosphere or a liquid is defined by the expression $a = g(1 - k^2v^2)$, where $g = 32.2$ ft/s^2, k is a constant, and v is velocity in ft/s. Initially, the position is $y_0 = 0$ and the velocity is $v_0 = 0$. Determine the velocity v as a function of the position y.

11–21 A projectile is shot in the vertical direction for atmospheric research. Determine the highest elevation reached by the projectile if its initial velocity is $v_0 = 500$ m/s and its downward acceleration is constant, $a = 9.81$ m/s^2.

11–22 In Prob. 11–21, determine the time of flight before the projectile hits the ground, and the velocity just before impact.

11-4 RELATIVE MOTION OF TWO PARTICLES

The motion of a body may be evaluated with respect to another moving body. There are numerous practical reasons for analyzing such relative motion. The most common cases are of vehicles whose motion may be monitored with respect to other vehicles (e.g., police using radar equipment). Some crashes between moving vehicles are in the same category. Another important area of relative motion is in structural mechanics. The relative motion of various parts of a given structure (which on the whole may or may not have a fixed location) is indicative of deformations and possible damage in the structural members. In simple cases the relative motions may be modeled by two particles moving along the same line. The particles may represent two separate bodies, or two distinct points of a deformable body.

Assume that two particles A and B are moving along the same straight line x in Fig. 11-7a, but that they are independent of each other. The positions, velocities, and accelerations of A and B are given in reference to a fixed point O as x_A, v_A, a_A, and x_B, v_B, a_B, respectively.

The *scalar relative position, velocity, and acceleration* of particle B with respect to A are defined using Fig. 11-7a as follows.

Relative position: $$x_{B/A} = x_B - x_A \quad (11\text{-}11)$$

Relative velocity $\left(\dfrac{dx_{B/A}}{dt}\right)$: $$v_{B/A} = v_B - v_A \quad (11\text{-}12)$$

Relative acceleration $\left(\dfrac{dv_{B/A}}{dt}\right)$: $$a_{B/A} = a_B - a_A \quad (11\text{-}13)$$

Note that according to Fig. 11-7 there are signs to consider. For example, $x_{A/B} = -x_{B/A}$, $v_{A/B} = -v_{B/A}$, and so on.

The significance of these equations varies depending on the particular problem. For example, in vehicle crashes the relative velocity is ultimately the most important, while in structural deformations the permanent changes in relative positions of chosen points indicate the extent of damage. The more general vector analyses of relative motion are presented in Chap. 12.

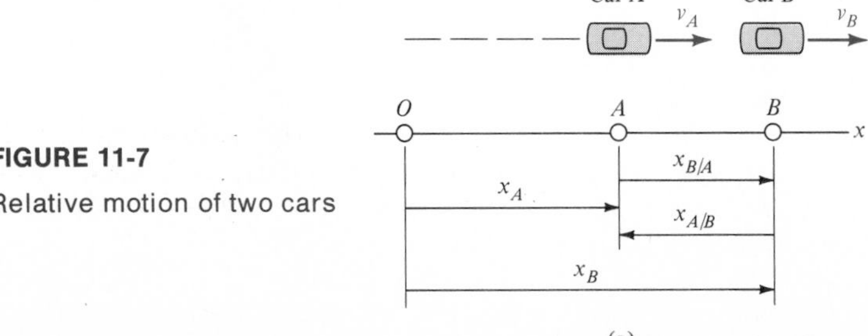

FIGURE 11-7
Relative motion of two cars

for $v_B > v_A$: $v_{A/B}$ (←) $v_{B/A}$ (→)

for $v_A > v_B$: $v_{A/B}$ (→) $v_{B/A}$ (←)

(b)

Dependent relative motions. Degrees of freedom. In some mechanical systems the position of a particle depends on the position of another particle or of several other particles. For example, a flexible tape of constant length is situated about a fixed peg O as in Fig. 11-8.

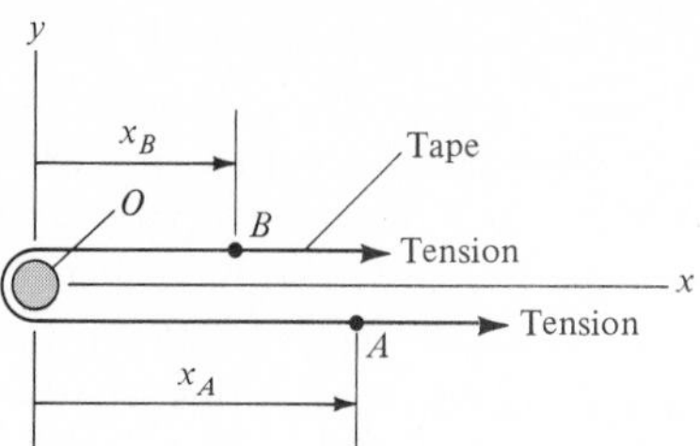

FIGURE 11-8

Tape wrapped around a fixed peg O

In this case only one of the two coordinates x_A and x_B (of particles A and B) may be chosen arbitrarily, so by inspection

$$x_A + x_B = \text{constant}$$

Any system in which only one coordinate is chosen arbitrarily is said to have *one degree of freedom.*

Consider a different tape transport system as in Fig. 11-9, where the tape is wrapped about fixed pegs O and O' and a peg P which is movable in the x direction. Here the coordinate x_A depends on coordinates x_B and x_P, both of which may be arbitrarily chosen, so this system has *two degrees of freedom.* To determine the relationship of the coordinates, consider the following facts: the tape AB has a constant length, and the lengths of tape in contact with the pegs are constant. Thus,

$$x_A + x_B + 2x_P = \text{constant}$$

The relationship of the particles' velocities or of their accelerations is obtained from the expression of their positions by differentiation with respect to time. In the case of the tape in Fig. 11-9,

$$v_A + v_B + 2v_P = 0$$

$$a_A + a_B + 2a_P = 0$$

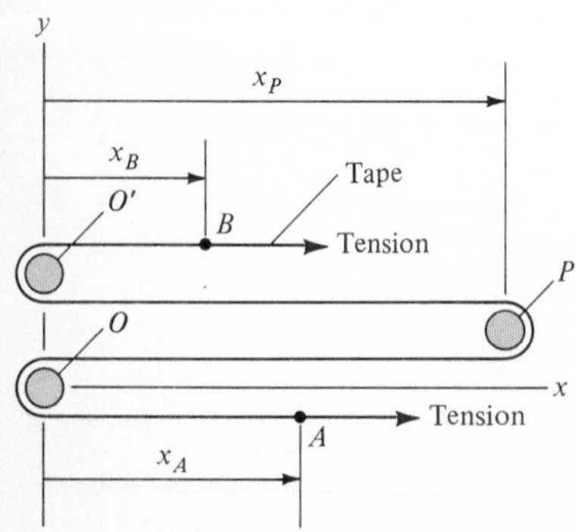

FIGURE 11-9

Tape wrapped around fixed pegs O and O' and movable peg P

EXAMPLE 11-5

Assume that police car P in Fig. 11-10 is equipped with an advanced radar that allows simultaneous tracking of two vehicles. At a particular instant $v_P = 40$ mph, $a_P = 10$ ft/s^2, $v_{A/P} = 25$ mph, $a_{A/P} = -8$ ft/s^2, $v_{B/P} = 20$ mph, $a_{B/P} = -13$ ft/s^2. Determine the absolute velocities v_A and v_B, and absolute accelerations a_A and a_B of vehicles A and B.

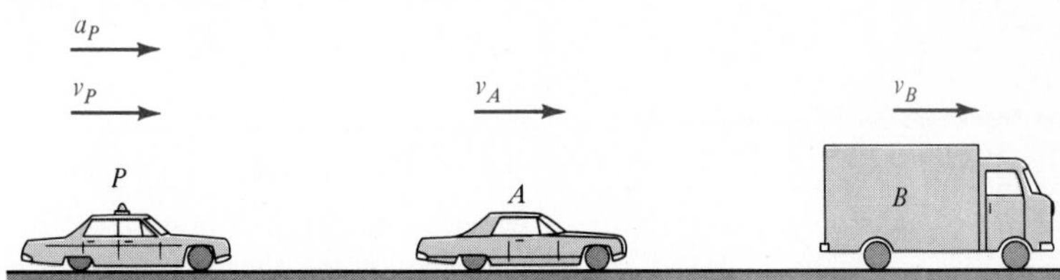

FIGURE 11-10
Relative motion of three vehicles on the same path

SOLUTION

From Eq. 11-12,

$$v_{A/P} = v_A - v_P$$

$$v_A = v_P + v_{A/P} = 40 \text{ mph} + 25 \text{ mph} = 65 \text{ mph}$$

$$v_B = v_P + v_{B/P} = 40 \text{ mph} + 20 \text{ mph} = 60 \text{ mph}$$

From Eq. 11-13,

$$a_{A/P} = a_A - a_P$$

$$a_A = a_P + a_{A/P} = 10 \text{ ft/s}^2 - 8 \text{ ft/s}^2 = 2 \text{ ft/s}^2 \quad \text{(increasing velocity)}$$

$$a_B = a_P + a_{B/P} = 10 \text{ ft/s}^2 - 13 \text{ ft/s}^2 = -3 \text{ ft/s}^2 \quad \text{(decreasing velocity)}$$

EXAMPLE 11-6

In an automobile road race the vehicles are started at 1-minute intervals. Assume that car A precedes car B, and that their accelerations are defined by $6e^{-t/\tau}$ m/s^2, where t is time in seconds and τ is a constant. The maximum velocity of car A is 165 km/h, of car B it is 170 km/h. Determine the relative position and velocity of car B with respect to A at $t = 3600$ s. Measure time from the start of car A.

SOLUTION

First, put the maximum velocities in units of meters and seconds. The velocity of car A is 45.83 m/s, of car B it is 47.22 m/s. The velocity and acceleration are determined using indefinite integrals (to show an alternative method to that used so far) according to Eqs. 11-1 and 11-2,

$$\int dv = \int a\,dt$$

$$= \int 6e^{-t/\tau}\,dt$$

$$v = -6\tau e^{-t/\tau} + C_1$$

and

$$\int dx = \int v\,dt$$

$$x = \int(-6\tau e^{-t/\tau} + C_1)\,dt$$

$$= 6\tau^2 e^{-t/\tau} + C_1 t + C_2$$

where C_1 and C_2 are constants of integration. They are determined using the initial conditions that at $t = 0$, $v = x = 0$. Thus,

$$C_1 = 6\tau$$

$$C_2 = -6\tau^2$$

From $v_{A\max} = 45.83$ m/s $= 6\tau_A$, and $v_{B\max} = 47.22$ m/s $= 6\tau_B$,

$$\tau_A = 7.638 \text{ s} \qquad \text{and} \qquad \tau_B = 7.870 \text{ s*}$$

The equations for velocity and displacement are

$$v_A = 45.83(1 - e^{-t/7.638}) \text{ m/s}$$

$$v_B = 47.22(1 - e^{-t/7.87}) \text{ m/s}$$

$$x_A = 45.83t + 350.0(e^{-t/7.638} - 1) \text{ m}$$

$$x_B = 47.22t + 371.6(e^{-t/7.87} - 1) \text{ m}$$

Using Eqs. 11-11 and 11-12, the relative position and velocity can be found. At $t = 3600$ s,

$$t_A = 3600 \text{ s} \qquad t_B = 3540 \text{ s}$$

$$v_A(3600) = 45.83 \text{ m/s} \qquad v_B(3540) = 47.22 \text{ m/s}$$

$$x_A(3600) = 164.64 \text{ km} \qquad x_B(3540) = 166.79 \text{ km}$$

$$v_{B/A} = 47.22 - 45.83 = 1.39 \text{ m/s}$$

$$x_{B/A} = 166.79 \text{ km} - 164.64 \text{ km} = 2.15 \text{ km}$$

JUDGMENT OF THE RESULTS

Car *A* covered about 2.5 km before car *B* was allowed to start. With a difference between the cars of 5 km/h, it is reasonable that car *B* should be a little over 2 km ahead of car *A* after an hour of racing. Note that the effect of the exponential decreases rapidly with increasing time.

* Note that these are approximations. Usually, $v_{\max}$ is determined using $dv/dt = 0$, but here $dv/dt = a = 6e^{-t/\tau}$, which only gradually approaches zero as t becomes large. Thus, $v_{\max} = -6\tau e^{-t/\tau} + 6\tau \simeq 6\tau$.

EXAMPLE 11-7

Two accelerometers A and B are mounted on an airplane fuselage for measurements in a structural test (two such transducers are shown in proximity in Fig. 11-11). A is located at $x_A = 10$ ft, and B at $x_B = 15$ ft. Thus, $x_{B/A} = 5$ ft initially. Starting at the same time, the acceleration outputs are $a_A = (\pi^2/5000)\sin(\pi t/10)$ ft/s^2 and $a_B = -(\pi^2/6000)\sin(\pi t/12)$ ft/s^2. Determine the value of $x_{B/A}$ at $t = 10$ s. Assume that all motions are along the x axis and that $v_A = v_B = 0$ at $t = 0$.

SOLUTION

Using Eq. 11-2, the displacement can be obtained by twice integrating the acceleration with respect to time. Using indefinite integrals gives

$$v_A = \int a_A\,dt = \int \frac{\pi^2}{5000}\sin\frac{\pi t}{10}\,dt = -\frac{\pi}{500}\cos\frac{\pi t}{10} + C_1$$

$$x_A = \int v_A\,dt = \int\left(-\frac{\pi}{500}\cos\frac{\pi t}{10} + C_1\right)dt$$

$$= -0.02\sin\frac{\pi t}{10} + C_1 t + C_2$$

Similarly,

$$v_B = \int a_B\,dt = \int -\frac{\pi^2}{6000}\sin\frac{\pi t}{12} = \frac{\pi}{500}\cos\frac{\pi t}{12} + C_3$$

$$x_B = \int v_B\,dt = \int\left(\frac{\pi}{500}\cos\frac{\pi t}{12} + C_3\right)dt$$

$$x_B = 0.024\sin\frac{\pi t}{12} + C_3 t + C_4$$

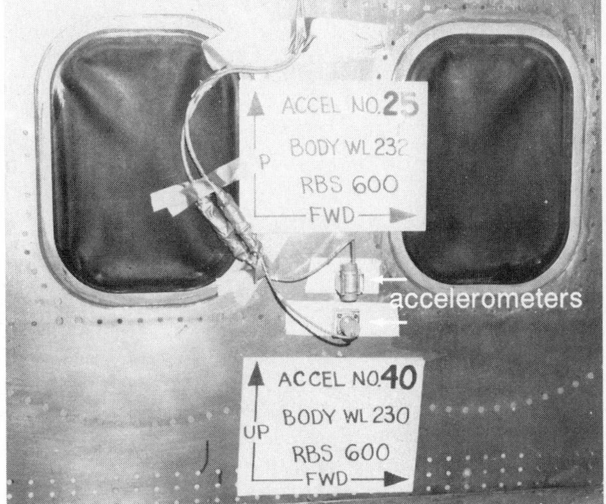

FIGURE 11-11

Two accelerometers on an airplane fuselage. The Boeing Company, Seattle, Washington

$C_1, C_2, C_3,$ and C_4 are constants of integration which can be determined by applying the initial conditions. At $t = 0$, $x_A = 10$ ft, $x_B = 15$ ft, $v_A = v_B = 0$. Therefore,

$$C_1 = \frac{\pi}{500} \text{ ft/s} \qquad C_3 = -\frac{\pi}{500} \text{ ft/s}$$

$$C_2 = 10 \text{ ft} \qquad C_4 = 15 \text{ ft}$$

The final displacement equations become

$$x_A = -0.02 \sin\frac{\pi t}{10} + \frac{\pi t}{500} + 10 \text{ ft}$$

$$x_B = 0.024 \sin\frac{\pi t}{12} - \frac{\pi t}{500} + 15 \text{ ft}$$

At $t = 10$ s, $x_A = 10.063$ ft and $x_B = 14.949$ ft, and $x_B - x_A$ is

$$x_{B/A} = 4.886 \text{ ft}$$

EXAMPLE 11-8

In the pulley system in Fig. 11-12, the motion of block B depends on that of block A. This is a case of dependent relative motion. Since only one of the coordinates y_A and y_B may be chosen arbitrarily if the rope *abc* is of constant length, this system has one degree of freedom. Assume that the pulleys have negligible diameter. Bars CE and BF are of constant length. Determine the acceleration of block B in terms of the acceleration of block A.

SOLUTION

First a relationship for the coordinates y_A and y_B must be established. The coordinates can be expressed as

$$y_A = a + l_1 \qquad \textbf{(a)}$$

$$y_B = c + l_2 = b + l_1 + l_2 \qquad \textbf{(b)}$$

where l_1 and l_2 are constants, and

$$a + b + c = \text{length of rope} = \text{constant} \qquad \textbf{(c)}$$

Substituting for the quantities a, b, and c from Eq. (a) and (b) in Eq. (c) yields

$$y_A - l_1 + y_B - l_1 - l_2 + y_B - l_2 = \text{constant}$$

Since l_1 and l_2 are constants,

$$y_A + 2y_B = \text{constant} \qquad \textbf{(d)}$$

The acceleration of block B is obtained by differentiating Eq. (d) twice,

$$v_A + 2v_B = 0 \qquad \textbf{(e)}$$

$$a_A + 2a_B = 0 \qquad \textbf{(f)}$$

$$a_B = -\frac{a_A}{2}$$

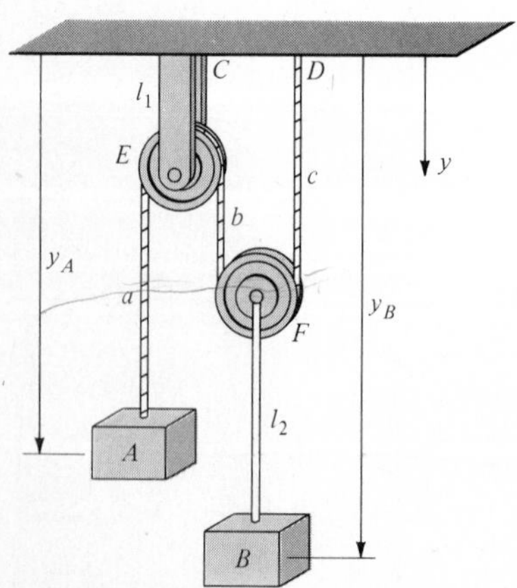

FIGURE 11-12

Relative motion of blocks A and B

11–23 The positions of two particles A and B are defined by $x_A = 3t^2 + 4$ and $x_B = 3t - 2$, where x is in meters and t is in seconds. Determine the relative velocity $v_{B/A}$ and relative acceleration $a_{B/A}$ at $t = 3$ s.

11–24 The positions of two particles A and B are defined by $x_A = t^3 - 2t$ and $x_B = 2t^3 + t^2 - 5$, where x is in feet and t is in seconds. Determine the relative velocity $v_{A/B}$ and relative acceleration $a_{A/B}$ at $t = 2$ s.

11–25 The velocities of two particles A and B are defined by $v_A = 4t^2 + t$ and $v_B = 3t - 4$, where v is in m/s and t is in seconds. Determine the relative position $x_{B/A}$ and relative acceleration $a_{B/A}$ at $t = 5$ s. When $t = 0$, $x_A = x_B$.

11–26 The motions of two particles A and B are defined by $x_A = 4t + 2$ and $v_B = 3t^2 + t$, where x is in feet, v is in ft/s, and t is in seconds. Determine $x_{B/A}$, $v_{B/A}$, and $a_{B/A}$ at $t = 10$ s. When $t = 0$, $x_B = 1$ ft.

11–27 The motions of two particles A and B are defined by $a_A = 20$ and $a_B = t + 2$, where a is in m/s^2 and t is in seconds. Determine $x_{A/B}$ and $v_{A/B}$ at $t = 5$ s. At $t = 0$, $x_A = x_B = v_A = v_B = 0$.

11–28 The motions of two particles A and B are defined by $a_A = 3t$ and $v_B = 2t^2$, where a is in ft/s^2, v is in ft/s, and t is in seconds. Determine $x_{B/A}$ and $a_{B/A}$ at $t = 3$ s. At $t = 0$, $x_A = 1$ ft, $x_B = 3$ ft, $v_A = 2v_B = 2$ ft/s.

11–29 Cars A and B are traveling in the same direction at constant velocities as shown. Determine the distance between the cars 10 s after the instant when they were 500 m apart.

FIGURE P11-29

11–30 Cars A and B are traveling toward each other at constant velocities as shown. Initially, $x_A = 0$ and $x_B = 1000$ ft. Determine when and where the cars would meet with respect to the initial configuration.

FIGURE P11-30

FIGURE P11-31

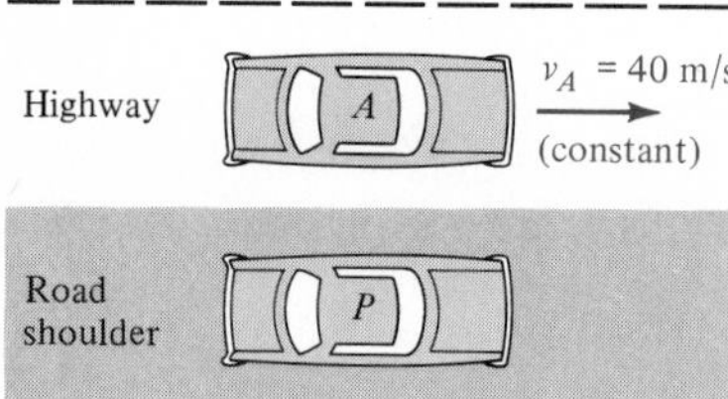

11–31 The police car P starts from rest when car A passes it. Car P accelerates at a constant rate of 1 m/s^2. Determine the speed of car P and the distance traveled where it overtakes car A.

FIGURE P11-32

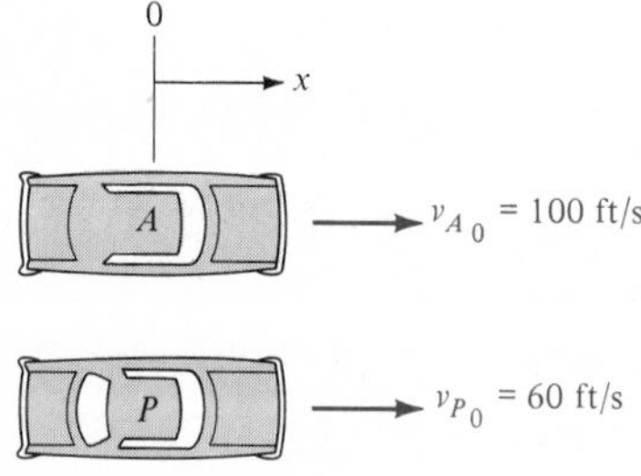

11–32 A motorist in car A passes an unmarked police car P at $x = 0$. At that instant car P accelerates at a constant rate of 3 ft/s^2 and A decelerates at a constant rate of 4 ft/s^2. Determine the time required for car P to overtake car A.

11–33 Airplanes A and B are 2 km apart and traveling in the same direction at the given velocities when B accelerates at a constant rate of 15 m/s^2. Determine the time required for B to overtake A.

FIGURE P11-33

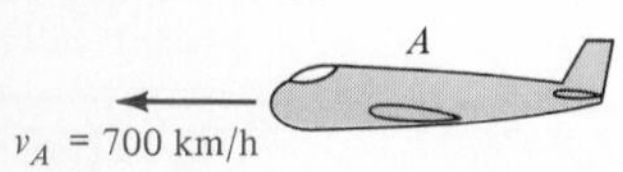

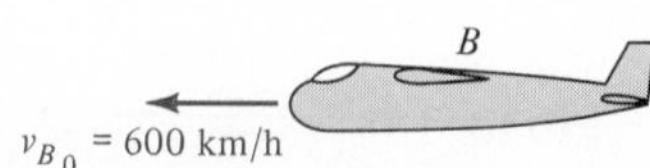

11–34 In a simulation of automobile crashes two cars A and B are started from rest toward each other from a distance of 400 m. The cars have constant accelerations, 6 m/s^2 and 5 m/s^2, respectively. Determine the relative velocity and position of the cars just before the impact.

FIGURE P11-34

11–35 Two drag racing cars A and B start side by side in a race to cover a quarter mile (1320 ft). Assume that each car can accelerate constantly over this distance. Their accelerations are $a_A = 50.00$ ft/s^2 and $a_B = 50.02$ ft/s^2. Determine the relative velocity and distance of the cars when car B crosses the finish line. Assume that there is no lateral distance between the cars.

11–36 What is the time Δt that car B can lose at the start of the race described in Prob. 11–35 to make the two cars even at the finish?

11–37 Blocks A and B are initially at rest when $y_A = 3$ m and $y_B = 2$ m. Block B is accelerated downward at a constant rate of 0.5 m/s^2. Determine the relative vertical position and velocity of the blocks 3 s after the motion begins.

FIGURE P11-37

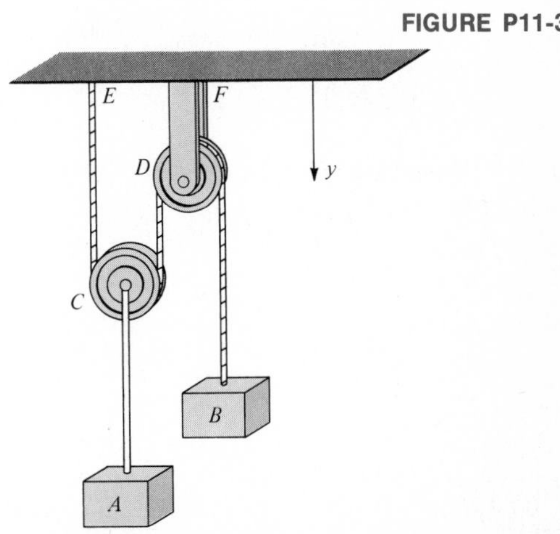

SUMMARY

Section 11-1

Rectilinear position, velocity, and acceleration

INSTANTANEOUS VELOCITY:

$$v = \lim_{\Delta t \to 0} \frac{\Delta x}{\Delta t} = \frac{dx}{dt} = \dot{x}$$

where x = position coordinate
t = time

AVERAGE VELOCITY:

$$v_{\text{ave}} = \frac{\Delta x}{\Delta t}$$

where Δx = net displacement
Δt = total time interval

INSTANTANEOUS ACCELERATION:

$$a = \lim_{\Delta t \to 0} \frac{\Delta v}{\Delta t} = \frac{dv}{dt} = \dot{v} = \ddot{x}$$

AVERAGE ACCELERATION:

$$a_{\text{ave}} = \frac{\Delta v}{\Delta t}$$

where Δv = net change in velocity
Δt = total time interval

Section 11-2

Measured and derived quantities as functions of time

See representative example in the figure on page 527.

TYPICAL FEATURES:

1. $v = 0$ is at a local maximum or minimum of the displacement–time curve.
2. $a = 0$ is at a local maximum or minimum of the velocity–time curve.
3. The area under the velocity–time curve in a specified time interval equals the net displacement in that interval.
4. The area under the acceleration–time curve in a specified time interval equals the net change in velocity in that interval.

EXAMPLE:

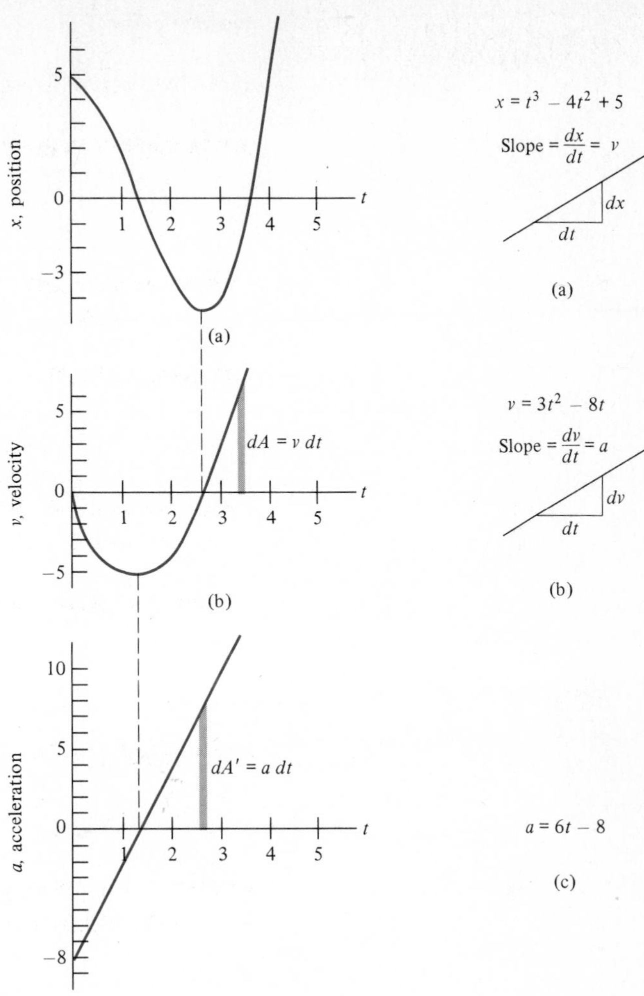

Section 11-3

Useful expressions based on acceleration

TIME AS A PARAMETER:

$$a = \frac{dv}{dt} \Rightarrow \int_{v_0}^{v} dv = \int_0^t a\,dt$$

DISPLACEMENT AS A PARAMETER:

$$v\,dv = a\,dx \Rightarrow \int_{v_0}^{v} v\,dv = \int_{x_0}^{x} a\,dx$$

Special cases

1. For constant acceleration a:

$$\int_{v_0}^{v} dv = a\int_0^t dt \quad \Rightarrow \quad v = at + v_0$$

$$\int_{v_0}^{v} dv = a\int_{x_0}^{x} dx \quad \Rightarrow \quad v^2 = 2a(x - x_0) + v_0^2$$

Position $x = \frac{1}{2}at^2 + v_0 t + x_0$.

2. For variable acceleration $a = f(t)$:

$$\int_{v_0}^{v} dv = \int_0^t f(t)\,dt \quad \Rightarrow \quad v = \int_0^t f(t)\,dt + v_0$$

$$\int_{x_0}^{x} dx = \int_0^t v\,dt \quad \Rightarrow \quad x = \int_0^t v\,dt + x_0$$

3. For variable acceleration $a = f(x)$:

$$\int_{v_0}^{v} v\,dv = \int_{x_0}^{x} f(x)\,dx \quad \Rightarrow \quad v^2 = 2\int_{x_0}^{x} f(x)\,dx + v_0^2$$

4. For variable acceleration $a = f(v)$:

$$a = \frac{dv}{dt} \quad \Rightarrow \quad dt = \frac{dv}{f(v)} \quad \Rightarrow \quad t = \int_0^t dt = \int_{v_0}^{v} \frac{dv}{f(v)}$$

$$a = f(v) = v\frac{dv}{dx} \quad \Rightarrow \quad x = \int_0^x dx = \int_{v_0}^{v} \frac{v\,dv}{f(v)}$$

Section 11-4

Relative motion of particles *A* and *B*

The positions, velocities, and accelerations of A and B are given with respect to a fixed point on the line of the particles as x_A, v_A, a_A, and x_B, v_B, a_B, respectively.

RELATIVE POSITION:

$$x_{B/A} = x_B - x_A$$

RELATIVE VELOCITY:

$$v_{B/A} = v_B - v_A$$

RELATIVE ACCELERATION:

$$a_{B/A} = a_B - a_A$$

Note: The following equations refer to position, velocity and accelerations.

$$x_{A/B} = -x_{B/A}$$

$$v_{A/B} = -v_{B/A}$$

etc.

The degrees of freedom of a mechanical system is the number of coordinates that may be arbitrarily chosen.

11–38 A package is released from a hovering helicopter. The downward acceleration of the package is estimated as $(32.2 - 0.02\,v^2)$ ft/s^2 where v is in ft/s. Determine the velocity of the package after it has fallen (a) 200 ft, and (b) 400 ft.

11–39 A particle falling toward a planet is accelerated by the gravity of that planet according to $a = -g_p(R^2/r^2)$, where g_p is the acceleration of gravity, measured at the surface, R is the radius of the planet, and r is the distance from the center of the spherical planet to the particle. Derive an equation for the velocity v at which a particle freely falling from a height h above the surface would hit the surface.

11–40 Consider the crash simulator shown in Fig. P11-40. The track is horizontal, and the sled is uniformly accelerated (a = const.) over a distance of 4 m from $v_0 = 0$ to $v = 100$ km/h. The sled is then uniformly decelerated and stopped in a distance of 1 m. Calculate the total time elapsed during the motion of the sled.

11–41 The system of three pulleys shown has *two degrees of freedom* since the coordinates of two blocks may be chosen arbitrarily. Write equations for the vertical velocities and accelerations of the three blocks.

11–42 The three blocks in Fig. P11-41 are initially at rest when $y_A = 5$ ft, $y_B = 6$ ft, and $y_C = 8$ ft. Blocks B and C are both accelerated upward at a constant rate of 3 ft/s^2. Determine the relative vertical position and velocity of blocks A and B at a time 2 s after the motion begins.

FIGURE P11-40

Test sled of crash simulator. Courtesy MTS Systems Corporation, Minneapolis, Minnesota

FIGURE P11-41

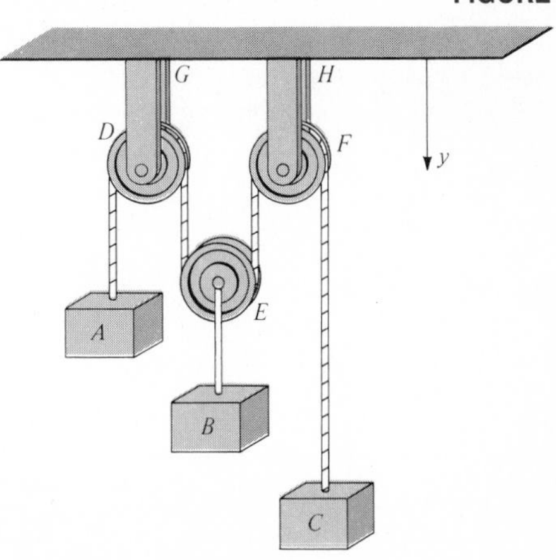

CHAPTER 12

KINEMATICS OF PARTICLES: VECTOR METHODS

OVERVIEW

Velocity and acceleration are vector quantities because they have magnitudes and directions. The vector methods that are developed in this chapter are useful in the analysis of a wide range of dynamics problems. These methods are most advantageous when the problems are complex. Many problems of two-dimensional and three-dimensional motion must be solved by using vector methods. Students are strongly advised to study thoroughly the basic principles and methods of vector analysis.

STUDY GOALS

SECTION 12-1 introduces the mathematical method of determining the velocity and acceleration vectors from the position vector of a particle which moves along a curved path. This analysis is the basis for studying all motions that are more complex than rectilinear motion. The concepts in Secs. 11-1 and 12-1 are readily correlated.

SECTION 12-2 presents an analysis of the physical aspects of a particle's motion along a curved path. Several concepts and definitions given here are essential for most work in dynamics.

SECTION 12-3 presents the concepts and analysis of tangential and normal components of the velocity and acceleration of a particle traveling on a curved path. These components are frequently used in dynamics, so they should be studied thoroughly. The vector expressions of the velocity and acceleration in curvilinear motion (Eqs. 12-8 and 12-11) and their implications are especially worthwhile to remember.

SECTION 12-4 extends the concepts of normal and tangential components of the velocity and acceleration to define radial and transverse components in polar coordinates. These coordinates are often convenient in solving problems of plane motion. The analysis in this section also provides useful practice with time derivatives of unit vectors.

SECTION 12-5 presents cylindrical coordinates which are used for the analysis of three-dimensional motion. The concept and application of these coordinates are based on polar coordinates. The additional mathematical complexity is minimal.

SECTION 12-6 presents the derivation of the velocity and acceleration of a particle using spherical coordinates. This derivation provides useful experience in three-dimensional vector mechanics, but it is quite difficult and is omitted from many courses. Nevertheless, the general expressions for velocity and acceleration in spherical coordinates can be used with an understanding of polar coordinates and the essence of the spherical system.

SECTION 12-7 describes the vector analysis of absolute and relative motions of two particles in translating rectangular coordinate systems. The radial and transverse components of velocity and acceleration studied in Sec. 12-4 using polar coordinates are applied in solving problems of relative motion.

12-1 CURVILINEAR MOTION. RECTANGULAR COORDINATES

The motion of a particle along a curved path is important to study because it is the basis for the analysis of all motions that are more complex than rectilinear motion. These are classified as *plane* (two-dimensional) or *space* (three-dimensional) *curvilinear motion.* The basic concepts for analyzing all such motions can be established by considering space motion of a particle. The mathematical method developed here is simple and immediately allows the solution of some problems of curvilinear motion. A more detailed analysis is presented in Sec. 12-2 to show the physical concepts of curvilinear motion that must be understood for solving a larger variety of problems.

Rectangular Coordinates

The fixed rectangular coordinate system is useful for analyzing curvilinear motion. Consider that the position vector **r** of a particle P is given in terms of rectangular components x, y, and z as in

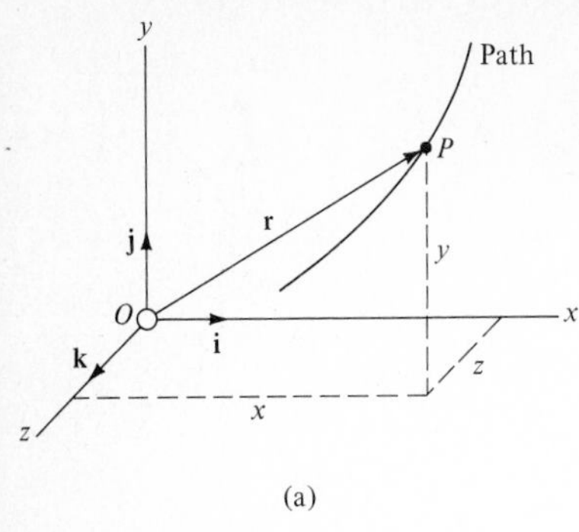

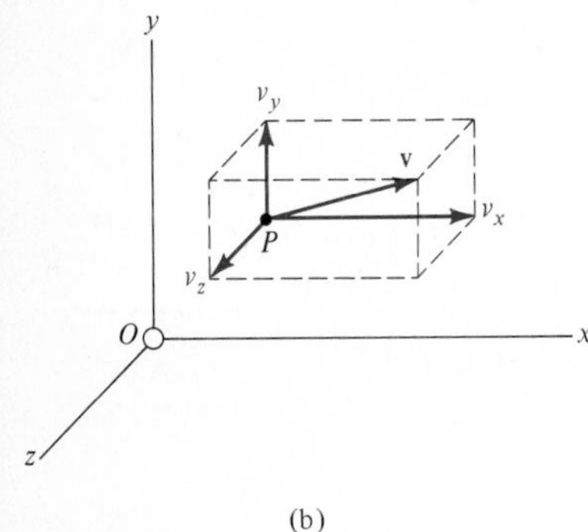

FIGURE 12-1

Position vector **r** in rectangular coordinates

Fig. 12-1a. With the unit vectors *i*, *j*, and *k*,

$$\mathbf{r} = x\mathbf{i} + y\mathbf{j} + z\mathbf{k} \tag{12-1}$$

Assume that x, y and z are continuous, differentiable functions of time t. In that case the velocity and the acceleration can be written as

$$\mathbf{v} = \frac{d\mathbf{r}}{dt} = \frac{dx}{dt}\mathbf{i} + \frac{dy}{dt}\mathbf{j} + \frac{dz}{dt}\mathbf{k} = \dot{x}\mathbf{i} + \dot{y}\mathbf{j} + \dot{z}\mathbf{k} \tag{12-2}$$

$$\mathbf{a} = \frac{d\mathbf{v}}{dt} = \frac{dv_x}{dt}\mathbf{i} + \frac{dv_y}{dt}\mathbf{j} + \frac{dv_z}{dt}\mathbf{k} = \ddot{x}\mathbf{i} + \ddot{y}\mathbf{j} + \ddot{z}\mathbf{k} \tag{12-3}$$

since the time derivatives of the constant unit vectors in the fixed xyz frame are zero; $v_x = \dot{x}$, $a_x = \ddot{x}$, and so on, are the scalar components of the corresponding vectors. The signs of these quantities denote the directions of the components in the positive or negative coordinate directions. Thus, curvilinear motion in general is the vector sum of orthogonal motions occurring simultaneously in the x, y, and z directions. For example, the absolute velocity **v** of particle P in Fig. 12-1b is the vector sum of velocities $v_x\mathbf{i}$, $v_y\mathbf{j}$, and $v_z\mathbf{k}$.

In a reverse procedure, the acceleration or velocity components may be individually integrated with respect to time to obtain velocity or position components, respectively.

12-2 PHYSICAL ASPECTS OF VELOCITY AND ACCELERATION IN CURVILINEAR MOTION

To understand the physical aspects of curvilinear motion, it is necessary to analyze the changes of the position and velocity vectors in detail. The effects of these changes on the acceleration of a particle are complex, but it is essential to study them in order to solve many problems in dynamics.

Consider a particle that is at position P at time t and at position P' at time $t + \Delta t$ as in Fig. 12-2a. The two positions are defined by vectors **r** and **r**′ with respect to the origin of a fixed reference frame. The vector $\Delta\mathbf{r}$ defines the change of the position vector **r** to **r**′, while Δs represents the actual displacement of the particle along the path of motion.

The *average velocity* of the particle over the time interval Δt is defined as the quotient $\Delta\mathbf{r}/\Delta t$. This is a vector in the direction of $\Delta\mathbf{r}$ since Δt is a scalar. The *instantaneous velocity* **v** of the particle at time t is defined at the limit as

$$\mathbf{v} = \lim_{\Delta t \to 0} \frac{\Delta \mathbf{r}}{\Delta t} = \frac{d\mathbf{r}}{dt} \qquad \boxed{12\text{-}4}$$

As Δt approaches zero, point P' approaches point P, and the vector $\Delta\mathbf{r}$ becomes tangent to the path of the particle. Consequently, the velocity vector $\mathbf{v}$ at any time t is tangent to the path of the particle as in Fig. 12-2b.

The magnitude v of the vector $\mathbf{v}$ is called the *speed* of the particle. It may also be defined by the quotient of the scalar quantities Δs and Δt at the limit,

$$v = \lim_{\Delta t \to 0} \frac{\Delta s}{\Delta t} = \frac{ds}{dt} \qquad \boxed{12\text{-}5}$$

where s is a length along the path of the particle.

The acceleration of the particle is analyzed by considering the velocity vectors at two nearby positions along the path as in Fig. 12-3a. For convenience, the vectors $\mathbf{v}$ and $\mathbf{v}'$ are redrawn with a common origin O' in Fig. 12-3b. The vector $\Delta\mathbf{v}$ defines the change of vector $\mathbf{v}$ to $\mathbf{v}'$.

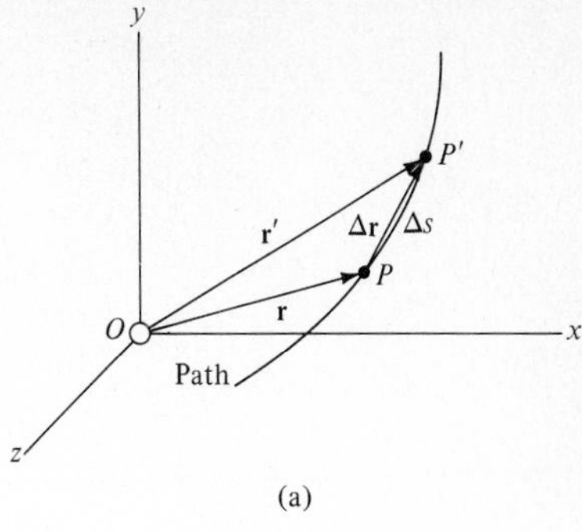

(a)

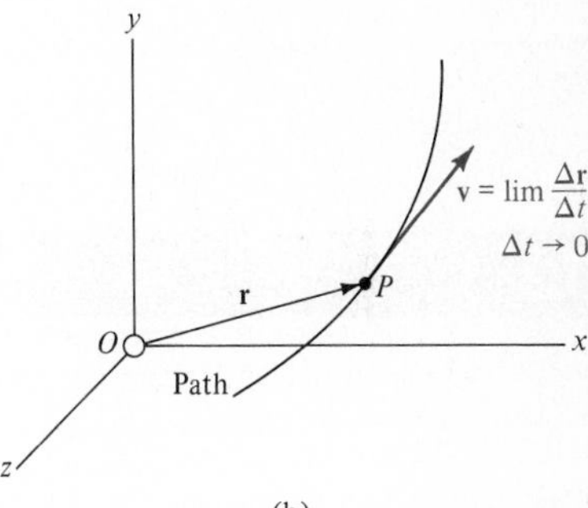

(b)

FIGURE 12-2

Definition of a particle's velocity

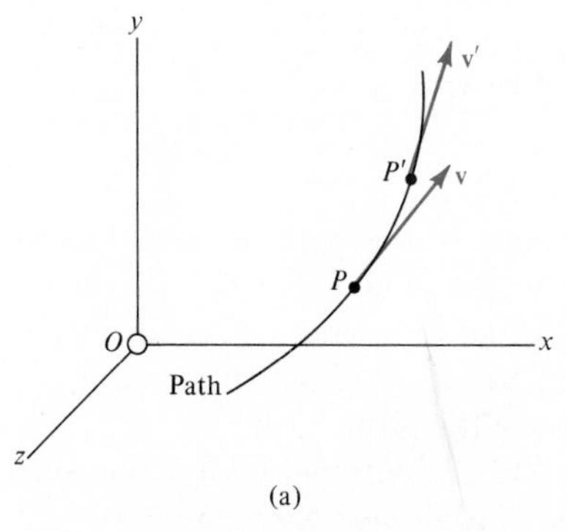

(a)

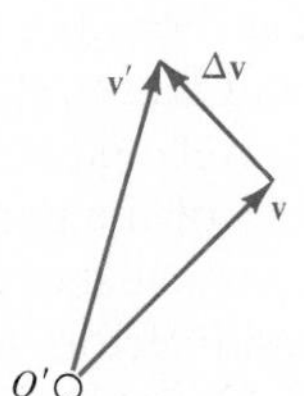

(b)

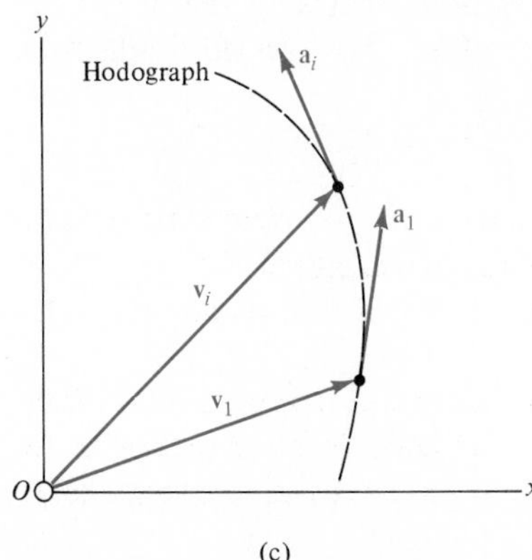

(c)

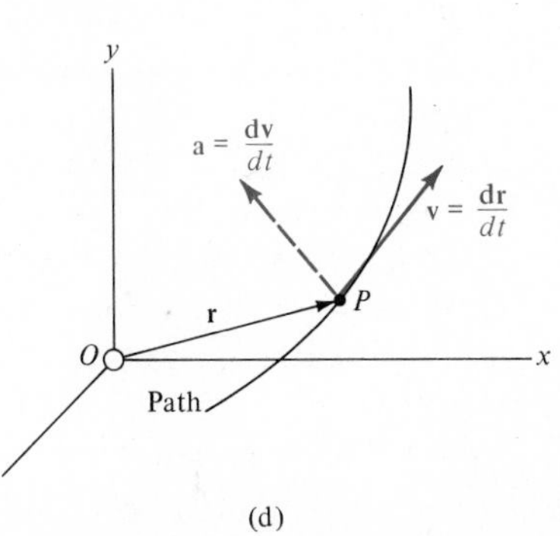

(d)

FIGURE 12-3

Definition of a particle's acceleration

The *average acceleration* of the particle over the time interval Δt is defined as the quotient $\Delta\mathbf{v}/\Delta t$. This is a vector in the direction of $\Delta\mathbf{v}$ since Δt is a scalar. The *instantaneous acceleration* $\mathbf{a}$ of the particle at time t is defined at the limit as

$$\mathbf{a} = \lim_{\Delta t \to 0} \frac{\Delta\mathbf{v}}{\Delta t} = \frac{d\mathbf{v}}{dt} \tag{12-6}$$

The acceleration vector is tangent to the curve described by the tip of the velocity vector $\mathbf{v}$. This curve, called the *hodograph* of the motion, is illustrated in Fig. 12-3c. Thus, the acceleration is not tangent to the path of the particle and is not collinear with the velocity vector in curvilinear motion. A general case is shown schematically in Fig. 12-3d.

12-3 TANGENTIAL AND NORMAL COMPONENTS

It is often convenient to define the velocity and acceleration of a particle using components along the tangent and the normal to the curved path, in the plane of that path. The concepts of this approach are established using plane motion for clarity, but they are valid for space motion, as well.

Curvilinear Plane Motion

The tangential and normal directions to the path are denoted by unity vectors $\mathbf{n}_t$ and $\mathbf{n}_n$, respectively, as in Fig. 12-4a. $\mathbf{n}_t$ is chosen in the direction of travel, and $\mathbf{n}_n$ is normal to $\mathbf{n}_t$ in the direction of the center of curvature of the path. These unit vectors rotate counterclockwise as the particle moves along the given path.

The tangential vectors $\mathbf{n}_t$ and $\mathbf{n}_t'$ are redrawn with a common origin O' in Fig. 12-4b. The vector $d\mathbf{n}_t$ defines the change of unit vector $\mathbf{n}_t$ to $\mathbf{n}_t'$ in a differential rotation $d\theta$. Using the arc-length approximation for the magnitude of $d\mathbf{n}_t$, $dn_t = n_t\, d\theta = d\theta$, since

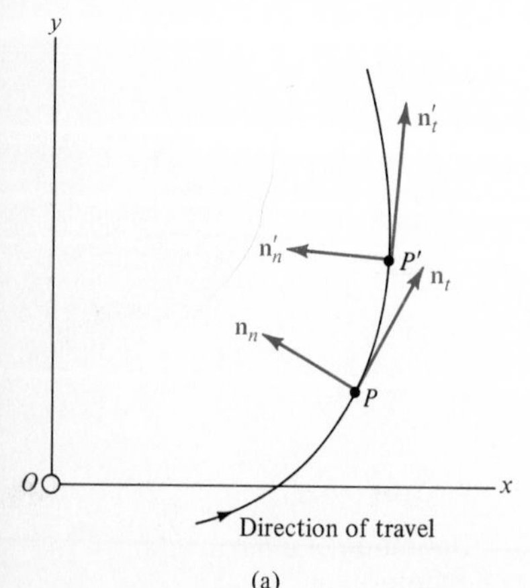

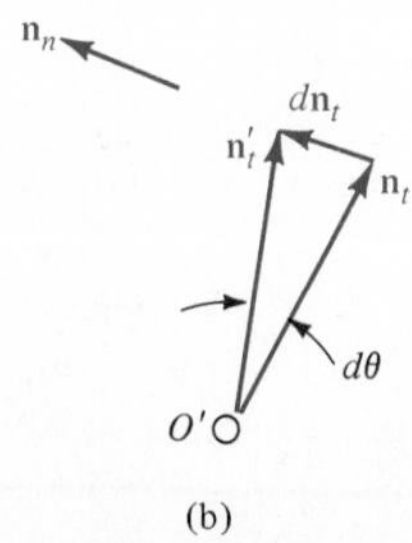

FIGURE 12-4
Unit vectors in curvilinear plane motion

$n_t = 1$. Next, it is noted that $d\mathbf{n}_t$ and $\mathbf{n}_n$ are in the same direction for an infinitesimal rotation. Thus, $d\mathbf{n}_t = d\theta\mathbf{n}_n$, or

$$\mathbf{n}_n = \frac{d\mathbf{n}_t}{d\theta} \qquad \text{12-7}$$

This result is useful in determining the acceleration in terms of tangential and normal components as follows.

The velocity $\mathbf{v}$ of the particle is always tangent to the path (Sec. 12-2), so it can be expressed as

$$\mathbf{v} = v\mathbf{n}_t \qquad \text{12-8}$$

The acceleration is obtained by differentiating Eq. 12-8 with respect to time t,*

$$\mathbf{a} = \frac{d\mathbf{v}}{dt} = \frac{dv}{dt}\mathbf{n}_t + v\frac{d\mathbf{n}_t}{dt} \qquad \text{12-9}$$

In an effort to eliminate the derivative of the unit vector $\mathbf{n}_t$, the following substitutions are found useful:

$$\frac{d\mathbf{n}_t}{dt} = \frac{d\mathbf{n}_t}{d\theta}\frac{d\theta}{ds}\frac{ds}{dt}$$

where $\dfrac{d\mathbf{n}_t}{d\theta} = \mathbf{n}_n$ from Eq. 12-7

$\dfrac{d\theta}{ds} = \dfrac{1}{\rho} =$ curvature of the path from Fig. 12-5 (ρ = radius of curvature)

$\dfrac{ds}{dt} = v$ from Eq. 12-5

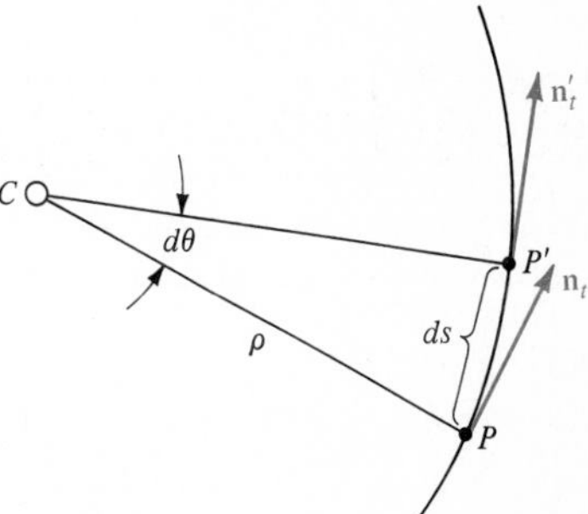

FIGURE 12-5

Quantities used in defining the curvature of the path

Thus,

$$\frac{d\mathbf{n}_t}{dt} = \frac{v}{\rho}\mathbf{n}_n \qquad \text{12-10}$$

and the acceleration can be expressed from Eq. 12-9 as

$$\boxed{\mathbf{a} = \frac{dv}{dt}\mathbf{n}_t + \frac{v^2}{\rho}\mathbf{n}_n} \qquad \text{12-11}$$

* The standard rule for the differentiation of products of scalar functions is extended to the product of a scalar function $A(t)$ and a vector function $\mathbf{B}(t)$ of the same variable t:

$$\frac{d(A\mathbf{B})}{dt} = \lim_{\Delta t\to 0}\frac{(A + \Delta A)(\mathbf{B} + \Delta\mathbf{B}) - A\mathbf{B}}{\Delta t}$$

$$= \lim_{\Delta t\to 0}\left(\frac{\Delta A}{\Delta t}\mathbf{B} + A\frac{\Delta\mathbf{B}}{\Delta t}\right) = \frac{dA}{dt}\mathbf{B} + A\frac{d\mathbf{B}}{dt}$$

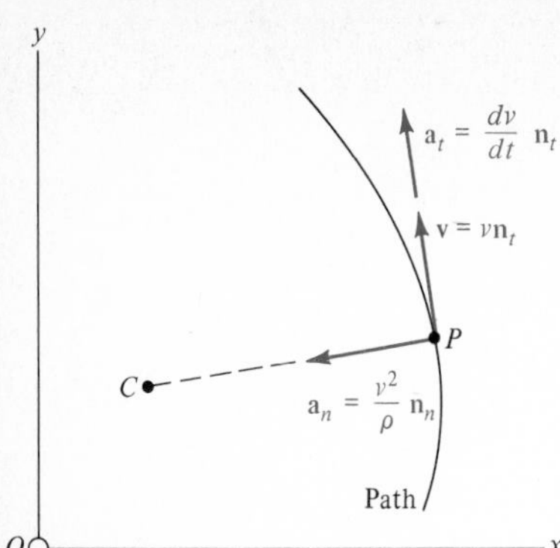

FIGURE 12-6

Normal and tangential accelerations

where the tangential and normal components of the acceleration are

$$a_t = \frac{dv}{dt} \qquad a_n = \frac{v^2}{\rho} \tag{12-12}$$

Note that a_t is equal to the rate of change of the speed of the particle and may be positive or negative, while a_n is always positive and $\mathbf{a}_n$ is directed toward the center of curvature of the path. These components are illustrated in Fig. 12-6.

Curvilinear Space Motion

The analysis of motion in space is made as an extension of the concepts used for motion in a plane. The connection between the two-dimensional and three-dimensional situations is based on the fact that at each point of interest on the path of the particle a plane can be drawn to include at least a small part of the path in that plane. Thus, the motion of the particle in that neighborhood can be analyzed using the methods established for plane motion.

It is necessary to define the directions of the unit vectors $\mathbf{n}_t$ and $\mathbf{n}_n$. As in plane motion, $\mathbf{n}_t$ is tangent to the path of the particle and is in the direction of travel. $\mathbf{n}_n$ is normal to $\mathbf{n}_t$ and is in the direction of the center of curvature of the small segment of path in the neighborhood of the particle. The plane containing the unit vectors $\mathbf{n}_t$ and $\mathbf{n}_n$ at a given position of the particle is called the *osculating plane*. Naturally, there is an infinite number of osculating planes as a particle moves along a three-dimensional path. Such a plane is illustrated in Fig. 12-7 for a particle moving upward along a helical path. Here the path segment and the unit vectors at P are drawn in the same color as the parallelogram representing the osculating plane.

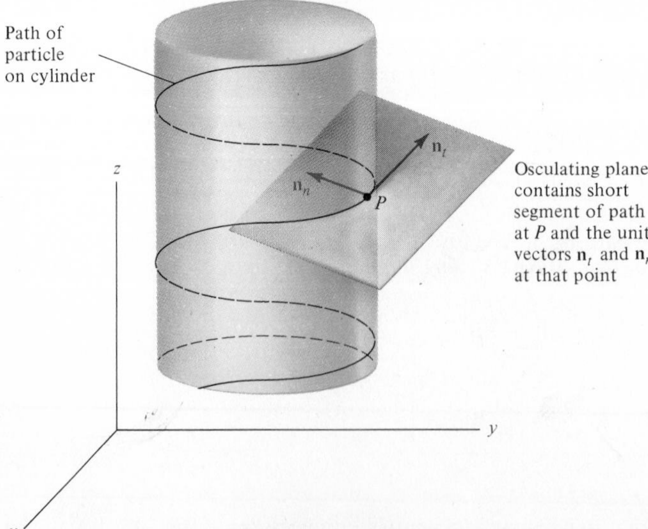

FIGURE 12-7

Definition of osculating plane

EXAMPLE 12-1

Supplies are dropped from an airplane to land at point A in Fig. 12-8. The velocity of the package at release from the airplane is $\mathbf{v}_0 = 200\mathbf{i}$ ft/s. The acceleration of the package is $\mathbf{a} = -32.2\mathbf{j}$ ft/s^2. (a) Determine the required distance x for the release of the package neglecting any effect of air on the package. (b) Next, change the initial velocity to $\mathbf{v}_0 = (200\mathbf{i} + 10\mathbf{j})$ ft/s.

SOLUTION

(a) The velocity in the x direction is constant throughout the fall since there is no acceleration in that direction. By finding the time for a vertical fall alone and multiplying it by the velocity in the x direction, the displacement in the x direction can be found. Using Eq. 11-10 for constant acceleration in the y direction,

$$y - y_0 = v_{0_y}t + \frac{1}{2}a_y t^2$$

$$0 - 300 = 0t + \frac{1}{2}(-32.2)t^2$$

$$\left(\frac{300}{16.1}\right)^{1/2} = t$$

$$4.32 \text{ s} = \text{t}$$

$$x - x_0 = v_{0_x}t + \frac{1}{2}a_x t^2$$

$$x - 0 = v_{0_x}t + 0$$

$$x = (200 \text{ ft/s})(4.32 \text{ s}) = 864 \text{ ft}$$

(b) For $v_{0_y} = 10$ ft/s,

$$y - y_0 = v_{0_y}t + \frac{1}{2}a_y t^2$$

$$0 - 300 = 10t + \frac{1}{2}(-32.2)t^2$$

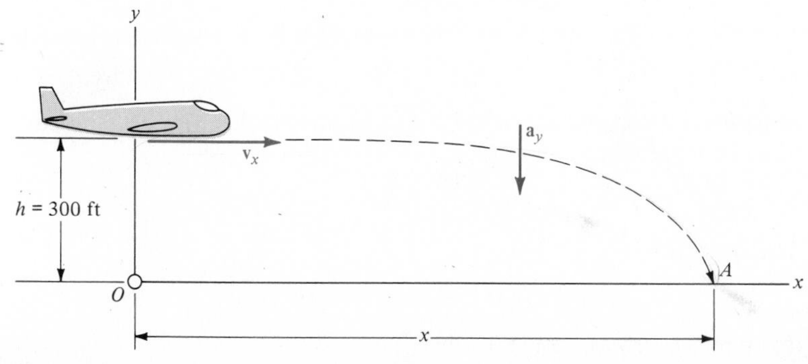

FIGURE 12-8
Motion of a package in free fall

$$16.1t^2 - 10t - 300 = 0$$

$$t = \frac{10 \pm \sqrt{100 + 19{,}300}}{32.2}$$

$$= \frac{10 \pm 139}{32.2}$$

$$= 4.63 \text{ s} \quad \text{and} \quad -4.02 \text{ s}$$

Using the positive root,

$$x = (4.63 \text{ s})(200 \text{ ft/s}) = 926 \text{ ft}$$

The negative root is not in the domain of the definition of the problem.

EXAMPLE 12-2

A racing car is traveling at a speed of 50 m/s on a circular track (Fig. 12-9). The driver, anticipating a straight section of the track, is accelerating the car at the rate of 2 m/s². Determine the magnitude and direction of the acceleration of the car at this instant.

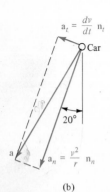

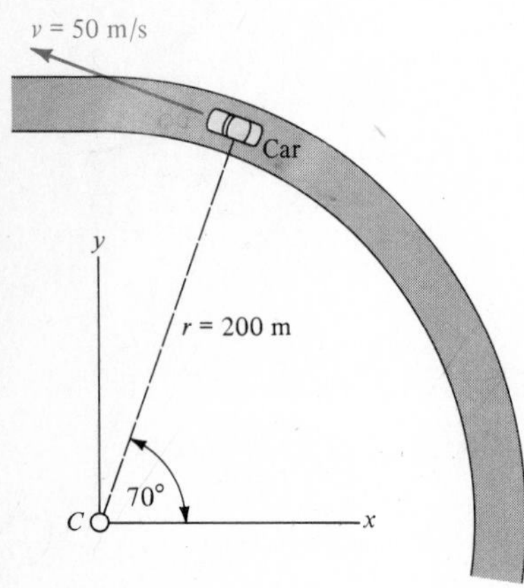

(a)

(b)

FIGURE 12-9
Motion of a car on a circular track

SOLUTION

TANGENTIAL ACCELERATION: The magnitude of this component is given as $a_t = 2 \text{ m/s}^2$. The vector expression is obtained by inspection,

$$\mathbf{a}_t = 2 \text{ m/s}^2 (-\cos 20^\circ\ \mathbf{i} + \sin 20^\circ\ \mathbf{j})$$
$$= (-1.88\mathbf{i} + 0.68\mathbf{j}) \text{ m/s}^2$$

NORMAL ACCELERATION: From Eq. 12-12, and by inspection of the direction of $\mathbf{n}_n$,

$$\mathbf{a}_n = \frac{v^2}{\rho}\mathbf{n}_n = \frac{50^2 \text{ (m/s)}^2}{200 \text{ m}}(-\cos 70^\circ\ \mathbf{i} - \sin 70^\circ\ \mathbf{j})$$
$$= (-4.28\mathbf{i} - 11.75\mathbf{j}) \text{ m/s}^2$$

MAGNITUDE AND DIRECTION OF ACCELERATION: The net acceleration in vector form is

$$\mathbf{a} = \mathbf{a}_t + \mathbf{a}_n = (-6.16\mathbf{i} - 11.07\mathbf{j}) \text{ m/s}^2$$

The magnitude is

$$a = \sqrt{6.16^2 + 11.07^2} = 12.67 \text{ m/s}^2$$

The results are shown in Fig. 12-9b.

EXAMPLE 12-3

In a packaging assembly operation small containers are deflected by a parabolic plate which is parallel to the ground (Fig. 12-10). What is the maximum speed v of the containers if their total acceleration caused by the deflector plate should not exceed 200 in./s^2? The plate does not change the speed of the containers.

SOLUTION

The magnitude of the total acceleration vector from Eq. 12-11 is

$$a = \left[\left(\frac{dv}{dt}\right)^2 + \left(\frac{v^2}{\rho}\right)^2\right]^{1/2}$$

Since all motion is occurring in a plane horizontal to the ground, the speed of each package is constant as it moves along the path, and $dv/dt = 0$. Therefore,

$$a_{\max} = \frac{v^2}{\rho} = 200 \text{ in./s}^2$$

Since v is constant, a has its largest magnitude when the radius of curvature ρ is minimum. The minimum radius of curvature for a parabola is at its vertex (point O in Fig. 12-10). The relation for the radius of curvature at a point is known from calculus,*

$$\rho = \frac{[1 + (dy/dx)^2]^{3/2}}{d^2y/dx^2}$$

Evaluating the appropriate derivatives at the point O,

$$\rho = \frac{[1 + 0]^{3/2}}{1/5} = 5 \text{ in.}$$

and

$$v^2 = (200 \text{ in./s}^2)(5 \text{ in.}) = 1000 \text{ in}^2/\text{s}^2$$

$$v = 31.6 \text{ in./s} = 2.64 \text{ ft/s}$$

* George B. Thomas, Jr., *Calculus and Analytic Geometry*, Alternate Edition (Reading, Mass.: Addison-Wesley Publishing Co., 1972), p. 601.

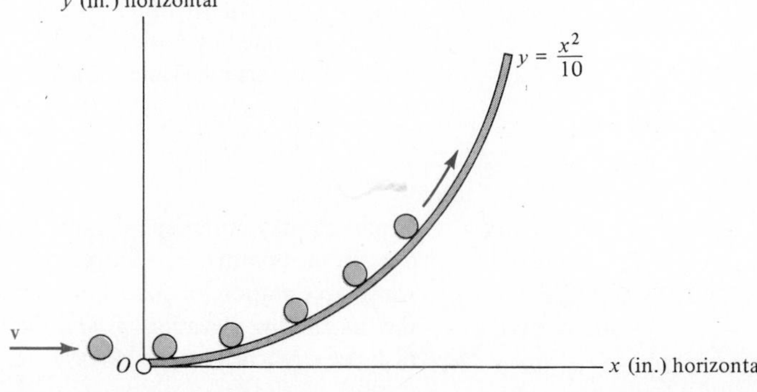

FIGURE 12-10

Motion of particles along a parabolic deflector

PROBLEMS

SECTIONS 12-1, 12-2, 12-3

12–1 The position vector of a particle in plane curvilinear motion is $\mathbf{r} = [(t^2 + 3t)\mathbf{i} + (3t - 2)\mathbf{j}]$ m, where t is in seconds. Determine the velocity and acceleration at $t = 3$ s.

12–2 The position of a particle is defined by $x = 2t^3$ and $y = 4t^2 + 3$, where x and y are in feet and t is in seconds. Determine the velocity and acceleration vectors at $t = 5$ s.

12–3 The velocity of a particle is defined by $v_x = 2t$ and $v_y = 3t^2 + 2$, where v_x and v_y are in m/s and t is in seconds. Determine the position and acceleration at $t = 2$ s if initially $x_0 = 2$ m and $y_0 = 5$ m.

12–4 The velocity vector of a particle is $\mathbf{v} = [20t\mathbf{i} - (8t + 3)\mathbf{j}]$ ft/s, where t is in seconds. Determine the position and acceleration vectors at $t = 10$ s if initially $x_0 = 10$ ft and $y_0 = 15$ ft.

12–5 The acceleration of a particle is defined by $a_x = 5\text{ m/s}^2$ and $a_y = -9\text{ m/s}^2$ at time $t_0 = 0$ when $x_0 = 0$, $y_0 = 100$ m, and $v_{x0} = v_{y0} = 0$. Determine the position of the particle at $t = 2$ s.

12–6 The acceleration vector of a particle is $\mathbf{a} = (10t\mathbf{i} - 30\mathbf{j})\text{ ft/s}^2$, where t is in seconds. Determine the velocity vector at $t = 3$ s if initially $x_0 = y_0 = 0$, and at $t = 1$ s, $x_1 = 40$ ft and $y_1 = 25$ ft.

12–7 The velocity of a particle in three-dimensional motion is defined by $\mathbf{v} = (2t\mathbf{i} + t^2\mathbf{j} - t\mathbf{k})$ m/s, where t is in seconds. Determine the position at $t = 10$ s if initially $x_0 = 0$, $y_0 = 200$ m, and $z_0 = 0$.

FIGURE P12-8

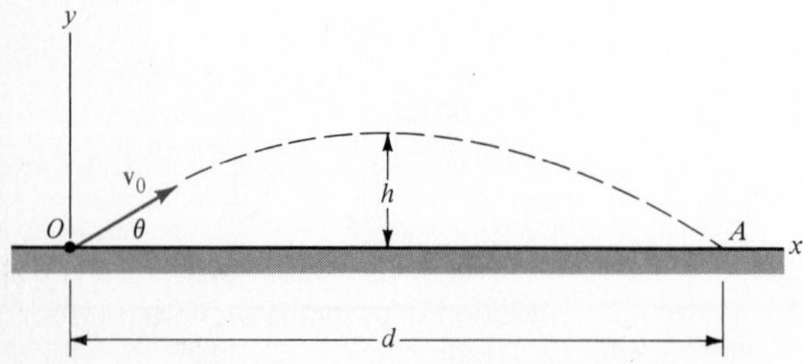

12–8 A particle has initial speed $v_0 = 100$ ft/s at $\theta = 30°$ at point O. It is subject to a downward acceleration of 32.2 ft/s^2. Determine the maximum elevation h and horizontal distance d reached by the particle.

12–9 For the same initial speed v_0 in Prob. 12–8, plot h vs. θ and d vs. θ as functions of $\theta = 30°$, 45°, and 60°.

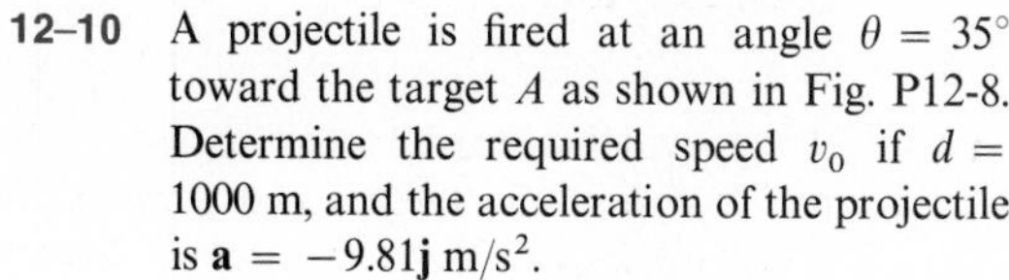

12–10 A projectile is fired at an angle $\theta = 35°$ toward the target A as shown in Fig. P12-8. Determine the required speed v_0 if $d = 1000$ m, and the acceleration of the projectile is $\mathbf{a} = -9.81\mathbf{j}$ m/s^2.

12–11 A helicopter at $x = 0$ and $y = 300$ ft has speed $v_{x_0} = 60$ ft/s when a package is ejected from it with a downward velocity of 5 ft/s. Determine the horizontal distance OA traveled by the package if it has a downward acceleration of 32.2 ft/s^2.

FIGURE P12-11

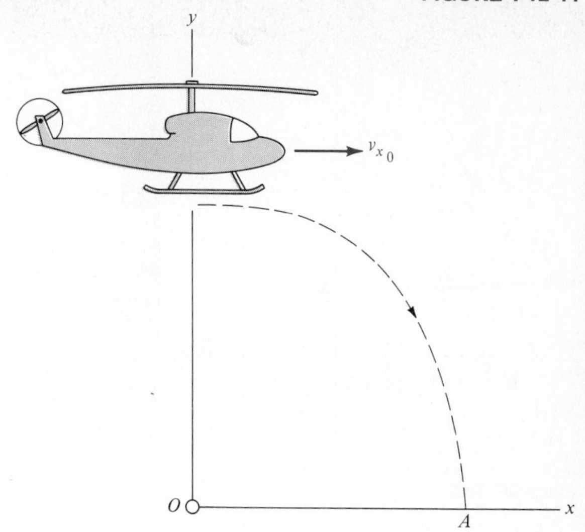

FIGURE P12-12

12–12 A person on a bluff horizontally throws a rock aimed at a target at point A. What should the initial speed v_0 of the rock be to hit the target if the rock has a downward acceleration of 9.81 m/s^2, $h = 40$ m, and $d = 30$ m?

12–13 A person in Fig. P12-12 horizontally throws a rock with an initial speed $v_0 = 30$ ft/s. The rock has a downward acceleration of 32.2 ft/s^2 and $h = 100$ ft. Determine the horizontal distance d and the velocity vector of the rock just before it lands at A.

FIGURE P12-14

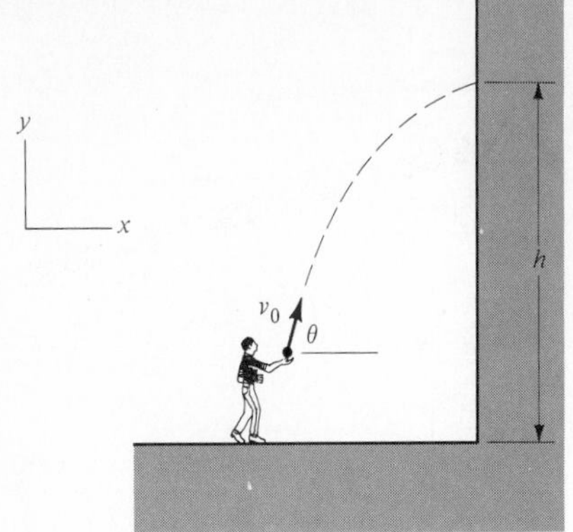

12–14 A person releases a ball 2 m above the ground and 7 m away from a wall. The ball has an initial velocity $v_0 = 20$ m/s at an angle $\theta = 70°$. Determine the height h where the ball hits the wall if the downward acceleration is 9.81 m/s^2.

12–15 In Prob. 12–14, determine the smallest speed v_0 with which the ball could land on the flat roof 18 m above the ground.

FIGURE P12-16

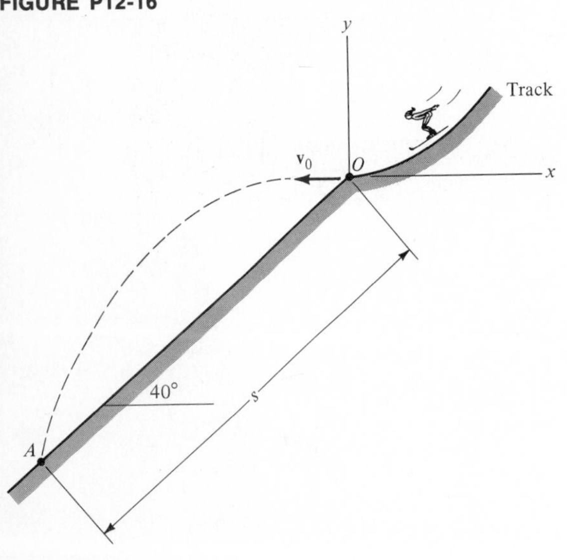

12–16 A ski jumper leaves the track at point O with a horizontal speed v_0. Determine v_0 if the skier touches down at $s = 150$ ft. The acceleration of gravity is 32.2 ft/s^2.

12–17 Assume that the takeoff velocity of the ski jumper in Fig. P12-16 is $\mathbf{v}_0 = (-20\mathbf{i} - 2\mathbf{j})$ m/s. Determine the distance s where the touchdown is if the acceleration of gravity is 9.81 m/s^2.

FIGURE P12-18

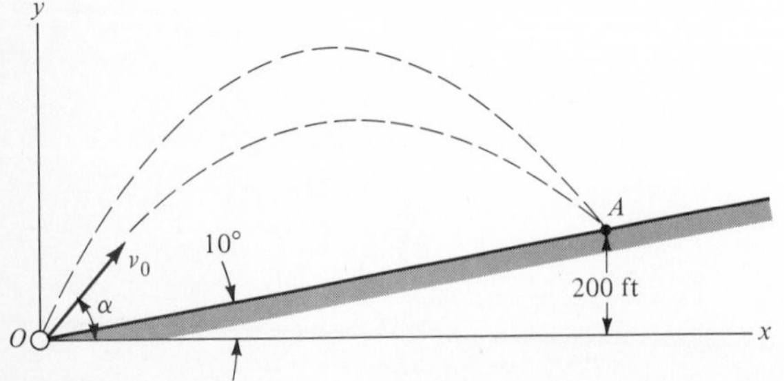

12–18 A projectile is fired with an initial speed $v_0 = 3000$ ft/s toward a target at point A. Determine the two firing angles α_1 and α_2 with which the projectile can reach the target. Also calculate the two times of flight. The acceleration of gravity is 32.2 ft/s^2.

12–19 A nozzle at point A discharges water which strikes the incline at point B. Determine the initial speed v_A of the water if the acceleration of gravity is 9.81 m/s^2.

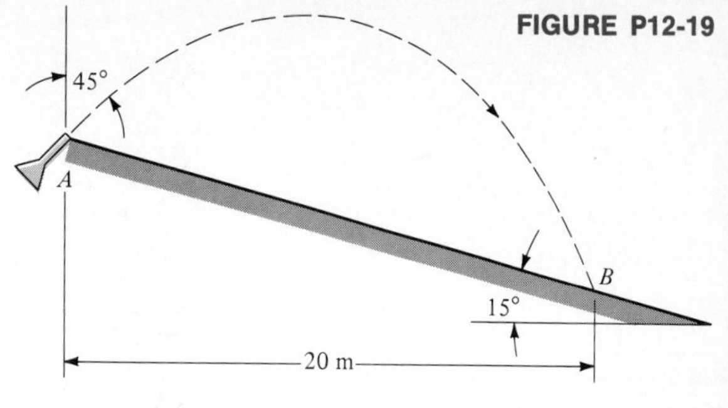

FIGURE P12-19

12–20 A projectile is thrown with an initial velocity $\mathbf{v}_0 = a\mathbf{i} + b\mathbf{j}$ from point A. The elevation h_{max} is known to be twice the elevation of the upper plateau with respect to A. (a) Determine the final velocity $\mathbf{v}_f$ of the projectile just before impact at point B. The acceleration of gravity is g. (b) Assume that the same projectile is thrown from point B with an initial velocity $-\mathbf{v}_f$. Determine the final velocity $\mathbf{v}_f'$ of the projectile just before impact at the lower level. Is this impact at point A?

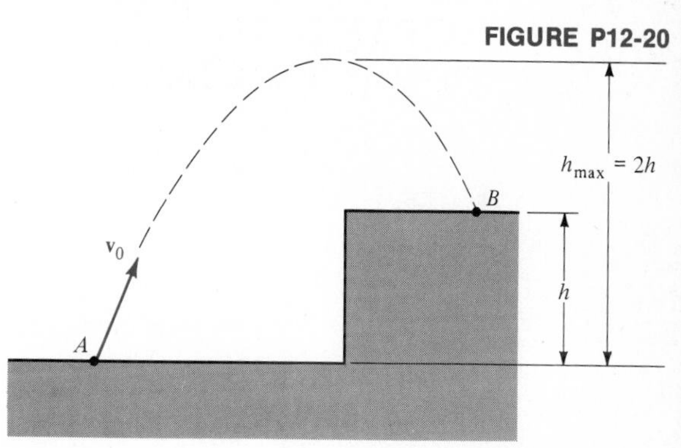

FIGURE P12-20

12–21 A particle P is moving with a constant speed $v = 10$ m/s on a circular track of radius $r = 0.2$ m. Determine the acceleration of the particle at $\theta = 0$ and 90°.

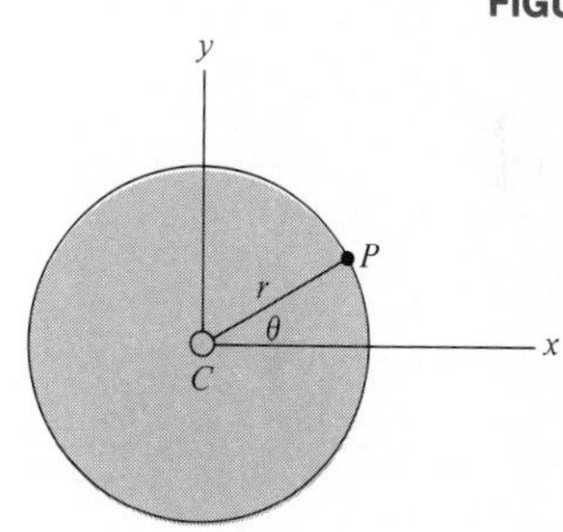

FIGURE P12-21

12–22 A particle P is moving clockwise on a circular track of radius $r = 1$ ft (Fig. P12-21). The speed of $v = 10$ ft/s is increasing at the rate of 2 ft/s^2 when $\theta = 30°$. Determine the acceleration vector of the particle.

12–23 A particle P starts from rest at $\theta = 0$ and travels counterclockwise on a circular track of radius $r = 0.5$ m (Fig. P12-21). With the speed increasing at a constant rate, the particle is at $\theta = 90°$ at time $t = 0.1$ s after the motion began. Determine the acceleration vectors at $\theta = 0$ and 90°.

12–24 A particle P moves with a constant speed $v = 10$ ft/s on a path defined by $y = 4x^2$ ft. Determine the radius of curvature ρ and the acceleration $\mathbf{a}$ of the particle at $x = 0$ and 2 ft.

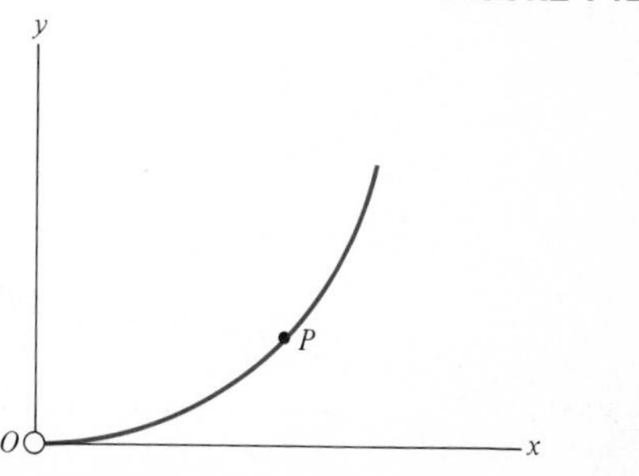

FIGURE P12-24

12–25 A particle P starts from rest at point O in Fig. P12-24. The path of motion is defined by $y = 3x^2$ m. The particle's speed increases at a constant rate and it reaches $x = 1$ m at time $t = 0.2$ s after the motion began. Determine the acceleration vectors at $x = 0$ and 1 m.

12–26 The position vector of a particle in plane curvilinear motion is $\mathbf{r} = [(4t^3 + 5)\mathbf{i} - t^2\mathbf{j}]$ m, with t in seconds. Calculate the radius of curvature ρ and the acceleration $\mathbf{a}$ of the particle at $t_1 = 1$ s and $t_2 = 2$ s.

12–27 The position vector of a particle is defined by $\mathbf{r} = [t^3\mathbf{i} + (5t^3 - t)\mathbf{j}]$ ft, where t is in seconds. Calculate the radius of curvature ρ and the acceleration $\mathbf{a}$ of the particle at $t_0 = 0$ and $t_1 = 5$ s.

12–28 A racing car travels at a constant speed v on a circular part of the track which has a radius of 100 m. Determine the maximum allowable speed assuming that the normal component of the acceleration must not exceed 7 m/s^2.

12–29 The car in Prob. 12–28 is traveling at a speed of $v = 20$ m/s when the driver wishes to accelerate forward. Determine the maximum rate at which v may be increased if the maximum total acceleration must not exceed 7 m/s^2.

FIGURE P12-30

12–30 An amusement ride has seats arranged in a circular pattern of 20 ft diameter and rotates in a horizontal plane. The speed of each person is 18 ft/s when the operator wishes to decrease the speed. Determine the maximum allowable rate of decrease of the speed if the total acceleration of each rider must not exceed 40 ft/s^2.

FIGURE P12-31

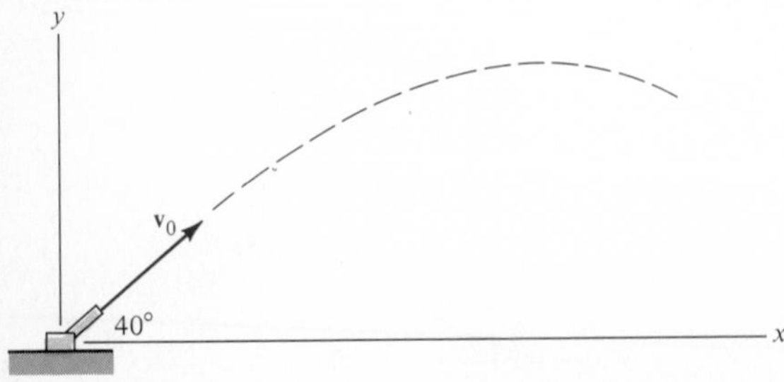

12–31 A projectile is fired with an initial speed $v_0 = 1000$ m/s. Determine the radius of curvature of the path of motion (a) immediately after the projectile leaves the muzzle, and (b) at the maximum elevation.

MOTION OF A PARTICLE IN POLAR COORDINATES

12-4

Sometimes the plane motion of a particle is convenient to analyze using polar coordinates. This is done with the aid of the methods developed in Sec. 12-3. According to common notation, the position, velocity, and acceleration of the particle are given in terms of the radial coordinate r and the angular coordinate θ. Of course, these can be readily converted to a rectangular system as shown in Fig. 12-11a, where $x = r\cos\theta$ and $y = r\sin\theta$.

The differential displacement $d\mathbf{r}$ is resolved into *radial* and *transverse components* $d\mathbf{r}_r$ and $d\mathbf{r}_\theta$, respectively, as in Fig. 12-11b. The magnitude of $d\mathbf{r}_r$ is dr, the change in length of $\mathbf{r}$. The approximation is made that the magnitude dr_θ is equal to the arc length

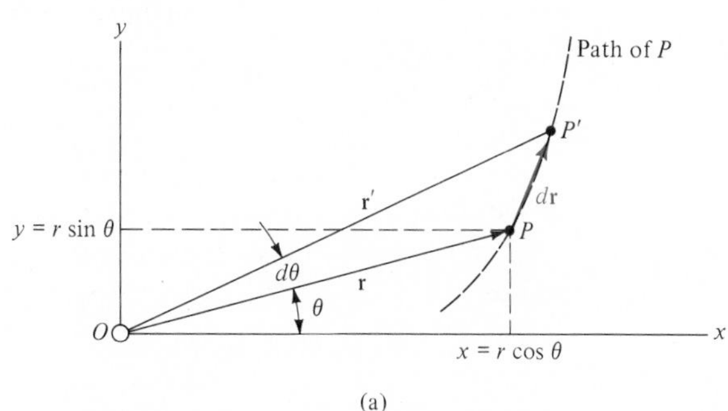

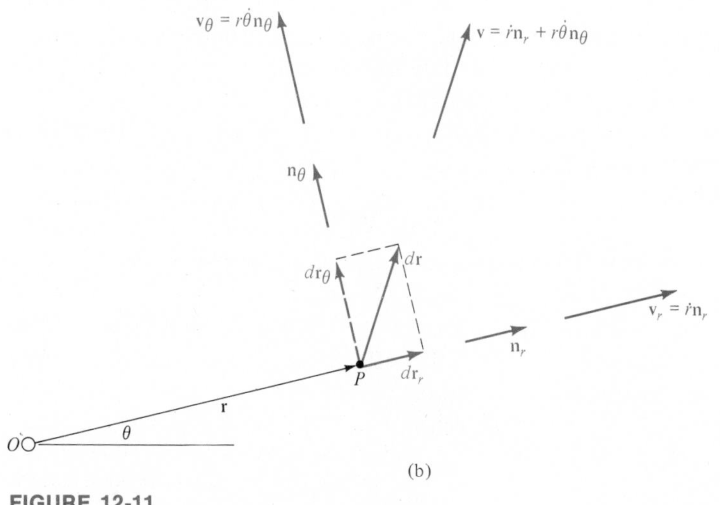

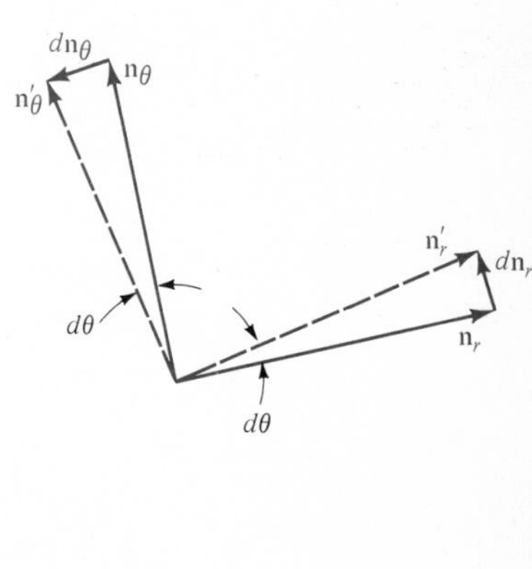

FIGURE 12-11

Analysis of motion using polar coordinates

$r\,d\theta$. With these, the radial and transverse scalar components of the velocity are

$$v_r = \frac{dr}{dt} = \dot{r} \tag{12-13}$$

$$v_\theta = r\frac{d\theta}{dt} = r\dot{\theta} = r\omega \tag{12-14}$$

where the symbols $\dot{\theta}$ and ω denote the *angular speed* of a radial line.* This scalar quantity is the time rate of change of the angle θ. Equations 12-13 and 12-14 can be combined vectorially using the unit vectors $\mathbf{n}_r$ and $\mathbf{n}_\theta$ (Fig. 12-11b),

$$\boxed{\mathbf{v} = \dot{r}\mathbf{n}_r + r\dot{\theta}\mathbf{n}_\theta} \tag{12-15}$$

Equation 12-15 is the expression for the velocity in polar coordinates. The units of angular speed are radians per second. The commonly used units of revolutions per minute (rpm) must be converted to rad/s for using Eq. 12-14 or 12-15.

The same result can be obtained in working directly with the derivative of the vector $\mathbf{r} = r\mathbf{n}_r$. This approach is lengthier than the pictorially advantageous method above, but it is so useful in some analyses that its details are presented here.

First express the orthogonal unit vectors $\mathbf{n}_r$ and $\mathbf{n}_\theta$ from Fig. 12-11 as

$$\mathbf{n}_r = \cos\theta\mathbf{i} + \sin\theta\mathbf{j} \qquad \mathbf{n}_\theta = -\sin\theta\mathbf{i} + \cos\theta\mathbf{j} \tag{12-16}$$

It is important to realize that the unit vectors $\mathbf{n}_r$ and $\mathbf{n}_\theta$ are constant only in magnitude but change directions as the particle moves to a new, nearby location. These changes are modeled as the rotations from $\mathbf{n}_r$ to $\mathbf{n}_r'$ and from $\mathbf{n}_\theta$ to $\mathbf{n}_\theta'$ in Fig. 12-11c. The differential changes of the unit vectors can be determined from this diagram using the arc-length approximations that $dn_r = n_r\,d\theta = d\theta$ and $dn_\theta = n_\theta\,d\theta = d\theta$. The time derivatives of the unit vectors can be written from these differentials or directly from the expressions for $\mathbf{n}_r$ and $\mathbf{n}_\theta$. For example,

$$\dot{\mathbf{n}}_r = \frac{d\mathbf{n}_r}{dt} = \frac{d\mathbf{n}_r}{d\theta}\frac{d\theta}{dt} = (-\sin\theta\mathbf{i} + \cos\theta\mathbf{j})\dot{\theta} = \dot{\theta}\mathbf{n}_\theta \tag{12-17}$$

The derivation of the expression for $\dot{\mathbf{n}}_\theta$ is left as an exercise for the reader. In summary, the differential changes and time derivatives of the unit vectors are tabulated as follows.

$$\text{Magnitudes:} \qquad \begin{aligned} dn_r &= n_r\,d\theta \\ &= d\theta \end{aligned} \qquad \begin{aligned} dn_\theta &= n_\theta\,d\theta \\ &= d\theta \end{aligned} \tag{12-18}$$

* The *angular velocity* vector $\boldsymbol{\omega}$ is defined using the right-hand rule.

$$\text{Vector derivatives:}\quad \dot{\mathbf{n}}_r = \frac{d\mathbf{n}_r}{dt} = \dot{\theta}\mathbf{n}_\theta \qquad \dot{\mathbf{n}}_\theta = \frac{d\mathbf{n}_\theta}{dt} = -\dot{\theta}\mathbf{n}_r \tag{12-19}$$

With these vectors, the velocity of the particle can be obtained by differentiating $\mathbf{r} = r\mathbf{n}_r$,

$$\mathbf{v} = \frac{d\mathbf{r}}{dt} = \frac{dr}{dt}\mathbf{n}_r + r\frac{d\mathbf{n}_r}{dt} = \dot{r}\mathbf{n}_r + r\dot{\theta}\mathbf{n}_\theta$$

which is identical to the result in Eq. 12-15.

The acceleration of the particle is obtained by differentiating Eq. 12-15 with respect to time,

$$\mathbf{a} = \dot{\mathbf{v}} = \ddot{r}\mathbf{n}_r + \dot{r}\dot{\mathbf{n}}_r + \dot{r}\dot{\theta}\mathbf{n}_\theta + r\ddot{\theta}\mathbf{n}_\theta + r\dot{\theta}\dot{\mathbf{n}}_\theta$$

After substituting the expressions for $\dot{\mathbf{n}}_r$ and $\dot{\mathbf{n}}_\theta$ and grouping the terms, the acceleration in polar coordinates becomes

$$\boxed{\mathbf{a} = (\ddot{r} - r\dot{\theta}^2)\mathbf{n}_r + (r\ddot{\theta} + 2\dot{r}\dot{\theta})\mathbf{n}_\theta} \tag{12-20}$$

Here the radial and transverse scalar components of the acceleration are

$$a_r = \ddot{r} - r\dot{\theta}^2 \tag{12-21}$$

$$a_\theta = r\ddot{\theta} + 2\dot{r}\dot{\theta} \tag{12-22}$$

It should be noted that $a_r \neq \dot{v}_r$ and $a_\theta \neq \dot{v}_\theta$ because the derivative of the radial velocity $\mathbf{v}_r$ has components in both the radial and transverse directions, and similarly for the derivative of the velocity $\mathbf{v}_\theta$. The quantity $\ddot{\theta} = \alpha$ is called the *angular acceleration* of a radial line, and has units of rad/s^2.*

An important special case of curvilinear motion is *circular motion* for which r = constant, $\dot{r} = \ddot{r} = 0$. In that case Eqs. 12-15 and 12-20 for velocity and acceleration become

$$\mathbf{v} = r\dot{\theta}\mathbf{n}_\theta \qquad \mathbf{a} = -r\dot{\theta}^2\mathbf{n}_r + r\ddot{\theta}\mathbf{n}_\theta \tag{12-23}$$

These agree, as they should, with Eqs. 12-8 and 12-11 when the differences in the sense of the vectors $\mathbf{n}_n$ and $\mathbf{n}_r$ are noted. Thus, for circular motion

$$v = r\dot{\theta} = r\omega \tag{12-24}$$

$$a_t = \frac{dv}{dt} = r\ddot{\theta} = r\alpha \tag{12-25}$$

$$a_n = \frac{v^2}{r} = r\dot{\theta}^2 = r\omega^2 \tag{12-26}$$

* The angular acceleration vector $\boldsymbol{\alpha}$ is defined using the right-hand rule.

12-5 CYLINDRICAL COORDINATES

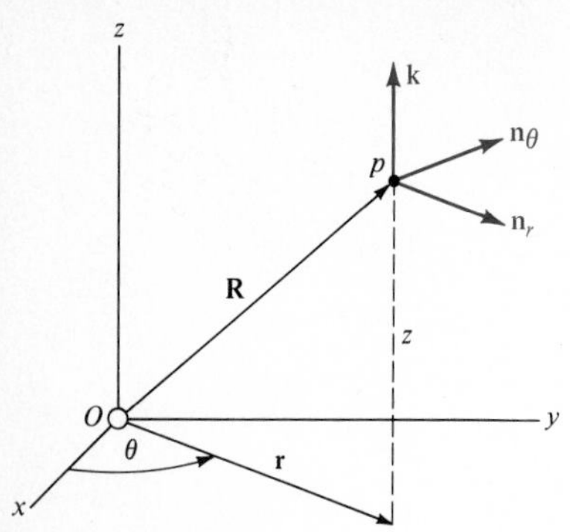

FIGURE 12-12

Unit vectors in cyclindrical coordinates. Note: $\mathbf{n}_r$ is parallel to r; $\mathbf{n}_\theta$ is parallel to xy plane; $\mathbf{k}$ is parallel to z.

The analysis of a particle's motion in three dimensions can be done in any one of three major ways. The use of rectangular coordinates was introduced in Sec. 12-1. Polar coordinates can be adapted to space motion as described in this section. The application of spherical coordinates is presented in Sec. 12-6.

The components of velocity and acceleration that were derived in Sec. 12-4 are also valid in three dimensions. It is only necessary to consider the additional motion which is perpendicular to the plane of the r and θ coordinates. When r and θ are defined in terms of the rectangular coordinates x and y as in Fig. 12-12, the third dimension is the z coordinate. The position vector $\mathbf{R}$ of particle P in cylindrical coordinates is defined according to Fig. 12-12 as

$$\mathbf{R} = r\mathbf{n}_r + z\mathbf{k} \tag{12-27}$$

The expressions for velocity and acceleration in cylindrical coordinates are as follows. From Eqs. 12-27 and 12-15, the velocity vector is

$$\boxed{\mathbf{v} = \dot{r}\mathbf{n}_r + r\dot{\theta}\mathbf{n}_\theta + \dot{z}\mathbf{k}} \tag{12-28}$$

where $\dot{z}$ is the scalar component of the velocity in the z direction, and $\mathbf{k}$ is the corresponding unit vector. The other terms were described in Sec. 12-4.

Similarly, the acceleration vector is obtained from Eqs. 12-28 and 12-20,

$$\boxed{\mathbf{a} = (\ddot{r} - r\dot{\theta}^2)\mathbf{n}_r + (r\ddot{\theta} + 2\dot{r}\dot{\theta})\mathbf{n}_\theta + \ddot{z}\mathbf{k}} \tag{12-29}$$

Note that the time derivatives of $z\mathbf{k}$ are $\dot{z}\mathbf{k}$ and $\ddot{z}\mathbf{k}$ because $\mathbf{k}$ is displaced only to parallel positions in cylindrical coordinates. Accordingly, $\dot{\mathbf{k}}$ is always zero.

12-6 SPHERICAL COORDINATES

The derivation of the equations of motion in spherical coordinates is very difficult for many people. The major concepts and steps of such a derivation are presented here for those who wish to obtain further experience in three-dimensional vector mechanics. A review of the derivatives of unit vectors in Sec. 12-3 and App. A is useful for this analysis. The concepts of polar coordinates presented in Sec 12-5 are also relevant.

The position of particle P on a spherical surface (Fig. 12-13a) is defined by the coordinates r, ϕ, and θ. Note that θ is not to the line OP but to its projection OQ in the xy plane (thus, θ is not the same angle as that defined for polar coordinates). The significance

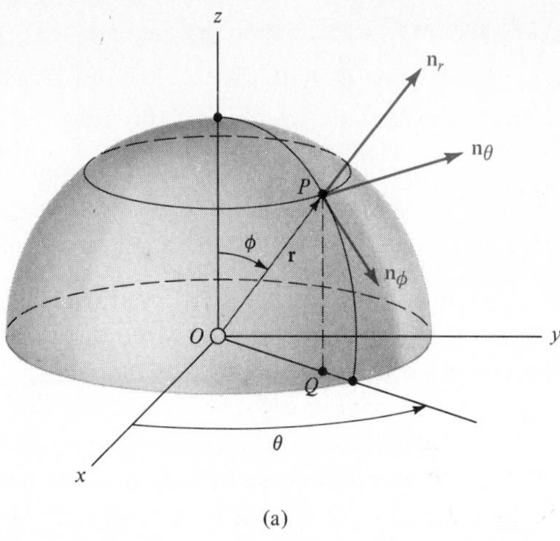

(a)

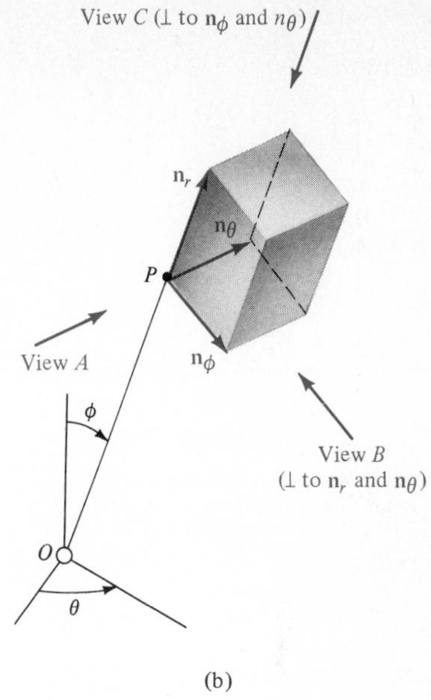

(b)

of this distinction can be appreciated by considering an arbitrary large rotation through an angle $\Delta\theta$. This rotation causes a large displacement of point P if it is located in the xy plane. The same rotation causes no displacement of point P if it is located on the z axis.

The derivation of the particle's velocity and acceleration is based on a triad of orthogonal unit vectors $\mathbf{n}_r$, $\mathbf{n}_\phi$, and $\mathbf{n}_\theta$ which satisfy the right-hand rule according to the vector product $\mathbf{n}_r \times \mathbf{n}_\phi = \mathbf{n}_\theta$. These unit vectors may change in directions, so their time derivatives must be included in the analysis. Pictorial views of these changing unit vectors are exemplified in Fig. 12-13c, which is typical of three orthogonal views A, B, C from Fig. 12-13b. Each of these views shows only two unit vectors as they change.

The complex rotations that are possible for the triad of unit vectors necessitate the use of vector notation for the angular velocities. To illustrate a convenient scheme, assume that the y and z axes rotate counterclockwise about the x axis with a common angular velocity ω_x. This angular velocity can be expressed as a vector $\boldsymbol{\omega}_x$ in the x direction, with its sense determined by the right-hand rule, $\boldsymbol{\omega}_x = \omega_x \mathbf{i}$. Similarly, the scheme can be applied to rotations about the y and z axes. A general rotational velocity $\boldsymbol{\omega}$ of a line such as OP in Fig. 12-13 can be expressed as the vector sum of three angular velocities,

$$\boldsymbol{\omega} = \omega_x \mathbf{i} + \omega_y \mathbf{j} + \omega_z \mathbf{k} \qquad \boxed{12\text{-}30}$$

In seeking relationships of unit vectors and angular velocities, a useful property is obtained from App. A. For any unit vector $\mathbf{n}$,

$$\dot{\mathbf{n}} = \boldsymbol{\omega}_n \times \mathbf{n} \qquad \boxed{12\text{-}31}$$

where $\boldsymbol{\omega}_n$ is the angular velocity of that unit vector. For example, $\dot{\mathbf{n}}_r = \dot{\phi}\mathbf{n}_\phi$ from Fig. 12-13c. Writing the vector product according to Eq. 12-31, with $\dot{\boldsymbol{\phi}} = \dot{\phi}\mathbf{n}_\theta$ (using the right-hand rule for $\dot{\boldsymbol{\phi}}$),

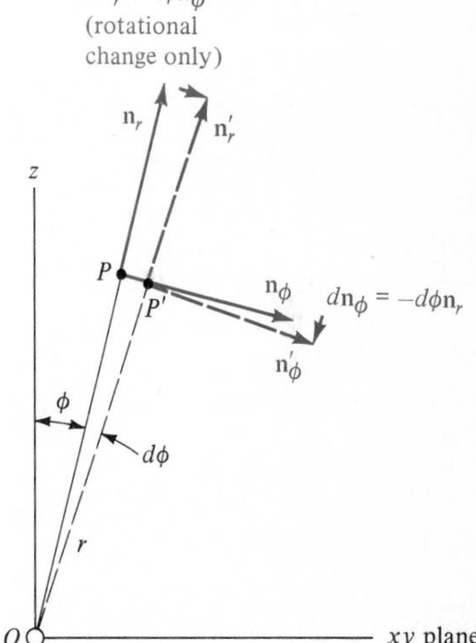

(c) View A from (b) (⊥ to $\mathbf{n}_r$ and $\mathbf{n}_\phi$)

FIGURE 12-13

Unit vectors in spherical coordinates

$$\dot{\boldsymbol{\phi}} \times \mathbf{n}_r = \begin{vmatrix} \mathbf{n}_r & \mathbf{n}_\phi & \mathbf{n}_\theta \\ 0 & 0 & \dot{\phi} \\ 1 & 0 & 0 \end{vmatrix} = \dot{\phi}\mathbf{n}_\phi = \dot{\mathbf{n}}_r$$

For the complete analysis of the particle's velocity and acceleration in the given spherical system, assume that its position $\mathbf{r}$ and the angles ϕ and θ are given as functions of time. In unit vector notation, the position is

$$\mathbf{r} = r\mathbf{n}_r \tag{12-32}$$

The total angular velocity $\boldsymbol{\omega}$ of vector $\mathbf{r}$ can be expressed in terms of components,

$$\boldsymbol{\omega} = \dot{\phi}\mathbf{n}_\theta + \dot{\theta}\mathbf{n}_z = \dot{\phi}\mathbf{n}_\theta + \dot{\theta}\cos\phi\mathbf{n}_r - \dot{\theta}\sin\phi\mathbf{n}_\phi \tag{12-33}$$

where Fig. 12-13c was used to convert from $\mathbf{n}_z$ to $\mathbf{n}_r$ and $\mathbf{n}_\phi$.

The velocity of the particle is obtained by differentiating Eq. 12-32 with respect to time,

$$\mathbf{v} = \dot{\mathbf{r}} = \dot{r}\mathbf{n}_r + r\dot{\mathbf{n}}_r \tag{12-34}$$

where $\dot{\mathbf{n}}_r$ can be written as

$$\dot{\mathbf{n}}_r = \boldsymbol{\omega} \times \mathbf{n}_r = \dot{\phi}\mathbf{n}_\phi + \dot{\theta}\sin\phi\mathbf{n}_\theta \tag{12-35}$$

Substituting Eq. 12-35 into Eq. 12-34, the general expression for the velocity is obtained,

$$\boxed{\mathbf{v} = \dot{r}\mathbf{n}_r + r\dot{\phi}\mathbf{n}_\phi + r\dot{\theta}\sin\phi\mathbf{n}_\theta} \tag{12-36}$$

The acceleration of the particle is derived by differentiating Eq. 12-36 with respect to time,

$$\begin{aligned} \mathbf{a} = \dot{\mathbf{v}} = {} & \ddot{r}\mathbf{n}_r + \dot{r}\dot{\mathbf{n}}_r + \dot{r}\dot{\phi}\mathbf{n}_\phi + r\ddot{\phi}\mathbf{n}_\phi \\ & + r\dot{\phi}\dot{\mathbf{n}}_\phi + \dot{r}\dot{\theta}\sin\phi\mathbf{n}_\theta + r\ddot{\theta}\sin\phi\mathbf{n}_\theta \\ & + r\dot{\theta}\cos\phi\dot{\phi}\mathbf{n}_\theta + r\dot{\theta}\sin\phi\dot{\mathbf{n}}_\theta \end{aligned} \tag{12-37}$$

This can be simplified by the following substitutions:

$$\dot{\mathbf{n}}_\phi = \boldsymbol{\omega} \times \mathbf{n}_\phi = -\dot{\phi}\mathbf{n}_r + \dot{\theta}\cos\phi\mathbf{n}_\theta \tag{12-38}$$

$$\dot{\mathbf{n}}_\theta = \boldsymbol{\omega} \times \mathbf{n}_\theta = -\dot{\theta}\sin\phi\mathbf{n}_r - \dot{\theta}\cos\phi\mathbf{n}_\phi \tag{12-39}$$

The resulting general expression for the acceleration is

$$\boxed{\begin{aligned} \mathbf{a} = {} & (\ddot{r} - r\dot{\phi}^2 - r\dot{\theta}^2\sin^2\phi)\mathbf{n}_r \\ & + (r\ddot{\phi} + 2\dot{r}\dot{\phi} - r\dot{\theta}^2\sin\phi\cos\phi)\mathbf{n}_\phi \\ & + (r\ddot{\theta}\sin\phi + 2\dot{r}\dot{\theta}\sin\phi + 2r\dot{\phi}\dot{\theta}\cos\phi)\mathbf{n}_\theta \end{aligned}} \tag{12-40}$$

where each parenthetical expression is a component of the total acceleration in the direction of the appropriate unit vector.

EXAMPLE 12-4

A wheel of radius $R = 0.4$ m is uniformly accelerated with an angular acceleration $\alpha = 30$ rad/s^2 until its angular speed reaches 100 revolutions per minute. Calculate and sketch the total acceleration of a particle on the outside of the wheel at $\omega_1 = 25$ rpm, $\omega_2 = 50$ rpm, and $\omega_3 = 100$ rpm (at the final velocity, $\alpha = 0$). Work with tangential and normal components based on the given angular acceleration and speeds.

SOLUTION

The normal and tangential components are obtained from Eqs. 12-25 and 12-26, which are applicable because the motion is circular. The total acceleration is expressed as

$$\mathbf{a} = \mathbf{a}_t + \mathbf{a}_n = r\alpha\mathbf{n}_t + r\omega^2\mathbf{n}_n$$

Before this equation can be used the values for ω must be converted from units of revolutions per minute (rpm) to radians per second. Recalling that an angle of 2π radians is traveled in one revolution,

$$\omega_1 = \left(\frac{25 \text{ rev}}{\text{min}}\right)\left(\frac{2\pi \text{ rad}}{\text{rev}}\right)\left(\frac{1 \text{ min}}{60 \text{ s}}\right) = 2.62 \text{ rad/s}$$

$$\begin{aligned}\mathbf{a}_1 &= 0.4 \text{ m}(30 \text{ rad/s}^2)\,\mathbf{n}_t + 0.4 \text{ m}(2.62 \text{ rad/s})^2\,\mathbf{n}_n \\ &= 12 \text{ m/s}^2\,\mathbf{n}_t + 2.74 \text{ m/s}^2\,\mathbf{n}_n\end{aligned}$$

$$\omega_2 = \left(\frac{50 \text{ rev}}{\text{min}}\right)\left(\frac{2\pi \text{ rad}}{\text{rev}}\right)\left(\frac{1 \text{ min}}{60 \text{ s}}\right) = 5.24 \text{ rad/s}$$

$$\begin{aligned}\mathbf{a}_2 &= 0.4 \text{ m}(30 \text{ rad/s}^2)\,\mathbf{n}_t + 0.4 \text{ m}(5.24 \text{ rad/s})^2\,\mathbf{n}_n \\ &= 12 \text{ m/s}^2\,\mathbf{n}_t + 11.0 \text{ m/s}^2\,\mathbf{n}_n\end{aligned}$$

$$\omega_3 = \left(\frac{100 \text{ rev}}{\text{min}}\right)\left(\frac{2\pi \text{ rad}}{\text{rev}}\right)\left(\frac{1 \text{ min}}{60 \text{ s}}\right) = 10.5 \text{ rad/s}$$

$$\begin{aligned}\mathbf{a}_3 &= 0\mathbf{n}_t + 0.4 \text{ m}(10.5 \text{ rad/s})^2\,\mathbf{n}_n \qquad (\text{since } \alpha = 0) \\ &= 43.9 \text{ m/s}^2\,\mathbf{n}_n\end{aligned}$$

Sketches of the acceleration vector of a general particle for each case are shown in Fig. 12-14.

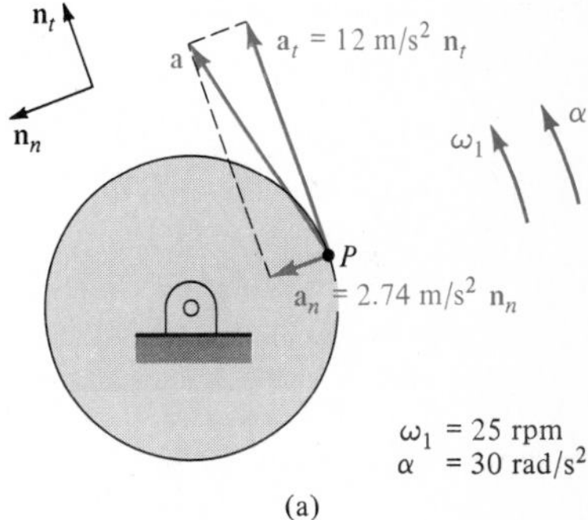

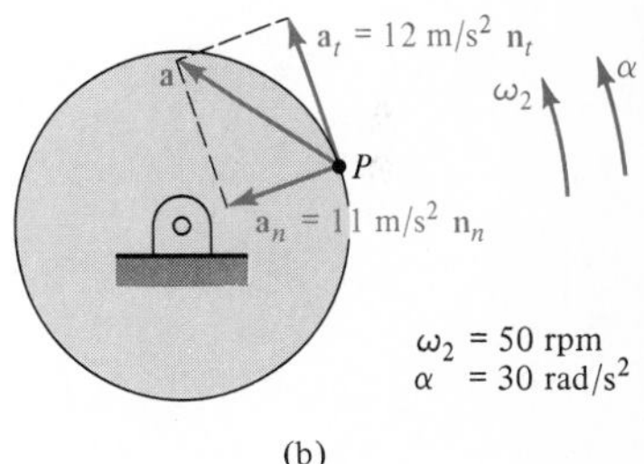

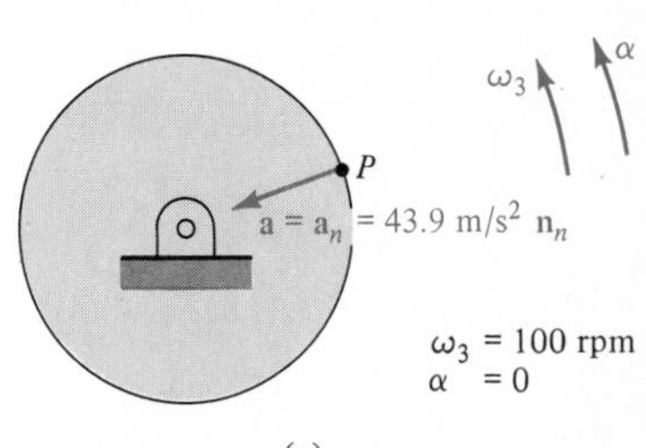

FIGURE 12-14
Acceleration vectors of particle P on a wheel

JUDGMENT OF THE RESULTS

The relative magnitudes of the acceleration components are reasonable because the magnitude a_n should become larger with respect to a_t under constant angular acceleration as the speed of rotation increases.

EXAMPLE 12-5

A radar station is tracking a rocket shortly after launch (Fig. 12-15). The data obtained at $\theta = 40°$ are $\dot{\theta} = 0.1$ rad/s, $\ddot{\theta} = 0.05$ rad/s^2, $r = 500$ ft, $\dot{r} = 40$ ft/s, and $\ddot{r} = 0$. (a) Determine the magnitudes of the velocity and acceleration of the rocket. (b) Obtain an expression for r in terms of θ, assuming that $\dot{r}$ = constant and $\dot{\theta}$ = constant for $0 < \theta < 40°$.

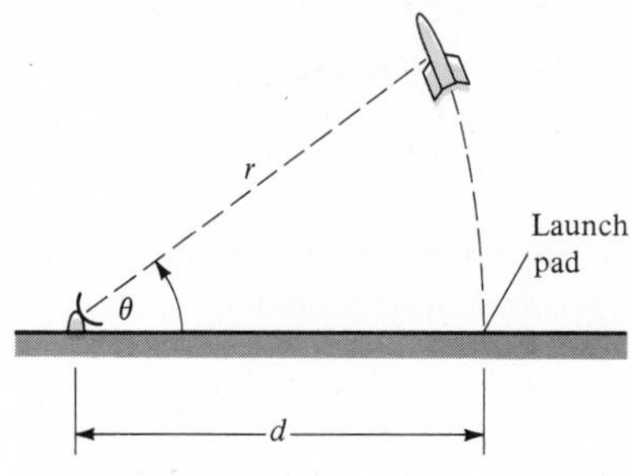

FIGURE 12-15
Tracking a rocket by radar

SOLUTION

(a) The magnitude of the velocity is obtained using Eq. 12-15, where $\mathbf{n}_r$ and $\mathbf{n}_\theta$ are orthogonal unit vectors,

$$\begin{aligned} v &= \sqrt{\dot{r}^2 + r^2\dot{\theta}^2} = \sqrt{40^2 + (500^2)(0.1^2)} \\ &= 64 \text{ ft/s} \end{aligned}$$

The magnitude of the acceleration is obtained using Eq. 12-20,

$$\begin{aligned} a &= \sqrt{(\ddot{r} - r\dot{\theta}^2)^2 + (r\ddot{\theta} + 2\dot{r}\dot{\theta})^2} \\ &= \sqrt{[0 - (500)(0.01)]^2 + [(500)(0.05) + 2(40)(0.1)]^2} \\ &= 33.4 \text{ ft/s}^2 \end{aligned}$$

Note that quantities with units of rad/s and rad/s^2 required no conversion in this solution.

(b) At all times t for motion in the given range of angles (since $\dot{r}$ = constant),

$$r = d + \dot{r}t$$

in which θ can be entered by applying (since $\dot{\theta}$ = constant)

$$\theta = \dot{\theta}t$$

Thus,

$$r = d + \frac{\dot{r}\theta}{\dot{\theta}}$$

EXAMPLE 12-6

The track of a proposed ski-jump tower is in the shape of a circular arc as shown in Fig. 12-16. The jumpers would start from rest at A and become airborne at B. Assume that they would have an acceleration $a_\theta = g \sin \theta$ tangential to the path from A to B, where $g = 9.81 \text{ m/s}^2$. Determine the takeoff speed v if $r = 60$ m and $\theta = 40°$.

SOLUTION

The takeoff velocity is horizontal by inspection. Since r is given, the problem is to determine $\dot{\theta}$ before calculating v using Eq. 12-14. Starting with the given relationship, a_θ is written as

$$a_\theta = g \sin \theta = \frac{dv}{dt}$$

Separating variables yields

$$g \sin \theta \, dt = dv$$

A problem arises here since it is not known how θ varies with time, so the equation cannot be integrated. Using transverse components from Eq. 12-22 gives

$$a_\theta = g \sin \theta = r\ddot{\theta} + 2\dot{r}\dot{\theta}$$

$$g \sin \theta = r \frac{d\dot{\theta}}{dt} + 0 \qquad \text{since } \dot{r} = 0$$

The chain rule is applied to the right side of this equation to allow for separation of variables,

$$g \sin \theta = r \frac{d\dot{\theta}}{dt} = r \frac{d\dot{\theta}}{d\theta}\frac{d\theta}{dt} = r \frac{d\omega}{d\theta}\,\omega$$

$$\omega \frac{d\omega}{d\theta} - \frac{g}{r} \sin \theta = 0$$

$$\omega \, d\omega - \frac{g}{r} \sin \theta \, d\theta = 0$$

Integrating yields

$$\frac{\omega^2}{2} + \frac{g}{r} \cos \theta = C = \text{constant}$$

Initial conditions require that $\omega = 0$ at $\theta = 0°$.

Hence

$$C = \frac{g}{r} \qquad \text{and} \qquad \omega = \sqrt{\frac{2g}{r}(1 - \cos \theta)}$$

For $\theta = 40°$,

$$\omega = \sqrt{\frac{2(9.81 \text{ m/s}^2)}{60 \text{ m}}(1 - \cos 40°)} = 0.277 \text{ rad/s}$$

From Eq. 12-14,

$$v = r\omega = (60 \text{ m})(0.277 \text{ rad/s}) = 16.6 \text{ m/s}$$

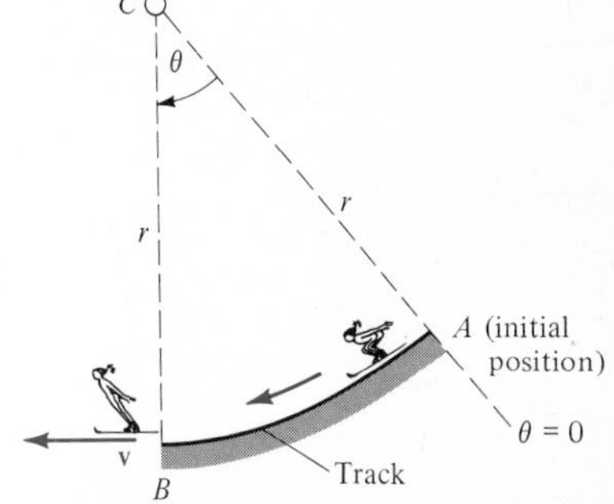

FIGURE 12-16
Motion of a ski-jumper

EXAMPLE 12-7

A satellite tracking antenna system includes a hemispherical body which rotates about its vertical axis as illustrated in cross-sectional view in Fig. 12-17a. (a) Determine the acceleration of P for a constant angle $\phi = 50°$ using spherical coordinates. Convert the result to rectangular coordinates. (b) Also determine the acceleration of P for the instant shown with the angle ϕ changing at a constant rate of $\dot{\phi} = -2$ rad/s.

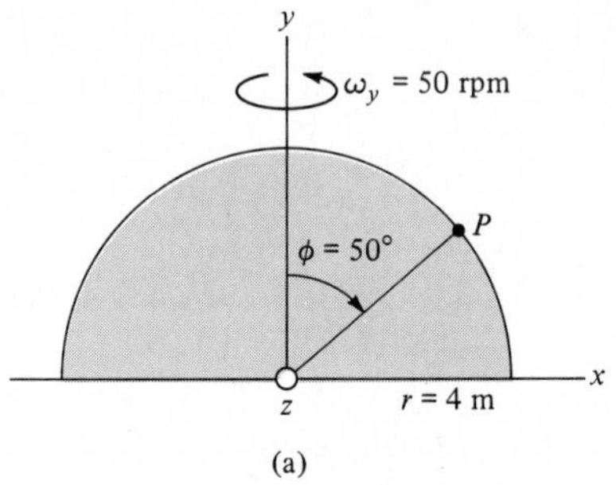

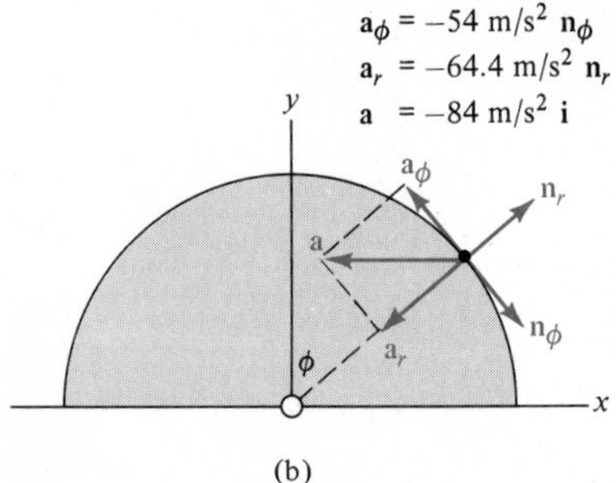

FIGURE 12-17

Accelerations of a particle using spherical coordinates

SOLUTION

(a) Applying Eq. 12-40 with $\dot{r} = \ddot{r} = \dot{\phi} = \ddot{\phi} = \ddot{\theta} = 0$, and

$$\dot{\theta} = \omega = 50 \text{ rpm} = \left(50 \frac{\text{rev}}{\text{min}}\right)\left(\frac{2\pi \text{ rad}}{\text{rev}}\right)\left(\frac{1 \text{ min}}{60 \text{ s}}\right)$$

$$= 5.24 \text{ rad/s}$$

yields

$$\begin{aligned}
\mathbf{a} &= -r\dot{\theta}^2 \sin^2 \phi \mathbf{n}_r - r\dot{\theta}^2 \sin \phi \cos \phi \mathbf{n}_\phi + 0\mathbf{n}_\theta \\
&= -4 \text{ m}(5.24 \text{ rad/s})^2 \sin^2 50^\circ \mathbf{n}_r \\
&\quad - 4 \text{ m}(5.24 \text{ rad/s})^2 \sin 50^\circ \cos 50^\circ \mathbf{n}_\phi \\
&= -64.4 \text{ m/s}^2 \mathbf{n}_r - 54.0 \text{ m/s}^2 \mathbf{n}_\phi
\end{aligned}$$

The spherical and rectangular components are sketched in Fig. 12-17, where

$$\begin{aligned}
a_x &= a_\phi \cos \phi + a_r \sin \phi \\
&= -54 \text{ m/s}^2 \cos 50^\circ - 64.4 \text{ m/s}^2 \sin 50^\circ
\end{aligned}$$

$$\mathbf{a}_x = -84 \text{ m/s}^2 \mathbf{i}$$

$$\begin{aligned}
a_y &= -a_\phi \sin \phi + a_r \cos \phi \\
&= 54 \text{ m/s}^2 \sin 50^\circ - 64.4 \text{ m/s}^2 \cos 50^\circ
\end{aligned}$$

$$\mathbf{a}_y = 0\mathbf{j}$$

(b) $\dot{\phi} = -2$ rad/s. From Eq. 12-40,

$$\mathbf{a} = (-r\dot{\phi}^2 - r\dot{\theta}^2 \sin^2 \phi)\mathbf{n}_r -- r\dot{\theta}^2 \sin \phi \cos \phi \mathbf{n}_\phi + 2r\dot{\phi}\dot{\theta} \cos \phi \mathbf{n}_\theta$$

Using the results that have already been obtained yields

$$\begin{aligned}
\mathbf{a} &= [-4\text{m}(-2 \text{ rad/s})^2 - 64.4 \text{ m/s}^2]\mathbf{n}_r - 54.0 \text{ m/s}^2 \mathbf{n}_\phi \\
&\quad + 2(4 \text{ m})(-2 \text{ rad/s})(5.24 \text{ rad/s}) \cos 50^\circ \mathbf{n}_\theta \\
&= -80.4 \text{ m/s}^2 \mathbf{n}_r - 54.0 \text{ m/s}^2 \mathbf{n}_\phi - 53.8 \text{ m/s}^2 \mathbf{n}_\theta
\end{aligned}$$

Realizing the relations between $\mathbf{a}_x$, $\mathbf{a}_y$ and $\mathbf{a}_r$, $\mathbf{a}_\phi$ stated above, and also that $\mathbf{n}_\theta = -\mathbf{k}$ (since $\mathbf{n}_\theta = \mathbf{n}_r \times \mathbf{n}_\phi$), $\mathbf{a}_z = -\mathbf{a}_\theta$ for the position shown.

$$\begin{aligned}
\mathbf{a} &= (a_\phi \cos \phi + a_r \sin \phi)\mathbf{i} + (-a_\phi \sin \phi + a_r \cos \phi)\mathbf{j} + a_\theta\mathbf{k} \\
&= (-54.0 \cos 50^\circ - 80.4 \sin 50^\circ) \text{ m/s}^2 \mathbf{i} + (54.0 \sin 50^\circ \\
&\quad - 80.4 \cos 50^\circ) \text{ m/s}^2 \mathbf{j} + 53.8 \text{ m/s}^2 \mathbf{k} \\
&= (-96.3\mathbf{i} - 10.3\mathbf{j} + 53.8\mathbf{k}) \text{ m/s}^2
\end{aligned}$$

JUDGMENT OF THE RESULTS IN PART (a)

Since the path of particle P is a circle in a horizontal plane, and the speed of P is constant, the only component of acceleration is the normal component parallel to the xz plane and directed toward the y axis. Thus,

$$\mathbf{a} = \mathbf{a}_n = \rho\omega^2\mathbf{n}_n = (4 \text{ m} \sin 50^\circ)(5.24 \text{ rad/s})^2 \mathbf{n}_n = 84 \text{ m/s}^2 \mathbf{n}_n$$

which agrees with the result obtained using spherical coordinates.

PROBLEMS

SECTIONS 12-4, 12-5, 12-6

FIGURE P12-32

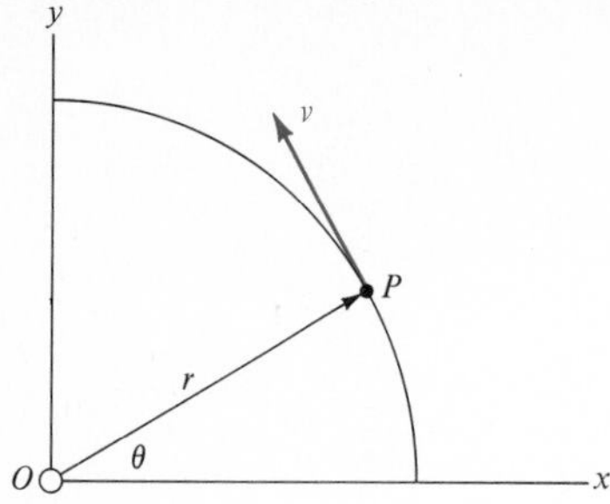

12–32 A particle P travels with a speed $v = 5$ m/s at $\theta = 30°$ on a circular path of radius $r = 2$ m. The speed is increasing at the rate of $3\ \text{m/s}^2$ at the given position. Determine the total acceleration of the particle using polar coordinates.

12–33 A particle P travels with a constant speed v on a circular track of radius $r = 2$ ft as in Fig. P12-32. Determine the maximum speed v if the total acceleration of the particle must not exceed $300\ \text{ft/s}^2$. Use polar coordinates.

FIGURE P12-34

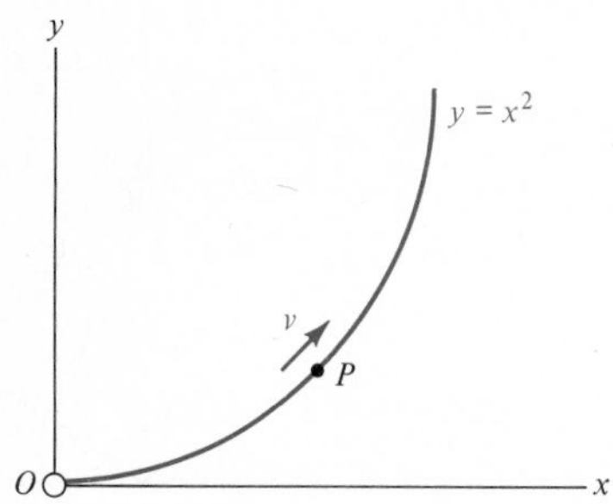

12–34 A particle P travels with a constant speed $v = 3$ m/s on a path defined by $y = x^2$. Determine the angular speed and angular acceleration of the line OP at $x = 2$ m with the angle θ measured from the x axis.

12–35 A particle P travels with a constant speed $v = 10$ ft/s as shown in Fig. P12-34. Determine the total acceleration of the particle at $x = 5$ ft using polar coordinates. The angle θ of line OP is measured from the x axis.

FIGURE P12-36

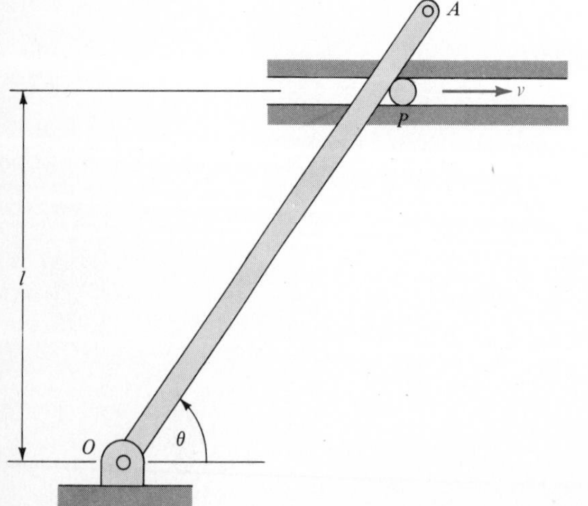

12–36 Peg P in a mechanism is to be pushed to the right with a constant speed $v = 8$ m/s. Determine the required angular speed and angular acceleration of the arm OA at $\theta = 60°$ if $l = 0.5$ m.

12–37 The acceleration of particle P in the horizontal groove in Fig. P12-36 must not exceed $100\ \text{ft/s}^2$. What are the limitations on the rotational motion of arm OA, which pushes the particle, if $l = 1$ ft and $\theta = 70°$? The arm starts from rest at $\theta = 90°$.

12–38 An aircraft is tracked by radar and the following data are obtained: $\theta = 80°$, $\dot{\theta} = 0.01$ rad/s (increasing θ), $\ddot{\theta} = 0$, $r = 3$ km, $\dot{r} = -100$ m/s, and $\ddot{r} = -5$ m/s^2. Determine the horizontal components of the velocity and acceleration of the aircraft.

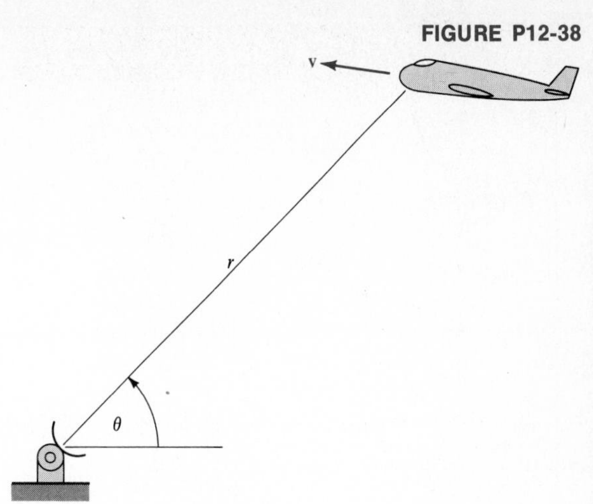

FIGURE P12-38

12–39 The following data are obtained by radar for the aircraft in Fig. P12-38: $\theta = 60°$, $\dot{\theta} = 0.03$ rad/s (increasing θ), $\ddot{\theta} = 0.01$ rad/s^2 (increasing $\dot{\theta}$), $r = 10{,}000$ ft, $\dot{r} = -400$ ft/s, and $\ddot{r} = -30$ ft/s^2. Determine the horizontal components of the velocity and acceleration of the aircraft.

12–40 The following data are obtained from an aircraft by radio: $\mathbf{v} = (200\mathbf{i} + 40\mathbf{j})$ m/s and $\mathbf{a} = 0$ at $x = 800$ m, $y = 3$ km. Determine $\dot{r}$, $\ddot{r}$, $\dot{\theta}$, and $\ddot{\theta}$ measured at that instant at a radar station located at point O.

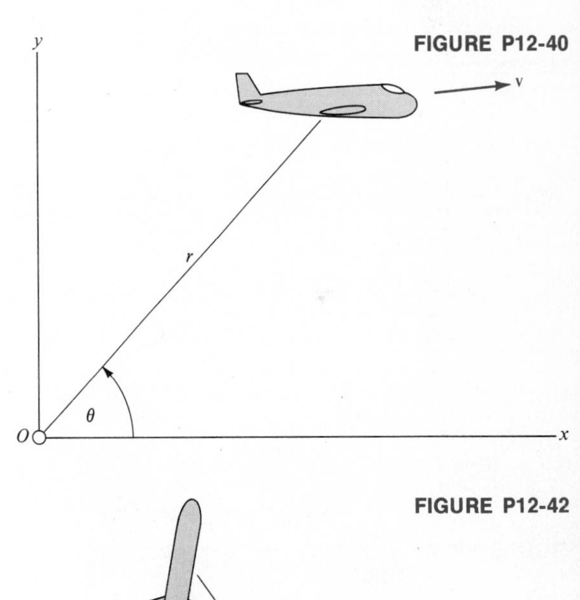

FIGURE P12-40

12–41 The following data are reported from an aircraft by radio: $\mathbf{v} = 400\mathbf{i}$ ft/s and $\mathbf{a} = 100\mathbf{i}$ ft/s^2 at $x = y = 6000$ ft. Determine $\dot{r}$, $\ddot{r}$, $\dot{\theta}$, and $\ddot{\theta}$ measured by radar at point O in Fig. P12-40.

12–42 A rocket is tracked by radar giving the data $\theta = 30°$, $\dot{\theta} = 0.02$ rad/s (increasing θ), $\ddot{\theta} = 0$, $r = 2$ km, $\dot{r} = 30$ m/s, and $\ddot{r} = 15$ m/s^2. Determine the vertical components of the velocity and the acceleration of the rocket.

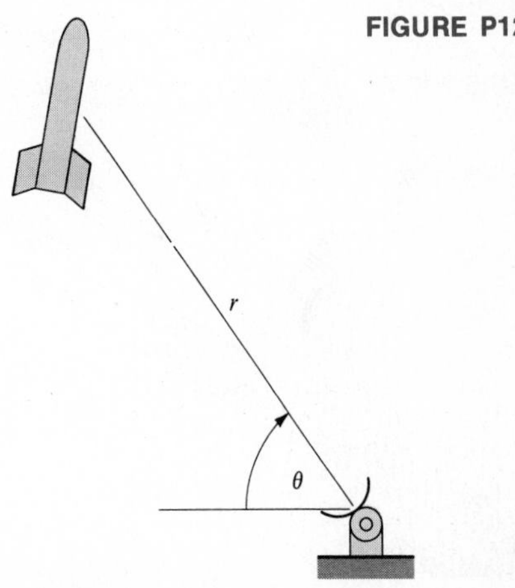

FIGURE P12-42

12–43 The following data are obtained by radar for the rocket in Fig. P12-42: $\theta = 60°$, $\dot{\theta} = 0.03$ rad/s (increasing θ), $\ddot{\theta} = 0.001$ rad/s^2 (decreasing $\dot{\theta}$), $r = 7000$ ft, $\dot{r} = 800$ ft/s, and $\ddot{r} = 50$ ft/s^2. Determine the magnitudes of the velocity and acceleration of the rocket.

FIGURE P12-44

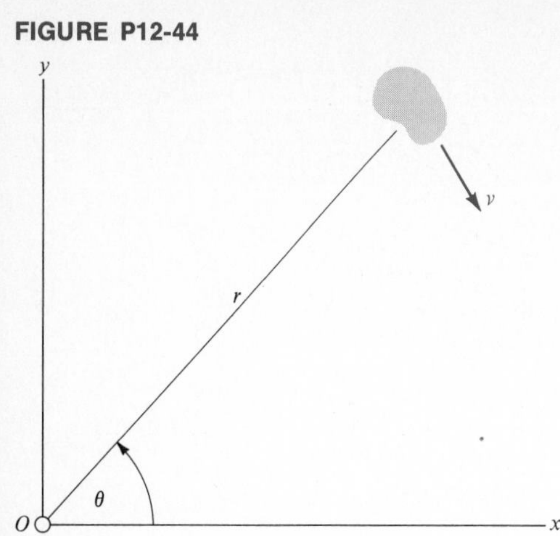

12–44 The data obtained from radar tracking of a meteor are as follows: $\theta = 60^\circ$, $\dot{\theta} = 3$ rad/s (decreasing θ), $\ddot{\theta} = 0.3$ rad/s^2 (decreasing $\dot{\theta}$), $r = 28$ km, $\dot{r} = -40$ m/s, and $\ddot{r} = 130$ m/s^2. Determine the velocity and acceleration vectors of the meteor.

FIGURE P12-45

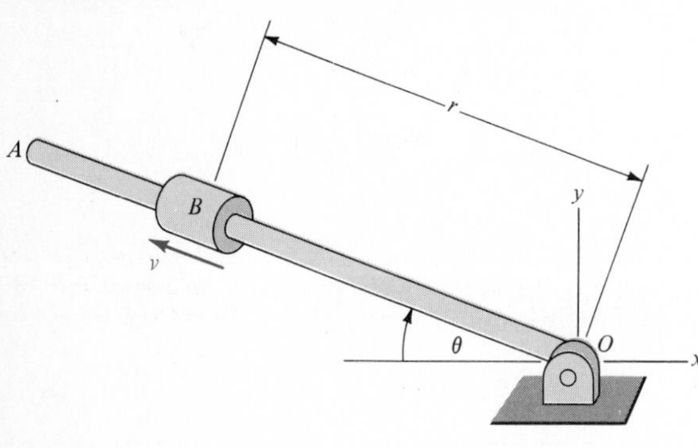

12–45 Block B slides along bar OA with a constant speed $v = 6$ ft/s while the bar is rotating with a constant angular speed $\omega = 120$ rpm. Determine the velocity and acceleration of block B at $r = 2$ ft and $\theta = 40^\circ$.

12–46 Arm A in Fig. P12-45 starts from rest at $\theta = 0$; its position is given by $\theta = 3t^2$, where θ is in radians and t is in seconds. The position of block B is given by $r = t^2/2 + 0.1$ m. Determine the total velocity $\mathbf{v}_B$ and acceleration $\mathbf{a}_B$ of the block at $t = 2$ s.

FIGURE P12-47

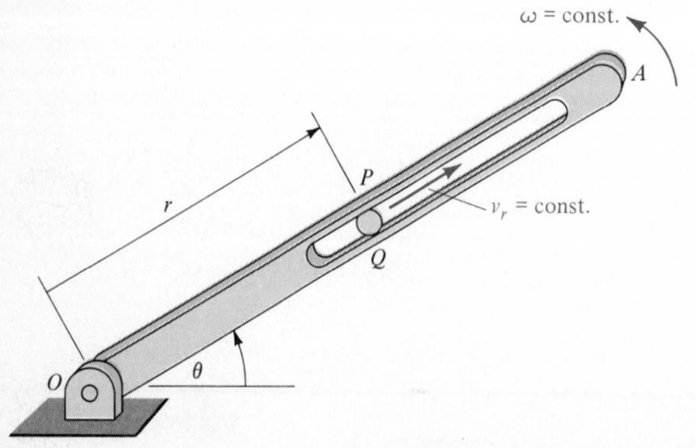

12–47 Peg P moves in the slotted bar with a constant speed v_r while the bar rotates with a constant angular speed ω. Write an expression for the total acceleration of the peg at position r using polar coordinates. Prove that peg P and a particle Q of the bar OA have different accelerations at the instant when P and Q are momentarily in contact.

12–48 A car travels with constant speed $v = 80$ ft/s on a circular part of a test track. The car's motion is monitored with radar located at point O. Determine the total acceleration of the car at $\theta = 70°$ if $R = 300$ ft. Use polar coordinates.

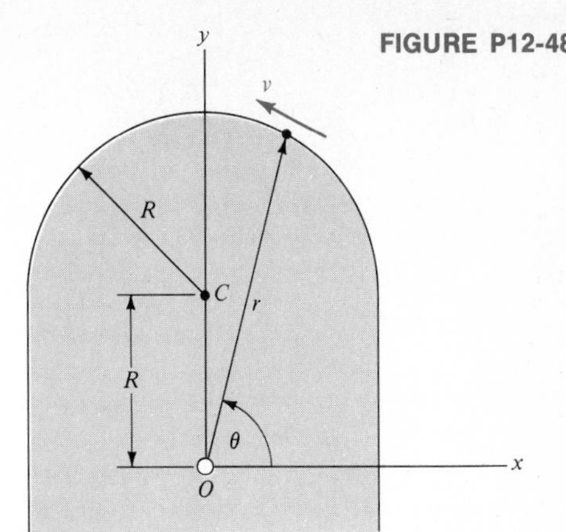

FIGURE P12-48

12–49 Small containers are moving on the helical path of an automated packaging machine. Calculate the acceleration of a particle at P if it has a constant speed $v = 2$ m/s, and $D = 0.6$ m.

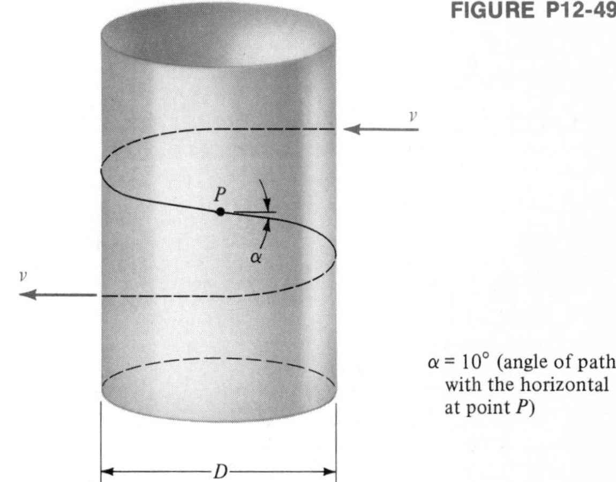

FIGURE P12-49

$\alpha = 10°$ (angle of path with the horizontal at point P)

12–50 Calculate the maximum allowable speed v of a container on the helical path shown in Fig. P12-49. The speed is constant, $D = 2$ ft, and the acceleration should not exceed 20 ft/s^2.

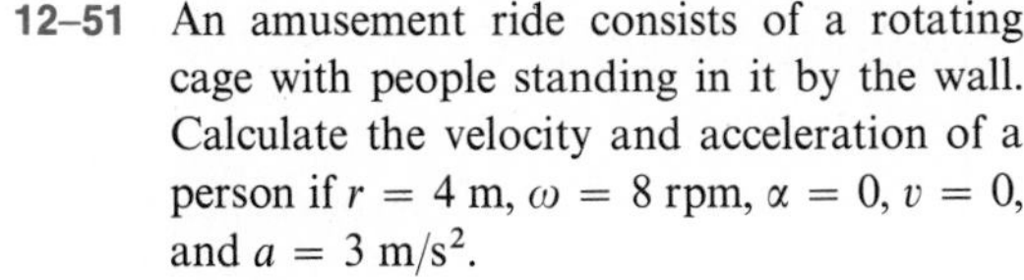

12–51 An amusement ride consists of a rotating cage with people standing in it by the wall. Calculate the velocity and acceleration of a person if $r = 4$ m, $\omega = 8$ rpm, $\alpha = 0$, $v = 0$, and $a = 3$ m/s^2.

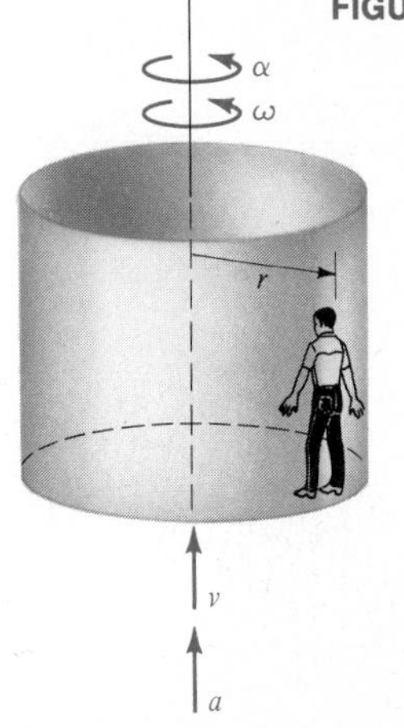

FIGURE P12-51

12–52 Calculate the velocity and acceleration of a person in the cage described in Prob. 12–51 if $r = 15$ ft, $\omega = 10$ rad/s, $\alpha = 2$ rad/s^2, $v = 5$ ft/s, and $a = 10$ ft/s^2.

FIGURE P12-53

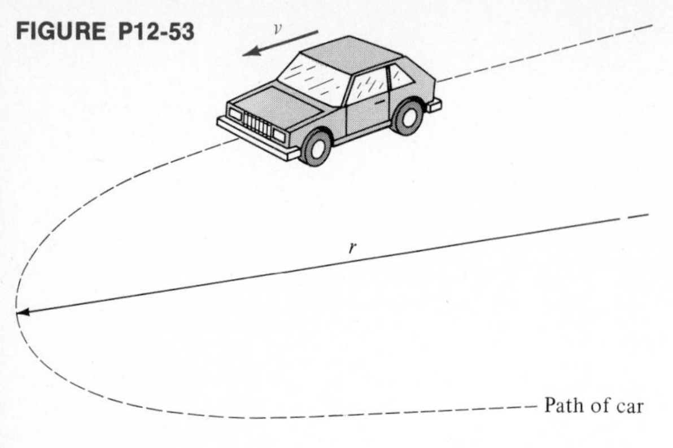

12–53 A car is descending on a curved road with a speed $v = 15$ m/s which is decreasing at the rate of 4 m/s^2. Calculate the acceleration of the car if $r = 200$ m and the road makes a 10° angle with the horizontal.

12–54 The loop of the amusement ride is approximately circular with a diameter of 74 ft. The lower parts A and B of the helical track are horizontally separated by a distance of 6 ft. Calculate the acceleration of the train at point C, where its speed along the track is 50 ft/s and increasing at the rate of 4 ft/s^2.

FIGURE P12-54

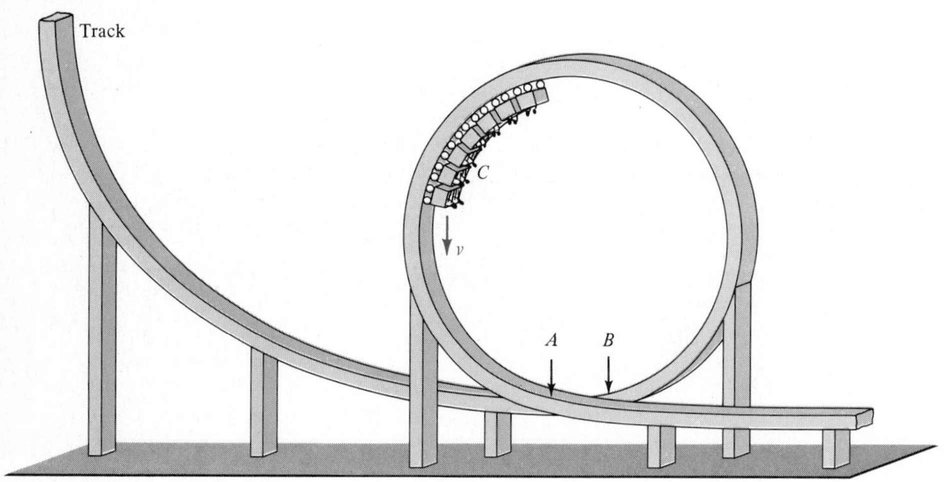

FIGURE P12-55

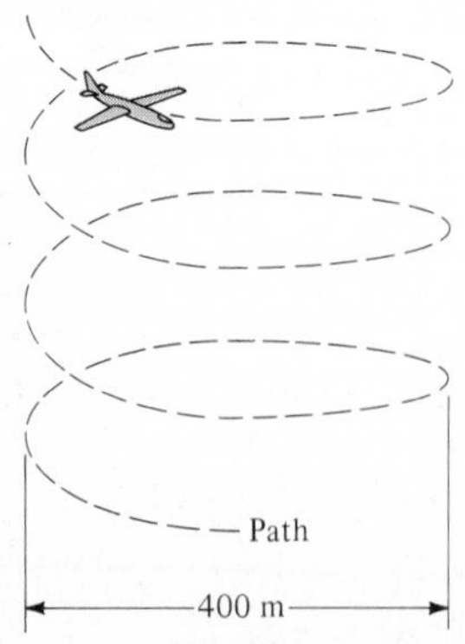

12–55 An airplane is descending in a helical pattern. It has a speed of 80 m/s horizontally, a downward speed of 5 m/s, and a downward acceleration of 2 m/s^2. Calculate the airplane's acceleration.

12–56 Small containers on a tray are lowered as the supporting screw rotates. The lead of the screw (advancement per revolution) is 0.3 in., and r is 10 in. Calculate the acceleration of a container if ω is 10 rad/s and decreasing at the rate of 1 rad/s^2.

FIGURE P12-56

12–57 Particles of fluid move at a constant speed $v = 5$ m/s in the rotating nozzle. Calculate the acceleration of the fluid at $r = 20$ cm.

FIGURE P12-57

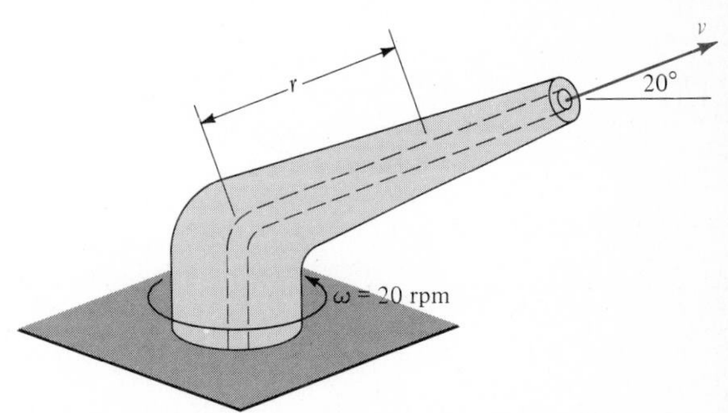

12–58 The train of the amusement ride has a helical track. The passengers travel on a 28-ft-diameter helix which makes a complete turn in a 50-ft horizontal distance. Calculate the acceleration of the riders when their speed along the track is 40 ft/s, increasing at the rate of 5 ft/s^2.

FIGURE P12-58

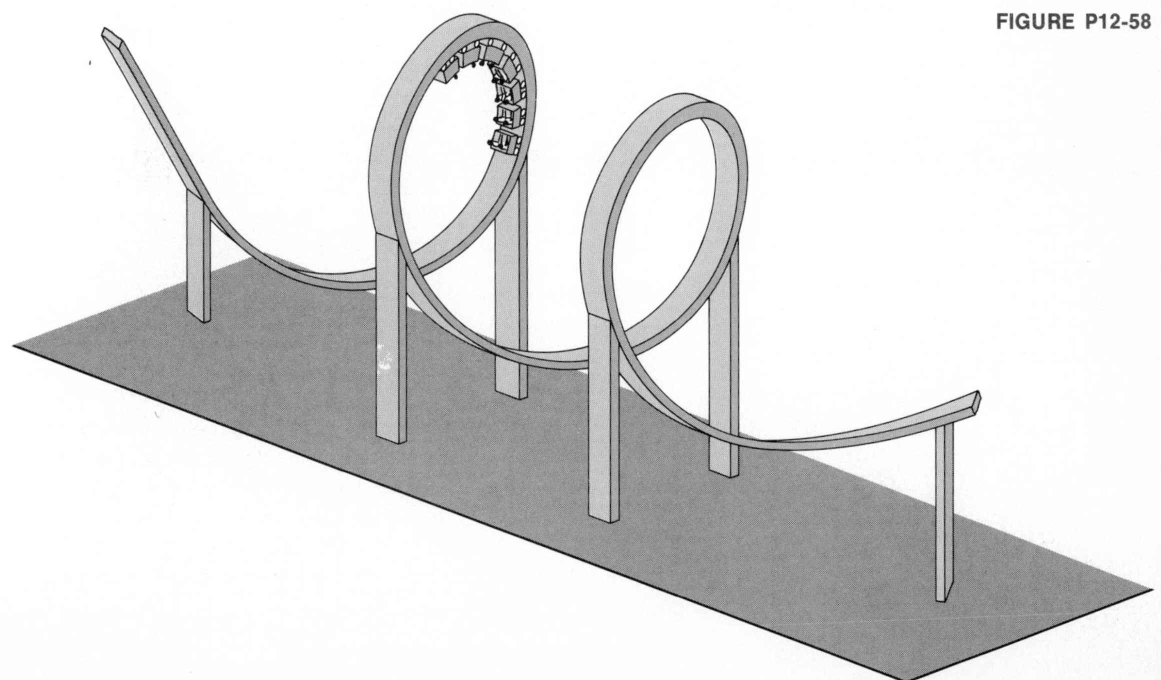

FIGURE P12-59

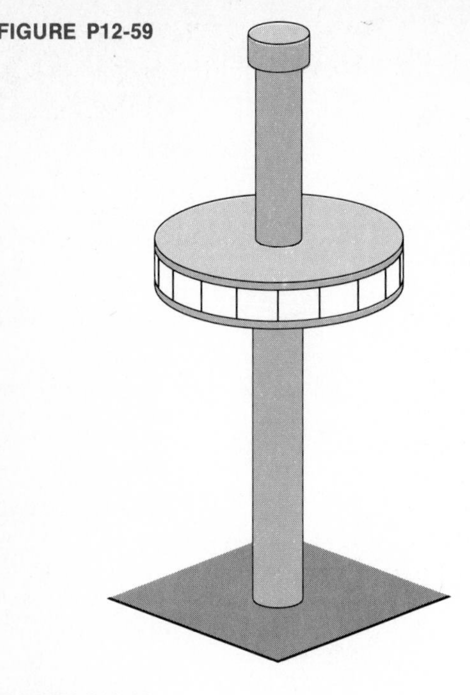

12–59 The rotary elevator on the outside of the tall tower has an angular speed $\omega = 0.5$ rad/s which increases at the rate $\alpha = 0.1$ rad/s^2. The elevator is moving downward at this instant with $v = 2$ ft/s and $a = 1$ ft/s^2. Calculate the acceleration of a person who is 12 ft from the center of the tower.

FIGURE P12-60

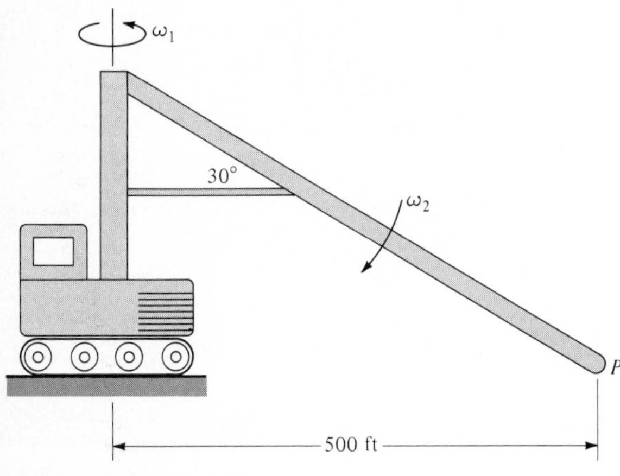

12–60 Consider the model of the large, high-speed earth mover used in strip mining. The long arm rotates about a vertical line with $\omega_1 = 0.01$ rad/s, while it also rotates about a horizontal line with $\omega_2 = 0.005$ rad/s. Calculate the total acceleration of a particle P at the end of the arm.

12–61 The hemispherical structure of a radar antenna is rotating about the z axis with $\omega_z = 2$ rad/s, which is increasing at the rate of 0.1 rad/s^2. Calculate the acceleration of a particle at P_1 if $R = 5$ m, $\phi_1 = 30°$, and $\dot{\phi}_1 = 3$ rad/s.

FIGURE P12-61

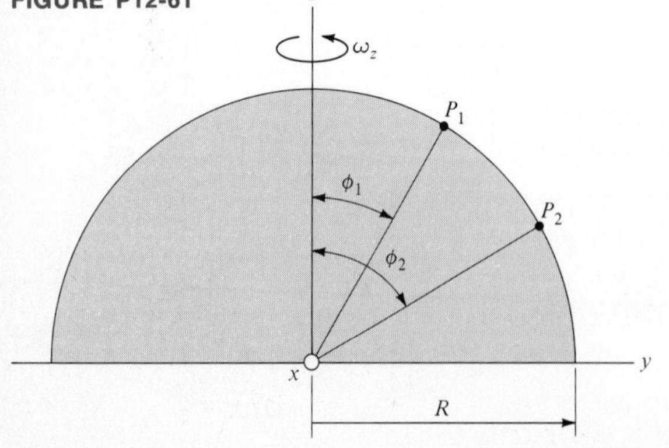

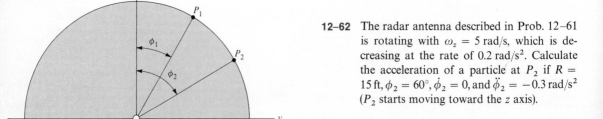

12–62 The radar antenna described in Prob. 12–61 is rotating with $\omega_z = 5$ rad/s, which is decreasing at the rate of 0.2 rad/s^2. Calculate the acceleration of a particle at P_2 if $R = 15$ ft, $\phi_2 = 60°$, $\dot{\phi}_2 = 0$, and $\ddot{\phi}_2 = -0.3$ rad/s^2 (P_2 starts moving toward the z axis).

RELATIVE MOTION OF TWO PARTICLES. TRANSLATION OF COORDINATES

12-7

The preceding analyses of curvilinear motion were made with respect to fixed coordinate systems. In some practical problems it is desirable to work with a moving coordinate system. This can be done by extending the concepts of relative motion that were introduced for rectilinear motion. The approach is to define the motion of a particle with respect to a convenient moving reference system, and define the motion of this system with respect to a fixed coordinate system. For example, the motion of a component in an aircraft engine can be analyzed with respect to a coordinate system fixed in the aircraft. The motion of this system can be described with respect to a coordinate system fixed on earth. Clearly, no reference system is fixed in an absolute sense. The coordinates fixed on earth move with respect to the sun, and so forth. In most engineering problems the fixed coordinate system is taken to be stationary with respect to the earth. The moving reference frame may or may not be rotating with respect to the fixed frame. The following analysis is for those cases where the moving reference frame is not rotating.

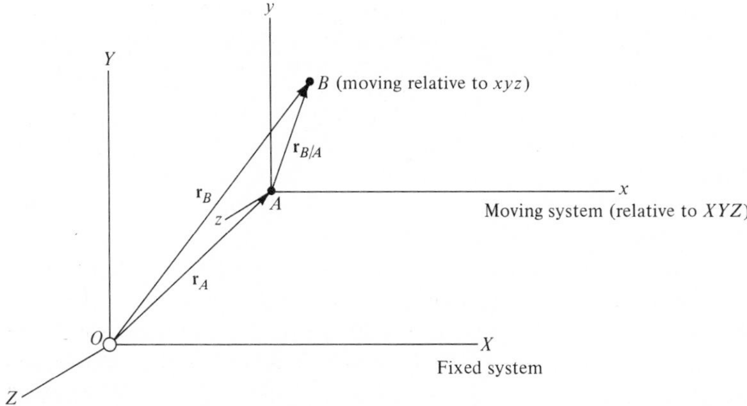

FIGURE 12-18

Translation of the xyz frame with respect to the XYZ frame

Translating Coordinate Systems

Consider Fig. 12-18 and assume that the XYZ coordinates are fixed and the xyz system is moving. This motion is called *translation* when the respective coordinates always remain parallel to each other. The moving coordinate system's position is defined by the vector $\mathbf{r}_A$, its velocity by $\mathbf{v}_A = \dot{\mathbf{r}}_A$, and its acceleration by $\mathbf{a}_A = \ddot{\mathbf{r}}_A$ with respect to the fixed coordinates X, Y, and Z.

Next assume that a particle B is moving with respect to the xyz coordinates. Its position vector with respect to point A is $\mathbf{r}_{B/A}$. It follows by vector addition that the position vector of particle B with respect to the fixed coordinates X, Y, and Z is completely specified by

$$\mathbf{r}_B = \mathbf{r}_A + \mathbf{r}_{B/A} \tag{12-41}$$

where the same set of unit vectors **i**, **j**, **k** is used in both reference frames. The velocity and acceleration of particle B in the XYZ frame is obtained by differentiating the position vector with respect to time,

$$\mathbf{v}_B = \dot{\mathbf{r}}_B = \dot{\mathbf{r}}_A + \dot{\mathbf{r}}_{B/A} = \mathbf{v}_A + \mathbf{v}_{B/A} \tag{12-42}$$

$$\mathbf{a}_B = \ddot{\mathbf{r}}_B = \ddot{\mathbf{r}}_A + \ddot{\mathbf{r}}_{B/A} = \mathbf{a}_A + \mathbf{a}_{B/A} \tag{12-43}$$

It should be noted in this derivation that the time derivatives of **i**, **j**, and **k** are zero since their magnitudes and directions are constant when the motion is translation.

The motion of particle B with respect to the moving reference frame at point A is called the *relative motion* of B and is denoted by the subscript B/A. The motion of particle B with respect to the fixed coordinate system at point O is called the *absolute motion* of B. This is denoted by the subscript B, the symbol for the particle. The choice of the subscripts is arbitrary in setting up a problem, but consistency is necessary in using the subscripts in Eqs. 12-41, 12-42, and 12-43. For example, interchange A and B in Fig. 12-18. Now Eq. 12-41 becomes $\mathbf{r}_A = \mathbf{r}_B + \mathbf{r}_{A/B}$, from which $\mathbf{r}_B = \mathbf{r}_A - \mathbf{r}_{A/B}$. Therefore, $\mathbf{r}_{B/A} = -\mathbf{r}_{A/B}$. The same is true for velocities and accelerations. Note that radial and transverse components are useful in analyzing relative motion as illustrated in the following example.

EXAMPLE 12-8

Ship A is monitoring the position and velocity of ship B by radar. The information is received at A in the form of the distance $r_{B/A}$ between the ships, the angle θ that the information beam makes with the north direction, and the rates at which these parameters change (Fig. 12-19). For the instant shown the absolute velocity of A is $\mathbf{v}_A = (-10\mathbf{i} + 40\mathbf{j})$ km/h, $\theta = 30^\circ$, $\dot{\theta} = -0.001$ rad/s, $r_{B/A} = 3$ km, and $\dot{r}_{B/A} = 2$ m/s. Determine the absolute velocity of B in units of m/s.

SOLUTION

From Eq. 12-42, $\mathbf{v}_B = \mathbf{v}_A + \mathbf{v}_{B/A}$. The velocity $\mathbf{v}_{B/A}$ must be calculated using radial and transverse components with respect to a translating reference system with its origin at A. From Eq. 12-15,

$$\begin{aligned} \mathbf{v}_{B/A} &= \dot{r}_{B/A}\mathbf{n}_r + r_{B/A}\dot{\theta}\mathbf{n}_\theta \\ &= 2\text{ m/s }\mathbf{n}_r + 3000\text{ m}(-0.001\text{ rad/s})\mathbf{n}_\theta \\ &= 2\text{ m/s }\mathbf{n}_r - 3\text{ m/s }\mathbf{n}_\theta \end{aligned}$$

Transforming from radial and transverse components to rectangular coordinates with the aid of Fig. 12-19b gives

$$\begin{aligned} (v_{B/A})_x &= -v_r \sin\theta + v_\theta \cos\theta \\ &= -2\text{ m/s} \sin 30^\circ + 3\text{ m/s} \cos 30^\circ \\ (v_{B/A})_y &= v_r \cos\theta + v_\theta \sin\theta \\ &= 2\text{ m/s} \cos 30^\circ + 3\text{ m/s} \sin 30^\circ \\ \mathbf{v}_{B/A} &= (1.6\mathbf{i} + 3.2\mathbf{j})\text{ m/s} \end{aligned}$$

Converting $\mathbf{v}_A$ to units of m/s yields

$$\begin{aligned} \mathbf{v}_A &= (-10\mathbf{i} + 40\mathbf{j})\text{ km/h}\left(\frac{1000\text{ m}}{\text{km}}\right)\left(\frac{1\text{ h}}{3600\text{ s}}\right) \\ &= (-2.8\mathbf{i} + 11.1\mathbf{j})\text{ m/s} \\ \mathbf{v}_B &= \mathbf{v}_A + \mathbf{v}_{B/A} = [(-2.8 + 1.6)\mathbf{i} + (11.1 + 3.2)\mathbf{j}]\text{ m/s} \\ &= (-1.2\mathbf{i} + 14.3\mathbf{j})\text{ m/s} \end{aligned}$$

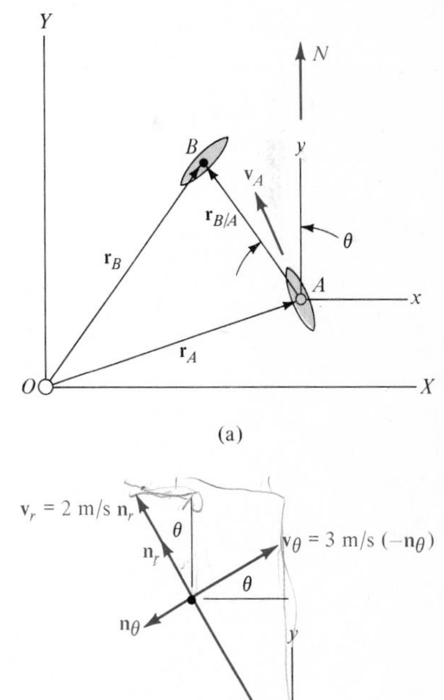

FIGURE 12-19
Relative motion of two ships

EXAMPLE 12-9

The radar equipment in a police aircraft during a calibration procedure is tracking a car as in Fig. 12-20a. Both the car and the airplane are traveling horizontally, the airplane in a transverse direction to the road. Calculate r and $\dot{r}$ for the given conditions.

SOLUTION

A translating coordinate system with its origin at P and its xyz axes parallel to the fixed XYZ axes is used as shown in Fig. 12-20b. The information requested is more accurately stated as $r_{C/P}$ and $v_{C/P}$. These scalars must be determined from vector quantities.

$$\mathbf{r}_{C/P} = \mathbf{r}_C - \mathbf{r}_P = 300m\mathbf{i} - 150m\mathbf{k} - 200m\mathbf{j} = 300m\mathbf{i} - 200m\mathbf{j} - 150m\mathbf{k}$$

$$r_{C/P} = \sqrt{300^2 + 200^2 + 150^2} = 391 \text{ m}$$

To calculate $\mathbf{v}_{C/P} = \dot{\mathbf{r}}_{C/P}$, v_C and v_P must be expressed in m/s.

$$v_C = \frac{100 \text{ km}}{\text{h}}\left(\frac{1000 \text{ m}}{\text{km}}\right)\left(\frac{1 \text{ h}}{3600 \text{ s}}\right) = 27.8 \text{ m/s}$$

$$v_P = \frac{150 \text{ km}}{\text{h}}\left(\frac{1000 \text{ m/km}}{3600 \text{ s/h}}\right) = 41.7 \text{ m/s}$$

$$\mathbf{v}_{C/P} = \dot{\mathbf{r}}_{C/P} = \mathbf{v}_C - \mathbf{v}_P = -27.8 \text{ m/s } \mathbf{k} - 41.7 \text{ m/s } \mathbf{i}$$

$$= (-41.7\mathbf{i} - 27.8\mathbf{k}) \text{ m/s}$$

$$v_{C/P} = \sqrt{-41.7^2 + 27.8^2} = 50 \text{ m/s}$$

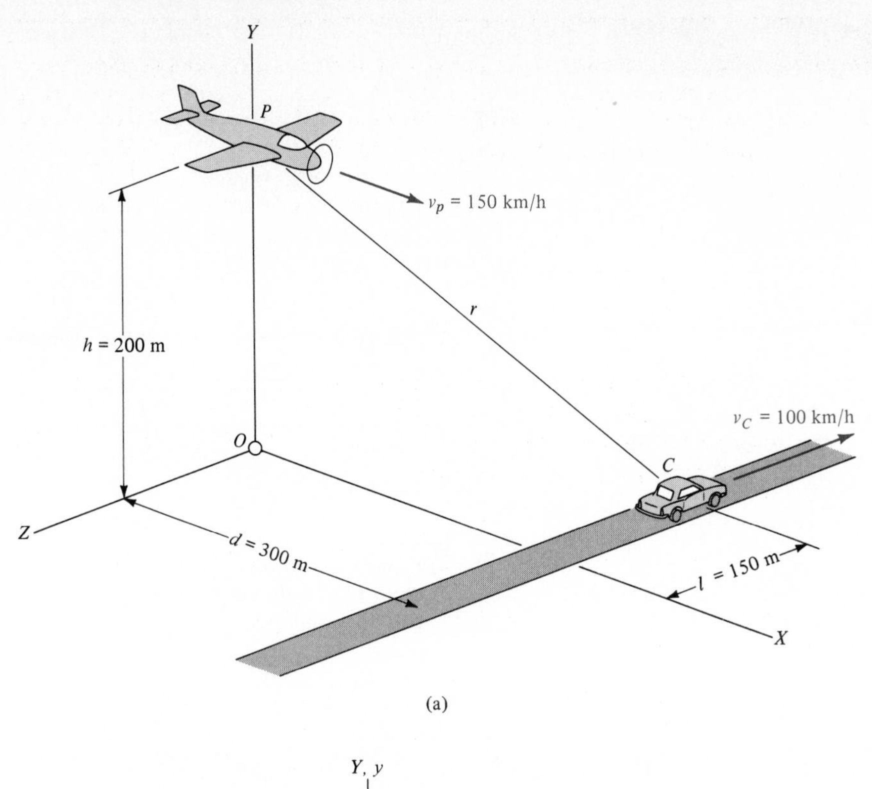

(a)

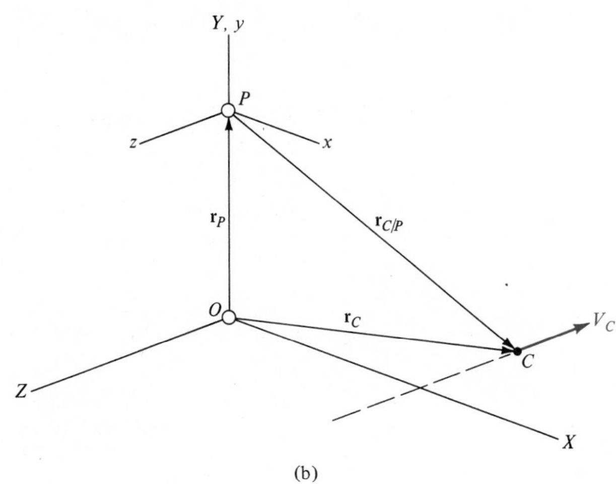

(b)

FIGURE 12-20

Relative motion of a car and an airplane

PROBLEMS

SECTION 12-7

12–63 The positions of particles A and B are defined in a fixed XY frame by $\mathbf{r}_A = 4t^2\mathbf{i} + 3t\mathbf{j}$ and $\mathbf{r}_B = 4t^2\mathbf{i} + 5t^2\mathbf{j}$, where r_A and r_B are in meters and t is in seconds. Determine the relative velocity $\mathbf{v}_{B/A}$ and the relative acceleration $\mathbf{a}_{B/A}$ at $t = 2$ s.

12–64 The positions of particles A and B are defined in a fixed XY frame by $\mathbf{r}_A = -3t^3\mathbf{i} - 2t^3\mathbf{j}$ and $\mathbf{r}_B = -4t^2\mathbf{i} + 3t^3\mathbf{j}$, where r_A and r_B are in feet and t is in seconds. Determine the relative velocity $\mathbf{v}_{A/B}$ and the relative acceleration $\mathbf{a}_{A/B}$ at $t = 4$ s.

12–65 Consider three particles A, B, and C in plane motion. Derive an equation for the relative velocity $\mathbf{v}_{A/C}$ in terms of $\mathbf{v}_{A/B}$ and $\mathbf{v}_{C/B}$.

12–66 Three particles A, B, and C are in plane motion. Derive an equation for the relative acceleration $\mathbf{a}_{B/A}$ in terms of $\mathbf{a}_{C/A}$ and $\mathbf{a}_{B/C}$.

FIGURE P12-67

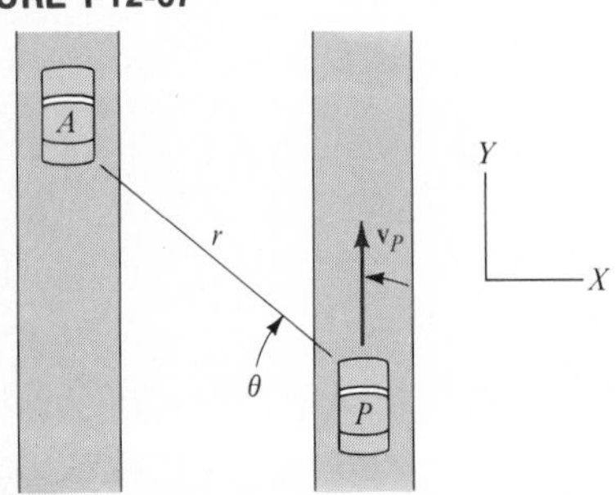

12–67 A police car P is traveling at velocity $\mathbf{v}_P = 100\mathbf{j}$ km/h. The data from the police car's radar indicate that $r = 200$ m, $\dot{r} = 30$ km/h, $\ddot{r} \approx 0$, $\theta = 0.1$ rad, $\dot{\theta} = -10^{-4}$ rad/s (θ decreasing), and $\ddot{\theta} \approx 0$. Determine $\mathbf{v}_A$.

12–68 In Prob. 12–67, $\mathbf{v}_P = 50\mathbf{j}$ mph, $r = 400$ ft, $\dot{r} = 30$ ft/s, $\ddot{r} = 6$ ft/s^2, $\theta = 0.12$ rad, $\dot{\theta} = -10^{-5}$ rad/s (θ decreasing), and $\ddot{\theta} \approx 0$. Determine the magnitudes of the velocity and acceleration of car A.

FIGURE P12-69

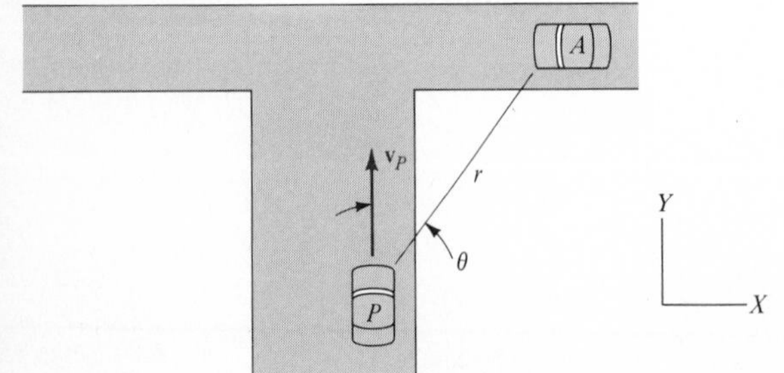

12–69 A police car P is traveling at velocity $\mathbf{v}_P = 20\mathbf{j}$ km/h when its radar gives the following data about car A: $r = 100$ m, $\dot{r} = -60$ km/h (r decreasing), $\ddot{r} \approx 0$, $\theta = 0.5$ rad, $\dot{\theta} = -10^{-3}$ rad/s (θ decreasing), and $\ddot{\theta} \approx 0$. Determine the velocity $\mathbf{v}_A$.

12–70 A police car P is traveling at velocity $\mathbf{v}_P = (30\mathbf{i} + 10\mathbf{j})$ mph when its radar shows that $r = 400$ ft, $\dot{r} = -100$ ft/s (r decreasing), $\ddot{r} \approx 0$, $\theta = 0.11$ rad, $\dot{\theta} = -10^{-5}$ rad/s (θ decreasing), and $\ddot{\theta} \approx 0$. Determine the velocity of car A.

FIGURE P12-70

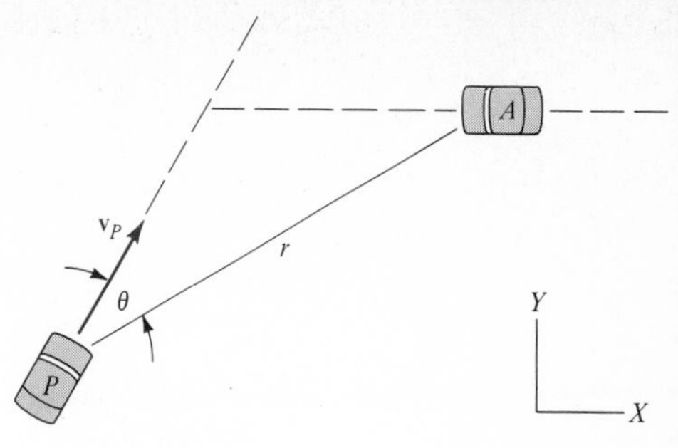

12–71 A police car's velocity is $\mathbf{v}_P = 40\mathbf{j}$ km/h when its radar gives the following data about car A: $r = 80$ m, $\dot{r} = -35$ m/s (r decreasing), $\ddot{r} = -3$ m/s^2, $\theta = 0.9$ rad, $\dot{\theta} = -0.2$ rad/s (θ decreasing), and $\ddot{\theta} \approx 0$. Determine the velocity and acceleration of car A.

FIGURE P12-71

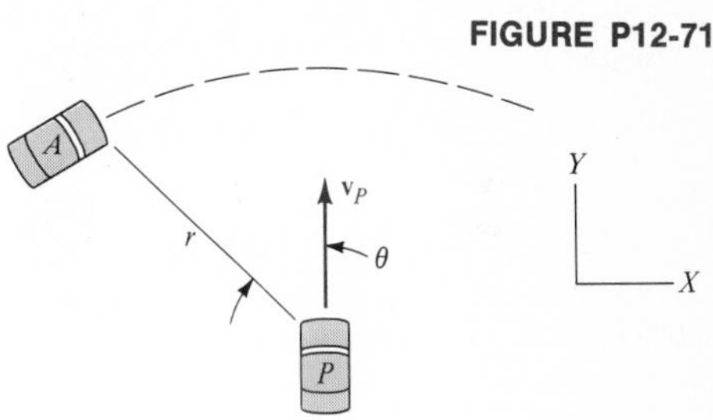

12–72 A coordinate system xy is in translational motion at a constant velocity with respect to fixed axes XY. Prove that a particle's acceleration is the same with respect to the xy and XY coordinate systems.

12–73 A mechanism consists of two rigid bars AB and BC which are pinned at A and B. $AB = BC = 1$ m, $\omega_1 = 0.3$ rad/s, $\alpha_1 = 0$ (in the XY frame), $\omega_2 = 0.2$ rad/s, and $\alpha_2 = 0$ (in the xy frame). Calculate and sketch the velocity $\mathbf{v}_C$ and acceleration $\mathbf{a}_C$ of point C in the XY and the xy frames.

FIGURE P12-73

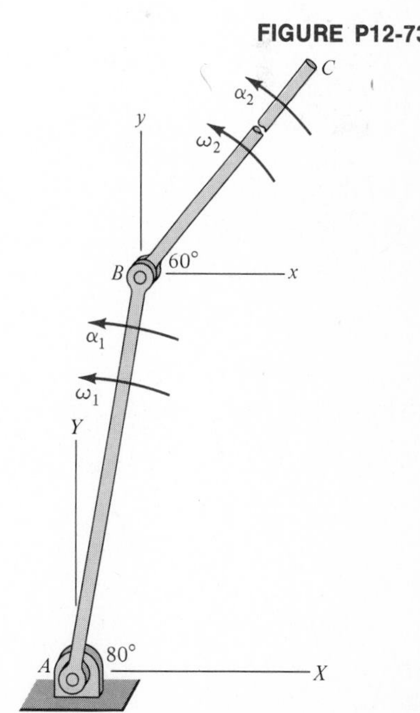

FIGURE P12-75

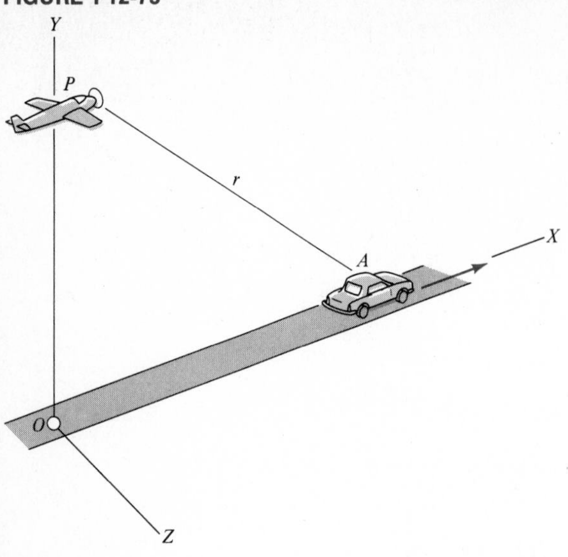

FIGURE P12-77

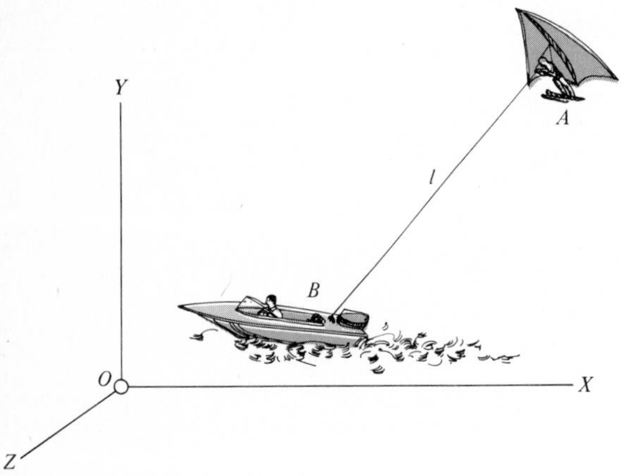

12–74 Solve Prob. 12–73 after changing the angular accelerations to $\alpha_1 = 0.3$ rad/s^2 and $\alpha_2 = -0.1$ rad/s^2 (decreasing ω_2).

12–75 A police aircraft is directly over point O at an altitude of 400 ft, flying with a velocity $\mathbf{v}_P = (200\mathbf{i} + 30\mathbf{k})$ ft/s. The car is at $X = 500$ ft and traveling at a velocity of $\mathbf{v}_A = 95\mathbf{i}$ ft/s. Determine the relative velocity $v_{P/A}$.

12–76 At the given instant in Prob. 12–75 the aircraft's acceleration is $\mathbf{a}_P = (15\mathbf{i} - 10\mathbf{j})$ ft/s^2 and the car's acceleration is $\mathbf{a}_A = -8\mathbf{i}$ ft/s^2. Calculate the relative acceleration $a_{A/P}$.

12–77 A hang glider A is being towed by boat B. The rope has a constant length $l = 30$ m. At a particular instant the velocities are defined by $\mathbf{v}_A = (-12\mathbf{i} + 4\mathbf{j} + 9\mathbf{k})$ m/s and $\mathbf{v}_B = (-10\mathbf{i} + 12\mathbf{k})$ m/s. Determine the angular speed ω of the rope.

12–78 The hang glider's acceleration in Prob. 12–77 is defined by $\mathbf{a}_A = (2\mathbf{i} + \mathbf{j})$ m/s^2 while the boat's acceleration is $\mathbf{a}_B = (1.5\mathbf{i} + 2\mathbf{k})$ m/s^2. Determine the angular acceleration of the rope at that instant.

SUMMARY

Section 12-1

Curvilinear motion in rectangular coordinates

POSITION VECTOR:

$$\mathbf{r} = x\mathbf{i} + y\mathbf{j} + z\mathbf{k}$$

VELOCITY VECTOR:

$$\mathbf{v} = \frac{d\mathbf{r}}{dt} = \frac{dx}{dt}\mathbf{i} + \frac{dy}{dt}\mathbf{j} + \frac{dz}{dt}\mathbf{k} = \dot{x}\mathbf{i} + \dot{y}\mathbf{j} + \dot{z}\mathbf{k}$$

ACCELERATION VECTOR:

$$\mathbf{a} = \frac{d\mathbf{v}}{dt} = \frac{dv_x}{dt}\mathbf{i} + \frac{dv_y}{dt}\mathbf{j} + \frac{dv_z}{dt}\mathbf{k} = \ddot{x}\mathbf{i} + \ddot{y}\mathbf{j} + \ddot{z}\mathbf{k}$$

The time derivatives of the unit vectors are zero when the xyz frame is fixed. The scalar components $\dot{x}$, $\ddot{x}$, and so on, may be analyzed by using appropriate equations from Sec. 11-1, 11-2, or 11-3. Each equation must include only those quantities which are relevant to the coordinate direction being considered.

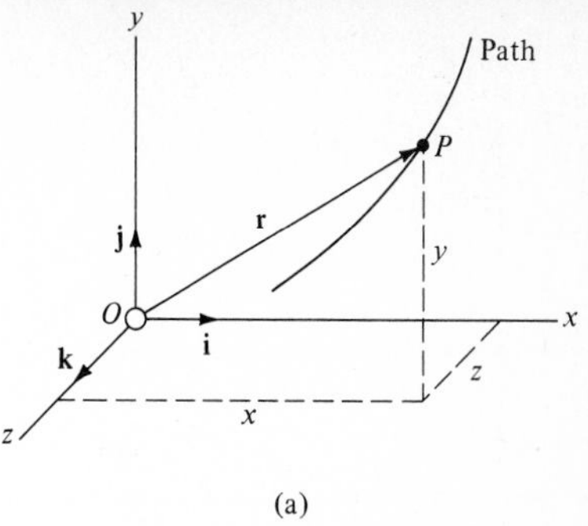

(a)

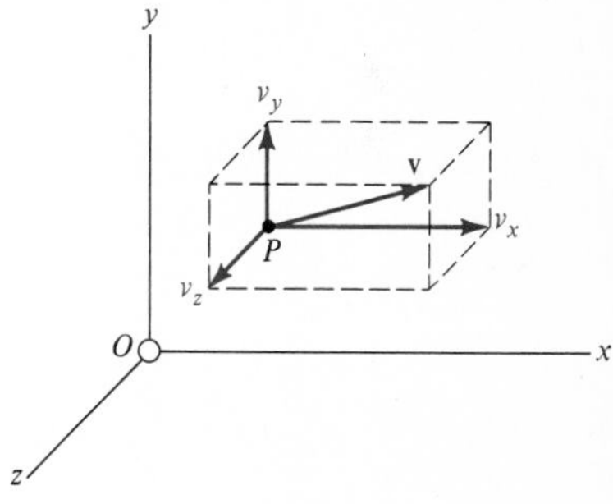

(b)

Section 12-2

Physical aspects of velocity and acceleration vectors

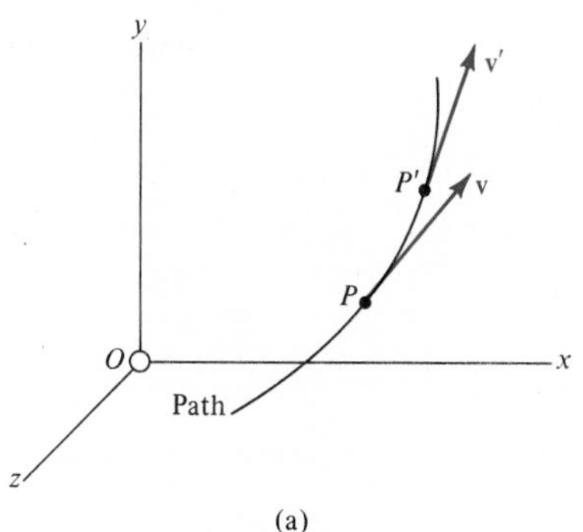

(a)

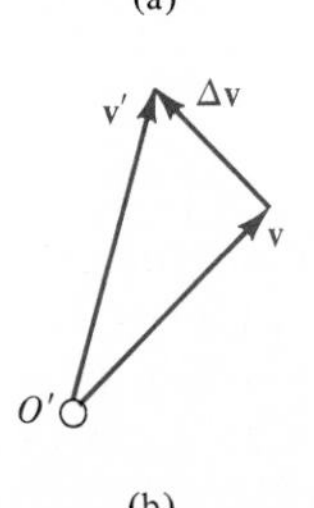

(b)

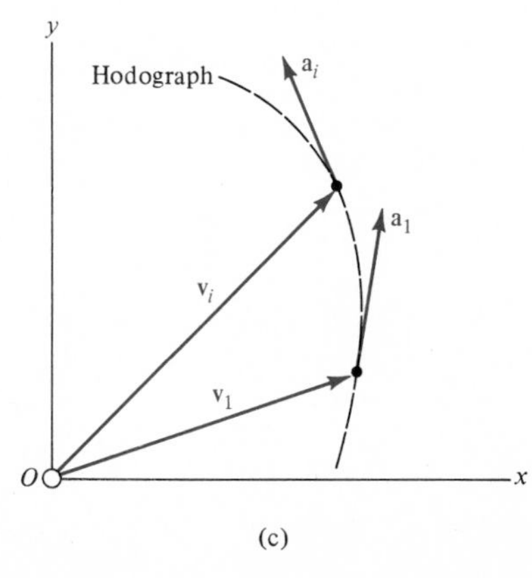

(c)

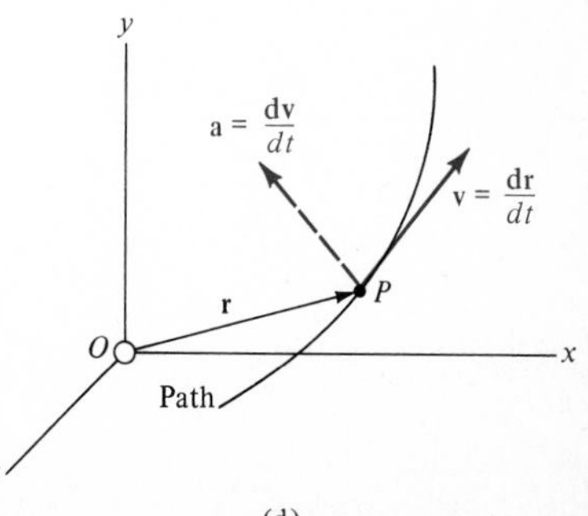

(d)

The instantaneous velocity **v** is always tangent to the path of the particle. Speed = magnitude v of vector **v**. The instantaneous acceleration **a** is in the direction of the infinitesimal change of the velocity vector.

The vector **a** is not tangent to the path of the particle and is not collinear with the velocity vector in curvilinear motion.

Section 12-3

Tangential and normal components

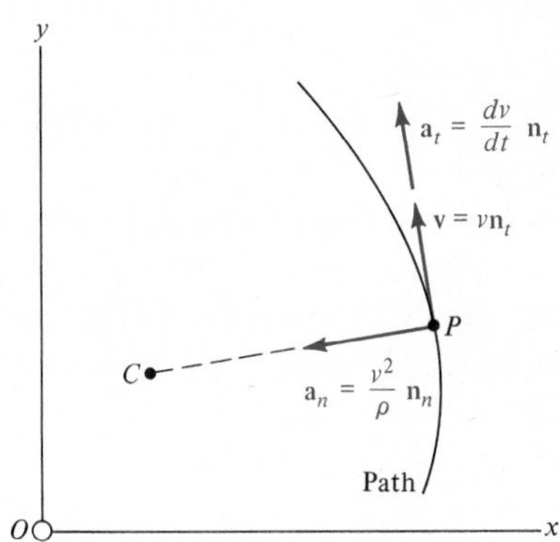

UNIT VECTORS:

$\mathbf{n}_t$ in direction of travel
$\mathbf{n}_n$ in direction of center of curvature of path

VELOCITY:

$\mathbf{v} = v\mathbf{n}_t$ always tangent to the path

ACCELERATION:

$$\mathbf{a} = a_t\mathbf{n}_t + a_n\mathbf{n}_n$$

$$a_t = \frac{dv}{dt} \qquad a_n = \frac{v^2}{\rho}$$

$$\rho = \frac{[1 + (dy/dx)^2]^{3/2}}{d^2y/dx^2}$$

For a circular path, $\rho = r =$ radius of circle.

Section 12-4

Polar coordinates

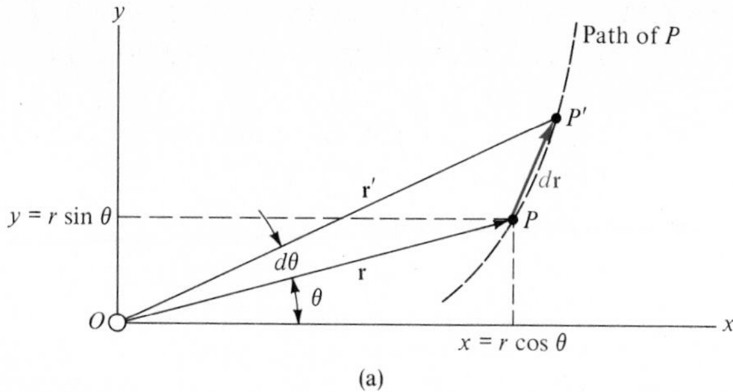

(a)

UNIT VECTORS:

$\mathbf{n}_r$ in radial direction (of position vector $\mathbf{r}$)
$\mathbf{n}_\theta$ in transverse direction to $\mathbf{r}$, in plane of path

VELOCITY:

$$\mathbf{v} = \dot{r}\mathbf{n}_r + r\dot{\theta}\mathbf{n}_\theta \qquad \text{always tangent to the path}$$

ANGULAR SPEED OF A RADIAL LINE:

$$\dot{\theta} = \omega, \text{ rad/s}$$

ACCELERATION:

$$\mathbf{a} = (\ddot{r} - r\dot{\theta}^2)\mathbf{n}_r + (r\ddot{\theta} + 2\dot{r}\dot{\theta})\mathbf{n}_\theta$$

ANGULAR ACCELERATION OF A RADIAL LINE:

$$\ddot{\theta} = \alpha, \text{ rad/s}^2$$

CIRCULAR MOTION:

$$\mathbf{v} = r\dot{\theta}\mathbf{n}_\theta$$

$$\mathbf{a} = -r\dot{\theta}^2\mathbf{n}_r + r\ddot{\theta}\mathbf{n}_\theta$$

Conversion to Tangential and Normal Coordinates:

$$v = r\dot{\theta} = r\omega$$

$$a_t = \frac{dv}{dt} = r\ddot{\theta} = r\alpha$$

$$a_n = \frac{v^2}{r} = r\dot{\theta}^2 = r\omega^2$$

Section 12-5

Cylindrical coordinates

UNIT VECTORS:

$\mathbf{n}_r$	in radial direction in xy plane
$\mathbf{n}_\theta$	in transverse direction to radial line in xy plane
$\mathbf{k}$	in z direction of xyz frame

VELOCITY:

$$\mathbf{v} = \dot{r}\mathbf{n}_r + r\dot{\theta}\mathbf{n}_\theta + \dot{z}\mathbf{k}$$

ACCELERATION:

$$\mathbf{a} = (\ddot{r} - r\dot{\theta}^2)\mathbf{n}_r + (r\ddot{\theta} + 2\dot{r}\dot{\theta})\mathbf{n}_\theta + \ddot{z}\mathbf{k}$$

These are based on the expressions for polar coordinates; r and θ are defined in the xy plane or in any plane parallel to xy.

Section 12-6

Spherical coordinates

UNIT VECTORS:

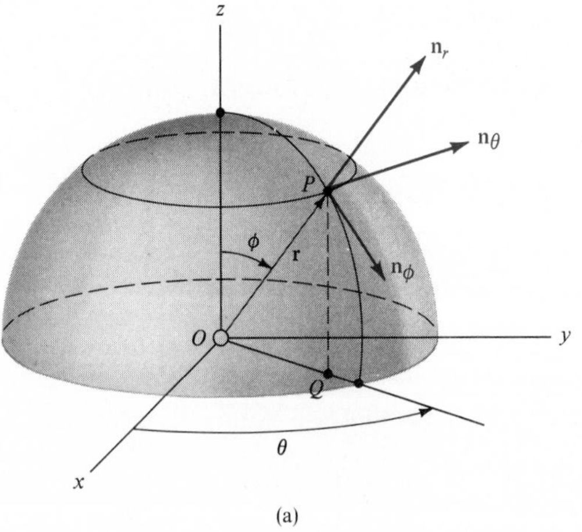

(a)

VELOCITY:

$$\mathbf{v} = \dot{r}\mathbf{n}_r + r\dot{\phi}\mathbf{n}_\phi + r\dot{\theta}\sin\phi\mathbf{n}_\theta$$

ACCELERATION:

$$\mathbf{a} = (\ddot{r} - r\dot{\phi}^2 - r\dot{\theta}^2 \sin^2 \phi)\mathbf{n}_r + (r\ddot{\phi} + 2\dot{r}\dot{\phi} - r\dot{\theta}^2 \sin \phi \cos \phi)\mathbf{n}_\phi + (r\ddot{\theta} \sin \phi + 2\dot{r}\dot{\theta} \sin \phi + 2r\dot{\phi}\dot{\theta} \cos \phi)\mathbf{n}_\theta$$

Section 12-7

Relative motion in translating coordinates

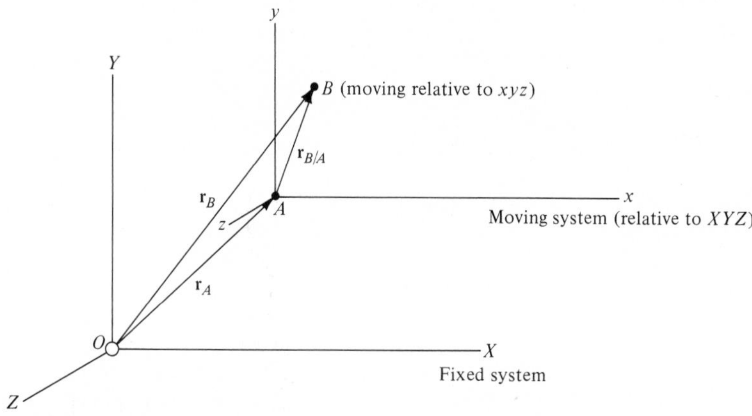

The motion of the xyz frame is translation when the respective coordinates of xyz and XYZ are always parallel to each other. The same unit vectors **i**, **j**, **k** are used in both reference frames. Absolute motions of particle B are defined with respect to point O of XYZ. Relative motions are defined with respect to point A of xyz.

POSITION:

$$\mathbf{r}_B = \mathbf{r}_A + \mathbf{r}_{B/A}$$

VELOCITY:

$$\mathbf{v}_B = \mathbf{v}_A + \mathbf{v}_{B/A}$$

ACCELERATION:

$$\mathbf{a}_B = \mathbf{a}_A + \mathbf{a}_{B/A}$$

Radial and transverse components (polar coordinates) are useful for determining the relative velocity $\mathbf{v}_{B/A}$ and relative acceleration $\mathbf{a}_{B/A}$. Subscripts are arbitrary but must be used consistently since

$$\mathbf{r}_{B/A} = -\mathbf{r}_{A/B} \qquad \mathbf{v}_{B/A} = -\mathbf{v}_{A/B} \qquad \mathbf{a}_{B/A} = -\mathbf{a}_{A/B}$$

REVIEW PROBLEMS

12–79 A roller coaster travels at speed $v_A = 60$ ft/s at point A where the radius of curvature is $r_A = 30$ ft. At this point v_A is decreasing at the rate of 3 ft/s². Determine the total acceleration of a person at point A.

FIGURE P12-79

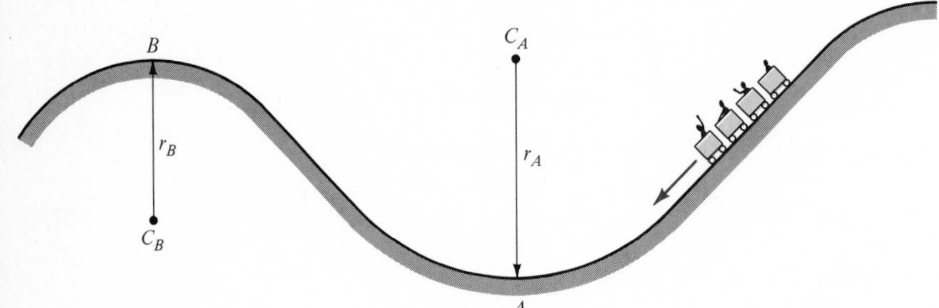

12–80 A roller coaster travels at speed $v_B = 5$ m/s at point B in Fig. P12-79 where the radius of curvature is $r_B = 8$ m. At this point v_B is decreasing at the rate of 1 m/s². Determine the total acceleration of a person at point B, noting that the train and its passengers are constrained to follow the track.

12–81 A prototype car moves on a complex test track. Recordings of the motion show that at time t the car's speed $v = 40$ ft/s was increasing at the rate of 3 ft/s², and its total acceleration was 25 ft/s². Determine the radius of curvature of the car's path at that instant.

FIGURE P12-82

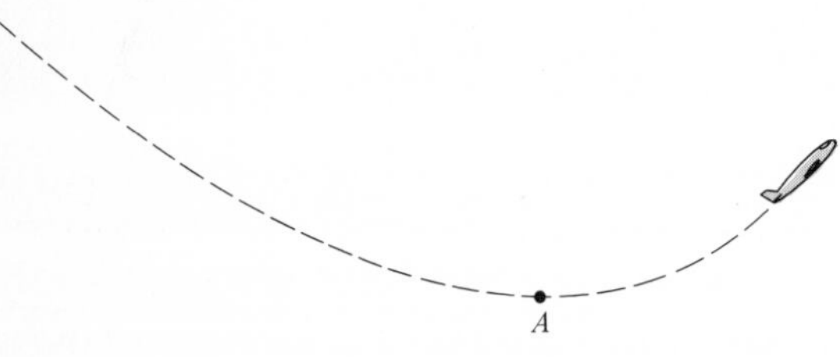

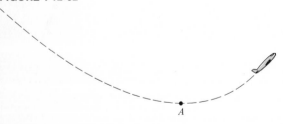

12–82 An airplane in a diving test has a speed $v_A = 200$ m/s, decreasing at the rate of 15 ft/s², at the bottom of its trajectory. Determine the radius of curvature of the trajectory at point A if the total acceleration of the airplane is 80 m/s² at point A.

FIGURE P12-83

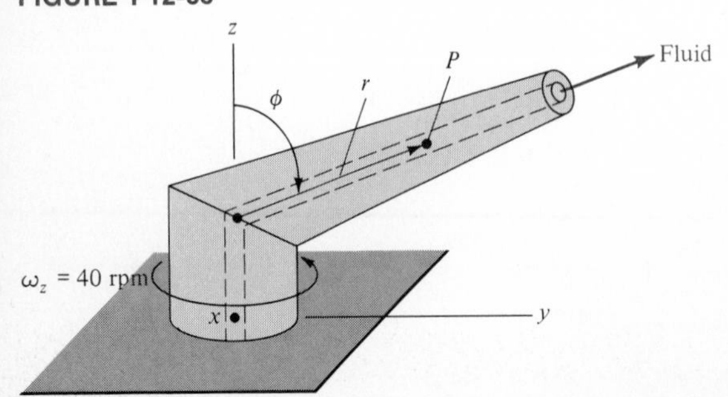

12–83 A fluid particle at point P in the rotating nozzle is moving with a constant radial speed $\dot{r} = 3$ m/s at $r = 15$ cm. Calculate the acceleration of the particle if $\phi = 60°$, $\dot{\phi} = 0.2$ rad/s, $\ddot{\phi} = 0.2$ rad/s².

12–84 In an amusement ride people sit in cages mounted on arms of constant length $r = 6.5$ m. The arms rotate at a constant angular speed $\omega_z = 2$ rad/s. A hydraulic actuator A controls the angle ϕ. Calculate the acceleration of a rider at $R = 8$ m from the pivot O if $\phi = 30°$, $\dot{\phi} = 0.2$ rad/s.

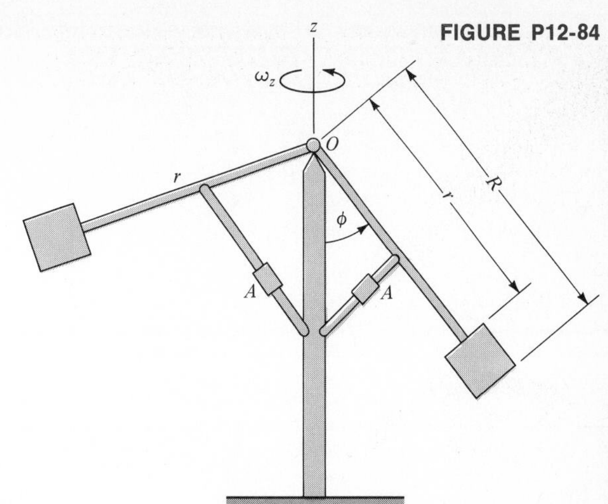

FIGURE P12-84

12–85 In a proposed amusement ride people sit in cages at a distance r from the pivot O. Hydraulic actuators A_1 and A_2 control the arm length r and the angle ϕ. Calculate the acceleration of a rider if $\omega_z = 5$ rpm, $r = 15$ ft, $\dot{r} = -2$ ft/s (r is decreasing), $\phi = 30°$, $\dot{\phi} = 0$, and $\ddot{\phi} = 0.1$ rad/s^2.

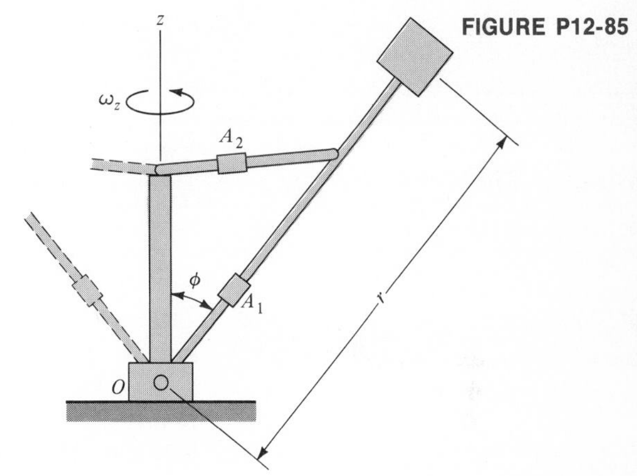

FIGURE P12-85

12–86 Consider the air traffic situation in the illustration. The controlling aircraft A is cruising at a steady speed v_A in a circular pattern at height $h = 15{,}000$ ft while another aircraft B is at height $l = 8000$ ft. Assume B and C are in the XY plane. Determine r and $\dot{r}$ for $\mathbf{v}_A = -300\mathbf{i}$ ft/s, $\mathbf{v}_B = -600\mathbf{k}$ ft/s, $R = 5000$ ft, and $d = 9000$ ft.

12–87 Determine the relative acceleration $\mathbf{a}_{B/A}$ in Prob. 12–86 assuming that the pilot of A is increasing the speed at the rate of 15 ft/s^2 while remaining in the same circular pattern.

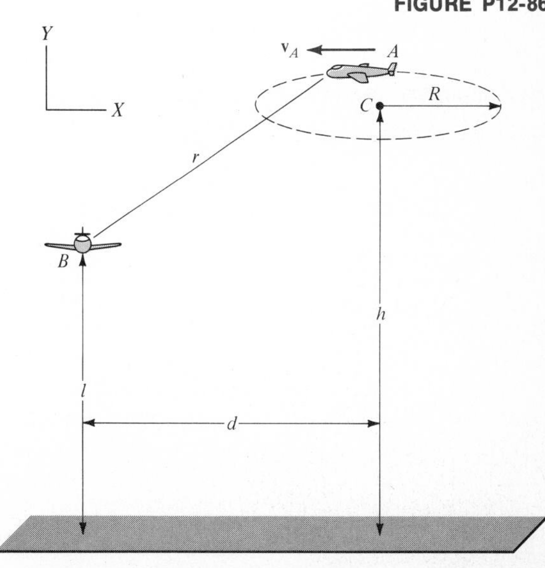

FIGURE P12-86

CHAPTER 13

KINETICS OF PARTICLES: NEWTON'S SECOND LAW

OVERVIEW

The methods developed for the kinematics analysis of particles are readily applied in the important area of kinetics of particles. In this area the relationship of forces, masses, and the acceleration of masses is investigated. The fundamental principles of kinetics were established by Isaac Newton. These principles are used in a wide range of kinetics analyses of particles, rigid bodies, deformable solids, and fluids.

STUDY GOALS

SECTION 13-1 defines Newton's second law of motion, the relationship of force, mass, and acceleration. The concepts and limitations of using this relationship should be remembered for future work in kinetics.

SECTION 13-2 presents definitions and important distinctions of engineering units that are used in all areas of kinetics. The student should note that the units and terminologies given in Sec. 13-2 are the correct ones for use in engineering work. Incorrect units are not acceptable even if they are commonly used in everyday life (these are mentioned in App. B to remove them from the mainstream of kinetics analyses).

SECTION 13-3 presents vector and scalar expressions of the equations of motion for the practical application of Newton's second law. These equations are indispensable for solving problems in kinetics. The appropriate equations of motion should be stated early in the solution of problems, immediately after the free-body diagrams are drawn.

13-1 NEWTON'S SECOND LAW OF MOTION

It is known from common experience that a body accelerates when a net force is acting on it. The acceleration is in the same direction as the direction of the resultant force. The magnitude of the acceleration depends on the magnitude of the force and mass of the body. The role of the mass is not clear to many people, as evidenced by those who believe erroneously that all heavy objects fall faster than light objects. Galileo was the first to demonstrate that the apparent differences caused by mass are only differences in the air friction of the objects; otherwise, heavy and light objects are accelerated at the same rate in the same gravitational field (e.g., in a given region of the earth's atmosphere).

The formal relation of force, mass, and acceleration of a particle is Newton's second law of motion, which may be stated as follows:

> The magnitude of the acceleration of a particle is directly proportional to the magnitude of the resultant force acting on it, and inversely proportional to its mass. The direction of the acceleration is the same as the direction of the resultant force.

This is a postulated law which can be proved only by experiment. For example, if a particle is subject to different forces $\mathbf{F}_1, \mathbf{F}_2, \mathbf{F}_3, \ldots$ at different times, the corresponding accelerations $\mathbf{a}_1, \mathbf{a}_2, \mathbf{a}_3, \ldots$ are determined using data of positions and times and kinematics relationships. Each acceleration is in the same direction as the force causing it, as shown in Fig. 13-1. Calculating the ratios of the magnitudes F_1/a_1, and so on, it is found that these ratios are constant for a given particle,

$$\frac{F_1}{a_1} = \frac{F_2}{a_2} = \frac{F_3}{a_3} = \cdots = \text{constant} = m$$

These constant ratios are defined as the *mass* of the particle, which is denoted by m. Mass is a property of the particle and is a measure of its *inertia*, or its *resistance to changes in velocity*. Newton's second law is expressed most concisely as

$$\boxed{\mathbf{F} = m\mathbf{a}} \qquad (13\text{-}1)$$

This equation accounts both for the directions of the force and acceleration vectors, and the proportions of the quantities involved. Equation 13-1 is easy to use, but caution is required in several particular situations as follows.

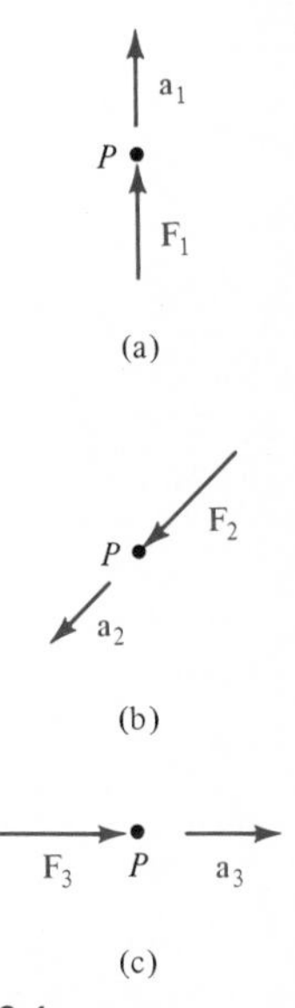

FIGURE 13-1
Acceleration of a particle

1. **F** must be the *resultant force* when two or more forces are acting on the particle.
2. Equation 13-1 is defined for a single particle. It may be used for systems of particles if **a** is taken as the acceleration of the mass center of the system (see Chap. 16).
3. The mass of the body may be changing significantly during its motion, in which case the mass must be treated as a function of time (e.g., rockets discharge fuel).
4. The motion of the particle is determined with respect to a "fixed," nonaccelerating reference frame. In most engineering problems the frame used is attached to the surface of the earth, which is not truly fixed. However, the results obtained are accurate enough for most applications. In some cases, other primary reference systems must be used (for example, one located at the sun). An ideal, fixed reference system is called a *Newtonian reference frame* or *inertial reference frame* (generally it is taken with respect to distant stars).
5. According to Einstein's theory of relativity, Eq. 13-1 is increasingly in error as the particle's velocity approaches the speed of light.

13-2 SYSTEMS OF UNITS

It is necessary to use a correct, internally consistent, system of units when applying Eq. 13-1 to solve problems. The following brief statements concern the two major systems in use, and are relevant to all areas of kinetics. These systems are the SI, which is adopted by international agreements, and the U.S. customary system, which is used by tradition.

SI Units

There are four units in Eq. 13-1, those of force, mass, length, and time. Any three of these may be arbitrarily chosen (they are independent of one another), while the fourth is dependent on the other three in order to satisfy $\mathbf{F} = m\mathbf{a}$. The base units of SI are the *kilogram* (kg) for mass, the *meter* (m) for length, and the *second* (s) for time. These three units are arbitrarily chosen and are dimensionally independent. The SI units form an absolute system, so the three base units are also independent of location in space. The unit of force is the *newton* (N), which is defined as the force that accelerates a 1-kg mass at the rate of 1 m/s^2. Its units are obtained from $F = ma$,

$$1\text{ N} = (1\text{ kg})(1\text{ m/s}^2) = 1\text{ kg}\cdot\text{m/s}^2$$

It is important to distinguish between the mass and the weight of a body whose motion is to be analyzed. Mass is a property of the body independent of its location or motion. Weight is a force that is caused by a particular gravitational field on the body. For any body near the earth, the weight of the body is the force of attraction of the earth on that body. This force W is calculated as

$$W = mg \tag{13-2}$$

where m is the mass of the body and g is the acceleration of gravity. g is approximately equal to 9.81 m/s^2 near the surface of the earth.

U.S. Customary Units

The base units in the U.S. customary system are the *pound* (lb) for force, the *foot* (ft) for length, and the *second* (s) for time. The pound is defined as the weight of a certain platinum cylinder that has a mass of about 0.453592 kg. The foot is defined as 0.3048 m. The units of time are the same in SI and in the U.S. customary system.

Since the pound is not an absolute unit (the weight slightly depends on location due to differences in elevation and the oblate shape of the earth), the U.S. customary units do not form an absolute system. It is called a *gravitational* system of units.

The mass is a derived quantity in the U.S. system of units, and is called a *slug*. From Eq. 13-1, $m = F/a$, so

$$1 \text{ slug} = \frac{1 \text{ lb}}{1 \text{ ft/s}^2} = 1 \frac{\text{lb}\cdot\text{s}^2}{\text{ft}}$$

In general, the mass m in slugs is calculated from

$$m = \frac{W}{g} \tag{13-3}$$

where W is in pounds and the acceleration of gravity g is about 32.2 ft/s^2.

Further comments about the applications of units are made in App. B.

EQUATIONS OF MOTION 13-3

In most situations there are several forces acting on a particle. Equation 13-1 can be expressed for an arbitrary number of concurrent forces $\mathbf{F}_1, \mathbf{F}_2, \mathbf{F}_3, \ldots$, as

$$\sum \mathbf{F}_i = m\mathbf{a} \tag{13-4}$$

This is called the *equation of motion*. It is the formal statement of the fact that the particle has only one acceleration no matter how many forces are acting on it simultaneously. The acceleration is caused by the resultant of the forces.

For solving problems it is convenient to express Eq. 13-4 in terms of scalar component equations. This is done using the most important coordinate systems in the following.

Rectangular coordinates. The forces and the acceleration can be resolved into x, y, and z components. In vector notation,

$$\sum (F_x\mathbf{i} + F_y\mathbf{j} + F_z\mathbf{k}) = m(a_x\mathbf{i} + a_y\mathbf{j} + a_z\mathbf{k})$$

The equations of motion in scalar form are

$$\boxed{\sum F_x = ma_x \qquad \sum F_y = ma_y \qquad \sum F_z = ma_z} \qquad \text{13-5a}$$

These may also be stated as differential equations of motion,

$$\boxed{\sum F_x = m\ddot{x} \qquad \sum F_y = m\ddot{y} \qquad \sum F_z = m\ddot{z}} \qquad \text{13-5b}$$

which can be integrated successively to express the velocity and position of the particle as functions of time.

Normal and tangential components. For plane curvilinear motion, the components of Eq. 13-4 may also be written using normal and tangential components. This is similar to using polar coordinates, and is particularly advantageous in circular motion where the radius of curvature ρ is constant. The scalar components of Eq. 13-4 are written on the basis of Eq. 12-12,

$$\sum F_n = ma_n = m\frac{v^2}{\rho} \qquad \text{13-6a}$$

$$\sum F_t = ma_t = m\dot{v} = mv\frac{dv}{ds} \qquad \text{13-6b}$$

where $\dot{v}$ is the time derivative of the magnitude of the velocity and s is distance along the path. The directions of these components should be noted. As shown in Fig. 13-2, the normal component $\sum F_n$ is always toward the center of curvature of the particle's path, while the tangential component $\sum F_t$, if not zero, is collinear with the velocity. There is always a normal component when the particle moves on a curved path, but the tangential component may be absent. The normal component $\sum F_n$ is called *centripetal force.**

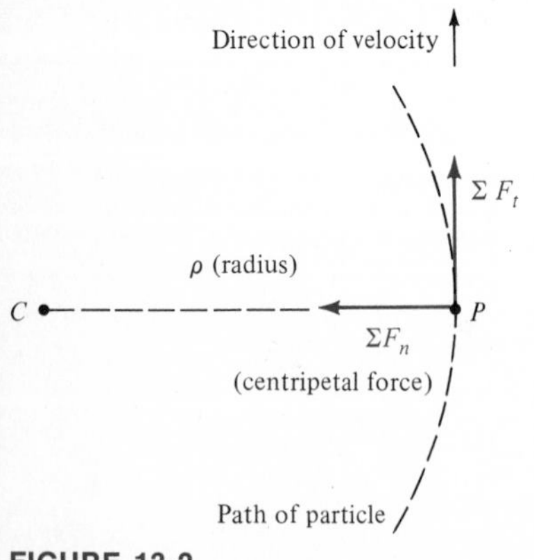

FIGURE 13-2

Normal and tangential forces in curvilinear motion

Polar coordinates. For plane curvilinear motion, the components of Eq. 13-4 are written in terms of r and θ according to Eqs. 12-21 and 12-22,

$$\sum F_r = ma_r = m(\ddot{r} - r\dot{\theta}^2) \qquad \text{13-7a}$$

$$\sum F_\theta = ma_\theta = m(r\ddot{\theta} + 2\dot{r}\dot{\theta}) \qquad \text{13-7b}$$

* A related term is *centrifugal force* as described in App. C. The concept of centrifugal force is from D'Alembert's definition of dynamic equilibrium for analyzing problems of dynamics as problems of statics. This method may be of interest to some people, but it should be omitted from the mainstream of kinetics topics at the elementary level for the reasons given in App. C.

EXAMPLE 13-1

The mass of each apple in a laboratory test (Fig. 13-3a) is 0.2 kg. Calculate the net force acting on the bottom of an apple which is accelerated upward at $a_y = 20 \text{ m/s}^2$ and horizontally at $a_x = 3 \text{ m/s}^2$, simultaneously.

SOLUTION

First the free-body diagram of an apple is sketched in Fig. 13-3b. The required force acting on the fruit is the resultant of the force components R_x and R_y. For rectangular coordinates the equations of motion are from Eq. 13-5a,

$$\sum F_x = ma_x \qquad \sum F_y = ma_y$$

$$R_x = ma_x \qquad R_y - W = (0.2 \text{ kg})(20 \text{ m/s}^2)$$

$$= (0.2 \text{ kg})(3 \text{ m/s}^2) \qquad R_y = 4 \text{ kg m/s}^2 + (0.2 \text{ kg})(9.81 \text{ m/s}^2)$$

$$= 0.6 \text{ kg}\cdot\text{m/s}^2 \qquad = 5.96 \text{ kg}\cdot\text{m/s}^2$$

$$= 0.6 \text{ N} \qquad = 5.96 \text{ N}$$

The net force at the bottom of the apple is

$$R = (R_x^2 + R_y^2)^{1/2} = [(0.6 \text{ N})^2 + (5.96 \text{ N})^2]^{1/2}$$
$$= 5.99 \text{ N} \quad \text{(at an angle of 84° from the horizontal)}$$

JUDGMENT OF THE RESULT

In static equilibrium the net force at the bottom of the fruit would be $R = R_y = W = 1.96$ N. Thus, the result is reasonable for the given accelerations a_x and a_y.

(a)

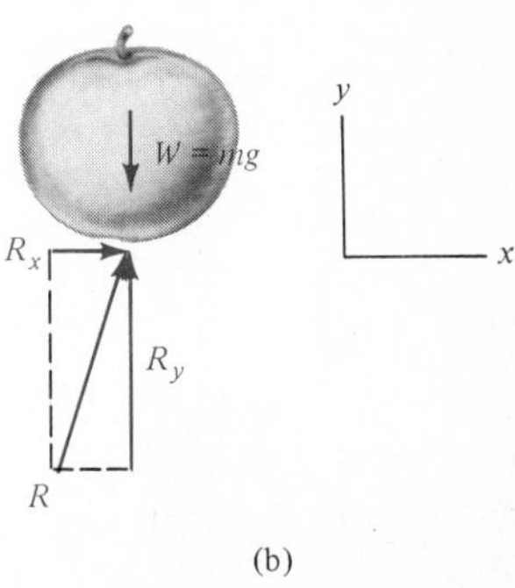

(b)

FIGURE 13-3
Dynamic testing of packaged fruit (Photograph courtesy MTS Systems Corporation, Minneapolis, Minnesota)

EXAMPLE 13-2

The total weight W of the elevator shown in Fig. 13-4a at rest is 5000 lb. Determine the tension T in the cable (a) during an upward acceleration $a_1 = 3\ \text{ft/s}^2$, and (b) during a downward acceleration of $a_2 = 4\ \text{ft/s}^2$.

SOLUTION

(a) The free-body diagram of the elevator is drawn in Fig. 13-4b. Next the mass of the elevator is calculated from Eq. 13-3,

$$m = \frac{W}{g} = \frac{5000\ \text{lb}}{32.2\ \text{ft/s}^2} = 155\,\frac{\text{lb}\cdot\text{s}^2}{\text{ft}} = 155\ \text{slug}$$

The equation of motion in the y direction is from Eq. 13-5a,

$$\sum F_y = ma_y$$

$$T - W = ma_y$$

$$T = W + ma_y = 5000\ \text{lb} + \left(155\frac{\text{lb}\cdot\text{s}^2}{\text{ft}}\right)\left(3\frac{\text{ft}}{\text{s}^2}\right)$$

$$= 5000\ \text{lb} + 465\ \text{lb} = 5465\ \text{lb}$$

(b) The free-body diagram of the elevator is drawn in Fig. 13-4c. The equation of motion is from Eq. 13-5a, as above,

$$\sum F_y = ma_y$$

$$T - W = ma_y$$

$$T = W + ma_y = 5000\ \text{lb} + \left(155\,\frac{\text{lb}\cdot\text{s}^2}{\text{ft}}\right)\left(-4\,\frac{\text{ft}}{\text{s}^2}\right)$$

$$= 4380\ \text{lb}$$

JUDGMENT OF THE RESULTS

In static equilibrium the tension in the cable is 5000 lb. Thus, the calculated small changes in this tension for the given small accelerations (in comparison with the acceleration of gravity) are reasonable.

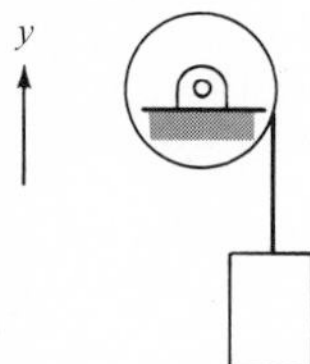

(a) Elevator at rest

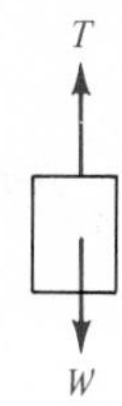

(b) Elevator accelerating upward

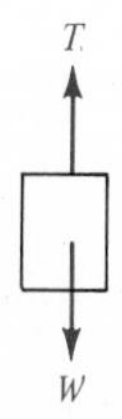

(c) Elevator accelerating downward

FIGURE 13-4

Analysis of an elevator

EXAMPLE 13-3

A ski-jump track has a circular part of radius $r = 40$ m at point A, where the skier has a horizontal speed $v = 20$ m/s (Fig. 13-5a). The 70-kg skier's speed is decreasing at point A at the rate of 1.3 m/s^2 because of air resistance. Determine the resultant force vector acting on the skier just before takeoff at A.

SOLUTION

The free-body diagram of the skier is drawn in Fig. 13-5b. The weight W is $mg = (70 \text{ kg})(9.81 \text{ m/s}^2) = 687$ N. The unknown forces R_x and R_y are determined using normal and tangential components according to Fig. 13-5b. The equations of motion are from Eq. 13-6,

$$\sum F_n = m\frac{v^2}{\rho} \qquad \sum F_t = m\dot{v}$$

$$R_y - W = m\frac{v^2}{\rho} \qquad -R_x = (70 \text{ kg})(-1.3 \text{ m/s}^2)$$

$$R_x = 91 \text{ N}$$

$$R_y = 687 \text{ N} + (70 \text{ kg})\left(\frac{400 \text{ m}^2/\text{s}^2}{40 \text{ m}}\right) = 1387 \text{ N}$$

The positive signs of R_x and R_y show that their directions were assumed correctly. The resultant force on the skier is

$$\mathbf{R} = R_x\mathbf{i} + (R_y - W)\mathbf{j} = 91 \text{ N}\,\mathbf{i} + 700 \text{ N}\,\mathbf{j}$$

JUDGMENT OF THE RESULTS

In static equilibrium the upward force on the skier would be $R_y = W = 687$ N. Since the center of curvature is above the skier, the normal acceleration caused by moving on a curved path is upward, so the increase in R_y is reasonable when compared with static equilibrium. The direction of $\mathbf{R}$ is reasonable by inspection.

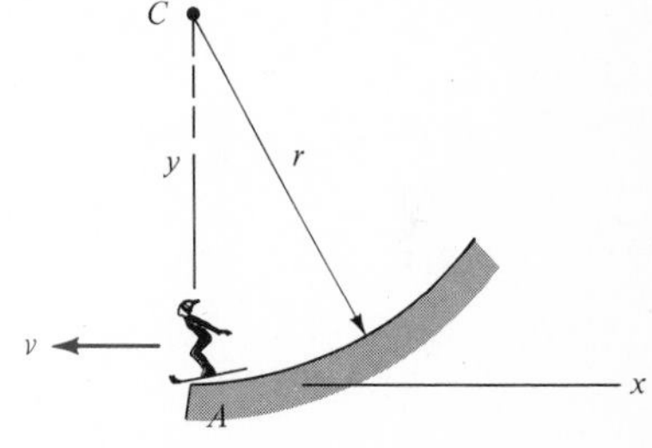

(a)

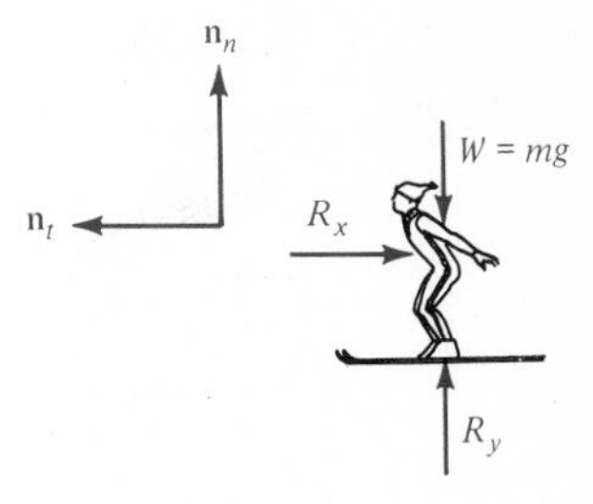

(b)

FIGURE 13-5
Analysis of a ski-jumper

EXAMPLE 13-4

Consider the amusement ride in Fig. 13-6a. A train of small cars with passengers travels inside the loop of the tracks, on a path of 70 ft diameter. Assume that the tracks can apply only normal forces P_n to the cars. The tracks and wheels are so constructed that the train cannot fall away from the tracks. Each car has a weight of $W = 700$ lb. (a) Determine the speed v_A of the train at point A at which there is no force from the tracks on a car at A. (b) At point B the train's speed is $v_B = 50$ ft/s. Determine the normal force on a car at point B.

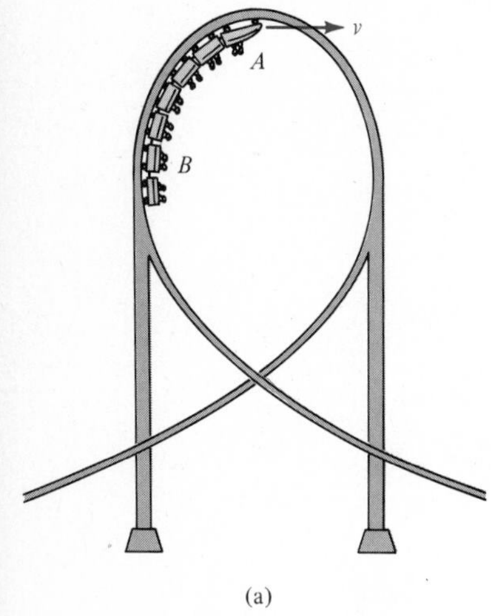

(a)

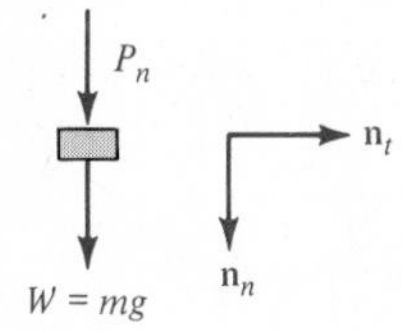

(b) Free-body diagram of a car at A

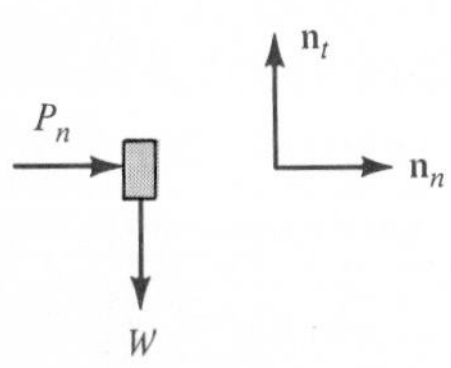

(c) Free-body diagram of a car at B

FIGURE 13-6

Analysis of an amusement ride with motion in a vertical plane

SOLUTION

(a) The free-body diagram of a car at A is drawn in Fig. 13-6b. Since the path has a constant radius of curvature and only normal forces have to be considered, the equation of motion is most convenient in terms of normal components. From Eq. 13-6,

$$\sum F_n = m\frac{v_A^2}{r}$$

From Fig. 13-6b,

$$P_n + W = m\frac{v_A^2}{r}$$

where P_n must be zero according to the statement of the problem. Using $W = mg$ the mass cancels out, and

$$v_A = (gr)^{1/2} = [(32.2 \text{ ft/s}^2)(35 \text{ ft})]^{1/2} = 33.57 \text{ ft/s}$$

(b) It is seen from the free-body diagram in Fig. 13-6c that W cannot affect the normal force in that particular position of the car. From Eq. 13-6,

$$P_n = m\frac{v_B^2}{r} = \left(\frac{W}{g}\right)\frac{v_B^2}{r}$$

$$= \left(\frac{700 \text{ lb}}{32.2 \text{ ft/s}^2}\right)\frac{(50 \text{ ft/s})^2}{35 \text{ ft}} = 1553 \text{ lb}$$

EXAMPLE 13-5

The large horizontal disk of a proposed amusement ride is modeled in Fig. 13-7a and b. The disk rotates at a constant angular speed $\omega = 0.5$ rad/s. Assume that a 150-lb person at point A ($r = 10$ ft) is walking outward at a constant speed of $v_r = 3$ ft/s. Calculate the radial and transverse forces that the disk exerts on the person at that instant.

SOLUTION

The free-body diagram of the person is drawn in Fig. 13-7c. The unknown radial and transverse forces are F_r and F_θ. From the equations of motion in polar coördinates (Eq. 13-7),

$$F_r = m(\ddot{r} - r\dot{\theta}^2)$$

$$F_\theta = m(r\ddot{\theta} + 2\dot{r}\dot{\theta})$$

where

$$\omega = \dot{\theta} = 0.5 \text{ rad/s}$$

$$r = 10 \text{ ft}$$

$$v_r = \dot{r} = 3 \text{ ft/s}$$

$$\ddot{\theta} = \ddot{r} = 0$$

Substituting these values gives

$$F_r = \frac{150 \text{ lb}}{32.2 \text{ ft/s}^2}\left[0 - (10 \text{ ft})\left(\frac{1}{2}\frac{\text{rad}}{\text{s}}\right)^2\right]$$

$$= -11.6 \text{ lb} \quad \text{(the force on the person is toward point } O)$$

$$F_\theta = \frac{150 \text{ lb}}{32.2 \text{ ft/s}^2}\left[(10 \text{ ft})(0) + 2\left(3\frac{\text{ft}}{\text{s}}\right)\left(\frac{1}{2}\frac{\text{rad}}{\text{s}}\right)\right]$$

$$= 14.0 \text{ lb}$$

(a)

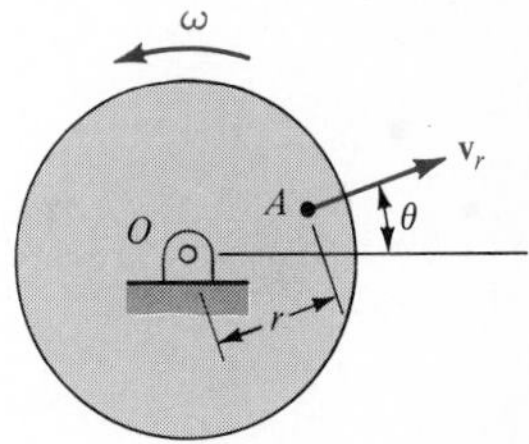

(b) Top view of disk

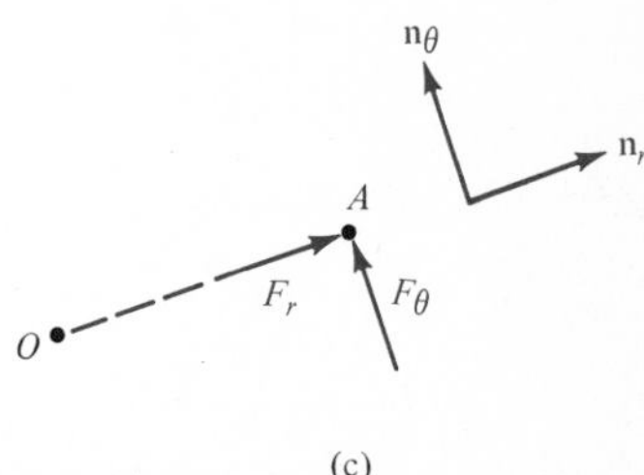

(c)

FIGURE 13-7

Analysis of an amusement ride with motion in a horizontal plane

EXAMPLE 13-6

An automobile of 1000-kg mass is traveling along a level road (Fig. 13-8a) at $v_0 = 100$ km/h when its brakes are applied locking the wheels. Calculate the stopping distance if the kinetic coefficient of friction of the wheels is 0.5. Neglect air resistance.

SOLUTION

The car is modeled as a block in Fig. 13-8b. There is no net force and consequently no acceleration in the y direction. The equation of motion for the x direction is

$$-\mathscr{F} = ma_x = m\frac{dv_x}{dt} = m\ddot{x}$$

where the friction force $\mathscr{F} = \mu N = \mu mg$. Since the distance traveled is of interest rather than time, x is the desired independent variable, not t. For the conversion, use

$$\frac{dv_x}{dt} = \frac{dv_x}{dx}\frac{dx}{dt} = v_x\frac{dv_x}{dx}$$

Thus, with $\mathscr{F}$ constant (since μ and N are constant),

$$-\mathscr{F} = mv_x\,dv_x/dx$$

After integration,

$$-\mathscr{F}x = \frac{1}{2}mv_x^2 + C$$

Taking $x = 0$ at $v_x = v_0$, $C = -\frac{1}{2}mv_0^2$. The stopping distance is at $x = x_s$, where $v_x = 0$, so

$$x_s = \frac{mv_0^2}{2\mathscr{F}} = \frac{mv_0^2}{2mg\mu} = \frac{10^{10}\text{ m}^2}{2(3600\text{ s})^2(9.81\text{ m/s}^2)(0.5)}$$

$$= 78.7\text{ m}$$

JUDGMENT OF THE RESULTS

The result is reasonable on the basis of common experiences. Note that the car's mass cancels out.

(a)

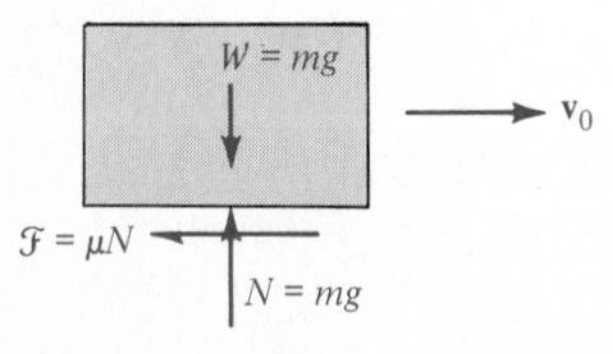

(b)

FIGURE 13-8
Analysis of a car's motion

PROBLEMS

13–1 Express Newton's second law of motion in terms of cylindrical component equations.

13–2 Write the scalar equations of motion using spherical coordinates.

13–3 An electronic instrument weighs 15 lb on the earth. Determine the mass of this device and its weight on the moon where the acceleration of gravity is 5.31 ft/s^2.

13–4 Determine the weight of a 70-kg person at sea level where $g = 9.81$ m/s^2, and at an elevation h_1 where $g_1 = 8$ m/s^2. Draw a free-body diagram of this person in a free fall at the elevation h_1.

13–5 A rocket is accelerating vertically at the rate of 80 ft/s^2 shortly after takeoff. Determine the net force from the seat on a 170-lb astronaut.

FIGURE P13-5

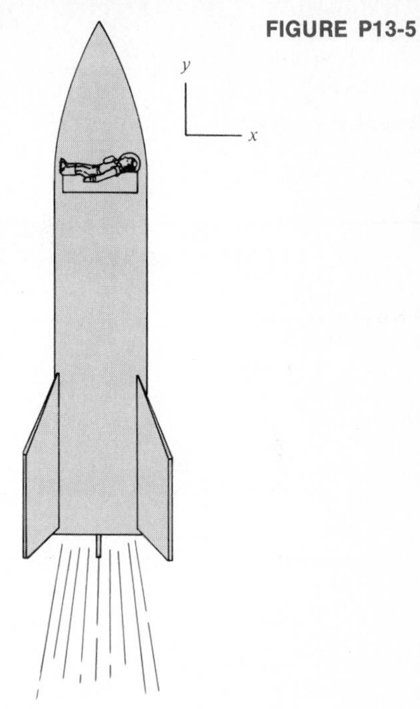

13–6 The acceleration of the rocket in Fig. P13-5 is $\mathbf{a} = (3\mathbf{i} + 25\mathbf{j})$ m/s^2 shortly after takeoff. Determine the vertical and horizontal forces from the seat on a 70-kg astronaut.

13–7 In a laboratory test of restraining belts for automobile passengers the initial horizontal speed of a dummy is $v_0 = 120$ ft/s. Its final speed is zero after a time interval of $\Delta t = 0.01$ s. Determine the net force that a 200-lb dummy applies to the belt in the crash test. All motion is horizontal.

FIGURE P13-7

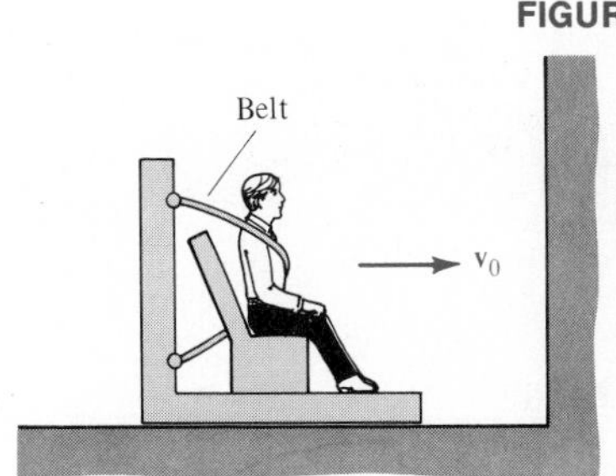

13–8 In a test of packaging materials a 10-kg electronic instrument is dropped on a cushioning pad. The speed of the device is changed by the pad from 15 m/s to zero in the time interval of $\Delta t = 0.002$ s. Calculate the force on the electronic device during its deceleration.

FIGURE P13-8

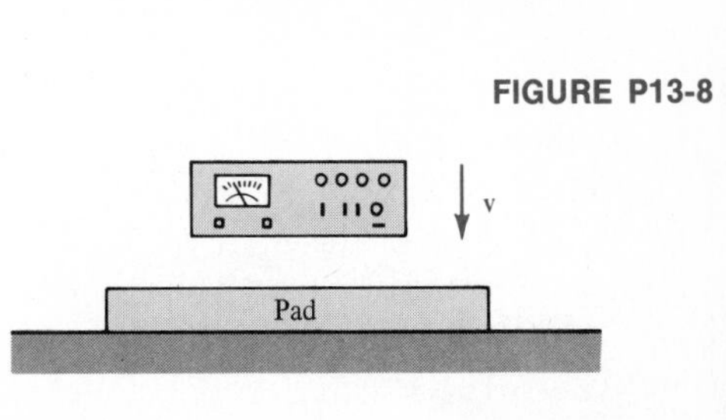

13–9 Assume that in Prob. 13–8 the time interval of deceleration is not known, but the deformation of the pad is measured. Determine the force on the device if the maximum deformation of the pad is 4 cm.

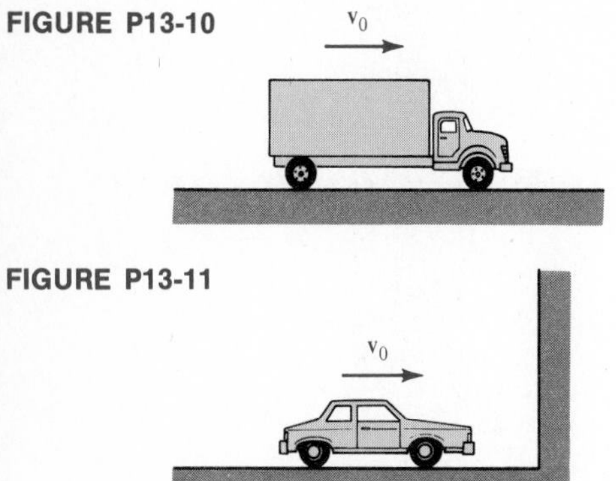

FIGURE P13-10

13–10 A truck travels horizontally at speed $v_0 =$ 60 ft/s when its brakes are applied locking the wheels. Determine the coefficient of kinetic friction if the truck stops after sliding 350 ft.

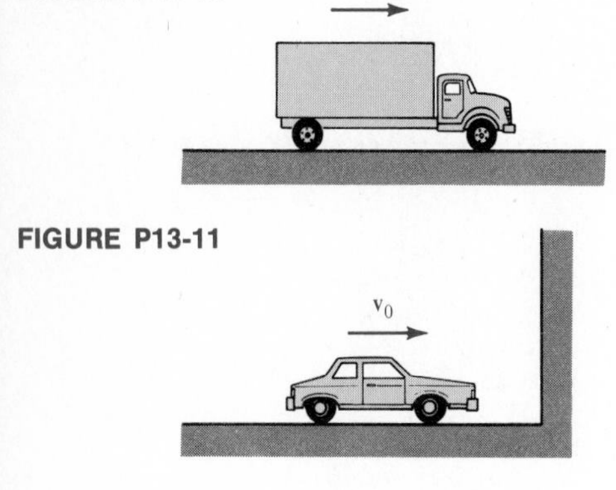

FIGURE P13-11

13–11 A car is traveling horizontally at speed $v_0 =$ 30 m/s when its brakes are applied locking the wheels. The car hits a structure after sliding 40 m. Determine the car's velocity just before impact if the coefficient of kinetic friction is 0.4.

13–12 For a car involved in an accident, it is estimated that its speed just before impact was $v_f = 40$ ft/s. Skid marks over a distance of $d = 150$ ft indicate the wheels were locked. Determine the initial speed v_0 of the car when the brakes were applied if the coefficient of kinetic friction was 0.5.

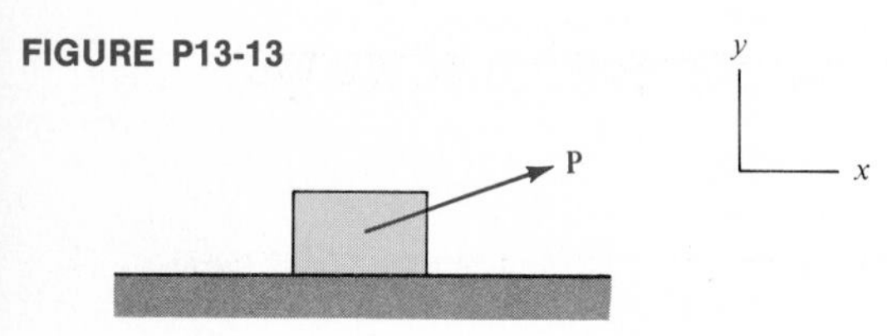

FIGURE P13-13

13–13 A 5-kg block is moved horizontally by a force $\mathbf{P} = (200\mathbf{i} + 25\mathbf{j})$ N. Determine the acceleration of the block if the coefficient of kinetic friction is 0.3.

13–14 Solve Prob. 13–13 assuming that $\mathbf{P} = (-200\mathbf{i} - 25\mathbf{j})$ N.

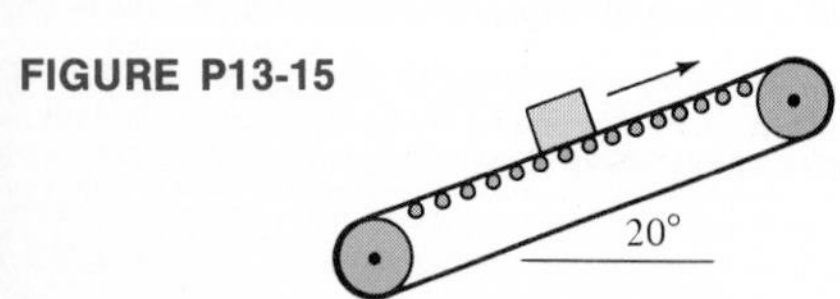

FIGURE P13-15

13–15 The coefficient of static friction between the 10-lb block and the conveyor belt is 0.6. Determine the maximum upward acceleration of the block along the incline that can be achieved without slippage of the block on the belt.

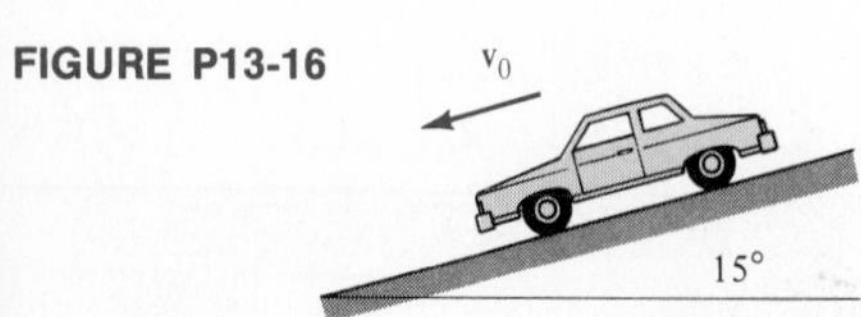

FIGURE P13-16

13–16 A 1000-kg car has initial speed $v_0 = 25$ m/s down the incline when its brakes are applied locking the wheels. Determine the stopping distance if the coefficient of kinetic friction is 0.7.

13–17 A small elevator A is connected to a counterweight B through a pulley of negligible weight and axle friction. (a) Determine the acceleration of A starting from rest. (b) Assume that the counterweight B is replaced by a constant-force spring which applies the same force F over the full range of displacements. Determine the acceleration of A in this case.

13–18 The blocks are released from rest. Determine the tensions in rope segments a, b, c, and d assuming that the pulleys have negligible mass and axle friction.

13–19 An experimental racing car of 2500-lb weight has initial velocity $\mathbf{v}_0 = 150\mathbf{i}$ ft/s when its brakes are applied locking the wheels. (a) Determine the stopping distance if the coefficient of kinetic friction is 0.6. (b) To assist in the braking, a jet applies a downward thrust $\mathbf{P} = -300\mathbf{j}$ lb throughout the deceleration. Calculate the stopping distance for this case.

13–20 An experimental car is equipped with a jet of thrust $\mathbf{P}$ (Fig. P13-19) to shorten the stopping distance of the car from its initial velocity $\mathbf{v}_0$. For any angle θ, $P = 0.2$ W. Plot the normalized stopping distance $d/(v_0^2/g)$ as a function of $\theta = 0°$, 45°, and 90° for a coefficient of friction $\mu = 0.2$.

13–21 Consider an X-Y plotter which records an electric signal Y as a function of another signal X. Bar A always remains vertical as shown but moves horizontally according to the X signal. Carriage B with the pen moves along the bar as controlled by the Y signal. The 0.2-kg bar and the 0.04-kg carriage move in the same vertical plane. Determine the net forces $\mathbf{F}_A$ on the bar and $\mathbf{F}_B$ on the carriage at an instant when $\mathbf{a}_A = 10\mathbf{i}$ m/s² and $\mathbf{a}_{B/A} = 25\mathbf{j}$ m/s².

13–22 A fork lift truck is lowering a 300-lb box at a downward acceleration of 4 ft/s². What is the maximum horizontal acceleration of the truck if the box is not to shift on the horizontal forks where the coefficient of static friction is 0.4?

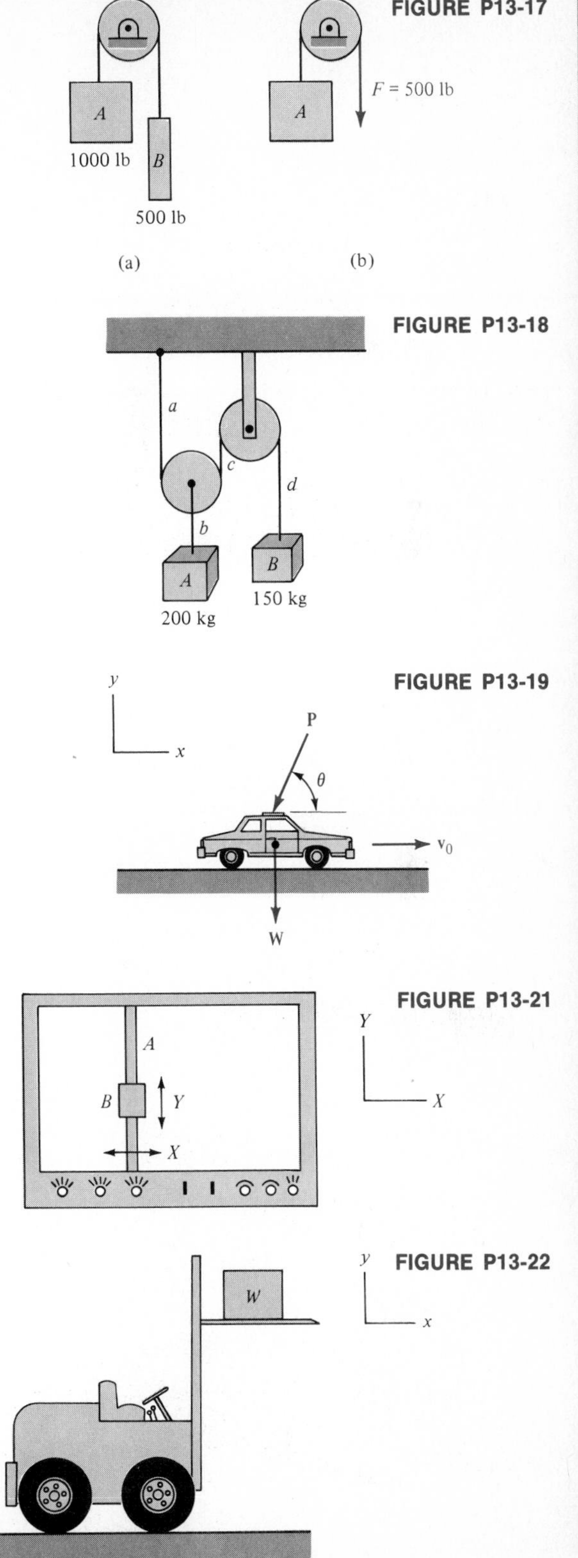

FIGURE P13-17

FIGURE P13-18

FIGURE P13-19

FIGURE P13-21

FIGURE P13-22

13–23 A 150-kg box on the horizontal forks of the lift truck in Fig. P13-22 is not to shift during motion of the truck. The forks are not moving vertically when the truck is slowing down from a speed of 15 km/h. What is the shortest stopping distance allowed if the coefficient of static friction is 0.3?

FIGURE P13-24

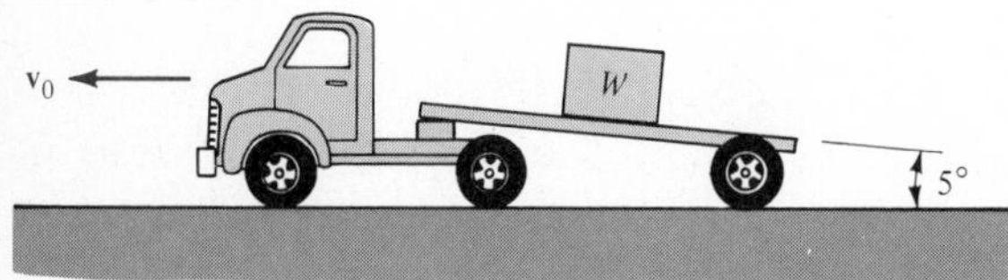

13–24 A truck is decelerating at the rate of 20 ft/s^2 when the 500-lb crate begins to shift forward on the sloping truck bed. Determine the coefficient of friction at the crate.

13–25 The truck in Fig. P13-24 travels at a speed of $v_0 = 90$ km/h when its brakes are applied. What is the shortest stopping distance allowed if the 200-kg crate is not to shift on the sloping truck bed? The coefficient of static friction at the crate is 0.25.

FIGURE P13-26

13–26 The locomotive A weighs 2×10^5 lb and the freight car B weighs 1.5×10^5 lb. They travel at a speed of $v_0 = 50$ mph when a 6000-lb braking force is applied to A and a 4000-lb braking force is applied to B. Determine the stopping distance and the force in the coupling between A and B.

FIGURE P13-28

13–27 In Prob. 13–26, an identical freight car C is coupled to B. The braking forces on B and C are 4000 lb each. Determine the stopping distance and the forces in the two couplings (A–B and B–C) if everything else is as in Prob. 13–26.

13–28 Consider the steam-driven pile driver. The 200-kg hammer is raised by steam pressure before each drop through a height of 2 m. The steam pressure can also be applied to cause a downward force on the falling hammer. Determine the constant downward force which would make the 2-m forced fall equivalent to a 4-m free fall for driving a pile.

13–29 A speed boat travels at $v = 35$ mph on a circular path of radius $r = 50$ ft. Determine the centripetal force on the 400-lb boat.

13–30 A speed skater travels at a constant speed of 15 m/s on a curved path of 30 m radius. Calculate the horizontal force on the 70-kg skater.

13–31 The road segment is horizontal and has a radius of curvature of 150 ft. The 2000-lb car travels at a constant speed of 40 mph. Determine the centripetal force on the car and the coefficient of friction necessary to keep the car on the circular path.

13–32 A 1000-kg car on a horizontal road of radius $r = 100$ m (Fig. P13-31) travels at speed $v = 30$ m/s, which is decreasing at the rate of 5 m/s^2. Determine the total horizontal force on the car.

13–33 A speed boat travels at a speed of 40 mph on a circular path of 50 ft radius. Determine the required angle θ of the boat with the horizontal if the passengers must not have any tendency to slide on the seats. For this, it is necessary to assume a coefficient of static friction of zero.

13–34 A speed boat travels at a speed of 60 km/h on a circular path of 30 m radius. Determine the required angle θ of the boat with the horizontal (Fig. P13-33) if the passengers must not slide on the seats, where the coefficient of static friction is 0.3.

13–35 The centerline of a road is in the horizontal plane and has a radius of curvature $r = 300$ ft. The legal speed at this location is planned to be 40 mph. Determine the appropriate banking angle θ of the road if a car traveling at the legal speed should not skid on the road where the coefficient of static friction is 0.2.

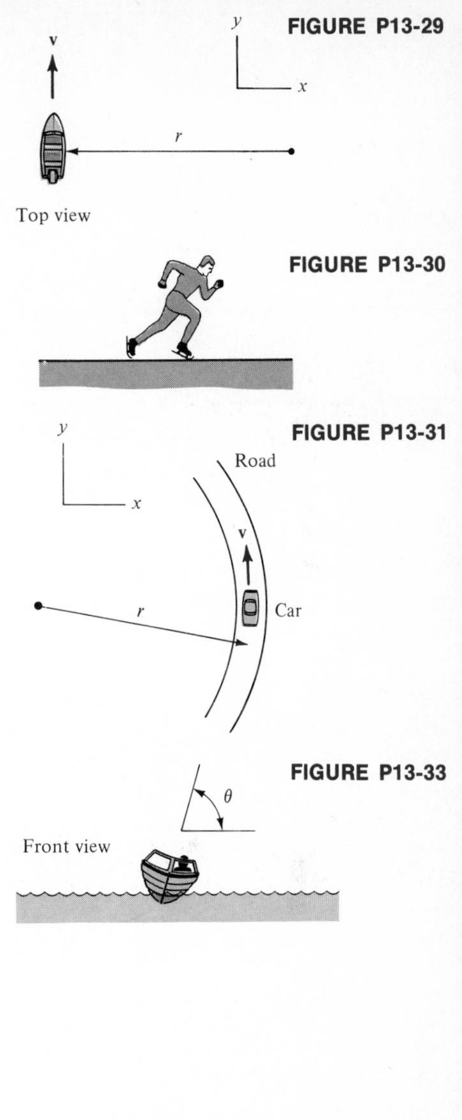

FIGURE P13-29

FIGURE P13-30

FIGURE P13-31

FIGURE P13-33

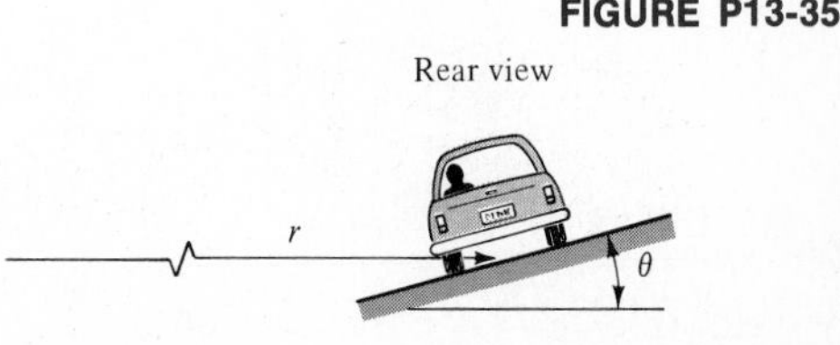

FIGURE P13-35

13–36 A car travels on a road of radius $r = 100$ m and banking angle $\theta = 10°$ (Fig. P13-35). Determine the maximum speed v of the car if it is not to skid on the road, where the coefficient of static friction is 0.4.

FIGURE P13-37

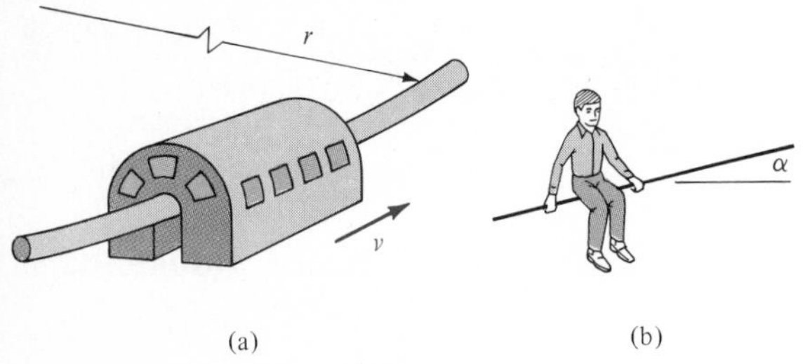

13–37 A high-speed train is proposed on a monorail track. Calculate the force on the track when a 50,000-lb coach travels at a speed of 200 ft/s on a track of 1000 ft radius.

13–38 Determine the required banking angle α of the seats in the coach in Prob. 13–37 if the passengers are not to slide on the seats, where the coefficient of static friction is 0.5.

FIGURE P13-39

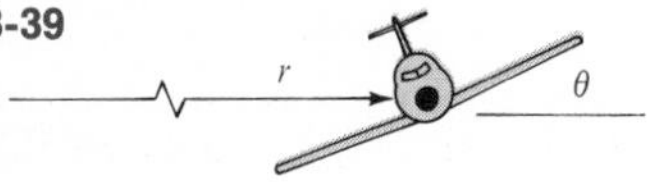

13–39 The airplane is flying on a horizontal circular path of 400 m radius at a speed of 100 m/s. Determine whether the pilot would slide on the seat where the coefficient of static friction is 0.5 if the banking angle is $\theta = 20°$. Assume that the seat is flat.

13–40 Solve Prob. 13–39 assuming that the speed of the airplane is decreasing at the rate of 5 m/s^2. Assume that the flat seat is not banked with respect to the direction of the tangential acceleration.

FIGURE P13-41

13–41 The track of the amusement ride has a radius of curvature $r = 40$ ft and a banking angle $\theta = 35°$ with the horizontal. Determine the speed v of the train if the passengers must not have any tendency to slide on the seats. Assume that the track is horizontal and the coefficient of friction at the seats is zero.

13–42 In Prob. 13–41, $v = 40$ ft/s and the coefficient of static friction at the seats is 0.4. Determine whether the passengers would slide on the seats.

FIGURE P13-43

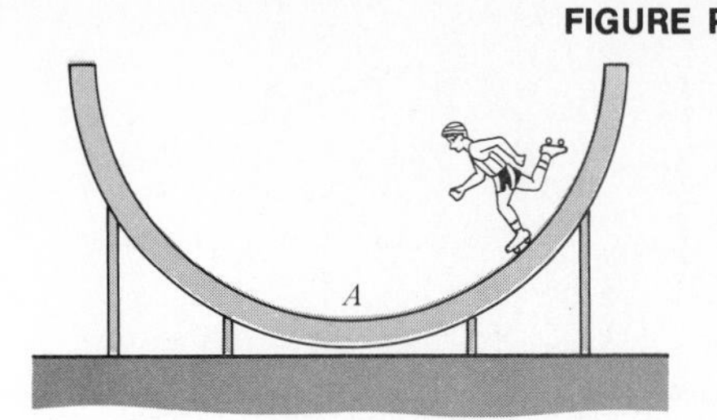

13–43 A 70-kg person on roller skates is doing a stunt. At the bottom of the circular track of 6 m radius the skater's speed is $v_A = 10$ m/s. Determine the vertical force on the skater at A.

FIGURE P13-44

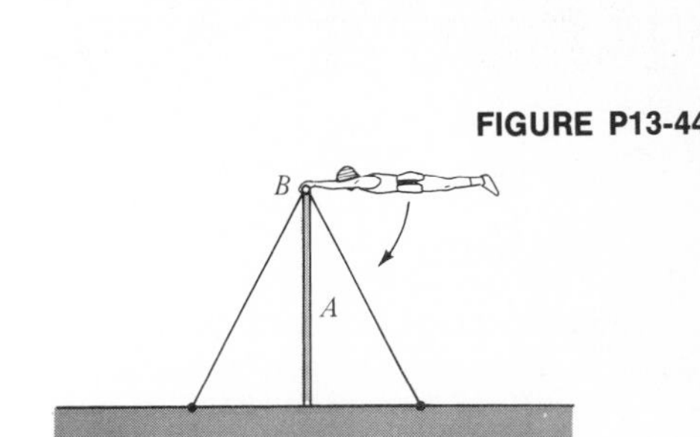

13–44 A gymnast swinging on a high bar is modeled as a 160-lb particle at a distance of 3.5 ft from bar B. Determine the net force on the gymnast's arms when he is at point A below the bar, the "particle" moving at a horizontal speed of $v_A = 15$ ft/s.

FIGURE P13-45

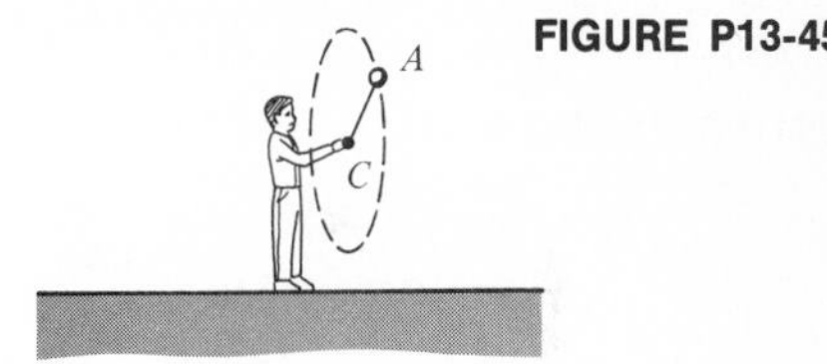

13–45 A child is twirling a 1-kg stone at the end of a 1-m-long rope at the rate of 200 rpm. Determine the maximum and minimum tension in the rope, assuming that the path is in the vertical plane and its center C is fixed.

13–46 Solve Prob. 13–45 assuming that the plane of the stone's path makes a 45° angle with the horizontal.

FIGURE P13-47

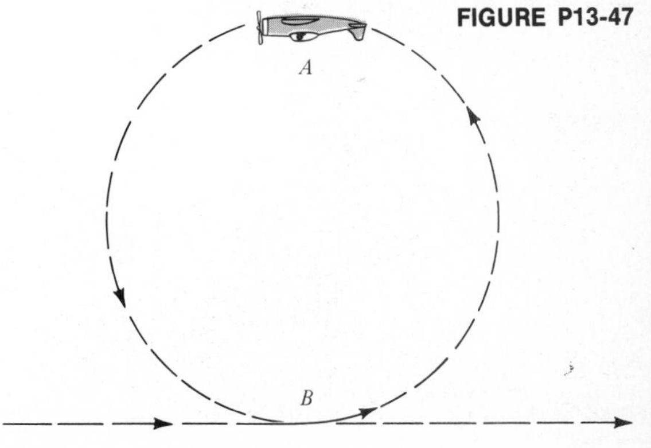

13–47 A 160-lb pilot is belted in the seat of the airplane which makes a vertical loop of 800-ft diameter. Determine the net force of the seat on the pilot at point A if the speed of the airplane is $v_A = 100$ mph and decreasing at the rate of 20 ft/s^2.

13–48 The airplane in Fig. P13-47 is flying at a speed of $v_B = 200$ m/s and accelerating at the rate of 15 m/s^2 at the low point B of a vertical loop of 400-m diameter. Determine the net force of the seat on the 65-kg pilot.

FIGURE P13-49

(a)

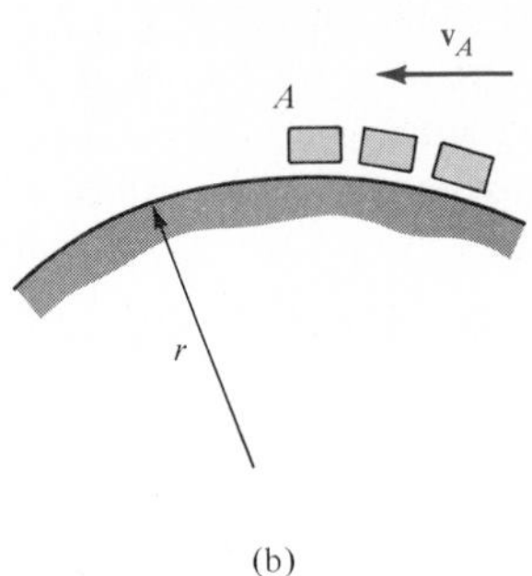

(b)

13–49 Passengers are belted in a roller coaster which travels in a vertical plane. At a high point A the speed of the train is $v_A = 8$ ft/s, decreasing at the rate of 10 ft/s^2, and $r = 25$ ft. Calculate the net force of the seat on a 200-lb passenger.

13–50 Determine the constant speed v_A in Prob. 13–49 at which the net force of the seat on the passenger is zero (the passenger experiences weightlessness).

FIGURE P13-51

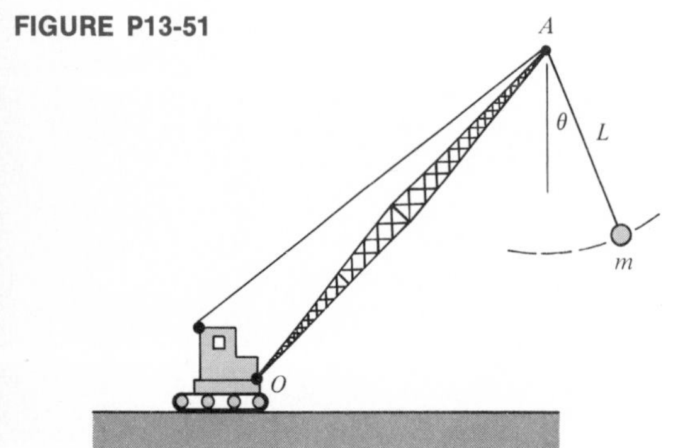

13–51 Consider a wrecker's ball suspended from a crane. Assume that point A is fixed. A proposed safety device would measure the tension in the cable at A and compute the velocity and acceleration of the ball as a function of θ. The ball swings in the vertical plane. Calculate the velocity and acceleration for $L = 10$ m and $\theta = 20°$ if the tension in the cable is twice the weight of the ball.

FIGURE P13-52

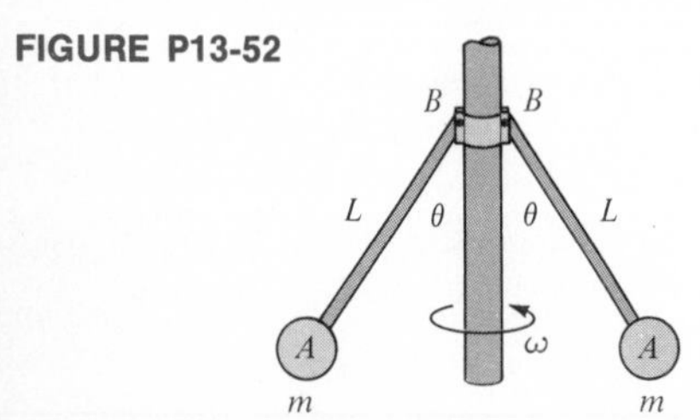

13–52 A speed-governing mechanism is modeled with two 0.2-kg balls A suspended by weightless rods hinged at B. Calculate the angular speed ω and the tension in the rods if $L = 10$ cm, and $\theta = 60°$. The hinges allow rotation of the rods only in a vertical plane. The given angle θ is constant in the present case.

13–53 Solve Ex. 13-3 using polar coordinates.

SUMMARY

Section 13-1

Newton's second law of motion

$$\mathbf{F} = m\mathbf{a}$$

The magnitude of the acceleration **a** of a particle is directly proportional to the magnitude of the resultant force **F** acting on it, and inversely proportional to its mass m. The direction of the acceleration is the same as the direction of the resultant force. The motion of the particle is determined with respect to a nonaccelerating (Newtonian) reference frame.

Section 13-2

SI units

FORCE:

$$F = ma$$

where F = newtons
m = kilograms
a = meters per second squared

WEIGHT:

$$W = mg$$

where W = newtons
m = kilograms
g = acceleration of gravity $\simeq 9.81\ \text{m/s}^2$

U.S. customary units

FORCE:

$$F = ma$$

where F = pounds
m = slug
a = feet per second squared

WEIGHT:

$$W = mg$$

where W = pounds
m = slug
g = acceleration of gravity $\simeq 32.2\ \text{ft/s}^2$

Equations of motion Basic equation of motion for any number of forces $\mathbf{F}_1, \mathbf{F}_2, \mathbf{F}_3, \ldots$, acting on one particle.

$$\sum \mathbf{F}_i = m\mathbf{a}$$

Scalar component equations of motion

RECTANGULAR COORDINATES:

$$\sum F_x = ma_x \qquad \sum F_y = ma_y \qquad \sum F_z = ma_z$$

NORMAL AND TANGENTIAL COORDINATES:

$$\sum F_n = ma_n = m\frac{v^2}{\rho} \qquad \text{(centripetal force)}$$

$$\sum F_t = ma_t = m\frac{dv}{dt} = mv\frac{dv}{ds}$$

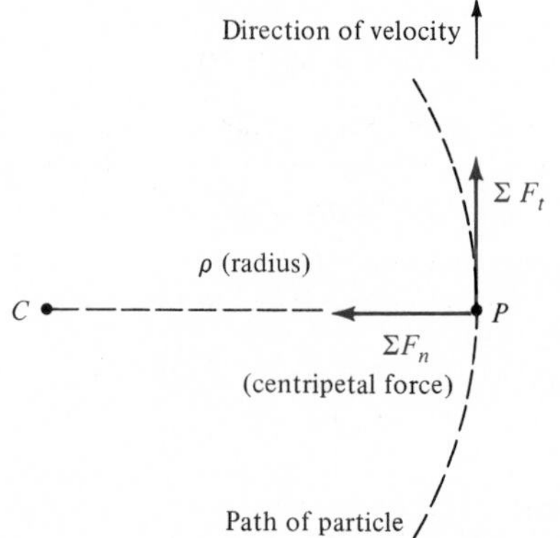

POLAR COORDINATES:

$$\sum F_r = ma_r = m(\ddot{r} - r\dot{\theta}^2)$$

$$\sum F_\theta = ma_\theta = m(r\ddot{\theta} + 2\dot{r}\dot{\theta})$$

Procedure for solving problems

1. Establish a convenient nonaccelerating reference frame.
2. Draw the free-body diagram of the particle showing known and unknown forces.
3. Apply the equations of motion which are appropriate for the chosen coordinate system to determine forces or accelerations.
4. If the velocity or position of the particle must be calculated, use the equations of kinematics after the acceleration has been determined.

REVIEW PROBLEMS

13–54 Solve Ex. 13-4 using polar coordinates.

13–55 An experimental car traveling on a horizontal road is tracked by radar located at point O. At a particular instant $r = 100$ m, $\dot{r} = 3$ m/s, $\ddot{r} = 0$, $\theta = 60^\circ$, $\dot{\theta} = 0.3$ rad/s, and $\ddot{\theta} = 0$. Calculate the net horizontal force on the 1300-kg car.

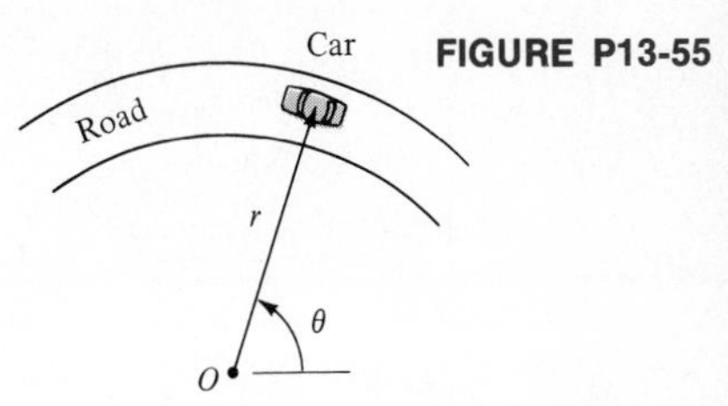

FIGURE P13-55

13–56 Radar data of a car traveling on a horizontal road (Fig. P13-55) indicate that $r = 250$ ft, $\dot{r} = 0$, $\ddot{r} = 10$ ft/s, $\theta = 45^\circ$, $\dot{\theta} = 0.6$ rad/s, $\ddot{\theta} = 0.01$ rad/s^2. Calculate the net horizontal force on the 2600-lb car.

FIGURE P13-57

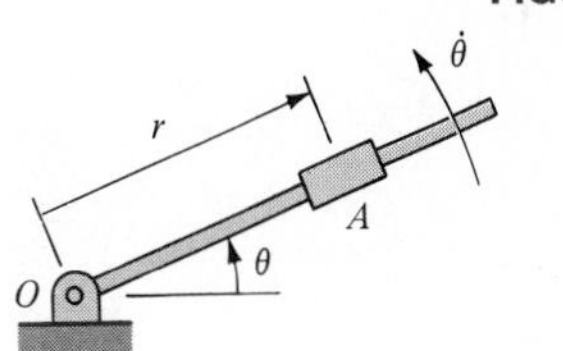

13–57 The 2-kg block A on the rotating arm is at $r = 0.4$ m when $\dot{r} = 2$ m/s, $\ddot{r} = 0$, $\dot{\theta} = 3$ rad/s, $\ddot{\theta} = 0.7$ rad/s^2. Determine the net force on the block.

13–58 Consider a wire-controlled model airplane which flies in a horizontal plane. Calculate the net horizontal force on the 10-lb airplane if $r = 30$ ft (in the plane of the path), $\dot{r} = 0$, $\ddot{r} = 0$, $\dot{\theta} = 1$ rad/s, and $\ddot{\theta} = 0.3$ rad/s^2.

FIGURE P13-58

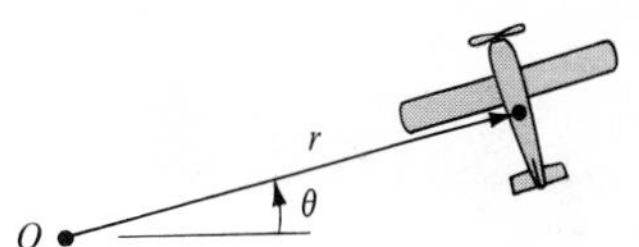

13–59 The 1000-kg wrecker's ball swings in the vertical plane. Determine the tension in the wire rope supporting the ball if point A is fixed, $L = 10$ m, $\dot{L} = 1$ m/s, $\ddot{L} = -0.5$ m/s^2 ($\dot{L}$ is decreasing), $\theta = 30^\circ$, $\dot{\theta} = 0.8$ rad/s, and $\ddot{\theta} = -0.1$ rad/s^2 ($\dot{\theta}$ is decreasing).

FIGURE P13-59

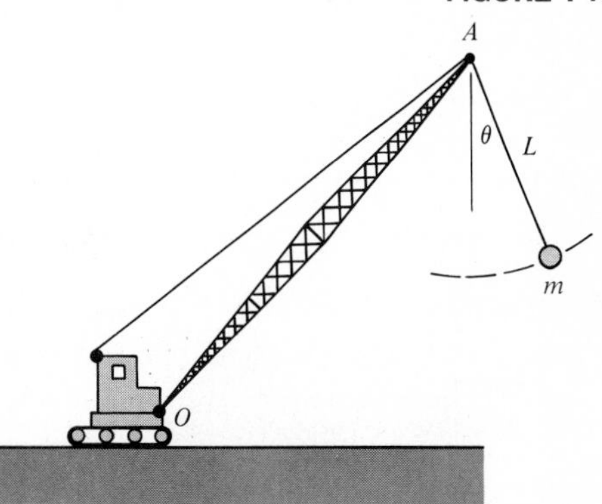

13–60 In the speed-governing mechanism $m = 0.2$ kg, $L = 10$ cm (rods AB have negligible mass), $\omega = 500$ rpm, $\theta = 40^\circ$, $\dot{\theta} = 0$. Determine $\ddot{\theta}$ at that instant and the tension in the rod.

FIGURE P13-60

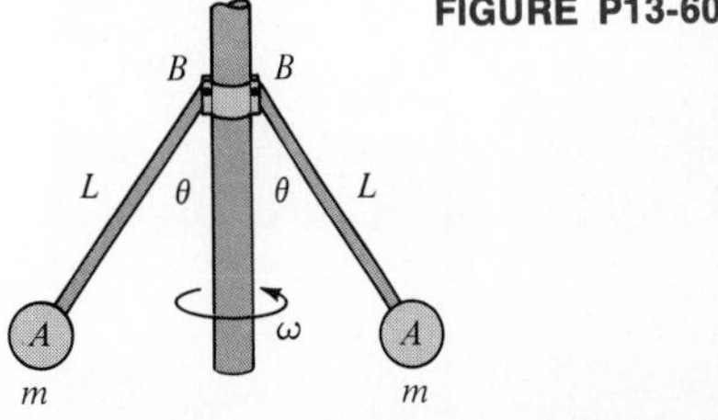

CHAPTER 14

KINETICS OF PARTICLES: WORK AND ENERGY METHODS

OVERVIEW

The fundamental method of analyzing the kinetics of particles is Newton's second law. The equation $\mathbf{F} = m\mathbf{a}$ also enables one to analyze the velocities and positions of a particle, but this is not always convenient in practice. For example, consider the mechanized hay-baling and transporting system shown in Fig. 14-1. The machine takes hay from the ground, makes bales of hay, and throws the bales to a higher level in the trailer. On the basis of kinematics, the initial velocity of the bales is the required quantity for attaining the elevation of the trailer. The acceleration of a bale in the launching phase of the motion is predetermined depending on the physical constraints of the machine. However, the bales are not identical in mass, so the forces propelling them are not identical either. For such a case, it is useful to develop supplementary methods for the analysis of forces, velocities, and positions of a particle, without the need to consider accelerations. The concepts of work and energy enable such a simplification of the basic method of kinetics.

FIGURE 14-1

Hay-baling machine in operation

STUDY GOALS

SECTION 14-1 presents the definition of work of a force acting on a particle. Work is a scalar quantity and is frequently used in engineering mechanics because it simplifies the solution of certain problems. A prior study of work in a statics course is useful but not essential for studying this material.

SECTION 14-2 describes the concept of potential energy for gravitational and elastic systems. Potential energy is frequently used in the analysis of moving bodies. The calculation of potential energy in most mechanical systems should be a routine procedure.

SECTION 14-3 presents the definition of kinetic energy of a particle on the basis of Newton's second law. All moving bodies have kinetic energy, and this quantity is very useful in the analysis of motion. The relationship of work and energy is especially important in dynamics analyses.

SECTION 14-4 introduces the concepts of power and efficiency which are important in the design of machines.

SECTION 14-5 illustrates the advantages and limitations of using energy methods in kinetics. Overall, the advantages are substantial, and it is important to acquire a working knowledge of these methods.

SECTION 14-6 defines the motion of particles in conservative systems in which friction does not oppose the motion. The elementary concepts of conservative systems are important for all students of dynamics. The advanced analysis at the end of Sec. 14-6 requires the use of partial derivatives, and for that reason it is often omitted from elementary dynamics courses.

SECTION 14-7 presents the principle of conservation of mechanical energy. The relationship of kinetic and potential energies in a conservative system is one of the most important and useful expressions in dynamics, and it should be studied with emphasis.

14-1 WORK OF A FORCE

The term *work* has an exact, quantitative meaning in the area of mechanics. Quantitatively, *work is the product of the force and the corresponding displacement of a body*. For example, the constant force **F** slowly moves the block from position 1 to 2 along a smooth plane in Fig. 14-2a. If **F** is parallel to the incline, the work U of the force is

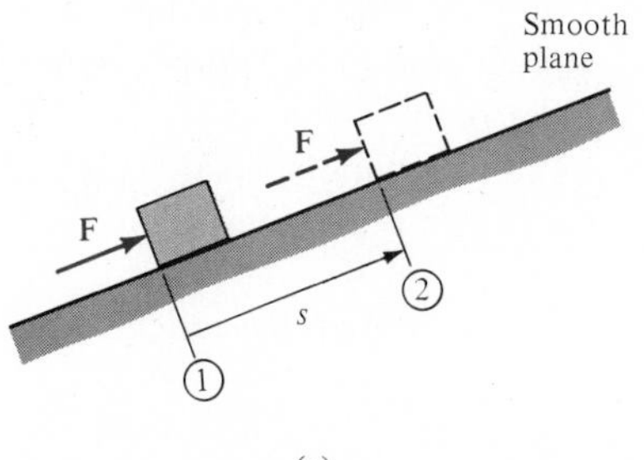

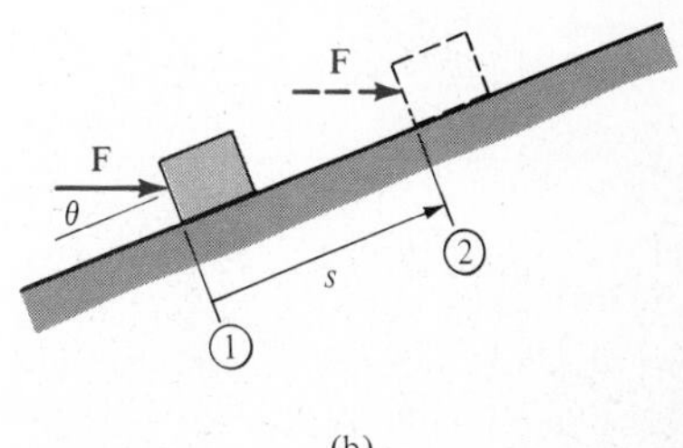

FIGURE 14-2
A force doing work on a body

$$U = Fs \quad \boxed{14\text{-}1}$$

If **F** is at an angle θ with the plane as in Fig. 14-2b,

$$U = F \cos \theta s \quad \boxed{14\text{-}2}$$

because only the component $F \cos \theta$ is associated with the displacement of the block if the inclined plane is fixed. The component $F \sin \theta$ does no work because there is no displacement of the block perpendicular to the fixed plane.

In general, the force **F** may change in magnitude and direction during the displacement. The work in that case is determined by considering the infinitesimal work dU of a force **F** during an infinitesimal displacement $d\mathbf{r}$ of particle P as in Fig. 14-3. The work dU is defined as the scalar product of the vectors **F** and $d\mathbf{r}$,

$$dU = \mathbf{F} \cdot d\mathbf{r} = F\,dr \cos \theta \quad \boxed{14\text{-}3}$$

The total work U_{12} in displacing particle P from point 1 to 2 is obtained by integrating Eq. 14-3. Using the rectangular components of **F** and $d\mathbf{r}$ gives

$$\boxed{\begin{aligned} U_{12} &= \int_1^2 \mathbf{F} \cdot d\mathbf{r} = \int_1^2 F \cos \theta\, dr \\ &= \int_1^2 (F_x\, dx + F_y\, dy + F_z\, dz) \end{aligned}} \quad \boxed{14\text{-}4}$$

Work is a scalar quantity with units of force times distance, N·m or lb·ft (dimensionally the same as for moments). Work may be positive or negative according to the cosine function in Eq. 14-3.

Forces that do no work are common. For example, a force applied to a fixed particle causes no displacement, so its work is zero. Similarly, a force acting perpendicular to the displacement has no component in the direction of the displacement, and its work is also zero.

Work can also be defined using Newton's second law, $\mathbf{F} = m\mathbf{a}$. From Eq. 14-4,

$$U_{12} = \int_1^2 \mathbf{F} \cdot d\mathbf{r} = \int_1^2 m\mathbf{a} \cdot d\mathbf{r} \quad \boxed{14\text{-}5}$$

Note that $\mathbf{a} \cdot d\mathbf{r} = a_t\, dr$, where a_t is the tangential acceleration of the particle and dr is approximately the infinitesimal displacement along the path.

The following special aspects of total work are easy to prove and are worth remembering:

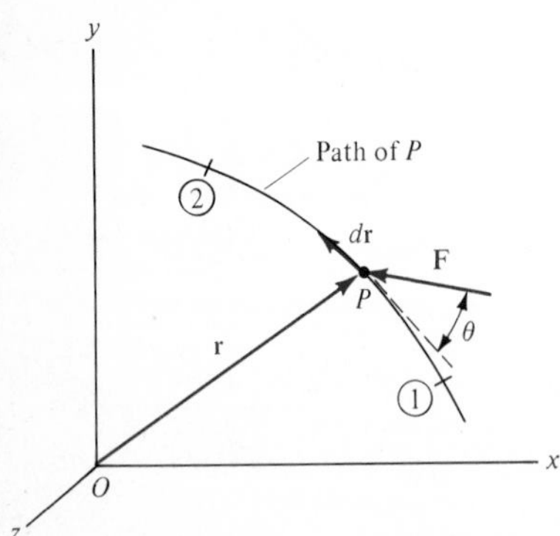

FIGURE 14-3

Analysis of a general displacement for the definition of work

1. The total work of concurrent forces is equal to the work of the resultant of those forces.
2. The work of internal forces in a rigid member cancels out.

Historically, the term *potential energy* comes from the perception that a weight W situated at any level could do useful work (drive a clock, pump water, etc.) if allowed to descend to a lower level with respect to the center of the earth. The formal expression of this concept is based on Fig. 14-4, where a body of weight W moves without friction between levels 1 and 2. In going from level 1 to 2, the total work of the constant force W according to Eq. 14-4 is

$$\underbrace{U_{12}}_{\text{work of } W \text{ from 1 to 2}} = \int_1^2 W\,dy = W(y_2 - y_1) = \underbrace{Wh}_{\text{potential energy of } W \text{ at 1 with respect to 2}} \qquad \text{14-6}$$

The potential of the weight W to do work because of its relative height h is defined as its *gravitational potential energy*. For further illustration of this concept, consider the potential energies at level 2 with respect to 3, at 1 with respect to 3, and at 2 with respect to 1 in Fig. 14-4. The results are, respectively,

$$U_{23} = \int_2^3 W\,dy = Wl \qquad \text{14-7a}$$

$$U_{13} = \int_1^3 W\,dy = W(h + l) \qquad \text{14-7b}$$

$$U_{21} = \int_2^1 -W\,dy = -Wh = -U_{12} \qquad \text{14-7c}$$

The minus sign in the last equation shows that gravity opposes the motion up.

It should be noted that Eq. 14-6 is valid only if $W = mg$ is essentially a constant over the displacement h.

Elastic Potential Energy

The potential energy of elastic members* is frequently important in mechanical systems. The background for analyzing any elastic member is the *theory of springs* discovered by Robert Hooke in 1642. This is illustrated in Fig. 14-5. A helical spring is loaded with progressively larger weights W, $2W$, and so on (Fig. 14-5a–d). The

* Ideal elastic members return to their original dimensions after the forces acting on them are removed. Elastic members are classified as linear or nonlinear depending on their diagrams of force vs. deformation. The term *inelastic behavior* means that the member does not return to its original shape after the forces on it are removed. Such behavior is analyzed in mechanics of materials.

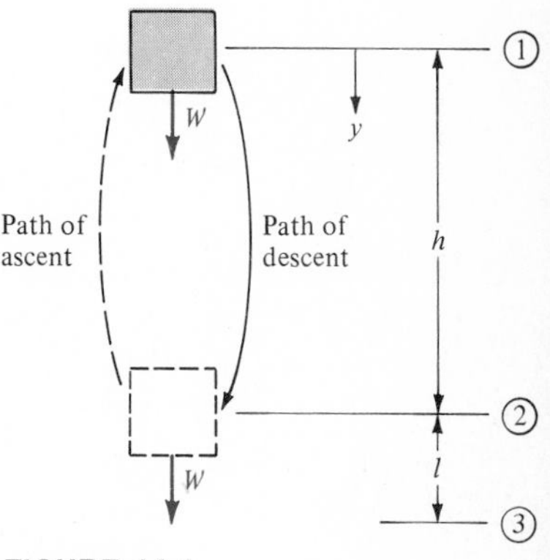

FIGURE 14-4

Model for defining gravitational potential energy

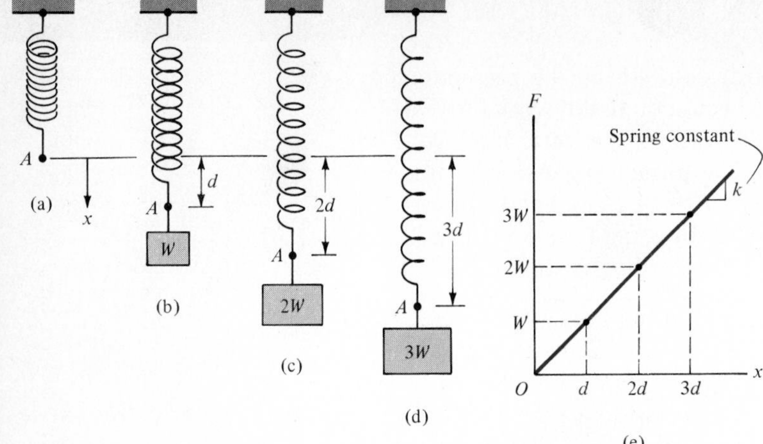

FIGURE 14-5

Model for the theory of elastic springs

deformations x are proportional to the applied force F for small deformations as shown by the straight-line plot in Fig. 14-5e. The slope k of this line is called the *spring constant* or *spring stiffness.* It has units of N/m or lb/in., and it is the property of each particular spring (depends on material, wire diameter, coil length, and coil diameter). From Fig. 14-5e,

$$F = kx \tag{14-8}$$

which is used in Eq. 14-4 to obtain the work of the external force F in stretching the spring from O to x_1,

$$U_1 = \int_O^{x_1} kx\,dx = \frac{1}{2}kx_1^2$$

This is equal to the shaded area in Fig. 14-6. In general, the work U of an *external force F* in *causing* a deformation x from the unstretched length of a spring is

$$\underbrace{U}_{\text{work of } F \text{ from } O \text{ to } x} = \underbrace{\frac{1}{2}kx^2}_{\text{potential energy of spring at } x \text{ with respect to } O} \tag{14-9a}$$

In the same deformation x, the work U' of the *force exerted by the spring* is

$$U' = -\frac{1}{2}kx^2 \tag{14-9b}$$

since the spring opposes the deformation which is caused by the external force F.

In general, the work done by an *external force* in deforming a linear, elastic spring from x_1 to x_2 (Fig. 14-6) is

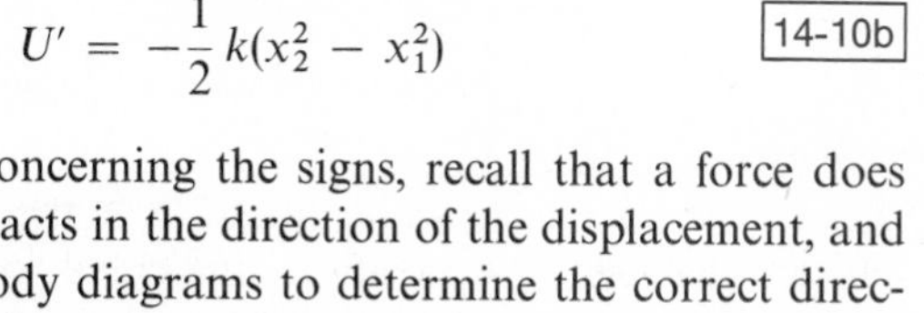

$$U = \int_{x_1}^{x_2} kx\,dx = \frac{1}{2}k(x_2^2 - x_1^2) \tag{14-10a}$$

while the work of the *force exerted by the spring* is

$$U' = -\frac{1}{2}k(x_2^2 - x_1^2) \tag{14-10b}$$

To avoid confusion concerning the signs, recall that a force does positive work when it acts in the direction of the displacement, and vice versa. Use free-body diagrams to determine the correct directions of forces and displacements. Thus, a spring does negative work while it is being deformed by an external force, and positive work when it returns to its undeformed position.

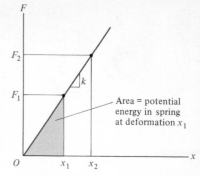

FIGURE 14-6

Definition of elastic potential energy

Notation for Potential Energy

The units of potential energy are the units of work, N·m or ft·lb, but it is useful to have a special notation for potential energy. A commonly used symbol for potential energy is V. The work U done on a mechanical system S by an external force is expressed as the negative of the resulting potential energy V of system S,

$$U = -V \tag{14-11}$$

For example, a machine raises a block of weight W. By the definition of work in Eq. 14-3, the machine does positive work on the block, while the weight W does negative work.

Equation 14-11 is valid for any ideal system based on gravity, springs, and so on, in which the work of a force depends only on the net change of position and not on the path of reaching that position. Such ideal systems are called *conservative*, and the forces involved are called *conservative forces*. These concepts are discussed in more detail in Sec. 14-6.

14-3 KINETIC ENERGY. PRINCIPLE OF WORK AND ENERGY

When a particle is in motion it has *kinetic energy* because of its velocity. To analyze this form of energy, consider a particle of mass m which moves on a curved path as in Fig. 14-7.

The kinetic energy associated with the velocity of the particle is defined on the basis of Newton's second law as follows. Denoting

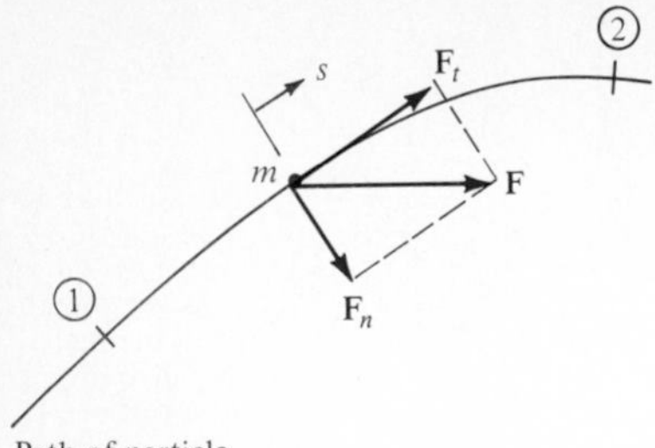

Path of particle

FIGURE 14-7

Quantities considered in the definition of kinetic energy

the displacement by s, and the magnitude of the velocity by v,

$$F_t = ma_t = m\frac{dv}{dt} = m\frac{dv}{ds}\frac{ds}{dt} = mv\frac{dv}{ds}$$

$$F_t\,ds = mv\,dv$$

This is integrated from position 1 to 2 where the position and velocity have values s_1, v_1, and s_2, v_2, respectively. For a constant mass m,

$$\int_{s_1}^{s_2} F_t\,ds = m\int_{v_1}^{v_2} v\,dv = \frac{1}{2}mv_2^2 - \frac{1}{2}mv_1^2 \qquad \text{14-12}$$

The integral $\int F\,ds$, a scalar quantity, is the work of the force on a particle between the positions indicated in Fig. 14-7. The scalar quantity $\frac{1}{2}mv^2$ is defined as the *kinetic energy* of the particle at the particular speed v. Kinetic energy is generally denoted by the symbol T,

$$\boxed{T = \frac{1}{2}mv^2} \qquad \text{14-13}$$

Kinetic energy is always positive, regardless of the direction of the velocity vector, since mass is positive and speed is squared. The units of kinetic energy are the units of work,

$$\frac{\text{kg}\cdot\text{m}^2}{\text{s}^2} = \left(\frac{\text{kg}\cdot\text{m}}{\text{s}^2}\right)\text{m}$$

$$= \text{N}\cdot\text{m (joule) in SI units and } \frac{\text{slug}\cdot\text{ft}^2}{\text{s}^2}$$

$$= \left(\frac{\text{slug}\cdot\text{ft}}{\text{s}^2}\right)\text{ft}$$

$$= \text{lb}\cdot\text{ft} \qquad \text{in U.S. customary units}$$

The relationship of work and kinetic energy as expressed by Eq. 14-12 can be written concisely as

$$\boxed{U_{12} = T_2 - T_1} \qquad \text{14-14}$$

where U_{12} is the work of the force on the particle as it moves from position 1 to 2, and T_1 and T_2 are the corresponding initial and final kinetic energies of the particle. Equation 14-14 is called the *principle of work and energy*. Since work may be positive or negative, a force may increase or decrease the kinetic energy of a particle as outlined in the following.

$$\textit{Positive work:}\quad T_2 = T_1 + U_{12}, \quad T_2 > T_1 \qquad \text{14-15}$$

In words, positive work always increases the kinetic energy.

Negative work: $T_2 = T_1 + U_{12}, \quad 0 \leq T_2 < T_1$ **14-16**

where it is noted that kinetic energy is always positive.

14-4 POWER AND EFFICIENCY

In the design of machines it is often necessary to consider how much work must be done by the whole machine in a given time. The *time rate* at which work is done is defined as *power*:

$$\text{power} = \frac{dU}{dt} \quad \text{14-17}$$

This can be manipulated to obtain an expression in terms of force and velocity which are relatively easily measured or known in advance in the design. Since $dU = \mathbf{F} \cdot d\mathbf{r}$ and $d\mathbf{r}/dt = \mathbf{v}$,

$$\text{power} = \frac{dU}{dt} = \frac{\mathbf{F} \cdot d\mathbf{r}}{dt} = \mathbf{F} \cdot \mathbf{v} \quad \text{14-18}$$

This gives a scalar quantity since both work and time are scalars. The units of power are the watt (W) in SI units, and ft·lb/s or *horsepower* (hp) in U.S. customary units. The details and relations of these units are as follows.

$$1 \text{ W} = 1 \text{ J/s} = 1 \text{ N}\cdot\text{m/s}$$

$$1 \text{ hp} = 550 \text{ ft}\cdot\text{lb/s} = 33{,}000 \text{ ft}\cdot\text{lb/min}$$

$$1 \text{ ft}\cdot\text{lb/s} = 1.356 \text{ J/s} = 1.356 \text{ W}$$

$$1 \text{ hp} = 550(1.356 \text{ W}) = 746 \text{ W} = 0.746 \text{ kW}$$

Mechanical Efficiency

Real mechanical systems always operate with frictional losses, so the useful work done by a machine is always less than the total work done by that machine. The *mechanical efficiency* η is defined as

$$\eta = \frac{\text{output work}}{\text{input work}} = \frac{\text{useful work}}{\text{total work required}} \quad \text{14-19}$$

Since both the output work and the input work are done in the same time interval, the efficiency can be defined in terms of power,

$$\eta = \frac{\text{power output}}{\text{power input}} = \frac{\text{useful power}}{\text{total power required}} \quad \text{14-20}$$

The mechanical efficiency of an ideal machine is $\eta = 1$, and of any real machine $\eta < 1$.

Overall Efficiency

Equation 14-20 is also valid when a system transforms energy from one form to another. The meaning of overall efficiency is that all losses of energy (electric, frictional, thermal, etc.) can be expressed in one consistent system of units and treated as loss in general. Of course, energy is conserved in a physical sense, and loss is simply the energy that must be expended in the process of doing useful work but not available for useful work.

14-5 OPPORTUNITIES AND LIMITATIONS OF USING ENERGY METHODS

The principle of work and energy (Eq. 14-14) is extremely useful in solving certain problems. For example, consider the initial steps in the design of the amusement ride shown in Fig. 14-8a. The train with passengers is to start from rest at point O. The train should accelerate over a short distance OA and make an inside loop at a height h with a radius of curvature r_B. The speed v_B at the top of the loop is specified. The known friction force $\mathscr{F}$ is constant along the track of given length s. It is required to determine the work U_{OA} of the starter machine acting on the motorless train so that it can roll from point A to B and have a speed v_B at B. This requirement leads to the following major steps in the analysis of an arbitrary mass m traveling from point O to B:

1. The required work of the starter machine is $U_{OA} = T_A - T_O$ from Eq. 14-14.
2. The kinetic energy at A is $T_A = \frac{1}{2}mv_A^2$ from Eq. 14-13. $T_O = 0$.
3. The kinetic energy at A should be sufficient for the train to rise to B and have kinetic energy $T_B = \frac{1}{2}mv_B^2$ at B.
4. The kinetic energies T_A and T_B are related according to Eq. 14-14, $U_{AB} = T_B - T_A$ where U_{AB} is the work done on the train from A to B.

FIGURE 14-8

Illustration for using the principle of work and energy

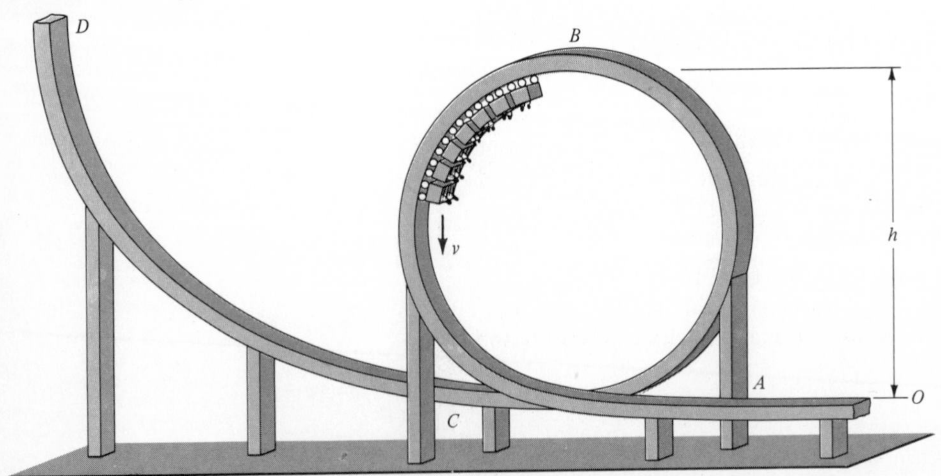

5. The work U_{AB} involves the work of gravity and the work of friction from A to B, so $U_{AB} = -mgh - \int_A^B \mathscr{F}\, ds$, where ds is measured along the path of the train.
6. Thus, the required work of the starter machine on the train is expressed in terms of the given quantities as

$$U_{OA} = T_A = T_B - U_{AB} = \frac{1}{2}mv_B^2 + mgh + \mathscr{F}s_{AB}$$

This schematic example shows the following advantages and limitations of using energy methods.

Advantages

1. The design can be made without determining any accelerations of the train.
2. Additional parts of the path are easy to include in the analysis. For example, making the train to continue from B to D and back all the way to O requires a simple modification of the procedure described above.
3. Only a summation of scalar quantities is necessary even if the path of the motion is complex.
4. Forces that do no work need not be considered. For example, the force of wind perpendicular to the vertical plane of the path O to D can do no work on the train and is ignored.

Limitations

1. The acceleration cannot be calculated directly from the quantities of work and energy. This may be necessary for other aspects of the problem besides getting a rudimentary solution. For example, the maximum initial (O to A) and final (A to O) accelerations of the train in Fig. 14-8 are important with respect to the comfort and safety of the passengers. The accelerations can be calculated but Newton's second law must be used directly to do this.*
2. Forces that do no work cannot be calculated from the quantities of work and energy. For example, the normal forces between the train and its track anywhere from O to D can be calculated only by using Newton's second law.*

Overall, it is worth noting that the limitations of using energy methods are negligible since Newton's second law is readily included in the analysis if necessary. The substantial advantages of these methods and their limitations are illustrated with the following examples.

* It should be remembered that the work–energy method involves no new principles, but is a reformulation of Newton's second law for convenience in solving certain problems.

EXAMPLE 14-1

The track of a ski-jump tower is in the shape of a circular arc as shown in Fig. 14-9a. The jumpers start at 1 and become airborne at 2. Calculate the takeoff speed and compare the solution with that in Ex. 12-6. Assume that the coefficient of friction is zero.

SOLUTION

From Eq. 14-14, $T_1 + U_{12} = T_2$. Since the skier starts from rest, $T_1 = \frac{1}{2}mv_1^2 = 0$. The work done on the skier is that done by gravity, $U_{12} = Wh = mgh$. This is positive according to the free-body diagram in Fig. 14-9b. The final kinetic energy is $T_2 = \frac{1}{2}mv_2^2$, and

$$0 + mgh = \frac{1}{2}mv_2^2$$

$$\begin{aligned} v_2 &= \sqrt{2gh} = \sqrt{2g(r - r\cos 40^\circ)} \\ &= \sqrt{2(9.81\ \text{m/s}^2)(60\ \text{m} - 60\ \text{m}\cos 40^\circ)} \\ &= 16.6\ \text{m/s} \end{aligned}$$

which agrees with the previous result. Note that Eq. 14-14 allowed a much faster solution.

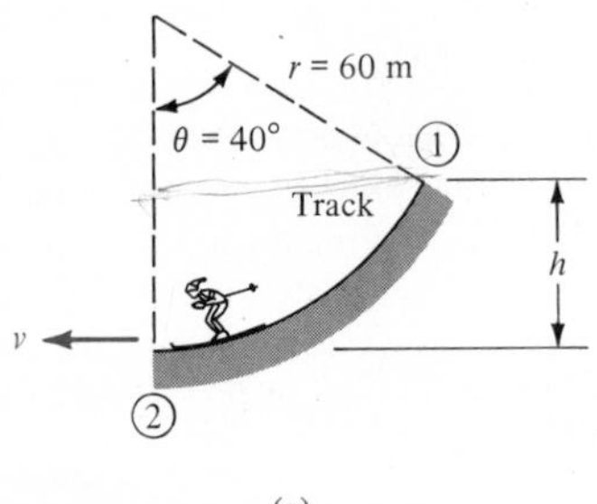

(a)

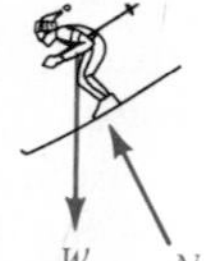

(b) Free-body diagram

FIGURE 14-9

Analysis of a ski-jumper's motion

EXAMPLE 14-2

Redo Ex. 13-6 using energy methods. To repeat the problem, a car of 1000 kg mass is traveling at $v_0 = 100$ km/h when its brakes are applied. Calculate the stopping distance if $\mu = 0.5$ and compare the solution with that of Ex. 13-6.

SOLUTION

From Eq. 14-14, $T_1 + U_{12} = T_2$. For this case,

$$T_1 = \frac{1}{2}mv_1^2 = \frac{1}{2}m_{\text{car}}\left[\left(\frac{100\text{ km}}{\text{h}}\right)\left(\frac{1000\text{ m}}{\text{km}}\right)\left(\frac{1\text{ h}}{3600\text{ s}}\right)\right]^2$$

$$T_1 = (385.8\text{ m}^2/\text{s}^2)m_{\text{car}}$$

$$U_{12} = -(\mathscr{F})\,\Delta x = -\mu N_{\text{car}}\,\Delta x = -\mu W_{\text{car}}\,\Delta x$$

$$= -(0.5)(m_{\text{car}}g)\,\Delta x = -(0.5)(9.81\text{ m/s}^2)m_{\text{car}}\,\Delta x$$

$$= -\left(4.9\,\frac{\text{m}}{\text{s}^2}\right)m_{\text{car}}\,\Delta x$$

$$T_2 = \frac{1}{2}mv_2^2 = \frac{1}{2}m_{\text{car}}(0)^2 = 0$$

Hence,

$$(385.8\text{ m}^2/\text{s}^2)m_{\text{car}} - (4.9\text{ m/s}^2)m_{\text{car}}\,\Delta x = 0$$

Dividing through by m_{car},

$$\Delta x = \frac{385.8\text{ m}^2/\text{s}^2}{4.9\text{ m/s}^2} = 78.7\text{ m}$$

which agrees with the previous result. Note that the work done by friction on the car was negative because it acted opposite to the direction of displacement of the car.

EXAMPLE 14-3

The starter spring of a pinball machine is modeled in Fig. 14-10a. The mass of the ball is $m = 0.1$ kg, the spring stiffness is $k = 200$ N/m, and the maximum deformation is $d = 10$ cm. Determine the speed of the ball as it leaves the spring at B. Neglect friction, and assume that the ball is a particle (which does not rotate).

SOLUTION

From Eq. 14-14, $T_B = T_A + U_{AB}$. At maximum deformation of the spring, $v_A = T_A = 0$. The total work done on the ball during its motion from A to B is expressed using the free-body diagram in Fig. 14-10b,

$$U_{AB} = \frac{1}{2}kd^2 - mgh$$

$$= \frac{1}{2}(200 \text{ N/m})(0.1 \text{ m})^2 - (0.1 \text{ kg})(9.81 \text{ m/s}^2)(0.1 \text{ m} \sin 5°)$$

$$= 0.991 \text{ N} \cdot \text{m}$$

The speed v_B is obtained from

$$\frac{1}{2}mv_B^2 = U_{AB}$$

$$v_B = \sqrt{\frac{2(0.991 \text{ N} \cdot \text{m})}{0.1 \text{ kg}}} = 4.45 \text{ m/s}$$

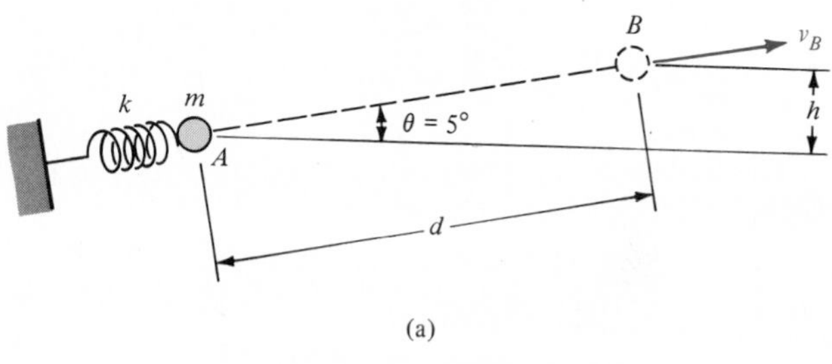

(a)

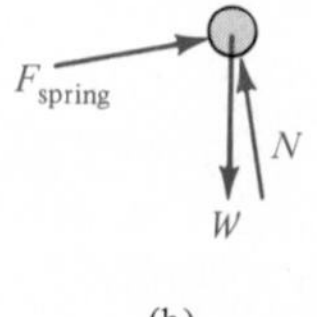

(b)

FIGURE 14-10

Starter mechanism of a pinball machine

EXAMPLE 14-4

The arresting mechanism of an aircraft carrier is modeled in Fig. 14-11a. The tensioning devices in the mechanisms at A and B (which let out the cable) apply a constant tension T to the cable throughout the braking process. Consider an airplane of weight $W = 15{,}000$ lb with a landing speed of $v_0 =$ 180 ft/s. What must the tension in the cable be if the airplane is to stop after rolling 700 ft horizontally?

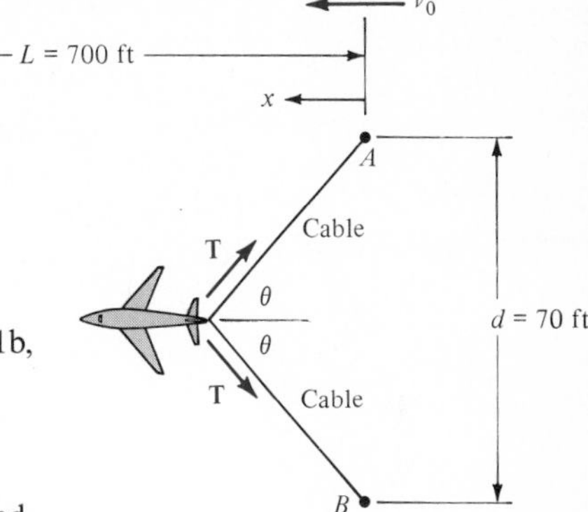

SOLUTION

From Eq. 14-14, $T_1 + U_{12} = T_2$. Using the free-body diagram in Fig. 14-11b,

$$\frac{1}{2}\left(\frac{W}{g}\right)v_0^2 - 2\int_0^{700\text{ ft}} T\cos\theta\,dx = 0$$

Since $\cos\theta$ varies with x, a relationship for it in terms of x must be found. From Fig. 14-11c,

$$\cos\theta = \frac{x}{(x^2 + 35^2)^{1/2}}$$

Therefore,

$$\frac{1}{2}\left(\frac{15{,}000\text{ lb}}{32.2\text{ ft/s}^2}\right)(180\text{ ft/s})^2 - 2\int_0^{700\text{ ft}} T\left[\frac{x}{(x^2 + 35^2)^{1/2}}\right]dx$$

Evaluating the integral yields, since T is constant,

$$7.55 \times 10^6\text{ lb}\cdot\text{ft} - 2T(x^2 + 35^2)^{1/2}\Big|_0^{700\text{ ft}} = 0$$

$$7.55 \times 10^6\text{ lb}\cdot\text{ft} - 2T\{[(700\text{ ft})^2 + (35\text{ ft})^2]^{1/2} - [0^2 + (35\text{ ft})^2]^{1/2}\} = 0$$

$$7.55 \times 10^6\text{ lb}\cdot\text{ft} - 2T(666\text{ ft}) = 0$$

$$T = 5670\text{ lb}$$

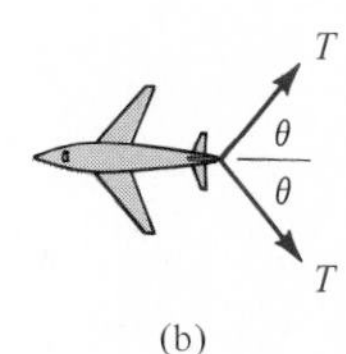

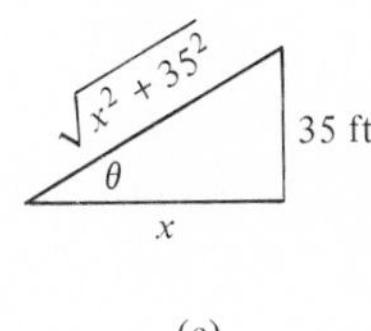

FIGURE 14-11
Arresting of an aircraft by a cable

JUDGMENT OF THE RESULTS

For a check, assume that $\cos\theta = 1$ during the whole motion. This is reasonable since θ is small for most of the displacement (for example, at $x = 200$ ft, $\cos\theta = 0.98$). Thus, assuming that $(x^2 + 35^2)^{1/2} \simeq 700$ ft,

$$7.55 \times 10^6\text{ lb}\cdot\text{ft} - 2T(700\text{ ft}) = 0$$

$$T = 5400\text{ lb}$$

so the more detailed solution appears correct.

EXAMPLE 14-5

Consider the amusement ride shown in Fig. 14-8. Assume that the mass of the train is $m = 3000$ kg, the height h is 25 m, and the minimum speed in the loop is $v_B = 2$ m/s (Fig. 14-12a). Calculate the required takeoff speed v_A and the maximum power input if the train is accelerated from rest at O to v_A at A over a distance of 15 m by a constant force F. (a) Solve the problem ignoring friction. (b) Assume that the system operates with 75% efficiency.

SOLUTION

(a) From Eq. 14-14, $T_A + U_{AB} = T_B$.

$$T_A = \frac{1}{2} m v_A^2 = \frac{1}{2}(3000 \text{ kg})(v_A)^2 = 1500 \text{ kg } v_A^2$$

Using a free-body diagram of the train as it moves up (Fig. 14-12b),

$$\begin{aligned} U_{AB} &= -mgh = -(3000 \text{ kg})(9.81 \text{ m/s}^2)(25 \text{ m}) \\ &= -7.36 \times 10^5 \text{ kg}\cdot\text{m}^2/\text{s}^2 \end{aligned}$$

$$T_B = \frac{1}{2} m v_B^2 = \frac{1}{2}(3000 \text{ kg})(2 \text{ m/s})^2 = 6000 \text{ kg}\cdot\text{m}^2/\text{s}^2$$

Hence,

$$1500 \text{ kg } v_A^2 - 7.36 \times 10^5 \text{ kg}\cdot\text{m}^2/\text{s}^2 = 6000 \text{ kg}\cdot\text{m}^2/\text{s}^2$$

$$v_A^2 = \frac{742\,000 \text{ kg}\cdot\text{m}^2/\text{s}^2}{1500 \text{ kg}} = 494.7 \text{ m}^2/\text{s}^2$$

$$v_A = 22.2 \text{ m/s}$$

To calculate the maximum power input Eq. 14-18 is used. Since F is constant,

$$P_{\max} = F v_{\max}$$

$v_{\max}$ occurs at point A, so the maximum power input must be at A. To find F, the principle of work and energy is used again,

$$T_O + U_{OA} = T_A$$

$$0 + Fd = \frac{1}{2} mv_A^2$$

$$F(15\text{ m}) = \frac{1}{2}(3000\text{ kg})(22.2\text{ m/s})^2$$

$$= 739 \times 10^3 \frac{\text{kg}\cdot\text{m}^2}{\text{s}^2} = 739\text{ kN}\cdot\text{m}$$

$$F = \frac{739\text{ kN}\cdot\text{m}}{15\text{m}} = 49.3\text{ kN}$$

Hence,

$$P_{\max} = Fv_{\max} = (49.3\text{ kN})(22.2\text{ m/s}) = 1094\text{ kN}\cdot\text{m/s}$$
$$= 1094\text{ kW} = 1466\text{ hp}$$

(b) From Eq. 14-20,

$$\text{total power required} = \frac{\text{useful power}}{\eta} = \frac{1094\text{ kW}}{0.75}$$
$$= 1459\text{ kW} = 1956\text{ hp}$$

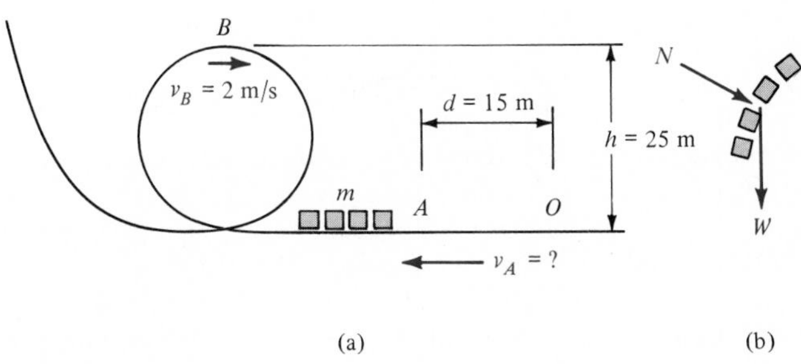

FIGURE 14-12
Analysis of an amusement ride

EXAMPLE 14-6

Consider the hay baler shown in Fig. 14-1 and assume that the machine and the trailer are stationary. Each bale weighs 30 lb and must travel a maximum of $h = 10$ ft vertically and $d = 15$ ft horizontally from the launching arm (Fig. 14-13). This arm starts from rest and moves over a maximum range of $\Delta y = 1.5$ ft vertically and $\Delta x = 1$ ft horizontally. Calculate the required launching velocity $\mathbf{v}$ and the maximum power input P if the system operates with a $\eta = 60\%$ efficiency, and air resistance is negligible.

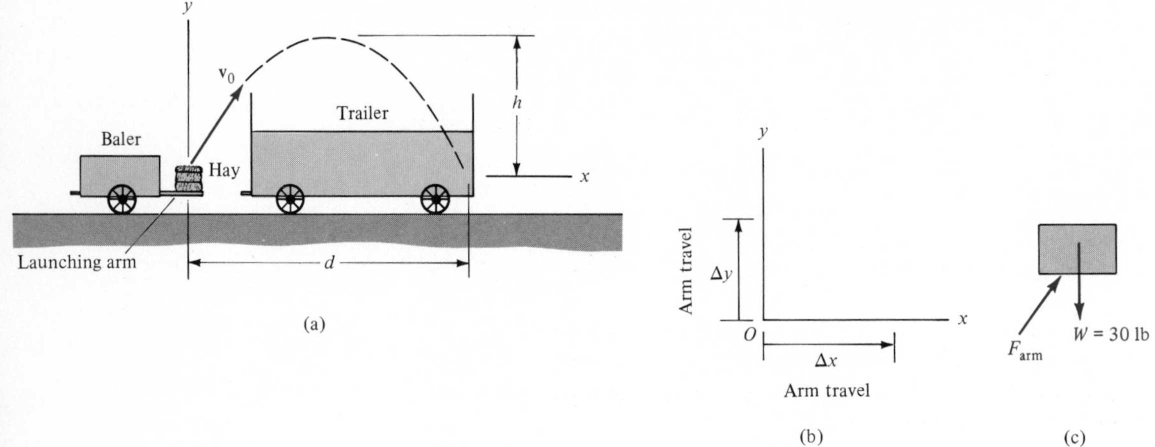

FIGURE 14-13

Analysis of the launching and flight of bales of hay

SOLUTION

The bale's aerial velocity in the x direction is assumed constant since there are no retarding forces in that direction. Gravity is the only force doing work on the bale in the air. In the analysis it is necessary to consider the launching phase and the flight phase of the motion separately.

LAUNCHING PHASE: From Eq. 14-14, $T_1 + U_{12} = T_2$, where 2 denotes the end of the launching phase. The force F_{arm} in the free-body diagram of Fig. 14-13c has components F_x and F_y. Thus,

$$0 + F_x\,\Delta x = \frac{1}{2}mv_{x_2}^2 \qquad \text{and} \qquad 0 + F_y\,\Delta y = \frac{1}{2}mv_{y_2}^2 \tag{a}$$

This cannot be solved yet since F_x, F_y, v_x, and v_y are all unknown.

FLIGHT PHASE: From Eq. 14-14, $T_2 + U_{23} = T_3$, where 3 denotes the peak height of the flight.

$$\frac{1}{2}m(v_x^2 + v_y^2)_2 - Wh = \frac{1}{2}mv_{x_3}^2$$

Note that $v_{y_3} = 0$. Also, $v_{x_2} = v_{x_3}$, so $\frac{1}{2}mv_{x_2}^2$ may be subtracted from both sides, yielding

$$\frac{1}{2}mv_{y_2}^2 - mgh = 0$$

$$v_{y_2} = \sqrt{2gh} = \sqrt{2(32.2 \text{ ft/s}^2)(10 \text{ ft})} = 25.4 \text{ ft/s}$$

Next, the time of flight t to the peak of the trajectory is obtained from Eq. 11-8.

$$v_y = v_{y0} - gt$$

$$0 = 25.4 \text{ ft/s} - (32.2 \text{ ft/s}^2)t$$

$$t = 0.789 \text{ s}$$

In this time the bale moves 7.50 ft in the x direction. Thus, v_x can be calculated from Eq. 11-10,

$$x = x_0 + v_{0_x}t$$

$$15 \text{ ft} = 0 + v_{0_x}(0.789 \text{ s})(2)$$

$$v_{0_x} = v_x = 9.51 \text{ ft/s}$$

$$\mathbf{v} = (9.51\mathbf{i} + 25.4\mathbf{j}) \text{ ft/s at takeoff}$$

LAUNCHING PHASE: The equations of work and energy for the x and y directions (Eq. a) can now be solved.

$$F_x \, \Delta x = \frac{1}{2}mv_{x_2}^2$$

$$F_x(1 \text{ ft}) = \frac{1}{2}\left(\frac{30 \text{ lb}}{32.2 \text{ ft/s}^2}\right)(9.51 \text{ ft/s})^2 = 42.1 \text{ lb}\cdot\text{ft}$$

$$F_x = 42.1 \text{ lb}$$

$$F_y \, \Delta y = \frac{1}{2}mv_{y_2}^2$$

$$F_y(1.5 \text{ ft}) = \frac{1}{2}\left(\frac{30 \text{ lb}}{32.2 \text{ ft/s}^2}\right)(25.4 \text{ ft/s})^2 = 300 \text{ lb}\cdot\text{ft}$$

$$F_y = 200 \text{ lb}$$

The maximum power is required when the speed is maximum (at the end of the launching phase). From Eq. 14-18,

$$\begin{aligned} P_{\text{max}} &= F_{\text{arm}}v_{\text{max}} = (F_x^2 + F_y^2)^{1/2}(v_x^2 + v_y^2)^{1/2} \\ &= (42.1^2 + 200^2)^{1/2} \text{ lb}(9.51^2 + 25.4^2)^{1/2} \text{ ft/s} \\ &= 5543 \text{ lb}\cdot\text{ft/s} \end{aligned}$$

From Eq. 14-20,

$$P_{\text{required}} = \frac{P_{\text{useful}}}{\eta} = \frac{5543}{0.6}\frac{\text{lb}\cdot\text{ft}}{\text{s}} = 16.8 \text{ hp}$$

PROBLEMS

SECTIONS 14-1, 14-2, 14-3, 14-4, 14-5

14–1 A particle of mass m freely falls from rest at a height $h = 20$ m. Calculate the speed v_f of the particle just before it hits the ground.

FIGURE P14-3

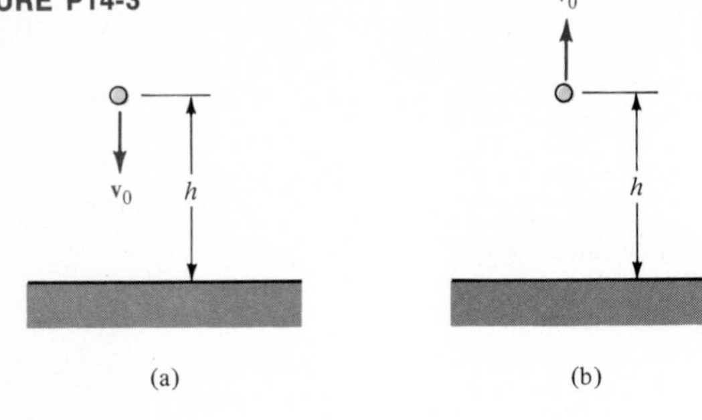

14–2 A 3-lb particle is dropped with no initial velocity from a height h. Its final speed just before impact is $v_f = 40$ ft/s. Determine h.

14–3 A ball of mass m is released at height h with an initial velocity $\mathbf{v}_0$. Determine the final velocity $\mathbf{v}_f$ at the ground for (a) the case when $\mathbf{v}_0$ is downward, and (b) the case when $\mathbf{v}_0$ is upward.

FIGURE P14-4

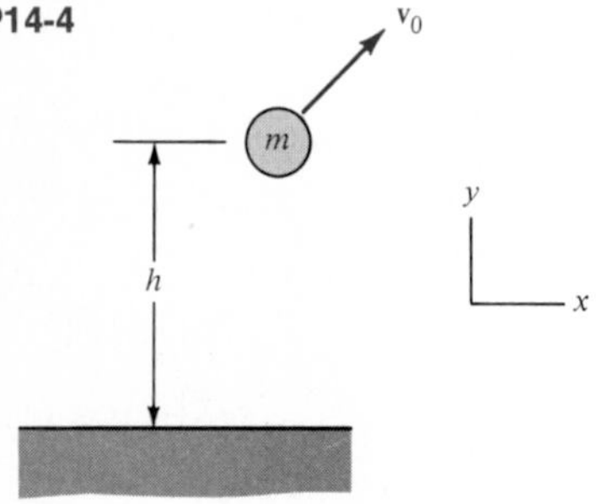

14–4 A 0.5-kg particle is projected at height $h = 40$ m with an initial velocity $\mathbf{v}_0 = (20\mathbf{i} + 15\mathbf{j})$ m/s. Determine its final velocity $\mathbf{v}_f$ just before impact at the ground.

FIGURE P14-5

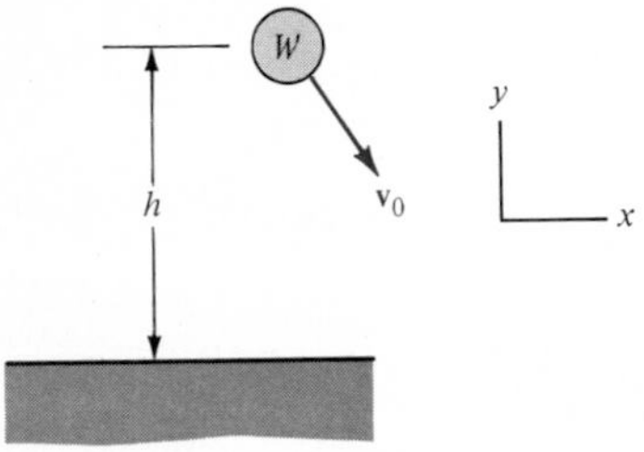

14–5 A 2-lb particle is released at height $h = 50$ ft with an initial velocity $\mathbf{v}_0 = (30\mathbf{i} - 10\mathbf{j})$ ft/s. Determine its final velocity $\mathbf{v}_f$ just before impact at the ground.

FIGURE P14-6

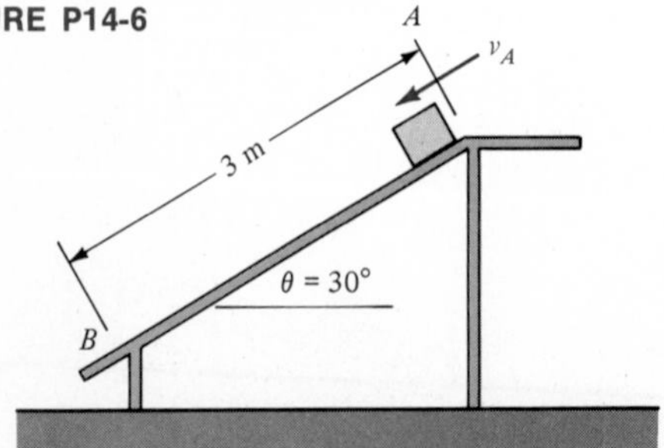

14–6 A 10-kg package has a speed $v_A = 8$ m/s at point A of the sloping board. Determine its speed at point B if the coefficient of kinetic friction is 0.2.

14–7 Using Fig. P14-6, determine the angle θ of the board that makes a 10-kg package fall off the board at point B if $v_A = 8$ m/s and $\mu_k = 0.6$.

14–8 A 20-lb box is pushed along the incline by a constant force F. The box starts from rest at point A and must have a speed of $v_B = 5$ ft/s after traveling 10 ft to point B. Determine the force F neglecting friction.

FIGURE P14-8

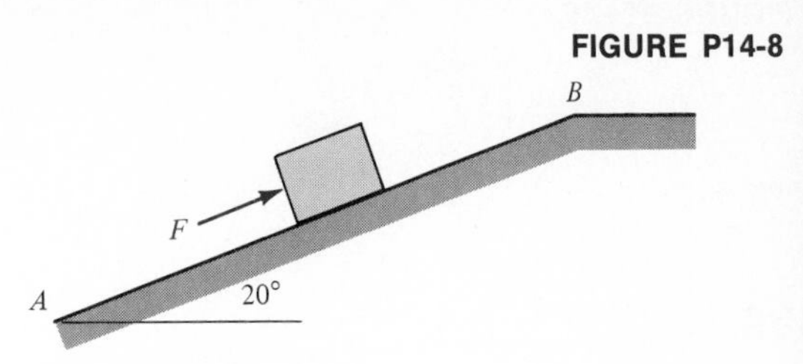

14–9 Determine the required force F in Prob. 14–8 assuming a coefficient of kinetic friction of 0.4.

14–10 A 5-kg block is sliding on a horizontal surface on which the coefficient of kinetic friction is 0.3. Determine the required initial speed v_0 to stop the block by friction after a travel of 4 m.

14–11 A ski slope has a straight section on which the coefficient of kinetic friction is 0.1. Where should the starting point A be located with respect to B if a 150-lb skier's speed at B should not exceed 30 ft/s? Assume that the poles are not used.

FIGURE P14-11

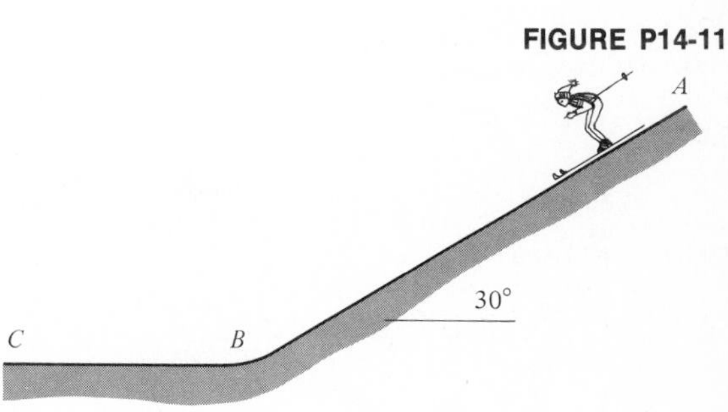

14–12 In Prob. 14–11, the path of the skiers from B to C is horizontal. Determine the maximum distance BC that a skier could glide without using the poles.

14–13 Skiers start from rest at point A, which has an elevation of 200 m above point B. The total path length is 800 m, and the net friction force is 10 N over the path. Determine the speed v_B of a 70-kg skier who is not using the poles for propulsion.

FIGURE P14-13

14–14 Determine the stopping distance of a 3000-lb car traveling up the incline on which the coefficient of kinetic friction is 0.5. The initial speed is $v_0 = 50$ mph when the brakes are applied locking the wheels.

FIGURE P14-14

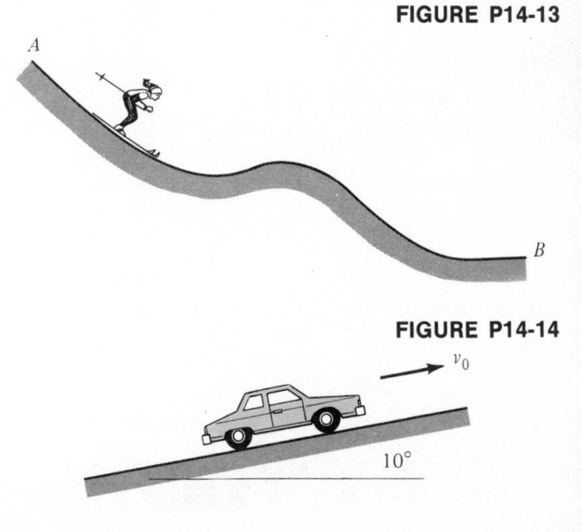

14–15 A 1000-kg car is traveling downhill at a speed of $v_0 = 20$ m/s when its brakes are applied locking the wheels. The coefficient of kinetic friction on the road is 0.3. Plot the stopping distance s as a function of $\theta = 5°$, 10°, and 15°. Estimate from the diagram the smallest incline on which the car would not stop by using the brakes.

FIGURE P14-15

14–16 A motor applies a constant, 200-lb force after the cart starts from rest. Determine the velocity of the cart after it has moved 30 ft up the incline. Also calculate the maximum power required for this motion.

FIGURE P14-16

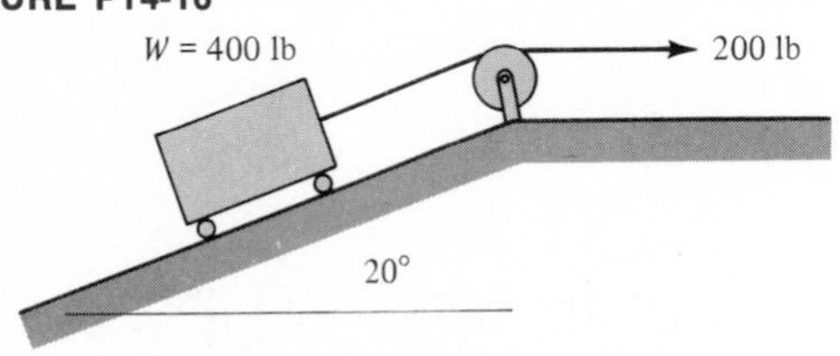

14–17 A bicycle racer and his bicycle have a total mass of 100 kg. He can develop 0.2 hp of power in a sustained effort. Determine his speed up the incline neglecting frictional losses.

FIGURE P14-17

14–18 A power winch is pulling a 100-lb crate up on the incline, where the coefficient of kinetic friction is 0.4. The crate starts from rest at A and has a speed of 5 ft/s at B. Calculate the required maximum power of the winch assuming a constant tension in the rope.

FIGURE P14-18

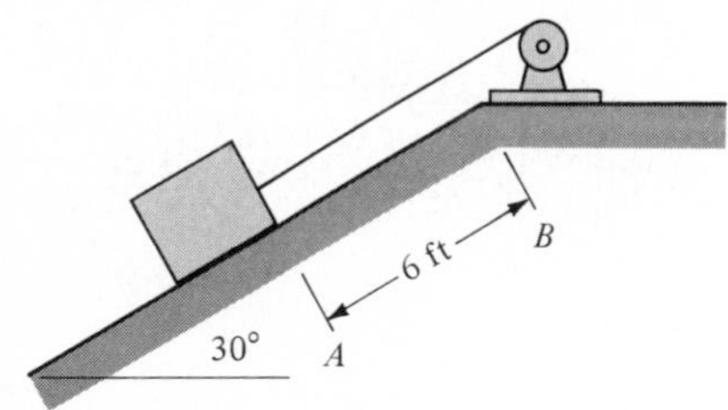

14–19 A 2-kg mass is suspended on a rigid rod of negligible mass and length $l = 50$ cm. The rod has a smooth hinge at O. The pendulum is released from rest at $\theta = 0$. Determine the speed of the mass at $\theta = 90°$ and 135° neglecting friction.

FIGURE P14-19

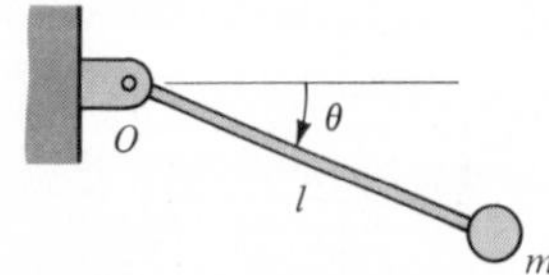

14–20 An amusement ride is modeled as a small block of weight W mounted on a weightless, rigid rod of length l. The rod rotates in a vertical plane about a frictionless hinge at O. The arm starts from rest at $\theta = 0$ under the action of a constant moment $M_O = 2Wl$. Determine the speed of the block at $\theta = 45°$ and 90°.

FIGURE P14-20

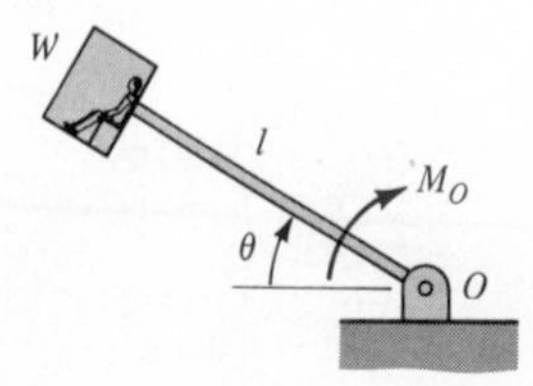

14–21 Calculate the maximum power required of the motor that drives the amusement ride described in Prob. 14–20. Consider the range of angles $0 \leq \theta \leq 90°$.

14–22 A 5000-lb elevator is being raised by a motor at a speed $v = 5$ ft/s. Determine the required power of the motor.

14–23 A 100-kg crate is to be raised by an electric motor (Fig. P14-22) which is rated at 60% efficiency and requires an electrical input of 1 kW. Calculate the constant speed v of the crate moving upward.

FIGURE P14-22

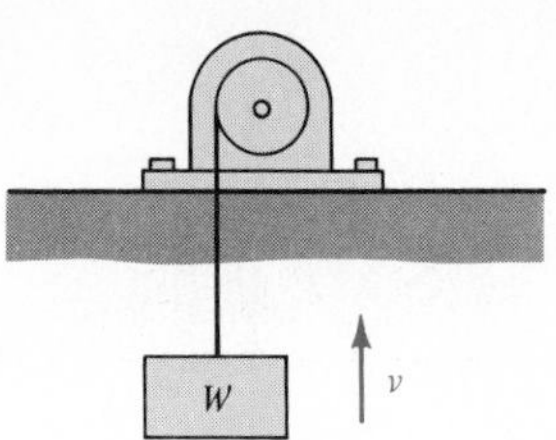

14–24 A machine partially shown in the photograph is designed to test electronic equipment E of mass $m = 10$ kg. The stroke (maximum travel) of the constant-force actuator A is $\Delta y = 5$ cm. Calculate the required maximum power to impart an upward velocity of 10 m/s to the equipment E.

FIGURE P14-24
Dynamic testing of a portable electronic instrument (Courtesy MTS Systems Corporation, Minneapolis, Minnesota)

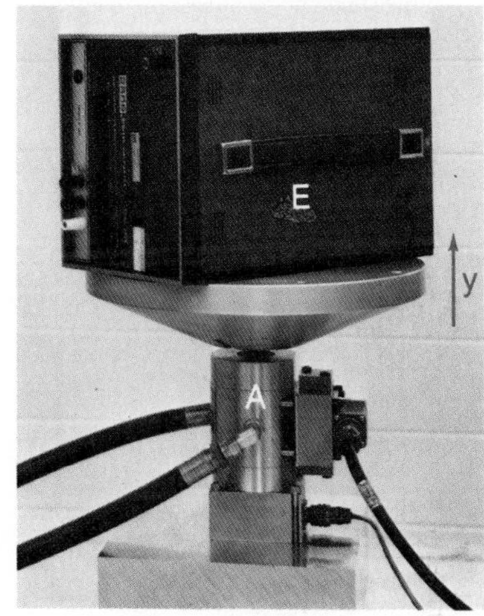

14–25 A 5-lb block is suspended from a spring which has a spring constant $k = 0.5$ lb/in. When not stretched, the spring has length $l_0 = 20$ in. The block is released from rest at $y = 0$ (the spring is not stretched initially). Determine (a) the speed of the spring at $y = 5$ in., and (b) the maximum displacement of the block.

FIGURE P14-25

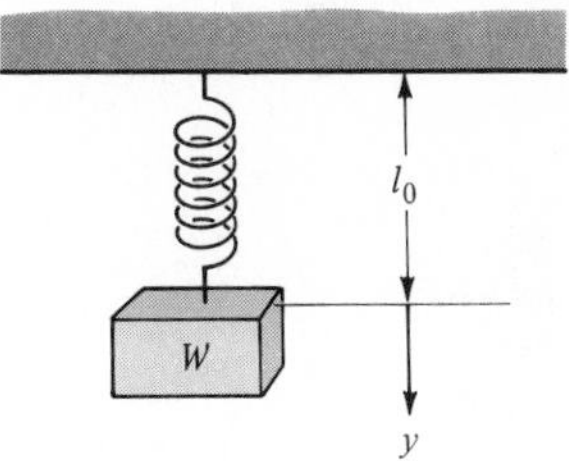

14–26 The block in Prob. 14–25 is displaced upward and released from rest with the spring at half its original length. Determine the lowest position of the block.

14–27 A 1200-kg car has a light bumper supported horizontally by two springs of stiffness $k =$ 15 kN/m. Determine the initial speed v of impact with a fixed wall that causes a 20-cm compression of the springs.

FIGURE P14-27

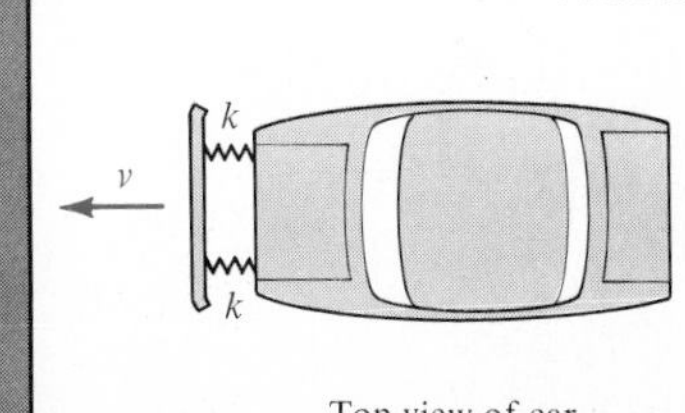

Top view of car

FIGURE P14-28

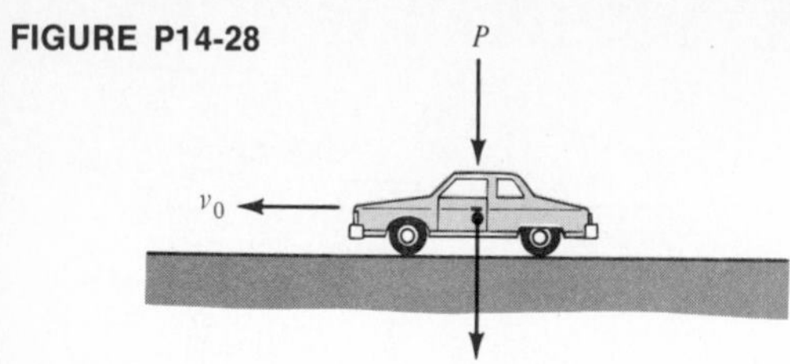

14–28 A 3000-lb experimental car can apply a downward thrust P in an emergency braking situation. Calculate the stopping distance d on a level road if the initial speed is $v_0 =$ 60 mph, $P = 500$ lb, and $\mu_k = 0.5$.

14–29 An interesting question concerning the vehicle in Prob. 14–28 is the direction of the thrust P for the shortest stopping distance. To obtain some insight, solve Prob. 14–28 as stated, and also with P acting not vertically but horizontally to oppose the motion directly.

FIGURE P14-30

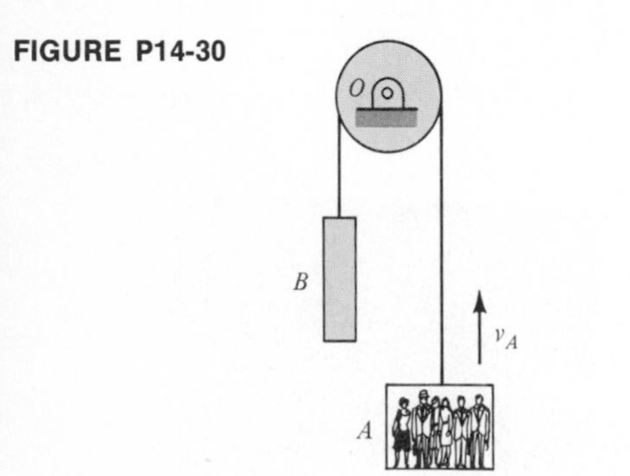

14–30 The elevator cab has a mass $m_A = 2000$ kg and the counterweight has a mass $m_B =$ 1500 kg. The system is moved by an electric motor at point O. Determine the required power of the motor to maintain a velocity $v_A = 2$ m/s upward.

FIGURE P14-31

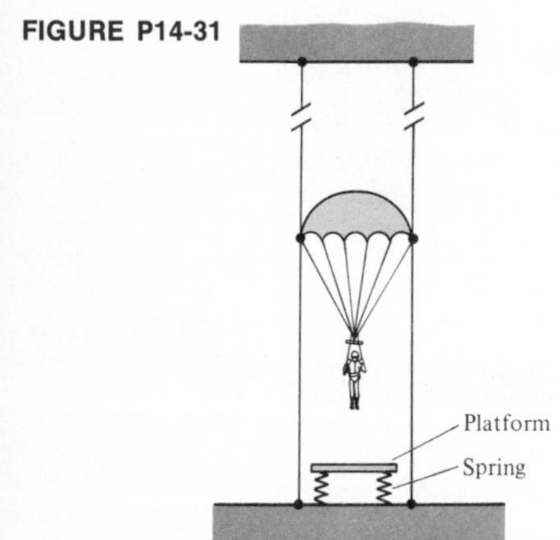

14–31 At an amusement park people may jump by parachutes which are guided to the ground by vertical wires. The jumpers land on a spring-supported platform at a speed of 6 ft/s. The effective spring constant (as if one spring supported each platform) is $k = 30$ lb/in. How much space should be allowed for the vertical displacement of each platform assuming that a 250-lb person may land on it?

FIGURE P14-32

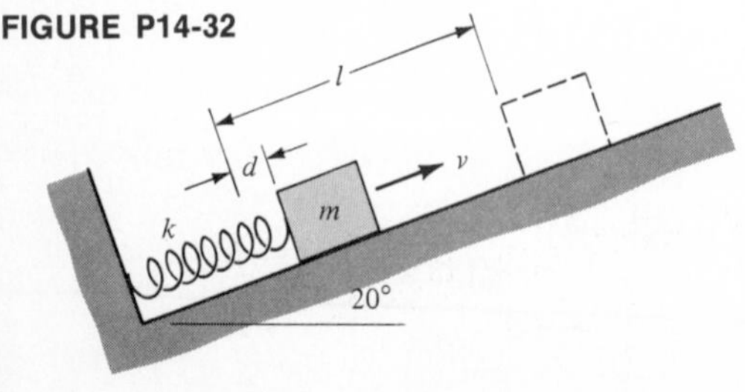

14–32 A spring of stiffness $k = 2$ kN/m is compressed through a distance $d = 5$ cm by a mechanism. The spring propels a 1-kg block along the incline when the mechanism is released. Determine the maximum distance l traveled by the block if friction is negligible. The block and the spring are not attached.

14–33 Determine the coefficient of kinetic friction in Prob. 14–32 if the block stops after traveling $l = 40$ cm.

14–34 The device shown is for measuring coefficients of friction. The fixture is slowly rotated counterclockwise until the 4-lb block starts sliding at $\theta = 10°$. Determine the coefficients of static and kinetic friction if the spring of stiffness $k = 5$ lb/in. deflects a maximum of 3 in.

FIGURE P14-34

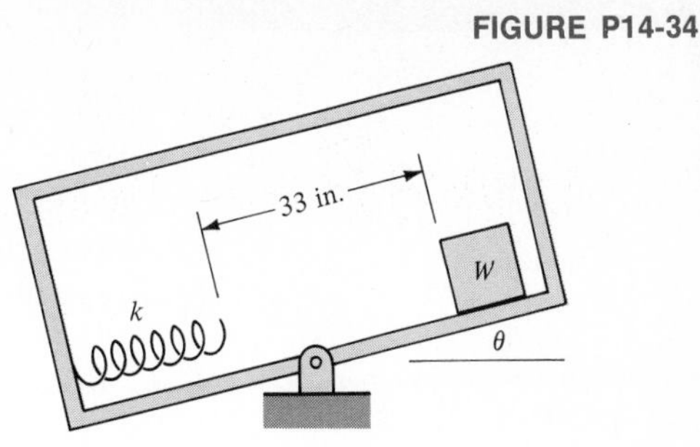

14–35 The bumper system of a tuned mass damper for a skyscraper (for a description of these, see Sec. 1-7 of the associated statics volume) is modeled as a mass m sliding at speed v_0 just before touching the bumper spring. Determine the rebound speed v_f of the block if the coefficient of kinetic friction is μ_k.

FIGURE P14-35

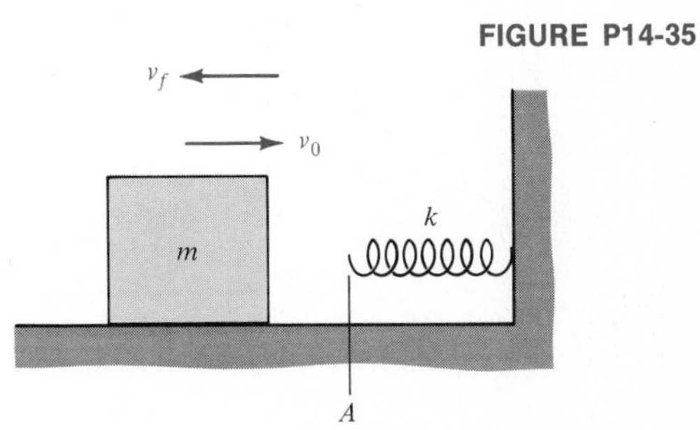

14–36 In Prob. 14–35, $m = 3 \times 10^5$ kg, $\mu_k = 0.05$, and $v_0 = 2$ m/s. Plot the ratio v_f/v_0 as a function of the spring stiffness, using $k =$ 200 kN/m, 400 kN/m, and 600 kN/m.

14–37 A 5000-lb aircraft is launched from a catapult which has a spring of stiffness $k = 100$ lb/in. The spring is compressed through a distance of 4 ft, then it is released to push the aircraft, whose engine applies a constant force $P =$ 1200 lb during the launch. Determine the takeoff speed v of the aircraft at the instant when the spring regains its original length. Neglect friction and assume that the aircraft is propelled by horizontal forces.

FIGURE P14-37

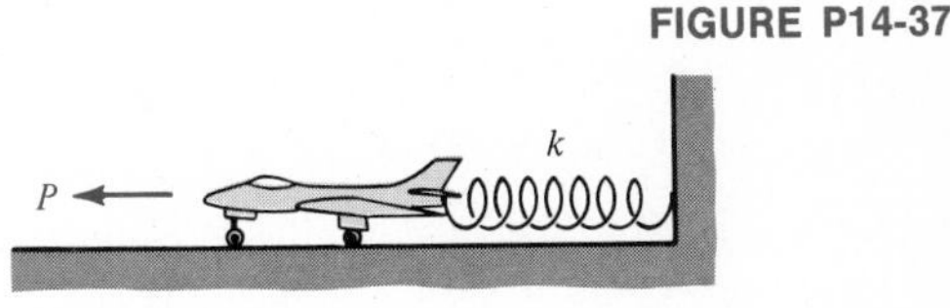

14–38 In Prob. 14–37, the launch angle θ is also important besides the takeoff speed. Plot the speed v as a function of $\theta = 0$, 15°, and 30°, assuming that the two propelling forces on the aircraft are acting parallel to the launching ramp.

FIGURE P14-38

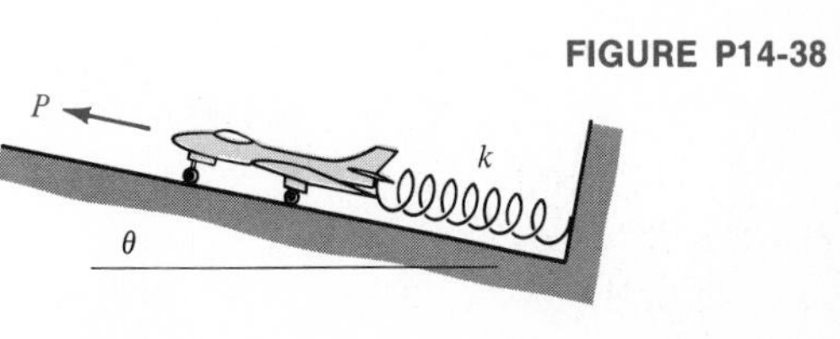

14-6 CONSERVATIVE SYSTEMS. POTENTIAL FUNCTIONS

The work of a force involved in a displacement of a particle between two specified points is independent of the path when the force is conservative (there is no friction opposing the motion). This is illustrated in Fig. 14-14, where the work of gravity is $U_{AB} = -Wh$ regardless of what path is followed if there is no friction.

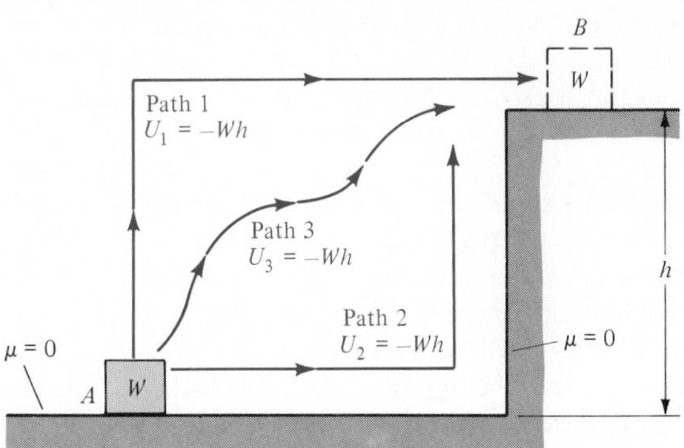

FIGURE 14-14

Work in a conservative system is independent of the path

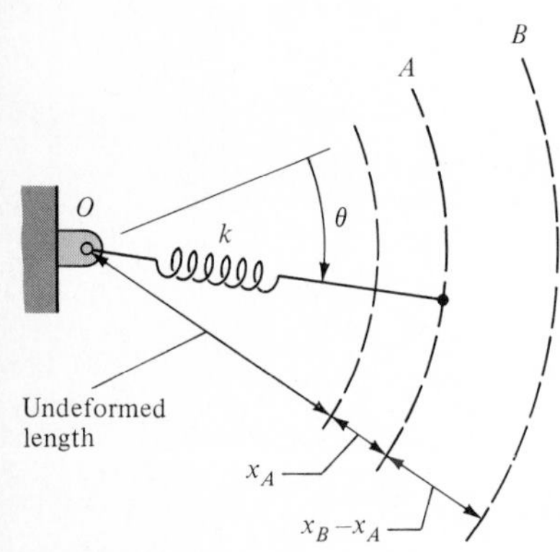

FIGURE 14-15

Conservative system with an elastic spring

The work of the force of an elastic spring is also independent of path for most practical purposes. For example, the work of the spring in the elongation from A to B in Fig. 14-15 is $U_{AB} = -\frac{1}{2}k(x_B^2 - x_A^2)$. This is independent of the orientation θ of the spring.

In general, a conservative system may involve motion in three dimensions. For example, a particle may move from point P_1 to P_2 while a conservative force **F** is acting on it as in Fig. 14-16. The potential energy $V(x, y, z)$ at any point x, y, z along a path is known as the *potential function* of the conservative force **F**. The work U_{12} of the force **F** between arbitrary points P_1 and P_2 is defined in terms of the corresponding potential functions,

$$U_{12} = V_1(x_1, y_1, z_1) - V_2(x_2, y_2, z_2) \qquad \text{14-21}$$

or, simply,

$$U_{12} = V_1 - V_2 \qquad \text{14-22}$$

which agrees in principle with the definition in Eq. 14-11.

An important special case of conservative systems is when the particle moves on a *closed path*, such as leaving from P_1 and finishing at P_1. One closed path starting from any point is called a *cycle* of displacement. In that case $V_1 = V_2$ in Eq. 14-22, and the total work in the cycle is zero. The detailed mathematical expression of this

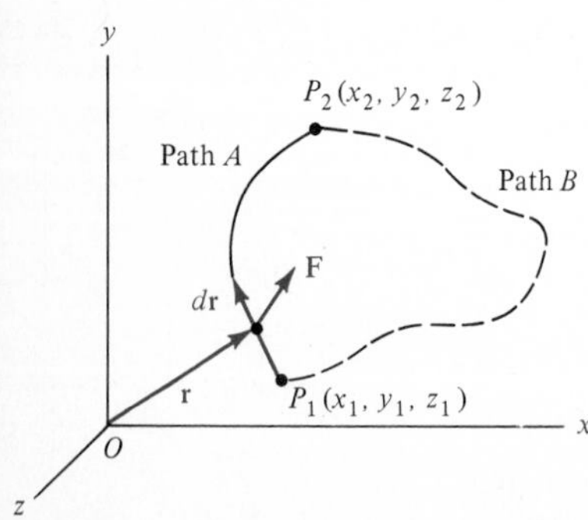

FIGURE 14-16

Analysis of a general conservative system

condition is made by using the line integral $\oint$ for the closed path,

$$U = \oint dU = \oint \mathbf{F} \cdot d\mathbf{r} = \oint (F_x\, dx + F_y\, dy + F_z\, dz) = 0 \qquad \boxed{14\text{-}23}$$

Advanced Analysis

In some cases it is useful to consider infinitesimal variations of the potential function V. Since the difference in potential energies at two positions is equal to the work of the force between these positions,

$$dU = V(x, y, z) - V(x + dx, y + dy, z + dz)$$

$$dU = -V(dx, dy, dz) = -dV(x, y, z) \qquad \boxed{14\text{-}24}$$

where dU is the elementary work of a conservative force* and $-dV$ is the corresponding change in potential energy. The differential change in V can be expressed with partial derivatives as the sum of independent changes along x, y, and z (the total derivative of V),

$$dV = \frac{\partial V}{\partial x} dx + \frac{\partial V}{\partial y} dy + \frac{\partial V}{\partial z} dz$$

From Eqs. 14-23 and 14-24,

$$F_x\, dx + F_y\, dy + F_z\, dz = -\frac{\partial V}{\partial x} dx - \frac{\partial V}{\partial y} dy - \frac{\partial V}{\partial z} dz \qquad \boxed{14\text{-}25}$$

Equating the terms that refer to the same axes,

$$F_x = -\frac{\partial V}{\partial x} \qquad F_y = -\frac{\partial V}{\partial y} \qquad F_z = -\frac{\partial V}{\partial z} \qquad \boxed{14\text{-}26}$$

These expressions can be written more concisely using vector notation. From Eq. 14-25,

$$\mathbf{F} = F_x\mathbf{i} + F_y\mathbf{j} + F_z\mathbf{k} = -\left(\frac{\partial V}{\partial x}\mathbf{i} + \frac{\partial V}{\partial y}\mathbf{j} + \frac{\partial V}{\partial z}\mathbf{k}\right)$$

The vector quantity on the right side is called the *gradient of the scalar function* V and is denoted by **grad** V. Another common notation uses the vector operator ∇ (called del), which is written as

$$\nabla = \frac{\partial}{\partial x}\mathbf{i} + \frac{\partial}{\partial y}\mathbf{j} + \frac{\partial}{\partial z}\mathbf{k}$$

* Equation 14-24 implies that dU is an *exact differential.*

With these two notations, the force and the potential function can be related as

$$\mathbf{F} = -\mathbf{grad}\, V = -\nabla V \qquad \boxed{14\text{-}27}$$

The essential feature of this relation is that a force is conservative when all of its components can be derived from a potential function based on position coordinates alone. This implies that the potential function is independent of the path used in reaching a given position and the potential energy associated with it. Conversely, when the force depends not only on position but frictional resistance (i.e., μ and length of path), a potential function does not exist, and the force is *nonconservative*. A force that is dependent on velocity (such as the drag force in a fluid) is also nonconservative.

14-7 CONSERVATION OF ENERGY

In any conservative system the work of a force on a particle can be converted to kinetic or potential energies of that particle. The resulting possibilities are illustrated using a particle of mass m in Fig. 14-17 and are explained as follows (these concepts are also valid for systems containing elastic members).

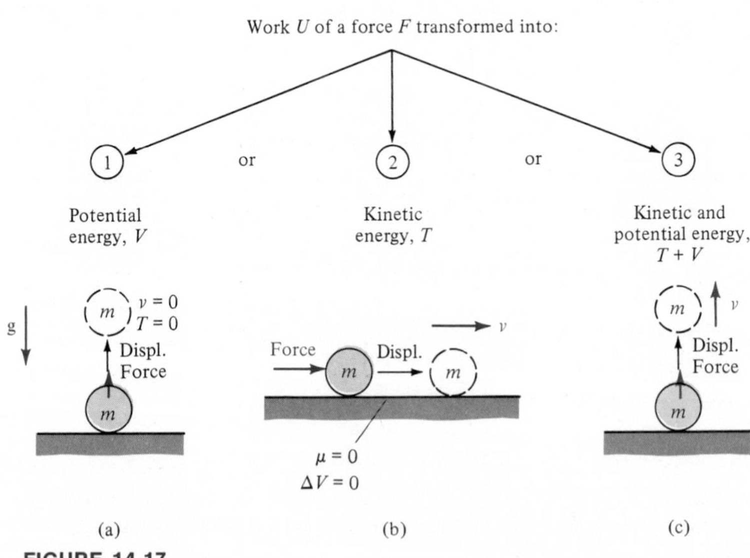

FIGURE 14-17

Transformation of work into potential energy or kinetic energy

1. All of the work of the force is transformed into potential energy at a new elevation where the velocity of the particle is zero (Fig. 14-17a).
2. The motion is in a horizontal plane. All of the work of the force is transformed into kinetic energy while there is no change in potential energy (Fig. 14-17b).
3. Some of the work of the force is transformed into potential energy, the rest of it is transformed into kinetic energy (Fig. 14-17c).

The work of conservative forces can be expressed entirely as a change in potential energy ($U_{12} = V_1 - V_2$ in Eq. 14-22) or as a change in kinetic energy ($U_{12} = T_2 - T_1$ in Eq. 14-14). Therefore, a relationship involving only potential and kinetic energies can be obtained for the general case illustrated in Fig. 14-17c. From Eqs. 14-22 and 14-14,

$$V_1 - V_2 = T_2 - T_1$$

where T and V have the same units. Rearranging yields

$$\boxed{T_1 + V_1 = T_2 + V_2} \qquad \text{14-28}$$

which means that the sum of the kinetic and potential energies is constant in any position of a particle that moves under the action of conservative forces. Equation 14-28 is a statement of the *conservation of mechanical energy*. In general, the *total mechanical energy* E of a particle at a particular position is $E = T + V$. At positions $1, 2, 3, \ldots, n$,

$$E_1 = E_2 = \cdots = E_n = \text{constant} \qquad \text{14-29}$$

In many systems there are other energies besides the two mechanical energies that should be considered. These include chemical, electrical, and thermal energies, among others. In a broad sense, total energy is conserved but its parts are constantly being transformed. Some of these energies can be relatively easily considered in modified versions of Eq. 14-28, but their experimental determination may be quite difficult. For example, velocity and position (T and V) are easily measured while the work of friction forces (dissipated as heat) is very difficult to measure.

Applications of conservation of energy. Equation 14-28 is most commonly used to determine the velocity or elevation of a particle, or the deformation of a spring. The calculations are simplified if reference levels are chosen to make the total energy at a position entirely kinetic energy or entirely potential energy.

EXAMPLE 14-7

Consider the swing shown in Fig. 14-18. Assume that the mass m of the child is 40 kg at the distance $L = 3$ m from point O. Calculate the maximum speed of the child if the maximum angle of the supporting chains with the vertical is $\theta = 70°$. Neglect frictional losses.

(a)

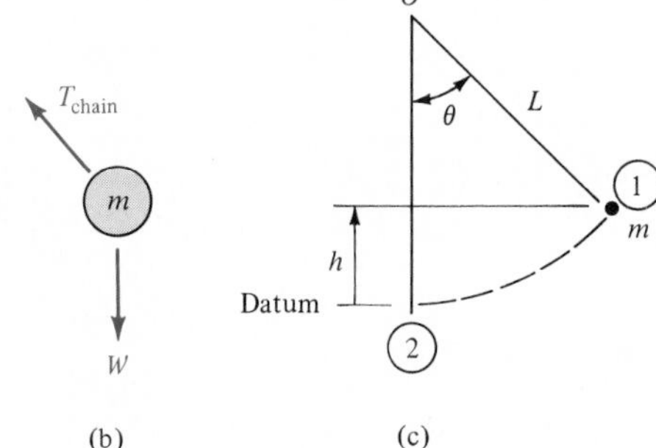

(b) (c)

FIGURE 14-18

Analysis of a swing

SOLUTION

A free-body diagram and a sketch of two extreme positions of the mass are drawn in Fig. 14-18b and c. Since $T + V = E =$ constant, T is a maximum when V is a minimum. A datum for measuring height h from is defined at the lowest position of the swing. Therefore, V_{min} occurs at the datum $h = 0$.

$$T_1 + V_1 = T_2 + V_2$$

$$T_1 = 0 \qquad (v_1 = 0 \text{ at the top of the path by inspection})$$

$$\begin{aligned} V_1 &= mgh_1 = mg(L - L\cos 70°) \\ &= 40\text{ kg}(9.81\text{ m/s}^2)(3\text{ m})(1 - \cos 70°) \\ &= 775\text{ kg}\cdot\text{m}^2/\text{s}^2 \end{aligned}$$

$$T_2 = \frac{1}{2}mv_2^2$$

$$V_2 = mgh_2 = mg(0) = 0$$

Therefore,

$$0 + 775\text{ kg}\cdot\text{m}^2/\text{s}^2 = \frac{1}{2}(40\text{ kg})(v_2)^2$$

$$v_2 = v_{max} = 6.22\text{ m/s}$$

EXAMPLE 14-8

A 3-kg collar slides with negligible friction along a vertical rod as shown in Fig. 14-19a. The spring is undeformed when the collar A is at the same elevation as point O. The collar is released from rest at $y_1 = 0.4$ m. Determine the velocities of the collar at $y_2 = 0$ and $y_3 = -0.4$ m.

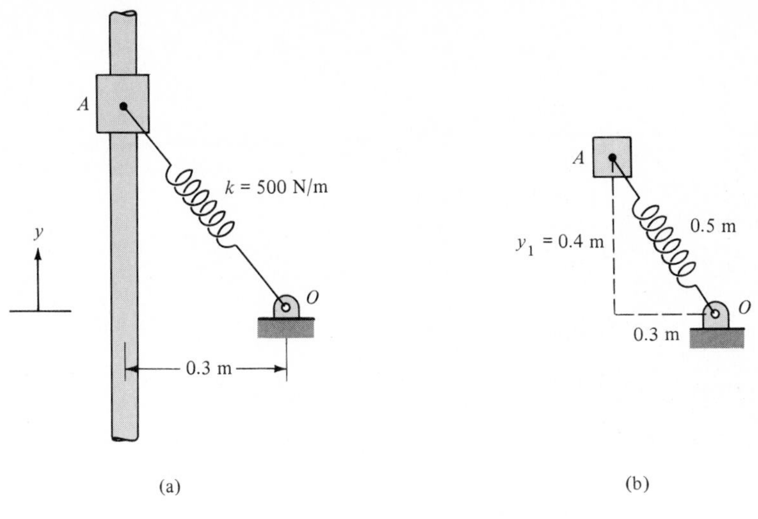

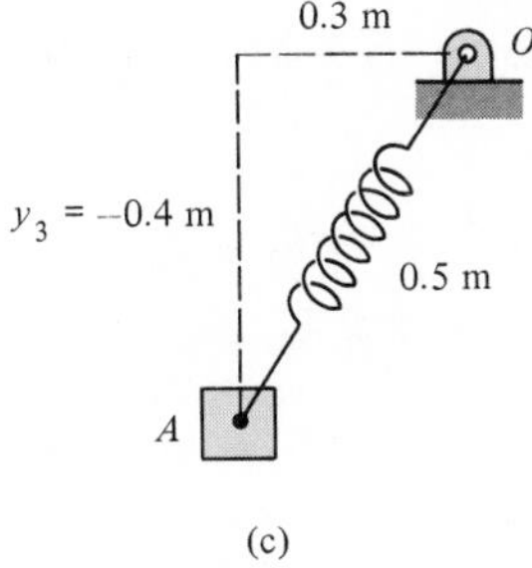

FIGURE 14-19

Sliding collar and spring mechanism

SOLUTION

For convenience, first the terms of Eq. 14-28 are evaluated at each position.

POSITION 1 (Fig. 14-19b): Potential energy: $V_1 = V_{g_1} + V_{e_1} = mgy_1 + \frac{1}{2}kd_1^2$, where d is the elongation of the spring.

$$V_1 = (3\text{ kg})(9.81\text{ m/s}^2)(0.4\text{ m}) + \frac{1}{2}(500\text{ N/m})(0.5\text{ m} - 0.3\text{ m})^2$$

$$= 21.8\text{ N}\cdot\text{m}$$

Kinetic energy: $T_1 = 0$ since $v_1 = 0$

POSITION 2: Potential energy:

$$V_2 = mgy_2 + \frac{1}{2}kd_2^2$$

$$= (3\text{ kg})(9.81\text{ m/s}^2)(0) + \frac{1}{2}(500\text{ N/m})(0)^2 = 0$$

Kinetic energy: $T_2 = \frac{1}{2}mv_2^2 = (1.5\text{ kg})v_2^2$

POSITION 3 (Fig. 14-19c): Potential energy:

$$V_3 = mgy_3 + \frac{1}{2}kd_3^2$$

$$V_3 = (3\text{ kg})(9.81\text{ m/s}^2)(-0.4\text{ m}) + \frac{1}{2}(500\text{ N/m})(0.5\text{m} - 0.3\text{ m})^2$$

$$= -1.8\text{ N}\cdot\text{m}$$

Kinetic energy: $T_3 = \frac{1}{2}mv_3^2 = (1.5\text{ kg})v_3^2$

The required speeds are calculated by applying the principle of conservation of energy (Eq. 14-28) first for positions 1 and 2, then for positions 1 and 3. To obtain the speed v_2,

$$T_1 + V_1 = T_2 + V_2$$

$$0 + 21.8\text{ N}\cdot\text{m} = (1.5\text{ kg})v_2^2 + 0$$

$$v_2 = \pm 3.8\text{ m/s}$$

By inspection, $\mathbf{v}_2$ is downward.

To obtain the speed v_3,

$$T_1 + V_1 = T_3 + V_3$$

$$0 + 21.8\text{ N}\cdot\text{m} = (1.5\text{ kg})v_3^2 - 1.8\text{ N}\cdot\text{m}$$

$$v_3 = \pm 4\text{ m/s}$$

The velocity $\mathbf{v}_3$ is downward from the following considerations: (a) the velocity is downward at position 2; (b) the downward velocity would increase from 2 to 3 in the absence of the spring; and (c) the changes in potential energy of the mass and spring from 2 to 3 are about the same in magnitude but opposite in sign.

EXAMPLE 14-9

Consider the sport of pole vaulting, in which tremendous improvements have been made by using poles that act as springs. Determine the additional height that a jumper can reach by using an elastic pole. To be specific, use the following data about a world-class athlete: weight = 170 lb, and maximum running speed = 30 ft/s. Assume that the speed of the vaulter in the horizontal direction is 7 ft/s during and after takeoff (Fig. 14-20a). To simplify this complex problem, assume that the vaulter is a particle that travels horizontally and compresses a coil spring, the spring rotates to the vertical position with the particle on top of it, and the spring propels the particle upward (Fig. 14-20b–f).

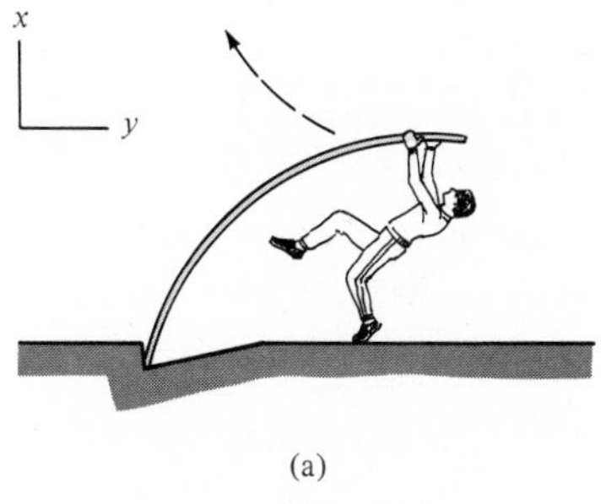

(a)

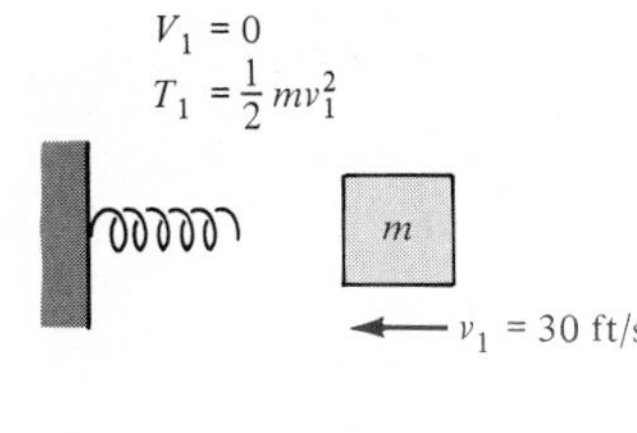

(b)

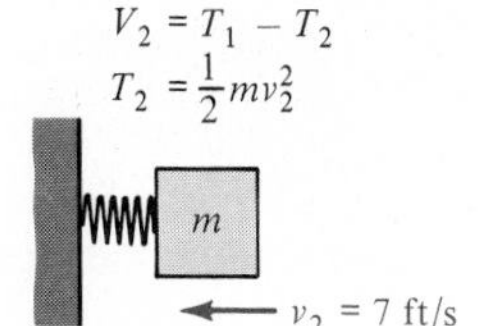

(c) Maximum compression of spring

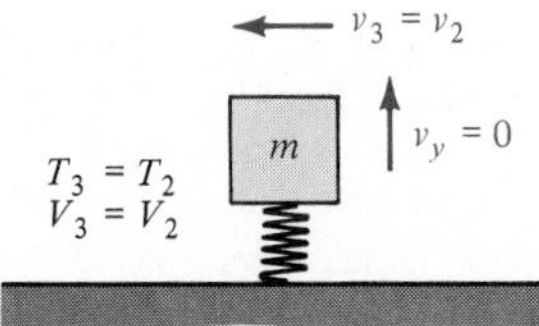

(d) Maximum compression of spring

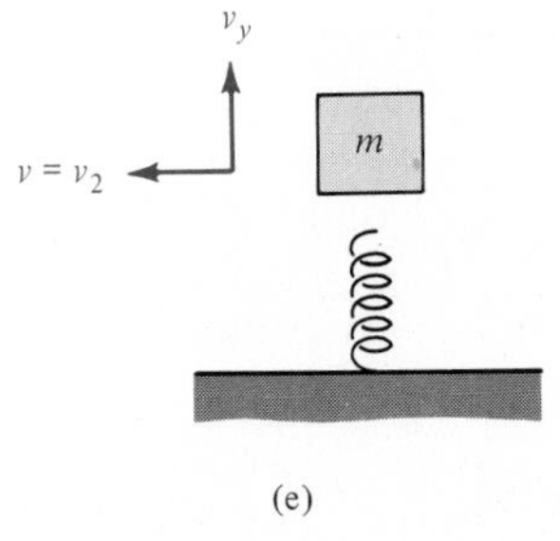

(e)

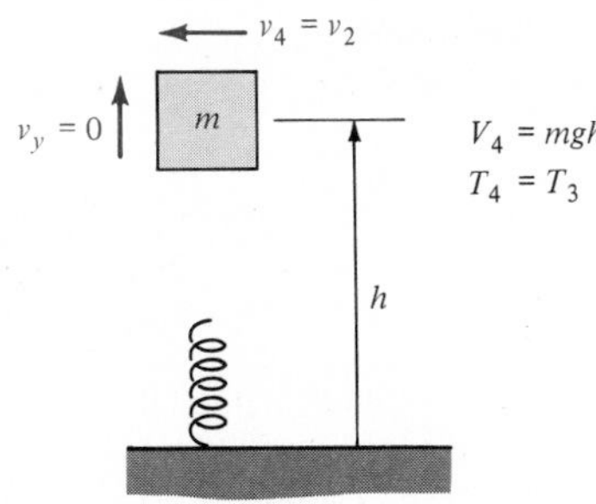

(f) Maximum elevation

FIGURE 14-20

Modeling the system configurations of pole-vaulting

SOLUTION

Assuming that the frictional loss in the pole is negligible, the energy stored in the pole can be calculated. Using the model of a mass and helical spring in Fig. 14-20b and c, and applying Eq. 14-28,

$$T_1 + V_1 = T_2 + V_2$$

$$T_1 = \frac{1}{2}mv_1^2 = \frac{1}{2}\left(\frac{170\text{ lb}}{32.2\text{ ft/s}^2}\right)(30\text{ ft/s})^2 = 2376\text{ ft}\cdot\text{lb}$$

$$V_1 = 0$$

$$T_2 = \frac{1}{2}mv_2^2 = \frac{1}{2}\left(\frac{170\text{ lb}}{32.2\text{ ft/s}^2}\right)(7\text{ ft/s})^2$$

$$= 129\text{ ft}\cdot\text{lb} \quad \text{(kinetic energy of the particle at maximum deformation of the spring)}$$

$$V_2 = \text{potential energy in the spring at its maximum deformation}$$

$$2376\text{ ft}\cdot\text{lb} + 0 = 129\text{ ft}\cdot\text{lb} + V_2$$

$$V_2 = 2247\text{ ft}\cdot\text{lb} \quad \text{(maximum potential energy in the pole from the kinetic energy of running horizontally)}$$

Eventually, the energy V_2 is released and the vaulter is projected upward by the pole, as modeled in Fig. 14-20d–f. The height reached by the mass m is calculated by using Eq. 14-28 again,

$$T_3 + V_3 = T_4 + V_4$$

where the subscripts 3 denote the initial conditions of Fig. 14-20d, and the subscripts 4 denote the quantities at the maximum elevation reached. Since the horizontal speed is 7 ft/s at positions 3 and 4, and the vertical velocity is zero at both 3 and 4, $T_3 = T_4$. Thus, with $V_3 = V_2$,

$$V_2 = V_4$$

$$2247\text{ ft}\cdot\text{lb} = mgh = (170\text{ lb})h$$

$$h = 13.2\text{ ft}$$

This is the additional height that can be reached, considering the pole as a spring and the vaulter as a particle during a jump. The actual height reached is greater than h because the athlete's center of gravity is initially above the ground by about 3.5 ft, he also jumps upward at takeoff, and he also uses his arms to push upward from the finally straightened pole. The performances of pole vaulters who clear bars set at heights over 18 ft can be explained using these considerations.

PROBLEMS

Problems 14–39 to 14–46 are intended only for those students who have worked with partial derivatives.

14–39 Prove that a force $\mathbf{F}(x, y, z) = F_x\mathbf{i} + F_y\mathbf{j} + F_z\mathbf{k}$ is conservative if

$$\frac{\partial F_x}{\partial y} = \frac{\partial F_y}{\partial x} \qquad \frac{\partial F_y}{\partial z} = \frac{\partial F_z}{\partial y} \qquad \frac{\partial F_z}{\partial x} = \frac{\partial F_x}{\partial z}$$

Show that these are necessary and sufficient conditions.

In Probs. 14–40 to 14–46, determine whether **F** acting on a particle is a conservative force. Also determine the associated potential function if it exists. Use the appropriate relationships given in Prob. 14–39.

14–40 $\mathbf{F} = x\mathbf{i} + 2y\mathbf{j}$

14–41 $\mathbf{F} = 3x^2\mathbf{i}$

14–42 $\mathbf{F} = -2x^2\mathbf{i} + 5y^3\mathbf{j}$

14–43 $\mathbf{F} = xy\mathbf{i} + 3z^2\mathbf{k}$

14–44 $\mathbf{F} = \dot{x}\mathbf{i} - z\mathbf{k}$

14–45 $\mathbf{F} = x^2y\mathbf{i} - xy^2\mathbf{j}$

14–46 $\mathbf{F} = y\dot{y}\mathbf{j} + 3z^2\dot{z}\mathbf{k}$

14–47 A 1-kg block starts from rest at point A and slides along a frictionless incline. Determine the speed of the block at point B, at a distance of 2 m from A.

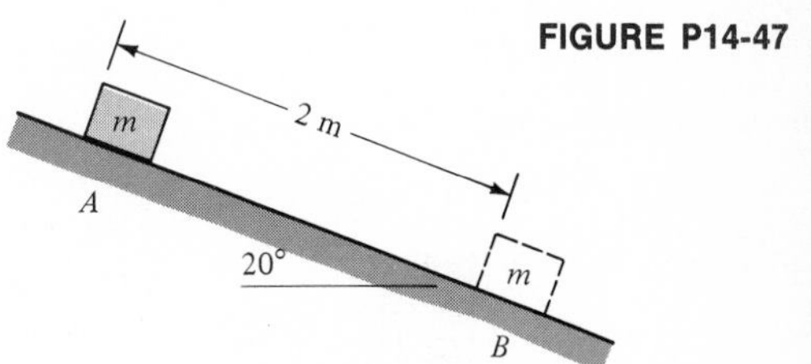

FIGURE P14-47

14–48 A 1-lb ball is thrown vertically upward at an initial speed of 30 ft/s. Determine the maximum elevation reached by the ball from its point of release.

FIGURE P14-49

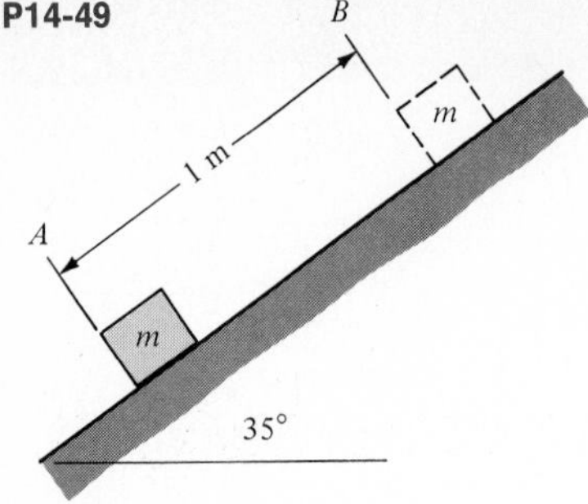

14–49 A 3-kg slider is moving up the smooth incline at a speed $v_A = 5$ m/s at point A. Calculate the speed at point B.

FIGURE P14-50

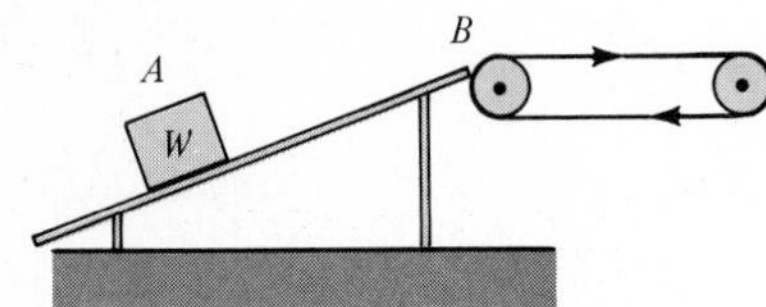

14–50 A 20-lb package is to be projected upward along a smooth incline toward a conveyor. Determine the required speed v_A at point A if the package should have a speed $v_B =$ 2 ft/s at point B, 6 ft above A.

FIGURE P14-51

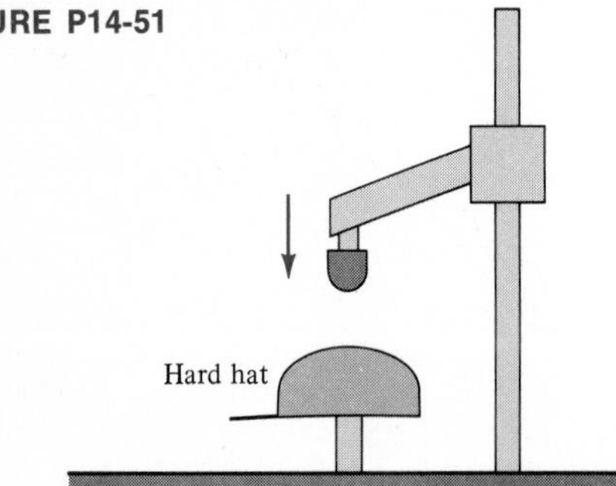

14–51 In the laboratory simulation of impacts on hard hats, a 4-lb plunger falls without resistance through a height of 6 ft. Determine the required initial downward speed of the plunger if its final speed should be 30 ft/s.

14–52 A high-jumper's center of gravity is 1.05 m above the ground. The jumper's maximum vertical speed at takeoff is 4.8 m/s. Estimate the elevation for setting the bar, which must be 15 cm below the maximum height reached by the jumper's center of gravity.

FIGURE P14-53

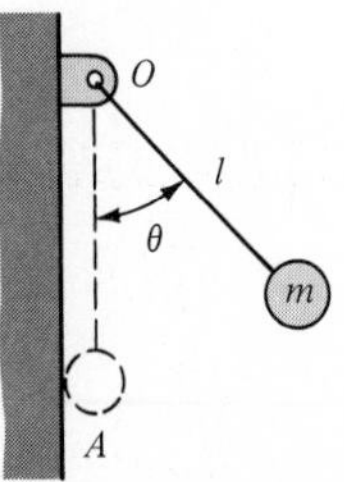

14–53 An impact tester is modeled as a pendulum of mass $m = 5$ kg on a thin wire of length $l = 0.8$ m. Determine the angle θ for releasing the ball from rest if its speed at point A should be 2.3 m/s.

14–54 A 160-lb gymnast is doing giant circles on a high bar, moving with negligible velocity in the position shown. Assume that his body may be modeled as a particle at the center of gravity, attached to a rigid, weightless rod of length $l = 3.5$ ft from the horizontal steel bar, and rotating without frictional resistance. Determine the speed of the gymnast's center of gravity at position A.

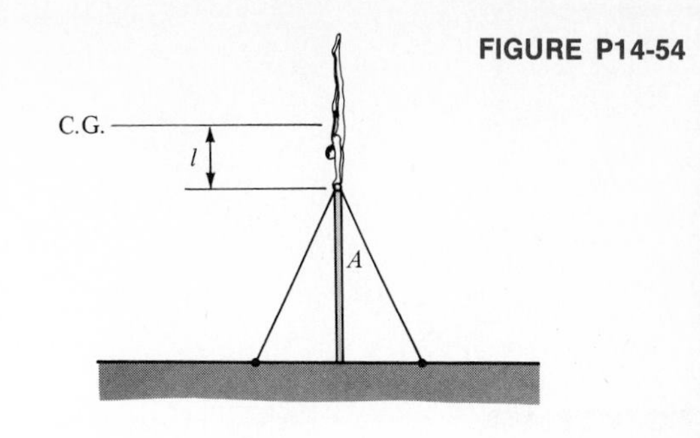

FIGURE P14-54

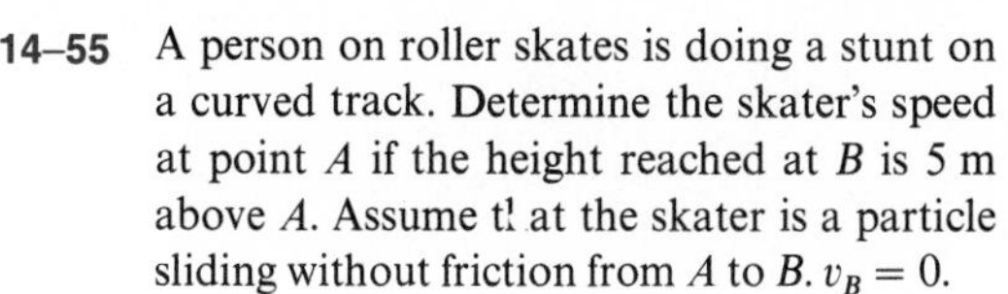

14–55 A person on roller skates is doing a stunt on a curved track. Determine the skater's speed at point A if the height reached at B is 5 m above A. Assume t! at the skater is a particle sliding without friction from A to B. $v_B = 0$.

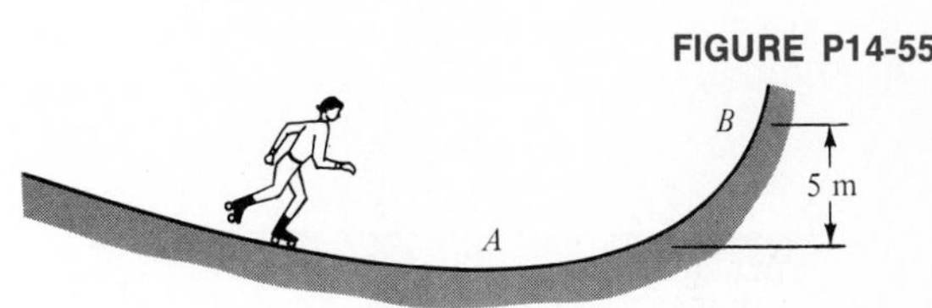

FIGURE P14-55

14–56 A ski jumper of mass $m = 70$ kg starts at the top of a tower that has a nearly straight track of 30 m vertical height and a 40° slope. The jumper acquires an initial speed of $v_0 = 2$ m/s by pulling on the railing. How much does this pulling improve the takeoff speed at the bottom of the track, assuming a frictionless slide?

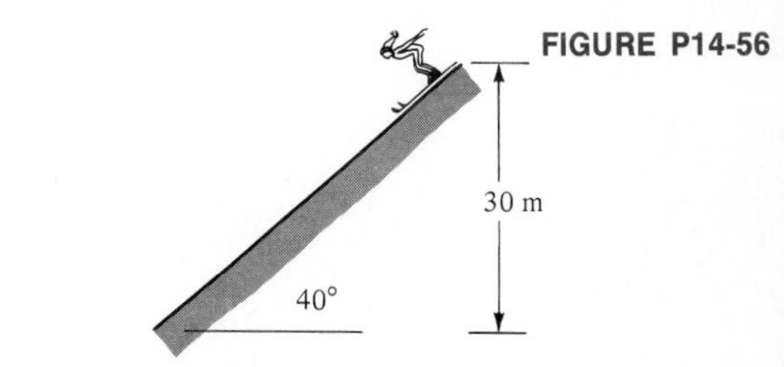

FIGURE P14-56

14–57 A ski jumper of weight $W = 160$ lb starts at the top of a tower that has a nearly straight track of 100 ft elevation at a 40° slope. Instead of pulling himself for an initial speed (as in Prob. 14–56), he decides to jump onto the track with a high, slightly forward jump. Assume that he can elevate his center of gravity 2 ft from a standing position, and that the track is frictionless. How much does this initial jump improve his takeoff speed at the bottom of the track?

14–58 A 5-kg block is attached to a spring of stiffness $k = 800$ N/m whose unstretched length is 0.3 m. The block is released from rest at (a) $y_1 = 0.3$ m, and (b) $y_2 = 0.25$ m. Determine the maximum deformation of the spring in each case.

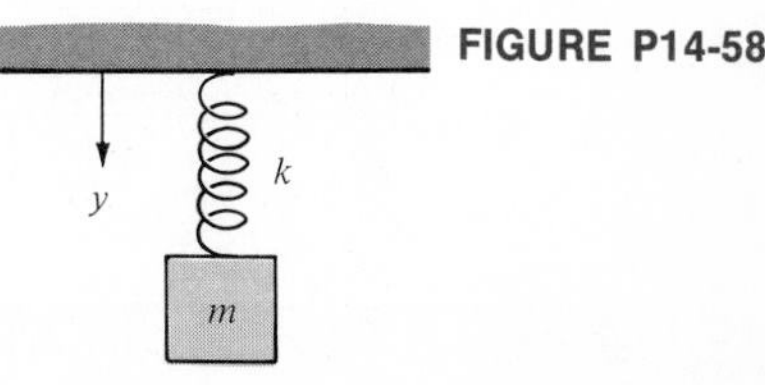

FIGURE P14-58

14–59 Assume that the block in Prob. 14–58 is released at an initial speed v_0 at $y = 0.4$ m. Calculate v_0 if the spring should be compressed to a length of 0.2 m by the moving block.

FIGURE P14-60

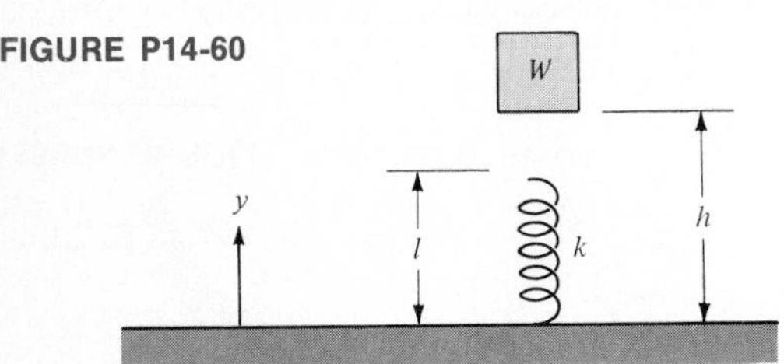

14–60 A 10-lb block is dropped from a height $h = 2$ ft on a spring of length $l = 10$ in. and spring constant k. Determine the required value of k if the spring should be compressed by the block to $y = 6$ in.

14–61 The block in Prob. 14–60 has an initial speed $v_0 = 7$ ft/s downward at height h. Determine the required stiffness in this case. Discuss the expected result if the same initial speed is applied upward.

FIGURE P14-62

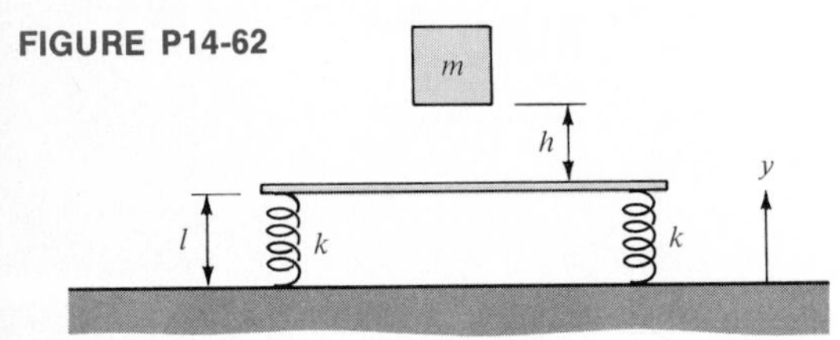

14–62 Assume that a rigid bar of negligible mass is supported by two identical springs of length $l = 0.2$ m and stiffness k. A 15-kg block is dropped on the bar from a height of $h = 0.5$ m so that the springs deform identically. Determine the required stiffness k of each spring if the lowest position of the bar is to be at $y = 0.1$ m.

FIGURE P14-63

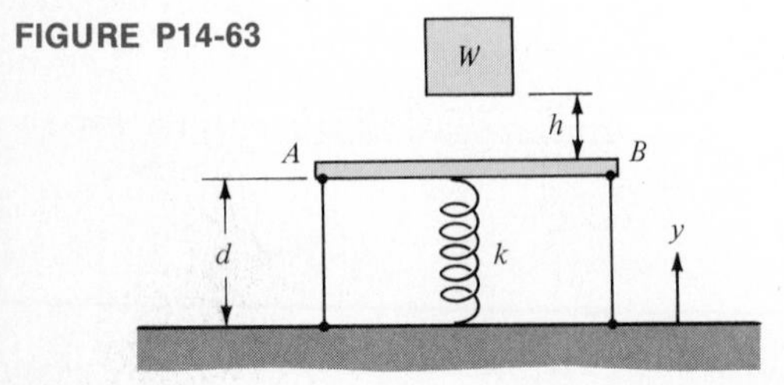

14–63 A rigid bar AB of negligible mass is used with flexible strings of constant length at A and B to compress a spring of stiffness $k = 10$ lb/in. and initial length $l = 14$ in. to a length $d = 12$ in. A 6-lb block is dropped on the bar from a height h. Determine h if the spring is compressed to $y = 9$ in. by the block.

14–64 The suspension system of a vehicle is modeled with springs of stiffness $k = 10$ kN/m and free lengths of $l = 0.3$ m. The maximum compression of each spring is to a length of 0.15 m. Determine the mass m that can be allowed to fall from rest on the spring from a height of $h = 0.2$ m so that the spring just bottoms out (its final length is 0.15 m).

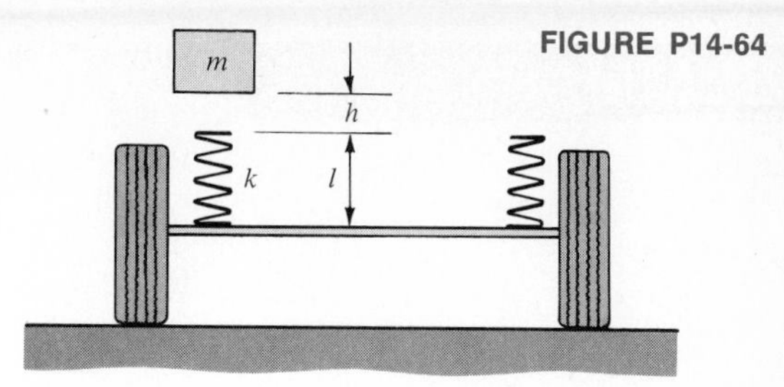

FIGURE P14-64

14–65 A 160-lb athlete exercises with a spring device to decrease the time between "explosive" jumps. A rigid bar AB of negligible weight is held on the shoulders and is connected by identical springs of stiffness $k = 2$ lb/in. to the ground. The jumper's vertical speed at takeoff is $v_0 = 13$ ft/s. At that instant $y = 5.3$ ft and the springs are not deformed. Determine the maximum force that the bar applies to the athlete.

FIGURE P14-65

14–66 Assume that in Prob. 14–65 the maximum force of the bar on the athlete is specified as 200 lb. Determine the stiffness k of each spring to achieve this force for the same speed v_0.

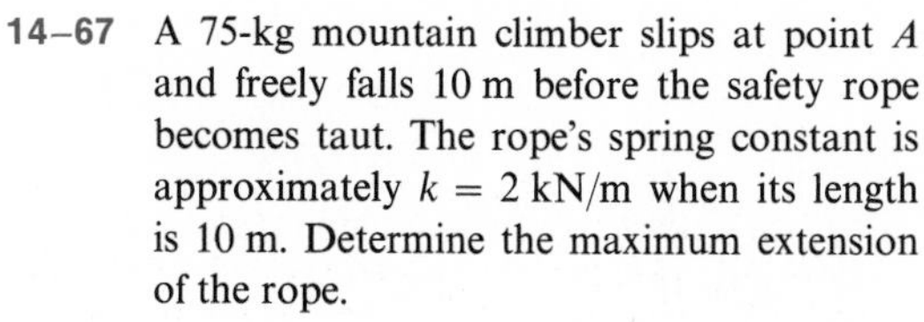

14–67 A 75-kg mountain climber slips at point A and freely falls 10 m before the safety rope becomes taut. The rope's spring constant is approximately $k = 2$ kN/m when its length is 10 m. Determine the maximum extension of the rope.

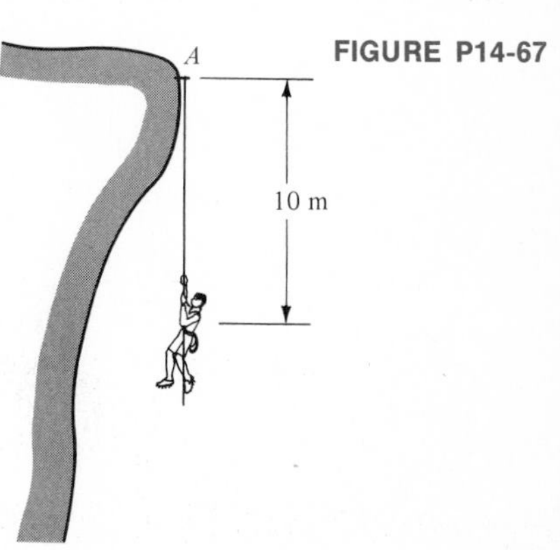

FIGURE P14-67

SUMMARY

Section 14-1

Work of a force

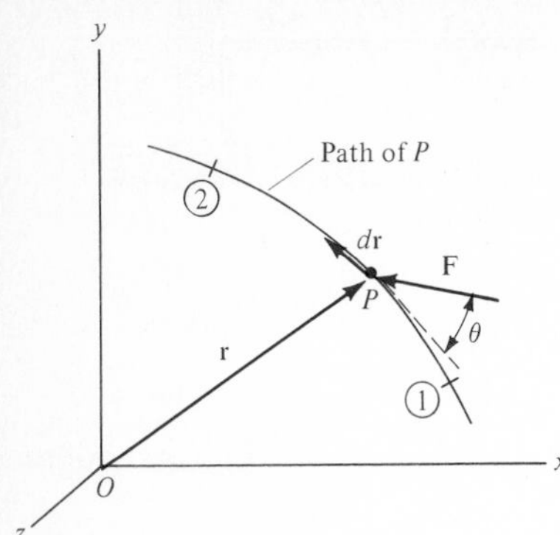

Total work of force **F** in displacing particle P from position 1 to position 2,

$$U_{12} = \int_1^2 \mathbf{F} \cdot d\mathbf{r} = \int_1^2 F \cos\theta \, dr$$

$$= \int_1^2 (F_x \, dx + F_y \, dy + F_z \, dz)$$

Section 14-2

Potential energy and work

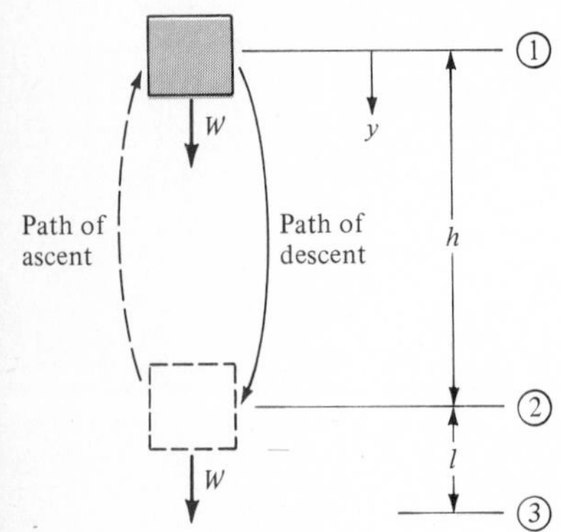

GRAVITATIONAL SYSTEM:

$$\underbrace{U_{12}}_{\text{work of } W \text{ from 1 to 2}} = \int_1^2 W \, dy = \underbrace{Wh}_{\text{potential energy of } W \text{ at 1 with respect to 2}}$$

$$U_{23} = Wl$$

$$U_{13} = W(h + l)$$

$$U_{21} = -Wh = -U_{12}$$

ELASTIC SYSTEM:

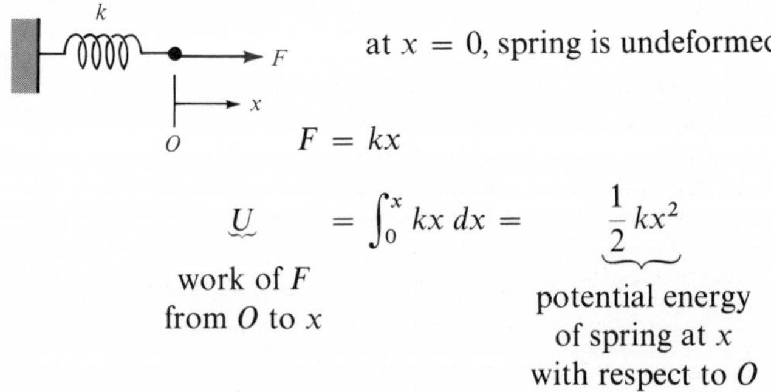

at $x = 0$, spring is undeformed

$$F = kx$$

$$\underbrace{U}_{\text{work of } F \text{ from } O \text{ to } x} = \int_0^x kx \, dx = \underbrace{\frac{1}{2} kx^2}_{\text{potential energy of spring at } x \text{ with respect to } O}$$

Section 14-3

Kinetic energy. Principle of work and energy

KINETIC ENERGY OF MASS *m* AT SPEED *v*:

$$T = \frac{1}{2} m v^2$$

PRINCIPLE OF WORK AND ENERGY:

$$U_{12} = T_2 - T_1$$

where U_{12} = work of a force on the particle from position 1 to 2
T_1 = kinetic energy of the particle at position 1 (initial kinetic energy)
T_2 = kinetic energy of the particle at position 2 (final kinetic energy)

Section 14-4

Power and efficiency

POWER:

$$\frac{dU}{dt} = \mathbf{F} \cdot \mathbf{v}$$

MECHANICAL EFFICIENCY:

$$\eta = \frac{\text{output work}}{\text{input work}} = \frac{\text{power output}}{\text{power input}}$$

Section 14-5

Opportunities and limitations using energy methods

ADVANTAGES:

1. Accelerations need not be determined.
2. Modifications of problems are easy to include in the analysis.
3. Scalars are summed even if the path of motion is complex.
4. Forces that do no work are ignored.

LIMITATIONS:

The quantities of work and energy cannot be used to determine accelerations or forces that do no work. In both cases Newton's second law must be used directly.

Section 14-6

Conservative systems. Potential functions

The work of a conservative force is independent of the path of displacement of a particle.

$$\text{Potential energy } V(x, y, z) = \text{potential function of a conservative force } \mathbf{F}$$

$$\underbrace{U_{12}}_{\substack{\text{work of } \mathbf{F} \\ \text{from 1 to 2}}} = \underbrace{V_1 - V_2}_{\substack{\text{difference of potential} \\ \text{energies at 1 and 2}}}$$

For a closed path, $U = \oint \mathbf{F} \cdot d\mathbf{r} = 0$

Analysis using partial derivatives:

$$\mathbf{F} = F_x\mathbf{i} + F_y\mathbf{j} + F_z\mathbf{k} = -\left(\frac{\partial V}{\partial x}\mathbf{i} + \frac{\partial V}{\partial y}\mathbf{j} + \frac{\partial V}{\partial z}\mathbf{k}\right)$$

Section 14-7

Conservation of mechanical energy

$$T_1 + V_1 = T_2 + V_2$$

where T = kinetic energy of a particle
V = potential energy of a particle or elastic member of the system

1 and 2 refer to two different positions of a particle. The equation is useful to determine the speed or elevation of a particle or the deformation of a spring. Choose reference levels to minimize the number of terms in the equation.

REVIEW PROBLEMS

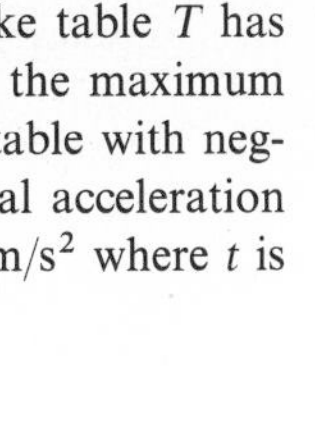

14–68 The horizontally moving shake table T has a mass of 300 kg. Determine the maximum power required to move the table with negligible friction if the horizontal acceleration is defined by $a = 0.5 \sin 5\pi t$ m/s^2 where t is in seconds.

FIGURE P14-68

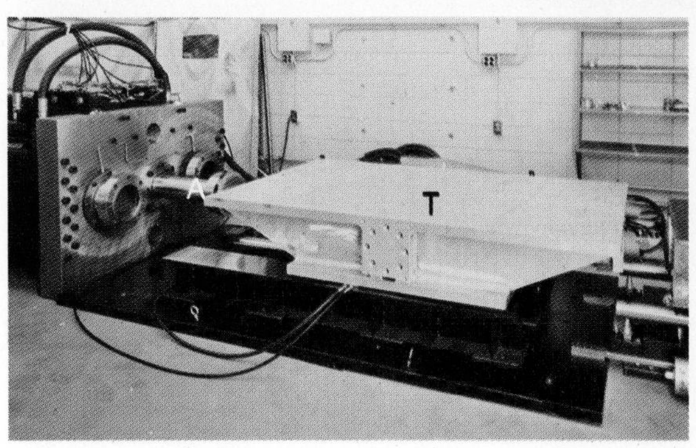

14–69 In Prob. 14–68 the total weight of the moving parts is 500 lb. Determine the maximum power if the horizontal displacement is defined by $d = 0.6 \sin 10\pi t$ in. where t is in seconds.

14–70 A 10 000-kg aircraft lands at $x = 0$ where its horizontal speed is $v_0 = 60$ m/s. The total retarding force on the aircraft is constant, $P = 120$ kN (the sum of braking force, reversed engine thrust, and force from the arresting cable). At $x = 50$ m the arresting cable breaks, and the retarding force becomes a constant $P_1 = 50$ kN. Determine the total stopping distance of the airplane.

FIGURE P14-70

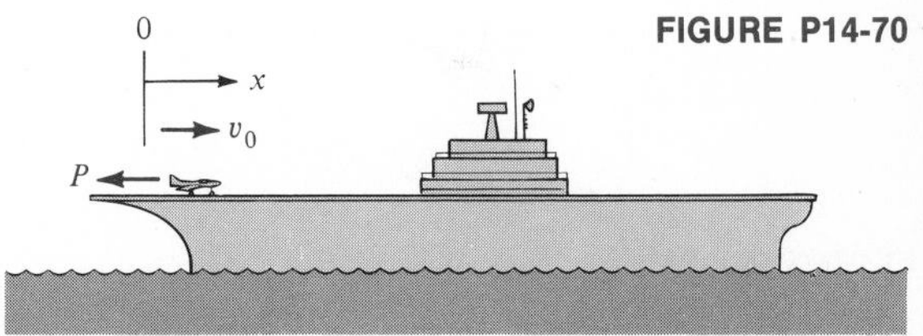

14–71 Assume that in Prob. 14–70 the brakes are automatically released when the cable breaks, and the reversed engine thrust is instantaneously changed to a constant forward thrust T. Determine the required force T if the aircraft must reach its takeoff speed of $v_t = 80$ m/s at $x = 200$ m.

FIGURE P14-72

(a)

14–72 The top car of the amusement ride is moving upward at point B at a speed v_B. The track makes a 60° angle with the horizontal. At point A, which has an elevation 90 ft above that of B, the top car contacts a spring of stiffness $k = 100$ lb/ft. The spring is compressed through a distance of $d = 8$ ft along the track. Determine the speed v_B of the 8000-lb train to cause the train to stop when the spring is deformed 8 ft.

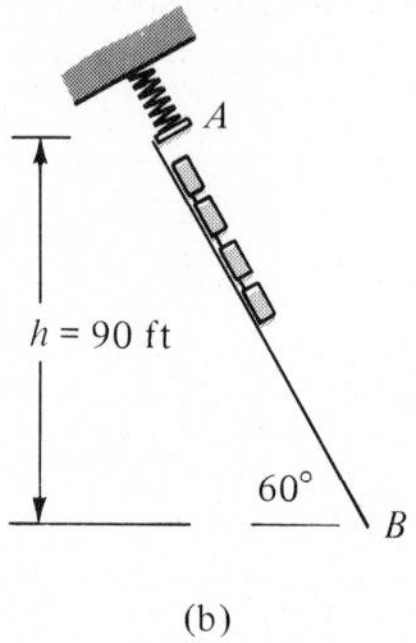

(b)

FIGURE P14-73

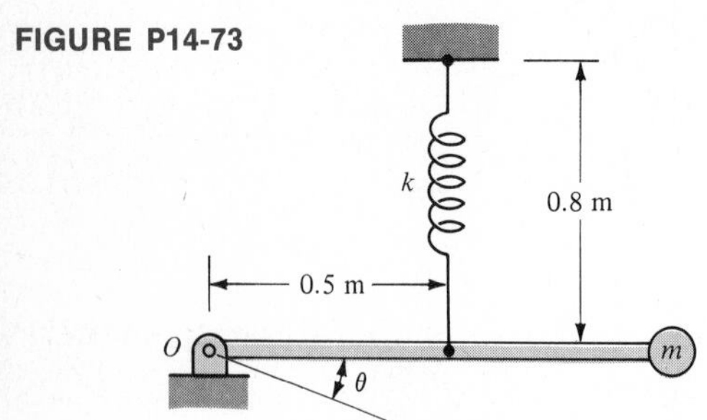

14–73 A 3-kg block is attached to a rigid bar of negligible mass which is pivoted at point O. The spring of stiffness $k = 700$ N/m is attached to the middle of the bar and is undeformed when the bar is released from rest in the horizontal position. Calculate the speed of the block at $\theta = 30°$.

FIGURE P14-74

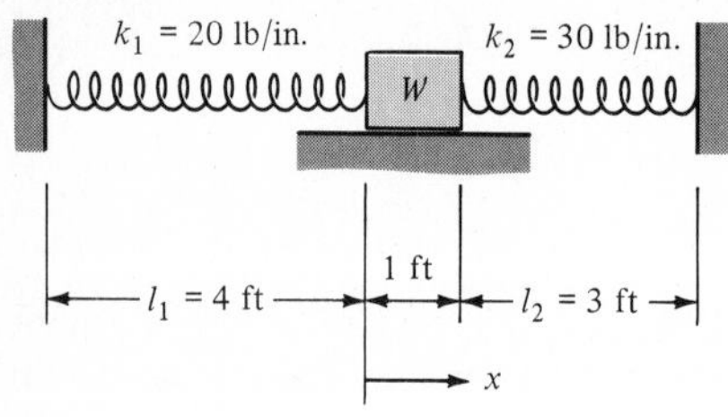

14–74 The 50-lb block slides horizontally with negligible friction. The springs are undeformed in the position shown. The block is displaced to $x = 1$ ft and is released from rest. Determine the speed at $x = 0$ and the maximum displacement to the left.

FIGURE P14-75

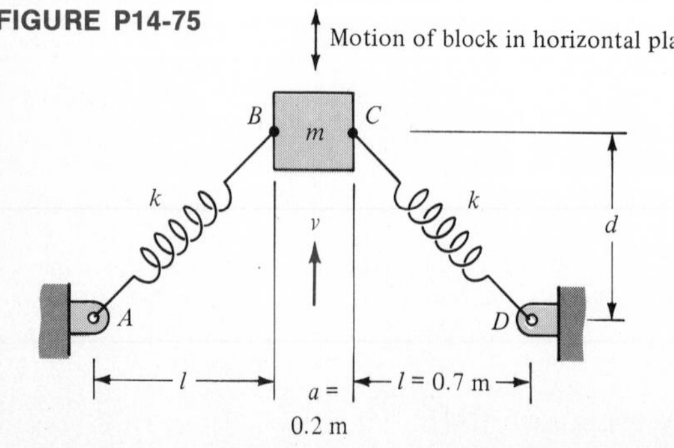

14–75 A vibration damper is modeled as a 10-kg block which slides horizontally with negligible friction. The springs apply no force to the block when A, B, C, and D are lined up. At that instant the speed of the block is $v = 5$ m/s. Determine the stiffness k of each spring if the maximum displacement should be $d = 0.6$ m.

CHAPTER 15

KINETICS OF PARTICLES: MOMENTUM METHODS

OVERVIEW

The foundation of all work in the area of kinetics is Newton's second law of motion (Eq. 13-1). It was demonstrated in Chap. 14 that special definitions and methods are useful in reformulating this law to facilitate the solution of problems where it is not essential to consider accelerations. In this chapter another approach is taken to reformulate Newton's second law and develop special methods that are advantageous in solving problems. The quantities of interest in this part of kinetics are force, mass, velocity, and time.

STUDY GOALS

SECTION 15-1 defines linear momentum by reformulating Newton's second law. The method of momentum is useful in solving common problems in kinetics, and it should be learned by all students of dynamics.

SECTIONS 15-2 AND 15-3 present the application of the method of momentum to curvilinear motion of a particle. The resulting concept of angular momentum is frequently used in dynamics. The method of angular momentum is convenient to use as a scalar method for analyzing circular motion in a plane, but it is particularly advantageous as a vector method for solving complex problems.

SECTION 15-4 presents Newton's law of universal gravitation, the mathematical expression of the gravitational force acting between two masses. This expression is used to define the acceleration of gravity g. The law of universal gravitation is most useful in aerospace engineering and scientific work.

SECTION 15-5 states Kepler's laws of planetary motion and the proofs of these laws based on Newton's second law. The method of momentum is used in the derivation. The initial steps in the analysis are of general interest, but the subsequent details are not important to the majority of students.

SECTIONS 15-6 THROUGH 15-8 present the method of impulse and momentum which is useful when time is involved in the analysis of motion of one or more particles. The definitions and major formulas of these sections are often used in dynamics analyses.

SECTIONS 15-9 THROUGH 15-11 extend the method of impulse and momentum to the analysis of deformable bodies in collisions. These sections introduce a few practical considerations and provide additional practice in solving problems of impulsive motion.

15-1 LINEAR MOMENTUM. TIME RATE OF CHANGE OF LINEAR MOMENTUM

Equation 13-1 can be expressed in a different form by noting that acceleration is the time rate of change of velocity. With this,

$$\boxed{\mathbf{F} = m\frac{d\mathbf{v}}{dt}} \tag{15-1}$$

If it can be assumed that the mass is constant during the motion,

$$\mathbf{F} = \frac{d}{dt}(m\mathbf{v}) \tag{15-2}$$

The vector $m\mathbf{v}$ (which is sometimes denoted by $\mathbf{G}$) is defined as the *momentum*, or, more specifically, the *linear momentum* of the particle. The direction of the momentum is in the direction of the velocity. Its magnitude is equal to the mass times the speed v of the particle. Thus, Newton's second law may be stated as

force = time rate of change of linear momentum

If several forces act simultaneously on a particle, the resultant force must be used in determining the change of linear momentum of the particle,

$$\boxed{\sum \mathbf{F} = \dot{\mathbf{G}} = \frac{d}{dt}(m\mathbf{v})} \tag{15-3}$$

where the mass m is assumed constant. When the resultant force on a particle is zero, the momentum vector $\mathbf{G}$ is constant in magnitude and direction. This is called *conservation of linear momentum*, which is a special definition of Newton's first law. The units of linear momentum are kg·m/s or N·s in SI, and lb·s in U.S. customary units.

ANGULAR MOMENTUM

15-2

A highly useful analytical method can be developed by considering various aspects of a particle's momentum with respect to a fixed center of motion. First assume that a particle P is moving at a constant speed v with respect to a circular path as in Fig. 15-1. At the particular instant considered, the linear momentum $m\mathbf{v}$ may be as shown (it is always tangent to the path). In circular motion the vector $m\mathbf{v}$ is perpendicular to the radius vector $\mathbf{r}$ to the particle. It is evident from Sec. 15-1 and Chap. 13 that both $m\mathbf{v}$ and $\mathbf{r}$ are important parameters in the complete description of the motion. Therefore, it is useful to combine them into a single parameter for a special characterization of the motion. The single parameter is formed by calculating the moment of the linear momentum vector about the fixed point O. This moment is called the *moment of linear momentum* or *angular momentum* about point O, and it is commonly denoted by $\mathbf{H}_O$. According to the definition of the moment of a vector,

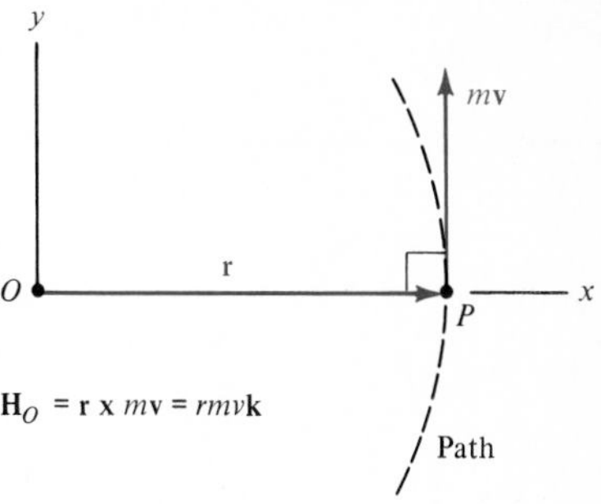

FIGURE 15-1

Definition of angular momentum of a particle

$$\mathbf{H}_O = \mathbf{r} \times m\mathbf{v} \tag{15-4}$$

which is perpendicular to the plane of motion (the xy axes in Fig. 15-1). $\mathbf{H}_O$ has a magnitude rmv for circular motion. The direction of $\mathbf{H}_O$ is determined by applying the right-hand rule. The units of angular momentum are kg·m²/s in SI, and ft·lb·s in U.S. customary units.

General Plane Motion

The concept of angular momentum is also applicable to general curvilinear motion of a particle. In the mathematical analysis any reference point moving in any arbitrary way is acceptable. For the analysis of angular momentum in general motion, consider Fig. 15-1, which shows plane curvilinear motion of a particle P. The vector $m\mathbf{v}$ is tangent to the path, and it can be resolved into radial and transverse components $m\mathbf{v}_r$ and $m\mathbf{v}_\theta$, respectively. Equation 15-4 is applicable to this configuration also. The magnitude of the cross product $\mathbf{r} \times m\mathbf{v}$ is $H_O = rmv \sin \phi$. (With $\phi = 90°$, this agrees with the result for circular motion.) It is seen from Fig. 15-2 that in general $v \sin \phi = v_\theta$, hence the magnitude of the angular momentum is $H_O = rmv_\theta$. This makes sense since v_r does not have a moment about point O. Therefore, in general the magnitude of the angular

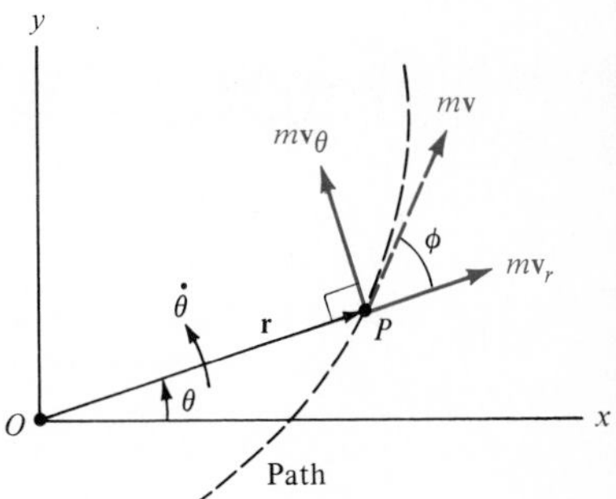

FIGURE 15-2

Angular momentum in general plane motion

momentum vector may be stated in several ways,

$$H_O = rmv \sin \phi = rmv_\theta = mr^2\dot{\theta} \tag{15-5}$$

The direction of $\mathbf{H}_O$ is again determined by applying the right-hand rule. These are shown in Fig. 15-2.

General Space Motion

The concepts discussed above are readily extended for the analysis of three-dimensional motion of a particle. Using rectangular coordinates, the position vector is $\mathbf{r} = x\mathbf{i} + y\mathbf{j} + z\mathbf{k}$ and the momentum vector is $m\mathbf{v} = mv_x\mathbf{i} + mv_y\mathbf{j} + mv_z\mathbf{k}$. With these, the most general form of the angular momentum can be written as follows:

$$\mathbf{H}_O = \mathbf{r} \times m\mathbf{v} = \begin{vmatrix} \mathbf{i} & \mathbf{j} & \mathbf{k} \\ x & y & z \\ mv_x & mv_y & mv_z \end{vmatrix} = m \begin{vmatrix} \mathbf{i} & \mathbf{j} & \mathbf{k} \\ x & y & z \\ v_x & v_y & v_z \end{vmatrix} \tag{15-6}$$

The angular momentum vector can also be expressed in terms of its rectangular components,

$$\mathbf{H}_O = H_x\mathbf{i} + H_y\mathbf{j} + H_z\mathbf{k} \tag{15-7}$$

where H_x is the angular momentum about the x axis, and so on. The magnitudes of the components in Eq. 15-7 can be expressed after evaluating the determinant in Eq. 15-6,

$$\begin{aligned} H_x &= m(yv_z - zv_y) \\ H_y &= m(zv_x - xv_z) \\ H_z &= m(xv_y - yv_x) \end{aligned} \tag{15-8}$$

In any plane motion two of the components of Eq. 15-7 are zero. For example, if the particle moves in the xz plane, $H_x = H_z = 0$ because $y = v_y = 0$.

15-3 TIME RATE OF CHANGE OF ANGULAR MOMENTUM

In the general case a net force is acting on the particle which changes the particle's angular momentum. The force and the resultant change in angular momentum can be related after evaluating the time derivative of the angular momentum vector.

$$\begin{aligned} \dot{\mathbf{H}}_O &= \frac{d}{dt}(\mathbf{r} \times m\mathbf{v}) = \dot{\mathbf{r}} \times m\mathbf{v} + \mathbf{r} \times m\dot{\mathbf{v}} \\ &= \mathbf{v} \times m\mathbf{v} + \mathbf{r} \times m\mathbf{a} = \mathbf{r} \times m\mathbf{a} \end{aligned}$$

since the vectors $\mathbf{v}$ and $m\mathbf{v}$ are collinear, so their vector product is zero. In the final term, $m\mathbf{a}$ is equal to the resultant $\mathbf{F}$ of all forces acting on the particle. Hence,

$$\dot{\mathbf{H}}_O = \mathbf{r} \times m\mathbf{a} = \mathbf{r} \times \mathbf{F} = \mathbf{M}_O$$

Thus, the term $\mathbf{r} \times m\mathbf{a}$ is equivalent to the moment $\mathbf{M}_O$ caused by the net force acting on the particle. It is concluded that *the rate of change of angular momentum of a particle is equal to the sum of the moments of forces acting on the particle.* This is expressed mathematically as

$$\boxed{\sum \mathbf{M}_O = \dot{\mathbf{H}}_O} \qquad \text{15-9}$$

It is often convenient to write Eq. 15-9 in terms of rectangular components,

$$\begin{aligned} \sum M_x &= \dot{H}_x \\ \sum M_y &= \dot{H}_y \\ \sum M_z &= \dot{H}_z \end{aligned} \qquad \text{15-10}$$

where M_x is the moment of all forces about the x axis, and so on. Equations 15-9 and 15-10 are very useful in more advanced dynamics analyses, especially in dealing with various systems of particles. In the special case when the moment $\mathbf{M}_O$ is zero, the angular momentum of the particle is constant. This is called *conservation of angular momentum.*

Motion under a Central Force

The net force acting on a particle that moves on a curved path is sometimes directed toward the same point O during a time interval. Such a force is called a *central force.* It is convenient to use the point O also as the reference point for the moments and angular momentum, as expressed in Eq. 15-9. In that case the line of action of the central force is through point O, $\sum \mathbf{M}_O = 0$, and

$$\dot{\mathbf{H}}_O = 0 \qquad \text{15-11}$$

This can be integrated with respect to time, resulting in

$$\mathbf{H}_O = \mathbf{r} \times m\mathbf{v} = \textbf{constant} \qquad \text{15-12}$$

Conservation of the angular momentum means that the particle's velocity and distance from the reference point tend to have an inverse relationship since the magnitude $H_O = r_1 m v_1 \sin \phi_1 = r_2 m v_2 \sin \phi_2 =$ constant. If the distance increases, the velocity must decrease, and vice versa. This is qualitatively analogous to the rapid, reversible changes in the rate of rotation of a figure skater who is spinning about a vertical axis. The rotation is slow when the arms are extended horizontally, and it becomes rapid when the arms are pulled close to the body. Of course, the skater is not a particle in such a case, even in the modeling for analysis.

EXAMPLE 15-1

A 5-kg block moves on a smooth horizontal xy plane (Fig. 15-3) under the action of a horizontal force **F** which varies with time. The velocity of the block is defined by $\mathbf{v} = (2t^2\mathbf{i} + 3t\mathbf{j})$ m/s, with time t given in seconds. Determine whether **F** and **v** are collinear at time $t = 3$ s.

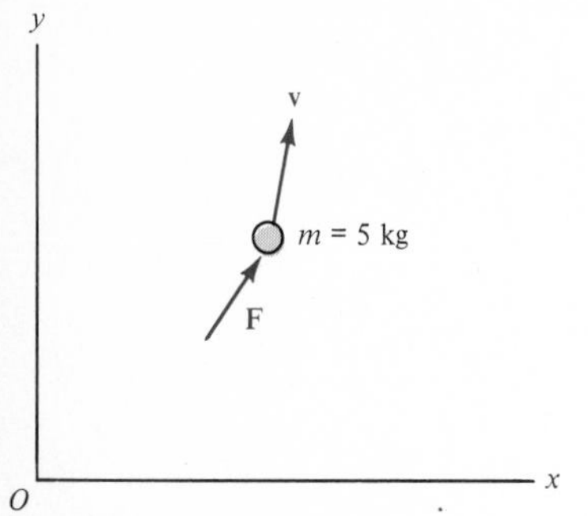

FIGURE 15-3

Analysis of a particle's motion using the method of linear momentum

SOLUTION

First the force is determined using Eq. 15-3,

$$\mathbf{F} = \dot{\mathbf{G}} = \frac{d}{dt}(m\mathbf{v}) = \frac{d}{dt}[(5)(2t^2)\mathbf{i} + (5)(3t)\mathbf{j}]$$

$$= 20t\mathbf{i} + 15\mathbf{j}$$

At $t = 3$ s,

$$\mathbf{F} = 60\mathbf{i} + 15\mathbf{j}$$

$$\mathbf{v} = 18\mathbf{i} + 9\mathbf{j}$$

By inspection, $F_x/F_y \neq v_x/v_y$, so **F** and **v** are vectors of different orientations at $t = 3$ s.

EXAMPLE 15-2

A pendulum consists of a particle of mass m suspended by an inextensible wire of negligible weight (Fig. 15-4). Write an expression for $\ddot{\theta}$ using the concept of angular momentum.

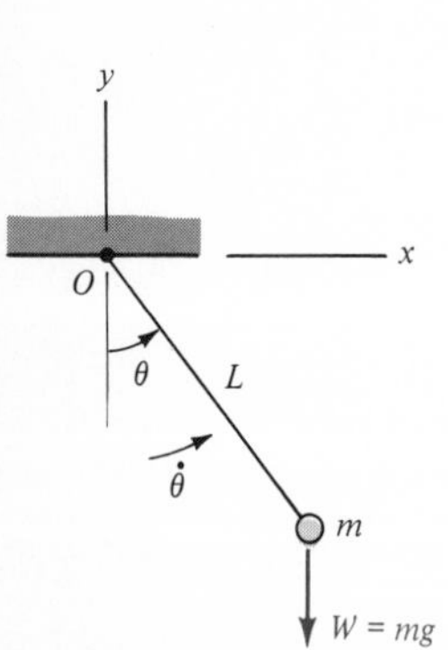

FIGURE 15-4

Analysis of a pendulum using the method of angular momentum

SOLUTION

In the general case when $\theta \neq 0$, the force of gravity has a moment about point O. Assuming that the pendulum moves in the xy plane, the moment of force W is about the z axis, so the last of Eq. 15-10 is used,

$$\sum M_z = \dot{H}_z$$

By inspection,

$$\sum M_z = -mgL \sin\theta$$

$$H_z = mrv$$

where $r = L$ and $v = L\dot{\theta}$. Thus,

$$\dot{H}_z = mL^2\ddot{\theta}$$

$$\ddot{\theta} = -\frac{g}{L}\sin\theta$$

EXAMPLE 15-3

A proposed amusement ride is modeled as two particles of weight $W = 600$ lb mounted on a weightless rod AB which rotates in a horizontal plane (Fig. 15-5). The rod is rotating at an angular speed $\omega_1 = 1$ rad/s when $r_1 = 20$ ft. An internal mechanism pulls both particles toward point O, to positions where $r_2 = 10$ ft. Determine the final angular speed ω_2. Assume that rod AB is freely rotating about the y axis while the particles are moved from one position to another.

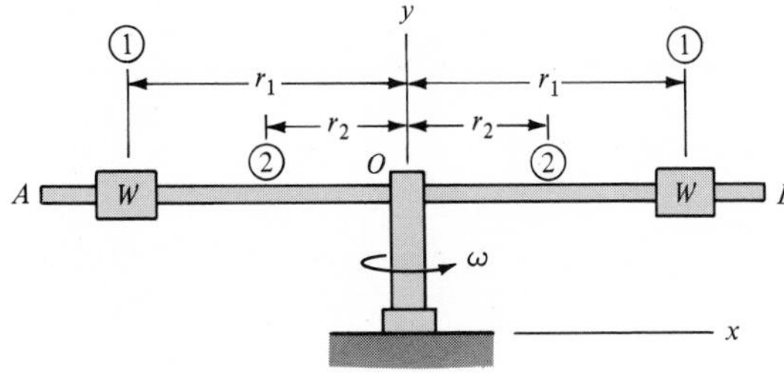

FIGURE 15-5

Conservation of angular momentum in an amusement ride

SOLUTION

Angular momentum is conserved according to the statement of the problem. Since all motion is in the same plane, the angular momentum does not change direction. Thus, the problem is solved using scalar quantities. The initial angular momentum is

$$H_O = rmv = 2\left(r_1 \frac{W}{g} r_1\omega_1\right)$$

which must be equal to the final angular momentum

$$H_O = 2\left(r_2 \frac{W}{g} r_2\omega_2\right)$$

Thus,

$$r_1^2\omega_1 = r_2^2\omega_2$$

The final angular speed is

$$\omega_2 = \frac{r_1^2}{r_2^2}\omega_1 = \frac{20^2}{10^2}(1) = 4 \text{ rad/s}$$

PROBLEMS

SECTIONS 15-1, 15-2, 15-3

FIGURE P15-1

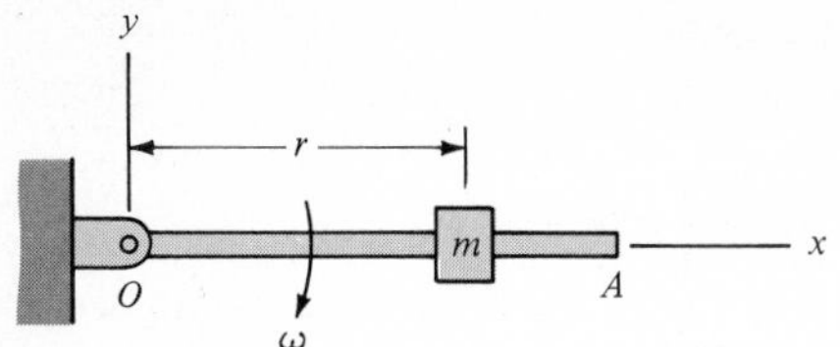

15–1 Bar OA of negligible mass is rotating in a horizontal plane about point O at an angular speed $\omega = 50$ rpm. A 3-kg block is at a constant distance $r = 0.4$ m from point O. Determine the linear momentum $\mathbf{G}$ and the angular momentum $\mathbf{H}_O$ of the block.

15–2 Solve Prob. 15–1 assuming that the block is moving away from the pin O at a constant speed of $\dot{r} = 6$ m/s.

FIGURE P15-3

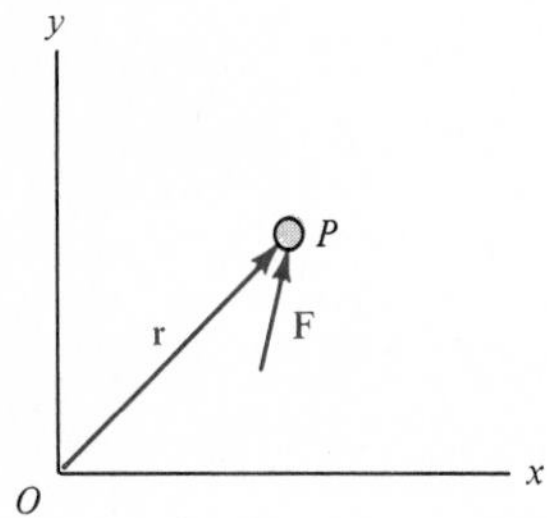

15–3 A 4-lb particle is moving in a horizontal xy plane under the action of force $\mathbf{F}$. The position of the particle is defined by $\mathbf{r} = (5\mathbf{i} + 2t^3\mathbf{j})$ ft, where t is in seconds. Determine the linear momentum $\mathbf{G}$ and the angular momentum $\mathbf{H}_O$ of the particle at $t = 3$ s.

15–4 Calculate the force $\mathbf{F}$ acting on particle P in Prob. 15–3 at $t = 3$ s.

FIGURE P15-5

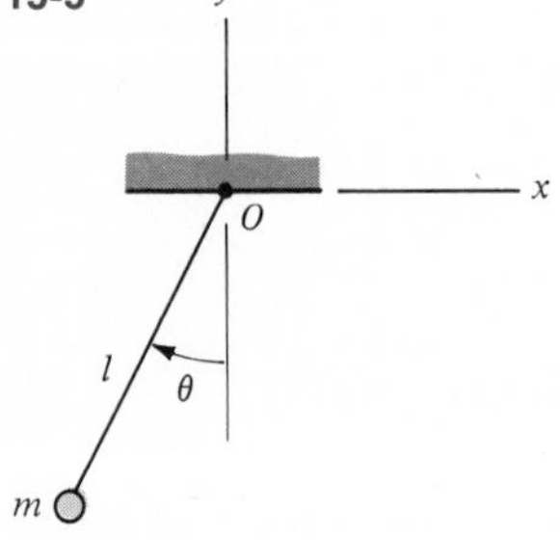

15–5 A 2-kg mass is suspended by a wire of length $l = 1$ m and negligible mass. At $\theta = 30°$ the angle is decreasing at the rate of 0.3 rad/s. Determine the linear momentum $\mathbf{G}$ and the angular momentum $\mathbf{H}_O$ of the mass using rectangular coordinates.

15–6 Calculate the tension in the wire in Prob. 15–5 using momentum methods.

FIGURE P15-7

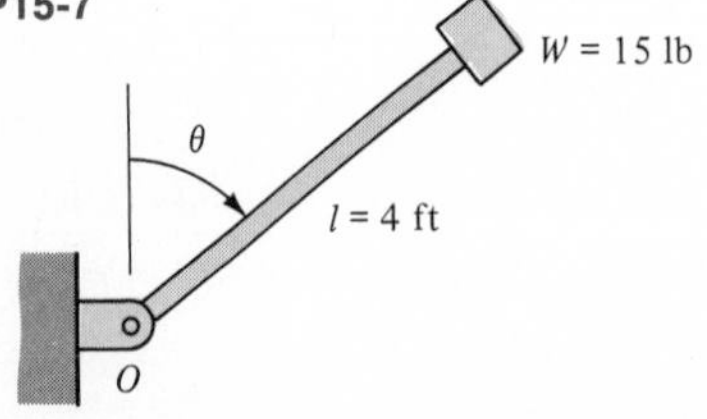

15–7 An impact tester is modeled as a particle on the end of a weightless, rigid rod. At $\theta = 60°$ the angle is increasing at the rate of 0.2 rad/s. Determine the axial force in the bar using momentum methods.

15–8 In Prob. 15–7, a moment M_O is applied to the bar by a mechanism to make the angular acceleration $\ddot{\theta} = 2$ rad/s^2 clockwise at $\theta = 60°$. Calculate M_O.

15–9 An astronaut during a space walk is on a tether of length $L = 20$ m which has an angular speed $\omega_y = 0.01$ rad/s (the motion is in the xz plane). The astronaut is pulled closer to the space vehicle by shortening the tether to $d = 4$ m. Determine the linear speed of the astronaut in the final position in the xyz frame neglecting the effects of gravity.

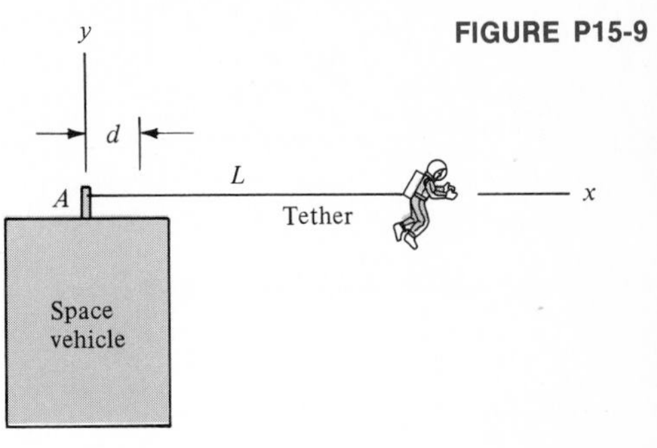

FIGURE P15-9

15–10 The astronaut shown in Fig. P15-9 pushes off the space vehicle at $d = 5$ ft at an initial velocity $\mathbf{v}_0 = (2\mathbf{i} - 3\mathbf{j})$ ft/s while the tether is loose. The tether becomes taut at length $L = 40$ ft. Determine the final linear velocity of the astronaut in the xyz frame neglecting the effects of gravity.

15–11 The unstretched length of the spring is 0.3 m, and its stiffness is 400 N/m. The 2-kg block attached to the spring moves on a smooth horizontal plane at velocity $\mathbf{v} = (3\mathbf{i} - 6\mathbf{j})$ m/s when $\theta = 10°$, $\mathbf{F} = (100\mathbf{i} - 70\mathbf{j})$ N, and $l = 0.5$ m. Determine the angular momentum $\mathbf{H}_O$ of the block and the rate of change of $\mathbf{H}_O$.

FIGURE P15-11

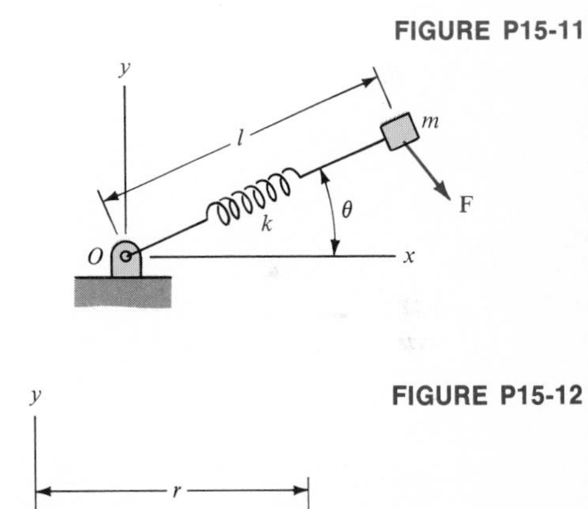

FIGURE P15-12

15–12 Rod OA is assumed weightless and that it rotates without friction in a horizontal plane. The rod is rotating at $\dot{\theta} = 100$ rpm. The 10-lb block B is at $r = 2$ ft and moving away from point O at a speed of $\dot{r} = 4$ ft/s. Determine the moment M that must act on the rod to maintain its angular speed.

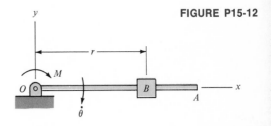

15–13 The rod of negligible mass in Fig. P15-12 is rotating in a vertical plane at $\dot{\theta} = 6$ rad/s. The 5-kg block B is at $r = 1$ m and moving toward point O at a speed of $\dot{r} = 2$ m/s. Determine the moment M that must act about point O to maintain the rod's angular speed. Neglect friction.

FIGURE P15-14

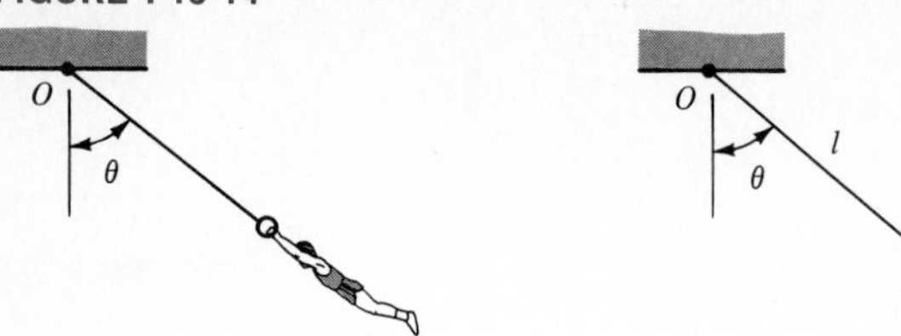

15–14 Assume a 160-lb gymnast on the flying rings is modeled as a particle P which can move along the supporting rope. At $\theta = 0, l = 10$ ft and P has a horizontal speed $v = 8$ ft/s. At that instant l is increased to 15 ft by the gymnast's own action. Calculate the new speed of P.

15–15 A 70-kg gymnast on the flying rings (Fig. P15-14) is modeled as a particle P. At $\theta = 60°$, $l = 6$ m, and P has no velocity. Assume that at $\theta = 0$ the gymnast applies a large force to suddenly make the length $l = 5.5$ m. Calculate the final velocity of P at $\theta = 0$.

FIGURE P15-16

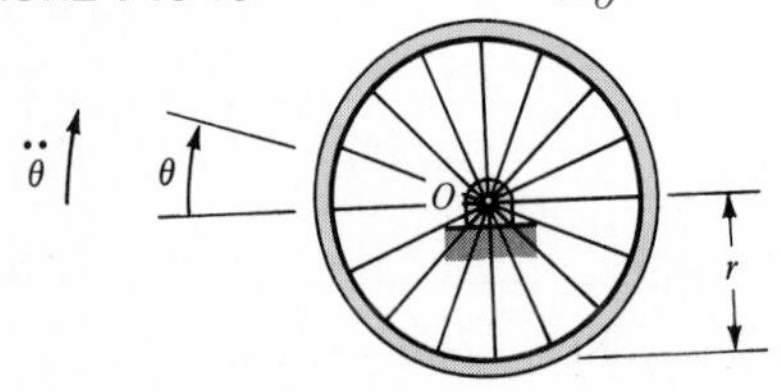

15–16 A bicycle wheel in a test is supported at point O and is initially at rest. A moment $M_O =$ 100 ft·lb is applied to the wheel. Determine the angular acceleration $\ddot{\theta}$ using momentum methods. Assume that the thin rim weighs 4 lb, the spokes are weightless, and $r = 13$ in. Model the rim as a particle at the distance r from point O.

FIGURE P15-17

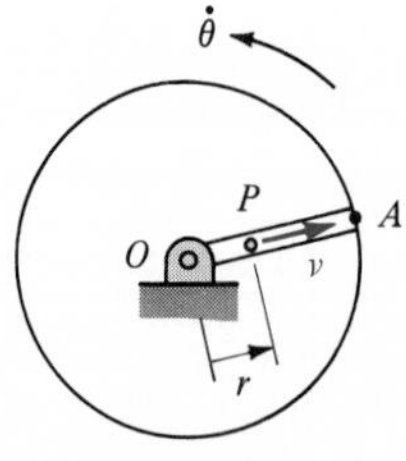

Top view

15–17 A proposed amusement ride consists of a flat disk rotating in a horizontal plane at angular speed $\dot{\theta} = 0.1$ rad/s. A 70-kg person P walks between safety rails from point O to A with a constant speed $v = 1$ m/s relative to the disk. Determine the moment M_O that must be applied to the disk to maintain its angular speed when the person is at $r_1 = 2$ m and at $r_2 = 3$ m, still walking toward point A. Neglect friction.

15–18 Determine the required moment M_O in Prob. 15–17 if $\dot{\theta} = 0.1$ rad/s must increase at the rate of 0.2 rad/s^2 when the person is at $r_1 = 2$ m. Consider only this position and neglect the mass of the disk.

15–19 A 12-lb block A is attached to a rigid bar of length $l = 4$ ft and negligible weight. A spring applies an upward force $F = 5$ lb at the middle of the bar when it starts from rest. The bar swings without frictional resistance in a vertical plane. Determine the initial acceleration of the block using momentum methods.

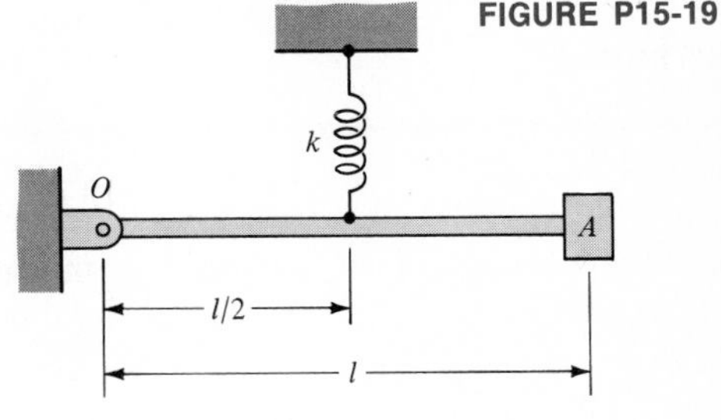

FIGURE P15-19

15–20 Solve Prob. 15–19 assuming that at the position shown block A moves at a constant speed $v = 2$ ft/s toward point O.

15–21 A ball of mass m is attached to a rigid rod of negligible mass and length r. The position ϕ of the rod is controlled by an internal mechanism. At $\phi = 0$, the rod rotates in a horizontal plane at angular speed $\dot{\theta}$. The mechanism adjusts the angle ϕ to 30° below the horizontal. Determine the new angular speed $\dot{\theta}_1$ assuming that the system is freely rotating about the vertical axis.

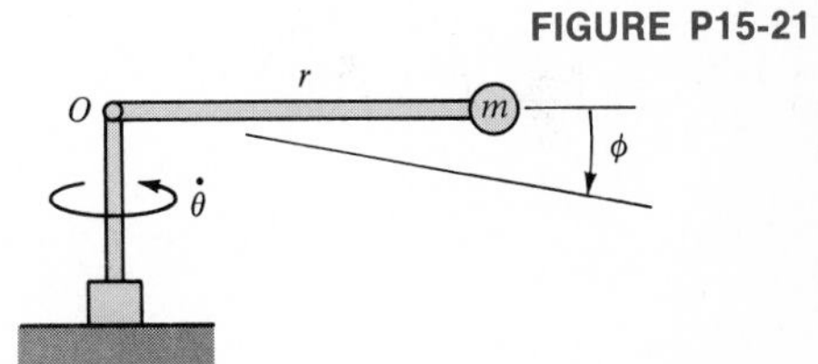

FIGURE P15-21

15–22 Assume that in Prob. 15–21 the original angular speed $\dot{\theta}$ must be maintained for all positions of the rod r. Determine the required moment to achieve this if the angle ϕ changes at a constant rate of $\dot{\phi} = 0.1\dot{\theta}$.

15-4 NEWTON'S LAW OF UNIVERSAL GRAVITATION

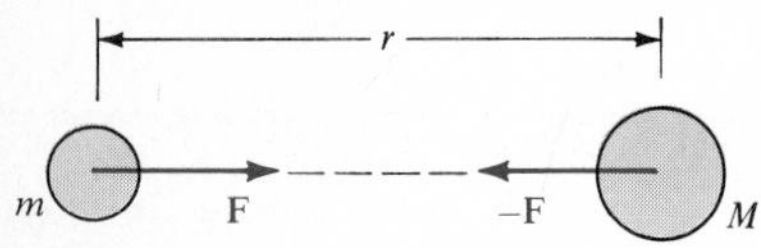

FIGURE 15-6

Mutual attraction of two masses

According to Newton's law of action and reaction, any central force acting on a particle must give rise to an equal and opposite force on the body that causes the force on the particle. This is illustrated in Fig. 15-6, where r is the distance between two masses M and m. Thus, the mass m moves under the central force $\mathbf{F}$, while M moves under the central force $-\mathbf{F}$, so that the forces are collinear. The mutual central forces have the same magnitude, but the resulting accelerations of the two bodies depend on their masses. Clearly, if M is much larger than m, the latter appears to move entirely with respect to M, while the force $-\mathbf{F}$ centered at m has a negligible effect on the larger mass. These differences depend on the relative masses and not on the magnitudes of the central forces between them. Of course, the magnitudes of the forces are important in analyzing the motion of each body in detail.

The mutual central force between two masses that are not in contact results from gravitational, electric, or magnetic effects. Of these, the gravitational force was first investigated theoretically and experimentally. Newton recognized that the mutual central force F between any two masses M and m is directly proportional to the product of the two masses and inversely proportional to the square of the distance r between them. The mathematical expression of Newton's *law of universal gravitation is*

$$\boxed{F = G\frac{Mm}{r^2}} \tag{15-13}$$

where G is a universal constant, the *constant of gravitation.* The value of this constant can be determined only experimentally. It was first measured by Cavendish in 1797, and many times since then to obtain a reliable value. This value in SI and U.S. customary units is

$$G = 6.673 \times 10^{-11}\ \text{m}^3/\text{kg}\cdot\text{s}^2$$

$$G = 3.442 \times 10^{-8}\ \text{ft}^4/\text{lb}\cdot\text{s}^4$$

The gravitational force F is significant only if at least one of the masses is very large because G is extremely small.

Equation 15-13 can be used to define the acceleration of gravity g and to partially explain why g is not a constant over the surface of the earth. Denoting the mass of the earth by M, its radius by R, and the mass of a body near or on the surface by m,

$$W = mg = G\frac{Mm}{R^2} \qquad g = \frac{GM}{R^2} \tag{15-14}$$

Here G and M are constant but R is not since the earth is not a perfect sphere. Other variations in g are caused by the spin of the earth and the fact that M is constant but it is not uniformly distributed. Nevertheless, the mass of the earth is much larger than the variations in the other quantities, so $g = 9.81 \text{ m/s}^2 = 32.2 \text{ ft/s}^2$ can be assumed in most engineering problems.

KEPLER'S LAWS OF PLANETARY MOTION 15-5

The methods of analyzing the motion of a particle under the action of a central force are ideally suitable in the area of celestial mechanics. When studying the motion of our solar system, the sun is often assumed immovable because its mass is 700 times the total mass of all its planets. Another simplifying assumption is that the planets do not affect one another in their orbits around the sun.

The first specific statements about planetary motion were made by the German astronomer Johannes Kepler (1571–1630), who analyzed astronomical data. The following two laws that Kepler discovered empirically are relevant to the concepts presented in this chapter.

KEPLER'S FIRST LAW (OR LAW OF SECTORIAL AREAS)

The radius drawn from a planet to the sun sweeps equal areas in equal times.

KEPLER'S SECOND LAW

The orbit of each planet is an ellipse, with the sun at one of its foci.

The mathematical proofs of these laws were provided by Newton. For a derivation consider the orbit of a planet as in Fig. 15-7. It is most convenient to use polar coordinates r and θ to describe the position and motion of the planet. The angle θ is measured from an arbitrary line of reference. Since the only force acting on

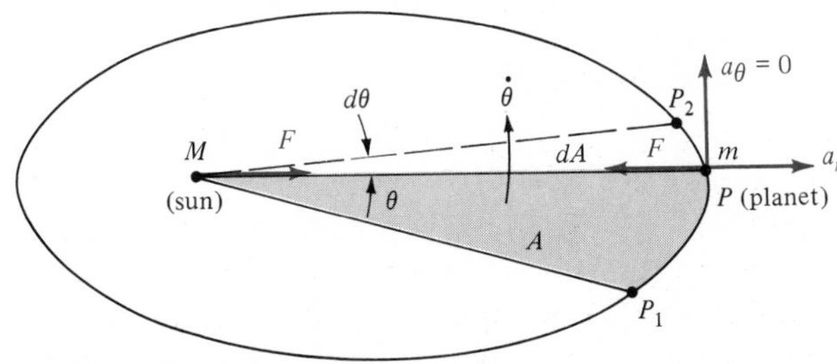

FIGURE 15-7

Illustration for proving Kepler's laws of planetary motion

the planet is the central force F directed toward the sun, the equations of motion are

$$F = -ma_r \qquad \text{and} \qquad a_\theta = 0 \tag{1}$$

From Eqs. 13-7 and 15-13,

$$F = G\frac{Mm}{r^2} = -m(\ddot{r} - r\dot{\theta}^2) \tag{2}$$

Since the angular momentum is constant, Eq. 15-5 can be written as

$$\dot{H}_O = \frac{d}{dt}(mr^2\dot{\theta}) = 0 \tag{3}$$

Thus, after integration of Eq. (3) with respect to time,

$$r^2\dot{\theta} = C \tag{4}$$

15-15

where C is a constant.

Equation (4) is the mathematical statement of Kepler's first law as can be seen from the following considerations:

1. A finite sectorial area A swept by the radius r is defined in Fig. 15-7 as the planet moves from point P_1 to P.
2. An infinitesimal sectorial area dA is swept by the radius as the planet moves from point P to P_2. Since the area dA is essentially a triangle of two sides r and a third side $r\,d\theta$, $dA \simeq \frac{1}{2}r^2\,d\theta$ for any orbital path.
3. Dividing dA by dt, and realizing that r is essentially constant during the time interval dt, $\dot{A} = \frac{1}{2}r^2(d\theta/dt)$. On the basis of Eq. (4), this is equivalent to the verbal statement of Kepler's first law.

   Kepler's second law can be proved by manipulating Eq. (2). Since the aim is to describe the shape of the orbit, time derivatives should be eliminated, and r should be expressed as a function of θ. Using Eq. (4), $\dot{\theta}$ is eliminated from Eq. (2),

$$\ddot{r} + \frac{GM}{r^2} - \frac{C^2}{r^3} = 0 \tag{5}$$

The following expressions can be used to eliminate $\ddot{r}$,

$$\dot{r} = \frac{dr}{dt} = \frac{dr}{d\theta}\frac{d\theta}{dt} = \frac{dr}{d\theta}\dot{\theta}$$

$$\ddot{r} = \frac{d\dot{r}}{dt} = \frac{d}{dt}\left(\frac{dr}{d\theta}\dot{\theta}\right) = \frac{d}{d\theta}\left(\frac{dr}{d\theta}\dot{\theta}\right)\frac{d\theta}{dt} = \dot{\theta}\frac{d}{d\theta}\left(\dot{\theta}\frac{dr}{d\theta}\right)$$

Once again, $\dot{\theta}$ can be eliminated by using Eq. (4), and

$$\ddot{r} = \frac{C^2}{r^2}\frac{d}{d\theta}\left(\frac{1}{r^2}\frac{dr}{d\theta}\right)$$

Thus, Eq. (5) becomes

$$\frac{C^2}{r^2}\frac{d}{d\theta}\left(\frac{1}{r^2}\frac{dr}{d\theta}\right) - \frac{C^2}{r^3} + \frac{GM}{r^2} = 0 \qquad (6)$$

This differential equation can be significantly simplified by making the substitution $q = 1/r$. The result is

$$\frac{d^2q}{d\theta^2} + q = \frac{GM}{C^2} \qquad (7)$$

The general solution of this equation is

$$q = \frac{GM}{C^2}[1 - K\cos(\theta - \alpha)] \qquad (8)$$

where K and α are constants. Now the two quantities of interest, r and θ, can be related in a usable form,

$$r = \frac{C^2/GM}{1 - K\cos(\theta - \alpha)} \qquad \boxed{15\text{-}16}$$

For repetitive orbital motion, r must be finite. Since $-1 < \cos(\theta - \alpha) < 1$, the constant K must be in the same range of values to obtain positive, noninfinite values of r. Thus, $-1 < K < 1$. It can be shown that under these conditions, Eq. 15-16 is the general equation of an ellipse. This proves Kepler's second law. In addition, three special cases of applying Eq. 15-16 are worth mentioning. For $K = 0$, $r =$ constant, and the orbit is circular. For $K = 1$, the orbit is parabolic. For $K > 1$, the orbit is hyperbolic. There are angles for which the radius r is infinite in the cases of parabolic and hyperbolic paths. The four distinct motions under a central force are shown schematically in Fig. 15-8.

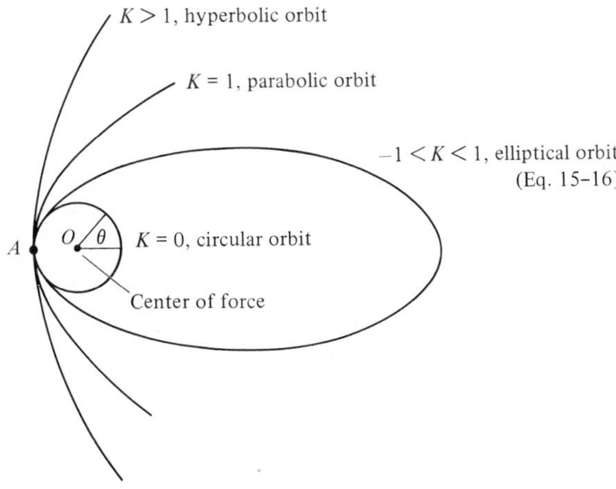

FIGURE 15-8
Four distinct planetary orbits

The foregoing considerations are also relevant to the flight of spacecraft. An important special aspect of such flights is the velocity required to allow a vehicle to escape from a celestial body. For a vehicle to escape from its dominant center of force, it must be moving on a parabolic or hyperbolic path with respect to that center of force. To understand the escape problem of spacecraft, consider a particle at point A in Fig. 15-8. The particle must have a minimum angular momentum with respect to point O to maintain a circular or elliptical orbit; otherwise, it will fall toward the center of force at point O. For escape, the angular momentum (thus, velocity) of the particle must be high enough to obtain a parabolic or hyperbolic path. Therefore, $K \geq 1$ in Eq. 15-16. Since it is most convenient to work with the particle at its closest approach to the center of force, r in Eq. 15-16 should be minimized. Letting $K = 1$ and $\cos(\theta - \alpha) = -1$,

$$2r_A = \frac{C^2}{GM}$$

The tangential velocity at point A is $v_A = r_A\dot{\theta}_A$. From Eq. 15-15, $C = r_A^2\dot{\theta}_A$; hence,

$$2r_A = \frac{v_A^2 r_A^2}{GM}$$

$$v_A = \sqrt{\frac{2GM}{r_A}} = v_{esc} \qquad \boxed{15\text{-}17}$$

This is the *escape velocity* since it allows the particle to move on a parabolic path away from the center of force. In the case of any celestial body the escape velocity can be expressed in two ways using Eq. 15-14, $GM = gR^2$,

$$v_{esc} = \sqrt{\frac{2GM_{cel}}{r_{esc}}} = \sqrt{\frac{2g_{cel}R_{cel}^2}{r_{esc}}} \qquad \boxed{15\text{-}18}$$

where G = universal constant of gravitation
M_{cel} = mass of celestial body from which the escape is made
g_{cel} = acceleration of gravity of celestial body
R_{cel} = radius of celestial body
r_{esc} = distance (from center of celestial body) at which the escape velocity is computed

Naturally, any particular escape is with respect to only one particular celestial body. For example, a vehicle that escapes from the earth may still remain a satellite of the sun.

The relationships of focal positions and characteristic dimensions of elliptic orbits are established using Fig. 15-9. The orbit has semi-major axis a and semi-minor axis b, and the foci are points B and C (of course, the major mass M is only at one of the foci). Consider a satellite of mass m at point A, the perigee of the orbit. It is seen that $2a = r_{\min} + r_{\max}$, so

$$a = \frac{1}{2}(r_{\min} + r_{\max}) \tag{15-19}$$

Next, it should be recalled that the sum of the distances from any point on an ellipse to the two foci is a constant for that ellipse. Therefore,

$$BD + CD = AB + AC = 2a$$

From the symmetry of the right triangles,

$$BD = CD = a = OC + r_{\min}$$

$$b^2 = (OD)^2 = (CD)^2 - (OC)^2 = a^2 - (a - r_{\min})^2$$

$$= r_{\min}(2a - r_{\min}) = r_{\min}r_{\max}$$

and

$$b = \sqrt{r_{\min}r_{\max}} \tag{15-20}$$

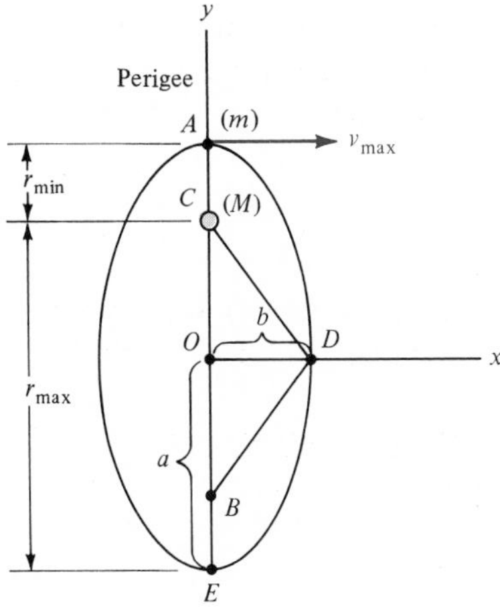

FIGURE 15-9

Geometry of elliptic orbits

EXAMPLE 15-4

The radar record of a comet indicates that it is moving at a speed of 26,000 mph in a direction perpendicular to the radial line r, and that r is 5000 mi from the center of the earth. Will this comet be captured by the earth?

SOLUTION

Since the velocity is perpendicular to the radial vector, the velocity is a maximum or a minimum at this point. Assume that the velocity is a maximum and compare it to the escape velocity. From Eq. 15-18,

$$v_{esc} = \sqrt{\frac{2g_{cel}R_{cel}^2}{r_{esc}}} = \sqrt{\frac{2(32.2\ \text{ft/s}^2)[(3960\ \text{mi})(5280\ \text{ft/mi})]^2}{5000\ \text{mi}\ (5280\ \text{ft/mi})}}$$

$$= 32{,}660\ \text{ft/s} = 22{,}270\ \text{mph}$$

The comet will escape.

EXAMPLE 15-5

An earth satellite of mass $m = 1000$ kg is planned to move in an elliptic orbit with semimajor axis $a = 3 \times 10^7$ m and semiminor axis $b = 2 \times 10^7$ m. Calculate the maximum and minimum velocities of the satellite. The earth is at focus C of the ellipse in Fig. 15-10a.

SOLUTION

Since the satellite is moving under a central force, from Eq. 15-5,

$$r_1 m v_1 \sin\phi_1 = r_2 m v_2 \sin\phi_2$$

The maximum and minimum velocities occur when $\sin\phi = \sin 90^\circ = 1$. The mass remains constant, hence the angular momentum relation can be written as

$$r_{max}v_{min} = r_{min}v_{max} \qquad \textbf{(a)}$$

This equation involves two unknown velocities. Therefore, one of the velocities must be calculated using another relationship. Considering v_{max}, the following equation is written using Fig. 15-10, Eq. 13-6, and Eq. 15-13,

$$\frac{GMm}{r_{min}^2} = m\frac{v_{max}^2}{\rho} \tag{b}$$

The radius of curvature ρ at point A in Fig. 15-10 is expressed according to a well-known formula as

$$\rho = \frac{[1 + (dx/dy)^2]^{3/2}}{|d^2x/dy^2|} \tag{c}$$

The general equation of an ellipse centered at the origin of the xy axes is

$$\frac{x^2}{a^2} + \frac{y^2}{b^2} = 1 = \frac{x^2}{(3 \times 10^7 \text{ m})^2} + \frac{y^2}{(2 \times 10^7 \text{ m})^2}$$

Taking the derivative twice with respect to y yields

$$\frac{2x}{a^2}\left(\frac{dx}{dy}\right) + \frac{2y}{b^2} = 0$$

$$\frac{dx}{dy} = -\frac{a^2}{b^2}\left(\frac{y}{x}\right)$$

$$\frac{d^2x}{dy^2} = -\frac{a^2}{b^2}\left(\frac{1}{x}\right)$$

At point A, $x = a$ and $y = 0$, so

$$\frac{dx}{dy} = 0 \qquad \frac{d^2x}{dy^2} = -\frac{a}{b^2}$$

$$\rho = \frac{[1 + (0)^2]^{3/2}}{a/b^2} = \frac{(2 \times 10^7 \text{ m})^2}{3 \times 10^7 \text{ m}} = 1.33 \times 10^7 \text{ m}$$

Note the difference between radius of curvature (ρ) and distance from the center of the earth (r) at any given point on the elliptical orbit. Solving Eq. (b) for v_{max} and substituting from Eq. 15-14 gives

$$v_{max} = \left(\frac{\rho GM}{r_{min}^2}\right)^{1/2} = \left(\frac{\rho g R^2}{r_{min}^2}\right)^{1/2}$$

$$= \left[\frac{(1.33 \times 10^7 \text{ m})(9.81 \text{ m/s}^2)(6.37 \times 10^6 \text{ m})^2}{(7.64 \times 10^6 \text{ m})^2}\right]^{1/2}$$

where $r_{min} = 7.64 \times 10^6$ m was determined from Eqs. 15-19 and 15-20. The result is

$$v_{max} = 9524 \text{ m/s} \qquad \text{at} \quad x = 3 \times 10^7 \text{ m} \quad \text{and} \quad y = 0$$

From Eq. (a),

$$v_{min} = \frac{r_{min}}{r_{max}} v_{max} = \left(\frac{7.64 \times 10^6 \text{ m}}{5.236 \times 10^7 \text{ m}}\right)(9524 \text{ m/s})$$

where $r_{max} = 5.236 \times 10^7$ m was determined from Eq. 15-19. Finally,

$$v_{min} = 1390 \text{ m/s} \qquad \text{at} \quad x = -3 \times 10^7 \text{ m} \quad \text{and} \quad y = 0$$

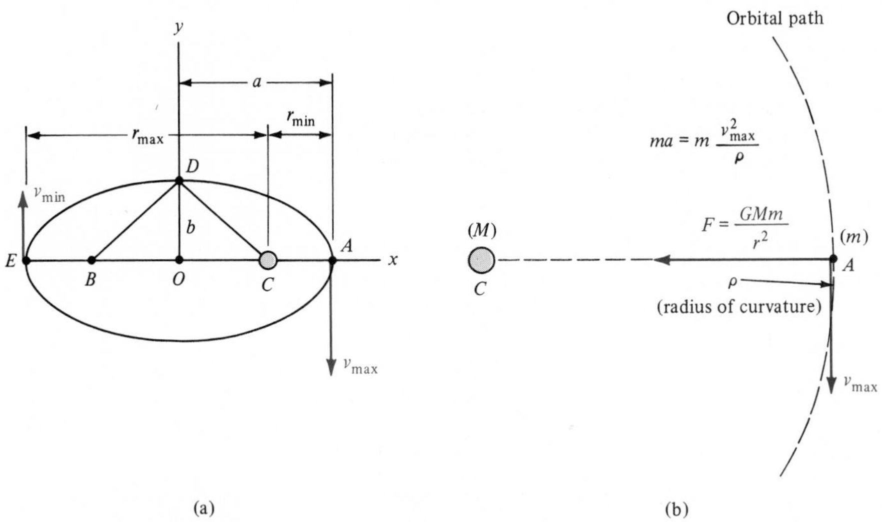

FIGURE 15-10
Elliptic orbit of an earth satellite

JUDGMENT OF THE RESULTS

The value of ρ is acceptable because it is less than a, the radius of a circle that could be drawn through point A while centered at O. The value of r_{min} might appear small, but it is reasonable for an elliptical orbit, and it allows the satellite to pass about 1300 km above the surface of the earth. The large velocities are possible in celestial motion.

PROBLEMS

15–23 Assume that the moon travels in a circular orbit of radius $r = 239{,}000$ mi around the earth. Calculate the time of a complete revolution of the moon.

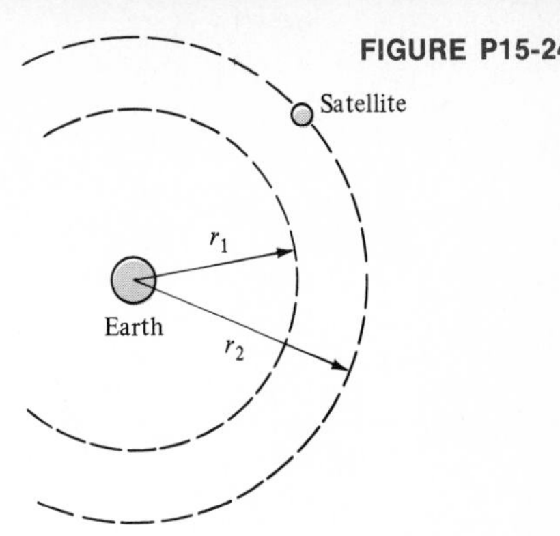

FIGURE P15-24

15–24 Two satellites revolve around the earth in concentric circular orbits. Determine the orbital speeds v_1 and v_2 for $r_1 = 2 \times 10^7$ m and $r_2 = 3 \times 10^7$ m, respectively.

15–25 Satellites in synchronous orbit always remain above the same point on the earth, as O, A, and B in Fig. P15-25 remain lined up. Calculate the radius r of the synchronous orbit. $R = 6370$ km.

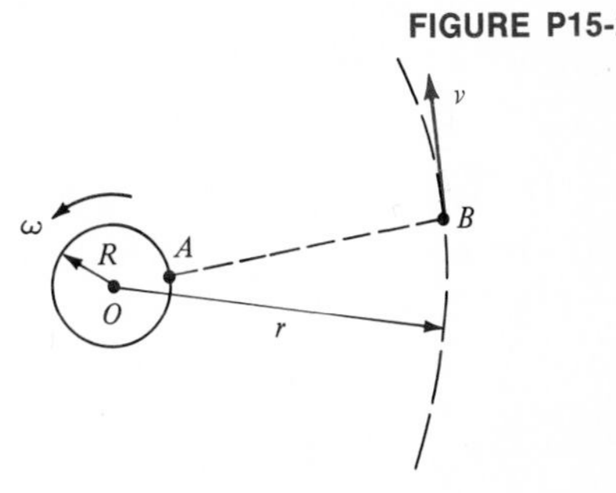

FIGURE P15-25

15–26 An earth satellite is moving in an elliptical orbit as shown. The closest approach is at point A, at $h = 1000$ km. Determine the dimension a of the orbit if $v_A = 30\,000$ km/h and $R = 6370$ km.

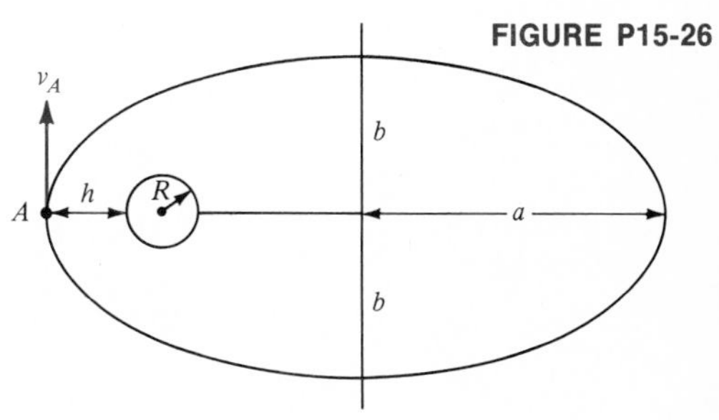

FIGURE P15-26

15–27 An earth satellite is planned to travel in an elliptical orbit whose major axis is $a =$ 10,000 mi (Fig. P15-26). Determine the required speed v_A of the satellite at point A if h must be 300 mi and $R = 3960$ mi.

15–28 The radar record of an earth satellite's orbit shows that $h = 700$ km and $a = 10\,000$ km (Fig. P15-26). Determine the maximum and minimum velocities of the satellite using $R = 6370$ km.

FIGURE P15-29

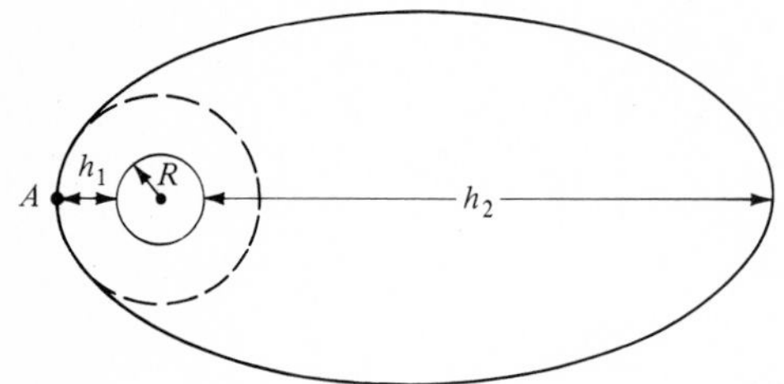

15–29 An earth satellite is in circular orbit at an altitude $h_1 = 400$ mi. Calculate the velocity in this orbit. Next, calculate the required change in velocity at point A to obtain a new orbit with a maximum altitude $h_2 = 30{,}000$ mi as shown. $R = 3960$ mi.

15–30 An earth satellite is in circular orbit at an altitude $h_1 = 500$ km and traveling at a speed v_1. At point A the original speed is increased by 25%. Determine the maximum altitude h_2 (Fig. P15-29) of the new orbit. $R = 6370$ km.

15–31 An earth satellite is in elliptical orbit with minimum altitude $h_1 = 300$ mi and maximum altitude $h_2 = 21{,}000$ mi (Fig. P15-29). Determine the velocity at point A.

15–32 In Prob. 15–31, determine the required change in velocity at point A to obtain a circular orbit of radius $r = h_1 + R$, where $R = 3960$ mi.

15–33 A space vehicle is in circular orbit of radius $r = 10\,000$ km around the earth. Calculate the escape velocity from the earth starting from this orbit.

15-6 IMPULSE AND MOMENTUM

The concept of momentum is especially useful for analyzing motion where time must be considered. For example, a force may act on a particle during a specified time interval. The final momentum in that case depends on the initial momentum, the force, and the time interval of the action of the force. The analysis of such a problem is facilitated by using a special method based on Newton's second law for what is called *impulsive motion* of a particle. This method is developed in the following.

Assume that a force **F** is acting on a particle of constant mass m. Newton's second law is written as

$$\mathbf{F} = \frac{d}{dt}(m\mathbf{v}) = \dot{\mathbf{G}}$$

where $\mathbf{G} = m\mathbf{v}$ is the linear momentum of the particle. Multiplying this equation by dt and integrating it from time t_1 to time t_2 yields

$$\mathbf{F}\,dt = d(m\mathbf{v})$$

$$\int_{t_1}^{t_2} \mathbf{F}\,dt = m\mathbf{v}_2 - m\mathbf{v}_1 \tag{15-21}$$

The integral on the left side in Eq. 15-21 is called the *linear impulse*, or simply the *impulse*. The impulse is said to be *of the force*, and acting *on the mass*. In simple terms, impulse is equal to the change in linear momentum of the particle. Thus, the symbol $\Delta\mathbf{G}$ is appropriate for impulse. The force **F** in Eq. 15-21 is sometimes called *impulsive force*.

In general, the force may change both in magnitude and direction, even during a very short time interval. To analyze such motion it is necessary to evaluate the integral of $\mathbf{F}\,dt$. It is often most convenient to use rectangular components of the force. For convenience, the impulse is written as the vector $\Delta\mathbf{G}$,

$$\Delta\mathbf{G} = \mathbf{G}_2 - \mathbf{G}_1 = m\mathbf{v}_2 - m\mathbf{v}_1$$

where the subscripts refer to time t_2 and time t_1. With this,

$$\Delta\mathbf{G} = \int_{t_1}^{t_2} \mathbf{F}\,dt \tag{15-22}$$

$$= \mathbf{i}\int_{t_1}^{t_2} F_x\,dt + \mathbf{j}\int_{t_1}^{t_2} F_y\,dt + \mathbf{k}\int_{t_1}^{t_2} F_z\,dt$$

The x, y, and z components of the change in linear momentum are set equal to the x, y, and z components of the impulse,

$$\Delta G_x = mv_{x_2} - mv_{x_1} = \int_{t_1}^{t_2} F_x\,dt$$

$$\Delta G_y = mv_{y_2} - mv_{y_1} = \int_{t_1}^{t_2} F_y\,dt$$

$$\Delta G_z = mv_{z_2} - mv_{z_1} = \int_{t_1}^{t_2} F_z\,dt$$

It must be kept in mind that impulse and momentum are vectors, and the scalar components are the most convenient to use in solving problems. A few major possibilities involving impulsive motion in a plane are shown in Fig. 15-11. Assume that in each case the initial momentum $m\mathbf{v}_1$ is known, a given force $\mathbf{F}$ acts on the particle, and the final momentum $m\mathbf{v}_2$ is to be determined. From Eq. 15-21,

$$\underbrace{m\mathbf{v}_1}_{\substack{\text{initial}\\ \text{momentum}}} + \underbrace{\int_{t_1}^{t_2} \mathbf{F}\,dt}_{\text{impulse}} = \underbrace{m\mathbf{v}_2}_{\substack{\text{final}\\ \text{momentum}}} \qquad \boxed{15\text{-}23}$$

The scalar or vector calculations with this equation are illustrated in Fig. 15-11. For any in-line impact the initial momentum and the

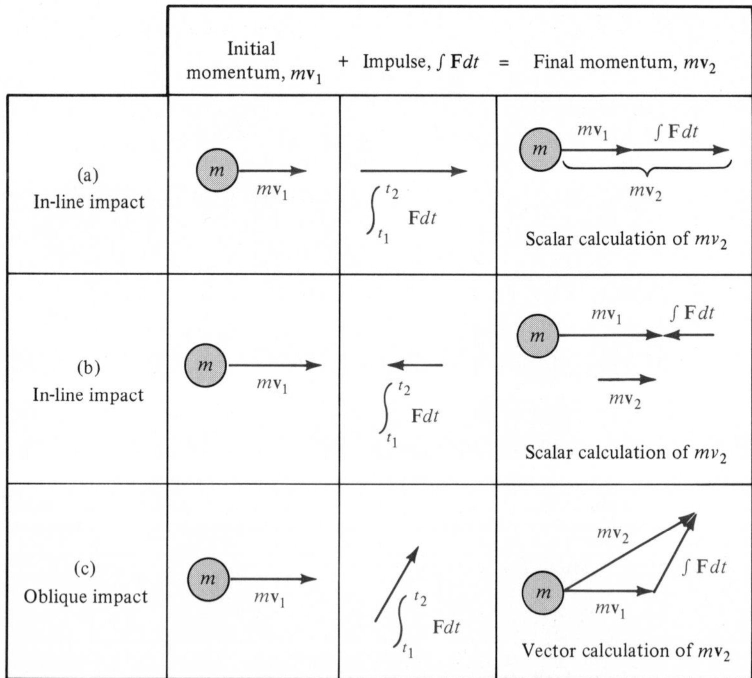

FIGURE 15-11

Major possibilities of impulsive motion of a particle

impulse are collinear and their magnitudes can be summed as scalars to obtain the magnitude of the final momentum. In all other cases they must be summed as vectors. These concepts are valid for all bodies that can be treated as particles that cannot deform or rotate for all practical purposes.

The units of impulse of a force are the same as those for linear momentum. In SI it is N·s = kg·m/s, and in U.S. customary units it is lb·s = slug·ft/s.

15-7 CONSERVATION OF TOTAL MOMENTUM OF PARTICLES

Consider the motion of two particles that temporarily interact. The mutual forces between the particles can be described using Newton's law of action and reaction. The force acting on each particle is equal in magnitude and opposite in direction to the force acting on the other particle. Both forces act in the same time interval. The resulting change in velocity of each particle depends on its mass. To analyze these changes in velocity it is useful to consider the changes in momentum of two interacting particles together.

Assume that two particles of mass m_A and m_B are at rest and touching a compressed spring as in Fig. 15-12a. The spring is suddenly released and the particles are propelled apart (Fig. 15-12b).

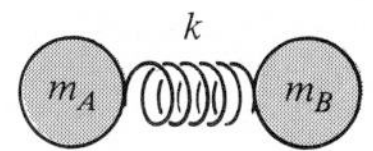

(a) Particles are at rest, spring is compressed

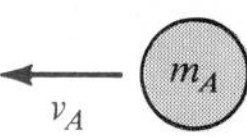

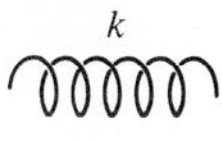

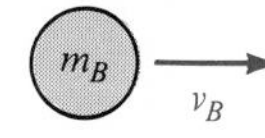

(b) Spring is released, particles are propelled apart

FIGURE 15-12

Conservation of total momentum of two interacting particles

The same force is acting on the two particles during their accelerations, so their changes in momentum can be written as

$$m_A\mathbf{v}_{A_2} - m_A\mathbf{v}_{A_1} = \int_{t_1}^{t_2} \mathbf{F}\, dt$$

$$m_B\mathbf{v}_{B_2} - m_B\mathbf{v}_{B_1} = -\int_{t_1}^{t_2} \mathbf{F}\, dt$$

Adding these two equations gives

$$m_A\mathbf{v}_{A_2} - m_A\mathbf{v}_{A_1} + m_B\mathbf{v}_{B_2} - m_B\mathbf{v}_{B_1} = 0$$

$$m_A\mathbf{v}_{A_1} + m_B\mathbf{v}_{B_1} = m_A\mathbf{v}_{A_2} + m_B\mathbf{v}_{B_2}$$

where the individual changes occur in the same time interval Δt. Note that it does not matter how the force varies during Δt. It is concluded that the sum of the momenta of the two particles A and B at any time is a constant,

$$m_A \mathbf{v}_A + m_B \mathbf{v}_B = \text{constant} \tag{15-24}$$

which means that *the total momentum of the interacting particles is conserved* when only a mutual force is acting on them.

The same result can be obtained by applying Eq. 15-23 to analyze the impulsive motion of two or more particles simultaneously. For each particle i the initial momentum is $m_i \mathbf{v}_{i_1}$, and the final momentum is $m_i \mathbf{v}_{i_2}$. After vectorially adding the momenta for all the particles and the impulses for all the forces acting on them, Eq. 15-23 can be written for n particles as

$$\sum_i^n (m_i \mathbf{v}_i)_1 + \sum_i^n \int_{t_1}^{t_2} \mathbf{F}_i \, dt = \sum_i^n (m_i \mathbf{v}_i)_2 \tag{15-25}$$

Here the integral represents the impulses of external forces acting on the particles *and* of interactive forces between the particles. The interactive forces always cancel out in pairs. Thus, in the absence of other external forces (such as friction) the final and initial momenta are the same,

$$\underbrace{\sum_i^n (m_i \mathbf{v}_i)_1}_{\substack{\text{total initial} \\ \text{momentum at} \\ \text{time } t_1}} = \underbrace{\sum_i^n (m_i \mathbf{v}_i)_2}_{\substack{\text{total final} \\ \text{momentum at} \\ \text{time } t_2}} \tag{15-26}$$

Equations 15-24 and 15-26 both express the conservation of total momentum for interactive particles that have only mutual conservative forces acting on them. It is worth noting that the change in velocity of a particle in mutual impulsive motion with another particle is inversely proportional to its mass.

15-8 APPLICATIONS OF PRINCIPLE OF IMPULSE AND MOMENTUM

Problems involving impulsive motion of two bodies can be grouped according to the significance of the velocity change for each body.

1. *The velocity change is significant only for one body.* This is the case when the mass of body A is much larger than the mass of body B. If the velocity of A does not change significantly (or even measurably), all of the change in momentum is assigned to body

B. In this case Eq. 15-23 is used and the conservation of total momentum cannot be applied. For example, a small meteor striking the earth, or an automobile hitting a large structure are in this category.

2. *The velocity change is significant for both bodies.* This is the situation when body *A* is not enormously larger in mass than body *B*. The conservation of total momentum (Eq. 15-24 or 15-26) is the ideal method for such problems. For example, the simplest modeling of vehicle collisions is in this category.

Approximations of Impulses

The ideal impulsive force has a square-wave shape as in Fig. 15-13a. In that case the impulse can be expressed as

$$\int_{t_1}^{t_2} F\,dt = F\,\Delta t \qquad \boxed{15\text{-}27}$$

In a real mechanical system the force varies relatively smoothly such as $F(t)$ in Fig. 15-13b. Nevertheless, in some cases the area $\int F\,dt$ under the curve $F(t)$ may be approximated by inspection using the equivalent quantities F_{eq} and Δt_{eq} in Eq. 15-27.

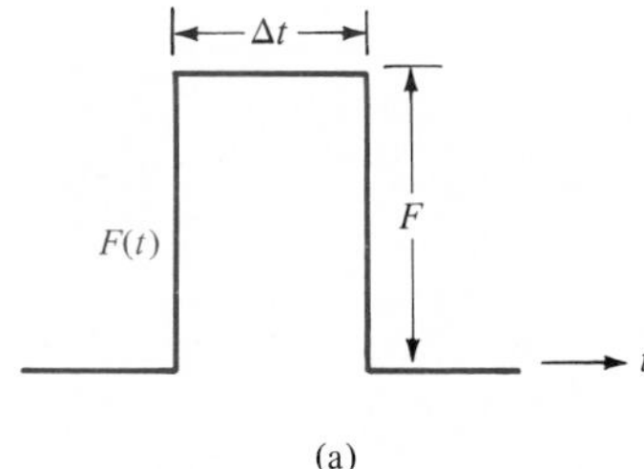

(a)

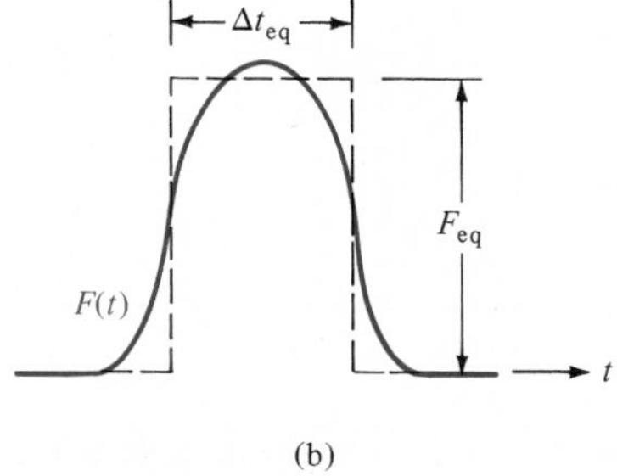

(b)

FIGURE 15-13

Ideal and realistic impulsive forces

EXAMPLE 15-6

A ballistics test consists of the setup shown in Fig. 15-14. The coefficient of friction between the 100-kg gun–platform system and the support is $\mu_k = 0.7$. Determine (a) the speed $(v_P)_1$ of the platform just after the gun fires a 0.05-kg bullet at a speed $v_B = 1000$ m/s, and (b) the distance Δx_p traveled by the platform before the bullet's impact at A.

SOLUTION

(a) The initial speed of the platform $(v_P)_1$ is found using Eq. 15-25,

$$\underbrace{\sum (m_i v_i)_0}_{0} + \underbrace{\sum \int F\, dt}_{0} = \sum (m_i v_i)_1$$

$$0 = (0.05 \text{ kg})(-1000 \text{ m/s}) + (100 \text{ kg})(v_p)_1$$

$$(v_P)_1 = 0.5 \text{ m/s}$$

(b) The distance Δx_p traveled by the platform is equal to $x - x_0$ in Eq. 11-9,

$$v^2 = v_0^2 + 2a(x - x_0)$$

Here $v_0 = (v_P)_1 = 0.5$ m/s, but $v = (v_P)_2^*$ and a must be calculated yet. This involves several intermediate steps as follows. The relative velocity of the bullet to the platform just after firing is

$$v_{B/P} = v_B - v_P = -1000 \text{ m/s} - (0.5 \text{ m/s}) = -1000.5 \text{ m/s}$$

Neglecting air resistance, the time Δt the bullet takes to travel the fixed distance of 2 m is

$$x_{B/P} = (x_{B/P})_0 + (v_{B/P})_0\, \Delta t$$

$$-2 \text{ m} = 0 + (-1000.5 \text{ m/s})\, \Delta t$$

$$\Delta t = 0.002 \text{ s}$$

Looking at the motion of the platform alone during this time interval, the friction force ($\mathscr{F} = \mu N = \mu W$) is an impulsive force. Thus,

* The different notations from those in Eq. 11-9 are useful to distinguish the various phases of the impulsive motion.

$$mv_1 + \mathcal{F}\,\Delta t = mv_2$$

$$(100 \text{ kg})(0.5 \text{ m/s}) - (0.7)(100 \text{ kg})(9.81 \text{ m/s}^2)(0.002 \text{ s}) = (100 \text{ kg})(v_P)_2$$

$$(v_P)_2 = 0.486 \text{ m/s} \qquad \text{(velocity of platform to the right just prior to impact)}$$

From Newton's second law, $-\mathcal{F} = ma$ ($\mathcal{F}$ is negative since friction opposes the motion). Thus, the acceleration a of the platform is

$$a = -\frac{\mathcal{F}}{m} = \frac{-(0.7)(100 \text{ kg})(9.81 \text{ m/s}^2)}{(100 \text{ kg})} = -0.7 \text{ g}$$
$$= -6.87 \text{ m/s}^2$$

The distance traveled is calculated using Eq. 11-9,

$$v^2 = v_0^2 + 2a(x - x_0)$$
$$(0.486 \text{ m/s})^2 = (0.5 \text{ m/s})^2 + 2(-6.87 \text{ m/s}^2)(\Delta x_p)$$
$$\Delta x_p = 0.001 \text{ m}$$

JUDGMENT OF THE RESULTS

The small initial speed of the platform is reasonable considering the masses of the bullet and the platform. The small distance traveled by the platform prior to the bullet's impact at A is reasonable considering the small average velocity of the platform and the short time of travel of the bullet in the air.

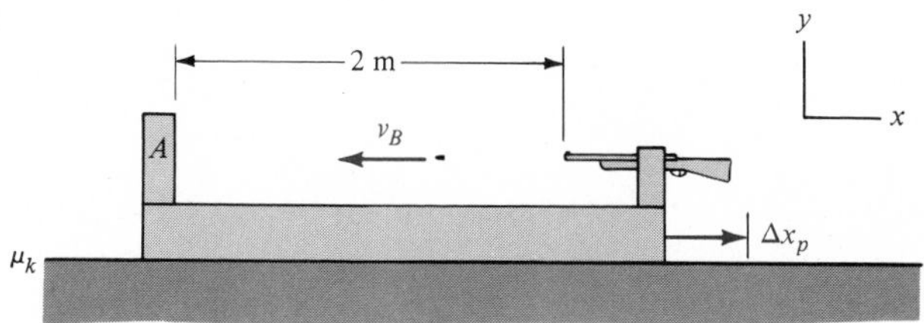

FIGURE 15-14

Gun–platform system for a ballistics test

EXAMPLE 15-7

(a)

Consider an athlete making vertical jumps on a force platform which electronically measures force (Fig. 15-15a). Figure 15-15b is the recording of force vs. time in a single jump from start to finish. Zero force corresponds to the person standing still, and the minimum force represents the person's weight (while the jumper is airborne). The takeoff velocity is the important factor in jumping. Determine the approximate takeoff velocity and height of the jump by estimating the area under the force vs. time plot in the relevant part of the motion. Each division of the vertical axis is 20 lb. The force recording is a graph of the *resultant* force $R_y = N - W$ acting on the person (N = normal force, W = weight).

SOLUTION

First, it is useful to analyze the major parts of the motion. While the person is standing still the normal force N exactly equals the weight W, so

$$R_y = \sum F = N - W = ma_y = 0$$

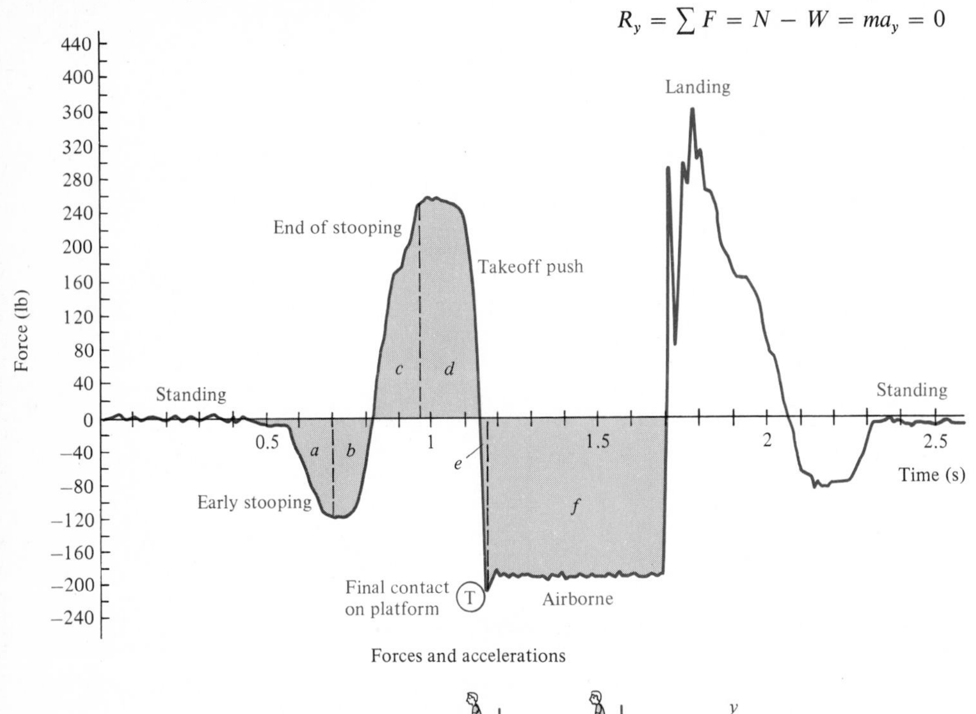

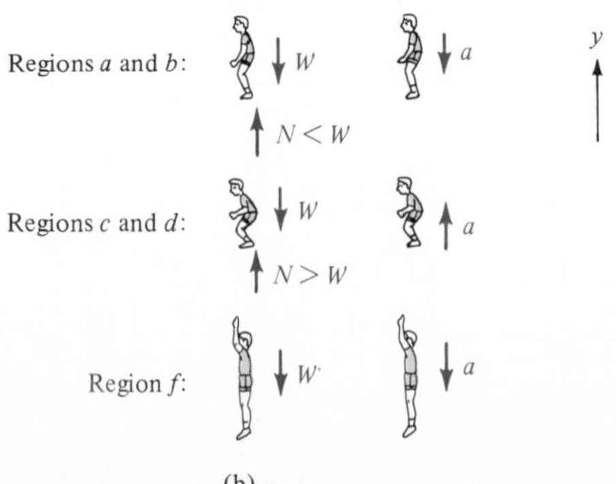

(b)

FIGURE 15-15

Test of an athlete on a force platform

At the beginning of the stooping process the acceleration of the center of gravity C is negative, hence a negative resultant force is registered,

$$R_y = ma_y = -ma_C \qquad \text{(i.e., } N < W\text{)}$$

Nearing the end of the stooping process, the person is decreasing a negative velocity which requires a positive force larger than the body weight,

$$R_y = ma_y = ma_C \qquad \text{(i.e., } N > W\text{)}$$

During the push-off process the person accelerates his center of gravity up (increasing a positive velocity), so the force registered is still positive. While in the air, the resultant force acting on the person is his weight (in the negative direction) and hence the force platform records a negative force of

$$R_y = \sum F = -W = -mg \qquad (N = 0)$$

Therefore, the weight of the jumper is seen to be about 180 lb. Considering the original problem, Eq. 15-23 is applicable:

$$mv_1 + \int R_y\, dt = mv_2$$

The integral $\int R_y\, dt$ is estimated as the area under the force vs. time curve. "Takeoff" occurs at point T. Breaking the area up into several component areas and approximating them as triangles and a rectangle,

$$\begin{aligned}\int R_y\, dA &= A_a + A_b + A_c + A_d + A_e \\ &= \frac{1}{2}(0.12\text{ s})(-120\text{ lb}) + \frac{1}{2}(0.11\text{ s})(-120\text{ lb}) \\ &\quad + \frac{1}{2}(0.14\text{ s})(240\text{ lb}) + (0.16\text{ s})(240\text{ lb}) \\ &\quad + \frac{1}{2}(0.02\text{ s})(-180\text{ lb}) \\ &= 41.4\text{ lb}\cdot\text{s}\end{aligned}$$

Therefore,

$$0 + 41.4\text{ lb}\cdot\text{s} = \left(\frac{180\text{ lb}}{32.2\text{ ft/s}^2}\right)v_2$$

$$v_2 = 7.41\text{ ft/s}$$

The height of the jump is calculated from the work–energy equation (Eq. 14-14),

$$T_1 + U_{12} = T_2$$

$$\frac{1}{2}(m)(7.41\text{ ft/s})^2 - mgh = 0$$

$$h = \frac{(7.41\text{ ft/s})^2}{2g} = 0.85\text{ ft (elevation of center of gravity)}$$

A more accurate analysis of the jump using the same plot is left as an exercise for the student. Note the negative signs of the areas A_a, A_b, and A_e in the summation for $\int R_y\, dA$. These represent that the resultant force allows the body to fall in those intervals.

PROBLEMS

SECTIONS 15-6, 15-7, 15-8

15–34 A 1-kg particle is initially at rest. A force $\mathbf{F} = 10\mathbf{i}$ N acts on the particle from time $t_1 = 0$ to $t_2 = 2$ s. Calculate the particle's final velocity.

15–35 A 4-lb particle is traveling at a velocity $\mathbf{v}_1 = 10$ ft/s $\mathbf{j}$ when a force $\mathbf{F} = 50$ lb $\mathbf{j}$ starts acting on it. Determine the particle's velocity $\mathbf{v}_2$ after a time interval of 0.5 s during which the force has acted.

15–36 A 10-kg block is traveling at a velocity $\mathbf{v}_1 = 20\mathbf{i}$ m/s at time $t_1 = 0$. A force $\mathbf{F} = -2\mathbf{i}$ kN acts on the block from time t_1 to $t_2 = 1$ s. Calculate the block's velocity $\mathbf{v}_2$ at times $t > t_2$.

15–37 A 5-lb block is traveling at a velocity $\mathbf{v}_1 = (-3\mathbf{j} + 5\mathbf{k})$ ft/s at time $t_1 = 2$ s. A force $\mathbf{F} = (400\mathbf{j} - 400\mathbf{k})$ lb acts on the block from time t_1 to $t_2 = 4$ s. Determine the block's velocity at time t_2.

15–38 A 0.5-kg particle is initially at rest. A force $\mathbf{F} = (50 + 10t)\mathbf{i}$ N acts on the particle in a time interval of 3 s. Calculate the final velocity of the particle if t is in seconds.

15–39 A 0.2-lb particle is traveling at a velocity $\mathbf{v}_1 = 20\mathbf{i}$ ft/s at time $t_1 = 1$ s. A force $\mathbf{F} = (30t + 40t^2)\mathbf{j}$ lb acts on the particle from time t_1 to $t_2 = 1.2$ s. Determine the particle's velocity at times $t > t_2$.

15–40 A highway divider is to be protected from crashing automobiles by a large, resilient bumper A. What is the average force acting on the bumper when it arrests a 2000-kg car traveling initially at 40 m/s? The car is stopped in a time interval of 0.2 s.

FIGURE P15-40

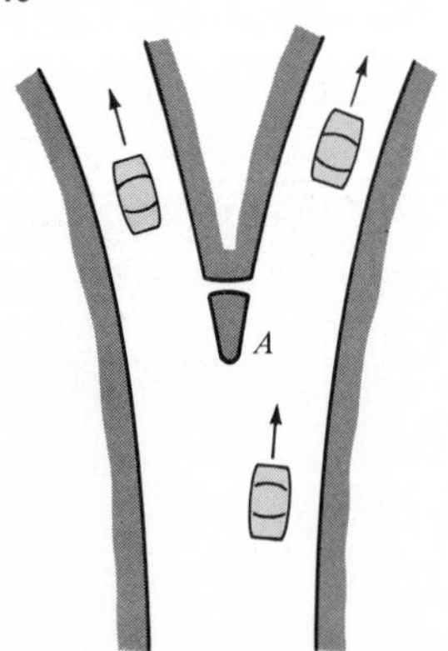

15–41 A new rifle is to be fired while firmly attached to a heavy bench. Calculate the average force applied to the vise A if a 0.05-lb bullet traverses the barrel in a time of 2 ms and exits at a speed of 2000 ft/s.

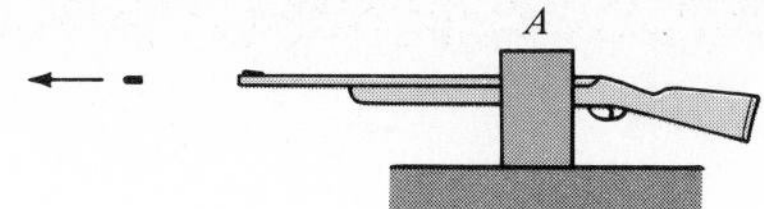

FIGURE P15-41

15–42 A 2000-kg car is traveling on a horizontal road at a speed of 100 km/h. Calculate the time required for the car to skid to a stop with the brakes applied if $\mu_k = 0.9$. Neglect air resistance.

15–43 A 3000-lb car is traveling up a 20% grade at a speed of 60 mph when the brakes are applied. Calculate the time required for the car to skid to a stop if $\mu_k = 0.5$, neglecting air resistance.

15–44 A 1500-kg car is traveling down a 15% grade at a speed of 70 km/h when the brakes are applied. Determine the time required for the car to skid to a stop if $\mu_k = 0.7$, neglecting air resistance.

15–45 A 2500-lb car is traveling down a 5% grade at 50 mph when the brakes are applied. Because of heat generated at the skidding tires, $\mu_k = 0.9 - 0.03t$, where t is in seconds. Calculate the time required to stop the car neglecting air resistance.

15–46 The seat belt for a car is assumed to be in the same plane as the velocity vector of the car. The belt should not break in restraining a 100-kg passenger when the car traveling at 100 km/h is stopped in 0.05 s. Calculate the required minimum strength of the belt.

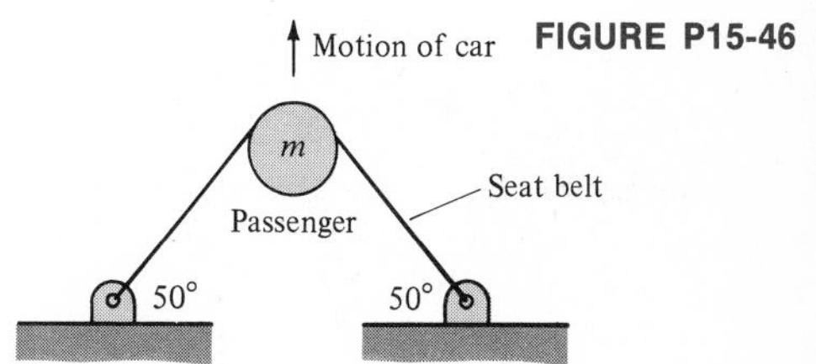

FIGURE P15-46

15–47 A 130-lb high jumper elevates her center of gravity by 2.5 ft. What is the average force exerted by her leg if it touches the ground for a time of 0.14 s during takeoff?

FIGURE P15-48

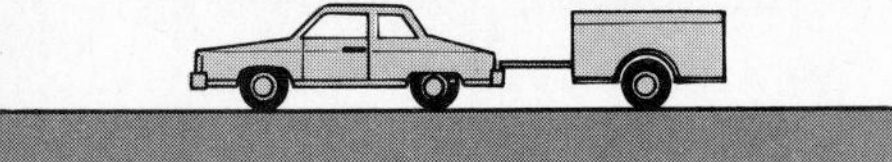

15–48 A 2000-kg car is pulling a 1000-kg trailer at a speed of 90 km/h. The brakes on the car (and none on the trailer) are applied causing skidding of its wheels. Calculate the horizontal force F between the car and the trailer during the initial time interval $\Delta t = 1$ s of the skidding if $\mu_k = 0.5$. Treat the car and the trailer as particles.

15–49 Consider the recording of force vs. time for a single vertical jump on a force platform. Estimate the initial impulse and check if it should make the jumper airborne for the length of time measured. Zero force corresponds to the person standing (or crouching) still on the platform.

FIGURE P15-49

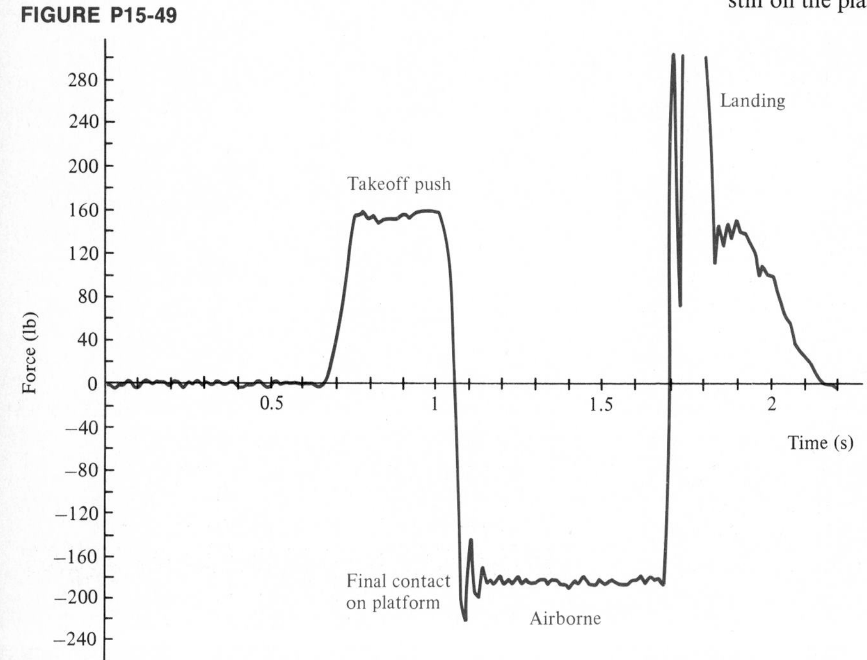

15–50 Reconsider Ex. 15-7 and Fig. 15-15. Compare and discuss the approximate total impulse of the jumper during takeoff with the total impulse during landing on the force platform.

15–51 Because of improper programming of the automatic recording equipment of a force platform, part of the data during landing after a jump was "clipped" as shown in Fig. P15-49. Estimate the part of the recording that was lost.

15–52 An important training method for athletes who need explosive movement by the legs is called *depth jumping*. In this exercise a person jumps from a bench to the floor and immediately rebounds vertically. Consider a 70-kg person who freely falls from a height of 0.6 m and rebounds to the same elevation. Calculate the average normal force acting on this person if the time interval of contact with the floor is 0.5 s.

15–53 A 0.05-kg bullet is fired from a 4-kg gun. The bullet leaves the barrel at a speed of 700 m/s. Calculate the recoil velocity of the gun.

15–54 A box that weighs 20 lb must be given an initial upward velocity of 20 ft/s by a spring-loaded mechanism. Calculate the recoil velocity of the 150-lb base structure. Assume that the box and the base structure are particles on opposite sides of the spring.

15–55 Reconsider Ex. 15-6 and Fig. 15-14. Determine the distance d that the platform travels after the bullet's impact at A. Assume that the bullet is stopped instantaneously upon impact.

15–56 Railroad car A of mass $m_A = 30\,000$ kg is moving at velocity $\mathbf{v}_A = 2\mathbf{i}$ km/h toward car B of mass $m_B = 15\,000$ kg, which moves at velocity $\mathbf{v}_B = 0.3\mathbf{i}$ km/h. The cars are coupled on contact and move together at velocity $\mathbf{v}_f$. Determine $\mathbf{v}_f$.

FIGURE P15-56

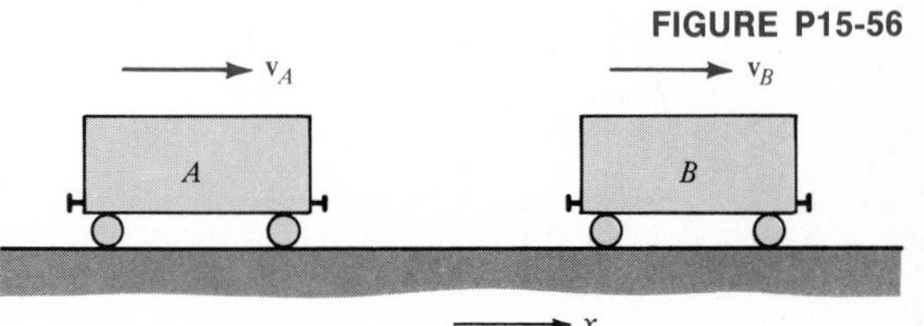

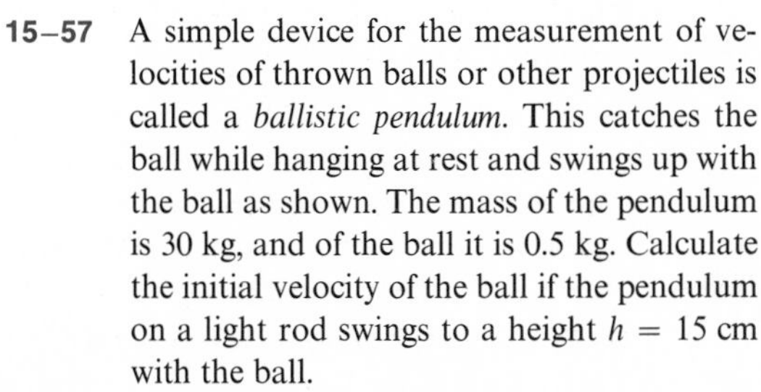
15–57 A simple device for the measurement of velocities of thrown balls or other projectiles is called a *ballistic pendulum*. This catches the ball while hanging at rest and swings up with the ball as shown. The mass of the pendulum is 30 kg, and of the ball it is 0.5 kg. Calculate the initial velocity of the ball if the pendulum on a light rod swings to a height $h = 15$ cm with the ball.

FIGURE P15-57

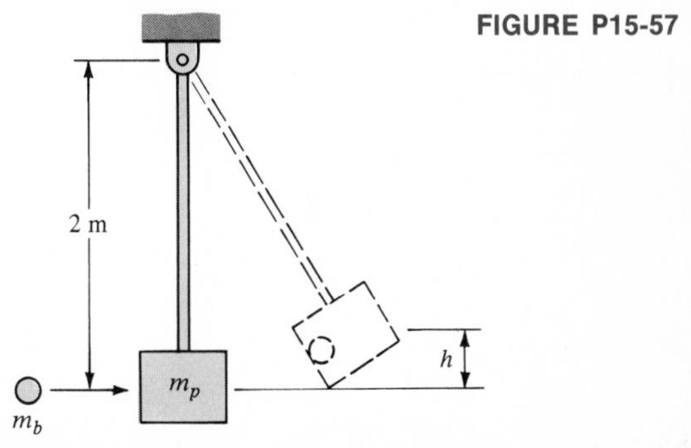

15-9 COLLISION OF DEFORMABLE BODIES

The principles of impulse and momentum were introduced in Secs. 15-6 through 15-8 for particles and for bodies that could be treated as particles in situations where the deformations and rotations of them could be ignored. At this stage the concepts developed for particles and particle-like bodies are extended to deal with deformable bodies that can be assumed to interact without rotations. The additional analysis of rotations is in the area of rigid body dynamics (Chap. 19).

Before considering the details of impulsive motion with deformations of the bodies, it is necessary to define the various possible relative motions at the instant of impact. These relative motions are described in reference to the line that is perpendicular to the plane of contacting surfaces of the bodies. This line is called the *line of impact* and is shown in Fig. 15-16. The relative motion is a *central impact* when the mass centers of the two bodies are on the line of impact during the impact. Otherwise, the motion involves *eccentric impact*. Only central impact is discussed in this chapter; it may occur in two ways as shown in Fig. 15-16. These are discussed in the following two sections.

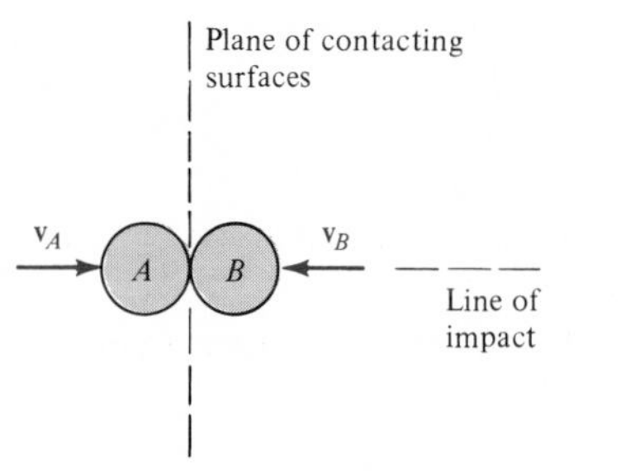

(a) Direct central impact

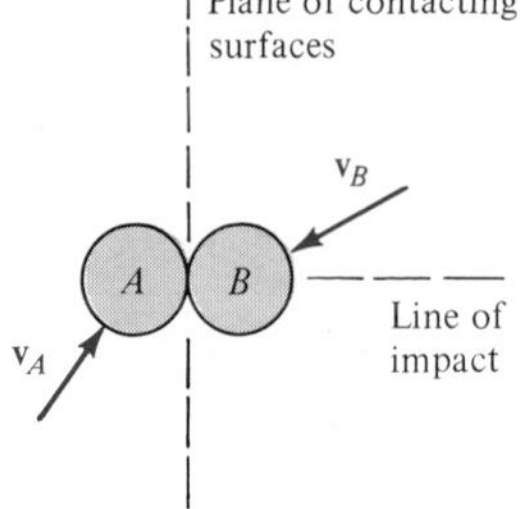

(b) Oblique central impact

FIGURE 15-16
Direct and oblique central impacts

15-10 DIRECT CENTRAL IMPACT

In the simple case when two masses are in collinear impact, as in Fig. 15-16a, the only matter of interest is the rebound of the bodies as it is affected by their deformations. It should be remembered that all bodies deform during an impact since all forces cause deformations. Each deformation has two components, an *elastic or recoverable deformation*, and a *plastic or permanent deformation*. Generally, both of these occur simultaneously, but either one may be dominant. They can also be equal.

The plastic deformation mainly dissipates energy in the form of heat, and it is this deformation that reduces the rebound velocities.

The ability of a body to deform plastically depends on the material, on the temperature, and the rate of deformation. In the impact of two bodies their deformations may be quite different overall and also in the proportions of elastic and plastic deformations. It is useful to understand this, but it is not necessary to analyze all of these details for each impact. What is important is the overall changes of velocities, regardless of which body deformed more than the other. If one body deforms plastically and the other one does not, it still affects the motion of both bodies.

The unified treatment of the impact problem is based on the principles of impulse and momentum. Assume that the bodies in impact are spheres of different masses, deformation characteristics, and velocities. It is worthwhile to consider four successive instants of the impact event as shown in Fig. 15-17.

1. The bodies before impact are traveling on a collision course. Their velocities may be in the same direction (pursuit), or in opposite directions heading toward each other. The momenta in this situation are the initial information for the analysis.
2. Contact is made, and both bodies deform. Their velocities rapidly change in this phase but are somewhat similar as the two bodies travel together for an instant. At maximum deformation $v'_A = v'_B = v_c$, where the common speed v_c is at the end of the phase.
3. The bodies are still in contact but are rebounding. At the beginning of this phase the particles still have the common speed $v''_A = v''_B = v_c$. The deformation is decreasing throughout this phase, which is called *restitution*.
4. If there is separation, the bodies travel at new velocities. The final speeds v_{Af} and v_{Bf} depend on the initial momenta and the energy lost during impact. v_{Af} may be positive or negative even if both v_A and v_B were positive.

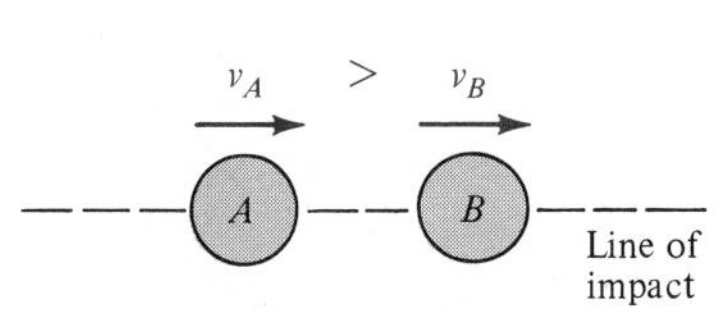

(a) Before impact (bodies approach)

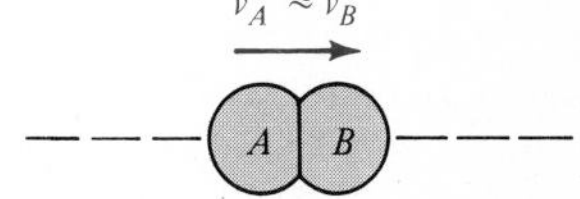

(b) Increasing deformation during contact

$v''_A \approx v''_B$

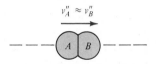

A B

(c) Decreasing deformation or restitution during contact

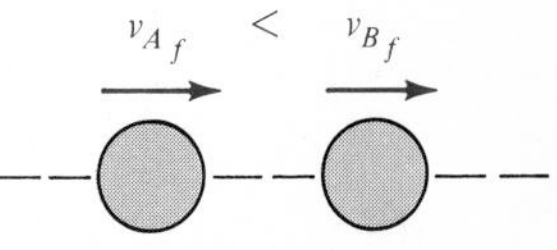

(d) After impact (bodies move apart; v_{Af} may be to the left)

FIGURE 15-17

Four successive events of a direct central impact

Assume that there are no other forces acting on the two bodies besides the mutual force between them. In this case the total momentum of the two bodies is conserved. Using Eq. 15-24 gives

$$m_A v_A + m_B v_B = m_A v_{Af} + m_B v_{Bf} \tag{15-28}$$

This equation contains two unknowns, v_{Af} and v_{Bf}. An additional relation can be obtained by considering the mutual impulses during contact. These impulses are expressed using Fig. 15-18. The impulses during increasing and decreasing deformations are caused by the mutual contact forces F_D and F_R, as shown in the figure. Using the equation of impulse and momentum for body A yields

$$\underbrace{m_A v_c}_{\substack{\text{momentum at}\\ \text{max. deformation}}} = \underbrace{m_A v_A}_{\substack{\text{initial}\\ \text{momentum}}} - \underbrace{\int_0^{t_1} F_D\,dt}_{\substack{\text{impulse of}\\ \text{contact force}\\ \text{during}\\ \text{increasing}\\ \text{deformation}}} \tag{15-29a}$$

and similarly for the restitution,

$$\underbrace{m_A v_{Af}}_{\substack{\text{final}\\ \text{momentum}}} = \underbrace{m_A v_c}_{\substack{\text{momentum}\\ \text{at max.}\\ \text{deformation}}} - \underbrace{\int_{t_1}^{t_2} F_R\,dt}_{\substack{\text{impulse of}\\ \text{contact force}\\ \text{during restitution}}} \tag{15-29b}$$

The ratio e of the restoration impulse and of the deformation impulse is written to represent the ability of the bodies to recover from the deformations.

$$e = \frac{\int_{t_1}^{t_2} F_R\,dt}{\int_0^{t_1} F_D\,dt} = \frac{m_A(v_c - v_{Af})}{m_A(v_A - v_c)} \tag{15-30}$$

For body B,

$$m_B v_c = m_B v_B + \int_0^{t_1} F_D\,dt \tag{15-31a}$$

$$m_B v_{Bf} = m_B v_c + \int_{t_1}^{t_2} F_R\,dt \tag{15-31b}$$

The ratio e for body B becomes

$$e = \frac{\int_{t_1}^{t_2} F_R\,dt}{\int_0^{t_1} F_D\,dt} = \frac{m_B(v_{Bf} - v_c)}{m_B(v_c - v_B)} \tag{15-32}$$

Since the mass cancels out in each equation, the common speed v_c at maximum deformation can be eliminated from the two ex-

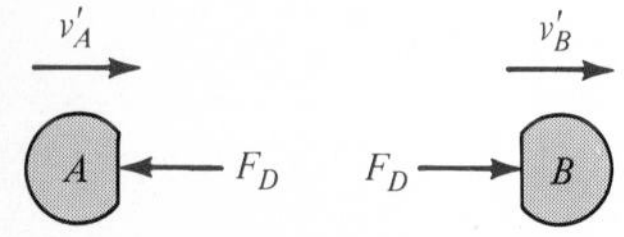

(a) Increasing deformation during contact from time 0 to t_1; velocity of A changes from v_A to v_c; $v'_A \approx v'_B$

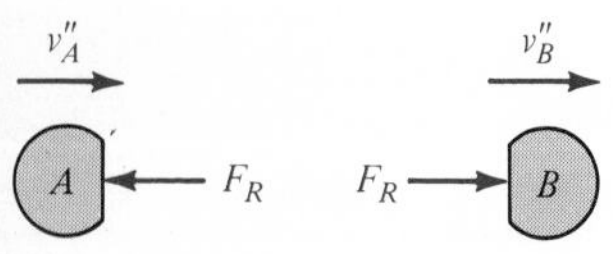

(b) Decreasing deformation or restitution during contact from time t_1 to t_2; velocity of A changes from v_c to v_{Af}; $v''_A \approx v''_B$

FIGURE 15-18

Mutual contact forces during direct central impact

pressions for e. Note that a quotient remains unchanged if an equal quotient is added to it by summing the numerators and denominators [for example, $a/b = (a + a)/(b + b)$]. Thus, from Eqs. 15-32 and 15-30,

$$e = \frac{v_{Bf} - v_c + v_c - v_{Af}}{v_c - v_B + v_A - v_c} = \frac{v_{Bf} - v_{Af}}{v_A - v_B} = \frac{\text{relative velocity of separation}}{\text{relative velocity of approach}} \qquad \boxed{15\text{-}33}$$

The ratio e is called the *coefficient of restitution.* Two extreme values of this coefficient are worth remembering.

1. *Perfectly elastic impact, e = 1.* Ideally, elastic materials sustain no permanent deformations during impact. The impulse during the increasing deformation equals the impulse during the rebound from maximum deformation. Consequently, $e = 1$, and the bodies separate with the same relative velocity as they were approaching each other (Eq. 15-33). The total kinetic energy of the bodies is conserved in this case.

2. *Perfectly plastic impact, e = 0.* Ideally, plastic materials are permanently deformed during impact and have no restoring force after the maximum deformation occurs. This means that $F_R = 0$ in Fig. 15-18b and $e = 0$ for both bodies when the ratio of the impulses is considered for the two phases of contact. Note that only one of the bodies has to be ideally plastic to obtain this condition. It is interesting that *the total momentum is conserved even though energy is lost* during the permanent deformation. The reason is that all forces are internal and cancel out in the isolated system of the two bodies. The conservation of momentum and having no rebound lead to the conclusion that the bodies move together after the impact with a common speed v_c. For initial speeds v_A and v_B,

$$m_A v_A + m_B v_B = (m_A + m_B) v_c \qquad \boxed{15\text{-}34}$$

It can be shown analytically that some of the total initial kinetic energy is lost during plastic impact. Most of this energy is converted into heat.

Real materials in impact. For real materials $0 < e < 1$ because forces cause elastic and plastic deformations simultaneously, even if one or the other is practically immeasurable. A complicating factor in this respect occasionally is the adhesive quality of the surfaces in contact, which may cause the bodies to stick together. Other effects on the coefficient of restitution are caused by the velocity of approach and temperature of the bodies. It is not possible to accurately determine e without experiments under the conditions of interest. Figure 15-19 shows the general tendencies of behavior that can be expected on the basis of experimental results.

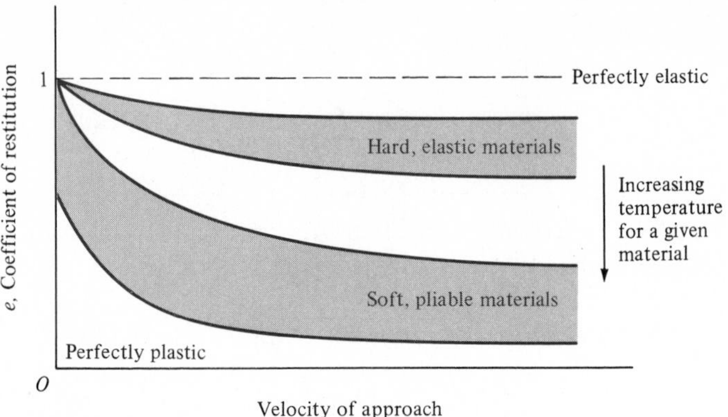

FIGURE 15-19

Schematic variation of coefficient of restitution for various materials and conditions

15-11 OBLIQUE CENTRAL IMPACT

The concepts of direct central impact can be readily extended to the case where the initial and final velocities are not collinear. This is called *oblique impact* and is illustrated in Fig. 15-20, where a coordinate system is conveniently positioned at the place of impact. The normally available initial information consists of the masses m_A and m_B, and their initial velocities $\mathbf{v}_A$ and $\mathbf{v}_B$. The final velocities $\mathbf{v}_{Af}$ and $\mathbf{v}_{Bf}$ are unknown in magnitude and direction, so four independent equations are required to determine the two components of each velocity.

The problem of oblique central impact is solved using the superposition of two motions if there is no mutual friction force between the bodies during contact. The two distinct motions are those along the x and y axes. The former is equivalent to two bodies moving along parallel lines without any chance of collision. The motion along the y axis is equivalent to direct central motion, which was analyzed in Sec. 15-10. With this approach the following four equations can be written (Fig. 15-20 is used for an example of applying each equation).

1. First consider mass A by itself (Fig. 15-20b). From Eq. 15-23,

$$m_A(v_A)_x + \int F_x\,dt = m_A(v_{Af})_x$$

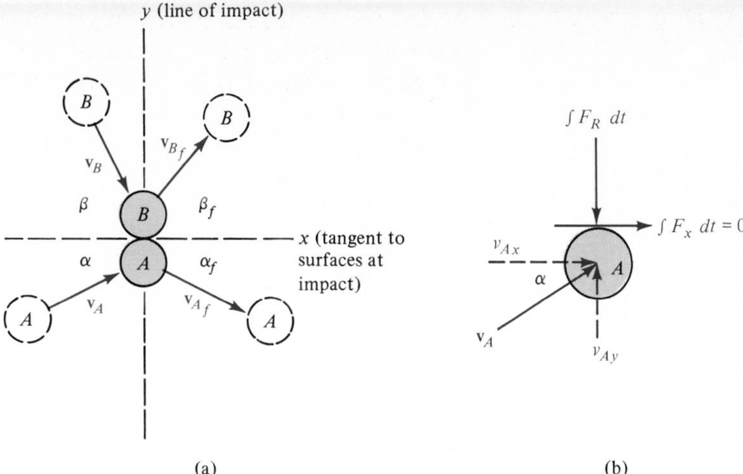

FIGURE 15-20
Oblique central impact

Since friction is assumed to be negligible, $F_x = 0$, and the x component of the momentum of body A is conserved,

$$(v_A)_x = (v_{Af})_x \qquad \boxed{15\text{-}35}$$

From Fig. 15-20a, $v_A \cos \alpha = v_{Af} \cos \alpha_f$.

2. Similarly, the x component of the momentum of body B is conserved,

$$(v_B)_x = (v_{Bf})_x \qquad \boxed{15\text{-}36}$$

From Fig. 15-20a,

$$v_B \cos \beta = v_{Bf} \cos \beta_f$$

3. The total momentum of the bodies in the y direction is conserved as in direct central impact,

$$m_A(v_A)_y + m_B(v_B)_y = m_A(v_{Af})_y + m_B(v_{Bf})_y \qquad \boxed{15\text{-}37}$$

From Fig. 15-20a,

$$m_A v_A \sin \alpha - m_B v_B \sin \beta = -m_A v_{Af} \sin \alpha_f + m_B v_{Bf} \sin \beta_f$$

4. The coefficient of restitution is applicable to the motion along the y axis (as in direct central impact). From Eq. 15-33,

$$e = \left(\frac{\text{relative velocity of separation}}{\text{relative velocity of approach}}\right)_{y \text{ direction}} \qquad \boxed{15\text{-}38}$$

From Fig. 15-20a,

$$e = \frac{v_{Bf} \sin \beta_f + v_{Af} \sin \alpha_f}{v_A \sin \alpha + v_B \sin \beta}$$

Equations 15-35 to 15-38 can be used to solve problems where the assumption of no friction during the collision is reasonable. Note that the detailed equations based on Fig. 15-20 are only for that example.

EXAMPLE 15-8

The energy-absorbing bumper of a 1500-kg car should be designed to allow for low-velocity impacts. Assume that the car travels at 10 km/h when it hits a stationary 2000-kg car which has a rigid bumper (Fig. 15-21) and is free to roll. Calculate the required coefficient of restitution of the bumper if car A is to continue at a speed of 3 km/h. Also calculate the energy that must be dissipated.

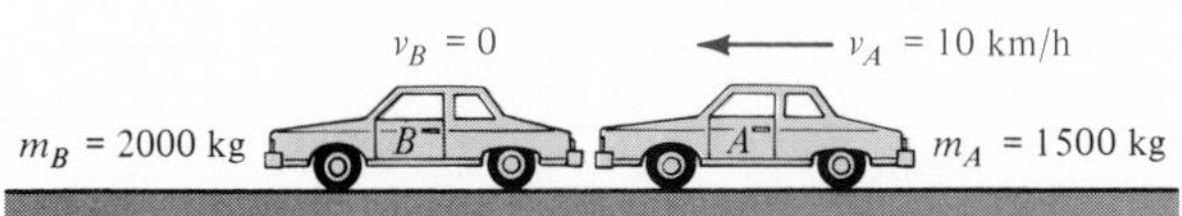

FIGURE 15-21
Collision of two cars

SOLUTION

Since no resultant external forces act on the cars, momentum of the system is conserved. A 3-km/h final speed means that the 1500-kg car travels in the same direction after impact. The final velocity of car B must be determined. From Eq. 15-24, and with positive taken to the left in Fig. 15-21,

$$m_A v_{A_1} + m_B v_{B_1} = m_A v_{A_2} + m_B v_{B_2}$$

$$(1500 \text{ kg})(10 \text{ km/h}) + 0 = (1500 \text{ kg})(3 \text{ km/h}) + (2000 \text{ kg})v_{B_2}$$

$$v_{B_2} = 5.25 \text{ km/h}$$

From Eq. 15-33,

$$e = \frac{v_{B_2} - v_{A_2}}{v_{A_1} - v_{B_1}} = \frac{5.25 - (3)}{10 - 0} = 0.225$$

The energy dissipated is calculated using Eq. 14-14,

$$T_1 + U_{12} = T_2$$

where U_{12} = energy dissipated (negative work).

$$\frac{1}{2} m_A v_A^2 + \frac{1}{2} m_B v_{B_1}^2 + U_{12} = \frac{1}{2} m_A v_{A_2}^2 + \frac{1}{2} m_B v_{B_2}^2$$

Converting the speeds to m/s (with 10 km/h = 2.78 m/s, 3 km/h = 0.833 m/s, and 5.25 km/h = 1.46 m/s),

$$\frac{1}{2}(1500 \text{ kg})(2.78 \text{ m/s})^2 + U_{12} = \frac{1}{2}(1500 \text{ kg})(0.833 \text{ m/s})^2$$

$$+ \frac{1}{2}(2000 \text{ kg})(1.46 \text{ m/s})^2$$

$$U_{12} = -3140 \text{ N}\cdot\text{m}$$

EXAMPLE 15-9

A Ping-Pong ball hits a table at an angle as shown in Fig. 15-22. Calculate v_f and θ neglecting friction if the coefficient of restitution is $e = 0.7$. Assume that the table is fixed.

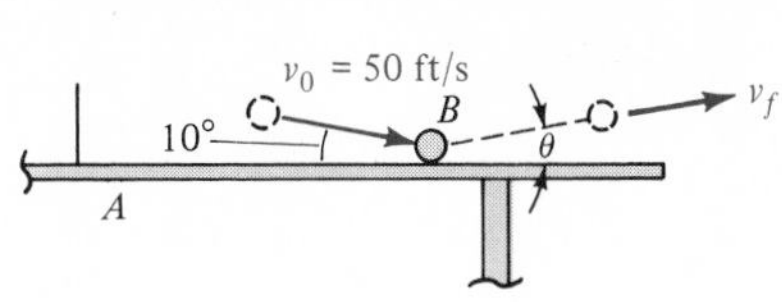

(a)

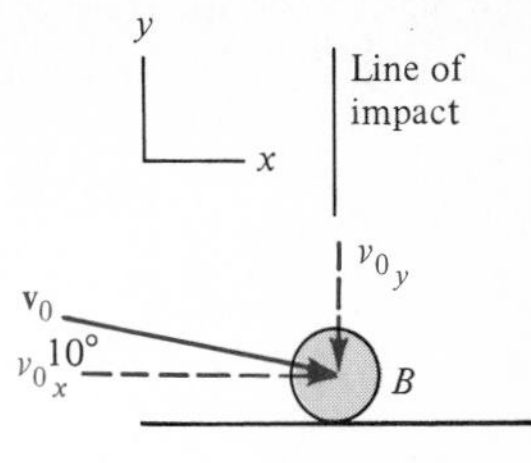

(b)

FIGURE 15-22
Rebound of a Ping-Pong ball

SOLUTION

The motion of the ball must be analyzed in terms of velocity components along and perpendicular to the line of impact. The line of impact in this case is defined by the mass center of the ball B and the point of impact on table A (Fig. 15-22b). Note that Eq. 15-33 is valid *only for velocity components along the line of impact*, and that the velocity of the table is zero both before and after the impact. Thus,

$$e = \frac{(v_{By})_f - (v_{Ay})_f}{v_{Ay} - v_{By}} = 0.7 = \frac{(v_{By})_f - 0}{0 - (-50 \text{ ft/s}) \sin 10^\circ}$$

$$(v_{By})_f = 0.7(50 \text{ ft/s})(\sin 10^\circ) = 6.08 \text{ ft/s}$$

Linear momentum is conserved along the line perpendicular to the line of impact for the Ping-Pong ball since no forces are exerted in that direction. Hence,

$$(v_{Bx})_f = (v_{Bx})_0 = (50 \text{ ft/s}) \cos 10^\circ = 49.2 \text{ ft/s}$$

$$v_f = [(v_{Bx})_f^2 + (v_{By})_f^2]^{1/2} = 49.6 \text{ ft/s}$$

$$\theta = \tan^{-1} \frac{(v_{By})_f}{(v_{Bx})_f} = \frac{6.08}{49.2}$$

$$= 7.0^\circ$$

Note that in this problem the principle of conservation of momentum cannot be used for the y direction. To appreciate this, try to apply Eq. 15-24 to the table A and the ball B,

$$m_A v_{Ay} + m_B v_{By} \stackrel{?}{=} m_A (v_{Ay})_f + m_B (v_{By})_f$$

It is clear that $v_{Ay} = (v_{Ay})_f = 0$ since the table does not move before or after the impact. Thus, Eq. 15-24 would require $v_{By} = (v_{By})_f$, which is not possible (the velocity of the ball changes in magnitude and direction). Equation 15-24 is not valid here because there is another impulsive force acting on the table (from the reactions at the legs) to maintain its zero velocity.

PROBLEMS

SECTIONS 15-9, 15-10, 15-11

FIGURE P15-58

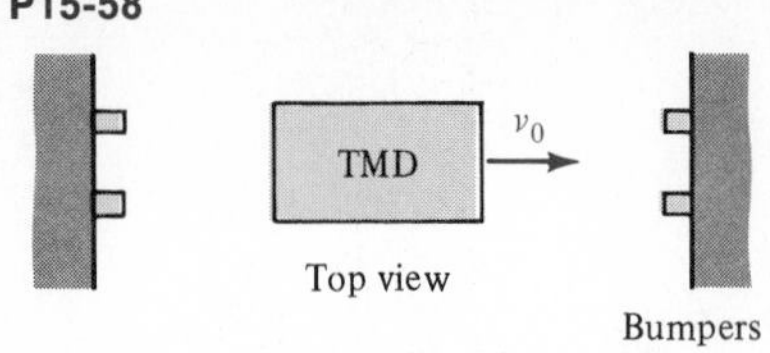

15–58 The 3×10^5-kg block of a tuned mass damper in a skyscraper (see Sec. 1-7 in the associated statics volume) is moving at $v_0 = 2$ m/s when it hits the overtravel bumpers. Calculate the rebound velocity if the coefficient of restitution is $e = 0.5$. Assume that friction is negligible and the bumpers are fixed.

15–59 Calculate the energy that the bumpers must dissipate during a single impact in the system described in Prob. 15–58.

15–60 Calculate the rebound velocity of the block in Prob. 15–58 after a total of three impacts between the left and right bumpers, starting with v_0. Plot the rebound velocity as a function of successive impacts. Estimate the number of bumps after which the rebound velocity would be about $0.01v_0$.

FIGURE P15-61

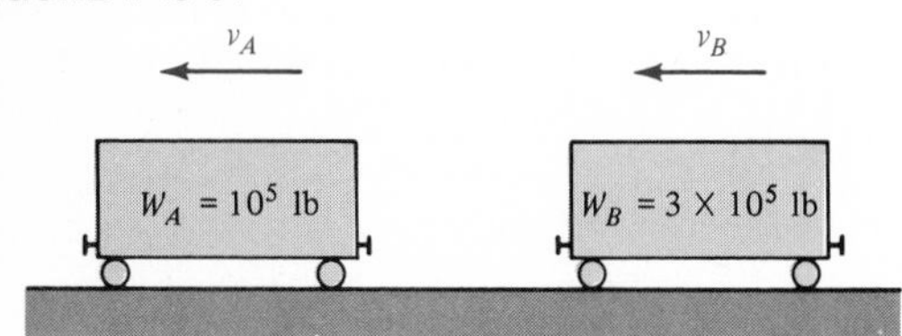

15–61 Freight car A is moving at $v_A = 1$ mph, and car B at $v_B = 2$ mph, as shown. Calculate their new velocities after an impact where the coefficient of restitution is $e = 0.4$.

15–62 Assume that cars A and B in Prob. 15–61 are coupled together at impact. Calculate their common velocity and the energy lost during the coupling.

FIGURE P15-63

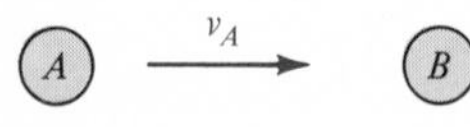

15–63 Particle A of mass $m_A = 1$ kg moves at speed v_A before hitting the stationary particle B of mass $m_B = 2$ kg. The speed of particle B after the impact is 5 m/s. Calculate the speed v_A if $e = 0.6$ and friction is negligible. Determine the energy lost during the collision.

15–64 The 300-kg ram of a pile driver is dropped on a 700-kg pile from a height of 2 m and drives it into soft soil. Determine the velocities of the ram and the pile after impact if the coefficient of restitution is $e = 0.5$.

FIGURE P15-64

(a)

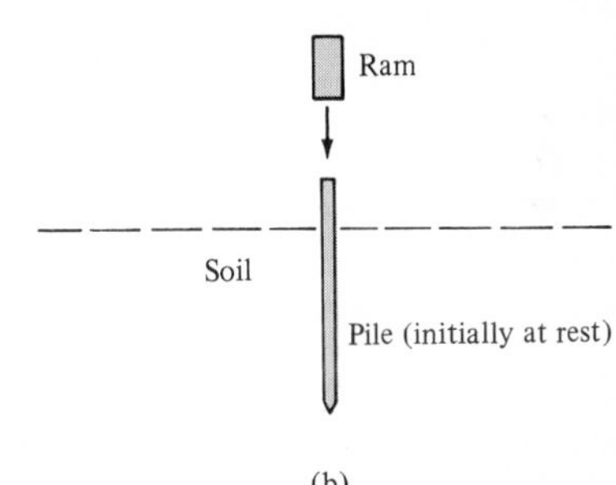

(b)

15–65 It is known that the ram of a pile driver (Fig. P15-64) stops momentarily after hitting an 800-lb pile in a free fall from a height of 6 ft. The coefficient of restitution is $e = 0.5$. Determine the weight of the ram.

15–66 Consider the rear-end collison of two cars whose initial velocities are collinear. Determine the final velocities if $m_A = 1000$ kg, $m_B = 2000$ kg, $v_A = 40$ km/h, $v_B = 80$ km/h, and $e = 0.5$.

FIGURE P15-66

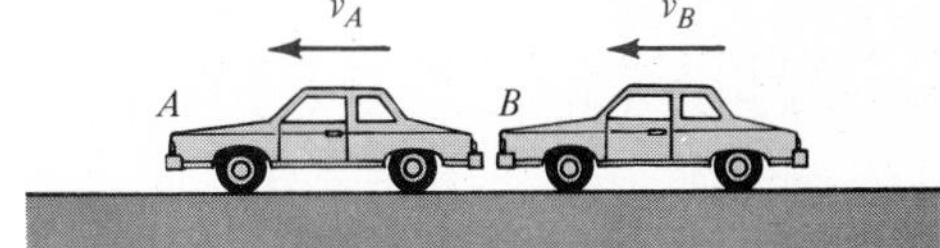

15–67 A 3000-lb car travels at $v_B = 60$ mph when it collides with a 2200-lb car that is moving in the same direction (Fig. P15-66). The speed of car B is 40 mph just after the impact. Determine the initial and final speeds of car A if $e = 0.4$.

FIGURE P15-68

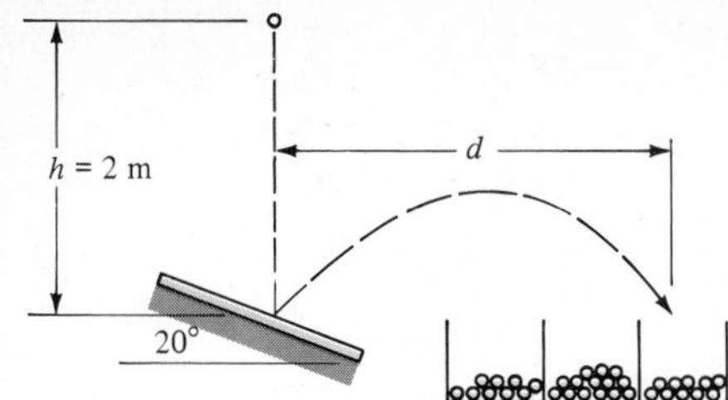

15–68 A manufacturer of tennis balls sorts the balls according to their rebounding quality as shown. Calculate the distance d of horizontal travel if the coefficient of restitution with the fixed racket is $e = 0.5$. Neglect friction.

15–69 Solve Prob. 15-68 using $e = 0.4$, 0.5, and 0.6. Plot d as a function of e and discuss the general tendency observed.

FIGURE P15-70

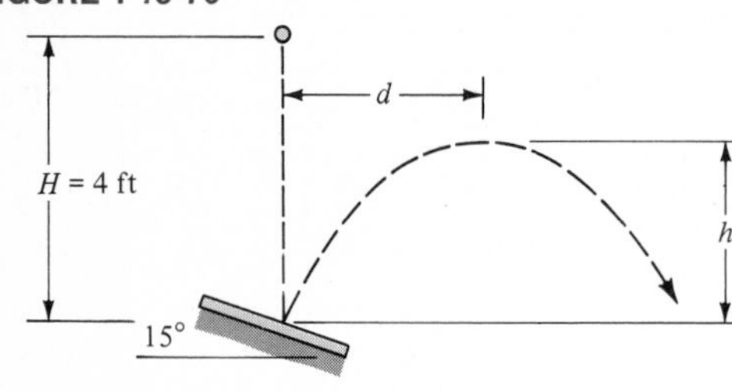

15–70 In the sorting of ball-bearing balls the rebound height h and horizontal distance d are to be electronically measured. Calculate h and d for the approximate positioning of the sensors if the platen is fixed and the coefficient of restitution is $e = 0.75$.

FIGURE P15-71

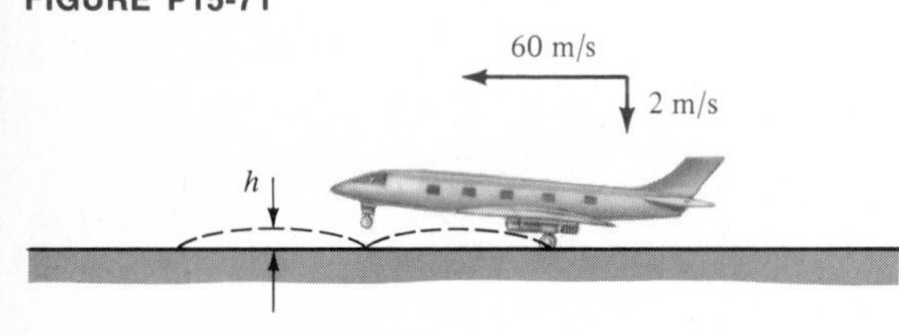

15–71 An 80 000-kg airplane lands at a horizontal speed of 60 m/s and a vertical descent of 2 m/s. Calculate the height h of the rebound after the first and second bounce on the runway if the coefficient of restitution is $e = 0.3$. Assume that there is no friction, aerodynamic lift, or engine thrust involved. Treat the airplane as a particle.

FIGURE P15-72

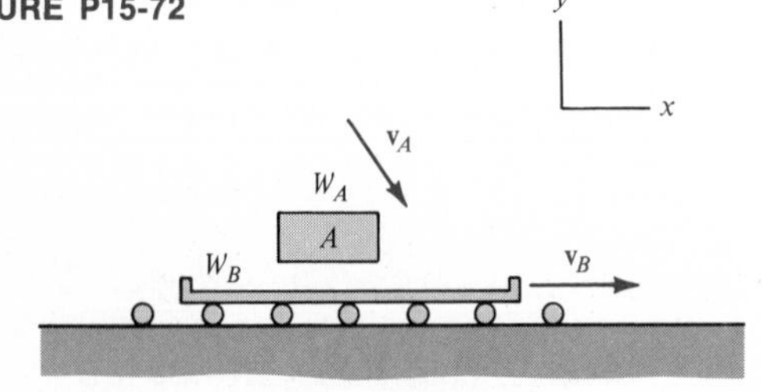

15–72 In an automated packaging operation a 20-lb box is dropped on a 40-lb tray that can freely roll horizontally. Calculate the velocity of the box just after the first impact if initially $\mathbf{v}_A = (3\mathbf{i} - 4\mathbf{j})$ ft/s, $\mathbf{v}_B = 0$, and the coefficient of restitution is $e = 0.2$. Neglect friction.

15–73 Assume a coefficient of kinetic friction of 0.3 between the box and the tray in Prob. 15–72. Estimate how far the box slides on the tray before coming to rest. Determine the final velocity of the tray after the box has stopped sliding.

15–74 A 20-ton aircraft is to land on a 60 000-ton aircraft carrier. Their initial velocities are $\mathbf{v}_A = -60\mathbf{i}$ km/h, and $\mathbf{v}_B = (-200\mathbf{i} - 3\mathbf{j})$ km/h. Calculate the rebound velocities of the aircraft assuming that (a) the ship cannot move up or down, and (b) the ship has no impulsive support reaction. The coefficient of restitution is $e = 0.4$. Neglect friction, engine thrust, aerodynamic lift, and treat the ship and aircraft as particles.

FIGURE P15-74

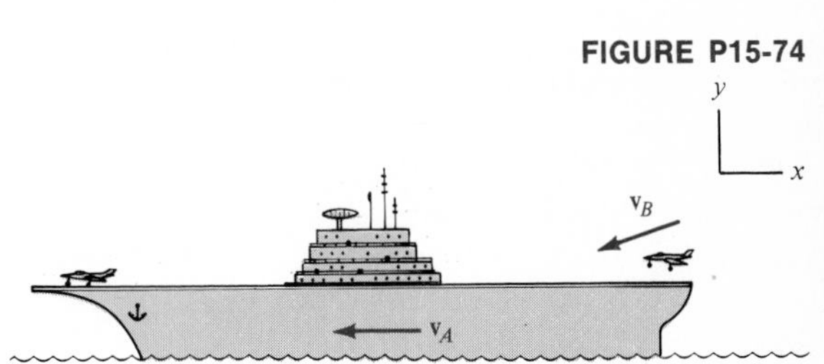

15–75 Two cars are colliding as they travel at initial velocities $\mathbf{v}_A = 50\mathbf{j}$ mph and $\mathbf{v}_B =$ unknown. The cars weigh $W_A = 2000$ lb and $W_B = 3000$ lb, and they become entangled during the impact. Their common velocity just after impact is estimated from skid marks as $\mathbf{v}_c = (10\mathbf{i} + 55\mathbf{j})$ mph. Determine the initial velocity $\mathbf{v}_B$.

FIGURE P15-75

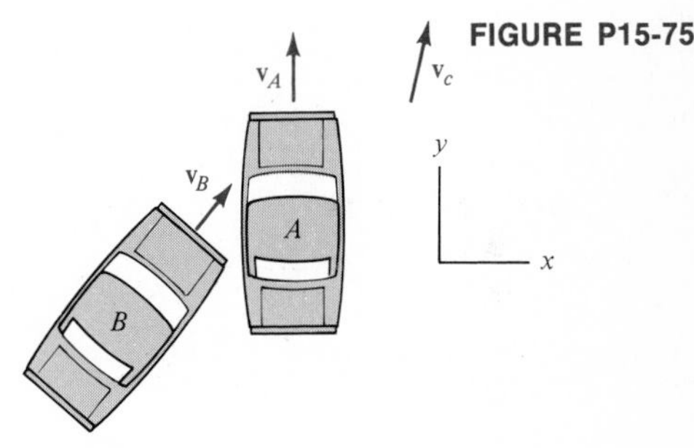

15–76 The block of a biaxial tuned mass damper of a skyscraper (see Sec. 1-7 in the associated statics volume) moves in the horizontal plane and may hit overtravel bumpers. Assume that this motion can be modeled by using a small particle bouncing from walls as shown. Determine v_f and ϕ in terms of the given initial quantities, assuming that the coefficient of restitution is e for each impact. Neglect friction.

FIGURE P15-76

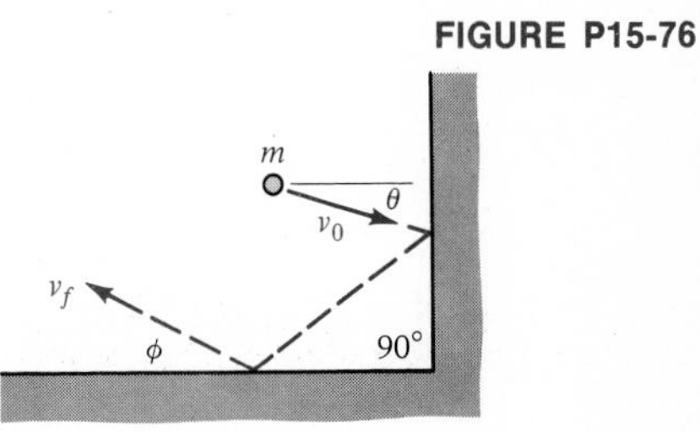

SUMMARY

Section 15-1

Linear momentum

$$\mathbf{G} = m\mathbf{v}$$

Time rate of change of linear momentum

$$\sum \mathbf{F} = \dot{\mathbf{G}} = \frac{d}{dt}(m\mathbf{v})$$

SPECIAL CASE:

$$\sum \mathbf{F} = 0 \qquad m\mathbf{v} = \text{constant} \qquad \text{(conservation of linear momentum)}$$

Section 15-2

Angular momentum

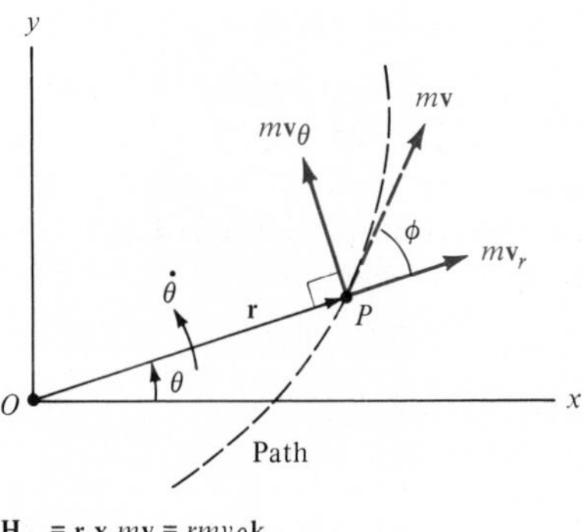

$\mathbf{H}_O = \mathbf{r} \times m\mathbf{v} = rmv_\theta\mathbf{k}$

Magnitude of angular momentum

GENERAL PLANE MOTION:

$$H_O = rmv \sin \phi = rmv_\theta = mr^2\dot{\theta}$$

GENERAL SPACE MOTION:

$$H_x = m(yv_z - zv_y)$$

$$H_y = m(zv_x - xv_z)$$

$$H_z = m(xv_y - yv_x)$$

Section 15-3

Time rate of change of angular momentum

$$\dot{\mathbf{H}}_O = \frac{d}{dt}(\mathbf{r} \times m\mathbf{v}) = \mathbf{r} \times \mathbf{F} = \mathbf{M}_O$$

$$\sum \mathbf{M}_O = \dot{\mathbf{H}}_O$$

$$\sum M_x = \dot{H}_x \qquad \sum M_y = \dot{H}_y \qquad \sum M_z = \dot{H}_z$$

SPECIAL CASE:

$$\sum \mathbf{M}_O = 0$$

$$\mathbf{H}_O = \mathbf{r} \times m\mathbf{v} = \text{constant} \qquad \text{(conservation of angular momentum)}$$

Section 15-4

Newton's law of universal gravitation

$$F = G\frac{Mm}{r^2}$$ for masses M and m at a distance r between them

ACCELERATION OF GRAVITY AT THE SURFACE OF THE EARTH (MASS *M*, RADIUS *R*):

$$g = \frac{GM}{R^2}$$

Section 15-5

Kepler's laws of planetary motion

FIRST LAW (LAW OF SECTORIAL AREAS)

The radius drawn from a planet to the sun sweeps equal areas in equal times.

SECOND LAW

The orbit of each planet is an ellipse, with the sun at one of its foci.

These laws are valid in any situation where the motion of a body is similar to planetary motion under the action of a central force.

Section 15-6

Impulse and momentum

LINEAR IMPULSE:

$$\int_{t_1}^{t_2} \mathbf{F}\,dt = \underbrace{m\mathbf{v}_2}_{\substack{\text{final}\\\text{momentum}}} - \underbrace{m\mathbf{v}_1}_{\substack{\text{initial}\\\text{momentum}}}$$

Section 15-7

Conservation of total momentum of particles
For n particles which move in such a way that only interactive forces are applied to them,

$$\underbrace{\sum_{i}^{n} (m_i\mathbf{v}_i)_1}_{\substack{\text{total initial}\\\text{momentum at time } t_1}} = \underbrace{\sum_{i}^{n} (m_i\mathbf{v}_i)_2}_{\substack{\text{total final momentum}\\\text{at time } t_2}}$$

Section 15-8

Applications of principle of impulse and momentum
Consider the impulsive motion of bodies A and B.

CASE 1: The velocity change is significant for body B, and negligible for body A ($m_A \gg m_B$).
Use

$$\underbrace{m\mathbf{v}_{B1}}_{\substack{\text{initial}\\\text{momentum}}} + \underbrace{\int_{t_1}^{t_2} \mathbf{F}\,dt}_{\text{impulse}} = \underbrace{m\mathbf{v}_{B2}}_{\substack{\text{final}\\\text{momentum}}}$$

Conservation of total momentum cannot be used in this case.

CASE 2: The velocity change is significant for both bodies (m_A and m_B are not enormously different). Use $m_Av_A + m_Bv_B =$ constant (total initial momentum $=$ total final momentum).

Sections 15-9 and 15-10

Direct central impact of deformable bodies

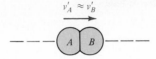

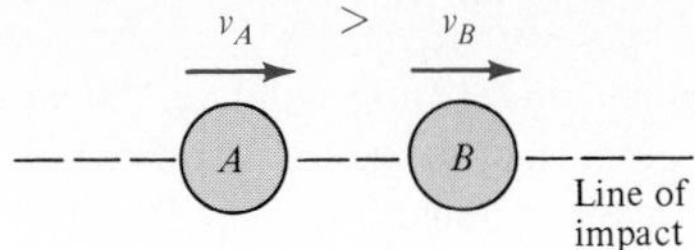

(a) Before impact (bodies approach)

$v'_A \approx v'_B$

(b) Increasing deformation during contact

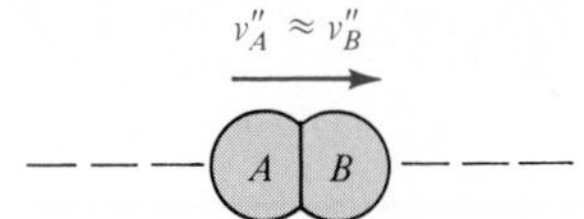

(c) Decreasing deformation or restitution during contact

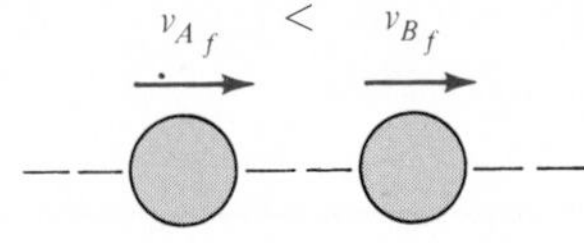

(d) After impact (bodies move apart; v_{A_f} may be to the left)

COEFFICIENT OF RESTITUTION:

$$e = \frac{v_{Bf} - v_{Af}}{v_A - v_B} = \frac{\text{relative velocity of separation}}{\text{relative velocity of approach}}$$

For real materials, $0 < e < 1$.

Section 15-11

Oblique central impact of deformable bodies

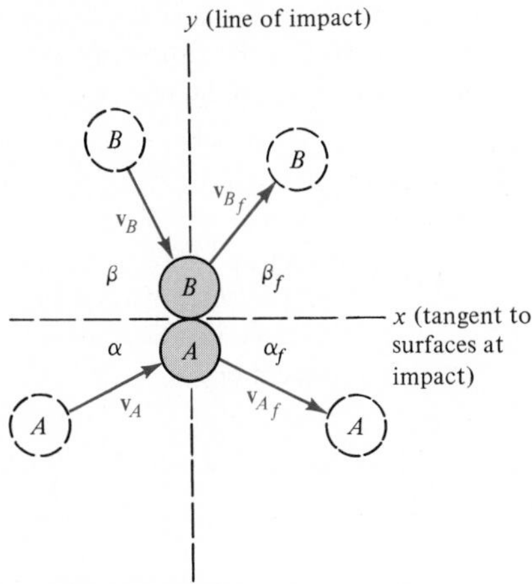

Assume that friction is negligible.

1. The x component of the momentum of body A is conserved.
2. The x component of the momentum of body B is conserved.
3. The total momentum of bodies A and B in the y direction is conserved.
4. The coefficient of restitution is applicable to the motion along the y axis.

REVIEW PROBLEMS

FIGURE P15-77

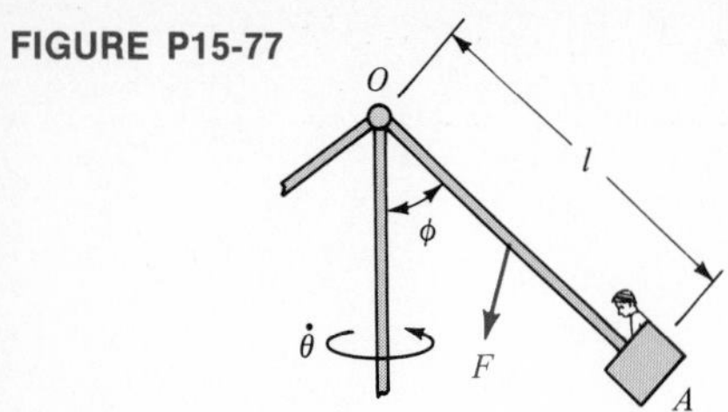

15–77 An amusement ride is modeled as a 300-kg particle at point A, attached to a rigid rod of negligible mass and length $l = 8$ m. The arm OA rotates about the vertical column at an angular speed $\dot{\theta} = 1.4$ rad/s. At the angle $\phi = 50°$ force F causes ϕ to decrease at the rate of 0.2 rad/s. Determine the required moment to maintain $\dot{\theta}$ constant, neglecting friction.

15–78 In Prob. 15–77, $\dot{\theta} = 1.4$ rad/s when $\phi = 50°$ Determine the new angular speed $\dot{\theta}_1$ after the angle ϕ is changed to 25°, assuming that the system is rotating about the vertical axis without any interference.

15–79 A space vehicle is in circular orbit of radius $r = 2200$ mi around a planet whose radius is $R = 2000$ mi. Determine the escape velocity from this planet starting from the given orbit if the acceleration of gravity is 10 ft/s^2 at the surface of the planet.

15–80 Sketch and discuss the orbits of a particle about a center of force for the borderline case of nearly identical maximum velocities but different orbits, one elliptical, the other parabolic.

FIGURE P15-81

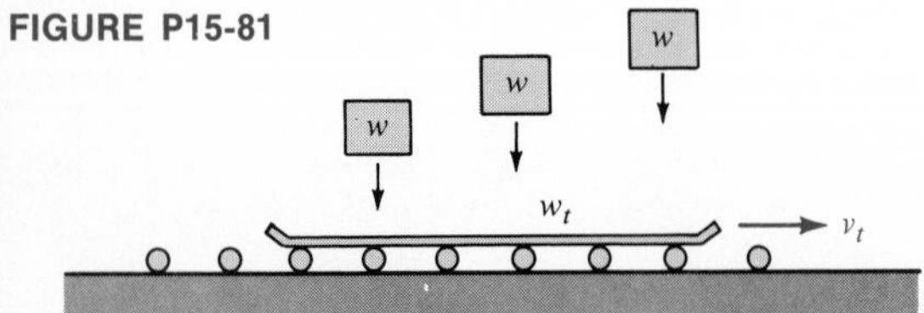

15–81 In an assembly operation a 20-lb empty tray is sliding on rollers with $v_t = 5$ ft/s. Three 5-lb objects are dropped on the tray in succession. Determine the new velocity of the tray after each object comes to rest on it. What would be the final velocity if all three objects were dropped simultaneously?

15–82 A 10 000-kg truck leaves a loosely moored ferry at a speed of 10 km/h relative to the ferry. Calculate the velocity imparted to the 20 times more massive ferry, assuming that both were at rest initially.

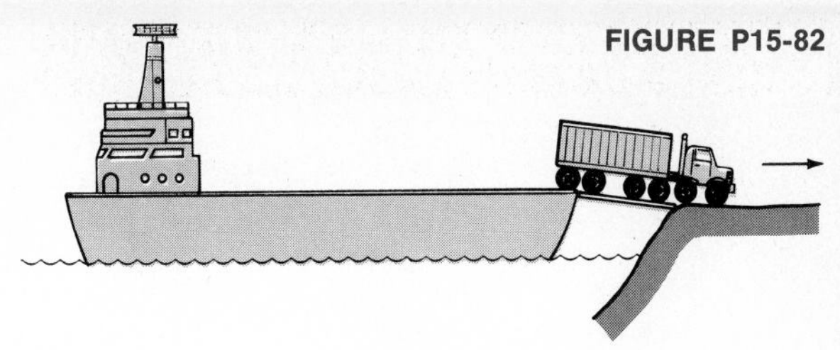

FIGURE P15-82

15–83 Two particles of mass m_A and m_B are at rest next to a compressed spring. Prove that the mass center remains fixed at any time after the spring is released sending the particles apart. Assume there is no friction.

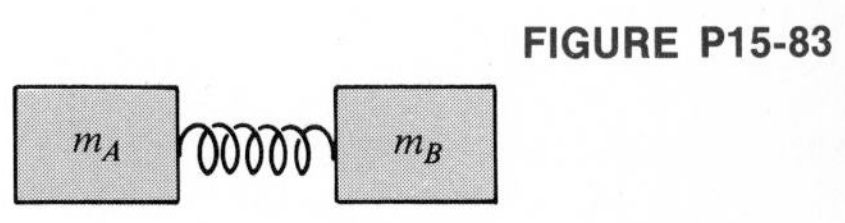

FIGURE P15-83

15–84 Consider two billiard balls of equal mass m and radius r. Ball A moves at velocity $\mathbf{v}_A$ and strikes ball B which is at rest. Determine the velocities of the balls after impact if $d = 0.7r$ and the coefficient of restitution is $e = 0.6$. Neglect friction.

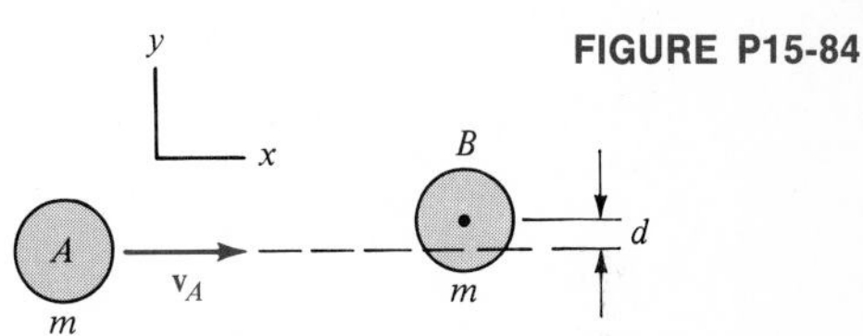

FIGURE P15-84

15–85 Solve Prob. 15–84 for $\mathbf{v}_B$ with $d = 0.5r$, $1.2r$, and $1.8r$. Draw the vectors $\mathbf{v}_B$ on the same xy diagram with a common origin for the three vectors. Estimate from the diagram the vector $\mathbf{v}_B$ relative to $\mathbf{v}_A$ for the situation when the two balls just graze each other.

Overview of Kinetics Analyses of Particles (Chaps. 13, 14, and 15)

The fundamental principle in kinetics is Newton's second law of motion. For a particle,

$$\sum \mathbf{F} = m\mathbf{a}$$

Several special methods are based on this equation and are developed for convenience in solving problems.

Energy Methods

$$U_{12} = \int_1^2 \mathbf{F} \cdot d\mathbf{r}$$

$$T = \frac{1}{2} mv^2$$

$$U_{12} = T_2 - T_1$$

$$U_{12} = V_1 - V_2$$

$$T_1 + V_1 = T_2 + V_2$$

Work and energy methods are useful when force, mass, position, and velocity are the primary quantities in the analysis.

Momentum Methods

$$\mathbf{G} = m\mathbf{v}$$

$$\sum \mathbf{F} = \dot{\mathbf{G}} = \frac{d}{dt}(m\mathbf{v})$$

$$\mathbf{H}_O = \mathbf{r} \times m\mathbf{v}$$

$$\sum \mathbf{M}_O = \dot{\mathbf{H}}_O = \frac{d}{dt}(\mathbf{r} \times m\mathbf{v})$$

$$\int_{t_1}^{t_2} \mathbf{F}\, dt = m\mathbf{v}_2 - m\mathbf{v}_1$$

$$\sum (m_i\mathbf{v}_i)_1 = \sum (m_i\mathbf{v}_i)_2$$

Momentum methods are useful when force, mass, velocity, and time are the primary quantities in the analysis.

CHAPTER 16

KINETICS OF SYSTEMS OF PARTICLES

OVERVIEW

Simple systems involving only a few particles have been analyzed on numerous occasions in previous sections. It is necessary to review and generalize the methods already developed to deal with a larger variety of problems than before. For this purpose it is useful to distinguish systems consisting of discrete particles from systems of continuously distributed particles.

1. ***Discrete particles.*** An example of this is the solar system. In certain studies it is reasonable to work with only two major masses such as the sun and the earth, but accurate analysis requires the consideration of numerous other interactions, as well. In this kind of a situation even the analysis of three interacting bodies is much more difficult than that of two bodies. Other discrete systems, such as certain exploding bodies, can be relatively easily analyzed.

2. ***Continuous particles in fluids.*** Even in small volumes of fluids the number of particles is enormous, so they cannot be individually analyzed. However, some macroscopic aspects of the motion of fluids can be modeled by lumping many particles in a simple system.

3. ***Continuous particles in rigid bodies.*** These bodies also contain innumerable particles for individual analysis, but the methods developed for the motion of particles can be extended to rigid bodies. Such generalizations of the methods are possible because in rigid bodies the geometric configuration of all particles is fixed.

The fundamentals of discrete systems and fluids are presented in this chapter. Much more information is available in other texts such as those dealing entirely with fluid mechanics. The motions of rigid bodies are more extensively covered in this text starting with Chap. 17. Chapter 16 serves, in a large part, as a bridge between the kinetics of particles and rigid bodies.

STUDY GOALS

SECTION 16-1 extends Newton's second law of motion to a system of particles. The analysis of external and internal forces acting on the system and the resulting governing equation of motion are relevant to further work in kinetics.

SECTION 16-2 presents important equations for the position, velocity, and acceleration of the center of mass of a system of particles. These equations must be understood by all students of dynamics.

SECTIONS 16-3 THROUGH 16-5 present the potential energy, kinetic energy, and work–energy relationship for systems of particles. The major equations derived are particularly useful in the dynamics analysis of rigid bodies.

SECTIONS 16-6 AND 16-7 are important for distinguishing and analyzing the linear and rotational motions of systems of particles.

SECTION 16-8 extends the results of Sec. 16-7 to the special case of using the center of mass for reference. Several useful expressions are derived, but the details of these derivations are not necessary to study in order to make substantial progress in elementary dynamics.

SECTION 16-9 presents the equations of conservation of linear and angular momentum for a system as a whole. These equations are useful for solving many problems and are worth remembering.

SECTION 16-10 describes the methods of analyzing impulsive motion of systems of particles. This special topic provides for additional practice in the application of momentum methods.

SECTIONS 16-11 THROUGH 16-13 present the concepts and basic methods of analyzing the flow of mass using the momentum methods developed for particles. These special topics are not essential for making progress in elementary dynamics, and may be omitted from the course syllabus. However, they are useful for practicing the solution of momentum problems. The students may also note that a large variety of interesting problems are available in this area.

NEWTON'S SECOND LAW APPLIED TO A SYSTEM OF PARTICLES 16-1

The analysis of a system consisting of many particles can be based on a system of only two particles. Assume that the forces acting on such a system are completely described in Fig. 16-1. The masses of

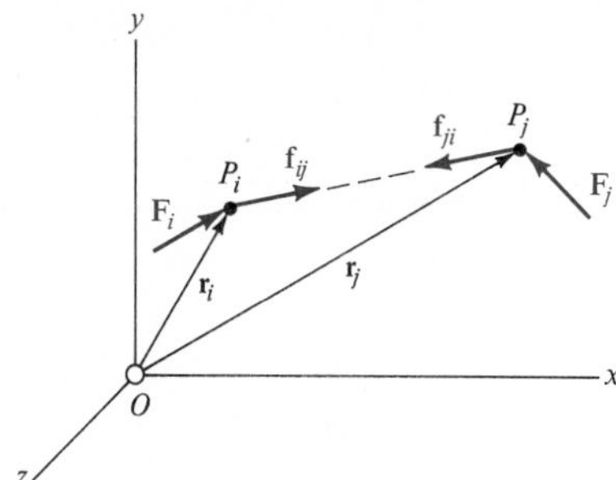

FIGURE 16-1

System of two interacting particles

particles P_i and P_j are m_i and m_j, respectively. The forces $\mathbf{f}_{ij}$ and $\mathbf{f}_{ji}$ are the mutual central forces between the particles. They are always collinear, equal in magnitude, and opposite in direction (attraction in the case of gravity, attraction, or repulsion in the cases of electricity and magnetism). It is tempting to cancel the *internal forces* $\mathbf{f}_{ij}$ and $\mathbf{f}_{ji}$ at once, but it is best to consider them individually in trying to relate the motion of each particle to the motion of the system as a whole. The forces $\mathbf{F}_i$ and $\mathbf{F}_j$ are *external forces* whether the particles are considered individually or together as a system. If there are only two particles in the system, Newton's second law can be written as

$$\mathbf{F}_i + \mathbf{f}_{ij} = m_i\mathbf{a}_i \qquad \text{for particle } P_i \tag{16-1a}$$

$$\mathbf{F}_j + \mathbf{f}_{ji} = m_j\mathbf{a}_j \qquad \text{for particle } P_j \tag{16-1b}$$

This can be extended to include a third particle P_k, which exerts mutual internal forces on each of the other particles in the system (Fig. 16-2). Summing all the forces acting on particle P_i, there are two internal forces now, and the external force $\mathbf{F}_i$. Therefore,

$$\mathbf{F}_i + \mathbf{f}_{ij} + \mathbf{f}_{ik} = m_i\mathbf{a}_i$$

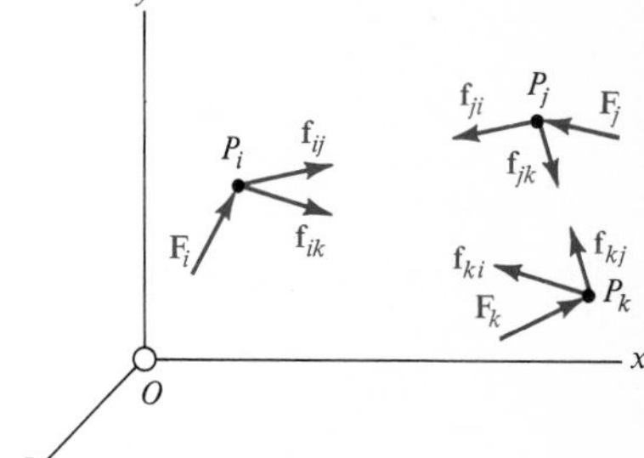

FIGURE 16-2

System of three interacting particles

Generalizing this to n particles in the system, the equation of motion for particle P_i is

$$\mathbf{F}_i + \sum_{j=1}^{n} \mathbf{f}_{ij} = m_i\mathbf{a}_i \tag{16-2}$$

The similar equation for particle P_j is

$$\mathbf{F}_j + \sum_{i=1}^{n} \mathbf{f}_{ji} = m_j\mathbf{a}_j$$

In both cases the forces $\mathbf{f}_{ii}$ and $\mathbf{f}_{jj}$ have no meaning and are defined to be zero.

The general situation is that each particle of n particles in the system has a resultant external force $\mathbf{F}$ on it, and also $n - 1$ internal forces $\mathbf{f}$ on it. Newton's second law for the whole system can be concisely written as

$$\sum_{i=1}^{n} \mathbf{F}_i + \sum_{i=1}^{n} \sum_{j=1}^{n} \mathbf{f}_{ij} = \sum_{i=1}^{n} m_i\mathbf{a}_i \tag{16-3}$$

The second term on the left side can be eliminated since each force $\mathbf{f}_{ij}$, and so on, has an equal and opposite force $\mathbf{f}_{ji}$, and so on, in the system. The resulting governing equation for the motion of the whole system is

$$\boxed{\sum_{i=1}^{n} \mathbf{F}_i = \sum_{i=1}^{n} m_i\mathbf{a}_i} \tag{16-4}$$

where $\sum \mathbf{F}_i$ is the vector sum of all forces acting on the system.

16-2 MOTION OF THE CENTER OF MASS

For a complete description of the motion of a system, it is useful to determine the position, velocity, and acceleration of the center C of the total mass. First, assume that $\mathbf{r}_i$ is the position vector of particle P_i in the chosen reference frame, and $\mathbf{r}_C$ is the position vector of the mass center of the system of particles in the same reference frame. For a system of n particles with a total mass m,

$$\boxed{m\mathbf{r}_C = \sum_{i=1}^{n} m_i\mathbf{r}_i} \tag{16-5}$$

by definition of the center of mass. This equation is valid regardless of any velocities and accelerations of the particles at a particular instant.

The velocities of the system as a whole and of the particles are obtained by differentiating Eq. 16-5 with respect to time,

$$m\mathbf{v}_C = \sum_{i=1}^{n} m_i\mathbf{v}_i \qquad \text{16-6}$$

It is assumed in this differentiation that the mass of each particle is constant. Note that the mass is not necessarily constant (e.g., when fuel is burned).

The accelerations of the system as a whole and of the particles are obtained by differentiating Eq. 16-6 with respect to time,

$$m\mathbf{a}_C = \sum_{i=1}^{n} m_i\mathbf{a}_i \qquad \text{16-7}$$

From Eqs. 16-4 and 16-7,

$$\sum \mathbf{F} = m\mathbf{a}_C \qquad \text{16-8}$$

where $\sum \mathbf{F}$ is the resultant of all forces acting on the system, m is the total mass of the system, and $\mathbf{a}_C$ is the acceleration of the center of mass.

On the basis of Eqs. 16-4 through 16-8, the following important remarks are made concerning the motion of a system of particles:

If there are only mutual forces acting among the particles (there are forces **f** but no forces **F** in Fig. 16-2), the center of mass of the system has no acceleration. Therefore, the center of mass of the system remains at rest if initially at rest, or it moves with a constant velocity regardless of the velocities of the constituent particles. These results agree with Newton's first and second laws for a single particle. In summary,

> The center of mass of a system of particles moves as if the total mass of the system were concentrated at the center of mass if there is no resultant force acting on the system.

These statements can be generalized for the situation where internal and external forces act on the particles (both forces **f** and **F** in Fig. 16-2). In that case,

> The center of mass of a system of particles moves under the action of internal and external forces as if the total mass of the system and all the external forces were at the center of mass.

Note that the last statement does not mean that the resultant of all the external forces must act at the center of mass, causing only a translation of the system. In general, the resultant force would not act through the center of mass, causing both a translation and a rotation (because of a moment about C) of the system. The derivations and statements above concern only the translation part of the total motion that is possible. The rotation will be considered in Sec. 16-7.

EXAMPLE 16-1

In the delicate docking maneuver of two spacecraft even small changes in their relative positions are important. Assume that the total weight of a space vehicle with crew is 4000 lb. What is the allowable displacement of a 160-lb astronaut in this vehicle to limit the resulting displacement of the vehicle's frame to 0.2 in.? The astronaut moves away from the center of mass of the vehicle as modeled in Fig. 16-3. The xy frame is fixed with respect to the center of mass C of the whole system, and the astronaut moves in the xy plane.

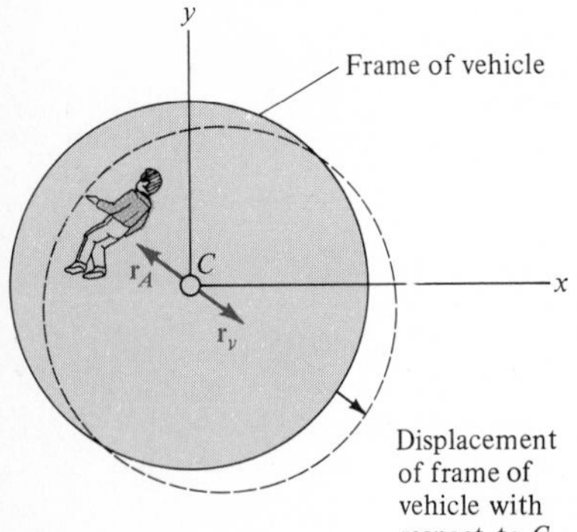

FIGURE 16-3
Motion of an astronaut in a spacecraft

SOLUTION

To determine the allowable displacement, apply Eq. 16-5,

$$m_T\mathbf{r}_C = \sum m_i\mathbf{r}_i = m_V\mathbf{r}_V + m_A\mathbf{r}_A$$

where m_T is the total mass of the system, m_A is the mass of the astronaut, and m_V is the mass of the vehicle without the moving astronaut, $m_V = m_T - m_A$. The corresponding position vectors $\mathbf{r}$ are defined with respect to the xy reference frame. Since the reference frame's origin is at the mass center, $\mathbf{r}_C = 0$. Thus,

$$0 = m_V\mathbf{r}_V + m_A\mathbf{r}_A$$

For a displacement $\mathbf{r}_A$ in any direction, $\mathbf{r}_A$ and $\mathbf{r}_V$ are collinear by inspection. Thus, a scalar solution can be used,

$$-\frac{3840 \text{ lb}}{32.2 \text{ ft/s}^2}(0.2 \text{ in.}) = \frac{160 \text{ lb}}{32.2 \text{ ft/s}^2}(r_A)$$

The allowable displacement r_A of the astronaut is

$$r_A = -4.8 \text{ in.}$$

Note the inverse proportionality of the displacements to the masses involved. The negative sign indicates that the astronaut and the vehicle move in opposite directions.

EXAMPLE 16-2

A spacecraft consists of a main cabin of mass $M = 6000$ kg and a nuclear power source of mass $m = 300$ kg (Fig. 16-4). The two main parts are treated as particles which are connected by a tube of negligible mass. An attitude-controlling jet applies a force $\mathbf{F} = (40\mathbf{i} + 20\mathbf{j})$ N to the main cabin. Determine the acceleration of the mass center.

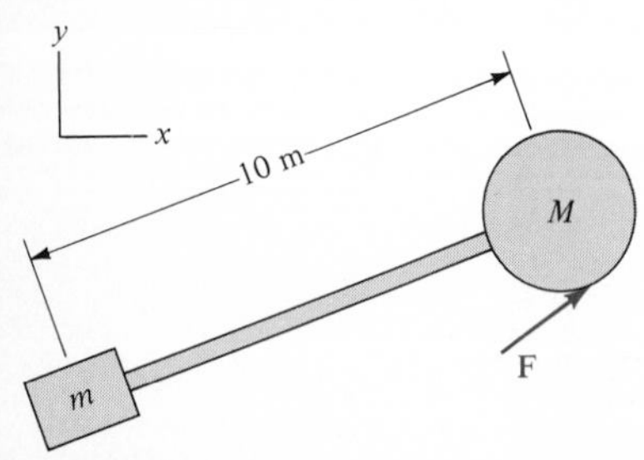

FIGURE 16-4
Model of a spacecraft

SOLUTION

The acceleration is obtained from Eq. 16-8,

$$\sum \mathbf{F} = m_T\mathbf{a}_C$$

where $m_T = M + m = 6300$ kg,

$$\mathbf{a}_C = \frac{40}{6300}\mathbf{i} + \frac{20}{6300}\mathbf{j} = (0.00635\mathbf{i} + 0.00317\mathbf{j}) \text{ m/s}^2$$

PROBLEMS

16–1 The masses and position vectors of three particles are $m_1 = 1$ kg, $\mathbf{r}_1 = (0.5\mathbf{i} + 0.8\mathbf{j})$ m, $m_2 = 2$ kg, $\mathbf{r}_2 = (0.9\mathbf{i} - 0.4\mathbf{j})$ m, $m_3 = 1.5$ kg, $\mathbf{r}_3 = (2\mathbf{i} + 0.7\mathbf{j})$ m. $\mathbf{v}_3 = 6\mathbf{i}$ m/s. Determine the position vector and velocity of the center of mass.

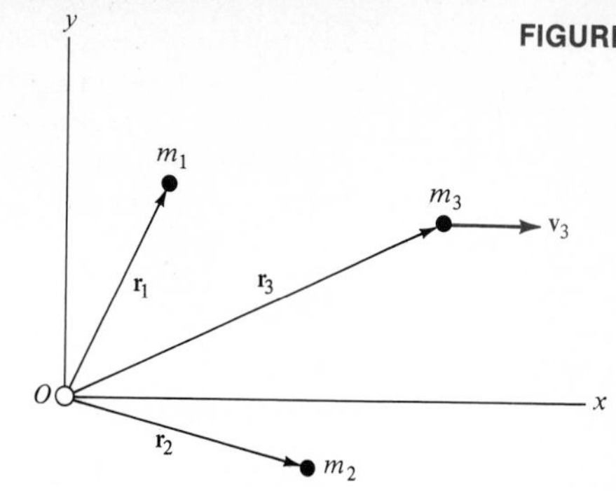

FIGURE P16-1

16–2 The position vectors of the three particles are $\mathbf{r}_1 = (2\mathbf{j} + 2\mathbf{k})$ ft, $\mathbf{r}_2 = (3\mathbf{i} + 2.5\mathbf{j} + 2\mathbf{k})$ ft, and $\mathbf{r}_3 = (4\mathbf{i} - 0.6\mathbf{j} - 2\mathbf{k})$ ft. Particle 2 has an acceleration $\mathbf{a}_2 = -8\mathbf{j}$ ft/s^2. Determine the position vector and acceleration of the center of mass.

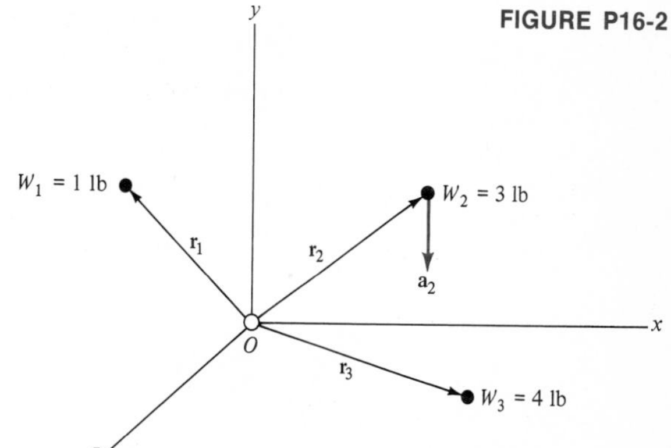

FIGURE P16-2

16–3 The three particles have identical mass $m = 2$ kg. Calculate the velocity of the center of mass.

16–4 The speed of each particle is increasing at the rate of 1 m/s^2 in Prob. 16–3. Determine the resultant force acting on the system, assuming rectilinear motion.

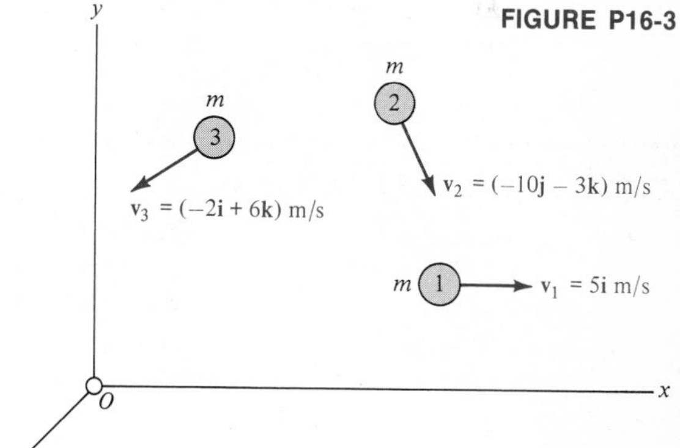

FIGURE P16-3

16–5 The three particles are released from rest and they all fall in a vertical plane. Force F acts horizontally on particle 3. Determine the acceleration of the center of mass of the system.

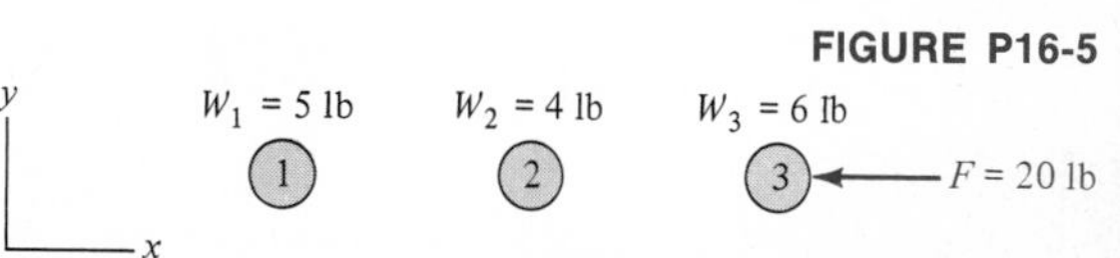

FIGURE P16-5

FIGURE P16-6

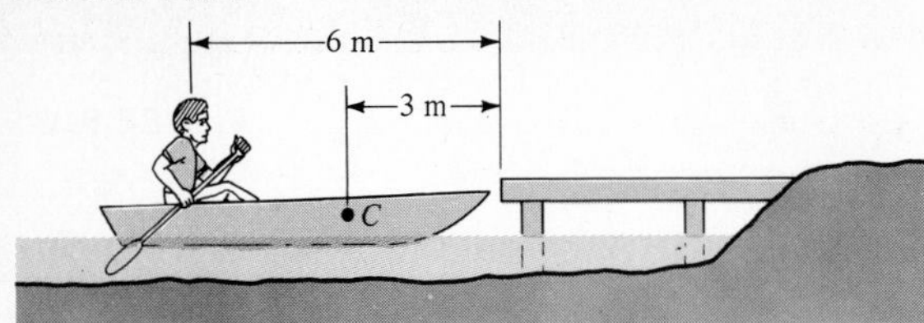

16–6 A 70-kg person steers a 120-kg boat to gently touch a dock. The person is 6 m away from the dock, gets up, and moves to the front of the boat to climb out. How far is the person from the dock after having moved to the front of the boat? C is the center of mass of the boat. Assume that there is no friction when the boat moves.

FIGURE P16-7

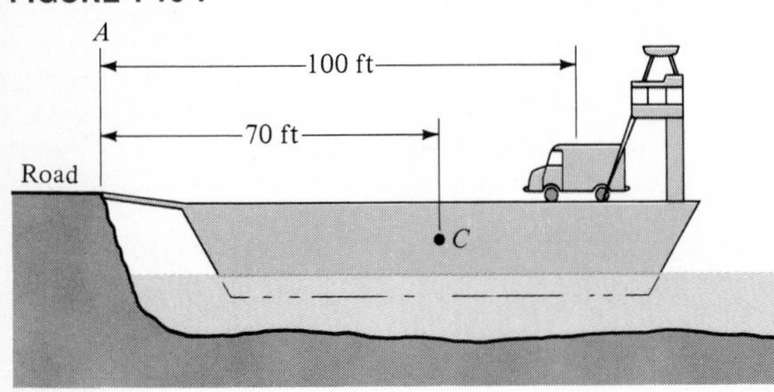

16–7 A 10-ton truck starts from the back of a 200-ton ferry. What would be the distance of the ferry from the road with the truck at A if the ferry initially touched land but was not tied? C is the center of mass of the ferry. Assume that there is no friction when the ferry moves.

FIGURE P16-8

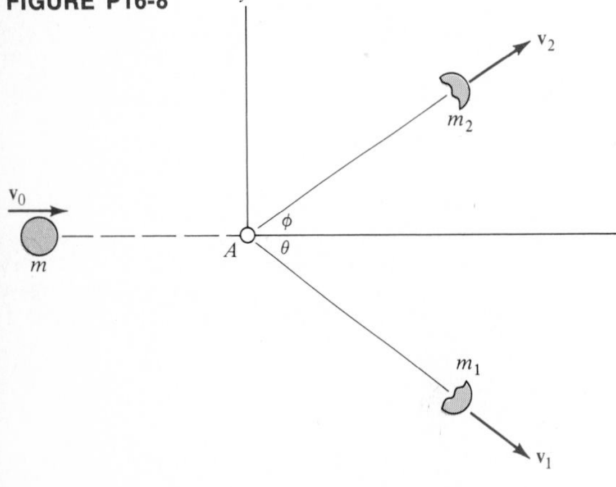

16–8 A 10-kg projectile traveling at $v_0 = 200$ m/s breaks into two parts at point A. Calculate v_1 and v_2 if $m_1 = 2$ kg, $\theta = 35°$, and $\phi = 30°$.

16–9 A meteor (tracked with radar) is traveling at $v_0 = 4000$ mph. It breaks into two parts at point A in Fig. P16-8. Calculate the weight W of the meteor if $\theta = 30°$, $W_1 = 400$ lb, $v_2 = 5110$ mph, and $\phi = 40°$. The weight W_1 is measured at the surface of the earth after the fragment is recovered.

16–10 A rocket (tracked with radar) is traveling at $v_0 = 8000$ km/h. It breaks into three parts at point A. The fragments move in the xy plane at $v_1 = 9000$ km/h, $\theta_1 = 45°$, $v_2 =$ 10 000 km/h, $\theta_2 = 0°$, $v_3 = 11\,000$ km/h, and $\theta_3 = 45°$. m_1 is recovered and found to be 1600 kg. Calculate m_2, m_3, and the total mass of the rocket.

FIGURE P16-10

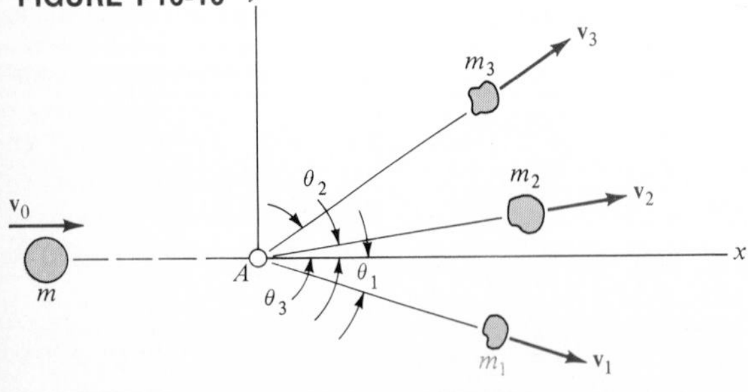

16–11 A rocket (tracked with radar) is traveling at $v_0 = 8000$ km/h. It breaks into three parts at point A as in Fig. P16-10. The fragments move in the xy plane at $v_1 = 9000$ km/h, $\theta_1 = 40°$, $v_2 = 10\,500$ km/h, $\theta_2 = 10°$, $v_3 =$ 14 000 km/h, and $\theta_3 = 45°$. m_1 is recovered and found to be 800 kg. Try to determine the masses m_2 and m_3 and show that the measured data are likely to be in error.

16-3 GRAVITATIONAL POTENTIAL ENERGY OF A SYSTEM OF PARTICLES

The gravitational potential energy of a system of particles is defined as the sum of the potential energies of the individual particles of the system. The individual potential energies may be measured with respect to a common reference level (for convenience) such as the x axis in Fig. 16-5. The potential energy V_i of a particle P_i is $m_i g y_i$, and the same factor g appears in every individual term of the total potential energy. Thus, the potential energy V of a system of n particles is expressed as

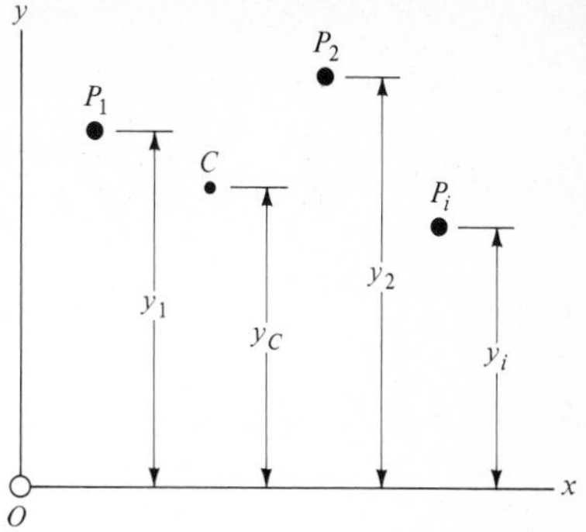

FIGURE 16-5

Schematic for the gravitational potential energy of a system of particles

$$V = g \sum_{i=1}^{n} m_i y_i = \sum_{i=1}^{n} W_i y_i \tag{16-9}$$

where m_i and W_i are the mass and weight of particle P_i, respectively. Equation 16-9 is valid regardless of any motion of the particles.

The potential energy can also be expressed as the product of the total mass m of the system, the position y_C of its center of mass, and g,

$$V = mgy_C = Wy_C \tag{16-10}$$

where W is the total weight of the system. In common engineering problems y_C represents both the center of mass and the center of gravity of the system.

16-4 KINETIC ENERGY OF A SYSTEM OF PARTICLES

The kinetic energy of a system of particles is defined as the sum of the kinetic energies of the individual particles of the system. The kinetic energy T_i of a particle P_i is $\frac{1}{2}m_i v_i^2$, where v_i is the magnitude of the particle's velocity in a fixed reference frame as in Fig. 16-6.

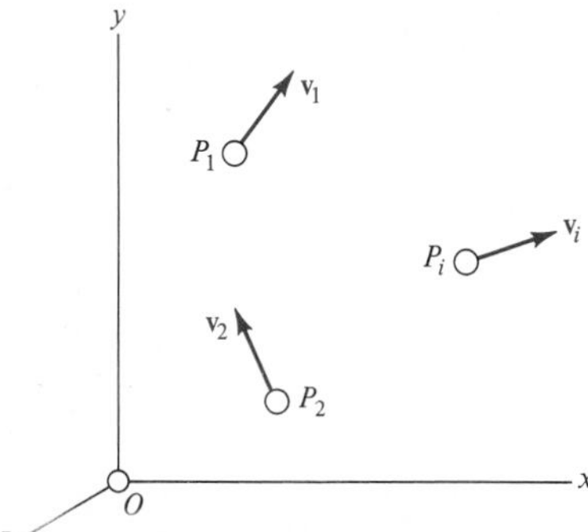

FIGURE 16-6

Schematic for the kinetic energy of a system of particles

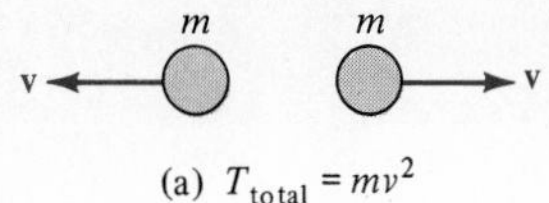

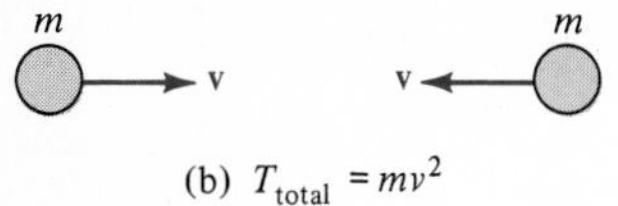

FIGURE 16-7

Two particles with identical kinetic energies in different motions

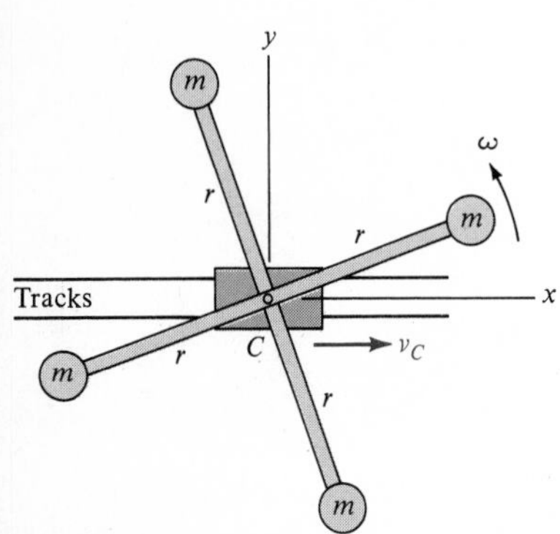

FIGURE 16-8

Amusement ride

The kinetic energy T of a system of n particles is given by

$$T = \frac{1}{2} \sum_{i=1}^{n} m_i v_i^2 \tag{16-11}$$

It should be remembered that the kinetic energy is always positive if there is motion, regardless of the number of particles and the directions of their velocities. For example, the two equal particles in Fig. 16-7a have equal and opposite velocities. Their total kinetic energy according to Eq. 16-11 is $T = 2(\frac{1}{2}mv^2) = mv^2$. The same is true for the system of two particles in Fig. 16-7b.

Reference Frame Translating with the Mass Center

There are many situations where it is desirable to determine the kinetic energy of a system of particles with respect to a given fixed frame, but it is not convenient to make the direct summation using Eq. 16-11. For example, consider the model of an amusement ride in Fig. 16-8. Each cage of mass m may be considered a particle in circular motion with respect to the carriage which translates with the xy frame. In such a case it is most convenient to superimpose the kinetic energy of the system with respect to its own center of mass (point C for the 4 cages in Fig. 16-8) and the kinetic energy of the total mass assumed to be concentrated at the center of mass. This concept is developed for the common special case of a translating reference frame as follows.

Assume that the xyz frame is at the center C of the total mass of a system of particles in Fig. 16-9. This frame is in *translation only* with respect to the XYZ frame at velocity $\mathbf{v}_C$. The particle P_i has velocity $\mathbf{v}_i$ and $\mathbf{v}_i'$ in the XYZ and xyz frames, respectively. $\mathbf{v}_i$ is the vector sum of $\mathbf{v}_C$ and $\mathbf{v}_i'$ as shown in Fig. 16-9,

$$\mathbf{v}_i = \mathbf{v}_C + \mathbf{v}_i' \tag{16-12}$$

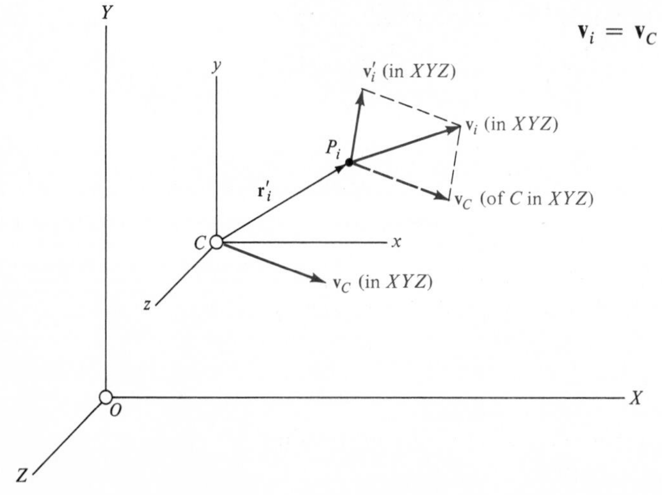

FIGURE 16-9

Schematic for using a translating reference frame located at the mass center

The total kinetic energy T of a system of n particles in the XYZ frame can be expressed using the scalar product $\mathbf{v}_i \cdot \mathbf{v}_i = \mathbf{v}_i^2$,

$$T = \frac{1}{2} \sum_{i=1}^{n} m_i v_i^2 = \frac{1}{2} \sum_{i=1}^{n} (m_i \mathbf{v}_i \cdot \mathbf{v}_i)$$

This is expanded using Eq. 16-12,

$$T = \frac{1}{2} \sum_{i=1}^{n} [m_i(\mathbf{v}_C + \mathbf{v}_i') \cdot (\mathbf{v}_C + \mathbf{v}_i')]$$

$$= \frac{1}{2}\left(\sum_{i=1}^{n} m_i\right) v_C^2 + \frac{1}{2} \sum_{i=1}^{n} m_i v_i'^2 + \sum_{i=1}^{n} (m_i \mathbf{v}_C \cdot \mathbf{v}_i')$$

The first term on the right side equals $\frac{1}{2}mv_C^2$, where m is the total mass of the system. Recognizing that $\mathbf{v}_C$ is the same for every term in the summation, the last term can be rewritten as

$$\sum_{i=1}^{n} (m_i \mathbf{v}_C \cdot \mathbf{v}_i') = \mathbf{v}_C \cdot \sum_{i=1}^{n} m_i \mathbf{v}_i' = \mathbf{v}_C \cdot \frac{d}{dt} \sum_{i=1}^{n} m_i \mathbf{r}_i' = 0$$

since $\sum_{i=1}^{n} m_i \mathbf{r}_i' = 0$ by definition of the center of mass. Thus, the total kinetic energy of the system in a fixed frame is

$$T = \underbrace{\frac{1}{2} m v_C^2}_{\substack{\text{motion of total} \\ \text{mass imagined to} \\ \text{be concentrated} \\ \text{at } C \text{ moving in} \\ \text{fixed } XYZ \text{ frame}}} + \underbrace{\frac{1}{2} \sum_{i=1}^{n} m_i v_i'^2}_{\substack{\text{motion of all} \\ \text{particles relative} \\ \text{to mass center } C \text{ at} \\ \text{origin of translating} \\ xyz \text{ frame}}} \tag{16-13}$$

Equation 16-13 is particularly useful in the dynamics analysis of rigid bodies. It is emphasized that Eq. 16-13 was obtained using a translating reference frame, and is not valid for a rotating frame. This is illustrated in Ex. 16-4.

16-5 WORK AND ENERGY. CONSERVATION OF ENERGY

The work–energy relationship for a single particle P_i is expressed as the sum of changes in the potential energy V_i and kinetic energy T_i,

$$U_i = \Delta V_i + \Delta T_i \tag{16-14}$$

where U_i is the work of all nonconservative forces acting on the particle.

To evaluate the work on a system of particles it is necessary to consider the work of the net external force $\mathbf{F}_i$ and of all internal forces $\mathbf{f}_{ij}$ on each particle P_i. These forces were discussed in detail in Sec. 16-1. The work of the equal and opposite internal forces $\mathbf{f}_{ij}$ and $\mathbf{f}_{ji}$ may or may not cancel out depending on the displacements of the particles P_i and P_j. These can be evaluated by inspection in many situations. Frequently, the pairs of displacements are identical in magnitude and only the external forces do work. The total work on a system of n particles and the resulting changes in potential and kinetic energies are expressed by

$$\sum_{i=1}^{n} U_i = \sum_{i=1}^{n} \Delta V_i + \sum_{i=1}^{n} \Delta T_i$$

which is commonly written in the concise form

$$\boxed{U = \Delta V + \Delta T} \tag{16-15}$$

The concept of a particle's potential energy can be generalized to include the internal elastic energy resulting from deformations. In that case the term ΔV may have two components, a gravitational and an elastic potential energy.

Conservation of Energy

There are many situations where it is reasonable to assume that all forces acting on a system of particles are conservative. This means that the work of external or internal friction forces must be zero or negligible in the entire system of particles. In that case Eq. 16-15 becomes

$$\boxed{\Delta V + \Delta T = 0} \tag{16-16}$$

which is the statement of the *conservation of total mechanical energy* in a system of particles. Equation 16-16 is frequently used to express the same total energy at two different configurations of a system and thus solve for an unknown quantity. This method is stated schematically for configurations 1 and 2,

$$\boxed{V_1 + T_1 = V_2 + T_2} \tag{16-17}$$

Equations 16-16 and 16-17 express the same principle of conservation of total mechanical energy. The applicability of the method is generally determined by judging the expected frictional losses.

Three acrobats simultaneously jump from a height $h = 3$ m onto a large trampoline as shown in Fig. 16-10. Their masses are $m_1 = 60$ kg, $m_2 = 70$ kg, and $m_3 = 50$ kg. (a) Calculate the total kinetic energy of the acrobats at the instant they touch the trampoline. (b) How much work is done during the rebound if the trampoline is 90% efficient? Assume that the acrobats are particles that return to 90% of their original height after rebound.

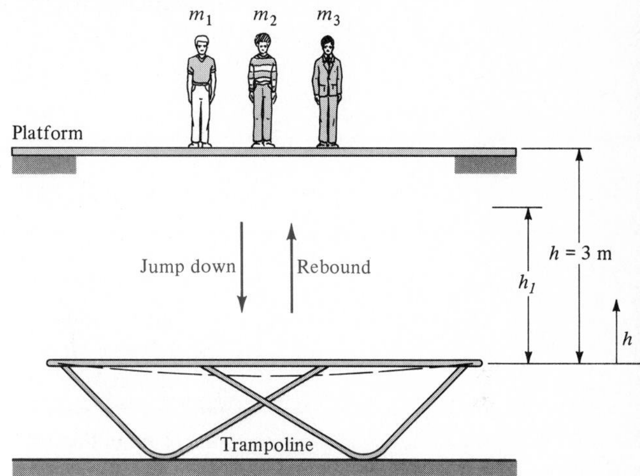

FIGURE 16-10
Acrobats jumping on a trampoline

SOLUTION

(a) Since energy is conserved in this system,

$$\Delta V + \Delta T = 0$$

$$(V_f - V_i) + (T_f - T_i) = 0$$

Because the acrobats start from rest, $T_i = 0$; because they make contact at zero height, $V_f = mgh = 0$.

$$-V_i + T_f = 0$$

$$T_f = V_i = (60 + 70 + 50)\text{kg}\,(9.81\ \text{m/s}^2)(3\ \text{m})$$
$$= 5297\ \text{N}\cdot\text{m}$$

This value is the required total kinetic energy because the potential energy is taken as zero at first contact with the trampoline.

(b) If the acrobats return to 90% of their original height after rebound,

$$h_1 = 0.90(3\ \text{m}) = 2.7\ \text{m}$$

The work U done during the rebound is determined using Eq. 16-15,

$$U = \Delta V + \Delta T$$

which is valid for the total motion through 3 m downward plus 2.7 m upward.

$$U = mg(\Delta h) + 0 \qquad \text{(since the kinetic energy is zero at the peak height after any rebound)}$$
$$= (60\ \text{kg} + 70\ \text{kg} + 50\ \text{kg})(9.81\ \text{m/s}^2)(3\ \text{m} - 2.7\ \text{m})$$
$$= 530\ \text{N}\cdot\text{m}$$

EXAMPLE 16-4

Consider the model of an amusement ride in Fig. 16-11. The carriage C of negligible mass moves at a constant speed v_C on a horizontal track that has straight sections and curved sections of radius R. The rigid arms have negligible mass and rotate at constant angular speed ω with respect to the carriage. The two cages of mass m each are assumed to be particles. Determine the total kinetic energy of the two cages (a) when the carriage is on the straight track, and (b) when the carriage is on the curved track.

SOLUTION

(a) The problem is solved using Eq. 16-13,

$$T = \frac{1}{2}(2m)v_C^2 + \frac{1}{2}(2)(mr^2\omega^2)$$

$$= m(v_C^2 + r^2\omega^2)$$

(b) It must be realized here that the carriage is rotating at an angular speed $\Omega = v_C/R$ with respect to the fixed XY frame. To use Eq. 16-13, the particles' motion relative to the translating $x'y'$ frame must be determined (and not just with respect to the carriage). Noting that the angular speeds ω and Ω are in the opposite sense to one another, the relative speed of a particle in the $x'y'$ frame is $r(\omega - \Omega)$. From Eq. 16-13,

$$T = \frac{1}{2}(2m)v_C^2 + \frac{1}{2}(2)(mr^2)(\omega - \Omega)^2$$

$$= mv_C^2 + mr^2(\omega - \Omega)^2$$

JUDGMENT OF THE RESULTS

The results are identical, as they should be, when $R \to \infty$ (a straight track). It is also noted that $v_i' = 0$ when $\omega = \Omega$, and $T = mv_C^2$ as can be visualized in the right part of Fig. 16-11.

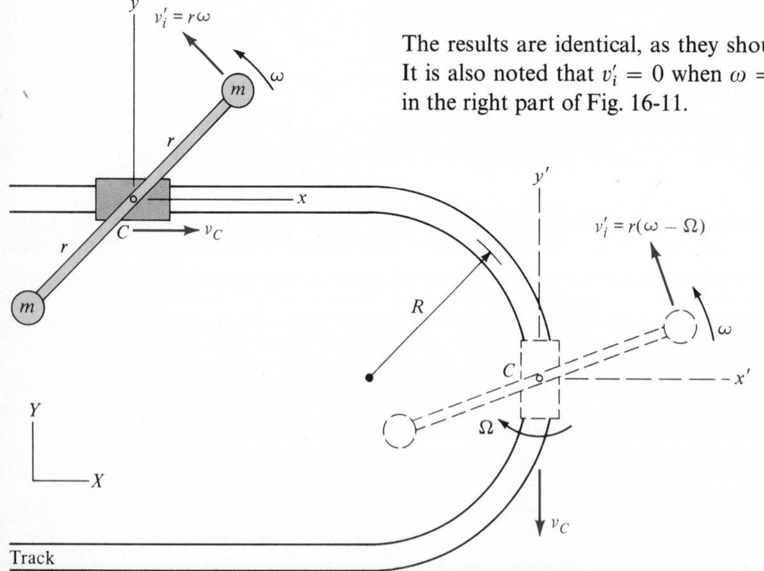

FIGURE 16-11

Schematic of an amusement ride

PROBLEMS

16–12 A roller coaster consists of five identical cars, each of 200 kg mass including passengers. The train of length $l = 8$ m has speed $v_A = 15$ m/s on a horizontal part of the track. Determine the elevation h reached by the front car if friction is negligible.

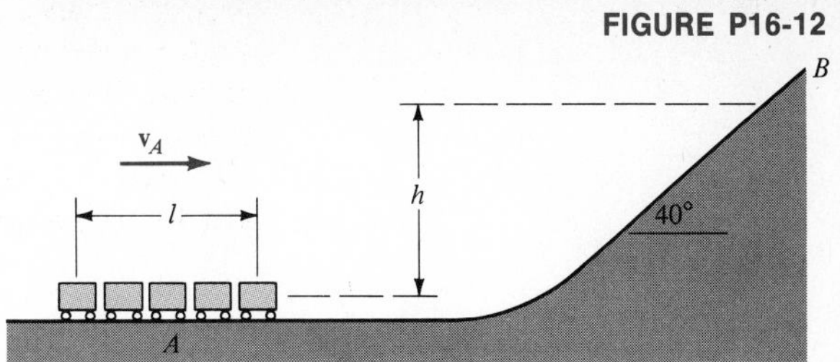

FIGURE P16-12

16–13 Each car of the roller coaster in Fig. P16-12 weighs 400 lb including passengers. The front car reaches an elevation $h = 50$ ft. Determine the speed v_A of the train to make it coast without friction to that height if $l = 20$ ft.

16–14 The track of a roller coaster has a circular part of radius $R = 50$ m to the center of mass of each 250-kg car. The maximum speed of the train is $v_A = 20$ m/s. Determine the speed v_B at a position $\theta = 40°$ of the front car. A typical distance between cars is $d = 2$ m. Neglect friction.

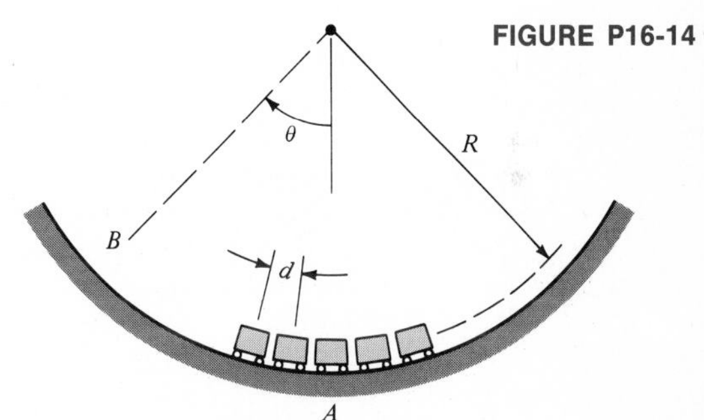

FIGURE P16-14

16–15 The front car of a roller coaster reaches a maximum displacement $\theta = 45°$ in Fig. P16-14. The radius of curvature to the center of mass of each car is $R = 180$ ft, each car weighs 400 lb, and a typical distance between cars is $d = 5$ ft. Determine the maximum speed v_A of the train to make it coast without friction to point B.

16–16 A Ferris wheel has three evenly spaced cages, which weigh 1000 lb with riders. Calculate the total kinetic energy of the cages if $\omega_A = 2$ rpm and $\omega_Y = 0$.

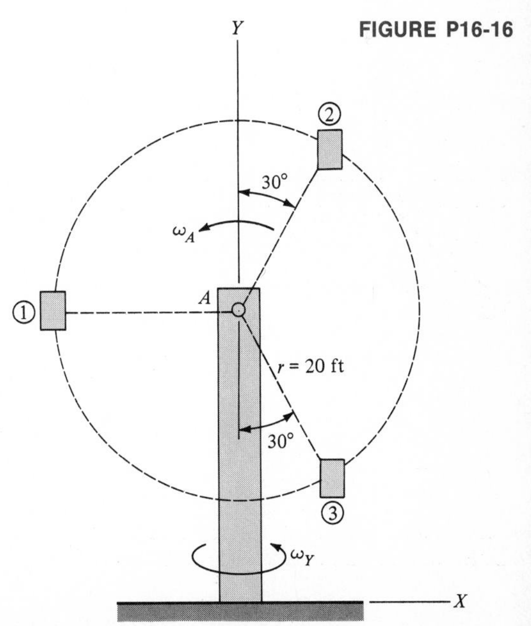

FIGURE P16-16

16–17 Calculate the total kinetic energy of the three cages in Prob. 16–16 assuming an additional motion at angular speed $\omega_Y = 1.5$ rpm for the position shown.

16–18 Determine the total work required to start the Ferris wheel of Prob. 16–16 from rest and bring it to the angular speeds $\omega_A = 2$ rpm and $\omega_y = 1.5$ rpm. Neglect friction and changes in elevation.

FIGURE P16-19

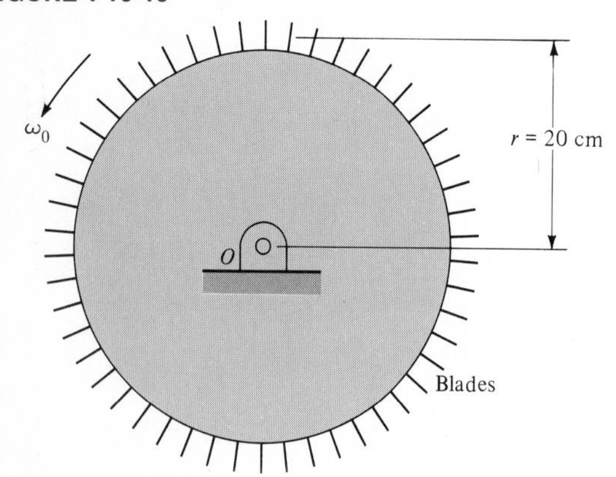

16–19 A turbine disk has 50 identical 0.15-kg blades of negligible dimensions. Assume that the disk itself has no mass and is rotating at $\omega_0 = 5000$ rpm. Determine the work of the braking mechanism that stops the rotation of the disk.

FIGURE P16-20

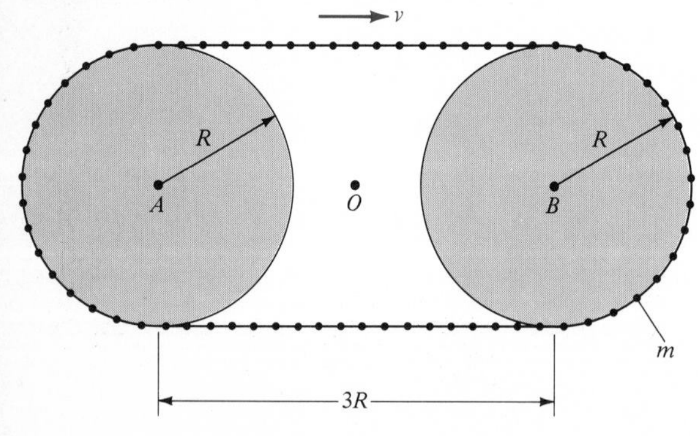

16–20 The chain-drive mechanism has 100 identical links of mass m. Write a formula for the kinetic energy of the chain.

16–21 The weight of each link is 0.05 lb and $R = 0.5$ ft in Prob. 16–20. Determine the average power requirement of the machine to start the chain from rest and bring it to a speed $v = 20$ ft/s in a time of 7 s.

16–22 An amusement ride is designed to rotate arm AOB about point O in the horizontal plane at $\omega_1 = 5$ rpm. The two cages with two riders in each are to rotate at $\omega_2 = 6$ rpm relative to the arm AOB. The average rider has a 70-kg mass and sits at $r = 1$ m from the center of the cage. Calculate the total kinetic energy of (a) riders 1 and 2, and (b) riders 3 and 4 for the position shown.

FIGURE P16-22

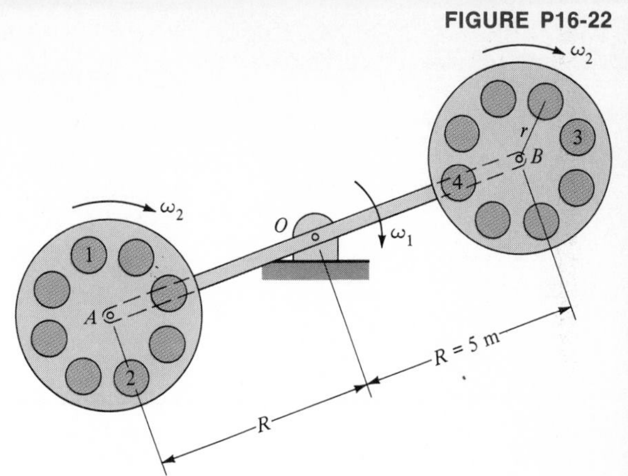

16–23 A proposed mechanism would change r from 1 m to 1.5 m during operation of the ride described in Prob. 16–22. Calculate the change in total kinetic energy and the required work to make the change. Ignore the masses of structural members. Assume that the change occurs instantaneously in the position shown.

16–24 The train of the amusement ride has a speed $v_A = 6$ ft/s when the train's center is at A. Each of the seven cars in the 40-ft long train weighs 800 lb with riders. Calculate the maximum elevation h_B of the train's front if $h_A = 70$ ft and friction is negligible as the train moves from A to B.

FIGURE P16-24

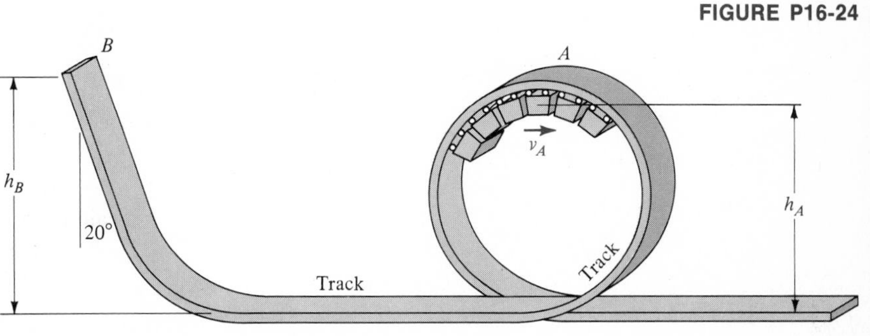

16-6 MOMENTS OF FORCES ON A SYSTEM OF PARTICLES

The analysis of forces in Sec. 16-1 can be readily extended to consider the moments of the forces about an arbitrary point of reference. This is done using Fig. 16-1. Assume that the moments of all forces acting on the system of particles should be evaluated with respect to point O. For generality, the moments of external and internal forces are included in the analysis. They are written for particle P_i as

$$\mathbf{r}_i \times \mathbf{F}_i + \mathbf{r}_i \times \mathbf{f}_{ij} = \mathbf{r}_i \times m_i\mathbf{a}_i$$

The moments of forces for a whole system of n particles can be written from this equation following the procedure used in Sec. 16-1,

$$\sum_{i=1}^{n} (\mathbf{r}_i \times \mathbf{F}_i) + \sum_{i=1}^{n} \sum_{j=1}^{n} (\mathbf{r}_i \times \mathbf{f}_{ij}) = \sum_{i=1}^{n} (\mathbf{r}_i \times m_i\mathbf{a}_i) \qquad \boxed{16\text{-}18}$$

The second term on the left side can be eliminated since the moments of the internal forces cancel in pairs for any number of interacting particles. The resulting governing equation for the motion of the system about the arbitrary point O caused by moments of forces is

$$\sum_{i=1}^{n} (\mathbf{r}_i \times \mathbf{F}_i) = \sum_{i=1}^{n} \mathbf{M}_{i_O} = \sum_{i=1}^{n} (\mathbf{r}_i \times m_i\mathbf{a}_i) \qquad \boxed{16\text{-}19}$$

16-7 LINEAR AND ANGULAR MOMENTA OF A SYSTEM OF PARTICLES

The linear momentum of a particle P_i is $\mathbf{G}_i = m_i\mathbf{v}_i$ by definition. The linear momentum of a system of n particles is, by definition,

$$\mathbf{G} = \sum_{i=1}^{n} m_i\mathbf{v}_i$$

Substituting Eq. 16-6 gives

$$\mathbf{G} = m\mathbf{v}_C \qquad \boxed{16\text{-}20}$$

which means that the linear momentum of a system of particles is equal to the total mass of the particles times the velocity of the center of mass. Equation 16-20 is valid if the mass of the system is constant.

Differentiating Eq. 16-20 with respect to time yields

$$\dot{\mathbf{G}} = m\dot{\mathbf{v}}_C$$

where $m\dot{\mathbf{v}}_C = m\mathbf{a}_C$. Using a simplified notation for the result obtained from Eqs. 16-4 and 16-7,

$$\sum \mathbf{F} = \dot{\mathbf{G}} \tag{16-21}$$

which means that *the resultant of the external forces on a system of particles equals the time rate of change of linear momentum of that system.* Equation 16-21 is an alternative expression of Eq. 16-4. These two equations are the generalized forms of Newton's second law of motion. It is assumed that the mass of the system is constant.

The angular momentum **H** of a system of particles is defined with respect to an arbitrary, fixed point of reference. Referring to point O in Fig. 16-1,

$$\mathbf{H}_O = \sum_{i=1}^{n} (\mathbf{r}_i \times m_i \mathbf{v}_i) \tag{16-22}$$

Differentiating this equation with respect to time gives

$$\dot{\mathbf{H}}_O = \sum_{i=1}^{n} (\dot{\mathbf{r}}_i \times m_i \mathbf{v}_i) + \sum_{i=1}^{n} (\mathbf{r}_i \times m_i \dot{\mathbf{v}}_i)$$

Since $\dot{\mathbf{r}}_i = \mathbf{v}_i$, $\dot{\mathbf{r}}_i \times m_i \mathbf{v}_i = \mathbf{v}_i \times m_i \mathbf{v}_i = 0$ (recall that the cross product of two collinear vectors is zero). Writing $\dot{\mathbf{v}}_i = \mathbf{a}_i$ yields

$$\dot{\mathbf{H}}_O = \sum_{i=1}^{n} (\mathbf{r}_i \times m_i \mathbf{a}_i) \tag{16-23}$$

It is seen from Eq. 16-19 that the right side of Eq. 16-23 is the sum of the moments of the external forces about point O for the whole system of particles. Denoting this sum by $\sum M_O$, the resulting equation is

$$\sum \mathbf{M}_O = \dot{\mathbf{H}}_O = \sum (\mathbf{r}_i \times m_i \mathbf{a}_i) \tag{16-24}$$

which means that *the resultant of the moments of the external forces on a system of particles equals the time rate of change of angular momentum of that system.* The moments and the angular momentum are defined with respect to a fixed point O. The mass of the system is assumed constant.

16-8 ANGULAR MOMENTUM ABOUT THE CENTER OF MASS

An important special case in using Eq. 16-24 is when the center of mass is taken as the point of reference for the moments of forces and the angular momentum. In this case it must be assumed that the center C of the mass may be moving with respect to the fixed coordinate system whose origin O was used in obtaining Eq. 16-24. The analysis of angular momentum is done using Fig. 16-12. The center C of the total mass of the particles moves in an arbitrary way with respect to the XYZ frame. A translating but not rotating reference

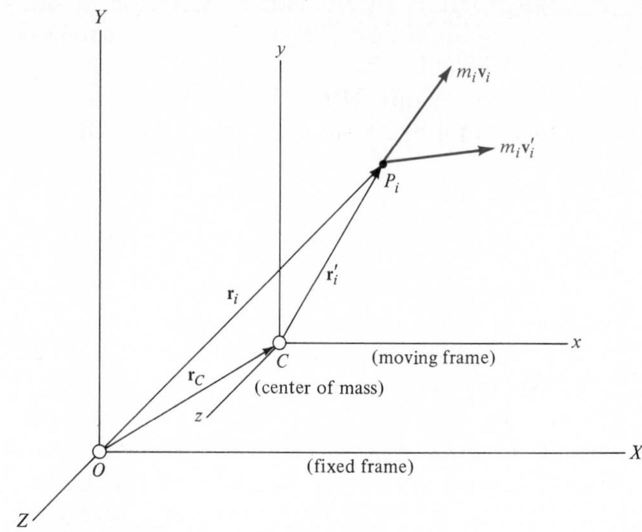

FIGURE 16-12

Motion of a particle with respect to the center of mass and a fixed reference frame

frame xyz is moving with point C. The position vectors, velocities, and accelerations of a particle P_i with mass m_i are represented according to the following scheme:

	FIXED XYZ FRAME	MOVING xyz FRAME
Position vector	$\mathbf{r}_i$	$\mathbf{r}_i'$
Velocity vector	$\mathbf{v}_i$	$\mathbf{v}_i'$
Acceleration vector	$\mathbf{a}_i$	$\mathbf{a}_i'$

The corresponding quantities for the mass center C moving in the XYZ frame are denoted by $\mathbf{r}_C$, $\mathbf{v}_C$, and $\mathbf{a}_C$.

The *absolute angular momentum* of the system of particles is defined with respect to point O of the XYZ frame using Eq. 16-22,

$$\mathbf{H}_O = \sum_{i=1}^{n} (\mathbf{r}_i \times m_i\mathbf{v}_i)$$

The angular momentum of the same system with respect to its own mass center C is

$$\mathbf{H}_C = \sum_{i=1}^{n} (\mathbf{r}'_i \times m_i\mathbf{v}_i) \qquad \boxed{16\text{-}25}$$

The same velocity $\mathbf{v}_i$ is used in Eqs. 16-22 and 16-25 because it is the same linear momentum $m_i\mathbf{v}_i$ that is being considered with two different position vectors to obtain the angular momenta in the two systems. From the expression of relative velocities,

$$\mathbf{v}_i = \mathbf{v}_C + \mathbf{v}'_i$$

Substituting for $\mathbf{v}_i$ in Eq. 16-25 gives

$$\mathbf{H}_C = \sum_{i=1}^{n} [\mathbf{r}'_i \times m_i(\mathbf{v}_C + \mathbf{v}'_i)]$$

$$= \sum_{i=1}^{n} (\mathbf{r}'_i \times m_i\mathbf{v}_C) + \sum_{i=1}^{n} (\mathbf{r}'_i \times m_i\mathbf{v}'_i) \qquad \boxed{16\text{-}26}$$

This equation can be simplified if it is recognized that the first term on the right side can be written as

$$\left(\sum_{i=1}^{n} \mathbf{r}'_i m_i\right) \times \mathbf{v}_C$$

since $\mathbf{v}_C$ is the same in each term of the summation and thus it can be factored out, and m_i is constant for each particle. Note that $\mathbf{v}'_i$ cannot be factored out since it is the velocity of only one particle in the most general case. By definition of the center of mass, $\sum_{i=1}^{n} \mathbf{r}'_i m_i = 0$ (this is the location of the center of mass with respect to itself). Thus,

$$\boxed{\mathbf{H}_C = \sum_{i=1}^{n} (\mathbf{r}'_i \times m_i\mathbf{v}'_i) = \sum_{i=1}^{n} (\mathbf{r}'_i \times m_i\mathbf{v}_i)} \qquad \boxed{16\text{-}27}$$

The last result also could be obtained if the xyz system were not moving and Eq. 16-22 were applied directly with respect to the center of mass. It is concluded that Eqs. 16-25 and 16-27 are the same. This means that *the angular momentum of a system of particles about its center of mass is the same whether it is observed from a newtonian reference frame or from the centroidal frame which may be translating but not rotating*. This result is not true for points other than the mass center.

The rate of change of angular momentum of a system of particles about its mass center depends on the moments of the external forces about the same point. This is shown by differentiating Eq. 16-27 with respect to time,

$$\dot{\mathbf{H}}_C = \sum_{i=1}^{n} (\dot{\mathbf{r}}'_i \times m_i\mathbf{v}'_i) + \sum_{i=1}^{n} (\mathbf{r}'_i \times m_i\dot{\mathbf{v}}'_i) \qquad \boxed{16\text{-}28}$$

With $\mathbf{v}'_i = \dot{\mathbf{r}}'_i$, the first term on the right side is zero since the cross product of two collinear vectors is zero. In the second term, $\dot{\mathbf{v}}'_i = \mathbf{a}'_i$. From the expression of relative accelerations,

$$\mathbf{a}_i = \mathbf{a}_C + \mathbf{a}'_i$$

The second term on the right side of Eq. 16-28 is expanded as

$$\sum_{i=1}^{n} [\mathbf{r}'_i \times m_i(\mathbf{a}_i - \mathbf{a}_C)] = \sum_{i=1}^{n} (\mathbf{r}'_i \times m_i\mathbf{a}_i) - \sum_{i=1}^{n} (\mathbf{r}'_i \times m_i\mathbf{a}_C)$$

where

$$\sum_{i=1}^{n} (\mathbf{r}'_i \times m_i\mathbf{a}_i) = \sum_{i=1}^{n} (\mathbf{r}'_i \times \mathbf{F}_i) = \sum_{i=1}^{n} \mathbf{M}_{iC}$$

from Eq. 16-8.

$$\sum_{i=1}^{n} (\mathbf{r}'_i \times m_i\mathbf{a}_C) = \left(\sum_{i=1}^{n} \mathbf{r}'_i m_i\right) \times \mathbf{a}_C = 0$$

as it was done with $\mathbf{v}_C$ in deriving Eq. 16-27. The resulting equation in the same concise form as that of Eq. 16-24 is

$$\boxed{\sum \mathbf{M}_C = \dot{\mathbf{H}}_C = \sum (\mathbf{r}'_i \times m_i\mathbf{a}_i) = \sum (\mathbf{r}'_i \times m_i\mathbf{a}'_i)} \qquad \text{16-29}$$

This means that *the resultant of the moments of the external forces about the mass center of a system of particles equals the time rate of change of the angular momentum about the mass center.* It is assumed that the total mass of the system is constant. Note that Eqs. 16-24 and 16-29 are similar in form, and the only difference between them is the point of reference. Both of these equations are useful in dynamics analysis. Remember in using Eq. 16-24 or 16-29 that moments may be taken only about a fixed point or the center of mass. If another point is used for reference, the equation of motion for rotation is in a more complicated form.

Useful Expressions

1. The angular momenta determined with respect to a fixed point O and the center of mass C are related as follows. From Eq. 16-22 and Fig. 16-5,

$$\mathbf{H}_O = \sum (\mathbf{r}_i \times m_i\mathbf{v}_i) = \sum (\mathbf{r}_C + \mathbf{r}'_i) \times m_i\mathbf{v}_i = \sum (\mathbf{r}_C \times m_i\mathbf{v}_i) + \sum (\mathbf{r}'_i \times m_i\mathbf{v}_i)$$

where $\mathbf{r}_C$ is the same for considering any particle of the system, and $\sum m_i\mathbf{v}_i = m\mathbf{v}_C$ from Eq. 16-6. The last term is $\mathbf{H}_C$ from Eq. 16-25. Thus,

$$\mathbf{H}_O = \mathbf{H}_C + \mathbf{r}_C \times m\mathbf{v}_C \qquad \text{16-30}$$

2. The moments about a fixed point O are related to the time rate of change of angular momentum with respect to the center of mass C as follows. From Eqs. 16-24 and 16-30,

$$\sum \mathbf{M}_O = \dot{\mathbf{H}}_O = \frac{d}{dt}(\mathbf{H}_C + \mathbf{r}_C \times m\mathbf{v}_C)$$

$$= \dot{\mathbf{H}}_C + \dot{\mathbf{r}}_C \times m\mathbf{v}_C + \mathbf{r}_C \times m\dot{\mathbf{v}}_C$$

Since $\dot{\mathbf{r}}_C = \mathbf{v}_C$, $\dot{\mathbf{r}}_C \times m\mathbf{v}_C = 0$. Writing $\dot{\mathbf{v}}_C = \mathbf{a}_C$ for the acceleration of the center of mass gives

$$\sum \mathbf{M}_O = \dot{\mathbf{H}}_C + \mathbf{r}_C \times m\mathbf{a}_C \qquad \text{16-31}$$

16-9 CONSERVATION OF MOMENTUM OF A SYSTEM OF PARTICLES

During any period of time when the resultant of the external forces acting on a system of particles is zero, $\dot{\mathbf{G}} = 0$ from Eq. 16-21. In that case,

$$\mathbf{G} = \text{constant} \qquad \text{16-32}$$

which means that the total linear momentum of the system is conserved. This does not mean that the linear momentum of any particle of the system must be conserved. For example, some particles could lose momentum while others gain an identical amount, thus leaving the total unchanged.

Similarly, the resultant moment of all external forces acting on a system of particles may be zero, either in reference to a fixed point O or to the mass center C. In the first case, $\dot{\mathbf{H}}_O = 0$ from Eq. 16-24, and

$$\mathbf{H}_O = \text{constant} \qquad \text{16-33}$$

In the second case, $\dot{\mathbf{H}}_C = 0$ from Eq. 16-29, and

$$\mathbf{H}_C = \text{constant} \qquad \text{16-34}$$

Equations 16-33 and 16-34 express the conservation of angular momentum. Note that the two constants may be different. Furthermore, either $\mathbf{H}_O$ or $\mathbf{H}_C$ may be constant while the other one is not constant.

16-10 IMPULSE AND MOMENTUM OF A SYSTEM OF PARTICLES

Equations 16-21 and 16-24 are the fundamental expressions relating forces and moments to the time rates of change of linear and angular momenta for systems of particles. However, these equations are not the best to analyze impulsive motion when the relationship of force and time is known. The concept of impulse which was developed for individual particles is also valid for systems of particles.

The time rate of change of linear momentum is expressed from Eq. 16-21 as

$$\sum \mathbf{F} = \frac{d}{dt}\mathbf{G}$$

This is integrated between time t_1 and t_2 during which the forces act and change the momentum of the system of particles from $\mathbf{G}_1$ to $\mathbf{G}_2$.

$$\sum_{i=1}^{n} \int_{t_1}^{t_2} \mathbf{F}_i \, dt = \mathbf{G}_2 - \mathbf{G}_1 = m\mathbf{v}_{C_2} - m\mathbf{v}_{C_1} \tag{16-35}$$

where the integral is the linear impulse of an external force F_i on a particle P_i of the system. The summation of these impulses over all particles equals the change in total linear momentum of the system of n particles.

A similar procedure is followed to express the angular impulse using Eqs. 16-24 and 16-22,

$$\begin{aligned} \sum \mathbf{M}_O &= \frac{d}{dt}\mathbf{H}_O \\ \sum_{i=1}^{n} \int_{t_1}^{t_2} \mathbf{M}_{i_O} \, dt &= \mathbf{H}_{O_2} - \mathbf{H}_{O_1} \\ &= \sum_{i=1}^{n} (\mathbf{r}_i \times m_i\mathbf{v}_i)_2 - \sum_{i=1}^{n} (\mathbf{r}_i \times m_i\mathbf{v}_i)_1 \end{aligned} \tag{16-36}$$

where the integral is the angular impulse of the moment $\mathbf{M}_{i_O} = \mathbf{r}_i \times \mathbf{F}_i$ on a particle P_i of the system. The summation of these impulses over all particles equals the change in total angular momentum of the system of n particles about the fixed reference point O. A similar expression can be written for the angular momentum about the center of mass.

EXAMPLE 16-5

A 1000-kg space vehicle is moving at a velocity $\mathbf{v}_0 = 2000\ \mathbf{k}$ km/h when spring-loaded mechanisms simultaneously eject two small probes in a plane normal to the parent vehicle's original velocity. The masses of the probes and their directions of motion are shown in Fig. 16-13. Determine the velocities of the three parts after the separation if $\mathbf{v}_f = (-0.56\mathbf{i} - 0.3\mathbf{j} + 0.77\mathbf{k})v_f$. In other words, the direction of the parent vehicle's final velocity is given, but its magnitude v_f and the magnitudes v_1 and v_2 are unknown.

SOLUTION

Since there are no external forces acting on this system of particles, the total linear momentum of the system is conserved,

$$\mathbf{G} = M\mathbf{v}_0 = m\mathbf{v}_f + m_1\mathbf{v}_1 + m_2\mathbf{v}_2$$

where M = total mass of the system = $m + m_1 + m_2$ = 1000 kg
m = mass of parent vehicle = 650 kg
$\mathbf{v}_0$ = original velocity of space vehicle
$\mathbf{v}_f$ = velocity of parent vehicle after firing

Substituting the given quantities yields

$$\begin{aligned}(1000\text{ kg})(2000\text{ km/h})\mathbf{k} &= (650\text{ kg})(-0.56\mathbf{i} - 0.3\mathbf{j} + 0.77\mathbf{k})v_f\\ &\quad + (200\text{ kg})(\cos 45^\circ\ \mathbf{i} + \sin 45^\circ\ \mathbf{j})v_1\\ &\quad + (150\text{ kg})v_2\mathbf{i}\end{aligned}$$

Equating vector components gives

$$x \text{ direction:}\quad 0 = (650)(-0.56)v_f + (200)(\cos 45^\circ)v_1 + (150)v_2$$

$$y \text{ direction:}\quad 0 = (650)(-0.3)v_f + (200)(\sin 45^\circ)v_1$$

$$z \text{ direction:}\quad (1000)(2000) = (650)(0.77)v_f$$

$$v_f = \frac{(1000)(2000)}{(650)(0.77)} = 4000\text{ km/h}$$

$$v_1 = \frac{(650)(0.3)(4000)}{(200)(\sin 45^\circ)} = 5515\text{ km/h}$$

$$v_2 = \frac{(650)(0.56)(4000) - (200)(\cos 45^\circ)(5515)}{150} = 4507\text{ km/h}$$

and

$$\mathbf{v}_f = (-2240\mathbf{i} - 1200\mathbf{j} + 3080\mathbf{k})\text{ km/h}$$

$$\mathbf{v}_1 = (3900\mathbf{i} + 3900\mathbf{j})\text{ km/h}$$

$$\mathbf{v}_2 = (4507\mathbf{i})\text{ km/h}$$

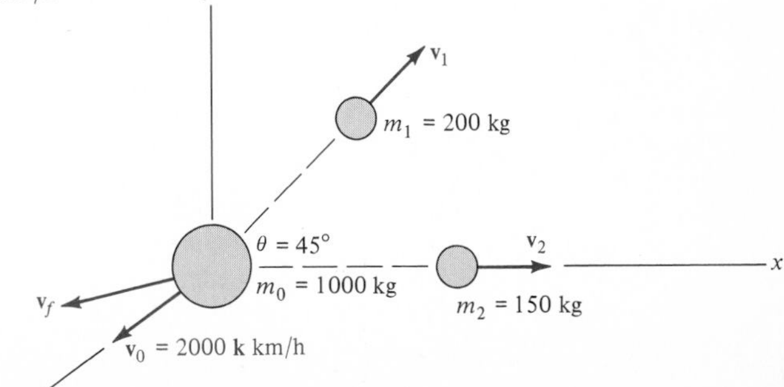

FIGURE 16-13
Motions of a spacecraft and its separated parts

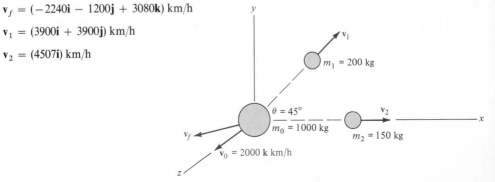

EXAMPLE 16-6

A proposed amusement ride consists of a large, horizontal turntable which should rotate with a constant angular speed $\omega = 0.3$ rad/s. Assume that a 100-kg and a 70-kg person are walking radially outward on the turntable as modeled in Fig. 16-14. Their positions, relative velocities on the turntable, and absolute accelerations are determined experimentally. At a particular instant $r_1 = 4$ m, $v_1 = 1$ m/s, $\mathbf{a}_1 = (-0.36\mathbf{i} + 0.6\mathbf{j})$ m/s^2, and $r_2 = 6$ m, $v_2 = 1.5$ m/s, $\mathbf{a}_2 = (0.54\mathbf{i} - 0.9\mathbf{j})$ m/s^2. Determine the torque M_O required to maintain the constant angular speed of the turntable at the given instant neglecting friction.

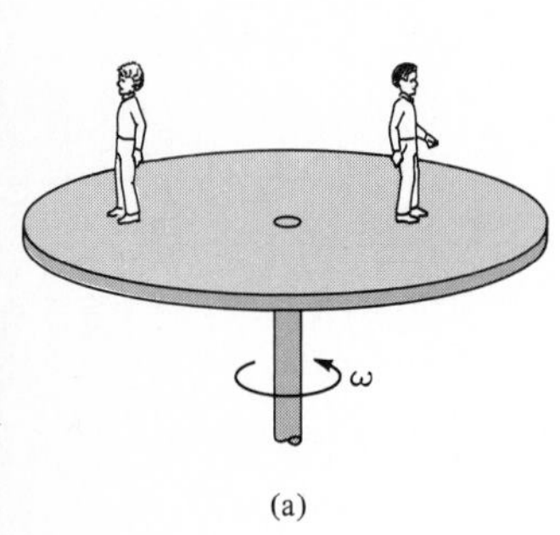

(a)

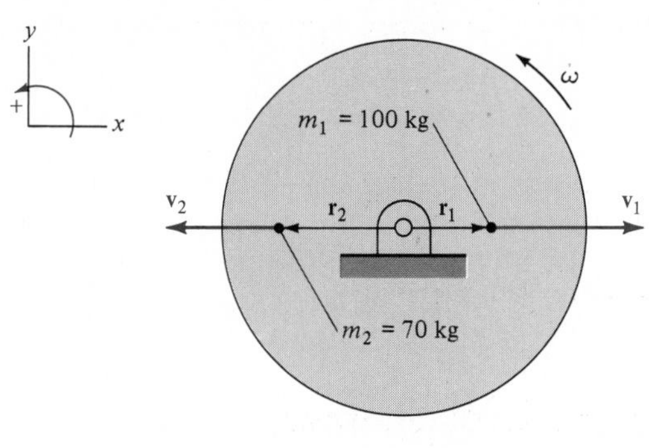

(b) Top view

FIGURE 16-14
Motion of a turntable

SOLUTION

From Eq. 16-15,

$$\begin{aligned}\mathbf{M}_O &= \dot{\mathbf{H}}_O = (\mathbf{r}_i \times m_i\mathbf{a}_i)\\ &= (4\mathbf{i}\text{ m}) \times 100(-0.36\mathbf{i} + 0.6\mathbf{j})\text{ N}\\ &\quad + (-6\mathbf{i}\text{ m}) \times 70(0.54\mathbf{i} - 0.9\mathbf{j})\text{ N}\\ &= 618\mathbf{k}\text{ N}\cdot\text{m}\end{aligned}$$

JUDGMENT OF THE RESULT

Another approach is to write the moment of momentum of the two masses, $H_O = H_1 + H_2 = m_1 r_1^2\omega + m_2 r_2^2\omega$:

$$M_O = \dot{H}_O = 2\omega\left(m_1 r_1 \frac{dr_1}{dt} + m_2 r_2 \frac{dr_2}{dt}\right)$$

$$\begin{aligned}M_O &= (0.6\text{ rad/s})[(100\text{ kg})(4\text{m})(1\text{ m/s})\\ &\quad + (70\text{ kg})(6\text{m})(1.5\text{ m/s})]\\ &= 618\text{ N}\cdot\text{m} \qquad \text{(counterclockwise)}\end{aligned}$$

In this scalar method special care must be taken to use the correct signs. Note that any mass moving outward on the turntable requires a positive moment according to the given convention.

EXAMPLE 16-7

Consider Ex. 16-3 and Fig. 16-10, in which three acrobats of masses $m_1 = 60$ kg, $m_2 = 70$ kg, and $m_3 = 50$ kg simultaneously jump from a height $h = 3$ m on a large trampoline below. Calculate the average force F_T acting on the trampoline if the acrobats are in contact with the net for a time of 0.4 s during a rebound which returns them to their starting level.

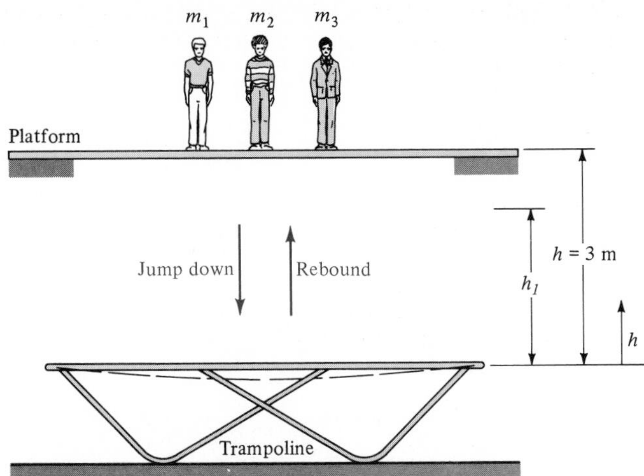

SOLUTION

The acrobats are in contact with the net for a given period of time, so the impulse-momentum method is advantageous to use. The force acts in the vertical direction. Apply Eq. 16-35 and assume that during the 0.4 s of contact the velocity changes from $-v_1$ to v_1 (where $-v_1$ is the initial velocity of acrobat 1 just before impact). Thus, $\Delta v_1 = v_1 - (-v_1) = 2v_1$. For the three people,

$$\sum_{i=1}^{n} \int_{t=0}^{t=0.4} F_i \, dt = G_2 - G_1 = m_1(2v_1) + m_2(2v_2) + m_3(2v_3)$$

Assuming the force F_T on the trampoline to be constant over this interval,

$$F_T(\Delta t) = 2m_1v_1 + 2m_2v_2 + 2m_3v_3$$

From Ex. 16-3, just before impact,

$$T_f = V_i$$

$$\frac{1}{2} m_i v_f^2 = m_i g h$$

$$v_f = \sqrt{2gh} = \sqrt{2(9.81 \text{ m/s}^2)(3 \text{ m})} = 7.67 \text{ m/s}$$

This is the speed of each of the three acrobats just before impact. Hence,

$$F_T = \frac{2}{0.4 \text{ s}}[60 \text{ kg} + 70 \text{ kg} + 50 \text{ kg}](7.67 \text{ m/s}) = 6.9 \text{ kN}$$

EXAMPLE 16-8

Consider Ex. 16-4 and Fig. 16-11 in which 2 cages of 1400 lb weight each are rotating about point C at $\omega = 0.5$ rpm. Determine the required average moment M_C about point C if $r = 12$ ft and the 2 cages are to be brought from $\omega = 0$ to 0.5 rpm in a time of 5 s. Assume that this is done while the carriage moves on the straight section of the track.

SOLUTION

From Eq. 16-36,

$$M_C\,\Delta t = H_{C_2} - H_{C_1} = \sum(\mathbf{r}_i \times m_i\mathbf{v}_i')_2 - \sum(\mathbf{r}_i \times m_i\mathbf{v}_i')_1$$

$$v_i' = r\omega = 12(0.5)\frac{2\pi}{60} = 0.63 \text{ ft/s}$$

$$M_C(5 \text{ s}) = 2\left[(12 \text{ ft})\left(\frac{1400 \text{ lb}}{32.2 \text{ ft/s}^2}\right)(0.63 \text{ ft/s}) - 0\right]$$

$$M_C = 131.5 \text{ ft}\cdot\text{lb}$$

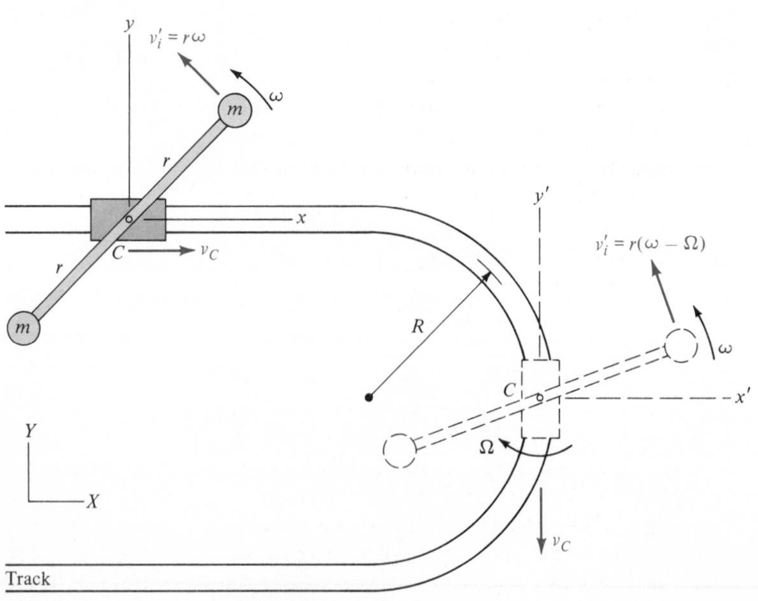

PROBLEMS

16–25 Two particles of mass $m_1 = 2$ kg and $m_2 = 5$ kg move in the fixed XY frame. Calculate the total angular momentum about point O and about the center of mass if $\mathbf{r}_1 = (2\mathbf{i} + \mathbf{j})$ m, $\mathbf{v}_1 = (0.4\mathbf{i} + 6\mathbf{j})$ m/s, $r_2 = (\mathbf{i} + 2.5\mathbf{j})$ m, and $\mathbf{v}_2 = (4\mathbf{i} + 3\mathbf{j})$ m/s.

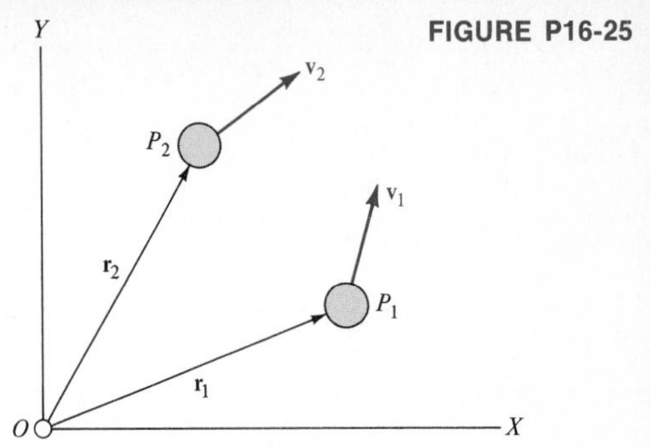

FIGURE P16-25

16–26 Two particles of weight $W_1 = 3$ lb and $W_2 = 4$ lb move in the fixed XY frame as shown in Fig. P16-25. Calculate the total angular momentum about point O and about the center of mass if $\mathbf{r}_1 = (5\mathbf{i} + 2\mathbf{j})$ ft, $\mathbf{v}_1 = (30\mathbf{i} + 30\mathbf{j})$ ft/s, $\mathbf{r}_2 = (3\mathbf{i} + 6\mathbf{j})$ ft, and $\mathbf{v}_2 = (-2\mathbf{i} + 10\mathbf{j})$ ft/s.

16–27 Two particles of mass $m_1 = 3$ kg and $m_2 = 1$ kg move in the fixed XY frame. (a) Determine the total linear momentum and its time rate of change. (b) Calculate the total angular momentum and its time rate of change about point O and about the center of mass if $\mathbf{r}_1 = (2\mathbf{i} + 3\mathbf{j})$ m, $\mathbf{v}_1 = (4\mathbf{i} + 5\mathbf{j})$ m/s, $\mathbf{F}_1 = (-20\mathbf{i} + 30\mathbf{j})$ N, $\mathbf{r}_2 = (\mathbf{i} + 10\mathbf{j})$ m, $\mathbf{v}_2 = (6\mathbf{i} + 2\mathbf{j})$ m/s, $\mathbf{F}_2 = 0$. Assume that $\mathbf{F}_1$ is the only force acting on the system.

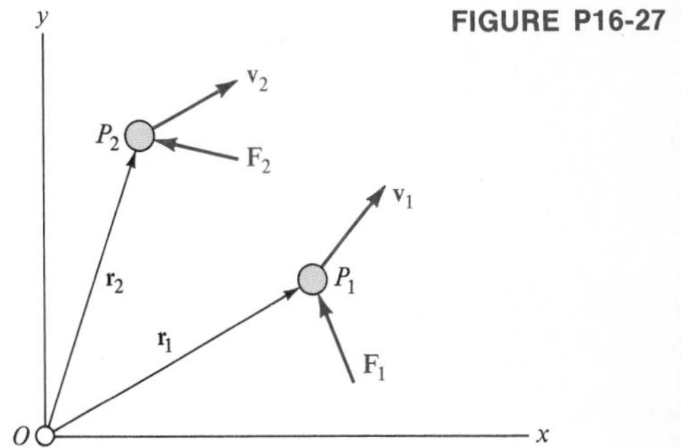

FIGURE P16-27

16–28 Two particles of weight $W_1 = 2$ lb and $W_2 = 5$ lb move in the vertical XY frame as shown in Fig. P16-27. (a) Calculate the total linear momentum and its time rate of change. (b) Calculate the total angular momentum and its time rate of change about point O and about the center of mass. $\mathbf{r}_1 = (5\mathbf{i} + 8\mathbf{j})$ ft, $\mathbf{v}_1 = (20\mathbf{i} + 25\mathbf{j})$ ft/s, $\mathbf{F}_1 = (100\mathbf{i} - 150\mathbf{j})$ lb, $\mathbf{r}_2 = (-3\mathbf{i} + 7\mathbf{j})$ ft, $\mathbf{v}_2 = 0$, and $\mathbf{F}_2 = (-100\mathbf{i} + 150\mathbf{j})$ lb. The particles are located near the surface of the earth.

16–29 The masses, positions, and velocities of three particles are given as $m_A = 80$ kg, $m_B = 70$ kg, $m_C = 100$ kg. $\mathbf{r}_A = (4\mathbf{i} + 3\mathbf{j})$ m, $\mathbf{r}_B = (2\mathbf{i} + \mathbf{j} + 3\mathbf{k})$ m, $\mathbf{r}_C = (3\mathbf{j} + 5\mathbf{k})$ m, $\mathbf{v}_A = (2\mathbf{j} - 0.5\mathbf{k})$ m/s, $\mathbf{v}_B = (0.5\mathbf{i} + 2\mathbf{j} + \mathbf{k})$ m/s, and $\mathbf{v}_C = (5\mathbf{i} + \mathbf{j})$ m/s. Determine the angular momentum of the system about its center of mass and about point O.

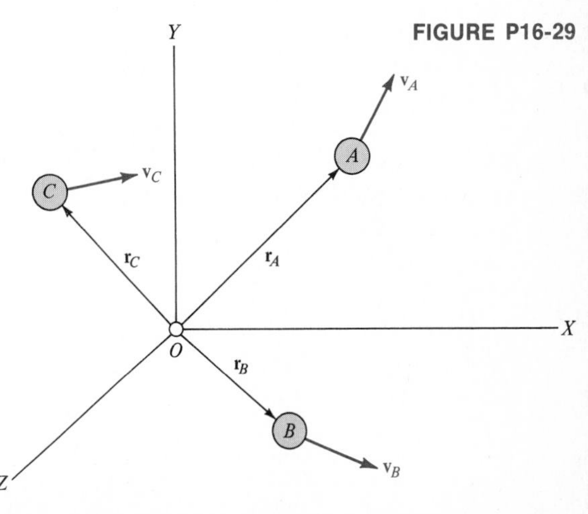

FIGURE P16-29

16–30 The three particles of Prob. 16–29 are located near the surface of the earth. Calculate the time rate of change of the angular momentum $\mathbf{H}_O$ of the system assuming that Y is a vertical axis.

FIGURE P16-31

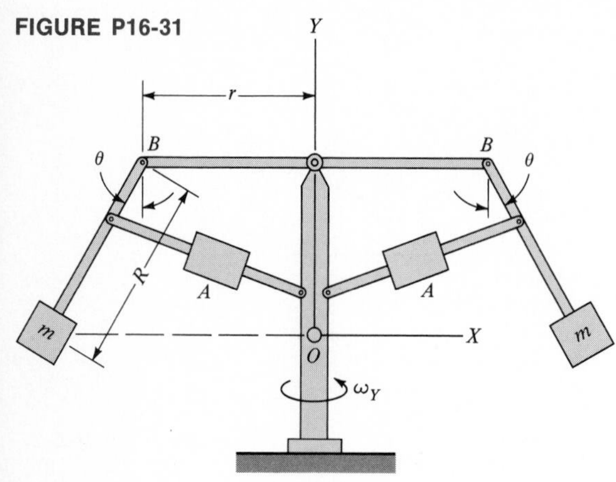

16–31 An amusement ride consists of 12 cages of mass $m = 500$ kg with riders. The cages are on rigid arms at distance $R = 5$ m from points B. The arms are pivoted at points B on a rigid horizontal platform of $r = 4$ m which rotates at $\omega_Y = 10$ rpm. The angle θ of each arm is controlled by a hydraulic actuator A. Calculate the total angular momentum H_O if $\theta = 20°$ for every arm.

16–32 Determine the required moment M_C about the center of mass of the 12 cages in Prob. 16–31 if a 50% increase in the angular speed ω_Y must be made in a time of 10 s. Neglect the mass of the supporting structure.

16–33 Consider the amusement ride described in Prob. 16–31. Assume that $\theta = 20°$ and $\dot{\theta} = 0$ for every other supporting arm R. For the remaining six cages, $\theta = 30°$ and increasing at the rate of 0.2 rad/s. Determine the total angular momentum $\mathbf{H}_O$ for the 12 cages.

FIGURE P16-34

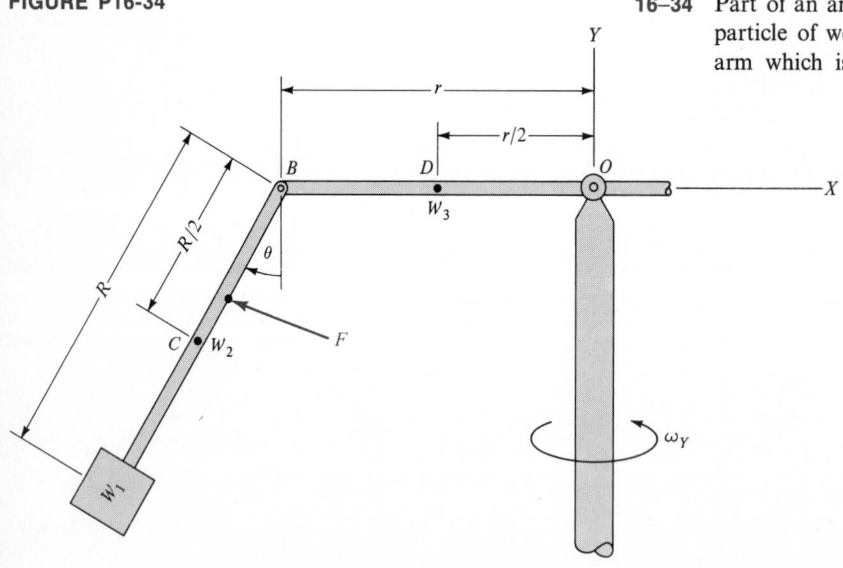

16–34 Part of an amusement ride is modeled as a particle of weight $W_1 = 1000$ lb on a rigid arm which is pivoted at B to rotate in a vertical plane. This arm is supported by a rigid member OB which rotates at $\omega_Y = 12$ rpm. For an approximation, the two rigid members are assumed to be particles of $W_2 = 150$ lb at point C, and $W_3 = 200$ lb at point D. The angle θ is controlled by force F. Calculate the angular momentum H_O of the three particles if $R = 16$ ft, $r = 12$ ft, and $\theta = 40°$.

16–35 Assume that a force $\mathbf{F} = (-500\mathbf{i} + 100\mathbf{j})$ lb is acting on the arm shown in Prob. 16–34. Determine the time rate of change of angular momentum about point O at that instant if ω_Y is constant and $\mathbf{F}$ acts 5 ft from point B.

16–36 Coordinates xyz are fixed at point A in a large space vehicle which has a translational velocity $\mathbf{v}_A = (10\,000\mathbf{i} + 2000\mathbf{j})$ km/h with respect to the fixed XYZ frame. Three small probes of mass $m_1 = 200$ kg, $m_2 = 300$ kg, and $m_3 = 250$ kg are moving relative to the xyz frame with velocities $\mathbf{v}_{1/A} = (500\mathbf{i} + 100\mathbf{j} - 50\mathbf{k})$ km/h, $\mathbf{v}_{2/A} = (600\mathbf{i} + 100\mathbf{k})$ km/h, and $\mathbf{v}_{3/A} = (700\mathbf{i} - 100\mathbf{j})$ km/h. (a) Calculate the total linear momentum of the three probes with respect to the moving and the fixed frames. (b) Determine the total angular momenta $\mathbf{H}_A$ and $\mathbf{H}_O$ of the three probes.

16–37 Generalize from Prob. 16–36 and derive a simple equation to express the total linear momentum $\mathbf{G}_O$ (in the fixed system) of any number of particles with total mass m. Write the equation in terms of the total momentum $\mathbf{G}_A$ (in the moving frame), the velocity $\mathbf{v}_A$ of the moving frame, and the total mass m.

16–38 A turbine disk has 50 identical 0.15-kg blades of negligible dimensions. Assume that the disk itself has no mass and is rotating at $\omega_0 = 5000$ rpm. Assume that one blade breaks off. Determine the total angular momentum H_O of the 50 blades just before and just after the fracture of the blade.

FIGURE P16-38

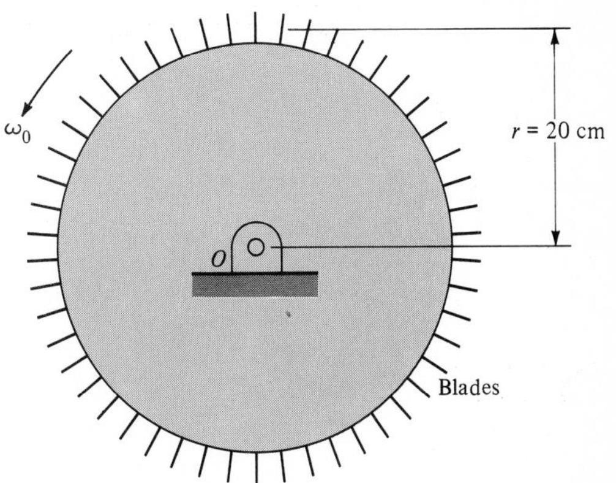

16–39 Determine the required average moment M_O to stop the rotation of the turbine disk described in Prob. 16–38. The braking of the intact disk is to be done in a time of 10 s. What is the time of stopping the disk after the fracture if the same moment M_O is applied? Assume that a constant moment $M_W = 10\ \text{N}\cdot\text{m}$ caused by the displaced weight opposes M_O.

FIGURE P16-40

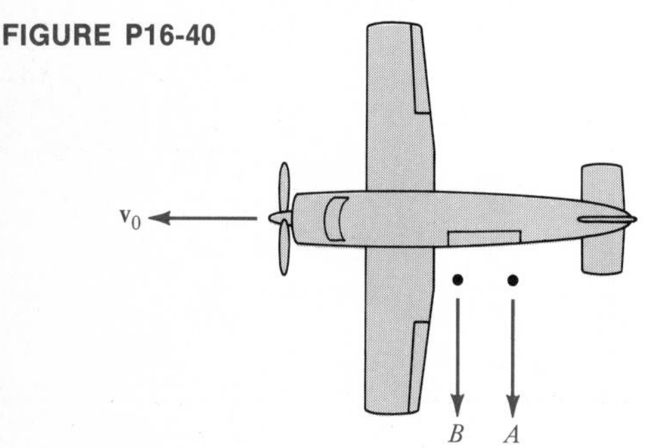

16–40 A small airplane is flying horizontally at $v_0 = 120$ mph when two parachutists jump out horizontally. Parachutist A weighs 180 lb and pushes against the airplane with a 200-lb force applied for 0.3 s. Parachutist B weighs 190 lb and jumps shortly after A, pushing with a 150-lb force for 0.2 s. Determine the final linear momentum of the airplane, which weighs 6500 lb without the two parachutists.

16–41 Solve Prob. 16–40 as stated and also assuming that the two people jump almost simultaneously within a total time of 0.3 s.

FIGURE P16-42

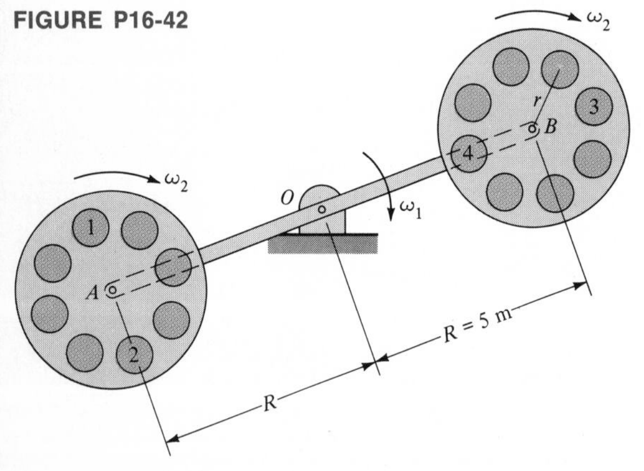

16–42 An amusement ride is designed to rotate arm AOB about point O in the horizontal plane at $\omega_1 = 5$ rpm. The two cages with two riders in each are to rotate at $\omega_2 = 6$ rpm relative to the arm AOB. The average rider has a 70-kg mass and sits at $r = 1$ m from the center of the cage. Calculate the total angular momentum of four riders with respect to point O for the position shown.

16–43 Assume that all rotations of the system described in Prob. 16–42 must be stopped in a time of 0.05 s. Determine the required braking moments $\mathbf{M}_A$, $\mathbf{M}_B$, and $\mathbf{M}_O$ to stop the motion of the riders.

16–44 An amusement ride moves up to eight 200-lb riders on a horizontal track at $\omega_Y = 7$ rpm, $R = 12$ ft, and $r = 5$ ft. The weight of the rigid arm and cage, $W = 1000$ lb, is assumed to be concentrated at $R/2$. Once in each revolution a hydraulic actuator applies a vertical force $F = 2500$ lb for a time of 1 s to the cage. Determine the maximum angular momentum H_O of the system if the arm is free to move up from the track.

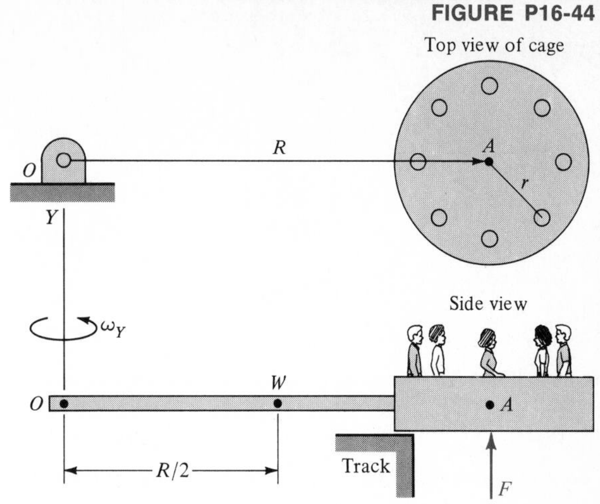

FIGURE P16-44

16–45 The chain-drive mechanism has 100 identical links of mass m. Determine the total angular momentum of the chain with respect to point O.

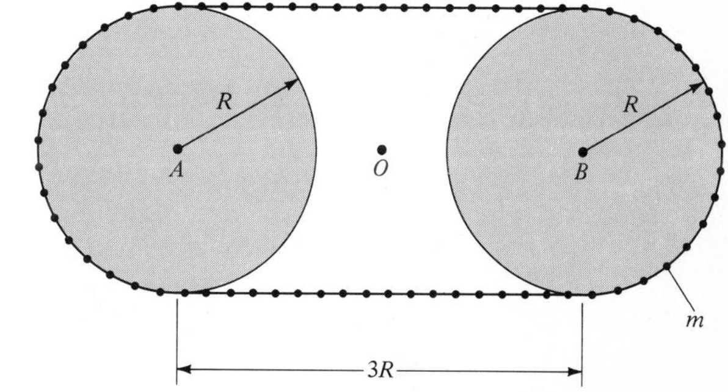

FIGURE P16-45

16–46 In Prob. 16–45, $m = 0.03$ kg, $R = 0.15$ m, and $v = 4$ m/s. Assume that a force $F = 100$ N is applied tangentially to the chain and evenly around the edge for a time of 5 s. Determine the new speed.

16–47 A Ferris wheel has three evenly spaced cages which weigh 1000 lb with riders. Calculate the total angular momentum of the cages if $\omega_A = 2$ rpm and $\omega_Y = 1.5$ rpm for the position shown.

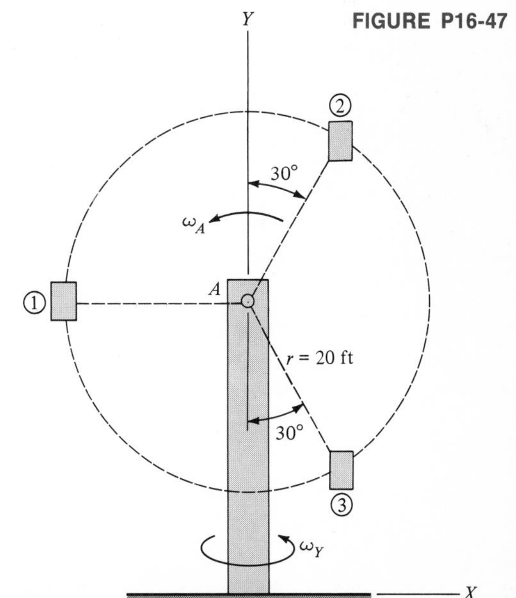

FIGURE P16-47

16–48 Consider the rotating system in Prob. 16–47. Moments $\mathbf{M}_A = -12{,}000\mathbf{k}$ ft·lb and $\mathbf{M}_Y = -5000\mathbf{j}$ ft·lb are applied for a time of 0.03 s. Determine the new ω_A and ω_Y, assuming that the system is still in the same position shown.

16-11 DEFINITIONS OF FLUID FLOW

The flow of fluids is readily analyzed at the elementary level using methods developed for particles. Two categories of fluid flow are distinguished. *Steady flow of mass* means that either no mass enters or leaves the system, or that the mass entering the system equals the mass leaving it in the same period of time. Typical examples of this are the flow of fluids in closed pipelines. *Variable mass* means that the system gains or loses mass. Rockets are examples of variable-mass systems.

The fluids involved in either steady-flow or variable-mass systems may be compressible or incompressible. Strictly speaking, all fluids are *compressible*, which means that they change volume without changing mass under compressive forces. However, liquids are much less compressible than gases, so they are considered *incompressible* in many engineering problems. Even gases may be assumed incompressible in some cases such as the flow of air around cars and low-speed aircraft (up to speeds of about 120 m/s).

16-12 STEADY FLOW OF MASS

The method of analyzing any flow of mass is based on the concept of free-body diagrams. A system of interest, which is commonly called *control volume*, is imagined to be isolated from its surroundings, and the masses entering and leaving the system are accounted for. The general case for steady flow of mass is illustrated in Fig. 16-15. Fluid flows at velocity $\mathbf{v}_1$ through area A_1 into the control volume, while fluid leaves the control volume at velocity $\mathbf{v}_2$ through area A_2. A net force $\sum \mathbf{F}$ acts on the fluid as the momentum of the fluid

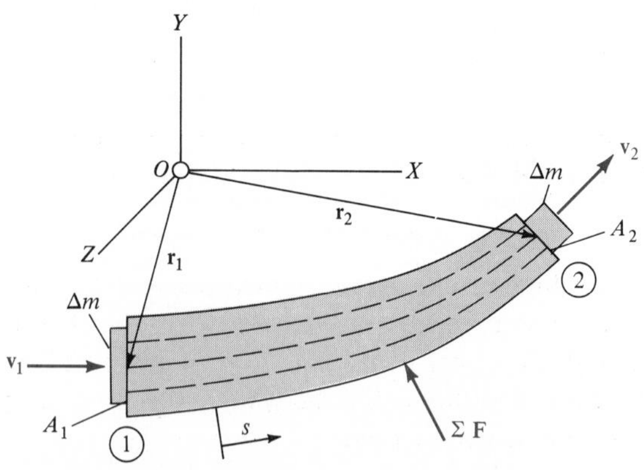

FIGURE 16-15

Control volume of a flowing fluid

changes in passing through the control volume. The flow of mass is described in more detail as follows.

Velocity Change of the Fluid

A useful approximation in the analysis of the flow is that at any cross section A all particles have the same velocity vector $\mathbf{v}$ ($\mathbf{v}$ is actually the average velocity of a nonuniform velocity distribution). Considering Fig. 16-15, assume that mass m_1 is entering the control volume at area A_1 in a unit of time. The exiting mass in the same time interval is $m_2 = m_1$ since mass is neither gained nor lost in steady flow. The density (mass per unit volume) of the fluid is denoted by ρ, and distance in the direction of flow is defined as the quantity s. An element of mass Δm moving through the control volume can be expressed as $\Delta m = m_1 = m_2$,

$$\Delta m = \rho_1 A_1 \Delta s_1 = \rho_2 A_2 \Delta s_2$$

Dividing by the time interval Δt, letting Δt approach zero at the limit, and substituting the speed v for ds/dt gives

$$\frac{dm}{dt} = \rho_1 A_1 v_1 = \rho_2 A_2 v_2 \qquad \text{16-37}$$

This is simplified in the case of incompressible fluids where $\rho_1 = \rho_2 = \rho$,

$$\boxed{\frac{1}{\rho}\frac{dm}{dt} = A_1 v_1 = A_2 v_2} \qquad \text{16-38}$$

The result $A_1 v_1 = A_2 v_2$ is useful in solving many common problems. The velocity vector is in the direction of the average flow at each cross section.

Momentum Change of the Fluid

The change in linear momentum $\Delta\mathbf{G}$ between sections 1 and 2 in Fig. 16-15 is expressed

$$\Delta\mathbf{G} = \Delta m\mathbf{v}_2 - \Delta m\mathbf{v}_1 = \Delta m\,\Delta\mathbf{v} \qquad \text{16-39}$$

The force causing the change in momentum is determined by considering the interval of time Δt during which an element of mass Δm of velocity $\mathbf{v}_1$ enters at section 1 while an equal mass Δm of velocity $\mathbf{v}_2$ exits at section 2. Dividing Eq. 16-39 by Δt and letting Δt approach zero at the limit yields

$$\dot{\mathbf{G}} = \frac{dm}{dt}\Delta\mathbf{v}$$

The time rate of change of momentum equals the resultant of all forces acting on the system of mass m in a unit of time, so from Eq. 16-21,

$$\sum \mathbf{F} = \frac{dm}{dt}\Delta\mathbf{v} = \frac{dm}{dt}(\mathbf{v}_2 - \mathbf{v}_1) \qquad \text{16-40}$$

Equation 16-40 has consistent SI units with forces expressed in newtons, dm/dt in kg/s, and the velocities in m/s. The corresponding units in the U.S. customary system are the lb, slug/s, and ft/s.

The methods described above can be extended to analyze the angular momentum of a steadily flowing mass. This is also illustrated using Fig. 16-15. Identical elements of mass Δm are considered at sections 1 and 2, and the angular momenta of these are evaluated with respect to an arbitrary fixed point O. For a general flow,

$$\mathbf{H}_1 = \Delta m(\mathbf{r}_1 \times \mathbf{v}_1) \qquad \mathbf{H}_2 = \Delta m(\mathbf{r}_2 \times \mathbf{v}_2)$$

The net moment of the external forces about point O equals the time rate of change of angular momentum. The equation of this is obtained in a way similar to that for linear momentum. Letting Δt approach zero at the limit gives

$$\sum \mathbf{M}_O = \dot{\mathbf{H}}_O = \frac{d}{dt}(\mathbf{H}_2 - \mathbf{H}_1) = \frac{dm}{dt}(\mathbf{r}_2 \times \mathbf{v}_2 - \mathbf{r}_1 \times \mathbf{v}_1) \qquad \text{16-41}$$

where dm/dt is the time rate of flow of mass through the system. Note that the angular momenta and the moments of the external forces are taken about the same point O.

Pressures in Flowing Fluids

It should be remembered that in Eqs. 16-40 and 16-41 all forces acting on the chosen element of fluid must be considered. This includes the distributed forces caused by pressure within the fluid and the forces applied by solid objects in contact with the fluid. In some cases the pressure can be ignored if the element is small and the pressure is constant over the element. In the general case such as the free-body diagram in Fig. 16-16, the pressures p_1 and p_2 are

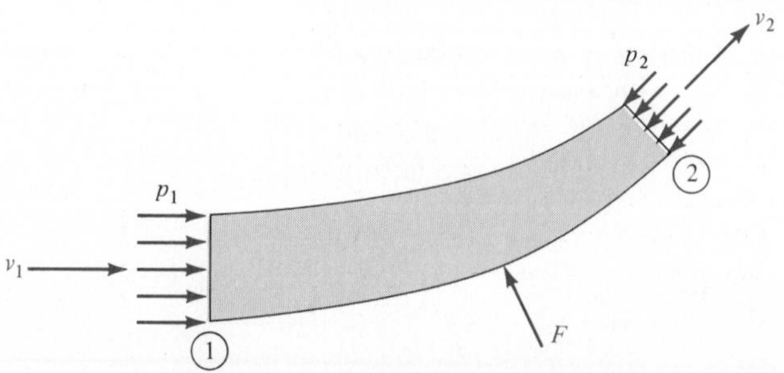

FIGURE 16-16

Pressures at two cross sections in a flowing fluid

not the same and they act on different areas. The forces caused by these must be evaluated by integrating the pressure over each appropriate area. With a constant pressure, force = (pressure) · (area).

Common Applications of Steady Flow

The principles of steady flow of mass are applied in many areas of engineering work. These are presented with a few comments to help in the solution of simple problems.

1. *Flow through pipes.* The major forces acting on fluids are caused by changing cross sections of pipes and by changing directions of flow. In long segments of a pipe the wall friction has to be considered in evaluating the change in pressure in the fluid.

2. *Flow through nozzles.* The major changes in flow velocity and fluid pressure occur within the nozzle. The free stream away from the nozzle maintains its linear momentum in the absence of external forces.

3. *Streams of fluid striking solid objects.* A stream of fluid striking a solid object exerts a force on that object. As a result, the linear momentum of the free stream changes during the contact with the solid object.

4. *Flow through propellers.* Air or water moving through fans and propellers is accelerated from one side of the blades to the other side. For fans it is often assumed that the air has no initial velocity as it enters the region of the blades, as from the left in Fig. 16-17. At exit, the *slipstream* moves at a high velocity. The major external force on the stream of air is the thrust of the blades. The air is considered incompressible (with atmospheric pressure throughout) in these situations.

The flow of fluid through the propellers of airplanes and boats is similar to the flow through fans. The major difference is that the fluid has an initial relative velocity $\mathbf{v}_0$ entering the region of the blades if the vehicle is in motion. The initial velocity is the velocity of the vehicle in the medium, and the velocity at exit is the velocity of the slipstream relative to the vehicle.

5. *Flow through jet engines.* Air entering a jet engine with a relative initial velocity $\mathbf{v}_0$ is accelerated by the burning of fuel in the engine, as modeled in Fig. 16-18. This is similar to the flow of air through propellers and can be analyzed in three ways. As a first assumption, the relatively small mass of the fuel added to the stream of air may be neglected. This allows treating the jet engine as a propeller system. Second, the mass of the fuel can be considered by isolating the engine from the fuel tanks. In this case air and fuel are

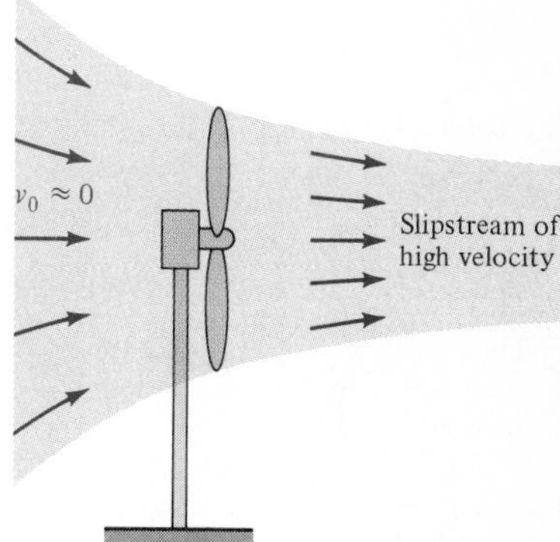

FIGURE 16-17
Fluid flow through a propeller

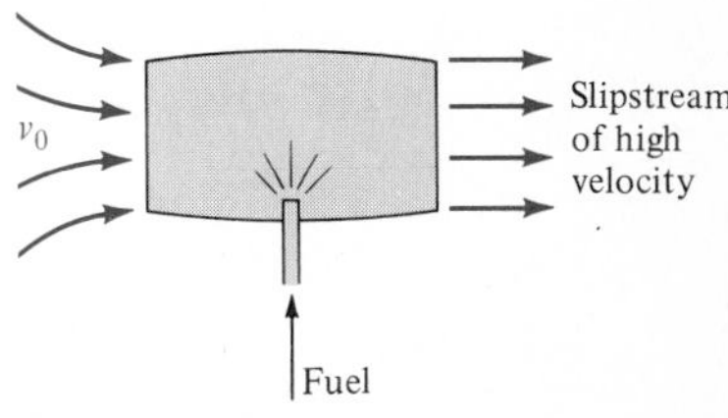

FIGURE 16-18
Air flow through a jet engine

taken as entering the system at right angles to each other and leaving the system in a common stream of the same total mass. In the third method the engine and its fuel supply are considered as the system of interest. Of course, this system loses mass, and it must be analyzed using the method presented in Sec. 16-13.

16-13 SYSTEMS WITH VARIABLE MASS

A system may continuously gain or lose mass. Examples of these are rockets and jet aircraft as whole. The analysis of such systems is based on a model that gains mass as in Fig. 16-19.

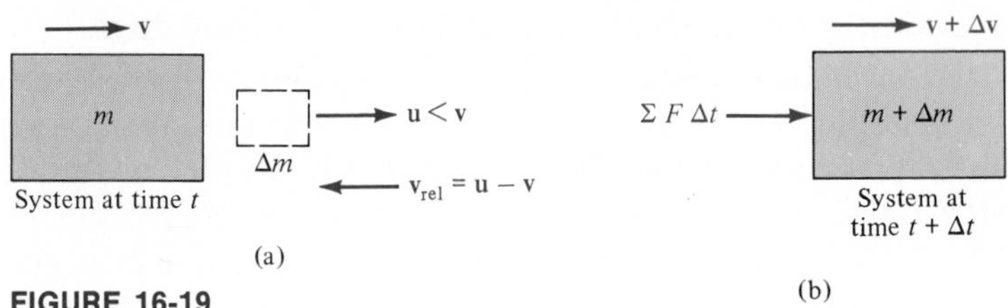

FIGURE 16-19

Moving system gaining mass

Assume that the system of mass m is moving with absolute velocity $\mathbf{v}$ at time t and is about to overtake and absorb mass Δm, which is moving with absolute velocity $\mathbf{u}$ in the same direction. The original relative velocity of Δm with respect to m is $\mathbf{v}_{rel} = \mathbf{u} - \mathbf{v}$. The mass m absorbs Δm in time Δt while external forces $\mathbf{F}$ may be acting on the system. The forces are included in the analysis for generality, but they exclude the mutual forces between m and Δm because these are internal to the new system of $m + \Delta m$. According to the principle of impulse and momentum, the final momentum equals the sum of the initial momentum and the impulse of the external forces,

$$(m + \Delta m)(\mathbf{v} + \Delta \mathbf{v}) = m\mathbf{v} + \Delta m\mathbf{u} + \sum \mathbf{F}\,\Delta t \qquad \text{16-42}$$

Expanding and rearranging gives

$$\sum \mathbf{F}\,\Delta t = m\,\Delta\mathbf{v} + \Delta m(\mathbf{v} - \mathbf{u}) + \Delta m\,(\Delta \mathbf{v})$$

This is simplified by substituting $-\mathbf{v}_{rel}$ for $\mathbf{v} - \mathbf{u}$ and neglecting the second-order term $\Delta m\,(\Delta\mathbf{v})$,

$$\sum \mathbf{F}\,\Delta t = m\,\Delta\mathbf{v} - \Delta m\mathbf{v}_{rel}$$

Dividing this equation by Δt and letting Δt approach zero at the limit yields

$$\sum \mathbf{F} = m\frac{d\mathbf{v}}{dt} - \frac{dm}{dt}\mathbf{v}_{\text{rel}} \tag{16-43}$$

where dm/dt is the time rate at which mass is absorbed by the original system of mass m.

Equation 16-43 is readily modified to obtain the governing equation for the motion of a system that loses mass. dm/dt is negative in that case, and

$$\sum \mathbf{F} = m\frac{d\mathbf{v}}{dt} + \frac{dm}{dt}\mathbf{v}_{\text{rel}} \tag{16-44}$$

It is interesting to compare Eqs. 16-43 and 16-44 with Newton's second law of motion for constant mass, $\sum \mathbf{F} = m(d\mathbf{v}/dt)$. The gain of mass is equivalent to an external force of magnitude $(v_{\text{rel}})dm/dt$ that opposes the original motion of m because $\mathbf{v}_{\text{rel}}$ and $\mathbf{v}$ are opposite in direction. Conversely, the expulsion of mass is equivalent to an external force in the direction of the velocity $\mathbf{v}$ of the system. This is the mechanism of propulsion by rockets.

Special Cases of Systems with Variable Mass

1. ***All external forces are zero.*** A rocket propelling itself in space encounters no air resistance. If gravity acting on it is also negligible, $\sum \mathbf{F} = 0$, and Eq. 16-44 becomes

$$m\frac{d\mathbf{v}}{dt} = \frac{dm}{dt}\mathbf{v}_{\text{rel}} \tag{16-45}$$

2. ***The system absorbs mass whose absolute initial velocity is zero.*** In this case $\mathbf{u} = 0$, $\mathbf{v}_{\text{rel}} = -\mathbf{v}$, so Newton's second law and Eq. 16-43 give the same result,

$$\sum \mathbf{F} = \frac{d}{dt}(m\mathbf{v}) = m\frac{d\mathbf{v}}{dt} + \frac{dm}{dt}\mathbf{v}$$

Newton's second law is not applicable to other systems that gain mass.

3. ***The system loses mass whose absolute final velocity is zero.*** In this case $\mathbf{u} = 0$, $\mathbf{v}_{\text{rel}} = \mathbf{v}$, so Newton's second law and Eq. 16-44 give the same result as in item 2 above. Newton's second law is not applicable to other systems that expel mass.

4. ***The system and the expelled mass move in the same direction.*** When the velocity $\mathbf{v}$ of the system is high, v may be larger than v_{rel}, so $\mathbf{v}$ and $\mathbf{u}$ have the same direction.

EXAMPLE 16-9

Water flows into a horizontal pipe of 0.4 m diameter at the rate of 50 kg/s. The flow is diverted into a temporary pipe of 0.2 m diameter as shown in Fig. 16-20. (a) Calculate the net horizontal and vertical forces acting on the water in the elbow region. (b) What are these forces after the temporary pipe is replaced by a 0.4-m-diameter pipe? The flow is in the horizontal xy plane.

SOLUTION

(a) The speed v_1 can be determined from Eq. 16-37,

$$\frac{dm}{dt} = A_1 v_1 \rho$$

$$v_1 = \frac{50 \text{ kg/s}}{(0.1257 \text{ m}^2)(1000 \text{ kg/m}^3)} = 0.4 \text{ m/s}$$

From Eq. 16-38,

$$v_2 = \frac{A_1}{A_2} v_1 = 1.6 \text{ m/s}$$

The unknown force is assumed to have positive rectangular components F_x and F_y. From Eq. 16-40,

$$F_x = \frac{dm}{dt}(v_{2_x} - v_{1_x})$$

$$= (50 \text{ kg/s})(1.6 \cos 50^\circ - 0.4) \text{ m/s} = 31.5 \text{ N}$$

$$F_y = \frac{dm}{dt}(v_{2_y} - v_{1_y})$$

$$= (50 \text{ kg/s})(1.6 \sin 50^\circ - 0) \text{ m/s} = 61.3 \text{ N}$$

(b) If the pipes are the same,

$$A_2 = A_1 \qquad v_2 = v_1$$

$$F_x = (50 \text{ kg/s})(0.4 \cos 50^\circ - 0.4) \text{ m/s} = -7.1 \text{ N}$$

$$F_y = (50 \text{ kg/s})(0.4 \sin 50^\circ - 0) \text{ m/s} = 15.3 \text{ N}$$

The signs of the force components should be interpreted remembering that the unknown forces F_x and F_y were assumed to be positive. The forces calculated are the resultants of forces from adjoining pipes, fixtures of the elbow joint, and pressure in the water.

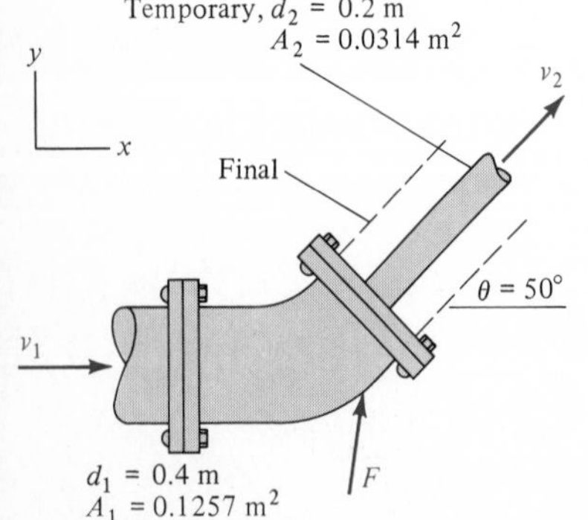

FIGURE 16-20
Water flowing through a bent pipe

EXAMPLE 16-10

A fixed nozzle of 0.05 m diameter emits a horizontal stream of water at the rate of 10 kg/s. The stream is diverted by a horizontally moving smooth vane whose absolute velocity is **u**. (a) Calculate the horizontal and vertical components of the force **F** that acts on the vane in this situation (Fig. 16-21). (b) Determine the power transmitted to the vane.

SOLUTION

(a) The flow speed at the nozzle is obtained from Eq. 16-37,

$$v = \frac{dm/dt}{\rho A} = \frac{10 \text{ kg/s}}{1000 \text{ kg/m}^3(\pi)(0.025 \text{ m})^2} = 5.1 \text{ m/s}$$

The relative speed of the water striking the vane at section 1 is

$$v_1 = v - u = \frac{v}{2}$$

and the water leaves the vane at section 2 without any change in the magnitude of the relative speed, so $v_2 = v_1$.

The forces are determined using Eq. 16-40, where the relative flow rate dm_1/dt must be used rather than the flow rate at the nozzle. Thus,

$$F_x = \frac{dm_1}{dt}(v_{2_x} - v_{1_x}) = (\rho A v_1)(v_1 \cos 60^\circ - v_1)$$

$$= (5 \text{ kg/s})(2.55 \text{ m/s})(\cos 60^\circ - 1) = -6.4 \text{ N}$$

This is the horizontal force acting on the water. The same force in the opposite direction is acting on the vane. Similarly, for the y direction,

$$F_y = \frac{dm_1}{dt}(v_{2_y} - v_{1_y}) = (\rho A v_1)(v_2 \sin 60^\circ - 0)$$

$$= (5 \text{ kg/s})(2.55 \text{ m/s})(\sin 60^\circ) = 11 \text{ N}$$

acting upward on the water and downward on the vane.

(b) Since the vane is moving only in the horizontal direction,

$$\text{power} = (\text{speed of vane})(F_x) = \frac{v}{2}F_x$$

$$= (2.55 \text{ m/s})(6.4 \text{ N}) = 16.3 \text{ W}$$

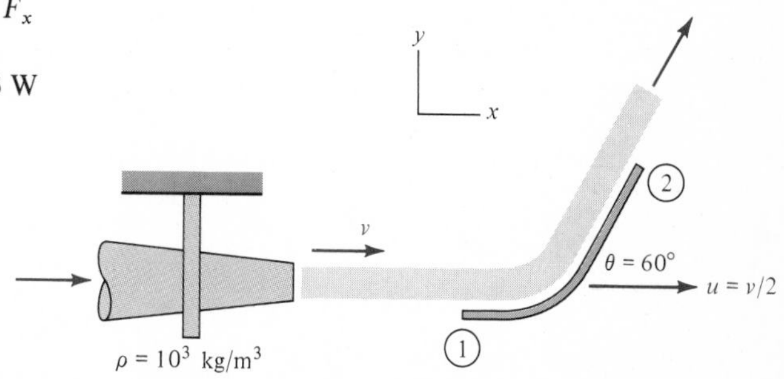

FIGURE 16-21
Water striking a moving vane

EXAMPLE 16-11

An escalator (Fig. 16-22) is designed to move people at a speed of 5 ft/s. (a) Assume that a group of people starts stepping on the empty escalator with a negligible relative speed. Calculate the magnitude of the required force to keep the escalator moving at its constant speed if weight is added on it at the rate of 200 lb/s. (b) What is the required force if the absolute initial velocity of each person is 3 ft/s parallel to the escalator? Neglect friction, the change in elevation of the people, and any changes in the direction of their velocity as they step on the escalator (i.e., their initial velocity is parallel to **v**).

FIGURE 16-22

People stepping on a moving escalator

SOLUTION

The system is gaining mass, so Eq. 16-43 can be used.

(a) In this case $dv/dt = 0$ and $v_{\text{rel}} = 0$; therefore, the system maintains its initial speed without requiring any force (noting the assumptions).

(b) Here $v_{\text{rel}} = u - v = 3 - 5 = -2$ ft/s, while $dv/dt = 0$. From Eq. 16-43,

$$F = \frac{-dm}{dt} v_{\text{rel}} = -\left(\frac{200 \text{ lb/s}}{32.2 \text{ ft/s}^2}\right)(-2 \text{ ft/s})$$

$$= 12.4 \text{ lb}$$

EXAMPLE 16-12

A 5000-kg space vehicle is at zero velocity in empty space (no air, no gravity) when its ion-propulsion engine is started (Fig. 16-23). The engine emits mass at a constant rate of $dm/dt = 10^{-6}$ kg/s with a relative speed of 6000 km/h. Calculate the speeds of the vehicle after the engine has operated for 1 h and 100 h.

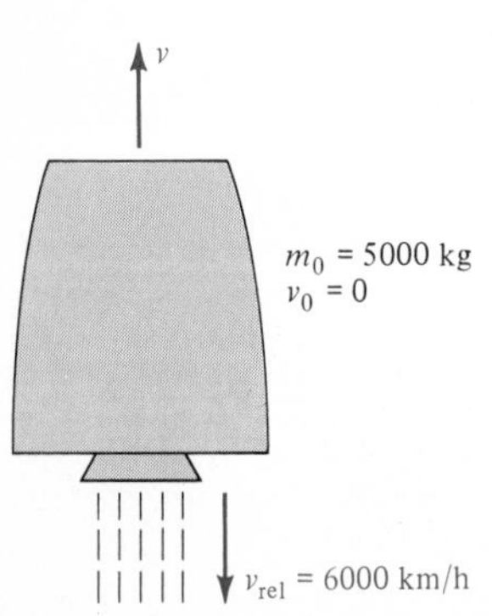

FIGURE 16-23

Propulsion of a spacecraft by ion emission

SOLUTION

Since the system is losing mass and there are no external forces, Eq. 16-45 is applicable,

$$m\frac{dv}{dt} = -\frac{dm}{dt} v_{\text{rel}}$$

With v_{rel} in the opposite direction as v,

$$(5000 \text{ kg})\left(\frac{dv}{dt}\right) = -(10^{-6} \text{ kg/s})(-6000 \text{ km/h})\left(\frac{3600 \text{ s}}{\text{h}}\right)$$

$$\frac{dv}{dt} = 0.00432 \text{ km/h}^2 = a$$

Because the space vehicle has zero initial velocity, the speed at time t is

$$v_t = at$$

$$v_{1\text{h}} = (0.00432 \text{ km/h}^2)(1 \text{ h}) = 0.00432 \text{ km/h}$$

$$v_{100\text{h}} = (0.00432 \text{ km/h}^2)(100 \text{ h}) = 0.432 \text{ km/h}$$

Note that the total mass emitted in 100 h is only 0.0036 kg.

PROBLEMS

16–49 Water flows through the horizontal pipe at the rate of 30 kg/s. Calculate the total horizontal force caused by the flowing water acting on flange 2 if $A_1 = 0.1\ \text{m}^2$, $A_2 = 0.06\ \text{m}^2$, $p_1 = 1$ kPa, and $p_2 = 920$ Pa.

FIGURE P16-49

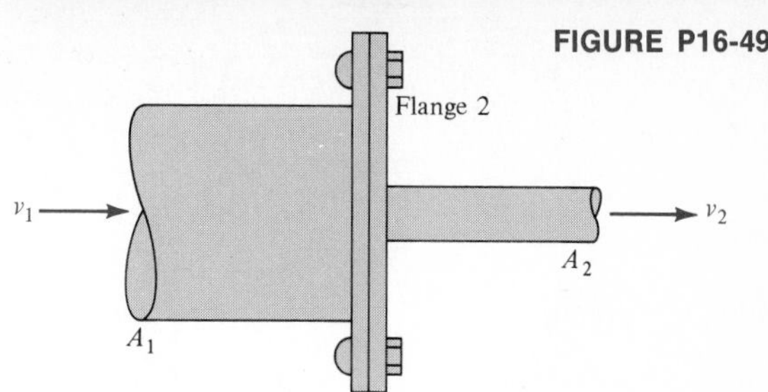

16–50 The jet engine on the test stand draws in air at the rate of 100 lb/s and expels it at a speed of 2000 ft/s. Calculate the thrust of the engine.

FIGURE P16-50

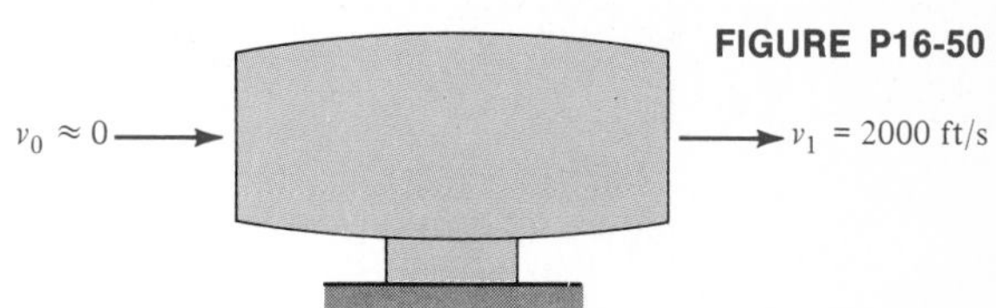

16–51 A fan accelerates air (density $= 1.21\ \text{kg/m}^3$) to a speed of 8 m/s in its slipstream. Calculate the force required to hold the fan.

FIGURE P16-51

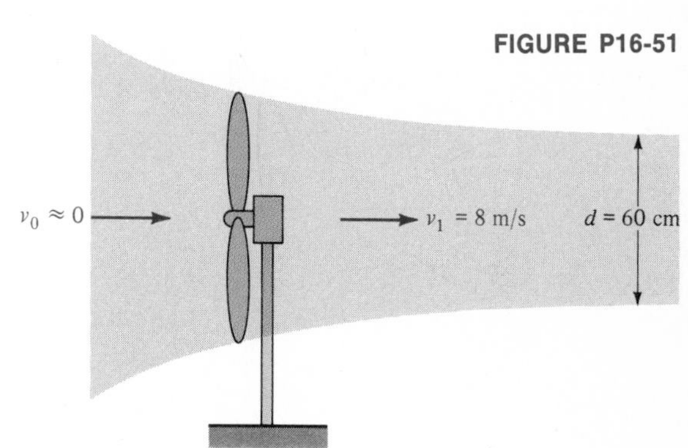

16–52 A small helicopter accelerates air (specific weight $= 0.076\ \text{lb/ft}^3$) to a maximum speed of 50 ft/s in its downward slipstream of 20 ft diameter. Determine the gross weight of the helicopter, which is hovering.

FIGURE P16-52

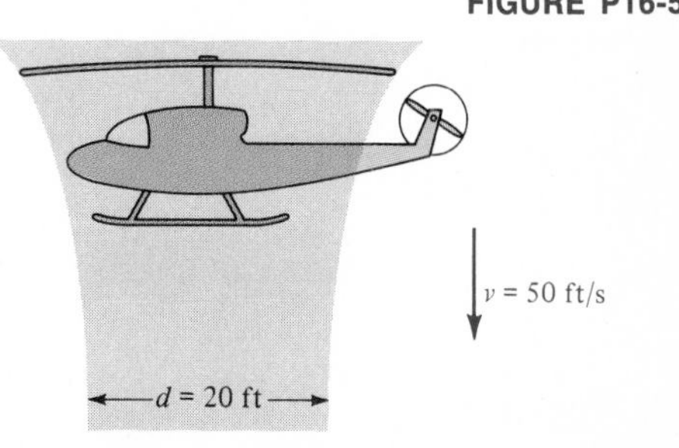

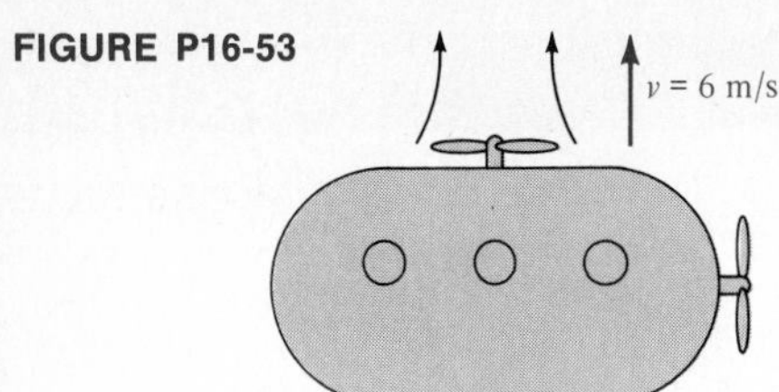

FIGURE P16-53

16–53 A research submarine should have a propeller with a downward thrust of 10 kN for floating at a given depth (the vehicle should rise to the surface if the motor fails). Assume that the proposed motor and blade can accelerate water (density $= 1000$ kg/m^3) from zero speed to 6 m/s in the slipstream. Estimate the diameter of the propeller circle.

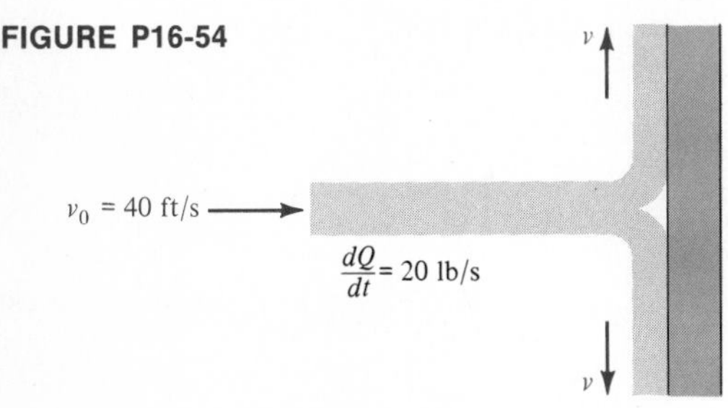

FIGURE P16-54

16–54 A stream of water is directed normal to a smooth, flat plate. The water flows away parallel to the plate. Calculate the required force to hold the plate.

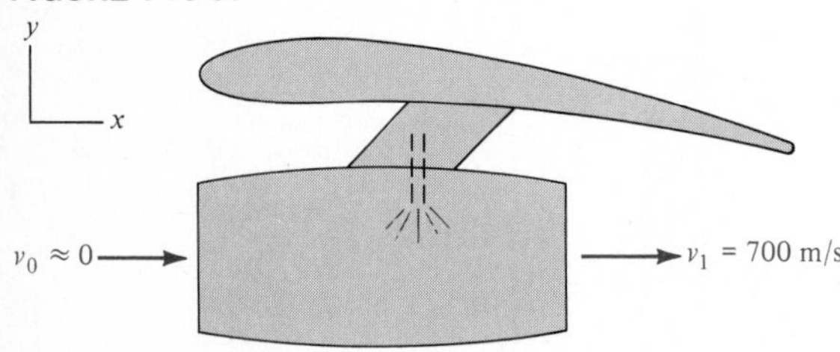

FIGURE P16-55

16–55 A jet engine draws in air at the rate of 70 kg/s and expels it at a speed of 700 m/s. The engine burns fuel at the rate of 1.2 kg/s. The fuel flows vertically to the engine at a speed of 5 m/s. Determine the x and y components of the force acting on the engine caused by the flowing fluids.

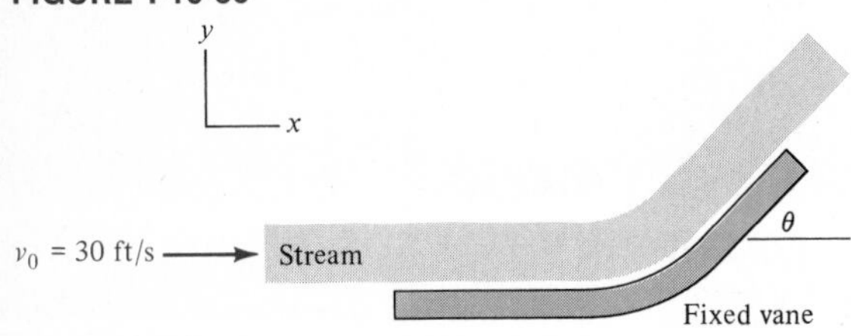

FIGURE P16-56

16–56 A stream of water flowing at the rate of 20 lb/s is deflected by a smooth, fixed vane. Determine the force required (x and y components) to hold the vane if $\theta = 60°$.

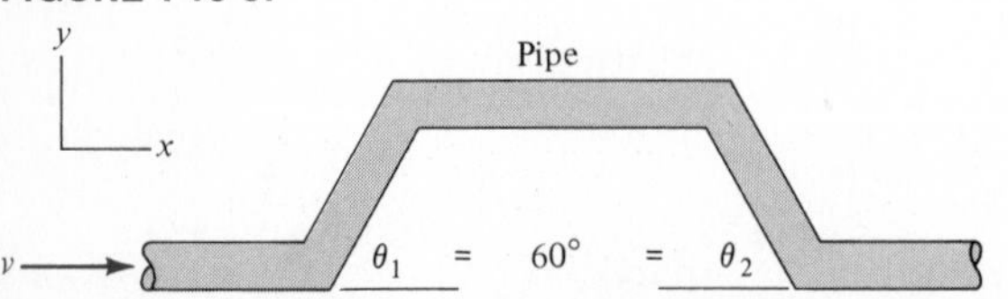

FIGURE P16-57

16–57 Oil flows in a 1-m-diameter pipe at a speed of 2 m/s. The symmetric expansion pipe is in the horizontal plane. (a) Calculate the force caused by the flowing oil on the first elbow. (b) Is there a net force acting on the system of the four elbows? The density of the oil is 970 kg/m^3.

16–58 Water flows into a branching pipe of equal cross sections. One branch carries water in the same direction as the original flow. Calculate the force acting on the control volume of water. The specific weight of water is 62.4 lb/ft^3.

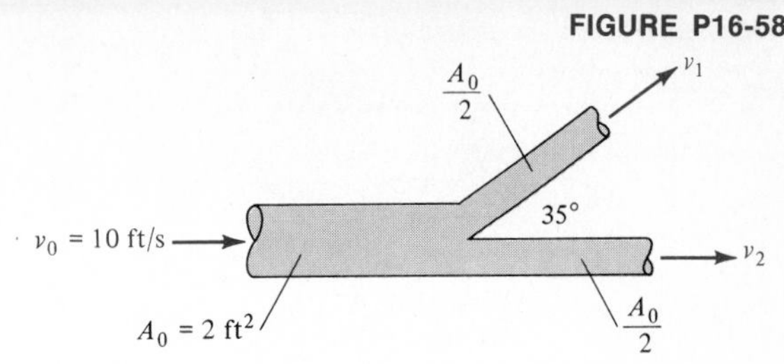

FIGURE P16-58

16–59 A jet engine in flight has an absolute speed of 1000 km/h. The engine draws in air at the rate of 100 kg/s and expels it at a relative speed of 750 m/s. Determine the thrust of the engine.

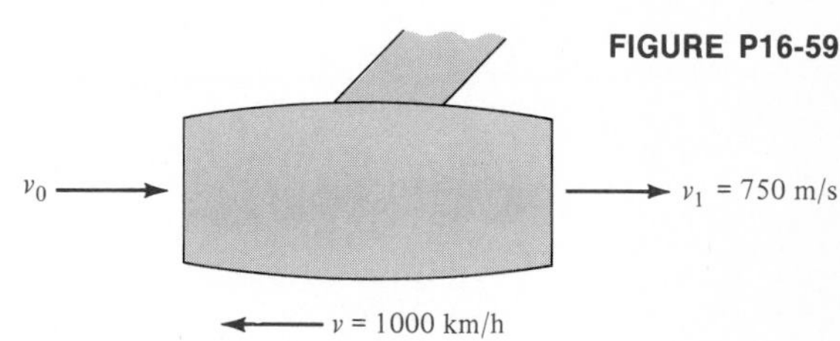

FIGURE P16-59

16–60 A fixed nozzle of 10 cm diameter emits a horizontal stream of water at the rate of 60 kg/s. The stream is to be diverted by a horizontally moving smooth vane whose absolute speed is $u = 5$ m/s. (a) Determine the horizontal component of the force acting on the vane if $\theta = 45°$. (b) Calculate the power transmitted to the vane.

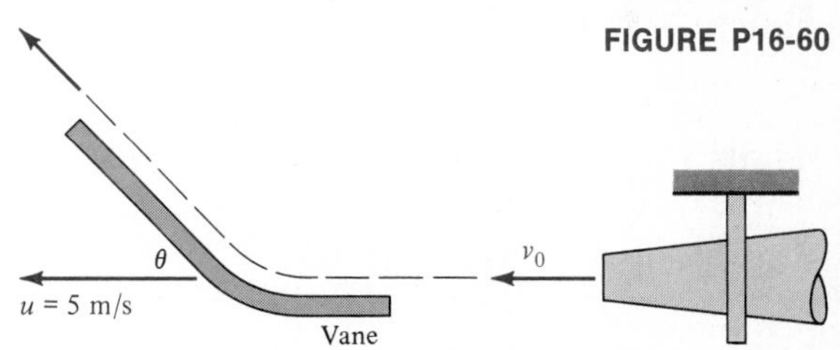

FIGURE P16-60

16–61 A jet engine is mounted at the tail section of the fuselage, with the air intake parallel to the exhaust. The engine draws in air at the rate of 180 lb/s and discharges it at a relative speed of 2200 ft/s. Determine the forces F_x and F_y acting on the tail section caused by the deflected and accelerated air flow.

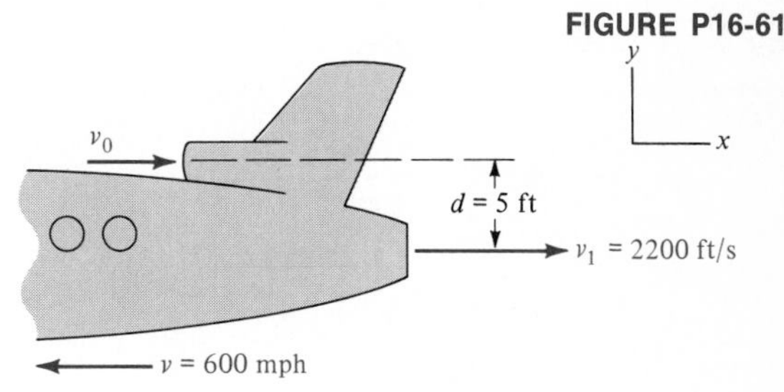

FIGURE P16-61

16–62 Each side scoop of the jet aircraft draws in air at the rate of 100 kg/s. The single engine burns fuel at the rate of 3 kg/s and discharges the gases at a relative speed of 900 m/s. Determine the thrust and the horsepower of the engine.

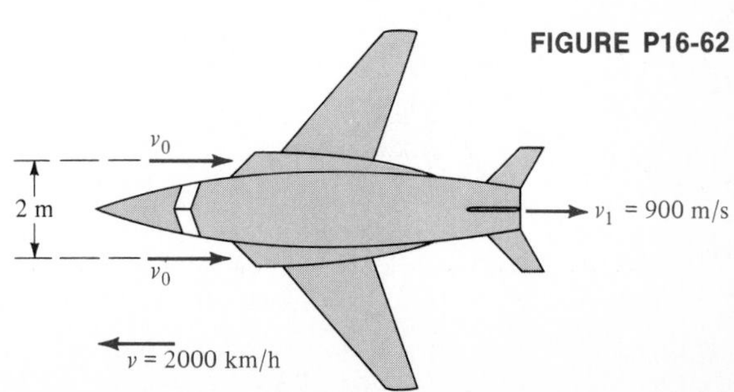

FIGURE P16-62

FIGURE P16-63

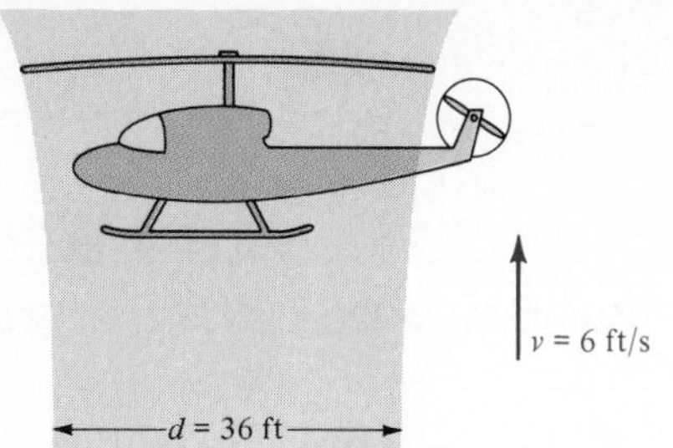

16–63 A 10,000-lb helicopter is designed to ascend at the maximum speed of 6 ft/s. What is the required relative speed of air in the slipstream if its diameter is estimated to be 36 ft? Air weighs 0.076 lb/ft^3.

FIGURE P16-64

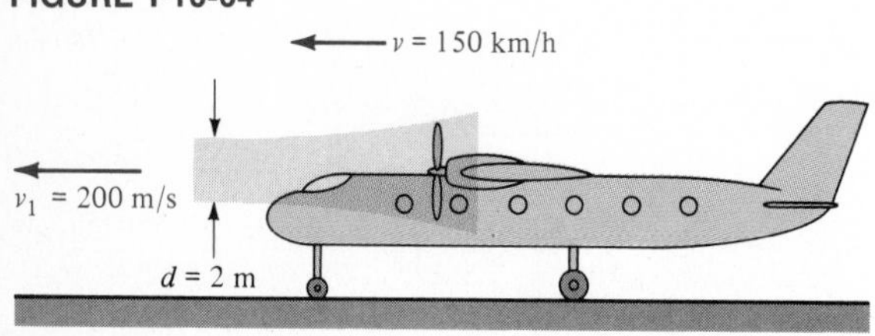

16–64 An airplane lands at a speed of 150 km/h when the airflow from the propellers is reversed for braking. The relative speed of the air with respect to the airplane is $v_1 = 200$ m/s in the 2-m-diameter slipstream. Calculate the reversed thrust of each engine for the given conditions. The density of air is 1.21 kg/m^3.

FIGURE P16-65

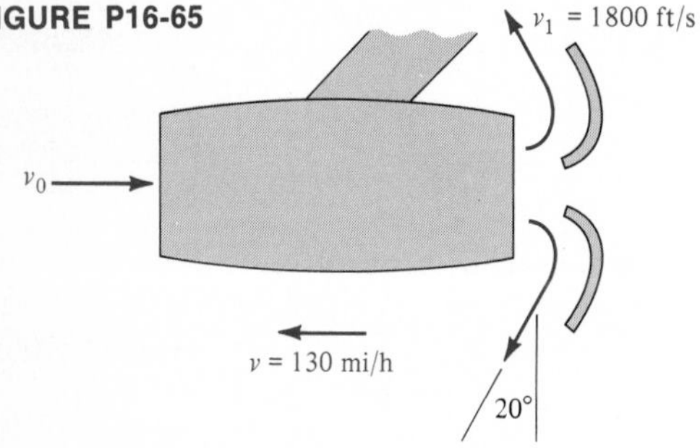

16–65 A jet aircraft lands at a speed of 130 mi/h. Each engine draws in air at the rate of 200 lb/s and exhausts it at a relative speed of 1800 ft/s. The thrust is reversed by baffles for purposes of braking the aircraft. Calculate the braking force of each engine for the given configuration.

FIGURE P16-66

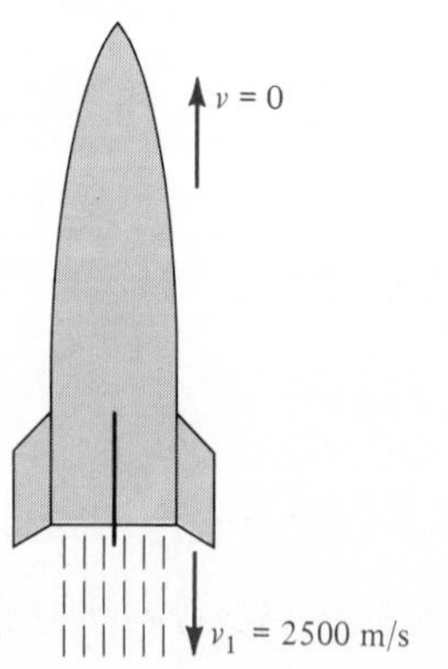

16–66 A rocket designed for atmospheric research is to hover briefly at its maximum elevation. The rocket's total mass at the beginning of this period is 4000 kg. Fuel is expelled at a constant speed of 2500 m/s. Find an expression for the rate dm/dt of the fuel consumption. Sketch this rate for a time of 20 s of operation. The acceleration of gravity is 8 m/s^2.

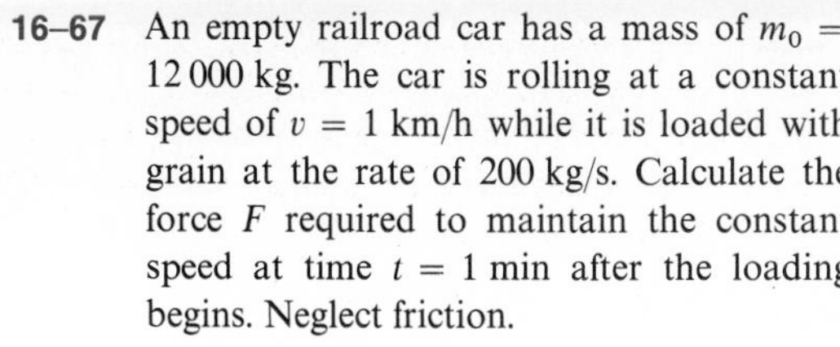

16–67 An empty railroad car has a mass of $m_0 =$ 12 000 kg. The car is rolling at a constant speed of $v = 1$ km/h while it is loaded with grain at the rate of 200 kg/s. Calculate the force F required to maintain the constant speed at time $t = 1$ min after the loading begins. Neglect friction.

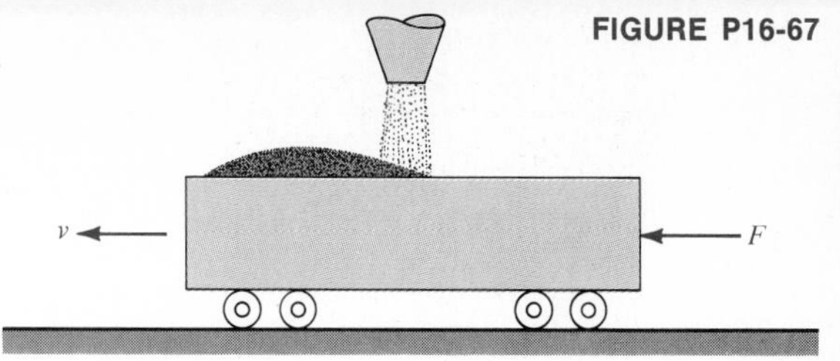

FIGURE P16-67

16–68 The railroad car described in Prob. 16–67 is rolling at an initial speed of $v_0 = 2$ km/h when the loading begins. Calculate its speed 20 s later assuming that no horizontal forces are acting during this time.

16–69 A pipeline-scrubber plug weighs 100 lb when empty. It accumulates scrapings at the rate of 0.02 lb/s while moving at a speed of 6 ft/s. Determine the force required to maintain the constant speed after 1 h of operation. Neglect friction.

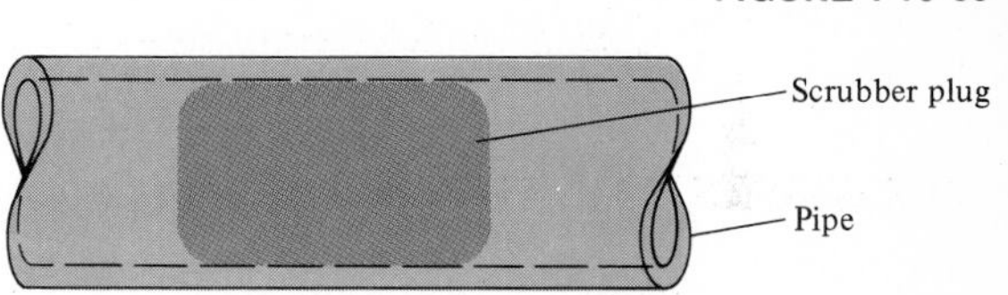

FIGURE P16-69

16–70 A balloon used for atmospheric research is to be tethered on a long wire rope that weighs 0.3 lb/ft. Plot the magnitude of the force between the balloon and the wire at point A as a function of time $t = 0$, 1 min, and 2 min. The balloon rises at a constant speed of $v = 5$ ft/s, and the wire uncoils at the ground without resistance.

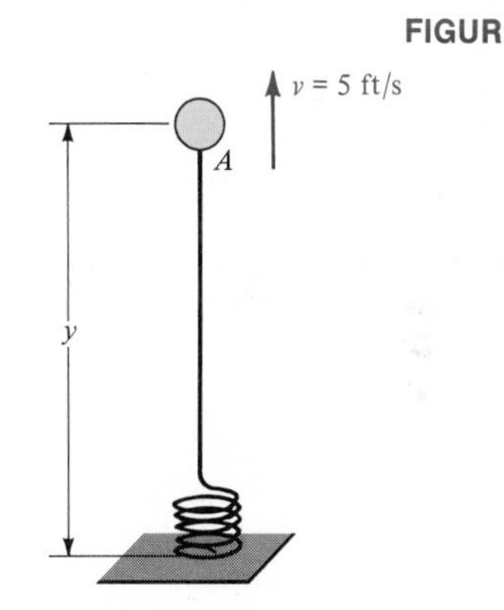

FIGURE P16-70

16–71 A moving sidewalk is designed to carry people horizontally at a speed of 5 ft/s. A group of people starts stepping on the empty mover, adding weight at the rate of 300 lb/s. (a) Calculate the force required to keep the machine moving at its constant speed if $v_{rel} = 0$. Neglect friction. (b) Calculate the force assuming that the people do not stand on the moving strip but walk forward on it at a speed of 4 ft/s relative to the strip.

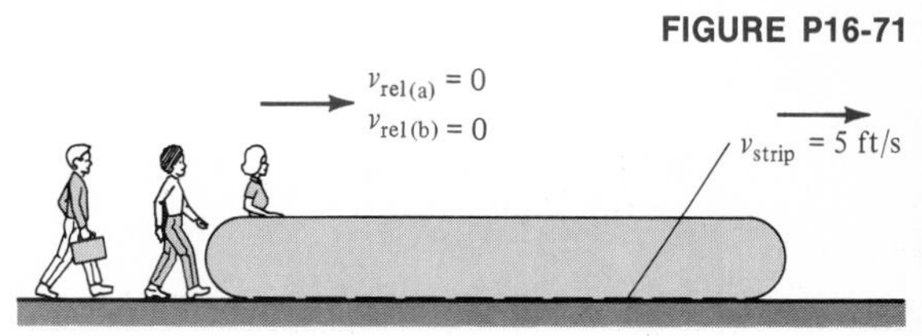

FIGURE P16-71

16–72 A spacecraft is designed with a maximum total thrust of 2 MN, which is to be obtained by ejecting propellant at a relative speed of 3000 m/s. Determine the required rate of fuel consumption.

16–73 A 4000-lb rocket is fired vertically. Fuel is burned at the rate of 40 lb/s and ejected at a relative speed of 10,000 ft/s. Sketch the acceleration and speed of the rocket from time $t = 0$ to 50 s.

FIGURE P16-74

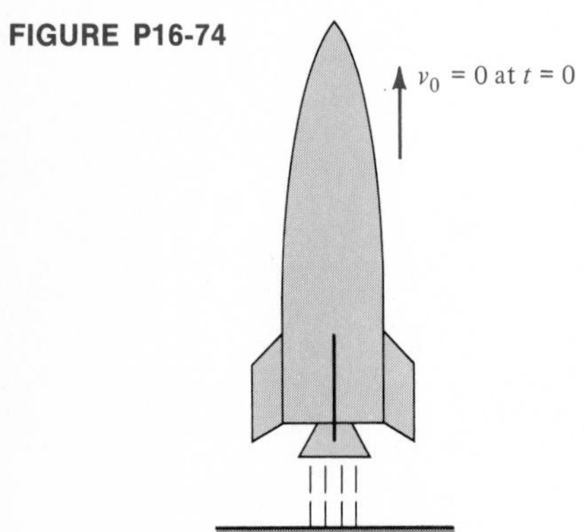

16–74 Determine the total initial mass of a rocket that begins to hover ($v_0 = 0$) in the vertical position immediately after ignition (time $t = 0$) if it burns fuel at the rate of 100 kg/s, ejecting it with a relative speed of 3400 m/s. Calculate the acceleration and speed of this rocket at $t = 20$ s.

FIGURE P16-75

16–75 A 5000-lb space vehicle is moving at a speed of 6000 ft/s when its retrorocket is fired to reduce the speed. The engine consumes fuel at the rate of 10 lb/s and expels it at a relative speed of 9000 ft/s. Determine the final speed of the vehicle if 100 lb of fuel can be used.

16–76 A 9000-kg spacecraft is moving in empty space at a speed of 2 km/s. The engine is fired for 10 s, burning fuel at the rate of 20 kg/s and ejecting it at a relative speed of 3500 m/s. Determine the maximum acceleration and speed of the vehicle during the 10-s period of using the engine.

SUMMARY

Section 16-1

Newton's second law for a system of n particles

$$\sum_{i=1}^{n} \mathbf{F}_i = \sum_{i=1}^{n} m_i \mathbf{a}_i$$

Section 16-2

Motion of the center of mass

POSITION:

$$m\mathbf{r}_C = \sum_{i=1}^{n} m_i \mathbf{r}_i$$

VELOCITY:

$$m\mathbf{v}_C = \sum_{i=1}^{n} m_i \mathbf{v}_i$$

ACCELERATION:

$$m\mathbf{a}_C = \sum_{i=1}^{n} m_i \mathbf{a}_i$$

$$\sum \mathbf{F} = m\mathbf{a}_C$$

Section 16-3

Gravitational potential energy

$$V = mgy_C = Wy_C$$

Section 16-4

Kinetic energy

$$T = \frac{1}{2}\sum_{i=1}^{n} m_i v_i^2$$ (v_i are with respect to a fixed frame)

$$T = \underbrace{\frac{1}{2} m v_C^2}_{\substack{\text{motion of total}\\ \text{mass imagined to}\\ \text{be concentrated at } C}} + \underbrace{\frac{1}{2}\sum_{i=1}^{n} m_i v_i'^2}_{\substack{\text{motion of all}\\ \text{particles relative}\\ \text{to mass center } C}}$$ (v' are with respect to a translating frame)

Section 16-5

Work and energy

$$\sum_{i=1}^{n} U_i = \sum_{i=1}^{n} \Delta V_i + \sum_{i=1}^{n} \Delta T_i$$

$$U = \Delta V + \Delta T$$

Conservation of energy

$$\Delta V + \Delta T = 0$$

For configurations 1 and 2: $V_1 + T_1 = V_2 + T_2$

Section 16-6

Moments of forces about point *O*

$$\sum_{i=1}^{n} (\mathbf{r}_i \times \mathbf{F}_i) = \sum_{i=1}^{n} \mathbf{M}_{i_O} = \sum_{i=1}^{n} (\mathbf{r}_i \times m_i\mathbf{a}_i)$$

Section 16-7

Linear and angular momenta

TRANSLATION:

$$\mathbf{G} = \sum_{i=1}^{n} m_i\mathbf{v}_i \qquad \sum \mathbf{F} = \dot{\mathbf{G}}$$

ROTATION:

$$\mathbf{H}_O = \sum_{i=1}^{n} (\mathbf{r}_i \times m_i\mathbf{v}_i) \qquad \text{(point } O \text{ is fixed)}$$

$$\sum \mathbf{M}_O = \dot{\mathbf{H}}_O = \sum (\mathbf{r}_i \times m_i\mathbf{a}_i)$$

Section 16-8

Angular momentum about the center of mass The angular momentum of a system of particles about its center of mass is the same whether it is observed from a fixed frame or from the centroidal frame which may be translating but not rotating.

USEFUL EXPRESSIONS:

$$\mathbf{H}_O = \mathbf{H}_C + \mathbf{r}_C \times m\mathbf{v}_C$$

$$\sum \mathbf{M}_O = \dot{\mathbf{H}}_C + \mathbf{r}_C \times m\mathbf{a}_C$$

where point O is fixed, C is the center of mass, and $\mathbf{r}_C$ is from O to C.

Section 16-9

Conservation of momentum

$$\left.\begin{aligned} \mathbf{G} &= \text{constant} \\ \mathbf{H}_O &= \text{constant} \\ \mathbf{H}_C &= \text{constant} \end{aligned}\right\} \quad \text{not the same constants in general}$$

Section 16-10

Impulse and momentum

TRANSLATION:

$$\sum_{i=1}^{n} \int_{t_1}^{t_2} \mathbf{F}_i \, dt = \mathbf{G}_2 - \mathbf{G}_1 = m\mathbf{v}_{C_2} - m\mathbf{v}_{C_1}$$

ROTATION:

$$\sum_{i=1}^{n} \int_{t_1}^{t_2} \mathbf{M}_{i_O} \, dt = \mathbf{H}_{O_2} - \mathbf{H}_{O_1}$$

Sections 16-11 and 16-12

Steady flow of mass

$$\frac{dm}{dt} = \rho_1 A_1 v_1 = \rho_2 A_2 v_2$$

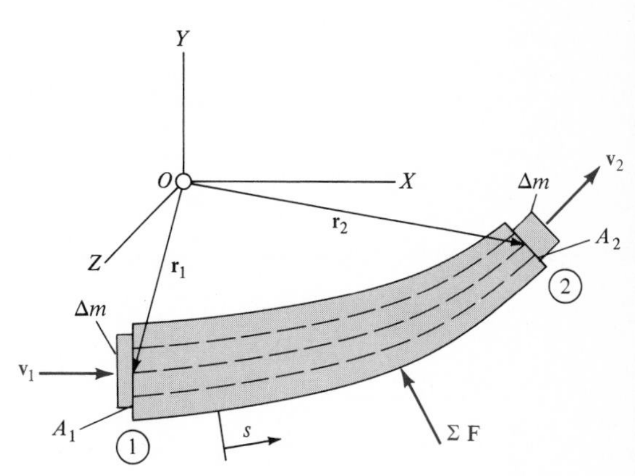

For an incompressible fluid, $\rho_1 = \rho_2 = \rho$,

$$\frac{1}{\rho}\frac{dm}{dt} = A_1 v_1 = A_2 v_2$$

$$\sum \mathbf{F} = \frac{dm}{dt}(\mathbf{v}_2 - \mathbf{v}_1)$$

$$\sum \mathbf{M}_O = \dot{\mathbf{H}}_O = \frac{dm}{dt}(\mathbf{r}_2 \times \mathbf{v}_2 - \mathbf{r}_1 \times \mathbf{v}_1)$$

Section 16-13

Systems with variable mass

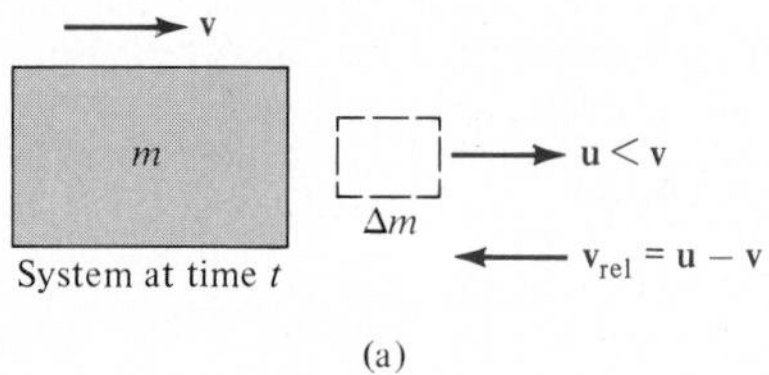

(a)

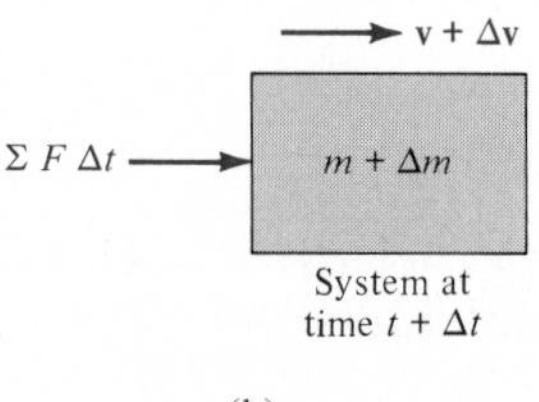

(b)

SYSTEM GAINING MASS:

$$\sum \mathbf{F} = m\frac{d\mathbf{v}}{dt} - \frac{dm}{dt}\mathbf{v}_{rel}$$

SYSTEM LOSING MASS:

$$\sum \mathbf{F} = m\frac{d\mathbf{v}}{dt} + \frac{dm}{dt}\mathbf{v}_{rel}$$

$$\frac{dm}{dt} = \text{time rate at which mass is absorbed or lost}$$

REVIEW PROBLEMS

16–77 Three 80-kg skydivers link up as shown. Assume they are falling vertically and that atmospheric forces are negligible. Diver C pushes away from the other two, acquiring a horizontal speed $v_C = 2$ m/s. Determine the distance between diver C and the others at $t = 5$ s after the separation if their common center of mass moves only vertically.

FIGURE P16-77

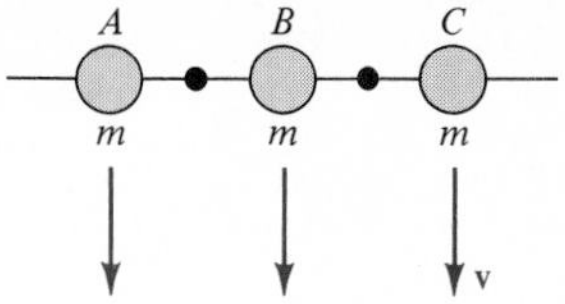

16–78 The amusement ride has 16 carriages of mass $m = 200$ kg which rotate at $\omega_A = 5$ rpm with respect to the arm OA. The carriages with riders move in the horizontal plane when $\theta = 0$. A hydraulic actuator at point A raises the arm with the rotating assembly. Calculate the kinetic energy of only two carriages with respect to point O when $\theta = 0$, $\dot{\theta} = 0.3$ rpm. One carriage is nearest to O, the other is farthest from O at the same instant.

FIGURE P16-78

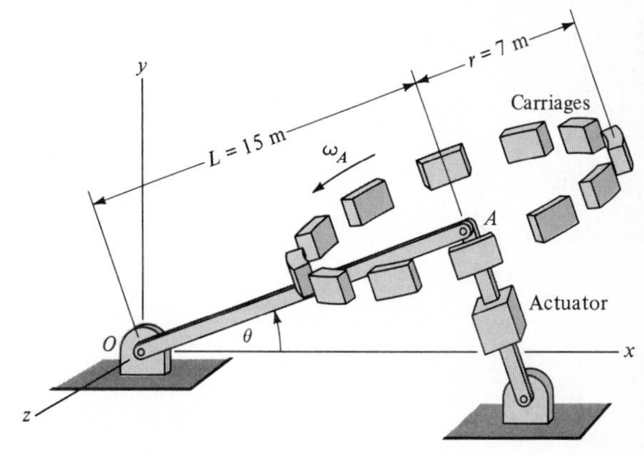

16–79 First solve Prob. 16–78 as stated, then for another pair of carriages. The latter two are equidistant from O at the same instant.

16–80 The amusement ride in Fig. P16-78 has 16 carriages of mass $m = 600$ kg which rotate at $\omega_A = 10$ rpm with respect to the arm OA. The carriages with riders move in the horizontal plane when $\theta = 0$. A hydraulic actuator at point A raises the arm with the rotating assembly. Calculate the angular momentum of two carriages with respect to point O when $\theta = 0$ and $\dot{\theta} = 0.3$ rpm. One carriage is nearest to O, the other is farthest from O at the same instant. The carriages and arm OA are always in the same plane for the purposes of this analysis.

16–81 Consider the rotating system of two masses in Prob. 16–80. Moments $\mathbf{M}_A = -80\mathbf{j}$ kN·m and $\mathbf{M}_O = 12\mathbf{k}$ kN·m are applied for a time of 0.01 s. Determine the new ω_A and $\dot{\theta}$, assuming that θ is still zero.

FIGURE P16-82

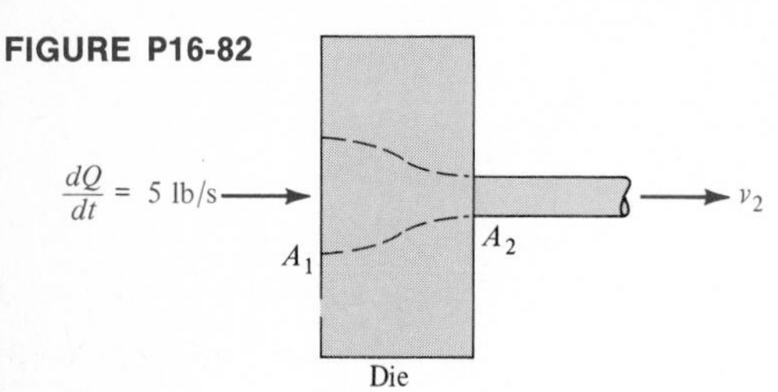

16–82 The die of a machine that extrudes plastic rods receives the liquid polymer at the rate of 5 lb/s. Calculate the exit speed v_2 of the rods and the force for holding the die if $A_1 = 10\ \text{in}^2$, $A_2 = 0.5\ \text{in}^2$, $\gamma = 124.8\ \text{lb/ft}^3$, $p_1 = 10$ psi and $p_2 = 8.21$ psi.

FIGURE P16-83

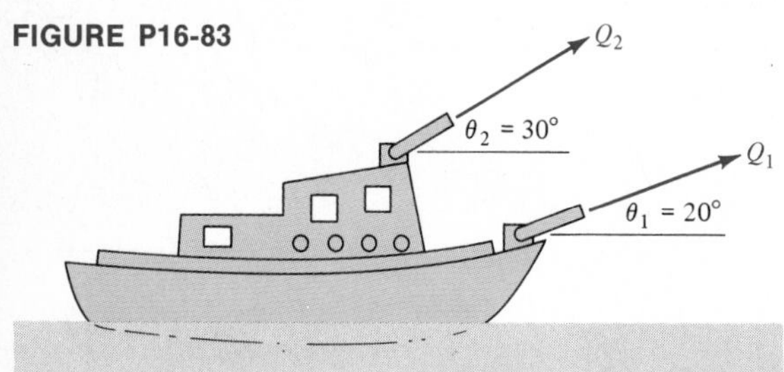

16–83 The fireboat discharges water from two nozzles at the rates of $Q_1 = 6000$ kg/min at $v_1 = 30$ m/s, and $Q_2 = 7000$ kg/min at $v_2 = 40$ m/s. Determine the required propeller thrust (horizontal) to hold the boat in a fixed position.

16–84 An orbiting 3000-lb spacecraft has a speed of 15,000 mi/h which must be increased to 20,000 mi/h to change the orbit. How much fuel is required if the engine burns fuel at the rate of 30 lb/s and expels it at a relative speed of 10,000 ft/s?

FIGURE P16-85

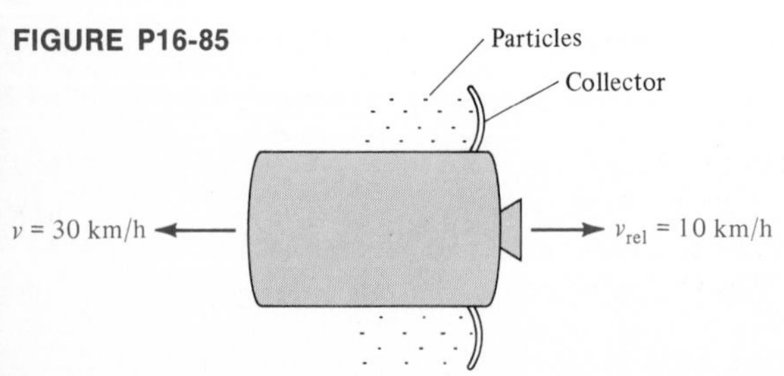

16–85 A 6000-kg vehicle is moving in space at a speed of $v = 30$ km/h. The vehicle collects interstellar particles at the rate of 10^{-9} kg/s. Determine the required rate of fuel consumption to keep the speed v constant if fuel is expelled at a relative speed of 10 km/h. The speed v is in a fixed reference frame in which the particles have negligible speed.

CHAPTER 17

KINEMATICS OF RIGID BODIES

OVERVIEW

A frequently used simplification in analyzing the motion of a body which cannot be treated as a particle is to assume that the body is rigid. By definition, all particles of a rigid body are in fixed positions with respect to each other. The basic problem in kinematics of rigid bodies is to relate the motions at specified locations in the body to the resulting motions at other arbitrary points. The concept of rigid body must be used with caution since nothing is perfectly rigid. Thus, there is always a difference between the actual and theoretical motions of a body when the assumption of rigid body is used in the analysis. The error depends on the deformability of the body. For example, pneumatic tires and springs are much less rigid than the frame of a car, which itself can bend or twist a great amount in comparison with a solid block of steel of the same overall dimensions. Nevertheless, the assumption of dealing with rigid bodies is a powerful concept in the analysis of numerous motions.

It is convenient to distinguish two motions of rigid bodies, translation and rotation. These may occur in pure form or in combination, which is considered as general motion. The various possibilities are described in the following categories and illustrated in Fig. 17-1.

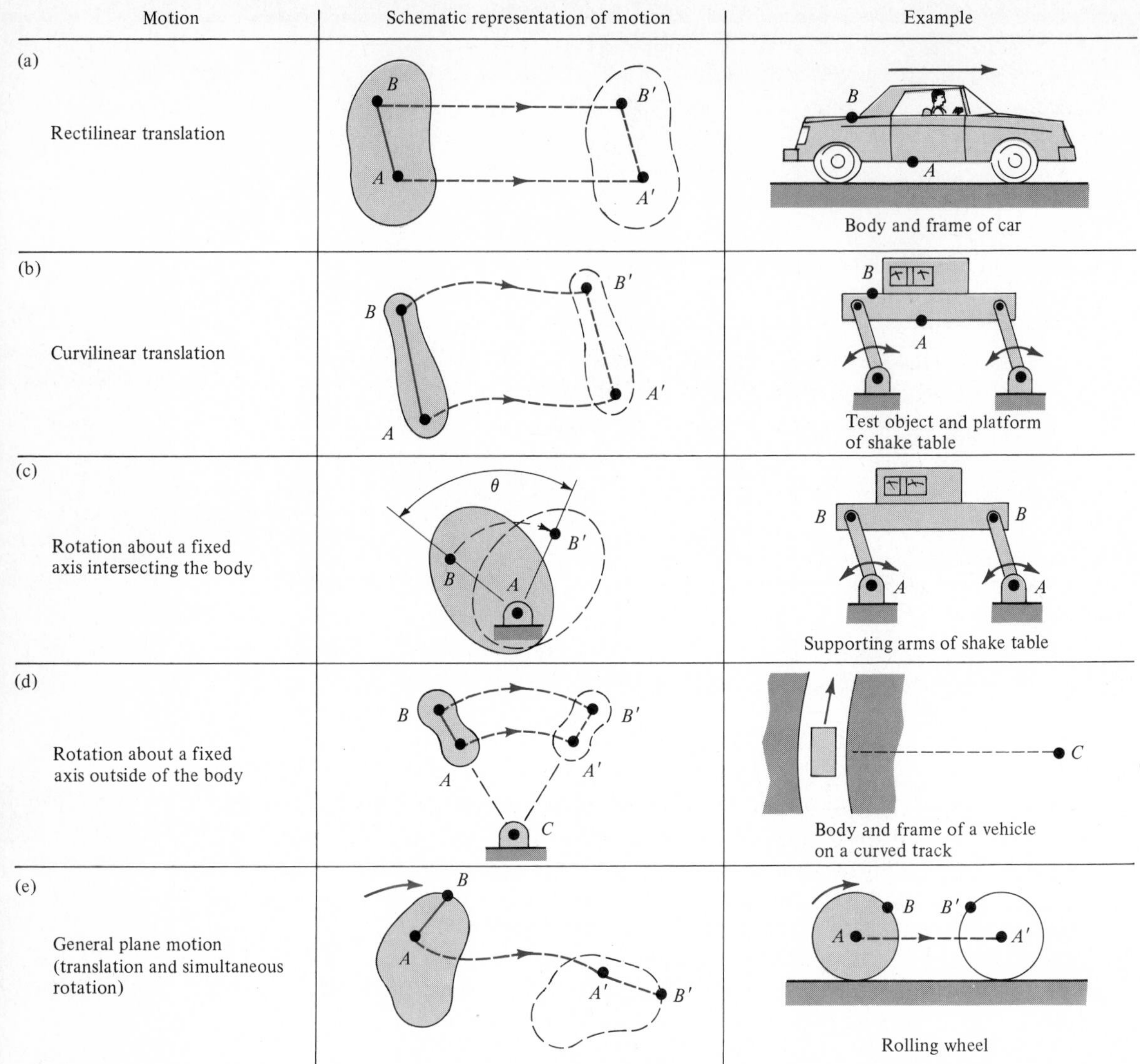

Motion
Schematic representation of motion
Example
(a)
Rectilinear translation
B
B'
A
A'
Body and frame of car
(b)
Curvilinear translation
Test object and platform
of shake table
(c)
Rotation about a fixed
axis intersecting the body
θ
Supporting arms of shake table
(d)
Rotation about a fixed
axis outside of the body
C
Body and frame of a vehicle
on a curved track
(e)
General plane motion
(translation and simultaneous
rotation)
Rolling wheel

FIGURE 17-1

Categories of plane motion of rigid bodies

1. ***Translation.*** A rigid body is in translational motion when every straight line in the body remains parallel to its original position throughout the motion. This implies that all particles of the body have parallel paths during translation. There are two subcategories of translation.

a. In *rectilinear translation* the paths of all particles of the body are parallel straight lines as in Fig. 17-1a. Fixed lines in the body retain their orientations. The motion of one point in the body completely specifies the motion of the whole body.

b. In *curvilinear translation* the paths of the particles of the body are curved lines as in Fig. 17-1b. Fixed lines in the body retain their orientations. The motion of one point in the body completely specifies the motion of the whole body.

2. ***Rotation about a fixed axis.*** A rigid body is in rotational motion about a fixed axis when all particles of the body move along circular paths whose centers are on the same fixed axis. The *axis of rotation* may or may not intersect the rigid body. If the axis intersects the body, particles on that axis have no velocity or acceleration (Fig. 17-1c). If the axis is outside the body, all particles of the body have velocity and acceleration (Fig. 17-1d). The orientation of a fixed line in a body changes during rotational motion which distinguishes rotation from curvilinear translation.

3. ***General plane motion.*** Any plane motion of a rigid body that is a combination of translation and rotation is general plane motion. The most common example of this is a rolling wheel (Fig. 17-1e).

4. ***General motion.*** All motions of rigid bodies which are not in one of the preceding three categories are called general motion. For example, the rolling wheel in Fig. 17-1e may be moving along a curved path on the ground. This means that the plane of rotation changes, and the motion is three-dimensional.

Sometimes a special category of three-dimensional motion is considered for rigid bodies that are attached at a fixed point. For example, the center A of the wheel in Fig. 17-1e may be fixed. This allows rotation of the wheel in a plane and also tilting of that plane about any line through point A.

STUDY GOALS

SECTIONS 17-1 AND 17-2 present the basic equations for analyzing the translation or rotation of a rigid body. The equations for the absolute velocity and acceleration of a particle in a rigid body should be studied thoroughly because they are frequently used in subsequent work.

SECTION 17-3 presents useful kinematics equations for those cases where the angular coordinate, or the angular speed, or the angular acceleration is known as a function of time. It is better to know how to derive these equations rather than memorize them.

SECTION 17-4 extends the methods of Sec. 17-2 and of relative motion to the analysis of velocities in general plane motion (translation and rotation). The basic procedure and the comments concerning solving problems are important for later work.

SECTION 17-5 describes the concept and important properties of the instantaneous center of rotation. This is one of the most illustrative and useful special methods in kinematics, which should be studied even though it does not introduce any new relationships.

SECTION 17-6 extends the methods of Sec. 17-4 to the analysis of accelerations in general plane motion. The basic procedure and the comments concerning practical applications are important for later work. It should be noted that in general the analysis of accelerations is more complex than that of velocities. Thus, it is important to devote sufficient time and effort to learning the physical significance and mathematical operations of the individual components of total acceleration.

SECTION 17-7 presents the generalization of the methods of planar kinematics (Secs. 17-4 and 17-6) to the analysis of three-dimensional motion. The equations for the absolute velocity and acceleration are valid for plane motion *and* for space motion. The terms of the equations are more complex to evaluate in the case of space motion; otherwise, the problems should not cause unusual difficulty to students who understand the results obtained in Sec. 17-6. In some elementary courses in dynamics progress can be achieved even if Sec. 17-7 is omitted.

SECTIONS 17-8 AND 17-9 present the method of analyzing velocities and accelerations using rotating reference frames. This method is advantageous in certain applications but is quite difficult for many students. For this reason, these sections are often omitted from elementary dynamics courses. Ambitious students are encouraged to study this topic to broaden their background and improve their skills in solving complex problems.

17-1 TRANSLATION

Assume that A and B are fixed particles in a rigid body which moves in rectilinear or curvilinear translation as in Fig. 17-2. The relative position vector $\mathbf{r}_{B/A}$ and the absolute position vectors $\mathbf{r}_A$ and $\mathbf{r}_B$ are related by the expression

$$\mathbf{r}_B = \mathbf{r}_A + \mathbf{r}_{B/A} \tag{17-1}$$

The relationship for the absolute and relative velocities is obtained

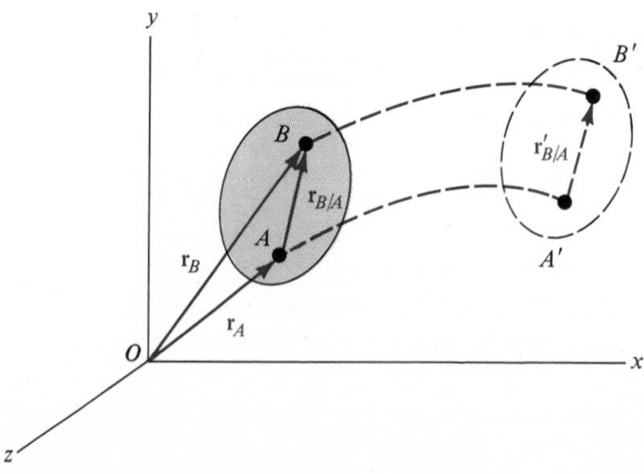

FIGURE 17-2
Translation of a rigid body

by differentiating Eq. 17-1 with respect to time. By the definition of translation, the vector $\mathbf{r}_{B/A}$ always has a constant magnitude and direction. Thus, $\dot{\mathbf{r}}_B = \dot{\mathbf{r}}_A + 0$, or

$$\mathbf{v}_B = \mathbf{v}_A \tag{17-2}$$

The relationship for the accelerations is obtained by differentiating Eq. 17-2 with respect to time,

$$\mathbf{a}_B = \mathbf{a}_A \tag{17-3}$$

Equations 17-2 and 17-3 are the mathematical statements of the fact that *when a rigid body is in translation, the motion of a single point completely specifies the motion of the whole body.* In rectilinear translation the velocity and acceleration have constant lines of action during the motion. In the case of curvilinear translation the directions of the velocity and acceleration vectors change throughout the motion depending on the curved path.

17-2 ROTATION ABOUT A FIXED AXIS

Assume that a rigid body of general shape is rotating about the fixed y axis as in Fig. 17-3. Note that in any situation a coordinate axis may be chosen to coincide with the axis of rotation for convenience. Consider an arbitrary point P and its position vector $\mathbf{r}$ in the body. The position of the body is defined by the *angular coordinate* θ in the plane of rotation of point P. θ is measured from an arbitrary line of reference such as z' to the line AP, where A is the projection of point P on the axis of rotation. The distance $\rho = r \sin \phi$ between points A and P is constant in a rigid body as point P moves on a circular path about the axis of rotation.

The velocity $\mathbf{v}$ of point P is obtained by differentiating the position vector $\mathbf{r}$ with respect to time,

$$\mathbf{v} = \frac{d\mathbf{r}}{dt}$$

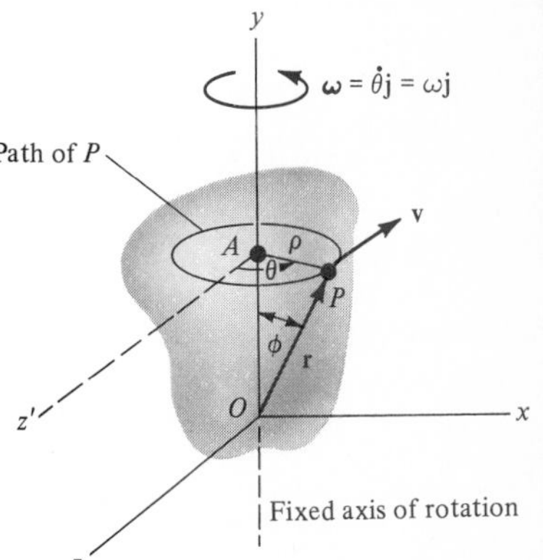

FIGURE 17-3

Analysis of velocities of a rigid body rotating about a fixed axis

For any vector $\mathbf{A}$ of constant magnitude, $\dot{\mathbf{A}} = \boldsymbol{\omega} \times \mathbf{A}$ as derived in App. A. Since the vector $\mathbf{r}$ has a constant magnitude for any given point P,

$$\boxed{\mathbf{v} = \dot{\mathbf{r}} = \boldsymbol{\omega} \times \mathbf{r}} \tag{17-4}$$

The angular velocity in Fig. 17-3 is $\boldsymbol{\omega} = \dot{\theta}\mathbf{j}$ according to the right-hand rule; $\boldsymbol{\omega}$ is the same for any line in a rigid body. The position vector is $\mathbf{r} = r_x\mathbf{i} + r_y\mathbf{j} + r_z\mathbf{k}$. Using these quantities in Eq. 17-4 and evaluating the determinant,

$$\mathbf{v} = \boldsymbol{\omega} \times \mathbf{r} = \dot{\theta}r_z\mathbf{i} - \dot{\theta}r_x\mathbf{k}$$

The magnitude of the velocity is expressed as

$$v = \dot{\theta}\sqrt{r_z^2 + r_x^2} = \dot{\theta}\rho = \dot{\theta}r \sin\phi \qquad \text{17-5}$$

according to Fig. 17-3.

Note that the angles θ and ϕ depend on the position of point P in the body while $\dot{\theta}$ is independent of that position. The velocity vector $\mathbf{v}$ of point P is tangent to the circular path of P and depends on θ; $\mathbf{v}$ is always in a plane that is perpendicular to the axis of rotation.

Plane motion is a special case of that illustrated in Fig. 17-3. For a thin slab parallel to the xz plane at point A, the velocity of any point P is obtained either from Eq. 17-4 or from the scalar expression $v = \rho\omega = \rho\dot{\theta}$ and the direction of ω. The vector method is most advantageous for analyzing three-dimensional motion.

The acceleration $\mathbf{a}$ of a particle P in a rigid body is obtained by differentiating Eq. 17-4 with respect to time,

$$\mathbf{a} = \frac{d\mathbf{v}}{dt} = \frac{d}{dt}(\boldsymbol{\omega} \times \mathbf{r})$$

Using the rule for the differentiation of a vector product gives

$$\mathbf{a} = \frac{d\boldsymbol{\omega}}{dt} \times \mathbf{r} + \boldsymbol{\omega} \times \frac{d\mathbf{r}}{dt}$$

Since $\mathbf{v} = \boldsymbol{\omega} \times \mathbf{r}$,

$$\mathbf{a} = \boldsymbol{\alpha} \times \mathbf{r} + \boldsymbol{\omega} \times \mathbf{v}$$

$$\boxed{\mathbf{a} = \boldsymbol{\alpha} \times \mathbf{r} + \boldsymbol{\omega} \times (\boldsymbol{\omega} \times \mathbf{r})} \qquad \text{17-6}$$

where $\boldsymbol{\alpha} = d\boldsymbol{\omega}/dt$ is the angular acceleration of the body as illustrated in Fig. 17-4. $\boldsymbol{\alpha}$ is the same for any line in a rigid body. The term $\boldsymbol{\alpha} \times \mathbf{r}$ in Eq. 17-6 is the tangential acceleration $\mathbf{a}_t$ of the particle P. The term $\boldsymbol{\omega} \times (\boldsymbol{\omega} \times \mathbf{r})$ is a vector triple product whose direction is always from point P to A on the axis of rotation (since $\boldsymbol{\omega}$ is along the axis of rotation and $\boldsymbol{\omega} \times \mathbf{r} = \mathbf{v}$ is tangent to the path of P). The term $\boldsymbol{\omega} \times (\boldsymbol{\omega} \times \mathbf{r})$ is the normal acceleration $\mathbf{a}_n$ of the particle P. The two components of the acceleration of P are shown in Fig. 17-4. Their magnitudes are

$$a_t = \rho\alpha \qquad a_n = \rho\omega^2 \qquad \text{17-7}$$

where ρ is applicable to a specific point P while ω and α are valid for any line in a rigid body at a given instant. These are important and useful results for the kinematics analysis of rigid bodies.

Plane motion is a special case of that illustrated in Fig. 17-4 concerning accelerations also. For a thin slab parallel to the xz plane at point A, the acceleration $\mathbf{a}$ of any particle P is obtained either from Eq. 17-6 or from the scalar expressions of Eq. 17-7 and the direction of α.

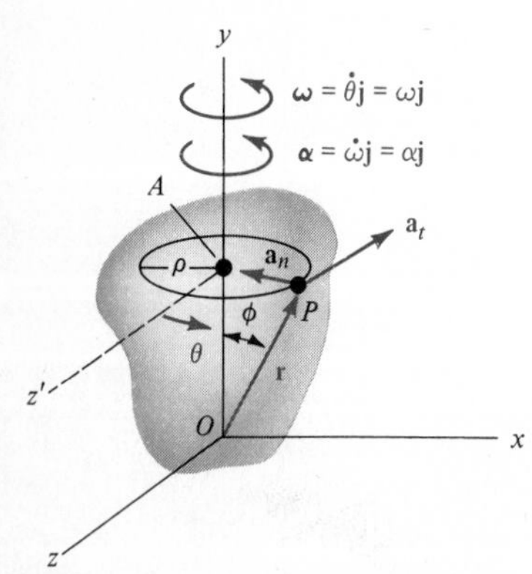

FIGURE 17-4

Analysis of accelerations of a rigid body rotating about a fixed axis

KINEMATICS EQUATIONS FOR ROTATING RIGID BODIES

17-3

The rotational motion of a rigid body (with or without a fixed axis) can be completely specified in magnitude if one parameter of that motion is known as a function of time. This parameter may be the angular coordinate θ, or the angular speed $\dot{\theta} = \omega$, or the angular acceleration $\ddot{\theta} = \alpha$. The time derivatives of the position coordinate are commonly written in one of the following scalar forms:

$$\omega = \frac{d\theta}{dt} = \dot{\theta} \qquad \boxed{17\text{-}8}$$

$$\alpha = \frac{d\omega}{dt} = \frac{d^2\theta}{dt^2} = \ddot{\theta} \qquad \boxed{17\text{-}9}$$

Sometimes it is convenient to eliminate dt as a variable. Using the chain rule yields

$$\alpha = \frac{d\omega}{dt} = \frac{d\omega}{d\theta}\frac{d\theta}{dt} = \omega\frac{d\omega}{d\theta} \qquad \boxed{17\text{-}10}$$

These three equations are analogous to those presented for rectilinear motion of a particle in Sec. 11-1, with θ, ω, and α replacing x, v, and a, respectively. They can be manipulated in the same way. In particular, two important special cases of rotation may be noted.

1. *Constant angular speed, ω.* In this case the angular acceleration is zero, and the angular position θ is given by

$$\theta = \theta_0 + \omega t \qquad \boxed{17\text{-}11}$$

where θ_0 is the initial coordinate for reference at time $t = 0$.

2. *Constant angular acceleration, α.* Integrating Eq. 17-8 and 17-10 with respect to time t and position θ, respectively, yields relationships analogous to those in Sec. 11-3,

$$\omega = \omega_0 + \alpha t \qquad \boxed{17\text{-}12a}$$

$$\theta = \theta_0 + \omega_0 t + \frac{1}{2}\alpha t^2 \qquad \boxed{17\text{-}12b}$$

$$\omega^2 = \omega_0^2 + 2\alpha(\theta - \theta_0) \qquad \boxed{17\text{-}12c}$$

where ω_0 is the initial angular speed corresponding to the initial position θ_0 at time $t = 0$.

If the angular acceleration is neither zero nor constant as in these special cases, Eqs. 17-8 through 17-10 must be used. It should be noted that in rotations about a fixed axis the positive direction is the same for θ, ω, and α.

EXAMPLE 17-1

The model of a tire tester in Fig. 17-5 shows a motor M which drives a large flywheel W. The diameter of the pulley on the motor is $d_1 = 1.1$ ft and on the flywheel it is $d_2 = 4$ ft. The flywheel has an outside diameter of $d_3 = 8$ ft. The motor starts from rest and rotates at a constant angular acceleration of $\alpha = 1\ \text{rad/s}^2$. Calculate the maximum speed at time $t = 20$ s of any particle in a 3-ft-diameter tire that is pressed against the flywheel. Assume that there is no slippage anywhere in the system, and that consequently particles in contact (of different members) have the same absolute velocity.

SOLUTION

The motion can be analyzed as planar motion. From Eq. 17-12a, using the subscript 1 for pulley 1,

$$\omega_1 = \omega_0 + \alpha_1 t$$

If the motor starts from rest, $\omega_0 = 0$. The angular speed at $t = 20$ s is

$$\omega_1 = \alpha_1 t = 1\ \text{rad/s}^2(t) = 1\ \text{rad/s}^2(20\ \text{s}) = 20\ \text{rad/s}$$

If there is no slipping, all particles at the perimeters of pulleys 1 and 2 have the same speed since all parts of the belt have the same speed. Thus, $v_1 = v_2$.

FIGURE 17-5

Model of a tire-testing system

From Eq. 17-4, $\mathbf{v} = \boldsymbol{\omega} \times \mathbf{r}$, and in two dimensions $v = \omega r$. Consequently, the angular speed of pulley 2 is

$$\omega_2 = \frac{\omega_1 r_1}{r_2} = \frac{(20 \text{ rad/s})(0.55 \text{ ft})}{2 \text{ ft}} = 5.5 \text{ rad/s}$$

Since the pulley 2 is connected to the flywheel 3,

$$\omega_3 = \omega_2 \qquad \text{and} \qquad v_3 = \omega_3 r_3 = 22 \text{ ft/s}$$

If there is no slipping, the speed v_3 is the same as the speed at the outside edge of the tire, which is the required maximum speed. Therefore, $v_A = v_3$, and

$$v_{\text{tire}} = v_3 = 22 \text{ ft/s}$$

EXAMPLE 17-2

Determine the maximum total acceleration of any particle in the tire described in Ex. 17-1. Use $t = 20$ s. Assume that particles in contact (of different members) have the same tangential acceleration if there is no slipping but in general have different normal accelerations.

SOLUTION

Both the tangential and normal components of the acceleration are maximum for particles at the outside edge of the tire according to Eq. 17-7. The tangential accelerations for pulley 1 and 2 are obtained using $a_t = r\alpha$ from Eq. 17-7,

$$a_{t_2} = a_{t_1} = r_1 \alpha_1 = \left(\frac{1.1}{2} \text{ ft}\right)(1 \text{ rad/s}^2) = 0.55 \text{ ft/s}^2$$

Thus, because pulley 2 is attached to the flywheel,

$$\alpha_3 = \alpha_2 = \frac{a_{t_2}}{r_2} = \frac{0.55 \text{ ft/s}^2}{2 \text{ ft}} = 0.275 \text{ rad/s}^2$$

$$a_{t_3} = \alpha_3 r_3 = 1.1 \text{ ft/s}^2$$

If there is no slipping, particles of the tire and the flywheel in contact at point A have the same tangential acceleration, so

$$a_{t_{\text{tire}}} = 1.1 \text{ ft/s}^2$$

The normal acceleration of a particle on the tire at point A is

$$a_n = \omega^2 r = \frac{v^2}{r} = \frac{(22 \text{ ft/s})^2}{1.5 \text{ ft}} = 323 \text{ ft/s}^2$$

The total acceleration for any point on the tire is obtained from Eq. 17-6,

$$\mathbf{a}_{\text{total}} = \mathbf{a}_t + \mathbf{a}_n$$

Both a_n and a_t are magnitudes of perpendicular vector quantities, so

$$a_{\text{total}} = \sqrt{a_t^2 + a_n^2} = \sqrt{(1.1)^2 + (323)^2} = 323 \text{ ft/s}^2$$

PROBLEMS

SECTIONS 17-1, 17-2, 17-3

FIGURE P17-1

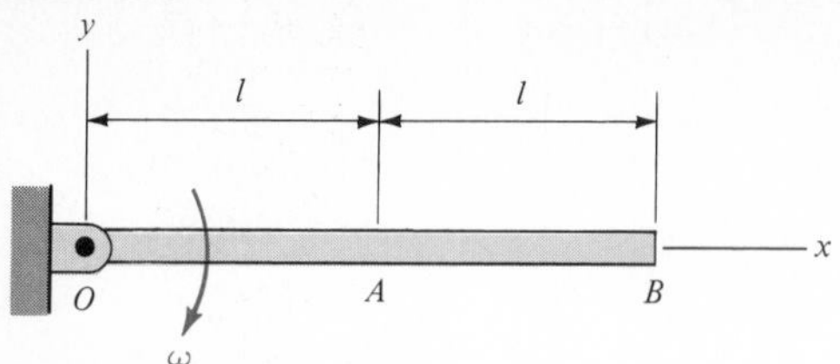

17–1 The rigid bar of length $2l = 1$ m is rotating about pin O at a constant rate of $\omega = 8$ rad/s. Determine the velocities and accelerations of particles at A and B for the position shown.

17–2 The angular speed of the bar described in Prob. 17–1 is increasing at the rate of $\alpha = 2\ \text{rad/s}^2$. Draw the acceleration vectors for particles at A and B on the same diagram of the bar.

FIGURE P17-3

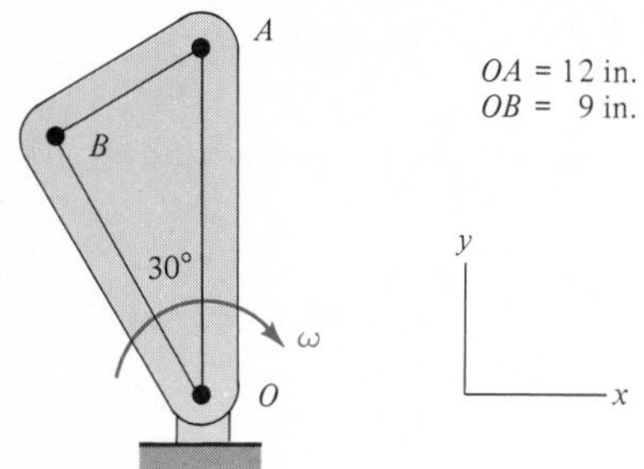

17–3 The triangular bracket is rotating about pin O at a constant rate of $\omega = 15$ rad/s. Determine the velocities and accelerations of pins A and B for the position shown.

17–4 The angular speed of the bracket described in Prob. 17–3 is decreasing at the rate of $\alpha = 3\ \text{rad/s}^2$. Draw the acceleration vectors for pins A and B on the same diagram of the bracket.

FIGURE P17-5

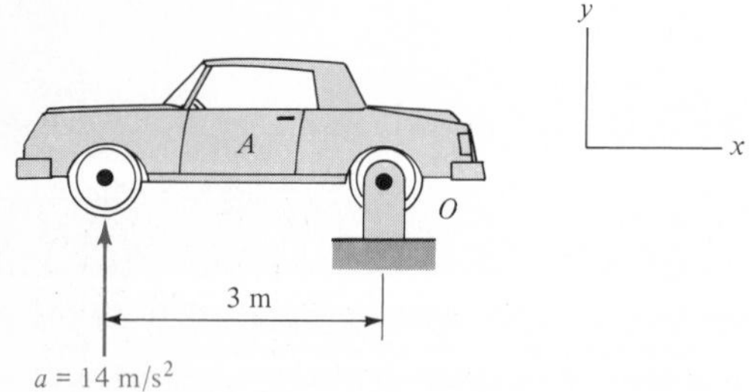

17–5 A car in a laboratory test is supported by hinges at the rear axles. The front wheels are simultaneously accelerated upward at the rate of $14\ \text{m/s}^2$. Calculate the angular acceleration of the car and the vertical acceleration of a point A midway between the wheels for the given position. Assume that the whole car is rigid.

17–6 The car described in Prob. 17–5 is rotating about hinges O at the rate of $\omega = 1$ rad/s counterclockwise when the front end is accelerated upwards as shown. Determine the total acceleration of a particle at A, midway between the wheels.

17–7 A 6-ft-tall electric console is tested on a shake table whose four supporting arms are 3 ft long. The arms are rotating at constant speeds of $\omega = 2$ rad/s, clockwise as viewed from the right of the picture, when the arms make 60° angles with the horizontal. Calculate the accelerations of parts A and B of the console. Assume that the platform is thin.

FIGURE P17-7

Electric console on a shake table (Courtesy MTS Systems Corporation, Minneapolis, Minnesota)

17–8 The angular velocities of the arms described in Prob. 17–7 are decreasing at the rate of 1 rad/s^2 in the given position. Draw the acceleration vectors at points A and B in the plane of rotation (as viewed from the z axis).

17–9 A radial line OA on a wheel has its position given by $\theta = 2t^3 - t^2 + 5$, where θ is in radians and t is in seconds. Determine the angular velocity, acceleration, and number of turns of the wheel at a time $t = 10$ s after starting from rest. How much time is required for the wheel to reach 5000 rpm?

FIGURE P17-9

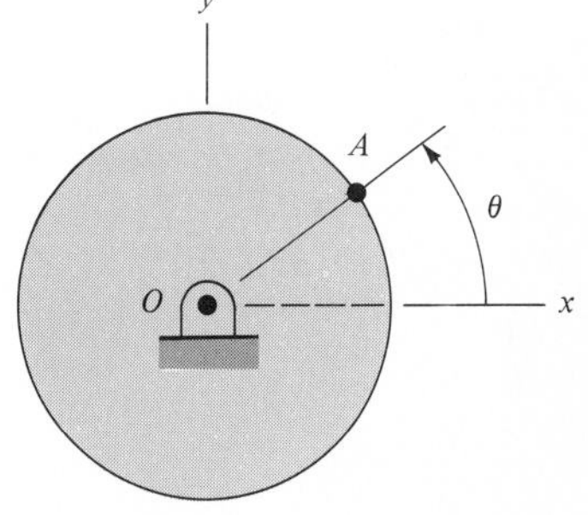

17–10 Draw the velocity and acceleration vectors of point A in Prob. 17–9 at time $t = 5$ s if the wheel's radius is 0.3 m.

17–11 The large drum on a fishing trawler is designed to take the rope up at a constant speed of $v = 6$ ft/s. Determine the angular velocity of the drum when the core of wound rope has a diameter of 2 ft. Calculate the required angular deceleration of the drum at that instant if the core diameter increases at the rate of 0.001 ft/s.

FIGURE P17-11

17–12 A flywheel is rotating at a speed of 800 rpm when it begins to decelerate at the rate of 0.02 rad/s^2. Calculate the time required to stop the wheel and the number of revolutions during the slowing process.

17–13 A turbine disk is rotating at a speed of 30,000 rpm. Calculate the angular acceleration required to stop the disk in a time of 100 s.

17–14 The rotor of a centrifugal pump starts from rest and reaches a speed of 1000 rpm in 300 revolutions. Calculate the time required for the 300 revolutions if the acceleration is at a constant rate.

FIGURE P17-15

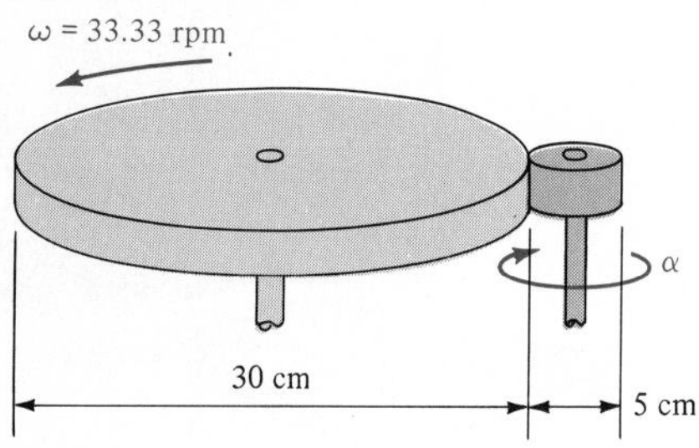

17–15 The turntable of a record player is to have a friction drive by a smaller wheel. What is the required angular acceleration of the small wheel if the large disk is to reach a speed of 33.33 rpm in a time of 5 s? How many revolutions does the large disk make in this time?

FIGURE P17-16

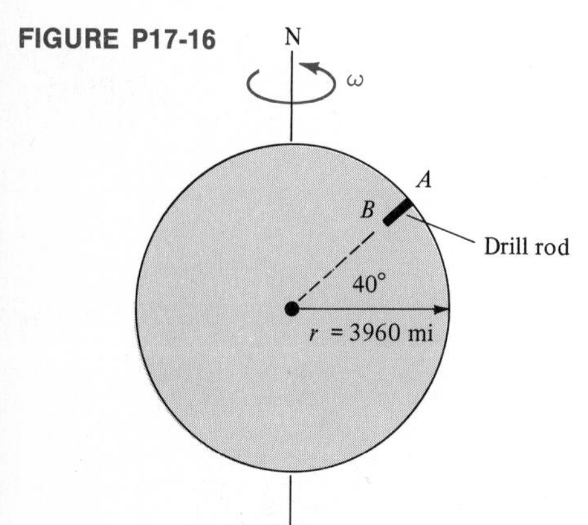

17–16 A drill rod for geologic exploration is 20,000 ft long, and it lies radially in the earth's crust. Calculate the linear velocities and accelerations of the ends A and B of the rod caused by rotation of the earth. The rod is located at 40° north latitude. Assume that the radius of the earth is 3960 mi, and that the earth makes one complete revolution about its axis in 24 h.

FIGURE P17-17

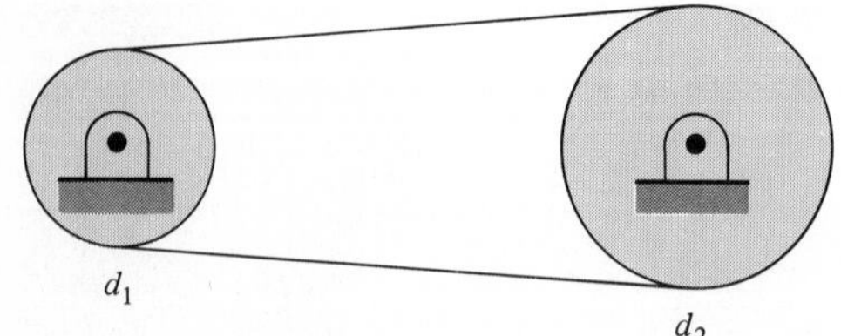

17–17 The sprockets of a chain drive have diameters $d_1 = 10$ cm and $d_2 = 20$ cm. The small sprocket is designed to reach a speed of 100 rpm in a time of 7 s. Determine the required angular acceleration of the large sprocket and the number of revolutions of both wheels during the acceleration.

17–18 The large wheel in Fig. P17-17 is rotating at speed of 200 rpm. Determine the angular acceleration of the small wheel to reduce the large wheel's speed to 100 rpm in a time of 5 s. $d_1 = 6$ in., $d_2 = 15$ in. There is no slipping.

17–19 A wheel is rotating at an angular speed ω. It is desirable to stop the wheel in one complete revolution of a radial line on it. Derive expressions for the required angular acceleration α and time t for such a braking.

17–20 A turntable has an inner wheel for friction drive as shown in the cross-sectional view. The small wheel has a diameter of 3 in., the large one of 12 in. (inside). Calculate the angular acceleration of the small wheel to increase ω_2 from 50 rpm to 100 rpm in a time of 3 s. Also calculate the number of revolutions of the large wheel during this time.

FIGURE P17-20

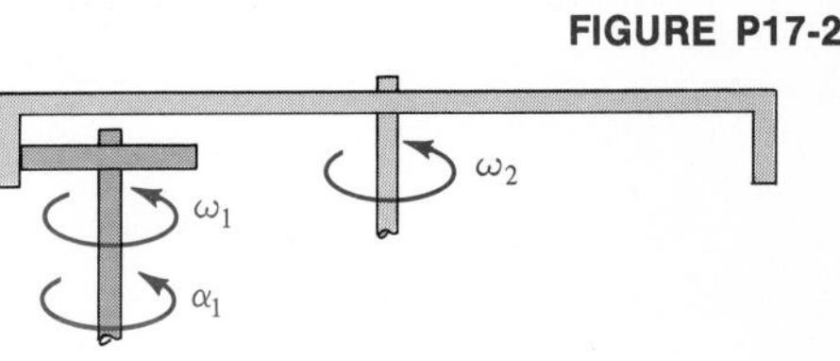

17–21 A hoist mechanism consists of wheel A ($r = 1$ m), gear B ($r = 0.5$ m), which is firmly attached to wheel A, and pinion gear C ($r = 0.1$ m), which drives gear B. The maximum allowed vertical acceleration of the load is 5 m/s^2. Its maximum velocity is 8 m/s. Determine the maximum angular velocity and angular acceleration of gear C, and the number of revolutions in which it reaches the maximum velocity starting from rest.

FIGURE P17-21

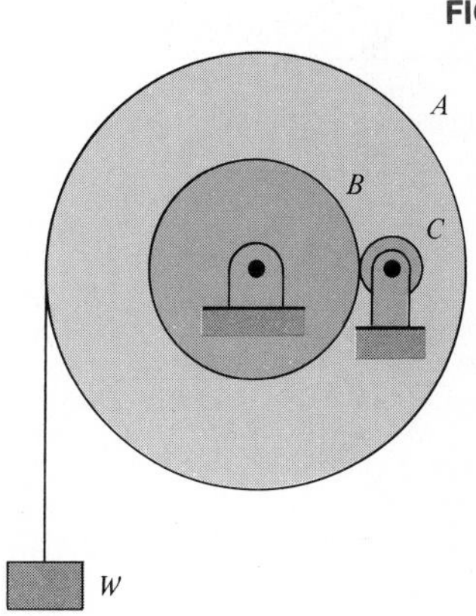

17–22 A failure in the hoist mechanism described in Prob. 17–21 allows the load to fall from rest at the rate of 7 m/s^2. Determine the angular acceleration of the gear C, and its angular velocity after the load has descended 20 m.

17-4 GENERAL PLANE MOTION. ANALYSIS OF VELOCITIES

A general plane motion of a rigid body can be considered a *simultaneous translation and rotation.* An example of this is a car moving on a plane, curved road as in Fig. 17-6a. The motion of the frame can be described as a combination of a pure curvilinear translation and a pure rotation shown in Fig. 17-6b and c. The absolute velocity of an arbitrary particle in such a rigid body is determined using the concept of relative motion of two particles.

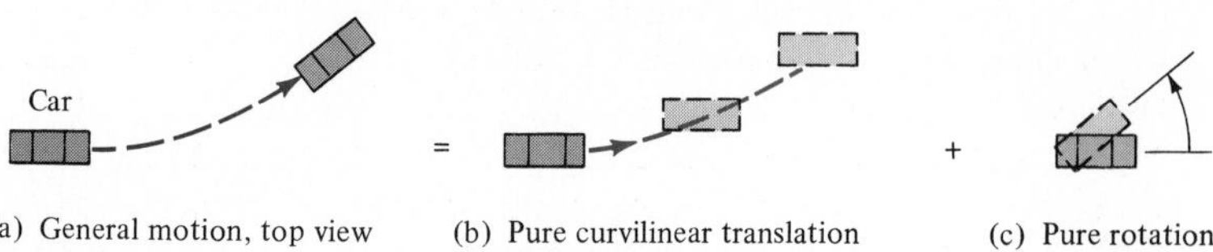

FIGURE 17-6

Schematic of general plane motion

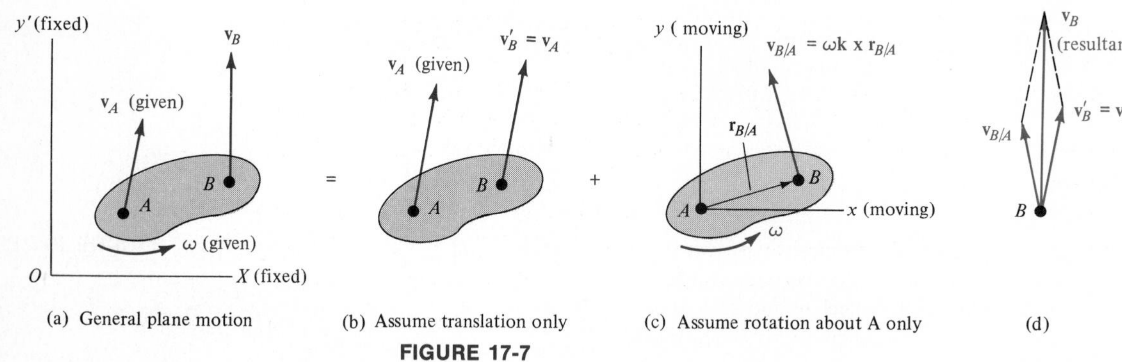

FIGURE 17-7

Analysis of velocities in general plane motion

Assume that the body in Fig. 17-7a is in general plane motion with respect to the fixed XY frame. Also assume that the absolute velocity $\mathbf{v}_A$ of particle A and the angular velocity $\boldsymbol{\omega}$ of the body are known, and the absolute velocity $\mathbf{v}_B$ of a given particle B is to be determined. The velocity $\mathbf{v}_B$ can be expressed (as for any two particles A and B) as

$$\boxed{\mathbf{v}_B = \mathbf{v}_A + \mathbf{v}_{B/A}} \qquad \text{17-13}$$

where $\mathbf{v}_{B/A}$ is the relative velocity of B with respect to A. Using the position vector $\mathbf{r}_{B/A}$ from point A to B, $\mathbf{v}_{B/A} = \dot{\mathbf{r}}_{B/A}$. From Eq. 17-4,

$$\mathbf{v}_{B/A} = \boldsymbol{\omega} \times \mathbf{r}_{B/A} \qquad v_{B/A} = r_{B/A}\omega \qquad \text{17-14}$$

$$\overset{>}{\mathbf{v}_B} = \mathbf{v}_A + \boldsymbol{\omega} \times \mathbf{r}_{B/A} \qquad \text{17-15}$$

since $\mathbf{r}_{B/A}$ has a constant magnitude in a rigid body. The velocity $\mathbf{v}_B$ is obtained as the vector sum of $\mathbf{v}_A$ and $\boldsymbol{\omega} \times \mathbf{r}_{B/A}$, where each quantity is known. This procedure is illustrated in Fig. 17-7 as the superposition of translation and rotation. In Fig. 17-7b all particles have the same velocity $\mathbf{v}_A$. The relative velocity $\mathbf{v}_{B/A}$ would be seen by a nonrotating observer located at point A (and moving with the body) as a pure rotation of B about point A as in Fig. 17-7c. The graphic vector summation of $\mathbf{v}_A$ and $\mathbf{v}_{B/A}$ gives the resultant $\mathbf{v}_B$ in Fig. 17-7d. Note that ω is the angular velocity of the body as a whole or of any line in the body.

It is instructive to interchange the roles assigned to the points A and B in the analysis, to determine the velocity of A in terms of $\mathbf{v}_B$. Assume that the rigid body in Fig. 17-8a has the same general plane motion as that in Fig. 17-7a. If the absolute velocity of B and the body's angular velocity ω are known, the analysis is made by using point B as an intermediate reference point, the origin of the moving reference frame. The translational component of the motion again causes points A and B to have identical velocities both in magnitude and in direction. Therefore, the translational velocity of point A is $\mathbf{v}'_A = \mathbf{v}_B$, as shown in Fig. 17-8b. Here the velocities are different from those in Fig. 17-7b even though the same body and total motion are analyzed in the two figures.

To consider rotation alone, point B is now assumed fixed, with point A moving on a circular path with respect to B as in Fig. 17-8c. Note that the angular velocities shown in Figs. 17-7c and 17-8c

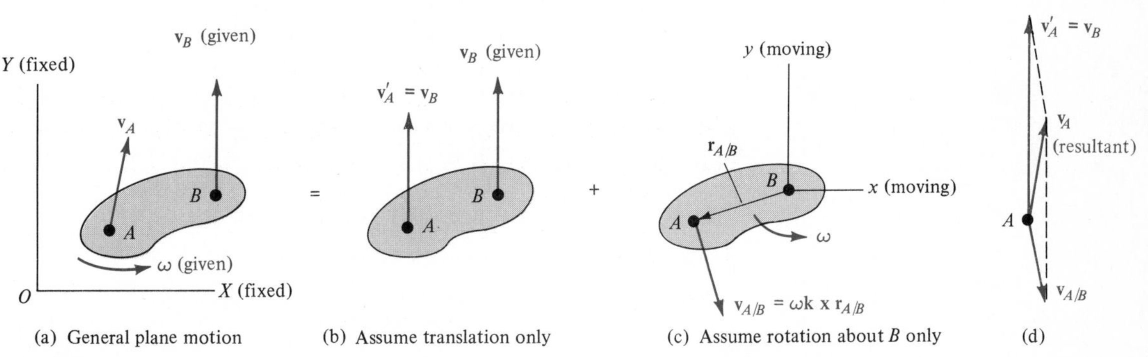

FIGURE 17-8
Alternative reference point for analyzing the motion illustrated in Fig. 17-7

represent the same rotation of the body.* The velocity vector of A caused by its rotation about point B is

$$\mathbf{v}_{A/B} = \boldsymbol{\omega} \times \mathbf{r}_{A/B} \qquad v_{A/B} = r_{A/B}\omega \tag{17-16}$$

It is seen from Eqs. 17-14 and 17-16 and Figs. 17-7c and 17-8c that $\mathbf{v}_{A/B}$ and $\mathbf{v}_{B/A}$ have parallel lines of action, equal magnitudes, and opposite directions.

The absolute velocity of point A is the vector sum of its translational velocity ($\mathbf{v}'_A = \mathbf{v}_B$) and its rotational velocity with respect to B, $\mathbf{v}_{A/B}$. This is shown in Fig. 17-8d, where $\mathbf{v}_A = \mathbf{v}_B + \boldsymbol{\omega} \times \mathbf{r}_{A/B}$. The resultant vector $\mathbf{v}_A$ is identical to $\mathbf{v}_A$ in Fig. 17-7, as it should be since the same total motion was analyzed in different ways.

The same procedure can be applied using any pair of points in a rigid body. These concepts can be applied for analyzing the motion of individual or interconnected rigid bodies. In solving problems it is useful to remember the following:

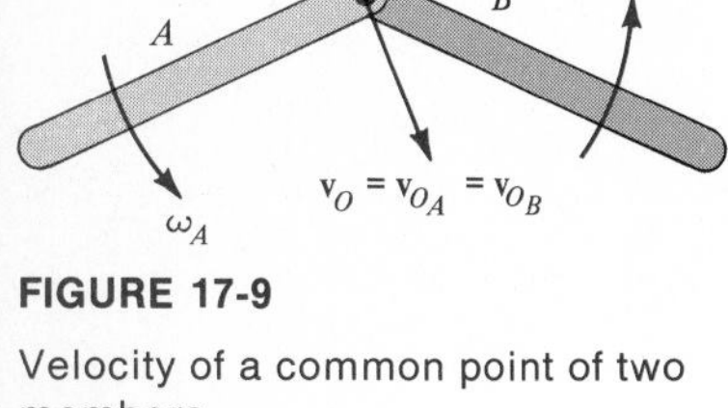

FIGURE 17-9

Velocity of a common point of two members

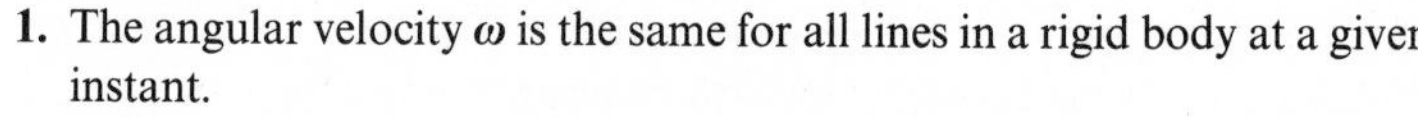

1. The angular velocity ω is the same for all lines in a rigid body at a given instant.
2. The common point of two or more pin-jointed members must have the same absolute velocity even though the individual members may have different angular velocities. For example, $\mathbf{v}_O = \mathbf{v}_{O_A} = \mathbf{v}_{O_B}$ in Fig. 17-9.

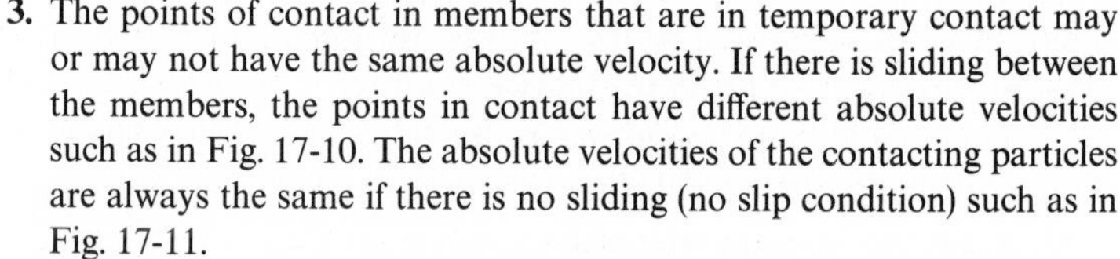

3. The points of contact in members that are in temporary contact may or may not have the same absolute velocity. If there is sliding between the members, the points in contact have different absolute velocities such as in Fig. 17-10. The absolute velocities of the contacting particles are always the same if there is no sliding (no slip condition) such as in Fig. 17-11.
4. Frequently, the angular velocity of a member is not known, but some points of the member move along defined paths (e.g., the end points of a piston rod). The geometric constraints of motion define the directions of the velocity vectors and are useful in the solution.

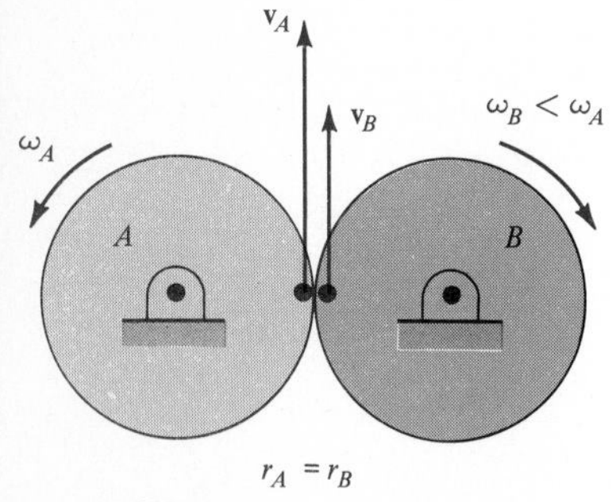

FIGURE 17-10

Two identical wheels slipping relative to one another

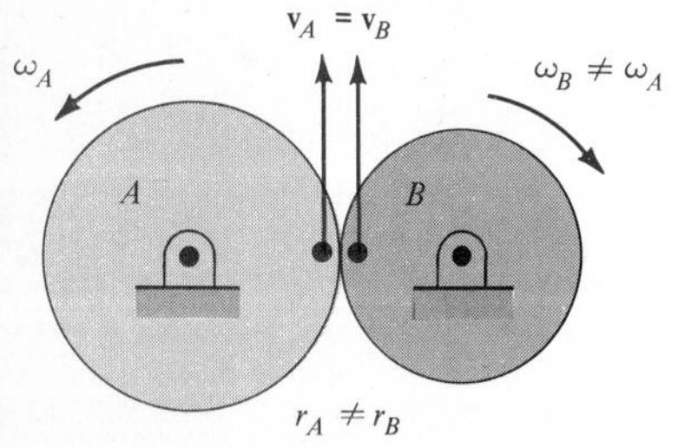

FIGURE 17-11

Two different wheels rotating without slipping relative to one another

* To check that ω is the same using A or B for a moving reference point, assume $\boldsymbol{\omega}_A$ for the angular velocity in Fig. 17-7c, and $\boldsymbol{\omega}_B$ in Fig. 17-8c. The two figures refer to the same body at the same instant. Thus,

$$\mathbf{v}_B = \mathbf{v}_A + \boldsymbol{\omega}_A \times \mathbf{r}_{B/A} \qquad \text{(Fig. 17-7c)}$$

$$\mathbf{v}_A = \mathbf{v}_B + \boldsymbol{\omega}_B \times \mathbf{r}_{A/B} \qquad \text{(Fig. 17-8c)}$$

With $\mathbf{r}_{B/A} = -\mathbf{r}_{A/B}$,

$$\mathbf{v}_B = \mathbf{v}_A + \boldsymbol{\omega}_A \times (-\mathbf{r}_{A/B}) = \mathbf{v}_A - \boldsymbol{\omega}_A \times \mathbf{r}_{A/B}$$

$$\mathbf{v}_A = \mathbf{v}_B + \boldsymbol{\omega}_A \times \mathbf{r}_{A/B}$$

$$\mathbf{v}_A = \mathbf{v}_A - \boldsymbol{\omega}_A \times \mathbf{r}_{A/B} + \boldsymbol{\omega}_B \times \mathbf{r}_{A/B}, \qquad \text{hence } \boldsymbol{\omega}_A = \boldsymbol{\omega}_B = \boldsymbol{\omega}$$

All lines in a rigid body have the same angular velocity at a given instant, and this is very useful in the analysis of velocities of the body. It is worthwhile to reconsider the common case of a body that is rotating about a fixed axis. In general this axis may or may not intersect the body.

For example, assume that the body in Fig. 17-12 is rotating at an angular velocity $\boldsymbol{\omega}$ about the fixed point C. This is enough information to determine the velocity vector of any particle such as A or B in the body if its distance from the axis of rotation is given. Each of these velocity vectors is normal to the line between the axis of rotation and the point of interest; $\mathbf{v}_A = \boldsymbol{\omega} \times \mathbf{r}_A$ and $\mathbf{v}_B = \boldsymbol{\omega} \times \mathbf{r}_B$. Their magnitudes are proportional to the angular velocity (which is common to all lines in that body) and the particular distances to the center of rotation, C. The proportionality of the magnitudes v_A and v_B can be illustrated by considering points A and B', with B' on line AC, at a distance r_B from point C (Fig. 17-12). Here $v_A = r_A\omega$ and $v_{B'} = v_B = r_B\omega$. The ratio of these speeds involves only the radial distances from the center of rotation, $v_A/v_B = r_A/r_B$, since $v_A/r_A = v_B/r_B = \omega$. These concepts are useful in analyzing the rotational motion of bodies that do not rotate about fixed axes.

The *general plane motion* (translation *and* rotation) of a rigid body may be analyzed if there are at least two points in the body whose velocity vectors are known. The need for knowing two velocity vectors is explained as follows.

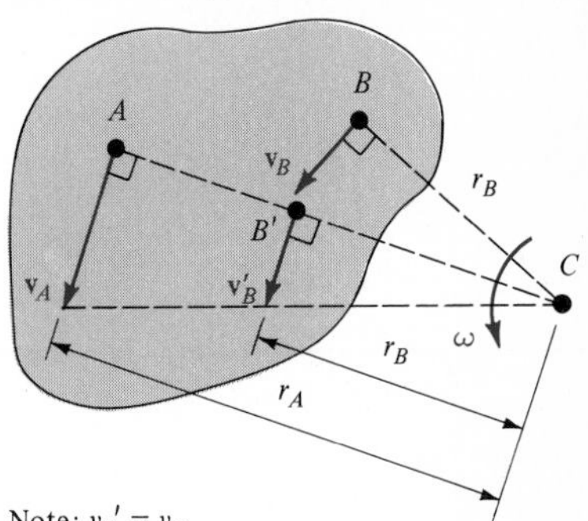

FIGURE 17-12

Schematic for the location of the instantaneous center of rotation

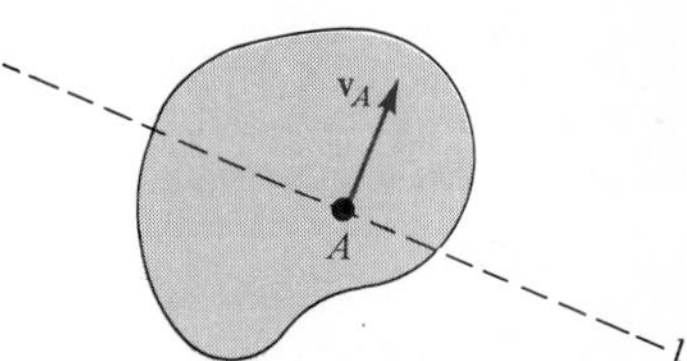

FIGURE 17-13

Indeterminate motion of a rigid body

1. *One velocity vector is specified.* Assume that a rigid body is in plane motion and that the velocity of one particle is given as in Fig. 17-13. The motion of the rigid body is indeterminate because the angular velocity of the body cannot be calculated from $\mathbf{v}_A$. The only statement that can be made about the total motion is that the body *may be* rotating at the given instant about a point on line l which is perpendicular to $\mathbf{v}_A$. Pure translation may be considered a rotation about a point on line l at an infinite distance from particle A. However, this motion cannot be analyzed with sufficient certainty.

2. *Two velocity vectors are specified.* Consider a rigid body for which the velocities of two particles are given as in Fig. 17-14. The different magnitudes and directions of these vectors indicate that the body is rotating. Assume that the rotation is about a point that is fixed at the instant considered. The velocity vector of any particle in the body must be perpendicular to its radial line drawn from the instantaneously fixed point, the center of rotation, as in rotation about a fixed axis. Referring to Fig. 17-14, the center of rotation

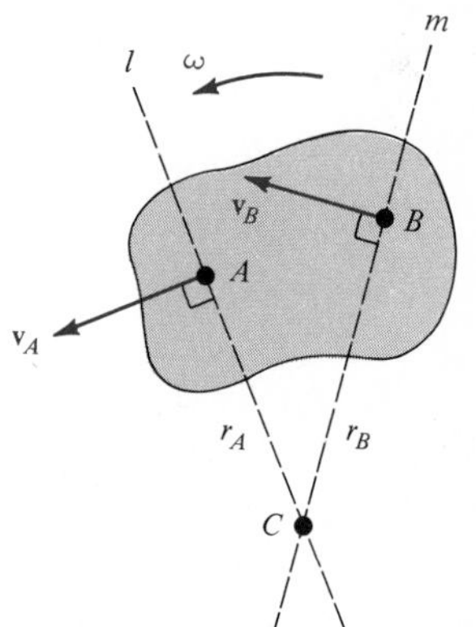

FIGURE 17-14

Two velocity vectors are sufficient to determine the instantaneous center of rotation

must be somewhere on line l for particle A, and somewhere on line m for particle B. To simultaneously satisfy both of these conditions, the body must be rotating instantaneously about point C, the intersection of lines that are normal to the given velocity vectors. This means that at the instant when v_A and v_B are specified the body is rotating about point C, which is called the *instantaneous center of rotation*. There are several properties of the instantaneous center of rotation that are useful or restrictive in the analysis of motion:

a. The angular velocity of the body can be determined from the velocity and radial distance of a single particle when the instantaneous center is located. From Fig. 17-14, $\omega = v_A/r_A$, or $\omega = v_B/r_B$. Since ω applies to all lines in the body, the velocity of all other particles in the body can be readily determined.

b. The instantaneous center of rotation has zero velocity at the given instant, and it may or may not be a point on the body.

c. The instantaneous center may be determined when the two given velocity vectors are parallel but of different magnitude. In that case the instantaneous center is obtained as point C in Fig. 17-12, where a triangular construction can be made using points A and B' and the parallel vectors $\mathbf{v}_A$ and $\mathbf{v}_{B'}$. Alternatively, the method shown in Fig. 17-14 may be used.

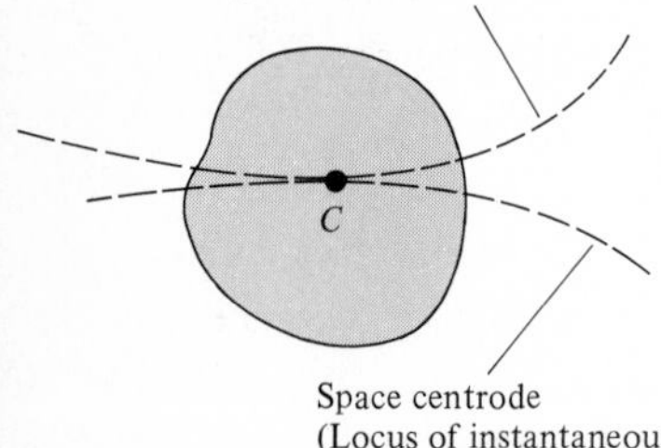

FIGURE 17-15

Body and space centrodes

d. Both the absolute position of the instantaneous center and its position relative to the body generally change in time. The instantaneous center can be imagined to be a fixed center of rotation only for the instant when the velocity vectors are valid. The locus of the absolute positions of the instantaneous center is called the *space centrode*. The locus of the positions of the instantaneous center relative to the body is called the *body centrode*. As shown schematically in Fig. 17-15, these curves are tangent at point C for the instant considered. The body centrode appears to roll on the space centrode as the body moves. For example, consider a wheel rolling on a flat surface as in Fig. 17-16.

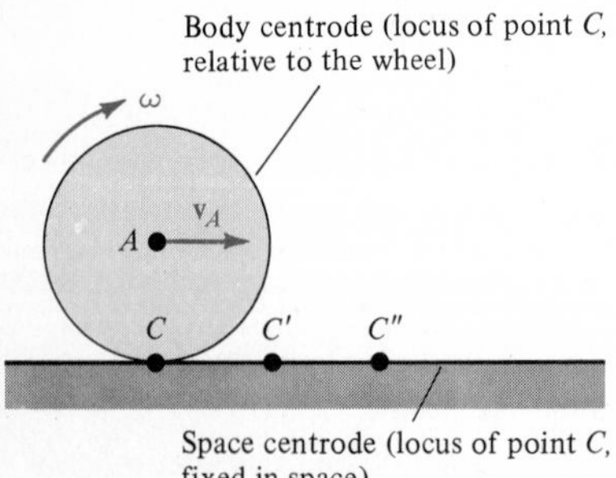

FIGURE 17-16

Body and space centrodes of a rolling wheel

e. The instantaneous center of rotation has zero velocity, but in general motion it has an acceleration. Thus, the instantaneous center of velocity is *not* an instantaneous center of zero acceleration, and cannot be used for convenient analyses of accelerations. This limitation of the concept of instantaneous center of rotation is important to remember. It should also be noted that the instantaneous center of rotation is introduced for convenience, but it implies no new relationship in kinematics.

EXAMPLE 17-3

A mechanism consists of two rigid bars AB and BC of length 2.4 m each (Fig. 17-17). Bar AB rotates about pin A while pin C moves in a horizontal guide. Determine the velocity of pin C and the angular speed $\dot{\phi}$ of bar BC if $\theta = 20^\circ$ to the right of the vertical and $\dot{\theta} = 0.8$ rad/s clockwise.

SOLUTION

From Eq. 17-13,

$$\mathbf{v}_C = \mathbf{v}_B + \mathbf{v}_{C/B}$$

$$\mathbf{v}_B = \mathbf{v}_A + \mathbf{v}_{B/A} \qquad \text{where } \mathbf{v}_A = 0$$

$$\mathbf{v}_B = \boldsymbol{\omega}_{AB} \times 2.4(\sin 20^\circ\,\mathbf{i} + \cos 20^\circ\,\mathbf{j})$$

$$= -0.8\mathbf{k} \times (0.821\mathbf{i} + 2.255\mathbf{j})$$

$$= (1.804\mathbf{i} - 0.657\mathbf{j})\text{ m/s}$$

Next, $\mathbf{r}_{C/B}$ is written in terms of vector components using the geometry of the two bars,

$$\mathbf{r}_{C/B} = (1.256\mathbf{i} + 2.045\mathbf{j})\text{ m}$$

$$\mathbf{v}_C = \mathbf{v}_B + \mathbf{v}_{C/B} = \mathbf{v}_B + \boldsymbol{\omega}_{BC} \times \mathbf{r}_{C/B}$$

Since pin C can move only horizontally,

$$v_C\mathbf{i} = 1.804\mathbf{i} - 0.657\mathbf{j} + \omega_{BC}\mathbf{k} \times (1.256\mathbf{i} + 2.045\mathbf{j})$$

$$= 1.804\mathbf{i} - \omega_{BC}(2.045)\mathbf{i} - 0.657\mathbf{j} + \omega_{BC}(1.256)\mathbf{j}$$

Comparing $\mathbf{j}$ components and $\mathbf{i}$ components gives

$$\omega_{BC}(1.256) = 0.657$$

$$\dot{\phi} = \omega_{BC} = 0.522\text{ rad/s} \qquad \text{(counterclockwise)}$$

$$v_C = 1.804 - \omega_{BC}(2.045)$$

$$= 0.735\text{ m/s} \qquad \text{(to the right)}$$

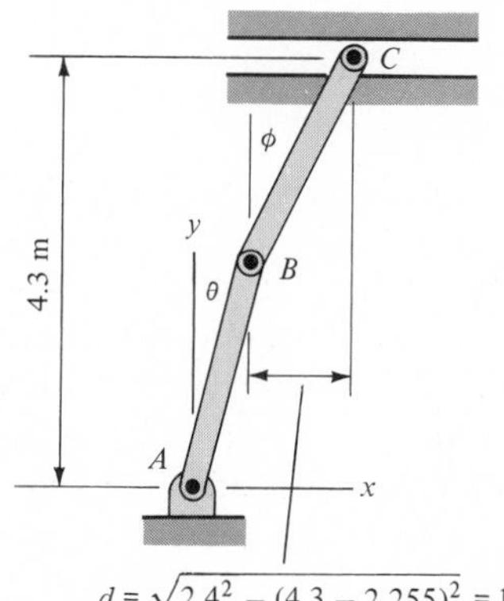

FIGURE 17-17
Analysis of a two-bar mechanism

EXAMPLE 17-4

The crank OA in Fig. 17-18 has an angular speed $\omega = 1200$ rpm. Determine the velocity of the piston B and the angular velocity of the piston rod AB at $\theta = 30°$, $\phi = 23.2°$.

SOLUTION

The given quantitites are expressed as

$$\omega_{AO} = 1200 \text{ rev/min} = 125.7 \text{ rad/s}$$

$$r_{A/O} = 5 \text{ in.} = 0.42 \text{ ft}, \qquad l = 11 \text{ in.} = 0.92 \text{ ft}$$

Using Eq. 17-13 yields

$$\mathbf{v}_B = \mathbf{v}_A + \mathbf{v}_{B/A} \qquad \text{and} \qquad \mathbf{v}_A = \mathbf{v}_O + \mathbf{v}_{A/O}$$

where $\mathbf{v}_0 = 0$, so

$$\mathbf{v}_A = \mathbf{v}_{A/O} = \boldsymbol{\omega}_{AO} \times \mathbf{r}_{A/O}$$

$$\mathbf{r}_{A/O} = 0.42(\sin\theta\mathbf{i} + \cos\theta\mathbf{j}) = 0.21\mathbf{i} + 0.362\mathbf{j}$$

$$\mathbf{v}_A = -125.7\mathbf{k} \times (0.21\mathbf{i} + 0.362\mathbf{j}) = -26.4\mathbf{j} + 45.5\mathbf{i}$$

$$\mathbf{v}_B = \mathbf{v}_A + \mathbf{v}_{B/A} = \mathbf{v}_A + \boldsymbol{\omega}_{AB} \times \mathbf{r}_{B/A}$$

where $\mathbf{r}_{B/A} = 0.846\mathbf{i} - 0.362\mathbf{j}$. Since v_B must be in the x direction, the vector equation becomes

$$\begin{aligned} v_B\mathbf{i} &= -26.4\mathbf{j} + 45.5\mathbf{i} + \omega_{AB}\mathbf{k} \times (0.846\mathbf{i} - 0.362\mathbf{j}) \\ &= 45.5\mathbf{i} - 26.4\mathbf{j} + 0.846\omega_{AB}\mathbf{j} + 0.362\omega_{AB}\mathbf{i} \end{aligned}$$

Comparing **j** and **i** components gives

$$0.846\omega_{AB} = 26.4$$

$$\omega_{AB} = 31.2 \text{ rad/s} \qquad \text{(counterclockwise)}$$

$$\begin{aligned} v_B &= 45.5 + 0.362\omega_{AB} \\ &= 56.8 \text{ ft/s} \qquad \text{(to the right)} \end{aligned}$$

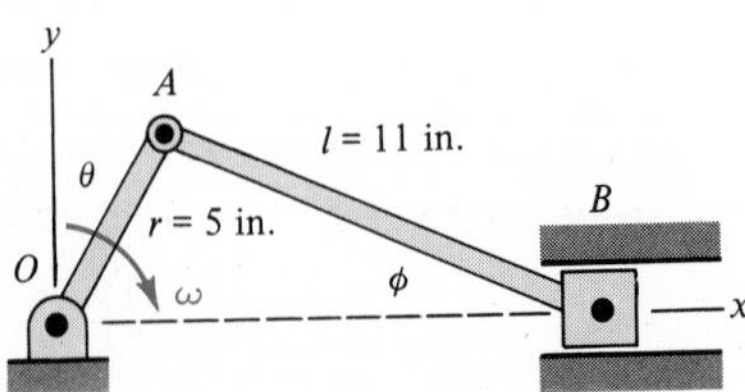

FIGURE 17-18

Analysis of a crank–piston mechanism

EXAMPLE 17-5

A wheel of 0.4 m diameter is rolling without slipping at a speed $v_A = 5$ m/s, as shown in Fig. 17-19a. Determine the speeds of points B and D using the instantaneous center of rotation.

SOLUTION

By inspection, $\mathbf{v}_B$ is perpendicular to line BC, and $\mathbf{v}_D$ is perpendicular to line DC. Since point C has zero velocity, the speeds are proportional to one another as in Fig. 17-19b. Thus,

$$v_B = \frac{r_{B/C}}{r_{A/C}} v_A = 2v_A = 10 \text{ m/s}$$

$$v_D = \frac{r_{D/C}}{r_{A/C}} v_A = \sqrt{2}v_A = 7.07 \text{ m/s}$$

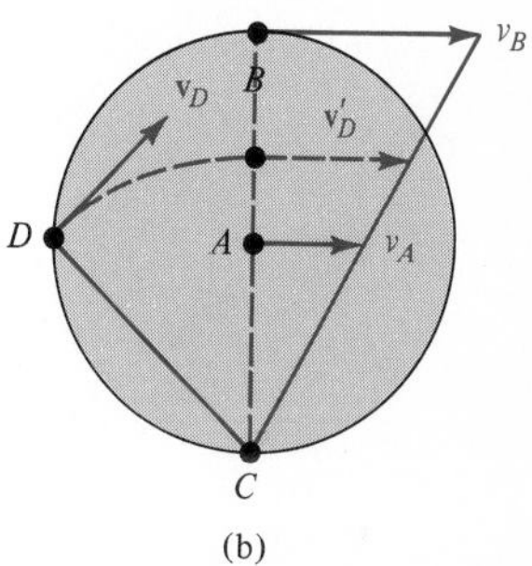

FIGURE 17-19

Analysis of a rolling wheel

EXAMPLE 17-6

Joint A of the link in Fig. 17-20a is sliding down the incline at a speed of $v_A = 2$ m/s. Determine the speed of joint B and angular velocity of the link AB for the position $\theta = 20°$. Use the instantaneous center of rotation.

FIGURE 17-20

Analysis of a constrained bar

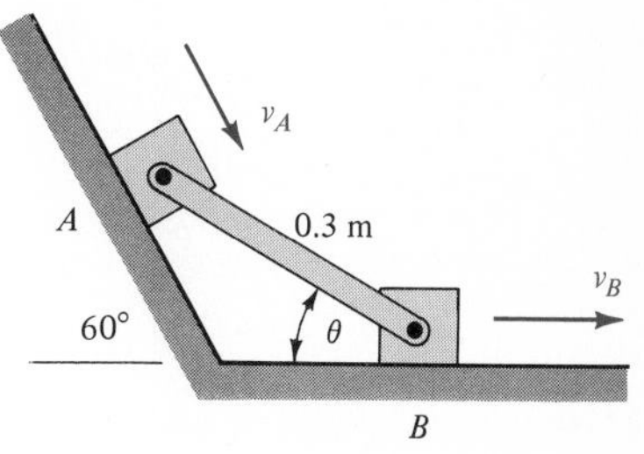

SOLUTION

The instantaneous center is determined from Fig. 17-20b using the law of sines,

$$\frac{\sin 60°}{0.3} = \frac{\sin 50°}{a} = \frac{\sin 70°}{b}$$

$$a = 0.265 \text{ m} \qquad b = 0.326 \text{ m}$$

Since ω_{AB} is the angular speed of line AB, line AC, and line BC at the given instant,

$$v_A = \omega_{AB} b \quad \text{so } \omega_{AB} = \frac{v_A}{b} = 6.13 \text{ rad/s} \quad \text{(counterclockwise)}$$

$$v_B = \omega_{AB} a = (6.13 \text{ rad/s})(0.265 \text{ m}) = 1.63 \text{ m/s}$$

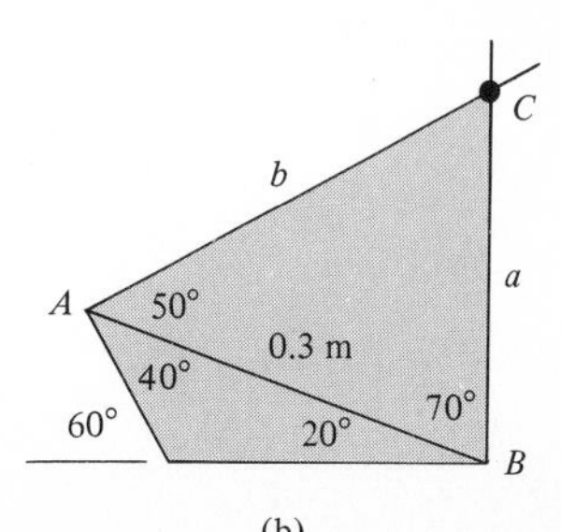

EXAMPLE 17-7

Determine the speed v_B in Ex. 17-4 using the instantaneous center of rotation.

SOLUTION

The instantaneous center C is located in Fig. 17-21b, where OA and AC are collinear, while BC is perpendicular to OB. The solution is obtained by relating v_A and v_B using the distances $r_{C/A}$ and $r_{C/B}$. From Eq. 17-4,

$$v_A = r\omega = (5 \text{ in.})(125.7 \text{ rad/s}) = 628.5 \text{ in./s} = 52.4 \text{ ft/s}$$

From Fig. 17-21b and the law of sines

$$\frac{\sin 60^\circ}{11} = \frac{\sin \phi}{5}$$

$$\phi = \sin^{-1}\left[\frac{(5)\sin 60^\circ}{11}\right] = 23.2^\circ$$

$$\alpha = 180^\circ - 60^\circ - 23.2^\circ = 96.8^\circ$$

$$\delta = 180^\circ - 96.8^\circ = 83.2^\circ$$

$$\beta = 90^\circ - \phi = 66.8^\circ$$

$$\gamma = \theta = 30^\circ$$

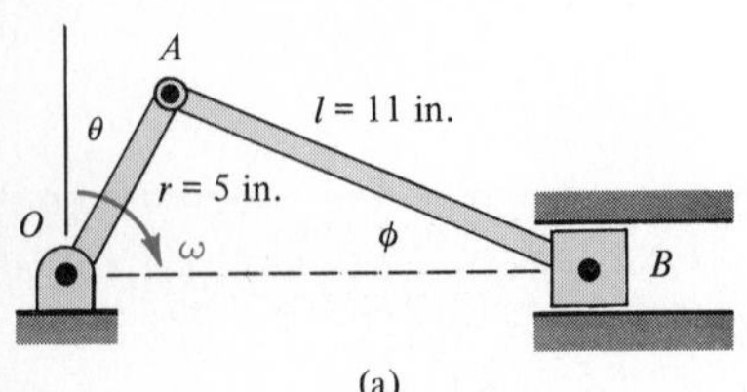

(a)

(b)

Again from the law of sines,

$$\frac{\sin 30^\circ}{11 \text{ in.}} = \frac{\sin 83.2^\circ}{r_{C/B}} = \frac{\sin 66.8^\circ}{r_{C/A}}$$

$$r_{C/B} = 22 \text{ in.} = 1.83 \text{ ft}$$

$$r_{C/A} = 20 \text{ in.} = 1.67 \text{ ft}$$

Since bar AB is rigid, its angular speed is

$$\omega_{AB} = \frac{v_A}{r_{C/A}} = \frac{v_B}{r_{C/B}}$$

$$v_B = \frac{v_A}{r_{C/A}}(r_{C/B}) = \frac{52.4(1.83)}{1.67} = 57.4 \text{ ft/s}$$

The results of Exs. 17-4 and 17-7 are slightly different because of round-off errors.

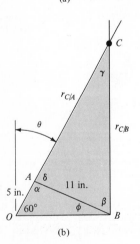

FIGURE 17-21
Analysis of a crank–piston mechanism

PROBLEMS

17–23 In the design of a motorcycle the absolute velocities of particles A and B on the tire are to be determined. Assume that $r = 40$ cm and $v_0 = 100$ km/h. The wheel rolls without slipping.

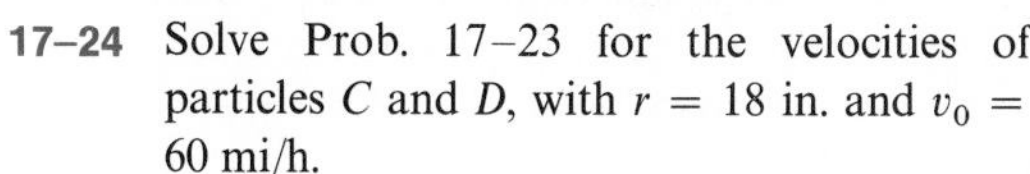

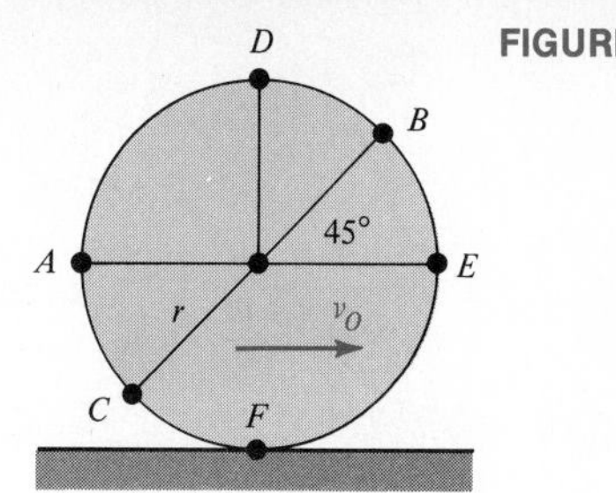

FIGURE P17-23

17–24 Solve Prob. 17–23 for the velocities of particles C and D, with $r = 18$ in. and $v_0 =$ 60 mi/h.

17–25 Solve Prob. 17–23 using the instantaneous center of rotation and sketch the velocity vectors $\mathbf{v}_A$ and $\mathbf{v}_B$ on the same diagram.

17–26 The wheel rolling to the left winds up a cable on a 20-cm-diameter drum. The diameter of the wheel is 40 cm. Calculate the absolute speed v of the cable if the wheel's angular speed is 2 rad/s.

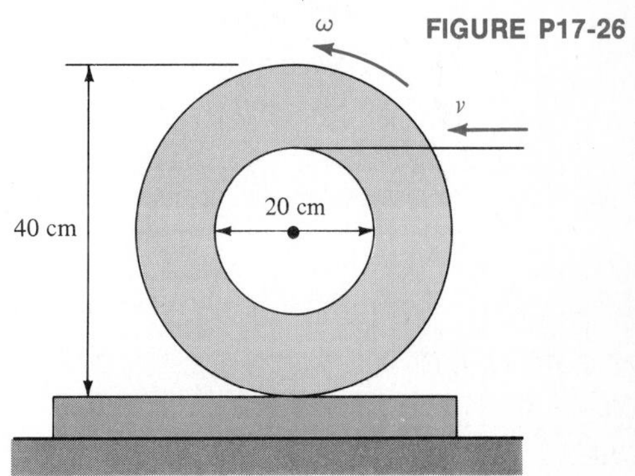

FIGURE P17-26

17–27 Solve Prob. 17–26 using the instantaneous center of rotation.

17–28 A cable drum of radius $R = 1$ ft is attached to rollers of radius $r = 3$ in. which roll on two horizontal tracks. Calculate the absolute speed of the cable if $\omega = 4$ rpm.

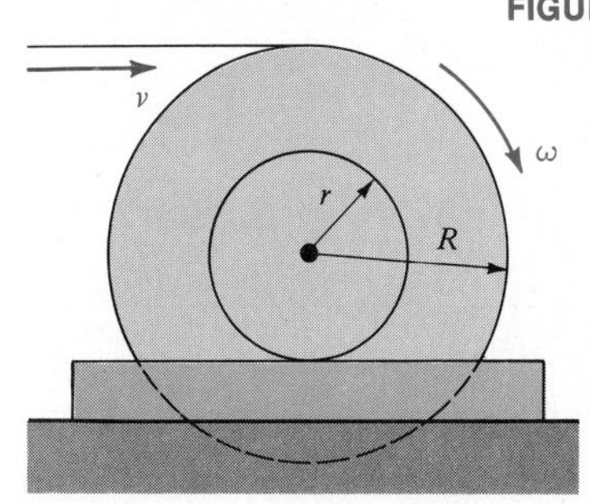

FIGURE P17-28

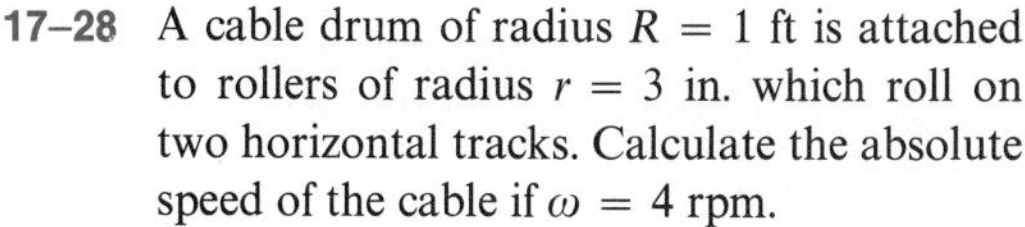

17–29 Solve Prob. 17–28 using the instantaneous center of rotation.

FIGURE P17-30

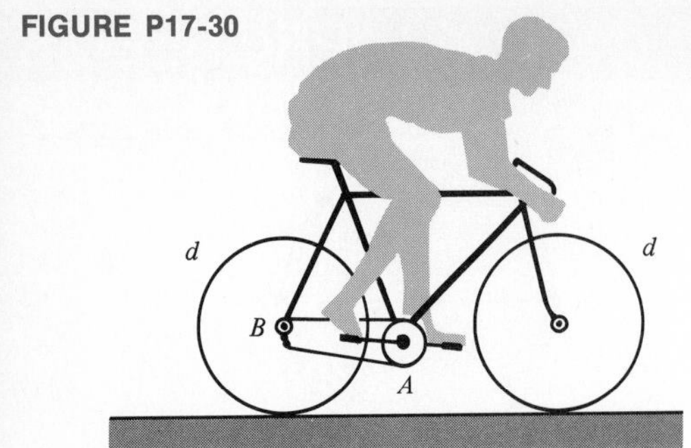

17–30 The bicycle has wheels of 70 cm diameter, sprockets A and B of 20-cm and 7-cm diameters, and a pedal circle of 30 cm diameter. Determine the required angular speed of the pedals to move the bicycle at a speed of 30 km/h if there is no slipping.

17–31 Plot the angular speed of sprocket A from Prob. 17–30 as a function of the sprocket diameter of B. Let $d_B = 5$ cm, 7 cm, and 9 cm.

FIGURE P17-32

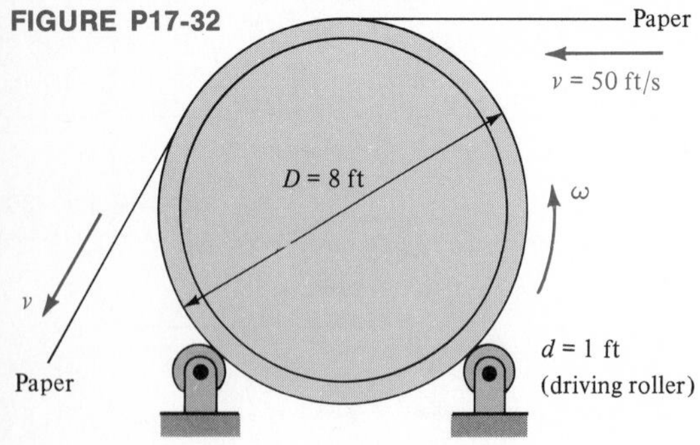

17–32 The large drum for drying paper moves the paper at a speed of 50 ft/s. Calculate the required angular speed of the small driving roller.

FIGURE P17-33

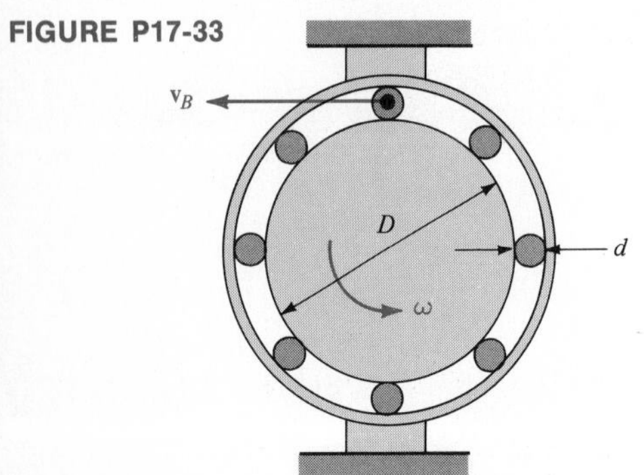

17–33 The outer case of a ball bearing is fixed. The shaft has a diameter of $D = 10$ cm, and the diameter of the balls is $d = 1$ cm. Calculate the angular speed of the shaft for which the speed v_B of the balls relative to the outer case is 10 m/s. The balls do not slip.

17–34 Determine the instantaneous center of rotation and the angular speed of a ball in the bearing described in Prob. 17–33.

FIGURE P17-35

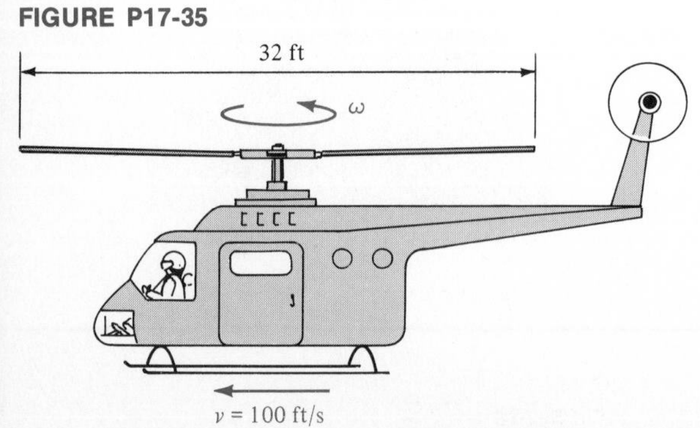

17–35 The air speed of a helicopter is $v = 100$ ft/s. Determine the maximum and minimum air speed at the tips of the large blades if they rotate at an angular speed $\omega = 160$ rad/s.

17–36 In an aerodynamic investigation of tennis balls a ball is given speeds v and ω as shown. Determine the maximum and minimum speeds of particles at the surface of a ball.

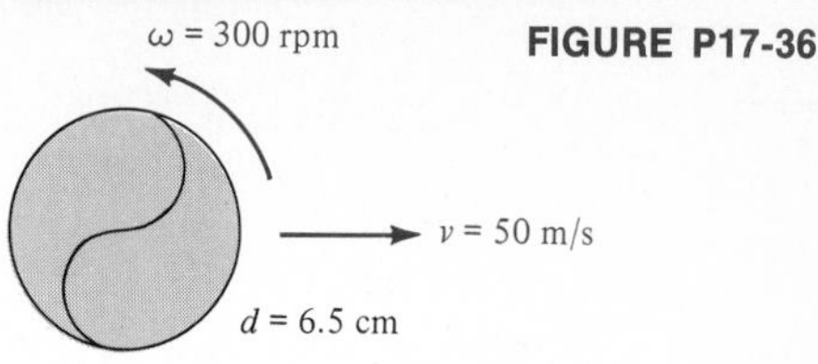

FIGURE P17-36

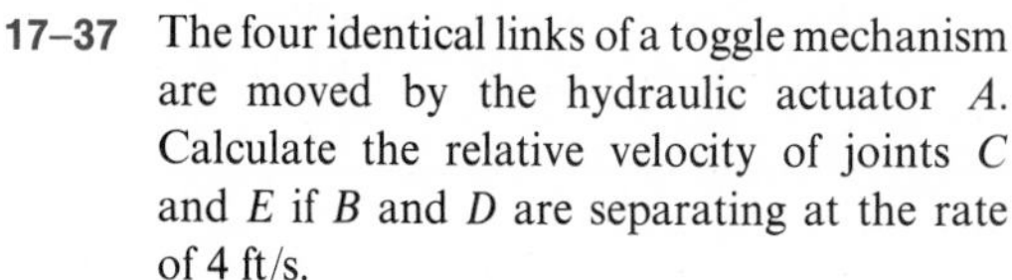

17–37 The four identical links of a toggle mechanism are moved by the hydraulic actuator A. Calculate the relative velocity of joints C and E if B and D are separating at the rate of 4 ft/s.

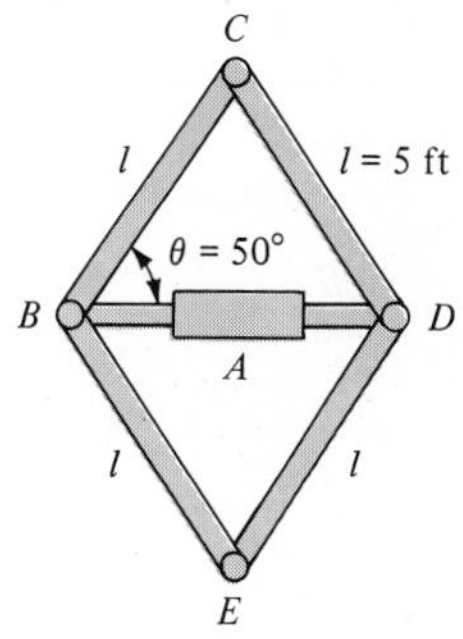

FIGURE P17-37

17–38 Determine the instantaneous center of rotation and the angular velocity of link CD in Prob. 17–37.

17–39 A horizontal platform is supported by 4-m-long bars that are pin-jointed at their ends and centers. Calculate the vertical speed of the platform if the actuator A moves wheels B and C apart at a relative velocity of $v_{B/C} = 1.4$ m/s.

17–40 Determine the instantaneous center of rotation and the angular velocity of member CD in Prob. 17–39.

17–41 Determine the velocity of joint B in the three-bar linkage if θ is zero and bar BC is horizontal.

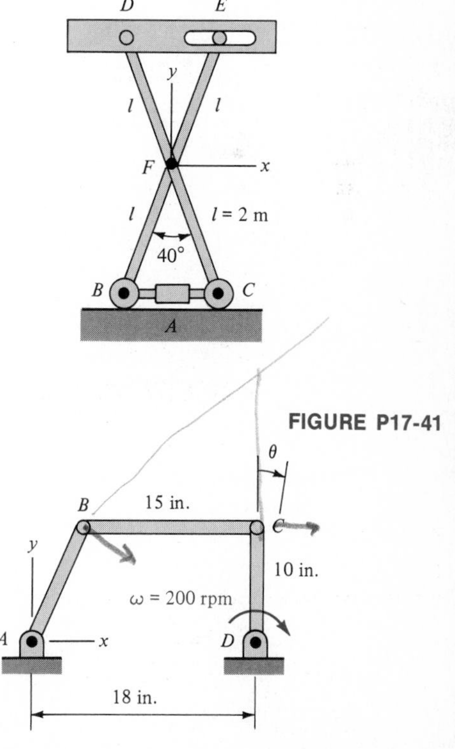

FIGURE P17-39

FIGURE P17-41

17–42 Sketch the velocity vectors of joint B in Prob. 17–41 for $\theta = 0, 15°$, and $30°$.

17–43 Solve Prob. 17–41 using the instantaneous center of rotation.

FIGURE P17-44

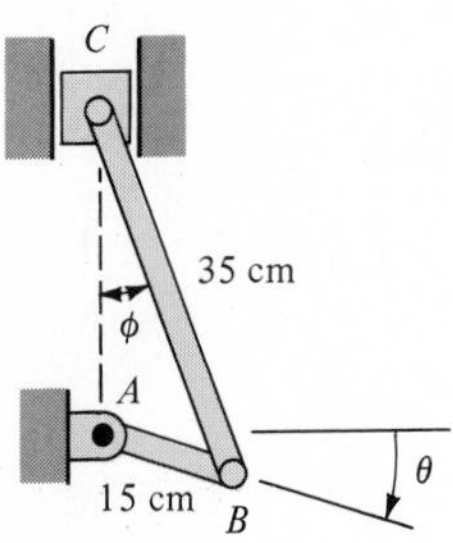

17–44 Determine the velocity of piston C for which the angular speed of the crank AB is maintained at 800 rpm at $\theta = 30°$.

17–45 Solve Prob. 17–44 using the instantaneous center of rotation.

FIGURE P17-46

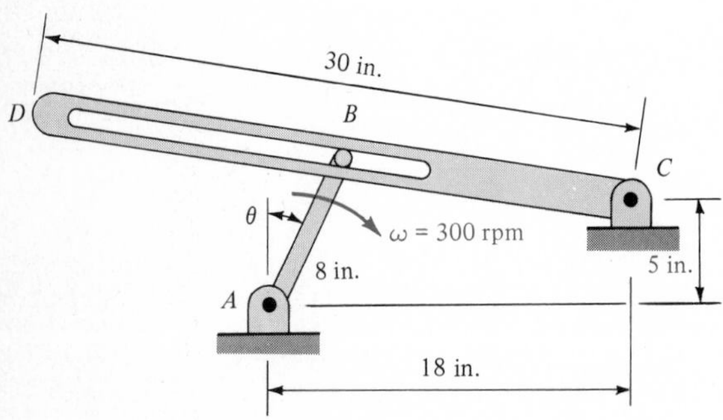

17–46 Calculate the angular velocity of the slotted arm CD for $\theta = 30°$.

17–47 Solve Prob. 17–46 using the instantaneous center of rotation.

FIGURE P17-49

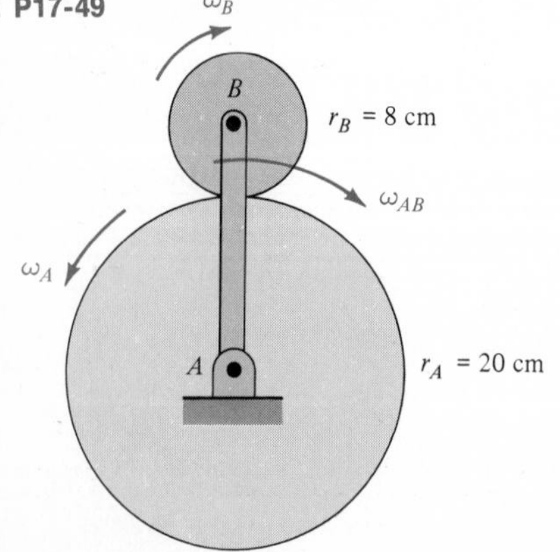

17–48 Sketch the velocity vectors of point D in Fig. P17-46 for $\theta = 0, 90°, 180°$, and $270°$. The arm CD can move past hinge A.

17–49 Wheel A rotates counterclockwise at $\omega_A =$ 200 rpm. Determine the angular velocity of wheel B for which the angular velocity of the arm AB is $\omega_{AB} = 50$ rpm clockwise.

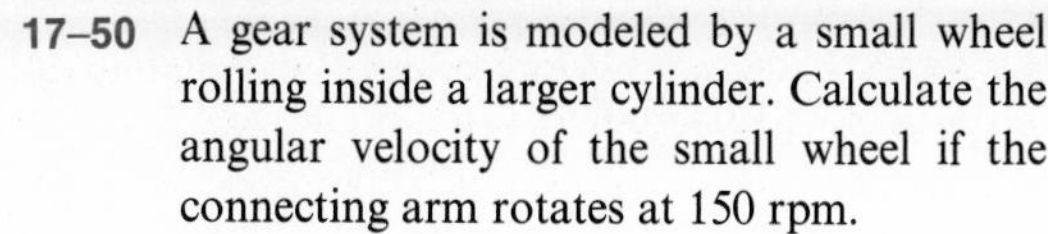
17–50 A gear system is modeled by a small wheel rolling inside a larger cylinder. Calculate the angular velocity of the small wheel if the connecting arm rotates at 150 rpm.

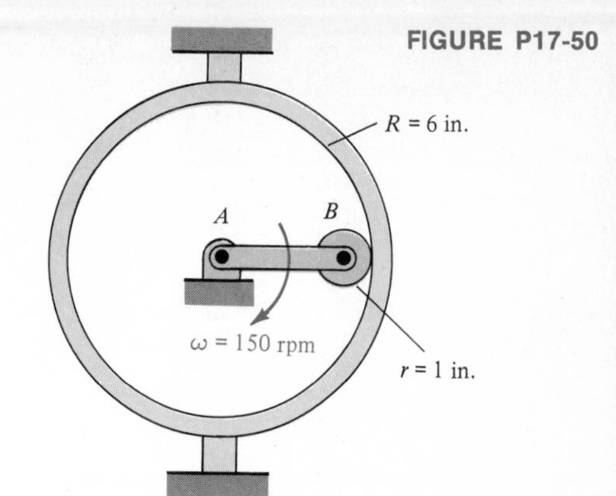

FIGURE P17-50

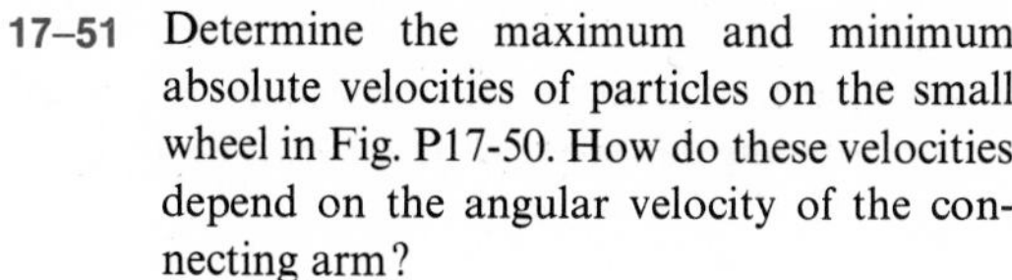
17–51 Determine the maximum and minimum absolute velocities of particles on the small wheel in Fig. P17-50. How do these velocities depend on the angular velocity of the connecting arm?

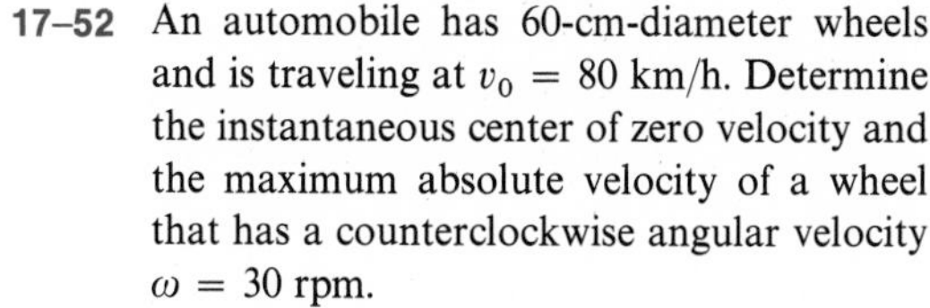
17–52 An automobile has 60-cm-diameter wheels and is traveling at $v_0 = 80$ km/h. Determine the instantaneous center of zero velocity and the maximum absolute velocity of a wheel that has a counterclockwise angular velocity $\omega = 30$ rpm.

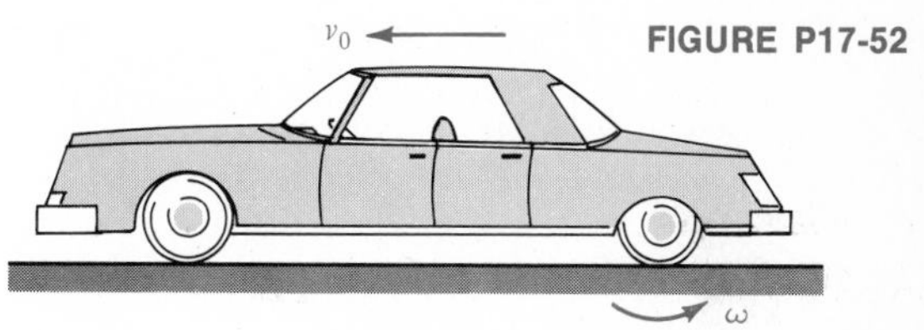

FIGURE P17-52

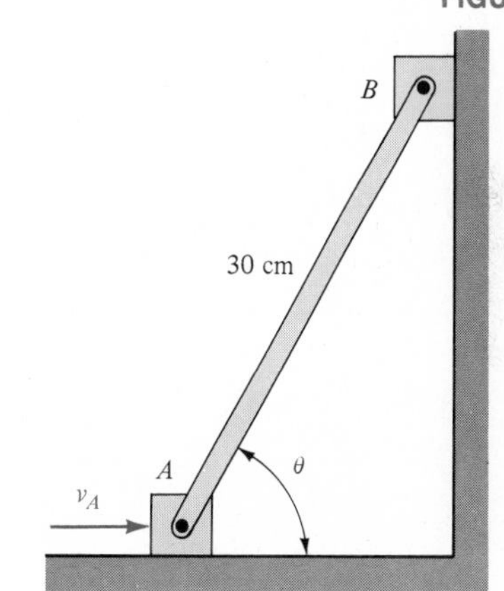

FIGURE P17-53

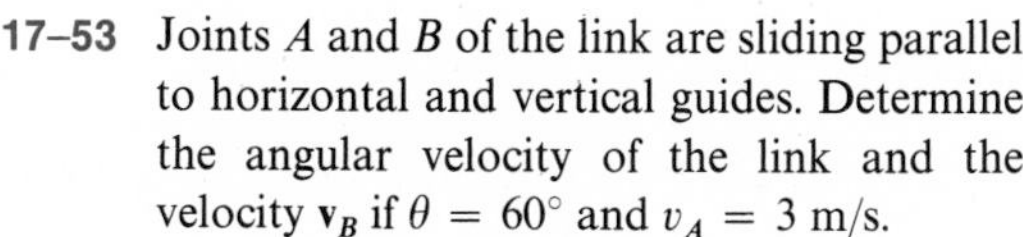
17–53 Joints A and B of the link are sliding parallel to horizontal and vertical guides. Determine the angular velocity of the link and the velocity $\mathbf{v}_B$ if $\theta = 60^\circ$ and $v_A = 3$ m/s.

17–54 A linear bearing has 0.4-in.-diameter rollers. Track A is moving to the left at $v_A = 4$ ft/s while track B is moving to the right at $v_B = 5$ ft/s. Determine the instantaneous center of rotation of a roller that moves without slipping. Also calculate the velocities of points 1 and 2. All linear velocities are with respect to the fixed xy frame.

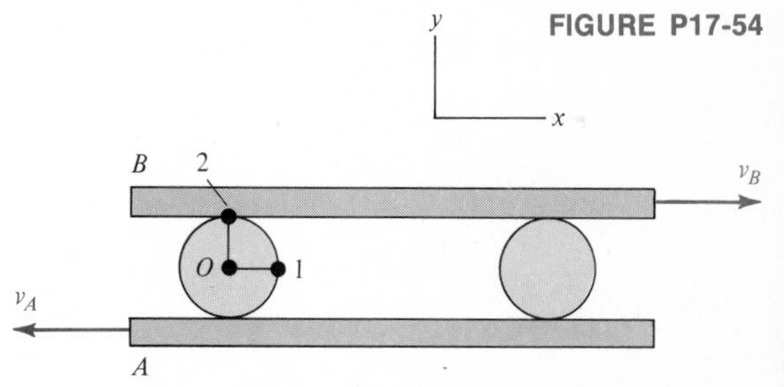

FIGURE P17-54

17-6 GENERAL PLANE MOTION. ANALYSIS OF ACCELERATIONS

The method developed for analyzing absolute and relative velocities of a rigid body in Sec. 17-4 can be extended for the analysis of accelerations. Differentiating Eq. 17-13 with respect to time results in the vector expression

$$\mathbf{a}_B = \mathbf{a}_A + \mathbf{a}_{B/A} \tag{17-17}$$

where $\mathbf{a}_B$ is the absolute acceleration of an arbitrary point B in the body, $\mathbf{a}_A$ is the absolute acceleration of a given reference point A in the body, and $\mathbf{a}_{B/A}$ is the relative acceleration of B with respect to A. This is the same expression as that for any two particles.

It is necessary to consider the relative acceleration in detail. Using Eqs. 17-14 and 17-6 gives

$$\begin{aligned} \mathbf{a}_{B/A} &= \frac{d}{dt}(\boldsymbol{\omega} \times \mathbf{r}_{B/A}) \\ &= \dot{\boldsymbol{\omega}} \times \mathbf{r}_{B/A} + \boldsymbol{\omega} \times \dot{\mathbf{r}}_{B/A} \\ &= \boldsymbol{\alpha} \times \mathbf{r}_{B/A} + \boldsymbol{\omega} \times (\boldsymbol{\omega} \times \mathbf{r}_{B/A}) \end{aligned} \tag{17-18}$$

Equation 17-17 is written in detail as

$$\mathbf{a}_B = \mathbf{a}_A + \boldsymbol{\alpha} \times \mathbf{r}_{B/A} + \boldsymbol{\omega} \times (\boldsymbol{\omega} \times \mathbf{r}_{B/A}) \tag{17-19}$$

Since terms in the form of $\boldsymbol{\alpha} \times \mathbf{r}$ and $\boldsymbol{\omega} \times (\boldsymbol{\omega} \times \mathbf{r})$ represent tangential and normal accelerations, respectively, Eq. 17-19 can be expressed as

$$\mathbf{a}_B = \underbrace{\mathbf{a}_A}_{\text{translation}} + \underbrace{(\mathbf{a}_{B/A})_t + (\mathbf{a}_{B/A})_n}_{\text{rotation}} \tag{17-20}$$

The magnitudes of the rotational components are written as

$$(a_{B/A})_t = r_{B/A}\alpha \tag{17-21a}$$

$$(a_{B/A})_n = r_{B/A}\omega^2 \tag{17-21b}$$

where the subscripts B/A are frequently omitted if the meaning of each quantity is clear. It should be noted that in these expressions $\boldsymbol{\omega}$ and $\boldsymbol{\alpha}$ are the angular velocity and acceleration of the body, and they are common to all lines in that body. On the other hand, the position vector $\mathbf{r}_{B/A}$ is valid only for the two points A and B.

Equation 17-20 can be appreciated from the schematic diagram in Fig. 17-22. Assume that a rigid body is in general plane motion and that the absolute accelerations of points A and B are known. One way to look at the details of this motion is to consider it as a simultaneous translation with acceleration $\mathbf{a}_A$ and a rotation about point A. The major complication in this analysis is that point B may have two components of acceleration as it rotates with respect to point A. The tangential component $(\mathbf{a}_{B/A})_t$ exists whenever ω is not constant. The normal component $(\mathbf{a}_{B/A})_n$ always exists when ω is other than zero, and it is always directed toward A. The vector sum of the translational and rotational components must equal the absolute acceleration of point B as in Fig. 17-22e.

It is worth noting that in general the absolute acceleration of the moving reference point (such as A in Fig. 17-22) is itself the resultant of normal and tangential components of acceleration. The motions of points A and B are still correctly related according to Eq. 17-18. In solving problems it is useful to remember the following:

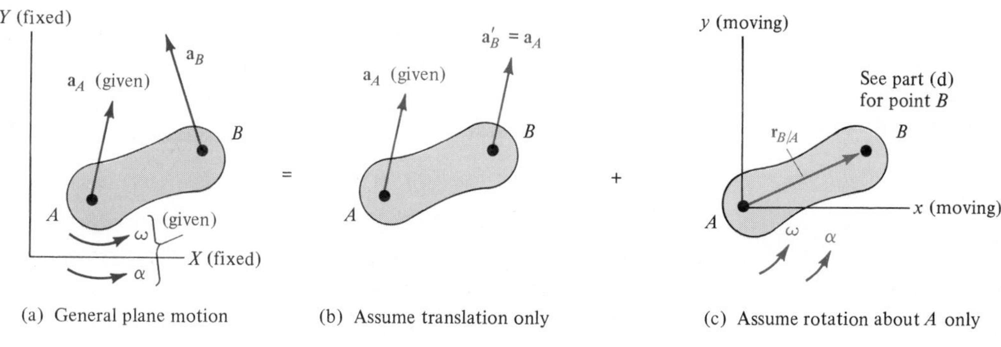

(a) General plane motion (b) Assume translation only (c) Assume rotation about A only

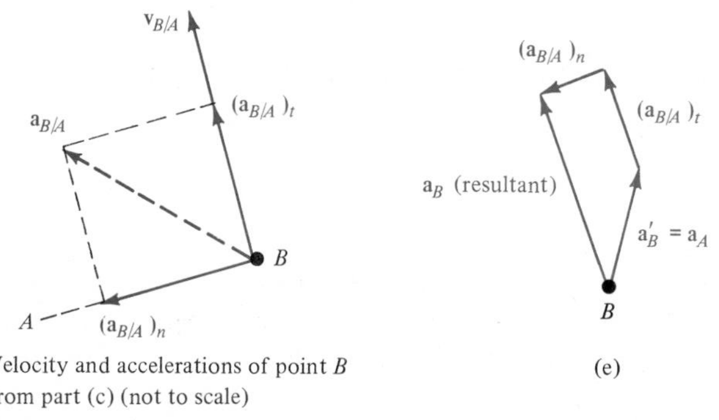

(d) Velocity and accelerations of point B from part (c) (not to scale)

(e)

FIGURE 17-22

Analysis of accelerations in general plane motion

1. The angular velocity $\boldsymbol{\omega}$ is the same for all lines, and the angular acceleration $\boldsymbol{\alpha}$ is the same for all lines in a rigid body at a given instant.

2. The common points of pin-jointed members must have the same absolute acceleration even though the individual members may have different angular velocities and angular accelerations such as in Fig. 17-23.

3. The points of contact in members that are in temporary contact generally do not have the same absolute acceleration. Even when there is no sliding between the members as in Fig. 17-24, only the tangential accelerations of the points in contact are the same, while the normal accelerations are frequently different in magnitude and direction.

4. The instantaneous center of zero velocity in general has an acceleration and should *not* be used as a reference point for accelerations unless its acceleration is known and included in the analysis.

5. Frequently, the angular acceleration of a member is not known, but some points of the member move along defined paths (e.g., the end points of a piston rod). The geometric constraints of motion can be used to define the directions of normal and tangential acceleration vectors and are useful in the solution.

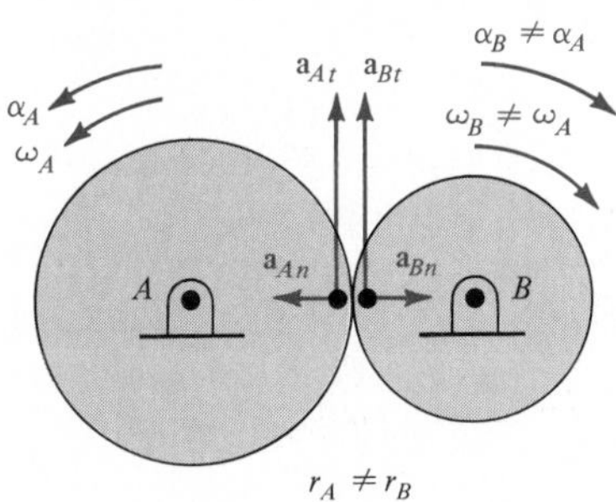

(a) No slipping at the point of contact

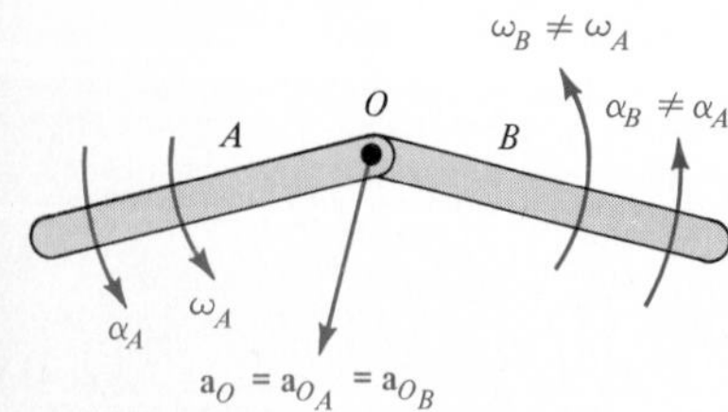

FIGURE 17-23

Acceleration of the common point of two members

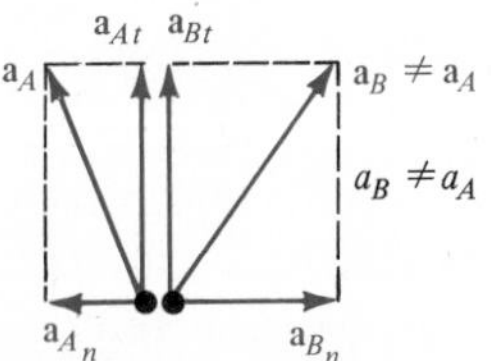

(b) Accelerations of points in contact on the two wheels

FIGURE 17-24

Accelerations of points in contact on two wheels

EXAMPLE 17-8

An automobile tire of $r = 0.3$ m is rolling without slipping. Determine the acceleration of points A and B in Fig. 17-25 if $v_O = 10$ m/s and $a_O = 4$ m/s^2.

SOLUTION

At the given instant point C is not moving, so from Eq. 17-14,

$$v_O = \omega r \qquad \omega = \frac{v_O}{r} = \frac{10 \text{ m/s}}{0.3 \text{ m}} = 33.3 \text{ rad/s}$$

$$\boldsymbol{\omega} = -33.3\mathbf{k} \text{ rad/s}$$

Since a_O is a tangential acceleration with respect to point C, from Eq. 17-21a,

$$a_O = \alpha r \qquad \alpha = \frac{a_O}{r} = \frac{4 \text{ m/s}^2}{0.3 \text{ m}} = 13.3 \text{ rad/s}^2$$

$$\boldsymbol{\alpha} = -13.3\mathbf{k} \text{ rad/s}^2$$

$$\mathbf{a}_O = a_O\mathbf{i} = 4\mathbf{i} \text{ m/s}^2$$

From Eq. 17-17,

$$\mathbf{a}_B = \mathbf{a}_O + \mathbf{a}_{B/O}$$

Using Eqs. 17-18 through 17-21b,

$$\mathbf{a}_{B/O} = (\mathbf{a}_{B/O})_n + (\mathbf{a}_{B/O})_t$$

$$(\mathbf{a}_{B/O})_n = \boldsymbol{\omega} \times (\boldsymbol{\omega} \times \mathbf{r}_{B/O}) = -33.3\mathbf{k} \times [-33.3\mathbf{k} \times (-0.3)\mathbf{i}]$$
$$= 333.3\mathbf{i} \text{ m/s}^2$$

$$(\mathbf{a}_{B/O})_t = \boldsymbol{\alpha} \times \mathbf{r}_{B/O} = -13.3\mathbf{k} \times (-0.3)\mathbf{i}$$
$$= 4\mathbf{j} \text{ m/s}^2$$

Combining these gives

$$\mathbf{a}_B = \mathbf{a}_O + \mathbf{a}_{B/O} = \mathbf{a}_O + (\mathbf{a}_{B/O})_n + (\mathbf{a}_{B/O})_t$$
$$= (4\mathbf{i} + 333.3\mathbf{i} + 4\mathbf{j}) \text{ m/s}^2$$
$$= (337.3\mathbf{i} + 4\mathbf{j}) \text{ m/s}^2$$

Solving for point A similarly yields

$$\mathbf{a}_A = \mathbf{a}_O + \mathbf{a}_{A/O} = \mathbf{a}_O + (\mathbf{a}_{A/O})_n + (\mathbf{a}_{A/O})_t$$
$$= 4\mathbf{i} + \boldsymbol{\omega} \times (\boldsymbol{\omega} \times \mathbf{r}_{A/O}) + \boldsymbol{\alpha} \times \mathbf{r}_{A/O}$$
$$= 4\mathbf{i} + (-33.3\mathbf{k}) \times (-33.3\mathbf{k} \times 0.3\mathbf{j}) + (-13.3\mathbf{k} \times 0.3\mathbf{j})$$
$$= (8\mathbf{i} - 333.3\mathbf{j}) \text{ m/s}^2$$

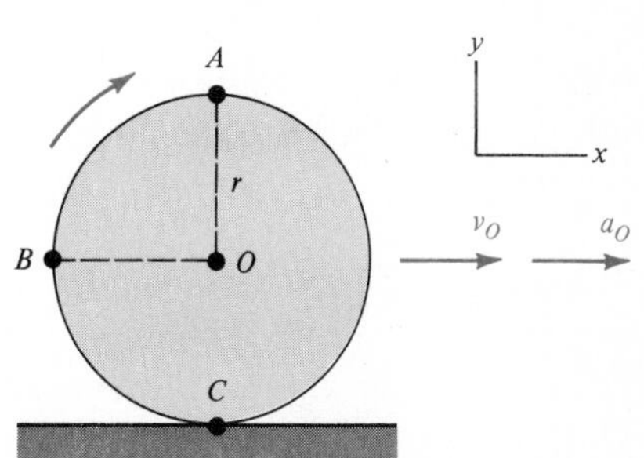

FIGURE 17-25

Analysis of accelerations of a rolling wheel

EXAMPLE 17-9

The crank OA in Fig. 17-26 is rotating at a constant angular speed of ω = 1500 rpm. Determine the acceleration of the piston B and the angular acceleration of the piston rod AB for the position when $\theta = 30°$.

SOLUTION

The angular velocity vector is

$$\omega = -157.1\mathbf{k} \text{ rad/s}$$

From Eq. 17-17,

$$\mathbf{a}_B = \mathbf{a}_A + \mathbf{a}_{B/A}$$

$$\mathbf{a}_A = \omega \times (\omega \times \mathbf{r})$$

$$= -157.1\mathbf{k} \times \left[-157.1\mathbf{k} \times \frac{4}{12}(\sin 30° \, \mathbf{i} + \cos 30° \, \mathbf{j})\right]$$

$$= (-4113\mathbf{i} - 7124\mathbf{j}) \text{ ft/s}^2$$

Using Eq. 17-18 through 17-21b gives

$$\mathbf{a}_{B/A} = \boldsymbol{\alpha}_{AB} \times \mathbf{r}_{B/A} + \omega_{AB} \times (\omega_{AB} \times \mathbf{r}_{B/A}) \quad \textbf{(a)}$$

Two equations can be distinguished here, the **i** direction equation and the **j** direction equation, but there are three unknowns ($\mathbf{a}_{B/A}$, $\boldsymbol{\alpha}_{AB}$, ω_{AB}), so it is necessary to solve for one unknown in another fashion. One choice is to solve for ω_{AB} using the instantaneous center of rotation of rod AB. Using Fig. 17-26b yields

$$\phi = \sin^{-1}\left[\frac{4(\sin 60°)}{12}\right] = 16.8°$$

$$\alpha = 180° - 60° - 16.8° = 103.2°$$

$$\delta = 180° - \alpha = 76.8°$$

$$\beta = 90° - \phi = 73.2°$$

$$\gamma = \theta = 30°$$

$$\frac{\sin 30°}{12} = \frac{\sin 73.2°}{r_{C/A}}$$

$$r_{C/A} = 23 \text{ in.} = 1.9 \text{ ft}$$

$$\omega_{AB} = \frac{v_A}{1.9} = \frac{\omega r_{A/O}}{1.9} = 27.56 \text{ rad/s}$$

$$\omega_{AB} = 27.56\mathbf{k} \text{ rad/s}$$

Now Eq. (a) may be solved knowing that $\mathbf{a}_B$ must have only an **i** component and that $\boldsymbol{\alpha}_{AB}$ must have only a **k** component.

$$
\begin{aligned}
a_B\mathbf{i} = &-4113\mathbf{i} - 7124\mathbf{j} + 27.56\mathbf{k} \\
&\times [27.56\mathbf{k} \times (\cos 16.8^\circ\,\mathbf{i} - \sin 16.8^\circ\,\mathbf{j})] \\
&+ \alpha_{AB}\mathbf{k} \times (\cos 16.8^\circ\,\mathbf{i} - \sin 16.8^\circ\mathbf{j}) \\
= &(-4113 - 727 + 0.289\alpha_{AB})\mathbf{i} \\
&+ (-7124 + 219.5 + 0.957\alpha_{AB})\mathbf{j}
\end{aligned}
$$

Using the **j** terms to solve for α_{AB} yields

$$
\begin{aligned}
0.957\alpha_{AB} &= 7124 - 219.5 \\
\alpha_{AB} &= 7215 \text{ rad/s}^2
\end{aligned}
$$

Using α_{AB} and the **i** terms gives

$$
\begin{aligned}
a_B &= -4113 - 723 + (7215)(0.289) \\
&= -2751 \text{ ft/s}^2
\end{aligned}
$$

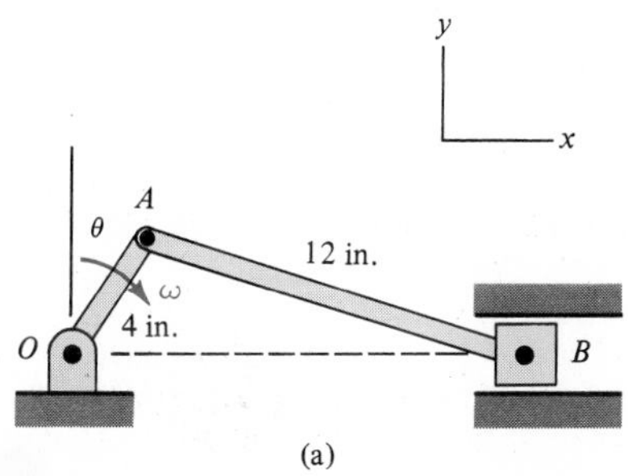

(a)

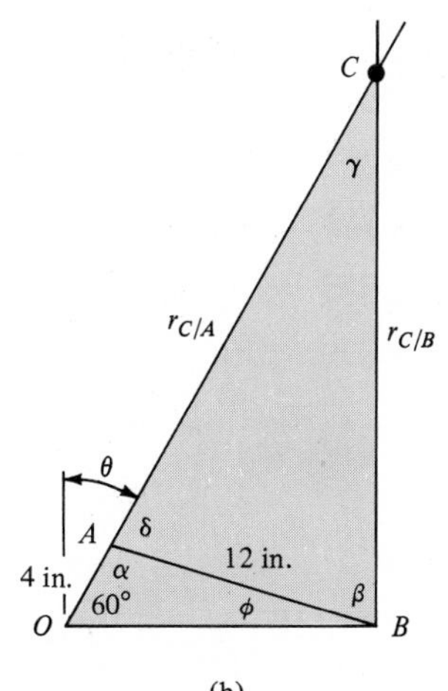

(b)

FIGURE 17-26

Analysis of a crank–piston mechanism

EXAMPLE 17-10

The mechanical hoe in Fig. 17-27 has the following dimensions: $OA = 2.3$ m, $AB = 1.7$ m. Assume that $\theta = 25°$, $\omega_{OA} = 0.5$ rad/s counterclockwise, $\alpha_{OA} = 0$, $\phi = 30°$, $\omega_{AB} = 0.8$ rad/s clockwise, and $\alpha_{AB} = 2$ rad/s^2 counterclockwise. Determine the acceleration of the bucket joint B for the given configuration.

SOLUTION

From Eq. 17-17,

$$\mathbf{a}_B = \mathbf{a}_A + \mathbf{a}_{B/A}$$

$$\begin{aligned}\mathbf{a}_A &= \boldsymbol{\omega}_{OA} \times (\boldsymbol{\omega}_{OA} \times \mathbf{r}_{A/O}) + \boldsymbol{\alpha}_{OA} \times \mathbf{r}_{A/O}\\ &= 0.5\mathbf{k} \times [0.5\mathbf{k} \times 2.3(\sin 25°\,\mathbf{i} + \cos 25°\,\mathbf{j})]\\ &= (-0.24\mathbf{i} - 0.52\mathbf{j})\ \text{m/s}^2\end{aligned}$$

$$\begin{aligned}\mathbf{a}_{B/A} &= \boldsymbol{\omega}_{AB} \times (\boldsymbol{\omega}_{AB} \times \mathbf{r}_{B/A}) + \boldsymbol{\alpha}_{AB} \times \mathbf{r}_{B/A}\\ &= -0.8\mathbf{k} \times [(-0.8\mathbf{k}) \times 1.7(\sin 30°\,\mathbf{i} - \cos 30°\,\mathbf{j})]\\ &\quad + 2\mathbf{k} \times 1.7(\sin 30°\,\mathbf{i} - \cos 30°\,\mathbf{j})\\ &= (-0.54\mathbf{i} + 0.94\mathbf{j} + 2.94\mathbf{i} + 1.7\mathbf{j})\ \text{m/s}^2\\ &= (2.4\mathbf{i} + 2.6\mathbf{j})\ \text{m/s}^2\end{aligned}$$

Combining the results for $\mathbf{a}_A$ and $\mathbf{a}_{B/A}$ gives

$$\begin{aligned}\mathbf{a}_B &= -0.24\mathbf{i} - 0.52\mathbf{j} + 2.4\mathbf{i} + 2.6\mathbf{j}\\ &= (2.16\mathbf{i} + 2.08\mathbf{j})\ \text{m/s}^2\end{aligned}$$

(a)

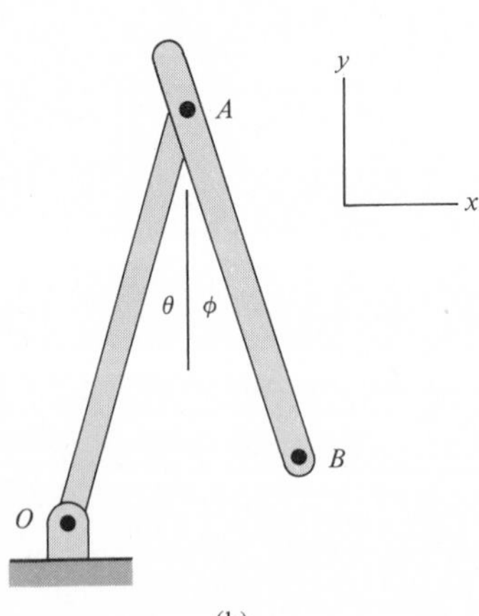

(b)

FIGURE 17-27
Analysis of a mechanical hoe

17–55 A wheel of radius $r = 50$ cm is rolling without slipping. Determine the accelerations of points A and B for $\mathbf{v}_O = 20\mathbf{i}$ m/s and $\mathbf{a}_O = 3\mathbf{i}$ m/s^2.

17–56 A wheel of radius $r = 2$ ft is rolling without slipping. Determine the accelerations of points C and D in Fig. P17-55 for $\mathbf{v}_O = 100\mathbf{i}$ ft/s and $\mathbf{a}_O = -20\mathbf{i}$ ft/s^2.

17–57 Sketch on a single diagram of a wheel the acceleration vectors of points A, B, C, and D in Fig. P17-55. Evaluate these using the values of r, $\mathbf{v}_O$, and $\mathbf{a}_O$ given in Prob. 17–55.

FIGURE P17-55

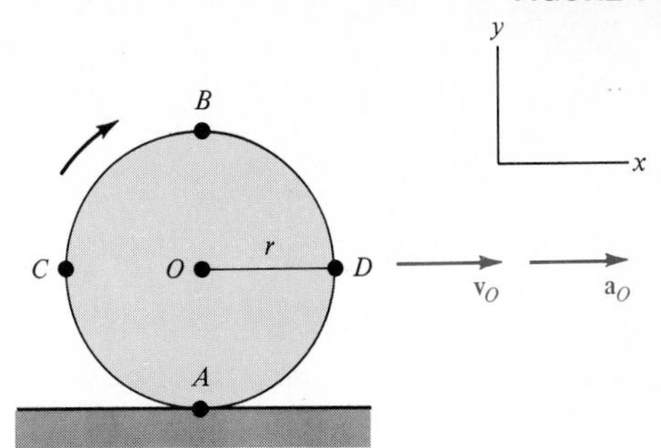

17–58 Block A is sliding down at a speed of $v_A = 2$ m/s and an acceleration of $a_A = 6$ m/s^2. Determine the acceleration of block B for $\theta = 40°$, $l = 40$ cm.

17–59 Block A in Fig. P17-58 has an upward velocity of 7 ft/s and a downward acceleration of 15 ft/s^2 when $\theta = 30°$. Calculate the acceleration of block B. $l = 18$ in.

FIGURE P17-58

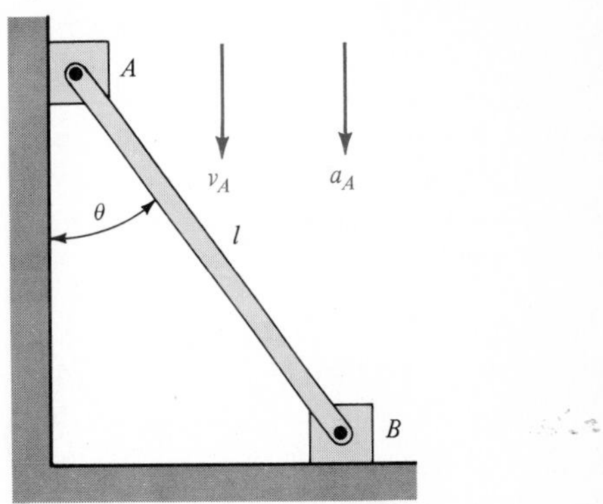

17–60 The tracks of the linear bearing move at $v_A = 1$ m/s and $v_B = 1.5$ m/s. Determine the accelerations of points 1 and 2 on the 2-cm-diameter roller.

FIGURE P17-60

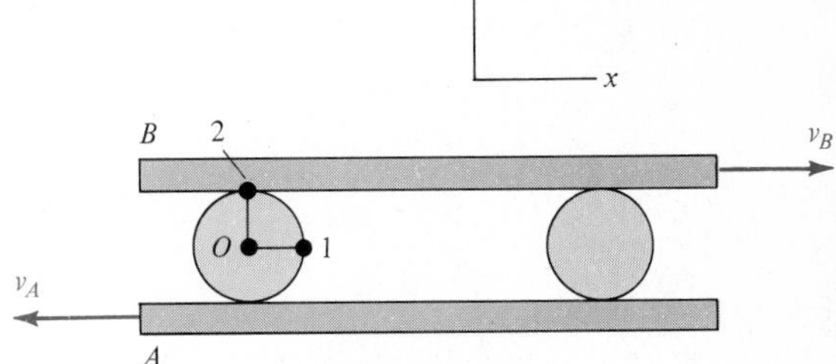

17–61 The outer case of the roller bearing is fixed. The shaft of diameter $D = 2$ in. is rotating with an angular speed of 5000 rpm. Determine the accelerations of both contacting points (at the shaft and at the outer case) of a 0.2-in.-diameter roller.

FIGURE P17-61

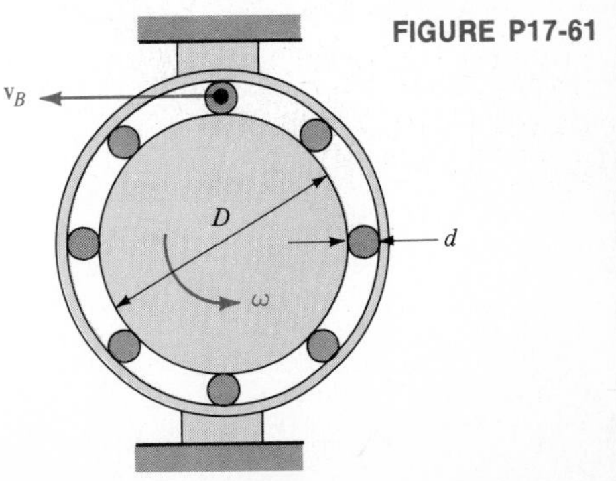

FIGURE P17-62

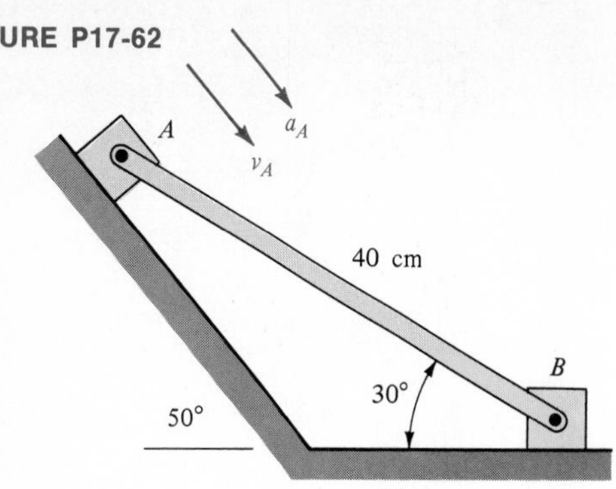

17–62 Determine the acceleration of block B for $v_A = 2$ m/s and $a_A = 50$ m/s^2.

17–63 Determine the vertical acceleration of joint C in Prob. 17–37.

FIGURE P17-64

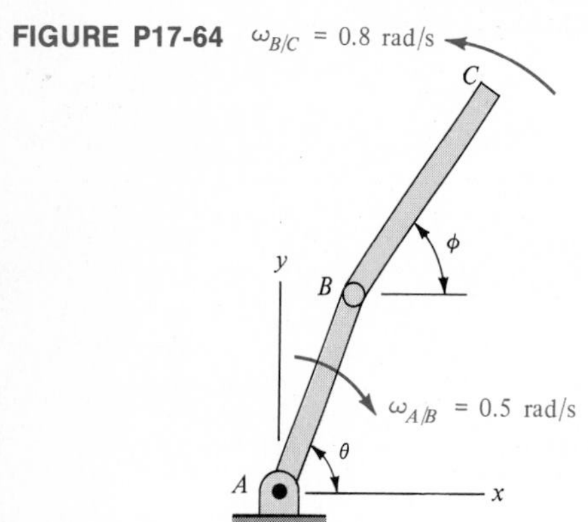

17–64 The rigid bars are pinned at A and B. The lengths of the bars are $AB = 1$ m and $BC = 1.2$ m. Determine the acceleration of point C for the given constant angular velocities if $\theta = 80°$ and $\phi = 60°$.

17–65 Solve Prob. 17–64 after adding the angular accelerations $\boldsymbol{\alpha}_{A/B} = 0.2\mathbf{k}$ rad/s^2 and $\boldsymbol{\alpha}_{B/C} = -0.3\mathbf{k}$ rad/s^2.

FIGURE P17-66

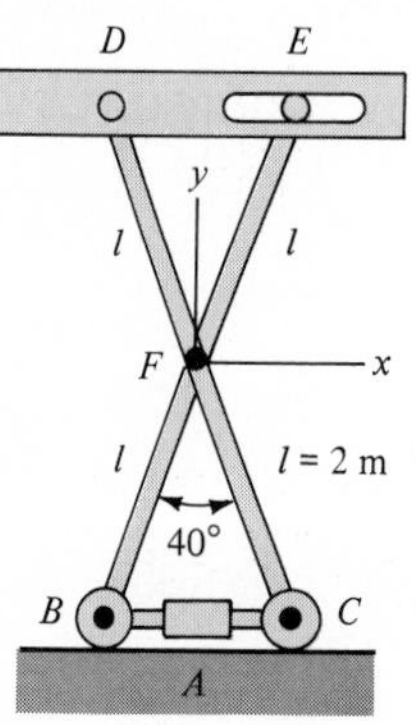

17–66 Determine the vertical acceleration of the horizontal platform in Prob. 17–39.

FIGURE P17-67

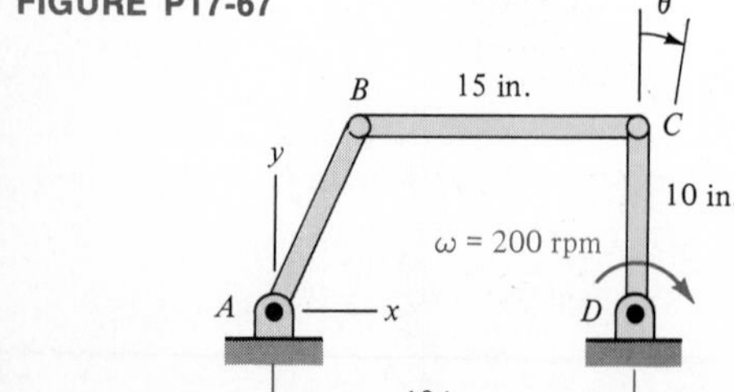

17–67 Determine the acceleration of joint B in Prob. 17–41 for $\theta = 30°$.

17–68 Sketch the acceleration vectors of point B in Fig. P17-67 for $\theta = 0$, 15°, and 30°.

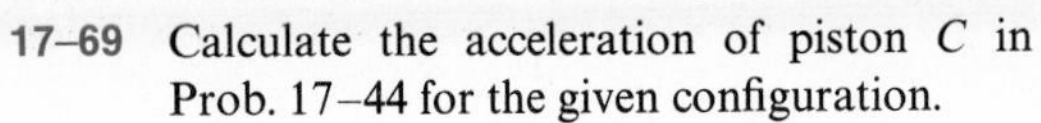

17–69 Calculate the acceleration of piston C in Prob. 17–44 for the given configuration.

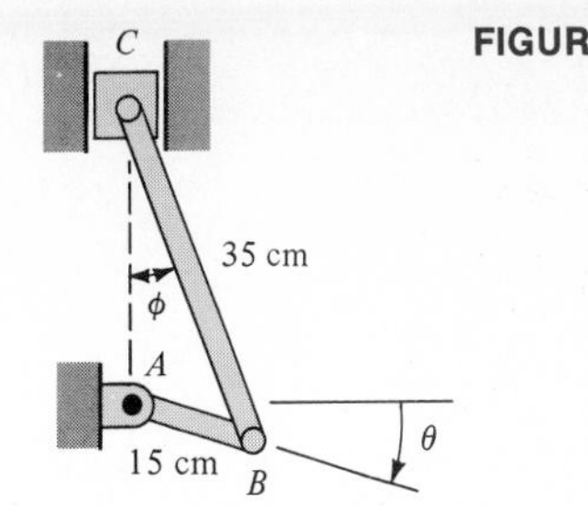

FIGURE P17-69

17–70 Calculate the acceleration of point D of the slotted arm for $\theta = 40°$.

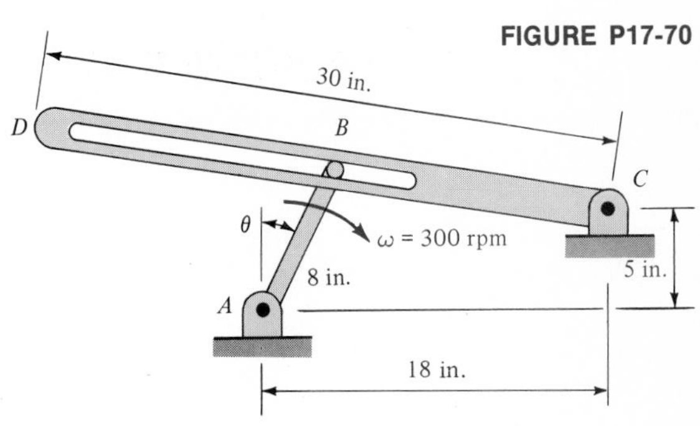

FIGURE P17-70

17–71 Determine the angle θ in Fig. P17-70 for which the acceleration of D is maximum.

17–72 Determine the acceleration of a particle on wheel B farthest from A if $\omega_A = 50$ rpm and $\omega_{AB} = 100$ rpm.

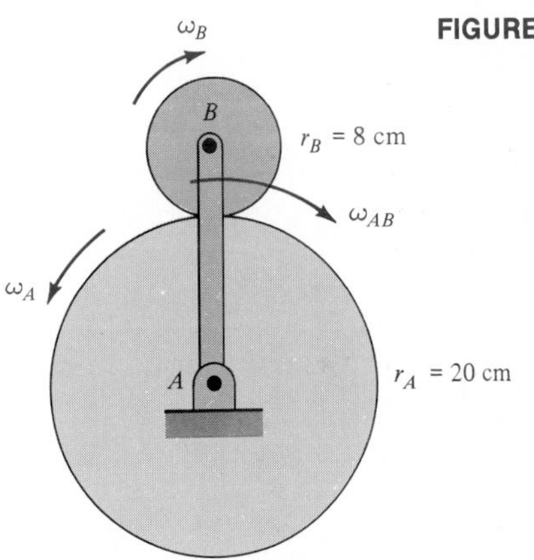

FIGURE P17-72

17–73 Calculate the acceleration of a particle closest to A on the small wheel which rolls without slipping in the cylinder.

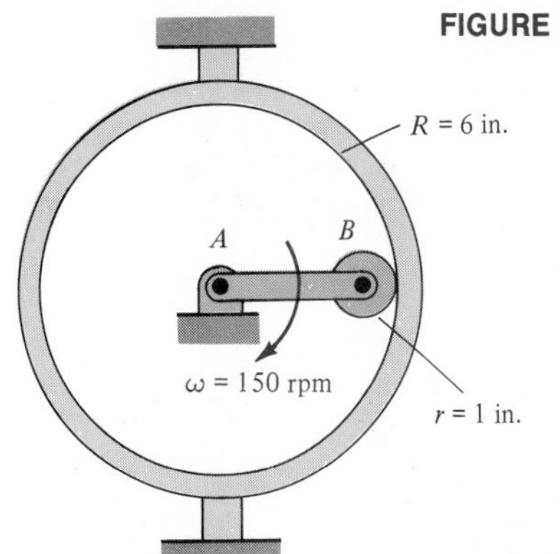

FIGURE P17-73

17-7 GENERAL MOTION

The three-dimensional motion of a rigid body is analyzed using the methods developed for the analysis of two-dimensional motion in Secs. 17-4 and 17-6. The difference between plane motion and space motion is in the relationship of the angular velocity and angular acceleration vectors, as illustrated in Fig. 17-28. Plane motion implies that, for any orientation of a position vector $\mathbf{r}_A$, the angular velocity vector $\boldsymbol{\omega}$ and the angular acceleration vector $\boldsymbol{\alpha}$ are collinear (Fig. 17-28a). In other words, $\boldsymbol{\omega}$ may change in magnitude but its direction is fixed in space, and all particles of the body move in fixed planes that are parallel to one another. In space motion both the magnitude and the direction of the angular velocity $\boldsymbol{\omega}$ may change, which means that the particles of the body do not move in fixed parallel planes, and $\boldsymbol{\omega}$ and $\boldsymbol{\alpha}$ are not collinear (Fig. 17-28b).

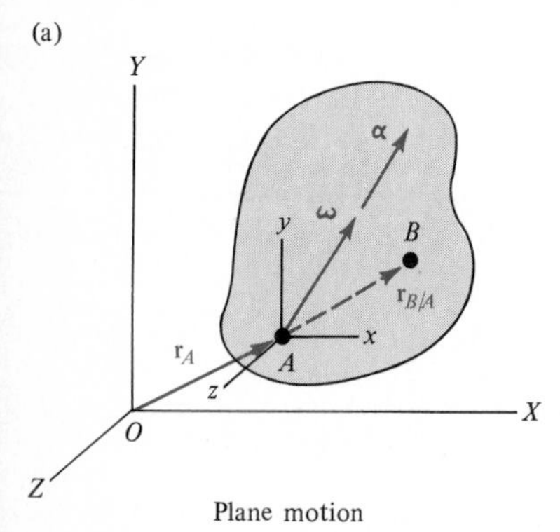

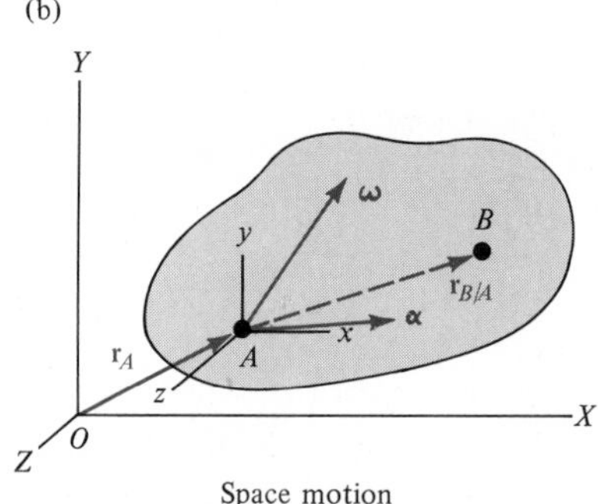

FIGURE 17-28
Schematic for general motion of a rigid body

For the analysis of the velocity and acceleration of an arbitrary point B in Fig. 17-28b, assume that the motion of point A and the vectors $\boldsymbol{\omega}$ and $\boldsymbol{\alpha}$ are known at a particular instant. The xyz frame is translating with respect to the XYZ frame. As in Eq. 17-15,

$$\mathbf{v}_B = \mathbf{v}_A + \boldsymbol{\omega} \times \mathbf{r}_{B/A} \qquad \text{17-22}$$

since $\mathbf{r}_{B/A}$ has a constant magnitude in a rigid body. From Eq. 17-19,

$$\mathbf{a}_B = \mathbf{a}_A + \boldsymbol{\alpha} \times \mathbf{r}_{B/A} + \boldsymbol{\omega} \times (\boldsymbol{\omega} \times \mathbf{r}_{B/A}) \qquad \text{17-23}$$

Equations 17-22 and 17-23 are valid for plane motion and for space motion since they are restatements of results from Secs. 17-4 and 17-6. The terms involving vector products are more complex to evaluate in the case of space motion than for plane motion. In both cases, the angular velocity and angular acceleration of a rigid body may vary with time, but they do not depend on what moving reference point is chosen.

Motion about a fixed point. A special case of three-dimensional motion is when a rigid body has a fixed point. This is in contrast to motion about a fixed axis, which is always two-dimensional motion. Assuming that the fixed point is A in Fig. 17-28b, $\mathbf{v}_A = 0$ in Eq. 17-22, and $\mathbf{a}_A = 0$ in Eq. 17-23.

Angular Acceleration

By definition, angular acceleration is the time derivative of angular velocity of a rigid body,

$$\boldsymbol{\alpha} = \frac{d\boldsymbol{\omega}}{dt} \tag{17-24}$$

This equation is valid in all situations, but it is useful to distinguish the following applications.

1. The direction of $\boldsymbol{\omega}$ is constant. This is plane motion and $\alpha = \dot{\omega}$ can be used in scalar solutions of problems.
2. Both the magnitude and direction of $\boldsymbol{\omega}$ change. This is space motion since all or some particles of the rigid body have three-dimensional paths. An example is the wheel of a car which accelerates on a curved path. Equation 17-24 must be used in this case.
3. The magnitude of $\boldsymbol{\omega}$ is constant but its direction changes. This is space motion since all or some particles of the rigid body have three-dimensional paths. An example is the wheel of a car which travels at a constant speed on a curved path. A useful expression for the angular acceleration in this case can be obtained as follows.

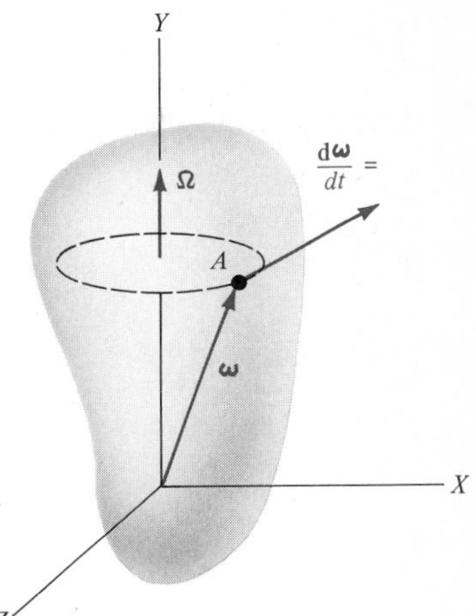

FIGURE 17-29

Schematic for the definition of angular acceleration when the magnitude ω of the angular velocity is constant

Assume that a rigid body is fixed at point O in Fig. 17-29, and the body has an angular velocity $\boldsymbol{\omega}$ of constant magnitude. Also assume that the tip A of the vector $\boldsymbol{\omega}$ moves on a circular path centered on the Y axis, and the motion of $\boldsymbol{\omega}$ about the Y axis is defined by the angular velocity vector $\boldsymbol{\Omega}$. Thus, $d\boldsymbol{\omega}/dt$ is tangent to the path of A. As for any vector from point O to A,

$$\frac{d\boldsymbol{\omega}}{dt} = \boldsymbol{\alpha} = \boldsymbol{\Omega} \times \boldsymbol{\omega} \tag{17-25}$$

This result is analogous in vector operation to the expression for the linear velocity of a particle: assume that $\mathbf{r}$ is the position vector from point O to A in Fig. 17-29; $\mathbf{v}_A = \dot{\mathbf{r}}_A = \boldsymbol{\Omega} \times \mathbf{r}$ from Eq. 17-4.

EXAMPLE 17-11

An automobile is traveling at a velocity $\mathbf{v} = 100\mathbf{i}$ km/h on a horizontal curved road of radius $R = 500$ m. Determine the velocity and acceleration of a particle B on the 0.6 m diameter wheel (Fig. 17-30a). Assume that the car is turning left as it travels on the curved road at a constant speed.

SOLUTION

Use the center of curvature O of the road for the fixed frame XYZ, and the axle A of the wheel for the translating frame xyz (Fig. 17-30b). Thus,

$$\mathbf{r}_{A/O} = (0.3\mathbf{j} + 500\mathbf{k})\ \text{m}$$

$$\mathbf{r}_{B/A} = 0.3\mathbf{i}\ \text{m}, \qquad \mathbf{r}_{A/C} = 0.3\mathbf{j}\ \text{m}$$

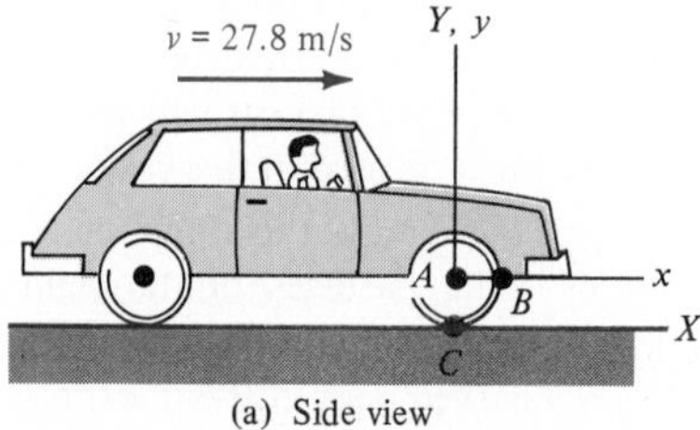

(a) Side view

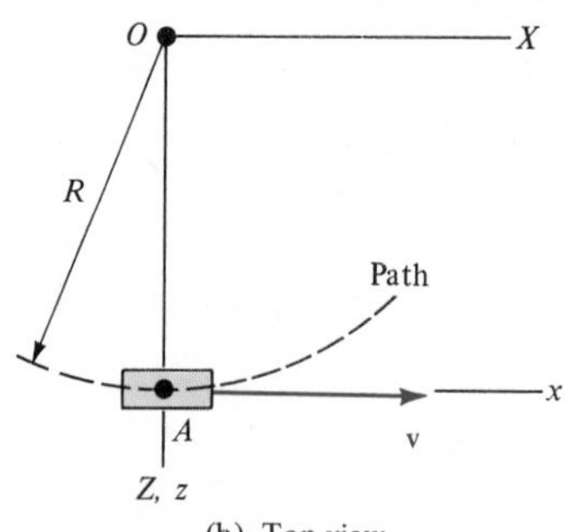

(b) Top view

FIGURE 17-30

Analysis of a particle's motion on a rolling wheel

From Eq. 17-22,

$$\mathbf{v}_B = \mathbf{v}_A + \boldsymbol{\omega} \times \mathbf{r}_{B/A} = \mathbf{v}_A + (\boldsymbol{\omega}_A + \boldsymbol{\omega}_O) \times \mathbf{r}_{B/A}$$

where $\mathbf{v}_A = 27.8\mathbf{i}$ m/s by inspection, $\boldsymbol{\omega}_A$ is the angular velocity of the wheel rotating about its own axis and $\boldsymbol{\omega}_O$ is the angular velocity of the wheel about point O. $\boldsymbol{\omega}_A$ and $\boldsymbol{\omega}_O$ are determined first.

$$\begin{aligned} \mathbf{v}_A &= \boldsymbol{\omega}_A \times \mathbf{r}_{A/C} & \mathbf{v}_A &= \boldsymbol{\omega}_O \times \mathbf{r}_{A/O} \\ 27.8\mathbf{i} &= \omega_A\mathbf{k} \times 0.3\mathbf{j} & 27.8\mathbf{i} &= \omega_O\mathbf{j} \times 500\mathbf{k} \\ 27.8 &= -\omega_A(0.3) & 27.8 &= 500\,\omega_O \\ \boldsymbol{\omega}_A &= -92.6\mathbf{k}\text{ rad/s} & \boldsymbol{\omega}_O &= 0.056\mathbf{j}\text{ rad/s} \end{aligned}$$

Thus,

$$\boldsymbol{\omega} \times \mathbf{r}_{B/A} = (0.056\mathbf{j} - 92.6\mathbf{k}) \times 0.3\mathbf{i} = (-27.8\mathbf{j} - 0.0168\mathbf{k})\text{ m/s}$$

Summing $\mathbf{v}_A$ and $\boldsymbol{\omega} \times \mathbf{r}_{B/A}$ gives

$$\mathbf{v}_B = (27.8\mathbf{i} - 27.8\mathbf{j} - 0.0168\mathbf{k})\text{ m/s}$$

From Eq. 17-23,

$$\mathbf{a}_B = \mathbf{a}_A + \dot{\boldsymbol{\omega}} \times \mathbf{r}_{B/A} + \boldsymbol{\omega} \times (\boldsymbol{\omega} \times \mathbf{r}_{B/A})$$

where $\boldsymbol{\omega} = \boldsymbol{\omega}_A + \boldsymbol{\omega}_O$ and

$$\begin{aligned} \mathbf{a}_A &= \dot{\boldsymbol{\omega}}_O \times \mathbf{r}_{A/O} + \boldsymbol{\omega}_O \times (\boldsymbol{\omega}_O \times \mathbf{r}_{A/O}) \\ \dot{\boldsymbol{\omega}}_O &= 0, \qquad \dot{\boldsymbol{\omega}} = \dot{\boldsymbol{\omega}}_A + \dot{\boldsymbol{\omega}}_O = \boldsymbol{\omega}_O \times \boldsymbol{\omega}_A \\ \mathbf{a}_A &= 0.056\mathbf{j} \times [0.056\mathbf{j} \times (0.3\mathbf{j} + 500\mathbf{k})] \\ &= -1.57\mathbf{k}\text{ m/s}^2 \end{aligned}$$

$$\begin{aligned} \dot{\boldsymbol{\omega}} \times \mathbf{r}_{B/A} &= (\boldsymbol{\omega}_O \times \boldsymbol{\omega}_A) \times \mathbf{r}_{B/A} = [0.056\mathbf{j} \times (-92.6\mathbf{k})] \times 0.3\mathbf{i} = 0 \\ \boldsymbol{\omega} \times (\boldsymbol{\omega} \times \mathbf{r}_{B/A}) &= (0.056\mathbf{j} - 92.6\mathbf{k}) \times [(0.056\mathbf{j} - 92.6\mathbf{k}) \times 0.3\mathbf{i}] \\ &= -2572\mathbf{i}\text{ m/s}^2 \end{aligned}$$

Combining these gives

$$\mathbf{a}_B = (-2572\mathbf{i} - 1.57\mathbf{k})\text{ m/s}^2$$

EXAMPLE 17-12

A helicopter is accelerating horizontally at the rate of $5\mathbf{i}$ ft/s². At $\theta = 0$ the blades are rotating at $\omega_O = 12$ rad/s and $\alpha_O = 2$ rad/s² (Fig. 17-31). A turbine disk of 4 in. diameter is rotating essentially in the vertical plane in the engine at the tip of each blade. At $\theta = 0$ the disk in engine A has an angular speed $\omega_C = 500$ rad/s and an angular acceleration $\alpha_C = 50$ rad/s², both clockwise as viewed from the rear of the helicopter. Determine the acceleration $\mathbf{a}_P$ of a particle P on the rim of the disk at 16 ft from the rotor O (this is the outermost particle on the disk with respect to point O).

SOLUTION

In this case point O accelerates with respect to the XYZ frame, point C with respect to point O, and particle P with respect to point C. Thus, it is necessary to use a set of equations based on Eqs. 17-17 and 17-23 as follows.

$$\mathbf{a}_P = \mathbf{a}_O + \mathbf{a}_{P/O} \qquad \mathbf{a}_{P/O} = \mathbf{a}_C + \mathbf{a}_{P/C}$$

where

$$\mathbf{a}_C = \boldsymbol{\alpha}_O \times \mathbf{r}_{C/O} + \boldsymbol{\omega}_O \times (\boldsymbol{\omega}_O \times \mathbf{r}_{C/O})$$

$$= -2\mathbf{j} \times \left(-16 - \frac{2}{12}\right)\mathbf{k} + (-12\mathbf{j}) \times \left[-12\mathbf{j} \times \left(-16 - \frac{2}{12}\right)\mathbf{k}\right]$$

$$= (32.3\mathbf{i} + 2328\mathbf{k}) \text{ ft/s}^2$$

$$\mathbf{a}_{P/C} = \boldsymbol{\alpha} \times \mathbf{r}_{P/C} + \boldsymbol{\omega} \times (\boldsymbol{\omega} \times \mathbf{r}_{P/C})$$

$$\boldsymbol{\alpha} = \boldsymbol{\alpha}_C + \boldsymbol{\omega}_O \times \boldsymbol{\omega}_C = 50\mathbf{i} + 6000\mathbf{k}$$

$$\boldsymbol{\alpha}_{P/C} = (50\mathbf{i} + 6000\mathbf{k}) \times \left(-\frac{2}{12}\right)\mathbf{k} + (500\mathbf{i} - 12\mathbf{j}) \times \left[(500\mathbf{i} - 12\mathbf{j}) \times \left(-\frac{2}{12}\right)\mathbf{k}\right]$$

$$= (8.3\mathbf{j} + 41{,}691\mathbf{k}) \text{ ft/s}^2$$

$$\mathbf{a}_O = 5\mathbf{i} \text{ ft/s}^2$$

Combining the intermediate results gives

$$\mathbf{a}_P = (37.3\mathbf{i} + 8.3\mathbf{j} + 44{,}019\mathbf{k}) \text{ ft/s}^2$$

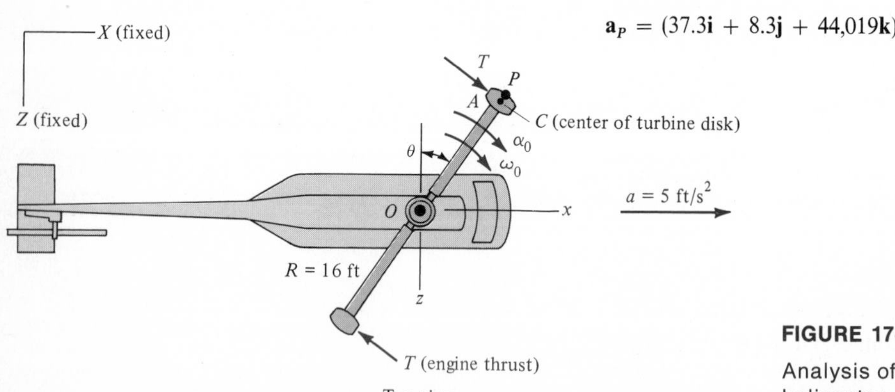

FIGURE 17-31

Analysis of a particle's motion in a helicopter

SECTION 17-7 PROBLEMS

17–74 A wheel of radius $r = 0.4$ m is always in a vertical plane as it travels on a horizontal track of radius $R = 100$ m at a constant speed $v = 20$ m/s. Determine the velocities of points B and D. The center of the track is on the Z axis.

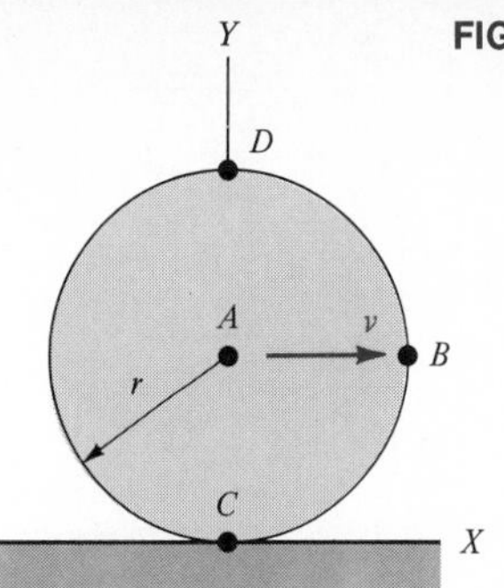

FIGURE P17-74

17–75 Determine the acceleration of particle D of the wheel described in Prob. 17–74.

17–76 Calculate the speeds and accelerations of particle D in Prob. 17–74 for two values of the track radius, $R_1 = 50$ m and $R_2 = 100$ m. Discuss the results.

17–77 The track of a roller coaster has a radius of curvature $r = 40$ ft in a vertical plane, and a radius of curvature $R = 70$ ft in the horizontal plane at the same point. The center of R is on the Z axis. Determine the angular acceleration of a car at point O if it has a constant speed $v_0 = 50$ ft/s.

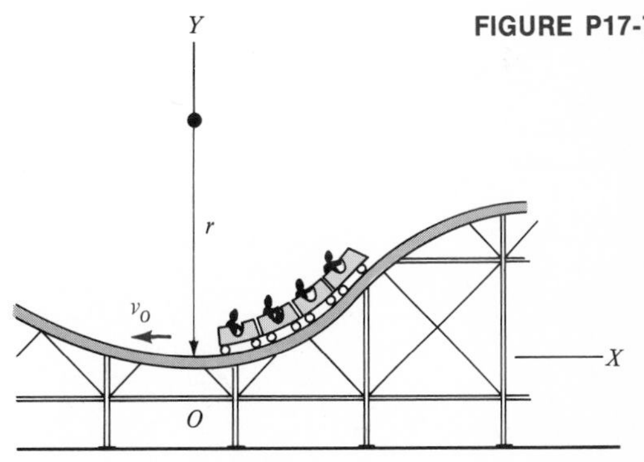

FIGURE P17-77

17–78 Determine the angular acceleration of a car in Prob. 17–77 if v_0 is decreasing at the rate of 15 ft/s^2.

17–79 The 40-m-long boom of the crane rotates about the vertical axis through point A at an angular speed $\omega_Y = 0.2$ rad/s. At the same time the boom is raised vertically at an angular speed $\omega_Z = 0.4$ rad/s when AB makes a 30° angle with the horizontal. Determine the velocity of point B.

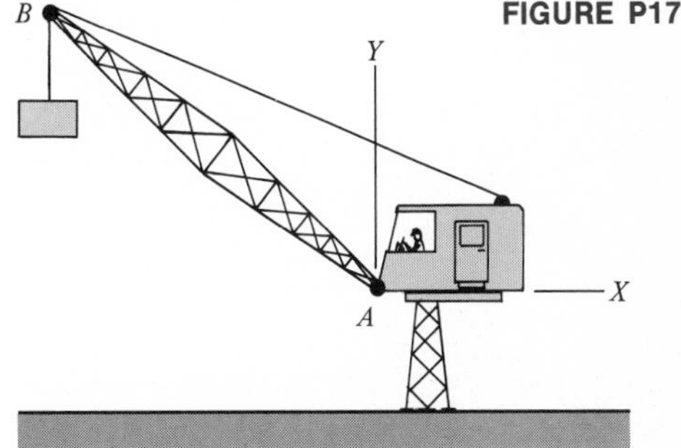

FIGURE P17-79

17–80 Determine the acceleration of point B in Prob. 17–79 assuming that the angular speeds are constant.

17–81 Determine the acceleration of point B in Prob. 17–79 assuming that ω_Y is increasing at the rate of 0.2 rad/s^2, and that ω_Z is decreasing at the rate of 0.1 rad/s^2.

FIGURE P17-82

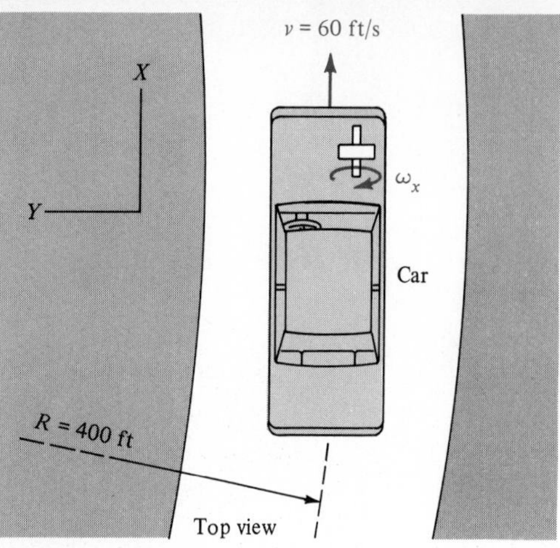

17–82 The rotor shaft of an alternator in a car is in the horizontal plane. It rotates at a constant angular speed $\omega_X = 1500$ rpm while the car travels at $v = 60$ ft/s on a horizontal road of 400 ft radius. Determine the magnitude of the angular acceleration of the rotor shaft if v increases at the rate of 8 ft/s^2.

FIGURE P17-83

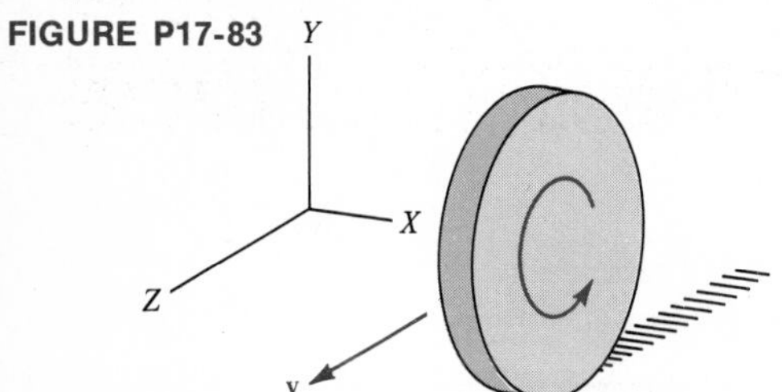

17–83 A wheel of 0.5 m radius is rolling at a velocity $\mathbf{v} = 30\mathbf{k}$ m/s. Calculate the angular acceleration of the wheel axle if it is turning at a constant angular velocity $\boldsymbol{\omega}_Y = 0.2\mathbf{j}$ rad/s.

17–84 Solve Prob. 17–83 with the additional angular velocity $\boldsymbol{\omega}_Z = -0.1\mathbf{k}$ rad/s of the wheel axle.

FIGURE P17-85

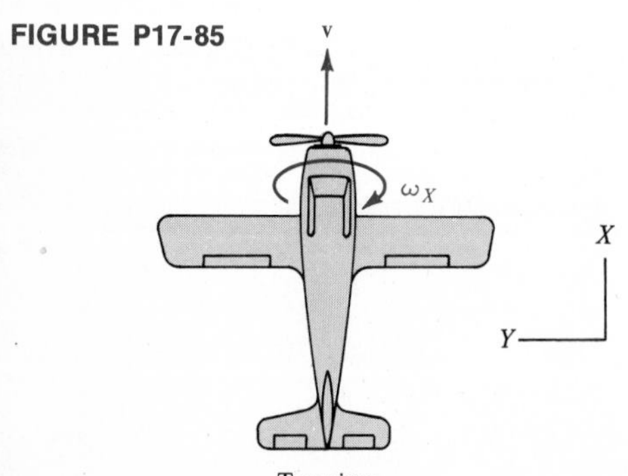

17–85 An airplane is flying horizontally at velocity $\mathbf{v} = 300\mathbf{i}$ ft/s when the angular speed of the propeller is $\omega_X = 2000$ rpm. Calculate the angular acceleration of the propeller when the airplane begins to turn right on a horizontal path of 1500 ft radius.

17–86 Determine the angular acceleration of the propeller in Prob. 17–85 if the airplane is simultaneously diving at the rate of $\boldsymbol{\omega}_Y = 0.8\mathbf{j}$ rad/s.

17–87 The tire rotates at a constant angular velocity $\boldsymbol{\omega}_Z = 100\mathbf{k}$ rad/s during a test. Determine the angular acceleration of the wheel axle if the wheel also rotates at angular velocity $\boldsymbol{\omega}_Y = 0.5\mathbf{j}$ rad/s.

FIGURE P17-87

Model of a tire-testing system. (Courtesy MTS Systems Corporation, Minneapolis, Minnesota)

17–88 Determine the angular acceleration of the wheel axle in Prob. 17–87 for the simultaneous angular velocities of $\boldsymbol{\omega}_X = -0.3\mathbf{i}$ rad/s, $\boldsymbol{\omega}_Y = 0.4\mathbf{j}$ rad/s, and $\boldsymbol{\omega}_Z = 120\mathbf{k}$ rad/s. Calculate the acceleration of particle P. The radius of the wheel is 1.4 ft.

17–89 A two-engine aircraft is flying horizontally at a constant speed. The two propellers are rotating at the same rate ω but in opposite directions. What are the directions of the angular accelerations of the propellers if the airplane begins a nosedive?

FIGURE P17-89

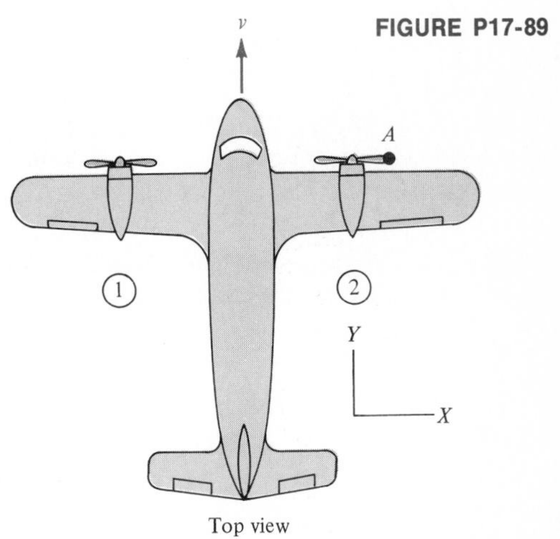

17–90 In Prob. 17–89, $\boldsymbol{\omega}_1 = 2300\mathbf{j}$ rpm and $\boldsymbol{\omega}_2 = -2300\mathbf{j}$ rpm. The airplane flies at $\mathbf{v} = 300\mathbf{j}$ km/h and starts turning to the left on a horizontal path of 1000 m radius. Determine the angular acceleration of the propellers caused by this maneuver. Calculate the acceleration of particle A. The length of the propeller is 1.5 m.

FIGURE P17-91

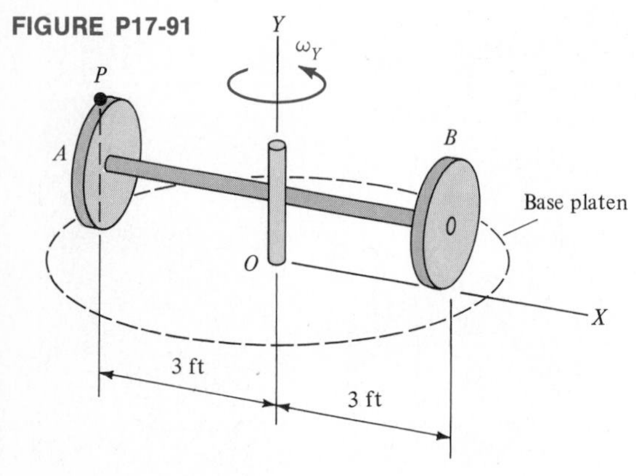

17–91 A food processor has wheels A and B of 1 ft radius mounted on a common arm which rotates in the horizontal plane at $\boldsymbol{\omega}_Y = 15\mathbf{j}$ rad/s. The wheels rotate without slipping on the base platen. Determine the angular accelerations of the wheels caused by the rotation. Calculate the acceleration of particle P.

FIGURE P17-92

17–92 The large amusement ride has two angular velocities at a given instant: the 12 small arms have $\boldsymbol{\omega}_B = (2\mathbf{i} + \mathbf{k})$ rpm, and the arm AB has $\boldsymbol{\omega}_Y = 0.5\mathbf{j}$ rpm. Determine the angular acceleration of a small arm that rotates about B.

17–93 Solve Prob. 17–92 with the additional angular velocity of $\boldsymbol{\omega}_X = 0.3\mathbf{i}$ rpm.

TIME DERIVATIVE OF A VECTOR USING A ROTATING FRAME 17-8

In some cases the velocity and acceleration of a particle or a rigid body are defined in a reference frame which is translating and rotating with respect to a fixed frame. For example, the motion of a turbine blade in a jet engine is conveniently defined with respect to the fuselage of the aircraft which translates and rotates with respect to the earth. The analysis of the absolute motion of the body in such a case is straightforward concerning translation of the moving frame, but the rotation of that frame requires additional considerations. These are discussed for an arbitrary vector **Q** as follows.

Assume that a fixed reference frame XYZ and a rotating frame xyz have a common origin O as in Fig. 17-32. Frame xyz is rotating at an angular velocity $\mathbf{\Omega}$ with respect to the XYZ frame. A vector **Q** is defined in terms of its rectangular components in the xyz frame,

$$\mathbf{Q} = Q_x\mathbf{i} + Q_y\mathbf{j} + Q_z\mathbf{k} \qquad \text{17-26}$$

The time rate of change of vector **Q** with respect to the xyz frame is obtained by differentiating Eq. 17-26 with respect to time. This is expressed as

$$(\dot{\mathbf{Q}})_{xyz} = \dot{Q}_x\mathbf{i} + \dot{Q}_y\mathbf{j} + \dot{Q}_z\mathbf{k} \qquad \text{17-27}$$

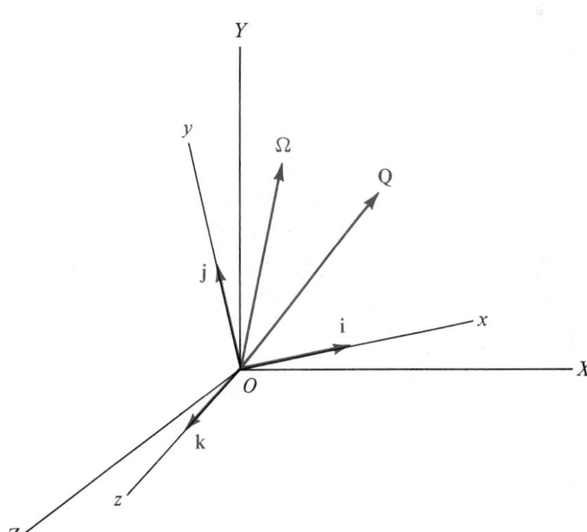

FIGURE 17-32

Fixed and rotating reference frames with a common origin

since $d\mathbf{i}/dt$, $d\mathbf{j}/dt$, and $d\mathbf{k}/dt$ are all zero (**i**, **j**, and **k** are constant in magnitude and have fixed directions in the xyz frame). The time rate of change of vector **Q** with respect to the XYZ frame is written from Eq. 17-26 as

$$(\dot{\mathbf{Q}})_{XYZ} = \dot{Q}_x\mathbf{i} + \dot{Q}_y\mathbf{j} + \dot{Q}_z\mathbf{k} + Q_x\frac{d\mathbf{i}}{dt} + Q_y\frac{d\mathbf{j}}{dt} + Q_z\frac{d\mathbf{k}}{dt} \tag{17-28}$$

where the time derivatives of the unit vectors are not zero since their directions change with respect to XYZ. The last three terms in Eq. 17-28 represent the velocity of the tip of vector **Q** if the magnitude Q is constant. Thus, from Eq. 17-25,

$$Q_x\frac{d\mathbf{i}}{dt} + Q_y\frac{d\mathbf{j}}{dt} + Q_z\frac{d\mathbf{k}}{dt} = \mathbf{\Omega} \times \mathbf{Q} \tag{17-29}$$

Using Eqs. 17-27 and 17-29, Eq. 17-28 is written as

$$\boxed{(\dot{\mathbf{Q}})_{XYZ} = (\dot{\mathbf{Q}})_{xyz} + \mathbf{\Omega} \times \mathbf{Q}} \tag{17-30}$$

where $(\dot{\mathbf{Q}})_{xyz}$ can be thought of as the time rate of change of the magnitude of vector **Q**, while $\mathbf{\Omega} \times \mathbf{Q}$ is the time rate of change of the direction of **Q**, both in the fixed frame. Equation 17-30 is useful in the analysis of velocities and accelerations when a rotating reference frame must be considered.

17-9 ANALYSIS OF VELOCITIES AND ACCELERATIONS USING A ROTATING FRAME

Assume that the position of a particle P is defined by the vector **r** with respect to the xyz frame in Fig. 17-33. The xyz frame is rotating at angular velocity $\mathbf{\Omega}$ with respect to the XYZ frame. The absolute velocity $\mathbf{v}_P$ of the particle in the XYZ frame is the time derivative of the position vector **r** with respect to the fixed frame. Using Eq. 17-30 and substituting the vector **r** for the arbitrary vector **Q** gives

$$\boxed{\mathbf{v}_P = (\dot{\mathbf{r}})_{XYZ} = (\dot{\mathbf{r}})_{xyz} + \mathbf{\Omega} \times \mathbf{r}} \tag{17-31}$$

where $(\dot{\mathbf{r}})_{xyz} = \mathbf{v}_{xyz}$, the velocity of the particle with respect to the rotating xyz frame. The term $\mathbf{\Omega} \times \mathbf{r}$ represents the absolute velocity $\mathbf{v}_{P'}$ of a particle P' which is fixed in the xyz frame and whose position

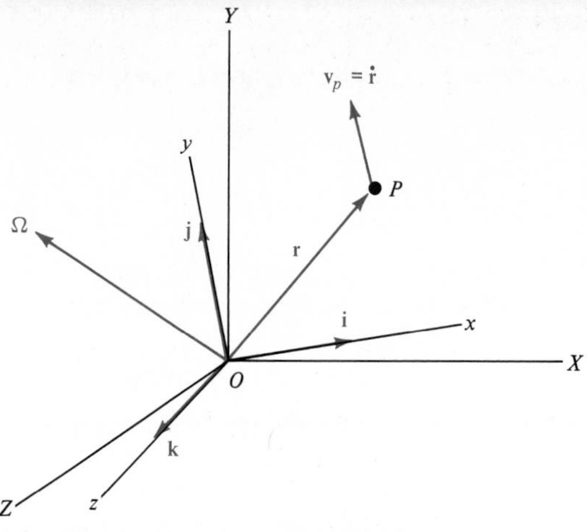

FIGURE 17-33

Analysis of a particle's motion with respect to a fixed frame and a rotating frame

coincides with the position of particle P at the given instant. Thus, Eq. 17-31 is written more concisely as

$$\boxed{\mathbf{v}_P = \mathbf{v}_{xyz} + \mathbf{v}_{P'}} \qquad \boxed{17\text{-}32}$$

where $\mathbf{v}_P$ = absolute velocity of particle P in the XYZ frame
$\mathbf{v}_{xyz}$ = velocity of P with respect to the moving xyz frame
$\mathbf{v}_{p'}$ = absolute velocity of particle P' fixed in the xyz frame; P and P' coincide at the given instant

Note that particle P has velocity $\mathbf{v}_{xyz}$ relative to P'.

The absolute acceleration $\mathbf{a}_P$ of particle P is obtained by differentiating Eq. 17-31 with respect to time,

$$\mathbf{a}_P = \dot{\mathbf{v}}_P = \frac{d}{dt}[(\dot{\mathbf{r}})_{xyz}] + \dot{\mathbf{\Omega}} \times \mathbf{r} + \mathbf{\Omega} \times \dot{\mathbf{r}} \qquad \boxed{17\text{-}33}$$

where $(\dot{\mathbf{r}})_{xyz}$ is with respect to the xyz frame, while $(d/dt)[(\dot{\mathbf{r}})_{xyz}]$, $\dot{\mathbf{\Omega}} \times \mathbf{r}$, and $\mathbf{\Omega} \times \dot{\mathbf{r}}$ are all defined with respect to the XYZ frame. Using Eq. 17-30 and substituting the vector $(\dot{\mathbf{r}})_{xyz}$ for the arbitrary vector $\mathbf{Q}$ gives

$$\frac{d}{dt}[(\dot{\mathbf{r}})_{xyz}] = (\ddot{\mathbf{r}})_{xyz} + \Omega \times (\dot{\mathbf{r}})_{xyz} \qquad \boxed{17\text{-}34}$$

The quantity $\dot{\mathbf{r}}$ in Eq. 17-33 is the absolute velocity $\mathbf{v}_P$ of particle P. and 17-40 gives

Thus, from Eq. 17-31,

$$\boldsymbol{\Omega} \times \dot{\mathbf{r}} = \boldsymbol{\Omega} \times (\dot{\mathbf{r}})_{xyz} + \boldsymbol{\Omega} \times (\boldsymbol{\Omega} \times \mathbf{r}) \tag{17-35}$$

Using Eqs. 17-34 and 17-35, Eq. 17-33 can be written as

$$\boxed{\mathbf{a}_P = \mathbf{a}_{xyz} + \dot{\boldsymbol{\Omega}} \times \mathbf{r} + \boldsymbol{\Omega} \times (\boldsymbol{\Omega} \times \mathbf{r}) + 2\boldsymbol{\Omega} \times \mathbf{v}_{xyz}} \tag{17-36}$$

where $\mathbf{a}_{xyz} = (\ddot{\mathbf{r}})_{xyz}$ = acceleration of particle P with respect to the moving xyz frame

$\dot{\boldsymbol{\Omega}} \times \mathbf{r}$ = tangential acceleration of P' because of angular acceleration of the xyz frame; it is always normal to $\mathbf{r}$

$\boldsymbol{\Omega} \times (\boldsymbol{\Omega} \times \mathbf{r})$ = normal acceleration of P' because of angular velocity of the xyz frame; it is always directed toward the origin of the vector $\mathbf{r}$ in the xyz frame

$\mathbf{v}_{xyz}$ = velocity of P with respect to the moving xyz frame

$2\boldsymbol{\Omega} \times \mathbf{v}_{xyz}$ = *Coriolis acceleration** = acceleration of particle P with respect to P', which is fixed in the xyz frame; it is always normal to $\mathbf{v}_{xyz}$ and thus to the path of P in xyz

Equation 17-36 is written more concisely as

$$\boxed{\mathbf{a}_P = \mathbf{a}_{xyz} + \mathbf{a}_{P'} + \mathbf{a}_{\text{Cor}}} \tag{17-37}$$

where $\mathbf{a}_P$ = absolute acceleration of particle P in the XYZ frame

$\mathbf{a}_{xyz}$ = acceleration of P with respect to the moving xyz frame

$\mathbf{a}_{P'} = \dot{\boldsymbol{\Omega}} \times \mathbf{r} + \boldsymbol{\Omega} \times (\boldsymbol{\Omega} \times \mathbf{r})$ = absolute acceleration of particle P' fixed in the xyz frame; P and P' coincide at the given instant

$\mathbf{a}_{\text{Cor}} = 2\boldsymbol{\Omega} \times \mathbf{v}_{xyz}$ = Coriolis acceleration of P with respect to P'

Translation and Rotation of the Moving Reference Frame

The general case of a moving reference frame is illustrated in Fig. 17-34a. The xyz frame rotates at angular velocity $\boldsymbol{\Omega}$ while its origin A translates with respect to the XYZ frame. The absolute position, velocity, and acceleration of particle P are defined,

* After the French mathematician DeCoriolis (1792–1843). The Coriolis acceleration is difficult to visualize, but the inclusion of this term (which may be significant) in the solution of problems causes little additional work. Those who wish to obtain further understanding of this acceleration should see App. D.

respectively, as

$$\mathbf{r}_P = \mathbf{r}_A + \mathbf{r}_{P/A} \tag{17-38}$$

$$\mathbf{v}_P = \dot{\mathbf{r}}_P = \mathbf{v}_A + \mathbf{v}_{P/A} \tag{17-39}$$

$$\mathbf{a}_P = \dot{\mathbf{v}}_P = \mathbf{a}_A + \mathbf{a}_{P/A} \tag{17-40}$$

The translation and rotation of frame xyz are distinguished using Fig. 17-34b, where $X'Y'Z'$ are parallel to XYZ. Since the velocity $\mathbf{v}_{P/A}$ can be determined by applying Eq. 17-31 to the $X'Y'Z'$ frame, Eq. 17-39 is written as

$$\boxed{\mathbf{v}_P = \mathbf{v}_A + (\dot{\mathbf{r}}_{P/A})_{xyz} + \boldsymbol{\Omega} \times \mathbf{r}_{P/A}} \tag{17-41}$$

where $\mathbf{v}_P$ and $\mathbf{v}_A$ are with respect to the XYZ frame.

The absolute acceleration $\mathbf{a}_P$ of particle P is obtained by differentiating Eq. 17-41 with respect to time. Using Eqs. 17-36 and 17-40 gives

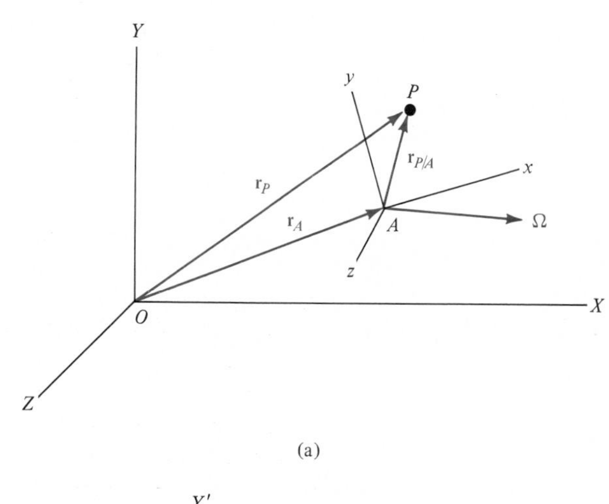

(a)

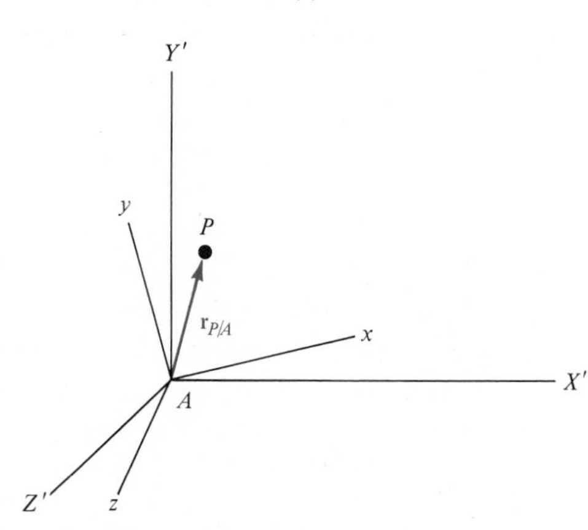

(b)

FIGURE 17-34

General analysis using a moving reference frame

$$\mathbf{a}_P = \mathbf{a}_A + \mathbf{a}_{xyz} + \dot{\mathbf{\Omega}} \times \mathbf{r}_{P/A} + \mathbf{\Omega} \times (\mathbf{\Omega} \times \mathbf{r}_{P/A}) + 2\mathbf{\Omega} \times \mathbf{v}_{xyz} \qquad \text{17-42}$$

where $\mathbf{a}_P$ and $\mathbf{a}_A$ are with respect to the XYZ frame, and the other terms are defined as for Eq. 17-36. Note that the vector $\mathbf{r}_{P/A}$ in Eqs. 17-41 and 17-42 is analogous to the vector $\mathbf{r}$ in Eqs. 17-31 and 17-36; it is a position vector to point P', which is fixed in the moving xyz frame and is coincident with the particle P at the instant considered. The fact that the temporarily coincident particles P and P' have different motions is especially important in the derivation of the kinematics relationship. In solving problems it is useful to remember the following:

1. Equations 17-31 through 17-42 are valid for individual particles or for particles that are parts of rigid bodies.
2. The angular velocity $\mathbf{\Omega}$ is the same for all lines, and the angular acceleration $\dot{\mathbf{\Omega}}$ is the same for all lines in a rigid body at a given instant.
3. The terms $\mathbf{v}_{xyz}$ and $\mathbf{a}_{xyz}$ are both zero when the rotating xyz frame is fixed in the rigid body whose motion is analyzed. The reason is that $\mathbf{v}_{xyz}$ and $\mathbf{a}_{xyz}$ represent the motion of a particle P relative to the xyz frame (and to point P' which is fixed in xyz). If xyz is fixed in the body, P and P' are always at the same location with respect to the xyz frame.
4. The common points of two or more pin-jointed members must have the same absolute velocity and the same absolute acceleration.
5. The points of contact in members that are in temporary contact generally have different absolute velocities and different absolute accelerations.
6. The geometric constraints of motion in a rotating frame are useful to define the directions of several components of the absolute velocity and acceleration. For an illustration of these, consider the plane motion of particle P with respect to the xy frame which rotates in the XY frame in Fig. 17-35.

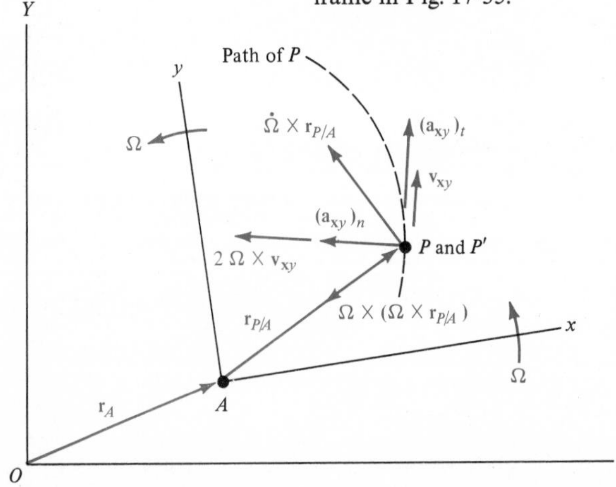

FIGURE 17-35

Analysis of a particle's plane motion using a rotating reference frame

EXAMPLE 17-13

Pin B of the crank r is moving in the slotted arm in Fig. 17-36. Assume that $l = 12$ in., $r = 5$ in., $\omega_1 = 10$ rad/s, and $\dot{\omega}_1 = 0$. Determine the angular velocity $\boldsymbol{\omega}_2$ of arm OC and the relative velocity $\mathbf{v}_{xyz}$ of pin B in the arm for $\theta = 20°$, $\phi = 55°$.

SOLUTION

The absolute velocity $\mathbf{v}_B$ of pin B can be expressed in two ways, using first point A, then point O as the fixed reference point. From Eq. 17-31,

$$\mathbf{v}_B = \boldsymbol{\omega}_1 \times \mathbf{r}_{B/A} \qquad \text{(reference point } A) \qquad \textbf{(a)}$$

$$\mathbf{v}_B = \mathbf{v}_{xyz} + \boldsymbol{\omega}_2 \times \mathbf{r}_{B/O} \qquad \text{(reference point } O) \qquad \textbf{(b)}$$

Equating these two expressions and using the two vector equations for **i** and **j** gives two equations with two unknowns which can be solved. In the configuration shown,

$$\mathbf{r}_{B/A} = \frac{5}{12}(\cos 55°\ \mathbf{i} + \sin 55°\ \mathbf{j})\ \text{ft}$$

$$= (0.24\mathbf{i} + 0.34\mathbf{j})\ \text{ft}$$

Using $\boldsymbol{\omega}_1 = -10\mathbf{k}$ rad/s yields

$$\mathbf{v}_B = \boldsymbol{\omega}_1 \times \mathbf{r}_{B/A} = (3.4\mathbf{i} - 2.4\mathbf{j})\ \text{ft/s} \qquad \textbf{(c)}$$

Using the law of sines to determine $\mathbf{r}_{B/O}$ gives

$$\mathbf{r}_{B/O} = \frac{8.4}{12}(\sin 20°\ \mathbf{i} - \cos 20°\ \mathbf{j})\ \text{ft}$$

$$= (0.24\mathbf{i} - 0.66\mathbf{j})\ \text{ft}$$

$$\mathbf{v}_{xyz} = v_{xyz}(0.34\mathbf{i} - 0.94\mathbf{j})\ \text{ft/s}$$

$$\boldsymbol{\omega}_2 \times \mathbf{r}_{B/O} = \omega_2\mathbf{k} \times (0.24\mathbf{i} - 0.66\mathbf{j})\ \text{ft/s}$$

Thus, Eq. (b) becomes

$$\mathbf{v}_B = v_{xyz}(0.34\mathbf{i} - 0.94\mathbf{j})\ \text{ft/s} + \omega_2(0.66\mathbf{i} + 0.24\mathbf{j})\ \text{ft/s} \qquad \textbf{(d)}$$

Equating the **i** components from Eqs. (c) and (d) yields

$$3.4 = v_{xyz}(0.34) + \omega_2(0.66)$$

Equating the **j** components from Eqs. (c) and (d) gives

$$-2.4 = v_{xyz}(-0.94) + \omega_2(0.24)$$

$$v_{xyz} = 3.4\ \text{ft/s} \qquad \mathbf{v}_{xyz} = (1.2\mathbf{i} - 3.2\mathbf{j})\ \text{ft/s}$$

$$\omega_2 = 3.4\ \text{rad/s} \qquad \boldsymbol{\omega}_2 = 3.4\mathbf{k}\ \text{rad/s}$$

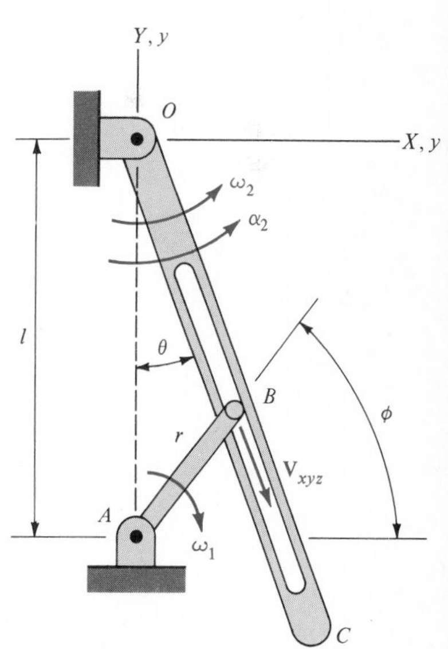

FIGURE 17-36

Analysis of a crank–slotted arm mechanism

EXAMPLE 17-14

The mechanism in Fig. 17-36 (Ex. 17-13) has dimensions $l = 0.3$ m and $r = 0.12$ m. Determine the angular acceleration $\boldsymbol{\alpha}_2$ of arm OC and the relative acceleration $\mathbf{a}_{xyz}$ of pin B in the arm for $\theta = 20°$, $\phi = 51.2°$, and a constant angular speed of $\omega_1 = 50$ rad/s.

SOLUTION

The absolute acceleration $\mathbf{a}_B$ of pin B can be expressed in two ways, using first point A, then point O as the fixed reference point. From Eq. 17-36,

$$\mathbf{a}_B = \boldsymbol{\omega}_1 \times (\boldsymbol{\omega}_1 \times \mathbf{r}_{B/A}) \qquad \text{(reference point } A) \qquad \textbf{(a)}$$

$$\mathbf{a}_B = \mathbf{a}_{xyz} + \boldsymbol{\alpha}_2 \times \mathbf{r}_{B/O} + \boldsymbol{\omega}_2 \times (\boldsymbol{\omega}_2 \times \mathbf{r}_{B/O}) + 2\boldsymbol{\omega}_2 \times \mathbf{v}_{xyz} \qquad \text{(reference point } O) \qquad \textbf{(b)}$$

Using a method of solution as in Ex. 17-13, $\boldsymbol{\omega}_2$ and $\mathbf{v}_{B/O}$ are found to be

$$\boldsymbol{\omega}_2 = 13.95\mathbf{k} \text{ rad/s}$$

$$\mathbf{v}_{xyz} = (1.75\mathbf{i} - 4.81\mathbf{j}) \text{ m/s}$$

From Eq. (a),

$$\mathbf{a}_B = -50\mathbf{k} \times [-50\mathbf{k} \times 0.12(\cos 51.2°\,\mathbf{i} + \sin 51.2°\,\mathbf{j})]$$
$$= (-188\mathbf{i} - 234\mathbf{j}) \text{ m/s}^2$$

From Eq. (b),

$$\mathbf{a}_{xyz} = a_{xyz}(\sin 20°\,\mathbf{i} - \cos 20°\,\mathbf{j})$$
$$= (0.34a_{xyz}\mathbf{i} - 0.94a_{xyz}\mathbf{j}) \text{ m/s}^2$$

For the given configuration, $\mathbf{r}_{B/O} = 0.22(\sin 20°\,\mathbf{i} - \cos 20°\,\mathbf{j})$, and

$$\boldsymbol{\alpha}_2 \times \mathbf{r}_{B/O} = \alpha_2\mathbf{k} \times [0.22(\sin 20°\,\mathbf{i} - \cos 20°\,\mathbf{j})]$$
$$= (0.075\alpha_2\mathbf{j} + 0.21\alpha_2\mathbf{i}) \text{ m/s}^2$$

$$\boldsymbol{\omega}_2 \times (\boldsymbol{\omega}_2 \times \mathbf{r}_{B/O}) = 13.95\mathbf{k} \times [13.95\mathbf{k} \times 0.22(\sin 20°\,\mathbf{i} - \cos 20°\,\mathbf{j})]$$
$$= (-14.6\mathbf{i} + 40.2\mathbf{j}) \text{ m/s}^2$$

$$2(\boldsymbol{\omega}_2 \times \mathbf{v}_{xyz}) = 2[13.95\mathbf{k} \times (1.75\mathbf{i} - 4.81\mathbf{j})]$$
$$= (134.2\mathbf{i} + 48.8\mathbf{j}) \text{ m/s}^2$$

Using these results provides two equations with two unknowns. Equating the **i** components from Eqs. (a) and (b) gives

$$-188 = 0.34a_{xyz} + 0.21\alpha_2 - 14.6 + 134.2$$

Equating the **j** components from Eqs. (a) and (b) yields

$$-234 = -0.94a_{xyz} + 0.075\alpha_2 + 40.2 + 38.8$$

These reduce to

$$-307.6 = 0.34a_{xyz} + 0.21\alpha_2$$

$$-323 = -0.94a_{xyz} + 0.075\alpha_2$$

$$a_{xyz} = 201 \text{ m/s}^2 \qquad \mathbf{a}_{xyz} = (68.3\mathbf{i} - 188.9\mathbf{j}) \text{ m/s}^2$$

$$\alpha_2 = -1790 \text{ rad/s}^2 \qquad \boldsymbol{\alpha}_2 = -1790 \text{ rad/s}^2\,\mathbf{k}$$

EXAMPLE 17-15

An amusement ride is modeled in Fig. 17-37. People sit in two circular cages that are mounted on a rigid rotating arm at A and B. The horizontal arm AB is rotating about point O at a constant angular speed $\omega_1 = 10$ rpm, and the cages rotate with respect to the arm AB at constant speed $\omega_2 = 3$ rpm. Calculate the absolute velocity and acceleration of a person at position 2.

SOLUTION

The first step is to choose coordinate systems, as shown in the figure. The absolute velocity of point 2 is determined using Eq. 17-41,

$$\mathbf{v}_2 = \mathbf{v}_A + \mathbf{v}_{2_{xyz}} + \boldsymbol{\omega}_1 \times \mathbf{r}_{2/A}$$

$\mathbf{v}_{2_{xyz}} = \boldsymbol{\omega}_2 \times \mathbf{r}_{2/A}$. The velocity of the center of the cage is expressed as

$$\mathbf{v}_A = \mathbf{v}_{A/D} + \boldsymbol{\omega}_1 \times \mathbf{r}_{A/O}$$

where $\mathbf{v}_{A/D} = 0$ because the center of the cage is fixed on the rotating arm.

Consequently,

$$\mathbf{v}_2 = \boldsymbol{\omega}_1 \times \mathbf{r}_{A/O} + \boldsymbol{\omega}_2 \times \mathbf{r}_{2/A} + \boldsymbol{\omega}_1 \times \mathbf{r}_{2/A}$$

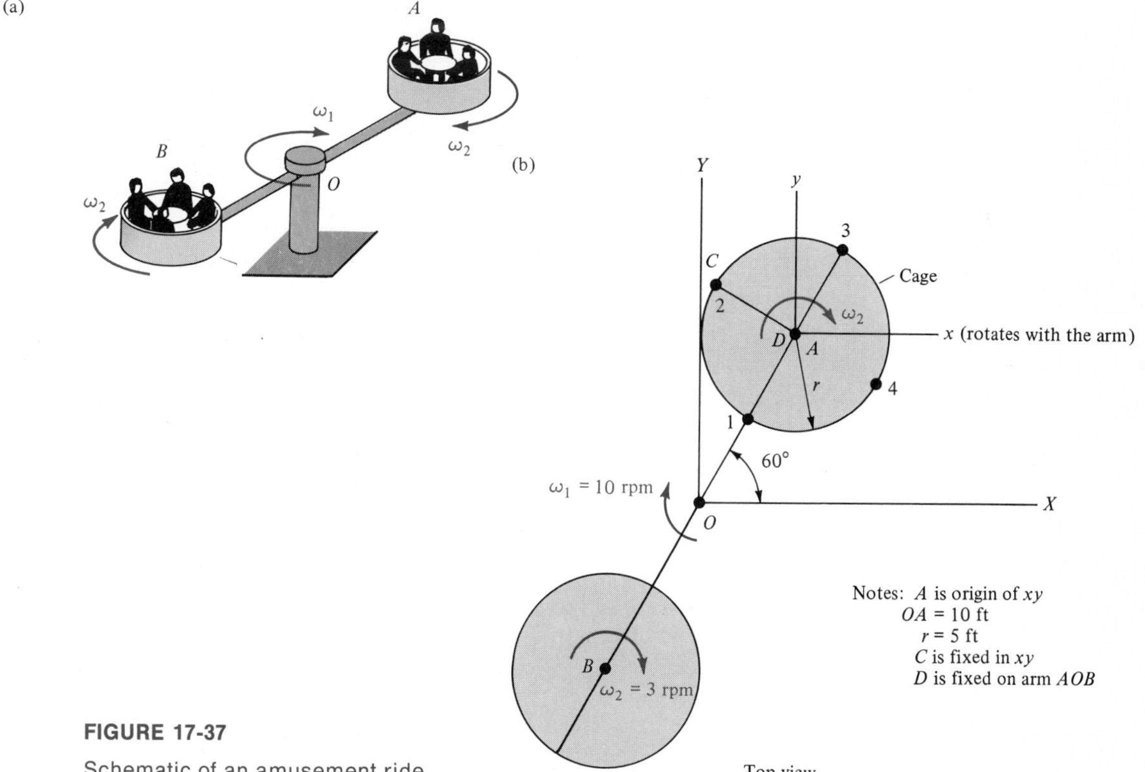

FIGURE 17-37

Schematic of an amusement ride

where

$$\mathbf{r}_{A/O} = (5\mathbf{i} + 8.66\mathbf{j}) \text{ ft}$$

$$\mathbf{r}_{2/A} = (-4.33\mathbf{i} + 2.5\mathbf{j}) \text{ ft}$$

The units of angular velocity should be in rad/s,

$$\begin{aligned}\omega_1 &= -10 \frac{\text{rev}}{\text{min}}\left(2\pi \frac{\text{rad}}{\text{rev}}\right)\left(\frac{1 \text{ min}}{60 \text{ s}}\right)\mathbf{k} \\ &= -1.047\mathbf{k} \text{ rad/s} \\ \omega_2 &= -0.314\mathbf{k} \text{ rad/s}\end{aligned}$$

Thus,

$$\begin{aligned}\mathbf{v}_2 &= (-1.047\mathbf{k}) \times (5\mathbf{i} + 8.66\mathbf{j}) \\ &\quad + (-0.314\mathbf{k}) \times (-4.33\mathbf{i} + 2.5\mathbf{j}) + (-1.047\mathbf{k}) \times (-4.33\mathbf{i} + 2.5\mathbf{j}) \\ &= (12.47\mathbf{i} + 0.659\mathbf{j}) \text{ ft/s}\end{aligned}$$

Solving for the acceleration using Eq. 17-42 gives

$$\begin{aligned}\mathbf{a}_2 = \mathbf{a}_A + \mathbf{a}_{2_{xyz}} + \dot{\omega}_1 \times \mathbf{r}_{2/A} + \omega_1 \\ \times (\omega_1 \times \mathbf{r}_{2/A}) + 2\omega_1 \times \mathbf{v}_{2_{xyz}}\end{aligned}$$

where $\mathbf{a}_A$ is recognized to be from pure rotation of A about the fixed point O. According to Eq. 17-36,

$$\mathbf{a}_A = \mathbf{a}_{A/D} + \dot{\omega}_1 \times \mathbf{r}_{A/O} + \omega_1 \times (\omega_1 \times \mathbf{r}_{A/O}) + 2\omega_1 \times \mathbf{v}_{A/D}$$

Here $\mathbf{a}_{A/D} = \mathbf{v}_{A/D} = \dot{\omega}_1 = 0$ since A is fixed on the arm, which rotates at a constant angular velocity. Thus,

$$\begin{aligned}\mathbf{a}_A &= \omega_1 \times (\omega_1 \times \mathbf{r}_{A/O}) \\ &= -1.047\mathbf{k} \times [-1.047\mathbf{k} \times (5\mathbf{i} + 8.66\mathbf{j})] \\ &= (-5.48\mathbf{i} - 9.49\mathbf{j}) \text{ ft/s}^2\end{aligned}$$

Also, $= \dot{\omega}_2 = 0$. Therefore,

$$\begin{aligned}\mathbf{a}_{2_{xyz}} &= \dot{\omega}_2 \times \mathbf{r}_{2/A} + \omega_2 \times (\omega_2 \times \mathbf{r}_{2/A}) \\ = \omega_2 \times (\omega_2 \times \mathbf{r}_{2/A}) &= -0.314\mathbf{k} \times [-0.314\mathbf{k} \times (-4.33\mathbf{i} + 2.5\mathbf{j})] \\ &= 0.416\mathbf{i} - 0.246\mathbf{j}\end{aligned}$$

$$\begin{aligned}\omega_1 \times (\omega_1 \times \mathbf{r}_{2/A}) &= -1.047\mathbf{k} \times [-1.047\mathbf{k} \times (-4.33\mathbf{i} + 2.5\mathbf{j})] \\ &= 4.75\mathbf{i} - 2.74\mathbf{j}\end{aligned}$$

$$\begin{aligned}2\omega_1 \times \mathbf{v}_{2_{xyz}} &= 2(-1.047\mathbf{k}) \times [-0.314\mathbf{k} \times (-4.33\mathbf{i} + 2.5\mathbf{j})] \\ &= 2.85\mathbf{i} - 1.64\mathbf{j}\end{aligned}$$

The person at point 2 has absolute acceleration

$$\mathbf{a}_2 = (2.54\mathbf{i} - 14.12\mathbf{j}) \text{ ft/s}^2$$

17–94 The two rigid bars of length $AB = BC = 1.5$ m are pinned at A and B. Determine the velocity $\mathbf{v}_C$ and acceleration $\mathbf{a}_C$ if $\omega_1 = 0.3$ rad/s, $\alpha_1 = 0$, $\omega_2 = 0.2$ rad/s, and $\alpha_2 = 0$.

FIGURE P17-94

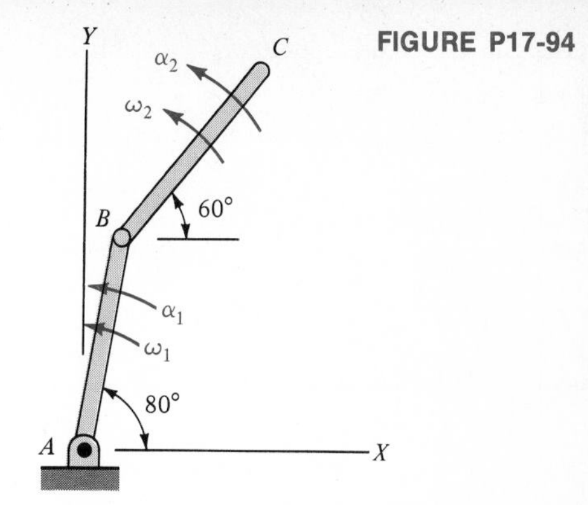

17–95 $AB = 3$ ft and $BC = 2$ ft in Fig. P17-94. $\omega_1 = 0.4$ rad/s, $\alpha_1 = 0.2$ rad/s^2 (increasing ω_1), $\omega_2 = 0.3$ rad/s, and $\alpha_2 = -0.2$ rad/s^2 (decreasing ω_2). Calculate the velocity $\mathbf{v}_C$ and acceleration $\mathbf{a}_C$.

FIGURE P17-96

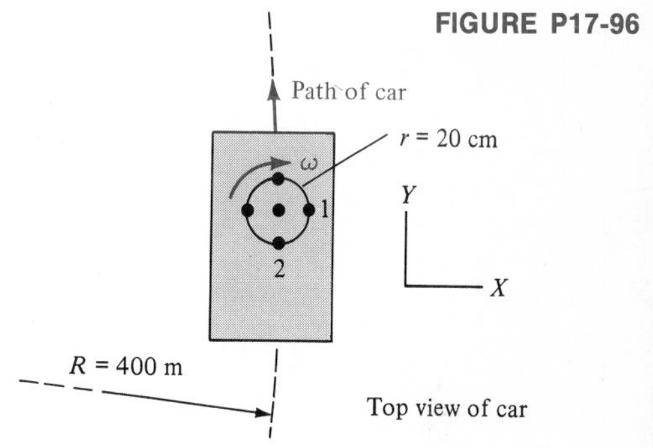

17–96 A flywheel is rotating in the horizontal plane in an experimental car. The car is traveling at a constant speed $v = 100$ km/h on a road of radius $R = 400$ m. Calculate the acceleration of a particle at positions 1 and 2 on the flywheel. The particle is 20 cm from the center of the wheel, which rotates at 5000 rpm.

FIGURE P17-97

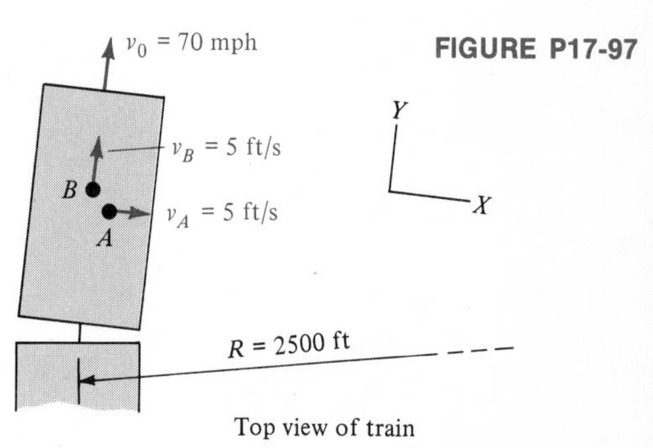

17–97 A train is traveling on a curved track at 70 mph when its brakes are applied causing a deceleration of 8 ft/s^2. At this instant persons A and B are walking in the train as shown. Calculate the accelerations of the two persons.

FIGURE P17-98

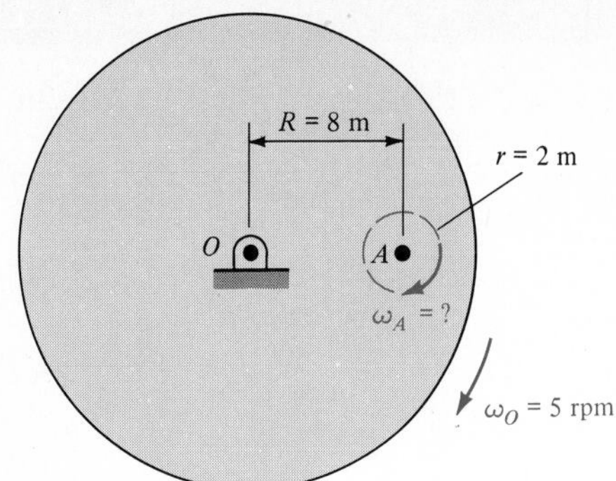

17–98 A proposed merry-go-round is modeled as shown. The two superimposed rotations about points O and A are in the horizontal plane. Calculate the maximum allowed angular speed ω_A if the maximum absolute acceleration of a person rotating 2 m away from A is 12 m/s^2. The angular speeds are constant.

17–99 Determine the absolute velocities $\mathbf{v}_1$ and $\mathbf{v}_3$ of persons at positions 1 and 3 in Fig. 17-37 (Ex. 17-15).

17–100 Calculate the absolute acceleration $\mathbf{a}_4$ of a person at position 4 in Fig. 17-37 (Ex. 17-15).

FIGURE P17-101

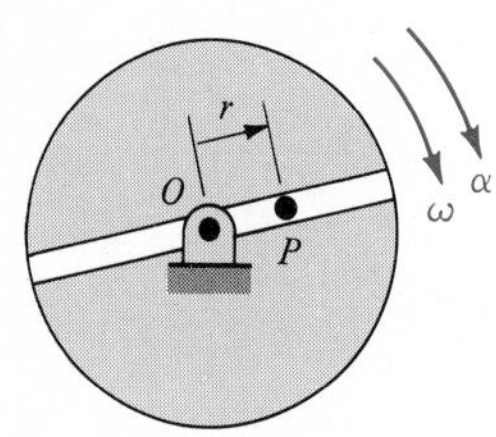

17–101 Particle P slides in the radial groove of the rotating disk. Determine the absolute velocity and acceleration of the particle using a rotating frame for $\omega = 10$ rad/s, $\alpha = 3$ rad/s^2, $r = 4$ in., $\dot{r} = 6$ in./s, and $\ddot{r} = 50$ ft/s^2.

17–102 Assume that $\ddot{r} = 0$ in Prob. 17–101. What is the maximum velocity $\dot{r}$ of the particle if its acceleration normal to the groove should not exceed 60 ft/s^2?

FIGURE P17-103

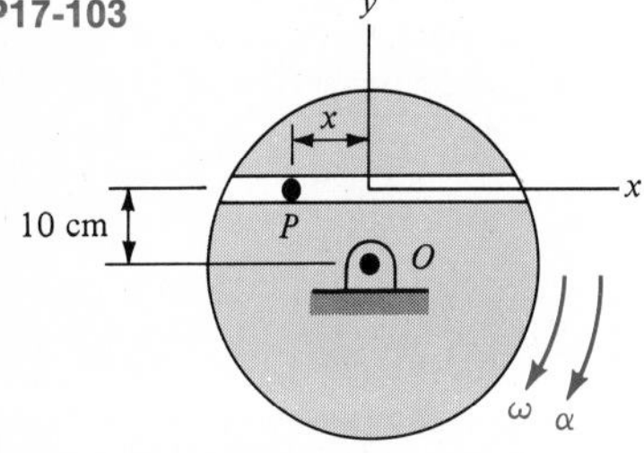

17–103 Particle P slides in the straight groove of the rotating disk. Determine the absolute velocity and acceleration of the particle for $\omega = 5$ rad/s, $\alpha = 4$ rad/s^2, $x = -20$ cm, $\dot{x} = 2$ m/s (increasing x), and $\ddot{x} = 1$ m/s^2 (increasing $\dot{x}$).

17–104 Assume that $\ddot{x} = 0$ in Prob. 17–103. What is the maximum velocity $\dot{x}$ of the particle if its acceleration normal to the groove should not exceed 20 m/s^2?

FIGURE P17-105

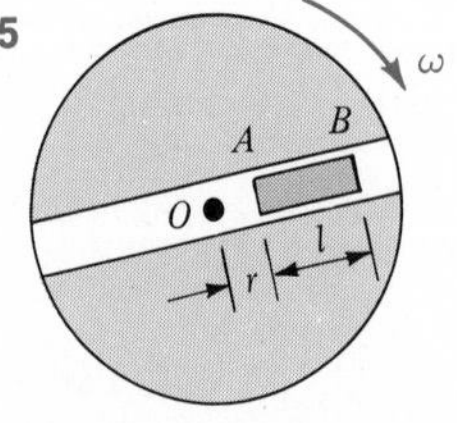

17–105 A slender body of length $l = 1.5$ ft is sliding in the radial groove of the rotating disk. Determine the radial and normal components of the accelerations of the ends A and B of the body for $\omega = 20$ rad/s, $r = 2$ ft, and $\dot{r} = 100$ ft/s (increasing r).

17–106 A slender body of length $l = 30$ cm is sliding in the straight groove of the rotating disk. Determine the x and y components of the accelerations of the ends A and B of the body for $\omega = 15$ rad/s, $\alpha = 2$ rad/s^2, $x_A = 1$ m, and $\dot{x}_A = 50$ m/s (increasing x_A).

FIGURE P17-106

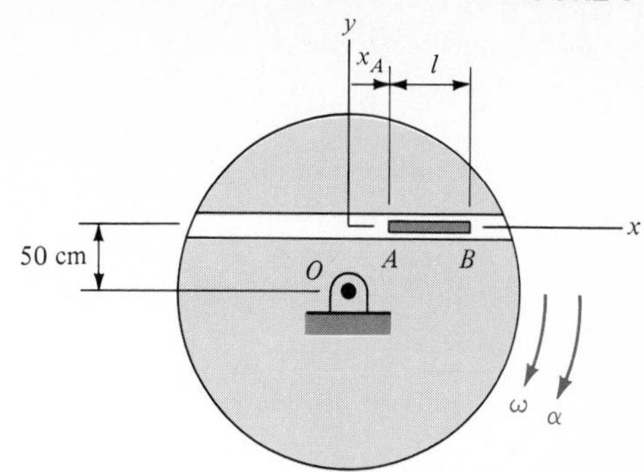

17–107 An amusement ride consists of arm OA rotating at $\Omega = 8$ rpm, and the circular cage at A rotating at $\omega = 5$ rpm relative to the arm OA. A person is moving away from A along the rail ABC at a speed of 0.5 m/s. Determine the person's acceleration at point B for $AB = 2$ m, $\theta = 60°$, and $\phi = 90°$. The angular speeds are constant.

FIGURE P17-107

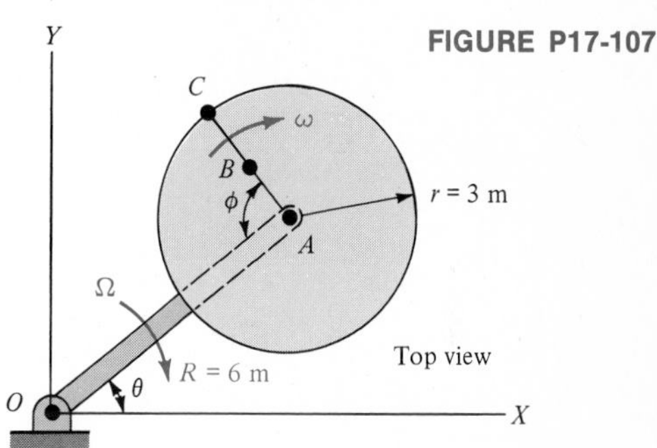

17–108 The crank arm BC rotates at $\omega = 200$ rpm, and pin C on the end of the crank slides in the slotted arm AD. Determine $\dot{\phi}$ and $\ddot{\phi}$ for $\theta = 30°$.

FIGURE P17-108

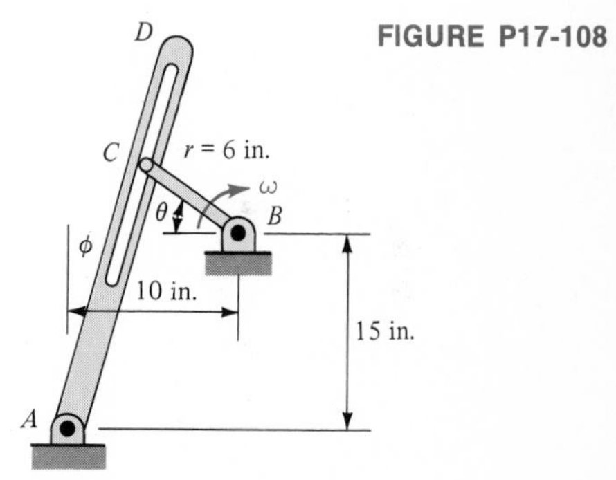

17–109 The Geneva mechanism of a mechanical counter converts the constant rotational motion of wheel A to intermittent rotation of wheel B. Pin C is mounted on A and slides in the slots of wheel B. Determine the angular velocity and angular acceleration of wheel B for $\theta = 30°$ and $\omega_A = 100$ rpm.

FIGURE P17-109

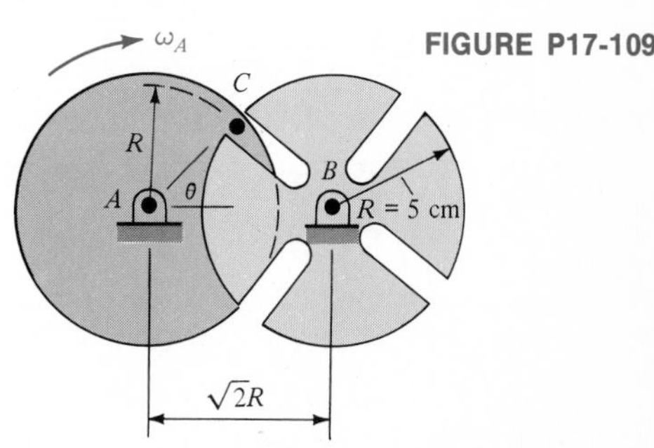

SUMMARY

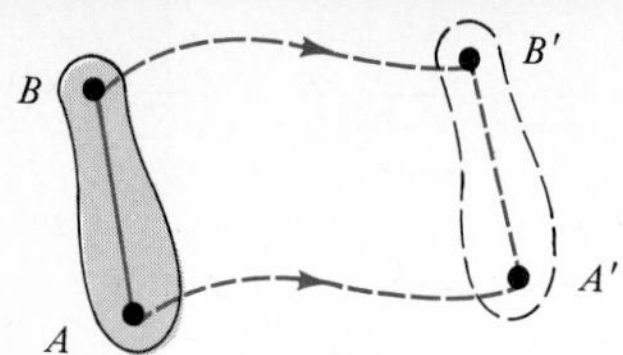

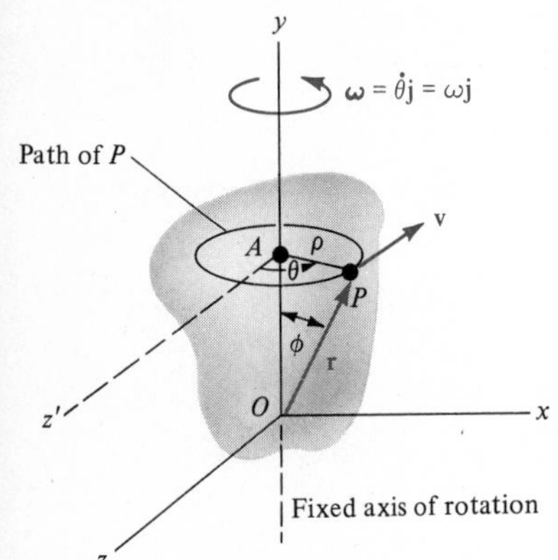

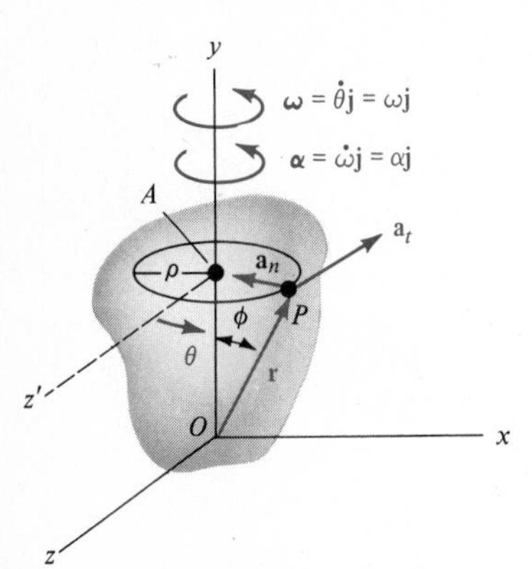

Section 17-1

Translation

$$\mathbf{v}_B = \mathbf{v}_A \qquad \mathbf{a}_B = \mathbf{a}_A$$

Section 17-2

Rotation about a fixed axis

$$\mathbf{v} = \boldsymbol{\omega} \times \mathbf{r}$$

$$\mathbf{a} = \underbrace{\boldsymbol{\alpha} \times \mathbf{r}}_{\mathbf{a}_t} + \underbrace{\boldsymbol{\omega} \times (\boldsymbol{\omega} \times \mathbf{r})}_{\mathbf{a}_n}$$

$$a_t = \rho\alpha \qquad a_n = \rho\omega^2$$

ω and α are valid for any line in a rigid body at a given instant.

Section 17-3

Kinematics equations for rotating rigid bodies. If θ (angular coordinate) is known as a function of time,

$$\omega = \frac{d\theta}{dt}$$

$$\alpha = \frac{d\omega}{dt} = \omega\,\frac{d\omega}{d\theta}$$

SPECIAL CASES:

1. $\omega = \text{constant}$
$\theta = \theta_0 + \omega t \qquad (\theta = \theta_0 \text{ at } t = 0)$

2. $\alpha = \text{constant}$
$\omega = \omega_0 + \alpha t \qquad (\omega = \omega_0 \text{ at } t = 0)$
$\theta = \theta_0 + \omega_0 t + \frac{1}{2}\alpha t$
$\omega^2 = \omega_0^2 + 2\alpha(\theta - \theta_0)$

Section 17-4

Velocities in general plane motion

$$\mathbf{v}_B = \underbrace{\mathbf{v}_A}_{\text{translation}} + \underbrace{\boldsymbol{\omega} \times \mathbf{r}_{B/A}}_{\text{rotation}}$$

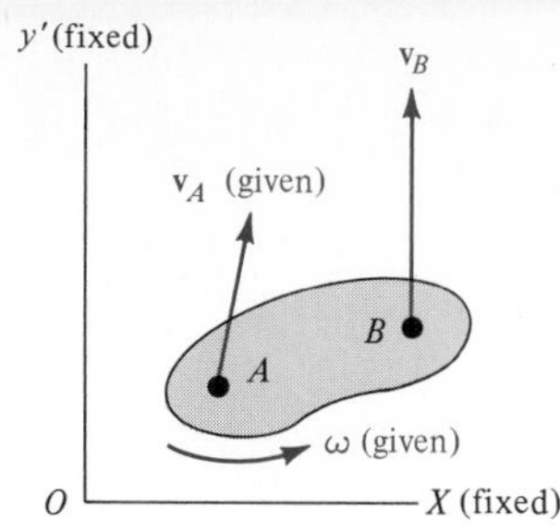

(a) General plane motion

=

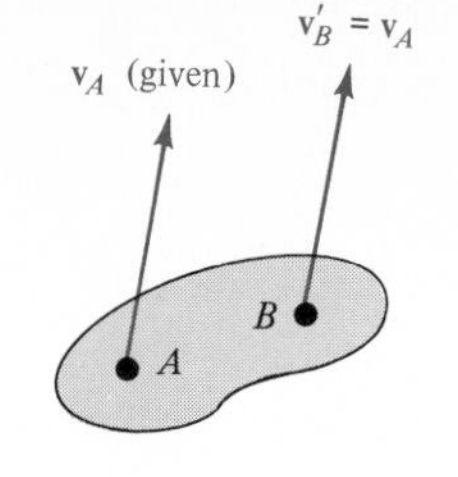

(b) Assume translation only

+

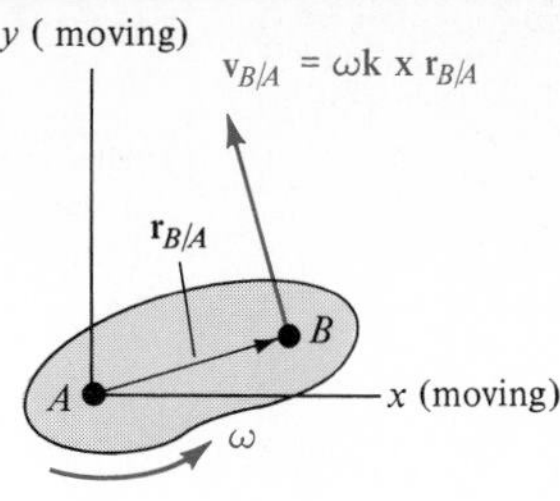

(c) Assume rotation about A only

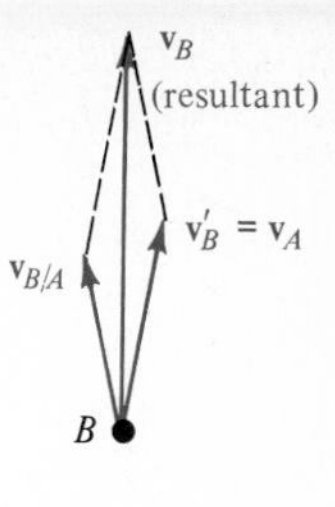

(d)

Section 17-5

Instantaneous center of rotation

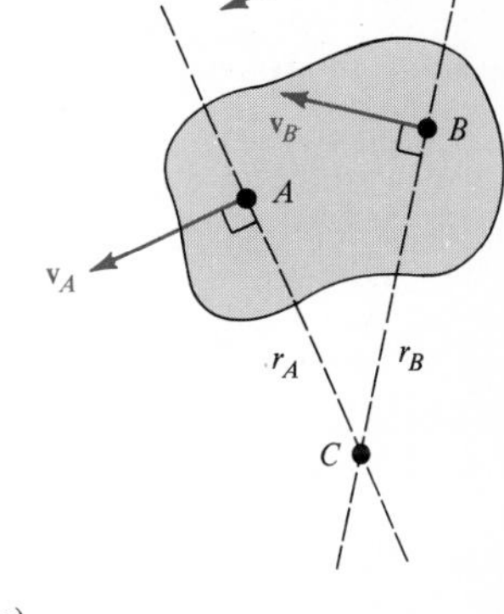

Velocities $\mathbf{v}_A$ and $\mathbf{v}_B$ are given; the body is rotating about point C at that instant. Point C has zero velocity at that instant, but in general it has an acceleration.

Section 17-6

Accelerations in general plane motion

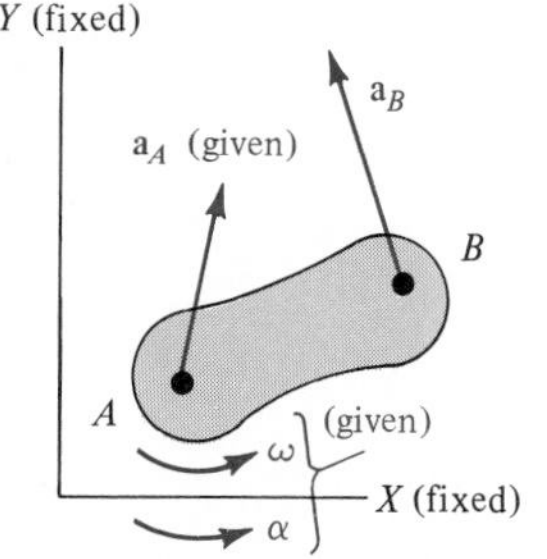

(a) General plane motion

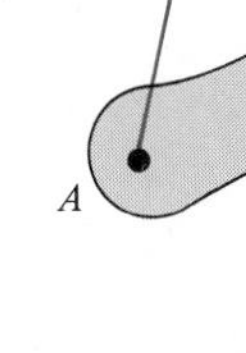

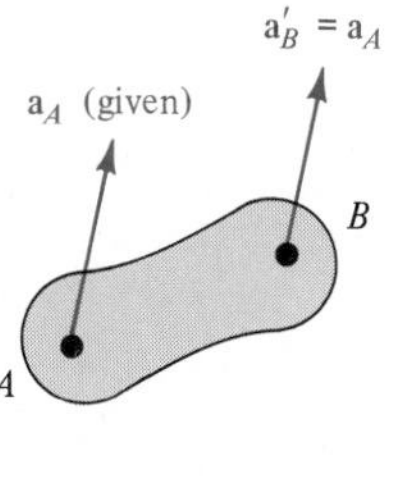

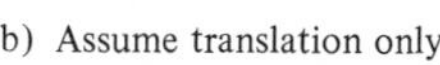

(b) Assume translation only

+

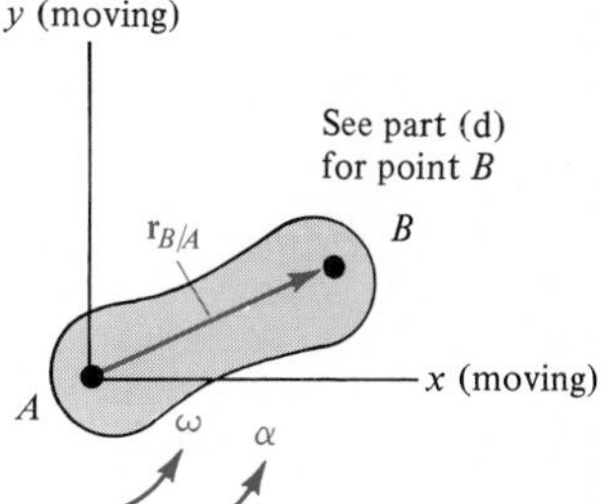

(c) Assume rotation about A only

$$\mathbf{a}_B = \mathbf{a}_A + \boldsymbol{\alpha} \times \mathbf{r}_{B/A} + \boldsymbol{\omega} \times (\boldsymbol{\omega} \times \mathbf{r}_{B/A})$$

$$\mathbf{a}_B = \underbrace{\mathbf{a}_A}_{\text{translation}} + \underbrace{(\mathbf{a}_{B/A})_t + (\mathbf{a}_{B/A})_n}_{\text{rotation}}$$

$$(a_{B/A})_t = r_{B/A}\alpha \qquad (a_{B/A})_n = r_{B/A}\omega^2$$

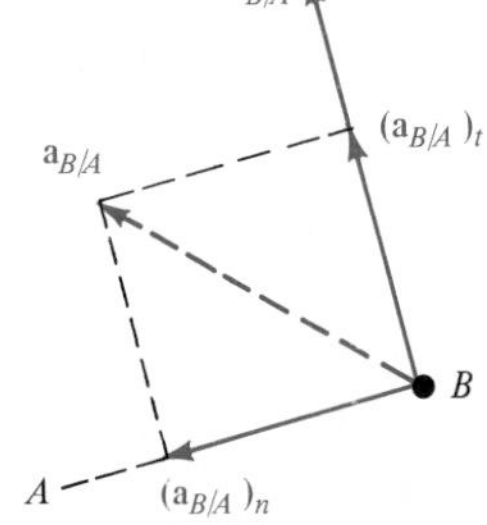

(d) Velocity and accelerations of point B from part (c) (not to scale)

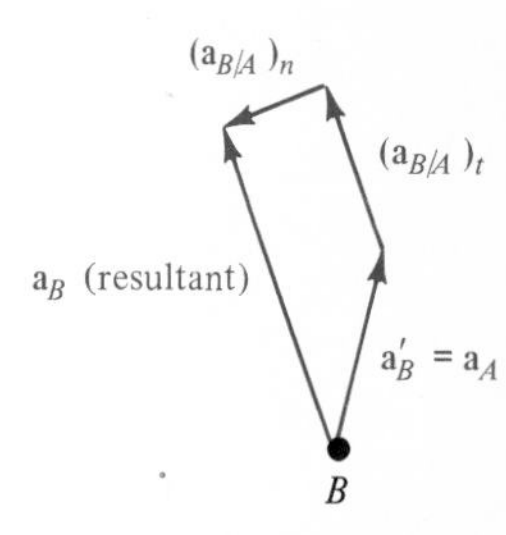

(e)

Section 17-7

General motion

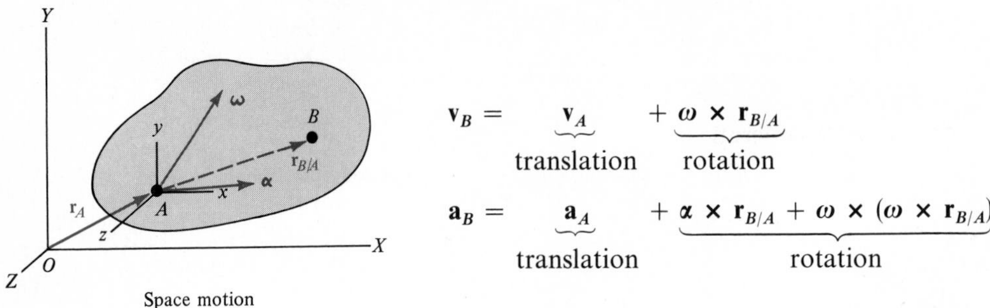

Space motion

$$\mathbf{v}_B = \underbrace{\mathbf{v}_A}_{\text{translation}} + \underbrace{\boldsymbol{\omega} \times \mathbf{r}_{B/A}}_{\text{rotation}}$$

$$\mathbf{a}_B = \underbrace{\mathbf{a}_A}_{\text{translation}} + \underbrace{\boldsymbol{\alpha} \times \mathbf{r}_{B/A} + \boldsymbol{\omega} \times (\boldsymbol{\omega} \times \mathbf{r}_{B/A})}_{\text{rotation}}$$

Angular acceleration. If ω has a constant magnitude,

$$\boldsymbol{\alpha} = \frac{d\boldsymbol{\omega}}{dt} = \boldsymbol{\Omega} \times \boldsymbol{\omega}$$

where $\boldsymbol{\Omega}$ is the angular velocity of $\boldsymbol{\omega}$.

Section 17-8

Time derivative of a vector using a rotating frame

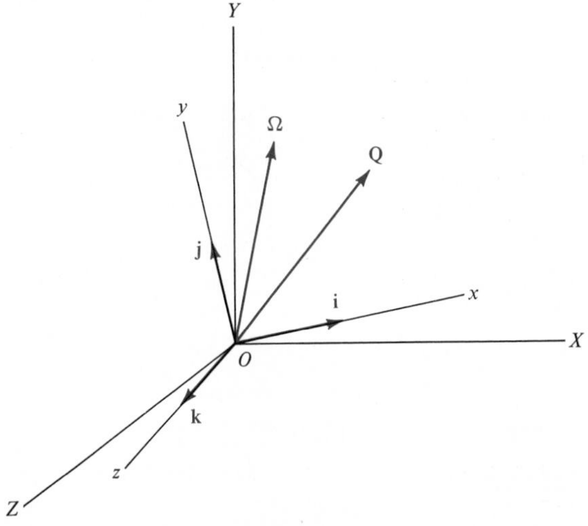

$$(\dot{\mathbf{Q}})_{XYZ} = (\dot{\mathbf{Q}})_{xyz} + \boldsymbol{\Omega} \times \mathbf{Q}$$

where $\mathbf{Q}$ is an arbitrary vector and $\boldsymbol{\Omega}$ is the angular velocity of the xyz frame with respect to XYZ.

Section 17-9

Analysis of velocities and accelerations using a rotating frame

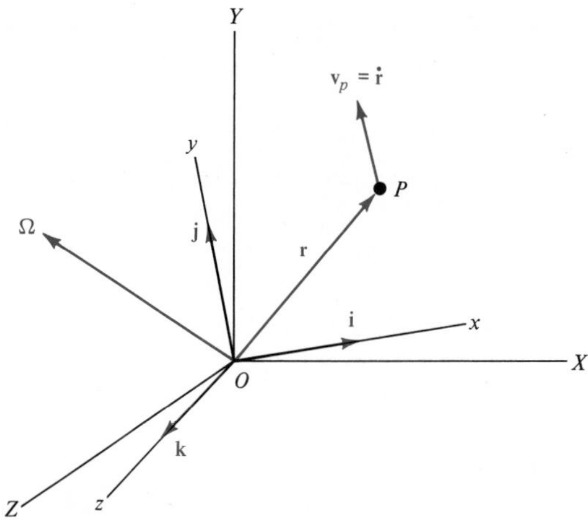

For common origin of XYZ and xyz:

$$\mathbf{v}_P = \mathbf{v}_{xyz} + \boldsymbol{\Omega} \times \mathbf{r}$$

$$\mathbf{a}_P = \mathbf{a}_{xyz} + \dot{\boldsymbol{\Omega}} \times \mathbf{r} + \boldsymbol{\Omega} \times (\boldsymbol{\Omega} \times \mathbf{r}) + 2\boldsymbol{\Omega} \times \mathbf{v}_{xyz}$$

For origin A of xyz translating with respect to XYZ:

$$\mathbf{v}_P = \mathbf{v}_A + (\dot{\mathbf{r}}_{P/A})_{xyz} + \boldsymbol{\Omega} \times \mathbf{r}_{P/A}$$

$$\mathbf{a}_P = \mathbf{a}_A + \mathbf{a}_{xyz} + \dot{\boldsymbol{\Omega}} \times \mathbf{r}_{P/A} + \boldsymbol{\Omega} \times (\boldsymbol{\Omega} \times \mathbf{r}_{P/A}) + 2\boldsymbol{\Omega} \times \mathbf{v}_{xyz}$$

where $\boldsymbol{\Omega}$ is the angular velocity of the xyz frame with respect to XYZ.

REVIEW PROBLEMS

FIGURE P17-110

17–110 The circular saw for road repairs has a diameter of 6 ft. Its chain drive has a 0.5 ft diameter sprocket at A, and a 0.8 ft diameter sprocket at B. Calculate the angular acceleration of sprocket A if the cutting teeth should reach a speed of 8 ft/s in a time of 5 s.

FIGURE P17-111

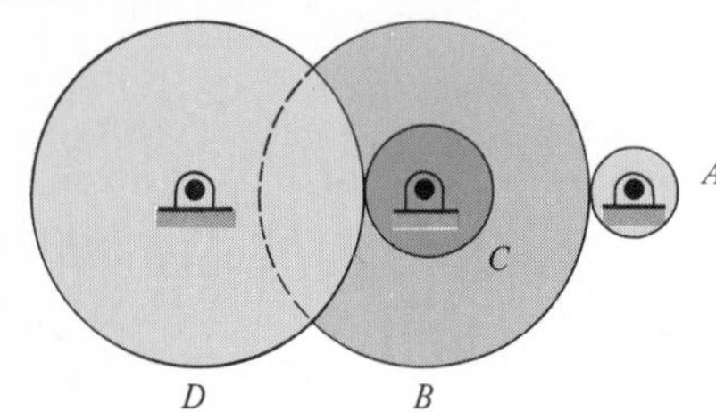

17–111 In a gearbox gear A drives B, and C which is firmly attached to B drives gear D. Determine the ratios of the angular velocities ω_A/ω_D, of the angular accelerations α_A/α_D, and of the number of revolutions n_A/n_D in terms of the diameters d_A, d_B, d_C, and d_D.

FIGURE P17-112

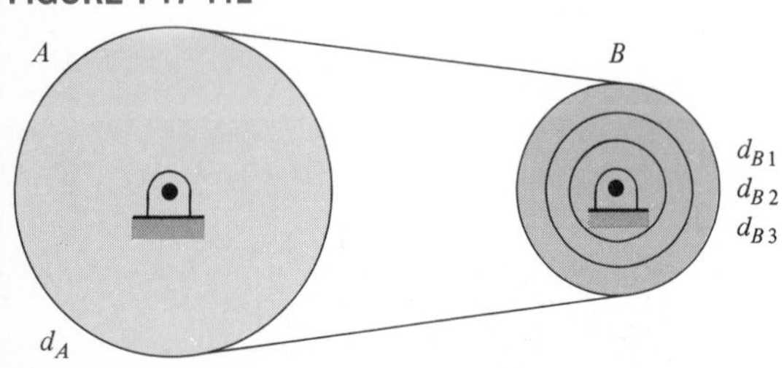

17–112 A chain drive consists of sprocket A with diameter $d_A = 26$ cm, and a composite sprocket B with diameters $d_{B_1} = 10$ cm, $d_{B_2} = 8$ cm, and $d_{B_3} = 6$ cm. The angular acceleration of A is 5 rad/s^2, starting from zero angular velocity. Determine the angular velocity of sprocket B and its number of turns after a complete revolution of sprocket A using each of the three diameters of B.

FIGURE P17-113

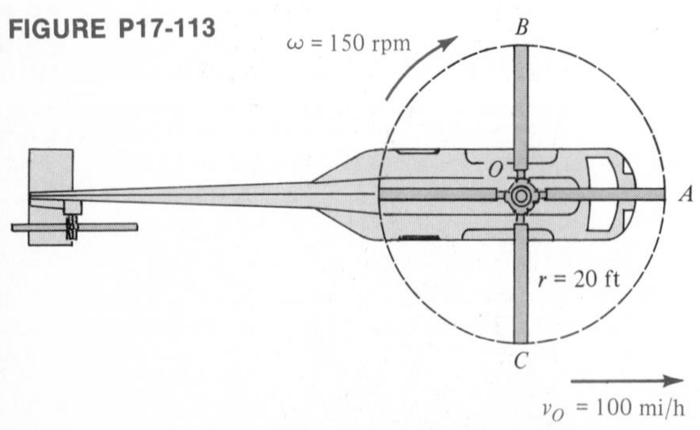

17–113 A helicopter shown in top view is moving horizontally at $v_0 = 100$ mi/h. Determine the instantaneous centers of rotation and the maximum air speeds for the blades A, B, and C in the positions shown.

17–114 A street sweeper is moving at speed $v = 6$ km/h while its 80-cm diameter brush is rotating at $\omega = 100$ rpm. Determine the instantaneous center of rotation of the brush and the bristles' speed against the ground at point A.

FIGURE P17-114

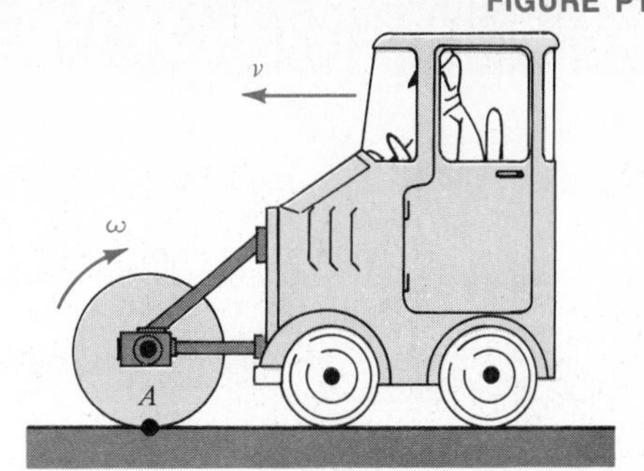

17–115 An amusement ride consists of circular cages rotating at $\omega_2 = 30$ rpm with respect to arm AB which rotates about point O at $\omega_1 = 20$ rpm. Determine the velocity of point A_1 for $\theta = 30°$ and $\phi = 45°$. Locate the instantaneous center of rotation of line AA_1 in the given position.

FIGURE P17-115

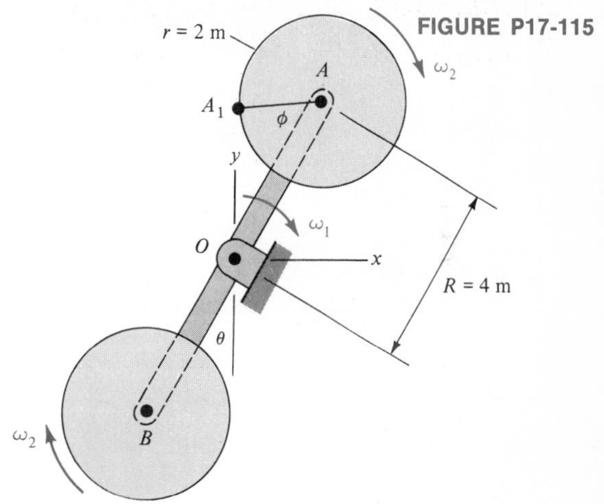

17–116 Tracks T of the model-testing equipment are inclined 10° from the vertical, and axis AA always moves parallel to these tracks. Assume that arm AB is vertical and rotating at a clockwise angular speed $\omega = 0.5$ rad/s, while axis AA has zero velocity and a downward acceleration of 3 m/s^2. Determine the acceleration of engine C which is fixed in the plane of axes AA and BB and is at a distance 1.3 m from axis BB.

17–117 Solve Prob. 17–116 with additional motion of the model aircraft. It rotates at a clockwise angular speed of 1.5 rad/s relative to the arm AB, and this speed is decreasing at the rate of 0.2 rad/s^2. Engine C is in the plane of AA and BB at the instant considered, but not fixed in that plane.

FIGURE P17-116

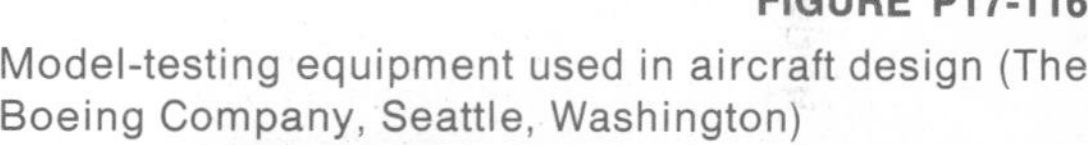
Model-testing equipment used in aircraft design (The Boeing Company, Seattle, Washington)

FIGURE P17-118

17–118 An airplane takes off at a velocity $\mathbf{v} = 220\mathbf{k}$ km/h. The wheels are 40 cm in diameter, and they are retracted soon after leaving the ground. The landing gear A has angular velocity $\omega_2 = 1\mathbf{k}$ rad/s. Determine the angular acceleration of the landing gear during retraction.

17–119 Determine the angular acceleration of both wheels A and B in Prob. 17–118. Calculate the acceleration of particle P assuming that it is directly below the pivot in the wing, at a distance $d = 1$ m from the pivot.

FIGURE P17-120

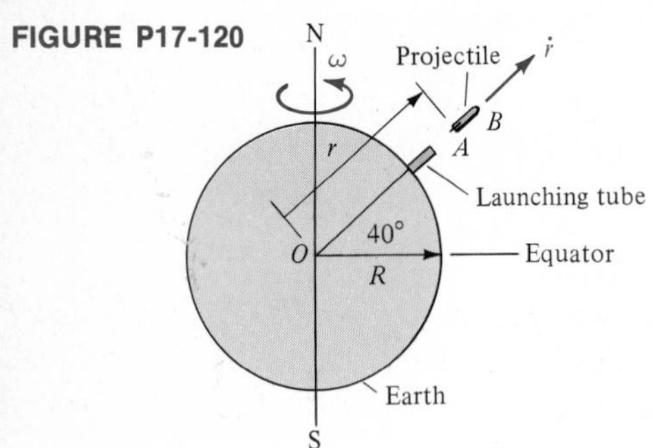

17–120 A 20 ft long projectile for atmospheric research is to be launched from a long vertical tube. $R = 3960$ mi, $r \approx R$, $\dot{r} = 2000$ ft/s, $\ddot{r} = 0$, $\omega = 0.729 \times 10^{-4}$ rad/s. Determine the accelerations of the projectile's head B and tail A in the tube, in the direction normal to the tube's axis.

17–121 Determine the maximum value of $\dot{r}$ in Prob. 17–120 if the acceleration normal to the tube's axis should not exceed 100 ft/s^2.

CHAPTER 18

KINETICS OF RIGID BODIES IN PLANE MOTION

OVERVIEW

In studying the kinetics of rigid bodies it is necessary to consider plane motion before space motion because the latter is quite complex and is often omitted from elementary courses. Also, planar kinetics is very important in engineering practice because many components of systems of rigid bodies have two-dimensional motions. The kinetics analysis of rigid bodies is based on Newton's second law of motion and the methods developed in Chap. 16.

STUDY GOALS

SECTION 18-1 presents the equation of motion for the translation of a rigid body. This equation is fundamental in analyzing the motion of rigid bodies and should be remembered for frequent application in dynamics.

SECTION 18-2 presents the derivation of the equation of motion for the rotation of a rigid body. This equation is also fundamental in analyzing the motion of rigid bodies and should be remembered. Memorizing the derivation is not necessary, but the resulting equation of motion is very important in dynamics.

SECTION 18-3 describes several practical aspects of applying the basic vector equations of motion, one for translation and another for rotation. Three scalar equations of motion are written from these for the convenient analysis of plane motion. Students should acquire skill in using these equations.

SECTION 18-4 presents the analysis of rigid bodies which rotate about fixed axes. This topic is important in machine design. The subtopic titled *center of percussion* is an interesting special case of rotation about a fixed axis, but it is not essential to make further progress in elementary dynamics.

SECTION 18-5 presents the methods of analyzing common problems of general plane motion, such as the rolling *and* sliding of wheels. These special problems provide for useful practice in applying the equations of motion.

18-1 EQUATION OF TRANSLATIONAL MOTION

To establish the methods of analysis, assume that a rigid body is a thin slab of arbitrary shape and is moving in such a way that all of its constituent particles move parallel to the fixed xy frame (Fig. 18-1a). The body consists of n particles of mass m_i with $i = 1, 2, \ldots, n$. Assume that the mass center C and all significant forces acting on the body are in the xy plane and are completely specified. In this case the system of forces can be replaced by its resultant force $\sum \mathbf{F}$ acting at C and by the resultant moment $\sum \mathbf{M}_C$ acting about C as in Fig. 18-1b.

The resultant moment is a couple, so it does not affect the translational motion of the body in any way. The translational motion depends only on the net force $\sum \mathbf{F}$ and the total mass m of the body. This can be expressed using Eq. 16-7, since the rigid body is a system of particles,

$$\boxed{\sum \mathbf{F} = \sum m_i \mathbf{a}_i = m\mathbf{a}_C} \qquad \text{18-1}$$

where $\mathbf{a}_C$ is the acceleration of the mass center. The resultant force and the acceleration vectors are collinear.

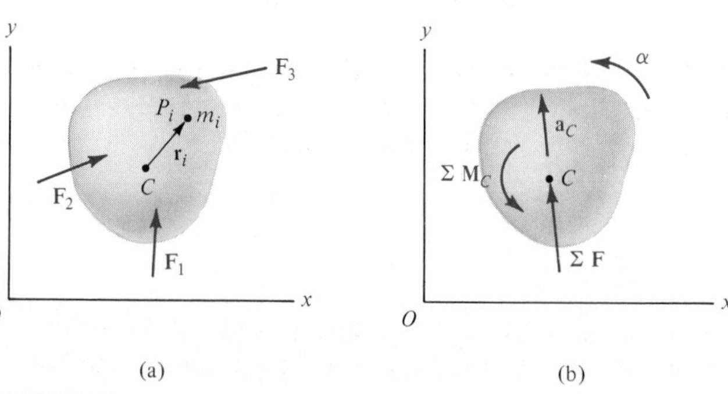

FIGURE 18-1
Plane motion of a rigid body

Equation 18-1 is fundamental in analyzing the motion of rigid bodies. It is simple to use, but certain aspects of it are worth noting as follows.

1. Newton's second law $\mathbf{F} = m\mathbf{a}$ is applicable only to individual particles having mass but no dimensions, or to the mass center of a system of discrete particles where $\mathbf{F}$ is the resultant of individual forces acting on individual particles. In the case of a rigid body it is assumed that the force acting on any one particle of the body is instantaneously acting on all particles of that body. This is modeled in Fig. 18-2, where a single force is acting on a body of six equal particles. The force is directly acting on particle 2, but the six particles are rigidly connected. According to Eq. 18-1, this is equivalent to having a single particle of mass 6 m, so the acceleration of the mass center is $\mathbf{a}_C = \mathbf{F}/6$ m.

2. The acceleration of the mass center depends only on the total mass of the body and the resultant force acting on the body. Thus, in Fig. 18-2 the force $\mathbf{F}$ may be moved to any particle without affecting the magnitude or direction of $\mathbf{a}_C$. This is independent of any rotation of the body, and consequently it is extremely useful in the analysis of motion.

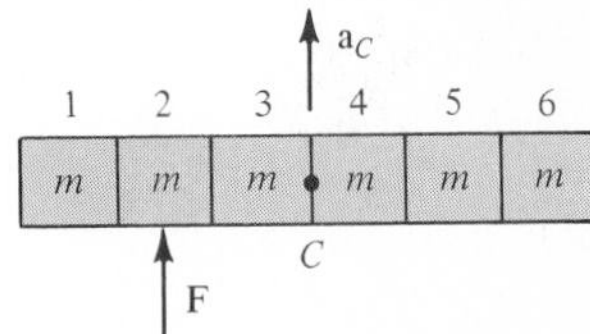

FIGURE 18-2

A rigid body modeled as a system of six identical particles

EQUATION OF ROTATIONAL MOTION 18-2

The resultant force $\sum \mathbf{F}$ in Fig. 18-lb fully accounts for the translational acceleration of a given body caused by the system of forces. The resultant moment of these forces about the center of mass C causes the body to rotate at an angular acceleration $\boldsymbol{\alpha}$ (independent of the linear acceleration $\mathbf{a}_C$). Using Eqs. 16-24 and 16-29 for a system of particles gives

$$\sum \mathbf{M}_C = \sum (\mathbf{r}_i \times m_i\mathbf{a}_i) = \dot{\mathbf{H}}_C \qquad \boxed{18\text{-}2}$$

where $\dot{\mathbf{H}}_C$ is the rate of change of the angular momentum $\mathbf{H}_C$ of the system of particles about its mass center. The angular momentum for the body in Fig. 18-1 is expressed according to Eq. 16-27 as

$$\mathbf{H}_C = \sum_{i=1}^{n} (\mathbf{r}_i \times m_i\mathbf{v}_i) \qquad \boxed{18\text{-}3a}$$

where $\mathbf{r}_i$ is the position vector* and $m_i\mathbf{v}_i = m_i\dot{\mathbf{r}}_i$ is the linear momentum of particle P_i with respect to the mass center of the body. With

* $\mathbf{r}_i$ here is equivalent to $\mathbf{r}'_i$ in Eq. 16-27. The change in notation is advantageous in the following analysis.

$\mathbf{v}_i = \boldsymbol{\omega} \times \mathbf{r}_i$ for any rigid body, where $\boldsymbol{\omega}$ is the angular velocity of the body,

$$\mathbf{H}_C = \sum_{i=1}^{n} [\mathbf{r}_i \times (\boldsymbol{\omega} \times \mathbf{r}_i)m_i] \tag{18-3b}$$

Since $\mathbf{r}_i$ and the velocity $\mathbf{v}_i$ are both parallel to the xy plane for every particle in a thin slab, $\mathbf{r}_i \times \mathbf{v}_i$ and hence the angular momentum vector $\mathbf{H}_C$ are perpendicular to the xy plane. Thus, for plane motion, $\mathbf{H}_C$ is in the same direction and has the same sense as $\boldsymbol{\omega}$ as found by using the right-hand rule. The magnitude of the angular momentum from Eq. 18-3b is $H_C = \omega \sum r_i^2 m_i$ since $\boldsymbol{\omega}$ is the same for all particles. Here each term $r_i^2 m_i$ is the moment of inertia of particle P_i about the center of mass. The summation of the moments of inertia for all infinitesimal particles of a rigid body is equivalent to writing $I_C = \int r^2\, dm$, the mass moment of inertia of the body about the centroidal axis perpendicular to the xy plane. Therefore, the concise expression for the angular momentum (for plane motion only) is

$$\boxed{\mathbf{H}_C = I_C \boldsymbol{\omega}} \tag{18-4}$$

Since I_C is a constant property of the body in plane motion, differentiating Eq. 18-4 with respect to time gives

$$\dot{\mathbf{H}}_C = I_C \dot{\boldsymbol{\omega}} = I_C \boldsymbol{\alpha} \tag{18-5}$$

where $\boldsymbol{\alpha}$ is the angular acceleration of the body. Substituting Eq. 18-5 in Eq. 18-2 yields

$$\boxed{\sum \mathbf{M}_C = I_C \boldsymbol{\alpha}} \tag{18-6}$$

This equation is equally important to Eq. 18-1 in analyzing the motion of rigid bodies. The following comments are worth noting.

1. Equation 18-2 is strictly applicable only to a system of discrete particles where individual forces are acting on individual particles. In the case of a rigid body it is assumed that the moment of a force acting on any one particle is instantaneously changing the angular momentum of all particles of the body. This is modeled in Fig. 18-3,

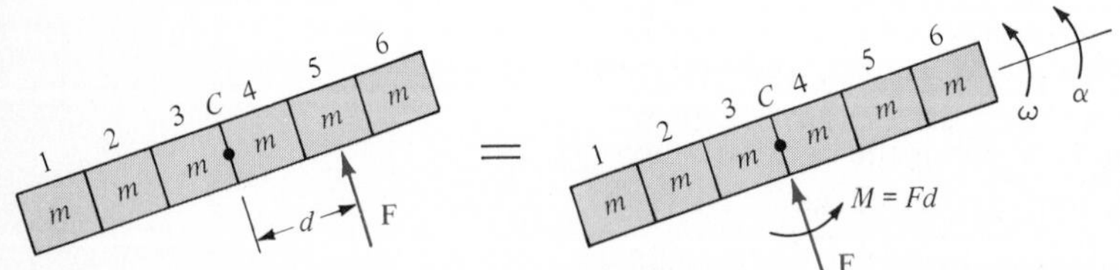

FIGURE 18-3

Rotation of a rigid body which is modeled as a system of six particles where a single force is acting on six rigidly connected particles that rotate about point C. Here $\sum \mathbf{M}_C = \mathbf{F}d = I_C\boldsymbol{\alpha}$, and the particles respond to the moment in unison, which means that ω and α are unique to the whole body at a given instant. This is equivalent to having a single particle of mass 6 m located on a massless link at a distance $r_g = \sqrt{I_C/6m}$ (r_g = radius of gyration) from point C. It appears that the angular motion of particle 5 is retarded by the other five particles. This effect depends on the square of the particles' distance from the mass center, so particle 6 has a much greater effect than particle 4 does on the motion of particle 5, even though they have the same mass.

2. The angular acceleration of the body depends only on its mass moment of inertia and the resultant moment acting on it. This is independent of any translational motion of the body.

3. An important restriction of Eq. 18-6 is that the body is assumed to be a thin slab with $\mathbf{r}_i$ and $\mathbf{v}_i$ parallel to the xy plane. This equation is also used for other bodies in special cases, when the xy plane is a plane of symmetry (with products of inertia being zero), or when all $\mathbf{v}_i$ are parallel to the xy plane because of constraints.

18-3 APPLICATION OF EQUATIONS OF MOTION

The fundamental equations (Eqs. 18-1 and 18-6) that govern the translation and rotation of a rigid body in plane motion are most convenient to use in scalar form. This results in a maximum of three equations. Defining the xy plane as the plane of motion,

$$\sum F_x = ma_{C_x} \qquad \sum F_y = ma_{C_y} \qquad \sum M_C = I_C\alpha \tag{18-7}$$

which can be integrated to obtain the velocities and positions of a body in terms of initial conditions.

To apply the equations of motion (Eq. 18-7), it is necessary to use proper free-body diagrams of the body or system of bodies. It is essential that the external force system acting on each body be represented by its equivalent resultant force and moment. The translation or rotation of a body is analyzed using the equations of motion as follows.

1. *Translation.* The resultant force in each coordinate direction is applied to the total mass as if the rigid body were a particle. There is no rotation of the body, so from Eq. 18-7,

$$\sum F_x = ma_{C_x} \qquad \sum F_y = ma_{C_y} \qquad \sum M_C = 0 \tag{18-8}$$

2. *Rotation.* The resultant moment about the center of mass causes rotation about that point. There is no translation of the center of mass, so from Eq. 18-7,

$$\sum F_x = 0 \qquad \sum F_y = 0 \qquad \sum M_C = I_C\alpha \tag{18-9}$$

Note that the equation of rotational motion was obtained using the center of mass for reference.

Constraints of Motion

The motions of rigid bodies are categorized according to the constraints affecting the motion:

1. *Unconstrained motion* is the general case which is exemplified by the motion of aircraft and rockets. In the case of unconstrained plane motion Eq. 18-7 can be directly applied with the three scalar equations being independent of one another.
2. *Constrained motion* is most commonly encountered in engineering work since the components of machines generally move in fixed patterns. When there is a physical constraint to the motion, a *kinematics analysis* has to be made first to determine the relation of linear and angular accelerations of the body. This is necessary because in constrained plane motion the three accelerations a_{C_x}, a_{C_y}, and α are interrelated, so the three equations of Eq. 18-7 are not independent of one another. The force and moment equations can be solved after the kinematics analysis. Two special cases of constrained motion are distinguished:
   a. *Point constraint* is typical of rotation about a fixed point or axis as in Fig. 18-4.
   b. *Line constraint* exists when the body moves along a fixed line or plane. An example of this is a wheel rolling along a road surface as in Fig. 18-5.

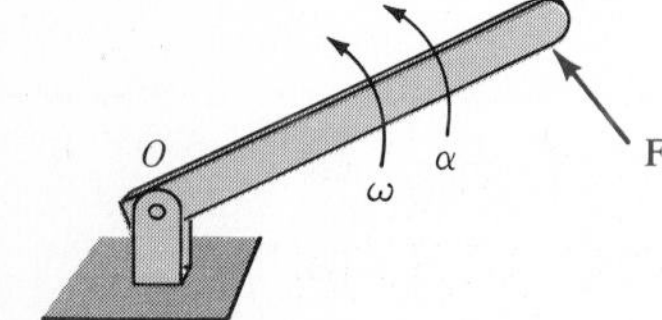

FIGURE 18-4

Axis of rotation is the point of constraint

Systems of Rigid Bodies

Frequently, a member's motion is constrained by other members that are also in motion. Such motions can be readily analyzed if the motions of the interconnected bodies can be determined kinematically. After the kinematics analysis, the force and moment equations can be solved using individual but interrelated free-body diagrams or using the free-body diagram for the entire system if convenient. It should be remembered that the required number of free-body diagrams depends on the number of unknowns in the problem. At most three equations of motion (Eq. 18-7) are available from a given free-body diagram for plane motion to solve for three unknowns. The equations of motion are first applied in examples of translational motion in the following.

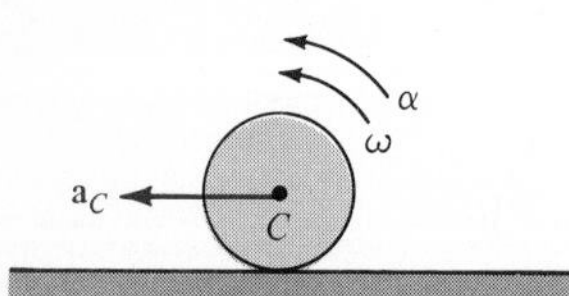

FIGURE 18-5

Line of constraint is on the supporting surface of the wheel

EXAMPLE 18-1

The combined mass of the motorcycle and its rider in Fig. 18-6a is 250 kg. The mass center is at C. The wheels are locked when the brakes are applied at a high speed. Determine the linear acceleration and the normal and tangential forces acting on each wheel assuming that the coefficient of kinetic friction is 0.5, and that the angular acceleration is zero.

SOLUTION

First the free-body diagram is sketched in Fig. 18-6b, showing the known and unknown forces externally applied to the body. The constraints of motion are that $a_{C_y} = 0$ and $\alpha = 0$. Using the three equations from Eq. 18-8 for pure translation,

$$\sum F_x = \mathscr{F}_A + \mathscr{F}_B = ma_{C_x} \quad \textbf{(a)}$$

$$\sum F_y = N_A + N_B - W = ma_{C_y} = 0 \quad \textbf{(b)}$$

$$\sum M_C = \mathscr{F}_A(0.7 \text{ m}) - N_A(0.8 \text{ m}) + \mathscr{F}_B(0.7 \text{ m}) + N_B(0.5 \text{ m}) = I_C\alpha = 0 \quad \textbf{(c)}$$

Since the wheels are locked and sliding, the friction forces are $\mathscr{F}_A = \mu_k N_A$ and $\mathscr{F}_B = \mu_k N_B$. Hence, from Eqs. (b) and (c),

$$N_A + N_B = W = (250 \text{ kg})(9.81 \text{ m/s}^2) = 2453 \text{ N}$$

$$\mu_k N_A(0.7 \text{ m}) - N_A(0.8 \text{ m}) + \mu_k N_B(0.7 \text{ m}) + N_B(0.5 \text{ m}) = 0$$

$$N_A(0.35 - 0.8) \text{ m} + N_B(0.35 + 0.5) \text{ m} = 0$$

Solving these simultaneously yields

$$N_A = 1604 \text{ N} \qquad N_B = 849 \text{ N}$$

$$\mathscr{F}_A = \mu_k N_A = 802 \text{ N} \qquad \mathscr{F}_B = \mu_k N_B = 425 \text{ N}$$

Hence, from Eq. (a),

$$a_{C_x} = \frac{1}{250 \text{ kg}}(802 + 425) \text{ N} = 4.91 \text{ m/s}^2$$

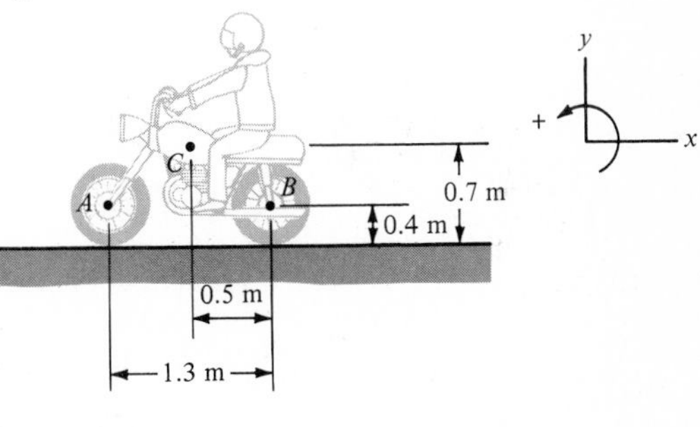

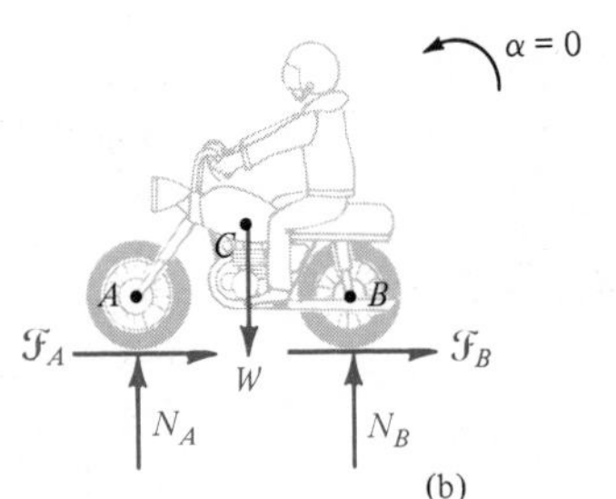

FIGURE 18-6

Motorcycle and its rider modeled as a rigid body

EXAMPLE 18-2

Assume that a small trailer of 70 kg mass is attached to the motorcycle described in Ex. 18-1. The draw bar is firmly attached to the trailer and is hinged at axle B (Fig. 18-7a). The trailer has a single wheel of negligible mass and no brake. Model the trailer as a rigid body which has no angular acceleration and which moves without friction. Determine the linear acceleration a_x of the system when the wheels of the motorcycle slide during braking.

SOLUTION

Free-body diagrams are drawn to show the external forces on the whole system (Fig. 18-7b), on the trailer alone (Fig. 18-7c), and on the motorcycle alone (Fig. 18-7d). The subscripts t and m are used for the trailer and motorcycle, respectively. Considering the system as a whole, there are four unknowns, N_A, N_B, N_t, and a_x, but only three equations of motion. Analyzing the trailer and the motorcycle separately, there are six independent equations of motion and six unknowns, N_A, N_B, N_t, a_x, R_x, and R_y. From the free-body diagrams:

Trailer:

$$\sum F_x = R_x = m_t a_x \qquad \textbf{(a)}$$

$$\sum F_y = R_y + N_t - W_t = m_t a_{t_y} = 0 \Rightarrow R_y = W_t - N_t \qquad \textbf{(b)}$$

$$\sum M_{C_t} = I_{C_t}\alpha_t = 0$$

$$R_x(0.1\text{ m}) - R_y(1.0\text{ m}) = 0 \Rightarrow R_x = 10R_y \qquad \textbf{(c)}$$

Motorcycle:

$$\sum F_x = \mathscr{F}_A + \mathscr{F}_B - R_x = m_m a_x \qquad \textbf{(d)}$$

$$\sum F_y = N_A + N_B - W_m - R_y = m_m a_{m_y} = 0 \qquad \textbf{(e)}$$

$$\sum M_{C_m} = \mathscr{F}_A(0.7\text{ m}) + \mathscr{F}_B(0.7\text{ m}) - N_A(0.8\text{ m}) + N_B(0.5\text{ m})$$
$$- (0.3\text{ m})R_x - (0.5\text{ m})R_y = I_{C_m}\alpha_m = 0 \qquad \textbf{(f)}$$

There are several alternative procedures to obtain a_x from these equations.

SOLUTION A

From Eq. (d),

$$\mathscr{F}_A + \mathscr{F}_B = R_x + m_m a_x$$

$$\mu_k(N_A + N_B) = R_x + m_m a_x \qquad \textbf{(d')}$$

From Eq. (e),

$$N_A + N_B = W_m + R_y$$

Dividing Eq. (d′) by μ_k and subtracting Eq. (e) from it,

$$0 = \frac{R_x}{\mu_k} + \frac{m_m a_x}{\mu_k} - (W_m + R_y)$$

$$R_x + m_m a_x - \mu_k W_m - \mu_k R_y = 0$$

Using $R_x = 10R_y$ from Eq. (c),

$$10R_y - \mu_k R_y + m_m a_x - \mu_k W_m = 0$$

$$9.5R_y + m_m a_x - \mu_k W_m = 0$$

Using Eqs. (a) and (c),

$$(0.1)(m_t)(a_x)(9.5) + m_m a_x = \mu_k W_m$$

$$a_x(66.5 + 250) = (0.5)(250)(9.81)$$

$$a_x = 3.87 \text{ m/s}^2 \qquad (3.8744)$$

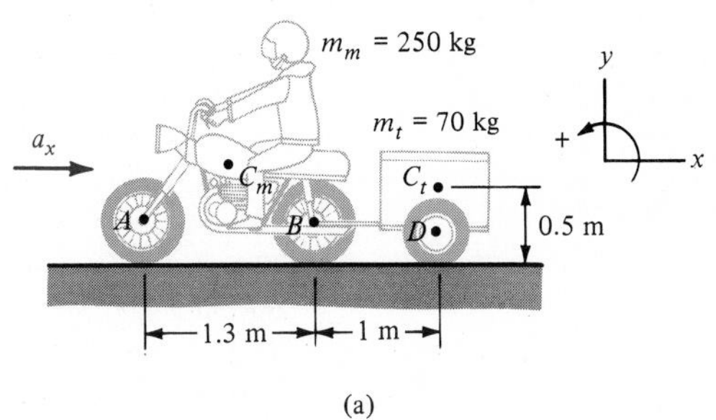

(a)

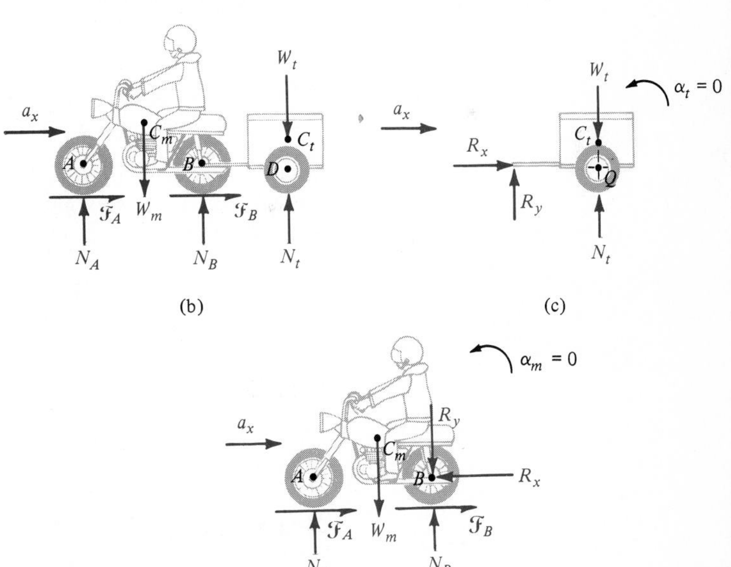

(b)

(c)

(d)

FIGURE 18-7

Analysis of a motorcycle and the attached trailer

SOLUTION B

Using Eqs. (a), (b), and (c),

$$m_t a_x = 10(W_t - N_t) \qquad \textbf{(g)}$$

From Eqs. (a) and (d),

$$\mathscr{F}_A + \mathscr{F}_B = (m_m + m_t)a_x \qquad \textbf{(h)}$$

From Eqs. (b), (c), and (e),

$$N_A + N_B = W_m + (0.1)m_t a_x \qquad \textbf{(i)}$$

From Eqs. (a), (b), (g), and (f),

$$(0.7\text{ m})(\mathscr{F}_A + \mathscr{F}_B) - N_A(0.8\text{ m}) + N_B(0.5\text{ m}) - (0.3\text{ m})m_t a_x - (0.5\text{ m})(0.1)m_t a_x = 0 \qquad \textbf{(j)}$$

Using $\mathscr{F}_A = \mu_k N_A$ and $\mathscr{F}_B = \mu_k N_B$, Eq. (h), (i), and (j) can be solved giving

$$a_x = 3.87\text{ m/s}^2$$

SOLUTION C

The moment equations may be simplified if, instead of using the mass centers, the reference points are carefully chosen to minimize the number of terms. Such a point for the motorcycle is point P at the intersection of forces $\mathscr{F}_B$ and N_B, eliminating the moments of $\mathscr{F}_A$, $\mathscr{F}_B$, N_B, and R_y. The moment equation using Eq. 16-31 and 18-5 is

$$\sum \mathbf{M}_P = I_C\boldsymbol{\alpha} + \mathbf{r}_{CP} \times m_m\mathbf{a}_C$$

where $\boldsymbol{\alpha} = 0$. Thus,

$$-N_A(1.3) + W_m(0.5) + R_x(0.4) = -(0.7)(m_m)(a_x)$$

For the trailer, point Q at the intersection of forces R_x and N_t is advantageous for reference of moments, giving (again from Eq. 16-31)

$$-R_y(1) = -(0.1)(m_t)(a_x)$$

The solution using the original force summations and the latter moment equations also yields

$$a_x = 3.87\text{ m/s}^2$$

JUDGMENT OF THE RESULT

The answer is also reasonable in comparison with $a_x = 4.91\text{ m/s}^2$ in Ex. 18-1 since the mass of the trailer moves without friction and its center of mass is directly above the trailer wheel (thus it has a small effect on the forces N_A and N_B). For another check, solve for N_A and N_B using $a_x = 3.8744\text{ m/s}^2$ and prove that this value of a_x is acceptable since it satisfies the equations of motion.

EXAMPLE 18-3

In Fig. 18-8a the hydraulic actuator A makes an angle θ with the horizontal and is attached to the horizontal platform in the plane of joints B and C. The uniform platform weighs 200 lb, is 8 in. thick, and 5 ft long between B and C. Each rocker arm (BE) is 2.5 ft long, is essentially weightless, and makes an angle ϕ with the horizontal. The test object weighs 250 lb and its center of mass is 2 ft above the platform. Determine the actuator force that would cause slipping of the test object while raising the platform starting at rest from $\theta = 45°$, $\phi = 60°$, if $\mu_s = 0.2$.

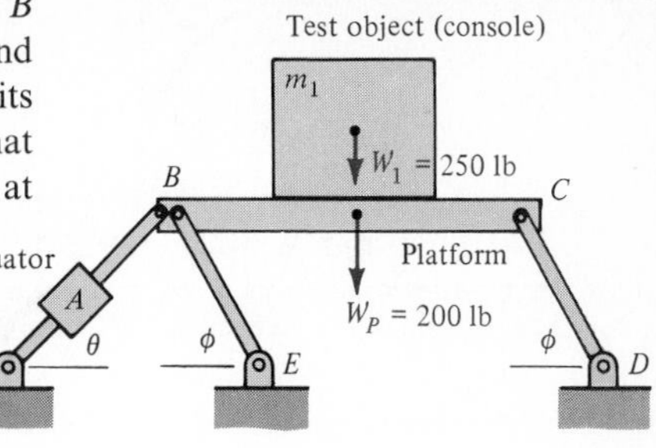

(a)

SOLUTION

First the free-body diagram of the platform and console is sketched in Fig. 18-8b. The directions of the normal and tangential components of the acceleration of the console and platform are known. Also, since this is curvilinear translation only, there is no rotational acceleration. At this stage there are four unknowns, forces F_A, F_{BE}, F_{CD}, and the tangential acceleration a_t. The actuator force necessary to cause slipping to impend is requested, which is enough information to determine the acceleration of the console a_t. A separate sketch of the console is shown in Fig. 18-8c. Since slipping is impending, $\mathscr{F} = \mu_s N_1$, which allows writing

$$F_x = \mu_s N_1 = m_1 a_x = m_1 a_t \cos 30° \qquad \textbf{(a)}$$

$$F_y = N_1 - W_1 = m_1 a_y = m_1 a_t \sin 30° \qquad \textbf{(b)}$$

From Eq. (a),

$$N_1 = \frac{1}{\mu_s}(m_1 \cos 30°)a_t = \frac{1}{0.2}\left(\frac{250 \text{ lb}}{32.2 \text{ ft/s}^2}\right)\cos 30° \; a_t$$

$$= \left(33.62\,\frac{\text{lb}\cdot\text{s}^2}{\text{ft}}\right)a_t$$

Substituting this into Eq. (b) gives

$$\left(33.62\,\frac{\text{lb}\cdot\text{s}^2}{\text{ft}}\right)a_t - 250 \text{ lb} = \left(\frac{250 \text{ lb}}{32.2 \text{ ft/s}^2}\right)\sin 30° \; a_t$$

which yields

$$a_t = 8.41 \text{ ft/s}^2 \qquad \text{in the } t \text{ direction}$$

This is the acceleration at which the console would begin slipping.

Referring again to Fig. 18-8b, forces are summed in the tangential direction,

$$\sum F_t = F_A \cos 15° - W_1 \sin 30° - W_p \sin 30° = (m_1 + m_p)a_t$$

$$= 0.97F_A - 125 \text{ lb} - 100 \text{ lb}$$

$$= \frac{(250 + 200) \text{ lb}}{32.2 \text{ ft/s}^2}(8.41 \text{ ft/s}^2)$$

$$F_A = 353.1 \text{ lb}$$

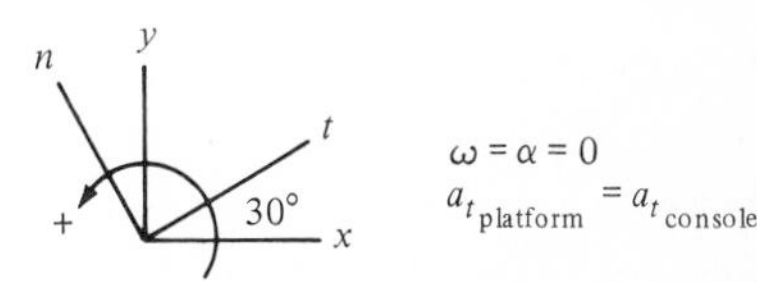

(b) Curvilinear translation

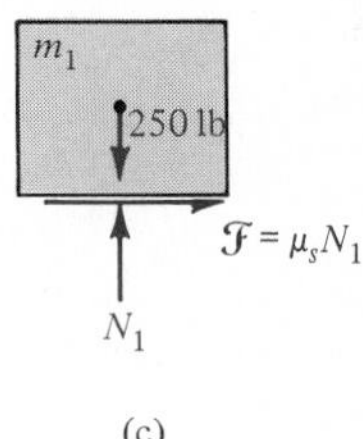

(c)

FIGURE 18-8

Schematic of a console tested on a shake table

PROBLEMS

SECTIONS 18-1, 18-2, 18-3

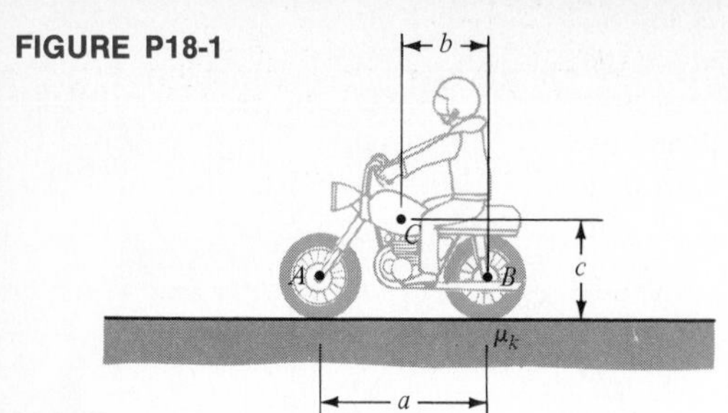

FIGURE P18-1

18–1 The mass of a motorcycle and its rider is 300 kg. It starts from rest with the driving wheel spinning on the ground. Determine the linear acceleration if $a = 1.2$ m, $b = 0.5$ m, $c = 0.6$ m, and $\mu_k = 0.8$. Assume that the wheels have negligible mass.

18–2 For the motorcycle in Prob. 18–1, determine the normal force on the front wheel.

18–3 Show that $N_A = 1548$ N and $N_B = 931$ N in Ex. 18-2, and that $a_x = 3.8744\ \text{m/s}^2$ is an acceptable result.

18–4 Show that $a_x = 5\ \text{m/s}^2$ is not an acceptable result in Ex. 18-2.

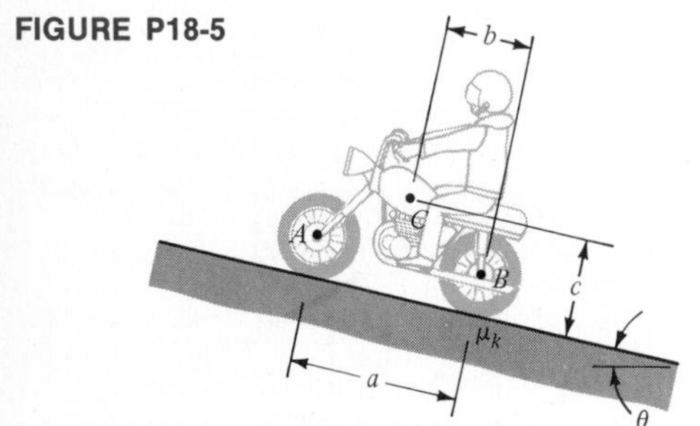

FIGURE P18-5

18–5 The weight of the motorcycle and its rider is 500 lb. It starts from rest going up on a 10° incline with the driving wheel spinning on the ground. Determine the normal and tangential forces on the wheels if $a = 4$ ft, $b = 1.7$ ft, $c = 2$ ft, and $\mu_k = 0.9$. Assume that the wheels have negligible mass.

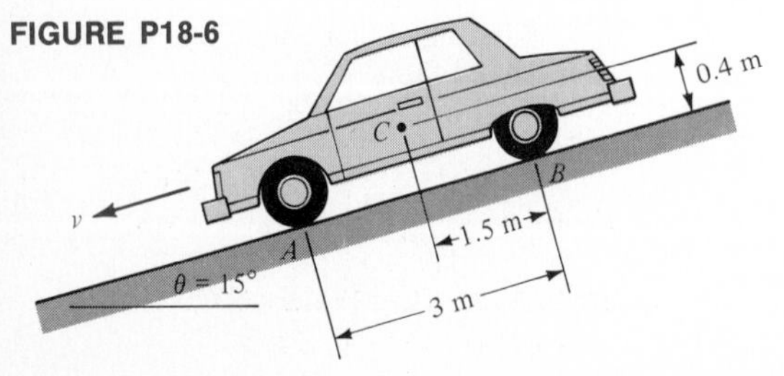

FIGURE P18-6

18–6 A 2000-kg car is traveling downhill when the brakes are applied locking all wheels. Calculate the normal and tangential forces on the wheels for $\mu_k = 0.5$, assuming symmetry between the left and right wheels.

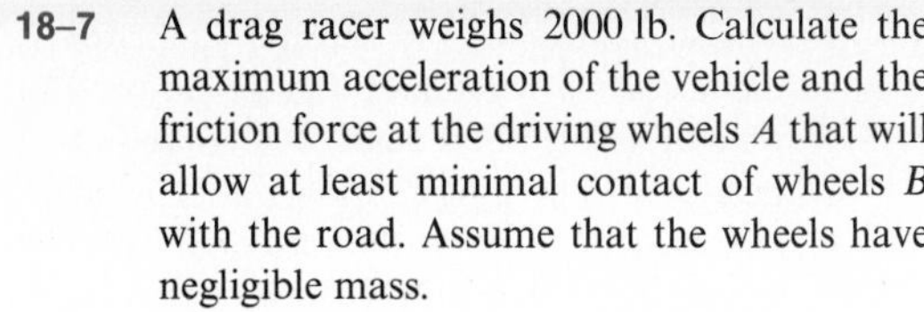

18–7 A drag racer weighs 2000 lb. Calculate the maximum acceleration of the vehicle and the friction force at the driving wheels A that will allow at least minimal contact of wheels B with the road. Assume that the wheels have negligible mass.

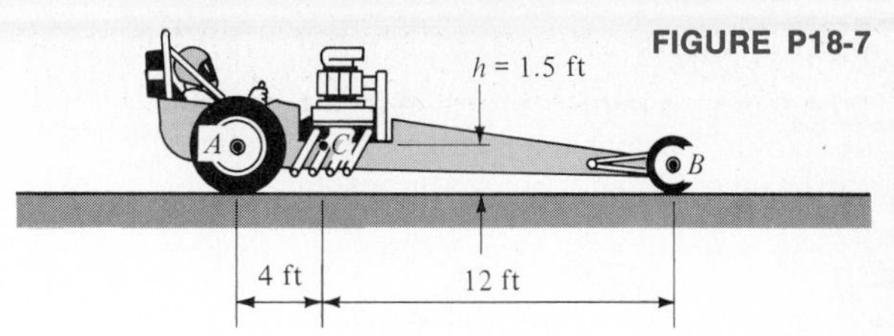

FIGURE P18-7

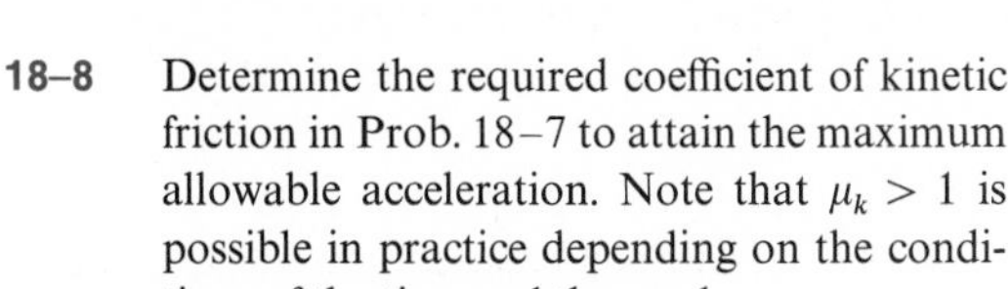

18–8 Determine the required coefficient of kinetic friction in Prob. 18–7 to attain the maximum allowable acceleration. Note that $\mu_k > 1$ is possible in practice depending on the conditions of the tires and the road.

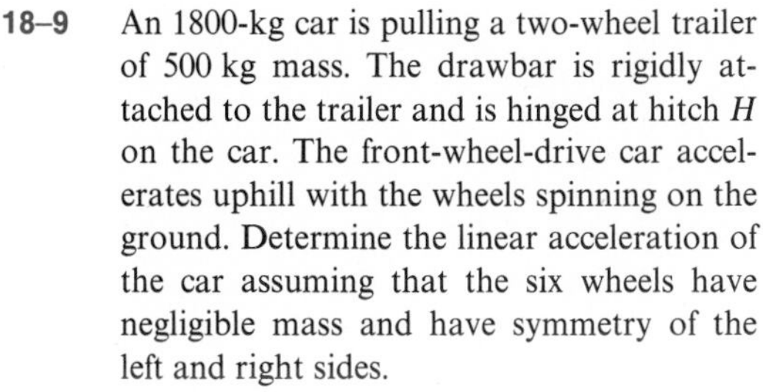

18–9 An 1800-kg car is pulling a two-wheel trailer of 500 kg mass. The drawbar is rigidly attached to the trailer and is hinged at hitch H on the car. The front-wheel-drive car accelerates uphill with the wheels spinning on the ground. Determine the linear acceleration of the car assuming that the six wheels have negligible mass and have symmetry of the left and right sides.

FIGURE P18-9

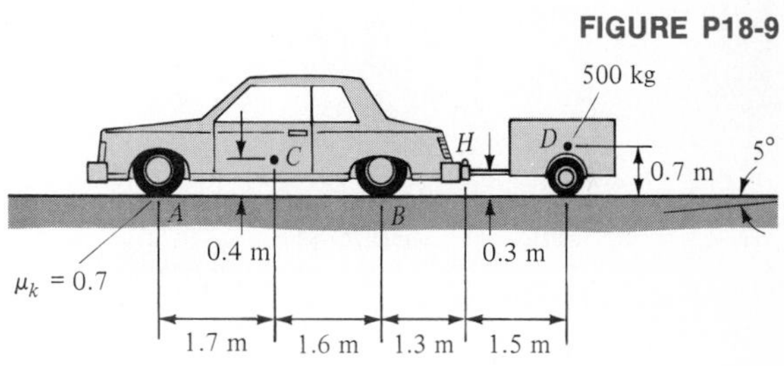

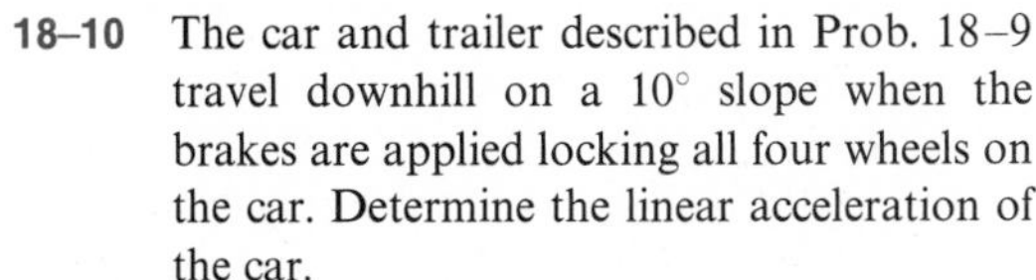

18–10 The car and trailer described in Prob. 18–9 travel downhill on a 10° slope when the brakes are applied locking all four wheels on the car. Determine the linear acceleration of the car.

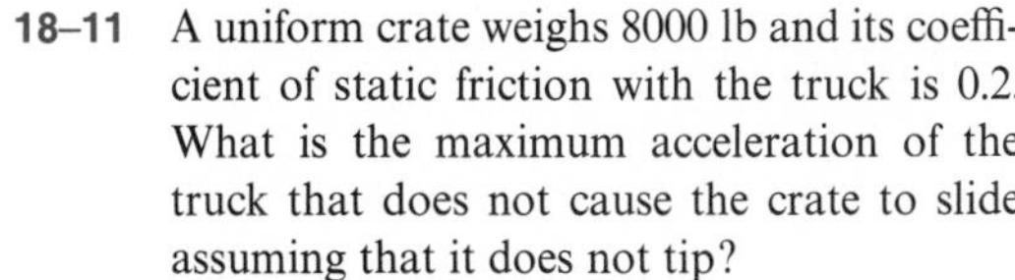

18–11 A uniform crate weighs 8000 lb and its coefficient of static friction with the truck is 0.2. What is the maximum acceleration of the truck that does not cause the crate to slide assuming that it does not tip?

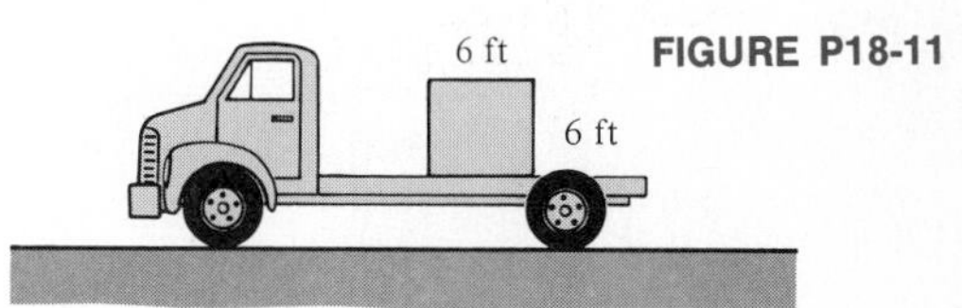

FIGURE P18-11

FIGURE P18-12

18–12 The crate in Prob. 18–11 is tied down with two cables that are in the same vertical plane with the center of mass. The truck is decelerating at the rate of 20 ft/s^2 and the forward cable becomes slack. Calculate the tension in the rear cable.

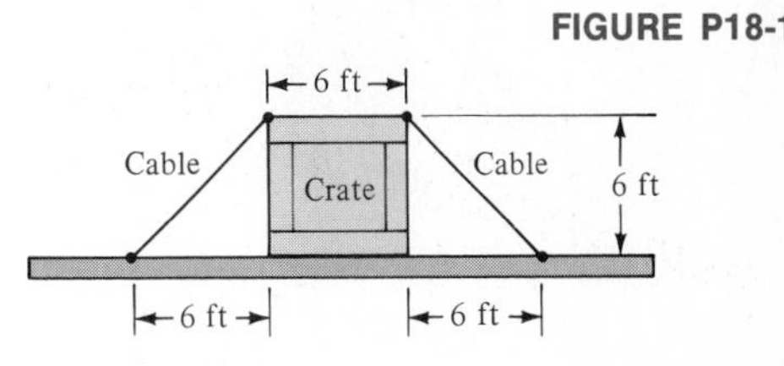

FIGURE P18-13

Model of a building frame tested on a seismic simulator (Photograph courtesy MTS Systems Corporation, Minneapolis, Minnesota)

(a)

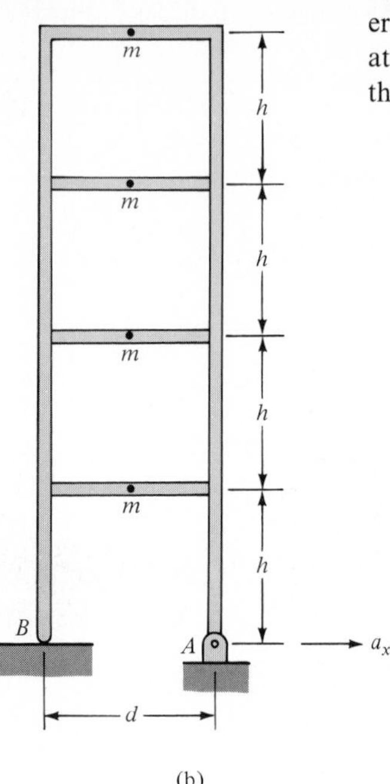

(b)

18–13 The structure on the seismic simulator is modeled as a two-dimensional rigid body of weightless members and four concentrated masses m. The hinge support at A is accelerated to the right at the rate of a_x. The leg at B is on a frictionless surface. Determine the forces acting on the legs at A and B.

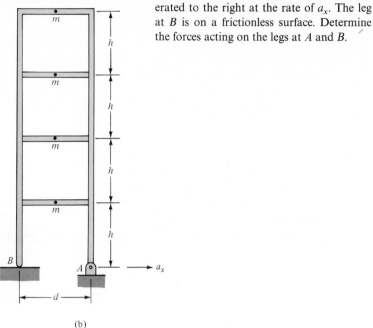

18–14 Determine the acceleration a_x in Prob. 18–13 at which leg B separates from its supporting surface.

FIGURE P18-15

Fixture *ABC* moving along horizontal tracks (Photograph courtesy MTS Systems Corporation, Minneapolis, Minnesota)

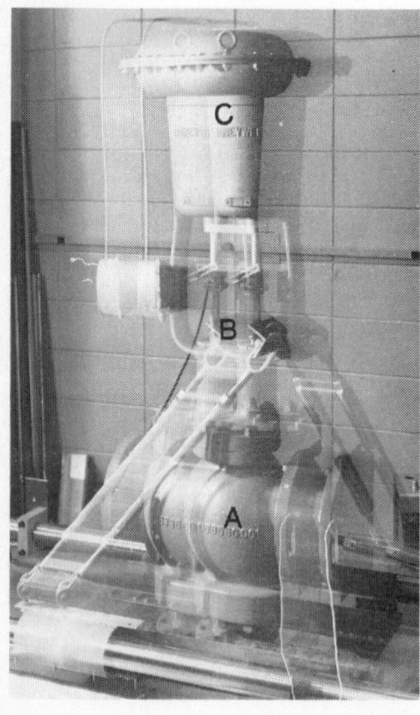

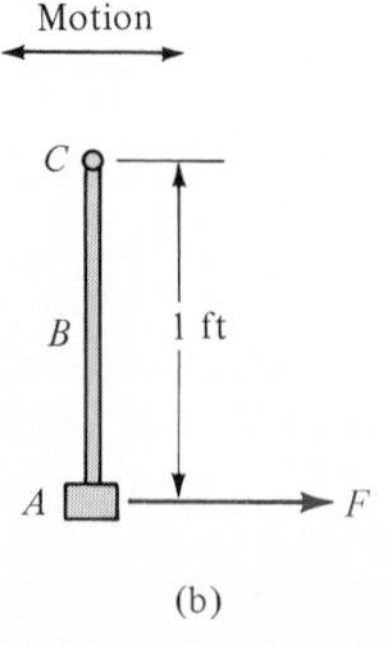

(b)

18–15 The fixture moving along horizontal tracks is modeled as three parts, A, B, and C, which remain in their vertical configuration during motion. Assume that A and C have concentrated masses as shown, and $W_A = 50$ lb, $W_C = 20$ lb. A force of 500 lb is applied horizontally to A and the fixture slides without friction. Calculate the maximum bending moment in the uniform rod B which has a weight of 40 lb.

18–16 A 100-kg console can be moved on casters whose friction may be neglected. Calculate the range of the height h at which a force $F = 400$ N may be applied without causing the console to overturn.

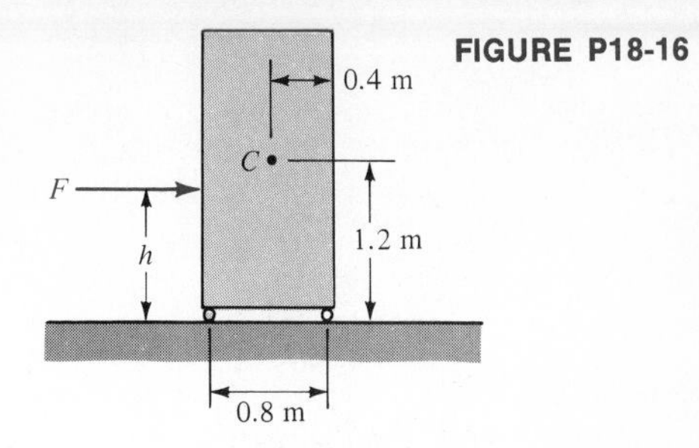

FIGURE P18-16

18–17 Determine the maximum force F in Prob. 18–16 that can be applied at a height of $h = 1.7$ m without causing the console to overturn. Assume that the casters have a coefficient of kinetic friction $\mu_k = 0.2$.

18–18 A 3000-lb vehicle may be driven either by its front or rear wheels. Calculate the maximum acceleration in each mode of operation without allowing slippage of the wheels if $\mu_s = 0.5$, the wheels have negligible mass, and $d = 4$ ft.

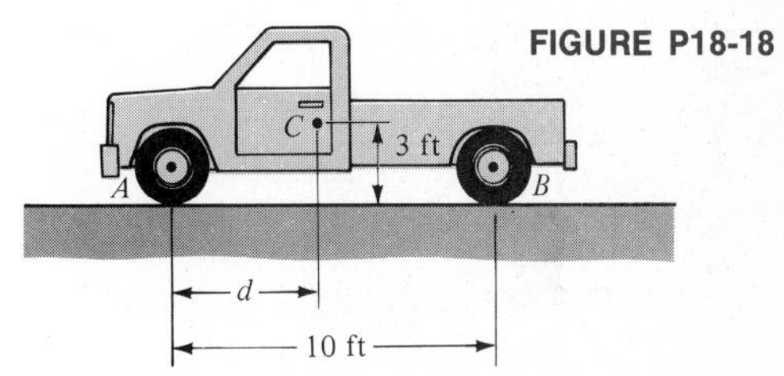

FIGURE P18-18

18–19 Determine the position d of the center of mass in Prob. 18–18 to equalize the tendency to slip of all wheels. Discuss why the value of d does or does not depend on the horizontal acceleration of the vehicle.

18–20 A 1000-kg trailer is pulled with a slanted draw bar. Determine the tension in the bar and the normal forces on wheels A and B if the truck accelerates at the rate of 3 m/s^2. The wheels have negligible mass.

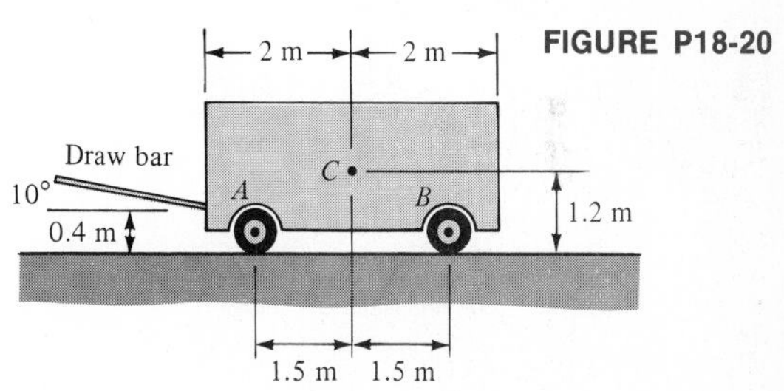

FIGURE P18-20

18–21 Calculate the force in the draw bar in Prob. 18–20 for a deceleration of the truck at the rate of 6 m/s^2. The wheels of the trailer are locked by the brakes and slide; $\mu_k = 0.3$.

18–22 Consider a 3000-lb car traveling on a curved, banked road at a speed of 80 ft/s. Determine the required bank angle θ for equalizing the normal forces on all wheels.

FIGURE P18-22

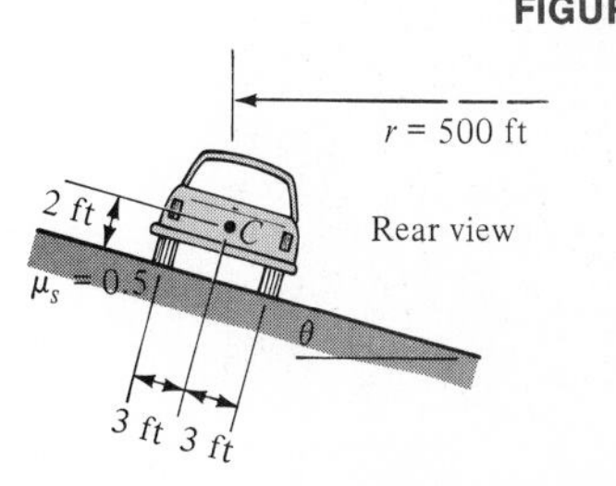

FIGURE P18-24

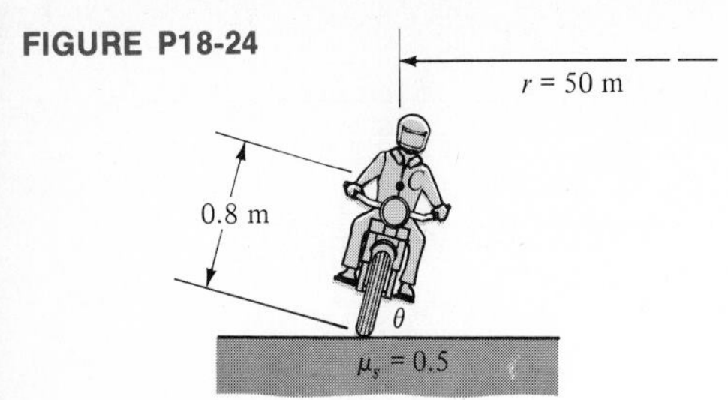

18–23 Determine the maximum speed of the car in Prob. 18–22 so that it does not slip or tip on the road with $\theta = 10°$.

18–24 A motorcycle and its rider have a total mass of 200 kg. Determine the maximum safe speed v and the required angle θ of leaning on a horizontal, curved road to prevent slipping or tipping ($\ddot{\theta}$ should be zero).

FIGURE P18-25

(a)

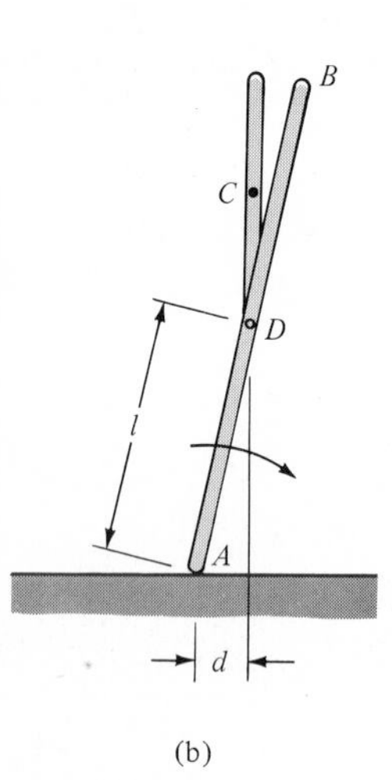

(b)

18–25 A person walking on stilts tries to stay upright even during a fall. This is modeled as a weightless bar AB (the stilts) starting from rest and rotating about A. A uniform bar representing the person is supported at D and remains vertical as AB rotates. Write an expression for the horizontal acceleration of the center of mass C assuming that the stilts start from rest with a small initial displacement d. What can be concluded concerning the stability of stilts as a function of their length l and the mass of the person?

FIGURE P18-26

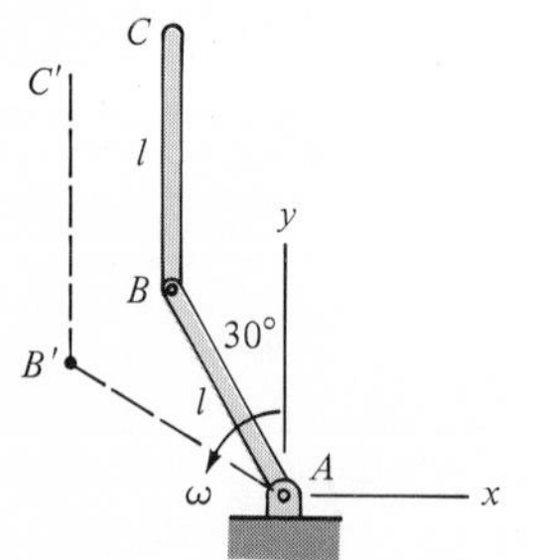

18–26 The uniform bars moving in the horizontal xy plane have length $l = 10$ in. and weight $W = 2$ lb each. AB is rotating at a constant angular speed $\omega = 50$ rad/s. Determine the required moment at B to constrain bar BC to remain parallel to the y axis as shown.

18–27 Calculate the required moment in Prob. 18–26 if the angular speed is decreasing at the rate of 10 rad/s^2.

ROTATION ABOUT A FIXED AXIS

18-4

The methods developed in the preceding three sections can be applied in the analysis of motion of bodies that rotate about fixed axes. This is an important area in plane motion of rigid bodies because of the numerous rotating shafts, wheels, gears, and linkages in essentially two-dimensional motion in almost every machine. For a general model of such rotational motions, consider the rigid body in Fig. 18-9a. This body is pinned at point O and moves in the vertical xy plane. Thus, it can only rotate about the fixed z axis at point O.

The free-body diagram of the body is shown in Fig. 18-9b, where $\mathbf{F}_R$ is the reaction of the pin on the body. The system of forces on the free body can be replaced by its resultant force $\sum \mathbf{F}$ acting at the center of mass C and by the resultant moment $\sum M_C$ acting about C as in Fig. 18-9c. These resultants cause a motion of the center of mass and a rotation of the body. For practical reasons, it is often best to use the fixed axis for a reference, and this is readily accomplished with the aid of Sec. 18-3. First, resolve the resultant force $\sum \mathbf{F}$ into normal and tangential components as shown in Fig. 18-9c. The normal component $\sum F_n$ intersects the fixed point O and causes the centripetal acceleration a_{Cn} of the center of mass,

$$\sum F_n = ma_{Cn} = mr_C\omega^2 = m\frac{v_C^2}{r_C} \qquad \boxed{18\text{-}10}$$

where r_C is the nearest distance between the fixed axis and the center of mass, ω is the angular speed of the body (of any line in the body), and $v_C = r_C\omega$.

The tangential force $\sum F_t$ causes the tangential acceleration a_{Ct},

$$\sum F_t = a_{Ct} = mr_C\alpha \qquad \boxed{18\text{-}11}$$

where α is the angular acceleration of the body (of any line in the body). The quantities m and r_C are the same as in Eq. 18-10.

The magnitude of the resultant moment $\sum \mathbf{M}_C$ on the body is obtained from Eq. 18-6,

$$\sum M_C = I_C\alpha \qquad \boxed{18\text{-}12}$$

The moment about the fixed reference point O in Fig. 18-9c is expressed using Eqs. 18-11 and 18-12,

$$\sum M_O = M_C + r_C \sum F_t = I_C\alpha + r_C m a_{Ct}$$

$$\sum M_O = (I_C + mr_C^2)\alpha \qquad \boxed{18\text{-}13}$$

where the quantity in parentheses is equal to I_O, the mass moment of inertia about O by the parallel-axis theorem as shown in Chap. 9

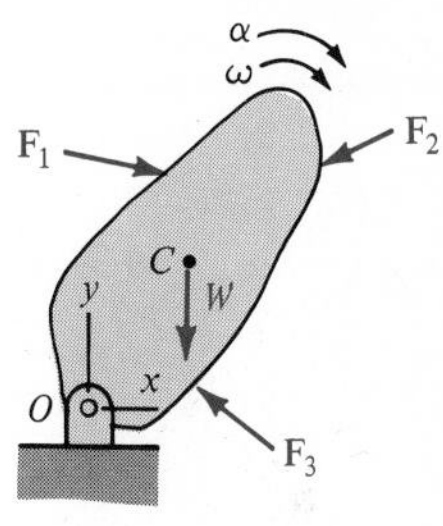

(a)

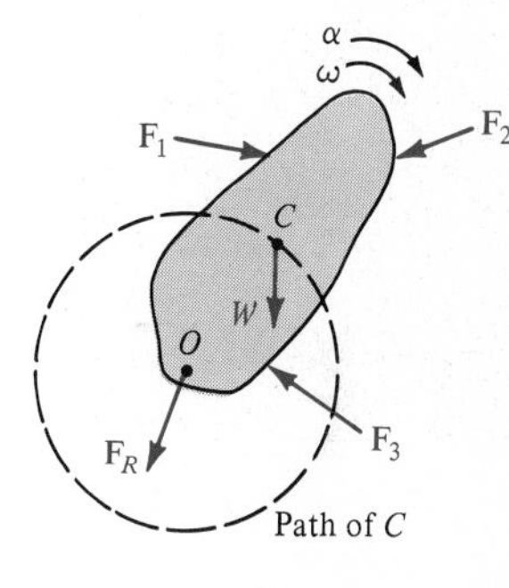

(b)

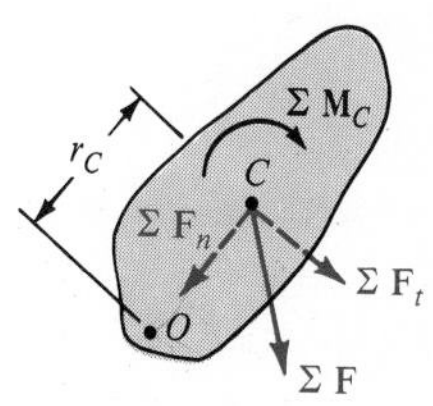

(c)

FIGURE 18-9

Rotation of a rigid body about a fixed axis

in the statics volume. Thus,

$$\sum M_O = I_O\alpha \tag{18-14}$$

where the angular acceleration α is independent of the point of reference.

In summary, the equations of motion for a rigid body rotating about a fixed axis O are

$$\begin{aligned} \sum F_n &= mr_C\omega^2 \\ \sum F_t &= mr_C\alpha \\ \sum M_O &= I_O\alpha \end{aligned} \tag{18-15}$$

where r_C is the distance between the fixed axis at O and the center of mass. The two force equations must include the n and t components of the reaction at the axis of rotation.

Rotation about the centroidal axis. Equation 18-15 may be used for rigid bodies that rotate about centroidal axes. In that special case $r_C = 0$, and $I_O = I_C$, so the equations of motion become

$$\begin{aligned} \sum F_n &= 0 \\ \sum F_t &= 0 \\ \sum M_C &= I_C\alpha \end{aligned} \tag{18-16}$$

where n and t may be taken as any pair of orthogonal directions in the plane of motion.

Center of Percussion

Another special case of rotation about a fixed axis is when the tangential force $\sum F_t$ alone accounts for both terms involving angular acceleration of the body. To illustrate this, consider the rotating body in Fig. 18-10, where again the fixed axis is at O, and the center of mass is at C. Assume that the resultants $\sum F_n$, $\sum F_t$, and $\sum M_C$ have been determined from the system of external forces acting on the body, and these are shown in Fig. 18-10a. It is possible to find a point P on the line OC where the forces $\sum F_n$ and $\sum F_t$ are equivalent to the original force-couple system at C, so $\sum M_C$ can be eliminated from consideration. The location of point P is such that the moment about C remains the same. The new location of the line of action of $\sum F_t$ at point P is at a distance p from the axis of rotation as shown in Fig. 18-10b. The value of p is calculated from the equality of moments in Fig. 18-10a and b, using point O for reference,

$$p \sum F_t = r_C \sum F_t + \sum M_C$$

From this equation of moments,

$$mr_C\alpha p = mr_C^2\alpha + I_C\alpha \qquad \boxed{18\text{-}17}$$

This can be simplified by using the radius of gyration r_g (with respect to the center of mass),

$$I_C = r_g^2 m$$

With this substitution, and canceling $m\alpha$,

$$p = r_C + \frac{r_g^2}{r_C}$$

Note that $r_C^2 + r_g^2 = r_O^2$, where r_O is the radius of gyration of the body about point O;

$$r_O^2 = \frac{I_O}{m} = \frac{I_C + r_C^2 m}{m} = \frac{(r_g^2 + r_C^2)m}{m}$$

Thus,

$$p = \frac{r_O^2}{r_C} \qquad \boxed{18\text{-}18}$$

The point P located in this manner is called the *center of percussion.* Any system of forces acting on the body can be represented by a single resultant force (or by its components $\sum F_n$ and $\sum F_t$) acting at the center of percussion. An interesting and important aspect of this concept is that there is no reaction at the supporting hinge of a body when a tangential force is applied to the body at its center of percussion.

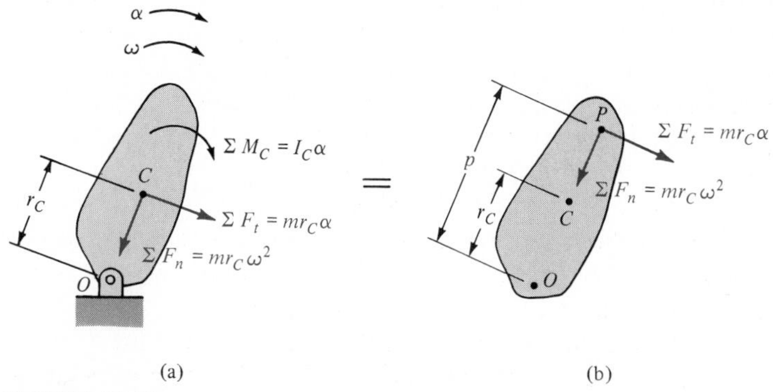

FIGURE 18-10

Schematic location of the center of percussion

EXAMPLE 18-4

The final step in a cable-making process is modeled in Fig. 18-11a. Assume that the cable drum is a homogeneous disk of 1 m radius and 1000 kg mass. The drum starts from rest and accelerates at a constant rate until the cable attains a velocity $\mathbf{v} = -2\mathbf{i}$ m/s in a period of 5 s. The drum winds up cable whose mass moving in the horizontal direction is 100 kg. Assume that the radius of the drum is constant in the given time interval and that the die applies a 5-kN friction force horizontally to the cable. Determine the required horizontal force and the moment at axle O.

SOLUTION

A sketch (Fig. 18-11b) of the cable alone allows for the calculation of the horizontal component T_{D_x} of the tension at the point of tangency to the disk. The acceleration of the cable is

$$a_{\text{cable}} = \frac{\Delta v}{\Delta t} = -\frac{2 \text{ m/s}}{5 \text{ s}} = -0.4 \text{ m/s}^2$$

$$\sum F_X = 5 \text{ kN} - T_{D_x} = m_{\text{cable}} a_{\text{cable}} = (100 \text{ kg})(-0.4 \text{ m/s}^2)$$

$$T_{D_x} = 5000 \text{ N} + 40 \text{ N} = 5040 \text{ N}$$

The free-body diagram of the drum alone is sketched in Fig. 18-11c. The angular acceleration α is calculated from the acceleration of the point of tangency D, $a_D = a_{\text{cable}} = r_{OD}\alpha$. Hence,

$$\alpha = \frac{a_{\text{cable}}}{r_{OD}} = \frac{0.4 \text{ m/s}^2}{1 \text{ m}} = 0.4 \text{ rad/s}^2$$

$$\sum F_x = 5040 \text{ N} - R_x = m_{\text{drum}} a_{\text{drum}} = 0 \qquad R_x = 5040 \text{ N}$$

$$\sum M_O = M_1 - 5040 \text{ N (1 m)} = I_O \alpha = \frac{m_{\text{drum}}}{2} r_{\text{drum}}^2 \alpha$$

$$M_1 = \left(\frac{1000 \text{ kg}}{2}\right)(1 \text{ m})^2(0.4 \text{ rad/s}^2) + 5040 \text{ N}\cdot\text{m}$$

$$= 5240 \text{ N}\cdot\text{m}$$

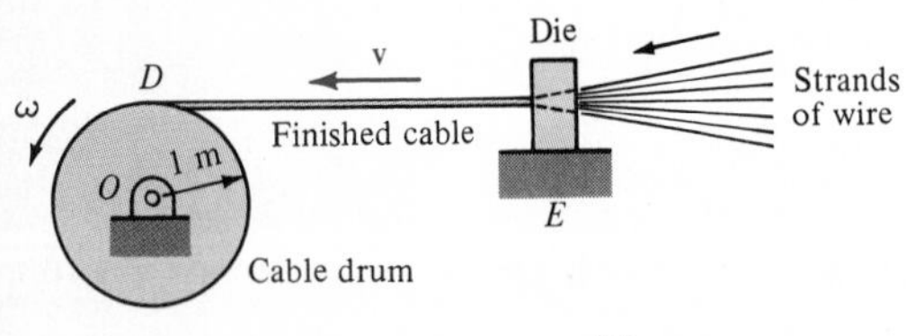

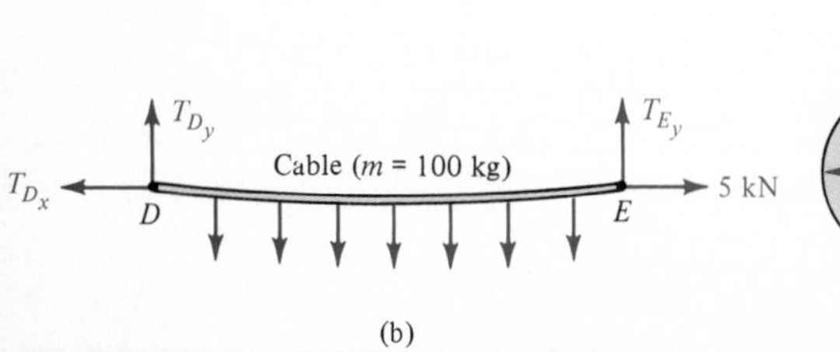

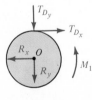

(c)

FIGURE 18-11
Schematic of cable manufacturing

EXAMPLE 18-5

A locomotive of 1 MN weight is tested in a laboratory. Assume that the mass is uniformly distributed in a rectangular volume 2.4 m × 2.4 m in cross section. Horizontal forces F are applied to the locomotive by hydraulic actuators as in Fig. 18-12a and b until the wheels are slightly off track B. The locomotive acts as if it were hinged at A when the actuator forces are suddenly decreased to zero. Determine the angular acceleration of the locomotive and the linear accelerations measured by the horizontal and vertical accelerometers H and V, respectively.

SOLUTION

This is a problem involving rigid body rotation about a fixed point. The appropriate free-body diagram is sketched in Fig. 18-12c. From Eqs. 18-13 and 18-14,

$$\sum M_A = W(0.65\text{ m}) = I_C\alpha + ma_{Ct}(r_C) = I_C\alpha + m(r_C\alpha)(r_C) = I_A\alpha$$

$$10^6\text{ N}(0.65\text{ m}) = \Bigg\{\underbrace{\frac{10^6\text{ N}}{9.81\text{ m/s}^2}\left(\frac{1}{12}\right)[(2.4\,m)^2 + (2.4\,m)^2]}_{I_C}$$

$$+\underbrace{\frac{10^6\text{ N}}{9.81\text{ m/s}^2}[(0.65\,m)^2 + (1.8\,m)^2]}_{mr_C^2}\Bigg\}\alpha$$

$$0.65\text{ m} = (0.47\text{ m}\cdot\text{s}^2)\alpha$$

$$\alpha = 1.38\text{ rad/s}^2 \qquad \text{(clockwise in Fig. 18-12)}$$

The acceleration at the point where the accelerometers are positioned is obtained by using Eq. 17-19,

$$\mathbf{a}_D = \mathbf{a}_A + \boldsymbol{\alpha}\times\mathbf{r}_{DA} + \boldsymbol{\omega}\times(\boldsymbol{\omega}\times\mathbf{r}_{DA})$$

Noting that $\boldsymbol{\omega} = 0$, $\boldsymbol{\alpha} = -1.38\text{ rad/s}^2\,\mathbf{k}$ (according to the right-hand rule and the sign convention in Fig. 18-12), and $\mathbf{r}_{DA} = 1.85\text{ m}\,\mathbf{i} + 0.8\text{ m}\,\mathbf{j}$,

$$\begin{aligned}\mathbf{a}_D &= 0 + (-1.38\text{ rad/s}^2\,\mathbf{k})\times(1.85\text{ m}\,\mathbf{i} + 0.8\text{ m}\,\mathbf{j}) + 0\\ &= 1.104\text{ m/s}^2\,\mathbf{i} - 2.553\text{ m/s}^2\,\mathbf{j}\end{aligned}$$

Hence,

$$a_H = 1.104\text{ m/s}^2 \qquad \text{to the right}$$

$$a_V = 2.553\text{ m/s}^2 \qquad \text{down}$$

(a)

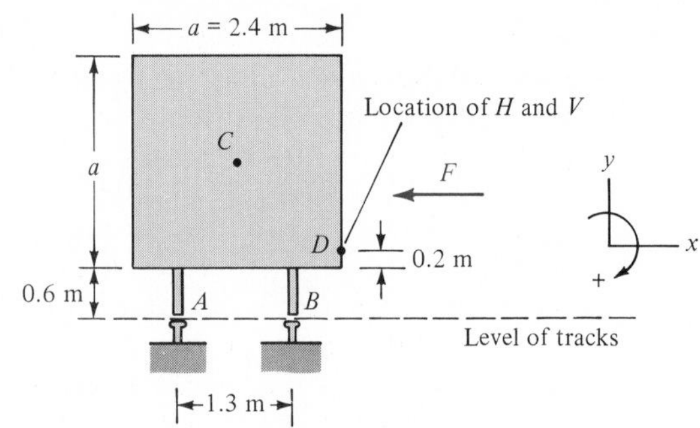

(b) Cross section of locomotive

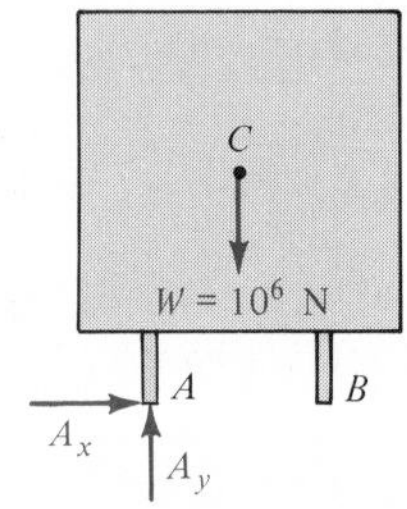

(c)

FIGURE 18-12

Laboratory test of a locomotive (Photograph courtesy MTS Systems Corporation, Minneapolis, Minnesota)

18–28 A 40-kg gear is represented as a 1-m-diameter uniform disk. Calculate the tangential force F for accelerating the gear at the rate of 2 rad/s^2.

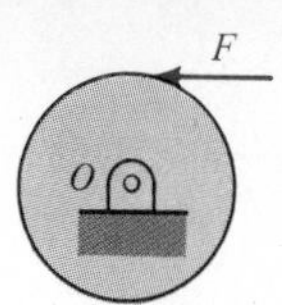

FIGURE P18-28

18–29 The small weight w is attached to the uniform disk of weight W. Calculate the required moment to accelerate the disk at the rate of 3 rad/s^2.

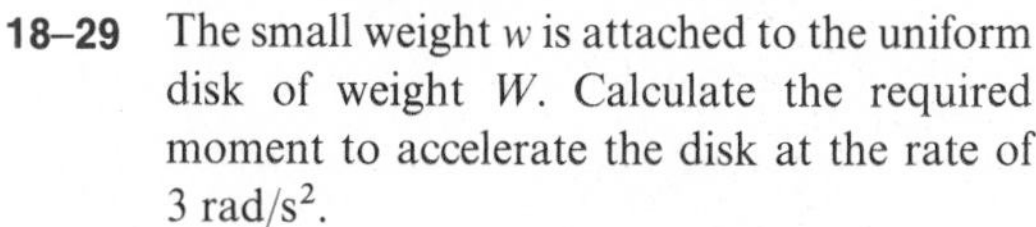

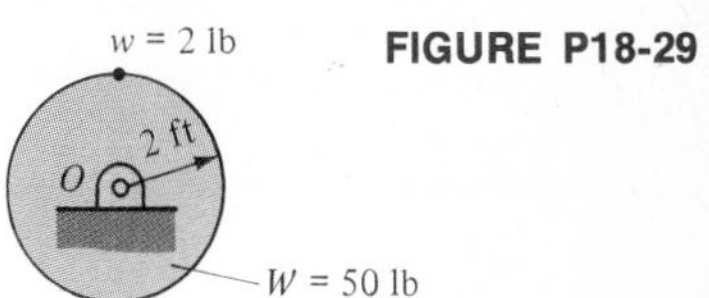

FIGURE P18-29

18–30 The turntable of a record player is a 2-kg, 0.3 m-diameter, uniform disk. Calculate the required moment to bring the turntable from rest to a speed of 33.33 rpm in a time of 5 s.

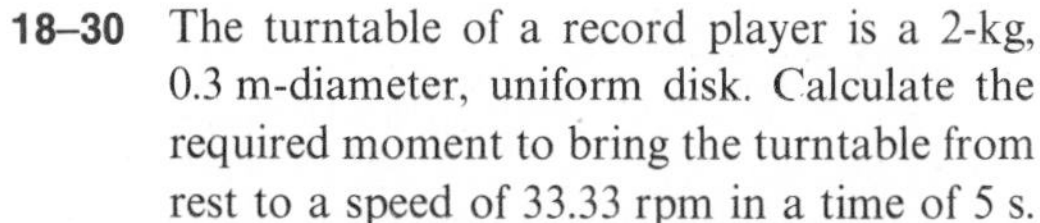

18–31 The pulley of an elevator is modeled as a 30-kg, 1-m-diameter, uniform disk. The elevator and its counterweight have masses $m_1 = 2000$ kg and $m_2 = 1800$ kg, respectively. Calculate the required torque on the pulley for accelerating the elevator upward at the rate of 1 m/s^2.

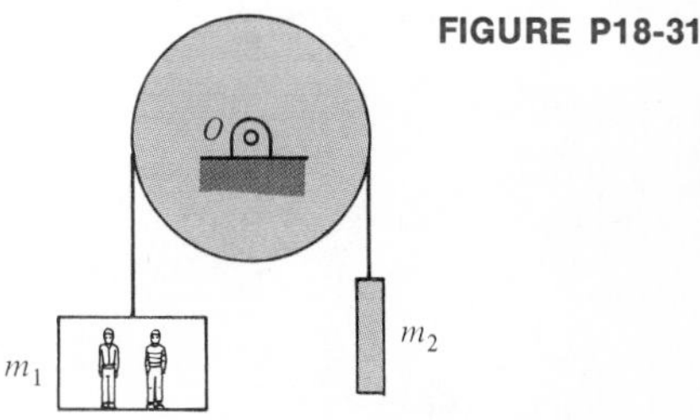

FIGURE P18-31

18–32 Solve Prob. 18–31 as stated and also after replacing the counterweight with a constant force. This force has the same magnitude as the static weight of m_2. Discuss the performances of the two systems for "assisting" the drive motor.

18–33 An amusement ride is a cylindrical ring of 20 ft outside diameter and 7 ft inside diameter. Assume that its 12,000-lb weight is uniformly distributed. Calculate the required tangential force at the inside circle to accelerate the ring at the rate of 0.5 rad/s^2.

FIGURE P18-34

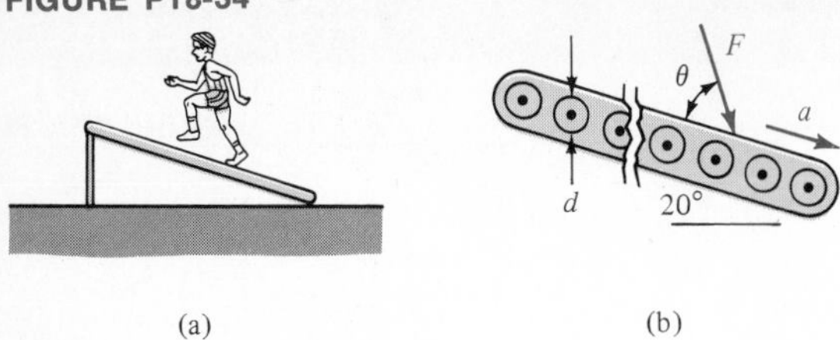

(a) (b)

18–34 The force F applied to an exercise treadmill is transmitted through a belt to 20 rollers that spin with negligible friction. Each roller has a mass of 2 kg and a diameter d of 0.08 m. Calculate the tangential acceleration a of the belt for $F = 1$ kN and $\theta = 60°$. The belt does not slip on the rollers.

FIGURE P18-35

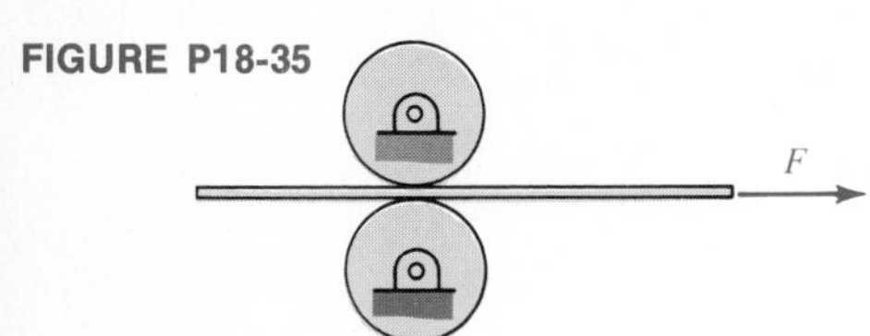

18–35 Determine the required force F to accelerate a thin sheet of metal at the rate of 5 ft/s^2 through the rollers that are free to rotate about their axes. Each uniform roller has a weight of 200 lb and a diameter of 1 ft. The sheet does not slip between the rollers.

FIGURE P18-36

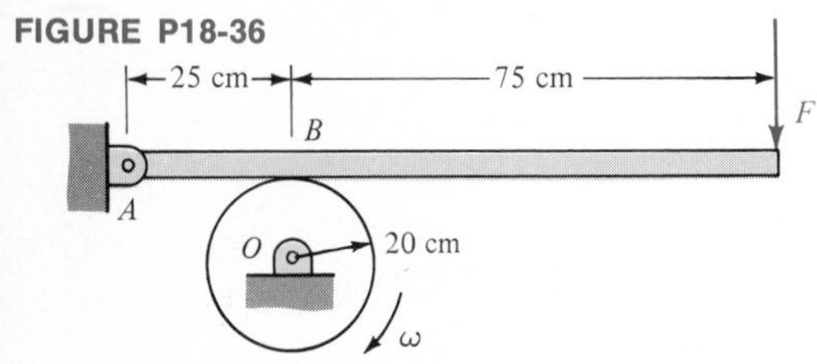

18–36 The uniform, 10-kg wheel is rotating at an angular speed $\omega = 500$ rpm when a braking force $F = 800$ N is applied to the hinged bar. The friction at O and A is negligible, while $\mu_k = 0.5$ at B. Calculate the time required to stop the wheel.

FIGURE P18-37

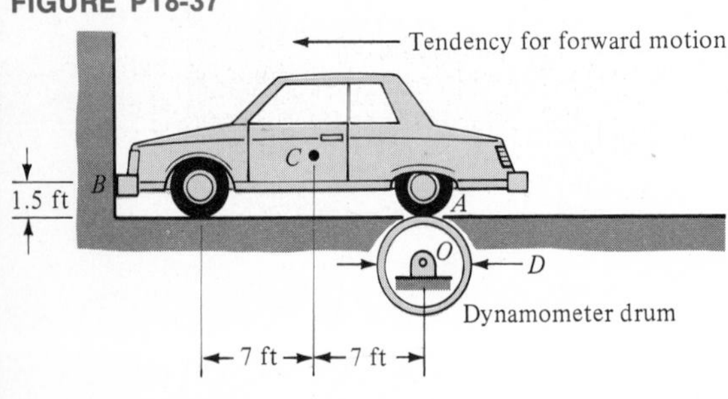

18–37 The driving wheels of a 3000-lb car are on the freely turning drum of a dynamometer. The drum is modeled as a thin-walled, open-ended cylinder of 300 lb weight and $D = 5$ ft diameter. What is the maximum angular acceleration of the dynamometer drum if the driving wheels have $\mu_s = 0.9$ on the drum?

FIGURE P18-38

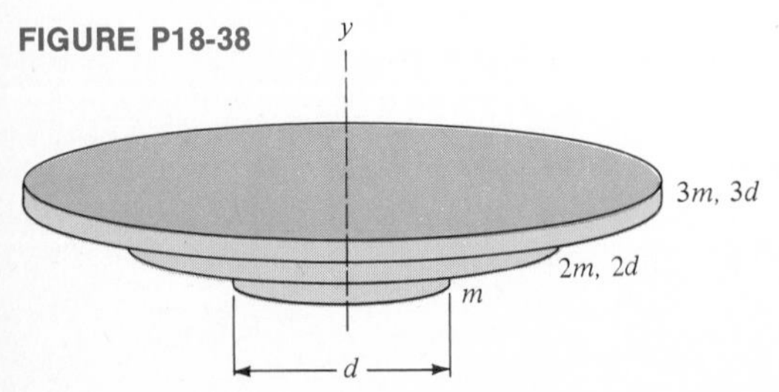

18–38 The base structure of a rotating restaurant is modeled as three uniform disks that rotate as a unit about the vertical axis y. The mass m is 8000 kg, and the diameter d is 7 m. Calculate the required moment to stop the disks in a time of 10 s from an angular speed of 0.1 rpm.

18–39 Assume that each propeller blade is a uniform, slender bar that weighs 20 lb. Calculate the moment that can accelerate the set of blades at the rate of 50 rad/s^2.

FIGURE P18-39

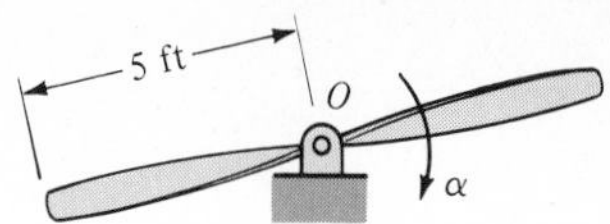

18–40 Each wing of a small airplane is modeled as a uniform, slender bar of 150-kg mass. Determine the angular acceleration of the airplane caused by the two forces F. Neglect the mass of the fuselage and air drag opposing the rotation.

FIGURE P18-40

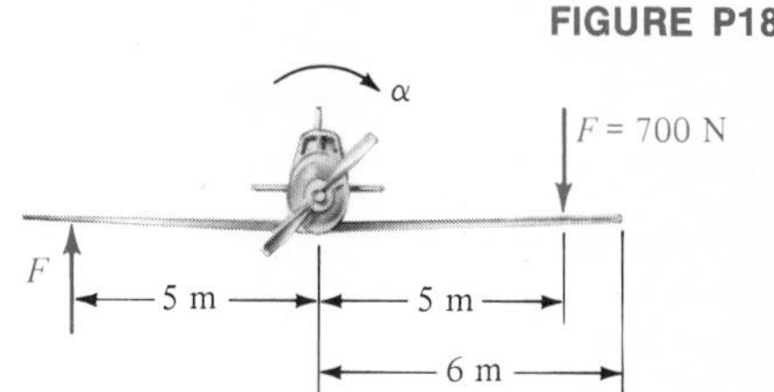

18–41 Solve Prob. 18–40 assuming that the fuselage is a uniform cylinder of 600-kg mass and 0.8 m diameter. Also assume that the center of the fuselage is fixed.

FIGURE P18-42

18–42 Assume the following for the vertical arm AC of the excavator: $AB = 3$ ft, $BC = 10.5$ ft; weight of member AC: 300 lb (uniformly distributed); weight of load at C: 350 lb. Determine the acceleration of the load at C caused by a force $\mathbf{F} = (-400\mathbf{i} - 70\mathbf{j})$ lb applied at joint A. Calculate the force at joint B.

18–43 The uniform, slender bar has a mass of 4 kg. It is released from rest at $\theta = 30°$. Calculate the angular acceleration of the bar and the x and y reactions at point O for the instant after release.

FIGURE P18-43

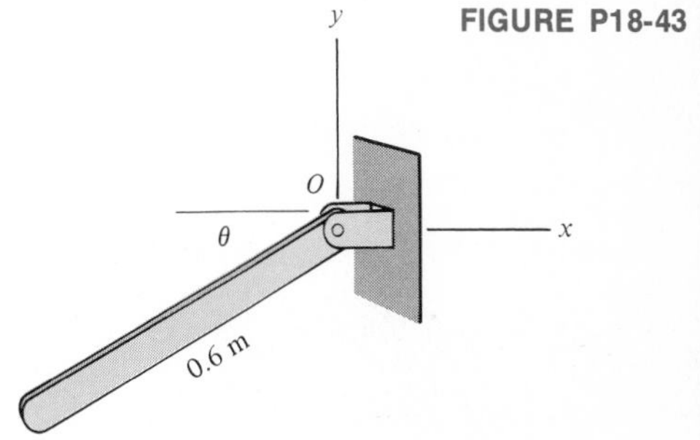

18–44 Solve Prob. 18–43 assuming that at $\theta = 30°$ the bar has an initial counterclockwise angular speed of $\dot{\theta} = 10$ rad/s.

FIGURE P18-45

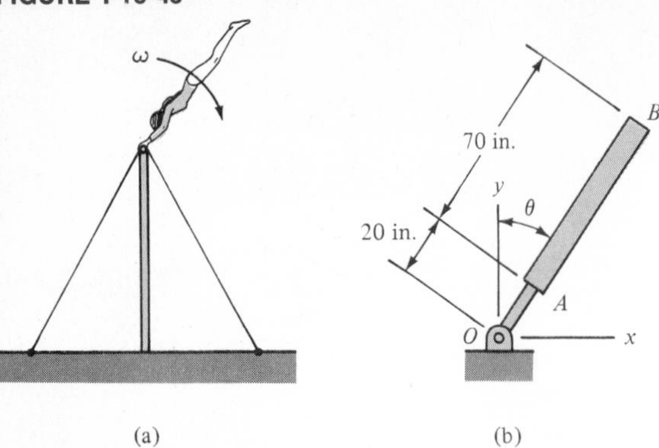

(a) (b)

18–45 Consider a gymnast doing giant circles on the high bar. Assume that her body can be modeled as a uniform rigid bar AB of 160 lb weight, firmly attached to a weightless rigid member OA. Assume that the angular speed is $\omega = 1$ rad/s at $\theta = 30°$. Determine the x and y reactions at the pivot O. Neglect friction of the hands on the bar.

FIGURE P18-46

Landing gear of an aircraft (Photograph, The Boeing Company, Seattle, Washington)

(a)

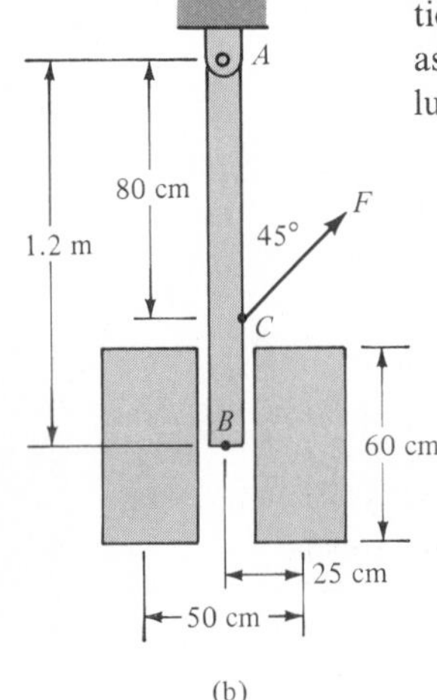

(b)

18–46 The landing gear of an aircraft is modeled as a uniform rod AB of 100 kg mass, and two uniform disks of 30 kg mass each. A force $F = 3$ kN is applied at C to retract the landing gear. Determine the angular acceleration of AB and the reaction at A caused by F, assuming that the mass of the wheels is lumped at point B.

18–47 Solve Prob. 18–46 by treating the wheels as uniform, thin disks located as in Fig. P18-46b.

FIGURE P18-48

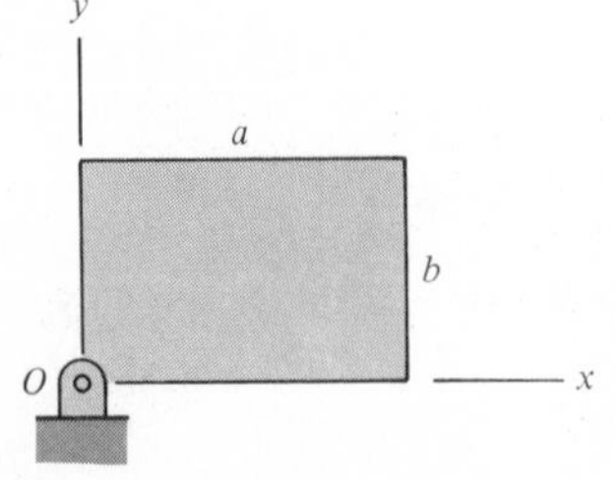

18–48 A uniform rectangular plate of mass m is hinged at point O and is released from rest when side a is horizontal. Determine the angular acceleration of the plate and the x and y reactions at O when the plate begins to swing.

18–49 A uniform ring of thickness t and density ρ is hinged at point O and is released from rest in the position shown. Determine the angular acceleration of the ring and the x and y reactions at O when the ring begins to swing. $r = 0.9R$.

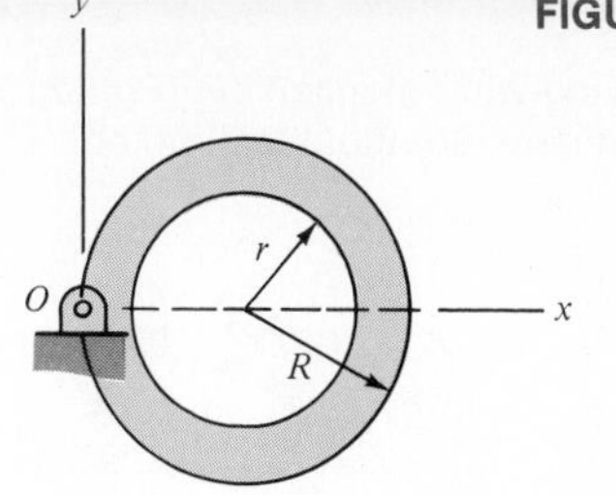

FIGURE P18-49

18–50 Plot the angular acceleration of the rigid body in Prob. 18–49 as a function of $r/R = 0$ (solid circular plate), 0.5, and 0.9.

18–51 The 20-ft-long swinging arm of an amusement ride is a uniform rod of 200 lb weight. The 500-lb weight of the riders and the cab may be assumed as concentrated at B, the end of the rod. At $\theta = 30°$, $\dot{\theta} = 0$. Determine the force F at $\phi = 45°$ and the reaction at O for causing an angular acceleration of 1 rad/s^2 of the rod.

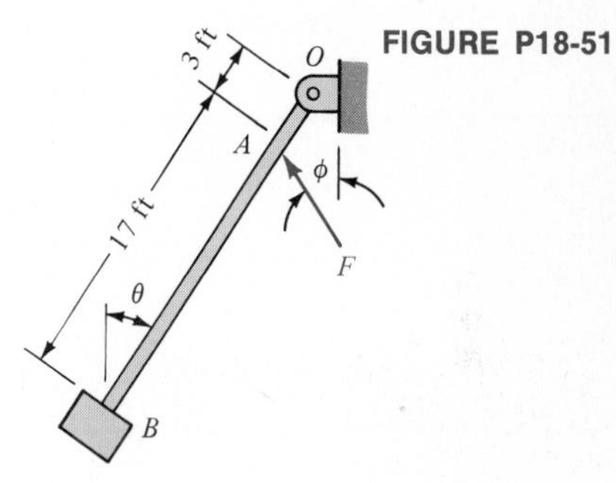

FIGURE P18-51

18–52 A 50-m-tall radio antenna tower is modeled as a uniform, rigid rod of 2000-kg mass, in a ball-and-socket joint at point O, and supported by four guy wires that are 90° apart (top view). The initial tension in each wire is 8 kN. Assume that one of these wires breaks. Calculate the initial angular acceleration of the tower and the reaction on the joint at O, assuming that the tension is constant in the other three wires for a short time after the breakage.

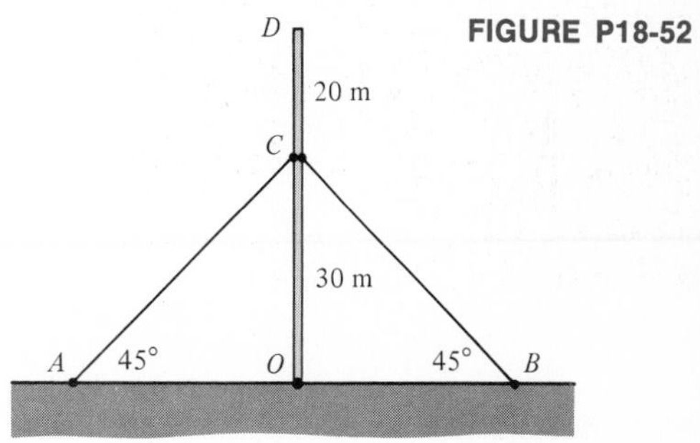

FIGURE P18-52

18–53 The crank arm OA is modeled as a uniform, slender bar of 4 lb weight. The uniform arm AB weighs 5 lb. A force $\mathbf{F} = 200\mathbf{j}$ lb is applied at A when $\theta = 45°$, and $\dot{\theta} = 0$. Determine the angular acceleration of arm OA and the x and y reactions at O. Neglect friction. The piston weighs 6 lb.

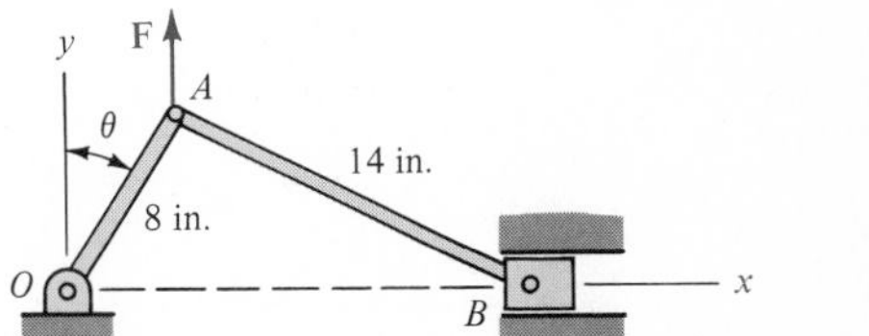

FIGURE P18-53

FIGURE P18-54

Test of a swing-wing mechanism (Photograph, The Boeing Company, Seattle, Washington)

(a)

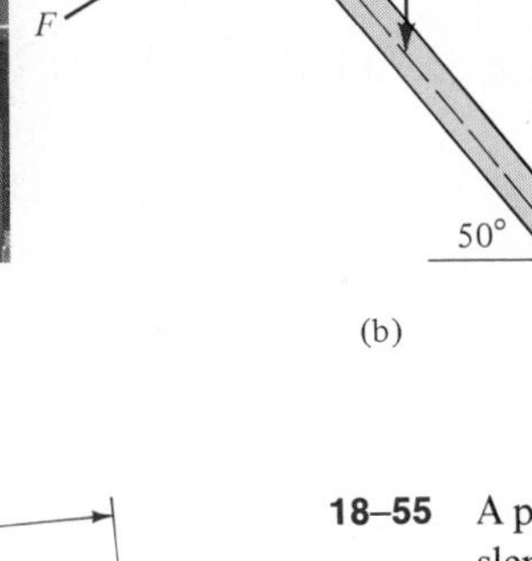

(b)

18–54 Consider the hinge mechanism A of a swing-winged supersonic transport in the testing machine. The test loads for the various angular positions of the hinge components are determined from the forces acting on the whole wing. Assume that the wing may be modeled as a uniform, slender bar of 12 000 kg mass and length $L = 20$ m. The drag force is $D = 300$ kN at a certain speed and altitude, and it acts at $L/2$. Determine the force F acting at $l = 2$ m to accelerate the wing counterclockwise at the rate of $\alpha = 0.2$ rad/s^2. Also calculate the x and y reactions at hinge A.

FIGURE P18-55

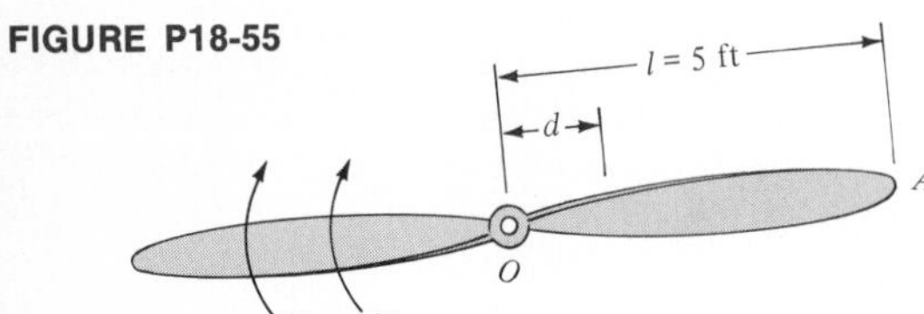

18–55 A propeller blade OA is modeled as a uniform, slender bar of 20-lb weight. Determine the axial and transverse forces in the blade at $d = 1$ ft for $\omega = 300$ rpm and $\alpha = 50$ rad/s^2.

FIGURE P18-56

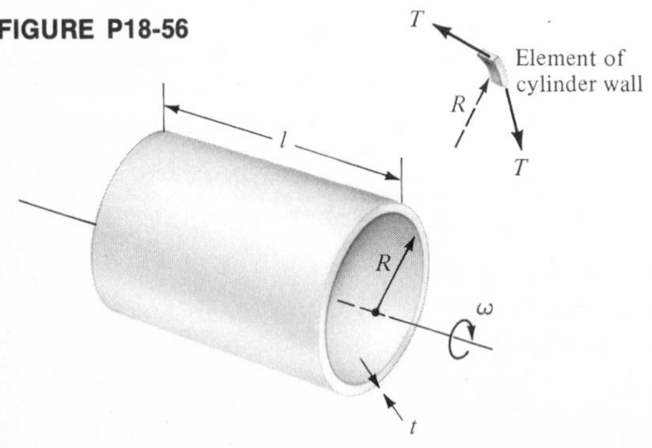

18–56 A thin-walled cylinder of mass m used for drying paper is rotating at a constant angular speed ω. $R \gg t$. Write an expression for the tension T in the cylinder caused by its rotation. This tension is normal to any radial plane through the axis of the cylinder.

A rigid body is said to be in general plane motion when it is translating and rotating. This combined motion is analyzed using Eq. 18-7. For a review, consider the free-body diagram of an arbitrary thin slab as in Fig. 18-13a. The basic set of resultants is shown in Fig. 18-13b, where $\sum \mathbf{F}$ is the resultant force and $\sum \mathbf{M}_C$ is the resultant moment about the center of mass C. The equations of motion using Fig. 18-13b are, as in Sec. 18-3,

$$\sum F_x = ma_{Cx} \qquad \sum F_y = ma_{Cy} \qquad \sum M_C = I_C\alpha \tag{18-7}$$

Note that the xy frame and the forces are parallel to the plane of the thin slab.

The resultant moment may also be determined with respect to an arbitrary, noncentroidal reference point as in Sec. 18-3. Assume that the desired reference point is point A in Fig. 18-13c. The general form of the resultant moment $\sum \mathbf{M}_A$ is

$$\sum \mathbf{M}_A = \sum \mathbf{M}_C + \mathbf{r} \times \sum \mathbf{F} \tag{18-19}$$

where $\sum \mathbf{M}_C$ is the resultant moment of the external forces about the center of mass C, and $\mathbf{r} \times \sum \mathbf{F}$ is the moment of the resultant force about the reference point A. The reference point A for moments should be selected to simplify the solution.

Constraints

In many problems of general plane motion the solution is facilitated if there is some constraint to the motion as shown in the following special cases of considerable practical significance.

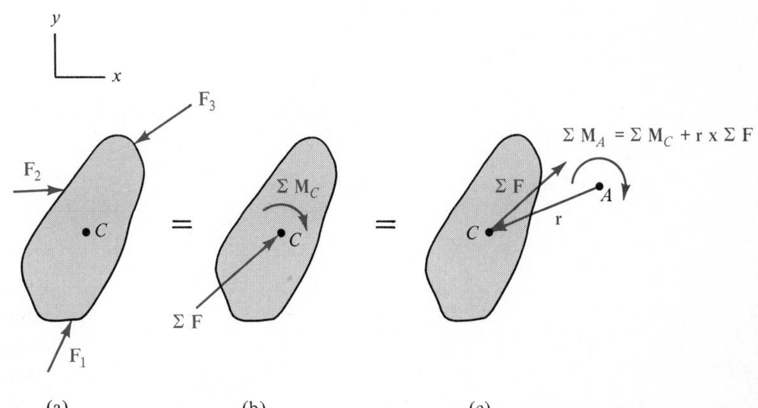

FIGURE 18-13

Resultant forces and moments with respect to the mass center or an arbitrary point

Pure rolling of wheels. Assume that a uniform wheel rolls without slipping on a flat surface at angular speed ω and angular acceleration α as in Fig. 18-14a. Here the center of mass C moves along a straight line and has acceleration $a_{Cx} = r\alpha$. (Note that the problem is more complex if the mass center is not at the geometric center since then the mass center is not in rectilinear translation.) The equation of motion for pure rolling is written using the free-body diagram in Fig. 18-14b. Summing moments about point A and using Eqs. 18-11 and 18-13,

$$\sum M_A = \text{Pr} = I_C\alpha + ma_{Cx}r = I_C\alpha + m(r\alpha)r$$
$$= (I_C + mr^2)\alpha$$

With $I_A = I_C + mr^2$,

$$\sum M_A = I_A\alpha \tag{18-20}$$

where I_A is the mass moment of inertia of the wheel about the point of contact. Equation 18-20 is valid only if the wheel rolls without slipping. Note that the friction force in this case is $\mathcal{F} \leq \mu_s N$, where μ_s is the coefficient of static friction.

UNBALANCED WHEELS: A special problem of practical interest is the plane motion of unbalanced wheels that roll without slipping. In such a case the mass center C and the geometric center G do not coincide, as in Fig. 18-15, and the relationship $a_C = r\alpha$ is not valid. Instead, one must use the expression

$$a_G = r\alpha \tag{18-21}$$

The acceleration of the mass center C is the vector sum of the acceleration of the geometric center G, and of the relative acceleration of C with respect to G,

$$\mathbf{a}_C = \mathbf{a}_G + \mathbf{a}_{C/G} = \mathbf{a}_G + (\mathbf{a}_{C/G})_n + (\mathbf{a}_{C/G})_t \tag{18-22}$$

where $(a_{C/G})_n = (CG)\omega^2$ and $(a_{C/G})_t = (CG)\alpha$.

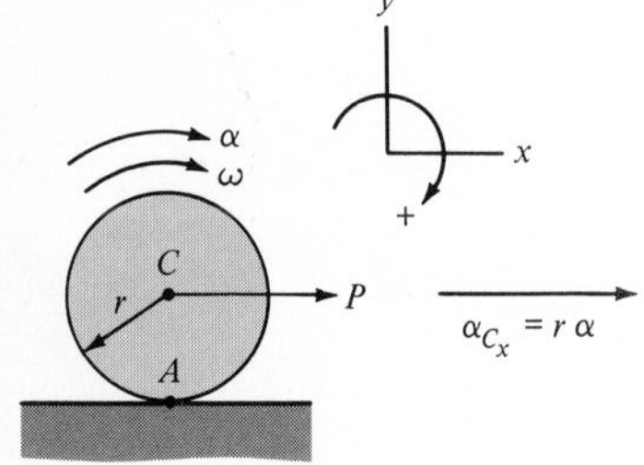

(a) No slipping at A

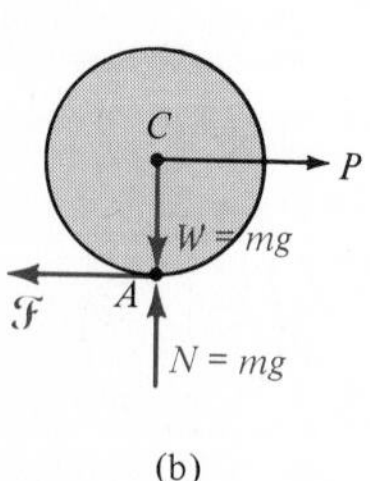

(b)

FIGURE 18-14

Pure rolling of a wheel

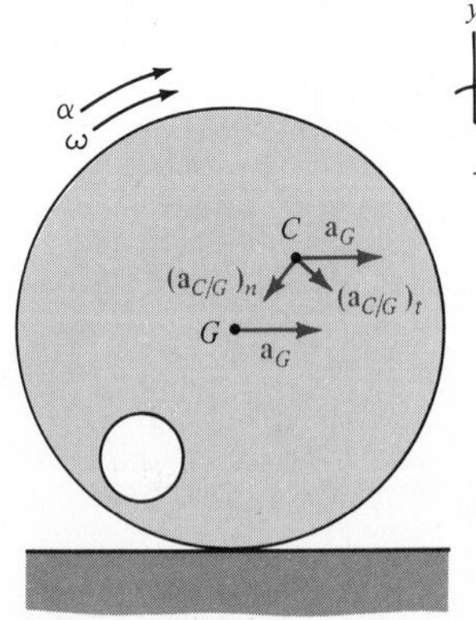

FIGURE 18-15

Rolling motion of an unbalanced wheel

Rolling and sliding of wheels. In the general case a wheel may be rolling while slipping at the point of contact. This means that the friction force at point A in Fig. 18-16 is known, $\mathscr{F} = \mu_k N$, where μ_k is the coefficient of kinetic friction. In this case, however, $a_{Cx} \neq \alpha r$ since the linear and angular accelerations are independent of one another.

Analysis of Rolling or Sliding Balanced Wheels

In general, there are four equations available for determining four unknowns in problems involving wheels in either pure rolling or in rolling *and* sliding. The equations are from force and moment summations and kinematic relationships. The method of solution is illustrated for the common case of a balanced wheel moving on a horizontal track under the action of a horizontal, centroidal force P as in Fig. 18-16.

		for no slipping	for slipping	
$\sum F_x = ma_{Cx}$	$P - \mathscr{F} = ma_{Cx}$	←	←	18-23
$\sum F_y = ma_{Cy}$	$N - mg = 0$	←	←	18-24
$\sum M_C = I_C\alpha$	$\mathscr{F}r = I_C\alpha$	←	←	18-25
	$a_C = \alpha r$	←		18-26
	$\mathscr{F} = \mu_k N$		←	18-27

If it is not certain whether the wheel is slipping or not, the condition for no slipping should be assumed and checked first. When the friction force is $\mathscr{F} \leq \mu_s N$, there is no slipping, and the assumption was correct. On the other hand, $\mathscr{F} > \mu_s N$ indicates that there is sliding, so the solution must be started over with $\mathscr{F} = \mu_k N$ now known but the accelerations a_C and α unrelated. Note that Eqs. 18-23 through 18-27 can be used with appropriate modifications for problems involving inclined tracks, or a force P which is not parallel to the track or not acting at the center of mass.

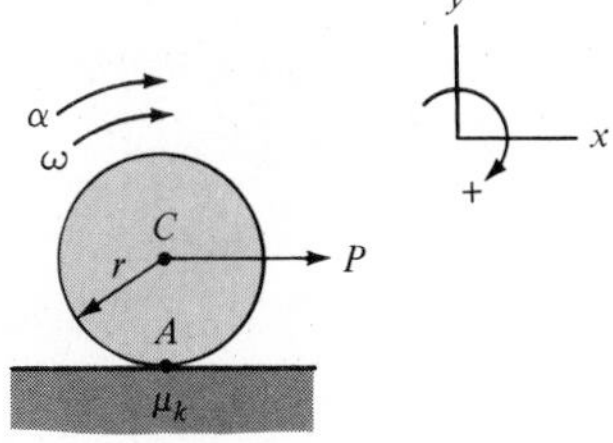

(a) Slipping at A is possible

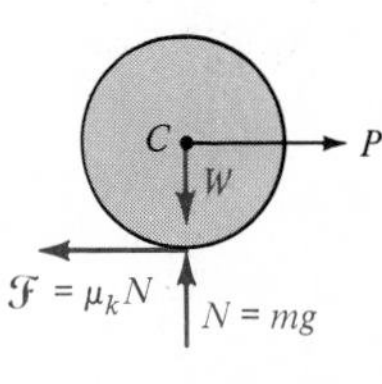

(b)

FIGURE 18-16

Schematic for a rolling and sliding wheel

EXAMPLE 18-6

A weight W of negligible dimensions is attached at point B to a uniform disk or radius r and weight $10W$ (Fig. 18-17a). The disk is released from rest in the position shown. Determine the acceleration of B if the disk does not slip.

SOLUTION

The free-body diagram is drawn in Fig. 18-17b. Since there is no slipping,

$$a_G = r\alpha$$

From Eqs. 18-19 and 18-20,

$$\sum M_A = (I_A\alpha)_{\text{disk}} + m_W r(\alpha_{Bx} + \alpha_{By})$$

where $m_W r(\alpha_{Bx} + \alpha_{By}) = 2m_W r^2\alpha$.

$$10Wr \sin 30^\circ + W(r \sin 30^\circ + r \cos 30^\circ)$$

$$= \left[\frac{1}{2}\frac{10W}{g}r^2 + \frac{10W}{g}r^2 + \frac{W}{g}(AB)^2\right]\alpha$$

Dividing by W and r,

$$5 + 0.5 + 0.866 = \left(\frac{5r}{g} + \frac{10r}{g} + \frac{2r}{g}\right)\alpha$$

$$\alpha = 0.37\frac{g}{r} \qquad \boldsymbol{\alpha} = -0.37\frac{g}{r}\mathbf{k}$$

The acceleration of point B is

$$\mathbf{a}_B = \mathbf{a}_G + (\mathbf{a}_{B/G})_n + (\mathbf{a}_{B/G})_t$$

$$= \boldsymbol{\alpha} \times r\mathbf{j} + 0 + \boldsymbol{\alpha} \times r\mathbf{i}$$

$$= 0.37g\mathbf{i} - 0.37g\mathbf{j}$$

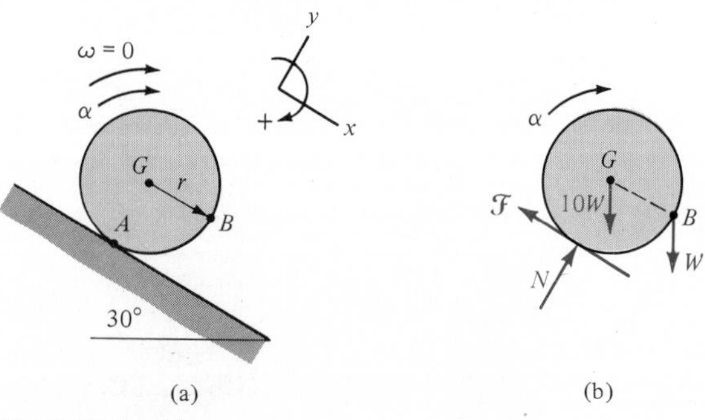

FIGURE 18-17

Analysis of a rolling, unbalanced wheel

EXAMPLE 18-7

A 20-kg uniform disk is at rest on a horizontal surface (Fig. 18-18a) when a horizontal force $F = 400$ N is applied at the center of the disk. Determine the acceleration of the mass center.

SOLUTION

The free-body diagram is drawn in Fig. 18-18b. It is not known whether the disk rolls without slipping or not. Hence, rolling without slipping is assumed and will be checked later.

$$\sum F_x = F - \mathscr{F} = ma_{Cx}$$

$$400 \text{ N} - \mathscr{F} = m(r\alpha) = (20 \text{ kg})(0.3 \text{ m})\alpha \qquad \text{(a)}$$

$$\sum F_y = N - W = ma_{Cy} = 0$$

$$N = W = (20 \text{ kg})(9.81 \text{ m/s}^2) = 196.2 \text{ N}$$

$$\sum M_A = F(r) = I_A\alpha$$

$$(400 \text{ N})(0.3 \text{ m}) = \left[\frac{1}{2}(20 \text{ kg})(0.3 \text{ m})^2 + (20 \text{ kg})(0.3 \text{ m})^2\right]\alpha$$

$$120 \text{ N}\cdot\text{m} = (2.70 \text{ kg}\cdot\text{m}^2)\alpha$$

$$\frac{44.4 \text{ N}\cdot\text{m}}{\text{kg}\cdot\text{m}^2} = 44.4 \text{ rad/s}^2 = \alpha$$

Substituting the value obtained for α into Eq. (a) yields

$$\mathscr{F} = 400 \text{ N} - (20 \text{ kg})(0.3 \text{ m})(44.4 \text{ rad/s}^2) = 133 \text{ N}$$

This is the friction force (to the left) necessary if rolling without slipping were to occur. A check of the maximum possible friction force shows that

$$\mathscr{F}_{\text{max}} = \mu_s N = (0.4)(196.2 \text{ N}) = 78.5 \text{ N}$$

Since $\mathscr{F}$ cannot exceed $\mathscr{F}_{\text{max}}$, the wheel must roll and slip. The solution must be started over, using the kinetic friction force $\mathscr{F}_k = (0.3)(196.2 \text{ N}) = 58.9$ N.

$$\sum F_x = F - \mathscr{F}_k = ma_{Cx}$$

$$= 400 \text{ N} - 58.9 \text{ N} = (20 \text{ kg})(a_{Cx})$$

$$a_{Cx} = 17.1 \text{ m/s}^2$$

$$\sum M_C = \mathscr{F}_k(r) = I_C\alpha$$

$$= (58.9 \text{ N})(0.3 \text{ m}) = \frac{1}{2}(20 \text{ kg})(0.3 \text{ m})^2\alpha$$

$$\alpha = 19.63 \text{ rad/s}^2$$

Note that if rolling and slipping had been assumed initially, there would have been no convenient check on the assumption available. Also notice that the sense of force $\mathscr{F}$ is not even known at first. A direction must be assumed and the solution will yield the correct sign.

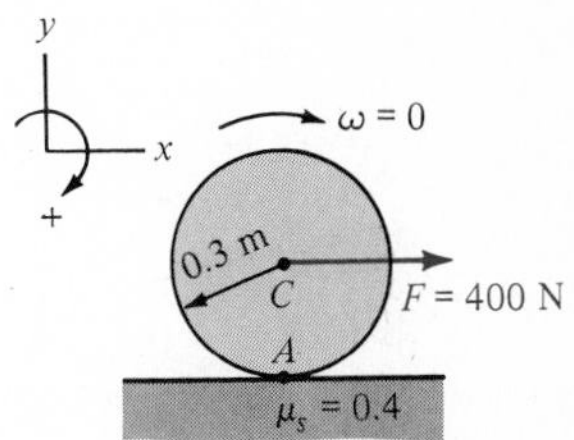

(a)

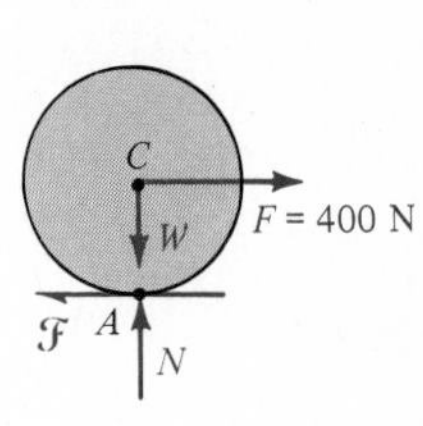

(b)

FIGURE 18-18

Analysis of a rolling wheel which may also be slipping

PROBLEMS

SECTION 18-5

FIGURE P18-57

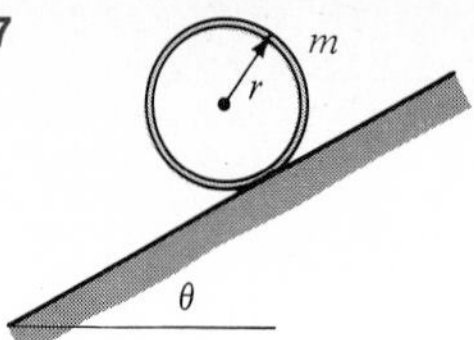

18–57 A thin hoop of mass m and radius r starts from rest on an inclined plane. The surfaces are rough and prevent slipping. Determine the hoop's linear acceleration for $m = 2$ kg, $r = 0.4$ m, and $\theta = 20°$.

18–58 Plot the acceleration of the hoop shown in Fig. P18-57 for the following conditions:

(a) m = constant, θ = constant, radius = r, $2r$, $3r$

(b) m = constant, r = constant, angle = θ, 2θ, 3θ

FIGURE P18-59

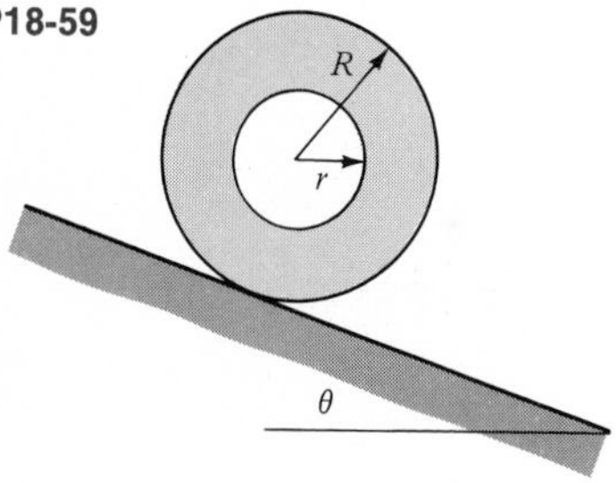

18–59 A 20-lb cylinder is released from rest. Determine the cylinder's linear acceleration if there is no slippage, $r = 6$ in., $R = 12$ in., and $\theta = 30°$.

18–60 Calculate the acceleration of the cylinder in Fig. P18-59 for $r = 3$ in. and $r = 9$ in. All other quantities are the same as in Prob. 18–59.

FIGURE P18-61

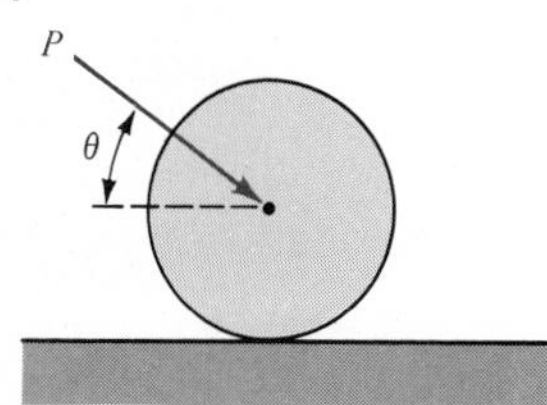

18–61 A force $P = 200$ N is acting at the mass center of a uniform disk of 10 kg mass and 0.4 m radius. Determine the linear acceleration of the disk if there is no slippage and $\theta = 25°$.

FIGURE P18-62

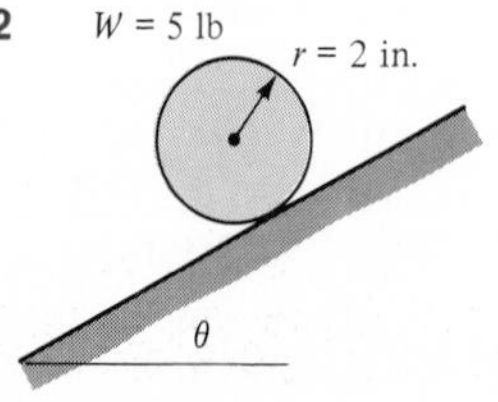

18–62 A uniform 5-lb cylinder is used in a simple experiment to demonstrate the acceleration of gravity. Assume that the cylinder rolls without slipping. Determine the angle θ for which the cylinder's linear acceleration along the incline is 30 ft/s^2.

18–63 A uniform cylinder of 20 kg mass and 0.3 m radius is supported by an inextensible band AB. Calculate the acceleration of the mass center and the tension in A for the instant immediately after part B of the band breaks. Assume that there is no slipping between the band and the cylinder.

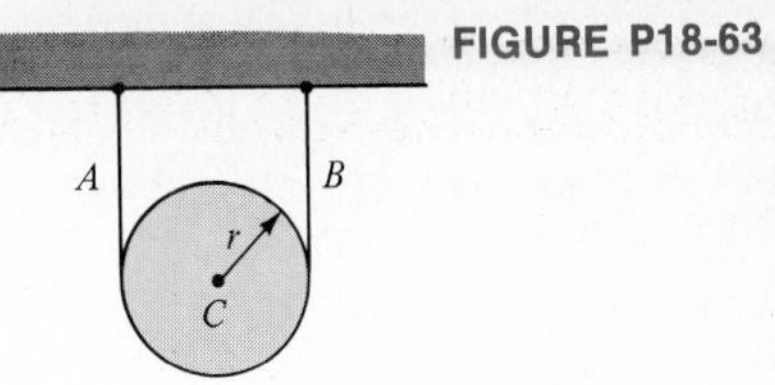

FIGURE P18-63

18–64 A 50-lb uniform wheel of 1 ft radius rolls downhill. Determine the wheel's linear acceleration, and the minimum value of μ_s that is sufficient to prevent slipping.

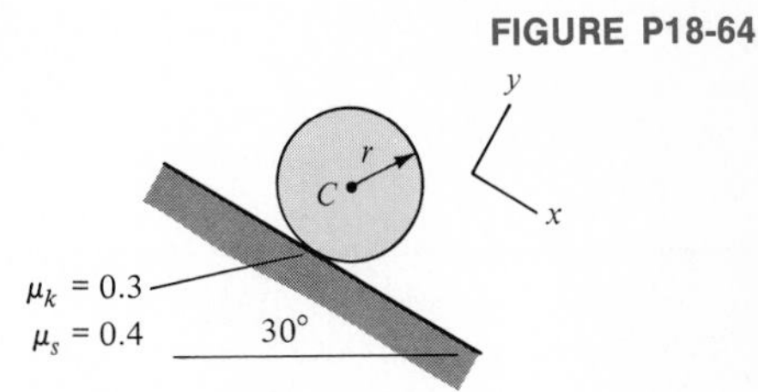

FIGURE P18-64

18–65 Add a force $\mathbf{P} = (20\mathbf{i} - 10\mathbf{j})$ lb acting at point C to the wheel in Prob. 18–64. Calculate the linear acceleration in this case.

18–66 A force $P = 500$ N is applied to a uniform disk of 30-kg mass and 0.5 m radius. Calculate the acceleration of the disk.

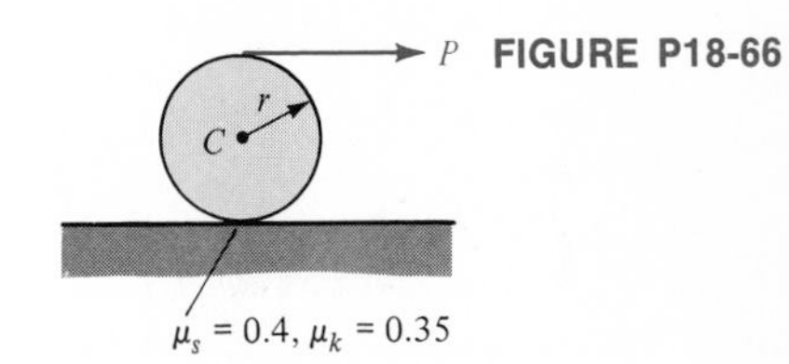

FIGURE P18-66

18–67 A 100-lb uniform wheel of 2-ft radius is initially at rest on an inclined plane. Calculate the linear acceleration of the disk caused by a force $P = 500$ lb.

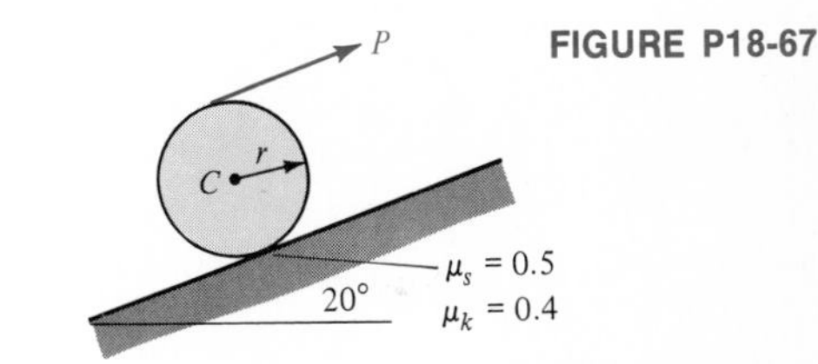

FIGURE P18-67

18–68 What is the maximum force P if the wheel in Prob. 18–67 is to be moved uphill without slipping?

18–69 The console has a mass of 250 kg which may be assumed as uniformly distributed. It rests on casters that roll freely in any direction. Determine the acceleration of the mass center caused by a force $\mathbf{F} = 800$ N $\mathbf{k}$ applied at point A.

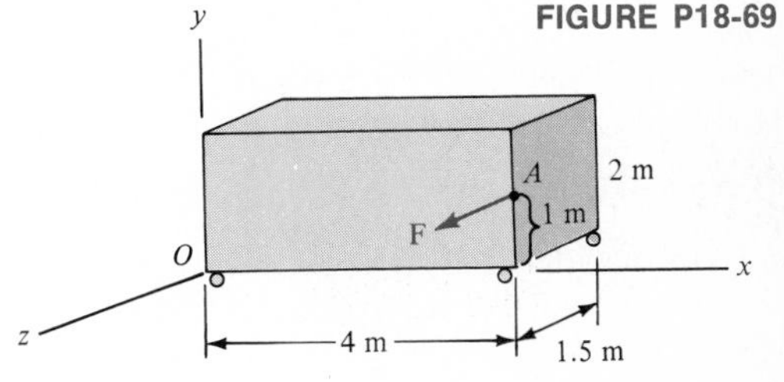

FIGURE P18-69

18–70 A truck is modeled as a 15,000-lb uniform box with a square cross section. Calculate the angular acceleration caused by a total vertical force $F = 10{,}000$ lb acting at wheels A. Assume that the springs at B cannot deflect any more.

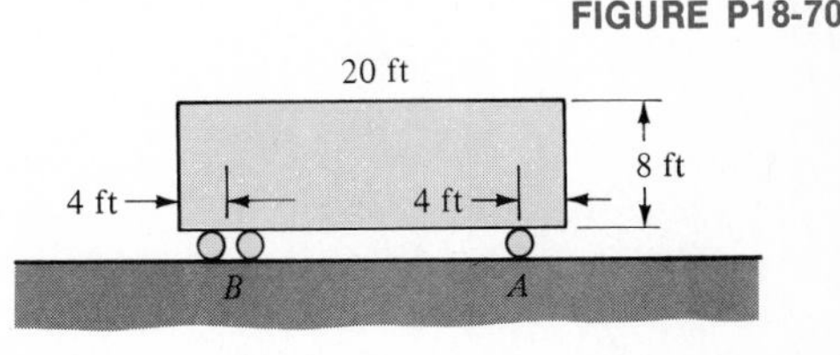

FIGURE P18-70

FIGURE P18-71

Spherical container for liquified natural gas (Photograph courtesy General Dynamics Quincy Shipbuilding Division)

(a)

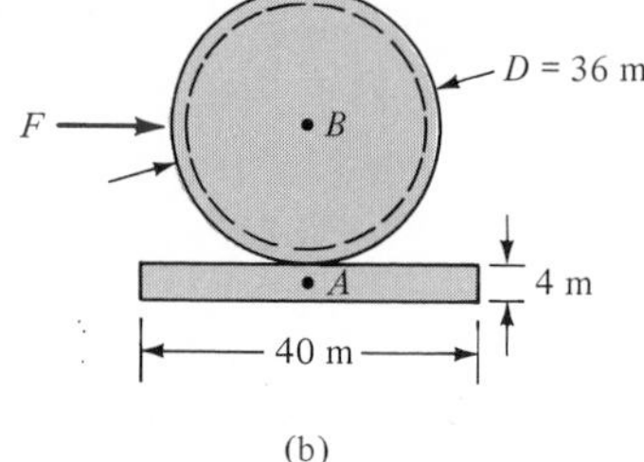

(b)

18–71 The barge carrying an empty spherical container for liquid natural gas is modeled as a rigid system in Fig. 18-71b. The thin-walled spherical shell has a mass of 800 t, and the barge is assumed to be a 40 m × 60 m × 4 m solid plate of 2000-t mass (1 t = 907.2 kg). Assume that the system is stable in the position shown and that center A of the barge always moves horizontally without frictional resistance. Determine the angular acceleration caused by a gust of wind that has a resultant $F = 50$ kN.

FIGURE P18-72

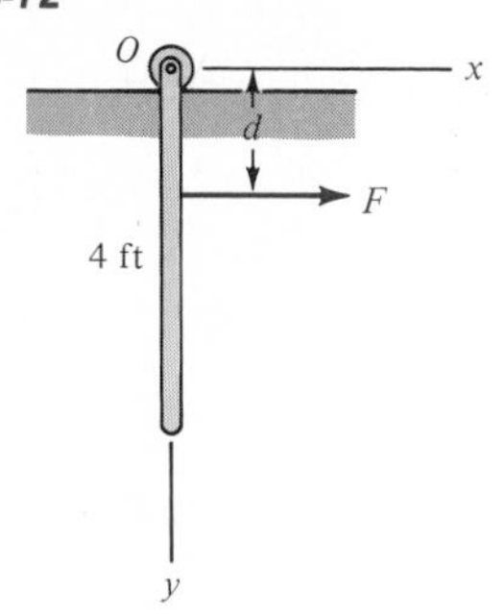

18–72 A uniform rod of 40 lb weight is hanging from the axle of a weightless, frictionless roller of negligible size. Determine the angular acceleration of the rod and the acceleration of the roller caused by a horizontal force $F =$ 20 lb at $d = 1$ ft.

FIGURE P18-73

Two-piece waveboard (Photograph courtesy MTS Systems Corporation, Minneapolis, Minnesota)

(a)

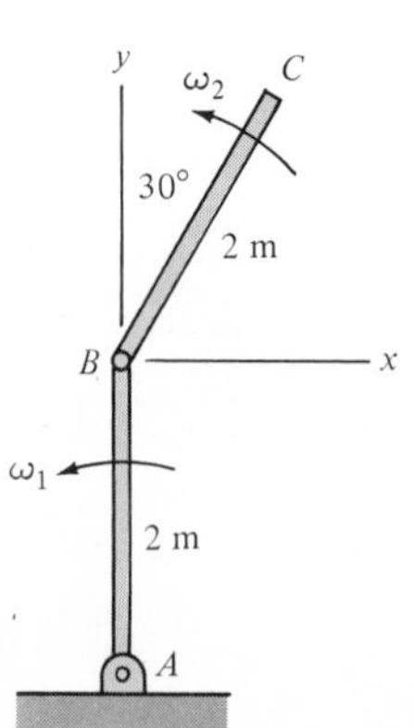

(b)

18–73 Consider the dry test of a two-piece waveboard (see Chap. 3 of the statics volume) which consists of two slender, uniform boards AB and BC, each of 2 m width and 300 kg mass. The lower board is stationary ($\omega_1 = 0$) when it is in the vertical position. At the same instant the upper board makes 30° with the vertical and has an angular speed of $\omega_2 =$ 0.3 rad/s, counterclockwise. ω_2 is increasing at the rate of 0.2 rad/s^2. Determine the moment about hinge B and the force acting on this hinge at the given instant.

18–74 Solve Prob. 18–73 assuming that $\omega_1 =$ 0.2 rad/s, constant.

18–75 A 60-lb ladder is assumed to be a uniform bar which starts sliding without frictional resistance. Determine the forces at A and B.

FIGURE P18-75

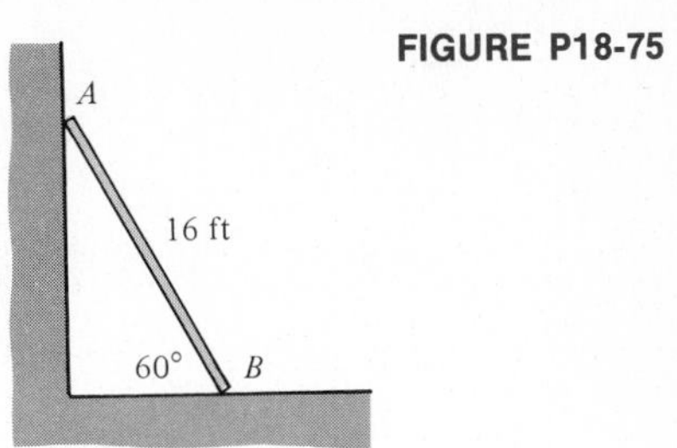

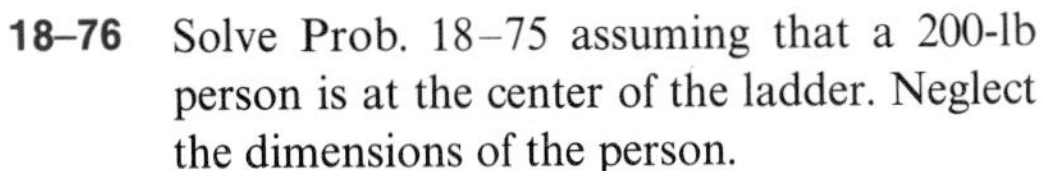

18–76 Solve Prob. 18–75 assuming that a 200-lb person is at the center of the ladder. Neglect the dimensions of the person.

FIGURE P18-77

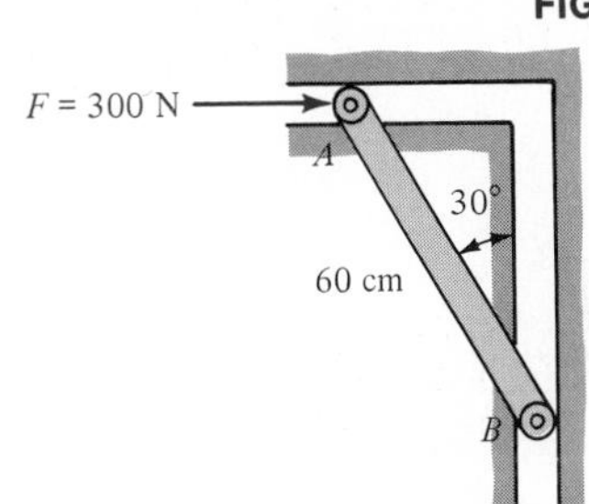

18–77 Pins A and B of the 5-kg, uniform bar have negligible mass and size, and they slide freely in their tracks. Calculate the acceleration of B and the forces between the pins and their guides caused by the force F.

FIGURE P18-78

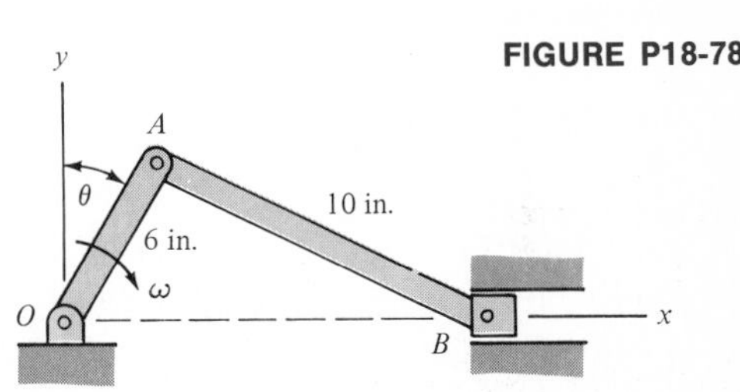

18–78 The 2-lb crank arm OA is rotating at a constant speed $\omega = 1500$ rpm. The uniform piston rod AB weighs 2.2 lb, and the piston B weighs 2.4 lb. Calculate the force acting on pin A for $\theta = 90°$, neglecting friction.

18–79 Determine the x and y components of the force acting on pin B in Prob. 18–78 for $\theta = 0$.

FIGURE P18-80

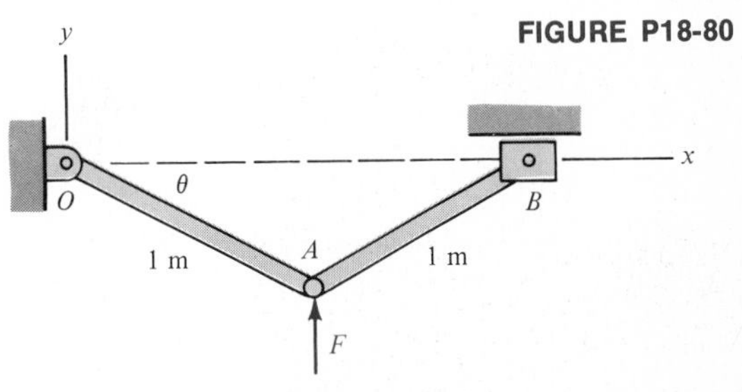

18–80 The toggle mechanism OAB moves in the horizontal xy plane. Each uniform bar has a mass of 12 kg, and block B has a mass of 6 kg. Calculate the x and y components of the force acting on pin B for $\theta = 30°$, and $F =$ 700 N. Friction is negligible. The system is released from rest at $\theta = 30°$.

FIGURE P18-81

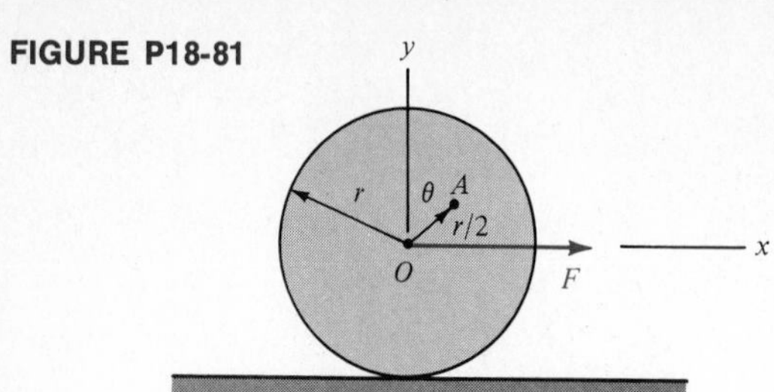

18–81 A uniform, 20-kg disk of 0.6 m radius has a 2-kg mass attached at point A. The disk is initially at rest when a force $F = 50$ N is applied at its geometric center O. Determine the acceleration of axis O for $\theta = 30°$ if there is no slippage.

18–82 Determine the acceleration of the mass center of the system in Prob. 18–81.

FIGURE P18-83

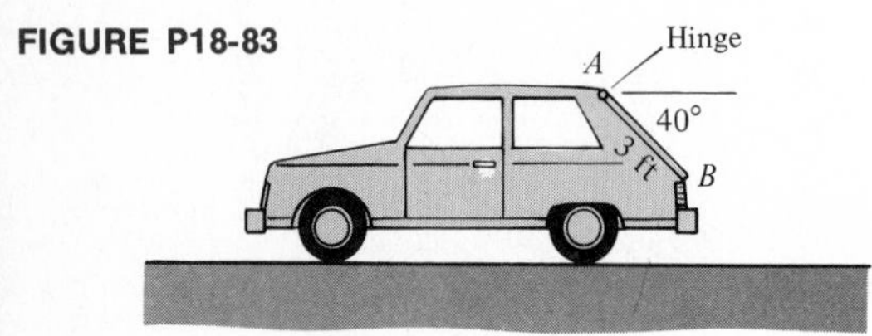

18–83 The hatch door AB of a car is modeled as a thin, uniform plate 3 ft long and 4 ft wide, and of 25 lb weight. Calculate the angular acceleration of the unlatched door if the car accelerates from rest at the rate of 20 ft/s^2. Neglect friction.

18–84 The door AB in Prob. 18–83 drops downward when released from rest at a 20° angle above the horizontal. Determine the acceleration of the car at which the door's angular acceleration is zero in its upper position.

FIGURE P18-85

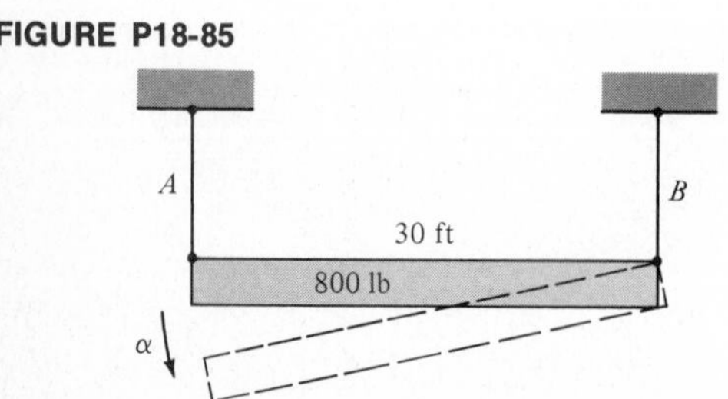

18–85 On April 20, 1979, cable A broke on a scaffolding used for window washing in Richmond, Virginia. Calculate the tension in cable B and the angular acceleration of the scaffolding immediately after the failure of cable A. Assume that the scaffolding is a uniform, slender bar.

18–86 The test sled is sliding on four linear bearings on two horizontal tracks. The dummy is restrained by a seat belt. Assume that the rigid torso (above the belt) has a 40-kg mass, a 35-cm radius of gyration about its own mass center, which is 30 cm from the pivot of the torso (at the seat belt). Determine the angular acceleration of an initially vertical torso if the sled is decelerating at the rate of 60 m/s^2.

FIGURE P18-86

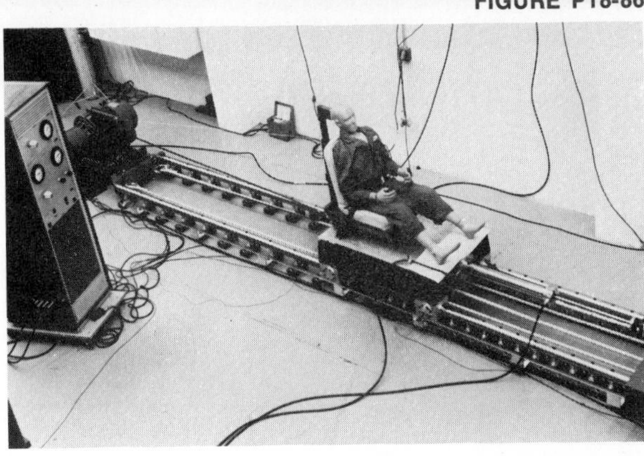

18–87 θ is the angle of the torso's initial recline (to the left) from the vertical in Fig. P18-86. Calculate the angular acceleration of the torso in Prob. 18–86 for $\theta = 30°$.

FIGURE P18-88

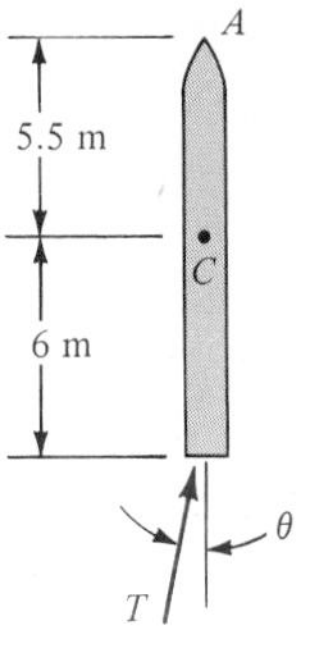

A
5.5 m
C
6 m
θ
T

18–88 The mass of a rocket is 6000 kg, and its radius of gyration about the mass center C is 3.5 m. Determine the thrust T that can cause an angular acceleration of 0.1 rad/s^2 when applied at $\theta = 5°$. Calculate the absolute acceleration of point A caused by this thrust.

18–89 A 200-kg spacecraft has two small attitude jets in the xz plane, the plane perpendicular to the vertical centerline (y direction) of the craft. The jets have the following thrusts and locations: $\mathbf{T}_1 = -0.5\ \text{N}\ \mathbf{k}$ at $x = 0.8$ m, and $\mathbf{T}_2 = 0.5\ \text{N}\ \mathbf{k}$ at $x = -0.8$ m. The spacecraft has an angular acceleration $\boldsymbol{\alpha} = 0.1\ \text{rad/s}^2\ \mathbf{j}$ when both jets are operating simultaneously. Determine the radius of gyration about the y axis. Assume that $\mathbf{T}_1$, $\mathbf{T}_2$, and the center of mass are in the same plane, and that the 20-m-long supporting wire applies only a vertical force to the spacecraft.

FIGURE P18-89

Spacecraft before a test in a vacuum chamber (The Boeing Company, Seattle, Washington)

18–90 Assume that only $\mathbf{T}_1$ is acting of the two jets in Prob. 18–89. Calculate the linear and angular accelerations of the spacecraft.

SUMMARY

Section 18-1

Equation of translational motion

$$\sum \mathbf{F} = m\mathbf{a}_C$$

where $\mathbf{a}_C$ is the acceleration of mass center C regardless of where the resultant force acts on the rigid body; this is independent of any rotation of the body.

Section 18-2

Equation of rotational motion

$$\sum \mathbf{M}_C = I_C\boldsymbol{\alpha}$$ (for a thin slab perpendicular to $\sum \mathbf{M}_C$, or special cases of symmetry or constraints)

where $\boldsymbol{\alpha}$ is the angular acceleration of the rigid body; this is independent of any translation of the body.

Section 18-3

Application of equations of motion. Three scalar equations of motion can be written for each rigid body in plane motion:

$$\sum F_x = ma_{Cx} \qquad \sum F_y = ma_{Cy} \qquad \sum M_C = I_C\alpha$$

TRANSLATION:

$$\sum F_x = ma_{Cx} \qquad \sum F_y = ma_{Cy} \qquad \sum M_C = 0$$

ROTATION:

$$\sum F_x = 0 \qquad \sum F_y = 0 \qquad \sum M_C = I_C\alpha$$

SYSTEMS OF RIGID BODIES: At most three equations of motion are available from each free-body diagram for plane motion to solve for three unknowns. The motions of interconnected bodies must be analyzed using interrelated free-body diagrams.

Section 18-4

Rotation about a fixed axis. The scalar equations of motion using normal and tangential components are

$$\sum F_n = mr_C\omega^2 \qquad \sum F_t = mr_C\alpha \qquad \sum M_O = I_O\alpha$$

where r_C is the nearest distance between the fixed axis O and the mass center C. $\sum F_n$ and $\sum F_t$ must include the reaction on the rigid body at the axis of rotation.

Section 18-5

General plane motion. The scalar equations of motion are

$$\sum F_x = ma_{Cx} \qquad \sum F_y = ma_{Cy} \qquad \sum M_C = I_C\alpha$$

PURE ROLLING OF WHEELS:

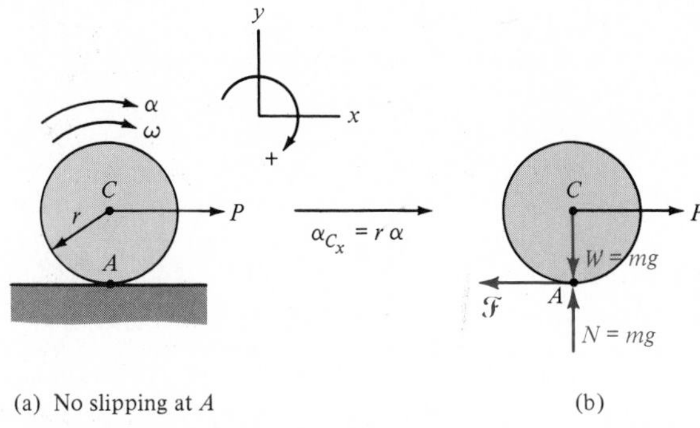

(a) No slipping at A (b)

If the mass center C coincides with the geometric center G (balanced wheel),

$$a_{Cx} = r\alpha \qquad \sum M_A = I_A\alpha$$

If C is not at G (unbalanced wheel),

$$a_C \neq r\alpha \qquad a_G = r\alpha$$

$$\mathbf{a}_C = \mathbf{a}_G + \mathbf{a}_{C/G} = \mathbf{a}_G + (\mathbf{a}_{C/G})_n + (\mathbf{a}_{C/G})_t$$

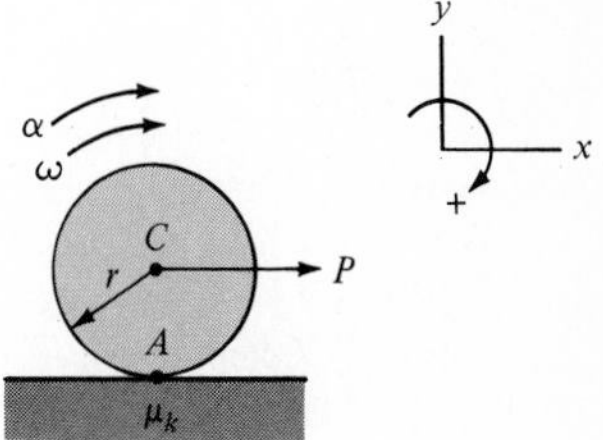

(a) Slipping at A is possible

ROLLING AND SLIDING OF BALANCED WHEELS:

$$\sum F_x = ma_{Cx} \qquad P - \mathscr{F} = ma_{Cx}$$

$$\sum F_y = ma_{Cy} \qquad N - mg = 0$$

$$\sum M_C = I_C\alpha \qquad \mathscr{F}r = I_C\alpha$$

$$a_C = \alpha r$$

$$\mathscr{F} = \mu_k N$$

(The first four equations are bracketed "for no slipping"; the first three together with $\mathscr{F} = \mu_k N$ are bracketed "for slipping".)

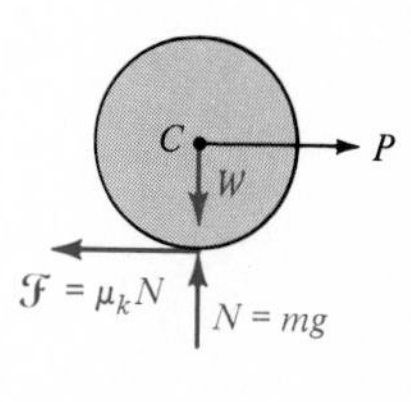

(b)

If slipping is not certain, assume no slipping and check whether $\mathscr{F} \leq \mu_s N$. $\mathscr{F} > \mu_s N$ is not possible; start the solution over, using $\mathscr{F} = \mu_k N$.

REVIEW PROBLEMS

FIGURE P18-91

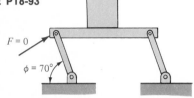

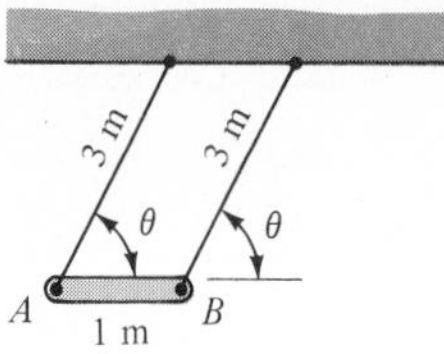

18–91 A playground swing is modeled as a 10-kg uniform bar on two wires. Determine the acceleration of the bar and the tension in the wires in the position with $\theta = 60^\circ$ and $\dot{\theta} = 0$.

18–92 A 30-kg child sits on the swing of Prob. 18–91. The coefficient of static friction between the bar and the child is 0.3. Determine the accelerations of the bar and the child at $\theta = 60^\circ$ and $\dot{\theta} = 0$.

FIGURE P18-93

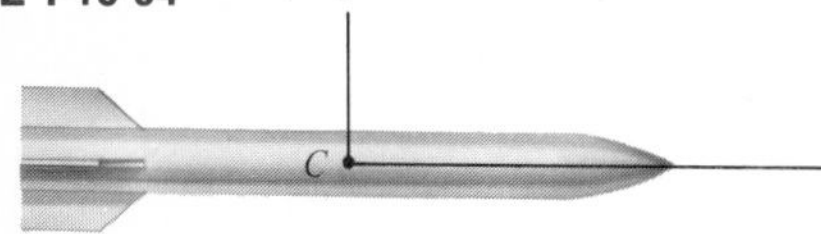

18–93 Consider the shake table described in Ex. 18-3. Assume that the actuator force is rapidly decreased to zero when $\phi = 70^\circ$ and $\dot{\phi} = 0$. Determine the accelerations of the test object and the platform assuming $\mu_s = 0$ between them.

FIGURE P18-94

18–94 Sometimes a rigid body is too complex for a direct calculation of its mass moment of inertia. An example of this is a small rocket which has numerous components in it. Describe a procedure for using experiments and calculations to determine the mass moment of inertia of the rocket about the transverse axis y through the centroid C. Assume that forces and accelerations can be accurately measured using electronic equipment.

FIGURE P18-95

Model aircraft in a laboratory test (Photograph The Boeing Company, Seattle, Washington)

(a)

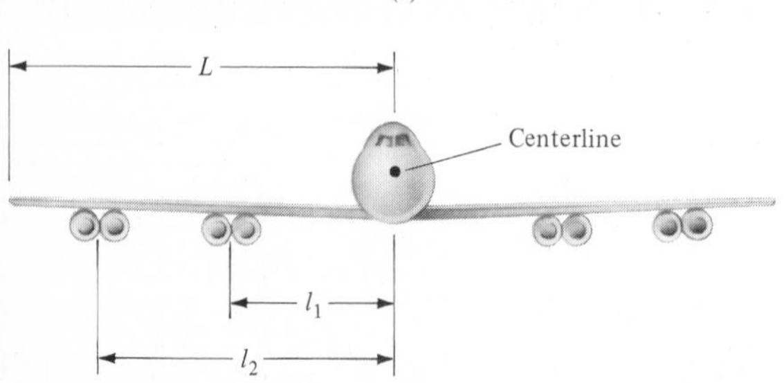

Note: Dimensions are in plane of paper, not along the swept wing

(b)

18–95 Assume that the wings and the engines of the model aircraft are the major parts that have mass moment of inertia about the longitudinal centerline of the fuselage. This centerline is a fixed axis in a test. Assume that each wing is a uniform, slender bar of length L and mass m, and each pair of engines is a concentrated mass m_1 at a distance l_1 and l_2 from the center. Write an expression for the angular acceleration α of the model caused by a moment M about the centerline. Neglect the differences in elevation of the various parts. The angle of sweep of the wings with the fuselage is θ.

18–96 A proposed impact tester consists of a 2-m long uniform bar OA of 16-kg mass. The position h of the 20-kg striker B can be adjusted before a test. Determine H so that there is no reaction at the pivot O caused by B striking the specimen S. The pendulum swings down with only its weight acting on it. Neglect the transverse dimensions of OA and B.

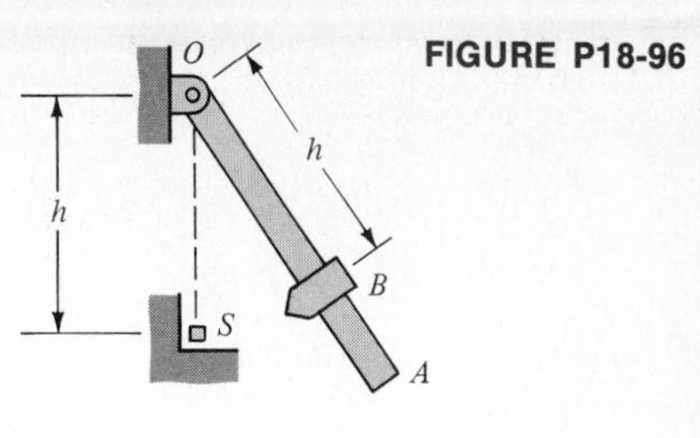

FIGURE P18-96

18–97 A 5-kg uniform disk is rotating at a constant speed $\omega = 30$ rad/s when it is lowered onto a horizontal surface and released. Determine the time elapsed until the disk begins to roll without slipping if $\mu_k = 0.3$.

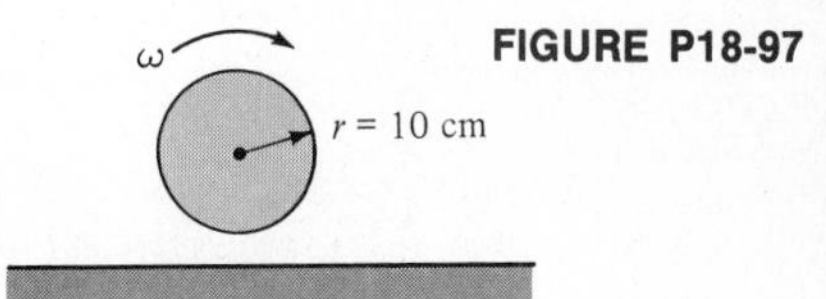

FIGURE P18-97

18–98 Determine the distance traveled by the disk in Prob. 18–97 during its motion involving skidding.

18–99 A linear bearing is modeled as a 5-lb bar on which a 10-lb force from a weightless spring mechanism is acting. The steel ball bearings have a coefficient of friction $\mu_s = 0.1$. Calculate the largest force F that can be applied without causing the balls to slip. Each bearing weighs 1/16 lb.

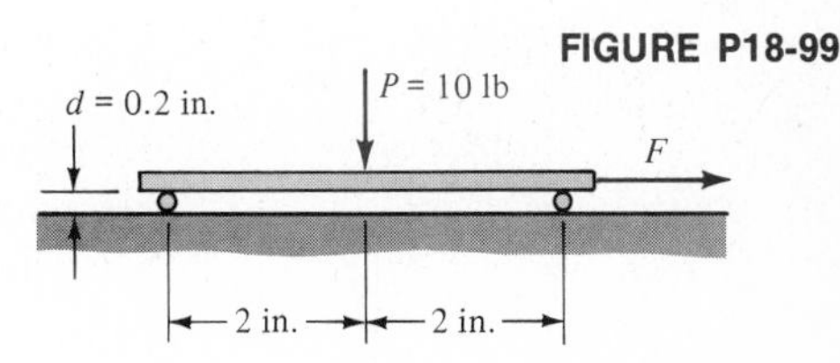

FIGURE P18-99

18–100 A conveyor belt starts from rest with an acceleration of 1 m/s^2. Determine the angular acceleration of the 2-kg sphere and the 3-kg cylinder. For both bodies, $\mu_s = 0.4$, and $\mu_k = 0.3$.

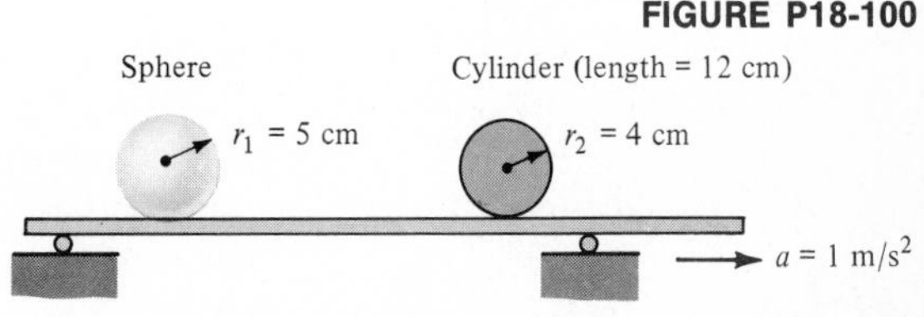

FIGURE P18-100

CHAPTER 19

ENERGY AND MOMENTUM METHODS FOR RIGID BODIES IN PLANE MOTION

OVERVIEW

It was shown in Chap. 14 that the methods of work–energy are very advantageous in solving a variety of problems in dynamics. These methods can be readily extended to analyze the motion of rigid bodies. For a glimpse of the usefulness of these methods, consider flywheel-powered cars which are currently investigated because they may be economical for urban transportation. A common feature of several proposed designs is the flywheel that can rotate at high speed about a vertical axis (Fig. 19-1). This disk may receive power periodically from an external source or an auxiliary engine in the car. The flywheel can be used to propel the car at the expense of gradually decreasing its own rotational speed. In turn, regenerative braking of the car can recharge the flywheel energy storage system. In such a case, there are repeated conversions of energy of the various components of the vehicle and some energy is dissipated by friction. The speed and the acceleration of the vehicle depend on the quantities of work and energy associated with the motion.

Momentum methods are also useful in analyzing the motion of rigid bodies. The major results of Chap. 15 are readily applied to rigid bodies. Since both the energy and momentum methods can be described concisely at this stage, they are presented together in one chapter.

STUDY GOALS

SECTION 19-1 presents the definition of the work of a force acting on a rigid body. The definition of work is also extended to systems of forces. This topic should be thoroughly studied because it is essential for applying the principle of work and energy to rigid bodies (Sec. 19-3).

SECTION 19-2 extends the concepts of kinetic energy of systems of particles to rigid bodies. The equation representing the kinetic energy due to translation and rotation of a body is frequently applied in dynamics analyses and is worth remembering.

SECTION 19-3 presents the principle of work and energy for analyzing the motion of a rigid body or system of rigid bodies. This method is particularly advantageous in solving problems where velocities and displacements are of interest.

SECTION 19-4 defines the conservation of energy in a conservative mechanical system. Even though an idealization, this is a very useful method for solving certain practical problems.

SECTION 19-5 extends the definition of power to translating and rotating rigid bodies. These concepts are used in machine design.

SECTIONS 19-6 AND 19-7 present the method of impules and momentum of a rigid body or system of rigid bodies. This special method is useful in the analysis of motion when time and velocities are of interest.

SECTION 19-8 defines the conservation of momentum for rigid bodies that have no external forces or moments acting on them. This method is useful in analyzing a mechanical system which changes configuration by internal means.

19-1 WORK OF A FORCE ON A RIGID BODY

The work of a force acting on a rigid body is defined as it is for a particle: work equals the magnitude of the force times the displacement along the line of action of the force. Formally, the work U_{12} of a force $\mathbf{F}$ during a displacement of its point of application from 1 to 2 is

$$U_{12} = \int_1^2 \mathbf{F} \cdot d\mathbf{r} \qquad \text{19-1}$$

where $d\mathbf{r}$ is the infinitesimal vector displacement of $\mathbf{F}$ along the path from 1 to 2 (see Sec. 14-1). In this equation the vectors $\mathbf{F}$ and

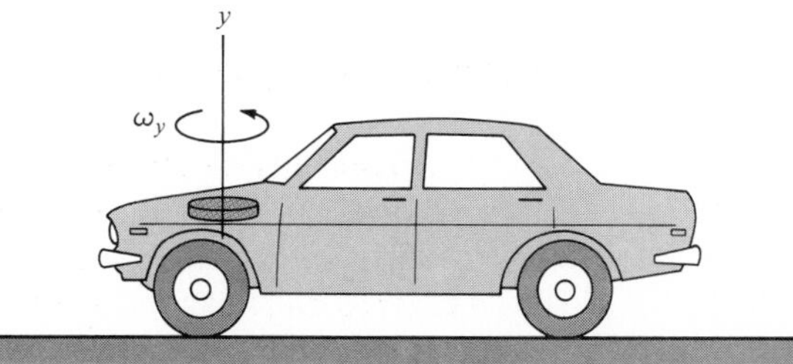

FIGURE 19-1

Schematic of a flywheel-powered car

$d\mathbf{r}$ are at any angle to each other since the dot product automatically accounts for the collinear components, the only ones that are involved in computing work. It is also useful to express the work of a force in terms of velocity. Since $\mathbf{v} = d\mathbf{r}/dt$,

$$U_{12} = \int_1^2 \mathbf{F} \cdot \mathbf{v}\, dt \tag{19-2}$$

where $\mathbf{v}$ is the velocity of the instantaneous point of application of $\mathbf{F}$.

The force $\mathbf{F}$ in Eqs. 19-1 and 19-2 may be the resultant of a system of forces. The work of these forces may be considered individually for a given rigid body. In that case the total work U of n forces is given by

$$U = \int \mathbf{F}_1 \cdot d\mathbf{r}_1 + \int \mathbf{F}_2 \cdot d\mathbf{r}_2 + \cdots + \int \mathbf{F}_n \cdot d\mathbf{r}_n \tag{19-3}$$

where it must be observed that no relative motion may occur between particles of a rigid body. In other words, the work of internal forces (equal but opposite forces in pairs) cancel for rigid bodies.

Work of a moment. The work of an applied moment acting on a rigid body may be developed from Eq. 19-3. For this it is desirable to establish a relationship involving the angular displacement caused by the moment. Consider an arbitrary rigid body with forces $\mathbf{F}$ and $-\mathbf{F}$ acting on it in the xy plane as in Fig. 19-2. For convenience, a reference frame is placed at point O such that line AB coincides with the x axis, and the y axis bisects the distance AB. For any translational motion of the body, the net work of forces $\mathbf{F}$ and $-\mathbf{F}$ is zero by inspection. In a rotational displacement $d\theta$, however, the resultant work of the two forces is positive. Working with points A and B, the total infinitesimal work dU can be expressed (for convenience, in scalar form) as

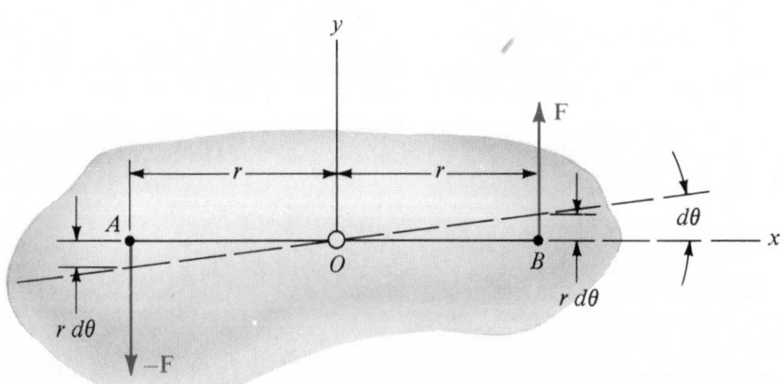

FIGURE 19-2

Schematic for defining the work of a couple

$$dU = dU_A + dU_B = -F(-r\,d\theta) + F(r\,d\theta)$$

$$dU = F(2r)\,d\theta = M\,d\theta$$

where M is the magnitude of the moment of the couple. The work U_{12} of the couple during a finite rotation of the body is the integral of this expression. The result is

$$U_{12} = \int_{\theta_1}^{\theta_2} M\,d\theta \tag{19-4}$$

which is limited to the common case where the moment vector **M** is perpendicular to the plane of motion.

The equations above for the work of a force or a moment are useful for solving many problems. There are several special cases that are worth noting as follows.

Work of a constant force. The force acting on the rigid body may have a constant magnitude F and a constant direction θ with respect to the path s of the body. The work U_{12} of the force F in moving the body from position s_1 to s_2 is written from Eq. 19-1 as

$$U_{12} = F\cos\theta\,(s_2 - s_1) \tag{19-5}$$

Work of a constant moment. When the moment M of the couple is constant, Eq. 19-4 becomes

$$U_{12} = M(\theta_2 - \theta_1) \tag{19-6}$$

Forces That Do No Work

Frequently, there are forces acting on moving rigid bodies without doing work on those bodies. The following list gives several such forces that are common.

1. Forces acting at fixed points on the body do no work. For example, the reaction at a fixed, frictionless pin does no work on the body that rotates about that pin.
2. A force which is always perpendicular to the direction of the motion does no work.
3. The weight of a body does no work when the body's center of gravity moves in a horizontal plane.
4. The friction force $\mathscr{F}$ at the point of contact on a body that rolls without slipping does no work. This is because the point of contact C is the instantaneous center of zero velocity ($v_C = 0$). From Eq. 19-2, $U = \int \mathscr{F} \cdot \mathbf{v}_C\,dt = 0$

19-2 KINETIC ENERGY OF A RIGID BODY

The system of particles comprising a rigid body has kinetic energy as a result of rotational motion as well as translational motion. These parts of the total kinetic energy are expressed using Fig. 19-3, where P_i is a representative particle of the body. This particle has mass m_i, absolute velocity $\mathbf{v}_i$ in the fixed XY frame, and relative velocity $\mathbf{v}_i'$ in the moving xy frame, which is attached at the center of mass C. Frame xy is translating with velocity $\mathbf{v}_C$ with respect to the fixed frame. The kinetic energy T_i of the particle P_i is written according to Sec. 16-4 as

$$T_i = \frac{1}{2} m_i v_i^2 = \frac{1}{2} m_i (v_C^2 + v_i'^2) \tag{19-7}$$

where $v_i' = r_i \omega$ may be substituted to represent the velocity of the particle relative to the xy frame. The total kinetic energy T is obtained by summing T_i over the n particles of the body,

$$T = \frac{1}{2} m v_C^2 + \frac{1}{2} \sum_{i=1}^{n} m_i v_i'^2$$

$$T = \frac{1}{2} m v_C^2 + \frac{1}{2} \left(\sum_{i=1}^{n} m_i r_i^2 \right) \omega^2 \tag{19-8}$$

Using an infinitesimal element of mass dm_i, and integrating over the whole body, the term in parentheses becomes the mass moment of inertia I_C about the centroidal axis through point C and normal to the plane of motion. This gives a concise expression for the kinetic energy,

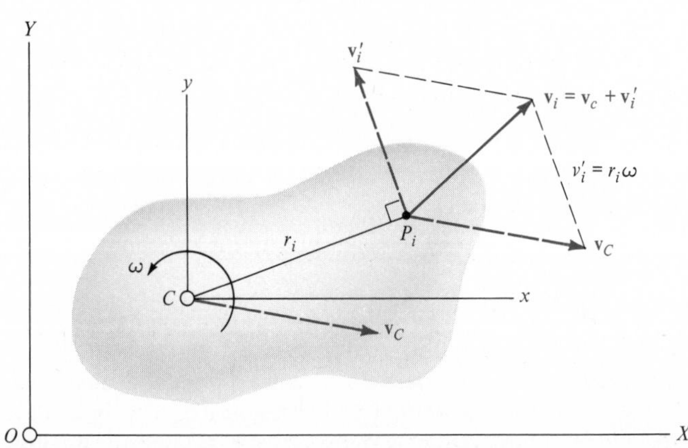

FIGURE 19-3

Schematic for defining the total kinetic energy of a rigid body

$$T = T_{\text{transl}} + T_{\text{rot}} = \frac{1}{2} m v_C^2 + \frac{1}{2} I_C \omega^2 \tag{19-9}$$

where $\frac{1}{2}mv_C^2$ represents the kinetic energy due to *translation*, and $\frac{1}{2}I_C\omega^2$ represents the kinetic energy due to *rotation* of the body, even though the two motions may be simultaneous. Note that the kinetic energy is always a positive quantity if the body has any motion. Equation 19-9 is valid for the general plane motion of any rigid body of arbitrary shape.

Rotation about an Arbitrary Fixed Axis

Equation 19-9 is valid for a body rotating about any fixed axis, but it can be further simplified for such a special case. Consider a body whose mass center is at C rotating about an axis at O as in Fig. 19-4. Here $v_C = r_C\omega$, and Eq. 19-9 can be written as

$$T = \frac{1}{2}(I_C + mr_C^2)\omega^2$$

where $I_C + mr_C^2 = I_O$, the mass moment of inertia about the axis O, by the parallel-axis theorem. Thus,

$$T = \frac{1}{2} I_O \omega^2 \tag{19-10}$$

which is equivalent to Eq. 19-9.

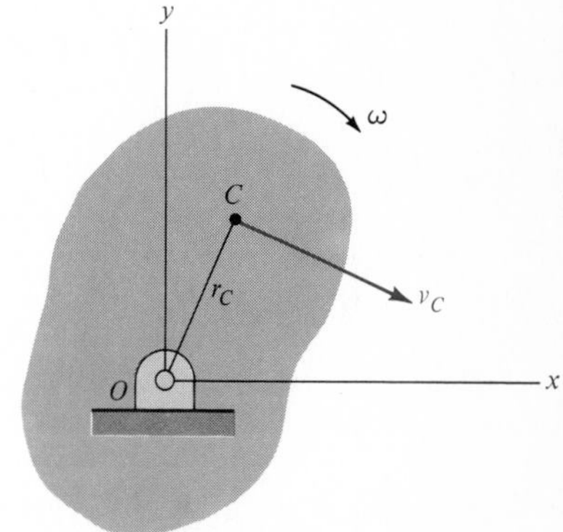

FIGURE 19-4

Rigid body rotating about an arbitrary fixed axis

19-3 PRINCIPLE OF WORK AND ENERGY

In a mechanical system work and energy are related for a rigid body as they are for a particle or a generalized system of particles. From Eq. 14-14,

$$T_2 = T_1 + U_{12} \tag{19-11}$$

where T_1 = initial kinetic energy of the body
T_2 = final kinetic energy of the body
U_{12} = work of all external forces and moments acting on the body during its motion from position 1 to 2

Equation 19-11 is particularly advantageous in solving problems of dynamics because it involves only scalar quantities and allows for direct calculations of velocities, distances, and rotations.

Systems of Rigid Bodies

Most mechanical systems consist of interconnected parts that do work on one another while energy is transferred among the parts. If such a system can be assumed to be internally conservative (i.e., the internal friction forces do no net work), the principle of work and energy expressed by Eq. 19-11 may be applied to the system as a whole. In that case

T_1 = sum of initial kinetic energies of all parts of the system

T_2 = sum of final kinetic energies of all parts of the system

U_{12} = work of all external forces and moments acting on the system as a whole; the internal forces (including friction forces) acting between various joints and parts of the system occur as pairs of equal and opposite forces, so the sum of their work is zero and need not be considered; there are no internal dissipative friction forces in the system

In analyzing an interconnected system of bodies, a free-body diagram of the whole system should be drawn to determine all significant external forces that are involved. All summations of kinetic energies and work are done arithmetically since these are scalar quantities.

19-4 CONSERVATION OF ENERGY

The work of conservative forces (such as weight of a body or force of a spring) cause changes in the potential energy of a mechanical system. In the case of a conservative system the work of the external forces may be expressed as a difference in potential energies, $U_{12} = V_1 - V_2$, at the initial and final positions. Thus, Eq. 19-11 can be written as

$$\boxed{T_1 + V_1 = T_2 + V_2} \qquad \text{19-12}$$

In words, *the sum of the kinetic and potential energies of a conservative system is constant.*

Note that in the general case the kinetic energy has translational and rotational parts. There is an interesting implication of this: a change from kinetic energy to potential energy, or vice versa, is not the only change possible in a conservative system of rigid bodies; it is also possible to convert translational kinetic energy to rotational kinetic energy, and vice versa. An example of the latter mechanism of energy conversion is a flywheel-powered automobile whose linear speed increases while the angular speed of its flywheel decreases.

The potential energy V of the system is most commonly in the form of gravitational potential energy or elastic potential energy. These quantities are evaluated as in the case of particles.

The *gravitational potential energy* V_g is a quantity that depends on the weight W of a system of bodies and on the relative vertical position h of the *center of gravity* with respect to a defined datum plane,

$$V_g = Wh \tag{19-13}$$

This energy must be taken as positive when h is above the datum plane (larger distance from the center of the earth), and negative for positions below that plane.

The *elastic potential energy* V_e of an elastic spring in linear deformation is given by

$$V_e = \frac{1}{2} kx^2 \tag{19-14}$$

where k is the spring constant and x is the change in length of the spring from its unstretched position. Note that a deformed spring does positive work on a body when the spring is returning to its undeformed position. This is true for either tensile or compressive deformations of the spring since in both cases the displacement and the force of the spring are in the same direction.

In general, the *total potential energy* V of a system in a particular configuration is expressed as the algebraic sum of the gravitational and elastic potential energies,

$$\boxed{V = V_g + V_e} \tag{19-15}$$

Note that conversions of gravitational potential energy to elastic potential energy, or vice versa, may occur as a system's configuration changes.

Limitations. The methods of work and energy described above are convenient to determine the speed of a body but not the reactions at joints and supporting surfaces. The equations of motion must be used to calculate forces.

19-5 POWER

The concept of power presented for particles in Sec. 14-4 is also valid for rigid bodies. However, it is necessary to distinguish analytically the power associated with translational and rotational motions. In both cases, power is the time rate at which work is done. For translation of a body under the action of a force **F**,

$$\text{power} = \frac{dU}{dt} = \mathbf{F} \cdot \mathbf{v} = m\mathbf{a} \cdot \mathbf{v} \tag{19-16}$$

where **v** is the velocity of any point in a translating body. Power can also be expressed as the time rate of change of kinetic energy,

$$\begin{aligned} \text{power} = \dot{T} &= \frac{d}{dt}\left(\frac{1}{2} m\mathbf{v} \cdot \mathbf{v}\right) \\ &= \frac{1}{2} m(\mathbf{a} \cdot \mathbf{v} + \mathbf{v} \cdot \mathbf{a}) \\ &= m\mathbf{a} \cdot \mathbf{v} \end{aligned}$$

which is equivalent to Eq. 19-16. For rotation of a body under the action of a constant moment **M** whose direction is parallel to the axis of rotation (a common case),

$$\text{power} = \frac{dU}{dt} = \frac{d}{dt}(M\, d\theta) = M \frac{d\theta}{dt} = M\omega \tag{19-17}$$

Equation 19-17 can also be obtained by taking the time derivative of the rotational kinetic energy; this is left as an exercise for the student.

The net power supplied to a system (or taken from it) is the algebraic sum of the powers required to change the kinetic and potential energies of the system in the same time interval. These can be expressed using the individual terms in concise notation as

$$\text{power} = \dot{T}_{\text{transl}} + \dot{T}_{\text{rot}} + \dot{V}_g + \dot{V}_e \tag{19-18}$$

The terms on the right side of this equation are obtained from Eqs. 19-9, 19-13, and 19-14. Each quantity is considered positive when it represents power supplied to the system, and negative when it is for power taken from the system. The units, watts or horsepower, must be consistent in the equation.

EXAMPLE 19-1

Each wheel of a vehicle is assumed to be a uniform disk of 15 kg mass and 0.8 m diameter. Determine the total kinetic energy of one wheel at a speed of 100 km/h of the vehicle (Fig. 19-5).

SOLUTION

A rotating wheel has kinetic energy due to both its translation and rotation. The angular speed of the wheel is calculated assuming that the wheel is rolling without slipping,

$$\omega = \frac{v_C}{R} = \frac{100 \text{ km}}{\text{h}}\left(\frac{1000 \text{ m/km}}{3600 \text{ s/h}}\right)\left(\frac{1}{0.4 \text{ m}}\right) = \frac{27.8 \text{ m/s}}{0.4 \text{ m}}$$
$$= 69.5 \text{ rad/s}$$

From Eq. 19-9,

$$T = \frac{1}{2}mv_C^2 + \frac{1}{2}I_C\omega^2$$
$$= \frac{1}{2}(15 \text{ kg})(27.8 \text{ m/s})^2$$
$$+ \frac{1}{2}\left[\frac{1}{2}(15 \text{ kg})(0.4 \text{ m})^2\right](69.5 \text{ rad/s})^2$$
$$= 8694 \text{ N·m}$$

JUDGMENT OF THE RESULT

Since the point in contact with the ground is an instantaneous center of rotation, Eq. 19-10 could also be used.

$$T = \frac{1}{2}I_O\omega^2 = \frac{1}{2}(I_C + mR^2)\omega^2$$
$$= \frac{1}{2}\left(\frac{1}{2}mR^2 + mR^2\right)\omega^2$$
$$= \frac{3}{4}mR^2\omega^2 = \frac{3}{4}mv_C^2 = \left(\frac{3}{4}\right)(15 \text{ kg})(27.8 \text{ m/s})^2$$
$$= 8694 \text{ N·m}$$

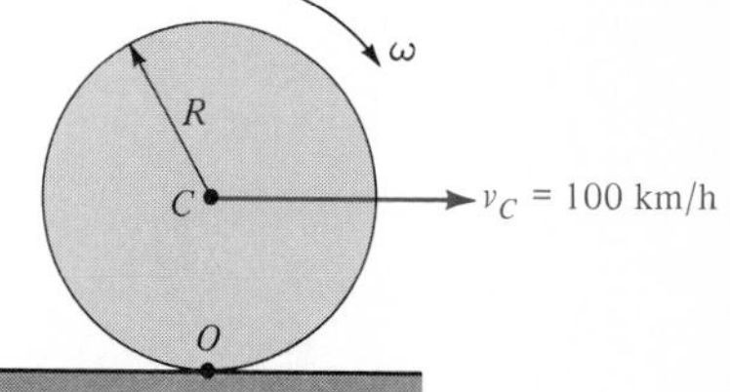

FIGURE 19-5
Analysis of a rolling wheel

EXAMPLE 19-2

Flywheel A of the tire tester in Fig. 19-6a drives two wheels B and C simultaneously. The flywheel has a weight of 500 lb, a 6-ft diameter, and a 2.5-ft radius of gyration. Wheels B and C weigh 35 lb each, have a 2.5-ft diameter, and may be considered as uniform disks. The wheels are at rest when a 2000-ft·lb counterclockwise moment is applied to wheel A. Determine the number of revolutions of wheel A before the angular speed of wheel C is what it would be in a car traveling at 150 mi/h. Neglect frictional losses at axles A, B, and C.

SOLUTION

The three wheels are considered as a system of rigid bodies. Since there is no slipping, there is no dissipative friction force in the system. Applying Eq. 19-11 gives

$$T_1 = 0 \qquad \text{(initially at rest)}$$

$$T_2 = \frac{1}{2} I_A \omega_A^2 + \frac{1}{2} m_A v_A^2 + \frac{1}{2} I_B \omega_B^2 + \frac{1}{2} m_B v_B^2 + \frac{1}{2} I_C \omega_C^2 + \frac{1}{2} m_C v_C^2$$

$$U_{12} = M_A \theta_A$$

For wheel C traveling (rolling without slipping) at 150 mi/h,

$$\omega_C = \frac{v_C}{R_C} = \frac{150 \text{ mi}}{\text{h}} \left(\frac{5280 \text{ ft/mi}}{3600 \text{ s/h}}\right)\left(\frac{1}{1.25 \text{ ft}}\right) = 176 \text{ rad/s}$$

The velocities of the points in contact between A and C must be the same if no slipping is occurring; hence,

$$v_C' = R_C \omega_C = R_A \omega_A$$

$$\omega_A = \left(\frac{R_C}{R_A}\right)\omega_C = \frac{1.25 \text{ ft}}{3 \text{ ft}}(176 \text{ rad/s}) = 73.3 \text{ rad/s}$$

Since wheel B is in contact with wheel A at the same radii as for wheels A and C, $\omega_B = \omega_C = 176$ rad/s (opposite direction). The translational velocities of the mass centers are always zero.

The only unknown now left in the work–energy equation is the angle that the applied moment acts through, θ_A. Hence,

$$T_1 + U_{12} = T_2$$

$$0 + 2000 \text{ ft}\cdot\text{lb}\,(\theta_A) = \frac{1}{2}\left[\frac{500 \text{ lb}}{32.2 \text{ ft/s}^2}(2.5 \text{ ft})^2\right]\left(73.3\,\frac{\text{rad}}{\text{s}}\right)^2$$

$$+\frac{1}{2}m_A(0)^2$$

$$+\frac{1}{2}\left[\frac{1}{2}\left(\frac{35 \text{ lb}}{32.2 \text{ ft/s}^2}\right)(1.25 \text{ ft})^2\right]\left(176\,\frac{\text{rad}}{\text{s}}\right)^2$$

$$+\frac{1}{2}m_B(0)^2$$

$$+\frac{1}{2}\left[\frac{1}{2}\left(\frac{35 \text{ lb}}{32.2 \text{ ft/s}^2}\right)(1.25 \text{ ft})^2\right]\left(176\,\frac{\text{rad}}{\text{s}}\right)^2$$

$$+\frac{1}{2}m_C(0)^2$$

$$\theta_A = \frac{(260{,}700 + 13{,}150 + 13{,}150) \text{ ft}\cdot\text{lb}}{2000 \text{ ft}\cdot\text{lb}}$$

$$= 143.5 \text{ rad}\left(\frac{1 \text{ rev}}{2\pi \text{ rad}}\right) = 22.8 \text{ rev}$$

FIGURE 19-6

Tire-testing equipment (Courtesy MTS Systems Corporation, Minneapolis, Minnesota)

EXAMPLE 19-3

Consider the system in Ex. 19-2 and determine the required normal force between the wheels to prevent slipping if $\mu_s = 0.7$.

SOLUTION

To calculate the minimum required normal force between the wheels, the friction force needed for the required motion must be calculated. Considering wheel A alone (Fig. 19-7), it is realized that the friction forces $\mathscr{F}_B$ and $\mathscr{F}_C$ do work *against* wheel A. The work of these friction forces was not included when the whole system was considered since the work of the equal but opposite friction forces acting on B and C canceled the work of $\mathscr{F}_B$ and $\mathscr{F}_C$ on A. The normal forces do no work since they have no component in the direction of the displacement. Hence, using the work energy equation for wheel A alone,

$$T_1 + U_{12} = T_2$$

$$0 + M_A\theta_A - \mathscr{F}_B(s_B) - \mathscr{F}_C(s_C) = \frac{1}{2}I_A(\omega_2)_A^2 + \frac{1}{2}m_A(v_2)_A^2$$

The arc lengths s_B and s_C which $\mathscr{F}_B$ and $\mathscr{F}_C$ act through are

$$s_B = s_C = R_A\theta_A$$

From symmetry,

$$\mathscr{F}_B = \mathscr{F}_C = \mathscr{F}$$

Substituting the values found in Ex. 19-2 yields

$$(2000 \text{ ft})(143.5 \text{ rad}) - 2\mathscr{F}(3 \text{ ft})(143.5 \text{ rad})$$

$$= \frac{1}{2}\left(\frac{500 \text{ lb}}{32.2 \text{ ft/s}^2}\right)(2.5 \text{ ft})^2\left(73.3\,\frac{\text{rad}}{\text{s}}\right)^2 + \frac{1}{2}m_A(0)^2$$

Solving for $\mathscr{F}$ gives

$$\mathscr{F} = 30.5 \text{ lb}$$

The minimum normal force N necessary is for the condition of impending slipping, $\mathscr{F} = \mathscr{F}_{\max} = \mu_s N$.

$$N = \frac{\mathscr{F}}{\mu_s} = \frac{30.5 \text{ lb}}{0.7} = 43.6 \text{ lb}$$

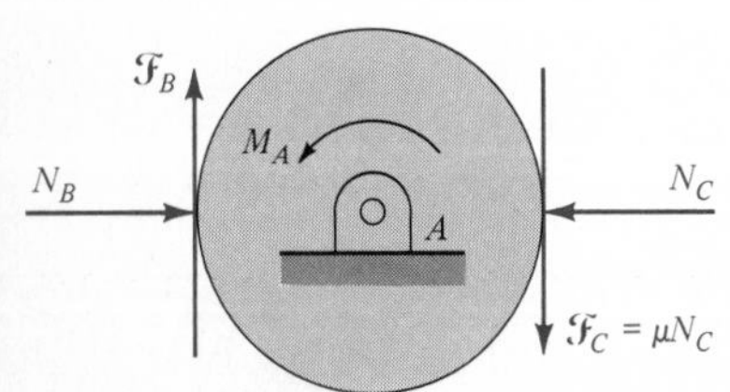

FIGURE 19-7
Flywheel A from Fig. 19-6

EXAMPLE 19-4

A uniform, 70-lb bar is released from rest with no force initially acting on it from the spring (Fig. 19-8). The ends of the bar slide along smooth surfaces. Calculate the speeds of ends A and B just before the bar strikes the horizontal surface.

SOLUTION

The bar undergoes both translation and rotation. The unknowns in this problem are the final linear speed v_C and angular speed ω. The work–energy principle allows for one equation. Another equation is obtained since v_C and ω are related. This relationship varies with the angle of inclination, but the relationship for the bar's nearly horizontal position is readily determined. At this instant, by inspection, $v_B = 0$, B is the instantaneous center of rotation,

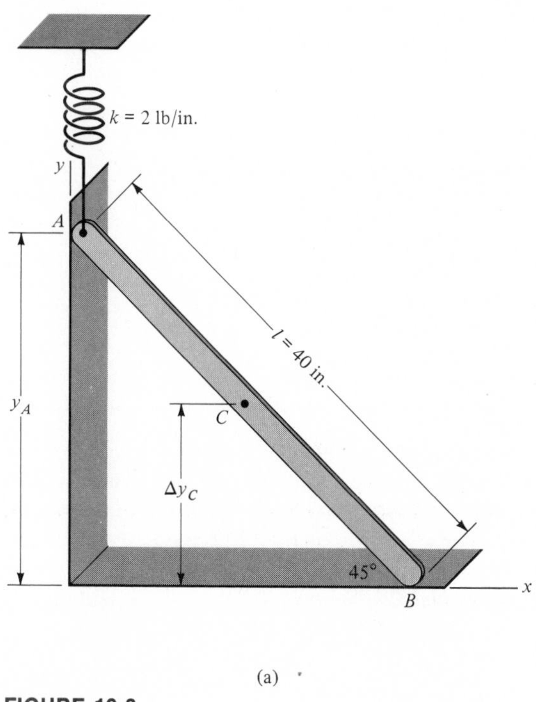

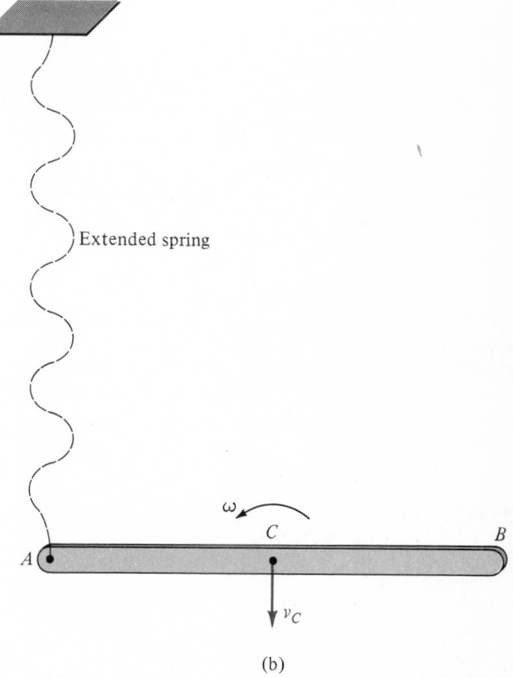

FIGURE 19-8
Analysis of a sliding bar

and $v_C = (l/2)\omega$. Work is done on the rod by gravity and the spring. The terms in the work–energy equation are

$$T_1 = 0 \text{ (released from rest)}$$

$$U_{12} = -\frac{1}{2}ky_A^2 + W_{\text{rod}}(\Delta y_C)$$

$$T_2 = \frac{1}{2}I_C\omega^2 + \frac{1}{2}mv_C^2$$

$$T_1 + U_{12} = T_2$$

$$0 - \frac{1}{2}(2\text{ lb/in.})(40\text{ in. }\sin 45^\circ)^2 + 70\text{ lb}(20\text{ in. }\sin 45^\circ)$$

$$= \frac{1}{2}\left[\frac{1}{12}\left(\frac{70\text{ lb}}{32.2\text{ ft/s}^2}\right)\frac{1\text{ ft}}{12\text{ in.}}(40\text{ in.})^2\right]\omega^2$$

$$+ \frac{1}{2}\left(\frac{70\text{ lb}}{32.2\text{ ft/s}^2}\right)\frac{1\text{ ft}}{12\text{ in.}}\left[\left(\frac{40\text{ in.}}{2}\right)\omega\right]^2$$

$$-800\text{ in}\cdot\text{lb} + 990\text{ in}\cdot\text{lb} = (12.08\text{ in}\cdot\text{lb}\cdot\text{s}^2)\omega^2$$

$$+ (36.2\text{ in}\cdot\text{lb}\cdot\text{s}^2)\omega^2$$

$$\omega^2 = \frac{190\text{ in}\cdot\text{lb}}{48.3\text{ in}\cdot\text{lb}\cdot\text{s}^2} = 3.93\frac{1}{\text{s}^2}$$

$$\omega = 1.98\text{ rad/s}$$

Since point B is the instantaneous center of rotation,

$$v_A = l\omega = (40\text{ in.})\left(1.98\frac{\text{rad}}{\text{s}}\right)$$

$$= 79.3\text{ in./s}$$

EXAMPLE 19-5

A high-speed tester for leaf springs (Fig. 19-9a) has a falling mass $m_A = 5$ kg which has a downward speed of 5 m/s when it contacts the undeformed spring. Assume that the spring constant is $k = 200$ kN/m. Determine (a) the speed of A when the spring's deflection is 0.01 m, and (b) the reactions at supports B and C at that instant. The supports can freely move horizontally. Neglect friction and the mass of the spring.

SOLUTION

(a) Since friction is not involved, the principle of conservation of energy can be used for the system as a whole. Defining the datum as the point where A first contacts the leaf spring,

$$T_1+V_1=T_2+V_2$$

$$\frac{1}{2}m_A(v_A)_1^2+0=\frac{1}{2}m_A(v_A)_2^2+\frac{1}{2}k(y_A)^2-m_A g y_A$$

$$\frac{1}{2}(5\text{ kg})(5\text{ m/s})^2=\frac{1}{2}(5\text{ kg})(v_A)_2^2+\frac{1}{2}(200\times10^3\text{ N/m})(0.01\text{ m})^2$$

$$-5\text{ kg}(9.81\text{ m/s}^2)(0.01\text{ m})$$

$$62.5\text{ kg}\cdot\text{m}^2/\text{s}^2=2.5\text{ kg}(v_A)_2^2+10\text{ N}\cdot\text{m}-0.49\text{ kg}\cdot\text{m}^2/\text{s}^2$$

$$\frac{53\text{ kg}\cdot\text{m}^2/\text{s}^2}{2.5\text{ kg}}=(v_A)_2^2$$

$$(v_A)_2=4.6\text{ m/s}$$

(b) The reactions at the supports do no work; hence, work–energy principles are of no use in solving for the reactions at A and B. The methods of kinetics (Chap. 18) must be used. The appropriate free-body diagrams are sketched in Fig. 19-9b. Summing forces acting on the spring (from symmetry, $R_B = R_C = R$) gives

$$F_y = R + R - P = m_{sp}a_{sp} = (0)(a_{sp}) = 0$$

The force P causes the deflection y_A of the spring, so

$$P = ky_A$$

$$2R = (200 \times 10^3\text{ N/m})(0.01\text{ m})$$

$$R = 1\ k\text{N}$$

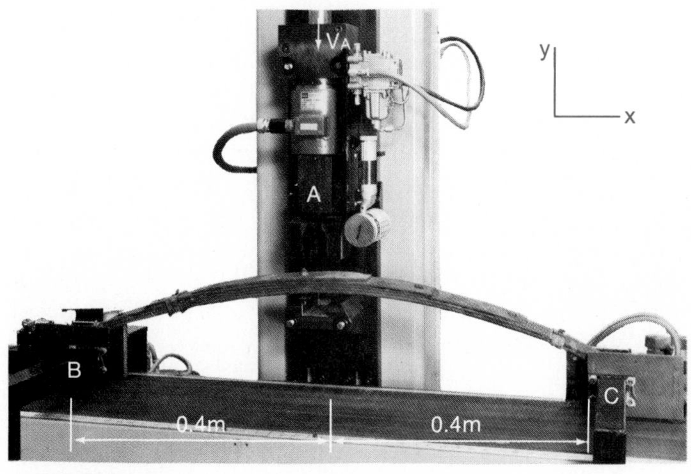

(a)

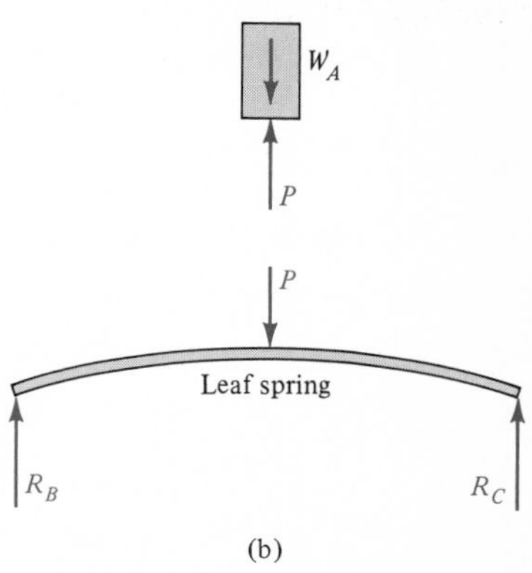

(b)

FIGURE 19-9

Leaf-spring-testing equipment (Photograph courtesy MTS Systems Corporation, Minneapolis, Minnesota)

PROBLEMS

SECTIONS 19-1, 19-2, 19-3, 19-4

FIGURE P19-1

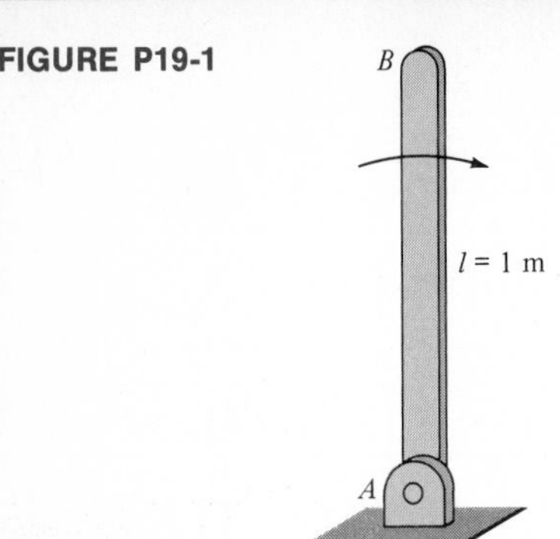

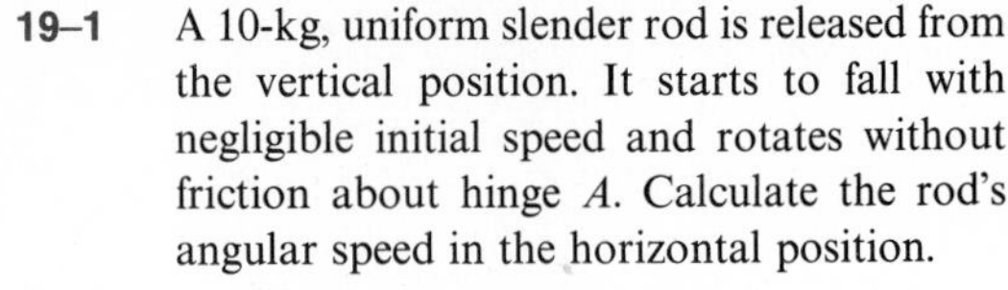

19–1 A 10-kg, uniform slender rod is released from the vertical position. It starts to fall with negligible initial speed and rotates without friction about hinge A. Calculate the rod's angular speed in the horizontal position.

19–2 Solve Prob. 19–1 assuming that an additional mass $m_B = 3$ kg of negligible dimensions is attached at point B, and that B has an initial speed $v_B = 1$ m/s when the rod is vertical.

FIGURE P19-3

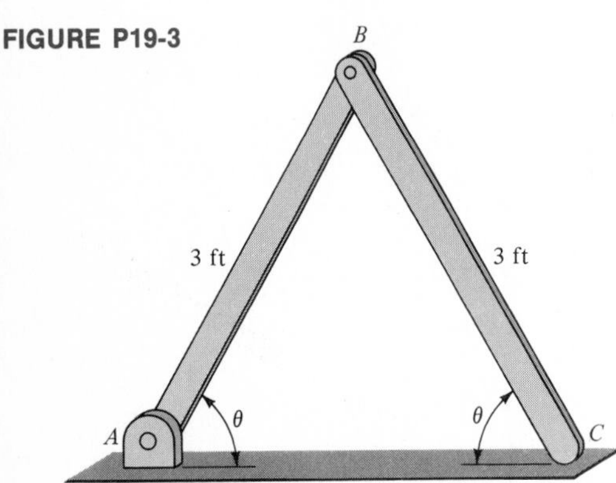

19–3 Two uniform slender rods are hinged at A and B, and part C slides on a smooth surface. The 10-lb rods are released from rest when $\theta = 60°$. Calculate the angular speed of BC when it strikes the surface.

19–4 Solve Prob. 19–3 assuming that $\dot{\theta} = 0.5$ rad/s at $\theta = 60°$ (decreasing angle).

FIGURE P19-5

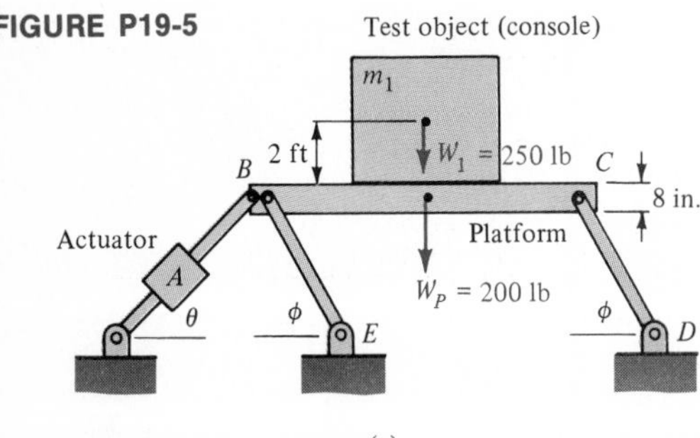

(a)

19–5 The test platform weighs 200 lb, and the console weighs 250 lb. There are four uniform rocker arms such as BE; each weighs 10 lb and makes an angle ϕ with the horizontal. The platform is released from rest at $\phi = 85°$. Calculate its speed when $\phi = 40°$. Assume that the actuator A has no mass and applies no force to the platform during this motion. $BC = 5$ ft, $BE = CD = 2.5$ ft.

19–6 A torsion spring attached at joint E in Fig. P19-5 applies no moment to arm BE when $\phi = 90°$. The platform slowly starts to swing downward at this angle, its motion only opposed by the spring. Calculate the platform's speed at $\phi = 50°$ if $k = 300$ ft·lb/rad of the single spring used. See Prob. 19–5 for further details.

19–7 A 30-kg flywheel is a uniform disk with a 0.2-m radius. Determine the moment that can increase the wheel's rotational speed from zero to 10,000 rpm in 100 revolutions.

19–8 A flywheel-powered, 1200-kg car begins regenerative braking when traveling at a speed of 70 km/h. The 15-kg flywheel of 0.2-m radius of gyration is rotating at a rate of ω_y = 1000 rpm at this instant. What is the final angular speed of the flywheel when the car is stopped if 30% of the car's translational kinetic energy can be used for recharging the power plant? Neglect the rotation of the car's wheels.

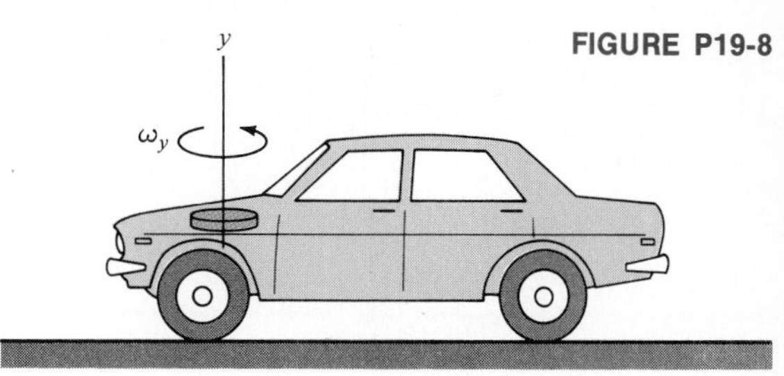

FIGURE P19-8

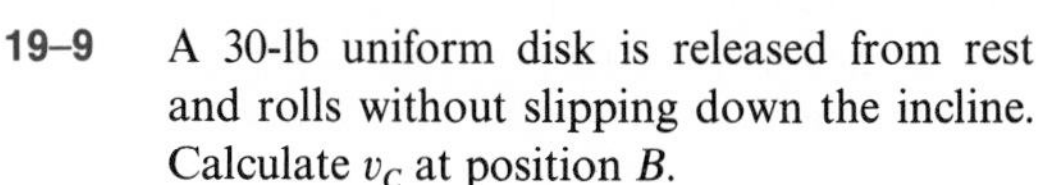

19–9 A 30-lb uniform disk is released from rest and rolls without slipping down the incline. Calculate v_C at position B.

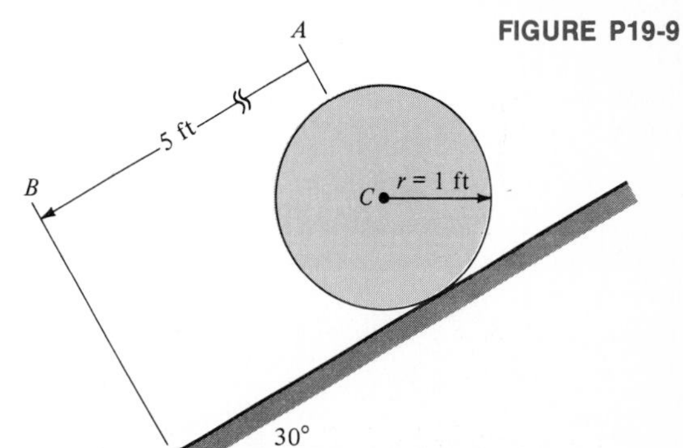

FIGURE P19-9

19–10 Solve Prob. 19–9 assuming that the disk has an initial speed v_C = 4 ft/s up the incline at position A.

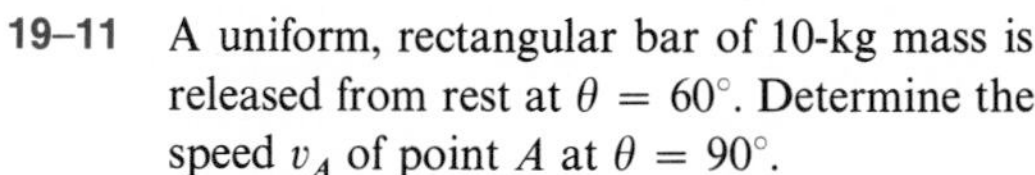

19–11 A uniform, rectangular bar of 10-kg mass is released from rest at $\theta = 60^\circ$. Determine the speed v_A of point A at $\theta = 90^\circ$.

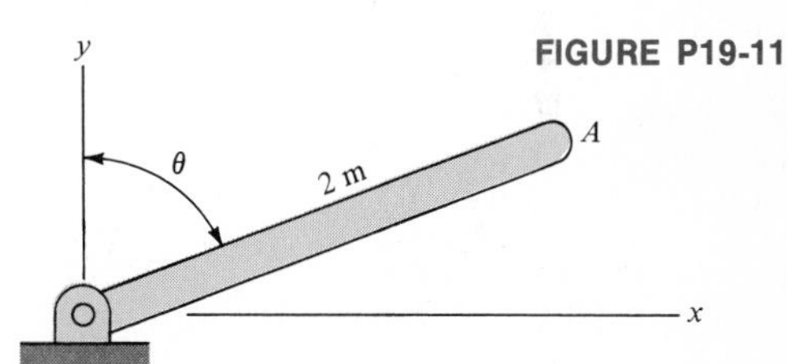

FIGURE P19-11

19–12 The bar in Prob. 19–11 has angular speed ω_O = 1 rad/s at $\theta = 0$. Determine the speed v_A at $\theta = 180^\circ$.

19–13 A uniform, rectangular plate of 80 lb weight is released from rest with edge OA horizontal. Determine the angular speed of the plate and the reaction at O when OA is vertical.

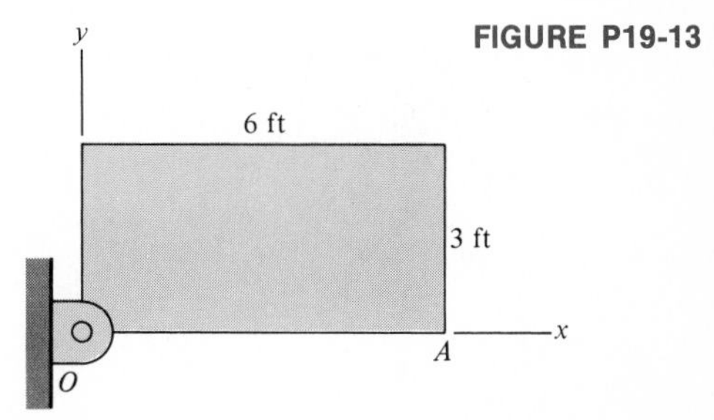

FIGURE P19-13

FIGURE P19-15

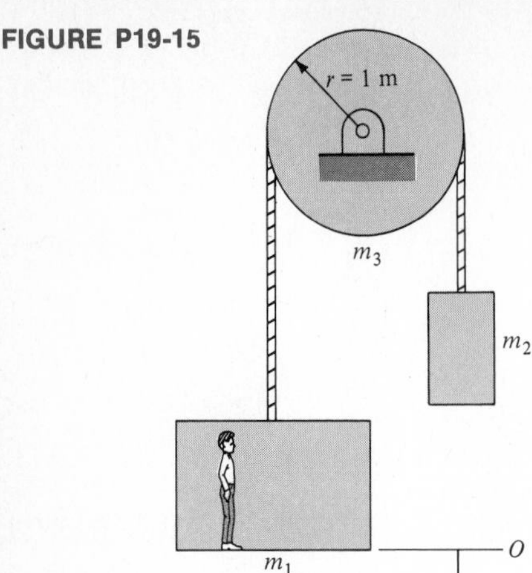

FIGURE P19-17

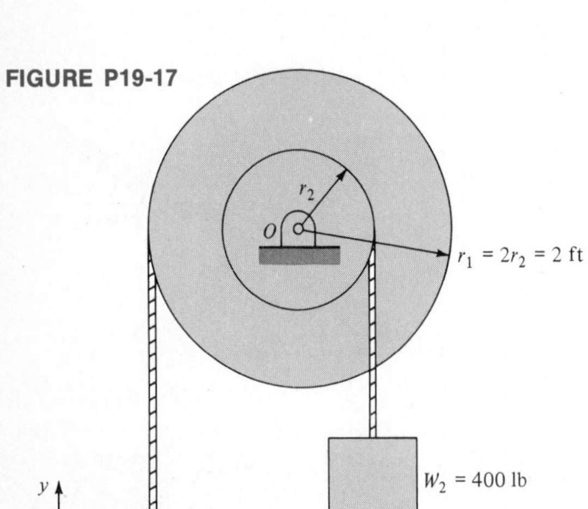

FIGURE P19-19

19–14 Solve Prob. 19–13 if $\mathbf{v}_A = 3$ ft/s $\mathbf{j}$ when point A is on the x axis.

19–15 An elevator system is modeled as $m_1 = 2000$ kg, $m_2 = 1800$ kg, and a uniform disk of $m_3 = 200$ kg. Assume that the system starts from rest and moves under the action of gravity, with negligible friction. Determine the speed of m_1 after it moves to $y = 30$ m.

19–16 Solve Prob. 19–15 assuming that m_1 has an initial upward speed of 2 m/s at $y = 0$.

19–17 The double pulley has a total weight of 50 lb and a radius of gyration of 1.6 ft. The system starts from rest. Calculate the speed of W_1 after a travel of 5 ft.

19–18 Determine the power of the system in Prob. 19–17 at a time of 3 s after the start.

19–19 The amusement ride of 12 cages is modeled as 12 concentrated masses of 700 kg each, on the ends of uniform, straight bars of 8 m length and 250 kg mass. The system is started from rest and moves in a plane. Calculate the required work input for the system to attain a speed of 4 rpm.

19–20 Determine the power requirement of the system in Prob. 19–19 if it must attain the speed of 4 rpm in a time of 5 s.

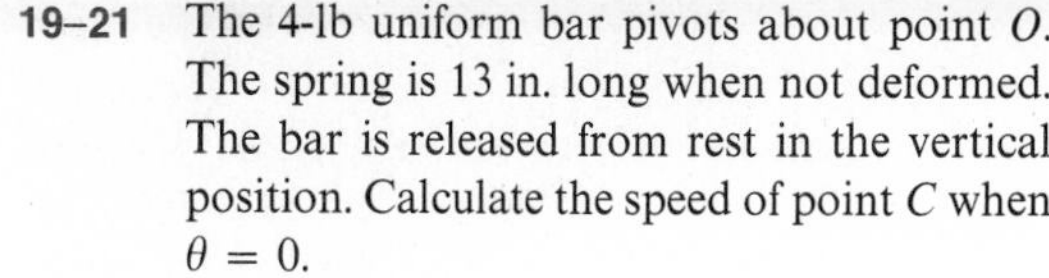

19–21 The 4-lb uniform bar pivots about point O. The spring is 13 in. long when not deformed. The bar is released from rest in the vertical position. Calculate the speed of point C when $\theta = 0$.

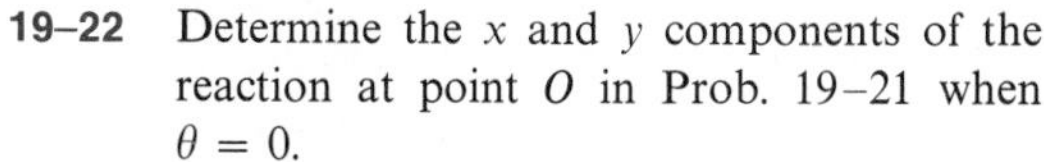

19–22 Determine the x and y components of the reaction at point O in Prob. 19–21 when $\theta = 0$.

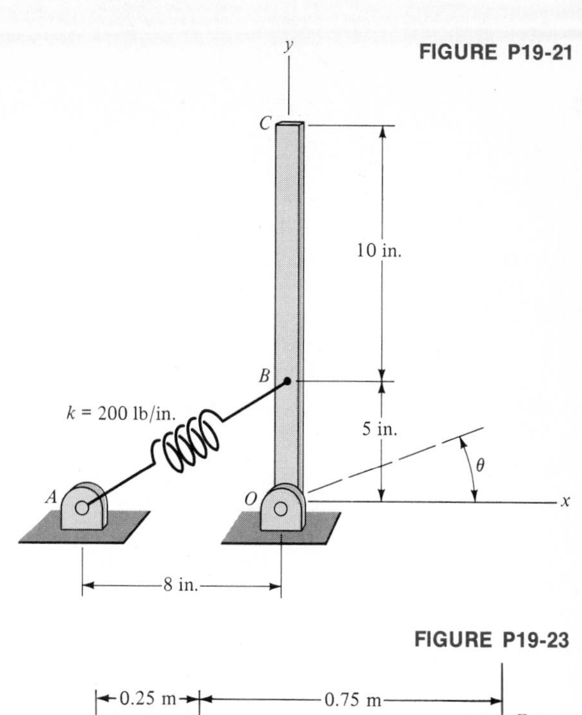

FIGURE P19-21

19–23 The uniform, 10-kg disk is rotating at an angular speed of 500 rpm when a braking force F is applied to the hinged bar. The friction at O and A is negligible, while $\mu_k = 0.5$ at B. Determine the force F for stopping the disk in 10 revolutions.

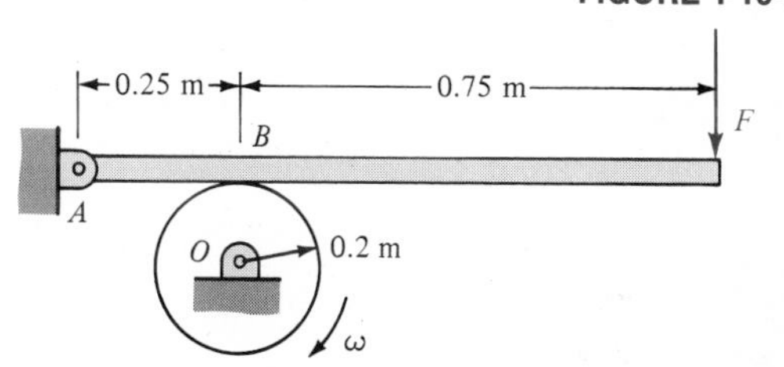

FIGURE P19-23

19–24 A moment is applied to the disk in Prob. 19–23 to accelerate it from rest, while the force F on the bar is 400 N. Determine the power input if the disk attains an angular speed of 500 rpm in a time of 20 s.

19–25 The base structure of a revolving restaurant is modeled as three uniform disks that rotate as a unit about the vertical axis. The mass m is 8000 kg, and the diameter d is 7 m. Determine the power input to accelerate the three disks from rest to $\omega = 0.1$ rpm in a time of 10 s.

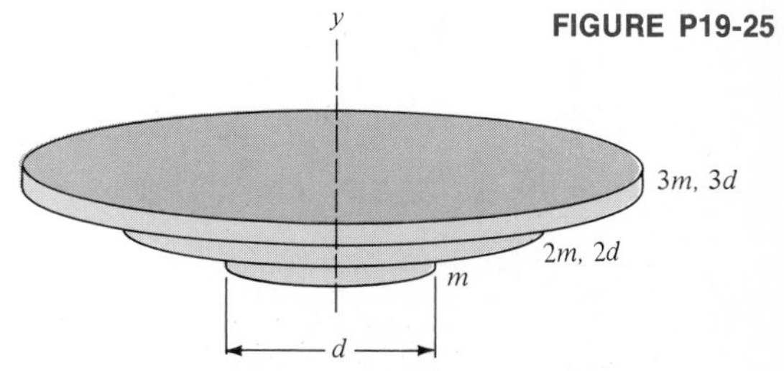

FIGURE P19-25

19–26 Show that in rotational motion power equals the time derivative of the rotational kinetic energy.

19–27 A front loader has a total weight of 10,000 lb. Each of its four wheels has a 350-lb weight, a 5-ft diameter, and a 1.5-ft radius of gyration. Calculate the required power to accelerate this machine from rest to a speed of 10 ft/s in a time of 5 s.

FIGURE P19-27

19–28 Assume that each wheel of a 1000-kg automobile has a 15-kg mass, 40-cm radius, and 20-cm radius of gyration. Plot the total kinetic energy of the four wheels and of the rest of the vehicle as functions of speed, using 20 km/h, 60 km/h, and 100 km/h.

FIGURE P19-29

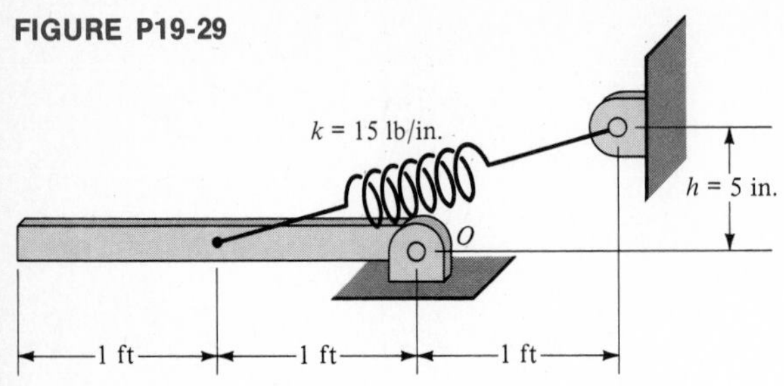

19–29 A uniform, 30-lb bar is hinged at point O. It is released from rest in the horizontal position. Calculate its angular speed in the vertical position if the spring applies no force in that position.

FIGURE P19-30

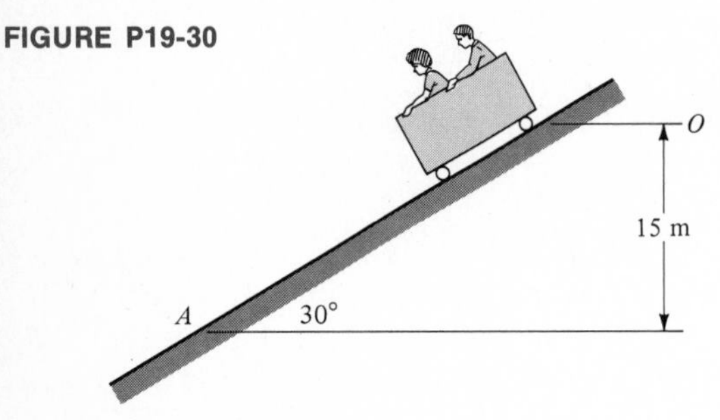

19–30 The total mass of a roller coaster is 300 kg. Each of its four wheels has a 3-kg mass, a 0.1-m radius, and a 0.07-m radius of gyration. Determine the linear speed of the coaster at A. It starts from rest at O. The wheels roll without slipping.

FIGURE P19-31

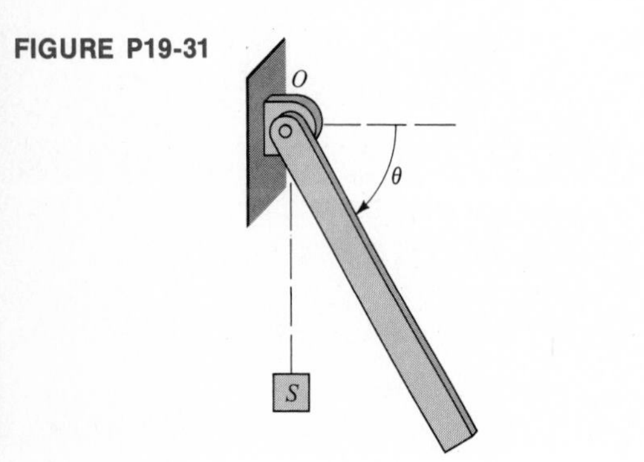

19–31 An impact tester is modeled as a pendulum of 110 lb weight and 2.5 ft radius of gyration about point O. It is released from rest at $\theta = 0$ to strike a specimen S at $\theta = 90°$. The specimen breaks absorbing 20 ft·lb of energy, and allows the pendulum to swing past the vertical position. Calculate the maximum angle θ from the horizontal position of the arm.

FIGURE P19-32

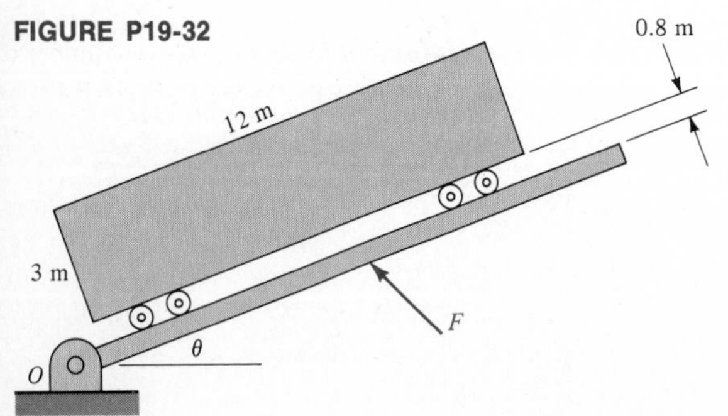

19–32 A 20 000-kg railroad car is modeled as a uniform bar of 3 m × 3 m cross section. It is firmly attached to the rigid rail bed OA, which is raised while pivoting about O for rapid unloading. Determine the maximum power required to elevate the car and the angle at which this occurs if $\dot{\theta} = 0.3$ rad/s (constant) from $\theta = 0$ to $\theta = 60°$.

19–33 A proposed conveyor system is modeled as a plate of mass M moving without slipping on small cylindrical rollers of mass m and radius r. Determine the speed of the plate after it had moved a distance l, starting from rest, along an inclined plane. Consider (a) free rollers, and (b) rollers freely rotating about their fixed axes. Assume the plate is on three rollers at all times as it moves a distance l.

FIGURE P19-33

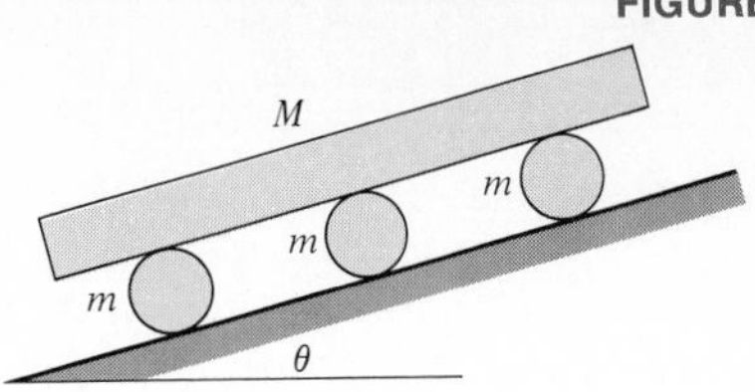

(a) Free rollers

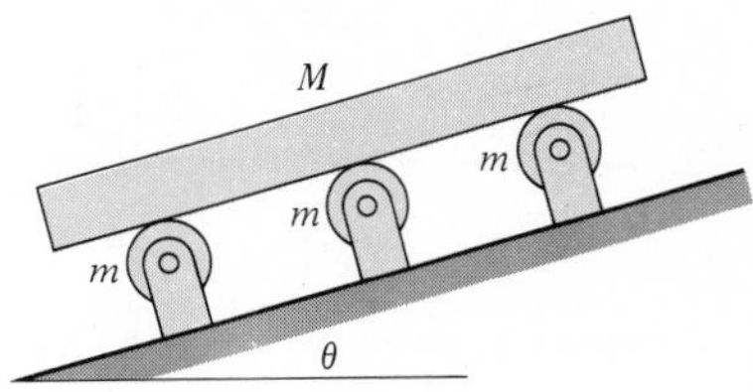

(b) Rollers on fixed axes

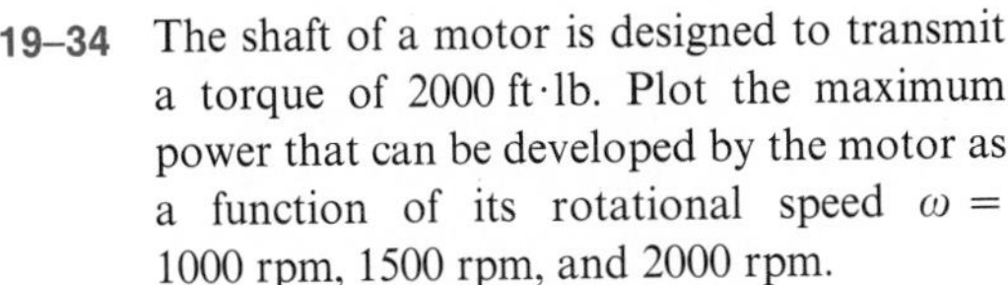
19–34 The shaft of a motor is designed to transmit a torque of 2000 ft·lb. Plot the maximum power that can be developed by the motor as a function of its rotational speed $\omega =$ 1000 rpm, 1500 rpm, and 2000 rpm.

FIGURE P19-35

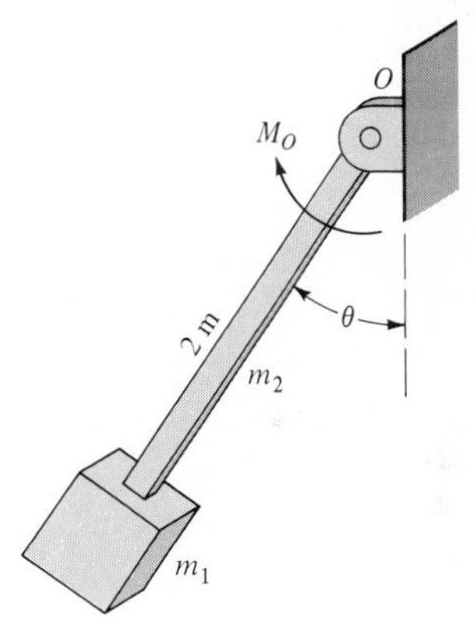

19–35 A concentrated mass $m_1 = 50$ kg is at the end of a uniform, hinged rod of mass $m_2 = 30$ kg. A moment M_O is applied about pivot O to raise the rod from $\theta = 0$ to 60° at a constant angular speed of 0.5 rad/s. Calculate the maximum power required during this motion.

19–36 Pins A and B of the uniform, 40-lb bar slide without friction in their guides. The spring applies no force to the bar when it is released from rest at $\theta = 90°$. Calculate the speed of pin A when $\theta = 45°$.

FIGURE P19-36

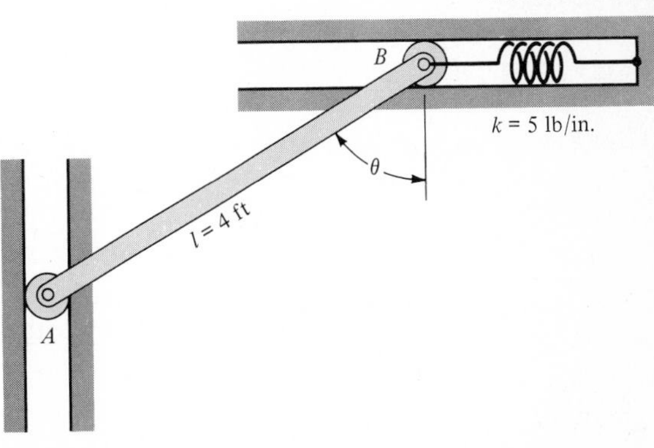

19–37 The bar in Prob. 19–36 is released from rest at $\theta = 135°$. Determine the smallest value of θ as the bar swings down.

19-6 IMPULSE AND MOMENTUM OF A RIGID BODY

The principles of impulse and momentum are particularly useful in the analysis of motion when time and velocities are involved in the problem as it was demonstrated in Chap. 15. These concepts are applicable to the systems of particles that constitute rigid bodies, and they are presented here for two-dimensional motion.

Linear Impulse and Momentum

The linear momentum **G** of a body is the vector sum of the linear momenta of all of its particles P_i of mass m_i and velocity $\mathbf{v}_i$. For n particles, from Eq. 16-20,

$$\mathbf{G} = \sum_{i=1}^{n} m_i \mathbf{v}_i = m\mathbf{v}_C \tag{19-19}$$

where m is the total mass and $\mathbf{v}_C$ is the velocity of the mass center.

Newton's equation of motion can be written as

$$\sum \mathbf{F} = m\mathbf{a}_C = m\frac{d\mathbf{v}_C}{dt} = \frac{d}{dt}(m\mathbf{v}_C) \tag{19-20}$$

since m is a constant. Multiplying this by dt and integrating from time t_1 to t_2,

$$\boxed{\int_{t_1}^{t_2} \sum \mathbf{F}\, dt = m(\mathbf{v}_{C_2} - \mathbf{v}_{C_1})} \tag{19-21}$$

where $\mathbf{v}_{C_2}$ and $\mathbf{v}_{C_1}$ are the velocities of the mass center at t_2 and t_1, respectively. The left side of this equation is the *impulse of the external forces* acting on the body from time t_1 to t_2. This impulse is equal to the change in linear momentum of the body in the given time interval. It should be noted that Eq. 19-21 is a vector equation and can be used in terms of scalar components for motion along conveniently selected x and y axes. In any case, the impulse term must include *all* external forces acting on the body (even those that do no work, because Newton's law does not allow such a distinction).

Angular Impulse and Momentum

The angular momentum **H** of a body is the vector sum of the angular momenta of all its particles. Angular momentum of a particle P_i is defined as the moment of its linear momentum with respect to a chosen reference point. Using the center of mass C in Fig. 19-10 for reference, and recalling from Eq. 18-4 that $\mathbf{H}_C = I_C\boldsymbol{\omega}$,

$$\mathbf{H}_C = \sum_{i=1}^{n} (\mathbf{r}_i \times m_i\mathbf{v}_i) = I_C\boldsymbol{\omega} \qquad \boxed{19\text{-}22}$$

where I_C is the centroidal mass moment of inertia about an axis perpendicular to the plane of motion, and $\boldsymbol{\omega}$ is the angular velocity of the body.

The equation of motion for rotation relates the external moments $\sum \mathbf{M}$ to the rate of change of angular momentum,

$$\sum \mathbf{M}_C = I_C\boldsymbol{\alpha} = I_C\frac{d\boldsymbol{\omega}}{dt} = \frac{d}{dt}(I_C\boldsymbol{\omega}) = \dot{\mathbf{H}}_C \qquad \boxed{19\text{-}23}$$

since I_C is a constant. Multiplying this by dt and integrating from time t_1 to t_2 gives

$$\boxed{\int_{t_1}^{t_2} \sum \mathbf{M}_C\, dt = I_C\boldsymbol{\omega}_2 - I_C\boldsymbol{\omega}_1 = \mathbf{H}_{C_2} - \mathbf{H}_{C_1}} \qquad \boxed{19\text{-}24}$$

where $\boldsymbol{\omega}_2$ and $\boldsymbol{\omega}_1$ are the angular velocities at t_2 and t_1, respectively. The left side of Eq. 19-24 is the *impulse of the external moments* acting on the body from time t_1 to t_2. This impulse equals the change in angular momentum of the body in the given time interval. The impulse term must include all external moments. For plane motion, Eq. 19-24 can be written in simplified scalar form if $\sum \mathbf{M}_C$ is parallel to $\boldsymbol{\omega}$,

$$\int_{t_1}^{t_2} \sum M_C\, dt = I_C(\omega_2 - \omega_1) \qquad \boxed{19\text{-}25}$$

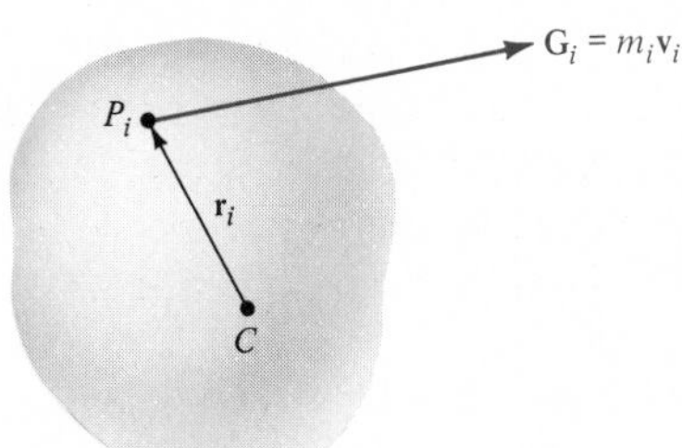

FIGURE 19-10

Schematic for defining angular momentum of a rigid body

Equations 19-22 through 19-25 are valid for any arbitrary fixed point of reference O (such as in Fig. 19-11) if the external moments and the mass moment of inertia are evaluated with respect to point O.

In that case the angular momentum about point O is the sum of the centroidal angular momentum $I_C\omega$, and the moment of the linear momentum about O, $r_C(mv_C)$. r_C is the distance between the reference point O and the centroid C, and v_C is the velocity of C perpendicular to the line OC. For plane motion, the angular impulse and momentum relationship can be written in scalar form if $\sum \mathbf{M}_O$ is parallel to $\boldsymbol{\omega}$,

$$\int_{t_1}^{t_2} \sum M_O \, dt = [I_C\omega + r_C(mr_C\omega)]_{t_1}^{t_2} \qquad \boxed{19\text{-}26}$$

$$= (I_C + mr_C^2)(\omega_2 - \omega_1)$$

$$= I_O(\omega_2 - \omega_1)$$

since $v_C = r_C\omega$.

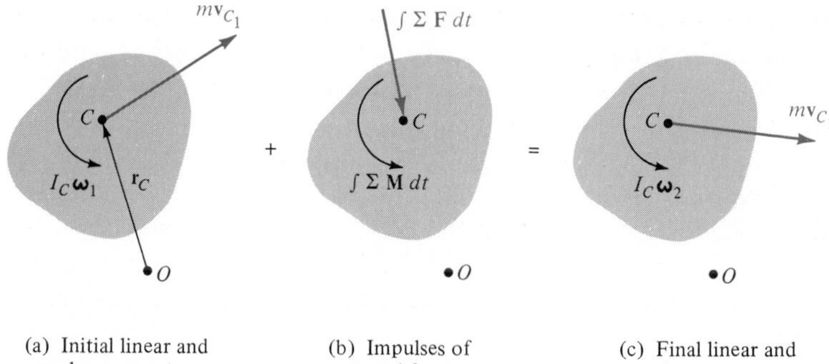

(a) Initial linear and angular momenta

(b) Impulses of external forces and moments

(c) Final linear and angular momenta

FIGURE 19-11

Analysis of impulse and momentum

19-7 IMPULSE AND MOMENTUM OF A SYSTEM OF RIGID BODIES

The methods developed in Sec. 19-6 can be applied to a system of rigid bodies (frequently a system of interconnected bodies) in two ways. First, Eqs. 19-21 and 19-24 (or 19-26) can be used to analyze the motion of each rigid member of the system separately. In this case the mutual forces acting between the members must be included in the formulation of the solution. The second approach is to apply the principle of impulse and momentum to the entire system of bodies. The advantage of this method is that the mutual forces acting between members of the system do not have to be included in the solution since they cancel in pairs.

Formulation of Problem Solutions

For a schematic illustration of common problems and the procedure for their solution, assume that the initial total linear and angular momenta are known for a given system of rigid bodies. The system may consist of one or more bodies, and the momenta may be entirely linear momentum, angular momentum, or both simultaneously. A system of external impulses is applied to the system of bodies, which thereby acquires a new, or final, set of momenta. This is shown schematically for a single rigid body in the *momentum and impulse diagrams* in Fig. 19-11.

The same procedure is used when the system consists of several bodies, but it is necessary to include in the formulation the linear and angular momenta and the external impulses for each moving member of the system. This is shown schematically for two interconnected bodies in Fig. 19-12.

The choice between the above two approaches is made according to the number of unknowns that must be determined. For any system of bodies in plane motion there are three scalar equations of impulse and momentum that can be written for a given configuration. These are obtained from Eqs. 19-21 and 19-26. Schematically,

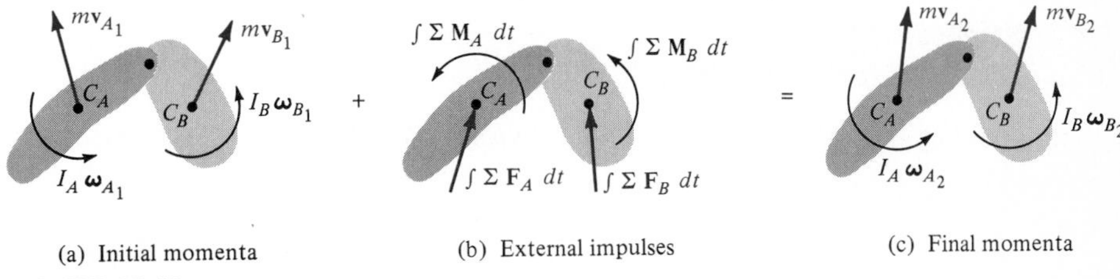

FIGURE 19-12

Impulse and momentum for two interconnected bodies

$$\sum\begin{pmatrix}\text{initial}\\ \text{linear}\\ \text{momentum}\end{pmatrix}_x + \sum\begin{pmatrix}\text{linear}\\ \text{impulse}\end{pmatrix}_x = \sum\begin{pmatrix}\text{final}\\ \text{linear}\\ \text{momentum}\end{pmatrix}_x \qquad \text{19-27a}$$

$$\sum\begin{pmatrix}\text{initial}\\ \text{linear}\\ \text{momentum}\end{pmatrix}_y + \sum\begin{pmatrix}\text{linear}\\ \text{impulse}\end{pmatrix}_y = \sum\begin{pmatrix}\text{final}\\ \text{linear}\\ \text{momentum}\end{pmatrix}_y \qquad \text{19-27b}$$

$$\sum\begin{pmatrix}\text{initial}\\ \text{angular}\\ \text{momentum}\end{pmatrix}_O + \sum\begin{pmatrix}\text{angular}\\ \text{impulse}\end{pmatrix}_O = \sum\begin{pmatrix}\text{final}\\ \text{angular}\\ \text{momentum}\end{pmatrix}_O \qquad \text{19-27c}$$

where x and y are the chosen coordinate directions and O is the reference point for the analysis of angular impulse and momentum. Point O must be the same *fixed reference point* for all parts of a given system to allow using one equation for the rotational motion of the entire system. Equations 19-27 make it possible to determine up to three unknowns working with the system as a whole. For more than three unknowns it is necessary to apply the principle of impulse and momentum to each member separately. In this case, three scalar equations can be written for each member but the forces between members must be included in the external impulses.

19-8 CONSERVATION OF MOMENTUM

The principles of conservation of momentum that were developed for particles (Sec. 16-9) are valid for rigid bodies that have no external forces or moments acting on them. The formal statements of the conservation principle are considered separately for linear and rotational motions.

Conservation of Linear Momentum

There is no linear impulse when $\sum \mathbf{F} = 0$, so Eqs. 19-19 and 19-21 become, respectively,

$$\Delta \mathbf{G} = 0 \qquad \text{19-28a}$$

$$m\mathbf{v}_{C_1} = m\mathbf{v}_{C_2} \qquad \mathbf{v}_{C_1} = \mathbf{v}_{C_2} \qquad \text{19-28b}$$

which express that the linear momentum is constant in the time interval t_1 to t_2. These are readily generalized for a system of interconnected bodies where the total linear momentum is conserved if there is no resultant linear impulse. This implies that some parts of the system may have changes in linear momentum while there is no net change for the whole system.

Conservation of Angular Momentum

The angular impulse is zero when $\sum \mathbf{M} = 0$, so Eqs. 19-22 and 19-24 become, respectively,

$$\Delta \mathbf{H}_C = 0 \qquad \text{19-29a}$$

$$I_O\omega_1 = I_C\omega_2 \qquad \omega_1 = \omega_2 \qquad \text{19-29b}$$

which express that the angular momentum is constant in the time interval t_1 to t_2. The same statements of the conservation principle can be obtained when working with an arbitrary reference point O, so $\Delta \mathbf{H}_O = 0$.

There are two noteworthy aspects of conservation of angular momentum. First, $\Delta \mathbf{H}_C = 0$ does not imply that $\Delta \mathbf{H}_O = 0$, or vice versa. The reason is that the angular impulse is defined with respect to a specific point as in Eqs. 19-24 and 19-26. Second, conservation of angular momentum does not require a simultaneous conservation of linear momentum. This means that there may be a linear impulse while the angular impulse is zero.

The net angular momentum of a system of interconnected bodies is conserved when there is no resultant angular impulse acting on the system. Again, some parts of the system may have changes in angular momentum while the system as a whole experiences no change. The same *fixed reference point O* must be used for all parts of the system.

EXAMPLE 19-6

A new pole for street lights is designed to shear off by a relatively small force when an automobile crashes into it (Fig. 19-13a). For modeling such an event, assume that the car is a uniform rectangular plate of 1200 kg mass. The car is traveling at $\mathbf{v}_0 = 100\mathbf{j}$ km/h prior to the impact. Assume that the force applied to the car by the pole is $\mathbf{F} = -5\mathbf{j}$ kN acting for a time of 0.1 s. Determine the motion of the car after the impact.

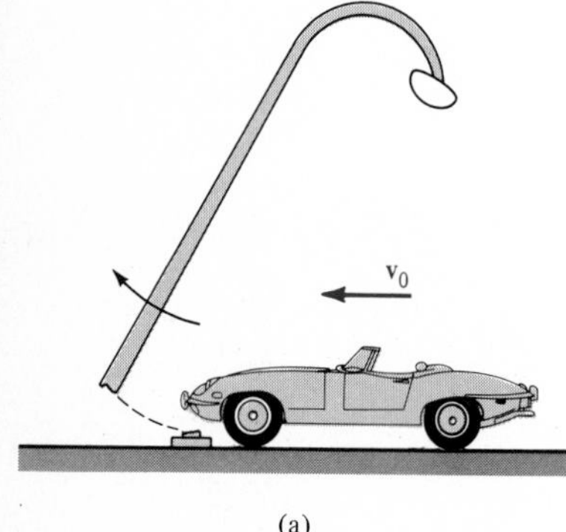

(a)

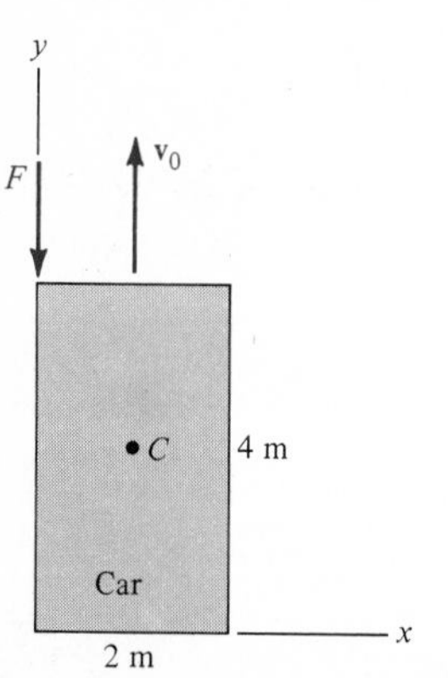

(b) Top view of car during impact

FIGURE 19-13

Motion of a car after an impact

SOLUTION

The impulse and momentum equations for a rigid body can be applied directly here. From Eq. 19-27 and Fig. 19-13b,

$$\sum\begin{pmatrix}\text{initial linear}\\ \text{momentum}\end{pmatrix}_x + \sum\begin{pmatrix}\text{linear}\\ \text{impulse}\end{pmatrix}_x = \sum\begin{pmatrix}\text{final linear}\\ \text{momentum}\end{pmatrix}_x$$

$$0 \quad + \quad 0 \quad = \quad 0$$

$$\sum(\text{initial linear momentum})_y + \sum(\text{linear impulse})_y$$

$$= \sum(\text{final linear momentum})_y$$

$$mv_{0y} + \sum F_y\,\Delta t = mv_{fy}$$

$$(1200\text{ kg})\left(100\,\frac{\text{km}}{\text{h}}\right)\left(\frac{1000\text{ m/km}}{3600\text{ s/h}}\right)$$

$$+ (-5000\text{ N})(0.1\text{ s}) = (1200\text{ kg})v_{fy}$$

$$27.78\text{ m/s} - \frac{500\text{ N}\cdot\text{s}}{1200\text{ kg}}\left(\frac{\text{kg}\cdot\text{m/s}^2}{\text{N}}\right) = v_f$$

$$v_f = 27.78\text{ m/s} - 0.42\text{ m/s} = 27.36\text{ m/s} = 98.5\text{ km/h}$$

$$\sum(\text{initial angular momentum})_C + \sum(\text{angular impulse})_C$$

$$= \sum(\text{final angular momentum})_C$$

$$I_C\omega_0 + \sum M_C\,\Delta t = I_C\omega_f$$

$$0 + (5000\text{ N})(1\text{ m})(0.1\text{ s}) = \frac{1}{12}(1200\text{ kg})[(2\text{ m})^2 + (4\text{ m})^2]\omega_f$$

$$\omega_f = 0.25\text{ rad/s}$$

EXAMPLE 19-7

The 50-lb wheel of an aircraft has a 1.5-ft radius and a 1-ft radius of gyration. Just prior to touchdown the wheel has a horizontal speed $v = 200$ ft/s and does not rotate (Fig. 19-14a). Assume that during touchdown a constant load $W = 20{,}000$ lb is acting on the wheel, v is constant, and there is no deformation or bouncing. Determine the distance traveled by the aircraft from the initial touchdown until the wheel can roll without sliding.

SOLUTION

Equations 19-27 are illustrated for this body in Fig. 19-14b during the interval of interest. Some qualitative observations must be made first, however. When considering motion in the x direction it must be remembered that the translational velocity of the wheel is the same as that of the aircraft. The horizontal velocity of the airplane will remain essentially constant until the brakes are applied (they are assumed "off" here). Therefore, when considering the linear momentum of the wheel it must be remembered that momentum of the whole aircraft is constant. On the other hand, the rotational velocity of the wheel may be considered separately from the rest of the aircraft since the wheel and aircraft rotate independently. Any change in motion in the y direction is ignored here; hence,

$$\sum (mv_y)_0 + N\,\Delta t - W\,\Delta t = \sum (mv_y)_f$$

$$N = W = 20{,}000 \text{ lb}$$

The problem then is to determine the time Δt after which the wheel rolls without slipping, when $v_C = r\omega = 200$ ft/s. During slipping, the friction force is $\mathscr{F} = \mathscr{F}_{\max} = \mu_k N$. From Eq. 19-27c,

$$\sum (\text{initial angular momenta})_C + \sum (\text{angular impulse})_C = \sum (\text{final angular momenta})_C$$

$$I_C(0) + \mathscr{F} r(\Delta t) = I_C\omega_f = m r_g^2 \left(\frac{v_C}{r}\right)$$

$$(0.4)(20{,}000 \text{ lb})(1.5 \text{ ft})(\Delta t) = \frac{50 \text{ lb}}{32.2 \text{ ft/s}^2}(1 \text{ ft})^2 \left(\frac{200 \text{ ft/s}}{1.5 \text{ ft}}\right)$$

$$\Delta t = \frac{207 \text{ lb}\cdot\text{ft}\cdot\text{s}}{12{,}000 \text{ lb}\cdot\text{ft}} = 0.017 \text{ s}$$

The distance x traveled by the aircraft until the wheel rolls without slipping is

$$x = v\,\Delta t = 200 \text{ ft/s } (0.017 \text{ s}) = 3.45 \text{ ft}$$

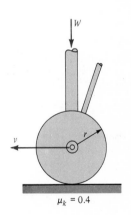

(a)

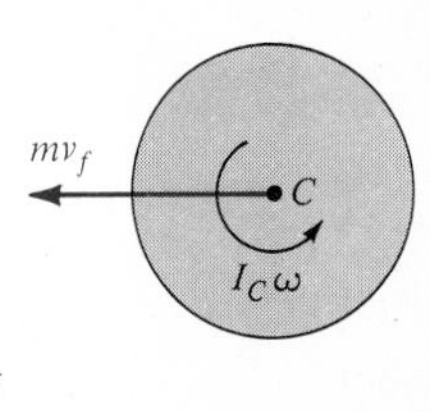

(b)

FIGURE 19-14

Motion of an airplane's wheel during landing

EXAMPLE 19-8

A system of two disks (Fig. 19-15a) is rotating without slipping. Disk A has a 2-kg mass, a 0.2-m radius, a 0.16-m radius of gyration, and is rotating at the rate of $\omega_A = 2000$ rpm. The uniform disk B has a 10-kg mass and a 0.4-m radius. Determine the required moment M_A on disk A to stop the rotation of the disks in a time of 20 s.

SOLUTION

The only unknown, at first glance, is the magnitude of the applied moment M_A. Therefore, one might be tempted to write the angular impulse-momentum equation for the system to simply solve for M_A (Fig. 19-15b). However, it must be remembered that stationary reaction impulsive forces, even though they do no work, must be included for impulse-momentum solution. Also, it must be recalled that one of the criteria for applying the angular impulse and momentum method was that moments must be taken *about the same point for the entire system*. Choosing point A as the reference point, it is seen that the vertical reaction component at B does cause an impulsive moment about A and therefore must be included. Therefore, each disk must be analyzed separately.

Looking first at disk A (Fig. 19-15c), the angular impulse-momentum equation is written,

$$I_A\omega_A + \mathscr{F}r_A(\Delta t) - M_A\,\Delta t = I_A(0) \tag{a}$$

The temptation to use $\mathscr{F} = \mathscr{F}_{max} = \mu N$ must be overcome here because slipping is not occurring and is not necessarily impending. Hence, both $\mathscr{F}$ and M_A are unknown and nothing can be solved for yet.

Looking at disk B (Fig. 19-15d), the angular impulse-momentum equation is written,

$$I_B\omega_B + \mathscr{F}r_B(\Delta t) = I_B(0) \tag{b}$$

The initial angular speed ω_B can be solved for,

$$r_A\omega_A = r_B\omega_B$$

$$\omega_B = (r_A/r_B)\omega_A = \left(\frac{0.2\text{ m}}{0.4\text{ m}}\right)\left(2000\,\frac{\text{rev}}{\text{min}}\right)\left(\frac{2\pi\text{ rad/rev}}{60\text{ s/min}}\right)$$

$$= 105\text{ rad/s}$$

Solving Eq. (b) for $\mathscr{F}$ (note the sign convention used) yields

$$\frac{1}{2}(10\text{ kg})(0.4\text{ m})^2(-105\text{ rad/s}) + \mathscr{F}(0.4\text{ m})(20\text{ s}) = 0$$

$$\mathscr{F} = 10.5\text{ N}$$

Substituting this into Eq. (a) and using $I_A = m_A r_g^2$ gives

$$(2\,\text{kg})(0.16\,\text{m})^2(210\,\text{rad/s}) + (10.5\,\text{N})(0.2\,\text{m})(20\,\text{s}) - M_A(20\,\text{s}) = 0$$

$$M_A = 2.64\ \text{N}\cdot\text{m}$$

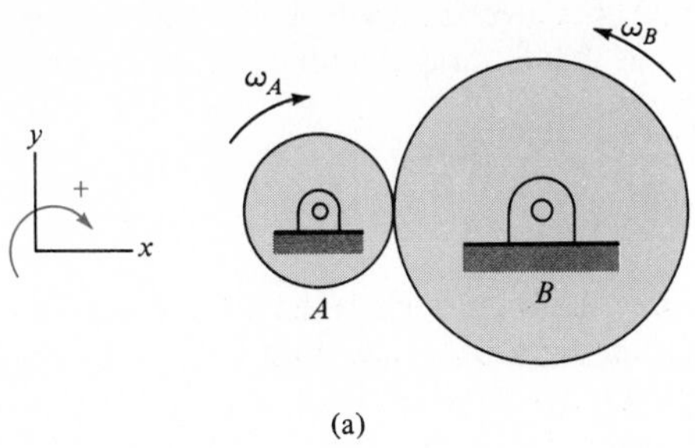

(a)

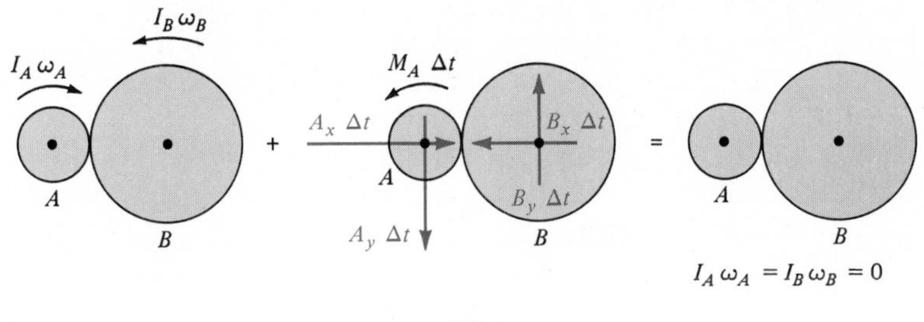

(b)

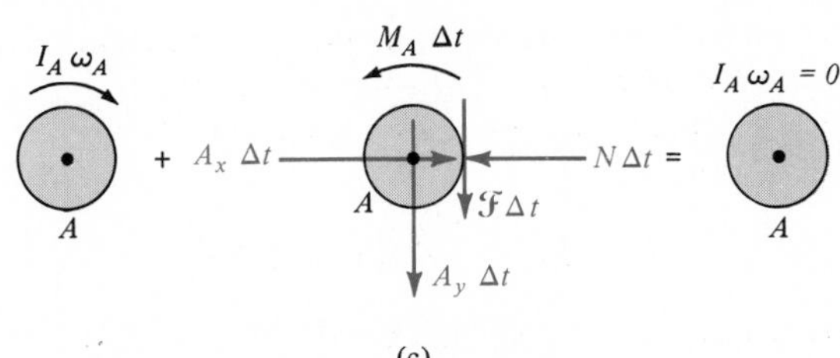

(c)

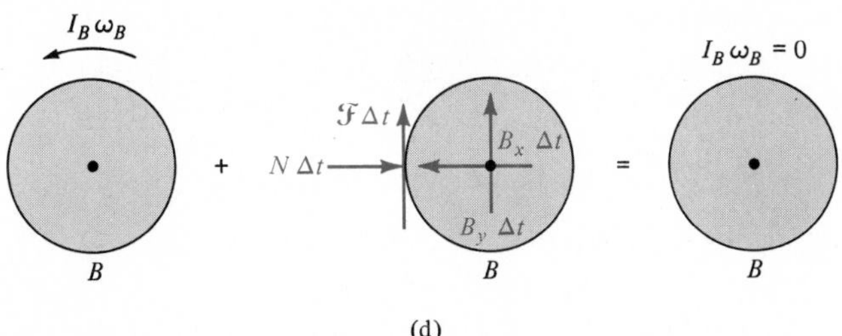

(d)

FIGURE 19-15

Analysis of a system of two wheels

EXAMPLE 19-9

A twin-engine jet aircraft (Fig. 19-16) has a total weight of 10^5 lb and a radius of gyration $r_z = 15$ ft about the z axis. Assume that the aircraft is in level flight at a constant velocity $\mathbf{v}_0$ in the y direction when engine B fails. It takes a time of 3 s for an automatic correction to cause an effect. Determine the change in v_0 and the angular speed ω_z that the airplane acquires in that time of 3 s. Assume that air drag is negligible for the rotation.

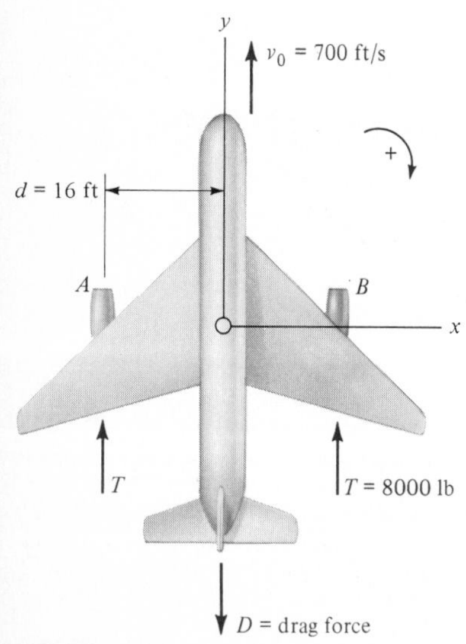

FIGURE 19-16

Top view of an airplane in level flight

SOLUTION

First the drag force D must be determined for inclusion in the linear momentum equation of the airplane. When both engines are producing thrusts of 8000 lb each, the airplane is flying at a constant velocity. Therefore, the drag force in the y direction must be $D = 16{,}000$ lb at velocity $\mathbf{v}_0$ ($\sum F_y = ma_y = 0$). Assuming the drag force to remain essentially constant during the 3-s time interval after the engine failure, the impulse-momentum equation is

$$(mv_0)_y + T(\Delta t) - D(\Delta t) = (mv_f)_y$$

$$\frac{10^5 \text{ lb}}{32.2 \text{ ft/s}^2}(700 \text{ ft/s}) + 8000 \text{ lb}(3 \text{ s}) - 16{,}000 \text{ lb}(3 \text{ s})$$

$$= \frac{10^5 \text{ lb}}{32.2 \text{ ft/s}^2}(v_f)_y$$

$$(v_f)_y = 700 \text{ ft/s} - \frac{(8000 \text{ lb})(3 \text{ s})(32.2 \text{ ft/s}^2)}{10^5 \text{ lb}} = 693 \text{ ft/s}$$

The linear momentum in the x direction does not change. The angular impulse-momentum equation is

$$I_C\omega_0 + \sum M_C \, \Delta t = I_C\omega_f$$

$$0 + (8000 \text{ lb})(16 \text{ ft})(3 \text{ s}) = \left(\frac{10^5 \text{ lb}}{32.2 \text{ ft/s}^2}\right)(15 \text{ ft})^2\omega_f$$

$$\omega_f = 0.55 \text{ rad/s} \qquad \text{(clockwise)}$$

PROBLEMS

19–38 A 3-kg uniform disk rolls without slipping on a horizontal surface. It comes to rest in a time of 20 s. Calculate the average friction force acting on the disk.

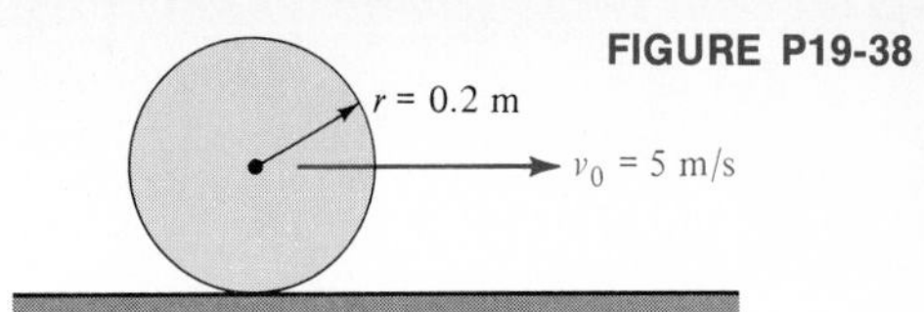

FIGURE P19-38

19–39 A 50-lb flywheel has a radius of gyration of 15 in. The wheel starts to idle in a vacuum chamber at 15,000 rpm and comes to rest in 30 days. Determine the moment caused by friction in the bearings.

19–40 A 30-kg flywheel is a uniform disk with a 0.25-m radius. Determine the moment that can increase the wheel's rotational speed from 1000 to 5000 rpm in a time of 100 s.

19–41 A 40-lb uniform wheel is at rest when a horizontal force F is applied at its axis. Calculate the horizontal speed of the wheel after a time of 5 s assuming that it rolls without slipping. Also calculate the minimum coefficient of friction to prevent slipping.

FIGURE P19-41

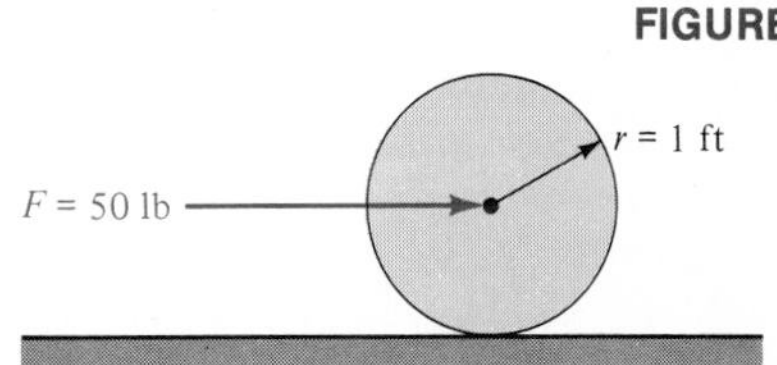

19–42 A thin hoop of 1 kg mass and 0.4-m radius starts to roll without slipping on an incline of $\theta = 30°$. Determine its speed after a time of 10 s and the minimum coefficient of friction to prevent slipping.

FIGURE P19-42

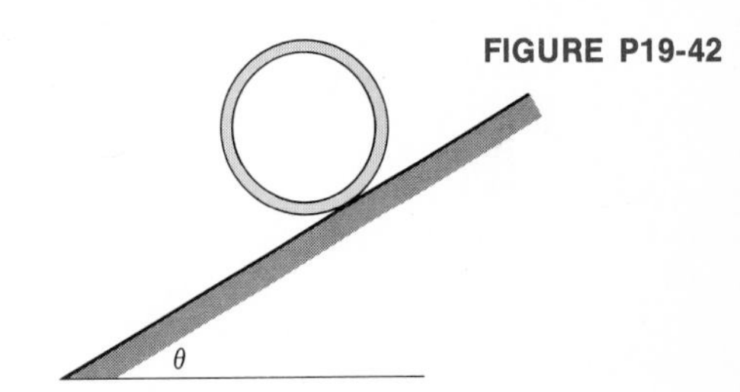

19–43 Solve Prob. 19–42 for a uniform disk of 10 lb weight and 1 ft radius.

19–44 A uniform 4-kg bar is released from rest at $\theta = 60°$. Determine its angular speed after a time of 0.05 s if the hinge at O is frictionless. Assume that θ is constant during Δt.

FIGURE P19-44

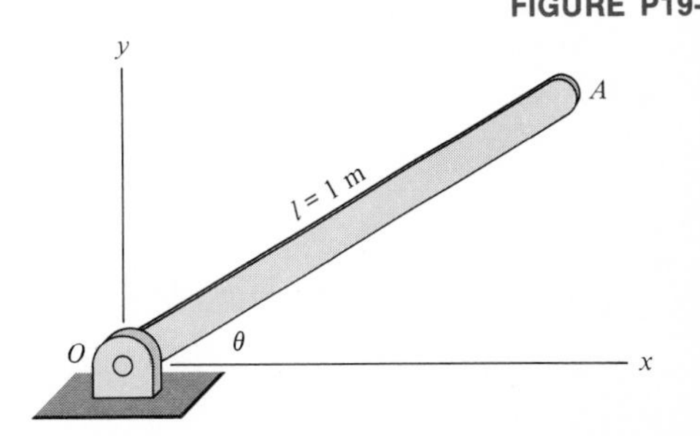

19–45 Determine the angular velocity of the bar in Prob. 19–44 assuming that it has an initial angular velocity $\omega_z = 0.5$ rad/s $\mathbf{k}$ at $\theta = 60°$.

19–46 Solve Prob. 19–44 assuming that a spring applies a constant downward force of 100 N at point A on the bar during the initial 0.05 s of the motion.

FIGURE P19-47

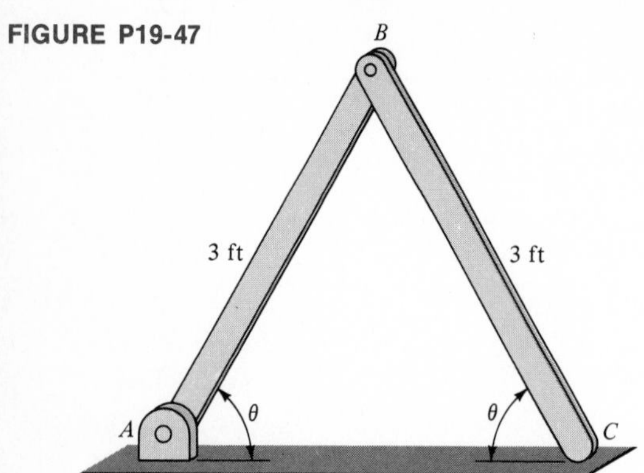

19–47 Two uniform, 10-lb rods are hinged at A and B, and part C slides on a smooth surface. The rods are released from rest when $\theta = 60°$. Calculate the vertical velocity of B after a time of 0.02 s.

19–48 Solve Prob. 19–47 assuming that a spring applies a constant horizontal force of 50 lb acting to the right at point C during the initial 0.02 s of the motion.

19–49 Determine the required normal force between the two disks in Ex. 19-8 to prevent slipping of the disks if $\mu_s = 0.6$.

FIGURE P19-51

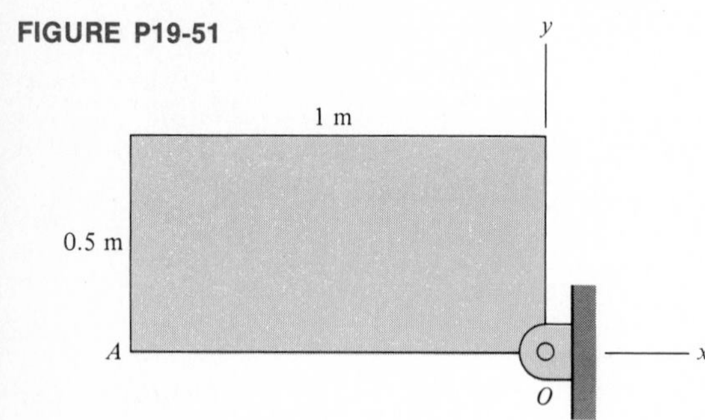

19–50 Consider the test setup in Prob. 19–5. Calculate the platform's velocity 0.1 s after it is released from rest.

19–51 A uniform plate of 40-kg mass is released from rest with edge OA horizontal. Determine the angular speed of the plate 0.05 s after its release.

19–52 Solve Prob. 19–51 assuming that the plate has an initial angular velocity $\boldsymbol{\omega}_z = -0.3$ rad/s **k** when OA is horizontal.

FIGURE P19-53

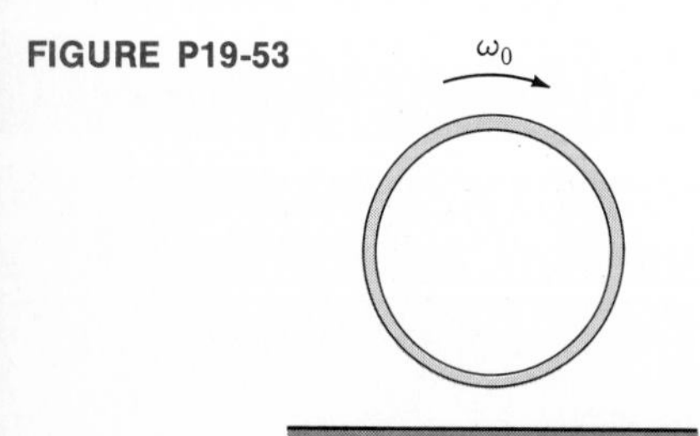

19–53 A thin hoop of 3 lb weight and 1 ft radius has an initial angular speed of $\omega_0 = 20$ rad/s when it is placed on a horizontal surface. The hoop does not bounce as it slides and rolls. Determine the time t when the hoop starts rolling without slipping if $\mu_k = 0.3$.

19–54 Solve Prob. 19–53 for a uniform sphere of 4-kg mass and 0.12-m radius. Also determine the horizontal speed of the sphere when it starts rolling without slipping.

19–55 The flywheel A of a tire tester (Fig. 19-6) has a 500-lb weight, a 6-ft diameter, and a 2.5-ft radius of gyration. It is freely rotating (no power input) at 100 rpm when a nonrotating wheel C is pressed against it with a normal force of 500 lb acting for a time $t = 1.5$ s. Wheel C is modeled as a uniform disk of 30 lb weight and 2.4 ft diameter. Determine the angular velocities of the two wheels after contact for the time of 1.5 s if $\mu_s = 0.7$ and $\mu_k = 0.6$.

19–56 An elevator system is modeled as $m_1 = 2000$ kg, $m_2 = 1800$ kg, and a uniform disk of $m_3 = 200$ kg. Calculate the velocity of m_1 at time $t = 10$ s after the system starts from rest.

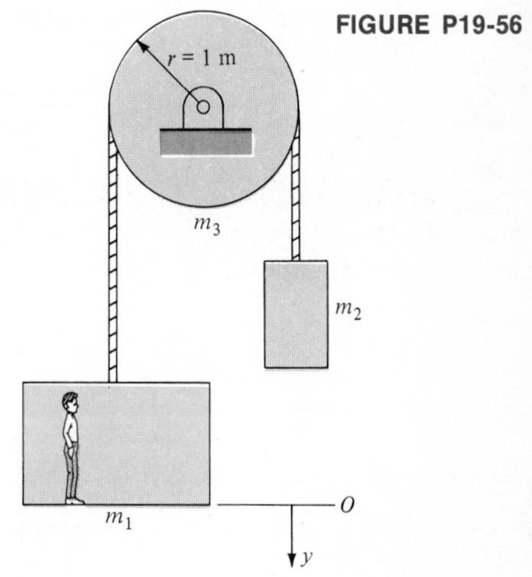

FIGURE P19-56

19–57 Determine the required braking moment acting on m_3 in Prob. 19–56 to bring it to a stop in a time of 3 s from an angular speed of 3 rad/s.

19–58 Consider Prob. 19–17. Calculate the required moment at 0 to give W_1 an upward velocity of 8 ft/s in a time of 2 s.

19–59 Determine the required moment in Prob. 19–19 to attain the angular speed of 4 rpm in a time of 5 s.

19–60 Determine the velocity of point C in Prob. 19–21 at time $t = 0.01$ s after the bar is released from rest as shown.

19–61 The pulley has a 0.15-m radius of gyration. Calculate the tension T to start the system from rest and attain an upward velocity of 8 m/s in a time of 5 s.

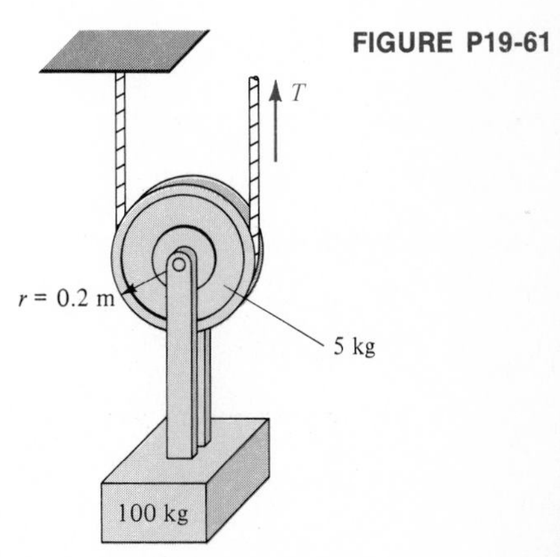

FIGURE P19-61

19–62 Consider the front loader in Prob. 19–27. Calculate the required torque applied to the two driving wheels in the first 5 s of motion.

19–63 Consider Prob. 19–32. Determine the required moment of the force F to attain $\dot{\theta} = 0.3$ rad/s in a time of 2 s starting from rest at $\theta = 0$.

19–64 Consider Prob. 19–33. Determine the velocity of the plate in the two cases, at a time t after it has started moving along the incline.

FIGURE P19-65

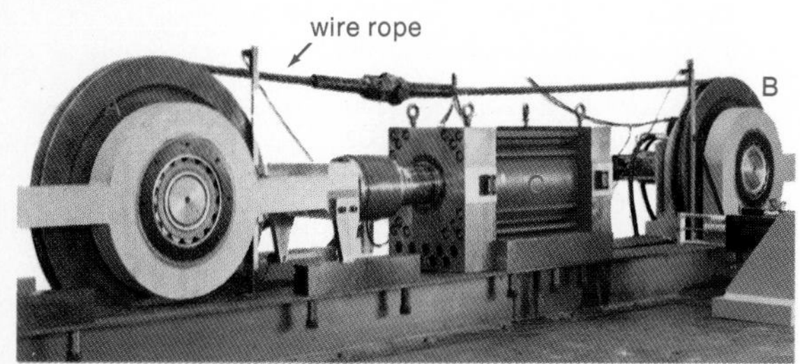

19–65 Pulleys A and B of the wire rope tester are essentially uniform disks of 300 lb weight and 2 ft radius. The wire rope makes a complete loop around the pulleys, which are pushed apart by the hydraulic actuator C applying a horizontal force of 100,000 lb. Determine the required moment at pulley B to attain a clockwise angular speed of 10 rad/s in a time of 0.1 s starting from rest. Also calculate the tension in the upper and lower parts of the wire rope assuming no slipping. Pulley A is free to rotate.

FIGURE P19-66

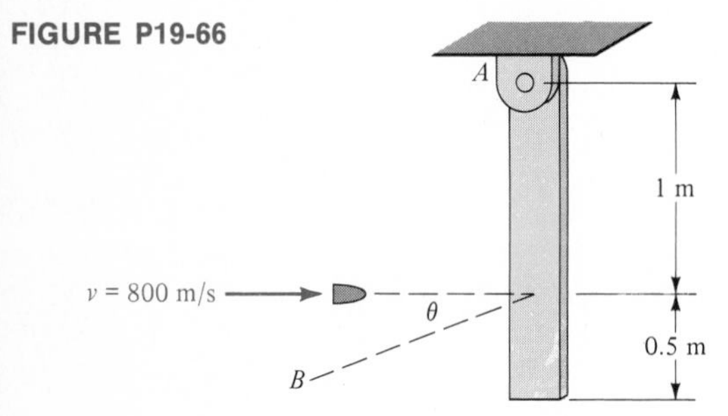

19–66 A 5-kg uniform bar is hinged at A. The bar is at rest when a 0.04-kg bullet strikes it horizontally and comes to rest in the bar. Calculate the initial angular speed of the bar with the bullet resting in it after the impact. Ignore direction B which is used in Prob. 19–67.

19–67 Solve Prob. 19–66 with the bullet coming from the direction of B at $\theta = 30°$. Also calculate the horizontal and vertical reactions at hinge A caused by the bullet during its arrest, which occurs in a time of 0.0015 s.

FIGURE P19-68

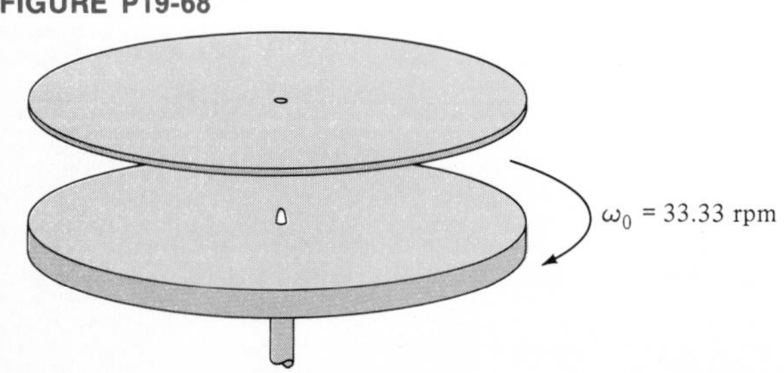

19–68 A 7-lb turntable has a 12-in. diameter and a 4-in. radius of gyration. The turntable is rotating at a speed of 33.33 rpm when a 0.08-lb, 12-in.-diameter, uniform, nonrotating disk is dropped on it. The disk stops slipping on the turntable in a time of 0.2 s. Determine the additional moment that the drive motor applies to the turntable during this 0.2-s interval to maintain the original speed ω_0.

Section 19-1

SUMMARY

Work of a force on a rigid body

$$U_{12} = \int_1^2 \mathbf{F} \cdot d\mathbf{r} = \int_1^2 \mathbf{F} \cdot \mathbf{v}\, dt$$

where $d\mathbf{r}$ = infinitesimal displacement of force $\mathbf{F}$ along the path from 1 to 2

$$\mathbf{v} = \frac{d\mathbf{r}}{dt}$$

Work of a moment

$$U_{12} = \int_{\theta_1}^{\theta_2} M\, d\theta$$

where $d\theta$ is the infinitesimal angular displacement.

Section 19-2

Kinetic energy of a rigid body

$$T = T_{\text{transl}} + T_{\text{rot}} = \frac{1}{2} m v_C^2 + \frac{1}{2} I_C \omega^2$$

where C denotes the center of mass.

Rotation about an arbitrary fixed axis at point O

$$T = \frac{1}{2} I_O \omega^2$$

Section 19-3

Principle of work and energy

$$T_2 = T_1 + U_{12}$$

where T_1 = initial kinetic energy of the body
T_2 = final kinetic energy of the body
U_{12} = work of all external forces and moments acting on the body from position 1 to 2

The equation is also valid for an internally conservative system of rigid bodies.

Section 19-4

Conservation of energy
In a conservative system,

$$T_1 + V_1 = T_2 + V_2$$

where T = kinetic energy
V = total potential energy = $V_g + V_e$
V_g = gravitational potential energy
V_e = elastic potential energy

Section 19-5

Power

TRANSLATION:

$$\text{Power} = \frac{dU}{dt} = \mathbf{F} \cdot \mathbf{v} = \dot{T}$$

ROTATION:

$$\text{Power} = \frac{dU}{dt} = M\omega \qquad \text{(for a constant moment } \mathbf{M} \text{ parallel to } \boldsymbol{\omega}\text{)}$$

Section 19-6

Impulse and momentum of a rigid body

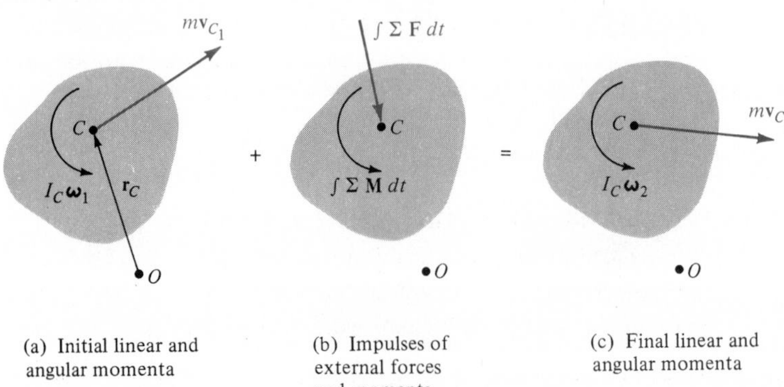

(a) Initial linear and angular momenta

(b) Impulses of external forces and moments

(c) Final linear and angular momenta

LINEAR IMPULSE AND MOMENTUM:

$$\int_{t_1}^{t_2} \sum \mathbf{F}\, dt = m(\mathbf{v}_{C_2} - \mathbf{v}_{C_1})$$

where t is time, C is the mass center, and $\sum \mathbf{F}$ includes *all* external forces.

ANGULAR IMPULSE AND MOMENTUM:

$$\int_{t_1}^{t_2} \sum \mathbf{M}_C \, dt = \mathbf{H}_{C_2} - \mathbf{H}_{C_1}$$

For plane motion, if $\sum \mathbf{M}$ is parallel to $\boldsymbol{\omega}$,

$$\int_{t_1}^{t_2} \sum M_C \, dt = I_C(\omega_2 - \omega_1)$$

$$\int_{t_1}^{t_2} \sum M_O \, dt = I_O(\omega_2 - \omega_1) \qquad \text{for an arbitrary reference point } O$$

Section 19-7

Impulse and momentum of a system of rigid bodies

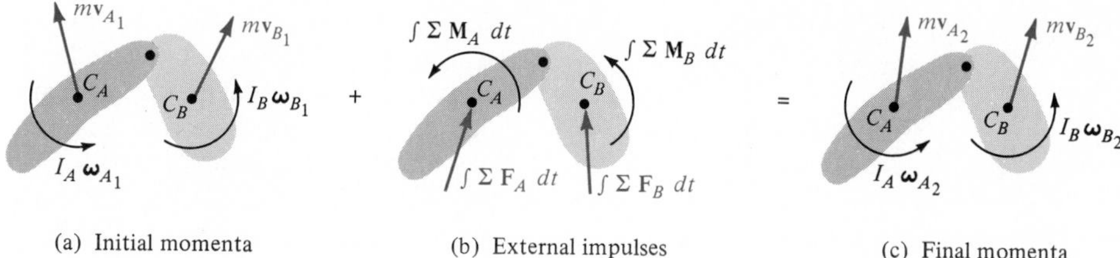

(a) Initial momenta (b) External impulses (c) Final momenta

PROCEDURES:

1. Apply the principle of impulse and momentum to each rigid member separately. The mutual forces acting between members must be included in the formulation of the solution; *or*
2. Apply the principle of impulse and momentum to the entire system of bodies, ignoring the mutual forces between members.

Section 19-8

CONSERVATION OF LINEAR MOMENTUM (time interval t_1 to t_2):

$$\sum \mathbf{F} = 0 \Rightarrow \Delta\mathbf{G} = 0$$

$$\mathbf{v}_{C_1} = \mathbf{v}_{C_2}$$

CONSERVATION OF ANGULAR MOMENTUM:

$$\sum \mathbf{M} = 0 \Rightarrow \Delta\mathbf{H}_C = 0$$

$$\boldsymbol{\omega}_1 = \boldsymbol{\omega}_2$$

For a system of rigid bodies, use the same *fixed reference point O* for all parts of the system:

$$\Delta\mathbf{H}_O = 0$$

REVIEW PROBLEMS

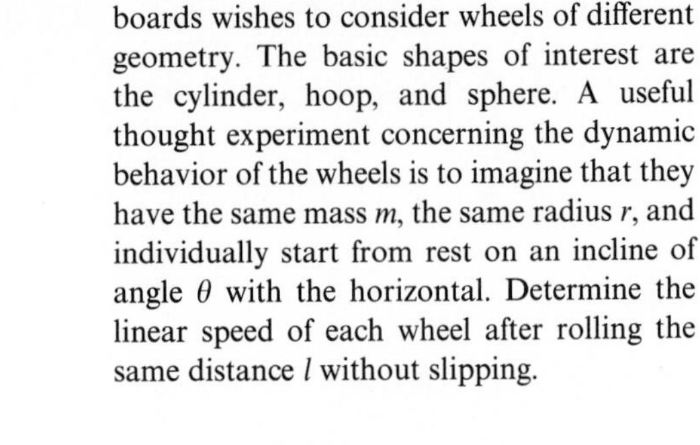

19–69 A designer of new roller skates and skateboards wishes to consider wheels of different geometry. The basic shapes of interest are the cylinder, hoop, and sphere. A useful thought experiment concerning the dynamic behavior of the wheels is to imagine that they have the same mass m, the same radius r, and individually start from rest on an incline of angle θ with the horizontal. Determine the linear speed of each wheel after rolling the same distance l without slipping.

FIGURE P19-70

19–70 The self-propelled, 4000-kg derrick at a dock raises a 20,000-kg container at a constant vertical speed of 0.3 m/s. At the same time, it accelerates horizontally at the rate of 0.6 m/s^2. Calculate the required power of the derrick in this motion, over a distance of 6 m.

FIGURE P19-71

19–71 A 70-kg person starts on a level road ($\theta = 0$) on a 12-kg bicycle. Each wheel is essentially a thin hoop of 1.5-kg mass and 74-cm diameter. What is the linear acceleration of the person if 0.4 hp is transmitted to the rear wheel which rolls without slipping? The linear speed at the given instant is 5 m/s.

19–72 Solve Prob. 19–71 with $\theta = 10°$ and the person heading uphill.

FIGURE P19-73

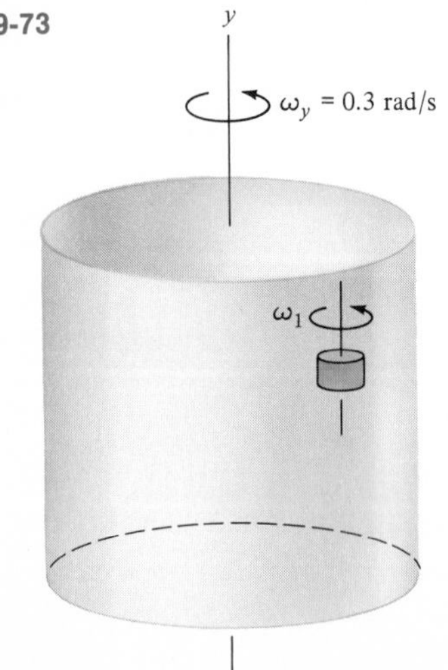

19–73 A 2000-kg spacecraft with radius of gyration of 0.3 m is rotating about its longitudinal axis at a constant speed of 0.3 rad/s. A 1-kg rotor of 0.04-m radius of gyration starts from rest (with respect to the spacecraft) and attains a speed of $\omega_1 = 12\,000$ rpm in a time of 40 s. The vectors $\boldsymbol{\omega}_y$ and $\boldsymbol{\omega}_1$ are parallel and in the same direction. Calculate the change in ω_y caused by ω_1 of the motor.

19–74 The total weight of a cylindrical spacecraft is 7,000 lb. Its radius of gyration is 2 ft when the four symmetrically arranged solar panels are parallel to the y axis; $\omega_y = 0.8$ rad/s in this configuration. Determine the final angular speed after the 20-lb uniform panels are dropped to their positions normal to the y axis.

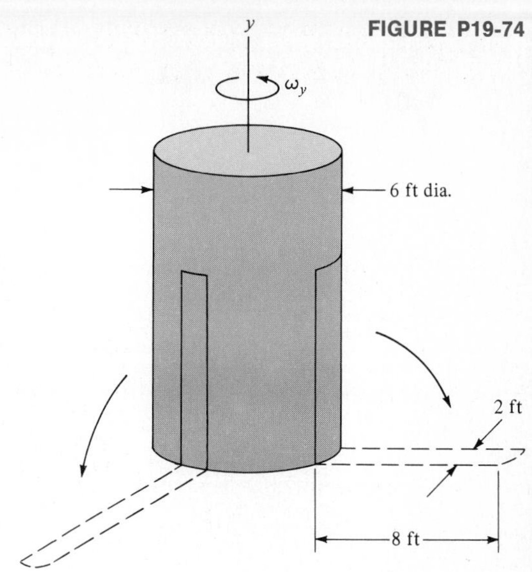

FIGURE P19-74

19–75 A truck skidding sideways and hitting a curb is modeled as a uniform square block of mass m and sides a. Its mass center has a speed v just before the impact which causes no rebound. Determine the angular speed of the block immediately after the impact. Assume that the curb applies a force at the bottom of the wheel.

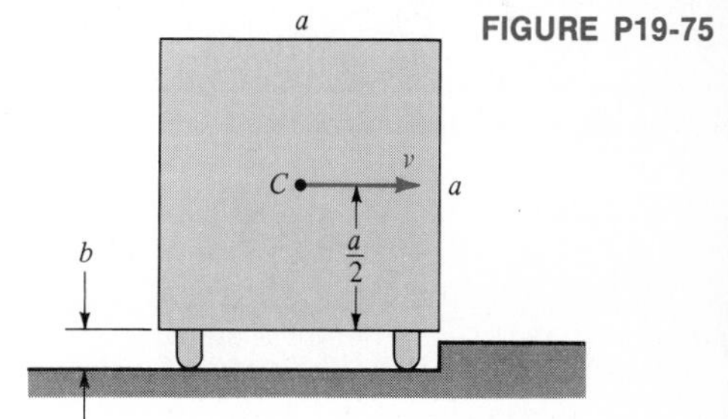

FIGURE P19-75

CHAPTER 20

KINETICS OF RIGID BODIES IN THREE DIMENSIONS

OVERVIEW

The forces and moments of forces acting on a rigid body may cause complex three-dimensional motion of that body. For a demonstration of these, experiment with toy tops or gyroscopes. Note that when an external force is applied to a rotating body, the resulting displacement may be in a different direction from that of the force. The reason for this phenomenon is that *the change in direction of the angular momentum vector involves an angular acceleration.* This concept was introduced in Sec. 17-7, dealing with the kinematics of rigid bodies, and is used here for analyzing the motion of bodies when the instantaneous axes of rotation change.

STUDY GOALS

All topics of spatial kinetics of rigid bodies are rather difficult for the average student and are often deferred to courses in advanced dynamics. However, the subject is important in practice, and ambitious students will find this chapter useful in further work in dynamics. The individual topics are described in perspective as follows.

SECTION 20-1 presents qualitative considerations and basic mathematical methods for the analysis of angular momentum in three dimensions. These methods are essential for studying the other topics in this chapter.

SECTION 20-2 extends the methods of impulse and momentum of systems of particles (Sec. 16-10) to rigid bodies. This application is straightforward after a thorough study of Sec. 20-1.

SECTION 20-3 presents the derivation of the expression for the kinetic energy of a rigid body on the basis of Sec. 16-4. This useful expression is also readily applied after studying Sec. 20-1.

SECTIONS 20-4 AND 20-5 present the equations of motion of a rigid body in three dimensions. The derivations are based on the results obtained in Secs. 16-7, 16-8, 17-8, and 20-1. These topics are intended only for the most ambitious students.

SECTION 20-6 describes the scalar governing equations for general three-dimensional motion of a rigid body. Relatively simple equations for special cases are also discussed. These are useful in machine design, but they have particular limitations.

SECTIONS 20-7 AND 20-8 apply the equations of three-dimensional motion to analyze gyroscopic motion. These important but rather difficult topics are intended only for the most ambitious and advanced students who wish to take a course in advanced dynamics.

20-1 ANGULAR MOMENTUM IN THREE DIMENSIONS

Qualitative Considerations

A little experimentation with a rapidly spinning top reveals the following remarkable facts. If the axis of the top is vertical, any arbitrary vertical displacement of the bearing support (O in Fig. 20-1) has no effect on the motion of the top. Thus, the angular momentum vector $\mathbf{H}$ appears to be constant. A more curious phenomenon is observed when the spinning top is placed on its bearing so that the axis of rotation is not vertical, as in Fig. 20-2a. In this case the shaft of the top appears to have a motion that would sketch out a cone which is called the *space cone.* The rotation of the angular velocity vector $\boldsymbol{\omega}$ about the y axis is called *precession.* The angular velocity of the precession of $\boldsymbol{\omega}$ (and of the vector $\mathbf{H}$) is denoted by the vector $\boldsymbol{\Omega}$ in Fig. 20-2a. Since the direction of the angular momentum vector $\mathbf{H}$ changes in the case of precession, $\mathbf{H}$ is not constant even if the magnitude ω for rotation about the top's own axis is a constant.

The angular velocity vector $\boldsymbol{\omega}$ also has a motion relative to the body. This motion would sketch out a cone which is called the *body cone*. For a rigid body with a fixed point, the instantaneous axis of rotation always passes through the fixed point of the body. This is shown schematically in Fig. 20-2b, where arbitrary space and body cones are also shown. The locus of the absolute positions of the instantaneous axis of rotation is the space cone. The locus of the positions of the instantaneous axis relative to the body is the body

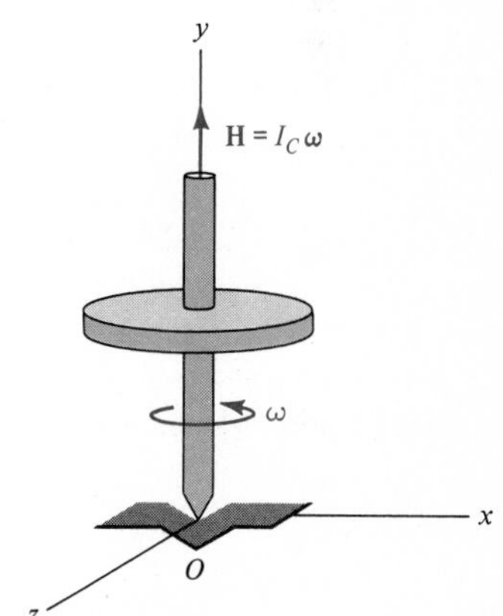

FIGURE 20-1
Top rotating about its axis of symmetry

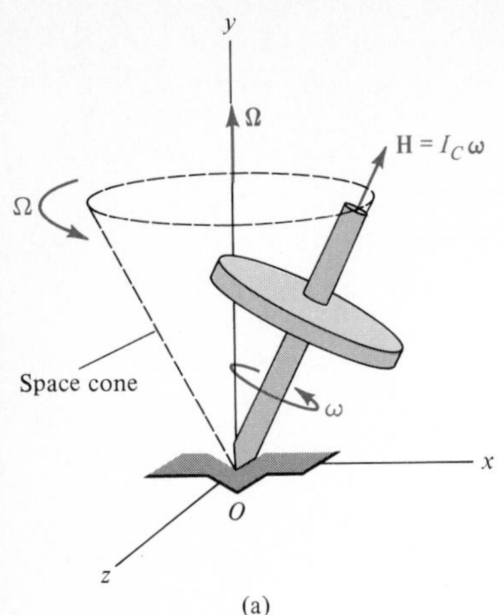

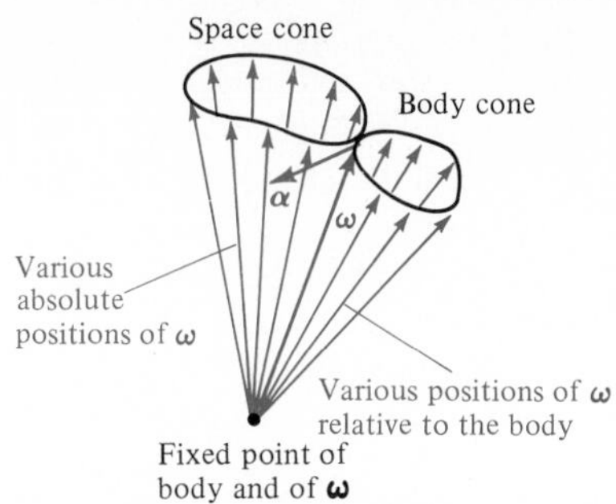

(b) General relation of body and space cones

FIGURE 20-2

Motion of an inclined, spinning top

cone. These cones may be circular or irregular cones, but they are tangent along $\boldsymbol{\omega}$ for the instant considered. The body cone appears to roll on the space cone analogously to the body and space centrodes introduced in Sec. 17-5.

It is useful to remember that $\boldsymbol{\alpha} = d\boldsymbol{\omega}/dt$. This means that, for $\boldsymbol{\omega}$ with a fixed point, $\boldsymbol{\alpha}$ is the velocity of the tip of the vector $\boldsymbol{\omega}$. Accordingly, $\boldsymbol{\alpha}$ is tangent to the path of the tip of $\boldsymbol{\omega}$ in the space cone as in Fig. 20-2b. $\boldsymbol{\alpha}$ is normal to $\boldsymbol{\omega}$ if the magnitude ω is constant.

Analysis of Angular Momentum

For the analysis of rotational motion in three dimensions, consider a rigid body with a reference frame xyz of fixed orientation attached at its mass center C (Fig. 20-3). The motion of the body is observed from point C, which may be moving with respect to the fixed frame XYZ (for details, see Sec. 16-8). The angular momentum of particle P_i of mass m_i, position vector $\mathbf{r}'_i$, and velocity $\mathbf{v}'_i$ is $\mathbf{H}_i = \mathbf{r}'_i \times m_i\mathbf{v}'_i$ (all with respect to the xyz axes which can only translate). Since the xyz axes are attached at the mass center C, the only possible motion of a point P_i relative to C is a rotation. Hence, $\mathbf{v}_i = \boldsymbol{\omega} \times \mathbf{r}'_i$ and $\mathbf{H}_i$ can be rewritten as $\mathbf{H}_i = \mathbf{r}'_i \times m_i(\boldsymbol{\omega} \times \mathbf{r}'_i)$. The total angular momentum of the body about C is

$$\mathbf{H}_C = \sum_{i=1}^{n} [\mathbf{r}'_i \times (\boldsymbol{\omega} \times \mathbf{r}'_i)m_i] \tag{20-1}$$

Equation 20-1 can be written as a determinant,

$$\mathbf{H}_C = \sum_{i=1}^{n} \begin{vmatrix} \mathbf{i} & \mathbf{j} & \mathbf{k} \\ r'_{i_x} & r'_{i_y} & r'_{i_z} \\ (\boldsymbol{\omega} \times \mathbf{r}'_i)_x & (\boldsymbol{\omega} \times \mathbf{r}'_i)_y & (\boldsymbol{\omega} \times \mathbf{r}'_i)_z \end{vmatrix} m_i \tag{20-2}$$

which can be expanded using $\mathbf{r}'_i = x_i\mathbf{i} + y_i\mathbf{j} + z_i\mathbf{k}$ and $\boldsymbol{\omega} = \omega_x\mathbf{i} +$

$\omega_y\mathbf{j} + \omega_z\mathbf{k}$. First, $\boldsymbol{\omega} \times \mathbf{r}_i'$ is expressed as $\boldsymbol{\omega} \times \mathbf{r}_i' = (\omega_y z_i - \omega_z y_i)\mathbf{i} + (\omega_z x_i - \omega_x z_i)\mathbf{j} + (\omega_x y_i - \omega_y x_i)\mathbf{k}$. With this, the x component of $\mathbf{H}_C$ is written in detail,

$$
\begin{aligned}
H_x &= \sum_{i=1}^{n} [y_i(\omega_x y_i - \omega_y x_i) - z_i(\omega_z x_i - \omega_x z_i)]m_i \\
&= \omega_x \sum_{i=1}^{n} (y_i^2 + z_i^2)m_i - \omega_y \sum_{i=1}^{n} x_i y_i m_i \\
&\quad - \omega_z \sum_{i=1}^{n} x_i z_i m_i
\end{aligned}
\qquad \text{20-3}
$$

Similar expressions can be obtained for H_y and H_z. Integrating over all particles of the body, the summation terms of Eq. 20-3 become

$$\int (y^2+z^2)\,dm = I_x \qquad \int xy\,dm = I_{xy} \qquad \int xz\,dm = I_{xz}$$

where I_x is the *centroidal mass moment of inertia* of the body about the x axis, and I_{xy} and I_{xz} are *centroidal mass products of inertia* with respect to the given axes.

The same procedure can be followed in completely evaluating Eq. 20-2. The resulting concise expressions for the three components of the total angular momentum are

$$
\begin{aligned}
H_x &= I_x\omega_x - I_{xy}\omega_y - I_{xz}\omega_z \\
H_y &= -I_{yx}\omega_x + I_y\omega_y - I_{yz}\omega_z \\
H_z &= -I_{zx}\omega_x - I_{zy}\omega_y + I_z\omega_z
\end{aligned}
\qquad \text{20-4}
$$

where all double subscripts denote products of inertia. The arrangement of the terms in Eq. 20-4 facilitates checking them for correctness by permutation of the subscripts. Note that Eq. 20-4 is valid for a particular position of the body; as the body rotates relative to xyz, the moments of inertia and products of inertia may in general change.

The mass moments of inertia and products of inertia in Eq. 20-4 are sometimes arrayed by themselves as

$$
\begin{pmatrix}
I_x & -I_{xy} & -I_{xz} \\
-I_{yx} & I_y & -I_{yz} \\
-I_{zx} & -I_{zy} & I_z
\end{pmatrix}
\qquad \text{20-5}
$$

which is called the *inertia matrix* or *inertia tensor*. By definition, $I_{xy} = I_{yx}$, and so on. This matrix depends on the xyz axes used in the analysis. However, the centroidal angular momentum $\mathbf{H}_C$ depends only on the angular velocity $\boldsymbol{\omega}$ for a given rigid body, so it must be independent of the choice of axes. This makes it reasonable to use axes that simplify the inertia matrix. The ideal set of centroidal axes is the *principal axes of inertia* $x'y'z'$ for which all products of

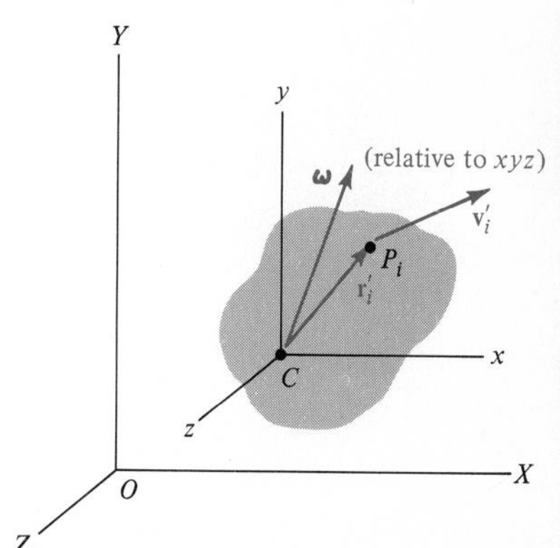

FIGURE 20-3
Schematic for the analysis of angular momentum

inertia of the given body are zero (see Chap. 9 in the statics volume). In that case the array in Eq. 20-5 becomes (for that instant only, in most cases)

$$\begin{pmatrix} I_{x'} & 0 & 0 \\ 0 & I_{y'} & 0 \\ 0 & 0 & I_{z'} \end{pmatrix} \qquad \boxed{20\text{-}6}$$

where $I_{x'}$, $I_{y'}$, and $I_{z'}$ are the *principal centroidal moments of inertia.* Two of these three values are the maximum and minimum mass moments of inertia of the given body. With these simplifications Eq. 20-4 becomes

$$H_{x'} = I_{x'}\omega_{x'} \qquad H_{y'} = I_{y'}\omega_{y'} \qquad H_{z'} = I_{z'}\omega_{z'} \qquad \boxed{20\text{-}7}$$

Note that in general the vectors $\boldsymbol{\omega}$ and $\mathbf{H}_C$ do not have the same directions. There are two special cases when these vectors are in the same direction. One is when $\omega_{x'}$, $\omega_{y'}$, and $\omega_{z'}$ are in the same ratios as $I_{x'}$, $I_{y'}$, and $I_{z'}$. The other special case is when $\boldsymbol{\omega}$ is collinear with a principal axis of inertia, which means that two of the angular velocity components in Eq. 20-7 are zero, as in Chapters 17 through 19.

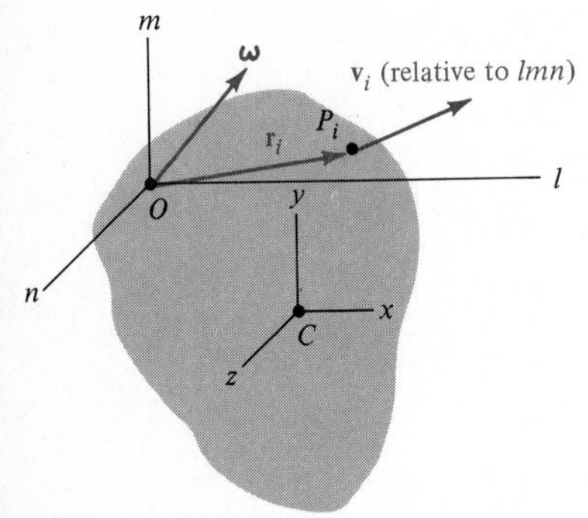

FIGURE 20-4

Motion of a rigid body about a fixed point

Angular Momentum about an Arbitrary Fixed Point

The concepts of analysis described above are also applicable when the rigid body is rotating about an arbitrary fixed point O as in Fig. 20-4. Here the frame lmn is attached at the fixed point which is the most convenient for reference. The total angular momentum of the body about point O is obtained similarly to Eq. 20-1,

$$\mathbf{H}_O = \sum_{i=1}^{n} [\mathbf{r}_i \times (\boldsymbol{\omega} \times \mathbf{r}_i)m_i] \qquad \boxed{20\text{-}8}$$

where $\mathbf{r}_i$ is the position vector of particle P_i in the fixed lmn frame. The lmn components of the total angular momentum $\mathbf{H}_O$ are obtained using the same procedure as for the centroidal reference frame xyz,

$$\boxed{\begin{aligned} H_l &= I_l\omega_l - I_{lm}\omega_m - I_{ln}\omega_n \\ H_m &= -I_{ml}\omega_l + I_m\omega_m - I_{mn}\omega_n \\ H_n &= -I_{nl}\omega_l - I_{nm}\omega_m + I_n\omega_n \end{aligned}} \qquad \boxed{20\text{-}9}$$

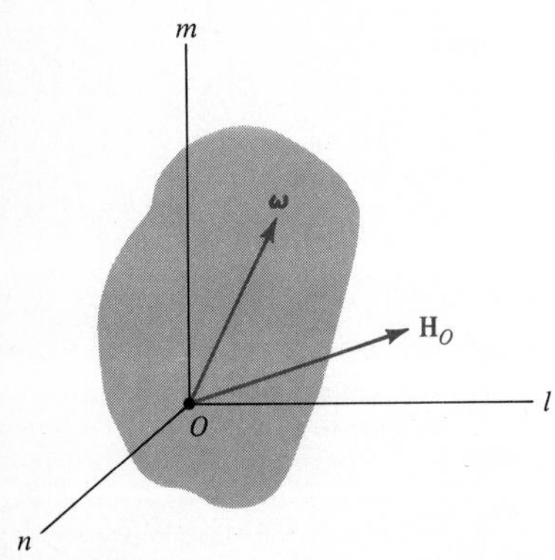

FIGURE 20-5

Generally the vectors ω and $\mathbf{H}_O$ are in different directions

The mass moments of inertia I_l, I_m, I_n, and the mass products of inertia I_{lm}, I_{mn}, I_{ln} are evaluated with respect to the lmn axes according to the definition of these quantities. Note that Eq. 20-9 is identical to Eq. 20-4 when the *lmn* axes are attached at the center of mass and the body rotates about that point. In the general case of rotation about an arbitrary fixed point O, the vectors ω and $\mathbf{H}_O$ are again possibly in different directions, as illustrated in Fig. 20-5.

20-2 IMPULSE AND MOMENTUM OF A RIGID BODY

The results obtained in Secs. 16-7 and 16-8 concerning the momenta of a system of particles are valid for a rigid body. The linear momentum of the body is given by

$$\mathbf{G} = m\mathbf{v}_C \qquad \text{20-10}$$

where m is the total mass of the body and $\mathbf{v}_C$ is the velocity of the center of mass. In a general case the total momentum of the body consists of linear and angular momenta, stated as

$$\text{system momenta} = \begin{cases} \text{linear momentum of mass center } (\mathbf{G}) \\ \text{angular momentum about mass center } (\mathbf{H}_C) \end{cases} \qquad \text{20-11}$$

Remember that G and H_C have different units. The vector $\mathbf{H}_C$ is determined from Eq. 20-4 or 20-7. The vectors $\mathbf{G}$ and $\mathbf{H}_C$ are analogous to a force and a couple, respectively, and this is advantageous in solving problems where the principle of impulse and momentum can be used. This is based on the concept that changes in momentum are easily determined when they are caused by external impulses. In all such cases the final total momentum of a system equals the initial total momentum plus the impulses during the interval considered. From time 1 to time 2,

$$\mathbf{G}_2 = \mathbf{G}_1 + (\text{external linear impulses})\Big|_1^2 \qquad \text{20-12a}$$

$$\mathbf{H}_{C_2} = \mathbf{H}_{C_1} + (\text{external angular impulses})\Big|_1^2 \qquad \text{20-12b}$$

These are illustrated for the simple case in Fig. 20-6, where the body is initially at rest and $\mathbf{F}$ acts in the y direction at point A of coordinates $(-a, -b, c)$.

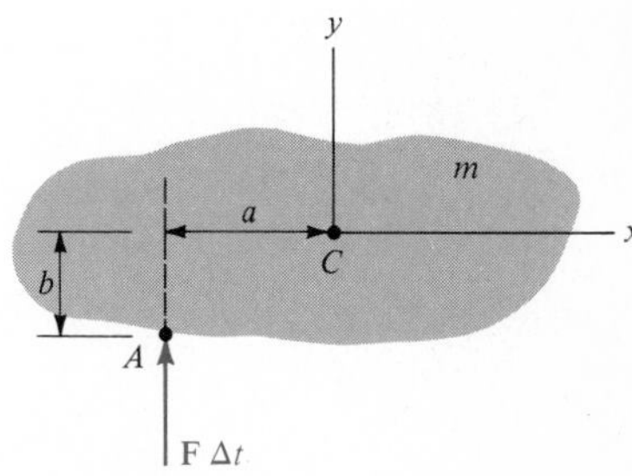

FIGURE 20-6

Example for the analysis of impulse and momentum

LINEAR IMPULSE AND MOMENTA

$$mv_{2_x} = mv_{1_x} + F_x\,\Delta t \qquad mv_{2_y} = mv_{1_y} + F_y\,\Delta t$$

$$mv_{2_x} = m(0) + (0)\,\Delta t = 0 \qquad mv_{2_y} = m(0) + F\,\Delta t = F\,\Delta t$$

$$H_{x_2} = H_{x_1} - F(c)\,\Delta t$$

$$H_{x_2} = 0 - F(c)\,\Delta t = -F(c)\,\Delta t$$ (the minus sign represents a moment of **F** in the $-\mathbf{i}$ direction since c is positive; check using the right-hand rule)

$$H_{y_2} = H_{y_1} + F(0)\,\Delta t = 0$$ (since **F** has no moment about the y axis, and $H_{y_1} = 0$)

$$H_{z_2} = H_{z_1} - F(a)\,\Delta t = -F(a)\,\Delta t$$

The angular velocity components can be determined from the angular momentum components using Eq. 20-4 and solving the equations above simultaneously. When a rigid body is rotating about a noncentroidal fixed point O, the impulse of the reaction must be included in the analysis if the center of mass is used for reference. It is generally more convenient to use the fixed point O for reference which requires taking moments of the momenta and of the impulses about O. The reaction at O has no moment about O and thus it is ignored. In this case Eq. 20-9 should be used to relate the angular velocities and momenta.

20-3 KINETIC ENERGY OF A RIGID BODY IN THREE-DIMENSIONAL MOTION

The kinetic energy of a rigid body in general motion is derived using the method developed in Sec. 16-4. The total kinetic energy T of a system of particles is given according to Eq. 16-13,

$$T = \underbrace{\frac{1}{2} m v_C^2}_{\text{motion of total mass imagined to be concentrated at mass center}} + \underbrace{\frac{1}{2} \sum_{i=1}^{n} m_i v_i'^2}_{\text{motion of all particles relative to mass center}} \tag{20-13}$$

where v_C is the speed of the mass center and v_i' is the speed of particle P_i in the xyz frame, which is attached at the center of mass and has a fixed orientation (Fig. 20-7). The translational part $(\frac{1}{2} m v_C^2)$ of the kinetic energy is in the simplest possible form, but the second part involving a summation over all particles can be further simplified for rigid bodies. This latter part of the kinetic energy can result only from the particles' rotation about the center of mass in the case of a rigid body. Therefore, $\mathbf{v}'_i$ may be replaced by $\boldsymbol{\omega} \times \mathbf{r}_i$, and the last term in Eq. 20-13 can be written as

$$T_{\text{rot}} = \frac{1}{2} \sum_{i=1}^{n} m_i v_i'^2 = \frac{1}{2} \sum_{i=1}^{n} m_i(\boldsymbol{\omega} \times \mathbf{r}'_i)^2 \qquad \boxed{20\text{-}14}$$

where $\boldsymbol{\omega} = \omega_x\mathbf{i} + \omega_y\mathbf{j} + \omega_z\mathbf{k}$ and $\mathbf{r}'_i = x_i\mathbf{i} + y_i\mathbf{j} + z_i\mathbf{k}$. Working with the rectangular components of the vector product and integrating over the whole body, Eq. 20-14 is restated as

$$T_{\text{rot}} = \frac{1}{2} \int [(\omega_y z - \omega_z y)^2 + (\omega_z x - \omega_x z)^2 + (\omega_x y - \omega_y x)^2]\, dm$$

After squaring and rearranging,

$$\begin{aligned} 2T_{\text{rot}} &= \omega_x^2 \int (y^2 + z^2)\, dm + \omega_y^2 \int (x^2 + z^2)\, dm \\ &\quad + \omega_z^2 \int (x^2 + y^2)\, dm - 2\omega_x\omega_y \int xy\, dm \\ &\quad - 2\omega_y\omega_z \int yz\, dm - 2\omega_z\omega_x \int zx\, dm \end{aligned}$$

The integrals are recognized as the centroidal mass moments of inertia and products of inertia used in deriving Eq. 20-4. Thus, the total kinetic energy of a rigid body can be expressed as

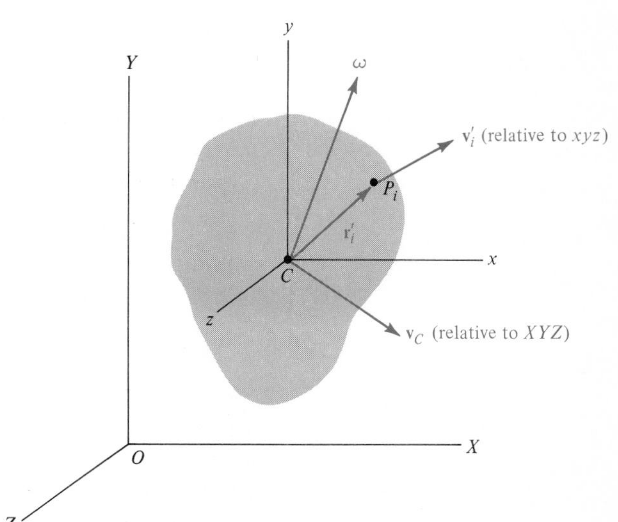

FIGURE 20-7
Schematic for the analysis of kinetic energy

$$T = \frac{1}{2}mv^2 + \frac{1}{2}(I_x\omega_x^2 + I_y\omega_y^2 + I_z\omega_z^2) - I_{xy}\omega_x\omega_y - I_{yz}\omega_y\omega_z - I_{zx}\omega_z\omega_x \tag{20-15}$$

When the chosen xyz frame coincides with the principal axes of inertia $x'y'z'$, Eq. 20-15 becomes

$$T = \frac{1}{2}mv_C^2 + \frac{1}{2}(I_{x'}\omega_{x'}^2 + I_{y'}\omega_{y'}^2 + I_{z'}\omega_{z'}^2) \tag{20-16}$$

where $\boldsymbol{\omega}$ is the same for any reference frame but its components $\omega_{x'}$, and so on, depend on the appropriate axes. An equivalent, concise form of Eqs. 20-15 and 20-16 is

$$\boxed{T = \frac{1}{2}mv_C^2 + \frac{1}{2}\boldsymbol{\omega} \cdot \mathbf{H}_C} \tag{20-17}$$

where $\mathbf{H}_C$ is determined from Eq. 20-4 or 20-7, as necessary.

Kinetic Energy of a Rigid Body Rotating about a Fixed Point

A special case of three-dimensional motion is when the body has a fixed point. This means that all particles of the body are in rotational motion about that point. Accordingly, the total kinetic energy of the body may be determined by evaluating only the rotational kinetic energy using the fixed point O for reference. Assume that a set of axes lmn is attached at the fixed point O. In this case Eq. 20-15 becomes

$$T = \frac{1}{2}(I_l\omega_l^2 + I_m\omega_m^2 + I_n\omega_n^2) - I_{lm}\omega_l\omega_m - I_{mn}\omega_m\omega_n - I_{nl}\omega_n\omega_l \tag{20-18}$$

With principal axes $l'm'n'$ at the fixed point O, Eq. 20-18 is further simplified to give

$$T = \frac{1}{2}(I_{l'}\omega_{l'}^2 + I_{m'}\omega_{m'}^2 + I_{n'}\omega_{n'}^2) \tag{20-19}$$

The most concise form of Eqs. 20-18 and 20-19 is

$$\boxed{T = \frac{1}{2}\boldsymbol{\omega} \cdot \mathbf{H}_O} \tag{20-20}$$

where care must be taken to evaluate $\mathbf{H}_O$ with respect to the fixed point of the rotating body.

EXAMPLE 20-1

The 15-kg wheel W in Fig. 20-8 has a 0.6-m diameter. Assume that the wheel is a homogeneous disk of very small thickness mounted on its axle at point A of coordinate 0.5 m $\mathbf{j}$. The axle is freely pivoted in a ball-and-socket joint at point O. The wheel has angular velocities $\omega_y = 15$ rad/s $\mathbf{j}$ and $\omega_z = 2$ rad/s $\mathbf{k}$. Determine (a) the linear and angular momenta of the wheel about point O, and (b) the kinetic energy of the wheel.

SOLUTION

(a) The linear momentum of the wheel is calculated from Eq. 20-10 involving only the velocity of the center of mass,

$$\mathbf{v}_A = \omega \times \mathbf{r}_{A/O} = \left(15\frac{\text{rad}}{\text{s}}\mathbf{j} + 2\frac{\text{rad}}{\text{s}}\mathbf{k}\right) \times (0.5 \text{ m } \mathbf{j})$$

$$= -1\frac{\text{m}}{\text{s}}\mathbf{i}$$

$$\mathbf{G} = m\mathbf{v}_A = 15 \text{ kg}\left(-1\frac{\text{m}}{\text{s}}\mathbf{i}\right) = -15\frac{\text{kg}\cdot\text{m}}{\text{s}}\mathbf{i}$$

The angular momentum of the wheel at this instant is calculated from Eq. 20-7,

$$\mathbf{H}_O = I_x\omega_x + I_y\omega_y + I_z\omega_z$$

Using $\omega_x = 0$, $I_y = \frac{1}{2}mr^2$, and $I_z = (mr^2/4) + m(OA)^2$ gives

$$\mathbf{H}_O = 0\mathbf{i} + \frac{1}{2}(15 \text{ kg})(0.3 \text{ m})^2(15 \text{ rad/s})\mathbf{j}$$

$$+ 15 \text{ kg}\left[\frac{(0.3 \text{ m})^2}{4} + (0.5 \text{ m})^2\right](2 \text{ rad/s})\mathbf{k}$$

$$= (10.1\mathbf{j} + 8.2\mathbf{k}) \text{ kg}\cdot\text{m}^2/\text{s}$$

FIGURE 20-8

Multiple-exposure photograph of a wheel in a complex test. (Courtesy MTS Systems Corporation, Minneapolis, Minnesota)

(b) Using the principal body axes (x', y', z') attached at A, which are parallel to the reference axes (x, y, z), Eq. 20-16 can be used to determine the kinetic energy,

$$T = \frac{1}{2}mv_O^2 + \frac{1}{2}(I_{x'}\omega_{x'}^2 + I_{y'}\omega_{y'}^2 + I_{z'}\omega_{z'}^2)$$

$$= \frac{1}{2}(15\text{ kg})(-1\text{ m/s})^2 + \frac{1}{2}\left[0 + \frac{1}{2}(15\text{ kg})(0.3\text{ m})^2(15\text{ rad/s})^2\right.$$

$$\left. + \frac{1}{4}(15\text{ kg})(0.3\text{ m})^2(2\text{ rad/s})^2\right]$$

$$= 7.5\frac{\text{kg}\cdot\text{m}^2}{\text{s}^2} + \frac{1}{2}(151.9 + 1.35)\text{ kg}\cdot\text{m}^2/\text{s}^2$$

$$= 84.13\frac{\text{kg}\cdot\text{m}^2}{\text{s}^2} = 84.13\text{ N}\cdot\text{m}$$

Alternatively, the xyz axes can be used as the reference frame, using Eq. 20-20,

$$T = \frac{1}{2}\omega\cdot\mathbf{H}_O = \frac{1}{2}(15\mathbf{j} + 2\mathbf{k})\text{ rad/s}\cdot(10.1\mathbf{j} + 8.2\mathbf{k})\text{ kg}\cdot\text{m}^2/\text{s}$$

$$= \frac{1}{2}(151.5 + 16.4)\text{ kg}\cdot\text{m}^2/\text{s}^2 = 83.95\text{ N}\cdot\text{m}$$

The difference between the two results for T is due to round-off errors in the two solutions.

EXAMPLE 20-2

In Ex. 20-1 a force $\mathbf{F} = 500\text{ N}\,\mathbf{j} + 800\text{ N}\,\mathbf{k}$ is applied to the wheel W at point A for a time of 0.03 s. $\mathbf{F}$ is acting in the yz plane. Determine the angular momentum of the wheel about point O just after the action of the force $\mathbf{F}$.

SOLUTION

Since the rotation is about a fixed, but noncentroidal point, the new angular momentum about point O can be calculated using Eq. 20-12. For this, the initial value of $\mathbf{H}_O$ calculated in Ex. 20-1 is used, and it is assumed that the change in orientation of the wheel is negligible during the short period considered ($I_{x'}, I_{y'}, I_{z'}$ remain constant).

$$\begin{aligned}\mathbf{H}_{O_2} &= \mathbf{H}_{O_1} + \mathbf{M}_O\,\Delta t\\ &= (10.1\mathbf{j} + 8.2\mathbf{k})\text{ kg}\cdot\text{m}^2/\text{s} + (0.5\text{ m}\,\mathbf{j})\\ &\quad\times(500\text{ N}\,\mathbf{j} + 800\text{ N}\,\mathbf{k})(0.03\text{ s})\\ &= (10.1\mathbf{j} + 8.2\mathbf{k})\text{ kg}\cdot\text{m}^2/\text{s} + 12\text{ N}\cdot\text{m}\cdot\text{s}\,\mathbf{i}\\ &= (12\mathbf{i} + 10.1\mathbf{j} + 8.2\mathbf{k})\text{ kg}\cdot\text{m}^2/\text{s}\end{aligned}$$

EXAMPLE 20-3

A crankshaft is modeled as a thin rod bent into the plane figure shown in Fig. 20-9a. It weighs 0.3 lb per linear inch and rotates freely in the bearings at A and B with $\omega_x = 1000$ rpm. Determine the angular momenta of the crankshaft about points A and B.

SOLUTION

First the relevant moments and products of inertia are determined using the parallel-axis theorem for a composite body. The crankshaft is analyzed in sections as shown in Fig. 20-9b. Using the subscript C for centroidal quantities yields

$$I_x = \sum (I_{C_x} + r_{C}^2 m)_i = \Bigg\{ (0)_{\mathrm{I}} + \left[\frac{1}{12}(6\text{ in.})^2 + (3\text{ in.})^2\right](1.8\text{ lb})_{\mathrm{II}}$$

$$+ [0 + (6\text{ in.})^2(1.2\text{ lb})]_{\mathrm{III}}$$

$$+ \left[\frac{1}{12}(3.6\text{ lb})(12\text{ in.})^2 + 0\right]_{\mathrm{IV}}$$

$$+ [0 + (-6\text{ in.})^2(1.2\text{ lb})]_{\mathrm{V}}$$

$$+ \left[\frac{1}{12}(6\text{ in.})^2 + (-3\text{ in.})^2\right](1.8\text{ lb})_{\mathrm{VI}}$$

$$+ (0)_{\mathrm{VII}} \Bigg\} \frac{1}{32.2\text{ ft/s}^2}$$

$$I_x = 5.36\text{ slug}\cdot\text{in}^2$$

$$(I_{xy})_A = \sum (I_{Cxy} + xym)_i = \{(0)_{\mathrm{I}} + [0 + (8\text{ in.})(3\text{ in.})(1.8\text{ lb})]_{\mathrm{II}}$$

$$+ [0 + (10\text{ in.})(6\text{ in.})(1.2\text{ lb})]_{\mathrm{III}}$$

$$+ (0 + (12\text{ in.})(0\text{ in.})(3.6\text{ lb})]_{\mathrm{IV}}$$

$$+ [0 + (14\text{ in.})(-6\text{ in.})(1.2\text{ lb}]_{\mathrm{V}}$$

$$+ [0 + (16\text{ in.})(-3\text{ in.})(1.8\text{ lb})]_{\mathrm{VI}}$$

$$+ (0)_{\mathrm{VII}} \Big\} \frac{1}{32.2\text{ ft/s}^2}$$

$$(I_{xy})_A = -2.24\text{ slug}\cdot\text{in}^2$$

Using Eq. 20-9 for the angular momentum about A with $\omega_x = 105$ rad/s, and $\omega_y = \omega_z = 0$ gives

$$H_x = I_x\omega_x - I_{xy}\omega_y - I_{xz}\omega_z$$

$$= (5.36\text{ slug}\cdot\text{in}^2)(105\text{ rad/s}) - 0 - 0$$

$$= 562\,\frac{\text{slug}\cdot\text{in}^2}{\text{s}}\left(\frac{1\text{ ft}^2}{144\text{ in}^2}\right) = 3.90\text{ lb}\cdot\text{ft}\cdot\text{s}$$

$$
\begin{aligned}
H_y &= -I_{xy}\omega_x + I_y\omega_y - I_{yz}\omega_z \\
&= -(-2.24\ \text{slug}\cdot\text{in}^2)(105\ \text{rad/s}) + 0 - 0 \\
&= 1.63 \frac{\text{slug}\cdot\text{ft}^2}{\text{s}} = 1.63\ \text{lb}\cdot\text{ft}\cdot\text{s}
\end{aligned}
$$

$$
H_z = 0 \qquad \text{(since } I_{xz} \text{ is zero by definition)}
$$

To calculate the angular momentum about B, the moments and products of inertia must be determined with respect to a reference system $x'y'z'$ with its origin at B. From symmetry, however,

$$
I_{x'} = I_x = 5.36\ \text{slug}\cdot\text{in}^2
$$

$$
I_{x'y'} = I_{xy} = -2.23\ \text{slug}\cdot\text{in}^2
$$

Thus,

$$
H_{x'} = 3.90\ \text{lb}\cdot\text{ft}\cdot\text{s}
$$

$$
H_{y'} = 1.63\ \text{lb}\cdot\text{ft}\cdot\text{s}
$$

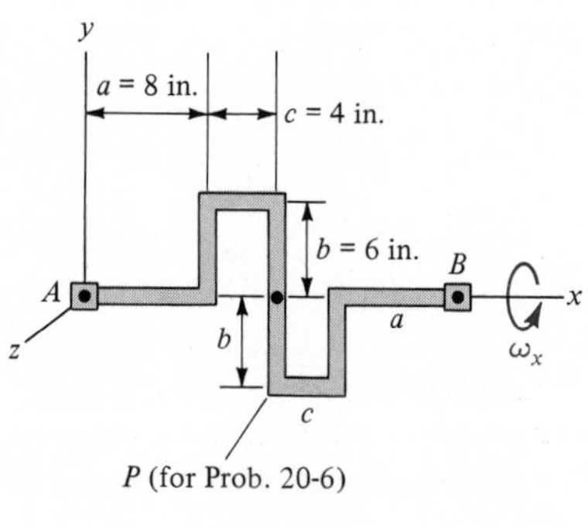

(a)

(b)

FIGURE 20-9

Model of a crankshaft

PROBLEMS

20–1 A slender rod l of 2 lb weight and 2 ft length is firmly attached to a vertical shaft that rotates in bearings A and B at $\omega_y = 20$ rad/s. Determine the angular momentum of the rod about point O for $\theta = 30°$.

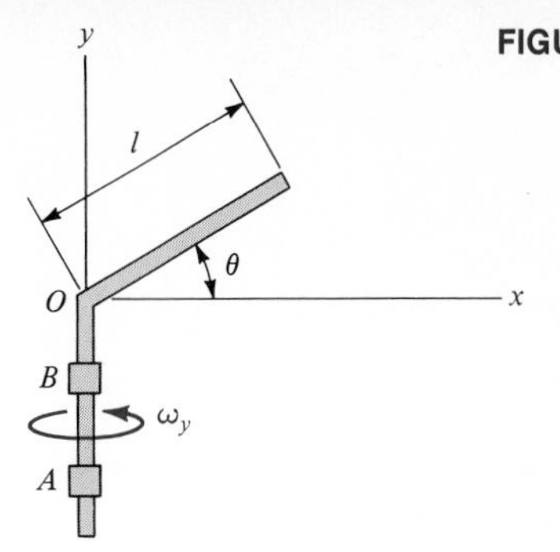

FIGURE P20-1

20–2 Determine the kinetic energy of the rod l in Prob. 20–1 for $\theta = 45°$.

20–3 A force $\mathbf{F} = 30$ lb $\mathbf{k}$ is acting for 0.01 s at the free end of the rod l in Prob. 20–1 when the rod is at rest in the xy plane. Determine the angular momentum of the rod about point O after the action of the force, assuming now that O is a ball-and-socket joint.

20–4 A slender rod of 1 kg mass and $2l = 1$ m length is firmly attached to a horizontal shaft that rotates in bearings A and B at $\omega_x = 10$ rad/s. Calculate the angular momentum of the rod about point O for $\theta = 30°$. θ is measured in the xy plane.

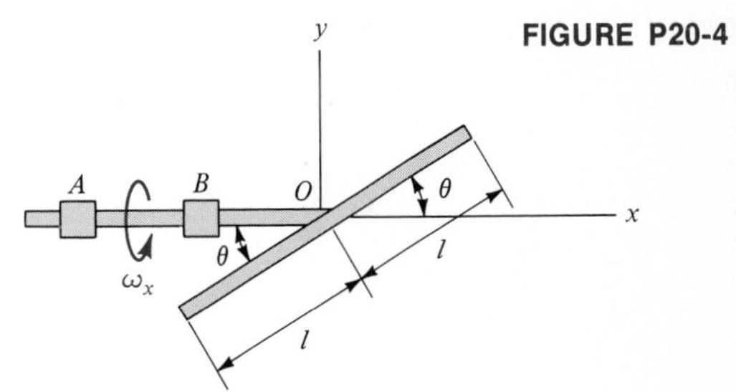

FIGURE P20-4

20–5 A force $\mathbf{F} = -2$ kN $\mathbf{k}$ is acting for 0.002 s at the upper free end of the rod in Prob. 20–4 when the rod is at rest in the xy plane. Calculate the final angular momentum of the rod about point O, assuming now that O is a ball-and-socket joint.

20–6 Assume that the crankshaft in Ex. 20-3 is changed by adding a 0.1-lb weight at point P. Determine the angular momenta of the system about points A and B.

FIGURE P20-7

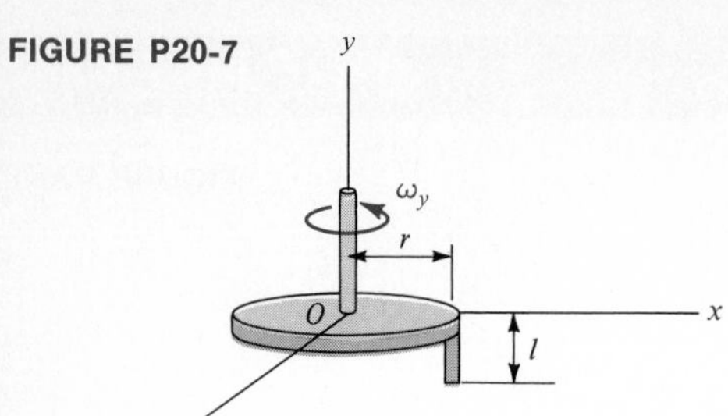

20–7 A thin peg of mass m and length l is mounted at a distance r from the center O of a uniform disk. The peg is parallel to the shaft of the disk, which is rotating at angular speed ω_y. Determine the angular momentum of the peg about point O when it is in the xy plane.

FIGURE P20-8

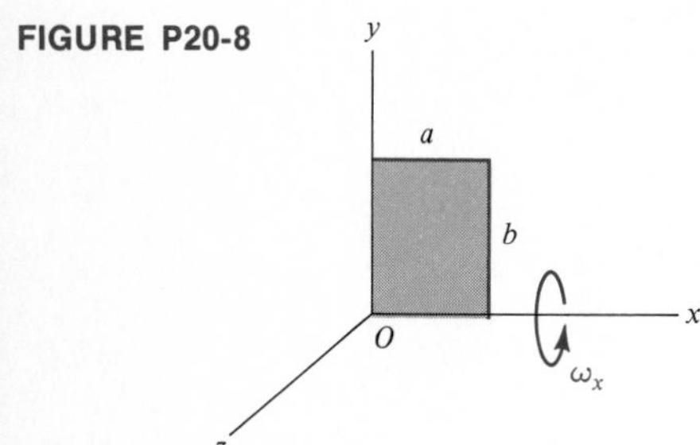

20–8 A uniform thin plate of mass m is rotating about the fixed x axis at angular speed ω_x. Calculate the angular momentum of the plate about point O and the kinetic energy of the plate.

FIGURE P20-9

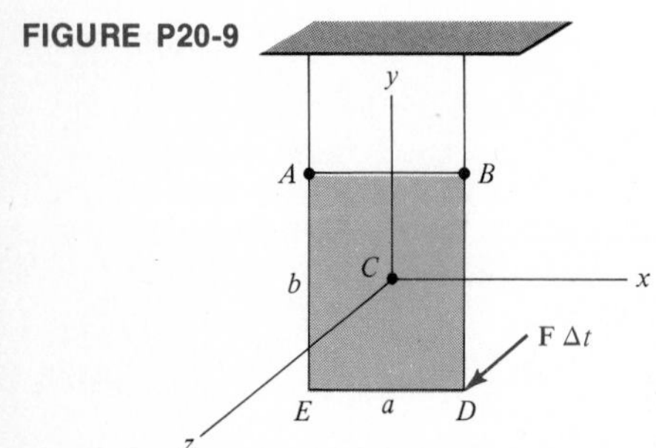

20–9 A uniform thin plate of mass m is suspended from two inextensible wires at A and B. An impulse $F\mathbf{k}\,\Delta t$ is applied at corner D. Determine the velocities of points C, D, and E immediately after the impact.

20–10 Calculate the tensions in wires A and B in Prob. 20–9 immediately after the impact. Note that all forces acting on the plate must be considered during the same time interval Δt.

20–11 Calculate the kinetic energy of the plate in Prob. 20–9 after the impact.

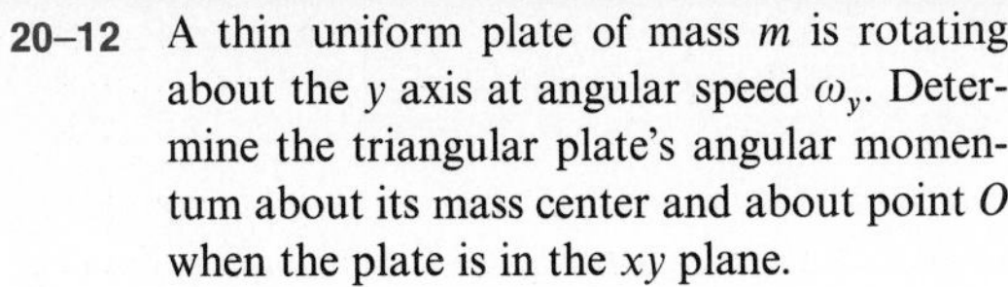

20–12 A thin uniform plate of mass m is rotating about the y axis at angular speed ω_y. Determine the triangular plate's angular momentum about its mass center and about point O when the plate is in the xy plane.

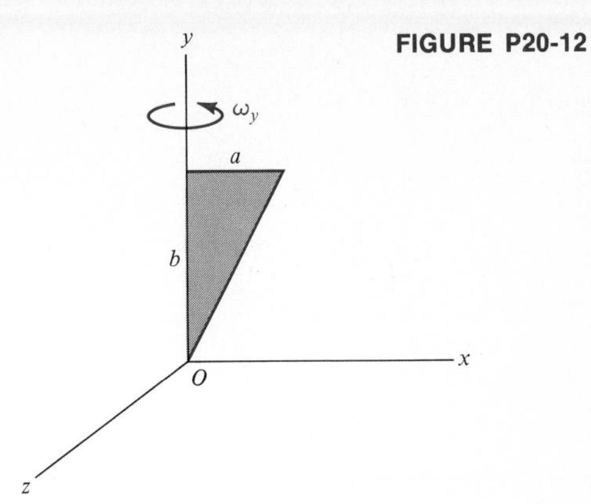

FIGURE P20-12

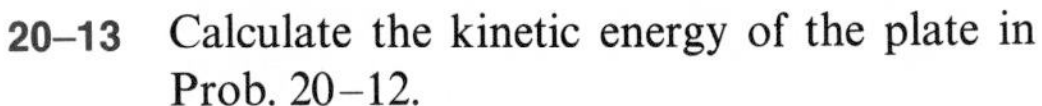

20–13 Calculate the kinetic energy of the plate in Prob. 20–12.

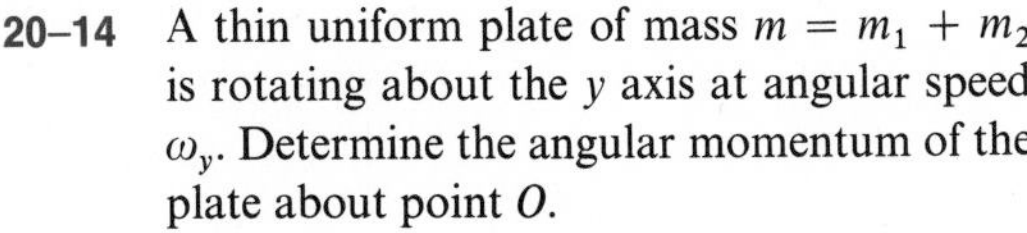

20–14 A thin uniform plate of mass $m = m_1 + m_2$ is rotating about the y axis at angular speed ω_y. Determine the angular momentum of the plate about point O.

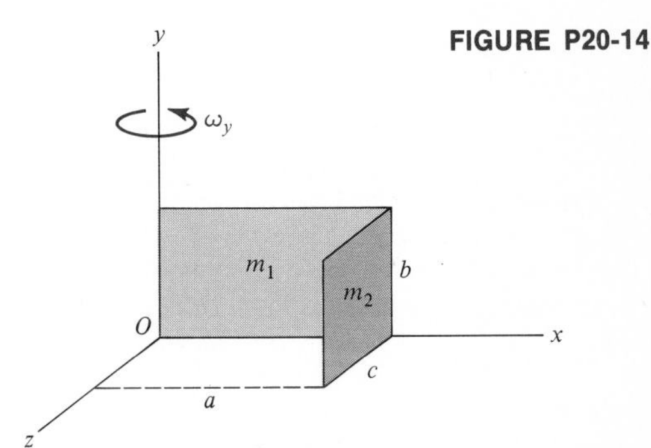

FIGURE P20-14

20–15 Calculate the kinetic energy of the bent plate in Prob. 20–14.

20–16 The photograph shows a working model of a retractable-bladed Savonius rotor for a modern windmill. Assume that one of the two blades breaks off, leaving a half-cylindrical shell rotating about the y axis at angular speed ω_y as sketched in Fig. 20-16b. The thin shell has a mass m, height h, and diameter d. Determine its angular momentum about point O for the position shown.

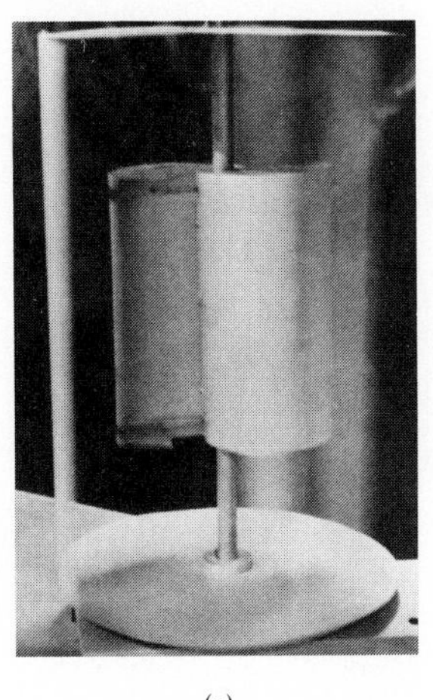

(a)

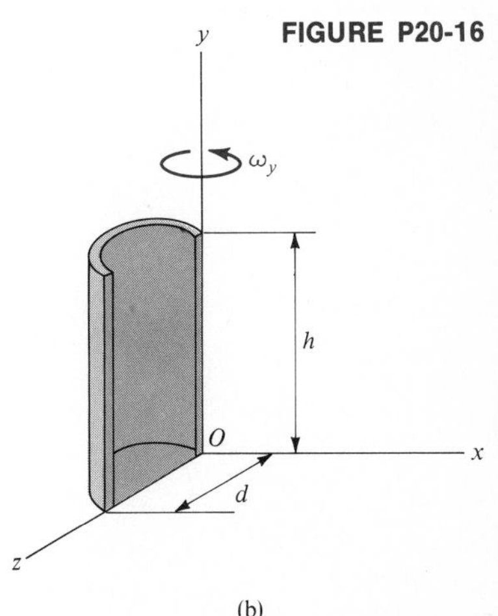

(b)

FIGURE P20-16

20–17 Calculate the kinetic energy of the single blade in Prob. 20–16.

FIGURE P20-18

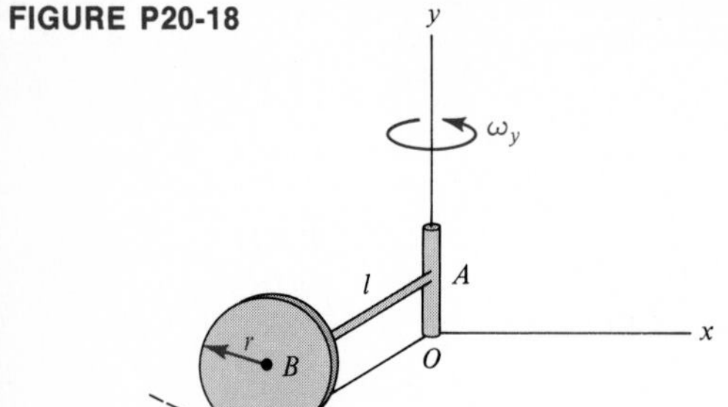

20–18 A thin uniform disk of mass m is rolling on the xz plane at angular speed ω_y about the y axis. Its shaft of length l is always parallel to the xz plane. Determine the angular momentum of the disk about point A.

20–19 Calculate the kinetic energy of the disk in Prob. 20–18.

FIGURE P20-20

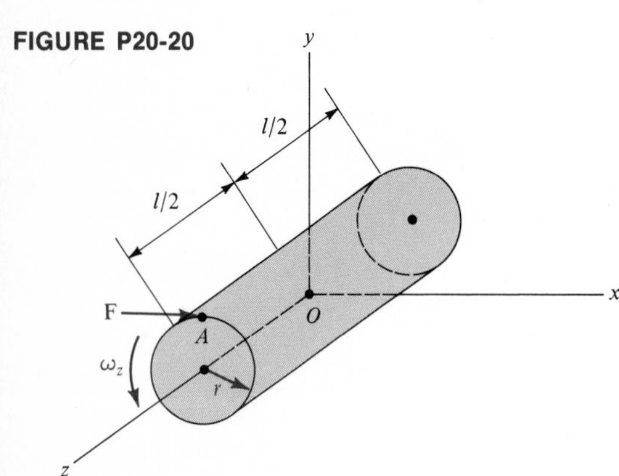

20–20 A 2000-kg spacecraft is modeled as a homogeneous cylinder of length $l = 6$ m and radius $r = 1$ m. It is spinning about the z axis at $\omega_z = 100$ rpm when a force $\mathbf{F} = 200$ N $\mathbf{i}$ is applied to it for 0.1 s by a control jet at point A in the yz plane. Determine the angular momentum of the cylinder after the impulse.

20–21 Calculate the final kinetic energy of the cylinder in Prob. 20–20.

FIGURE P20-22

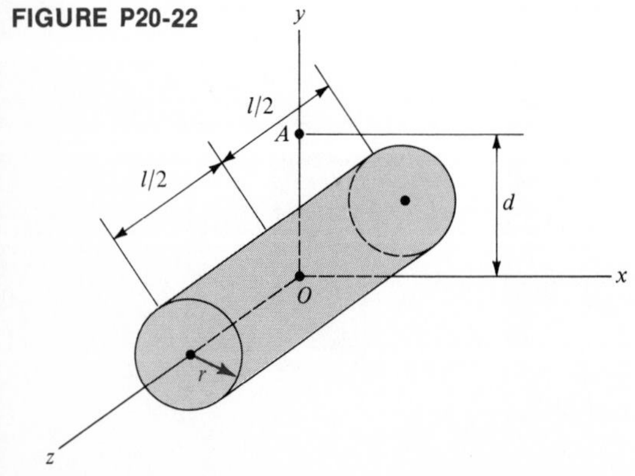

20–22 A rotor of a jet engine is modeled as a uniform cylinder of mass m, length l, and radius r. It has angular speeds ω_x, ω_y, and ω_z about the respective axes in the position shown. Determine the angular momentum about point A.

EQUATIONS OF MOTION IN THREE DIMENSIONS

The fundamental equations of motion given by Eqs. 16-20 and 16-29 are valid for any motion of a rigid body, so these are restated here:

$$\sum \mathbf{F} = \dot{\mathbf{G}} = m\mathbf{a}_C \qquad (20\text{-}21)$$

$$\sum \mathbf{M}_C = \dot{\mathbf{H}}_C \qquad (20\text{-}22)$$

where $\mathbf{a}_C$ is the acceleration of the center of mass and $\mathbf{H}_C$ is the angular momentum of the body about its center of mass. Equation 20-22 may also be used with any fixed point O for a reference if both $\sum \mathbf{M}$ and $\mathbf{H}$ are evaluated about that point.

The greatest difficulty in using these apparently simple equations is in obtaining derivatives of the angular momentum $\mathbf{H}$. Recall that $\mathbf{H}$ was defined for a particular position of the body with respect to axes of fixed orientation. At a different time and position, the body's mass moments and products of inertia may be different with respect to the same fixed frame. Thus, differentiation of $\mathbf{H}$ is possible only if it is given as a function of time. This problem is solved using the following approach.

1. Attach axes XYZ at the mass center C of the body. These axes have fixed orientations in space regardless of the body's motion. The origin of XYZ remains at C (Fig. 20-10) during the motion.
2. Attach axes xyz at the mass center of the body and in fixed orientations with respect to the body. It is often simplest to align the xyz axes with the principal inertia axes of the body.
3. Establish $\boldsymbol{\omega}$, the angular velocity of the body and of the xyz axes (since they are fixed in the body) with respect to the XYZ frame. This $\boldsymbol{\omega}$ has components ω_x, ω_y, and ω_z along the xyz axes.
4. The angular momentum vector $\mathbf{H}_C$ is defined here with respect to the

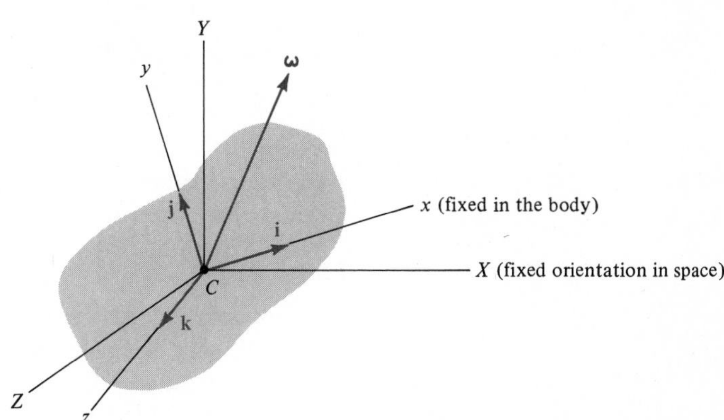

FIGURE 20-10

Schematic for deriving the equations of motion

fixed XYZ frame (similar to the derivation in Sec. 20-1, where xyz is fixed) and is determined using Eq. 20-4. The vector $\mathbf{H}_C$ can be resolved into components along the rotating xyz axes in Fig. 20-10 using **i**, **j**, and **k** unit vectors in the xyz frame.

5. The time derivative of the vector $\mathbf{H}_C$ in the rotating xyz frame using the corresponding unit vectors is

$$(\dot{\mathbf{H}}_C)_{xyz} = \dot{H}_x\mathbf{i} + \dot{H}_y\mathbf{j} + \dot{H}_z\mathbf{k} \tag{20-23}$$

6. The time derivatives of the angular momentum vector $\mathbf{H}_C$ in the fixed XYZ and rotating xyz frames are related to one another using Eq. 17-30,

$$(\dot{\mathbf{H}}_C)_{XYZ} = (\dot{\mathbf{H}}_C)_{xyz} + \boldsymbol{\omega} \times \mathbf{H}_C \tag{20-24a}$$

In more detail, using

$$\begin{aligned}\dot{\mathbf{H}}_C &= \frac{d}{dt}(H_x\mathbf{i} + H_y\mathbf{j} + H_z\mathbf{k}) \\ &= \dot{H}_x\mathbf{i} + \dot{H}_y\mathbf{j} + \dot{H}_z\mathbf{k} + H_x\dot{\mathbf{i}} + H_y\dot{\mathbf{j}} + H_z\dot{\mathbf{k}}\end{aligned}$$

gives

$$\underbrace{(\dot{\mathbf{H}}_C)_{XYZ}}_{\substack{\text{rate of}\\ \text{change in}\\ \text{magnitude}\\ \text{and}\\ \text{direction}\\ \text{of } \mathbf{H}_C}} = \underbrace{(\dot{H}_x\mathbf{i} + \dot{H}_y\mathbf{j} + \dot{H}_z\mathbf{k})}_{\substack{\text{rate of change in}\\ \text{magnitude of } \mathbf{H}_C}} + \underbrace{\boldsymbol{\omega} \times (H_x\mathbf{i} + H_y\mathbf{j} + H_z\mathbf{k})}_{\substack{\text{rate of change in}\\ \text{direction of } \mathbf{H}_C}} \tag{20-24b}$$

where $\boldsymbol{\omega} = \omega_x\mathbf{i} + \omega_y\mathbf{j} + \omega_z\mathbf{k}$ is the angular velocity of the body and of the xyz frame fixed in the body. Note that **i**, **j**, and **k** are unit vectors along the instantaneous xyz axes.

Equation 20-22 can be written using Eq. 20-24 as

$$\boxed{\sum \mathbf{M}_C = (\dot{\mathbf{H}}_C)_{xyz} + \boldsymbol{\omega} \times \mathbf{H}_C} \tag{20-25}$$

In expanded form,

$$\begin{aligned}\sum \mathbf{M}_C = &(\dot{H}_x + \omega_y H_z - \omega_z H_y)\mathbf{i} \\ &+ (\dot{H}_y + \omega_z H_x - \omega_x H_z)\mathbf{j} \\ &+ (\dot{H}_z + \omega_x H_y - \omega_y H_x)\mathbf{k}\end{aligned} \tag{20-26}$$

In certain problems it may be more advantageous to express these terms using components along a reference frame other than the xyz or XYZ frame. Assume that the new reference frame has an angular velocity $\boldsymbol{\Omega}$, and $\mathbf{M}_C$, $\mathbf{H}_C$, and $\dot{\mathbf{H}}_C$ are expressed using **i**, **j**, and **k** components along the new reference axes. The terms in the unit vector derivatives $\dot{\mathbf{i}}$, $\dot{\mathbf{j}}$, and $\dot{\mathbf{k}}$ above can be expressed as $\boldsymbol{\Omega} \times \mathbf{H}_C$, and Eq. 20-25 becomes

$$\sum \mathbf{M}_C = (\dot{\mathbf{H}}_C)_{xyz} + \mathbf{\Omega} \times \mathbf{H}_C \qquad \text{20-27}$$

where $\mathbf{\Omega}$ is the angular velocity of the rotating xyz frame relative to the XYZ frame. $(\dot{\mathbf{H}}_C)_{xyz}$ is the rate of change of $\mathbf{H}_C$ in the rotating xyz frame, and is evaluated using Eqs. 20-4 and 20-23. Note that both Eqs. 20-25 and 20-27 may be used in reference to an arbitrary fixed point O besides the mass center C, if done consistently.

20-5 EULER'S EQUATIONS OF MOTION

The xyz axes fixed in the body and rotating with it may be chosen to coincide with the principal axes of inertia $x'y'z'$ with respect to the mass center C or a fixed point O. In that case the inertia matrix contains only $I_{x'} = I_x$, $I_{y'} = I_y$, and $I_{z'} = I_z$ according to Eq. 20-6. With these and Eq. 20-7, Eq. 20-26 can be written as three scalar equations,

$$\begin{aligned} \sum M_x &= I_x\dot{\omega}_x - (I_y - I_z)\omega_y\omega_z \\ \sum M_y &= I_y\dot{\omega}_y - (I_z - I_x)\omega_z\omega_x \\ \sum M_z &= I_z\dot{\omega}_z - (I_x - I_y)\omega_x\omega_y \end{aligned} \qquad \text{20-28}$$

These expressions are known as *Euler's equations* after Leonhard Euler (1707–1783), a Swiss mathematician. Note that all moments, moments of inertia, and angular velocities in Eq. 20-28 must be evaluated with respect to the appropriate principal axes of the given body.

20-6 SOLUTION OF BASIC PROBLEMS IN THREE-DIMENSIONAL MOTION

The maximum number of independent scalar equations of motion of a rigid body is six. These are presented here as a group of governing equations for convenience.

$$\sum F_x = ma_{C_x} \quad \sum F_y = ma_{C_y} \quad \sum F_z = ma_{C_z} \qquad \text{20-29}$$

$$\begin{aligned} \sum M_x &= \dot{H}_x + \omega_y H_z - \omega_z H_y \\ \sum M_y &= \dot{H}_y + \omega_z H_x - \omega_x H_z \\ \sum M_z &= \dot{H}_z + \omega_x H_y - \omega_y H_x \end{aligned} \qquad \text{20-30}$$

where a_C refers to the linear acceleration of the center of mass, and the angular momenta are determined according to Eq. 20-4. These equations are valid *in general.* The following categories of problems are stated with a few comments to facilitate the analysis in common situations.

Unconstrained motion. The six governing equations, Eqs. 20-29 and 20-30, should be used with xyz axes attached at the center of mass of the body.

Motion of a body about a fixed point. The governing equations are valid for a body rotating about a noncentroidal fixed point O. The reference axes xyz must go through that fixed point to allow working with a set of moment equations (Eq. 20-30) that do not involve the unknown reaction at O.

Motion of a body about a fixed axis. This is the generalized form of plane motion of an arbitrary rigid body. Such a motion is very common since it includes the plane motion of unbalanced wheels and shafts. For the purpose of simplifying the governing equations, consider a body that is constrained to rotate about the fixed Z axis as in Fig. 20-11. Also, the body cannot slide along Z. The xyz axes are fixed in the body (with $z = Z$, but with point O not necessarily at the mass center) and are rotating with the body at an angular velocity $\boldsymbol{\omega} = \omega\mathbf{k}$. Thus, $a_{C_z} = 0$, and $\omega_x = \omega_y = 0$ for this situation. The xyz components of the angular momentum are obtained from Eq. 20-4,

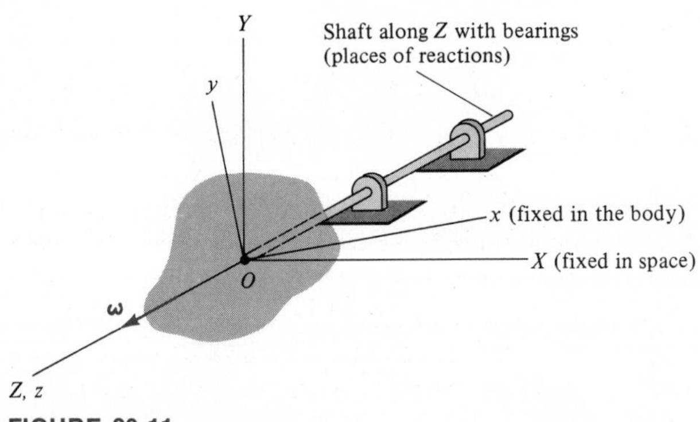

FIGURE 20-11

Rotation about a fixed axis

$$H_x = -I_{xz}\omega_z \qquad H_y = -I_{yz}\omega_z \qquad H_z = I_z\omega_z$$

Consequently, the governing equations of Eq. 20-29 and 20-30 are simplified to

$$\sum F_x = ma_{C_x} \qquad \sum F_y = ma_{C_y} \qquad \sum F_z = 0$$

$$\sum M_x = -I_{xz}\dot{\omega}_z + I_{yz}\omega_z^2$$

$$\sum M_y = -I_{yz}\dot{\omega}_z - I_{xz}\omega_z^2 \qquad \boxed{20\text{-}31}$$

$$\sum M_z = I_z\dot{\omega}_z$$

where $\dot{\omega}_z = \alpha_z$, the angular acceleration about the fixed axis. It is emphasized that Eqs. 20-31 are valid for the special case of motion about a fixed axis only.

In practice it is frequently possible to determine $\dot{\omega}_z$ directly from the last moment equation, and ω_z by integration of $\dot{\omega}_z$. With these, the bearing reactions on the fixed axis of rotation can be calculated.

1. *Dynamically balanced bodies.* The last set of governing equations is further simplified *if the mass distribution of the body is symmetrical with respect to the xy plane.* In that case $I_{xz} = I_{yz} = 0$, and the moment equations become

$$\sum M_x = 0 \qquad \sum M_y = 0 \qquad \sum M_z = I_z\dot{\omega}_z \qquad \boxed{20\text{-}32}$$

This means that, for ω_z = constant, no moments are involved with the rotation of the body in any direction. The body is said to be *dynamically balanced.* (Static balance requires the absence of moments when $\omega_z = 0$.)

2. *Dynamically unbalanced bodies.* A common problem concerning rotating rigid bodies is when $I_{xz} \neq 0$ and $I_{yz} \neq 0$, even with ω_z = constant according to the illustrative situation in Fig. 20-11. In that case,

$$\sum M_x = I_{yz}\omega_z^2 \quad \sum M_y = -I_{xz}\omega_z^2 \quad \sum M_z = 0 \qquad \boxed{20\text{-}33}$$

The mass products of inertia are constant in the rotating *xyz* reference frame, but they are constantly changing in the fixed *XYZ* frame as the body rotates. This creates vibratory reactions in the stationary supporting structure (transmitted through bearings) of the rotating member. It is worth remembering that the peak magnitudes of these reactions are directly proportional to ω^2 according to Eq. 20-33. Dynamic unbalance of machine parts is normally undesirable (requiring corrective measures to rearrange the mass distribution), but sometimes it is a design feature of machines that must produce vibrations.

EXAMPLE 20-4

The turbine disk of a jet engine is modeled as a uniform thin disk of 5 kg mass and 0.2 m radius. It is rotating at constant angular velocity $\omega_z = 3000$ rad/s $\mathbf{k}$ and $\boldsymbol{\omega}_x = -0.5$ rad/s $\mathbf{i}$ when it is in the XY plane (Fig. 20-12). Determine the moment required to maintain the axle of the disk in the YZ plane.

SOLUTION

Using xyz axes coincident with XYZ, the moments of inertia are calculated,

$$I_x = I_y = \frac{1}{4}mr^2 = \frac{1}{4}(5\text{ kg})(0.2\text{ m})^2 = 0.05\text{ kg}\cdot\text{m}^2$$

$$I_z = \frac{1}{2}mr^2 = \frac{1}{2}(5\text{ kg})(0.2\text{ m})^2 = 0.1\text{ kg}\cdot\text{m}^2$$

Since xyz are principal axes of inertia, Euler's equations of motion (Eq. 20-28) can be used. For this problem $\dot{\omega}_x = \dot{\omega}_y = \dot{\omega}_z = \omega_y = 0$; therefore,

$$\sum M_x = I_x\dot{\omega}_x - (I_y - I_z)\omega_y\omega_z = 0 - 0 = 0 \quad \textbf{(a)}$$

$$\begin{aligned}\sum M_y &= I_y\dot{\omega}_y - (I_z - I_x)\omega_z\omega_x \quad \textbf{(b)}\\ &= 0 - (0.1 - 0.05)\text{ kg}\cdot\text{m}^2(3000\text{ rad/s})(-0.5\text{ rad/s})\\ &= 75\text{ kg}\cdot\text{m}^2/\text{s}^2 = 75\text{ N}\cdot\text{m}\end{aligned}$$

$$\sum M_z = I_z\dot{\omega}_z - (I_x - I_y)\omega_x\omega_y = 0 - 0 = 0 \quad \textbf{(c)}$$

Note that even though rotations are occurring about the x and z axes, it is a moment about the Y axis that is necessary for the desired motion! If this moment (Eq. b) were not supplied by forces normal to the shaft in the bearings, the turbine would experience an angular acceleration about the Y axis ($\dot{\omega}_y$).

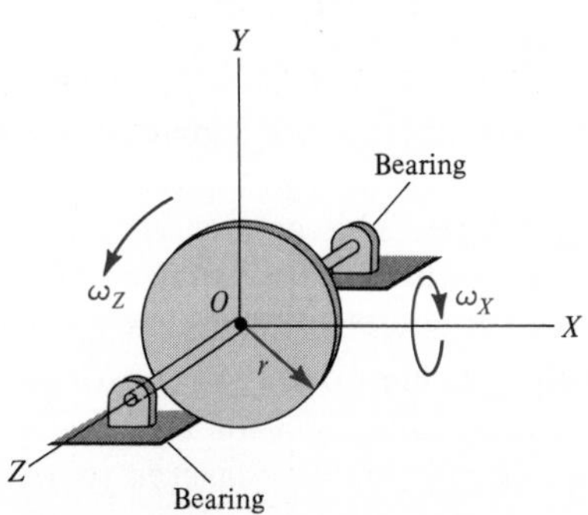

FIGURE 20-12

Model of a turbine disk. The bearings are fixed in the engine, not in the XYZ frame

EXAMPLE 20-5

Determine the bearing reactions at A and B for the crankshaft in Ex. 20-3.

SOLUTION

This problem involves motion of a body about a fixed axis (here the x axis). From Ex. 20-3,

$$H_x = 3.90\ \text{lb}\cdot\text{ft}\cdot\text{s}$$

$$H_y = 1.63\ \text{lb}\cdot\text{ft}\cdot\text{s}$$

$$H_z = 0$$

Applying Eqs. 20-25 and 20-26 gives

$$\sum \mathbf{M}_A = \dot{\mathbf{H}}_A + \boldsymbol{\omega} \times \mathbf{H}_A \quad \text{(point } A \text{ may be used since it is fixed)}$$

$$= (\dot{H}_x + \omega_y H_z - \omega_z H_y)_A \mathbf{i} + (\dot{H}_y + \omega_z H_x - \omega_x H_z)_A \mathbf{j} + (\dot{H}_z + \omega_x H_y - \omega_y H_x)_A \mathbf{k}$$

$$= (0+0-0)\mathbf{i} + (0+0-0)\mathbf{j} + \left[0 + 105\frac{\text{rad}}{\text{s}}(1.63\ \text{lb}\cdot\text{ft}\cdot\text{s})\mathbf{k} - 0\right]$$

$$= 171\ \text{lb}\cdot\text{ft}\ \mathbf{k}$$

From Fig. 20-13, the only reaction component that can cause a moment about the z axis is the force B_y.

$$\sum M_z = B_y(2\ \text{ft}) - W(1\ \text{ft}) = 171\ \text{lb}\cdot\text{ft}$$

$$B_y = \frac{1}{2\ \text{ft}}[171\ \text{lb}\cdot\text{ft} + 14.4\ \text{lb}\ (1\ \text{ft})]$$

$$= 92.7\ \text{lb} \quad \text{(the direction assumed in Fig. 20-13 is correct)}$$

$$\sum F_y = ma_y = 0$$

$$A_y + B_y - W = 0$$

$$A_y = 14.4\ \text{lb} - 92.7\ \text{lb} = -78.3\ \text{lb} \quad (A_y \text{ should be downward in Fig. 20-13})$$

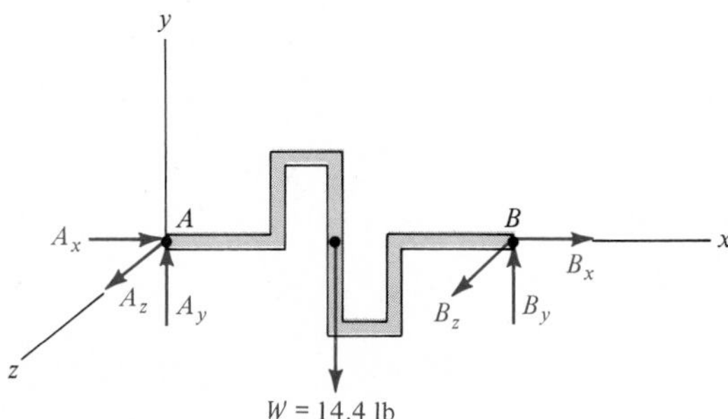

FIGURE 20-13
Model of a crankshaft

PROBLEMS

SECTIONS 20-4, 20-5, 20-6

FIGURE P20-23

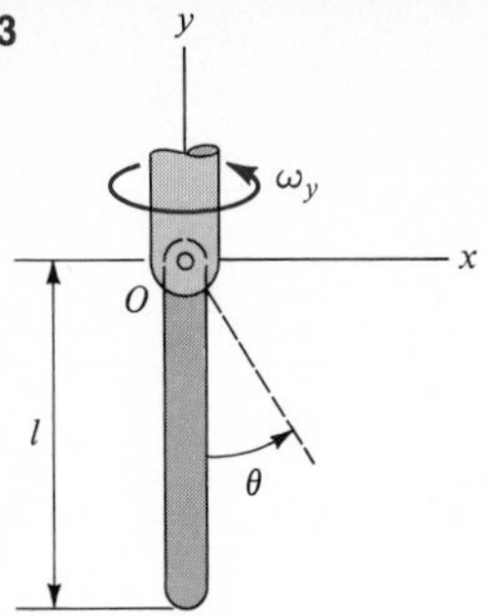

20–23 A well-drilling rod is modeled as a slender rod of mass m and length l moving without resistance. The rod rotates at angular speed ω_y about the y axis and can freely pivot about the horizontal z axis. Determine the value of ω at which the rod would start to swing away from a nearly vertical position (from $\theta \approx 0$ to $\theta > 0$).

20–24 Add a concentrated mass M to the tip of the rod in Prob. 20–23. Write an expression for the angle θ of the rod. Does M allow an increased operating speed ω?

FIGURE P20-25

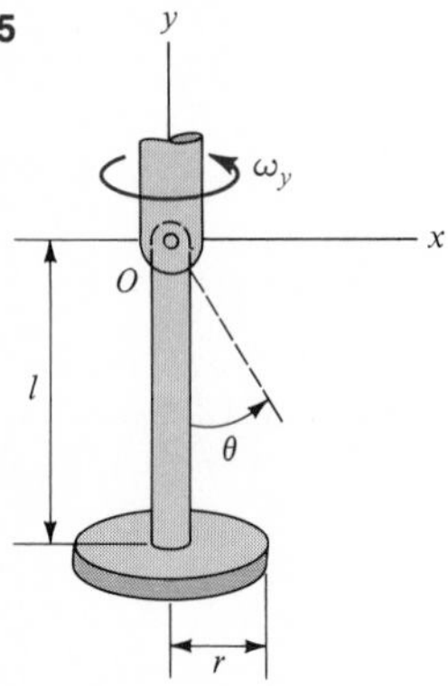

20–25 A proposed amusement ride is modeled as a thin, uniform disk of mass m and radius r, firmly attached to a slender rod of length l and negligible mass. The rod rotates at angular speed ω_y about the y axis and can freely pivot about the horizontal z axis. Write an expression for the angle θ of the rod.

20–26 Solve Prob. 20–25 including the mass M of the rod.

20–27 Determine the rate of change $\dot{\mathbf{H}}_O$ of the angular momentum of the rod in Prob. 20–4.

20–28 Determine the reactions at bearings A and B in Prob. 20–4 for the position shown. $AB = 0.3$ m, $BO = 0.2$ m.

20–29 Determine the rate of change $\dot{\mathbf{H}}_O$ of the angular momentum of the peg in Prob. 20–7.

20–30 The thin plate of mass m is attached to a shaft with bearings A and B at $x = 0$ and $2a$, respectively. It rotates about the fixed x axis at angular speed ω_x. Determine the reactions at A and B for the position shown.

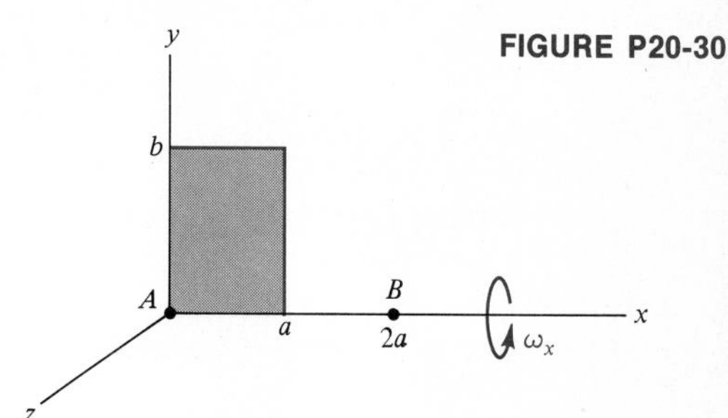

FIGURE P20-30

20–31 The thin plate of mass m is attached to a shaft with bearings at A and B. It rotates about the fixed y axis at angular speed ω_y. Determine the reactions at A and B for the position shown.

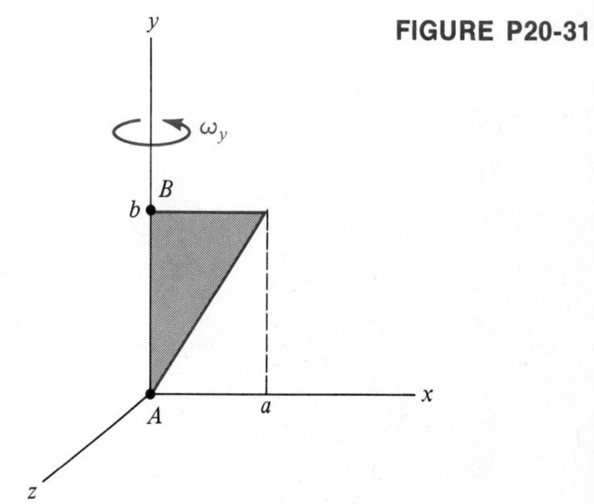

FIGURE P20-31

20–32 The thin bent plate of mass $m_1 + m_2$ is rotating about the fixed y axis at angular speed ω_y. Determine the rate of change $\dot{\mathbf{H}}_A$ of the angular momentum for ω_y = constant and also for ω_y changing at the rate of $\dot{\omega}_y$.

20–33 The bent plate in Prob. 20–32 is connected to bearings at A and B. Determine the reactions at A and B for the position shown and ω_y = constant.

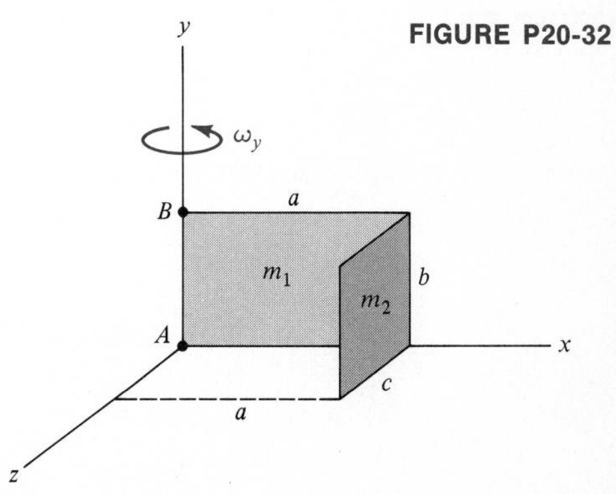

FIGURE P20-32

FIGURE P20-34

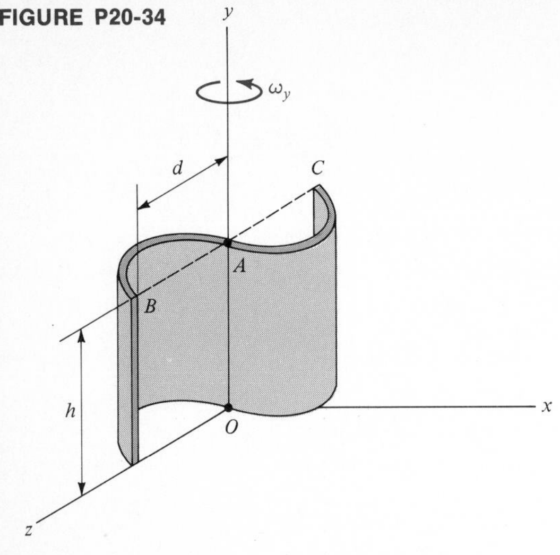

20–34 Consider the semicylindrical blades of a Savonius rotor (Prob. 20–16) of total mass m which are rigidly connected to a shaft at the y axis. Corners A, B, and C (and identical lower ones) are in the yz plane. Determine the rate of change $\dot{\mathbf{H}}_O$ of the angular momentum of the pair of blades if $\dot{\omega}_y = 0$.

20–35 Calculate the reactions at bearings O and A for the blades in Prob. 20–34.

FIGURE P20-36

Complex test of a tire (Courtesy MTS Systems Corporation, Minneapolis, Minnesota)

20–36 The 50-kg wheel is modeled as a uniform disk of 0.6 m radius and 0.4 m thickness. Its center is located at $x = -30$ cm, $y = z = 0$, and pivoted about O ($x = y = z = 0$). The wheel is parallel to the yz plane when its angular velocity is $\boldsymbol{\omega} = (10\mathbf{i} - 3\mathbf{k})$ rad/s. Determine the rate of change $\dot{\mathbf{H}}_O$ of the angular momentum of the wheel.

20–37 Calculate the required moment M_O about point O in Prob. 20–36 to keep the wheel's axle in the xy plane.

FIGURE P20-39

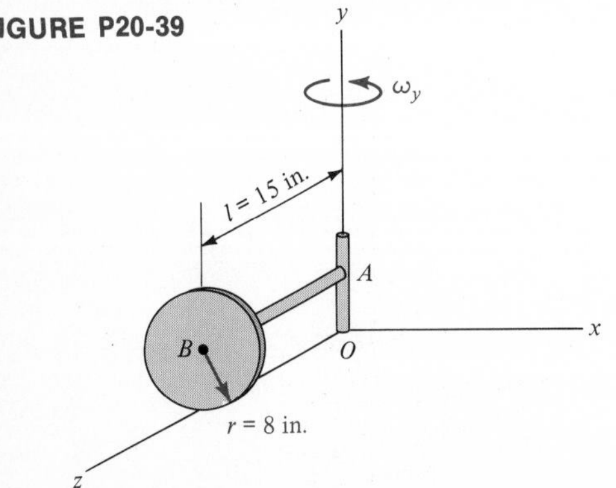

20–38 A 5-lb, thin, uniform disk of 8 in. radius is rotating about a 15-in. arm which is pivoted at the fixed point A. Arm AB of negligible weight is always parallel to the xz plane and rotates about the y axis at $\omega_y = 200$ rpm as the disk rolls on the xz plane. Calculate the rate of change $\dot{\mathbf{H}}_A$ of the angular momentum of the disk.

20–39 Determine the normal force exerted by the disk on the floor in Prob. 20–38.

20–40 A thin plate of mass m has bearings at points O and A. A hole of radius r is drilled at $x = 3a/4$ and $y = b/2$. Determine the reactions at O and A in the position shown for an angular speed ω_x about the x axis.

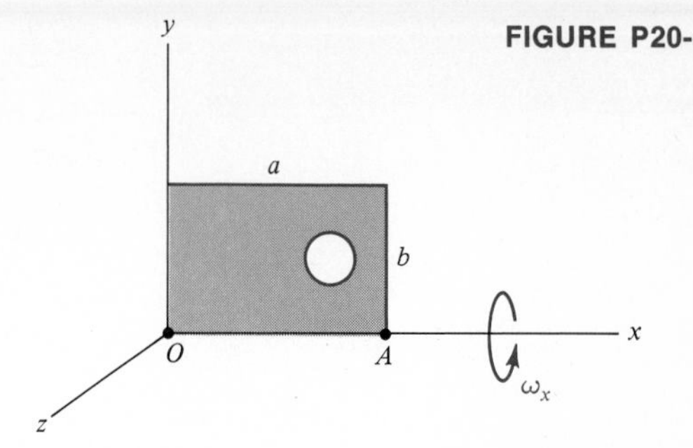

FIGURE P20-40

20–41 A thin rod of mass m per unit length is bent into the shape of a semicircle and firmly attached to the shaft which is in bearings A and B. Determine the rate of change $\dot{\mathbf{H}}_O$ of the angular momentum when the angular speed about the x axis is ω_x.

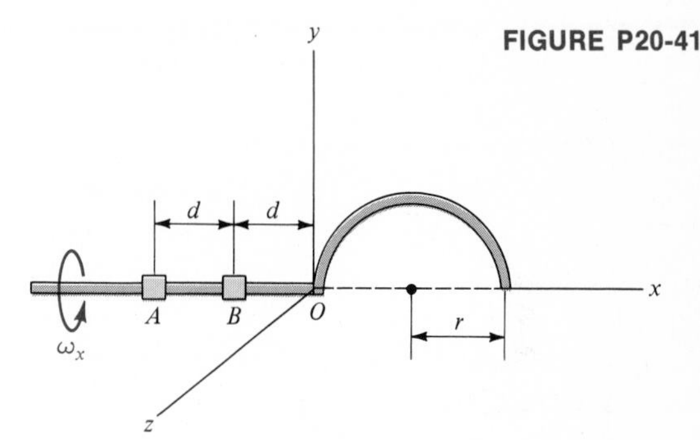

FIGURE P20-41

20–42 Determine the bending moment in the rod at point O in Prob. 20–41. Also determine the reactions at A and B when the member is in the xy plane.

20–43 A two-bladed propeller of an airplane has a total mass of 12 kg and a radius of gyration of 0.8 m. The propeller is rotating at a speed of 1400 rpm when the airplane is flying at a 300-km/h speed in a circular path of 200 m radius. Determine the moment acting on the propeller shaft because of these motions.

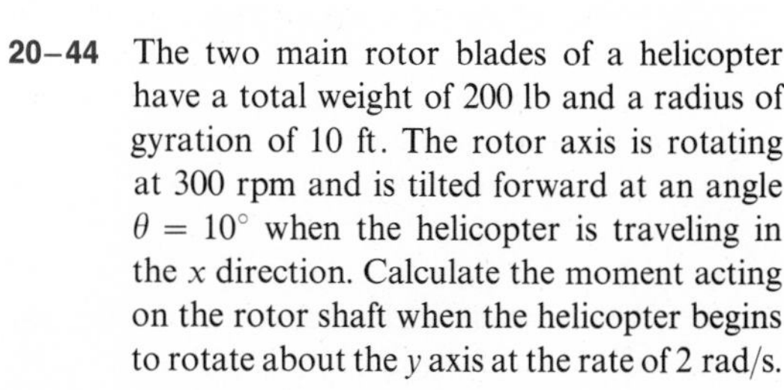

20–44 The two main rotor blades of a helicopter have a total weight of 200 lb and a radius of gyration of 10 ft. The rotor axis is rotating at 300 rpm and is tilted forward at an angle $\theta = 10^\circ$ when the helicopter is traveling in the x direction. Calculate the moment acting on the rotor shaft when the helicopter begins to rotate about the y axis at the rate of 2 rad/s.

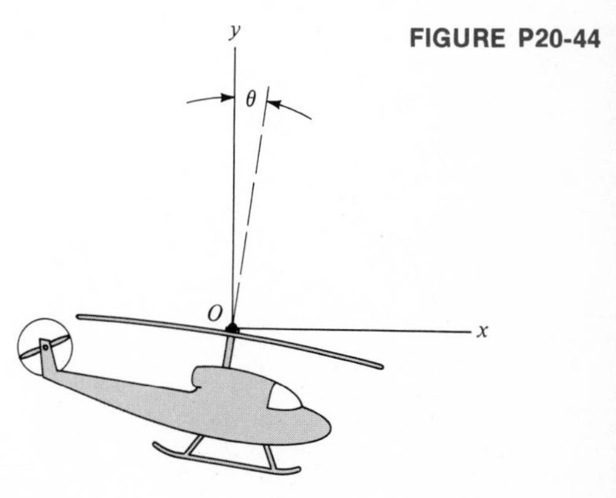

FIGURE P20-44

20-7 GYROSCOPIC MOTION

The basic concepts of analyzing the three-dimensional motion of a rigid body were presented in the preceding sections. These concepts are extended here for introductory analysis of gyroscopes which include navigational devices in aircraft and stabilizing machinery in ships.

Precession

It is worthwhile to review qualitatively the appreciable differences that are possible between two-dimensional and three-dimensional motions of a rigid body. Consider first a uniform disk on a rigid axle at rest. Assume that a moment **M** about the x axis is applied to the shaft in Fig. 20-14. With the disk at rest initially, the shaft rotates about the x axis only. It can be said that the system changes its angular momentum (in this case, from $\mathbf{H}_O = 0$) toward the applied moment **M**. This is analogous to the change in linear momentum of a particle in the direction of the force acting on it.

Next, assume that the disk has an initial angular velocity $\boldsymbol{\omega}_z$ about the z axis as in Fig. 20-15a. This velocity is called the spin velocity, and it results in the initial angular momentum $\mathbf{H}_O$ about point O. A moment **M** about the x axis causes this angular momentum vector to rotate toward the x axis as shown in Fig. 20-15b. The resulting angular velocity $\boldsymbol{\omega}_y$ of **H** is called the precession velocity. The precession should be viewed as a change in angular momentum in the direction of the applied moment vector **M** (again analogous to the linear motion of a particle). The applied moment and the rate of precession are related according to Eq. 20-26. Precession is always slow relative to the spin rate. This particular motion may be readily experienced by gradually tilting the axis of a high-speed rotary tool while *very carefully* holding it in the hand.

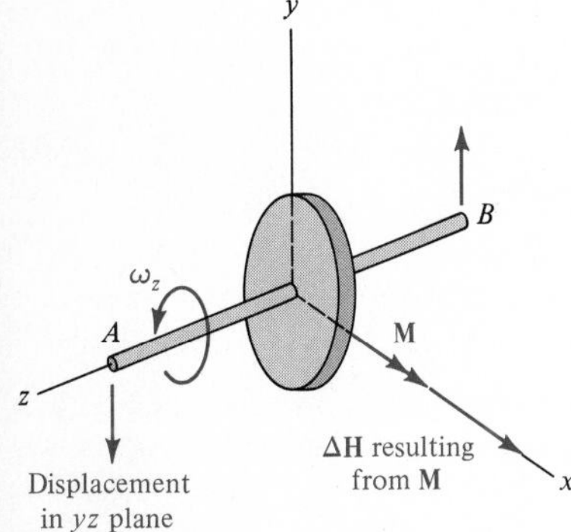

FIGURE 20-14
Analysis of a nonspinning disk

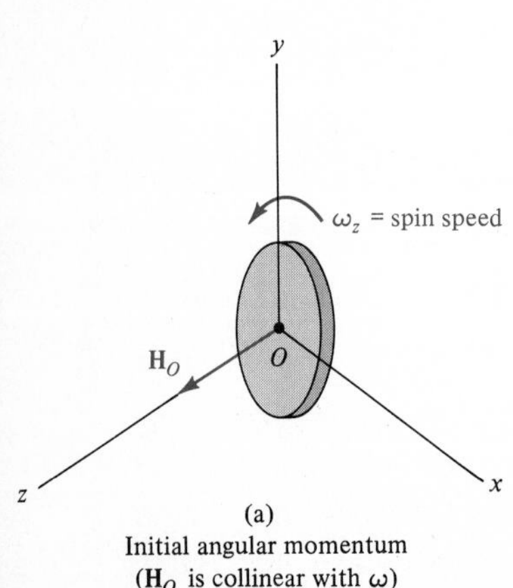

(a)
Initial angular momentum
($\mathbf{H}_O$ is collinear with ω)

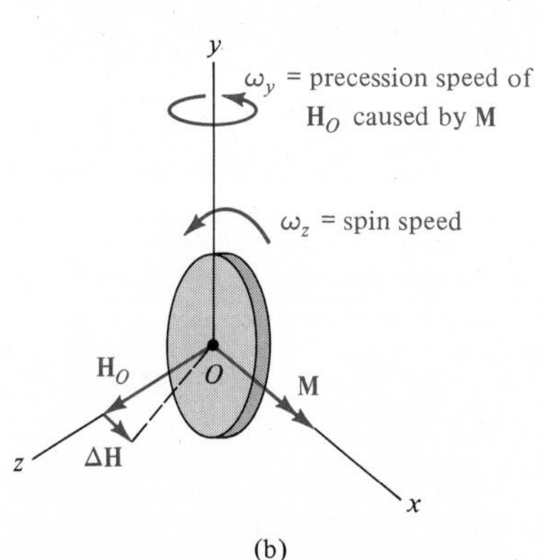

(b)
M causes $\mathbf{H}_O$ to rotate toward the x axis

FIGURE 20-15
Analysis of a spinning disk

The precession axis of a rotating body may itself change orientation in general. Analysis of the motion including this additional complication is made using the model of a gyroscope in Fig. 20-16. In

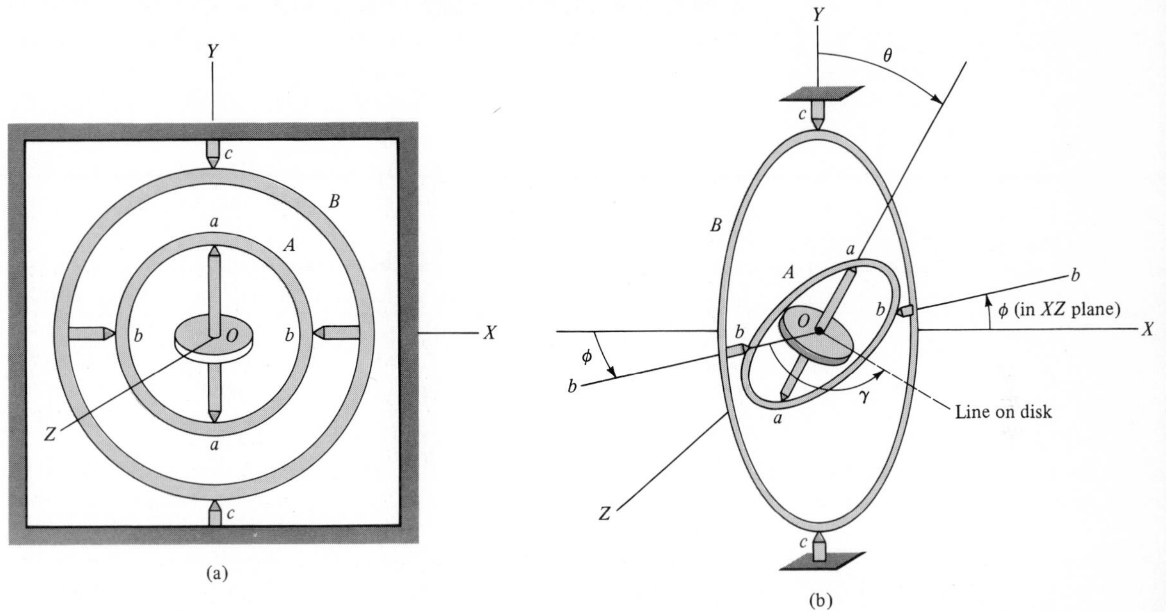

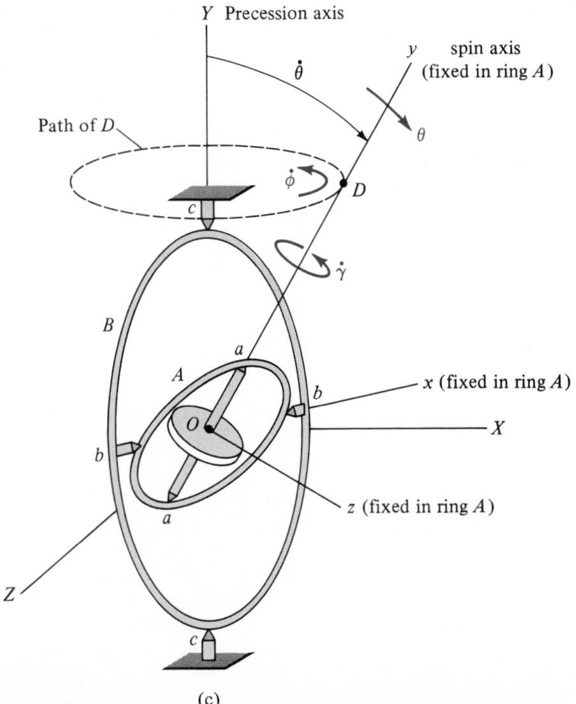

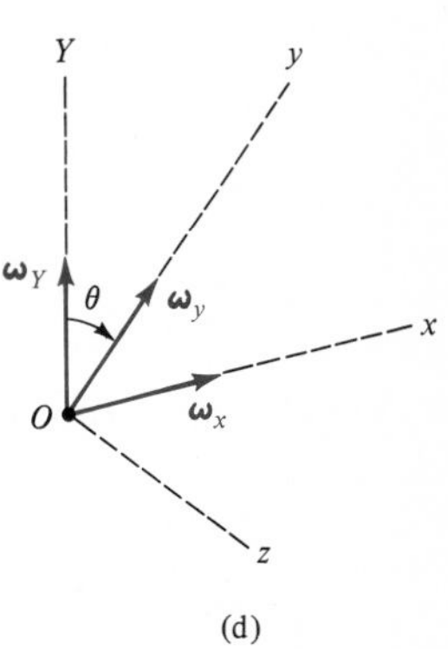

FIGURE 20-16
Model of a gyroscope

part (a) of the figure the shaft of a disk is pivoted at *aa* to spin freely in the *inner gimbal* ring A, which is pivoted at *bb* to rotate freely with respect to the *outer gimbal* ring B. The latter is pivoted at *cc* to rotate freely about the Y axis of a fixed reference frame XYZ. A general position of these parts is shown in Fig. 20-16b. Angle ϕ specifies the position of line *bb* from the fixed X axis (line *bb* always moves in the XZ plane). Angle θ represents the position of line *aa* of the inner gimbal from the fixed Y axis. Angle γ gives the position of a radial line on the disk measured from line *bb* (γ is measured in the XZ plane when $\theta = 0$). The angles ϕ, θ, and γ are known as the *Euler angles*. These angles completely define the position of the gyroscope at a given instant. Their time derivatives are defined as follows: $\dot{\phi}$ = rate of *precession*, $\dot{\theta}$ = rate of *nutation*, and $\dot{\gamma}$ = rate of *spin*.

For further analysis of the angular velocities, consider a rotating reference frame xyz which is attached to the inner gimbal A as in Fig. 20-16c. Thus, at the instant indicated, the disk spins about the y axis at angular speed $\dot{\gamma}$, while the y axis itself has two motions. The y axis is rotating away from the fixed Y axis at angular speed $\dot{\theta}$, and it is precessing about the Y axis at angular speed $\dot{\phi}$ (the dashed line shows the circular path of point D on the y axis caused by the precession).

The absolute angular velocity $\boldsymbol{\omega}$ of the gyroscope is the vector sum of its three angular velocities. These are shown schematically in Fig. 20-16d. Considering the fixed unit vectors **I**, **J**, and **K**, and the **i**, **j**, and **k** unit vectors of the rotating xyz frame,

$$\boldsymbol{\omega} = \boldsymbol{\omega}_Y + \boldsymbol{\omega}_x + \boldsymbol{\omega}_y = \dot{\phi}\mathbf{J} + \dot{\theta}\mathbf{i} + \dot{\gamma}\mathbf{j} \tag{20-34}$$

This equation should be transformed to use entirely either the fixed or the rotating frame. The latter is more convenient (both here and eventually when dealing with moments), and this is accomplished realizing that the Y, y, and z axes are in the same plane, so

$$\mathbf{J} = \cos\theta\mathbf{j} - \sin\theta\mathbf{k} \tag{20-35}$$

With this, Eq. 20-34 becomes

$$\boldsymbol{\omega} = \dot{\theta}\mathbf{i} + (\dot{\gamma} + \dot{\phi}\cos\theta)\mathbf{j} - \dot{\phi}\sin\theta\mathbf{k} \tag{20-36}$$

where all components are defined in the rotating frame.

Equations of Motion

The effects of external moments on the motion of a gyroscope are determined using Eq. 20-27. Here the center of mass C is at the origin O of the rotating frame. The concise form of the equation of motion is

$$\sum \mathbf{M}_O = (\dot{\mathbf{H}}_O)_{xyz} + \mathbf{\Omega} \times \mathbf{H}_O \qquad \text{20-37}$$

where $\mathbf{\Omega}$ is the angular velocity of the xyz frame. For the configuration in Fig. 20-16c and d,

$$\mathbf{\Omega} = \omega_Y + \omega_x = \dot{\theta}\mathbf{i} + \dot{\phi}\cos\theta\mathbf{j} - \dot{\phi}\sin\theta\mathbf{k} \qquad \text{20-38}$$

The angular momentum vector $\mathbf{H}_O$ and its derivative are evaluated (neglecting the mass of the two gimbals) using Eqs. 20-4 and 20-23. A considerable simplification can be made here since the xyz axes are principal axes of inertia of the gyroscope, which is always symmetrical with respect to its spin axis. Thus, only two moments of inertia have to be used, I_1 about the spin axis and I_2 about any centroidal line transverse to the spin axis. With these, the angular momentum is expressed using Eqs. 20-4 and 20-36 as

$$\mathbf{H}_O = I_2\dot{\theta}\mathbf{i} + I_1(\dot{\gamma} + \dot{\phi}\cos\theta)\mathbf{j} - I_2\dot{\phi}\sin\theta\mathbf{k} \qquad \text{20-39}$$

After performing the mathematical manipulations indicated in Eq. 20-37 (noting that **i**, **j**, and **k** are fixed in the xyz frame), the xyz components of the moment are written as

$$\begin{aligned}
\sum M_x &= I_1\dot{\phi}\sin\theta(\dot{\gamma} + \dot{\phi}\cos\theta) \\
&\quad + I_2(\ddot{\theta} - \dot{\phi}^2\sin\theta\cos\theta) \\
\sum M_y &= I_1(\ddot{\gamma} + \ddot{\phi}\cos\theta - \dot{\phi}\dot{\theta}\sin\theta) \\
\sum M_z &= I_1\dot{\theta}(\dot{\gamma} + \dot{\phi}\cos\theta) \\
&\quad - I_2(\ddot{\phi}\sin\theta + 2\dot{\phi}\dot{\theta}\cos\theta)
\end{aligned} \qquad \text{20-40}$$

The details of the derivation are left as an exercise for the student.

Equations 20-40 are the differential equations of motion of a gyroscope. Note in the above analysis that the xyz frame is attached at the mass center of the gyroscope, where it is effectively supported by gimbals of negligible mass. These equations are also valid for a noncentroidal xyz frame which is attached at any fixed point O located on the spin axis of the gyroscope. The reason is that the Euler angles can be identically defined for centroidal and noncentroidal rotating frames, and the symmetry of the body allows using only two moments of inertia in either case. The position of point O with respect to the mass center does not affect the value of I_1 in Eq. 20-40, but it affects I_2 (the parallel-axis theorem must be used).

20-8 SPECIAL CASES OF GYROSCOPIC MOTION

The solution of the nonlinear equations of motion (Eq. 20-40) are very complex in general. In a few particular cases they are relatively simple and can be considered at the introductory level to gain further insight into three-dimensional motion.

Steady Precession

Assume that a gyroscope precesses under the following conditions:

$$\text{angular speed of spin } \dot{\gamma} = \text{constant}, \ddot{\gamma} = 0$$
$$\text{angular speed of precession } \dot{\phi} = \text{constant}, \ddot{\phi} = 0$$
$$\text{angle } \theta \text{ to spin axis} = \text{constant}, \dot{\theta} = \ddot{\theta} = 0$$

In this case, Eqs. 20-40 reduce to

$$\begin{aligned} \sum M_x &= I_1\dot{\phi} \sin \theta(\dot{\gamma} + \dot{\phi} \cos \theta) - I_2\dot{\phi}^2 \sin \theta \cos \theta \\ \sum M_y &= 0 \\ \sum M_z &= 0 \end{aligned} \qquad \text{20-41}$$

The same result can be obtained directly from Eq. 20-37 after realizing that H_O is constant in steady precession when observed from the rotating xyz frame, so its time derivative is zero. Equation 20-37 reduces to

$$\sum \mathbf{M}_O = \mathbf{\Omega} \times \mathbf{H}_O \qquad \text{20-42}$$

which is equivalent to Eq. 20-41 after the vector product is evaluated.

An important observation concerning the direction of the external moment acting on a steadily precessing gyroscope can be made from Fig. 20-16c and Eq. 20-41. It is noted that *the moment vector should be perpendicular to the spin axis and the precession axis to maintain steady precession of a gyroscope.* This is illustrated in Fig. 20-17a for the general case of steady precession. A spinning top with its weight causing the moment is in this category.

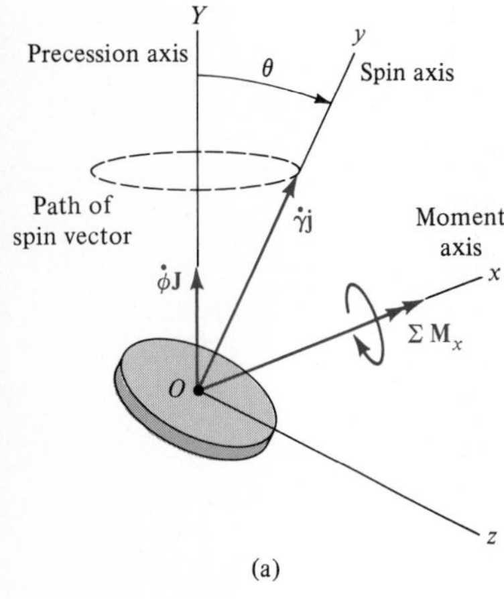

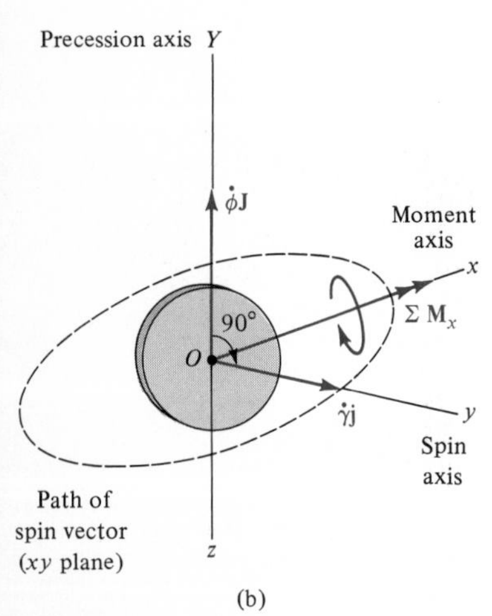

FIGURE 20-17

Special case of a gyroscope in steady precession

Precession and Spin Axes Are at a Right Angle

In many engineering problems involving steady precession of a gyroscope the precession axis is normal to the spin axis. This allows a further simplification of Eq. 20-41 since $\theta = 90°$. The resulting single equation is

$$\sum M_x = I_1 \dot{\phi} \dot{\gamma} \qquad \boxed{20\text{-}43}$$

which is illustrated in Fig. 20-17b. The general meaning of this specific situation can be summarized as follows:

1. The gyroscope precesses about an axis perpendicular to both the spin axis and the moment axis.
2. The rate of precession $\dot{\phi}$ is inversely proportional to the rate of spin $\dot{\psi}$ for a given moment acting on the gyroscope.
3. Large moments are required if the orientation of a rapidly spinning gyroscope is to be changed.

Steady Precession of Axisymmetric Bodies in the Absence of Moments

Projectiles and spacecraft are commonly given spin velocities to achieve stabilization. Such motions may be analyzed using the methods developed so far.

Assume that the body is an axisymmetric rigid body spinning about its longitudinal axis, and that no external moments are acting on it. In this case $\sum \mathbf{M}_C = \dot{\mathbf{H}}_C = 0$ from Eq. 20-22. Thus, the angular momentum $\mathbf{H}_C$ is constant in magnitude and direction. It is convenient to set up the fixed frame XYZ with respect to this angular momentum vector. In particular, the axis of precession Y is chosen to coincide with the vector $\mathbf{H}_C$ as in Fig. 20-18a. Axis y is the spin axis of the body as before, while x is chosen to be in the Yy plane.

The total angular velocity $\boldsymbol{\omega}$ is the vector sum of the spin velocity $\dot{\gamma}\mathbf{j}$ and the precession velocity $\dot{\phi}\mathbf{J}$. Since the angle θ between the Y and y axes is generally not a right angle (but θ = constant as shown below*), components of $\boldsymbol{\omega}$ are different from the precession and spin velocities, respectively. These distinctions are shown in Fig. 20-18b. The xyz components of $\mathbf{H}_C$ and $\boldsymbol{\omega}$ depend on the angles θ and α according to Fig. 20-18,

$$H_x = -H_C \sin\theta \qquad H_y = H_C \cos\theta \qquad H_z = 0 \qquad \boxed{20\text{-}44}$$

$$\omega_x = \omega \sin\alpha \qquad \omega_y = \omega \cos\alpha \qquad \omega_z = 0 \qquad \boxed{20\text{-}45}$$

* First assume that in the fixed XYZ frame $\boldsymbol{\omega} = \omega \sin\alpha\mathbf{i} + \omega\cos\alpha\mathbf{j} + \dot{\theta}\mathbf{k}$. However,

$$\mathbf{H}_C = H_C \sin\theta\mathbf{i} + H_C\cos\theta\mathbf{j} + 0\mathbf{k}$$

since $\mathbf{H}_C$ is in the xy plane. Hence, $\dot{\theta} = 0$ and θ = constant.

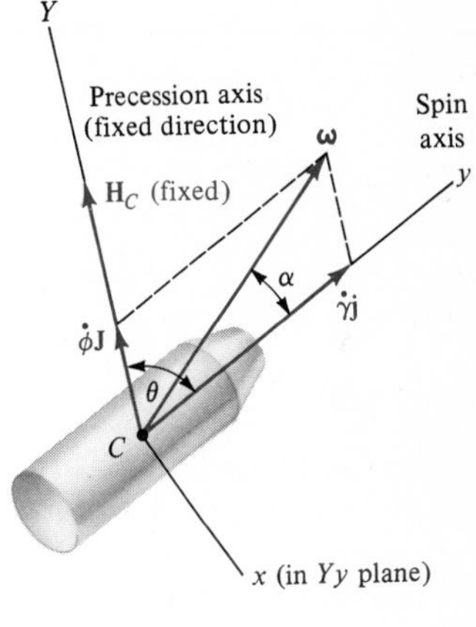

(a)

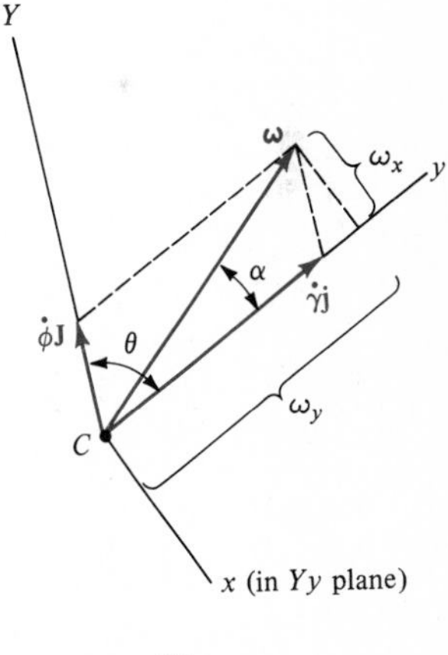

(b)

FIGURE 20-18

Special case of a steadily precessing axisymmetric body

where the magnitudes H_C and ω are constant. Note that there are no components in the z direction, which is perpendicular to the plane of $\mathbf{H}_C$ and $\boldsymbol{\omega}$. The components of angular momentum and angular velocity are simply related to one another since the xyz axes are principal axes of inertia,

$$H_x = I_2\omega_x \qquad H_y = I_1\omega_y \qquad H_z = 0 \tag{20-46}$$

where I_1 and I_2 are the moments of inertia about the spin axis and a transverse axis through C, respectively. From Eqs. 20-44 and 20-46,

$$\omega_x = -\frac{H_C \sin\theta}{I_2} \qquad \omega_y = \frac{H_C \cos\theta}{I_1} \qquad \omega_z = 0 \tag{20-47}$$

Since H_C, I_1, I_2, and θ are all constant, ω_x and ω_y are constant. These are the mathematical expressions of *steady precession* of the body about the Y axis, with a constant angle θ to the spin axis y (since $\omega_z = 0$).

The angle α between the spin axis and the vector $\boldsymbol{\omega}$ can be determined from Fig. 20-18b and Eq. 20-47, from which

$$\boxed{\tan\alpha = -\frac{\omega_x}{\omega_y} = \frac{I_1}{I_2}\tan\theta} \tag{20-48}$$

Thus, α is a constant angle in this motion.

The rate of precession is obtained from Eq. 20-41. With $\sum \mathbf{M} = 0$,

$$\boxed{\dot{\phi} = \frac{I_1\dot{\gamma}}{(I_2 - I_1)\cos\theta}} \tag{20-49}$$

The occurrence of precession depends on the initial rotational motion of the body. The direction of precession depends on the relative magnitudes of I_1 and I_2 according to Eq. 20-49. Four particular cases are worth noting as outlined in the following.

1. The initial spin is entirely about the axis of symmetry as in Fig. 20-19a. There is no precession since ω_x, H_x, and $\sum \mathbf{M}$ are all zero.

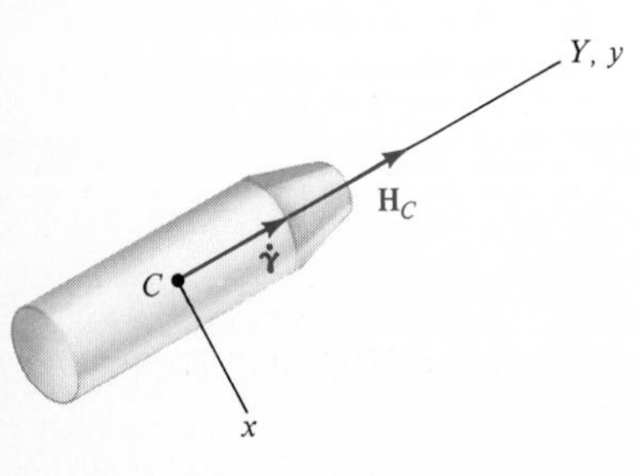

(a) No precession

(b) No precession

FIGURE 20-19

Special cases involving no precession

2. The initial spin is entirely about a transverse axis (with respect to the axis of symmetry) as in Fig. 20-19b. There is no precession since ω_y, H_y, and $\sum \mathbf{M}$ are all zero.
3. There is initial spin and precession as illustrated in Figs. 20-20a and 20-21a for two distinct possibilities.
   a. $I_2 > I_1$; the body is elongated along the spin axis y. $\theta > \alpha$ at all times according to Eq. 20-48. The body cone (Sec. 20-1) appears to roll on the outside of the space cone as shown in Fig. 20-20b. This is called *direct precession* because the spin and the precession have the same sense when viewed from the positive Y or y axes.
   b. $I_1 > I_2$; the body is short along the spin axis y. $\theta < \alpha$ at all times according to Eq. 20-48. This is possible if the spin and precession vectors have components in opposite directions as in Fig. 20-21a. The space cone is interior to the body cone, and the spin and the precession have opposite senses when viewed from the positive Y or y axes as in Fig. 20-21b. This precession is said to be *retrograde*.

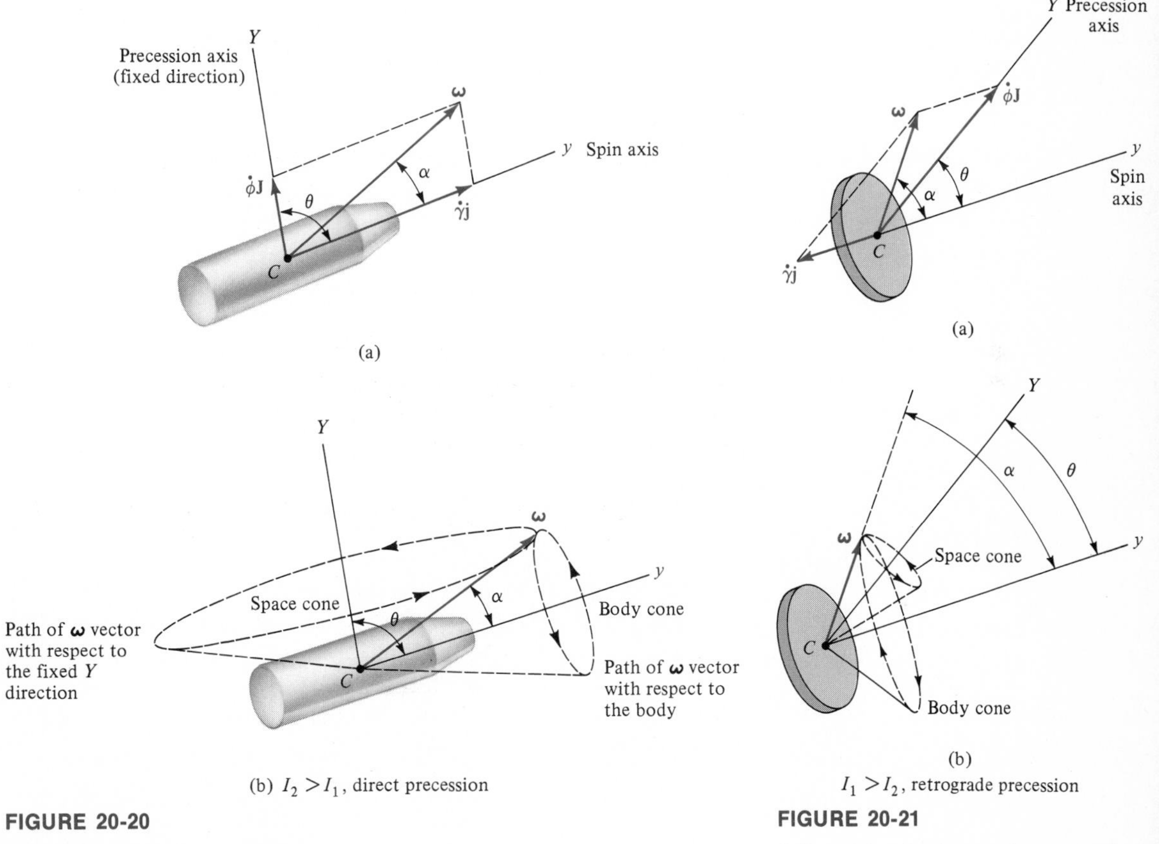

FIGURE 20-20
Schematic of direct precession

FIGURE 20-21
Schematic of retrograde precession

EXAMPLE 20-6

A 2-kg disk of 0.3 m diameter is attached to a light axle of length $l = 0.05$ m. It is released with l in the XY plane and $\beta = 60^\circ$ (Fig. 20-22a), and spinning about l at the rate of $\dot{\gamma} = 800$ rpm $= 83.8$ rad/s. Determine the rate of precession $\dot{\phi}$ of the disk.

SOLUTION

Ignoring the weight of the axle, the only force acting on the disk is its weight, which causes a moment about the z axis. By Eq. 20-42 and using the xyz frame as the reference system (Fig. 20-22b),

$$\sum \mathbf{M}_o = \boldsymbol{\Omega} \times \mathbf{H}_o$$

where

$$\sum \mathbf{M}_o = Wl \cos \beta \mathbf{k}$$

$$\boldsymbol{\Omega} = \Omega_x \mathbf{i} + \Omega_y \mathbf{j} + \Omega_z \mathbf{k}$$

$$= (\dot{\phi} \cos \beta)\mathbf{i} + (\dot{\phi} \sin \beta)\mathbf{j} \qquad \text{(angular velocity of } xyz \text{ system)}$$

$$\mathbf{H}_o = I_x \omega_x \mathbf{i} + I_y \omega_y \mathbf{j} + I_z \omega_z \mathbf{k} \qquad (\boldsymbol{\omega} = \text{angular velocity of the body})$$

$$= I_x \dot{\phi} \cos \beta \mathbf{i} + I_y(\dot{\gamma} + \dot{\phi} \sin \beta)\mathbf{j}$$

Evaluating the cross product yields

$$Wl \cos \beta \mathbf{k} = (I_y \Omega_x \omega_y - I_x \Omega_y \omega_x)\mathbf{k}$$
$$= I_y \dot{\phi} \cos \beta(\dot{\gamma} + \dot{\phi} \sin \beta)\mathbf{k} - I_x(\dot{\phi} \sin \beta)(\dot{\phi} \cos \beta)\mathbf{k}$$

Dividing through by $\cos \beta$, and realizing that moments of inertia must be calculated with respect to point O using the parallel-axis theorem, gives

$$mg(0.05\text{m}) = \frac{1}{2} m(0.15\text{m})^2 \left(83.8 \frac{\text{rad}}{\text{s}} \dot{\phi} + 0.866\dot{\phi}^2\right)$$
$$- \left[\frac{1}{4} m(0.15\text{m})^2 + m(0.05\text{m})^2\right](0.866)\dot{\phi}^2$$

Dividing through by the mass m, a quadratic in $\dot{\phi}$ is obtained. Using the quadratic formula, the solutions are

$$\dot{\phi} = 0.4 \text{ rad/s} \qquad \text{and} \qquad -56.7 \text{ rad/s}$$

Of these, the positive root satisfies the condition for the change of angular momentum to be in the direction of the external moment.

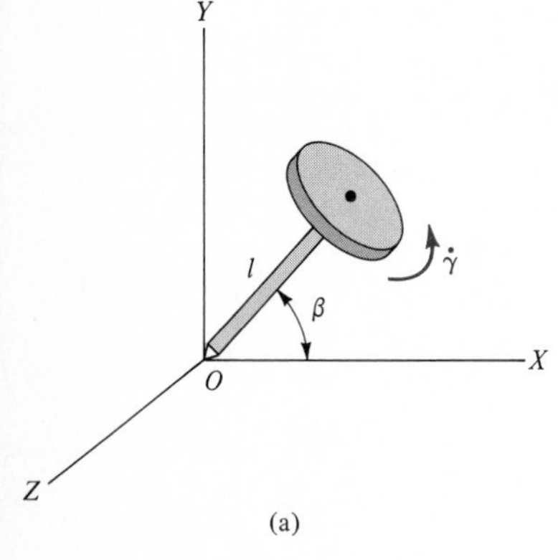

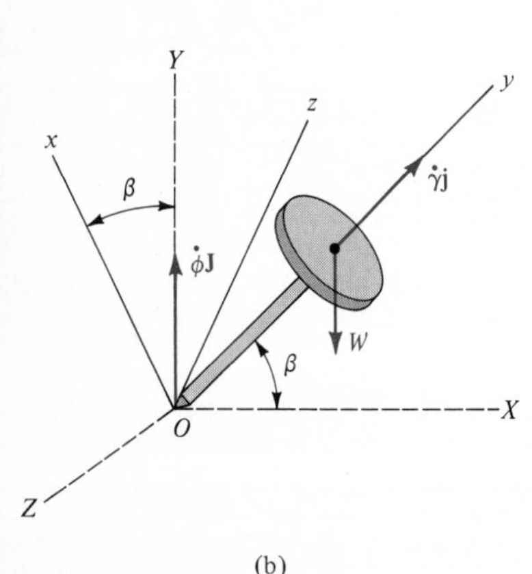

FIGURE 20-22
Analysis of a spinning disk

A 2000-lb spacecraft is modeled as a uniform cylinder of 6 ft diameter and 10 ft length (Fig. 20-23a). It is spinning at the rate of $\dot{\gamma} = 1$ rad/s. Small jets apply a moment $\mathbf{M}_x = 30$ ft·lb $\mathbf{i}$ for a time of 2 s to the cylinder. Determine the resulting precession and sketch the body and space cones for the motion of the spacecraft after the burn.

SOLUTION

Since the reference axes are principal inertial axes and initially $\omega_x = \omega_z = 0$, the initial angular momentum is simply

$$\mathbf{H}_1 = I_y\omega_y = \left(\frac{1}{2}mr^2\right)\omega_y$$

$$= \frac{1}{2}\left(\frac{2000 \text{ lb}}{32.2 \text{ ft/s}^2}\right)(3 \text{ ft})^2\left(1\,\frac{\text{rad}}{\text{s}}\right)\mathbf{j}$$

$$= 279.5 \text{ ft·lb·s } \mathbf{j}$$

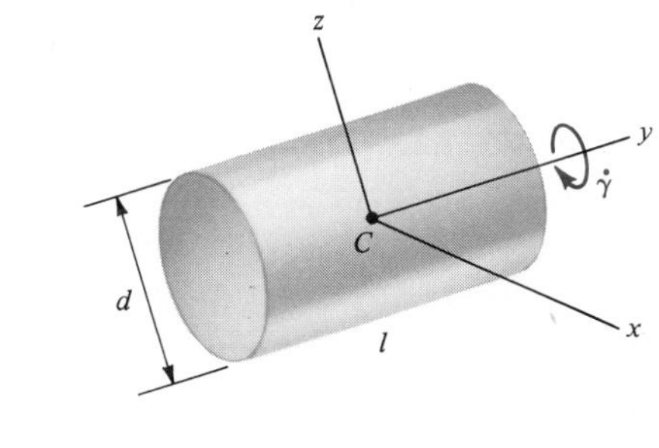

(a)

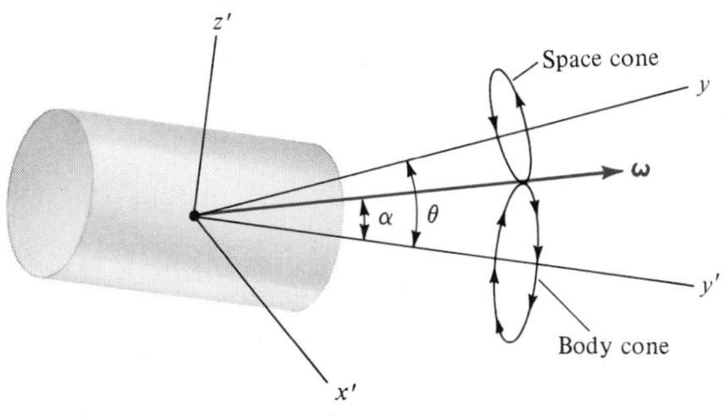

(b)

FIGURE 20-23
Model of a spinning spacecraft

From impulse-momentum, Eq. 20-12,

$$\begin{aligned}\mathbf{H}_2 &= \mathbf{H}_1 + \sum \mathbf{M}_C\,\Delta t\\ &= 279.5\ \text{ft}\cdot\text{lb}\cdot\text{s}\,\mathbf{j} + (30\ \text{ft}\cdot\text{lb})(2\ \text{s})\mathbf{i}\end{aligned}$$

where **i** and **j** are unit vectors in the rotating xyz reference frame (M_x is always about the same axis, and moments of inertia remain the same and products of inertia are still zero). Setting components equal, from Eq. 20-4, yields

$$\omega_{y_2} = \omega_{y_1} = 1\ \text{rad/s}$$

$$I_x\omega_{x_2} = \frac{1}{12}m(3a^2 + L^2)\omega_{x_2} = 60\ \text{ft}\cdot\text{lb}\cdot\text{s}$$

$$\begin{aligned}\omega_{x_2} &= \frac{60\ \text{ft}\cdot\text{lb}\cdot\text{s}}{\dfrac{1}{12}\left(\dfrac{2000\ \text{lb}}{32.2\ \text{ft/s}^2}\right)[3(3\ \text{ft})^2 + (10\ \text{ft})^2]}\\ &= \frac{60\ \text{ft}\cdot\text{lb}\cdot\text{s}}{657.4\ \text{ft}\cdot\text{lb}\cdot\text{s}^2}\\ &= 0.0913\ \text{rad/s}\end{aligned}$$

From Eq. 20-48,

$$\tan\alpha = -\frac{\omega_x}{\omega_y} = -\frac{0.0913}{1}$$

$$\alpha = -5.2^\circ$$

$$\tan\theta = \frac{I_2}{I_1}\tan\alpha = \frac{657.4}{279.5}(-0.0913)$$

$$\theta = -12.1^\circ$$

The rate of precession is obtained from Eq. 20-49,

$$\begin{aligned}\dot{\phi} &= \frac{I_1\dot{\gamma}}{(I_2 - I_1)\cos\theta} = \frac{279.5(1\ \text{rad/s})}{(657.4 - 279.5)\cos(-12.1^\circ)}\\ &= 0.756\ \text{rad/s}\end{aligned}$$

The body and space cones are sketched in Fig. 20-23b.

PROBLEMS

20–45 Plot the rate of precession of the disk in Ex. 20-6 for $\beta = 0, 30^\circ$, and 60°.

20–46 Plot the rate of precession of the disk in Ex. 20-6 for $\dot{\gamma} = 600$ rpm, 800 rpm, and 1000 rpm.

20–47 The disk in Ex. 20-6 is precessing at $\dot{\boldsymbol{\phi}} = -0.1\mathbf{J}$ rad/s. Determine the rate of spin $\dot{\gamma}$.

20–48 A 3-lb disk of 8 in. diameter is attached to a light rod which freely rotates in the ball-and-socket joint at O. Determine the rate of precession if the rate of spin $\dot{\gamma}$ about OA is 1000 rpm and $\beta = 45^\circ$.

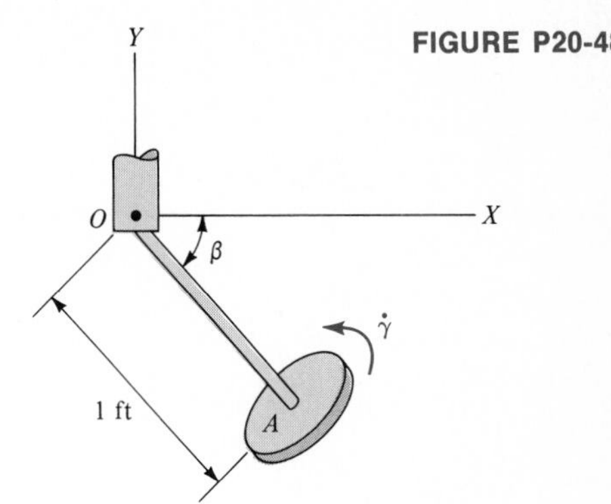

FIGURE P20-48

20–49 Solve Prob. 20–48 assuming that the member OA is a uniform 2-lb rod of 0.6 in. diameter which rotates with the disk.

20–50 A thin-walled tube of mass m, length l, and diameter d is spinning at angular speed $\dot{\gamma}$ about a smooth weightless rod OA which is hinged at O. Write an expression for the rate of precession $\dot{\phi}$ of the rod if it is released in the horizontal position.

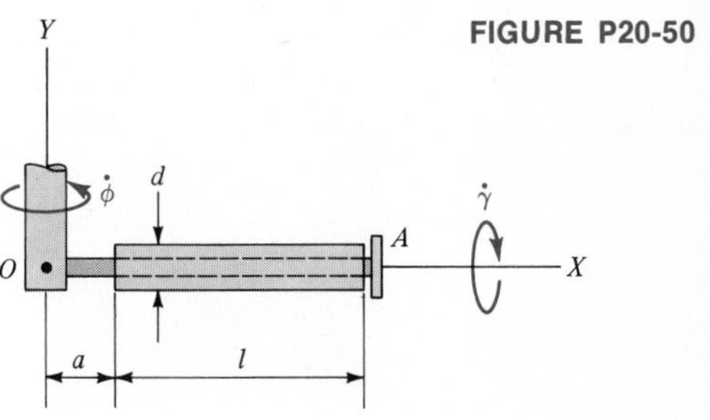

FIGURE P20-50

20–51 The tube in Prob. 20–50 has $m = 0.5$ kg, $l = 0.3$ m, $d = 0.04$ m, and $a = 0.1$ m. Determine $\dot{\gamma}$ if the tube is precessing at $\dot{\boldsymbol{\phi}} = 0.2\mathbf{J}$ rad/s.

FIGURE P20-52

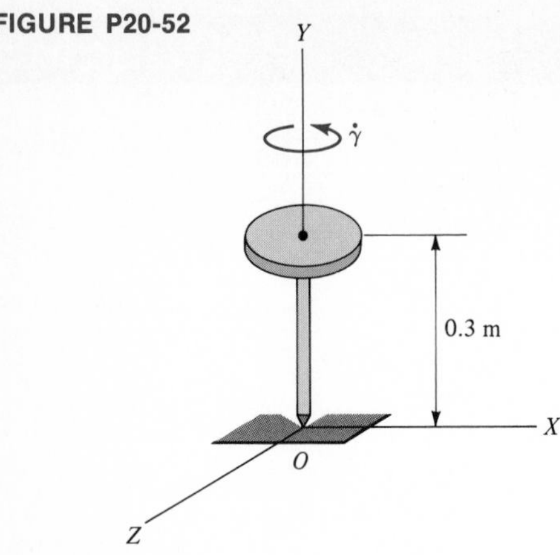

20–52 The disk of 0.1 m diameter and 1 kg mass supported on a light rod is spinning in the upright position at $\dot{\gamma} = 1200\mathbf{J}$ rad/s. A force $\mathbf{F} = 20\,\mathbf{K}$ N is applied to the disk for a time of 0.1 s. Determine the resultant rate of precession if the spin is unchanged.

FIGURE P20-53

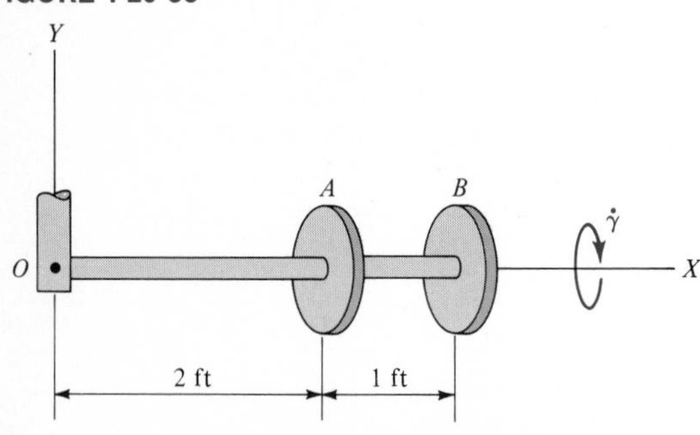

20–53 Two identical 5-lb disks of 1 ft diameter are spinning at the same rate of $\dot{\gamma} = 1000$ rpm about a light shaft which is supported at a ball-and-socket joint at O. Determine the precession of the shaft if it is released in the horizontal position.

FIGURE P20-54

20–54 The takeoff speed of an aircraft is 60 m/s. The wheels freely spin at that time. Each 20-kg wheel has a 0.4 m radius and a 0.2 m radius of gyration about its axle. The main strut is pivoted at O, 1.1 m above the wheel axle. Determine the torsional moment on the main strut caused by gyroscopic action when the landing gear is retracted (soon after takeoff) at the rate of $\omega = 1$ rad/s.

FIGURE P20-55

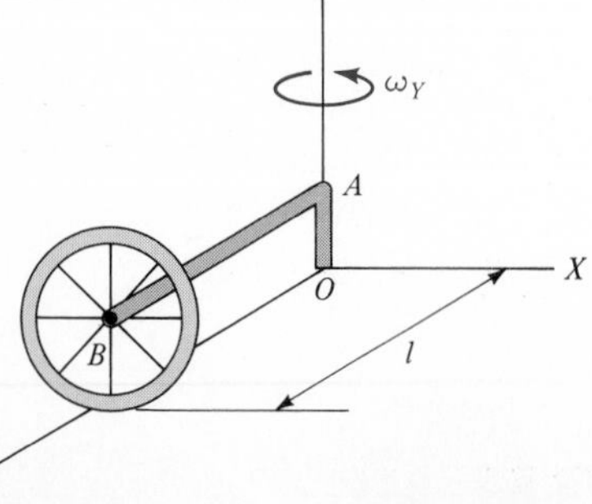

20–55 A wheel in a test is modeled as a thin hoop of 10 lb weight and 1 ft radius. The light guiding arm AB has a length $l = 6$ ft and is pivoted at A. The wheel rolls on the horizontal plane at the rate of $\omega_Y = 50$ rpm. Calculate the normal force between the wheel and the surface.

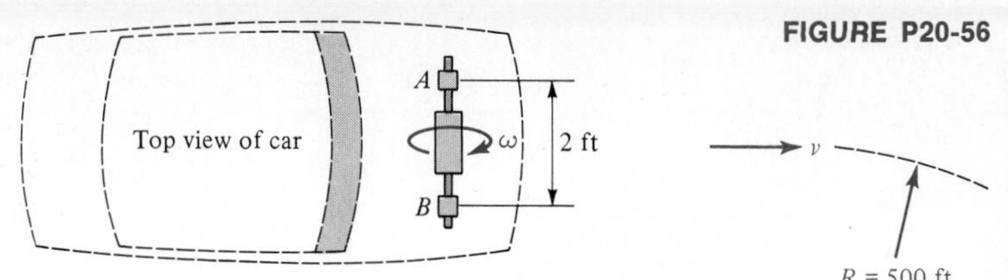

FIGURE P20-56

20–56 An electric car has a transverse-mounted rotor of 12 lb weight and 3 in. radius of gyration about its axle. The rotor is turning at the rate of $\omega = 1800$ rpm when the car enters a 500-ft-radius curve on a level road at a speed of 80 ft/s. Determine the horizontal and vertical reactions at bearings A and B caused by the motion of the vehicle.

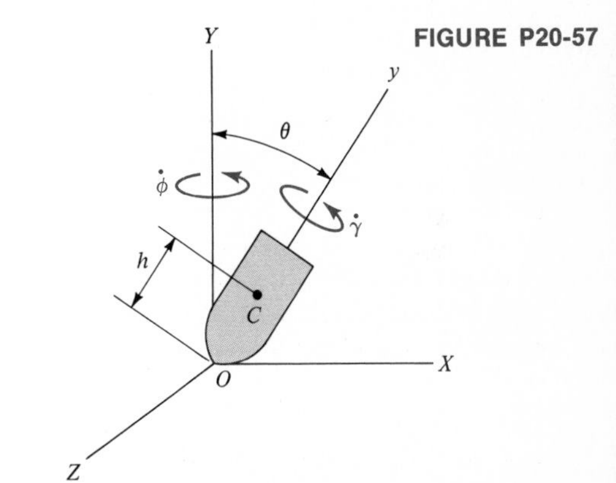

FIGURE P20-57

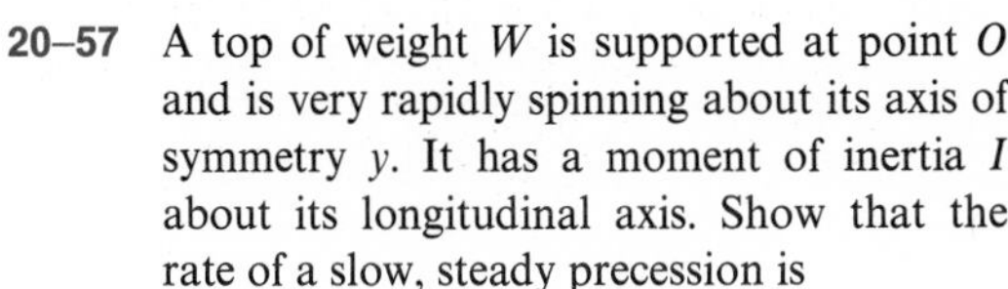

20–57 A top of weight W is supported at point O and is very rapidly spinning about its axis of symmetry y. It has a moment of inertia I about its longitudinal axis. Show that the rate of a slow, steady precession is

$$\phi \simeq \frac{Wh}{I\dot{\gamma}}$$

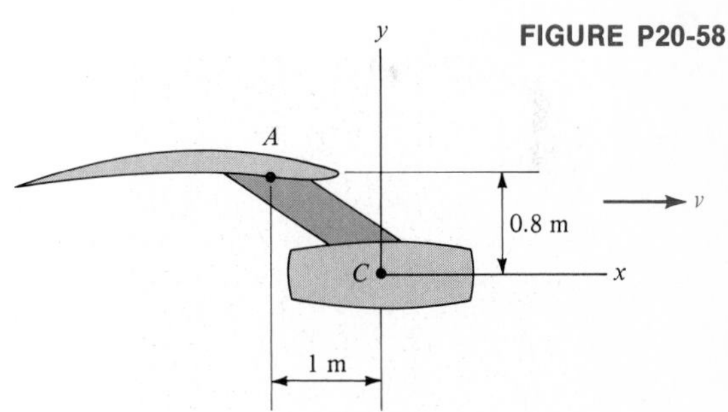

FIGURE P20-58

20–58 The 50-kg rotor of a jet engine has a 0.17 m radius of gyration about its axle and rotates at $\boldsymbol{\omega} = 35\,000\mathbf{i}$ rpm. A maneuver causes a rotation of $\boldsymbol{\Omega} = 1\mathbf{k}$ rad/s. Determine the moment reaction caused by these motions at point A of the supporting pylon.

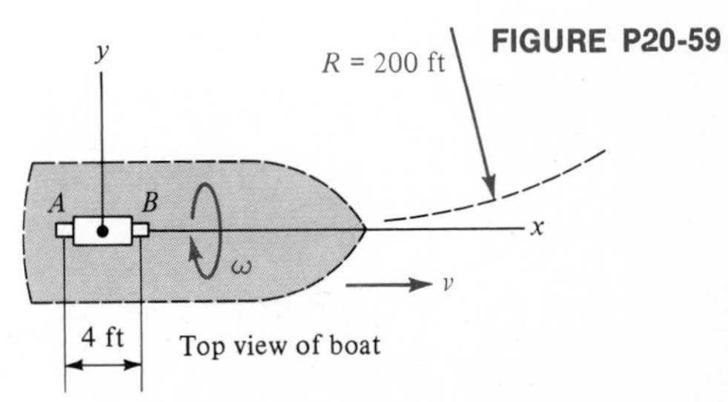

FIGURE P20-59

20–59 The 20-lb turbine rotor of a hydroplane boat has a 5-in. radius of gyration about its axle and is rotating at $\boldsymbol{\omega} = 20{,}000\mathbf{i}$ rpm. The boat is turning left on a circular path of 200 ft radius while traveling at 100 ft/s. Determine the horizontal and vertical reactions at bearings A and B caused by these motions.

FIGURE P20-60

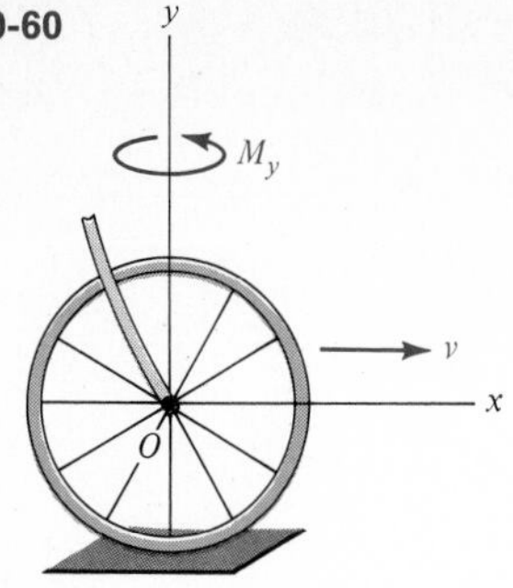

20–60 The wheels of a motorcycle are modeled as thin hoops of 6 kg mass and 0.7 m diameter. The motorcycle is traveling at 100 km/h when the rider applies a moment $\mathbf{M}_y = 10 \text{ N}\cdot\text{m } \mathbf{j}$ to the front wheel. Determine the resulting precession assuming that the front wheel is a free body.

FIGURE P20-61

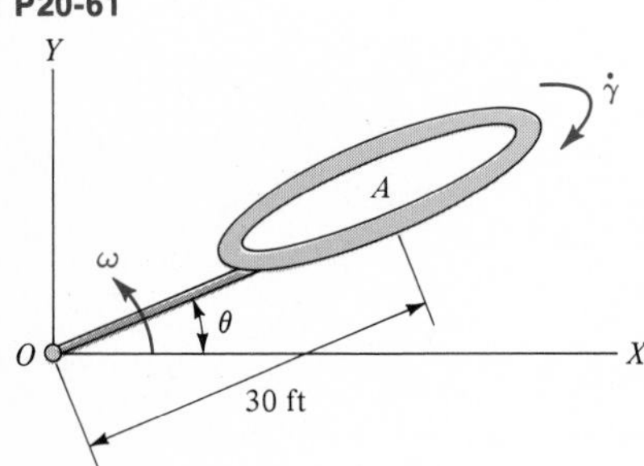

20–61 An amusement ride is modeled as a thin ring of 12,000 lb weight and 20 ft radius. It is spinning at the rate of $\dot{\gamma} = 1.5$ rad/s about A on the rigid arm OA, which is pivoted at O. The arm and the ring are tilted upward about the Z axis through O at the rate of $\omega =$ 0.4 rad/s. Determine the moment $\mathbf{M}_O$ caused by gyroscopic action when $\theta = 30°$.

FIGURE P20-62

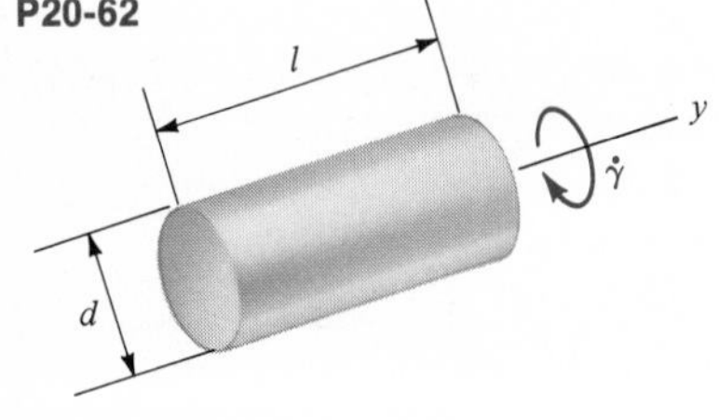

20–62 A uniform, 2-kg cylinder is spinning at the rate of $\dot{\gamma} = 300$ rpm about its axis of symmetry y, which itself has a slight precession. Determine if the precession is direct or retrograde for $l = 2d = 0.2$ m.

20–63 Determine the ranges of l/d in Prob. 20–62 for direct and retrograde precessions.

FIGURE P20-64

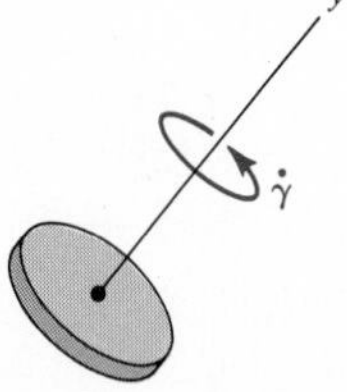

20–64 A thin disk of mass m and radius r is spinning about its axis of symmetry y at the rate $\dot{\gamma}$. The y axis has a slight precession at the rate of $\dot{\phi} = 10$ rpm. Calculate $\dot{\gamma}$.

Section 20-1

Angular momentum in three dimensions

PRECESSION: Rotation of angular velocity vector $\boldsymbol{\omega}$ ($\boldsymbol{\Omega}$ is angular velocity of $\boldsymbol{\omega}$ and $\mathbf{H}$).

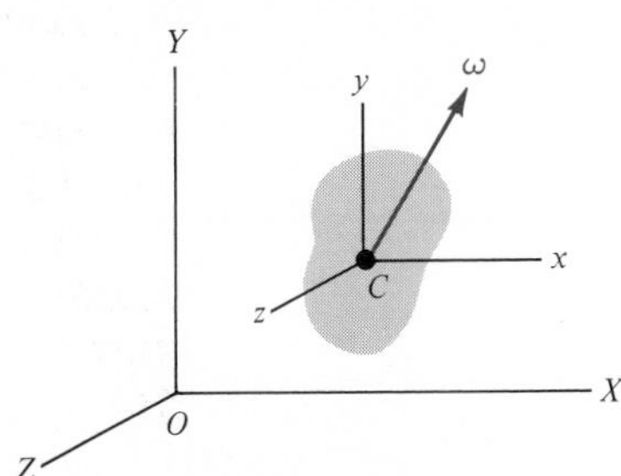

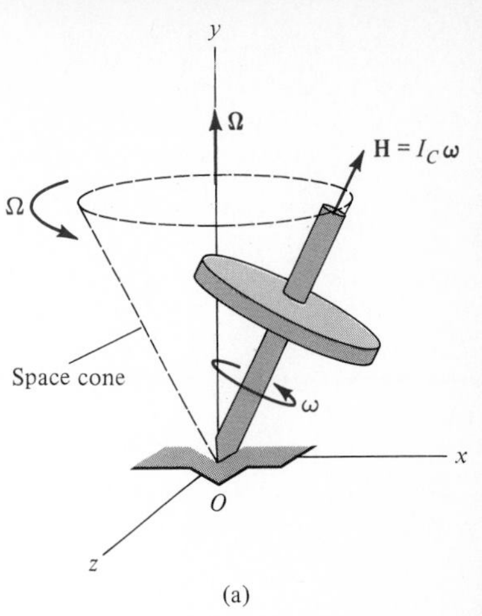

(a)

The xyz frame is attached at the mass center C and has fixed orientation. The body rotates relative to xyz.

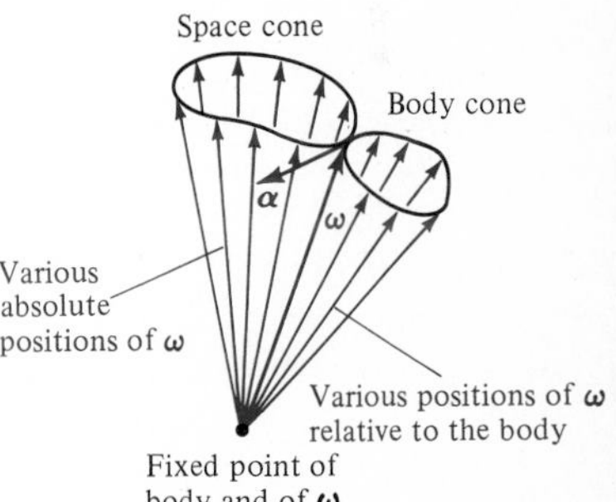

(b) General relation of body and space cones

RECTANGULAR COMPONENTS OF TOTAL ANGULAR MOMENTUM $\mathbf{H}_C$:

$$H_x = I_x\omega_x - I_{xy}\omega_y - I_{xz}\omega_z$$
$$H_y = -I_{xy}\omega_x + I_y\omega_y - I_{yz}\omega_z$$
$$H_z = -I_{zx}\omega_x - I_{zy}\omega_y + I_z\omega_z$$

These are valid for a particular position of the body.

For an arbitrary point O of a fixed lmn frame, $\mathbf{H}_O$ is determined in a similar procedure (l, m, and n replace x, y, and z, respectively), but all I must be evaluated with respect to the lmn frame.

Section 20-2

Impulse and momentum of a rigid body in three-dimensional motion

$$\text{System momenta} = \begin{cases} \text{linear momentum of mass center } (\mathbf{G}) \\ \text{angular momentum about mass center } (\mathbf{H}_C) \end{cases}$$

APPLICATION OF PRINCIPLE OF IMPULSE AND MOMENTUM FROM TIME t_1 TO TIME t_2:

$$\mathbf{G}_2 = \mathbf{G}_1 + (\text{external linear impulses})\big|_1^2$$
$$\mathbf{H}_{C_2} = \mathbf{H}_{C_1} + (\text{external angular impulses})\big|_1^2$$

Section 20-3

Kinetic energy of a rigid body in three-dimensional motion

$$T = \underbrace{\frac{1}{2}mv_C^2}_{\substack{\text{translation}\\ \text{of mass center}}} + \underbrace{\frac{1}{2}\omega \cdot \mathbf{H}_C}_{\substack{\text{rotation about}\\ \text{mass center}}}$$

For a rigid body that has a fixed point O,

$$T = \frac{1}{2}\omega \cdot \mathbf{H}_O$$

Section 20-4

Equations of motion in three dimensions

$$\sum \mathbf{F} = m\mathbf{a}_C$$

$$\sum \mathbf{M}_C = \dot{\mathbf{H}}_C = (\dot{\mathbf{H}}_C)_{xyz} + \mathbf{\Omega} \times \mathbf{H}_C$$

where $\mathbf{a}_C$ = acceleration of mass center
$\mathbf{H}_C$ = angular momentum of the body about its mass center
xyz = frame fixed in the body with origin at the mass center
$\mathbf{\Omega}$ = angular velocity of the xyz frame with respect to a fixed XYZ frame

An arbitrary fixed point O may be used for reference (rather than point C) if used consistently.

Section 20-5

Euler's equations of motion

If the xyz axes (fixed in the body) coincide with the principal axes of inertia of the body,

$$\sum M_x = I_x\dot{\omega}_x - (I_y - I_z)\omega_y\omega_z$$

$$\sum M_y = I_y\dot{\omega}_y - (I_z - I_x)\omega_z\omega_x$$

$$\sum M_z = I_z\dot{\omega}_z - (I_x - I_y)\omega_x\omega_y$$

where all quantities must be evaluated with respect to the appropriate principal axes.

Section 20-6

Solution of basic problems in three-dimensional motion

Apply six independent scalar equations to analyze the motion of an unconstrained rigid body:

$$\sum F_x = ma_{C_x} \qquad \sum F_y = ma_{C_y} \qquad \sum F_z = ma_{C_z}$$

$$\sum M_x = \dot{H}_x + \omega_y H_z - \omega_z H_y$$

$$\sum M_y = \dot{H}_y + \omega_z H_x - \omega_x H_z$$

$$\sum M_z = \dot{H}_z + \omega_x H_y - \omega_y H_x$$

The equations of motion are simplified if the body has a fixed point or a fixed axis.

Section 20-7

Gyroscopic motion

EQUATION OF MOTION:

$$\sum \mathbf{M}_O = (\dot{\mathbf{H}}_O)_{xyz} + \mathbf{\Omega} \times \mathbf{H}_O$$

where $\mathbf{\Omega}$ is the angular velocity of the xyz frame, which is fixed in the inner gimbal of the gyroscope.

Section 20-8

Special cases of gyroscopic motion

STEADY PRECESSION:

$$\sum \mathbf{M}_O = \mathbf{\Omega} \times \mathbf{H}_O$$

PRECESSION AND SPIN AXES ARE AT A RIGHT ANGLE IN STEADY PRECESSION:

$$\sum M_x = I_1 \dot{\phi} \dot{\gamma}$$

where x = axis fixed in inner gimbal, perpendicular to the spin axis
I_1 = moment of inertia about the spin axis
$\dot{\phi}$ = rate of precession
$\dot{\gamma}$ = rate of spin

REVIEW PROBLEMS

FIGURE P20-65

Model aircraft in a wind tunnel (The Boeing Company, Seattle, Washington)

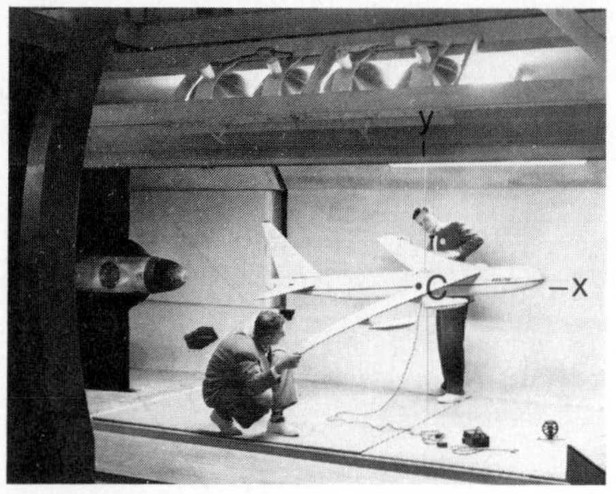

20–65 A model aircraft is suspended at its center of mass C by taut vertical wires. The fuselage is assumed to be a uniform cylinder of mass M, radius r, and length l, with the nose of the craft at $x = 0.5l$. The horizontal wings are assumed to be uniform slender rods of mass m and length $0.6l$, attached to the fuselage at $x = 0.1l$ and $y = r$, and swept back at 45°. The model is nosing over at the rate of ω when it is in the horizontal position. Determine its angular momentum about point C.

20–66 The model in Prob. 20–65 is at rest when it receives an impulse $F\,\Delta t$ in the positive x direction. F acts at $x = -0.2l$, $y = 0$, and $z = 0.2l$. Determine the resulting angular velocity of the model about point C.

20–67 Calculate the final kinetic energy of the model in Prob. 20–66.

FIGURE P20-68

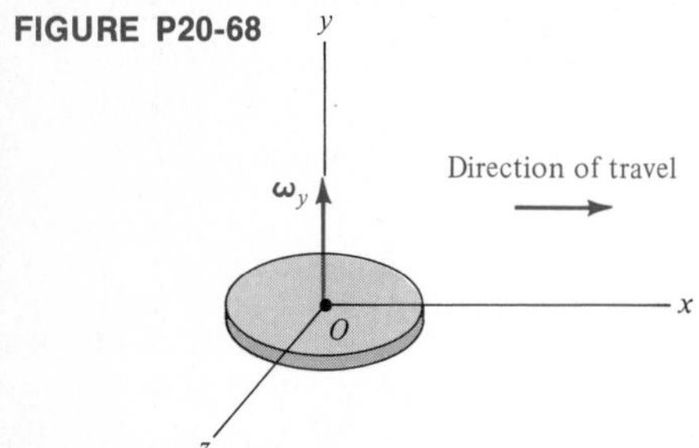

20–68 The flywheel of an experimental car is modeled as a uniform disk of 15-kg mass and 0.3 m radius. It is rotating at $\omega_y = 12\,000$ rpm when braking of the car tilts $\boldsymbol{\omega}_y$ toward the x axis at the rate of 1 rad/s. Determine the moment acting on the disk's shaft caused by the braking.

FIGURE P20-69

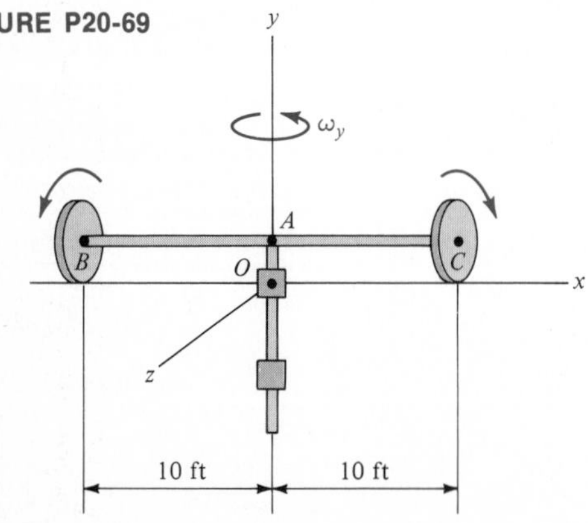

20–69 A tire tester has two identical 30-lb wheels in a symmetric arrangement. Axles AB and AC have negligible weight and rotate at a constant angular speed $\omega_y = 100$ rpm. The wheels are modeled as uniform disks rolling on the horizontal xz plane. Determine the vertical reactions on each wheel if joint A allows rotations of the axles in vertical planes.

20–70 A spacecraft is modeled as a uniform cylinder of 2000-kg mass, 3-m length, and 2-m diameter. It is rotating about its axis of symmetry y at $\omega_y = 30$ rpm. A 0.1-kg meteorite is traveling in the xy plane, in the negative x direction, at $v = 3000$ m/s relative to the spacecraft. The meteorite is embedded in the cylinder at point A. Determine the angular velocity and angular momentum vectors of the spacecraft immediately after the impact.

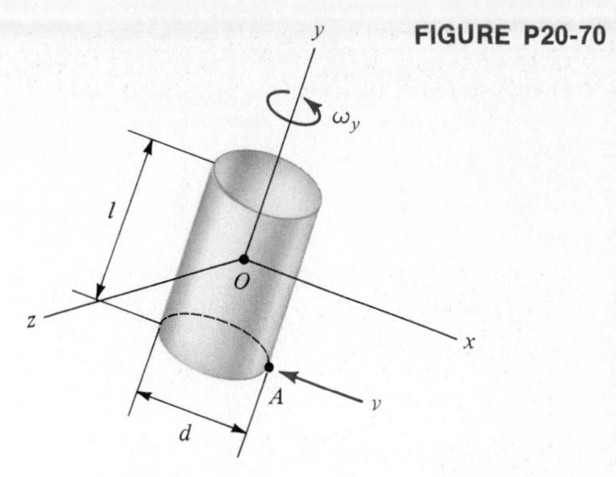

FIGURE P20-70

20–71 A space station is modeled as a thin ring of mass m and average radius $r = 200$ ft. It is rotating about its axis of symmetry y at $\omega_y = 2$ rpm. A meteorite of mass $m_1 = m/10^5$ is traveling in the xy plane, in the negative y direction, at $v = 20{,}000$ ft/s relative to the station. The meteorite is embedded in the ring at A. Determine the angular velocity and angular momentum vectors of the space station immediately after the impact.

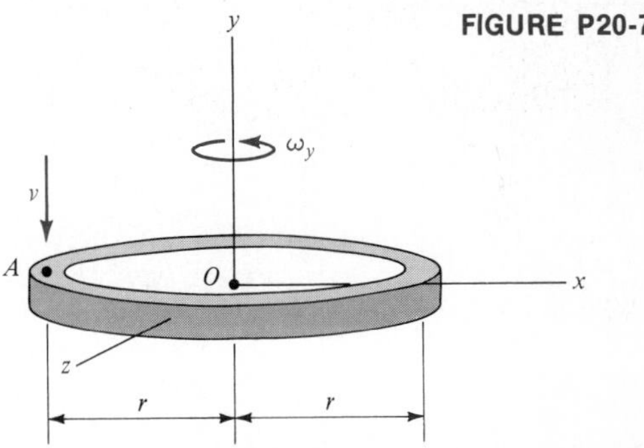

FIGURE P20-71

CHAPTER 21

VIBRATIONS

OVERVIEW

A rigid body or a system of bodies is in vibration when its motion involves repeated oscillations about an equilibrium position. The oscillation may be repeated uniformly, or change with time. Vibrations in machines and structures are quite common and undesirable in most cases. Their undesirable effects may be classified with respect to human characteristics and damage to engineering structures. An extreme example is a slender skyscraper whose wind-induced oscillations are entirely safe for the structure yet unpleasant to the occupants on the upper floors. At the other extreme, certain vibrations in an airplane may be unnoticeable to the passengers yet cause damage (fatigue) with catastrophic consequences. Naturally, the simultaneous occurrence of unpleasant and structurally damaging vibration, such as those in cars and trucks, is the most common. Vibration is useful in certain testing machines, compactors, and pile drivers.

STUDY GOALS

SECTIONS 21-1 AND 21-2 present the basic characteristics and definitions of vibrating systems. The governing differential equation of motion of a simple system is described in detail. The student of vibrations should acquire skill in setting up and solving a variety of governing differential equations of motion, so it is essential to learn the basic method of analysis.

SECTION 21-3 extends the concepts of vibrating particles to vibrating rigid bodies. It should be noted that the governing equation for a rigid body depends not only on the mass of that body and the elastic members of the system, but also on the configuration of the particular system.

SECTION 21-4 describes the analysis of undamped vibrating systems using the principle of conservation of energy. This method is advantageous when the maximum displacement and velocity of a body are involved in the analysis.

SECTION 21-5 presents forced vibrations without damping, a useful approach in machine design and structural engineering work. The condition of resonance is a particularly important topic of study. Readers of this section may note that analyzing various vibration problems requires solving differential equations with different forcing functions.

SECTIONS 21-6 THROUGH 21-8 present the analyses of vibrating systems with viscous damping, which frequently must be considered in engineering practice. The student should note the similarities and differences between damped and undamped vibrating systems of otherwise identical components.

SECTION 21-9 describes electrical analogues of mechanical vibrations. This topic is not essential for the mechanics analysis of vibrations, but it is relevant to design practice, and it shows the value of a broad engineering education.

SIMPLE HARMONIC MOTION 21-1

The basic characteristics of all vibrating systems are introduced using a simple system that consists of a mass m and a light spring of stiffness k. Figure 21-1a shows this system in equilibrium where the spring exerts no force on the block. Assume that the block is displaced to position x on the frictionless surface by a *disturbing force* P (also called *shaking force* or *exciting force*). This force is opposed by the *restoring force* F, as shown in the free-body diagram in Fig. 21-1b. The restoring force is always directed toward the equilibrium position. When the block is released in a position of displacement, the disturbing force P becomes zero and the restoring force F accelerates the block toward its equilibrium position. The equation of motion is written according to Newton's second law, with the positive direction taken to the right,

$$\boxed{\begin{aligned} \sum F_x &= ma_x \\ -kx &= m\ddot{x} \end{aligned}} \qquad \boxed{21\text{-}1}$$

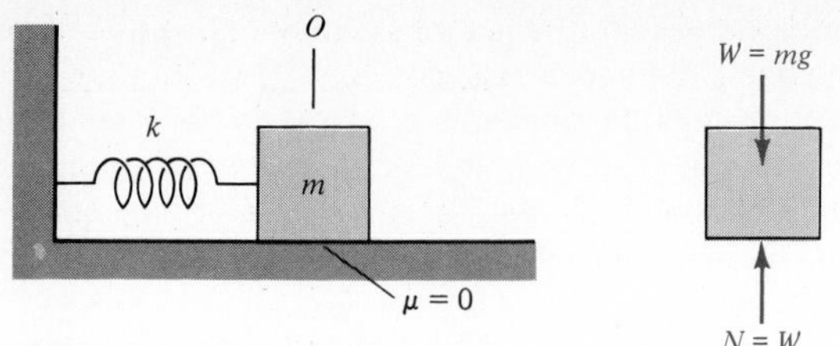

(a) Equilibrium position (force of spring is zero here)

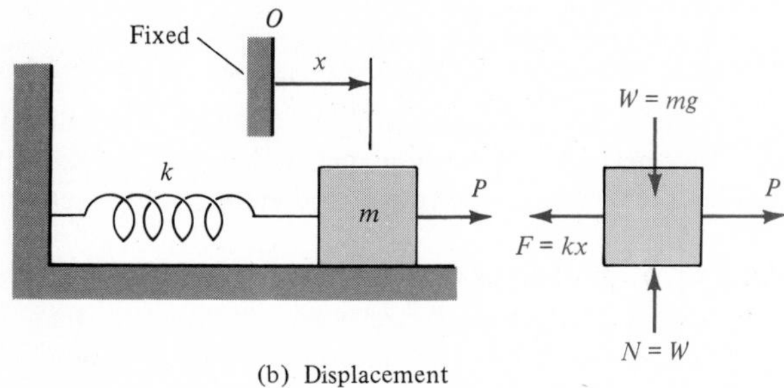

(b) Displacement

FIGURE 21-1

Model of a vibrating system

This indicates that the block in Fig. 21-1b accelerates to the left when $P = 0$, as expected.

As the block is pulled to the left by the stretched spring, it has kinetic energy upon reaching the equilibrium position O, overshoots this position, and compresses the spring. Now the restoring force is to the right, and the block accelerates to the right. Such changes in the direction of the restoring force cause the repeated oscillations of the block.

For a detailed analysis of the motion, Eq. 21-1 is written as

$$\ddot{x} + \omega^2 x = 0 \tag{21-2}$$

where

$$\omega = \sqrt{k/m} \tag{21-3}$$

is called the *natural circular frequency* of the vibration. A motion for which Eq. 21-2 is valid is called *simple harmonic motion.* The same equation may also be obtained by considering a mass suspended by a spring and oscillating in the vertical direction. In that case the spring has an initial deformation in the equilibrium position.

Equation 21-2 is a linear differential equation of the second order with constant coefficients. The general solution of this equation is

$$x = C_1 \sin \omega t + C_2 \cos \omega t \tag{21-4}$$

where t is time and C_1 and C_2 are arbitrary constants. The speed v and acceleration a of the vibrating mass are obtained by differentiating Eq. 21-4,

$$v = \dot{x} = C_1\omega \cos \omega t - C_2\omega \sin \omega t \tag{21-5}$$

$$a = \ddot{x} = -C_1\omega^2 \sin \omega t - C_2\omega^2 \cos \omega t \tag{21-6}$$

Substituting from Eqs. 21-4 and 21-6 into Eq. 21-2 is used to prove that Eq. 21-4 is the solution of the equation of motion. The constants C_1 and C_2 can be determined from the *initial conditions* of the motion.

For convenience, Eq. 21-4 is expressed as a single sine function by introducing new constants A and ϕ such that $C_1 = A \cos \phi$ and $C_2 = A \sin \phi$. With these, Eq. 21-4 becomes

$$x = A \cos \phi \sin \omega t + A \sin \phi \cos \omega t$$

This is simplified by using the formula $\sin(\alpha + \beta) = \sin \alpha \cos \beta + \cos \alpha \sin \beta$, resulting in

$$\boxed{x = A \sin(\omega t + \phi)} \tag{21-7}$$

The plot of this equation as x vs. ωt in Fig. 21-2 is the typical graphic representation of vibratory motion. The maximum displacement of the vibrating mass from its equilibrium position ($x = 0$) is A, the amplitude of the vibration. The constant ϕ is called the *phase angle*, which allows defining the initial displacement $A \sin \phi$ of the mass at time $t = 0$. The time τ required to complete a *cycle* of the motion is the *period* of the motion. A complete cycle occurs when $\omega t = 2\pi$, so

$$\boxed{\text{period} = \tau = \frac{2\pi}{\omega} = 2\pi\sqrt{\frac{m}{k}} \quad \text{(seconds per cycle)}} \tag{21-8}$$

after substituting for ω from Eq. 21-3. An important characteristic

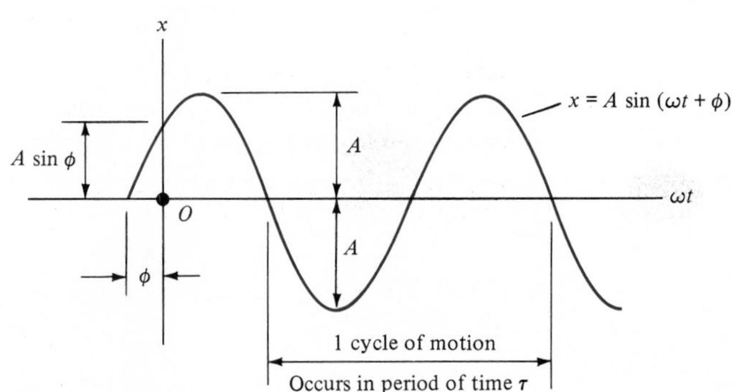

FIGURE 21-2

Graph of simple harmonic displacements

of a vibrating system is the rate at which complete cycles of the motion are repeated. This is defined by the *frequency* of the motion which represents the number of cycles that occur in a unit of time. The frequency f is the reciprocal of the period,

$$\text{frequency} = f = \frac{1}{\tau} = \frac{\omega}{2\pi} = \frac{1}{2\pi}\sqrt{\frac{k}{m}} \tag{21-9}$$

The frequency f is expressed in cycles per second (cps), with units of 1/s. Frequency is denoted by *hertz* (Hz) in the SI system of units. The circular frequency ω has units of rad/s.

Graphic Recordings of Simple Harmonic Motion

It is illustrative to consider two experiments for obtaining graphic vibration patterns. First, assume that a mass m is suspended by a spring of stiffness k (Fig. 21-3). A strip of paper can be moved near the mass to continually record the position of the mass. Assume that the mass is released at position $-x_0$ with the spring stretched when time $t = 0$. The paper moves at a constant speed v, and the oscillations of the mass are recorded as in Fig. 21-3. The amplitude of the vibration is x_0, and time is measured to the right on the paper (as by an observer moving with the paper). It should be noted that the recorded amplitude is not dependent on the chart speed v, but the spacing of the oscillations is dependent on v. Thus, v must be known for measuring the frequency from the paper. The recorded pattern for simple harmonic motion is always a sine wave, even though this may not be obvious at very low or very high chart speeds.

For another illustration, assume that a particle P is moving

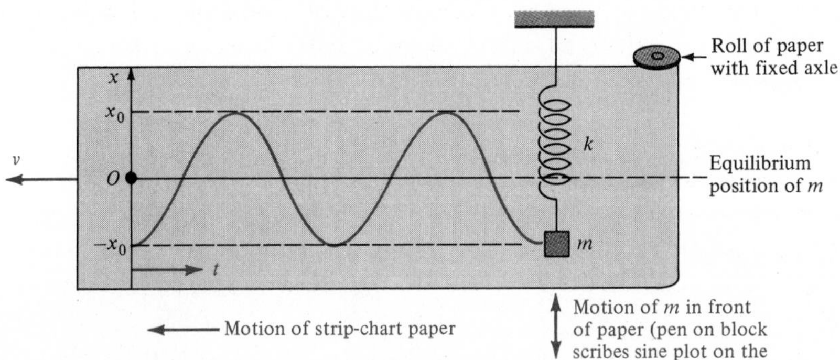

FIGURE 21-3

Schematic of a vibration experiment with recording of displacements

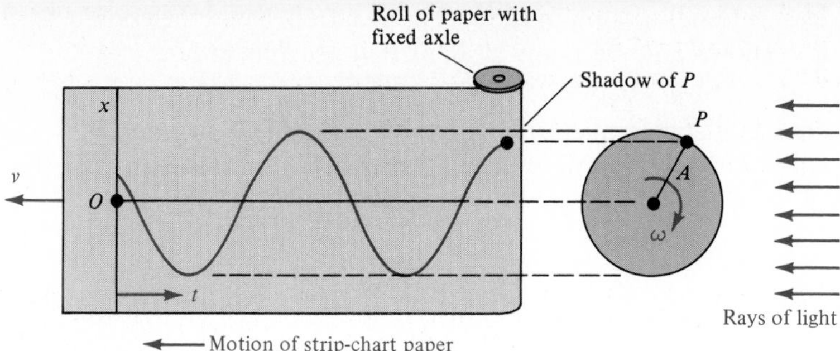

FIGURE 21-4
Illustration of simple harmonic motion

at a constant angular speed ω on a fixed circular path as in Fig. 21-4. The vertical position of P is recorded by its shadow on a horizontally moving chart. The recorded trace is a sine wave of amplitude A (radius of the circle) when ω and v are constant. This shows (in comparison with Fig. 21-3) that uniaxial vibrations may be associated with rotating members. This model also provides an interpretation for the meaning of the circular frequency ω since a complete cycle on the chart corresponds to the radius A sweeping through an angle of 2π radians.

FREE VIBRATIONS OF PARTICLES WITHOUT DAMPING

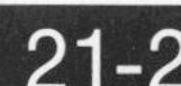

The basic concepts of analyzing a vibrating system that were developed in Sec. 21-1 are of a particular class of problems. The following list gives the characteristics of these problems and some practical considerations.

1. *One degree of freedom.* The motion of a mass m is defined by a single coordinate as a function of time (such as x in Eqs. 21-2 and 21-7). Such a system is said to have *one degree of freedom.*

2. *Free vibration.* Equation 21-2 is valid when a disturbing force is applied only once to give the mass an initial displacement (as in plucking a taut string). The mass is in *free vibration* when only two kinds of forces are acting on it: (a) the elastic restoring force within the system, and (b) gravitational or other constant forces that cause no displacement from the equilibrium position of the system.

3. *Undamped vibration.* In the absence of dissipative forces acting on a vibrating mass, the amplitude of the vibration is constant, and the motion is said to be *undamped.* This is an idealization because all engineering systems have some dissipative forces (at least internal friction in the spring material). Nevertheless, Eq. 21-2 is useful to analyze many systems that oscillate in air.

4. *Natural frequency.* Each mass–spring system vibrates at a characteristic frequency in free vibration. This is known as the *natural frequency* of the system. It is readily calculated using Eq. 21-9,

$$f = \frac{1}{2\pi}\sqrt{\frac{k}{m}}$$

5. *Lumped parameters.* Strictly speaking, Eq. 21-2 is valid only for a particle of mass m and a spring of no mass and spring constant k. In practice the mass of a translating rigid body is assumed to be concentrated in a particle, and the mass of the spring (whose mass center also moves) is often ignored resulting in a small error. Occasionally, several springs are connected to a single vibrating body. It is generally possible to represent these by a single equivalent spring constant.

21-3 FREE VIBRATIONS OF RIGID BODIES

Rigid bodies may be vibrating in rotational motion about a fixed axis. In such a case the particles of the body are in nonuniform motion, and it is necessary to derive an equation of motion which is appropriate for the particular case. For example, consider a concentrated mass m_1 on a uniform bar of mass m_2 which is hinged at point A and supported by a spring at point B as shown in Fig. 21-5a. The system is oscillating about the horizontal axis at A. The free-body diagram of the system in static equilibrium is given in Fig. 21-5b assuming that the bar is horizontal in equilibrium. δ is the static deflection of the spring. The force $k\delta$ is obtained after summing moments about A and writing $\sum M_A = 0$,

$$k\,\delta = \frac{3}{2}m_1 g + m_2 g \qquad \boxed{21\text{-}10}$$

The free-body diagram for an arbitrary position of the bar during its oscillations is shown in Fig. 21-5c. If θ is always small, $\sin\theta \simeq \theta$ (in radians), and the force of the spring may be written as $k(\delta + l\theta)$.

The equation of motion is

$$\sum M_A = I_A\alpha = I_A\ddot{\theta} \qquad \boxed{21\text{-}11}$$

Assume that clockwise is positive in Fig. 21-5c. After substituting for the moments M_A and the moments of inertia I_A,

$$\underbrace{-k(\delta + l\theta)}_{\text{force}}\underbrace{l}_{\text{arm}} + \underbrace{m_1 g}_{\text{force}}\underbrace{\left(\frac{3}{2}l\right)}_{\text{arm}} + \underbrace{m_2 g}_{\text{force}}\;\underbrace{(l)}_{\text{arm}}$$

$$= \underbrace{\left[m_2 \frac{4l^2}{3}\right.}_{(I_{\text{bar}})_A} + \underbrace{\left. m_1 \left(\frac{3}{2} l \right)^2 \right]}_{(I_{m_1})_A} \ddot{\theta}$$

But from Eq. 21-10,

$$\left(\frac{3}{2} m_1 g + m_2 g \right) l - (k\,\delta) l = 0$$

Hence,

$$\left(\frac{9}{4} m_1 + \frac{4}{3} m_2 \right) \ddot{\theta} + k\theta = 0 \qquad \boxed{21\text{-}12}$$

which is similar in form to Eq. 21-1. The term in parentheses in Eq. 21-12 is called the *effective mass* of the system. In general there may also be an *effective spring constant* which is numerically different from k. The details of these effective coefficients depend on the configuration of the system. The natural frequency of the system is determined by solving Eq. 21-9 using the effective mass and spring constant.

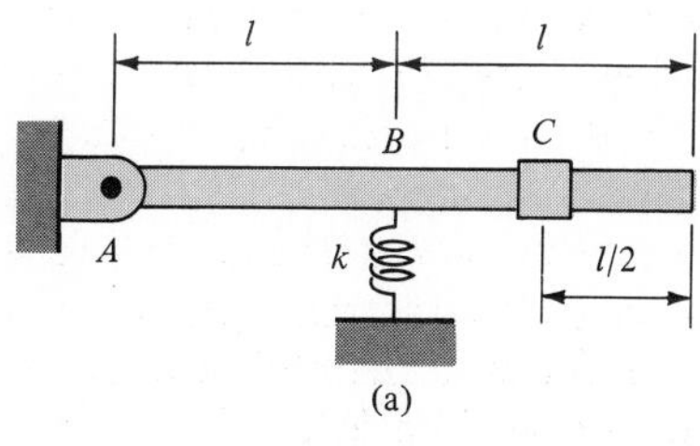

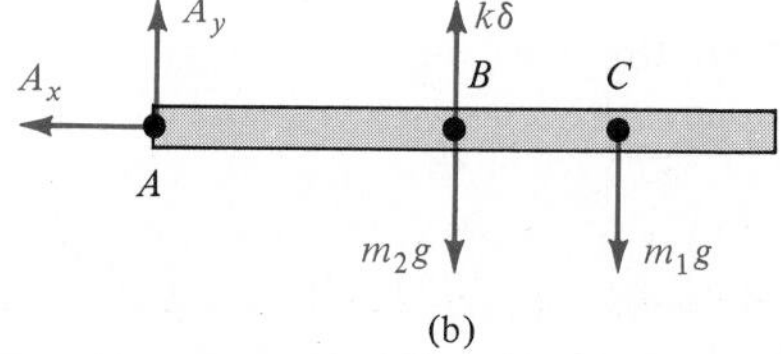

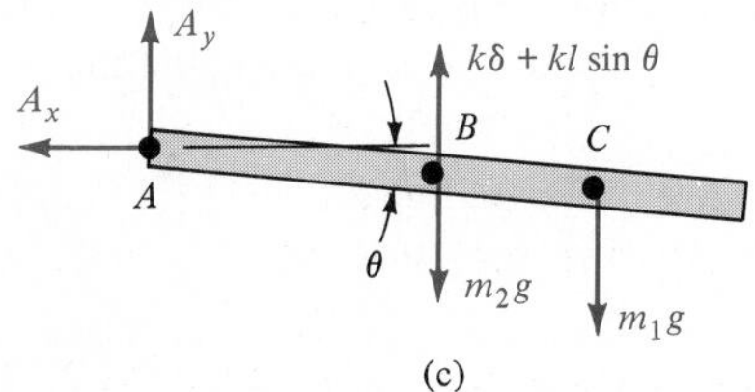

FIGURE 21-5

Model of a vibrating rigid body

EXAMPLE 21-1

A demonstration model for power-line galloping (wind-induced vibrations of aerial electric wires) consists of a uniform bar of mass $m = 1$ kg and four springs of stiffness $k = 100$ N/m arranged as in Fig. 21-6a. Write the equation of motion for free vertical vibration of this system and determine its natural frequency.

SOLUTION

The static equilibrium position is shown in Fig. 21-6b. With a static deformation δ of each spring,

$$\sum F_y = -4k\,\delta + mg = 0$$

It is seen here that the four springs have a resultant stiffness equivalent to a single spring of (effective) stiffness $K = 4k$. For a positive displacement y (down) as shown in Fig. 21-6c,

$$\sum F_y = ma_y$$

$$-4k(y + \delta) + mg = m\ddot{y}$$

From the equilibrium equation, $-4k\,\delta + mg = 0$; hence, the equation of motion becomes

$$m\ddot{y} + 4ky = 0$$

or

$$\ddot{y} + \frac{4k}{m}y = 0$$

From Eq. 21-2,

$$\omega = \sqrt{\frac{4k}{m}} = \sqrt{\frac{4(100\text{ N/m})}{1\text{ kg}}} = \sqrt{400\left(\frac{\text{kg}\cdot\text{m}}{\text{s}^2}\right)\left(\frac{\text{m}}{\text{kg}}\right)}$$

$$= 20\text{ rad/s}$$

$$f = \frac{1\text{ cycle}}{2\pi\text{ rad}}\,\omega = 3.18\text{ cycles/s}$$

Note that the equilibrium deformation force $k\,\delta$ and weight mg are again not involved in the final equation of motion. For simple harmonic motion and small deformations this is always true. Therefore, such forces can be (carefully) ignored when writing the equation of motion.

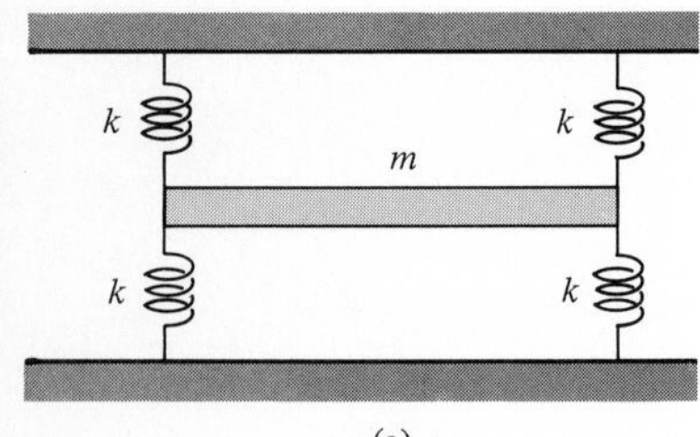

(a)

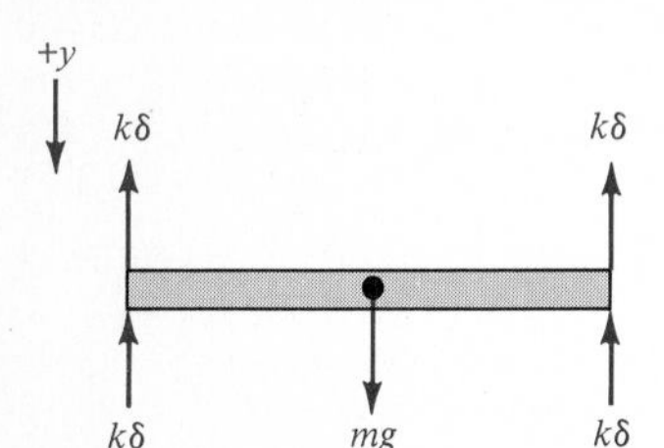

(b) Equilibrium position

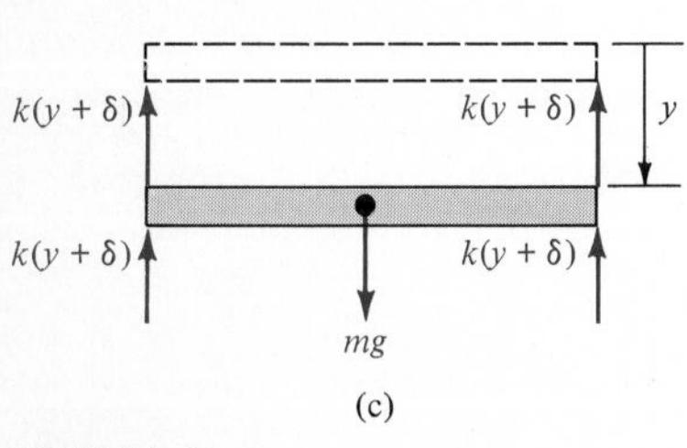

(c)

FIGURE 21-6
Model for power-line galloping

EXAMPLE 21-2

A simple pendulum consists of a concentrated mass m suspended on a light string of length l (Fig. 21-7a). Write the equation of motion of the pendulum for small amplitudes of displacement θ. Determine how the period of oscillations depends on m and l.

SOLUTION

The appropriate free-body diagram is sketched in Fig. 21-7b. Here the weight is not involved in any initial deflections and therefore must be included in the equation of motion (see Ex. 21-1).

$$\sum F_t = ma_t$$

$$-W \sin \theta = ma_t = m(l\ddot{\theta})$$

$$ml\ddot{\theta} + mg \sin \theta = 0$$

But, $\sin \theta \simeq \theta$ (in radians) is often used up to approximately 30° or 0.5 rad. Therefore, the equation of motion becomes

$$\ddot{\theta} + \left(\frac{g}{l}\right)\theta = 0$$

$$\omega = \sqrt{\frac{g}{l}}$$

$$\tau = \frac{2\pi}{\omega} = 2\pi\sqrt{\frac{l}{g}}$$

The period depends only on the length l (and g) and not on the mass. The period increases with increasing l, which agrees with common experience.

It might be noted that the equation of motion could have also been obtained by summing moments about the support A,

$$\sum M_A = I_A\alpha$$

$$-mgl \sin \theta = (ml^2)\ddot{\theta}$$

$$\ddot{\theta} + \left(\frac{g}{l}\right)\theta = 0$$

which agrees with the previous result.

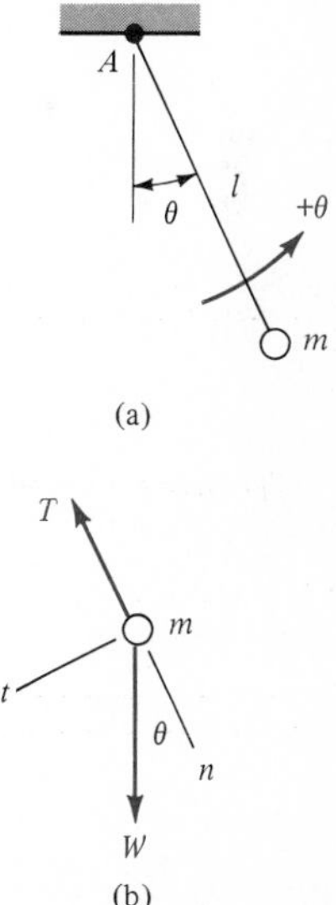

FIGURE 21-7
Analysis of an oscillating pendulum

EXAMPLE 21-3

The torsion pendulum of a satellite-tracking radar station is modeled as a 0.1-lb uniform disk of 2 in. diameter suspended on a thin rod of torsional stiffness k (Fig. 21-8a). Determine k in units of in·lb/rad if the natural frequency of the system should be 5 cycles per second.

SOLUTION

First the equation of motion is written in general terms. A torsion spring resists twists with a torque that depends on the angle turned. Hence, $M_{sp} = k\theta$, and from Fig. 21-8b,

$$\sum M_A = I_A \alpha$$

$$-k\theta = I_A \ddot{\theta}$$

$$\ddot{\theta} + \left(\frac{k}{I_A}\right)\theta = 0$$

$$\omega = \sqrt{\frac{k}{I_A}}$$

$$f = \frac{1}{2\pi}\sqrt{\frac{k}{I_A}}$$

$$k = (2\pi f)^2 I_A = \left(10\pi \frac{1}{s}\right)^2 I_A$$

$$= \left(100\pi^2 \frac{1}{s^2}\right)\left[\frac{1}{2}\left(\frac{0.1 \text{ lb}}{32.2 \text{ ft/s}^2}\right)\left(\frac{1 \text{ ft}}{12 \text{ in.}}\right)(1 \text{ in}^2)\right]$$

$$= 0.128 \text{ in} \cdot \text{lb/rad}$$

JUDGMENT OF THE RESULTS

The expression for the frequency appears reasonable since f increases with k and decreases with I. A stiffer spring will apply a larger torque for a given rotation hence it will cause a large angular acceleration and therefore more cycles per time period. An increase in the inertia I can be considered as an increase in resistance to rotation. In other words, the system will be more "sluggish," resulting in less cycles per time period for an increase in I.

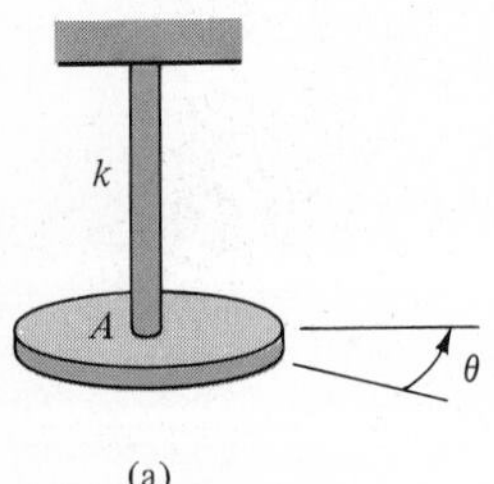

(a)

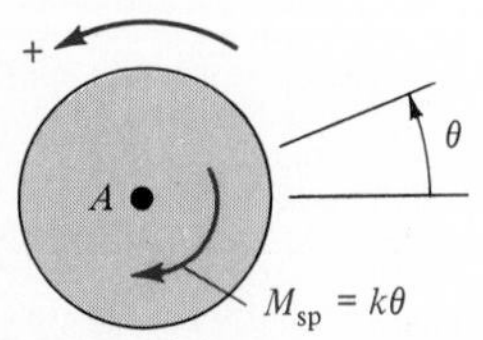

(b) Top view of disk

FIGURE 21-8

Analysis of an oscillating torsion pendulum

PROBLEMS

21–1 A 1-kg mass is suspended from a spring of stiffness $k = 2$ kN/m. Determine the period of oscillations of the mass.

21–2 A 2-lb block is supported by a spring of stiffness k. Determine the required stiffness to obtain the frequency of 5 s^{-1}.

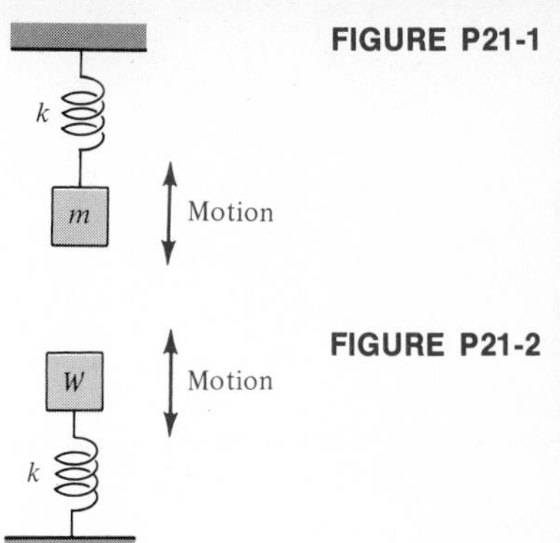

FIGURE P21-1

FIGURE P21-2

21–3 Calculate the maximum velocity and the maximum acceleration of the block in Prob. 21–2.

21–4 Determine the natural frequencies of the system in Prob. 21–1 for the mass moving without friction in the (a) vertical direction, and (b) horizontal direction.

21–5 A tuned mass damper has a mass of 300 000 kg and should have a 7-s period in free vibration. Calculate the required spring constant of the system (for a description of these, see Sec. 1-7 in the statics volume).

FIGURE P21-5

Tuned mass damper in motion (Courtesy MTS Systems Corporation, Minneapolis, Minnesota)

21–6 It is found that the average fruit freely vibrating in its container moves with an amplitude of 0.5 in. and a period of 0.3 s. Assuming that the motion is simple harmonic, calculate the maximum vertical acceleration of a fruit.

21–7 A uniform, 10-kg bar is suspended by two springs whose spring constants are $k = 5$ kN/m. Calculate the natural frequency of the bar in uniform vertical motion.

FIGURE P21-7

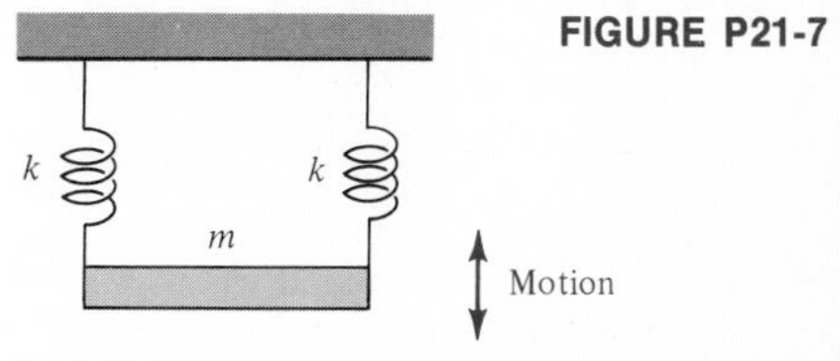

FIGURE P21-8

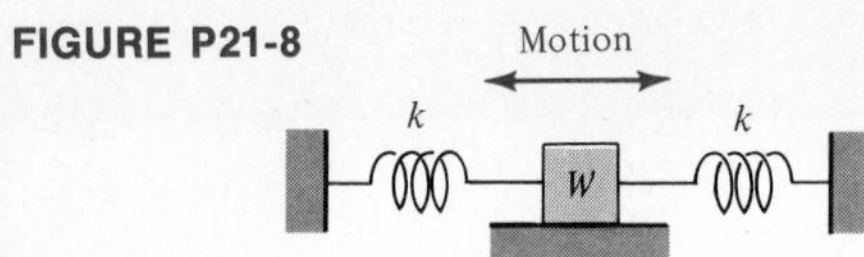

21–8 A 5-lb block slides without friction on a horizontal surface. The two springs of spring constant $k = 500$ lb/in. are attached to the block but apply no force to it in the equilibrium position. Determine the natural frequency of the block.

FIGURE P21-9

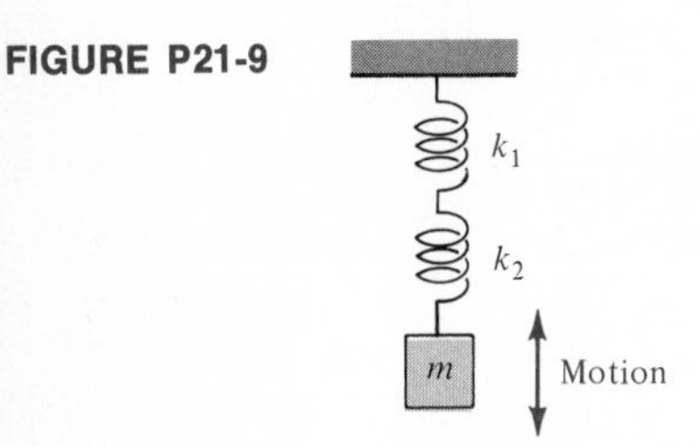

21–9 Derive the equation for the natural frequency of the mass supported by two springs in series.

FIGURE P21-10

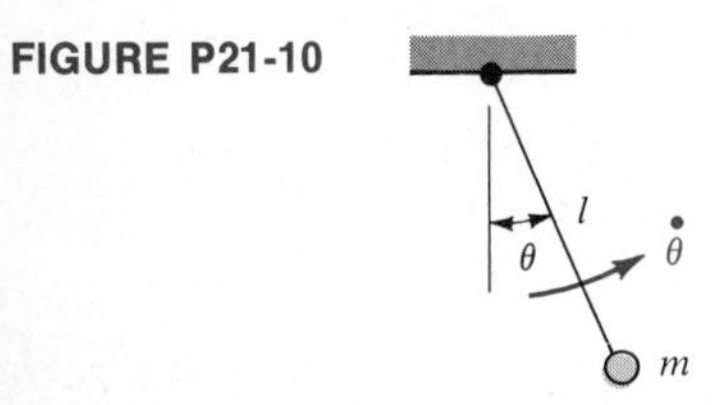

21–10 The pendulum has a length $l = 1$ m. Its angular speed is $\dot{\theta} = 0$ at $\theta_0 = 0.1$ rad. Calculate the period of the pendulum and write the equation for θ as a function of time if time is zero at θ_0.

FIGURE P21-11

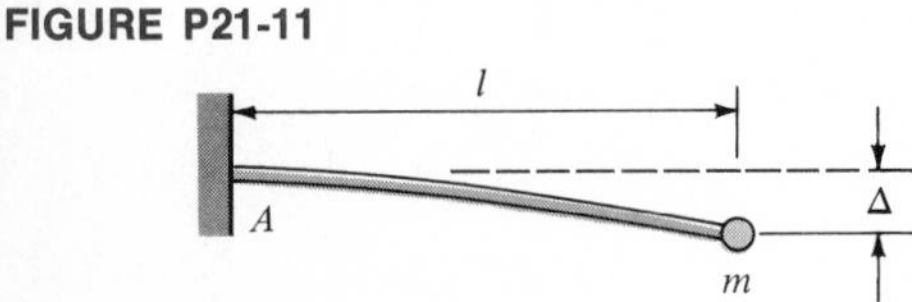

21–11 A bar of negligible mass with a concentrated mass m on its free end is firmly gripped at A. The static deflection of the free end is Δ. Derive the equation for the natural frequency of the system.

FIGURE P21-12

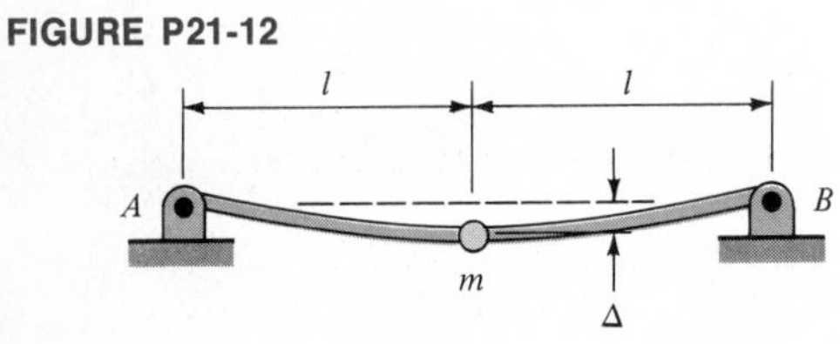

21–12 A bar of negligible mass with a concentrated mass m on it is hinged at the ends. The static deflection of the bar is Δ. Derive the equation for the natural frequency of the system.

FIGURE P21-13

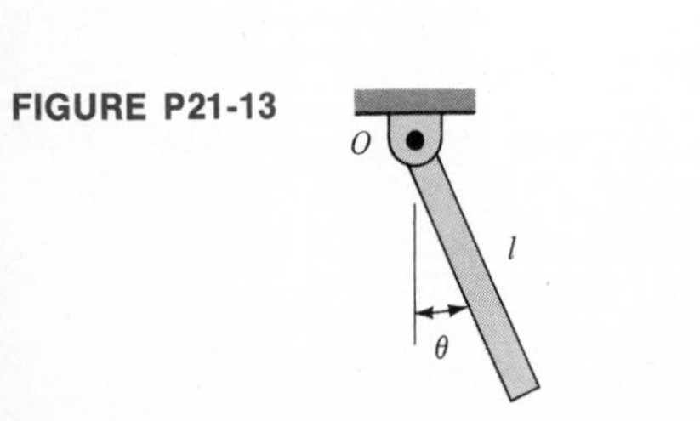

21–13 A uniform, slender bar of 3 kg mass and 1 m length is hinged at O. Calculate the period of small oscillations of the bar.

21–14 Determine the maximum angular speed $\dot{\theta}$ and angular acceleration $\ddot{\theta}$ of the bar in Prob. 21–13 if the angular amplitude is 0.1 rad.

21–15 A uniform, slender bar of weight $W_1 = 6$ lb and length $l = 4$ ft is hinged at O. A small block of weight $W_2 = 10$ lb is attached to the bar at $d = 3$ ft. Determine the required angular amplitude of this pendulum if the maximum speed of W_2 should be 1 ft/s.

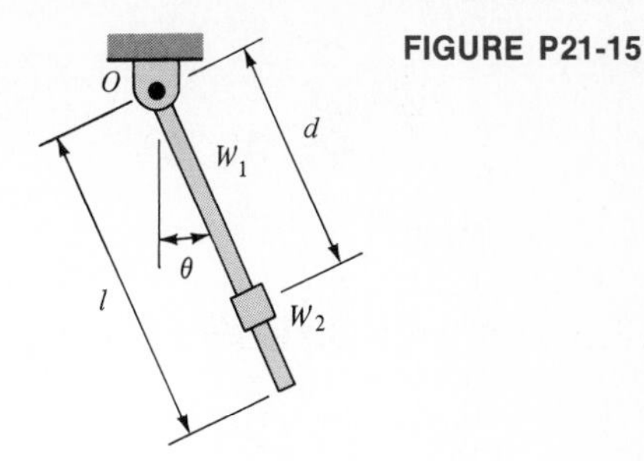

FIGURE P21-15

21–16 A uniform, slender bar of 4 kg mass and 1 m length is hinged at O and supported by a spring of spring constant $k = 5$ kN/m at $d = 0.75$ m. Determine the natural frequency of the system assuming that the bar is attached to the spring.

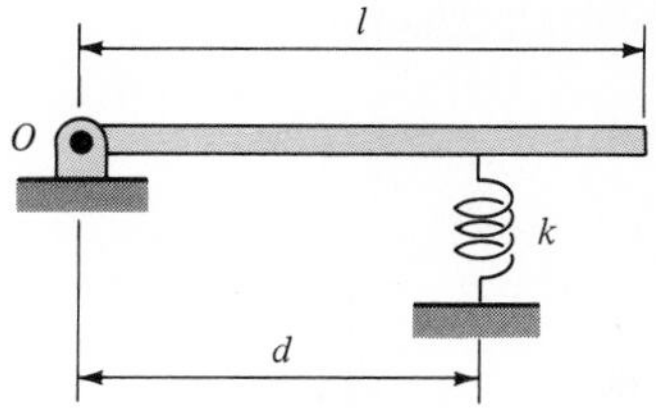

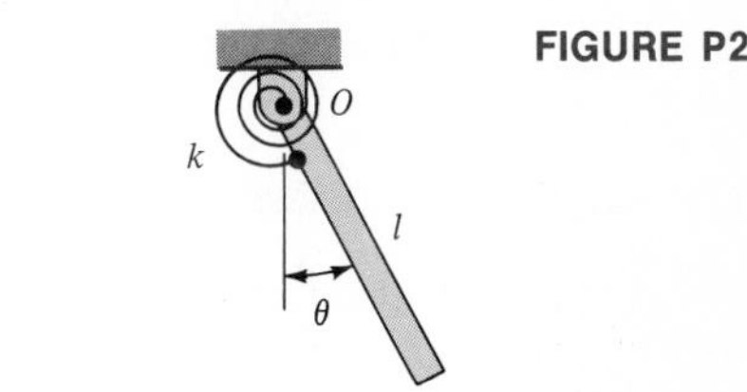

FIGURE P21-16

21–17 A uniform, slender bar of 10 lb weight and 3 ft length is hinged at O. A torsion spring of stiffness $k = 500$ ft·lb/rad is attached between the bar and its support. The spring applies no moment to the bar when $\theta = 0$. Calculate the natural frequency of the system.

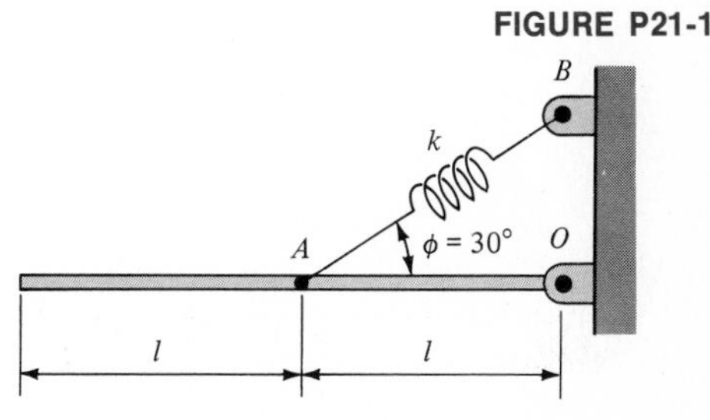

FIGURE P21-17

21–18 A uniform, slender bar of 10 kg mass and length $2l = 1.6$ m is hinged at O and is supported in the horizontal position by a spring of spring constant $k = 4$ kN/m. Determine the natural frequency of the system.

FIGURE P21-18

21–19 A uniform, thin plate of mass m is hinged at O. Determine the period of small oscillations of the plate.

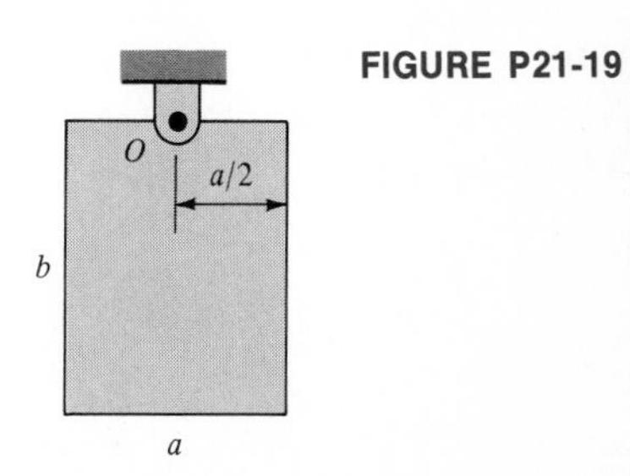

FIGURE P21-19

FIGURE P21-20

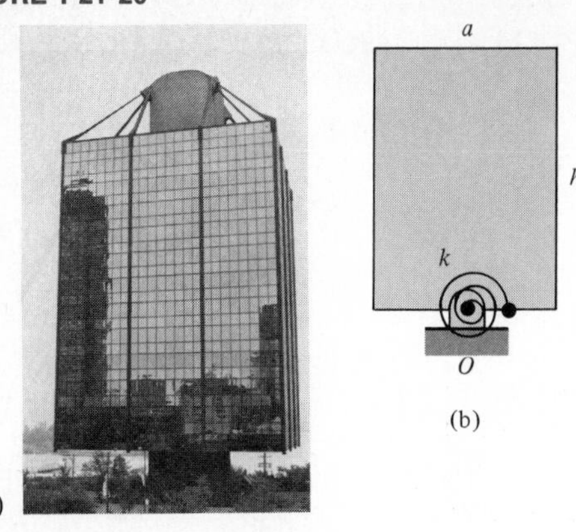

21–20 The modern building is modeled as a uniform, rigid bar of mass m, cross section $a \times a$, and height h. It is assumed to be hinged at O with a torsion spring of stiffness k restraining the building motions. Write the equation for the natural frequency of the system.

FIGURE P21-21

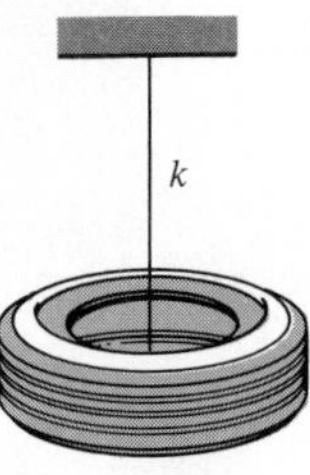

21–21 A wheel is suspended by a wire of torsional stiffness $k = 3$ in·lb/rad. The period of torsional vibration is 12 s. Calculate the moment of inertia of the wheel.

FIGURE P21-22

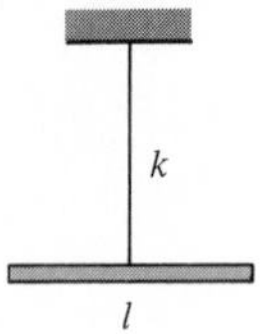

21–22 A slender rod of 1 kg mass and 0.4 m length is suspended at its center by a wire. Determine the torsional spring constant k of the wire if the period of torsional vibrations is 2 s.

FIGURE P21-23

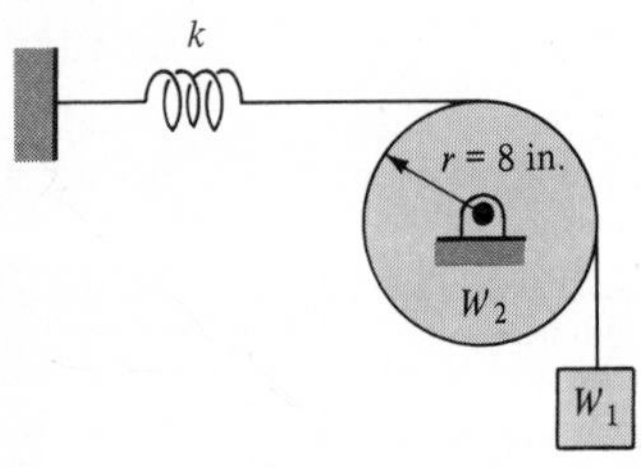

21–23 A 5-lb block is suspended from a wire that passes over a uniform disk of 6 lb weight. The disk rotates freely about its axle, and the spring constant is $k = 100$ lb/in. Determine the natural frequency of the system if the wire does not slip on the disk.

FIGURE P21-24

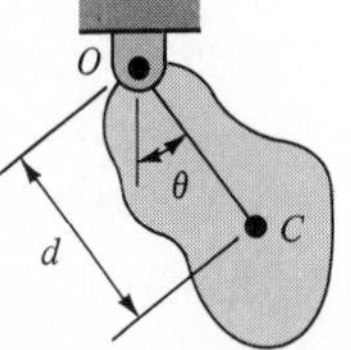

21–24 An arbitrary body of mass m and mass center at C is suspended by the hinge at O. The radius of gyration about C is r_g. Determine the period of small oscillations.

21-4 ENERGY METHODS IN VIBRATION ANALYSIS

Systems that vibrate without damping are conservative systems that can be analyzed using the principle of conservation of energy. This means that the total mechanical energy of the system is constant, total energy = kinetic energy + potential energy = constant,

$$T + V = \text{constant}$$

The principle can be used to determine the equation of vibratory motion and the natural frequency of a system as illustrated in the following. Consider a mass–spring system where the mass can move without friction on a horizontal surface (Fig. 21-9). The total energy of this system with the mass m at position x (measured from the equilibrium position) and moving at speed $\dot{x}$ is

$$T + V = \frac{1}{2}m\dot{x}^2 + \frac{1}{2}kx^2 = \text{constant} \qquad \text{21-13}$$

Differentiating this equation with respect to time gives

$$m\dot{x}\ddot{x} + kx\dot{x} = 0$$

which may be divided by the speed $\dot{x}$ since $\dot{x}$ is not zero when the mass moves between its extreme positions. The result is

$$\boxed{m\ddot{x} + kx = 0} \qquad \text{21-14}$$

which agrees with Eq. 21-1. The circular frequency is $\omega = \sqrt{k/m}$, as in Eqs. 21-3 and 21-9.

The energy method of vibration analysis is particularly advantageous when those positions of the mass are considered where the total energy consists entirely of potential energy or of kinetic energy.

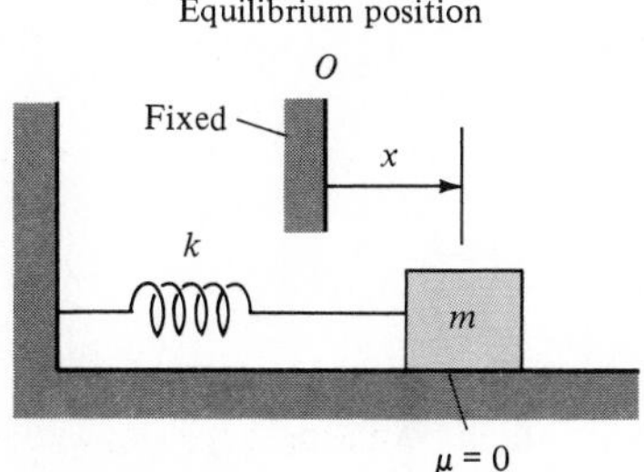

FIGURE 21-9

Model of a vibrating system

These positions are shown schematically in Fig. 21-10 and further explained as follows.

1. At every peak displacement A the kinetic energy T is zero as $\dot{\mathbf{x}}$ changes direction. At the same instants, the potential energy V is maximum and represents the total energy of the system.
2. At every instant B when the mass passes through its equilibrium position $x = 0$, the potential energy V is zero. At the same instants, the kinetic energy T is maximum since the speed $\dot{x}$ is maximum. This value of T represents the total energy of the system.

The analysis of the motion is facilitated by using the simple expressions that can be written for the maximum displacement x_{max} and speed $\dot{x}_{max}$. From Eq. 21-7,

$$x = A \sin(\omega t + \phi)$$

$$x_{max} = A \tag{21-15}$$

$$\dot{x} = A\omega \sin(\omega t + \phi)$$

$$\dot{x}_{max} = A\omega \tag{21-16}$$

Using these results gives

$$V_{max} = \frac{1}{2} kA^2 \quad \text{(the total energy of the system)} \tag{21-17}$$

$$T_{max} = \frac{1}{2} mA^2\omega^2 \quad \text{(also the total energy of the system)} \tag{21-18}$$

Since $V_{max} = T_{max}$, the circular frequency is

$$\omega = \sqrt{\frac{k}{m}}$$

which agrees with Eq. 21-3.

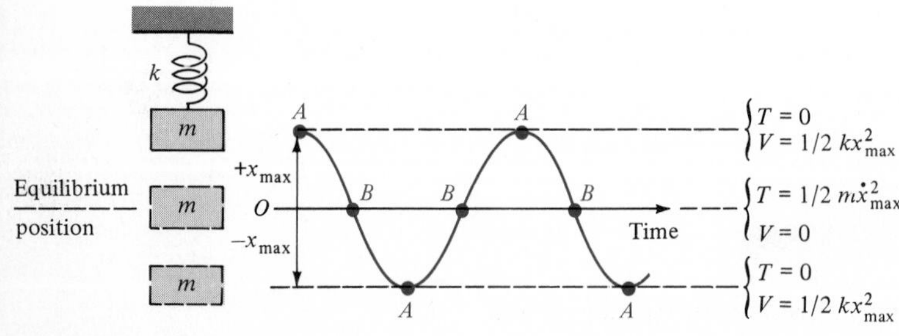

FIGURE 21-10

Kinetic and potential energies at equilibrium and peak displacements

EXAMPLE 21-4

A rocker arm is modeled as a uniform slender rod of mass m hinged at point O as in Fig. 21-11a. The spring touches the rod but applies no force to it when the rod is vertical. Derive an expression for the circular frequency ω of the system for small displacements.

SOLUTION

Two positions of the bar are of interest. Assume that position 1 in Fig. 21-11b represents a maximum displacement where the bar has no velocity. In position 2 the velocity of the bar is maximum but the system's potential energy is chosen to be zero. Thus,

$$T_1 = 0$$

$$V_1 = \frac{1}{2}k(2l)^2\theta_{max}^2 + \frac{1}{2}Wl\theta_{max}^2$$

where the last term is obtained using

$$1 - \cos\theta_{max} = 2\sin^2\frac{\theta_{max}}{2} \simeq 2\left(\frac{\theta_{max}}{2}\right)^2 = \frac{\theta_{max}^2}{2}$$

$$V_2 = 0$$

$$T_2 = \frac{1}{2}I_O\dot{\theta}_{max}^2 \qquad \text{(note: } \omega \neq \dot{\theta} \text{ here)}$$

Using $V_{max} = T_{max}$ yields

$$\frac{1}{2}k(2l)^2\theta_{max}^2 + \frac{1}{2}Wl\theta_{max}^2 = \frac{1}{2}I_O\dot{\theta}_{max}^2$$

From Eqs. 21-15 and 21-16,

$$x_{max} = A = 2l\theta_{max} \qquad \theta_{max} = \frac{A}{2l}$$

$$\dot{x}_{max} = A\omega = 2l\dot{\theta}_{max} \qquad \dot{\theta}_{max} = \frac{A\omega}{2l}$$

With $I_O = \frac{1}{3}m(2l)^2$,

$$\omega = \sqrt{\frac{2k}{m} + \frac{g}{2l}}$$

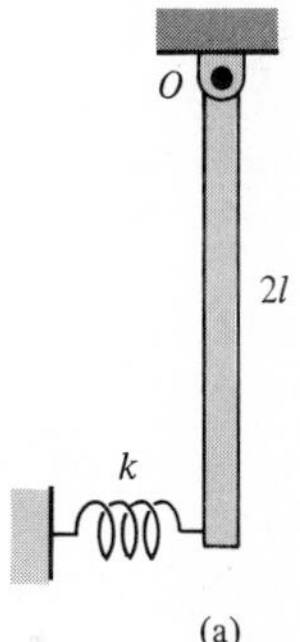

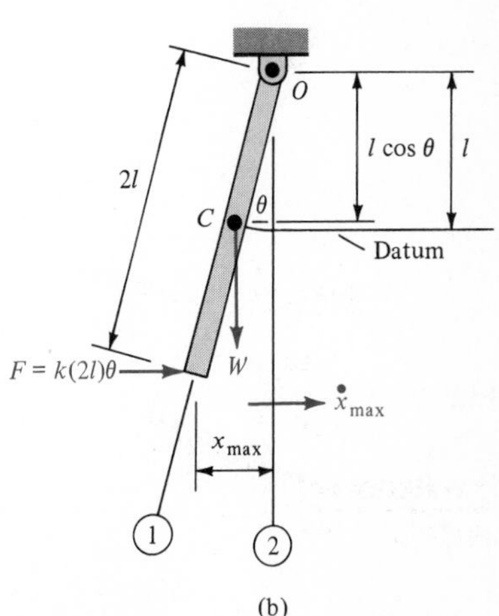

FIGURE 21-11
Model of a rocker-arm system

JUDGMENT OF THE RESULT

The quantities k, m, g, and l are all in reasonable places in the equation according to previous results for oscillating systems. For comparison, see Exs. 21-1 and 21-2.

PROBLEMS

SECTION 21-4

In Probs. 21–25 to 21–36, solve the indicated problem using energy methods.

21–25 Prob. 21–1.

21–26 Prob. 21–2.

21–27 Prob. 21–7.

21–28 Prob. 21–8.

21–29 Prob. 21–13.

21–30 Prob. 21–15.

21–31 Prob. 21–16.

21–32 Prob. 21–17.

21–33 Prob. 21–18.

21–34 Prob. 21–19.

21–35 Prob. 21–23.

21–36 Prob. 21–75.

FIGURE P21-37

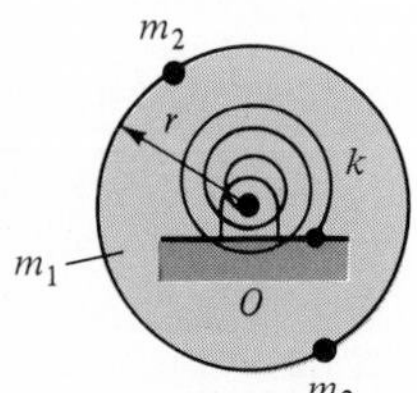

21–37 A uniform disk of mass $m_1 = 5$ kg and radius $r = 0.2$ m is supported at its axle O. A torsion spring of stiffness $k = 800$ N·m/rad causes angular oscillations of the disk. Two concentrated masses $m_2 = 1$ kg are placed diametrically on the disk. Calculate the period of vibrations of the system.

21–38 The axle of a 10-lb uniform disk is connected to a spring of stiffness $k = 50$ lb/in.; $r = 8$ in. Determine the period of vibrations of the system for $\theta = 0$. The disk rolls without slipping.

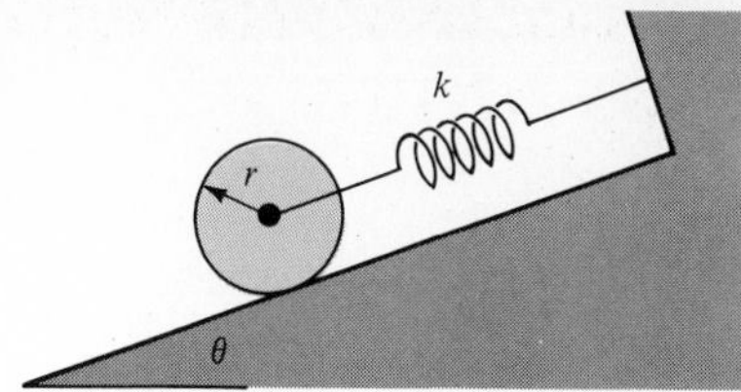

FIGURE P21-38

21–39 Determine the natural frequency of the system in Prob. 21–38 for $\theta = 30°$.

21–40 A thin, circular hoop of mass m and radius r is suspended at O. Determine the natural frequency of small oscillations in the xy plane.

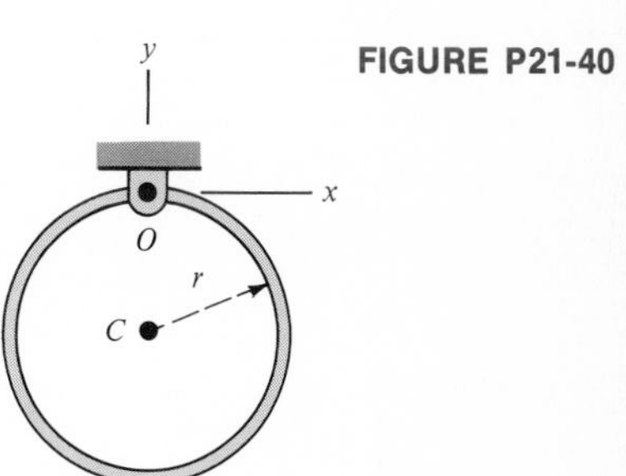

FIGURE P21-40

21–41 Solve Prob. 21–40 for oscillations of the hoop's center C in the yz plane.

21–42 Fluid of mass m and mass density ρ occupies a segment of length l of the U-tube. The column of fluid is displaced to x_0 from the equilibrium position and starts to oscillate with negligible friction. Determine the period of oscillations.

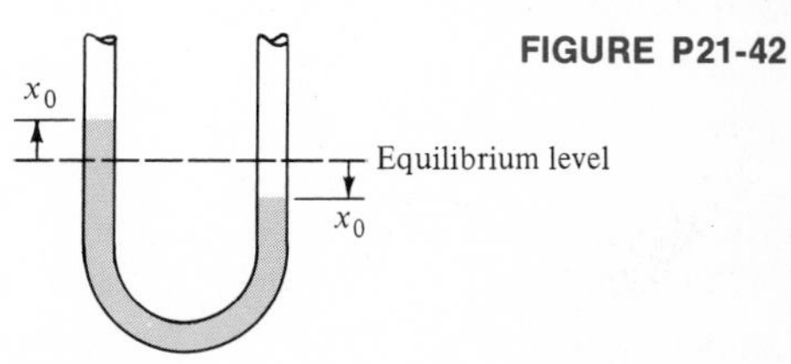

FIGURE P21-42

21-5 FORCED VIBRATIONS WITHOUT DAMPING

In many important vibration problems encountered in engineering work the exciting force is applied periodically during the motion. These are called *forced vibrations*. The most common periodic force is a *harmonic function of time* such as

$$P = P_0 \sin \Omega t \tag{21-19}$$

where P_0 is a constant, Ω is the *forcing frequency*, and t is time. The motion is analyzed using Fig. 21-12, where the force P is generally not equal to the restoring force F of the spring. Summing forces in the horizontal direction with positive x to the right, the equation of motion in terms of the forcing function P is

$$\boxed{m\ddot{x} + kx = P_0 \sin \Omega t} \tag{21-20}$$

Considering different problems, a forced vibration may be described in terms of the displacement d of a foundation or primary mass M to which the vibrating system is attached. This is illustrated in Fig. 21-13, where d is measured with respect to a fixed frame A. Here the restoring force F depends on the difference of the displacements x and d. The equation of motion is written according to Newton's second law with the positive direction to the right. From $F = m\ddot{x}$,

$$m\ddot{x} + k(x - d_0 \sin \Omega t) = 0$$

$$m\ddot{x} + kx = kd_0 \sin \Omega t \tag{21-21}$$

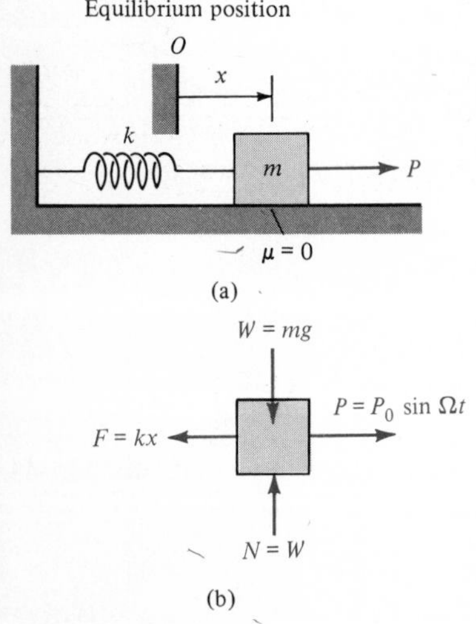

FIGURE 21-12

Model for analyzing forced vibrations

where d_0 is the amplitude of vibration of the primary mass M, Ω is the frequency of motion of M, and t is time. Equations 21-20 and 21-21 are of the same form, and they have the same solution if $P_0 = kd_0$.

The *general solution* of Eq. 21-20 or 21-21 (nonhomogeneous, second-order differential equations) consists of two parts, $x = x_c + x_p$, where x_c is the *complementary solution* and x_p is the *particular solution*. The complementary solution is obtained by setting the right side of Eq. 21-20 or 21-21 equal to zero. The solution of this equation is Eq. 21-4 for a freely vibrating system,

$$x = C_1 \sin \omega t + C_2 \cos \omega t \tag{21-22}$$

where $\omega = \sqrt{k/m}$ and C_1 and C_2 are arbitrary constants.

For the particular solution, it is reasonable to assume a harmonic function since the motion is harmonic. Let

$$x_p = A \sin \Omega t \tag{21-23}$$

where A is the amplitude, a constant. The value of this amplitude is determined by substituting Eq. 21-23 into Eq. 21-20,

$$-m\Omega^2 A \sin \Omega t + kA \sin \Omega t = P_0 \sin \Omega t$$

From this, and $\omega = \sqrt{k/m}$,

$$A = \frac{P_0}{k - m\Omega^2} = \frac{P_0/k}{1 - (\Omega/\omega)^2} \qquad \boxed{21\text{-}24}$$

where ω is the natural circular frequency of the system of m and k, and Ω is the forcing frequency.

The *general solution* of the forced vibration without damping is $x = x_c + x_p$,

$$x = \underbrace{C_1 \sin \omega t + C_2 \cos \omega t}_{\substack{\text{free vibration} \\ \text{(transient)}}} + \underbrace{A \sin \Omega t}_{\substack{\text{forced vibration} \\ \text{(steady-state)}}} \qquad \boxed{21\text{-}25}$$

The two terms of free vibration are dependent only on properties m and k of the system and on the initial conditions that affect the constants C_1 and C_2. These define the initial natural vibration of the system (e.g., after the forcing function suddenly begins or stops). This is also called *transient vibration* because in a real system it is damped out by friction.*

The single term representing forced vibration in Eq. 21-25 depends on the amplitude of the applied force or displacement, on the forcing frequency Ω, and on the circular frequency ω of the system (because A depends on Ω/ω according to Eq. 21-24). This is also called *steady-state* vibration since it is the motion of the system after a transient vibration is dissipated.

Resonance

A particularly important aspect of forced vibrations is that the amplitude of the oscillations depends on the frequency ratio Ω/ω. According to Eq. 21-24, the amplitude becomes infinite when $\Omega = \omega$. This condition is called *resonance*. Of course, the amplitude does not become infinite in practice because of damping or physical constraints, but the condition is often a dangerous one (it can cause fractures). Dangerously large amplitudes may also occur at other frequency ratios near the resonant (natural) frequency. The various conditions can be readily compared graphically, as shown in the following.

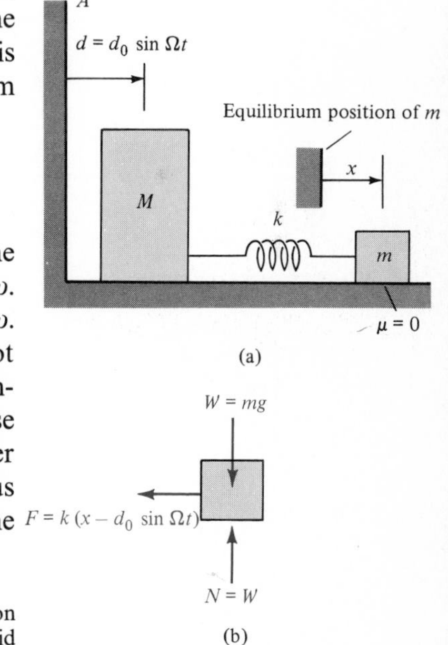

FIGURE 21-13

Forced vibration in terms of displacement of a primary mass

* Some readers may notice the contradiction that the equation of free vibration was obtained assuming negligible damping, while here the transient vibration is said to be damped out by friction. This is a difficult subject to discuss, but here it may suffice to say that even small friction has a cumulative effect in reducing the amplitudes of the free vibration. Thus, the latter eventually becomes negligible *in comparison* with the forced component of the vibration.

Consider the ratio of the amplitude A of the steady-state vibration to the maximum static deflection P_0/k that the exciting force P would cause. This ratio is called the *magnification factor* and can be expressed in three ways using Eq. 21-20, 21-21, and 21-24:

$$\text{magnification factor} = \frac{A}{P_0/k} = \frac{A}{d_0} = \frac{1}{1-(\Omega/\omega)^2} \tag{21-26}$$

This quantity is plotted as a function of the frequency ratio Ω/ω in Fig. 21-14. Several items are of particular interest in this diagram:

1. *Static loading.* $\Omega = 0$; $A/(P_0/k) = 1$. The same is true for $\Omega \neq 0$ if $\Omega \ll \omega$.

2. *Resonance.* $\Omega = \omega$; $A/(P_0/k) = \infty$

3. *High-frequency excitation.* $\Omega \gg \omega$; $A/(P_0/k) \simeq 0$; the mass remains essentially stationary because of its inertia (the system cannot respond quickly enough).

4. *Phase relationships.* The sign of the magnification factor indicates whether the direction of motion of the vibrating mass is the same as that of the exciting force or displacement. The vibration is *in phase* for $\Omega < \omega$, and it is 180° *out of phase* for $\Omega > \omega$.

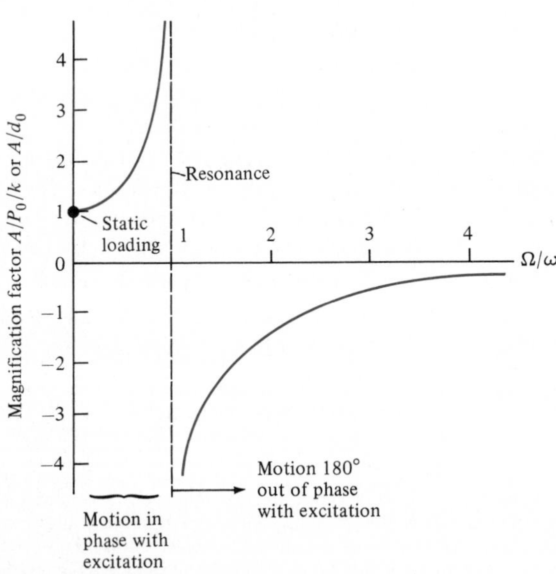

FIGURE 21-14

Magnification factor in forced vibrations

EXAMPLE 21-5

A 1500-kg truck cab is assumed to be a uniform block symmetrically supported by four springs, each with a stiffness of 120 kN/m, on top of four hydraulic actuators A (Fig. 21-15). Determine the resonance frequency of the cab in units of Hz, and the amplitude of the vibration if the displacement input of each actuator is $d = 0.05 \sin 6t$ m.

SOLUTION

For this motion, with x measured from the static equilibrium position,

$$\sum F_x = ma_x$$

$$-4k(x - d) = m\ddot{x}$$

$$m\ddot{x} + 4kx = 4k(0.05 \sin 6t) \text{ m}$$

This is now in the form of Eq. 21-21, with an effective stiffness, $k_{\text{eff}} = 4k$.

$$\omega = \sqrt{\frac{k_{\text{eff}}}{m}} = \sqrt{\frac{4k}{m}}$$

$$f = \frac{\omega}{2\pi} = \frac{1}{2\pi}\sqrt{\frac{4(120 \times 10^3 \text{ N/m})}{1500 \text{ kg}}} = 2.85 \text{ s}^{-1}$$

From Eq. 21-24,

$$\frac{P_0}{k_{\text{eff}}} = (4k)\frac{d_0}{4k} = d_0 = 0.05 \text{ m}$$

$$A = \frac{d_0}{1 - (\Omega/\omega)^2} = \frac{0.05 \text{ m}}{1 - (6/17.9)^2}$$

$$= 0.0563 \text{ m}$$

FIGURE 21-15

Road-simulation test of a truck cab (Courtesy MTS Systems Corporation, Minneapolis, Minnesota)

PROBLEMS

SECTION 21-5

FIGURE P21-43

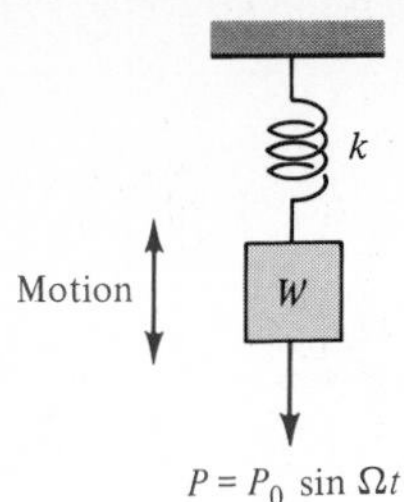

21–43 A periodic force $P = 20 \sin 30t$ lb is acting on a 10-lb block which is supported by a spring of stiffness $k = 150$ lb/in. Calculate the amplitude of vibration of the block.

21–44 In Prob. 21–43, determine the value of Ω for which the amplitude of the vibration is 0.2 in.

FIGURE P21-45

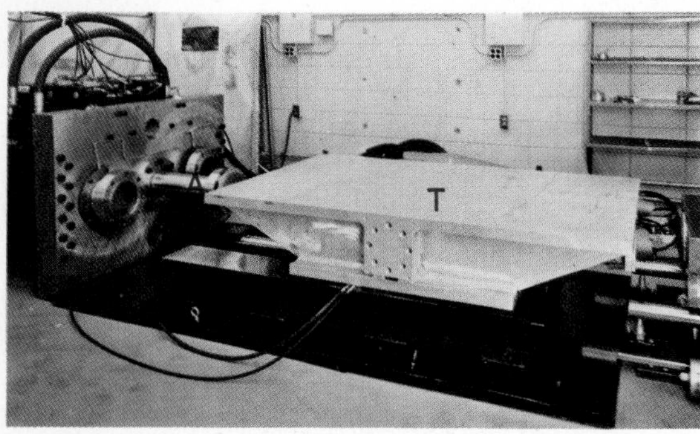

21–45 A 200-kg shake table T moves horizontally with negligible friction on its tracks. The actuator rod A moves under displacement control defined by $d = 0.15 \sin 12t$ m. Determine the amplitude of vibration of the table if it is connected to rod A through a spring of stiffness $k = 1$ MN/m.

21–46 Plot the amplitude of the vibration in Prob. 21–45 for $k = 10$ kN/m, 100 kN/m, and 1 MN/m.

FIGURE P21-47

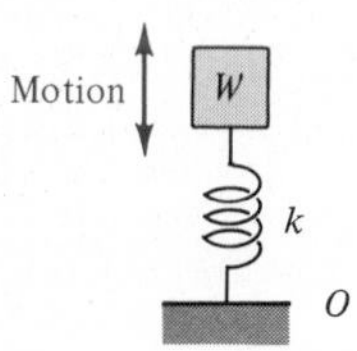

21–47 An accelerometer consists of a 0.01-lb block on a spring of stiffness k. Determine k if the amplitude A of the block's vibration should be 0.1 in. when the base O has vertical cyclic displacements of $d = 0.005 \sin 500t$ in.

21–48 Determine the amplitude of the block's vibration in Prob. 21–47 for the forcing frequency reduced by 50%, using the value of k calculated in Prob. 21–51.

FIGURE P21-49

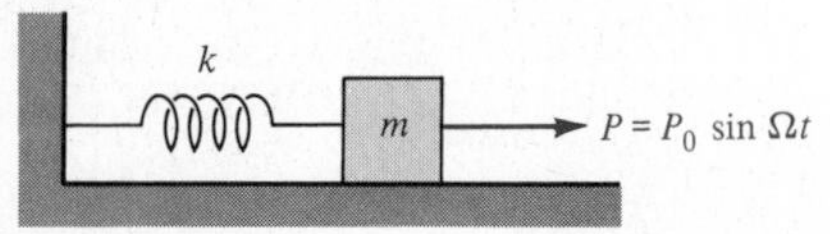

21–49 Determine the range of values of the forcing frequency Ω for which the amplitude of the vibration is less than 1% of the static deflection caused by P_0, a constant force.

21–50 The machine on the shake table has a component that is modeled as a 0.1-kg mass on the end of a light cantilever beam. In a test the machine is vibrated vertically with displacements of $d = 0.03 \sin 40t$ m. The resulting amplitude A of the small mass is 2 mm. Calculate the spring constant k of the beam.

21–51 The shake table in Fig. P21-50 consists of a 50-kg platform with the test piece and a 1200-kg base on four air springs. It is desired to apply a vertical force $P = \sin 15t$ kN to the platform while limiting the amplitude of vibration of the base to 1 mm. Determine the required spring constant of the air springs. There is no spring between the upper platform and the heavy base.

FIGURE P21-50

Vibration test of a gardening implement (Photograph courtesy MTS Systems Corporation, Minneapolis, Minnesota)

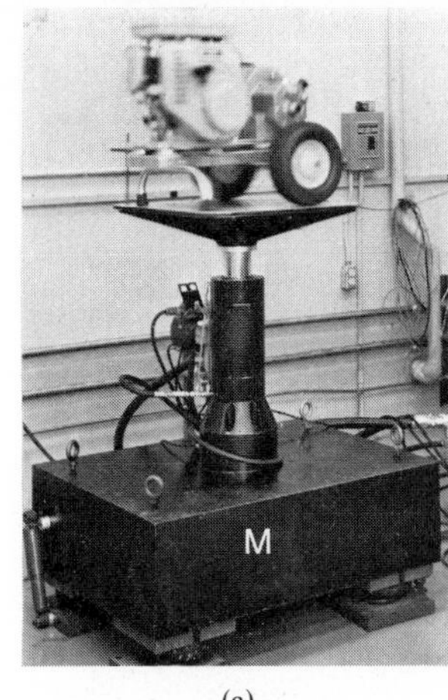

(a)

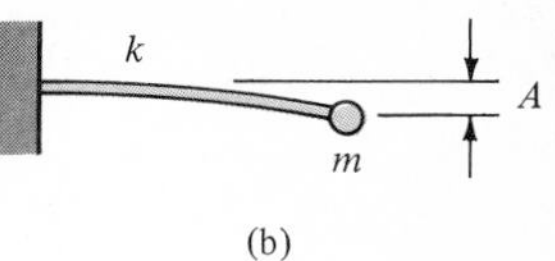

(b)

21–52 An interesting problem of tuned-mass dampers of skyscrapers (see Sec. 1-7 in the statics volume) is how to test them mechanically. The best method is to wait for a windless day and force the heavy block to oscillate with respect to the building. This eventually causes substantial vibrations of the building if the system is properly designed and constructed. Consider a 600,000-lb block with a spring of stiffness $k = 1500$ lb/in. Determine the required force P of the hydraulic actuator to move the block with an amplitude of 50 in. and a period of 7 s in a perfectly rigid building.

FIGURE P21-52

Model of a tuned-mass damper system (Courtesy MTS Systems Corporation, Minneapolis, Minnesota)

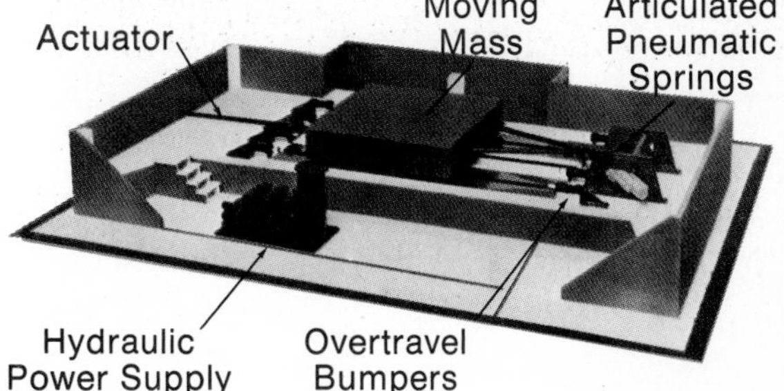

21–53 A 70-kg astronaut is belted to a chair of stiffness $k = 10$ kN/m. The vertical acceleration of the astronaut should not exceed 3 g. Determine the allowable amplitude d_0 of the massive platform at a frequency of 10 Hz.

FIGURE P21-53

NASA–Langley three-degree-of-freedom (vertical, horizontal, roll) human factors research motion simulator with 7 ft × 7 ft welded aluminum table (Courtesy MTS Systems Corporation, Minneapolis, Minnesota)

21-6 DAMPING OF VIBRATIONS

The ever-present damping of vibrating systems is one of the most easily demonstrated phenomena in dynamics when used in free vibrations. The decaying amplitudes of plucked strings and small cantilever beams show that the initial mechanical energy of the system (potential and kinetic energy) is gradually dissipated. The mechanical energy may be directly converted into heat by *dry friction* (also called *Coulomb friction*) between rigid bodies, or by *internal friction* during deformation of a solid material. The mechanical energy of a vibrating system may also be dissipated by *fluid friction* in a gas or a liquid which temporarily acquires some kinetic energy. For all practical purposes, the mechanical energy in vibrations is ultimately converted into heat.

The damping caused by fluid friction is called *viscous damping*. This is quite common in engineering work and is relatively easy to analyze using the idealization that *viscous damping is directly proportional to the speed* of the body in the fluid at low speeds. The presence of this damping is always modeled by a *dashpot*, which consists of a piston A moving in a cylinder B as illustrated in Fig. 21-16a. The relative motion of the piston with respect to the cylinder is caused by the external force F and resisted by the fluid friction between the moving parts. The elemental friction force f is uniformly distributed on the circumference of the piston as implied in Fig. 21-16b. As stated previously, the resultant friction force is proportional to velocity and is denoted by $c\dot{x}$ in Fig. 21-16c, where $\dot{x}$ is the relative velocity of the piston in the cylinder, and the constant c is the *coefficient of viscous damping*. The value of c depends on the fluid and its temperature (inverse dependence on temperature) and on the construction of the dashpot. The units of the coefficient c are N·s/m or lb·s/ft. Dashpots are real devices such as the shock absorbers of automobiles, but sometimes they are also used in the analytical modeling of systems that do not have distinct damping devices. The mass of the dashpot is generally neglected.

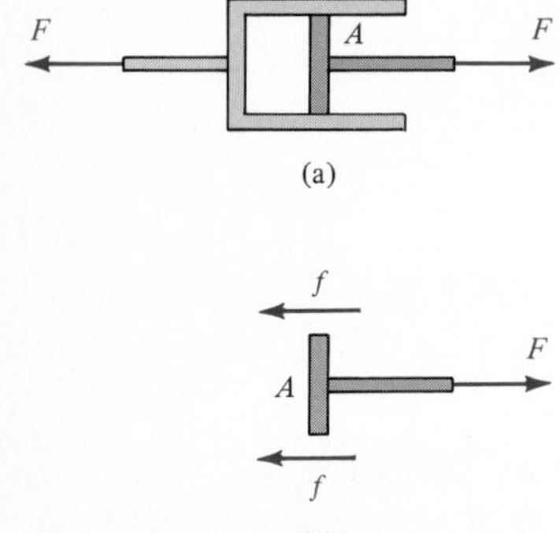

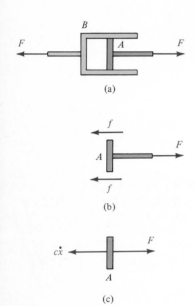

FIGURE 21-16
Model of a dashpot

21-7 DAMPED FREE VIBRATIONS

The analysis of a damped, freely vibrating system is illustrated in Fig. 21-17. Assume that the mass m is moving to the right so that the friction force $c\dot{x}$ is to the left. With x, $\dot{x}$, and $\ddot{x}$ taken as positive to the right, the equation of motion of the mass from the free-body diagram in Fig. 21-17b is

$$-c\dot{x} - kx = m\ddot{x}$$

according to Newton's second law. This is rewritten as

$$m\ddot{x} + c\dot{x} + kx = 0 \qquad \boxed{21\text{-}27}$$

which is a linear, second-order, homogeneous, differential equation. It has solutions of the form

$$x = e^{\lambda t} \qquad \boxed{21\text{-}28}$$

where e is the base of the natural logarithm, λ is a constant, and t is time. Substituting this solution in Eq. 21-27 and dividing the resulting equation by $e^{\lambda t}$,

$$m\lambda^2 + c\lambda + k = 0 \qquad \boxed{21\text{-}29}$$

which has two roots according to the quadratic formula,

$$\lambda = \frac{-c}{2m} \pm \sqrt{\left(\frac{c}{2m}\right)^2 - \frac{k}{m}} \qquad \boxed{21\text{-}30}$$

Thus, the general solution of Eq. 21-27 is the sum of two exponentials with λ_1 and λ_2 defined by Eqs. 21-28 and 21-30, as discussed shortly.

An important parameter for describing free vibrations is obtained when the term under the radical in Eq. 21-30 is zero. The value of c that satisfies this condition is called the *critical damping coefficient* c_c. It is determined from Eq. 21-30 that

$$\left(\frac{c_c}{2m}\right)^2 = \frac{k}{m} \quad \text{or} \quad c_c = 2m\sqrt{\frac{k}{m}} = 2m\omega \qquad \boxed{21\text{-}31}$$

where ω is the circular frequency of the system without damping. There are three special cases of damping that can be distinguished with respect to the critical damping coefficient.

1. *Overdamped system.* $c > c_c$. The roots λ_1 and λ_2 of Eq. 21-29 are real and distinct. The general solution of Eq. 21-27 is

$$x = Ae^{\lambda_1 t} + Be^{\lambda_2 t} \qquad \boxed{21\text{-}32}$$

where A and B are constants. The exponents λ_1 and λ_2 are both negative, so x decreases from its initial value $x_0 = A + B$ toward

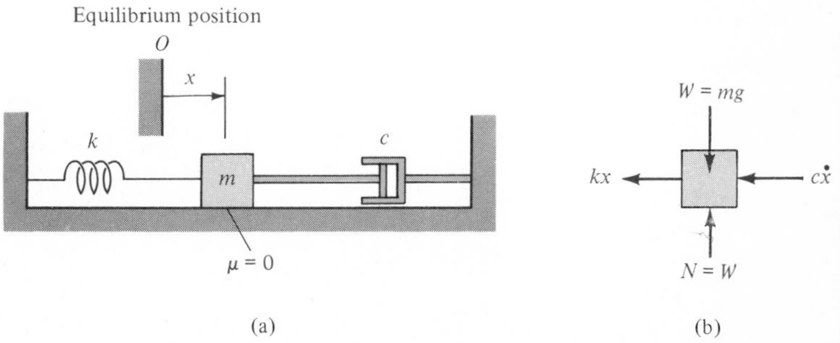

FIGURE 21-17

Model of a damped vibrating system

zero as time t increases. This motion is *nonvibratory* (or *aperiodic*), as illustrated in Fig. 21-18.

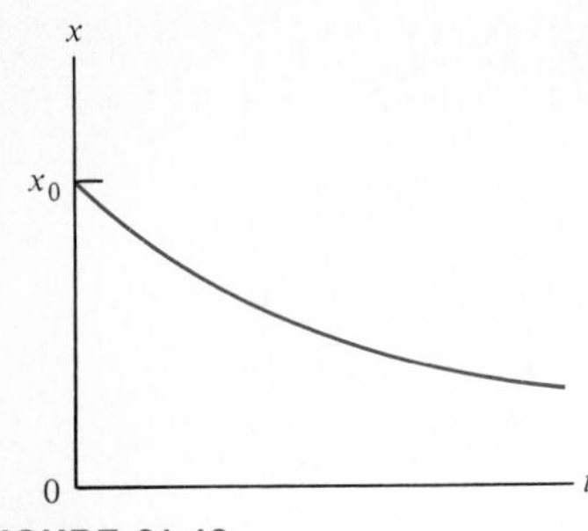

FIGURE 21-18

Nonvibratory motion of an overdamped system

2. *Critically damped system.* $c = c_c$. In this case $\lambda_1 = \lambda_2 = -c_c/2m = -\omega$. The general solution of Eq. 21-27 is

$$x = (A + Bt)e^{-\omega t} \tag{21-33}$$

This motion is also nonvibratory but it is of special interest because x decreases at the fastest possible rate without oscillation of the mass.

3. *Underdamped system.* $c < c_c$. The roots λ_1 and λ_2 of Eq. 21-29 are complex numbers. It can be shown that the general solution of Eq. 21-27 is

$$x = Ae^{-(c/2m)t} \sin (\omega_d t + \phi) \tag{21-34}$$

where A and ϕ are constants depending on the initial conditions of the problem (similar to those in Eq. 21-7). The constant ω_d is defined as the *damped natural frequency* of the system and is expressed as

$$\omega_d = \omega \sqrt{1 - \left(\frac{c}{c_c}\right)^2} \tag{21-35}$$

where $\omega = \sqrt{k/m}$, the natural frequency for undamped vibration. The ratio c/c_c is called the *damping factor* ζ.

Equation 21-34 defines harmonic oscillations of diminishing amplitude as shown in Fig. 21-19. The amplitude is $Ae^{-(c/2m)t}$.

The *period of the damped vibration* τ_d is defined as the time interval between successive humps where they touch one of the exponential envelopes of the oscillations. It is interesting to note that the period is constant even though the amplitude decreases,

$$\tau_d = \frac{2m}{\omega_d} \tag{21-36}$$

The period of damped vibration is always larger than the period of

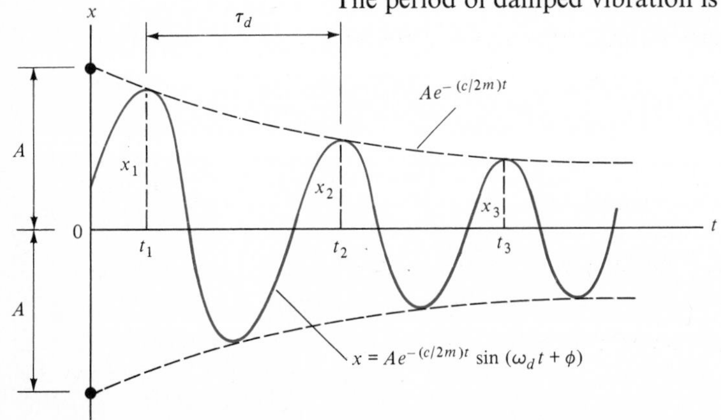

FIGURE 21-19

Oscillations of an underdamped system

the same system without damping, according to Eqs. 21-35 and 21-36.

21-8 DAMPED FORCED VIBRATIONS

Assume that a periodic force $P = P_0 \sin \Omega t$ is acting horizontally on mass m in Fig. 21-20. The appropriate equation of motion in this case becomes

$$m\ddot{x} + c\dot{x} + kx = P_0 \sin \Omega t \tag{21-37}$$

A similar equation is applicable when a support of the system moves periodically with displacement $d = d_0 \sin \Omega t$ as in Sec. 21-5. That requires replacing P_0 in Eq. 21-37 by kd_0.

Equation 21-37 is a nonhomogeneous differential equation, so its general solution consists of a complementary solution x_c, and a particular solution x_p. The complementary solution is the same as for Eq. 21-27, and it is given by Eq. 21-32, 21-33, or 21-34, depending on the damping of the system. These are *transient vibrations* and they rapidly diminish in most cases. The particular solution that applies to the *steady-state vibration* of the system should be a harmonic function of time, such as

$$x_p = A \sin (\Omega t - \phi) \tag{21-38}$$

where A and ϕ are constants. Substituting x_p into Eq. 21-37 gives

$$-m\Omega^2 A \sin (\Omega t - \phi) + c\Omega A \cos (\Omega t - \phi) + kA \sin (\Omega t - \phi) = P_0 \sin \Omega t$$

The constant coefficients A and ϕ in this equation can be determined by first setting $\Omega t - \phi = 0$ and then setting $\Omega t - \phi = \pi/2$, giving

$$c\Omega A = P_0 \sin \phi \tag{21-39}$$

$$(k - m\Omega^2)A = P_0 \cos \phi \tag{21-40}$$

The *phase angle* ϕ (phase difference between the applied force and the resulting vibration) is determined from the ratio of Eqs. 21-39 and 21-40,

$$\tan \phi = \frac{c\Omega}{k - m\Omega^2} \tag{21-41}$$

The sine and cosine functions can be eliminated by summing the squares of Eqs. 21-39 and 21-40,

$$A^2[(c\Omega)^2 + (k - m\Omega^2)^2] = P_0^2 \tag{21-42}$$

The amplitude A is expressed as

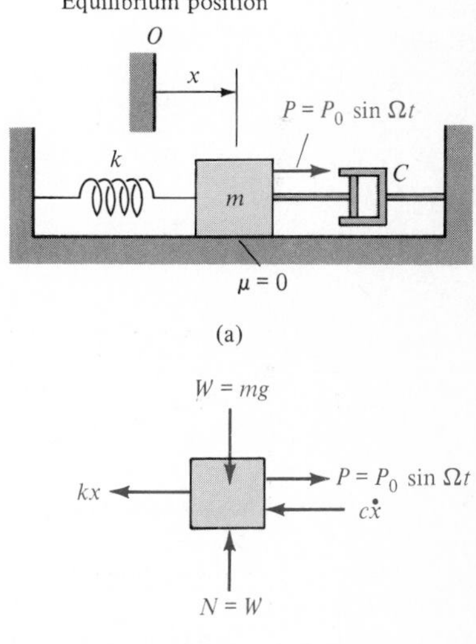

FIGURE 21-20

Model for forced vibrations of a damped system

$$A = \frac{P_0}{\sqrt{(c\Omega)^2 + (k - m\Omega^2)^2}} \qquad \text{21-43}$$

A magnification factor for the amplitude of the vibration can be defined as in the case of undamped forced vibration (Eq. 21-26). Using $c_c = 2m\omega$ and $\omega = \sqrt{k/m}$ gives

$$\text{magnification factor} = \frac{A}{P_0/k} = \frac{A}{d_0} = \frac{1}{\sqrt{[2(c/c_c)(\Omega/\omega)]^2 + [1-(\Omega/\omega)^2]^2}} \qquad \text{21-44}$$

This quantity is plotted as a function of the frequency ratio Ω/ω for different damping factors in Fig. 21-21. The following items are of interest in this diagram:

1. ***Static loading.*** $\Omega = 0$; $A/(P_0/k) = 1$, independent of the damping (agrees with Eq. 21-24).

2. ***Resonance.*** $\Omega = \omega$; the amplitude is magnified substantially when the coefficient of viscous damping c is low.

3. ***High-frequency excitation.*** $\Omega \gg \omega$; $A/(P_0/k) \simeq 0$; the mass remains essentially stationary because of its inertia, regardless of any damping of its motion.

4. ***Large coefficient of viscous damping.*** The amplitude of the vibration is reduced at all values of Ω/ω as the coefficient of viscous damping c is increased in a particular system.

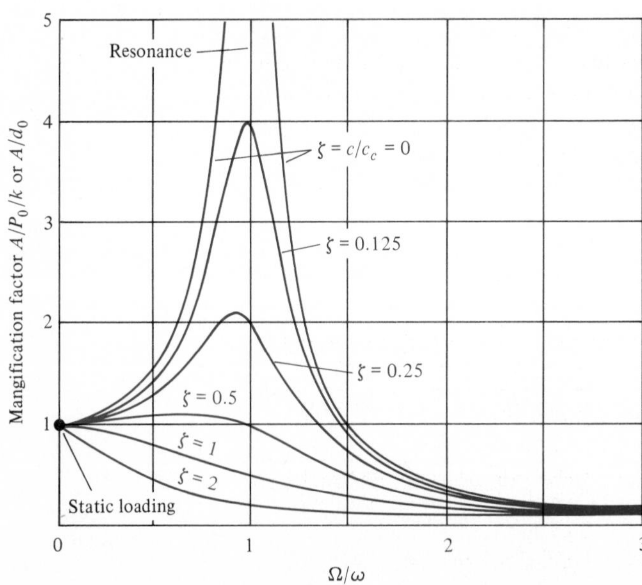

FIGURE 21-21

Magnification factor in damped forced vibrations

The phase angle ϕ can also be determined using the same parameters as in Eq. 21-44,

$$\tan \phi = \frac{2(c/c_c)(\Omega/\omega)}{1 - (\Omega/\omega)^2} \tag{21-45}$$

21-9 ELECTRICAL ANALOGUES OF MECHANICAL VIBRATIONS

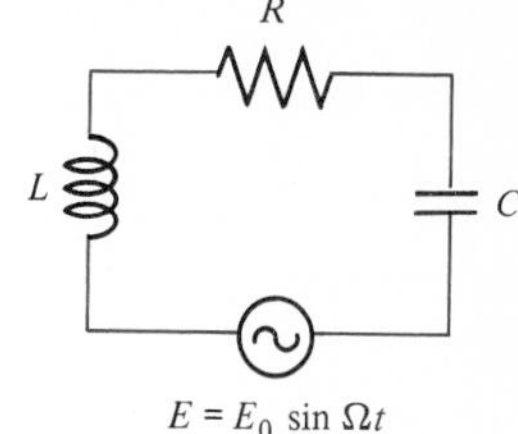

FIGURE 21-22

Series electric circuit

Electrical circuits with transient or steadily oscillating currents are common. Interestingly, for each mechanical system there is an electrical circuit analogy. The oscillations of the two analogous systems can be analyzed using the same differential equation. This is of practical value because an electrical circuit is generally more easily constructed and modified than the corresponding mechanical system. Thus, a mechanical system can be designed, evaluated, and refined using an appropriate electrical circuit.

The analogy between electrical and mechanical systems is illustrated using the simple series circuit in Fig. 21-22. The source of alternating voltage $E = E_0 \sin \Omega t$ generates a current $i = dq/dt$ in the circuit, where q denotes electric charge. The drop in electric potential (voltage) around the circuit is $L(di/dt)$ across the inductor, Ri across the resistor, and q/C across the capacitor. The sum of the voltage drops around the circuit equals the applied voltage according to Kirchhoff's second law,

$$L\ddot{q} + R\dot{q} + \frac{1}{C}q = E_0 \sin \Omega t \tag{21-46}$$

which is identical in form to Eq. 21-37 since L, R, and C are constants. The mechanical–electrical analogues can be established by comparing these two equations. The results are given in Table 21-1. These analogues allow the student to perform at least simple experiments to simulate the behavior of mechanical systems using electrical circuits.

TABLE 21-1

Mechanical–Electrical Analogues

MECHANICAL SYSTEM		ELECTRICAL SYSTEM	
Mass	m	Inductance	L
Coefficient of viscous damping	c	Resistance	R
Spring constant	k	Reciprocal of capacitance	$1/C$
Displacement	x	Charge	q
Velocity	$\dot{x}$	Current	i
Applied force	P	Applied voltage	E

EXAMPLE 21-6

An automobile tested in a laboratory is modeled as a 1500-kg mass on a spring of stiffness $k = 200$ kN/m. The system has a damping factor of $c/c_c = 0.4$. Assume that the spring and dashpot are attached to the base whose vertical displacements are defined by $d = 0.04 \sin 6t$ m (Fig. 21-23a). Write the equation of motion of m for steady-state vibration. Determine the magnification factor of the amplitude of vibration, the amplitude A, and the phase angle ϕ.

SOLUTION

Summing forces on the automobile, and noting that the weight and static spring deflection force cancel, gives

$$\sum F_y = ma_y$$

$$-k(y - d) - c\dot{y} = m\ddot{y}$$

$$m\ddot{y} + c\dot{y} + k(y - d_0 \sin \Omega t) = 0$$

$$m\ddot{y} + c\dot{y} + ky = kd_0 \sin \Omega t$$

This equation is of the form of Eq. 21-37.

$$\omega = \sqrt{\frac{k}{m}} = \sqrt{\frac{200 \times 10^3 \text{N/m}}{1500 \text{ kg}}}$$

$$= 11.5 \text{ s}^{-1}$$

From Eq. 21-44,

$$\text{magnification factor} = \frac{A}{d_0} = \frac{1}{\sqrt{(2c\Omega/c_c\omega)^2 + [1 - (\Omega/\omega)^2]^2}}$$

$$= \frac{1}{\sqrt{[(2)(0.4)(6/11.5)]^2 + [1 - (6/11.5)^2]^2}}$$

$$= 1.19$$

$$A = (1.19)d_0 = (1.19)(0.04 \text{ m}) = 0.0476 \text{ m}$$

$$c = 0.4c_c = 0.4(2m\omega) = \frac{13\,860 \text{ N}\cdot\text{s}}{\text{m}}$$

From Eq. 21-41,

$$\phi = \tan^{-1}\left(\frac{c\Omega}{k - m\Omega^2}\right)$$

$$= \tan^{-1}\left[\frac{13\,860 \dfrac{\text{N}\cdot\text{s}}{\text{m}}(6 \text{ s}^{-1})}{200 \times 10^3 \text{ N/m} - 1500 \text{ kg}\,(6 \text{ s}^{-1})^2}\right]$$

$$= \tan^{-1}(0.57)$$

$$= 29.7^\circ$$

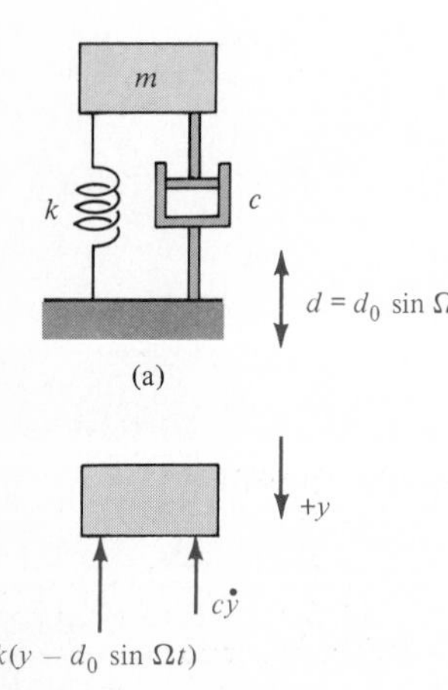

FIGURE 21-23
Model of a damped system in forced vibrations

PROBLEMS

21–54 A 2-kg mass is in series with a spring and a dashpot. The spring stiffness is $k = 500$ N/m and the coefficient of viscous damping is $c = 100$ N·s/m. The mass is released from rest at an initial displacement $x_0 = 0.05$ m. Determine the general solution of the vibrations.

21–55 A 5-lb block is in series with a spring and a dashpot. The spring has stiffness $k = 100$ lb/in. and the coefficient of viscous damping is $c = 20$ lb·s/ft. The block has initial speed $\dot{x}_0 = 3$ ft/s at the initial displacement $x_0 = 0$. Determine the general solution of the vibrations.

21–56 Determine the period of damped vibrations in Prob. 21–55 for $k = 100$ lb/in. and 500 lb/in.

21–57 A 100-kg instrument is resting on four springs whose static deflection is 0.02 m. The mass is deflected an additional 0.03 m by an external force and released. The damping factor is $c/c_c = 0.1$. Determine the general solution of the vibrations.

21–58 Calculate the time t_1 in which the amplitude of vibrations in Prob. 21–57 diminishes to 1% of its initial value.

21–59 A 2000-lb vehicle is designed with an energy-absorbing, self-recovering bumper. The light bumper is positioned by two springs of stiffness $k = 600$ lb/in. Two shock absorbers of viscous damping c also resist the deflection of the springs in an impact. Determine the coefficient c that allows the most rapid compression of the springs without tending to cause oscillations. Assume that the bumper becomes wedged in a massive body during impact.

FIGURE P21-59

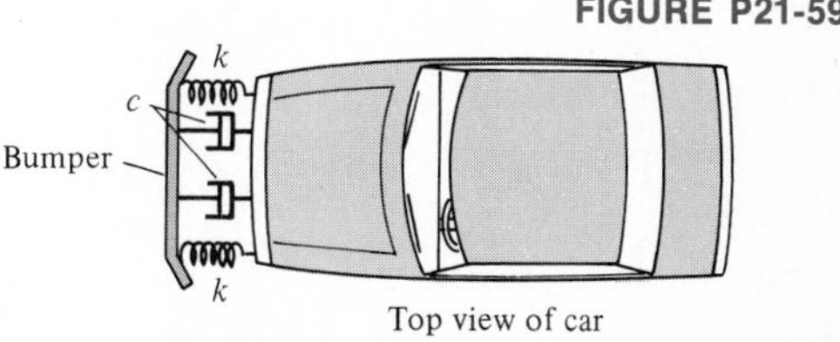

21–60 Consider an overdamped system of a mass m, spring stiffness k, and damping coefficient $c\,(c > c_c)$. Show that m never passes its equilibrium position $x = 0$ after being released from rest at an arbitrary deflection x_0.

21–61 Show that the result of Prob. 21–60 is also valid if m has an arbitrary initial speed $\dot{x}_0$ at its equilibrium position $x = 0$.

21–62 Consider an underdamped system of mass m, spring stiffness k, and damping coefficient $c\ (c < c_c)$. Figure 21-19 applies in this case. Assume that $x_1, x_2, \ldots$ measured to the exponential envelope represent maximum displacements. Show that $x_1/x_2 = x_2/x_3 =$ constant, and that for the nth cycle,

$$D = \ln \frac{x_n}{x_{n+1}} = \frac{2\pi(c/c_c)}{\sqrt{1 - (c/c_c)^2}}$$

The quantity D is known as the *logarithmic decrement*.

21–63 Prove the alternative expression for the logarithmic decrement in Prob. 21–62,

$$D = \frac{1}{N} \ln \frac{x_a}{x_b}$$

where N is the number of cycles between cycle a and cycle b.

21–64 Use the result of Prob. 21–63 to determine the damping factor c/c_c of a system consisting of a 10-kg mass and a spring of stiffness $k = 5$ kN/m. The amplitude of the vibration decreases from 2 cm to 1 cm in 100 cycles.

21–65 Solve Prob. 21–43 assuming that a dashpot of damping coefficient $c = 3$ lb·s/ft is connected parallel with the spring.

21–66 A 20-kg mass is shaken by a force $P = 30 \sin 40t$ N. $k = 500$ N/m, and $c = 10$ N·s/m. Determine the amplitude of the steady-state vibrations.

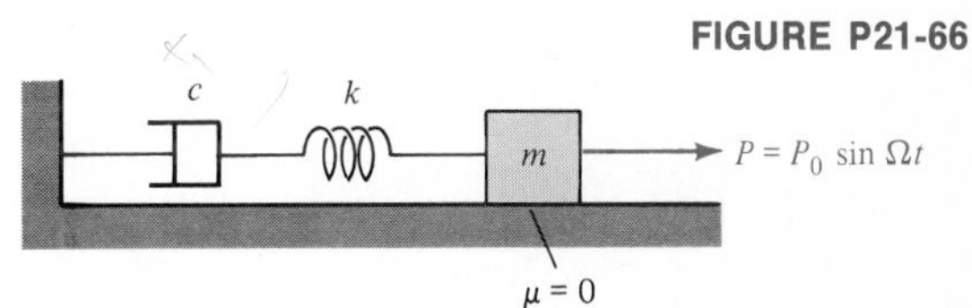

FIGURE P21-66

21–67 Solve Prob. 21–53 assuming that the chair has viscous damping of $c = 100$ N·s/m.

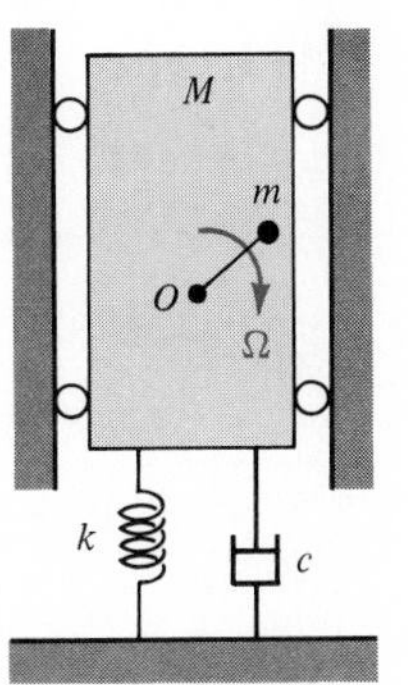

FIGURE P21-68

21–68 A simple fatigue testing machine is modeled as a mass $M = 20$ kg which is constrained to move vertically. A mass $m = 1$ kg is rotating at 1800 rpm on a 20-cm-diameter path centered at point O on block M. Calculate the amplitude of vibration of M if $k = 20$ kN/m and $c = 300$ N·s/m.

In Probs. 21–69 to 21–73 determine the analogous electrical circuit and write the differential equation for it.

FIGURE P21-69

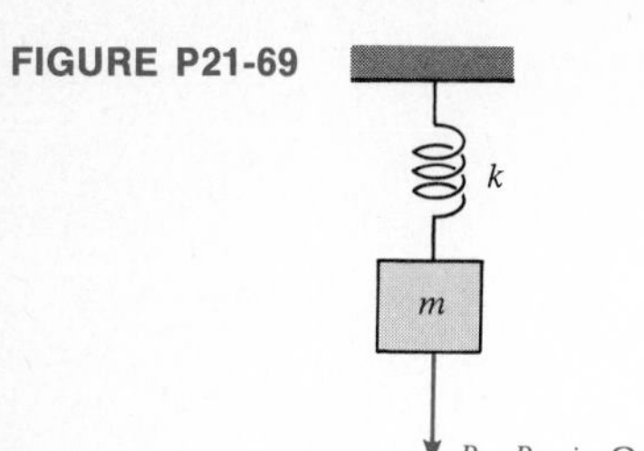

21–69

FIGURE P21-70

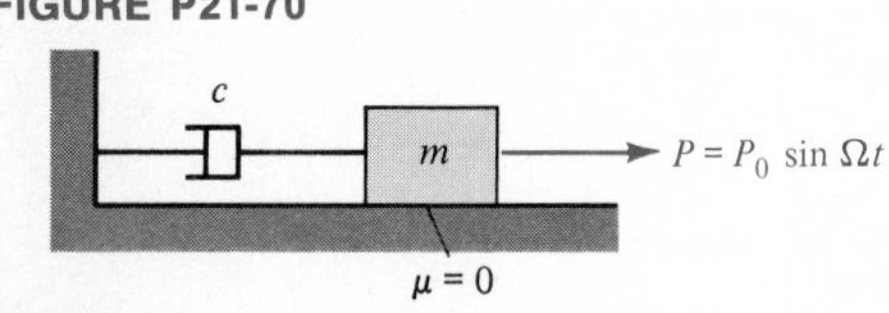

21–70

FIGURE P21-71

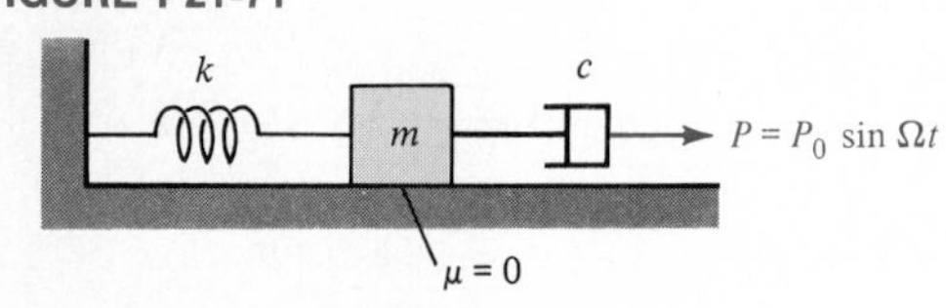

21–71

FIGURE P21-72

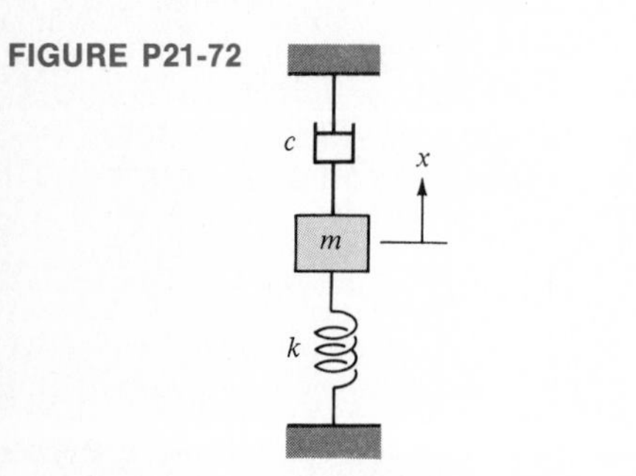

21–72

FIGURE P21-73

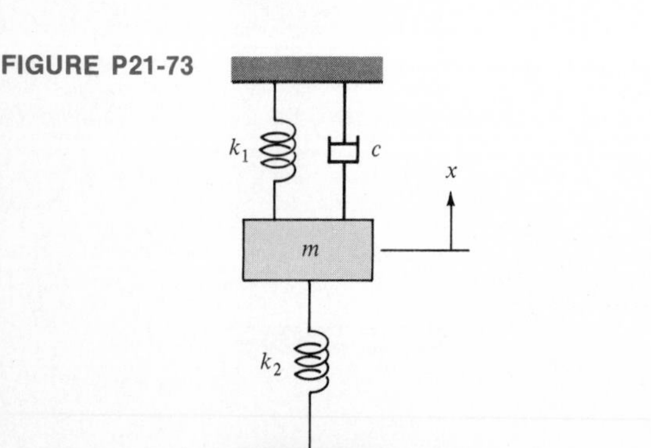

21–73

For a system of a particle (mass m) and spring (spring constant k) unless otherwise noted.

SUMMARY

Section 21-1

Simple harmonic motion

NATURAL CIRCULAR FREQUENCY:

$$\omega = \sqrt{\frac{k}{m}}$$

PERIOD:

$$\tau = \frac{2\pi}{\omega} = 2\pi\sqrt{\frac{m}{k}}$$

FREQUENCY:

$$f = \frac{1}{\tau} = \frac{\omega}{2\pi} = \frac{1}{2\pi}\sqrt{\frac{k}{m}}$$

Section 21-2

Free vibrations of particles without damping. Free vibration occurs when only two kinds of forces act on a mass: (a) elastic restoring force, and (b) constant force.

NATURAL FREQUENCY:

$$f = \frac{1}{2\pi}\sqrt{\frac{k}{m}}$$

Section 21-3

Free vibrations of rigid bodies. The differential equation of motion for a rigid body depends on the mass of that body, the elastic members of the system, *and* on the configuration of the particular system.

Section 21-4

Energy methods in vibration analysis. For an undamped system,

$$T + V = \text{constant}$$

DISPLACEMENT:

$$x = A \sin(\omega t + \phi)$$

$$x_{\max} = A$$

$$\dot{x}_{\max} = A\omega$$

$$V_{\max} = \frac{1}{2} kA^2 \qquad T_{\max} = \frac{1}{2} mA^2\omega^2 \qquad \omega = \sqrt{k/m}$$

Section 21-5

Forced vibrations without damping

PERIODIC FORCE:

$$P = P_0 \sin \Omega t$$

PERIODIC DISPLACEMENT OF PRIMARY MASS:

$$d = d_0 \sin \Omega t$$

where Ω is the forcing frequency.

AMPLITUDE:

$$A = \frac{P_0}{k - m\Omega^2} = \frac{P_0/k}{1 - (\Omega/\omega)^2}$$

where $\omega = \sqrt{k/m}$.

MAGNIFICATION FACTOR:

$$\frac{A}{P_0/k} = \frac{A}{d_0} = \frac{1}{1 - (\Omega/\omega)^2}$$

RESONANCE:

$$\Omega = \omega$$

Section 21-6

Damping of vibrations

Coefficient of viscous damping $= c$
Friction force in viscous damping $= c\dot{x}$

Section 21-7

Damped free vibrations

Critical damping coefficient $= c_c$

DAMPING FACTOR:

$$\zeta = \frac{c}{c_c}$$

OVERDAMPED SYSTEM:

$$c > c_c \qquad \text{(nonvibratory motion)}$$

CRITICALLY DAMPED SYSTEM:

$$c = c_c \qquad \text{(nonvibratory motion)}$$

UNDERDAMPED SYSTEM:

$$c < c_c \qquad \text{(vibratory motion)}$$

DAMPED NATURAL FREQUENCY:

$$\omega_d = \omega \sqrt{1 - \left(\frac{c}{c_c}\right)^2}$$

Section 21-8

Damped forced vibrations

AMPLITUDE:

$$A = \frac{P_0}{\sqrt{(c\Omega)^2 + (k - m\Omega^2)^2}}$$

MAGNIFICATION FACTOR:

$$\frac{A}{P_0/k} = \frac{A}{d_0} = \frac{1}{\sqrt{[2(c/c_c)(\Omega/\omega)]^2 + [1 - (\Omega/\omega)^2]^2}}$$

RESONANCE:

$$\Omega = \omega$$

For phase angle ϕ,

$$\tan \phi = \frac{c\Omega}{k - m\Omega^2} = \frac{2(c/c_c)(\Omega/\omega)}{1 - (\Omega/\omega)^2}$$

Section 21-9

Electrical analogues of mechanical vibrations. Compare Eq. 21-37 (mechanical system) and Eq. 21-45 (electrical system). Use Table 21-1 to establish an analogous system.

21–74 A 1000-kg rocket is suspended by a hinge at O. Determine the centroidal moment of inertia of the rocket if its period of small oscillations is 8 s.

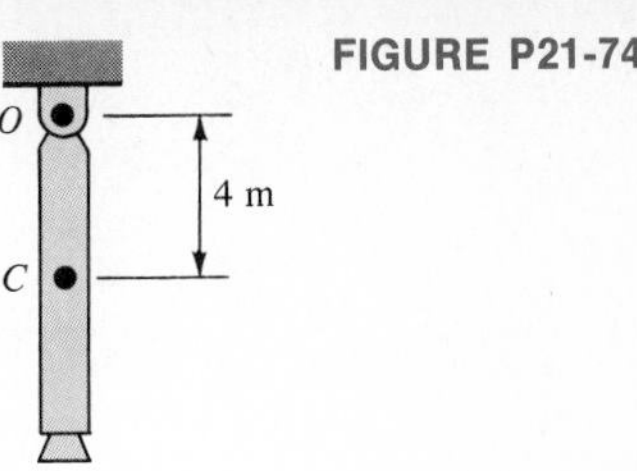

FIGURE P21-74

21–75 A 4000-lb helicopter is suspended by a hinge at O. Determine the centroidal moment of inertia (transverse axis) of the helicopter if its period of small oscillations is 12 s.

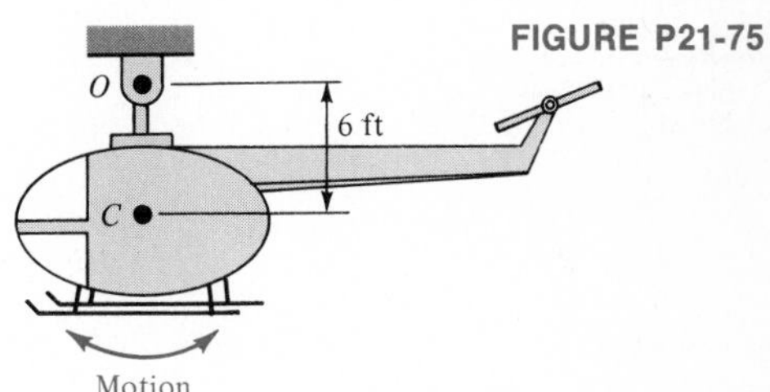

FIGURE P21-75

21–76 An amusement ride is modeled as a small block of mass m sliding inside a smooth surface of radius R. Determine the period of small oscillations. Use energy methods.

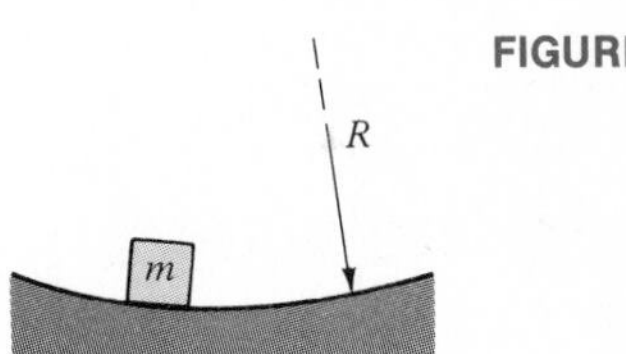

FIGURE P21-76

21–77 An amusement ride is modeled as a uniform disk of mass m and radius r rolling without slipping inside a curved track of radius R. Determine the period of small oscillations. Use energy methods.

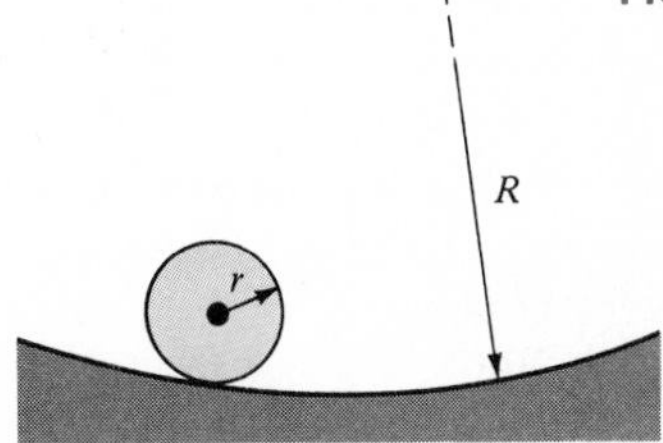

FIGURE P21-77

FIGURE P21-78

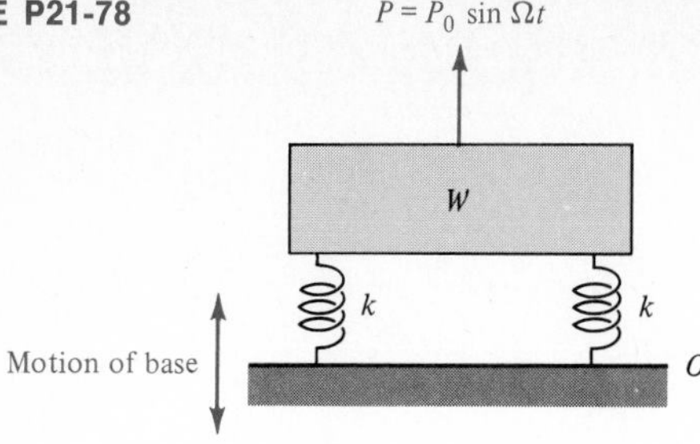

21–78 A vibrating pile driver is modeled as a 200-lb block on two springs of stiffness $k = 400$ lb/in. The desired peak displacement of the base is $d_0 = 0.05$ in. when the device is operated at 10 cycles per second. Determine the amplitude of vibrations of the block.

21–79 Determine the required forcing amplitude P_0 in Prob. 21–78.

FIGURE P21-80

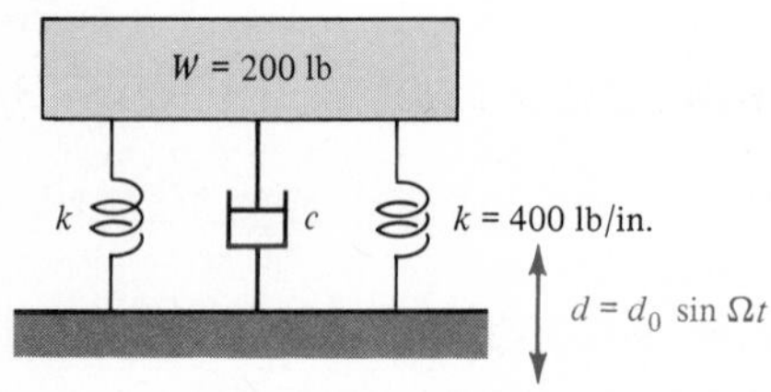

21–80 The pile driver in Prob. 21–78 has a coefficient of viscous damping $c = 30$ lb·s/ft. Determine the amplitude of vibrations of the block.

FIGURE P21-81

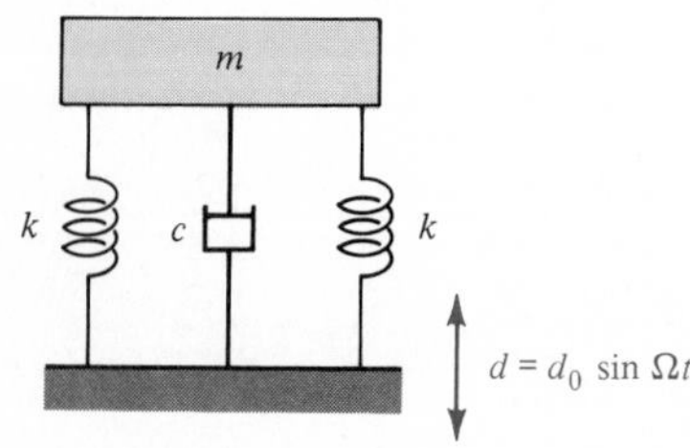

21–81 Determine the analogous electrical circuit of the system and write the differential equation for it.

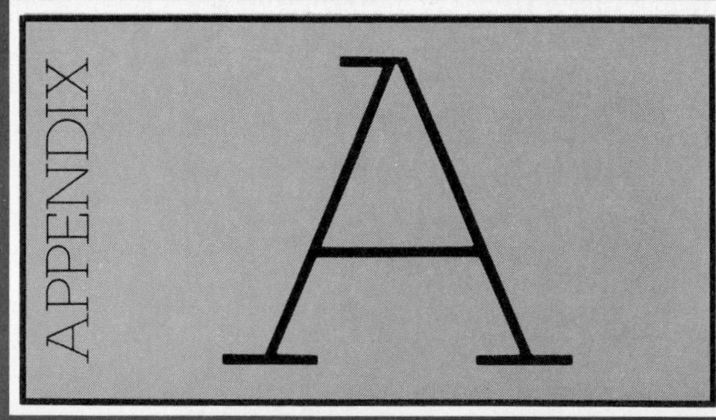

VECTOR DERIVATIVES

The student of dynamics will find the following vector analysis of general interest. Consider a vector **A** defined at time t as

$$\mathbf{A} = A\mathbf{n}_A$$

where A is the magnitude of the vector and $\mathbf{n}_A$ is a unit vector in the direction of **A**. Assume that the vector **A** changes to $\mathbf{A} + \Delta\mathbf{A}$ in the time interval Δt, as shown in Fig. A-1. The unit vector $\mathbf{n}_n$ is normal to $\mathbf{n}_A$ and lies in the plane of **A** and $\mathbf{A} + \Delta\mathbf{A}$. $\Delta\theta$ is the angular change of vector **A** in this plane. At the limit, the rotation $\Delta\theta$ is represented by a vector $d\boldsymbol{\theta}$ normal to $\mathbf{n}_A$ and $\mathbf{n}_n$ according to the right-hand rule. The time derivative of vector **A** is expressed using the chain rule as

$$\dot{\mathbf{A}} = \frac{d\mathbf{A}}{dt} = \frac{d}{dt}(A\mathbf{n}_A) = \dot{A}\mathbf{n}_A + A\dot{\mathbf{n}}_A \qquad \text{A-1}$$

This can be expressed in a more convenient form after noting the following:

$$\mathbf{n}_A = \frac{d\mathbf{n}_A}{dt} = \lim_{\Delta t \to 0}\left[\frac{(\mathbf{n}_A + \Delta\mathbf{n}_A) - \mathbf{n}_A}{\Delta t}\right] = \lim_{\Delta t \to 0}\left[\frac{\Delta\mathbf{n}_A}{\Delta t}\right] \qquad \text{A-2}$$

$$|\mathbf{n}_A| = |\mathbf{n}_A + \Delta\mathbf{n}_A| = 1 \qquad \text{by definition of the unit vector} \tag{A-3}$$

$$|\Delta\mathbf{n}_A| \simeq |\mathbf{n}_A|\,\Delta\theta = \Delta\theta \tag{A-4}$$

Since $\Delta\mathbf{n}_A$ is perpendicular to $\mathbf{n}_A$ at the limit,

$$\dot{\mathbf{n}}_A = \lim_{\Delta t \to 0}\left[\frac{\Delta\theta}{\Delta t}\right]\mathbf{n}_n = \dot{\theta}\mathbf{n}_n \tag{A-5}$$

Denoting the total instantaneous angular velocity of the vector **A** by ω_A gives

$$\dot{\mathbf{n}}_A = \dot{\theta}\mathbf{n}_n = \boldsymbol{\omega}_A \times \mathbf{n}_A \tag{A-6}$$

With these considerations, Eq. A-1 is written as

$$\dot{\mathbf{A}} = \underbrace{\dot{A}\mathbf{n}_A}_{\text{magnitude change}} + \underbrace{\boldsymbol{\omega}_A \times \mathbf{A}}_{\text{direction change}} \tag{A-7}$$

where all quantities are defined in the same arbitrary reference frame. Equation A-7 is one of the most important and most generally useful equations in dynamics.

Special Cases

1. Vector **A** has constant direction. $\omega_A = 0$, and

$$\dot{\mathbf{A}} = \dot{A}\mathbf{n}_A \tag{A-8}$$

2. Vector **A** has constant magnitude. $\dot{A} = 0$, and

$$\dot{\mathbf{A}} = \boldsymbol{\omega}_A \times \mathbf{A} \tag{A-9}$$

3. Arbitrary unit vector **n**. Since the magnitude of **n** is always constant,

$$\dot{\mathbf{n}} = \boldsymbol{\omega}_n \times \mathbf{n} \tag{A-10}$$

where ω_n is the angular velocity of **n**.

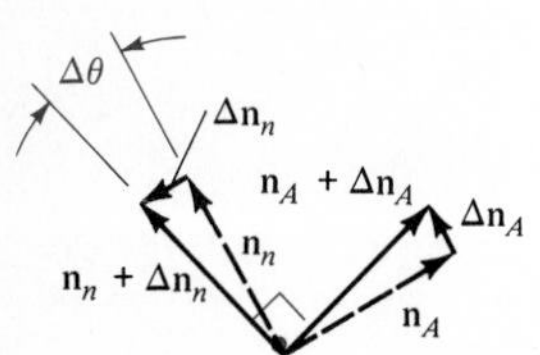

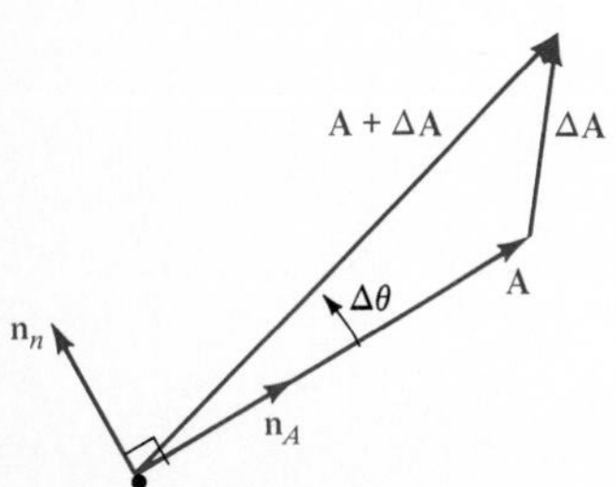

FIGURE A-1

APPENDIX B

COMMENTS ON THE USE OF UNITS

The two correct systems of units to be used in kinetics are described in Sec. 13-2. In addition, the reader should note that several other systems of units have been widely used until recently, and these may be encountered from time to time. In that case it is necessary to seek clarification of the precise meaning of each unit and use appropriate conversion factors. An even greater problem arises in commercial practice, as discussed in the following.

The units of mass and weight are often not distinguished in commercial practice. This is true throughout the world, regardless of the officially adopted technical system of units. Thus, the kilogram is widely used as a measure of weight in many countries, whereas the pound is the commonly used unit of mass in the United States. In the study of dynamics it is necessary to understand the difference between mass and weight and use the appropriate units and conversion factors if necessary. An overview of the relation between mass and weight in the two systems is shown in Fig. B-1. Parts a and b of this figure show the *correct* units and terminologies to be used in engineering work. Parts c and d show the technically *incorrect* systems that have widespread usage in everyday life. Kilogram as force is sometimes denoted by kgf, and pound as mass is sometimes denoted by lbm. Unfortunately, even these clarifying notations are often omitted.

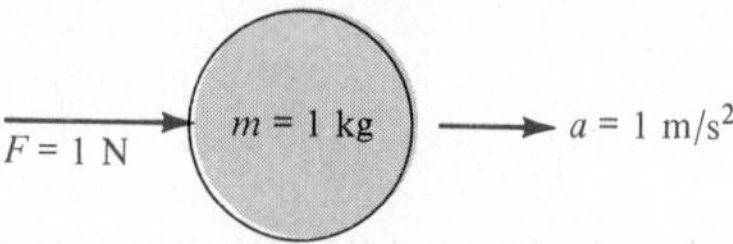

(a) SI units

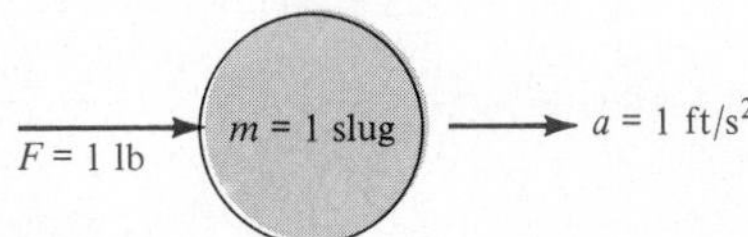

(b) U.S. customary units

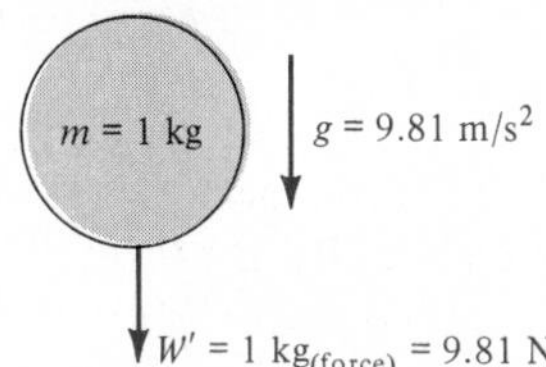

(c) kg as a commercial unit of weight, not to be used in technical work

$m' = 1\ \text{lb}_{(\text{mass})} = \frac{1}{32.2}$ slug

m'

$g = 32.2\ \text{ft/s}^2$

$W = 1$ lb

(d) lb as a commercial unit of mass, not to be used in technical work

FIGURE B-1

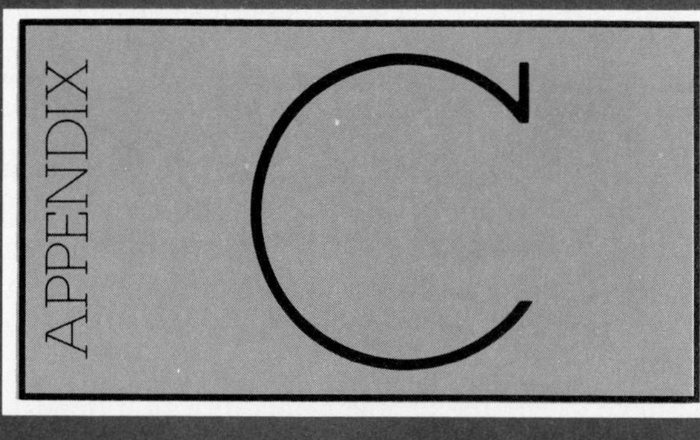

INERTIA FORCE AND DYNAMIC EQUILIBRIUM

The French philosopher D'Alembert (1717–1783) suggested an intriguing concept in dynamics by making an apparently trivial restatement of Newton's second law. He wrote from $\mathbf{F} = m\mathbf{a}$,

$$\mathbf{F} - m\mathbf{a} = 0 \qquad \text{C-1}$$

The implication of this is that if $-m\mathbf{a}$ is considered a force and added to all of the other forces acting on the particle, the analysis is reduced to one of statics. This is illustrated using the force triangle in Fig. C-1. The triangle is closed when the fictitious force $-m\mathbf{a}$ is taken into account, creating the dynamic equivalent of static equilibrium, called *dynamic equilibrium.*

The quantity $-m\mathbf{a}$ is called *inertia force.* Its direction is always opposite to the direction of the particle's acceleration. The philosophical merits of this concept are controversial since cause and effect are debatable (the inertia force is the reaction to the force which is causing the acceleration of the body). The concept of the inertia force is further illustrated in the following two examples.

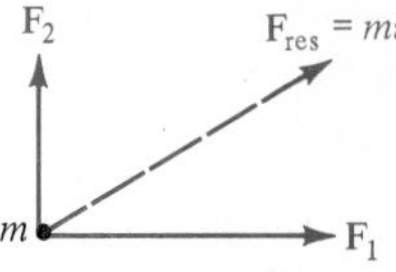

(a) m accelerates ($\mathbf{F}_{res} = \mathbf{F}_1 + \mathbf{F}_2$)

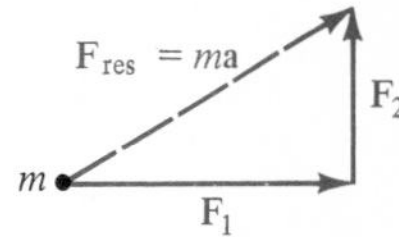

(b) m accelerates

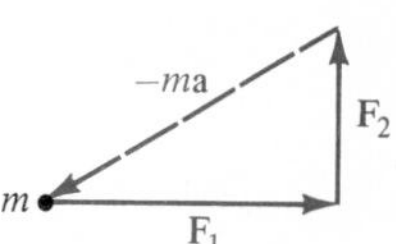

(c) Dynamic equilibrium

FIGURE C-1

*1. **Inertia force in linear motion.*** Consider a vehicle traveling on a straight path and accelerating. The passengers feel that they are forced in a direction opposite to the direction of the acceleration of the vehicle.

*2. **Centrifugal force.*** Consider a vehicle traveling at a constant speed on a circular path. The force that keeps each passenger on the curved path is the *centripetal force*, which is directed toward the center of the path. The passengers feel that they are forced away from the center of the path. This fictitious force is called *centrifugal force*.

Again, the inertia force felt by the people is in a direction opposite to the acceleration of the vehicle, which is entirely toward the center of the path if the angular velocity is constant (Fig. C-2). In spite of such human sensations, the fictitious inertia forces and dynamic equilibrium should not be used in elementary dynamics for several reasons:

1. The application of dynamic equilibrium is confusing to most students because it is fictitious.
2. Imagining dynamic equilibrium is not necessary to solve any problems.
3. The method of dynamic equilibrium becomes increasingly cumbersome (even mind-boggling) as the complexity of the problems increases.
4. Most dynamicists prefer to use the basic method of drawing free-body diagrams with real forces and applying the equations of motion according to these drawings. This is the simplest method in solving the majority of problems.

The method of dynamic equilibrium is illustrated in solving Ex. 18-1 as follows.

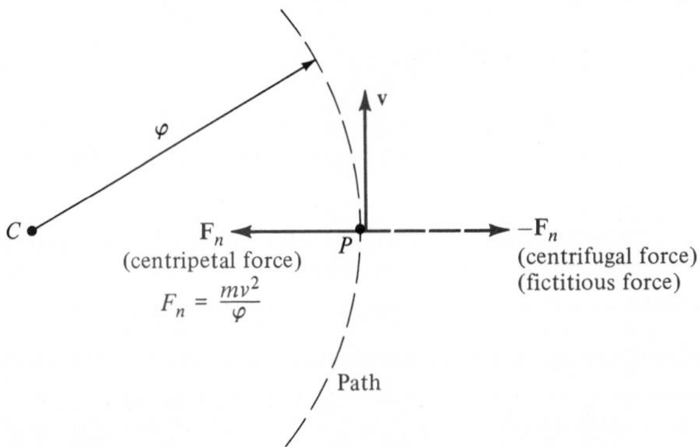

FIGURE C-2

The combined mass of the motorcycle and its rider in Fig. C-3a is 250 kg. The mass center is at C. The wheels are locked when the brakes are applied at a high speed. Determine the linear acceleration and the normal and tangential forces acting on each wheel assuming that the coefficient of kinetic friction is 0.5, and that the angular acceleration is zero.

EXAMPLE C-1

SOLUTION

First two free-body diagrams are sketched, one showing only actual forces and moments externally applied to the body, and the other showing the inertial forces and moments acting at the center of mass (Fig. C-3b). The constraints of motion are that $a_{Cy} = 0$ and $\alpha = 0$. Using the three equations from Eq. 18-8 for pure translation,

$$\sum F_x = \mathscr{F}_A + \mathscr{F}_B = ma_{Cx} \quad \textbf{(a)}$$

$$\sum F_y = N_A + N_B - W = ma_{Cy} = 0 \quad \textbf{(b)}$$

$$\sum M_C = \mathscr{F}_A(0.7\text{ m}) - N_A(0.8\text{ m}) + \mathscr{F}_B(0.7\text{ m}) + N_B(0.5\text{ m}) = I_C\alpha = 0 \quad \textbf{(c)}$$

Since the wheels are locked and sliding, the friction forces are $\mathscr{F}_A = \mu_k N_A$ and $\mathscr{F}_B = \mu_k N_B$. Hence, from Eqs. (b) and (c),

$$N_A + N_B = W = (250\text{ kg})(9.81\text{ m/s}^2) = 2453\text{ N}$$

$$\mu_k N_A(0.7\text{ m}) - N_A(0.8\text{ m}) + \mu_k N_B(0.7\text{ m}) + N_B(0.5\text{ m}) = 0$$

$$N_A(0.35 - 0.8)\text{ m} + N_B(0.35 + 0.5)\text{ m} = 0$$

Solving these simultaneously yields

$$N_A = 1604\text{ N} \qquad N_B = 849\text{ N}$$

$$\mathscr{F}_A = \mu_k N_A = 802\text{ N} \qquad \mathscr{F}_B = \mu_k N_B = 425\text{ N}$$

Hence, from Eq. (a),

$$a_{Cx} = \frac{1}{250\text{ kg}}(802 + 425)\text{ N} = 4.91\text{ m/s}^2 \qquad \text{as in Ex. 18-1}$$

It should be noted that for the moment equation moments could have been summed about one of the contact points instead of the mass center C. For example,

$$\sum M_A = N_B(1.3\text{ m}) - W(0.8\text{ m}) = -ma_{Cx}(0.7\text{ m})$$

The negative sign of the moment of the inertia force ma_{Cx} is due to the assumed direction of a_{Cx} and the positive moment convention chosen (counterclockwise). It is essential that all forces and moments, applied or inertial, be summed into equations *according to the same sign convention and on the correct sides of the equation* (applied forces and moments on one side and inertia forces and moments on the other side).

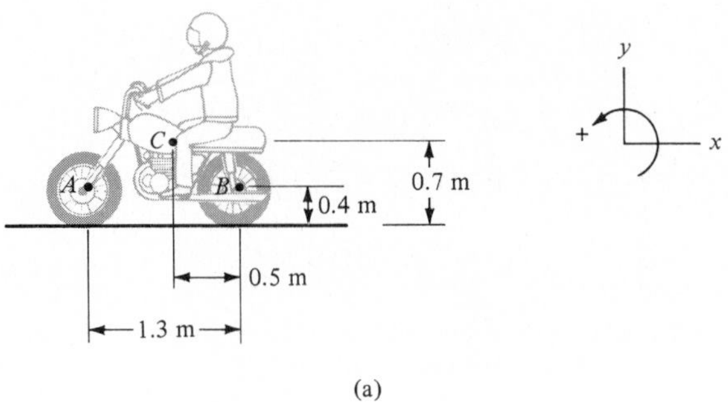

(a)

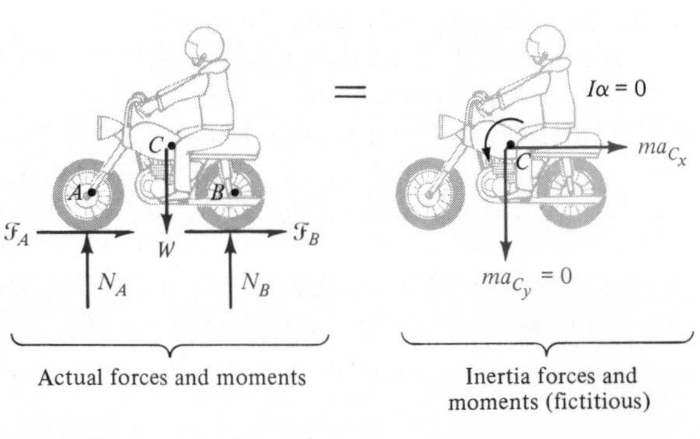

(b)

FIGURE C-3

APPENDIX D

CORIOLIS EFFECTS

For a qualitative understanding of the Coriolis acceleration, consider Fig. D-1. Assume that persons *A* and *B* are standing on a disk which rotates at a constant angular speed Ω in a horizontal plane. Person *A* throws a ball perfectly aimed in the direction of person *B*, as shown in the top view of the disk in Fig. D-1b. In this situation the ball would miss person *B* for two reasons.

First, the ball has not only the relative velocity $\mathbf{v}_{rel}$ from *A* to *B* at the instant of release, but also a rotational velocity $\mathbf{v}_{rot}$ because *A* is rotating with the disk. Thus, the actual direction of the ball is determined by the vector sum $\mathbf{v}_{abs} = \mathbf{v}_{rel} + \mathbf{v}_{rot}$. The resulting error in the ball's path to the position B_1 of person *B* at the instant of release is the *Coriolis drift* 1. An additional error, Coriolis drift 2, occurs since *B* moves into a new position B' while the ball flies (Fig. D-1b).

The Coriolis drift can be eliminated if the ball moves in a groove or along a rail from *A* to *B*. Such constraints accelerate the ball to correct its path, and it will reach person *B*.

Note that the two components of the Coriolis drift discussed above are meant to clarify the physical situation, while the factor 2 in $\mathbf{a}_{Cor} = 2\boldsymbol{\Omega} \times \mathbf{v}_{rel}$ (such as in Eq. 17-36) is obtained from the mathematical derivation using a rotating reference frame. However, it should be clear that a larger Ω would increase the drift and would necessitate a larger acceleration to correct the path. Also, a larger v_{rel} means a larger magnitude of the velocity vector $\mathbf{v}_{abs}$; hence, a larger acceleration necessary to change its direction toward B', the actual target (Fig. D-1b).

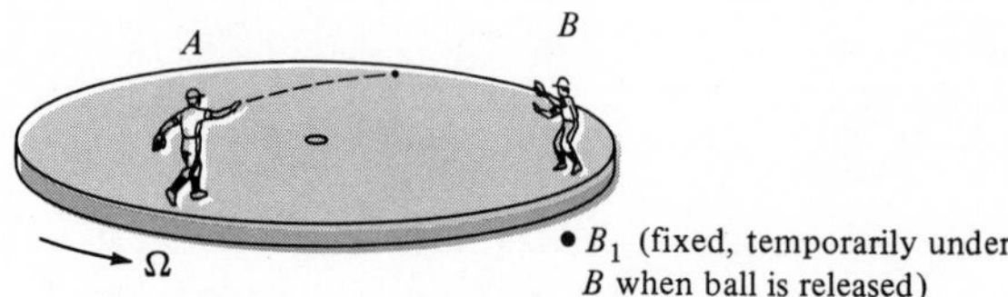

(a) A and B stand on rotating disk; A throws a ball in direction of B

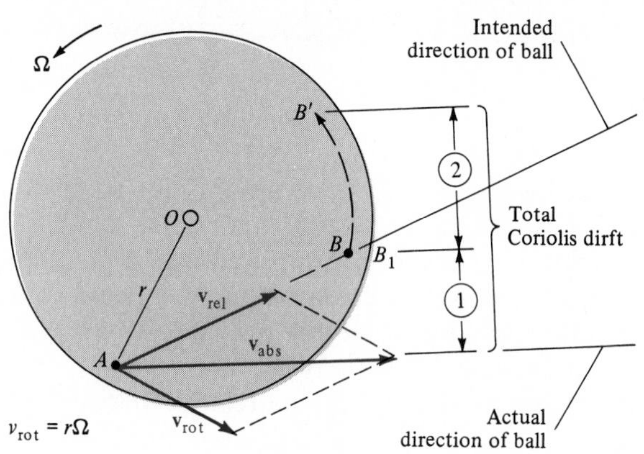

(b) Top view of disk

Notes: (1) = Coriolis drift 1 (ball would miss fixed point B_1 because of $\mathbf{v}_{rot}$ of A at instant ball is released)

(2) = Coriolis drift 2 (ball would further lag behind B because B moves into new position B' while ball flies)

In general, (1) ≠ (2)

FIGURE D-1

EXAMPLE D-1

Assume that person A in Fig. D-1 is standing at the center of the disk while throwing the ball. Discuss the Coriolis drift in this situation.

SOLUTION

Consider two reference systems: the fixed XY frame, and the xy axes which are painted on the rotating disk. Let $\mathbf{v}_{rel} = v_{rel}\mathbf{i}$ when the ball is released by A, as shown in Fig. D-2a. In this case $\mathbf{v}_{rel} = \mathbf{v}_{abs}$, and the path of the ball with respect to the rotating xy frame is the curved line in Fig. D-2b. Note that in this special case the Coriolis drift 1 is zero since $r = 0$ in $v_{rot} = r\Omega$ (Fig. D-1b). Thus, the entire drift consists of Coriolis drift 2: the target B moves to B' during the flight of the ball to the fixed point B_1 (Fig. D-2b).

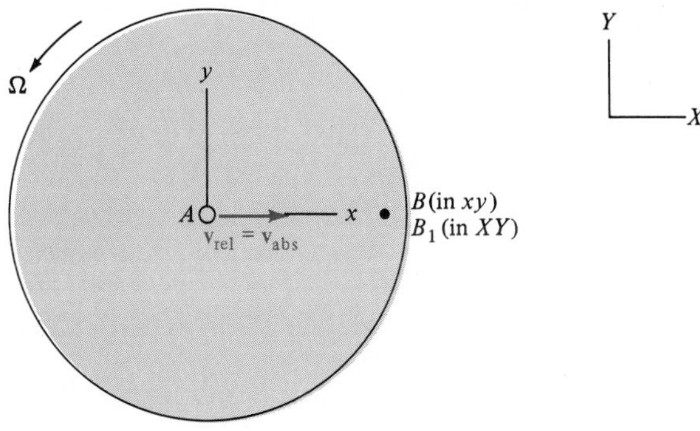

(a) Top view of disk

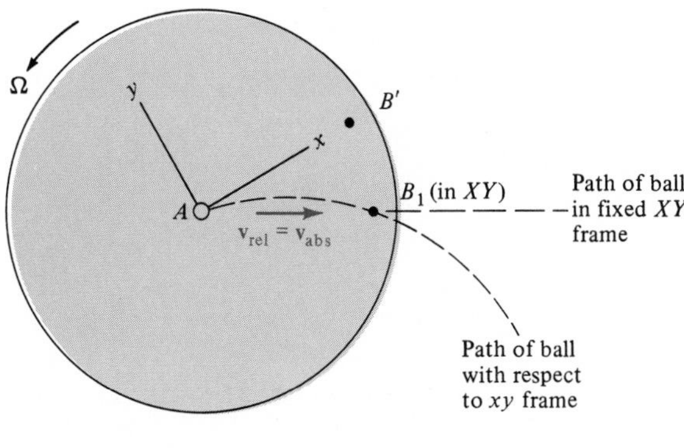

(b)

FIGURE D-2

ANSWERS TO SELECTED PROBLEMS

Chapter 2

2–1 (a) $R = 5.6$ kN↓
(b) $R = 5.6$ kN↓

2–2 $R = 1130$ lb 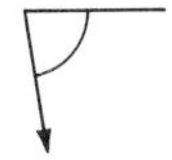 82°

2–4 $F_2 = 490$ lb; $R = 750$ lb

2–5 $R = 4.42$ kN

2–7 $262.8^\circ \geq \theta \geq 97.2^\circ$

2–8 $R = 1900$ N

2–10 $R = 12.1$ kN  72.4°

2–11 $R = 1043$ lb 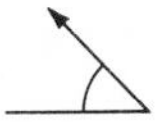 45.8°

2–13 $P = 9.35$ MN 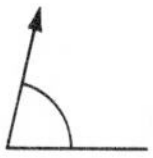 76.8°

2–14 $R = 2.24 \times 10^5$ lb 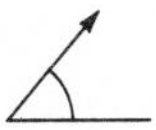 49.8°

2–16 $F_H = 19{,}319$ lb, $F_V = 5176$ lb

2–17 $F_H = 4.7$ kN; $F_V = 1.7$ kN; $F_{AB} = 4.9$ kN

2–18 For 8-kN force: $F_x = 6.93$ kN←; $F_y = 4$ kN↓
For 10-kN force: $F_x = 6.43$ kN→; $F_y = 7.66$ kN↓

2–19 $F_{AB} = 772.7$ lb  75°;
$F_{AC} = 772.7$ lb  75°

2–21 $\mathbf{R} = -5.64\mathbf{j}$ kN

2–22 $\mathbf{R} = (150\mathbf{i} - 1126\mathbf{j})$ lb

2–24 $F_2 = 490$ lb

2–25 $\mathbf{R} = (4.03\mathbf{i} + 1.24\mathbf{j})$ kN

2–28 $\mathbf{R} = (1811.5\mathbf{i} - 439.4\mathbf{j})$ N

2–29 $\mathbf{F}_4 = 307.5\mathbf{j}$ lb

2–31 $\mathbf{F}_4 = 2780.2\mathbf{j}$ lb

2–32 $\mathbf{R} = (-45\mathbf{i} + 31\mathbf{j})$ kN

2–34 $\mathbf{P} = -2.7\mathbf{i}$ MN

2–35 $\mathbf{R} = (-0.40\mathbf{i} + 2.22\mathbf{j}) \times 10^5$ lb

2–37 $\mathbf{R} = (1035.9\mathbf{i} + 500.9\mathbf{j})$ lb

2–39 $\mathbf{n} = \dfrac{a\mathbf{i} + b\mathbf{j} + c\mathbf{k}}{\sqrt{a^2 + b^2 + c^2}}$

2–41 $n_x = -0.423; n_y = 0.906; n_z = 0$
2–42 $n_x = 0.643; n_y = -0.766; n_z = 0$
2–44 $n_x = 0; n_y = 0.515; n_z = 0.858$
2–45 $\mathbf{n} = 0.421\mathbf{i} + 0.337\mathbf{j} + 0.842\mathbf{k}$
2–47 $\theta_x = 138^\circ; \theta_y = 56.1^\circ; \theta_z = 68.2^\circ$
2–48 $\mathbf{n}_{AB} = -0.743\mathbf{i} + 0.558\mathbf{j} + 0.372\mathbf{k}$
2–50 $\mathbf{R} = (-407\mathbf{i} + 232.6\mathbf{j})$ lb
2–51 $\mathbf{F}_H = \mathbf{F}_y = -1.08\mathbf{j}$ kN;
$\mathbf{F}_V = \mathbf{F}_z = -3.58\mathbf{k}$ kN
2–53 $\mathbf{R}_{xy} = -0.3F\mathbf{i} + 0.05F\mathbf{j}$
2–55 $\mathbf{R}_H = (a - c - f)\mathbf{i} + (e - h)\mathbf{k}$;
$\mathbf{R}_V = -(b + d + g)\mathbf{j}$
2–56 $F_{4x} = -R_x = -a + c + f$
$F_{4y} = -R_y = b + d + g$
$F_{4z} = -R_z = -e + h$
2–57 $\theta = 9.9^\circ$
2–58 $F' = 1.98$ kN
2–60 $\phi = 79.6^\circ$
2–61 $F' = 200$ lb
2–63 $F' = 1.25$ kN
2–64 $\theta = 48.9^\circ$
2–65 $F'_{BA} = 394.3$ N; $F'_{CA} = 352.7$ N
2–67 $\phi = \cos^{-1} \dfrac{-b}{\sqrt{a^2 + b^2 + c^2}}$
2–68 $F' = \dfrac{-Fc}{\sqrt{a^2 + b^2 + c^2}}$
2–70 (a) $F' = 76.82$ lb; (b) $F' = 76.82$ lb
2–71 $T = 1$ kN
2–73 $T = 333.3$ N
2–74 $F_{BC} = 666.6$ N
2–76 $F_{BC} = 3.33$ kN; $F_{DE} = 3.33$ kN
2–77 $T = 1.33$ kN; $F_{BC} = 2.67$ kN;
$F_{BD} = 5.33$ kN
2–79 $F_{DA} = 1200$ N; $F_{BC} = 800$ N;
$F_{BE} = 1600$ N
2–80 $T = 100$ lb
2–82 $T_{CA} = 228.5$ lb; $T_{CB} = 233.1$ lb
2–84 $R_x = 216.5$ N; $R_y = 125$ N
2–86 $C_x = -221.6$ N; $C_y = -896.6$ N
2–88 $T_B = 1.48$ kN
2–89 $R_x = -3789.3$ lb; $R_y = 1668.6$ lb
2–90 $\mathbf{F} = (28.2\mathbf{i} - 48.8\mathbf{j})$ N; $\mathbf{P} = -3.15\mathbf{i}$ kN
2–92 $\mathbf{R} = (20.3\mathbf{j} + 823.7\mathbf{k})$ lb
2–93 $P = 5.29$ kN; $W = 5.29$ kN
2–94 $P = \dfrac{Wl}{3h}$
2–95 $T = \dfrac{3}{8}W$
2–97 $F = 714.3$ lb
2–98 $\theta \simeq 55^\circ$
2–102 $\mathbf{R} = (72.3\mathbf{i} - 15.4\mathbf{j})$ kN
2–103 $\mathbf{R} = (350\mathbf{i} - 1056.2\mathbf{j})$ lb
2–104 $R_V = 4F_y = 5.29$ kN
$F_H = 0$ [symmetrical loadings]
2–106 $\theta = 27.7^\circ$

Chapter 3

3–1 $A_x = 9.40$ kN; $A_y = 3.42$ kN;
$M_A = 8.66$ kN·m ↙
3–2 $M_O = 1500$ ft·lb ↗
3–5 $M_A = 8.19$ kN·m ↙; $M_B = 9.96$ kN·m ↙
3–7 $M_A = 11.28$ kN·m ↙;
$M_C = 13.65$ kN·m ↙
3–8 $\theta = 0$
3–10 $\alpha = 45^\circ$
3–11 $M_1 = 1.73$ kN·m ↙; $M_2 = 2.95$ kN·m ↙;
$M_A = 4.69$ kN·m ↙
3–13 $M_W = 800$ N·m ↗; $M_O = 700$ N·m ↙
3–14 $M_{F_O} = 662$ ft·lb; $M_{W_O} = 350$ ft·lb
3–16 $M_{FA} = 1503.5$ ft·lb; $M_{WA} = 80$ ft·lb;
$F_R = 210.7$ lb ↘ 20.8°;
$M_R = 1583.5$ ft·lb ↙
3–17 $\sum M_A = -0.34(W_1 AB + W_2 AC) +$
$0.98FAD$ ↙
3–20 $\text{Area}_{\theta=90^\circ} = 6$; $\text{area}_{\theta=30^\circ} = 3$;
$\text{area}_{\theta=150^\circ} = 3$
3–21 $\mathbf{M}_O = 250\mathbf{k}$ N·m
3–23 $\mathbf{M}_A = -14.5\mathbf{k}$ in·lb
3–24 $\sum \mathbf{M}_A = 116.6\mathbf{k}$ N·m
3–26 $\mathbf{M}_O = (90\mathbf{i} - 360\mathbf{j} - 42\mathbf{k})$ in·lb
3–27 $\mathbf{M}_A = -8.66\mathbf{k}$ kN·m
3–29 $\mathbf{M}_B = -9.96\mathbf{k}$ kN·m
3–30 $\mathbf{M}_C = -2676.8\mathbf{k}$ ft·lb
3–32 $\mathbf{M}_O = (-600\mathbf{i} - 280\mathbf{j} + 350\mathbf{k})$ ft·lb
3–33 $\mathbf{M}_O = 0$
3–35 $\mathbf{M}_{WA} = -1.52\mathbf{k}$ kN·m;
$\mathbf{M}_{FA} = -3.45\mathbf{k}$ kN·m
3–36 $\mathbf{M}_O = (1200\mathbf{i} + 300\mathbf{j} + 1400\mathbf{k})$ ft·lb
3–38 $\mathbf{M}_C = (80\mathbf{i} + 360\mathbf{k})$ ft·kip
3–39 $\mathbf{M}_O = (24\mathbf{i} + 40\mathbf{j} - 40\mathbf{k})$ N·m;
$\mathbf{M}_A = (12\mathbf{i} + 10\mathbf{j} - 40\mathbf{k})$ N·m;
$\mathbf{M}_B = (4\mathbf{i} + 10\mathbf{j})$ N·m
3–41 $\mathbf{M}_O = 1.5\mathbf{k}$ kN·m
3–42 $\mathbf{M}_{O_1} = -1500\mathbf{k}$ ft·lb;
$\mathbf{M}_{O_2} = -1000\mathbf{k}$ ft·lb

3–43 $\mathbf{V} = -16ab^2\mathbf{i} - 12a^2b\mathbf{j} + (60a^2c - 136b^2c)\mathbf{k}$

3–45 $\mathbf{A}\cdot(\mathbf{B}\times\mathbf{C}) = -14$

3–46 $\mathbf{A}\cdot(\mathbf{B}\times\mathbf{C}) = -4070$

3–48 $M_x = -320$ ft·lb; $M_y = 0$; $M_z = -400$ ft·lb

3–49 $M_x = 20$ N·m; $M_y = -20$ N·m; $M_z = -14$ N·m

3–51 $M_x = 0$; $M_y = 0$; $M_z = -114$ N·m

3–52 $M_x = -520$ in·lb; $M_y = 0$; $M_z = 650$ in·lb

3–54 $M_x = -600$ in·lb
$M_y = -1500$ in·lb
$M_z = -8500$ in·lb

3–55 $M_x = -4.2$ kN·m; $M_y = 0.615$ kN·m; $M_z = 7.5$ kN·m

3–57 $|F_{min}| = 360$ ft·lb

3–58 $M_x = 94.1$ N·m

3–60 $M_x = 4$ N·m; $M_y = 48$ N·m; $M_z = -48$ N·m

3–61 $M_l = 36.8$ N·m

3–62 $\mathbf{M}_A = -Fd\mathbf{k}$

3–64 $\mathbf{M}_{net} = -(F + P)c\mathbf{j}$

3–65 $\mathbf{M}_{net} = -(P + \frac{F}{2})c\mathbf{i}$

3–67 $\mathbf{M}_{net} = Fc\mathbf{j} - Pa\mathbf{k}$

3–68 $P = 1000$ N; $Q = 1500$ N; $F = 250$ N

3–70 $F = 428.6$ lb

3–71 $F_{min} = 509.9$ N

3–73 $F_{eq} = 200$ lb; $M_{eq} = 590.9$ ft·lb↺

3–74 $F_{eq_1} = 500$ N; $M_{eq_1} = 281.9$ N↻

3–76 $d = 1.88$ ft

3–77 $P = 34.5$ kN

3–78 $\mathbf{F}_{eq} = (200\mathbf{i} - 30\mathbf{j} - 20\mathbf{k})$ N
$\mathbf{M}_{eq} = (-68\mathbf{i} + 80\mathbf{j} - 80\mathbf{k})$ N·m

3–80 $P = 50$ lb

3–81 $\mathbf{F}_{eq} = (-3\mathbf{i} + 0.5\mathbf{j})$ kN
$\mathbf{M}_{eq} = 2.38\mathbf{k}$ kN·m

3–83 $\mathbf{F}_R = -(0.996F + P + W)\mathbf{j} + 0.087F\mathbf{k}$

3–84 $\mathbf{F}_R = (729.8\mathbf{i} + 192.8\mathbf{j})$ N
$\mathbf{M}_R = -451.0\mathbf{k}$ N·m

3–86 $\mathbf{F}_R = F\mathbf{i} - W\mathbf{j}$; $\mathbf{M}_R = (Wd - Fh)\mathbf{k}$

3–87 $\mathbf{F}_R = (F - P)\mathbf{i}$; $\mathbf{M}_R = [Pa - F(a + b)]\mathbf{k}$

3–89 $P = 2435$ lb

3–90 $\mathbf{F}_R = -(F + P)\mathbf{j}$;
$\mathbf{M}_R = -(Qb + Pa + 3Fa)\mathbf{k}$

3–91 $\mathbf{F}_R = 8500\mathbf{j}$ lb; $\mathbf{M}_R = 95{,}000\ \mathbf{k}$ ft · lb

3–92 $\mathbf{F}_R = 8500\mathbf{j}$ lb;
$\mathbf{M}_R = 61{,}065\mathbf{i} + 72{,}774\mathbf{k}$ ft·lb

3–94 $\mathbf{F}_R = (-400\mathbf{j} - 300\mathbf{k})$ N; $d = 0.48$ m

3–95 $x = 1.052$ m

3–97 $d = 0.47$ ft (left of origin)

3–98 $\mathbf{F}_R = (50\mathbf{i} + 750\mathbf{j} + 40\mathbf{k})$ lb
$\mathbf{M}_R = (-68\mathbf{i} + 100\mathbf{j} - 270\mathbf{k})$ ft·lb

3–99 $\mathbf{F}_R = 40\mathbf{j}$ kN; $\mathbf{M}_R = (52\mathbf{i} - 100\mathbf{k})$ kN·m

3–100 $M_x = 59$ kN·m; $M_y = 20$ kN·m; $M_z = -100$ kN·m

3–101 $\sum M_O = 200$ ft·lb↻

3–102 $M_{1x} = 10$ kN·m; $M_{1y} = 0$; $M_{1z} = 0$
$M_{2x} = 3.54$ kN·m; $M_{2y} = 0$; $M_{2z} = -3.54$ kN·m
$M_{3x} = 0$; $M_{3y} = 0$; $M_{3z} = -10$ kN·m
$R_x = 2.71$ kN; $R_y = 0$; $R_z = 2.71$ kN
$M_{Rx} = 13.54$ kN·m; $M_{Ry} = 0$; $M_{Rz} = -13.54$ kN·m

3–104 $M_O = \dfrac{-Frl\sin\theta}{\sqrt{l^2 + r^2 - 2lr\cos\theta}}$

3–106 $(M_x)_{max} = -4500$ ft·lb

3–108 $\mathbf{F}_{eq} = (300\mathbf{i} + 2000\mathbf{j} - 300\mathbf{k})$ lb
$\mathbf{M}_{eq} = (1200\mathbf{i} + 300\mathbf{j} + 3200\mathbf{k})$ ft·lb

Chapter 4

4–1 $R_x = T_1(\sin\theta - \sin\phi)$; $R_y = T_1(\cos\theta + \cos\phi)$

4–2 $T_2 = T_1 = 5$ kN; $R_x = 4.70$ kN; $R_y = 6.71$ kN

4–4 $B_y = 3.67$ kN↑; $A_x = 0$; $A_y = 3.33$ kN↑

4–5 $R = 1142.9$ lb↑; $F_1 = 285.7$ lb↓

4–7 Statically indeterminate (three unknowns with two equations)

4–8 $A_x = 1.53$ kN←; $A_y = 1.29$ kN↑; $M_A = 7.66$ kN·m↺

4–10 $T = 8.77$ kN; $A_y = 0$; $A_x = 8.24$ kN→

4–12 Statically indeterminate (four unknowns with three equations)

4–13 $T_B = 3.67$ kN; $A_y = 1.10$ kN↓; $A_x = 6.90$ kN→

4–15 $O_y = 0.866T$↑; $O_x = 0.5T$→

4–16 $R_B = 8$ kN↑; $R_A = 4$ kN↑

4–18 $\mathbf{F} = (5.6\mathbf{i} - 2.0\mathbf{j})$ kN

4–19 $\mathbf{F} = (491.7\mathbf{i} - 178.9\mathbf{j})$ lb; $R_x = 8.3$ lb→; $R_y = 528.9$ lb↑

4–21 For $d = 0.3$: $R_x = 2.91$ kN→; $R_y = 5.06$ kN↓
For $d = 1.0$: $R_x = 0.86$ kN→; $R_y = 1.52$ kN↓

4–24 $A_x = 43.45$ kN→; $A_y = 105.2$ kN↑

4–26 $F = 1334$ lb

4-27 From front view: $A_x = 0$; $A_y = 25$ lb↑

4-30 $F_{CB} = 75$ N; $F_{DA} = 25$ N; partially constrained

4-31 $F_{BC} = 141.4$ lb; $F_{AD} = 141.4$ lb; partially constrained

4-34 $F_{FA} = 500$ N↑; $F_{BC} = 1250$ N←; $F_{ED} = 1250$ N→; completely constrained

4-35 $A_y = 0$; partially constrained

4-37 $F_{DE} = 0$; completely constrained (statically indeterminate)

4-38 $F_{AE} = 800$ N↑; partially constrained (improperly)

4-41 $F_C = 8$ kN↗; $F_A = 6.93$ kN←

4-42 $F_C = 1766$ lb↗; $F_A = 1850.8$ lb←

4-44 $F_A = 500$ lb; $F_B = 0$; $C_x = 0$; $C_y = 500$ lb

4-45 $A_x = F \tan\theta$→; $B_y = F$↓; $B_x = F \tan\theta$←

4-47 $A_y = \frac{M}{r}$↓; $B_y = \frac{M}{r}$↑

4-48 $M_C = 240$ N·m; $C_y = 6000$ N↓

4-50 $F_A = -\frac{M}{l \sin 2\theta}$ (direction at angle θ);

$C_x = \frac{M \cos\theta}{l \sin 2\theta}$←; $C_y = \frac{M \sin\theta}{l \sin 2\theta}$↑

4-51 $F_B = F_A = \frac{W}{\sin\theta}$

4-53 $A_x = 300$ lb→; $A_y = 116.1$ lb↓; $F_A = 321.7$ lb

4-54 $O_x = 400$ lb←; $O_y = 154.8$ lb↑; $M_O = 2119.1$ ft·lb

4-55 $F_D = 502$ lb; $A_x = 43.6$ lb→; $A_y = 500$ lb↓; $M_B = 7150.4$ in·lb

4-56 $N_C = 6.05$ kN↑; $N_D = 11.28$ kN↑; $N_E = 4.67$ kN↑; $N_F = 2.33$ kN↑

4-58 $A_x = 0$; $A_y = W$; $M_A = W(l \sin\theta - d \cos\phi)$; $B_x = 0$; $B_y = W$

4-59 $\mathbf{C} = (-8.5\mathbf{i} - 0.44\mathbf{j})$ kN

4-61 $F_{CD} = 241.6$ N; $A_z = 255$ N↓; $B_z = 165$ N↑; $A_x = 38.2$ N→; $B_x = 10.4$ N←

4-62 $A_x = 30$ kN←; $A_y = 10$ kN↑; $B_x = 30$ kN→

4-64 $A_x = 0$; $A_y = 3300$ N↑; $M_A = 600$ N·m↺

4-65 $A_x = 0$; $A_y = 3300$ N↑; $A_z = 500$ N↙; $M_{A_z} = 600$ N·m; $M_{A_x} = 3500$ N·m; $M_{Ay} = -1000$ N·m

4-67 $O_x = 200$ lb→; $O_y = 29.6$ lb↑; $O_z = 538.4$↙; $M_{0x} = 0$; $M_{0y} = 4400$ ft·lb; $M_{0z} = -1000$ ft·lb

4-68 $T = 5.14$ kN; $N_A = 1.69$ kN; $N_B = 4.43$ kN

4-70 $A_x = 240$ lb→; $A_y = 1020$ lb↑; $M_A = 8080$ ft·lb↺

4-71 $O_x = 3576$ lb→; $O_y = 150$ lb↑; $O_z = 0$

4-74 $A_z = B_z = 18{,}750$ lb; $D_z = 12{,}500$ lb

4-75 $A_x = 0$; $A_y = 40.5$ kN↑; $M_A = 300.8$ kN·m↺

4-77 $O_x = 7.8$ kN→; $O_y = 41.6$ kN↑; $A_y = 12.3$ kN↓

4-78 $\sum F_y = L - W + F_{Ay} = 0$; $\sum F_z = F_{Az} + P = 0$; $\sum M_A = M_{A_x} + Ph = 0$

4-79 $T_C = 225.2$ lb; $T_A = 450.4$ lb

4-80 $M_y = M + Fd$; $M_x = Fh$; $O_z = F$

4-81 $\Delta A_x = \Delta B_x = 22$ lb→; $\Delta A_y = 0$

4-82 $A_y = 1200$ N↑; $A_z = -480$ N

4-83 $A_y = 360$ N↑; $B_y = 120$ N↑; $A_z = -415.7$ N; $B_z = 623.5$ N

Chapter 5

Note: The signs of the shearing force (V) and the bending moment (M) are consistent with the accepted positive orientations discussed in the text.

5-1 At $x = 0$: $V_O = 500$ N; $M_O = 500$ N·m
At $x = 2$ m: $V = 500$ N; $M = -400$ N·m
At $x = 4$ m: $V = 500$ N; $M = -300$ N·m

5-2 At $y = 0$: $V_O = -200$ lb; $M_O = 600$ ft·lb
At $y = 1$ ft: $V = -200$ lb; $M = 400$ ft·lb
At $y = 2$ ft: $V = -200$ lb; $M = 200$ ft·lb

5-4 At $x = 0$: $V_O = 500$ lb; $M_O = -1300$ ft·lb
At $x = 1$ ft: $V = 500$ lb; $M = -800$ ft·lb
At $x = 2.5$ ft: $V = 300$ lb; $M = -150$ ft·lb

5-5 At $y = 0$: $V_O = 2$ kN; $M_O = -14$ kN·m
At $y = 1$ m: $V = 2$ kN; $M = -12$ kN·m
At $y = 1.3$ m: $V = 2$kN; $M = -11.4$ kN·m

5-7 At $x_C = 1$ m, $M = -7$ kN·m
At $x_C = 2$ m, $M = -8$ kN·m
At $x_C = 4$ m, $M = -10$ kN·m

5-8 At $x_C = 3$ ft, $M_O = -1500$ ft·lb
At $x_C = 6$ ft, $M_O = -3000$ ft·lb
At $x_C = 9$ ft, $M_O = -4500$ ft·lb

5-10 At $y = 8$ in: $V_D = 60$ lb; $M_D = 120$ in·lb
At $y = 2$ in: $V_E = -40$ lb; $M_E = 80$ in·lb
$M_A = 0$; $M_B = 0$; $M_C = 240$ in·lb

5–11 At $x_C = 1$ m, $M = 2.4$ kN·m
At $x_C = 2.5$ m, $M = 3.75$ kN·m
At $x_C = 4$ m, $M = 2.4$ kN·m

5–13 $M_C = 3.1$ kN·m; $M_D = 4.75$ kN·m
$V_E = -1.9$ kN; $M_E = 1.9$ kN·m

5–14 $M_C = 600$ ft·lb
$M_D = 600$ ft·lb
$M_E = 600$ ft·lb

5–16 At $x_C = 4$ ft: $V_C = 75$ lb;
$M_C = -1425$ ft·lb
At $x_D = 6$ ft: $V_D = 75$ lb;
$M_D = -1275$ ft·lb

5–17 $V_A = 5$ kN; $M_A = -8$ kN·m
$V_D = 5$ kN; $M_D = -3$ kN·m

5–19 $V_C = -2.5$ kN; $M_C = 6.5$ kN·m

5–20 $M_B = 16{,}400$ ft·lb; $M_D = 16{,}867$ ft·lb
$V_C = 233.3$ lb; $M_C = 16{,}633$ ft·lb

5–22 $M_A = \frac{5}{12} FL$; $V_B = -F/3$;
$M_B = FL/2$; $M_C = FL/3$

5–23 Case 1: $V_B = 0$; $M_B = 2$ kN·m;
$V_C = 0$; $M_C = 2$ kN
Case 2: $V_B = -3$ kN
$M_B = 1.25$ kN · m; $V_C = 0$;
$M_C = 2$ kN·m

5–25 For $x_C = 50$ ft: $N_E = 23{,}325$ lb;
$V_E = 3750$ lb; $M_E = 18{,}750$ ft·lb
For $x_C = 80$ ft: $N_E = 37{,}320$ lb;
$V_E = 0$; $M_E = 0$
For $x_C = 110$ ft: $N_E = 51{,}315$ lb;
$V_E = -3750$ lb; $M_E = -18{,}750$ ft·lb

5–26 $V_C = 0.66$ kN; $M_C = 2.0$ kN·m
$V_D = -0.66$ kN; $M_D = 2.0$ kN · m

5–28 Maximum shear $= \pm 10{,}000$ lb
Maximum moment $= -12{,}500$ ft·lb

5–29 $V = \pm 10{,}000$ lb; $M = -150{,}000$ in·lb

5–31 $V_E = -9526$ lb; $M_E = -1.1 \times 10^6$ ft·lb
$V_F = -866$ lb; $M_F = 125{,}000$ ft·lb

5–32 $V_E = -F \sin(\phi - \theta)$;
$M_E = Fd \sin(\phi - \theta)$

$$V_D = F\frac{d}{l}\sin(\phi - \theta);$$

$$M_D = Fd\sin(\phi - \theta)$$

5–34 $V_O = W_1 + W_2$;

$$M_O = -\left[\frac{W_1 R}{2} + W_2 R + M_A\right]$$

$$V_B = W_1 + W_2;$$

$$M_B = -\left[\frac{W_1 R}{4} + \frac{3W_2 R}{4} + M_A\right]$$

5–35 $V_1 = 0.9W$; $M_1 = [0.45a - 0.1b]\,W$
$V_2 = -0.1W$; $M_2 = 0.05bW$

5–36 $V(x) = 500$ N;
$M(x) = (-500 + 500x)$ N·m

5–37 $V(y) = -200$ lb;
$M(y) = (600 - 200y)$ ft·lb

5–39 For $2 > x > 0$: $V(x) = 500$ lb;
$M(x) = (500x - 1300)$ ft·lb
For $3 > x > 2$: $V(x) = 300$ lb;
$M(x) = (300x - 900)$ ft·lb

5–40 For $2 > y > 0$: $V(y) = 2$ kN;
$M(y) = (2y - 14)$ kN·m
For $4 > y > 2$: $V(y) = 5$ kN;
$M(y) = (5y - 20)$ kN·m

5–42 For $1.9 > x > 0$: $V(x) = 4$ kN;
$M(x) = (4x - 7.9)$ kN·m
For $2.0 > x > 1.9$: $V(x) = 3$ kN;
$M(x) = (3x - 6.0)$ kN·m

5–43 For $4 > x > 0$: $V(x) = 500$ lb;
$M(x) = (500x - 2500)$ ft·lb
For $6 > x > 4$: $V(x) = 250$ lb;
$M(x) = (250x - 1500)$ ft·lb

5–45 For $4 > y > 0$: $V(y) = -40$ lb;
$M(y) = 40y$ in·lb
For $10 > y > 6$: $V(y) = 60$ lb;
$M(y) = (-60y + 600)$ in·lb

5–46 For $4.8 > x > 0$: $V(x) = 0.12$ kN;
$M(x) = 0.12x$ kN·m
For $5 > x > 4.8$: $V(x) = -2.88$ kN;
$M(x) = (-2.88x + 14.4)$ kN·m

5–48 For $3 > x > 0$: $V(x) = 200$ lb;
$M(x) = 200x$ ft·lb
For $7 > x > 3$: $V(x) = 0$; $M(x) = 600$ ft·lb
For $10 > x > 7$: $V(x) = -200$ lb;
$M(x) = (-200x + 2000)$ ft·lb

5–49 For $0.8 > x > 0$: $V(x) = -2$ kN;
$M(x) = -2x$ kN·m
For $1.8 > x > 0.8$. $V(x) = 2.3$ kN;
$M(x) = (2.3x - 3.44)$ kN·m
For $2.8 > x > 1.8$: $V(x) = -0.7$ kN;
$M(x) = (-0.7x + 1.96)$ kN·m

5–51 For $2 > x > 0$: $V(x) = 5$ kN;
$M(x) = (5x - 8)$ kN·m
For $3 > x > 2$: $V(x) = 0$; $M(x) = 2$ kN·m

5–52 For $1.5 > x > 0$: $V(x) = 200$ lb;
$M(y) = (200y - 600)$ ft·lb

5–56 For $6 > x > 0$: $V(x) = 2733.3$ lb;
$M(x) = 2733.3x$ ft·lb
For $8 > x > 6$: $V(x) = 233.3$ lb;
$M(x) = (233.3x + 15{,}000)$ ft·lb
For $15 > x > 8$: $V(x) = -2266.7$ lb;
$M(x) = (-2266.7x + 35{,}000)$ ft·lb

5–57 For $80 > x > 0$: $V(x) = -5000$ lb;
$M(x) = -5000x$ ft·lb
For $120 > x > 80$: $V(x) = 10{,}000$ lb;
$M(x) = (10{,}000x - 1{,}200{,}000)$ ft·lb

5–59 For $15 > x > 0$: $V(x) = -10{,}000$ lb;
$M(x) = -1000x$ in·lb
For $30 > x > 15$: $V(x) = 10{,}000$ lb;
$M(x) = (10{,}000x - 300{,}000)$ in·lb

5–60 Point A: $V = -W$; $M = 0$
Point B: $V = -0.943\ W$; $M = -Wd$
Point C: $V = -0.745\ W$; $M = -2Wd$
Point D: $V = 0$; $M = -3\ Wd$

5–62 For $6 > x > 0$: $V(x) = -1.74$ kN;
$M(x) = 1.74x$ kN · m
For $8 > x > 6$: $V(x) = -5.21$ kN;
$M(x) = (-5.21x + 41.7)$ kN·m

5–63 For $1 > x > 0$: $V(x) = -58.82$ lb;
$M(x) = -58.82x$ ft·lb
For $2.2 > x > 1$: $V(x) = -58.82$ lb;
$M(x) = (-58.82x + 200)$ ft·lb
For $3.2 > x > 2.2$: $V(x) = -58.82$ lb;
$M(x) = (-58.82x - 200)$ ft·lb
For $4.0 > x > 3.2$: $V(x) = -58.82$ lb;
$M(x) = -58.82x$ ft·lb
For $5.1 > x > 4.0$: $V(x) = -58.82$ lb;
$M(x) = (-58.82x + 300)$ ft·lb

5–65 For $1 > x > 0$: $V(x) = 500$ lb;
$M(x) = (500x - 1500)$ ft·lb
For $2 > x > 1$: $V(x) = 400$ lb;
$M(x) = (400x - 1400)$ ft·lb
For $3 > x > 2$: $V(x) = 300$ lb;
$M(x) = (300x - 1200)$ ft·lb
For $4 > x > 3$: $V(x) = 200$ lb;
$M(x) = (200x - 900)$ ft·lb
For $5 > x > 4$: $V(x) = 100$ lb;
$M(x) = (100x - 500)$ ft·lb

5–67 For $a > x > 0$: $V(x) = 0.9W$;
$M(x) = W(0.9x - 0.9a + 0.1b)$
For $(a + b) > x > a$: $V(x) = -0.1W$;
$M(x) = W[-0.1x + 0.1(a + b)]$

5–69 For $5 > x > 0$: $V(x) = -1900$ kN;
$M(x) = (-1900x + 17{,}200)$ kN·m
For $9 > x > 5$: $V(x) = -1950$ kN;
$M(x) = (-1950x + 17{,}450)$ kN·m
For $11 > x > 9$: $V(x) = 50$ kN;
$M(x) = (50x - 550)$ kN·m

5–71 For $a > x > 0$: $V(x) = P$;
$M(x) = Px - [M_O + P(a + b)]$
For $b > x > a$; $V(x) = P$;
$M(x) = Px - P(a + b)$

5–72 For $a > x > 0$: $V(x) = \dfrac{Pc + M_O}{a + b + c}$;

$$M(x) = \left[\frac{Pc + M_O}{a + b + c}\right]x$$

For $(a + b) > x > a$: $V(x) = \dfrac{Pc + M_O}{a + b + c}$;

$$M(x) = \left[\frac{Pc + M_O}{a + b + c}\right]x - M_O$$

For $(a + b + c) > x > (a + b)$:

$$V(x) = -\frac{P(a + b) + M_O}{a + b + c};$$

$$M(x) = \left[-\frac{P(a + b) + M_O}{a + b + c}\right]x + P(a + b) - M_O$$

5–75 $\mathbf{V} = (10\mathbf{i} - 2\mathbf{j})$ N;
$\mathbf{M} = (0.08\mathbf{i} + 0.40\mathbf{j})$ N·m

5–76 $\mathbf{V}_J = 10\mathbf{i}$ N;
$M_J = (-0.32\mathbf{i} - 2.50\mathbf{k})$ N·m

5–78 $M(y) = 10y$ N·m

5–79 $M_x(z) = 2z$ N·m; $M_y(z) = 10z$ N·m

Chapter 6

6–1 $AC = 2.81$ kN T

6–2 $BC = 1322$ lb C
$AC = 986$ lb T
$AB = 1055$ lb T

6–4 $AD = BD = 1170$ lb T
$AC = BC = 0$
$AB = 1099$ lb C

6–5 $CD = AD = 0$
$BC = 2$ kN C
$AB = 10$ kN C
$BD = 10.2$ kN T

6–7 $AB = BC = AD = 0$
$CD = 3F$ lb C
$AC = 3.16F$ lb T

6–8 $CD = 0$
$AD = BC = F$ lb T
$AB = 3F$ lb T
$BD = 3.16F$ lb C

6–10 $BC = 470$ lb T
$CD = 171$ lb T
$AB = 2144$ lb T
$AD = 670$ lb T
$BD = 2247$ lb C

6-11 $AB = 10$ kN T
$BF = 7.07$ kN C
$FE = 5$ kN C
$BE = 5$ kN T
6-13 Nine independent equations
6-14 $BD = 1$ kN T
$AB = 12.86$ kN T
$BC = 6.98$ kN T
$BE = 8.72$ kN C
6-16 $FG = 5150$ lb C
$AF = 2758$ lb T
$BF = 1700$ lb C
$EF = 3200$ lb T
6-17 $AH = 43.3$ kN T
6-19 $AB = CD = BC = 3.46$ kN C
$AE = DE = 1.73$ kN T
$BE = CE = 3.46$ kN T
6-20 $AB = CD = 4.62$ kN C
$AE = DE = 2.31$ kN T
$BE = CE = 3.47$ kN T
$BC = 4.05$ kN C
6-22 $CJ = EJ = DJ = 0$;
$JK = IJ = 25$ kN T
6-23 $BL = 5000$ lb T
$AL = KL = 10{,}020$ lb T
$CJ = EJ = 3205$ lb T
$CK = 2500$ lb C
$BK = 9610$ lb T
6-26 $F_{AB} = 5.01$ kN
6-27 $F_{AB} = 1055$ lb
$F_{BC} = 1323$ lb
6-29 $F_{AB} = 1099$ lb
6-33 $F_{AB} = 2144$ lb
6-34 $F_{BC} = 5$ kN
6-37 $F_{AB} = 3000$ lb
6-39 $AC = 3.16F$ lb T
$CD = 3F$ lb C
6-40 $CD = 18$ kN C
$AB = 1.5$ kN C; $AC = 18.98$ kN T
6-42 $AB = 12$ kN T
$BF = 7.21$ kN C
$EF = 18$ kN C
6-43 $EF = 840$ lb C
$BE = 600$ lb T
$BC = 840$ lb T
6-45 $AB = 12.86$ kN T
$BE = 8.72$ kN C
$DE = 5.72$ kN C
6-46 $EF = 3000$ lb C
$BE = 2121$ lb T
$BC = 1500$ lb C
6-48 $BC = 3.46$ kN C
$CE = 3.46$ kN T
$DE = 1.73$ kN T
6-50 $CD = 5410$ lb C
$JK = 4500$ lb T
6-52 $DE = 27.9$ kN C
$IJ = 25$ kN T
6-54 $CD = 18$ kips C
$CJ = 3200$ lb T
$JK = 16$ kips T
6-56 $GH = 195.8$ kN T
$CH = 150.5$ kN T
$BC = 341$ kN C
6-57 $F_{8-9} = 100$ kN T
$F_{5-9} = 40$ kN T
$F_{9-10} = 115$ kN T
$F_{4-9} = 21.2$ kN C
6-59 $F_{1-9} = 26$ kN T
$F_{4-6} = 1$ kN T
$F_{2-8} = 4.67$ kN T
$F_{3-7} = 2.5$ kN T
6-60 $F_{17-18} = 25.5$ kips T
$F_{5-18} = 16$ kips C
$F_{3-18} = 13$ kips T
$F_{4-18} = 10$ kips T
6-63 $F_{BD} = 4540$ lb T
$C_y = 1440$ lb$\downarrow$
6-64 $F_{BD} = 10$ kN C
$E_y = 5$ kN$\uparrow$
6-66 $BE = 12$ kN T
6-67 $CD = 3610$ lb C
$BE = 4680$ lb T
6-69 $F_{BE} = 457.6$ lb
6-71 $C_x = 16.15$ lb
$C_y = 0$
6-72 $C_x = 27$ lb
$C_y = 14.7$ lb
6-74 $C_x = 48.75$ lb
$C_y = 18.75$ lb
6-75 $C_x = 20$ N
$C_y = 64.7$ N
6-77 $F_{BC} = 6.75$ lb
$D_y = 4.78$ lb
6-78 $D_x = 7.78$ lb
$D_y = 4.78$ lb
6-79 Force on bolt = 59.5 kN
6-81 $F_y = 13.3$ kN
6-83 $F_x = 26$ kN
$F_y = 52$ kN
6-85 $T_{AB} = T_{CD} = 8.06$ kN
$T_{BC} = 8$ kN

6–86 $y_B = 2.4$ ft
$T_{BC} = 1671$ lb
6–88 $y_C = 3.83$ ft
$y_D = 2.26$ ft
6–90 $T_{max} = 2$ kN
6–91 $T_{DE} = 481$ lb
$T_{CD} = 455$ lb
6–93 $T_{BC} = 570$ lb
6–95 $F_{3-18} = 13$ kips T
$F_{4-18} = F_{5-17} = 10$ kips T
$F_{5-18} = F_{6-17} = 16$ kips C
$F_{17-18} = 25.5$ kips T
$F_{16-17} = 38$ kips T
6–96 $F_{AE} = 1.73$ kN
$F_{BC} = 3.46$ kN
6–97 $F_{1-2} = 31.75$ kN T
$F_{8-9} = 35.75$ kN C
$F_{2-3} = 15.52$ kN T
6–98 $F_{15-13} = 10{,}390$ lb T
$F_{4-14} = 2594$ lb C
6–99 $C_x = -666$ lb
$C_y = 0$
6–100 $E_x = 22.6$ kN
$E_y = 30.7$ kN
6–102 At $y_C = 3a$, $T_{AB} = 2.24$ F, $T_{CD} =$ F

Chapter 7

7–1 $\bar{x} = 0$ (by symmetry); $\bar{y} = \dfrac{2ab + \dfrac{b^2}{2}}{2a + b}$

7–2 $\bar{x} = \dfrac{a^2}{2(a+b)}; \bar{y} = \dfrac{ab + \dfrac{b^2}{2}}{a + b}$

7–4 $\bar{x} = \dfrac{ab + \dfrac{a^2}{2}}{a + 3b}; \bar{y} = \dfrac{5b^2}{2(a + 3b)}$

7–5 $\bar{x} = \dfrac{c^2 - a^2}{2(a + b + c)}; \bar{y} = \dfrac{\dfrac{b^2}{2} + bc}{a + b + c}$

7–7 $\bar{x} = 0.921r; \bar{y} = 0.335r$

7–8 $\bar{x} = \bar{y} = 0.636r$

7–10 $\bar{x} = \dfrac{7}{18}b; \bar{y} = \dfrac{2}{9}h$

7–11 $\bar{x} = \dfrac{(b - c)d^2 + ca^2}{2[(b - c)d + ca]};$

$\bar{y} = \dfrac{b^2d + c^2(a - d)}{2[(b - c)d + ca]}$

7–13 $\bar{x} = \dfrac{b}{6}; \bar{y} = \dfrac{\sqrt{3}b}{6}$

7–14 $\bar{x} = \dfrac{a^2(b + 2c)}{3a(b + c)}; \bar{y} = \dfrac{a(b^2 + bc + c^2)}{3a(b + c)}$

7–16 $\bar{x} = \dfrac{65}{132}a; \bar{y} = \dfrac{53}{99}b$

7–17 $\bar{x} = 0; \bar{y} = \dfrac{4(r_o^3 - r_i^3)}{3\pi(r_o^2 - r_i^2)}$

7–19 $\bar{x} = \dfrac{2r \sin\theta}{3\theta}; \bar{y} = 0$ (by symmetry)

7–20 $\bar{x} = \dfrac{4a}{5}; \bar{y} = \dfrac{3b}{8}$

7–22 $\bar{x} = \dfrac{3a}{4}; \bar{y} = \dfrac{3b}{10}$

7–23 $\bar{x} = \dfrac{3a}{8}; \bar{y} = \dfrac{3b}{10}$

7–25 $\bar{x} = 0$ (by symmetry); $\bar{y} = -0.00884r_o$

7–26 $\bar{x} = \dfrac{a^2 + \dfrac{\pi}{2}ab + \dfrac{2}{3}b^2}{2a + \dfrac{\pi b}{2}}; \bar{y} = b$

7–27 $\bar{x} = \bar{z} = 0$ (by symmetry); $\bar{y} = 0.781r$

7–29 $\bar{x} = \bar{z} = 0$ (by symmetry); $\bar{y} = \dfrac{3h}{4}$

7–31 $\bar{x} = \dfrac{6l^2 + 4dl + d^2}{4(3l + d)}$

$\bar{y} = \bar{z} = 0$ (by symmetry)

7–32 $\bar{y} = \dfrac{3(h^2 + 8hd + 4d^2)}{4(2h + d)}; \bar{x} = \bar{z} = 0$

7–34 $V = 2\pi^2 r^2(R + r)$

7-35 $A = 2\pi r(\frac{2}{3}\pi R + r)$
7-37 $A = \pi[3r^2 + Rr(\pi + 2)]$

7-38 $V = \frac{\pi^2 r^2}{2}\left(R + \frac{4r}{3\pi}\right)$

7-40 $A = \pi r(r + \sqrt{h^2 + r^2})$; $V = \frac{\pi r^2 h}{3}$

7-41 $A = \pi[r^2 + 2rl + r\sqrt{r^2 + d^2}]$

7-43 $A = 13.32\ \text{m}^2$
7-44 $V = 17.76\ \text{ft}^3$
7-46 $V = 30.8\ \text{cm}^3$
7-47 $A = 4710\ \text{m}^2$

7-49 $V(x) = wx$; $M(x) = -\frac{wx^2}{2}$

7-51 $V(x) = w\left(\frac{L}{2} - x\right)$; $M(x) = \frac{wx}{2}(L - x)$

7-53 For $0 < x < l$:

$$V(x) = \frac{w}{2}\left(l - \frac{x^2}{l}\right);\ M(x) = \frac{wx}{2}\left(l - \frac{x^2}{3l}\right)$$

7-55 $V(x) = -\frac{wx^2}{2L}$; $M(x) = -\frac{wx^3}{6L}$

7-57 $V(x) = w(a + b - c - x)$

$$- \frac{w[(b - c)^2 - a^2]}{2b}$$

$$M(x) = w\left[(a + b - c) - \frac{(b - c)^2 - a^2}{2b}\right](x - a)$$

7-59 $V(x) = \frac{w(a + b - x)^2}{2b}$

$$- \frac{w}{a + b}\left[\frac{a^2}{2} + \frac{b^2}{6} + \frac{ab}{2}\right]$$

$$M(x) = C_y(a + b - x) - \frac{w(a + b - x)^3}{6b}$$

7-61 $V(x) = w(2L + x)$;

$$M(x) = wx\left(-2L - \frac{x}{2}\right)$$

7-63 $V(x) = \frac{w}{30}\left[59L - \frac{15x^2}{L}\right]$;

$$M(x) = \frac{wx}{30}\left[59L - \frac{5x^2}{L}\right]$$

7-65 $V(x) = -wx$; $M(x) = w\left(3L^2 - \frac{x^2}{2}\right)$

7-67 $V(x) = \frac{-w}{a + b}\left[\frac{3a^2}{2} - \frac{b^2}{2} - ab\right] - wx$

$$M(x) = \frac{-w}{a + b}\left[\frac{3a^2}{2} - \frac{b^2}{2} - ab\right]x - \frac{wx^2}{2}$$

7-69 $V(x) = \frac{-w_B}{L}(2L^2 - 3Lx + x^2)$

$$M(x) = \frac{w_B}{6}\left(5L^2 - 12Lx - \frac{2}{L}x^3 + 9x^2\right)$$

7-71 For $2 > \overline{x} > 0$: $M(x) = -150x^2\ \text{N}\cdot\text{m}$
7-72 For $2 > x > 0$: $V(x) = (-0.3x - 3)\ \text{kN}$; $M(x) = (0.15x^2 - 3x)\ \text{kN}\cdot\text{m}$
For $4 > x > 2$: $V(x) = (4.2 - 0.3x)\ \text{kN}$; $M(x) = (-0.15x^2 + 4.2x - 14.4)\ \text{kN}\cdot\text{m}$
7-74 $T_{max} = 67.3\ \text{kN}$
7-75 $T_{max} = 1.044 \times 10^5\ \text{lb}$; $y_{x=L/4} = 3.75\ \text{ft}$
7-77 $T_A = 330\ \text{kN}$; $T_B = 342\ \text{kN}$
7-78 $T_B = 227\ \text{kips}$;
$T_A = 239\ \text{kips}$
7-80 $T = 860\ \text{N}$
7-81 $T_{min} = 550\ \text{lb}$;
$T_{max} = 586\ \text{lb}$; $y = 4.46\ \text{ft}$
7-83 $T_{max} = 635.5\ \text{N}$
7-84 $T_{max} = 268.9\ \text{kN}$
7-85 $P_x = 58.8\ \text{kN}$; $P_y = 21.4\ \text{kN}$;
$\overline{x} = 0.485\ \text{m}$; $\overline{y} = 1.33\ \text{m}$
7-87 $P_x = 10{,}140\ \text{lb}$; $\overline{x} = 2.82$ ft from top of plate
7-88 $P = 15.42\ \text{kN}$
7-90 $d = 3.6$ ft below top of line AB
7-91 $F_x = 600.3\ \text{lb}$; $\overline{y} = 1.17\ \text{m}$; $F_y = 457.7\ \text{lb}$;
$\overline{x} = 0.75\ \text{m}$
7-93 $\sum M_B$ is counterclockwise; therefore, the dam will not overturn
7-94 $V_A = 1920\ \frac{\text{lb}}{\text{ft}}$

$$M_A = 4608\ \frac{\text{ft}\cdot\text{lb}}{\text{ft}}$$

7-96 $W = 1190\ \text{lb}$

7–98 $F = 944.4$ kN
7–99 $d = 3.43$ in.
7–101 $h = 0.43R$
7–102 Maximum h is obtained when r approaches R
7–105 $\tau_\theta = 0; \sigma_\theta = 5000$ psi
7–106 $\tau_\theta = 0; \sigma_\theta = -20$ MPa
7–108 $\tau_\theta = 35$ MPa; $\sigma_\theta = 65$ MPa
7–109 $\tau_\theta = -190$ psi; $\sigma_\theta = 851$ psi
7–111 $\tau_\theta = -939.7$ psi; $\sigma_\theta = -342$ psi
7–112 $\tau_\theta = 29.9$ MPa; $\sigma_\theta = -698$ MPa
7–115 $\bar{x} = 8.78$ cm; $\bar{y} = 1.99$ cm
7–116 $\bar{x} = 0.582$ in.; $\bar{y} = 0.479$ in.
7–117 $A = 2\pi\left[6\left(\frac{r}{2} + R\right)\sqrt{l^2 + (R - r)^2} + r^2\right]$
7–118 $V = 2\pi l[r^2 + Rr + R^2]$
7–119 $V(x) = W + w_c\left(4x - \frac{x^2}{2a} - \frac{15a}{2}\right)$

$$M(x) = W(x - a) + w_c\left(\frac{-x^3}{6a} + 2x^2 - \frac{15ax}{2} + 9a^2\right)$$

7–121 $T_{min} = T_o \simeq 230$ lb; $T_{max} = 356.5$ lb
7–123 $P = \frac{\gamma_w \pi}{6}(R^2L + RrL + r^2L)$;

$$d = \frac{L(R^2 + 2Rr + 3r^2)}{4(R^2 + Rr + r^2)}$$

7–124 $P = 134.7$ lb
7–125 $\tau_\theta = 100$ MPa; $\sigma_\theta = 200$ MPa
7–126 $\tau_\theta = -295$ psi; $\sigma_\theta = -206.8$ psi

Chapter 8

8–1 The block slides
8–2 $\theta_c =$ constant
8–4 The block slides
8–5 The block slides
8–7 $\theta = 21°$
8–9 $P = 0.67$ W
8–10 $W = 6.19$ P
8–12 $\theta \geq 63.4°$
8–13 The board is at rest
8–15 $F = 102.7$ N
8–16 The beam slides
8–18 The boxes slide
8–19 $F = 237$ N
8–21 $F = 78.57$ N
8–22 $M_C = 143$ N·m
8–25 $M_C = 333$ N·m
8–26 $M_C = 1176$ lb·in.
8–29 $P = \mu W/\cos\phi$
8–31 $M_C = \frac{1}{2} Wr \sin 2\phi/\cos\theta$
8–32 $T = 39.2$ kN
8–34 $F = 1.093$ kN
8–35 $F = -827$ N
8–37 The force F is not sufficient
8–38 Block A is self-locking
8–40 $P = 2(N - W)$
8–41 $F = 10$ kN $\cos\theta/2 + 20$ kN $\sin\theta/2$
8–43 $F = 6.94$ kN
8–44 $F = 939$ lb
8–46 $M = 1.137$ kN·m
8–48 $M_t = 188.9$ ft·lb
8–49 $T = 42.6$ kN
8–51 $M = 10$ N·m
8–52 $p = 1.885$ cm
8–54 $F_B = 125$ lb
8–55 $P = 58.2$ kN
8–57 $M = 67.3$ N·m
8–59 $D = 8$ in.
8–60 $P = 46.4$ kN
8–62 $M = 4.8$ N·m
8–63 $M_{total} = 888$ in·lb
8–65 $M = 35$ ft·lb
8–66 $M = 2$ N·m
8–68 $M = 560$ in·lb
8–69 $M = 2.25$ N·m
8–71 $M = 37.5$ lb·in.
8–72 $P = 2.13$ kN
8–74 $T_B = 197$ lb
$T_C = 203$ lb
8–75 $T_A \simeq 198$ lb
$T_B = 203$ lb
$T_C = 197$ lb
8–77 $M_{total} = 38.48$ N·m
8–79 $\theta = 0.286°$
8–82 $F = 40$ lb/wheel
8–83 $a = 2$ cm
8–85 $T_2 = 451$ lb
8–86 $T_1 = 1.151$ kN
8–88 $T = 513$ lb
8–90 $T_2 = 0.204$ lb

8–93 $T_2 = 14.65\ T_1$
8–96 For clockwise rotation,
$T_A = 190.8$ N
$T_B = 670$ N
8–98 $T_1 = 2$ kN
$T_2 = 8.67$ kN
8–99 $F_x = F_y = 4.22$ kN
8–102 There is slippage
8–103 $P = 288.6$ lb
8–105 $W_1 = 8.91$ kN
8–106 $M = 281$ in·lb
8–108 The mechanism does not unwind
8–109 $M = 118.5$ N·m
8–111 $F = 400$ lb
8–112 $\mu = 3$; friction is inadequate
8–113 $T_1 = 81.2$ lb
$T_2 = 660$ lb

Chapter 9

9–1 $I_z = \dfrac{ML^2}{3}$

9–2 $I_{y'} = \dfrac{ML^2}{12}$

$r_x = a;\ r_y = \dfrac{L}{\sqrt{3}}$

9–4 $I_x = \dfrac{MR^2}{2};\ I_y = MR^2$

9–5 $I_x = MR^2$

9–7 $I_x = \dfrac{a}{3}[b^3 - 3b^2d + 3bd^2];$

$I_y = \dfrac{b}{3}[a^3 + 3a^2c + 3ca^2];$

$r_x = \left[\dfrac{b^2 - 3bd + 3d^2}{3}\right]^{1/2}$

9–8 $J_C = \dfrac{1}{12}ab(a^2 + b^2)$

9–10 $I_x = \dfrac{\delta tab^3}{12}$

9–11 $I_y = \dfrac{ba^3}{12}$

9–13 $I_y = \dfrac{\delta tba^3}{4}$

9–14 $I_x = \dfrac{ab^3}{12};\ r_x = \dfrac{b}{\sqrt{6}}$

9–16 $I_x = \dfrac{\pi r^4}{8};\ I_y = \dfrac{5\pi r^4}{8}$

9–17 $I_y = \dfrac{5\pi\delta t}{8}$

9–19 $I_x = \dfrac{\pi r^4}{4};\ I_y = \dfrac{\pi r^4}{4}$

9–20 $J_O = \dfrac{\pi r^4}{2}$

9–22 $I_y = \dfrac{1}{3}[(\mathbf{i}^3 - g^3)e + (c^3 - a^3)d];$

$r_y = \left[\dfrac{(\mathbf{i}^3 - g^3)e + (c^3 - a^3)d}{3(bd + he)}\right]^{1/2}$

9–23 $I_z = \dfrac{ML^2}{3}$

9–25 $I_x = \dfrac{1}{3}a(b^3 - 3b^2d + 3bd^2);$

$I_y = \dfrac{1}{3}b(a^3 + 3a^2c + 3ca^2)$

9–26 $J_0 = \dfrac{ab}{3}(a^2 + b^2 - 3bd + 3ac + 3d^2 + 3c^2)$

9–28 $I_{y_1} = 2MR^2$
9–29 $I_{x_1} = 2MR^2$

9–31 $J_O = \dfrac{1}{12}ab(a^2 + b^2)$

9–32 $I_{x'} = 0.11r^4$
9–33 $J_O = 2.07r^4$

9–35 $I_{x_1} = \dfrac{\pi r^2}{2}$

9–37 $I_l = \dfrac{160}{27}Mc^2$

9–38 $I_x = \dfrac{1}{3}ac^3 - a(c - b)\left[\dfrac{b^2}{12} + \dfrac{bc}{6} + \dfrac{c^2}{4}\right]$

9–40 $I_x = \frac{1}{3}hf^3 - \frac{1}{3}(h - b)d^3$

9–41 $I_y = \frac{1}{12}eh^3 + (eh)\left(g + \frac{h}{2}\right)^2 + \frac{1}{12}db^3 + (bd)\left(a + \frac{b}{2}\right)^2$

9–43 $I_x = 327.3$
9–44 $I_y = 723.3$
9–46 $I_{xy} = \delta a^2 b^2 t$
9–47 $I_{uv} = -0.144ab(a^2 - b^2) + 0.125a^2b^2$
9–49 $I_{xy} = -196.9a^4$
9–50 $I_{xy} = -196.9\delta ta^4$
9–52 $I_{uv} = -264.9\delta ta^4$
9–54 $I_{xy} = 1090a^4$
9–55 $I_{uv} = 2710\delta ta^4$
9–57 $I_{uv} = 0.041[cb(b^2 - c^2) + a(c - 2d)((c - 2d)^2 - a^2)]$
9–58 $I_{uv} = 0.036\delta t[cb(b^2 - c^2) + a(c - 2d)((a - 2d)^2 + a^2)]$
9–60 $I_{uv} = 1163.4\delta t$
9–61 $I_{xy} = -26.75\ \text{in}^4$
9–63 $I_{uv} = 5.46\ \text{in}^4$
9–64 $I_x = 0.888\ \text{kg}\cdot\text{cm}^2$; $I_y = 2.00\ \text{kg}\cdot\text{cm}^2$; $I_z = 1.162\ \text{kg}\cdot\text{cm}^2$
9–66 $I_x = 4.79 \times 10^{-3}\ \text{slug}\cdot\text{in}^2$; $I_y = 5.94 \times 10^{-3}\ \text{slug}\cdot\text{in}^2$; $I_z = 4.85 \times 10^{-3}\ \text{slug}\cdot\text{in}^2$

9–68 $I_x = \frac{ab^3}{21}$; $r_x = \frac{b}{\sqrt{7}}$

9–69 $J_0 = ab\left[\frac{a^2}{5} + \frac{b^2}{21}\right]$

9–70 $I_y = \frac{ba^3}{4}$

9–72 $I_l = \frac{7}{5}MR^2$

9–73 $I_{uv} = 0$ (symmetric about a point)
9–74 $I_{xy} = 0$

Chapter 10

10–1 $U_{AB} = -Wl \sin\theta$
10–2 $U_{AB} = -2067\ \text{N}\cdot\text{m}$
10–4 $U_{AB} = 500\ \text{N}\cdot\text{m}$
10–5 $U_{AB} = 1000\ \text{N}\cdot\text{m}$
10–7 $U_{AB} = Wl[\mu_1 \cos 20^\circ - \sin 20^\circ + \mu_2]$

10–8 $\text{Ratio} = \frac{(\mu \cos 20^\circ - \sin 20^\circ + \mu)}{(\mu \cos 20^\circ + \sin 20^\circ + \mu)}$

10–10 $U_{AC} = 1697\ \text{lb}\cdot\text{ft}$; $U_{AB} = 4510\ \text{lb}\cdot\text{ft}$
10–11 $U_{ABA} = 44{,}300\ \text{lb}\cdot\text{ft}$
10–13 $U = 132.6\ \text{kN}\cdot\text{m}$

10–14 $U = \frac{Wl}{2}$

10–16 $U = 251\ \text{N}\cdot\text{m}$
10–17 $U = 502\ \text{N}\cdot\text{m}$
10–19 $U_{\text{total}} = 60{,}600\ \text{in}\cdot\text{lb}$
10–20 $U = 1.17\ \text{kN}\cdot\text{m}$
10–22 $U = 2830\ \text{lb}\cdot\text{in}$

10–23 $P = \frac{F}{3}$

10–25 $P = (0.1986)F$

10–26 $M_O = \frac{Fd}{2}$

10–28 $F = \frac{W}{2}$

10–29 $P = \frac{1}{2}F \cot\theta$

10–35 $F_B = 141.8\ \text{lb}$
10–41 $\eta_{BA} = 0.549$

10–42 $\eta = \frac{\sin 20^\circ}{\mu_k \cos 20^\circ - \sin 20^\circ + \mu_k}$

10–44 Frictional work = 19.9 kN·m
10–46 $\Delta U = 7.5\ \text{N}\cdot\text{m}$
10–47 $U = 1440\ \text{in}\cdot\text{lb}$
10–49 $F = 145.4\ \text{N}$
10–50 $\Delta U = 2.28\ \text{N}\cdot\text{m}$
10–51 $\theta = 80.5^\circ$
10–53 $\theta = 63.4^\circ$; stable
10–54 $x = 26.25$ cm; stable
10–56 $\theta = 38.0^\circ$; stable
10–57 $\theta = 27.5^\circ$
10–59 $\theta = 52.3^\circ$
10–60 $\theta = 0.477^\circ$
10–65 $U = 2560\ \text{lb}\cdot\text{ft}$
10–66 $F = 17{,}320\ \text{lb}$
10–67 $45^\circ \le \theta \le 90^\circ$
10–68 $\eta = 0$

10–69 $k = 382\ \frac{\text{N}\cdot\text{m}}{\text{rad}}$

Chapter 11

11–1 $v = 3$ m/s; $a = 0$ m/s^2
11–2 $v = 12$ ft/s; $a = 4$ ft/s^2
11–4 $v_{t_1} = -4.161$ in./s; $v_{t_2} = -6.536$ in./s; $a_{t_1} = -18.186$ in./s^2; $a_{t_2} = 15.136$ in./s^2
11–5 $v = -3.355$ m/s
11–7 $y = 10.368$ m; $a = 1.632$ m/s^2
11–8 $x = 67.167$ in.; $a = 13$ in./s^2
11–10 $a = 12{,}800$ ft/s^2
11–11 $v = 30$ m/s; $x = 75$ m
11–13 $v_{\max} = 8$ m/s
11–14 $v = 25$ in./s
11–16 $v = 10.44$ ft/s
11–17 $x = 500$ m
11–18 (a) $t = 0.358$ s; (b) $t = \infty$
11–20 $v = \frac{1}{k}\sqrt{(1 - e^{-2k^2gy})}$
11–22 $t_f = 102$ s; $v_f = 500$ m/s
11–23 $v_{B/A} = -15$ m/s; $a_{B/A} = -6$ m/s^2
11–25 $x_{B/A} = -161.7$ m; $a_{B/A} = -38$ m/s^2
11–26 $x_{B/A} = 1009$ ft; $v_{B/A} = 306$ ft/s; $a_{B/A} = 61$ ft/s^2
11–28 $x_{B/A} = 3.5$ ft; $a_{B/A} = 3$ ft/s^2
11–29 $x_{B/A} = 450$ m
11–31 $v_P = t = 80$ m/s; $x_P = 3200$ m
11–32 $t = 11.4$ s
11–34 $x_A = 218.2$ m; $v_{A/B} = 93.5$ m/s
11–35 $x_{B/A} = 0.527$ ft; $v_{B/A} = 0.145$ ft/s
11–37 $x_{B/A} = 2.375$ m$\downarrow$; $v_{B/A} = 2.25$ m/s$\downarrow$
11–38 (a) $v = 40.1$ ft/s; (b) $v = 40.1$ ft/s
11–40 $t_{\text{total}} = 0.36$ s
11–41 $v_A + 2v_B + v_C = 0$; $a_A + 2a_B + a_C = 0$

Chapter 12

12–1 $\mathbf{v} = (9\mathbf{i} + 3\mathbf{j})$ m/s; $\mathbf{a} = 2\mathbf{i}$ m/s^2
12–2 $\mathbf{v} = (150\mathbf{i} + 40\mathbf{j})$ ft/s; $\mathbf{a} = (60\mathbf{i} + 8\mathbf{j})$ ft/s^2
12–4 $\mathbf{r} = (1010\mathbf{i} - 415\mathbf{j})$ ft; $\mathbf{a} = (20\mathbf{i} - 8\mathbf{j})$ ft/s^2
12–5 $\mathbf{r} = (10\mathbf{i} + 82\mathbf{j})$ m
12–7 $\mathbf{r} = (100\mathbf{i} + 533.3\mathbf{j} - 50\mathbf{k})$ m
12–8 $h = 38.82$ ft; $d = 269.3$ ft
12–10 $v_0 = 102.17$ m/s
12–11 $d = 249.84$ ft
12–13 $d = 74.7$ ft; $\mathbf{v} = (-30\mathbf{i} - 80.18\mathbf{j})$ ft/s
12–14 $h = 16.1$ m
12–16 $v_0 = 46.95$ ft/s
12–17 $s = 78.61$ m
12–19 $v_0 = 16.37$ m/s
12–20 (a) $\mathbf{v}_f = a\mathbf{i} - \sqrt{b^2 - 2gh}\mathbf{j}$; (b) $\mathbf{v}'_f = -a\mathbf{i} - b\mathbf{j}$; yes, it is at point A
12–22 $\mathbf{a} = (-85.6\mathbf{i} - 51.73\mathbf{j})$ ft/s^2
12–23 $\mathbf{a}_1 = 157.1\mathbf{j}$ m/s^2; $\mathbf{a}_2 = (-157.1\mathbf{i} - 493\mathbf{j})$ m/s^2
12–25 $\mathbf{a} = (-293\mathbf{i} + 209\mathbf{j})$ m/s^2
12–26 At t_1: $\mathbf{a} = (24\mathbf{i} - 2\mathbf{j})$ m/s^2; $\rho = 79.57$ m
At t_2: $\mathbf{a} = (48\mathbf{i} - 2\mathbf{j})$ m/s^2; $\rho = 1165.83$ m
12–28 $v_{\max} = 26.46$ m/s
12–29 $a_t = 5.74$ m/s^2
12–31 (a) $\rho = 133\,069$ m; (b) $\rho = 59\,812$ m
12–32 $\mathbf{a} = (-12.5\mathbf{n}_r + 3\mathbf{n}_\theta)$ m/s^2
12–34 $\dot{\theta} = 0.15$ rad/s; $\ddot{\theta} = -0.139$ rad/s^2
12–35 $\mathbf{a} = -0.0192\mathbf{n}_r + 0.196\mathbf{n}_\theta$
12–37 $\dot{\theta} = -7.54$ rad/s; $\ddot{\theta} = -46.9$ rad/s^2
12–38 $v_{\text{hor}} = 46.9$ m/s; $a_{\text{hor}} = -1.05$ m/s^2
12–40 $\dot{r} = 90.1$ m/s; $\dot{\theta} = 0.06$ rad/s; $\ddot{r} = 11.18$ m/s^2; $\ddot{\theta} = 0.003$ rad/s^2
12–41 $\dot{\theta} = 0.033$ rad/s; $\dot{r} = 282.8$ ft/s; $\ddot{\theta} = 0.011$ rad/s^2; $\ddot{r} = 79.9$ ft/s^2
12–43 $v = 827.1$ ft/s; $a = 59.9$ ft/s$^2\uparrow$
12–44 $\mathbf{v} = (-40\mathbf{n}_r - 84{,}000\mathbf{n}_\theta)$ m/s; $\mathbf{v} = (72\,726\mathbf{i} - 42\,034\mathbf{j})$ m/s; $\mathbf{a} = (-251\,870\mathbf{n}_r - 8160\mathbf{n}_\theta)$ m/s^2; $\mathbf{a} = (-118\,695\mathbf{i} - 222\,296\mathbf{j})$ m/s^2
12–46 $\mathbf{v}_B = (2\mathbf{n}_r + 25.2\mathbf{n}_\theta)$ m/s; $\mathbf{a}_B = (-301.4\mathbf{n}_r + 60.6\mathbf{n}_\theta)$ m/s^2
12–47 $\mathbf{a}_Q = -r\omega^2\mathbf{n}_r$
12–49 $\mathbf{a} = -12.95\mathbf{n}_r$
12–50 $v_{\max} = 20.6$ ft/s
12–52 $\mathbf{v} = (150\mathbf{n}_\theta + 5\mathbf{k})$ ft/s; $\mathbf{a} = (-1500\mathbf{n}_r + 30\mathbf{n}_\theta + 10\mathbf{k})$ ft/s^2
12–53 $a = 4.15$ m/s^2
12–55 $a = 45.4$ m/s^2
12–56 $\mathbf{a} = (-1000\mathbf{n}_r - 10\mathbf{n}_\theta + 0.048\mathbf{k})$ in./s^2
12–58 $a = 81.5$ ft/s^2
12–59 $\mathbf{a} = (-3\mathbf{n}_r + 1.2\mathbf{n}_\theta - 1\mathbf{k})$ ft/s^2
12–61 $\mathbf{a}_{p_1} = (-50\mathbf{n}_r - 8.66\mathbf{n}_\phi + 52.21\mathbf{n}_\theta)$ m/s^2
12–62 $\mathbf{a}_{p_2} = (-281.25\mathbf{n}_r - 166.9\mathbf{n}_\phi - 2.6\mathbf{n}_\theta)$ ft/s^2
12–64 $\mathbf{v}_{A/B} = (-112\mathbf{i} - 240\mathbf{j})$ ft/s; $\mathbf{a}_{A/B} = (-64\mathbf{i} - 120\mathbf{j})$ ft/s^2
12–65 $\mathbf{v}_{A/C} = \mathbf{v}_{A/B} - \mathbf{v}_{C/B}$
12–67 $v_A = 36.1$ m/s $= 130$ km/h
12–68 $v_A = 103$ ft/s $= 70.4$ mph; $a_A = 6$ ft/s^2
12–70 $v_A = 54.2$ ft/s
12–71 $v_A = 37.4$ m/s; $a_A = 15.31$ m/s^2
12–73 $\mathbf{v}_C = (-0.468\mathbf{i} + 0.152\mathbf{j})$ m/s; $\mathbf{a}_C = (-0.036\mathbf{i} - 0.124\mathbf{j})$ m/s^2
12–74 $\mathbf{a}_B = (-0.311\mathbf{i} - 0.0365\mathbf{j})$ m/s^2; $\mathbf{a}_C = (-0.378\mathbf{i} - 0.1215\mathbf{j})$ m/s^2

12–76 $a_{A/P} = 25.1 \text{ ft/s}^2$
12–77 $\omega = 0.18$ rad/s
12–79 $\mathbf{a}_T = (-3\mathbf{i} + 120\mathbf{j}) \text{ ft/s}^2$
12–80 $\mathbf{a}_B = (-1\mathbf{i} - 3.125\mathbf{j}) \text{ m/s}^2$
12–82 $\rho = 509$ ft
12–83 $\mathbf{a} = (-1.981\mathbf{n}_r + 0.060\mathbf{n}_\phi + 21.898\mathbf{n}_\theta) \text{ m/s}^2$
12–85 $\mathbf{a} = (-1.01\mathbf{n}_r - 0.26\mathbf{n}_\phi - 1.04\mathbf{n}_\theta) \text{ ft/s}^2$
12–86 $\dot{r} = -459$ ft/s
12–87 $\mathbf{a}_{B/A} = (-15\mathbf{i} + 18\mathbf{k}) \text{ ft/s}^2$

Chapter 13

13–1 $F_r\mathbf{n}_r + F_\theta\mathbf{n}_\theta + F_z\mathbf{k} = m\{(\ddot{r} - r\dot{\theta}^2)\mathbf{n}_r + (2\dot{r}\dot{\theta} + r\ddot{\theta})\mathbf{n}_\theta + \ddot{z}\mathbf{k}\}$
13–2 $F_r = m(\ddot{r} - r\dot{\phi}^2 - r\dot{\theta}^2 \sin^2\phi)$
$F_\phi = m(r\ddot{\phi} + 2\dot{r}\dot{\phi} - r\dot{\theta}^2 \sin\phi \cos\phi)$
$F_\theta = m(r\ddot{\theta}\sin\phi + 2\dot{r}\dot{\theta}\sin\phi + 2r\dot{\phi}\dot{\theta}\cos\phi)$
13–4 $W = 686.7$ N; $W_1 = 560$ N
13–5 $R_s = 592.4$ lb↑
13–7 $F_H = 74{,}534$ lb
13–8 $R = 75\,098$ N↑
13–10 $\mu = 0.16$
13–11 $v_f = 24.21$ m/s→
13–13 $a = 38.56 \text{ m/s}^2$→
13–14 $a = 35.56 \text{ m/s}^2$←
13–16 $(x - x_0) = 76.41$ m
13–17 (a) $a_A = 10.73 \text{ ft/s}^2$↓; $a_A = 16.1 \text{ ft/s}^2$↓
13–19 (a) $x - x_0 = 582.3$ ft; (b) $x - x_0 = 519.9$ ft
13–21 $\mathbf{F}_A = 2\mathbf{i}$ N; $\mathbf{F}_B = (0.4\mathbf{i} + 1\mathbf{j})$ N
13–22 $a_x = 11.28 \text{ ft/s}^2$
13–23 $x - x_0 = 2.96$ m
13–25 $x = 92.4$ m
13–26 $x - x_0 = 2{,}922.7$ ft; $F = 285.7$ lb←
13–28 $P = 1964$ N
13–29 $F = 653.8$ lb
13–31 $F = 1427$ lb; $\mu = 0.713$
13–32 $F = 10.296$ kN, $\theta = 29°$
13–34 $\theta = 63.23°$
13–35 $\theta = 8.28°$
13–37 $\theta = 38.8°$; $F = 79{,} 736$ lb
13–38 $\alpha = 49.65°$
13–40 He will slide. $F_{\text{Tot}} > F_{\text{Act}}$ in Prob. 13–39
13–41 $v = 30.03$ ft/s
13–43 $N = 1853.4$ N↑
13–44 $N = 479$ lb
13–46 $T_{\min} = 431.5$ N (top)
$T_{\max} = 445.4$ N (bottom)
13–47 $\theta = 47.16°$, $R = 146.2$ lb
13–49 $R = 194.3$ lb, $\theta = 71.4°$
13–50 $v = 28.4$ ft/s
13–52 $T = 3.92$ N, $\omega = 14$ rad/s
13–53 $R = 705.9$ N, $\theta = 82.6°$
13–55 $F_R = -11.7$ kN, $F_\theta = 2340$ N,
$\mathbf{F}_{\text{hor}} = (-7876.5\mathbf{i} - 8962.5\mathbf{j})$ N
13–56 $\mathbf{F}_{\text{hor}} = (-4711\mathbf{i} - 4425\mathbf{j})$ lb
13–58 $F_{\text{hor}} = 9.7$ lb
13–59 $T = 17.86$ kN, $\alpha = 79.7°$

Chapter 14

14–1 $v_f = 19.8$ m/s↓
14–2 $h = 24.8$ ft
14–4 $\mathbf{v}_f = (20\mathbf{i} - 31.7\mathbf{j})$ m/s
14–5 $\mathbf{v}_f = (30\mathbf{i} - 57.7\mathbf{j})$ ft/s
14–7 $\theta \simeq -38.25°$
14–8 $F = 7.6$ lb
14–10 $v_0 = 4.8$ m/s
14–11 $h = 33.8$ ft
14–13 $v_f = 60.8$ m/s
14–14 $d = 125.4$ ft
14–16 $v = 17.5$ ft/s; $P = 3494$ ft·lb/s = 6.4 hp
14–17 $v = 0.86$ m/s
14–19 $v_{\theta=90°} = 3.1$ m/s; $v_{\theta=135°} = 2.6$ m/s
14–20 $v_{\theta=45°} = 1.31\sqrt{gl}$; $v_{\theta=90°} = 2.07\sqrt{gl}$
14–22 $P = 45.5$ hp
14–23 $v = 6$ m/s
14–25 (a) $v_f = 4.49$ ft/s; (b) $y = 20$ in.
14–26 $x = 24.14$ in.
14–28 $d = 206.14$ ft
14–29 $d = 180.37$ ft
14–31 $\delta = 21.8$ in.
14–32 $l = 0.745$ m
14–34 $\mu = 0.017$
14–35 $v_f^2 = v_0^2 - (2\mu g)^2\left(\frac{m}{k}\right)\left\{\sqrt{1 + \left(\frac{v_0}{\mu g}\right)^2\left(\frac{k}{m}\right)} - 1\right\}$
14–37 $v_f = 13.62$ ft/s
14–40 $v_1 = v_2$ or $v = -\left(\frac{1}{2}x^2 + y^2\right) + C$
14–41 $v = -x^3 + C$
14–42 $v_1 = v_2 = v = \frac{2}{3}x^3 - \frac{5}{4}y^4 + C$
14–43 $\frac{\partial F_x}{\partial y} = x$, $\frac{\partial F_y}{\partial x} = 0$. $\mathbf{F}$ is not conservative.
14–44 $\mathbf{F}$ is a function of velocity and is not conservative.

14–47 $v = 3.66$ m/s
14–48 $h = 13.98$ ft
14–49 $v_B = 6.02$ m/s
14–51 $v_1 = 22.66$ ft/s
14–52 elevation = 2.07 m
14–54 $v_2 = 21.23$ ft/s
14–55 $v_1 = 9.90$ m/s
14–58 (a) $y_{max} = 0.123$ m; (b) $y_{max} = 0.173$ m
14–59 $v_0 = 1.98$ m/s
14–61 $k = 33.91$ lb/in.
14–62 $k = 8829$ N/m
14–64 $m = 32.77$ kg
14–65 $F = 96.96$ lb
14–67 $\Delta y = 2.71$ m
14–68 $P_m = 2.39$ W
14–70 $x_T = 290$ m
14–71 $T = 133$ N
14–73 $v_2 = 1.03$ m/s
14–74 (a) $v = 19.66$ ft/s; (b) $x_{max} = 1$ ft

Chapter 15

15–1 $\mathbf{H}_O = -2.51\mathbf{k}$ kg·m²/s
15–2 $\mathbf{H}_O = -2.51\mathbf{k}$ kg·m²/s
15–4 $\mathbf{F} = 0.75\mathbf{j}$ lb
15–5 $\mathbf{H} = 0.60\mathbf{k}$ kg·m²/s
15–7 $T = 7.42$ lb
15–8 $\mathbf{M}_O = -37.05\mathbf{k}$ ft·lb
15–10 $\mathbf{v}_f = (-0.289\mathbf{i} - 0.239\mathbf{j})$ ft/s
15–11 $\mathbf{H}_O = -6.43\mathbf{k}$ kg·m²/s; $\dot{\mathbf{H}}_O = \mathbf{M}_O = 43.13\mathbf{k}$ N·m
15–13 $M_O = 120$ N·m
15–14 $v = 5.33$ ft/s
15–16 $\ddot{\theta} = 685.93$ rad/s²
15–17 (a) $M_O = 28$ N·m; (b) $M_O = 42$ N·m
15–19 $a = 25.5$ ft/s²↓
15–20 $a = 25.5$ ft/s²↓
15–22 $\mathbf{M} = 0.5r^2m\dot{\theta}^2 \sin 2\phi \mathbf{n}_\theta + 0.2r^2m\dot{\theta}^2 \cos \phi \mathbf{n}_\phi$
15–23 $t = 660$ h = 27.5 days
15–25 $r = 42\,230$ km
15–26 $a = 10\,317$ km
15–28 $v_A = 30\,717$ km/h; $v_B = 16\,800$ km/h
15–29 $v_A = 16{,}856$ mph; $\Delta v = 5584$ mph
15–31 $v_A = 22{,}289$ mph
15–32 $\Delta v = 5237$ mph decrease
15–34 $\mathbf{v}_2 = 20$ **i**m/s
15–35 $\mathbf{v}_2 = 211.25\mathbf{j}$ ft/s
15–37 $\mathbf{v}_2 = (5149\mathbf{j} - 5147\mathbf{k})$ ft/s
15–38 $\mathbf{v}_2 = 390$ m/s**i**
15–40 $F_{avg} = 400$ kN
15–41 $F_{avg} = 1553$ lb
15–43 $\Delta t = 3.98$ s
15–44 $\Delta t = 3.64$ s
15–46 $F = 36.362$ kN
15–47 $F = 496.2$ lb
15–49 $v_{requ.} = 9.7$ ft/s $\simeq v_{meas.}$
15–50 $F_L\,\Delta t = 47$ lb·s
15–52 $N = 1647$ N
15–53 $v_G = 8.75$ m/s
15–55 $x = 14 \times 10^{-6}$ m
15–56 $v_f = 1.43$ km/h
15–58 $v_f = 1$ m/s
15–59 $\Delta T = 4.5 \times 10^5$ J
15–61 $v'_B = 1.65$ mph; $v'_A = 2.05$ mph
15–62 $\Delta T = -1165$ ft · lb
15–64 $v'_P = 4.07$ m/s; $v'_R = 0.94$ m/s
15–65 $W_R = 400$ lb
15–67 $v_A = 26.23$ mph; $v'_A = 53.5$ mph
15–68 $d = 1.24$ m
15–70 $h = 1.6$ ft; $d = 2.25$ ft
15–71 $h_1 = 0.018$ m; $h_2 = 1.65 \times 10^{-3}$ m
15–73 $\mathbf{v}'_A = (1.56\mathbf{i} + 0.8\mathbf{j})$ ft/s; $\mathbf{v}'_B = 0.72\mathbf{i}$ ft/s
15–75 $\mathbf{v}_B = (16.7\mathbf{i} + 58.3\mathbf{j})$ mph
15–76 $\mathbf{v}_f = -e\mathbf{v}_0$; $\phi = \theta$
15–77 $\mathbf{M} = 1{,}378{,}200\mathbf{n}_\theta - 6911\mathbf{n}_\phi$
15–79 $v_{esc} = 9448$ mph
15–81 $v_{f_1} = 4$ ft/s; $v_{f_2} = 3.3$ ft/s; $v_{f_3} = 2.86$ ft/s; $v_f = 2.86$ ft/s
15–82 $v_f = 0.48$ km/h
15–84 $v'_A = 0.377v_A$ ↘ 47.6°; $v'_B = 0.796v_A$ ↗ 20.5°

Chapter 16

16–1 $\mathbf{v}_G = 2\mathbf{i}$ m/s
16–2 $\mathbf{a}_G = -3\mathbf{j}$ ft/s²
16–4 $\mathbf{F} = (1.36\mathbf{i} - 1.92\mathbf{j} + 1.32\mathbf{k})$ N
16–5 $\mathbf{a}_G = (-42.9\mathbf{i} - 32.2\mathbf{j})$ N
16–7 $r_T = 4.76$ ft
16–8 $\mathbf{v}_1 = (452.3\mathbf{i} - 316.7\mathbf{j})$ m/s; $\mathbf{v}_2 = (137.0\mathbf{i} + 79.1\mathbf{j})$ m/s
16–10 $m_3 = 1309.1$ kg; $m_2 = 1436$ kg; $m_T = 4345$ kg
16–11 $W = 40{,}800$ lb
16–13 $v = 52.97$ ft/s
16–14 $v_f = 13.12$ m/s
16–16 $T = 818$ ft · lb
16–17 $T = 1047$ ft·lb
16–19 $U_{1-2} = -41{,}123$ ft·lb
16–20 $T = \frac{1}{2}(100)(m)v^2$

16–22 (a) $T = 501$ N·m; (b) $T = 501$ N·m
16–23 $T_2 - T_1 = 71.2$ N·m; $U = T_2 - T_1 = 71.2$ N·m
16–25 $\mathbf{H}_O = -11.8\mathbf{k}$ kg·m²/s
16–26 $\mathbf{H}_C = 8.94\mathbf{k}$ slug·ft²/s
16–28 (a) $\mathbf{G} = (1.24\mathbf{i} + 1.55\mathbf{j})$ slug·ft/s; $\sum \mathbf{F} = -7\mathbf{j}$ lb;

(b) $\mathbf{H}_O = -2.18\mathbf{k}$ slug·ft²/s; $\frac{d}{dt}(\mathbf{H}_O) = -1295\mathbf{k}$ ft·lb; $\mathbf{H}_G = 5.78\mathbf{k}$ slug·ft²/s

$\frac{d}{dt}(\mathbf{H}_G) = -1290\mathbf{k}$ ft · lb
16–29 $\mathbf{H}_O = (-970\mathbf{i} + 2590\mathbf{j} - 615\mathbf{k})$ kg·m²/s; $\mathbf{H}_G = (-2033\mathbf{i} + 1126\mathbf{j} - 45.2\mathbf{k})$ kg · m²/s
16–31 $\mathbf{H}_O = 2.05 \times 10^5\,\mathbf{k}$ kg·m²/s
16–32 $\frac{d}{dt}(\mathbf{H}_C) = 10{,}278\mathbf{k}$ ft·lb
16–34 $\mathbf{H}_O = (-11\,276\mathbf{i} + 21\,392\mathbf{j})$ kg·m²/s
16–35 $\dot{\mathbf{H}}_O = 21{,}435\mathbf{k}$ ft · lb
16–37 $\mathbf{G}_O = m_T\mathbf{v}_A + \mathbf{G}_A$
16–38 before fracture: $\mathbf{H}_O = 157.08\mathbf{k}$ kg·m²/s; after fracture: $\mathbf{H}_O = 153.9\mathbf{k}$ kg·m²/s
16–40 $m_p\mathbf{v}_{pf} = (35{,}528\mathbf{i} + 88.8\mathbf{j})$ slug·ft/s
16–41 $m_p\mathbf{v}_{pf} = (35{,}528\mathbf{i} + 88.8\mathbf{j})$ slug·ft/s
16–43 $\mathbf{M}_A = 1764$ N·m $\mathbf{k}$; $\mathbf{M}_B = 7280$ N·m $\mathbf{k}$; $\mathbf{M}_O = 76\,328$ N·m $\mathbf{k}$
16–44 $\mathbf{H}_O = (6065\mathbf{j} + 4800\mathbf{i})$ ft·lb·s
16–46 $v_f = 170.67$ m/s
16–47 $\mathbf{H}_O = (7807.5\mathbf{k} + 2925.5\mathbf{j})$ ft·lb·s
16–49 $F = 38.8$ N→
16–50 Thrust = 6211 lb←
16–52 $W = F = 1854$ lb
16–53 $d = 0.595$ m
16–55 $\mathbf{F} = (-5740\mathbf{i} + 6\mathbf{j})$ N
16–56 $\mathbf{P} = (-9.32\mathbf{i} + 16.14\mathbf{j})$ lb
16–58 $\mathbf{F} = A_0(158.8\mathbf{i} + 111.2\mathbf{j})$ lb
16–59 Thrust of engine = 47.2 kN→
16–61 Force on tail section = 7379 lb←
16–62 Thrust of engine = 37.1 kN→; Horsepower = 27,650 hp
16–64 Reverse thrust = 120.4 kN→
16–65 Braking force = 5008 lb→
16–67 $F_x = 55.6$ N
16–68 $v_x = 1.501$ km/h
16–69 $F = 0.00373$ lb
16–71 (a) machine will move with constant speed with no force required; (b) $F = 37.3$ lb
16–72 $\frac{dm}{dt} = 667$ kg/s
16–74 $W_0 = 340$ kN; $a = 0.588$ m/s²; $v = 5.64$ m/s
16–75 $v_f = 5818$ ft/s
16–76 $a = 7.95$ m/s²; $v = 2.08$ km/s
16–78 $T = 2701$ N·m
16–80 $\mathbf{H}_O = (61\,740\,\mathbf{j} + 10\,193\mathbf{k})$ kg · m²/s
16–81 $\omega_A = 9.94$ rpm; $\dot{\theta} = 0.3$ rpm
16–83 $T = 6985$ N
16–84 $M_f = 100.2$ kg

Chapter 17

17–1 $\mathbf{v}_A = -4\mathbf{j}$ m/s; $\mathbf{v}_B = -8\mathbf{j}$ m/s; $\mathbf{a}_A = -32\mathbf{i}$ m/s²; $\mathbf{a}_B = -64\mathbf{i}$ m/s²
17–2 $\mathbf{a}_B = (-2\mathbf{j} - 64\mathbf{i})$ m/s²; $\mathbf{a}_A = (-\mathbf{j} - 32\mathbf{i})$ m/s²
17–4 $\mathbf{a}_A = (-3\mathbf{i} - 225\mathbf{j})$ ft/s²; $\mathbf{a}_B = (82.426\mathbf{i} - 147.3\mathbf{j})$ ft/s²
17–5 $\mathbf{a}_A = 7\mathbf{j}$ m/s²; $\boldsymbol{\alpha} = -4.67$ r/s² $\mathbf{k}$
17–7 $\mathbf{a}_C = (6\mathbf{i} - 10.39\mathbf{j})$ ft/s²; $\mathbf{a}_A = \mathbf{a}_B = \mathbf{a}_C = (6\mathbf{i} - 10.39\mathbf{j})$ ft/s²
17–8 $\mathbf{a}_C = (3.4\mathbf{i} - 11.9\mathbf{j})$ ft/s²
17–10 $|\mathbf{v}_A| = 42$ m/s
17–11 $\alpha = -0.006$ rad/s²; $\omega = 6$ rad/s
17–13 $\alpha = 31.41$ rad/s² (deceleration)
17–14 $t = 36$ s
17–16 $\mathbf{v}_A = -787.96\mathbf{k}$ mph; $\mathbf{a}_B = -0.109$ ft/s² $\mathbf{n}_r$; $\mathbf{a}_A = -0.109$ ft/s² $\mathbf{n}_r$
17–17 $\alpha_2 = 0.75$ rad/s²; $\Delta\theta_1 = 5.85$ rev; $\Delta\theta_2 = 2.92$ rev
17–19 $t = 4\pi$ s
17–20 $\alpha_1 = 6.98$ rad/s²; $\Delta\theta_2 = 3.76$ rev
17–22 $\omega_{fc} = 83.67$ rad/s; $\alpha_C = 35$ rad/s²
17–23 $\mathbf{v}_A = (27.78\mathbf{i} + 27.78\mathbf{j})$ m/s; $\mathbf{v}_B = (47.42\mathbf{i} - 19.64\mathbf{j})$ m/s
17–25 $\mathbf{v}_A = (27.78\mathbf{i} + 27.78\mathbf{j})$ m/s; $\mathbf{v}_B = (47.42\mathbf{i} - 19.64\mathbf{j})$ m/s
17–26 $\mathbf{v}_C = -0.6\mathbf{i}$ m/s
17–28 $\mathbf{v}_A = 0.525\mathbf{i}$ ft/s
17–29 $v_A = 0.525$ ft/s→
17–30 $\omega_A = 8.33$ rad/s
17–32 $\omega_{\text{roll}} = 100$ rad/s
17–34 $\omega = 2000$ rad/s
17–35 $v_{\max} = 2660$ ft/s; $v_{\min} = -2460$ ft/s
17–37 $v_{C/E} = 3.36$ ft/s
17–38 $\omega_C = 0.523$ rad/s ↰
17–40 $\omega_{CD} = 0.19$ rad/s
17–41 $\mathbf{v}_B = (209.4\mathbf{i} - 62.82\mathbf{j})$ in./s

17–43 $v_B = 218.6$ in./s
17–44 $\mathbf{v}_C = -8.34$ m/s $\mathbf{j}$
17–46 $\omega_{BC} = 6.7$ rad/s ↺
17–47 $\omega_{CB} = 6.7$ rad/s ↺
17–49 $\omega_B = 70.69$ rad/s ↻
17–50 $\omega_B = 78.55$ rad/s
17–52 $r = 7.08$ m (below center of wheel)
17–53 $\omega_{AB} = 0.115$ rad/s ↺; $v_B = 1.73$ cm/s ↑
17–55 $\mathbf{a}_A = 800\mathbf{j}$ m/s^2; $\mathbf{a}_B = [6\mathbf{i} - 800\mathbf{j}]$ m/s^2
17–56 $\mathbf{a}_C = (4980\mathbf{i} - 20\mathbf{j})$ ft/s^2;
$\mathbf{a}_D = (-5020\mathbf{i} + 20\mathbf{j})$ ft/s^2
17–58 $\mathbf{a}_B = -30.52\mathbf{i}$ m/s^2
17–59 $\mathbf{a}_B = -235.34\mathbf{i}$ ft/s^2
17–61 $\mathbf{a}_O = -57{,}116\mathbf{i}$ ft/s^2; $\mathbf{a}_I = 57{,}116\mathbf{i}$ ft/s^2
17–62 $\mathbf{a}_B = 45.22\mathbf{i}$ m/s^2
17–64 $\mathbf{a}_C = (-0.427\mathbf{i} - 0.911\mathbf{j})$ m/s^2
17–65 $\mathbf{a}_C = (-0.312\mathbf{i} - 1.05\mathbf{i})$ m/s^2
17–67 $\mathbf{a}_B = (55.24\mathbf{i} - 1392.42\mathbf{j})$ ft/s^2
17–68 $\mathbf{a}_B = (-21.95\mathbf{i} + 337.53\mathbf{j})$ ft/s^2, $\theta = 0$
17–70 $\mathbf{a}_D = (-4.57\mathbf{i} - 90.14\mathbf{j})$ ft/s^2
17–71 $\theta = 9.84°$
17–73 $\mathbf{a}_C = 411.34\mathbf{i}$ ft/s^2
17–74 $\mathbf{v}_B = (20\mathbf{i} - 20\mathbf{j} + 0.08\mathbf{k})$ m/s;
$\mathbf{v}_D = 40\mathbf{i}$ m/s
17–76 $\mathbf{a}_D = (24\mathbf{k} - 1000\mathbf{j})$ m/s^2
17–77 $\boldsymbol{\alpha} = -0.893\mathbf{i}$ rad/s^2
17–79 $\mathbf{v}_B = (8\mathbf{i} + 13.86\mathbf{j} + 6.93\mathbf{k})$ m/s
17–80 $\mathbf{a}_B = (6.93\mathbf{i} - 3.2\mathbf{j} - 3.2\mathbf{k})$ m/s^2
17–82 $\boldsymbol{\alpha} = (23.6\mathbf{j} + 0.02\mathbf{k})$ ft/s^2
17–83 $\boldsymbol{\alpha} = -12\mathbf{k}$ rad/s^2
17–85 $\boldsymbol{\alpha} = -41.9\mathbf{j}$ rad/s^2
17–86 $\boldsymbol{\alpha} = (-41.9\mathbf{j} - 167.5\mathbf{k})$ rad/s^2
17–88 $\mathbf{a}_P = (0.168\mathbf{i} + 140{,}000\mathbf{j} - 123.2\mathbf{k})$ ft/s^2;
$\boldsymbol{\alpha} = (48\mathbf{i} + 36\mathbf{j})$ rad/s^2
17–90 $\mathbf{a}_A = -187\,600\mathbf{i}$ m/s^2
17–91 $\boldsymbol{\alpha}_A = -75\mathbf{k}$ rad/s^2; $\boldsymbol{\alpha}_B = 75\mathbf{k}$ rad/s^2;
$\mathbf{a}_P = (825\mathbf{i} - 25\mathbf{j})$ *ft/s*^{2}
17–92 $\boldsymbol{\alpha} = (0.00549\mathbf{i} - 0.01095\mathbf{k})$ rad/s^2
17–94 $\mathbf{v}_C = (-0.703\mathbf{i} + 0.228\mathbf{j})$ m/s;
$\mathbf{a}_C = (-0.0534\mathbf{i} - 0.1849\mathbf{j})$ m/s^2
17–95 $\mathbf{v}_C = (-1.702\mathbf{i} + 0.508\mathbf{j})$ m/s;
$\mathbf{a}_C = (-0.417\mathbf{i} - 0.724\mathbf{j})$ m/s^2
17–97 $\mathbf{a}_A = (4.22\mathbf{i} - 8.41\mathbf{j})$ ft/s^2;
$\mathbf{a}_B = (4.63\mathbf{i} - 8\mathbf{j})$ ft/s^2
17–98 $\boldsymbol{\omega}_A = -1.69\mathbf{k}$ or $2.74\mathbf{k}$ rad/s
17–100 $\mathbf{a}_4 = (-20.14\mathbf{i} + 17.96\mathbf{j})$ ft/s^2
17–101 $|\mathbf{v}_P| = 3.37$ ft/s; $|\mathbf{a}_P| = 31.2$ ft/s^2
17–103 $\mathbf{v}_P = (-1.5\mathbf{i} + 1\mathbf{j})$ m/s;
$\mathbf{a}_P = (8\mathbf{i} + 14.3\mathbf{j})$ m/s^2
17–104 $\dot{x} = 2.71$ m/s
17–106 $\mathbf{a}_A = (-224\mathbf{i} - 1615\mathbf{j})$ m/s^2;
$\mathbf{a}_B = (-292\mathbf{i} - 1615\mathbf{j})$ m/s^2
17–107 $\mathbf{a}_B = (1.781\mathbf{i} - 4.33\mathbf{j})$ m/s^2
17–109 $\omega_B = 40.6$ rpm; $\alpha_B = 255$ rad/s^2
17–110 $\alpha_A = 2.56$ rad/s^2
17–112 $\Delta\theta_{B_1} = 2.6$ rev.; $\Delta\theta_{B_2} = 3.25$ rev.;
$\Delta\theta_{B_3} = 4.33$ rev.;
$\omega_{B_1} = 20.61$ rad/s; $\omega_{B_2} = 25.76$ rad/s;
$\omega_{B_3} = 34.35$ rad/s
17–113 $v_A = 846.8$ ft/s
17–115 $l_A = 1.598$ m
17–116 $\mathbf{a}_C = (0.52\mathbf{i} - 3.125\mathbf{j})$ m/s^2
17–118 $\boldsymbol{\alpha} = 0.0611\mathbf{j}$ m/s^2
17–119 $\mathbf{a}_p = (-3.7\mathbf{i} - 18\,720\mathbf{j})$ m/s^2
17–121 $\dot{r} = 685{,}871$ ft/s

Chapter 18

18–1 $\mathbf{a}_x = -7.63$ m/s^2 $\mathbf{i}$
18–2 $N_A = 81.75$ N
18–5 $N_A = 0$; $N_B = 492.4$ lb; $F_B = 443.16$ lb
18–6 $N_B = 8204$ N; $N_A = 10.75$ kN;
$F_A = 5373$ N; $F_B = 4102$ N
18–7 $a_x = 85.87$ ft/s^2
18–8 $\mu_k = 2.67$
18–10 $a_x = 3.82$ m/s^2
18–11 $a_x = 6.44$ ft/s^2
18–13 $B = 2$ mg↑; $A_y = 2$ mg↑; $A_x = 4$ ma$_x$→
18–14 $a_x = 0.2\,g\left(\dfrac{d}{h}\right)$
18–16 $0.22 \text{ m} < h < 2.18$ m
18–17 $F = 313.9$ N
18–19 $d = 3.5$ ft
18–20 $T = 3046.3$ N ↘
18–22 $\theta = 21.68°$
18–23 $v_{\text{safe}} = 74.52$ mph
18–25 $a_x = g\cdot d/l$
18–26 $M = 26.96$ ft·lb
18–28 $F = 20$ N
18–29 $M = 10.06$ ft · lb ↻
18–31 $T = 2888.5$ N·m
18–32 $T = 1989$ N · m
18–34 $a_B = 25$ m/s^2
18–35 $F = 31.06$ lb
18–37 $\alpha = 64.14$ rad/s
18–38 $M = 1847$ N · m
18–40 $\alpha = 1.94$ rad/s^2
18–41 $\alpha = 1.92$ rad/s^2

18–43 $F_x = 12.7\ \text{N}\rightarrow;\ F_y = 22.1\ \text{N}\downarrow$

18–44 $F_x = 116.8\ \text{N}\rightarrow;\ F_y = 38\ \text{N}\uparrow$

18–46 $\alpha = 12.63\ \text{rad/s}^2;\ A_y = 551.72\ \text{N}\downarrow;$
$A_x = 454.1\ \text{N}\leftarrow$

18–47 $\alpha = 12.56\ \text{rad/s}^2;\ A_x = 514.9\ \text{N}\leftarrow;$
$A_y = 551.72\ \text{N}\downarrow$

18–49 $O_x = 0;\ O_y = 0.09\rho t g \pi R^2$

18–51 $F = 4496\ \text{lb};\ O_X = 2857\ \text{lb}\rightarrow;$
$O_y = 2293\ \text{lb}\uparrow$

18–52 $\alpha = 0.10\ \text{rad/s}^2;\ O_y = 25280\ \text{N}\uparrow;$
$O_x = 656.8\ \text{N}\rightarrow$

18–53 $\alpha_{AO} = 129.6\ \text{rad/s}^2 \circlearrowright;\ O_x = 143.8\ \text{lb}\leftarrow;$
$O_y = 139.7\ \text{lb}\downarrow$

18–55 $F_A = 1471.6\ \text{lb}\leftarrow;\ F_t = 58.5\ \text{lb}\downarrow$

18–56 $$T = \frac{mg}{4}\left[1 + \frac{R\omega^2}{g}(0.424)\right]$$

18–57 $a = 1.68\ \text{m/s}^2 \swarrow$

18–59 $a = 9.91\ \text{ft/s}^2 \searrow$

18–61 $a = 12.1\ \text{m/s}^2\rightarrow$

18–63 $a = 6.54\ \text{m/s}^2\downarrow;\ T = 65.4\ \text{N}\uparrow$

18–64 $a = 10.73\ \text{ft/s}^2\searrow;\ \mu = 0.19$

18–65 $a = 20.3\ \text{ft/s}^2$

18–67 $a = 162.09\ \text{ft/s}^2\nearrow$

18–68 $P = 106.1\ \text{lb}$

18–70 $\alpha = 1.468\ \text{rad/s}^2$

18–71 $\alpha = 0.0013\ \text{rad/s}^2 \circlearrowright$

18–73 $M = 2198\ \text{N}\cdot\text{m} \circlearrowleft$

18–74 $M = 2501\ \text{N}\cdot\text{m} \circlearrowleft$

18–76 $N_B = 210.7\ \text{lb}\uparrow;\ N_A = 111.7\ \text{lb}\rightarrow$

18–77 $A = 154.8\ \text{N}\downarrow;\ B = 39.4\ \text{N}\leftarrow$

18–79 $B_x = 689.8\ \text{lb};\ B_y = 654.6\ \text{lb}$

18–80 $B_x = 151.23\ \text{N};\ B_y = 140\ \text{N}$

18–82 $\mathbf{a}_G = (-168\mathbf{i} + 0.04\mathbf{j})\ \text{m/s}^2$

18–83 $\alpha = 18.76\ \text{rad/s}^2 \circlearrowleft$

18–85 $\alpha = 1.61\ \text{rad/s}^2 \circlearrowleft$

18–88 $T = 14055\ \text{N};\ \mathbf{a}_A = (-0.346\mathbf{i} - 7.48\mathbf{j})\ \text{m/s}^2$

18–89 $r_g = 0.2\ \text{m}$

18–90 $\alpha = 0.05\ \text{rad/s}^2 \circlearrowleft;\ a_z = 2.5 \times 10^{-3}\ \text{m/s}^2$

18–91 $a = 4.91\ \text{m/s}^2;\ T_1 = T_2 = 42.5\ \text{N}$

18–93 $a_p = 21.61\ \text{ft/s}^2 \ \angle 70^\circ;\ a_c = 7.39\ \text{ft/s}^2\downarrow$

18–95 $$\alpha = \frac{M}{\left\{2mL^2\left(\frac{1}{12} + \frac{\sin^2\theta}{4}\right) + 2m(l_1^2 + l_2^2)\right\}}$$

18–96 $h = 0.133\ \text{m}$

18–97 $t = 0.34\ \text{s}$

18–99 $F = 194.8\ \text{lb}$

18–100 $\alpha_c = 16.67\ \text{rad/s}^2 \circlearrowleft;\ \alpha_{sp} = 0.714\ \text{rad/s}^2 \circlearrowleft$

Chapter 19

19–1 $\omega_f = 5.42\ \text{rad/s}$

19–2 $\omega_f = 5.08\ \text{rad/s}$

19–4 $\omega_f = 5.42\ \text{rad/s}$

19–5 $v_p = 5.3\ \text{ft/s}$

19–7 $M = 523.6\ \text{N}\cdot\text{m}$

19–8 $\omega_f = 4657\ \text{rpm}$

19–10 $\omega_f = 8.97\ \text{rad/s}$

19–11 $v_A = 5.42\ \text{m/s}\downarrow$

19–13 $\omega = 4.4\ \text{rad/s};\ O_x = 96.15\ \text{lb}\leftarrow:$
$O_y = 212.3\ \text{lb}\uparrow$

19–14 $\omega = 4.49\ \text{rad/s};\ O_x = 98.04\ \text{lb}\leftarrow;$
$O_y = 216.32\ \text{lb}\uparrow$

19–16 $v_1 = 5.85\ \text{m/s}$

19–17 $v_1 = 6.69\ \text{ft/s}\downarrow$

19–19 $U = 53.06\ \text{kN}\cdot\text{m}$

19–20 $P = 21.22\ \text{kN}\cdot\text{m/s}$

19–22 $O_x = 329.8\ \text{lb}\leftarrow:\ O_y = 0.97\ \text{lb}\uparrow$

19–23 $F = 10.91\ \text{N}$

19–25 $P = 19.4\ \text{N}\cdot\text{m/s}$

19–27 $P = 6524\ \text{ft}\cdot\text{lb/s}$

19–29 $\omega = 7.89\ \text{rad/s}$

19–30 $v_C = 17\ \text{m/s}$

19–32 $P_{\max} = 353.16\ \text{kN}\cdot\text{m/s at } \theta = 0$

19–33 $$\text{(a) } v^2 = \frac{\left(1 + 3\frac{m}{M}\right)2gl\sin\theta}{\left(1 + \frac{9}{8}\frac{m}{M}\right)};$$

$$\text{(b) } v^2 = \frac{2gl\sin\theta}{1 + \frac{3}{2}\frac{m}{M}}$$

19–35 $P = 31.04\ \text{W}$

19–36 $v_A = 6.09\ \text{ft/s}$

19–37 $\theta \simeq 23.8^\circ$

19–39 $M \simeq 0.0015\ \text{ft}\cdot\text{lb}$

19–40 $M = 3.93\ \text{N}\cdot\text{m}$

19–41 $\mu = 0.4$

19–43 $v = 107.3\ \text{ft/s} \quad \mu = 0.19$

19–44 $\omega_f = 0.37\ \text{rad/s} \circlearrowright$

19–46 $\omega_f = 2.25\ \text{rad/s} \circlearrowright$

19–47 $v_{B_y} = 0.044\ \text{ft/s}\downarrow$

19–49 $N = 17.5\ \text{N}$

19–50 $\omega = 0.84\ \text{rad/s} \circlearrowleft$

19–52 $\omega_f = 0.29\ \text{rad/s} \circlearrowleft$

19–53 $\Delta t = 1.04\ \text{s}$

19–55 $\omega_{f_A} = 5.09\ \text{rad/s} \circlearrowleft;\ \omega_{f_C} = 12.73\ \text{rad/s} \circlearrowright$

19–56 $v_1 = 5.03\ \text{m/s}\downarrow$

19–58 $M = 307.4 \text{ ft}\cdot\text{lb}$
19–59 $M = 50.534 \text{ kN}\cdot\text{m}$
19–61 $T = 935.7 \text{ N}\uparrow$
19–62 $T = 19.6 \text{ ft}\cdot\text{lb}$

19–64 (a) $v_p = \dfrac{\left[1 + 3\left(\dfrac{m}{M}\right)\right] g\,\Delta t \sin\theta}{\dfrac{3}{2}\dfrac{m}{M} + 1}$;

(b) $v_p = \left(3\dfrac{m}{M} + 1\right) g\,\Delta t \sin\theta$

19–65 $T_U = 49{,}813 \text{ lb}$; $T_L = 50{,}186 \text{ lb}$
19–67 $\omega_f = 7.31 \text{ rad/s}$↰; $O_y = 10\,617 \text{ N}\downarrow$; $O_x = 5.3 \text{ N}\leftarrow$
19–68 $M = 0.021 \text{ ft}\cdot\text{lb}$
19–70 $P_{\text{Tot}} = 97.5 \text{ kN}\cdot\text{m/s}$
19–73 $\Delta\omega = 1.1 \times 10^{-6} \text{ rad/s}$
19–74 $\omega_f = 0.705 \text{ rad/s}$

Chapter 20

20–1 $\mathbf{M}_O = (-0.717\mathbf{i} + 1.242\mathbf{j}) \text{ lb}\cdot\text{ft}\cdot\text{s}$
20–2 $T = 8.28 \text{ ft}\cdot\text{lb}$
20–4 $\mathbf{H}_O = (-0.362\mathbf{i} + 0.624\mathbf{j}) \text{ kg}\cdot\text{m}^2/\text{s}$
20–5 $\mathbf{H}_{O_2} = (-1\mathbf{i} + 1.732\mathbf{j}) \text{ kg}\cdot\text{m}^2/\text{s}$

20–7 $\mathbf{H}_O = \dfrac{1}{2} lrm\omega_y\mathbf{i} + mr^2\omega_y\mathbf{j}$

20–8 $T = \dfrac{1}{6} mb^2\omega_x^2$

20–10 $T_A = T_B = \dfrac{mg}{2}$

20–11 $T = \dfrac{7}{2}\dfrac{F^2(\Delta t)^2}{M}$

20–13 $T = \dfrac{1}{12} ma^2\omega_y^2$

20–14 $\mathbf{H}_O = -\dfrac{1}{4} m_2ab\omega_y\mathbf{i} + \left[\dfrac{1}{3} m_1a^2 + m_2\left(a^2 + \dfrac{1}{3}c^2\right)\right]\omega_y j - \dfrac{1}{4} m_2bc\omega_y\mathbf{k}$

20–16 $\mathbf{H}_O = \dfrac{md\omega_y}{2}\left(\dfrac{h}{\pi}\mathbf{i} + d\mathbf{j} - \dfrac{h}{2}\mathbf{k}\right)$

20–17 $T = \dfrac{1}{4} md^2\omega_y^2$

20–19 $T = \dfrac{1}{4} m\omega_y^2\left(\dfrac{r^2}{2} + 3l^2\right)$

20–20 $\mathbf{H}_{O_2} = (60\mathbf{j} + 10\,450\mathbf{k}) \text{ N}\cdot\text{m}\cdot\text{s}$
20–21 $T = 54.6 \text{ kN}\cdot\text{m}$

20–23 $\omega_{\min} = \sqrt{\dfrac{3g}{2l}}$

20–25 $\theta = \cos^{-1}\left[\dfrac{gl}{\omega^2\left(l^2 - \dfrac{r^2}{4}\right)}\right]$

20–26 $\theta = \cos^{-1}\left[\dfrac{mgl + \dfrac{1}{2}Mgl}{\left(ml^2 + \dfrac{1}{3}Ml^2 - \dfrac{1}{4}mr^2\right)\omega^2}\right]$

20–28 $\mathbf{B}_z = \mathbf{A}_z = 0$; $\mathbf{B}_y = 74.5\mathbf{j} \text{ N}$; $\mathbf{A}_y = -42.3\mathbf{j} \text{ N}$

20–29 $\dot{\mathbf{H}}_O = -\dfrac{1}{2} lrm\omega_y^2\mathbf{k}$

20–31 $\mathbf{B}_x = -\dfrac{ma}{b}\left(\dfrac{g}{3} + \dfrac{b\omega_y^2}{4}\right)\mathbf{i}$; $\mathbf{A}_z = 0$;

$\mathbf{A}_x = \left(\dfrac{mag}{3b} - \dfrac{ma\omega_y^2}{12}\right)\mathbf{i}$; $\mathbf{A}_y = mg\mathbf{j}$

20–32 $\dot{\mathbf{H}}_A = -\dfrac{b}{4}(m_1a\dot{\omega}_y + m_2c\omega_y^2)\mathbf{i} + \left[\dfrac{1}{3}m_1a^2 + m_2\left(a^2 + \dfrac{1}{3}c^2\right)\right]\dot{\omega}_y\mathbf{j} + \dfrac{b}{4}(m_1a\omega_y^2 - m_2c\dot{\omega}_y)\mathbf{k}$

20–34 $\dot{\mathbf{H}}_O = \omega_y\mathbf{j} \times \dfrac{1}{3} md^2\omega_y\mathbf{j} = 0$
20–35 $A_x = A_z = O_x = O_z = 0$
20–37 $\mathbf{M}_O = 0$
20–38 $\dot{\mathbf{H}}_A = -28.3\mathbf{i} \text{ ft}\cdot\text{lb}$
20–40 $\mathbf{O}_z = 0$;

$\mathbf{A}_y = \left\{ mg\left[\dfrac{\dfrac{a}{2} - \dfrac{3\pi r^2}{b}}{1 - \dfrac{\pi r^2}{ab}}\right] - \dfrac{3}{8}\pi r^2 m\omega_x^2 + \dfrac{1}{4} mab\,\omega_x^2 \right\}\mathbf{j}$;

$$\mathbf{O}_y = -\left\{\frac{mb}{2}\omega_x^2 - mg + mg\left[\frac{\frac{a}{2} - \frac{3\pi r^2}{b}}{1 - \frac{\pi r^2}{ab}}\right]\right.$$

$$\left. - \frac{3}{8}\pi r^2 m\omega_x^2 + \frac{1}{4} mab\,\omega_x^2\right\}\mathbf{j}$$

20–41 $\dot{\mathbf{H}}_O = -2mr^3\omega_x^2\mathbf{k}$
20–43 $\mathbf{M}_O = 469\mathbf{k}$ N·m
20–44 $\dot{\mathbf{H}}_O = -1250\mathbf{k}$ lb · ft
20–47 $\dot{\gamma} = -436$ rad/s
20–48 $\dot{\phi} = -52.5$ rad/s

20–50 $\dot{\phi} = -\dfrac{4g\left(a + \dfrac{l}{2}\right)}{d^2\dot{\gamma}}$

20–51 $\gamma = 30{,}656\mathbf{l}$ rad/s
20–52 $\dot{\phi} = 18.08$ rad/s
20–53 $\dot{\phi} = -6.15$ rad/s
20–55 $N = 18.53$ lb
20–56 $\mathbf{A}_z = -0.3513\mathbf{k}$ lb; $\mathbf{B}_z = 0.3513\mathbf{k}$ lb; $(A_y + B_y)\mathbf{j} = -4.77\mathbf{j}$ lb; $A_x\mathbf{i} = -B_x\mathbf{i}$
20–58 $\mathbf{M}_A = (5296\mathbf{j} - 490.5\mathbf{k})$ N·m
20–59 $\mathbf{A}_y = \mathbf{B}_y = -15.53\mathbf{j}$ lb
20–61 $\mathbf{M}_O = (77{,}458\mathbf{i} + 44{,}721\mathbf{j} - 207{,}850\mathbf{k})$ lb · ft
20–62 $I_2 > I_1$, direct precession
20–64 $\dot{\gamma} = -5\mathbf{j}$ rad/s
20–65 $\mathbf{H}_C = \{(0.333M + 0.551m)l^2 + (0.25\,M + 2\,m)r^2\}\omega\mathbf{k}$
20–67 $T = 0.01l^2F^2\,(\Delta t)^2$
20–68 $\mathbf{M}_O = 848\mathbf{i}$ N·m
20–70 $\boldsymbol{\omega} = (3.14\mathbf{j} - 0.225\mathbf{k})$ rad/s; $\mathbf{H}_O = (3140\mathbf{j} - 450\mathbf{k})$ N·m·s
20–71 $(\mathbf{H}_O)_2 = [191{,}400\mathbf{j} + 40m\mathbf{k}]$ ft·lb/s; $\boldsymbol{\omega} = [0.209\mathbf{j} + 0.002\mathbf{k}]$ rad/s

Chapter 21

21–1 $\tau = 0.14$ s
21–2 $k = 1.55$ lb/ft
21–4 (a) $\omega = 44.72$ rad/s; (b) $\omega = 44.72$ rad/s
21–5 $k = 76.94$ kN/m
21–7 $\omega = 31.62$ rad/s; $A = 40$ cm
21–8 $\omega = 278$ rad/s
21–10 $\tau = 2.01$ s; $\theta = 0.1\sin\left(3.132t + \dfrac{\pi}{2}\right)$
21–11 $\omega = \sqrt{\dfrac{g}{\Delta}}$
21–13 $\tau = 1.34$ s
21–14 $\dot{\theta}_{max} = 0.38$ rad/s; $\ddot{\theta}_{max} = 1.47$ rad/s^2
21–16 $\omega = 45.93$ rad/s
21–17 $\omega = 23.5$ rad/s
21–19 $\tau = 2\pi\sqrt{\dfrac{2\left(\dfrac{b^2}{3} + \dfrac{a^2}{12}\right)}{bg}}$
21–20 $\omega = 2.45\sqrt{\dfrac{-gh + \dfrac{2k}{m}}{4h^2 + a^2}}$
21–22 $k = 0.13$ N/m
21–23 $\omega = 69.5$ rad/s
21–25 $\tau = 0.14$ s
21–26 $k = 1.55$ lb/ft
21–28 $\omega = 278$ rad/s
21–29 $\tau = 1.34$ s
21–31 $\omega = 45.93$ rad/s
21–32 $\omega = 23.5$ rad/s
21–34 $\tau = 2\pi\sqrt{\dfrac{4b^2 + a^2}{6bg}}$
21–35 $\omega = 69.5$ rad/s
21–37 $\tau = 0.09$ s
21–38 $\tau = 0.175$ s
21–40 $\omega = \sqrt{\dfrac{g}{2r}}$
21–41 $\omega = \sqrt{\dfrac{3g}{4r}}$
21–43 $A = 0.158$ in.
21–44 $\Omega = 43.95$ rad/s
21–45 $A \simeq 15$ cm
21–47 $k = 81.73$ lb/ft
21–49 $0 < \Omega < 10.45\sqrt{\dfrac{k}{m}}$
21–50 $k = 10$ N/m
21–51 $k_i = 250$ kN/m
21–53 $d_0 = 5.49$ m
21–54 $x = 0.56e^{44.4t} + 4.44e^{5.6t}$
21–56 $\tau_{d_1} = 0.0052$ s; $\tau_{d_2} = 0.0017$ s

21–57 $x = 0.043e^{-22.15t}[\sin(22.04t + 0.783)]$
21–59 $c = 945.73$ lb·s/ft
21–64 $c/c_c = 0.00133$
21–65 $A = 1.034 \times 10^{-4}$ in.
21–66 $A = 0.095$ cm
21–67 $d_0 = 5.49$ m
21–68 $A = 30.0\ \mu$m
21–74 $I_O = 63\,613$ kg·m^2
21–75 $I_G = 83{,}069$ slug·ft^2
21–76 $\tau = 2\pi\sqrt{\dfrac{R}{g}}$
21–77 $\tau = 2\pi\sqrt{\dfrac{3(R - r)}{2g}}$
21–78 $A = -0.012$ m
21–79 $P_0 = 20$ lb
21–80 $A = 0.016$ in.

LIST OF SYMBOLS

QUANTITY REPRESENTED	SYMBOL
Acceleration	$\mathbf{a}$
Distance; magnitude of acceleration; semi-major axis of ellipse	a
Acceleration of mass center	$\mathbf{a}_C$
Acceleration of B relative to frame at A in translation	$\mathbf{a}_{B/A}$
Coriolis acceleration	$\mathbf{a}_{\text{Cor}}$
Points in space or on rigid body	$A, B, C, \ldots$
Planar area; amplitude of vibration	A
Width; distance; semi-minor axis of ellipse	b
Distance; coefficient of viscous damping	c
Critical damping coefficient	c_c
Constant; center of mass; instantaneous center; capacitance	C
Distance; diameter; periodic displacement	d
Constant	d_O
Drag force	D
Natural logarithm base; coefficient of restitution	e
Total mechanical energy; voltage	E
Frequency	f
Force (vector form)	$\mathbf{F}$
Magnitude of force	F
Friction force	$\mathscr{F}$
Gravitational acceleration	g
Linear momentum	$\mathbf{G}$
Constant of gravitation; geometric center; magnitude of linear momentum	G
Height	h
Angular momentum about mass center	$\mathbf{H}_C$

SYMBOL	QUANTITY REPRESENTED
$\mathbf{H}_O$	Angular momentum about point O
$\dot{\mathbf{H}}_O$	Time rate of change of angular momentum about point O with respect to fixed XYZ frame
$(\dot{\mathbf{H}}_O)_{xyz}$	Time rate of change of angular momentum about point O with respect to rotating xyz frame
$\mathbf{i}, \mathbf{j}, \mathbf{k}$	Unit vectors along coordinate axes x, y, z, respectively
i	Current
I_x, I_y, I_z	Mass moment of inertia about x, y, z, axes, respectively
I_C	Centroidal mass moment of inertia
$I_{x'}, I_{y'}, I_{z'}$	Principal centroidal moments of inertia
I_1	Principal moment of inertia about spin axis
I_2	Mass moment of inertia about centroidal axis transverse to spin axis
I_{xy}, I_{yz}, I_{xz}	Mass products of inertia
J	Polar mome..t of inertia
k	Spring constant
K	Constant
l	Length
l, m, n	Direction cosines
L	Inductance
m	Mass
$\mathbf{M}$	Moment (vector form)
M	Magnitude of moment; mass
$\mathbf{M}_O$	Moment about point O
$\mathbf{M}_{AB}$	Moment about axis AB
$\mathbf{n}$	Unit vector
$\mathbf{n}_n, \mathbf{n}_t$	Unit vectors in normal, tangential directions
$\mathbf{n}_r, \mathbf{n}_\theta$	Unit vectors in radial, transverse directions
N	Force normal to a surface
O	Origin of coordinate system
p	Pressure; distance from fixed axis to center of percussion
$\mathbf{P}$	Force
P	Point in space; power; center of percussion
P_O	Constant
q	Electric charge
Q	Quantity of matter in fluid flow
$\mathbf{Q}$	Vector
$(\dot{\mathbf{Q}})_{XYZ}$	Time rate of change of vector $\mathbf{Q}$ with respect to fixed XYZ frame
$(\dot{\mathbf{Q}})_{xyz}$	Time rate of change of vector $\mathbf{Q}$ with respect to rotating xyz frame
$\mathbf{r}$	Position vector
$\mathbf{r}_C$	Position vector of mass center
r	Radius; radial coordinate
r_C	Nearest distance between fixed axis and mass center
r_g	Radius of gyration about mass center

QUANTITY REPRESENTED	SYMBOL
Radius of gyration about point *O*	r_O
Resultant force, reaction	$\mathbf{R}$
Radius of earth; resistance	R
Length of curve; displacement	s
Time; thickness	t
Kinetic energy; tension; thrust	T
Velocity	$\mathbf{u}$
Work	U
Velocity	$\mathbf{v}$
Velocity of mass center	$\mathbf{v}_C$
Velocity of *B* with respect to frame at *A* in translation	$\mathbf{v}_{B/A}$
Potential energy; shearing force	V
Elastic potential energy	V_e
Gravitational potential energy	V_g
Weight per unit length	w
Weight	$\mathbf{W}$, W
Rectangular coordinates; distances	x, y, z
Principal axes of inertia	x', y', z'
Rectangular coordinates of fixed frame	X, Y, Z
Angular acceleration	$\boldsymbol{\alpha}$, α
Angle	α (alpha)
Angle	β (beta)
Specific weight; angle; Eulerian angle	γ (gamma)
Rate of spin	$\dot{\gamma}$
Change in length; static deflection of spring; angle	δ (delta)
Damping factor	ζ (zeta)
Efficiency	η (eta)
Angular coordinate; angle; Eulerian angle	θ (theta)
Angular speed; rate of nutation	$\dot{\theta}$
Constant	λ (lambda)
Coefficient of friction	μ (mu)
Coefficient of kinetic friction	μ_k
Coefficient of static friction	μ_s
Mass density; radius of curvature	ρ (rho)
Angle; Eulerian angle; phase angle	ϕ (phi)
Angular speed; rate of precession	$\dot{\phi}$
Period	τ (tau)
Period of damped vibration	τ_d
Angular velocity of rigid body	$\boldsymbol{\omega}$
Angular speed; natural circular frequency	ω (omega)
Damped natural frequency	ω_d
Angular velocity of rotating coordinate system	$\boldsymbol{\Omega}$
Angular speed; forcing frequency	Ω (omega)

INDEX